POLYMERS
A PROPERTY DATABASE
SECOND EDITION

POLYMERS

A PROPERTY DATABASE

SECOND EDITION

EDITED BY

BRYAN ELLIS

RAY SMITH

CRC Press
Taylor & Francis Group
Boca Raton London New York

CRC Press is an imprint of the
Taylor & Francis Group, an **informa** business

CRC Press
Taylor & Francis Group
6000 Broken Sound Parkway NW, Suite 300
Boca Raton, FL 33487-2742

First issued in paperback 2019

© 2009 by Taylor & Francis Group, LLC
CRC Press is an imprint of Taylor & Francis Group, an Informa business

No claim to original U.S. Government works

ISBN-13: 978-0-8493-3940-0 (hbk)
ISBN-13: 978-0-367-38651-1 (pbk)

This book contains information obtained from authentic and highly regarded sources. Reasonable efforts have been made to publish reliable data and information, but the author and publisher cannot assume responsibility for the validity of all materials or the consequences of their use. The authors and publishers have attempted to trace the copyright holders of all material reproduced in this publication and apologize to copyright holders if permission to publish in this form has not been obtained. If any copyright material has not been acknowledged please write and let us know so we may rectify in any future reprint.

Except as permitted under U.S. Copyright Law, no part of this book may be reprinted, reproduced, transmitted, or utilized in any form by any electronic, mechanical, or other means, now known or hereafter invented, including photocopying, microfilming, and recording, or in any information storage or retrieval system, without written permission from the publishers.

For permission to photocopy or use material electronically from this work, please access www.copyright.com (http://www.copyright.com/) or contact the Copyright Clearance Center, Inc. (CCC), 222 Rosewood Drive, Danvers, MA 01923, 978-750-8400. CCC is a not-for-profit organization that provides licenses and registration for a variety of users. For organizations that have been granted a photocopy license by the CCC, a separate system of payment has been arranged.

Trademark Notice: Product or corporate names may be trademarks or registered trademarks, and are used only for identification and explanation without intent to infringe.

Library of Congress Cataloging-in-Publication Data

Polymers : a property database / editors, Bryan Ellis, Ray Smith. -- 2nd ed.
   p. cm.
  Includes bibliographical references and index.
  ISBN 978-0-8493-3940-0 (hardcover : alk. paper)
  1. Polymers--Dictionaries. 2. Polymerization--Dictionaries. I. Ellis, Bryan, 1926- II. Smith, Ray, 1955- III. Title.

QD380.3.P656 2009
668.903--dc22

2008039585

Visit the Taylor & Francis Web site at
http://www.taylorandfrancis.com

and the CRC Press Web site at
http://www.crcpress.com

Table of Contents

Preface	vii
Editorial Board	ix
Introduction	xi
Dictionary Entries	
A	1
B	51
C	109
D	135
E	139
F	161
G	175
H	179
I	197
J	199
K	201
L	203
M	217
N	243
O	325
P	333
R	1029
S	1031
T	1063
U	1067
V	1075
W	1081
X	1083
Z	1085
Name and Synonym Index	1089

Preface

The latest edition of *Polymers: A Property Database* is an improved version of an indispensable polymer reference work. The database is a comprehensive and in-depth collection of properties for a very wide range of polymers, both synthetic and natural; and where polymers exist with subtle variations in structure, these are also covered in some detail. Processing grades are also described along with typical applications for each major polymer.

Each polymer has a general properties section, where very useful information such as chemical structure, synonyms, CAS registration numbers, and monomers used in the polymerisation is given. The mechanical properties are particularly well presented and all source material is meticulously referenced.

The database is well indexed (even more comprehensively in the online version) and extremely easy to use making this a very desirable addition to any researcher's library. Polymers are indexed by common name allowing ease of access even to those with only a basic knowledge of polymers; industrial/commercial names are also used where these exist. This is particularly useful for those working in industry or in multidisciplinary fields where polymers are used.

Editorial Board

Editors
Bryan Ellis
Formerly Sheffield University, UK
Dr. Ray Smith
University of London, UK

Advisory Editor
Dr. Howard Barth
Formerly DuPont, US

Executive Editors
Dr. Fiona Macdonald
CRC Press, Boca Raton, US
Dr. Matt Griffiths
CRC Press, London, UK

Contributing Editors
Dr. Paul Tindal
Dr. Martin Edwards
Dr. Philip Drachman
Additional data collected by contributors to the *CRC Chemical Database*.

Introduction

Those of us who work with polymers and polymeric materials rely heavily on data regarding solution and bulk properties, and manufacturing procedures. These parameters usually can be found spread over different handbooks, encyclopedias, and the Internet. However, *Polymers: A Property Database* is one-stop shopping, whereby this information is now available from a single source. Entries range from a few lines for research polymers to encyclopedic submissions for more common polymers.

To produce this comprehensive *Database*, which conveniently can be used as a desk reference book, a double-column format was used with small, but easy to read font with spaced-out tables. Because main-chain or common polymer names are used as headers, arranged alphabetically in dictionary style, the *Database* is practical to use. (IUPAC approved nomenclature is given under a separate heading). To quickly locate a polymer, a comprehensive polymer name index is available.

The *Database* contains a listing of polymer properties that are, for the most part, associated with polymer manufacturing, processing, and applications. As such, the *Database* contains other useful information in addition to polymer properties, not found in other source books; an example of this is given for nylon 6,6, a well-studied commercial polymer:

Structural Formulae	Elastic Modulus
Additives	Poisson Ratio
Morphology	Tensile Strength Yield
Density	Compression Strength
Thermal Expansion Coefficient	Impact Strength
Latent Heat Crystallization	Viscoelastic Behavior
Thermal Conductivity	Hardness
Specific Heat Capacity	Failure Properties
Glass Transition Temperature	Fracture Mechanical Properties
Melting Temperature	Friction Abrasion and Resistance
Deflection Temperature	Electric Properties
Brittleness Temperature	Dielectric Permittivity
Surface Properties	Dielectric Strength
Solubility	Dissipation Power
Transport Properties	Magnetic Properties
Melt Flow Index	Optical Properties
Intrinsic Viscosity	Refractive Index
Polymer Melts	Molar Refraction
Permeability of Gases	Polymer Stability
Permeability of Liquids	Thermal Stability
Water Content	Upper-Use Temperature
Water Absorption	Decomposition
Gas Permeability	Flammability
Mechanical Properties	Environmental Stress
Tensile Modulus	Chemical Stability
Flexural Modulus	Hydrolytic Stability
Tensile Strength Break	Biological Stability
Flexural Strength at Break	Applications and Selected References

The *Database* also lists many types of polysaccharides, modified cellulosics, and other important biopolymers. Different types of polymers and polymeric structures are presented, such as inorganic polymers, blends, block copolymers, graft polymers, ionomers, elastomers, fibers, hydrogels, interpenetrating networks, structural foams, polymer composites, polysiloxanes, resins, and natural rubbers.

For this second edition, an introductory chapter has been added that reviews polymer complexity as it relates to polymer properties. Furthermore, tabulated lists of polymer properties are given to serve as a guide in selecting appropriate test procedures.

Polymer property data are now available on the Internet in a variety of tabular forms. The advantages and convenience of having a desk reference book of the magnitude of *Database* cannot be overstated. No other reference handbooks contain the caveats, descriptions, and explanations that are found in *Polymers: A Property Database* as exemplified in the above table.

A. IMPORTANCE OF POLYMER PROPERTIES

Because of their high molecular mass, polymers, as compared to small molecules, have unique properties that are often difficult to predict. As such, some background knowledge of the physical chemistry of polymers is desirable for dealing with polymers and polymeric materials.

Polymer properties, like solubility behavior, are used as a guide on a laboratory scale when analyzing or characterizing polymers or when determining structure-property relationships. On an industrial scale, properties such as melt viscosity or heat capacity are important for establishing polymerization and processing conditions. A listing of properties is required for selecting polymers to meet specific applications.

Polymers are ubiquitous as they are used in all applications, from consumer products to high-temperature industrial use to medical devices, under a wide-range of conditions. In modern polymer science and engineering, more complex structures, such as multilayer films, nanomaterials, electro-optical, and electronic devices, are being developed that require more specialized and complex testing for end-use performance evaluation. Furthermore, from knowledge of structure-property relations of polymers and polymeric materials, one can begin to design and tailor-make polymers and complex polymeric structures to meet specific end-use performance requirements.

It is sometimes difficult to accurately predict end-use performance characteristics of the final product using tabulated data of individual components. As a result, accurate measurements are those made on the final product itself, rather than using *model* polymers or components. In these cases, empirically derived measurements using the actual product, verified with authentic samples, may be the best option. It should be noted that most empirically derived data are trade secrets, and, as such, not available. Nevertheless, compilations of properties are still valuable.

B. POLYMER COMPLEXITY

Because of polymer complexity, property variability must be taken into consideration. In this section, we will discuss possible sources of polymer inconsistency and offer suggestions to recognize and reduce these errors.

Chemical or compositional *heterogeneity* refers to the chemical or structural difference among chains of the same polymer. Thus, a measured property of a chemically *heterogeneous* sample will be an *averaged* value dependent upon sample source. For chemically *homogeneous* samples, property variability will not be a concern. In a similar fashion, polymers that are *polydisperse* in molecular weight have averaged property values, while *monodisperse* samples will give accurate data. Obviously, samples that are both chemically *homogeneous* and *monodisperse* will give the most accurate and precise values.

Compared to synthetic polymers, almost all nucleic acids and mammalian proteins are compositionally (chemically) *homogeneous* and *monodisperse*. If not, there would be no life; biopolymers carry highly specific and selective information. Mammalian polysaccharides, for the most part, are also compositionally *homogeneous*, but are *polydisperse* in molecular weight; whereas plant polysaccharides are polydisperse. Chemically modified cellulose (cellulosics) are typically both compositionally heterogeneous and polydisperse in molecular weight. Starches (α-amylose and amylopectin), another major class of polysaccharides, are highly polydisperse in molecular weight, but quite compositionally homogeneous. In addition, amylopectin and many other polysaccharides are highly branched, which may further complicate listed property values.

Synthetic polymers can be quite complex and, as such, tabulated and measured property data must be interpreted with care. *Homogeneous* synthetic polymers are those produced from condensation polymerization reactions, in which all polymer chains are *chemically* indistinguishable from another. Even though these types of polymers show a finite polydispersity of two, accuracy and precision will not be compromised because all samples (and reference standards) will have the same degree of polydispersity. Lastly, synthetic polymers produced by addition polymerization (i.e., ionic, complex coordination catalytic or free-radical copolymerization) will have the greatest amount of compositional heterogeneity and, with the exception of anionically polymerized samples, will also have a large molecular weight polydispersity. For these polymers, tabulated data must be interpreted with caution, unless users establish their own data sets with reference polymers obtained from the same polymerization conditions.

Sequence distribution or polymer microstructure is the next higher level of complexity in which the average *arrangement* of monomers along a chain is considered. The polymerization mechanism and reactivity ratios of monomers dictate this parameter. Monomers can be randomly arranged along chains in the case of statistic or random copolymers or, in the extreme form, as block copolymers. In any event, the microstructure of reference polymers should be defined when properties are listed.

Next in line of complexity is macromolecular *architecture*, or polymer *configuration*, in which the topological nature of the chain is of interest. Thus polymer branching can take on a wide range of configurations including short- and long-chain branching, and comb, star, and dendritic structures with or without comonomer segregation or blockiness. Because of the strong influence of polymer configuration on properties, this parameter needs to be defined and care taken when comparing tabulated data to those of actual samples.

In summary, polymers may have up to two or more distributed characteristics depending on the number of different monomers used in the polymerization, the type of polymerization mechanism, and whether or not the sample was fractionated during isolation. As a rough estimate, polymer "complexity" increases exponentially with the number of distributive properties, making it more difficult to measure accurate polymer properties.

Some polymers are modified *after* polymerization; however, this process can be somewhat difficult to control because polymer chain segments can influence the chemistry of a neighboring groups. Chemical modifications are done mainly on cellulosics and other polysaccharides to tailor-make specific property characteristics. Thus tabulated property data given for cellulosics and polysaccharides represent average values of the entire sample ensemble of polymer chains that differ in composition. To complicate matters further, insoluble gels, comprised of three-dimensional networks, may form if chains are allowed to chemically or physically (via hydrogen bonding) react with one another, either during or after polymerization.

Post-polymerization processes are also accomplished via vulcanization, irradiation, or through the addition of a low molecular weight cross-linking agent. The resulting polymer (i.e., rubber, elastomer, resin, or gel) in essence, is one super or giant molecule approaching infinite molecular weight. These *viscoelastic* materials have wonderful consumer, industrial, and aerospace end-use applications when properly formulated.

The next level of polymer complexity is polymer blends and multicomponent systems. To adjust the glass-transition temperature, plasticizers are added, often times at high concentrations. To increase polymer strength, reinforced polymeric materials are used that consist of added inorganic material, the most common being carbon black or glass fibers. Laminated structures are also produced for increased material strength.

High-value added, specialty products with controlled molecular weight, branching, or architecture are being developed for high-technology industries, most notably electronic and optical devices, printing inks, and coatings in the aerospace industry. Because of their specialized uses, most of these polymeric materials are not listed in this compilation.

C. REGULATORY AGENCIES

Most industries issue testing protocols and polymer property specifications to the trade. To ensure uniformity, national regulatory agencies have formed to deal with standardized methods and testing approaches. In the United States, ASTM is the most prominent independent agency supported by industry with about 100 test methods in place specifically for polymers and polymeric materials. API specializes in the development of procedures for petroleum products, some of which are polymeric. In Britain, BSI is the key agency for testing, while in Europe, DIN procedures are followed. Many of these agencies are overseen by ISO, a federation of national regulatory bodies. (See Table 1 for complete names and acronyms.)

Governmental departments of commerce, defense, and military are also involved in issuing protocols and specifications. For example, the FDA is responsible for establishing acceptable limits of extractable components from polymeric materials in contact with food and drugs.

Table 1. Key agencies involved in standardized testing of polymers and polymeric materials under the umbrella of ISO

Abbreviation	Organization
API	American Petroleum Institute
ASTM	American Society for Testing and Materials
BSI	British Standards Institution
DIN	Deutsches Institut fur Normung
FDA	Food and Drug Agency
ISO	International Organization for Standardization*

*Global federation of national standards bodies representing 100 countries

D. REFERENCE POLYMERS AND SPECIALTY MATERIALS

Sources of reference polymer standards that can be used for instrument calibration and validating methods are listed in Table 2. In the United States NIST is responsible for distributing a number of well-characterized polymer standards.

These standards have well-defined chemical composition and molecular weight, and are also suitable for formulating materials for R&D. All reference standards and polymeric material should come with certificates of analysis. (Because water content in polymers, especially hydrophilic ones and polysaccharides, may affect properties, it is advisable to vacuum dry and properly store them to prevent moisture buildup and degradation.)

Table 2. Sources of polymer standards used for instrument calibration, method development and verification, and formulating R&D samples

American Polymer Standards Corp.	USA
Gearing Scientific Ltd.	UK
National Institute of Standards and Technology	USA
Polymer Laboratories Varian	UK
Polymer Source Inc.	Canada
Polymer Standards Service (PSS)	Germany
Pressure Chemical Co.	USA
Putus Macromolecular	China
Sigma-Aldrich	USA
Tosoh Corporation	Japan
Waters Corporation	USA

E. POLYMER PROPERTIES

In this section we discuss and list polymer properties that are included in data tables of this book. Some properties reviewed in this section are not listed in this text, but they are included for completeness. Specific properties for certain classes of polymers are not given, especially those used for optical, electronic and magnetic devices.

Much of this section and the book's content is based on van Krevelen's (1976) property schemes, with modification. His book should be consulted for more detailed discussions. Other books of interest are listed at the end of this chapter.

Basic information that characterizes polymers is listed in Table 3. These properties can be estimated from the expected outcome of the polymerization, measured, or calculated from group contributions (see van Krevelen,

Table 3. Basic polymer information

Property measured	Remarks
CAS registration number	
Physical state at rt	
Chemical composition of repeat units	
Structural formula of repeat group	
Comonomer molar ratios	For copolymers
Molar substitution	For cellulosics
Molecular weight of repeat unit	
Statistical average molecular weights	M_n, M_v, M_w, M_z, and polydispersity
Percent added inorganic or carbon filler or plasticizer	Polymer additives used to impart selected performance
Polymer additives, e.g., antioxidants, UV stabilizers, etc.	
Moisture level	
Crosslinking density	If applicable
Branching, degree (frequency) and extent (length)	Short- or long-chain branching if applicable, as estimated
Polymer architecture (topology), other than linear or branched, if applicable	graft, star, comb, or dendritic
Crystallinity	
Tacticity	
Microstructure, i.e., monomer sequence distribution	Block, random, or alternate
Toxicity and stability	Should be determined or at least estimated from structure of corresponding comonomers
Environmental impact	Must be known or at least estimated for safe disposal

1976). Methods for measuring these properties can be found in the reference list (for example, see Barth and Mays 1991; Brady 2003; Wu 1995). Some of the more important properties will be considered here.

The most useful average molecular weights are the number- (M_n), weight- (M_w), and z-averages (M_z). These averages are easily determined from the molecular weight distribution obtained using size exclusion chromatography (Mori and Barth 2001). Oftentimes just the viscosity-average molecular weight (M_v) is available, which can be conveniently determined from the measured intrinsic viscosity of the polymer in a given solvent at a specified temperature using tabulated Mark–Houwink coefficients. Alternatively, M_w can be determined from light scattering and M_n from osmometry.

Branching, molecular topology, and comonomer sequence distribution along the chain are more difficult to estimate; these properties are best estimated by the chemistry of the polymerization procedure, with support from NMR measurements. Polymer toxicity and stability must be known or at least estimated from functional group and comonomer chemistry. It should be realized that polymer toxicity, to a first approximation, is lower than the corresponding comonomer toxicity; because of the low polymer diffusion coefficient, macromolecules cannot readily pass through biomembranes, thus have limited bioavailability.

The effect of molecular weight of a polymer in solution on its colligative properties, summarized in Table 4, is a well-established phenomenon. These properties are dependent on the *number* of macromolecules in solution, which independent on molecular weight and chemical composition. In fact, the number-average molecular weight of a polymer can be determined by measuring one of its colligative properties.

Table 4. Colligative polymer properties

Property measured	Remarks
Freezing point depression	MW dependent
Boiling point elevation	MW dependent
Vapor pressure depression	MW dependent
Osmotic pressure elevation	MW dependent

Table 5 lists volumetric properties of polymers in the liquid or solid state as a function of temperature; these properties are related to the compactness of chains and the interaction of comonomers within and among neighboring chains. These properties are more dependent on chemical composition, than molecular weight. Volumetric properties also depend on factors influenced by comonomer sequence distribution, such as tacticity, branching, and polymer crystallinity.

Table 5. Volumetric properties

Property measured	Remarks
Specific volume (reciprocal of specific density)	Depends on polymer state
Molar volume (reciprocal of molar density)	Depends on polymer state
Specific thermal expansivity	Depends on polymer state
Molar thermal expansivity	Depends on polymer state
Specific melt expansivity	Applicable to crystalline polymers
Molar melt expansivity	Applicable to crystalline polymers

Table 6 lists thermodynamic and calorimetric attributes of a polymer, while Table 7 deals with polymer solubility and cohesive energy. Except for molar entropy, all these properties depend mainly of chemical composition, rather than molecular weight. Furthermore, polymer crystallinity, in addition to the chemical nature of a polymer, plays a major role in dictating solubility behavior. In order to effect solubility in the case of crystalline or semicrystalline polymers, the solution must be heated near or above its melting point to break up crystalline regions.

Light scattering and inherent viscosity measurements made at infinite dilution are used to determine polymer size parameters, conformation, second virial coefficient, weight-average molecular weight, and long-chain branching parameters (Table 8). These are fundamental parameters that allow us to probe structural features of polymer molecules. These properties are dependent on molecular mass and shape, rather than polymer composition.

Table 6. Calorimetric and thermodynamic properties including transition temperatures

Property measured	Remarks
Molar entropy	
Molar enthalpy	
Molar heat capacity	
Latent heat of crystallization	
Thermal conductivity	
Melting temperature, T_m	Disappearance of polymeric crystalline phase
Glass-transition temperature, T_g	Onset of extensive macromolecular motion
Secondary transition temperatures	Other than T_m and T_g
Deflection temperature (heat distortion)	Highest continuous temperature material will withstand
Vicat softening point	Temperature at which a needle penetrates material
Brittleness temperature	

Table 7. Cohesive properties and solubilities

Property measured	Remarks
Cohesive energy	
Cohesive energy density	Related to the "internal pressure" of a polymer in solution
Surface and interfacial energy	
Solubility parameter	Equal to the square root of the cohesive energy density
Good Solvency	Good solvent imparts solubility via polymer solvation
Nonsolvency	Poor solvent cannot solvate polymer
Theta temperature	The temperature at which polymer–polymer, polymer–solvent, and solvent–solvent interactions are equal
Theta solvent	A solvent in which polymer–polymer, polymer–solvent, and solvent–solvent interactions are equal

Table 8. Dilute solution properties

Property measured	Remarks
Intrinsic viscosity	Measured quantity related to the hydrodynamic shape and molecular volume of a polymer in solution
Mark–Houwink coefficients	Coefficients related to the shape of macromolecules in solution.
Molecular conformation	Molecular shape parameter
Specific refractive index	Parameter needed for calculating M_w from light scattering data
Polymer-solvent 2nd virial coefficient	Determined from light scattering measurements
Radius of gyration	Macromolecular size parameter
End-to-end distance	Macromolecular size parameter
Hydrodynamic volume	Macromolecular volume parameter

Melt index and viscosity are critical parameters needed for polymer processing. These and other polymer transport properties are listed in Table 9. As in the case of other viscosity measurements, these properties depend mainly on higher statistical molecular weight averages, such as M_w and M_z.

Table 9. Transport properties

Property measured	Remarks
Melt viscosity	Depends on molecular weight and chain entanglement
Melt index	Inversely proportional to viscosity
Gas permeability across a polymer film or membrane	Usually water vapor, oxygen, nitrogen, or carbon dioxide, or specialty gases
Diffusion coefficient	Diffusion of polymer in a given solvent at defined conditions
Water absorption	Water content taken up at specified relative humidity and temperature

Tables 10 to 13 list polymer characteristics directly involved with end-use properties: mechanical properties (Table 10), electric and magnetic properties (Table 11), optical properties (Table 12), and polymer stability (Table 13). (A more complete discussion of these properties is given in selected references at the end of this chapter.)

Table 10. Mechanical properties

Property measured	Remarks
Adhesion (tackiness)	
Ball indentation hardness	
Bulk modulus (reciprocal of compressibility)	
Coefficient of friction	
Compression strength	Force needed to rupture material
Tensile creep	Shape change of material caused by suspended weight
Damping	Absorption or dissipation of vibrations
Dynamic mechanical behavior	
Elastic modulus	
Elongation	
Fatigue	Number of cycles required for fracture
Flexural stiffness	
Flexural strength at break	Amount of stress needed to break material
Fracture mechanical properties	Fracture energy, fatigue resistance, fatigue crack growth, void coalescence
Friction abrasion and resistance	
Hardness	Resistance to compression, indentation, and scratch
Impact strength	Energy absorbed by sample prior to fracture
Indention hardness	
Load deformation	
Mar resistance	
Mold shrinkage	
Poisson's ratio	
Scratch resistance	
Shear strength	Maximum load to produce a fracture by shearing
Surface abrasion resistance	
Tear resistance	
Tensile strength break (yield)	See Young's modulus
Toughness	Amount of energy to break a material (area under stress-strain curve)
Ultimate strength	
Viscoelastic behavior	
Young's modulus	
(Tensile strength)	Modulus of elasticity or tensile modulus

Table 11. Electrical and magnetic properties

Property measured	Remarks
Arc resistance	Time needed for current to make material surface conductive because of carbonization
Dielectric constant	Ability of material to store electric energy for capacitor application
Dielectric permittivity	
Dielectric strength	Voltage required to break down or arc material
Dissipation power factor (loss tangent)	Watts (power) lost in material used as insulator
Insulation resistance	
Magnetic susceptibility	
Resistivity	
Volume resistivity	

Table 12. Optical properties

Property measured	Remarks
Color	Physiological response; measured using three parameters: lightness, chroma, and delta
Luminous transmittance	Measure of plastic haze or clarity
Molar refraction	
Percent transmission	Transparency
Refractive index	
Specular gloss	Surface "flatness"; mirror "finish"
Total internal reflectance	
UV-visible absorbance spectrum	

Table 13. Polymer stability

Property measured	Remarks
Accelerated aging studies	
Biological stability	Stability in the presence of microorganisms
Burning rate	
Chemical resistance	Hydrolytic stability (extreme pH conditions), exposure to chemicals and solvents
Flammability	Flame resistance
Flash ignition temperature	
Long-term immersion	
Permeability	Amount of gas or liquid penetrating film
Recyclability	
Resistance to cold	
Self-extinguishing temperature	
Stress cracking	Caused by weathering
Thermal stability	In addition to temperature, decomposition products are measured
UV resistance	Color fading, pitting, crumbling, surface cracking, crazing, brittleness
Weathering (environmental stress)	Color and gloss change, cracks, crazing, weakening (see UV resistance)

CONCLUSIONS

Polymer science can be viewed as an applied branch of chemistry based on deliverable properties. It is of interest to note that most of these properties depend on just four attributes: (1) polymer molecular weight, (2) crystallinity, (3) chemical composition, and (4) macromolecular topology or architecture; furthermore, these parameters interact with one another in a complex manner. By varying these parameters, polymers can be tailor-made to fit a list of desirable characteristics.

It is hoped that this polymer property database will serve as a guideline to help pave the way for the development of newer materials of improved characteristics.

SELECTED REFERENCES

Billmeyer, F. W., *Textbook of Polymer Science*, 3rd ed., Interscience Publishers, 1984 (classic book with excellent treatment of polymer properties).

Barth, H. G. and Mays, J. W., Eds., *Modern Methods of Polymer Characterization*, New York: Wiley, 1991 (covers latest developments at the time of most methods).

Brady, Jr., R. F., Ed., *Comprehensive Desk Reference of Polymer Characterization and Analysis*, American Chemical Society-Oxford, 2003 (survey of characterization and analytical methods).

Brandrup, J., Immergut, E. H., Grulke, E. A., Abe, A, and Bloch, D. R., Eds., *Polymer Handbook*, 4th ed., New York: John Wiley & Sons, 2005 (premier handbook of polymer science, listing virtually all polymer characteristics for most polymers).

Brydson, J. A., *Plastics Materials*, Butterworth Heinemann, 2000 (comprehensive treatment of plastics, their synthesis, properties, and applications).

Bueche, F., *Physical Properties of Polymers*, Krieger Publishing, 1979 (emphasis is on polymer physics).

Cowie, J.M.G. and Arrighi, V., *Polymers: Chemistry and Physics of Modern Materials*, 3rd ed., CRC Press, 2008 (excellent discussion of physical properties and applications).

Heimenz, P.C. and Lodge, T. P., *Polymer Chemistry*, 2nd ed., CRC Press, 2007 (comprehensive treatment of polymer chemistry—synthesis and physical chemistry).

Mark, J.E., Allcock, H. R., and West, R., *Inorganic Polymers*, Oxford, 2005 (physical chemistry and properties of inorganic polymers).

Mark, J. E., Ed., *Polymer Data Handbook*, Oxford, 1999 (compilation of major classes of polymers and their physical properties).

Mori, S. and Barth, H. G., *Size Exclusion Chromatography*, Springer-Verlag, 1999 (comprehensive treatment of SEC, theory and applications).

Munk, P. and Aminabhavi, T. M., *Introduction to Macromolecular Science*, 2nd ed., New York: John Wiley & Sons, 2002 (emphasis on polymer physical chemistry).

Nielsen, L. E., *Polymer Rheology*, Marcel Dekker, 1977 (introductory text on polymer rheology).

Richardson, T. L. and Lokensgard, E., *Industrial Plastics: Theory and Applications*, Delmar, 1996 (practical overview of some important properties and polymer processing).

Carraher, Jr., C. E., *Seymour/Carraher's Polymer Chemistry*, 7th ed., CRC Press, 2007 (popular introduction to polymer chemistry).

Seymour, R. B., *Engineering Polymer Sourcebook*, McGraw-Hill, 1990 (good overview of physical properties of engineering polymers).

Sperling L. H., *Introduction to Physical Polymer Science*, 2nd ed., Wiley-Interscience, 1992 (good treatment of polymer physics and properties).

van Krevelen, D. W., *Properties of Polymers*, 3rd ed., Elsevier, 1990 (in-depth treatment of polymer properties, best resource available).

Whistler, R., *Industrial Gums*, 2nd ed., Academic Press, 1973 (although outdated, gives solid background on the chemistry and properties of cellulosics and polysaccharides).

Wu, C. S., Ed., *Handbook of Size Exclusion Chromatography*, 2nd ed., Marcel Dekker, 2003 (covers all aspects of this important technique).

ABS, Blow moulding

Monomers: Acrylonitrile, 1,3-Butadiene, Styrene
Material class: Thermoplastic, Copolymers
Polymer Type: ABS
CAS Number:

CAS Reg. No.
9003-56-9

Applications/Commercial Products:

Trade name	Supplier
Cycolac MSB	General Electric Plastics
Cycolac MSC	General Electric Plastics
Cycolac MSD	General Electric Plastics
Cycolac MSL	General Electric Plastics

ABS, Carbon reinforced

Synonyms: *Acrylonitrile-butadiene-styrene copolymer. Acrylonitrile-butadiene-styrene terpolymer. Poly(2-propenenitrile-co-1,3-butadiene-co-ethenylbenzene)*
Related Polymers: ABS
Monomers: Acrylonitrile, 1,3-Butadiene, Styrene
Material class: Composites, Copolymers, Thermoplastic
Polymer Type: ABS
CAS Number:

CAS Reg. No.
9003-56-9

Molecular Formula: $[(C_8H_8)(C_4H_6)(C_3H_3N)]_n$
Fragments: C_8H_8 C_4H_6 C_3H_3N
Additives: 15% Carbon fibre
General Information: An amorph. thermoplastic that combines high performance with design versatility and easy processability

Volumetric & Calorimetric Properties:

Density:

No.	Value	Note
1	d 1.12 g/cm^3	ASTM D792

Thermal Expansion Coefficient:

No.	Value	Note	Type
1	3.6×10^{-5} K^{-1}	ASTM D696 [3]	L

Melting Temperature:

No.	Value	Note
1	210–260°C	[4]

Glass-Transition Temperature:

No.	Value	Note
1	100–110°C	[4]

Deflection Temperature:

No.	Value	Note
1	80°C	1.8 MPa, ASTM D648 [3]
2	90°C	0.45 MPa, ASTM D648 [3]

Surface Properties & Solubility:

Solvents/Non-solvents: Sol. polar solvents, esters, ketones and some chlorinated hydrocarbons [5]

Transport Properties:

Transport Properties General: Exhibits viscoelastic behaviour in both the melt and solid states. Like other thermoplastics, it may be described as exhibiting non-Newtonian behaviour; viscosity decreases due to increasing shear rate [2]

Water Absorption:

No.	Value	Note
1	0.3 %	24h, ASTM D570 [3]
2	0.75 %	saturation, ASTM D570 [3]

Mechanical Properties:

Mechanical Properties General: Has high impact resistance over a wide temp. range, with very good rigidity and toughness. Its tensile and flexural strength, tensile and flexural moduli are all higher than those of unreinforced ABS. Has very low notched Izod impact strength compared to that of unreinforced ABS grades. Elongation at break 1–2% (ASTM D638). [3]

Tensile (Young's) Modulus:

No.	Value	Note
1	12409–13788 MPa	1800–2000 kpsi, ASTM D638 [4]

Flexural Modulus:

No.	Value	Note
1	6000 MPa	ASTM D790 [3]

Tensile Strength Yield:

No.	Value	Note
1	80 MPa	ASTM D638 [3]

Flexural Strength Yield:

No.	Value	Note
1	100 MPa	ASTM D790 [3]

Compressive Strength:

No.	Value	Note
1	75–117 MPa	ASTM D695 [3]

Impact Strength: Izod 60 J m^{-1} (notched, ASTM D256 [3], 280 J m^{-1} (unnotched, ASTM D256) [3]
Hardness: Rockwell M90 (ASTM D785) [3]
Fracture Mechanical Properties: Shear strength 45 MPa (ASTM D732) [3]. Coefficient of friction 0.2 (static) (LNP SOP) 0.15, (dynamic) (LNP SOP) [3]
Friction Abrasion and Resistance: Coefficient of friction 0.2 (static, (LNP SOP), 0.15 (dynamic, LNP SOP) [3]

Izod Notch:

No.	Value	Notch	Note
1	60 J/m	Y	ASTM D256
2	280 J/m	N	ASTM D256 [3]

Electrical Properties:
Electrical Properties General: Is electrically conductive. Surface resistivity 10^3–10^6 Ω cm (ASTM D257) [3]; Volume resistivity 10^4–10^6 Ω cm (ASTM D257) [3]

Polymer Stability:
Thermal Stability General: Has a relatively flat stress-temp. response due to its amorph. character. Retains its props. at low temps. [2,5]

Upper Use Temperature:

No.	Value	Note
1	60°C	[3]
2	60°C	UL 746B [3]

Flammability: Flammability rating HB (3.2 mm, UL94) [3]
Environmental Stress: Has poor weather and uv resistance unless protected. Undergoes stress cracking in contact with certain chemicals. The rubber phase in ABS oxidises more rapidly than the rigid component. Oxidation of polybutadiene results in embrittlement of the rubber [2,5]
Chemical Stability: Resistant to hydrocarbon solvents, mineral and vegetable oils, waxes and related household and commercial materials. Has high resistance to staining agents in household applications. Attacked by conc. nitric and sulfuric acids. Unaffected by weak acids at room temp. and alkalis below 65°. Has poor resistance to most organic solvents [1,2,5]
Recyclability: Scrap and reject material can be reworked at an 80:20 (virgin:regrind) ratio if kept clean and segregated by colour and grade [4]
Stability Miscellaneous: Processing conditions can influence resultant props. by chemical and physical means. Degradation of the rubber and matrix phases occurs under very severe conditions. Evidence of morphological changes is shown by agglomeration of dispersed rubber particles during injection-moulding at higher temps. Physical effects such as orientation and moulded-in stress can affect mechanical props. [10]

Applications/Commercial Products:
Processing & Manufacturing Routes: Prod. by grafting styrene and acrylonitrile directly on to the polybutadiene latex in a batch or continuous emulsion polymerisation process. The degree of grafting is a function of the 1,2-vinyl content of the polybutadiene, the monomer concentration, the extent of conversion, temp. and mercaptan concentration. The emulsion polymerisation process involves two steps; (i) production of a rubber latex; (ii) subsequent polymerisation of styrene and acrylonitrile in the presence of the rubber latex to produce an ABS latex. This latex is then processed to isolate the ABS resin. The polymerisation ingredients include the monomers, an emulsifier, a polymerisation initiator and usually a chain transfer agent for MW control. [1,6,7,8,9] Processed on standard reciprocating-screw injection moulding machines at 210–216°. [4] Should be dried before melt processing.
Applications: Uses where high strength and stiffness are required

Trade name	Supplier
AS-15CF/000	Compounding Technology
RTP 682 FR	RTP Company
Stat-Kon AC-1003	LNP Engineering
Styvex	Ferro Corporation
Thermocomp (AC-1003)	LNP Engineering

Bibliographic References
[1] *Kirk-Othmer Encycl. Chem. Technol.*, Vol. 1, 4th edn., (ed. J.I. Kroschwitz), Wiley Interscience, 1991, 391
[2] *Encycl. Polym. Sci. Eng.*, Vol. 1, 2nd edn., (ed. J.I. Kroschwitz), John Wiley & Sons, 1985, 388
[3] LNP Engineering Plastics, *Product Data Book*, LNP Engineering Plastics, (technical datasheet)
[4] Guide to Plastics, *Property and Specification Charts*, (ed. W.A. Kaplan), McGraw-Hill, 1994, **70**, 12
[5] Harper, C.A., *Handb. Plast. Elastomers*, (ed. C.A. Harper), McGraw-Hill, 1975
[6] Hayes, R.A. and Futamara, S., *J. Polym. Sci., Polym. Chem. Ed.*, 1981, **19**, 985
[7] Brydon, A., Burnett, G.M. and Cameron, C.G., *J. Polym. Sci., Polym. Chem. Ed.*, 1974, **12**, 1011
[8] Allen, P.W., Ayrey, G. and Moore, C.G., *J. Polym. Sci.*, 1959, **36**, 55
[9] Odian, G., *Principles of Polymerization*, McGraw-Hill, 1970
[10] Casale, A. and Salvatore, O., *Polym. Eng. Sci.*, 1975, **15**, 286

ABS, Extrusion, Unfilled

Synonyms: *Acrylonitrile-butadiene-styrene copolymer. Acrylonitrile-butadiene-styrene terpolymer. Poly(2-propenenitrile-co-1,3-butadiene-co-ethenylbenzene)*
Related Polymers: ABS
Monomers: Acrylonitrile, 1,3-Butadiene, Styrene
Material class: Thermoplastic, Copolymers
Polymer Type: ABS
CAS Number:

CAS Reg. No.
9003-56-9

Molecular Formula: $[(C_8H_8)(C_4H_6)(C_3H_3N)]_n$
Fragments: C_8H_8 C_4H_6 C_3H_3N
General Information: An amorph. thermoplastic grade which has good heat resistance, very good processability, chemical resistance and toughness. It is highly suitable for thermoforming

Volumetric & Calorimetric Properties:
Density:

No.	Value	Note
1	d 1.06 g/cm^3	ISO 1183 [3]

Thermal Expansion Coefficient:

No.	Value	Note	Type
1	9×10^{-5}–0.0001 K^{-1}	23–55°, ASTM E831 [3]	L

Thermal Conductivity:

No.	Value	Note
1	0.17 W/mK	DIN 52612 [4]

Melting Temperature:

No.	Value	Note
1	230–250°C	[3,4]

Glass-Transition Temperature:

No.	Value	Note
1	88–120°C	[8]

ABS, Extrusion, Unfilled

Deflection Temperature:

No.	Value	Note
1	94–97°C	80°, 1.8 MPa, 4h, ISO 75
2	98–102°C	80°, 0.45 MPa, 4h, ISO 75

Vicat Softening Point:

No.	Value	Note
1	92–101°C	50 K h^{-1}, 50 N, ISO 306 [3]

Surface Properties & Solubility:
Solvents/Non-solvents: Sol. polar solvents, esters, ketones, some chlorinated hydrocarbons [5]

Transport Properties:
Transport Properties General: ABS exhibits viscoelastic behaviour in both the melt state and the solid state. Like other thermoplastics, it may be described as exhibiting non-Newtonian behaviour; viscosity decreases due to increasing shear rate [2].

Melt Flow Index:

No.	Value	Note
1	1–8 g/10 min	220°, 10 kg, ISO 1133 [3,4]

Polymer Melts: Storage shear modulus and loss shear modulus as a function of angular frequency, and shear stress as a function of shear rate depend strongly on grafting degree in the long-time region associated with particle-particle interactions. As the grafting degree increases, the viscoelastic functions first decrease and then increase. The minima in the functions occurs at the grafting degree of about 0.45 for ABS having a rubber particle size of 170 nm [6]

Water Absorption:

No.	Value	Note
1	0.4 %	23°, 24h, DIN 53495 [4]

Mechanical Properties:
Mechanical Properties General: Has high impact resistance over a wide temp. range down to -40°, with very good rigidity and toughness and a high yield strength. Elongation at break 15–25% (5 mm min^{-1}, ISO 527) [3]

Tensile (Young's) Modulus:

No.	Value	Note
1	1600–2700 MPa	1 mm min^{-1}, ISO 527 [3,4]

Flexural Modulus:

No.	Value	Note
1	1800–2400 MPa	2 mm min^{-1}, ISO 178 [3]

Compressive Modulus:

No.	Value	Note
1	1030–2690 MPa	ASTM D695 [5]

Elastic Modulus:

No.	Value	Note
1	850–900 MPa	ISO 537 [4]

Tensile Strength Yield:

No.	Value	Note
1	35–45 MPa	50 mm min^{-1}, ISO 527 [3]

Flexural Strength Yield:

No.	Value	Note
1	52–85 MPa	5 mm min^{-1}, ISO 178 [3,4]

Compressive Strength:

No.	Value	Note
1	36–70 MPa	ASTM D695 [5]

Miscellaneous Moduli:

No.	Value	Note	Type
1	1250–1400 MPa	≤0.5% elongation, 1000h, ISO/IEC 899 [4]	tensile creep modulus

Viscoelastic Behaviour: Stress-strain behaviour, creep, stress and relaxation and fatigue have been reported [7]
Hardness: Ball indentation (H358/30) 75–110 MPa (ISO 2039) [4]. Rockwell R75–R115 (ASTM D785) [5]
Friction Abrasion and Resistance: Has good abrasion resistance [3]

Izod Area:

No.	Value	Notch	Note
1	19–42 kJ/m^2	Notched	23°, ISO 180/1A [3,4]
2	10–25 kJ/m^2	Notched	-30°, ISO 180/1A [3,4]

Electrical Properties:
Electrical Properties General: Electrical props. are fairly constant over a wide range of frequencies and are unaffected by temp. or humidity [2]

Surface/Volume Resistance:

No.	Value	Note	Type
1	0.01–1 × 10^{15} Ω.cm	IEC 93 [3,4]	S

Dielectric Permittivity (constant):

No.	Value	Frequency	Note
1	3–3.4	100 Hz	
2	2.9–3	1 MHz	IEC 250 [3,4]

Dielectric Strength:

No.	Value	Note
1	34–85 kV.mm^{-1}	IEC 243-1 [3,4]

Arc Resistance:

No.	Value	Note
1	50–90s	[2]

Dissipation (Power) Factor:

No.	Value	Frequency	Note
1	0.009	100 Hz	
2	0.007–0.01	1 MHz	IEC 250 [3,4]

Optical Properties:
Transmission and Spectra: Transmittance 33.3% (ASTM D1003) [5]
Volume Properties/Surface Properties: Haze 100% [5]. Gloss 75–86% (measuring angle 20°) [3]. An extensive range of colours is available [3]. Inherent colour opaque [3]

Polymer Stability:
Polymer Stability General: Has very good dimensional stability and chemical resistance. Undergoes autoxidation resulting in embrittlement of the rubber, and is liable to photooxidative degradation [2,5]
Thermal Stability General: The polymer has a relatively flat stress-temp. response due to its amorph. character. The change in tensile, compressive and flexural props. is small below 80%. Retains its props. at low temps. [2,5]
Upper Use Temperature:

No.	Value	Note
1	60–90°C	[5]

Flammability: Flammability rating HB (1.6 mm thick, UL 94) [4]
Environmental Stress: Has poor weather and uv resistance unless protected. The rubber phase in ABS oxidises more rapidly than the rigid component. Oxidation of polybutadiene results in embrittlement of the rubber. [2] May yellow slightly in sunlight [2,5]
Chemical Stability: Is resistant to hydrocarbon solvents, mineral and vegetable oils, waxes and related household and commercial materials. Is highly resistant to staining agents in household applications. It is attacked by conc. nitric and sulfuric acids but is unaffected by weak acids at room temp., or alkalis (up to 65°). Attacked by organic solvents (C_6H_6, toluene, xylene, ethers, esters, chlorinated aromatic and aliphatic hydrocarbons, amines, ketones and hot alcohols. Certain chemicals cause stress cracking [1,2,5]
Hydrolytic Stability: Stable to neutral salt solns. [1,5]
Biological Stability: Is biologically inert
Recyclability: Scrap and reject material can be reworked at an 80:20 (virgin:regrind) ratio if kept clean and segregated by colour and grade [8]

Applications/Commercial Products:
Processing & Manufacturing Routes: Prod. by grafting styrene and acrylonitrile directly onto the polybutadiene latex in a batch or continuous emulsion polymerisation process. The degree of grafting is a function of the 1,2-vinyl content of the polybutadiene, the monomer concentration, the extent of conversion, temp. and mercaptan concentration. The emulsion polymerisation process involves two steps; (i) production of a rubber latex; (ii) subsequent polymerisation of styrene and acrylonitrile in the presence of the rubber latex to produce an ABS latex. This latex is then processed to isol. the ABS resin. The polymerisation ingredients include the monomers, an emulsifier, a polymerisation initiator and, usually, a chain transfer agent for MW control [1,9,10,11,12]. Processed on conventional single or twin-screw extruders into pipe, sheet, blow moulded shapes or profiles. Extruded sheet can be thermoformed
Applications: Uses in automotive exterior applications including extruded/thermoformed fascias for large trucks; in refrigerator doors, luggage and as tank liners. Other uses in large structural components in transportation and pipes. Industrial uses include protective covers

Trade name	Details	Supplier
ABS		John Gibson (Plastics) Ltd.
Cycolac		General Electric Plastics
Lucky ABS	various grades	Standard Polymers
Lustran		Bayer Inc.
Magnum		Dow
Novodur	various grades	Bayer Inc.
Sinkral		Enichem America
Terluran		BASF

Bibliographic References
[1] *Kirk-Othmer Encycl. Chem. Technol.*, Vol. 1, (eds. J.I. Kroschwitz and M. Howe-Grant), Wiley Interscience, 1993, 391
[2] *Encycl. Polym. Sci. Eng.*, Vol. 1, 2nd edn., (ed. J.I. Kroschwitz), John Wiley & Sons, 1985, 388
[3] Lustran ABS, Novadur ABS, *Application Technology Information*, Bayer, 1997, (technical datasheet)
[4] Terluran, BASF Plastics, (technical datasheet)
[5] Harper, C.A., *Handb. Plast. Elastomers*, (ed. C.A. Harper), McGraw-Hill, 1975
[6] Aoki, Y., *Macromolecules*, 1987, **20**, 2208, (dynamic viscoelastic and steady flow props)
[7] Pillichody, C.T. and Kelley, P.D., *Handbook of Plastic Materials and Technology*, (ed. I.I. Rubin), Wiley Interscience, 1990, (stress-strain behaviour, creep, stress relaxation, fatigue)
[8] Guide to Plastics, *Property and Specification Charts*, (ed. W.A. Kaplan), McGraw-Hill, 1994, **70**
[9] Hayes, R.A., Futamura, S., *J. Polym. Sci., Polym. Chem. Ed.*, 1981, **19**, 985
[10] Brydon, A., Burnett, G.M. and Cameron, C.G., *J. Polym. Sci., Polym. Chem. Ed.*, 1974, **12**, 1011
[11] Allen, P.W., Ayrey, G. and Moore, C.G., *J. Polym. Sci.*, 1959, **36**, 55
[12] Odian, G., *Principles of Polymerization*, McGraw-Hill, 1970

ABS Flame retardant

Synonyms: Acrylonitrile-butadiene-styrene copolymer. Acrylonitrile-butadiene-styrene terpolymer. Poly(2-propenenitrile-co-1,3-butadiene-co-ethenylbenzene)
Related Polymers: ABS
Monomers: Acrylonitrile, 1,3-Butadiene, Styrene
Material class: Thermoplastic, Copolymers
Polymer Type: ABS
CAS Number:

CAS Reg. No.
9003-56-9

Molecular Formula: $[(C_8H_8)(C_4H_6)(C_3H_3N)]_n$
Fragments: C_6H_8 C_4H_6C C_3H_3N
Additives: Flame retardants such as halogenated additives. Additive can be brominated, chlorinated or both
General Information: Thermoplastic, amorph. polymer that combines high performance with design versatility and easy processability. Has better fire resistance than standard grades of ABS. Good resistance to abrasion, chemicals and stress ensures a long lasting attractive surface finish. Has excellent flow props. [3,4]

Volumetric & Calorimetric Properties:
Density:

No.	Value	Note
1	d 1.2 g/cm^3	ISO 1183 [6]

ABS Flame retardant

Thermal Expansion Coefficient:

No.	Value	Note	Type
1	9×10^{-5}–0.0001 K^{-1}	DIN 53732, 23–55°, ASTM E831 [4, 6]	L

Thermal Conductivity:

No.	Value	Note
1	0.19 W/mK	DIN 52612 [4]

Melting Temperature:

No.	Value	Note
1	200–230°C	[4, 5]

Glass-Transition Temperature:

No.	Value	Note
1	110–125°C	[3]

Deflection Temperature:

No.	Value	Note
1	71–92°C	1.8 MPa, 80°, 4h, ISO 75 [4, 5, 6]
2	90°C	0.45 MPa, 80°, 4h, ISO 75 [4, 5, 6]

Vicat Softening Point:

No.	Value	Note
1	84–96°C	50N, 50° h^{-1}, ISO 306B [4, 5]

Surface Properties & Solubility:
Solvents/Non-solvents: Sol. polar solvents, esters, ketones, chlorinated hydrocarbons [8]

Transport Properties:
Transport Properties General: ABS exhibits viscoelastic behaviour in the melt solid state. Exhibits non-Newtonian behaviour, by the decrease in viscosity resulting from an increase in shear rate [2]

Melt Flow Index:

No.	Value	Note
1	15–48 g/10 min	220°, 10 kg, ISO 1133 [4, 6]

Polymer Melts: Viscoelastic functions, in terms of storage shear modulus and loss shear modulus as a function of angular frequency and shear stress as a function of shear rate, depend strongly on grafting degree in the long-time region accociated with particle-particle interactions. As the grafting degree increases, the viscoelastic functions first decrease and then increase. The minima in the functions occurs at the grafting degree of about 0.45 for ABS having rubber particle size of 170 nm [9]

Water Absorption:

No.	Value	Note
1	0.2–0.4 %	24h, 23°, DIN 53495 [4, 7]

Mechanical Properties:
Mechanical Properties General: High impact resistance over a wide temp. range down to -40°

Tensile (Young's) Modulus:

No.	Value	Note
1	2050–2600 MPa	1 mm $(min)^{-1}$, ISO 527 [4, 5]

Flexural Modulus:

No.	Value	Note
1	1950–2400 MPa	2 mm $(min)^{-1}$, ISO 178 [5, 6]

Tensile Strength Break:

No.	Value	Note
1	6–9 MPa	ISO 527 [4,7]

Compressive Modulus:

No.	Value	Note
1	896–1448 MPa	ASTM D695 [8]

Tensile Strength Yield:

No.	Value	Note
1	37–46 MPa	50 mm $(min)^{-1}$, ISO 527 [5,6]

Flexural Strength Yield:

No.	Value	Note
1	55–68 MPa	5 mm $(min)^{-1}$, ISO 178 [5,6]

Compressive Strength:

No.	Value	Note
1	45–52 MPa	ASTM D695 [8]

Viscoelastic Behaviour: Stress-strain behaviour, creep, stress relaxation and fatigue in ABS materials have been reported [10]
Hardness: Ball indentation hardness 110 N mm^{-2} (ISO 2039-1) [6]. Rockwell R90-R107 (ASTM D785) [8]
Friction Abrasion and Resistance: Has good abrasion resistance [4]
Izod Area:

No.	Value	Notch	Note
1	7–15 kJ/m^2	Notched	23°, ISO 180-1A [4]
2	5–8 kJ/m^2	Notched	-30°, ISO 180-1A [4,6]

Electrical Properties:
Electrical Properties General: Electrical props. are fairly constant over a wide range of frequencies and are unaffected by temp. or humidity [2]
Surface/Volume Resistance:

No.	Value	Note	Type
1	$>1 \times 10^{15}$ Ω.cm	min., IEC 93 [4]	S

Dielectric Permittivity (constant):

No.	Value	Frequency	Note
1	2.8–3.2	50 Hz	IEC 250 [4, 5]
2	2.7–3.1	1 MHz	IEC 250 [4, 5]

ABS Flame retardant

Dielectric Strength:

No.	Value	Note
1	24–25 kV.mm^{-1}	IEC 243-1 [4, 6]

Arc Resistance:

No.	Value	Note
1	50–90s	[2]

Dissipation (Power) Factor:

No.	Value	Frequency	Note
1	0.005	50 Hz	IEC 250 [4, 5]
2	0.004–0.011	1 MHz	IEC 250 [4, 5]

Optical Properties:
Volume Properties/Surface Properties: Supplied in either standard or custom colours, or as natural pellet for colouring [4]. An extensive range of colours is available [6]

Polymer Stability:
Polymer Stability General: Offers very good dimensional stability and chemical resistance. Undergoes autooxidation resulting in embrittlement of the rubber. Liable to photooxidative degradation [2, 8]
Thermal Stability General: Has a relatively flat stress-temp. response due to its amorph. character. The change in tensile, compressive and flexural props. is small up to 80°. Props. retained at low temps. [2, 8]
Upper Use Temperature:

No.	Value	Note
1	70°C	[7]
2	50–80°C	[7, 8]

Decomposition Details: Thermal decomposition of the fire retardant additives can occur at temps. in excess of 240° or during long residence times. In addition to the constituent monomers, antimony compounds, brominated hydrocarbons and hydrogen bromide might also be released [4]
Flammability: Flammability rating V0 (1.57–3.2 mm thick, UL 94) [4, 6]; 5VA (2.5–3.2 mm thick, UL 94) [4, 6]. Glow wire test, wire applied during 30 seconds, extinguishing time ≤ 5 s (1.5–3.2 mm, 960°, IEC 695-2-1/1) [4]. Limiting oxygen index 27–30% (ASTM D2863, ISO4589 IV) [4, 6]
Environmental Stress: Resistance to weathering and Uv is poor unless protected. Undergoes stress cracking when in contact with certain chemical agents under stress. ABS polymeric components differ in oxidative stability. The rubber phase in ABS oxidises more rapidly than the rigid component. Oxidation of polybutadiene results in the embrittlement of the rubber [2]. May discolour in sunlight [2, 8]
Chemical Stability: Undergoes stress cracking when in contact with certain chemical agents. Resistant to mineral and vegetable oils, waxes and related household and commercial materials. High resistance to staining agents in household applications. Attacked by conc. nitric acid, sulfuric acid. Unaffected by weak acids, weak and strong bases up to 65°. Poor resistance to most organic solvents [2,8]
Biological Stability: Is biologically inert [4]
Recyclability: Can be recycled. The scrap should be clean, uncontaminated, and free from degradation. Normally the content of the reground material should not exceed 20–25 % [4]
Stability Miscellaneous: Sufficiently stable during processing provided that processing temp. is kept as low as conveniently possible and residence times are kept to an absolute minimum [4]

Applications/Commercial Products:
Processing & Manufacturing Routes: Produced by grafting styrene and acrylonitrile directly upon the polybutadiene latex in a batch or continuous emulsion polymerisation process. The degree of grafting is a function of the 1,2-vinyl content of the polybutadiene, monomer concentration, extent of conversion, temp. and mercaptan concentration. The emulsion polymerisation process involves two steps; production of a rubber latex and subsequent polymerisation of styrene and acrylonitrile in the presence of the rubber latex to produce an ABS latex. This latex is then processed to isolate the ABS resin. The polymerisation ingredients include the monomers, an emulsifier, a polymerisation initiator and, usually, a chain transfer agent for MW control. [1,11,12,13,14]. Processed on standard reciprocating-screw injection moulding machines. Processing characteristics are short cycle time and low energy consumption giving an excellent price/performance ratio. Processing temps. must be minimised when moulding with flame-retardant grades due to the nature of the self-extinguishing mechanism. This is based on the thermal decomposition of additives. As they decompose, these additives release large volumes of flame-suppressant gases. A high processing temp., which may trigger the gas release, must be avoided. Melt temp. must never exceed 250°. Drying at 80–85° (2–4h) is required prior to use in moulding machines [4, 5]. Recommended melt temp. 200–230° [4, 6]. Recommended mould temp. 40–70° [4, 6]. Injection velocity 240 mm s^{-1} (ISO 294) [6]
Mould Shrinkage (%):

No.	Value	Note
1	0.4–0.7%	ASTM D955 [5]

Applications: Transformer housings, switches, personal computer housings, television cabinets, oven doors, electric tool housings, institutional furniture and shelving, air conditioner housings, carpet cleaner housings, smoke detectors [6, 7]

Trade name	Details	Supplier
Ashlene		Ashley Polymers
Comalloy ABS		ComAlloy International
Cycolac	CKM1, CKM2	General Electric Plastics
Cycolac	K25, KJL	General Electric Plastics
GPC Delta D-1000		H Simex International
Lastilac		LATI Industria Thermoplastics S.p.A.
Lustran		Bayer Inc.
Lustran ABS	911, 914HM	Monsanto Chemical
Magnum		Dow
Novodur	L3FR, M3FR	Bayer Inc.
Novodur		Bayer Inc.
RESIN RX1-550FR		Resin Exchange
RXI-500		Resin Exchange
Ronfalin		DSM
Sinkral PDFRI-UV		Enichem America
Starflam ABS		Ferro Corporation
Terluran		BASF
Thermofil ABS		Thermofil
Toyolac		Toray
VP II		Diamond Polymers

Bibliographic References

[1] *Kirk-Othmer Encycl. Chem. Technol.*, Vol. 1, 4th edn., (eds. J.I. Kroschwitz and M. Howe-Grant), Wiley Interscience, 1993, 391
[2] *Encycl. Polym. Sci. Eng.*, Vol. 1, 2nd. edn., (ed. J.I. Kroschwitz), John Wiley and Sons, 1985, 388
[3] Guide to Plastics, *Property and Specification Charts*, McGraw-Hill, 1994, **70**, 12
[4] *DSM Polymers*, Edition 01/95, (technical datasheet)
[5] *Cycolac ABS Resins*, GE Plastics, (technical datasheet)
[6] Lustran ABS, Novadur ABS, *Application Technology Information*, Bayer, 1997, (technical datasheets)
[7] The Plastics Compendium, *Key Properties and Sources*, Vol. 1, (ed. R. Dolbey), RAPRA, 1995, 1
[8] *Handb. Plast. Elastomers*, (ed. C.A. Harper), McGraw-Hill, 1975
[9] Aoki, Y., *Macromolecules*, 1987, **20**, 2208, (dynamic viscoelastic props)
[10] Pillichody, C.T. and Kelley, P.D., *Handb. Plast. Elastomers*, (ed. I.I. Rubin), John Wiley and Sons, 1990, (stress-strain behaviour, creep, stress relaxation and fatigue)
[11] Hayes, R.A. and Futamura, S., *J. Polym. Sci., Polym. Chem. Ed.*, 1981, **19**, 985
[12] Brydon, A., Burnett, G.M. and Cameron, C.G., *J. Polym. Sci., Polym. Chem. Ed.*, 1974, **12**, 1011
[13] Allen, P.W., Ayrey, G. and Moore, C.G., *J. Polym. Sci.*, 1959, **36**, 55
[14] Odian, G., *Principles of Polymerization*, McGraw-Hill, 1970

ABS General purpose

Synonyms: *Acrylonitrile-butadiene-styrene copolymer. Acrylonitrile-butadiene-styrene terpolymer. Poly(2-propenenitrile-co-1,3-butadiene-co-ethenylbenzene)*
Related Polymers: ABS
Monomers: Acrylonitrile, 1,3-Butadiene, Styrene
Material class: Thermoplastic, Copolymers
Polymer Type: ABS
CAS Number:

CAS Reg. No.
9003-56-9

Molecular Formula: $[(C_8H_8)(C_4H_6)(C_3H_3N)]_n$
Fragments: C_8H_8 C_4H_6 C_3H_3N
General Information: Thermoplastic amorph. polymer that combines high performance with design versatility and easy processability. Offers a combination of good mechanical props. and heat resistance. Excellent stiffness, flow characteristics and dimensional stability enable the production of consistent mouldings. Good resistance to abrasion, chemicals and stress ensure a long-lasting attractive surface finish

Volumetric & Calorimetric Properties:

Density:

No.	Value	Note
1	d 1.05 g/cm^3	ISO 1183 [5, 6]

Thermal Expansion Coefficient:

No.	Value	Note	Type
1	8.4×10^{-5}–9.6×10^{-5} K^{-1}	DIN 53752 [4, 6]	L

Thermal Conductivity:

No.	Value	Note
1	0.2 W/mK	DIN 52612 [4]

Specific Heat Capacity:

No.	Value	Note	Type
1	1.3 kJ/kg.C	[7]	P

Melting Temperature:

No.	Value	Note
1	220–260°C	[5, 6]

Glass-Transition Temperature:

No.	Value	Note
1	100°C	[7]

Deflection Temperature:

No.	Value	Note
1	89–95°C	1.8 MPa, 80°, 4 h, ISO 75
2	94–96°C	0.45 MPa, 80°, 4 h, ISO 75 [4, 6]
3	95–100°C	[7]

Vicat Softening Point:

No.	Value	Note
1	93–100°C	49 N, 50° h^{-1}, ISO 306/B50 [4]

Surface Properties & Solubility:

Solvents/Non-solvents: Sol. polar solvents, esters, ketones, chlorinated hydrocarbons [8]

Transport Properties:

Transport Properties General: ABS Exhibits viscoelastic behaviour in the melt and solid state. Exhibits non-Newtonian behaviour, by the decrease in viscosity resulting from an increase in shear rate [2]
Melt Flow Index:

No.	Value	Note
1	9–45 g/10 min	220°, 10 kg, ISO 1133 [5, 6]

Polymer Melts: Viscoelastic functions, in terms of storage shear modulus and loss shear modulus as a function of angular frequency and shear stress as a function of shear rate, depend strongly on grafting degree in the long-time region associated with particle-particle interactions. As the grafting degree increases the viscoelastic functions first decrease and then increase. The minima in the functions occurs at the grafting degree of about 0.45 for ABS having rubber particle size of 170 nm [15]
Permeability and Diffusion General: Permeability decreases with increasing acrylonitrile content
Water Absorption:

No.	Value	Note
1	0.2–0.45 %	1.8" thick, 24h, ASTM D570 [8]

Gas Permeability:

No.	Gas	Value
1	O_2	79–102 cm^3 mm/(m^2 day atm)
2	N_2	9.8–13.8 cm^3 mm/(m^2 day atm)
3	CO_2	3.54–4.72 cm^3 mm/(m^2 day atm)

Mechanical Properties:

Mechanical Properties General: High impact resistance over a wide temp. range down to -40°. Excellent toughness and rigidity. Deforms in a ductile manner over a broad temp. range and at high strain rates. [2, 4] Elongation at break 13–26 % (50 mm min^{-1}, ISO 527) [4]

ABS General purpose

Tensile (Young's) Modulus:

No.	Value	Note
1	2000–2700 MPa	1 mm min^{-1}, ISO 527 [4, 6]

Flexural Modulus:

No.	Value	Note
1	2400–2600 MPa	ISO 178 [5, 6]

Tensile Strength Break:

No.	Value	Note
1	13–26 MPa	50 mm min^{-1}, ISO 527 [4]

Compressive Modulus:

No.	Value	Note
1	830–1380 MPa	ASTM D695 [8]

Tensile Strength Yield:

No.	Value	Note
1	44–50 MPa	50 mm min^{-1}, ISO 527 [4, 5]

Flexural Strength Yield:

No.	Value	Note
1	68–81 MPa	ISO 178 [5, 6]

Compressive Strength:

No.	Value	Note
1	35–76 MPa	ASTM D695 [8]

Impact Strength: Impact strength is dependent on grafted rubber content. The ability of the rubber domains to promote craze formation and shear yielding accounts for improved impact props. [1, 2]
Viscoelastic Behaviour: Stress-strain behaviour, creep, stress relaxation and fatigue have been reported [9]
Hardness: Rockwell R110 (ISO 2039/2) [5]. Ball indentation hardness 105–120 N mm^{-2} (ISO 2039-1) [6]
Fracture Mechanical Properties: Shear strength 28 MPa [7]
Friction Abrasion and Resistance: Exhibits good abrasion resistance. Coefficient of friction 0.30-0.35
Izod Area:

No.	Value	Notch	Note
1	10–28 kJ/m^2	Notched	23°, ISO 180/1A, samples dry as moulded/accelerated conditioning according to ISO 1110, 70°, 63% humidity [4,6]
2	4–20 kJ/m^2	Notched	-30°, ISO 180/1A samples dry as moulded/accelerated conditioning according to ISO 1110, 70°, 63% humidity [4,6]

Electrical Properties:
Electrical Properties General: Exhibits electrical props. that are fairly constant over a wide range of frequencies and are unaffected by temp. or humidity [2]

Surface/Volume Resistance:

No.	Value	Note	Type
1	$>1 \times 10^{15}$ Ω.cm	min., IEC 93 [4]	S

Dielectric Permittivity (constant):

No.	Value	Frequency	Note
1	3	50 Hz	IEC 250 [4, 6]
2	2.9–3	1 MHz	IEC 250 [4, 6]

Dielectric Strength:

No.	Value	Note
1	25–33 kV.mm^{-1}	IEC 243-1 [4, 6]

Arc Resistance:

No.	Value	Note
1	50–90s	[2]

Dissipation (Power) Factor:

No.	Value	Frequency	Note
1	0.004	50 Hz	IEC 250 [4, 6]
2	0.007–0.008	1 MHz	IEC 250 [4, 6]

Optical Properties:
Transmission and Spectra: Poor Uv resistance unless protected by the use of stabilizing additives, pigments and proctective coatings and film. Continued exposure to strong Uv causes embrittlement [2]
Volume Properties/Surface Properties: Gloss, untextured, >90 (60°, ASTM D523) [5]. Supplied in either standard or custom colours or as a natural pellet for colouring. Available in most colours [4]. Haze 100 % (ASTM D1003) [8]

Polymer Stability:
Polymer Stability General: Offers very good dimensional stability and chemical resistance. Undergoes autoxidation resulting in embrittlement of the rubber. Liable to photooxidative degradation
Thermal Stability General: Has a relatively flat stress-temp. response due to its amorph. character. The change in tensile, compressive and flexural props. is small up to 80° [2]
Upper Use Temperature:

No.	Value	Note
1	75–85°C	[7]

Flammability: Flammability rating HB (1.6 mm thick, UL94) [4]. Limiting oxygen index 18% (ISO 4589, IV) [4]. Glow wire test, wire applied during 30 seconds extinguishing time 5 s (max., 2 mm thick, 650°, IES 695-2-1/1) [4]
Environmental Stress: Has poor Uv resistance unless protected by the use of stabilising additives, pigments and protective coatings and film. Continued exposure to strong uv radiation causes embrittlement. [2] Poor weather resistance unless protected. ABS polymeric components differ in oxidative stability. The rubber phase in ABS oxidises more rapidly than the rigid component. Oxidation of polybutadiene results in embrittlement of the rubber [2]. May show slight yellow discoloration in sunlight [2, 8]
Chemical Stability: Undergoes stress cracking when in contact with certain chemical agents. Resistant to mineral and vegetable oils, waxes and related household and commercial materials. High resistance to staining agents in household applications. Attacked

by concentrated nitric acid, sulfuric acid; unaffected by weak acids, weak and strong bases. Poor resistance to most organic solvents. Commercial grades may contain Uv absorbers [2, 8]
Biological Stability: Is biologically inert [4]
Recyclability: Can be recycled. The scrap should be clean, uncontaminated and free from degradation. Normally the content of the reground material should not exceed 20 to 25% [9]
Stability Miscellaneous: Processing conditions can influence resultant props. by chemical and physical means. Degradation of the rubber and matrix phases occurs under very severe conditions. Morphological changes can be evident during injection moulding (at higher temps.) as agglomeration of dispersed rubber particles. Physical effects such as orientation and moulded-in stress affect mech. props. [10]

Applications/Commercial Products:

Processing & Manufacturing Routes: Made by grafting styrene and acrylonitrile directly on to the polybutadiene latex in a batch or continuous emulsion polymerisation process. Grafting is achieved by the free-radical copolymerisation of styrene and acrylonitrile monomers in the presence of polybutadiene. The degree of grafting is a function of the 1,2-vinyl content of the polybutadiene, monomer concentration, extent of conversion, temp. and mercaptan concentration. The emulsion polymerisation process involves two steps; production of a rubber latex and subsequent polymerisation of styrene and acrylonitrile in the presence of the rubber latex to produce an ABS latex. This latex is then processed to isolate the ABS resin. The polymerisation ingredients include the monomers, an emulsifier, a polymerisation initiator and, usually, a chain transfer agent for MW control [1,11,12,13,14]. Processed on standard reciprocating-screw injection moulding machines. ABS is slightly hygroscopic and requires drying prior to use in moulding machines. Drying temp./time 80–95° (2–4h) [4,5]. Recommended melt temp. 220–260° [4,5,6]. Recommended mould temp. 40–70° [4,5,6,]. Injection velocity 240 mm s^{-1} (ISO 294) [6] The superior flow props. of ABS mean that large products can be moulded with thinner walls - giving savings in product weight and shorter cycle times. Low energy consumption and short cycle times give an excellent price/performance ratio [4]
Applications: Housings for electrical domestic appliances, telecommunications equipment, audio and video equipment and business machines. Automotive interior and exterior components such as instrument panels, consoles, radiator grilles and headlight housings. Electrical powered tools and garden equipment. Other uses in furniture, toys, leisure articles

Trade name	Details	Supplier
ABS		Diamond Polymers
ABS 101		John Gibson (Plastics) Ltd.
Abson A.B.S.	230, 300, 500	BF Goodrich Chemical
Cyclolac	various grades	General Electric Plastics
JSR-21		Japan Synthetic
Lubricomp A	Filled and unfilled grades	LNP Engineering
Lucky ABS	HF-350, HF-380	Standard Polymers
Lustran ABS	248, 448, PG-298, PG-299	Monsanto Chemical
Magnum		Dow
Magnum 541	graft	Dow
Novodur		Bayer Inc.
Ronfalin	various grades	DSM (UK) Ltd.
Sinkral	A112, A123, B133P, B423	Enichem America
Thermocomp A-1000		LNP Engineering
Urtal		Montedison

Bibliographic References

[1] *Kirk-Othmer Encycl. Chem. Technol.*, 4th edn. (eds. J.I. Kroschwitz and M. Howe-Grant), Wiley Interscience, 1993, 391
[2] *Encycl. Polym. Sci. Eng.*, 2nd edn., (ed. J.I. Kroschwitz), John Wiley and Sons, 1985, 388
[3] Ulrich, H., *Introduction to Industrial Polymers*, Hanser, 1982
[4] *DSM Polymers*, Edition 01/95, (technical datasheet)
[5] *Cycolac ABS Resins*, GE Plastics, (technical datasheet)
[6] Lustran ABS, Novadur ABS, *Application Technology Information*, Bayer, 1997, (technical datasheet)
[7] Daniels, C.A., *Polymers: Structure and Properties*, Technomic, 1989
[8] *Handb. Plast. Elastomers*, (ed. C.A. Harper), McGraw-Hill, 1975
[9] Pillachody, C.T. and Kelly, P.D., *Handbook of Plastic Materials and Technology*, (ed. I.I. Rubin), John Wiley and Sons, 1990, **C3**, (stress-strain behaviour, creep, stress relaxation, fatigue)
[10] Casale, A. and Salvatore, O., *Polym. Eng. Sci.*, 1975, **15**, 286
[11] Hayes, R.A. and Futamura, S., *J. Polym. Sci.*, *Polym. Chem. Ed.*, 1981, **19**, 985
[12] Brydon, A., Burnett, G.M. and Cameron, C.G., *J. Polym. Sci.*, *Polym. Chem. Ed.*, 1974, **12**, 1011
[13] Allen, P.W., Ayrey, G. and Moore, C.G., *J. Polym. Sci.*, 1959, **36**, 55
[14] Odian, G., *Principles of Polymerization*, McGraw-Hill, 1970
[15] Aoki, Y., *Macromolecules*, 1987, **20**, 2208, (dynamic viscoelastic steady-flow props)

ABS, Glass fibre reinforced

Synonyms: *Acrylonitrile-butadiene-styrene, Glass fibre reinforced*
Related Polymers: ABS
Monomers: Acrylonitrile, 1,3-Butadiene, Styrene
Material class: Thermoplastic, Copolymers
Polymer Type: ABS

Applications/Commercial Products:

Trade name	Supplier
ABS-G1FG-2	Washington Penn
AS-10GF	Compounding Technology
Novodor PTGV	Bayer Inc.
RTP 601	RTP Company
Styvex	Ferro Corporation
Thermocomp AF	LNP Engineering

ABS, Heat Resistant

Synonyms: *Acrylonitrile-butadiene-styrene copolymer. Acrylonitrile-butadiene-styrene terpolymer. Poly(2-propenenitrile-co-1,3-butadiene-co-ethenylbenzene)*
Related Polymers: ABS
Monomers: Acrylonitrile, 1,3-Butadiene, Styrene, α-Methylstyrene
Material class: Thermoplastic, Copolymers
Polymer Type: ABS
CAS Number:

CAS Reg. No.
9003-56-9

Molecular Formula: $[(C_8H_8)(C_4H_6)(C_3H_3N)(C_9H_{10})]_n$
Fragments: C_8H_8 C_4H_6 C_3H_3N C_9H_{10}
Additives: α-Methylstyrene
General Information: An amorph. thermoplastic that combines high performance with design versatility and easy processability. Addition of α-methylstyrene gives a heat distortion temp. higher than that of conventional grades of ABS. Good resistance to abrasion, chemicals and stress ensures a long lasting attractive surface finish. [2,3,4,5]

ABS, Heat Resistant

Volumetric & Calorimetric Properties:

Density:

No.	Value	Note
1	d 1.06 g/cm^3	ISO 1183 [5]

Thermal Expansion Coefficient:

No.	Value	Note	Type
1	7.8×10^{-5}–9.1×10^{-5} K^{-1}	DIN 53752, ASTM E831 [4,5]	L

Thermal Conductivity:

No.	Value	Note
1	0.2 W/mK	DIN 52612 [4]

Specific Heat Capacity:

No.	Value	Note	Type
1	1.3–1.7 kJ/kg.C	[7]	P

Melting Temperature:

No.	Value	Note
1	240°C	[4,5]

Glass-Transition Temperature:

No.	Value	Note
1	110–125°C	[15]

Deflection Temperature:

No.	Value	Note
1	94–102°C	80°, 1.8 MPa, 4h, ISO 75
2	105–108°C	80°, 0.45 MPa, 4h, ISO 75 [4,5]

Vicat Softening Point:

No.	Value	Note
1	100–115°C	50 K h^{-1}, 50N, ISO 306 [4,5]

Surface Properties & Solubility:

Solvents/Non-solvents: Sol. polar solvents, esters, ketones and some chlorinated hydrocarbons [7]

Transport Properties:

Transport Properties General: Exhibits viscoelastic behaviour in both the melt and solid states. Like other thermoplastics, it may be described as exhibiting non-Newtonian behaviour; viscosity decreases due to increasing shear rate

Melt Flow Index:

No.	Value	Note
1	2–18 g/10 min	220°, 10 kg, ISO 1133 [5]

Polymer Melts: Storage shear modulus and loss shear modulus as a function of angular frequency, and shear stress as a function of shear rate depend strongly on grafting degree in the long-time region associated with particle-particle interactions. As the grafting degree increases the viscoelastic functions first decrease and then increase. The minima in the functions occurs at the grafting degree of about 0.45 for ABS having rubber particles of 170 nm [16]

Water Absorption:

No.	Value	Note
1	0.2–0.45 %	23°, 24h, DIN 53495 [4,8]

Mechanical Properties:

Mechanical Properties General: Has high impact resistance over a wide temp. range down to -40°. Has good toughness and rigidity and high yield strength. Can deform in a ductile manner over a broad temp. range and at high strain rates [2,5]. Its creep resistance is higher than that of other grades of ABS. Little change in tensile, compressive and flexural props. below 80° [2,7]

Tensile (Young's) Modulus:

No.	Value	Note
1	2400–3000 MPa	1 mm min^{-1}, ISO 527 [4,5]

Flexural Modulus:

No.	Value	Note
1	2500–2800 MPa	2 mm min^{-1}, ISO 178 [5]

Tensile Strength Break:

No.	Value	Note	Elongation
1	15–23 MPa	5 mm min^{-1}, ISO 527 [4,5]	Elongation at break

Compressive Modulus:

No.	Value	Note
1	1310–1650 MPa	ASTM D695 [7]

Elastic Modulus:

No.	Value	Note
1	950 MPa	ISO 537 [8]

Tensile Strength Yield:

No.	Value	Elongation
1	44–48 MPa	50 mm min^{-1}, ISO 527 [4]

Flexural Strength Yield:

No.	Value	Note
1	75–85 MPa	5 mm min^{-1}, ISO 178 [5]

Compressive Strength:

No.	Value	Note
1	50–69 MPa	ASTM D695 [7]

ABS, Heat Resistant

Miscellaneous Moduli:

No.	Value	Note	Type
1	2400–2500 MPa	1h, ISO 899-1	tensile creep modulus
2	1900 MPa	1000h, ISO 899-1 [5]	

Viscoelastic Behaviour: Stress-strain behaviour, creep, stress relaxation and fatigue have been reported [14]
Hardness: Ball indentation 110-120 N mm^{-2} (ISO 2039-1) [5]. Rockwell R100–R115 (ASTM D785) [7]
Friction Abrasion and Resistance: Has good abrasion resistance [4]
Izod Area:

No.	Value	Notch	Note
1	12–23 kJ/m^2	Notched	23°, ISO 180-1A
2	3–10 kJ/m^2	Notched	-30°, ISO 180-1A [4,5]

Electrical Properties:

Electrical Properties General: Electrical props. are fairly constant over a wide range of frequencies and are unaffected by temp. or humidity [2]
Surface/Volume Resistance:

No.	Value	Note	Type
1	0.1–1 × 10^{15} Ω.cm	IEC 93 [5]	S

Dielectric Permittivity (constant):

No.	Value	Frequency	Note
1	2.7–3	100 Hz	
2	2.7–3.1	1 MHz	IEC 250 [5]

Dielectric Strength:

No.	Value	Note
1	25–33 kV.mm^{-1}	IEC 243-1 [4,5]

Arc Resistance:

No.	Value	Note
1	50–90s	[2]

Dissipation (Power) Factor:

No.	Value	Frequency	Note
1	0.005–0.009	100 Hz	
2	0.005–0.008	1 MHz	IEC 250 [5]

Optical Properties:

Transmission and Spectra: Transmittance 33.3% (ASTM D1003) [7]. Commercial grades may contain uv absorber
Volume Properties/Surface Properties: Haze 100% [7]. Gloss 76–87% (measuring angle 20°) [5]. May be supplied in either standard or custom colours or as natural pellet for colouring [4]

Polymer Stability:

Polymer Stability General: Combines high heat resistance with dimensional stability at higher temps.
Thermal Stability General: Has a relatively flat stress-temp. response due to its amorph. character [2,7]

Upper Use Temperature:

No.	Value	Note
1	80–110°C	[7]

Flammability: Flammability rating HB (1.6 mm, UL94) [4]. Oxygen index 18% (ISO 4589, specimen IV) [4]. Glow wire test, extinguishing time 55 (min), extinguishing temp. 650° (30s application, 2 mm thick, IES 695-2-1/1) [4]
Environmental Stress: Has poor weather and uv resistance unless protected. Undergoes stress cracking in contact with certain chemicals. The rubber phase in ABS oxidises more rapidly than the rigid component. Oxidation of polybutadiene results in embrittlement of the rubber. Yellows slightly in sunlight [2,4]
Chemical Stability: Is resistant to hydrocarbon solvents, mineral and vegetable oils, waxes and related household and commercial materials. Has high resistance to staining agents in household applications. Attacked by conc. nitric acid and sulfuric acids. Unaffected by weak acid at room temp. and alkalis below 65°. Poor resistance to most organic solvents [2,7]
Hydrolytic Stability: Stable to solns. of acid, alkaline and neutral salts
Biological Stability: Is biologically inert [4]
Recyclability: Can be recycled. The scrap should be clean, uncontaminated, and free from degradation. Normally the content of the reground material should not exceed 20-25% [4]
Stability Miscellaneous: Processing conditions can influence resultant props. by chemical and physical means. Degradation of the rubber and matrix phases occurs under very severe conditions. Evidence of morphological changes is shown by agglomeration of dispersed rubber particles during injection-moulding at higher temps. Physical effects such as orientation and moulded-in stress can affect mechanical props. [9]

Applications/Commercial Products:

Processing & Manufacturing Routes: Prod. by grafting styrene and acrylonitrile directly on to the polybutadiene latex in a batch or continuous emulsion polymerisation process. The degree of grafting is a function of the 1,2-vinyl content of the polybutadiene, the monomer concentration, the extent of conversion, temp. and mercaptan concentration. The emulsion polymerisation process involves two steps; (i) production of a rubber latex; (ii) subsequent polymerisation of styrene and acrylonitrile in the presence of the rubber latex to produce an ABS latex. This latex is then processed to isolate the ABS resin. The polymerisation ingredients include the monomers, an emulsifier, a polymerisation initiator and usually a chain transfer agent for MW control [1,10,11,12,13]. α-Methylstyrene is added during the polymerisation step [15]. Processed on standard reciprocating-screw injection-moulding machines. Drying is required prior to use in moulding machines
Mould Shrinkage (%):

No.	Value	Note
1	0.4–0.7%	ISO 2577 [4,5]

Applications: Used in automotive components (radiator grilles, airvents and interior trim); in business machine housings; in domestic appliances (e.g., electric iron handles); in optical equipment and in instrument panel sections

Trade name	Supplier
ABS 700	John Gibson (Plastics) Ltd.
Ashlene	Ashley Polymers
Comalloy ABS	ComAlloy International
Cycolac X-15	General Electric Plastics
Dow ABS	Dow
Lastilac	LATI Industria Thermoplastics S.p.A.

Lucky ABS HR-420	Standard Polymers
Lustran	Bayer Inc.
Novodur	Bayer Inc.
Sinkral	Enichem
Terluran	BASF
Thermofil ABS	Thermofil
Toyolac	Toray Industries

Bibliographic References

[1] *Kirk-Othmer Encycl. Chem. Technol.*, Vol. 1, 4th edn., (ed. J.I.Kroschwitz), Wiley Interscience, 1991, 391
[2] *Encycl. Polym. Sci. Eng.*, Vol. 1, 2nd edn. (ed. J.I. Kroschwitz), John Wiley & Sons, 1985, 388
[3] Brydson, J.A., *Plast. Mater.*, 5th edn., Butterworths, 1988
[4] *DSM Polymers*, DSM, 1995, (technical datasheet)
[5] Lustran ABS, Novadur ABS, *Application Technology Information*, Bayer, 1997, (technical datasheet)
[6] The Plastics Compendium, *Key Properties and Sources*, (ed. R. Dolbey), Rapra Technology Ltd., 1995, **1**
[7] Harper, C.A., *Handb. Plast. Elastomers*, (ed. C.A. Harper), McGraw-Hill, 1975
[8] *Terluran*, BASF Plastics, (technical datasheet)
[9] Casale, A. and Salvatore, O., *Polym. Eng. Sci.*, 1975, **15**, 286
[10] Hayes, R.A. and Futamura, S., *J. Polym. Sci., Polym. Chem. Ed.*, 1981, **19**, 985
[11] Brydon, A., Burnett, G.M. and Cameron, C.G., *J. Polym. Sci., Polym. Chem. Ed.*, 1974, **12**, 1011
[12] Allen, P.W., Ayrey, G. and Moore, C.G., *J. Polym. Sci.*, 1959, **36**, 55
[13] Odian, G., *Principles of Polymerization*, McGraw-Hill, 1970
[14] Pillichody, C.T. and Kelley, P.D., *Handbook of Plastic Materials and Technology*, (ed. I.I.Rubin), Wiley Interscience, 1990, (stress-strain behaviour, creep, stress relaxation, fatigue)
[15] Guide to Plastics, *Property and Specification Charts*, (ed W.A. Kaplan) McGraw-Hill, 1991, **68**, 11
[16] Aoki, Y., *Macromolecules*, 1987, **20**, 2208, (dynamic viscoelastic and steady-flow props)

ABS, High impact

Synonyms: *Acrylonitrile-butadiene-styrene copolymer. Acrylonitrile-butadiene-styrene terpolymer. Poly(2-propenenitrile-co-1,3-butadiene-co-ethenylbenzene)*
Related Polymers: ABS
Monomers: Acrylonitrile, 1,3-Butadiene, Styrene
Material class: Thermoplastic, Copolymers
Polymer Type: ABS
CAS Number:

CAS Reg. No.
9003-56-9

Molecular Formula: $[(C_8H_8)(C_4H_6)(C_3H_3N)]_n$
Fragments: C_8H_8 C_4H_6 C_3H_3N
General Information: An amorph. thermoplastic that combines high performance with design versatility and easy processability

Volumetric & Calorimetric Properties:
Density:

No.	Value	Note
1	d 1.03 g/cm^3	ISO 1183 [3]

Thermal Expansion Coefficient:

No.	Value	Note	Type
1	9.5×10^{-5}–0.00011 K^{-1}	ASTM D696, DIN 53732 [1,3]	L

Thermal Conductivity:

No.	Value	Note
1	0.2 W/mK	DIN 52612 [3]

Melting Temperature:

No.	Value	Note
1	230–270°C	[3,5]

Glass-Transition Temperature:

No.	Value	Note
1	91–110°C	[14]

Deflection Temperature:

No.	Value	Note
1	89–101°C	120 K h^{-1}, 1.8 MPa, ISO 75/A [3,4]
2	98°C	0.45 MPa [5]

Vicat Softening Point:

No.	Value	Note
1	91–100°C	50 K h^{-1}, 50 N, ISO 306B [3,4]

Surface Properties & Solubility:
Solvents/Non-solvents: Sol. polar solvents, esters, ketones and some chlorinated hydrocarbons [6]

Transport Properties:
Transport Properties General: Exhibits viscoelastic behaviour in both the melt state and the solid state. Like other thermoplastics, it may be described as exhibiting non-Newtonian behaviour; viscosity decreases due to increasing shear rate [2]
Melt Flow Index:

No.	Value	Note
1	4.5–17 g/10 min	220°, 10 kg, ISO 1133 [3,4]

Polymer Melts: Storage shear modulus and loss shear modulus as a function of angular frequency, and shear stress as a function of shear rate depend strongly on grafting degree in the long-time region associated with particle-particle interactions. As the grafting degree increases, the viscoelastic functions first decrease and then increase. The minima in these functions occurs at the grafting degree of about 0.45 for ABS having a rubber particle size of 170 nm [8]
Water Absorption:

No.	Value	Note
1	0.2–0.3 %	23°, 24h, DIN 53495 [3,5]

Mechanical Properties:
Mechanical Properties General: Has high impact resistance over a wide temp. range down to -40°. Has very good rigidity, high yield strength and a higher notched Izod impact strength than other types of ABS [3,5]. The change in tensile, compressive and flexural props. is small below 80°

Tensile (Young's) Modulus:

No.	Value	Note
1	1900–2250 MPa	1mm min^{-1}, ISO 527 [4,7]

ABS, High impact

Flexural Modulus:

No.	Value	Note
1	2100–2200 MPa	ASTM D790M [4,5]

Tensile Strength Break:

No.	Value	Note
1	10–40 MPa	ISO 527 [3,4]

Compressive Modulus:

No.	Value	Note
1	965–1380 MPa	ASTM D695 [6]

Elastic Modulus:

No.	Value	Note
1	800–845 MPa	23°, ISO 537 [4,7]

Tensile Strength Yield:

No.	Value	Note
1	38–42 MPa	50 mm min^{-1}, ISO 527 [3,4]

Flexural Strength Yield:

No.	Value	Note
1	65 MPa	ISO 178 [4]

Compressive Strength:

No.	Value	Note
1	31–55 MPa	ASTM D695 [6]

Miscellaneous Moduli:

No.	Value	Note	Type
1	1150 MPa	100h, ≤0.5% elongation, ISO/IEC 899 [7]	tensile creep modulus

Viscoelastic Behaviour: Stress-strain behaviour, creep, stress relaxation and fatigue have been reported [9]
Hardness: Ball indentation (H358/30) 65-90 MPa (ISO 2039) [4,7]. Rockwell R80–R105 (ASTM D785) [1,4]
Friction Abrasion and Resistance: Has good abrasion resistance [3]
Izod Notch:

No.	Value	Notch	Note
1	35–38 J/m	Y	23°, ISO 180/1A [3,7]
2	17–26 J/m	Y	-30°, ISO 180/1A [3,7]

Izod Area:

No.	Value	Notch	Note
1	35–38 kJ/m^2	Notched	-30°, ISO 180/1A [3,7]
2	17–26 kJ/m^2	Notched	23°, ISO 180/1A [3,7]

Electrical Properties:

Electrical Properties General: Electrical props. are fairly constant over a wide range of frequencies and are unaffected by temp. or humidity [2]
Surface/Volume Resistance:

No.	Value	Note	Type
1	$>1\times10^{15}$ Ω.cm	min., IEC 93 [3]	S

Dielectric Permittivity (constant):

No.	Value	Frequency
1	2.9	50 Hz
2	2.9	1 MHz IEC 250 [3]

Dielectric Strength:

No.	Value	Note
1	16–31 kV.mm^{-1}	ASTM D149 [1]

Arc Resistance:

No.	Value	Note
1	50–90s	[2]

Dissipation (Power) Factor:

No.	Value	Frequency	Note
1	0.01	50 Hz	[3]
2	0.008	1 MHz	IEC 250 [3]

Optical Properties:

Transmission and Spectra: Transmittance 28% (ASTM D10003) [6]. Commercial grades may contain uv absorber [6]
Volume Properties/Surface Properties: Haze 100% [6]. Colour supplied in either standard or custom colours, or as natural pellet for colouring [3]

Polymer Stability:

Polymer Stability General: Undergoes autooxidation resulting in embrittlement of the rubber and is liable to photooxidative degradation [2,6]
Thermal Stability General: Has a relatively flat stress-temp. response due to its amorph. character. Retains its props. at low temps. [2,6]
Upper Use Temperature:

No.	Value	Note
1	60–100°C	[6]

Flammability: Flammability rating HB (1.6 mm thick, UL 94) [3]. Oxygen index 18% (ISO 4589, specimen N) [3]. Glow wire test extinguishing time 5 s (max), extinguishing temp. 650°
Environmental Stress: Has poor weather and uv resistance unless protected. The rubber phase in ABS oxidises more rapidly than the rigid component. Oxidation of polybutadiene results in embrittlement of the rubber [2]
Chemical Stability: Resistant to hydrocarbon solvents, mineral and vegetable oils, waxes and related household and commercial materials. Is highly resistant to staining agents in household applications. Attacked by conc. nitric and sulfuric acids but is unaffected by weak acids at room temp. and most alkalis up to 65°. Has poor resistance to most organic solvents [1,2,6]. Certain chemicals cause cracking under applied stress

Hydrolytic Stability: Is resistant to neutral salt solns [1,6]
Biological Stability: Is biologically inert [3]
Recyclability: Can be recycled. The scrap should be clean, uncontaminated, and free from degradation. Normally the content of the reground material should not exceed 20–25% [3]
Stability Miscellaneous: Processing conditions can influence resultant props. by chemical and physical means. Degradation of the rubber and matrix phases occurs under very severe conditions. Evidence of morphological changes is shown by agglomeration of dispersed rubber particles during injection-moulding at higher temps. Physical effects such as orientation and moulded-in stress can affect mechanical props. [15]

Applications/Commercial Products:
Processing & Manufacturing Routes: Prod. by grafting styrene and acrylonitrile directly on to the polybutadiene latex in a batch or continuous emulsion polymerisation process. The degree of grafting is a function of the 1,2-vinyl content of the polybutadiene, the monomer concentration, the extent of conversion, temp. and mercaptan concentration. The emulsion polymerisation process involves two steps; (i) production of a rubber latex; (ii) subsequent polymerisation of styrene and acrylonitrile in the presence of the rubber latex to produce an ABS latex. This latex is then processed to isolate the ABS resin. The polymerisation ingredients include the monomers, an emulsifier, a polymerisation initiator and, usually, a chain transfer agent for MW control. [1,10,11,12,13]. Processed on standard reciprocating-screw injection-moulding machines

Mould Shrinkage (%):

No.	Value	Note
1	0.4–0.7%	DSM [3]

Applications: Used in automotive industry (interior and exterior trims); in pipes, luggage, refrigerator inner liners, power tools, boat hulls, ski-boots, furniture and in agricultural parts

Trade name	Details	Supplier
ABS		Diamond Polymers
Ashlene		Ashley Polymers
Bapolan		Bamberger
Comalloy ABS		ComAlloy International
Cycolac		General Electric Plastics
GPC Delta		H Simex International
Lastilac		LATI Industria Thermoplastics S.p.A.
Lucky ABS		Standard Polymers
Lustran	ABS 114-8, Ultra HDX	Monsanto Chemical
Lustran Ultra HX		Monsanto Chemical
Magnum		Dow
Novodur		Bayer Inc.
Polyman ABS		A. Schulman Inc.
RXI	101, 105	Resin Exchange
Ronfalin		DSM
Sinkral		Enichem
Styvex	40007 BKL2	Ferro Corporation
Terluran		BASF
Thermofil ABS		Thermofil
Toyolac		Toray Industries

Bibliographic References

[1] *Kirk-Othmer Encycl. Chem. Technol.*, Vol. 1, 4th edn., (eds. J.I. Kroschwitz), Wiley Interscience, 1993, 391
[2] *Encycl. Polym. Sci. Eng.*, Vol. 1, 2nd. edn., (ed. J.I. Kroschwitz), John Wiley & Sons, 1985, 388
[3] *DSM Polymers*, DSM, 1995, (technical datasheet)
[4] *Magnum ABS Resins*, Dow, (technical datasheet)
[5] Dolbey, R., The Plastics Compendium, *Key Properties and Sources*, RAPRA Technology, 1995, **1**
[6] Harper, C.A., *Handb. Plast. Elastomers*, (ed. C.A. Harper), McGraw-Hill, 1975
[7] *Terluran*, BASF, (technical datasheet)
[8] Aoki, Y., *Macromolecules*, 1987, **20**, 2208, (dynamic viscoelastic and steady-flow props)
[9] Pillichody, C.T. and Kelley, P.D., *Handbook of Plastic Materials and Technology*, (ed. I.I. Rubin), Wiley Interscience, 1990, (stress-strain behaviour, creep, stress relaxation, fatigue)
[10] Hayes, R.A. and Futamara, S., J. Polym. Sci., *Polym. Chem. Ed.*, 1981, **19**, 985
[11] Brydon, A., Burnett, G.M. and Cameron, C.G., J. Polym. Sci., *Polym. Chem. Ed.*, 1974, **12**, 1011
[12] Allen, P.W., Ayrey, G. and Moore, C.G., *J. Polym. Sci.*, 1959, **36**, 55
[13] Odian, G., *Principles of Polymerization*, McGraw-Hill, 1970
[14] Guide to Plastics, *Property and Specification Charts*, (ed. W.A. Kaplan), McGraw-Hill, 1994, **70**
[15] Casale, A. and Salvatore, O., *Polym. Eng. Sci.*, 1975, **15**, 286

ABS, Injection moulding, Glass fibre reinforced

Synonyms: *Acrylonitrile-butadiene-styrene copolymer. Acrylonitrile-butadiene-styrene terpolymer. Poly(2-propenenitrile-co-1,3-butadiene-co-ethenylbenzene)*
Related Polymers: ABS
Monomers: Acrylonitrile, 1,3-Butadiene, Styrene
Material class: Composites, Copolymers
Polymer Type: ABS
CAS Number:

CAS Reg. No.
9003-56-9

Molecular Formula: $[(C_8H_8)(C_4H_6)(C_3H_3N)]_n$
Fragments: C_8H_8 C_4H_6 C_3H_3N
Additives: 20% Glass fibre
General Information: An amorph. thermoplastic that combines high performance with design versatility and easy processability

Volumetric & Calorimetric Properties:
Density:

No.	Value	Note
1	d 1.2 g/cm^3	ISO 1183 [3]

Thermal Expansion Coefficient:

No.	Value	Note	Type
1	3.4×10^{-5}–5.4×10^{-5} K^{-1}	23–55°, DIN 53732, ASTM E831 [3,5]	L

Volumetric Properties General: Fibre reinforcement reduces the coefficient of thermal expansion compared with that of standard ABS grades
Thermal Conductivity:

No.	Value	Note
1	0.22–0.24 W/mK	DIN 52612, ASTM C177 [3,4]

ABS, Injection moulding, Glass fibre reinforced

Melting Temperature:

No.	Value	Note
1	240°C	[5]

Glass-Transition Temperature:

No.	Value	Note
1	100–110°C	[12]

Deflection Temperature:

No.	Value	Note
1	100–103°C	120 K h^{-1}, 1.8 MPa, ISO 75/A
2	104–105°C	0.45 MPa, ASTM D648 [3,4,5]

Vicat Softening Point:

No.	Value	Note
1	103–104°C	50 K h^{-1}, 50 N, ISO 306B [3,5]

Surface Properties & Solubility:
Solvents/Non-solvents: Sol. polar solvents, esters, ketones and some chlorinated hydrocarbons [7]

Transport Properties:
Transport Properties General: Exhibits viscoelastic behaviour both in the melt state and the solid state. Like other thermoplastics, it may be described as exhibiting non-Newtonian behaviour; viscosity decreases due to increasing shear rate [2]. Melt flow index 2.5–9.2 cm^3 10 min^{-1}, (220°, 10 kg, ISO 1133) [3,5]

Water Absorption:

No.	Value	Note
1	0.2–0.3 %	23°, 24h, DIN 53495 [3,4]
2	0.7 %	saturation, ASTM D570 [4]

Mechanical Properties:
Mechanical Properties General: Exhibits high impact resistance over a wide temp. range down to -40°. Has very good rigidity and toughness with tensile strength and modulus higher than those of standard grades of ABS. Elongation at break 1–2% (50 mm min^{-1}, ISO 527). [3,5] Notched Izod impact strength is very low compared with that of ordinary grades of ABS [3]

Tensile (Young's) Modulus:

No.	Value	Note
1	5500–6700 MPa	1mm min^{-1}, ISO 527 [3,5]

Flexural Modulus:

No.	Value	Note
1	4500–5500 MPa	2mm min^{-1}, ISO 178 [4,5]

Compressive Modulus:

No.	Value	Note
1	5515 MPa	ASTM D695 [12]

Tensile Strength Yield:

No.	Value	Note
1	70–75 MPa	50mm min^{-1}, ISO 527 [3,5]

Flexural Strength Yield:

No.	Value	Note
1	95–100 MPa	ISO 178 [4,5]

Compressive Strength:

No.	Value	Note
1	65–96 MPa	ASTM D695 [4,12]

Hardness: Ball indentation 130 N mm^{-2} (ISO 2039-1) [5]. Rockwell M90 (ASTM D785) [4]
Failure Properties General: Shear strength 40 MPa (ASTM D732) [4]
Friction Abrasion and Resistance: Has good abrasion resistance. [3] Coefficient of friction 0.25 (static) (LNP S0P), 0.3 (dynamic) (LNP S0P) [4]

Izod Area:

No.	Value	Notch	Note
1	7 kJ/m^2	Notched	23°, ISO 180/1A [3,5]
2	6 kJ/m^2	Notched	-30°, ISO 180/1A [3,5]

Electrical Properties:
Electrical Properties General: Electrical props. are fairly constant over a wide range of frequencies and are unaffected by temp. or humidity [2]

Surface/Volume Resistance:

No.	Value	Note	Type
1	0.1–1 × 10^{15} Ω.cm	IEC 93 [3,5]	S

Dielectric Permittivity (constant):

No.	Value	Frequency	Note
1	3	50 Hz	
2	3	1 MHz	IEC 250 [3]

Dielectric Strength:

No.	Value	Note
1	25–32 kV.mm^{-1}	IEC 243-1 [3]

Arc Resistance:

No.	Value	Note
1	50–90s	[2]

Dissipation (Power) Factor:

No.	Value	Frequency	Note
1	0.004	50 Hz	
2	0.008	1 MHz	IEC 250 [3]

– ABS, Injection moulding, Unfilled   A-10 – A-10

Optical Properties:
Volume Properties/Surface Properties: Inherent colour is opaque. An extensive range of colours is available [5]

Polymer Stability:
Polymer Stability General: Has very good dimensional stability and high heat stability. Undergoes autooxidation resulting in embrittlement of the rubber and is liable to photooxidative degradation
Thermal Stability General: Has a relatively flat stress-temp. response due to its amorph. character. Retains its props. at low temps. [2,6]
Upper Use Temperature:

No.	Value	Note
1	70°C	[6]

Flammability: Flammability rating HB (1.6 mm, UL 94). [3] Oxygen index 18% (ISO 4589, Specimen N). [3] Glow wire test extinguishing time 5 s (min), extinguishing temp. 650°, (2mm thick, wire applied for 30 s, IEC 695-2-1/1) [3]
Environmental Stress: Has poor weather and uv resistance unless protected. The rubber phase in ABS oxidises more rapidly than the rigid component. Oxidation of polybutadiene results in embrittlement of the rubber [2]
Chemical Stability: Is resistant to hydrocarbon solvents, mineral and vegetable oils, waxes and related household and commercial materials. Has high resistance to staining agents in household applications. Attacked by conc. nitric and sulfuric acids and organic solvents. Unaffected by weak acids at room temp. and most alkalis up to 65° [1,2,7]
Biological Stability: Is biologically inert [3]
Recyclability: Can be recycled. The scrap should be clean, uncontaminated, and free from degradation. Normally the content of the reground material should not exceed 20-25% [3]
Stability Miscellaneous: Processing conditions can influence resultant props. by chemical and physical means. Degradation of the rubber and matrix phases occurs under very severe conditions. Morphological changes can be evident as agglomeration of dispersed rubber particles during injection moulding at higher temps. Physical effects such as orientation and moulded-in stress can affect mechanical props. [13]

Applications/Commercial Products:
Processing & Manufacturing Routes: Prod. by grafting styrene and acrylonitrile directly on to the polybutadiene latex in a batch or continuous emulsion polymerisation process. The degree of grafting is a function of the 1,2-vinyl content of the polybutadiene, the monomer concentration, the extent of conversion, temp. and mercaptan concentration. The emulsion polymerisation process involves two steps; (i) production of a rubber latex; (ii) subsequent polymerisation of styrene and acrylonitrile in the presence of the rubber latex to produce an ABS latex. This latex is then processed to isolate the ABS resin. The polymerisation ingredients include the monomers, an emulsifier, a polymerisation initiator and usually a chain transfer agent for MW control. [1,8,9,10,11] Processed on standard reciprocating-screw injection-moulding machines
Mould Shrinkage (%):

No.	Value	Note
1	0.15–0.5%	glass fibre reinforcement lowers mould shrinkage compared with standard ABS grades, ISO 2577 [3,5]

Applications: Applications for automotive components such as instrument panel supports, rear light housings. Industrial ventilator housings and structural parts for office equipment (e.g. chassis)

Trade name	Details	Supplier
Ashlene		Ashley Polymers
Comalloy ABS		ComAlloy International
Cycolac		General Electric
D-10 FG-0100		Thermofil
Dow ABS		Dow
Lastilac		LATI Industria Thermoplastics S.p.A.
Lustran G6220		Monsanto Chemical
Mabs-FG10		Modified Plastics
Novodur		Bayer Inc.
Polyman ABS		A. Schulman Inc.
RTP	601, 601 FR	RTP Company
RTP 605		Vigilant Plastics
RTP 605		RTP Company
Ronfalin		DSM (UK) Ltd.
Terluran		BASF
Thermocomp	A-1000 series	LNP Engineering
Thermofil ABS		Thermofil
Toyolac		Toray

Bibliographic References
[1] *Kirk-Othmer Encycl. Chem. Technol.*, 4th edn., (ed. J.I. Kroschwitz), Wiley Interscience, 1993, **1**, 391
[2] *Encycl. Polym. Sci. Eng.*, 2nd edn., (ed. J.I. Kroschwitz), John Wiley and Sons, 1985, **1**, 388
[3] *DSM Polymers*, DSM, 1995, (technical datasheet)
[4] *LNP Engineering Plastics*, Product Data Book, Product Information, (technical datasheet)
[5] Lustran ABS, Novadur ABS, *Application Technology Information*, Bayer, 1997, (technical datasheet)
[6] Dolbey, R., The Plastics Compendium, *Key Properties and Sources*, Vol. 1, RAPRA Technology, 1995
[7] Harper, C.A., *Handb. Plast. Elastomers*, (ed. C.A. Harper), McGraw-Hill, 1975
[8] Hayes, R.A. and Futamara, S, *J. Polym. Sci.*, *Polym. Chem. Ed.*, 1981, **19**, 985
[9] Brydon, A., Burnett, G.M. and Cameron, C.G., *J. Polym. Sci.*, *Polym. Chem. Ed.*, 1974, **12**, 1011
[10] Allen, P.W., Ayrey, G. and Moore, C.G., *J. Polym. Sci.*, 1959, **36**, 55
[11] Odian, G., *Principles of Polymerization*, McGraw-Hill, 1970
[12] Guide to Plastics, *Property and Specification Charts*, (ed. W.A. Kaplan), McGraw-Hill, 1994, **70**
[13] Casale, A. and Salvatore, O., *Polym. Eng. Sci.*, 1975, **15**, 286

ABS, Injection moulding, Unfilled   A-10
Related Polymers: ABS
Monomers: Acrylonitrile, 1,3-Butadiene, Styrene
Material class: Thermoplastic, Copolymers
Polymer Type: ABS
CAS Number:

CAS Reg. No.
9003-56-9

Applications/Commercial Products:

Trade name	Supplier
ABS	John Gibson (Plastics) Ltd.
Abson A.B.S. 213	BF Goodrich Chemical

Cycolac CTB	General Electric Plastics
Cycolac DH	General Electric Plastics
Cycolac GSM	General Electric Plastics
Cycolac GTM 5300	General Electric Plastics
Cycolac X-11	General Electric Plastics
Lustran ABS	Monsanto Chemical
Magnum	Dow
Novodur P2H-AT	Bayer Inc.
Novodur P2K-AT	Bayer Inc.
Novodur P2L	Bayer Inc.
Novodur P2L-AT	Bayer Inc.
Ronfalin CP	DSM (UK) Ltd.
Ronfalin FF	DSM (UK) Ltd.
Terluran	BASF

ABS/Nylon blends, Injection moulding A-11

Synonyms: *Acrylonitrile-butadiene-styrene terpolymer/polyamide blends. ABS/PA. ABS/Polyamide Alloy. Nylon/ABS Alloy*
Related Polymers: ABS, Nylon 6, Nylon 6,6
Monomers: Acrylonitrile, 1,3-Butadiene, Styrene
Material class: Blends, Copolymers
Polymer Type: ABS/nylon, ABSpolyamide
CAS Number:

CAS Reg. No.	Note
9003-56-9	ABS

General Information: A semicrystalline thermoplastic blend of ABS and Nylon 6,6 or Nylon 6

Volumetric & Calorimetric Properties:
Density:

No.	Value	Note
1	d 1.06 g/cm^3	ASTM D792 [1]

Thermal Expansion Coefficient:

No.	Value	Note	Type
1	0.0001 K^{-1}	6–20°, ASTM D696 [1]	L
2	0.00017 K^{-1}	20–80°, ASTM D696 [1]	L
3	0.00038 K^{-1}	80–140°, ASTM D696 [1]	L

Thermal Conductivity:

No.	Value	Note
1	0.29 W/mK	ASTM C177 [1]

Specific Heat Capacity:

No.	Value	Note	Type
1	2.5 kJ/kg.C	260°, ASTM C351	P

Melting Temperature:

No.	Value	Note
1	260–274°C	[3]

Deflection Temperature:

No.	Value	Note
1	85°C	1.8 MPa, ASTM D648 [3]
2	94°C	0.45 MPa, ASTM D648 [3]

Vicat Softening Point:

No.	Value	Note
1	197°C	1 kg, ASTM D1525 [1]
2	110°C	5 kg, ASTM D1525 [1]
3	225°C	B120, ISO 306 [2]

Surface Properties & Solubility:
Solvents/Non-solvents: Shows good resistance to solvents

Transport Properties:
Melt Flow Index:

No.	Value	Note
1	1.5 g/10 min	ASTM D1238 [1]

Polymer Melts: Apparent viscosity 16000 P (250°, 200 s^{-1}, Dayton rheometer) [1]; 8900 P (250°, 500 s^{-1}, Dayton rheometer) [1]; 5600 P (250°, 1000 s^{-1}, Dayton rheometer) [1]
Water Absorption:

No.	Value	Note
1	1.1 %	23°, 90d, ASTM D570 [1]
2	1.6 %	23°, equilib., ASTM D570 [1]

Mechanical Properties:
Mechanical Properties General: Exhibits excellent impact resistance over a wide range of temps. Possesses higher elongation at break and notched Izod impact strength than those of unmodified ABS or unmodified PA. Elongation at break 270%, (23°, ASTM D638) [1,3]

Tensile (Young's) Modulus:

No.	Value	Note
1	1900 MPa	23°, 1 mm min^{-1}, ASTM D638, ISO 527 [1]

Flexural Modulus:

No.	Value	Note
1	2070–2140 MPa	23°, ASTM D790 [1,3]

Poisson's Ratio:

No.	Value	Note
1	0.38	[1]

Tensile Strength Yield:

No.	Value	Elongation	Note
1	46 MPa	6%	23°, ASTM D638 [1,3]

Impact Strength: Tensile impact strength 545 kJ m^{-2} (23°, ASTM D1822) [1]

– ABS, Plating

Hardness: Rockwell R95 (23°, ASTM D785) [1]. Shore D75 (23°, ASTM D676) [1]
Friction Abrasion and Resistance: Has good abrasion resistance
Izod Notch:

No.	Value	Notch	Note
1	850 J/m	Y	23°, 1/8" bar, ASTM D256 [1]

Electrical Properties:
Electrical Properties General: Exhibits lower dielectric strength than polyamide and ABS
Surface/Volume Resistance:

No.	Value	Note	Type
2	8.3×10^{15} Ω.cm	ASTM D257 [1]	S

Dielectric Permittivity (constant):

No.	Value	Frequency	Note
1	3.9	100 Hz	ASTM D150 [1]
2	3.8	1 kHz	ASTM D150 [1]
3	3.5	100 kHz	ASTM D150 [1]

Dielectric Strength:

No.	Value	Note
1	0.425 kV.mm^{-1}	short time, ASTM D149 [1]
2	0.35 kV.mm^{-1}	step by step, ASTM D149 [1]

Arc Resistance:

No.	Value	Note
1	108s	3.2 mm, ASTM D495

Static Electrification: Comparative tracking index 600 V (ASTM D3638) [1]. High voltage arc tracking rate 0.4 in min^{-1} (UL 746A)
Dissipation (Power) Factor:

No.	Value	Frequency	Note
1	0.015	100 Hz	ASTM D150 [1]
2	0.014	1 kHz	ASTM D150 [1]
3	0.024	100 kHz	ASTM D150 [1]

Polymer Stability:
Thermal Stability General: Stable at high temps.
Upper Use Temperature:

No.	Value	Note
1	70°C	[3]

Flammability: Flammability rating HB (1.6 mm, UL 94) [1]. Limiting oxygen index 20% (ASTM D2863) [1]
Chemical Stability: Good resistance to solvents; has better resistance compared to ABS

Applications/Commercial Products:
Processing & Manufacturing Routes: ABS and polyamide (Nylon 6) blends may be compatibilised with imidised acrylic polymers as well as with styrenic copolymers containing maleic anhydride. The compatibilising copolymer is miscible with the SAN matrix in ABS. The resulting blend has a well-defined ABS dispersion in the Nylon 6 matrix owing to the grafting reaction between maleic anhydride of the compatibiliser and the primary amine end group in Nylon 6. [4] Processed in conventional injection-moulding equipment. Various finishing techniques including painting, flocking, hot stamping, and some welding are possible. The blend should be dried before processing [1]
Mould Shrinkage (%):

No.	Value	Note
1	1.3–1.6%	ASTM D955 [2,3]

Applications: Housings for power tools, electro-technical components. Lawn and garden equipment, sporting goods, gears and impellers

Trade name	Details	Supplier
Cycloy	EHA	General Electric Plastics
Gelon	polyamide resins	General Electric Plastics
Staloy N		DSM
Triax	1000, 1120, 1125	Monsanto Chemical

Bibliographic References
[1] *Polym. Blends Alloys,* Technomic Inc., USA, 1988
[2] Gelon, *Polyamide Resins,* GE Plastics, (technical datasheet)
[3] Dolbey, R., The Plastics Compendium, *Key Properties and Sources,* Vol. 1, RAPRA Technology, 1995
[4] *Kirk-Othmer Encycl. Chem. Technol.,* 4th edn., (ed. J.I. Kroschwitz), Wiley Interscience, 1993, **19**, 862

ABS, Plating A-12
Synonyms: *Acrylonitrile-butadiene-styrene copolymer. Acrylonitrile-butadiene-styrene terpolymer. Poly(2-propenenitrile-co-1,3-butadiene-co-ethenylbenzene)*
Related Polymers: ABS
Monomers: Acrylonitrile, 1,3-Butadiene, Styrene
Material class: Thermoplastic, Copolymers
Polymer Type: ABS
CAS Number:

CAS Reg. No.
9003-56-9

Molecular Formula: $[(C_8H_8)(C_4H_6)(C_3H_3N)]_n$
Fragments: C_8H_8 C_4H_6 C_3H_3N
Additives: Nickel, copper and chrome
General Information: An amorph. thermoplastic that is electroplated to give parts with the more decorative finish of metals such as chrome, gold, silver and brass. The metal layer screens radio waves and conducts electricity. The adhesion of electroplated metal is excellent

Volumetric & Calorimetric Properties:
Density:

No.	Value	Note
1	d 1.04 g/cm^3	ISO 1183 [4]

Thermal Expansion Coefficient:

No.	Value	Note	Type
1	0.0001 K^{-1}	23–55°, ASTM E831 [4]	L

– **ABS, Plating**

Melting Temperature:

No.	Value	Note
1	240–285°C	[3,4]

Glass-Transition Temperature:

No.	Value	Note
1	100–110°C	[5]

Deflection Temperature:

No.	Value	Note
1	92°C	80°, 1.8 MPa, 4h, ISO 75 [4]
2	96°C	80°, 0.45 MPa, 4h, ISO 75 [4]

Vicat Softening Point:

No.	Value	Note
1	95°C	50 K h^{-1}, 50 N, ISO 306 B [4]

Surface Properties & Solubility:
Solvents/Non-solvents: Sol. polar solvents (esters, ketones and some chlorinated hydrocarbons) [6]

Transport Properties:
Transport Properties General: Exhibits viscoelastic behaviour in both the melt and solid states. Like other thermoplastics, it may be described as exhibiting non-Newtonian behaviour; viscosity decreases due to increasing shear rate [2]

Melt Flow Index:

No.	Value	Note
1	25 g/10 min	220°, 10 kg, ISO 1133/1991 [4]

Polymer Melts: Storage shear modulus and loss shear modulus as a function of angular frequency, and shear stress as a function of shear rate depend strongly on grafting degree in the long-time region associated with particle-particle interactions. As the grafting degree increases the viscoelastic functions first decrease and then increase. The minima in the functions occurs at the grafting degree of about 0.45 for ABS having rubber particle size of 170 nm [13]

Water Absorption:

No.	Value	Note
1	0.3 %	24 h [3]

Mechanical Properties:
Mechanical Properties General: Has high impact resistance over a wide temp. range down to -40°. Has very good rigidity and toughness. Elongation at break 8–30% (5 mm min^{-1}, ISO 527). [2,4] Little change in tensile, compressive and flexural props. below 80° [2,6]

Tensile (Young's) Modulus:

No.	Value	Note
1	2200–2600 MPa	1 mm min^{-1}, ISO 527 [4,5]

Flexural Modulus:

No.	Value	Note
1	2100–2700 MPa	2 mm min^{-1}, ISO 178 [4,5]

Tensile Strength Yield:

No.	Value	Note
1	40–46 MPa	50 mm min^{-1}, ISO 527 [4,5]

Flexural Strength Yield:

No.	Value	Note
1	60–80 MPa	5 mm min^{-1}, ISO 178 [4,5]

Viscoelastic Behaviour: Stress-strain behaviour, creep, stress relaxation and fatigue have been reported [7]
Hardness: Ball indentation 90 N mm^{-2} (ISO 2039-1) [4]. Rockwell R103–R110 (ASTM D785) [2]
Friction Abrasion and Resistance: The plating can increase scratch resistance

Izod Notch:

No.	Value	Notch	Note
1	23 J/m	Y	23°, ISO 180-1A
2	10 J/m	Y	-30°, ISO 180-1A

Izod Area:

No.	Value	Notch	Note
1	23 kJ/m^2	Notched	23°, ISO 180-1A
2	10 kJ/m^2	Notched	-30°, ISO 180-1A

Electrical Properties:
Electrical Properties General: Electrical props. are fairly constant over a wide range of frequencies and are unaffected by temp. or humidity [2]

Surface/Volume Resistance:

No.	Value	Note	Type
2	1×10^{15} Ω.cm	IEC 93 [4]	S

Dielectric Permittivity (constant):

No.	Value	Frequency	Note
1	2.8	100 Hz	
2	2.7	1 MHz	IEC 250 [4]

Dielectric Strength:

No.	Value	Note
1	34 kV.mm^{-1}	IEC 243-1 [4]

Dissipation Power Factor: tan δ 0.005 (100 Hz); 0.007 (1 MHz, IEC 250) [4]

Arc Resistance:

No.	Value	Note
1	50–90s	[2]

Dissipation (Power) Factor:

No.	Value	Frequency	Note
1	0.005	100 Hz	
2	0.007	1 MHz	IEC 250 [4]

Polymer Stability:

Polymer Stability General: Undergoes autooxidation, resulting in embrittlement of the rubber and is liable to photooxidative degradation [2,6]

Thermal Stability General: Has a relatively flat strss-temp. response due to its amorph. character. Retains props at low temps.[2,6]

Upper Use Temperature:

No.	Value	Note
1	70°C	[3]

Flammability: Flammability rating HB (UL94) [3] Oxygen index 19%

Chemical Stability: Is resistant to hydrocarbon solvents, mineral and vegetable oils, waxes and related household and commercial materials. Has high resistance to staining agents in household applications. Attacked by conc. nitric and sulfuric acids. Unaffected by weak acids at room temp. and most alkalis up to 65°. Has poor resistance to most organic solvents [1,2,6]

Hydrolytic Stability: Resistant to solns. of acid, alkaline and neutral salts [1,6]

Recyclability: Regrind should not be used for items to be electroplated because of possible flaws in the surface finish [3]

Stability Miscellaneous: Processing conditions can influence resultant props. by chemical and physical means. Degradation of the rubber and matrix phases occurs under very severe conditions. Evidence of morphological changes is shown by agglomeration of dispersed rubber particles during injection-moulding at higher temps. Physical effects such as orientation and moulded-in stress can affect mechanical props. [12]

Applications/Commercial Products:

Processing & Manufacturing Routes: Prod. by grafting styrene and acrylonitrile directly on to the polybutadiene latex in a batch or continuous emulsion polymerisation process. The degree of grafting is a function of the 1,2-vinyl content of the polybutadiene, the monomer concentration, the extent of conversion, temp. and mercaptan concentration. The emulsion polymerisation process involves two steps; (i) production of a rubber latex; (ii) subsequent polymerisation of styrene and acrylonitrile in the presence of the rubber latex to produce an ABS latex. This latex is then processed to isolate the ABS resin. The polymerisation ingredients include the monomers, an emulsifier, a polymerisation initiator and usually a chain transfer agent for MW control. [1,8,9,10,11] Processed on standard reciprocating-screw injection-moulding machines. Must be dried before melt processing. Temp. profile and mould surface temps. need to be kept higher than for standard ABS grades. This is in order to ensure good surface finish for subsequent electroplating and to minimise moulded-in stresses. Mould release agents should not be used as they affect plating behaviour [3]

Applications: Used in the automotive industry (radiator grilles, wheel covers, headlight bezels, interior and exterior trim and knobs). Other uses in sanitary ware, in marine hardware, in parts for radios and TVs, in shavers, light fixtures and in appliance housings

Trade name	Details	Supplier
Ashlene		Ashley Polymers
Blendex 131		General Electric Speciality
Blendex 338	graft	General Electric Speciality
Comalloy ABS		ComAlloy International
Cycolac		General Electric Plastics
Dow ABS		Dow
Novodur	(P2MC)	Bayer Inc.
Polyman ABS		A. Schulman Inc.
Sinkral		Enichem
Terluran		BASF
Thermofil ABS		Thermofil
Toyolac		Toray

Bibliographic References

[1] *Kirk-Othmer Encycl. Chem. Technol.*, Vol. 1, 4th edn., (ed. J.I. Kroschwitz), Wiley Interscience, 1993, 391
[2] *Encycl. Polym. Sci. Eng.*, Vol. 1, 2nd edn., (ed. J.I. Kroschwitz), John Wiley & Sons, 1985, 388
[3] Dolbey, R., The Plastics Compendium, *Key Properties and Sources*, RAPRA Technology, 1995
[4] Lustran ABS, Novadur ABS, *Application Technology Information*, Bayer, 1997, (technical datasheet)
[5] Guide to Plastics, *Property and Specification Charts*, (ed. W.A. Kaplan), McGraw-Hill, 1994, **70**, 12
[6] Harper, C.A., *Handb. Plast. Elastomers*, (ed. C.A. Harper), McGraw-Hill, 1975
[7] Pillichody, C.T. and Kelley, P.D., *Handbook of Plastic Materials and Technology*, (ed. I.I.Rubin), Wiley Interscience, 1990, (stress-strain behaviour, creep, stress relaxation, fatigue)
[8] Hayes, R.A. and Futamura, S., J. Polym. Sci., *Polym. Chem. Ed.*, 1981, **19**, 985
[9] Brydon, A., Burnett, G.M. and Cameron, C.G., J. Polym. Sci., *Polym. Chem. Ed.*, 1974, **12**, 1011
[10] Allen. P.W., Ayrey, G. and Moore, C.G., *J. Polym. Sci.*, 1959, **36**, 55
[11] Odian, G., *Principles of Polymerization*, McGraw-Hill, 1970
[12] Casale, A. and Salvatore, O., *Polym. Eng. Sci.*, 1975, **15**, 286
[13] Aoki, Y., *Macromolecules*, 1987, **20**, 2208, (dynamic viscoelastic and steady-flow props)

ABS/Polycarbonate blend, General purpose injection moulding A-13

Synonyms: *Acrylonitrile-butadiene-styrene terpolymer/polycarbonate blend. ABS/PC blend. ABS/Polycarbonate alloy. PC/ABS blend. PC/ABS alloy. Polycarbonate/ABS blend. Polycarbonate/ABS alloy*

Related Polymers: ABS, Bisphenol A polycarbonate, ABS/Polycarbonate blends, Flame retardant

Monomers: Acrylonitrile, 1,3-Butadiene, Styrene, Bisphenol A

Material class: Blends, Copolymers

Polymer Type: ABS, polycarbonate alloy blendsABS/PC

CAS Number:

CAS Reg. No.	Note
9003-56-9	ABS
80-05-7	Bisphenol A

Molecular Formula: $[(C_8H_8)(C_4H_6)(C_3H_3N)]_n$ and $(C_{16}H_{14}O_3)_n$
Fragments: C_8H_8 C_4H_6 C_3H_3N $C_{16}H_{14}O_3$
Additives: May also contain PTFE
General Information: An amorph. thermoplastic blend of ABS resin and polycarbonate, which has high heat distortion and high dimensional stability with excellent impact strength at low temps. It has great design flexibility, fast part production rates and high yields with consistent physical props. Has attractive cost/performance characteristics

Volumetric & Calorimetric Properties:

Density:

No.	Value	Note
1	d 1.1 g/cm^3	ISO 1183 [2,3]

Thermal Expansion Coefficient:

No.	Value	Note	Type
1	$8 \times 10^{-5} - 9 \times 10^{-5}$ K^{-1}	4–80 mm thick, DIN 53752, ASTM D696 [2,3]	L

ABS/Polycarbonate blend, General purpose injection moulding

Thermal Conductivity:

No.	Value	Note
1	0.2 W/mK	ASTM C177 [3]

Melting Temperature:

No.	Value	Note
1	250–280°C	[3]

Transition Temperature General: Has higher deflection temps. under load than have ABS resins. The ductile-brittle transition temp. decreases with increasing ABS content up to 50% ABS [8]

Deflection Temperature:

No.	Value	Note
1	96–110°C	1.8 MPa
2	105–130°C	0.45 MPa, ISO 75, DIN 53461 [2,3]

Vicat Softening Point:

No.	Value	Note
1	112–131°C	VST/B120, DIN ISO 306 [2]

Transport Properties:
Melt Flow Index:

No.	Value	Note
1	2.5 g/10 min	230°, 3.8 kg, ISO 1133 [3]
2	8.8–17.6 g/10 min	260°, 5 kg, ISO 1133 [2]

Polymer Melts: The melt viscosity of PC/ABS blends relative to PC is significantly lower, even lower than that of pure ABS in some compositions. The presence of ABS, even at values as low as 20%, reduces the melt viscosity of PC significantly. The presence of ABS (35% or higher) in the blends results in viscosity close to or even lower than that of ABS. An increase in the MW of PC in PC/ABS blends results in low temp. fracture toughness improvement but higher melt viscosity. The selection of PC in PC/ABS blends must be a compromise between the toughness advantages of higher PC MW and the disadvantages of higher melt viscosity. A ratio of PC:ABS of 65:35 is usually ideal. [8] Viscosity 5100 P (510 Pa s) (270°, 100 s^{-1}); 1600 P (160 Pa s) (270°, 1000 s^{-1}); 1050 P (105 Pa s) (270°, 2000 s^{-1}) [3]

Water Absorption:

No.	Value	Note
1	0.2–0.6 %	ISO R62, DIN 53495-1L [2,3]

Mechanical Properties:
Mechanical Properties General: Has excellent impact strength down to -50° and improved stiffness and notched Izod impact strength compared with conventional high impact ABS. Its notched Izod impact resistance is lower than that of polycarbonate

Tensile (Young's) Modulus:

No.	Value	Note
1	2000–2200 MPa	ISO R527, DIN 53457-t [2,3]

Flexural Modulus:

No.	Value	Note
1	2000–2250 MPa	ISO 178, DIN 53457-B3 [2,3]

Tensile Strength Break:

No.	Value	Note	Elongation
1	40–50 MPa	5 mm min^{-1}, ISO R527, DIN 53455 [2,3]	80%

Tensile Strength Yield:

No.	Value	Elongation	Note
1	44–55 MPa	4–5.5%	5 mm min^{-1}, ISO R527, DIN 53455 [2,3]

Flexural Strength Yield:

No.	Value	Note
1	70 MPa	ISO 178 [3]

Miscellaneous Moduli:

No.	Value	Note	Type
1	800 MPa	0°	Shear modulus
2	700 MPa	100°	Shear modulus
3	100 MPa	120°	Shear modulus

Viscoelastic Behaviour: Has good creep resistance [6]
Hardness: Ball indentation 80–97 MPa (ISO 2039, DIN 53459) [2,3]

Izod Notch:

No.	Value	Notch	Note
1	500 J/m	Y	23°, ISO R180 [3]
2	300 J/m	Y	-40°, ISO R180 [3]

Electrical Properties:
Electrical Properties General: Has good electrical insulation props.
Surface/Volume Resistance:

No.	Value	Note	Type
1	$>0.1 \times 10^{15}$ Ω.cm	min., IEC 93, DIN VDE 0303, part 3 [2]	S

Dielectric Permittivity (constant):

No.	Value	Frequency	Note
1	2–3	50 Hz	IEC 250, DIN VDE 0303, part 4 [2]
2	2.9–3	1 MHz	IEC 250, DIN VDE 0303, part 4 [2]

Dielectric Strength:

No.	Value	Note
1	24 kV.mm^{-1}	IEC 243, DIN VDE 0303, part 2 [2]

Static Electrification: Comparative tracking index Grade 250–500 (IEC 112, DIN VDE 0303, part 1) [2]

– ABS/Polycarbonate blends, extrusion unreinforced

Dissipation (Power) Factor:

No.	Value	Frequency	Note
1	0.004	50 Hz	IEC 250, DIN VDE 0303, part 4 [2]
2	0.007	1 MHz	IEC 250, DIN VDE 0303, part 4 [2]

Optical Properties:
Volume Properties/Surface Properties: Transparency not transparent. Colour many opaque colours. Surface finish good mould surface reproduction [2]

Polymer Stability:
Polymer Stability General: Has high dimensional accuracy and stability [2]. Retains props., allowing for long-term durability [7]
Thermal Stability General: Has good low temp. resistance [2]. Has high thermal stability during processing [3]
Upper Use Temperature:

No.	Value	Note
1	70°C	[5]
2	70°C	[6]

Flammability: Flammability rating HB (1.6 mm, UL94, IEC 707) [2,3]. Oxygen index 21-24% (ASTM D2863) [4]
Environmental Stress: Has stress crack resistance slightly better than that of PC [2]
Chemical Stability: Is resistant to oils and fats but has poor resistance to solvents
Recyclability: Trim or scrap can be reground and reused with virgin material in proportions of up to 30%, with little loss of physical props., provided materials are free of contaminants [3]

Applications/Commercial Products:
Processing & Manufacturing Routes: ABS and polycarbonate are alloyed in a compounding operation using a range of equipment including batch or continuous melt mixers, and both single and twin-screw extruders. In the compounding step more than one type of ABS may be used (i.e., emulsion and mass produced) to obtain an optimum balance of props. for a specific application [1]. ABS/PC blends have a very wide processing window and can endure longer residence times in injection-moulding machines. This is due to the lower viscosities of ABS/PC blends. The resins absorb small amounts of moisture and must be dried (100–110°, 3-4h) before processing. Standard or vented injection moulded machines are used for processing [3]
Mould Shrinkage (%):

No.	Value	Note
1	0.3–0.7%	ASTM D955 [3]

Applications: Has applications in internal and external automotive components; housings for business machines and in electrical and electronic components. Typical automotive interior applications include air vents, central consoles, decorative trim, glove compartment doors, grilles, electrical housings, instrument frames and panels, buffer strips, steering column shrouds, wheel nut caps, window surrounds and mirror housings. Exterior applications include cowl vent grilles, body side mouldings, wheel covers, tail-light housings, headlamp surrounds, logo plates, spoilers and body components. May also be used in certain food and medical applications

Trade name	Details	Supplier
Bayblend	non exclusive	Bayer Inc.
Bayblend T	44, 45 M_n (opaque), T95, T65 M_n, 88	Mobay
Bayblend T	84 M_n, 64, 65 M_n (opaque)	Bayer Inc.
Cycoloy MC	8100 (extrusion blow moulding grade)	General Electric Plastics
Cycoloy C	1000, 1100 EHA	General Electric Plastics
Pulse	600, 700, 759, 1310, 1725 (flame retardant)	Dow
Pulse 880BG		Dow
Pulse B250		Dow
Pulse EXT 20		Dow
Triax 2000		Monsanto Chemical

Bibliographic References

[1] *Kirk-Othmer Encycl. Chem. Technol.*, 4th edn., (ed. J.I. Kroschwitz), Wiley Interscience, 1991, **1**, 391
[2] *Bayblend Product Range*, Bayer, 1993, (technical datasheet)
[3] *Pulse engineering resins*, Dow Plastics, (technical datasheet)
[4] *Guide to Plastics, Property and Specification Charts*, McGraw-Hill, 1994, **70**
[5] Dolbey, R., The Plastics Compendium, *Key Properties and Sources*, RAPRA Technology, 1995, **1**
[6] Harper, C.A., *Handb. Plast. Elastomers*, (ed. C.A. Harper), McGraw-Hill, 1975
[7] *Guide to Plastics, Property and Specification Charts*, McGraw-Hill, 1990, **66**
[8] Wu, J.S., Shen, S.C. and Chang, F.C., *J. Appl. Polym. Sci.*, 1993, **50**, 1379

ABS/Polycarbonate blends, extrusion unreinforced A-14

Synonyms: *Acrylonitrile-butadiene-styrene terpolymer/polycarbonate blends. ABS/PC blends. ABS/Polycarbonate alloys*
Related Polymers: ABS/Polycarbonate blends, ABS, Bisphenol A polycarbonate
Monomers: Acrylonitrile, 1,3-Butadiene, Styrene, Bisphenol A
Material class: Thermoplastic, Blends, Copolymers
Polymer Type: ABS, polycarbonate alloy blends
CAS Number:

CAS Reg. No.	Note
9003-56-9	ABS
80-05-7	Bisphenol A

General Information: Amount of polycarbonate in the blend brings a compromise to the blend props.; ABS confers better melt processing ability, PC confers better mech. props. and rigidity

Volumetric & Calorimetric Properties:
Density:

No.	Value	Note
1	d 1.15 g/cm³	ASTM D792 [2]

Thermal Expansion Coefficient:

No.	Value	Note	Type
1	6.2×10^{-5}–7.2×10^{-5} K^{-1}	ASTM D696 [2,3]	L

Melting Temperature:

No.	Value	Note
1	230–260°C	[2]

Transition Temperature General: Deflection temps. are higher than those of ABS resins. The ductile-brittle transition temp. decreases with increasing ABS content up to a max. content of 50% [4]

ABS/Polycarbonate blends, extrusion unreinforced

Deflection Temperature:

No.	Value	Note
1	104°C	1.8 MPa [3]
2	107°C	0.45 MPa, ASTM D648 [3]

Vicat Softening Point:

No.	Value	Note
1	113°C	Rate B, ASTM D1525 [2]

Surface Properties & Solubility:
Solvents/Non-solvents: Has poor resistance to solvents

Transport Properties:
Melt Flow Index:

No.	Value	Note
1	2.5 g/10 min	230°, 3.8 kg, ASTM D1238 [2]

Polymer Melts: The melt viscosity of PC/ABS blends relative to PC is significantly lower, even lower than that of pure ABS in some compositions. The presence of ABS, even at amounts as low as 20%, reduces the melt viscosity of PC significantly. The presence of ABS in the blends in amounts of 35% or higher results in viscosity close to or even lower than that of ABS. An increase in the MW of PC in PC/ABS blends improves low-temp. fracture toughness but increases melt viscosity. Therefore the amount of PC in PC/ABS blends must be selected to allow a compromise between these props. Typically, a ratio of PC:ABS of 65:35 is ideal [4]

Water Absorption:

No.	Value	Note
1	0.2–0.45 %	24h, ASTM D570 [2]

Mechanical Properties:
Mechanical Properties General: Has excellent impact strength down to -50°, and has high stiffness and toughness. Its notched Izod impact resistance is better than that of high impact ABS

Tensile (Young's) Modulus:

No.	Value	Note
1	2400–2600 MPa	ASTM D638 [2]

Flexural Modulus:

No.	Value	Note
1	2000–2600 MPa	ASTM D790 [2]

Tensile Strength Break:

No.	Value	Note	Elongation
1	50 MPa	ASTM D638 [3]	65%

Tensile Strength Yield:

No.	Value	Elongation	Note
1	58 MPa	5%	ASTM D638 [3]

Flexural Strength Yield:

No.	Value	Note
1	84 MPa	ASTM D790 [3]

Viscoelastic Behaviour: Has good creep resistance [5]
Hardness: Rockwell R95–R120 (ASTM D785) [3]
Izod Notch:

No.	Value	Notch	Note
1	538 J/m	Y	3.2 mm, 23°, ASTM D256 [3]

Electrical Properties:
Electrical Properties General: Has good electrical insulation props.
Dielectric Permittivity (constant):

No.	Value	Frequency	Note
1	2.9	60 Hz	
2	2.8	1 MHz	ASTM D150 [2]

Dielectric Strength:

No.	Value	Note
1	>17 kV.mm^{-1}	min., ASTM D149 [3]

Arc Resistance:

No.	Value	Note
1	>50s	min., ASTM D495 [3]

Static Electrification: Comparative tracking index rating 300–550
Dissipation (Power) Factor:

No.	Value	Frequency	Note
1	0.004	60 Hz	ASTM D150 [3]
2	0.008	1 MHz	ASTM D150 [3]

Optical Properties:
Volume Properties/Surface Properties: Not transparent; available in many opaque colours [6]

Polymer Stability:
Polymer Stability General: Good dimensional stability; has good long-term property retention [2]
Thermal Stability General: Has good low-temp. resistance [6]
Flammability: Flammability rating HB (1.6 mm, UL 94) [2]
Environmental Stress: Resistance to stress cracking is slightly better than that of PC [6]
Chemical Stability: Is resistant to oils and fats, but has poor resistance to solvents

Applications/Commercial Products:
Processing & Manufacturing Routes: ABS and polycarbonate are alloyed in a compounding operation using a range of equipment including batch or continuous melt mixers, and both single and twin-screw extruders. In the compounding step more than one type of ABS may be used (i.e., emulsion and mass produced) to obtain an optimum balance of props. for a specific application. [1] The resins must be dried at 100° for 3–4h before processing. Processed by extrusion [2]

Mould Shrinkage (%):

No.	Value	Note
1	0.5–0.7%	ASTM D955 [3]

ABS/Polycarbonate blends, Flame retardant

Applications: Uses in internal and external automotive components; in housings for business machines, electrical and electronic components; in certain food and medical applications

Trade name	Details	Supplier
Bayblend	MJ-2500	Mobay
Bayblend		Bayer Inc.
Cycoloy	MC8100	General Electric Plastics

Bibliographic References

[1] *Kirk-Othmer Encycl. Chem. Technol.*, 4th edn., Vol. 1, (ed. J.I. Kroschwitz), Wiley Interscience, 1993, 391
[2] Guide to Plastics, *Property and Specification Charts*, (ed. W.A. Kaplan), McGraw-Hill, 1990, **60,** No.1
[3] *Polym. Blends Alloys*, Technomic, 1988
[4] Wu, J.S., Shen, S.C. and Chang, F.C., *J. Appl. Polym. Sci.*, 1993, **50,** 1379, (MW effects)
[5] Harper, C.A., *Handb. Plast. Elastomers*, (ed. C.A. Harper), McGraw-Hill, 1975
[6] *Engineering Thermoplastics: Properties and Applications*, Product ranges and characteristics, processing and fabrication, Bayer, (technical datasheet)

ABS/Polycarbonate blends, Flame retardant   A-15

Synonyms: *Acrylonitrile-butadiene-styrene terpolymer/polycarbonate blends, Flame retardant. ABS/PC blends, Flame retardant. ABS/Polycarbonate alloys, Flame retardant*
Related Polymers: ABS, Bisphenol A polycarbonate, ABS/Polycarbonate blends
Monomers: Acrylonitrile, 1,3-Butadiene, Styrene, Bisphenol A
Material class: Blends, Copolymers
Polymer Type: ABS, polycarbonate alloy blendsABS/PC
CAS Number:

CAS Reg. No.	Note
9003-56-9	ABS
80-05-7	Bisphenol A

Molecular Formula: $[(C_8H_8)(C_4H_6)(C_3H_3N)]_n$ and $(C_{16}H_{14}O_3)_n$
Fragments: C_8H_8 C_4H_6 C_3H_3N $C_{16}H_{14}O_3$
Additives: An additive free from antimony, bromine and chlorine
General Information: An amorph. thermoplastic blend of ABS and polycarbonate which has a good basis for flame-retardant formulations owing to its high polycarbonate content. If the components are appropriately distributed, carbonisation processing of the aromatic polymer component can be used to build up heat-shielding zones which inhibit the pyrolysis of the aliphatic components. Thus it is possible to produce flame-retardant systems containing little or no halogen. Blends have outstanding characteristics including heat resistance (between that of ABS and PC), good toughness and low temp. resistance, high stiffness and dimensional stability

Volumetric & Calorimetric Properties:
Density:

No.	Value	Note
1	d 1.17 g/cm^3	ISO 1183, DIN 53479 [2]

Thermal Expansion Coefficient:

No.	Value	Note	Type
1	7.6×10^{-5} K^{-1}	DIN 53752 [2]	L

Melting Temperature:

No.	Value	Note
1	220–270°C	[3,4]

Transition Temperature General: The ductile-brittle transition temp. decreases with increasing ABS content up to a max. of 50% [6]
Deflection Temperature:

No.	Value	Note
1	80–90°C	1.8 MPa [2]
2	90–100°C	0.45 MPa, ISO 75, DIN 53461 [2]

Vicat Softening Point:

No.	Value	Note
1	90–110°C	B120, DIN ISO 306 [2]

Transport Properties:
Melt Flow Index:

No.	Value	Note
1	16–31 g/10 min	240°, 5 kg, DIN ISO 1133 [2]

Polymer Melts: The melt viscosity of PC/ABS blends relative to PC is significantly lower, even lower than that of pure ABS in some compositions. The presence of ABS, even at values as low as 20%, reduces the melt viscosity of PC significantly. The presence of ABS ($\geq$ 35%) in the blends results in viscosity close to or even lower than that of ABS. An increase in the MW of PC in PC/ABS blends results in low-temp. fracture toughness improvement but higher melt viscosity. The selection of PC in PC/ABS blends must be a compromise between the toughness advantages of higher PC MW and the disadvantages of higher melt viscosity. A component ratio of PC:ABS of 65:35 is ideal [6]
Water Absorption:

No.	Value	Note
1	0.2 %	DIN 53495-1L [2]

Mechanical Properties:
Mechanical Properties General: Has high impact strength down to -50° and high stiffness

Tensile (Young's) Modulus:

No.	Value	Note
1	2600–2700 MPa	DIN 53457-t [2]

Flexural Modulus:

No.	Value	Note
1	2300–2700 MPa	DIN 53457-B3 [2]

Tensile Strength Break:

No.	Value	Note	Elongation
1	45 MPa	ISO 527, DIN 53455 [2]	50%

Compressive Modulus:

No.	Value	Note
1	1586 MPa	ASTM D695 [3]

Tensile Strength Yield:

No.	Value	Elongation	Note
1	60 MPa	4%	ISO 527, DIN 53455 [2]

Flexural Strength Yield:

No.	Value	Note
1	87–96 MPa	ASTM D790 [4]

Compressive Strength:

No.	Value	Note
1	76–78 MPa	ASTM D695 [3]

Impact Strength: Charpy no break (23°, ISO 179, DIN 53453) [2]; 20–30 kJ m^{-2} (notched, 23°, ISO 179, DIN 53453) [2]
Hardness: Ball indentation H30 121–122 N mm^{-2} (ISO 2039, DIN 53456) [2]. Rockwell R115–R119 (ASTM D985) [3]
Charpy:

No.	Value	Notch	Note
1	20–30	Y	23°, ISO 179, DIN 53453 [2]

Electrical Properties:
Electrical Properties General: Has good electrical insulation props.
Surface/Volume Resistance:

No.	Value	Note	Type
2	0.1–10 × 10^{15} Ω.cm	min., IEC 93, DIN VDE 0303, part 3	S

Dielectric Permittivity (constant):

No.	Value	Frequency	Note
1	3	50 Hz	IEC 250, DIN VDE 0303, part 4 [2]
2	2.9	1 MHz	IEC 250, DIN VDE 0303, part 4 [2]

Dielectric Strength:

No.	Value	Note
1	30 kV.mm^{-1}	IEC 243, DIN VDE 0303, part 2 [2]

Static Electrification: Comparative tracking index Grade 300 (IEC 112, DIN VDE 0303, part 1) [2]
Dissipation (Power) Factor:

No.	Value	Frequency	Note
1	0.003–0.005	50 Hz	IEC 250, DIN VDE 0303, part 4 [2]
2	0.007–0.008	1 MHz	IEC 250, DIN VDE 0303, part 4 [2]

Optical Properties:
Volume Properties/Surface Properties: Transparency not transparent. Colour many opaque colours. Surface finish good mould surface reproduction [2]
Polymer Stability:
Polymer Stability General: Has high dimensional stability and retains props., allowing for long-term durability [2,5]
Thermal Stability General: Has good low temp. resistance [2]
Flammability: Flammability rating V0 (1.6 mm, UL94, IEC 707) [2]. Oxygen index 30% (ASTM D2863) [3]
Environmental Stress: Stress crack resistance is slightly better than that of PC [2]
Chemical Stability: Is resistant to oils and fats but has poor resistance to solvents

Applications/Commercial Products:
Processing & Manufacturing Routes: ABS and polycarbonate are alloyed in a compounding operation using a range of equipment including batch or continuous melt mixers, and both single and twin-screw extruders. In the compounding step more than one type of ABS may be used (i.e., emulsion and mass produced) to obtain an optimum balance of props. for a specific application [1]. Because of their lower viscosity ABS/PC blends have a very wide processing window and can endure longer residence times in injection-moulding machines. Predrying (80–90°, 3–4h) prior to processing is essential to ensure desired appearance and property performance. Standard or vented injection-moulding machines are used for processing
Mould Shrinkage (%):

No.	Value	Note
1	0.4–0.6%	DIN 53464 [2]

Applications: Has applications in business machines (personal and point-of-sale computer housings typewriter housings; general office equipment). Electrical/electronic applications include general wiring devices, plugs and receptacles, control housings; telecom system connectors

Trade name	Details	Supplier
Bayblend	FR grades	Bayer Inc.
Bayblend	DP-2 grades	Mobay
Cycoloy	C 2800, C 2950	General Electric Plastics

Bibliographic References

[1] *Kirk-Othmer Encycl. Chem. Technol.*, 4th Edn., Vol. 19, (ed. J.I. Kroschwitz), Wiley Interscience, 1993, 391, 602
[2] *Bayblend Product Range*, Bayer, 1993, (technical datasheet)
[3] Guide to Plastics, *Property and Specification Charts*, McGraw-Hill, 1994, **70**
[4] *Polym. Blends Alloys*, Technomic Co. Inc., USA, 1988
[5] Guide to Plastics, *Property and Specification Charts*, McGraw-Hill, 1990, **66**
[6] Wu, J.S., Shen, S.C. and Chang, F.C., *J. Appl. Polym. Sci.*, 1993, **50**, 1379

ABS/Polycarbonate blends, Glass fibre reinforced

Synonyms: *Acrylonitrile-butadiene-styrene terpolymer/polycarbonate blends. ABS/PC blends. ABS/Polycarbonate alloys*
Related Polymers: ABS/Polycarbonate blends, ABS, Bisphenol A polycarbonate
Monomers: Acrylonitrile, Bisphenol A, 1,3-Butadiene, Styrene
Material class: Blends, Copolymers
Polymer Type: ABS, polycarbonate alloy blendsABS/PC
CAS Number:

CAS Reg. No.	Note
9003-56-9	ABS
80-05-7	Bisphenol A

ABS/Polycarbonate blends, Glass fibre reinforced

Additives: 20% Glass fibre
General Information: An amorph. thermoplastic blend of ABS and polycarbonate

Volumetric & Calorimetric Properties:

Density:

No.	Value	Note
1	d 1.25 g/cm^3	ISO 1183, DIN 53479 [2]

Thermal Expansion Coefficient:

No.	Value	Note	Type
1	2.6×10^{-5} K^{-1}	DIN 53752 [2]	L

Volumetric Properties General: The coefficient of thermal expansion is reduced compared with the unreinforced ABS/PC blend due to glass fibre reinforcement

Melting Temperature:

No.	Value	Note
1	240–280°C	[3]

Transition Temperature General: Deflection temps. are higher than those of ABS resins. The ductile-brittle temp. decreases with increasing ABS content up to a max. of 50% ABS [5]

Deflection Temperature:

No.	Value	Note
1	115°C	1.8 MPa
2	130°C	0.45 MPa, ISO 75, DIN 53461 [2]

Vicat Softening Point:

No.	Value	Note
1	134°C	B120, DIN ISO 306 [2]

Transport Properties:

Polymer Melts: The melt viscosity of PC/ABS blends is significantly lower than that of PC and is even lower than that of pure ABS in some compositions. The presence of ABS, even at amounts as low as 20%, is able to reduce the melt viscosity of PC significantly. The presence of ABS (35% or greater) in the blends results in viscosity close to or even lower than that of ABS. An increase in the MW of PC in PC/ABS blends improves low-temp. fracture toughness but gives higher melt viscosity. A component ratio of PC/ABS of 65:35 can be considered as a compromise composition in terms of the product toughness (due to PC component), cost and the processing melt viscosity [5]

Water Absorption:

No.	Value	Note
1	0.2 %	DIN 53495-1L [2]

Mechanical Properties:

Mechanical Properties General: Has high impact strength down to -50°, and very good rigidity. Tensile and flexural moduli are both higher than unreinforced ABS/PC blends

Tensile (Young's) Modulus:

No.	Value	Note
1	6000 MPa	DIN 53457-t [2]

Flexural Modulus:

No.	Value	Note
1	6000 MPa	DIN 53457-B3 [2]

Tensile Strength Break:

No.	Value	Note
1	75 MPa	ISO 527, DIN 53455 [2]

Tensile Strength Yield:

No.	Value	Elongation	Note
1	75 MPa	2%	ISO 527, DIN 53455 [2]

Hardness: Ball indentation H30 (125 N mm^{-2}, ISO 2039, DIN 53456 [2]

Charpy:

No.	Value	Notch	Note
1	20	N	23°, ISO 179, DIN 53453 [2]
2	8	Y	23°, ISO 179, DIN 53453 [2]

Electrical Properties:

Electrical Properties General: Has good electrical insulation props.

Surface/Volume Resistance:

No.	Value	Note	Type
2	$>0.1 \times 10^{15}$ Ω.cm	min., IEC 93, DIN VDE 0303, part 3 [2]	S

Dielectric Permittivity (constant):

No.	Value	Frequency	Note
1	3	50 Hz	IEC 250, DIN VDE 0303, part 4 [2]
2	3.2	1 MHz	IEC 250, DIN VDE 0303, part 4 [2]

Dielectric Strength:

No.	Value	Note
1	30 kV.mm^{-1}	IEC 243, DIN VDE 303, part 2 [2]

Static Electrification: Comparative tracking index Grade 200 (IEC 112, DIN VDE 0303, part 1) [2]

Dissipation (Power) Factor:

No.	Value	Frequency	Note
1	0.002	50 Hz	IEC 250, DIN VDE 0303 [2]
2	0.009	1 MHz	IEC 250, DIN VDE 0303 [2]

Optical Properties:

Volume Properties/Surface Properties: Not transparent, has many opaque colours available. Provides good mould surface reproduction [2]

Polymer Stability:

Polymer Stability General: Has high dimensional stability and accuracy and long term prop. retention [3]
Thermal Stability General: Exhibits good low temp. resistance [2]
Flammability: Flammability rating HB (1.6 mm, UL 94) [2]. Limiting oxygen index 24% (ASTM D2863) [3]

Environmental Stress: Stress crack resistance is slightly better than that of PC [2]
Chemical Stability: Is resistant to oils and fats, but has poor resistance to solvents

Applications/Commercial Products:
Processing & Manufacturing Routes: ABS and polycarbonate are alloyed in a compounding operation using a range of equipment including batch or continuous melt mixers, and both single and twin-screw extruders. In the compounding step more than one type of ABS may be used (i.e., emulsion and mass produced) to obtain an optimum balance of props. for a specific application [1]. The resins absorb small amounts of moisture and must be dried at 100–110° for 3–4h before processing. Processed by injection-moulding [4]

Mould Shrinkage (%):

No.	Value	Note
1	0.2–0.4%	DIN 53464 [2]

Applications: Has applications in internal and external automotive components; in housings for business machines; in electrical and electronic components

Trade name	Details	Supplier
Bayblend	T88-4N, 20% glass fibre	Bayer Inc.
Bayblend	T88-2N, 10% glass fibre	Bayer Inc.

Bibliographic References
[1] *Kirk-Othmer Encycl. Chem. Technol.*, 4th edn., (ed. J.I. Kroschwitz), Wiley Interscience, 1991, **1**,19, 391, 602
[2] Bayblend, Mechanical, Thermal, *Electrical and Other Properties*, Bayer Engineering Plastics, 1993, (technical datasheet)
[3] Guide to Plastics, *Property and Specification Charts*, (ed. W.A. Kaplan), McGraw-Hill, 1990, **66**, No.11
[4] *Pulse engineering resins*, Dow Plastics, (technical datasheet)
[5] Wu, J.S., Shen, S.C. and Chang, F.C., *J. Appl. Polym. Sci.*, 1993, **50**, 1379

ABS Polymers

$$\{CH_2-CH(CN)-CH_2CH=CHCH_2CH_2-CH(Ph)\}_n$$

Synonyms: *Acrylonitrile-butadiene-styrene copolymer. Poly(1-butenylene-graft-1-phenylethylene-co-cyanoethylene). Poly(acrylonitrile-co-butadiene-co-styrene). Poly(2-propenenitrile-co-1,3-butadiene-co-ethenylbenzene). Acrylonitrile-butadiene-styrene terpolymer*
Related Polymers: ABS - general purpose, ABS - injection moulding unfilled, ABS - extrusion unfilled, ABS - high impact, ABS - flame retardent, ABS - Heat resistant, ABS - Modifier, ABS - Blow moulding, ABS - Injection moulding glass fibre filled, ABS - Glass reinforced, ABS/Polycarbonate blends, Polyacrylonitrile, ABS - stainless steel filled, ABS - carbon reinforced, ABS - plating
Monomers: Acrylonitrile, 1,3-Butadiene, Styrene
Material class: Thermoplastic, Copolymers
Polymer Type: ABS
CAS Number:

CAS Reg. No.	Note
9003-56-9	
108146-73-2	block
106677-58-1	graft copolymer

Molecular Formula: $[(C_8H_8)(C_4H_6)(C_3H_3N)]_n$
Fragments: C_8H_8 C_4H_6 C_3H_3N

Molecular Weight: MW is instrumental in determining the physical props. of ABS. As MW increases, the strength of the material increases while its processability declines
Additives: Glass fibre, carbon fibre, stainless-steel fibre. Fire retardants to improve flame retardancy. α-Methylstyrene to raise heat distortion temp. Uv resistant additives, films, laminates or paint coatings to retard colour, gloss and property value changes. Methyl methacrylate may replace acrylonitrile in order to improve transparency
General Information: ABS terpolymers are elastomeric and thermoplastic composites formed from three different monomers. Acrylonitrile contributes heat resistance, high strength and chemical resistance; butadiene contributes impact strength, toughness and low temp. property retention; styrene contributes gloss, processability and rigidity. Typical ABS contains 20% rubber, 25% acrylonitrile and 55% styrene. The elastomeric component polybutadiene or butadiene copolymer is dispersed as a grafted particulate phase in the thermoplastic matrix of styrene and acrylonitrile copolymer (SAN). Props. of ABS can be modified by varying the relative proportions of the basic components, the degree of grafting and the MW. ABS is normally opaque as it is a two-phase system, and each phase has a different refractive index. Very versatile material whose props. are tailored to meet specific product requirements. Speciality products include electroplating grades, high temp. resistant grades, fire retardant grades and ABS blends with other polymers such as polycarbonates and PVC. ABS is sold as an unpigmented powder, unpigmented pellets, precoloured pellets and 'salt and pepper' blends of ABS and colour concentrates. Reinforced ABS grades are available using glass, carbon and steel fibres
Morphology: Similar to high-impact polystyrene, but the dispersed rubber particles are much smaller (1–10 μm) and contain styrene-acrylonitrile copolymer inclusions. The average particle size and particle size distribution within the rubber phase have a significant effect on the overall balance of props. including strength, toughness and appearance. Large rubber particles typically enhance toughness while lowering gloss. The rigid phase chain length, as correlated to MW, can also have a significant effect on ABS resin props. Generally, the longer the polymer chain, the higher the strength including impact/ductility, but the lower the flow. The ratio of the rigid SAN phase to the rubber phase affects the flow/impact balance of the ABS resin. An increase in the rubber content will increase the impact/toughness of an ABS material, generally at the expense of the flow. The balance between the flow and impact props. of an ABS material is a basic product characteristic

Applications/Commercial Products:
Processing & Manufacturing Routes: Most ABS is made by grafting styrene and acrylonitrile directly on to the polybutadiene latex in a batch or continuous emulsion polymerisation process. In all manufacturing processes, grafting is achieved by the free-radical copolymerisation of styrene and acrylonitrile monomers in the presence of an elastomer. Ungrafted styrene-acrylonitrile is formed during graft polymerisation and/or added afterwards. Other commercial processes used for manufacturing ABS are mass and mass suspension. Each manufacturing method results in the formation of a material composed of discrete rubber particles dispersed in a rigid matrix. Can be processed by all the techniques commonly used for thermoplastic materials; compression and injection moulding, extrusion, calendering and blow-moulding. Post-processing operations include cold forming, thermoforming, metal plating, painting, hot stamping, ultrasonic, spin and vibrational welding and adhesive bonding. ABS materials have a broad processing window of operating conditions and favourable shear thinning flow characteristics

Mould Shrinkage (%):

No.	Value	Note
1	0.5–0.9%	[1]

Applications: Is the biggest selling engineering thermoplastic. The principal market worldwide is in transportation. Uses include both interior and exterior applications. General purpose and high heat grades are used for automotive instrument panels, consoles, doorpost covers and other interior trim parts. Exterior applications include radiator grilles, headlight housings and extruded/thermoformed fascias for large trucks. ABS plating grades include applications such as knobs, light bezels, mirror housings, grilles and decorative trim. The second largest market is in appliances and applications including injection-moulded housings for kitchen appliances, power tools, vacuum cleaners, sewing machines, hair dryers, business machines and other electric and electronic equipment. Another significant market is pipe and fittings. Other uses include consumer and industrial applications such as luggage, toys, medical devices, furniture and bathroom fixtures

Trade name	Details	Supplier
ABS		Mobil
ABS	various grades	Diamond Polymers
ABS	various grades	Multibase
ABS	various grades	Engineered Polymer
ABS 3333/3500		Richard Group/RRT
ABS 400/600/700		Michael Day
ABS BK		TP Composites
ABS General black		IPC/GWC
ABS-G1FG-2		Washington Penn
API 800/900		American Polymers
AVP	various grades	Polymerland Inc.
Abson ABS		BF Goodrich Chemical
Ashlene	various grades	Ashley Polymers
Bapolan	various grades	Bamberger Polymers
Blendex		General Electric Speciality
CTI AS	various grades	CTI
Celstran ABSS	various grades	Hoechst Celanese
Cevian SA/SHE/SER	various grades	Daicel Chemical
Claradex	various grades	Synthetic Rubber
Comalloy 220		Comalloy Intl. Corp.
Conductomer		Synthetic Rubber
Conductor 700/790		Japan Synthetic
Cyclolac		General Electric Plastics
D-10/D-12/D-13/D-18	various grades	Thermofil Inc.
D-20/D-30/D-40/D-50	various grades	Thermofil Inc.
Denka	various grades	Denka
Electrafil	various grades	DSM Engineering
Emiclear	EC	Toshiba Chem. Prod.
FPC	various grades	Federal Plastics
Faradex	various grades	DSM Engineering
Fiberfil	various grades	DSM Engineering
Fiberstran	various grades	DSM Engineering
GPC DELTA D	various grades	Grand Pacific
JSR ABS/AES/NC/NF/XT	various grades	Japan Synthetic
Lubricomp AFL/AL	various grades	LNP Engineering
Lucky ABS	various grades	Lucky
Lustran ABS		Bayer Inc.
Lustran ABS	various grades	Monsanto Chemical
Lustran Elite		Monsanto Chemical
Lustran Ultra		Monsanto Chemical
Lustran Ultra ABS		Bayer Inc.
Magnum		Dow
Marbon resins		General Electric Plastics
Maxloy AK/BK/HK/HN	various grades	Japan Synthetic
Miwon ABS	various grades	Miwon Petrochem
NP ABS 6201F		Network Polymers
Nitriflex	various grades	A.Schulman Inc.
Novatemp 88		Novatec
Novodur		Bayer Inc.
Novodur PMTM		Bayer Inc.
Nyloy	various grades	Nytex Composites
Nytron		Nytex Composites
PKF-600		Ferro Corporation
PRL-ABS AB		Polymer Resources
Perma Stat		RTP Company
Polyfabs	various grades	A.Schulman Inc.
Polyflam		A.Schulman Inc.
Polylac PA	various grades	Chi Mei Industrial
Proquigel	various grades	Proquigel
RAB	various grades	Ferro Corporation
RTP	various grades	RTP Company
RX1/SC1	various grades	Spartech
Ravikral	various grades	Enichem America
Ronfalin		DSM
Shinko-Lac	various grades	Mitsubishi Rayon
Shuman 700/7000 series		Shuman Plastics
Sicoflex		Mazzucchelli Celuloide SpA
Sinkral	various grades	Enichem America
Stat-Kon A/AC/AS		LNP Engineering
Stat-Loy		LNP Engineering
Styvex	various grades	Ferro Corporation
Taitalac	various grades	Taita Chem. Co.
Terluran		BASF
Terluran		Hydro Polymers
Terluran B46L		BASF
Thermocomp A-1000	several grades	LNP Engineering
Toydax	various grades	Toray Industries
Ultrastyr 2003		Enichem America
Urtal	various grades	Enichem America

Bibliographic References
[1] *Kirk-Othmer Encycl. Chem. Technol.*, Vol. 1, 4th edn., (eds. J.I. Kroschwitz and M. Howe-Grant), Wiley Interscience, 1993, 391
[2] *Encycl. Polym. Sci. Eng.*, Vol. 1, 2nd edn., (ed. J.I. Kroschwitz), John Wiley and Sons, 1985, 388

[3] Ulrich, H., *Introduction to Industrial Polymers*, Hanser, 1982
[4] *Handb. Plast. Elastomers*, (ed. C.A. Harper), McGraw-Hill, 1975
[5] Brydson, J.A., *Plast. Mater.*, 5th edn., Butterworths, 1988
[6] Odian, G., *Principles of Polymerization*, McGraw-Hill, 1970
[7] Guide to Plastics, *Property and Specification Charts*, McGraw-Hill, 1990, **66**, 11
[8] Alger, M.S.M., *Polymer Science Dictionary*, Elsevier Science, 1990

ABS/polyurethane blends, General purpose A-18

Monomers: Acrylonitrile, 1,3-Butadiene, Styrene
Material class: Blends, Copolymers
Polymer Type: ABS, ABS/polyurethanepolyurethane

Applications/Commercial Products:

Trade name	Supplier
Prevail 3050	Dow
Prevail 3100	Dow
Prevail 3150	Dow

ABS/Polyvinylchloride blends A-19

Synonyms: *Acrylonitrile-butadiene-styrene terpolymer and polyvinylchloride blends. ABS/PVC Blends. ABS/PVC Alloy. PVC/ABS Blends*
Monomers: Acrylonitrile, 1,3-Butadiene, Styrene, Vinyl chloride
Material class: Blends, Copolymers
Polymer Type: PVC, ABSABS/PVC
CAS Number:

CAS Reg. No.	Note
9003-56-9	ABS

General Information: ABS content of blends is at least 40% up to 80%. Blends combine rigidity and impact props. of ABS with inherent flame retardancy of PVC. Flame retardant grades available.

Volumetric & Calorimetric Properties:
Density:

No.	Value	Note
1	d 1.25 g/cm^3	ASTM D792 [2,3]

Thermal Expansion Coefficient:

No.	Value	Note	Type
1	$4.6 \times 10^{-5} - 8 \times 10^{-5}$ K^{-1}	ASTM D696 [3]	L

Deflection Temperature:

No.	Value	Note
1	70°C	1.8 MPa, ASTM D648 [2]
2	75°C	0.45 MPa, ASTM D648 [2]

Surface Properties & Solubility:
Solvents/Non-solvents: Sol. ethers, esters

Transport Properties:
Melt Flow Index:

No.	Value	Note
1	1.9 g/10 min	ASTM 1238 [3]

Water Absorption:

No.	Value	Note
1	0.2 %	ASTM D570 [2]

Mechanical Properties:
Mechanical Properties General: Has good rigidity and impact resistance. Its notched Izod impact strength is higher than that of unmodified PVC, but lower than that of medium impact ABS

Tensile (Young's) Modulus:

No.	Value	Note
1	2200–2600 MPa	ASTM D638 [3]

Flexural Modulus:

No.	Value	Note
1	2200–2750 MPa	ASTM D790 [3]

Tensile Strength Break:

No.	Value	Note	Elongation
1	40–45 MPa	ASTM D638, ISO R527 [3]	20%

Flexural Strength at Break:

No.	Value	Note
1	55–70 MPa	ASTM D790 [3]

Tensile Strength Yield:

No.	Value	Elongation	Note
1	30–45 MPa	8%	ASTM D638, ISO R527 [2,3]

Hardness: Rockwell R100–R109 (ASTM D785) [2,3]
Izod Notch:

No.	Value	Notch	Note
1	100 J/m	Y	ASTM D256A [2]

Electrical Properties:
Electrical Properties General: Has good electrical insulation props.
Dielectric Permittivity (constant):

No.	Value	Frequency	Note
1	3	1 kHz	ASTM D150 [2]

Dielectric Strength:

No.	Value	Note
1	23 kV.mm^{-1}	ASTM D149 [2]

Dissipation (Power) Factor:

No.	Value	Frequency	Note
1	0.04	1 kHz	ASTM D150 [2]

ABS/PSU blend

Polymer Stability:
Thermal Stability General: The thermal stability of the blends is lower than that of the individual components [4]
Upper Use Temperature:

No.	Value	Note
1	60°C	[2]

Decomposition Details: Each component of the blend may undergo thermal degradation. The polybutadiene rubber phase degrades most rapidly [4]
Flammability: Flammability rating V0 (UL 94) [2]. Limiting oxygen index 28% (ASTM D2863) [2]. Is flame resistant owing to the presence of PVC. Has self-extinguishing props.
Environmental Stress: PVC component enhances stability to uv
Chemical Stability: Has good resistance to solvents, oils and fats, except ethers and esters
Stability Miscellaneous: The presence of PVC lowers the processing stability

Applications/Commercial Products:
Processing & Manufacturing Routes: ABS and PVC are alloyed in a compounding operation using a range of equipment including batch or continuous melt mixers, and both single and twin-screw extruders. In the compounding step more than one type of ABS may be used (i.e. emulsion and mass produced) to obtain an optimum balance of props. for a specific application [1]. Processed in conventional injection-moulding equipment. The resin should be dried before processing [2]
Mould Shrinkage (%):

No.	Value	Note
1	0.45%	lower than unmodified ABS, ASTM D955 [2]

Applications: 80:20 ABS/PVC blends are used to prod. fire-retarding ABS-type materials. 10:90 Blends are used as impact-modified forms of unplasticised PVC. Other uses in blends with plasticised PVC to give a crashpad sheet material; in housings for domestic appliances and computers; in printer and plumbing components; in cash register bases

Trade name	Details	Supplier
832-1000		ComAlloy International
Cyclovin		General Electric Plastics
Lustran ABS	600	Monsanto Chemical
Novaloy 9000		Novatec
Triax CBEI		Monsanto Chemical
Trosiplast PVC/ABS		Hüls AG

Bibliographic References
[1] *Kirk-Othmer Encycl. Chem. Technol.*, 4th edn., (ed. J.I. Kroschwitz), Wiley Interscience, 1993, **1**, 391
[2] Dolbey, R., The Plastics Compendium, *Key Properties and Sources*, Vol. 1, Rapra Technology, 1995
[3] Guide to Plastics, *Property and Specification Charts*, (ed. W.A. Kaplan), McGraw-Hill, 1994, **70**, No.12
[4] *Polymer Science and Technology*, (eds. D. Klempner and K.C. Frisch), Vol. 11, Plenum Press, (thermal degradation)
[5] Brydson, J.A., *Plast. Mater.*, 4th edn., Butterworth Heinemann, 1982

ABS/PSU blend A-20

Synonyms: *ABS/PSO blend. Acrylonitrile-butadiene-styrene/polysulfone blend*
Related Polymers: Polysulfone, ABS
Monomers: 4,4'-Dichlorodiphenyl sulfone, Bisphenol A

Material class: Thermoplastic
Polymer Type: Polysulfone
CAS Number:

CAS Reg. No.	Note
25135-51-7	Udel

Molecular Formula: $(C_{27}H_{22}O_4S)_n$
Fragments: $C_{27}H_{22}O_4S$
General Information: Unique modified polysulfone formulated for maximum cost-effectiveness while maintaining key props. Is resistant to hot water and has dimensional stability, high heat deflection temp. and toughness. Can withstand temp. 38° higher than ABS at an increased cost compared to ABS [1]

Volumetric & Calorimetric Properties:
Density:

No.	Value	Note
1	d 1.13 g/cm^3	ASTM D792 [1]

Thermal Expansion Coefficient:

No.	Value	Note	Type
1	6.5×10^{-5} K^{-1}	ASTM D696 [1]	L

Volumetric Properties General: Has higher coefficient of linear thermal expansion compared to unreinforced and reinforced polysulfone grades
Melting Temperature:

No.	Value	Note
1	280–310°C	[1]

Deflection Temperature:

No.	Value	Note
1	150°C	1.8 MPa, ASTM D648 [2]

Transition Temperature:

No.	Value	Note	Type
1	150°C	[2]	High heat deflection

Transport Properties:
Water Absorption:

No.	Value	Note
1	0.25 %	3.2 mm, 24h, 23°, ASTM D570 [1]

Mechanical Properties:
Mechanical Properties General: Has lower tensile strength at yield compared with unreinforced PSU. High tensile and flexural modulus. Lower tensile impact strength compared with unreinforced PSU. Better toughness than polysulfone [2]

Tensile (Young's) Modulus:

No.	Value	Note
1	2120 MPa	ASTM D638 [1]

– ABS, Stainless steel filled, Flame retardant

Flexural Modulus:

No.	Value	Note
1	2180 MPa	ASTM D790 [1]

Tensile Strength Break:

No.	Note	Elongation
1	ASTM D638 [1]	75%

Tensile Strength Yield:

No.	Value	Note
1	>50 MPa	min., ASTM D638 [1]

Flexural Strength Yield:

No.	Value	Note
1	83 MPa	ASTM D790 [1]

Impact Strength: Tensile impact strength 189 kJ m^{-2} (ASTM D1822) [1]
Izod Notch:

No.	Value	Notch	Break	Note
1	374 J/m	Y		ASTM [1]
2		N	Y	ISO [1]

Izod Area:

No.	Value	Notch	Note
1	24.6 kJ/m^2	Notched	ISO [1]

Electrical Properties:
Electrical Properties General: Has excellent electrical props. which are maintained over a wide temp. range and after immersion in water [1]. Volume resistivity has been reported [2]
Dielectric Permittivity (constant):

No.	Value	Frequency	Note
1	3.31	60 Hz	ASTM D150 [1]
2	3.28	1 kHz	ASTM D150 [1]
3	3.25	1 MHz	ASTM D150 [1]

Dielectric Strength:

No.	Value	Note
1	17 kV.mm^{-1}	3.2 mm, ASTM D149 [1]

Dissipation (Power) Factor:

No.	Value	Frequency	Note
1	0.0052	60 Hz	ASTM D150 [1]
2	0.0055	1 kHz	ASTM D150 [1]
3	0.013	1 MHz	ASTM D150 [1]

Optical Properties:
Volume Properties/Surface Properties: Opaque [1]

Polymer Stability:
Upper Use Temperature:

No.	Value	Note
1	125°C	[2]

Flammability: Flammability rating V2 (UL 94) [2]. Limiting oxygen index 22% [2]
Hydrolytic Stability: Very good hydrolytic stability and steam resistance [1]

Applications/Commercial Products:
Processing & Manufacturing Routes: Melt temp. range is 280–310° and mould temp. range is 70–120° in injection moulding. Prior to melt processing, the resin must be dried for 3h at 120°. Has improved melt flow over polysulfone. Can be electroplated in conventional ABS preplate systems [1,2]
Mould Shrinkage (%):

No.	Value	Note
1	0.66%	ASTM D955 [1]

Applications: Food contact application has FDA recognition. Used in automotive parts (painted and plated) and plumbing fixtures. Mindel A-670 is chromium-platable for automotive interior lighting fixtures

Trade name	Details	Supplier
Mindel A	modified polysulfone	Amoco Performance Products

Bibliographic References
[1] *Engineering Plastics for Performance and Value,* Amoco Performance Products, Inc., (technical datasheet)
[2] *The Plastics Compendium, Key Properties and Sources,* (ed. R. Dolbey), Rapra Technology Ltd., 1995, **1**
[3] *Kirk-Othmer Encycl. Chem. Technol.,* Vol. 19, 4th edn., (eds. J.I. Kroschwitz and M. Howe-Grant), Wiley Interscience, 1993, 945

ABS, Stainless steel filled, Flame retardant A-21
Synonyms: *Acrylonitrile-butadiene-styrene copolymer. Acrylonitrile-butadiene-styrene terpolymer. Poly(2-propenenitrile-co-1,3-butadiene-co-ethenylbenzene)*
Related Polymers: ABS, Stainless Steel filled
Monomers: Acrylonitrile, 1,3-Butadiene, Styrene
Material class: Composites, Copolymers
Polymer Type: ABS
CAS Number:

CAS Reg. No.
9003-56-9

Molecular Formula: $[(C_8H_8)(C_4H_6)(C_3H_3N)]_n$
Fragments: C_8H_8 C_4H_6 C_3H_3N
Additives: Stainless steel fibres. Brominated or chlorinated compounds added as fire retardants
General Information: An amorph. thermoplastic

Volumetric & Calorimetric Properties:
Density:

No.	Value	Note
1	d 1.27 g/cm^3	ASTM D792 [3]

ABS, Stainless steel filled, Flame retardant

Thermal Expansion Coefficient:

No.	Value	Note	Type
1	7.5×10^{-5} K^{-1}	ASTM D696 [3]	L

Melting Temperature:

No.	Value	Note
1	230–245°C	[10]

Glass-Transition Temperature:

No.	Value	Note
1	100–110°C	[5]

Deflection Temperature:

No.	Value	Note
1	80°C	1.8 MPa, ASTM D648 [3]
2	90°C	0.45 MPa, ASTM D648 [3]

Vicat Softening Point:

No.	Value	Note
1	92°C	50° h^{-1}, 49 N, unannealed, DIN 53480 [10]

Surface Properties & Solubility:
Solvents/Non-solvents: Sol. polar solvents, esters, ketones and some chlorinated hydrocarbons [4]

Transport Properties:
Transport Properties General: Exhibits viscoelastic behaviour in both the melt and the solid states. Like other thermoplastics, it may be described as exhibiting non-Newtonian behaviour; viscosity decreases due to increasing shear rate [2]

Water Absorption:

No.	Value	Note
1	0.35 %	24h, ASTM D570 [3]
2	0.85 %	saturation, ASTM D570 [3]

Mechanical Properties:
Mechanical Properties General: Has high impact resistance over a wide temp. range down to -40°. Has very good rigidity and very low notched Izod impact strength compared with that of ordinary grades of ABS. Elongation at break 4–6% (ASTM D638) [3]

Tensile (Young's) Modulus:

No.	Value	Note
1	2700 MPa	50 mm min.$^{-1}$, ISO R527 [10]

Flexural Modulus:

No.	Value	Note
1	2900 MPa	ASTM D790 [3]

Tensile Strength Yield:

No.	Value	Note
1	40 MPa	ASTM D638 [3]

Flexural Strength Yield:

No.	Value	Note
1	65 MPa	ASTM D790 [3]

Hardness: Rockwell M75 (ASTM D785) [3]
Friction Abrasion and Resistance: Has good abrasion resistance [2]
Izod Area:

No.	Value	Notch	Note
1	6 kJ/m^2	Notched	23°, ISO 179/2C [10]
2	16 kJ/m^2	Unnotched	23°, ISO 179/2C [10]

Electrical Properties:
Electrical Properties General: Is electrically conductive. Volume Resistivity 100–10000 Ω cm (ASTM D257) [3]; Surface Resistivity 10–1000 Ω (ASTM D257) [3]
Static Electrification: Shielding effectiveness 40-55 dB (field 3 mm, ASTM D4935) [10]

Polymer Stability:
Thermal Stability General: Has a relatively flat stress-temp. response due to its amorph. character. Retains its props. at low temps [2,4]
Upper Use Temperature:

No.	Value	Note
1	60°C	UL 746B [3]

Flammability: Flammability rating V0 (3.2 mm, UL94). [3] Limiting oxygen index 28% (ASTM 2883) [10]
Environmental Stress: Has poor weather and uv resistance unless protected. Undergoes stress cracking in contact with certain chemicals. The rubber phase in ABS oxidises more rapidly than the rigid component. Oxidation of polybutadiene results in embrittlement of the rubber [2]
Chemical Stability: Resistant to hydrocarbon solvents, mineral and vegetable oils, waxes and related household and commercial materials. Is highly resistant to staining agents in household applications. Attacked by conc. nitric and sulfuric acids. Unaffected by weak acids at room temp. and alkalis below 65°. Has poor resistance to most organic solvents [1,2,4]
Hydrolytic Stability: Resistant to solns. of acid, alkaline and neutral salts [1,4]
Biological Stability: Is biologically inert
Recyclability: Scrap and reject material can be reworked at an 80:20 (virgin:regrind) ratio if kept clean and segregated by colour and grade [5]
Stability Miscellaneous: Processing conditions can influence resultant props. by chemical and physical means. Degradation of the rubber and matrix phases occurs under very severe conditions. Evidence of morphological changes is shown by agglomeration of dispersed rubber particles during injection-moulding at higher temps. Physical effects such as orientation and moulded-in stress can affect mechanical props. [11]

Applications/Commercial Products:
Processing & Manufacturing Routes: Prod. by grafting styrene and acrylonitrile directly on to the polybutadiene latex in a batch or continuous emulsion polymerisation process. The degree of grafting is a function of the 1,2-vinyl content of the polybutadiene, the monomer concentration, the extent of conversion, temp. and mercaptan concentration. The emulsion polymerisation process involves two steps; (i) production of a rubber latex; (ii) subsequent polymerisation of styrene and acrylonitrile in the presence of the rubber latex to produce an ABS latex. This latex is then processed to isolate the ABS resin. The polymerisation ingredients include the monomers, an emulsifier, a polymerisation initiator and, usually, a chain transfer agent for MW control.[1,6,7,8,9] Processed on standard reciprocating-screw

injection-moulding machines. Recommended processing temp. 230–245° [10]

Mould Shrinkage (%):

No.	Value	Note
1	0.3–0.5%	3mm section, ASTM D955 [3]

Applications: Suitable for applications where protection against electrostatic discharges is important. Applications include all devices that have to comply with CISPR and FCC regulations regarding electromagnetic interference (EMI) shielding

Trade name	Supplier
Faradex (XA 611)	DSM Engineering Plastics
PDX-84357	LNP Engineering
PDX-A-89777	LNP Engineering

Bibliographic References

[1] *Kirk-Othmer Encycl. Chem. Technol.*, Vol. 1, 4th edn., (ed. J.I. Kroschwitz), Wiley Interscience, 1993, 391
[2] *Encycl. Polym. Sci. Eng.*, Vol. 1, 2nd edn., (ed. J.I. Kroschwitz), John Wiley & Sons, 1985, 388
[3] LNP Engineering Plastics, *Product Data Book,* LNP Engineering, (technical datasheet)
[4] Harper, C.A., *Handb. Plast. Elastomers,* (ed. C.A. Harper), McGraw-Hill, 1975
[5] Guide to Plastics, *Property and Specification Charts,* (ed. W.A. Kaplan), McGraw-Hill, 1994, **70,** 12
[6] Hayes, R.A. and Futamara, S., *J. Polym. Sci., Polym. Chem. Ed.,* 1981, **19,** 985
[7] Brydon, A., Burnett, G.M. and Cameron, C.G., *J. Polym. Sci., Polym. Chem. Ed.,* 1974, **12,** 1011
[8] Allen, P.W., Ayrey, G. and Moore, C.G., *J. Polym. Sci.,* 1959, **36,** 55
[9] Odian, G., *Principles of Polymerization,* McGraw-Hill, 1970
[10] Faradex DSM Engineering Plastics,(technical datasheet)
[11] Casale, A. and Salvatore, O., *Polym. Eng. Sci.,* 1975, **15,** 286

ABS, Stainless steel filled, General purpose   A-22

Synonyms: *Acrylonitrile-butadiene-styrene copolymer. Acrylonitrile-butadiene-styrene terpolymer. Poly(2-propenenitrile-co-1,3-butadiene-co-ethenylbenzene)*
Monomers: Acrylonitrile, 1,3-Butadiene, Styrene
Material class: Composites, Copolymers
Polymer Type: ABS
CAS Number:

CAS Reg. No.
9003-56-9

Molecular Formula: $[(C_8H_8)(C_4H_6)(C_3H_3N)]_n$
Fragments: C_8H_8 C_4H_6 C_3H_3N
Additives: 7-10% Stainless steel fibres
General Information: An amorph. thermoplastic. The fibres provide shielding against electromagnetic interference and protection against electrostatic discharges

Volumetric & Calorimetric Properties:
Density:

No.	Value	Note
1	d 1.15 g/cm^3	ASTM D792 [3]

Thermal Expansion Coefficient:

No.	Value	Note	Type
1	8×10^{-5} K^{-1}	ASTM D696 [3]	L

Melting Temperature:

No.	Value	Note
1	240–255°C	[6]

Glass-Transition Temperature:

No.	Value	Note
1	100–110°C	[5]

Deflection Temperature:

No.	Value	Note
1	80°C	1.8 MPa, ASTM D648 [3]
2	90°C	0.45 MPa, ASTM D648 [3]

Vicat Softening Point:

No.	Value	Note
1	92°C	50° h^{-1}, 49 N, unannealed, DIN 53480 [6]

Surface Properties & Solubility:
Solvents/Non-solvents: Sol. polar solvents, esters, ketones and some chlorinated hydrocarbons [4]

Transport Properties:
Transport Properties General: Exhibits viscoelastic behaviour in both the melt and solid states. Like other thermoplastics, it may be described as exhibiting non-Newtonian behaviour; viscosity decreases due to increasing shear rate [2]

Water Absorption:

No.	Value	Note
1	0.35 %	24h, ASTM D570 [3]
2	0.85 %	saturation, ASTM D570 [3]

Mechanical Properties:
Mechanical Properties General: Has high impact resistance over a wide temp. range down to -40°. Has very good rigidity and very low notched Izod impact strength compared with that of ordinary grades of ABS. Elongation at break 3–6% (ASTM D638) [3,6]

Tensile (Young's) Modulus:

No.	Value	Note
1	2758 MPa	400 kpsi, ASTM D638 [5]

Flexural Modulus:

No.	Value	Note
1	3000 MPa	ASTM D790 [3]

Tensile Strength Yield:

No.	Value	Note
1	50 MPa	ASTM D638 [3]

Flexural Strength Yield:

No.	Value	Note
1	75 MPa	ASTM D790 [3]

Hardness: Rockwell M 77 (ASTM D785) [3]
Friction Abrasion and Resistance: Has good abrasion resistance [2]
Izod Notch:

No.	Value	Notch	Note
1	9 J/m	Y	23°, ISO 180/1A [6]
2	23 J/m	N	23°, ISO 180/1A [6]

Izod Area:

No.	Value	Notch	Note
1	9 kJ/m^2	Notched	23°, ISO 180/1A [6]
2	23 kJ/m^2	Unnotched	23°, ISO 180/1A [6]

Electrical Properties:
Electrical Properties General: Is electrically conductive. Surface Resistivity 10–1000 Ω cm (ASTM D257) [3]; Volume Resistivity 100–10000 Ω cm (ASTM D257) [3]
Static Electrification: Shielding effectiveness 40–55 dB (field 3 mm, ASTM D4935) [6]

Polymer Stability:
Thermal Stability General: Has a relatively flat stress-temp. response due to its amorph. character. [2,4] Retains its props at low temps.[2,4]
Upper Use Temperature:

No.	Value	Note
1	60°C	UL 746B [3]

Flammability: Flammability rating HB (3.2 mm, UL94). [3] Limiting oxygen index 18% (ASTM 2883) [6]
Environmental Stress: Has poor weather and uv resistance unless protected. Undergoes stress cracking in contact with certain chemicals. The rubber phase in ABS oxidises more rapidly than the rigid component. Oxidation of polybutadiene results in embrittlement of the rubber [2]
Chemical Stability: Resistant to hydrocarbon solvents, mineral and vegetable oils, waxes and related household and commercial materials. Is highly resistant to staining agents in household applications. Attacked by conc. nitric and sulfuric acids. Unaffected by weak acids at room temp. and alkalis below 65°. Has poor resistance to most organic solvents [2,4]
Hydrolytic Stability: Resistant to solns. of acid, alkaline and neutral salts [1,4]
Biological Stability: Is biologically inert
Recyclability: Scrap and reject material can be reworked at an 80:20 (virgin:regrind) ratio if kept clean and segregated by colour and grade [5]
Stability Miscellaneous: Processing conditions can influence resultant props. by chemical and physical means. Degradation of the rubber and matrix phases occurs under very severe conditions. Evidence of morphological changes is shown by agglomeration of dispersed rubber particles during injection-moulding at higher temps. Physical effects such as orientation and moulded-in stress can affect mechanical props. [11]

Applications/Commercial Products:
Processing & Manufacturing Routes: Prod. by grafting styrene and acrylonitrile directly on to the polybutadiene latex in a batch or continuous emulsion polymerisation process. The degree of grafting is a function of the 1,2-vinyl content of the polybutadiene, the monomer concentration, the extent of conversion, temp. and mercaptan concentration. The emulsion polymerisation process involves two steps: (i) production of a rubber latex; (ii) subsequent polymerisation of styrene and acrylonitrile in the presence of the rubber latex to produce an ABS latex. This latex is then processed to isolate the ABS resin. The polymerisation ingredients include the monomers, an emulsifier, a polymerisation initiator and, usually, a chain transfer agent for MW control. [1,7,8,9,10] Processed on standard reciprocating-screw injection-moulding machines. Should be dried before melt processing. Recommended processing temp. 240–255° [6]
Applications: Applications in the protection of equipment against electrostatic discharge, and against electromagnetic interference, so that equipment complies with CISPR and FCC regulations

Trade name	Details	Supplier
A5-7SS/006		Compounding Technology
Electrafil G1204/SS/ 7/BK-2089		DSM Engineering Plastics
Faradex (XA 111)		DSM Engineering Plastics
Stat-Kon AS		LNP Engineering
Thermocomp (PDX-A-88057)	(also used to refer to other materials)	LNP Engineering

Bibliographic References
[1] *Kirk-Othmer Encycl. Chem. Technol.*, Vol. 1, 4th edn., (ed. J.I.Kroschwitz), Wiley Interscience, 1993, 391
[2] *Encycl. Polym. Sci. Eng.*, Vol. 1, 2nd edn., (ed. J.I. Kroschwitz), John Wiley & Sons, 1985, 388
[3] LNP Engineering Plastics, *Product Data Book,* LNP Engineering Plastics, (technical datasheet)
[4] Harper, C.A., *Handb. Plast. Elastomers,* (ed. C.A. Harper), McGraw-Hill, 1975
[5] Guide to Plastics, *Property and Specification Charts,* (ed. W.A. Kaplan), McGraw-Hill, 1994, **70**, 12
[6] *Faradex,* DSM Engineering Plastics, (technical datasheet)
[7] Hayes, R.A. and Futamara, S., *J. Polym. Sci., Polym. Chem. Ed.,* 1981, **19**, 985
[8] Brydon, A., Burnett, G.M. and Cameron, C.G., *J. Polym. Sci., Polym. Chem. Ed.,* 1974, **12**, 1011
[9] Allen, P.W., Ayrey, G. and Moore, C.G., *J. Polym. Sci.,* 1959, **36**, 55
[10] Odian, G., *Principles of Polymerization,* McGraw-Hill, 1970
[11] Casale, A. and Salvatore, O., *Polym. Eng. Sci.,* 1975, **15**, 286

Acemannan   A-23
Synonyms: *Carrisyn. Polymannoacetate*
Material class: Polysaccharides
CAS Number:

CAS Reg. No.
110042-95-0

Molecular Weight: >10000 Dalton
Morphology: Long-chain polymer consisting of linear (1→4)-D-mannopyranosyl units, randomly acetylated (0.8 Ac per monomer)
Miscellaneous: Constit. of leaf juice of *Aloe barbadensis* and *Aloe arborescens* (Liliaceae). Biol. active component of carrisyn extract (73-90%). Potent inducer of Interleukin 1 and Prostaglandin E_2 production by human peripheral blood adherent cells *in vitro*. Influences activity and production of T-lymphocytes, fibroblasts, B-lymphocytes and endothelial cells. Adjuvant and immunoenhancer. Increases antibody production, stimulates phagocytosis. Shows specific antineoplastic activity against sarcoid tumours in horse. Used in the treatment of inflammatory bowel disease

Bibliographic References
[1] *Pat. Coop. Treaty (WIPO),* (Carrington Lab), 1987, 0 052, (extraction)
[2] Womble, D., *Int. J. Immunopharmacol.,* 1988, **10**, 967, (activity)
[3] *Pat. Coop. Treaty (WIPO),* (Carrington Lab), 1990, 01 253, (prep; activity)
[4] Fogleman, R.W., *Vet. Hum. Toxicol.,* 1992, **34**, 144; 201, (tox)
[5] Martindale, The Extra Pharmacopoeia, 30th edn., *Pharmaceutical Press,* 1993, 536

Acetal copolymer A-24

Synonyms: *Poly(oxymethylene-co-ethylene oxide). Polyacetal copolymer. Ethylene oxide-trioxane copolymer. Poly(1,3,5-trioxane-co-oxirane)*
Related Polymers: Polyacetal, Polyoxymethylene
Monomers: Trioxane, Ethylene oxide
Material class: Thermoplastic, Copolymers
Polymer Type: Acetal, Polyethylene oxide
CAS Number:

CAS Reg. No.
24969-25-3

Molecular Formula: $[(CH_2O).(C_2H_4O)]_n$
Fragments: CH_2O C_2H_4O
Additives: Antioxidants, UV stabilisers, pigments, glass fibre, lubricants, PTFE
Miscellaneous: The exact composition of many available copolymers is undisclosed. Many are thought to be copolymers with ethylene oxide in the main chain

Applications/Commercial Products:
Processing & Manufacturing Routes: Processed using conventional techniques such as extrusion, injection moulding and rotational moulding
Applications: Consumer goods, e.g. kettles. Plumbing such as fittings and valves, fuel caps, pump housings, gears, bearings, air flow valves and filter housings

Trade name	Details	Supplier
AC/AH	various grades	Michael Day
ACX		United Composites
Acetal	AC range	Albis Corp.
Acetron		DSM Engineering
Ashlene	R190/180/190/195	Ashley Polymers
CP/HP	various grades	Plastic Materials
CTI	AT series	CTI
Celcon		Celanese Plastics (Hoechst)
Celstran	ACG40-0104 / ACK30-01-2	Polymer Composites
Delrin	various grades	DuPont
Duracon		Polyplastics
Electrafil	J-80	DSM Engineering
Fiberfil	J-80	DSM Engineering
Fulton	400 series	LNP Engineering
G-10/G-20/G-9900/ G-1/G6	various grades	Thermofil Inc.
Hostaform		Hoechst Celanese
Iupital	various grades	Mitsubishi Gas
J-82	various grades	DSM Engineering
Kematal		Amcel
Kematal		Celanese Plastics
Kemlex	10000/12000 series	Ferro Corporation
Kepital	F10/F20/F30	Korea Engineering
Lubricomp	K series	LNP Engineering
Lubriplas	POM-1100 series	Bay Resins
Lucet	various grades	Lucky
MAC/MAT	various grades	Modified Plastics
POM CO	various grades	TP Composites
PRL-Acetal	AC range	Polymer Resources
Perma Stat 800		RTP Company
Plaslube	AC/J ranges	DSM Engineering
RTP	800 series	RTP Company
Stat-Kon	KC-1002/KCL-4022	LNP Engineering
Tenac	various grades	Asahi Chemical
Tenac-C	various grades	Asahi Chemical
Thermocomp KF		LNP Engineering
Ultraform		BASF

Acrylamide copolymer emulsion A-25

Related Polymers: Polyacrylamide
Monomers: Acrylamide, Acrylic acid
Material class: Emulsions and Gels, Copolymers
Polymer Type: Acrylic
CAS Number:

CAS Reg. No.	Note
9003-06-9	polymer with 2-propenoic acid
121236-26-8	polymer with 2-propenoic acid, block
25987-30-8	polymer with 2-propenoic acid, sodium salt
59326-44-2	polymer with 2-propenoic acid, potassium salt
70559-15-8	polymer with 2-propenoic acid, magnesium salt
55554-55-7	polymer with 2-propenoic acid, ammonium salt
137295-46-6	polymer with 2-propenoic acid, calcium sodium salt
26100-47-0	polymer with 2-propenoic acid, ammonium salt
32843-78-0	polymer with 2-propenoic acid, lithium salt
31212-13-2	polymer with 2-propenoic acid, potassium salt
25085-02-3	polymer with 2-propenoic acid, sodium salt
115528-83-1	polymer with 2-propenoic acid, sodium salt, block
119547-32-9	polymer with 2-propenoic acid, aluminium salt

General Information: One of the most useful groups of water-soluble polymers
Tacticity: The cotacticity parameters of the copolymer are very similar to those of the homopolymers in the range 0.4–0.5 [10]
Morphology: The copolymers prepared by either alkaline hydrol. or direct copolymerisation, are heterogeneous. Acid hydrol. produces blocks of acrylate [10,12]. The microstruct. of copolymer hydrogels varies with the amount of acrylate. Multifractal theory has been used to study and characterise the micromophologies of copolymer hydrogels [20]. Pattern formation during the shrinking of a constrained gel in Me_2CO aq. has been reported [26]
Miscellaneous: Can be cross-linked to form gels which are very highly water-absorbing [5]

Volumetric & Calorimetric Properties:
Equation of State: Mark-Houwink parameters have been reported [1]
Glass-Transition Temperature:

No.	Value	Note
1	96.6–99.8°C	1.5–5% acrylamide [5]
2	105–113.5°C	cross-linked copolymer; 1.5–5% acrylamide [5]

Surface Properties & Solubility:
Solubility Properties General: The solubility of copolymer in water can be affected to different degrees by the presence of salts. The acid copolymers are more susceptible to adverse effects than are the partially neutralised copolymers. A soln. of a copolymer with 37 mol% acrylic acid will become milky in the presence of $CaCl_2$ at a level of 0.08 mol l^{-1} and the copolymer is precipitated at 0.4 mol l^{-1} [15]

Transport Properties:
Polymer Solutions Dilute: The rheological and flow props. of copolymer solns. are adversely affected by the presence of CO_2. This is due to a combination of effects: chemical degradation of the copolymer and, more importantly, modification of the soln. struct. [6]. Zero shear intrinsic viscosity, molecular expansion factor and polymer size all vary with pH, showing a maximum in the region of pH 8 [4]. The viscoelastic props. and rheological behaviour of copolymer solns. can be seriously affected by the presence of salts. The presence of multivalent cations causes complex formation with the copolymer which in turn affects the viscosity behaviour. Univalent cations affect the viscosity but di- and trivalent cations affect chain conformation and solubility as well. The intrinsic viscosity increases with increasing temp. [7,13,18,19]. Two types of negative thixotropy have been reported in dilute solns. of partially hydrolysed polyacrylamide in water/glycerol mixtures. The character of the effect depends on amount of glycerol, shear rate and polymer concentration. The critical shear rate at which type I negative thixotropy occurs increases in the presence of NaCl, whereas critical shear rate of type II negative thixotropy is unaffected [17]. Intrinsic viscosity data have been reported [2]

Mechanical Properties:
Mechanical Properties General: The mechanical props. of the gels depend very much on the conc., degree of swelling and network density [11]. Di- and trivalent cations give rise to ionic cross-links which change the elastic modulus of the gel [9]

Optical Properties:
Optical Properties General: Refractive index increment 0.183-0.191 ml g^{-1} (dependent on amount) [2]. Aq. solns. of the copolymer show a flow-induced negative birefringence [14]
Transmission and Spectra: Ir and C-13 nmr spectral data have been reported [5,10,13]

Polymer Stability:
Decomposition Details: Decomposition starts at approx. 220° with some prior loss of moisture. Second decomposition starts at approx. 320° and ends at approx. 400° when a third stage begins. The maximum decomposition rate occurs during the third stage at approx. 425°. The main reaction in stage 1 is imidisation with loss of ammonia; in stage 2 degradation of carboxylate groups and in stage 3 breakdown of backbone. Mechanistic details have been reported [16]
Hydrolytic Stability: The hydrol. of the copolymer as a function of pH and composition has been studied [8]
Stability Miscellaneous: The copolymer is subject to shear degradation, the intrinsic viscosity falling as the degradation proceeds. Degradation is reduced for low soln. concentrations, in solns. of high ionic strength and with higher levels of ionic charge in the copolymer [7]

Applications/Commercial Products:
Processing & Manufacturing Routes: The copolymer can be synth. by hydrol. of polyacrylamide or by the copolymerisation of acrylamide and acrylic acid. The hydrol. can be carried out under alkaline (or sometimes acid) conditions at elevated temps. (0.1M polyacrylamide, 0.25M NaOH, 0.1 M NaCl, at 30° or 50°). The copolymerisation can be performed by the same methods used for acrylamide, e.g. radical copolymerisation (using persulfate initiator and $K_2S_2O_8/Na_2S_2O_5$ redox system in water, or 0.1M NaCl soln. at 60°) or inverse emulsion copolymerisation [10,12,13,21]. Hydrol. of a homopolymer gel gives a copolymer gel with unchanged topography [11]

Applications: Used in the oil industry (drilling fluids and enhanced oil recovery), as paper filter retention aids and effluent water clarification aids

Trade name	Details	Supplier
HPAM		Calgon Corp.
HPAM		BASF
Pusher		Dow
Separan		Dow
Texipol	63, 67 series	Scott Bader

Bibliographic References
[1] Wu, X.Y., Hunkeler, D., Hamielec, A.E., Pelton, R.H. and Woods, D.R., *J. Appl. Polym. Sci.*, 1991, **42**, 2081, (MW)
[2] McCarthy, K.J., Burkhardt, C.W. and Parazak, D.P., *J. Appl. Polym. Sci.*, 1987, **33**, 1699, (dilute soln. props.)
[3] McCarthy, K.J., Burkhardt, C.W. and Parazak, D.P., *J. Appl. Polym. Sci.*, 1987, **33**, 1683, (conformn.)
[4] Tam, K.C., and Tiu, C., *Polym. Commun.*, 1990, **31**, 102
[5] Liu, Z.S. and Rempel, G.L., *J. Appl. Polym. Sci.*, 1997, **64**, 1345, (synth.)
[6] Lakatos, I. and Lakatos-Szabo, J., *Colloid Polym. Sci.*, 1996, **274**, 959, (rheological props.)
[7] Henderson, J.M. and Wheatley, A.D., *J. Appl. Polym. Sci.*, 1987, **33**, 669
[8] Kheradmand, H., Francois, J. and Plazanet, V., *Polymer*, 1988, **29**, 860, (hydrol.)
[9] Ben Jar, P.-Y. and Wu, Y.S., *Polymer*, 1997, **38**, 2557
[10] Truong, N.D., Galin, J.C., Francois, J. and Pham, Q.T., *Polymer*, 1986, **27**, 459, (struct.)
[11] Opperman, W. Rose, S. and Rehage, G., *Br. Polym. J.*, 1985, **17**, 175, (elastic behaviour)
[12] Halverson, F., Lancaster, J.E. and O'Connor, M.N., *Macromolecules*, 1985, **18**, 1139
[13] Rangaraj, A., Vangani, V. and Rakshit, A.K., *J. Appl. Polym. Sci.*, 1997, **66**, 45, (synth.)
[14] Farinato, R.S., *Polymer*, 1988, **29**, 2182, (birefringence)
[15] Muller, G. Laine, J.P. and Fenyo, J.C., *J. Polym. Sci. Part A*, 1979, **17**, 659, (conformn.)
[16] Leung, W.M., Axelson, D.E. and Van Dyke, J.D., *J. Polym. Sci. Part A*, 1987, **25**, 1825, (thermal degradation)
[17] Bradna, P., Quadrat, O. and Dupuis, D., *Colloid Polym. Sci.*, 1995, **273**, 421
[18] Tam, K.C. and Tiu, C., *Colloid Polym. Sci.*, 1990, **268**, 911
[19] Tam, K.C. and Tiu, C., *Colloid Polym. Sci.*, 1994, **272**, 516, (viscoelastic props.)
[20] Shan, J., Chen, J., Liu, Z. and Zhang, M., *Polym. J. (Tokyo)*, 1996, **28**, 886
[21] Banda, D; Snupcrek, J. and Cermak, V., *Eur. Polym. J.*, 1997, **33**, 1345, (synth.)

Acrylic fibres
Related Polymers: Polyacrylonitrile
Monomers: Acrylamide, Acrylonitrile, Butyl acrylate, Vinyl acetate, Methylenebutanedioic acid, Ethylhexyl acrylate, Methyl acrylate
Material class: Thermoplastic, Fibres and Films, Copolymers
Polymer Type: Acrylic, Acrylonitrile and copolymers, Acrylic copolymers

Molecular Weight: M_n 40000-60000; MW 90000-140000
Additives: Dyes. Triclosan (as an antibacterial agent). Flame retardants (tris(dibromopropyl phosphate)
General Information: Acrylic fibre is available in a variety of forms including: coloured, uv-resistant, bi-component, abrasion resistant and various specialist forms (e.g. antibacterial). It is used in both staple and tow forms. Fibres are available with strengths of 1.2-15 den; 3 den is probably the normal strength. Acrylic microfibre is also available. Fibres are manufactured with a minimum of 85% acrylonitrile monomer: the comonomer(s) chosen to give a final product with the desired props. Perhaps the commonest forms of acrylic fibre are ternary copolymers containing 89-95% acrylonitrile, 4-10% of a non-ionogenic

Acrylic fibres

monomer and 0–1% of an ionogenic monomer. 100% Acrylonitrile fibres are produced for industrial use. Much of the interest in PAN fibres in recent years has been in its application in the manufacture of carbon fibres

Tacticity: In a series of acrylonitrile/methyl acrylate copolymers the tacticities are virtually the same regardless of the copolymer composition [21]

Morphology: PAN and acrylonitrile copolymer fibres are paracrystalline. The conform. of PAN chains is mainly planar zig-zag. Packing of the chains is pseudohexagonal with sequences of average periodicity of approx. 2.4Å. [9,10] The shape of the fibre's cross-section depends on composition, spinning process and conditions, e.g. a dumbbell shape indicates a dry-spun fibre and a kidney shape a wet-spun fibre. Props. of the fibre, such as a lustre, moisture regain and some mech. props., are affected by the shape of its cross-section [1,12]

Miscellaneous: Acrylic fibre is wool-like in nature and is the only synthetic fibre with an uneven surface even when extruded from a round-hole spinneret. Acrylic fibres have excellent wickability, are quick-drying, resistant to uv degradation, oil, chemicals and moth-attack. They are readily dyeable to most colours with excellent fastness. They are washable and retain their shape after washing

Volumetric & Calorimetric Properties:

Density:

No.	Value	Note
1	d 1.14–1.19 g/cm^3	[2]

Melting Temperature:

No.	Value	Note
1	320°C	approx., PAN; decomposes
2	150°C	approx., 10% H$_2$O [34]

Glass-Transition Temperature:

No.	Value	Note
1	97°C	PAN
2	80°C	approx., commercial fibres [3]

Transition Temperature General: Measurement of T_m of PAN fibres is difficult owing to its decomposition. The presence of moisture lowers T_g by 35–50° and can lower T_m to approx. 150° [2,3,34]. Studies on the thermal transitions between 22° and 200° have been reported [11,13]

Surface Properties & Solubility:

Plasticisers: PAN is plasticised by H$_2$O and other hydrogen-bonding solvents of low MW [3,34]

Solvents/Non-solvents: Sol. DMF, DMSO, dimethyl acetamide, ethylene carbonate. Insol. H$_2$O, C$_6$H$_6$, mineral oil, CCl$_4$, aliphatic hydrocarbons, aliphatic alcohols, aliphatic esters, aliphatic ketones [1]

Wettability/Surface Energy and Interfacial Tension: Contact angle 48° (H$_2$O, advancing) [3]

Transport Properties:

Transport Properties General: The presence of water increases the melt flow index. Addition of ethylene carbonate to the hydrated PAN melt increases the melt flow index but replacement of some of the water with ethylene carbonate gives rise to significant decrease in melt flow index [34]

Melt Flow Index:

No.	Value	Note
1	0 g/10 min	approx., 170°, 1.5 MPa, 17% H$_2$O [34]
2	2 g/10 min	approx., 170°, 1.5 MPa, 20% H$_2$O [34]
3	10 g/10 min	approx., 170°, 1.5 MPa, 23% H$_2$O [34]
4	20 g/10 min	approx., 170°, 1.5 MPa, 26% H$_2$O [34]

Polymer Solutions Dilute: Intrinsic viscosity 1.67 dl g^{-1} (DMF, 30°, acrylonitrile, MW 370000); 1.71 dl g^{-1} (DMF, 30°, acrylonitrile, ethylhexyl acrylate/itaconic acid (98/1.5/0.5), MW 372000); 1.61 dl g^{-1} (DMF, 30°, acrylonitrile/ ethylhexyl acrylate/itaconic acid (93/6.5/0.5), MW 331000); 0.98 dl g^{-1} (DMF, 30°, acrylonitrile/methyl acrylate/itaconic acid, MW 165000); 1.88 dl g^{-1} (DMF, 30°, acrylonitrile/methyl acrylate/itaconic acid, MW 429000) [13]

Permeability of Gases: Acrylonitrile/methyl acrylate copolymer has a high intrinsic selectivity and a low intrinsic permeability. Acrylic fibres do not swell: articles made from them remain permeable to air even when wet [1,23]

Water Content: Moisure regain can be enhanced by incorporation of hydrophilic comonomers

Water Absorption:

No.	Value	Note
1	1.5–2.5 %	[2]
2	2.1–2.3 %	65% relative humidity, acrylonitrile/methyl acrylate [25]
3	7.2–7.5 %	65% relative humidity, acrylonitrile/acrylamido methylpropanesulfonic acid [25]

Mechanical Properties:

Mechanical Properties General: Mech. props. are affected by the crystallite sizes and orientation of the fibres and by varying composition and MW. [13,14] Wet conditions do not severely affect the mech. props. but hot, wet conditions adversely affect some props. Acrylic fibres are better than wool but similar to cotton in terms of their tensile strength and abrasion resistance. Gel-spinning can be used to produce high strength fibres from high MW polymer [1,2] As temps. increases elongation at break increases, slowly initially and then rapidly as T_g is approached. Creep rate also increases with temp., the rate of increase becoming greater nearer T_g. The effects of stretching a fibre at temps. below the stabilisation range (at 185°) have been reported. As the stretch ratio increases fibre diameter decreases and Young's modulus increases. Tensile strength initially decreases but then increases as stretch ratio is taken above zero; similarly, elongation initially increases slightly and then decreases. The mechanism of tensile extension/recovery has been reported. [4,8,17] Tenacity 0.09–0.33 N tex^{-1} (35–55% elongation, dry); 0.14–0.24 N tex^{-1} (40–60% elongation, wet); 0.09–0.3 N tex^{-1} (loopknot); 2 N tex^{-1} (7–10% elongation, gel-spun fibre); 0.7–0.8 N tex^{-1} (9–11% elongation, asbestos replacement grade) [2]; 0.22–0.3 N tex^{-1} (acrylonitrile/vinyl acetate (91/9), 17.4–20.7% elongation); 0.14–0.29 N tex^{-1} (polyacrylonitrile/vinyl acetate (91/9), flame retardant); 3.4–4.4 g den^{-1} (acrylonitrile/itaconic acid (97/3), 7.6–9.7% elongation, MW 131000; high temp. drawing); 1.8–3.1 g den^{-1} (acrylonitrile/itaconic acid (97/3), 11.1–11.8% elongation, MW 131000, boiling water drawing) [26]; average modulus 0.44–0.62 N tex^{-1} (dry). Tensile modulus 114–144g den^{-1} (acrylonitrile/itaconic acid (97/3), MW 131000, high temp. drawing); 78–90 g den^{-1} (acrylonitrile/itaconic acid (97/3), MW 131000, boiling water drawing) [26]

Tensile (Young's) Modulus:

No.	Value	Note
1	16000 MPa	hollow fibre, MW 307000, 10% doped [5,6]
2	21000 MPa	hollow fibre, MW 307000, 20% doped [5,6]
3	47000 MPa	hollow fibre, MW 326000 [5,6]

Tensile Strength Break:

No.	Value	Note
1	470 MPa	hollow fibre, MW 307000, 10% doped [5,6]
2	600 MPa	hollow fibre, MW 30700, 20% doped [5,6]
3	800 MPa	hollow fibre, MW 326000 [5,6]

Viscoelastic Behaviour: Elastic recovery 99% (2% stretch). Sonic modulus 95–207 g den^{-1} (acrylonitrile/itaconic acid (97/3), MW 131000, varying drawing conditions) [26]
Mechanical Properties Miscellaneous: Macromolecular entanglements in PAN and PAN-copolymer fibres and the effects of various thermal treatments on the texture of fibres have been reported [19]
Failure Properties General: Crease recovery is an important factor for fibres with a flat or rectangular cross-section. The effect of flatability (ratio of cross-sectional length to width) and modulus on crease recovery has been reported. Crease recovery of acrylonitrile/vinyl acetate (85/15) fibres improves with increasing modulus and flatability but reaches a maximum when the modulus reaches approx. 8.6 g den^{-1} [16]
Friction Abrasion and Resistance: Increasing the draw ratio gives fibres with an increased molecular orientation, a smoother surface and as a result, an increased fibre-to fibre coefficient of friction [32]

Electrical Properties:
Electrical Properties General: Fibres have high electrical resistance [2]. Electrically conducting acrylic fibres can be made by binding copper sulfide onto the surface of fibres which have pendant groups with an affinity for Cu(II) [33]
Static Electrification: Show moderate static build-up. Treatment of fibres or fabric with cationic/anionic surfactant mixtures can have a dramatic effect on the surface conductivity, greatly reducing the discharge time [2,29]. Can be made with antistatic props. by a treatment which results in the formation of polyaniline within and at the surface of the fibres. Antistatic fibres made in this way have a much lower max. static charge and half-life than the untreated material [30]

Optical Properties:
Optical Properties General: Birefringence -0.00042 (Dralon) [31]
Refractive Index:

No.	Value	Note
1	1.512	546.1 nm, Dralon, parallel to fibre axis [31]
2	1.5162	546.1 nm, Dralon, perpendicular to fibre axis [31]

Total Internal Reflection: Index of birefringence 0.1 [2]

Polymer Stability:
Thermal Stability General: Fibres have good colour stability below 120°. [2] PAN fibres shrink by up to 9.4% when heated to 220° [15]
Decomposition Details: If heated above 200° PAN darkens, resulting in a black solid in which the nitrile groups have cyclised to give a conjugated ladder polymer with high thermal stability. The temp. at which cyclisation begins is dependent on the comonomers present. Further heating at higher temps. under the correct conditions can yield carbon fibres. The formation of the cyclised product, the ladder polymer, is of great interest because of its influence on the props. of the subsequently produced carbon fibres [7,21,24]
Flammability: On burning acrylic fibres melt, ignite and then burn with a sooty, yellow flame leaving a dark, brittle residue. [1] Fibres are moderately flammable. [2] Limiting oxygen index 18% [2] 33.3% (acrylonitrile/vinyl acetate, (91/9) 38.5–39.6% (acrylonitrile/vinyl acetate, (91/9), 25% tributyl phosphate) 43.3% (acrylonitrile/vinyl acetate, (91/9), 25% aluminium hydroxide) 43.3–50.0% (acrylonitrile/vinyl acetate, (91/9), 25% tris(dibromopropyl) phosphate) [8] The flame resistance of PAN fibres can be improved by copolymerisation, blending and by surface treatment. Copolymerisation and blending can lead to significant changes in props., however, and surface treatments are often readily removed. Surface modification of PAN fibres by plasma polymerisation using hexamethyldisiloxane and ethyldichlorophosphate has been reported. [6] Stabilisation of PAN-based fibres, I.e cyclisation by heating above 200°, improves flame resistance [7]
Environmental Stress: Has excellent resistance to weathering and is lightfast. These props. are better than those of natural or other synthetic products. Discoloration brought about by weathering is very much a surface effect. A decrease in MW accompanies chemical change, evidenced by an increase in the oxygen content of the fibre and by changes in its ir spectrum [1,2,19]
Chemical Stability: Acrylic fibre has good chemical resistance. Is resistant to weak alkali, oxidising agents and moderately concentrated mineral acids. Dissolved by concentrated nitric acid or sulfuric acid. Fibres have good resistance to smoke, soot and industrial gases [1,2]
Hydrolytic Stability: PAN is resistant to hydrol. and can be used in humid atmospheres at temps. up to 140° [1]
Biological Stability: Acrylic fibre does not rot and is not attacked by bacteria, fungus or insects. Antibacterial props. can be imparted to PAN fibres by the incorporation of carboxylic groups into the polymer and subsequent impregnation with antibiotics [1,22]
Stability Miscellaneous: The effect of γ-irradiation on optical props. has been reported [31]

Applications/Commercial Products:
Processing & Manufacturing Routes: The polymers or copolymers used for acrylic fibre production can be made by anionic methods but radical polymerisation is more common. Precipitation polymerisation and soln. polymerisation are used to prepare the parent copolymer; soln. polymerisation has the advantage in that the polymer soln. can be spun with no further preparation step. Aq. media are the most widely used for acrylic fibre polymerisations. Commonly used non-ionogenic comonomers include methyl methacrylate, methyl acrylate, vinyl acetate and acrylamide. Ionogenic comonomers are generally monomers with sulfonic acid groups (e.g. styrenesulfonic acid) or basic compounds such as the vinylpyridine and dialkylaminoalkyl methacrylates. The composition of the copolymer determines the dyeing props. of the fibres: the use of different types of dye is made possible by use of the appropriate comonomers. Acrylic fibres can be dry spun or wet spun and can be used in staple or tow form. Polyacrylonitrile decomposes below its melting point and therefore cannot be spun from the melt. Mixtures of polyacrylonitrile and water, however, can melt to give a single phase without decomposition of the polymer but spinning from this has not been carried out commercially. The presence of water in excess of its saturation level gives rise to foaming during melt spinning. Fibres spun with foaming have a greatly reduced density and inferior mech. props. Spinning in the supercooled state can give foam-free fibres. The shape of the fibre's cross section depends on composition, MW, spinning process and conditions e.g. a dumbbell shape indicates a dry-spun fibre and a kidney shape a wet-spun product. Props. of the fibre, such as lustre, moisture regain and some mech. props. are affected by the shape of its cross-section. Fibres are spun to give the cross-section shape most appropriate for the intended use, e.g. triangular fibres where a good lustre is required and flat fibres for carpets, blankets and synthetic fur. Hollow fibres can be spun using an appropriate spinneret. After spinning, the fibres are put through several further stages including: washing, drawing, finish application, crimping, drying, cutting, steaming and packing. Dyeing of the fibre can be carried out before (spin-dyeing), during or immediately after (gel-dyeing) or during the after treatment of the spun fibre [1,2,5,12,13,25,23]. Acrylic fibres can be modified in many ways to achieve the required props. The incorporation of hydrophilic monomers can improve the moisture-absorbing props. of acrylic fibres as can increasing their porosity. Abrasion resistance can be improved by increasing fibre density. This may be achieved by adjustment of spinning conditions; the incorporation of comonomers with smaller molar volumes (e.g. vinylidene chloride); the incorporation of hydrophilic comonomers to reduce

void content (e.g. sulfonated or acrylamide derivatives). Pilling problems can be reduced by producing a more brittle fibre which may be achieved by adjustment of spinning conditions, annealing at lower temps., by using less comonomer or by adding an ionic comonomer [2]. Antistatic acrylic fibre can be prepared by exposing fibre impregnated with oxidising agent to aniline vapour [31]

Applications: One of the major uses of acrylic fibre is as a precursor for carbon fibre. Other uses include clothing (sportswear, knitwear, etc.); furnishings (upholstery, blankets, carpets, garden furnishings, awnings, etc.); in brake and clutch linings, as an asbestos replacement, and in concrete reinforcement. Stabilised PAN-based fibres (heated to 200–300°) find application in garments where non-burning fabric is required

Trade name	Details	Supplier
Acrilan		Solutia
Biofresh	antibacterial fibre	Sterling
Cashmilon		Asahi Chemical
Courtelle		Courtaulds
Dolan		Hoechst Celanese
Dolanit		Hoechst Celanese
Dralon		Bayer Inc.
Dralon T		Bayer Inc.

Bibliographic References

[1] Nozaj, A. and Suliag, G., *Ullmanns Encykl. Ind. Chem.*, **A10**, 629
[2] Knorr, R.S., *Encyclopaedia of Chemical Technology*, 4th edn., 1993, **10**, 559
[3] Henrici-Olive, G. and Olive, S., *Adv. Polym. Sci.*, 1979, **32**, 123
[4] Rosenbaum, S., *J. Appl. Polym. Sci.*, 1965, **9**, 2071, (mech props.)
[5] Yang, M-C. and Yu, D.-G., *J. Appl. Polym. Sci.*, 1998, **68**, 1331, (props)
[6] Akovali, G and Gundogan, G., *J. Appl. Polym. Sci.*, 1990, **41**, 2011, (flammability)
[7] Ko, T.-H., *J. Appl. Polym. Sci.*, 1993, **47**, 707, (flammability)
[8] Tsai, J.-S., *J. Mater. Sci.*, 1993, **28**, 1161, (additives)
[9] Liu, X.D. and Ruland, W., *Macromolecules*, 1993, **26**, 3030, (struct)
[10] Rizzo, P., Auriemma, F., Guerra, G., Petraccone, V. and Corradini, P., *Macromolecules*, 1996, **29**, 8852, (conformn)
[11] Hu, X.P. and Hsieh, Y.-L., *Polymer*, 1997, **38**, 1491, (struct)
[12] Tsai, J.-S. and Su, W.-C., *J. Mater. Sci. Lett.*, 1991, **10**, 1253
[13] Tsai, J.-S. and Lin, C.-H., *J. Appl. Polym. Sci.*, 1991, **42**, 3039, 3045, (props)
[14] Tsai, J.-S. and Lin, C.-H., *J. Mater. Sci.*, 1991, **26**, 3996, (props)
[15] Sarvaranta, I., *J. Appl. Polym. Sci.*, 1995, **56**, 1085, (shrinkage)
[16] Tsai, J.-S., *J. Mater. Sci. Lett.*, 1995, **14**, 66
[17] Wang, P.H. Liu, J. and Li, R.Y., *J. Appl. Polym. Sci.*, 1994, **52**, 1667, (mech props)
[18] Howell, H.E. and Patil, A.S., *J. Appl. Polym. Sci.*, 1992, **44**, 1523, (weathering)
[19] Qian, B.; Hu, P.; He, J.; Zhao, J.X. and Wu, C., *Polym. Eng. Sci.*, 1992, **32**, 1290
[20] Qian, B.J., Qin, J., Wu, Z.Q., Wu, C.X. et al, *J. Appl. Polym. Sci.*, 1992, **45**, 871
[21] Bang, Y.H., Lee, S. and Cho, H.H., *J. Appl. Polym. Sci.*, 1998, **68**, 2205
[22] Buchenska, J., *J. Appl. Polym. Sci.*, 1997, **65**, 1955
[23] Shilton, S.J., Bell, G. and Ferguson J., *Polymer*, 1994, **35**, 5327, (gas permeability)
[24] Zhu, Y., Wilding, M.A. and Mukhapodhay, S.K., *J. Mater. Sci.*, 1996, **31**, 3831
[25] Oh, Y.S. Lee, S., Min, S.K., Shin, Y.J. and Kim, B.K., *J. Appl. Polym. Sci.*, 1997, **64**, 1937
[26] Jain, M.K., Balassubramanian, M., Desai, P. and Abhiraman, A.S., *J. Mater. Sci.*, 1987, **22**, 301
[27] Hellsten, K.M., Klingberg, A.W., Karlsson, B.T.G., *J. Am. Oil Chem. Soc.*, 1989, **66**, 1381
[28] Park, Y.H., Kim, Y.K., Nam, S.W., *J. Appl. Polym. Sci.*, 1991, **43**, 1307, (synth., electrical props.)
[29] Hamza, A.A., Ghander, A.M., Oraby, A.H., Mabrouk, M.A., Guthrie, J.T., *J. Phys. D: Appl. Phys.*, 1986, **19**, 2443, (optical props., irradiation)
[30] Gupta, B.S., El-Mogahzy, Y.E., Selivansky, D., *J. Appl. Polym. Sci.*, 1989, **38**, 899, (coefficient of friction)
[31] Hudson, M.J., Galer, J.M., *Solid State Ionics*, 1994, **73**, 175
[32] Min, B.G., Son, T.W., Kim, B.C., Jo, W.H., *Polym. J. (Tokyo)*, 1992, **24**, 841, (plasticisers)

Acrylic/methacrylic copolymers A-27

$$\left[\left\{CH_2C(CH_3)\text{–}\underset{O}{\overset{C}{|}}\text{–}OR\right\}_x \left\{CH_2C(CH_3)\text{–}\underset{O}{\overset{C}{|}}\text{–}OR'\right\}_y\right]_n$$

R = Me, Et, $CH_2CH_2CH_2CH_3$ (butyl), $CH_2(CH_2)_8CH_3$ (lauryl)

R' = Et, $CH_2CH_2CH_2CH_3$ (butyl), $CH_2(CH_2)_8CH_3$ (dodecyl)

= $\underset{CH_2}{\overset{O}{\triangle}}$

= $CH_2CH(CH_3)_2$

= $PhC(CH_3)_3$

Synonyms: *Poly(methyl methacrylate-co-ethyl methacrylate). Poly(methyl methacrylate-co-butyl methacrylate). Poly(methyl methacrylate-co-dodecyl methacrylate). Poly(methyl methacrylate-co-tert-butylphenyl methacrylate). Poly(methyl methacrylate-co-methyl acrylate). Poly(methyl methacrylate-co-ethyl acrylate). Poly(methyl methacrylate-co-butyl acrylate). Poly(ethyl methacrylate-co-ethyl acrylate). Poly(butyl methacrylate-co-isobutyl methacrylate). Poly(lauryl methacrylate-co-glycidyl methacrylate). Poly(methyl methacrylate-co-ethyl methacrylate-co-ethyl acrylate)*

Related Polymers: Polymethacrylates, General, Hydroxyalkyl methacrylate copolymers, Methacrylic ionomers, Poly(methyl methacrylate), Poly(ethyl methacrylate), Poly(butyl methacrylate), Poly(isobutyl methacrylate), Poly(*tert*-butyl methacrylate), Poly(dodecyl methacrylate), Poly(methyl acrylate), Poly(ethyl acrylate), Poly(butyl acrylate)

Monomers: Methyl methacrylate, Ethyl methacrylate, Butyl methacrylate, Lauryl methacrylate, *tert*-Butylphenyl methacrylate, Glycidyl methacrylate, Methyl acrylate, Ethyl acrylate, Butyl acrylate

Material class: Thermoplastic
Polymer Type: Acrylic, Acrylic copolymers
CAS Number:

CAS Reg. No.	Note
25685-29-4	MMA-*co*-ethyl methacrylate
106923-27-7	MMA-*co*-ethyl methacrylate, block
128440-91-5	MMA-*co*-ethyl methacrylate, graft
128948-52-7	MMA-*co*-ethyl methacrylate, graft, isotactic
117307-83-2	MMA-*co*-ethyl methacrylate, block, isotactic
123673-59-6	MMA-*co*-ethyl methacrylate, syndiotactic
117307-84-3	MMA-*co*-ethyl methacrylate, syndiotactic, block
25608-33-7	MMA-*co*-butyl methacrylate
107404-23-9	MMA-*co*-butyl methacrylate, block
112966-33-3	MMA-*co*-butyl methacrylate, graft
130196-30-4	MMA-*co*-butyl methacrylate, block, isotactic
30795-64-3	MMA-*co*-dodecyl methacrylate
9010-88-2	MMA-*co*-ethyl acrylate
25852-37-3	MMA-*co*-butyl acrylate
9011-53-4	butyl methacrylate-*co*-isobutyl methacrylate

– Acrylic/methacrylic copolymers

Volumetric & Calorimetric Properties:
Density:

No.	Value	Note
1	d 1.18 g/cm³	MMA-co-ethyl methacrylate [1]
2	d 1.043–1.151 g/cm³	MMA-co-butyl methacrylate, ASTM D1475 [2]
3	d 1.079 g/cm³	MMA-co-dodecyl methacrylate, 50/50 copolymer [2]
4	d 1.083 g/cm³	butyl methacrylate-co-isobutyl methacrylate [2]

Equation of State: Equation of state information for the MMA/butyl methacrylate copolymer has been reported [3]

Glass-Transition Temperature:

No.	Value	Note
1	101°C	MMA-co-ethyl methacrylate, 75% MMA, 25% ethyl methacrylate, MW 1690000 [5]
2	91.3°C	MMA-co-ethyl methacrylate, 50% MMA, 50% ethyl methacrylate, MW 2040000 [5]
3	29°C	MMA-co-ethyl methacrylate, 1:1 isotactic random copolymer, MW 31800 [6]
4	19°C	MMA-co-ethyl methacrylate, 1:1 isotactic block copolymer, MW 12500 [6]
5	79.4°C	MMA-co-ethyl methacrylate, 25% MMA, 75% ethyl methacrylate, MW 2070000 [5]
6	36–80°C	MMA-co-butyl methacrylate, ASTM D3418 [2]
7	89°C	MMA-co-butyl methacrylate, 78.3% MMA, 21.7% butyl methacrylate, MW 1590000 [5]
8	64.8°C	MMA-co-butyl methacrylate, 52.3% MMA, 47.7% butyl methacrylate, MW 2210000 [5]
9	-22–120°C	MMA-co-butyl methacrylate, random copolymers, varying tacticities [6]
10	54–74°C	MMA-co-butyl methacrylate [7]
11	90°C	MMA-co-dodecyl methacrylate, 50/50 copolymer [2]
12	-60–120°C	MMA-co-dodecyl methacrylate, syndiotactic random copolymers [6]
13	35°C	butyl methacrylate-co-isobutyl methacrylate [2]
14	60°C	ethyl methacrylate-co-ethyl acrylate [7]
15	43°C	MMA-co-ethyl acrylate [7]
16	65°C	MMA-co-ethyl methacrylate-co-ethyl acrylate [7]

Transition Temperature General: Generally T_g increases as the amount of MMA in the copolymer increases [4]

Vicat Softening Point:

No.	Value	Note
1	135°C	MMA-co-ethyl methacrylate [7]
2	150–190°C	MMA-co-butyl methacrylate [7,8]
3	43°C	MMA-co-ethyl acrylate [7]
4	135°C	MMA-co-ethyl methacrylate-co-ethyl acrylate [7]
5	143°C	ethyl methacrylate-co-ethyl acrylate [7]

Surface Properties & Solubility:
Cohesive Energy Density Solubility Parameters: δ 19.22 $J^{1/2}$ $cm^{-3/2}$ (9.4 $cal^{1/2}$ $cm^{-3/2}$) [1]

Solvents/Non-solvents: MMA-co-butyl methacrylate sol. chlorohydrocarbons, esters, ketones, toluene, xylene, acetonitrile, nitroparaffins, DMF, THF, ethoxyethanol. Insol. alcohols, formamide, hydrocarbons, vegetable oils, Et$_2$O, diisopropyl ether, turpentine. MMA-co-ethyl acrylate sol. aromatics, esters, ketones, chlorinated hydrocarbons. MMA-co-ethyl methacrylate-co-ethyl acrylate sol. aromatics, esters, ketones, chlorinated hydrocarbons, alcohols

Transport Properties:
Polymer Solutions Dilute: Theta temps. (lauryl methacrylate-co-glycidyl methacrylate, 3% glycidyl methacrylate, 97% lauryl methacrylate) -24° (EtOH/heptane, 69:31), 1° (EtOH/heptane, 62:38), 24° (EtOH/heptane, 56:44), 40° (EtOH/heptane, 49:51), 1° (propanol/heptane, 44:56), 24° (propanol/heptane, 30:70) [9]. Viscosity (MMA-co-butyl methacrylate, 25°) [2] 2.30–2.75 St (toluene, 40% solids), 0.50–0.6 St (toluene, 30% solids), 0.65–1.3 St (2-butanone, 40% solids), 0.15–0.31 St (2-butanone, 30% solids), 2.75–6.27 St (isopropyl acetate, 40% solids), 0.40–0.65 St (isopropyl acetate, 30% solids), 17.6–10.66 St (ethoxyethanol, 40% solids), 2.30–8.9 St (ethoxyethanol, 30% solids)

Water Absorption:

No.	Value	Note
1	0.35–0.5 %	max., MMA-co-butyl methacrylate [2]
2	<0.5 %	max., MMA-co-dodecyl methacrylate [2]
3	<0.4 %	max., butyl methacrylate-co-isobutyl methacrylate [2]

Mechanical Properties:
Tensile (Young's) Modulus:

No.	Value	Note
1	2910 MPa	23°, in air, MMA-co-ethyl methacrylate, 75% MMA, 25% ethyl methacrylate [5]
2	2420 MPa	37°, in H$_2$O, MMA-co-ethyl methacrylate, 75% MMA, 25% ethyl methacrylate [5]
3	2680 MPa	23°, in air, MMA-co-ethyl methacrylate, 50% MMA, 50% ethyl methacrylate [5]
4	2170 MPa	37°, in H$_2$O, MMA-co-ethyl methacrylate, 50% MMA, 50% ethyl methacrylate [5]
5	2270 MPa	23°, in air, MMA-co-ethyl methacrylate, 25% MMA, 75% ethyl methacrylate [5]
6	1890 MPa	37°, in H$_2$O, MMA-co-ethyl methacrylate, 25% MMA, 75% ethyl methacrylate [5]
7	2790 MPa	23°, in air, MMA-co-butyl methacrylate, 78.3% MMA, 21.7% butyl methacrylate [5]
8	2340 MPa	37°, in H$_2$O, MMA-co-butyl methacrylate, 78.3% MMA, 21.7% butyl methacrylate [5]
9	2080 MPa	23°, in air, MMA-co-butyl methacrylate, 52.3% MMA, 47.7% butyl methacrylate [5]
10	1680 MPa	37°, in H$_2$O, MMA-co-butyl methacrylate, 52.3% MMA, 47.7% butyl methacrylate [5]

Tensile Strength Break:

No.	Value	Note	Elongation
1	11–16 MPa	23°, 50% relative humidity, MMA-co-butyl methacrylate [2]	0.5–2%
2	15 MPa	23°, butyl methacrylate-co-isobutyl methacrylate [2]	175%

Hardness: Knoop 3.5–13 (23°, 50% relative humidity, MMA-co-butyl methacrylate, Tukon, 1.6 mm thick, 25 g load) [2], 16 (23°, MMA-co-dodecyl methacrylate, Tukon, 50/50 copolymer) [2], 4 (butyl methacrylate-co-isobutyl methacrylate, Tukon) [2]

Fracture Mechanical Properties: Fracture toughness 1.03 MPa m$^{1/2}$ (23°, in air, MMA-co-ethyl methacrylate, 75% MMA, 25% ethyl methacrylate) [5], 1.45 MPa m$^{1/2}$ (37°, in H$_2$O, MMA-co-ethyl methacrylate, 75% MMA, 25% ethyl methacrylate) [5], 0.98 MPa m$^{1/2}$ (23°, in air, MMA-co-ethyl methacrylate, 50% MMA, 50% ethyl methacrylate) [5], 1.24 MPa m$^{1/2}$ (37°, in H$_2$O, MMA-co-ethyl methacrylate, 50% MMA, 50% ethyl methacrylate) [5], 0.85 MPa m$^{1/2}$ (23°, in air, MMA-co-ethyl methacrylate, 25% MMA, 75% ethyl methacrylate) [5], 0.98 MPa m$^{1/2}$ (37°, in H$_2$O, MMA-co-ethyl methacrylate, 25% MMA, 75% ethyl methacrylate) [5], 0.99 MPa m$^{1/2}$ (23°, in air, MMA-co-butyl methacrylate, 78.3% MMA, 21.7% butyl methacrylate) [5], 1.1 MPa m$^{1/2}$ (37°, in H$_2$O, MMA-co-butyl methacrylate, 78.3% MMA, 21.7% butyl methacrylate) [5], 0.73 MPa m$^{1/2}$ (23°, in air, MMA-co-butyl methacrylate, 52.3% MMA, 47.4% butyl methacrylate) [5], 0.75 MPa m$^{1/2}$ (37°, in H$_2$O, MMA-co-butyl methacrylate, 52.3% MMA, 47.7% butyl methacrylate) [5]

Optical Properties:

Transmission and Spectra: C-13 nmr spectral data have been reported [10]

Applications/Commercial Products:

Processing & Manufacturing Routes: Stereoregular MMA/ethyl acrylate copolymers can be synth. using *tert*-BuLi-AlEt$_3$ to produce syndiotactic copolymers or *tert*-BuMgBr to produce isotactic copolymers [6]

Applications: Has uses in adhesives, coatings, paints, as a binder in inks; in cosmetics, dentistry, drug-delivery systems, imaging, optical lenses and as moulding additives. Copolymers, with N-methylolacrylamide as cross-linking agent, are used as auxiliary agents in the textile industry

Trade name	Details	Supplier
Acryloid	MMA-co-ethyl methacrylate	Rohm and Haas
Acryloid	MMA-co-ethyl acrylate	Rohm and Haas
Acryloid (Paraloid)	MMA-co-butyl methacrylate (various grades)	Rohm and Haas
Elvacite	MMA-co-butyl methacrylate (various grades)	ICI, UK
Elvacite 2046	butyl methacrylate-co-isobutyl methacrylate	ICI, UK
Elvacite 2552	MMA-co-dodecyl methacrylate	ICI, UK
Neocryl	MMA-co-butyl methacrylate (various grades)	Zeneca
Neocryl	MMA-co-ethyl acrylate	Zeneca
Neocryl	ethyl methacrylate-co-ethyl acrylate	Zeneca
Neocryl	MMA-co-ethyl methacrylate-co-ethyl acrylate	Zeneca
Plexigum MB	MMA-co-ethyl acrylate	Rohm

Bibliographic References

[1] *Paraloid Range*, Rohm and Haas, 1996, (technical datasheet)
[2] *Elvacite*, ICI, 1996, (technical datasheet)
[3] Shiomi, T., Tohyama, M., Endo, M., Sato, T. and Imai, K., *J. Polym. Sci., Part B: Polym. Phys.*, 1996, **34**, 2599, (equation of state)
[4] Fernandez-Garcia, M., Lopez-Garcia, M.M.C., Barrales-Rienda, J.M., Madruga, E.L. and Arias, C., *J. Polym. Sci., Part B: Polym. Phys.*, 1994, **32**, 1191
[5] Johnson, J.A. and Jones, D.W., *J. Mater. Sci.*, 1994, **29**, 870, (mechanical props.)
[6] Kityama, T., Ute, K. and Hatada, K., *Br. Polym. J.*, 1990, **23**, 5, (synth.)
[7] *Neocryl Range*, Zeneca, 1997, (technical datasheet)
[8] Ash, M. and Ash, I., Handbook of Plastic Compounds, *Elastomers and Resins*, (eds. M.B. Ash and I.A. Ash), Wiley-VCH, 1992
[9] Napper, D.H., *Trans. Faraday Soc.*, 1968, **64**, 1701, (theta temps.)
[10] Moad, G. and Willing, R.I., *Polym. J. (Tokyo)*, 1991, **23**, 1401, (C-13 nmr)

Acrylonitrile-butadiene-isoprene terpolymer A-28

Synonyms: Poly(acrylonitrile-co-butadiene-co-isoprene). Poly(2-propenenitrile-co-1,3-butadiene-co-2-methyl-1,3-butadiene)
Monomers: Acrylonitrile, 1,3-Butadiene, Isoprene
Material class: Synthetic Elastomers, Copolymers
Polymer Type: Polydienes, Acrylonitrile and copolymers, Polybutadiene
CAS Number:

CAS Reg. No.
25135-90-4

Applications/Commercial Products:

Trade name	Supplier
Nipol DN-1201	Nippon Zeon

Acrylonitrile-methyl acrylate copolymer A-29

Synonyms: Barex. Poly(acrylonitrile-co-methyl acrylate). Poly(2-propenenitrile-co-(methyl 2-propenoate)
Related Polymers: Acrylic fibres, Acrylonitrile-methyl acrylate copolymer - injection moulding, Polyacrylonitrile
Monomers: Acrylonitrile, Methyl acrylate
Material class: Thermoplastic
Polymer Type: Acrylonitrile and copolymers, Acrylic copolymers
CAS Number:

CAS Reg. No.
24968-79-4

General Information: Also contains some acrylonitrile-butadiene copolymer
Tacticity: Copolymers of varying composition synth. by aq. redox polymerisation similar tacticity [7]
Morphology: Morphology has been reported [10]

Volumetric & Calorimetric Properties:

Equation of State: Mark-Houwink-Sakurada equation of state parameters and second virial coefficients have been reported [9]
Glass-Transition Temperature:

No.	Value	Note
1	58°C	acrylonitrile: methyl acrylate 1:1 [1]
2	53°C	acrylonitrile: methyl acrylate 1:2 [1]
3	37°C	acrylonitrile: methyl acrylate 1:3 [1]
4	102°C	10% methyl acrylate [1,4,5]
5	73°C	50% methyl acrylate [1,4,5]
6	30°C	90% methyl acrylate [1,4,5]

Transition Temperature General: T_g decreases with increasing methyl acrylate content [1,4,5]

Surface Properties & Solubility:
Solvents/Non-solvents: Swollen by nitromethane. Swelling causes a partial conformational change in the paracrystalline regions from hexagonal to orthorhombic [10]

Transport Properties:
Polymer Solutions Dilute: Intrinsic viscosity data 0.294 dl g^{-1} (DMF, 30°, acrylonitrile:methyl acrylate (1:0.86), 0.266 dl g^{-1} (DMF, 45°, acrylonitrile:methyl acrylate (1:0.86), 0.69 dl g^{-1} (DMF, 30°, acrylonitrile:methyl acrylate (91.5:8.5), MW 47800), 3.83 dl g^{-1} (DMF, 30°, acrylonitrile:methyl acrylate (91.5:8.5), MW 526000), 0.43 dl g^{-1} (82.5% aq. ethylene carbonate 30°, acrylonitrile:methyl acrylate (91.5:8.5), MW 71400), 1.12 dl g^{-1} (82.5% aq. ethylene carbonate, 30°, acrylonitrile:methyl acrylate (91.5:8.5, MW 526000), 0.48 dl g^{-1} (51% aq. nitric acid, acrylonitrile:methacrylate (91.5:8.5), MW 47800), 1 dl g^{-1} (80% aq. nitric acid, acrylonitrile:methyl acrylate (91.5:8.5), MW 47800 [6,9]

Optical Properties:
Transmission and Spectra: Ir, H-1 nmr and C-13 nmr spectral data have been reported [1,3,4,10]

Polymer Stability:
Thermal Stability General: The effect of composition on cyclisation behaviour under heating has been reported [7]. Three stages in the behaviour of the copolymer on annealing have been reported [10]
Decomposition Details: 51% Weight loss for a 1:1 copolymer occurs at 360-500°, with decomposition complete (97%) at 510–800°. 62% Weight loss for a 1:2 (acrylonitrile:methylacrylate) copolymer occurs at 360–510° with decomposition complete (98%) at 540–700°. 70% Weight loss for a 1:3 (acrylonitrile:methylacrylate) copolymer occurs at 420–620° with decomposition complete (94%) at 700–850° [1]

Applications/Commercial Products:
Processing & Manufacturing Routes: Synth. by bulk polymerisation using benzoyl peroxide initiator; the kinetics of polymerisation have been reported [2,3,4]. Electro-copolymerisation of methyl acrylate and acrylonitrile onto graphite fibres has been reported [8]
Applications: Used in the manufacture of carbon fibres

Bibliographic References
[1] Joseph, R., Devi, S. and Rakshit, A.K., *J. Appl. Polym. Sci.*, 1993, **50**, 173, (synth.)
[2] Czajlik, I., Foldes-Berezsnich, T., Tudos, F. and Mader-Vertes, E., *Eur. Polym. J.*, 1997, **15**, 236
[3] Takenawa, M., Johnson, A.F. and Kamide, K., *Polymer*, 1994, **35**, 3908, (synth.)
[4] Brar, A.S. and Sunita, J. Polym. Sci. Part A, 1992, **30**, 2549, (struct., T_g, C-13 nmr)
[5] Penzel, E., Rieger, J. and Schneider, H.A., *Polymer*, 1997, **38**, 325, (T_g)
[6] Joseph, R., Devi, S. and Rakshit, A.K., *Polym. Int.*, 1991, **26**, 89, (viscosity)
[7] Bang, Y.H., Lee, S. and Cho, H.H., *J. Appl. Polym. Sci.*, 1998, **68**, 2205
[8] Chang, J., Bell, J.P. and Shkolnik, S., *J. Appl. Polym. Sci.*, 1987, **34**, 2105, (synth.)
[9] Kamide, K., Miyazaki, Y., and Kobayashi, H., *Polym. J. (Tokyo)*, 1982, **14**, 591, (soln. props.)
[10] Grobelny, J., Sokol, M. and Turska, E., *Polymer*, 1989, **30**, 1187, (nmr)

Acrylonitrile-methyl acrylate copolymer, Injection moulding A-30
Synonyms: *Barex 210*
Related Polymers: Acrylonitrile-methyl acrylate copolymer
Monomers: Acrylonitrile, Methyl acrylate
Material class: Thermoplastic
Polymer Type: Acrylonitrile and copolymers, Acrylic copolymers

Applications/Commercial Products:

Trade name	Supplier
Barex 210 resins	BP Chemicals

AES, Impact modified A-31
Monomers: Acrylonitrile, Styrene, Ethylene, Propylene
Material class: Synthetic Elastomers
Polymer Type: Polyolefins, Acrylonitrile and copolymers, Polystyrene

Applications/Commercial Products:

Trade name	Supplier
Royaltuf 372	Uniroyal Chemical

Agar A-32

Synonyms: *Agarose. Agar-Agar*
Material class: Polysaccharides
Polymer Type: Carbohyrate
CAS Number:

CAS Reg. No.
9012-36-6

Molecular Formula: $(C_{12}H_{17}O_{13}S)_n$
Fragments: $C_{12}H_{17}O_{13}S$
General Information: Colloidal polysaccharide isol. from the red algae species *gelidium* and *gracilaria*. Agar samples are a mixture of agarose (28–80%) and agaropectin, itself a mixture of related galactose-based polymers

Surface Properties & Solubility:
Solvents/Non-solvents: Sol. urea, DMSO [1]

Transport Properties:
Polymer Solutions Concentrated: Forms strong gels at 40° (approx.) [3]. Gels are formed at concentrations $\geq 1\%$ at 20° [2]
Water Content: Absorbs 20 times its weight of cold water with swelling [3]. A hot 1.5% soln. gels at 32–39° and is liquefied by heating above 85°

Mechanical Properties:
Mechanical Properties General: Variation of Young's modulus with concentration has been reported in graphical form [2]

Bibliographic References
[1] Watanabe, M. and Nishimari, K., *Polym. J. (Tokyo)*, 1988, **20**, 1125
[2] *Polymeric Materials Encyclopedia*, (ed. J.C. Salamone), CRC Press, 1996, **1**, 137
[3] *Hawley's Condensed Chemical Dictionary*, 11th edn., (eds. N.I. Sax and R.J. Lewis), Van Nostrand Reinhold, New York, 1987

Alginic acid A-33

Synonyms: *Alginate. Polymannuronic acid. Polyguluronic acid*
Related Polymers: Cellulose
Monomers: Mannuronic acid, Guluronic acid
Material class: Polysaccharides
Polymer Type: Carbohyrate
CAS Number:

CAS Reg. No.	Note
9005-32-7	mixture of monomers

Molecular Formula: $(C_6H_8O_6)_n$
Fragments: $C_6H_8O_6$
Molecular Weight: 30000-200000 (DP 186-930)
General Information: White-yellow powder isol. from brown seaweed (e.g. *Laminarae*) by extraction with alkali. Strongly hydrophilic with good colloidal props. Calcium alginate fibres have been manufactured for specialist uses. The arrangement and proportions of the two main acidic monomers vary widely with the source of the polymer

Surface Properties & Solubility:
Solubility Properties General: Polymer and fibre are sol. alkali
Solvents/Non-solvents: Insol. organic solvents [1]

Transport Properties:
Water Content: 10% [2]. Monovalent cationic salts of alginic acid can absorb 300 times their weight of water [2]

Optical Properties:
Total Internal Reflection: $[\alpha]_D^{20}$ -113--148 [2]

Polymer Stability:
Flammability: Fibres made from metal alginates are flame-resistant [1]
Hydrolytic Stability: Has pK_a 2.95 at pH 2.8-6 [2]

Applications/Commercial Products:
Processing & Manufacturing Routes: Dried *Laminaria* (brown seaweed) is milled and treated with sodium carbonate/sodium hydroxide. The viscous soln. is bleached and sterilised with sodium hypochlorite. Acidification gives the free acid. Fibres are made by spinning solns. of sodium alginate into dilute hydrochloric acid/calcium chloride to give calcium alginate
Applications: Applications of the calcium salt include fireproof and dissolving fibres. Used as thickener and emulsifier in foods, cosmetics, agrochemical and pharmaceutical formulations, paper and textile coating, concrete waterproofing and drilling muds

Trade name	Supplier
Alginate fibre	Courtaulds

Bibliographic References
[1] Moncrieff, R.W., *Man-Made Fibers,* 6th edn., Newnes-Butterworth, 1975
[2] *Encyclopaedia of Marine Resources,* (ed. F.E. Firth), Van Nostrand Reinhold, 1969
[3] McDowell, R.H., *Properties of Alginates,* Alginate Ind. Ltd., 1955

Amorphous Nylon A-34

Synonyms: *Poly(1,3-benzenedicarboxylic acid-co-1,4-benzenedicarboxylic acid-co-1,6-hexanediamine)*
Monomers: 1,3-Benzenedicarboxylic acid, 1,4-Benzenedicarboxylic acid, 1,6-Hexanediamine
Material class: Thermoplastic
Polymer Type: Polyamide
CAS Number:

CAS Reg. No.
25750-23-6

Molecular Formula: $(C_{14}H_{18}N_2O_2)_n$
Fragments: $C_{14}H_{18}N_2O_2$
General Information: An amorph. polyamide with good O_2 and CO_2 barrier props. especially at high moisture levels. 20% Blends with Nylon 6 and Nylon 6,6 give amorph. products with excellent barrier props. in refrigerated conditions. Wide ranging chemical resistance to dil. acids and bases, aliphatic and aromatic hydrocarbons, esters and ketones. When blended with Nylon 6, Selar improves O_2 barrier, transparency and glass thermoformability as well as processing of Nylon 6

Volumetric & Calorimetric Properties:
Density:

No.	Value	Note
1	d 1.19 g/cm^3	[1]

Glass-Transition Temperature:

No.	Value	Note
1	125°C	[1]

Surface Properties & Solubility:
Solvents/Non-solvents: Insol. aliphatic hydrocarbons, aromatic hydrocarbons, ketones and esters

Transport Properties:
Melt Flow Index:

No.	Value	Note
1	15 g/10 min	225°, 2.16 kg pressure

Polymer Solutions Dilute: η_{inh} 0.83 dl g^{-1} (*m*-cresol)
Polymer Solutions Concentrated: Apparent viscosity 1000 Pa s (approx., shear rate 50-100 s^{-1}, 280°) [1], 500 Pa s (approx., shear rate 50-100 s^{-1}, 300°)
Water Content: 1%
Gas Permeability:

No.	Gas	Value	Note
1	O_2	1.23 cm^3 mm/(m^2 day atm)	18.7 × 10^{-13} cm^2 (s cmHg)$^{-1}$, 0% relative humidity [1]
2	O_2	0.4 cm^3 mm/(m^2 day atm)	6.1 × 10^{-13} cm^2 (s cmHg)$^{-1}$, 100% relative humidity [1]

Mechanical Properties:
Tensile Strength Break:

No.	Value	Note	Elongation
1	68 MPa	ASTM D882 [1]	20%

Impact Strength: Spencer impact strength 4.5 J mm^{-1} (ASTM D3420, film) [1]
Failure Properties General: Elmendorf tear 16 g mil^{-1} (ASTM D1922) [1]

Optical Properties:
Refractive Index:

No.	Value	Note
1	1.597	measured on 25 m film

Transmission and Spectra: Uv transmission >80% (λ > 320 nm). Light transmission 70% (approx., 25 μm monolayer blown film, ASTM D1003)
Volume Properties/Surface Properties: Transparency transparent. Haze 0.9% (20°, cast or blown film)

Polymer Stability:
Chemical Stability: Stable to hydrocarbons, dilute mineral acids and bases, ketones, esters and higher alcohols. Poor resistance to EtOH, isopropanol, ethylene glycol and AcOH

Applications/Commercial Products:
Processing & Manufacturing Routes: Similar to Trogamid T (Nylon 6(3)6). Packaged at <0.1% moisture content for immediate use. May be dry blended with Nylon 6 before extrusion. Blending temp. of 240-250° in extrusion processes
Applications: Modifier for semi-crystalline Polyamides to improve clarity, water vapour and oxygen impermeability. Thin extruded films used for packaging. Has FDA approval for food packaging applications. Also used as modifier for HDPE when suitably dispersed to provide platelets which form a 'tortuous path' for gas molecules. Selar PA is used in bottles, jars and other rigid structs. that require good physical props. (low weight and impact resistance) combined with the clarity and high oxygen barrier props. of glass. Often used in co-extrusion coatings or other multilayer-structs.

Trade name	Details	Supplier
Selar PA	Several grades	DuPont

Bibliographic References
[1] Selar PA. Property and Extrusion Guide DuPont, 1993, (technical datasheet)
[2] *Eur. Pat. Appl.*, BP Chemicals, 1991, 411 791, (permeability)

Amylopectin   A-35
Related Polymers: Starch, Amylose
Monomers: Base monomer unit glucose
Material class: Polysaccharides
Polymer Type: Cellulosics
CAS Number:

CAS Reg. No.
9037-22-3

Molecular Formula: $(C_6H_{10}O_5)_n$
Fragments: $C_6H_{10}O_5$
Molecular Weight: MW 65000000-500000000 (degree of polymerisation 400000-3000000), 160000-1600000 (degree of polymerisation 1000-10000), 36000000-25000000 (degree of polymerisation 200000-1500000)
General Information: One of the largest natural polymers, its MW range is variously quoted in the range 1-500000000, depending on source [1,4]. Branches occur at every tenth-twelfth glucose residue, usually by reaction at C-6, though C-3 bonds are also found. Each side-chain comprises 20-30 glucose residues, which may also be non-linear. Thus the struct. of amylopectin is variable and complex

Surface Properties & Solubility:
Solvents/Non-solvents: Sol. H_2O, butanol [5]. Insol. EtOH, Me_2CO, Et_2O [5]
Surface Tension:

No.	Value	Note
1	35 mN/m	20°, contact angle method [3]

Polymer Stability:
Hydrolytic Stability: Hydrolysed by mineral acids
Biological Stability: Non-branched sections hydrolysed by α-amylase and β-amylase leaving a highly branched core unit (the limit dextrin)

Bibliographic References
[1] Banks, W. and Greenwood, C.T., *Starch and its Components*, Edinburgh University Press, 1975
[2] *Handbook of Starch Hydrolysis Products and Their Derivatives*, (eds. M.W. Kearsley and S.Z. Dziedzic), Blackie Academic and Professional, 1995
[3] Ray, B.R., Anderson, J.R. and Scholz, J.J., *J. Phys. Chem.*, 1958, **62**, 1220
[4] Zimm, B.H. and Thurmond, C.D., *J. Am. Chem. Soc.*, 1952, **74**, 1111
[5] Meyer, K.H., *Angew. Chem., Int. Ed. Engl.*, 1951, **63**, 155

Amylose   A-36
Synonyms: *Poly [(1→4)-α-D-glucose]*. α-*Amylose*
Related Polymers: Starch, Amylopectin
Monomers: Base monomer unit glucose
Material class: Polysaccharides
Polymer Type: Carbohyrate
CAS Number:

CAS Reg. No.
9005-82-7

Molecular Formula: $(C_6H_{10}O_5)_n$
Fragments: $C_6H_{10}O_5$
Molecular Weight: MW 160000-320000 (degree of polymerisation 200-2000), 160000-700000 (degree of polymerisation 200-4300)
General Information: This component of starch is usually found as a linear polymer, though occasional branching is found in samples from some sources. Amylose usually comprises 18-28% of the mixture in starch, though in some seeds it appears to make up approx. 85% of the total
Morphology: In parallel with starch, amylose has been identified in four forms, A, B, C and V, though detailed conformational information is scarce. Polymer chains appear to have a left-handed helical struct., containing six glucose residues per turn [7]. This struct. accounts for the ability of amylose to absorb water (up to 10%) and to form clathrates with a wide range of molecules. Complexation with butanol is important in its isolation and purification. In soln. amylose is thought to form random coils, consisting of helical sections separated by disordered areas of the polymer [8]

Surface Properties & Solubility:
Solvents/Non-solvents: Sol. hot H_2O, ethylenediamine, nitromethane, DMSO, aq. potassium hydroxide, aq. potassium chloride [5,6]. Insol. butanol, Et_2O [5,6]
Surface Tension:

No.	Value	Note
1	37 mN/m	20°, contact angle method [3]

Polymer Stability:
Hydrolytic Stability: Hydrolysed by mineral acids
Biological Stability: Hydrolysed by α-amylase and β-amylase

Arabinan A-37

→ 5)-α-L-Ara*f*-(1 → 5)-α-L-Ara*f*-(1 →
                    3
                    ↑
                    1
               α-L-Ara*f*

Portion of idealised structure

Synonyms: *Araban*
Material class: Polysaccharides
CAS Number:

CAS Reg. No.
11078-27-6
9060-75-7

Molecular Formula: $(C_5H_8O_4)n$
Fragments: $C_5H_8O_4$
General Information: Minimum formula given
Morphology: A polymer of linearly α-(1→5)-linked L-arabinofuranose units with single L-arabinofuranose α-(1→3)-linked to the main chain at intervals.
Miscellaneous: Arabinans devoid of other sugars have been isol. from mustard seeds and maritime pine. Heteroarabinans have been found in sugar beet and apples

Optical Properties:
Optical Properties General: $[\alpha]_D$ -157; -114; -129 (H_2O)

Bibliographic References
[1] Hirst, E.L., *J. Chem. Soc.*, 1939, 452; 454; 1865
[2] Hirst, E.L., *Adv. Carbohydr. Chem.*, 1947, **2**, 235, (rev)
[3] Hirst, E.L., *Biochem. J.*, 1965, **95**, 453, (isol)
[4] Roudier, A.J., *Bull. Soc. Chim. Fr.*, 1965, 460
[5] Jones, J.K.N., *Methods Carbohydr. Chem.*, 1965, **5**, 74, (synth)
[6] Aspinal, G.O., The Carbohydrates (Pigman W. etal, Ed.), Academic Press, 1970, **2B**, 517
[7] Radha, A., *Carbohydr. Res.*, 1997, **298**, 105, (cryst struct; conform)

Aramids A-38

Related Polymers: Poly(*p*-phenylene terephthalamide), Poly(*m*-phenylene isophthalamide)
Material class: Polymer liquid crystals
Polymer Type: Polyamide
General Information: Aramids are defined as fibre forming substances comprising long-chain polyamides with >85% of amide groups attached to aromatic rings. These are synthetic polymers with the qualities of stiffness and heat resistance as well as high tenacity, which make them competitive substitutes for steel wire or glass fibres for the production of ropes, cables, hoses and coated fabrics etc. The three best known aramids are all manufactured by DuPont and include poly (*p*-phenyleneterephthalamide), poly *p*-benzamide see and poly *m*-phenylene isophthalamide. The chain stiffness of these materials has a large positive effect upon the initial modulus, which can be further increased by orientation due to drawing the fibre to several times its original length, and by crystallinity. Fibres spun from anisotropic solns. of rigid rods do not need to be hot drawn to attain high strength [1]. Anisotropic doped solns. permit higher spinning rates because they have much lower viscosity. Aramid fibres can survive temps. up to 500° and do not melt. Upper use temps. depend on Tg and are in range 250–400°. They are characterised by medium to ultrahigh tensile strength, medium to low elongation and moderate to ultrahigh modulus. Aramids with a high proportion of meta rings have superior heat and flame resistance to the ultrahigh modulus rigid rod types like Kevlar. Tensile props. are not only superior to Nylon 6,6 but at 250° are the equal of conventional textile fibres. Aramid fibres retain useful props. for up to two weeks of heat ageing in air at 300°. Aramid fibres amy be used for flame resistant applications, e.g. protective apparel, are highly resistant to most chemicals but are susceptible to UV degradation. High breakdown voltages suit aramids for dielectrics in transformers and motors. The stress-strain curves for ultrahigh modulus aramid fibres resemble those of glass and steel, on a specific basis, they are actually stronger and stiffer than steel or glass, and find use in composites
Morphology: Aramids exhibit lyotrpoic liquid crystalline behaviour, enhancing chain orientation and phys. props.

Applications/Commercial Products:
Processing & Manufacturing Routes: Aramids are prepared commercially from the reaction of the aromatic diacid chlorides with the diamine in *N*-methyl pyrrolidone (NMP) containing some calcium chloride which helps to keep the growing polymer in soln. The polymer is prepared under high shear mixing conditions below 95°. It is dried and dissolved in sulfuric acid to provide a dope for spinning [2]. Fibres are spun from sulfuric acid soln. because the acid forms a low melting complex with the aramid enabling higher levels to be spun [3]. Filament yarn is then crimped and cut to provide staple which can be blended and processed into fabrics and felts using conventional textile technology. Aromatic polyamides may also be prepared by the interfacial polym. method in which an acid chloride is dissolved in a water immiscible solvent and added to aq. diamine. High MW polymer is best obtained by stirring in the presence of an emulsifying agent at close to 0°. Direct polycondensation from the acid and amine can be achieved in soln.
Applications: The major areas where aramid fibres are employed commercially include ballistics, rope and cable, fibre optics, protective apparel, mechanical rubber goods and composites. The polymers are solid in the form of yarn, roving, staple, flock, pulp and selected spun yarns and fabrics [12]. Nomex (MPD-I) has been marketed since 1965 and is used in continuous filament yarn for reinforcement in rubber goods, radiator hose, conveyor belts, protective apparel and hot gas filtration. It is converted from spun yarn to paper used for wrapping tape, electric sheets or insulation for motors and transformers. Spun yarn is also converted to needle felts (used in hot gas filtration, laundry and office machine parts), pressboard (spacers, gaskets, electrical motor/transformer insulation), fabric (personal protective clothing, wrappers). Floc is converted to a honeycomb sheet for aerospace applications, ground and marine transportation, and sports equipment. The first and foremost application is in personal protective garments. Fire protective garments are produced for oil workers, firemen, and motor racing drivers. Needle felts are used for filter media to separate dust from flue gases continuously up to 200°. Nomex paper and pressboard are made by heat and pressure consolidation of fibre floc. Pressboard is used routinely for spacers and support in high performance electromechanical equipment. Nomex honeycomb technology finds applications in skis, tailfins of aircraft and helicopter blades. *Kevlar (PPD-T)* was originally designed at DuPont as a replacement for steel wire in radial tyres and has been marketed since 1974. As fabric, it gives excellent ballistic performance at low weight for body armour. Light weight military helmets offer superior head protection. Kevlar reinforced phenolics are employed as spall liners for armoured cars. Kevlar light weight knitted or woven gloves are widely used for protection against heat, chemical attack and/or cuts. Forestry workers use Kevlar protective clothing designed to protect against chain saw chaps. In composite form it is used in motorcycle racing to reinforce bodywork, tyres brake discs and helmets. It has found

many other uses in composites, e.g. to save weight in many aircraft parts while maintaining desired levels of strength, in boat hulls and canoes for its lightness and impact resistance as well as skis where its vibration damping props. are advantageous. It has replaced asbestos in brake and clutch linings, both on safety and performance criteria. Kevlar has many reinforced rubber applications, e.g. high performance tyres for racing, puncture resistant cycle tyres, truck and aircraft tyres. It is also used in conveyor belt construction and for car and other high pressure hydraulic hoses. There are many applications where Kevlar ropes provide the strength of steel at 5% of the weight, especially in marine moorings or anchor lines for ships and off-shore platforms. Trawl nets for fishing are now often made from Kevlar to reduce drag through the water. The automobile industry was seen as the major client for replacement of asbestos with aramid fibre [5], with uses in bullet proofing also important. The industrial applications of aramids has been assisted by the development of pultrusion processes for impregnating continuous fibres with thermoplastic resins, leading to pelleted products with up to twenty times as long fibred incorporated [6]. A fibre foam coining process was described as a means to reduce the tendency of long fibres to orient in the direction of matrix flow in moulding by ICI [7,8]. Their higher strength, stiffness and impact resistance made them suitable for a wide variety of markets in aerospace automotive and chemical sectors [9]. Moulded pellets made by melt coating coutinuous aramid fibres have been shown to have higher wear resistance than glass reinforced materials. Kevlar/nylon 6,6 compds. are being applied to wear plates, bushings and bearings [10] and a new composite material of Nylon 6,6 with Kevlar fibre has been used for car transmission components [11,12]

Bibliographic References

[1] U.S. Pat., 1973, 3 767 756
[2] Encycl. Polym. Sci. Technol., 2nd edn., John Wiley and Sons, 1986, **11**, 400
[5] Plastics (London), 1989, **20**, 65
[6] J. Adv. Compos. Bull., 1987, **1**, 3
[7] J. Usine Nouv., 1985, **23**, 91
[9] J. Adv. Mater., 1985, **7**, 4
[10] J. Adv. Mater., 1987, **9**, 2
[11] J. High Perform. Plast., 1985, **2**, 4

ASA

A-39

$$\left[CH_2C(CH_3) \atop O\ \ OH \right]_x \left[CH_2CH \atop Ph \right]_y \left[CH_2CH \atop CN \right]_z$$

Synonyms: *ASA terpolymer. AAS. Acrylonitrile-styrene-acrylate terpolymer. Acrylic-styrene-acrylonitrile terpolymer. Poly(2-propenoic acid-co-ethenylbenzene-co-2-propenenitrile)*
Related Polymers: Polyacrylonitrile, ASA - Injection moulding, Unfilled, ASA - Extrusion, Unfilled, ASA - High Impact, ASA - Blow moulding, ASA and PC Blend, ASA and PVC Blend, Polystyrene
Monomers: Acrylonitrile, Styrene, 2-Propenoic acid
Material class: Thermoplastic, Copolymers
Polymer Type: ASA
CAS Number:

CAS Reg. No.	Note
26299-47-8	butyl ester
24980-16-3	
108554-70-7	graft, butyl ester
112965-31-8	block, butyl ester

Volumetric & Calorimetric Properties:
Density:

No.	Value	Note
1	d 1.07 g/cm^3	ISO 1183 [17]

Thermal Expansion Coefficient:

No.	Value	Note	Type
1	9.5×10^{-5} K^{-1}	DIN 53752 [17]	L
2	8×10^{-5}–0.00011 K^{-1}		L

Thermal Conductivity:

No.	Value
1	0.17 W/mK

Deflection Temperature:

No.	Value	Note
1	96°C	1.8 MPa, ISO 75 [17]
2	102°C	67% styrene, 32% acrylonitrile, 1% acrylic acid
3	120°C	55% styrene, 25% acrylonitrile, 20% acrylic acid [16]

Vicat Softening Point:

No.	Value	Note
1	92°C	50N, ISO 306 [17]

Surface Properties & Solubility:
Plasticisers: Tris(2,6-dimethylphenyl)phosphate

Transport Properties:
Melt Flow Index:

No.	Value	Note
1	8 g/10 min	200°, 21.6 kg, ISO 1133 [17]

Permeability and Diffusion General: Permeability of gases depends on processing and moulding conditions
Water Content: Water absorption data have been reported [9]
Water Absorption:

No.	Value	Note
1	1.65 %	ISO 62A [17]

Gas Permeability:

No.	Gas	Value	Note
1	N_2	7.6–10.1 cm^3 mm/(m^2 day atm)	0.1 mm film, 23°, DIN 53380 [17]
2	O_2	50.7–55.7 cm^3 mm/(m^2 day atm)	0.1 mm film, 23°, DIN 53380 [17]
3	CO_2	203–233 cm^3 mm/(m^2 day atm)	0.1 mm film, 23°, DIN 53380 [17]

Mechanical Properties:

Tensile (Young's) Modulus:

No.	Value	Note
1	2200 MPa	ISO 527 [17]
2	2200–2900 MPa	20°

Flexural Modulus:

No.	Value	Note
1	2500 MPa	20°

Tensile Strength Break:

No.	Value	Elongation
1	47–66 MPa	10–20%

Elastic Modulus:

No.	Value	Note
1	3516 MPa	510 kpsi, graft acrylic terpolymer, ASTM D747-50 [15]

Tensile Strength Yield:

No.	Value	Elongation	Note
1	568.8 MPa	5.3%	5800 kg cm^{-2}, graft acrylic terpolymer [15]
2	47 MPa	20%	ISO 527 [17]

Impact Strength: Impact strength 20.8 J m^{-1} (0.39 ft lb in^{-1}) (55% styrene, 25% acrylonitrile, 20% acrylic acid), 21.9 J m^{-1} (0.41 ft lb in^{-1}) (70% styrene, 10% acrylonitrile, 20% acrylic acid) [16]

Viscoelastic Behaviour: Dynamic viscoelastic behaviour has been reported [5]. Creep of compression-moulded compounds is due to crazing and may be reduced by hot-drawing parallel to the draw direction [14]

Hardness: Penetration depth measurements have been reported [6]

Fracture Mechanical Properties: Crazing study has been reported [13]

Izod Notch:

No.	Value	Notch	Note
1	100 J/m	Y	20°

Charpy:

No.	Value	Notch	Note
1	270	N	23°, ISO 179/1D [17]
2	125	N	-30°, ISO 179 [17]
3	25	Y	23°, ISO 179/1A [17]
4	3	Y	-30°, ISO 179 [17]

Electrical Properties:

Surface/Volume Resistance:

No.	Value	Note	Type
2	0.01 × 10^{15} Ω.cm	ISO 1325 [17]	S

Dielectric Permittivity (constant):

No.	Value	Frequency	Note
1	3.8	100 Hz	IEC 250 [17]
2	3.4	1 MHz	[17]

Dielectric Strength:

No.	Value
1	22 kV.mm^{-1}

Static Electrification: Comparative tracking index 600 V (IEC 112A) [17]

Dissipation (Power) Factor:

No.	Value	Frequency	Note
1	0.009	100 Hz	ISO 1325 [17]
2	0.034	1 MHz	[17]

Optical Properties:

Transmission and Spectra: Ir [2,3] and H-1 nmr [7] spectral data have been reported

Volume Properties/Surface Properties: Clarity clear to opaque, depending upon composition [16]

Polymer Stability:

Flammability: Flammability rating HB (0.81 mm, UL 94) [17]

Environmental Stress: Weathering data have been reported [8,10]. Excellent weathering resistance is due to the acrylic elastomer content

Chemical Stability: Acrylic terpolymers are resistant to gasoline and CCl_4 [16]

Hydrolytic Stability: Stable to boiling H_2O and to cold alkalis [16]

Applications/Commercial Products:

Processing & Manufacturing Routes: Terpolymer containing butyl acrylate may be synth. by a two-stage emulsion polymerisation [1]. Block terpolymer can be synth. by initial polymerisation of butyl acetate with a peroxide initiator at 68° under N_2. Acrylonitrile and styrene are added, and polymerisation continued at 110° [4]. Graft terpolymer can be synth. by grafting styrene and acrylonitrile onto butyl acrylate rubber in aq. media in the presence of anionic emulsifiers [11]. The mechanism of graft polymerisation has been reported [12]

Mould Shrinkage (%):

No.	Value	Note
1	0.5%	linear

Applications: Applications due to excellent weather resistance, mechanical props. and mouldability in automobile industry (door mirror and lamp housings, radiator grilles, bumpers); in building (gutters, shutters, baths); in air conditioner parts, lamps, sockets, surfboards, garden furniture, swimming pool pump and filter housings

Trade name	Details	Supplier
ABS/PC	S-10/WR-2	Diamond Polymers
ASA	V range	Hitachi
Centrex	800 series	Monsanto Chemical
Geloy		General Electric Plastics
Geloy	BG/GY/XP	General Electric
Luran S		BASF

| Luran S | various grades | BASF |
| Terblend | S KR 2861/1 | BASF |

Bibliographic References

[1] Kweon, H.Y. and Park, Y.H., *Pollimo*, 1996, **20**, 877, (synth.)
[2] Mi, P., Yang, J., Xue, X., Chen, H.et al, *Hecheng Xiangjiao Gongye*, 1996, **19**, 43, (ir)
[3] Ryabikova, V.M., Zigel, A.N., Agnivtseva, T.G., Nesterova, V.M. and Tumina, S.D., *Plast. Massy*, 1990, 76, (ir)
[4] *Jpn. Pat.*, 1989, 01 121 312, (synth.)
[5] Pavlyuchenko, V.N., Kholodnova, L.V., Kanuan, S.K., Byrdina, N.A. et al, *Vysokomol. Soedin.*, Ser. A, 1989, **31**, 922, (viscoelastic behaviour)
[6] Kratschmann, K. and Dengel, D., *Z. Werkstofftech.*, 1985, **16**, 246, (hardness)
[7] Bergmann, K., Schmeidberger, H. and Unterforsthuber, K., *Colloid Polym. Sci.*, 1984, **262**, 283, (H-1 nmr)
[8] Kudryavtseva, T.V., Fratkina, G.P., Kirillova, E.I., Egorova, E.I. et al, *Plast. Massy*, 1979, 28, (weathering)
[9] Knoll, M., *Plastverarbeiter*, 1978, **29**, 75, (water absorption)
[10] Stangl, M., Binder, K. and Tschamler, H., Chem., *Kunstst. Aktuell*, 1977, **31**, 221, 225
[11] *U.S.S.R. Pat.*, 1976, 525 716, (synth.)
[12] Gaylord, N.G., Polym. Prepr. (Am. Chem. Soc., Div. Polym. Chem.), 1972, **13**, 505
[13] Bucknall, C.B., Page, C.J. and Young, V.O., Am. Chem. Soc., Div. Org. Coat. Plast. Chem., *Pap*, 1974, **34**, 288
[14] Bucknall, C.B., Page, C.J. and Young, V.O., *Adv. Chem. Ser.*, 1976, **154**, 179
[15] *U.S. Pat.*, 1967, 3 322 734, (mechanical props.)
[16] *U.S. Pat.*, 1956, 2 772 252, (props.)
[17] Luran S ASA Product Line, Properties, *Processing*, BASF, 1990, (technical datasheet)

ASA, Blow moulding

Related Polymers: ASA
Monomers: Acrylic acid, Styrene, Acrylonitrile
Material class: Thermoplastic, Copolymers
Polymer Type: ASA

Applications/Commercial Products:

Trade name	Supplier
Geloy 1020	General Electric Plastics
Geloy BG 10	General Electric Plastics

ASA, Extrusion unfilled

Related Polymers: ASA
Monomers: Acrylic acid, Acrylonitrile, Styrene
Material class: Thermoplastic, Copolymers
Polymer Type: ASA

Applications/Commercial Products:

Trade name	Supplier
Centrex 381	Monsanto Chemical
Centrex 833	Monsanto Chemical
Geloy 1020	General Electric Plastics
Geloy 1120	General Electric Plastics
Luran S 776SE	BASF
Luran S 797SE	BASF

ASA, High impact

Related Polymers: ASA
Monomers: Acrylic acid, Acrylonitrile, Styrene
Material class: Thermoplastic, Copolymers
Polymer Type: ASA

Applications/Commercial Products:

Trade name	Supplier
Geloy 1220	General Electric Plastics
Geloy 1221	General Electric Plastics

ASA, Injection moulding, Unfilled

Related Polymers: ASA
Monomers: Acrylonitrile, Styrene, Acrylic acid
Material class: Thermoplastic, Copolymers
Polymer Type: ASA
CAS Number:

CAS Reg. No.
24980-16-3

Applications/Commercial Products:

Trade name	Supplier
Centrex 811	Monsanto Chemical
Centrex 821	Monsanto Chemical
Geloy XP 1001	General Electric Plastics

ASA/PVC blend

Synonyms: *PVC/ASA blend. Polyvinylchloride/acrylic-styrene-acrylonitrile terpolymer blend*
Monomers: Styrene, Acrylonitrile, Acrylic acid, Vinyl chloride
Material class: Blends, Copolymers
Polymer Type: PVC, ASA

Volumetric & Calorimetric Properties:
Density:

No.	Value	Note
1	d 1.21 g/cm^3	[1]

Thermal Expansion Coefficient:

No.	Value	Note	Type
1	8.46×10^{-5} K^{-1}	[1]	1

Deflection Temperature:

No.	Value	Note
1	73°C	1.82 MPa, 0.25" thick [1]
2	82°C	0.45 MPa, 0.25" thick [1]

Transport Properties:
Melt Flow Index:

No.	Value	Note
1	4.5 g/10 min	200°, 3.8 kg [1]

Water Absorption:

No.	Value	Note
1	0.11 %	24h [1]

– Atactic Polypropylene

Mechanical Properties:

Flexural Modulus:

No.	Value	Note
1	2069.2 MPa	21100 kg cm^{-2}, 0.25" thick [1]

Tensile Strength Break:

No.	Elongation
1	40% elongation [1]

Tensile Strength Yield:

No.	Value	Note
1	46.2 MPa	471 kg cm^{-2} [1]

Flexural Strength Yield:

No.	Value	Note
1	68.25 MPa	696 kg cm^{-2}, 0.25" thick [1]

Izod Notch:

No.	Value	Notch	Note
1	960.8 J/m	y	18 ft Ib in^{-1}, room temp. [1]

Applications/Commercial Products:

Mould Shrinkage (%):

No.	Value	Note
1	0.4%	[1]

Trade name	Supplier
Geloy XP2003	General Electric Plastics

Bibliographic References

[1] *Geloy Range,* GE Plastics, (technical range)

Atactic Polypropylene A-45

Synonyms: *PP-atactic. Polypropene-atactic. Poly(1-propene) atactic*
Related Polymers: Polypropylene
Monomers: Propylene
Material class: Thermoplastic
Polymer Type: Polypropylene
CAS Number:

CAS Reg. No.
9003-07-0

Molecular Formula: $(C_3H_6)_n$
Fragments: C_3H_6
General Information: Data given here are specific to atactic polypropylene. More general data, including the effect of the atactic content on commercial material, can be found under Polypropylene
Morphology: Soft and rubbery

Volumetric & Calorimetric Properties:

Latent Heat Crystallization: Entropy of polymerisation ΔS_{polym} 116 J K^{-1} (25°, monomer liq. phase, polymer condensed amorph. phase). [19,20,21] (See also polypropylene, isotactic polypropylene)
Specific Heat Capacity:

No.	Value	Note	Type
1	2.34 kJ/kg.C	[2,26]	P

Glass-Transition Temperature:

No.	Value	Note
1	-7.16°C	[3]

Transition Temperature General: Increase in M_n causes increase in Tg; non linear deviation if $M_n < 10000$ [3]
Transition Temperature:

No.	Value	Type
1	-80°C	γ

Surface Properties & Solubility:

Internal Pressure Heat Solution and Miscellaneous: Heat of mixing at infinite dilution has been reported [22,24]
Solvents/Non-solvents: Sol. hydrocarbons, chlorinated hydrocarbons, isoamyl acetate, Et$_2$O. Insol. more polar organic solvents [16,17]

Transport Properties:

Polymer Solutions Dilute: Theta temps. have been reported [6,7,8,9,10,11,12,13]. Intrinsic viscosity and Huggins coefficients have been reported [18,27]
Polymer Melts: Viscosity has been reported [4]
Permeability of Gases: [He] 3.3×10^{-6}; [H$_2$] 1.5×10^{-6} cm^2 (s Pa)$^{-1}$ [5]

Optical Properties:

Volume Properties/Surface Properties: Concentration dependence of refractive index has been reported [14,15]

Polymer Stability:

Chemical Stability: More stable to sulfuric acid than isotactic polypropylene

Applications/Commercial Products:

Trade name	Details	Supplier
APP	Amorphous not truly atactic	Elf Atochem (UK) Ltd.

Bibliographic References

[1] Antsiburov, K.A., Balakhonov, E.G., Burmistrova, T.I., Kal'yanova, S.P. and Berzin, V.I., *CA,* 1988, **109**, 110988e
[2] Quirk, R.P. and Alsamarraie, M.A.A., *Polym. Handb.,* 3rd edn., (eds. J. Brandrup and E.H. Immergut), Wiley Interscience, 1989, 378
[3] Cowie, J.M.G., *Eur. Polym. J.,* 1973, **9**, 1041
[4] Wasserman, S.H. and Graessley, W.W., *Polym. Eng. Sci.,* 1996, **36**, 852
[5] Stannett, V. and Yasuda, H., *High Polymers,* (ed. K.W. Doak), John Wiley and Sons, 1964, **20**
[6] Elias, H.G., *Polym. Handb.,* 3rd edn., (eds. J. Brandrup and E.H. Immergut), Wiley Interscience, 1989, 209
[7] Moraglio, G. and Brzezinski, J., *J. Polym. Sci., Part B: Polym. Lett.,* 1964, **2**, 1105
[8] Takamizawa, K., Kambara, Y. and Oyama, T., *Rep. Prog. Polym. Phys. Jpn.,* 1967, **10**, 85
[9] Dietschy, H., PhD Thesis ETH. Zurich, 1966
[10] Shiokawa, Y., Takeo, H., Takamizawa, K. and Oyama, T., *Rep. Prog. Polym. Phys. Jpn.,* 1964, **7**, 49
[11] Kinsinger, J.B. and Hughes, R.E., *J. Phys. Chem.,* 1963, **67**, 1922

[12] Kinsinger, J.B. and Wessling, R.A., *J. Am. Chem. Soc.*, 1959, **81**, 2908
[13] Cowie, J.M. and McEwan, I.J., *J. Polym. Sci., Polym. Phys. Ed.*, 1974, **12**, 441
[14] Huglin, M.B., *Polym. Handb.*, 3rd edn., (eds. J. Brandrup and E.H. Immergut), Wiley Interscience, 1989, 409
[15] Horska, J., Stejskal, J. and Kratochvil, P., *J. Appl. Polym. Sci.*, 1983, **28**, 3873
[16] Fuchs, O, *Polym. Handb.*, 3rd edn., (eds. J. Brandrup and E.H. Immergut), Wiley Interscience, 1989, 379
[17] Kurata, M. and Stockmayer, W.H., *Adv. Polym. Sci.*, Springer-Verlag, 1963, **3**, 196
[18] Stickler, M. and Sütterlin, N., *Polym. Handb.*, 3rd edn., (eds. J. Brandrup and E.H. Immergut), Wiley Interscience, 1989, 183
[19] Busfield, W.K., *Polym. Handb.*, 3rd edn., (eds. J. Brandrup and E.H. Immergut), Wiley Interscience, 1989, 295
[20] Dainton, F.S., Evans, D.M., Hoare, F.E. and Melia, T.-P., *Polymer*, 1962, **3**, 277
[21] Passaglia, E. and Kevorkian, H.K., *J. Appl. Phys.*, 1963, **34**, 90
[22] Phuong-Nguyen, H. and Delmas, G., *Macromolecules*, 1979, **12**, 740, 746
[24] Ochiai, H., Ohashi, T., Tadokoro, Y. and Murackami, I., *Polym. J. (Tokyo)*, 1982, **14**, 457
[25] Delmas, G. and Tancrède, P., *Eur. Polym. J.*, 1973, **9**, 199
[26] Wilski, H. and Grewes, T., *J. Polym. Sci.*, 1964, **C6**, 33, (heat capacity)
[27] Moraglio, G. and Danusso, F., *Ann. Chim. (Rome)*, 1959, **49**, 902, (Huggins coefficients)

Avimid K-III

A-46

Synonyms: *Poly[(5,7-dihydro-1,3,5,7-tetraoxobenzo[1,2-c:4,5-c']dipyrrole-2,6(1H,3H)-diyl)-1,4-phenyleneoxy(2,6-dichloro-1,4-phenylene)(1-methylethylidene)(3,5-dichloro-1,4-phenylene)oxy-1,4-phenylene]. Poly[1H,3H-benzo[1,2-c:4,5-c']difuran-1,3,5,7-tetrone-alt-4,4'-[(1-methylethylidene)bis[(2,6-dichloro-4,1-phenylene)oxy]]bis[benzenamine]]*
Monomers: 1*H*,3*H*-Benzo[1,2-c:4,5-c']difuran-1,3,5,7-tetrone, 4,4'-[(1-Methylethylidene)bis[(2,6-dichloro-4,1-phenylene)oxy]]-bis[benzenamine]
Material class: Thermoplastic
Polymer Type: Polyimide
CAS Number:

CAS Reg. No.	Note
94148-53-5	
94148-82-0	source based

Molecular Formula: $(C_{37}H_{20}Cl_4N_2O_6)_n$
Fragments: $C_{37}H_{20}Cl_4N_2O_6$

Volumetric & Calorimetric Properties:
Latent Heat Crystallization: ΔH_m 26.9 J g^{-1} [2]
Melting Temperature:

No.	Value	Note
1	351°C	[2]
2	372°C	[3]

Glass-Transition Temperature:

No.	Value	Note
1	240°C	[3]

Transition Temperature General: T_g reported not to be measurable [2]

Surface Properties & Solubility:
Solubility Properties General: May be blended with compounds containing both ketone and ether functional groups e.g. 1,4-bis (4-phenoxybenzoyl)benzene, to improve processability [2]

Applications/Commercial Products:
Processing & Manufacturing Routes: Avimid K may be autoclave-moulded owing to its excellent tack and drape characteristics [1]. Synth. by reaction of benzenetetracarboxylic acid with 2,2-bis. (3,5-dichloro-4-(4-aminophenoxy)phenyl)propane [3]
Applications: Applications due to high temp. resistance in aircraft engine parts

Trade name	Details	Supplier
Avimid K	discontinued material	DuPont

Bibliographic References
[1] Baker, E.T. and Gesell. T.L., *Int. SAMPE Symp. Exhib.*, 1990, **35**, 979
[2] *Eur. Pat. Appl.*, 1992, 501 436
[3] *Eur. Pat. Appl.*, 1984, 122 060

Bis(2,3-epoxycyclopentyl)ether resin   B-1

Synonyms: *Poly[2,2'-oxybis(6-oxabicyclo[3.1.0]hexane)]*
Monomers: Bis(2,3-epoxycyclopentyl)ether
Material class: Thermosetting resin
Polymer Type: Epoxy
CAS Number:

CAS Reg. No.	Note
25322-94-5	homopolymer
119280-14-7	copolymer with *m*-phenylenediamine
75260-76-3	copolymer with 1,2-propanediol
29439-76-7	copolymer with ethylene glycol and methylenedianiline
26010-47-9	copolymer with ethylene glycol and *m*-phenylenediamine

Molecular Formula: $(C_{10}H_{14}O_3)_n$
Fragments: $C_{10}H_{14}O_3$
Additives: Glass fibre, carbon fibre
General Information: Props. of cured resin depend on curing agent, stoichiometry, cure time and cure temp. Uncured ERRA-0300 is a solid whereas uncured ERLA 0400 is an amber liq. [11]

Volumetric & Calorimetric Properties:
Thermodynamic Properties General: The linear thermal expansion coefficient of ERLA 4617 and of its carbon fibre reinforced analogue increases with increasing temp. A discontinuity in α_L occurs in the range 119–131° due to an approach to the glass transition [10]

Melting Temperature:

No.	Value	Note
1	40–60°C	ERRA-0300, uncured [11]

Deflection Temperature:

No.	Value	Note
1	115°C	ERLA 4300/4305, *m*-phenylenediamine cured, ASTM D648-56 [4,5]
2	175°C	ERLA 4617, *m*-phenylenediamine cured, ASTM D648-56 [4,5]
3	190°C	ERRA 0300, *m*-phenylenediamine cured [11]
4	180°C	ERLA 0400, *m*-phenylenediamine cured [11]
5	151°C	ERLA 0400, Tonox cured [11]
6	187°C	1.82 MPa, ERLA 0400, *m*-phenylenediamine cured [12]

Transport Properties:
Polymer Solutions Dilute: Viscosity 38 cps (25°, ERLA 0400, uncured) [11]
Water Content: Water absorption is dependent upon the nature of the curing agent used [6]. Water absorption of cured ethylene glycol copolymers, reinforced with carbon fibre, increases with increasing temp. [7]

Water Absorption:

No.	Value	Note
1	2.95 %	24h, 100°, ERRA 4205, methylenedianiline cured [6]
2	5.92 %	24h, 100°, ERRA 4205, diaminodiphenylsulfone cured [6]
3	3.12 %	24h, 100°, ERLA 4617, methylenedianiline cured [6]
4	5.23 %	24h, 100°, ERLA 4617, *m*-phenylenediamine cured [6]
5	7.4 %	24h, 100°, copolymer with 1,2-propanediol, *m*-phenylenediamine cured [6]
6	4.53 %	24h, 100°, copolymer with 1,2-propanediol, methylenedianiline cured [6]

Mechanical Properties:
Mechanical Properties General: The flexural props. of ERRA 0300 (Tonox cured, glass reinforced) deteriorate with increasing temp. [11]. Increasing cure time of ERRA 0300 or ERLA 0400 with *m*-phenylenediamine has little effect on mechanical props.

Tensile (Young's) Modulus:

No.	Value	Note
1	4481 MPa	650 kpsi, ethylene glycol/methylenedianiline cured [2]
2	5377 MPa	780 kpsi, ethylene glycol/*m*-phenylenediamine cured [2]
3	5805 MPa	842 kpsi, ERLA-4300, *m*-phenylenediamine cured, ASTM D638-64T [4,5]
4	6008 MPa	871.5 kpsi, ERLA-4305, *m*-phenylenediamine cured, ASTM D638-64T [4,5]
5	5398 MPa	783 kpsi, ERLA-4617, *m*-phenylenediamine cured, ASTM D638-64T [4,5]
6	4688 MPa	680 kpsi, ERRA 0300, *m*-phenylenediamine cured [11]
7	4895 MPa	710 kpsi, ERLA 0400, *m*-phenylenediamine cured [11]
8	19924 MPa	2890 kpsi, ERRA 0300, Tonox cured, glass reinforced [11]

Flexural Modulus:

No.	Value	Note
1	4653 MPa	675 kpsi, ethylene glycol/methylenedianiline cured [2]
2	5515 MPa	800 kpsi, ethylene glycol/*m*-phenylenediamine cured [2]
3	5998 MPa	870 kpsi, ERLA-4300, *m*-phenylenediamine cured, ASTM D790-66 [4,5]
4	6273 MPa	910 kpsi, ERLA-4305, *m*-phenylenediamine cured, ASTM D790-66 [4,5]
5	5619 MPa	815 kpsi, ERLA-4617, *m*-phenylenediamine cured, ASTM D790-66 [4,5]
6	24818 MPa	3600 kpsi, 23°, ERRA 0300, Tonox cured, glass reinforced [11]

– Bis(2,3-epoxycyclopentyl)ether resin

Tensile Strength Break:

No.	Value	Note	Elongation
1	121 MPa	17.5 kpsi, ethylene glycol/methylenedianiline or ethylene glycol/m-phenylenediamine cured [2]	2-9%–5.1%
2	94 MPa	13.65 kpsi, ERLA-4300, m-phenylenediamine cured [4,5]	
3	108 MPa	15.65 kpsi, ERLA-4305, m-phenylenediamine cured [4,5]	
4	132 MPa	19.2 kpsi, ERLA-4617, m-phenylenediamine cured [4,5]	
5	345 MPa	50 kpsi, ERRA 0300, Tonox cured, glass reinforced [11]	
6	131 MPa	19 kpsi, ERRA 0300, m-phenylenediamine cured [11]	
7	121–124 MPa	17.5-18 kpsi, ERLA 0400, m-phenylenediamine cured [11,12]	

Flexural Strength at Break:

No.	Value	Note
1	172 MPa	25 kpsi, ethylene glycol/methylenedianiline cured [2]
2	217 MPa	31.5 kpsi, ethylene glycol/m-phenylenediamine cured [2]
3	160 MPa	23.25 kpsi, ERLA-4300, m-phenylenediamine cured [4,5]
4	164 MPa	23.8 kpsi, ERLA-4305, m-phenylenediamine cured [4,5]
5	214 MPa	31 kpsi, ERLA-4617, m-phenylenediamine cured [4,5]
6	503 MPa	73 kpsi, 23°, ERRA 0300, Tonox cured, glass reinforced [11]
7	190 MPa	27.6 kpsi, ERLA 0400, m-phenylenediamine cured [12]

Compressive Modulus:

No.	Value	Note
1	4736 MPa	687 kpsi, ethylene glycol/methylenedianiline cured [2]
2	5846 MPa	848 kpsi, ethylene glycol/m-phenylenediamine cured [2]
3	6791 MPa	985 kpsi, ERLA-4300, m-phenylenediamine cured, ASTM D695-63T [4,5]
4	6808 MPa	987.5 kpsi, ERLA-4305, m-phenylenediamine cured, ASTM D695-63T [4,5]
5	6136 MPa	890 kpsi, ERLA-4617, m-phenylenediamine cured, ASTM D695-63T [4,5]
6	4757 MPa	690 kpsi, ERRA 0300, m-phenylenediamine cured [11]
7	4826 MPa	700 kpsi, ERLA 0400, m-phenylenediamine cured [11]

Poisson's Ratio:

No.	Value	Note
1	0.35	ERRA 0300, m-phenylenediamine cured [11]

Compressive Strength:

No.	Value	Note
1	176 MPa	25.6 kpsi, ethylene glycol/methylenedianiline cured [2]
2	227 MPa	32.9 kpsi, ethylene glycol/m-phenylenediamine cured [2]
3	256 MPa	37.15 kpsi, ERLA-4300, m-phenylenediamine cured [4,5]
4	246 MPa	35.65 kpsi, ERLA-4305, m-phenylenediamine cured [4,5]
5	226 MPa	32.8 kpsi, ERLA-4617, m-phenylenediamine cured [4,5]
6	207 MPa	30 kpsi, ERRA 0300 and ERLA 0400, m-phenylenediamine cured [11]

Miscellaneous Moduli:

No.	Value	Note	Type
1	4736 MPa	687 kpsi, ERLA 0400, m-phenylenediamine cured [12]	tensile tangent modulus
2	4715 MPa	684 kpsi, ERLA 0400, m-phenylenediamine cured [12]	compressive tangent modulus
3	5012 MPa	727 kpsi, ERLA 0400, m-phenylenediamine cured [12]	flexural tangent modulus

Hardness: Barcol 35-40 (methyl nadic anhydride cured) [8]
Izod Notch:

No.	Value	Notch	Note
1	28.8 J/m	Y	0.54 ft lb in^{-1}, ERRA 0300, m-phenylenediamine cured [11]
2	13.9 J/m	Y	0.26 ft lb in^{-1}, ERRA 0300, Tonox cured [11]
3	15.5 J/m	Y	0.29 ft lb in^{-1}, ERLA 0400, Tonox cured [11]

Polymer Stability:
Decomposition Details: 10% Weight loss for m-phenylenediamine cured resin occurs after 2h heating in air at 212°; 20% weight loss occurs at 257° [3]. Activation energy for 10–15% weight loss 111 kJ mol^{-1} (m-phenylenediamine cured), 97 kJ mol^{-1} for 15–20% weight loss (m-phenylenediamine cured) [3]
Flammability: ERLA 4617 (carbon fibre filled) does not meet the requirements of the FAR standard [9]. Not self-extinguishing (ERLA 4617, carbon fibre filled, FAA standard) [9]

Applications/Commercial Products:
Processing & Manufacturing Routes: Curing with m-phenylenediamine is probably a two-step process [1]. ERLA-4300 and ERLA 4305 are synth. by reaction of bis(2,3-epoxycyclopentyl)ether with H$_2$O (1 mol) in the presence of benzyldimethylamine catalyst for 7h at 100° [4]. Strong, tough polymers may be synth. by curing bis(2,3-epoxycyclopentyl)ether with methyl nadic anhydride [8] or m-phenylenediamine [11] using a tin octoate catalyst

Applications: In prepreg systems. Potential applications in military filament winding applications (missiles, submarines) and in environments exposed to cryogenic temps.

Trade name	Details	Supplier
ERLA	4617 (ethylene glycol copolymer)/4300/4305	Union Carbide
ERLA	4617 (ethylene glycol copolymer)	Bakelite Xylonite Ltd
ERNA	4205 (65% solids)	Union Carbide
ERR	0300/4205	Union Carbide
ERRA	4205 (65% solids; discontinued material)	Union Carbide
ERRA	0300/0400	Union Carbide

Bibliographic References
[1] Huang, J., Wang, D., Zhang, B. and Lu, Y., *Gaodeng Xuexiao Huaxue Xuebao*, 1982, **3**, 418, (cure)
[2] Soldatos, A.C., Burhans, A.S. and Cole, L.F., *SPE J.*, 1969, **25**, 61, (mechanical props.)
[3] Knight, G.J. and Wright, W.W., *Br. Polym. J.*, 1989, **21**, 303, (thermal stability)
[4] Soldatos, A.C. and Burhans, A.S., *Ind. Eng. Chem. Proc. Des. Dev.*, 1967, **6**, 205, (synth., mechanical props.)
[5] Soldatos, A.C. and Burhans, A.S., Am. Chem. Soc., Div. Org. Coat. Plast. Chem., *Pap*, 1967, **27**, 282, (synth., mechanical props.)
[6] Eckstein, B.H., *Org. Coat. Plast. Chem.*, 1978, **38**, 503, (water absorption)
[7] Judd, N.C.W., *Br. Polym. J.*, 1977, **9**, 36, (water absorption)
[8] U.S. Pat., 1964, 3 117 099, (synth.)
[9] Hognat, J., *J. Fire Flammability*, 1977, **8**, 506, (flammability)
[10] Pirgon, O., Wostenholm, G.H. and Yates, B., *J. Phys. D: Appl. Phys.*, 1973, **6**, 309, (thermal expansion)
[11] Madden, J.J. Burhans, A.S. and Pitt, C.F., *Adhesives Age*, 1965, **8**, 22, (mechanical props.)
[12] Cole, L.F. and Mulvaney, W.P., *Mod. Plast.*, 1967, **44**, 151, (mechanical props.)

1-(*N,N*-Bis(2,3-epoxypropyl)amino)-4-(2,3-epoxyprop-1-oxy)benzene) resin   B-2

Synonyms: *Poly[N-[4-(oxiranylmethoxy)phenyl]-N-(oxiranylmethyl)oxiranemethanamine]. Poly(triglycidyl-p--aminophenol). PTGAP*
Monomers: Triglycidyl *p*-aminophenol
Material class: Thermosetting resin
Polymer Type: Epoxy
CAS Number:

CAS Reg. No.	Note
31305-88-1	homopolymer
115632-64-9	copolymer with 1,3-benzenediamine
119670-44-9	copolymer with aniline

Molecular Formula: ($C_{15}H_{19}NO_4$)
Fragments: $C_{15}H_{19}O_4$
Molecular Weight: M_n 360 (uncured)
General Information: Props. of cured resin depend on curing agent, stoichiometry, cure time and cure temp.

Volumetric & Calorimetric Properties:
Glass-Transition Temperature:

No.	Value	Note
1	200°C	diaminodiphenylmethane:aniline (75:25) cured [8]
2	168°C	diaminodiphenylmethane:aniline (50:50) cured [8]
3	141°C	diaminodiphenylmethane:aniline (25:75) cured [8]
4	117°C	aniline cured [8]
5	192°C	*N*-tetraethyldiaminodiphenylmethane cured [7]
6	123°C	aniline cured [7]
7	227°C	diaminodiphenylmethane cured [7]
8	249°C	diaminodiphenylsulfone cured [7]

Deflection Temperature:

No.	Value	Note
1	135°C	diaminodiphenylamine:aniline (25:75) cured; 0.18 MPa, ASTM D648 [8]
2	118°C	aniline cured; 0.18 MPa, ASTM D648 [8]
3	197°C	diaminodiphenylamine:aniline (75:25) cured; 0.18MPa, ASTM D648
4	167°C	diaminodiphenylamine:aniline (50:50) cured; 0.18MPa, ASTM D648 [8]

Transition Temperature:

No.	Value	Note	Type
1	-63--23°C	aniline cured [8]	T_β
2	47-67°C	aniline cured [8]	T'_β

Surface Properties & Solubility:
Cohesive Energy Density Solubility Parameters: δ 22.2–25.9 MPa$^{1/2}$ (*N*-tetraethyldiaminodiphenylmethane cured) [7], 23.0–27 MPa$^{1/2}$ (aniline cured) [7], 23.5–26.7 MPa$^{1/2}$ (diaminodiphenylmethane cured) [7], 25.6–27.7 MPa$^{1/2}$ (diaminodiphenylsulfone cured) [7]

Transport Properties:
Polymer Solutions Dilute: Viscosity 2500–5000 cP (25°, uncured)
Permeability of Liquids: Diffusivity [H_2O] 27×10^{-9} cm^2 s^{-1} (diaminodiphenylmethane:aniline (75:25) cured) [8], 28×10^{-9} cm^2 s^{-1} (diaminodiphenylmethane:aniline (50:50) cured) [8], 33×10^{-9} cm^2 s^{-1} (diaminodiphenylmethane:aniline (25:75) cured) [8], 47×10^{-9} cm^2 s^{-1} (aniline cured) [8]
Water Content: The equilibrium concentration of absorbed H_2O in diaminodiphenylmethane-aniline cured systems decreases with the weight fraction of aniline [8]
Water Absorption:

No.	Value	Note
1	5.47 %	48h, 100°, 95% relative humidity, diaminodiphenylmethane:aniline (75:25) cured [8]
2	4.98 %	48h, 100°, 95% relative humidity, diaminodiphenylmethane:aniline (50:50) cured [8]
3	4.92 %	48h, 100°, 95% relative humidity, diaminodophenylmethane:aniline (25:75) cured [8]
4	5.29 %	48h, 100°, 95% relative humidity, aniline cured [8]

Mechanical Properties:
Tensile (Young's) Modulus:

No.	Value	Note
1	3530 MPa	20°, diaminodiphenylmethane:aniline (75:25) cured
2	3660 MPa	20°, diaminodiphenylmethane:aniline (50:50) cured
3	3810 MPa	20°, diaminodiphenylmethane:aniline (25:75) cured
4	3910 MPa	20°, aniline cured [8]

Bisphenol A Diglycidyl ether resin

Poisson's Ratio:

No.	Value	Note
1	0.372	diaminodiphenylmethane:aniline (75:25) cured [8]
2	0.374	diaminodiphenylmethane:aniline (50:50) cured [8]
3	0.366	diaminodiphenylmethane:aniline (25:75) cured [8]

Miscellaneous Moduli:

No.	Value	Note	Type
1	1600 MPa	20°, 0.3 Hz, diaminodiphenylmethane:aniline (75:25) cured [8]	dynamic modulus
2	1940 MPa	20°, 0.3 Hz, diaminodiphenylmethane:aniline (50:50) cured) [8]	dynamic modulus
3	1820 MPa	20°, 0.3 Hz, diaminodiphenylamine:aniline (25:75) cured [8]	dynamic modulus
4	1750 MPa	20°, 0.3 Hz, aniline cured [8]	

Viscoelastic Behaviour: Longitudinal sound velocity 2983 m s^{-1} (20°, diaminodiphenylmethane:aniline (75:25) cured), 2953 m s^{-1} (20°, diaminodiphenylmethane:aniline (50:50) cured), 2905 m s^{-1} (20°, diaminodiphenylmethane:aniline (25:75) cured) [8]

Optical Properties:
Transmission and Spectra: H-1 nmr spectral data have been reported [3]

Polymer Stability:
Decomposition Details: Decomposition characteristic depend upon the nature of the curing agent. For example, 10% weight loss in air occurs at 285° for resin cured with diaminodiphenylsulfone but at 240° when cured with diethylenetriamine [4]. Char yield 8.3% (diaminodiphenylsulfone cured), 3.8% (diethylenetriamine cured) [4]
Hydrolytic Stability: Exposure of uncured ERL 0500 to 96% relative humidity at 60° results in loss of 20% of the epoxide groups after approx. 4.5 d [9]

Applications/Commercial Products:
Processing & Manufacturing Routes: Kinetic studies on the curing process with acid anhydrides have been reported [1]. Curing with aromatic diamines follows first order kinetics [2]. Other details of curing behaviour have been reported [5]. Kinetics of cure with diaminodiphenylsulfone have been reported [6]
Applications: Uncured resin has applications due to its low viscosity and low temp. reactivity. Increases heat resistance and cure speed of bisphenol A epoxy resins; other uses in adhesives, as toolings compounds and as laminating system. Molecularly distilled resin has applications as a binder for solid propellants and in military flares

Trade name	Details	Supplier
ERL	0500/0510	Ciba-Geigy Corp.

Bibliographic References
[1] Patel, R.D., Patel, R.G. and Patel, V.S., *Angew. Makromol. Chem.*, 1987, **155**, 57
[2] Patel, R.D., Patel, R.G. and Patel, V.S., *J. Therm. Anal.*, 1988, **34**, 1283
[3] Jagannathan, N.R. and Herring, F.G., *J. Polym. Sci., Part A: Polym. Chem.*, 1987, **25**, 897, (H-1 nmr)
[4] Patel, R.D., Patel, R.G. and Patel, V.S., *Thermochim. Acta*, 1988, **128**, 149, (thermal degradation)
[5] Lin, S. and Lu, C., *Reguxing Shuzhi*, 1994, **9**, 5
[6] Varley, R.J., Hodgkin, J.H., Hawthorne, D.G. and Simon, G.P., *J. Appl. Polym. Sci.*, 1996, **60**, 2251
[7] Bellenger, V., Morel, E. and Verdu, J., *J. Appl. Polym. Sci.*, 1989, **37**, 2563, (solubility parameters)
[8] Morel, E., Bellenger, V., Bocquet, M. and Verdu, J., *J. Mater. Sci.*, 1988, **23**, 4244, (T_g, water sorption, elastic props)
[9] Pearce, P.J., Davidson, R.G. and Morris, C.E.M., *J. Appl. Polym. Sci.*, 1981, **26**, 2363, (hydrolytic stability)

Bisphenol A Diglycidyl ether resin

Synonyms: *Poly(4,4'-(1-methylethylidene)bisphenol-co-(chloromethyl)oxirane). Diglycidyl ether of Bisphenol A. BADGE. DGEBPA*
Monomers: Bisphenol A, (Chloromethyl)oxirane
Material class: Thermosetting resin
Polymer Type: Epoxy
CAS Number:

CAS Reg. No.	Note
40364-42-9	copolymer with methylenedianiline
56727-50-5	copolymer with hexamethylenediamine
76397-91-6	copolymer with tetrahydromethylisophthalic anhydride
71745-12-5	copolymer with 4,4-diaminodiphenylsulfone
25068-38-6	

Molecular Formula: $C_{18}H_{20}O_3$
Fragments: $C_{18}H_{20}O_3$
Molecular Weight: MW 370 (approx., commercial liq. resin, M_n 900-3750 (n=2-12, solid resin), MW 9454 (Epikote 1009), M_n 4746 (Epikote 1009), M_n 1400 (Epikote 1004)
Additives: Carbon fibre, glass fibre, aramid fibre, aluminium borate to improve mechanical props. Copper acrylate to improve electrical props. and thermal and chemical resistance. TiO_2 (30–40%) or graphite to improve thermomechanical props. Silica, talc, calcium silicate, mica, clay, calcium carbonate to reduce shrinkage and improve dimensional stability. Reactive diluents (e.g. glycidyl ethers) to reduce viscosity
General Information: MW of the commercial product (liq. resin with degree of polymerisation approx. zero) may be increased by reaction of epichlorohydrin and bisphenol A in a mole ratio as close to 1:1 as possible. Typically, however, an excess (10:1) of epichlorohydrin is preferred. Solid resins have a degree of polymerisation of 2-30 [1]
Identification: Uncured liq. resin is characterised by hydrolysable chlorine content [1,16]

Volumetric & Calorimetric Properties:
Density:

No.	Value	Note
1	d 1.16 g/cm^3	DER 331, BF_3.monoethylamine cured [49]
2	d 1.16 g/cm^3	Epon 828, cured [50]
3	d^{27} 1.15 g/cm^3	BF_3-amine complex cured [4]
4	d^{27} 1.19 g/cm^3	diethylenetriamine cured [4]
5	d^{27} 1.05 g/cm^3	m-phenylenediamine cured [4]
6	d 1.24 g/cm^3	tetrahydro-5-methylisophthalic anhydride cured [5]
7		4,4'-methylenedianiline cured [5]
8	d^{35} 1.18 g/cm^3	[33]

– Bisphenol A Diglycidyl ether resin

Thermal Expansion Coefficient:

No.	Value	Note	Type
1	0.00047 K^{-1}	BF$_3$-amine complex cured [4]	1
2	0.00041 K^{-1}	diethylenetriamine cured [4]	1
3	0.00044 K^{-1}	m-phenylenediamine cured [4]	1

Equation of State: Equation of state information for liq. resin has been reported [12]

Specific Heat Capacity:

No.	Value	Note	Type
1	2.093 kJ/kg.C	0.5 cal g$^{-1\circ}$ C^{-1} Epon 828, uncured [50]	p

Melting Temperature:

No.	Value	Note
1	65–75°C	solid resin, M$_n$ 900
2	95–105°C	solid resin, M$_n$ 1400
3	125–135°C	solid resin, M$_n$ 2900
4	145–155°C	solid resin, M$_n$ 3750 [1]

Glass-Transition Temperature:

No.	Value	Note
1	124°C	BF$_3$-amine complex cured [4]
2	122°C	diethylenetriamine cured [4]
3	132°C	m-phenylenediamine cured [4]
4	102°C	tetrahydro-5-methylisophthalic anhydride cured [5]
5	145°C	4,4'-methylenedianiline cured [5]
6	130°C	polyamide cured [26]
7	82°C	Epikote 1009 [25]
8	51°C	Epikote 1004 [27]
9	-24°C	Epon 828 [32]
10	61°C	Epon 828, BF$_3$.ethylamine complex cured, 0.045 equiv. curing agent [32]
11	94°C	[33]
12	96°C	PKHH [35]
13	145°C	methylenedianiline cured, 150°, 4h, 0.1 MPa pressure [37]
14	117°C	diaminodiphenylsulfone cured, 81% cured [38]
15	55°C	diaminodiphenylsulfone cured, 58% cured [38]
16	187°C	AER-331, 4,4'-sulfonyldianiline cured [39]
17	-28°C	ED-22, tetrahydromethylisophthalic anhydride cured, 30% curing agent [42]
18	-41°C	ED-22, tetrahydromethylisophthalic anhydride cured, 64% curing agent [42]
19	47°C	dry, room temp. cured with Versamid 125 [43]
20	51°C	dry, cured 50°, Versamid 125 [43]
21	37°C	wet, cured 50°, Versamid 125 [43]
22	47°C	dry, cured 80°, Versamid 125 [43]

Transition Temperature General: T$_g$ decreases with increasing cure pressure [37]. T$_g$ of Epon 828 cured with BF$_3$.ethylamine complex increases amount of curing agent used [32]. T$_g$ of tetrahydromethylisophthalic anhydride cured resin varies linearly with the composition of resin and curing agent in the composite [42]

Deflection Temperature:

No.	Value	Note
1	80–120°C	aliphatic polyamine cured
2	142°C	aromatic polyamine cured
3	40–60°C	polyamine cured [1]
4	120°C	cured with hexahydrophthalic anhydride [1]
5	111°C	triethylenetetramine cured [2]
6	160°C	4,4'-methylenedianiline cured [2]
7	101°C	Versamide cured [2]
8	163°C	1.82 MPa, DER 331, BF$_3$.ethylamine cured, unannealed [49]
9	96°C	1.82 MPa, DER 331, diaminodiphenylsulfone cured, unannealed [49]
10	120°C	1.82 MPa, BF$_3$-amine complex cured, ASTM D648-45T [4]
11	91°C	1.82 MPa, diethylenetriamine cured, ASTM D648-45T [4]
12	131°C	1.82 MPa, m-phenylenediamine cured, ASTM D648-45T [4]
13	160°C	1.82 MPa, DER 331, methylenediamine cured, unannealed [49]
14	160°C	Epon 828, methylenedianiline cured [50]
15	320°C	Epon, cured with norbornyl dianhydride siloxane [36]
16	175°C	Epon 828, diaminodiphenylsulfone cured [50]
17	170°C	Epon 828, BF$_3$.ethylamine cured [50]

Vicat Softening Point:

No.	Value	Note
1	125°C	BF$_3$-amine complex cured, modified ASTM D1525-70 [4]
2	93°C	diethylenetriamine cured, modified ASTM D1525-70 [4]
3	136°C	m-phenylenediamine cured, modified ASTM D1525-70 [4]

Surface Properties & Solubility:

Solubility Properties General: High MW resins form miscible blends with poly(ether sulfone) [10] and poly (N-vinyl pyrrolidone) [35]. Miscible with poly(tetrahydrofurfuryl methacrylate) [14]. Incompatible with epoxidised natural rubber [31]

Solvents/Non-solvents: Solubility data have been reported [17]. Uncured liq. resin sol. nitric acid, sulfuric acid [26], insol., C$_6$H$_6$, toluene, dioxane, hydrochloric acid, MeOH, sodium hydroxide, H$_2$O, DMF [26]

Wettability/Surface Energy and Interfacial Tension: Studies on contact angle have been reported [11]. Contact angle 39° (22°, equilib., ED-20 on untreated OT-4 (titanium) alloy) [41]. Contact angle decreases upon treatment of the alloy surface with chelated transition metal compounds [41]

Surface Tension:

No.	Value	Note
1	32.9 mN/m	diethylaminopropylamine cured [3]
2	44 mN/m	diethylenetriamine cured [3]

Transport Properties:

Transport Properties General: Transport props. have been reported [19]

Polymer Solutions Dilute: η 11000–15000 mPa s (25°, liq. resin, commercial grade); 3000 mPa s (min., 25°, cured with aliphatic polyamide), 10000–20000 mPa s (25°, cured with aromatic polyamide), 10000 mPa s (25°, cured with polyamine) [1]. Viscosity of the cured resin can be lowered by the addition of diluents such as glycidyl ethers and dibutyl phthalate. Certain mechanical props., however, may be sacrificed [1]

Polymer Melts: Theta temp. 63° (1,2-dichloroethane) [35]. The temp. dependence of melt viscosity, itself related closely to dielectric relaxation time, may be described by the Williams-Landel-Fern equation [48]

Permeability of Gases: Sorption by amorph. films of N_2, CH_4 and CO_2 is low [33]. Diffusion coefficients [CO_2] $0.129-0.206 \times 10^8$ $cm^2\ s^{-1}$ (35°, 1.7 mil thick); [N_2] $0.316-0.322 \times 10^8\ cm^2\ s^{-1}$ (1.7 mil thick, 35°); [CH_4] $0.063-0.069 \times 10^8\ cm^2\ s^{-1}$ (35°, 1.7 mil thick) [3]

Water Absorption:

No.	Value	Note
1	0.5 %	1h 100°, cured with hexahydrophthalic anhydride [1]

Gas Permeability:

No.	Gas	Value	Note
1	[H_2O]	11163 cm^3 mm/(m^2 day atm)	$1.7 \times 10^{-8}\ cm^2$ (s cm Hg)$^{-1}$, 40° [34]
2	[CO_2]	26.5 cm^3 mm/(m^2 day atm)	$0.403 \times 10^{-10}\ cm^2$ (s cm Hg)$^{-1}$, 35°, 10 atm, 1.7 mil thick [45]
3	[CH_4]	1.25 cm^3 mm/(m^2 day atm)	$0.019 \times 10^{-10}\ cm^2$ (s cm Hg)$^{-1}$, 35°, 10 atm, 1.7 mil thick [45]
4	[He]	164.2 cm^3 mm/(m^2 day atm)	$2.5 \times 10^{-10}\ cm^2$ (s cm Hg)$^{-1}$, 35°, 10 atm, 1.7 mil thick [45]
5	[O_2]	6.6 cm^3 mm/(m^2 day atm)	$0.1 \times 10^{-10}\ cm^2$ (s cm Hg)$^{-1}$, 35°, 10 atm, 1.7 mil thick [45]

Mechanical Properties:

Mechanical Properties General: Mechanical props. are dependent upon the type of curing agent used. Mechanical props. of resin cured with hexamethylenediamine have been reported [20]. Tensile props. of Epon 828 have been reported [24]. Mechanical props. decrease with increasing cure pressure [37]. Tensile modulus is increased by addition of fillers to cured resin. The increase is especially marked with the addition of graphite [40]. Tensile strength of Epikote 828 is improved by post-curing at up to 100° [43]. Mech. props. of cured DER 331 may be improved by incorporation of poly(4-phenyleneterephthalimide) anion [46]. Epon 828 cured with polyamide, aliphatic or aromatic amine undergoes plastic yield under compression and brittle failure under tension. Compressive strength increases with increasing resin density and increasing T_g [47]

Tensile (Young's) Modulus:

No.	Value	Note
1	3600 MPa	Epon 828, cured [30]
2	2230 MPa	methylenedianiline cured, 200°, 1.5h, 0.1 MPa pressure [37]
3	1390 MPa	methylenedianiline cured, 200°, 1.5h, 1 kPa pressure [37]
4	3400 MPa	25°, cured with hexahydrophthalic anhydride [1]
5	20.2 MPa	equilib., ED-20, polyethylene polyamine cured [40]
6	2000 MPa	DER 331, m-phenylenediamine cured [46]
7	27.7 MPa	283 kg cm^{-2}, DER 331, BF_3.ethylamine cured [49]
8	2157 MPa	22000 kg cm^{-2}, DER 331, diaminodiphenylsulfone cured [49]
9	1510 MPa	154 kg mm^{-2}, AER 331, 4, 4′-sulfonyl dianiline cured) [39]

Flexural Modulus:

No.	Value	Note
1	2795 MPa	28500 kg cm^{-2}, DER 331, BF_3.ethylamine cured [49]
2	3400 MPa	25°, cured with hexahydrophthalic anhydride [1]
3	3050 MPa	triethylenetetramine cured [2]
4	2700 MPa	4,4′-methylenedianiline cured [2]
5	2140 MPa	Versamide cured [2]
6	3177 MPa	32400 kg cm^{-2}, DER 331, diaminodiphenylsulfone cured [49]
7	2687 MPa	27400 kg cm^{-2}, DER 331, methylenediamine cured [49]
8	3491 MPa	356 kg mm^{-2}, AER-331, 4,4′-sulfonyldianiline cured [39]

Tensile Strength Break:

No.	Value	Note	Elongation
1	40–70 MPa	aliphatic polyamine cured [1]	
2	83 MPa	aromatic polyamine cured [1]	
3	35–55 MPa	polyamide cured [1]	
4	65 MPa	25°, cured with hexahydrophthalic anhydride [1]	
5	28.5 MPa	295 kg cm^{-2}, DER 331, BF_3.ethylamine cured [49]	1%
6	79 MPa	triethylenetetramine cured [2]	
7	70.4 MPa	4,4′-methylenedianiline cured [2]	
8	57.3 MPa	Versamide cured [2]	
9	20.6–27.5 MPa	3-4 kpsi, Epon cured with norbornyl dianhydride siloxane [36]	4%
10	98.4 MPa	methylenedianiline cured, 200°, 1.5h, 0.1 MPa pressure [37]	13.8%
11	18.5 MPa	methylenedianiline cured, 200°, 1.5h, 1 kPa pressure [37]	1.45%
12	89 MPa	9.08 kg mm^{-2}, AER-331, 4,4′-sulfonyldianiline cured [39]	
13	42 MPa	DER 331, m-phenylenediamine cured [46]	
14	78.5 MPa	801 kg cm^{-2}, DER 331, diaminodiphenylsulfone cured [49]	5%

– Bisphenol A Diglycidyl ether resin

Flexural Strength at Break:

No.	Value	Note
1	104–124 MPa	aliphatic polyamine cured [1]
2	152 MPa	aromatic polyamine cured [1]
3	48–62 MPa	polyamide cured [1]
4	131 MPa	25°, cured with hexahydrophthalic anhydride [1]
5	96 MPa	triethylenetetramine cured [2]
6	93 MPa	4,4′-methylenedianiline cured [2]
7	67 MPa	Versamide cured [2]
8	157 MPa	16 kg mm^{-2}, AER-331, 4,4′-sulfonyldianiline cured [39]
9	130 MPa	tetrahydro-5-methylisophtalic anhydride cured [5]
10	116.7 MPa	1190 kg cm^{-2}, DER 331, BF$_3$.ethylamine cured [49]
11	17.9 MPa	1830 kg cm^{-2}, DER 331, diaminodiphenylsulfone cured [49]
12	116 MPa	1180 kg cm^{-2}, DER 331, methylenedianiline cured [49]

Compressive Modulus:

No.	Value	Note
1	3050 MPa	triethylenetetramine cured [2]
2	2600 MPa	4,4′-methylenedianiline or Versamide cured [2]

Poisson's Ratio:

No.	Value	Note
1	0.4	Epon 828, cured [30]

Tensile Strength Yield:

No.	Value	Elongation	Note
1	82–97 MPa		Epon 828, cured [30]
2	37.2 MPa	5.4%	5.4 kpsi, Epon cured with norbornyl anhydride siloxane [36]

Flexural Strength Yield:

No.	Value	Note
1	85.5 MPa	872 kg cm^{-2}, DER 331, BF$_3$.ethylamine cured [49]
2	163.8 MPa	1670 kg cm^{-2}, DER 331, diaminodiphenyl-sulfone cured [49]
3	93 MPa	947 kg cm^{-2}, DER 331, methylenedianiline cured [49]

Compressive Strength:

No.	Value	Note
1	138–227 MPa	aliphatic polyamine cured [1]
2	207 MPa	aromatic polyamine cured [1]
3	62–76 MPa	polyamide cured [1]
4	124 MPa	25°, cured with hexahydrophthalic anhydride [1]
5	112 MPa	triethylenetetramine cured [2]
6	116 MPa	4,4′-methylenedianiline cured [2]
7	85.6 MPa	Versamide cured [2]
8	126 MPa	4,4′-methylenedianiline cured [2]

Viscoelastic Behaviour: Acoustic emission behaviour under tensile loading has been reported [30]
Hardness: Pencil hardness 2H (aliphatic polyamine cured), F (polyamide cured) [1]. Scratch hardness 1000g (film, polyamide cured) [26]. Rockwell M115 (DER 331, BF$_3$.ethylamine cured), M80 (DER 331, diaminodiphenylsulfone cured), M106 (DER 331, methylenedianiline cured) [49]
Failure Properties General: Tensile modulus is increased by addition of fillers to cured resin. The increase is especially marked with the addition of graphite [40]. Tensile strength of Epikote 828 is improved by post-curing at up to 100° [43]. Mechanical props. of cured DER 331 may be improved by incorporation of poly p-phenyleneterephthalamide) anion [46]. Epon 828 cured with polyamide, aliphatic or aromatic amine undergoes plastic yield under compression and brittle failure under tension. Compressive yield strength increases with increasing resin density and increasing T_g [47]
Fracture Mechanical Properties: K_{IC} 0.567–1.074 MPa m$^{1/2}$ (Epon 828, cured) [30]. The dependence of crack growth on curing agent has been reported [21]. Shear strength 41 MPa (max., Epon 828, unspecified cure) [50]. K_{IC} 0.62 M$_n$ m$^{-3/2}$ (AER-331) 4,4′-sulfonyldianiline cured) [39]
Izod Notch:

No.	Value	Notch	Note
1	12.8 J/m	Y	0.24 ft lb^{-1}, room temp., DER 331, BF$_3$.ethylamine cured [49]
2	30.9 J/m	Y	0.58 ft lb^{-1}, room temp., DER 331, methylenedianiline cured [49]

Electrical Properties:

Electrical Properties General: Electrical props. are dependent upon the curing agent used. Dielectric loss per mole of Epon 828 is increased upon changing from toluene to THF as a solvent [23]. Dielectric relaxation time of Epikote 1009 has temp. dependency that fits the Williams-Landel-Ferry equation [25]. Dielectric constant of resin cured with diaminodiphenylsulfone increases with increasing temp. but decreases with increasing degree of cure. Dielectric loss factor of diaminodiphenylsulfone cured resin increases with increasing degree of cure at low temps. [38]
Surface/Volume Resistance:

No.	Value	Note	Type
1	780×10^{15} Ω.cm	25°, triethylenetetramine cured [2]	S
2	$>790 \times 10^{15}$ Ω.cm	min., 25°, 4,4′-methylenedi-aniline cured [2]	S
3	550×10^{15} Ω.cm	25°, Versamide cured [2]	S
4	$>7.85 \times 10^{15}$ Ω.cm	min., DER 331, BF$_3$.ethylamine or methylenedianiline cured [49]	S
5	0.0067×10^{15} Ω.cm	43°, polyamide cured, conductivity value [26]	S

Dielectric Permittivity (constant):

No.	Value	Frequency	Note
1	4–5	50 Hz	25°, aliphatic and aromatic polyamine cured [1]
2	3.2–3.6	50 Hz	25°, polyamide cured [1]
3	3.4	60 Hz	cured wth hexahydrophthalic anhydride [1]
4	3.9	1 kHz	triethylenetetramine cured [2]
5	4.06	1 kHz	4,4'-methylenedianiline cured [2]
6	3.19	1 kHz	Versamide cured [2]
7	3.15	1 MHz	DER 331, BF_3.ethylamine cured [49]
8	4.89	1 kHz	DER 331, diaminodiphenylsulfone cured [49]
9	3.44	1 MHz	DER 331, methylenedianiline cured [49]

Dielectric Strength:

No.	Value	Note
1	14.1 kV.mm^{-1}	DER 331, diaminodiphenylsulfone cured [49]

Dissipation (Power) Factor:

No.	Value	Frequency	Note
1	0.006	60 Hz	cured with hexahydrophthalic anhydride [1]
2	0.02	1 kHz	cured triethylenetetramine [2]
3	0.015	1 kHz	cured 4,4'-methylenedianiline [2]
4	0.007	1 kHz	cured with Versamide [2]
5	0.023	1 MHz	DER 331, BF_3.ethylamine cured [49]
6	0.022	1 kHz	DER 331, diaminodiphenylsulfone cured [49]
7	0.033	1 MHz	DER 331, methylenedianiline cured [49]
8	0.002–0.02		room temp., Epon 828, anhydride or amine cured [50]

Optical Properties:
Refractive Index:

No.	Value	Note
1	1.5695	liq., uncured [26]
2	1.573	25°, Epon 828, uncured [50]

Transmission and Spectra: Uv [13], ir [13,22], ms [28], C-13 nmr [22] and H-1 nmr [22] spectral data have been reported
Volume Properties/Surface Properties: Colour pale yellow (Epon 828, uncured) [32]

Polymer Stability:
Thermal Stability General: Thermal stability [18] is dependent upon the nature of the curing agent used; thermal props. are a measure of the degree of cure [1]. Heat resistance is decreased by addition of V_2O_5 [40]

Upper Use Temperature:

No.	Value	Note
1	205°C	AER-331, 4,4'-sulfonyldianiline cured [39]

Decomposition Details: Decomposition temp. 392° (cured with hexahydrophthalic anhydride), 420° (cured with 2-methylimidazole) [1]. Shows 6.8% weight loss (triethylenetetramine cured), 5.5% weight loss (4,4'-methylenedianiline cured), 5% weight loss (Versamide cured) after heating at 210° for 300h [2]. Epon 828 shows 10% weight loss in N_2 atmosphere at 402° (cured with 4,4'-diaminodiphenylsulfone); at 375° (cured with 4,4'-diaminodiphenylether). It shows 50% weight loss in N_2 at 427° and 398° for the same respective curing agents. Char yield at 600° 21.8% (N_2, 4,4'-diaminodiphenylsulfone cured), 19.2% (N_2, 4,4'-diaminodiphenylether cured) [6], 2.4% (N_2, 700°, Epon 828, uncured) [32]. Epon 828 shows 10% weight loss in air at 392° (4,4'-diaminodiphenylsulfone cured); at 368° (4,4'-diaminodiphenylether cured); shows 50% weight loss at approx. 420° for both curing agents. Char yield at 600° in air 3.1% (4,4'-diaminodiphenylsulfone cured); 16.7% (4,4'-diaminodiphenylether cured) [6]; char yield at 700° 11.5–19.6% (in N_2, Epon 828, BF_3.ethylamine complex cured, 0.0045 equiv. curing agent, cure 120–140°) [32]. 5% Weight loss in air at 420° (Epon, norbornyl dianhydride siloxane cured) [36]
Flammability: Epon 828 has no fl. p. at 249° [50]
Chemical Stability: Resin cured with triethylenetetramine, 4,4'-methylenedianiline or Versamide is swollen by 50% sodium hydroxide, 30% sulfuric acid, Me_2CO, toluene and H_2O [2]. Epon 828 is highly resistant to sodium hydroxide, acids, fuels and solvents [50]
Hydrolytic Stability: Hydrolysed by H_2O, leading to degradation of mech. props. [8]
Stability Miscellaneous: The effect of γ-irradiation upon dissipation factor depends on the nature of the curing agent and the radiation dose [44]

Applications/Commercial Products:
Processing & Manufacturing Routes: Synth. commercially by reacting excess epichlorohydrin with bisphenol A under alkaline conditions [22]. The resin comprises approx. 80% of the monomeric material and is a viscous liq. Solid resins may be synth. by two processes; the taffy process using epichlorohydrin and bisphenol A with a stoichiometric amount of sodium hydroxide; chain-extension involving reaction of bisphenol A with crude liq. resin. Depending upon the desired use, a curing agent such as a primary or secondary aliphatic polyamine is used to cross-link the crude resin to obtain optimum props. The use of dicyandiamide in combination with a urone solvent as a curing agent has been reported [7]. Temp. must be greater than 145° for dicyandiamide to act as a curing agent [27]. Change in struct. during cure has been reported [9]. Product yield studies have been reported [15]. Mechanism of epoxidation has been reported [29]
Applications: Liq. resin used in casting and tooling; in adhesives; in coatings; in the synth. of modified resins and high MW epoxy resins. Use of cured resin depends upon the type of curing agent employed. Typical applications include castings, adhesives; coatings, electrical laminates; in flooring, mouldings, filament wiring and structural laminates. Epon 828 when appropriately formulated and cured is passed by FDA regulations for food contact applications

Trade name	Details	Supplier
AER	331 (liq. resin)	Asahi Kasei
DER	331	Dow Chemical
ED-20	Russian tradename	
Epikote	1009	Yuka-Shell
Epikote	828/1004/1001/1007	Shell Chemicals

Epon	828	Shell Chemicals
PKHH		Union Carbide

Bibliographic References

[1] McAdams, L.V., and Gannon, J.A., *Encycl. Polym. Sci. Eng.*, 2nd edn., Vol. 6 (ed. J.I. Kroschwitz), 1985, 322, (rev)
[2] Muskopf, J.W. and McCollister, S.B., *Ullmanns Encykl. Ind. Chem.*, 5th edn, Vol A9 (ed. W. Gerhartz), 1987, 547, (rev.)
[3] Schonhorn, H. and Sharpe, L.H., *J. Polym. Sci., Part A-1*, 1965, **3**, 3087, (surface tension)
[4] Kaelble, D.H., Moacanin, J. and Gupta, A., *Epoxy Resins: Chemistry and Technology*, 2nd edn. (ed. C.A. May), Marcel Dekker, 1988, 603, (props.)
[5] Frier, M., *Actes Colloq.-IFREMER*, 1988, **7**, 357, (props.)
[6] Wang, T.-S., Yeh, J.-F. and Shau, M.-D., *J. Appl. Polym. Sci.*, 1996, **59**, 215
[7] Guthner, T. and Hammer, B., *J. Appl. Polym. Sci.*, 1993, **50**, 1453, (curing agent)
[8] Paz Abuin, S. and Pazos Pollin, M., *Quim. Ind. (Madrid)*, 1990, **36**, 423, (hydrolysis)
[9] Prudkai, P.A., Voloshkin, A.F. and Tytyuchenko, V.S., *Plast. Massy*, 1988, 13, (cure)
[10] Singh, V.B., *Polym. Sci.*, 1991, **2**, 912, (miscibility)
[11] Wagner, H.D., Wiesel, E. and Gallis, H.E., *Mater. Res. Soc. Symp. Proc.*, 1990, **170**, 141, (contact angle)
[12] Dee, G.T. and Walsh, D.J., *Macromolecules*, 1988, **21**, 811, (eqn of state)
[13] Nechitailo, L.G., Gerasimov, I.G., Palii, A.I., Reznikova, M.Z. et al, *Zh. Prikl. Spektrosk.*, 1987, **46**, 236, (uv, ir)
[14] Goh, S.H., *Polym. Bull. (Berlin)*, 1987, **17**, 221, (miscibility)
[15] Trivedi, M.K. and Dedhia, S., *Paintindia*, 1987, **37**, 23
[16] Ishimura, H., Akiyama, S. and Higasikibaru, M., *Kotingu Jiho*, 1986, 3, (props.)
[17] Shvets, T.M., Melnichenko, Z.M. and Kushchevskaya, N.F., *Lakokras. Mater. Ikh Primen.*, 1990, 21, (solubility)
[18] Pet'ko, I.P., Batog, A.E. and Zaitsev, Yu.S., *Kompoz. Polim. Mater.*, 1987, **34**, 10, (thermal stability)
[19] Aleman, J.V., Garcia-Fierro, J.L., Legross, R. and Lesbats, J.P., *Crosslinked Epoxies, Proc. Discuss. Conf. 9th*, 1986, 61, (transport props)
[20] Takahashi, K., Yamamoto, T., Itoh, S., Harakawa, K. and Kajiwara, M., *Zairyo*, 1988, **37**, 454, (mechanical props.)
[21] Babaerskii, P.G., Kulik, S.G., Pavlenko, A.A. and Borovko, V.V., *Mekh. Kompoz. Mater. (Zinatne)*, 1987, 10
[22] Braun, D. and Vargha, V., *Angew. Makromol. Chem.*, 1989, **73**, 219, (synth.)
[23] Dallas, G. and Ward, T.C., *Polym. Mater. Sci. Eng.*, 1992, **67**, 288, (dielectric loss)
[24] Guess, T.R., Wischmann, K.B. and Stavig, M.E., *CA*, 1993, **121**, 302043n, (tensile props.)
[25] Koike, T., *J. Appl. Polym. Sci.*, 1992, **45**, 901, (T_g)
[26] Anand, M. and Srivastava, A.K., *High Performance Polym.*, 1992, **4**, 97, (synth., props.)
[27] Topic, M., Mogus-Milankovic and Katovic, Z., *Polymer*, 1991, **32**, 2892, (T_g)
[28] Treverton, J.A., Paul, A.J. and Vickerman, J.C., *Surf. Interface Anal.*, 1993, **20**, 449, (ms)
[29] Oyanguren, P.A. and Williams, R.J.J., *Polymer*, 1992, **33**, 2376
[30] Hollmann, K. and Hahn, H.T., *Polym. Eng. Sci.*, 1989, **29**, 513, (acoustic emission)
[31] Kallitsis, J.K. and Kalfoglou, N.K., *J. Appl. Polym. Sci.*, 1989, **37**, 453, (miscibility)
[32] Chen, C.S. and Pearce, E.M., *J. Appl. Polym. Sci.*, 1989, **37**, 1105
[33] Barbari, T.A., Koros, W.J. and Paul, D.R., *J. Polym. Sci., Part B: Polym. Phys.*, 1988, **26**, 709, 729, (gas transport)
[34] Swinyard, B.T., Sagoo, P.S., Barrie, J.A. and Ash, R., *J. Appl. Polym. Sci.*, 1990, **41**, 2479, (water sorption)
[35] Iribarren, J.I., Iriarte, M., Uriarte, C. and Iruin, J.J., *J. Appl. Polym. Sci.*, 1989, **37**, 3459, (soln props, miscibility)
[36] Eddy, V.J., Hallgren, J.E. and Colborn, R.E., *J. Polym. Sci., Part A: Polym. Chem.*, 1990, **28**, 2417
[37] Nakamae, K., Nishino, T., Airu, X. and Takatsuka, K., *Polym. J. (Tokyo)*, 1991, **23**, 1157, (cure behaviour)
[38] Chen, Y.P., Pollard, J.F., Graybeal, J.D. and Ward, T.C., Polym. Prepr. (Am. Chem. Soc., Div. Polym. Chem.), 1988, **29**, 207, (dielectric behaviour)
[39] Iijima, T., Hiraoka, H., Tomoi, M. and Kakiuchi, H., *J. Appl. Polym. Sci.*, 1990, **41**, 2301, (props)
[40] Zakordonskii, V.P., *Zh. Prikl. Khim. (Leningrad)*, 1995, **68**, 1532, (thermochemical props.)
[41] Bibik, E.E., Popova, E.A. and Skok, G.S., *Zh. Prikl. Khim. (Leningrad)*, 1987, **60**, 454, (contact angle)
[42] Kozhevnikova, I.I., Vereshkina, L.V., Sazhin, B.I. and Gusev, A.V., *Zh. Prikl. Khim. (Leningrad)*, 1989, **62**, 148
[43] Armstrong, K.B., *Adhesion (London)*, 1989, **13**, 237
[44] Fouracre, R.A., Banford, H.M., Tedford, D.J. et al, *Radiat. Phys. Chem.*, 1991, **37**, 581, (γ-irradiation)
[45] Barbari, T.A., Koros, W.-J. and Paul, D.R., *J. Membr. Sci.*, 1989, **42**, 69, (gas permeability)
[46] Moore, D.R. and Mathias, L.J., *Polym. Compos.*, 1988, **9**, 144
[47] Kozey, V., and Kumar, S., *Int. SAMPE Tech. Conf.*, 1994, **26**, 96, (compressive behaviour)
[48] Koike, T., *Plast. Eng. (N.Y.)*, 1996, **31**, 217, (melt viscosity)
[49] *DER*, Dow Chemical Co., 1996, (technical datasheet)
[50] *Epon Resin 828*, sheet no. SC, Shell Chemical Co., 235-91, 828, (technical datasheet)

Bisphenol A-isophthalic acid polyester carbonate B-4

Synonyms: *Bisphenol A-1,3-benzenedicarboxylic acid polyester carbonate. Poly[4,4'-(1-methylethylidene)bis[phenol]-co-1,3-benzenedicarboxylic acid-co-carbonic acid]. PCI*
Related Polymers: Poly[2,2-propanebis(4-phenyl)carbonate], Polyester carbonates
Monomers: Carbonic dichloride, Bisphenol A, 1,3-Benzenedicarboxylic acid, Carbonic acid
Material class: Thermoplastic, Copolymers
Polymer Type: polyester, bisphenol A polycarbonate
CAS Number:

CAS Reg. No.	Note
31133-79-6	source: carbonic acid
67536-57-6	source: carbonic dichloride

Molecular Formula: $[(C_{16}H_4O_3)_x \cdot (C_{23}H_{18}O_4)_y]_n$
Fragments: $C_{16}H_{14}O_3$ $C_{23}H_{18}O_4$
General Information: The polymers have both carbonate and ester linkages, which confer props. intermediate between those of polycarbonate and the polyester. Common usage of the form PCI(X), where I refers to the isophthalate ester moiety and X to the mole fraction of ester (relative to total ester + carbonate). The molar ratio of Bisphenol:(ester + carbonate) residues is 1:1
Morphology: Polyester carbonates with ester:carbonate molar ratios up to 4.9 (PCI (0.83)) are amorph. [2,3]; ordering is not observed (cf. terephthalates) [2]. Isophthalate polyester carbonates are less easily crystallised than the terephthalates; at ester:carbonate ratios of 6 (PCI (0.86)) and above the cast films are about 10% cryst. [3]

Volumetric & Calorimetric Properties:
Density:

No.	Value	Note
1	d 1.1958 g/cm^3	PCI (0.5) [4]
2	d 1.24 g/cm^3	15.6°, ester:carbonate 3.3, PCI (0.77) [3]
3	d 1.245 g/cm^3	15.6°, ester:carbonate 4.9, PCI (0.83) [3]
4	d 1.248 g/cm^3	15.6°, ester:carbonate 6.7, PCI (0.87) [3]

Specific Heat Capacity:

No.	Value	Note	Type
1	0.17–0.21 kJ/kg.C	ester:carbonate 3.3, PCI (0.77) [3]	p
2	0.18–0.22 kJ/kg.C	ester:carbonate 4.9, PCI (0.83) [3]	p
3	0.22–0.26 kJ/kg.C	ester:carbonate 6.7, PCI (0.87) [3]	p

– **Bisphenol A polycarbonate**

Glass-Transition Temperature:

No.	Value	Note
1	160°C	PCI (0.1) [1]
2	168°C	PCI (0.3) [1]
3	178°C	PCI (0.5) [1]
4	159°C	[4]
5	166°C	PCI (0.67) [4]
6	184–188°C	ester:carbonate 3.3, PCI (0.77) [3]
7	188–192°C	ester:carbonate 4.9, PCI (0.83) [3]
8	184–188°C	ester:carbonate 6.7, PCI (0.87) [3]

Mechanical Properties:
Tensile Strength Break:

No.	Value	Note	Elongation
1	47 MPa	23°, 40% relative humidity, ester:carbonate:3.3, PCI (0.77) [3]	32%
2	53 MPa	23°, 40% relative humidity, ester:carbonate 4.9, PCI (0.83) [3]	56%
3	52 MPa	23°, 40% relative humidity, ester:carbonate 6.7, PCI (0.87) [3]	39%

Tensile Strength Yield:

No.	Value	Elongation	Note
1	56.5–57.5 MPa	8%	23°, 40% relative humidity, ester:carbonate 3.3, PCI (0.77) [3]
2	58.3–59.3 MPa	9%	23°, 40% relative humidity, ester:carbonate 4.9, PCI (0.83) [3]
3	58.7–59.7 MPa	8%	23°, 40% relative humidity, ester:carbonate 6.7, PCI (0.87) [3]

Complex Moduli: PCI (0.5) tan δ 0.016 (80°) [4]
Viscoelastic Behaviour: Some tensile stress-strain curves have been reported [3]. There is a viscoelastic region up to the yield point, followed by a stress drop and then some necking behaviour until the break point is reached

Optical Properties:
Transmission and Spectra: Ir spectral data have been reported [2]
Volume Properties/Surface Properties: Ester:carbonate molar ratios up to 4.9 (PCI (0.83)) are transparent, but at 6 (PCI (0.86)) and above the cast films are white due to increasing crystallinity [3]

Polymer Stability:
Decomposition Details: Decomposition curve (0-800°) in air has been reported [5]. The isophthalate polyester carbonates are more stable to thermo-oxidation than the corresponding terephthalates [5]. An additional major breakdown mechanism is also found above 390° for the high ester content isophthalate polyester carbonates [5]. Initial weight loss temp. 360° (air, ester:carbonate 4.2, PCI (0.81)) [5], 370° (air, ester:carbonate 6.7, PCI (0.87)) [5]. Weight loss temp. 415°, 505°, 540° (PCI (0.81)); [5] 480°, 545° (PCI (0.87)) [5]
Stability Miscellaneous: Good stability to thermal ageing [4]

Applications/Commercial Products:
Processing & Manufacturing Routes: Polyester carbonates may be synth. in soln. in two stages or by interfacial polycondensation in one stage [6,7]. Isophthaloyl chloride in CH_2Cl_2 is added to bisphenol A in anhydrous pyridine under N_2 at 25° with *p-tert*-butyl phenol as a chain capping agent - this results in oligomer formation. Phosgenation is then carried out by adding phosgene to the vapour space of the reaction vessel [3,6]. Interfacial polycondensation route; bisphenol A is dissolved in aq. NaOH soln., to which is added a mixture of isophthaloyl chloride and phosgene in an organic solvent (e.g. CCl_4) at room temp. [2,7] tertiary amines, quaternary ammonium or phosphonium salts are used as catalysts; phenols, phenyl chlorocarbonates and aromatic carbonyl halides are used as chain terminators [7]

Applications:

Trade name	Supplier
Lexan KL1-9306	General Electric Plastics

Bibliographic References
[1] Prevorsek, D.C. and De Bona, B.T., *J. Macromol. Sci. Pt B*, 1981, **19**, 605, (nomenclature, T_g)
[2] Kolesnikov, G.S., Smirnova, O.V. and Samsoniya, S.A., *Polym. Sci. USSR (Engl. Transl.)*, 1967, **9**, 1127, (morphology, ir, synth.)
[3] Bosnyak, C.P., Parsons, I.W., Hay, J.N. and Haward, R.N., *Polymer*, 1980, **21**, 1448, (props., synth.)
[4] Bubeck, R.A. and Bales, S.E, *Polym. Eng. Sci.*, 1984, **24**, 1142, (density, T_g, complex moduli, ageing)
[5] Bosnyak, C.P., Knight, G.J. and Wright, W.W., *Polym. Degrad. Stab.*, 1981, **3**, 273, (thermal stability)
[6] Prevorsek, D.C., De Bona, B.T. and Kesten, Y., *J. Polym. Sci., Polym. Chem. Ed.*, 1980, **18**, 75, (synth.)
[7] Encycl. Polym. Sci. Eng. 2nd edn. vol 11, (ed. J.I. Kroschwitz), Wiley-Interscience, 1987, 692, (synth.)

Bisphenol A polycarbonate

Synonyms: *Poly[2,2-propanebis(4-phenyl)carbonate]. Poly[oxycarbonyloxy-1,4-phenylene(1-methylethylidene)-1,4-phenylene]. Poly(4,4'-dihydroxydiphenyl-2,2-propane-co-carbonic dichloride). Polycarbonate fast cycling resins. Easy flow polycarbonate. Polycarbonate, Uv stabilised. Polycarbonate, Flame retardant. Extrusion polycarbonate. Polycarbonate film*
Related Polymers: Bisphenol A Polycarbonate, Structural foam, Bisphenol A Polycarbonate, Glass fibre reinforced, Bisphenol A Polycarbonate, PTFE lubricated, Bisphenol A Polycarbonate, Carbon fibre reinforced, Bisphenol A Polycarbonate, Glass filled, Bisphenol A Polycarbonate, Metal filled, Polycarbonate/polyester alloys, Polycarbonate/polyethylene alloy, Polycarbonate/polyurethane alloy, Poly[2,2-propanebis(4-phenyl)carbonate]-*block*-poly(ethylene oxide), Poly[2,2-propanebis(4-phenyl)carbonate]-*block*-polysulfone, Poly[2,2-propanebis(4-phenyl)carbonate]-*block*-poly(dimethylsiloxane), Poly[2,2-propanebis(4-phenyl)carbonate]-*block*-poly(methyl methacrylate), Polyester carbonates
Monomers: Bisphenol A, Carbonic dichloride
Material class: Thermoplastic, Fibres and Films
Polymer Type: bisphenol A polycarbonate
CAS Number:

CAS Reg. No.	Note
24936-68-3	
25037-45-0	source: carbonic acid
25971-63-5	source: carbonic dichloride

Molecular Formula: $(C_{16}H_{14}O_3)_n$
Fragments: $C_{16}H_{14}O_3$

– Bisphenol A polycarbonate

Molecular Weight: 18000–22000 (high flow), 74000–91000 (film), 32000 (extrusion grade), 32000–60000 (extruded film), 60000 (min., film cast from soln.)

Additives: May contain impact modifiers (such as graft copolymers based on butadiene and acrylate rubbers); uv stabilisers (benzotriazoles); coloured terminal groups; mould release agents (<1% esters of long-chain carboxylic acids); heat stabilisers (<0.5% phosphites, phosphonites or phosphines if necessary in combination with epoxide compounds, or also organosilicon compounds); for easy-flow, modification is with higher branched alkyl phenyl end groups. Flame retardants include small amounts of salt-like compounds e.g. $C_4F_9SO_3K$ or sodium 2,4,5-trichlorobenzenesulfonate; chlorinated organic sulfonates; PTFE

General Information: High flow, uv stabilised, flame retardant, high impact and extrusion versions are all available. Flame retardant grades defined in this entry are only those having a UL94 flammability rating of V0 or 5V. Films may have surface finishes of matt, gloss, velvet, polished, suede, dazzle-free, light diffusing and scratch resistant. Film thicknesses of 125–475 μm are available depending on the surface props. required. Very thin films (<10 μm) are also available

Volumetric & Calorimetric Properties:
Density:

No.	Value	Note
1	d^{25} 1.196 g/cm^3	amorph. [2,3]
2	d 1.316 g/cm^3	cryst., specific volume 0.76 cm^3 g^{-1} [4]

Thermal Expansion Coefficient:

No.	Value	Note	Type
1	3.9×10^{-5} K^{-1}	0-60°, ASTM D696 [1]	L
2	7×10^{-5} K^{-1}	0-100° [5]	L
3	6.4×10^{-5} K^{-1}	0° [6]	L
4	0.00024–0.00026 K^{-1}	87°, glass, 9.8 MPa (100 kg cm^{-2}) [7]	V
5	0.00051–0.00071 K^{-1}	310°, melt [7]	V

Volumetric Properties General: C_p depends on the degree of crystallinity [15,16,17], and $C_p(T)$ for amorph. and crystalline samples may be calculated

Equation of State: Equation of state information has been reported [8,9,10]

Thermodynamic Properties General: ΔC_p (C_p amorph. glass - C_p melt) 0.29 kJ kg^{-1} °C^{-1} (146°) [17]. ΔC_p (C_p crystalline - C_p melt) 0.03 kJ kg^{-1} °C^{-1} (146°) [16,17]

Latent Heat Crystallization: Heat of fusion varies with degree of crystallinity [15,23]; extrapolation gives ΔH_f for the crystalline solid. ΔH_f 109.7 kJ kg^{-1} (26.2 cal g^{-1}) [23]; 127.3 kJ kg^{-1} (30.4 cal g^{-1}) [15]. ΔS_f 0.066 E.U. g^{-1} [18]

Thermal Conductivity:

No.	Value	Note
1	0.18 W/mK	0.00043 cal (cm s °C)$^{-1}$, 20°, thin film [11]
2	0.04 W/mK	-263° [13]
3	0.15 W/mK	-173° [13]
4	0.115 W/mK	-213° [13]

Thermal Diffusivity: α 0.00015 m^2 s^{-1} (1.5 cm^2 s^{-1}, 20°) [12]; α 0.00013 m^2 s^{-1} (1.3 cm^2 s^{-1}, 100°) [12]

Specific Heat Capacity:

No.	Value	Note	Type
1	1.81 kJ/kg.C	solid, 20° [14]	P
2	0.803 kJ/kg.C	cryst., 0.192 cal g^{-1} °C^{-1}, 20°, calc. [17]	P
3	1.178 kJ/kg.C	amorph., 0.281 cal g^{-1} °C^{-1}, 20°, calc. [17]	P

Melting Temperature:

No.	Value	Note
1	233°C	1 atm [18]
2	177–247°C	450–520K, from C_p measurements [16]

Glass-Transition Temperature:

No.	Value	Note
1	145°C	[13,17,19]

Transition Temperature General: Polycarbonate displays a number of secondary (sub-T_g) dynamic relaxations, designated β, γ and δ (in order of decreasing temp.; the glass transition is designated α). These relaxations have been associated with changes in physical props. [24], and have been identified [24,25,26] with molecular motions. T_g is a function of MW [15]

Deflection Temperature:

No.	Value	Note
1	138°C	280°F, 1.82 MPa, ASTM D648 [1]
2	143°C	290°F, 0.45 MPa, ASTM D648 [1]

Vicat Softening Point:

No.	Value	Note
1	157.1°C	[20]

Brittleness Temperature:

No.	Value	Note
1	-140°C	[21]
2	-130°C	[22]

Transition Temperature:

No.	Value	Note	Type
1	50–80°C	[25,27,28]	β
2	-100°C	approx. [25,26,28]	γ

Surface Properties & Solubility:
Solubility Properties General: Most of the commercial polymer is prod. and characterised in soln. [1], the preferred solvent being CH_2Cl_2. Polycarbonate can be combined with a number of other polymers to form blends [29]. Commercial blends include the bulk polymers PC/ABS, PC/PBT and PC/PET, and also the blends PC/acrylic, PC/polyethylene, PC/styrene copolymer and PC/PU. Polycarbonate forms blends, miscible in all proportions, with polymers such as tetramethylpolycarbonate [30], "Kodar" (a copolyester formed from 1,4-cyclohexanedimethanol and a mixture of iso- and terephthalic acids) [31] and poly(vinylidene chloride-co-vinyl chloride) (13 wt.% vinyl chloride) [32].

Bisphenol A polycarbonate

Polycarbonate forms some partially miscible blends, including those with polystyrene [33], PET [34], PBT [35] and polymethylmethacrylate [36]. The polymer-polymer compatibility depends on various factors including composition [33,36,37], temp. [37,38,39], the preparation process [40] and the presence [30] and nature of solvents [40] used. A further compatibilising effect is that of transesterification at high temps. [41,42,43]. Immiscible blends include PC/PVC [32], PC/poly(vinylidene fluoride) [32], PC/ABS [44] and PC/polyethylene [45,46]. Despite their immiscibility, PC/ABS blends display good mechanical props. resulting from strong interfacial adhesion between phases [47,48]. The compatibility of PC/PE blends can be improved by block copolymers [46]. Flory-Huggins interaction parameters [36], spreading coefficients in the melt [49] and interfacial adhesion measurements [48] for pairs of polymers can all be related to miscibility

Cohesive Energy Density Solubility Parameters: Polymer-solvent interaction parameters (Flory-Huggins), 0.42 (CH_2Cl_2, 230°) [50]; 0.28 ($CHCl_3$, 250°) [50]; 1.33 (n-decane, 250°) [50]; 0.03 (cyclohexanol, 250°) [50]. Cohesive energy, E_{coh} (563 J g^{-1}) (-273°) [51]; E_{coh} 92400 J mol^{-1} [52,53]. δ 39.8–44.4 J$^{1/2}$ cm$^{-3/2}$ (9.5–10.6 cal$^{1/2}$ cm$^{-3/2}$) (poorly H bonded solvents); δ 39.8–41.9 J$^{1/2}$ cm$^{-3/2}$ (9.5–10 cal$^{1/2}$ cm$^{-3/2}$) (moderately H bonded solvents) [54]; δ 40.2–41 J$^{1/2}$ cm$^{-3/2}$ (9.6–9.8 cal$^{1/2}$ cm$^{-3/2}$) [69]; δ_d 39.8 J$^{1/2}$ cm$^{-3/2}$ (9.5 cal$^{1/2}$ cm$^{-3/2}$) [55]; δ_a 12.6 J$^{1/2}$ cm$^{-3/2}$ (3 cal$^{1/2}$ cm$^{-3/2}$) [55].

Internal Pressure Heat Solution and Miscellaneous: Excess partial molar heats of mixing [50] -4.98 kJ mol^{-1} (-1.19 kcal mol^{-1}, 200–230°, CH_2Cl_2); 4.65 kJ mol^{-1} (1.11 kcal mol^{-1}, 200–250°, n-decane); -0.88 kJ mol^{-1} (-0.21 kcal mol^{-1}, 200–250°, C_6H_6); 14.24 kJ mol^{-1} (3.4 kcal mol^{-1}, 200–250°, cyclohexanol).

Plasticisers: Commercial polycarbonate does not usually include plasticisers. Polycarbonate may form plasticised (T_g lowered, elastic modulus lowered) or antiplasticised (T_g lowered, elastic modulus raised) systems with other species. Plasticisers include dibutyl succinate [56] (25°, 0-25 wt.%), dibutyl phthalate [56,57] (25°, greater than 20 wt.%), dioctyl phthalate [58] (0.8 wt.%) and 1,3- and 1,4-dichlorobenzene [58] (0-5 wt.%). Antiplasticisers include diphenyl phthalate [56,57] (25°, 0-25 wt.%) and dinitrobiphenyl (25°, 0-25 wt.%) [56]. Some species show both plasticising and antiplasticising props., depending on the concentration and temp. of the additives [59,60,61,62]. For example di-n-butyl phthalate [56,60] is an antiplasticiser at concentrations of 5-20 wt.% and a plasticiser at concentrations greater than 20 wt.%. Plasticisers also accelerate the rate of crystallisation of polycarbonate [63,64]

Solvents/Non-solvents: Sol. CH_2Cl_2, $CHCl_3$, 1,1,2,2-tetrachloroethane, 1,1,2-trichloroethane, cis-1,2-dichloroethene [1]. Slightly sol. dioxane, cyclohexanone, DMF, nitrobenzene [1]. Swelling effects are observed with C_6H_6, chlorobenzene, Me_2CO, EtOAc, CCl_4, acetonitrile and 1,1-dichloroethane [1]. Insol. H_2O, aliphatic hydrocarbons, cycloaliphatic hydrocarbons, ethers, carboxylic acids [1]

Wettability/Surface Energy and Interfacial Tension: Surface free energy of solid, γ_s 40.4 mJ m^{-2} (40.4 erg cm^{-2}, 25°) [65]; γ_s 49.5 mJ m^{-2} (49.5 erg cm^{-2}) [69]; γ_s 40.8 mJ m^{-2} (20°) [49]. Dispersion component of surface free energy of solid, γ_s^d 33.6 mJ m^{-2} (33.6 erg cm^{-2}, 25°) [65]; γ_s^d 44.3 mJ m^{-2} (44.3 erg cm^{-2}) [69]. Polar component of surface free energy of solid, γ_s^p 6.8 mJ m^{-2} (6.8 erg cm^{-2}, 25°) [65]; γ_s^p 5.69 mJ m^{-2} (5.69 erg cm^{-2}) [69]. Critical surface tension, γ_c 34.5 mJ m^{-2} (34.5 dyne cm, 25°) [66]. Temp. dependence of solid surface tension 0.06 mJ m^{-2} K^{-1} (0.06 dyne cm^{-1}) [66]; 0.065 mJ m^{-2} K^{-1} (estimated from similar polymers) [49]. Surface free energy of melt, γ_m 24.6 mJ m^{-2} (calc., 270°) [49]. Dispersion component of surface free energy of melt, γ_m^d 18 mJ m^{-2} (calc., 270°) [49]. Polar component of surface free energy of melt, γ_m^p 6.6 mJ m^{-2} (calc., 270°) [49]. Work of adhesion between polycarbonate and water 97 mJ m^{-2} (pH 1-4, room temp.) [67]; 86 mJ m^{-2} (pH 10-14, room temp.) [67]; W^d 86 mJ m^{-2} (calc., room temp.) [67]; $W^{acid-base}$ 11 mJ m^{-2} (calc., room temp.) [67]. Work of adhesion between polycarbonate and ice, W 105 mJ m^{-2} (-15°) [68] Wettability may be improved by irradiation with argon ions [251]

Transport Properties:
Melt Flow Index:

No.	Value	Note
1	1.2 g/10 min	ASTM D1238, 2160 g load, MW 48100, 280° [70]
2	25 g/10 min	ASTM D1238, 2160 g load, MW 26100, 280° [70]
3	12.3 g/10 min	ASTM D1238, MW 26300 [71]
4	4 g/10 min	ASTM D1238, MW 34400 [71]

Polymer Solutions Dilute: Theta temp. θ 170° (n-butyl benzyl ether) [72]; θ 25° (63.9 wt.% dioxane, 36.1% cyclohexane) [72]. Mark-Houwink constant, K 0.000204 (1,2-dichloroethane, 25°, MV 10900–69700) [73]; 0.000142 (1,2-dichloroethane, 25°, MV 27600–88000) [74]; 0.000134 (tetrachloroethane, 25°, MV 10900–69700) [73]; 0.000389 (THF, 25°, MV 10900–69700) [73]; 0.00049 (THF, 25°, MV 27600–88000) [74]; 0.00012 ($CHCl_3$, 25°, MV 10900-69700) [73]; 0.000112 ($CHCl_3$, 25°, MV 27600–88000) [74]; 0.000119 (CH_2Cl_2, 25°, MW 4400–75800) [72]; 0.000123 (CH_2Cl_2, 25°, MV 27600–88000) [74]; 0.000111 (CH_2Cl_2, 25°, MW 13000–80000) [75]; 0.0021 (n-butyl benzyl ether, 170°, MW 40000–800000) [72]; 0.0021 (63.9 wt.% dioxane/36.1% cyclohexane, 25°, MW 40000–800000) [72]; 0.00214 (63.9 wt.% dioxane/36.1% cyclohexane, 25°, MV 27600-88000) [74]; 0.001 (63.9% wt.% dioxane/36.1% cyclohexane, 25°, MW 4000–40000) [72]; 0.001 (n-butyl benzyl ether, 170°, MW 4000–40000) [72]; 0.000776 (cyclohexanone, 25°, MV 27600–88000) [74]; 0.000309 (dioxane, 25°, MV 27600–88000) [74]. Mark-Houwink a values 0.76 (1,2-dichloroethane, 25°, MV 10900–69700) [73]; 0.76-0.8 (1,2-dichloroethane, 25°, MV 27600–88000) [74]; 0.82 (tetrachloroethane, 25°, MV 10900–69700) [73]; 0.7 (THF, 25°, MV 10900–69700) [73]; 0.65–0.69 (THF, 25°, MV27600–88000) [74]; 0.82 ($CHCl_3$, 25°, MV 10900–69700) [73]; 0.80–0.84 ($CHCl_3$, 25°, MV 27600–88000) [74]; 0.8 (CH_2Cl_2, 25°, MW 4400–75800) [72]; 0.79–0.81 (CH_2Cl_2, 25°, MV 27600–88000) [74]; 0.82 (CH_2Cl_2, 25°, MW 13000–80000) [75]; 0.5 (n-butyl benzyl ether, 170°, MW 40000–800000) [72]; 0.56 (n-butyl benzyl ether, 170°, MW 4000–40000) [72]; 0.5 (63.9 wt.% dioxane/36.1% cyclohexane, 25°, MW 40000–800000) [72]; 0.5 (63.9 wt.% dioxane/36.1% cyclohexane, 25°, MV 27600–88000) [74]; 0.56 (63.9 wt.% dioxane/36.1% cyclohexane, 25°, MW 4000–40000) [72]; 0.71 (dioxane, 25°, MV 27600–88000) [74]; 0.62 (cyclohexanone, 25°, MV 27600–88000) [74]. Temp. dependence of intrinsic viscosity, d log [η]/dT, 0.0015 ($CHCl_3$, 25–48°, MV 10900-69700) [73]; 0.0018 (tetrachloroethane, 25–56°, MV 10900–69700) [73]; 0.0016 (1,2-dichloroethane, 25–56°, MV 10900-69700) [73]; 0.0019 (THF, 25–48°, MV 10900-69700) [73]

Polymer Melts: Power law parameters have been reported [75,76]

Permeability of Gases: Permselectivity for CO_2/CH_4 19 (35°, 10 atm) [77,78]; 24.2 (35°, 5 atm, measured for mixture of gases) [79]; 23.3 (35°, 20 atm) [81]; 22.8 (35°, 2 atm) [83]; 20 (35°, 5 atm) [85]; 3.7 (175°, 5 atm) [85]. Permselectivity for O_2/N_2: 4.3 (35°, 2 atm) [78]; 4.8 (35°, 2 atm) [77]; 5.13 (35°, 1 atm) [81]. Permselectivity for He/CH_4: 35 (35°, 10 atm) [78]; 32.3 (35°, 2 atm) [83]; 53 (35°, 20 atm) [81]. Permselectivity for He/N_2: 45 (35°, 5 atm) [85]; 11 (175°, 5 atm) [85]. Permselectivity for CO_2/O_2: 4.3 (25°, 1 atm, 0% relative humidity) [80]. Pre-exponential permeability factor, P_o [CO_2] 7.5 × 10^{-11} cm^3 cm^{-1} s^{-1} Pa^{-1} (1000 Barrers, 35–120°, 5 atm) [85]; [CH_4] 6.3 × 10^{-10} cm^3 cm^{-1} s^{-1} Pa^{-1} (8400 Barrers, 35–120°, 5 atm); [He] 8.25 × 10^{-10} cm^3 cm^{-1} s^{-1} Pa^{-1} (11000 Barrers, 35–125°, 5 atm) [85]; [O_2] 4.1 × 10^{-10} cm^3 cm^{-1} s^{-1} Pa^{-1} (5400 Barrers, 35–125°, 5 atm) [85], 7 × 10^{-10} cm^3 cm^{-1} s^{-1} Pa^{-1} (9.3 × 10^{-8} cm^3 cm^{-1} s^{-1} (cmHg)$^{-1}$, 35–65°, 1 atm) [82]; [N_2] 3.8 × 10^{-10} cm^3 cm^{-1} s^{-1} Pa^{-1} (5000 Barrers, 35–120°, 5 atm) [85], 1.01 × 10^{-10} cm^3 cm^{-1} s^{-1} Pa^{-1} (13.42 × 10^8 cm^3 cm^{-1} s^{-1} (cmHg)$^{-1}$, 35–65°, 1 atm) [82]. Gas permeability decreases with increasing gas pressure up to about 20 atm [87,88,89,90]. However, for CO_2 at pressures above about

– Bisphenol A polycarbonate

50 atm (800 psi) an increase in pressure leads to a rapid increase in permeability, due to the onset of plasticisation [89]. Subsequent depressurisation from about 60 atm (900 psi) shows hysteresis. This effect is known as "conditioning", and can be achieved by exposing the polycarbonate film to CO_2 at high pressures for about 2–7 days [89,91,92]. Hysteresis occurs for CO_2 at pressures of 40 atm (600 psi) and 61 atm (900 psi), but not 20 atm (300 psi) [92], and does not appear to occur with other gases [89]. Changes induced by CO_2 conditioning include increased permeability to other gases [84,89], without significant loss in permselectivity [84], which gives improvements in separation membrane props.

Permeability of Liquids: Liq. transport in polycarbonate is complicated by solvent induced crystallisation [94,95,96,97,98,99]. One consequence of crystallisation by common solvents (such as Me_2CO, EtOAc or xylene) is a loss of mech. props. [100,101]. Diffusion coefficient, D [Me_2CO] 74.2×10^{-8} cm^2 s^{-1} (20°) [96], 6.3×10^{-7} cm^2 s^{-1} (25°) [102], 5.5×10^{-7} cm^2 s^{-1} (24–26°) [100], 4×10^{-7} cm^2 s^{-1} (25°) [94]; [CCl_4] 1.9×10^{-9} cm^2 s^{-1} (25°) [102]; [toluene] 13.8×10^{-7} cm^2 s^{-1} (25°) [102]; [MeOH] 4.3×10^{-10} cm^2 s^{-1} (22°) [97]; [xylene] 3.6×10^{-8} cm^2 s^{-1} (20°) [96]; [methyl isobutyl ketone] 3×10^{-8} cm^2 s^{-1} (20°) [96]; [1,2-dichloroethane] 4.6×10^{-7} cm^2 s^{-1} (25°) [104]; [CH_2Cl_2] 1.4×10^{-7} cm^2 s^{-1} (25°) [104]; [H_2O] 6.5×10^{-8} cm^2 s^{-1} (23°) [103], 6×10^{-7} cm^2 s^{-1} (80°) [105], 1×10^{-6} cm^2 s^{-1} (96°) [103], 8.7×10^{-7} cm^2 s^{-1} (100°) [105]

Volatile Loss and Diffusion of Polymers: Volatile loss of 2.4-3.05 wt.% occurs upon heating in air at 350° [83,106]

Permeability and Diffusion General: Diffusity, D [He] 5×10^{-6} cm^2 s^{-1} (35°, 10 atm) [78], 7×10^{-6} cm^2 s^{-1} (35°) [86], 64.5×10^{-7} cm^2 s^{-1} (35°, 20 atm) [82], 5.5×10^{-6} cm^2 s^{-1} (35°, Henry) [88], 74.4×10^{-6} cm^2 s^{-1} (35°, Langmuir) [88]; [CO_2] 3.2×10^{-8} cm^2 s^{-1} (35°, 10 atm) [77,78], 30.9×10^{-9} cm^2 s^{-1} (35°, 20 atm) [81,82], 62.2×10^{-9} cm^2 s^{-1} (35°, Henry) [82,87,88], 48.5×10^{-9} cm^2 s^{-1} (35°, Langmuir) [82,87,88]; [O_2] 5×10^{-8}–6×10^{-8} cm^2 s^{-1} (25°, 0.03 atm) [93], 5.8×10^{-8} cm^2 s^{-1} (35°, 2 atm) [78], 56.1×10^{-9} cm^2 s^{-1} (35°, 1 atm) [81,82]; [Ar] 3.3×10^{-8} cm^2 s^{-1} (35°, Henry) [88], 59.4×10^{-10} cm^2 s^{-1} (35°, Langmuir) [88]; [N_2] 18.1×10^{-9} cm^2 s^{-1} (35°, 1 atm) [81,82], 17.6×10^{-9} cm^2 s^{-1} (35°, Henry) [82,87,88], 50.9×10^{-9} cm^2 s^{-1} (35°, Langmuir) [82,88], 51.4×10^{-10} cm^2 s^{-1} (35°, Langmuir) [87]; [CH_4] 4.8×10^{-9} cm^2 s^{-1} (35°, 20 atm) [81,82], 10.9×10^{-9} cm^2 s^{-1} (35°, Henry) [82,87], 1.3×10^{-9} cm^2 s^{-1} (35°, Langmuir) [82,87]

Water Content: Water is absorbed when polycarbonate is immersed. At lower temps. (0-60°) the water molecules are thought to form hydrogen bonds with the carbonate residue [109]. Polycarbonate immersed in water at about 100° absorbs water, the distribution of which changes with time. Initially (20-24h), the water is evenly distributed throughout the polymer matrix [110], and after about 10 days pools of aggregated water or clusters also form g, cluster formation is thought to be due to hydrol. [111,112]. The formation of clusters has an activation energy of 155 kJ mol^{-1} (37 kcal mol^{-1}) [111]

Water Absorption:

No.	Value	Note
1	0.1–0.2 %	room temp. and humidity [107]
2	0.1 %	24.5°, 100h, 0% relative humidity [108]
3	0.2 %	24.5°, 100h, 50% relative humidity [108]
4	0.3 %	100°, 140h, 56% relative humidity [105]
5	0.35 %	100°, 100h, 100% relative humidity [108]
6	0.6 %	100°, 140h, 100% relative humidity [105]
7	0.5 %	80°, 100h [105]
8	0.5 %	80° [111]
9	0.6 %	100°, 100h [105]
10	0.63 %	96°, 30h [103]
11	0.3–0.4 %	100°, 24h [110]
12	0.5–0.8 %	100°, 30–40 days [110]
13	0.63 %	97°, <240h [111]
14	1.25 %	97°, 900h [111]
15	0.41 %	23° [103]
16	0.4 %	room temp., 150–750h [109]
17	0.35 %	25° [111]
18	0.73 %	125°, 20 psi steam, <3.5h [111]
19	1.98 %	125°, 20 psi steam, 35h [111]

Gas Permeability:

No.	Gas	Value	Note
1	CO_2	446.5 cm^3 mm/(m^2 day atm)	5.1×10^{-13} cm^3 (cm s Pa)$^{-1}$, 35°, 10 atm [77,78]
2	CO_2	464 cm^3 mm/(m^2 day atm)	5.3×10^{-13} cm^3 (cm s Pa)$^{-1}$, 35°, 5 atm [79]
3	CO_2	376–411 cm^3 mm/(m^2 day atm)	4.3–4.7×10^{-13} cm^3 (cm s Pa)$^{-1}$, 25°, 1 atm, 0% relative humidity [80]
4	CO_2	394 cm^3 mm/(m^2 day atm)	6×10^{-10} cm^2 (s cmHg)$^{-1}$, 35°, 20 atm [81,82]
5	CO_2	495.5 cm^3 mm/(m^2 day atm)	5.66×10^{-13} cm^3 (cm s Pa)$^{-1}$, 35°, 2 atm [83]
6	CH_4	23.6 cm^3 mm/(m^2 day atm)	2.7×10^{-14} cm^3 (cm s Pa)$^{-1}$, 35°, 10 atm [78,79]
7	CH_4	16.9 cm^3 mm/(m^2 day atm)	0.257×10^{-10} cm^2 (s cmHg)$^{-1}$, 35°, 20 atm [81,82]
8	CH_4	13.6 cm^3 mm/(m^2 day atm)	1.56×10^{-14} cm^3 (cm s Pa)$^{-1}$, 35°, 61 atm [84]
9	O_2	105 cm^3 mm/(m^2 day atm)	1.2×10^{-13} cm^3 (cm s Pa)$^{-1}$, 35°, 2 atm [77,78]
10	O_2	85.8–95.4 cm^3 mm/(m^2 day atm)	0.98–1.09×10^{-13} cm^3 (cm s Pa)$^{-1}$, 25°, 1 atm, 0% relative humidity [80]
11	O_2	97.4 cm^3 mm/(m^2 day atm)	1.113×10^{-13} cm^3 (cm s Pa)$^{-1}$, 35°, 1–2 atm [81,82,83]
12	N_2	25.4 cm^3 mm/(m^2 day atm)	2.9×10^{-14} cm^3 (cm s Pa)$^{-1}$, 35°, 2 atm [78]
13	N_2	19 cm^3 mm/(m^2 day atm)	2.17×10^{-14} cm^3 (cm s Pa)$^{-1}$, 35°, 1 atm [81,82]
14	N_2	21 cm^3 mm/(m^2 day atm)	2.4×10^{-14} cm^3 (cm s Pa)$^{-1}$, 35°, 10 atm [78]
15	He	892.9 cm^3 mm/(m^2 day atm)	1.02×10^{-12} cm^3 (cm s Pa)$^{-1}$, 35°, 20 atm [81,82]
16	He	866.7 cm^3 mm/(m^2 day atm)	9.9×10^{-13} cm^3 (cm s Pa)$^{-1}$, 35°, 2–10 atm [78,83]

Mechanical Properties:

Mechanical Properties General: Tensile stress-strain curve at 25° [113,114,115,116]. Stress-strain behaviour is very sensitive to the testing temp.. The yield stress is less affected by changes in the strain rate [113,114,117,120,121,122,123,124]. Tensile stress-strain curve shows a viscoelastic region up to the yield point, then a stress drop, then necking occurs and increase in stress until the break point [114,116,117,118]

– Bisphenol A polycarbonate

Tensile (Young's) Modulus:

No.	Value	Note
1	2300 MPa	room temp., strain rate 0.00083 s^{-1} [125]
2	2200 MPa	DIN 53 454 [126]
3	2160 MPa	21°, 100 mm min^{-1} [127]
4	2410 MPa	23°, ASTM D638 [122]
5	2068–2275 MPa	300-330 kpsi, ASTM D689-44 [62]

Flexural Modulus:

No.	Value	Note
1	2585 MPa	375 kpsi, ASTM D790 [128]
2	2209 MPa	21°, calc. [127]
3	2360 MPa	ambient temp. [129]
4	2151 MPa	312 kpsi [130]

Tensile Strength Break:

No.	Value	Note	Elongation
1	121 MPa	strain rate 10% min^{-1} [122]	85%
2	120 MPa	strain rate 0.0017 s^{-1} [126]	65%
3	100.7 MPa	strain rate 5.08 mm min^{-1} [121]	74%
4	60.2–61.8 MPa	strain rate 5% min^{-1} [125]	106%
5	69 MPa	[135]	107%

Compressive Modulus:

No.	Value	Note
1	1990 MPa	289 kpsi, 22° [131]
2	1660 MPa	241 kpsi, ASTM D695 [128]
3	2030 MPa	20.3 kbar, room temp. [132]

Elastic Modulus:

No.	Value	Note
1	1740 MPa	253 kpsi, 22° [131]
2	4940 MPa	4.94 × 10^{10} dyn cm^{-2}, 25° [136]
3	420 MPa	23°, shear <0.15 [126]

Poisson's Ratio:

No.	Value	Note
1	0.31	22° [131]
2	0.38	room temp. [132]
3	0.36	room temp. [133]
4	0.3043–0.3135	room temp. [134]

Tensile Strength Yield:

No.	Value	Elongation	Note
1	68 MPa	6.9%	strain rate 10% min^{-1} [122]
2	72 MPa		strain rate 0.0017 s^{-1} [126]
3	64.81 MPa		strain rate 5.08 mm min^{-1} [121]
4	58–59.2 MPa	6.6–7.0%	strain rate 5% min^{-1} [125]
5	65 MPa		[135]

Flexural Strength Yield:

No.	Value	Note
1	98.6 MPa	14.3 kpsi [129]
2	76–90 MPa	11–13 kpsi, ASTM D790 [128]

Compressive Strength:

No.	Value	Note
1	75.8 MPa	11 kpsi, ASTM D695 [128]
2	72.52 MPa	10.52 kpsi, 22°, 6.9% strain [131]

Miscellaneous Moduli:

No.	Value	Note	Type
1	2200 MPa	-268° [142]	Complex shear storage modulus
2	1900 MPa	-196° [142]	Complex shear storage modulus
3	1300 MPa	-190° [140]	Complex shear storage modulus
4	1000 MPa	20° [129]	Complex shear storage modulus
5	740 MPa	20° [140]	Complex shear storage modulus
6	700 MPa	140° [129]	Complex shear storage modulus
7	550 MPa	140° [140]	Complex shear storage modulus

Complex Moduli: Tensile storage modulus decreases slowly with increasing temp. [137,138]. The degree of crystallinity is also found to be important: 2300 MPa (23°, 110 Hz, 27% crystallinity), 1500 MPa (23°, 110 Hz, 0% crystallinity) [138]. Above T_g, E' falls rapidly with increasing temp. [139]. Above T_g, E' also increases with frequency [139]. The complex shear storage modulus decreases with temp., with a more rapid decrease occurring above about 130° [116,129,140] and below about -100° [140,142]. Loss tangent has been reported [116,138,140,141]

Impact Strength: Izod impact strength varies with notch depth. Dart drop impact strength 2013–2753 J m^{-2} [147]. Charpy impact energy 3100 J m^{-2} (sharp notch 25 μm, ASTM D256) [148], 12 kJ m^{-2} (blunt notch, 250 μm, ASTM D256) [148], 9800 J m^{-2} (2500 μm) [135]

Viscoelastic Behaviour: A typical creep-time curve has been reported [149,150,151,152]. Below the T_g, the creep rate increases with increasing temp. [150] and with increasing applied stress [152,153]. Stress relaxation curves have been reported [152,153,154]. Time-temp. shift factors have been reported [139,157,158,159]

Hardness: Rockwell M65 (ASTM D785) [144], Rockwell M70 (ASTM D785) [128], Rockwell M80 (ASTM D785) [130], Rockwell R118 (ASTM D785) [128], Rockwell R115 (ASTM D785) [143]

Failure Properties General: Polycarbonate yields before fracture for a wide range of strain rates (0.01-2000 s^{-1}) and over a wide range of temps. (-100-200°) [117]. The mode of failure (brittle/ductile) depends on temp. [146,160], and sample thickness [121,160,161,162,163]. At room temp., thin samples (approx. 3 mm) show ductile failure and thicker specimens show brittle failure. At temps. below -100°, samples show brittle failure. K_I 3.0–3.3 MPa m$^{1/2}$ (23°) [164], 3.7 MPa m$^{1/2}$ (room temp.) [148], 3.3 MPa m$^{1/2}$ (20°) [165], 3.6 MPa m$^{1/2}$ (25°) [165], 3.15–3.31 MPa m$^{1/2}$ (ASTM E399-83) [167]

– Bisphenol A polycarbonate

Fracture Mechanical Properties: The shape of the fatigue fracture lifetime curve depends on temp. and stress: above 75° typical decay behaviour is seen, but below 75° S-shaped fatigue fracture lifetime curves are observed [168,169]. There are two types of fatigue failure: shear fracture cracking and craze growth, with the latter predominating at low temps. [168]. For a fixed temp. crazing occurs at low stresses, with shear fracture predominating at higher stresses and necking occurring at still higher stresses [168,170]. Other factors include the frequency of the applied stress, and the history of the sample. Within these constraints the following fatigue strengths can be considered 7.8 MPa (10^7 cycles, 20°, JIS K7119, 30 Hz) [171], 4.9 MPa (10^7 cycles, 50°, JIS K7119, 30 Hz) [171]. Other values have been reported [168,169]. Strain rate sensitivity index, 0.0102–0.028 [172], 0.028 [119], 0.027 [126]. Shear strength at break, 63.4 MPa (9.2 kpsi, ASTM D732) [128], 56 MPa (23°, shear rate 0.003 s^{-1}) [126]

Friction Abrasion and Resistance: Dynamic friction coefficient, μ_{dy}, 0.46 (27°, polymer/hardened steel, 10.7 cm s^{-1}) [173], 0.45 (polymer/steel, 173–202 cm s^{-1}) [128], 0.35 (polymer/steel, 1 cm s^{-1}) [128], 0.38–0.42 (23°, polymer\cold rolled steel, 0.08 cm s^{-1}, ASTM D1894) [174]. Abrasion resistance 24 mg (1000 cycles)$^{-1}$ (ASTM D1044, Taber CS-17 test) [128]

Izod Notch:

No.	Value	Notch	Note
1	107–160 J/m	Y	2-3 ft lb in^{-1}, 0.5 in., ASTM D256 [130,143]
2	160 J/m	Y	3 ft lb in^{-1}, 0.245 in., ASTM D256-56 [144]
3	640–854 J/m	Y	12-16 ft lb in^{-1}, 0.125 in., ASTM D256 [1,140,143,145,146]

Electrical Properties:

Electrical Properties General: Polycarbonate conducts electricity at moderate to high fields by two mechanisms: the Richardson-Schottky mechanism of charge transport at moderate fields (50 to 1000 V) and space charge-limited conduction at high fields (>100 V) [175,176]. Dielectric permittivity is approx. independent of temp. in the range -30° to 125° [128,177], but increases with increasing temp. above 125° [177]. It has a small inverse dependence on frequency. Dielectric strength varies with temp. and sample thickness. tan δ depends on temp. and frequency [1,128,180,181]

Surface/Volume Resistance:

No.	Value	Note	Type
4	0.9–1.1 × 10^{15} Ω.cm	500 V [179]	S

Dielectric Permittivity (constant):

No.	Value	Frequency	Note
1	3.47	800 Hz	30°, from graph [177]
2	3.63	800 Hz	160° [177]
3	3.8	800 Hz	170° [177]
4	3.17	60 Hz	23°, ASTM D150 [1,128]
5	2.96	1 MHz	23°, ASTM D150 [1,128]
6	2.74	10 GHz	23°, ASTM D150 [1]

Dielectric Strength:

No.	Value	Note
1	153.9 kV.mm^{-1}	3910 V mil^{-1}, 25°, 1.5 mil, ASTM D149 [128]
2	44.5 kV.mm^{-1}	1130 V mil^{-1}, 25°, 23 mil, ASTM D149 [128]
3	26.9 kV.mm^{-1}	683 V mil^{-1}, 23°, 55 mil [178]
4	15.7 kV.mm^{-1}	400 V mil^{-1}, 25°, 125 mil, ASTM D149 [128]
5	23.6 kV.mm^{-1}	600 V mil^{-1}, 100°, 125 mil, ASTM D149 [128]

Arc Resistance:

No.	Value	Note
1	10–11s	23°, stainless-steel electrodes, ASTM D495 [1,128]
2	12s	ASTM D495 [143]
3	120s	23°, tungsten electrodes, ASTM D495 [1]

Complex Permittivity and Electroactive Polymers: The variation of complex relative permittivity, as a function of frequency and temp., has been reported [128,180,182,183,184]. Imaginary relative permittivity has been reported [180,182,183,184].

Magnetic Properties: Magnetic susceptibility, -0.981 × 10^{-5} – -0.931 × 10^{-5} (room temp., 0.3–0.6 T) [185]. Details of magnetic birefringence have been reported [186]

Dissipation (Power) Factor:

No.	Value	Frequency	Note
1	0.001–0.03	0.1–1 MHz	-200–140° [1,128,180]
2	0.03–25	0.1–1 MHz	140–190° [1,180]

Optical Properties:
Refractive Index:

No.	Value	Note
1	1.61097	435.8 nm [187]
2	1.5995	486.1 nm [188]
3	1.5976–1.6006	514.5 nm [189]
4	1.59006	546.1 nm [187]
5	1.58545	587.6 nm [187]
6	1.58071	643.8 nm [187]
7	1.58	656.3 nm [188]
8	1.4237	435.8 nm, CH_2Cl_2 [72]
9	1.4445	435.8 nm, $CHCl_3$ [72]

Transmission and Spectra: 75–84% transmittance (400–450 nm, 5.08 nm thick film) [190]. 84–89% transmittance (450–900 nm, 5.08 nm thick film) [190]. 90% transmittance (400–800 nm, 100 μm thick film) [1]. Transmission of x-rays 91% (5.37 Å, 2 μm thick film) [191], 85% (7.12 Å, 2 μm thick film) [191], 78% (8.34 Å, 2 μm thick film) [191], 56% (11.91 Å, 2 μm thick film) [191]. Uv [177,187,192,193], fluorescence [192,193] and ir [1,175,187,194] spectral data have been reported

Molar Refraction: Concentration dependence of refractive index, dn/dc 0.162 ($CHCl_3$, 546.1 nm) [73], 0.156 ($CHCl_3$, 546.1 nm) [74], 0.177 (CH_2Cl_2, 435.8 nm) [72], 0.188 (dioxane, 546 nm) [195]

Total Internal Reflection: Intrinsic birefringence 0.189 (546 nm) [196], 0.132 (514.5 nm) [189], 0.186–0.198 (546 nm) [197]. Stress optical coefficient 0.089 × 10^{-9} Pa^{-1} (room temp.) [198], 0.087 × 10^{-9} Pa^{-1} (87 × 10^{-13} $dyne^{-1}$ cm^2, 150°) [199], 0.035 × 10^{-9} Pa^{-1} (166–170°) [139], 5 × 10^{-9} Pa^{-1} (166–170°) [139], 5.9 × 10^{-9} Pa^{-1} (157°) [200], 3.5 × 10^{-9} Pa^{-1} (200°, melt) [201], 3.45 × 10^{-9} Pa^{-1} (201–233°, melt) [198]

Volume Properties/Surface Properties: Front surface reflectance 5.476% (435.8 nm) [187], 5.190% (546.1 nm) [187], 5.063% (643.8 nm) [187]. Rayleigh light scattering loss, $\alpha_{anisotropic}$ 0.106 db m^{-1}

— **Bisphenol A polycarbonate**

(633 nm) [202], $\alpha_{isotropic}$ 0.06 db m^{-1} (633 nm) [202]. Clarity transparent [143]

Polymer Stability:
Polymer Stability General: Shows good stability under a wide range of conditions
Thermal Stability General: Has high degree of thermal stability. Remains essentially intact up to 400°, although degradation begins above 300°, and rearrangement above T_g [203]
Upper Use Temperature:

No.	Value	Note
1	100–125°C	air [207]
2	135–140°C	[1]
3	135°C	[1]

Decomposition Details: At temps. between 300 and 400°, mechanism of decomposition depends on the conditions [208,209]. In a sealed evacuated environment, viscosity decreases; a continuously evacuated flask results in gel formation. Mechanism is via an ionic ester interchange route, or by free radical scission [205,211] giving cyclic dimers, cross-linked species and CO_2. Above 420°, a complex series of degradation reactions occur, resulting in branching, cross-linking, formation of ester, ether and unsaturated compds. and other products [204,206]. Most functionality of the original polymer is removed by 490°; above 500° aromatisation, and above 540° extensive carbonisation. Maximum rate of decomposition temp. 470° (N_2) [210]; 458–462° (N_2, or vacuum) [205]. Temp. 35% wt. loss 400° (N_2) [203]; temp. 75% wt. loss 42–52° (He) [206]. Polymer appears optically black at 490° [204]. Char yield 30% (800°, no fire retardant) [212]. Char yield 27% (800°, N_2, fire retardant) [212]
Flammability: Self extinguishing (ASTM D635) [1,128]. Flammability rating HB (1.57 mm), V2 (1.57 mm, 3.18 mm, UL 94) [213]; V0 (0.8 mm, 3.2mm, flame retardant, UL 94); 5V (3.2 mm, flame retardant, UL 94) [210]. Limiting oxygen index 32.0% (3 mm, no flame retardant, ASTM D2863) [212]; 33.5% (3 mm, flame retardant, ASTM D2863) [212]
Environmental Stress: Processed polycarbonate gradually becomes more glass-like with age, resulting in a decrease in Young's modulus [214,215]. Elongation at break also decreases over time as does loss tangent [215]. Exhibits good resistance to atmospheric oxygen between -30-100° [1,216]; however, ozonolysis occurs on exposure to ozone [217]. Non uv-stabilised polycarbonate outdoors for several years becomes yellow, and cross-linked, with evidence of pitting and cracking [218]. Photochemical reactions taking place vary in type with wavelength of radiation [207,218,219]. Photo-oxidation predominates in natural weather conditions, in more humid conditions mech. props. are affected as well as yellowing; polymer loses ductility and becomes brittle [221,222]. The effects are reduced by the use of stabilisers [222]. Critical strain, below which cracking or crazing does not occur on exposure to liquids, increases with T_g of swollen polymer. Value decreases as solubility increases [95,223] and with variation of solvent, e.g. as chain length increases [69,95,223]
Chemical Stability: Commercial grades may contain uv stabilisers or flame retardants. Solvent resistance is limited; unaffected by aliphatic hydrocarbons and alcohols except MeOH; swollen or dissolved by most other solvents [1,69,95,128]. Also fully resistant to aq. solns. of soap, detergents, bleaches, oils, waxes, fruit juices, photographic chemicals, foodstuffs including beverages and dyes and pigments [1,128]. Stable to dilute and conc. oxidising agents and weak alkali but degraded by ammonia and amines [1,230]. Rapidly attacked by strong alkali solns. at room temp. Good resistance to salt solns. at room temp. [1]. Degraded by organic salts at melt temp.
Hydrolytic Stability: Very stable to hydrol. at room temp. [1,103,222] but at higher temp. results in breakdown of mech. props. [105,112,222]. Long term use above 60° in water is not recommended. Exposure to water vapour results in increasing brittleness, [236,237] and rate of hydrol. is dependent on temp. and relative humidity [236,238]. Hydrol. mechanism involves chain scission [239,240,241] and decarboxylation [240] but at higher temps. annealing occurs [239]. This results in a self-catalytic hydrol. process [241,243]. Presence of glass enhances hydrol. [105,242]
Biological Stability: Soil burial shows no real effects from soil corrosion agents [1,244], microorganisms or insects. Non toxic to animals [1,245]. Has demonstrated oestrogenic effects in rats [1]
Stability Miscellaneous: Annealing increases yield stress [124,167,214,215,246,247,248,249,250], decreases Young's modulus [248] and increases brittleness [167,247,249], reduces loss tangent [215], and reduces fatigue lifetime [250]. Annealing effect increases with increasing temp. [124,248]

Applications/Commercial Products:
Processing & Manufacturing Routes: Prod. by interfacial polymerisation of Bisphenol A disodium salt in aq. alkaline soln. or suspension via reaction with phosgene in the presence of an inert organic solvent. The oligocarbonates formed are condensed to high MW polycarbonates over catalysts such as tertiary amines. For processing there is a need to keep the materials dry, so product stored in vacuum sealed tins heated for several hours at 110° prior to use. Processing by injection moulding, extrusion, compression moulding, cold forming, and casting. Processing temp. between 230° and 300°. Blow moulded articles, films, sheets, profiles, tubes and rods are formed by conventional processing methods. Melt temp. at die outlet 230–236°; moisture content must be below 0.01%. Films may be cast from soln. or extruded (blow moulded). Extremely thin films (<10 μm) are cast on to a carrier film from which they are not separated before the final manufacturing step (e.g. winding capacitors)
Mould Shrinkage (%):

No.	Value
1	0.5–0.7%
2	0.3–0.8%

Applications: Electrical applications (for insulation) as voltage carrying components, covers for electrical equipment, coil formers. Household and consumer articles (special grades conform to food laws) as housing for razors; in kitchen sinks and appliances. Automotive applications as light coverings; in motorcycle windscreens, ventilation and radiator grilles, aircraft windows, motorcycle safety helmets, safety glasses, helmet visors. Photographic and optical equipment in camera housings, binoculars, slide projectors, microscope components. Sporting goods (ski sticks, surfboards). Used in construction for windows, double sheets for glazing greenhouses and industrial windows, dome construction. Computer and office machines. Medical applications as blood filters, dialysers, pacemaker components, sterilising equipment. Used in packaging (transparent bottles and containers, re-usable packaging, e.g. baby bottles, food containers, milk pails, electronic packaging). Laser-optical data storage systems such as CDs. Telecommunications in telephone switching mechanisms

Trade name	Supplier
Apec HT DP	Mobay
Apec HT DP	Bayer Inc.
Bapolan	Bamberger
Bayfol	Bayer Inc.
Calibre	Dow
Cyrolon ZX	CYRO Industries
Fiberfil	Akzo Engineering Plastics
Iupilon	Mitsubishi Gas Chem.
Jupilon	Mitsubishi Gas Chem.
Karlex	Ferro Corporation
Lexan	General Electric Plastics

Makrofol	Bayer Inc.
Makrolon	Mobay
Makrolon	Bayer Inc.
Marc PC	MRC
Margard sheet	General Electric Plastics
Merlon	Mobay
Novarex	Mitsubishi Chemical
Orgalan	ATO Chimie
PC-1100	Bay Resins
PDX	LNP Engineering
Panlite	Teijin Chem. Ltd.
Poly-glaz	Sheffield Plastics
Polypenco Polycarbonate	Polymer Corp.
Silaprene	Uniroyal Chemical
Sinvet	Anic
Stat-Kon	LNP Engineering
Thermocomp	LNP Engineering
Tuffak	General Electric Plastics
Upilon	Mitsubishi Gas Chem.
Xantar	Xantar Polycarbonates
Zelux	Westlake Polymers

Bibliographic References

[1] Schnell, H., *Polym. Rev.*, 1964, **9**, 1
[2] Adams, G.W. and Farris, R.J., *Polymer*, 1989, **30**, 1829, (refractive index)
[3] Ali, M.S. and Sheldon, R.P., *J. Appl. Polym. Sci.*, 1970, **14**, 2619, (refractive index)
[4] Bonart, R., *Makromol. Chem.*, 1966, **92**, 149, (refractive index)
[5] Möginger, B. and Fritz, U., *Polym. Int.*, 1991, **26**, 121, (thermal expansion)
[6] Roe, J.M. and Simha, R., *Int. J. Polym. Mater.*, 1974, **3**, 193, (thermal expansion)
[7] Matheson, R.R., *Macromolecules*, 1987, **20**, 1851, (thermal expansion)
[8] Zoller, P., *J. Polym. Sci., Polym. Phys. Ed.*, 1982, **20**, 1453, (eqn of state)
[9] Wohlfarth, C., *J. Appl. Polym. Sci.*, 1993, **48**, 1923, (eqn of state)
[10] Hartmann, B. and Haque, M.A., *J. Appl. Polym. Sci.*, 1985, **30**, 1553, (eqn of state)
[11] Steere, R.C., *J. Appl. Polym. Sci.*, 1966, **10**, 1673
[12] Chen, F.C., Poon, Y.M. and Choy, C.L., *Polymer*, 1977, **18**, 129, (thermal diffusivity)
[13] Choy, C.L. and Greig, D., *J. Phys.: Solid State Phys.*, 1977, **10**, 169
[14] Dainton, F.S., Evans, D.M., Hoare, F.E. and Melia, T.P., *Polymer*, 1962, **3**, 263, (heat capacity)
[15] Adam, G.A., Hay, J.N., Parsons, I.W. and Haward, R.N., *Polymer*, 1976, **17**, 51, (heat capacity)
[16] O'Reilly, J.M., Karasz, F.E. and Bair, H.E., *J. Polym. Sci., Part C: Polym. Lett.*, 1963, **6**, 109, (heat capacity)
[17] Wissler, G.E. and Gist, B., *J. Polym. Sci., Polym. Phys. Ed.*, 1980, **18**, 1257, (heat capacity)
[18] Jones, L.D. and Karasz, F.E., *J. Polym. Sci., Polym. Lett. Ed.*, 1966, **4**, 803, (T_m)
[19] Frosini, V., Butta, E. and Calamia, M., *J. Appl. Polym. Sci.*, 1970, **11**, 527, (T_g)
[20] *Jpn. Pat.*, 1992, 04 252 263, (transition temps)
[21] Vincent, P.I., *Polymer*, 1972, **13**, 558, (transition temp)
[22] Hartmann, B. and Lee, G.F., *J. Appl. Polym. Sci.*, 1979, **23**, 3639, (transition temp)
[23] Mercier, J.P. and Legras, R., *J. Polym. Sci., Polym. Lett. Ed.*, 1970, **8**, 643, (enthalpies)
[24] Bicerano, J., *J. Polym. Sci., Polym. Phys. Ed.*, 1991, **29**, 1329, (relaxation)
[25] Yee, A.F. and Smith, S.A., *Macromolecules*, 1981, **14**, 54, (relaxation)
[26] Legrand, D. and Erhardt, P.F., *J. Appl. Polym. Sci.*, 1969, **13**, 1707, (relaxation)
[27] Bernès, A., Chatain, D. and Lacabanne, C., *Polymer*, 1992, **33**, 4682, (relaxation)
[28] Hong, J. and Brittain, J.O., *J. Appl. Polym. Sci.*, 1981, **26**, 2459, (relaxation)
[29] Utracki, L.A., *Polym. Eng. Sci.*, 1995, **35**, 2, (PC blends, rev)
[30] Hellmann, E.H., Hellmann, G.P. and Rennie, A.R., *Colloid Polym. Sci.*, 1991, **269**, 343, (miscibility, solvent effect)
[31] Mohn, R.N., Paul, D.R., Barlow, J.W. and Cruz, C.A., *J. Appl. Polym. Sci.*, 1979, **23**, 575, (miscibility)
[32] Woo, E.M., Barlow, J.W. and Paul, D.R., *J. Appl. Polym. Sci.*, 1985, **30**, 4243, (miscibility)
[33] Kim, W.N. and Burns, C.M., *J. Appl. Polym. Sci.*, 1987, **34**, 945, (PC/PS blends)
[34] Murff, S.R., Barlow, J.W. and Paul, D.R., *J. Appl. Polym. Sci.*, 1984, **29**, 3231, (PC/PET blends)
[35] Wahrmund, D.C., Paul, D.R. and Barlow, J.W., *J. Appl. Polym. Sci.*, 1978, **22**, 2155, (PC/PBT blends)
[36] Kim, W.N. and Burns, C.M., *J. Appl. Polym. Sci.*, 1990, **41**, 1575, (compatibility, phase diagrams)
[37] Perreault, F. and Prud'homme, R.E., *Polym. Eng. Sci.*, 1995, **35**, 34, (PC/PMMA phase diagrams)
[38] Kyu, T. and Saldanha, J.M., *Macromolecules*, 1988, **21**, 1021, (PC/PMMA blends, effects of temp)
[39] Kyu, T. and Saldanha, J.M., *J. Polym. Sci., Polym. Lett. Ed.*, 1988, **26**, 33, (PC/PMMA blends)
[40] Chiou, J.S., Barlow, J.W. and Paul, D.R., *J. Polym. Sci., Polym. Phys. Ed.*, 1987, **25**, 1459, (PC/PMMA blends)
[41] Pilati, F., Marianucci, E. and Berti, C., *J. Appl. Polym. Sci.*, 1985, **30**, 1267, (PC/PET blends)
[42] Devaux, J., Godard, P. and Mercier, J.P., *J. Polym. Sci., Polym. Phys. Ed.*, 1982, **20**, 1875, 1881, 1895, 1901, (PC/PBT blends)
[43] Zhang, R., Luo, X. and Ma, D., *J. Appl. Polym. Sci.*, 1995, **55**, 455, (PC/copolyester blends)
[44] Paul, D.R., *Macromol. Symp.*, 1994, **78**, 83, (PC/ABS blends)
[45] Min, K., White, J.L. and Fellers, J.F., *Polym. Eng. Sci.*, 1984, **24**, 1327, (PC/PE blends)
[46] Endo, S., Min, K., White, J.L. and Kyu, T., *Polym. Eng. Sci.*, 1986, **26**, 45, (PC/PE blends)
[47] Greco, R. and Dong, L.S., *Macromol. Symp.*, 1994, **78**, 141, (PC/ABS blends)
[48] Keitz, J.D., Barlow, J.W. and Paul, D.R., *J. Appl. Polym. Sci.*, 1984, **29**, 3131, (PC/styrene-acrylonitrile blends)
[49] Hobbs, S.Y., Dekkers, M.E.J. and Watkins, V.H., *Polymer*, 1988, **29**, 1598, (spreading coefficients)
[50] Dipaola-Baranyi, G., Hsiao, C.K., Spes, M., Odell, P.G. and Burt, R.A., *J. Appl. Polym. Sci.: Appl. Polym. Symp.*, 1992, **51**, 195, (inverse gas chromatography)
[51] Arends, C.B., *J. Appl. Polym. Sci.*, 1993, **49**, 1931, (cohesive energy parameters)
[52] Seitz, J.T., *J. Appl. Polym. Sci.*, 1993, **49**, 1331, (cohesive energy parameters)
[53] Fedors, R.F., *Polym. Eng. Sci.*, 1974, **14**, 147, 472, (loss tangent)
[54] Barton, A.F.M., *Chem. Rev.*, 1975, **75**, 731, (loss tangent)
[55] Jacques, C.H.M. and Wyzgoski, M.G., *J. Appl. Polym. Sci.*, 1979, **23**, 1153, (critical strain)
[56] Belfiore, L.A., Henrichs, P.M., Massa, D.J., Zumbulyadis, N. *et al*, *Macromolecules*, 1983, **16**, 1744, (plasticisers, antiplasticisers)
[57] Belfiore, L.A. and Cooper, S.L., *J. Polym. Sci., Polym. Phys. Ed.*, 1983, **21**, 2135, (plasticisers, antiplasticisers)
[58] Wang, C.H., Xia, J.L. and Yu, L., *Macromolecules*, 1991, **24**, 3638, (plasticisers)
[59] Suvorova, A.I. and Hannanova, E.G., *Makromol. Chem.*, 1990, **191**, 993, (plasticisers, antiplasticisers)
[60] Liu, Y., Roy, A.K., Jones, A.A., Inglefield, P.T. and Ogden, P., *Macromolecules*, 1990, **23**, 968, (plasticisation, antiplasticisation)
[61] Petrie, S.E.B., Moore, R.S. and Flick, J.R., *J. Appl. Phys.*, 1972, **43**, 4318, (plasticisers, antiplasticisers)
[62] Jackson, W.J. and Caldwell, J.R., *J. Appl. Polym. Sci.*, 1967, **11**, 211, 227, (plasticisers, antiplasticisers)
[63] Gallez, F., Legras, R. and Mercier, J.P., *Polym. Eng. Sci.*, 1976, **16**, 276, (plasticisers, crystallisation)
[64] Legras, R. and Mercier, J.P., *J. Polym. Sci., Polym. Phys. Ed.*, 1979, **17**, 1171, (plasticisers, crystallisation)
[65] van der Valk, P., van Pelt, A.W.J., Busscher, H.J., de Jong, H.P. *et al*, *J. Biomed. Mater. Res.*, 1983, **17**, 807, (contact angles)
[66] Petke, F.D. and Ray, B.R., *J. Colloid Interface Sci.*, 1969, **31**, 216, (contact angles)
[67] Hüttinger, K.J., Höhmann, S. and Seiferling, M., *Carbon*, 1991, **29**, 449, (work of adhesion)
[68] Murase, H., Nanishi, K., Kogure, H., Fujibayashi, T. *et al*, *J. Appl. Polym. Sci.*, 1994, **54**, 2051, (work of adhesion)
[69] Henry, L.F., *Polym. Eng. Sci.*, 1974, **14**, 167, (contact angles)
[70] Dobkowski, Z., *Rheol. Acta*, 1986, **25**, 195, (melt flow index)
[71] Pryde, C.A., Kelleher, P.G., Hellman, M.Y. and Wentz, R.P., *Polym. Eng. Sci.*, 1982, **22**, 370, (melt flow index)

[72] Berry, G.C., Nomura, H. and Mayhan, K.G., J. Polym. Sci., Polym. Phys. Ed., 1967, **5**, 1, (dilute soln props)

[73] Sitaramaiah, G., J. Polym. Sci., Part A: Polym. Chem., 1965, **3**, 2743, (dilute soln props)

[74] Moore, W.R. and Uddin, M., Eur. Polym. J., 1969, **5**, 185, (dilute soln props)

[75] Baumann, G.F. and Steingiser, S., J. Polym. Sci., Part A: Polym. Chem., 1963, **1**, 3395, (dilute soln props, melt rheology)

[76] Dobkowski, Z., Eur. Polym. J., 1982, **18**, 1051, (melt rheology)

[77] Hellums, M.W., Koros, W.J., Paul, D.R. and Husk, G.R., AIChE Symp. Ser., 1989, **85**, 6, (gas permeability)

[78] Hellums, M.W., Koros, W.J., Husk, G.R. and Paul, D.R., J. Membr. Sci., 1989, **46**, 93, (gas permeability)

[79] Costello, L.M. and Koros, W.J., Ind. Eng. Chem. Res., 1993, **32**, 2277, (gas permeability)

[80] Schmidhauser, J.C. and Longley, K.L., J. Appl. Polym. Sci., 1990, **39**, 2083, (gas permeability)

[81] Muruganandam, N. and Paul, D.R., J. Membr. Sci., 1987, **34**, 185, (gas permeability)

[82] Muruganandam, N., Koros, W.J. and Paul, D.R., J. Polym. Sci., Polym. Phys. Ed., 1987, **25**, 1999, (gas permeability)

[83] Aguilar-Vega, M. and Paul, D.R., J. Polym. Sci., Polym. Phys. Ed., 1993, **31**, 1599, (gas permeability)

[84] Jordan, S.M. and Koros, W.J., J. Membr. Sci., 1990, **51**, 233, (gas permeability)

[85] Costello, L.M. and Koros, W.J., J. Polym. Sci., Polym. Phys. Ed., 1994, **32**, 701, (gas permeability)

[86] Gusev, A.A., Suter, U.W. and Moll, D.J., Macromolecules, 1995, **28**, 2582, (gas permeability)

[87] Barbari, T.A., Koros, W.J. and Paul, D.R., J. Polym. Sci., Polym. Phys. Ed., 1988, **26**, 709, (gas permeability)

[88] Koros, W.J., Chan, A.H. and Paul, D.R., J. Membr. Sci., 1977, **2**, 165, (gas permeability)

[89] Jordan, S.M., Koros, W.J. and Fleming, G.K., J. Membr. Sci., 1987, **30**, 191, (gas permeability)

[90] Jordan, S.M., Koros, W.J. and Beasley, S.K., J. Membr. Sci., 1989, **43**, 103, (gas permeability)

[91] Fleming, G.K. and Koros, W.J., Macromolecules, 1986, **19**, 2291, (gas permeability)

[92] Jordan, S.M., Fleming, G.K. and Koros, W.J., J. Polym. Sci., Polym. Phys. Ed., 1990, **28**, 2305, (gas permeability)

[93] Gao, Y., Baca, A.M., Wang, B. and Ogilby, P.R., Macromolecules, 1994, **27**, 7041, (gas permeability)

[94] Wu, T., Lee, S. and Chen, W., Macromolecules, 1995, **28**, 5751, (liq permeability)

[95] Kambour, R.P., Gruner, C.L. and Romagosa, E.E., Macromolecules, 1974, **7**, 248, (liq permeability)

[96] Turska, E. and Benecki, W., J. Appl. Polym. Sci., 1979, **23**, 3489, (liq permeability)

[97] Ware, R.A., Tirtowidjojo, S. and Cohen, C., J. Appl. Polym. Sci., 1981, **26**, 2975, (liq permeability)

[98] Mercier, J.P., Groeninckx, G. and Lesne, M., J. Polym. Sci., Polym. Symp., 1967, **16**, 2059, (liq permeability)

[99] Sheldon, R.P. and Blakey, P.R., Nature (London), 1962, **195**, 172, (liq permeability)

[100] Grinsted, R.A. and Koenig, J.L., Macromolecules, 1992, **25**, 1229, (liq permeability)

[101] Ercken, M., Adriaensens, P., Vanderzande, D. and Gelan, J., Macromolecules, 1995, **28**, 8541, (liq permeability)

[102] Miller, G.W., Visser, S.A.D. and Morecroft, A.S., Polym. Eng. Sci., 1971, **11**, 73, (liq permeability)

[103] Robeson, L.M. and Crisafulli, S.T., J. Appl. Polym. Sci., 1983, **28**, 2925, (liq permeability, water absorption)

[104] Titow, W.V. and Braden, M., Plast. Polym., 1973, **41**, 94, (liq permeability)

[105] Golovoy, A. and Zinbo, M., Polym. Eng. Sci., 1989, **29**, 1733, (liq permeability, water content, water absorption)

[106] Levantovskaya, I.I., Dralyuk, G.V., Pshenitsyna, V.P., Smirnova, O.V. et al, Polym. Sci. USSR (Engl. Transl.), 1968, **10**, 1892, (volatile loss)

[107] Lee, P.L., Xiao, C., Wu, J., Yee, A.F. and Schaefer, J., Macromolecules, 1995, **28**, 6477, (water content)

[108] Ito, E. and Kobayashi, Y., J. Appl. Polym. Sci., 1978, **22**, 1143, (water content)

[109] Allen, G., McAinsh, J. and Jeffs, G.M., Polymer, 1971, **12**, 85, (water absorption)

[110] Fyfe, C.A., Randall, L.H. and Burlinson, N.E., Chem. Mater., 1992, **4**, 267, (water absorption)

[111] Bair, H.E., Johnson, G.E. and Merriweather, R., J. Appl. Phys., 1978, **49**, 4976, (water absorption)

[112] Narkis, M., Nicolais, L., Apicella, A. and Bell, J.P., Polym. Eng. Sci., 1984, **24**, 211, (water absorption)

[113] G'Sell, C., Hiver, J.M., Dahoun, A. and Souahi, A., J. Mater. Sci., 1992, **27**, 5031, (tensile stress-strain curve)

[114] Kitagawa, M. and Zhou, D., Polym. Eng. Sci., 1995, **35**, 1725, (tensile stress-strain curve)

[115] Nied, H.F., Stokes, V.R. and Ysseldyke, D.A., Polym. Eng. Sci., 1987, **27**, 101, (tensile stress-strain curve)

[116] G'Sell, C., El Bari, H., Perez, J., Cavaille, J.Y. and Johari, G.P., Mater. Sci. Eng., A, 1989, **110**, 223, (tensile shear stress-shear curve, complex moduli)

[117] Fleck, N.A., Stronge, W.J. and Liu, J.H., Proc. R. Soc. London A, 1990, **429**, 459, (tensile shear stress-shear curve, fracture)

[118] Grenet, J. and G'Sell, C., Polymer, 1990, **31**, 2057, (tensile shear stress-shear curve)

[119] Nazarenkos, S., Bensason, S., Hiftner, A. and Baer, E., Polymer, 1994, **35**, 3883, (stress-strain curves, strain rate sensitivity)

[120] Bauwens-Crowet, C. and Bauwens, J.C., Polymer, 1983, **24**, 921, (stress-strain curves)

[121] Landes, J.D. and Zhou, Z., Fracture, 1993, **63**, 383, (stress-strain curves, fracture)

[122] Wert, M.J., Saxena, A. and Ernst, H.A., J. Test. Eval., 1990, **18**, 1, (stress-strain curves, tensile strength, fracture toughness)

[123] Droh, N.N, Leevers, P.S. and Williams, J.G., Polymer, 1993, **34**, 4230, (stress-strain curves)

[124] Bauwens-Crowet, C. and Bauwens, J.-C., Polymer, 1982, **23**, 1599, (stress-strain curves)

[125] Kau, C., Tse, A., Hiltner, A. and Baer, E., J. Mater. Sci., 1993, **28**, 529, (Young's modulus, tensile strength)

[126] G'Sell, C. and Gopez, A.J., J. Mater. Sci., 1985, **20**, 3462, (Young's modulus, tensile strength, shear stress-shear curves)

[127] Tsou, A.H., Greener, J. and Smith, G.D., Polymer, 1995, **36**, 949, (Young's modulus, flexural modulus)

[128] Thompson, R.J. and Goldblum, K.B., Mod. Plast., April, 1958, **35**, 131, (props)

[129] Kambour, R.P., Caraher, J.M., Fasoldt, C.C., Seeger, G.T. et al, J. Polym. Sci., Polym. Phys. Ed., 1995, **33**, 425, (flexural modulus, complex moduli)

[130] Gruenwald, G., Mod. Plast. Jan, 1960, **38**, 137, (flexural modulus)

[131] Whitney, W. and Andrews, R.D., J. Polym. Sci., Part C: Polym. Lett., 1967, **16**, 2981, (Poisson's ratio, moduli)

[132] Pampillo, C.A. and Davis, L.A., J. Appl. Phys., 1971, **42**, 4674, (Poisson's ratio, compressive modulus)

[133] Raftopoulos, D.D., Karapanos, D. and Theocaris, P.S., J. Phys. D: Appl. Phys., 1976, **9**, 869, (Poisson's ratio)

[134] Krbecek, H., Krüger, H. and Pietralla, M., J. Polym. Sci., Polym. Phys. Ed., 1993, **31**, 1477, (Poisson's ratio)

[135] Quayum, M.M. and White J.R., J. Appl. Polym. Sci., 1991, **43**, 129, (tensile strength)

[136] Warfield, R.W. and Barnet, F.R., Angew. Makromol. Chem., 1972, **27**, 215, (bulk modulus)

[137] Adams, G.W. and Farris, R.J., Polymer, 1989, **30**, 1829, (complex moduli)

[138] Wyzgoski, M.G. and Yeh, G.S.Y., Int. J. Polym. Mater., 1974, **3**, 133, (complex moduli)

[139] Hwang, E.J., Inoue, T. and Osaki, K., Polymer, 1993, **34**, 1661, (complex moduli, stress-optical coefficient)

[140] Robeson, L.M. and Faucher, J.A., J. Polym. Sci., Polym. Lett. Ed., 1969, **7**, 35, (complex moduli, Izod impact strength)

[141] Trznadel, M., Pakuta, T. and Kryszewski, M., Polymer, 1988, **29**, 619, (complex moduli)

[142] Ahlborn, K., Cryogenics, 1988, **28**, 234, (complex moduli)

[143] Mod. Plast. Jan, 1960, **37**, 164, (impact strength, electrical props, hardness)

[144] Broutman, L.J. and Patil, R.S., Polym. Eng. Sci., 1971, **11**, 165, (impact strength, hardness)

[145] Turley, S.G., Polym. Prepr. (Am. Chem. Soc., Div. Polym. Chem.), 1967, **8**, 1524, (impact strength)

[146] Legrand, D.G., J. Appl. Polym. Sci., 1969, **13**, 2129, (impact strength, fracture)

[147] Fried, J.R., Zhang, C. and Liu, H.-C., J. Polym. Sci., Polym. Lett. Ed., 1990, **28**, 7, (impact energy)

[148] Seldén, R., Polym. Test., 1987, **7**, 209, (impact energy)

[149] Bauwens-Crowet, J.M. and Bauwens, J.-C., J. Mater. Sci., 1974, **9**, 1197, (creep)

[150] Challa, S.R. and Progelhof, R.C., Polym. Eng. Sci., 1995, **35**, 546, (creep)

[151] Garbarski, J., Polym. Eng. Sci., 1992, **32**, 107, (creep)

[152] Malkin, A.Y.A., Teishev, A.E. and Kutsenko, M.A., J. Appl. Polym. Sci., 1992, **45**, 237, (creep, stress relaxation)

[153] Grzywinski, G.G. and Woodford, D.A., Polym. Eng. Sci., 1995, **35**, 1931, (creep, stress relaxation)

[154] Higuchi, H., Jamieson, A.M., Simha, R. and McGervey, J.D., J. Polym. Sci., Polym. Phys. Ed., 1995, **33**, 2295, (stress relaxation)

[155] Yee, A.F., Bankert, R.J., Ngai, K.L. and Rendell, R.W., J. Polym. Sci., Polym. Phys. Ed., 1988, **26**, 2463, (stress relaxation)

[156] Mattioli, T.K. and Quesnel, D.J., Polym. Eng. Sci., 1987, **27**, 848, (stress relaxation, fatigue)

[157] Arisawa, K., Hirose, H., Minoru, I., Harada, T. and Wada, Y., *Jpn. J. Appl. Phys.*, 1963, **2**, 695, (creep)
[158] Mendelson, R.A., *Polym. Eng. Sci.*, 1983, **23**, 79
[159] Alegría, A., Macho, E. and Colmenero, J., *Macromolecules*, 1991, **24**, 5196
[160] Parvin, M. and Williams, J.G., *J. Mater. Sci.*, 1975, **10**, 1883, (fracture)
[161] Parvin, M. and Williams, J.G., *Int. J. Fract.*, 1975, **11**, 963, (fracture)
[162] Bahadur, S. and Henkin, A., *Polym. Eng. Sci.*, 1973, **13**, 422, (fracture)
[163] Bubeck, R. and Bales, S.E., *Polym. Eng. Sci.*, 1984, **24**, 1142, (fracture, shear yield stress)
[164] Adams, G.C., Bender, R.G., Crouch, B.A. and Williams, J.G., *Polym. Eng. Sci.*, 1990, **30**, 241, (fracture)
[165] Maekawa, Z., Hamada, H., Iwamoto, M. and Ukai, M., *Polym. Eng. Sci.*, 1993, **33**, 996, (fracture)
[166] Glover, A.P., Johnson, F.A. and Radon, J.C., *Polym. Eng. Sci.*, 1974, **14**, 420, (fracture)
[167] Hill, A.J., Heater, K.J. and Agrawal, C.M., *J. Polym. Sci., Polym. Phys. Ed.*, 1990, **28**, 387, (fracture)
[168] Takemori, M.J., *Polym. Eng. Sci.*, 1988, **28**, 641, (fatigue)
[169] Liu, L.B., Yee, A.F. and Lewis, J.C., *J. Non-Cryst. Solids*, 1991, **131-133**, 492, (fatigue)
[170] Matsumoto, D.S. and Gifford, S.K., *J. Mater. Sci.*, 1985, **20**, 4610, (fatigue)
[171] Furue, H. and Shimamura, S., *Int. J. Fract.*, 1980, **16**, 553, (fatigue)
[172] Goble, D.L. and Wolff, E.G., *J. Mater. Sci.*, 1993, **28**, 5986, (strain rate sensitivity)
[173] Chung, C.I., Hennessey, W.J. and Tusim, M.H., *Polym. Eng. Sci.*, 1977, **17**, 9, (friction coefficient)
[174] Bayer, R.G., Engel, P.A. and Sacher, E., *Wear*, 1975, **32**, 181, (friction coefficient)
[175] Kalkar, A.K., Kundagol, S., Chand, S. and Chandra, S., *Radiat. Phys. Chem.*, 1992, **39**, 435, (electrical conductivity, ir)
[176] Chand, S., Agarwal, J.P. and Mehendru, P.C., *Thin Solid Films*, 1983, **99**, 351, (electrical conductivity)
[177] Amin, M., Darwish, K.A. and Mounir, M., *Angew. Makromol. Chem.*, 1987, **150**, 81, (electrical conductivity, dielectric constant, uv)
[178] Swanson, J.W. and Dall, F.C., *IEEE Trans. Electr. Insul.*, 1977, **12**, 142, (dielectric props, volume resistivity)
[179] Hatfield, L.L., Leiker, G.R. and Kristiansen, M., *IEEE Trans. Electr. Insul.*, 1988, **23**, 57, (surface resistivity)
[180] Pratt, G.J. and Smith, M.J.A., *Br. Polym. J.*, 1986, **18**, 105, (tan δ)
[181] Allen, G., McAinsh, J. and Jeffs, G.M., *Polymer*, 1971, **12**, 85, (tan δ)
[182] Watts, D.C. and Perry, E.P., *Polymer*, 1978, **19**, 248, (complex relative permittivity)
[183] Monsour, A.A. and Madbouly, S.A., *Polym. Int.*, 1995, **36**, 269, (complex relative permittivity)
[184] Yianakopoulos, G., Vanderschueren, J., Niezette, J. and Thielen, A., *IEEE Trans. Electr. Insul.*, 1990, **25**, 693, (complex relative permittivity)
[185] Keyser, P.T. and Jefferts, S.R., *Rev. Sci. Instrum.*, 1989, **60**, 2711, (magnetic susceptibility)
[186] Champion, J.V., Desson, R.A. and Meeten, G.H., *Polymer*, 1974, **15**, 301, (magnetic birefringence)
[187] Philipp, H.R., Legrand, D.G., Cole, H.S. and Liu, Y.S., *Polym. Eng. Sci.*, 1987, **27**, 1148, (refractive index)
[188] Waxter, R.M., Horowitz, D. and Feldman, A., *Appl. Opt.*, 1979, **18**, 101, (refractive index)
[189] Peetz, L., Krüger, J.K. and Pietralla, M., *Colloid Polym. Sci.*, 1987, **265**, 761, (intrinsic birefringence)
[190] Lytle, J.D., Wilkerson, G.W. and Jaramillo, J.G., *Appl. Opt.*, 1979, **18**, 1842, (transmittance)
[191] Brown, G. and Kanaris-Sotiriou, R., *J. Phys. E: Sci. Instrum.*, 1969, **2**, 551, (transmission of x-rays)
[192] Pankasem, S., Kuczynski, J. and Thomas, J.K., *Macromolecules*, 1994, **27**, 3773, (uv, fluorescence spectrum)
[193] Itagaki, H. and Umeda, Y., *Polymer*, 1995, **36**, 29, (uv, fluorescence spectrum)
[194] Kulczycki, A., *Spectrochim. Acta*, 1985, **41**, 1427, (ir)
[195] Fabre, M.J., Tagle, L.H., Gargallo, L., Radue, D. and Hernandez-Fuentes, I., *Eur. Polym. J.*, 1989, **25**, 1315
[196] Vogt, V.-D., Dettenmaier, M., Spies, H.W. and Pietralla, M., *Colloid Polym. Sci.*, 1990, **268**, 22, (intrinsic birefringence)
[197] Wu, M.-S.S., *J. Appl. Polym. Sci.*, 1986, **32**, 3263, (intrinsic birefringence)
[198] Wimberger-Friedl, R. and Hendriks, R.D.H.M., *Polymer*, 1989, **30**, 1143, (stress-optical coefficient)
[199] Lee, S., de la Vega, J. and Bogue, D.C., *J. Appl. Polym. Sci.*, 1986, **31**, 2791, (stress-optical coefficient)
[200] Muller, R. and Pesce, J.J., *Polymer*, 1994, **35**, 734, (stress-optical coefficient)
[201] Wimberger-Friedl, R., *Rheol. Acta*, 1991, **30**, 329, (stress-optical coefficient)
[202] Takezawa, Y., Taketani, N., Tanno, S. and Ohara, S., *J. Appl. Polym. Sci.*, 1992, **46**, 2033
[203] Rufus, I.B., Shah, H. and Hoyle, C.E., *J. Appl. Polym. Sci.*, 1994, **51**, 1549, (thermal stability)
[204] Politou, A.S., Morterra, C. and Low, M.J.D., *Carbon*, 1990, **28**, 529, (thermal stability)
[205] McNeill, I.C. and Rincon, A., *Polym. Degrad. Stab.*, 1991, **31**, 163, (thermal stability)
[206] Ball, G.L. and Boettner, E.A., *J. Appl. Polym. Sci.*, 1972, **16**, 855, (thermal stability)
[207] Factor, A., *Angew. Makromol. Chem.*, 1995, **232**, 27, (rev)
[208] Davis, A. and Golden, J.H., *Nature (London)*, 1965, **206**, 397, (thermal stability)
[209] Kuroda, S.-I., Terauchi, K., Nogami, K. and Mita, I., *Eur. Polym. J.*, 1989, **25**, 1, (thermal stability)
[210] Foti, S., Giuffrida, M., Maravigna, P. and Montaudo, G., *J. Polym. Sci., Polym. Chem. Ed.*, 1983, **21**, 1567, (thermal stability)
[211] McNeill, I.C. and Rincon, A., *Polym. Degrad. Stab.*, 1993, **39**, 13, (thermal stability)
[212] Kourtides, D.A. and Parker, J.A., *Polym. Eng. Sci.*, 1978, **18**, 855, (thermal stability, flammability)
[213] Ash, M. and Ash, I., Handbook of Plastic Compounds, *Elastomers and Resins*, (eds. M.B. Ash and I.A. Ash), Wiley-VCH, 1992, 218, (commercial polycarbonate props)
[214] Bauwens, J.-C., *Plast. Rubber Process. Appl.*, 1987, **7**, 143, (annealing)
[215] Othmezouri-Decerf, J., *Polymer*, 1988, **29**, 641, (annealing)
[216] Szlezyngier, W., *Ochr. Koroz.*, 1978, **21**, 51, (oxidation)
[217] Clark, D.T. and Munro, H.S., *Polymer*, 1984, **25**, 826, (ozonolysis)
[218] Factor, A., Ligon, W.V. and May, R.J., *Macromolecules*, 1987, **20**, 2461, (weathering)
[219] Clark, D.T. and Munro, H.S., *Polym. Degrad. Stab.*, 1984, **8**, 195, (weathering)
[220] Rivaton, A., *Polym. Degrad. Stab.*, 1995, **49**, 163, (weathering)
[221] Sherman, E.S., Ram, A. and Kenig, S., *Polym. Eng. Sci.*, 1982, **22**, 457, (weathering)
[222] Ram, A., Zilber, O. and Kenig, S., *Polym. Eng. Sci.*, 1985, **25**, 535, (weathering, hydrolytic stability)
[223] Jacques, C.H.M. and Wyzgoski, M.G., *J. Appl. Polym. Sci.*, 1979, **23**, 1153, (environmental stress cracking)
[224] Wyzgoski, M.G. and Jacques, C.H.M., *Polym. Eng. Sci.*, 1977, **17**, 854, (environmental stress cracking)
[225] Iannone, M., Nicolais, L., Nicodemo, L. and Di Benedetto, A.T., *J. Mater. Sci.*, 1982, **17**, 81, (environmental stress cracking)
[226] Priori, A., Nicolais, L. and Di Benedetto, A.T., *J. Mater. Sci.*, 1983, **18**, 1466, (environmental stress cracking)
[227] Lee, W., Nobile, M.R., Di Benedetto, A.T. and Nicolais, L., *Int. J. Polym. Mater.*, 1989, **12**, 185, (environmental stress cracking)
[228] Kirloskar, M.A. and Donovan, J.A., *Polym. Eng. Sci.*, 1987, **27**, 124, (environmental stress cracking)
[229] Kovapatnitskii, A.M., Klinov, I.Y. and Bokshitskii, M.N., *Tr. Mosk. Inst. Khim. Mashinostr.*, 1971, **37**, 144, (acid stability)
[230] Gaines, G.L., *Polym. Degrad. Stab.*, 1990, **27**, 13, (base stability)
[231] Bukreeva, T.V., Smirnova, O.V., Paulov, N.N., Ivanova, L.V. and Serzhantova, N.A., *Tr. Inst. - Mosk. Khim.-Tekhnol. Inst. im. D.I. Mendeleeva*, 1980, **110**, 28, (base stability)
[232] Bailly, C., Daumerie, M., Legras, R. and Mercier, J.P., *Makromol. Chem.*, 1986, **187**, 1197, (salt stability)
[233] Gorelov, Y.P. and Miller, V.B., *Polym. Sci. USSR (Engl. Transl.)*, 1979, **20**, 2134, (thermal oxidation)
[234] Dobkowski, Z. and Rudnik, E., *J. Therm. Anal.*, 1992, **38**, 2211, (thermal oxidation)
[235] Lee, L.-H., *J. Polym. Sci., Part A: Polym. Chem.*, 1964, **2**, 2859, (thermal oxidation)
[236] Gardner, R.J. and Martin, J.R., *J. Appl. Polym. Sci.*, 1979, **24**, 1269, (hydrolytic stability)
[237] Pryde, C.A., Kelleher, P.G., Hellman, M.Y. and Wentz, R.P., *Polym. Eng. Sci.*, 1982, **22**, 370, (hydrolytic stability)
[238] Zinbo, M. and Golovoy, A., *Polym. Eng. Sci.*, 1992, **32**, 786, (hydrolytic stability)
[239] Ghorbel, I., Thominette, F., Spiteri, P. and Verdu, J., *J. Appl. Polym. Sci.*, 1995, **55**, 163, (hydrolytic stability)
[240] Ghorbel, I., Akele, N., Thominette, F., Spiteri, P. and Verdu, J., *J. Appl. Polym. Sci.*, 1995, **55**, 173, (hydrolytic stability)
[241] Schilling, F.C., Ringo, W.M., Sloane, N.J.A. and Bovey, F.A., *Macromolecules*, 1981, **14**, 532, (hydrolytic stability)
[242] Bair, H.E., Falcone, D.R., Hellman, M.Y., Johnson, G.E. and Kelleher, P.G., *J. Appl. Polym. Sci.*, 1981, **26**, 1777, (hydrolytic stability)
[243] Pryde, C.A. and Hellman, M.Y., *J. Appl. Polym. Sci.*, 1980, **25**, 2573, (hydrolytic stability)
[244] Miner, R.J., *Bell Syst. Tech. J.*, 1972, **51**, 23, (biological stability)
[245] Guess, W.L. and Aution, J., *J. Oral Ther. Pharmacol.*, 1966, **3**, 116, (toxicology)
[246] Allen, G., Morley, D.C.W. and Williams, T., *J. Mater. Sci.*, 1973, **8**, 1449, (annealing)

[247] Adam, G.A., Cross, A. and Haward, R.N., *J. Mater. Sci.*, 1975, **10**, 1582, (annealing)
[248] Neki, K. and Geil, P.H., *J. Macromol. Sci., Phys.*, 1973, **8**, 295, (annealing)
[249] Risch, B.G. and Wilkes, G.L., *J. Appl. Polym. Sci.*, 1995, **56**, 1511, (annealing)
[250] Liu, L.B., Yee, A.F. and Gidley, D.W., *J. Polym. Sci., Polym. Phys. Ed.*, 1992, **30**, 221, (annealing, fatigue)
[251] Cho, J.S., Choi, W.-K., Jung, H.-J., Koh, S.-K. and Yoon, K.H., *J. Mater. Res.*, 1997, **12**, 277, (wettability)

Bisphenol A polycarbonate, Carbon fibre reinforced B-6

Synonyms: *Poly[2,2-propanebis(4-phenyl)carbonate], Carbon fibre reinforced*
Related Polymers: Poly[2,2-propanebis(4-phenyl)carbonate], Polycarbonate, carbon fibre reinforced, PTFE lubricated, Glass fibre reinforced and carbon fibre reinforced polycarbonate
Monomers: Bisphenol A, Carbonic dichloride
Material class: Thermoplastic, Composites
Polymer Type: Bisphenol A polycarbonate
CAS Number:

CAS Reg. No.
24936-68-3
25037-45-0

Molecular Formula: $(C_{16}H_{14}O_3)_n$
Fragments: $C_{16}H_{14}O_3$
Additives: 10–40% Carbon fibre
General Information: The props. of carbon fibre reinforced polycarbonate depend on a range of factors including fibre type and treatment, processing conditions, fibre orientation and fibre concentration. Often the material tested is in the form of a single coated fibre, rather than a composite of fibres and polymer matrix. Flame retardant and nickel coated versions available

Volumetric & Calorimetric Properties:
Density:

No.	Value	Note
1	d 1.23 g/cm³	10% carbon fibre [1]
2	d 1.24 g/cm³	10% carbon fibre [1]
3	d 1.29 g/cm³	20% carbon fibre [1]
4	d 1.34 g/cm³	30% carbon fibre [1]
5	d 1.33 g/cm³	30% PAN based carbon fibre [1]
6	d 1.35 g/cm³	30% pitch carbon fibre [1]
7	d 1.38 g/cm³	40% carbon fibre [1]
8	d 1.28 g/cm³	
9	d 1.37 g/cm³	

Thermodynamic Properties General: The longitudinal thermal conductivity increases linearly with increasing fibre content up to about 20% volume fraction. Above 20% volume fraction the conductivity is constant [2] and is about 50 times that of unreinforced polycarbonate and is similar to that of metal alloys such as stainless steel [2]. The transverse thermal conductivity is much less than the longitudinal value. From 5–30% volume carbon fibre the thermal conductivity increases approx. linearly with carbon fibre level without reaching saturation point
Thermal Conductivity:

No.	Value	Note
1	1.4–2.1 W/mK	longitudinal, 5% pitch carbon fibre [2]
2	3.3–4.9 W/mK	longitudinal, 10% vol. pitch carbon fibre [2]
3	8.6–10.4 W/mK	longitudinal, 20% vol. pitch carbon fibre [2]
4	0.33 W/mK	transverse, 10% vol. pitch carbon fibre [2]
5	0.49 W/mK	transverse, 20% vol. pitch carbon fibre [2]
6	0.63 W/mK	transverse, 30% vol. pitch carbon fibre [2]

Melting Temperature:

No.	Value	Note
1	288–316°C	20–40% carbon fibre [1]

Glass-Transition Temperature:

No.	Value	Note
1	138°C	0% carbon fibre [3]
2	136°C	10% wt. carbon fibre [3]
3	134°C	20% wt. carbon fibre [3]
4	133°C	30% wt. carbon fibre [3]
5	133°C	40% wt. carbon fibre [3]

Transition Temperature General: T_g of carbon fibre composites depends on the formation conditions [3], and decreases with increasing carbon fibre content [3]
Deflection Temperature:

No.	Value	Note
1	138–143°C	280–290°F, 1.8 MPa, 10% carbon fibre [1]
2	146°C	294°F, 1.8 MPa, 20% carbon fibre [1]
3	146–149°C	295–300°F, 1.8 MPa, 30% carbon fibre [1]
4	149°C	300°F, 1.8 MPa, 30% PAN based carbon fibre [1]
5	143°C	290°F, 30% pitch carbon fibre [1]
6	147–149°C	296°F, 1.8MPa, 40% carbon fibre [1]

Surface Properties & Solubility:
Wettability/Surface Energy and Interfacial Tension: Generally, the adhesion between carbon fibres and the polycarbonate matrix is relatively weak, with failure occurring at the fibre-matrix interface, rather than cohesive failure of the matrix [4,5,7]. SEM studies show single fibres pulled out from the composite in tensile tests [5,7,8,9]. Adhesion, however, does depend on the nature of the carbon fibre [4,7], and can be improved by oxidation of the fibres. For example, treatment with RF plasma in oxygen [6] or oxygen at 400° for 30 mins. [5] gives improved adhesion; still better results are achieved using 0.75% ozone at 100° for 1 min. [5]. In the case of ozone, SEM shows bundles of fibres pulled out from the composite - the fibres are linked together with polymer [5]. Adhesion also improves with increased processing temp. [5]

Transport Properties:
Water Absorption:

No.	Value	Note
1	0.1 %	10% carbon fibre [1]
2	0.1 %	20% carbon fibre [1]
3	0.04–0.08 %	30% carbon fibre [1]
4	0.1 %	40% carbon fibre [1]

– Bisphenol A polycarbonate, Carbon fibre reinforced

Mechanical Properties:

Mechanical Properties General: Carbon fibre reinforced polycarbonate shows greatly increased tensile and flexural moduli and increased tensile and flexural strengths relative to the unreinforced polycarbonate; these values increase with increasing carbon fibre content. However, the improvement in strength is achieved at the expense of toughness. The composites are harder than their unreinforced counterparts, but display brittle fracture. The mech. props. vary with factors such as fibre type and orientation as well as concentration. Elongation at break 4% (20% wt. carbon fibre) [1], 3% (30% wt. carbon fibre) [1], 3% (40% wt. carbon fibre) [1]

Tensile (Young's) Modulus:

No.	Value	Note
1	7310 MPa	1060 kpsi, 10% vol. graphite, random fibre orientation [10]
2	11240 MPa	1630 kpsi, 10% vol. graphite, unidirectional fibre orientation [10]
3	8550 MPa	1240 kpsi, 20% vol. graphite, random fibre orientation [10]
4	12580 MPa	1825 kpsi, 20% vol. graphite, unidirectional fibre orientation [10]
5	11030 MPa	1600 kpsi, 30% vol. graphite, random fibre orientation [10]
6	28950 MPa	4200 kpsi, 30% vol. graphite, unidirectional fibre orientation [10]
7	10620 MPa	1540 kpsi, 40% vol. graphite, random fibre orientation [10]
8	23920 MPa	3470 kpsi, 40% vol. graphite, unidirectional fibre orientation [10]
9	109000–121000 MPa	57-62% vol. PAN based carbon fibre, commercial oxidation treatment, unidirectional fibre orientation [5]

Flexural Modulus:

No.	Value	Note
1	42000 MPa	50% vol. unidirectional carbon fibre, longitudinal [11]
2	8000 MPa	50% vol. unidirectional carbon fibre, transverse [11]

Tensile Strength Break:

No.	Value	Note	Elongation
1	64.1 MPa	9.3 kpsi, 10% vol. graphite, random fibre orientation [10]	1.1%
2	97.9 MPa	14.2 kpsi, 10% vol. graphite, unidirectional fibre orientation [10]	1.4%
3	68.9 MPa	10 kpsi, 20% vol. graphite, random fibres [10]	0.84%
4	176.5 MPa	25.6 kpsi, 20% vol. graphite, unidirectional fibres [10]	1.66%
5	77.9 MPa	11.3 kpsi, 30% vol. graphite, random fibres [10]	0.64%
6	261 MPa	37.8 kpsi, 30% vol. graphite, unidirectional fibres [10]	1.17%
7	78.6 MPa	11.4 kpsi, 40% vol. graphite, random fibre orientation [10]	0.83%
8	345 MPa	40% vol. graphite, unidirectional fibres [10]	1–40%

Flexural Strength at Break:

No.	Value	Note
1	138–152 MPa	20–22 kpsi, 10% carbon fibre [1]
2	220–221 MPa	31.9 kpsi, 20% carbon fibre [1]
3	247–262 MPa	35.9–38 kpsi, 30% carbon fibre [1]
4	248 MPa	36 kpsi, PAN based carbon fibre [1]
5	117 MPa	17 kpsi, pitch carbon fibre [1]
6	251–254 MPa	36.4 kpsi, 40% carbon fibre [1]
7	1435.8–1571.2 MPa	57–62% vol. PAN based carbon fibre, commercially oxidised fibres, unidirectional orientation, DIN 29971 [5]

Tensile Strength Yield:

No.	Value	Elongation	Note
1	103.4 MPa	1–4%	15 kpsi, 10% wt. carbon fibre [1]
2	110.3 MPa		16 kpsi, 10% wt. carbon fibre [1]
3	159 MPa		20% wt. carbon fibre [1]
4	157 MPa		22.8 kpsi, 20% wt. carbon fibre [1]
5	171.1 MPa		30% wt. carbon fibre [1]
6	169–172 MPa		24.5–25 kpsi, 30% wt. carbon fibre [1]
7	165 MPa		24 kpsi, 30% wt. PAN based carbon fibre [1]
8	80 MPa		11.6 kpsi, 30% wt. pitch carbon fibre [1]
9	176 MPa		40% wt. carbon fibre [1]
10	179 MPa		26 kpsi, 40% wt. carbon fibre [1]

Miscellaneous Moduli:

No.	Value	Note	Type
1	3767–5613 MPa	53% vol. graphite [12]	shear modulus

Hardness: Rockwell M85 (20% carbon fibre) [1]; M85 (30% carbon fibre) [1]; R110 (30% carbon fibre) [1]; M90 (40% carbon fibre) [1]

Failure Properties General: Generally, in tensile tests of carbon fibre reinforced polycarbonate the mode of failure is primarily fibre-matrix debonding [7,8,13] followed by fibre cracking and pull-out [13]. However, the failure mode is influenced by the fibre type and any treatment (e.g. oxidation) [4,5,6]. These features are reflected in the shear strength values.

Fracture Mechanical Properties: Shear strength 18 MPa (single filament, PAN based graphite fibre) [8]; 62.2–90.6 MPa (single filament, high strength carbon fibre) [7]; 39.9 MPa (20% vol. unidirectional graphite fibre) [9]. Engineering shear strength 79.5–101.1 MPa (53% vol. graphite) [12]. Shear strength 60.7 MPa (57–62% vol. unidirectional PAN based carbon fibre, no oxidation treatment, ASTM D2344) [5]; 62.6 MPa (57–62%

vol. unidirectional PAN based carbon fibre, oxidation treatment, ASTM D2344) [5]; 66.3 MPa (57–62% vol. unidirectional PAN based carbon fibre, ozone treated, ASTM D2344) [5]. Carbon fibre reinforced polycarbonate displays brittle fracture [9,11]. Fracture energy (longitudinal) 60 kJ m^{-2} (50% vol. carbon fibre, unidirectional) [11]. Fracture energy (transverse) 3 kJ m^{-2} (50% vol. carbon fibre, unidirectional) [11], 2.9–3.9 kJ m^{-2} (34–40% wt. PAN based carbon fibre) [14]. Under fatigue testing, cohesive failure of the polycarbonate matrix occurs as well as polymer-fibre debonding [7]. Fatigue strength 35 MPa (43000 cycles, 37°, 1 Hz) [7]

Friction Abrasion and Resistance: Carbon fibre at levels of about 15% reduces the friction coefficient of unreinforced polycarbonate by a factor of 2.1 [15]

Izod Notch:

No.	Value	Notch	Note
1	85–96 J/m	Y	1.6-1.8 ft lb in^{-1}, ¼" thick, 10% carbon fibre [1]
2	107 J/m	Y	2 ft lb in^{-1}, 10% carbon fibre [1]
3	96.2 J/m	Y	20% carbon fibre [1]
4	91 J/m	Y	1.7 ft lb in^{-1}, ⅛" thick, 20% carbon fibre [1]
5	96 J/m	Y	1.8 ft lb in^{-1}, ⅛" thick, 30% carbon fibre [1]
6	101–107 J/m	Y	1.9-2 ft lb in^{-1}, 30% carbon fibre [1]
7	96 J/m	Y	1.8 ft lb in^{-1}, 30% PAN based carbon fibre [1]
8	53 J/m	Y	1 ft lb in^{-1}, 30% pitch carbon fibre [1]
9	101.6 J/m	Y	40% carbon fibre [1]
10	80 J/m	Y	1.5 ft lb in^{-1}, ⅛" thick, 40% carbon fibre [1]

Electrical Properties:

Electrical Properties General: The presence of carbon fibres gives composites with greatly reduced electrical resistivity, changing by 5 orders of magnitude between 3 and 20% volume fraction. Volume resistivity 1000–20000 Ω cm (3% vol. pitch carbon fibre) [2]; 8–30 Ω cm (5% vol. pitch carbon fibre) [2]; 0.5–0.8 Ω cm (10% vol. pitch carbon fibre) [2]; 0.07 Ω cm (20% vol. pitch carbon fibre) [2]; 0.04 Ω cm (30% vol. pitch carbon fibre) [2]

Polymer Stability:

Thermal Stability General: Heat resistance decreases as the carbon fibre content increases [15]

Decomposition Details: The temp. at which autocatalytic thermal oxidation of polycarbonate occurs is lower in carbon fibre reinforced polycarbonate than in unreinforced [15]

Flammability: Flammability rating V0 (10% carbon fibre, UL94) [1]; V1 (30% PAN based carbon fibre, UL94) [1]; V1 (30% pitch carbon fibre, UL94) [1]

Applications/Commercial Products:

Mould Shrinkage (%):

No.	Value
1	0.1–0.25%

Trade name	Details	Supplier
EMI-X	DC-1008 (40%)	LNP Engineering
Karlex BKC	13000 series (20%, 30%, 40%)	Ferro Corporation
PC-1100	H30 (30%)	Bay Resins
PC-CF	flame retardant and nickel coated available	Compounding Technology
Stat-Kon D	carbon powder	LNP Engineering
Stat-Kon DC	1006 (30%), 1002 FR (10%, flame retardant), 1003 FR (15%, flame retardant)	LNP Engineering
Thermocomp DC	1006 (30%)	LNP Engineering

Bibliographic References

[1] Ash, M. and Ash, I., Handbook of Plastic Compounds, *Elastomers and Resins*, (eds. M.B. Ash and I.A. Ash), Wiley-VCH, 1992, 325, (props)
[2] Demain, A. and Issi, J.-P., *J. Compos. Mater.*, 1993, **27**, 668, (thermal conductivity, electrical resistivity)
[3] Yurkevich, O.R., *Polym. Eng. Sci.*, 1996, **36**, 1087, (T_g)
[4] Bascom, W.D., Yon, K.-J., Jensen, R.M. and Cordner, L., *J. Adhes.*, 1991, **34**, 79, (adhesion)
[5] Krekel, G., Zielke, U.J., Hüttinger, K.J. and Hoffman, W.P., *J. Mater. Sci.*, 1994, **29**, 2968, 3461, 3984, (adhesion, tensile and flexural props)
[6] Bascom, W.D. and Chen, W.-J., *J. Adhes.*, 1991, **34**, 99, (adhesion)
[7] Latour, R.A., Black, J. and Miller, B., *J. Compos. Mater.*, 1992, **26**, 256, (adhesion, fatigue)
[8] Schadler, L.S., Laird, C., Melanitis, N., Galiotis, C. and Figueroa, J.C., *J. Mater. Sci.*, 1992, **27**, 1663, (adhesion)
[9] Blumentritt, B.F., Vu, B.T. and Cooper, S.L., *Composites*, 1975, **6**, 105, (tensile props, fracture, adhesion)
[10] Blumentritt, B.F., Vu, B.T. and Cooper, S.L., *Polym. Eng. Sci.*, 1974, **14**, 633, (tensile props)
[11] Stori, A.A. and Magnus, E., *Proc. 2nd Int. Conf. on Composite Structures*, (ed. I.H. Marshall), Applied Science, 1983, 332, (flexural moduli, fracture)
[12] Weinberg, M., *Composites*, 1987, **18**, 386, (shear modulus, shear strength)
[13] Krey, J., Friedrich, K. and Schwalbe, K.-H., *J. Mater. Sci. Lett.*, 1987, **6**, 851, (failure props)
[14] Brady, R.L., Porter, R.S. and Donovan, J.A., Interfaces Polym., Ceram., Met. Matrix Compos., *Proc. Int. Conf. Compos. Interfaces*, 2nd, 1988, 463, (fracture)
[15] Burya, A.I., Levi, A.G. and Skylar, T.A., *Plast. Massy*, 1989, 54, (friction coefficient, thermal oxidation)

Bisphenol A Polycarbonate, Glass fibre reinforced

Synonyms: *Poly[2,2-propanebis(4-phenyl)carbonate], Glass fibre reinforced*
Related Polymers: Bisphenol A polycarbonate, Polycarbonate foam, Glass fibre reinforced, Polycarbonate, Glass fibre and carbon fibre reinforced, Polycarbonate, Glass fibre reinforced, PTFE lubricated
Monomers: Bisphenol A, Carbonic dichloride
Material class: Thermoplastic
Polymer Type: Polycarbonates (miscellaneous), Bisphenol A polycarbonate
CAS Number:

CAS Reg. No.	Note
24936-68-3	
25037-45-0	source: carbonic acid

Molecular Formula: $(C_{16}H_{14}O_3)_n$
Fragments: $C_{16}H_{14}O_3$
Additives: 10–50% Glass fibre. May include easy mould release agents. Adhesion promoters such as aminosilanes or epoxysilanes are added. Alkali metal salts are added for flame retardancy

– Bisphenol A Polycarbonate, Glass fibre reinforced

General Information: Uv stabilised, flame retardant and extrusion versions available

Volumetric & Calorimetric Properties:
Density:

No.	Value	Note
1	d 1.26 g/cm^3	10 wt.% glass fibre [1,2]
2	d 1.27 g/cm^3	10 wt.% glass fibre or bead [2]
3	d 1.34–1.35 g/cm^3	20% glass fibre [2]
4	d 1.43–1.44 g/cm^3	30% glass fibre [2]

Thermal Expansion Coefficient:

No.	Value	Note	Type
1	3.5×10^{-5} K^{-1}	10 wt.% glass fibre, estimated [3]	L
2	3.2×10^{-5} K^{-1}	10 wt.% glass fibre [1]	L
3	2.5×10^{-5} K^{-1}	20 wt.% glass fibre, estimated [3]	L
4	1.9×10^{-5} K^{-1}	30 wt.% glass fibre, estimated [3]	L

Thermal Conductivity:

No.	Value	Note
1	0.66 W/mK	4.6 Btu h^{-1} ft^{-2} °F^{-1}, 10% glass fibre [1]

Melting Temperature:

No.	Value	Note
1	271–316°C	10% glass fibre [2]
2	274–316°C	20% glass fibre [2]
3	293–316°C	30% glass fibre [2]
4	293–316°C	40% glass fibre [2]

Deflection Temperature:

No.	Value	Note
1	138–141°C	1.82 MPa, 10% glass fibre [2]
2	146–149°C	1.8 MPa, 20% glass fibre [2]
3	143–149°C	1.8 MPa, 30% glass fibre [2]
4	146–152°C	1.8 MPa, 40% glass fibre [2]
5	149–154°C	1.8 MPa, 50% glass fibre [2]
6	147–150°C	1.81 MPa, 10% glass staple fibre [2]
7	139–147°C	1.8 MPa, 10% glass fibre [2]
8	129°C	265°F, 1.82 Mpa, 10% glass bead [2]
9	135–147°C	1.81 MPa, 20% glass staple fibre [2]
10	140–147°C	1.8 MPa, 20% glass fibre [2]
11	139°C	1.81 MPa, 30% glass staple fibre [2]
12	142–150°C	1.8 MPa, 30% glass fibre [2]
13	141°C	285°F, 264 psi, 10% glass fibre [1]
14	146°C	1.8 MPa, 30% glass fibre [1]
15	149–154°C	300–310°F, 264 psi, 30–50% long fibre glass [1]

Vicat Softening Point:

No.	Value	Note
1	146–166°C	20% glass fibre [2]
2	151–166°C	30% glass fibre [2]
3	150°C	35% glass fibre [2]
4	166°C	40% glass fibre [2]
5	150–153°C	10% glass staple fibre, B [2]
6	150°C	20-30% glass staple fibre, B [2]

Surface Properties & Solubility:

Surface and Interfacial Properties General: Examination by SEM of glass fibre containing composites shows poor adhesion between fibres and matrix [4,5]. This is improved by silane coupling agents (e.g. aminopropyltriethoxysilane, glycidoxypropyltriethoxysilane, methacryloxypropyltriethoxysilane, aminophenyltriethoxysilane, because of better wetting (but these are susceptible to hydrol.) [4]. Annealing of bare glass fibre composites at 275° for 1h gives a high quality interface. It is thought that sub-microscopic ordered domains form in the PC matrix melt above 265°, and nucleate on the glass surface. SEM shows extensive plastic deformation near the fibre surface [4]. In addition the surface is significantly rougher for glass fibre composites compared with the pure polycarbonate [7]

Transport Properties:
Melt Flow Index:

No.	Value	Note
1	5 g/10 min	10% glass fibre [2]
2	5.4 g/10 min	300°, 20 wt.% short glass fibre, 1.2 kg load, ASTM D1238, estimated [6]
3	5–9 g/10 min	20% glass fibre [2]
4	5–10 g/10 min	30% glass fibre [2]
5	5 g/10 min	40% glass fibre [2]

Polymer Melts: Melt flow rate as a function of MW is found to deviate from the theoretical power law expectations, and the deviation increases with increasing glass fibre content [5]. Melt viscosity as shear rate approaches zero 1200 Pa s (290°, 10 wt.% glass fibre, estimated) [7], 3800 Pa s (290°, 20 wt.% glass fibre, estimated) [7], 1600 Pa s (290°, 30 wt.% glass fibre, estimated) [7], 2000 Pa s (290°, 40 wt.% glass fibre, estimated) [7]

Water Absorption:

No.	Value	Note
1	0.12–0.14 %	10% glass fibre [2]
2	0.15 %	10% glass fibre [1,2]
3	0.09–0.16 %	20% glass fibre [2]
4	0.07–0.14 %	30% glass fibre [2]
5	0.08 %	40% glass fibre, ASTM D570 [16]
6	0.06–0.12 %	40% glass fibre [2]
7	0.07 %	50% glass fibre [2]
8	0.14 %	10% glass bead [2]
9	0.13 %	20% glass fibre [2]
10	0.11 %	30% glass fibre [2]

– Bisphenol A Polycarbonate, Glass fibre reinforced

Mechanical Properties:

Mechanical Properties General: The effect of glass fibre reinforcement (relative to the parent polycarbonate matrix) on mech. props. is to give increased rigidity, greater hardness, greater tensile and flexural strengths, less creep tendency, less fatigue, less shrinkage, less thermal expansion and a greater heat deflection temp. On the other hand, glass fibre reinforced polycarbonate is less tough, has some reduction in impact strength, does not extend by neck formation and displays brittle failure. Many of these property values are largely unchanged if the temp. is lowered to -40°. Examples of tensile stress-strain curves for different glass fibre contents have been reported [17,18]. Both the fibre length and orientation can make significant differences to the mech. props. For example both longer and undirectional glass fibres increase the tensile strength and modulus. It should be noted that for some commercial data, the orientation information is not available (or not known). Some improvement in general props. can be achieved by annealing at 275° for 1h

Tensile (Young's) Modulus:

No.	Value	Note
1	2700 MPa	10% wt. glass fibre, estimated [3]
2	4200 MPa	20% wt. glass fibre, estimated [3]
3	3240 MPa	470 kpsi, =20 wt.% glass fibre, random fibre orientation [18]
4	5410 MPa	785 kpsi, =20 wt.% glass fibre, undirectional fibre orientation [17]
5	6500 MPa	30 wt.% glass fibre, estimated [3]
6	3940 MPa	572 kpsi, =35 wt.% glass fibre, random fibre orientation [18]
7	7540 MPa	1093 kpsi, =35 wt.% glass fibre, undirectional fibre orientation [17]
8	11700 MPa	1700 kpsi, 40% glass fibre, ASTM D638 [16]
9	8800 MPa	40 wt.% glass fibre [3]
10	6740 MPa	977 kpsi, =50 wt.% glass fibre, random fibre orientation [18]
11	11770 MPa	1707 kpsi, =50 wt.% glass fibre, undirectional fibre orientation [17]
12	10300–15200 MPa	1500–2200 kpsi, 30–50 wt.% long glass fibre [1]

Flexural Modulus:

No.	Value	Note
1	4100 MPa	600 kpsi, 10% glass fibre [1]
2	3170–3310 MPa	20 wt.% short glass fibre [6]
3	7200 MPa	30% glass fibre [1]
4	8030 MPa	30 wt.% glass fibre [8]
5	8300 MPa	1200 kpsi, 40 wt.% glass fibre, ASTM D790 [16]
6	3800–12200 MPa	=40 wt.% glass fibre [10]
7	18300–19300 MPa	=50 wt.% glass fibre [10]
8	26000 MPa	=50 wt.% glass fibre [9]
9	4900 MPa	=50 wt.% glass fibre [9]
10	9000–14500 MPa	1300–2000 kpsi, 30–50 wt.% long glass fibre [1]

Tensile Strength Break:

No.	Value	Note	Elongation
1	48–63 MPa	7000-9100 psi, 10% glass fibre [2]	4–8%
2	50.64–58.44 MPa	20 wt.% short glass fibre [6]	
3	100–110 MPa	20% glass fibre [2]	3–6%
4	48.3 MPa	7 kpsi, 10 vol.% (=20 wt.%) glass fibre, random fibre orientation [18]	
5	88.6 MPa	10 vol.% (=20 wt.%) glass fibre, unidirectional fibre orientation [17]	
6	130–141 MPa	30% glass fibre [2]	3–5%
7	135 MPa	35% glass fibre [2]	3–8%
8	61.4 MPa	8.9 kpsi, 35 wt.% glass fibre, random fibre orientation [18]	
9	114 MPa	16.54 kpsi, 35 wt.% glass fibre, unidirectional fibre orientation [17]	
10	160 MPa	40% glass fibre [2]	3–5%
11	124 MPa	40% glass fibre, ASTM D638 [16]	
12	93.1 MPa	13.5 kpsi, =50 wt.% glass fibre, random fibre orientation [18]	
13	160 MPa	23.17 kpsi, =50 wt.% glass fibre, unidirectional fibre orientation [17]	
14	159.7 MPa	23.1 kpsi, 40% long glass fibre [2]	1.7%
15	173–176 MPa	25.1–25.5 kpsi, 50% long glass fibre [2]	1.5–4%
16	159–173 MPa	23.0–25.1 kpsi, 30-50% long glass fibre [1]	
17	70 MPa	10% glass staple fibre [2]	7–8%
18	70–80 MPa	10% glass staple fibre [2]	4–6%
19	55–100 MPa	20% glass staple fibre [2]	3.8–7%
20	90–100 MPa	20% glass staple fibre [2]	3–5%
21	70 MPa	30% glass staple fibre [2]	3.5%
22	100–130 MPa	30% glass staple fibre [2]	3–4%

Flexural Strength at Break:

No.	Value	Note
1	110 MPa	16 kpsi, 10% glass fibre [2]
2	128–165 MPa	24 kpsi, 20% glass fibre, upper limit [2]
3	141.4–193 MPa	28 kpsi, 30% glass fibre, upper limit [2]
4	166 MPa	30% glass fibre [1]
5	151 MPa	=30 wt.% glass fibre [8]
6	241 MPa	35 kpsi, 30% long glass fibre [2]
7	160 MPa	35% glass fibre [2]
8	179–221 MPa	26 kpsi, 40% glass fibre, lower limit [2]

– Bisphenol A Polycarbonate, Glass fibre reinforced

9	186 MPa	27 kpsi, 40% glass fibre, ASTM D790 [16]
10	251 MPa	36.4 kpsi, 40% long glass fibre [2]
11	248–346 MPa	40 wt.% long glass fibre [10]
12	281–290 MPa	40.8-42 kpsi, 50% long glass fibre [2]
13	286–362 MPa	50 wt.% long glass fibre [10]
14	241–281 MPa	35.0-40.8 kpsi, 50% long glass fibre [1]
15	130 MPa	10% glass staple fibre [2]
16	110–120 MPa	10% glass staple fibre [2]
17	52 MPa	7500 psi, 10% glass bead [2]
18	120–160 MPa	20% glass staple fibre [2]
19	130–150 MPa	20% glass staple fibre [2]
20	130 MPa	30% glass staple fibre [2]
21	140–190 MPa	30% glass staple fibre [2]

Tensile Strength Yield:

No.	Value	Elongation	Note
1	66 MPa		10% glass fibre [2]
2	59.7–64.64 MPa		20 wt.% short glass fibre [6]
3	107 MPa		20% glass fibre [2]
4	120.7 MPa		30% glass fibre [2]
5	159 MPa		40% glass fibre [2]
6	58 MPa	1-4% elongation	8400 psi, 10% glass bead [2]

Flexural Strength Yield:

No.	Value	Note
1	169.8–172.8 MPa	20% short glass fibre [6]

Compressive Strength:

No.	Value	Note
1	96.5 MPa	14 kpsi, 10% glass fibre [1]
2	128 MPa	18.5 kpsi, 40% glass fibre, ASTM D690 [16]
3	205–225 MPa	29.8-32.7 kpsi, 30–50% long glass fibre [1]

Miscellaneous Moduli:

No.	Value	Note
1	3500 MPa	37 wt.% short glass fibre, highly oriented [12]
2	6700 MPa	37 wt.% highly oriented short glass fibre [12]

Complex Moduli: The storage modulus for glass-fibre reinforced polycarbonate depends on the angle between the measurement direction and that of the fibres. It is also expected to depend on temp. and frequency

Hardness: Rockwell M88 (10% glass fibre) [2], R118 (10% glass fibre) [2], M92 (20% glass fibre) [2]; R118–R122 (20% glass fibre), M85–M90 (10% glass fibre), M90–M95 (20–30% glass fibre) [2], M93–M95 (30% glass fibre) [2], R119–R120 (30% glass fibre) [2], M95–M97 (40% glass fibre) [2], R119–R120 (40% glass fibre) [2], M97 (50% long glass fibres) [2]. Ball indentation hardness 150 MPa (20% glass fibre) [2], 155 MPa (35% glass fibre), 125 MPa, (10% glass staple fibre), 140 MPa (20% glass staple fibre), 145 MPa (30% glass staple fibre) [2]

Failure Properties General: Glass fibre reinforced polycarbonate displays brittle failure (compare Bisphenol A polycarbonate) [11,19]. However, the composites show a decreased rate of fatigue crack propagation relative to the pure matrix, because of crack stopping by fibre bundles (if present) [14,15]. The failure mechanism involves matrix cracking, followed by fibre-matrix debonding, then fibre cracking in the bundle and finally crack propagation in the pure matrix [14,15,19]

Fracture Mechanical Properties: Fracture toughness K_c 4.18–4.28 MPa m$^{1/2}$ (parallel to direction of melt flow, 20% short glass fibre) [6]; 4.74–5.06 MPa m$^{1/2}$ (transverse direction, 20% short glass fibre) [6]. The crack stopping props. of the polycarbonate-glass fibre composite lead to an increase in fatigue lifetime by several orders of magnitude relative to the pure matrix [14,15]. Fatigue strength 32.4 MPa (20°, flexure, =30 wt.% glass fibre, 10^7 cycles, 30 Hz, JIS K7119) [8]; 30.4 MPa (50°, flexure, 17 vol.% (=30 wt.%) glass fibre, 10^7 cycles, JIS K7119, 30 Hz) [8]; 75.7 MPa (7.7 kg mm^{-2}, tension, 23°, 30% glass fibre, 10^4 cycles, 10 Hz) [11]; 67 MPa (20°, estimated, flexure, 17 vol.% (=30 wt.%) glass fibre, 10^4 cycles, 30 Hz, JIS K7119) [8]

Izod Notch:

No.	Value	Notch	Note
1	105 J/m	Y	10% glass staple fibre [2]
2	43–64 J/m	Y	0.8-1.2 ft lb in^{-1}, ¼ in., 10% glass bead [2]
3	80–120 J/m	Y	30% glass staple fibre [2]
4	80 J/m	Y	30% glass staple fibre [2]
5	106–139 J/m	Y	2-6 ft lb in^{-1}, 10% glass fibre [2]
6	117.4 J/m	Y	10% glass fibre [1]
7	75–123 J/m	Y	1.4-2.3 ft lb in^{-1}, 20% glass fibre [2]
8	91–133 J/m	Y	1.7-2.5 ft lb in^{-1}, 30% glass fibre [2]
9	251 J/m	Y	30% long glass fibre [2]
10	117–160 J/m	Y	40% glass fibre [2]
11	267 J/m	Y	5 ft lb in^{-1}, 40% long glass fibre [2]
12	352 J/m	Y	6.6 ft lb in^{-1}, 50% long glass fibre [2]

Izod Area:

No.	Value	Notch	Note
1	12 kJ/m^2	Notched	min., 10% glass staple fibre [2]
2	14 kJ/m^2	Notched	min., 20% glass staple fibre [2]
3	17 kJ/m^2	Notched	min., 30% glass staple fibre [2]

Charpy:

No.	Value	Notch	Note
1	5.8	N	tip radius 250 μm, 20% short glass fibre, estimated [6]
2	6	N	35% glass fibre [13]

– Bisphenol A Polycarbonate, Glass fibre reinforced

Electrical Properties:

Dielectric Permittivity (constant):

No.	Value	Frequency	Note
1	3.17	60 Hz	20% glass fibre [2]
2	3.35	60 Hz	30% glass fibre [2]
3	3.32	1 kHz	30% glass fibre [2]
4	3.53	60 Hz	40% glass fibre [2]
5	3.8	60 Hz	40% glass fibre, ASTM D150 [16]
6	3.58	1 MHz	40% glass fibre, ASTM D150 [16]
7	3.1	60 Hz	10% glass fibre [2]
8	3.25	60 Hz	20% glass fibre [2]
9	3.4	60 Hz	30% glass fibre [2]

Dielectric Strength:

No.	Value	Note
1	17.3–19.6 kV.mm^{-1}	upper limit 490 V mil^{-1}, 10% glass fibre [2]
2	18.8–19.6 kV.mm^{-1}	upper limit 490 V mil^{-1}, 20% glass fibre [2]
3	16 kV.mm^{-1}	1/8" thick, 30% glass fibre [2]
4	18.7–19.2 kV.mm^{-1}	upper limit 480 V mil^{-1}, 30% glass fibre [2]
5	17.7–18.9 kV.mm^{-1}	40% glass fibre [2]
6	18.7 kV.mm^{-1}	475 V mil^{-1}, 40% glass fibre, ASTM D149 [16]
7	30 kV.mm^{-1}	min., 10–30% glass staple fibre [2]
8	28–30 kV.mm^{-1}	10–30% glass staple fibre [2]

Dissipation (Power) Factor:

No.	Value	Frequency	Note
1	0.006	60 Hz	40% glass fibre, ASTM D150 [16]
2	0.007	1 MHz	40% glass fibre, ASTM D150 [16]

Optical Properties:
Transmission and Spectra: Ir spectral data have been reported [6]
Volume Properties/Surface Properties: Not transparent [11]

Polymer Stability:
Thermal Stability General: The thermal stability of glass-reinforced polycarbonate is similar to that of bisphenol A polycarbonate. At 650° the glass fibres devitrify and become brittle [6]
Flammability: Flammability rating V1 [1,2]; V0 (3.2 mm) [2] (10% glass fibre); V0 (10% glass fibre, flame retardant) [2]; V1 (1.47 mm, 20% glass fibre) [2]; V0 (1.6–3.2 mm, 20% glass fibre) [2]; V1 (1.47 mm, 30% glass fibre) [2]; V0 (3.2 mm, 30% glass fibre) [2]; V0 (1.57 mm, 30% glass fibre, flame retardant) [2]; V1 (1.47 mm, 30% glass fibre) [2]; V0 (3.05–3.2 mm, 40% glass fibre) [2] V2 (1.47 mm, 10% glass fibre); V0 (10% glass staple fibre, with flame retardant); V2 (1.47–1.6 mm, 20% glass staple fibre); V1 (20% glass staple fibre); V2 (1.6 mm, 30% glass staple fibre); V2 (1.47 mm, 30% glass staple fibre) [2]
Hydrolytic Stability: Exposure to water at 45° for up to 3 months gives no statistically significant changes in flexure props. [10]. Glass-reinforced polycarbonate is fairly resistant to stress relaxation; at deflections up to 50° and submersion in water at 37° for 60 days, 90% of the original bending moment is retained [10]. Immersion in water at 85° for 100h increases the rate of hydrol. for glass fibre reinforced polycarbonate relative to the parent matrix [5,16]. Hydrol. results in a reduction in MW, but the tensile strength and longitudinal flexural strength and modulus are not significantly affected [5,9]. The transverse flexural modulus and strength, however, show significant deterioration, thought to be caused by water absorption and then degradation at the matrix-fibre interface [9]. In the presence of silane coupling agents to aid adhesion, significant deterioration of the interface occurs after exposure to water at 85° [4]. Resistance to water is improved by annealing at 275° for 1h, which decreases the permeability of water at the interface [9]
Recyclability: Reprocessing by grinding and re-moulding results in severe physical degradation of the glass fibres, resulting in a decrease in relative mean fibre length, increase in melt flow index [6] and decrease in Charpy impact strength for 20% short glass fibre filled polymer [6]. The tensile strength at yield and break are not significantly changed on recycling and the ir spectrum is unchanged [9]. Changes in props. following recycling are caused by fibre shortening and some reduction in the polymer chain length [9]
Stability Miscellaneous: Annealing the polycarbonate-glass fibre composite for 1h at 275° improves the adhesion between the two components, resulting in an increased flexural modulus and strength, and improved water resistance [9]. Exposure to higher shear rates, such as during processing (extrusion, injection), results in degradation of the glass fibres (reduced length) [6,7]

Applications/Commercial Products:
Processing & Manufacturing Routes: Glass fibres are usually chopped strands or short fibres (typically 6 mm long, 8–12 μm diameter). Incorporation reduces the mean fibre length to 0.1–0.3 mm. The fibre is E-glass-alkali free aluminium borosilicates containing <0.8% alkali metal oxides

Mould Shrinkage (%):

No.	Value	Note
0	0.05–0.5%	depending on glass fibre content
2	0.1–0.7%	

Applications: Housings of electrical components, transportation, electronics, sporting gear, connectors, switches, structural applications, office machine parts

Trade name	Details	Supplier
Calibre	510 (flame retardant)	Dow
Celstran	SF7070 (flame retardant)	Polymer Composites
Karlex NA	20%, 30%, 40%	Ferro Corporation
Karlex NAFR	12018NAFR. 10%, flame retardant	Ferro Corporation
Korlex	12003NA (10%)	Ferro Corporation
Lexan	20%, 30%, 40%	General Electric Plastics
Makrolan	8000 series; 20%, 35%	Bayer Inc.
Novarex	7025G (10–40% glass fibre reinforced, flame retardant)	Mitsubishi Chemical
PC	10GF/000FR (10%, flame retardant)	Compounding Technology
PC	10GB/000 10%	Compounding Technology
PC-1100G		Bay Resins

PC-1700	G10FR (10% flame retardant)	Bay Resins
RTP	301 (10%), 303 (20%), 305 (30%), 307 (40%)	RTP Company
Stat-Kon DF-FR	flame retarded	LNP Engineering
Thermo-Comp DF	10%, 30%, 40%	LNP Engineering
Verton	DF-700-10 (50% long fibre)	LNP Engineering
Xantar	19S(R), 19V (uv stabilised)	Xantar Polycarbonates
	PCG (30–50% glass fibre)	Polymer Composites
	-40GM/000 (40%, milled glass)	Compounding Technology

Bibliographic References

[1] Murphy, J., *Handbook of Reinforced Plastics*, Elsevier, 1994, (props)
[2] Ash, M. and Ash, I., Handbook of Plastic Compounds, *Elastomers and Resins*, (eds. M.B. Ash and I.A. Ash), Wiley-VCH, 1992, 218, 325, (commercial sources, props)
[3] Kircher, K., *Kyoka Purasuchikkusu*, 1981, **25**, 382, (props)
[4] Jancar, J. and Di Benedetto, A.T., *J. Mater. Sci.: Mater. Med.*, 1993, **4**, 555, (adhesion, hydrol)
[5] Kelleher, P.G., Wentz, R.P., Hellman, M.Y. and Gilbert, E.H., *Polym. Eng. Sci.*, 1983, **23**, 537, (adhesion, hydrolysis)
[6] Chrysostomou, A. and Hashemi, S., *J. Mater. Sci.*, 1996, **31**, 1183, (melt flow index, flexural props, impact strength, recycling, ir)
[7] Knutsson, B.A. and White, J.L., *J. Appl. Polym. Sci.*, 1981, **26**, 2347, (polymer melt, surface morphology)
[8] Furue, H. and Shimamura, S., *Int. J. Fract.*, 1980, **16**, 553, (flexural props, fatigue)
[9] Jancar, J., Di Benedetto, A.T. and Goldberg, A.J., *J. Mater. Sci.: Mater. Med.*, 1993, **4**, 562, (flexural props, hydrol)
[10] Goldberg, A.J., Burstone, C.J., Hadjinikolaou, I. and Jancar, J., *J. Biomed. Mater. Res.*, 1994, **28**, 167, (flexural props, hydrol)
[11] Misaki, T. and Kishi, J., *J. Appl. Polym. Sci.*, 1978, **22**, 2063, (tensile strength, fatigue)
[12] Haidar, B., Interfacial Phenom. Compos. Mater. '91, Proc. Int. Conf., 2nd, 1991, 145, (complex modulus)
[13] Kircher, K., *Kyoka Purasuchikkusu*, 1981, **25**, 420, (props)
[14] Krey, J., Friedrich, K. and Schwalbe, K.-H., ICCM & ECCM, *Sixth Int. Conf. Compos. Mater. Second Eur. Conf. Compos. Mater.*, 1987, **3**, 3439, (fracture, fatigue)
[15] Krey, J., Friedrich, K. and Schwalbe, K.-H., *J. Mater. Sci. Lett.*, 1987, **6**, 851, (fracture)
[16] Mohr, J.G., Oleesky, S.S., Shook, G.D. and Meyer, L.S., *SPI Handbook of Technology and Engineering of Reinforced Plastics/Composites*, 2nd edn., (ed. J.G. Mohr), Van Nostrand Reinhold, 1973, 327, (props)
[17] Blumentritt, B.F., Vu, B.T. and Cooper, S.L., *Polym. Eng. Sci.*, 1974, **14**, 633, (tensile props)
[18] Blumentritt, B.F., Vu, B.T. and Cooper, S.L., *Polym. Eng. Sci.*, 1975, **15**, 428, (tensile props)
[19] Blumentritt, B.F., Vu, B.T. and Cooper, S.L., *Composites*, 1975, **6**, 105, (tensile props, fracture)

Bisphenol A polycarbonate, Glass filled   B-8

Synonyms: *Poly[2,2-propanebis(4-phenyl)carbonate], Glass filled*
Related Polymers: Poly[2,2-propanebis(4-phenyl)carbonate]
Monomers: Bisphenol A, Carbonic dichloride
Material class: Thermoplastic
Polymer Type: Polycarbonates (miscellaneous)
CAS Number:

CAS Reg. No.
24936-68-3
25037-45-0

Molecular Formula: $(C_{16}H_{14}O_3)_n$
Fragments: $C_{16}H_{14}O_3$
Additives: 10–35% Glass filled. May include easy mould release agents
General Information: UV stabilised and flame retardant versions available. Extrusion version available

Applications/Commercial Products:
Mould Shrinkage (%):

No.	Value
1	0.1–0.7%

Trade name	Details	Supplier
Makrolan	8000 series	Bayer Inc.
PC-10GB/000	10%	Compounding Technology
Xantar	19S(R), 19V (uv stabilised)	Xantar Polycarbonates

Bisphenol A polycarbonate, Metal filled   B-9

Synonyms: *Poly[2,2-propanebis(4-phenyl)carbonate], Metal filled*
Related Polymers: Poly[2,2-propanebis(4-phenyl)carbonate]
Monomers: Bisphenol A, Carbonic dichloride
Material class: Thermoplastic
Polymer Type: Bisphenol A polycarbonate
CAS Number:

CAS Reg. No.	Note
24936-68-3	
25037-45-0	source: carbonic acid

Molecular Formula: $(C_{16}H_{14}O_3)_n$
Fragments: $C_{16}H_{14}O_3$
Additives: Nickel, aluminium and stainless steel used as fillers
General Information: Flame retardant versions available for aluminium and stainless steel filled grades

Volumetric & Calorimetric Properties:
Density:

No.	Value	Note
1	d 1.54 g/cm^3	40 % Al [1]
2	d 1.4 g/cm^3	25 % Ni [1]
3	d 1.26–1.29 g/cm^3	Stainless steel [1]

Deflection Temperature:

No.	Value	Note
1	143°C	[1]
2	148°C	[1]
3	141°C	[1]

Transport Properties:
Water Absorption:

No.	Value	Note
1	0.13–0.15 %	Stainless steel [1]

– **Bisphenol A polycarbonate, PTFE lubricated**

Mechanical Properties:
Tensile Strength Break:

No.	Value	Elongation
1	44.1 MPa	2 %
2	103 MPa	3.5 %
3	55–69 MPa	5.0-6 %

Flexural Strength at Break:

No.	Value	Note
1	86.1 MPa	[1]
2	151 MPa	[1]
3	83–96 MPa	[1]

Polymer Stability:
Flammability: V0 (40 % Al) V1 (25 % Ni) HB stainless steel, without flame retardant V0 (Stainless steel, with flame retardant) [1]

Applications/Commercial Products:
Mould Shrinkage (%):

No.	Value
1	0.5%

Trade name	Details	Supplier
EMI	DA-30 (30% Al, flame retardant), DA-40 (40% Al, flame retardant)	LNP Engineering
PDX	83393 (25% Ni)	LNP Engineering
Stat-Kon	DS (stainless steel, flame retardant)	LNP Engineering
	84356 (stainless steel, flame retardant)	LNP Engineering

Bisphenol A polycarbonate, PTFE lubricated B-10
Synonyms: *Poly[2,2-propanebis(4-phenyl)carbonate], PTFE lubricated*
Related Polymers: Poly[2,2-propanebis(4-phenyl)carbonate], Glass fibre reinforced PTFE lubricated Polycarbonate, Carbon fibre reinforced PTFE lubricated Polycarbonate
Monomers: Bisphenol A, Carbonic dichloride, Tetrafluoroethylene
Material class: Thermoplastic
Polymer Type: Bisphenol A polycarbonate
CAS Number:

CAS Reg. No.	Note
24936-68-3	
25037-45-0	source: carbonic acid

Molecular Formula: $(C_{16}H_{14}O_3)_n$
Fragments: $C_{16}H_{14}O_3$
Additives: Lubricated with 5–20% PTFE

Volumetric & Calorimetric Properties:
Density:

No.	Value	Note
1	d 1.23 g/cm^3	5% PTFE [1]
2	d 1.26 g/cm^3	10% PTFE, 15% PTFE [1,2]
3	d 1.29 g/cm^3	15% PTFE [1]
4	d 1.32 g/cm^3	20% PTFE [1]
5	d 1.22 g/cm^3	
6	d 1.28 g/cm^3	

Thermal Expansion Coefficient:

No.	Value	Note	Type
1	7×10^{-5} K^{-1}	10% PTFE, ASTM D696 [2]	L

Thermal Conductivity:

No.	Value	Note
1	0.2 W/mK	10% PTFE, ASTM C177 [2]

Deflection Temperature:

No.	Value	Note
1	130°C	1.81 MPa, 10% PTFE, ASTM D648 [2]
2	135°C	0.45 MPa, 10% PTFE, ASTM D648 [2]
3	127–132°C	260–270°F, 1.8 MPa, lower limit, 5% PTFE [1]
4	127–132°C	260–270°F, 1.8 MPa, 10% PTFE [1]
5	132–135°C	270–275°F, 1.8 MPa, 15% PTFE [1]
6	132–135°C	270–275°F, 1.8 MPa, 20% PTFE [1]

Vicat Softening Point:

No.	Value	Note
1	152–157°C	305–315°F, 5% PTFE [1]

Transport Properties:
Water Absorption:

No.	Value	Note
1	0.2 %	saturation, 10% PTFE, ASTM D570 [2]
2	0.14–0.15 %	5% PTFE [1]
3	0.13–0.14 %	10% PTFE [1,2]
4	0.12–0.13 %	15% PTFE [1]
5	0.11–0.12 %	20% PTFE [1]

– Bisphenol A polycarbonate, PTFE lubricated

Mechanical Properties:
Flexural Modulus:

No.	Value	Note
1	2100 MPa	10% PTFE, ASTM D790 [2]

Tensile Strength Break:

No.	Value	Note	Elongation
1	55 MPa	5% and 10% PTFE [1,2]	75%
2	59–61 MPa	8500–8800 psi, 5% PTFE [1]	

Flexural Strength at Break:

No.	Value	Note
1	85 MPa	5% PTFE [1]
2	65 MPa	9400 psi, 10% PTFE [1]

Compressive Strength:

No.	Value	Note
1	70 MPa	10% PTFE, ASTM D695 [2]

Miscellaneous Moduli:

No.	Value	Note	Type
1	50 MPa	10% PTFE, ASTM D732 [2]	Shear strength

Impact Strength: 139 J m^{-1} (2.6 ft lb in^{-1}, ¼" thick, 5% PTFE) [1]; 139–160 J m^{-1} (2.6 ft lb in^{-1}, lower limit, 5% PTFE) [1]; 133 J m^{-1} (2.5 ft lb in^{-1}, ¼" thick, 10% PTFE) [1]; 133–137 J m^{-1} (2.50–2.56 ft lb in^{-1}, ¼" thick, 15% PTFE) [1]; 133–160 J m^{-1} (2.5–2.6 ft lb in^{-1}, lower limit, 15% PTFE) [1]; 133 J m^{-1} (2.5 ft lb in^{-1}, ¼" thick, 20% PTFE) [1]
Hardness: M78 (5% PTFE) [1]; M74-M75 (15% PTFE) [1]; Rockwell M75 (10% PTFE, ASTM D785) [2]
Friction Abrasion and Resistance: Has excellent wear and frictional characteristics. Even 5% PTFE offers improved surface wear and reduced static and dynamic coefficients of friction [1]. Wear factor 85 (10% PTFE, LNP SOP) [2]. Coefficients of friction 0.14 (static) and 0.2 (dynamic) (10% PTFE, LNP SOP) [2]
Izod Notch:

No.	Value	Notch	Note
1	120 J/m	Y	10% PTFE, ASTM D256 [2]

Electrical Properties:
Surface/Volume Resistance:

No.	Value	Note	Type
2	1×10^{15} Ω.cm	10% PTFE, ASTM D257 [2]	S

Dielectric Permittivity (constant):

No.	Value	Frequency	Note
1	3	60 Hz	10% PTFE, ASTM D150 [2]
2	2.9	1 MHz	10% PTFE, ASTM D150 [2]

Dielectric Strength:

No.	Value	Note
1	15 kV.mm^{-1}	15 MV m^{-1}, 10% PTFE, ASTM D149 [2]
2	15.6 kV.mm^{-1}	390 V mil^{-1}, 10% PTFE [1]
3	15.6–18.1 kV.mm^{-1}	390-460 V mil^{-1}, 15% PTFE [1]

Arc Resistance:

No.	Value	Note
1	120s	10% PTFE, ASTM D495 [2]

Complex Permittivity and Electroactive Polymers: Tracking resistance 275 (10% PTFE, DIN IEC 112) [2]
Dissipation (Power) Factor:

No.	Value	Frequency	Note
1	0.0009	60 Hz	10% PTFE, ASTM D150 [2]
2	0.01	1 MHz	10% PTFE, ASTM D150 [2]

Polymer Stability:
Upper Use Temperature:

No.	Value	Note
1	120°C	continuous service, 10% PTFE, UL 746B [2]

Flammability: Flammability rating HB (1.57 mm, 5% PTFE, UL94) [1]; V1 (15% PTFE, UL94) [1]; V2 (3.2 mm, 10% PTFE, UL94) [2]

Applications/Commercial Products:
Mould Shrinkage (%):

No.	Value	Note
1	0.5–0.7%	
2	0.1–0.3%	3 mm section, 10% PTFE, ASTM D955 [2]

Trade name	Details	Supplier
Lexan	WR 1210	General Electric Plastics
Lubricomp	DL-4030 (15%)	LNP Engineering
Migralube	DL-4030	ICI Advanced Materials
PC-000/T	5%, 10%, 15%, 20%	Compounding Technology
PC-1100	TF15 (15%)	Bay Resins
Plaslube	PC-50/TF/15 (15%)	Akzo
Thermocomp DL	-4010 (5%), -4020 (10%), -4030 (15%), -4040 (20%)	LNP Engineering

Bibliographic References
[1] Ash, M. and Ash, I., Handbook of Plastic Compounds, *Elastomers and Resins*, (eds. M.B. Ash and I.A. Ash), Wiley-VCH, 1992, 325
[2] *LNP Engineering Plastics Product Information*, LNP Engineering Plastics, 1996, (technical datasheet)

Bisphenol A polycarbonate, Structural Foam B-11

Synonyms: *Polycarbonate, Structural foam*
Related Polymers: Bisphenol A polycarbonate, Glass fibre reinforced polycarbonate foam, Poly[2,2-propanebis(4-phenyl)-carbonate]
Monomers: Bisphenol A, Carbonic dichloride
Material class: Thermoplastic
Polymer Type: Bisphenol A polycarbonate
CAS Number:

CAS Reg. No.
24936-68-3

Molecular Formula: $(C_{16}H_{14}O_3)_n$
Fragments: $C_{16}H_{14}O_3$
Molecular Weight: MW 28000-32000
Additives: Flame retardants; blowing agents e.g. 5-phenyltetrazole
General Information: Foams are classified by density; 0.96 g cm^{-3}, 20% volume foam; 0.84 g cm^{-3} 30% volume foam and 0.72 g cm^{-2} 40% volume foam. Foam densities increased with fillers present

Volumetric & Calorimetric Properties:
Density:

No.	Value	Note
1	d 1.14–1.2 g/cm^3	5% vol. foam [2]
2	d 1.09–1.21 g/cm^3	10% vol. foam [4]
3	d 1.08–1.2 g/cm^3	10% vol. foam [4]
4	d 0.97–1.2 g/cm^3	19% vol. foam [2]
5	d 0.98–1.232 g/cm^3	20% vol. foam, 5% glass fibre [3]
6	d 0.9–1.23 g/cm^3	30% vol. foam, 5% glass fibre [4]
7	d 1.13–1.25 g/cm^3	10% vol. foam, 7% glass fibre [4]
8	d 1.12–1.25 g/cm^3	10% vol. foam, 7% glass fibre [4]
9	d 0.95–1.27 g/cm^3	25% vol. foam, 10% glass fibre [4]
10	d 1.19–1.32 g/cm^3	10% vol. foam, 20% glass fibre [4]
11	d 1.1–1.32 g/cm^3	17% vol. foam, 20% glass fibre [4]
12	d 1.29–1.43 g/cm^3	10% vol. foam, 30% glass fibre [4]
13	d 1.23 g/cm^3	[8]
14	d 1.2 g/cm^3	Lexan [8]

Thermal Expansion Coefficient:

No.	Value	Note	Type
1	5.04×10^{-5} K^{-1}	[8]	L
2	6.66×10^{-5} K^{-1}	Lexane [8]	L

Volumetric Properties General: Polycarbonate microcellular foams have average bubble diameters of about 10μm and a bubble volume fraction of 10–40% [1]. The type of foam achieved depends on the foaming temp. and foaming time [1,2]

Deflection Temperature:

No.	Value	Note
1	127°C	1.8 MPa [8]
2	137°C	0.4 Mpa, Lexan [8]
3	138°C	0.45 MPa [8]
4	127°C	1.82 Mpa, 10% vol. foam [4]
5	127°C	1.81 Mpa, 30% vol. foam, 5% glass fibre, DIN 53461 [4,5]
6	137°C	0.45 Mpa, 30% vol. foam, 5% glass fibre, DIN 53461 [5]
7	132–133°C	1.82 Mpa, 10% vol. foam, 7% glass fibre [4]
8	134°C	1.82 Mpa, 25% vol. foam, 10% glass fibre [4]
9	133°C	1.82 Mpa, 10% vol. foam, 20% glass fibre [4]
10	136°C	1.82 Mpa, 17% vol. foam, 20% glass fibre [4]
11	139°C	1.82 Mpa, 10% vol. foam, 30% glass fibre [4]

Vicat Softening Point:

No.	Value	Note
1	140°C	30% vol. foam, 5% glass fibre, DIN 53460 [5]

Transport Properties:
Melt Flow Index:

No.	Value	Note
1	7 g/10 min	[8]
2	5.5 g/10 min	[8]

Water Absorption:

No.	Value	Note
1	0.13–0.2 %	10% vol. foam [4]
2	0.15 %	30% vol. foam, 5% glass fibre [4,8]
3	0.14–0.16 %	10% vol. foam, 7% glass fibre [4]
4	0.15 %	25% vol. foam, 10% glass fibre [4]
5	0.14 %	24h, Lexan [8]
6	0.14 %	10% vol. foam, 20% glass fibre [4]
7	0.15 %	17% vol. foam, 20% glass fibre [4]
8	0.12 %	10% vol. foam, 30% glass fibre [4]

Mechanical Properties:
Mechanical Properties General: The values of the tensile props. are generally reduced by foaming, and are further reduced as the foam content increases. [7] Foamed polycarbonate is less tough (break strain 5–8%) than unfoamed (break strain 80–110%). The relative tensile toughness decreases with foaming time. [2,6] The presence of foam improves impact resistance props. and the failure mode is ductile, unlike unfoamed polycarbonate which displays brittle failure under some conditions. [1] Foamed polycarbonate shows an increased tendency to creep compared with unfoamed, but creep rupture occurs at higher strains. [2] The fatigue lifetime appears to depend on the foam level. [7] Extension to break 5.2–7.8% (10% vol. foam) [4]; 4.8% (10% vol. foam, 7% glass fibre). [4]; 3.6% (10% vol. foam, 20% glass fibre) [4]; 3% (10% vol. foam, 30% glass fibre) [4]

Tensile (Young's) Modulus:

No.	Value	Note
1	1800 MPa	261 kpsi, 20% vol. foam, 5% glass fibre [3]
2	2000 MPa	30% vol. foam, 5% glass fibre, DIN 53457 [5]
3	2069 MPa	21100 kg cm^{-2} [8]
4	2059 MPa	21000 kg cm^{-2} [8]
5	2098 MPa	21400 kg cm^{-2}, Lexan [8]

– Bisphenol A polycarbonate, Structural Foam

Flexural Modulus:

No.	Value	Note
1	2300 MPa	23°, 30% vol. foam, 5% glass fibre, DIN 53457 [5]
2	2000 MPa	60°, 30% vol. foam, 5% glass fibre, DIN 53457 [5]

Tensile Strength Break:

No.	Value	Note	Elongation
1	49 MPa	3% foam [7]	
2	43 MPa	17% foam [7]	
3	36 MPa	33% foam [7]	
4	28 MPa	46% foam [7]	
5	42–43 MPa	30% vol. foam, 5% glass fibre [4]	
6	43 MPa	25% vol. foam, 10% glass fibre [4]	
7	45 MPa	17% vol. foam, 20% glass fibre [4]	
8	38.6 MPa	394 kg cm^{-2} [8]	5%
9	42.1 MPa	429 kg cm^{-2} [8]	
10	54.4 MPa	555 kg cm^{-2}, Lexan [8]	9%
11	50–55 MPa	10% vol. foam [4]	
12	50 MPa	10% vol. foam, 7% glass fibre [4]	
13	60 MPa	10% vol. foam, 20% glass fibre [4]	
14	70 MPa	10% vol. foam, 30% glass fibre [4]	

Flexural Strength at Break:

No.	Value	Note
1	75–85 MPa	10% vol. foam [4]
2	70 MPa	23°, 5% strain, 30% vol. foam, 5% glass fibre, DIN 53452 [5]
3	60 MPa	60°, 5% strain, 30% vol. foam, 5% glass fibre, DIN 53452 [5]
4	90–95 MPa	10% vol. foam, 7% glass fibre [4]
5	83 MPa	5% strain, 25% vol. foam, 10% glass fibre [4]
6	105 MPa	10% vol. foam, 20% glass fibre [4]
7	97 MPa	5% strain, 17% vol. foam, 20% glass fibre [4]

Poisson's Ratio:

No.	Value	Note
1	0.33–0.35	20% vol. foam, 5% glass fibre [3]

Tensile Strength Yield:

No.	Value	Note
1	38–43 MPa	5000-6300 psi, 20% vol. foam, 5% glass fibre [3]
2	50.3 MPa	513 kg cm^{-2}, Lexan [8]

Flexural Strength Yield:

No.	Value	Note
1	74–76 MPa	5% strain, 30% vol. foam, 5% glass fibre [4]
2	74.4 MPa	759 kg cm^{-2} [8]
3	75.8 MPa	773 kg cm^{-2} [8]
4	82.6 MPa	879 kg cm^{-2}, Lexan [8]

Compressive Strength:

No.	Value	Note
1	66.2 MPa	675 kg cm^{-2}, Lexan [8]

Impact Strength: Charpy: 40 kJ m^{-2} (23°, 30% foam, 5% glass fibre, DIN 53453) [5]; 40 kJ m^{-2} (-20°, 30% foam, 5% glass fibre, DIN 53453) [5]
Hardness: Rockwell M85 (30% vol. foam, 5% glass fibre) [4]
Failure Properties General: Unfoamed polycarbonate displays brittle failure in thicker films or sharp-notched specimens. Foamed polycarbonate, however, shows ductile failure for sharp-notched samples, and the fracture energy is about 16 times higher for foamed relative to unfoamed polycarbonate. [1] Blunt-notched samples, which show ductile failure in unfoamed polycarbonate, remain ductile when foamed. [1] The bubbles are thought to act as impact modifiers, and this effect appears to be optimised for 40μm diameter bubbles and 28% vol. fraction [1]
Fracture Mechanical Properties: The fatigue lifetime depends on the stress, and also on the foam level; the fatigue lifetime for 3% vol. foam is significantly higher than that of unfoamed polycarbonate, but higher foam levels show progressively shorter lifetimes. Average fatigue lifetime 28500 cycles (48.9 MPa, 0% vol. foam). [7], 132100 cycles (46.5 MPa, 10 Hz, 3% vol. foam) [7], 30000 cycles (46.5 MPa, 10 Hz, 17% vol. foam) [7], 10000 cycles (46.5 MPa, 10 Hz, 32% vol. foam) [7], 7000 cycles (46.5 MPa, 10 Hz, 48% vol. foam) [7]

Izod Notch:

No.	Value	Notch	Note
1	690–720 J/m	N	unnotched, 10% vol. foam [4]
2	350 J/m	Y	20°, 30% foam, 5% glass fibre, ASTM D256 [5]
3	350 J/m	Y	-20°, 30% foam, 5% glass fibre, ASTM D256 [5]
4	2668 J/m	Y	50 ft lb in^{-1}; notched, room temp., Lexan [8]
5	850 J/m	N	unnotched, 10% vol. foam, 7% glass fibre [4]
6	430 J/m	N	unnotched, 10% vol. foam, 20% glass fibre [4]
7	480 J/m	N	unnotched, 10% vol. foam, 30% glass fibre [4]

Electrical Properties:
Dielectric Permittivity (constant):

No.	Value	Frequency	Note
1	2.22–2.48	100 Hz	10% vol. foam [4]
2	2.59–3	60 Hz	30% vol. foam, 5% glass fibre [4]
3	2.4–2.45	100 Hz	10% vol. foam, 7% glass fibre [4]
4	3.1	60 Hz	25% vol. foam, 10% glass fibre [4]
5	2.52	100 Hz	10% vol. foam, 20% glass fibre [4]

| 6 | 3.1 | 60 Hz | 17% vol. foam, 20% glass fibre [4] |
| 7 | 2.7 | 100 Hz | 10% vol. foam, 30% glass fibre [4] |

Dielectric Strength:

No.	Value	Note
1	8–13 kV.mm^{-1}	10% vol. foam [4]
2	10 kV.mm^{-1}	10% vol. foam, 7% glass fibre [4]
3	12 kV.mm^{-1}	10% vol. foam, 20% glass fibre [4,8]
4	11 kV.mm^{-1}	10% foam, 30% glass fibre [4]
5	8.22 kV.mm^{-1}	[8]
6	15.7 kV.mm^{-1}	[8]

Polymer Stability:
Flammability: Flammability rating V0 5V (3-18 mm 10% vol. foam) [4]; V0/5V (6.10-6.27mm 10% vol. foam) [4]; (V0 6.25 mm 10% vol. foam) [4]; V0 (6.25 mm 30% vol. foam, 5% glass fibre UL94) [5]; 5V/V0 (6.4 mm 30% vol. foam, 5% glass fibre, flame retardant grade) [4]; V0/5V (3.8 mm 10% vol. foam, 7% glass fibre) [4]; V0/5V (6.29 mm 10% vol. foam, 7% glass fibre) [4]; 5V/V0 (4.4 mm 25% vol. foam, 10% glass fibre, flame retardant grade [4]; 5V/V0 (6.29 mm 10% vol. foam, 20% glass fibre) [4]; 5V/V0 (6.4 mm 17% vol. foam, 20% glass fibre, flame retardant grade) [4]; 5V/V0 (6.29 mm 10% vol. foam, 30% glass fibre) [4]

Applications/Commercial Products:
Processing & Manufacturing Routes: Blowing agent concentrates (e.g. 5-phenyltetrazole, phenyldihydrooxadiazinone) are incorporated into the melt and decompose to give gas. Mouldings are prod. by a low pressure process.

Mould Shrinkage (%):

No.	Value	Note
1	0.3–0.8%	
2	0.6%	[8]
3	0.7%	[8]

Applications: Housings for office equipment and lamps, roofs for jeeps, telecommunications, office machines

Trade name	Details	Supplier
Fiberfil F	30% foam level	Akzo Engineering Plastics
Lexan FL400		General Electric Plastics
Makrolan SF		Bayer Inc.
Makrolon SF		Mobay

Bibliographic References
[1] Collias, D.I., Baird, D.G. and Borggreve, R.J.M., *Polymer*, 1994, **35**, 3978, (fracture props)
[2] Wing, G., Pasricha, A., Tuttle, M. and Kumar, V., *Polym. Eng. Sci.*, 1995, **35**, 673, (creep and general props)
[3] Hobbs, S.Y., *J. Appl. Phys.*, 1977, **48**, 4052, (tensile props)
[4] Ash, M. and Ash, I., Handbook of Plastic Compounds, *Elastomers and Resins*, (eds. M.B. Ash and I.A. Ash), Wiley-VCH, 1992, 325, (props)
[5] *Encycl. Polym. Sci. Eng.*, (ed. J.I. Kroschwitz), Vol.11, Wiley-Interscience, 1985, 648, (props)
[6] Collias, D.I. and Baird, D.G., Soc. Plast. Eng., *Tech. Pap.*, 50th, 1992, **38**, 1532, (tensile toughness)
[7] Seeler, K.A. and Kumar, V., Soc. Plast. Eng., *Tech. Pap.*, 50th, 1992, **38**, 1497, (creep)
[8] *Lexan*, G.E. Plastics, 1996, (technical datasheet)

Bisphenol A-terephthalic acid-isophthalic acid polyester carbonate

Synonyms: *2,2-Propanebis[4-phenol]-1,4-benzenedicarboxylic acid-1,3-benzenedicarboxylic acid polyester carbonate. Poly[4,4'-(1-methylethylidene)bis[phenol]-co-1,4-benzenedicarboxylic acid-co-1,3-benzenedicarboxylic acid-co-carbonic acid]. Poly[4,4'-(1-methylethylidene)bis[phenol]-co-1,4-benzenedicarbonyldichloride-co-1,3-benzenedicarbonyldichloride-co-carbonic dichloride]*
Related Polymers: Poly[2,2-propanebis(4-phenyl)carbonate], Polyester carbonates
Monomers: Bisphenol A, 1,3-Benzenedicarboxylic acid, 1,4-Benzenedicarboxylic acid, Carbonic acid
Material class: Thermoplastic, Copolymers
Polymer Type: Polyester, Bisphenol A polycarbonate
CAS Number:

CAS Reg. No.	Note
89001-40-1	source: carbonic dichloride and diacids
31133-80-9	source: carbonic acid and diacids
71519-80-7	source: carbonic dichloride and acid chlorides

Molecular Formula: $[(C_{23}H_{18}O_4).(C_{16}H_{14}O_3)]_n$
Fragments: $C_{23}H_{18}O_4$ $C_{16}H_{14}O_3$
General Information: Contains approx. equal amounts of terephthalic acid and isophthalic acid in random distribution with Bisphenol A carbonate. The polymers have both carbonate and ester linkages, conferring props. intermediate between those of polycarbonate and the polyesters

Volumetric & Calorimetric Properties:
Density:

No.	Value	Note
1	d 1.2 g/cm^3	approx. [2,3]
2	d 1.2 g/cm^3	bisphenol A:terephthalate:isophthalate 1:0.29:0.06 [2]
3	d 1.2 g/cm^3	bisphenol A:terephthalate:isophthalate 1:0.12:0.28 [2]
4	d 1.2 g/cm^3	bisphenol A:terephthalate:isophthalate 1:0.25:0.25 [2]
5	d 1.197 g/cm^3	bisphenol A:terephthalate:isophthalate 1:0.54:0.16 [3]
6	d 1.202 g/cm^3	bisphenol A:terephthalate:isophthalate 1:0.69:0.17 [3]

Thermal Expansion Coefficient:

No.	Value	Note	Type
1	7.2×10^{-5} K^{-1}	DIN 53 752 [4]	1

Glass-Transition Temperature:

No.	Value	Note
1	170°C	bisphenol A:terephthalate:isophthalate 1:0.25:0.25 [2,5,6]
2	184°C	bisphenol A:terephthalate:isophthalate 1:0.55:0.1 [1]

Deflection Temperature:

No.	Value	Note
1	163°C	1.82 Mpa, bisphenol A:terephthalate:isophthalate 1:0.29:0.06
2	152°C	1.82 Mpa, bisphenol A:terephthalate:isophthalate 1:0.12:0.28
3	145°C	1.8 Mpa, bisphenol A:terephthalate:isophthalate 1:0.25:0.25 [2]

Vicat Softening Point:

No.	Value	Note
1	159°C	bisphenol A:terephthalate:isophthalate 1:0.0925:0.0025 [7]
2	170°C	bisphenol A:terephthalate:isophthalate 1:0.25:0.25 [2,4]
3	182°C	bisphenol A:terephthalate:isophthalate 1:0.4:0.4 [4]
4	159°C	bisphenol A:terephthalate:isophthalate 1:0.15:0.15 [4]

Surface Properties & Solubility:
Solubility Properties General: Cohesive energy density 398 J cm^{-3} (bisphenol A: terephthalate:isophthalate 1:0.55:0.1) [1]

Transport Properties:
Permeability of Gases: Diffusivity values and temp. dependence of permeability have been reported [6]

Gas Permeability:

No.	Gas	Value	Note
1	CO_2	595 cm^3 mm/(m^2 day atm)	6.8×10^{-13} cm^2 (s Pa)$^{-1}$, 35°, 4.4 atm, bisphenol A:terephthalate:isophthalate 1:0.25:0.25 [6]
2	O_2	122 cm^3 mm/(m^2 day atm)	1.39×10^{-13} cm^2 (s Pa)$^{-1}$, 35°, 4.4 atm, bisphenol A:terephthalate:isophthalate 1:0.25:0.25 [6]
3	N_2	25.4 cm^3 mm/(m^2 day atm)	2.9×10^{-14} cm^2 (s Pa)$^{-1}$, 35°, 4.4 atm, bisphenol A:terephthalate:isophthalate 1:0.25:0.25 [6]
4	H_2	963 cm^3 mm/(m^2 day atm)	1.1×10^{-12} cm^2 (s Pa)$^{-1}$, 35°, 4.4 atm, bisphenol A:terephthalate:isophthalate 1:0.25:0.25 [6]

Mechanical Properties:
Tensile (Young's) Modulus:

No.	Value	Note
1	2400 MPa	bisphenol A:terephthalate:isophthalate 1:0.15:0.15, DIN 53457 [4]
2	2300 MPa	bisphenol A:terephthalate:isophthalate 1:0.25:0.25 [4]
3	2200 MPa	bisphenol A:terephthalate:isophthalate 1:0.4:0.4 [4]

Tensile Strength Break:

No.	Value	Note	Elongation
1	72 MPa	bisphenol A:terephthalate:isophthalate 1:0.29:0.06 [2]	122%
2	66 MPa	bisphenol A:terephthalate:isophthalate 1:0.25:0.25 [2,4]	80%
3	60 MPa	bisphenol A:terephthalate:isophthalate 1:0.4:0.4 [4]	50%
4	70 MPa	bisphenol A:terephthalate:isophthalate 1:0.15:0.15 [4]	100%
5	78 MPa	bisphenol A:terephthalate:isophthalate 1:0.29:0.06 [2]	78%

Flexural Strength at Break:

No.	Value	Note
1	66 MPa	bisphenol A:terephthalate:isophthalate 1:0.4:0.4 [4]
2	77 MPa	bisphenol A:terephthalate:isophthalate 1:0.15:0.15 [4]
3	97 MPa	bisphenol A:terephthalate:isophthalate 1:0.29:0.06 [2]
4	96 MPa	bisphenol A:terephthalate:isophthalate 1:0.12:0.28 [2]
5	71 MPa	bisphenol A:terephthalate:isophthalate 1:0.25:0.25 [2,4]

Tensile Strength Yield:

No.	Value	Note
1	59.8 MPa	bisphenol A:terephthalate:isophthalate 1:0.56:0.1 [1]

Impact Strength: Tensile impact strength 251 kJ m^{-2} (bisphenol A:terephthalate:isophthalate 1:0.12:0.28 ASTM D1822) [8]. Dart drop impact strength 100 J (bisphenol A:terephthalate:isophthalate 1:0.12:0.28) [8]
Mechanical Properties Miscellaneous: Loss factor values have been reported [3]
Hardness: Ball indentation 100 MPa (bisphenol A:terephthalate:isophthalate 1:0.25:0.25) [2]. Rockwell R127 (bisphenol A:terephthalate:isophthalate 1:0.29:0.06); R122 (bisphenol A:terephthalate:isophthalate 1:0.12:0.28) [2]
Failure Properties General: Fatigue lifetime 250000-550 000 cycles (bisphenol A: terephthalate:isophthalate 1:0.55:0.1) [9]

Izod Notch:

No.	Value	Notch	Note
1	706 J/m	Y	13.23 ft lb in^{-1}, bisphenol A:terephthalate:isophthalate 1:0.025:0.025 [7]
2	530 J/m	Y	bisphenol A:terephthalate:isophthalate 1:0.29:0.06 [2]
3	530 J/m	Y	bisphenol A:terephthalate:isophthalate 1:0.12:0.28 [2]
4	548 J/m	Y	3.2 mm, bisphenol A:terephthalate:isophthalate 1:0.25:0.25 [2]

– Bisphenol A-terephthalic acid polyester carbonate

Charpy:

No.	Value	Notch	Note
1	35	Y	23°, bisphenol A: terephthalate:isophthalate 1:0.15:0.15 [4]
2	16	Y	-40°, bisphenol A: terephthalate:isophthalate 1:0.15:0.15 [4]
3	32	Y	23°, bisphenol A: terephthalate:isophthalate 1:0.25:0.25 [4]
4	16	Y	-40°, bisphenol A: terephthalate: isophthalate1:0.25:0.25 [4]
5	28	Y	23°, bisphenol A: terephthalate:isophthalate 1:0.4:0.4 [4]
6	18	Y	-40°, bisphenol A: terephthalate: isophthalate 1:0.4:0.4 [4]

Electrical Properties:
Dielectric Permittivity (constant):

No.	Value	Frequency	Note
1	3.27	60 Hz	bisphenol A:terephthalate:isophthalate 1:0.29:0.06 [2]
2	3.15	60 Hz	bisphenol A:terephthalate:isophthalate 1:0.12:0.28 [2]

Dielectric Strength:

No.	Value	Note
1	20.1 kV.mm^{-1}	bisphenol A:terephthalate:isophthalate 1:0.29:0.06 [2]
2	20.3 kV.mm^{-1}	bisphenol A:terephthalate:isophthalate 1:0.12:0.28 [2]
3	45.5 kV.mm^{-1}	bisphenol A:terephthalate:isophthalate 1:0.25:0.25 [2]

Optical Properties:
Refractive Index:

No.	Value	Note
1	1.6	[2]

Transmission and Spectra: Transmittance 85% [2]
Molar Refraction: Haze 1% [2]. Transparent (bisphenol A: terephthalate:isophthalate 1:0.25:0.25) [2]

Polymer Stability:
Decomposition Details: Studies of thermal decomposition have been reported [10]. Temp. of initial weight loss 335–350° [10]. Max. rate of weight loss 390–395°, 490–500°, 520–530° [10]
Flammability: Flammability rating HB (bisphenol A:terephthalate: isophthalate 1:0.29:0.06) [2], V2 (bisphenol A:terephthalate: isophthalate 1:0.12:0.28) [2]
Hydrolytic Stability: Has poor resistance to water immersion (96°) and steam sterilisation [8]. Mech. props. are greatly reduced, MW decreases and microcavities are formed [8]
Stability Miscellaneous: Has good resistance to embrittlement on ageing at 120°

Applications/Commercial Products:
Processing & Manufacturing Routes: Synth. in soln. in two stages or by interfacial polycondensation in one stage [4,11]. Two stage polymerisation; a mixture of terephthaloyl and isophthaloyl chlorides in CH$_2$Cl$_2$ is added to bisphenol A in anhydrous pyridine under N$_2$ at 25° with *p-tert*-butyl phenol as a chain capping agent- this results in oligomer formation. Phosgenation is then carried out by adding phosgene to the vapour space of the reaction vessel [7,11,12]

Trade name	Details	Supplier
Lexan	4501,4701, 3250	General Electric Plastics

Bibliographic References
[1] Kambour, R.P., *Polym. Commun.*, 1983, **24**, 292, (cohesive energy density, T$_g$, tensile props.)
[2] Ash, M. and Ash, I., Handbook of Plastic Compounds, *Elastomers and Resins*, 1992, 218, (props.)
[3] Bubeck, R.A. and Bales, S.E., *Polym. Eng. Sci.*, 1984, **24**, 1142, (density, complex moduli, ageing)
[4] Encycl. Polym. Sci. Eng. 2nd edn., (ed. J.I. Kroschwitz), Wiley-Interscience, 1987, **11**, 692, (props., synth.)
[5] Pinnau, I. and Koros, W.J., *J. Appl. Polym. Sci.*, 1992, **46**, 1195, (T$_g$)
[6] Pinnau, I., Hellums, M.W. and Koros, W.J., *Polymer*, 1991, **32**, 2612, (T$_g$, gas permeability)
[7] *Eur. Pat. Appl.*, 1982, 50847, (Vicat, impact strength, synth.)
[8] Robeson, L.M., Dickinson, B.L. and Crisafulli, S.T., *J. Appl. Polym. Sci.*, 1986, **32**, 5965, (props.)
[9] Takemori, M.T., Kambour, R.P. and Matsumoto, D.S., *Polym. Commun.*, 1983, **24**, 297, (fatigue)
[10] Bosnyak, C.P., Knight, G.J. and Wright, W.W., *Polym. Degrad. Stab.*, 1981, **3**, 273, (thermal stability)
[11] Prevorsek, D.C., De Bona, B.T. and Kesten, Y., J. Polym. Sci., *Polym. Chem. Ed.*, 1980, **18**, 75, (synth.)
[12] Bosnyak, C.P., Parsons, I.W., Hay, J.N. and Haward, R.N., *Polymer*, 1980, **21**, 1448, (synth.)

Bisphenol A-terephthalic acid polyester carbonate

Synonyms: *Bisphenol A-1,4-benzenedicarboxylic acid polyester carbonate. Poly[4,4'-(1-methylethylidene)bis[phenol]-co-1,4-benzenedicarboxylic acid-co-carbonic acid]*. PCT
Related Polymers: Poly[2,2-propanebis(4-phenyl)carbonate], Polyester carbonates
Monomers: Bisphenol A, 1,4-Benzenedicarboxylic acid, Carbonic acid
Material class: Thermoplastic, Copolymers
Polymer Type: Polyester, Bisphenol A polycarbonate
CAS Number:

CAS Reg. No.	Note
31133-78-5	source: carbonic acid
73951-28-7	source: carbonic dichloride

Molecular Formula: [(C$_{16}$H$_{14}$O$_3$).(C$_{23}$H$_{18}$O$_4$)]$_n$
Fragments: C$_{16}$H$_{14}$O$_3$ C$_{23}$H$_{18}$O$_4$
General Information: The polymers have both carbonate and ester linkages, which confer props. intermediate between polycarbonate and the polyester. Common usage of the form PCT(X), where T refers to the terephthalate ester moiety and X to the mole fraction of ester (relative to total ester + carbonate). The molar ratio of Bisphenol: (ester + carbonate) residues is 1:1
Morphology: Terephthalate polyester carbonates show some molecular ordering (X-ray analysis; cf. isophthalates) [1]; and are more easily crystallised than isophthalates. [2] At ester levels of PCT (0.82) [ester:carbonate ratio 4.5] and above the cast films are about 15–30% cryst. [2]

Volumetric & Calorimetric Properties:
Density:

No.	Value	Note
1	d 1.1957 g/cm^3	0.1 mol terephthalate [5]
2	d 1.196 g/cm^3	0.2 mol terephthalate [5]

– Bisphenol A-terephthalic acid polyester carbonate

3	d 1.199 g/cm^3	0.5 mol terephthalate [3]
4	d 1.2 g/cm^3	0.75 mol terephthalate [5]
5	d 1.217 g/cm^3	0.15 mol terephthalate [2]

Specific Heat Capacity:

No.	Value	Note	Type
1	0.12–0.15 kJ/kg.C	0.64 mol terephthalate; ester: carbonate 18 [2]	p
2	0.14–0.17 kJ/kg.C	0.8 mol terephthalate; ester: carbonate 4 [2]	p
3	0.09–0.11 kJ/kg.C	0.9 mol terephthalate; ester carbonate: 8.8 [2]	P

Glass-Transition Temperature:

No.	Value	Note
1	154°C	0.05 mol terephthalate [4]
2	153–158°C	0.1 mol terephthalate [4,5,6]
3	189–192°C	0.5 mol terephthalate [4,6]
4	177–178°C	0.5 mol terephthalate [3,7]
5	190°C	0.75 mol terephthalate [3]
6	225–229°C	0.85 mol terephthalate; ester: carbonate 4 [2]

Transition Temperature:

No.	Value	Note
1	-95–-85°C	γ transition, 0.05 mol terephthalate [7]

Surface Properties & Solubility:
Solubility Properties General: Forms miscible blends with PET in the range 0–67% PET (0.5 mol terephthalate); above 67% crystallisation of PET occurs [3]
Cohesive Energy Density Solubility Parameters: Solubility parameter 19.7 J$^{1/2}$ cm$^{-3/2}$ (9.61 cal$^{1/2}$ cm$^{-3/2}$, 0.5 mol terephthalate, calc.) [3]
Solvents/Non-solvents: V. sol. chlorinated organic solvents, tetrachloroethane [1,3], CH$_2$Cl$_2$ (less so) [1]; insol. Me$_2$CO, H$_2$O [4], MeOH, 2-propanol [1]

Transport Properties:
Polymer Solutions Dilute: Mark Houwink constants have been reported (0.5 mol terephthalate) [4]; K 0.000466 (CH$_2$Cl$_2$); K 0.000216 (CH$_2$Cl$_2$, M$_n$ 6100-5000); α 0.71 (CH$_2$Cl$_2$); α 0.83 (CH$_2$Cl$_2$, M$_n$ 6100-5000)

Mechanical Properties:
Tensile (Young's) Modulus:

No.	Value	Note
1	2030 MPa	0.5 mol terephthalate [3]

Tensile Strength Break:

No.	Value	Note	Elongation
1	59.4 MPa	0.5 mol terephthalate [3]	
2	56 MPa	23°, 40% relative humidity, ester: carbonate 1.8, 0.64 mol terephthalate [2]	30%
3	56 MPa	23°, 40% relative humidity, ester: carbonate 4.5, 0.82 mol terephthalate [2]	60%
4	60 MPa	23°, 40% relative humidity, ester: carbonate 8.3, 0.89 mol terephthalate [2]	60%

Tensile Strength Yield:

No.	Value	Elongation	Note
1	63.5 MPa		0.5 mol terephthalate [3]
2	55.5–56.5 MPa	14%	23°, 40% relative humidity, ester: carbonate 1.8, 0.64 mol terephthalate [2]
3	56–57 MPa	14%	23°, 40% relative humidity, ester: carbonate 4.5, 0.82 mol terephthalate [2]
4	55.5–56.5 MPa	14%	23°, 40% relative humidity, ester: carbonate 8.3, 0.89 mol terephthalate [2]

Complex Moduli: The variations of tensile storage and loss moduli and tan δ with temp. (-150–200°, 1 Hz) have been reported [5,7,8]
Impact Strength: Impact strength decreases on thermal ageing
Viscoelastic Behaviour: Tensile stress-strain curves have been reported for ester/carbonate ratios of > 1.8. There is a viscoelastic region up to the yield point, followed by a stress drop, and then necking behaviour (depending on the ester/carbonate ratio) until the break point is reached [2]
Failure Properties General: Polyester carbonate (PCT 0.5) displays ductile failure [9]

Optical Properties:
Refractive Index:

No.	Value	Note
1	1.601	0.5 mol terephthalate [3]

Transmission and Spectra: Ir spectral data have been reported [1]
Volume Properties/Surface Properties: Ester:carbonate ratios above 4.2 cause whitening of the polymer due to increasing crystallinity [2]

Polymer Stability:
Decomposition Details: Decomposition curves in N$_2$ have been reported [10]. Above 70% wt loss some stabilisation occurs. Initial wt loss temp. increases with increase in terephthalate content 280° (ester:carbonate 1.8, 0.64 mol terephthalate) [10]; 350° (ester:carbonate 5.6, 0.85 mol terephthalate) [10]. Max wt loss 400–450° (0.64 mol terephthalate); 475° (0.85 mol terephthalate) [10]
Environmental Stress: Shows good resistance to embrittlement at 120° [5]

Applications/Commercial Products:
Processing & Manufacturing Routes: Polyester carbonates may be synth. in soln. in two stages or by interfacial polycondensation in one stage [4,11]. Terephthaloyl chloride in CH$_2$Cl$_2$ is added to bisphenol A in anhydrous pyridine under N$_2$ at 25° with *p-tert*-butyl phenol as a chain capping agent: this results in oligomer formation. Phosgenation is then carried out by adding phosgene to the vapour space of the reaction vessel [2,4,12]. Interfacial polycondensation route; bisphenol A is dissolved in aq. NaOH soln., to which is added a mixture of terephthaloyl chloride and phosgene in an organic solvent (e.g. CCl$_4$) at room temp. [1,11]. Tertiary amines, quaternary ammonium or phosphonium salts are used as catalysts; phenols, phenyl chlorocarbonates and aromatic carbonyl halides are used as chain terminators [11]
Applications: Lighting engineering, electronics and electrical engineering applications

– Bisphenol AF polycarbonate   B-14 – B-14

Trade name	Supplier
Apec KL1-9308	Bayer Inc.
Copec	Mitsubishi Chemical

Bibliographic References
[1] Kolesnikov, G.S., Smirnova, O.V. and Samsoniya, S.A., *Polym. Sci. USSR (Engl. Transl.)*, 1967, **9**, 1127, (morphology, solvents, ir, synth)
[2] Bosnyak, C.P., Parsons, I.W., Hay, J.N. and Haward, R.N., *Polymer*, 1980, **21**, 1448, (synth., props.)
[3] Aharoni, S.M., *J. Macromol. Sci. Pt B*, 1983-4, **22**, 813, (synth., props.)
[4] Prevorsek, D.C., De Bona, B.T. and Kesten Y., *J. Polym. Sci., Polym. Chem. Ed.*, 1983-4, **22**, 813, (solvents, T_g, dilute soln. props., synth.)
[5] Bubeck, R.A and Bales, S.E., *Polym. Eng. Sci.*, 1984, **24**, 1142, (density, T_g, complex moduli, ageing)
[6] Prevorsek, D.C. and De Bona, B.T., *J. Macromol. Sci. Pt B*, 1981, **19**, 605, (T_g, nomenclature)
[7] Khanna, Y.P., *J. Therm. Anal.*, 1985, **30**, 153, (T_g, T_γ, complex moduli)
[8] Smith, P.B, *Macromolecules*, 1988, **21**, 2058, (complex moduli)
[9] Aharoni, S.M., *Macromolecules*, 1985, **18**, 2624, (failure props.)
[10] Bosnyak, C.P., Knight, G.J. and Wright, W.W., *Polym. Degrad. Stab.*, 1981, **3**, 273, (thermal stability)
[11] Encycl. Polym. Sci. Eng. 2nd edn., vol 11, (ed. J.I. Kroschwitz), Wiley-Interscience, 1987, 692, (synth.)
[12] *U.S. Pat.*, 1982, 4, 388 455, (synth.)

Bisphenol AF polycarbonate   B-14

Synonyms: *Poly[1,1-(2,2,2-trifluoro-1-(trifluoromethyl)ethane bis(4-phenyl)carbonate]. Poly[oxycarbonyloxy-1,4-phenylene[2,2,2-trifluoro-1-(trifluoromethyl)ethylidene]-1,4-phenylene]. Poly(4,4'-[2,2,2-trifluoro-1-(trifluoromethyl)ethylidene]bisphenol-co-carbonic dichloride)*
Monomers: Bisphenol AF, Carbonic dichloride
Material class: Thermoplastic
Polymer Type: Polycarbonates (miscellaneous)
CAS Number:

CAS Reg. No.
32291-26-2

Molecular Formula: $(C_{16}H_8F_6O_3)_n$
Fragments: $C_{16}H_8F_6O_3$

Volumetric & Calorimetric Properties:
Density:

No.	Value	Note
1	d 1.478 g/cm^3	[4,5,6,7]

Equation of State: Flory/Van der Waals equation of state information for polymer melt has been reported [8]. Tait, information and Sanchez-Lacombe equation of state information has been reported [1]
Glass-Transition Temperature:

No.	Value	Note
1	158°C	[9]
2	148°C	[2]
3	172°C	[4]
4	171°C	[11]
5	172–176°C	[6,8,10]
6	169°C	[1]

Surface Properties & Solubility:
Solubility Properties General: Forms miscible blends with poly methyl methacrylate, and immiscible blends with Bisphenol A polycarbonate, and dichloroethylene polycarbonate
Cohesive Energy Density Solubility Parameters: Solubility parameter 20.5 $J^{1/2}$ cm$^{3/2}$ (10 cal$^{1/2}$ cm$^{3/2}$, calc.) [1]
Solvents/Non-solvents: Very sol. Me$_2$CO, THF, N-methylpyrrolidinone, DMSO, pyridine [2,3], chlorinated hydrocarbons, C$_6$H$_6$, m-cresol, cyclohexanone, dioxane, EtOAc, polar aprotic solvents [3]. Sol. on heating to 60° DMF, CHCl$_3$, 1,4-dioxane, phenol [2]. Insol. MeOH [2]
Wettability/Surface Energy and Interfacial Tension: Contact angle 93° (H$_2$O in air, 25°, 65% relative humidity) [2]; 91° (H$_2$O in air, 25°) [3]
Surface Tension:

No.	Value	Note
1	20 mN/m	20 dyne cm^{-1}, 25° [3]

Transport Properties:
Permeability of Gases: The permeability of CO_2 decreases with pressure up to 20 atm [4,6,7] after which plasticisation occurs and permeability increases; the permeability-pressure curve shows hysteresis [7]. Conditioning at 450 psi (30 atm) for 7 days leads to an increase in permeability [7]. Gas selectivity and diffusivity have been reported [6,12]
Gas Permeability:

No.	Gas	Value	Note
1	CO$_2$	1838 cm^3 mm/(m^2 day atm)	2.1 × 10^{-12} cm^2 (s Pa)$^{-1}$, 35°, 5 atm [4]
2	CO$_2$	1575 cm^3 mm/(m^2 day atm)	1.8 × 10^{-12} cm^2 (s Pa)$^{-1}$, 35°, 10 atm [6]
3	CO$_2$	1488 cm^3 mm/(m^2 day atm)	1.7 × 10^{-12} cm^2 (s Pa)$^{-1}$, 35°, 20 atm [4,6]
4	CH$_4$	788 cm^3 mm/(m^2 day atm)	9 × 10^{-14} cm^2 (s Pa)$^{-1}$, 35°, 3.5 atm [4]
5	CH$_2$	68 cm^3 mm/(m^2 day atm)	7.77 × 10^{-14} cm^2 (s Pa)$^{-1}$, 35°, 10 atm [6,7]
6	O$_2$	414 cm^3 mm/(m^2 day atm)	4.72 × 10^{-13} cm^2 (s Pa)$^{-1}$, 25°, 1 atm [9]
7	O$_2$	455 cm^3 mm/(m^2 day atm)	5.2 × 10^{-13} cm^2 (s Pa)$^{-1}$, 35°, 2 atm [6,12]
8	N$_2$	113.8 cm^3 mm/(m^2 day atm)	1.3 × 10^{-13} cm^2 (s Pa)$^{-1}$, 35°, 2 atm [6]
9	N$_2$	105 cm^3 mm/(m^2 day atm)	1.2 × 10^{-13} cm^2 (s Pa)$^{-1}$, 35°, 10 atm [6]
10	He	3939 cm^3 mm/(m^2 day atm)	4.5 × 10^{-12} cm^2 (s Pa)$^{-1}$, 35°, 10 atm [6]

Mechanical Properties:
Complex Moduli: The variations of E', E'' and tan δ with temp. (-150–200°, 110 Hz) have been reported [14]

Optical Properties:
Refractive Index:

No.	Value	Note
1	1.426	[4]
2	1.539	calc. [12]

Transmission and Spectra: Ir spectral data has been reported [3]
Volume Properties/Surface Properties: Colourless and transparent film [1,4]

Polymer Stability:
Thermal Stability General: Hexafluoropolycarbonate has good thermal stability, aided by the presence of fluorine
Decomposition Details: Temp. for 10% weight loss 443° (N_2) [2]. Temp. for 10% weight loss 460° (air) [3]. Residual mass at 500° 45.1% (N_2) [2]. Residual mass at 500° 57% (air) [3]

Applications/Commercial Products:
Processing & Manufacturing Routes: Synth. by transesterification, in which hexafluorobisphenol A and hexafluorodiphenyl dicarbonate are mixed in the melt (150–300°) in the absence of O_2 with elimination of phenol. Reaction rate is increased by catalysts (*e.g.* zinc acetate) and the use of a medium-high vacuum in the final stages [3]. Can also be synth. by interfacial polycondensation where trichloromethylchloroformate is introduced into a soln. or suspension of hexafluorobisphenol A in aq. NaOH at 20° in an inert solvent (*e.g.* 1,2-dichloroethane). Quaternary ammonium salt catalysts are used to achieve high MW [3]

Bibliographic References
[1] Kim, C.K. and Paul, D.R., *Polymer*, 1992, **33**, 4929, 4941, (props, solubility parameter)
[2] Liaw, D.-J. and Chang, P., *Polymer*, 1996, **37**, 2857, (props, synth)
[3] Saegusa, Y., Kuriki, M., Kawai, A. and Nakamura, S., *J. Polym. Sci., Polym. Chem. Ed.*, 1990, **28**, 3327, (props, synth)
[4] Jordan, S.M. and Koros, W.J., *J. Membr. Sci.*, 1990, **51**, 233, (density, T_g, gas permeability)
[5] Fleming, G.K. and Koros, W.J., *J. Polym. Sci., Polym. Phys. Ed.*, 1990, **28**, 1137, (density, T_g)
[6] Hellums, M.W., Koros, W.J., Husk, G.R. and Paul, D.R., *J. Membr. Sci.*, 1989, **46**, 93, (density, T_g, gas permeability)
[7] Jordan, S.M., Fleming, G.K. and Koros, W.J., *J. Polym. Sci., Polym. Phys. Ed.*, 1990, **28**, 2305, (density, T_g, gas permeability)
[8] Brannock, G.R. and Sanchez, I.C., *Macromolecules*, 1993, **26**, 4970, (equation of state)
[9] Schmidhauser, J.C. and Longley, K.L., *J. Appl. Polym. Sci.*, 1990, **39**, 2083, (T_g, gas permeability)
[10] Garfield, L.J., *J. Polym. Sci., Polym. Symp.*, 1970, **30**, 551, (T_g)
[11] Kim, C.K. and Paul, D.R., *Macromolecules*, 1992, **25**, 3097, (T_g, refractive index)
[12] Koros, W.J., Coleman, M.R. and Walker, D.R.B., *Annu. Rev. Mater. Sci.*, 1992, **22**, 47, (gas permeability)
[13] McHattie, J.S., Koros, W.J. and Paul, D.R., *J. Polym. Sci., Polym. Phys. Ed.*, 1991, **29**, 731, (complex moduli)

Bisphenol F diglycidyl ether resin   B-15

Synonyms: *Poly(bisphenol F bis(oxiranylmethyl) ether)*
Monomers: Bisphenol F, (Chloromethyl)oxirane
Material class: Thermosetting resin
Polymer Type: Epoxy

CAS Number:

CAS Reg. No.	Note
96141-20-7	(homopolymer)
28064-14-4	

Molecular Formula: $(C_{16}H_{16}O_3)_n$
General Information: Blended with Bisphenol A diglycidyl ether resin to reduce viscosity and improve crystallisation resistance

Volumetric & Calorimetric Properties:
Density:

No.	Value	Note
1	d 0.987 g/cm^3	9.9 lb gal^{-1}, EPON 862 [1]
2	d 1.23 g/cm^3	15% triethylenetetramine cured, 25°, ASTM D792 [1]
3	d 1.22 g/cm^3	4% 2-ethyl-4-methylimidazole cured, 25°, ASTM D792 [1]

Glass-Transition Temperature:

No.	Value	Note
1	121°C	15% triethylenetetramine cured, rheometrics dynamical tester [1]
2	105°C	15% triethylenetetramine cured, TMA [1]
3	151°C	4% 2-ethyl-4-methylimidazole cured, rheometrics peak [1]
4	125°C	4% 2-ethyl-4-methylimidazole cured, TMA [1]

Deflection Temperature:

No.	Value	Note
1	107°C	15% triethylenetetramine cured, ASTM D648
2	134°C	4% 2-ethyl-4-methylimidazole cured, ASTM D648 [1]

Transport Properties:
Polymer Solutions Dilute: Viscosity 5000–7000 cP (250°, uncured); 775 cP (25°, 15% triethylenetetramine cured, 25°) [1]; 3670 cP (4% 2-ethyl-4-methylimidazole cured, 25°) [1]

Mechanical Properties:
Tensile (Young's) Modulus:

No.	Value	Note
1	3240 MPa	470 kpsi, 15% triethylenetetramine cured, 23° [1]
2	3033 MPa	440 kpsi, 2-ethyl-4-methylimidazole cured, 23° [1]

Flexural Modulus:

No.	Value	Note
1	3433 MPa	498 kpsi, 15% triethylenetetramine cured, 23° [1]

– **Bisphenol G polycarbonate**

Tensile Strength Break:

No.	Value	Note	Elongation
1	82 MPa	11.9 kpsi, 15% triethylenetetramine cured, 23° [1]	7.5%
2	60 MPa	8.7 kpsi, 4% 2-ethyl-4-methylimidazole cured, 23° [1]	2.9%

Flexural Strength at Break:

No.	Value	Note
1	127.8 MPa	15% triethylenetetramine cured, 23°, ASTM D790
2	166.6 MPa	5% elongation, 16.9 kpsi, 4% 2-ethyl-4-methylimidazole cured, 23°, ASTM D790 [1]

Tensile Strength Yield:

No.	Value	Elongation	Note
1	82.7 MPa	6.6% elongation	12 kpsi, 15% triethylenetetramine cured, 23°, ASTM D638 [1]

Compressive Strength:

No.	Value	Note
1	105.5 MPa	18.3 kpsi, 8%, 15% triethylenetetramine cured, 23°, ASTM D695 [1]
2	111 MPa	16.1 kpsi, 8.9%, 4% triethylenetetramine cured, 23°, ASTM D690 [1]

Fracture Mechanical Properties: Fracture toughness 2.64 $MPa^{1/2}$ (1015 $psi^{1/2}$, 15% triethylenetetramine cured, 23°, ASTM D399) [1]; 2.03 $MPa^{1/2}$ (600 $psi^{1/2}$, 4% 2-ethyl-4-methylimidazole cured, 23°, ASTM D399) [1]

Electrical Properties:
Surface/Volume Resistance:

No.	Value	Note	Type
2	1.6×10^{15} Ω.cm	15% triethylenetetramine cured, 23°, ASTM D257 [1]	S

Dielectric Permittivity (constant):

No.	Value	Frequency	Note
1	3.57	1 MHz	15% triethylenetetramine cured, 23°, ASTM D150 [1]
2	3.35	1 MHz	4% 2-ethyl-4-methylimidazole cured, 23°, ASTM D150 [1]

Dielectric Strength:

No.	Value	Note
1	22.87 $kV.mm^{-1}$	581 V mil^{-1}, 15% triethylenetetramine cured, 23°, ASTM D149 [1]
2	22.88 $kV.mm^{-1}$	582 V mil^{-1}, 4% 2-ethyl-4-methylimidazole cured, 23°, ASTM D149 [1]

Dissipation (Power) Factor:

No.	Value	Frequency	Note
1	0.028	1 MHz	15% triethylenetetramine cured, 23°, ASTM D150 [1]
2	0.022	1 MHz	4% 2-ethyl-4-methylimidazole cured, 23°, ASTM D150 [1]

Polymer Stability:
Chemical Stability: Has good resistance to chemicals. [1] Similar resistance to bisphenol A diglycidyl ether resins. Methyl tetrahydrophthalic anhydride cured material absorbs 1.25% MeOH after 180d exposure, and a similar amount of fuel oil at 25°. Unstable to Me_2CO at -25°
Hydrolytic Stability: Stable to inorganic acids. Large weight loss in NaOH at 80°

Applications/Commercial Products:

Trade name	Details	Supplier
Epon	862	Shell Chemicals

Bibliographic References
[1] *EPON Resin Epoxy Bisphenol F BPF Resin 862 SC 772:95*, Shell Chemical, 1996, (technical datasheet)

Bisphenol G polycarbonate B-16

Synonyms: *Poly(oxycarbonyloxy[2-(1-methylethyl)-1,4-phenylene](1-methylethylidene)[3-(1-methylethyl)-1,4-phenylene]). Poly[2,2-propane bis[4-(2-(1-methylethyl)phenyl)]carbonate]. Poly(4,4'-(1-methylethylidene)bis[2-(1-methylethyl)phenol]-co-carbonic dichloride)*
Monomers: Bisphenol G, Carbonic dichloride
Material class: Thermoplastic
Polymer Type: Polycarbonates (miscellaneous)
CAS Number:

CAS Reg. No.
115775-53-6

Molecular Formula: $(C_{22}H_{26}O_3)_n$
Fragments: $C_{22}H_{26}O_3$

Volumetric & Calorimetric Properties:
Glass-Transition Temperature:

No.	Value	Note
1	80°C	[1]

Transport Properties:
Gas Permeability:

No.	Gas	Value	Note
1	O_2	0.0899 cm^3 mm/(m^2 day atm)	1.37×10^{-13} cm^2 (s Pa)$^{-1}$, 25° [1]

Bibliographic References

[1] Schmidhauser, J.C. and Longley, K.L., *J. Appl. Polym. Sci.*, 1990, **39**, 2083, (T_g, gas permeability)

Bisphenol P polycarbonate   B-17

Synonyms: *Poly[1,4-phenylene bis[(2,2-propane)(4-phenyl)]carbonate]. Poly[oxycarbonyloxy-1,4-phenylene(1-methylethylidene)-1,4-phenylene(1-methylethylidene)-1,4-phenylene]. Poly(4,4'-[1,4-phenylenebis(1-methylethylidene)]bisphenol-co-carbonic dichloride)*
Monomers: Bisphenol P, Carbonic dichloride
Material class: Thermoplastic
Polymer Type: Polycarbonates (miscellaneous)
CAS Number:

CAS Reg. No.	Note
25036-51-5	
25038-79-3	source: carbonic dichloride

Molecular Formula: $(C_{25}H_{24}O_3)_n$
Fragments: $C_{25}H_{24}O_3$

BPDA-ODA   B-19

Synonyms: *Poly(3,3',4,4'-biphenyltetracarboxylic dianhydride-alt-4,4'-oxydianiline). Poly(3,3',4,4'-biphenyltetracarboxylic dianhydride-alt-bis(4-aminophenyl)ether). Poly(4,4'-oxydiphenylene biphenyltetracarboximide). Poly[(1,1',3,3'-tetrahydro-1,1',3,3'-tetraoxo[5,5'-bi-2H-isoindole]-2,2'-diyl)-1,4-phenyleneoxy-1,4-phenylene]. Poly(4,4'-oxybis[benzenamine]-alt-[5,5'-biisobenzofuran]-1,1',3,3'-tetrone). Poly([5,5'-biisobenzofuran]-1,1',3,3'-tetrone-alt-4,4'-oxybisbenzenamine)*
Monomers: [5,5'-Biisobenzofuran]-1,1',3,3'-tetrone, 4,4'-Oxybis[benzenamine]
Material class: Semithermoplastic
Polymer Type: polyimide
CAS Number:

CAS Reg. No.	Note
26615-45-2	
26298-81-7	source based

Molecular Formula: $(C_{28}H_{14}N_2O_5)_n$
Fragments: $C_{28}H_{14}N_2O_5$

Volumetric & Calorimetric Properties:
Density:

No.	Value	Note
1	d 1.366 g/cm^3	amorph. film [4,8]
2	d 1.398 g/cm^3	Upilex R film

Thermal Expansion Coefficient:

No.	Value	Note	Type
1	5.2×10^{-5} K^{-1}	50-250° [1]	L

Volumetric Properties General: Fractional free volumes 0.121 [4], 0.1 (Upilex R) [5]
Glass-Transition Temperature:

No.	Value	Note
1	270°C	[4,5,8]
2	302°C	265-370°, broad transition [3]

Surface Properties & Solubility:
Cohesive Energy Density Solubility Parameters: CED 834 J cm^{-3} [2], δ 32.3 J$^{1/2}$ cm$^{3/2}$ [4,5]

Transport Properties:
Water Absorption:

No.	Value	Note
1	1.27 %	50h, 25°, 75% relative humidity [1]

Gas Permeability:

No.	Gas	Value	Note
1	H_2	241.7 cm^3 mm/(m^2 day atm)	3.68×10^{-10} cm^2 (s cmHg)$^{-1}$, 35°, 10 atm. [4]
2	CO	2.4 cm^3 mm/(m^2 day atm)	0.036×10^{-10} cm^2 (s cmHg)$^{-1}$, 35°, 10 atm. [4]
3	CO_2	42.2 cm^3 mm/(m^2 day atm)	0.642×10^{-10} cm^2 (s cmHg)$^{-1}$, 35°, 10 atm. [4]
4	CO_2	4728–13987 cm^3 mm/(m^2 day atm)	$72-213 \times 10^{-10}$ cm^2 (s cmHg)$^{-1}$, 50-110°, 20 atm., amorph. film [8]
5	CO_2	781–3677 cm^3 mm/(m^2 day atm)	$11.9-56 \times 10^{-10}$ cm^2 (s cmHg)$^{-1}$, 35-110°, 20 atm., Upilex R film [8]
6	CH_4	6.5 cm^3 mm/(m^2 day atm)	0.0099×10^{-10} cm^2 (s cmHg)$^{-1}$, 35°, 10 atm. [4]
7	H_2O	51219.5 cm^3 mm/(m^2 day atm)	7.8×10^{-8} cm^2 (s cmHg)$^{-1}$ [5]

Mechanical Properties:
Tensile (Young's) Modulus:

No.	Value	Note
1	3000 MPa	[7]
2	3400 MPa	room temp., 2 mm min^{-1} [3]
3	3500–3600 MPa	[2]
4	5800 MPa	cold drawn film, parallel to drawing direction [7]
5	2500 MPa	cold drawn film, perpendicular to drawing direction [7]

Tensile Strength Break:

No.	Value	Note	Elongation
1	119 MPa	room temp., 2 mm min^{-1} [3]	41%
2	140 MPa	[7]	
3	230 MPa	cold drawn film, parallel to drawing direction [7]	
4	110 MPa	cold drawn film, perpendicular to drawing direction [7]	

Tensile Strength Yield:

No.	Value	Elongation	Note
1	117 MPa	7%	room temp., 2 mm min^{-1} [3]

Optical Properties:
Total Internal Reflection: Birefringence Δn ($\eta_{TE}-\eta_{TM}$) 0.0004–0.0045 (depends on cure conditions) [8], 0.0056 (Upilex R) [8]

Polymer Stability:
Environmental Stress: Good resistance to γ radiation [6]. Good resistance to ionising radiation in air for doses over 50 MGy. In oxygen, deterioration of mechanical props. occurs at doses 10–20% of those in air [9]

Applications/Commercial Products:

Trade name	Supplier
Upilex R	Ube Industries Ltd.

Bibliographic References

[1] Numata, S., Fujisaki, K. and Kinjo, N., *Polymer*, 1987, **28**, 2282, (thermal expansion)
[2] Numata, S. and Miwa, T., *Polymer*, 1989, **30**, 1170, (thermal expansion)
[3] Kim, Y., Ree, M., Chang, T., Ha, C.S. *et al*, *J. Polym. Sci., Part B: Polym. Phys.*, 1995, **33**, 2075, (blends, general props)
[4] Tanaka, K., Kita, H., Okano, M. and Okamoto, K.-I., *Polymer*, 1992, **33**, 585, (gas permeability)
[5] Okamoto, K.-I., Tanihara, N., Watanabe, H., Tanaka, K. *et al*, *J. Polym. Sci., Part B: Polym. Phys.*, 1992, **30**, 1223, (water sorption)
[6] Hegazy, E.-S.A., Sasuga, T., Nishii, M. and Seguchi, T., *Polymer*, 1992, **33**, 2897, (radiation resistance)
[7] Kochi, M., Yokota, R., Iizuka, T. and Mita, I., *J. Polym. Sci., Part B: Polym. Phys.*, 1990, **28**, 2463, (mech props)
[8] Okamoto, K.-I., Tanaka, K., Kita, H., Nakamura, A. and Kusuki, Y., *J. Polym. Sci., Part B: Polym. Phys.*, 1989, **27**, 1221, (CO_2 sorption)
[9] Sasuga, T., *Polymer*, 1988, **29**, 1562, (radiation resistance)

BPDA-*p*PDA   B-20

Synonyms: *Poly(3,3',4,4'-biphenyltetracarboxylic dianhydride-alt-p-phenylenediamine). Poly(p-phenylene biphenyltetracarboximide). Poly[(1,1',3,3'-tetraoxo[5,5'-bi-2H-isoindole]-2,2'-diyl)-1,4-phenylene]. Poly(1,4-benzenediamine-alt-[5,5'-biisobenzofuran]-1,1',3,3'-tetrone). Poly[5,5'-biisobenzofuran]-1,1',3,3'-tetrone-alt-1,4-benzenediamine)*
Monomers: [5,5'-Biisobenzofuran]-1,1',3,3'-tetrone, 1,4-Benzenediamine
Material class: Semithermoplastic
Polymer Type: Polyimide
CAS Number:

CAS Reg. No.	Note
32197-39-0	
29319-22-0	source base

Molecular Formula: $(C_{22}H_{10}N_2O_4)_n$
Fragments: $C_{22}H_{10}N_2O_4$

Volumetric & Calorimetric Properties:
Density:

No.	Value	Note
1	d 1.488 g/cm^3	Upilex S film [9]
2	d^{30} 1.391 g/cm^3	[1]

Thermal Expansion Coefficient:

No.	Value	Note	Type
1	3×10^{-6}–2.4×10^{-5} K^{-1}	in plane, parallel to optic axis, 50–250°, depends on orientation [9]	L
2	3×10^{-6}–2.5×10^{-5} K^{-1}	in plane, perpendicular to optic axis, 50–250°, depends on orientation [9]	L
3	8.4×10^{-5}–0.000148 K^{-1}	out of plane, 50–250°	L
4	0.000125–0.000154 K^{-1}	50–250° [9]	V

Equation of State: Tait equation of state parameters have been reported [9]

Glass-Transition Temperature:

No.	Value	Note
1	352°C	broad transition, 320–470°, spin cast film
2	>500°C	min., Upilex S film

Surface Properties & Solubility:
Cohesive Energy Density Solubility Parameters: CED 930 J cm^{-3} [3]

Transport Properties:
Permeability and Diffusion General: Permeability of water 0.86–7.5 $\times 10^{-10}$ cm^2 s^{-1} (25°) [4]. Permeability is dependent on relative humidity and sample

Water Absorption:

No.	Value	Note
1	1.37 %	50h, 25°, 75% relative humidity [1]
2	0.12–2.4 %	saturation, 25°, depends on relative humidity, film thickness and sample preparation [4]

Mechanical Properties:
Tensile (Young's) Modulus:

No.	Value	Note
1	5300 MPa	[10]
2	5900 MPa	25° [8]
3	10200 MPa	25° [8]
4	4000–55000 MPa	depends on orientation [3]
5	59100 MPa	cold drawn film, parallel to drawing direction [10]
6	2400 MPa	cold drawn film, perpendicular to drawing direction [10]

Tensile Strength Break:

No.	Value	Note	Elongation
1	210 MPa	25°	32%
2	240 MPa	[10]	
3	570 MPa	[5]	45%
4	1220 MPa	cold drawn film, parallel to drawing direction [10]	
5	90 MPa	cold drawn film, perpendicular to drawing direction [10]	

Tensile Strength Yield:

No.	Value	Elongation	Note
1	170 MPa	4%	25° [8]
2	280 MPa	8%	

Viscoelastic Behaviour: Dynamic mech. props. have been reported [5,8]. Residual stress 6.7 MPa (room temp., film cast and cured on Si) [5]

Electrical Properties:
Dielectric Permittivity (constant):

No.	Value	Frequency	Note
1	2.98–3.05	100 Hz–1 MHz	room temp., out-of-plane
2	3.4	12000 GHz	approx., in plane [7]

Optical Properties:
Total Internal Reflection: Δ_n 0.2333–0.1776 (632.8 nm, room temp., depends on thickness) [4]. Δ_n 0.262–0.217 (543.5–1064.2 nm, room temp.) [7]

Polymer Stability:
Environmental Stress: Good resistance to γ radiation, low gas evolution for doses up to 25 MGy [6]. Good resistance to ionising radiation in air for doses over 50 MGy. In oxygen, deterioration of mech. props. occurs at 10–20% of that in air [2]

Applications/Commercial Products:

Trade name	Details	Supplier
Pyralin	PI-2610, PI-2611, PI-581D	DuPont
Upilex S		Ube Industries Ltd.

Bibliographic References
[1] Numata, S., Fujisaki, K. and Kinjo, N., *Polymer*, 1987, **28**, 2282, (thermal expansion)
[2] Sasuga, T., *Polymer*, 1988, **29**, 1562, (radiation resistance)
[3] Numata, S. and Miwa, T., *Polymer*, 1989, **30**, 1170, (thermal expansion)
[4] Ree, M., Han, H. and Gryte, C.C., *J. Polym. Sci., Part B: Polym. Phys.*, 1995, **33**, 505, (water sorption)
[5] Ree, M., Nunes, T.L. and Rex Chen, K.-J., *J. Polym. Sci., Part B: Polym. Phys.*, 1995, **33**, 453, (general props)
[6] Hegazy, E.-S.A., Sasuga, T., Nishii, M. and Seguchi, T., *Polymer*, 1992, **33**, 2897, (radiation resistance)
[7] Boese, D., Lee, H., Yoon, D.Y., Swalen, J.D. and Rabolt, J.F., *J. Polym. Sci., Part B: Polym. Phys.*, 1992, **30**, 1321, (optical props, dielectric props)
[8] Kim, Y., Ree, M., Chang, T., Ha, C.S., et al, *J. Polym. Sci., Part B: Polym. Phys.*, 1995, **33**, 2075, (blends, general props)
[9] Pottiger, M.T., Coburn, J.C. and Edman, J.R., *J. Polym. Sci., Part B: Polym. Phys.*, 1994, **32**, 825, (thermal expansion)
[10] Kochi, M., Yokota, R., Iizuka, T. and Mita, M., *J. Polym. Sci., Part B: Polym. Phys.*, 1990, **28**, 2463, (mech props)

Bromobutyl rubber

Synonyms: *Brominated isobutylene-isoprene elastomer. BIIR. Butyl rubber brominated*
Related Polymers: Butyl rubber
Monomers: Isobutylene, Isoprene
Material class: Synthetic Elastomers
Polymer Type: Polydienes

Additives: Stabiliser 1.5% epoxidised soybean oil
General Information: 2% Bromine. Brominated butyl rubber can be vulcanised more quickly than butyl rubber, and is more compatible with other rubbers
Morphology: Mainly composed of 1,4 linked isoprene units [7]

Volumetric & Calorimetric Properties:
Transition Temperature General: Thermal characteristics of blends of bromobutyl rubber with SBR, natural rubber and EPM in N_2 and O_2 for pure gum and sulfur vulcanisates by DSC and DTG have been reported [9,10]

Surface Properties & Solubility:
Solubility Properties General: Solubility props. of bromobutyl rubber are very similar to those of butyl rubber, see butyl rubber. Bromination improves compatibility with other rubbers [16]

Transport Properties:
Permeability of Gases: Has low permeability to gases
Permeability of Liquids: Water permeation data have been reported for bromobutyl rubber with different loadings of carbon black and for blends of bromobutyl rubber with natural rubber [13]

Mechanical Properties:
Mechanical Properties General: Bromination of butyl rubber speeds up the rate of curing and improves adhesion to metals. [16] Addition of bromobutyl rubber to natural rubber causes a decrease in tensile strength, storage modulus and loss factor [2]. Blends of bromobutyl rubber, natural rubber and polyisobutylene have also been reported [3,6]
Tensile Strength Break:

No.	Value	Note
1	14 MPa	50% natural rubber [2]

Failure Properties General: Blending with natural rubber reduces tear strength, 26 kN m^{-1} (50% natural rubber). Fatigue life 88000 cycles (70% natural rubber). [2]. Heat ageing also decreases mechanical props. [2]
Fracture Mechanical Properties: The rate of fatigue crack growth for bromobutyl rubber has been reported as similar to natural rubber under most conditions, though bromobutyl rubber does not undergo strain crystallisation. [8] Fatigue crack propagation measurements have been reported for blends of bromobutyl rubber with natural rubber and EPDM over a range of tear energies [8]. Propagation is lower by a factor of 15 compared with natural rubber

Optical Properties:
Transmission and Spectra: Nmr studies have identified the structure of brominated groups in the polymer [1]

Polymer Stability:
Thermal Stability General: Heat resistance data have been reported [4]
Decomposition Details: Bromobutyl rubber has been found to cross-link extensively under thermo-oxidative conditions [17]. Max. weight loss at 386°. [11] Weight loss studies for bromobutyl rubber blends have been reported [10]
Flammability: Flammability rating 5.6 (Match test); 1.4 (Mol 417 test) [15]
Environmental Stress: More resistant to ozone than are hydrocarbon diene rubbers [14]
Stability Miscellaneous: High energy radiation causes cross-linking rather than chain scission [12]

Applications/Commercial Products:

Processing & Manufacturing Routes: Prod. by bromination of butyl rubber

Applications: Used in tyres

Trade name	Supplier
Bromobutyl	Exxon Chemical
Polysar bromobutyl	Polysar

Bibliographic References

[1] Chu, C.Y., Watson, R.N., Vukov, R., *Rubber Chem. Technol.*, 1987, **60**, 636
[2] Lemieux, M.A., Killgour, P.C., *Rubber Chem. Technol.*, 1984, **57**, 792
[3] Mazich, K.A., Samus, M.A., Killgour, P.C., Plummer, H.K., *Rubber Chem. Technol.*, 1986, **59**, 623
[4] Timar, J., Edwards, W.S., *Rubber Chem. Technol.*, 1979, **52**, 319
[5] Young, D.G., Danik, J.A., *Rubber Chem. Technol.*, 1994, **67**, 137
[6] Hess, W.M., Herd, C.R., Vegvari, P.C., *Rubber Chem. Technol.*, 1993, **66**, 329
[7] Cheng, D.M., Gardner, I.J., Wang, H.C., Frederick, C.B., et al, *Rubber Chem. Technol.*, 1990, **63**, 265
[8] Young, D.G., *Rubber Chem. Technol.*, 1985, **58**, 785
[9] Sircar, A.K., *Rubber Chem. Technol.*, 1977, **50**, 71
[10] Sircar, A.K., Lamond, T.G., *Rubber Chem. Technol.*, 1975, **48**, 301, 640, 653
[11] Brazier, D.W., Nickel, G.H., *Rubber Chem. Technol.*, 1975, **48**, 319
[12] Hill. D.J.T., O'Donnell, J.H., Perera, M.C.S., Pomery, P.J., *Polymer*, 1995, **36**, 4185
[13] Cassidy, P.E., Aminabhavi, T.M., Thompson, C.H., *Rubber Chem. Technol.*, 1983, **56**, 594
[14] Keller, R.W., *Handbook of Polymer Science and Technology*, (ed. N.P. Cheremisinoff), Marcel Dekker, 1989, 165
[15] Trexler, H.E., *Rubber Chem. Technol.*, 1973, **47**, 1114
[16] Fusco, J.V., *Rubber Chem. Technol.*, 1982, **55**, 103
[17] Jipa, S., Giurginea, M., Setnescu, T., Setnescu, R., et al, *Polym. Degrad. Stab.*, 1996, **54**, 1

BTDA-4-BDAF B-22

Synonyms: *Poly(3,3',4,4'-benzophenonetetracarboxylic dianhydride-alt-2,2-bis[4-(4-aminophenoxy)phenyl]hexafluoropropane). Poly[(1,3-dihydro-1,3-dioxo-2H-isoindole-2,5-diyl)carbonyl(1,3-dihydro-1,3-dioxo-2H-isoindole-5,2-diyl)-1,4-phenyleneoxy-1,4-phenylene[2,2,2-trifluoro-1-(trifluoromethyl)ethylidene]-1,4-phenyleneoxy-1,4-phenylene]. Poly(4,4'-[[2,2,2-trifluoro-1-(trifluoromethyl)ethylidene]bis(4,1-phenyleneoxy)]bisbenzenamine-alt-5,5'-carbonylbis[1,3-isobenzofurandione])*

Monomers: 5,5'-Carbonylbis[1,3-isobenzofurandione], 4,4'-[2,2,2-Trifluoro-1-(trifluoromethyl)ethylidene]bis(4,1-phenyleneoxy)bis[benzenamine]

Material class: Thermoplastic
Polymer Type: Polyimide
CAS Number:

CAS Reg. No.	Note
69572-62-9	
69577-65-7	source based

Molecular Formula: $(C_{44}H_{22}F_6N_2O_7)_n$
Fragments: $C_{44}H_{22}F_6N_2O_7$

Volumetric & Calorimetric Properties:

Density:

No.	Value	Note
1	d^{30} 1.39 g/cm^3	[1]
2	d 1.384 g/cm^3	[3]

Thermal Expansion Coefficient:

No.	Value	Note	Type
1	5.59×10^{-5} K^{-1}	50–250° [1]	L

Volumetric Properties General: Fractional free volume 0.163 [3]

Glass-Transition Temperature:

No.	Value	Note
1	232°C	[3]
2	236°C	[2]

Surface Properties & Solubility:

Cohesive Energy Density Solubility Parameters: δ 27 J$^{1/2}$ cm$^{-3/2}$ [3]

Solvents/Non-solvents: Insol. *N*-methyl-2-pyrrolidone, *N,N*-dimethylacetamide, DMSO, pyridine, THF [2]

Transport Properties:

Gas Permeability:

No.	Gas	Value	Note
1	H_2	2252 cm^3 mm/(m^2 day atm)	34.3×10^{-10} cm^2 (s cmHg)$^{-1}$, 35°, 10 atm. [3]
2	CO	20.6 cm^3 mm/(m^2 day atm)	0.313×10^{-10} cm^2 (s cmHg)$^{-1}$, 35°, 10 atm. [5]
3	CO_2	287 cm^3 mm/(m^2 day atm)	4.37×10^{-10} cm^2 (s cmHg)$^{-1}$, 35°, 10 atm. [5]
4	O_2	74.9 cm^3 mm/(m^2 day atm)	1.14×10^{-10} cm^2 (s cmHg)$^{-1}$, 35°, 10 atm. [5]
5	N_2	12.8 cm^3 mm/(m^2 day atm)	0.195×10^{-10} cm^2 (s cmHg)$^{-1}$, 35°, 10 atm. [5]

Mechanical Properties:

Tensile (Young's) Modulus:

No.	Value	Note
1	2000 MPa	[2]

Tensile Strength Break:

No.	Value	Note	Elongation
1	98 MPa	[2]	10%

Electrical Properties:

Surface/Volume Resistance:

No.	Value	Note	Type
1	0.303×10^{15} Ω.cm	conductivity value [2]	S

Dielectric Permittivity (constant):

No.	Value	Frequency	Note
1	2.7	10 Hz	24° [2]

Polymer Stability:
Decomposition Details: Under N_2 atmosphere, single step decomposition occurs. Initial decomposition at 512°; 10% wt. loss at 540° with heat rate 10° min^{-1} [2]

Applications/Commercial Products:

Trade name	Details	Supplier
Eymyd	L20, L20N	Ethyl Corporation

Bibliographic References

[1] Numata, S., Fujisaki, K. and Kinja, N., *Polymer,* 1987, **28**, 2282, (thermal expansion)
[2] Negi, Y.S., Suzuki, Y.-I., Kawamura, I., Hagiwara, T. *et al,* *J. Polym. Sci., Part A: Polym. Chem.,* 1992, **30**, 2281, (general props)
[3] Tanaka, K., Kita, H., Okano, M. and Okamoto, K.-I., *Polymer,* 1992, **33**, 585, (gas permeability)
[4] Fusaro, R.L. and Hady, W.F., *ASLE Trans.,* 1985, **28**, 542, (friction, wear)

BTDA-diaminophenylindane   B-23

Synonyms: *Poly(3,3',4,4'-benzophenonetetracarboxylic dianhydride-alt-diaminophenylindane). BTDA-DAPI. Poly(1-(4-aminophenyl)-2,3-dihydro-1,3,3-trimethyl-1H-inden-5-amine-alt-5,5'-carbonylbis[1,3-isobenzofurandione]). Poly(3-(4-aminophenyl)-2,3-dihydro-1,1,3-trimethyl-1H-inden-5-amine-alt-5,5'-carbonylbis(1,3-isobenzofurandione)*
Monomers: 5,5'-Carbonylbis[1,3-isobenzofurandione], Diaminophenylindane
Material class: Thermoplastic
Polymer Type: Polyimide
CAS Number:

CAS Reg. No.	Note
62929-02-6	source based: XU series
104983-64-4	source based: Matrimid series

Molecular Formula: $(C_{35}H_{24}N_2O_5)_n$
Fragments: $C_{35}H_{24}N_2O_5$
Molecular Weight: M_n 11000 (Matrimid 5218). MW 80000 (Matrimid 5218)
Additives: Graphite fibre

Volumetric & Calorimetric Properties:
Density:

No.	Value	Note
1	d 1.2 g/cm^3	Matrimid 5218 [1]
2	d 1.2 g/cm^3	XU-218 [5]

Thermal Expansion Coefficient:

No.	Value	Note	Type
1	5.04×10^{-5} K^{-1}	Matrimid 5218 [1]	L
2	8.2×10^{-6} K^{-1}	approx., Probimide 293 [4]	L
3	5×10^{-5} K^{-1}	XU-218 [5]	L
4	8×10^{-6} K^{-1}	Probimide 293 [6]	L

Glass-Transition Temperature:

No.	Value	Note
1	320°C	Matrimid 5218 [3]
2	320–330°C	XU-218 [5]
3	280°C	dry, XU-218 [8]
4	265°C	wet, XU-218 [8]
5	320°C	[9]

Surface Properties & Solubility:
Solubility Properties General: Blends with poly(ether sulfone) have been reported [2]. Matrimid 5218 forms miscible blends with poly(ether sulfone) when cast as films or precipitated from soln. [3]
Solvents/Non-solvents: XU-218 sol. *N,N*-dimethylacetamide [5,8,9], DMF, *N*-methylpyrrolidone [5,8,9], DMSO [5], CH_2Cl_2, ethylene chloride, $CHCl_3$ [9] tetrachloroethane, THF [9] dioxane, [9] acetophenone, cyclohexanone, [9] *m*-cresol [9], γ-butyrolactone [8]. Insol. 2-methoxyethyl acetate, glyme, diglyme, isophorone, 2-butanone, cellosolve, xylene [9]

Transport Properties:
Polymer Solutions Dilute: η 75 P (γ-butyrolactone, Probimide 298) [4]. $[\eta]_{inh}$ 0.55 dl g^{-1} (*N*-methylpyrrolidone, 25°)
Permeability and Diffusion General: Highly porous membranes may be formed from low concentration casting solns. [5]

Mechanical Properties:
Tensile (Young's) Modulus:

No.	Value	Note
1	2893 MPa	29500 kg cm^{-1}, Matrimid 5218 [1]
2	13400 MPa	Probimide 293 [4]
3	2900 MPa	25°, XU-218 [5]
4	13000 MPa	55–215°, Probimide 293 [6]

Tensile Strength Break:

No.	Value	Note	Elongation
1	85.4 MPa	871 kg cm^{-2}, Matrimid 5218 [1]	48%
2	87 MPa	25°, XU-218 [9]	48.6%

Failure Properties General: Short beam shear strength 75.8 MPa (11 kpsi, XU-218) [8]
Fracture Mechanical Properties: G_{Ic} 4.1 in in^{-2} [8]

Electrical Properties:
Dielectric Strength:

No.	Value	Note
1	220 kV.mm^{-1}	Matrimid 5218 [1]

Optical Properties:
Total Internal Reflection: dn/dc 0.189 (Matrimid 5218), 0.194 (XU-218 HP) [7]

Polymer Stability:
Thermal Stability General: XU-218 has good stability up to 315° and is more stable than its graphite fibre composite [8]

Upper Use Temperature:

No.	Value	Note
1	285°C	XU-218, compression moulded [8]

Decomposition Details: Probimide 293 films (prebaked in N_2) show a 15–18% weight loss between 80° and 230° and decompose above 500° [4]. XU-218 shows a max. weight loss of 0.4% after heating in air for 13 weeks at 232° [8]. Shows 10% weight loss at 485° in air and 492° in N_2 [9]

Chemical Stability: Resistance to solvents may be improved by heating XU-218 composites at 315° for more than 16 h followed by heating at 371° for 30–40 mins. [8] Resistant to hexane, xylene, Freon 113, diacetone alcohol, butyl cellosolve, amyl ethyl ketone. Attacked by methyl cellosolve, 2-butanone, Me_2CO, acetonitrile [9]

Applications/Commercial Products:
Processing & Manufacturing Routes: Synth. by addition of benzophenonetetracarboxylic acid dianhydride to a 15–20% soln. of diaminophenylindane in N-methylpyrrolidone. The mixture is reacted at room temp. for 18 h, with imidisation achieved by addition of Ac_2O/pyridine. The polymer is insol. by precipitation into H_2O [9]. Soln. casting may be used for production of ultrafiltration membranes

Applications: Coatings for high temp. applications. Potential application as an ultrafiltration membrane

Trade name	Details	Supplier
Matrimid 5218		Ciba-Geigy Corporation
Probimide 293	New name for XU-293	Ciba-Geigy Corporation
XU-218		Ciba-Geigy Corporation
XU-293		Ciba-Geigy Corporation

Bibliographic References
[1] 1996, Ciba-Geigy, (technical datasheet)
[2] Blizard, K.G., Druy, M.A. and Karasz, F.E., *Annu. Tech. Conf. - Soc. Plast. Eng.*, 1992, **1**, 582, (miscibility)
[3] Liang, K., Grebowitz, J., Valles, E., Karasz, F.E. and MacKnight, W.J., *J. Polym. Sci., Part B: Polym. Phys.*, 1992, **30**, 465, (miscibility)
[4] Elsner, G., *J. Appl. Polym. Sci.*, 1987, **34**, 815
[5] Sarbolouki, M.N., *J. Appl. Polym. Sci.*, 1984, **29**, 743
[6] Elsner, G., Kempf, J., Bartha, J.W. and Wagner, H.H., *Thin Solid Films*, 1990, **185**, 189
[7] Segudovic, N., Karasz, F.E. and MacKnight, W.J., *J. Liq. Chromatogr.*, 1990, **13**, 2581
[8] Cobuzzi, C.A. and Chaudhari, M.A., *Natl. SAMPE Tech. Conf.*, 1985, **17**, 318, (props)
[9] Bateman, J.H., Geresy, W. and Neiditch, D.S., Am. Chem. Soc., Div. Org. Coat. Plast. Chem., *Pap.*, 1975, **35**, 77, (synth, props)

BTDA-MDA

Synonyms: *Poly(3,3′,4,4′-benzophenonetetracarboxylic dianhydride-alt-4,4′-methylenedianiline). Poly(3,3′,4,4′-benzophenonetetracarboxylic dianhydride-alt-bis(4-aminophenyl)methane). Poly[(1,3-dihydro-1,3-dioxo-2H-isoindole-2,5-diyl)carbonyl(1,3-dihydro-1,3-dioxo-2H-isoindole-5,2-diyl)-1,4-phenylenemethylene-1,4-phenylene]. Poly[5,5′-carbonylbis(1,3-isobenzofurandione)-alt-4,4′-methylene(bisbenzenamine)]. Poly(4,4′-methylenebisbenzenamine-alt-5,5′-carbonylbis[1,3-isobenzofurandione]). Poly[(1,3-dioxo-2,5-isoindolinediyl)carbonyl(1,3-dioxo-5,2-isoindolinediyl)-p-phenylenemethylene-p-phenylene]*

Monomers: 5,5′-Carbonylbis[1,3-isobenzofurandione], 4,4′-Methylenebis[benzenamine]

Material class: Semithermoplastic
Polymer Type: Polyimide
CAS Number:

CAS Reg. No.	Note
26913-87-1	
25038-84-0	source based

Molecular Formula: $(C_{30}H_{16}N_2O_5)_n$
Fragments: $C_{30}H_{16}N_2O_5$

Volumetric & Calorimetric Properties:
Density:

No.	Value	Note
1	d 1.05–1.08 g/cm^3	Skybond 705 [1]
2	d 1.15–1.18 g/cm^3	Skybond 703 [1]
3	d 1.332 g/cm^3	[5]
4	d 1.36991 g/cm^3	film, bifix cured [14]
5	d 1.36904 g/cm^3	film, free cured [14]

Thermal Expansion Coefficient:

No.	Value	Note	Type
1	4.5×10^{-5} K^{-1}	[8,14]	L
2	4.9×10^{-5} K^{-1}	film, free cured [14]	L

Glass-Transition Temperature:

No.	Value	Note
1	296°C	[5]
2	240°C	[7]
3	270°C	imidised, in air, 200° [17]
4	306°C	imidised, in air, 300° [17]
5	290°C	imidised *in vacuo*, 300° [17]
6	290°C	imidised in N_2, 300° [17]
7	284°C	[19]
8	285–300°C	imidised sample heated to 350° [20]

Transition Temperature:

No.	Value	Note	Type
1	105°C	[19]	Tβ

Surface Properties & Solubility:
Solubility Properties General: The relationship between struct. and solubility has been reported [11]

Solvents/Non-solvents: V. sol. sulfuric acid, fuming nitric acid [13]. Insol. DMF, C_6H_6, $CHCl_3$, Et_2O, EtOH, Me_2CO [13], fuming nitric acid (air-imidised film) [17]

Transport Properties:
Polymer Solutions Dilute: η 1.1–2.6 Pa s (Skybond 705) [1], 1.15–1.18 (Skybond 703) [1], 0.65 dl g^{-1} (dimethylacetamide, 30°) [15], $[\eta]_{inh}$ 0.629 dl g^{-1} (conc. sulfuric acid, 30°) [13], 1.1768 dl g^{-1} (conc. sulfuric acid, 30°) [13], 0.6324 dl g^{-1} (conc. sulfuric acid, 30°) [16], 0.96 dl g^{-1} (dimethylacetamide, 35°, 24 h) [17]
Permeability of Liquids: Diffusion of DMF has been reported [10]. Permeability of water 96.6 g (cm s cmHg)$^{-1}$ (25°, 75% humidity) [14], 178.3 g (m^2 atm)$^{-1}$ (37°, 100% relative humidity, 25 μm thick) [9]

Water Absorption:

No.	Value	Note
1	3.25 %	50 h, 25°, 75% relative humidity [14]

Gas Permeability:

No.	Gas	Value	Note
1	H_2	236.4 cm^3 mm/(m^2 day atm)	3.6×10^{-10} cm^2 (s cm Hg)$^{-1}$, 30°, 10 atm
2	H_2	682.9 cm^3 mm/(m^2 day atm)	10.4×10^{-10} cm^2 (s cm Hg)$^{-1}$, 100°, 10 atm
3	O_2	11 cm^3 mm/(m^2 day atm)	0.168×10^{-10} cm^2 (s cm Hg)$^{-1}$, 30°, 10 atm
4	O_2	38.8 cm^3 mm/(m^2 day atm)	0.591×10^{-10} cm^2 (s cm Hg)$^{-1}$, 100°, 10 atm
5	N_2	0.92 cm^3 mm/(m^2 day atm)	0.014×10^{-10} cm^2 (s cm Hg)$^{-1}$, 30°, 10 atm
6	N_2	5.45 cm^3 mm/(m^2 day atm)	0.0834×10^{-10} cm^2 (s cm Hg)$^{-1}$, 100°, 10 atm
7	O_2	19400 cm^3 mm/(m^2 day atm)	485.5 cm^3 m^{-2} d^{-1} atm^{-1}, room temp., 0.0025 cm thick [9]
8	CO_2	44840 cm^3 mm/(m^2 day atm)	1121 cm^3 m^{-2} d^{-1} atm^{-1}, room temp, 0.0025 cm thick [9]

Mechanical Properties:
Tensile (Young's) Modulus:

No.	Value	Note
1	1194 MPa	cast film [4]
2	2516 MPa	365 ksi, 1 mil film, N_2 imidised [17]

Tensile Strength Break:

No.	Value	Note	Elongation
1	113.4 MPa	cast film [4]	9.2%
2	0.09 MPa	13.3 psi, 1 mil film, N_2 imidised [17]	

Tensile Strength Yield:

No.	Value	Note
1	0.05 MPa	7.3 psi, 1 mil film, N_2 imidised [17]

Mechanical Properties Miscellaneous: Adhesion to steel and to PA7 aluminium alloy has been reported [3]

Friction Abrasion and Resistance: Dynamic coefficient of friction μ_{dyn}, 0.26 (steel, 2000 cycles, 5 N load, 22°), 0.32 (steel, 16000 cycles, 5 N load, 22°), 0.19 (steel, 2000 cycles, 10 N load, 22°)[4]. Wear rate 171 μm^2 (1000 cycles)$^{-1}$ [4]

Electrical Properties:
Dielectric Permittivity (constant):

No.	Value	Frequency	Note
1	3	1 kHz	[13]
2	2.95	10 kHz	[13]

Strong Field Phenomena General: Ac field breakdown strength 370 V m^{-1} (26–60 μm film); dc field breakdown strength 534 V m^{-1} (55 μm film) [6]
Dissipation (Power) Factor:

No.	Value	Frequency	Note
1	0.002	1 kHz	[13]
2	0.011	10 kHz	[13]

Optical Properties:
Transmission and Spectra: Ir spectral data have been reported [18]
Volume Properties/Surface Properties: Opaque (film) [13]

Polymer Stability:
Polymer Stability General: Skybond has a shelf life of 3 months at 5° [1]
Thermal Stability General: Details of thermal stability have been reported [2,12]
Upper Use Temperature:

No.	Value	Note
1	340°C	max., Skybond 703 [1]

Decomposition Details: Degradation temp. 500° (min., N_2) [7]. 1.25% Weight loss in air at 100°, 57.5% weight loss in air at 700°, 50% weight loss in air at 800° [13]
Flammability: Non-flammable [13]
Environmental Stress: Is stable to sunlight [13]
Chemical Stability: Is resistant to organic solvents [13]. Attacked by alkali and strong acids [13]

Applications/Commercial Products:
Processing & Manufacturing Routes: Synth. by reaction of benzophenonetetracarboxylic acid dianhydride with diaminophenylmethane [13,15] in 10% dimethylacetamide at 0° for 3 h under a N_2 atmosphere. The poly(amic acid) is then imidised at up to 250° over 4 h [7]. Also synth. by reacting benzophenonetetracarboxylic acid with bis(4-isocyanophenyl)methane in dimethylacetamide at 0°, followed by heating up to 130° over a 6.5 h period [13,16].
Applications: Used in high-temp. applications; in coatings for wires, as an adhesive for high performance composites in the aerospace industry

Trade name	Details	Supplier
Skybond	703, 705	Monsanto Chemical

Bibliographic References
[1] Structural Adhesives, *Directory and Databook,* Chapman and Hall, 1996, 240
[2] Mathur, A.B., Srivastava, S.K., Singh, P.K. and Mathur, G.N., Therm. Anal., *Proc. Int. Conf.,* 7th, 1982, **2**, 1071, (thermal stability)
[3] Zurakowska-Orszagh, J., Kurzela, M. and Gora, M., *Polimery (Warsaw),* 1977, **22**, 80, (adhesion)
[4] Jones, J.W. and Eiss, N.S., *ACS Symp. Ser.,* 1985, **287**, 135, (wear)
[5] Li, Y., Wang, X., Ding, M. and Xu, J., *J. Appl. Polym. Sci.,* 1996, **61**, 741, (gas permeability)

[6] Bjellheim, P. and Helgee, B., *J. Appl. Polym. Sci.*, 1993, **48**, 1587, (electric strength)
[7] Jin, Q., Yamashita, T., Horie, K., Yokota, R. and Mita, I., *J. Polym. Sci., Part A: Polym. Chem.*, 1993, **31**, 2345, (synth, thermal props)
[8] Numata, S., Kinjo, N. and Makino, D., *Polym. Eng. Sci.*, 1988, **28**, 906
[9] Sykes, G.F. and St Clair, A.K., *J. Appl. Polym. Sci.*, 1986, **32**, 3725, (gas permeability)
[10] Chartoff, R.P. and Chiu, T.W., *Org. Coat. Plast. Chem.*, 1978, **39**, 129
[11] St Clair, T.L., St Clair, A.K. and Smith, E.N., Struct.-Solubility Relat. Polym., *(Proc. Symp.)*, 1976, 199, (solubility)
[12] Gillham, J.K., Hallock, K.D. and Stadnicki, S.J., Soc. Plast. Eng., *Tech. Pap.*, 1972, **18**, 229, (thermal anal)
[13] Khune, G.D., *J. Macromol. Sci., Chem.*, 1980, **14**, 687, (synth, props)
[14] Numata, S., Fujisaki, K. and Kinjo, N., *Polymer*, 1987, **28**, 2282, (thermal expansion, moisture permeability)
[15] Yang, C.-P. and Chen, Y.-H., *Angew. Makromol. Chem.*, 1988, **160**, 91, (use)
[16] Ghatge, N.D. and Dandge, D.K., *Angew. Makromol. Chem.*, 1976, **56**, 163, (synth, degradation)
[17] Bell, V.L., Stump, B.L. and Gager, H., *J. Polym. Sci., Polym. Chem. Ed.*, 1976, **14**, 2275, (mech props)
[18] Crandall, E.W., Johnson, E.L. and Smith, C.H., *J. Appl. Polym. Sci.*, 1975, **19**, 897, (ir)
[19] Gillham, J.K. and Gillham, H.C., Am. Chem. Soc., Div. Org. Coat. Plast. Chem., *Pap*, 1973, **33**, 201
[20] Coulehan, R.E., *Rheol. Acta*, 1974, **13**, 149, (rheological props)

BTDA-MPD/MDA   B-25

where x : y = 1 : 4

Synonyms: *Poly(3,3′,4,4′-benzophenonetetracarboxylic dianhydride-co-2-methyl-1,3-phenylenediamine-co-4,4′-methylenedianiline). Poly[4,4′-carbonylbis-1,2-benzenedicarboxylic acid-co-1,3-diisocyanatomethylbenzene-co-1,1′-methylenebis(4-isocyanatobenzene)]. Poly[5,5′-carbonylbis(1,3-isobenzofurandione)-co-1,3-diisocyanatomethylbenzene-co-1,1′-methylenebis(4-isocyanatobenzene)]*

Monomers: 5,5′-Carbonylbis[1,3-isobenzofurandione], 4,4′-Carbonylbis[1,2-benzenedicarboxylic acid], 1,3-Diisocyanato-2-methylbenzene, 1,1′-Methylenebis[4-isocyanatobenzene]
Material class: Thermoplastic
Polymer Type: Polyimide
CAS Number:

CAS Reg. No.	Note
9046-51-9	source based: dianhydride and diisocyanates
62181-46-8	source based: tetraacid and diisocyanates

Molecular Formula: $[(C_{24}H_{12}N_2O_5).(C_{30}H_{16}N_2O_5)]_n$
Fragments: $C_{24}H_{12}N_2O_5$ $C_{30}H_{16}N_2O_5$
Molecular Weight: MW 30100-47100 (increasing with increasing annealing temp.). M_n 9900-11100 (decreases with increasing annealing temp.)
Additives: Silica to improve mouldability
Morphology: Annealing of PI2080 at temps. close to T_g gives a homogeneous, amorph. polymer [9]

Volumetric & Calorimetric Properties:
Density:

No.	Value	Note
1	d 1.3507 g/cm^3	film, PI2080 [5]
2	d 1.337 g/cm^3	PI2080 [9]

Thermal Expansion Coefficient:

No.	Value	Note	Type
1	5.9×10^{-5} K^{-1}	PI2080 [10]	L

Volumetric Properties General: Density is relatively unaffected by annealing [9]
Glass-Transition Temperature:

No.	Value	Note
1	315°C	[5]
2	310°C	[7]
3	303°C	PI2080, DSC method [8,9]
4	307°C	PI2080, TMA method [8]
5	310–315°C	PI2080 [10]

Transition Temperature General: T_g of PI2080 increases with increasing annealing temp.
Deflection Temperature:

No.	Value	Note
1	270–280°C	PI2080 [10]
2	292°C	PI2080 [9]

Surface Properties & Solubility:
Solubility Properties General: PI2080 is miscible with poly(2,2′-(*m*-phenylene)-5,5′-bibenzimidazole) in all proportions [7]. PI2080 forms blends with *N,N*′-(methylenedi-*p*-phenylene)bismaleimide which may be polymerised above 180° to give hard films [13]
Plasticisers: *N*-Cyclohexyl-2-pyrrolidone [3]
Solvents/Non-solvents: PI2080 sol. dimethylacetamide [7,10], DMF [8,10], *N*-methylpyrrolidone [10], DMSO [10]

Transport Properties:
Polymer Solutions Dilute: η 0.5 dl g^{-1} (approx., PI2080 film) [9]
Permeability of Gases: The separation factor of cast PI2080 membranes increases with increasing annealing time and shows optimal performance when the casting solution is composed of 25 wt% PI2080, 37.5 wt% DMF and 37.5 wt% dioxane [4].
Permeability of Liquids: Permeability: MeOH 5.02×10^{-8}; 1.07×10^{-8}; EtOH 6.73×10^{-8}; C_6H_6 1.15×10^{-8}; CHCl$_3$ 4.79×10^{-9} mol (m^2 Pa)$^{-1}$ [6]. Other permeability values have been reported [6]
Permeability and Diffusion General: Long-time diffusion coefficients have been reported [5,8]

Mechanical Properties:
Tensile (Young's) Modulus:

No.	Value	Note
1	1300 MPa	25°, PI2080 [10]
2	660 MPa	285°, PI2080 [10]
3	6200 MPa	6.2×10^{10} dyne cm^{-2} PI2080) [9]

Flexural Modulus:

No.	Value	Note
1	3300 MPa	25°, PI2080 [10]
2	1100 MPa	285°, PI2080 [10]

– BTDA-ODA

Tensile Strength Break:

No.	Value	Note	Elongation
1	120 MPa	25°, PI2080 [10]	10%
2	29 MPa	285°, PI2080 [10]	
3	170 MPa	1.7×10^9 dyne cm^{-2}, PI2080 [9]	7%

Flexural Strength at Break:

No.	Value	Note
1	200 MPa	25°, PI2080 [10]
2	34 MPa	285°, PI2080 [10]

Compressive Modulus:

No.	Value	Note
1	2000 MPa	25°, PI2080 [10]

Compressive Strength:

No.	Value	Note
1	200 MPa	25°, PI2080 [10]

Viscoelastic Behaviour: Compression creep 0.04% (24 h, 25°, 1.87 GPa), 0.47% (24 h, 312°) [10]

Optical Properties:
Optical Properties General: The effect of curing on optical props. has been reported [12]
Molar Refraction: dn/dc 0.208 (PI2080) [11]

Polymer Stability:
Thermal Stability General: Thermal stability in a flow of N_2 has been reported [2]. PI2080 undergoes thermal and thermooxidative degradation causing cross-linking which is initiated by the formation of radicals from cleavage of methylene-phenyl bond [8]
Decomposition Details: Decomposition temp. 523–534° (annealed sample) [1]
Flammability: Oxygen index 44% (PI2080) [10]

Applications/Commercial Products:

Trade name	Supplier
Lenzing P84	Lenzing AG
PI2080	Dow
PI2080	Mitsubishi Kasei
Upjohn Polyimide 2080	Upjohn

Bibliographic References

[1] Kokugan T., Narabe, H., Shibamoto, A., Kitamura, A et al, *Sen'i Gakkaishi*, 1989, **45**, 95, (thermal anal)
[2] Yokota, R., Sakino, T. and Mita, I., *Kobunshi Ronbunshu*, 1990, **47**, 207, (thermal stability)
[3] *Eur. Pat. Appl.*, 1985, 148 719
[4] Yanagishita, H., Maejima, C., Kitamoto, D. and Nakane, T., *J. Membr. Sci.*, 1994, **86**, 231
[5] Toi, K., Ito, T., Shirakawa, T. and Ikemoto, I., J. Polym. Sci., *Part B: Polym. Phys.*, 1992, **30**, 549, (gas diffusion)
[6] Feng, X., Sourirajan, S., Tezel, H. and Matsuura, T., *J. Appl. Polym. Sci.*, 1991, **43**, 1071, (vapour permeability)
[7] Stankovic, S., Guerra, G., Williams, D.J., Karasz, F.E. and MacKnight, W.J., *Polym. Commun.*, 1988, **29**, 14, (miscibility)
[8] Kuroda, S.-I. and Mita, I., *Eur. Polym. J.*, 1989, **25**, 611, (degradation)
[9] Kochi, M., Mita, I. and Yokota, R., *Polym. Eng. Sci.*, 1984, **24**, 1021, (mech props)
[10] Sarbolouki, M.N., *J. Appl. Polym. Sci.*, 1984, **29**, 743
[11] Segudovic, N., Karasz, F.E. and MacKnight, W.J., *J. Liq. Chromatogr.*, 1990, **13**, 2581
[12] Franke, H. and Crow, J.D., *Proc. SPIE-Int. Soc. Opt. Eng.*, 1986, **651**, 102
[13] Yamamoto, Y., Satoh, S. and Etoh, S., *SAMPE J.*, 1985, **21**, 6

BTDA-ODA B-26

Synonyms: Poly(3,3',4,4'-benzophenonetetracarboxylic dianhydride-alt-4,4'-oxydianiline). Poly(3,3',4,4'-benzophenonetetracarboxylic dianhydride-alt-bis(4-aminophenyl) ether). Poly(4,4'-oxydiphenylene benzophenonetetracarboximide). Poly[(1,3-dihydro-1,3-dioxo-2H-isoindole-2,5-diyl)carbonyl(1,3-dihydro-1,3-dioxo-2H-isoindole-5,2-diyl)-1,4-phenyleneoxy-1,4-phenylene]. Poly[5,5'-carbonylbis-1,3-isobenzofurandione-alt-4,4'-oxybis(benzenamine)]. Poly[4,4'-oxybis(benzenamine)-alt-5,5'-carbonylbis-1,3-isobenzofurandione]
Monomers: 5,5'-Carbonylbis[1,3-isobenzofurandione], 4,4'-Oxybis[benzenamine]
Material class: Semithermoplastic
Polymer Type: Polyimide
CAS Number:

CAS Reg. No.	Note
24991-11-5	
24980-39-0	source based

Molecular Formula: $(C_{29}H_{14}N_2O_6)_n$
Fragments: $C_{29}H_{14}N_2O_6$

Volumetric & Calorimetric Properties:
Density:

No.	Value
1	d^{30} 1.373 g/cm^3

Thermal Expansion Coefficient:

No.	Value	Note	Type
1	5.52×10^{-5} K^{-1}	50-250° [1]	L

Volumetric Properties General: Fractional free volume 0.124 [2]
Glass-Transition Temperature:

No.	Value	Note
1	266°C	[2]

Surface Properties & Solubility:
Cohesive Energy Density Solubility Parameters: δ 32.7 J$^{1/2}$ cm$^{-3/2}$ [2]

Transport Properties:
Permeability of Liquids: Water permeability 41.2×10^{12} g (cm s cmHg)$^{-1}$
Water Absorption:

No.	Value	Note
1	2.35 %	50h, 25°, 75% relative humidity [1]

Gas Permeability:

No.	Gas	Value	Note
1	H$_2$	314.5 cm^3 mm/(m^2 day atm)	4.79×10^{-10} cm^2 (s cmHg)$^{-1}$, 35°, 10 atm. [2]
2	CO	3.3 cm^3 mm/(m^2 day atm)	0.0499×10^{-10} cm^2 (s cmHg)$^{-1}$, 35°, 10 atm. [2]

BTDA-mPDA

3	CO_2	41 cm³ mm/(m² day atm)	0.625×10^{-10} cm² (s cmHg)$^{-1}$, 35°, 10 atm. [2]
4	CH_4	0.72 cm³ mm/(m² day atm)	0.0109×10^{-10} cm² (s cmHg)$^{-1}$, 35°, 10 atm. [2]
5	O_2	12.5 cm³ mm/(m² day atm)	0.191×10^{-10} cm² (s cmHg)$^{-1}$, 35°, 2 atm. [2]
6	N_2	1.5 cm³ mm/(m² day atm)	0.0236×10^{-10} cm² (s cmHg)$^{-1}$, 35°, 2 atm. [2]

Mechanical Properties:
Tensile (Young's) Modulus:

No.	Value	Note
1	4300 MPa	[3]
2	12100 MPa	cold drawn film, parallel to drawing direction [3]
3	3300 MPa	cold drawn film, perpendicular to drawing direction

Tensile Strength Break:

No.	Value	Note
1	190 MPa	[3]
2	310 MPa	cold drawn film, parallel to drawing direction [3]
3	160 MPa	cold drawn film, perpendicular to drawing direction [3]

Applications/Commercial Products:

Trade name	Supplier
Pyralin Pl 2550	DuPont

Bibliographic References

[1] Numata, S., Fujisaki, K. and Kinjo, N., *Polymer,* 1987, **28**, 2282, (thermal expansion)
[2] Tanaka, K., Kita, H., Okano, M. and Okamoto, K.-I., *Polymer,* 1992, **33**, 585, (gas permeability)
[3] Kochi, M., Yokota, R., Iizuka, T. and Mita, I., *J. Polym. Sci., Part B: Polym. Phys.,* 1990, **28**, 2465, (mechanical props)

BTDA-mPDA B-27

Synonyms: *Poly(3,3',4,4'-benzophenonetetracarboxylic dianhydride-alt-m-phenylenediamine). Poly[(1,3-dihydro-1,3-dioxo-2H-isoindole-2,5-diyl)carbonyl(1,3-dihydro-1,3-dioxo-2H-isoindole-5,2-diyl)-1,3-phenylene]. Poly[5,5'-carbonylbis(1,3-isobenzofurandione)-alt-1,3-benzenediamine]*
Monomers: 5,5'-Carbonylbis[1,3-isobenzofurandione], 1,3-Benzenediamine
Material class: Semithermoplastic
Polymer Type: Polyimide

CAS Number:

CAS Reg. No.	Note
25868-65-9	
25038-83-9	source based

Molecular Formula: $(C_{23}H_{10}N_2O_5)_n$
Fragments: $C_{23}H_{10}N_2O_5$
Additives: Boron fibre, glass fibre. Aromatic phosphorus-containing compounds added to improve fire resistance

Volumetric & Calorimetric Properties:
Density:

No.	Value	Note
1	d 1.15–1.18 g/cm³	Skybond 700 [1]
2	d 1.3896–1.495 g/cm³	[4]
3	d 1.40844 g/cm³	film, bifix cured [5]
4	d 1.41039 g/cm³	film, free cured [5]

Thermal Expansion Coefficient:

No.	Value	Note	Type
1	2.94×10^{-5} K^{-1}	film, bifix cured [5,9]	L
2	3.67×10^{-5} K^{-1}	film, free cured [5]	L

Melting Temperature:

No.	Value	Note
1	520°C	[3]

Glass-Transition Temperature:

No.	Value	Note
1	307°C	[10]
2	300°C	[11]

Surface Properties & Solubility:
Solubility Properties General: The relationship between solubility and struct. has been reported [8]

Transport Properties:
Polymer Solutions Dilute: η 3-7 Pa s (Skybond 700) [1], 0.43 dl g^{-1} [4]
Permeability of Liquids: Permeability of water 38.5×10^{-12} g (cm s cmHg)$^{-1}$, (25°, 75% relative humidity) [5]
Water Absorption:

No.	Value	Note
1	3.15 %	50 h, 25°, 75% relative humidity [5]

Gas Permeability:

No.	Gas	Value	Note
1	O_2	608 cm³ mm/(m² day atm)	106.4 cm³ m^{-2}, room temp., atm. pressure, 0.0025 cm film [11]
2	CO_2	2394 cm³ mm/(m² day atm)	304.7 cm³ m^{-2} d^{-1} room temp., atm. pressure, 0.0025 cm film [11]

Mechanical Properties:
Mechanical Properties General: Films retain their strength and flexibility after annealing at 300° [3]
Mechanical Properties Miscellaneous: Bonds aluminium to aluminium with a lap shear strength of 3.51 MPa (510 psi, ASTM D1002-72). Commercial Skybond resin bonds aluminium to aluminium with a lap shear strength of 2.83 MPa (410 psi) [4]

Electrical Properties:
Electrical Properties General: The effect of moisture on electrical props. has been reported [12]

Optical Properties:
Volume Properties/Surface Properties: Colour light yellow (freshly cast film) [3]

Polymer Stability:
Polymer Stability General: Skybond has a shelf life of 3 months at 5° [1]
Thermal Stability General: Thermal stability may be decreased upon addition of aromatic phosphorus compounds used as fire retarding agents [10]
Upper Use Temperature:

No.	Value	Note
1	370°C	max., Skybond 700 [1]

Decomposition Details: Decomposition temp. 550° (in N_2) [3]. 10% weight loss 475°; 50% weight loss 790° (N_2) [4]. Char yield 49% (in N_2) [4]; 63% (700, N_2) [10] 1% (700°, air) [10]. Activation energy of decomposition 124 kJ mol^{-1} (N_2) [4]. The major products of decomposition are CO and CO_2. Shows approx. 40% weight loss after heating in air at 400° for approx. 30 h. The max. rate of weight loss is 1.74% h^{-1} which occurs after 14 h heating in air at 400° [7]. Max. decomposition temp. 631° (in N_2) [10]
Flammability: Limiting oxygen index 43.9% (ASTM D2863-74) [10]
Hydrolytic Stability: Effect of hot geothermal brine has been reported [2]

Applications/Commercial Products:
Processing & Manufacturing Routes: Synth. by reaction of benzophenonetetracarboxylic acid and *m*-phenylenediamine in dry DMF below 30°, followed by thermal imidisation [7] or chemical imidisation using an Ac_2O/pyridine mixture [10]
Applications: Used in high temp. applications and as an adhesive for high performance composites in the aerospace industry

Trade name	Supplier
Skybond 700	Monsanto Chemical

Bibliographic References
[1] Structural Adhesives, *Directory and Databook*, Chapman & Hall, 1996, 240
[2] Lorensen, L.E., Walkup, C.M. and Mones, E.T., Natl. Sci. Found., Res. Appl. Natl. Needs, *(Rep.) NSF/RA (U.S.)*, 1976, 1725
[3] Preston, J., Dewinter, W.F. and Black, W.B., *J. Polym. Sci., Part A-1*, 1969, **7**, 283, (film props)
[4] Varma, I.K. and Rao, B.S., *J. Polym. Sci., Polym. Lett. Ed.*, 1983, **21**, 545, (degradation, adhesion)
[5] Numata, S., Fujisaki, K. and Kinjo, N., *Polymer*, 1987, **28**, 2282, (thermal expansion, moisture permeability)
[6] Jewell, R.A., *J. Appl. Polym. Sci.*, 1971, **15**, 1717, (degradation)
[7] Dine-Hart, R.A. and Wright, W.W., *Makromol. Chem.*, 1972, **153**, 237, (degradation)
[8] St Clair, T.L., St Clair, A.K. and Smith, E.N., Struct.-Solubility Relat. Polym., *(Proc. Symp.)*, 1976, 199, (solubility)
[9] Numata, S., Kinjo, N. and Makino, D., *Polym. Eng. Sci.*, 1988, **28**, 906, (thermal expansion)
[10] Mikroyannidis, J.A., *J. Polym. Sci., Polym. Chem. Ed.*, 1984, **22**, 1065, (thermal stability, additives)
[11] Sykes, G.F. and St Clair, A.K., *J. Appl. Polym. Sci.*, 1986, **32**, 3725, (gas permeability)
[12] Subramanian, R., Pottiger, M.T., Morris, J.H. and Curilla, J.P., *Mater. Res. Soc. Symp. Proc.*, 1991, **227**, 147

BTDA-*m*PDA/ODA   B-28

Synonyms: *Poly(3,3',4,4'-benzophenonetetracarboxylic dianhydride-co-m-phenylene-4,4'-oxydianiline). Poly(4,4'-oxybisbenzenamine-co-1,3-benzenediamine-co-5,5'-carbonylbis[1,3-isobenzofurandione]). Poly(5,5'-carbonylbis[1,3-isobenzofurandione]-co-1,3-benzenediamine-co-4,4'-oxybisbenzenamine). Poly(1,3-benzenediamine-co-5,5'-carbonylbis[1,3-isobenzofurandione]-co-4,4'-oxybisbenzenamine)*
Monomers: 5,5'-Carbonylbis[1,3-isobenzofurandione], 1,3-Benzenediamine, 4,4'-Oxybis[benzenamine]
Material class: Semithermoplastic
Polymer Type: Polyimide
CAS Number:

CAS Reg. No.	Note
31942-21-9	source based

Molecular Formula: $[(C_{23}H_{10}N_2O_5).(C_{29}H_{14}N_2O_6)]_n$
Fragments: $C_{23}H_{10}N_2O_5$ $C_{29}H_{14}N_2O_6$

Volumetric & Calorimetric Properties:
Thermal Expansion Coefficient:

No.	Value	Note	Type
1	4.7×10^{-5} K^{-1}	50–250°, unoriented film [1]	L
2	0.00015 K^{-1}	50–250°, unoriented film [1]	V

Equation of State: Tait equation of state parameters have been reported [1]

Applications/Commercial Products:

Trade name	Supplier
Pyralin Pl-2525	DuPont
Pyralin Pl-2555	DuPont
Pyralin Pl-2556	DuPont

Bibliographic References
[1] Pottiger, M.T., Coburn, J.C. and Edman, J.R., *J. Polym. Sci., Part B: Polym. Phys.*, 1994, **32**, 825, (thermal expansion)

Butadiene-acrylonitrile copolymers   B-29

Synonyms: *Poly(butadiene-co-acrylonitrile)*
Related Polymers: Polybutadiene, Nitrile rubber
Monomers: Acrylonitrile, 1,3-Butadiene
Material class: Synthetic Elastomers
Polymer Type: Acrylonitrile and copolymers, Polybutadiene
CAS Number:

CAS Reg. No.
9003-18-3

Applications/Commercial Products:

Trade name	Supplier
Krynac	Polysar
Nipol N	Nippon Zeon
Nysyn	Copolymer Rubber

Butadiene-pentadiene copolymers   B-30

Synonyms: *Poly(1,3-butadiene-co-1,3-pentadiene)*
Related Polymers: Polybutadiene
Monomers: 1,3-Butadiene, Pentadiene
Material class: Synthetic Elastomers
Polymer Type: Polybutadiene
CAS Number:

CAS Reg. No.	Note
25102-53-8	
107194-69-4	block

Butadiene-styrene-vinylidene chloride terpolymer   B-31

$$-[CH_2CCl_2]_x-[CH_2CH=CHCH_2]_y-[CH_2CH(Ph)]_z-_n$$

Synonyms: *Poly(1,3-butadiene-co-1,1-dichloroethene-co-ethenylbenzene). Poly[(1,1-dichloroethene)-co-(1,4-butadiene)-co-ethenylbenzene]*
Related Polymers: Poly(vinylidene chloride), Vinylidene chloride copolymers, Polybutadiene, Polystyrene
Monomers: 1,3-Butadiene, Styrene, Vinylidene chloride
Material class: Copolymers, Synthetic Elastomers
Polymer Type: Polybutadiene, Polystyrenepolyvinylidene chloridepolyhaloolefins
CAS Number:

CAS Reg. No.
31669-55-3

Molecular Formula: $[(C_2H_2Cl_2)_x.(C_4H_6)_y.(C_8H_8)_z]_n$
Fragments: $C_2H_2Cl_2$ C_4H_6 C_8H_8
General Information: Typical compositions vinylidene chloride:-butadiene:styrene, 5:70:25-35:36:27. Some commercial products are carboxylated
Miscellaneous: Sulfur-free vulcanised materials can be produced by using metal oxides, amines and thiuram. [1] Binder saturant formulations of carboxylated terpolymer (52–53% total solids) are available commercially. [10] Formulations containing ≤40% vinylidene chloride are acceptable under US Federal Food, Drug, and Cosmetic Act in certain food-contact applications [3]

Volumetric & Calorimetric Properties:
Density:

No.	Value	Note
1	d_1 0.928 g/cm^3	9.1–9.3 lb gal^{-1}, binder saturant formulation of carboxylated terpolymer; 52–53% total solids [10]

Surface Properties & Solubility:
Surface Tension:

No.	Value	Note
1	40 mN/m	binder saturant formulation of carboxylated terpolymer; 52–53% total solids [10]

Transport Properties:
Transport Properties General: High-solids latex formulations display good high-shear rheology and good low-shear viscosity [4]
Polymer Solutions Dilute: Viscosity 40–50 cP (binder saturant formulation of carboxylated terpolymer; 52–53% total solids) [10]

Mechanical Properties:
Mechanical Properties General: Physicomechanical props. of terpolymer vulcanisates are similar to those of styrene/butadiene copolymer rubbers. Inferior tear strength and superior cold resistance have been reported [1]
Friction Abrasion and Resistance: Carboxylated terpolymer displays good abrasion resistance [10]

Polymer Stability:
Flammability: Carboxylated terpolymer displays good flame resistance [10]
Chemical Stability: Vulcanised material has similar resistance to petroleum as styrene/butadiene copolymer rubbers

Applications/Commercial Products:
Processing & Manufacturing Routes: Prepared by emulsion polymerisation of mixture of corresponding monomers (free-radical initiated) [1]; can be vulcanised in presence of metal oxides, amines and thiuram [1]
Applications: Bindings in textile coverings (e.g. nylon carpets); smoulder-resistant back coatings to cotton fabrics; paper coatings (high-solids latexes); industrial microbiocide research (as latex); quick-set latex adhesive compositions; indirect food additives, e.g. as components of paper/paperboard coatings for aq. and fatty food contact

Trade name	Details	Supplier
Darex	529 L, 535 L non-exclusive; carboxylated binder saturant formulations (52–53% total solids)	W R Grace
Gen Flo 8520	non-exclusive	General Tyre and Rubber Co.
SKS-25KhV-5ARK	5/70/25 (VDC/butadiene/styrene) terpolymer	AB Svensk Konstsilke

Bibliographic References
[1] Epshtein, V.G., Chekanova, A.A., Tsouilingol'd, V.L., and Nazarova, A.N., *Uch. Zap. Yaroslav. Tekhnol. Inst.*, 1971, **22**, 159, (synth, processing, vulcanisation, stability, mechanical props.)
[2] Donaldson, D.J., Mard, H.H.S., Harper, R.J., *Text. Res. J.*, 1979, **49**, 185, (uses)
[3] *Fed. Regist.*, 21 Jan 1983, **48**, 2749, (uses, haz.)
[4] *U.S. Pat.*, 1984, 4 474 860, (uses, transport props)
[5] *U.S. Pat.*, 1986, 4 567 099, (uses, transport props.)
[6] *Eur. Pat. Appl.*, 1987, 226 111, (uses)
[7] *U.S. Pat.*, 1988, 4 725 611, (uses)
[8] *U.S. Pat.*, 1988, 4 725 612, (uses)
[9] *Eur. Pat. Appl.*, 1988, 294 027, (uses)
[10] Ash, M. and Ash, I., *Handb. Plast. Elastomers*, VCH, 1992, 618, (volumetric props, uses, stability, mechanical props, transport props, surface props)

1,4-Butanediol diglycidyl ether resin B-32

Synonyms: *Poly(butylene glycol diglycidyl ether). Poly(2,2'-[1,4-butanediylbis(oxymethylene)]bisoxirane). PDGEBD*
Monomers: 1,4-Butanediol diglycidyl ether
Material class: Thermosetting resin
Polymer Type: Epoxy
CAS Number:

CAS Reg. No.	Note
29611-97-0	homopolymer
69777-27-1	copolymer with 1,2-ethanediamine
69777-29-3	copolymer with diaminodiphenylmethane
104559-45-7	copolymer with 4-ethylpyridine
105692-42-0	copolymer with butylamine and 1,8-octanediamine
105692-54-4	copolymer with 1,10-decanediamine

Molecular Formula: $(C_{10}H_{18}O_4)_n$
Fragments: $C_{10}H_{18}O_4$
Molecular Weight: M_n 220
General Information: Props. of cured resin depend on curing agent, stoichiometry, cure time and cure temp.

Volumetric & Calorimetric Properties:
Density:

No.	Value	Note
1	d^{20} 1.05 g/cm^3	uncured [4]
2	d 1.2863 g/cm^3	diaminodiphenylsulfone cured [5]
3	d 1.073 g/cm^3	butylamine/1,8-octanediamine cured, zero cross-linking density [6]
4	d 1.182 g/cm^3	1,10-decanediamine cured, cross-linking density 4.27 mol kg^{-1} [6]

Volumetric Properties General: Density of diaminodiphenylsulfone cured resin increases with ageing below T_g [5]. Density of resin cured with butylamine and 1,8-octanediamine, or with 1,10-decanediamine, increases with increasing cross-linking density [6]
Glass-Transition Temperature:

No.	Value	Note
1	78°C	diaminodiphenylsulfone cured [1]
2	82°C	diaminodiphenylmethane cured [1]
3	7°C	1,2-ethanediamine cured [1]
4	77°C	diaminodiphenylsulfone cured [5]
5	-33°C	butylamine/1,8-octanediamine cured [6]
6	0°C	1,10-decanediamine cured, cross-linking density 4.27 mol kg^{-1} [6]
7	76°C	diaminodiphenylmethane cured [7]

Transition Temperature General: T_α of diaminodiphenylsulfone cured resin, but not T_β or T_γ, is affected by ageing below T_g [5]. T_g of resin cured with butylamine and 1,8-octanediamine increases with increasing cross-linking density [6]
Transition Temperature:

No.	Value	Note	Type
1	-55°C	approx., 1,2-ethanediamine, diaminodiphenylmethane cured [3]	T_β
2	64°C	diaminodiphenylsulfone cured [5]	T_α

Surface Properties & Solubility:
Cohesive Energy Density Solubility Parameters: δ 22.2 MPa$^{1/2}$ (uncured) [4]
Surface Tension:

No.	Value	Note
1	28.7 mN/m	50°, uncured [4]

Mechanical Properties:
Mechanical Properties General: Dynamic loss tangent is halved, at temps. between 60° and 80°, by ageing below T_g in resin cured by diaminodiphenylsulfone [5]
Viscoelastic Behaviour: Resin cured with diaminodiphenylsulfone shows a yield point at 6.2% strain. Specimens compressed to half their original length may be recovered by heating. Yield point may be improved by ageing below T_g [5]

Applications/Commercial Products:
Processing & Manufacturing Routes: A rubber-like solid may be obtained by heating 4-ethylpyridine with 1,4-butanediol diglycidyl ether for 5 min. at 140°. The aromaticity of the pyridine ring is lost [2]
Applications: Reactive diluent for epoxy resins. Used in adhesives, coatings

Trade name	Supplier
RD-2	Ciba-Geigy Corp.

Bibliographic References
[1] Bellenger, V., Verdu, J. and Morel, E., *J. Polym. Sci., Part B: Polym. Phys.*, 1987, **25**, 1219, (T_g)
[2] Xue, G., Ishida, H. and Koenig, J.L., *Polymer*, 1986, **27**, 1134
[3] Charlesworth, J.M., *J. Polym. Sci., Polym. Phys. Ed.*, 1979, **17**, 329, (mechanical relaxation)
[4] Chen, D., Pascault, J.P. and Sage, D., *Makromol. Chem.*, 1991, **192**, 883, (surface tension)
[5] Chang, T.-D. and Brittain, J.O., *Polym. Eng. Sci.*, 1982, **22**, 1221
[6] Charlesworth, J.M., *J. Macromol. Sci., Phys.*, 1987, **26**, 105
[7] Cukierman, S., Halary, J.-L. and Monnerie, L., *J. Non-Cryst. Solids*, 1991, **131-133**, 898

Butyl glycoyl ether resin B-33

Synonyms: *Poly((butoxymethyl) oxirane). Poly(1-butoxy-2,3-epoxypropane)*
Monomers: Butyl glycidyl ether
Material class: Thermosetting resin
Polymer Type: Epoxy
CAS Number:

CAS Reg. No.	Note
25610-58-6	homopolymer

Molecular Formula: $(C_7H_{14}O_2)_n$

– para-tert-Butyl Phenolic resins

Volumetric & Calorimetric Properties:
Density:

No.	Value	Note
1	d 0.982 g/cm^3	[5]

Melting Temperature:

No.	Value	Note
1	27°C	[5]

Glass-Transition Temperature:

No.	Value	Note
1	-79°C	[5]

Transport Properties:
Polymer Solutions Dilute: $[\eta]_{inh}$ 2.4 dl g^{-1} (toluene, 30°) [5]

Optical Properties:
Refractive Index:

No.	Value	Note
1	1.458	30° [5]

Applications/Commercial Products:
Processing & Manufacturing Routes: Synth. by anionic polymerisation using potassium hydride [1] or potassium tert-butoxide [2] in the presence of 18-crown-6; by ring-opening polymerisation using an aryl silyl ether [3]; by photopolymerisation initiated by an aluminium alkyl/aryl silyl ether [4] or by benzylsulfonium salts [6]; by soln. polymerisation in C_6H_6 at 50° using a diethylzinc-sulfur catalyst [5]
Applications: Reactive diluent for epoxy resins

Bibliographic References
[1] Stolarzewicz, A., Neugebauer, D. and Grobelny, J., *Macromol. Rapid Commun.*, 1996, **17**, 787, (synth)
[2] Stolarzewicz, A., Neugebauer, D. and Grobelny Z., *Macromol. Chem. Phys.*, 1995, **196**, 1295, (synth)
[3] Nambu, Y. and Endo, T., *Macromolecules*, 1991, **24**, 2127, (synth)
[4] Hayase, S., Onishi, Y., Suzuki, S. and Wada, M., *Macromolecules*, 1985, **18**, 1799, (synth)
[5] Lal, J. and Trick, G.S., *J. Polym. Sci., Part A-1*, 1970, **8**, 2339, (props)
[6] Hamazu, F., Akashi, S., Koizumi, T., Takata, T. and Endo, T., *J. Photopolym. Sci. Technol.*, 1992, **5**, 247, (synth)

para-tert-Butyl Phenolic resins   B-34

Synonyms: *4-(1,1-Dimethylethyl)phenol based resins*
Monomers: 4-*tert*-Butylphenol, Formaldehyde
Material class: Thermoplastic
Polymer Type: Phenolic
CAS Number:

CAS Reg. No.	Note
25085-50-1	
28963-94-2	copolymer with formaldehyde and hexa

Molecular Formula: $[(C_{10}H_{14}O).(CH_2O)]_n$
Fragments: $C_{10}H_{14}O$ CH_2O
Molecular Weight: MW 530-542 (novolak, uncured), 1260 (novolak, uncured); M_n 1050 (novolak, uncured); MW 460 (Yarresin B), 520 (101 K); M_n 1085 (all-*ortho*-novolak, uncured)
General Information: The resins do not crosslink as the phenol is difunctional (the *para*-position is blocked)
Morphology: Adopts a cyclic conformn. in non-polar solvents [2]

Volumetric & Calorimetric Properties:
Density:

No.	Value	Note
1	d 1.2 g/cm^3	novolak, uncured film [14]

Equation of State: Mark-Houwink-Sakurada equation of state information has been reported [9]
Melting Temperature:

No.	Value	Note
1	400–401°C	recrystallised cyclic resole, uncured [6]
2	60–95°C	all-*ortho* novolak, uncured, dependent on synth. [7]
3	84–86°C	novolak, uncured [10]
4	80–100°C	Yarresin B [15]
5	101°C	novolak, uncured [20]
6	157–171°C	all-*ortho* novolak, uncured [21]

Glass-Transition Temperature:

No.	Value	Note
1	51°C	novolak, uncured [8]
2	65°C	Bakelite 92370 [16]
3	85°C	PN430 [16]

Vicat Softening Point:

No.	Value	Note
1	120°C	novolak, uncured [20]

Surface Properties & Solubility:
Solubility Properties General: Low MW novolak resin forms miscible blends with atactic poly(methyl methacrylate) and bisphenol A polycarbonate [11]
Solvents/Non-solvents: Uncured novolak sol. Me_2CO, EtOH, $CHCl_3$, C_6H_6, petrol, turpentine, linseed oil, tung oil, 2-butanone, toluene, CCl_4, EtOAc [10,20]

Transport Properties:
Water Absorption:

No.	Value	Note
1	1.43–1.52 %	24h, 100% relative humidity, novolak, uncured [10]
2	4.12–4.3 %	144h, 100% relative humidity, novolak, uncured [10]

Electrical Properties:
Electrical Properties General: Has semiconductor props.; conduction dependency on temp. has been reported [5]

Optical Properties:
Transmission and Spectra: Mass [1], C-13 nmr [3,18] and H-1 nmr [18] spectral data have been reported
Volume Properties/Surface Properties: Colour 6 (Gardner, novolak, uncured) [20]

Polymer Stability:
Chemical Stability: Uncured novolak has excellent compatibility at 250° with linseed oil, tung oil, castor oil and double-boiled linseed oil [20]
Stability Miscellaneous: Electron beam irradiation *in vacuo* causes cross-linking [14]

Applications/Commercial Products:
Processing & Manufacturing Routes: May be synth. by a continuous process involving reaction of hexa and *p-tert*-butylphenol in a melt at 110-120° followed by addition of paraformaldehyde; optimum paraformaldehyde: *p-tert*-butylphenol 0.7:1 [4]. Reaction of p-*tert*-butylphenol with formaldehyde in the presence of aq. base produces a mixture of cyclic oligomers, ie. with calixarene struct. [6,13,15]. The solid resole is compounded with neoprene rubber and metallic oxides in solvent to make a contact adhesive. The solid novolak is compounded with kaolin and styrene-butadiene rubber latex with minor amounts of calcium carbonate, colloidal silica and hydroxyethyl starch to make a coating for carbonless paper. All-*ortho*novolak resin may be synth. by reaction of paraformaldehyde with 4-*tert*-butylphenoxymagnesium bromide in C_6H_6 or toluene [7,19,21]. Other synth. by acetal reaction has been reported [17]
Applications: All-*ortho* novolaks have potential applications as antioxidants in lard and as antimicrobial agents. Other uses include contact adhesives for electrical insulation varnishes, as coatings for metals, ceramics and some plastic surfaces. The resole tackifier is compounded to make a Neoprene rubber-phenolic contact adhesive. This finds application in the furniture, automotive, construction and footware industries as a rubber adhesive. The Novolak is part of the coated front (CF) sheet formulation which receives the image in the pressure-sensitive record sheet of the carbonless paper system. All-*ortho* novolaks have potential applications as antioxidants in lard and as antimicrobial agents. Other uses include contact adhesives for electrical insulation varnishes, as coatings for metals ceramics and some plastic surfaces

Trade name	Details	Supplier
101K	resole (NaOH catalysed); Russian tradename, Sirfen	SIR
92370	novolak	Bakelite
Durez 26799		Hooker/Durez
PN 430	novolak	Hoechst
Rutaphen KA		Bakelite
Yarresin B	resole (NH_3 catalysed); Russian tradename, Alresen	Huls America

Bibliographic References
[1] Saito, J.Toda, S. and Tanaka, S., *Netsu Kokasei Jushi*, 1981, **2**, 72, (ms)
[2] Ishida, S., Wakaki, S. and Nakamoto, Y., *Kobunshi Ronbunshu*, 1984, **41**, 565, (struct., conformn.)
[3] Mukoyama, Y., Tanno, T., Yokokawa, H. and Fleming, J., *J. Polym. Sci., Polym. Chem. Ed.*, 1973, **11**, 3193, (C-13 nmr)
[4] Shalangovskaya, T.M. and Arkhipov, M.I., *Lakokras. Mater. Ikh Primen.*, 1974, 7, (synth.)
[5] Kornev, E.V. and Negrobov, A.F., *Sb. Nauch. Rab. Aspir. Voronezh. Gos. Univ.*, 1968, 138, (electrical conductivity)
[6] Gutsche, C.D. and Muthukrishnan, R., *J. Org. Chem.*, 1978, **43**, 4905, (synth.)
[7] Casiraghi, G., Cornia, M., Ricci, G., Balduzzi, G. et al, *Makromol. Chem.*, 1983, **184**, 1363, (synth.)
[8] Kalkar, A.K. and Parkhi, P.S., *J. Appl. Polym. Sci.*, 1995, **57**, 233, (miscibility)
[9] Sue, H., Ueno, E., Nakamoto, Y. and Ishida, S.-I., *J. Appl. Polym. Sci.*, 1989, **38**, 1305, (equation of state)
[11] Kalkar, A.K. and Roy, N.K., *Eur. Polym. J.*, 1993, **29**, 1391, (miscibility)
[12] Pennacchia, J.R., Pearce, E.M., Kwei, T.K., Bulkin, B.J. and Chen, J.-P., *Macromolecules*, 1986, **19**, 973, (miscibility)
[13] Osawa, Z. and Tsurumi, K., *Polym. Degrad. Stab.*, 1989, **26**, 151, (synth., use)
[14] Tanigaki, K. and Iida, Y., *Makromol. Chem., Rapid Commun.*, 1986, **7**, 485, (synth., use)
[15] Semenov, S.A., Valkina, E.M. and Reznik, A.M., *Zh. Neorg. Khim.*, 1994, **39**, 670, (use)
[16] Weill, A., Dechenaux, E. and Paniez, P., *Microelectron. Eng.*, 1986, **4**, 285, (T_g)
[17] Zhang, P. and Zhao, M., *Tuliao Gongye*, 1993, 22, (synth.)
[18] Novikov, N.A., Mozoleva, A.P., Uvarov, A.V., Alekseeva, S.G. and Slonim, I.Ya., *Vysokomol. Soedin., Ser. A*, 1986, **28**, 1064, (H-1 nmr, C-13 nmr)
[19] Abe, Y., Matsumura, S., Asakura, and Kasama, H., *Yukagaku*, 1986, **35**, 751, (synth., use)
[20] Ahisanuddin, Saksena, S.C., Panda, H. and Rakhshinda, *Paintindia*, 1982, **32**, 3, (synth., solubility)
[21] Uchibori, T., Kawada, K., Watanabe, S., Asakura, K. et al, *Bokin Bobai*, 1990, **18**, 215, (synth., use)

Butyl rubber
Synonyms: GR-I. Poly(isobutylene-co-isoprene). IIR
Related Polymers: Bromobutyl rubber, Chlorobutyl rubber
Monomers: Isobutylene, Isoprene
Material class: Synthetic Elastomers
Polymer Type: Polydienes
CAS Number:

CAS Reg. No.
9006-49-9

Additives: Can be vulcanised by sulfur, zinc oxide and thiazole. Zinc dibutyl dithiocarbamate added as stabiliser, zinc stearate as an antioxidant. Carbon black added as a filler
Morphology: Is a random copolymer containing 1–3% isoprene. Methyl substituents create steric hindrance to rotation about the polymer backbone

Volumetric & Calorimetric Properties:
Density:

No.	Value	Note
1	d 0.917 g/cm^3	unvulcanised [1]
2	d 0.933 g/cm^3	pure gum vulcanisate [2]
3	d 0.93–0.97 g/cm^3	pure gum vulcanisate
4	d 1.13 g/cm^3	vulcanisate, 50 phv carbon black [2]

Thermal Expansion Coefficient:

No.	Value	Note	Type
1	0.00075 K^{-1}	unvulcanised [3]	V
2	0.00056 K^{-1}	pure gum vulcanisate [4]	V
3	0.00046 K^{-1}	vulcanisate, 50 phv carbon black	V
4	0.000194 K^{-1}	[48]	L

Thermodynamic Properties General: Studies of thermophysical props. of butyl rubber loaded with 100 phv of different types of carbon black and also barium titanate have been reported at temps. ranging from 30 to 150°C. The props. include thermal conductivity, thermal diffusivity and specific heat capacity [67]

Butyl rubber

Thermal Conductivity:

No.	Value	Note
1	0.13 W/mK	pure gum vulcanisate [9]
2	0.23 W/mK	vulcanisate, 50 phv carbon black [9]

Thermal Diffusivity: 7×10^{-8} m^2 s^{-1} [85]

Specific Heat Capacity:

No.	Value	Note	Type
1	1.95 kJ/kg.C	unvulcanised [6,7,8]	P
2	1.85 kJ/kg.C	pure gum vulcanisate [7]	P

Melting Temperature:

No.	Value	Note
1	1.5°C	equilibrium, unvulcanised [3]

Glass-Transition Temperature:

No.	Value	Note
1	-71°C	unvulcanised [3]
2	-63°C	pure gum vulcanisate [5]
3	-75--67°C	[61]

Surface Properties & Solubility:
Solubility Properties General: The compatibility of butyl rubber with other elastomers, cured and uncured, with and without carbon black has been reported [56,88]
Cohesive Energy Density Solubility Parameters: Solubility parameter δ 31.4 J$^{1/2}$ cm$^{-3/2}$ [26]
Solvents/Non-solvents: Sol. cyclohexane, C_6H_6. Insol. dioxane
Surface and Interfacial Properties General: Adhesion is affected by amount of carbon black filler [65]
Surface Tension:

No.	Value	Note
1	35.6 mN/m	24° [30,31]
2	27 mN/m	critical surface tension of spreading [82]

Transport Properties:
Polymer Solutions Dilute: θ Temp. 21° (isoamyl isovalerate); 22° (C_6H_6); 23.7° (isoamyl benzyl ether); 46.2° (butyl butyrate); 55.5° (3-methyl-5-heptanone). [84] Bulk viscosity at zero shear has been reported. [26] Mark-Houwink parameters have been reported. [26] MW determinations have been reported [56]
Polymer Melts: Relationships between melt viscosity and limiting viscosity have been reported [33]
Permeability of Gases: Relative gas permeabilities and diffusivities have been reported for a range of gases. [22,24] Permeability (O_2) 0.44 mol (m s Pa)$^{-1}$ [23]. Diffusion coefficients for H_2 and N_2 in butyl rubber have been reported over a range of temps. and compared with other elastomers. [49] The coefficients for butyl rubber are lower than those of other hydrocarbon elastomers by an order of magnitude
Permeability of Liquids: Water permeability is low
Volatile Loss and Diffusion of Polymers: Self-diffusion constants over a range of MW have been reported for butyl rubber and other elastomers [81]
Permeability and Diffusion General: Has low permeability as steric hindrance of methyl groups on the polymer backbone slows chain motion. [32] Diffusion coefficients have been reported for aliphatic and aromatic hydrocarbons at concentrations up to 2.4% in butyl rubber and compared with SBR and BR [87]

Gas Permeability:

No.	Gas	Value	Note
1	air	36.4 cm^3 mm/(m^2 day atm)	24° [94]
2	N_2	21.6 cm^3 mm/(m^2 day atm)	21° [94]

Mechanical Properties:
Mechanical Properties General: Steric hindrance between the methyl groups in butyl rubber gives it low resilience (and high damping) at low or ambient temp. but high resilience at 100° [26,32]. Hydrostatic pressure increases ultimate props. [51]. Has similar tensile props. to styrene-butadiene rubber, but with poorer resistance to abrasion and tearing. Carbon black improves tear resistance but not tensile props. [32,37,75,83] and its effect on cure has been reported [94]. Poisson's ratio has been measured for butyl rubber with different types of carbon black [91]

Tensile (Young's) Modulus:

No.	Value	Note
1	1 MPa	pure gum vulcanisate [5,17]
2	3-4 MPa	vulcanisate, 50 phv carbon black [16]
3	7.7 MPa	25° [78]
4	1.2 MPa	190° [78]
5	0.74 MPa	[53]

Tensile Strength Break:

No.	Value	Note	Elongation
1	18-21 MPa	pure gum vulcanisate [15,16]	750-950%
2	18 MPa	atmospheric pressure, 22° [51]	
3	24 MPa	200 MPa pressure, 22° [51]	310%
4	18-21 MPa	vulcanisate, 50 phv carbon black [15,16]	650-850%
5	10 MPa	pure gum vulcanisate [32]	800%
6	12-20 MPa	vulcanisate, 50 phv carbon black [43]	490-700%

Elastic Modulus:

No.	Value	Note
1	1970 MPa	pure gum vulcanisate [13]
2	2170 MPa	vulcanisate, 50 phv carbon black [12]

Miscellaneous Moduli:

No.	Value	Note	Type
1	0.33 MPa	pure gum vulcanisate [5,17]	shear modulus
2	1.8 MPa	vulcanisate, 50 phv carbon black [18]	shear modulus
3	2.1-8.6 MPa	vulcanisate, 50 phv carbon black [43]	300% modulus

Complex Moduli: Storage and loss moduli have been reported for vulcanised and unvulcanised rubber [20,25,45,64]
Mechanical Properties Miscellaneous: Compressibility 508 MPa^{-1} (pure gum vulcanisate) [4]; 460 MPa^{-1} (vulcanisate, 50 phv carbon black) [12]. Longitudinal bulk wave velocity 1485 m s^{-1} (pure gum vulcanisate) [12]; 1510 m s^{-1} (vulcanisate, 50 phv carbon black)

— Butyl rubber

[12]; longitudinal strip wave velocity 100 m s^{-1} (1 kHz, pure gum vulcanisate); 210 m s^{-1} (1 kHz, vulcanisate, 50 phv carbon black) [14]. Oxidised carbon black gives enhanced adhesion, lower tan δ, improved ultimate props. and abrasion resistance. [43] Peak longitudinal absorption 450 dB cm^{-1} (10 MHz) [27]. Resilience 13–16% (pure gum vulcanisate) [16,21]; 14% (vulcanisate, 50 phv carbon black) [16]. The effect of carbon black loading on dynamic mech. props. has been reported. [37,44,45,64,89] Interaction with carbon black is weaker than for other elastomers. [52] Heat build up and resilience are affected by curing temp. and system and carbon black fillers [94]

Hardness: Shore A40-A80 [32]; A53-A64 (vulcanisate, 50 phv carbon black) [43]

Fracture Mechanical Properties: Tearing energy has been reported for blends of butyl rubber with polyisobutylene over a range of composition from -30 to 100° [46]

Friction Abrasion and Resistance: Unlike other rubbers, butyl rubber causes more wear by sliding on stainless steel in air than in an inert atmosphere, owing to the stability of the peroxy radical. [47] Has poor abrasion resistance. Abrasion props., flex cracking, fatigue and cut-growth rate with respect to other elastomers have been reported. [86] Carbon black fillers affect these props. [92]

Electrical Properties:

Electrical Properties General: Has excellent electrical insulation props. [78] Resistivity, dielectric props. and current-voltage relationships are affected by amount of carbon black filler [37,72,73,74,76]

Dielectric Permittivity (constant):

No.	Value	Frequency	Note
1	2.38	1 kHz	unvulcanised [11]
2	2.42	1 kHz	pure gum vulcanisate [11]

Dissipation (Power) Factor:

No.	Value	Frequency	Note
1	0.003	1 kHz	unvulcanised [11]
2	0.0054	1 kHz	pure gum vulcanisate [11]

Optical Properties:

Optical Properties General: FTIR and nmr spectral data show that isoprene units have 1,4 structure [41,54]

Refractive Index:

No.	Value	Note
1	1.5081	unvulcanised [10]
2	1.5105	[88]

Volume Properties/Surface Properties: SIMS characterisation shows saturated hydrocarbon with no significant evidence of oxidation [55]

Polymer Stability:

Polymer Stability General: Butyl rubber is more resistant to attack by oxygen and ozone than diene rubbers because it has no tertiary hydrogen atoms in the isobutene segments and only has a small proportion of unsaturated monomer segments

Thermal Stability General: Phenol formaldehyde (resole) resins give thermally stable cross-links for butyl rubber. [57] Stability depends on such factors as thermal and oxidative conditions, mechanical stresses, catalyst residues, ozone and radiation [68]

Upper Use Temperature:

No.	Value	Note
1	120°C	[89]
2	200°C	[33]

Decomposition Details: Maximum weight loss temp. at 386°. [62] Pyrolysis begins at approx. 300° and becomes very rapid at 400°. The main product is isobutylene. [38,70] The photochemical breakdown of butyl rubber has been reported. [38] Under vacuum no breakdown occurred at any wavelength. Breakdown in air occurs only at wavelengths shorter than 460 nm. [34] Volatilisation within 30 min: 0% (280°), 50% (340°), 100% (370°). [63] Monomer yield 35–38% (500°), 65–73% (800°), 11–15% (1200°). [63] Butyl rubber degrades under thermo-oxidative conditions mainly by chain scission [90]

Flammability: Release of flammable isobutylene occurs at 500-600°. [58] Heat of combustion ΔH 47 kJ g^{-1} [63]

Environmental Stress: Has good resistance to vegetable oils and ester based lubricants. [32] Surprisingly low radiation stability, useful props. are destroyed at 10^5 Gy [28]. High energy radiation gives chain scission [40,60] and cross-linking. [71] Good weathering and ozone resistance [32], much more resistant to ozone cracking than other diene rubbers [80]. Crack propagation critical stress 12 MPa, slower than for natural rubber [53]

Chemical Stability: Suitable for use with inorganic chemicals such as metal salts, acids and alkalis up to 100°. [32] Useful life for butyl rubber (resin cured) with 50 phv carbon black and 5 phv oil at 170° is 280 h [32]

Hydrolytic Stability: The effect of steam ageing on the mechanical properties of butyl and chlorobutyl rubbers has been compared. [32]

Stability Miscellaneous: The resistance of butyl rubber to thermo-oxidation and to gas permeation has been confirmed in accelerated reacting as a coating to improve the stability of natural rubber. [59] The effect of silane coupling agent on the thermo-oxidative stability of butyl rubber has been investigated [66]

Applications/Commercial Products:

Processing & Manufacturing Routes: Manufactured by a slurry process using $AlCl_3$ as catalyst at low temp. (-90 –-98°). Mixture is isobutene with a small amount of diolefin, usually isoprene (1–3%). Butyl rubber does not have vulcanisable groups in the monomer, so the curable isoprene is incorporated. Curing rate is slow

Mould Shrinkage (%):

No.	Value	Note
1	2.36%	cured, without filler [48]
2	1.6%	cured, 20 phv filler [48]

Applications: Used in applications where low air permeability is important, such as tyre inner linings and pneumatic devices. Also used in adhesives, caulks, sealants, lining material, electrical cable sheathing for its good durability. Used in high temp. hoses, conveyor belts and in load bearing and mounting structures due to good damping props.

Trade name	Supplier
Exxon Butyl 1077	Exxon Chemical
Kalar 264, 5245	Hardman
Kalene 800, 1300	Hardman
Polysar Butyl	Bayer Inc.

Bibliographic References

[1] Wood, L.A., Bekkedahl, N., Roth, F.L., *Rubber Chem. Technol.*, 1943, **16**, 244
[2] Bekkedahl, N., *J. Res. Natl. Bur. Stand. (U.S.)*, 1949, **43**, 145
[3] Kell, R.M., Bennett, B., Stickney, P.B, *Rubber Chem. Technol.*, 1958, **31**, 499
[4] Price, C., Padget, J., Kirkham, M.C., Allen, G., *Polymer*, 1969, **10**, 495
[5] Wood, L.A., Roth, F.L., *Rubber Chem. Technol.*, 1963, **36**, 611
[6] Furukawa, G.T., Reilly, M.L., *J. Res. Natl. Bur. Stand. (U.S.)*, 1956, **56**, 285
[7] Hamill, W.H., Mrowca, B.A., Anthony, R.L., *Rubber Chem. Technol.*, 1946, **19**, 622

[8] Wood, L.A., Bekkedahl, N., *J. Polym. Sci., Part B: Polym. Lett.*, 1967, **5**, 169
[9] Schilling, H., *Kautsch. Gummi Kunstst.*, 1963, **16**, 84
[10] Wood L.A., *Synthetic Rubbers*, (ed. G.S. Whitby), John Wiley, 1954
[11] McPherson, A.T., *Rubber Chem. Technol.*, 1963, **36**, 1230
[12] Cramer, W.S. and Silver, I., Feb. 1957, U.S. Naval Ordnance Lab.
[13] Rands, R.D, Ferguson, W.J., Prather, J.L., *J. Res. Natl. Bur. Stand. (U.S.)*, 1944, **33**, 63
[14] Witte, R.S., Mrowca, B.A., Guth, E., *Rubber Chem. Technol.*, 1950, **23**, 163
[15] Ball, J.M., Maasen, G.C., ASTM, *Symp. Appl. Synth. Rubbers*, 1944, 27
[16] Boomstra, B.B.S.T., Elastomers. Their Chemistry, *Physics and Technology*, (ed. R. Houwink), Elsevier, 1948
[17] Martin, G.M., Roth, F.L., Stiehler, R.D., *Rubber Chem. Technol.*, 1957, **30**, 876
[18] Prettyman, I.B., *Handbook of Chemistry and Physics*, CRC Press, 1962, 1564
[19] Schmieder, L., Wolk, K., *Kolloidn. Zh.*, 1953, **134**, 149
[20] Dillon, J.H., Prettyman, I.B., Hall, G.L., *Rubber Chem. Technol.*, 1944, **17**, 597
[21] Boonstra, B.B.S.T., *Rubber Chem. Technol.*, 1951, **24**, 199
[22] Cabasso, I., *Encycl. Polym. Sci. Eng.*, 2nd edn., Vol. 9, (ed. J.I.Kroschwitz), John Wiley & Sons, 1987, 562
[23] Koros, W.J., Hellums, M.W., *Encycl. Polym. Sci. Eng.: Supplement Volume*,742
[24] Stannett, V., Crank, J., Park, G.S., *Diffusion in Polymers*, Academic Press, 1968
[25] Ferry, J.D., *Viscoelastic Properties of Polymers*, 3rd edn., Wiley, 1980, 606
[26] Kresge, E.N., Schatz, R.H., Wang, H.-C., *Encycl. Polym. Sci. Eng.*, 2nd edn., Vol. 8, (ed. J.I.Kroschwitz), John Wiley & Sons, 1987, 433
[27] Hartmann, B., *Encycl. Polym. Sci. Eng.*, 2nd edn., Vol. 1, (ed. J.I.Kroschwitz), John Wiley & Sons, 1985, 148
[28] Clough, R., *Encycl. Polym. Sci. Eng.*, 2nd edn., Vol. 13, (ed. J.I. Kroschwitz), 1985, 687
[29] Sanders, J.F., Ferry, J.D., *Macromolecules*, 1974, **7**, 681
[30] Le Grand, D.G., Gaines, G.L., *J. Colloid Interface Sci.*, 1969, **31**, 162
[31] Wu, S., *J. Colloid Interface Sci.*, 1969, **31**, 153
[32] *The Materials Selector*, 2nd edn., (eds. N.A. Waterman and M.F. Ashby), Chapman & Hall, 1997
[33] Baldwin, *J. Am. Chem. Soc.*, 1950, **72**, 1833
[34] Buckley, D.J., *Rubber Chem. Technol.*, 1959, **32**, 1487
[35] Gehman, *Rubber Chem. Technol.*, 1957, **30**, 1202
[36] Stambaugh, *Phys. Rev. Lett.*, 1944, **65**, 7
[37] Gessler, *Rubber Age (London)*, 1953, **74**, 397
[38] Madorsky, Strauss, Thompson, Williamson, *J. Res. Natl. Bur. Stand. (U.S.)*, 1949, **42**, 499
[39] Buckley, Robison, *J. Polym. Sci.*, 1956, **19**, 145
[40] Charlesky, *Trans., Inst. Rubber Ind.*, 1958, **34**, 175
[41] Werstler, D.D., *Rubber Chem. Technol.*, 1980, **53**, 1191
[42] Dolezal, P.T., Johson, P.S., *Rubber Chem. Technol.*, 1980, **53**, 252
[43] Medalia, A.I., Laube, S.G., *Rubber Chem. Technol.*, 1978, **51**, 89
[44] McKenna, G.B., Zapas, L.J., *Rubber Chem. Technol.*,15
[45] Ferry, J.D., Fitzgerald, E.R., *Rubber Chem. Technol.*, 1982, **55**, 1403
[46] Hamed, G.R., Ogrimi, F., *Rubber Chem. Technol.*, 1983, **56**, 1111
[47] Gent, A.N., Pulford, C.T.R., *Rubber Chem. Technol.*, 1980, **53**, 176
[48] Beatty, J.R., *Rubber Chem. Technol.*, 1978, **51**, 1044
[49] Young, D.G., Danik, J.A., *Rubber Chem. Technol.*, 1994, **67**, 137
[50] Hess, W.M., Herd, C.R., Vegvari, P.C., *Rubber Chem. Technol.*, 1993, **66**, 329
[51] Pai, P.C.H, Meier, D.J., *Rubber Chem. Technol.*, 1992, **65**, 396
[52] Ayala, J.A., Hess, W.M., Kistler, F.D., Joyce, G.A., *Rubber Chem. Technol.*, 1991, **64**, 19
[53] Layer, R.W., Lattimer, R.P., *Rubber Chem. Technol.*, 1990, **63**, 426
[54] Cheng, D.M., Gardner, I.J., Wang, H.C., Frederich, C.B. et al, *Rubber Chem. Technol.*, 1990, **63**, 265
[55] Van Ooli, W.J., Nahmias, M., *Rubber Chem. Technol.*, 1989, **62**, 658
[56] Lobos, Z.J., Tang, H., *Rubber Chem. Technol.*, 1989, **62**, 623
[57] Lattimer, R.P., Kinsey, R.A., Layer, R.W., *Rubber Chem. Technol.*, 1989, **62**, 107
[58] Lawson, D.F., *Rubber Chem. Technol.*, 1986, **59**, 455
[59] Stenberg, B., Peterson, L.-O., Flink, P., Bjork, F., *Rubber Chem. Technol.*, 1986, **59**, 70
[60] Bohm, G.G.A., Tveekram, J.O., *Rubber Chem. Technol.*, 1982, **55**, 575
[61] Brazier, D.W., *Rubber Chem. Technol.*, 1980, **53**, 437
[62] Brazier, D.W., Nickel, G.H., *Rubber Chem. Technol.*, 1975, **48**, 661
[63] Fabris, H.J., Sommer, J.G., *Rubber Chem. Technol.*, 1977, **50**, 523
[64] Studebaker, M.L., Beatty, J.R., *Rubber Chem. Technol.*, 1974, **47**, 803
[65] Lawandy, S.N., Younan, A.F., Darwish, N.A., Yousef, Y., Mounir, A., *J. Adhes. Sci. Technol.*, 1997, **11**(3), 327
[66] Adhikary, A., Mukhopadhyay, R., Dean, A.S., *J. Mater. Sci.*, 1995, **30**, 4112
[67] Nasr, G.M., Badawy, M.M., Gwaily, S.E., Shash, N.M., Hassan, H.H., *Polym. Degrad. Stab.*, 1995, **48**, 237, 391
[68] Dubey, V., Pandey, S.K., Rao, N.B.S.N., *J. Anal. Appl. Pyrolysis*, 1995, **34**, 111
[69] Liao, F.S., Su, A.C., Hsu, T.C.J., *Polymer*, 1994, **35**, 2579
[70] Vaidyanathaswamy, R., *J. Anal. Appl. Pyrolysis*, 1993, **27**, 207
[71] Hill, D.J.T., O'Donnell, J.H., Perera, M.C.S., Pomery, P.J., *Radiat. Phys. Chem.*, 1992, **40**, 127
[72] Abohashem, A., *J. Appl. Polym. Sci.*, 1992, **45**, 1733
[73] Abdelnour, K.N., Hanna, F.F., Abelmessich, S.L., *Polym. Degrad. Stab.*, 1992, **35**, 121
[74] Hashem, A.A., Ghani, A.A., Eatah, A.I., *J. Appl. Polym. Sci.*, 1991, **42**, 1081
[75] Eatah, A.I., Ghani, A.A., Hashem, A.A., *Polym. Degrad. Stab.*, 1990, **27**, 75
[76] Hakim, I.K., Bishai, A.M., Saad, A.L., *J. Appl. Polym. Sci.*, 1988, **35**, 1123
[77] Richards, J.R., Mancke, R.G., Ferry, J.D., *J. Polym. Sci., Part B: Polym. Lett.*, 1964, **2**, 197
[78] Gent, A.N., Hindi, M., *Rubber Chem. Technol.*, 1988, **61**, 892
[79] McKenna, G.B., Zapar, L.J., *Rubber Chem. Technol.*, 1981, **54**, 718
[80] *Handbook of Polymer Science and Technology*, (ed. N.P. Cheremisinoff), Marcel Dekker, 1989, 244
[81] Hamed, G.R., *Rubber Chem. Technol.*, 1981, **54**, 576
[82] Loran, A.Y., Patel, R., *Rubber Chem. Technol.*, 1981, **54**, 91
[83] Medalia, A.I., *Rubber Chem. Technol.*, 1978, **51**, 437
[84] Tsuji, T., Fujita, H., *Rubber Chem. Technol.*, 1975, **48**, 765
[85] Frensdorff, H.K., *Rubber Chem. Technol.*, 1974, **47**, 849
[86] Corish, B., Powell, B.D.W., *Rubber Chem. Technol.*, 1974, **47**, 481
[87] Corman, B.G., Deviney, M.L., Whittington, L.E., *Rubber Chem. Technol.*, 1972, **45**, 278
[88] Callan, J.E., Hess, W.M., Scott, C.E., *Rubber Chem. Technol.*, 1971, **44**, 814
[89] Sircar, A.K., Voet, A., Cook, F.R., *Rubber Chem. Technol.*, 1971, **44**, 185
[90] Jipa, S., Giurginea, M., Setnescu, T., Setnescu, R. et al, *Polym. Degrad. Stab.*, 1996, **54**, 1
[91] Fajdiga, B., Susteric, Z., *Kautsch. Gummi Kunstst.*, 1993, **46**, 225
[92] Heinrich, G., *Kautsch. Gummi Kunstst.*, 1993, **45**, 173
[93] Emmons, H.W., *J. Appl. Polym. Sci.*, 1981, **26**, 2447
[94] *Permeability and other Film Properties of Plastics and Elastomers*, Plastic Design Library, 1995, 437

Butyl Rubber Chlorinated B-36

Synonyms: *CIIR. Chlorobutyl rubber. Chlorinated isobutylene-isoprene copolymer/elastomer*
Related Polymers: Butyl rubber
Monomers: Isobutylene, Isoprene
Material class: Synthetic Elastomers
Polymer Type: Polydienes
Additives: Sulfur and zinc oxide as vulcanising agents
General Information: 1% Chlorine. Chlorinated butyl rubber can be vulcanised more quickly than butyl rubber, and is more compatible with other rubbers

Volumetric & Calorimetric Properties:
Density:

No.	Value	Note
1	d 1.043 g/cm^3	pure gum vulcanisate [4]
2	d 1.132–1.253 g/cm^3	vulcanisate, carbon black filled [4]

Thermal Conductivity:

No.	Value	Note
1	0.136 W/mK	40° [8]

Surface Properties & Solubility:
Solubility Properties General: The compatibility of cured and uncured rubber with other elastomers has been reported. [18] Compatibility is increased compared to butyl rubber. Partially miscible with EPDM, immiscible with styrene-butadiene rubber, natural rubber, polybutadiene and polychloroprene [12]

– Butyl Rubber Chlorinated

Wettability/Surface Energy and Interfacial Tension: Chlorobutyl rubber gives good adhesion to fillers and highly unsaturated elastomers. This alleviates the difficulty of vulcanising butyl rubber blends with diene rubbers. [11] Chlorination of butyl rubber improves the rate of curing, and adhesion to metals [3]

Transport Properties:
Transport Properties General: Transport props. are very similar to those of butyl rubber

Mechanical Properties:
Tensile Strength Break:

No.	Value	Note	Elongation
1	1.26 MPa	pure gum vulcanisate [4]	290%
2	4–11.5 MPa	vulcanisate, carbon black filled [4]	290%

Mechanical Properties Miscellaneous: Energy loss is higher than that of other polymers. Storage and loss moduli and tan δ over a range of frequencies for pure gum vulcanisate and for carbon black filled vulcanisate over a range of loadings have been reported [4, 9]
Hardness: Shore A24; A44-A56 (carbon black filled) [4]
Failure Properties General: The phase behaviour of chlorobutyl rubber blends with other elastomers and its relationship to tread wear, skid resistance, flex cracking, fatigue and cut-growth rate has been reported. [11] Tear strength 8.75 kN m^{-1} (pure gum vulcanisate); 23–43 kN m^{-1} (vulcanisate, carbon black filled) [4]
Fracture Mechanical Properties: Fatigue crack propagation has been reported over a range of strain rates (1–20 s^{-1}) and temps. (0–75°) in N$_2$, air and 0.5 ppm O$_2$. The rate of crack propagation does not increase abruptly at high tearing energy unlike butyl rubber and natural rubber [7]
Friction Abrasion and Resistance: Has higher skid resistance on wet ice than other rubbers [14]

Optical Properties:
Refractive Index:

No.	Value	Note
1	1.5105	[18]

Transmission and Spectra: Nmr spectral data have identified the structure of chlorinated groups in the polymer [2]

Polymer Stability:
Thermal Stability General: Thermal characteristics of a range of chlorobutyl rubber blends have been reported [6,9]
Decomposition Details: High temp. degradation is similar to butyl rubber, the main decomposition product being isobutylene. [1,9] Wt. loss onset at 352°; 50% wt. loss at 395° [6]; max. wt. loss at 386° [10]. Cross-linking is extensive under thermo-oxidative conditions [19]
Flammability: Flammability rating 5.4 (filled match test); 2 (MIL 417 test) [17]
Environmental Stress: Has better ozone resistance than other diene rubbers. [15] Useful life 230h (50 phv carbon black, 5 phv oil, 170°). [16]
Hydrolytic Stability: Steam ageing effects have been reported [16]
Stability Miscellaneous: High energy radiation causes cross-linking rather than chain scission. [13]

Applications/Commercial Products:
Processing & Manufacturing Routes: Prod. by chlorination of butyl rubber
Applications: Used in tyres

Trade name	Supplier
Chlorobutyl	Exxon Chemical
Polysar Chlorobutyl	Polysar

Bibliographic References
[1] Buckley, D.J., *Rubber Chem. Technol.*, 1959, **32**, 1475
[2] Chu, C.Y., Watson, K.N., Vukov, R., *Rubber Chem. Technol.*, 1987, **60**, 636
[3] Fusco, J.V., *Rubber Chem. Technol.*, 1982, **55**, 1584
[4] Capps, R.N., *Rubber Chem. Technol.*, 1986, **59**, 103
[5] Hess, W.M., Herd, C.R., Vegvari, P.C., *Rubber Chem. Technol.*, 1993, **66**, 329
[6] Sircar, A.K., *Rubber Chem. Technol.*, 1977, **50**, 71
[7] Young, D.G., *Rubber Chem. Technol.*, 1985, **58**, 785
[8] Sircar, A.K., Wells, J.A., *Rubber Chem. Technol.*, 1982, **55**, 191
[9] Sircar, A.K., Lamond, T.G., *Rubber Chem. Technol.*, 1972, **45**, 3291
[10] Brazier, D.W., Nickel, G.H., *Rubber Chem. Technol.*, 1975, **48**, 661
[11] Corish, P.J., Powell, B.D.W., *Rubber Chem. Technol.*, 1974, **47**, 481
[12] Gardiner, J.B., *Rubber Chem. Technol.*, 1970, **43**, 370
[13] Hill, D.J.T., O'Donnell, J.H., Perera, M.C.S., Pomery, P.J., *Polymer*, 1995, **36**, 4185
[14] Ahagon, A., Kobayashi, T., Misawa, M., *Rubber Chem. Technol.*, 1988, **61**, 14
[15] Keller, R.W., *Handbook of Polymer Science and Technology*, (ed. N.P. Cheremisinoff), Marcel Dekker, 1989, 165
[16] *The Materials Selector*, 2nd edn., (eds. N.A Waterman and M.F. Ashby), Chapman & Hall, 1977, 504
[17] Trexler, H.E., *Rubber Chem. Technol.*, 1973, **46**, 1114
[18] Callan, J.E., Hess, W.M., Scott, C.E., *Rubber Chem. Technol.*, 1971, **44**, 814
[19] Jipa, S., Giurginea, M., Setnescu, T., Setnescu, R., et al, *Polym. Degrad. Stab.*, 1996, **54**, 1

Carboxymethylhydroxyethyl cellulose C-1

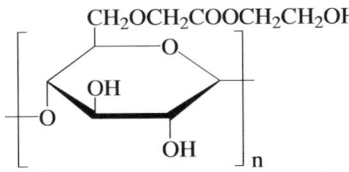

Synonyms: *CMHEC. 6-Carboxymethyl-2'-hydroxyethyl cellulose*
Related Polymers: Cellulose, Sodium carboxymethyl cellulose
Monomers: Base monomer unit 6-carboxymethyl-(2'-hydroxyethyl) glucose
Material class: Polysaccharides
Polymer Type: cellulosics
CAS Number:

CAS Reg. No.
9004-30-2

Molecular Formula: $(C_{10}H_{16}O_8)_n$
Fragments: $C_{10}H_{16}O_8$
General Information: The degree of substitution for commercial material is 0.3 carboxymethyl groups per base monomer unit and 0.7 hydroxyethyl groups per carboxymethyl substituted unit [1,2].

Volumetric & Calorimetric Properties:
Density:

No.	Value	Note
1	d 0.6 g/cm^3	[2]

Transition Temperature:

No.	Value	Note	Type
1	>230°C	min. [1,2]	Discoloration

Transport Properties:
Water Absorption:

No.	Value	Note
1	11 %	25°, 10% relative humidity [1,2]
2	15 %	25°, 50% relative humidity [1,2]
3	31 %	25°, 90% relative humidity [1,2]

Mechanical Properties:
Tensile Strength Break:

No.	Value	Note
1	69 MPa	film, 50% relative humidity [2]

Viscoelastic Behaviour: Brookfield viscosity 23–85 (25°, 2% soln.) [2]

Optical Properties:
Refractive Index:

No.	Value	Note
1	1.53	film [2]

Applications/Commercial Products:

Trade name	Supplier
CMHEC 37L	Hercules
Tylose CHR	Hoechst

Bibliographic References
[1] *Hercules CMHEC-37L*, Technical Bulletin No. VC-402C, Hercules Inc., Wilmington, Delaware, U.S.A., (technical datasheet)
[2] Just, E.K. and Majewicz, T.G., *Encycl. Polym. Sci. Eng.*, 2nd edn., (eds. H.F. Mark, N.M. Bikales, C.G. Overberger and G. Menges), John Wiley and Sons, 1985, **3**, 247

Cardanol Resins C-2

Synonyms: *Cashew nut shell liquid resins. CNSL resins*
Monomers: Cardanol, Formaldehyde
Material class: Thermosetting resin
Polymer Type: phenolic
CAS Number:

CAS Reg. No.	Note
37311-73-2	polymer with formaldehyde
130572-32-6	polymer with formaldehyde, hexa cured

Molecular Formula: $[(C_{21}H_{32}O).(CH_2O)]_n$
Fragments: $C_{21}H_{32}O$ CH_2O
Additives: Glass fibre
General Information: Cardanol has a *meta*-substituted alkenyl chain. The *meta*-position enhances the reactivity of the monomer towards the addition of formaldehyde. The resin hardens by polyaddition and polymerisation to yield a solid infusible product which in powdered form is Friction Dust

Volumetric & Calorimetric Properties:
Glass-Transition Temperature:

No.	Value	Note
1	126°C	resol, cured [5]
2	128°C	novolac, hexa cured [5]

Surface Properties & Solubility:
Solubility Properties General: Both novolac and resol-type resins form semi-interpenetrating networks with poly(methyl methacrylate) [5]
Solvents/Non-solvents: Uncured novolac sol. hexane, MeOH [4]

Mechanical Properties:
Mechanical Properties General: Tensile strength of hexa cured novolac decreases with increasing hexa content [5]

Tensile Strength Break:

No.	Value	Note	Elongation
1	5.3 MPa	novolac, hexa cured [1]	
2	4.4 MPa	novolac, hexa (3.5%) cured [5]	19.4%
3	1.87 MPa	novolac, hexa (7.0%) cured [5]	4.0%
4	2.63 MPa	resol, cured [5]	

Hardness: Shore A87.5 (novolac, hexa (3.5%) cured) [5], A92 (novolac, hexa (7.0%) cured); D45 (novolac, hexa (3.5%) cured) [5], D46 (novolac, hexa (7.0%) cured) [5] ; Shore A92 (resol, cured), D57.5 (resol, cured) [5]

Optical Properties:
Transmission and Spectra: Uv and ir spectral data have been reported [4]

Polymer Stability:
Polymer Stability General: Stability may be improved by grafting on polysiloxanes [3]
Upper Use Temperature:

No.	Value	Note
1	93°C	200°F

Decomposition Details: Details of thermal degradation have been reported [6]
Flammability: Limiting oxygen index 21% (novolac, hexa cured, ASTM D3863-77) [1]. Fire retardant props. may be improved by phosphorylation of the prepolymer [1]
Chemical Stability: Resistant to acids and alkalis but attacked by aromatic hydrocarbons and chlorinated hydrocarbons

Applications/Commercial Products:
Processing & Manufacturing Routes: Cardanol is obtained from the fibrous shell of cashew nuts. (Cashew nut shell liq. results from a special heat treatment and decarboxylation). The reacted cashew nut shell liq. resin hardening reaction includes polym. and polyaddition yields a solid infusible product which in powdered form (Friction Dust) retains high binding power at raised temps. Kinetics of the alkaline catalysed reaction have been reported [2]. Novolak-type resin may be synth. by reaction of cardanol with aq. formaldehyde (mole ratio 1: 0.9) in the presence of oxalic acid catalyst [3]. Adipic acid also used as catalyst [4]. Use of succinic acid catalyst [6] yields "high-*ortho*" novolaks [7]. Resol-type resin synth. by reaction of cardanol and aq. formaldehyde in the presence of barium hydroxide catalyst. May be heat cured if desired [5]
Applications: Cardanol Friction Dust is used (5–15%) in brake lining formulations. Oil modification of phenolic resins may be achieved with Cashew nut shell liq. or other oils, leading to a softer braking action. Has other uses in inks and in anticorossive coatings and as a cement hardener

Trade name	Supplier
Carditol	3M Industrial Chemical Products

Bibliographic References
[1] Pillai, C.K.S, Prasad, V.S., Sudha, J.D., Bera, S.C. and Menon, A.R.R., *J. Appl. Polym. Sci.*, 1990, **41**, 2487
[2] Misra, A.K. and Pandey, G.N., *J. Appl. Polym. Sci.*, 1985, **30**, 969, 979
[3] Pillot, J-P., Dunogues, J., Gerval, J., The M.D. and Thanh, M.V., *Eur. Polym. J.*, 1989, **25**, 285, (synth.)
[4] Sathiyalekshmi, K. and Kumaresan, S., *Indian J. Technol.*, 1993, **31**, 702, (synth., uv, ir)
[5] Manjula, S., Pavithran, C., Pillai, C.K.S. and Kumar, V.G., *J. Mater. Sci.*, 1991, **26**, 4001, (synth., mechanical props.)
[6] Lekshmi, K.S., *Polym. Sci.*, (ed. I.S. Bhardway), 1994, **2**, 862, (synth., thermal degradation)
[7] Sathiyalekshmi, K., *Bull. Mater. Sci.*, 1993, **16**, 137, (synth.)
[8] Attanasi, Passalenti, B. and Errigo, U., *Congr. FATIPEC*, 1986, **1B**, 511, (uses)

Carrageenan

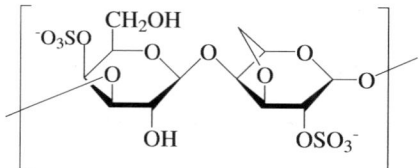

K - Carrageenan (idealised)

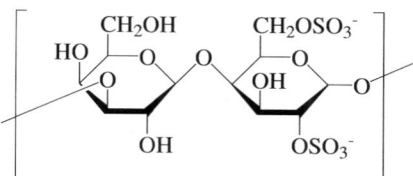

X - Carrageenan (idealised)

Monomers: D-Galactose
Material class: Polysaccharides
CAS Number:

CAS Reg. No.	Note
11114-20-8	κ carrageenan
9064-57-7	λ carrageenan
9062-07-1	ι carrageenan
51311-95-6	ε carrageenan
51311-96-7	μ carrageenan

Molecular Weight: 100000–1000000 (commercial)
General Information: Isol. from red algae spp. including *Chondrus, Furcellaria* and *Euchena* by hot water treatment at alkaline pH, followed by alcohol precipitation. At least six discrete polymers have been identified (prefixed ε, λ, μ, ν, ι, κ). κ-Carrageenan is the most regular polymer. All components are derived from D-galactose, 3,6-anhydrogalactose and related sulfates [1]. Gels result from complex interactions of polymer chains, giving coiled regions that are further strengthened by the presence of cations

Applications/Commercial Products:
Processing & Manufacturing Routes: Dried seaweed is macerated and extracted with hot water at alkaline pH. The polymer soln. is conc. before the polymer is precipitated with alcohol [1]
Applications: ι, κ and λ Carrageenans are of commercial significance. They are used to prod. high-strength gels with milk, and in other food applications as a viscosity modifier. Also used in cosmetics and pharmaceutical formulations

Bibliographic References
[1] *Polysaccharides*, (ed. G.O. Aspinall), Academic Press, 1983, **2**, 447

Casein
Synonyms: *Calcium caseinate. Milk protein*
Related Polymers: Proteins, Casein Formaldehyde
Monomers: Amino acids
Material class: Proteins and polynucleotides

– Casein Formaldehyde

CAS Number:

CAS Reg. No.
9005-46-3

Molecular Weight: 33000–375000 Casein complex. Component protein MW's α 21000, β 24000, γ 11-20000, κ 19000
General Information: Casein, as its calcium salt, forms 3% of milk. It is a mixture of several related and interacting proteins. Fractions obtained are designated α, β, γ, and κ [1]. Its amino acid content is similar to that of wool, though its sulfur content is only 20% of that of wool

Volumetric & Calorimetric Properties:
Density:

No.	Value	Note
1	d 1.25 g/cm^3	[1]
2	d 1.29 g/cm^3	Aralac fibre [3]

Surface Properties & Solubility:
Solubility Properties General: Low solubility in H_2O at neutral pH. Forms colloidal suspensions in acidic and alkaline soln. and in the presence of salts. Casein can also form colloids in some organic solvents [1]. Treatment with formaldehyde renders material waterproof and insol.
Solvents/Non-solvents: Spar. sol. H_2O [1]. Polymer and fibres insol. organic solvents [3]
Surface and Interfacial Properties General: Diameter of colloidal particles 20–140 nm [1]

Transport Properties:
Water Content: 8% (25°, 20% relative humidity), 12% (25°, 50% relative humidity) [1]
Water Absorption:

No.	Value	Note
1	14 %	standard conditions [3]

Electrical Properties:
Electrical Properties General: Isoelectric point of aq. soln. at pH 4.6 [1]
Dielectric Permittivity (Constant):

No.	Value	Note
1	8	[1]

Applications/Commercial Products:
Processing & Manufacturing Routes: Wool like fibres are made by formaldehyde cross-linking of spun protein or by chrome tanning
Applications: Used to make plastics, paints and coatings. Other uses in food processing, paper coating and adhesives.

Trade name	Details	Supplier
Fibrolane Bx/Bc	Wool-like fibres	Courtaulds
Lamital	Wool-like fibre	Snia
Mermova	Blending fibre	Snia

Bibliographic References
[1] *Encycl. Polym. Sci. Eng.*, (eds. H.F. Mark, N.M. Bikales, C.G. Overberger and G. Menges), John Wiley and Sons, 1985, **2**
[2] Davey, P.T., Houchin, M.R. and Winter, G., *J. Chem. Technol. Biotechnol.*, 1983, **33**, 164
[3] Moncrieff, R.W., *Man-Made Fibers*, 5th edn., Newnes Butterworth, 1975
[4] Wormell, R.J., *J. Text. Inst.*, 1953, **44**, 258

Casein Formaldehyde   C-5
Synonyms: *Casein Plastic. Artificial Horn. Galalith. CF. CS*
Related Polymers: Casein
Material class: Thermosetting resin
Polymer Type: polyamide
General Information: A polymer originating from natural proteins in sources such as milk, horn, soya beans, wheat etc. The first synth. was described in 1897, with commercial production starting in France in 1900, in Britain in 1913 and in the USA in 1920. Worldwide production reached 10000 tons by 1930. It was quickly superseded by synthetic compounds such as Bakelite and its uses were further restricted by its poor moulding quality and its low resistance to moisture. [1] Uses for Galalith have been reported [6].
Identification: Gives a pink colour when subjected to the biuret protein test. Gives a yellow precipitate when heated with conc. nitric acid (xanthoproteic test) [1]. Hydrol. with aq. sulfuric acid produces a cheese-like odour [3]

Volumetric & Calorimetric Properties:
Density:

No.	Value	Note
1	d 1.35 g/cm^3	ASTM D792 [1]
2	d 1.34 g/cm^3	Erinoid [9]

Thermal Expansion Coefficient:

No.	Value	Note	Type
1	$4.1 \times 10^{-5} - 6.8 \times 10^{-5}$ K^{-1}	ASTM D696 [1]	L

Deflection Temperature:

No.	Value	Note
1	149°C	ASTM D648 [1]

Vicat Softening Point:

No.	Value	Note
1	93.3°C	[1]

Surface Properties & Solubility:
Solvents/Non-solvents: Galalith sol. boiling 5% ammonium chloride; sl. sol. 1% potassium hydroxide, 1% ammonium oxalate; spar. sol. aq. ammonium sulfate [3]

Transport Properties:
Water Content: Water absorption may be reduced by addition of coal-tar pitch [4] or by coating with soluble quinones or tannic acid soln. [5]. Red-coloured Erinoid absorbs 1.43g H_2O after 400h exposure, whereas blue-coloured Erinoid absorbs 1.26g H_2O after the same exposure period [9]
Water Absorption:

No.	Value	Note
1	7–14 %	24h, $\frac{1}{8}$" thick, ASTM D570 [1]

– Cellulose   C-6 – C-6

Mechanical Properties:

Tensile (Young's) Modulus:

No.	Value	Note
1	3516–3930 MPa	510–570 kpsi, ASTM D638, ASTM D651 [1]

Tensile Strength Break:

No.	Value	Note	Elongation
1	51.7–68.9 MPa	7.5–10 kpsi, ASTM D638, ASTM D651 [1]	2.5%

Flexural Strength at Break:

No.	Value	Note
1	69–124 MPa	10–18 kpsi, ASTM D790 [1]

Compressive Strength:

No.	Value	Note
1	186–365 MPa	27–53 kpsi, ASTM D695 [1]

Hardness: Rockwell M70–M100 (ASTM D785) [1]. Brinell 23 (2.5 mm ball, 25 kg load) [1]

Izod Notch:

No.	Value	Notch	Note
1	53.4 J/m	Y	1 ft lb in^{-1}, ASTM D256 [1]

Electrical Properties:

Electrical Properties General: Dielectric constant shows non-uniformity at wavelength of approx. 800 m. [7] Erinoid has good electrical insulating props. when dry [9]

Dielectric Permittivity (Constant):

No.	Value	Frequency	Note
1	6.1–6.8	1 MHz	ASTM D150 [1]

Dielectric Strength:

No.	Value	Note
1	15.7–27.5 kV.mm^{-1}	400–700 V mil^{-1}, ASTM D149 [1]

Dissipation (Power) Factor:

No.	Value	Frequency	Note
1	0.052	1 MHz	ASTM D150 [1]

Optical Properties:

Volume Properties/Surface Properties: Clarity: transparent to opaque. May be coloured with both deep and pastel shades [1,9]

Polymer Stability:

Upper Use Temperature:

No.	Value	Note
1	135°C	[1]

Flammability: Has low flammability. [1,9] Burns very slowly with a smoky flame (ASTM D635); the ash does not glow upon removal of the flame [1]

Environmental Stress: Coloured product may be faded by sunlight [1]

Chemical Stability: Resistant to weak acids [9] and organic solvents but is decomposed by alkalis and strong acids (ASTM D543) [1]. Erinoid disintegrates after immersion in olive oil at 200° [9]

Hydrolytic Stability: Erinoid is softened by boiling H_2O [9]

Biological Stability: Undergoes very slow decomposition in soil [8]

Applications/Commercial Products:

Processing & Manufacturing Routes: First synth. in 1897 by the enzyme-catalysed reaction of whey with formaldehyde. The modern process involves dry extrusion. Rennet casein (obtained by the action of rennin on fresh skimmed milk) is wetted with sufficient water to produce a hand-mouldable mass. Any desired additives may be added at this stage. The mass is extruded under pressure into rods, ribbons or tubes. The extruded product is then cured by immersion in 5% formaldehyde soln. at room temp. for a period of days or months depending upon the thickness of the extruded material. Curing may be accelerated by the addition of alum, ammonium chloride or Werner amine complex. The cured material is dried in circulating hot air. [1] Galalith may be synth. by the reaction of powdered casein with 1–4% aq. paraformaldehyde for 2–3 h. Moulding under pressure follows, with the moulded product being cured at 30–35° for 2–4 d [2]

Applications: Used to make articles of simple contour that are unharmed by minor dimensional change due to moisture absorption: in coat buttons, fountain pens, cigarette cases, umbrella handles and radio cabinets. Other applications in dress ornaments, knitting needles and poker chips

Trade name	Details	Supplier
Ameroid		American Plastics Corp.
Erinoid	British tradename	Erinoid Ltd.

Bibliographic References

[1] Salzberg, H.K., Encycl. Polym. Sci. Technol. (ed. N.M. Bikales), John Wiley & Sons, Inc., 1969, **11**, 696, (rev)
[2] *Span. Pat.*, 1967, 326 498, (synth)
[3] Cocosinschi, Al.St., Z. Anal. Chem., 1936, 107, 197; 1939, **117**, 103 (anal, solubility)
[4] Panova, E.V and D'yachenko, P.F., *Org. Chem. Ind. (USSR)*, 1940, **7**, 110, (additives)
[5] *Brit. Pat.*, 1946, 577753
[6] Guth, H.J., Chem. Zentralbl., 1950, **1**, 2166, (uses)
[7] Kühlewein, H., Z. Tech. Phys., 1929, **10**, 280, (dielectric constant)
[8] Blanck, E., Landwirtsch. Vers.-Stn., 1917, **90**, 17
[9] Allen, R.G., Sci. Proc. R. Dublin Soc., 1918, **15**, 331, 405, (stability, water absorption)

Cellulose   C-6

Related Polymers: Cellulose I, Cellulose II
Monomers: Base monomer unit glucose
Material class: Polysaccharides
Polymer Type: cellulosics
CAS Number:

CAS Reg. No.
9004-34-6

Molecular Formula: $(C_6H_{10}O_5)_n$
Fragments: $C_6H_{10}O_5$

General Information: Cellulose is a polysaccharide polymer, containing a linear array of anhydroglucose units that are assembled into complex multi-stranded fibres and films. The polymer is the major structural component of plant cell walls and is particularly important in cells that form stems and wood. It is

Cellulose I

also the major constit. of fibres such as cotton, flax, jute, hemp and ramie. Certain green algae and bacteria also produce the polymer. Chemical processing can rearrange the macromolecular struct. of the polymer, producing fibres such as rayon. The struct. and props. of cellulose can be described at several levels 1. The single polymer chain. 2. The arrays of multiple overlapping chains of the polymer that make elementary fibrils and microfibrils, the basic building blocks found in natural cellulose-containing structs. 3. Assemblies of microfibrils that combine to make fibrils, the intermediate building blocks in construction of components of plants, such as the framework for cell walls, the skeletons of complex structs. such as stems, wood etc. and valuable fibres such as cotton. 4. Reconstituted fibres produced commercially by re-precipitating cellulose polymer chains after solubilisation as salts of derivs.

Bibliographic References

[1] *Encycl. Polym. Sci. Eng.*, 2nd edition, (eds. H.F. Mark, N.M. Bikales, C.G. Overberger and G. Menges), John Wiley and Sons, 1985
[2] Elias, H.-G., *Macromolecules*, Synthesis, Materials and Technology, Plenum Press, N.Y., 1984
[3] Manley, R. St.J, *Nature (London)*, 1964, **204**, 1155
[4] Lee, J.H. et al, *Proc. Natl. Acad. Sci. U.S.A.*, 1994, **91**, 7425
[5] Cousins, S.K. and Brown, R.M., *Polymer*, 1995, **36**, 3885
[6] Heiner, A.P. et al, *Carbohydr. Res*, 1995, **273**, 207 (struct)

Cellulose I C-7

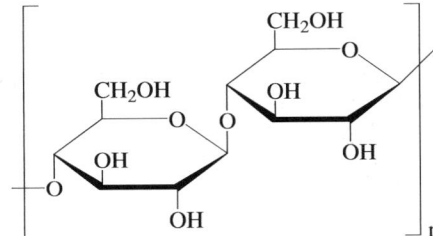

Synonyms: *Natural cellulose. α-Cellulose. Cellulon*
Related Polymers: Cellulose II, Cellulose
Monomers: Base monomer unit glucose
Material class: Polysaccharides
Polymer Type: cellulosics
CAS Number:

CAS Reg. No.
9004-34-6

Molecular Formula: $(C_6H_{10}O_5)_n$
Fragments: $C_6H_{10}O_5$
Molecular Weight: MW 331100 (cotton), 366100 (softwood), 339700 (hardwood). DP (of crystalline regions): 200–250 (purified cotton), 300–350 (ramie), 250–400 (unbleached sulfite wood pulp), 2000 (cotton). DP of whole fibres: 2000–14000 (cotton fibre), 9550 (bast fibres), 26500 (*Valonia*), 2000–3700 (*A. xylinium*), 8200–8450 (wood)
Additives: Flame retardants
General Information: Two different forms of crystalline Cellulose I named 1α and 1β have been recognised recently. The 1α form can be converted to 1β at high temps. in alkali. The 1β form is thought to predominate in most fibre and wood celluloses of commercial importance. Main component of higher plant cell walls. Also found in some bacteria, algae, fungi and tunicates. Secondary cell walls of fibres such as cotton are >90% cellulose, while in bast fibres it is *ca.* 70% and in woods the content is *ca.* 40–50%. Cellulose is the most abundant, continually synthesised chemical on earth, wth *ca.* 8×10^{10} t/yr produced and degraded. Pulping and processing gives the rearranged polymer, Cellulose II

Morphology: Degree of crystallinity: cotton 82–87% (acid hydrol.), 70% (X-ray), 60% (density), 60% (deuterium exchange), 72% (formylation); Ramie 95% (acid hydrol.), 70% (X-ray), 60% (density); wood pulp 65% (X-ray), 65% (density), 45–50% (deuterium exchange), 53–65% (formylation). A struct. based on a modified two-phase (crystalline-amorph.) model has been reported [27]
Miscellaneous: Crystalline average length: 144 nm (cotton), 120 nm (ramie), 153 nm (wood); width: 50 nm (cotton), 35 nm (ramie), 37 nm (wood); thickness 64 nm (cotton), 40 nm (ramie), 45 nm (wood)

Volumetric & Calorimetric Properties:

Density:

No.	Value	Note
1	d 1.59–1.63 g/cm^3	crystalline, x-ray detn.
2	d 1.482–1.489 g/cm^3	amorph. x-ray detn. [1]
3	d 1.582–1.63 g/cm^3	[11]
4	d 1.545–1.585 g/cm^3	cotton [11]
5	d 1.55 g/cm^3	ramie [11]
6	d 1.535–1.547 g/cm^3	wood pulps [11,22]

Thermodynamic Properties General: Heat of combustion 17.43 kJ kg^{-1} K^{-1} (cotton) [1,4]. Heat of crystallisation 121.8 kJ kg^{-1} [11]
Thermal Conductivity:

No.	Value	Note
1	0.071 W/mK	cotton, density 0.5 g cm^{-1} [11]
2	0.029–0.17 W/mK	papers, various [11]

Specific Heat Capacity:

No.	Value	Note	Type
1	1.214–1.357 kJ/kg.C	cotton [4]	P
2	1.365 kJ/kg.C	ramie [4]	P
3	1.327–1.353 kJ/kg.C	hemp [4]	P

Glass-Transition Temperature:

No.	Value	Note
1	220–245°C	[11]
2	243–433°C	[11]

Transition Temperature:

No.	Value	Note
1	>290°C	min. Ignition temp. [1]
2	475°C	Cotton [4]
3	19–23°C	Secondary transition temp. [13]

Surface Properties & Solubility:

Solvents/Non-solvents: Sol. DMF, chloral, pyridine, [7], conc. mineral acids (with degradation) [1], inorganic salts (e.g. ZnCl$_2$, LiCl, Ca(SCN)$_2$), strong alkalis, metal complexes (e.g. [Cu(NH$_3$)$_4$](OH)$_2$ (Cuoxam), [Cd(en)$_3$](OH)$_2$, (Cadoxen)) [11]. Swollen by: liq. NH$_3$, hydrazine [11]. Liq. cryst. with *N*-methylmorpholine-*N*-oxide [15]

Cellulose I

Surface Tension:

No.	Value	Note
1	42 mN/m	20% cotton [17]
2	36–42 mN/m	20°, wood pulp [17]

Transport Properties:
Water Absorption:

No.	Value	Note
1	7–8 %	cotton [16]
2	0.6–11.7 %	amorph. cellulose, 11–97% relative humidity [16]
3	8–14 %	cellulose, 20°, 60% relative humidity [21]

Mechanical Properties:
Mechanical Properties General: Moduli of rigidity have been reported [6]

Tensile Strength Break:

No.	Value	Note	Elongation
1	900 MPa	dry, ramie [1]	2.3%
2	1060 MPa	wet, ramie [1]	2.4%
3	200–800 MPa	dry, cotton [1]	12–16%
4	200–800 MPa	wet, cotton [1]	6–13%
5	824 MPa	dry, flax [1]	1.8%
6	863 MPa	wet, flax [1]	2.2%

Elastic Modulus:

No.	Value	Note
1	78–108 MPa	flax [1]
2	59–78 MPa	hemp
3	48–69 MPa	ramie [1]

Electrical Properties:
Electrical Properties General: Insulating value 500 kV cm^{-1} [10]

Dielectric Permittivity (Constant):

No.	Value	Frequency	Note
1	1.67	10 kHz	25°, cotton linters [4]
2	2.86	200 kHz	25°, cotton cellulose [4]
3	2.42	10 MHz	25°, cotton cellulose [4]
4	2.2–2.3		pulp sheets [1]
5	5.7		crystalline portions [1]

Dielectric Strength:

No.	Value	Note
1	50 kV.mm^{-1}	Native cellulose fibre [11]
2	30–50 kV.mm^{-1}	50Hz, cellophane [4]
3	7.7–9.2 kV.mm^{-1}	50Hz, insulating paper [11]

Strong Field Phenomena General: Zeta potential 21.1mV (cotton), 18mV (unbleached sulfite pulp, water), 9mV (unbleached sulfite pulp, 20°, water) [11]

Dissipation (Power) Factor:

No.	Value	Frequency	Note
1	0.02	1 kHz	20° [11]
2	0.07	100 MHz	20° [11]

Optical Properties:
Refractive Index:

No.	Value	Note
1	1.618	parallel to fibre axis, cellulose 1 [11]
2	1.544	perpendicular to fibre axis cellulose 1 [11]
3	1.576–1.595	parallel, cotton [11,14]
4	1.527–1.534	perpendicular, cotton [11,14]
5	1.595–1.601	parallel, ramie
6	1.525–1.534	perpendicular, ramie [11,14]

Polymer Stability:
Thermal Stability General: Decomposition temp. 200–270° [11], 150° (cotton) [12]

Decomposition Details: Decomposes at 250–397° giving H_2O, CO_2, CO and tar. Major tar component is laevoglucosan [9]

Flammability: Fl. p. 361° (cotton) [4]. Limiting Oxygen Index 18.4 % [11]. Mechanism of pyrolysis and the use of flame retardants have been reported [28]

Hydrolytic Stability: Susceptible to acid hydrol., with amorph. regions reacting faster than crystalline sections. In alkali the polymer dissolves at elevated temps. undergoing depolymerisation by β-elimination of H_2O

Biological Stability: Degraded by cellulases prod. by bacteria such as *Trichoderma* sp.

Applications/Commercial Products:

Trade name	Details	Supplier
Cellulon	Bacterial cellulose	Weyerhauses

Bibliographic References

[1] *Encycl. Polym. Sci. Eng.*, 2nd edn. (eds. H.F. Mark, N.M. Bikales, C.G. Overberger and G. Menges), John Wiley and Sons, 1985, **3**
[2] Elias, H.-G., Macromolecules Volume 2: Synthesis, *Materials and Technology*, Plenum Press, 1984, 1068
[3] *Cellulose Sources and Exploitation*, (eds. J.F. Kennedy, G.O. Phillips and P.A. Williams), Ellis Horwood, 1990
[4] *Handbook of Physical and Mechanical Testing of Paper and Paperboard*, (ed. R.E. Mark), Marcel Dekker, 1984
[5] *Cellulose and Cellulose Derivatives*, (eds. E. Oh, H.M. Spurlun and M.W. Grafflin), Interscience, 1954
[6] *High Polymers*, (eds. Bikales, N.M. and Segal, L.), Wiley Interscience, 1971
[7] Cellulose Chemistry, *Biochemistry and Material Aspects*, (eds. J.F. Kennedy, G.O. Phillips and P.A. Williams), Ellis Horwood, 1993
[8] *Cellul. Its Deriv.*, (eds. J.F. Kennedy, G.O. Phillips and P.A. Williams), Ellis Horwood, 1985
[9] Madorsky, S.L., Hart, V.E. and Straus, S., J. Res. Natl. Bur. Stand., Sect. A, 1958, **60**, 343
[10] *Concise Encyclopedia of Polymer Science and Engineering*, (eds. H.F. Mark, N.M. Bikales, C.G. Overberger and G. Menges), John Wiley and Sons, 1990
[11] *Polym. Handb.*, 3rd edn., (eds. J. Brandrup and E.H. Immergut), Wiley Interscience, 1989
[12] Morton, W.E. and Hearle, J.W.S., *Physical Properties of Textile Fibres*, Butterworth, 1962
[13] Ramiah, M.V. and Goring, D.A., J. Polym. Sci., Part C: Polym. Lett., 1965, **11**, 27
[14] Frey-Wyssling, A., Helv. Chim. Acta, 1936, **19**, 981
[15] Chanzy, H. and Pegny, A., J. Polym. Sci., *Polym. Phys. Ed.*, 1980, **18**, 1137
[16] Zeronian, S.H., Coole, M.L., Alger, K.W. and Chandler, J.M., *J. Appl. Polym. Sci.: Appl. Polym. Symp.*, 1983, **37**, 1053

[17] Luner, P. and Sandell, M., *J. Polym. Sci., Part C: Polym. Lett.*, 1969, **28**, 115
[18] Sugiyama, J., Vuong, R. and Chanzy, H., *Macromolecules*, 1991, **24**, 4168
[19] Sugiyama, J., Okano, T., Yamamoto, H. and Horii, F., *Macromolecules*, 1990, **23**, 3196
[20] Atalla, R.H. and Vanderhart, D.L., Science (Washington, *D.C.*), 1984, **223**, 1465
[21] *Ullmanns Encykl. Ind. Chem.*, 5th edn., (ed. W. Gerhartz), VCH, 1986, **A5**
[22] Bremer, F.C., Frilette, J. and Mark, H., *J. Am. Chem. Soc.*, 1948, **70**, 8177
[23] Marx-Figini, M., *J. Polym. Sci., Part C: Polym. Lett.*, 1969, **28**, 57
[24] Palma, A., Buldt, G. and Jovanovic, S.M., *Makromol. Chem.*, 1976, **177**, 1063
[25] Marx-Figini, M. and Pion, B.G., *Biochim. Biophys. Acta*, 1974, **338**, 382
[26] Goring, D.A.I. and Timell, T.E., *Tappi*, 1962, **45**, 454
[27] Ioelovitch, M. and Gordeev, M., *Acta Polym.*, 1994, **45**, 121
[28] Kandola, B.K., Horrocks, A.R., Price, D. and Coleman, G.V., *J. Macromol. Sci., Rev. Macromol. Chem. Phys.*, 1996, **36**, 721, (pyrolysis, flame retardants)

Cellulose II

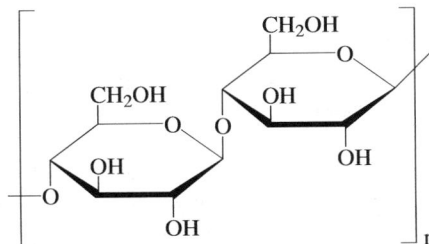

Synonyms: *Regenerated cellulose. Mercerised cellulose. Viscose. Rayon. Cellophane*
Related Polymers: Cellulose I, Cellulose
Monomers: Base monomer unit glucose
Material class: Polysaccharides
Polymer Type: cellulosics
CAS Number:

CAS Reg. No.	Note
9004-34-6	
9005-81-6	Cellophane

Molecular Formula: $(C_6H_{10}O_5)_n$
Fragments: $C_6H_{10}O_5$
Molecular Weight: DP (of microcrystals): 40–60 (Fortisan); 30–50 (Textile yarns); 15–30 (Tyre-yarns); DP (of fibres): 350–400 (viscose); 600–650 (polynosic fibres)
Miscellaneous: Average fibre crystallinity: 40% (by x-ray), 60% (acid hydrol.), 37% (deuterium exch.), 83% (Fortisan-acid hydrol.), 68% (Textile rayon-acid hydrol.). Crystalline-Rayon fibre: P2, space group, a. = 0.793, b. = 0.918, c. = 1.034, γ = 117.3°. Crystallite average length: 33.4 nm (Fortisan), 24 nm (Viscose), width: 2.7 nm (Fortisan), 2.3 nm (Viscose), thickness: 5 nm (Fortisan/viscose)

Volumetric & Calorimetric Properties:
Density:

No.	Value	Note
1	d 1.5–1.53 g/cm^3	viscose [2,6,18]
2	d 1.15 g/cm^3	viloft [6]
3	d 1.534 g/cm^3	cellophane [11]
4	d 1.59–1.63 g/cm^3	crystalline, x-ray detn [4]
5	d 1.482–1.489 g/cm^3	amorph., x-ray detn [4]

Thermodynamic Properties General: Heat of formation 5900 kJ kg^{-1} [21]. Heat of combustion 14732 kJ kg^{-1} [36,19]. Heat of crystallisation 134.8 kJ kg^{-1} [19]
Thermal Conductivity:

No.	Value	Note
1	0.054–0.07 W/mK	rayon [19]
2	0.8 W/mK	sulfite pulp, wet
3	0.067 W/mK	sulfite pulp, dry

Specific Heat Capacity:

No.	Value	Note	Type
1	1.357 kJ/kg.C	viscose fibre [1]	p
2	1.776 kJ/kg.C	isotopic rayon [1]	p
3	1.403 kJ/kg.C	oriented rayon [1]	p
4	1.357–1.595 kJ/kg.C	viscose [19]	P

Melting Temperature:

No.	Value	Note
1	235–245°C	[3]

Transition Temperature:

No.	Value	Note
1	450°C	Ignition temp. (rayon) [1]
2	350–420°C	[17]
3	-20°C	Secondary transition temp. [23]
4	15°C	Secondary transition temp. [23]

Surface Properties & Solubility:
Solubility Properties General: Solubility: 20–30% (6.5% NaOH soln. 20°, viscose) [1]
Solvents/Non-solvents: Sol. DMF, chloral, pyridine, conc. mineral acids (with hydrolytic degradation), inorganic salts (e.g. $ZnCl_2$, LiCl, $Ca(SCN)_2$), strong alkalis, metal complexes (e.g. [Cu$(NH_3)_4$](OH)$_2$ (Cuoxam), [Cd(en)$_3$](OH)$_2$, (Cadoxen)) [19,27]. Swollen by H_2O liq. NH_3, hydrazine [19], alkylamines. Liq. cryst. formed with: *N*-methylmorpholine-*N*-oxide, [26], dimethyl acetamide/LiCl [25]
Surface Tension:

No.	Value	Note
1	45.4 mN/m	20°, polarity 0.344, cellophane [19]

Transport Properties:
Polymer Solutions Dilute: Viscosity ranges in water: 20 000 mPa s (1%, type HH, 25°, ASTM 2364), 10 000 mPa s (2%, type M, 25°, ASTM 2364), 200 mPa s (5%, type L, 25°, ASTM 2364) [3]
Permeability of Gases: Gas permeability of cellophane at 25° [He] 0.005 × 10^{-10} cm^2 (5.cm Hg)$^{-1}$), [N$_2$] 0.0032 × 10^{-10}, [H$_2$] 0.0065 × 10^{-10}, [O$_2$] 0.0021 × 10^{-10}, [CO$_2$] 0.0047 × 10^{-10}, [H$_2$O] 1900 × 10^{-10}.
Permeability and Diffusion General: Gas and liq. permeabilities have been reported [14]
Water Content: Water retention 90–110% (regular rayon); 60–70% (polynosic rayon); 60–80% (HWM rayon); 130% (viloft); 100% (viscose) [16]

– Cellulose II

Water Absorption:

No.	Value	Note
1	11–13 %	24°, 65% relative humidity, rayon [16]
2	1–11 %	9–100%, relative humidity, viscose fibre [19]
3	45–115 %	cellophane [34]

Mechanical Properties:
Tensile (Young's) Modulus:

No.	Value	Note
1	7239 MPa	extrapolated [12]

Tensile Strength Break:

No.	Value	Note	Elongation
1	200–400 MPa	dry [1]	8–26%
2	100–200 MPa	wet [1]	13–43%
3	610 MPa	oriented viscose, dry [2,4]	19%
4	520 MPa	oriented viscose, wet [2,4]	9%
5	80 MPa	cellophane [34]	30%

Elastic Modulus:

No.	Value	Note
1	33000 MPa	oriented rayon [2,4]
2	80000 MPa	mercerised ramie [2]

Miscellaneous Moduli:

No.	Value	Note
1	1100 MPa	calc. modulus of rigidity, Fortisan H [3]
2	70000–90000 MPa	$7-9 \times 10^{11}$ dyne cm^{-2} Fortisan H [13]

Impact Strength: 35–58 J m^{-1} (cellophane) [34]
Mechanical Properties Miscellaneous: Relative wet/dry strength 50% (viscose) [4], 86% (high, oriented viscose) [4]
Failure Properties General: Strength lost 150° (rayon) [16]. Tear strength 0.8–8 N mm^{-2} (cellophane) [34]. Burst strength (Mullen) 30–50 [34]

Electrical Properties:
Dielectric Permittivity (Constant):

No.	Value	Frequency	Note
1	2.67	1 kHz	[1]
2	2.2–2.7		paper [19]
3	7.3	10 kHz	25°, cellophane [30]
4	4.04	3000 MHz	cellophane [30]
5	3.5	100 kHz	0% relative humidity, viscose
6	8.4	1 kHz	65% relative humidity, viscose [20,19]

Dielectric Strength:

No.	Value	Note
1	79.99 kV.mm^{-1}	Cellophane [36]

Strong Field Phenomena General: Zeta potential: 30mV (filter paper) [33], 26mV (viscose rayon), NaBH$_4$-reduced sample, water) [19]

Dissipation (Power) Factor:

No.	Value	Frequency	Note
1	0.0026–0.0028		30°, 0.006–0.18% water, Kraft paper [31]
2	0.009	60 Hz	cellophane [32]

Optical Properties:
Refractive Index:

No.	Value	Note
1	1.58	Parallel to fibre axis [15,19]
2	1.52	Perpendicular to fibre axis [15,19]
3	1.529–1.547	Parallel, viscose fibres [19,24]
4	1.512–1.52	Perpendicular viscose fibres [19,24]

Total Internal Reflection: Birefringence (by refractive index) 0.018 (regular viscose); 0.036–0.057 (high tenacity rayon); 1.040–1.045 (polynosic rayon) [2]
Volume Properties/Surface Properties: Haze 3.5% (cellophane) [34]

Polymer Stability:
Polymer Stability General: Crosslinking of cellulose chains is important for production of crease-resistant and crease-retaining materials. This process also prevents swelling of the polymer in alkali and hence reduces degradation
Thermal Stability General: Decompostion temp. 175–240° (rayon) [16], 180° (viscose) [20]. Minimum use temp. -17° (cellophane) [34]

Upper Use Temperature:

No.	Value	Note
1	177°C	cellophane [34]

Decomposition Details: Limiting oxygen index 19 % [17], 19.7 % [19]. Heat of combustion 18 J kg^{-1} [17]
Flammability: Ignition temp. 450° (rayon) [1]. Fl. p. 322° (rayon) [1]

Applications/Commercial Products:
Processing & Manufacturing Routes: Cellulose from trees is obtained by pulping (high-grade material is from spruce and eucalyptus). Pulp is dissolved in sodium hydroxide soln. and the alkali-cellulose converted to cellulose xanthate (CS$_2$). The resulting syrup is extruded into a soln. of sulfuric acid. Filaments are spun into a continuous tow, which can be cut to produce staple fibre
Applications: Textiles: woven and knitted fabrics, blends with polyesters, wool, mohair and silk, used in knitting yarns, furnishings and upholstery. Stitch-bonded non-wovens in drapes, ticking, cloth and footwear. Needle-punched non-wovens in blankets and wall/floor coverings. Medical and hygiene uses as dressings, swabs and sanitary towels. Drylaid non-wovens: wipes, interlinings, filters. Wetlaid non-wovens: food casings, apparel. Flock: wall coverings, apparel, shoes, packaging

Trade name	Details	Supplier
Avisco	general name for products	American Viscose Corporation
Cellophane		DuPont
Dutermatt	Extra-dull Viscose staple fibre	Courtaulds
Evla	Coarse crumped Viscose staple	Courtaulds
Fibro	Standard Viscose staple fibre	Courtaulds
Sarille	Crimped Viscose staple fibre	Courtaulds
Transparent P	film (uncoated)	Wolff
Viloft	deatex viscose fibre	Courtaulds
Viscose	fibre	Courtaulds

Bibliographic References

[1] *Handbook of Physical and Mechanical Testing of Paper and Paperboard*, (ed. R.E. Mark), Marcel Dekker, 1984
[2] *Concise Encyclopedia of Polymer Science and Engineering*, (eds. H.F. Mark, N.M. Bikales, C.G. Overberger and G. Menges), John Wiley and Sons, 1990
[3] *High Polymers*, (eds. Bikales, N.M. and Segal, L.), Wiley Interscience, 1971
[4] *Encycl. Polym. Sci. Eng.*, (eds. H.F. Mark, N.M. Bikales, C.G. Overberger and G. Menges), John Wiley and Sons, **3**
[5] Elias, H.-G., Macromolecules Volume 2: Synthesis, *Materials and Technology*, Plenum Press, 1984, 1068
[6] *ACS Symp. Ser.*, 1977, **58**, 3
[7] Kolpak, K.J. and Blackwell, J., *Macromolecules*, 1976, **9**, 273
[8] *Cellulose and Cellulose Derivatives*, (eds. E. Ott, H.M. Spurlin and M.W. Grafflin), Interscience, 1954
[9] Philippe, J.H., Nelson, M.L. and Ziifle, H.M., *Text. Res. J.*, 1947, **17**, 585
[10] Moorhead, F.F., *Text. Res. J.*, 1950, **20**, 549
[11] Venkeswaran, A. and Vanden Akker, J.A., *J. Appl. Polym. Sci.*, 1965, **9**, 1149
[12] Wellisch, E., Marker, L. and Sweeting, O.T., *J. Appl. Polym. Sci.*, 1961, **5**, 647
[13] Mann, J. and Roldan-Gonzales, L., *Polymer*, 1962, **3**, 549
[14] Dean, J.A., *Lange's Handbook of Chemistry*, 14th edn., McGraw-Hill, 1992
[15] Moncrieff, R.W., *Man-Made Fibers*, Newnes-Butterworth, 1975
[16] Rodriguez, F., *Principles of Polymer Systems*, 3rd edn., Hemisphere Publishing Corp, 1989, 596
[17] *Chemistry of the Textiles Industry*, (ed. C.M. Carr), Blackie Academic and Professional, 1995
[18] Davidson, G.F., *J. Text. Inst.*, 1927, **18**, 1175
[19] *Polym. Handb.*, 3rd edn., (eds. J. Brandrup and E.H. Immergut), Wiley Interscience, 1989
[20] Morton, W.E. and Heale, J.W.S, *Physical Properties of Textile Fibres*, Butterworth, 1962
[21] Jessup, R.S. and Proser, E.I., *J. Res. Natl. Bur. Stand. (U.S.)*, 1950, **44**, 385
[22] Han, S.T. and Ulmanen, T., *Tappi*, 1958, **41**, 185
[23] Kubat, J. and Pattyranie, C., *Nature (London)*, 1967, **215**, 390
[24] Rose, L. and Griffiths, J.D., *J. Text. Inst.*, 1948, **39**, 265
[25] Conio, C., Corazza, P., Bianchi, E., Tealdi, A. and Ciferri, A., *J. Polym. Sci., Polym. Lett. Ed.*, 1984, **22**, 273
[26] Chanzy, H. and Pegny, A., *J. Polym. Sci., Polym. Phys. Ed.*, 1980, **18**, 1137
[27] *Cellulose Chemistry, Biochemistry and Material Aspects*, (eds. J.F. Kennedy, G.O. Phillips and P.A. Williams), Ellis Horwood, 1993
[28] Hergh, S.P. and Montgomery, D.J., *Text. Res. J.*, 1953, **22**, 805
[29] Church, H.F., *J. Soc. Chem. Ind., London*, 1947, **66**, 221
[30] Stoops, W.N., *J. Am. Chem. Soc.*, 1934, **56**, 1480
[31] Clark, F.M., *Ind. Eng. Chem.*, 1952, **44**, 887
[32] Reddish, W., *Trans. Faraday Soc.*, 1950, **46**, 459
[33] Rabinov, G. and Heymann, E., *J. Chem. Phys.*, 1943, **47**, 655
[34] *Kirk-Othmer Encycl. Chem. Technol.*, 4th edn. (ed. J.I. Kroschwitz), Wiley Interscience, 1993, **10**, 766, 781
[35] *Ullmanns Encykl. Ind. Chem.*, 5th edn. (ed. W. Gerhartz) VCH, 1986, **A5**, 375
[36] Birky, M.M. and Yeh, K.N., *J. Appl. Polym. Sci.*, 1973, **17**, 239

Cellulose acetate

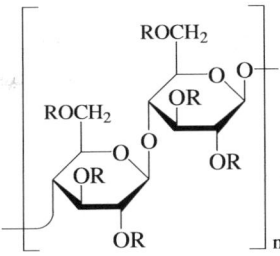

R = Ac, H

Synonyms: *Acetate*
Related Polymers: Cellulose II, Cellulose acetate propionate, Cellulose acetate butyrate, Cellulose triacetate
Monomers: Base monomer unit glucose
Material class: Polysaccharides
Polymer Type: Cellulosics
CAS Number:

CAS Reg. No.
9004-35-7

Molecular Formula: $(C_{11-12}H_{15-16}O_{7.5-8})_n$
General Information: Thermoplastic polymer with excellent optical props.

Volumetric & Calorimetric Properties:

Density:

No.	Value	Note
1	d 1.28 g/cm^3	[1]
2	d 1.3–1.31 g/cm^3	[2]
3	d 1.26 g/cm^3	varies with degree of acetylation [8]

Thermal Expansion Coefficient:

No.	Value	Note	Type
1	0.0001–0.00015 K^{-1}	sheet [3]	L
2	8×10^{-5}–0.00018 K^{-1}	mould [3]	L
3	0.00014 K^{-1}	[4]	L
4	0.00012 K^{-1}	ASTM D696 [8]	L

Thermodynamic Properties General: Heat of combustion 17179 kJ kg^{-1}
Thermal Conductivity:

No.	Value	Note
1	0.17–0.34 W/mK	[3,8]

Specific Heat Capacity:

No.	Value	Note	Type
1	1.26–1.67 kJ/kg.C	23°, ASTM C177 [8]	P
2	1.26–2.09 kJ/kg.C	sheet [1,9]	P
3	1.26–1.76 kJ/kg.C	moulding [1,9]	P

– **Cellulose acetate**

Melting Temperature:

No.	Value	Note
1	230°C	[3]
2	235–255°C	[2]

Deflection Temperature:

No.	Value	Note
1	49–98°C	0.455 MPa
2	44–91°C	1.82 MPa, moulding [3]
3	61°C	0.0383 Mpa
4	72°C	0.0096 MPa, ASTM D648 [4]
5	70°C	0.0455 MPa
6	57°C	1.82 MPa, ASTM D648 [8]
7	55–118°C	0.455 MPa, ASTM D648 [1]
8	44–113°C	1.82 MPa, ASTM D648 [1]

Vicat Softening Point:

No.	Value	Note
1	70°C	[5]

Surface Properties & Solubility:
Plasticisers: Diglycerol ethers
Solvents/Non-solvents: Sol. ketones, esters and alcohols (ASTM D543) [5]. Liq. cryst. trifluoroacetic acid/CH_2Cl_2
Surface Tension:

No.	Value	Note
1	45.9 mN/m	20°, polarity 0.296 [13,14]

Transport Properties:
Water Absorption:

No.	Value	Note
1	3–8.5 %	[2]

Gas Permeability:

No.	Gas	Value	Note
1	He	893 cm^3 mm/(m^2 day atm)	13.6×10^{-10} cm^2 (s cmHg)$^{-1}$ [3]
2	N_2	18.4 cm^3 mm/(m^2 day atm)	0.28×10^{-10} cm^2 (s cmHg)$^{-1}$, 30° [3]
3	H_2	229.8 cm^3 mm/(m^2 day atm)	3.5×10^{-10} cm^2 (s cmHg)$^{-1}$, 20° [3]
4	O_2	51.2 cm^3 mm/(m^2 day atm)	0.78×10^{-10} cm^2 (s cmHg)$^{-1}$, 20° [3]
5	H_2O	361163 cm^3 mm/(m^2 day atm)	5500×10^{-10} cm^2 (s cmHg)$^{-1}$, 20° [3]
6	H_2S	229.8 cm^3 mm/(m^2 day atm)	3.5×10^{-10} cm^2 (s cmHg)$^{-1}$, 30° [3]
7	Ethylene oxide	1116.3 cm^3 mm/(m^2 day atm)	17×10^{-10} cm^2 (s cmHg)$^{-1}$, 0° [3]
8	Bromomethane	446.5 cm^3 mm/(m^2 day atm)	6.8×10^{-10} cm^2 (s cmHg)$^{-1}$, 60° [3]
9	C_6H_6	33621 cm^3 mm/(m^2 day atm)	512×10^{-10} cm^2 (s cmHg)$^{-1}$, 35° [3]
10	Hexane	183.9 cm^3 mm/(m^2 day atm)	2.8×10^{-10} cm^2 (s cmHg)$^{-1}$, 35° [3]
11	CCl_4	245.6 cm^3 mm/(m^2 day atm)	3.74×10^{-10} cm^2 (s cmHg)$^{-1}$, 35° [3]
12	EtOH	195685 cm^3 mm/(m^2 day atm)	2980×10^{-10} cm^2 (s cmHg)$^{-1}$, 35° [3]
13	EtOAc	236069 cm^3 mm/(m^2 day atm)	3595×10^{-10} cm^2 (s cmHg)$^{-1}$, 35° [3]

Mechanical Properties:
Tensile (Young's) Modulus:

No.	Value	Note
1	2174 MPa	ASTM D638 [4]
2	2800–3000 MPa	$2.8–3 \times 10^{10}$ dyne cm^{-2}, DS 1.87–2.99 [6]

Flexural Modulus:

No.	Value	Note
1	1749 MPa	ASTM D790 [4]
2	1449 MPa	ASTM D790 [4]

Tensile Strength Break:

No.	Value	Note	Elongation
1	13.5–58.6 MPa	23°, ASTM D638 [1,9]	
2	5.5–41.4 MPa	70°, ASTM D638 [1,9]	50–56%
3	75 MPa	film [2]	15–55%

Elastic Modulus:

No.	Value	Note
1	620.7–1793 MPa	ASTM D790 [1,9]
2	1310 MPa	ASTM D790 [8]

Tensile Strength Yield:

No.	Value	Elongation	Note
1	13.8–48.3 MPa		73°, ASTM D638 [1,9]
2	15.2–51 MPa		sheet [3]
3	28.3–52.4 MPa		mould [3]
4	22.8 MPa	30%	ASTM D638 [8]

Flexural Strength Yield:

No.	Value	Note
1	13.8–110.3 MPa	ASTM D790 [1,10]
2	33.1 MPa	ASTM D790 [8]
3	41–68.9 MPa	sheet [3]
4	13.8–110 MPa	mould [3]

– Cellulose acetate

Compressive Strength:

No.	Value	Note
1	13.1–64.1 MPa	ASTM D695 [1,10]

Hardness: Rockwell R39–R120 (ASTM D785) [1], R100–R123 (moulding) [3], R85–R120 (sheeting) [3], R82 (ASTM D785) [4]
Failure Properties General: Deformation under load 40–1% (13.79 MPa), 15–1% (6.9 MPa, ASTM D621) [1]. Tear strength 1.6–3.9 N mm^{-1} (film) [2]. Burst strength (Mullen) 30-60 [2]
Izod Notch:

No.	Value	Notch	Note
1	132.7–136.6 J/m	N	23°, ASTM D256 [1,9]
2	1.9–14.3 J/m	N	-40°, ASTM D758 [1,9]
3	107–454 J/m	N	23°, sheet [3]
4	53–214 J/m	N	23°, mould [3]
5	235 J/m	Y	23°, ASTM D256 [8]
6	59 J/m	Y	-40°, ASTM D256 [8]

Electrical Properties:
Surface/Volume Resistance:

No.	Value	Note	Type
3	0.68×10^{15} Ω.cm	ASTM D257 [8]	S

Dielectric Permittivity (Constant):

No.	Value	Frequency	Note
1	3.2–7	1 MHz	ASTM D150 [5]
2	3.4–7.4	60 Hz	sheet [3]
3	3.2–7	1 MHz	sheet [3]
4	3.5–7.5	60 Hz	mould [3]
5	3.2–7	1 MHz	mould [3]
6	3.5	1 MHz	ASTM D150 [8]

Dielectric Strength:

No.	Value	Note
1	0.25–0.365 kV.mm^{-1}	0.125 cm thick [1,9]
2	13.4 kV.mm^{-1}	ASTM D149 [4]
3	11–24 kV.mm^{-1}	sheet [3]
4	9–24 kV.mm^{-1}	mould [3]
5	14.5 kV.mm^{-1}	ASTM D149 [8]
6	126–197 kV.mm^{-1}	sheet [2]

Complex Permittivity and Electroactive Polymers: Zeta potential 52 mV (cellulose acetate fibre, H$_2$O) [12]

Optical Properties:
Optical Properties General: Uv light screening >99% absorbed (ASTM E308) [8]
Refractive Index:

No.	Value	Note
1	1.46–1.5	25°, ASTM D542 [1,9]
2	1.49–1.5	sheet [3]
3	1.46–1.5	mould [3]
4	1.46–1.49	ASTM D542 [8]

Total Internal Reflection: $[\alpha]_D^{20}$ +12.6 (MeCN), +5.15 (Me$_2$CO), -10 (dioxan), -26 (pyridine) [7]
Volume Properties/Surface Properties: Light transmittance 88% (sheet) [3], >90% (sheet, 1.52 mm thick, ASTM E308) [8]. Haze <1% (sheet) [3]. Haze <8.5% (sheet, 1.52 mm thick, ASTM D1003) [8]

Polymer Stability:
Polymer Stability General: Aq. potassium or calcium iodide stabilises the polymer against thermal degradation and discoloration during processing
Upper Use Temperature:

No.	Value	Note
1	79°C	film [2]
2	-26°C	film [2]

Decomposition Details: Weight loss in accelerated ageing of 0.4–12.0% (82°, 72h, ASTM D706) [1,2]
Flammability: 1.27–5.08 cm min^{-1} (ASTM D635) [1]
Chemical Stability: Stable to chlorine bleach and to dilute acids and alkalis. Decomposed by strong mineral acids and alkalis, particularly at higher temps. [1]

Applications/Commercial Products:
Processing & Manufacturing Routes: Cellulose fibres (or cellulose dissolved in N,N-dimethylacetamide-LiCl) is treated with Ac$_2$O in AcOH, using sulfuric acid as catalyst. Water is added to terminate the reaction. Partial hydrol. of acetyl groups can also be achieved to give a range of products. The product is collected by precipitation into water or dil. AcOH
Mould Shrinkage (%):

No.	Value	Note
1	0.2–0.6%	ASTM D955

Applications: Fibres, films, plastics, liquid crystals, cigarette filters, packaging

Trade name	Supplier
Acety	Diacel Chem. Ind.
CA700	Albis Corp.
Cellidor S	Bayer Inc.
H/H2/H3/H4	Rotuba
M/MH/MS	Rotuba
Tenite acetate	Eastman Chemical Company
	Hoechst Celanese

Bibliographic References
[1] *Encycl. Polym. Sci. Eng.*, 2nd edn., (eds. H.F. Mark, N.M. Bikales, C.G. Overberger and G. Menges), John Wiley and Sons, 1985, **3**
[2] *Kirk-Othmer Encycl. Chem. Technol.*, 4th edn., (ed. J.I. Kroschwitz), Wiley Interscience, 1993, **5**
[3] Dean, J.A., *Lange's Handbook of Chemistry*, 14th edn., McGraw-Hill, 1992
[4] *Handbook of Plastic Materials and Technology*, (ed. I.I. Rubin), Wiley Interscience, 1990, 55
[5] Crompton, T.R., *Practical Polymer Analysis*, Plenum Press, 1993, 716
[6] Richards, G.N., *J. Appl. Polym. Sci.*, 1961, **5**, 545
[7] Fort, R.J., Moore, W.R. and Tidswell, B.M., *Chem. Ind. (London)*, 1964, 108
[8] *Tenite Cellulose Plastics*, Pub. No. PPC-100B, Eastman Chemical Co., Kingsport, TN, USA, 1995, (technical datasheet)

[9] Guide to Plastics, *Property and Specification Charts*, (ed. J. Agronoff), McGraw Hill Inc., 1982, **59**, 470
[10] Physical and Chemical Properties of Triacetate Filaments, *Yarns and Staple Fibres*, Tech. Bull. TBT 30, Celanese Fibres Mktg. Co., Charlotte, NC, USA, 1974, (technical datasheet)
[11] Birky, M.M. and Yeh, K.N., *J. Appl. Polym. Sci.*, 1973, **17**, 239
[12] Gröbe, A, *Polym. Handb.*, 3rd edn., (eds. J. Brandrup and E.H. Immergut), Wiley Interscience, 1989, V137
[13] Kaelble, D.H. and Moacanin, J., *Polymer*, 1977, **18**, 475
[14] Busscher, H.J. and Arends, J., *J. Colloid Interface Sci.*, 1981, **81**, 75

Cellulose acetate butyrate   C-10

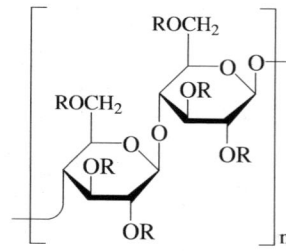

R = Ac, $CH_2CH_2CH_2CH_3$

Related Polymers: Cellulose II, Cellulose acetate, Cellulose acetate propionate
Monomers: Base monomer unit glucose
Material class: Polysaccharides
Polymer Type: Cellulosics
CAS Number:

CAS Reg. No.
9004-36-8

General Information: Composition by weight: Ac 13–15%, Butyrate 36–38%, H 1–2%. Commercial grades: Ac 6–29.5%, Butyrate 17–48%, H 1–2.5%. Commercial grades (lacquers, plastics and coatings): Degree of substitution Ac: Bu: H = 0.5–2.1: 0.7–2.3: 0.2–0.5

Volumetric & Calorimetric Properties:
Density:

No.	Value	Note
1	d 1.15–1.22 g/cm^3	ASTM D792 [1]
2	d 1.19 g/cm^3	ASTM D792 [2,5]

Thermal Expansion Coefficient:

No.	Value	Note	Type
1	0.00011–0.00016 K^{-1}	ASTM D696 [1]	L
2	0.00012 K^{-1}	ASTM D696 [5]	L

Equation of State: Equation of state information has been reported [9]
Thermal Conductivity:

No.	Value	Note
1	0.16–0.33 W/mK	ASTM C177 [1,6]

Specific Heat Capacity:

No.	Value	Note	Type
1	1.2–1.6 kJ/kg.C	ASTM C177 [1,6]	P

Melting Temperature:

No.	Value	Note
1	140°C	[4]

Deflection Temperature:

No.	Value	Note
1	65°C	0.038 MPa [2]
2	72°C	0.0096 MPa, ASTM D648 [2]
3	74°C	1.82 MPa [5]
4	85°C	0.455 MPa, ASTM D648 [5]
5	59.5–112°C	0.455 MPa [1]
6	50–99°C	1.82 MPa, ASTM D648 [1]

Vicat Softening Point:

No.	Value	Note
1	70°C	[6]

Surface Properties & Solubility:
Solubility Properties General: Compatible with polyester, acrylic, vinyl and alkyd resins [1]
Solvents/Non-solvents: Sol. ketones, esters and alcohols (ASTM D543) [6]
Surface Tension:

No.	Value	Note
1	34 mN/m	20° [10,11]

Transport Properties:
Water Absorption:

No.	Value	Note
1	1–4 %	24h, ASTM D570 [1]
2	1.5 %	24h, ASTM D570 [2]

Gas Permeability:

No.	Gas	Value	Note
1	O_2	320.7 cm^3 mm/ (m^2 day atm)	4.61 × 10^{-10} cm^2 (s cmHg)$^{-1}$ [7]

Mechanical Properties:
Tensile (Young's) Modulus:

No.	Value	Note
1	1725 MPa	ASTM D638 [2]

Flexural Modulus:

No.	Value	Note
1	482.8–1379 MPa	ASTM D790 [1]
2	1449 MPa	ASTM D790 [2]

– Cellulose acetate propionate

Tensile Strength Break:

No.	Value	Note	Elongation
1	13.8–51.7 MPa	23°, ASTM D638 [1,3]	
2	8.3–39.3 MPa	70°, ASTM D638 [1,3]	
3	34.5 MPa	ASTM D638 [2]	50%

Elastic Modulus:

No.	Value	Note
1	1379 MPa	ASTM D790 [5]

Tensile Strength Yield:

No.	Value	Note
1	10.3–48.3 MPa	73°, ASTM D638 [1,3]

Flexural Strength Yield:

No.	Value	Note
1	10.3–64.1 MPa	ASTM D790 [1,3]
2	27.6–62 MPa	sheet [4]
3	12.4–64.1 MPa	moulding [4]

Compressive Strength:

No.	Value	Note
1	7.6–52.4 MPa	ASTM D695 [1,3]

Hardness: Rockwell 29–117 (ASTM D785) [1]. Rockwell R75 (ASTM D785) [2]
Failure Properties General: Deformation under load 40–1% (13.79 MPa), 15–1% (6.9 MPa, ASTM D621) [1]
Izod Notch:

No.	Value	Notch	Note
1	187 J/m	Y	ASTM D256 [2]
2	133–288 J/m	Y	23°, sheet [4]
3	53–582 J/m	Y	23°, moulding [4]
4	240 J/m	Y	23°, ASTM D256 [5]
5	96 J/m	Y	-40°, ASTM D256 [5]

Electrical Properties:
Surface/Volume Resistance:

No.	Value	Note	Type
1	14×10^{15} Ω.cm	ASTM D257 [5]	S

Dielectric Strength:

No.	Value	Note
1	0.25–0.4 kV.mm^{-1}	0.125 cm thick, ASTM D149 [1,3]
2	16.6 kV.mm^{-1}	ASTM D149 [5]

Optical Properties:
Refractive Index:

No.	Value	Note
1	1.46–1.49	25°, ASTM D542 [1,3]

Volume Properties/Surface Properties: Light transmittance >90% (sheet, 1.52 mm thick, ASTM E308) [5]. Haze <8.5% (sheet, 1.52 mm thick, ASTM D1003) [5]
Polymer Stability:
Decomposition Details: Weight loss in accelerated ageing of 0.1–4.0% at 82° over 72h (ASTM D706) [1]
Flammability: 1.27–3.81 cm mm^{-1} (ASTM D635) [1,3]
Chemical Stability: Decomposed by strong acids and alkalis; slightly decomposed by weak acids and alkalis (ASTM D543) [6]
Applications/Commercial Products:
Mould Shrinkage (%):

No.	Value	Note
1	0.2–0.4%	ASTM D955

Trade name	Supplier
Cellidor B	Bayer Inc.
Tenite butyrate	Eastman Chemical Company

Bibliographic References

[1] *Encycl. Polym. Sci. Eng.*, 2nd edn., (ed. J.I. Kroschwitz), John Wiley and Sons, 1985, **3**
[2] *Handbook of Plastic Materials and Technology*, (ed. I.I. Rubin), Wiley Interscience, 1990
[3] Physical and Chemical Properties of Triacetate Filaments, *Yarns and Staple Fibres*, Tech. Bull. TBT 30, Celanese Fibres Mktg. Co., Charlotte, NC, USA, 1974, (technical datasheet)
[4] Dean, J.A., *Lange's Handbook of Chemistry*, 14th edn., McGraw-Hill, 1992
[5] *Tenite Cellulose Plastics*, Pub. No. PPC-100B, Eastman Chemical Co., Kingsport, TN, USA, 1995, (technical datasheet)
[6] Crompton, T.R., *Practical Polymer Analysis*, Plenum Press, 1993, 716
[7] Yang, W.H., Smolen, V.F. and Peppas, N.A., *J. Membr. Sci.*, 1981, **9**, 53
[8] Saunders, J.K., *Org. Polym. Chem.*, Chapman and Hall, 1973, 265
[9] Spenser, R.S. and Gilmore, G.D., *J. Appl. Phys.*, 1950, **21**, 523
[10] Wu, S., *Polymer Interface and Adhesion*, Marcel Dekker, 1982, (surface tension)
[11] Gröbe, A., *Polym. Handb.*, 3rd edn., (eds. J. Brandrup and E.H. Immergut), Wiley Interscience, 1989, **V 137**, (surface tension, general data)

Cellulose acetate propionate

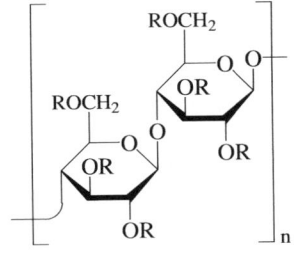

R = Ac, CH$_2$CH$_2$CH$_3$

Related Polymers: Cellulose acetate, Cellulose II
Monomers: Base monomer unit glucose
Material class: Lyotropic, Polysaccharides

Cellulose acetate propionate

Polymer Type: cellulosics
CAS Number:

CAS Reg. No.
9004-39-1

General Information: Composition by weight Ac 1.5–3.5%; propionate 43–47%; H 2–3% (commercial grades)

Volumetric & Calorimetric Properties:

Density:

No.	Value	Note
1	d 1.17–1.24 g/cm^3	[4]
2	d 1.19 g/cm^3	[3]
3	d 1.19–1.23 g/cm^3	ASTM D792 [2,6]
4	d 1.2 g/cm^3	ASTM D792 [5]

Thermal Expansion Coefficient:

No.	Value	Note	Type
1	0.00012–0.00016 K^{-1}	ASTM D696 [2,6]	L
2	0.00011 K^{-1}	ASTM D696 [5]	L

Thermal Conductivity:

No.	Value	Note
1	0.16–0.33 W/mK	ASTM C177 [2,6]

Specific Heat Capacity:

No.	Value	Note	Type
1	1.2–1.6 kJ/kg.C	[2,6]	P

Melting Temperature:

No.	Value	Note
1	190°C	[4]

Deflection Temperature:

No.	Value	Note
1	72°C	0.038 MPa [3]
2	80°C	0.0096 MPa, ASTM D648 [3]
3	75°C	1.82 MPa [5]
4	83°C	0.455 MPa, ASTM D648 [5]

Vicat Softening Point:

No.	Value	Note
1	100°C	ASTM D1525 [3]

Transport Properties:

Water Absorption:

No.	Value	Note
1	1.3 %	24h, ASTM D570 [3]
2	1–3 %	24h, ASTM D570 [2,6]

Mechanical Properties:

Tensile (Young's) Modulus:

No.	Value	Note
1	414–1480 MPa	[4]
2	1725 MPa	ASTM D638 [3]

Flexural Modulus:

No.	Value	Note
1	689.7–1931 MPa	ASTM D790 [2,6]
2	1449 MPa	ASTM D790 [3,5]

Tensile Strength Break:

No.	Value	Note	Elongation
1	35 MPa	ASTM D638 [3]	60%
2	13.8–51.7 MPa	23°, ASTM D638 [2,6]	
3	6.7–44.8 MPa	70°, ASTM D638 [2,6]	

Tensile Strength Yield:

No.	Value	Note
1	10.3–48.3 MPa	73°, ASTM D638 [2,6]
2	31.7 MPa	ASTM D638 [5]

Flexural Strength Yield:

No.	Value	Note
1	20.7–75.9 MPa	Moulding compound [2,6]
2	41.4 MPa	ASTM D790 [5]

Hardness: Rockwell R10–R122 (moulding compound) [4], R20–R120 [2], R70 (ASTM D785) [3]

Izod Notch:

No.	Value	Notch	Break	Note
1	27 J/m	N	Y	[4]
2	411 J/m	Y		ASTM D256 [3]
3	416 J/m	Y		ASTM D256 [5]

Electrical Properties:

Surface/Volume Resistance:

No.	Value	Note	Type
3	39 × 10^{15} Ω.cm	ASTM D257 [5]	S

Dielectric Permittivity (Constant):

No.	Value	Frequency	Note
1	3100000	1 MHz	ASTM D150 [3]
2	3300000		ASTM D150 [5]

Dielectric Strength:

No.	Value	Note
1	0.3–0.45 kV.mm^{-1}	0.125 cm thick, ASTM D149 [2,6]

– Cellulose nitrate

| 2 | 13.4 kV.mm^{-1} | ASTM D149 [3] |
| 3 | 15.9 kV.mm^{-1} | ASTM D149 [5] |

Dissipation (Power) Factor:

No.	Value	Frequency	Note
1	0.03	1 MHz	ASTM D150 [3,5]

Optical Properties:
Refractive Index:

No.	Value	Note
1	1.46–1.48	25°, ASTM D542 [2]

Volume Properties/Surface Properties: Light transmission 90° (min., sheet, 1.52 mm thick, ASTM E308) [5]. Haze 8% (max., sheet, 1.52 mm thick, ASTM D1003) [5]

Applications/Commercial Products:

Trade name	Supplier
Cellidor CP	Bayer Inc.
Tenite propionate	Eastman Chemical Company

Bibliographic References
[1] Crompton, T.R., *Practical Polymer Analysis*, Plenum Press, 1993
[2] *Encycl. Polym. Sci. Eng.*, 2nd edn., (eds. H.F. Mark, N.M. Bikales, C.G. Overberger and G. Menges), John Wiley and Sons, 1985, **3**, 158
[3] *Handbook of Plastic Materials and Technology*, (ed. I.I. Rubin), Wiley Interscience, 1990, 55
[4] Dean, J.A., *Lange's Handbook of Chemistry*, 14th edn., McGraw-Hill, 1992
[5] *Tenite Cellulose Plastics*, Pub. No. PPC-100B, Eastman Cehmical Co., Kingsport, TN, USA, 1995, (technical datasheet)
[6] *Physical and Chemical Properties of Triacetate Filaments*, *Yarns and Staple Fibres*, Tech. Bull. TBT 30, Celanese Fibres Mktg. Co., Charlotte, NC, USA, 1974, (technical datasheet)

Cellulose nitrate C-12

R = ONO$_2$

Synonyms: *Nitrocellulose. Collodion cotton. Collodion wool. Colloxylin. Celloidin. Gun cotton. Pyrocellulose. Celluloid*
Related Polymers: Cellulose I, Cellulose II
Monomers: Base monomer unit glucose
Material class: Polysaccharides
Polymer Type: Cellulosics
CAS Number:

CAS Reg. No.	Note
9004-70-0	Cellulose nitrate
8050-88-2	Celluloid

Molecular Weight: Trinitrate unit molar mass 297. Dinitrate unit molar mass 252. Mononitrate unit molar mass 207. MW 630000, M_n 6900
Additives: H_2O, EtOH, isopropanol, butanol as damping agents
Identification: Trinitrate Nitrogen content 14.14%. Dinitrate Nitrogen content 11.12%. Pyrocellulose Nitrogen content 12.6 ± 0.10%. Gun cotton Nitrogen content >13.35%

Volumetric & Calorimetric Properties:
Density:

No.	Value	Note
1	d 1.58–1.65 g/cm^3	[2]
2	d 1.65–1.66 g/cm^3	[3]

Thermal Conductivity:

No.	Value	Note
1	0.23 W/mK	[4]

Specific Heat Capacity:

No.	Value	Note	Type
1	3.1–4.1 kJ/kg.C	[4]	P
2	1.26–1.67 kJ/kg.C	[4]	P

Melting Temperature:

No.	Value	Note
1	271°C	[4]

Glass-Transition Temperature:

No.	Value	Note
1	326°C	[8,9,10]
2	339°C	[8,9,10]

Deflection Temperature:

No.	Value	Note
1	60–71°C	1.82 MPa [4]

Transition Temperature:

No.	Value	Note	Type
1	155–220°C	[1]	Parr softening point

Surface Properties & Solubility:
Solubility Properties General: Compatible with most resins and plasticisers, but incompatible with waxes and tars. Also compatible with ethyl cellulose, cellulose acetate and ethylhydroxyethyl cellulose [1]. Wide compatibility with adipates, phthalates, phosphates and vegetable oils [11]
Plasticisers: Dibutyl phthalate, tricresyl phosphate
Solvents/Non-solvents: Sol. esters, ketones, Et_2O/EtOH mixtures [1]
Surface Tension:

No.	Value	Note
1	38 mN/m	20° [7,8]

– Cellulose triacetate

Transport Properties:

Water Absorption:

No.	Value	Note
1	1 %	24h, 21°, 80% relative humidity [1]

Gas Permeability:

No.	Gas	Value	Note
1	He	453 cm^3 mm/(m^2 day atm)	6.9×10^{-10} cm^2 (s cmHg)$^{-1}$ [4]
2	N$_2$	7.87 cm^3 mm/(m^2 day atm)	0.12×10^{-10} cm^2 (s cmHg)$^{-1}$ [4]
3	H$_2$	131.3 cm^3 mm/(m^2 day atm)	2×10^{-10} cm^2 (s cmHg)$^{-1}$, 20° [4]
4	O$_2$	128 cm^3 mm/(m^2 day atm)	1.95×10^{-10} cm^2 (s cmHg)$^{-1}$ [4]
5	CO$_2$	139.2 cm^3 mm/(m^2 day atm)	2.12×10^{-10} cm^2 (s cmHg)$^{-1}$ [4]
6	SO$_2$	115.6 cm^3 mm/(m^2 day atm)	1.76×10^{-10} cm^2 (s cmHg)$^{-1}$ [4]
7	NH$_3$	3749 cm^3 mm/(m^2 day atm)	57.1×10^{-10} cm^2 (s cmHg)$^{-1}$ [4]
8	H$_2$O	413039 cm^3 mm/(m^2 day atm)	6290×10^{-10} cm^2 (s cmHg)$^{-1}$ [4]

Mechanical Properties:

Tensile (Young's) Modulus:

No.	Value	Note
1	1310–1520 MPa	[4]

Tensile Strength Break:

No.	Value	Note	Elongation
1	62–110.3 MPa	23°, 50% relative humidity [1]	13–14%
2	48.3–55.2 MPa	[4]	40–45%

Hardness: Sward 90% of glass [1]. Rockwell R95–R115 [4]

Electrical Properties:

Electrical Properties General: Specific resistance 10^{11}–10^{12} Ω cm [11]

Dielectric Permittivity (Constant):

No.	Value	Frequency	Note
1	7–7.5	60 Hz	25–30° [6]
2	7	1 kHz	25–30° [1]
3	6	1 MHz	25–30° [1]

Dissipation (Power) Factor:

No.	Value	Frequency	Note
1	0.09–0.12	50–60 Hz	[11]
2	0.06–0.09	1 MHz	[11]

Optical Properties:

Refractive Index:

No.	Value	Note
1	1.51	[1]
2	1.49–1.51	[4]

Volume Properties/Surface Properties: Lower limit of substantially complete transmission 313 nm [1]

Polymer Stability:

Flammability: Fl. p. 13° [5]. Autoignition temp. 160–170° [5]
Chemical Stability: Has excellent resistance to water, hydrocarbons and mineral oils, and good resistance to aromatic hydrocarbons but has poor resistance to strong acids and alkalis [1]

Applications/Commercial Products:

Processing & Manufacturing Routes: Batch process using nitric acid and 2–5% sulfuric acid as catalyst. The reaction mixture is poured into water, washed and centrifuged
Applications: Lacquers and coatings

Trade name	Supplier
Walsroder Nitrocellulose	Wolff
	Hercules
	SNPE (France)
	ICI, UK

Bibliographic References

[1] *Encycl. Polym. Sci. Eng.*, 2nd edn., (ed. J.I. Kroschwitz), John Wiley and Sons, 1985, **3**
[2] *Kirk-Othmer Encycl. Chem. Technol.*, 4th edn., (ed. J.I. Kroschwitz), Wiley Interscience, 1993, **5**
[3] Quinchon, J. and Tranchant, J., *Properties of Nitrocelluloses*, Ellis Horwood, 1989, 63
[4] Dean, J.A., *Lange's Handbook of Chemistry*, 14th edn., McGraw-Hill, 1992
[5] *Dictionary of Organic Compounds*, 6th edn., (eds. J. Buckingham and F. Macdonald), Chapman and Hall, 1995
[6] Pillonnet, A., Asch, G., Dufour, J. and Lasjaunias, J.C., C.R. Seances Acad. Sci., Ser. A, 1973, **276**, 639
[7] Wu, S., *Polymer Interface and Adhesion*, Marcel Dekker, 1982, (surface tension)
[8] Gröbe, A., *Polym. Handb.*, 3rd edn., (eds. J. Brandrup and E.H. Immergut), Wiley Interscience, 1989, **V 137**
[9] Wiley, F.E., *Ind. Eng. Chem.*, 1942, **34**, 1052, (glass transition temps)
[10] Clash, R.F., Runkieweiz, L.M., *Ind. Eng. Chem.*, 1944, **36**, 279, (glass transition temps)
[11] *Ullmanns Encykl. Ind. Chem.*, 5th edn., (ed. W. Gerhartz), VCH, 1986, **5A**

Cellulose triacetate

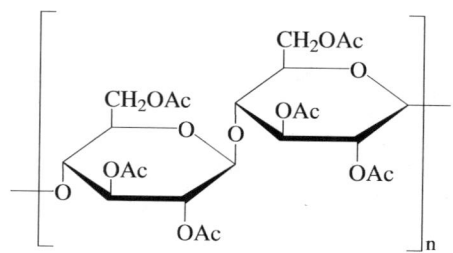

Monomers: Base monomer unit glucose
Material class: Polysaccharides
Polymer Type: Cellulosics

– Cellulose triacetate

CAS Number:

CAS Reg. No.
9012-09-3

Molecular Formula: $(C_{12}H_{16}O_8)_n$
Fragments: $C_{12}H_{16}O_8$
Molecular Weight: DP 300 (fibres). Degree of substitution ≥ 2.8

Volumetric & Calorimetric Properties:
Density:

No.	Value	Note
1	d 1.27–1.29 g/cm^3	[1]
2	d 1.3 g/cm^3	fibres [1]

Thermodynamic Properties General: Heat of combustion 17598 kJ kg^{-1} [7]
Melting Temperature:

No.	Value	Note
1	265–295°C	[3]

Glass-Transition Temperature:

No.	Value	Note
1	49–478°C	[9]

Transition Temperature:

No.	Value	Note	Type
1	310–315°C	[1]	Decomposition temp.
2	195°C	[4]	Crystalline rearrangement
3	130°C	in steam [4]	Crystalline rearrangement

Surface Properties & Solubility:
Solvents/Non-solvents: Sol. CH_2Cl_2, formic acid, glacial AcOH. Slightly sol. dioxan, Me_2CO. Insol. hydrocarbons, C_6H_6, toluene, CCl_4, tetrachloroethene [4]. Liq. crystal sol. trifluoroacetic acid/water [8]. Swollen by H_2O [1], dichloroethene, trichloroethene [4]

Transport Properties:
Water Content: Water retention capacity 16–17% [1], 10% (after heat setting) [4]
Water Absorption:

No.	Value	Note
1	4–4.5 %	20°, 65% relative humidity [1]
2	2.5 %	after heat setting [4]

Mechanical Properties:
Tensile Strength Break:

No.	Value	Note	Elongation
1	137.39–245.2 MPa	14–25 kg mm^{-2}, fibres [1]	
2	117.78–235.3 MPa	12–24 kg mm^{-2}, foil, longitudinal [1]	
3	98.1–117.7 MPa	10–12 kg mm^{-2}, foil, transverse [1]	
4	86 MPa		10–50% elongation [10]

Viscoelastic Behaviour: Viscoelastic behaviour has been reported [3]
Failure Properties General: Tear strength 1.6–11.8 N mm^{-1} [10]. Burst strength (Mullen) 50–70 [10]

Electrical Properties:
Electrical Properties General: Specific resistance 10^{13}–10^{15} Ω cm [1]
Dielectric Permittivity (Constant):

No.	Value	Frequency	Note
1	3–4.5	50–60 Hz	[1]
2	4		[10]

Dielectric Strength:

No.	Value	Note
1	1.46 kV.mm^{-1}	[10]

Complex Permittivity and Electroactive Polymers: Zeta potential 37 mV (H_2O) [9]
Dissipation (Power) Factor:

No.	Value	Frequency	Note
1	0.01–0.02	50–60 Hz	[1]
2	0.016		[10]

Optical Properties:
Refractive Index:

No.	Value	Note
1	1.469	fibres, along axis [1]
2	1.472	fibres, transverse to axis [1]

Total Internal Reflection: $[\alpha]_D$ -22.5 ($CHCl_3$) [2]. Double refraction -0.003 [1]
Volume Properties/Surface Properties: Appearance yellowish flakes [2]

Polymer Stability:
Thermal Stability General: Crystallinity increases after heating at 240° for 1 min. [6]
Upper Use Temperature:

No.	Value	Note
1	175°C	[10]

Decomposition Details: Thermal decomposition range 230–320°. Decomposes to give gaseous products together with acetylated derivatives of D-glucose [5]
Chemical Stability: Fibres are resistant to slightly acid/alkaline conditions at ambient temp. Resistant to chlorine bleaches. Decomposed by strong mineral acids [6]
Biological Stability: Resistant to microorganisms (similar to polyesters and nylon) [6]

Bibliographic References
[1] *Ullmanns Encykl. Ind. Chem.*, 5th edn., (ed. W. Gerhartz), VCH, 1985, **A5**, 444
[2] *Handbook of Chemistry and Physics*, 63rd edn., (eds. R.C. Weast and M.J. Astle), CRC Press, 1983
[3] *Kirk-Othmer Encycl. Chem. Technol.*, 3rd edn., (ed. M. Grayson), Wiley Interscience, New York, 1979, **5**, 89
[4] Moncrieff, R.W., *Man-Made Fibers*, 6th edn., Newnes-Butterworth, 1975, 257
[5] Brown, W.P. and Tipper, C.F.H., *J. Appl. Polym. Sci.*, 1978, **22**, 1459

[6] *Concise Encyclopedia of Polymer Science and Engineering*, (eds. H.F. Mark, N.M. Bikales, C.G. Overberger and G. Menges), John Wiley and Sons, 1990
[7] Birley, M.M. and Yeh, K.N., *J. Appl. Polym. Sci.*, 1973, **17**, 239
[8] Meeten, G.H. and Navard, P., *Polymer*, 1983, **24**, 815
[9] Gröbe, A., *Polym. Handb.*, 3rd edn., (eds. J. Brandrup and E.H. Immergut), Wiley Interscience, 1989, V155, (zeta potential)
[10] *Kirk-Othmer Encycl. Chem. Technol.*, Vol. 10, 4th edn., (ed. J.I. Kroschwitz), Wiley Interscience, 1993

Chicle   C-14

Related Polymers: *cis*-1,4-Polyisoprene, *trans*-1,4-Polyisoprene, Jelutong, Balata, Gutta-percha
Monomers: 2-Methyl-1,3-butadiene
Material class: Thermoplastic, Gums and resins
Polymer Type: polybutadiene
Molecular Weight: MW and MW distribution have been determined for chicle and related polymers. M_n 91000 (*cis*), M_n 16000 (*trans*)
Additives: Contains large amounts of resin, (75–80%) starch, salts, sugars and ethers
General Information: Contains both *cis* and *trans* polyisoprene in a ratio of 1:3 [7,8]. Is gummy rather than hard and rubbery due to low MW
Miscellaneous: Physical and chemical props. have been reported [2,3]. The occurrence, location, props. and biosynth. of natural isoprene polymers, including chicle, have been reported [9]

Optical Properties:
Transmission and Spectra: Nmr spectra for chicle have been analysed, revealing structural details [4,5,6] including terminal OH groups

Applications/Commercial Products:
Processing & Manufacturing Routes: Isol. from *Achras sapota* [9]
Applications: Used in chewing gum

Bibliographic References
[1] Hager, T., MacArthur, J., McIntyre, D. and Suger, R., *Rubber Chem. Technol.*, 1979, **52**, 693
[2] Wren, W.G., *Annu. Rep. Prog. Rubber Technol.*, 1948, **12**, 6, (rev)
[3] Blow, C.M., *Annu. Rep. Prog. Rubber Technol.*, 1949, **13**, 14, (rev)
[4] Tanaka, Y., *J. Appl. Polym. Sci.: Appl. Polym. Symp.*, 1989, **44**, 1
[5] Tanaka, Y., Mori, M., Takei, A., Boochathum, P. and Sato, Y., *J. Nat. Rubber Res.*, 1990, **5**, 241
[6] Tanaka, Y. and Sato, H., *Polymer*, 1976, **17**, 113
[7] Schlesinger, W. and Leeper, H.M., *Science (Washington, D.C.)*, 1950, **112**, 51
[8] Archer, B.L. and Audley, B.G., *Phytochemistry*, 1973, **2**, 310
[9] Schlesinger, W. and Leeper, H.M., *J. Polym. Sci.*, 1953, **11**, 203

Chitin   C-15

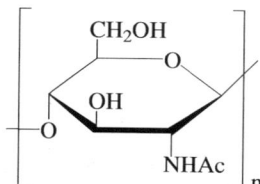

Synonyms: *Poly[(1→4)-β-2-acetamido-2-deoxy-D-glucose]*. *Poliglusam*
Related Polymers: Chitosan
Monomers: Base monomer unit 2-acetamido-2-deoxyglucose
Material class: Polysaccharides
Polymer Type: Carbohyrate
CAS Number:

CAS Reg. No.
1398-61-4
26023-36-9

Molecular Formula: $(C_8H_{13}NO_5)_n$
Fragments: $C_8H_{13}NO_5$
Molecular Weight: MW 1000000–2000000 [9]
General Information: A white, amorph., translucent solid. Chitin is estimated to be the world's second most abundant natural polymer after cellulose. It is a linear polymer predominantly made of (1,4)-β linked 2-acetamido-2-deoxy-D-glucose (*N*-acetyl glucosamine), though approx. one amine moiety in six appears to be unacetylated. Chitin does not usually occur as a pure polymer but is generally associated with proteins, in what is thought to be a non-covalent interaction that may involve the free amine groups. It is found in this form in many species of fungi, algae, protozoa, bryozoa, molluscs and arthropods. In the hard cuticle of insects the polymer can comprise 30–45% of the total mass. In a few instances, such as the spines of certain diatoms, chitin does appear to occur as a pure, crystalline polymer. Here the polymer appears to be fully acetylated [1]
Morphology: Analysis of the assembly of chitin polymer chains in insect cuticle shows that they form microfibrils, which are of almost uniform size within particular samples but which vary widely between species. Microfibril diameters can vary from 2.5 nm [2] to 300 nm [3]. A survey of thirteen arthropod species gave an average microfibril diameter of 2.8 nm [2], suggesting that each microfibril contains eighteen chitin chains arranged in three sheets, though this interpretation may underestimate the dimensions [4]. The microfibrils form lamellae which build up in around five layers to make the thin insect epicuticle which is <1μm thick. Chitin is isolated by removal of associated protein with dilute alkali. Purified chitin is generally fibrous in appearance but contains crystalline regions that have been found in three forms: α-, β- and γ-. In the most stable α-form, found in insect cuticle, chains are aligned in an antiparallel array [5]. The β-form contains chains arranged in a parallel array [6]. For the γ-form, it has been suggested that a unit contains two parallel chains and one antiparallel chain, [7] though this interpretation has been challenged. A distorted form of one of the α- or β- form may be involved [8]. The α-chains are assembled via three-dimensional hydrogen bonding. This permits formation of long fibrils of high tensile strength, which do not swell in water. β Crystallites appear to contain mainly two-dimensional hydrogen bonds, allowing them to absorb water and form crystalline hydrates. Treatment with cold 6M hydrochloric acid is reported to convert this form to the α-form [7]

Surface Properties & Solubility:
Solvents/Non-solvents: Sol. concentrated mineral acids [15] including formic acid, dichloroacetic acid and trichloroacetic acid [11]. Sol. in concentrated solutions of salts such as lithium thiocyanate, lithium chloride/*N*-methylpyrrolidinone etc. [14]. α-Chitin dissolves in ≥98% formic acid. β-Chitin dissolves slowly in 88–90% formic acid [12]. Insol. organic solvents

Transport Properties:
Permeability and Diffusion General: [O_2] 30–35 cm^3 m^{-2} day^{-1} (loligo pen) [13,18]. [H_2O] 1.69–2.17 cm$_3$ cm^{-2} s^{-1} (6100–7800 g cm^{-2} h^{-1}, loligo pen)
Water Content: 2–10% [20]

Mechanical Properties:
Tensile (Young's) Modulus:

No.	Value	Note
1	8900–11300 MPa	(Loligo pen) [13,17]
2	5100 MPa	(Crab shell) [13,17]
3	4300 MPa	(non-woven sheet) [13,17]

Mechanical Properties Miscellaneous: Breaking length 6.3–6.9 km (Loligo pen), 3 km (Crab shell) [13,16]. Bursting factor 5.3–7.4 (Loligo pen) [13,16]. Tearing factor 37–40 (Loligo pen) [13,16]. Viscosity 200–3000 MPa s (1% soln. in 1% AcOH) [20]

Optical Properties:
Optical Properties General: Amorph. form is semi-transparent
Total Internal Reflection: $[\alpha]_D^{18}$ +22 (c. 1 in methanesulfonic acid) [19]

Polymer Stability:
Chemical Stability: Hydrolysed by strong mineral acids
Biological Stability: Degraded by chitinase-producing bacteria
Stability Miscellaneous: Dissociation constant Ka 6.0-7 [19]

Bibliographic References
[1] Fall, M., Smith, D.G., McLachlan, J. and McInnes, A.G., *Can. J. Chem.*, 1966, **44**, 2269
[6] Blackwell, J., *Biopolymers*, 1969, **7**, 281
[7] Rudall, K.M., *Adv. Insect Physiol.*, 1963, **1**, 257
[8] Blackwell, J., *Methods Enzymol.*, 1988, **161**, 435
[9] Brine, C.J. and Austin, P.R., Comp. Biochem. Biophys. Photosynth., *Pap. Conf.*, 1981, **B69**, 283
[10] *Concise Encyclopedia of Polymer Science and Engineering*, (eds. H.F. Mark, N.M. Bikales, C.G. Overberger and G. Menges), John Wiley and Sons, 1990, 417
[11] Roberts, G.A.F., *Chitin Chemistry*, McMillan, 1992
[12] Austin, P.R., Chitin, Chitosan, Relat. Enzymes, (Proc. Jt U.S.-Jpn Semin. Adv. Chitin, Chitosan, *Relat. Enzymes)*, (ed. J.P. Zikakis), Academic Press, 1984, 227
[13] Takai, M., Shimizu, Y., Hayashi, J., Uraki, Y. *et al*, Chitin Chitosan, *Proc. Int. Conf.*, (eds. G. Skjak-Braek, T. Anthonsen and P. Sandford), Elsevier Applied Science, 1989, 475
[14] Rutherford, F.A. and Austin, P.R., Proc. Int. Conf. Chitin/Chitosan, *1st*, (eds. R.A.A. Muzzarelli and E.R. Panser), 1978, 182
[15] Hackman, R.H., *Aust. J. Biol. Sci.*, 1962, **15**, 526
[16] Hirano, S., Ohe, Y., and Kendo, S., *Polymer*, 1976, **47**, 315
[17] Hirano, S., Ohe, Y., and Ono, H., *Carbohydr. Res.*, 1976, **47**, 315
[18] Hirano, S. and Yamaguchi, R., *Biopolymers*, 1976, **15**, 1685
[19] *Ullmanns Encykl. Ind. Chem.*, (eds. B. Elvers, S. Hankins, and W. Russey), VCH, 1994
[20] *Encycl. Polym. Sci. Eng.*, 2nd edn., (eds. H.F. Mark, N.M. Bikales, C.G. Overberger and G. Menges), John Wiley and Sons, 1985, **3**

Chitosan   C-16

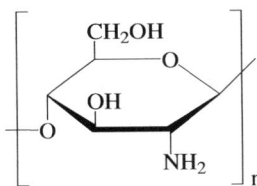

Synonyms: *Poly[(1→4)-2-amino-2-deoxy-β-D-glucose]*
Related Polymers: Chitin
Monomers: Base monomer unit glucosamine
Material class: Polysaccharides
Polymer Type: Carbohyrate
CAS Number:

CAS Reg. No.
9012-76-4

Molecular Formula: $(C_6H_{11}NO_4)_n$
Fragments: $C_6H_{11}NO_4$
Molecular Weight: MW 90000–1140000. (Depending on method of preparation MW can be as low as 3000)
Morphology: Though some data on fibre struct. have been obtained, the unit cell dimensions and chain alignment in chitosan have not yet been determined. Fibres contain crystalline areas, allowing preparation of microcrystalline material in a manner analogous to that of cellulose. The average MW of this material is 420000, with loss of 62% of the original *N*-acetyl groups [12].
Miscellaneous: The name chitosan is also applied to some partially de-acetylated samples of chitinous material, which is rendered sol. in dilute acids. **Chitosan - sources and props.** Chitosan is a cationic linear polymer of 2-amino-2-deoxy-D-glucose (glucosamine), which is prepared by alkaline deacetylation of chitin. Chain lengths vary widely with the method of preparation and are usually less than 50% of those of the parent chitin polymers. Reported values range from >2250 [13], to as low as 20 residues [3]

Surface Properties & Solubility:
Solvents/Non-solvents: Sol. dilute acids such as HCl, HNO_3, $HClO_4$; insol. H_2SO_4. Forms H_2O sol. salts with many organic acids [4,5]

Transport Properties:
Polymer Solutions Concentrated: Viscosity 73–76 cps (commercial Protasan) [6]. Viscosity has been reported [12]
Water Content: Water retention value 700% [12]
Water Absorption:

No.	Value	Note
1	2–10 %	[11]

Mechanical Properties:
Tensile Strength Break:

No.	Value	Note	Elongation
1	132.39 MPa	13.5 kg mm^{-2}, dry [9]	14.1%
2	31.38 MPa	3.2 kg mm^{-2}, wet [9]	78.0%

Optical Properties:
Total Internal Reflection: $[\alpha]_D^{11}$ -3--10 (c, 0.5 in AcOH aq.) [10]

Polymer Stability:
Chemical Stability: Stable in soln. at pH 4.6–5.1
Stability Miscellaneous: Dissociation constant Ka 6.0-7 [12]

Applications/Commercial Products:
Processing & Manufacturing Routes: Prod. in the form of solns., powders, beads and fibres
Applications: Used as a cationic flocculant, humectant, viscosifier and selective chelator. Also used in water clarification and soilds concentration. Absorbs heavy metals from industrial wastes. Used as an assistant in dyeing and in photographic emulsions

Trade name	Details	Supplier
Chitosan		Katacura
Chitosan		Kypro Co.
Protasan	87% deacetylated chitin	Protan Inc.

Bibliographic References
[1] Roberts, G.A.F. and Domszy, J.G., *Int. J. Biol. Macromol.*, 1982, **4**, 374
[2] Muzzarelli, R.A.A., Chitin Chemistry Pergamon Press, 1977
[3] Horton, D. and Lineback, D., *Methods Carbohydr. Chem.*, 1965, **5**, 401
[4] Austin, P.R., Chitin, Chitosan, Relat. Enzymes, (Proc. Jt U.S.-Jpn Semin. Adv. Chitin, Chitosan, *Relat. Enzymes)*, (ed. J.P. Zikakis), Academic Press, 1984, 227
[5] Chitin Nat. Technol., *(Proc. Int. Conf. Chitin Chitosan)*, (eds. R.A.A. Muzzarelli, C. Jeuniaux and G.W. Gooday), Plenum Press, 1986
[6] Skjak-Braek, G., Anthonsen, T. and Sandford, P., Chitin Chitosan, Proc. Int. Conf. Elsevier Applied Science, 1989, 52
[7] Gummow, B. and Roberts, G.A.F., *Makromol. Chem.*, 1986, **187**, 995
[8] Maghami, G.G. and Roberts, G.A.F., *Makromol. Chem.*, 1988, **189**, 2239
[9] Kim, J.M. and Lee, Y.M., *Polymer*, 1993, **34**, 1952
[10] *Ullmanns Encykl. Ind. Chem.*, (eds. B. Elvers, S. Hankins, R.A.A. Muzzarelli and W. Russey), VCH, 1994, **A6**, 231
[11] *Encycl. Polym. Sci. Eng.*, (eds. H.F. Mark, N.M. Bikales, C.G. Overberger and G. Menges), John Wiley and Sons, 1985, 431
[12] Strucszczyk, H., *J. Appl. Polym. Sci.*, 1987, **33**, 177
[13] Muzzarelli, R.A.A., Tanfani, F., Scarpini, G. and Laterza, G., *Biophys. Methods*, 1990, **2**, 299

CN-PPV6   C-17

Synonyms: *Poly[[2,5-bis(hexyloxy)-1,4-phenylene](1-cyano-1,2-ethenediyl) [2,5-bis(hexyloxy)-1,4-phenylene] (2-cyano-1,2-ethenediyl)]*
Monomers: 2,5-Bis(hexyloxy)terephthalaldehyde, 2,5-Bis(hexyloxy)benzene-1,4-diacetonitrile
Polymer Type: poly(arylene vinylene)
CAS Number:

CAS Reg. No.
151897-69-7

Molecular Formula: $(C_{42}H_{60}N_2O_4)_n$
Fragments: $C_{42}H_{60}N_2O_4$
Molecular Weight: M_n 4000–12000
General Information: Is a bright powder [1,2]
Miscellaneous: XPS, UPS and optical absorption spectroscopy have been used to study both doped and undoped polymer [16]. Other related cyano-substituted polymers are also available (CN-PPPV1, CN-PPV1A, CN-PPV6, CN-PPVV1,8)

Volumetric & Calorimetric Properties:
Melting Temperature:

No.	Value	Note
1	225°C	CN-PPV6 [3]

Transition Temperature:

No.	Value	Type
1	32°C	CN-PPV6

Surface Properties & Solubility:
Solvents/Non-solvents: CN-PPV6 sol. $CHCl_3$, toluene [1,17]
Surface and Interfacial Properties General: Interface formation with aluminium has been reported [11]

Optical Properties:
Optical Properties General: The electronic struct. in both the ground state and the excited state and photo/electroluminescence props. have been reported [13,14]. Luminescence from both intermolecular and intramolecular photoexcitation has been reported [16,18]. Photoluminescence efficiency 0.35–0.46 (CN-PPV6; 488 nm excitation; 0.48 (CN-PPV1,8, 488 nm excitation) [12]
Transmission and Spectra: H-1 nmr [1], photoluminescence [3,8,9,10], uv [4] and absorption spectral data [5] for CN-PPV6 have been reported

Applications/Commercial Products:
Processing & Manufacturing Routes: Synth. by Knoevenagel condensation polymerisation of 2,5-bis(hexyloxy)terephthaldehyde and 2,5-bis(hexyloxy)benzene-1,4-diacetonitrile [1,2,16].
In situ doping has been reported [16]
Applications: Used in electro-optical devices and photoresponsive devices. Potential application in LEDS

Supplier
Uniax Corp.

Bibliographic References
[1] Greenham, N.C., Moratti, S.C., Bradley, D.D.C., Friend, R.H. and Holmes, A.B., *Rubber Age (London)*, 1993, **365**, 628, (synth.)
[2] Moratti, S.C., Bradley, D.D.C., Friend, R.H., Greenham, N.C. and Holmes, A.B., *Polym. Prepr. (Am. Chem. Soc., Div. Polym. Chem.)*, 1994, **35**, 214, (synth.)
[3] Moratti, S.C., Bradley, D.D.C., Cervini, R., Friend, R.H. et al, *Proc. SPIE-Int. Soc. Opt. Eng.*, **2144**, 108, (photoluminescence spectroscopy)
[4] Fahlman, M., Brëdas, J.L. and Salaneck, W.R., *Synth. Met.*, 1996, **78**, 237, (uv photoelectron spectroscopy)
[5] Samuel, I.D.W., Rumbles, G., Collisan, C.J., Crystall, B. et al, *Synth. Met.*, 1996, **76**, 15, (absorption spectrum)
[6] Staring, E.G.J., Demandt, R.C.J.E., Braun, D., Rikken, G.L.J. et al, *Synth. Met.*, 1995, **71**, 2179, (synth.)
[7] *PCT Int. Appl.*, 1994, 94 29 883, (synth., applications)
[8] *PCT Int. Appl.*, 1996, 96 16 449, (applications)
[9] Hayes, G.R., Samuel, I.D.W. and Phillips, R.T., *Phys. Rev. B: Condens. Matter*, 1997, **54**, R 8301, (photoluminescence)
[10] Samuel, I.D.W., Rumbles, G. and Collison, C.J., *Phys. Rev. B: Condens. Matter*, 1995, **52**, R 11573, (photoluminescence)
[11] Fahlman, M., Salaneck, W.R., Moratti, S.C., Holmes, A.B. and Bredas, J.L., *Chem. Eur. J.*, 1997, **3**, 286, (surface props.)
[12] Greenham, N.C., Samuel, I.D.W., Hayes, G.R., Phillips, R.T. et al, *Chem. Phys. Lett.*, 1995, **241**, 89, (photoluminescence)
[13] Cornil, J., Dos Santos, D.A., Beljonne, D. and Bredas, J.L., *J. Phys. Chem.*, 1995, **99**, 5604, (struct.)
[14] Yu, G. and Heeger, A.J., *J. Appl. Phys.*, 1995, **78**, 4510
[15] Dos Santos, D.A., Beljonne, D., Cornil, J., Bredas, J.L., *Chem. Phys.*, 1998, **227**, 1
[16] Fahlman, M., Broms, P., Dos Santos, D.A. Moratti, S.C. et al, *J. Chem. Phys.*, 1995, **102**, 8167, (struct.)
[17] Samuel, I.D.W., Rumbles, G., Collison, C.J.Moratti, S.C. and Holmes A.B., *Chem. Phys.*, 1998, **227**, 75
[18] Wu, M. and Conwell, E.M., *Chem. Phys.*, 1998, **227**, 11

Collagen   C-18

where X and Y = proline, hydroxyproline, alanine or arginine

Related Polymers: Proteins, Gelatin
Monomers: Amino acids
Material class: Proteins and polynucleotides
Molecular Weight: 95000 (monomer strand), 1700000 (polymer)
General Information: This protein occurs in all but the most primitive organisms, providing extracellular support elements and, in higher animals, tendons, ligaments and other connective tissues. It is the most abundant protein in most animals. The polymer forms 3-chain helical structs., high in proline and hydroxyproline

Volumetric & Calorimetric Properties:
Density:

No.	Value	Note
1	d 1.34 g/cm^3	[2]

Melting Temperature:

No.	Value	Note
1	230°C	[6]

Glass-Transition Temperature:

No.	Value	Note
1	120°C	[5]

Transition Temperature:

No.	Value	Note	Type
1	37–40°C	[1]	Denaturation temp.

Transport Properties:
Polymer Solutions Dilute: [η] 62–70 dl g^{-1} (invertebrate collagen) [8]

Mechanical Properties:
Tensile (Young's) Modulus:

No.	Value	Note
1	2600 MPa	2.6 × 10^{10} dynes cm^{-2}, acid or enzyme-solubilised collagen [3]

Tensile Strength Break:

No.	Value	Note	Elongation
1	2–7 MPa	2–7 × 10^7 dyn cm^{-2}, nature collagen [4]	15%

Optical Properties:
Total Internal Reflection: [α] -350–-400 [1]. [α]$_D$ -390 (invertebrate) [8]

Polymer Stability:
Thermal Stability General: Denatures above 40° [1]

Bibliographic References
[1] Balian, G. and Boures, J.H., *Food Sci. Technol.*, (eds. A.G. Ward and A. Courts), Academic Press, 1977
[2] Chien, J.C.W., *J. Macromol. Sci.*, *Rev. Macromol. Chem. Phys.*, 1975, **12**, 1
[3] Chien, J.C.W. and Chang, E.P., *Biopolymers*, 1973, **12**, 2045
[4] Kaplan, D. and Bettelheim, F.A., *Biochim. Biophys. Acta*, 1972, **279**, 92
[5] Marshall, A.S. and Petrie, S.E.B., *J. Photogr. Sci.*, 1980, **28**, 128
[6] Jolley, J.E., *Photogr. Sci. Eng.*, 1970, **14**, 169
[7] Miller, E.J. and Gay, S., *Methods Enzymol.*, 1982, **82**, 3
[8] Murray, L.W., Waite, J.H., Tanzer, M.L. and Hauschka, P.V., *Methods Enzymol.*, 1982, **82**, 65

Cresol-formaldehyde Resins C-19
Related Polymers: Cresole Resoles, Cresol Novolaks
Monomers: *o*-Cresol, *m*-Cresol, *p*-Cresol, Formaldehyde
Material class: Thermosetting resin
Polymer Type: Phenolic
Molecular Formula: [(C$_7$H$_8$O).(CH$_2$O)]$_n$
Fragments: C$_7$H$_8$O CH$_2$O
General Information: The cresol isomers have different functionalities in the condensation reaction with formaldehyde. This influences the degree of crosslinking possible, with *ortho*-cresol and *para*-cresol being difunctional while *meta*-cresol is trifunctional

Applications/Commercial Products:
Processing & Manufacturing Routes: Cresol-formaldehyde resins are made by the condensation reaction of molten cresols with formaldehyde, generally 37% aq. soln. With basic catalyst and a molar excess of formaldehyde, a resole is formed. With an acidic catalyst and excess cresol, a Novolak resin is formed. Cresol Resoles may be used as solids or in soln. They cure by application of heat or acid. Cresol Novolaks may be used as solid powdered resin, or in alcoholic solvents. They require a curing agent, generally hexamethylenetetramine
Applications: See Phenol-aldehyde resins for a general discussion of the uses of Cresol-formaldehyde resins

Cresol Resoles C-20
Synonyms: *Methylphenol-formaldehyde Resoles. Cresol-formaldehyde Resoles. Cresylic Resoles*
Related Polymers: Cresol-formaldehyde Novolaks, Cresol-formaldehyde resins
Monomers: *o*-Cresol, *m*-Cresol, *p*-Cresol, Formaldehyde
Material class: Thermosetting resin
Polymer Type: phenolic
CAS Number:

CAS Reg. No.	Note
25053-96-7	*o*-cresol-formaldehyde copolymer
25086-36-6	*m*-cresol-formaldehyde copolymer
25053-88-7	*p*-cresol-formaldehyde copolymer

Molecular Formula: [(C$_7$H$_8$O).(CH$_2$O)]$_n$
General Information: Of the three isomers, *m*-cresol is the most desirable material as it is the only isomer with three reactive positions necessary for cross-linking

Optical Properties:
Transmission and Spectra: C-13 nmr [1], H-1 nmr [1], Raman [3] and esr [4] spectral data have been reported

Applications/Commercial Products:
Processing & Manufacturing Routes: Synth. by the condensation of molten cresols with formaldehyde with basic catalyst. Cresol resoles may be used as solids or in soln. Cured by application of heat or acid. Effect of synthetic procedure on yield and viscosity has been reported [2]
Applications: Solid cresol resoles may be used to make friction materials, which are used as brake and clutch linings. Other applications in electrical engineering as varnishes in the manufacture of laminates in coatings and as hardeners for epoxy resins

Bibliographic References
[1] Tong, S.N., Park, K.Y. and Harwood, H.J., Polym. Prepr. (Am. Chem. Soc., Div. Polym. Chem.), 1983, **24**, 196, (C-13 nmr, H-1 nmr)
[2] Ciernik, J., Linhartova, M. and Bartaskova, P., *Chem. Prum.*, 1978, **28**, 517, (synth)
[3] Chow, S. and Chow, Y.L., *J. Appl. Polym. Sci.*, 1974, **18**, 735, (Raman)
[4] Toyoda, S., Sugawara, S., Furuta, T. and Honda, H., *Carbon*, 1970, **8**, 473, (esr)

Cresylic Novolaks C-21
Synonyms: *Cresol-formaldehyde Novolaks. Cresol Novolaks. Methylphenol-formaldehyde Novolaks*
Related Polymers: Cresol-Resoles
Monomers: *o*-Cresol, *m*-Cresol, *p*-Cresol, Formaldehyde
Material class: Thermosetting resin
Polymer Type: Phenolic
CAS Number:

CAS Reg. No.	Note
25053-96-7	*o*-cresol-formaldehyde copolymer
25086-36-6	*m*-cresol-formaldehyde copolymer
25053-88-7	*p*-cresol-formaldehyde copolymer

Molecular Formula: [(C$_7$H$_8$O).(CH$_2$O)]$_n$
Fragments: C$_7$H$_8$O CH$_2$O
Molecular Weight: MW 2790 (*o*-cresol), 702 (*p*-cresol). M$_n$ 1468 (*o*-cresol), 595 (*p*-cresol)
General Information: Of the three isomers, *m*-cresol is the most desirable as it contains the three reactive positions necessary for cross-linking
Morphology: *Ortho*-linked *p*-cresol resins have a highly ordered structure owing to intramolecular hydrogen bonding which may favour formation of a cyclic tetramer structure [13]

Cresylic Novolaks

Volumetric & Calorimetric Properties:

Density:

No.	Value	Note
1	d 1.3 g/cm^3	o and p-cresol, uncured, film [4]

Equation of State: Mark-Houwink-Sakurada equation of state information for o- and p- cresol resins has been reported [1]

Melting Temperature:

No.	Value	Note
1	65°C	p-cresol, uncured [5]
2	104–140°C	m-cresol, uncured, dependent on synth. [5]
3	60°C	o-cresol, uncured [5]
4	71–78°C	all ortho p-cresol, uncured [6]
5	128–144°C	all ortho m-cresol, uncured [6]
6	85°C	m-cresol, uncured [15]
7	90°C	p-cresol, uncured [15]

Glass-Transition Temperature:

No.	Value	Note
1	93°C	o-cresol, uncured [4]
2	81°C	p-cresol, uncured [4]

Transition Temperature General: Softening temp. of uncured m-cresol orthonovolac decreases with a decrease in formaldehyde content [3]

Vicat Softening Point:

No.	Value	Note
1	115–117°C	m-cresol orthonovolac, uncured, formaldehyde mole ratio 1 [3]
2	69–72°C	m-cresol orthonovolac, uncured, formaldehyde mole ratio 0.6 [3]
3	80°C	p-cresol, uncured [5]
4	125–160°C	m-cresol, uncured, dependent on synth. [5]
5	75°C	o-cresol, uncured [5]

Surface Properties & Solubility:

Solvents/Non-solvents: Uncured o-, m- and p-cresols sol. Me$_2$CO, 2-butanone, EtOAc, EtOH [5,15]; insol. petrol. Uncured p-cresol sol. C$_6$H$_6$, turpentine, toluene, CCl$_4$, CHCl$_3$ [5], MeOH [15]. Uncured o and m-cresols insol. C$_6$H$_6$, turpentine, toluene, CCl$_4$ [5]. Uncured m-cresol insol. CHCl$_3$ [5], CCl$_4$, toluene [15]. Uncured o-cresol sol. CHCl$_3$ [5]

Transport Properties:

Polymer Solutions Dilute: Reduced viscosity 0.044–0.049 dl g^{-1} (DMF/H$_2$O (90:10), 35°, p-cresol, concentration 1.15–2.14%) [14], 0.086–0.075 dl g^{-1} (pyridine, 35°, p-cresol, concentration 2.14–1.15%) [14], 0.108–0.141 dl g^{-1} (piperidine, 35°, p-cresol, concentration 2.14–1.15% [14]. Viscosity 0.065–0.072 dl g^{-1} (DMF/H$_2$O (90:10), 35°, m-cresol) [14], 0.028–0.036 dl g^{-1} (DMF/H$_2$O (90:10), 35°, o-cresol) [14], 0.043–0.048 dl g^{-1} (2-butanone, 35°, o-cresol) [14], 0.029–0.04 dl g^{-1} (2-butanone, 35°, p-cresol) [14]. Theta temp. 119.2° (1,2-dichlorobenzene, o-cresol [16])

Optical Properties:

Transmission and Spectra: C-13 nmr [7,9,17,19], H-1 nmr [10], ms [11] and ir [17] spectral data have been reported

Volume Properties/Surface Properties: Colour 7 (o and p-cresol, uncured, Gardner scale), 12 (m-cresol, uncured, Gardner scale) [5]

Polymer Stability:

Thermal Stability General: Thermal stability of high MW o and p-cresol resins has been reported [17]

Decomposition Details: Uncured m-cresol orthonovalac resins decompose at 149–181° in air depending upon the amount of formaldehyde used in synth.; decomposition temp. increases with increasing formaldehyde content. 10% Weight loss in air of uncured orthonovolacs occurs at 311-369° depending on amount of formaldehyde used in synth.; temp. increases with increasing formaldehyde content [3]. Char yield (800° in N$_2$) 30% (m-cresol, uncured), 33% (p-cresol, uncured), 51% (m-cresol, trioxane cured), 34% (m-cresol, terephthaloyl chloride cured), 39% (p-cresol, terephthaloyl chloride cured) [15]. Pyrolysis of p-cresol resin at 600° yields p-cresol and methylxanthenes among the products; pyrolysis of the o-cresol resin at 600° yields various methyl-substituted derivatives of o-cresol whereas m-cresol resin yields dimers [12,13]. Uncured m-cresol resin degrades at 270° whereas uncured p-cresol resin degrades at 350° [15]

Flammability: Oxygen index 33% (m-cresol, trioxane cured), 23% (m-cresol, terephthaloyl chloride cured), 25% (p-cresol, terephthaloyl chloride cured) [15]

Chemical Stability: Uncured o and p-cresols have excellent compatibility at 250° with linseed oil, tung oil, castor oil and double-boiled linseed oil. Under the same conditions, the m-isomer is compatible with double-boiled linseed oil and with castor oil, provided hydrochloric acid and not oxalic acid has been used as the catalyst in synth. [5]

Stability Miscellaneous: Electron beam irradiation *in vacuo* causes cross-linking of o and p-cresol resins [4]

Applications/Commercial Products:

Processing & Manufacturing Routes: Synth. by the condensation reaction of molten cresols with formaldehyde with acidic catalyst. May be used as solids or in soln. They require a curing agent, generally hexamethylenetetramine. Synth. in the presence of water-miscible organic solvents gives products of higher MW [2,7]. All-*ortho*-linked novolaks may be synth. by reaction of bromomagnesium cresol derivative with paraformaldehyde in C$_6$H$_6$ [8]. Resins with narrow MW distribution (M$_w$/M$_n$ 1.25) may be synth. using hydroxycarboxylic acids (e.g. citric acid) as catalysts [18]

Applications: Cresol novolaks are used as host material for positive photoresists in microelectronics and in printing plates. They are reacted with a photoreactive compound (e.g. 5-substituted diazonaphthalene) and when exposed to uv radiation, the resist is transformed into an indene carboxylic acid which is sol. in buffered alkaline soln. Other uses in coatings and as hardeners for epoxy resins. All-*ortho* resins have potential applications as antimicrobial agents and antioxidants in lard

Bibliographic References

[1] Sue, H., Ueno, E., Nakamoto, Y. and Ishida, S.-I., *J. Appl. Polym. Sci.*, 1989, **38**, 1305, (equation of state)
[2] Miloshev, St., Novakov, P., Dimitrov, V.L. and Gitsov, I., *Polymer*, 1991, **32**, 3067, (synth)
[3] Chetan, M.S. Ghadage, R.S., Rajan, C.R., Gunjikar, V.G. and Ponrathnam, S., *Thermochim. Acta*, 1993, **228**, 261, (thermal degradation)
[4] Tanigaki, K. and Iida, Y., *Makromol. Chem., Rapid Commun.*, 1986, **7**, 485, (irradiation)
[5] Ahisanuddin, Saksena, S.C., Panda, H. and Rakhshinda, *Paintindia*, (synth., solubility)
[6] Uchibori, T., Kawada, K., Watanabe, S., Asakura, K. et al, *Bokin Bobai*, 1990, **18**, 215, (synth., use)
[7] Carothers, J.A., Gipstein, E., Fleming, W.W. and Tompkins, T., *J. Appl. Polym. Sci.*, 1982, **27**, 3449, (C-13 nmr)
[8] Abe, Y., Matsumura, S., Asakura, K. and Kasama, H., *Yukagaku*, 1986, **35**, 751, (synth., use)
[9] De Breet, A.J.J., Dankelman, W., Huysmans, W.G.B. and De Wit, J., *Angew. Makromol. Chem.*, 1977, **62**, 7, (C-13 nmr)
[10] Yoshikawa, T. and Kumanotani, J., *Makromol. Chem.*, 1970, **131**, 273, (H-1 nmr)
[11] Eichhoff, H.J., Kaemmerer, H. and Weller, D., *Makromol. Chem.*, 1970, **132**, 163, (ms)

- Crystalline modified polysulfone with glass fibre

[12] Blazsó, M. and Tóth, T., *J. Anal. Appl. Pyrolysis*, 1986, **10**, 41, (thermal degradation)
[13] Tobiason, F.L., *J. Polym. Sci., Polym. Chem. Ed.*, 1979, **17**, 949, (configuration)
[14] Suthar, B.P., *J. Inst. Chem. (India)*, 1986, **58**, 53, (viscosity)
[15] Zaks, Y., Lo, J., Raucher, D. and Pearce, E.M., *J. Appl. Polym. Sci.*, 1982, **27**, 913, (flammability)
[16] Nakamoto, Y., Morita, T., Kumagai, H., Sue, H. and Ishida, S., *Netsu Kokasei Jushi*, 1989, **10**, 79, (theta temp.)
[17] Nakano, Y., *Nippon Setchaku Kyokaishi*, 1989, **25**, 194, (synth., uses, thermal stability, ir, C-13 nmr)
[18] *Jpn. Pat.*, 1996, 08 003 257, (synth)
[19] Khadim, M.A., Rahman, M.D. and Durham, D.L., *Proc. SPIE-Int. Soc. Opt. Eng.*, **1672**, 347, (C-13 nmr)

Crystalline modified polysulfone with glass fibre C-22

Synonyms: *Poly[oxy-1,4-phenylenesulfonyl-1,4-phenyleneoxy-1,4-phenylene-(1-methylethylidene)-1,4-phenylene]*
Related Polymers: Polysulfone
Monomers: 4,4′-Dichlorodiphenyl sulfone, Bisphenol A
Material class: Thermoplastic
Polymer Type: Polysulfone
CAS Number:

CAS Reg. No.	Note
25135-51-7	Udel

Molecular Formula: $(C_{27}H_{22}O_4S)_n$
Fragments: $C_{27}H_{22}O_4S$
Additives: Glass fibre
General Information: Unique modified polysulfone formulated for maximum cost-effectiveness while maintaining key props. The resin is a blend of amorph. polysulfone and a proprietary semi-crystalline polymer which combines key features of both the warp-free props. of polysulfone and the chemical resistance of the semi-crystalline material. Levels of the latter are selected to meet end-use thermal requirements

Volumetric & Calorimetric Properties:
Density:

No.	Value	Note
1	d 1.47 g/cm³	ASTM D792 [1]

Thermal Expansion Coefficient:

No.	Value	Note	Type
1	2.7×10^{-5} K^{-1}	ASTM D696 [1]	L

Melting Temperature:

No.	Value	Note
1	270–315 °C	[1]

Deflection Temperature:

No.	Value	Note
1	160 °C	1.8 MPa, ASTM D648 [1]

Transition Temperature:

No.	Value	Note	Type
1	160 °C	[1]	High heat deflection

Transport Properties:
Melt Flow Index:

No.	Value	Note
1	6.5 g/10 min	0.3 MPa, 275°, ASTM D1238 [1]

Water Absorption:

No.	Value	Note
1	0.14 %	3.2 mm, 24h, 23°, ASTM D570 [1]

Mechanical Properties:
Mechanical Properties General: Has higher tensile strength at yield and flexural modulus, but lower tensile impact strength compared with unreinforced PSU. Tough and rigid
Flexural Modulus:

No.	Value	Note
1	6900 MPa	ASTM D730 [1]

Tensile Strength Break:

No.	Value	Note	Elongation
1	121 MPa	17600 psi, ASTM D638 Mindel B-430 [1]	2.5%

Tensile Strength Yield:

No.	Value	Note
1	103 MPa	ASTM D638 [1]

Flexural Strength Yield:

No.	Value	Note
1	159 MPa	ASTM D790 [1]

Impact Strength: Tensile impact strength 84 kJ m^{-2} (ASTM D1822) [1]
Izod Notch:

No.	Value	Notch	Note
1	53 J/m	Y	ASTM [1]

Izod Area:

No.	Value	Notch	Note
1	7 kJ/m²	Notched	ISO [1]
2	31.7 kJ/m²	Unnotched	ISO [1]

Electrical Properties:
Electrical Properties General: Has excellent electrical props., which are maintained over a wide temp. range and after immersion in water [1]
Dielectric Permittivity (Constant):

No.	Value	Frequency	Note
1	3.7	60 Hz	[1]
2	3.7	1 kHz	[1]
3	3.7	1 MHz	ASTM D150 [1]

– Cyclomethicone

Dielectric Strength:

No.	Value	Note
1	20 kV.mm^{-1}	3.2 mm, ASTM D145 [1]

Dissipation (Power) Factor:

No.	Value	Frequency	Note
1	0.002	60 Hz	ASTM D150 [1]
2	0.003	1 kHz	ASTM D150 [1]
3	0.009	1 MHz	ASTM D150 [1]

Optical Properties:
Volume Properties/Surface Properties: Opaque [1]
Polymer Stability:
Polymer Stability General: Stable over a wide temp. range. Has the warp resistance of polysulfone and the chemical resistance of the semi-crystalline material [1]
Thermal Stability General: Very good high temp. performance
Upper Use Temperature:

No.	Value	Note
1	160°C	UL 746 [1]

Flammability: Flammability rating V0 (0.79 mm thick, UL 94) [1]
Chemical Stability: Good chemical resistance [1]
Hydrolytic Stability: Very good hydrolytic stability and steam resistance [1]

Applications/Commercial Products:
Processing & Manufacturing Routes: Melt temp. range is 270–315° and mould temp. range is 65–100° for injection moulding. Prior to melt processing, the resin must be dried for 4h at 150° [1]
Mould Shrinkage (%):

No.	Value	Note
1	0.3–0.5%	ASTM D955 [1]

Applications: Food contact applications, has FDA approval. Used in electrical/electronic components, connectors, snap-fit assemblies. Glass-filled Mindel A-322 is used for miniature relays which are warp and shrink free, and have the necessary electrical props. It combines the best props. of crystalline and amorph. polymers for high-density connectors

Trade name	Details	Supplier
Mindel B		Amoco Performance Products
Mindel B-322	crystalline modified with 22% glass fibre various grades available	Amoco Performance Products

Bibliographic References
[1] *Engineering Plastics for Performance and Value*, Amoco Performance Products, Inc., (technical datasheet)
[2] *Kirk-Othmer Encycl. Chem. Technol.*, 4th edn., (eds. J.I. Kroschwitz and M. Howe-Grant), Wiley Interscience, 1993, **13**, 945

Cyclomethicone C-23
Synonyms: *Cyclic dimethyl siloxane. Cyclic dimethyl polysiloxane. Cyclosiloxane. Hexamethylcyclotrisiloxane. Octamethylcyclotetrasiloxane. Decamethylcyclopentasiloxane. Dodecamethylcyclohexasiloxane*
Related Polymers: Polysiloxanes

Monomers: Dichlorodimethylsilane, Hexamethylcyclotrisiloxane, Octamethylcyclotetrasiloxane, Decamethylcyclopentasiloxane, Dodecamethylcyclohexasiloxane
Material class: Fluids, Oligomers
Polymer Type: dimethylsilicones
CAS Number:

CAS Reg. No.	Note
69430-24-6	
556-67-2	n = 4
541-05-9	n = 3
540-97-6	n = 6
541-02-6	n = 5

Molecular Formula: $[C_2H_6OSi]_n$
Fragments: C_2H_6OSi
Molecular Weight: MW 300–400 (specifically 222, 296, 370, 444)

Volumetric & Calorimetric Properties:
Density:

No.	Value	Note
1	d 0.959 g/cm^3	[1,2,3]

Melting Temperature:

No.	Value	Note
1	17°C	primarily D4 [1,2,3]
2	-40°C	primarily D5 [1,2,3]

Transition Temperature General: Boiling point 175° (D4); 190–210° (D5)

Surface Properties & Solubility:
Solvents/Non-solvents: Insol. H$_2$O. Misc. alcohols, esters, ethers, ketones, aliphatic hydrocarbons, aromatic hydrocarbons, chlorinated hydrocarbons [2]
Transport Properties:
Polymer Melts: Viscosity values have been reported [1,2,3,5]. $\eta^{25°}$ 0.023 St (primarily D4), $\eta^{25°}$ 0.039 St (primarily D5), $\eta^{38°}$ 0.017 St (primarily D4), $\eta^{38°}$ 0.03 St (primarily D5), Viscosity temp. coefficient 0.6. Viscosity more temp. dependent than for linear siloxanes of the same MW

Optical Properties:
Refractive Index:

No.	Value	Note
1	1.394	[1,2,3]

Polymer Stability:
Thermal Stability General: Usable from -40–190° (primarily D4) [1,2,3].
Upper Use Temperature:

No.	Value	Note
1	190°C	primarily D5

Flammability: Fl. p. 55° (primarily D4). Fl. p. 82° (primarily D5) [1,2]
Hydrolytic Stability: Stable to H$_2$O at room temp. At higher temps., particularly in acid or alkaline conditions, it can break down into linear polysiloxanes
Biological Stability: Non-biodegradable [4]

Applications/Commercial Products:

Processing & Manufacturing Routes: Formed as a major product of the hydrolysis and equilibration of dichlorodimethylsilane

Applications: Sold as a lubricant and emollient. Used industrially as intermediate in manufacture of long linear polydimethylsiloxanes

Trade name	Details	Supplier
Dow Corning 344	fluid	Dow Corning STI
SF	1200 series, 1173	General Electric Silicones

Bibliographic References

[1] Meals, R.N. and Lewis, F.M., Silicones Reinhold, 1959
[2] Ash, M. and Ash, I., Handbook of Plastic Compounds, *Elastomers and Resins*, (eds. M.B. Ash and I.A. Ash), Wiley-VCH, 1992
[3] Noll, W., *Chemistry and Technology of Silicones*, Academic Press, 1968
[4] Hardman, B. and Torkelson, A., *Encycl. Polym. Sci. Eng.*, Vol. 15, 2nd edn., (ed. J.I. Kroschwitz), John Wiley and Sons, 1985, 204
[5] Hardy, D.V.N. and Megson. N.J.L., Q. Rev., *Chem. Soc.*, 1948, **2**, 25

Deoxyribonucleic acid D-1

Synonyms: *DNA. Thymus nucleic acid*
Material class: Polynucleotides
Molecular Weight: DNA from viruses have molecular weights *ca.* $1-130 \times 10^6$ whereas bacterial and animal DNA has much higher molecular weights with chains easily severed during isoln.
General Information: A polynucleotide constructed from chains of 2-deoxy-D-ribose purine and pyrimidine units linked by phosphate diester bonds between the 3′- and 5′-hydroxyls of adjacent sugars.
Morphology: The adenine and thymine contents are equal and the guanine and cytosine contents also equal, regardless of source. In the secondary structure two polymer chains form right-handed helices about a common axis, with the sequence of atoms running in opposite directions in each strand. The bases are inside the helix and the adenine of one chain is hydrogen bonded to the thymine of the other and similarly with the guanine and cytosine
Miscellaneous: Present in all life forms except some viruses. An essential component of chromosomes in cell nuclei which carries genetic information by containing a chemical code in its structure, which is exactly reproducible. The linear sequence of bases in one strand determines the sequence in the other. Thus each strand can serve as a template for replication of the original DNA molecule. DNA also serves as a template for the formation of ribonucleic acids

Optical Properties:
Transmission and Spectra: nmr [9], ir [8], Raman [8] and ms [7] have been reported

Bibliographic References
[1] Watson, J.D., *Nature (London)*, 1953, **171**, 737; 964, (struct)
[2] Chargaff, E., The Nucleic Acids Academic Press, Vol. 1, 1955, (rev)
[3] Brown, D.M., *Compr. Biochem.*, 1963, **8**, 157, (rev)
[4] Crick, F., *Nature (London)*, 1970, **227**, 561
[5] Davidson, J.N., The Biochemistry of Nucleic Acids Academic Press, 7th Ed., 1972
[6] Narang, S.A., *Tetrahedron*, 1983, **39**, 3, (rev)
[7] Jankowski, K., *Adv. Heterocycl. Chem.*, 1986, **39**, 79, (rev)
[8] Taillandier, E., *J. Mol. Struct.*, 1989, **214**, 185
[9] Wemmer, D.E., *Biol. Magn. Reson.*, 1992, **10**, 195

Dichloropoly(methylphenylsilane) D-2

$$\text{Cl}-\underset{\text{Ph}}{\overset{\text{Me}}{\text{Si}}}-\left[\underset{\text{Me}}{\overset{\text{Me}}{\text{Si}}}\right]_n-\underset{\text{Me}}{\overset{\text{Me}}{\text{Si}}}-\text{Cl}$$

Synonyms: α,ω-*Dichloropolymethylphenylsilane. Dichloropolyphenylmethylsilane. Dichloropolyphenylmethylsilylene*
Related Polymers: Poly(methylphenylsilane), More general information (polysilanes)
Monomers: Dichloromethylphenylsilane
Material class: Thermoplastic
Polymer Type: Polysilanes
CAS Number:

CAS Reg. No.	Note
76188-55-1	polymethylphenylsilane

Molecular Formula: $(C_7H_8Si)_n$
Fragments: C_7H_8Si
Molecular Weight: MW 20000; M_n 6000

Volumetric & Calorimetric Properties:
Melting Temperature:

No.	Value	Note
1	220°C	[5]

Surface Properties & Solubility:
Solvents/Non-solvents: Sol. aromatic hydrocarbons, chlorinated hydrocarbons, THF. Mod. sol. ethers, aliphatic hydrocarbons. Slightly sol. acetonitrile. Insol. alcohols [1,2,3,7]

Optical Properties:
Transmission and Spectra: Strong absorption 338-341 nm [4,6,8]. Fluorescence 360 nm

Polymer Stability:
Polymer Stability General: Functional end groups react readily with other chemicals [2,9]
Hydrolytic Stability: The SiCl end groups are hydrolysed to SiOH see polymethylphenylsilane [1,2]

Applications/Commercial Products:
Processing & Manufacturing Routes: Prepared by Wurtz coupling of dichloromethylphenylsilane using metallic sodium in toluene at around 110°. Water must be excluded to prevent hydrol. of ClSi end groups to hydroxyl groups [1,2,7]
Applications: Precursor to polysilane-polystyrene block copolymers. Potentially useful electronic and non-linear optical props [9,10,11,12,13,14,15,16]

Bibliographic References
[1] Jones, R.G., Benfield, R.E., Cragg, R.H., Swain, A.C. and Webb, S.J., *Macromolecules*, 1993, **26**, 4878
[2] Demoustier-Champagne, S., Marchand-Brynaert, J. and Devaux, J., *Eur. Polym. J.*, 1996, **32**, 1037
[3] West, R. and Maxka, J., *ACS Symp. Ser.*, 1988, **360**, 6
[4] Yu-Ling, H., Banovetz, J.P. and Waymouth, R.M., *ACS Symp. Ser.*, 1994, 55
[5] West, R., *Comprehensive Organometallic Chemistry*, (eds. G. Wilkinson, F.G.A. Stone and E.W. Abel), Pergamon Press, 1982, **2**, 365
[6] Trefonas, P., *Encycl. Polym. Sci. Eng.*, Vol. 13, 2nd edn., (ed. J.I. Kroschwitz), John Wiley and Sons, 1985, 162
[7] Trefonas, P., Djvrovich, P.I., Zhang, X.-H., West, R., *et al*, J. Polym. Sci., *Polym. Lett. Ed.*, 1983, **21**, 819
[8] Trefonas, P., West, R., Miller, R.D. and Hofer, D., J. Polym. Sci., *Polym. Lett. Ed.*, 1983, **21**, 823
[9] Demoustier-Champagne, S., de Mahieu, A.-F., Devaux, J., Fayt, R. and Teyssie, Ph., J. Polym. Sci., *Part A: Polym. Chem.*, 1993, **31**, 2009
[10] Eckhardt, A. and Schnabel, W., *J. Inorg. Organomet. Polym.*, 1996, **6**, 95
[11] Eckhardt, A., Nespurek, S. and Schnabel, W., *Ber. Bunsen-Ges. Phys. Chem.*, 1994, **98**, 1325, (photoconductivity)
[12] Klingensmith, K., Downing, J.W., Miller, R.D. and Michl, J., *J. Am. Chem. Soc.*, 1986, **108**, 7438, (photoconductivity)
[13] Fujino, M., *Chem. Phys. Lett.*, 1986, **136**, 451, (general photoconductivity)
[14] Stolka, M., Yuh, H.-J., McGrane, K. and Pai, D.M., J. Polym. Sci., *Polym. Chem. Ed.*, 1987, **25**, 823, (photoconductivity)
[15] Abkowitz, M.A., Stolka, M., Weagley, R.J., McGrane, K.M. and Knier, F.E., *Adv. Chem. Ser.*, 1990, **224**, 467, (photoconductivity)

Dicyclopentadiene dioxide resin D-3

Synonyms: *Poly(octahydro-2,4-methano-2H-indeno[1,2-b:5,6-b′]bisoxirene). Poly(1,2:5,6-diepoxyhexahydro-4,7-methanoindan)*
Monomers: Dicyclopentadiene dioxide
Material class: Thermosetting resin

Polymer Type: epoxy
CAS Number:

CAS Reg. No.	Note
29987-76-6	homopolymer
71855-14-6	copolymer with phenylphosphonic dichloride

Molecular Formula: $(C_{10}H_{12}O_2)_n$
Fragments: $C_{10}H_{12}O_2$
Additives: Glass fibre

Applications/Commercial Products:
Processing & Manufacturing Routes: Synth. of the resin from unsaturated precursor by the action of peroxyacetic acid has been reported [2]
Applications: Uncured resin (in presence of Ba/Ca ions) has potential application as stabiliser for PVC. Resin cured with phenylphosphonic dichloride has potential application as a fireproofing agent for polyester fibres. May also be used as a reactive diluent for epoxy resin compositions

Trade name	Details	Supplier
EP	207	Union Carbide

Bibliographic References
[1] *Jpn. Pat.*, 1979, 74 896, (use)
[2] Brojer, Z., Penczek, P. and Penczek, S., *Przemysl Chem.*, 1962, **41**, 684, (synth, use)

Diglycidyl aniline resin   D-4

Synonyms: *Poly[N-(oxiranylmethyl)-N-phenyloxiranemethanamine]*
Monomers: Diglycidyl aniline
Material class: Thermosetting resin
Polymer Type: Epoxy
CAS Number:

CAS Reg. No.	Note
30999-33-8	
119391-54-7	copolymer with aniline
129888-35-3	copolymer with 1,3-diaminobenzene
120468-21-5	copolymer with tetrahydro-5-methylisophthalic acid anhydride
110430-27-8	copolymer with diaminodiphenylmethane
120468-22-6	copolymer with 1,3-benzenedimethanamine

Molecular Formula: $(C_{12}H_{15}NO_2)_n$
Fragments: $C_{12}H_{15}NO_2$
Miscellaneous: The props. of the cured resin depend on hardener, stoichiometry, cure time and cure temp.

Volumetric & Calorimetric Properties:
Density:

No.	Value	Note
1	d 1.24 g/cm^3	tetrahydro-5-methylisophthalic acid anhydride cured [5]
2	d 1.21 g/cm^3	diaminodiphenylmethane cured [5]
3	d 1.22 g/cm^3	1,3-benzenedimethanamine [5]

Glass-Transition Temperature:

No.	Value	Note
1	155°C	1,3-diaminobenzene cured [2]
2	105°C	tetrahydro-5-methylisophthalic acid anhydride cured [5]
3	123°C	diaminodiphenylmethane cured [5]
4	74°C	1,3-benzenedimethanamine cured [5]

Transport Properties:
Permeability of Liquids: Average diffusion coefficient [H_2O] 0.32×10^{-9} cm^2 s^{-1} (20°, 1,3-diaminobenzene cured, water-aged, 70°) [2]
Water Content: Polymer only partially cured may be fully cured by immersion in H_2O at 70° [2]

Applications/Commercial Products:
Processing & Manufacturing Routes: The mechanism and kinetics of cure with aniline have been reported [1]. Diglycidyl aniline undergoes anionic polymerisation with 10% potassium *tert*-butoxide to give a soluble polymer. The polymer is a mixture of six and seven-membered rings [3]. The influence of cyclisation and functional group reactivity on the formation of networks in diaminodiphenylmethane cured systems has been reported [4]

Trade name	Supplier
STF-5	Reichhold Chemie

Bibliographic References
[1] Matejka, L. and Dusek, K., *Macromolecules*, 1989, **22**, 2911
[2] Johncock, P., *J. Appl. Polym. Sci.*, 1990, **41**, 613
[3] Johncock, P. and Cunliffe, A.V., *Polymer*, 1993, **34**, 1933
[4] Matejka, L. and Dusek, K., *Polymer*, 1991, **32**, 3195
[5] Frier, M., *Actes Colloq.-IFREMER*, 1088, **7**, 357, (props.)

Dimethiconol   D-5

Synonyms: α-*Hydro-ω-hydroxypoly[oxy(dimethylsilylene)]. Poly(dimethylsiloxane)diol. Dimethicone polyol*
Related Polymers: Polydimethylsiloxane, Methylsilicone rubber, Room temp. vulcanised
Monomers: Dichlorodimethylsilane, Hexamethylcyclotrisiloxane
Material class: Fluids
Polymer Type: Dimethylsilicones
CAS Number:

CAS Reg. No.	Note
31692-79-2	Dimethiconol
63394-02-5	elastomer
9006-65-9	PDMS fluid
9016-00-6	PDMS fluid
63148-62-9	PDMS fluid

– Dimethiconol

Molecular Formula: $[C_2H_6OSi]_n$
Fragments: C_2H_6OSi
Molecular Weight: MW 300–100000

Volumetric & Calorimetric Properties:
Density:

No.	Value	Note
1	d 0.96 g/cm^3	[6]

Thermal Expansion Coefficient:

No.	Value	Note	Type
1	0.001 K^{-1}	[10,11]	V

Specific Heat Capacity:

No.	Value	Note		Type
1	1.5 kJ/kg.C	0.36 cal (g°C)$^{-1}$	[7,9,10]	P

Transition Temperature General: Transition temps. have been reported [1,3]

Surface Properties & Solubility:
Cohesive Energy Density Solubility Parameters: δ 15.5 J$^{1/2}$ cm$^{-3/2}$ (7.5–7.6 cal$^{1/2}$ cm$^{-3/2}$) [4,5]
Solvents/Non-solvents: Sol. aliphatic hydrocarbons, chlorinated hydrocarbons, aromatic hydrocarbons. Mod. sol. ketones, esters, ethers. Insol. alcohols, H$_2$O [4,5,6]
Surface and Interfacial Properties General: The mixture of polar hydrophilic end groups and the non-polar hydrophobic repeat sequence gives dimethiconol valuable surfactant props.

Transport Properties:
Polymer Melts: Viscosity has been reported [2,7]

Optical Properties:
Refractive Index:

No.	Value	Note
1	1.4	[11]

Transmission and Spectra: Transparent above 280 nm [8]

Polymer Stability:
Polymer Stability General: The reactive SiOH groups make dimethiconol less stable than PDMS fluid [9]

Chemical Stability: Liable to undergo condensation reactions in acid conditions [9]
Hydrolytic Stability: More liable to undergo condensation reactions than depolymerisation in H$_2$O and aq. solns. at room temp.
Biological Stability: Non-biodegradable

Applications/Commercial Products:
Processing & Manufacturing Routes: Short chain polysiloxanes are produced by the hydrol. of dichlorodimethylsilane in weak acid. In the absence of chain terminating groups dimethiconol is formed. High MW dimethiconol is formed by polymerisation of cyclic dimethysiloxanes at 150–200° with alkaline catalysis in the presence of small quantities of water
Applications: Used as a wetting agent and surfactant. High MW dimethiconol is used to manufacture methylsilicone rubber

Trade name	Details	Supplier
Eccofoam SIL	elastomer	Emerson & Cuming
Eccosil	elastomer	Emerson & Cuming
Masil SFR	fluid	PPG/Specialty Chem
Norsil RTV	elastomer	RH Carlson
RTF 762	elastomer	General Electric Silicones
RTV	various grades, not exclusive	General Electric Silicones
Unisil SF-R	fluid	UPI

Bibliographic References
[1] Clarson, S.J. and Semlyen, J.A., *Siloxane Polymers*, 1993
[2] Meals, R.N., *Ann. N. Y. Acad. Sci.*, 1964, **125**, 137
[3] Clarson, S.J., Dodgson, K. and Semlyen, J.A., *Polymer*, 1985, **26**, 930, (transition temps)
[4] Yerrick, K.B. and Beck, H.N., *Rubber Chem. Technol.*, 1964, **37**, 261, (solubility)
[5] Baney, R.H., Voigt, C.E. and Mentele, J.W., Struct.-Solubility Relat. Polym., *(Proc. Symp.)*, (eds. F.W. Harris and R.B. Seymour), Academic Press, 1977, 225, (solubility)
[6] Roff, W.J. and Scott, J.R., Fibres, Films, *Plastics and Rubbers*, Butterworths, 1971
[7] Stark, F.O., Fallender, J.R. and Wright, A.P., *Comprehensive Organometallic Chemistry*, (eds. G. Wilkinson, F.G.A. Stone and E.W. Abel), Pergamon Press, 1982, **2**, 305
[8] Hardman, B. and Torkelson, A., *Encycl. Polym. Sci. Eng.*, 2nd edn., (ed. J.I. Kroschwitz), John Wiley and Sons, 1985, **15**, 204
[9] Noll, W., *Chemistry and Technology of Silicones*, Academic Press, 1968
[10] Bates, O.K., *Ind. Eng. Chem.*, 1949, **41**, 1966, (thermal props)
[11] Hardy, D.V.N. and Megson, N.J.L., *Q. Rev. Chem. Soc.*, 1948, **2**, 25

E-Bonite

Synonyms: *Hard rubber. Vulcanite*
Related Polymers: Natural rubber
Monomers: 2-Methyl-1,3-butadiene
Material class: Thermosetting resin
Polymer Type: polybutadiene
Additives: Additives include vulcanising sulfur, plasticisers, amines. Typical rubber-sulfur ratio 68:32
General Information: Obtained by vulcanising natural rubber with sulfur. One of the earliest synthetic polymers

Volumetric & Calorimetric Properties:
Density:

No.	Value	Note
1	d 1.08–1.2 g/cm^3	[1]
2	d 1.15 g/cm^3	[3]

Thermal Expansion Coefficient:

No.	Value	Note	Type
1	0.0002–0.00024 K^{-1}	[1]	V

Thermal Conductivity:

No.	Value	Note
1	2.9–3.8 W/mK	[1]

Specific Heat Capacity:

No.	Value	Note	Type
1	1.38–1.42 kJ/kg.C	[1]	P

Transition Temperature General: Yield temp. 85° [4]

Surface Properties & Solubility:
Solvents/Non-solvents: Swollen by aromatic hydrocarbons, chlorinated hydrocarbons [4]

Transport Properties:
Permeability and Diffusion General: (H$_2$O) 2–10 × 10^{-16} mol (m s Pa)$^{-1}$

Water Absorption:

No.	Value	Note
1	0.25 %	equilibrium [4]

Gas Permeability:

No.	Gas	Value	Note
1	H$_2$	2 × 10^{-5} cm^3 mm/(m^2 day atm)	50 × 10^{-17} mol (cm s Pa)$^{-1}$
2	He	4 × 10^{-5} cm^3 mm/(m^2 day atm)	100 × 10^{-17} mol (cm s Pa)$^{-1}$
3	N$_2$	4 × 10^{-7} cm^3 mm/(m^2 day atm)	max., 1 × 10^{-17} mol (cm s Pa)$^{-1}$

Mechanical Properties:
Mechanical Properties General: Mechanical props. depend on composition and processing

Tensile (Young's) Modulus:

No.	Value	Note
1	>500 MPa	min. [1]
2	2700 MPa	[3]

Flexural Modulus:

No.	Value	Note
1	315–900 MPa	hard resin rubber [1]

Tensile Strength Break:

No.	Value	Note	Elongation
1	8–44 MPa	[2]	
2	62 MPa	[4]	3%

Flexural Strength at Break:

No.	Value	Note
1	8–16 MPa	hard resin rubber [1]

Poisson's Ratio:

No.	Value	Note
1	0.39	static, unloaded [1]
2	0.46	10 MHz [1]
3	0.2–0.3	static, loaded [1]

Miscellaneous Moduli:

No.	Value	Note	Type
1	970 MPa	[3]	Shear modulus
2	4100 MPa	[3]	Bulk modulus

Impact Strength: Impact strength 1–7 kJ m^{-2} (ebonite) [1], 25 kJ m^{-2} (hard resin rubber) [1]
Hardness: Shore A100 [1]; Shore D70–D85 [2]
Izod Notch:

No.	Value	Notch	Note
1	26.6 J/m	Y	0.5 ft lb in^{-1} [4]

Electrical Properties:
Electrical Properties General: Resistivity decreases on exposure to light
Surface/Volume Resistance:

No.	Value	Note	Type
1	1000 × 10^{15} Ω.cm	[4]	S

Dielectric Permittivity (Constant):

No.	Value	Frequency	Note
1	2.7	1 kHz	25–75° [4]
2	3	1 MHz	25–75° [4]

Ecolyte

Optical Properties:
Refractive Index:

No.	Value
1	1.6–1.66

Polymer Stability:
Polymer Stability General: Stability depends on composition and processing
Thermal Stability General: Prolonged exposure to temps. above 70° increases rigidity and lowers impact strength [1]
Chemical Stability: Has good chemical resistance
Hydrolytic Stability: Has high resistance to acid and to other inorganic liquids

Applications/Commercial Products:
Processing & Manufacturing Routes: Prod. by vulcanising with high levels of sulfur to increase cross-linking. May also contain thermoplastic resins or is compounded with anthracite
Applications: Uses include car battery cases (with anthracite), water meters, water pipes, shoe heels

Trade name	Supplier
Dexonite	Dexine Rubber Co. Ltd

Bibliographic References
[1] *Concise Encyclopedia of Polymer Science and Engineering,* (eds. H.F. Mark, N.M. Bikales, C.G. Overberger and G. Menges), John Wiley and Sons, 1990
[2] Cooper, D.L., *Rubber Prod. Manuf. Technol.,* (eds. A.K. Bhowmick, M.M. Hall and H.A. Benarey), Marcel Dekker, 1994
[3] Clark, E.S., *Encycl. Polym. Sci. Eng.,* 2nd edn., (ed. J.I. Kroschwitz), John Wiley and Sons, 1985, **5**, 370
[4] Brydson, J.A., *Plast. Mater.,* 6th edn., Butterworth Heinemann, 1995, 863, (props., rev.)

Ecolyte   E-2

$$\left[\left[CH_2CH \atop R \right]_x \left[CH_2CR' \atop \underset{R''}{O=} \right]_y \right]_n$$

R = aryl
R' = R'' = alkyl

Related Polymers: Ecolyte PS
Monomers: Ethylene, Propylene, Styrene
Material class: Copolymers
Polymer Type: polyethylene, polypropylenepolyolefins
CAS Number:

CAS Reg. No.	Note
52682-94-7	Ecolyte PS
25191-48-4	Ecolyte PS102, Ecolyte PS108
36343-62-1	3-methyl-3-buten-2-one polymer with ethene
53859-04-4	Ecolyte P
53801-30-2	Ecolyte PP
86403-35-2	Ecolyte E
53801-29-9	Ecolyte PT
27340-61-0	Ecolyte S

General Information: The ecolyte process imparts photodegradable props. to the base polymers by copolymerisation with 3-methyl-3-buten-2-one. The ecolyte compositions contain less than 1% by weight of carbonyl groups. The physical props. are essentially identical to those of the base polymers, e.g. polyethylene, polypropylene, and polystyrene

Applications/Commercial Products:

Trade name	Supplier
Ecolyte E	Ecolyte Atlantic Inc.
Ecolyte P	Ecolyte Atlantic Inc.
Ecolyte PE	Ecolyte Atlantic Inc.
Ecolyte PP	Ecolyte Atlantic Inc.
Ecolyte PS	Ecolyte Atlantic Inc.
Ecolyte S	Ecolyte Atlantic Inc.

Bibliographic References
[1] Cooney, J.D. and Wilnes, D.M., *ACS Symp. Ser.,* 1976, **25**, 307, (degradation)
[2] Jones, P.H., Prasad, D., Heskins, M., Morgan, M.H. and Guillet, J.E., *Environ. Sci. Technol.,* 1974, **8**, 919, (biodegradability)
[3] Rabek, J.F., *Polymer Photodegradation: Mechanisms and Experimental Methods,* Chapman and Hall, 1995, (rev)
[4] Guillet, J.E., *Degrad. Polym.,* (eds. G. Scott and D. Gilead), Chapman and Hall, 1995, 216, (rev)
[5] Degradable Materials: Perspectives, *Issues and Opportunities,* (eds. S.A. Barenberg, J.L. Brash, R. Narayan and A.E. Redpath), CRC Press, 1990, 62, 70, 593, (rev)

Ecolyte PS   E-3

$$\left[\left[CH_2CH \atop Ph \right]_x \left[CH_2C(CH_3) \atop \underset{OH}{O=} \right]_y \right]_n$$

Synonyms: *Poly(3-methyl-3-buten-2-one-co-ethenylbenzene). Poly(methyl isopropenyl ketone-co-styrene). PSE. SMiPK*
Related Polymers: Ecolyte
Monomers: Styrene, 3-Methyl-3-buten-2-one
Material class: Copolymers
Polymer Type: polyolefins, polystyrene
CAS Number:

CAS Reg. No.	Note
52682-94-7	Ecolyte PS
25191-48-4	3-methyl-3-buten-2-one polymer with styrene

Molecular Formula: $(C_5H_8O.C_8H_8)_n$
Fragments: C_5H_8O C_8H_8
Additives: Ecolyte E, Ecolyte PE added to polystyrene to assist photodegradability
General Information: A photodegradable polymer prod. by the Ecolyte process
Miscellaneous: Copolymerisation constants for styrene-methyl isopropenyl ketone r_1 0.32, r_2 0.66 [5]

Volumetric & Calorimetric Properties:
Transition Temperature General: Transition temp. and activation energy of phenyl group motion have been reported [8]
Transition Temperature:

No.	Value	Note
1	-90°C	phenyl group motion

Transport Properties:
Polymer Solutions Dilute: Viscosity values have been reported [4] in the study of polymer blends and copolymers

Mechanical Properties:
Tensile Strength Break:

No.	Value	Note
1	63 MPa	approx., 1.5 mm thick [1]
2	38 MPa	approx., 3.2 mm thick [1]

Optical Properties:
Transmission and Spectra: Phosphorescence, [8] ir [3,6], emission [6] and absorbance [3] spectral data have been reported

Polymer Stability:
Polymer Stability General: Ecolyte PE and Ecolyte E degrade readily in uv light. [4] Ecolyte is used as a prodegradant when added to polystyrene. 10% Ecolyte E added to polystyrene increases the photodegradability of polystyrene 2.5 fold [12]
Environmental Stress: Weathering studies have been reported (ASTM D1435-85; ASTM D638-87b) [1]. Degradation depends on level of Ecolyte in sample. Embrittlement times have been reported [9]
Biological Stability: Degrades in approx. one year in natural soils [10,11]
Stability Miscellaneous: Stability to synchroton radiation is similar to the polymer's stability to uv radiation [2,3]

Bibliographic References
[1] May, S.A., Fuentes, E.C. and Sato, N., *Polym. Degrad. Stab.*, 1991, **32**, 357, (weathering)
[2] Guillet, J.E., Li, S.K.L., MacDonald, S.A. and Wilson, C.G., Polym. Prepr. (Am. Chem. Soc., *Div. Polym. Chem.*), 1984, **25**, 296, (irradiation)
[3] Guillet, J.E., Li, S.K.L., MacDonald, S.A. and Wilson, C.G., *ACS Symp. Ser.*, 1984, **266**, 389, (photochem)
[4] Heskins, M., McAneney, T.B. and Guillet, J.E., *ACS Symp. Ser.*, 1976, **25**, 281, (photodegradation)
[5] Nenkov, G., Georgieva, T., Stoyanov, A. and Kabaivanov, V., *Angew. Makromol. Chem.*, 1980, **91**, 69, (preparation and photodestruction)
[6] Hrdlovič, P., Lukáč, I., Zvara, I., Kuličková, M. and Berets, D., *Eur. Polym. J.*, 1980, **16**, 651, (degradation)
[7] Erben, F. and Veselý, R., *Angew. Makromol. Chem.*, 1983, **114**, 161, (weathering)
[8] Somersall, A.C., Dan, E. and Guillet, J.E., *Macromolecules*, 1974, **7**, 233, (phosphorescence)
[9] Cooney, J.D. and Wilnes, D.M., *ACS Symp. Ser.*, 1976, **25**, 307, (degradation)
[10] Jones, P.H., Prasad, D., Heskins, M., Morgan, M.H. and Guillet, J.E., *Environ. Sci. Technol.*, 1974, **8**, 919, (biodegradability)
[11] Guillet, J.E., Regulski, T.W. and McAneney, T.B., *Environ. Sci. Technol.*, 1974, **8**, 923, (biodegradability of Ecolyte PS polystyrene)
[12] Degradable Materials: Perspectives, *Issues and Opportunities*, (eds. S.A. Barenberg, J.L. Brash, R. Narayan and A.E. Redpath), CRC Press, 1990, 62, 70, 593, (degradation)
[13] Hanner, M.J., McKelvy, M.L., Sikkema, L. and Priddy, D.B., *Polym. Degrad. Stab.*, 1993, **39**, 235, (degradation)

ECTFE E-4

$$-[CH_2CH_2]_x-[CF_2CFCl]_y-$$

Synonyms: *Poly(chlorotrifluoroethene-co-ethylene). Chlorotrifluoroethylene-ethylene copolymer*
Related Polymers: Polychlorotrifluoroethylene, Polyethylene
Monomers: Chlorotrifluoroethylene, Ethylene
Material class: Thermoplastic, Copolymers
Polymer Type: polyethylene, PCTFEpolyhaloolefins

CAS Number:

CAS Reg. No.	Note
25101-45-5	
110872-65-6	alternating

Molecular Formula: $[(C_2H_4).(C_2ClF_3)]_n$
Fragments: C_2H_4 C_2ClF_3
Additives: Stabilisers, glassfibre
General Information: Copolymers are typically 1:1 alternating with an extended zig-zag chain. Crystallinity is typically 50–55% depending on the method of preparation
Morphology: Chain repeat distance 0.502 nm (approx.). Unit cell contains three molecules and occupies a volume of 0.324 m^3 [1]

Volumetric & Calorimetric Properties:
Density:

No.	Value	Note
1	d 1.7 g/cm^3	[3]

Thermal Expansion Coefficient:

No.	Value	Note	Type
1	8×10^{-5} K^{-1}		L
2	0.0002 K^{-1}	[3]	L

Thermal Conductivity:

No.	Value
1	0.16 W/mK

Melting Temperature:

No.	Value	Note
1	240°C	[1]
2	260–282°C	[6]

Vicat Softening Point:

No.	Value	Note
1	76°C	1.8 MPa, ASTM D648 [3]
2	115°C	0.45 MPa, ASTM D648 [3]

Brittleness Temperature:

No.	Value	Note
1	<-76°C	max. [1,6]

Transition Temperature:

No.	Value	Note	Type
1	90°C	motion in amorph. phase [1]	β transition
2	140°C	motion in cryst. phase	α transition
3	-65°C	motion in cryst. phase	γ transition

Surface Properties & Solubility:

Plasticisers: None
Solvents/Non-solvents: Insol. all solvents. Some hot polar solvents are absorbed
Surface and Interfacial Properties General: Coefficient of friction 0.7 (ASTM D1894) [7]

Transport Properties:

Transport Properties General: Has excellent barrier props. against water vapour and other gases
Polymer Melts: Heat seal temp. 246–260° [7]
Water Absorption:

No.	Value	Note
1	0.01 %	max., 24h, 3 mm thick [5]

Gas Permeability:

No.	Gas	Value	Note
1	O_2	0.00135 cm^3 mm/(m^2 day atm)	3278 m mm (h MPa)$^{-1}$, 25° [1]
2	N_2	0.000899 cm^3 mm/(m^2 day atm)	2186 m mm (h MPa)$^{-1}$, 25° [1]
3	H_2O	0.061 cm^3 mm/(m^2 day atm)	148310 m mm (h MPa)$^{-1}$, 25° [1]
4	Cl_2	0.00019 cm^3 mm/(m^2 day atm)	468 m mm (h MPa)$^{-1}$, 25° [1]
5	HCl	0.00086 cm^3 mm/(m^2 day atm)	2100 m mm (h MPa)$^{-1}$, 25° [1]
6	H_2S	0.00128 cm^3 mm/(m^2 day atm)	3120 m mm (h MPa)$^{-1}$, 25° [1]

Mechanical Properties:

Mechanical Properties General: Shows mechanical props. similar to nylon. Has good impact resistance and is tougher than PTFE. Shows very good retention of props. after heat ageing up to 175°
Flexural Modulus:

No.	Value	Note
1	2000 MPa	[1]

Tensile Strength Break:

No.	Value	Note	Elongation
1	46–47 MPa	[1]	110–160%

Tensile Strength Yield:

No.	Value	Elongation	Note
1	32–35 MPa	6–10%	[1]

Flexural Strength Yield:

No.	Value	Note
1	48 MPa	Flexural stress [2]

Miscellaneous Moduli:

No.	Value	Note	Type
1	620 MPa	10 MPa, 250h [2]	Creep modulus

Impact Strength: Impact strength no break (room temp.) [2]
Hardness: Rockwell R93 [2]
Friction Abrasion and Resistance: Armstrong abrasion volume loss 0.3 cm^3 [2]. Coefficient of friction 0.7 (ASTM D1894) [7]
Izod Notch:

No.	Value	Notch	Note
1	1060 J/m	Y	min., room temp. [3]
2	64.1 J/m	Y	[6]

Electrical Properties:

Electrical Properties General: Widely used in wire and cable insulation due to good electrical props.
Surface/Volume Resistance:

No.	Value	Note	Type
1	14–20 × 10^{15} Ω.cm	ASTM D257 [1]	S

Dielectric Permittivity (Constant):

No.	Value	Frequency	Note
1	2.52	100 Hz	ASTM D150 [1]
2	2.56	1 kHz	ASTM D150 [1]
3	2.6	1 MHz	ASTM D150 [1]

Dielectric Strength:

No.	Value	Note
1	18.9 kV.mm^{-1}	480 V mil^{-1}, 3 mm, short time, ASTM D149 [4]

Arc Resistance:

No.	Value	Note
1	124s	[1]

Complex Permittivity and Electroactive Polymers: Insulation resistance >10^{15} Ω [2]
Dissipation (Power) Factor:

No.	Value	Frequency	Note
1	0.005	100 Hz	ASTM D150, Halar 300 [1]
2	0.004	1 kHz	ASTM D150, Halar 300 [1]
3	0.012	100 kHz	ASTM D150, Halar 300 [1]
4	0.016	1 MHz	ASTM D150, Halar 300 [1]

Optical Properties:

Transmission and Spectra: Light transmittance 94–96% (ASTM E424, 1 mil) [7]
Volume Properties/Surface Properties: Gloss 70–90% (ASTM D2457) [7]. Haze 2% (ASTM D1003) [7]

Polymer Stability:

Polymer Stability General: Has good thermal and oxidative stability, non-flammability and good weatherability
Thermal Stability General: No significant deterioration in props. up to 1000h between 150–175° [1]

– ECTFE, Carbon fibre filled

Upper Use Temperature:

No.	Value	Note
1	140–180°C	Halar grades [2]
2	180–200°C	[1]
3	165–180°C	[1]

Decomposition Details: Decomposes at 350° in N_2 and at lower temps. in air [1]
Flammability: Passes the UL Steiner tunnel (UL910) smoke and flame spread test [1]. Flammability rating V0 (UL94, vertical, 0.18mm thick) vertical, 0.18 mm thick = 94 V-O (SE-O) Chars, but does not melt or drip) [1,2]. Oxygen index 48–60% (ASTM D2863) [1,7].
Environmental Stress: Subject to stress cracking above 150–170° (Federal specification L-P-390C, class H) [1]
Chemical Stability: Inert to acids and most bases, strong oxidising agents and other chemicals. May absorb small amounts of hot polar organic solvents [1,2]
Hydrolytic Stability: Has excellent dimensional stability in H2O and salt solns. [1,2]
Stability Miscellaneous: Resistance to radiation is excellent. Useful props. are maintained when exposed to dosages as high as 5 MGy (500 Mrad) [1]

Applications/Commercial Products:
Processing & Manufacturing Routes: Manufactured by aq. suspension, low temp. polymerisation and radiation induced polymerisation. Processed by powder coating, roto moulding, roto lining, injection moulding and extrusion
Mould Shrinkage (%):

No.	Value	Note
1	0.05%	[6]

Applications: Wire coating, film, tubing, fibres, pumps, valves, pipe coating, linings

Trade name	Details	Supplier
ECTFE		Furon
Fluoromelt FP-C		LNP Engineering
Fluoromelt FP-CF		LNP Engineering
Halar		Allied Chemical
Halar		Ausimont
Korton ECTFE		Norton
Lubricomp FP-C		LNP Engineering
Thermocomp FP	various grades	LNP Engineering

Bibliographic References
[1] *Encycl. Polym. Sci. Eng.*, (ed. J.I. Kroschwitz), Vol. 3, Wiley Interscience, 1989, 481
[2] Skrypa, M.J., Robertson, A.B. and Toelcke, G.A., Soc. Plast. Eng., *Tech. Pap.*, 1972, **18**, 659
[3] *Encyclopedia of Advanced Materials*, (eds. D. Bloor, R.J. Brook, M.C. Flemings, S. Mahajan and R.W. Cahn), Pergamon Press, 1994, **3**, 863
[4] Khanna, Y.P. and Taylor, T.J., *J. Appl. Polym. Sci.*, 1989, **38**, 135, (dielectric props.)
[5] *Kirk-Othmer Encycl. Chem. Technol.*, Vol. 13, Wiley Interscience, 1980, 574
[6] *Hylar 500*, Ausimont, (technical datasheet)
[7] *Korton ECTFE* Norton, (technical datasheet)

ECTFE, Carbon fibre filled E-5

Synonyms: *Poly(ethylene-co-chlorotrifluoroethylene), Carbon fibre filled. Ethylene-chlorotrifluoroethylene copolymer, Carbon fibre filled*
Related Polymers: ECTFE, Glass fibre filled
Monomers: Ethylene, Chlorotrifluoroethylene
Material class: Thermoplastic, Copolymers, Composites
Polymer Type: polyethylene, PCTFEfluorocarbon (polymers)

Volumetric & Calorimetric Properties:
Density:

No.	Value	Note
1	d 1.7 g/cm^3	20% fillled [1]

Deflection Temperature:

No.	Value	Note
1	143°C	1.82 MPa, 20% filled [1]

Mechanical Properties:
Flexural Modulus:

No.	Value	Note
1	9.9 MPa	101 kg cm^{-2}, 20% filled [1]

Tensile Strength Break:

No.	Value	Note	Elongation
1	75.8 MPa	773 kg cm^{-2}, 20% filled [1]	2%

Flexural Strength Yield:

No.	Value	Note
1	65.5 MPa	668 kg cm^{-2}, 20% filled [1]

Izod Notch:

No.	Value	Notch	Note
1	267 J/m	Y	5 ft lb in^{-1}, room temp., 20% filled [1]

Applications/Commercial Products:
Processing & Manufacturing Routes: Processed by injection moulding
Mould Shrinkage (%):

No.	Value	Note
1	0.35%	20% filled [1]

Trade name	Supplier
Fluoromelt-CC-1004	LNP Engineering
Thermocomp FP-CC-1004	LNP Engineering

Bibliographic References
[1] 1996, LNP Engineering, (technical datasheet)

ECTFE, Glass fibre filled

E-6

Synonyms: *Poly(ethylene-co-chlorotrifluoroethylene), Glass fibre filled. Ethylene-chlorotrifluoroethylene copolymer, Glass fibre filled*
Related Polymers: ECTFE, Carbon fibre filled
Monomers: Ethylene, Chlorotrifluoroethylene
Material class: Copolymers, Composites, Thermoplastic
Polymer Type: polyethylene, PCTFEfluorocarbon (polymers)
General Information: 30% Filled material has excellent tensile strength and flexural modulus [1]

Volumetric & Calorimetric Properties:
Density:

No.	Value	Note
1	d 1.8 g/cm^3	20% filled [1]
2	d 1.87 g/cm^3	30% filled [1]
3	d 1.85 g/cm^3	[2]

Thermal Expansion Coefficient:

No.	Value	Note	Type
1	3.6×10^{-5} K^{-1}	30% filled [1]	L
2	2.9×10^{-5} K^{-1}	[2]	L

Thermal Conductivity:

No.	Value	Note
1	0.24 W/mK	0.000586 cal (cm s°C)$^{-1}$, 30% filled [1]

Melting Temperature:

No.	Value	Note
1	240°C	20–30% filled [1]

Deflection Temperature:

No.	Value	Note
1	135°C	1.82 MPa, 20% filled [1]
2	210°C	1.82 MPa, 30% filled [1]
3	200°C	1.8 MPa [2]

Transport Properties:
Water Absorption:

No.	Value	Note
1	0.01 %	24h [2]

Mechanical Properties:
Flexural Modulus:

No.	Value	Note
1	5236 MPa	53400 kg cm^{-2}, 20% filled [1]
2	5854 MPa	59700 kg cm^{-2}, 30% filled [1]
3	6500 MPa	[2]

Tensile Strength Break:

No.	Value	Note	Elongation
1	60 MPa	611 kg cm^{-2}, 20% filled [1]	3%
2	69 MPa	703 kg cm^{-2}, 30% filled [1]	3%
3	82 MPa	[2]	10% [2]

Flexural Strength Yield:

No.	Value	Note
1	96.5 MPa	984 kg cm^{-2}, 20% filled [1]
2	103 MPa	1050 kg cm^{-2}, 30% filled [1]

Hardness: Shore R100 [2]
Izod Notch:

No.	Value	Notch	Note
1	261 J/m	Y	4.9 ft lb in^{-1}, room temp., 20% filled [1]
2	240 J/m	Y	4.5 ft lb in^{-1}, ¼", room temp., 30% filled [1]
3	370 J/m	Y	[2]

Electrical Properties:
Dielectric Permittivity (Constant):

No.	Value	Frequency	Note
1	3	1 kHz	30% filled [1]
2	2.9	1 kHz	[2]

Dielectric Strength:

No.	Value	Note
1	4.33 kV.mm^{-1}	30% filled [1]
2	35 kV.mm^{-1}	[2]

Dissipation (Power) Factor:

No.	Value	Frequency	Note
1	0.12	1 MHz	30% filled [1]
2	0.004	1 kHz	dry [2]

Polymer Stability:
Upper Use Temperature:

No.	Value	Note
1	148°C	continuous service, 30% filled [1]
2	130°C	continuous service [2]

Applications/Commercial Products:
Processing & Manufacturing Routes: Processed by extrusion or injection moulding
Mould Shrinkage (%):

No.	Value	Note
1	0.65%	20% filled
2	0.5%	30% filled [1]
3	0.04%	[2]

Applications: Has applications in electronic parts

Trade name	Details	Supplier
Fluoromelt FP-CF-1005		LNP Engineering
Thermocomp FP-CF-1004	20% filled	LNP Engineering
Thermocomp FP-CF-1006	30% filled; discontinued material	LNP Engineering

Bibliographic References

[1] *Thermocomp Range*, LNP Engineering, 1996, (technical datasheet)
[2] *The Materials Selector*, 2nd edn., (eds. N.A. Waterman and M.F. Ashby), Chapman & Hall, 1997, **3**, 233

Epoxidised natural rubber E-7

Synonyms: *ENR*
Related Polymers: Natural rubber
Monomers: 2-Methyl-1,3-butadiene
Material class: Natural elastomers
Polymer Type: polybutadiene
CAS Number:

CAS Reg. No.	Note
9006-04-6	Rubber

Additives: Phosphorus-containing compounds to improve flame retardancy. Carbon black, silica, wood flour and rice husk ash
General Information: Is an elastic material with a higher density and T_g than natural rubber. Epoxidation allows improvement of some props. of natural rubber (e.g. oil resistance, gas permeability) and hence widens the range of applications. it also provides potential for further chemical modification. Commercial products are available with varying levels of epoxidation (25% or 50%)
Morphology: Will strain crystallise at epoxidation levels up to 95%. The degree of crystallinity diminishes at epoxidation levels above 50%. The lattice parameters vary with level of epoxidation [8]. (25 mol% epoxidation). a 1.254nm, b 0.950nm, c 0.827nm; a 1.172nm, b 1.038nm, c 0.852nm (95 mol% epoxidation) [8]
Miscellaneous: Can be vulcanised by conventional methods and can also be cross-linked by other means, e.g. peroxides or dibasic acids

Volumetric & Calorimetric Properties:
Density:

No.	Value	Note
1	d 0.907 g/cm^3	approx., 10% epoxidation [4]
2	d 0.94 g/cm^3	10% epoxidation [22]
3	d 0.97 g/cm^3	25% epoxidation [22]
4	d 1.03 g/cm^3	50% epoxidation [22]
5	d 0.99 g/cm^3	approx., 50% epoxidation [4]

Specific Heat Capacity:

No.	Value	Note	Type
1	0.46 kJ/kg.C	0.111 cal g^{-1}°C^{-1} [4]	P

Glass-Transition Temperature:

No.	Value	Note
1	-60°C	10% epoxidation [22]
2	-48--45°C	25% epoxidation [8,22]
3	-25.7°C	47% epoxidation [4]
4	-20°C	50% epoxidation [22]
5	-9°C	75% epoxidation [8]
6	18°C	fully epoxidsed [4]
7	26°C	50% epoxidation; amine-cured; 30% carbon filled [5]
8	16°C	50% epoxidation; sulfur-cured; 30% carbon filled [5]

Transition Temperature General: T_g is increased by epoxidation, rising linearly by approx. 1° for each mol% epoxidation. Only one T_g is reported for all epoxidation levels [1,4]. Curing raises the T_g; amine-cured polymer has higher T_g than the sulfur-cured polymers [5]

Surface Properties & Solubility:
Solubility Properties General: Compatible with chlorinated polyolefins and phenolic resins (Novolaks). Is semi-compatible with resoles and incompatible with cured epoxy resins on Bisphenol A [3]
Cohesive Energy Density Solubility Parameters: 18 J$^{1/2}$ cm$^{-3/2}$ [20]
Plasticisers: Aromatic oils, phthalate esters, paraffinic oil [19]

Transport Properties:
Polymer Solutions Dilute: Mooney viscosity has been reported [22]
Permeability of Gases: Gas permeability is decreased by epoxidation [1,4,10,11]. CO_2 29.1 × 10^{-4} cm^2 (s cm Hg)$^{-1}$, 25% epoxidation; CO_2 998 × 10^{-4} cm^2 (s cm Hg)$^{-1}$, 50% epoxidation; CH_4

Transport Properties:
Permeability of Liquids: The permeability of vulcanised polymer to a solvent varies with the solvent, the temp. and the vulcanisation method. The variation has been ascribed to differences in cross-link density and, to a lesser extent, the flexibility of cross-link bonds. Permeability to C_6H_6, toluene, xylene and mesitylene has been reported [12]

Mechanical Properties:
Mechanical Properties General: Tensile strength and elongation increase with epoxidation levels of 10–15% and then decrease significantly [4]. Amine-cured polymer is more rigid, less stretchable and has lower tensile strength than sulfur-cured polymer [5]. For diamine-cured rubbers the tensile strength and 100% modulus increase, and the elongation decreases, with increasing amine concentration (at same curing conditions). A similar trend is observed for increasing cure time [5]. The presence of fillers generally improves the mechanical props. of cured polymer, the improvement depending on the type and level of filler used [15]. Dunlop resilience 81.1% (5.5% epoxidation; cross-linked), 79.6% (10.5% epoxidation; cross-linked), 79.5% (15.4% epoxidation; cross-linked), 74.3% (20.4% epoxidation; cross-linked) [4]
Tensile Strength Break:

No.	Value	Note	Elongation
1	7.6 MPa	5.5% epoxidation; cross-linked [4]	380%
2	8 MPa	15.4% epoxidation; cross-linked [4]	380%
3	2 MPa	20.4% epoxidation; cross-linked [4]	180%
4	1 MPa	30% epoxidation; cross-linked [4]	200%
5	15.5–16.7 MPa	158–170 kg cm^{-2}, 28.31% epoxidation; cured and filled [7]	600–650%

Hardness: Shore A46–A48 (28–31% epoxidation, cured and filled) [7]. Other hardness values have been reported [4,15]

Failure Properties General: The addition of fillers reduces the failure life of vulcanised polymers. At a constant filler level the fatigue life decreases in the order silica >white rice husk ash >carbon black. The use of a coupling agent improves interfacial props. and dispersion of filler particles thus improving the fatigue life. Failure mode in filled vulcanised polymers has a dual nature [16]

Optical Properties:
Transmission and Spectra: Ir and nmr spectral data have been reported [2,4,6]

Polymer Stability:
Polymer Stability General: Hydrogenation can improve ageing resistance [21]
Thermal Stability General: Thermal degradation occurs in a single stage in nitrogen and multi-stages in air [9]
Chemical Stability: Swollen by petrol, C_6H_6, butanol and DMF. Degree of swelling is dependent upon the level of epoxidation [4]. Degraded by phenylhydrazine/O_2 [6]

Applications/Commercial Products:
Processing & Manufacturing Routes: Natural rubber can be epoxidised in the latex state or in soln. Peracids are commonly used either preformed or generated *in situ*. Methods employing a bromohydrin intermediate and also hydrogen peroxide-catalysed reactions have been reported but are not as efficient as peracetic acid epoxidation. The latex concentration, up to 40–50%, has little effect on the epoxidation and product. Epoxidation in acid soln. can lead to the formation of by-products by side reactions. Reaction of a latex stabilised using a nonionic surfactant such as Vulcastab LW and neutralised with AcOH prior to the addition of peracetic acid gives efficient epoxidation with reduced side reactions. [4,7]. Can be cross-linked using *p*-phenylenediamine or dicumyl peroxide but the conventional method is based on sulfur. The curing temp. for sulfur reactions is generally higher (180° compared to 150° for amine-cure) and other components added to the vulcanisation mixture include: activators (e.g. zinc oxide and stearic acid), antioxidant, and accelerator e.g. 2-morpholinobenzothiazole. [5,8,12]. May be cured with uv in the presence of a cationic initiator (e.g. triaryl sulfonium salts) [6]
Applications: Applications include engineering components, oil seals, inner lining of tubeless tyres, tyre treads and adhesives

Trade name	Supplier
ENR	Gutherie Latex

Bibliographic References
[1] *Encycl. Polym. Sci. Technol.*, 1985, **14**, 768
[2] Bradbury, J.H. and Perera, M.C.S., *J. Appl. Polym. Sci.*, 1995, **58**, 2057, (nmr)
[3] Kallitsis, J.K., and Kafoglou, N.K., *J. Appl. Polym. Sci.*, 1989, **37**, 453, (compatibility)
[4] Burfield, D.R., Lim, K.L. and Law, K.-S., *J. Appl. Polym. Sci.*, 1984, **29**, 1661, (synth., props.)
[5] Hashim, A.S. and Kohjiva, S., *J. Polym. Sci., Part A-1*, 1994, **32**, 1149
[6] Decker, C., Xuan, H.L. and Viet, N.T., *J. Polym. Sci., Part A-1*, 1995, **33**, 2759, (synth)
[7] Vernekar, S.P., Sabne, S.D., Patil, S.D., Patil, A.S. et al, *J. Appl. Polym. Sci.*, 1992, **44**, 2107
[8] Davies, C.K.L., Wolfe, S.V., Gelling, I.R. and Thomas, A.G., *Polymer*, 1983, **24**, 107
[9] Li, S.-D., Chen, Y., Zhou, J., Li, P.-S. et al, *J. Appl. Polym. Sci.*, 1998, **67**, 2207, (thermal degradation)
[10] Barrie, J.A., Becht, M., and Campbell, D.S., *Polymer*, 1992, **33**, 2450, (gas permeability)
[11] Fitch, W., Koros, W.J., Nolen, R.L. and Carnes, J.R., *J. Appl. Polym. Sci.*, 1993, **47**, 1033, (gas permeability)
[12] Johnson, T. and Thomas, S., *J. Macromol. Sci., Phys.*, 1997, **36**, 401, (liq. permeability)
[13] Nasir, M. and Choo, C.H., *Eur. Polym. J.*, 1989, **25**, 355, (mechanical props.)
[14] Roychoudhury, A. and De, P.P., *J. Appl. Polym. Sci.*, 1993, **50**, 181
[15] Ishak, Z.A.M., Bakar, A.A., *Eur. Polym. J.*, 1995, **31**, 259, (additives)
[16] Ishak, Z.A.M., Bakar, A.A., Ishiaku, U.S., Hashim, A.S. and Azahari, B., *Eur. Polym. J.*, 1997, **33**, 73, (additives)
[17] Ismail, H., Rozman, H.D., Jaffri, R.M. and Mohd Ishak, Z.A., *Eur. Polym. J.*, 1997, **33**, 1627, (additives)
[18] Varughes, S. and Tripathy, D.K., *J. Appl. Polym. Sci.*, 1992, **44**, 1847, (additives)
[19] Varughes, S. and Tripathy, D.K., *J. Elastomers Plast.*, 1993, **25**, 343
[20] Ng, S.C. and Chee, K.K., *Eur. Polym. J.*, 1997, **33**, 749, (solubility parameters)
[21] Roy, S., Bhattacharjee, S. and Gupta, B.R., *J. Appl. Polym. Sci.*, 1993, **49**, 375
[22] Poh, B.T. and Tan, B.K., *J. Appl. Polym. Sci.*, 1991, **42**, 1407
[23] Poh, B.T., Chen, M.F. and Ding, B.S., *J. Appl. Polym. Sci.*, 1996, **60**, 1569
[24] Poh, B.T. and Tang, W. L., *J. Appl. Polym. Sci.*, 1995, **55**, 537
[25] Poh, B.T., Kwok, C.P. and Lim, G.H., *Eur. Polym. J.*, 1995, **31**, 223

Epoxidised SBS   E-8

Synonyms: *Epoxidised styrene-butadiene-styrene block copolymer. Styrene-epoxidised butadiene-styrene block copolymers. Epoxidised poly(styrene-block-butadiene-block-styrene) triblock copolymer. ESBS*
Related Polymers: SBS
Monomers: Styrene, 1,3-Butadiene
Material class: Copolymers
Polymer Type: polybutadiene, polystyrene
Additives: Carbon black, zinc oxide, stearic acid
General Information: A chemically modified form of SBS which is suitable for specific applications. Such modifications are possible because of the presence of unsaturated polybutadiene units. Polystyrene is more resistant to oxidation than is polybutadiene, so oxidation (epoxidation) largely results in degradation of polybutadiene
Morphology: The physical props. and mech. props. of block copolymers are not adversely affected by epoxidation, indicating that such chemical modification does not interfere with specific domain morphology which is of such importance to these thermoplastic elastomers [5,6]. Even at extreme (80%) levels of epoxy group content, the microphase separated struct. is still preserved; epoxidised polybutadienes are essentially incompatible with the polystyrene segments [5]. Conformational changes of polymer chains, including chain extension, occurs during epoxidation. Hydrodynamic volume of polymer chain depends on solvent used and epoxy group content of polymer [5,10]
Identification: Oxirane content of epoxidised copolymers can be determined by direct titration with tetraethylammonium bromide or tetrabutylammonium iodide, using crystal violet as indicator [6,11]. Total oxygen content can be determined by neutron activation anal. [6]
Miscellaneous: Detailed studies of the kinetics and mechanism of the epoxidation reaction have been reported for systems using *o*-monoperoxyphthalic acid [5,10] and peracetic acid [8,9] as oxidation agents: epoxidation reaction in dilute soln. follows elementary second order kinetics [8,10]. Rates depend on composition and microstruct. of block copolymers and experimental conditions such as concentration of oxidant, reaction temp. and solvent [5,8,9,10]. The solvent chosen influences conformn. which in turn influences reaction rate [5,10]. The rate of epoxidation is reduced as styrene content of block copolymer increases [9]. Epoxidised copolymers can be further modified by reaction with dicarboxylic acids or aldehydes to give thermosetting epoxy resins which are suitable for coatings and castings [7]. Hydrol. with perchloric acid removes approx. 50% of epoxide groups to give a material that may swell in H_2O or other hydrogen-bonding solvents, but which retains its tensile strength even after prolonged contact with solvents [12]. Epoxidation of linear and tapered block copolymers with MWs greater than 60 000 using *in-situ* peroxyformic acid as oxidising agent can lead to gelation when carried out in certain solvents, e.g. cyclohexane. Such problems can be alleviated by use of mixed epoxidation agents, i.e. peroxyformic/peroxyacetic acid(s); there are no adverse effects on tensile props. or oil resistance of the resulting polymers [6]

Epoxy Resins

Volumetric & Calorimetric Properties:
Transition Temperature General: T_g increases with epoxidation by approx. 8° per 10% increase in epoxy-group content [5]

Surface Properties & Solubility:
Solvents/Non-solvents: Sol. toluene, cyclohexane, $CHCl_3$, dioxane. Insol. 2-propanol, EtOH. Swollen by H_2O or other hydrogen-bonding solvents (50% epoxidised polymer, after hydrol. with $HClO_4$) [12]

Transport Properties:
Transport Properties General: Epoxidised SBS (20/60/60) displays higher melt flow than parent copolymer [6]
Melt Flow Index:

No.	Value	Note
1	5.6 g/10 min	180°, 5 kg, uncompounded SBS (20/60/20) [6]

Polymer Solutions Dilute: Reduced and intrinsic viscosities in dioxane and in $CHCl_3$ (plus corresponding mixed solvent system) have been reported [5,10]

Mechanical Properties:
Mechanical Properties General: Tensile storage modulus and loss tangent have been reported [5]. Part epoxidised material (approx. 50% conversion) after hydrol. with $HClO_4$ gives polymer which swells in H_2O or other hydrogen-bonding solvents. Tensile strength is retained even in presence of H_2O, with only approx. 10% of initial tensile strength lost after approx. 14 days [12]. Mech. props of epoxidised SBS (20/60/20) block copolymers (using in-situ peroxyformic acid as oxidant) have been reported for both uncompounded and compounded (containing carbon black, zinc oxide and stearic acid) materials. Uncompounded systems show only slight differences in tensile props. when compared to parent copolymers, but epoxidised polymers display lower set at break, lower shore hardness and higher melt flow [6]. Improvements in tensile props. of epoxidised polymers are observed for compounded systems [6]

Tensile Strength Break:

No.	Value	Note	Elongation
1	34.3 MPa	epoxidised SBS (20/60/20) [6]	690%
2	23.4 MPa	epoxidised SBS (20/60/20) carbon black filled [6]	240%

Miscellaneous Moduli:

No.	Value	Note
1	3 MPa	300% Modulus, epoxidised SBS (20/60/20) [6]
2	19 MPa	200% Modulus, epoxidised SBS (20/60/20) carbon black filled [6]

Viscoelastic Behaviour: Set at break 17% (epoxidised SBS (20/60/20) [6]
Hardness: Shore A86 (epoxidised SBS (20/60/20); A88 (epoxidised SBS (20/60/20) carbon black filled) [6]
Fracture Mechanical Properties: Crescent tear strength 51.8 kg cm^{-1} (290 lb in^{-1}, epoxidised SBS (20/60/20) carbon black filled) [6]

Electrical Properties:
Electrical Properties General: Dielectric constant and dielectric dissipation factor have been reported [5]

Optical Properties:
Transmission and Spectra: H-1 nmr spectral data have been reported [5]

Polymer Stability:
Chemical Stability: Resistant to chemical degradation during epoxidation up to approx. 54% epoxy group content [1,7]. At higher levels of epoxidation, or after prolonged storage with epoxidising agent (o-monopereoxyphthalic acid) and its residue (phthalic acid), degradation occurs by chain scission process involving polybutadiene units [1]; which is accelerated by increasing temp. Epoxidised polymer (using in situ peroxyformic acid as oxidant) displays increased resistance to ASTM oils nos. 1, 2 and 3 and to ASTM fuel A. Resistance to ASTM fuels B and C is poor [2]. Epoxidised copolymer (using in situ peroxyformic acid as oxidant) compound woth carbon black, zinc oxide and stearic acid exhibits low swelling in vinegar, brake fluid, ASTM oils nos. 1, 2, MeOH and mineral oil but relatively high swelling in 2-butanone and dibutyl phthalate. Resistance to swelling by ASTM fuel A, vegetable oil and ASTM oil is improved relative to parent copolymer [2]

Applications/Commercial Products:
Processing & Manufacturing Routes: Epoxidation can be carried out by using either preformed peroxy acids or by in situ methods using H_2O_2 and lower aliphatic acids. The most efficient conversion of double bonds to oxirane groups, with fewer unwanted side reactions is achieved using preformed peroxy acids including peracetic acid [3,4,5], perbenzoic acid [6] and o-monoperoxyphthalic acid [1,7]. The use of peroxyformic acid (produced in situ from formic acid and H_2O_2) has also been reported [2,3,4]. Mixed formic acid/AcOH/peroxide systems also used for epoxidation of high MW (above 60000) block copolymers (in situ method) [2]
Applications: Potential applications in electronic industry, e.g. as components in circuit board fabrications

Bibliographic References

[1] *Macromolecules*, 1991, **24**, No. 14, 4010-4016
[2] *J. Polym. Sci., Part A: Polym. Chem.*, 1991, **29**, 1183
[3] *J. Polym. Sci., Part A: Polym. Chem.*, 1990, **28**, 3761
[4] *J. Polym. Sci., Part A: Polym. Chem.*, 1990, **28**, 1867
[5] Huang, W.-K., Hsiue, G.-H. and Hou, W.-H., *J. Polym. Sci., Polym. Chem. Ed.*, 1988, **26**, 1867, (synth., transport props., H-1 nmr, chemical stability, morphology, thermal props, electrical props., dynamic mechanical props)
[6] Udipi, K., *J. Appl. Polym. Sci.*, 1979, **23**, 3301, (synth., mechanical props., chemical stability)
[7] *U.S. Pat.*, 1958, 2 829 130, (synth., modification, mechanical props.)
[8] Poluteltou, P.T., Gonsouskaya, T.B., Ponomarev, F.G. and Gusev, Yu.K., *Vysokomol. Soedin., Ser. A*, 1973, **15**, 606, (synth.)
[9] Bukanova, E.F., Tutorskii, I.A. and Boikacheva, E.G., *Vysokomol. Soedin., Ser. A*, 1976, **18**, 2223, (synth., transport props., chemical stability, ordering)
[10] Huang, W.-K., Hsiue, G.-H. and Hou, W.-H., *J. Chin. Inst. Chem. Eng.*, 1987, **18**, 289, (synth., transport props., chemical stability, ordering)
[11] Jay, R.R., *Anal. Chem.*, 1964, **36**, 667, (anal.)
[12] *U.S. Pat.*, 1971, 3555112, (synth, modification, mech. props)

Epoxy Resins

Synonyms: *Epoxy Resins, General Information*
Material class: Thermosetting resin
Polymer Type: epoxy
General Information: Epoxy resins first appeared in the 1940s, having been developed for moulding and coating applications. Since then these materials have been widely developed, though coatings remain the major application. The structure of the resins is based around the chemistry of the epoxy group, with the 'cured' resin consisting of a pre-polymer which is cross-linked with a curing agent. The pre-polymers are usually molecules with 2 epoxy groups spaced with an aromatic or aliphatic backbone. The first reported, and still most widely used resin is that formed when epichlorohydrin is treated with Bisphenol A under basic conditions to give a diglycidyl ether. The reaction may continue with several additions of Bisphenol A units opening the epoxy rings, to give short chains of up to 20 repeat units , terminated with epoxy groups. For commercial resins, the number of repeat units is often only slightly greater than 0, e.g. EPON 828, n=0.13 Other phenol derivatives are commonly used to form the aromatic backbone, such as Bisphenol F, resorcinol, brominated bisphenols, and more highly functionalised molecules such as tetrakisphenylolethane.

Other alcohols, amines and carboxylic acids may also be combined with epichlorohydrin to give a range of diglycidyl ether based epoxy resins. Non aromatic, commercially available epoxies are produced by peracid epoxidation of alkenes and dienes, such as vinyl cyclohexene and esters of cyclohexane carboxylic acids While the composition of the prepolymer and number of repeat units have an effect on properties, the prepolymers do not in themselves have useful properties until they are cured. Curing occurs when the prepolymer is treated with a suitable multi-functional cross linking agent such as a polyamine. These react with the epoxy end-groups of several short chains, linking them together, ultimately to form a 3 dimensional network of polymer chains. The curing process is complicated, beginning in most cases with a liquid resin and hardener, passing through a gel phase and continuing until virtually all the initial components have been combined in the solid cross linked network. The properties of the cured resin depend on a number of factors; the structure of the prepolymer, the curing agent, stoichiometry of resin and hardener, cure time and cure temperature. Subtle variation of any of these factors can have a profound effect on the props. of the cured resin. In some cases prepolymer resins have high viscosities which are detrimental for certain applications and processing techniques. Certain diluents may be added, such as hydrocarbon solvents, to reduce viscosity, but these can adversely affect props of the cured resin as they are not incorporated into the polymer network during cure. Reactive diluents are more often used, which are mainly low MW mono- or difunctional epoxy compounds. Both of these types of diluent are incorporated into the polymer network, though monofunctional diluents effectively terminate the chain, and so reduce cross linking and thus physical and mechanical props. Difunctional epoxies, which may be cured as epoxy resins in their own right, maybe completely incorporated into the polymer network and offer other advantages over non-reactive diluents and props are little affected, especially at elevated temp where other diluents often fail, and may even cause enhancement of certain props. In general, not withstanding the dependence of the props of cured epoxy resins on type and amount of hardener, cure time and temp., epoxy resins display high strength and modulus, wear resistance and cracking resistance. Electrical properties are also good, with high resistivity, good dielectric strength and track resistance especially at high humidity. High temp. stability and ageing props are good. Epoxy resins may be processed by a range of thermosetting processing techniques, with low cure shrinkage and low evolution of volatiles on cure. Cure, however, may be highly exothermic especially on a large scale. Many of these props. may be enhanced using suitable fillers, e.g. metals and carbon black to improve conductivity, minerals, glass and fire retardants to modify thermal and electrical props. Epoxy resins, in general have good resistance to chemicals and good dimensional stability; however aromatic based resins are susceptible to degradation in UV light and other weathering conditions and so cycloaliphatic resins are preferred for outdoor applications. The reactions to form the epoxy prepolymer chains and in the curing process give rise to highly hydroxyl functionalised polymers which display excellent adhesion to all but the most non-polar surfaces, thus the widespread usage as coating materials and adhesives and in composites.

Applications/Commercial Products:

Processing & Manufacturing Routes: Epoxy resins are produced by mixing the prepolymer with a defined quantity of curing agent and then allowing reaction to take place between epoxy and curing agent for a set time at a set temp. Despite a relatively short pot life, the mixtures can be injection moulded (easily handleable moulding powders are available) and cast. No volatiles are produced on cure, so curing can be carried out at normal pressures, though the process is exothermic, especially on a large scale. Cure shrinkage is also generally low, especially for filled grades. Laminates are produced by drawing the material through a solution of epoxy and curing agent, cure begins as the solvent begins to evaporate. Solid powder coatings consisting of a mixture of resin and hardener are applied by spraying directly onto a heated surface, rapid cure begins as the solids melt. Coatings are also applied by spraying and dipping.

Applications: Major use for Epoxy resins is as a surface coating, due to high chemical resistance, mechanical strength and adhesion. Variation of the prepolymer and cure agent gives a huge range of coatings which may be applied as liquids or solids. Used in household appliances such as washing machines, furniture, tools, pipes, food cans, tubing, as marine coatings, in chemical plant pipelines, automotive parts. Coatings are also applied to many surfaces for excellent wear and chemical resistance, such as laboratory benches, bridge and road surfaces. Electronic components, resistors, transistors, capacitors etc. are encapsulated with epoxy coatings, for insulation and protection, also used for fibre optics and printed circuit boards. Moulded resins can be used for pump housings and impellers, pipe fittings, valves, tools and tooling patterns and fixtures. Cast products include tools, insulation, electrical components, switchgear, work surfaces . Used as an adhesive capable of bonding most surfaces, effective at high temp. Used in aircraft for bonding wing and fuselage parts, stiffeners for rudders, helicopter rotors. Also used extensively in car parts, and also for repair to cracks in bridges, dams as well as in domestic usage. Laminates and other composite materials are used for printed circuit boards, in electronic equipment, tubes and panels in aircraft.

Bibliographic References

[1] Lee, H. and Neville, K., *Handbook of Epoxy Resins*, McGraw Hill, 1982, (manuf, processing)
[2] May, C.A. and Tanaka, Y., Epoxy Resins: Chemistry and Technology Marcel Dekker, 1987
[3] Bhatnagar, M.S., Epoxy Resins Universal Book, 1996
[4] *Encycl. Polym. Sci. Eng.*, John Wiley, 1986, **6**, 322, (rev)
[5] Ullmanns Encykl. Ind. Chem. VCH, 1987, **A9**
[6] *Kirk-Othmer Encycl. Chem. Technol.*, (eds. J.I. Kroschwitz and M. Howe-Grant), 4th edn., Wiley, vol.9, 730, (rev)
[7] *Encyclopedia of Chemical Process and Design*, Marcel Dekker, 1983, **19**
[8] Bhatnagar, M.S., *Polym.-Plast. Technol. Eng.*, 1993, **32**, 53
[9] Petrosyan, V.A. *et al*, *Arin. Khim. Zh.*, 1989, **42**, 54
[10] Polyakov, V.A. *et al*, *Mekh. Kompoz. Mater. (Zinatne)*, 1989, **2**, 218
[11] Kharakha, Zh.G. *et al*, *Polimery (Warsaw)*, 1989, **34**, 204
[12] Czvikovsky, T., *Kem. Kozl.*, 1989, **6**, 4
[13] Singh, J., *Paintindia*, 1989, **4**, 1
[14] Bhatnagar, M.S., *Pop. Plast.*, 1987, **32**, 20, 29
[15] Marris, J.J. and Tremin, S.C., *J. Appl. Polym. Sci.*, 1966, **10**, 523, (manuf, processing)
[16] Morgan, R.J. and Mones, E.T., *J. Appl. Polym. Sci.*, 1987, **33**, 999, (manuf, processing)
[17] Arnold, R.J., *Mod. Plast.*, 1964, **41**, 149, (manuf, processing)

2-(3,4-Epoxycyclohexyl-5,5-spiro-3,4-epoxy)cyclohexane metadioxide resin   E-10

Synonyms: *Poly[2-(7-oxabicyclo[4.1.0]hept-3-yl)spiro[1,3-dioxane-5,3'-[7]oxabicyclo[4.1.0]heptane]]*
Monomers: 2-(3,4-Epoxycyclohexyl-5,5-spiro-3,4-epoxy)cyclohexane metadioxide
Material class: Thermosetting resin
Polymer Type: Epoxy
CAS Number:

CAS Reg. No.	Note
26616-47-7	homopolymer

Molecular Formula: $(C_{14}H_{20}O_4)_n$
Fragments: $C_{14}H_{20}O_4$
General Information: Props. of cured resin depend on curing agent, stoichiometry, cure time and cure temp.

Volumetric & Calorimetric Properties:
Density:

No.	Value	Note
1	d^{25} 1.18 g/cm^3	ERL-4234, uncured [8]

Transition Temperature General: Bp 250° (min., ERL-4234, uncured) [8]

Transport Properties:
Polymer Solutions Dilute: Viscosity 7000–17000 cP (38°, ERL-4234, uncured) [8]
Water Content: Water sorption can be modified by altering the amount of cross-linking agent and/or modifier added to the resin [3]

Electrical Properties:
Electrical Properties General: Volume resistivity of an aromatic triamine cured sample decreases upon heating or exposure to 95% relative humidity [1]. Decreasing the amount of isomethyltetrahydrophthalic anhydride curing agent in UP 612 resin decreases the dielectric loss tangent and the dielectric constant [5]
Dielectric Permittivity (Constant):

No.	Value	Note
1	3.36	UP 612, aromatic triamine cured, aged, 230°, 100h [1]

Optical Properties:
Volume Properties/Surface Properties: Colour 2 (max., 1933 Gardner, ERL-4234, uncured) [8]

Polymer Stability:
Polymer Stability General: Improved heat resistance may be obtained by curing with Leucoparafuchsine (an aromatic amine) [4]
Thermal Stability General: UP 612 cured with aromatic triamine is stable up to 320° [1]. Details of thermal ageing in air of cross-linked resin have been reported [6]
Decomposition Details: The kinetics and mechanism of thermal decomposition have been reported [7]
Environmental Stress: Uv irradiation causes cross-linking of free epoxy groups, oxidation and polymerisation [2]

Applications/Commercial Products:

Trade name	Details	Supplier
Araldite	CY 175	Ciba-Geigy Corp.
ERL	4234	Union Carbide

Bibliographic References
[1] Podzolko, Yu. G., Esikov, Yu. G., Grigorovich, I.V. and Solov'eva, E.V., *Tr. Mosk. Energy Inst.*, 1982, **586**, 23, (electrophysical props.)
[2] Leyrer, R., *Muanyag Gumi*,(uv irradiation)
[3] Anufrieva, G.P., Trizno, M.S., Limasov, A.I. and Nikolaev, A.F., *Plast. Massy*, 1979, 23, (water sorption)
[4] Läpitskii, V.A., Grigorovich, I.V. and Kritsuk, A.A., *Mekh. Kompoz. Mater. (Zinatne)*, 1981, **552**
[5] Gladehenko, V., Ya. and Mikhailus, Yu, V., *Vestn. Khark Politekh. Inst.*, 1974, **87**, 67, (electrical props.)
[6] Pet'ko, I.P. and Batog, A.E., *Plast. Massy*, 1976, 39, (thermal ageing)
[7] Gnitsevich, S.E., Lazareva, S.Ya., Mal'kov, Yu. E. and Vilesova, M.S., *Zh. Prikl. Khim. (Leningrad)*, 1991, **64**, 403, (thermal degradation)
[8] ERL-4234, Union Carbide, (technical datasheet)

3',4'-Epoxycyclohexylmethyl-3,4-epoxycyclohexane carboxylate resin   E-11

Synonyms: *Poly(7-oxabicyclo[4.1.0]hept-3-ylmethyl 7-oxabicyclo[4.1.0] heptane-3-carboxylate)*
Related Polymers: Epoxy resins
Monomers: 3'-Cyclohexenemethyl-3-cyclohexenecarboxylate, Peracetic acid
Material class: Thermosetting resin
Polymer Type: Epoxy

CAS Number:

CAS Reg. No.	Note
34010-59-8	copolymer with isophthalic acid
25085-98-7	
120468-26-0	copolymer with tetrahydro-5-methylisophthalic anhydride

Molecular Formula: $C_{14}H_{20}O_4$
Fragments: $C_{14}H_{20}O_4$
Miscellaneous: Props. are dependent upon the type of curing agent and/or method of cure used

Volumetric & Calorimetric Properties:
Density:

No.	Value	Note
1	d 1.23 g/cm^3	tetrahydro-5-methylisophthalic anhydride cured [16]

Latent Heat Crystallization: Enthalpy of cure -149 J g^{-1} (cured with methyl tetrahydroisophthalic anhydride) [11], 13.1 J g^{-1} (uv cured, triphenylsulfonium hexafluoroantimonate initiator [8], 27.5 J g^{-1} (heat cured, triphenylsulfonium hexafluoroantimonate initiator [8]), 34 J g^{-1} (heat cured, aliphatic sulfonium initiator [8]
Glass-Transition Temperature:

No.	Value	Note
1	200°C	approx., uv-cured, ceramic filled [7]
2	172°C	methyl tetrahydroisophthalic anhydride cured [11]
3	128°C	tetrahydro-5-methylisophthalic anhydride cured [16]

Transition Temperature General: T_g may be reduced by the presence of photoinitiators in the cured polymer, which have a plasticisation effect [8]
Deflection Temperature:

No.	Value	Note
1	150°C	hexahydrophthalic anhydride cured [14]
2	49°C	benzylsulfonium tetrafluoroborate cured, ASTM D648 [9]
3	85°C	benzylsulfonium hexafluorophosphate cured, ASTM D648 [9]
4	171°C	benzylsulfonium hexafluoroarsenate cured, ASTM D648 [9]
5	172°C	benzylsulfonium hexafluoroantimonate cured, ASTM D648 [9]

Vicat Softening Point:

No.	Value	Note
1	85°C	copolymer with isophthalic acid [6]

Transition Temperature:

No.	Value	Note	Type
1	-80°C	approx., uv cured, triphenylsulfonium hexafluoroantimonate initiator [8]	T_β

Transport Properties:
Polymer Solutions Dilute: Viscosity 350–450 cP (25°, ERL 4221)

Water Absorption:

No.	Value	Note
1	2.29 %	5h, 100°, benzyltetramethylene sulfonium tetrafluoroborate, cured [9]
2	1.97 %	5h, 100°, benzyltetramethylene sulfonium hexafluorophosphate cured [9]
3	2.22 %	5h, 100°, benzyltetramethylene sulfonium hexafluoroarsenate cured [9]
4	1.31 %	5h, 100°, benzyltetramethylene sulfonium hexafluoroantimonate cured [9]
5	0.4 %	1h, 100°, hexahydrophthalic anhydride cured [14]

Mechanical Properties:

Mechanical Properties General: The dependence of bending strength and impact strength on struct. of polymer cured with tetrahydromethylisobenzofurandione has been reported [2]. The effect of method of synth. on tensile strength of tetrahydromethylisobenzofurandione cured material has been reported [3]. Elastic modulus of ceramic filled material shows a gradual decrease with temp. [7]. Storage modulus shows a gradual decrease with increasing temp. until 100°, when the decrease is more pronounced. Dynamic mechanical props. are affected by the method of cure [8]

Tensile (Young's) Modulus:

No.	Value	Note
1	3300 MPa	25°, hexahydrophthalic anhydride cured [14]

Flexural Modulus:

No.	Value	Note
1	3000 MPa	25°, hexahydrophthalic anhydride cured [14]

Tensile Strength Break:

No.	Value	Note
1	68 MPa	25°, hexahydrophthalic anhydride cured [14]

Flexural Strength at Break:

No.	Value	Note
1	89 MPa	25°, hexahydrophthalic anhydride cured [14]

Compressive Strength:

No.	Value	Note
1	151 MPa	25°, hexahydrophthalic anhydride cured [14]
2	163 MPa	tetrahydro-5-methylisophthalic anhydride cured [16]

Impact Strength: Falling-ball impact resistance 5 cm (1.8 kg ball, uv and heat cured, triphenylsulfonium hexafluoroantimonate initiator, ASTM D3451), 40 cm (1.8 kg ball, heat cured, aliphatic sulfonium initiator, ASTM D3451) [8]

Electrical Properties:

Electrical Properties General: Electrical props. may be improved by curing with Al-alkoxysilane complex [5]

Dielectric Permittivity (Constant):

No.	Value	Frequency	Note
1	9.33	10 MHz	23°, ERL 4221 [15]
2	6.23	100 MHz	23°, ERL 4221 [15]
3	4.72	200 MHz	23°, ERL 4221 [15]
4	3.3	60 Hz	hexahydrophthalic anhydride cured [14]

Dissipation (Power) Factor:

No.	Value	Frequency	Note
1	0.005	60 Hz	hexahydrophthalic anhydride cured [14]

Optical Properties:

Transmission and Spectra: C-13 nmr [1,8] and H-1 nmr [1] spectral data have been reported

Polymer Stability:

Chemical Stability: Details of chemical stability have been reported [4]. Polymer cured with tetrafluoroborate or hexafluorophosphate salts of benzyltetramethylene sulfonium is decomposed by 2-butanone whereas polymer cured with hexafluoroarsenate or antimonate salts is merely swollen [9]
Hydrolytic Stability: Details of irradiation at low temps. have been reported [12]

Applications/Commercial Products:

Processing & Manufacturing Routes: May be cured [5] with uv [7] or by electron beam irradiation in the presence of inorganic salts [10]. May be synth. by cationic polymerisation using silver hexafluoroantimonate [13] or benzyl sulfonium salts. Depending on the nature of the counterion, improvements in shelf-life and gel time can be achieved [9]. Synth. comly by epoxidation of 3'-cyclohexenylmethyl-3-cyclohexenecarboxylate with 25.5% peracetic acid in Me_2CO for 2h at room temp. The mixture is cooled to -11° for 16h, followed by addition to ethylbenzene at 40–45° to allow for removal of by-products. Product yield is typically 86% [14]
Applications: Used in the manufacture of electrical insulators; in adhesives and coatings; in printing inks. Potential application as an encapsulant for electroluminescent devices. Used as a general purpose casting resin, as a plasticiser acid scavenger and in filament winding

Trade name	Details	Supplier
CY-179		Ciba-Geigy Corp.
Cyracure	UVR 6110	Union Carbide
ERL	4221	Union Carbide

Bibliographic References

[1] Udagawa, A., Yamamoto, Y. and Chujo, R., *Polymer*, 1990, **31**, 2425, (nmr)
[2] Pet'ko, I.P., Batog, A.E. and Zaitsev, Yu. S., *Kompoz. Polim. Mater.*, 1987, **34**, 10
[3] Voloskov, G.A., Morozov, V.N., Lipskaya, V.A. and Kovriga, V.V, *Mekh. Kompoz. Mater. (Zinatne)*, 1987, 585
[5] Hayase, S., Onishi, Y., Suzuki, S., Kurokawa, T., and Wada, M., *Kobunshi Ronbunshu*, 1984, **41**, 581, (curing agent)
[6] *Ger. Pat.*, 1971, 2 117 563
[7] Maruno, T. and Murata, N., *J. Adhes. Sci. Technol.*, 1995, **9**, 1343
[8] Udagawa, A., Yamamoto, Y., Inoue, Y. and Chujo, R., *Polymer*, 1991, **32**, 2779, 2947, (C-13 nmr dynamic mech. props.)
[9] Morio, K., Murase, H. and Tsuchiya, H., *J. Appl. Polym. Sci.*, 1986, **32**, 5727, (synth., props.)
[10] Davidson, R.S. and Wilkinson, S.A., *J. Photochem. Photobiol. A*, 1991, **58**, 123
[11] Kretzschmar, K. and Hoffmann, K.W., *Thermochim. Acta*, 1985, **94**, 105, (cure enthalpy)
[12] Bonjour, E., Brauns, P., Lagnier. and Van de Voorde, M., *CERN (Rep.)*, 1977, **77-30**, 51, (irradiation)

[13] Yagci, Y., Fischer, C.-H., and Schnabel, W., *Makromol. Chem., Rapid Commun.*, 1989, **10**, 137, (synth.)
[14] McAdams, L.V., and Gannon, J.A., *Encycl. Polym. Sci. Eng.*, 2nd edn., Vol.6 (ed J.I.Kroschwitz), 1985, 322, (rev.)
[15] Snell, J.B. and Olyphant, M., *IEEE Trans. Electr. Insul.*, 1975, **EI10**, 54, (dielectric constant)
[16] Frier, M., *Actes Colloq.-IFREMER*, 1988, **7**, 357, (props.)

ETFE

$$\left[\left[CF_2CF_2 \right]_x \left[CH_2CH_2 \right]_y \right]_n$$

Synonyms: *Tetrafluoroethylene-ethylene copolymer. ETFE copolymer. Poly(tetrafluoroethylene-co-ethylene). Poly(ethylene-co-tetrafluoroethylene)*
Related Polymers: Polytetrafluoroethylene, Poly(tetrafluoroethylene-*co*-ethylene), Glass fibre filled, Polyethylene
Monomers: Tetrafluoroethylene, Ethylene
Material class: Thermoplastic, Copolymers
Polymer Type: PTFE, ETFE copolymers
CAS Number:

CAS Reg. No.	Note
25038-71-5	
111939-51-6	alternating
132201-85-5	block

Molecular Formula: $[(C_2F_4).(C_2H_4).(C_4H_4F_4)]_n$
Fragments: C_2F_4 C_2H_4 $C_4H_4F_4$
Additives: Glass fibre, thermally stable pigments, stabiliser and antioxidants
General Information: The two monomers combine readily into a 1:1 alternating struct. The copolymers are sometimes modified with a third monomer such as perfluoroalkylvinyl or vinylidene compds., perfluoroalkyl ethylenes and perfluoroalkoxy vinyl compds. to improve high temp. props. Commercial material is usually 50–60% crystalline

Volumetric & Calorimetric Properties:
Density:

No.	Value	Note
1	d 1.7 g/cm^3	[3]

Thermal Expansion Coefficient:

No.	Value	Note	Type
1	0.00055 K^{-1}	ASTM D864-52 [7]	V
2	4.2×10^{-5} K^{-1}	[3]	L
3	7.9×10^{-5}–0.000141 K^{-1}	[6]	L

Thermodynamic Properties General: Heat of combustion 3507.2 kJ mol^{-1} (13.7 MJ kg^{-1}) [6]
Thermal Conductivity:

No.	Value	Note
1	0.238 W/mK	[1]

Specific Heat Capacity:

No.	Value	Note	Type
1	1.05 kJ/kg.C	[8]	P

Melting Temperature:

No.	Value	Note
1	275°C	ASTM D2117-64 [7]

Glass-Transition Temperature:

No.	Value	Note
1	110°C	cubic thermal expansion [7]

Deflection Temperature:

No.	Value	Note
1	150–180°C	0.45 MPa, ASTM D648 [3]
2	70°C	1.8 MPa, ASTM D648 [3]

Brittleness Temperature:

No.	Value	Note
1	-80°C	ASTM D746-64T [7]
2	-100°C	ASTM D746 [1]

Transition Temperature:

No.	Value	Note	Type
1	110–135°C	[1,9]	α relaxation
2	-25°C	[1,9]	β relaxation
3	-120°C	[1,9]	γ relaxation

Surface Properties & Solubility:
Solvents/Non-solvents: Insol. most organic and inorganic solvents, including acids and bases

Transport Properties:
Transport Properties General: Permeability characteristics, with respect to oxygen, nitrogen, carbon dioxide and moisture, are similar to those of high-density polyethylene
Melt Flow Index:

No.	Value	Note
1	7 g/10 min	ASTM D1238-15T [8]
2	8–45 g/10 min	ASTM D1238-15T [1]

Polymer Melts: Melt viscosity 7000-100000 Poise (700-10000 Pa s, ASTM D3159, capillary rheometer) [6]. Critical shear rate 12000 s^{-1} (Tefzel 210), 4800 s^{-1} (Tefzel 200), 4500 s^{-1} (Tefzel 280) [1]
Water Absorption:

No.	Value	Note
1	<0.02 %	max., ASTM D570 [2]
2	0.03 %	[3]

Gas Permeability:

No.	Gas	Value	Note
1	O$_2$	7000 cm^3 mm/(m^2 day atm)	
2	N$_2$	2000 cm^3 mm/(m^2 day atm)	
3	CO$_2$	34000 cm^3 mm/(m^2 day atm)	flat film, 20μm thick, ASTM D1434-66 [7]

– ETFE

Mechanical Properties:

Mechanical Properties General: Excellent mech. props., at both low and high temps. ETFE is tougher, stiffer and has a higher tensile strength and creep resistance than PTFE. Flexural strength no break (ASTM D790) [10]

Tensile (Young's) Modulus:

No.	Value	Note
1	830 MPa	23° [6]

Flexural Modulus:

No.	Value	Note
1	1380 MPa	[3]

Tensile Strength Break:

No.	Value	Note	Elongation
1	49 MPa	500 kg cm^{-2}, 23°, ASTM D1708-59T [7]	100–300%

Compressive Modulus:

No.	Value	Note
1	860 MPa	[6]

Elastic Modulus:

No.	Value	Note
1	1370 MPa	14000 kg cm^{-2}, 23°, ASTM D638-67T [7]
2	1650 MPa	Tensile elastic modulus [3]

Compressive Strength:

No.	Value	Note
1	17 MPa	ASTM D695 [8]
2	48.9 MPa	ASTM D695 [1]

Mechanical Properties Miscellaneous: Creep 0.6% (30 kg cm^{-2}, 100h), 1.75% (60 kg cm^{-2}, 100h), 4.25% (90 kg cm^{-2}, 100h), (ASTM D674-56T, 23°) [7]
Hardness: Shore R50, Shore D75 [3]. Shore D68 (ASTM D676-49T) [7]
Friction Abrasion and Resistance: Abrasion resistance 10 mg (1000 cycles)$^{-1}$ (1000 g load, Taber abraser, CS-17) [7]. Static coefficient of friction 0.2 (1 mm thick, polished aluminium sheet) [7]. Abrasion resistance 4 (sand, ASTM D968) [6]. Dynamic coefficient of friction 0.4 (689 kPa, >3 m min^{-1}, mating material AISI 1018 steel, Rc 20, 16AA) [1]

Electrical Properties:

Electrical Properties General: Has excellent electrical props. Dielectric constant and loss factor are low and are practically independent of temp. and frequency [7]
Surface/Volume Resistance:

No.	Value	Note	Type
2	0.5 × 10^{15} Ω.cm	ASTM D257 [1]	S
3	>0.1 × 10^{15} Ω.cm	min., [6]	S

Dielectric Permittivity (Constant):

No.	Value	Frequency	Note
1	240	50 Hz	23° [7]

Dielectric Strength:

No.	Value	Note
1	25.9 kV.mm^{-1}	1.7 mm thick, short time, ASTM D149 [7]

Arc Resistance:

No.	Value	Note
1	72s	ASTM D495 [3]

Dissipation (Power) Factor:

No.	Value	Frequency	Note
1	0.0002	50 Hz	23° [7]

Optical Properties:
Refractive Index:

No.	Value	Note
1	1.4	[3]

Volume Properties/Surface Properties: Clarity: transparent in thin sections [3]

Polymer Stability:
Polymer Stability General: Good thermal stability and resistance to weathering
Thermal Stability General: Ageing at 180° in an air circulating oven has no effect on physical props. after 1500h [7]
Upper Use Temperature:

No.	Value	Note
1	148–198°C	[3]
2	>200°C	min. [1]

Decomposition Details: Weight loss in air 0.3% (330°) [7]
Flammability: Burning rate self extinguishing (ASTM D635-63) [7]. Oxygen index 30–32% (ASTM D2863) [8]. Flammability rating V0 (UL 94) [1]. Average time of burning 5 s (max.), average length of burn 10 mm (ASTM D635, bar thickness 2.9 mm) [1]
Environmental Stress: Shows no change in tensile props. after 2000h simulated weathering [7]
Chemical Stability: Excellent chemical resistance with no significant changes in weight or physical props. [7] after exposure to a range of organic and inorganic solvents. Sensitive to attack from amines [6]
Hydrolytic Stability: Stable to boiling water over long periods [1]

Applications/Commercial Products:
Processing & Manufacturing Routes: Copolymerised in aq. and non-aq. media with free radical initiators using emulsion and suspension techniques. Processed by injection moulding, compression moulding, blow moulding, transfer moulding, rotational moulding, extrusion and coating. Films can be thermoformed and heat sealed. Articles can be formed below the melting temp. using conventional metal-forming techniques. Articles can be readily machined

– ETFE, Carbon fibre filled   E-13 – E-13

Mould Shrinkage (%):

No.	Value	Note
1	3–4%	[3]

Applications: Cable coating, wiring for chemical plants, sockets, connectors, switch components, seals, corrugated tubing, fasteners, pump vanes, linings, coatings, valve components and injection moulded articles

Trade name	Details	Supplier
Aflon COP		Asahi Glass
Halon ET		Ausimont
Hostaflon ET		Hoechst Celanese
Nesflon EP		Daikin Kogyo
Stat-Kon	FP-E	LNP Engineering
Tefzel		DuPont
Tefzel	200 series	DuPont
Thermocomp	FP series	LNP Engineering

Bibliographic References

[1] *Encycl. Polym. Sci. Eng.*, 2nd edn., (ed. J.I. Kroschwitz), Wiley Interscience, 1989, **16**
[2] Brydson, J.A., *Plast. Mater.*, 5th edn., Butterworth Heinemann Ltd., 1989
[3] *The Materials Selector*, 2nd edn., (eds. N.A. Waterman and M.F. Ashby), Chapman and Hall, 1997, **3**
[4] *Encyclopedia of Advanced Materials*, (eds. D. Bloor, R.J. Brook, M.C. Flemings, S. Mahajan and R.W. Cahn), Pergamon Press, 1994, **3**
[5] *Kirk-Othmer Encycl. Chem. Technol.*, Vol 11, 4th edn., (ed. J.I. Kroschwitz), Wiley Interscience, 1980
[6] Kerbow, D.L. and Sperati, C.A., *Polym. Handb.*, 4th edn., (eds. J. Brandrup, E.H. Immergut and E.A. Grulke), John Wiley and Sons, 1999, **V**, 31
[7] Garbuglio, C., Modena, M., Valera, M. and Ragazzini, M., *Eur. Polym. J.*, 1974, **10**, 91
[8] *Tefzel 200*, DuPont, (technical datasheet)
[9] Starkweather, H.W., *J. Polym. Sci., Polym. Phys. Ed.*, 1973, **11**, 587
[10] Basics of Design Engineering: Materials: Plastics, *Mach. Des.*, 1995, **67**, 79

ETFE, Carbon fibre filled   E-13

Synonyms: Poly(ethylene-co-tetrafluoroethylene), Carbon fibre filled. Ethylene-tetrafluoroethylene copolymer, Carbon fibre filled
Related Polymers: ETFE, Glass fibre filled
Monomers: Ethylene, Tetrafluoroethylene
Material class: Thermoplastic, Copolymers, Composites
Polymer Type: polyethylene, ETFE copolymersfluorocarbon (polymers)
General Information: 10% Filled material combines good heat and chemical resistance with low wear [2]

Volumetric & Calorimetric Properties:
Density:

No.	Value	Note
1	d 1.7 g/cm^3	20% filled [1,2]
2	d 1.75 g/cm^3	10% filled [2]
3	d 1.72 g/cm^3	20% filled [2]
4	d 1.73 g/cm^3	30% filled [2,3]

Thermal Expansion Coefficient:

No.	Value	Note	Type
1	5.04×10^{-5} K^{-1}	10% filled [2]	L
2	3.96×10^{-5} K^{-1}	15% filled [2]	L
3	1.72×10^{-5} K^{-1}	20% filled [2]	L
4	1.45×10^{-5} K^{-1}	30% filled [3]	L

Thermal Conductivity:

No.	Value	Note
1	0.87 W/mK	0.00207 cal (cm s°C)$^{-1}$, 20% filled [2]

Melting Temperature:

No.	Value	Note
1	270°C	30% filled [2]

Deflection Temperature:

No.	Value	Note
1	213°C	1.82 MPa, 30% filled [2]
2	191°C	1.82 MPa, 20% filled [1]
3	198°C	1.82 MPa, 10% filled [2]
4	241°C	1.8 MPa, 30% filled [3]
5	218°C	1.82 MPa, 15% filled [2]
6	223°C	1.82 MPa, 20% filled [2]
7	213°C	1.82 MPa, 30% filled [2]

Transport Properties:
Water Absorption:

No.	Value	Note
1	0.02 %	24h, 10–20% filled [1,2]

Mechanical Properties:
Tensile (Young's) Modulus:

No.	Value	Note
1	7855 MPa	80100 kg cm^{-2}, 20% filled [1]

Flexural Modulus:

No.	Value	Note
1	8885 MPa	90600 kg cm^{-2}, 20% filled [1]
2	3442 MPa	35100 kg cm^{-2}, 10% filled [2]
3	6894 MPa	70300 kg cm^{-2}, 15% filled [2]
4	7924 MPa	80800 kg cm^{-2}, 20% filled [2]
5	11278 MPa	115000 kg cm^{-2}, 30% filled [2]
6	11400 MPa	30% filled [3]

Tensile Strength Break:

No.	Value	Note	Elongation
1	60.6 MPa	618 kg cm^{-2}, 20% filled [1]	1%
2	58.5 MPa	597 kg cm^{-2}, 10% filled [2]	10%

– ETFE, Glass fibre filled

3	75.8 MPa	773 kg cm^{-2}, 15% filled [2]	
4	91.7 MPa	935 kg cm^{-2}, 20% filled [2]	
5	103 MPa	1050 kg cm^{-2}, 30% filled [2,3]	5%

Flexural Strength Yield:

No.	Value	Note
1	86 MPa	878 kg cm^{-2}, 10–20% filled [1,2]
2	99 MPa	1010 kg cm^{-2}, 20% filled [2]
3	137 MPa	1400 kg cm^{-2}, 30% filled [2]

Compressive Strength:

No.	Value	Note
1	89.5 MPa	913 kg cm^{-2}, 20% filled [1]
2	55 MPa	562 kg cm^{-2}, 10% filled [2]

Izod Notch:

No.	Value	Notch	Note
1	96 J/m	Y	1.8 ft lb in^{-1}, room temp., 20% filled [1]
2	320 J/m	Y	6 ft lb in^{-1}, room temp., 10% filled [2]
3	165 J/m	Y	3.1 ft lb in^{-1}, room temp., 15% filled [2]
4	208 J/m	Y	3.9 ft lb in^{-1}, room temp., 20% filled [2]
5	240 J/m	Y	4.5 ft lb in^{-1}, room temp., 30% filled [2,3]

Applications/Commercial Products:
Processing & Manufacturing Routes: Processed by injection moulding
Mould Shrinkage (%):

No.	Value	Note
1	0.385%	20% filled [1]
2	0.55%	10% filled [2]
3	0.5%	20% filled [2]
4	0.35%	30% filled [2]
5	0.15–0.25%	30% filled [3]

Trade name	Details	Supplier
Electrafil J-1400/CF/20	20% PAN carbon fibre filled; discontinued material	DSM Engineering Plastics
Fluoromelt FP-EC-1004		LNP Engineering
Fluoromelt FP-EF-1006		LNP Engineering
Thermocomp FP-EC-1002	10% filled	LNP Engineering
Thermocomp FP-EC-1003	15% filled	LNP Engineering
Thermocomp FP-EC-1004	20% PAN carbon fibre filled	LNP Engineering
Thermocomp FP-EC-1006	30% filled	LNP Engineering

Bibliographic References
[1] 1996, DSM Engineering Plastics, (technical datasheet)
[2] *Thermocomp Range*, LNP Engineering, 1996
[3] *The Materials Selector*, 2nd edn., (eds N.A. Waterman and M.F. Ashby), Chapman and Hall, 1997, **3**, 233

ETFE, Glass fibre filled

Synonyms: *Poly(ethylene-co-tetrafluoroethylene), Glass fibre filled. Ethylene-tetrafluoroethylene copolymer, Glass fibre filled*
Related Polymers: ETFE, Carbon fibre filled
Monomers: Ethylene, Tetrafluoroethylene
Material class: Thermoplastic, Copolymers, Composites
Polymer Type: polyethylene, ETFE copolymersfluorocarbon (polymers)

Volumetric & Calorimetric Properties:
Density:

No.	Value	Note
1	d 1.75 g/cm^3	10% filled [1]
2	d 1.79 g/cm^3	15% filled [1]
3	d 1.82 g/cm^3	20% filled [1]
4	d 1.89 g/cm^3	30% filled [1,2]

Thermal Expansion Coefficient:

No.	Value	Note	Type
1	5.04×10^{-5} K^{-1}	10% filled [1]	L
2	4.5×10^{-5} K^{-1}	15% filled [1]	L
3	3.6×10^{-5} K^{-1}	20% filled [1]	L
4	3.1×10^{-5} K^{-1}	30% filled [2]	L

Melting Temperature:

No.	Value	Note
1	270°C	30% filled [1]

Deflection Temperature:

No.	Value	Note
1	199°C	1.82 MPa, 10% filled [1]
2	204°C	1.82 MPa, 15-20% filled [1]
3	237°C	1.82 MPa, 30% filled [1]
4	238°C	1.8 MPa, 30% filled [2]

Transport Properties:
Water Absorption:

No.	Value	Note
1	0.02 %	24h, 10-30% filled [1,2]

Mechanical Properties:
Flexural Modulus:

No.	Value	Note
1	7200 MPa	30% filled [2]
2	3452 MPa	35200 kg cm^{-2}, 10% filled [1]

– Ethylcellulose

No.	Value	Note
3	4138 MPa	42200 kg cm^{-2}, 15% filled [1]
4	5168 MPa	52700 kg cm^{-2}, 20% filled [1]
5	7237 MPa	73800 kg cm^{-2}, 30% filled [1]

Tensile Strength Break:

No.	Value	Note	Elongation
1	58.6 MPa	598 kg cm^{-2}, 10% filled [1]	
2	67.6 MPa	689 kg cm^{-2}, 15% filled [1]	
3	75.8 MPa	773 kg cm^{-2}, 20% filled [1]	7%
4	96.5 MPa	984 kg cm^{-2}, 30% filled [1]	5%
5	96 MPa	room temp., 30% filled [2]	4–5%

Flexural Strength Yield:

No.	Value	Note
1	131 MPa	1340 kg cm^{-2}, 30% filled [1]

Hardness: Rockwell R74 (30% filled) [1,2]
Izod Notch:

No.	Value	Notch	Note
1	320 J/m	Y	6 ft lb in^{-1}, room temp., 10% filled [1]
2	347 J/m	Y	6.5 ft lb in^{-1}, room temp., 15% filled [1]
3	374 J/m	Y	7 ft lb in^{-1}, room temp., 20% filled [1]
4	400 J/m	Y	7.5 ft lb in^{-1}, room temp., 30% filled [1,2]

Electrical Properties:
Dielectric Permittivity (Constant):

No.	Value	Frequency	Note
1	3.4	1 MHz	30% filled [1]

Dielectric Strength:

No.	Value	Note
1	16.1 kV.mm^{-1}	30% filled [1]

Dissipation (Power) Factor:

No.	Value	Frequency	Note
1	0.005	1 MHz	30% filled [1]

Optical Properties:
Volume Properties/Surface Properties: Clarity: opaque (30% filled) [2]

Polymer Stability:
Upper Use Temperature:

No.	Value	Note
1	177°C	continuous service, 30% filled [1]

Applications/Commercial Products:
Processing & Manufacturing Routes: Processed by injection moulding

Mould Shrinkage (%):

No.	Value	Note
1	0.55%	20–30% filled [1]
2	0.2–0.3%	30% filled [2]

Trade name	Details	Supplier
Fluoromelt FP-EF-1004		LNP Engineering
Thermocomp FP-EF-1002	10% filled	LNP Engineering
Thermocomp FP-EF-1003	15% filled	LNP Engineering
Thermocomp FP-EF-1004	20% filled	LNP Engineering
Thermocomp FP-EF-1006	30% filled	LNP Engineering

Bibliographic References
[1] *Thermocomp Range,* LNP Engineering, 1996, (technical datasheet)
[2] *The Materials Selector,* 2nd edn., (eds. N.A. Waterman and M.F. Ashby), Chapman and Hall, 1997, **3**, 233

Ethylcellulose   E-15
Synonyms: *Perethylcellulose*
Related Polymers: Cellulose II
Monomers: Base monomer unit glucose
Material class: Polysaccharides
Polymer Type: Cellulosics
CAS Number:

CAS Reg. No.
9004-57-3

General Information: Approx 2.3 ethyl groups per monomer for films; 2.4–25 for lacquer

Volumetric & Calorimetric Properties:
Density:

No.	Value	Note
1	d 1.14 g/cm^3	[1,3]
2	d 1.09–1.17 g/cm^3	[6]

Thermal Expansion Coefficient:

No.	Value	Note	Type
1	0.0001–0.0002 K^{-1}	[6]	L

Thermal Conductivity:

No.	Value	Note
1	0.16–0.3 W/mK	[6]

Melting Temperature:

No.	Value	Note
1	135°C	crystalline [6]

Glass-Transition Temperature:

No.	Value	Note
1	316°C	[11,13]

Ethylcellulose

Deflection Temperature:

No.	Value	Note
1	45–88°C	1.82 MPa [6]

Vicat Softening Point:

No.	Value	Note
1	152–162°C	[1,3]

Surface Properties & Solubility:
Solubility Properties General: Solubility 4–8% NaOH 0.5–0.7 [4]. Cold water 0.8–1.3 [4]. Organic solvents 2.3–2.6 [4]
Solvents/Non-solvents: Liq. crystals sol. $CHCl_3$ [9], dioxan [11], AcOH [10], DMSO [8]
Surface Tension:

No.	Value	Note
1	32 mN/m	20° [11,12]

Transport Properties:
Polymer Solutions Dilute: Intrinsic viscosity 11 dl g^{-1} (MW 625000, DP 2650); 7.75 dl g^{-1} (MW 335000, DP 1420); 3.95 dl g^{-1} (MW 190000, DP 805) [5]
Water Absorption:

No.	Value	Note
1	2 %	[1]
2	0.8–1.8 %	24h [6]

Gas Permeability:

No.	Gas	Value	Note
1	H_2O	0.06675 cm^3 mm/(m^2 day atm)	890 g m^{-2} day^{-1}, 75 μm film, ASTM E96-66 [1]
2	He	26266 cm^3 mm/(m^2 day atm)	400×10^{-10} cm^2 $(s\ cmHg)^{-1}$, 30°
3	SO_2	13396 cm^3 mm/(m^2 day atm)	204×10^{-10} cm^2 $(s\ cmHg)^{-1}$
4	N_2	5516 cm^3 mm/(m^2 day atm)	84×10^{-10}, 30°
5	Ethylene oxide	27580 cm^3 mm/(m^2 day atm)	420×10^{-10} cm^2 $(s\ cmHg)^{-1}$, 0°
6	H_2	5713 cm^3 mm/(m^2 day atm)	87×10^{-10} cm^2 $(s\ cmHg)^{-1}$, 20°
7	NH_3	46294 cm^3 mm/(m^2 day atm)	705×10^{-10} cm^2 $(s\ cmHg)^{-1}$ [6]
8	O_2	1740 cm^3 mm/(m^2 day atm)	26.5×10^{-10}, 30°
9	CO_2	2692 cm^3 mm/(m^2 day atm)	41×10^{-10} cm^2 $(s\ cmHg)^{-1}$, 30°
10	H_2O	787992 cm^3 mm/(m^2 day atm)	12000×10^{-10} cm^2 $(s\ cmHg)^{-1}$, 20°

Mechanical Properties:
Tensile Strength Break:

No.	Value	Note
1	46–72 MPa	dry, 75 μm film [2]

Mechanical Properties Miscellaneous: Wet tensile strength is 80–85% of dry film value [1,3]. Flexibility, folding endurance 160–2000 (dry, 75 μm film) [1,3]
Hardness: Sward 52–61 (75 μm film) [1,3]. Rockwell R50–R115 [6]
Izod Notch:

No.	Value	Notch	Note
1	21 J/m	N	23° [6]

Electrical Properties:
Dielectric Permittivity (Constant):

No.	Value	Frequency	Note
1	2.8–3.9	1 MHz	25° [1,3]
2	3–4.1	1 kHz	25° [1,3]
3	2.5–4	60 Hz	25° [1,3]

Dielectric Strength:

No.	Value	Note
1	600 $kV.mm^{-1}$	1500 V in^{-1}, ASTM D149-64 [1,3]
2	13.8–19.7 $kV.mm^{-1}$	[6]

Dissipation (Power) Factor:

No.	Value	Frequency	Note
1	0.002–0.02	1 kHz	25° [1,3]
2	0.005–0.02	60 Hz	25° [1,3]

Optical Properties:
Refractive Index:

No.	Value	Note
1	1.47	cast film [2]

Volume Properties/Surface Properties: Light transmission at 310–400 nm is practically complete [1,3] and at 280–310 nm it is >50% complete [1,3]. Discolouration by sunlight is very slight [1,3]

Polymer Stability:
Decomposition Details: Decomposes at 306°. Products mainly gases; other products include acetaldehyde, aliphatic compounds and furans

Applications/Commercial Products:
Processing & Manufacturing Routes: Reaction of alkali cellulose with ethyl chloride at high pressure and temp.
Applications: Lacquers and varnishes, printing inks and adhesives. Also used in food and food packaging applications

Supplier
Dow
Aqualon

Bibliographic References
[1] *Encycl. Polym. Sci. Eng.*, 2nd edn., (ed. J.I. Kroschwitz), John Wiley and Sons, 1985, **3**
[2] *Kirk-Othmer Encycl. Chem. Technol.*, 4th edn., (ed. J.I. Kroschwitz), Wiley Interscience, 1993, **5**
[3] *Chemical and Physical Properties of Hercules Ethyl Cellulose*, Hercules Inc., 1982, (technical datasheet)
[4] *Concise Encyclopedia of Wood and Wood-based Materials*, (ed. A.P. Schneiwind), Pergamon Press, 1989

[5] *High Polymers*, Part 4, (eds. N.M. Bikales and L. Segal), Wiley Interscience, 1971
[6] Dean, J.A., *Lange's Handbook of Chemistry*, 14th edn., McGraw-Hill, 1992
[7] Brown, W.P. and Tipper, C.F.H., *J. Appl. Polym. Sci.*, 1978, **22**, 1459
[8] Patton, P.A. and Gilbert, R.D., J. Polym. Sci., *Polym. Phys. Ed.*, 1983, **21**, 515, (liq crysts)
[9] Suto, S., J. Polym. Sci., *Polym. Phys. Ed.*, 1984, **22**, 637, (liq cryst)
[10] Bheda, J., Fellers, J.F. and White, J.L., *Colloid Polym. Sci.*, 1980, **258**, 1335, (liq cryst)
[11] Gröbe, A., *Polym. Handb.*, 3rd edn., (eds. J. Brandrup and E.H. Immergut), Wiley Interscience, 1989, V137
[12] Wu, S., *Polymer Interface and Adhesion*, Marcel Dekker, 1982, (surface tension)
[13] Wiley, F.E., *Ind. Eng. Chem.*, 1942, **34**, 1052, (glass transition temps)

Ethylene copolymers with acrylates   E-16

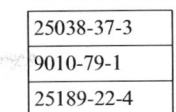

R = CH_3 methyl acrylate copolymer
= C_2H_5 ethyl acrylate copolymer
= C_4H_9 butyl acrylate copolymer

Polymer Stability:
Thermal Stability General: Has good low temp. props. down to -65°
Environmental Stress: Excellent environmental stress crack resistance
Biological Stability: FDA approved grades available

Ethylene-propylene elastomers   E-17

Synonyms: *EPR. EPM. EPDM. Ethylene-propylene rubber. EPM Rubber. EPDM Rubber. Poly(ethylene-propylene monomer) rubber. Poly(ethylene-propylene-diene monomer). Polyolefin rubber, synthetic. Ethylene-propylene terpolymer. Poly(ethylene-ethylidenenorbornene-propene) rubber. Poly(ethylene-methylidenenorbornene-propene) rubber. Poly(dicyclopentadiene-ethylene-propene) rubber. Poly(ethylene-hexadiene-propene) rubber. EPT. Ethylene-propylene copolymer. Ethylene-propylene-diene polymer. Ethylene-propylene-diene rubber*
Related Polymers: Propylene ethylene copolymer, Blends with polypropylene, Blends with polyethylene, Polypropylene copolymers, Polypropylene oxide elastomers, Polypropylene/rubber blends
Monomers: Ethylene, Propylene, 5-Ethylidene-2-norbornene, Methylidenenorbornene, Dicyclopentadiene, 1,4-Hexadiene
Material class: Synthetic Elastomers, Copolymers
Polymer Type: polyethylene, polypropylenepolyolefins
CAS Number:

CAS Reg. No.
25038-36-2
25034-71-3
25038-37-3
9010-79-1
25189-22-4

Molecular Formula: $[(C_2H_4).(C_3H_6)]_n$
Fragments: C_2H_4 C_3H_6
Molecular Weight: Most probable distribution MZ:MW:MN 3:2:1; $M_w/M_n \leq 50$; MZ/MW ≤ 20. MV/MW can be as low as 0.85
Additives: Fillers include carbon black, also calcined clay and calcium carbonate. Plasticisers - usually paraffinic oils with addition of naphthalenic oil for high temp. usage
Morphology: Most commercial grades amorph. at 20°, some grades may attain up to 30% crystallinity. If ethylene content >40 mol. % crystallinity will develop below ambient temp. if annealed for several hours. If ethylene content ≤ 40 mol. % polymer remains amorph. at all temps. Crystallinity observed is the polyethylene orthorhombic cell, polypropylene crystallinity is not a controlling factor [2]

Volumetric & Calorimetric Properties:
Density:

No.	Value	Note
1	d 0.854 g/cm³	40-60% ethylene, amorph. [2]
2	d 0.86-0.87 g/cm³	[3]

Thermal Expansion Coefficient:

No.	Value	Note	Type
1	0.00018 K⁻¹	[3]	L
2	0.00075 K⁻¹	40-60% ethylene, 23°	L
3	0.00088 K⁻¹	[2]	V
4	0.00025 K⁻¹	40-60% ethylene, 23° [2]	L

Thermal Conductivity:

No.	Value	Note
1	0.335 W/mK	[3]

Thermal Diffusivity: 1.9×10^{-5} m² s⁻¹ [3]
Specific Heat Capacity:

No.	Value	Note	Type
1	2.22 kJ/kg.C	40-60% ethylene, 23°, amorph. [2]	P

Glass-Transition Temperature:

No.	Value	Note
1	-60--45°C	[3]

Brittleness Temperature:

No.	Value	Note
1	-60°C	[2]

Surface Properties & Solubility:
Cohesive Energy Density Solubility Parameters: δ 16 J^{1/2} cm^{-3/2} [2]
Internal Pressure Heat Solution and Miscellaneous: Heats of mixing at infinite dilution have been reported for a range of polymer ratios in aliphatic solvents [7]

Ethylene-propylene elastomers

Plasticisers: Paraffinic oil is most widely used, esp. of low volatility. May be blended with naphthenic oils but these interfere with peroxide curing. Aromatic oils are not recommended [3]
Solvents/Non-solvents: Sol. aliphatic hydrocarbons (C_5 and above), low polarity halogenated and non-halogenated aromatic and naphthenic hydrocarbons, THF. Insol. aldehydes, ketones, carboxylic acids, esters, dioxane [2]

Transport Properties:
Polymer Solutions Dilute: Theta temp. θ 80° (approx., ethyl phenyl ether) [3,5]; θ 40° (approx., 1-chloronaphthalene) [3]. Polymer-solvent interaction parameters have been reported [2,5]. Mark-Houwink relationships have been reported [2]. Specific refractive index increment 0.104 cm^3 g^{-1} (1,2,4-trichlorobenzene, 135°, 633 nm) [2]
Permeability of Liquids: Permeability of liquids has been reported [2]. D_o 6.1 × 10^{-8} (23°, C_6H_6); 2.8 × 10^{-7} cm^2 s^{-1} (30°, CH_2Cl_2)
Water Content: Water content <1% [3]

Mechanical Properties:
Mechanical Properties General: Mech. props. depend considerably on structural characteristics and the type and amount of filler. Elastic props. are comparable to most synthetic rubbers but not as good as NR, or SBR

Tensile Strength Break:

No.	Value	Note	Elongation
1	10 MPa	[2]	500%

Tensile Strength Yield:

No.	Value	Note
1	20 MPa	reinforced [2]
2	>50 MPa	min. 20°, uncured gum elastomer [2]
3	10 MPa	20°, amorph.

Mechanical Properties Miscellaneous: Tear strength 15–45 kN m^{-1} (vulcanised) [2]
Hardness: Shore A35–A90 (vulcanised)

Electrical Properties:
Electrical Properties General: Electrical props. are generally excellent, especially at high temp., after heat ageing or in water submersion. They are however strongly dependant on the fillers used [3]

Dielectric Permittivity (Constant):

No.	Value	Frequency	Note
1	2.5	10 kHz	[2]

Dissipation (Power) Factor:

No.	Value	Frequency	Note
1	0.2	10 kHz	[2]

Optical Properties:
Refractive Index:

No.	Value	Note
1	1.474	23°
2	1.4524	90°
3	1.4423	125° [2]

Transmission and Spectra: H-1 nmr [6], C-13 nmr [7,8] and ir [1,2] (ASTM D3900) spectral data have been reported
Volume Properties/Surface Properties: Glassy white in appearance

Polymer Stability:
Thermal Stability General: Decomposes at or above 300° [2]
Upper Use Temperature:

No.	Value	Note
1	150°C	no significant hardening after 1000h, better than NR or SBR [3]
2	150–175°C	[2]

Flammability: Flammable only above thermal decomposition temp. [2]
Environmental Stress: More resistant to oxidation than other rubbers, natural rubber and styrene-butadiene rubber
Chemical Stability: Resistant to ozone, dilute acids, alkalis and alcohols. Swollen considerably by aliphatic, aromatic or chlorinated hydrocarbons [3]
Hydrolytic Stability: Has good stability

Applications/Commercial Products:
Processing & Manufacturing Routes: Manufacture is highly proprietary, many variations from large number of patents. Soln. or slurry processes are used; hexane soln. is the most common. Catalyst has two components: transition metal halide ($TiCl_4$, VCl_4, $VOCl_3$, etc., - $VOCl_3$ most widely used), and an alkyl component ((C_2H_5)$_2$AlCl, $C_2H_5AlCl_2$, or a mixture of these). After reaction the catalyst is destroyed/removed by water washing. Resulting polymer crumb is dewatered and dried. Unreacted monomer is recycled (except ethylene where initial reaction rate is very high). A similar Ziegler-Natta process is used for Polypropylene [2,3]. MW affects filler dispersion (especially Carbon black). A narrow MW distribution for average dispersion, broad MW distribution [2] for excellent dispersion. If ethylene content is high, form should be chosen to avoid excessive mixing times. Generally, extrusion, calendering, curing and injection moulding are carried out on common, commercial rubber factory equipment [2,3]
Applications: Due to good chemical resistance, especially to ozone, and good toleration of high concentrations of filler and oil whilst retaining good physical props., ethylene-propylene elastomers are widely used in cost effective rubber compounding. Over 150 grades are available. Main applications are automotive profiles, hoses, seals; building and construction profiles, roofing foil, seals; electrical cable insulation and jacketing; general moulded articles; as blends in ozone resistant tyre side walls, and as an impact modifier for polypropylene, polyamides, polystyrenes. In lubrication used as an additive to oil. Potential use in biomedical applications (heart-valve leaflets)

Trade name	Details	Supplier
Buna		Bunawerke Hüls GmbH
Buna		DSM
Epcar		Bayer Inc.
Epcar		Polysar
Epsyn		Copolymer Rubber
Esprene		Sumitomo Chem Co
Hostalen		Hoechst (UK) Ltd. Polymer Division
JSR		Japan Synthetic
Kelton	K740, 778, 4802	DSM
Mitsui		Mitsui Petrochemicals
Moplen		Enimont Iberica SA
Nordel		E.I. DuPont de Nemours & Co., Inc.
Nordel		Enichem
Polysar EPDM		Polysar

Polysar EPM		Polysar
Profax		Hercules
Royalene		Uniroyal Chemical
Tenite		Eastman Chemical Company
Vistalon		Exxon Chemical

Bibliographic References

[1] Phuong-Nguyen, H. and Delmas, G., *Macromolecules*, 1979, **12**, 740, 746
[2] *Encycl. Polym. Sci. Eng.*, 2nd edn., (ed. J.I. Kroschwitz), John Wiley and Sons, 1986
[3] *Kirk-Othmer Encycl. Chem. Technol.*, 4th edn., (eds. J.I. Kroschwitz and M. Howe-Grant), Wiley Interscience, 1993
[4] Bruckner, S., Allegra, G., Gianotti, G. and Moraglio, G., *Eur. Polym. J.*, 1974, **10**, 347
[5] Baldwin, F.P. and Ver Strate, G., *Rubber Chem. Technol.*, 1972, **45**, 709
[6] Gardner, I.J., Cozewith, C. and Ver Strate, G., *Rubber Chem. Technol.*, 1971, **44**, 1015
[7] Smith, W.V., *J. Polym. Sci., Polym. Phys. Ed.*, 1980, **18**, 1573, 1587
[8] Hikichi, K., Hiraoki, T., Takemura, S., Ohuchi, M. and Nishioka, A., *ACS Symp. Ser.*, 1984, **247**, 119
[9] Mast, F., Hoschtitzky, J.A.R.D., van Blitterswijk, C.A. and Huysmans, H.A., *J. Mater. Sci.: Mater. Med.*, 1997, **8**, 5, (use)

Ethylhydroxyethyl cellulose — E-18

Synonyms: EHEC. Cellulose ethyl-2-hydroxyethyl ether
Monomers: Base monomer unit glucose, Ethylene oxide
Material class: Polysaccharides
Polymer Type: Cellulosics
CAS Number:

CAS Reg. No.
9004-58-4

Volumetric & Calorimetric Properties:
Density:

No.	Value	Note
1	d 1.12 g/cm^3	film [1,2]

Transition Temperature:

No.	Value	Note	Type
1	175°C	[2]	unplasticised flow temp.

Mechanical Properties:
Tensile Strength Break:

No.	Value	Note	Elongation
1	34–41 MPa	film [1,2]	6–10%

Mechanical Properties Miscellaneous: Flexibility 500–900 (MIT double folds) [1,2]

Optical Properties:
Refractive Index:

No.	Value	Note
1	1.47	[2]

Volume Properties/Surface Properties: Colour 0.9 (ASTM D1209, soln.) [1,2]

Applications/Commercial Products:
Processing & Manufacturing Routes: Reaction of alkali cellulose with a limited quantity of ethylene oxide, followed by reaction with an excess of ethyl chloride at high temp. and pressure
Applications: Lacquers and varnishes, printing inks, adhesives

Trade name	Supplier
Bermocoll	Berol Kemi AB
EHEC	Hercules
	Dow

Bibliographic References

[1] *Chemical and Physical Properties of Hercules Ethyl Cellulose*, Hercules Inc., 1981, (technical datasheet)
[2] *Encycl. Polym. Sci. Eng.*, 2nd edn., (ed. J.I. Kroschwitz), John Wiley and Sons, 1985, 3

Eucommia ulmodia gum — E-19

Synonyms: *trans-1,4-Polyisoprene*
Monomers: 2-Methyl-1,3-butadiene
Material class: Thermoplastic, Natural elastomers
Polymer Type: polybutadiene
Molecular Formula: [C$_5$H$_8$]$_n$
Fragments: C$_5$H$_8$
General Information: Isol. from *Eucommia ulmodia* olive tree, found in China, N. America and Europe. Has similar props. and struct. to gutta-percha, and balcita. Cross-linking disrupts crystallinity to give rubbery props.

Volumetric & Calorimetric Properties:
Density:

No.	Value
1	d 0.93–0.97 g/cm^3

Melting Temperature:

No.	Value
1	64°C

Glass-Transition Temperature:

No.	Value
1	-70°C

Brittleness Temperature:

No.	Value
1	-54°C

Surface Properties & Solubility:
Solvents/Non-solvents: Sol. aromatic hydrocarbons, halohydrocarbons at low temp., aliphatic hydrocarbons at high temp. Insol. alcohols, ketones

– Eucommia ulmodia gum

Mechanical Properties:
Tensile Strength Break:

No.	Value	Note
1	16.1–33 MPa	
2	20 MPa	cross-linked

Tensile Strength Yield:

No.	Elongation	Note
1	335–450%	
2	500%	cross-linked

Impact Strength: Yield strength 120 MPa
Hardness: Shore A97-A98. Shore A70-A100 (cross-linked)

Electrical Properties:
Surface/Volume Resistance:

No.	Value	Type
1	0.001×10^{15} Ω.cm	S

Dielectric Permittivity (Constant):

No.	Value
1	3.034

Dielectric Strength:

No.	Value
1	27.2 kV.mm^{-1}

Polymer Stability:
Chemical Stability: Stable in HF and conc. HCl, unstable in nitric acid and hot, conc. sulfuric acid

Applications/Commercial Products:
Processing & Manufacturing Routes: Isol. by solvent extraction of leaves of *Eucommia ulmodia*, although present throughout the tree. Not isol. by bark slashing. Low energy requirements for processing due to low softening temp.
Applications: Has a range of potential applications due to rubbery props.

Bibliographic References
[1] Yan, R.-F., *Polymeric Materials Encyclopedia,* (ed. J.C. Salamone), CRC Press, 1996, 2291

6FDA-4-BDAF F-1

Synonyms: *Poly(2,2'-bis(3,4-dicarboxyphenyl)hexafluoropropane dianhydride-alt-2,2-bis[4-(4-aminophenoxy)phenyl]hexafluoropropane). Poly(5,5'-[2,2,2-trifluoro-1-(trifluoromethyl)ethylidene]-bis-1,3-isobenzofurandione-alt-2,2-bis[4-(4-aminophenoxy)phenyl]hexafluoropropane). Poly(4,4'-(hexafluoroisopropylidene)-bis(phthalic anhydride)-alt-2,2-bis[4-(4-aminophenoxy)phenyl]-hexafluoropropane). Poly[(1,3-dihydro-1,3-dioxo-2H-isoindole-2,5-diyl)[2,2,2-trifluoro-1-(trifluoromethyl)ethylidene](1,3-dihydro-1,3-dioxo-2H-isoindole-5,2-diyl)-1,4-phenyleneoxy-1,4-phenylene[2,2,2-trifluoro-1-(trifluoromethyl)ethylidene]-1,4-phenyleneoxy-1,4-phenylene]*
Monomers: 5,5'-[2,2,2-Trifluoro-1-(trifluoromethyl)ethylidene]bis-1,3-isobenzofurandione, 4,4'-[2,2,2-Trifluoro-1-(trifluoromethyl)ethylidene]bis(4,1-phenyleneoxy)bis[benzenamine]
Material class: Thermoplastic
Polymer Type: polyimide
CAS Number:

CAS Reg. No.	Note
87186-94-5	
87182-96-5	source based

Molecular Formula: $(C_{46}H_{22}F_{12}N_2O_6)_n$
Fragments: $C_{46}H_{22}F_{12}N_2O_6$

Volumetric & Calorimetric Properties:
Density:

No.	Value	Note
1	d 1.429 g/cm^3	[6,7]
2	d 1.432 g/cm^3	[3]
3	d 1.438 g/cm^3	[5]

Volumetric Properties General: Fractional free volume 0.1783, [3] 0.181 [6,7]. Specific free volume 0.1245 [3]
Glass-Transition Temperature:

No.	Value	Note
1	232°C	[4]
2	234°C	[1]
3	243°C	[6,7]
4	257°C	[2]
5	262°C	[5]
6	263°C	[3]

Surface Properties & Solubility:
Cohesive Energy Density Solubility Parameters: δ 21.7 J$^{1/2}$ cm$^{3/2}$ (10.6 cal$^{1/2}$ cm$^{3/2}$) [2], 24 J$^{1/2}$ cm$^{3/2}$ [6,7]
Solvents/Non-solvents: Sol. *N*-methyl-2-pyrrolidone, *N,N*-dimethylacetamide, DMSO, pyridine, EtOAc, CHCl$_3$, Me$_2$CO, 1,4-dioxane, *m*-cresol [1,2]. Insol. THF, hexane, methylcellulose [1,2]

Transport Properties:
Polymer Solutions Dilute: η_{inh} 0.54 dl g^{-1} (135°, 0.5 wt.%) [6]

Gas Permeability:

No.	Gas	Value	Note
1	CO_2	1241 cm^3 mm/(m^2 day atm)	18.9 × 10^{-10} cm^2 (s cmHg)$^{-1}$, 35°, 6.8 atm. [3,5]
2	CO_2	1254.2 cm^3 mm/(m^2 day atm)	19.1 × 10^{-10} cm^2 (s cmHg)$^{-1}$, 35°, 10 atm. [6]
3	CH_4	33.5 cm^3 mm/(m^2 day atm)	0.51 × 10^{-10} cm^2 (s cmHg)$^{-1}$, 35°, 6.8 atm. [3]
4	He	3021 cm^3 mm/(m^2 day atm)	46 × 10^{-10} cm^2 (s cmHg)$^{-1}$, 35°, 6.8 atm. [3]
5	O_2	337 cm^3 mm/(m^2 day atm)	5.13 × 10^{-10} cm^2 (s cmHg)$^{-1}$, 35°, 2 atm. [6]
6	O_2	354.6 cm^3 mm/(m^2 day atm)	5.4 × 10^{-10} cm^2 (s cmHg)$^{-1}$, 35°, 6.8 atm. [5]
7	H_2	3021 cm^3 mm/(m^2 day atm)	46 × 10^{-10} cm^2 (s cmHg)$^{-1}$, 35°, 6.8 atm. [5]
8	H_2	3113 cm^3 mm/(m^2 day atm)	47.4 × 10^{-10} cm^2 (s cmHg)$^{-1}$, 35°, 10 atm. [6]
9	CO	99.2 cm^3 mm/(m^2 day atm)	1.51 × 10^{-10} cm^2 (s cmHg)$^{-1}$, 35°, 10 atm. [6]
10	N_2	64.4 cm^3 mm/(m^2 day atm)	0.981 × 10^{-10} cm^2 (s cmHg)$^{-1}$, 35°, 10 atm. [6]

Mechanical Properties:
Tensile (Young's) Modulus:

No.	Value	Note
1	1900 MPa	[1]

Tensile Strength Break:

No.	Value	Note	Elongation
1	98 MPa	[1]	6%

Electrical Properties:
Dielectric Permittivity (Constant):

No.	Value	Frequency	Note
1	2.96	100 Hz	room temp. [4]
2	2.9–2.99	100 Hz	100-300° [4]

Optical Properties:
Transmission and Spectra: Uv spectral data have been reported [2]

Polymer Stability:
Decomposition Details: Under N_2 atmosphere, single step decomposition occurs; initial decomposition at 511°; 10% wt. loss occurs at 532° at heat rate of 10° min^{-1} [1]. Extrapolated end of wt. loss at 635°; residue at 600° 66% [4]

Applications/Commercial Products:

Trade name	Supplier
Eymyd HP40	Ethyl Corporation

Bibliographic References
[1] Negi, Y.S., Suzuki, Y.-I., Kawamura, I., Hagiwara, T. *et al*, *J. Polym. Sci., Part A: Polym. Chem.*, 1992, **30**, 2281, (general props)
[2] Chun, B.-W., *Polymer*, 1994, **35**, 4203, (solubility)

[3] Zoia, G., Stern, S.A., StClair, A.K. and Pratt, J.R., *J. Polym. Sci., Part B: Polym. Phys.*, 1994, **32**, 53, (gas permeability)
[4] Misra, A.C., Tesoro, G., Hougham, G. and Pendharkar, S.M., *Polymer*, 1992, **33**, 1078, (thermal props)
[5] Stern, S.A., Mi, Y., Yamamoto, H. and StClair, A.K., *J. Polym. Sci., Part B: Polym. Phys.*, 1989, **27**, 1887, (gas permeability)
[6] Tanaka, K., Kita, H., Okano, M. and Okamoto, K.-I., *Polymer*, 1992, **33**, 585, (gas permeability)
[7] Tanaka, K., Kita, H. and Okamoto, K.-I., *J. Polym. Sci., Part B: Polym. Phys.*, 1993, **31**, 1127, (CO_2 sorption)

6FDA-durene   F-2

Synonyms: *Poly(2,2'-bis(3,4-dicarboxyphenyl)hexafluoropropane dianhydride-alt-2,3,5,6-tetramethyl-1,4-benzenediamine. Poly(5,5'-[2,2,2-trifluoro-1-(trifluoromethyl)ethylidene]bis-1,3-isobenzofurandione-alt-2,3,5,6-tetramethyl-1,4-benzenediamine). Poly(4,4'-(hexafluoroisopropylidene)bis(phthalic anhydride)-alt2,3,5,6-tetramethyl-1,4-benzenediamine. 6FDA-pTeMPD. Poly[(1,3-dihydro-1,3-dioxo-2H-isoindole-2,5-diyl)[2,2,2-trifluoro-1-(trifluoromethyl)ethylidene](1,3-dihydro-1,3-dioxo-2H-isoindole-5,2-diyl)(2,3,5,6-tetramethyl-1,4-phenylene)]*
Monomers: 5,5'-[2,2,2-Trifluoro-1-(trifluoromethyl)ethylidene]bis-1,3-isobenzofurandione, 2,3,5,6-Tetramethyl-1,4-benzenediamine
Material class: Thermoplastic
Polymer Type: polyimide
CAS Number:

CAS Reg. No.	Note
112870-05-0	
112869-57-5	source based

Molecular Formula: $(C_{29}H_{18}F_6N_2O_4)_n$
Fragments: $C_{29}H_{18}F_6N_2O_4$

Volumetric & Calorimetric Properties:
Density:

No.	Value	Note
1	d^{25} 1.326 g/cm^3	[2]

Volumetric Properties General: Fractional free volume 0.184 (25°), 0.189 (25°, CO_2 conditioned) [2]
Glass-Transition Temperature:

No.	Value	Note
1	420°C	[2]

Surface Properties & Solubility:
Cohesive Energy Density Solubility Parameters: δ 26.1 J$^{1/2}$ cm$^{-3/2}$ [1]

Transport Properties:
Polymer Solutions Dilute: η_{inh} 0.99 dl g^{-1} (35°, 0.5 wt% in dimethylacetamide)
Gas Permeability:

No.	Gas	Value	Note
1	CO_2	48593 cm^3 mm/(m^2 day atm)	740 × 10^{-10} cm^2 (s cmHg)$^{-1}$, 35°, 2 atm. [2]
2	CO_2	87992 cm^3 mm/(m^2 day atm)	1340 × 10^{-10} cm^2 (s cmHg)$^{-1}$, CO_2 conditioned film, 35°, 2 atm. [2]
3	CO_2	28893 cm^3 mm/(m^2 day atm)	440 × 10^{-10} cm^2 (s cmHg)$^{-1}$, 35°, 10 atm. [1]
4	CH_4	1852 cm^3 mm/(m^2 day atm)	28.2 × 10^{-10} cm^2 (s cmHg)$^{-1}$, 35°, 10 atm. [1]
5	H_2	36051 cm^3 mm/(m^2 day atm)	549 × 10^{-10} cm^2 (s cmHg)$^{-1}$, 35°, 10 atm. [1]
6	O_2	8011 cm^3 mm/(m^2 day atm)	122 × 10^{-10} cm^2 (s cmHg)$^{-1}$, 35°, 2 atm. [2]

Applications/Commercial Products:

Trade name	Supplier
Sixef Durene	Ethyl Corporation

Bibliographic References
[1] Tanaka, K., Okano, M., Toshino, H., Kita, H. and Okamoto, K.-I, *J. Polym. Sci., Part B: Polym. Phys.*, 1992, **30**, 907, (gas permeability)
[2] Tanaka, K., Ito, M., Kita, H., Okamoto, K.-I. and Ito, Y., *Bull. Chem. Soc. Jpn.*, 1995, **68**, 3011, (CO_2 permeability)
[3] Tanaka, K., Kita, H. and Okamoto, K.-I., *J. Polym. Sci., Part B: Polym. Phys.*, 1993, **31**, 1127, (CO_2 sorption)

6FDA-3,3'-6FDA   F-3

Synonyms: *Poly(5,5'-[2,2,2-trifluoro-1-(trifluoromethyl)ethylidene]bis-1,3-isobenzofurandione-alt-3,3'-[2,2,2-trifluoro-1-(trifluoromethyl)ethylidene]bisbenzenamine). Poly[(1,3-dihydro-1,3-dioxo-2H-isoindole-2,5-diyl)[2,2,2-trifluoro-1-(trifluoromethyl)ethylidene](1,3-dihydro-1,3-dioxo-2H-isoindole-5,2-diyl)[2,2,2-trifluoro-1-(trifluoromethyl)ethylidene]-1,3-phenylene]*
Related Polymers: 6FDA-4,4'-6FDA
Monomers: 5,5'-[2,2,2-Trifluoro-1-(trifluoromethyl)ethylidene]bis-1,3-isobenzofurandione, 3,3'-[2,2,2-Trifluoro-1-(trifluoromethyl)ethylidene]bisbenzenamine
Material class: Thermoplastic
Polymer Type: polyimide
CAS Number:

CAS Reg. No.	Note
94322-31-3	
94289-79-9	source based

Molecular Formula: $(C_{34}H_{14}F_{12}N_2O_4)_n$
Fragments: $C_{34}H_{14}F_{12}N_2O_4$

Volumetric & Calorimetric Properties:
Density:

No.	Value	Note
1	d^{25} 1.493 g/cm^3	[1]

Volumetric Properties General: Fractional free volume 0.225 [1]

– 6FDA-4,4'-6FDA

Glass-Transition Temperature:

No.	Value	Note
1	254°C	[1]

Transition Temperature:

No.	Value	Note	Type
1	149°C	[1]	Sub T_g

Transport Properties:
Gas Permeability:

No.	Gas	Value	Note
1	He	3152 cm³ mm/(m² day atm)	48×10^{-10} cm² (s cmHg)$^{-1}$, 35°, 10 atm. [2]
2	He	249531 cm³ mm/(m² day atm)	3800×10^{-10} cm² (s cmHg)$^{-1}$, 35-275° [1]
3	CO_2	335 cm³ mm/(m² day atm)	5.1×10^{-10} cm² (s cmHg)$^{-1}$, 35°, 10 atm. [2]
4	CO_2	13790 cm³ mm/(m² day atm)	210×10^{-10} cm² (s cmHg)$^{-1}$, (35-275°) [1]
5	O_2	118.2 cm³ mm/(m² day atm)	1.8×10^{-10} cm² (s cmHg)$^{-1}$, 35°, 2 atm. [2]
6	O_2	21013 cm³ mm/(m² day atm)	320×10^{-10} cm² (s cmHg)$^{-1}$ (35-275°) [1]

Optical Properties:
Total Internal Reflection: n_{TE} 1.5466, n_{TM} 1.5432, Birefringence Δn (n_{TE}- n_{TM}) 0.0034 (film cured at 300°, depends on cure temp.) [3]

Applications/Commercial Products:

Trade name	Supplier
Sixef 33	Ethyl Corporation

Bibliographic References

[1] Costello, L.M. and Koros, W.J., *J. Polym. Sci., Part B: Polym. Phys.*, 1995, **33**, 135, (gas permeability)
[2] Coleman, M.R. and Koros, W.J., *J. Polym. Sci., Part B: Polym. Phys.*, 1994, **32**, 1915, (gas permeability)
[3] Reuter, R., Franke, H. and Feger, C., *Appl. Opt.*, 1988, **27**, 4565, (optical props)

6FDA-4,4'-6FDA   F-4

Synonyms: *Poly(5,5'-[2,2,2-trifluoro-1-(trifluoromethyl)ethylidene]bis-1,3-isobenzofurandione-alt-4,4'-[2,2,2-trifluoro-1-(trifluoromethyl)ethylidene]bisbenzenamine). Poly[(1,3-dihydro-1,3-dioxo-2H-isoindole-2,5-diyl)[2,2,2-trifluoro-1-(trifluoromethyl)]ethylidene](1,3-dihydro-1,3-dioxo-2H-isoindole-5,2-diyl)[2,2,2-trifluoro-1-(trifluoromethyl)ethylidene]-1,4-phenylene]*

Related Polymers: 6FDA-3,3'-6FDA
Monomers: 5,5'-[2,2,2-Trifluoro-1-(trifluoromethyl)ethylidene]bis-1,3-isobenzofurandione, 4,4'-[2,2,2-Trifluoro-1-(trifluoromethyl)ethylidene]bis[benzenamine]
Material class: Thermoplastic
Polymer Type: polyimide
CAS Number:

CAS Reg. No.	Note
32036-79-6	
29896-40-0	source based

Molecular Formula: $(C_{34}H_{14}F_{12}N_2O_4)_n$
Fragments: $C_{34}H_{14}F_{12}N_2O_4$

Volumetric & Calorimetric Properties:
Density:

No.	Value	Note
1	d^{25} 1.466 g/cm³	[2]
2	d 1.48 g/cm³	[3]

Volumetric Properties General: Fractional free volume 0.184 (25°) [1], 0.182 (room temp.) [3,4], 0.272 [2], 0.194 (CO_2 conditioned, 25°)[1]
Glass-Transition Temperature:

No.	Value	Note
1	305°C	[1]
2	320°C	[2,3]

Transition Temperature General: 25°, 118°, 217°. Sub T_g transition temps. [2]

Surface Properties & Solubility:
Cohesive Energy Density Solubility Parameters: δ 24.1 J$^{1/2}$ cm$^{-3/2}$ [3,4]

Transport Properties:
Polymer Solutions Dilute: η_{inh} 0.95 dl g^{-1} (35°) [4]
Gas Permeability:

No.	Gas	Value	Note
1	CO_2	3349 cm³ mm/(m² day atm)	51×10^{-10} cm² (s cmHg)$^{-1}$, 35°, 10 atm. [1]
2	CO_2	5516 cm³ mm/(m² day atm)	84×10^{-10} cm² (s cmHg)$^{-1}$, CO_2 conditioned film, 35°, 10 atm. [1]
3	CO_2	4196 cm³ mm/(m² day atm)	63.9×10^{-10} cm² (s cmHg)$^{-1}$, 35°, 10 atm. [6]
4	CO_2	13790 cm³ mm/(m² day atm)	210×10^{-10} cm² (s cmHg)$^{-1}$, 35–120° [2]
5	CO_2	27579 cm³ mm/(m² day atm)	420×10^{-10} cm² (s cmHg)$^{-1}$, 121–325° [2]

Optical Properties:
Total Internal Reflection: n_{TE} 1.5405, n_{TM} 1.5275, Birefringence Δ_n 0.013 (film cured at 300°, depends on cure temp.) [5]

Applications/Commercial Products:

Trade name	Supplier
Sixef 44	Ethyl Corporation

– 6FDA-ODA

Bibliographic References
[1] Tanaka, K., Ito, M., Kita, H., Okamoto, K.-I. and Ito, Y., *Bull. Chem. Soc. Jpn.*, 1995, **68**, 3011, (CO_2 sorption)
[2] Costello, L.M. and Koros, W.J., *J. Polym. Sci., Part B: Polym. Phys.*, 1995, **33**, 135, (gas permeability)
[3] Tanaka, K., Kita, H. and Okamoto, K.-I., *J. Polym. Sci., Part B: Polym. Phys.*, 1993, **31**, 1127, (CO_2 sorption)
[4] Tanaka, K., Kita, H., Okano, M. and Okamoto, K.-I., *Polymer*, 1992, **33**, 585, (gas permeability)
[5] Reuter, R., Franke, H. and Feger, C., *Appl. Opt.*, 1988, **27**, 4565, (optical props)
[6] Coleman, M.R. and Koros, W.J., *J. Polym. Sci., Part B: Polym. Phys.*, 1994, **32**, 1915, (gas permeability)

6FDA-ODA   F-5

Synonyms: *Poly(2,2'-bis(3,4-dicarboxyphenyl)hexafluoropropane dianhydride-4,4'-oxydianiline). Poly(2,2'-bis(3,4-dicarboxyphenyl)hexafluoropropane dianhydride bis(4-aminophenyl)ether). Poly(5,5'-[2,2,2-trifluoro-1-(trifluoromethyl)ethylidene]bis-1,3-isobenzofurandione-alt-4,4'-oxybisbenzenamine). Poly(4,4'-oxybis[benzenamine]-alt-5,5'-[2,2,2-trifluoro-1-(trifluoromethyl)ethylidene]bis-1,3-isobenzofurandione). Poly[(1,3-dihydro-1,3-dioxo-2H-isoindole-2,5-diyl)[2,2,2-trifluoro-1-(trifluoromethyl)ethylidene](1,3-dihydro-1,3-dioxo-2H-isoindole-5,2-diyl)-1,4-phenyleneoxy-1,4-phenylene]*
Monomers: 5,5'-[2,2,2-Trifluoro-1-(trifluoromethyl)ethylidene]bis-1,3-isobenzofurandione, 4,4'-Oxybis[benzenamine]
Material class: Thermoplastic
Polymer Type: polyimide
CAS Number:

CAS Reg. No.	Note
39940-16-4	
32240-73-6	source based

Molecular Formula: $(C_{31}H_{14}F_6N_2O_5)_n$
Fragments: $C_{31}H_{14}F_6N_2O_5$

Volumetric & Calorimetric Properties:
Density:

No.	Value	Note
1	d 1.432 g/cm^3	[2,5]

Volumetric Properties General: Fractional free volume 0.165 [2,5]
Glass-Transition Temperature:

No.	Value	Note
1	287°C	[5]
2	288°C	[7]
3	299°C	[2,6]
4	285–295°C	depends on cure temp. [1]
5	272–315°C	depends on cure temp. [3]

Surface Properties & Solubility:
Cohesive Energy Density Solubility Parameters: δ 27.2 $J^{1/2}$ cm$^{-3/2}$ [2]
Solvents/Non-solvents: Sol. *N*-methyl-2-pyrrolidone, conc. sulfuric acid. Slightly sol. THF [1]

Transport Properties:
Polymer Solutions Dilute: η_{inh} 1.1–1.9 dl g^{-1}, (30°, 0.5 g dl^{-1} in *N*-methylpyrrolidone) [1]. Viscosity values depend on cure conditions
Gas Permeability:

No.	Gas	Value	Note
1	H_2	2673 cm^3 mm/(m^2 day atm)	40.7 × 10^{-10} cm^2 (s cmHg)$^{-1}$, 35°, 10 atm. [6]
2	H_2	3447 cm^3 mm/(m^2 day atm)	52.5 × 10^{-10} cm^2 (s cmHg)$^{-1}$, 35°, 6.8 atm. [7]
3	CO	76 cm^3 mm/(m^2 day atm)	1.16 × 10^{-10} cm^2 (s cmHg)$^{-1}$, 35°, 10 atm. [6]
4	CH_4	22.4 cm^3 mm/(m^2 day atm)	0.341 × 10^{-10} cm^2 (s cmHg)$^{-1}$, 35°, 10 atm. [6]
5	CH_4	34.8 cm^3 mm/(m^2 day atm)	0.53 × 10^{-10} cm^2 (s cmHg)$^{-1}$, 35°, 6.8 atm. [7]
6	O_2	254.8 cm^3 mm/(m^2 day atm)	3.88 × 10^{-10} cm^2 (s cmHg)$^{-1}$, 35°, 2 atm. [6]
7	O_2	331.6 cm^3 mm/(m^2 day atm)	5.05 × 10^{-10} cm^2 (s cmHg)$^{-1}$, 35°, 6.8 atm. [7]
8	H_2O	141839 cm^3 mm/(m^2 day atm)	21.6 × 10^{-8} cm^2 (s cmHg)$^{-1}$, 50° [5]

Mechanical Properties:
Tensile (Young's) Modulus:

No.	Value	Note
1	2200 MPa	room temp., 1 mm min^{-1} [1]

Tensile Strength Yield:

No.	Value	Elongation	Note
1	101 MPa	20.7%	room temp., 1 mm min^{-1} [1]

Viscoelastic Behaviour: Dynamic mech. props. have been reported [1]

Optical Properties:
Total Internal Reflection: n_{TE} 1.5949, n_{TM} 1.5870, birefringence Δ_n ($n_{TE}- n_{TM}$) 0.0079 (film cured at 500°, depends on cure temp.) [8]

Polymer Stability:
Decomposition Details: 5% Wt. loss occurs at 505°; 10% wt. loss occurs at 524° [1]
Environmental Stress: Uv photolabile. Chain cleavage occurs above 300 nm; below 300 nm in air, photooxidative ablation occurs with liberation of gases. In the absence of oxygen there is little cleavage and no ablation [3,4]

Applications/Commercial Products:

Trade name	Details	Supplier
NR 150A2	Series	DuPont
Pyralin Pl-2566		DuPont

Bibliographic References

[1] Takeichi, T., Ogura, S. and Takayama, Y., *J. Polym. Sci., Part A: Polym. Chem.,* 1994, **32**, 579, (thermal props, mech props, solubility)
[2] Tanaka, K., Kita, H. and Okamoto, K.-I., *J. Polym. Sci., Part B: Polym. Phys.,* 1993, **31**, 1127, (CO_2 sorption)
[3] Hoyle, C.E. and Anzures, E.T., *J. Polym. Sci., Part A: Polym. Chem.,* 1992, **30**, 1233, (photodegradation)
[4] Hoyle, C.E., Creed, D., Nagarajan, R., Subramanian, P. and Anzures, E.T., *Polymer,* 1992, **33**, 3162, (photodegradation)
[5] Okamoto, K.-I., Tanihara, N., Watanabe, H., Tanaka, K. *et al*, *J. Polym. Sci., Part B: Polym. Phys.,* 1992, **30**, 1223, (water sorption)
[6] Tanaka, K., Kita, H., Okano, M. and Okamoto, K.-I., *Polymer,* 1992, **33**, 585, (gas permeability)
[7] Stern, S.A., Mi, Y., Yamamoto, H. and StClair, A.K., *J. Polym. Sci., Part B: Polym. Phys.,* 1989, **27**, 1887, (gas permeability)
[8] Reuter, R., Franke, H. and Feger, C., *Appl. Opt.,* 1988, **27**, 4565, (optical props)

6FDA-mPDA/pPDA F-6

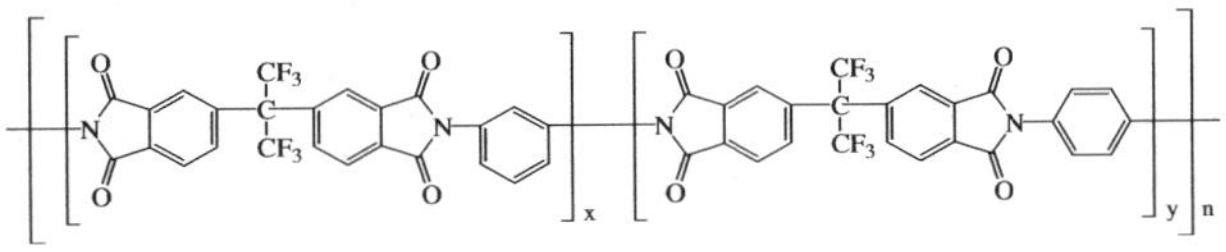

Synonyms: 2,2'-Bis(3,4-dicarboxyphenyl)hexafluoropropane dianhydride-co-m-phenylenediamine-co-p-phenylenediamine. Poly(5,5'-[2,2,2-trifluoro-1-(trifluoromethyl)ethylidene]bis-1,3-isobenzofurandione-co-m-phenylenediamine-co-p-phenylenediamine. Poly(4,4'-(hexafluoroisopropylidene)bis(phthalic anhydride)-co-m-phenylenediamine-co-p-phenylenediamine). Poly[(1,3-dihydro-1,3-dioxo-2H-isoindole,2,5-diyl)[2,2,2-trifluoro-1-(trifluoromethyl)ethylidene](1,3-dihydro-1,3-dioxo-2H-isoindole-5,2-diyl)phenylene]
Monomers: 4,4'-[2,2,2-Trifluoro-1-(trifluoromethyl)ethylidene]bis[1,2-benzenedicarboxylic acid], 1,3-Benzenediamine, 1,4-Benzenediamine
Material class: Thermoplastic
Polymer Type: polyimide
CAS Number:

CAS Reg. No.	Note
63849-68-3	CRU, phenylene bonding unspecified
70850-07-6	source based

Molecular Formula: $(C_{25}H_{10}F_6N_2O_4)_n$
Fragments: $C_{25}H_{10}F_6N_2O_4$
Additives: Carbon fibre, glass fibre
Miscellaneous: A review of applications as an adhesive has been reported [1]

Volumetric & Calorimetric Properties:
Density:

No.	Value	Note
1	d 1.44 g/cm^3	Avimid N [2,4]
2	d 1.53 g/cm^3	Avimid N, ½" graphite fibre [5]
3	d 1.58 g/cm^3	Avimid N, ¼" Celion G-40 fibre
4	d 1.57 g/cm^3	Avimid N, ⅛" Celion G-40 fibre [5]
5	d 1.4 g/cm^3	NR-150B2 [7]

Thermal Expansion Coefficient:

No.	Value	Note	Type
1	5.6×10^{-5} K^{-1}	Avimid N [2]	L

Glass-Transition Temperature:

No.	Value	Note
1	370°C	Avimid N [2]
2	333°C	dry, carbon fibre laminate [3]
3	340–370°C	Avimid N [4]
4	332–352°C	Avimid N, fibre length dependent [5]
5	340°C	NR-150B2 [7]

Transition Temperature General: T_g decreases in carbon fibre laminates containing 1.6% H_2O when cycled under hot/wet conditions. There is no change in T_g in dry samples under the same conditions [3]. T_g decreases with increasing fibre length in laminates [5]

Surface Properties & Solubility:
Plasticisers: H_2O [3]

Transport Properties:
Water Content: Water absorption increases with an increase in T_g [3]. Water absorption in carbon fibre laminates is greater when the fibres are aligned perpendicular to the longest edge [6]

Mechanical Properties:
Tensile (Young's) Modulus:

No.	Value	Note
1	4000 MPa	Avimid N [4]
2	42053 MPa	6100 kpsi, Avimid N, ½" Magnamite graphite fibre [5]
3	4137 MPa	600 kpsi, NR-150B2 [7]
4	53773 MPa	7800 kpsi, Avimid N, ¼" Celion G-40 fibre [5]
5	46879 MPa	6800 kpsi, Avimid N, ⅛" Celion G-40 fibre [5]

Flexural Modulus:

No.	Value	Note
1	4130 MPa	Avimid N [2]

Tensile Strength Break:

No.	Value	Note
1	110 MPa	Avimid N [2,4]
2	213 MPa	31 kpsi, Avimid N, ½" Magnamite graphite fibre [5]
3	134 MPa	19.5 kpsi, Avimid N, ¼" Celion G-40 fibre [5]
4	169 MPa	24.5 kpsi, Avimid N, ⅛" Celion G-40 fibres [5]
5	110 MPa	16 kpsi, NR-150B2 [7]

Mechanical Properties Miscellaneous: Interlaminar shear strength of a carbon fibre laminate decreases by 80% upon heating from room temp. to 315°. This loss of strength may be minimised by post-curing to high T_gs [3]. Yield stress 61 MPa (0.2% offset method, Avimid N) [4]
Failure Properties General: NR-150B2 reinforced with Celion 6000 fibres retains 47% of its shear strength and 57% of its flex strength after ageing at 316° [7]
Fracture Mechanical Properties: Avimid N and its composites have excellent fracture toughness props. [4] G_{Ic} 2 kJ m^{-2} [4]. Stress intensity factor 3.05 MPa m (3.2 mm thick), 2.98 MPa m (6.2 mm thick) [4]

– 6FDA-TPE Addition cured (acetylenic end-caps)

Izod Notch:

No.	Value	Notch	Note
1	42 J/m	Y	Avimid N [2]
2	42.7 J/m	Y	0.8 ft lb in^{-1}, NR-150B2 [7]

Electrical Properties:
Dielectric Permittivity (Constant):

No.	Value	Frequency	Note
1	2.7	1 kHz	Avimid N [2]

Polymer Stability:
Upper Use Temperature:

No.	Value	Note
1	316°C	NR-150B2 [7]

Decomposition Details: Weight loss at 343° of carbon fibre laminates is independent of fibre length up to 580 h. After this time, laminates containing Celion G-40 fibres are more stable with only 1.9% weight loss after 1243 h [5]. NR-150B2 shows a max. 10.7% weight loss after heating in air at 316° for 4122 h [7]

Applications/Commercial Products:
Processing & Manufacturing Routes: Synth. by treating 2,2′-bis (3,4-dicarboxyphenyl)hexafluoropropane dianhydride with *m*- and *p*- phenylenediamine at 150° for 1.25 h in a solvent such as dimethylacetamide or EtOH. The poly(amic acid) precursor is then imidised in an autoclave at 360° for 16 h [7]
Applications: Adhesive. Potential applications in jet engines; in ablative heat shields; in syntactic foams and honeycomb sandwich constructions. Applications due to electromagnetic transparency and good electrical props. as insulators and in radar domes

Trade name	Supplier
Avimid N	DuPont
NR-150B2 Series	DuPont

Bibliographic References
[1] *Secchaku*, 1979, **23**, 205, (uses, rev)
[2] Takekoshi, T., *Kirk-Othmer Encycl. Chem. Technol.*, 4th edn., (ed. J.I. Kroschwitz), 1991, **19**, 832, (props)
[3] Cornelia, D., *Int. SAMPE Symp. Exhib.*, 1994, **39**, 917
[4] Wedgewood, A.R., Su, K.B. and Narin, J.A., *SAMPE J.*, 1988, **24**, 41, (fracture toughness)
[5] Gibbs, H.H. and Myrick, D.E., *Int. SAMPE Symp. Exhib.*, 1988, **33**, 1473, (mech props)
[6] Paplham, W.P., Brown, R.A., Salin, I.M. and Seferis, J.C., *J. Appl. Polym. Sci.*, 1995, **57**, 123, (water absorption)
[7] Scola, D.A. and Vontell, J.H., *CHEMTECH*, 1989, **19**, 112, (props)
[8] Gibbs, H.H., *Natl. SAMPE Exhib.*, (*Proc.*), 1976, **21**, 592, (uses, additives)

6FDA-TPE Addition cured (acetylenic end-caps) F-7

Synonyms: *Poly(2,2′-bis(3,4-dicarboxyphenyl)hexafluoropropane dianhydride-alt-1,3-bis(3-aminophenoxy)benzene) Addition cured (acetylenic end-caps). Poly(5,5′-[2,2,2-trifluoro-1-(trifluoromethyl)ethylidene]bis-1,3-isobenzofurandione-alt-1,3-bis(3-aminophenoxy)benzene) Addition cured (acetylenic end-caps). Poly(4,4′-(hexafluoroisopropylidene)bis(phthalic anhydride)-1,3-isobenzofurandione-alt-1,3-bis(3-aminophenoxy)benzene) Addition cured (acetylenic end-caps). α-(3-Ethynylphenyl)-ω-[5-[1-[2-(3-ethynylphenyl)-2,3-dihydro-1,3-dioxo-1H-isoindol-5-yl]-2,2,2-trifluoro-1-(trifluoromethyl)ethylidene]-1,3-dihydro-1,3-dioxo-2H-isoindol-2-yl]poly[(1,3-dihydro-1,3-dioxo-2H-isoindole-2,5-diyl)[2,2,2-trifluoro-1-(trifluoromethyl)ethylidene](1,3-dihydro-1,3-dioxo-2H-isoindole-5,2-diyl)-1,3-phenyleneoxy-1,3-phenyleneoxy-1,3-phenylene]*
Monomers: 1,3-Bis(3-aminophenoxy)benzene, 2,2′-Bis(3,4-dicarboxyphenyl)hexafluoropropane dianhydride
Material class: Thermosetting resin
Polymer Type: polyimide (end capped)
CAS Number:

CAS Reg. No.	Note
67297-78-3	CRU with end-groups

Molecular Formula: $(C_{37}H_{18}F_6N_2O_6)_n$
Fragments: $C_{37}H_{18}F_6N_2O_6$
Molecular Weight: MW 1343 M_n 1329, 1500, 2600
General Information: Isol. as an amorph. beige powder [9]

Volumetric & Calorimetric Properties:
Melting Temperature:

No.	Value	Note
1	160–180°C	[5]

Glass-Transition Temperature:

No.	Value	Note
1	105°C	[6]
2	205°C	after heating at 275° for 20 min. [6]

Transition Temperature General: T_g increases with an increase in either curing time or curing temp. but increases with increasing amounts of plasticiser [6]
Vicat Softening Point:

No.	Value	Note
1	175°C	[9]

Surface Properties & Solubility:
Solubility Properties General: Forms optically transparent blends with poly(diethyl 4,4′-oxydiphenylene pyromellitamate) provided the soln. composition is not a 50:50 mixture or a 30:70 blend (70% 6FDA-TPE) [3]. Addition of 10% (w/w) of Ultem 1000 to Thermid FA-700 causes an increase in the peak exotherm temp. whereas further addition results in a decrease in this temp. [4]
Plasticisers: Bis(2-(4-ethynylphenoxy)ethyl) ether [6]
Solvents/Non-solvents: Uncured oligomer sol. CH_2Cl_2, Me_2CO, THF, 2-butanone, glyme, DMF, cyclohexanone, 1-cyclohexyl-2-pyrrolidone (28 wt% or lower), diglyme (44 wt% or lower), *N,N*-dimethylacetamide, DMSO (44 wt% or lower), *N*-methylpyrrolidone, butyrolactone. Insol. cyclohexane, 2-propanol, ethyl glyme, 2-methoxyethanol, xylene, ethyl diglyme [9]

Transport Properties:
Polymer Solutions Dilute: [η] 0.21 Pa s (diglyme, 22°, 53% solids) [9]
Permeability of Liquids: Diffusion coefficient [H_2O] 2.8×10^{-8} cm^2 s^{-1} (1.5 μm film, cured) [7]. Films absorb water rapidly at greater than 10% relative humidity [7]

Mechanical Properties:
Mechanical Properties Miscellaneous: Peel strength decreases with increasing temp. [2]
Fracture Mechanical Properties: Inherently brittle owing to high degree of cross-linking. Brittleness may be overcome by forming a semi-interpenetrating network with high performance thermoplastic polyimides [8]

Electrical Properties:
Dielectric Permittivity (Constant):

No.	Value	Frequency
1	2.902	100 Hz, cold compacted [9]
2	2.8	30 MHz, cold compacted [9]
3	2.93	30 MHz, cured and post-cured [9]
4	2.983	100 Hz, cured [9]
5	2.9	30 MHz, cured [9]
6	3.071	100 Hz, cured and post-cured [9]

Dissipation (Power) Factor:

No.	Value	Frequency	Note
1	0.00176	100 Hz	cold compacted [9]
2	0.0048	30 MHz	cold compacted [9]
3	0.00373	100 Hz	cured [9]
4	0.0041	30 MHz	cured [9]
5	0.00191	100 Hz	cured and post-cured [9]
6	0.0044	30 MHz	cured and post-cured [9]

Optical Properties:
Transmission and Spectra: Ir spectral data have been reported [5]

Polymer Stability:
Thermal Stability General: May be cross-linked upon heating with a max. conversion of 81.49% after heating for 1 h at 215° [5]
Upper Use Temperature:

No.	Value	Note
1	316°C	[1]

Decomposition Details: Decomposition temp. 510° (in N_2) 549° (in air). The max. rate of weight loss at 560° (in N_2), 583.5° (in air). Char yield 61.7% (750°, in N_2) [4]. 5% Weight loss in moist air occurs at 535°; 50% weight loss at 608° [6]. Heating in circulating air at 316° causes a max. weight loss of 21.6% after 723 h [6].

Applications/Commercial Products:
Processing & Manufacturing Routes: Synth. by treating 4,4'-hexafluoroisopropylidene(phthalic anhydride) with 1,3-bis(3-aminophenoxy)benzene and 3-ethynylaniline in the mole ratio 2:1:2 followed by thermal imidisation in a toluene/N-methylpyrrolidone mixture [9]
Applications: Potential application in adhesives, in the aerospace industry; in electrical and electronic applications. Possible use as encapsulants for *in vitro* biosensors

Trade name	Supplier
Thermid EL 5512	National Starch & Chemicals
Thermid FA 700	National Starch & Chemicals

Bibliographic References
[1] *Thermid FA700*, National Starch & Chemicals, 199, (technical datasheet)
[2] Fujimoto, M., *Nippon Setchaku Gakkaishi*, 1993, **29**, 421
[3] Ree, M. and Yoon, D.Y., *Adv. Chem. Ser.*, 1994, **239**, 247, (miscibility)
[4] Choudhary, V. and Pearce, E.M., *J. Appl. Polym. Sci.*, 1994, **51**, 1823
[5] Huang, W.X. and Wunder, S.L., *J. Polym. Sci., Part B: Polym. Phys.*, 1994, **32**, 2005
[6] Harris, F.W., Sridhar, K., and Das, S., *Polym. Prepr. (Am. Chem. Soc., Div. Polym. Chem.)*, 1984, **25**, 110
[7] Chang, L.-H. and Tompkins, H., *Appl. Phys. Lett.*, 1991, **59**, 2278, (moisture diffusion)
[8] Pater, R.H., *SAMPE J.*, 1990, **26**, 19
[9] Capo, D.J. and Schoenberg, J.E., *SAMPE J.*, 1987, **23**, 35, (props., uses)

FEP, Carbon fibre filled   F-8

Synonyms: *Tetrafluoroethylene-hexafluoropropylene copolymer, Carbon fibre filled. Fluorinated ethylene-propylene copolymer, Carbon fibre filled. Poly(tetrafluoroethylene-co-hexafluoropropylene), Carbon fibre filled. Poly(tetrafluoroethene-co-hexafluoropropene), Carbon fibre filled*
Related Polymers: Filled FEP, FEP, PTFE filled, FEP, Glass fibre filled
Monomers: Tetrafluoroethylene, Hexafluoropropene
Material class: Composites, Copolymers
Polymer Type: PTFE, fluorocarbon (polymers)
General Information: Has superior chemical resistance and thermal props.

Volumetric & Calorimetric Properties:
Density:

No.	Value	Note
1	d 2.08 g/cm^3	15% filled [1]
2	d 2.1 g/cm^3	10% filled [1]

Thermal Expansion Coefficient:

No.	Value	Note	Type
1	3.96×10^{-5} K^{-1}	15% filled [1]	L
2	1.44×10^{-5} K^{-1}	10% filled [1]	L

Thermal Conductivity:

No.	Value	Note
1	0.26 W/mK	0.000621 cal (s cm °C)$^{-1}$ [1]
2	0.29 W/mK	0.00069 cal (s cm °C)$^{-1}$ [1]

Melting Temperature:

No.	Value	Note
1	268°C	15% filled [1]
2	271°C	10% filled [1]

Deflection Temperature:

No.	Value	Note
1	90°C	1.82 MPa, 15% filled [1]
2	118°C	1.82 MPa, 10% filled [1]

FEP, Filled — FEP, Glass fibre filled

FEP, Filled — F-9

Synonyms: *Tetrafluoroethylene-hexafluoropropylene copolymer, Filled. Poly(tetrafluoroethylene-co-hexafluoropropylene), Filled. Fluorinated ethylene-propylene copolymer, Filled*
Related Polymers: FEP, Glass fibre filled FEP, Carbon fibre filled FEP, PTFE filled FEP
Monomers: Tetrafluoroethylene, Hexafluoropropene
Material class: Composites, Copolymers
Polymer Type: PTFE, fluorocarbon (polymers)

Transport Properties:
Water Absorption:

No.	Value	Note
1	0.027%	24h, 10% filled [1]

Mechanical Properties:
Flexural Modulus:

No.	Value	Note
1	4688 MPa	47800 kg cm^{-2}, 15% filled [1]
2	3099 MPa	31600 kg cm^{-2}, 10% filled [1]

Tensile Strength Break:

No.	Value	Note	Elongation
1	17.8 MPa	182 kg cm^{-2}, 15% filled [1]	4%
2	33 MPa	337 kg cm^{-2}, 10% filled [1]	

Flexural Strength Yield:

No.	Value	Note
1	26.2 MPa	267 kg cm^{-2}, 15% filled [1]
2	50.3 MPa	513 kg cm^{-2}, 10% filled [1]

Izod Notch:

No.	Value	Notch	Note
1	256 J/m	Y	4.8 ft lb in^{-1}, room temp., 15% filled [1]
2	309 J/m	Y	5.8 ft lb in^{-1}, room temp., 10% filled [1]

Polymer Stability:
Upper Use Temperature:

No.	Value	Note
1	204°C	continuous service 15% filled [1]

Applications/Commercial Products:
Processing & Manufacturing Routes: Processed by extrusion or injection moulding
Mould Shrinkage (%):

No.	Value	Note
1	1.5%	[1]

Applications: Used in housings and valves

Trade name	Details	Supplier
Thermocomp FP-FC-1002	10% filled discontinued material	LNP Engineering
Thermocomp FP-FC-1003	15% filled	LNP Engineering

Bibliographic References
[1] *Thermocomp Range*, LNP Engineering, 1996, (technical datasheet)

FEP, Glass fibre filled — F-10

Synonyms: *Tetrafluoroethylene-hexafluoropropylene copolymer, Glass fibre filled. Fluorinated ethylene-propylene copolymer, Glass fibre filled. Tetrafluoroethylene-hexafluoropropylene copolymer, Carbon fibre filled. Poly(tetrafluoroethylene-co-hexafluoropropylene), Glass fibre filled. Poly(tetrafluoroethylene-co-hexafluoropropene), Glass fibre filled*
Related Polymers: Filled FEP, FEP, Carbon fibre filled, FEP, PTFE filled
Monomers: Tetrafluoroethylene, Hexafluoropropene
Material class: Composites, Copolymers
Polymer Type: PTFE, fluorocarbon (polymers)

Volumetric & Calorimetric Properties:
Density:

No.	Value	Note
1	d 2.2 g/cm^3	20% filled [1]
2	d 2.21 g/cm^3	

Thermal Expansion Coefficient:

No.	Value	Note	Type
0	4.32 × 10^{-5} K^{-1}	[2]	L
1	5 × 10^{-5} K^{-1}	20% filled [1]	L

Deflection Temperature:

No.	Value	Note
1	158°C	1.8 MPa, 20% filled [1]
2	176°C	1.82 MPa [2]

Transport Properties:
Water Absorption:

No.	Value	Note
1	0.01%	24h, 20% filled [1,2]

Mechanical Properties:
Flexural Modulus:

No.	Value	Note
1	5500 MPa	20% filled [1]
2	5511 MPa	56200 Kg cm^{-2} [2]

Tensile Strength Break:

No.	Value	Note	Elongation
1	40 MPa	room temp., 20% filled [1]	2.5%
2	41.4 MPa	42.2 Kg cm^{-2} [2]	

– **FEP, PTFE filled**

Hardness: Shore R65 (20% filled) [1], D63 [2]
Izod Notch:

No.	Value	Notch	Note
1	200 J/m	Y	20% filled [1]
2	427 J/m	Y	8 ft lb in^{-1}, room temp.

Electrical Properties:
Dielectric Permittivity (Constant):

No.	Value	Frequency	Note
1	2.5	1 KHz	20% filled [1]
2	2.52	1 MHz	[2]

Dielectric Strength:

No.	Value	Note
1	13 kV.mm^{-1}	20% filled [1]
2	18.7 kV.mm^{-1}	[2]

Dissipation (Power) Factor:

No.	Value	Frequency	Note
1	0.0005	1 KHz	dry, 20% filled [1]
2	0.0002	1 MHz	[2]

Polymer Stability:
Upper Use Temperature:

No.	Value	Note
1	150°C	continuous service, 20% filled [1]

Applications/Commercial Products:
Processing & Manufacturing Routes: Processed by injection moulding
Mould Shrinkage (%):

No.	Value	Note
1	0.4%	20% filled [1]

Trade name	Supplier
Fluoromelt FP-FF-1002	LNP Engineering
Fluoromelt FP-FF-1004M	LNP Engineering
Thermocomp FP-FF-1003M/1004/1004M	LNP Engineering
Thermocomp LF	LNP Engineering

Bibliographic References
[1] *The Materials Selector*, 2nd edn., (eds. N.A. Waterman and M.F. Ashby), Chapman & Hall, 1997, **3**, 231

FEP, PTFE filled   F-11

Synonyms: *FEP, Polytetrafluoroethylene filled. Poly(tetrafluoroethene-co-hexafluoropropene), PTFE filled. Poly(tetrafluoroethene-co-hexafluoropropylene), PTFE filled*
Related Polymers: Filled FEP, FEP, Carbon fibre filled, FEP, Glass fibre filled
Monomers: Tetrafluoroethylene, Hexafluoropropene
Material class: Composites, Copolymers
Polymer Type: PTFE, fluorocarbon (polymers)
General Information: Has superior chemical resistance and thermal props.

Volumetric & Calorimetric Properties:
Density:

No.	Value	Note
1	d 2.15 g/cm^3	10% filled [1]

Thermal Expansion Coefficient:

No.	Value	Note	Type
1	0.000117 K^{-1}	10% filled [1]	L

Thermal Conductivity:

No.	Value	Note
1	0.25 W/mK	0.000586 cal (cm s°C)$^{-1}$ [p], 10% filled [1]

Melting Temperature:

No.	Value	Note
1	268°C	10% filled [1]

Deflection Temperature:

No.	Value	Note
1	71°C	1.82 MPa, 10% filled [1]

Mechanical Properties:
Flexural Modulus:

No.	Value	Note
1	689 MPa	7030 kg cm^{-2}, 10% filled [1]

Tensile Strength Break:

No.	Value	Note	Elongation
1	13.7 MPa	140 kg cm^{-2}, 10% filled [1]	10%

Flexural Strength Yield:

No.	Value	Note
1	20.6 MPa	210 kg cm^{-2}, 10% filled [1]

Izod Notch:

No.	Value	Notch	Note
1	800 J/m	Y	15 ft lb in^{-1}, room temp., 10% filled [1]

Electrical Properties:
Dielectric Permittivity (Constant):

No.	Value	Frequency	Note
1	2.1	1 MHz	10% filled [1]

Dielectric Strength:

No.	Value	Note
1	19.6 kV.mm^{-1}	10% filled [1]

Dissipation (Power) Factor:

No.	Value	Frequency	Note
1	0.0003	1 MHz	10% filled [1]

Polymer Stability:
Upper Use Temperature:

No.	Value	Note
1	204°C	continuous service, 10% filled [1]

Applications/Commercial Products:
Processing & Manufacturing Routes: Processed by extrusion or injection moulding
Mould Shrinkage (%):

No.	Value	Note
1	3.5%	10% filled [1]

Applications: Used in housings and valves

Trade name	Details	Supplier
Fluoromelt FP-FL-4020		LNP Engineering
Teflon	B/30/100/140/160	DuPont
Teflon	FEP	Furon
Thermocomp	FP series	LNP Engineering
Thermocomp FP-FL-4020	discontinued	LNP Engineering

Bibliographic References
[1] *Thermocomp Range*, LNP Engineering, 1996, (technical range)

Fluoroalkoxyphosphazene elastomers F-12

Synonyms: *Alkylfluoroalkoxyphosphazene elastomers. Arylfluoroalkoxyphosphazene elastomers*
Monomers: Cyclic phosphazene trimers, Cyclic phosphazene tetramers
Material class: Synthetic Elastomers
Polymer Type: polyphosphazenes
Molecular Formula: $[(C_4H_4F_6NOP)(C_3H_5F_3NOP)]_n$
Fragments: $C_4H_4F_6NOP$ $C_3H_5F_3NOP$
Molecular Weight: MW 300000-1200000; M_n 30000-200000

Volumetric & Calorimetric Properties:
Glass-Transition Temperature:

No.	Value	Note
1	-65--44°C	depends on exact polymer [1]

Transition Temperature General: No crystalline melting transition temp. [1]

Surface Properties & Solubility:
Solvents/Non-solvents: Sol. Me$_2$CO, ketones, aldehydes. Insol. H$_2$O, aliphatic hydrocarbons [1]
Surface and Interfacial Properties General: Good adhesive props. [1]

Mechanical Properties:
Mechanical Properties General: Good elastomeric props. [1]
Viscoelastic Behaviour: Very slow viscous flow at room. temp. [1]

Polymer Stability:
Thermal Stability General: Good low temp. props. [1]
Flammability: Non-flammable [1]
Environmental Stress: Resistant to oxidation [1]
Chemical Stability: Resistant to oil and many organic solvents [1]
Recyclability: Not recyclable

Applications/Commercial Products:
Processing & Manufacturing Routes: Phosphazene cyclic trimers and tetramers with halogen, alkyl or aryl side groups are heated at 250–300°, either alone or with $(NPCl_2)_3$ for a day or more to prod. long-chain polymers. These polymers are treated with excess sodium trifluoroethoxide in THF at over 65° for several hours [1]
Applications: Possible uses as rubbers with good solvent and oxidation resistance and for non-burning foam insulation

Bibliographic References
[1] Allcock, H.R., McDonnell, G.S. and Desorcie, J.L., *Macromolecules*, 1990, **23**, 3873

Fluoromethyl silicone rubber, Heat vulcanised F-13

Synonyms: *Fluorosilicone rubber*
Related Polymers: Methylsilicone rubber, Heat vulcanised, Fluorosilicone oil
Monomers: Dichloroethenylmethylsilane, Dichloromethyl(3,3,3-trifluoropropyl)silane
Material class: Synthetic Elastomers, Copolymers
Polymer Type: silicones
CAS Number:

CAS Reg. No.
152425-27-9

Molecular Formula: $[(C_4H_7F_3OSi).(C_3H_6OSi)]_n$
Fragments: $C_4H_7F_3OSi$ C_3H_6OSi
Additives: Finely divided silica is incorporated as a reinforcer to improve mech. props.
General Information: Vinyl groups are present at less than 1%. They cause the gum to cure more readily with lower activity peroxides

Volumetric & Calorimetric Properties:
Density:

No.	Value	Note
1	d 1.45 g/cm^3	[10]

Thermal Expansion Coefficient:

No.	Value	Note	Type
1	0.0008 K^{-1}	[8]	V

Glass-Transition Temperature:

No.	Value	Note
1	-75°C	[12]

Brittleness Temperature:

No.	Value	Note
1	-68°C	-90°F [3]

Fluoromethyl silicone rubber, Room temp. vulcanised

Transition Temperature:

No.	Value	Note	Type
1	-61°C	-78°F [3]	Stiffening

Surface Properties & Solubility:
Cohesive Energy Density Solubility Parameters: δ 19.6 $J^{1/2}$ $cm^{-3/2}$ (9.6 $cal^{1/2}$ $cm^{-3/2}$) [6,7,8], δ_d 15.8, δ_p 6.6, δ_h 5.6 $J^{1/2}$ $cm^{-3/2}$
Solvents/Non-solvents: Vulcanised rubber is swollen by common organic solvents rather than dissolved, e.g. ketones prod. high volume swell (2-butanone Volume Swell 760) [8]. Sol. ketones, esters, ethers [4,6,8]. Mod. sol. chlorinated hydrocarbons, aromatic hydrocarbons. Insol. aliphatic hydrocarbons, alcohols, H_2O
Surface and Interfacial Properties General: Exhibits generally low surface adhesion [5]

Transport Properties:
Transport Properties General: The substitution of trifluoropropyl for methyl substantially reduces gas permeability [2]
Polymer Melts: Kinematic viscosity 100000 St (approx., pre-cure gum) [4]
Permeability of Gases: [O_2] 0.0085 (25°); [N_2] 0.0036 cm^2 s^{-1} GPa^{-1} (25°)

Mechanical Properties:
Mechanical Properties General: Requires vulcanisation with a reinforcing filler (usually silica) in order to possess acceptable mech. props. [3]
Tensile Strength Break:

No.	Value	Note	Elongation
1	6.9 MPa	1000 psi, 25°, filled [3]	120%

Viscoelastic Behaviour: Compression set 22% (22h, 150°, filled) [9]
Hardness: Durometer 45-70 (25°) [10]
Fracture Mechanical Properties: Tear strength 1.3 J cm^{-2} (filled) [9]

Electrical Properties:
Electrical Properties General: Fluorosilicone rubber is not as good an insulator as normal silicone rubber [6]
Dielectric Permittivity (Constant):

No.	Value	Note
1	6-7	[6,13]

Dielectric Strength:

No.	Value	Note
1	20 $kV.mm^{-1}$	[6]

Polymer Stability:
Polymer Stability General: Very resistant material used where resistance to chemicals is important. Commercially the most important, solvent resistant, low temp. material [1,9]
Thermal Stability General: Usable from -68° to 232° [1,9]
Upper Use Temperature:

No.	Value	Note
1	232°C	[9]

Decomposition Details: Decomposes at temps. of over 300° with emission of toxic vapour [5]
Environmental Stress: Resistant to oxidative degradation, water, weathering, ozone and corona. Uv radiation has no effect [4,5]
Chemical Stability: Not swollen by non-polar organic solvents, H_2O and alcohols. Swollen by organic solvents such as esters, ethers and ketones. Damaged by strong acids and alkalis [4,11]
Hydrolytic Stability: Stable in H_2O and aq. solns. at room temp. Damaged by high pressure steam
Biological Stability: Non-biodegradable. If properly vulcanised it is not subject to biological attack
Recyclability: Can be recycled by depolymerisation using steam or acid catalysis [5]

Applications/Commercial Products:
Processing & Manufacturing Routes: Methyl trifluoropropyl and methylvinyl cyclic polysiloxanes are polymerised and equilibrated together at 150-200° with an alkaline catalyst. There is a large excess of fluorosiloxane producing a linear polymer with less than 1% of methyl vinylsiloxane. Traces of hexamethyldisiloxane are added to terminate the siloxane chains. Elastomer is readily cured by organic peroxides (typically 1-2 parts per 100 rubber) at temps. of over 110°. Several hours post cure treatment is required to remove residual peroxide
Applications: Used as an encapsulant and coating; used for diaphragms and O-rings. It is preferred where solvent resistance is particularly important and is especially valuable where a solvent resistant material with good low temp. props. is required

Trade name	Details	Supplier
Sylon FX	11200 series, 11300 series	3M Industrial Chemical Products

Bibliographic References
[1] Pierce, O.R., *Silicone Technology*, (ed. P.F. Bruins), John Wiley and Sons, 1970, 7
[2] Robb, W.L., *Ann. N. Y. Acad. Sci.*, 1968, **146**, 119, (gas permeability)
[3] Pierce, O.R., Holbrook, G.W., Johannson, O.K., Saylor, J.C. and Brown, E.D., *Ind. Eng. Chem.*, 1960, **52**, 783
[4] Roff, W.J. and Scott, J.R., Fibres, Films, *Plastics and Rubbers*, Butterworths, 1971
[5] Noll, W., *Chemistry and Technology of Silicones*, Academic Press, 1968
[6] Stark, F.O., Fallender, J.R. and Wright, A.P., *Comprehensive Organometallic Chemistry*, (eds. G. Wilkinson, F.G.A. Stone and E.W. Abel), Pergamon Press, 1982, **2**, 305
[7] Yerrick, K.B. and Beck, H.N., *Rubber Chem. Technol.*, 1964, **37**, 261, (solubility)
[8] Baney, R.H., Voigt, C.E. and Mentele, J.W., Struct.-Solubility Relat. Polym., *(Proc. Symp.)*, (eds. F.W. Harris and R.B. Seymour), Academic Press, 1977, 225, (solubility)
[9] Hardman, B. and Torkelson, A., *Encycl. Polym. Sci. Eng.*, Vol. 15, 2nd edn., (ed. J.I. Kroschwitz), John Wiley and Sons, 1985, 204
[10] Ash, M. and Ash, I., Handbook of Plastic Compounds, *Elastomers and Resins*, (eds. M.B. Ash and I.A. Ash), Wiley-VCH, 1992
[11] Freeman, G.G., Silicones Plastics Institute, 1962
[12] Borisov, A.F., *Soviet Rubber Technology*, 1966, **25**, 5, (transition temps)
[13] Vincent, G.A., Feuron, F.W.G. and Orbech, T., Annu. Rep., *Conf. Electr. Insul. Dielectr. Phenom.*, 1972, 17, (dielectric constant)

Fluoromethyl silicone rubber, Room temp. vulcanised

Synonyms: *Methyltrifluoropropyl silicone rubber, Room temp. vulcanised. Fluorosilicone rubber, Room temp. vulcanised*
Related Polymers: Methylsilicone rubber, Room temp. vulcanised
Monomers: Dichlorodimethylsilane, Dichloromethyl(3,3,3-trifluoropropyl)silane, Hexamethylcyclotrisiloxane, Cyclic methyltrifluoropropylsiloxane
Material class: Synthetic Elastomers, Copolymers
Polymer Type: silicones, dimethylsilicones
CAS Number:

CAS Reg. No.
156395-52-7

Molecular Formula: $[(C_4H_7F_3OSi).(C_2H_6OSi)]_n$

Fragments: $C_4H_7F_3OSi$ C_2H_6OSi
Additives: Finely divided silica is usually incorporated as a reinforcer to improve mech. props.
General Information: Silanol groups are present in the pre-cure gum. During cure they form cross links through condensation. Fully cured elastomer has few free silanol groups

Volumetric & Calorimetric Properties:
Density:

No.	Value	Note
1	d 1.25 g/cm^3	[8]

Thermal Expansion Coefficient:

No.	Value	Note	Type
1	0.0009 K^{-1}	[5]	V

Glass-Transition Temperature:

No.	Value	Note
1	-100°C	20% trifluoropropyl, 80% methyl; increases with increase in trifluoropropyl substitution [6]

Surface Properties & Solubility:
Cohesive Energy Density Solubility Parameters: δ 17–18 J$^{1/2}$ cm$^{-3/2}$ (8.3–8.8 cal$^{1/2}$ cm$^{-3/2}$) (increases with trifluoropropyl substitution) [3,4,5]
Solvents/Non-solvents: Sol. ethers, esters, ketones. Mod. sol. aliphatic hydrocarbons, aromatic hydrocarbons, chlorinated hydrocarbons. Insol. alcohols, H$_2$O [4,5,7]. Vulcanised rubber swollen by solvents rather than dissolved
Surface and Interfacial Properties General: Exhibits generally low surface adhesion [2]

Transport Properties:
Permeability of Gases: Permeability of gases has been reported. [1] Gas permeability depends on composition; it decreases as trifluoropropyl substitution increases

Mechanical Properties:
Mechanical Properties General: Requires reinforcing filler to possess acceptable mech. props. [12,13]

Electrical Properties:
Dielectric Permittivity (Constant):

No.	Value	Note
1	4–5	increases with increase in trifluoropropyl substitution [4,11]

Dielectric Strength:

No.	Value	Note
1	20 kV.mm^{-1}	[4]

Polymer Stability:
Polymer Stability General: A solvent resistant, room temp. vulcanising rubber
Thermal Stability General: Usable from -68–200° [9,10]
Upper Use Temperature:

No.	Value	Note
1	200°C	[9]

Environmental Stress: Resistant to oxidative degradation, H$_2$O, weathering and ozone. Unaffected by uv radiation [2]
Chemical Stability: Swollen by esters, ethers and ketones; relatively unaffected by toluene, aliphatic hydrocarbons and CCl$_4$. Disintegrates in conc. sulfuric acid. Relatively unaffected by weak alkalis, weak acids and alcohols [2,7]
Hydrolytic Stability: Stable in H$_2$O and aq. solns. at room temp. Damaged by high pressure steam [2]
Biological Stability: Non-biodegradable. If properly vulcanised it is not subject to biological attack [2]

Applications/Commercial Products:
Processing & Manufacturing Routes: Dimethyl- and methyl trifluoropropyl-cyclic polysiloxanes are polymerised together at 150–200° with alkaline catalysis in the presence of small quantities of water. The gum cures at room temp. with alkoxysilane cross-linking agents and metal salt catalyst. Cross-linking requires moisture
Applications: Potential applications due to good solvent resistance as a sealant and coating

Bibliographic References
[1] Robb, W.L., *Ann. N. Y. Acad. Sci.*, 1968, **146**, 119, (gas permeability)
[2] Noll, W., *Chemistry and Technology of Silicones*, Academic Press, 1968
[3] Yerrick, K.B. and Beck, H.N., *Rubber Chem. Technol.*, 1964, **37**, 261, (solubility)
[4] Stark, F.O., Fallender, J.R. and Wright, A.P., *Comprehensive Organometallic Chemistry*, (eds. G. Wilkinson, F.G.A. Stone and E.W. Abel), Pergamon Press, 1982, **2**, 305
[5] Baney, R.H., Voigt, C.E. and Mentele, J.W., *Struct.-Solubility Relat. Polym.*, (*Proc. Symp.*), (eds. F.W. Harris and R.B. Seymour), Academic Press, 1977, 225, (solubility)
[6] Borisov, M.F., *Soviet Rubber Technology*, 1966, **25**, 5, (transition temps)
[7] Roff, W.J. and Scott, J.R., *Fibres, Films, Plastics and Rubbers*, Butterworths, 1971
[8] Ash, M. and Ash, I., Handbook of Plastic Compounds, *Elastomers and Resins*, (eds. M.B. Ash and I.A. Ash), Wiley-VCH, 1992
[9] Meals, R.N., *Ann. N. Y. Acad. Sci.*, 1965, **125**, 137
[10] Hardman, B. and Torkelson, A., *Encycl. Polym. Sci. Eng.*, 2nd edn., (ed. J.I Kroschwitz), John Wiley and Sons, 1985, **15**, 204
[11] Vincent, G.A., Fearon, F.W.G. and Orbech, T., Annu. Rep., Conf. Electr. Insul. Dielectr. Phenom., 1972, 17, (dielectric constant)
[12] *Ger. Pat.*, 1977, 2 644 555
[13] *Ger. Pat.*, 1977, 2 644 551

Fluorosilicone oil

Synonyms: *Fluoromethylsilicone fluid. Trifluoropropylmethyl silicone fluid. Poly[oxy(methyl(3,3,3-trifluoropropyl)silylene)]*
Related Polymers: PDMS fluid
Monomers: Dichloromethyl(3,3,3-trifluoropropyl)silane, Cyclic methyltrifluoropropylsiloxane
Material class: Fluids
Polymer Type: silicones
CAS Number:

CAS Reg. No.
25791-89-3
156395-51-6

Molecular Formula: $[C_4H_7F_3OSi]_n$
Fragments: $C_4H_7F_3OSi$

Volumetric & Calorimetric Properties:
Density:

No.	Value	Note
1	d 1.25 g/cm^3	[2]

Thermal Expansion Coefficient:

No.	Value	Note	Type
1	0.0008 K^{-1}	[4]	V

Transition Temperature:

No.	Value	Note
1	-48°C	Pour Point [2]

Surface Properties & Solubility:
Cohesive Energy Density Solubility Parameters: δ 19.5 J$^{1/2}$ cm$^{-3/2}$ (9.6 cal$^{1/2}$ cm$^{-3/2}$); δ_d 15.8; δ_p 6.6; δ_h 5.6 J$^{1/2}$cm$^{-3/2}$ [3,4]
Solvents/Non-solvents: Sol. ethers, esters, ketones. Insol. aliphatic hydrocarbons, chlorinated hydrocarbons, aromatic hydrocarbons, alcohols, H$_2$O [2,4,5]
Surface and Interfacial Properties General: Better lubricating props. than dimethyl silicone fluid [9,10,12]
Wettability/Surface Energy and Interfacial Tension: γ_c 24 mJ m^{-2} (24 mN m^{-1}) [1]

Transport Properties:
Polymer Melts: Viscosity 3 St (25°) [2,11]. Viscosity is very pressure dependent due to the bulky alkyl chain as well as the fluorine substitution. A rise in pressure of 1–1000 atm. increases viscosity by a factor of 15. Viscosity temp. coefficient 0.85 [4,9]

Electrical Properties:
Dielectric Permittivity (Constant):

No.	Value	Note
1	6.8	[5,11]

Dielectric Strength:

No.	Value	Note
1	15 kV.mm^{-1}	[15]

Dissipation (Power) Factor:

No.	Value	Note
1	0.001	[5]

Polymer Stability:
Polymer Stability General: Stable material which is particularly resistant to solvents
Thermal Stability General: Usable from -48–170° [2,7]
Upper Use Temperature:

No.	Value
1	170°C

Flammability: Fl. p. 200° [9]. Fire point 315°
Environmental Stress: Stable to H$_2$O, air, oxygen and other substances. Props. deteriorate due to cross-linking when exposed to radiation [8]
Chemical Stability: Chemically inert but decomposed by conc. mineral acids. Very thin films liable to degradation [2,5,8]
Hydrolytic Stability: Stable to H$_2$O and aq. solns. Very thin films spread on water; liable to hydrolytic degradation [2,5,8]
Biological Stability: Non-biodegradable [2,5]
Recyclability: Can be reused. No large scale recycling

Applications/Commercial Products:
Processing & Manufacturing Routes: For low viscosity fluids methyltrifluoropropyldichlorosilane is hydrolysed and the hydrolysate equilibrated under acid catalysis at 180° with small amounts of hexamethyldisiloxane. For high viscosity fluids cyclic methyltrifluoropropylsiloxanes are polymerised together under alkaline catalysis with traces of chain terminator at 150°
Applications: Used in lubrication where chemical solvents may be encountered

Bibliographic References
[1] Owen, M.J., *J. Appl. Polym. Sci.*, 1988, **35**, 895, (surface tension)
[2] Hardman, B. and Torkelson, A., *Encycl. Polym. Sci. Eng.*, 2nd edn. (ed. J.I. Kroshwitz), John Wiley and Sons, 1985, **15**, 204
[3] Yerrick, K.B. and Beck, H.N., *Rubber Chem. Technol.*, 1964, **37**, 261, (solubility)
[4] Barney, R.H., Voigt, C.E. and Mentele, J.W., Struct.-Solubility Relat. Polym., *(Proc. Symp.)*, (eds. F.W. Harris and R.B. Seymour), Academic Press, 1977, 225, (solubility)
[5] Stark, F.O., Fallender, J.R. and Wright, A.P., *Comprehensive Organometallic Chemistry*, (eds. G. Wilkinson, F.G.A. Stone and E.W. Abel), Pergamon Press, 1982, **2**, 305
[6] Pierce, O.R., *Silicone Technology*, (ed. P.F. Bruins), John Wiley and Sons, 1970, 9
[7] Roff, W.J. and Scott, J.R., Fibres, Films, *Plastics and Rubbers*, Butterworths, 1971
[8] Noll, W., *Chemistry and Technology of Silicones*, Academic Press, 1968
[9] Schiefer, H.M. and Van Dyke, J., *ASLE Trans.*, 1964, **7**, 32
[10] Tabor, D. and Winer W.O., *ASLE Trans.*, 1965, **8**, 69, (pressure dependence of viscosity)
[11] Buch, R.R., Klimisch, H.M. and Johannson, O.K., J. Polym. Sci., *Part A-2*, 1967, **7**, 563
[12] Clarson, S.J. and Semlyen, J.A., *Siloxane Polymers*, Prentice Hall, 1993

Furan resins, Foundry applications F-16
Related Polymers: Furan Resins, Phenolic Resoles, Phenolic Novolaks
Monomers: Furfuryl alcohol, Urea, Formaldehyde
Material class: Thermosetting resin
Polymer Type: furan
Additives: Addivtives depend on the foundry moulding process. Latent curing agents are needed for heat cure materials and sulfonic acids are needed for cure at lower temps.
General Information: Furan resins cover a wide range of chemical compositions. Some are well-condensed materials with little or no free furfuryl alcohol while other resins are urea-formaldehyde condensates with furfuryl alcohol

Applications/Commercial Products:
Processing & Manufacturing Routes: In the hot box process, quartz sand is batch mixed with phenolic resole or urea-modified furan resin (1.4–2%) and a latent curing agent such as ammonium nitrate (0.3–0.5%). The mix is shaped and precured at 200–260° for up to 2 min in core boxes. Use of modified furan resins (0.9–1.2%) with curing agents (0.2–0.4%, usually sulfonic acids) enables a warm-box process for binding sand at 140–200°. Furan and phenolic resins with special curing agent may be employed in a vacuum warm box process where the corebox is heated at 70–100°. For a no-bake process, sand and phenolic resole or furan resin (0.7–1.3%) and an acidic curing agent (0.2–0.6%) are cured at ambient temps. for 20 mins to several hours
Applications: The processing and manufacturing routes are all used to prepare sand-based moulds for casting of various metals in the foundry environment

Trade name	Details	Supplier
Rütaphen GA	foundry resin-amine cold box	Bakelite
Rütaphen GC	foundry resin-shell moulding	Bakelite
Rütaphen GH	foundry resin-hot box	Bakelite
Rütaphen GK	foundry resin-cold curing	Bakelite
Rütaphen GN	foundry resin-no bake	Bakelite
Rütaphen GS	foundry resin	Bakelite
Rütaphen GW	foundry resin-warm box	Bakelite
	sand mould resin	Neste

Gelatin G-1

Synonyms: *Hydrolysed collagen*
Related Polymers: More general information, Source of Gelatin, Collagen
Monomers: Amino acids
Material class: Proteins and polynucleotides
CAS Number:

CAS Reg. No.
9000-70-8

Molecular Weight: 65000–300000, 100000–500000, 90000
General Information: Vitreous solid obtained by hydrol. of collagen. Acid hydrol. gives type A gelatin, alkaline hydrol. gives type B gelatin. Both are largely amorph. solids, though they can contain up to 17% crystalline regions [13]. Gelatin is a mixture of proteins, and is pale-coloured, odourless and tasteless

Volumetric & Calorimetric Properties:
Density:

No.	Value	Note
1	d 1.34 g/cm^3	dry [3]

Melting Temperature:

No.	Value	Note
1	230°C	dry [7]

Glass-Transition Temperature:

No.	Value	Note
1	217°C	dry [6]
2	200°C	[15]

Surface Properties & Solubility:
Solubility Properties General: Polymer granules are hydrated in cold water, dissolving fully in hot water [1]. Gels are formed at concentrations ≥0.5%, below 35° [2]
Solvents/Non-solvents: Sol. ethylene glycol, glycerol, DMSO, trifluoroacetic acid [4,9], warm H$_2$O [11]. Spar. sol. EtOH, Me$_2$CO [8]. Insol. EtOH, CCl$_4$, Me$_2$CO, C$_6$H$_6$, petrol [1]
Wettability/Surface Energy and Interfacial Tension: Contact angle 120–126° (H$_2$O, 15% gel) [11]
Surface Tension:

No.	Value	Note
1	44–48 mN/m	10% soln., 50° [11]

Transport Properties:
Polymer Solutions Concentrated: Viscosity 20–70 mP (Type A) [2], 20–75 mP (Type B) [2], 5–10 mPa s (6.67 wt%, 60°) [4], 3–10 mPa s (40°) [4], 7.3–8.46 mPa s [14]
Water Content: 9–13% [1], 9–15% (50% relative humidity) [4], 44% (90% relative humidity, pH 6.10, 25°) [10]. Absorbs less than ten times its weight of water [1]

Electrical Properties:
Electrical Properties General: Amphoteric isoelectric points for type A pH 7–9 and for type B pH 4.7–5.4 [2]

Optical Properties:
Refractive Index:

No.	Value	Note
1	1.54	dry [5]

Total Internal Reflection: $[\alpha]_D$ -350--120 (0–40°) [11]. $[\alpha]_D$ -130--140 [4]

Polymer Stability:
Polymer Stability General: Amorph. solid formed if dried above T$_m$

Applications/Commercial Products:

Supplier
Eastman Chemical Company

Bibliographic References
[1] *Kirk-Othmer Encycl. Chem. Technol.*, 4th edn., (ed. J.I. Kroschwitz), John Wiley and Sons, 1994, **12**
[2] *Microencapsulation - Processes and Applications*, (ed. J.E. Vandegaer), Plenum Press, 1974, 21
[3] Chein, J.C.W., *J. Macromol. Sci., Rev. Macromol. Chem.*, 1975, **12**, 1
[4] *Encycl. Polym. Sci. Eng.*, (eds. H.F. Mark, N.M. Bikales, C.G. Overberger and G. Menges), John Wiley and Sons, 1985, **7**
[5] Sklar, E., *Photogr. Sci. Eng.*, 1969, **13**, 29
[6] Marshall, A.S. and Petrie, S.E.B., *J. Photogr. Sci.*, 1980, **28**, 128
[7] Godard, P., Biebuyck, J.J., Daumerie, M. Naveau, H. and Mercier, J.P., *J. Polym. Sci., Polym. Phys. Ed.*, 1978, **16**, 1817
[8] *Molecular Biology*, Academic Press, 1964, **5**
[9] Umberger, J.Q., *Photogr. Sci. Eng.*, 1967, **11**, 385
[10] Mason, C.M. and Silcox, H.E., *Ind. Eng. Chem.*, 1943, **35**, 726
[11] *Sci. Technol. Gelatin*, (eds. A.G. Ward and A. Courts), Academic Press, 1977
[12] Piez, K.A., *Treatise Collagen*, (ed. G.N. Ramachandran), Academic Press, **1**, 207
[13] Chien, J.C.W., *J. Macromol. Sci., Rev. Macromol. Chem. Phys.*, 1975, **12**, 1

Groundnut protein fibre G-2

$$-[NHCHC(=O)]_n-$$
          $|$
          R

Synonyms: *Peanut protein fibre*
Related Polymers: Proteins, Zein
Material class: Amino acid
Polymer Type: protein
General Information: Composed of various amino acids

Volumetric & Calorimetric Properties:
Density:

No.	Value
1	d 1.31 g/cm^3

Surface Properties & Solubility:
Internal Pressure Heat Solution and Miscellaneous: Heat of wetting 111.4 J g^{-1} (26.6 cal g^{-1}) [1], 114.7 J g^{-1} (27.4 cal g^{-1}) [2]
Solvents/Non-solvents: Sol. NaOH. Insol. organic solvents, acid soln.

Transport Properties:
Water Absorption:

No.	Value	Note
1	12–15 %	moisture regain; expands slightly when wet [1]

Mechanical Properties:
Mechanical Properties General: Tenacity 6.2–8 cN tex^{-1} [1]
Tensile Strength Break:

No.	Value	Note
1	78.5–117.7 MPa	8–12 kg mm^{-2} [1]

– Guayule rubber

Tensile Strength Yield:

No.	Value	Note
1	40–60 MPa	dry
2	80 MPa	wet [1]

Complex Moduli: Modulus of torsional rigidity 1300 MPa (1.3×10^{10} dynes cm^{-2}) [1]

Optical Properties:
Refractive Index:

No.	Value	Note
1	1.53	[1]

Polymer Stability:
Thermal Stability General: Does not soften or melt [1]
Flammability: Is less flammable than wool [1]
Chemical Stability: Is resistant to organic solvents [1]
Hydrolytic Stability: Is insol. acidic sols. Degrades in alkali [1]

Applications/Commercial Products:
Processing & Manufacturing Routes: The hexane extract of groundnut (peanut) meal is dissolved in dilute sodium hydroxide, allowed to mature (uncoiling the globular protein chains) and is then spun into dilute acid soln. The fibre is hardened by treatment with formaldehyde [2]
Applications: Used in fabrics as a blend with wool, cotton or rayon

Trade name	Details	Supplier
Ardil	discontinued	ICI, UK
Fibrolane C	discontinued	Courtaulds

Bibliographic References
[1] Gordon Cook, J., *Handbook of Textile Fibers*, 5th edn., Merrow, 1984, **2**
[2] Moncrieff, R.W., *Man-Made Fibers*, 6th edn., Newnes-Butterworth, 1975

Guayule rubber G-3

Synonyms: *Polyisoprene. cis-1,4-Polyisoprene*
Related Polymers: Natural rubber
Monomers: 2-Methyl-1,3-butadiene
Material class: Natural elastomers
Polymer Type: polydienes
Molecular Formula: $(C_5H_8)_n$
Fragments: C_5H_8
Molecular Weight: MW 200000, MW 100000–2500000, M_n 60000–800000, MW 2200000, M_n 920000, MZ/MN 1.8
Additives: Antioxidants, vulcanising agents, carbon black, normally added in practical applications
General Information: Rubber from *Guayule* shrubs differs from rubber from *Hevea* trees mainly with respect to non-rubber components. Guayule rubber contains less non-rubber and is more linear and gel-free; gives better processing characteristics [1]. Contains 100% *cis*-1,4-polyisoprene, with resins, rubber sol. triglycerides and higher terpenes. Has little or no branching and contains no natural oxidant. Evidence for bimodality when grown in wet conditions [5], normally unimodal

Volumetric & Calorimetric Properties:
Specific Heat Capacity:

No.	Value	Note	Type
1	1.67 kJ/kg.C	13° [3]	P

Melting Temperature:

No.	Value	Note
1	3°C	[3]

Glass-Transition Temperature:

No.	Value	Note
1	-68°C	[3]

Surface Properties & Solubility:
Solubility Properties General: Swelling index 3.44 (C_6H_6) [3], 3 (toluene) [6]
Solvents/Non-solvents: Swollen by C_6H_6, toluene

Transport Properties:
Polymer Solutions Dilute: Critical MW for entanglement 13000 [3]

Mechanical Properties:
Mechanical Properties General: Is slower to vulcanise than *Hevea* natural rubber and has lower moduli and tensile props. [2]
Tensile Strength Break:

No.	Value	Note	Elongation
1	11–21 MPa	[2]	640–850%
2	7.24 MPa	[3]	635%
3	18.89 MPa	vulcanised [6]	435%

Miscellaneous Moduli:

No.	Value	Note	Type
1	1.93 MPa	[6]	100% Modulus

Hardness: Shore A54 [3]. Other hardness values have been reported [6]
Failure Properties General: Tear strength 31.2 kN m^{-1} (vulcanised) [8]
Friction Abrasion and Resistance: Abrasion resistance has been reported [3]

Optical Properties:
Refractive Index:

No.	Value	Note
1	1.5205	25°
2	1.5415	25°

Transmission and Spectra: Ir spectrophotometric data have been reported [8]

Polymer Stability:
Polymer Stability General: Storage of *Guayule* plants results in partial degradation of the rubber [7]
Thermal Stability General: Degradation behaviour of guayule rubber, hevea rubber and synthetic, isoprene rubber is very similar. Natural antioxidants in hevea rubber give it slightly better stability. Unsaturated fatty acids (15% w/w) in guayule accelerate degradation [1]
Decomposition Details: Accelerated thermal and uv oxidative degradation cause changes in surface composition with the appearance of new functional groups, including hydroxyl, ether, carbonyl and carboxyl. Polynuclear aromatics are formed under extreme conditions [1]

Applications/Commercial Products:
Processing & Manufacturing Routes: Isol. from *Guayule* shrub by maceration and solvent extraction

Bibliographic References

[1] Lin, S.-S., *Rubber Chem. Technol.*, 1989, **62**, 315
[2] Ramos-De Valle, L.F., *Rubber Chem. Technol.*, 1981, **54**, 24
[3] Eagle, F.A., *Rubber Chem. Technol.*, 1981, **54**, 662
[4] Estibi, A. and Hamerstrand, G.E., *Rubber Chem. Technol.*, 1989, **62**, 635
[5] Backhaus, R.A. and Nakayama, F.S., *Rubber Chem. Technol.*, 1988, **61**, 78
[6] Sloan, J.M., Magliochetti, M.J. and Zukar, W.X., *Rubber Chem. Technol.*, 1986, **59**, 800
[7] Black, L.T., Swanson, C.L. and Hamerstrand, G.E., *Rubber Chem. Technol.*, 1986, **59**, 123
[8] Barigan, T.F., Verbisear, A.J. and Oda, T.A., *Rubber Chem. Technol.*, 1982, **55**, 407

Gutta-percha G-4

Synonyms: *Gutta. trans-1,4-Polyisoprene. trans-Polyisoprene*
Related Polymers: *trans*-1,4-Polyisoprene, Chicle, Balata, Jelutong, Guayule rubber
Monomers: 2-Methyl-1,3-butadiene
Material class: Natural elastomers, Gums and resins
Polymer Type: polybutadiene
CAS Number:

CAS Reg. No.
9000-32-2

Molecular Weight: MW 30000. MW determination has been reported
Additives: Contains large quantities of resin (esters of phytosterols)
General Information: Exists in several cryst. forms. α Crystalline form is stable, β-form is metastable [19]. First described in 1845. Isol. from *Palaquim, Isonandra* and *Payana* spp. (*Sapotacea* family) in S.E. Asia. Isol. has been reviewed [25]

Volumetric & Calorimetric Properties:
Density:

No.	Value	Note
1	d^{20} 0.945–0.955 g/cm^3	[3]

Thermal Expansion Coefficient:

No.	Value	Note	Type
1	2.3×10^{-5} K^{-1}	unstressed, 5–23° [10]	V
2	3.3×10^{-5} K^{-1}	stressed by quenching [10]	V

Thermodynamic Properties General: Heat of fusion 12.9 kJ mol^{-1} (α-form), 6.5 kJ mol^{-1} (β-form) [4]
Melting Temperature:

No.	Value	Note
1	65°C	α-form [3]
2	56°C	β-form [3]

Glass-Transition Temperature:

No.	Value	Note
1	-68°C	midpoint [1]
2	-71°C	[11]
3	-58°C	[9]

Transition Temperature General: α and β transitions depend on rate of cooling and thermal history [8]

Surface Properties & Solubility:
Solvents/Non-solvents: Sol. hot petrol, C$_6$H$_6$, CHCl$_3$, carbon disulfide [2]. Insol. EtOH, H$_2$O [2]

Transport Properties:
Permeability of Gases: Rate of gas permeation and diffusion increases sharply at 45° [14]. 1.3×10^{-8} g (h cm mmHg)$^{-1}$ (25°, H$_2$O) [13]

Mechanical Properties:
Mechanical Properties General: Mech. props. of gutta-percha have been reported in a study of semicrystalline polymers [5]
Tensile Strength Break:

No.	Value	Note
1	20 MPa	[3]
2	1 MPa	90% *trans* isomer [3]

Mechanical Properties Miscellaneous: The resilience of gutta-percha and balata has been measured from -70–200°. A rapid decrease occurs above 30° [21]

Optical Properties:
Transmission and Spectra: Ir and nmr spectral data have been reported [3,15,22,23,24]

Polymer Stability:
Decomposition Details: Pyrolysis under vacuum yields 96% isoprenes and pentenes [6]. From 50–400° isoprene, dipentene and mixed large molecules are the main pyrolysis products [12]
Environmental Stress: Degrades rapidly on exposure to ozone, air and light

Applications/Commercial Products:
Processing & Manufacturing Routes: Produced from isolates of *Palaquium* and *Payena* spp. by extraction into warm petroleum. Moulding temp. must be 60–100° to avoid degradation. Softens in hot water to allow processing [18]. Must not dry out completely
Applications: Previously been used for cable insulation, including underwater cables, golf ball covers and dental applications. No longer of commercial significance, has been superseded by wholly synthetic materials

Bibliographic References

[1] Burfield, D.R. and Lim, K., *Macromolecules*, 1983, **16**, 1170
[2] Roff, W.J., Fibres, Films, *Plastics and Rubbers*, Academic Press, 1956
[3] Barnard, D., Bateman, L., Cunneen, J.I. and Smith, J.F., *The Chemistry and Physics of Rubber-Like Substances*, (ed. L. Bateman), Applied Science Publishers, 1963
[4] Mandelkern, L., Quinn, F.A. and Roberts, D.E., *Rubber Chem. Technol.*, 1956, **29**, 1181
[5] Kargin, V.A., Sogolova, T.I. and Nadareiskivli, L.J., *Polym. Sci. USSR (Engl. Transl.)*, 1954, **6**, 1404
[6] Svab, V., *Tehnika (Belgrade)*, 1966, **21**, 2174
[7] Kirchof, F., Gummi, Asbest, *Kunstst.*, 1963, **16**, 302
[8] Cooper, W. and Smith, R.K., *J. Polym. Sci., Part A-1*, 1963, 159
[9] Butta, E. and Frosine, V., *Chim. Ind. (Milan)*, 1963, **45**, 703
[10] Feldman, R.I. and Sokolov, S.I., *Kolloidn. Zh.*, 1958, **20**, 388
[11] Marei, A.I., *Vysokomol. Soedin.*, 1952, 274
[12] Straus, S. and Madorsky, S.L., *J. Res. Natl. Bur. Stand. (U.S.)*, 1953, **50**, 165
[13] Heering, H., Puell, H. and Drewitz, I., *Kunststoffe*, 1948, **38**, 49
[14] van Amerongen, G.J., *J. Polym. Sci.*, 1947, **2**, 381
[15] Kishore, K. and Pandey, H.K., *Prog. Polym. Sci.*, 1986, **12**, 155
[16] Wren, W.G., *Annu. Rep. Prog. Rubber Technol.*, 1948, **12**, 6, (rev)
[17] Blow, C.M., *Annu. Rep. Prog. Rubber Technol.*, 1949, **13**, 14, (rev)
[18] Stern, H.J., Rubber, *Natural and Synthetic*, 2nd edn., Maclaren and Sons, 1967, 12
[19] Heinisch, K.F., *Dictionary of Rubber*, Applied Science Publishers, 1966
[20] Hager, T., MacArthur, A., McIntyre, D. and Suger, R., *Rubber Chem. Technol.*, 1979, **52**, 693
[21] Fujimoto, K., *CA*, 1962, **56**, 3609, (resilience)
[22] Tanaka, Y., *J. Appl. Polym. Sci.: Appl. Polym. Symp.*, 1989, **44**, 1
[23] Tanaka, Y., Mori, M., Takei, A., Boochathum, P. and Sato, Y., *J. Nat. Rubber Res.*, 1990, **5**, 2415
[24] Tanaka, Y. and Sato, H., *Polymer*, 1976, **17**, 113
[25] Archer, B.L. and Audley, B.G., *Phytochemistry*, 1973, **2**, 310

Heparin H-1

Synonyms: *Heparinic acid*
Monomers: Base monomer units include glucose, Base monomer units include L-iduronic acid, Base monomer units include glucosamine, Base monomer units include glucuronic acid
Material class: Polysaccharides
Polymer Type: Carbohyrate
CAS Number:

CAS Reg. No.
9005-49-6

Molecular Formula: $(C_{12}H_{19}NS_2)_n$ (average for free acid form)
Fragments: $C_{12}H_{19}NS_2$
Molecular Weight: Isolated MW 5000–25000; natural source MW in proteoglycan 600000–1000000 [1,4]
General Information: Heparin is a complex polysaccharide from mammalian tissues consisting of derivatives of glucosamine, glucuronic acid and other sugars, carrying sulfonic acid, sulfate and sulfonamide residues. It is isol. as an amorph. powder. Occurs in the blood, skin, aorta, umbilical cord and most tissues, supposedly manufactured by the most cells. Possesses anticoagulant and antithrombolytic props. Some heparin chains are linked to serine or small peptides and contain the linkage → GlcUA → Gal → Gal → Xyl → Ser, as found in other connective tissue polysaccharides, suggesting *in vivo* synth. as a proteoglycan [5]

Surface Properties & Solubility:
Solvents/Non-solvents: Sol. H_2O; insol. EtOH, $CHCl_3$, Me_2CO, C_6H_6

Transport Properties:
Polymer Solutions Concentrated: Mark Houwink viscosity relationships have been reported [2,3]

Optical Properties:
Total Internal Reflection: $[\alpha]_D$ +44–+70 (H_2O)

Polymer Stability:
Chemical Stability: Stable in soln. of pH 5.0–7.5

Bibliographic References
[1] Robinson, H.C., Horner, A.A., Hook, M., Ogren, S. and Lindahl, U., *J. Biol. Chem.*, 1978, **253**, 6687
[2] Liberti, P.A. and Sturala, S.S., *Arch. Biochem. Biophys.*, 1967, **119**, 510
[3] Liberti, P.A. and Sturala, S.S., *J. Polym. Sci.*, 1966, **134**, 137
[4] Nieduszynski, I., *Heparin,* (eds. D.A. Lane and U. Lindahl), Edward Arnold, 1989, 51
[5] Roden, L., *Biochem. Glycoproteins Proteoglycans*, (ed. W.J. Lennarz), Plenum Press, 1980, 321, (rev)

High-impact polystyrene H-2

Synonyms: *HIPS. Impact polystyrene. Impact resistant polystyrene. Rubber modified polystyrene. Rubber reinforced polystyrene. Polystyrene, High impact*
Related Polymers: Polystyrene - General Information, Polybutadiene, Styrene-butadiene block copolymer, Noryl (HIPS/PPO blend)
Monomers: Styrene, 1,3-Butadiene
Material class: Composites, Thermoplastic
Polymer Type: polystyrene
CAS Number:

CAS Reg. No.	Note
9003-53-6	high impact
9003-55-8	rubber modified

Molecular Formula: $(C_8H_8O)_n$
Fragments: C_8H_8O
Additives: A uv stabiliser is usually added to commercial products to prevent photooxidation of the rubber phase. Examples of uv stabilisers used are carbon black, zinc oxide, benzophenones and benzotriazoles (Tinuvin P). A fire retardant is usually added and similar compds. to those that are efficient for types of polystyrene are used. Typically a halogen bearing organic compd. e.g. decabromodiphenyl oxide (11–13%) is combined with antimony oxide (3–4%) [3,4]. Processing aids (plasticisers) and mould-release agents are often added for extrusion and injection-moulding applications
General Information: High-Impact Polystyrene polymer products are opaque and not transparent. High-impact polystyrene has the following improved physical props. compared to unmodified polystyrene polymer: increased elongation, ductility and environmental stress crack resistance and increased impact resistance. The physical props. of HIPS depend strongly on the type of rubber used and its props., the concentration ratio of styrene and butadiene, as well as the polymerisation and processing conditions, including the additives used. HIPS is a composite material consisting of discrete rubber particles (3–10 wt%) dispersed in a polystyrene matrix. The rubber used is usually grafted to the polystyrene to improve compatibility of the polystyrene and rubber phases at the phase interface. Rubbers used include pure polybutadiene (this is the most important commercially); styrene-butadiene copolymers (particularly those of 1,4-polybutadiene) and ethylene-propylene-diene terpolymer (EPDM rubber) which may be used to impart better thermal and uv light stability leading to impact resistant polymers with more resistance to outdoor ageing. Other rubbers which have been reported in the literature include acrylates, polyisoprene and polyethylene but these have had limited commercial success because of their low chemical reactivity. The rubber phase volume can be increased to 10-40% (w/v) by filling the rubber particles with polystyrene occlusions. Most commercial HIPS resins have particle sizes ranging from about 1 to 10 μm. However if the particle diameters are much larger than 5–10 μm, the toughness decreases and the finished part has a low surface gloss. The rubber particle size distribution does not have to be monodisperse. A bimodal particle size distribution containing some particles of rubbers with a diameter less than 1 μm has increased toughness props.
Rubber Cross-Linking Cross-linking of the rubber phase is important in forming a stable rubber particle during the polymerisation process. The vinyl content (pendant vinyl group formed by 1.2 polymerisation) of the rubber is important in determining the final cross-link density. The physical props. of HIPS are affected by the cross-link density. Grafting occurs during the polymerisation of styrene when some of the free radicals react with the rubber. Grafting is important because the graft copolymer concentrates at the interface of the polystyrene and rubber phases and strongly affects its particle size, morphology and toughness
Morphology: The morphology of high-impact polystyrene has been studied by electron microscopy [9,11,12]. Differences in morphology among the high-impact polystyrene resins are a result of different production processes, types of rubber and rubber concentrations. For high-impact polystyrene prod. using pure polybutadiene rubber, the soln. polymerisation process forms

spherical, cellular particles. This struct. has been called a "salami" or "honeycomb" struct. The mass suspension polymerisation process poduces a similar morphology, however blending yields irregularly shaped particles. Other morphologies can be obtained when styrene-butadiene block copolymers are used including cylindrical core-shell, lamellar and cellular structs. The morphology of HIPS is a non-equilibrium morphology because it depends on the shear rate applied during its synthesis [13]

Volumetric & Calorimetric Properties:
Density:

No.	Value	Note
1	d 0.98–1.1 g/cm^3	[5]
2	d 1.05 g/cm^3	ASTM D792 [1,2]

Thermal Expansion Coefficient:

No.	Value	Note	Type
1	3.4×10^{-5}–0.00021 K^{-1}	[5]	L

Volumetric Properties General: Density may be affected by additives e.g. fire retardants and uv stabilisers [2,5]
Thermodynamic Properties General: The ratio of the change in specific heat at T_g for one component in the blend to the change in specific heat for the pure material gives the relative amount of that component in the blend [43]

Thermal Conductivity:

No.	Value	Note
1	0.04–0.12 W/mK	[5]

Transition Temperature General: The Vicat softening point, deflection temp. under load, and service temp. are all lower than those of polystyrene [2,5]. Two T_g values are observed: one for the rubber phase and the other for the polystyrene phase. As the rubber and polystyrene phases are immiscible, there is generally no interaction and the presence of one phase does not greatly affect the T_g of the other [2]. The T_g of the rubber phase [2] can be affected by the degree of cross-linking of the rubber [67] and the presence of additives [3]. The T_g of the rubber phase may be used to identify the type of rubber in the polymer [73]. An additional transition may be observed just above the T_g of the rubber phase which corresponds to the T_g of the styrene-rubber copolymer [73]. T_m is lower than that of polystyrene due to the lack of a regular struct. [87]. Heat distortion temp. decreases with increasing rubber concentration and decreases sharply with increasing plasticiser concentration. It is relatively unaffected by rubber particle size, cross-link density and grafting. There is a small increase with increasing MW [2]. It is similar to that of polystyrene [68,73]

Deflection Temperature:

No.	Value	Note
1	81–96°C	1.8 MPa, ASTM D648 [2]
2	64–93°C	1.8 MPa [23]

Vicat Softening Point:

No.	Value	Note
1	78–100°C	[5]
2	94–104°C	ASTM D1525 [2]
3	95°C	[1]
4	100°C	[60]

Transport Properties:
Melt Flow Index:

No.	Value	Note
1	3–13.5 g/10 min	ASTM D1238, test condition G [2]
2	1 g/10 min	190°, 21.6 kg [11]
3	13 g/10 min	190°, 21.6 kg, rubber phase modified by copolymer [11]

Polymer Solutions Concentrated: Rheological behaviour during polymerisation has been reported [2,15,94]
Polymer Melts: The melt flow index decreases with increasing rubber concentration and MW; it decreases slightly as the amount of grafting increases. Melt flow index increases with increasing rubber particle size (up to a limiting value) and with increasing plasticiser concentration. It is relatively unaffected by cross-link density. The introduction of rubber particles into polystyrene causes only minor changes in the melt rheological behaviour [2]. Melt viscosity and melt elasticity and their comparison to those of polystyrene have been reported [88]. Viscosity-shear rate dependence and the melt elasticity have been reported [2]. High-impact polystyrene samples have slightly higher melt viscosities but lower melt elasticity than does general purpose polystyrene [89]. The temp. and MW dependence of melt viscosity are identical to those of polystyrene unmodified [88,90]. Melt rheology props. [91] are non-Newtonian [88]; pseudoplastic behaviour occurs at 220°. The melt viscosity and primary normal stress coefficients of mineral oil plasticised material as a function of shear rate and mineral oil concentration have been reported [93]. The concentration dependence is lower than expected because of the higher solubility of mineral oil in the rubber phase [2]. Shear viscosity and primary normal stress as a function of shear rate for various temps. have been reported [98]

Water Absorption:

No.	Value	Note
1	0.1–0.6 %	24h [5]

Mechanical Properties:
Mechanical Properties General: Modification by rubber reduces both the tensile strength and the modulus compared with polystyrene [2,5,68] but impact strength and elongation at break are both significantly increased [2,5,68]. Higher tensile props. are achieved by reduced rubber concentration, decreased rubber particle size, increased rubber cross-link density, increased matrix MW, reduced plasticiser concentration and low amount of grafting [2]. Elongation is increased by increased rubber particle size, increased percentage of rubber, decreased rubber cross-link density, or by the addition of plasticiser [2]. The addition of plasticisers can increase impact strength and processability [2,3]. The mechanical props. are strongly affected by the nature of the rubber phase including the rubber composition and concentration, rubber phase volume, particle size distribution, degree of grafting and cross-linking, and the MW and MW distribution of the matrix polystyrene [68]. Antioxidants, plasticisers, flame retardants and other additives also affect the mechanical props. [68,74]. The variables that affect the mechanical props. are not independent and the optimisation of one prop. may be detrimental to other props. [2]. Elongation at rupture 58% [1]. The tensile yield decreases with increasing particle size for a given particle type [60]. The loss modulus and storage modulus decrease with increasing temp. [98]. Plateau modulus 0.04 40000 N m^{-2} [98]

Tensile (Young's) Modulus:

No.	Value	Note
1	900–3500 MPa	[5]
2	1500–2500 MPa	rubber phase, styrene-butadiene copolymer [69]

3	2140 MPa	[1]
4	1500 MPa	[75]
5	1655–2241 MPa	ASTM D638 [2]
6	1600 MPa	ASTM D638 [68,73]
7	1300–3200 MPa	40–0% volume fraction rubber [59,60]

Flexural Modulus:

No.	Value	Note
1	1930 MPa	[60]
2	1793–2276 MPa	ASTM D790 [2]
3	1000–3400 MPa	20° [5]

Tensile Strength Break:

No.	Value	Note	Elongation
1	12–62 MPa	20° [5]	7–60%
2	15.9–23 MPa	ASTM D638 [2]	20–30%
3	21 MPa	[68,73]	40%
4	15–40 MPa	styrene-butadiene rubber [69]	15–60%
5	18 MPa	[60]	40%
6	17.35–18.63 MPa	177–190 kg cm^{-2}, copolymer modified rubber phase [11]	37–48%
7	17.35–18.24 MPa	177–186 kg cm^{-2}, copolymer modified rubber phase [11]	14–63%

Flexural Strength at Break:

No.	Value	Note
1	27–69 MPa	20° [5]
2	32.4–51.7 MPa	ASTM D790 [2]

Tensile Strength Yield:

No.	Value	Elongation	Note
1	13.8–25.5 MPa		ASTM D638 [2]
2	29.6 MPa		[1]
3	17.5 MPa	2%	[68,73]
4	20 MPa		[75]
5	10–40 MPa		40.0% volume fraction rubber [59,60]

Compressive Strength:

No.	Value	Note
1	27–62 MPa	[5]

Miscellaneous Moduli:

No.	Value	Note
1	0.04 MPa	Plateau modulus 0.04 40000 N m^{-2} [98]

Complex Moduli: The dynamic storage modulus decreases as the rubber particle volume fraction increases; values of the dynamic storage modulus converge at temps. below the T_g of the rubber [59]

Impact Strength: Impact strength generally increases with rubber phase volume. However, the rubber content is limited to 14% (w/w) for commercial resins owing to the high viscosity of the feed solns. in the soln. polymerisation process [2]. The rubber phase volume can be increased to 10–40% (w/v) by filling the rubber particles with polystyrene occlusions. At a constant rubber level, increasing the rubber particle size improves the impact strength with a similar effect observed when the polystyrene occlusion content is increased [2,71,72]. For particle sizes greater than 5–10 μm the impact strength decreases [2]. The particle size distribution also affects impact strength, with impact strength improved by the use of a bimodal distribution [2,17]. Gardner impact strength 29–173 cm kg (5 × 0.32 cm sample) [2]. Dart-drop impact strength medium to high [1]. Izod impact values may also be affected by processing conditions, e.g. temp. [2,11,75]. The notched Charpy impact strength depends strongly on the testing temp. [58,59]. Charpy impact strength 26–36 kg cm^2 cm^{-2} (unnotched) [11], 16–30 kg cm^2 cm^{-2} (unnotched) [11]. At temps. above the T_g of the rubber the notched Charpy impact strength increases with increasing rubber volume fraction [58,59]

Viscoelastic Behaviour: Viscoelastic behaviour has been reported [88,90,96]. The compliance of the rubber-modified melt is slightly lower than that of the polystyrene matrix [2]. The hard-elastic behaviour of high-impact polystyrene is similar to that of crystalline polymers [97]. The recovery process of elastic high-impact polystyrene includes an instantaneous component and a time-dependent mechanism [97]

Hardness: Rockwell M35–M70 [5], L65–L88 [2], M25–M65 (ASTM D785) [2]. Hardness is lower than that of polystyrene [5]

Failure Properties General: Crazing is the dominant mechanism in fatigue [68] and causes large increases in hysteresis and a decrease in modulus [68,80,81]. Although high-impact polystyrene is more fatigue resistant than polystyrene, it fails in a brittle fashion under cyclic loading [2]. Fatigue resistance is inferior to that of ABS [82]. The effect of stress and strain history upon mechanical props. and creep testing and repeated tensile tests has been reported [83,84]. Fatigue and creep response is affected by environment and solvents which do not show any effect on unstressed polymer do affect the stressed polymer [57]. High-impact polystyrene protected from the environment exhibits ductile behaviour at all pressures. However, specimens exposed to silicone oil exhibit two transitions as the pressure is increased: ductile-to-brittle transition followed by a brittle-to-ductile transition [85]. Is brittle at temps. below the T_g of the rubber phase [58,59].

Fracture Mechanical Properties: Fracture toughness K_{Ic} 1–2 J cm^{-3} m$^{1/2}$ [69]. The role of rubber particles in the toughening of polystyrene has been reported [2,6,9,17,60,68,71,72,73]. Rubber particles both initiate and control the craze growth. In the presence of rubber particles a large number of small crazes are formed prior to failure. Energy is absorbed upon formation of the craze. Provided the stress concentration does not become catastrophic (leading to the formation of a crack) the material will retain its structural integrity. At temps. approaching T_g, the crazed material can recover after the stress is removed [2]. Craze formation [2] is increased by good rubber matrix adhesion [74], large particle diameter (1–6 μm) [71,72], high rubber concentration [9], low rubber T_g [74], spherical rubber particles [74], high testing temp. [74] and high rubber thermal expansion coefficient [74]. The polystyrene MW has a negligible effect on craze formation, but stabilises the crazes once formed [2,76]. The mechanism of craze formation has been reported [2,6,77,78,79]

Izod Notch:

No.	Value	Notch	Note
1	27–400 J/m	Y	20° [5]
2	37.4–122.8 J/m	Y	23°, ASTM D256 [2]
3	134 J/m	Y	[1]
4	96 J/m	Y	[68,73]
5	80 J/m	Y	[60]
6	50–400 J/m	Y	styrene-butadiene rubber [69]

| 7 | 130–160 J/m | N | 25% w/v polybutadiene rubber [70] |
| 8 | 370–430 J/m | N | 40% w/v polybutadiene rubber melt/styrene-butadiene-styrene [70] |

Electrical Properties:
Dielectric Permittivity (Constant):

No.	Value	Frequency	Note
1	2.4–4.8	60 Hz	[5]
2	2.4–4.5	1 kHz	[5]
3	2.4–3.8	1 MHz	[5]
4	2.8	1 kHz	uv stabilised and fire retardant grade [5]

Dielectric Strength:

No.	Value	Note
1	12–24 kV.mm^{-1}	[5]
2	15 kV.mm^{-1}	uv stabilised and fire retardant grade [5]

Arc Resistance:

No.	Value	Note
1	20–140s	[5]

Dissipation (Power) Factor:

No.	Value	Frequency	Note
1	0.0004–0.002	60–1 MHz	[5]
2	0.0006	1 kHz	uv stabilised and fire retardant grade [5]

Optical Properties:
Optical Properties General: Optical props. are affected by the discrete rubber particles dispersed in the polystyrene phase. These rubber particles scatter light, with max. scattering at 0.75 μm particle size. The amount of scattering is less for smaller or larger rubber particles [2]. Transparent grades can be prod. by matching the refractive index of the rubber with the polystyrene phase [2,44,45,47]. However, the refractive index of the polystyrene and rubber phases must have the same temp. dependence or the amount of haze will also be temp. dependent [2,46,47]. Optical props. depend on the polystyrene-rubber composition, type of rubber used, rubber particle size, MW, cross-linking and grafting [2]. If the rubber particle size is reduced to less than 0.1 μm, which is shorter than the wavelength of visible light, transparent grades may be prod. [47]

Volume Properties/Surface Properties: Haze 30% (min.) (opaque/translucent). Gloss high or low depending on the rubber particles [2]. The gloss decreases with increasing rubber concentration, particle size, MW and grafting [2]. The gloss increases with increasing cross-link density and is little affected by plasticiser [2]. Gloss 91–99% (0.6–1 μm rubber particles) [2,48]. The gloss in bimodal rubber particle size systems is reduced by the larger-sized particles [2,48]. Transparency translucent-opaque [5]

Polymer Stability:
Thermal Stability General: The thermal degradation of blends of polybutadiene and polyisoprene rubbers with polystyrene in the absence of O_2 has been reported [2,49,50]. Thermal degradation of the polystyrene phase is dramatically slowed by the presence of the rubber phase. This is shown by the total weight loss and the appearance of volatile products [49]. The degradation products of rubber that can readily donate hydrogen, such as 4-vinylcyclohexene from polybutadiene and dipentene from polyisoprene, deactivate polystyryl radicals as they form, inhibiting the formation of volatile degradation products from polystyrene [2,49]. The presence of rubber particles, however, accelerates property loss on thermal ageing [2], compared to polystyrene. The polybutadiene component breaks down first [49]. The heat stability and softening temp. of high-impact polystyrene are inferior to those of polystyrene [5]

Upper Use Temperature:

No.	Value	Note
1	<90°C	max. continuous service [5]

Decomposition Details: Volatile decomposition products are evolved at approx. 445–465° [49]. Decomposition products from polybutadiene include butadiene, 4-vinylcyclohexene, propene, pentene, ethane and ethene; polystyrene decomposes to give styrene [49]

Flammability: Flammability rating SB [5]. Rubber-modified polystyrene ignites and burns readily yielding a thick, sooty smoke, with the characteristic odour of burning rubber [51]. Fire retardants may be added [3,4,52,53,54,61,62]. Oxygen index 17.6 (without flame retardants). Data for the percentage of char have been reported [54]

Environmental Stress: Resistance to environmental stress-cracking can be improved by increasing rubber particle size, the rubber cross-link density, the MW of the matrix, or by reducing plasticiser concentration [2,55]. Increasing grafting increases environmental stress-crack resistance initially until a plateau value is reached [2]. Rubber modification also gives improved environmental stress-crack resistance compared to unmodified polystyrene [2]. The environmental stress-cracking caused by hexadecane [56,57], various alcohols, and cottonseed-oleic acid has been reported [55]. Environmental stress-cracking is induced by the same chemicals as cause stress-cracking in polystyrene [5]. Yellows upon exposure to sunlight which also causes brittleness and cracking [2]. The effect of outdoor weathering is accelerated compared to polystyrene because of the degradation of the rubber phase [2,63,64]. Weathering reduces the impact strength dramatically. Unprotected high-impact polystyrene is therefore not suitable for long-term outdoor applications [2]. Improved weatherability may be obtained by using unsaturated rubbers with greater uv resistance, e.g. ethylene-propylene-diene rubber [2]. Uv stabilisers may be added [1,2,65,66]

Chemical Stability: Has chemical stability similar to that of polystyrene

Hydrolytic Stability: Has excellent hydrolytic stability

Biological Stability: Has biological stability similar to that polystyrene

Recyclability: Tensile modulus, tensile strength and elongation at yield are unaffected by recycling at 290° [75]. However, rubber particles lose their sphericity and the ratio of weight to M_n of the polystyrene component increases. Impact strength is reduced by recycling at 290°, changes from ductile to brittle and the mode of failure [75]

Applications/Commercial Products:
Processing & Manufacturing Routes: In mass prod. rubber modified PS, the rubber is dissolved in styrene monomer in a prepolymerisation stage [14]. This is followed by free radical polymerisation of the styrene monomer, typically using a peroxide catalyst. As polymerisation of the styrene proceeds two phases are formed; a rubber rich phase and a polystyrene rich phase. At the beginning of the reaction, the rubber rich phase is the continuous phase with the polystyrene rich phase as the dispersed phase. As the polymerisation proceeds the volume fraction of the polystyrene rich phase increases. When the volume fractions of the rubber and polystyrene phases are approx. equal, phase inversion begins to occur and polystyrene becomes the continuous phase containing small domains of rubber. The phase inversion generally occurs when the amount of PS is 2–3 times the initial amount of rubber [2]. Formation of styrene-rubber graft copolymer occurs throughout the polymerisation. This graft copolymer concentrates at the phase boundary between rubber and polystyrene and

High-impact polystyrene

stabilises the rubber particles. After phase inversion, the particle boundaries are fixed and polymerisation continues both inside the rubber particles and in the continuous phase. As conversion nears completion the temp. is increased to approx. 200° to cross-link the rubber chain to protect the rubber particles from the very high shear fields during fabrication. Although the manufacturing process for HIPS is similar to that for general purpose polystyrene, the apparatus used needs to be adapted [1,14] to meet additional demands like dissolving approx. 5–10% w/w polybutadiene in styrene which may take 10–20h. (The rubber content of commercial HIPS resins is limited to about 14% w/w). Additionally shear conditions need to be controlled to achieve phase inversion and the desired particle size. If shear agitation is stopped or insufficient, phase inversion does not occur, and the product is a quasi interpenetrating network of cross-linked rubber with polystyrene [2]. The shear agitation rate can also affect the particle morphology formed. A minimum shear rate was found to be required to achieve phase inversion and this was determined for a range of rubber concentrations [15]. HIPS is mostly prepared commercially by a soln. (or bulk soln.) continuous process using a linear flow reactor or a continuous stirred tank reactor. The processability of medium impact PS is similar to that of the homopolymer, general purpose grades. Mould shrinkage and dimensional stability are also similar. The high impact grades may be processed by injection-moulding, extrusion, blow moulding and vacuum forming (thermoforming). No special pretreatment is needed for printing and decorating. In processing these materials it is necessary to avoid overheating or long residence times which cause yellowing. Resins with high melt viscosities are most suitable for extrusion. A similar extrusion process is used for all styrene-based plastics. Most rubber-modified polystyrenes are fabricated into sheet by extrusion primarily for subsequent thermoforming operations. HIPS resins with low melt viscosity are used for injection moulding applications. Processing conditions are similar to those for general purpose polystyrene. Injection moulding can also be used to produce HIPS structural foams. Thermoforming of HIPS extruded sheet is a process of considerable commercial importance. Thermoforming is usually accomplished by heating a thermoplastic sheet above its glass-transition temp. and forcing it against a mould by appling vacuum, air or mechanical pressure. On coding, the contour of the mould is reproduced in detail. Carefully determined conditions for the plastic sheet temp. (heating time and mould temp.) must be maintained. For styrene plastics the normal forming temp. is 149° with an upper processing limit of 182°. Blow moulding is a multistep fabrication process used for manufacturing hollow, symmetrical objects. The granules are melted and a parison is obtained by extrusion or injection moulding. The parison is then enclosed by the mould, and pressure or vacuum is applied to force the material to assume the contour of the mould. After sufficient cooling, the object is ejected. Blow moulding is used for rubber-modified PS and for ABS to form speciality products. Injection blow moulding is used extensively for PS and styrene copolymers. HIPS may be used to form high-density structural foam. This foam is produced by injection moulding or profile extrusion methods. A decomposable chemical blowing agent that releases nitrogen or carbon dioxide is usually used e.g. sodium bicarbonate or azodicarbonamides

Applications: Injection moulded HIPS is used to make thin wall products for food packaging, closures and container lids, toys and novelties. The largest market for impact PS is food packaging. Vacuum forming of extruded HIPS sheet is used to produce refrigerator door liners, food-container liners, containers for dairy products and luggage. Refrigerator parts include large-scale applications as well as domestic freezers and other refrigerator fittings. Typical injection moulded HIPS products are air-conditioner grills, radio and TV cabinets, tape-cartridges and video cassettes. Injection moulded HIPS products are also used for kitchen/bathroom cabinets, toilet seats and tanks, instrument control knobs and housings for electrical appliances. Extruded HIPS products include furniture drawers and extruded panelling. Blow moulded HIPS is used for speciality bottles, containers and furniture parts. Other applications of HIPS include drain pipes, personal care products, lighting features, disposable-service ware, wall coverings, signs, plumbing and bath fixtures. HIPS structural foam parts are used to replace wood, metals or solid plastics. They are widely used in appliances, automobiles, furniture and in the construction industry. HIPS can be coextruded with other polymers to produce containers, dinner-ware, drink cups and packaging materials. Products containing HIPS blended with other polymers are used for many applications including containers for electrophotographic toner, recyclable trash bags, foam cushioning material, table tennis balls, radioactive shielding, wood-free pencils, heat resistant utensils, multilayer containers for microwave cooking, tyre treads. An application of HIPS is the formation of the HIPS/PPO (polypropylene oxide) blend used in the electrical and electronic applications. Applications are described in more detail in a separate entry

Trade name	Details	Supplier
Estyrene	high impact	Nippon Steel Chem. Co.
Kayson	high impact	Polysar
Lacqrene	high impact and general purpose	ATO Chimie
Lustrex	high impact	Monsanto Chemical
Noryl	HIPS-PPO blend	General Electric
Novacor		Novacor Chem. Int.
Shell	high impact	Shell Chemicals
Styron	high impact and general purpose	Dow
Styrowood	imitation wood, high impact	Sordelli
Styvarene	impact resistant	Plastugil
Toporex	high impact	Japan Synthetic
Vestyron	polystyrene and styrene butadiene	Hüls AG
	high impact	Amoco Chemical Company
	high impact	Gulf
	impact polystyrene	BASF
	HIPS containing bimodal styrene-butadiene rubber particles	Asahi Chem. Industry Co.
	fire-proof HIPS	Firestone
	transparent HIPS	Rohm and Haas
	HIPS wood composite	Daicel Chem.
	HIPS	Neste
	HIPS foam	Sumitomo Chem Co
	thermoplastic resin containing HIPS	Nissan Chemical Ind. Ltd.
	HIPS containing bimodal particles	Fina Chemicals (Division of Petrofina SA)
	intumescent HIPS	Polymer Co. Ltd.
	injection moulded HIPS	Showa Denko
	HIPS filled with calcium carbonate	United Sterling Corp. Ltd.
	impact polystyrene	Svenska Polystyren Fabriken AB
	high impact	Huntsman Chemical
	high impact	America Petrofina Inc.
	high impact	Mobil

Bibliographic References

[1] *Kirk-Othmer Encycl. Chem. Technol.*, 3rd edn., (ed. M. Grayson), Wiley Interscience, 1978, **21**, 811, 827
[2] *Encycl. Polym. Sci. Eng.*, 2nd edn., (ed. J.I. Kroschwitz), John Wiley and Sons, 1985, **16**, 1
[3] Chang, E.P., Kirsten, R. and Slagowski, E.L., *J. Appl. Polym. Sci.*, 1979, **21**, 2167, (flame retardants)
[4] Kuryla, W. and Papa, A., *Flame Retard. Polym. Mater.*, Marcel Dekker, 1973
[5] *The Materials Selector*, 2nd edn., (eds. N.A. Waterman and M.F. Ashby), Chapman and Hall, 1997, **3**, (uses, processing, tradenames)
[6] Molau, G.E. and Keskkula, H., *J. Polym. Sci., Part A-1*, 1966, **4**, 1595, (struct)
[7] Keskkula, H., *Appl. Polym. Symp.*, 1970, **15**, 51, (struct)
[8] Craig, T.O., Quick, R.M. and Jenkins, T.E., *J. Polym. Sci., Polym. Chem. Ed.*, 1977, **15**, 433, 441
[9] Michler, G.H., *Polymer*, 1986, **27**, 323, (morphology)
[10] Riess, G., Schlienger, M. and Marti, S., *J. Macromol. Sci., Phys.*, 1980, **17**, 355, (morphology)
[11] Abramov, V.V., Yenalyev, V.D., Akutin, M.S., Tchalaya, N.M. *et al*, *Org. Coat. Plast. Chem.*, 1981, **45**, 191, (morphology, mech props)
[12] Yenalyev, V.D., Noshova, N.A., Melnichenko, V.I., Bovkunenko, O.P. and Bulatova, V.M., *Org. Coat. Plast. Chem.*, 1981, **45**, 285, (morphology)
[13] Eastmond, G.C., *Polym. Eng. Sci.*, 1984, **24**, 541, (morphology)
[14] Amos, J.L., *Polym. Eng. Sci.*, 1974, **14**, 1
[15] Freeguard, G.F. and Karmarkar, M., *J. Appl. Polym. Sci.*, 1971, **15**, 1657, (synth)
[16] Whitmore, M.D. and Noolandi, J., *Macromolecules*, 1985, **18**, 657, (MW)
[17] Hobbs, S.Y., *Polym. Eng. Sci.*, 1986, **26**, 74
[18] *Eur. Pat. Appl.*, 1982, 48 389
[19] *Encycl. Polym. Sci. Eng.*, 2nd edn., (ed. J.I. Kroschwitz), John Wiley and Sons, 1985, **3**, 38
[20] *Kirk-Othmer Encycl. Chem. Technol.*, 3rd edn., (ed. M. Grayson), Wiley Interscience, 1978, **18**, 475, (uses)
[21] *Kirk-Othmer Encycl. Chem. Technol.*, 3rd edn., (ed. M. Grayson), Wiley Interscience, 1978, **11**, 86
[22] Gardner, W., Cooke, E. and Cooke, R., *Chemical Synonyms and Tradenames*, 8th edn., Technical Press Ltd., 1978
[23] *Chemical Industry Directory*, Tonbridge, Benn Business Information Services, 1988
[24] *UK Kompass Register*, Reed Information Services, 1996
[25] *New Horizons in Plastics*, (ed. J. Murphy), WEKA, 1991
[26] *Jpn. Pat.*, 1987, 62 218 428
[27] *Jpn. Pat.*, 1981, 10 177
[28] *Jpn. Pat.*, 1976, 25 389
[29] *Jpn. Pat.*, 1993, 05 04 274, (use)
[30] *U.S. Pat.*, 1993, 5 219 628, (use)
[31] *Brit. Pat.*, 1990, 88 06 731, (use)
[32] *Jpn. Pat.*, 1991, 03 176 354, (use)
[33] *Jpn. Pat.*, 1980, 80 110 136, (use)
[34] *Jpn. Pat.*, 1983, 58 31 749, (use)
[35] *U.S. Pat.*, 1986, 4 596 670, (use)
[36] *Jpn. Pat.*, 1994, 06 230 669, (use)
[37] *U.S. Pat.*, 1994, 5 334 657, (use)
[38] *PCT Int. Appl.*, 1993, 93 13 168, (use)
[39] Tung, H.C., Nonnemacher, G.S. and Hollandsworth, D., *Res. Discl.*, 1993, **350**, 431, (use)
[40] *Ger. Pat.*, 1994, 4 403 017, (use)
[41] *Jpn. Pat.*, 1994, 06 157 846, (use)
[42] *Can. Pat.*, 1992, 1 057 790, (use)
[43] *Analytical Calorimetry*, (eds. R.S. Porter and J.F. Johnson), Vol. 2, Plenum Press, 1970, 51, (calorimetry)
[44] Baum, B., Holly, W.H., Stiskin, H., White, R.A. *et al*, *Adv. Chem. Ser.*, 1976, **154**, 263, (haze)
[45] Gesner, B.D., *J. Appl. Polym. Sci.*, 1967, **11**, 2499, (haze)
[46] Rosen, S.L., *Polym. Eng. Sci.*, 1967, **7**, 115, (haze)
[47] Conaghan, B.F. and Rosen, S.L., *Polym. Eng. Sci.*, 1972, **12**, 134, (haze)
[48] *U.S. Pat.*, 1980, 4 528 327, (gloss, mech props)
[49] McNeill, I.C., Ackerman, L. and Gupta, S.N., *J. Polym. Sci., Polym. Chem. Ed.*, 1978, **16**, 2169, (thermal degradation)
[50] McNeill, I.C. and Gupta, S.N., *Polym. Degrad. Stab.*, 1980, **2**, 95, (thermal degradation)
[51] Birley, A.W., Haworth, B. and Batchelor, J., *Physics of Plastics: Processing, Properties and Materials Engineering*, Hanser/Oxford University Press, 1991
[52] *U.S. Pat.*, 1980, 4 214 056, (flame retardants)
[53] Sprenkle, W.E. and Southern, J.H., *J. Appl. Polym. Sci.*, 1981, **26**, 2229, (flame retardants)
[54] Granzow, A. and Savides, C., *J. Appl. Polym. Sci.*, 1980, **25**, 2195, (flame retardants)
[55] Bubeck, R.A., Arends, C.B., Hall, E.L. and van der Sande, J.B., *Polym. Eng. Sci.*, 1981, **21**, 624, (stress-cracking)
[56] McCammond, D. and Hoa, S.V., *Polym. Eng. Sci.*, 1980, **17**, 869, (stress-cracking)
[57] Hoa, S.V., *Polym. Eng. Sci.*, 1980, **20**, 1157, (stress-cracking)
[58] Bucknall, C.B., *Makromol. Chem., Makromol. Symp.*, 1988, **16**, 209, (transition temps, mech props)
[59] Bucknall, C.B., Cote, F.F.P. and Partridge, I.K., *J. Mater. Sci.*, 1986, **21**, 301, (transition temps, mech props)
[60] *Polymer Toughening*, (ed. C.B. Arends), Marcel Dekker, 1996, (mech props)
[61] Horn, W.E., Stinson, J.M. and Smith, D.R., *Annu. Tech. Conf. - Soc. Plast. Eng.*, 1992, **50**, 2020, (flame retardants)
[62] *U.S. Pat.*, 1992, 5 137 937, (flame retardants)
[63] Davis, A. and Sims, D., *Weathering of Polymers*, Applied Science, 1983, 208, (weathering)
[64] Bair, H.E., Boyle, D.J. and Kelleher, P.G., *Polym. Eng. Sci.*, 1980, **20**, 995, (weathering)
[65] Loelinger, H. and Gilg, B., *Angew. Makromol. Chem.*, 1985, **137**, 163, (uv stabilisers)
[66] Virt, J., Rosik, L., Kovarova, J. and Popisil, J., *Eur. Polym. J.*, 1980, **16**, 247, (uv stabilisers)
[67] Chen, C.C., Habibullah, M. and Sauer, H.A., *J. Appl. Polym. Sci.*, 1983, **28**, 391, (transition temp)
[68] *Encycl. Polym. Sci. Eng.*, 2nd edn., (ed. J.I. Kroschwitz), John Wiley and Sons, 1985, **12**, 436, (blends)
[69] *Physical Properties of Polymers Handbook*, (ed. J.E. Mark), AIP Press, 1996
[70] Wu, S., *Polym. Eng. Sci.*, 1990, **30**, 753, (mech props)
[71] Donald, A.M. and Kramer, E.J., *J. Appl. Polym. Sci.*, 1982, **27**, 3729, (mech props)
[72] Donald, A.M. and Kramer, E.J., *J. Mater. Sci.*, 1982, **17**, 2351, (mech props)
[73] Bucknall, C.B., *Toughened Plastics*, Applied Science, 1977, (mech props)
[74] *MMI Press Symp. Ser.*, 1982, **2**, 323, (blends)
[75] Kalfoglou, N.K. and Chaffey, C.E., *Polym. Eng. Sci.*, 1979, **19**, 552, (mech props, recycling)
[76] Fellers, J.F. and Kee, B.F., *J. Appl. Polym. Sci.*, 1974, **18**, 2355, (mech props)
[77] Bucknall, C.B., *Adv. Polym. Sci.*, Springer-Verlag, 1978, 121, (mech props)
[78] Bubeck, R.A., Buckley, D.J., Kramer, E.J. and Brown, H.R., *J. Mater. Sci.*, 1991, **26**, 6249, (mech props)
[79] Clieslinski, R.C., *J. Mater. Sci. Lett.*, 1992, **11**, 813, (mech props)
[80] Bucknall, C.B. and Stevens, W.W., *J. Mater. Sci.*, 1980, **15**, 2950, (mech props)
[81] Sauer, J., Habibullah, M. and Chen, C.C., *J. Appl. Phys.*, 1981, **52**, 5970, (mech props)
[82] Bucknall, C.B. and Stevens, W.W., Toughening Plast., Int. Conf., (Prepr.), Plastics and Rubber Institute, 1978, (mech. props)
[83] Bucknall, C.B., Clayton, D. and Weast, W.E., *J. Mater. Sci.*, 1973, **8**, 514, (mech props)
[84] Castellani, L. and Maestrini, C., *Polymer*, 1990, **31**, 2278, (mech props)
[85] Trent, J.S., Miles, M.J. and Baer, E., *J. Mater. Sci.*, 1979, **14**, 789, (mech props)
[86] Ghaffer, A., Scott, A. and Scott, G., *Eur. Polym. J.*, 1975, **11**, 271, (uv stability)
[87] Seymour, R.B. and Carraher, C.E., Structure-Property Relationships in Polymers Plenum Press, 1984, (transition temp)
[88] Kruse, R.L. and Southern, J.H., *J. Rheol. (N.Y.)*, 1980, **24**, 755, (rheology)
[89] Hwang, E.J. and Kim, K.U., *Pollimo*, 1986, **10**, 27, (rheology)
[90] Santamaria, A. and Guzman, G.M., *Polym. Eng. Sci.*, 1982, **22**, 365, (rheology)
[91] Pena, J.J., Guzman, G.M. and Santamaria, A., *Polym. Eng. Sci.*, 1981, **21**, 307, (rheology)
[92] Hyun, K.S. and Karam, H.J., *Trans. Soc. Rheol.*, 1969, **12**, 335, (rheology)
[93] Kruse, R.L. and Southern, J.H., *Polym. Eng. Sci.*, 1979, **19**, 815, (rheology)
[94] Song, Z., Yuan, H. and Pen, Z., *J. Appl. Polym. Sci.*, 1982, **32**, 3349, (rheology)
[95] Freeguard, G.F. and Karmarkar, M., *J. Appl. Polym. Sci.*, 1971, **15**, 1649, (rheology)
[96] Masuda, T., Kitamura, M. and Onogi, S., *J. Rheol. (N.Y.)*, 1981, **25**, 453, (rheology)
[97] Moet, A., Palley, I. and Baer, E., *J. Appl. Phys.*, 1980, **51**, 5175, (rheology)
[98] Schmidt, L.R., *J. Appl. Polym. Sci.*, 1979, **23**, 2463, (rheology)

High-impact polystyrene/poly(phenylene oxide) blends

H-3

Synonyms: *HIPS/PPO blends. HIPS/DMPPO blends. PS/PPO blends. Polystyrene/poly(phenylene oxide) blends, High impact. High-impact polystyrene/poly(2,6-dimethyl-1,4-phenylene oxide) blends. High-impact polystyrene/poly(phenylene ether) blends. HIPS/PPE blends. PS/PPE blends. High-impact polystyrene/polyoxyphenylene blend*
Related Polymers: High-impact polystyrene, Poly(phenylene oxide)
Monomers: Styrene, 1,3-Butadiene, 2,6-Dimethylphenol
Material class: Blends, Thermoplastic, Composites
Polymer Type: polyalkylene ether, polystyrene
CAS Number:

CAS Reg. No.	Note
9003-53-6	polystyrene
25134-01-4	PPO

Molecular Weight: It is not possible to give a definite MF due to the variety of rubber compounds which may be used to make high-impact polystyrene. The MW of the HIPS/PPO blend depends on the MW of the PPO, PS and rubber used, the ratio of HIPS to PPO, and the amount of rubber used

Additives: About 1% (w/w) low-density polyethylene is added to commercial HIPS/PPO resins as a processing lubricant [3,10]. Flame retardants may be added if required [3]. As PPO has an inherent flame resistance halogenated additives are not used. HIPS/PPO blends may be glass reinforced for applications requiring greater rigidity and dimensional stability [10]. Coloured HIPS/PPO blends may be prod. by using suitable additives. Other additives which may be used include proprietary stabilisers and impact modifiers

General Information: HIPS and PPO are compatible across the entire range of compositions and therefore the blends each have a single glass-transition temp. and form a clear film [7]. The available blends comprise a large family of products with a wide range of props. [1]. HIPS/PPO blends were designed to fill the significant price/property void between commodity materials such as HIPS or ABS with other engineering plastics [2]. Blending PPO with HIPS lowers the cost and improves some physical props. [3]. HIPS/PPO blends are tough, rigid materials that maintain excellent mechanical props. over a wide range of temps. such as good impact resistance and high tensile modulus. They have good dimensional stability with low creep; high heat distortion temp.; excellent electrical props. that are little affected by temp., humidity and frequency; good hydrolytic props. being resistant to water, detergents, acids, bases and steam; and low density [1,2,3,4]. The addition of HIPS improves the processability by enabling melt processing at lower temps. [3]. The lower melt viscosity of HIPS/PPO blends (compared to that for PPO) means that they can be injection moulded

Miscellaneous: Rubber types used in HIPS/PPO blends include high *cis* and low *cis* polybutadiene, EPDM rubber and styrene-butadiene block copolymers. The most common type is high *cis* polybutadiene. Optimum impact props. occur with an average rubber particle size of approx. 2μm or less. A core-shell morphology and the addition of very small rubber particles (0.3μm) improve the props. Most of the HIPS that is used for PPO blends is designed specifically for this purpose [3]. Typical compositions for HIPS/PPO blends contain 40–85% (w/w) PPO [3,11]. The ratio of PPO to PS in blends may be determined from the intensity of the PPO peak at 854cm^{-1} and the PS peak at 700cm^{-1} in the IR spectrum [22]. Many grades of HIPS/PPO blends are available commercially. Some grades are designed for specific processing conditions such as foaming, profile extrusion and electroplating. Some speciality grades are designed for a single application e.g. electrical connectors (high heat grade) and exterior automotive parts (high-impact mineral-filled grade) [3]. Although HIPS/PPO blends may be prod. as sheet and for vacuum forming, the greatest volume use is in pellets for injection moulding

Volumetric & Calorimetric Properties:

Density:

No.	Value	Note
1	d 1.06 g/cm^3	injection moulded/extrusion [1]
2	d 1.1 g/cm^3	foamable-injection moulded [1]
3	d 1.21–1.29 g/cm^3	20–30% glass reinforced, ASTM D792 [23]
4	d 1.29 g/cm^3	solid [44]
5	d 0.91 g/cm^3	[44]

Thermal Expansion Coefficient:

No.	Value	Note	Type
1	$5.4 \times 10^{-5} – 7.4 \times 10^{-5}$ K^{-1}	ASTM D696 [1]	L
2	$2.5 \times 10^{-5} – 3.6 \times 10^{-5}$ K^{-1}	20–30% glass reinforced blends, ASTM D696 [23]	L
3	6.8×10^{-5} K^{-1}	Foamable, ASTM D696 [1]	L

Volumetric Properties General: The density changes slightly for different polymer grades and processing methods. [1] Fire-retardant additives may increase the density. [23] Glass reinforcement increases the density. [23]

Thermodynamic Properties General: Thermal expansion coefficient varies with grade and composition but it is very low compared with other polymers [23]. Calorimetry has shown a negative heat of mixing across the complete composition range [1,29]. Typically ΔH_{mix} 40 J mol^{-1} (approx.) [3,13]

Thermal Conductivity:

No.	Value	Note
1	0.22–0.23 W/mK	unreinforced, ASTM C177 [23]
2	0.16–0.17 W/mK	glass reinforced, ASTM C177 [23]
3	0.12 W/mK	foamable, ASTM C177 [1]
4	0.26 W/mK	[44]
5	0.24 W/mK	solid [44]

Specific Heat Capacity:

No.	Value	Note	Type
1	1.2 kJ/kg.C	foamable [1]	P
2	2.07 kJ/kg.C	[45]	P
3	1.46 kJ/kg.C	solid [45]	V

Melting Temperature:

No.	Value	Note
1	180°C	[44]

Transition Temperature General: The glass-transition temp. depends upon the composition of the blend. A linear relationship has been demonstrated between Tg and the percentage of PPO in the blend by plotting a graph of results. Each blend composition has a single Tg because PPO and HIPS are compatible polymer systems. Tm and Tg have intermediate values depending on the amount of each of the constituents. Other thermal props. are also dependent on the blend composition. The heat deflection temp. is high and also depends on the blend composition [1,2]. Blends containing large amounts of PPO have the highest heat deflection temp. [24,25]. An equation relating the Tg of the blend and the mass fractions of HIPS and PPO has been reported. [26].

High-impact polystyrene/poly(phenylene oxide) blends

A thermodynamic relationship which relates the Tg of HIPS-PPO blends to composition and heat capacities has also been reported. [27]. A theory has been developed and used to predict the compositional variation of Tg for PPO-HIPS polymer blends [28]

Deflection Temperature:

No.	Value	Note
1	137.2°C	0.455 MPa [5]
2	129.4°C	1.82 MPa [5]
3	96–140°C	0.45 MPa, ASTM D648 [1,23]
4	88–133°C	1.8 MPa, ASTM D648 [1,23]
5	142–150°C	0.45 MPa, 20–30% glass reinforced ASTM D648 [23]
6	135–150°C	1.8 MPa, 20–30% glass reinforced, ASTM D648 [23]
7	82°C	0.46 Mpa, foamable, ASTM D648 [1]
8	96°C	1.82 Mpa, foamable, ASTM D648 [1]

Surface Properties & Solubility:

Solubility Properties General: Studies show that PS and PPO are miscible and form a compatible polymer blend. [1,13,40]. The values for the Flory Huggins interaction parameter for PS-PPO blends have been reported [29,42,43]

Transport Properties:

Polymer Melts: The rheological behaviour of HIPS-PPO blends is intermediate between the behaviour of the separate polymers. The melt viscosity of a particular blend is a function of the MW of PPO and PS, the ratio of PPO to HIPS, the amount of rubber and the rubber particle size in the HIPS, and the amount of plasticizers or fillers [1]. The melting behaviour of one HIPS-PPO commercial blend has been studied in detail and rheological constants have been reported [44]. A plot of log (melt viscosity) against log (shear rate) at 250° shows the zero shear viscosity plateau region. The transition region between the zero shear viscosity and power law asymptotes is broader for PPO-HIPS than for amorph. polymers [44]. Viscosity 11400 kPa s (Tm). The intrinsic viscosity of a HIPS-PPO blend has been measured and is intermediate between the corresponding values for HIPS and PPO [48]. Equation constants have also been reported. [48]. Graphs of the shear viscosity and primary normal stress difference as a function of shear rate have been plotted for this blend at temps. in the range 200–300°. Steady state shear viscosity data was included for graphs at 260° [48]. The intrinsic viscosity of a HIPS-PPO blend has been measured and is intermediate between the corresponding values for HIPS and PPO [48]. Equation constants have also been reported. [48]. Graphs of the shear viscosity and primary normal stress difference as a function of shear rate have been plotted for this blend at temps. in the range 200–300°. Steady state shear viscosity data was included for graphs at 260° [48]

Permeability of Gases: The permeability and diffusivity of He, CH_4 and CO_2 in PS-PPO blends has been found to give concave-upward curves characteristic of a strongly interacting miscible blend. Ne shows more complex behaviour. [36] The transport props. of He, CH_4 and CO_2 in PS-PPO blends at 35° have been studied as a function of pressure and absorption isotherms produced. [37] The permeability of CO_2 in these blends decreases as the pressure increases. [37] The effect of mineral oil plasticizer on the sorption data has been studied [37]. The separation factors for He-CH_4 and CO_2-CH_4 are larger than in the pure component polymer

Water Content: Polymer has lowest water absorption rate for engineering thermoplastics [13]

Water Absorption:

No.	Value	Note
1	0.066–0.08 %	24h, 23°, ASTM D570 [1,23]
2	0.06 %	24h, 23°, foamable, ASTM D570 [1]
3	0.06–0.07 %	24h, 23°, 20–30% glass reinforced, ASTM D570 [23]
4	0.14–0.21 %	equilibrium, 23°, [23]
5	0.3–0.45 %	equilibrium, 100°, [23]
6	0.12–0.22 %	equilibrium, 23°, 20–30% glass reinforced [23]
7	0.3–0.33 %	equilibrium, 100°, 20–30% glass reinforced [23]
8	0.07 %	equilibrium, 23°, foamable [1]

Mechanical Properties:

Mechanical Properties General: HIPS-PPO resins have good mechanical props. over a broad temp. range from -40°–105° [23]. The impact strength is high, which is retained at -40° [23]. HIPS-PPO resins have excellent creep resistance [23] and are tough, rigid materials [2]. Tensile strength and modulus of PPO and PS reach a maximum in a compostion containing 80% w/w PPO, but most props. of blends are close to the weighted average for the two polymers [13]. Blends with HIPS also have intermediate property values [13,17] but the impact strength is higher than either HIPS or PPO alone. [2] The effect of rubber particle concentration on the mechanical props. has been reported in the [45,46,47]. Other flexural modulus values have been reported [4]. The influence of composition, MW and MW distribution on the tensile modulus has been investigated. The tensile modulus decreases with increasing PPO composition.

The modulus at each blend composition is higher than that calculated using a simple rule of mixtures [47]. The effect of rubber particle concentration on the tensile elongation at break has been reported [45]

Tensile (Young's) Modulus:

No.	Value	Note
1	2500 MPa	23°, ASTM D638 [23]
2	1600 MPa	95°, ASTM D638 [23]
3	2448 MPa	23° [5]
4	1586 MPa	93° [5]
5	1620 MPa	foamable, ASTM D638 [1]
6	650–8400 MPa	23°, 20–30% glass reinforced, ASTM D638 [23]
7	5250–7800 MPa	95°, 20–30% glass reinforced, ASTM D638 [23]
8	2200–2700 MPa	[30]

Flexural Modulus:

No.	Value	Note
1	2500 MPa	23°, ASTM D790 [2,23]
2	1800 MPa	95° [5,23]
3	2200–2500 MPa	ASTM D790 [1,5]
4	1800 MPa	foamable, ASTM D790 [1]
5	2620 MPa	-18° [5]
6	4000–7000 MPa	95°, 20–30% glass reinforced [23]
7	5300–7700 MPa	23°, 20–30% glass reinforced, ASTM D790 [13]
8	2414–7586 MPa	all grades, ASTM D790 [1,4]

High-impact polystyrene/poly(phenylene oxide) blends

Tensile Strength Break:

No.	Value	Note	Elongation
1	49–61 MPa	23°, ASTM D638/D651 [23]	50–60%
2	45–76 MPa	ASTM D638 [1,13]	60% [1]
3	23 MPa	foamable, ASTM D638 [1]	6%
4	100–118 MPa	23°, 20–30% glass reinforced, ASTM D638 [23]	4–6%
5	66 MPa	23° [5]	20–30%
6	45 MPa	93° [5]	30–40%
7	48–117 MPa	all grades, ASTM D635 [4]	

Flexural Strength at Break:

No.	Value	Note
1	88.2–93.1 MPa	23°, ASTM D790 [23]
2	56–104 MPa	23°, ASTM D790 [1,2]
3	47 MPa	foamable, ASTM D790 [1]
4	110 MPa	-18 [5]
5	93.1 MPa	23° [5]
6	50.3 MPa	93° [5]
7	51 MPa	95° [23]
8	132–142 MPa	23° [23] 20–30% glass reinforced, ASTM D790
9	92–126 MPa	95°, 20–30% glass reinforced [23]
10	86–138 MPa	all grades, ASTM D790 [4]

Tensile Strength Yield:

No.	Value	Elongation	Note
1	49–64 MPa	7%	23°, ASTM D638 [23]
2	45 MPa		95°, ASTM D638 [23]
3	45–48 MPa		23°, ASTM D638 [1]
4	45–140 MPa	4–6%	23°, 20–30% glass reinforced, ASTM D638 [23]
5	74–94 MPa		95°, 20–30% glass reinforced, ASTM D638 [23]

Compressive Strength:

No.	Value	Note
1	103–123 MPa	all grades, ASTM D695 [4]
2	36 MPa	10% deformation foamable, ASTM D695 [1]

Impact Strength: Optimum impact props. occur with an average rubber particle size of 2 μm or less, which is lower than the corresponding particle size for HIPS. [1,34]. The impact strength of PPO-PS blends decreases with increasing PS composition [3,4]. The use of HIPS substantially increases the impact strength of the blends [3]. The Izod impact strength is a maximum at compositions near 50% w/w PPO [3]. Gardner impact 15–34 J (23°, ASTM D256) [1]; 28 J (35% rubber phase) [46]; 4–13 J (-40°) [1]; 5.7 J (26% rubber phase) [46]. Puncture impact strength (Dynatup test) 51 J (23°); 40 J (-40°, ASTM D3763) [1]. Falling ball impact 24 J (ASTM GE, foamable). [1] Graphs of the Izod impact strength have been reported [4,34]

Viscoelastic Behaviour: The loss modulus G-q has been measured as a function of frequency for a HIPS-PPO blend at various temps. from 193–300°. G-q decreases as the temp. increases [48]. The viscoelastic response of the blend at 260° is found to be intermediate to the HIPS and PPO values but closer to that of HIPS, the major component in the blend. Similar trends are observed for the storage modulus G-q and the relaxation-time spectra [48]. Plateau modulus G-q 60,000 Nm^{-2} (entanglement molecular weight Me 20,700) [48]. These values are intermediate between those of HIPS and PPO, as above [48]. The viscoelastic response of HIPS-PPO has been studied as a function of composition, frequency and temp. The relaxation processes of a blend exhibit the same temp. dependence but the viscoelastic response is a function of composition. The HIPS-PPO system therefore forms a rheologically compatible blend. Graphs of the loss modulus G-q and relaxation time τ_{12} are plotted as a function of weight fraction PPO and reduced frequency. Graphs of the storage modulus G-q as a function of reduced frequency and weight fraction of PPO have been reported [49]

Mechanical Properties Miscellaneous: Deformation under load 0.3% (13.8 MPa, 50°) [5]

Hardness: Rockwell R115–R119 (injection moulded) [1,23]. Rockwell R113 (extrusion) [1]. Rockwell L106–L109 (20–30% glass reinforced) [1,23]

Failure Properties General: Shear strength 72.4 MPa [5], 31 MPa (foamable, ASTM D732) [1]. The fracture behaviour and morphology of HIPS-PPO polymer blends under monotonic and cyclic loading conditions has been studied [33]. The HIPS particles fail by particle fracture during both fatigue and fast fracture. When PE is used as an additive, this affects the failure mechanism [33]. The effect of rubber particle size and characteristic ratio (chain flexibility) on the toughening mechanism of HIPS-PPO blends has been reported [34]. Shear yielding is enhanced for small (0.2 μm diameter) rubber particles and relatively low characteristic ratio. Larger rubber particles (4.9 μm diameter) rarely induce shear yielding and instead crazing occurs resulting in crack formation [34]. In blends containing less than 50% PPO crazing is the dominant failure mechanism, whereas in blends containing more than 50% PPO shear yielding is dominant [1]. If the tensile strength of PPO-PS blend is plotted as a function of blend composition, two maxima in tensile or yield strength are noted at both high and low PPO compositions [3,7]. The transition from ductile to brittle modes of deformation depends on the blend composition [3,45]. These failure mechanisms have been widely reported [6,7,8,9,31,32,33,34,35]

Fracture Mechanical Properties: Flexural fatigue endurance limit 2–17 MPa, (23° 2500 MHz); 27–32 MPa (20–30% w/w glass reinforced). Creep 0.75% (300h, 23°, 13.8 MPa) [1,5]. Creep 0.6–0.8% (300h, 13.8 MPa) [13]. Creep 0.275% (1000h, 23°, 3.4 MPa foamable) [1]. Creep curves showing the apparent modulus of HIPS-PPO resins at various temps., initial stress levels and times have been reported [23]

Friction Abrasion and Resistance: Taber abrasion resistance 20 mg (1000 cycles)$^{-1}$ CS-17, 20–30% glass reinforced, ASTM D1044). 35 mg (1000 cycles)$^{-1}$ (CS-17, 20–30% glass reinforced, ASTM D1044) [24]

Izod Notch:

No.	Value	Notch	Note
1	210–270 J/m	Y	23°, ASTM D256) [23]
2	1.4 J/m	Y	-40° [23]
3	200–350 J/m	Y	[31]
4	215–530 J/m	Y	23°, ASTM D256 [1,3]
5	130–190 J/m	Y	-40° [1,3]
6	90 J/m	Y	23°, 20–30% glass reinforced, ASTM D256 [23]
7	0.7 J/m	Y	-40°, 20–30% glass reinforced, [23]
8	75 J/m	Y	-40° [5]

9	96 J/m	Y	23° [5]
10	213.5 J/m	Y	35% rubber phase [46]
11	2.24 J/m	Y	93° [5]
12	149.5 J/m	Y	26% rubber phase [46]

Electrical Properties:

Electrical Properties General: High volume resistivity for all blend compositions [23]. The electrical props. are relatively unaffected by temp. and frequency [4,13] and humidity [13, 23]

Surface/Volume Resistance:

No.	Value	Note	Type
2	110×10^{15} Ω.cm	ASTM D257, foamable grade	S

Dielectric Permittivity (Constant):

No.	Value	Frequency	Note
1	2.65–2.69	60 Hz	20°, ASTM D150 [23]
2	2.65–2.78	60 Hz	23°, ASTM D150 [1]
3	2.27	100 Hz	23°, foamable, ASTM D150 [1]
4	2.86–3.15	60 Hz	20°, 20–30% glass reinforced, ASTM D150 [23]
5	2.85–3.11	1 MHz	20°, 20–30% glass reinforced, ASTM D150 [23]
6	2.64–2.68	1 MHz	20°, ASTM D150 [23]

Dielectric Strength:

No.	Value	Note
1	16–22 kV.mm^{-1}	3mm, short time, ASTM D149 [23]
2	16–24 kV.mm^{-1}	3mm short time, 20–30% glass reinforced, ASTM D149 [23]
3	16–25 kV.mm^{-1}	ASTM D149 [1]
4	7.49 kV.mm^{-1}	foamable, ASTM D149 [1]

Arc Resistance:

No.	Value	Note
1	75s	tungsten, ASTM D495 [23]
2	110–120s	tungsten, 20–30% glass reinforced, ASTM D495 [23]
3	67s	foamable, ASTM D495 [1]

Strong Field Phenomena General: High voltage arc tracking rate 22 cm min^{-1} (foamable) (UL746A) [1]. High ampere arc ignition 43 arcs (foamable) (UL746A) [1]

Dissipation (Power) Factor:

No.	Value	Frequency	Note
1	0.0004–0.0007	60 Hz	20°, ASTM D150 [23]
2	0.0004–0.003	60 Hz	23°, ASTM D150 [1]
3	0.0047	100 Hz	23°, foamable, ASTM D150 [1]
4	0.0008–0.002	60 Hz	20°, ASTM D150 glass reinforced [23]
5	0.0009–0.0024	1MHz	20°, ASTM D150 [23]
6	0.0039	1MHz	23°, foamable, ASTM D150 [1]
7	0.0014–0.0021	1MHz	20°, 20–30%, glass reinforced, ASTM D150 [23]

Optical Properties:

Optical Properties General: The optical props. of HIPS-PPO blends are not reported in the literature as these props. are not significant commercially. Light scattering from the rubber particles affects the refractive index and other optical props.

Transmission and Spectra: C-13 nmr spectral data have been obtained for HIPS-PPO blends. [38] Ir spectral data have also been used to determine the blend composition [22]. Other nmr spectral data have been reported [41].

Volume Properties/Surface Properties: Gloss 65.7% (35% rubber phase, rubber particle size 0.3 µm) [39]; 52.9% (26% rubber phase, rubber particle diameter 0.3µm) [39].

Polymer Stability:

Polymer Stability General: The stability of HIPS-PPO blends is expected to be intermediate between the stability of high-impact polystyrene and polyphenylene oxide, depending on the blend composition and processing conditions.

Upper Use Temperature:

No.	Value	Note
1	85–110°C	mechanical and electrical without impact, UL746B [1,23]
2	85–105°C	mechanical with impact, UL746B [1,23]
3	90–110°C	mechanical and electrical without impact, 20–30% glass reinforced [23]
4	90–105°C	mechanical with impact, 20–30% glass reinforced [23]

Flammability: HIPS-PPO are classified as slow burning materials and fire retardant additives may be used for self-extinguishing grades. [23] As the PPO component has inherent fire resistance, blends can be made flame retardant without the use of halogenated additives that tend to lower impact strength and melt stability in other polymers. [12] Flammability rating V0-V1 (UL94) [1,23]; SE1 (UL94) [23]; V0/5V (foamable, UL94) [1].

Environmental Stress: Prone to environmental stress cracking. [23] Prolonged outdoor exposure will cause parts made in standard grade to discolour, craze and lose props; can be protected by organic or metal coatings. [23]

Chemical Stability: HIPS-PPO resins have good resistance to some organic solvents, but are attacked by aromatic or chlorinated aliphatic compds. [13,23]

Hydrolytic Stability: HIPS-PPO resins are completely stable to hydrolysis. [13] Unaffected by aq. media including detergents, weak or strong acids and bases [5,23] and have good resistance to steam [4]

Stability Miscellaneous: Has good resistance to wear [23] and outstanding dimensional stability, particularly under stress at elevated temps. and in humid conditions. [23]

Applications/Commercial Products:

Processing & Manufacturing Routes: HIPS/PPO blends can be processed by injection moulding, foam moulding, blow moulding, extrusion and thermoforming. In Injection Moulding [1,12] conventional injection moulding machines are used. The melt temp. varies for different grades of product (blend composition) but approximates to the heat deflection temp. under load plus 127°. The melt temp. may also be affected by the length of melt flow, part configuration, mould temp. and other factors. The most commonly used melt temps. are in the range 77–95°. Temps. below 65° may cause edge cracking and excessive stresses in the moulded part. Typical clamping pressures are in the range 41–69 MPa. The low moisture absorption of HIPS/PPO blends means that for many applications, the pellets can be moulded without drying. For optimum surface appearance, the material must be dried for 2–4h at 100–120°. In foam moulding HIPS/PPO blends can occur via several different processes. HIPS/PPO blends are used in blow moulding [1,21] because of their high melt strength, excellent melt elasticity, and high blow ratio capacity. The blow moulding process requires different equipment than that used for blow moulding polyolefin-based resins. Pressures for shaping the part are in the vicinity of 0.4 MPa and the melt temp. used is approximately 21–27° below that employed for injection moulding of similar blends. Blow moulding is useful for the production of large, low volume parts, since tooling costs are relatively low and complex designs can be incorporated in the parts. HIPS/PPO blends are used for extrusion because of their excellent melt strength, good melt stability and wide processing range. The extrusion melt temp. can be approximated by adding 82° to the heat deflection temp. under load of the resin being processed. Drying the resin improves surface appearance and prevents porosity, particularly in parts over 0.2cm thick. Drying temps. vary from 88° to 115° depending on the resin grade. Extruded shapes are cooled and shaped in water (minimum temp. 38°) or air. Using the latter minimises stresses and reduces warping. Most extruded HIPS/PPO blend sheet material is thermoformed into large, complex shapes with draw ratios in normal parts of 1:4. Deeper draw ratios are feasible, because of the excellent melt strength of HIPS/PPO blends. The heating temp. depends on the grade of material. Spray mist cooling is not recommended for thermoforming HIPS/PPO resins. A draft angle of 3° is commonly used, but this angle may have to be increased for very complex shapes. HIPS/PPO resin parts may be finished by common techniques such as painting, hot-stamping, sputtering or plating. Articles can be bonded by adhesives or ultrasonic welding. The precise details of how the HIPS and PPO polymers are blended are protected by patents [11,20]. The conditions used may be varied for different polymer blend compositions and when different additives are used. However HIPS/PPO polyblends are generally formed by a melt-blending process

Mould Shrinkage (%):

No.	Value	Note
0	5–7%	injection moulding [1,12]
1	0.0006–0.008%	foam moulding [1]

Applications: HIPS/PPO blends are used in many industrial applications such as in computer housings, television components (e.g. housings, deflection yoke parts, sockets, tuner strips, high voltage cups and connectors), keyboard frames, interface boxes and structural parts for telecommunication applications. Special HIPS/PPO composites are replacing metal in chassis or frames of electrical and electronic equipment because of the high stiffness and dimensional stability of these materials and the reduced weight of the product Foam-moulded HIPS/PPO blends are used for large housings and panels (air conditioner and business computers). They are also used for sound-deadening applications (office partitions) and for other large articles such as hospital and office furniture and storm doors. Blow moulded instrument panels permit functionalised assemblies, wire and cable management, integrated ductwork and structural ribbing. HIPS/PPO blends are used in automobile/automotive applications including instrument panels, interior trim, speaker grilles, consoles, glove compartments, exterior trim, cowl tops, wheel covers, headlight housings, mirror cases, electrical connectors and fuse boxes. These materials are used because their low specific gravity reduces the total vehicle weight without sacrificing the cost-performance ratio or the ease of processing. Large exterior automotive parts such as doors and fenders are being developed. HIPS/PPO blends are used for fluid handling applications because of their low water absorption props. and dimensional stability. These applications include pump housings, pump impellers, valves and pipes and other cold and warm water applications including shower heads and water sprinklers. HIPS/PPO resins are used in small and large home appliances. The excellent electrical props. of these materials permit functional parts to serve also as electrical insulators. Injection moulded HIPS/PPO may be used in small home appliances such as hair dryers, coffee makers and steam irons. HIPS/PPO materials may also be used in larger home appliances e.g. refrigerators and ranges, in control panels, door liners, handles, and freezer parts. Other applications of HIPS/PPO resins include electrical construction such as lighting, outlet boxes and smoke detectors and they are also used as electrical insulators in electrical connectors, wiring and cable covers, wiring ducts and protective devices. These blends are also used for electrical applications in automobiles as well as for domestic and industrial appliances. Speciality applications of HIPS/PPO materials include aircraft, material handling trays, recreational equipment, heat-resistant utensils and polymer electret films. The polymer film acts as an electret because its dielectric props. allow the film to have a stable charge.

Trade name	Details	Supplier
Laril	unfilled, fire-retardant and glass filled HIPS/PPO blends	Freeman
Luranyl	HIPS/PPO blends for injection moulding	BASF
Luranyl	HIPS/PPO blends (coatings for wires)	Hitachi Cable Co Ltd
Luranyl	HIPS/PPO automobile parts	General Electric Plastics
Luranyl	HIPS/PPO injection mouldings	Mitsubishi Rayon Co Ltd
Luranyl	HIPS/PPO compounds	Idemitsu Kosan Co Ltd
Noryl	family of HIPS/PPO blends with many different grades of material	General Electric
Noryl	HIPS/PPO blends	Asahi Chem. Industry Co.
Prevex	unfilled, fire-retardant and glass filled HIPS/PPO blends	Borg Warner
Thermocomp	various grades of HIPS/PPO blends unfilled, glass filled, glass bead filled, mineral filled and PTFE filled	LNP Engineering

Bibliographic References

[1] *Encycl. Polym. Sci. Eng.*, Vol. 13, 2nd edn. (eds. N.M. Bikales, H.F. Mark, G. Menges and C.G.Overberger), John Wiley & Sons, 1989, 18, 16, 81
[2] *Kirk-Othmer Encycl. Chem. Technol.*, 3rd edn. (eds. M. Grayson and D.Eckreth), Wiley Interscience, 1984, **18**, 474, 603, (HIPS/PPO blends)
[3] Macknight, W.J., Karasz, F.E. and Fried, J.R., *Polym. Blends*, (eds. D.R. Paul and S. Newman), Academic Press, 1978, **1**, 185, (rev, polymer blends)
[4] Kramer, M., *Appl. Polym. Symp.*, 1971, **15**, 227, (HIPS/PPO blends)
[5] Hay, A.S., *Polym. Eng. Sci.*, 1976, **16**, 1, (rev, PPO and blends)
[6] Bucknall, C.B., *Toughened Plastics*, Applied Science Publishers Ltd, 1977, 193, (rev, polymer blends)
[7] Yee, A.F., Polym. Prepr. (Am. Chem. Soc., Div. Polym. Chem.), 1976, **17**, 145, (PPO HIPS blends, props, mechanical props.)
[8] Bucknall, C.B., Clayton, D. and Weast, W.E., *J. Mater. Sci. Lett.*, 1972, **7**, 1443, (PPO HIPS blends, mechanical props.)
[9] Bair, H.E., *Polym. Eng. Sci.*, 1970, **10**, 247, (HIPS/PPO composition)
[10] U.S. Pat., 1968, 3 383 435, (HIPS/PPO manufacture)
[11] *Engineering Thermoplastics: Properties and Applications*, (ed. J.M. Margoplis), Marcel Dekker, 1985, 123, (rev)
[12] Jpn. Pat., 1990, 02 295 007
[13] Eur. Pat. Appl., 1994, 618 267, (automobile parts)
[14] Jpn. Pat., 1991, 03 176 354, (heat resistant utensils)
[15] Ger. Pat., 1993, 4 209 031, (injection mouldings)
[16] Eur. Pat. Appl., 1990, 381 239, (injection mouldings)
[17] Jpn. Pat., 1995, 07 278 325, (polymer electret films)
[18] PCT Int. Appl., 1994, 94 20 571, (HIPS/PPO manufacture)
[19] Eur. Pat. Appl., 1992, 476 366, (blow moulding)
[20] White, D.M. and Hallgreen, J.E., *J. Polym. Sci., Polym. Chem. Ed.*, 1983, **21**, 2921, (ir)
[21] Yee, A.F., *Polym. Eng. Sci.*, 1977, **17**, 213, (PPO HIPS blends, mechanical props.)
[22] *The Materials Selector*, (eds. N.A. Waterman & M.F. Ashby), Chapman & Hall, The Materials Selecter, 1997, **3**, 1816, (trade-names, physical props)

[23] *Encycl. Polym. Sci. Eng.*, Vol. 12, 2nd edn. (eds. N.M. Bikales, H.F. Mark, G. Menges & C.G. Overberger, John Wiley & Sons, 1989, **443**, (PPO blends)
[24] Cooper, G.D., Lee, G.F., Katchman, A. and Shank, C.P., *Mater. Technol.*, 1981, **12**, (HIPS-PPO blends)
[25] Schultz, A.R. and Gendron, B.M., *J. Appl. Polym. Sci.*, 1972, **16**, 461, (thermal props)
[26] Couchman, P.R., *Macromolecules*, 1978, **11**, 1156, (thermal props)
[27] Kwon, I.H. and Jo. W.H, *Polym. J. (Tokyo)*, 1992, **24**, 625, (thermodynamic props)
[28] Weeks, N.E., Karasz, F.E. and Macknight, W.J., *J. Appl. Phys.*, 1977, **48**, 4068, (calorimetry)
[29] *Physical Properties of Polymers Handbook,* (ed. J.E. Mark), AIP Press, 1996, (rev)
[30] Wellinghoff, S. and Baer, E., *Org. Coat. Plast. Chem.*, 1976, **36**, 140, (mech props)
[31] Bucknall, C.B., Clayton, D. and Keast, W.E., *J. Mater. Sci.*, 1973, **8**, 514, (mech props)
[32] Rimnac, C.M., Hertzberg, R.W. and Manson, J.A., *Polymer*, 1982, **23**, 1977, (mech props)
[33] Okamoto, Y., Miyagi, H., Uno, T. and Amemiya, Y., *Polym. Eng. Sci.*, 1993, **33**, 1606, (mech props)
[34] Stallings, R.L., Hopfenberg, H.B. and Stannett, V., J. Polym. Sci., *Polym. Symp.*, 1973, **41**, 23, (diffusion)
[35] Morel, G. and Paul, D.R., *J. Membr. Sci.*, 1982, **10**, 273, (diffusion)
[36] Maeda, Y. and Paul, D.R., *Polymer*, 1985, **26**, 2055, (diffusion)
[37] Stejskal, E.O., Schaefer, J., Sefcik, M.D. and McKay, R.A., *Macromolecules*, 1981, **14**, 275, (C-13 nmr)
[38] *U.S. Pat.*, 1980, 4 528 327, (gloss)
[39] Karasz, F.E. and MacKnight, W.J., *Stud. Phys. Theor. Chem.*, 1980, **10**, 109, (compatibility)
[40] Feng, H., Feng, Z., Ruan, H. and Shen, L., *Chemtracts: Macromol. Chem.*, 1993, **4**, 104
[41] Kwei, T.K. and Frisch, H.L., *Macromolecules*, 1978, **11**, 1267, (interaction parameter)
[42] Schultz, A.R. and McCullough, C.R., J. Polym. Sci., Part A-2, 1972, **10**, 307, (interaction parameter)
[43] Poslinski, A.J. and Stokes, V.K., *Polym. Eng. Sci.*, 1992, **32**, 1147, (melting props)
[44] van der Sanden, M.C.M., Meijer, H.E.H., and Lemstra, P.J., *Prog. Colloid Polym. Sci.*, 1993, **92**, 120, (mech props)
[45] *U.S. Pat.*, 1980, 4 528 327, (mech props)
[46] Kleiner, L.W., Karasz, F.E. and MacKnight, W.J., *Polym. Eng. Sci.*, 1979, **19**, 519, (mech props)
[47] Schmidt, L.R., *J. Appl. Polym. Sci.*, 1979, **23**, 2463, (melt rheology)
[48] Prest, W.M. and Porter, R.S., J. Polym. Sci., Part A-2, 1972, **10**, 1639, (viscoelastic props)
[49] *Jpn. Pat.*, 1989, 01 313 806, (electric insulators and wire coatings)

High styrene resins H-4

Synonyms: *High styrene rubbers. Poly(styrene-butadiene). Poly(-ethenylbenzene-co-1,3-butadiene)*
Related Polymers: Styrene-butadiene copolymers, Styrene-butadiene rubber, K-resins
Monomers: Styrene, 1,3-Butadiene
Material class: Copolymers, Synthetic Elastomers
Polymer Type: polybutadiene, polystyrene
CAS Number:

CAS Reg. No.
9003-55-8

Additives: Vulcanised by sulfur
General Information: Around 51%–85% styrene to 49%–15% butadiene. Are white, essentially odourless, free-flowing powders or granular solids [1]. They are random copolymers of styrene and butadiene containing relatively higher levels (greater than 40%) of styrene compared with conventional SBR (approx. 23.5%) (see styrene-butadiene rubber). They are leather-like in behaviour and appearance for 50% styrene level but rigid thermoplastics at 70% styrene level [2]

Volumetric & Calorimetric Properties:
Density:

No.	Value	Note
1	d 1.04 g/cm^3	commercial resins, 82.5 wt% styrene [1]

Melting Temperature:

No.	Value	Note
1	42–52°C	commercial resins, 82.5 wt% styrene [1]

Glass Transition Temperature:

No.	Value	Note
1	<50°C	max.

Transition Temperature General: Copolymers containing approx. 70% styrene have low softening points (behave like rigid thermoplastics) [2]

Mechanical Properties:
Mechanical Properties General: A high styrene content results in improved hardness, stiffness and abrasion resistance [1]

Applications/Commercial Products:
Processing & Manufacturing Routes: Usually prepared by emulsion polymerisation, using similar process to that used for SBR
Applications: Used as reinforcing agents, e.g. for rubbers (increase in hardness, stiffness and abrasion resistance); often blended with hydrocarbon rubbers such as NR or SBR. Have found applications in shoe soles and various mouldings (although in competition with both conventional thermoplastics and thermoplastic elastomers). Also used in tyres where increased hardness gives improved road grip

Trade name	Details	Supplier
Arlatex 43 DA	85 wt% styrene	Housemex
Pliolite S	various grades; non-exclusive; also used for other styrene, butadiene, and copolymer rubbers; 82.5 wt % styrene	Goodyear

Bibliographic References
[1] Ash, M. and Ash, I., Handbook of Plastic Compounds, Elastomers and Resins VCH, 1992, 446, 621, (thermal props., volumetric props, mech. props, uses)
[2] Brydson, J.A., Plast. Mater. 6th edn, Butterworth-Heinemann, 1995, 282, (physical props, thermal props, uses)

Hydroxyalkyl methacrylate copolymers H-5

Synonyms: *Poly(hydroxyethyl methacrylate-co-methacrylic acid). Poly(hydroxypropyl methacrylate-co-methacrylic acid). Poly(hydroxyethyl methacrylate-co-methyl methacrylate). Poly(hydroxyethyl methacrylate-co-dimethylaminoethyl methacrylate). Poly(hydroxyethyl methacrylate-co-hydroxyethoxyethyl methacrylate). Poly(hydroxyethyl methacrylate-co-butyl methacrylate). Poly(hydroxyethyl methacrylate-co-acrylic acid). Poly(hydroxyethyl methacrylate-co-butyl acrylate). Poly(hydroxyethyl methacrylate-co-hydroxypropyl acrylate). Poly(2-hydroxyethyl methacrylate-co-styrene). Poly(2-hydroxyethyl methacrylate-graft-propylene). Poly(2-hydroxyethyl methacrylate-co-N-vinylpyrrolidone). Poly(hydroxyethyl methacrylate-co-ethyl methacrylate). Poly(hydroxyethyl methacrylate-co-ethylene glycol dimethacrylate). Poly(2-hydroxyethyl methacrylate-co-N,N-dimethylacrylamide)*
Related Polymers: Polymethacrylates - General, Poly(hydroxyalkyl methacrylates), Polymethacrylate ionomers, Polystyrene, Polypropylene, Poly*N*-vinylpyrrolidone), Poly(*N,N*-dimethylacrylamide)
Monomers: Hydroxyethyl methacrylate, Hydroxypropyl methacrylate, Methacrylic acid, Methyl methacrylate, Ethyl methacrylate, Butyl methacrylate, Dimethylaminoethyl methacrylate, Ethylene glycol dimethacrylate, Hydroxyethoxyethyl methacrylate, Acrylic acid, Butyl acrylate, Hydroxypropyl acrylate, Styrene, Propylene, *N*-Vinylpyrrolidone, *N,N*-Dimethylacrylamide
Material class: Thermoplastic, Copolymers
Polymer Type: acrylic

Hydroxyalkyl methacrylate copolymers

CAS Number:

CAS Reg. No.	Note
31693-08-0	hydroxyethyl methacrylate-co-methacrylic acid
25053-81-0	hydroxyethyl methacrylate-co-ethylene glycol dimethacrylate
26355-01-1	hydroxyethyl methacrylate-co-methyl methacrylate
119182-44-4	hydroxyethyl methacrylate-co-methyl methacrylate, block
117650-87-0	hydroxyethyl methacrylate-co-methyl methacrylate, graft
26335-61-5	hydroxyethyl methacrylate-co-ethyl methacrylate
25702-92-5	hydroxyethyl methacrylate-co-butyl methacrylate
131004-73-4	hydroxyethyl methacrylate-co-butyl methacrylate, graft
26010-51-5	2-hydroxyethyl methacrylate-co-styrene
108648-39-1	2-hydroxyethyl methacrylate-co-styrene, block
106173-87-9	2-hydroxyethyl methacrylate-co-styrene, graft
29612-57-5	hydroxyethyl methacrylate-co-N-vinylpyrrolidone

Volumetric & Calorimetric Properties:

Density:

No.	Value	Note
1	d 1.149–1.205 g/cm^3	hydroxymethyl methacrylate-co-ethyl methacrylate, 25–67% hydroxyethyl methacrylate [1]
2	d 1.109–1.179 g/cm^3	hydroxymethyl methacrylate-co-butyl methacrylate, 31–69% hydroxymethyl methacrylate [1]
3	d 1.271–1.273 g/cm^3	hydroxymethyl methacrylate-co-butyl acrylate, 75–95% hydroxymethyl methacrylate [2]
4	d 1.274–1.275 g/cm^3	hydroxymethyl methacrylate-co-hydroxypropyl acrylate, 75–95% hydroxymethyl methacrylate [2]

Glass Transition Temperature:

No.	Value	Note
1	90–155°C	hydroxyethyl methacrylate-co-methacrylic acid, 10.1–90.1% methacrylic acid [3,5]
2	11–100°C	hydroxyethyl methacrylate-co-hydroxyethoxyethyl methacrylate [6]
3	98–145°C	hydroxypropyl methacrylate-co-methacrylic acid, 51.3–78.1% methacrylic acid [5]

Transition Temperature General: The presence of H$_2$O in the hydrogel results in a lowering of the T_g. The reduction in T_g is dependent on the amount of H$_2$O and the pH [3,4]. The plasticising action of some additives, e.g. PEG, also lowers T_g [3]

Surface Properties & Solubility:

Cohesive Energy Density Solubility Parameters: Solubility parameters, δ, of (hydroxyethyl methacrylate-co-ethyl methacrylate) [1] 22.09 J$^{1/2}$ cm$^{-3/2}$ (25% hydroxyethyl methacrylate), 22.81 J$^{1/2}$ cm$^{-3/2}$ (45% hydroxyethyl methacrylate), 22.4 J$^{1/2}$ cm$^{-3/2}$ (58% hydroxyethyl methacrylate), 23.63 J$^{1/2}$ cm$^{-3/2}$ (56% hydroxyethyl methacrylate), 24.14 J$^{1/2}$ cm$^{-3/2}$ (67% hydroxyethyl methacrylate). Solubility parameters, δ (hydroxyethyl methacrylate-co-butyl methacrylate) [1] 21.74 J$^{1/2}$ cm$^{-3/2}$ (31% hydroxyethyl methacrylate), 21.27 J$^{1/2}$ cm$^{-3/2}$ (41% hydroxyethyl methacrylate), 22.61 J$^{1/2}$ cm$^{-3/2}$ (54% hydroxyethyl methacrylate), 22.81 J$^{1/2}$ cm$^{-3/2}$ (66% hydroxyethyl methacrylate), 23.63 J$^{1/2}$ cm$^{-3/2}$ (69% hydroxyethyl methacrylate)

Solvents/Non-solvents: (Hydroxyethyl methacrylate-co-methyl methacrylate, 0.7-1 mol. fraction methyl methacrylate) [7]. Insol. H$_2$O, MeOH, EtOH, Me$_2$CO, CHCl$_3$, CCl$_4$, CH$_2$Cl$_2$, 2-butanone, THF

Wettability/Surface Energy and Interfacial Tension: Hamilton contact angle (hydroxyethyl methacrylate-co-methacrylic acid) 135–146° (25–75 mol % hydroxyethyl methacrylate), 131–140° (hydroxyethyl methacrylate-co-acrylic acid, 25–75 mol % hydroxyethyl methacrylate), 93–126° (hydroxyethyl methacrylate-co-styrene, 25–75 mol % hydroxyethyl methacrylate), 130–154° (hydroxyethyl methacrylate-co-N-vinylpyrrolidone, 50–75 mol % hydroxyethyl methacrylate) [8]. Surface energy, γ, (hydroxyethyl methacrylate-co-methacrylic acid) [8] 52.6 mN m^{-2} (γ^p 15.9; γ^d 36.7) (75 mol % hydroxyethyl methacrylate), 55.4 mN m^{-2} (γ^p 20.1; γ^d 35.3) (50 mol % hydroxyethyl methacrylate), 56.4 mN m^{-2} (γ^p 21.5; γ^d 35.0) (25 mol % hydroxyethyl methacrylate); (hydroxyethyl methacrylate-co-acrylic acid) [8] 51.9 mN m^{-2} (γ^p 14.4; γ^d 37.5) (75 mol % hydroxyethyl methacrylate), 51.8 mN m^{-2} (γ^p 15.4; γ^d 36.4) (50 mol % hydroxyethyl methacrylate), 52.1 mN m^{-2} (γ^p 16.3; γ^d 35.8) (25 mol % hydroxyethyl methacrylate); (hydroxyethyl methacrylate-co-methylmethacrylate) [4] 49.4 mN m^{-2} (γ^p 12.7; γ^d 36.7) (75 mol % hydroxyethyl methacrylate), 49.2 mN m^{-2} (γ^p 7.9; γ^d 41.3) (50 mol % hydroxyethyl methacrylate), 46.9 mN m^{-2} (γ^p 6.9; γ^d 39.5) (25 mol % hydroxyethyl methacrylate); (hydroxyethyl methacrylate-co-styrene) [4] 50.5 mN m^{-2} (γ^p 18.8; γ^d 31.7) (90 mol % hydroxyethyl methacrylate), 47 mN m^{-2} (γ^p 14.9; γ^d 32.1) (50 mol % hydroxyethyl methacrylate), 45.1 mN m^{-2} (γ^p 12.3; γ^d 32.8) (30 mol % hydroxyethyl methacrylate), 42.6 mN m^{-2} (γ^p 7.3; γ^d 35.3) (10 mol % hydroxyethyl methacrylate); (hydroxypropyl methacrylate-co-methacrylic acid) [4] 48.5 mN m^{-2} (γ^p 13.8; γ^d 34.7) (50 mol % hydroxypropyl methacrylate); (hydroxyethyl methacrylate-co-N-vinylpyrrolidone) [8] 45.5 mN m^{-2} (γ^p 6.2; γ^d 39.3) (90 mol % hydroxyethyl methacrylate), 43.4 mN m^{-2} (γ^p 3.4; γ^d 40.0) (70 mol % hydroxyethyl methacrylate), 43.9 mN m^{-2} (γ^p 4.1; γ^d 39.8) (50 mol % hydroxyethyl methacrylate); 46.5 mN m^{-2} (γ_p 9.5, γ_d 37.0) (hydroxyethyl methacrylate-co-N-vinylpyrrolidone, dry, 95 mol% hydroxyethyl methacrylate); 43.9 mN m^{-2} (γ_p 4.1, γ_d 39.8) (hydroxyethyl methacrylate-co-N-vinylpyrrolidone, dry, 50 mol% hydroxyethyl methacrylate)[8,13]; 43.8 mN m^{-2} (γ_p 8.8, γ_d 35.0) (hydroxyethyl methacrylate-co-N,N-dimethylacrylamide, dry, 60 mol% hydroxyethyl methacrylate); 46.3 mN m^{-2} (γ_p 15.6, γ_d 30.7 (hydroxyethyl methacrylate-co-N,N-dimethylacrylamide, dry, 95 mol% hydroxyethyl methacrylate) [13]

Transport Properties:

Polymer Solutions Dilute: Intrinsic viscosity (hydroxyethyl methacrylate-co-methacrylic acid) [5] 0.93–2.9 dl g^{-1} (36.8–76.5% methacrylic acid); (hydroxypropyl methacrylate-co-methacrylic acid) [5] 0.41–1.09 dl g^{-1} (51.3–78.1% methacrylic acid); (hydroxyethyl methacrylate-co-ethyl methacrylate) [1] 0.99–1.96 dl g^{-1} (25–67% hydroxyethyl methacrylate, M$_n$ 151000–304000); (hydroxyethyl methacrylate-co-butyl methacrylate) [1] 1.25–1.5 dl g^{-1} (31–69% hydroxyethyl methacrylate, M$_n$ 151000–900000). Theta temps. (hydroxyethyl methacrylate-co-ethyl methacrylate) [1]: 29.5° (DMF, 25 mol % hydroxyethyl methacrylate, M$_n$ 151000); (hydroxyethyl methacrylate-co-butyl methacrylate) [1]: 29.5° (DMF, 31 mol % hydroxyethyl methacrylate, M$_n$ 181000)

Permeability of Gases: The permeability of the copolymers to gases is dependent on the equilibrium H$_2$O content as well as the composition [8] The permeability coefficient of O$_2$ increases with increasing degree of hydration [12]

Water Content: The incorporation of carboxyl and nitrogen-containing monomers into hydroxyalkyl methacrylate copolymers gives hydrogels with equilibrium H$_2$O contents in the range 30–90% [8]. Thermal anal. of some copolymers suggests that there is both free and bound H$_2$O in (hydroxyethyl methacrylate-co-methacrylic acid) hydrogel. The water may be defined as 'freezing' or 'non-freezing', the former having a plasticising effect on the

Hydroxyalkyl methacrylate copolymers

polymer and the latter little effect [3,4]. Equilibrium water content (hydroxyethyl methacrylate-co-methyl methacrylate) [4] 12.3–32.5% (50–90 mol % hydroxyethyl methacrylate); (hydroxyethyl methacrylate-co-butyl acrylate) [2] 19.7–34.4% (75–95 mol % hydroxyethyl methacrylate, value dependent upon amount of cross-linking agent added); (hydroxyethyl methacrylate-co-hydroxypropyl acrylate) [2] 33.8–40.9% (75–95 mol % hydroxyethyl methacrylate, value dependent on amount of cross-linking yet added); (hydroxyethyl methacrylate-co-styrene) [4] 16.7–25.2% (80–90 mol % hydroxyethyl methacrylate); 40.4–85.4% (hydroxyethyl methacrylate-co-N-vinylpyrrolidone, 30–95% hydroxyethyl methacrylate cross-linked) [9]; 64.3–38.7% (N-vinylpyrrolidone copolymer, 50–95% hydroxyethyl methacrylate) [13]; 41.3–70.5% (N,N-dimethylacrylamide copolymer, 95–60% hydroxyethyl methacrylate) [13]

Mechanical Properties:

Mechanical Properties General: The mechanical props. of the copolymers vary tremendously with both composition, degree of cross-linking and equilibrium H_2O content [8]. Tensile modulus of hydroxyethyl methacrylate-co-N-vinylpyrrolidone 0.08–2.99 MN m^{-2} depending on amount of hydroxyethyl methacrylate and added cross-linking agent [9] Tensile modulus and tensile strength at break of hydrogels vary with copolymer composition [13]

Tensile (Young's) Modulus:

No.	Value	Note
1	0.22 MPa	hydroxyethyl methacrylate-co-methyl methacrylate, 90 mol % hydroxyethyl methacrylate
2	0.33 MPa	hydroxyethyl methacrylate-co-methyl methacrylate, 80 mol % hydroxyethyl methacrylate
3	0.33 MPa	hydroxyethyl methacrylate-co-methyl methacrylate, 70 mol % hydroxyethyl methacrylate
4	22.71 MPa	hydroxyethyl methacrylate-co-methyl methacrylate, 60 mol% hydroxyethyl methacrylate
5	64.14 MPa	hydroxyethyl methacrylate-co-methyl methacrylate, 50 mol% hydroxyethyl methacrylate-co-methyl methacrylate
6	0.37 MPa	hydroxyethyl methacrylate-co-styrene, 90 mol % hydroxyethyl methacrylate
7	2.44 MPa	hydroxyethyl methacrylate-co-styrene, 80 mol % hydroxyethyl methacrylate
8	0.2–0.76 MPa	hydroxyethyl methacrylate-co-N-vinylpyrrolidone [13]
9	0.16–0.7 MPa	hydroxyethyl methacrylate-co-N,N-dimethylacrylamide [13]

Tensile Strength Break:

No.	Value	Note	Elongation
1	0.35 MPa	hydroxyethyl methacrylate-co-methyl methacrylate, 90 mol % hydroxyethyl methacrylate	160%
2	0.8 MPa	hydroxyethyl methacrylate-co-methyl methacrylate, 80 mol % hydroxyethyl methacrylate	241%
3	0.92 MPa	hydroxyethyl methacrylate-co-methyl methacrylate, 70 mol % hydroxyethyl methacrylate	282%
4	13.17 MPa	hydroxyethyl methacrylate-co-methyl methacrylate, 60 mol % hydroxyethyl methacrylate	58%
5	13.3 MPa	hydroxyethyl methacrylate-co-methyl methacrylate, 50 mol % hydroxyethyl methacrylate [4]	21%
6	1.32 MPa	hydroxyethyl methacrylate-co-styrene, 90 mol % hydroxyethyl methacrylate	401%
7	7.84 MPa	hydroxyethyl methacrylate-co-styrene, 80 mol % hydroxyethyl methacrylate [4]	324%
8	0.2–0.5 MPa	hydroxyethyl methacrylate-co-N-vinylpyrrolidone [13]	74–198%
9	0.2–0.48 MPa	hydroxyethyl methacrylate-co-N,N-dimethylacrylamide [13]	101–186%

Optical Properties:

Optical Properties General: The refractive index of hydrated copolymers varies markedly with temp., decreasing with increasing temp. [4] Refractive index decreases with increasing degree of hydration [12]

Transmission and Spectra: Ir spectral data have been reported [1]

Polymer Stability:

Decomposition Details: Decomposition temp. (for 60% weight loss) of hydroxyethyl methacrylate-co-methacrylic acid 402–407° (36.8–76.5% methacrylic acid) [5]. Decomposition temp. (hydroxypropyl methacrylate-co-methacrylic acid) 370–390° (51.3–78.1% methacrylic acid) [5]. Decomposition prod. monomers by an unzipping process. Cross-linking competes with the unzipping process and there is more cross-linking with increasing amounts of hydroxyethyl methacrylate [10]

Biological Stability: Poly(hydroxymethyl acrylate) hydrogels adsorb proteins more readily at higher temps. (20° and above) with the rate of adsorption increasing with increasing time. The incorporation of cross-linking agent (e.g., ethylene glycol dimethacrylate) or styrene into the copolymer can have varying effects on the adsorption of proteins: the rate of fibrinogen absorption decreases as the level of cross-linking or amount of styrene increases whereas there is comparatively little effect on the adsorption of albumin [4]

Stability Miscellaneous: The MW of hydroxyethyl methacrylate/methyl methacrylate copolymers decreases with increasing dosage of irradiation [7]

Applications/Commercial Products:

Processing & Manufacturing Routes: Acrylic copolymers are synth. by radiation-induced bulk polymerisation [5]. The polypropylene graft copolymer is synth. at 50° in H_2O or aq. MeOH using benzoin ethyl ether as initiator [11]

Applications: Biomedical uses (transdermal drug delivery systems, contact lenses). Hydroxyethyl methacrylate/propylene graft copolymers have improved wettability and dyeing props. compared to polypropylene

Bibliographic References

[1] Choudhary, M.S. and Varma, I.K., *J. Polym. Sci., Polym. Chem. Ed.*, 1985, **23**, 1917
[2] Perera, D.I. and Shanks, R.A., *Polym. Int.*, 1995, **37**, 133
[3] Verhoeven, J., Schaeffer, R., Bouwstra, J.A. and Junginger, H.E., *Polymer*, 1989, **30**, 1946
[4] Barnes, A., Corkhill, P.H. and Tighe, B.J., *Polymer*, 1988, **29**, 2191
[5] Ramakrishna, M., Deshparda, D.D. and Babu, G.N., *J. Polym. Sci., Part A: Polym. Chem.*, 1988, **26**, 445
[6] Svetlik, J. and Pouchly, J., *Eur. Polym. J.*, 1976, **12**, 123
[7] Parsonage, E.E. and Peppas, N.A., *Br. Polym. J.*, 1987, **19**, 469
[8] Baker, D.A., Corkhill, P.H., Ng, C.O., Skelly, P.J. and Tighe, B.J., *Polymer*, 1988, **29**, 691

- Hydroxyethyl cellulose

[9] Davis, T.P. and Huglin, M.B., *Macromolecules,* 1989, **22**, 2824
[10] Choudhary, M.S. and Lederer, K., *Eur. Polym. J.,* 1982, **18**, 1021, (thermal degradation)
[11] Shukla, S.R. and Athalye, A.R., *J. Appl. Polym. Sci.,* 1994, **51**, 1567, (synth.)
[12] Kossmehl, G. and Volkheimer, J, *Makromol. Chem.,* 1989, **190**, 1253
[13] Corkhill, P.H. and Tighe, B.J., *Polymer,* 1990, **31**, 1526

Hydroxyethyl cellulose   H-6

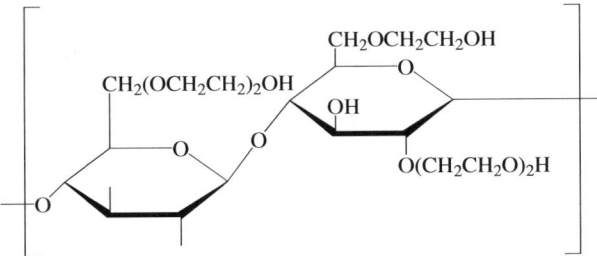

Monomers: Base monomer unit glucose, Ethylene oxide
Material class: Polysaccharides
Polymer Type: cellulosics
CAS Number:

CAS Reg. No.
9004-62-0

Molecular Weight: MW 90000-1300000
General Information: Substitution *ca.* 0.15 per monomeric unit (film)

Volumetric & Calorimetric Properties:
Density:

No.	Value	Note
1	d 0.62 g/cm^3	[1]
2	d 0.7 g/cm^3	[3]
3	d 1.0033 g/cm^3	2% soln. [1,2]

Melting Temperature:

No.	Value	Note
1	135–140°C	[1]

Vicat Softening Point:

No.	Value	Note
1	135–140°C	Softening range [1]

Transition Temperature:

No.	Value	Note	Type
1	205–210°C	[1]	Browning range

Surface Properties & Solubility:
Plasticisers: Glycerol, triethanolamine, triethylene glycol [3]
Solvents/Non-solvents: Sol. 30% EtOH aq. (25–60°), DMSO, ethylene chlorohydrin. Mod. sol. 70% EtOH aq. (25–60°). Insol. MeOH, EtOH, methylcellosolve, glacial AcOH, toluene, hexane, mineral spirits (25–60°). Swollen by ethylene glycol, propylene glycol and glycerol [3]
Surface and Interfacial Properties General: Interfacial tension 25.5 mN m^{-1} (0.001%, MS 2.5), 23.7 mN m^{-1} (0.001%, MS 1.8) [1,2]

Surface Tension:

No.	Value	Note
1	66.8 mN/m	0.1%, MS 2.5 [1,2]
2	67.3 mN/m	0.001%, MS 2.5 [1,2]
3	66.7 mN/m	0.1%, MS 1.8 [1,2]
4	69.8 mN/m	0.001%, MS 2.5 [1,2]

Transport Properties:
Polymer Solutions Dilute: Limits of viscosity in water 100000 mPa s (2%, 25°, ASTM 2364, Type HH), 6000 mPa s (2%, 25°, ASTM 2364, Type M), 15 mPa s (2%, 25°, ASTM 2364, Type L) [3]
Water Content: 5% (max.) [3]

Mechanical Properties:
Tensile Strength Break:

No.	Value	Note
1	14–22 MPa	film, cationic [4]

Flexural Strength at Break:

No.	Value	Note
1	60–70 MPa	[4]

Impact Strength: Impact strength 6–8 kg cm^{-2} (film) [5]
Failure Properties General: Tearing strength 2–4 g (film) [5]

Optical Properties:
Refractive Index:

No.	Value	Note
1	1.336	2% soln. [1]
2	1.51	film [1]

Polymer Stability:
Environmental Stress: Non-toxic

Applications/Commercial Products:
Processing & Manufacturing Routes: Reaction of alkali cellulose with ethylene oxide in a stirred autoclave under pressure
Applications: Thickener for coatings and cements, binder for ceramics, stabiliser in foods and health care products

Trade name	Details	Supplier
Cellosize		Union Carbide Corporation
Natrosol	Biostable and pharmaceutical grades	Aqualon
Nexton	Blends of hydrophobically modified hydroxyethyl cellulose	Aqualon
Tylose H		Hoechst Celanese
		Fuji Chemical

Bibliographic References
[1] *Encycl. Polym. Sci. Eng.,* 2nd edn., (ed. J.I. Kroschwitz), John Wiley and Sons, 1985, **3**
[2] *Natrosol® Hydroxyethylcellulose,* Hercules Inc., 1978, (technical datasheet)
[3] *Natrosol® Hydroxyethylcellulose,* 1994, Aqualon Div. Hercules Inc., (technical datasheet)

[4] *Kirk-Othmer Encycl. Chem. Technol.*, 4th edn., (ed. J.I. Kroschwitz), John Wiley and Sons, 1993, **5**
[5] *Cellulose and Cellulose Derivatives,* Part 5, (eds. N.M. Bikales and L. Segal), Wiley Interscience, 1971

Hydroxypropyl cellulose

H-7

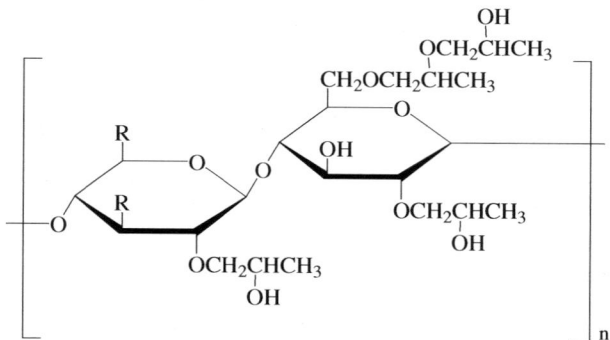

Monomers: Base monomer unit glucose, Propylene oxide
Material class: Polysaccharides, Thermoplastic
Polymer Type: Lyotropic
CAS Number:

CAS Reg. No.
9004-64-2

Molecular Weight: MW 416000 (degree of substitution 4), MW 200000, MW 80000-1150000
General Information: Thermoplastic polymer, showing liq. cryst. props. at very high concentrations in soln.

Volumetric & Calorimetric Properties:
Density:

No.	Value	Note
1	d 1.1 g/cm^3	[1,4]
2	d 1.17 g/cm^3	film [3]
3	d 1.088 g/cm^3	amorph., 24° [3]

Melting Temperature:

No.	Value	Note
1	235–245°C	[3]

Transition Temperature:

No.	Value	Note	Type
1	130°C	[1,4]	Softening temp.
2	100–150°C	[6]	Softening temp.
3	450–500°C	[1,4]	Burn out temp.
4	40–45°C	2% aq. soln. [1]	Cloud point

Surface Properties & Solubility:
Solubility Properties General: Specific gravity 1.01 (2% soln., 30°) [5]. Bulking value in soln. 0.334 l kg^{-1} [5]
Solvents/Non-solvents: Liq. cryst. sol. H$_2$O, MeOH, DMF [7,8,9]

Surface and Interfacial Properties General: Interfacial tension 12.5 mN m^{-2} (0.1% in H$_2$O vs refined mineral oil) [5]
Surface Tension:

No.	Value	Note
1	43.6 mN/m	0.1% soln. [5]

Transport Properties:
Polymer Solutions Dilute: Viscosity 1500–3000 cP (1500–3000 mPa s, 1%, Type H, MW 1150000), 150–700 cP (150–700 mPa s, 10%, Type E, MW 80000) (H$_2$O) [5], 1000–4000 cP (1000–4000 mPa s, 1%, Type H, MW 1150000), 150–700 cP (150–700 mPa s, 10%, Type E, MW 80000) (EtOH) [5]
Water Content: Equilibrium moisture content 4% (23°, 50% relative humidity), 12% (23°, 84% relative humidity) [5]
Water Absorption:

No.	Value	Note
1	4 %	film, 23°, 50% relative humidity [1,4]

Gas Permeability:

No.	Gas	Value	Note
1	O$_2$	9685 cm^3 mm/(m^2 day atm)	0.05 × 10^{-12} mol (m s Pa)$^{-1}$ [1,4]
2	N$_2$	1162 cm^3 mm/(m^2 day atm)	0.006 × 10^{-12} mol (m s Pa)$^{-1}$ [1,4]
3	CO$_2$	65857 cm^3 mm/(m^2 day atm)	0.34 × 10^{-12} mol (m s Pa)$^{-1}$ [1,4]

Mechanical Properties:
Tensile (Young's) Modulus:

No.	Value	Note
1	430 MPa	[2]

Tensile Strength Break:

No.	Value	Note	Elongation
1	13.8 MPa	[1,4]	50%
2	14 MPa	[2]	

Elastic Modulus:

No.	Value	Note
1	414 MPa	film [1,4]

Mechanical Properties Miscellaneous: Folding endurance flexibility 10000 (MIT Double folds (50 μm film)) [1,2]

Electrical Properties:
Dielectric Permittivity (Constant):

No.	Value	Note
1	9.07	[2]

Dissipation (Power) Factor:

No.	Value	Note
1	0.0706	[2]

− Hydroxypropyl cellulose

Optical Properties:
Refractive Index:

No.	Value	Note
1	1.337	2% aq. soln. [1,4,5]
2	1.559	film [1,4]
3	1.48	film [3]

Applications/Commercial Products:
Processing & Manufacturing Routes: Alkali cellulose is treated with a limiting quantity of propylene oxide in a stirred autoclave
Applications: Polymerisation control (PVC), binder for pharmaceutical tablets, thickener for paints, stabiliser for foods and ceramics

Trade name	Supplier
Klucel	Aqualon
Klucel	Hercules

Bibliographic References

[1] *Encycl. Polym. Sci. Eng.,* 2nd edn., (ed. J.I. Kroschwitz), John Wiley and Sons, 1985, **3**
[2] Elias, H.-G., *New Commercial Polymers,* Gordon and Breach, 1969-75, 1176
[3] Samuels, R.J., J. Polym. Sci., *Part A-2,* 1969, **7**, 1197
[4] Klucel®, *Hydroxypropyl Cellulose,* Chemical and Physical Properties, Hercules Inc., 1976, (technical datasheet)
[5] Klucel®, *Physical and Chemical Properties,* Tech. Bull., 1995, Aqualon Division, Hercules Inc., (technical datasheet)
[6] *Kirk-Othmer Encycl. Chem. Technol.,* 4th edn., (ed. J.I. Kroschwitz), John Wiley and Sons, 1993, **5**
[7] Onogi, Y., White, J.L. and Fellers, J.F., J. Polym. Sci., *Polym. Phys. Ed.,* 1980, **18**, 663
[8] Gröbe, A., *Polym. Handb.,* 3rd edn., (eds. J. Brandrup and E.H. Immergut), Wiley Interscience, 1989, V155
[9] Werbowyj, R.S. and Gray, D.G., *Mol. Cryst. Liq. Cryst.,* Letts., 1976, **34**, 97

Isotactic Polypropylene I-1

Synonyms: *PP-isotactic. Polypropene-isotactic. Poly(1-propene) isotactic*
Related Polymers: Polypropylene, Filled polypropylenes, Polypropylene copolymers, Polypropylene Blends, Polypropylene Films, Polypropylene Fibres
Monomers: Propylene
Material class: Thermoplastic
Polymer Type: polypropylene
CAS Number:

CAS Reg. No.
25085-53-4

Molecular Formula: $(C_3H_6)_n$
Fragments: C_3H_6
General Information: The most practically important form of polypropylene. Synth. using a Ziegler-Natta type catalyst, the most common now normally based on $TiCl_4$ supported by $MgCl_2$. Since commercial polypropylene is best described as 'mainly isotactic', cross references to normal variants are included here (in addition to general information). Data given here are known to be specific to isotactic polypropylene. For more general data where details are primarily for commercial, mainly isotactic polypropylene, but reference to other forms is also made.
Tacticity: Isotactic
Morphology: Morphological props. vary a little with MW distribution, and more so with draw temp. In general, as distribution broadens, birefringence may increase and amorph. orientation may increase (>165°), long period spacing decreases slightly; as draw temp. increases, crystalline fraction, crystalline orientation and especially long period spacing increase but birefringence and amorph. orientation decrease [5]
Miscellaneous: By the very nature of blends, great variation in props. occurs as a result of proportions of components, and the exact nature of components, method of mixing, as well as all the usual causes of variation. It follows that where props. are given, ranges may be very wide and not necessarily exhaustive

Volumetric & Calorimetric Properties:
Density:

No.	Value	Note
1	d^{25} 0.94 g/cm^3	calc.
2	d^{25} 0.939 g/cm^3	calc. [5]
3	d 0.932–0.943 g/cm^3	crystalline [1]
4	d^{25} 0.85 g/cm^3	amorph. [5]

Thermodynamic Properties General: Thermodynamic data including C_v, C_p, $H_T^C - H_0^C$, S_T^C, $G_T^C - H_C^C$ have been reported [29]
Latent Heat Crystallization: ΔH_{fusion} 8.79 kJ mol^{-1} (209 kJ kg^{-1}) [1]. Entropy of polymerisation $\Delta S°$ 205 J K^{-1} mol^{-1} (25°, monomer gas phase, polymer crystalline/partly crystalline)[26]. $\Delta S°$ 116 J K^{-1} mol^{-1} (25°, monomer liquid phase, polymer condensed amorph. phase) [26,27]. $\Delta S°$ 136 J K^{-1} mol^{-1} (25°, monomer liquid phase, polymer 100% crystalline) [26,27,28]
Specific Heat Capacity:

No.	Value	Note	Type
1	1.8–1.92 kJ/kg.C	[28]	P

Melting Temperature:

No.	Value	Note
1	186°C	crystalline, monoclinic [1]
2	165°C	[7]

Glass-Transition Temperature:

No.	Value	Note
1	-165°C	[9]

Transition Temperature General: T_g varies with increasing M_n if more than 10000 [10]

Surface Properties & Solubility:
Solvents/Non-solvents: Sol. hydrocarbons, halogenated hydrocarbons, higher aliphatic esters and ketones, diamyl ether (above 80°). Insol. all common organic solvents (room temp.), polar organic solvents, inorganic solvents [24]

Transport Properties:
Polymer Solutions Dilute: Calculations for dynamic viscosity have been reported [3]. Theta temp. 70° (amyl acetate) [10,11]; 183.2 (dibenzyl ether) [10,12]; 125.0° (biphenyl) [11,13,14]; 142–148° (diphenyl ether) [10,12,13,14]
Permeability and Diffusion General: Effects of ageing on transport props. of cryst. isotactic polymer have been reported [30], leading to a decrease in value of D
Gas Permeability:

No.	Gas	Value	Note
1	He	9849900 cm^3 mm/(m^2 day atm)	1.5×10^{-5} cm^2 (s cmHg)$^{-1}$ [3]
2	H$_2$	2626640 cm^3 mm/(m^2 day atm)	4×10^{-6} cm^2 (s cmHg)$^{-1}$ [3]

Mechanical Properties:
Mechanical Properties General: Mech. props. often determined on commercial polypropylene
Hardness: Microhardness 111 MN m^{-2} (α phase, crystallinity 0.721, MW 255000, 97% isotactic). Microhardness 90 MN m^{-2} (β phase, crystallinity 0.619, MW 255000, 97% isotactic) [5]

Optical Properties:
Transmission and Spectra: Ir absorption 790 cm^{-1} (amorph.), 809 cm^{-1} (crystalline) [1,2], 841–842 cm^{-1}
Molar Refraction: Concentration dependence of refractive index has been reported [16,17,18,19,20,21,22,23]

Polymer Stability:
Chemical Stability: Decomposed by sulfuric acid above 100° [7,8]

Applications/Commercial Products:

Trade name	Supplier
"Standard Samples"	Polysciences Inc.
"Standard Samples"	Pressure Chemical Co.
Appryl	Elf Atochem (UK) Ltd.
Carlona	Shell Chemicals
Daplen	PCD Polymere GmbH.
Eltex P	Solvay
Hostaten PP	Hoechst (UK) Ltd. Polymer Division
Isplen	Repsol Quimica
Moplen	Novamont
Napryl	Naphthachimie SA
Novolen	BASF
Polypro	Chisso Corporation
Profax	Hercules
Propathene	ICI Acrylics

Shell	Shell Chemicals
Tatren	Slovnaft Joint Stock Company
Tenite	Eastman Chemical Company
Tipplen	Tiszai Vegyi Kombinát
Trovidor 500	Anbus Pares-Parfonry SA
Vestolen	Chemische Werke Huels AG

Bibliographic References

[1] Quirk, R.P. and Alsamarraie, M.A.A., *Polym. Handb.*, 3rd edn., (eds. J. Brandrup and E.H. Immergut), Wiley Interscience, 1989
[2] *Wunderlich Macromolecular Physics*, Academic Press, 1973
[3] Stannett, V. and Yasuda, H., *High Polymers*, (ed. K.W. Doak), Wiley, 1964, **20**
[4] Wasserman, S.H. and Graessley, W.W., *Polym. Eng. Sci.*, 1996, **36**, 852
[5] Balta, C.F.J., Martinez, S.J. and Asano, T.J., *J. Mater. Sci. Lett.*, 1988, **7**, 165
[6] Dole, M. and Wunderlich, B., *Makromol. Chem.*, 1959, **34**, 29
[7] *Encycl. Polym. Sci. Eng.*, 2nd edn., (ed. J.I. Kroschwitz), John Wiley and Sons, 1985
[8] Garton, A., Carlsson, D.J. and Wiles, D.M., *Dev. Polym. Photochem.*, 1980, **1**, 93
[9] Cowie, J.M.G., *Eur. Polym. J.*, 1973, **9**, 1041
[10] Elias, H.G., *Polym. Handb.*, 3rd edn., (eds. J. Brandrup and E.H. Immergut), Wiley Interscience, 1989
[11] Inagaki, H., Miyamoto, T. and Ohta, S., *J. Phys. Chem.*, 1966, **70**, 3420
[12] Nakajima, A. and Saijyo, H., J. Polym. Sci., *Part A-2*, 1968, **6**, 735
[13] Moraglio, G. and Brzezinski, J., J. Polym. Sci., *Part B: Polym. Lett.*, 1964, **2**, 1105
[14] Kinsinger, J.B. and Hughes, R.E., *J. Phys. Chem.*, 1963, **67**, 1922
[15] Kinsinger, J.B. and Wessling, R.A., *J. Am. Chem. Soc.*, 1959, **81**, 2908
[16] Hughes, M.B., *Polym. Handb.*, 3rd edn., (eds. J. Brandrup and E.H. Immergut), Wiley Interscience, 1989
[17] Weston, W.E. and Billmeyer, F.W., *J. Phys. Chem.*, 1961, **65**, 576
[18] Chiang, R., *J. Polym. Sci.*, 1958, **28**, 235
[19] Westerman, L., J. Polym. Sci., *Part A: Polym. Chem.*, 1963, **1**, 411
[20] Parrini, P.P., Sebastiano, F. and Messina, G., *Makromol. Chem.*, 1960, **38**, 27
[21] Carpenter, D.K., J. Polym. Sci., *Part A-2*, 1966, **4**, 923
[22] Miyamoto, T. and Inagaki, H., J. Polym. Sci., *Part A-2*, 1969, **7**, 963
[23] Horska, J., Stejskal, J. and Kratochv@eil, P., *J. Appl. Polym. Sci.*, 1983, **28**, 3873
[24] Fuchs, O., *Polym. Handb.*, 3rd edn., (eds. J. Brandrup and E.H. Immergut), Wiley Interscience, 1989
[25] Busfield, W.K., *Polym. Handb.*, 3rd edn., (eds. J. Brandrup and E.H. Immergut), Wiley Interscience, 1989
[26] Dainton, F.S., Evans, D.M., Hoare, F.E. and Metia, T.P., *Polymer*, 1962, **3**, 277
[27] Passaglia, E. and Kevorkian, H.K., *J. Appl. Phys.*, 1963, **34**, 90
[28] Bu, H.S., Aycock, W. and Wunderlich, B., *Polymer*, 1987, **28**, 1165
[29] Grebowicz, J., Lau, S.-F. and Wunderlich, B., J. Polym. Sci., *Polym. Symp.*, 1984, **71**, 19
[30] Vittoria, V., *Polym. Commun.*, 1987, **28**, 199

Jelutong

Related Polymers: *cis*-1,4-Polyisoprene, Gutta-percha, Balata, Chicle
Monomers: 2-Methyl-1,3-butadiene
Material class: Gums and resins, Natural elastomers
Polymer Type: polybutadiene
Additives: Contains resin (75–80%) which is extractable with Me_2CO
General Information: Contains mainly *cis*-1,4-polyisoprene. Physical and chemical props. have been reported [1,2]

Optical Properties:
Transmission and Spectra: Ir and nmr spectral data of samples of *Achras* and *Dyera* spp. latex indicate presence of both *cis* and *trans* isomers [3]

Applications/Commercial Products:
Processing & Manufacturing Routes: Isol. from latex from *Dyera costulata*
Applications: Used in chewing gum

Bibliographic References
[1] Wren, W.G., *Annu. Rep. Prog. Rubber Technol.*, 1948, **12**, 6, (rev)
[2] Blow, C.M., *Annu. Rep. Prog. Rubber Technol.*, 1949, **13**, 14, (rev)
[3] Tessier, A.M., Nolot, P., Hoffelt, J., Gaugain, B. and Delaveau, P., *Ann. Pharm. Fr.*, 1977, **35**, 37

K-Resin

Synonyms: *Styrene-butadiene polymer. Styrene-butadiene block copolymer*
Related Polymers: Styrene-butadiene block copolymers
Monomers: Styrene, 1,3-Butadiene
Material class: Thermoplastic
Polymer Type: polybutadiene, polystyrene
General Information: Composition generally 75% styrene: 25% butadiene

Applications/Commercial Products:

Trade name	Supplier
K Resin	Phillips 66

LARC-CPI

Synonyms: *Langley Research Center Crystalline Polyimide. BTDA-1,3-BABB. Poly(3,3',4,4'-benzophenonetetracarboxylic dianhydride-1,3-bis(4-aminophenoxy-4'-benzoyl)benzene. Poly[(1,3-dihydro-1,3-dioxo-2H-isoindole-2,5-diyl)carbonyl(1,3-dihydro-1,3-dioxo-2H-isoindole-5,2-diyl)-1,4-phenyleneoxy-1,4-phenylenecarbonyl-1,3-phenylenecarbonyl-1,4-phenyleneoxy-1,4-phenylene]. Poly(5,5'-carbonylbis[1,3-isobenzofurandione]-alt-1,3-phenylenebis[[4-(4-aminophenoxy)phenyl]methanone])*
Monomers: 5,5'-Carbonylbis[1,3-isobenzofurandione], 1,3-Phenylenebis[4-(4-aminophenoxy)phenyl]methanone
Material class: Thermoplastic
Polymer Type: polyimide
CAS Number:

CAS Reg. No.	Note
103320-42-9	
107194-52-5	source based

Molecular Formula: $(C_{49}H_{26}N_2O_9)_n$
Fragments: $C_{49}H_{26}N_2O_9$

Volumetric & Calorimetric Properties:
Density:

No.	Value	Note
1	d 1.335 g/cm^3	amorph. [1,5]
2	d 1.507 g/cm^3	crystalline [5]

Thermal Expansion Coefficient:

No.	Value	Note	Type
1	5.5×10^{-5}–0.000105 K^{-1}	150–300°, bulk [5]	L
2	0.000117 K^{-1}	25–275°, unit cell expansion a axis [1]	L
3	7.5×10^{-5} K^{-1}	25–275°, unit cell expansion b axis [1]	L
4	0.000192 K^{-1}	25–275° [1]	V
5	5.1×10^{-5}–0.000123 K^{-1}	150–300°, amorph. phase [5]	L
6	6.4×10^{-5} K^{-1}	25–275°, crystalline phase [1]	L

Latent Heat Crystallization: ΔH_f 92 J g^{-1}, 125 J g^{-1} (100% crystalline)
Melting Temperature:

No.	Value	Note
1	350°C	[1,2,3]

Glass-Transition Temperature:

No.	Value	Note
1	222°C	[1,2,3]

Surface Properties & Solubility:
Solvents/Non-solvents: Insol. *N*-methyl-2-pyrrolidone, *m*-cresol, CHCl$_3$, *N,N*-dimethylacetamide, ethylene glycol, jet fuel, tricresyl phosphate (hydraulic fluid) [2,3]

Mechanical Properties:
Tensile (Young's) Modulus:

No.	Value	Note
1	3000 MPa	25°, 10 kg [1]
2	1450–3630 MPa	526–210 kpsi, 25–232°, ASTM D882 [2]
3	1650–2770 MPa	402–239 kpsi, 25–177°, ASTM D882, quenched film [3]

Tensile Strength Break:

No.	Value	Note	Elongation
1	32–137 MPa	20.1–4.6 kpsi, 25–232°, ASTM D882 [2]	6.5–75.1%
2	32–66 MPa	9.6–4.7 kpsi, 25–177°, ASTM D882, quenched film [3]	3.3–9.1%

Tensile Strength Yield:

No.	Value	Note
1	21–84 MPa	12.2–3 kpsi, 25–232°, ASTM D882 [2]
2	24.1–28.3 MPa	3.5–4.1 kpsi, 25–177°, ASTM D882, quenched film [3]

Fracture Mechanical Properties: Critical strain energy release rate (G_{Ic}) 5.5 kJ m^{-2} (32.9 in lb in^{-2}), 6.6 kJ m^{-2} (37.8 in lb in^{-2}) (powder moulding samples, 2770 MPa (402 kpsi), ASTM E399) [2,3]

Polymer Stability:
Decomposition Details: Exhibits complex decomposition kinetics. 0.5% Weight loss occurs at 100° (H$_2$O), then main weight loss occurs at 425°, 5% weight loss at 530°. Residue at 750° 42% [5]. At 300° in air, 1.6% weight loss after 48h, then becomes dark and brittle after 240h (2.4% wt. loss), 5.9% weight loss after 500h [2]

Applications/Commercial Products:

Trade name	Supplier
LARC-CPI	Mitsui Toatsu Chemicals

Bibliographic References
[1] Brillhart, M.V. and Cebe, P., *J. Polym. Sci., Part B: Polym. Phys.*, 1995, **33**, 927, (thermal expansion)
[2] Hergenrother, P.M., Beltz, M.W. and Havens, S.J., *J. Polym. Sci., Part A: Polym. Chem.*, 1989, **27**, 1161, (synth, props)
[3] Hergenrother, P.M., Wakelyn, N.T. and Havens, S.J., *J. Polym. Sci., Part A: Polym. Chem.*, 1987, **25**, 1093
[4] Muellerleile, J.T., Risch, B.G., Rodrigues, D.E., Wilkes, G.L. and Jones, D.M., *Polymer*, 1993, **34**, 789, (crystallisation and morphology)
[5] Xin Lu, S., Cebe, P. and Capel, M., *Polymer*, 1996, **37**, 2999, (thermal expansion)

LaRC-TPI   L-2

Synonyms: *Langley Research Center Thermoplastic Polyimide. BTDA-3,3'-DABP. Poly[(1,3-dihydro-1,3-dioxo-2H-isoindole-2,5-diyl)carbonyl(1,3-dihydro-1,3-dioxo-2H-isoindole-5,2-diyl)-1,3-phenylenecarbonyl-1,3-phenylene]. Poly(5,5'-carbonylbis[1,3-isobenzofurandione]-alt-bis(3-aminophenylmethanone)*
Related Polymers: LaRC-I-TPI
Monomers: 5,5'-Carbonylbis[1,3-isobenzofurandione], Bis(3-aminophenyl)methanone
Material class: Thermoplastic
Polymer Type: polyimide
CAS Number:

CAS Reg. No.	Note
28827-74-9	source based
51518-44-6	

Molecular Formula: $(C_{30}H_{14}N_2O_6)_n$
Fragments: $C_{30}H_{14}N_2O_6$
Morphology: Crystallisation behaviour has been reported [2,27,28]. Approx. 45–48% crystalline [13]. Is semi-crystalline if annealed below 330° but is amorph. if annealed above this temp. [12]. Crystallisation may be induced by annealing at 320° for 3h, air quenching and thermal cycling [25]
Miscellaneous: Grades with improved melt flow props. have been reported [29]. A review of applications as an adhesive has been reported [7]

Volumetric & Calorimetric Properties:
Density:

No.	Value	Note
1	d 1.37 g/cm^3	51µm film [11,19]
2	d 1.37 g/cm^3	Durimid film [20]

Thermal Expansion Coefficient:

No.	Value	Note	Type
1	3.5×10^{-5} K^{-1}	Durimid [20]	L

Melting Temperature:

No.	Value	Note
1	295°C	[25]
2	355°C	[25]

Glass-Transition Temperature:

No.	Value	Note
1	250°C	approx. [9]
2	242°C	[10,12]
3	260°C	[11]
4	267°C	[15]
5	246°C	amorph. [18]
6	265°C	[19]
7	256°C	Durimid [20]
8	240°C	amorph. [21]
9	259°C	[22]
10	216°C	cured in air, 200° [23]
11	264°C	cured in air or *in vacuo*, 300° [23]
12	257°C	preheated to 350° [24]
13	245°C	[25]

Transition Temperature General: T_m not detected for the amorph. form [18,21]. T_g is dependent upon the thermal history of the sample [24]
Transition Temperature:

No.	Value	Note
1	125°C	Tβ Preheated to 350° [24]

Surface Properties & Solubility:
Solubility Properties General: May be blended with PMR-15 to form semi-interpenetrating networks which improve the fracture toughness of PMR-15 [9]. Forms blends with Xydar liq. cryst. polymer to improve processability [10] and with poly(2,2'-(*m*-phenylene)-5,5'-bibenzimidazole [15]
Cohesive Energy Density Solubility Parameters: δ 33.1 J$^{1/2}$ cm$^{3/2}$ [11]
Solvents/Non-solvents: Insol. *N*-methylpyrrolidone, DMF, dimethylacetamide [9]. Sol. dimethylacetamide [15]

Transport Properties:
Polymer Solutions Dilute: [η]$_{inh}$ 0.16 dl g^{-1} (*m*-cresol, 35°, annealed) [13], 0.49 dl g^{-1} (*p*-chlorophenol/phenol (9:1), 35°, amorph.) [18,21], 0.83 dl g^{-1} (dimethylacetamide, 35°) [23], 0.39 dl g^{-1} (*m*-chlorophenol/phenol (95%: 5%), 25°) [25]. Inherent viscosity increases with increasing annealing time [13]. Zero shear viscosity 2000–100000 Pa s (350°) [12]
Gas Permeability:

No.	Gas	Value	Note
1	H_2	131.3 cm^3 mm/(m^2 day atm)	2×10^{-10} cm^2 (s cm Hg)$^{-1}$, 35° [4]
2	H_2O	20356 cm^3 mm/(m^2 day atm)	3.1×10^{-8} cm^2 (s cm Hg)$^{-1}$, 50°, 51 mm thick [11]
3	H_2O	23180 cm^3 mm/(m^2 day atm)	353×10^{-10} cm^2 (s cm Hg)$^{-1}$, 50° [26]
4	O_2	3024 cm^3 mm/(m^2 day atm)	75.6 cm^3 m^{-2} d^{-1} 0.0025 cm thick, room temp [17]
5	CO_2	11152 cm^3 mm/(m^2 day atm)	278.8 cm^3 m^{-2} d^{-1} 0.0025 cm thick, room temp [17]

Mechanical Properties:
Tensile (Young's) Modulus:

No.	Value	Note
1	3520 MPa	[14]
2	3516 MPa	film [19]
3	3450 MPa	Durimid film [20]
4	3723 MPa	540 ksi, 25°, film [22]
5	1476 MPa	214 kpsi, 246°, film [22]
6	3723 MPa	540 kpsi, air cured [23]

Flexural Modulus:

No.	Value	Note
1	4278 MPa	Durimid [20]

Tensile Strength Break:

No.	Value	Note	Elongation
1	114.5 MPa	[19]	
2	138 MPa	Durimid [20],	5%
3	136 MPa	19.7 kpsi, 25°, film [22,23]	

Flexural Strength at Break:

No.	Value	Note
1	158.6 MPa	[19]

Compressive Modulus:

No.	Value	Note
1	3588 MPa	Durimid [20]

Poisson's Ratio:

No.	Value	Note
1	0.36	Durimid [20]

Tensile Strength Yield:

No.	Value	Note
1	119 MPa	17.3 kpsi, 25°, film
2	16.5 MPa	2.4 kpsi, 246°, film [22]

Compressive Strength:

No.	Value	Note
1	236 MPa	Durimid [20]

Complex Moduli: Complex moduli decrease with increasing temp. [13]
Mechanical Properties Miscellaneous: Lap shear strength 12 MPa (titanium alloy, room temp., ASTM D100 amorph.) [18]. Lap shear strength at constant bonding pressure increases with increasing temp. [18]
Fracture Mechanical Properties: G_{Ic} 1800 J M^{-2} [9]. Has good toughness, with crazing apparent in a notched sample when stress is applied [14]. K_c 1.98 MPa m$^{1/2}$ (25 µm thick), 2.79 MPa m$^{1/2}$ (45 µm thick) [14]
Izod Notch:

No.	Value	Notch	Note
1	53 J/m	Y	Durimid [20]
2	770 J/m	N	Durimid [20]
3	200 J/m	N	LaRC-TPI [20]
4	21 J/m	Y	LaRC-TPI [20]

Electrical Properties:
Dielectric Permittivity (Constant):

No.	Value	Frequency	Note
1	3.4	1 kHz	Durimid film, ASTM D150-64T [20]

Dissipation (Power) Factor:

No.	Value	Frequency	Note
1	0.003	1 kHz	Durimid film, ASTM D150-64T [20]

Optical Properties:
Transmission and Spectra: Charge-transfer emission [1] and ir [6,9] spectral data have been reported
Volume Properties/Surface Properties: Colour yellow (amorph.) [18]

Polymer Stability:
Thermal Stability General: Thermal stability in a N_2 atmosphere has been reported [5]. Oxidative ageing between 300 and 400° causes a colour change from bright yellow to dark brown, and embrittlement. Toughness decreases by 50% after 200 h at 340° [8]
Upper Use Temperature:

No.	Value	Note
1	300°C	[19]

Decomposition Details: Shows 10% weight loss after 600 h and 20% weight loss after 1000 h upon heating in air at 316°. [19] Durimid shows 10% weight loss in air above 550° [20]. Amorph. polymer shows 5% weight loss after 100 h at 350° [21]
Flammability: Flammability rating V0 (Durimid film, UL94) [20]
Chemical Stability: Stability of films to aq. alkali has been reported [3]. Durimid film is resistant to 2-propanol, 2-butanone, Freon TF and toluene [20]. Exposure of moulded Durimid resin to chemicals generally causes deterioration in mech. props. [20]
Stability Miscellaneous: Electron-beam irradiation causes little change in tensile strength or elongation at break. The shear modulus above T_g of LaRC-TPI 1500 increases with increasing radiation dose [16]

Applications/Commercial Products:
Processing & Manufacturing Routes: Synth. by reaction of 3,3′ 4,4′-benzophenonetetracarboxylic dianhydride with 3,3′-diaminobenzophenone in dimethylacetamide at room temp. The polyamic acid product is then imidised at 300° [17]
Applications: As an adhesive in high temp. environments; in insulating fibres; in composite matrices

Trade name	Details	Supplier
Durimid	100	Rogers Corporation
LaRC-TPI		Mitsui Toatsu Chemicals
LaRC-TPI-1500		Mitsui Toatsu Chemicals

Bibliographic References
[1] Hasegawa, M., Mita, I., Kochi, M., Yokota, R., *J. Polym. Sci., Polym. Chem. Ed.*, 1989, **27**, 263
[2] Hou, T.H. and Bai, J.M., *NASA Contract. Rep.*, 1990, 41
[3] Croall, C.I., *NASA Tech. Memo.*, 1990, 43, (alkaline degradation)
[4] Tanaka, K., Kita, H. and Okamoto, K., *Kobunshi Ronbunshu*, 1990, **47**, 945, (gas permeability)
[5] Yokota, R., Sakino, T. and Mita, I., *Kobunshi Ronbunshu*, 1990, **47**, 207, (thermal stability)
[6] Young, P.R. and Chang, A.C., *SAMPE J.*, 1986, **22**, 70, 162, (ir)

[7] Esaki, S. and Ohta, M., *Kino Zairyo*, 1985, **5**, 22, (rev, uses)
[8] Hinkley, J.A. and Yue, J.J., *J. Appl. Polym. Sci.*, 1995, **57**, 1539
[9] Tai, H.J., Jang, B.Z. and Wang, J.B., *J. Appl. Polym. Sci.*, 1995, **58**, 2293
[10] Blizard, K.G. and Haghighat, R.R., *Polym. Eng. Sci.*, 1993, **33**, 799
[11] Okamoto, K.-I., Tanihara, N., Watanabe, H., Tanaka, K. et al, *J. Polym. Sci., Part B: Polym. Phys.*, 1992, **30**, 1223, (water sorption)
[12] Hou, T.H. and Bai, J.M., *High Performance Polym.*, 1989, **1**, 191
[13] Hou, T.H., Bai, J.M. and St Clair, T.L., *J. Appl. Polym. Sci.*, 1988, **36**, 321
[14] Hinkley, J.A. and Mings, S.L., *Polymer*, 1990, **31**, 75, (fracture toughness)
[15] Guerra, G., Williams, D.J., Karasz, F.E. and MacKnight, W.J., *J. Polym. Sci., Part B: Polym. Phys.*, 1988, **26**, 301
[16] Sasuga, T., *Polymer*, 1991, **32**, 1539, (irradiation)
[17] Sykes, G.F. and St Clair, A.K., *J. Appl. Polym. Sci.*, 1986, **32**, 3725, (gas permeability)
[18] Progar, D.J. and St Clair, T.L., *J. Adhes. Sci. Technol.*, 1994, **8**, 67, (adhesion)
[19] Scola, D.A. and Vontell, J.H., *CHEMTECH*, 1989, **19**, 112, (props)
[20] Sherman, D.C., Chen, C.-Y. and Cercena, J.L., *Int. SAMPE Symp. Exhib.*, 1988, **33**, 134
[21] Tamai, S., Yamashita, W. and Yamaguchi, A., *Int. SAMPE Tech. Conf.*, 1994, **26**, 365
[22] Bell, V.L., *J. Polym. Sci., Polym. Chem. Ed.*, 1976, **14**, 225, (mech props)
[23] Bell, V.L., Stump, B.L. and Gager, H., *J. Polym. Sci., Polym. Chem. Ed.*, 1976, **14**, 2275, (mech props)
[24] Gillham, J.K. and Gillham, H.C., *Am. Chem. Soc., Div. Org. Coat. Plast. Chem., Pap*, 1973, **33**, 201
[25] Mullerleile, J.T. and Wilkes, G.L., *Polym. Prepr. (Am. Chem. Soc., Div. Polym. Chem.)*, 1990, **31**, 637
[26] Kita, H., Tanaka, K., Okamoto, K.-I. and Yamamoto, M., *Chem. Lett.*, 1987, 2053
[27] Maffezzoli, A., *Mater. Sci. Forum*, 1996, **203**, 25, (cryst behaviour)
[28] Theil, M.H. and Gangal, P.D., *NASA Contract. Rep.*, 1992, 36
[29] Progar, D.J., St Clair, T., Burks, H., Gautreaux, C. et al, *SAMPE J.*, 1990, **26**, 53

LaRC-I-TPI   L-3

Synonyms: *IDPA-mPDA. Poly(4,4'-isophthaloyldiphthalic anhydride-alt-m-phenylenediamine). Langley Research Center Isomeric Thermoplastic Polyimide. Poly[(1,3-dihydro-1,3-dioxo-2H-isoindole-5,2-diyl)-1,3-phenylene(1,3-dihydro-1,3-dioxo-2H-isoindole-2,5-diyl)carbonyl-1,3-phenylenecarbonyl]. Poly[5,5'-(1,3-phenylenedicarbonyl)bis-1,3-isobenzofurandione-alt-1,3-benzenediamine]*
Related Polymers: LaRC-TPI
Monomers: 5,5'-(1,3-Phenylenedicarbonyl)bis-1,3-isobenzofurandione, 1,3-Benzenediamine
Material class: Thermoplastic
Polymer Type: polyimide
CAS Number:

CAS Reg. No.	Note
115967-80-1	
115949-39-8	source based

Molecular Formula: $(C_{30}H_{14}N_2O_6)_n$
Fragments: $C_{30}H_{14}N_2O_6$
Additives: Graphite fibre to improve flexural strength
Morphology: Amorph. [4]
Miscellaneous: Calculation of bulk props. has been achieved via computational methods [6]

Volumetric & Calorimetric Properties:
Glass-Transition Temperature:

No.	Value	Note
1	255°C	[1]
2	259°C	[2,4]
3	269°C	[3]

Surface Properties & Solubility:
Solvents/Non-solvents: Insol. refluxing CH_2Cl_2, dimethylacetamide, *m*-cresol [4]

Transport Properties:
Polymer Melts: $[\eta]_{inh}$ 0.5–1.5 dl g^{-1} (*m*-cresol, 35°) [1], 0.45 dl g^{-1} (dimethylacetamide, 35°) [3], 0.47 dl g^{-1} (dimethylacetamide, 35°) [4], 0.52 dl g^{-1} (dimethylacetamide, 35°) [3]. Melt viscosity 2200 Pa s [1]

Mechanical Properties:
Tensile (Young's) Modulus:

No.	Value	Note
1	3500–4140 MPa	[1]
2	3723.3 MPa	53 mm thick film [4]

Tensile Strength Break:

No.	Value	Note	Elongation
1	131 MPa	25° [4]	4.9%
2	58 MPa	200° [4]	13%

Elastic Modulus:

No.	Value	Note
1	1000 MPa	20° [1]
2	50 MPa	350° [1]

Tensile Strength Yield:

No.	Value	Elongation	Note
1	96.53 MPa	4.9%	25° [4]
2	50.33 MPa	13%	200° [4]

Miscellaneous Moduli:

No.	Value	Note	Type
1	3723.3 MPa	25° [4]	Tan modulus
2	2620.1 MPa	200° [4]	Tan modulus

Mechanical Properties Miscellaneous: Lap shear strength 18.5 MPa (titanium, room temp., bonding strength, 2.07 MPa, ASTM D1002) [3], 33 MPa (titanium, room temp., 204° for 1000 h) [4]
Failure Properties General: Regarded as a creasable film, one that can withstand a 180° fold followed by a reverse 360° fold [2]
Fracture Mechanical Properties: K_{Ic} 3.8–3.9 mPa m$^{1/2}$ (ASTM E399, D5045) [1]; K_c 26.6 MPa cm (53 mm thick film) [4]; G_{Ic} 3000–3800 J m^2 [1]. G_c 5.2 in-lb in^{-2} (53 mm thick film) [4]

Electrical Properties:
Electrical Properties General: Dielectric props. have been reported [5]

Polymer Stability:
Decomposition Details: Cured film shows max. weight loss of 3.1% after heating for 350 h at 300° [2]

Applications/Commercial Products:
Processing & Manufacturing Routes: Synth. by reaction of 4,4′-isophthaloyldiphthalic anhydride with *m*-phenylenediamine in dimethylacetamide [3], diglyme-dimethylacetamide, or *m*-cresol soln. Imidisation is achieved by heating at 200° for 2 h. Processability may be improved by blending with a bisnadimide [1]
Applications: Has adhesive props.

Bibliographic References
[1] Durand, V., Senneron, M. and Sillion, B., *Polym. Prepr. (Am. Chem. Soc., Div. Polym. Chem.)*, 1994, **35**, 365, (synth, blends)
[2] Pratt, J.R., Blackwell, D.A., St Clair, T.L. and Allphin, N.L., *Polym. Prepr. (Am. Chem. Soc., Div. Polym. Chem.)*, 1988, **29**, 128
[3] Progar, D.J., St Clair, T.L. and Pratt, J.R., *Polyimides: Mater., Chem. Charact., Proc. Int. Conf. Polyimides*, 3rd, 1988, 151, (synth, adhesion)
[4] Pratt, J.R. and St Clair, T.L., *SAMPE J.*, 1990, **26**, 29, (rev, props)
[5] Sillion, B, Rabilloud, G., Garapon, J., Gain, O. and Vallet, J., *Mater. Res. Soc. Symp. Proc.*, 1995, **381**, 93, (dielectric props)
[6] Collantes, E.R., Gahimer, T., Welsh, W.J., *Comput. Theor. Polym. Sci.*, 1996, **6**, 29

LDPE L-4

Synonyms: *Low-density polyethylene. PE-LD. High pressure polyethylene. HPPE. Polythene*
Related Polymers: Polyethylene, Polyethylene films, Polyethylene fibres, Use of comonomers (inc. polar ones) in the high pressure process
Monomers: Ethylene
Material class: Thermoplastic
Polymer Type: polyethylene, polyolefins
CAS Number:

CAS Reg. No.	Note
9002-88-4	all density types

Molecular Formula: $(C_2H_4)_n$
Fragments: C_2H_4
Molecular Weight: 500–60000
Additives: Antioxidants (*e.g.* 2,6-di-*tert*-butyl-*p*-cresol (BHT); octadecyl 3,5-di-*tert*-butyl-4-hydroxyhydrocinnamate; tetrakis [methylene(3,5-di-*tert*-butyl-4-hydroxycinnamate)]methane), uv stabilisers (*e.g.* benzophenones), antiblock agents (*e.g.* diatomaceous silica particle size approx. 5–15 µm), slip agents (*e.g.* erucamide; oleamide; steramide), photooxidation inhibitor (carbon black approx. 10 µm particle size, approx. 2–3%)
General Information: Originally referred to homopolymers of density 0.915–0.94 g cm^{-3}. However, use of Ziegler-type catalysts in low pressure processes produces copolymers of density 0.900–0.964 g cm^{-3}, and use of these catalysts in the high pressure processes can achieve densities of 0.870–0.96 g cm^{-3}. The MW, MW distribution and frequency and distribution of long and short chain branching all affect the physical and mech. props.
Morphology: Unbranched PE is 70–90% crystalline with an orthorhombic unit cell; the presence of short chain branches [1] disrupts the chains from folding into the flat zig-zag formation reducing crystallinity to approx. 40–60%. Effects of long-chain branching have been reported [20]
Identification: Can be differentiated from LLDPE, by using DSC to fractionate and then measuring the melt peaks (typically four peaks for LDPE, seven for LLDPE). Temp. range of melt peaks is related to the comonomer [19]
Miscellaneous: Owing to the customisable nature of polyethylenes, and differences between manufacturing processes, data given here tend to be overall ranges. Specific manufacturer's grades will normally have closer tolerances - or, with certain processes, even be available outside the ranges quoted

Volumetric & Calorimetric Properties:
Density:

No.	Value	Note
1	d^{25} 0.91–0.955 g/cm^3	[1,2,3,4]
2	d 0.916–0.93 g/cm^3	ASTM D1505 [1,2,3,4]

Thermal Expansion Coefficient:

No.	Value	Note	Type
1	0.0001–0.00022 K^{-1}	ASTM D696 [1,4]	L

Volumetric Properties General: Data given here are restricted to more common polymers *e.g.* with melt index 0.2–150 or 100000–20000 Mpa s viscosity range. Greases, waxes, highly cross-linked and other speciality polymers are excluded unless specifically stated

Thermal Conductivity:

No.	Value	Note
1	0.34 W/mK	0.0008 cal (s cm °C)$^{-1}$, ASTM C177 [4]

Melting Temperature:

No.	Value	Note
1	80–115°C	[3,4]
2	105–115°C	45–55% cryst. [1,2]

Glass-Transition Temperature:

No.	Value	Note
1	<-40°C	max. [2]

Transition Temperature General: T_g related to T_β and T_γ [2,34]
Deflection Temperature:

No.	Value	Note
1	40–44°C	104–112°F, 0.455 MPa (66 psi), ASTM D648 [4]
2	40–50°C	0.455 MPa, ASTM D648 [2]

Vicat Softening Point:

No.	Value	Note
1	86–102°C	[11]

Brittleness Temperature:

No.	Value	Note
1	>76°C	min., 50% failure, ASTM D746 [1]

Transition Temperature:

No.	Value	Note	Type
1	20–80°C	molecular motion [2]	T_α
2	-30--20°C	onset of branch point motion [2]	T_β
3	-125--120°C	segmental motion [2]	T_g

Surface Properties & Solubility:

Solubility Properties General: Solubility increases with increasing melt flow index but decreases with increasing density and crystallinity [16]

Solvents/Non-solvents: Sol. above 50° hydrocarbons, chlorinated hydrocarbons [16], higher aliphatic esters and aliphatic ketones [17]. Insol. below 50° all common solvents

Transport Properties:

Melt Flow Index:

No.	Value	Note
1	0.15–150 g/10 min	commercial LDPE excludes waxes, greases, cross-linked speciality polymers

Polymer Solutions Concentrated: Props. are affected by long-chain branching; molecules less extended than those of linear PE. Mark-Houwink equation parameters K 1.05×10^{-3}, a 0.63 (high MW); K 1.35×10^{-3}, a 0.63 (MW to 76000) have been reported [2]

Polymer Melts: Highly viscous; viscosity depends on MW, MW distribution (chain branching), shear rate and shear stress. Pseudoplastic; apparent viscosity decreases with increasing shear stress. At 200° melt viscosity falls two orders of magnitude for a four order of magnitude increase in shear rate. Under practical extrusion conditions viscosity of LDPE is much lower than that of LLDPE of the same melt flow index [2]. Effects of shear rate, etc. vs. viscosity have been reported [12,13,14,15]. Molten LDPE shows elastic props. e.g. die swell which increases with shear rate

Permeability of Gases: Other gas permeability values have been reported [5,9]

Water Content: 12–23 gm^{-2} (24h, 38°, 90% relative humidity, 25 μm) [4,11]

Water Absorption:

No.	Value	Note
1	<0.01 %	max., 24h, 1/8 in thick, ASTM D570 [4]

Gas Permeability:

No.	Gas	Value	Note
1	H_2	647 cm^3 mm/(m^2 day atm)	7.4×10^{-13} cm^2 (s Pa)$^{-1}$, 25° [5,6]
2	He	3294 cm^3 mm/(m^2 day atm)	3.7×10^{-13} cm^2 (s Pa)$^{-1}$, 25° [5,7]
3	Ar	184 cm^3 mm/(m^2 day atm)	2.1×10^{-13} cm^2 (s Pa)$^{-1}$, 25° [5,7]
4	Xe	350 cm^3 mm/(m^2 day atm)	4.01×10^{-13} cm^2 (s Pa)$^{-1}$, 25° [5,6]
5	CO	96.3 cm^3 mm/(m^2 day atm)	1.1×10^{-13} cm^2 (s Pa)$^{-1}$, 25° [5,7]
6	H_2	767 cm^3 mm/(m^2 day atm)	1950 cm^3 mil (100 in^2 day atm)$^{-1}$, 25° [4]
7	O_2	197 cm^3 mm/(m^2 day atm)	500 cm^3 mil (100 in^2 day atm)$^{-1}$, 25° [4,5,7]
8	Ne	42 cm^3 mm/(m^2 day atm)	4.8×10^{-14} cm^2 (s Pa)$^{-1}$, 25° [5,6]
9	CO_2	1063 cm^3 mm/(m^2 day atm)	2700 cm^3 mil (100 in^2 day atm)$^{-1}$, 25° [4]
10	N_2	70.8 cm^3 mm/(m^2 day atm)	180 cm^3 mil (100 in^2 day atm)$^{-1}$, 25° [5,7]
11	CH_4	192.6 cm^3 mm/(m^2 day atm)	2.2×10^{-13} cm^2 (s Pa)$^{-1}$, 25° [5,7]
12	ethane	446 cm^3 mm/(m^2 day atm)	5.1×10^{-13} cm^2 (s Pa)$^{-1}$, 25° [5,7]
13	NH_3	1838 cm^3 mm/(m^2 day atm)	2.1×10^{-12} cm^2 (s Pa)$^{-1}$, 20° [5,8]
14	H_2O	5953 cm^3 mm/(m^2 day atm)	6.8×10^{-12} cm^2 (s Pa)$^{-1}$, 25° [5,10]

Mechanical Properties:

Mechanical Properties General: Data refer to typical commercial polymers with melt index 0.2–150 or viscosity range 100000–20000 Mpa s. Elongation, tear strength and tensile strength compared with LLDPE, have been reported [13]

Tensile (Young's) Modulus:

No.	Value	Note
1	100–300 MPa	25–41 kpsi, ASTM D638, ISO 527, DIN 53457 [4,11]

Flexural Modulus:

No.	Value	Note
1	240–330 MPa	35–38 kpsi, 23° (73°F), ASTM D790 [4]

Tensile Strength Break:

No.	Value	Note	Elongation
1	8.3–34 MPa	1200–4550 psi, ASTM D638, ISO 527, ISO 1184 [2,4,11]	100–800%

Elastic Modulus:

No.	Value	Note
1	120–330 MPa	ISO 537, DIN 53445 [11]

Tensile Strength Yield:

No.	Value	Elongation	Note
1	6–15 MPa	100–800%	870–2200 psi, ASTM D638, ISO 1184 [2,4,11]

Impact Strength: Dart drop impact strength 50–300 g (25 μm)$^{-1}$ (38 μm thickness, ASTM D1709) [1,2,11] decreases as density and melt index increase

Hardness: Rockwell D41-D60 (ASTM D785) [1,2]. Shore D44-D54 (ASTM D2240, ISO 868, DIN 53505) [4,11]

Failure Properties General: Failure is ductile except at very low temp. Stress cracking, however, gives brittle failure. Resistance to stress cracking increases as density decreases, MW increases, MW distribution narrows and temp. increases

Electrical Properties:

Electrical Properties General: Outstanding electrical props. make LDPE the dielectric of choice for many applications including high voltage and high, very high and ultra high frequency. Carbon black can be added in sufficient quantity to produce a semiconducting composite

Dielectric Permittivity (Constant):

No.	Value	Frequency	Note
1	2.3	60 Hz	ASTM D150, IEC 250/VDE 0303 T4 [1,2,11]
2	2.2	1 kHz-1 GHz	ASTM D150 [4]

– LDPE

Dielectric Strength:

No.	Value	Note
1	196 kV.mm^{-1}	5000 V mil^{-1}, 1 mil thick [4]
2	118 kV.mm^{-1}	3000 V mil^{-1}, 5 mil thick [4]
3	17.7–39.4 kV.mm^{-1}	450–1000 V mil^{-1}, $\frac{1}{8}$ in thick [4]

Dissipation (Power) Factor:

No.	Value	Note
1	0.0001–0.0003	ASTM D150 [4,11]

Optical Properties:

Refractive Index:

No.	Value	Note
1	1.51–1.52	25°, ASTM D542 [1,2]

Transmission and Spectra: C-13 nmr spectral data have been reported [26]
Volume Properties/Surface Properties: Haze 15% (max., ASTM D1003) [11]. Gloss 45–75 (45°, ASTM D2457); 10–100 (20°, DIN 67530) [11]

Polymer Stability:

Polymer Stability General: Generally stable but subject to photooxidation. Low environmental impact because of presence of carbon and hydrogen only
Thermal Stability General: Melts at 70–130°; above 225° fumes may be evolved, decomposition commences about 300° [11]. -2% Change in linear dimensions at 100° (30 min.) [4]
Upper Use Temperature:

No.	Value	Note
1	60°C	max. recommended storage temp. [11]
2	88°C	max. film use temp. [35]

Decomposition Details: Above 300° oxidative pyrolysis gives CO, H_2O and smaller amounts of hydrocarbons, acids and aldehydes [11,25]. Ignition of evolved gases accelerates pyrolysis. Depending on oxygen level, products may contain carbon (soot), CO_2, CO, acrolein, or other aldehydes [11,26]
Flammability: Self ignition temp. 350° (small scale, ASTM D1929-91A). Burning may be accompanied by release of flaming molten droplets. LDPE powder/dust constitutes a dust explosion hazard (Group (a) flammable dust) [11,27,29]
Environmental Stress: Stress in the presence of wetting agents, silicone oils, or polar substances may lead to environmental stress cracking. Resistance (ESCR) increases with decreasing density, decreasing melt flow index and increasing comonomer content [11]. Photooxidation occurs between 260-420 nm. Abiotic degradation may be enhanced by addition of starch and a pro-oxidant (SBS/manganese stearate) [36]
Chemical Stability: Non polar and predominantly inert. May be swollen by some organic solvents (e.g. hydrocarbon oils) [4,11]
Biological Stability: Biologically inert. Modification of biological/environmental degradation has been reported [21,30,31,32]
Recyclability: Recycling of sorted waste is no problem and produces high quality parts. Composites may be more problematic but can be suitable for incineration or feedstock recycling [22]
Stability Miscellaneous: Microwave irradiation causes limited cross-linking [37]

Applications/Commercial Products:

Processing & Manufacturing Routes: Polymerisation of ethene in stirred autoclave cylindrical chamber, at 207 MPa and 220–300°. Conversion per pass 22%. In a tubular reactor, at 202 MPa (pulsed periodically to approx. 150–180 MPa) and 300–325° (peak) or 250–275° (during partial cooling). Conversion per pass 35% [1,2]. Model for high pressure polymerisation has been reported [33]. May be blow moulded, injection moulded or rotationally moulded. Films are blown or extruded

Mould Shrinkage (%):

No.	Value	Note
1	1.5–5%	[4]

Applications: Used in household containers, closures, pharmaceutical products, wire and cable insulation (the material of choice for high frequency use), adhesives and sealants [23]. Sheet is blow moulded (for small volume containers - the high purity of LDPE makes it excellent for medical and pharmaceutical use), extrusion (usually cross-linked using peroxides, normally only used where high flexibility is required), compounding with other polyethylenes. Also used in production of EVA (accounts for 50% of PE production) and in blends

Trade name	Details	Supplier
Alcudia	also includes HDPE, MDPE	Repsol Quimica
Bralen	also used for EVA and other products	Slovnaft Joint Stock Company
Cabot PE	masterbatches and compounds in an LDPE carrier	Cabot Plastics Ltd.
Corthene		Royalite Plastics Ltd.
Empee	also HDPE, PP, PS (flame retardant)	Monmouth Plastics
Eucalene	also HDPE	Van Raaijen Kunststoffen
Finathene	also HDPE	Fina Chemicals (Division of Petrofina SA)
LDPE ***		Dow
Lacqtene	aslo HDPE, LLDPE, linear HDPE	Elf Atochem (UK) Ltd.
Lacqtene		Ashland
Lupolen	includes most types of PE as well as several other polymers	BASF
Luwax	wax, also includes those from HDPE, oxidised PE and ethylene copolymers	BASF
Marlex	also HDPE, MDPE, butene-ethene copolymer (inc. films)	Philips Petroleum International
Microthene	also includes a blend with vinyl acetate	Quantum Chemical
Mitsubishi	also HDPE, LLDPE (inc. films)	Mitsubishi Kasei
Novapol	also HDPE, LLDPE, linear MDPE and butene-ethene copolymer	Novacor

Novex	homo and copolymers	BP Chemicals
Novex		KD Thermoplastics
Okiten		Ina-Oki
PE ***	homo and copolymers also EVA, PVC and others	Chevron
PE-25	homo and copolymers, also EVA, PVC and others	Washington Penn
Petrothene	also LLDPE, HDPE, MDPE, EVA, PP, EPR	Quantum Chemical
Plastotrans	film	4P Folie Forchheim GmbH.
Polyethylene LD	others also available	Montell Polyolefins
Polyter	drinking water grade	Meridionale des Plastiques SA
Repsol	also HDPE and PP	Commercial Quimica Insular SA
Rexene	also includes EVA, iPP, EPR	Rexene Products
Riblene	also EVA and ethene-butene copolymer	Enichem
Riblene		Ashland
Sabic	also HDPE	NV Chemach SA
Sarmatene	LDPE based master-batch	Sandoz Chemicals Ltd.
Stamylan	covers a range of olefin polymers	DSM
Stamylan		Macrodan A/S
Stamylan		Ashland Plastics
Statoil	includes HDPE, also iPP, ethene-propene copolymers and others	Northern Industrial Plastics Ltd.
Sumiplene	includes HDPE	Sumica SA
Suntec	includes HDPE also EVA	Asahi Chem. Co Ltd
Tipolen		Tiszai Vegyi Kombinát

Bibliographic References

[1] Pebsworth, L.W., *Kirk-Othmer Encycl. Chem. Technol.*, 4th edn., (eds. J.I. Kroschwitz and M. Howe-Grant), Wiley Interscience, 1996, **17**, 707
[2] Doak, K.W., *Encycl. Polym. Sci. Eng.*, 2nd edn., (ed. J.I. Kroschwitz), Wiley Interscience, 1986, **6**, 386
[3] Alger, M., *Polymer Science Dictionary*, 2nd edn., Chapman and Hall, 1997
[4] Guide to Plastics, *Property and Specification Charts*, McGraw-Hill, 1987
[5] Pauly, S., *Polym. Handb.*, 4th edn., (eds. J. Brandrup, E.H. Immergut and E.A. Grulke), John Wiley and Sons, 1989, **VI**, 543
[6] Miyake, H., Matsuyama, M., Ashida, K. and Watanabe, J., *Zhenkong Kexue Yu Jishu*, 1983, **A1**, 3, 1447
[7] Michaels, A.S. and Bixler, H.J., *J. Polym. Sci.*, 1961, **50**, 413
[8] Braunisch, V.L. and Lenhart, H., *Kolloid Z. Z. Polym.*, 1961, **177**, 24
[9] Waack, R., Alex, N.H., Frisch, H.L., Stannett, V. and Szwarc, M., *Ind. Eng. Chem.*, 1955, **47**, 2524
[10] Myers, A.W., Meyer, J.A., Rogers, C.E., Stannett, V. and Szwarc M., *Tappi*, 1961, **44**, 58
[11] *Luwax*, BASF, (technical datasheet)
[12] Kwack, T.H. and Han, C.D., *J. Appl. Polym. Sci.*, 1983, **28**, 3419
[13] Attalla, G. and Bertinotti, F., *J. Appl. Polym. Sci.*, 1983, **28**, 3503
[14] Han, C.D., Kim, Y.J., Chung, H.-K, and Kwack, T.H., *J. Appl. Polym. Sci.*, 1983, **28**, 3435
[15] Kuhn, R. and Kromer H., *Colloid Polym. Sci.*, 1982, **260**, 1083
[16] Roff, W.J. and Scott, J.R., Fibres, Films, *Plastics and Rubbers*, Butterworths, 1971
[17] Bloch, D.R., *Polym. Handb.*, 4th edn., (eds. J. Brandrup, E.H. Immergut and E.A. Grulke), John Wiley and Sons, 1999, **VII**, 500
[18] Butcher, J., *Plast. World*, 1996, **54**, 50
[19] Wolf, B., Kenig, S., Klopstock, J. and Miltz, J., *J. Appl. Polym. Sci.*, 1996, **62**, 1339
[20] Sukhadia, A.M., Annu. Tech. Conf. - Soc. Plast. Eng., 54th, 1996, 1157
[21] Xingzhan, H., *Polym. Degrad. Stab.*, 1997, **55**, 131
[22] Hartzmann, G., Jagdmann, S. and Klimesch, R.G., *Kunststoffe*, 1996, **86**, 9
[23] *Adhesives and Sealants Industry*, Feb., 1997, **4(1)**, 26
[24] Smedberg, A., Hjertberg, T. and Gustafsson, B., *Polymer*, 1997, **38**, 4127
[25] Mitera, J., Michal, J., Kubat, J. and Kubelka, V., *Z. Anal. Chem.*, 1976, **281**, 23
[26] Morikawa, T., *J. Combust. Toxicol.*, 1976, **3**, 135
[27] Field, P., *Handbook of Powder Technology*, Elsevier, 1982, **4**
[28] O'Mara, M.M., *J. Fire Flammability*, 1974, **5**, 34
[29] HSE Health and Safety at Work Booklet No. 22, "Dust explosions in factories", HMSO, 1976
[30] Graff, G., *Modern Plast. Intern.*, 1996, **26**, 27
[31] Lin, Y., *J. Appl. Polym. Sci.*, 1997, **63**, 811
[32] Erlandsson, B., Karlsson, S. and Albertsson, A.C., *Polym. Degrad. Stab.*, 1997, **55**, 237
[33] Brandolin, A., Lacunza, M.H., Ugrin, P.E. and Capicti, N.J., *Polym. React. Eng.*, 1996, **4**, 193
[34] Dechter, J.J., Axelson, D.E., Dekmezion, A., Glotin, M. and Mandelkern, L., *J. Polym. Sci., Polym. Phys. Ed.*, 1982, **20**, 641
[35] Mackenzie, K.J., *Kirk-Othmer Encycl. Chem. Technol.*, 4th edn., (eds. J.I. Kroschwitz and M. Howe-Grant), Wiley Interscience, 1993, **10**, 761
[36] Albertsson, A.-C., Barenstedt, C. and Karlsson, S., *Acta Polym.*, 1994, **45**, 97
[37] Hindeleh, A.M. et al, *Acta Polym.*, 1996, **47**, 105, (microwave irradiation)

LLDPE   L-5

Synonyms: *Linear low-density polyethylene.* PE-LLD
Related Polymers: Polyethylene, Polyethylene film, HDPE, MDPE, ULDPE
Monomers: Ethylene, 1-Butene (co-monomer), 1-Hexene (co-monomer), 1-Octene (co-monomer), 4-Methyl-1-pentene (co-monomer), Cyclopentene (speciality co-monomer), Norbornene (speciality co-monomer)
Material class: Thermoplastic
Polymer Type: polyethylene, polyolefins
CAS Number:

CAS Reg. No.	Note
9002-88-4	all density types
25087-34-7	1-butene co-monomer
25213-02-9	1-hexene co-monomer
26221-73-8	1-octene co-monomer
25213-96-1	4-methyl, 1-pentene co-monomers
60785-11-7	1-hexene, 1-butene co-monomers
74746-95-5	1-butene, 4-methyl-1-pentene co-monomers
32536-03-1	cyclopentene co-monomer
26007-43-2	norbornene co-monomer
72238-82-5	norbornene, 1-butene co-monomers
172501-09-6	norbornene, 1-decene co-monomers

Molecular Formula: $(C_2H_4)_n$
Molecular Weight: 50000–200000; M_w/M_n 2.5–4.5 (typical commodity grades); 10–35 (chromium oxide catalysed)

LLDPE

Additives: Antislip agents (e.g. erucamide), antiblock agents (e.g. talc), PPA, zinc oxide may be used. Antioxidant normally used

General Information: LLDPE designation is usually applied to ethylene copolymers in the density range 0.915–0.925 g cm^{-3}, with a co-monomer (normally butene, hexene or octene) content of approx. 2.5–3.5 mol%. For lower co-monomer content and for higher co-monomer content. As with LDPE, props. are affected by MW, MW distribution and frequency type, and distribution of branches (co-monomer). The structural regularity and narrow MW distribution achieved by the use of metallocene catalysts is having a major impact, a 'new' non-metallocene LLDPE has also been reported [16]

Morphology: Primarily CH_2CH_2 repeating units with occasional branches derived from a co-monomer such as 1-butene, 1-hexene, 1-octene, 4-methyl-1-pentene, or cyclopentene and norbornene. Normally these branches are short but soln. process, metallocene catalysed LLDPE may contain approx. 0.002 long branches per 100 ethylene units. Chain ends are normally methyl and either methyl (hydrogen as chain transfer agent), vinyl or vinylidene. The resulting chains crystallise principally in the orthorhombic form with cell parameters a 0.77 nm; b 0.5 nm; c (the direction of the polymer chain) 0.2534 nm - cf. Degree of crystallinity depends (primarily) on α-olefin content and is typically less than 45%. Spherulites, 1–5μm diameter, are formed from flat crystallites (lamellae) of folded chains, interconnected by chains passing from one to another and providing mech. strength. The remaining approx. 60% is amorph., giving flexibility. Short chains inhibit chain folding, decreasing lamella thickness and increasing the number of interlamellar tie molecules. High compositional uniformity gives poor crystallisation, very thin lamellae, and high flexibility. Non-uniformity gives more rigidity [1,2]

Identification: DSC can be used to fractionate sample and then measure melt peaks, the number of which (6 or 7 for LLDPE, 4 for LDPE) and temp. range is related to the co-monomer [28]

Volumetric & Calorimetric Properties:

Density:

No.	Value	Note
1	d 0.9 g/cm^3	uniform [1]
2	d 0.912–0.915 g/cm^3	non-uniform, 4 mol % [1]
3	d 0.908–0.912 g/cm^3	uniform [1]
4	d 0.92–0.922 g/cm^3	non-uniform, 3 mol % [1]
5	d 0.918–0.92 g/cm^3	uniform [1]
6	d 0.927–0.93 g/cm^3	non-uniform, 2 mol % [1]

Volumetric Properties General: Density varies with α-olefin content and branching uniformity [1]

Melting Temperature:

No.	Value	Note
1	146–147°C	equilibrium, linear unbranched [5]
2	133–138°C	
3	100–120°C	commercial uniform LL (or mLL) DPE [1]
4	122–128°C	[1,4]

Transition Temperature General: T_m decreases with increasing branching [1]

Brittleness Temperature:

No.	Value	Note
1	-140–-100°C	[1]

Surface Properties & Solubility:

Solvents/Non-solvents: Sol. binary systems comprising carbon disulfide with xylene, cyclohexane, C_6H_6, toluene, CCl_4, THF, or $CHCl_3$, [7] at room temp.; aromatic, aliphatic and halogenated hydrocarbons at elevated temp. (80–100°C) [1,2]. Insol. all known single solvents at room temp. [1,2]. High branching content and low MW fractions sol. hexane (50°)

Transport Properties:

Melt Flow Index:

No.	Value	Note
1	>25 g/10 min	min., cat. 1, ASTM D1248-84
2	10–25 g/10 min	cat. 2, ASTM D1248-84
3	1–10 g/10 min	cat. 3, ASTM D1248-84
4	0.4–1 g/10 min	cat. 4, ASTM D1248-84
5	<0.4 g/10 min	max., cat. 5, ASTM D1248-84 [1]
6	0.1–5 g/10 min	typical commercial [1,2]

Polymer Melts: Melt viscosity 5000–70000 Pa. s. [1] (Effect of MW and temp. described by Arrhenius equation [7]). Shear thinning occurs [8] - strongly MW distribution dependent. Metallocene catalysed LLDPE (narrow MW distribution) has nearly constant melt viscosity [9]. Non-uniform resins can reduce viscosity by approx.. 6-8 times at high processing speeds and resins from chromium oxide based catalysts (very broad MW distribution) by up to 100 times [1]. Processability of mLLDPE can be improved to that of LDPE [9,10,11]

Mechanical Properties:

Mechanical Properties General: Increasing α-olefin content reduces modulus, tensile strength and increases elongation at break. [1,2,13,14] Mech. props. improve with increasing chain length of the α-olefin for a given density and crystallinity. Increasing uniformity of branching increases elasticity and strain recovery and decreases tensile modulus. Increasing MW (including HMW fraction in broad MW distribution resins) improves mech. props. especially dart impact strength of films [1,2,14]. Effect of orientation has been reported [2,15]

Tensile (Young's) Modulus:

No.	Value	Note
1	75–500 MPa	[1,2,4]

Flexural Modulus:

No.	Value	Note
1	276–724 MPa	40–105 kpsi, 23° (73°F), ASTM D790 [4]

Tensile Strength Break:

No.	Value	Note	Elongation
1	10–30 MPa	ASTM D638, ISO 527 [1,2,4]	100–970%

Tensile Strength Yield:

No.	Value	Note
1	7–19 MPa	ASTM D638, ISO 527 [4]

Miscellaneous Moduli:

No.	Value	Note	Type
1	50 MPa	ISO 6721-2, mLLDPE, density 0.903 g. cm^{-3}	Torsional

Impact Strength: Dart drop impact strength (film) 65–1500g (ASTM D1709:ISO 7765-1); 65–350g (LLDPE non-uniform branching); 230–1500g (mLLDPE, uniform branching) [1]

Hardness: Shore D55-D56 (LLDPE ASTM D2240); D43 (93A, ISO 868 mLLDPE)

Fracture Mechanical Properties: Nature and extent of morphological changes under shear or crazing have been reported [27]

Friction Abrasion and Resistance: Coefficient of friction 1.1–1.3 (ISO 8295). Good abrasion resistance

Izod Notch:

No.	Value	Notch	Note
1	54 J/m	Y	1 ft lb in^{-1} [4]

Electrical Properties:

Electrical Properties General: Excellent electrical insulator with high resistance to the effects of temp. and mech. damage

Optical Properties:

Optical Properties General: Uniform branching distribution gives high transparency (haze $\geq 3\%$). Non-uniform branching distribution gives larger crystalline lamellae and hence more opacity (haze 10–15%) [1]

Volume Properties/Surface Properties: Haze 1.3–28% (mLLDPE); 6–14% (LLDPE, ASTM D1003, film). Gloss 35–97 units (45°, ASTM D2457, film). Clarity 33–95% (ASTM D1746, film)

Polymer Stability:

Polymer Stability General: Stability is generally high as expected for a saturated, branched hydrocarbon. Unstable to thermo-oxidation and photo-oxidation

Thermal Stability General: Thermal degradation onset 250°, thermo-oxidative degradation begins above 150° [1]

Upper Use Temperature:

No.	Value	Note
1	60°C	max. recommended storage temp.

Decomposition Details: At 250°C MW reduction, double bond formation; 450° pyrolysis to isoalkanes and olefins (absence of oxygen). At 150° thermo-oxidation commences with production of hydroxyl and carboxyl groups; above 300° oxidative pyrolysis gives CO, water, hydrocarbons, acids, aldehydes, ketones and alcohols. [1,5] Product toxicity depends on temp., heating rate and sample size [1,22,23,24,25]

Flammability: Self-ignition temp. 350°, (ASTM D1929-91A). At lower temps. decomposition products may ignite, accelerating the process and releasing flaming molten droplets. Resulting smoke contains CO, acrolein and other aldehydes. [18,19] Dust explosion hazard [19,20]

Chemical Stability: Most reactive sites are tertiary CH bonds and chain-end double bonds. Forms sulfo-compounds in 70% H_2SO_4 (elevated temp.), can be nitrated with conc. HNO_3. Stable to alkaline and salt solns. [1]

Biological Stability: Food use grades readily available

Recyclability: Recycled LLDPE can be compatibilised with virgin LLDPE by solid state shear extrusion pulverisation [26]

Applications/Commercial Products:

Processing & Manufacturing Routes: Three catalyst classes used in manufacture [1,2]. Titanium-based (Ziegler) catalysts have the market share, comprising a Ti/organoaluminium/$MgCl_2$ mixture. Used heterogeneously, at 2–0.2% and 70–100°. MW is controlled by hydrogen transfer, diversity of active sites gives broad MW distribution. [10,29,30,31,32,33,34,35,36] Metallocene catalysts include Kaminsky methylaluminoxane Zr metallocenes [30,37,38,39,40,41,42], used at 1–0.1% and give narrow MW distribution. Especially used for speciality resins. [43,44] Exposure to hydrogen leads to large MW reduction. Metallocenes may also be used in combination with perfluorinated aromatic boron compounds, but are highly susceptible to poisoning. [45,46] Constrained geometry (Dow) catalysts are known, where aluminoxanes are used with a monocyclopentadienyl Ti or Zr complex [47,48,49,84]. Produces a highly uniform composition, though long-chain branching can occur. [9] Chromium oxide (Phillips) [50] catalysis produces a broad MW distribution, using allyl boron or aluminium co catalyst. Processes include gas phase [56,57] (economic and flexible), heavy solvent [44,58] (metallocene catalysis), high pressure polymer melt and slurry (oldest) process. [50,82,83] Polymer may be blow moulded [61,65,66,81], rotationally moulded, injection moulded [61,63,65] or extruded [61,65,69] or processed by most other techniques [61,62,63,64,65,88]

Mould Shrinkage (%):

No.	Value	Note
1	2–2.2%	[4]

Applications: Major use (50% of all LLDPE) is for production of film - as a single material/layer, as a blend or as a co-extrusion. LLDPE film can be produced at high rates (because of low strain hardening, low extensional viscosity). LLDPE is tending to replace vinyl acetate and EVA films, especially mLLDPE because of its high clarity and good oxygen barrier prop. - especially important for food and medical use. Appreciable quantities of LLDPE (especially compositionally uniform) are used in to upgrade the mech. props. of largest application (10% of consumption) is injection moulding especially housewares and industrial containers (replacing EVA). Compositionally uniform LLDPE is used for flexible lids, flexible parts for small appliances, disposable oxygen masks. Blow moulded and rotationally moulded articles - for which the ESCR of LLDPE is superior to that of Pipe/tube accounts for approx. 1% LLDPE use where the high flexibility, high burst strength, high ESCR and high heat distortion temp. are superior. Compositionally uniform LLDPE is replacing PVC in medical tubing. LLDPE (espesially mLLDPE) is used for wire and cable insulation

Trade name	Details	Supplier
Alidathene	blended linear PE	Alida Packaging
Dowlex	large MW distribution	Dow
Elite	metallocene catalysed	Dow
Evolue	metallocene catalysed, easy processing	Mitsui Petrochemicals
Exact	metallocene catalysed	Exxon Chemical
Exceed	metallocene catalysed	Exxon Chemical
Finacene PE	metallocene catalysed	Fina Chemicals (Division of Petrofina SA)
Flexirene	VLDPE	Enichem
Flexirene	"Clearflex" film	Ashland Plastics
Fusabond	anhydride modified	DuPont
Innovex		BP Chemicals
Laclene	non exclusive	Sirypetrol
Laclene		NV Chemach SA
Lacqtene	non exclusive	Elf Atochem (UK) Ltd.
Lacqtene		Ashland Plastics
Luflexen	metallocene catalysed	BASF
Lupolen	non exclusive	BASF
Lupolen		Northern Industrial Plastics Ltd.
Lupolex		BASF
Marlex	metallocene catalysed	Philips Petroleum International

Microlin	linear PE powder	Micropol
Mitsubishi	non exclusive	Mitsubishi Kasei
Novapol	non exclusive	Novacor
Petrothene	non exclusive	Quantum Chemical
Petrothene	non exclusive	USI
Polybond	maleic anhydride modified	BP Performance Polymers Inc
Polyethylene mLL & LL	other density/types available	Montell Polyolefins
Quickwrap	stretch film	Centurion Packaging
Siclair	new process	Nova Chemicals
Stamylan	non exclusive	DSM
Stamylan		Macrodan A/S
Stamylan		Ashland Plastics
Stamylex		DSM
Stamylex		Ashland Plastics
Tuflin	film	Union Carbide
Xtrathene	film	Alida Packaging

Bibliographic References

[1] Kissin, Y.V., Kirk-Othmer Encycl. Chem. Technol. 4th edn., (eds. J.I. Kroschwitz and M. Howe-Grant) Wiley Interscience, 1996, **17**, 756
[2] James, D.E., Encycl. Polym. Sci. Eng. 2nd edn., Vol. 6, (ed. J.I. Kroschwitz) Wiley Interscience, 1986, 429
[3] Alger, M., *Polymer Science Dictionary*, 2nd edn., Chapman and Hall, 1997
[4] *Modern Plastics Encyclopedia*, McGraw-Hill, 1987
[5] Hay, J.N., *Polymer*, 1981, **22**, 718
[6] Vidyarthi, N., Palit, S.R., *J. Polym. Sci., Polym. Chem. Ed.*, 1980, **18**, 3315
[7] Woo, L., Ling, T.K. and Westphal, S.P, *Proceedings of SPE Polyolefins: VIII International Conference*, 1993, 242
[8] Nowlin, T.E., Kissin, Y.V. and Wagner, K.P., *J. Polym. Sci., Part A: Polym. Chem.*, 1988, **26**, 755
[9] Knight, G.W. and Lai, S., *Proceedings of SPE Polyolefins: VIII International Conference*, 1993, 226
[10] *PCT Int. Appl.*, 1993, 93 082 21
[11] *Eur. Pat. Appl.*, 1991, 416 815 A2
[12] Sehanobish, K., Patel, R.M., Croft, B.A., Chu, S.P. and Kao, C.I., *J. Appl. Polym. Sci.*, 1994, **51**, 887
[13] Belov, G.P., Belova, V.N., Raspopov, L.N., Kissin, Y.V. et al, *Polym. J. (Tokyo)*, 1972, **3**, 681
[14] Crotty, V.J.; Firdaus, V.; Hagerty, R.O., *Proceedings of SPE Polyolefins: VIII International Conference*, 1993, 192
[15] Butcher, J., *Plast. World*, 1996, **54(12)**, 50
[16] Schutt, J.H., *Plast. World*, 1997, **55(1)**, 8
[17] Mitera, J., Michal, J., Kubat, J. and Kubelka, Z., *Z. Anal. Chem.*, 1976, **281**, 23
[18] Morikawa, T., *J. Combust. Toxicol.*, 1976, **3**, 135
[19] Field, P., Handbook of Powder Technology Vol.4, Elsevier, 1982
[20] Palmer, K.N., *Dust Explosions and Fires*, Chapman and Hall, 1973
[21] Harding, G., *Acta Derm.-Venereol.*, 1969, **49**, 147
[22] Hilado, C.J., Smouse, K.Y., Kourtides, D.A. and Parker, J.A., *J. Combust. Toxicol.*, 1976, **3**, 305
[23] Hilado, C.J., Solis, A.N., Marcussen, W.H. and Furst, A., *J. Combust. Toxicol.*, 1976, **3**, 381
[24] Hilado, C.J., Soriano, J.A. and Kosala, K.J., *J. Combust. Toxicol.*, 1977, **4**, 533
[25] Hilado, C.J. and Huttlinger, N.V., *J. Combust. Toxicol.*, 1978, **5**, 81 361
[26] Nesarikar, A.R., Carr, S.H., Khait, K. and Mirabella, F.M., *J. Appl. Polym. Sci.*, 1997, **63**, 1179
[27] Glutton, E., Rose, V. and Capaccio, G., *J. Macromol. Sci., Phys.*, 1996, **35**, 629
[28] Wolf, B., Kenig, S., Klopstock, J. and Miltz, J., *J. Appl. Polym. Sci.*, 1996, **62**, 1339
[29] Nowlin, T.E., *Prog. Polym. Sci.*, 1985, **11**, 29
[30] Sinn, H. and Kaminsky, W., *Adv. Organomet. Chem.*, 1980, **18**, 99
[31] Chien, J.C.W and Hsieh, J.T.T., *J. Polym. Sci., Polym. Chem. Ed.*, 1976, **14**, 1915
[32] Zakharov, V.A. and Yermakov, Y.I., *Catal. Rev.*, 1979, **19**, 67
[33] Dyachkovsky, F.S. and Pomogailo, A.D., *J. Polym. Sci., Polym. Symp.*, 1980, **68**, 97
[34] Munoz-Escalona, A., *MMI Press Symp. Ser.*, (ed. R.P. Quirk), Harwood, 1983, **4**, 323
[35] Kissin, Y.V., *MMI Press Symp. Ser.*, (ed. R.P. Quirk), Harwood, 1983, **4**, 597
[36] *Eur. Pat. Appl.*, 699 944 A1
[37] Sinn, H., Kaminsky, W., Vollmer, H.-J. and Woldt, R., *Angew. Chem., Int. Ed. Engl.*, 1980, **19**, 390
[38] Chien, J.C.W. and Wang, B.-P., *J. Polym. Sci., Part A: Polym. Chem.*, 1988, **26**, 3089
[39] *Eur. Pat. Appl.*, 1991, 447 722 A2
[40] *Eur. Pat. Appl.*, 1991, 422 703
[41] *U.S. Pat.*, 4 258 475
[42] *Eur. Pat. Appl.*, 323 716 A1
[43] Kaminsky, W. and Spiehl, R., *Makromol. Chem.*, 1989, **190**, 515
[44] Kaminsky, W., Proceedings of the Worldwide Metallocene Conference, MetCon '93, Houston, Texas, 1993, 246
[45] Jordan, R.F., *Adv. Organomet. Chem.*, 1991, **32**, 325
[46] Yang, X, Stern, C.L. and Marks, T.J., *J. Am. Chem. Soc.*, 1991, **113**, 3623
[47] *Eur. Pat. Appl.*, 416 815 A2
[48] Story, B., Proceedings of the Worldwide Metallocene Conference, MetCon '93, Houston, Texas, 1993, 111
[49] Stevens, J.C., Proceedings of the Worldwide Metallocene Conference, MetCon '93, Houston, Texas, 1993, 158
[50] *U.S. Pat.*, 5 208 309
[51] *U.S. Pat.*, 400 371 2
[52] *U.S. Pat.*, 401 138 2
[53] *U.S. Pat.*, 425 554 2
[54] Burdett, I.D., *Chem. Technol.*, 1992, **10**, 616
[55] Choi, K.-Y. and Ray, W.H., *J. Macromol. Sci., Rev. Macromol. Chem. Phys.*, 1985, **25**, 57
[56] *U.S. Pat.*, 540 592 2
[57] Miller, B., *Plast. World*, 1992, **May**, 46
[58] Akimoto, C. and Yano, A., Proceedings of Worldwide Metallocene Conference, MetCon '95, Houston, Texas, 1995
[59] Forsman, J.P., *Hydrocarbon Process.*, 1972, **51**, 132
[60] Swogger, K.W. and Kao, C.-I., *Proceedings of SPE Polyolefins: VIII International Conference*, 1993, 13
[61] Tanner, R.I., *Engineering Rheology*, Clarendon Press, 1986
[62] Pearson, J.R.A., *Mechanics of Polymer Processing*, Elsevier Applied Science, 1985
[63] Injection Molding Handbook (eds. D.V. Rosato and D.V. Rosato), Van Nostrand Reinhold, 1986
[64] Blow Molding Handbook (eds. D.V. Rosato and D.V. Rosato), Hanser, 1989
[65] Encycl. Polym. Sci. Eng. 2nd edn., (ed. J.I. Kroschwitz), Wiley Interscience, 1985
[66] Bristol, J., *Plastic Films and Packaging*, Longman, 1983
[67] Goffreda, F., Proceedings of Worldwide Metallocene Conference, MetCon '95, Houston, Texas, 1995
[68] Wu, P.C., Proceedings of Worldwide Metallocene Conference, MetCon '95, Houston, Texas, 1995
[69] *Plastics Extrusion Technology Handbook*, (ed. S. Levy), Industrial Press, 1981
[70] Smock, D., *Plast. World*, 1994, **Jan**, 16
[71] Yu, T.C. and Wagner, G.J., *Proceedings of SPE Polyolefins: VIII International Conference*, 1993, 539
[72] *Plast. World*, 1984, **42**, Oct., 108
[73] *Plast. World*, 1983, **41**, Mar., 37
[74] Schut, J.H., *Plast. Technol.*, 1994, **Feb.**, 19
[75] *U.S. Pat.*, 524 783
[76] Lancaster, G., Proceedings of Worldwide Metallocene Conference, MetCon '95, Houston, Texas, 1995
[77] Kirschner, E., *Chem. Eng. News*, 1997, **75(10)**, 12
[78] *Packaging Week*, Benn Publications, 1997, **12(38)**, 2
[79] *Packaging Week*, Benn Publications, 1997, **12(35)**, 1
[80] Stinson, S., *Chem. Eng. News*, 1997, **75(10)**, 31
[81] Colvin, R., *Modern Plast. Intern.*, 1996, **26(12)**, 30
[82] *Eur. Chem. News*, 67(**No. 1761**), 24
[83] Rotman, D., *Chem. Week*, 1997, **159(10)**, 8
[84] *Packaging Week*, Benn Publications, 1997, **12(38)**, 16
[85] Hartzmann, G., Jagdmann, S. and Klimesch, R.G., *Kunststoffe*, 1996, **86(10)**, 9
[86] Nowlin, T.E., Kissin, Y.V. and Wagner, K.P., *J. Polym. Sci., Part A: Polym. Chem.*, 1988, **26**, 755
[87] *Plast. Rubber Wkly.*, **No. 1684**, 7
[88] Lee, M., *Eur. Plast. News*, 1997, **24(5)**, 23

Long-chain Poly(alkyl methacrylates) L-6

$$\left[\begin{array}{c} CH_2C(CH_3) \\ | \\ O^{\diagdown C \diagup OR} \end{array} \right]_n$$

Synonyms: *Poly(alkyl 2-methyl-2-propenoate). Poly(dodecyl methacrylate). Poly(tetradecyl methacrylate). Poly(hexadecyl methacrylate). Poly(octadecyl methacrylate). Poly(lauryl methacrylate). Poly(myristyl methacrylate). Poly(cetyl methacrylate). Poly(stearyl methacrylate). Poly(palmityl methacrylate). Poly(3,5,5-trimethylhexyl methacrylate). Poly(isodecyl methacrylate). Poly(tridecyl methacrylate). Poly(docosyl methacrylate)*
Related Polymers: Polymethacrylates General
Monomers: 3,5,5-Trimethylhexyl methacrylate, 1-Methylnonyl methacrylate, Dodecyl methacrylate, Tridecyl methacrylate, Tetradecyl methacrylate, Hexadecyl methacrylate, Octadecyl methacrylate, Docosyl methacrylate
Material class: Thermoplastic
Polymer Type: acrylic
CAS Number:

CAS Reg. No.	Note
29320-53-4	(decyl)
25719-52-2	(docecyl)
30525-99-6	(tetradecyl)
25986-80-5	(hexadecyl)
25639-21-8	(octadecyl)
116697-32-6	(octadecyl, isotactic)
27252-90-0	(docosyl)

Molecular Formula: $(C_{13}H_{24}O_2)_n$
Morphology: Hexagonal a 4.73, b 4.73, c 30.2, 1 monomer per unit cell (octadecyl, syndiotactic) [1]
Identification: Contains only C, H and O

Volumetric & Calorimetric Properties:
Density:

No.	Value	Note
1	d 0.929 g/cm^3	dodecyl [2]
2	d 0.97 g/cm^3	octadecyl [3]

Equation of State: Mark-Houwink constants for the decyl, dodecyl, tridecyl, hexadecyl, octadecyl and docosyl polymers have been reported [16,19,21]
Melting Temperature:

No.	Value	Note
1	-34°C	dodecyl [2,6]
2	-2°C	tetradecyl, side-chain crystallisation [6]
3	22°C	hexadecyl, side-chain crystallisation; bulk polymerisation [7]
4	37.5°C	octadecyl, uv polymerisation [4]
5	36°C	octadecyl [2]
6	34°C	octadecyl, side-chain crystallisation [6]

Glass-Transition Temperature:

No.	Value	Note
1	1°C	trimethylhexyl [8]
2	-70°C	decyl [9]
3	-60°C	decyl [10]
4	-44°C	decyl [10]
5	-41°C	decyl [11]
6	-55°C	dodecyl, bulk polymerisation [7,12]
7	-65°C	dodecyl, uv polymerisation [4]
8	-43°C	dodecyl [8]
9	-43°C	tridecyl [10]
10	-46°C	tridecyl [11]
11	-72°C	tetradecyl, bulk polymerisation [7]
12	-9°C	tetradecyl, birefringent below T_m [5]
13	15°C	hexadecyl [13]
14	-100°C	octadecyl [6,10]

Transition Temperature General: No transitions above -55°, decyl [5]. No T_g for bulk polymerised poly(hexadecyl methacrylate) [7]
Vicat Softening Point:

No.	Value	Note
1	38°C	octadecyl, crystalline melting point [3]

Brittleness Temperature:

No.	Value	Note
1	-30°C	decyl, refractometric method [5]
2	-5°C	tetradecyl, refractometric method [5]
3	36°C	octadecyl [2]

Surface Properties & Solubility:
Solubility Properties General: The phase behaviour of the dodecyl system has been reported [15]
Cohesive Energy Density Solubility Parameters: Cohesive energy 77500 J mol^{-1} (dodecyl) [12]. Solubility parameters δ 16.8 J$^{1/2}$ cm$^{3/2}$ (8.2 cal$^{1/2}$ cm$^{3/2}$, dodecyl), δ 16 J$^{1/2}$ cm$^{3/2}$ (7.8 cal$^{1/2}$ cm$^{3/2}$, octadecyl) [15]
Solvents/Non-solvents: Sol. butanol; CCl$_4$, 2-butanone (octyl); cyclohexane, propylacetate (stearyl); EtOAc, THF (tridecyl); C$_6$H$_6$ (octyl, cetyl), amyl acetate, isooctane (decyl), n-pentanol (>29°, isopropyl acetate (decyl) [16,20]. Insol. MeOH (dodecyl), EtOAc (dodecyl) [16]
Wettability/Surface Energy and Interfacial Tension: Contact angle 75° (decyl, advancing, H$_2$O); 89° (decyl, advancing, H$_2$O); 77° (tridecyl, advancing, H$_2$O); 84° (octadecyl, advancing, H$_2$O) [10]. Surface tension 32.8 mN m^{-1} (dodecyl, 20°) [17]; 36.3 mN m^{-1} (octadecyl, 20°) [17]. Critical surface tension 33 mN m^{-1} (33 dyne cm^{-1}, decyl) [18]

Transport Properties:
Polymer Solutions Dilute: Molar volumes at 25° 168.6 cm^3 mol^{-1} (dodecyl, amorph., van der Waal's), 273.8 cm^3 mol^{-1} (dodecyl, amorph., rubbery polymer) [2]. Intrinsic viscosity [19] 1.05 dl g^{-1} (30°, MW 923000, THF, decyl), EtOAc 320 dl g^{-1} (11°, EtOAc, decyl, MW 923000), 1.1 dl g^{-1} (30°, THF, MW 1380000, tridecyl), 0.37 dl g^{-1} (27°, EtOAc, MW 1380000, tridecyl), 0.94 dl g^{-1} (30°, THF, MW 934000, octadecyl), 0.36 dl g^{-1} (36°, propyl acetate, MW 934000, octadecyl), 0.86 dl g^{-1} (30°, THF, MW 1350000,

docosyl), 0.47 dl g^{-1} (31°, amyl acetate, MW 1350000, docosyl). Intrinsic viscosity parameters 0.000357 (g MW)$^{1/2}$, decyl) [19]; 0.032–0.035 cm$^{3/2}$ mol$^{1/2}$ g$^{-1/2}$ (dodecyl) [2,25,27]; 0.0332 (tridecyl) [19]; 0.048–0.075 (hexadecyl) [2,25,27]. Second virial coefficients for the decyl [19], dodecyl [28], tridecyl [19], hexadecyl [28], octadecyl [19] and docosyl [19] polymers have been reported. Theta temps. 11° (EtOAc, decyl); 9.6° (n-pentanol, decyl) [19,22]; 29.5° (n-amyl alcohol, dodecyl, MW 270000–2400000); 13° (isopropyl acetate, dodecyl, MW 260000–3600000); 29.5° (pentanol, dodecyl, MW 270000–2400000) [23,24,25]; 27° (EtOAc, tridecyl) [19]; 21° (n-heptane, hexadecyl) [25]; 10.5° (n-butyl acetate, octadecyl) [19,26]; 36° (n-propyl acetate, octadecyl)
Permeability of Liquids: Diffusion coefficient [H$_2$O] 120.4 × 10^{-8} cm^2 s^{-1} (decyl, 37°); 160.8 × 10^{-8} cm^2 s^{-1} (decyl, 37°); 771.1 × 10^{-8} cm^2 s^{-1} (tridecyl, 37°); 529.9 × 10^{-8} cm^2 s^{-1} (octadecyl, 37°) [10]. The diffusion coefficient of hexadecyl polymer in heptane has been reported [28]
Water Absorption:

No.	Value	Note
1	<4.1 %	max., decyl
2	<6.4 %	max., decyl
3	<5.6 %	max., tridecyl [10]

Mechanical Properties:
Viscoelastic Behaviour: Williams-Landel-Ferry constants for temp. superposition of dodecyl polymer have been reported [29].

Optical Properties:
Refractive Index:

No.	Value	Note
1	1.474	30°, dodecyl, bulk polymerisation [2,7]
2	1.4746	30°, tetradecyl bulk polymerisation [7]
3	1.475	30°, hexadecyl, bulk polymerisation [7]

Total Internal Reflection: Intrinsic segmental anisotropy -160 × 10^{-25} cm^3 (hexadecyl, C$_6$H$_6$) [30]

Polymer Stability:
Polymer Stability General: The thermal stability of the decyl polymer can be increased by the incorporation of styrene, α-methyl styrene, indene, tetradecene or dicyclopentadiene units [31]

Applications/Commercial Products:

Trade name	Details	Supplier
Standard samples	Dodecyl, octadecyl	Aldrich
Standard samples	Dodecyl	Polysciences Inc
Standard samples	Dodecyl, octadecyl	Scientific Polymer Products Inc.

Bibliographic References

[1] Alihaud, H., Gallot, Y. and Skoulois, A., *C.R. Seances Acad. Sci., Ser. C*, 1968, **267**, 139, (morphology)
[2] Van Krevelen, D.W., *Properties of Polymers: Their Correlation with Chemical Structure*, Elsevier, 1976, 79, (density)
[3] Hoff, E.A.W., Robinson, D.W. and Willbourn, A.H., *Polym. Sci.*, 1955, **18**, 161, (density)
[4] Rogers, S.S. and Mandelkern, L., *J. Phys. Chem.*, 1957, **61**, 985, (thermal expansion)
[5] Wiley, R.H. and Brauer, G.M., *J. Polym. Sci.*, 1948, **3**, 647, (transition temp)
[6] Nielsen, L.E., *Mechanical Properties of Polymers*, Reinhold, 1962
[7] Lal, J. and Trick, G.S., J. Polym. Sci., *Part A: Polym. Chem.*, 1964, **2**, 4559, (transition temp)
[8] Williams, G. and Watts, D.C., *Trans. Faraday Soc.*, 1971, **67**, 2793
[9] *Perspex Acrylic Sheet Bulletin*, ICI, (technical datasheet)
[10] Kalachandra, S. and Kusy, R.P., *Polymer*, 1991, **32**, 2428
[11] Hopfinger, A.J., Koehler, M.G., Pearlstein, R.A. and Tripathy, S.K., *J. Polym. Sci., Part B: Polym. Phys.*, 1988, **26**, 2007, (transition temps)
[12] Porter, D., *Group Interaction Modelling of Polymer Properties*, Dekker, 1995, 283
[13] Koehler, M.G. and Hopfinger, A.J., *Polymer*, 1989, **30**, 116
[14] Hughes, L.J. and Britt, G.E., *J. Appl. Polym. Sci.*, 1961, **5**, 337, (solubility parameters)
[15] Shine, A.D., *Physical Properties of Polymers Handbook*, (ed. J.E. Mark), AIP Press, 1996, 249
[16] Kurata, M. and Stockmayer, W.H., *Adv. Polym. Sci.*, Springer-Verlag, 1963, **3**, 196, (solvents)
[17] Wu, S., *Polymer Interface and Adhesion*, Marcel Dekker, 1982
[18] Lee, L.H., *CA*, 1967, **61**, 154th
[19] Xu, Z., Hadjichristidis, N. and Fetters, L.J., *Macromolecules*, 1984, **17**, 2303, (soln props)
[21] Hadjichristidis, N., Touloupis, L. and Fetters, J., *Macromolecules*, 1981, **14**, 128
[22] Herold, F.K. and Wolf, B.A., *Makromol. Chem.*, 1981, **184**, 2539, (theta temp)
[23] Chinai, S.N., *J. Polym. Sci.*, 1959, **41**, 475
[24] Lee, T.H. and Levi, D.W., *J. Polym. Sci.*, 1960, **47**, 449
[25] Tsvetkov, V.N., Andreyeva, L.N., Korneyeva, Ye.V., Lavrenko, P.N. et al, *Polym. Sci. USSR (Engl. Transl.)*, 1972, **14**, 1944
[26] Ricker, M. and Schmidt, M., *Abstr. Pap.-Am. Chem. Soc.*, 247
[27] Chee, K.K., *J. Appl. Polym. Sci.*, 1987, **33**, 1067
[28] Tsvetkov, V.N. et al, *Vysokomol. Soedin., Ser. A*, 1970, **11**, 349
[29] Ngai, K.L. and Plazek, D.J., *Physical Properties of Polymers Handbook*, (ed. J.E. Mark), AIP Press, 1996, 341
[30] Tsvetkov, V.N., Andreeva, L.N., Korneeva, E.V. and Lavrenko, P.N., *Dokl. Akad. Nauk SSSR*, 1972, **205**, 895
[31] Akhmedov, A.I., *Khim. Tekhnol. Topl. Masel*, 1987, **3**, 33, (thermal stability)
[32] Aharoni, S.M., *Macromolecules*, 1983, **16**, 1722, (viscoelastic props.)

Low cis-1,4-polybutadiene

Synonyms: Low cis-1,4-poly(1,3-butadiene). Butadiene rubber. BR
Related Polymers: Polybutadiene, cis-1,4-Polybutadiene, 1,2-Polybutadiene
Monomers: 1,3-Butadiene
Material class: Synthetic Elastomers
Polymer Type: polybutadiene
CAS Number:

CAS Reg. No.
9003-17-2

Molecular Formula: (C$_4$H$_6$)$_n$
Fragments: C$_4$H$_6$
Molecular Weight: MW 10000-840000, 110000
Additives: Reinforced carbon black
General Information: Contains less than 1,4-cis isomer. Exhibits high resilience and low hysteresis

Volumetric & Calorimetric Properties:
Density:

No.	Value	Note
1	d 0.89 g/cm^3	40% 1,4-cis, 33% 1,4-trans, 27% 1,2 [1]
2	d 0.895 g/cm^3	44% 1,4-cis, 42% 1,4-trans, 14% 1,2 [1]

Thermal Expansion Coefficient:

No.	Value	Note	Type
1	0.00075 K^{-1}	43% 1,4-cis, 50% 1,4-trans, 7% 1,2 [2]	V

Surface Properties & Solubility:
Solubility Properties General: Miscibility with other rubbers has been reported [32]

Solvents/Non-solvents: Sol. hydrocarbons, THF, higher ketones, higher aliphatic esters. Insol. alcohol, lower ketones and esters, nitromethane, propionitrile, H_2O [3,4,5]. Theta solvents dioxane 26.5° (36% 1,4-*cis*, 57% 1,4-*trans*, 7% 1,2) [6]

Transport Properties:
Polymer Solutions Dilute: Williams-Landel-Ferry equation parameters have been reported [18]. Unperturbed dimensions have been reported [8,9,10,11,12,13,14,15,16,17]. [η] 1.32 (toluene, 25°, 46% 1,4-*cis*, 47% 1,4-*trans*, 8% 1,2, MW 110000) [23]. Other viscosity values have been reported for a range of compositions [37]. [η] 1.26-6.51 (toluene, 25°, MW 10000-84000, 40-44% 1,4-*cis*, 31-44% 1,4-*trans*, 9-29% 1,2) [22]

Mechanical Properties:
Mechanical Properties General: A range of mech. props. data for low *cis* mixtures has been reported [11,24,27,28,31,33,] including storage moduli and loss moduli. Mech. props. related to use in tyres has been reported [29,30]

Optical Properties:
Transmission and Spectra: Ir spectral data have been reported [7]

Polymer Stability:
Polymer Stability General: Oxidative degradation of polybutadiene leads to cross-linking and to hardening of vulcanisates [36]
Decomposition Details: With increasing *trans* content, 1,4-polybutadiene yields much less 4-vinyl-1-cyclohexene dimer, much more cyclopentene, less 1,3-butadiene and more 1,3-cyclohexadiene in the first 15% of weight loss by pyrolysis [25]. Thermal degradation analysis has been reported [38,39,40]
Flammability: Classed as high-volatile-loss, low-char polymers [34]
Biological Stability: 40% Assimilated by *Moraxella* strain isol. from soil (MW 2350). No assimilation at greater MW [26]
Stability Miscellaneous: High energy radiation increases radical concentration linearly with radiation up to 100 Mrad. Unsaturation decreases markedly. *Cis-trans* isomerisation occurs. Some double bonds are formed. The yield of cross-linking is higher for *cis*-1,4 than for *trans*-1,4 but lower than for 1,2. Stress-strain props. of radiation cross-linked polybutadiene have been reported [35]

Applications/Commercial Products:
Processing & Manufacturing Routes: Produced by anionic polymerisation using lithium initiators. Polar additives increase 1,2 content
Applications: Used in tyres and as toughening agents for polystyrene

Trade name	Supplier
Afdene BR	Lehmann and Voss
Buna CB 35R, 35NF	Bayer Inc.
Buna CB 45NF/CB 55N	Bayer Inc.
Calprene 248	Repsol Quimica
Finaprene 250	Petrofina
Intene 50/50A	Enichem

Bibliographic References

[1] Natta, G., Porri, L., Corradini, P. and Morero, D., *Chim. Ind. (Milan)*, 1958, **40**, 366
[2] Valentine, R.H., Ferry, J.D., Homma, T. and Ninomiya, K., *J. Polym. Sci., Part A-2*, 1968, **6**, 479
[3] Dexheimer, H. and Fuchs, O., *Struktur und Physikaleches Verkalten der Kunststoffe*, (eds. R. Nitsche and K.A.Wolf), Springer Verlag, 1963, **1**
[4] Kurata, M. and Stockmayer, W.H., *Adv. Polym. Sci.*, Springer Verlag, 1963, **3**, 196
[5] Roff, W.J., *Fibres, Films, Plastics and Rubbers*, Academic Press, 1956
[6] Hadjichristidis, N., Xu, Z., Fetters, L.J. et al, *J. Polym. Sci., Polym. Phys. Ed.*, 1982, **20**, 743
[7] Noda, I., Downey, A.E. and Marcott, C., *Physical Properties of Polymers Handbook*, (ed. J.E. Mark), AIP Press, 1996, 293
[8] Fetters, L.J., Lohse, D.J. and Colby, R.H., *Physical Properties of Polymers Handbook*, (ed. J.E. Mark), AIP Press, 1996, 338
[9] Carrella, J.M., Graessley, W.W. and Fetters, L.J., *Macromolecules*, 1984, **17**, 2775
[10] Roovers, J., *Polym. J. (Tokyo)*, 1986, **18**, 153
[11] Struglinski, M.J. and Graessley, W.W., *Macromolecules*, 1985, **18**, 2630
[12] Mancke, R.G., Dickie, R.A. and Ferry, J., *J. Polym. Sci., Part A-2*, 1968, **6**, 1783
[13] Rochefort, W.E., Smith, G.G., Rachapudy, H., Raju, V.R. and Graessley, W.W., *J. Polym. Sci., Part B: Polym. Phys.*, 1979, **17**, 1197
[14] Colby, R.H., Fetters, L.J. and Graessley, W.W., *Macromolecules*, 1987, **20**, 2940
[15] Cheng, P.-L., Barney, C.V. and Cohen, R.H., *Makromol. Chem.*, 1989, **190**, 589
[16] Sakurai, S., Hasegawa, H. and Han, C.C., *Polym. Commun.*, 1990, **31**, 99
[17] Roovers, J. and Martin, J.E., *J. Polym. Sci., Part B: Polym. Phys.*, 1989, **27**, 2513
[18] Ngai, K.L. and Plazek, D.J., *Physical Properties of Polymers Handbook*, (ed. J.E. Mark), AIP Press, 1996, 341
[19] Maekawa, E., Mancke, R.G. and Ferry, J.D., *J. Phys. Chem.*, 1965, **69**, 2811
[20] Taylor, C.R., *Viscoelastic Properties of Polymers*, 3rd edn., Wiley, 1980, 320
[21] Colby, R.H. and Graessley, W.W., *Macromolecules*, 1987, **20**, 2226
[22] Baumgaertel, M., De Rosa, M.E., Machado, J. and Winter H., *Rheol. Acta*, 1992, **31**, 75
[23] Yanovsky, Y.G., Vinogradov, G.V. and Ivanova, L.I., *Chem. Eng. Commun.*, 1985, **32**, 219
[24] Vinogradov, G.V., Dzyura, E.A., Malkin, A.Ya. and Grechanovskii, V.A., *J. Polym. Sci., Part A-2*, 1971, **9**, 1153
[25] Tamura, S. and Gilham, J.K., *J. Appl. Polym. Sci.*, 1978, **22**, 1867
[26] Tsuchii, A., Suzuki, T. and Fukuoka, S., *Agric. Biol. Chem.*, 1984, **48**, 621
[27] Morin, G., Peyrelasse, J. and Mouge, Ph., *Rheol. Acta*, 1983, **22**, 476
[28] Vinogradov, G.V., Dreval, V.E. and Yanovsky, Yu.G., *Rheol. Acta*, 1985, **24**, 574
[29] Vinogradov, G.V., Malkin, A.Ya., Yanovskii, Yu.G., Borisenkova, E.K., Yarlykov, B.V. and Berezhnaya, G.V., *J. Polym. Sci., Part A-2*, 1972, **10**, 1061
[30] Short, J.N., Kraus, G., Zelinski, R.P. and Naylor, F.E., Soc. Plast. Eng., *Tech. Pap.*, Paper No. 3, 1959, **5**, 8
[32] Roland, C.M., *Rubber Chem. Technol.*, 1989, **62**, 456
[33] Rahalkar, R.R. and Tang, H., *Rubber Chem. Technol.*, 1988, **61**, 812
[34] Lawson, D.F., *Rubber Chem. Technol.*, 1986, **59**, 455
[35] Bohm, G.G.A. and Tveekrem, J.O., *Rubber Chem. Technol.*, 1982, **55**, 575
[36] Gent, A.N., *Rubber Chem. Technol.*, 1982, **55**, 525
[37] Luxton, A.R., *Rubber Chem. Technol.*, 1981, **54**, 596
[38] Brazier, D.W., *Rubber Chem. Technol.*, 1980, **53**, 437
[39] Sircar, A.K. and Lamond, T.G., *Rubber Chem. Technol.*, 1972, **45**, 329
[40] Sircar, A.K. and Lamond, T.G., *Thermochim. Acta*, 1973, **7**, 287
[41] Sharaf, M.A., *Rubber Chem. Technol.*, 1994, **67**, 88

MDPE

M-1

$$\left[\left[CH_2CH_2\right]_x\left[branch\right]_y\left[CH_2CH_2\right]_z\right]_n$$

Synonyms: *Medium-density polyethylene. PE-MD*
Related Polymers: Polyethylenes, LLDPE, HDPE
Monomers: Ethylene, 1-Butene, 1-Hexene, 1-Octene
Material class: Thermoplastic
Polymer Type: polyethylene, polyolefins
CAS Number:

CAS Reg. No.	Note
9002-88-4	all density types
25087-34-7	1-butene comonomer
25213-02-9	1-hexene comonomer
26221-73-8	1-octene comonomer
60785-11-7	1-hexene 1-butene comonomer

Molecular Formula: $(C_2H_4)_n$
Fragments: C_2H_4
General Information: MDPE designation is applied to ethylene polymers containing sufficient comonomer to give a density range of 0.925–0.94 g cm^{-3}. Typical comonomer content is 1–2 mol%. MDPE can be produced by the low pressure process [1] or a modified high pressure process. [2] Props are generally intermediate between those of HDPE and LLDPE, as is the struct.
Morphology: Structural unit consists of polymer backbone interspersed with short branched systems of comonomer. Contains less branched portions than for LLDPE. Due to the customisable nature of polyethylenes and differences in manuf. process, props highly variable.

Volumetric & Calorimetric Properties:
Density:

No.	Value	Note
1	d 0.925–0.94 g/cm^3	[1,3,4]

Thermal Conductivity:

No.	Value	Note
1	0.35 W/mK	[4]

Thermal Diffusivity: K 2.7×10^{-7} m^2 s^{-1} 0° [5]
Specific Heat Capacity:

No.	Value	Note	Type
1	1.9–2.14 kJ/kg.C	50° [2,5]	P

Melting Temperature:

No.	Value	Note
1	110–130°C	DSC, ISO 3146 [4,5]

Deflection Temperature:

No.	Value	Note
1	38–41°C	A, 1.8 MPa [4]
2	42–69°C	B, 0.45 MPa [4]

Vicat Softening Point:

No.	Value	Note
1	119–120°C	Rate A, DIN/VDE/ISO/IEC 306 [4]
2	62–65°C	Rate B, DIN/VDE/ISO/IEC 306 [4]

Transport Properties:
Melt Flow Index:

No.	Value	Note
1	0.15–2.2 g/10 min	190°, 2.16 kg [4]
2	0.8–8 g/10 min	190°, 5 kg [4]
3	20–110 g/10 min	190°, 21.6 kg [4]

Polymer Solutions Concentrated: Viscosity number 260 ml g^{-1} (soln 0.001 g m^{-1} decalin, ISO/IEC 1628) [4]
Water Absorption:

No.	Value	Note
1	<0.01 %	max, 23°, saturation [4]

Gas Permeability:

No.	Gas	Value	Note
1	H$_2$	0.3 cm^3 mm/(m^2 day atm)	ASTM, E96(E), 37° [3]

Mechanical Properties:
Tensile (Young's) Modulus:

No.	Value	Note
1	200–700 MPa	[2,4]

Tensile Strength Break:

No.	Value	Note	Elongation
1	10–30 MPa	[1,3,4]	50–1000%

Elastic Modulus:

No.	Value	Note
1	500–900 MPa	[1]

Tensile Strength Yield:

No.	Value	Elongation	Note
1	10–20 MPa	8–600%	[2,4]

Miscellaneous Moduli:

No.	Value	Note	Type
1	500 MPa	[4]	shear modulus

Impact Strength: Double V notched tensile impact strength 200 kJ m^{-2} (23°); 140–160 kJ m^{-2} (-30°) [4]
Hardness: Shore D45–D60 (durometer) [2]. Ball indentation H132/30 (34 MPa) [4]
Fracture Mechanical Properties: A non-linear-elastic thermal decohesion model for impact fracture, which correlates well with notched Charpy specimens, has been reported [5]

Izod Notch:

No.	Value	Notch	Note
1	27–850 J/m	N	[2]

Electrical Properties:
Surface/Volume Resistance:

No.	Value	Note	Type
2	$>0.1 \times 10^{15}$ Ω.cm	min [4]	s

Dielectric Permittivity (Constant):

No.	Value	Frequency	Note
1	2.4	100 Hz–1 MHz	DIN/VBE 0303-4, ISO/IEC 250 [4]
2	2.2	1 kHz–1 GHz	ASTM D150 [3]

Dielectric Strength:

No.	Value	Note
1	150 kV.mm^{-1}	3750 Vmil^{-1}, 1mm thick DIN/VDE 0303-21, ISO/IEC 243/1 [4]
2	197 kV.mm^{-1}	5000 Vmil^{-1}, thick [3]
3	118 kV.mm^{-1}	3000 Vmil^{-1}, 0.125mm thick, ASTM D149 [3]

Dissipation (Power) Factor:

No.	Value	Frequency	Note
1	<-0.0002	100 Hz–1 MHz	max, DIN/VDE 0303-4, ISO/IEC 250 [4]
2	0.0003	1 kHz–1 GHz	ASTM D150 [3]

Optical Properties:
Refractive Index:

No.	Value	Note
1	1.52–1.53	25° [2]

Volume Properties/Surface Properties: Haze 4.5–6.0% (25–31 mm thick film) [3]

Applications/Commercial Products:
Processing & Manufacturing Routes: Manufactured by modified high pressure process (LDPE) or Ziegler-Natta process (esp. slurry). (LLDPE). Processed by blow moulding, injection moulding or extrusion (temp. range 170–240°) [4]
Applications: Used for bottles, in linens, profiles, sheet, cable insulation, pipes, film

Trade name	Details	Supplier
Alcudia	non exclusive	Repsol Quimica
Alcudia	non exclusive	Petroplas
Finathene	non exclusive	Fina Chemicals (Division of Petrofina SA)
Fortiflex	non exclusive	Solvay
Hostalen	non exclusive	Hoechst
Hostalen	non exclusive	Hoechst
Hostalen	non exclusive	Tinto Plastics
Lupolen	non exclusive	BASF
Marlex	non exclusive	Philips Petroleum International
Novapol	non exclusive	Novacor
Petrothene	non exclusive	Quantum Chemical
Warin		Barflo-LCA

Bibliographic References
[1] Kissin, Y.V., *Kirk-Othmer Encycl. Chem. Technol.*, 4th edn. (eds. M. Howe-Grant, and J.I. Kroschwitz), Wiley-Interscience, 1996, 756, (rev)
[2] Dook, K.W., *Encycl. Polym. Sci. Eng.*, 2nd edn. (ed. J.I. Kroschwitz Wiley), Interscience, 1986, 386, (rev)
[3] Guide to Plastics, *Property and Specification Charts*, McGraw-Hill, 1987, 74, 92, 150
[4] *Lupolen High Density Polyethylene*, BASF, (technical datasheet)
[5] Leevers, P.S. and Morgan, R.E., *Eng. Fract. Mech.*, 1995, **52**, 999

Melamine formaldehyde resins M-2
Synonyms: *Melamine resins. Melamines*
Related Polymers: Melamine Resins, Wood applications, Urea resins, Wood applications
Monomers: Melamine, Formaldehyde, Paraformaldehyde
Material class: Thermosetting resin
Polymer Type: melamine formaldehyde
Miscellaneous: Melamine is a reactive hexa-functional monomer which readily forms hexa-methylol derivs. This intermediate crosslinks forming methylene ether bridges. Melamine can also be co-condensed with urea to form a range of melamine urea formaldehyde (MUF) resins

Applications/Commercial Products:
Processing & Manufacturing Routes: Melamine-formaldehyde resins are made by the condensation reaction of melamine with formaldehyde, generally 37% aq. soln. The condensation generally proceeds with basic catalyst and a molar excess of formaldehyde. The resins are used in soln. generally aq., but may also be used if alcoholic solvents added. They cure by application of heat or acid. Melamine is co-condensed with urea to form a range of melamine urea formaldehyde resins
Applications: Melamine formaldehyde resins have various applications. See the relevant cross references for further details

Trade name	Details	Supplier
Blagden MF	MF resin	Blagden
Dynomel	MF, MUF resins	Dyno
Melmex	melamine materials	BIP
Melolam	MF, MUF resins	Dyno
Melurex	MUF resin	Neste

Melamine resin, Moulding powders, Cellulose filled M-3
Related Polymers: Urea Formaldehyde Resins, Melamine Formaldehyde Resins, Urea resins, Moulding Compounds, Cellulose Filled
Monomers: Urea, Melamine, Formaldehyde
Material class: Thermosetting resin
Polymer Type: melamine formaldehyde
Additives: Additives include cellulose filler, pigments and a curing catalyst

– Meroxapol

Volumetric & Calorimetric Properties:
Density:

No.	Value	Note
1	d 1.47–1.52 g/cm^3	ASTM D792 [1]

Thermal Expansion Coefficient:

No.	Value	Note	Type
1	$2 \times 10^{-5} - 5.7 \times 10^{-5}$ K^{-1}	ASTM D896 [1]	L

Thermal Conductivity:

No.	Value	Note
1	0.00293–0.00423 W/mK	ASTM C177 [1]

Deflection Temperature:

No.	Value	Note
1	182°C	1.8 MPa, ASTM D648 [1]

Transport Properties:
Water Absorption:

No.	Value	Note
1	0.1–0.6 %	24h, 3.2 mm thick, ASTM D590

Mechanical Properties:
Tensile (Young's) Modulus:

No.	Value	Note
1	9300 MPa	ASTM D638 [1]

Flexural Modulus:

No.	Value	Note
1	7600 MPa	ASTM D7900 [1]

Tensile Strength Break:

No.	Value	Note
1	48–90 MPa	ASTM D638 [1]

Flexural Strength at Break:

No.	Value	Note
1	83–104 MPa	ASTM D790 [1]

Tensile Strength Yield:

No.	Elongation	Note
1	0.6–0.9%	ASTM D638 [1]

Hardness: Rockwell M120 (ASTM D785) [1]

Electrical Properties:
Dielectric Strength:

No.	Value	Note
1	0.03–0.083 kV.mm^{-1}	60 Hz, ASTM D149 [1]
2	0.036–0.05 kV.mm^{-1}	1 kHz, ASTM D149 [1]

Arc Resistance:

No.	Value	Note
1	125–136s	ASTM D495 [1]

Polymer Stability:
Upper Use Temperature:

No.	Value
1	99°C

Flammability: Flammability rating V0 (UL 94) [1]

Applications/Commercial Products:
Processing & Manufacturing Routes: Melamine moulding materials are made from colourless thermosetting resin with cellulose fillers, pigments, catalyst amd other additives. For the compression moulding process preheating of the resin powder pellets or granules is required. Injection moulding grades are preheated efficiently both by the injection machine and by the frictional heat generated as the material flows through the sprue, runner or gate. Resin-impregnated paper foils are used as decorative finishes for melamine items. The process involves partial cure of the moulding, opening the press, inserting a specially made foil and continuing the moulding cycle. Decoration is also achieved by applying special glaze resin to the moulding, by painting, printing or vacuum metallising methods

Mould Shrinkage (%):

No.	Value	Note
1	0.58–0.65%	compression moulding grades
2	0.75%	injection moulding grades

Applications: Melamine materials give mouldings with a superior surface performance to urea-formaldehyde materials. The low formaldehyde emission and excellent stain resistance props. make them suitable for tableware and kitchenware. The heat resistance of melamine materials enables their use as ashtrays. Injection moulded melamine is used where surface appearance is important and especially for electrical accessories

Trade name	Details	Supplier
Melmex	melamine materials	BIP
Melmex SMX	modified melamine	BIP

Bibliographic References
[1] *Amino Moulding Powders*, BIP Plastics, (technical datasheet)

Meroxapol

$$HO{-}[CH(CH_3)CH_2O]_x{-}[CH_2CH_2O]_y{-}[CH_2CH(CH_3)O]_z{-}H$$

$7 < x < 21 \qquad 4 < y < 163 \qquad 7 < z < 21$

Synonyms: *Polyoxypropylene-polyoxyethylene block copolymer. Polypropylene polyethylene glycol. Oxirane polymers with methyloxirane (block, triblock)*

Methacrylate amide/imide copolymers

Related Polymers: Polyoxypropylene, Poloxamers
Monomers: Propylene oxide, Oxirane
Material class: Copolymers
Polymer Type: polyalkylene ether
CAS Number:

CAS Reg. No.	Note
9003-11-6	generic
106392-12-5	block copolymer

Molecular Formula: [$(C_3H_6O).(C_2H_4O).(C_3H_6O)$]
Fragments: C_3H_6O C_2H_4O C_3H_6O
Molecular Weight: M_n 1500–9500

Applications/Commercial Products:

Trade name	Supplier
Hodag Nonionic	Hodag
Macol	PPG-Mazer
Pluronic R	BASF

Bibliographic References

[1] *The Thesaurus of Chemical Products*, 2nd edn., (eds. M. Ash and I. Ash), Edward Arnold, 1992, **1**
[2] Šimek, L., Petřík, S., Hadobaš, F. and Bohdanecký, M., *Eur. Polym. J.*, 1990, **26**, 371, 375

Methacrylate amide/imide copolymers   M-5

R = Me, Et or CH$_2$CH$_2$CH$_2$CH$_3$ (butyl)
X = H (acrylamide)
X = CH$_3$ (methacrylamide)
X = Ph (N-phenmaleimide)
X = (N-o-tolylmaleimide)
X = (N-p-tolylmaleimide)

Synonyms: *Poly(ethyl methacrylate-co-acrylamide). Poly(butyl methacrylate-co-methacrylamide). Poly(methyl methacrylate-co-N-phenylmaleimide). Poly(methyl methacrylate-co-N-tolylmaleimide). Poly(methyl methacrylate-co-N,N-dimethylacrylamide). Poly(lauryl methacrylate-co-N,N-dimethylacrylamide)*
Related Polymers: Polymethacrylates, General, Poly(methyl methacrylate), Poly(ethyl methacrylate), Poly(butyl methacrylate), Polyacrylamide, Polymethacrylamide, Poly(lauryl methacrylate), Poly(N,N-dimethylacrylamide)
Monomers: Acrylamide, Methacrylamide, N-Phenylmaleimide, Methyl methacrylate, Ethyl methacrylate, Butyl methacrylate, Lauryl methacrylate, N,N-Dimethylacrylamide
Material class: Thermoplastic
Polymer Type: acrylic copolymers

Volumetric & Calorimetric Properties:
Glass-Transition Temperature:

No.	Value	Note
1	134°C	80% maleimide, MMA-co-N-tolylmaleimide [1]
2	176°C	36% maleimide, MMA-co-N-tolylmaleimide [1]
3	130–220°C	MMA-co-N-phenylmaleimide [2]
4	76–88°C	0.8–48% butyl methacrylate, butyl methacrylate-co-methacrylamide [3]
5	76–90°C	80–60% ethyl methacrylate, ethyl methacrylate-co-acrylamide [3]

Transition Temperature General: T_g is dependent upon the composition of the copolymer, generally increasing with decreasing content of the methacrylate component

Surface Properties & Solubility:
Cohesive Energy Density Solubility Parameters: δ (ethyl methacrylate-co-acrylamide) [3]: 18.62 $J^{1/2}$ cm$^{-3/2}$ (9.1 cal$^{1/2}$ cm$^{-3/2}$) (THF), 19.01 $J^{1/2}$ cm$^{-3/2}$ (9.3 cal$^{1/2}$ cm$^{-3/2}$) (CHCl$_3$), 19.84 $J^{1/2}$ cm$^{-3/2}$ (9.7 cal$^{1/2}$ cm$^{-3/2}$) (CH$_2$Cl$_2$), 20.67 $J^{1/2}$ cm$^{-3/2}$ (10.1 cal$^{1/2}$ cm$^{-3/2}$) (AcOH), 21.06 $J^{1/2}$ cm$^{-3/2}$ (10.3 cal$^{1/2}$ cm$^{-3/2}$) (Ac$_2$O), 21.69 $J^{1/2}$ cm$^{-3/2}$ (10.6 cal$^{1/2}$ cm$^{-3/2}$) (DMF)

Wettability/Surface Energy and Interfacial Tension: Surface energy 47 mN m^{-2} (γ_p 12.0, γ_d 35.0) (methyl methacrylate-co-N,N-dimethylacrylamide, dry, 10 mol% methyl methacrylate) [4]; 46.5 mN m^{-2} (γ_p 9.8, γ_d 36.7) (methyl methacrylate-co-N,N-dimethylacrylamide, dry, 20 mol% methyl methacrylate) [4]; 49 mN m^{-2} (γ_p 15.6, γ_d 33.4) (lauryl methacrylate-co-N,N-dimethylacrylamide, dry, 10 mol% lauryl methacrylate) [4]; 47.1 mN m^{-2} (γ_p 12.8, γ_d 34.3) (lauryl methacrylate-co-N,N-dimethylacrylamide, dry, 20 mol% lauryl methacrylate) [4]; 46.7 mN m^{-2} (γ_p 11.4, γ_d 35.3) (lauryl methacrylate-co-N,N-dimethylacrylamide, dry, 30 mol% lauryl methacrylate) [4]; 46.6 mN m^{-2} (γ_p 10.6, γ_d 36.0) (lauryl methacrylate-co-N,N-dimethylacrylamide, dry, 40 mol% lauryl methacrylate)[4]

Transport Properties:
Polymer Solutions Dilute: Intrinsic viscosity 0.53–0.59 dl g^{-1} (ethyl methacrylate-co-acrylamide, range of solvents) [3]. η 1.865–2.51 dl g^{-1} (30°, CHCl$_3$, MMA-co-N-phenylmaleimide, 2.91–9.92% N-phenylmaleimide) [1]. η 0.164–0.423 dl g^{-1} (30°, CHCl$_3$, MMA-co-N-(o-tolylmaleimide), 35.3–7.9% maleimide) [1], 0.202–0.41 dl g^{-1} (30°, CHCl$_3$, MMA-co-N-(p-tolylmaleimide), 36.6–13.4% maleimide) [1]

Water Content: The equilib. water content of N,N-dimethylacrylamide copolymer hydrogels with both methyl and lauryl methacrylate decreases with increasing methacrylate content [4]

Mechanical Properties:
Mechanical Properties General: Tensile stress, tensile modulus, plateau modulus and flexural strength of MMA-co-N-phenylmaleimide are dependent upon the mole fraction of the maleimide component. Tensile modulus and tensile strength at break of N,N-dimethylacrylamide copolymer hydrogels with both methyl and lauryl methacrylate increase with decreasing equilib. water content [4]

Tensile (Young's) Modulus:

No.	Value	Note
1	1177–1672 MPa	MMA-co-N-phenylmaleimide, 0–9.92% maleimide [1]
2	0.09–0.18 MPa	MMA-co-N,N-dimethylacrylamide hydrogel, 5–20% MMA [4]
3	0.17–0.85 MPa	lauryl methacrylate-co-N,N-dimethylacrylamide hydrogel, 10–40% lauryl methacrylate [4]

Tensile Strength Break:

No.	Value	Note	Elongation
1	48–60 MPa	MMA-co-N-phenyl-maleimide, 0–10% maleimide [1]	4.2–7.2%
2	0.13–0.16 MPa	MMA-co-N,N-dimethyl-acrylamide hydrogel, 5–20% MMA [4]	130–170%

| 3 | 0.21–0.53 MPa | lauryl methacrylate-co-N,N-dimethylacrylamide hydrogel, 10–40% lauryl methacrylate [4] | 105–317% |

Flexural Strength at Break:

No.	Value	Note
1	96–114 MPa	MMA-co-N-phenylmaleimide, 0–9.92% maleimide [1]

Miscellaneous Moduli:

No.	Value	Note	Type
1	0.014–0.42 MPa	140000–4200000 dyne cm^{-2}, MMA-co-N-phenylmaleimide, 100% maleimide [2]	Plateau modulus

Electrical Properties:
Dielectric Permittivity (Constant):

No.	Value	Note
1	3.77	butyl methacrylate-co-methacrylamide, 61% butyl methacrylate
2	3.89	butyl methacrylate-co-methacrylamide, 55% butyl methacrylate
3	3.54	butyl methacrylate-co-methacrylamide, 48% butyl methacrylate [3]

Optical Properties:
Transmission and Spectra: Ir and nmr spectra data have been reported [1,3]

Polymer Stability:
Decomposition Details: Initial decomposition temp. 220° (MMA-co-methacrylamide, 61% MMA), 230.4° (MMA-co-methacrylamide, 55% MMA), 237.1° (MMA-co-methacrylamide, 48% MMA) [3]. MMA-co-N-(o-tolylmaleimide) undergoes a two-stage decomposition process. Initial decomposition temps. for the first stage range from 365–380° and for the second stage from 456–480°. Decomposition temps. depend on the mole fraction of maleimide. Initial decomposition temp. (MMA-co-N-(p-tolylmaleimide)) 358–373° depending on mole fraction of maleimide [1]. Decomposition temp. (ethyl methacrylate-co-acrylamide) 239–260° depending on the mole fraction of ethyl methacrylate [3]

Applications/Commercial Products:
Processing & Manufacturing Routes: Synth. by mixing monomers in various ratios with initiator at 60–80° [1,3]

Bibliographic References
[1] Bharel, R., Choudhary, V. and Varma, I.K., *J. Appl. Polym. Sci.*, 1993, **49**, 31, (thermal behaviour)
[2] Wu, S. and Beckerbauer, R., *Polymer*, 1992, **33**, 509
[3] Srinivasulu, B., Rao, P.R., Sundaram, E.V., Sathaiah, G. and Sirdeshmukh, L., *J. Appl. Polym. Sci.*, 1991, **43**, 1521
[4] Corkhill, P.H. and Tighe, B.J., *Polymer*, 1990, **31**, 1526

Methacrylate/olefin copolymers   M-6

Synonyms: Poly(alkyl methacrylate-co-ethene). Poly(alkyl methacrylate-co-ethylene). Poly(methyl methacrylate-co-ethylene). Poly(methyl methacrylate-co-butadiene). Poly(methyl methacrylate-graft-natural rubber). Poly(ethyl methacrylate-co-chloroprene). Poly(glycidyl methacrylate-co-chloroprene). Poly(methyl methacrylate-co-tetrafluoroethylene). Poly(methyl methacrylate-co-EPDM-graft-styrene). Poly(methyl methacrylate-co-EPDM-co-acrylonitrile). Poly(methyl methacrylate-co-EPDM-graft-2-vinylnaphthalene). Poly(methyl methacrylate-co-vinylidene fluoride). Poly(butyl methacrylate-co-vinylidene chloride)
Related Polymers: Poly(tert-butyl methacrylate), Polymethacrylates, General, Poly(methyl methacrylate), Poly(ethyl methacrylate), Poly(sec-butyl methacrylate), Natural rubber, Polychloroprene, Polybutadiene, EPDM, Poly(vinylidene chloride), Poly(vinylidene fluoride)
Monomers: Methyl methacrylate, Ethyl methacrylate, Butyl methacrylate, Ethylene, Butadiene, Isoprene, Chloroprene, Vinylidene fluoride, Vinylidene chloride
Material class: Thermoplastic
Polymer Type: acrylic copolymers
CAS Number:

CAS Reg. No.	Note
25101-13-7	methyl methacrylate-co-ethylene
132076-25-6	methyl methacrylate-co-ethylene, block
110807-37-9	methyl methacrylate-co-ethylene, graft
25232-40-0	methyl methacrylate-co-butadiene
106399-43-3	methyl methacrylate-co-butadiene, block
107439-29-2	methyl methacrylate-co-butadiene, graft

Molecular Weight: MW 200000–2000000 (methyl methacrylate-graft-natural rubber), 200000–330000 (main rubber chains), 2000–5000 (grafted side chains)
General Information: The props. of MMA-graft-natural rubber vary with MMA content. The copolymer can be vulcanised to give strong, hard compounds that can be coloured and translucent, or almost transparent. The rubbers are hard and can be blended with natural rubber in all proportions. The periodic A_2B MMA-co-ethylene copolymer is a soft, tough, partially crystalline solid with good elasticity [1]. The AB copolymer is a transparent, soft and rubbery material [1]. Generally, the methacrylate-co-ethylene periodic A_2B copolymers are amorph., soft and extensible [2]. The MMA-vinylidene fluoride-methyl perfluoroacrylate terpolymer is a translucent solid [3]
Morphology: In MMA-graft-natural rubber copolymers the PMMA chains are attached either at the α-methylene carbon of the polyisoprene or at a double bond at a chain end. The ethylene copolymers with alkyl methacrylates can be graft, statistical (unordered or random) or periodic, i.e., sequence ordered, with repeat units (ethylene-ethylene-alkyl methacrylate)$_n$, or (ethylene-alkyl methacrylate)$_n$ denoted as A_2B and AB systems respectively. The random ethylene copolymers are amorph.

Volumetric & Calorimetric Properties:
Density:

No.	Value	Note
1	d^{20} 1.052 g/cm^3	ethylene-ethylene-MMA periodic copolymer [2]
2	d^{20} 1.02 g/cm^3	MMA-graft-natural rubber [4]

Melting Temperature:

No.	Value	Note
1	90°C	MMA-co-ethylene, periodic A_2B [1,2]
2	47°C	benzyl methacrylate-co-ethylene, A_2B system [2]
3	280°C	MMA-co-vinylidene fluoride, vinylidene fluoride blocks [3]

Glass-Transition Temperature:

No.	Value	Note
1	−4.2–92.7°C	MMA-*co*-ethylene, 36.6–90.0% MMA [2]
2	−6°C	MMA-*co*-ethylene, A$_2$B, 33.3% MMA [2]
3	45°C	MMA-*co*-ethylene, A$_2$B, 33.3% MMA [2]
4	19°C	MMA-*co*-ethylene, AB, 50% MMA [2]
5	15.7–16.9°C	butyl methacrylate-*co*-vinylidene chloride, semicontinuous emulsion polymerisation
6	23.2–24.2°C	butyl methacrylate-*co*-vinylidene chloride, batch process [6]
7	−81–105°C	MMA-*co*-ethylene
8	−32°C	ethyl methacrylate-*co*-ethylene, A$_2$B system
9	−26°C	isopropyl methacrylate-*co*-ethylene, A$_2$B system
10	14°C	phenyl methacrylate-*co*-ethylene, A$_2$B system [1,2,7]
11	−65–−35°C	MMA-*co*-butadiene, block; MMA-*co*-butadiene, triblock
12	114–129°C	MMA-*co*-butadiene, block; MMA-*co*-butadiene, triblock
13	−58°C	ethyl methacrylate-*co*-butadiene, block; ethyl methacrylate-*co*-butadiene, triblock
14	84–90°C	ethyl methacrylate-*co*-butadiene, block; ethyl methacrylate-*co*-butadiene, triblock
15	−50°C	isobutyl methacrylate-*co*-butadiene, block; isobutyl methacrylate-*co*-butadiene, triblock
16	116°C	isobutyl methacrylate-*co*-butadiene, block; isobutyl methacrylate-*co*-butadiene, triblock
17	−55°C	isobornyl methacrylate-*co*-butadiene, block; isobornyl methacrylate-*co*-butadiene, triblock [5]
18	202°C	isobornyl methacrylate-*co*-butadiene, block; isobornyl methacrylate-*co*-butadiene, triblock [5]
19	−42.3°C	ethyl methacrylate-*co*-chloroprene, graft, MW 140000
20	−34.5°C	ethyl methacrylate-*co*-chloroprene, random, MW 71000 [7]
21	−29.7°C	glycidyl methacrylate-*co*-chloroprene, graft, MW 128000
22	−16.4°C	glycidyl methacrylate-*co*-chloroprene, random, MW 74200 [7]
23	138°C	MMA-*co*-vinylidene fluoride, block [3]
24	15–25°C	butyl methacrylate-*co*-vinylidene chloride [6]

Transition Temperature General: T_g varies with composition [2,5] and synth. method [6]. T_m is not observed for poly(MMA-*co*-ethylene AB system) [1,2]

Surface Properties & Solubility:
Cohesive Energy Density Solubility Parameters: δ (MMA-*co*-butadiene) 5.9 J$^{1/2}$ cm$^{-3/2}$ 17.2 (MPa)$^{1/2}$ (30% MMA), 6 J$^{1/2}$ cm$^{-3/2}$ 17.6 (MPa)$^{1/2}$ (50% MMA) [8]. Cohesive energy (MMA-*co*-butadiene) 295.4 J cm^{-3} (70.56 cal cm^{-3}, 30% MMA) [8], 310.7 J cm^{-3} (74.25 cal cm^{-3}, 50% MMA) [8]

Solvents/Non-solvents: MMA-*co*-ethylene AB system sol. CHCl$_3$, C$_6$H$_6$, Me$_2$CO, THF, DMF. Natural rubber graft copolymers are generally sol. solvents which will dissolve both natural rubber and PMMA, *e.g.* C$_6$H$_6$, cyclohexanone, CCl$_4$, amyl acetate. Insol. EtOH, Me$_2$CO, 2-butanone [4]. MMA-*co*-vinylidene fluoride, sol. *o*-(bistrifluoromethyl)benzene (hot); insol. 2-butanone [3]

Wettability/Surface Energy and Interfacial Tension: Contact angle (MMA-*co*-vinylidene fluoride) 94° (advancing; H$_2$O), 37° (advancing; hexadecane) [3]

Transport Properties:
Polymer Solutions Dilute: Poly(alkyl methacrylates) with olefin copolymer side chains have excellent low temp. viscosity and good thickening efficiency [9]

Mechanical Properties:
Mechanical Properties General: The presence of diblocks in triblock MMA/butadiene copolymers gives tensile props. that are worse than those of pure triblocks [5]

Tensile (Young's) Modulus:

No.	Value	Note
1	98.1 MPa	10000 kg cm^{-2}, MMA-*co*-ethylene, A$_2$B system
2	1 MPa	10 kg cm^{-2}, ethyl methacrylate-*co*-ethylene, A$_2$B system
3	1 MPa	10 kg cm^{-2}, isopropyl methacrylate-*co*-ethylene, A$_2$B system
4	1.5 MPa	15 kg cm^{-2}, phenyl methacrylate-*co*-ethylene, A$_2$B system
5	24.5 MPa	250 kg cm^{-2}, benzyl methacrylate-*co*-ethylene, A$_2$B system [1,2]
6	689–966 MPa	100–140 kpsi, MMA-*co*-vinylidene fluoride [3]
7	3.4–12.7 MPa	butyl methacrylate-*co*-vinylidene chloride, semicontinuous emulsion polymerisation [6]
8	230.1–296.3 MPa	butyl methacrylate-*co*-vinylidene chloride, batch process [6]

Flexural Modulus:

No.	Value	Note
1	1750 MPa	254 kpsi, 25°, MMA-*co*-vinylidene fluoride [3]
2	357 MPa	51.8 kpsi, 100°, MMA-*co*-vinylidene fluoride [3]
3	12 MPa	1800 psi, 150°, MMA-*co*-vinylidene fluoride [3]

Tensile Strength Break:

No.	Value	Note	Elongation
1	2 MPa	syndiotactic MMA-*co*-butadiene triblock, 28% MMA	320%
2	35 MPa	syndiotactic MMA-*co*-butadiene triblock, 72% MMA [5]	140%
3	2.2 MPa	125°, MMA-*co*-butadiene, 38% MMA [5]	1086%
4	32 MPa	25°, MMA-*co*-butadiene, 38% MMA [5]	835%
5	1.4–18.7 MPa	90–25°, ethyl methacrylate-*co*-butadiene, 31% ethyl methacrylate [5]	530–1120%
6	19 MPa	25°, ethyl methacrylate-*co*-butadiene, 21% ethyl methacrylate [5]	1200%

7	1.8–24 MPa	90–25°, tert-butyl methacrylate-co-butadiene, 29% tert-butyl methacrylate [5]	570–1080%
8	2.2–35 MPa	150–25°, isobornyl methacrylate-co-butadiene, 42% isobornyl methacrylate [5]	600–650%
9	7.23 MPa	syndiotactic MMA-isobutylene-syndiotactic MMA triblock, 20% PMMA [10]	525%
10	6.52 MPa	syndiotactic MMA-isobutylene-syndiotactic MMA triblock, 20% MMA, annealed	550%
11	11.82 MPa	syndiotactic MMA-isobutylene-syndiotactic MMA triblock, 27% PMMA	550%
12	14.52 MPa	syndiotactic MMA-isobutylene-syndiotactic MMA triblock, 34% PMMA	650%
13	7.69 MPa	syndiotactic MMA-isobutylene-syndiotactic MMA triblock, 27% PMMA	25%
14	25.5 MPa	260 kg cm^{-2}, MMA-co-ethylene, periodic A_2B	300%
15	0.2 MPa	2 kg cm^{-2}, ethyl methacrylate-co-ethylene, A_2B system	500%
16	0.3 MPa	3 kg cm^{-2}, isopropyl methacrylate-co-ethylene, A_2B system	1100%
17	4.9 MPa	50 kg cm^{-2}, phenyl methacrylate-co-ethylene, A_2B system	600%
18	14.7 MPa	150 kg cm^{-2}, benzyl methacrylate-co-ethylene, A_2B system [1,2]	500%
19	27.5 MPa	2.8 kg mm^{-2}, natural rubber graft copolymer, 23% MMA	520–560%
20	17.7 MPa	1.8 kg mm^{-2}, natural rubber graft copolymer, 49% MMA [4]	215%
21	2–35 MPa	MMA-co-butadiene, triblock	140–1300%
22	1.4–19 MPa	ethyl methacrylate-co-butadiene	580–1280%
23	1.8–24 MPa	tert-butyl methacrylate-co-butadiene	570–1270%
24	2.2–35 MPa	isobornyl methacrylate-co-butadiene [5]	600–1080%
25	6.5–14.5 MPa	MMA-co-isobutylene [10]	25–650%
26	21.4–24.8 MPa	vinylidene fluoride-co-MMA, 3.1–3.6 kpsi [3]	10–40%
27	1.3–2.9 MPa	butyl methacrylate-co-vinylidene chloride, semicontinuous emulsion polymerisation [6]	468–588%
28	7.23 MPa	butyl methacrylate-co-vinylidene chloride, batch process [6]	74–87%
29	38.3 MPa	390 kg cm^{-2}, MMA-EPDM-styrene graft copolymer [7]	3.5%
30	33.8 MPa	345 kg cm^{-2}, MMA-EPDM-(2-vinylnaphthalene) graft copolymer [11]	1.8%

Tensile Strength Yield:

No.	Value	Note
1	1.7–4.6 MPa	125–25°, MMA-co-butadiene, 38% MMA
2	2.3–4 MPa	70–25°, MMA-co-butadiene, 31% MMA
3	2.3 MPa	25°, ethyl methacrylate-co-butadiene, 21% ethyl methacrylate
4	1.9–9.6 MPa	150–25°, isobornyl methacrylate-co-butadiene, 42% isobornyl methacrylate [5]
5	9.8 MPa	100 kg cm^{-2}, MMA-co-ethylene, periodic A_2B [1]
6	0.3 MPa	3.5 kg cm^{-2}, ethyl methacrylate-co-ethylene, A_2B system [2]
7	0.3 MPa	3.5 kg cm^{-2}, isopropyl methacrylate-co-ethylene, A_2B system [2]
8	0.5 MPa	5 kg cm^{-2}, phenyl methacrylate-co-ethylene, A_2B system [2]
9	2.9 MPa	30 kg cm^{-2}, benzyl methacrylate-co-ethylene, A_2B system [2]
10	3–42 MPa	syndiotactic MMA-co-butadiene
11	2.3–4 MPa	ethyl methacrylate-co-butadiene
12	1.9–9.6 MPa	isobornyl methacrylate-co-butadiene [5]

Miscellaneous Moduli:

No.	Value	Note	Type
1	1681 MPa	244 kpsi, 1h [3]	Tensile creep modulus
2	1530 MPa	222 kpsi, 10h [3]	Tensile creep modulus
3	861 MPa	125 kpsi, 300h [3]	Tensile creep modulus

Hardness: IRHD natural rubber graft copolymers 72–75 (23% MMA), 96 (49% MMA) [4]

Electrical Properties:
Surface/Volume Resistance:

No.	Value	Note	Type
2	>0.0009 × 10^{15} Ω.cm	min., natural rubber graft, vulcanised, 49% MMA [4]	S

Dielectric Permittivity (Constant):

No.	Value	Note
1	2.9	natural rubber graft, vulcanised, 49% MMA [4]

Dielectric Strength:

No.	Value	Note
1	16 kV.mm^{-1}	natural rubber graft, vulcanised, 49% MMA [4]

Dissipation (Power) Factor:

No.	Value	Note
1	0.02	natural rubber graft, vulcanised, 49% MMA [4]

Optical Properties:
Transmission and Spectra: Ir and nmr spectral data have been reported [1,2,5,6,7,11,12,13,14]

Polymer Stability:
Thermal Stability General: The MMA-*co*-ethylene periodic A_2B copolymer decomposes slowly above 200° in air [1]. The MMA-EPDM-acrylonitrile graft copolymer is more stable than either ABS or poly(acrylonitrile-*co*-MMA). Variations in the method of preparation have little effect on the thermal stability [13]

Upper Use Temperature:

No.	Value	Note
1	160°C	isobornyl methacrylate-*co*-butadiene [5]

Flammability: MMA-*co*-vinylidene fluoride burns slowly, and drips. Limiting oxygen index 17.7% [3]

Environmental Stress: The graft copolymer of MMA-EPDM-acrylonitrile weathers better than ABS [13]. MMA-EPDM-(2-vinylnaphthalene) graft copolymer has better resistance to light than ABS [11]. MMA-EPDM-*graft*-styrene copolymer has good weathering props. and good resistance to light [7]

Applications/Commercial Products:
Processing & Manufacturing Routes: Natural rubber graft copolymers are synth. by transfer of the free-radical followed by monomer polymerisation. The reaction is initiated by γ-radiation (most efficient), uv photosensitisers, redox systems or peroxides. Graft copolymers can be made by the reaction of azodicarboxylate substituted poly(alkyl methacrylates) with natural rubber or synthetic polyisoprenes [15]. The grafting of MMA onto radiation-vulcanised natural rubber by a radiation-induced process has been reported. The thermoplastic elastomer thus produced can be processed satisfactorily but improvements are gained by blending with the graft copolymer of non-vulcanised material [16]. The graft copolymer has been prepared in toluene using benzoyl peroxide at 80°. At a constant level of peroxide the grafting efficiency and level decrease and MW increases as the level of MMA increases. At a constant level of MMA the graft level decreases to a constant level as the amount of peroxide is increased [17]. Processed and vulcanised in a similar manner to natural rubber. MMA-*co*-ethylene periodic AB copolymer is synth. by reduction of poly(MMA-*alt*-vinyl bromide) [1]. The periodic A_2B copolymers may be synth. by the hydrogenation of poly(methacrylate-*alt*-butadiene) [1,2]. A statistical ethylene-MMA copolymer can be synth. by reduction of the corresponding statistical copolymer of MMA and vinyl chloride [14]. Triblock alkyl methacrylate/butadiene copolymers can be synth. by sequential anionic polymerisation using dilithium initiators [5]. Butyl methacrylate/vinylidene chloride copolymer is synth. by batch polymerisation at 25° using a redox catalyst to give crystalline heterogeneous product. Seeded semicontinuous emulsion polymerisations give amorph. polymers with a more uniform composition [6]. MMA-*co*-vinylidene fluoride block copolymer can be synth. by termonomer-induced copolymerisation as the monomers themselves will not copolymerise. A suitable comonomer is methyl perfluoroacrylate [3]

Applications: MMA-*co*-ethylene is used in composite articles and polymer alloys. Poly(alkyl methacrylates) with olefinic copolymer side-chains are used as viscosity modifiers in modern multigrade oils. Natural rubber graft copolymers are used in the manufacture of impact-resistant articles (cutting boards) and rigid mouldings (golf ball covers, roller-skates, castor wheels, electric plugs); as reinforcing agents; as adhesives (in soln. and latex form) to bond rubber to a variety of materials, *e.g.* metal or textiles

Trade name	Details	Supplier
Nucrel	MMA-*co*-ethylene	DuPont
PE	MMA-*co*-ethylene	Chevron

Bibliographic References

[1] Yokata, K. and Hirabayashi, T., *Macromolecules*, 1981, **14**, 1613, 1615, (synth.)
[2] Yokata, K., Kougo, T. and Hirabayashi, T., *Polym. J. (Tokyo)*, 1983, **15**, 349, (synth., props.)
[3] Weise, J.K., *Polym. Prepr. (Am. Chem. Soc., Div. Polym. Chem.)*, 1971, **12**, 512
[4] Roff, W.J. and Scott, J.R., *Fibres, Films, Plastics and Rubbers*, Butterworths, 1971, 332
[5] Yu, J.M., Dubois, P. and Jerome, R., *Macromolecules*, 1996, **29**, 6090, 7316, 8362, (synth., props.)
[6] Lee, K., El-Aasser, M.S. and Vanderhof, J.W., *J. Appl. Polym. Sci.*, 1991, **42**, 3133
[7] Park, C.K., Ha, C.S., Lee, J.K. and Cho, W.J., *J. Appl. Polym. Sci.*, 1994, **53**, 763, 967, (synth., chloroprene copolymers)
[8] Makarova, L.V., Shvarts, A.G., Zakharov, N.D. and Priborets, A.M., *Polym. Sci. USSR (Engl. Transl.)*, 1965, **7**, 1168
[9] Omeis, J. and Pennewiss, H., *Abstr. Pap.-Am. Chem. Soc.*, 1994, **207**, 335
[10] Kennedy, J.P., Price, J.L. and Koshimura, K., *Macromolecules*, 1991, **24**, 6567
[11] Park, J.-Y., Park, D.-J., Ha, C.S. and Cho, W.J., *J. Appl. Polym. Sci.*, 1994, **51**, 1303, (synth., props.)
[12] Barnard, D., *J. Polym. Sci.*, 1956, **22**, 213
[13] Beldie, C., Onu, A., Bourceanu, G. and Poinescu, I.C., *Eur. Polym. J.*, 1985, **21**, 321
[14] Yokata, K., Miwa, M., Hirabayashi, T. and Insi, Y., *Macromolecules*, 1992, **25**, 5821
[15] Kiew, W.A., Campbell, D.S. and Tinker, A.J., *Polymer*, 1987, **28**, 2157, 2161
[16] Razzak, M.T., Yoshi, F., Makuuchi, K. and Ishigaki, I., *J. Appl. Polym. Sci.*, 1991, **43**, 883
[17] Enyiegbulam, M.E. and Aloka, I.U., *J. Appl. Polym. Sci.*, 1992, **44**, 1841

Methacrylate-urethane polymers and copolymers

M-7

PEG: $HO-[CH_2CH_2O]_n-H$

SES: $HO-[CH_2CH_2OCH_2CH_2OCCH_2CH_2CO]_n-CH_2CH_2OCH_2CH_2OH$

UES: $HO-[CH_2CH_2OCH_2CH_2OCCH=CHCO]_n-CH_2CH_2OCH_2CH_2OH$

SPU: $HO-[CH_2CH_2OCH_2CH_2OCNH(CH_2)_6NHCO]_n-CH_2CH_2OCH_2CH_2OH$

Synonyms: *Poly(urethane methacrylate), Poly(methacrylourethane)*
Material class: Thermoplastic
Polymer Type: acrylic copolymers
General Information: There is a wide variety of possible methacrylate/urethane polymers and copolymers. This entry contains data on some examples. Poly(urethane methacrylates) designated UMPEG, UMSES, UMUES and UMSPU [1]. Methacrylate-urethane copolymer from 1,4-butanediol/diphenylmethane diisocyanate/poly(tetramethylene ether glycol) with incorporated double bonds and methyl methacrylate designated DP. Methacrylate-urethane copolymer from hexamethane diisocyanate/tetraethylene glycol/hydroxyethyl methacrylate and methyl methacrylate, designated HP. Methacrylate-urethane copolymer from toluene diisocyanate/tetraethylene glycol/hydroxyethyl methacrylate and methyl methacrylate, designated TP

Volumetric & Calorimetric Properties:
Thermal Expansion Coefficient:

No.	Value	Note	Type
1	$5.7 \times 10^{-5} - 7.5 \times 10^{-5}$ K^{-1}	TP; $T > T_g$ [2]	1
2	0.000157 – 0.000181 K^{-1}	TP; $T < T_g$ [2]	1
3	0.000115 – 0.000126 K^{-1}	HP, $T > T_g$ [2]	1
4	0.000101 – 0.000166 K^{-1}	HP, $T > T_g$ [2]	1

Glass-Transition Temperature:

No.	Value	Note
1	-10.5°C	UMPEG, cured [1]
2	-0.5°C	UMSES, cured [1]
3	4°C	UMUES, cured [1]
4	5.5°C	UMSPU, cured [1]
5	133–137°C	TP, varying methyl methacrylate content [2]
6	114–136°C	TP, varying tetraethylene glycol/hydroxyethyl methacrylate content [2]
7	64–82°C	HP, varying tetraethylene glycol/hydroxyethyl methacrylate content [2]

Surface Properties & Solubility:
Solvents/Non-solvents: Sol. Me_2CO, $CHCl_3$, dioxane, THF. Insol. C_6H_6, toluene, xylene, alcohols, cyclohexane, CCl_4 [1]

Transport Properties:
Polymer Solutions Dilute: Viscosity 232.5 P (40°, UMPEG, M_n 2060), 257 P (40°, UMSES, M_n 2470), 281 P (40°, UMUES, M_n 2610), 355 P (40°, UMSPU, M_n 2210) [1]

Water Absorption:

No.	Value	Note
1	0.17 %	TP [2]
2	0.87 %	HP [2]

Mechanical Properties:
Tensile (Young's) Modulus:

No.	Value	Note
1	250–390 MPa	DP, MW 43500000 [3]
2	210–370 MPa	DP, 21% unreacted polyurethane [3]
3	120–260 MPa	DP, MW 13000000 [3]
4	360–420 MPa	DP, 21% unreacted polyurethane [3]
5	260–264 MPa	DP, MW 414500 [3]
6	370–450 MPa	DP, MW 2850000 [3]
7	580–640 MPa	DP, 12% unreacted polyurethane [3]

Tensile Strength Break:

No.	Value	Note	Elongation
1	3–11 MPa	MW 43500000 [3]	30–36%
2	8.4–9.8 MPa	DP, 21% unreacted polyurethane [3]	10–14%
3	6–12 MPa	DP, MW 13000000 [3]	7–11%
4	8–12 MPa	DP, 21% unreacted polyurethane [3]	8–12%
5	9–13 MPa	DP, MW 414500 [3]	5–7%
6	11–13 MPa	DP, MW 2850000 [3]	16.8–17.8%
7	9–11 MPa	DP, 12% unreacted polyurethane [3]	43–47%

Tensile Strength Yield:

No.	Value	Note
1	12–18 MPa	DP, MW 43500000 [3]
2	8.4–9.8 MPa	DP, 21% unreacted polyurethane [3]
3	15–19 MPa	DP, MW 2850000 [3]
4	16.6–17 MPa	DP, 12% unreacted polyurethane [3]

Impact Strength: Falling dart impact strength 0.417 m s^{-1} (DP, MW 43500000); 0.427 m s^{-1} (DP, MW 13000000); 0.428 m s^{-1} (DP, MW 2850000). Isolated graft copolymers of DP have not been tested by the falling dart technique only those polymers with unreacted PU present [3]. Impact strength of TP and HP increases (11.2–16.4 N m^{-2}) increases with the number of hydroxyl groups contributed by tetraethylene glycol/hydroxyethyl methacrylate. Increasing methyl methacrylate content increases with strength of TP whereas in HP it causes a decrease

Hardness: Pencil hardness B (UMPEG, 91.7% gel content), HB (UMSES, 94.0% gel content), 2H (UMUES, 90.4% gel content), H (UMSPU, 90.0% gel content) [1]. Surface hardness of TP generally increases with tetraethylene glycol/hydroxyethyl methacrylate mole ratio whereas little change is observed for HP. Surface hardness of TP generally decreases with increasing methyl methacrylate content whereas it increases for HP [2]

Optical Properties:
Refractive Index:

No.	Value	Note
1	1.55	TP [2]

Transmission and Spectra: Ir and uv spectral data have been reported [1,2,4]

Volume Properties/Surface Properties: Colour: transparent to yellowish-brown

Polymer Stability:
Thermal Stability General: All copolymers are stable at 120°; UMUES gels at 140° and all copolymers gel to various degrees at 160° and above

Stability Miscellaneous: Exposure to uv causes rapid cure of all copolymers. Irradiation of UMPEG, UMSES, UMUES and UMSPU generally increases tensile modulus and strength [1]

Applications/Commercial Products:
Processing & Manufacturing Routes: DP may be synth. by reacting polyurethane, 1,4-butanediol/4,4'-diphenylmethane diisocyanate and polyethylene ether glycol with glyceryl monomethacrylate or α,α-dimethyl-m-isopropenyl benzyl isocyanate to give incorporated unsaturated diol or unsaturated isocyanate groups. This product is then copolymerised with methyl methacrylate by a radical mechanism. Typical conditions may be mole ratio polyurethane/glycerol monomethacrylate:methyl methacrylate 1.5:4 by weight in DMF at 80°. The type and level of initiator is varied along with the reaction time. The product contains a mixture of polyurethane and DP depending on the reaction conditions [3,4]. Poly(methacrylate) methyl/polyurethane inter-penetrating networks can be synth. by γ-irradiation [5] Oligomers can be thermally cured (above 140°) or cured by uv photolysis, electron-beam irradiation or γ-irradiation [1]

Applications: Used in photoimaging, and as interfacial agents in blending

Bibliographic References
[1] Chiang, W-Y. and Chen, S.-C., *J. Appl. Polym. Sci.*, 1989, **37**, 1669, (synth., props).
[2] Zhang, L., Pan, M., Tai, H. and Li, X., *J. Appl. Polym. Sci.*, 1994, **54**, 1847, (synth., props., morphology)
[3] Wilson, D. and George, M.H., *Polymer*, 1992, **33**, 3723, (mechanical props.)
[4] Wilson, D. and George, M.H., *Polym. Commun.*, 1990, **31**, 90, 355, (synth.)
[5] Wang, G.-C., Cui, H.-W., Fang, B. and Zhang, Z.-P., *J. Appl. Polym. Sci.*, 1992, **44**, 1165, (synth.)

Methacrylates, General M-8

Synonyms: *Acrylics*
Related Polymers: Poly(ethyl methacrylate), Poly(propyl methacrylate), Poly(isopropyl methacrylate), Poly(octyl methacrylate), Poly(pentyl methacrylate), Poly(aryl methacrylate), Poly(hydroxyalkyl methacrylate), Poly methacrylates with side-chains containing heteroatoms, Poly(methyl methacrylate) General, Poly(butyl methacrylate), Poly(isobutyl methacrylate), Poly(*sec*-butyl methacrylate), Poly(*tert*-butyl methacrylate), Hexyl methacrylates, Long-chain poly(alkyl methacrylates), Poly(haloalkyl methacrylate), Poly(glycol methacrylates)
Material class: Thermoplastic
Polymer Type: acrylic
General Information: There is a tremendous range of methacrylate homopolymers, copolymers and polymer blends; the principal commercial product is poly(methyl methacrylate). The products are available as sheets, rods, tubes, moulding and extrusion products, emulsions, dispersions and solns. Generally, the methacrylates are harder and have a higher tensile strength and lower elongation than the corresponding polyacrylates. Poly(methyl methacrylate) is a hard rigid polymer whereas *n*-hexyl methacrylate and polymers with longer side-chains are rubber-like. *n*-Dodecyl methacrylate is brittle owing to crystallisation of the side-chain. Fragments of polymerisation reagents such as initiators, solvents and chain-transfer reagents can be incorporated into the polymer chain as end-groups during manufacture. These groups, although at a low concentration, can have an effect on some polymer props., *e.g.* thermal stability [1]. Cross-linked methacrylate polymers and copolymers have important applications as hydrogels. Perhaps the most important is poly(2-hydroxyethyl methacrylate) which when cross-linked with, *e.g.* ethylene diacrylate, gives pliable, soft hydrogels which have a wide variety of applications, particularly in medicine
Tacticity: Normally atactic. Syndiotactic and isotactic forms are known
Morphology: There is some local order in the amorph. polymers although the behaviour of poly(methyl methacrylate) is atypical [2]
Identification: Fragments of polymerisation reagents that have been incorporated into the polymer struct. as end groups during manufacture may be identified by pyrolysis [1,3]. Soln. polymers are usually characterised by solids' content, viscosity and solvent as well as resin composition and MW. Emulsion polymers are classified on the basis of the type of surfactant used in their preparation: anionic, cationic or nonionic. Emulsion polymers are then characterised in a similar way to soln. polymers with the additional parameter of particle size

Volumetric & Calorimetric Properties:

Volumetric Properties General: For a series of linear alkyl methacrylates the density decreases with the length of the alkyl chain
Transition Temperature General: T_g decreases as the length of the alkyl side-chain increases (disregarding tacticity). T_g increases with increasing branching of the alkyl chain. T_g varies markedly with thermal history and residual solvent [5]. Methacrylate polymers show side-chain transitions at about $20°$ and below $-150°$ due to the relaxation of the carbomethoxy side-chains and the aliphatic ester group respectively [6]. A T_β transition occurs owing to rotation of the ester group about its bond to the backbone chain. The length of the side group has no effect on the steric hindrance to the rotation. However, when the group is closer to the backbone both the rigidity of the backbone and the hindrance to the rotation can increase [7,8,9]. A γ-relaxation for larger alkyl methacrylates (*e.g.* butyl) at low temps. and a δ-transition for ethyl and propyl methacrylates have been reported [8]

Surface Properties & Solubility:

Solubility Properties General: The nature of the side-chain affects the solubility. Short side-chain polymers tend to be sol. polar solvents but as the length of the side-chain increases the polymers become more sol. non-polar solvents
Surface and Interfacial Properties General: The monolayer behaviour of the polymethacrylates depends on tacticity. Isotactic polymers form expanded monolayers whereas syndiotactic polymers give more condensed layers [12]

Wettability/Surface Energy and Interfacial Tension: For linear alkyl methacrylates the receding contact angle increases from methyl to butyl but decreases from hexyl to dodecyl. The advancing contact angles, however, increase with the length of the alkyl group [13]. There is a linear relationship between the wt % oxygen content of the linear alkyl methacrylates and the advancing contact angle between the polymer and H_2O [14]. The receding contact angles of methyl methacrylate copolymers containing hydrophilic or charged groups decrease relative to the homopolymer whereas the advancing angles are virtually the same. In the case of hydroxyethyl copolymers, however, the receding contact angles are similar to those of the homopolymer and the advancing angles are larger [13]

Mechanical Properties:

Mechanical Properties General: The mechanical props. of the methacrylates tend to improve with increasing MW. However, above a certain value (the critical MW) the rate of change of improvement tends to level off. The critical MW for an amorph. polymer is typically 100000-200000. The impact resistance of methacrylic polymers may be improved by the incorporation of modifiers such as butadiene or other acrylic materials
Viscoelastic Behaviour: The sound velocity of methacrylates shows dispersion when the temp. is greater than T_g [16]

Optical Properties:

Optical Properties General: The refractive index changes with the bulk of the side-chain. Refractive index measurements have been used to monitor extent of reaction in the polymerisation of methyl methacrylate polymers and copolymers [21]
Transmission and Spectra: Ir and nmr spectral data have been reported [3]

Polymer Stability:

Polymer Stability General: Polymethacrylates are more stable to photodegradation than are acrylics. Large doses of ionising radiation can cause brittleness, loss of strength and loss of desirable props. Polymethacrylates exhibit predominantly main-chain fracture during irradiation and deterioration of props. is due to lowering of MW. Irradiation of dodecyl, octadecyl and other long-chain esters causes cross-linking in preference to scission and yields infinite networks. The effect of γ-irradiation on polymethacrylates depends on the alkyl substituent. Side-chain scission is the predominant mechanism. Long-chain alkyl methacrylates undergo both scission and cross-linking. Addition of the benzyl group stabilises the polymer against radiation damage [17]
Decomposition Details: Methacrylates generally decompose readily above $300°$ to give the monomer at a yield of greater than 90%
Flammability: Flame resistance of polymethacrylates can be improved by copolymerisation with bromoethyl methacrylate
Hydrolytic Stability: Methacrylic polymers are more stable to hydrolysis in both acidic and alkaline media than are polyacrylates
Stability Miscellaneous: Electron beam and γ-irradiation cause such rapid polymer degradation that poly(alkyl methacrylates) can be used as positive-working electron beam photoresists. Some silicon-containing methacrylates undergo both scission and cross-linking when irradiated [18]. Poly(methacrylates) undergo shear-induced degradation in soln., particularly in poor solvents. The degradation is minimal for long-chain alkyl methacrylate polymers, however [19]

Applications/Commercial Products:

Processing & Manufacturing Routes: Free-radical polymerisation is the primary commercial method. Also synth. by bulk (main commercial route to sheet, rod, tube and moulding products), soln. (important for polymers to be used in coatings, adhesives and laminates), suspension (moulding powders), emulsion (paint, textile and leather applications) or anionic polymerisation. Initiators such as azo compounds, diacyl peroxides and *tert*-alkyl peroxides are used, with water-sol. persulfates and redox initiators and oxidants used particularly in emulsion polymerisation. Radiation curing is used for coating and thin film systems and oxygen inhibitors are used in these and most other acrylic systems

Applications: Polymethacrylate sheet has applications in glazing (windscreens, diffusers, mirrors, aircraft canopies); in meter cases, furniture, sanitary ware and in protective casing. Rods, tubes and fibres have uses in medical instruments, optics, and in electrical parts. Other uses of soln., and emulsion polymerised products include coatings; as a modifier (oils, impact modifier, to improve flow); in inks and in coating and processing applications. Many methacrylic homo- and copolymers have applications in the imaging industry, *e.g.*, in litho coatings and other photoresists. A major use for methacrylic polymers is the medical industry (manufacture of contact lenses, artificial joints, dental products, bone cements; and in instrument and equipment parts). Long-chain methacrylates are used as performance-improving additives to lubricating oils and hydraulic fuels. Solns. of methacrylic polymers are used in ceramic transfer lacquers; in coatings (for concrete, fibrous cement, household appliances, non-ferrous metals, plastics, rubber, wood); in coil coatings, heat sealing, masonry paints, primers for building structs., corrosion protective coatings, PVC finishing, sterilisable coatings, stoving lacquers. Also used in the production of enstatite-free forsterite for the laser industry

Bibliographic References

[1] Ohtani, H., Ishiguro, S., Tanaka, M. and Tsuge, S., *Polym. J. (Tokyo)*, 1989, **21**, 41
[2] Miller, R.L. and Boyer, R.F., *J. Polym. Sci., Polym. Phys. Ed.*, 1984, **22**, 2021, (morphology)
[3] Wang, F.C., Gerhart, B. and Smith, C.G., *Anal. Chem.*, 1995, **67**, 3681, (pyrolysis)
[4] Siakali-Kioulafa, E., Hadjichristidis, N. and Mays, J.W., *Macromolecules*, 1990, **23**, 3530, (glass transition temp.)
[5] Brostrom, L.R., Coleman, D.L., Gregonis, D.E. and Andrade, J.D., *Makromol. Chem., Rapid Commun.*, 1980, **1**, 341, (thermal anal.)
[6] Billmeyer, F.W., *Textbook of Polymer Science*, Wiley Interscience, 1962
[7] Ribelles, J.L.G. and Calleja, R.D., *J. Polym. Sci., Polym. Phys. Ed.*, 1985, **23**, 1297, (transition temps.)
[8] Plazek, D.J. and Ngai, K.L., *Physical Properties of Polymers Handbook*, (ed. J.E. Mark), AIP Press, 1996, 139, (transition temps.)
[9] Struik, L.C.E., *Polymer*, 1987, **28**, 1869, (transition temps.)
[10] Kurata, M. and Stockmayer, W.H., *Adv. Polym. Sci.*, Springer-Verlag, 1963, **3**, 196, (solubility)
[11] Strain, D.E., Kennelly, R.G. and Dittmar, H.R., *Ind. Eng. Chem.*, 1939, **31**, 382, (solubility)
[12] Brinkhuis, R.H.G. and Schouten, A.J., *Langmuir*, 1992, **8**, 2247
[13] Hogt, A.H., Gregonis, D.E., Andrade, J.D., Kim, S.W. *et al*, *J. Colloid Interface Sci.*, 1985, **106**, 289
[14] Kalachandra, S. and Kusy, R.P., *Polymer*, 1991, **32**, 2428
[15] Glanvill, A.B., *Plastic Engineers' Databook*, Machinery Publishing Co., 1971
[16] Li, B.Y., Jiang, D.Z., Fytas, G. and Wang, C.H., *Macromolecules*, 1986, **19**, 778, (acoustic props.)
[17] Dong, L., Hill, D.J.T., O'Donnell, J.H., Pomery, P.J. and Hatada, K., *J. Appl. Polym. Sci.*, 1996, **59**, 589, (irradiation)
[18] Dawes, K. and Glover, L.C., *Physical Properties of Polymers Handbook*, (ed. J.E. Mark), AIP Press, 1996, 557, (irradiation)
[19] Herold, F.K., Schultz, G.V. and Wolf, B.A., *Polym. Commun.*, 1986, **27**, 59, (degradation)
[20] *Handbook of Industrial Materials*, 2nd edn., Elsevier Advanced Technology, 1992, 393
[21] Bahr, D. and Pinto, J.C., *J. Appl. Polym. Sci.*, 1991, **42**, 2795

Methacrylic acid copolymers   M-9

Synonyms: *Methacrylic acid ionomers. Poly(methacrylic acid-co-styrene). Poly(methacrylic acid-co-N-vinylpyrrolidone). Poly(methacrylic acid-co-urethane)*
Related Polymers: Polymethacrylates-General, Poly(methacrylic acid), Poly(methacrylic acid salts), Polystyrene, Poly(N-vinylpyrrolidone), Polyurethane, Poly(methacrylic acid salt-*co*-ethylene), Poly(methacrylate ionomers)
Monomers: Methacrylic acid, Styrene, *N*-Vinylpyrrolidone
Material class: Thermoplastic
Polymer Type: acrylic
CAS Number:

CAS Reg. No.	Note
88215-93-4	methacrylic acid-*co*-styrene, ammonium salt
37382-81-3	methacrylic acid-*co*-styrene, lead salt
33970-45-5	methacrylic acid-*co*-styrene, Na salt
174204-73-0	methacrylic acid-*co*-styrene, europium salt
174204-74-1	methacrylic acid-*co*-styrene, holmium salt
131151-50-3	methacrylic acid-co-styrene, neodymium salt
174204-72-9	methacrylic acid-*co*-styrene, samarium salt
174204-75-2	methacrylic acid-*co*-styrene, strontium salt

General Information: This entry combines data on the methacrylic acid copolymers other than those with ethylene and methacrylates.
Morphology: Ionomers generally behave according to a multiple-cluster model with ionic groups aggregating to form multiplets which are surrounded by regions of restricted mobility. At very low ion contents only multiplets are present; as the ion content increases the areas of restricted mobility begin to overlap and form contiguous regions of restricted mobility. These areas are called clusters but do not appear when the degree of neutralisation is less than 50% [1,2,3]
Miscellaneous: Forms complexes with basic polymers, e.g. poly(4-vinylpyridine) [11]

Volumetric & Calorimetric Properties:
Glass-Transition Temperature:

No.	Value	Note
1	121°C	methacrylic acid-*co*-styrene
2	132–211°C	methacrylic acid-*co*-styrene, 5.5% Na ion
3	142–218°C	methacrylic acid-*co*-styrene, 8.2% Na ion [1,3]
4	127°C	methacrylic acid-*co*-styrene, 0.04% H_2O [3]
5	129°C	methacrylic acid-*co*-styrene, 0.16% H_2O [3]
6	100°C	methacrylic acid-*co*-N-vinylpyrrolidone, 65% methacrylic acid
7	122°C	methacrylic acid-*co*-N-vinylpyrrolidone, 81.4% methacrylic acid
8	-0.6–45.2°C	methacrylic acid-*co*-urethane, 10% Li salt
9	12.7°C	methacrylic acid-*co*-urethane, 10% methacrylic acid, lithium salt
10	29.8°C	methacrylic acid-*co*-urethane, 30% methacrylic acid, lithium salt [5]

Transition Temperature General: Some ionomer systems (8-14 mol % ions) show two glass transitions which correspond to an ion-poor matrix and an ion-rich cluster. Water plasticises the ionic cores but very small amounts do not affect the cluster glass transition mechanism [1,2,3]

Surface Properties & Solubility:
Solubility Properties General: In aq. media the styrene segments form the core and the methacrylic segments form the shell of each micelle. In aq. dioxan the copolymers form micelles whose props. depend on the ratio of dioxan:H_2O when the amount of H_2O is greater then 5% [5]
Wettability/Surface Energy and Interfacial Tension: γ_t 48.6; γ_p 11.0; γ_d 37.6 (methacrylic acid-*co*-N-vinylpyrrolidone, 100% *N*-vinylpyrrolidone). γ_t 49.4; γ_p 9.9; γ_d 39.5 (methacrylic acid-*co*-N-vinylpyrrolidone, 75% *N*-vinylpyrrolidone). γ_t 47.9; γ_p 11.3; γ_d 36.6 (methacrylic acid-*co*-N-vinylpyrrolidone, 50% *N*-vinylpyrrolidone). γ_t 53.0; l_p 14.5; γ_d 38.5 (methacrylic acid-*co*-N-vinylpyrrolidone, 25% *N*-vinylpyrrolidone) [7]

Transport Properties:
Polymer Solutions Dilute: Intrinsic viscosity 2.06 dl g^{-1} (methacrylic acid-*co*-N-vinylpyrrolidone, 65.8% methacrylic acid); 3.85 dl g^{-1} (methacrylic acid-*co*-N-vinylpyrrolidone, 81.4% methacrylic acid) [4]

Water Content: 58.5% (methacrylic acid-*co*-*N*-vinylpyrrolidone 75% wt. methacrylic acid); 56.5% (methacrylic acid-*co*-*N*-vinylpyrrolidone 25% wt. methacrylic acid); 50.0% (methacrylic acid-*co*-*N*-vinylpyrrolidone 50% wt. methacrylic acid)

Mechanical Properties:
Mechanical Properties General: Rigidity modulus of methacrylic acid [7]: 4.2×10^{-7} mN m^{-2} (75% wt. methacrylic acid), 5.5×10^{-7} mN m^{-2} (25% wt. methacrylic acid), 10×10^{-7} mN m^{-2} (min.) (50% wt. methacrylic acid)
Viscoelastic Behaviour: Williams-Landel-Ferry constants for the methacrylic acid-*co*-styrene diblock copolymer have been reported [8]

Electrical Properties:
Electrical Properties General: Conductivity increases linearly with salt concentration [9]
Dielectric Permittivity (Constant):

No.	Value	Note
1	2.7–3.2	room temp., methacrylic acid-*co*-styrene, 2–9% Na salt [9]

Dissipation (Power) Factor:

No.	Value	Note
1	0.0014–0.0126	high temp., methacrylic acid-*co*-styrene, 2–9% salt content [9]
2	0.0025–0.0051	low temp., methacrylic acid-*co*-styrene, 2–9% salt content [9]

Polymer Stability:
Decomposition Details: 60% weight loss for methacrylic acid-*co*-*N*-vinylpyrrolidone occurs at 452° (81.4% methacrylic acid) and at 457° (65.8% methacrylic acid) [4]

Applications/Commercial Products:
Processing & Manufacturing Routes: Methacrylic acid-*co*-styrene copolymer salts (random and block) are synth. by hydrolysing the corresponding esters to give acids which are then neutralised to produce Na, K, Cu, and Zn salts [10] Methacrylic acid/*N*-vinylpyrrolidone copolymers are synth. by radiation-induced bulk-polymerisation [4]

Bibliographic References
[1] Ma, X., Sauer, J.A. and Hara, M., *Macromolecules*, 1995, **28**, 3953, (morphology, dynamic mechanical props.)
[2] Kim, J.S., Jackman, R.J. and Eisenberg, A., *Macromolecules*, 1994, **27**, 2789
[3] Kim, J.S. and Eisenberg, A., *J. Polym. Sci., Part B: Polym. Phys.*, 1995, **33**, 197, (dynamic mechanical props.)
[4] Ramakrishna, M., Deshpanda, D.D. and Babu, G.N., *J. Polym. Sci., Part A: Polym. Chem.*, 1988, **26**, 445
[5] Wu, C., Zhang, R., Liu, W., Yang, F. and Zhang, G., *Polym. Int.*, 1996, **40**, 13
[6] Qin, A., Tian, M., Ramireddy, C., Webber, S.E. and Munk, P., *Macromolecules*, 1994, **27**, 120
[7] Baker, D.A., Corkhill, P.H., Ng, C.O., Skelly, P.J. and Tighr, B.J., *Polymer*, 1988, **29**, 691
[8] Yoshikawa, K., Desjardins, A., Dealy, J.M. and Eisenberg, A., *Macromolecules*, 1996, **29**, 1235
[9] Hodge, I.M. and Eisenberg, A., *Macromolecules*, 1978, **11**, 283, (dielectric props.)
[10] Leung, L.M. and Ng, C.W., *Abstr. Pap.-Am. Chem. Soc.*, 1987, **194**, 163, (synth.)

Methyl hydrogen rubber, Heat vulcanised M-10
Synonyms: *Methyl hydrogen based silicone elastomer*
Related Polymers: Methylsilicone rubber, Heat vulcanised
Monomers: Dichlorodimethylsilane, Dichloromethylsilane, Hexamethylcyclotrisiloxane, Cyclic methylhydrogensiloxane
Material class: Synthetic Elastomers, Copolymers
Polymer Type: silicones, dimethylsilicones

Molecular Formula: $[(C_2H_6OSi).(CH_4OSi)]_n$
Fragments: C_2H_6OSi CH_4OSi
Additives: Finely divided silica incorporated as a reinforcer to improve mech. props.

Volumetric & Calorimetric Properties:
Density:

No.	Value	Note
1	d 1.1 g/cm^3	[3]

Specific Heat Capacity:

No.	Value	Note	Type
1	1.3 kJ/kg.C	0.3 Btu lb^{-1}°F^{-1} [3]	P

Glass-Transition Temperature:

No.	Value	Note
1	-130°C	[2,10]

Surface Properties & Solubility:
Cohesive Energy Density Solubility Parameters: δ 15.5 J$^{1/2}$ cm$^{-3/2}$ (7.5–7.6 cal$^{1/2}$ cm$^{-3/2}$) [6,7]. For more details see [methyl silicone rubber (heat vulcanised)]
Solvents/Non-solvents: Sol. aliphatic hydrocarbons, chlorinated hydrocarbons, aromatic hydrocarbons. Mod. sol. ketones, esters, ethers. Insol. alcohols, H_2O [6,7,8]
Surface and Interfacial Properties General: Strong adhesion to surfaces, particularly metals [1,5]

Transport Properties:
Polymer Melts: Viscosity 100000 St (approx., pre-cure gum) [8]

Mechanical Properties:
Mechanical Properties General: Requires silica filler for acceptable mech. props. [1]

Electrical Properties:
Dielectric Permittivity (Constant):

No.	Value	Note
1	3	[4,9]

Dielectric Strength:

No.	Value	Note
1	15–20 kV.mm^{-1}	[4]

Polymer Stability:
Polymer Stability General: See [methylsilicone rubber (heat vulcanised)]. The presence of Si-H bonds reduces stability, particularly in the presence of water

Applications/Commercial Products:
Processing & Manufacturing Routes: Dimethyl- and methyl hydrogen-cyclic polysiloxanes are equilibrated together under acid catalysis at 100–110° with a large excess of dimethylsiloxanes and trace amounts of hexamethyldisiloxane to terminate the siloxane chains. Alkaline conditions or temps. much over 110° must be avoided due to the reactivity of the Si-K bond. Organic peroxides, typically benzoyl peroxide, are used at 1–2 parts per 100 rubber to cure the gum. The rubber is cured at 110° under pressure for less than an hour
Applications: Used where tight adhesion to metals is important

Bibliographic References
[1] *Belg. Pat.*, 1963, 632 587
[2] Clarson, S.J. and Semlyen, J.A., *Siloxane Polymers*, Prentice Hall, 1993

[3] Ash, M. and Ash, I., *Handbook of Plastic Compounds, Elastomers and Resins*, (eds. M.B. Ash and I.A. Ash), Wiley-VCH, 1992
[4] Stark, F.O., Fallender, J.R. and Wright, A.P., *Comprehensive Organometallic Chemistry*, (eds. G. Wilkinson, F.G.A. Stone and E.W. Abel), Pergamon Press, 1982, **2**, 305
[5] Noll, W., *Chemistry and Technology of Silicones*, Academic Press, 1968
[6] Yerrick, K.B. and Beck, H.N., *Rubber Chem. Technol.*, 1964, **37**, 261, (solubility)
[7] Barney, R.H., Voigt, C.E. and Mentele, J.W., Struct.-Solubility Relat. Polym., *(Proc. Symp.)*, (eds. F.W. Harris and R.B. Seymour), Academic Press, 1977, 225
[8] Roff, W.J. and Scott, J.R., Fibres, Films, *Plastics and Rubbers*, Butterworths, 1971
[9] Vincent, G.A., Fearon, F.W.G. and Orbech, T., Annu. Rep., *Conf. Electr. Insul. Dielectr. Phenom.*, 1972, 17, (dielectric constant)
[10] Lee, C.L. and Haberland, G.G, *J. Polym. Sci., Part B: Polym. Lett.*, 1965, **3**, 883, (transition temps)

Methyl phenyl silicone resin M-11

$$\left[-O-\underset{\underset{R}{|}}{\overset{\overset{Me}{|}}{Si}}- \right]_n$$

R = Me, Ph

Synonyms: *Methyl phenyl siloxane resin*
Related Polymers: Methyl silicone resin, Glass reinforced silicone resin, Room temp. vulcanised flexible silicone resin
Monomers: Trichloromethylsilane, Trichlorophenylsilane, Dichlorodimethylsilane, Dichloromethylphenylsilane
Material class: Gums and resins, Copolymers
Polymer Type: silicones, dimethylsilicones
CAS Number:

CAS Reg. No.
159968-66-8

Molecular Formula: $[(C_2H_6OSi)(C_7H_8OSi)]_n$
Fragments: C_2H_6OSi C_7H_8OSi
General Information: The functionality of a silicone resin is the ratio of oxygen to silicon atoms. For commercial resins it varies from 1.2 to 1.6. High functionality resins are highly crosslinked

Volumetric & Calorimetric Properties:
Density:

No.	Value	Note
1	d 1.1 g/cm^3	[6]

Thermal Conductivity:

No.	Value	Note
1	0.15 W/mK	0.00035 cal (cm s °C)$^{-1}$ [6]

Specific Heat Capacity:

No.	Value	Note	Type
1	1.5 kJ/kg.C	0.36–0.37 cal^{-1} g^{-1}°C^{-1} [6]	P

Brittleness Temperature:

No.	Value	Note
1	-70°C	low functionality [6]

Surface Properties & Solubility:
Solvents/Non-solvents: Sol. chlorinated hydrocarbons, aromatic hydrocarbons, ketones, esters, ethers. Mod. sol. alcohols, aliphatic hydrocarbons. Insol. H_2O [1]. (Solubility data refers to pre-cure resin)
Surface and Interfacial Properties General: Adhesive power 4 MPa (40 kg cm^{-2}, 18°). Adhesive power, 0.1 MPa (1 kg cm^{-2}, 150°) [1]

Transport Properties:
Polymer Solutions Concentrated: Viscosity values have been reported 0.5–2.5 Poise (pre-cure resin, toluene) [10]

Mechanical Properties:
Mechanical Properties General: Below 30° resin is much stiffer. High functionality resins are hard and stiff over whole temp. range. Methyl phenyl resins are softer and more flexible than methyl resins of the same functionality

Tensile (Young's) Modulus:

No.	Value	Note
1	3.5 MPa	0.35 kgf mm^{-2}, low functionality resin [6]

Tensile Strength Break:

No.	Value	Note	Elongation
1	6 MPa	0.6 kgf mm^{-2}, low functionality resin [6]	125%

Hardness: Shore A35–A40 (low functionality resin; hardness decreases as temp. increases) [1,6]

Electrical Properties:
Dielectric Permittivity (Constant):

No.	Value	Note
1	2.9	[2,3]

Dielectric Strength:

No.	Value	Note
1	30 kV.mm^{-1}	800 V mil^{-1}, 0.13 mm thick [3,4,5]

Dissipation (Power) Factor:

No.	Value	Note
1	0.005	[2,3]

Polymer Stability:
Thermal Stability General: Usable from -100–250° [1,3,6]
Upper Use Temperature:

No.	Value
1	250°C

Decomposition Details: At high temp. decomposition to C_6H_6 and cyclic siloxanes occurs [9]
Environmental Stress: Outstanding weather resistance. Unaffected by sunlight or water; excellent radiation resistance [1,8]
Chemical Stability: Attacked by strong acids; unaffected by alkali or weak acid. Degraded by chlorinated hydrocarbons, aromatic hydrocarbons, ketones and esters. Mod. effect of alcohols and aliphatic hydrocarbons [1,8]
Hydrolytic Stability: Fully stable to cold water; poor resistance to steam [1]
Biological Stability: Non-biodegradable [6]
Recyclability: Not recyclable

– **Methyl silicone resin** M-12 – M-12

Applications/Commercial Products:

Processing & Manufacturing Routes: Dichlorodimethylsilane, trichloromethylsilane [6,7], trichlorophenylsilane and dichloromethylphenylsilane are dissolved in toluene and hydrolysed to form a branched copolymer which remains dissolved. Further polym. occurs on heating with zinc octanoate at 120–200°. After application of the resin, cure occurs with a catalyst such as zinc octanoate usually at elevated temp.

Applications: Silicone resins can be copolymerised with other polymers, particularly polyesters, to prod. resins with improved hardness, adhesive props. and solvent resistance, but which maintain the heat resistance and water repellency of silicones. Used in heat resistant coatings, its miscibility with paint makes it important in the paint industry. Mixed with glass fibre it can be used in moulding compds. or as a laminate with thin glass sheets. It can be modified and cured with acyloxy silanes to form room temp. vulcanising flexible resins. Has important use as an insulator and dielectric in electric industry

Trade name	Details	Supplier
Amicon SC	RTV resins	Emerson & Cuming
ECC		General Electric Silicones
Eccosil	RTV resins	Emerson & Cuming
Elastosil	various grades not exclusive	Wacker Silicones
Norcast A-4000		RH Carlson
SC	filled resins	Bacon
Sylgard	RTV resins	Dow Corning STI
Thermoset	RTV resins	Thermoset

Bibliographic References

[1] Noll, W., *Chemistry and Technology of Silicones*, Academic Press, 1968
[2] Dakins, T.W. and Works, C.N., *J. Appl. Phys.*, 1947, **18**, 786, (dielectric constant)
[3] Rochow, E.G., An Introduction to the Chemistry of the Silicones, 2nd edn., John Wiley and Sons, 1951
[4] Freeman, G.G., *Silicones*, The Plastics Institute, 1962
[5] U.S. Pat., 1941, 2 258 222
[6] Roff, W.J. and Scott, J.R., Fibres, Films, *Plastics and Rubbers*, Butterworths, 1971
[7] Hardman, B. and Torkelson, A., *Encycl. Polym. Sci. Eng.*, Vol. 15, 2nd edn., (ed. J.I. Kroschwitz), John Wiley and Sons, 1985, 204
[8] McGregor, R.R., *Silicones and their Uses*, McGraw-Hill, 1954
[9] Doyle, C.D., *J. Polym. Sci.*, 1958, **31**, 95, (thermal stability)
[10] Meals, R.N. and Lewis, F.M., *Silicones*, Reinhold, 1959

Methyl silicone resin   M-12

Synonyms: *Methyl siloxane resin*
Related Polymers: Methyl phenyl silicone resin, Glass reinforced silicone resin, Flexible silicone resin, Room temp. vulcanised
Monomers: Dichlorodimethylsilane, Trichloromethylsilane
Material class: Gums and resins, Copolymers
Polymer Type: silicones, dimethylsilicones
CAS Number:

CAS Reg. No.
159011-85-5

Molecular Formula: $(C_2H_6OSi)_x$
Fragments: C_2H_6OSi
General Information: The functionality of a silicone resin is the ratio of oxygen to silicon atoms. For commercial resins it varies from 1.2 to 1.6. High functionality resins are highly crosslinked. Silicone resins can be copolymerised with other polymers, particularly polyesters, to prod. resins with improved hardness, adhesive props. and solvent resistance but which maintain the heat resistance and water repellency of silicones

Volumetric & Calorimetric Properties:

Density:

No.	Value	Note
1	d 1.06 g/cm^3	low functionality
2	d 1.2 g/cm^3	high functionality [1,2,3]

Thermal Expansion Coefficient:

No.	Value	Note	Type
1	0.00011 K^{-1}	[4,10]	L

Thermal Conductivity:

No.	Value	Note
1	0.15 W/mK	0.00035 cal cm^{-1} S^{-1} °C^{-1} [4]

Specific Heat Capacity:

No.	Value	Note	Type
1	1.5 kJ/kg.C	0.36-0.37 cal^{-1} g$^{°-1}$ [4]	P

Brittleness Temperature:

No.	Value	Note
1	-70°C	low functionality [4]

Surface Properties & Solubility:

Solvents/Non-solvents: Sol. chlorinated hydrocarbons, aromatic hydrocarbons, ketones, esters. Mod. sol. ethers, alcohols. Insol. H_2O, aliphatic hydrocarbons [6] (solubility data refers to pre-cure resin)

Transport Properties:

Polymer Solutions Concentrated: Viscosity 0.5–2.5P (toluene, pre-cure gum) [9]

Water Absorption:

No.	Value	Note
1	0.2 %	24h [4]

Mechanical Properties:

Mechanical Properties General: High functionality resins are hard and stiff. Resins become much softer and elastic as functionality falls. Methyl resins are harder and stiffer than methyl phenyl resins of the same functionality

Tensile (Young's) Modulus:

No.	Value	Note
1	400 MPa	40 kgf.mm^{-2}, high functionality [4]

Tensile Strength Break:

No.	Value	Note	Elongation
1	30 MPa	3 kgf.mm^{-2}, high functionality [4]	7.5%

Flexural Strength at Break:

No.	Value	Note
1	600 MPa	60 kgf.mm^{-2}, high functionality [4]

– Methyl silicone rubber, One part room temp. vulcanised

Compressive Strength:

No.	Value	Note
1	800 MPa	80 kgf.mm^{-2}, high functionality [4]

Impact Strength: 7.5 J/25.4 mm (5 ftlb.in^{-1}, high functionality [4]

Electrical Properties:
Electrical Properties General: Very good insulator
Dielectric Permittivity (Constant):

No.	Value	Note
1	2.9	[5,7]

Dielectric Strength:

No.	Value	Note
1	60 kV.mm^{-1}	1500 Vmil^{-1}, 0.25 mm thickness [5,7]

Dissipation Power Factor: tan δ 0.002 [5,7]

Optical Properties:
Refractive Index:

No.	Value	Note
1	1.418	low functionality
2	1.425	high functionality [1,2,3]

Polymer Stability:
Polymer Stability General: Stable at high temps.
Thermal Stability General: Usable from -100–250° [6,10]
Upper Use Temperature:

No.	Value
1	250°C

Flammability: Inherently self-extinguishing
Environmental Stress: Outstanding weather resistance. Unaffected by sunlight or H$_2$O. Excellent radiation resistance [6,10]
Chemical Stability: Attacked by strong acids. Unaffected by alkali or weak acid. Degraded by chlorinated hydrocarbons, aromatic hydrocarbons, ketones and esters. Alcohols and aliphatic hydrocarbons have little effect [6,10]
Hydrolytic Stability: Fully stable to cold H$_2$O; poor resistance to steam [6]
Biological Stability: Non-biodegradable [9]
Recyclability: Not recyclable

Applications/Commercial Products:
Processing & Manufacturing Routes: Dichlorodimethylsilane and trichloromethylsilane in toluene are hydrolysed to form a branched copolymer which remains dissolved. Further polym. occurs on heating with zinc octoate at 150–200°. After application of the resin, cure occurs, with a catalyst such as zinc octoate, usually at elevated temp. [4,8]
Applications: Used for heat resistant enamels, varnishes and coatings. Mixed with glass fibre it can be used in moulding compds. or as a laminate with thin glass sheets. It can be modified and cured with acyloxysilanes to form room temp. vulcanising flexible resins. Important use as insulator and dielectric in electric industry

Trade name	Details	Supplier
Amicon SC	RTV resins	Emerson & Cuming
Eccosil	RTV resins	Emerson & Cuming
Elastoril	various grades not exclusive	Wacker Silicones
SC	filled resins	Bacon
Sylgard	RTV resins	Dow Corning STI

Bibliographic References
[1] Roehow, E.G. and Gilliam, W.G., *J. Am. Chem. Soc.*, 1941, **63**, 798
[2] Roehow, E.G., An Introduction to the Chemistry of the Silicones, 2nd edn., John Wiley and Sons, 1951
[3] Hardy, D.V.N. and Megson, N.J.L, *Q. Rev., Chem. Soc.*, 1948, **2**, 25
[4] Roff, W.J. and Scott, J.R., Fibres, Films, *Plastics and Rubbers*, Butterworths, 1971
[5] Freeman, G., *Silicones*, The Plastics Institute, 1962
[6] Noll, W., *Chemistry and Technology of Silicones*, Academic Press, 1968
[7] Johnson, P., *Silicones*, (ed. S. Fordham), 1960, 213
[8] Hardman, B. and Torkelson, A., *Encycl. Polym. Sci. Eng.*, 2nd edn., (ed. J.I. Kroschwitz), John Wiley and Sons, 1985, **15**, 204
[9] Meals, R.N. and Lewis, F.M., Silicones Reinhold, 1959
[10] McGregor, R.R., Silicones and their Uses McGraw-Hill, 1954

Methyl silicone rubber, One part room temp. vulcanised
M-13

Synonyms: *Silicone elastomer, One part room temp. vulcanised. One part room temp. vulcanised methyl silicone rubber*
Related Polymers: Methylsilicone rubber, Room temp. vulcanised, Methylsilicone rubber, Two part room temp. vulcanised, Dimethiconol
Monomers: Dichlorodimethylsilane, Hexamethylcyclotrisiloxane
Material class: Synthetic Elastomers
Polymer Type: dimethylsilicones
CAS Number:

CAS Reg. No.	Note
63394-02-5	elastomer
31900-57-9	PDMS
31692-79-2	dimethiconol

Molecular Formula: [C$_2$H$_6$OSi]
Fragments: C$_2$H$_6$OSi
Additives: Finely divided silica is usually incorporated as a reinforcer to improve mech. props.
General Information: Silanol groups are present in pre-cure gum. During cure they form cross links through condensation. Fully cured elastomer has few free silanol groups

Volumetric & Calorimetric Properties:
Density:

No.	Value	Note
1	d 1.18–1.48 g/cm^3	filler dependent [3]

Thermal Expansion Coefficient:

No.	Value	Note	Type
1	0.00035 K^{-1}	[5]	L
2	0.0008 K^{-1}	[2]	V

Thermal Conductivity:

No.	Value	Note
1	0.3 W/mK	0.0007 cal (cm s °C)$^{-1}$ [2]

Specific Heat Capacity:

No.	Value	Note	Type
1	1.3 kJ/kg.C	0.3 Btu (lb °F)$^{-1}$ [4]	P

– **Methylcellulose**

Surface Properties & Solubility:
Cohesive Energy Density Solubility Parameters: δ 15.5 J$^{1/2}$ cm$^{-3/2}$ (7.5–7.6 cal$^{1/2}$ cm$^{-3/2}$) [6,7]
Solvents/Non-solvents: Sol. aliphatic hydrocarbons, chlorinated hydrocarbons, aromatic hydrocarbons. Mod. sol. ketones, esters, ethers. Insol. alcohols, H$_2$O [6,7,8]. Vulcanised rubber swollen by solvents rather than dissolved
Surface and Interfacial Properties General: Good adhesion to many surfaces. Can be used as adhesive peel; adhesion can be as high as 2.5 J cm^{-2} (140 ppi) [1,5,9]

Mechanical Properties:
Mechanical Properties General: Mech. props. are inferior to those of heat vulcanised methyl silicone rubber see methyl silicone rubber, heat vulcanised
Tensile Strength Break:

No.	Value	Note	Elongation
1	2.4 MPa	350 psi, 25° [1]	400

Hardness: Shore A33 [1]
Fracture Mechanical Properties: Tear strength 0.9 J cm^{-2} (50 ppi, 25°) [1]

Electrical Properties:
Electrical Properties General: See methyl silicone rubber, room temp. vulcanised

Optical Properties:
Optical Properties General: See methyl silicone rubber, room temp. vulcanised

Applications/Commercial Products:
Processing & Manufacturing Routes: Dimethiconol gum is cured at room temp. with alkoxysilane cross-linking agent and metal salt catalyst. It is supplied mixed with cross-linking agent and catalyst and undergoes cure only when exposed to atmospheric moisture
Mould Shrinkage (%):

No.	Value
1	0.6%

Applications: Used as sealant, encapsulant, coating and adhesive valued for the ease of vulcanisation. It is important in the construction and electronics industries

Trade name	Details	Supplier
Eccosil 1776		Emerson & Cuming
RTV	100 series	General Electric Silicones
RTV 6424		General Electric Silicones
Sylgard		Dow Corning STI

Bibliographic References
[1] Hamilton, S.B., *Silicone Technology*, (ed. P.F. Bruins), John Wiley and Sons, 1970, 17
[2] Hainline, A.N., *Silicone Technology*, (ed. P.F. Bruins), John Wiley and Sons, 1970, 51
[3] Meals, R.N., *Ann. N. Y. Acad. Sci.*, 1965, **125**, 137
[4] Ash, M. and Ash, I., Handbook of Plastic Compounds, *Elastomers and Resins*, (eds. M.B. Ash and I.A. Ash), Wiley-VCH, 1992
[5] Hardman, B. and Torkelson, A., *Encycl. Polym. Sci. Eng.*, 2nd edn., (ed. J.I. Kroschwitz), John Wiley and Sons, 1985, **15**, 204
[6] Yerrick, K.B. and Beck, H.N., *Rubber Chem. Technol.*, 1964, **37**, 261, (solubility)
[7] Baney, R.H., Voigt, C.E. and Mentele, J.W., Struct.-Solubility Relat. Polym., *(Proc. Symp.)*, (eds. F.W. Harris and R.B. Seymour), Academic Press, 1977, 225
[8] Roff, W.J. and Scott, J.R., Fibres, Films, *Plastics and Rubbers*, Butterworths, 1971
[9] Noll, W., *Chemistry and Technology of Silicones*, Academic Press, 1968

Methylcellulose

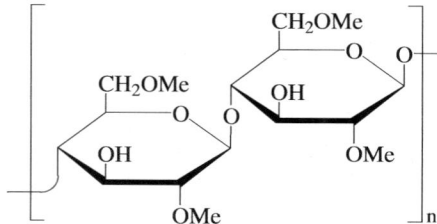

Synonyms: *Permethylcellulose*
Related Polymers: Cellulose II
Monomers: Base monomer unit glucose
Material class: Polysaccharides
Polymer Type: cellulosics
CAS Number:

CAS Reg. No.
9004-67-5

Volumetric & Calorimetric Properties:
Density:

No.	Value	Note
1	d 1.39 g/cm^3	film [1,2]

Melting Temperature:

No.	Value	Note
1	290–305°C	film [1,2]

Glass-Transition Temperature:

No.	Value	Note
1	150°C	[7]

Transition Temperature:

No.	Value	Note	Type
1	48°C	soln. [1,2]	Gelation temp.

Surface Properties & Solubility:
Solubility Properties General: Solubility - cold 4–8% NaOH: 0.1–0.4 [4]; hot 4–8% NaOH: 0.4–0.6 [4]; cold water: 1.3–2.6 [4]; organic solvents: 2.5–3 [4]
Surface and Interfacial Properties General: Interfacial tension 19–23 mN m^{-1} (25°, 1% soln.) [1,2]
Surface Tension:

No.	Value	Note
1	47–53 mN/m	(25°, 1% soln.) [1,2]

Transport Properties:
Polymer Solutions Dilute: Viscosity 10000–15000 MPa s (soln.) [6]
Permeability of Gases: Transmission rate (film) [H$_2$O 540 nmol m^{-2} s^{-1} (37°, 90–100% relative humidity) [1,2]; [O$_2$] 200 nmol m^{-2} s^{-1} (24°) [1,2]
Water Absorption:

No.	Value	Note
1	2–6.5 %	24h, ASTM D570 [1,2]

Mechanical Properties:

Tensile (Young's) Modulus:

No.	Value	Note
1	3300 MPa	3.3×10^{10} dyn cm^{-2} [3]

Tensile Strength Break:

No.	Value	Note	Elongation
1	58.7–78.6 MPa	24°, 50% relative humidity, film [1,2]	10–15%

Optical Properties:

Refractive Index:

No.	Value	Note
1	1.49	film [1,2]

Transmission and Spectra: 55% transmission (400 nm), 49% transmission (290 nm), 26% transmission (210 nm) [1,2]

Polymer Stability:

Flammability: Charring temp. 290–305° (film) [1,2]
Environmental Stress: The film is stable to uv after 500h exposure [1,2]
Chemical Stability: Excellent resistance to oils and solvents [1]

Applications/Commercial Products:

Processing & Manufacturing Routes: Alkali cellulose is alkylated with methyl chloride
Applications: Thickener for cements, foods, cosmetics, and coatings. Binder and emulsifier for pharmaceutical preps., adhesives and binder for ceramics

Trade name	Details	Supplier
Culminal MC		Shim Etsu
Methocel	registered trademark	Dow
Tylose M		Hoechst Celanese
Walocel M		Wolff

Bibliographic References

[1] *Encycl. Polym. Sci. Eng.*, 2nd edn., (ed. J.I. Kroschwitz), John Wiley and Sons, 1985
[2] *Methocel®*, Dow Chemical Co., Midland, Michigan, U.S.A., 1978, (technical datasheet)
[3] Richards, G.N., *J. Appl. Polym. Sci.*, 1961, **5**, 545
[4] *Concise Encyclopedia of Wood and Wood-based Materials*, (ed. A.P. Schniewind), Pergamon Press, 1989
[5] *Kirk-Othmer Encycl. Chem. Technol.*, 3rd edn., (ed. M. Grayson), Wiley Interscience, 1979, **5**
[6] *Kirk-Othmer Encycl. Chem. Technol.*, 4th edn., (ed. J.I. Kroschwitz), Wiley Interscience, 1993, **5**
[7] Gröbe, A., *Polym. Handb.*, 3rd edn., (eds. J. Brandrup and E.H. Immergut), Wiley Interscience, 1989, V155, (glass transition temps)

Methylphenylsilicone rubber, Heat vulcanised M-15

Synonyms: *Phenyl methyl silicone rubber, Heat vulcanised. Methyl phenyl based silicone elastomer*
Related Polymers: Methylsilicone rubber, Heat vulcanised, Vinyl methyl phenyl silicone rubber, Heat vulcanised
Monomers: Dichlorodimethylsilane, Dichloromethylphenylsilane, Hexamethylcyclotrisiloxane
Material class: Synthetic Elastomers, Copolymers
Polymer Type: silicones, dimethylsilicones

CAS Number:

CAS Reg. No.
156048-35-0

Molecular Formula: $[(C_2H_6OSi).(C_7H_8OSi)]_n$
Fragments: C_2H_6OSi C_7H_8OSi
Additives: Finely divided silica is incorporated as a reinforcer to improve mech. props.

Volumetric & Calorimetric Properties:

Density:

No.	Value	Note
1	d 1.15 g/cm^3	[5]

Thermal Expansion Coefficient:

No.	Value	Note	Type
1	0.0009 K^{-1}	90% dimethyl siloxane, 10% methyl phenyl siloxane [1]	V
2	0.00075 K^{-1}	60% dimethyl siloxane, 40% methyl phenyl siloxane	V

Thermal Conductivity:

No.	Value	Note
1	0.25–0.3 W/mK	0.0006–0.0007 cal (cm s °C)$^{-1}$ [12,13]

Specific Heat Capacity:

No.	Value	Note	Type
1	1.3 kJ/kg.C	0.3 Btu lb^{-1}°F^{-1} [5]	P

Melting Temperature:

No.	Value	Note
1	-77°C	60% dimethyl siloxane, 40% methyl phenyl siloxane [1,10]

Glass-Transition Temperature:

No.	Value	Note
1	-112°C	90% dimethyl siloxane, 10% methyl phenyl siloxane [1]
2	-96°C	60% dimethyl siloxane, 40% methyl phenyl siloxane [1]
3	-84°C	60% dimethyl siloxane, 40% methyl phenyl siloxane [10]

Transition Temperature General: The addition of small amounts of methyl phenyl siloxane as a copolymer with dimethyl siloxane sharply lowers the stiffening temp. and improves low temp. elastic props. This involves loss of a clearly defined T_m. Increases in methyl phenyl siloxane content cause a deterioration of low temp. props. either by restoration of T_m or sharp rise in T_g [1,10]. For transitions of high phenyl content silicone rubber see phenyl silicone rubber (HV)

Transition Temperature:

No.	Value	Note	Type
1	-109°C	90% dimethyl siloxane, 10% methyl phenyl siloxane [1]	Stiffening

– **Methylsilicone rubber, Heat vulcanised** M-16 – M-16

| 2 | -70°C | 60% dimethyl siloxane, 40% methyl phenyl siloxane [1] | Stiffening |

Surface Properties & Solubility:
Cohesive Energy Density Solubility Parameters: δ 16–17 $J^{1/2}$ $cm^{-3/2}$ (7.7–8.2 $cal^{1/2}$ $cm^{-3/2}$) [7]. Value increases with amount of phenyl substitution
Solvents/Non-solvents: Sol. aliphatic hydrocarbons, chlorinated hydrocarbons, aromatic hydrocarbons, ketones, esters, ethers. Insol. alcohols, H_2O
Surface and Interfacial Properties General: Exhibits generally low surface adhesion [4]

Transport Properties:
Transport Properties General: Gas permeability falls sharply as amount of phenyl substitution rises [2]
Polymer Melts: Kinematic viscosity 100000 St (approx. pre-cure gum) [9]
Permeability of Gases: [O_2] 0.034 (25°, 5% phenyl); [N_2] 0.15 (25°, 5% phenyl); [O_2] 0.0095 (25°, 20% phenyl); [CO_2] 0.055 cm^2 s^{-1} GPa^{-1} (25°, 20% phenyl) [2]

Mechanical Properties:
Mechanical Properties General: Mech. props. deteriorate at high temp. Silicone rubber [4] containing phenyl for use at high temps. must incorporate some methyl vinyl groups see vinyl methyl phenyl silicone rubber (HV)
Tensile Strength Break:

No.	Value	Note	Elongation
1	6.9 MPa	1000 psi, 25°, filled [3,5]	500%

Viscoelastic Behaviour: Compression set at 150° and 22h may be as high as 50% for filled high phenyl rubber [6]
Hardness: Shore A50 [5]
Fracture Mechanical Properties: Tear strength 2 J cm^{-2} (110 psi, 25°, filled) [5]

Electrical Properties:
Electrical Properties General: Exhibits excellent insulating props. but is inferior to dimethyl silicone rubber
Dielectric Permittivity (Constant):

No.	Value	Note
1	3	approx. [8,11]

Dielectric Strength:

No.	Value	Note
1	20 $kV.mm^{-1}$	500 V mil^{-1} [5]

Optical Properties:
Transmission and Spectra: No transmission below 290 nm [4,6]

Polymer Stability:
Polymer Stability General: Retains elastic props. down to low temp. [6]
Thermal Stability General: Usable from -100° to 260°
Upper Use Temperature:

No.	Value
1	260°C

Decomposition Details: Oxidation temp. in dry air 300–350° (rises with rise in phenyl substitution) [6]
Environmental Stress: Resistant to oxidative degradation, water, weathering, ozone and corona. Uv radiation has no effect. Very resistant to high energy radiation (radioactivity) [4]

Chemical Stability: Vulcanised rubber is swollen rather than dissolved e.g. swollen by toluene, aliphatic hydrocarbons and CCl_4. Disintegrates in conc. sulfuric acid. Relatively unaffected by weak alkali, weak acids and alcohols [9]
Hydrolytic Stability: Stable in H_2O and aq. solns. at room temp. Damaged by high pressure steam
Biological Stability: Non-biodegradable. If properly vulcanised it is not subject to biological attack
Recyclability: Can be recycled by depolymerisation using steam or acid catalysis [4]
Stability Miscellaneous: Stability to solvents reduced by phenyl incorporation [4]

Applications/Commercial Products:
Processing & Manufacturing Routes: Dimethyl- and methylphenyl-cyclic polysiloxanes are polymerised together at 150–200° with an alkaline catalyst and traces of hexamethyldisiloxane. Organic peroxide, typically benzoyl peroxide, is used at 1–2 parts per 100 rubber to cure the gum to an elastomer. The rubber is cured at 110–200° under pressure, typically for less than an hour. It is often post cured for several hours at temps. of 200–250°
Applications: Used for low temp. applications. Used for coatings, pads and belts

Trade name	Details	Supplier
COHR lastic	several grades	CHR Industries

Bibliographic References
[1] Polmanteer, K.E. and Hunter, M.J., *J. Appl. Polym. Sci.*, 1959, **1**, 3, (transition temps)
[2] Robb, W.L., *Ann. N. Y. Acad. Sci.*, 1968, **146**, 119, (gas permeability)
[3] Meals, R.N. and Lewis, F.M., Silicones Reinhold, 1959
[4] Noll, W., *Chemistry and Technology of Silicones*, Academic Press, 1968
[5] Ash, M. and Ash, I., Handbook of Plastic Compounds, *Elastomers and Resins*, (eds. M.B. Ash and I.A. Ash), Wiley-VCH, 1992
[6] Hardman, B. and Torkelson, A., *Encycl. Polym. Sci. Eng.*, Vol. 15, 2nd edn., (ed. J.I. Kroschwitz), John Wiley and Sons, 1985, 20
[7] Yerrick, K.B. and Beck, H.N., *Rubber Chem. Technol.*, 1964, **37**, 261, (solubility)
[8] Stark, F.O., Fallender, J.R. and Wright, A.P., *Comprehensive Organometallic Chemistry*, (eds. G. Wilkinson, F.G.A. Stone and E.W. Abel), Pergamon Press, 1982, **2**, 305
[9] Roff, W.J. and Scott, J.R., Fibres, Films, *Plastics and Rubbers*, Butterworths, 1971
[10] Borisov, M.F., *Soviet Rubber Technology*, 1966, **25**, 5, (transition temps)
[11] Vincent, G.A., Feuron, F.W.G. and Orbech, T., Annu. Rep., *Conf. Electr. Insul. Dielectr. Phenom.*, 1972, 17, (dielectric constant)
[12] Ames, J., *J. Sci. Instrum.*, 1958, **35**, 1
[13] Jamieson, O.T. and Irving, J.B., *Proc. Int. Conf. Therm. Conduct.*, 1975, 279, (thermal conductivity)

Methylsilicone rubber, Heat vulcanised M-16
Synonyms: *Silicone elastomer. Polydimethylsiloxane rubber. Poly[oxy(dimethylsilylene)]*
Related Polymers: Polydimethylsiloxane gum, Silicone rubber, Glass fibre filled, Methylphenylsilicone rubber, Heat vulcanised, Vinyl methylsilicone rubber, Heat vulcanised, Vinyl methyl phenyl silicone rubber, Heat vulcanised, Phenylsilicone rubber, Heat vulcanised, Fluoromethyl silicone rubber, Methyl hydrogen rubber, Heat vulcanised, Silicone rubber, Conductive
Monomers: Dichlorodimethylsilane, Hexamethylcyclotrisiloxane
Material class: Synthetic Elastomers
Polymer Type: dimethylsilicones
CAS Number:

CAS Reg. No.	Note
9016-00-6	
63394-02-5	elastomer

Molecular Formula: $[C_2H_6OSi]_n$
Fragments: C_2H_6OSi

– Methylsilicone rubber, Heat vulcanised

Additives: Usually filled with finely divided silica incorporated as a reinforcer to improve mech. props.

Volumetric & Calorimetric Properties:
Density:

No.	Value	Note
1	d 1.1 g/cm^3	[17]

Thermal Expansion Coefficient:

No.	Value	Note	Type
1	0.0003 K^{-1}	[8]	L
2	0.00095 K^{-1}	[14]	V

Thermal Conductivity:

No.	Value	Note
1	0.3 W/mK	0.0007 cal (cm s °C)$^{-1}$ [8]

Specific Heat Capacity:

No.	Value	Note	Type
1	1.3 kJ/kg.C	0.3 Btu (lb °F)$^{-1}$ [17]	P

Melting Temperature:

No.	Value	Note
1	-54°C	[14]

Glass-Transition Temperature:

No.	Value	Note
1	-123°C	[14,21]

Brittleness Temperature:

No.	Value	Note
1	-65°C	-85°F [6]

Transition Temperature:

No.	Value	Note
1	-38°C	Stiffening temp. related to T_m rather than T_g [14]

Surface Properties & Solubility:
Cohesive Energy Density Solubility Parameters: δ 15.5 J$^{1/2}$ cm$^{3/2}$ (7.5–7.6 cal$^{1/2}$ cm$^{-3/2}$; δ_d 15.9; δ_p 0.0; δ_h 4.1 J$^{1/2}$ cm$^{3/2}$ [2,10,12,15]
Solvents/Non-solvents: Sol. chlorinated hydrocarbons, aromatic hydrocarbons, aliphatic hydrocarbons. Mod. sol. ketones, esters, ethers. Insol. alcohols, H$_2$O [2,10,11,12]. Vulcanised rubber swollen by solvents rather than dissolved [12]
Surface and Interfacial Properties General: Generally low surface adhesion [3]

Transport Properties:
Transport Properties General: Silicone rubber is ten times more permeable to gases than other rubbers. The solubility of gases in polysiloxanes is similar to that for other polymers but the diffusion coefficients are much higher [13,16,22]
Polymer Melts: Viscosity 100000 St (approx., pre-cure gum) [2]
Permeability of Gases: Other gas permeability values have been reported [13]

Water Absorption:

No.	Value	Note
1	35 %	7 days, 25° [2,3]

Gas Permeability:

No.	Gas	Value	Note
1	O$_2$	61266 cm^3 mm/(m^2 day atm)	9.33 × 10^{-8} cm^2 (s cmHg)$^{-1}$ [24]
2	CO$_2$	298977 cm^3 mm/(m^2 day atm)	45.53 × 10^{-8} cm^2 (s cmHg)$^{-1}$ [24]

Mechanical Properties:
Mechanical Properties General: Silicone rubber displays commercially acceptable tensile strength only when vulcanised and mixed with a reinforcing filler (usually finely divided silica). The mech. props. of silicone rubber are generally inferior to those of other rubbers. Silicone elastomers are used when their other advantages outweigh this [4,8,9]

Tensile (Young's) Modulus:

No.	Value	Note
1	1 MPa	25°, 150 psi, filled [6]

Tensile Strength Break:

No.	Value	Note	Elongation
1	0.35 MPa	0.035 kgf mm^{-2}, unfilled [2]	
2	6.2 MPa	25°, 900 psi, filled [5,8,18]	500%

Viscoelastic Behaviour: Compression set 40% at 175° and 22h compression (filled). Compression set at room temp. <10%. Compression set 20% at 175° and 22h compression (filled low shrinkage rubber) [4]
Hardness: Shore A50–A60 [17]
Fracture Mechanical Properties: Tear strength 1.6 J cm^{-2} (90 psi, 25°, filled) [6]

Electrical Properties:
Electrical Properties General: Silicone rubber has excellent insulating props. [3,7]
Surface/Volume Resistance:

No.	Value	Note	Type
1	0.1 × 10^{15} Ω.cm	dry, filled [8]	S

Dielectric Permittivity (Constant):

No.	Value	Note
1	3	filled [3]
2	2.7	unfilled [2]

Dielectric Strength:

No.	Value	Note
1	20 kV.mm^{-1}	500 V mil^{-1} [7,11]

Arc Resistance:

No.	Value	Note
1	200–250s	[18]

Dissipation (Power) Factor:

No.	Value	Frequency	Note
1	0.003	60 Hz	filled [7,8]
2	0.00024	60 Hz	unfilled [2]

Optical Properties:
Refractive Index:

No.	Value	Note
1	1.404	[2]

Transmission and Spectra: Absorption maximum at 290 nm [3]

Polymer Stability:
Polymer Stability General: Stable material [1]
Thermal Stability General: Usable from -60–260°
Upper Use Temperature:

No.	Value
1	260°C

Decomposition Details: Oxidation temp. 290° (dry air), CO produced [1,23]
Flammability: Limiting oxygen index 20% (approx.) [1]
Environmental Stress: Resistant to oxidative degradation, H_2O, weathering, ozone and corona. Unaffected by uv radiation
Chemical Stability: Swollen by toluene, aliphatic hydrocarbons and CCl_4. Disintegrates in conc. H_2SO_4. Relatively unaffected by weak alkali, weak acids, alcohols and oils (except silicone oil)
Hydrolytic Stability: Stable in H_2O and aq. solns. at room temp. Damaged by high pressure steam [3]
Biological Stability: Non-biodegradable. If properly vulcanised it is not subject to biological attack
Recyclability: Can be recycled by depolymerisation using steam or by acid catalysis [3]

Applications/Commercial Products:
Processing & Manufacturing Routes: An organic peroxide, typically benzoyl peroxide, is used at 1–2 parts per 100 rubber to cure the gum to an elastomer. The rubber is cured at 100–200° under pressure, typically for less than an hour, often post-cured for several hours at temps. of 200–250°. Cyclic dimethylpolysiloxanes are polymerised at 150–200° with an alkaline catalyst. Small quantities of hexamethyldisiloxane are added to terminate the siloxane chains. Siloxane rubbers can be cured by exposure to high energy electrons (e.g. from a β-radiation source, ca. 10 Mrep). Curing will occur at room temp. leaving no residual catalyst. To reduce mould shrinkage, low MW siloxanes can be removed by solvent extraction or thermal treatment. In solvent extraction volatile siloxanes are removed by washing in MeOH. In thermal treatment the gum is steam or air stripped at 250°, reducing the volatile fraction from around 10% by weight to less than 2%

Mould Shrinkage (%):

No.	Value	Note
1	2–4%	standard
2	0.5%	low shrinkage

Applications: Used where good temp. props. and stability are important. Used as encapsulants, coatings, electrical cable insulation, pads and belts, in electronic and plastic fabrication and in insulating tapes. Used in aircraft engines, gas turbines, autoclaves, irons and ovens. Important use as electrical insulant. Used in medical tubing and for vacuum seals. Important in aircraft and aerospace industry. Used in cosmetic surgery but there are doubts over safety. Radiation curing produces more stable cross links than peroxide vulcanisation and a product free from peroxide residues. The process has not been widely adopted commercially and is used in the laboratory to vulcanise PDMS without chemicals. Low shrinkage gums are used in the manufacture of precision gaskets and other applications where low shrinkage is important.

Trade name	Details	Supplier
B-2		Wacker Silicones
B-210		Wacker Silicones
COHR lastic	various grades not exclusive	CHR Industries
Rhodorsil	RS 40 series, RS 50	Rhone-Poulenc
SE	various grades not exclusive	General Electric Silicones
Silastic GP		Dow Corning STI
Silopren HV		Bayer Inc.

Bibliographic References
[1] Hardman, B. and Torkelson, A., *Encycl. Polym. Sci. Eng.*, 2nd edn., (ed. J.I. Kroschwitz), John Wiley and Sons, 1985, **15**, 204
[2] Roff, W.J. and Scott, J.R., Fibres, Films, *Plastics and Rubbers*, Butterworths, 1971
[3] Noll, W., *Chemistry and Technology of Silicones*, Academic Press, 1968
[4] Ames, J., *Silicones*, (ed. S. Fordham), 1960, 154
[5] Meals, R.N. and Lewis, F.M., Silicones Reinhold, 1959
[6] Hamilton, S.B., *Silicone Technology*, (ed. P.F. Bruins), John Wiley and Sons, 1970, 17
[7] Freeman, G.G., Silicones The Plastics Institute, 1962
[8] Ames, J., *J. Sci. Instrum.*, 1958, **35**, 1
[9] Warrick, E.L. and Lauterbur, P.C., *Ind. Eng. Chem.*, 1955, **47**, 486
[10] Yerrick, K.B. and Beck, H.N., *Rubber Chem. Technol.*, 1964, **37**, 261, (solubility)
[11] Stark, F.O., Fallender, J.R. and Wright, A.P., *Comprehensive Organometallic Chemistry*, (eds. G. Wilkinson, F.G.A. Stone and E.W. Abel), Pergamon Press, 1982, **2**, 305
[12] Baney, R.H., Voigt, C.E. and Mentele, J.W., Struct.-Solubility Relat. Polym., *(Proc. Symp.)*, (eds. F.W. Harris and R.B. Seymour), Academic Press, 1977, 225, (solubility)
[13] Robb, W.L., *Ann. N. Y. Acad. Sci.*, 1968, **146**, 119, (gas permeability)
[14] Polmanteer, K.E. and Hunter, M.J., *J. Appl. Polym. Sci.*, 1959, **1**, 3, (transition temps)
[15] Hauser, R.L., Walker, C.A. and Kilbourne, F.L., *Ind. Eng. Chem.*, 1956, **48**, 1202, (solubility)
[16] Kammermeyer, K., *Ind. Eng. Chem.*, 1957, **49**, 1685, (gas permeability)
[17] Ash, M. and Ash, I., Handbook of Plastic Compounds, *Elastomers and Resins*, (eds. M.B. Ash and I.A. Ash), Wiley-VCH, 1992
[18] McGregor, R.R., Silicones and their Uses McGraw Hill, 1954
[19] Drake, J., Peters, D.P. and McGuire, S., *Silicones*, (ed. S. Fordham), 1960, 117
[20] *Brit. Pat.*, 1957, 773 325, (low shrinkage gum)
[21] Yim, A. and St Pierre, L.E., *J. Polym. Sci., Part B: Polym. Lett.*, 1969, **7**, 237, (transition temps)
[22] Barrer, R.M. and Chio, H.T., *J. Polym. Sci., Part C: Polym. Lett.*, 1965, **10**, 111, (gas permeability)
[24] Stern, S.A., Shah, V.M. and Hardy, B.J., *J. Polym. Sci., Polym. Phys. Ed.*, 1987, **25**, 1263, (permeability, gas transport)

Methylsilicone rubber, Room temp. vulcanised M-17

Synonyms: Silicone elastomer, Room temp. vulcanised. α-Hydro-w-hydroxy-poly[oxy(dimethylsilylene)]
Related Polymers: Dimethiconol, Methylsilicone rubber, Two part room temp. vulcanised, Methylsilicone rubber, One part room temp. vulcanised, Nitrilemethylsilicone rubber, Room temp. vulcanised, Fluoromethylsilicone rubber, Room temp. vulcanised
Monomers: Hexamethylcyclotrisiloxane, Dichlorodimethylsilane
Material class: Synthetic Elastomers
Polymer Type: dimethylsilicones

– Methylsilicone rubber, Room temp. vulcanised M-17 – M-17

CAS Number:

CAS Reg. No.	Note
63394-02-5	elastomer
31900-57-9	PDMS
31692-79-2	dimethiconol

Molecular Formula: [C_2H_6OSi]
Fragments: C_2H_6OSi
Molecular Weight: MW 10000–100000 (pre-cure gum)
Additives: Finely divided silica is usually incorporated as a reinforcer to improve mech. props.
General Information: Silanol groups are present in the pre-cure gum. During cure they form cross links through condensation and fully cured elastomer has few free silanol groups

Volumetric & Calorimetric Properties:
Density:

No.	Value	Note
1	d 1.18–1.48 g/cm^3	filler dependent [1]

Thermal Expansion Coefficient:

No.	Value	Note	Type
1	0.00035 K^{-1}	[9]	L
2	0.00018 K^{-1}	[11]	V

Thermal Conductivity:

No.	Value	Note
1	0.3 W/mK	0.0007 cal (cm s °C)$^{-1}$ [9,11,14]

Specific Heat Capacity:

No.	Value	Note	Type
1	1.3 kJ/kg.C	0.3 Btu (lb °F)$^{-1}$ [8]	P

Melting Temperature:

No.	Value	Note
1	-54°C	[10]

Glass-Transition Temperature:

No.	Value	Note
1	-123°C	[10]

Transition Temperature General: T_m and T_g are nearly constant for dimethylsilicone rubbers. Brittleness and stiffening temps. vary with precise mech. props.
Brittleness Temperature:

No.	Value	Note
1	-65°C	-85°F [12]

Transition Temperature:

No.	Value	Note	Type
1	-38°C	[10]	S

Surface Properties & Solubility:
Cohesive Energy Density Solubility Parameters: δ 15.5 J$^{1/2}$ cm$^{-3/2}$ (7.5–7.6 cal$^{1/2}$ cm$^{-3/2}$); δ_d 15.9; δ_p 0.0; δ_h 4.1 J$^{1/2}$ cm$^{-3/2}$
Internal Pressure Heat Solution and Miscellaneous: Swelling agents have been reported [2]
Solvents/Non-solvents: Sol. aliphatic hydrocarbons, chlorinated hydrocarbons, aromatic hydrocarbons. Mod. sol. ketones, esters, ethers. Insol. alcohols, H_2O [4,5,7]. Vulcanised rubber swollen by solvents rather than dissolved

Transport Properties:
Transport Properties General: Silicone rubber is ten times more permeable than other rubbers to gases. The solubility of gases in polysiloxanes is similar to that for other polymers but the diffusion coefficients are much higher [3]
Polymer Melts: Viscosity has been reported [1]
Permeability of Gases: Permeability of gases has been reported [3]

Mechanical Properties:
Mechanical Properties General: As with heat vulcanised silicone rubber, good mech. props. depend on a reinforcing filler (usually silica) [1]. Under laboratory conditions room temp. vulcanised silicone rubber can have a tensile strength comparable to heat vulcanised silicone rubber. Commercial room temp. vulcanised silicone rubber, particularly one component room temp. vulcanised, has a lower strength [1]
Tensile Strength Break:

No.	Value	Note	Elongation
1	0.08 MPa	0.8 kg cm^{-2}, 30°, unfilled [2]	320%
2	4 MPa	40 kg cm^{-2}, high filler loading [2]	120%

Electrical Properties:
Electrical Properties General: Room temp. vulcanised silicone rubber has excellent insulating props. nearly as good as heat vulcanised rubber see methylsilicone rubber (heat vulcanised)
Dielectric Permittivity (Constant):

No.	Value	Note
1	3.6	[1,12,13]

Dielectric Strength:

No.	Value	Note
1	20 kV.mm^{-1}	500 V mil^{-1} [1,8,12,13]

Arc Resistance:

No.	Value	Note
1	90–125s	[11]

Dissipation (Power) Factor:

No.	Value	Note
1	0.015	[1,12]

Optical Properties:
Refractive Index:

No.	Value	Note
1	1.404	[7]

Transmission and Spectra: Absorption maximum at 290 nm [1]

Polymer Stability:

Polymer Stability General: Stable, but less so than heat vulcanised silicone rubber
Thermal Stability General: Usable from -60–260° [8,9]
Upper Use Temperature:

No.	Value
1	200°C
2	260°C

Decomposition Details: Oxidation temp. 290° (dry air), CO is produced [9,14]
Flammability: Limiting oxygen index 20% (approx.) [9]
Environmental Stress: Resistant to oxidative degradation, H_2O, weathering, ozone and corona. Unaffected by uv radiation
Chemical Stability: Swollen by toluene, aliphatic hydrocarbons and CCl_4. Disintegrates in conc. sulfuric acid. Relatively unaffected by weak alkali, weak acids, alcohols and oils (except silicone oil) [7]
Hydrolytic Stability: Stable in H_2O and aq. solns. at room temp. Damaged by high pressure steam
Biological Stability: Non-biodegradable. If properly vulcanised it is not subject to biological attack
Recyclability: Reclaimable in theory by depolymerisation; not reclaimed in practice [6]

Applications/Commercial Products:

Processing & Manufacturing Routes: Dimethiconol chains are formed by polymerisation of cyclic dimethyl siloxane at 150–200° with alkaline catalysis in the presence of small quantities of water. Dimethiconol gum cures at room temp. with alkoxysilane cross-linking agent and metal salt catalysis. Cross-linking requires moisture

Mould Shrinkage (%):

No.	Value
1	0.6%

Applications: Used as a sealant, encapsulant and coating. Important as flexible rubber mould. Of use in construction, aerospace and automobile industries. Valued for ease of vulcanisation. Important in electronics industry

Trade name	Details	Supplier
Eccofoam SIL		Emerson & Cuming
Eccosil		Emerson & Cuming
Norsil RTV		RH Carlson
RTF 762		General Electric Silicones
RTV	various grades, not exclusive	General Electric Silicones
Sylgard		Dow Corning STI

Bibliographic References

[1] Meals, R.N., *Ann. N. Y. Acad. Sci.*, 1965, **125**, 137
[2] Bajaj, P., Babu, G.N., Khanna, D.N. and Vorshney, S.K., *J. Appl. Polym. Sci.*, 1979, **23**, 3505, (solubility)
[3] Robb, W.L., *Ann. N. Y. Acad. Sci.*, 1968, **146**, 119, (gas permeability)
[4] Yerrick, K.B. and Beck, H.N., *Rubber Chem. Technol.*, 1964, **37**, 261, (solubility)
[5] Baney, R.H., Voigt, C.E. and Mentele, J.W., Struct.-Solubility Relat. Polym., *(Proc. Symp.)*, (eds. F.W. Harris and R.B. Seymour), Academic Press, 1977, 225
[6] Noll, W., *Chemistry and Technology of Silicones*, Academic Press, 1968
[7] Roff, W.J. and Scott, J.R., Fibres, Films, *Plastics and Rubbers*, Butterworths, 1971
[8] Ash, M. and Ash, I., Handbook of Plastic Compounds, *Elastomers and Resins*, (eds. M.B. Ash and I.A. Ash), Wiley-VCH, 1992
[9] Hardman, B. and Torkelson, A., *Encycl. Polym. Sci. Eng.*, 2nd edn., (ed. J.I. Kroschwitz), John Wiley and Sons, 1985, **15**, 204
[10] Polmanteer, K.E. and Hunter, M.J., *J. Appl. Polym. Sci.*, 1959, **1**, 3, (transition temps)
[11] Hainline, A.N., *Silicone Technology*, (ed. P.F. Bruins), John Wiley and Sons, 1970
[12] Hamilton, S.B., *Silicone Technology*, (ed. P.F. Bruins), John Wiley and Sons, 1970, 17
[13] Stark, F.O., Fallender, J.R. and Wright, A.P., *Comprehensive Organometallic Chemistry*, (eds. G. Wilkinson, F.G.A. Stone and E.W. Abel), Pergamon Press, 1982, **2**, 305

Methylsilicone rubber, Room temp. vulcanised M-18

Synonyms: *Phenylmethylsilicone rubber, Room temp. vulcanised. Methylphenyl rubber, Room temp. vulcanised*
Related Polymers: Methylsilicone rubber, Heat vulcanised
Monomers: Dichlorodimethylsilane, Dichloromethylphenylsilane, Hexamethylcyclotrisiloxane, Cyclic methylphenylsiloxane
Material class: Synthetic Elastomers, Copolymers
Polymer Type: silicones, dimethylsilicones
CAS Number:

CAS Reg. No.
156048-35-0

Molecular Formula: $[(C_2H_6OSi).(C_7H_8OSi)]_n$
Fragments: C_2H_6OSi C_7H_8OSi
Additives: Finely divided silica is usually incorporated as a reinforcer to improve mech. props.
General Information: Silanol groups are present in the pre-cure gum. During the cure they form cross links through condensation. Fully cured elastomer has few free silanol groups

Volumetric & Calorimetric Properties:
Density:

No.	Value	Note
1	d 0.97 g/cm^3	[3]

Thermal Expansion Coefficient:

No.	Value	Note	Type
1	0.0009 K^{-1}	90% dimethylsiloxane, 10% methylphenylsiloxane [2]	V
2	0.00075 K^{-1}	60% dimethylsiloxane, 40% methylphenylsiloxane [2]	V

Specific Heat Capacity:

No.	Value	Note	Type
1	1.3 kJ/kg.C	0.3 Btu (lb °F)$^{-1}$ [3]	P

Melting Temperature:

No.	Value	Note
1	-77°C	60% dimethylsiloxane, 40% methylphenylsiloxane [2,8]

Glass-Transition Temperature:

No.	Value	Note
1	-112°C	90% dimethylsiloxane, 10% methylphenylsiloxane [2]
2	-96°C	60% dimethylsiloxane, 40% methylphenylsiloxane [2]
3	-84°C	60% dimethylsiloxane, 40% methylphenylsiloxane [8]

Transition Temperature General: For transitions of low phenyl content silicone rubber see methylphenylsilicone rubber (heat vulcanised). For transitions of high phenyl content silicone rubber see phenylsilicone rubber heat vulcanised
Transition Temperature:

No.	Value	Note
1	-109°C	Stiffening temp. (90% dimethylsiloxane, 10% methylphenylsiloxane) [2]
2	-70°C	Stiffening temp. (60% dimethylsiloxane, 40% methylphenylsiloxane) [2]

Surface Properties & Solubility:
Cohesive Energy Density Solubility Parameters: δ 16–17 $J^{1/2}$ $cm^{-3/2}$ (7.7–8.2 $cal^{1/2}$ $cm^{-3/2}$ (increases with phenyl substitution) [7]
Solvents/Non-solvents: Sol. aliphatic hydrocarbons, chlorinated hydrocarbons, aromatic hydrocarbons, ketones, esters, ethers [5]. Insol. alcohols, H_2O [6]. Vulcanised rubber swollen by solvents rather than dissolved [7]
Surface and Interfacial Properties General: Exhibits generally low surface adhesion [4]

Transport Properties:
Polymer Melts: Viscosity has been reported [3]
Permeability of Gases: Gas permeability falls sharply as phenyl substitution rises [1]
Gas Permeability:

No.	Gas	Value	Note
1	H_2	12256 cm^3 mm/(m^2 day atm)	0.014 cm^2 (s GPa)$^{-1}$, 25°, 20% methylphenylsiloxane, 80% dimethylsiloxane [1]
2	O_2	8317 cm^3 mm/(m^2 day atm)	0.0095 cm^2 (s GPa)$^{-1}$, 25°, 20% methylphenylsiloxane, 80% dimethylsiloxane [1]
3	N_2	3239 cm^3 mm/(m^2 day atm)	0.0037 cm^2 (s GPa)$^{-1}$, 25°, 20% methylphenylsiloxane, 80% dimethylsiloxane [1]
4	O_2	2977 cm^3 mm/(m^2 day atm)	0.034 cm^2 (s GPa)$^{-1}$, 25°, 5% methylphenylsiloxane, 95% dimethylsiloxane [1]
5	N_2	13131 cm^3 mm/(m^2 day atm)	0.015 cm^2 (s GPa)$^{-1}$, 25°, 5% methylphenylsiloxane, 95% dimethylsiloxane [1]

Mechanical Properties:
Mechanical Properties General: Soft elastomer, when used for protecting delicate instruments, requires reinforcing filler to possess acceptable mech. props. [3]
Tensile Strength Break:

No.	Value	Note	Elongation
1	6.2 MPa	63 kg cm^{-2} [3]	150%

Hardness: Shore A45 [3]
Fracture Mechanical Properties: Tear strength 0.4 $J\ cm^{-2}$ (4 $kg\ cm^{-1}$) [3]

Electrical Properties:
Electrical Properties General: Has excellent insulating props.
Dielectric Permittivity (Constant):

No.	Value	Frequency	Note
1	2.8–3.1	1 kHz	[3,9,10]

Dielectric Strength:

No.	Value	Note
1	19.7 $kV.mm^{-1}$	[3,9]

Optical Properties:
Transmission and Spectra: No transmission below 290 nm [4,6]
Polymer Stability:
Polymer Stability General: Has good low temp. props.
Thermal Stability General: Usable from -100–260° [6]
Upper Use Temperature:

No.	Value
1	260°C

Decomposition Details: Oxidation temp. 300–350° (dry air; increases with increase in phenyl substitution) [6]
Environmental Stress: Resistant to oxidative degradation, H_2O, weathering, ozone and corona. Unaffected by uv radiation. Very resistant to high energy radiation (radioactivity) [4]
Chemical Stability: Swollen by toluene, aliphatic hydrocarbons and CCl_4. Disintegrates in conc. H_2SO_4. Relatively unaffected by weak alkali, weak acids and alcohols [4,5]
Hydrolytic Stability: Stable in H_2O and aq. solns. at room temp. Damaged by high pressure steam [5]
Biological Stability: Non biodegradable. If properly vulcanised it is not subject to biological attack
Recyclability: Reclaimable in theory by depolymerisation. Not reclaimed in practice [4]

Applications/Commercial Products:
Processing & Manufacturing Routes: Dimethyl- and methylphenyl-cyclic polysiloxanes are polymerised together at 150–200° with alkaline catalysis in the presence of small quantities of water. Phenyl dimethiconol gum cures at room temp. with alkoxysilane cross-linking agent and metal salt catalysts. Cross-linking requires moisture
Applications: Used as an encapsulant particularly in low temp. applications. Used particularly in the electronics industry

Trade name	Details	Supplier
RTV	500 series, 655 series, 6110 series	General Electric Silicones

Bibliographic References
[1] Robb, W.L., *Ann. N. Y. Acad. Sci.*, 1968, **146**, 119, (gas permeability)
[2] Polmanteer, K.E. and Hunter, M.J., *J. Appl. Polym. Sci.*, 1959, **1**, 3, (transition temps)
[3] Ash, M. and Ash, I., Handbook of Plastic Compounds, *Elastomers and Resins*, (eds. M.B. Ash and I.A. Ash), Wiley-VCH, 1992
[4] Noll, W., *Chemistry and Technology of Silicones*, Academic Press, 1968
[5] Roff, W.J. and Scott, J.R., Fibres, Films, *Plastics and Rubbers*, Butterworths, 1971
[6] Hardman, B. and Torkelson, A., *Encycl. Polym. Sci. Eng.*, 2nd edn. (ed. J.I. Kroschwitz), John Wiley and Sons, 1985, **15**, 204
[7] Yervick, K.B. and Beck, H.N., *Rubber Chem. Technol.*, 1964, **37**, 261, (solubility)
[8] Borisov, M.F., *Soviet Rubber Technology*, 1966, **25**, 5, (transition temps)
[9] Stark, F.O., Fallender, J.R. and Wright, A.P., *Comprehensive Organometallic Chemistry*, (eds. G. Wilkinson, F.G.A. Stone and E.W. Abel), Pergamon Press, 1982, **2**, 305
[10] Vincent, G.A., Feuron, F.W.G. and Orbech, T., Annu. Rep., *Conf. Electr. Insul. Dielectr. Phenom.*, 1972, 17, (dielectric constant)

Methylsilicone rubber, Two part room temp. vulcanised

Synonyms: *Silicone elastomer, Two part room temp. vulcanised. Two part room temp. vulcanised methylsilicone rubber.* α-Hydro-ω-hydroxypoly[oxy(dimethylsilylene)]

Related Polymers: Methylsilicone rubber, Room temp. vulcanised, Methylsilicone rubber, One part room temp. vulcanised, Dimethiconol, Methylsilicone rubber, Heat vulcanised
Monomers: Dichlorodimethylsilane, Hexamethylcyclotrisiloxane
Material class: Synthetic Elastomers
Polymer Type: dimethylsilicones
CAS Number:

CAS Reg. No.	Note
63394-02-5	elastomer
31900-57-9	PDMS
31692-79-2	dimethiconol

Molecular Formula: $[C_2H_6OSi]_n$
Fragments: C_2H_6OSi
Additives: Finely divided silica is usually incorporated as a reinforcer to improve mech. props.
General Information: Silanol groups are present in the pre-cure gum. During cure cross links are formed through condensation. Fully cured elastomer has few free silanol groups

Volumetric & Calorimetric Properties:
Density:

No.	Value	Note
1	d 1.18–1.48 g/cm^3	filler dependent [5]

Thermal Expansion Coefficient:

No.	Value	Note	Type
1	0.00035 K^{-1}	[3]	L
2	0.0008 K^{-1}	[4]	V

Thermal Conductivity:

No.	Value	Note
1	0.3 W/mK	0.0007 cal (cm s °C)$^{-1}$

Specific Heat Capacity:

No.	Value	Note	Type
1	1.3 kJ/kg.C	0.3 Btu (lb °F)$^{-1}$ [6]	P

Transition Temperature General: See methylsilicone rubber, room temp. vulcanised

Surface Properties & Solubility:
Cohesive Energy Density Solubility Parameters: δ 15.5 J$^{1/2}$ cm$^{-3/2}$ (7.5–7.6 cal$^{1/2}$ cm$^{-3/2}$) [1,2]
Solvents/Non-solvents: Sol. aliphatic hydrocarbons, chlorinated hydrocarbons, aromatic hydrocarbons. Mod. sol. ketones, esters, ethers. Insol. alcohols, H$_2$O [1,2,8]. Vulcanised rubber swollen by solvents rather than dissolved
Surface and Interfacial Properties General: Low adhesion to most surfaces. Some commercial forms have primer added to increase adhesion [3,9]

Mechanical Properties:
Tensile Strength Break:

No.	Value	Note	Elongation
1	4.2 MPa	42.2 kg cm^{-2}, 25° [3,7]	400%

Hardness: Shore A30–A60 [3,7]
Fracture Mechanical Properties: Tear strength 0.3–2.5 J cm^{-2} (3–26 kg cm^{-1}) [3,7]

Electrical Properties:
Electrical Properties General: See methyl silicone ruuber, room temp. vulcanised

Optical Properties:
Optical Properties General: See methyl silicone rubber, room temp. vulcanised

Applications/Commercial Products:
Processing & Manufacturing Routes: Dimethiconol gum is cured at room temp. with alkoxysilane cross-linking agent and metal salt catalyst. It is supplied lacking cross-linking agent and/or catalyst. Undergoes cure when missing ingredients are added
Mould Shrinkage (%):

No.	Value
1	0.6%

Applications: Used as sealant, encapsulant and coating. It is important in the construction industry and as a flexible rubber mould

Trade name	Details	Supplier
RTF 762		General Electric Silicones
RTV	00 series, 500 series, 600 series	General Electric Silicones

Bibliographic References
[1] Yerrick, K.B. and Beck, H.N., *Rubber Chem. Technol.*, 1964, **37**, 261, (solubility)
[2] Baney, R.H., Voigt, C.E. and Mentele, J.W., Struct.-Solubility Relat. Polym., *(Proc. Symp.)*, (eds. F.W. Harris and R.B. Seymour), Academic Press, 1977, 225, (solubility)
[3] Hardman, B. and Torkelson, A., *Encycl. Polym. Sci. Eng.*, 2nd edn., (ed. J.I. Kroschwitz), John Wiley and Sons, 1985, **15**, 204
[4] Hainline, A.N., *Silicone Technology*, (ed. P.F. Bruins), John Wiley and Sons, 1970, 51
[5] Meals, R.N., *Ann. N. Y. Acad. Sci.*, 1965, **125**, 137
[6] Ash, M. and Ash, I., Handbook of Plastic Compounds, *Elastomers and Resins*, (eds. M.B. Ash and I.A. Ash), Wiley-VCH, 1992
[7] Cold Cure Silicone Rubber for Mould Making Alec Tiranti, Reading, 1979
[8] Roff, W.J. and Scott, J.R., Fibres, Films, *Plastics and Rubbers*, Butterworths, 1971
[9] Noll, W., *Chemistry and Technology of Silicones*, Academic Press, 1968

Modacrylic fibres M-20
Related Polymers: Polyacrylonitrile, Acrylic fibres
Monomers: Vinylidene chloride, Acrylonitrile, Vinyl chloride, Vinyl bromide
Material class: Fibres and Films, Thermoplastic, Copolymers
Polymer Type: acrylic, acrylonitrile and copolymers
Additives: Antimony compounds, tin halides and phosphorus compounds as flame retardants.
General Information: Modacrylic fibres are manufactured using 35–85% acrylonitrile monomer. The commonest comonomers include vinyl chloride, vinylidene chloride and vinyl bromide
Morphology: Crystallinity and crystal size vary with copolymer composition and also with spinning method and conditions [5,6]
Miscellaneous: Modacrylic fibres are soft, resilient, easily dyed, abrasion resistant, flame resistant, resistant to acid/alkali attack, shape retentive and quick drying

Volumetric & Calorimetric Properties:
Density:

No.	Value	Note
1	d 1.28–1.37 g/cm^3	[2]

Modacrylic fibres

Transition Temperature General: The presence of moisture lowers T_g [2]

Transport Properties:
Water Absorption:

No.	Value	Note
1	1.5–3.5 %	[2]

Mechanical Properties:
Mechanical Properties General: Wet conditions do not severely affect mech. props. The effect of drawing rate, polymer content, and composition on mech. props. of wet-spun acrylonitrile/vinylidene chloride copolymers has been reported. [3,5,6] Tenacity 0.13–0.25 N tex^{-1} (45–60% elongation, dry), 0.11–0.23 N tex^{-1} (45–65% elongation, wet), 0.11–0.19 N tex^{-1} (loopknot), [2] 0.13–0.32 N tex^{-1} (acrylonitrile/vinylidene chloride (50/50), 13.7–14.2% elongation), 0.17–0.3 N tex^{-1} (acrylonitrile/vinylidene chloride (50/50), 7% antimony oxide, 13.5–19.7% elongation) [3], 0.13–0.33 N tex^{-1} (acrylonitrile/vinylidene chloride), 27% chlorine content, MW 640000, 13–18.2% elongation). [5] Average modulus 0.34 N tex^{-1} (dry) [2]
Viscoelastic Behaviour: Elastic recovery 99–100% (2% stretch); 95% (10% stretch)
Failure Properties General: Crease recovery is an important factor for fibres with a flat or rectangular cross-section. Flat fibres generally have the best crease recovery but this is a difficult property to control for modacrylic fibres and so variation of denier is used to give the required crease recovery [6]

Electrical Properties:
Electrical Properties General: Fibres have high electrical resistance [2]
Static Electrification: Fibres have moderate static build-up [2]

Polymer Stability:
Decomposition Details: The first weight loss, occurring at 285–308°, is due to dehydrohalogenation [4]
Flammability: Has low flammability [2]. The halogenated copolymers generally used for modacrylic fibre have good flame-retardant props. When burnt modacrylic fibres melt to give a dark, hard material. Some burning takes place but ceases when the applied flame is removed. [1,3,5] Finer fibres have a higher limiting oxygen index. The incorporation of antimony oxide can improve the limiting oxygen index. Limiting oxygen index 27% [2], 43.8% (acrylonitrile/vinylidene chloride (50/50)) 44.7–50.0% (acrylonitrile/vinylidene chloride (50/50), 7% antimony oxide) [3], 49–50% (acrylonitrile/vinylidene chloride, 33% chlorine content), 41–42% (acrylonitrile/vinylidene chloride, 22.1% chlorine content), 37–38% (acrylonitrile/vinylidene chloride, 8.6% chlorine content) [5]
Environmental Stress: Has excellent resistance to sunlight [2]
Chemical Stability: Has excellent resistance to chemical attack [2]

Applications/Commercial Products:
Processing & Manufacturing Routes: The polymers or copolymers used for modacrylic fibre production can be made by anionic methods but radical polymerisation is more commonly used. Modacrylic fibres can be dry spun or wet spun and can be used in staple or tow form. The effect of polymer content, MW and composition on the selection of spinneret has been investigated for wet-spun acrylonitrile/vinylidene chloride copolymers. Their low softening temp. permits them to be stretched, moulded or embossed. Modacrylic fibres can lose lustre on heating in a moist atmosphere, perhaps during dyeing. Improvements in spinning and aftertreatment have helped lustre stability and lustre stabilisers can be added. Relustering can be achieved by dry heat treatment [1,5,6]
Applications: Applications due to low flammability in clothing (simulated fur, deep-pile coats), in trims and linings, wigs, knitwear, children's wear; in furnishings (blankets, carpets, awnings), in filters, fabrics, paint rollers and stuffed toys

Trade name	Supplier
Acrilan	Solutia

Bibliographic References
[1] Nozaj, A and Suliag, G, *Ullmanns Encykl. Ind. Chem.*, **A10**, 629
[2] Knorr, R.S., Encyclopaedia of Chemical Technology 4th edn., 1993, **10**, 559
[3] Tsai, J.-S., *J. Mater. Sci.*, 1993, **28**, 1161, 4841, (flame retardant, viscoelastic props)
[4] Tsai, J.-S.; Ho, D.-L. and Hung, S.-C., *J. Mater. Sci. Lett.*, 1991, **10**, 881
[5] Tsai, J.-S., *J. Mater. Sci. Lett.*, 1992, **11**, 953, 1017, (flammability, struct)
[6] Tsai, J.-S. and Wu, C.-H., *J. Mater. Sci. Lett.*, 1993, **12**, 523

Natural balata N-1

Synonyms: *trans*-1,4-Polyisoprene. *trans*-Polyisoprene. Balata rubber
Related Polymers: *trans*-1,4-Polyisoprene, Chicle, Jelutong, Gutta percha
Monomers: 2-Methyl-1,3-butadiene
Material class: Thermoplastic, Gums and resins
Polymer Type: polybutadiene
CAS Number:

CAS Reg. No.
9005-99-6

Molecular Weight: MW and MW distribution have been determined for balata and related polymers
Additives: Contains 35–50% resin
General Information: Isol. from S. American trees, *Mimusop* and *Eulinisu* sp. Produced as a substitute for gutta-percha. Has inferior props. to pure *trans*-1,4-polyisoprene, due to larger quantities of resin contaminants. Isol., production and usage have been reported [4,5,6,9]
Morphology: Two cryst. forms known, α and β

Volumetric & Calorimetric Properties:
Melting Temperature:

No.	Value	Note
1	64°C	β-form
2	74°C	α-form

Glass-Transition Temperature:

No.	Value	Note
1	-70°C	[1]
2	-58°C	raw polymer [15]
3	-55°C	vulcanised [15]

Transition Temperature General: α and β transition temps. depend on rate of cooling and sample thermal history [8]

Transport Properties:
Polymer Solutions Dilute: Limiting viscosity number 1.1 (100 ml g^{-1} toluene) [10]

Mechanical Properties:
Tensile Strength Break:

No.	Value	Note
1	24.6 MPa	3580 psi [10]

Failure Properties General: Resilience has been reported from -70–200°; decreases rapidly above 30° [11]

Optical Properties:
Transmission and Spectra: Ir and nmr spectral data have been reported [2,15]

Polymer Stability:
Decomposition Details: Pyrolysis under vacuum yields 50% isoprene, 13% dipentene, 13% methane, 6% cyclopentadiene, 4% propene, 4% ethane and 7% cyclopentadiene [7]. Other pyrolysis studies have been reported [12]

Applications/Commercial Products:
Processing & Manufacturing Routes: Isol. from latex taken by bark slashing of *Mimusop balata* or *Eulinisa* sp.
Applications: Used for golf ball covers

Bibliographic References
[1] Burfield, D.R. and Lim, K., *Macromolecules*, 1983, **16**, 1170
[2] Kishore, K. and Pandey, H.K., *Prog. Polym. Sci.*, 1986, **12**, 155
[3] Hager, T., MacArthur, A., McIntyre, D. and Suger, R., *Rubber Chem. Technol.*, 1979, **52**, 693
[4] Wren, W.G., *Annu. Rep. Prog. Rubber Technol.*, 1948, **12**, 6, (rev)
[5] Blow, C.M., *Annu. Rep. Prog. Rubber Technol.*, 1949, **13**, 14, (rev)
[6] Cook, L., *Rubber Age (London)*, 1955, **77**, 872
[7] Wall, L.A., *J. Res. Natl. Bur. Stand. (U.S.)*, 1948, **41**, 315
[8] Cooper, W. and Smith, R.K., *J. Polym. Sci.*, Part A-1, 1963, 159
[9] Kirchof, F., Gummi, Asbest, *Kunstst.*, 1963, **16**, 4
[10] Forbes, W.G. and McLeod, L.A., *Trans., Inst. Rubber Ind.*, 1958, **34**, 154
[11] Fujimoto, K., *CA*, 1962, **56**, 3609c, (resilience)
[12] Galin-Vacherot, M., *Eur. Polym. J.*, 1971, **7**, 1455
[13] Vacherot, M., *J. Gas Chromatogr.*, 1967, **5**, 155
[14] Vacherot, M. and Marchal, J., C.R. Seances Acad. Sci., Ser. C, 1966, **263**, 210
[15] Clark, J.K., *Appl. Spectrosc.*, 1968, **22**, 204

Natural rubber N-2

Synonyms: *cis*-1,4-Polyisoprene. Hevea guianensis. Tree rubber. Poly(2-methyl-1,3-butadiene). Poly(1,3-butadiene-2-methyl). Rubber
Related Polymers: *cis*-1,4-Polyisoprene, Epoxidized natural rubber, Ebonite, Rubber hydrochloride, Guayule rubber
Monomers: 2-Methyl-1,3-butadiene
Material class: Natural elastomers
Polymer Type: polybutadiene
CAS Number:

CAS Reg. No.
9006-04-6

Additives: Vulcanised by sulfur, organic peroxides, urethane
General Information: Natural rubber has nearly 100% *cis*-1,4 microstruct. can be isomerised, cyclised, hydrogenated, halogenated and alkylhalogenated. Contains antioxidants, carbohydrates, triglycerides, proteins and resins
Morphology: Typically 30–35% cryst. [70]. Effects of cross-linking on crystallisation have been reported [72]. Faster crystallisation rate, increased by presence of additives such as stearic acid [92]
Miscellaneous: Bimodal

Volumetric & Calorimetric Properties:
Density:

No.	Value	Note
1	d 0.913 g/cm^3	unvulcanised [1]
2	d 0.95 g/cm^3	0°, unvulcanised [33]
3	d 0.934 g/cm^3	20°, unvulcanised [33]
4	d 0.95 g/cm^3	vulcanised [48]
5	d 0.97 g/cm^3	vulcanised [17]

Thermal Expansion Coefficient:

No.	Value	Note	Type
1	0.00067 K^{-1}	unvulcanised [1]	V
2	0.00062 K^{-1}	unvulcanised [33]	V
3	0.0066 K^{-1}	vulcanised [1]	V
4	0.0067 K^{-1}	vulcanised [48]	V

– Natural rubber

Volumetric Properties General: Density values have been calculated [9]
Equation of State: Tait equation of state parameters have been reported [9]
Thermodynamic Properties General: Heat of combustion 45.2 kJ g^{-1} (unvulcanised) [1,33]; ΔH_{fusion} 4.64 kJ mol^{-1} (unvulcanised) [6]; 44.4 kJ g^{-1} (vulcanised) [1]
Thermal Conductivity:

No.	Value	Note
1	0.134 W/mK	unvulcanised [1]
2	0.13 W/mK	unvulcanised
3	0.151 W/mK	vulcanised [113]
4	0.153 W/mK	vulcanised [20,21]
5	0.15 W/mK	vulcanised [48]

Thermal Diffusivity: 7×10^{-8} m^2 s^{-1} (vulcanised) [22]
Specific Heat Capacity:

No.	Value	Note	Type
1	1.905 kJ/kg.C	unvulcanised [34]	P
2	0.502 kJ/kg.C	unvulcanised [33]	P
3	1.828 kJ/kg.C	vulcanised [19,48]	P

Melting Temperature:

No.	Value	Note
1	35°C	unvulcanised, cryst. [5]
2	40°C	vulcanised [23]

Glass-Transition Temperature:

No.	Value	Note
1	-72°C	unvulcanised [2,34]
2	-70°C	unvulcanised [32]
3	-70°C	vulcanised [48]
4	-63°C	unvulcanised [48]
5	-68°C	dry, vulcanised [78]
6	-78°C	swollen, vulcanised [78]

Transition Temperature General: Melting temp. for unvulcanised rubber varies with crystallisation temp. [68,69]. 1% Cross-linking depresses melt temp. by 20° [71]

Surface Properties & Solubility:
Cohesive Energy Density Solubility Parameters: Cohesive energy density 266.5 J cm^{-3} (unvulcanised) [35]. Solubility parameter δ 15.2 J$^{1/2}$ cm$^{-3/2}$ (calc., unvulcanised) [36], 16.2 J$^{1/2}$ cm$^{-3/2}$ (unvulcanised) [37], 16.6-16.7 J$^{1/2}$ cm$^{-3/2}$ (unvulcanised) [38,39], 17.1 J$^{1/2}$ cm$^{-3/2}$ (unvulcanised) [40]. Swelling of vulcanisates has been reported [98]
Solvents/Non-solvents: Unvulcanised sol. aliphatic hydrocarbons, aromatic hydrocarbons, chlorinated hydrocarbons, esters, ketones, Et$_2$O, carbon disulfide [66]; insol. aniline, phenol, H$_2$O, lower esters, alcohols [66]
Wettability/Surface Energy and Interfacial Tension: Contact angles have been reported [88] for ethylene glycol 59.5° (unvulcanised, cut); 63.7° (unvulcanised, cast); 61° (vulcanised); 45.7° (vulcanised, peroxide cure); 81.5° (carbon black)
Adhesion General: Has good adhesion to metals and fabrics [67]

Surface Tension:

No.	Value	Note
1	35 mN/m	unvulcanised, calc. [62]
2	31 mN/m	unvulcanised, Zisman critical surface tension [62]

Transport Properties:
Polymer Solutions Dilute: Mark-Houwink constants have been reported [41,42,43,51,52]
Polymer Melts: Melt viscosity 20 MPa s, higher than synthetic polyisoprene, attributed to long-chain branching [115]
Permeability of Gases: Other gas permeability values have been reported [63,64,65]
Gas Permeability:

No.	Gas	Value	Note
1	O$_2$	1540 cm^3 mm/(m^2 day atm)	1.76×10^{-12} cm^2 (s Pa)$^{-1}$, unvulcanised [24]
2	Ar	1505 cm^3 mm/(m^2 day atm)	1.72×10^{-12} cm^2 (s Pa)$^{-1}$, unvulcanised [24]
3	CO$_2$	10067 cm^3 mm/(m^2 day atm)	1.15×10^{-11} cm^2 (s Pa)$^{-1}$, unvulcanised [24]
4	CO	1033 cm^3 mm/(m^2 day atm)	1.18×10^{-12} cm^2 (s Pa)$^{-1}$, unvulcanised [24]
5	N$_2$	622 cm^3 mm/(m^2 day atm)	7.11×10^{-13} cm^2 (s Pa)$^{-1}$, unvulcanised [24]
6	CH$_4$	1987 cm^3 mm/(m^2 day atm)	2.27×10^{-12} cm^2 (s Pa)$^{-1}$, unvulcanised [24]
7	H$_2$O	150577 cm^3 mm/(m^2 day atm)	1.72×10^{-10} cm^2 (s Pa)$^{-1}$, unvulcanised [24]

Mechanical Properties:
Mechanical Properties General: Compressibility has been reported [9]. Storage and loss moduli have been reported [14]. Props. of filled, vulcanised materials have been reported [34]. Hysteresis props. have been reported [74,93]. Has good green strength [111]

Tensile (Young's) Modulus:

No.	Value	Note
1	1.3 MPa	vulcanised [26,27]
2	2 MPa	vulcanised [48]

Tensile Strength Break:

No.	Value	Note	Elongation
1	30 MPa	20°, vulcanised [114]	
2	5 MPa	140°, vulcanised [114]	
3	17–25 MPa	vulcanised [25,64]	750-850% elongation
4	20 MPa	vulcanised [67]	
5	18.8 MPa	vulcanised [110]	550% elongation

Flexural Strength at Break:

No.	Value	Note
1	17–30 MPa	650–900% elongation, vulcanised [50]
2	14–28 MPa	450–600% elongation, vulcanised, 50 phv carbon black [50]

– **Natural rubber**

Elastic Modulus:

No.	Value	Note
1	0.43 MPa	vulcanised [18,27]
2	2000 MPa	vulcanised [48]

Poisson's Ratio:

No.	Value	Note
1	0.4999	vulcanised [9,28,29,48,80]

Miscellaneous Moduli:

No.	Value	Note	Type
1	1.94 MPa	[9]	Isothermal bulk modulus
2	2.27 MPa	[9]	Adiabatic bulk modulus

Viscoelastic Behaviour: Viscoelastic response behaviour by the Williams-Landel-Ferry equation has been reported [55,56,57,58,59]. Creep 2% (vulcanised) [18,26,108]. Velocity of sound 54 m s^{-1} (vulcanised) [48].
Mechanical Properties Miscellaneous: Stress relaxation moduli have been reported [106]. Resilience 75–77% (unvulcanised) [15,16]; 75–84% (vulcanised) [16,31]. Dynamic mech. props. have been reported for filled and blends [76,97,102]. Tear strength 58 kN m^{-1} (vulcanised) [73].
Hardness: Shore A30–A90 (vulcanised) [67]
Failure Properties General: Stress relaxation has been reported (50–115°) [86]
Friction Abrasion and Resistance: Coefficient of friction 2.3–3.5 (vulcanised, smooth glass) [77,79], 0.53 (chlorinated, vulcanised) [79]. Wear and crack resistance have been reported [103]. Abrasion resistance 29 mg (1000 cycles)$^{-1}$ (Tabor CS-17, ASTM D1044, vulcanised) [17]. Fatigue life in air is improved by antioxidants [81]. Annealing, and other fatigue factors have been reported [85,100]. Wet skid resistance is related to chain flexibility [87]

Electrical Properties:
Electrical Properties General: Addition of carbon black decreases resistivity to 0.02 Ω [94]
Dielectric Permittivity (Constant):

No.	Value	Frequency	Note
1	0.14–0.2		unvulcanised [33]
2	2.37–2.45	1 kHz	unvulcanised [1,8]
3	2.5–3	1 kHz	vulcanised [1,8,98]

Dielectric Strength:

No.	Value	Note
1	0.0039 kV.mm^{-1}	unvulcanised [33]

Dissipation (Power) Factor:

No.	Value	Frequency	Note
1	0.001–0.003	1 kHz	unvulcanised [8]
2	0.002–0.004	1 kHz	vulcanised [8]

Optical Properties:
Refractive Index:

No.	Value	Note
1	1.5191	unvulcanised [7]
2	1.5264	vulcanised [1]

Transmission and Spectra: Ir spectral data have been reported [49,54]

Polymer Stability:
Thermal Stability General: Slightly more resistant to degradation than guayule rubber, and synthetic, isoprene rubber due to the presence of natural antioxidants [83]
Upper Use Temperature:

No.	Value	Note
1	100°C	vulcanised [67]
2	80°C	[61]

Decomposition Details: Undergoes thermo-oxidative degradation by free-radical processes to give main chain scission and low MW products [84]. Accelerated thermal and uv oxidative degradation causes changes in surface composition with the appearance of functional groups, including hydroxyl, ether, carbonyl and carboxyl. Polynuclear aromatics are formed under extreme conditions [83]
Flammability: Limiting oxygen index 17.2% (vulcanised, foam) [60]
Environmental Stress: Has good resistance to oxidation and radiation
Chemical Stability: Is stable to oxygenated hydrocarbons
Hydrolytic Stability: Has good resistance to water and dilute acids (vulcanised). Has fair resistance to steam and conc. base (vulcanised)
Stability Miscellaneous: Intermittent stress relaxation measurements at elevated temps. in oxygen correlate with conventional ageing [109]

Applications/Commercial Products:
Processing & Manufacturing Routes: Isol. as latex from *Hevea brasiliensis* by bark tapping. Coagulated to produce solid material. Hardens during storage and may 'freeze' at low temp. Used either as latex, or dried and smoked to produce sheet, or wasted and milled to produce crepe
Applications: Latex used in adhesives, gloves, contraceptives, carpet backing and tubing. Sheet and crepe used in tyres, belts, flexible mountings, footwear, hoses, seals, matting. May also be blended with other elastomers

Trade name	Supplier
DPNR	H.A. Astlett
Dynatex GTZ	Gutherie Latex
GPNS, GPS	Muehlstein
Hyflo NS, S	H.A. Astlett
Itex TSR	Intertex
PA	Akrochem
Revertex	Diversified Compounders
Unitex	Gutherie Latex

Bibliographic References
[1] Wood, L.A., *Rubber Chem. Technol.*, 1939, **12**, 130
[2] Wood, L.A., *Rubber Chem. Technol.*, 1940, **13**, 861
[3] Chang, S.S. and Bestul, A.B., *J. Res. Natl. Bur. Stand. (U.S.)*, 1971, **75A**, 113
[4] Wood, L.A. and Bekkedahl, N., *J. Polym. Sci., Part B: Polym. Lett.*, 1967, **5**, 169

[5] Dalai, E.N., Taylor, K.D. and Phillips, P.J., *Polymer*, 1983, **24**, 1623
[6] Kim, H.G. and Mandelkern, L., *J. Polym. Sci., Part A-2*, 1972, **10**, 1125
[7] Wood, L.A. and Tilton L.W., *J. Res. Natl. Bur. Stand. (U.S.)*, 1949, **43**, 57
[8] McPherson, A.T., *Rubber Chem. Technol.*, (Rubber Rev.), 1963, **36**, 1230
[9] Wood, L.A. and Martin, G.M., *Rubber Chem. Technol.*, 1964, **37**, 850
[10] Payne, A.R. and Scott, J.R., *Engineering Design with Rubber*, Interscience, 1960
[11] Fletcher, W.P. and Gent, A.N., *Br. J. Appl. Phys.*, 1957, **8**, 194
[12] Ferry, J.D., *Viscoelastic Properties of Polymers*, Wiley, 1961, 458
[13] Schneider, L. and Wolk, K., *Kolloid Z. Z. Polym.*, 1953, **134**, 149
[14] Zapas, L.J., Shufler, S.L. and de Witt, T.W., *Rubber Chem. Technol.*, 1956, **29**, 725
[15] Boomstra, B.B.S.T., *Properties of Elastomers*, (ed. R. Houwink), Elsevier, 1948, 3
[16] Boomstra, B.B.S.T., *Rubber Chem. Technol.*, 1951, **24**, 199
[17] Wildschut, A.J., *Technological and Physical Investigations on Natural and Synthetic Rubbers*, Elsevier, 1946
[18] Wood, L.A. and Roth, F.L., *Rubber Chem. Technol.*, 1963, **36**, 611
[19] Hamill, W.H., Mrowca, A. and Anthony, R.L., *Rubber Chem. Technol.*, 1946, **19**, 622
[20] Caruile, L.C. and Hoge, H.J., *Rubber Chem. Technol.*, 1966, **39**, 126
[21] Pillsworth, M.N., Hoge, H.J. and Robinson, H.E., *J. Mater. Sci.*, 1972, **7**, 550
[22] Hands, D., *Rubber Chem. Technol.*, (Rubber Rev.), 1977, **50**, 480
[23] Furukawa, G.T. and Reilly, M.L., *J. Res. Natl. Bur. Stand. (U.S.)*, 1956, **56**, 285
[24] Michaels, A.S. and Bixler, H.J., *J. Polym. Sci.*, 1961, **50**, 413
[25] Boomstra, B.B.S.T., *Elastomers. Their Chemistry, Physics and Technology*, (ed. R. Houwink), Elsevier, 1948, 3
[26] Martin, G.M., Roth, F.L. and Stiehler, R.D., *Rubber Chem. Technol.*, 1957, **30**, 876
[27] Roth, F.L., Bullman, G.W. and Wood, L.A., *Rubber Chem. Technol.*, 1966, **39**, 397
[28] Holownia, B.P., *Rubber Chem. Technol.*, 1975, **48**, 246
[29] Rightmire, G.K., *J. Lubrication Technol.*, 1970, 381
[30] Perry, J.D., Mancke, R.G., Maekawa, E., Oyanagi, Y. and Dickie, R.A., *J. Phys. Chem.*, 1964, **68**, 3414
[31] Dillon, J.H., Prettyman, J.B. and Hall, G.L., *J. Appl. Phys.*, 1944, **15**, 309
[32] Dannis, M.L., *J. Appl. Polym. Sci.*, 1959, **1**, 121
[33] Bristow, G.M. and Sears, A.G., *NR Technol.*, 1982, **13**, 73
[34] *The Materials Selector*, 2nd edn., (eds. N.A. Waterman and M.F. Ashby), Chapman and Hall, 1991
[35] Orwell, R.A. and Arnold, P.A., *Physical Properties of Polymers Handbook*, (ed. J.E. Mark), AIP Press, 1996, 187
[36] Di Benedetto, A.T., *J. Polym. Sci., Part A-1*, 1963, 3459
[37] Gee, G., Trans., *Inst. Rubber Ind.*, 1943, **18**, 206
[38] Mangaraj, D., Bhatnagav, S.K. and Rath, S.B., *Makromol. Chem.*, 1963, **67**, 75
[39] Mangaraj, D., Patra, S. and Rath, S.B., *Makromol. Chem.*, 1963, **65**, 39
[40] Mark, H. and Tobolsky, A.V., *Physical Chemistry of High Polymers*, Interscience, 1950, 263
[41] Wagner, H.L. and Flory, P.J., *J. Am. Chem. Soc.*, 1952, **74**, 195
[42] Altgelt, R. and Schultz, G.V., *Makromol. Chem.*, 1960, **36**, 209
[43] Carter, W.C., Scott, R.L. and Magat, M., *J. Am. Chem. Soc.*, 1946, **68**, 1480
[44] Poddubnyi, I.Y. et al, *Vysokomol. Soedin.*, 1964, **5**, 1588
[45] Abe, M. et al, *Kogyo Kagaku Zasshi*, 1969, **72**, 2313
[46] Poddubnyi, I.Y. and Ehrenberg, E.G., *J. Polym. Sci.*, 1962, **57**, 545
[47] Balsara, N.P., *Physical Properties of Polymers Handbook*, (ed. J.E. Mark), AIP Press, 1996, 266
[48] Lindley, P.B., *Engineering Design with Natural Rubber MRPRA Tech. Bulletin*, Malaysian Rubber Producers Assoc., 1978
[49] Noda, I., Dowrey, A.E. and Marcott, C., *Physical Properties of Polymers Handbook*, (ed. J.E. Mark), AIP Press, 1996, 293
[50] Brostow, W., Kubát, J. and Kubát, M., *Physical Properties of Polymers Handbook*, (ed. J.E. Mark), AIP Press, 1996, 333
[51] Roovers, J., *Polym. J. (Tokyo)*, 1986, **18**, 153
[52] Ansora, F.N., Revuelta, L.M., Guzman, G.M. and Irwin, J.J., *Eur. Polym. J.*, 1982, **18**, 9
[53] Sanders, J.F., Ferry, J.D. and Valentine, R.H., *J. Polym. Sci., Part A-2*, 1968, **6**, 967
[54] Noda, I., Dowrey, A.E. and Marcott, C., *Physical Properties of Polymers Handbook*, (ed. J.E. Mark), AIP Press, 1996, 293
[55] Payne, A.R., *Rheology of Elastomers*, (eds. P. Mason and N. Wookey), Pergamon, 1958
[56] Dickie, R.A. and Ferry, J.D., *J. Phys. Chem.*, 1966, **70**, 2594
[57] Schwartzl, F.R. and Zahradnik, F., *Rheol. Acta*, 1980, **19**, 137
[58] Roland, C.M. and Ngai, K.L., *Macromolecules*, 1992, **24**, 5315
[59] Allegria, A., Colmenero, J., Ngai, K.L. and Roland, C.M., *Macromolecules*, 1994, **27**, 4486
[60] Tewarson, A., *Physical Properties of Polymers Handbook*, (ed. J.E. Mark), AIP Press, 1996, 587
[61] Billmeyer, F.W., *Textbook of Polymer Science*, Wiley Interscience, 1984, 136, 361
[62] Johnson, K.L., Kendall, K. and Roberts, A.D.., *Proc. R. Soc. London A*, 1971, **324**, 301
[63] Koros, W.J. and Hellums, M.W., *Encycl. Polym. Sci. Eng.*, (eds. H.F. Mark, N.M. Bikales, C.G. Overberger, G. Menges and J.I. Kroschwitz), Wiley Interscience, 1989, **Supplement**, 742
[64] van Amerongen, G.J., *J. Appl. Phys.*, 1946, **17**, 972
[65] Cabasso, J., *Encycl. Polym. Sci. Eng.*, (eds. H.F. Mark, N.M. Bikales, C.G. Overberger, G. Menges and J.I. Kroschwitz), Wiley Interscience, 1987, **9**, 562
[66] Orwell, R.A., *Encycl. Polym. Sci. Eng.*, (eds. H.F. Mark, N.M. Bikales, C.G. Overberger, G. Menges and J.I. Kroschwitz), Wiley Interscience, 1987, **15**, 381
[67] *The Materials Selector*, 2nd edn., (eds. N.A. Waterman and M.F. Ashby), Chapman and Hall, 1996
[68] Wood, L.A. and Bekkedahl, N., *J. Appl. Phys.*, 1946, **17**, 362
[69] Dalal, E.N., Taylor, K.D. and Phillips, P.J., *Polymer*, 1983, **24**, 1623
[70] Mandelkern, L., *Rubber Chem. Technol.*, 1993, **66**, 661
[71] Mandelkern, L., Roberts, D.E., Halpin, J.C. and Price, F.P., *J. Am. Chem. Soc.*, 1960, **82**, 46
[72] Gent, A.N., *Trans. Faraday Soc.*, 1954, **50**, 521
[73] *Natsyn Polyisoprene Rubber Bulletin*, Goodyear Chemical Division, Oct., 1988, (technical datasheet)
[74] Hess, W.M., Herd, C.R. and Vegvari, P.C., *Rubber Chem. Technol.*, 1993, **66**, 329
[75] Lee, M.-P. and Moet, A., *Rubber Chem. Technol.*, 1993, **66**, 304
[76] Coran, A.Y. and Donnet, J.-B., *Rubber Chem. Technol.*, 1992, **65**, 1016
[77] Roberts, A.D., *Rubber Chem. Technol.*, 1992, **65**, 673
[78] Jackson, C.L. and McKenna, G.B., *Rubber Chem. Technol.*, 1991, **64**, 760
[79] Extrand, C.W., Gent, A.N. and Kaang, S.Y., *Rubber Chem. Technol.*, 1991, **64**, 108
[80] Kuglev, H.P., Stacer, R.G. and Steimle, C., *Rubber Chem. Technol.*, 1990, **63**, 473
[81] Gent, A.N. and Hindi, M., *Rubber Chem. Technol.*, 1990, **63**, 123
[82] Qin, C., Yin, J. and Huang, B., *Rubber Chem. Technol.*, 1990, **63**, 71
[83] Lin, S.-S., *Rubber Chem. Technol.*, 1989, **62**, 315
[84] Sambhi, M.S., *Rubber Chem. Technol.*, 1989, **62**, 779
[85] Rowland, C.M. and Sobieski, J.W., *Rubber Chem. Technol.*, 1989, **62**, 683
[86] Björk, F., Dickwan, O. and Stenberg, B., *Rubber Chem. Technol.*, 1989, **62**, 387
[87] Goldberg, A., Lesuer, D.R. and Patt, J., *Rubber Chem. Technol.*, 1989, **62**, 288
[88] Extrand, C.W. and Gent, A.N., *Rubber Chem. Technol.*, 1988, **61**, 688
[89] Akhtar, S., *Rubber Chem. Technol.*, 1988, **61**, 577
[90] Chen, Y.-J., White, J.L., Min, K., Nakajima, N. and Weissert, F.C., *Rubber Chem. Technol.*, 1988, **61**, 324
[91] Shimomura, Y., White, J.L. and Spruiell, J.E., *J. Appl. Polym. Sci.*, 1982, **27**, 3553
[92] White, J.L., *J. Polym. Eng.*, 1985, **5**, 235
[93] Coran, A.Y., *Rubber Chem. Technol.*, 1988, **61**, 281
[94] Siswanto, M.G., Bo, N.P., Parangtoko, Neubacker, H. and Burton, L.C., *Rubber Chem. Technol.*, 1988, **61**, 269
[95] Harris, J.A., *Rubber Chem. Technol.*, 1987, **60**, 870
[96] Lee, R.F. and Donovan, J.A., *Rubber Chem. Technol.*, 1987, **60**, 674
[97] Akhtar. S. and Bhagawan, S.-S., *Rubber Chem. Technol.*, 1987, **60**, 591
[98] Oikawa, H. and Murakami, K., *Rubber Chem. Technol.*, 1987, **60**, 579
[99] Hamed, G.R. and Shieh, C.-H., *Rubber Chem. Technol.*, 1986, **59**, 883
[100] Young, D.G., *Rubber Chem. Technol.*, 1986, **59**, 809
[101] Harris, J. and Stevenson, A., *Rubber Chem. Technol.*, 1986, **59**, 740
[102] Mazich, K.A., Samur, M.A., Killgoar, P.C. and Plummer, H.K., *Rubber Chem. Technol.*, 1986, **59**, 623
[103] Dannenberg, E.M., *Rubber Chem. Technol.*, 1986, **59**, 497
[104] Ahagon, A., *Rubber Chem. Technol.*, 1986, **59**, 187
[105] McKenna, G.B. and Zapas, L.J., *Rubber Chem. Technol.*, 1986, **59**, 130
[106] Campbell, D.S. and Fuller, K.N.G., *Rubber Chem. Technol.*, 1984, **57**, 104
[107] Montes, S. and White, J.L., *Rubber Chem. Technol.*, 1982, **55**, 1354
[108] Thiele, J.L. and Cohen, R.E., *Rubber Chem. Technol.*, 1983, **56**, 465
[109] Clamroth, R. and Ruetz, L., *Rubber Chem. Technol.*, 1983, **56**, 31
[110] Young, D.G., *Rubber Chem. Technol.*, 1985, **58**, 785
[111] Barnard, D., Baker, C.S.L. and Wallace, I.R., *Rubber Chem. Technol.*, 1985, **58**, 740
[112] Lindley, P.B. and Stevenson, A., *Rubber Chem. Technol.*, 1982, **55**, 337
[113] Sircar, A.K. and Walls, J.L., *Rubber Chem. Technol.*, 1982, **55**, 191
[114] Bell, C.L.M., Stimson, D. and Thomas, A.G., *Rubber Chem. Technol.*, 1982, **55**, 66

[115] Senyek, M., *Encycl. Polym. Sci. Eng.*, (eds. H.F. Mark, N.M. Bikales, C.G. Overberger, G. Menges and J.I. Kroschwitz), Wiley Interscience, 1987, **8**, 567

[116] Ball, J.M. and Maassen, G.C., ASTM, *Symp. Appl. Synth. Rubbers*, 1944, 27

Natural rubber/poly(methyl methacrylate) inter-penetrating network   N-3

Synonyms: *Natural rubber/PMMA blend. Natural rubber/PMMA/graft-natural rubber blend*
Related Polymers: Natural rubber, Poly(methyl methacrylate), General
Material class: Blends
Polymer Type: polydienes, acrylicpolyisoprene
Morphology: The two polymers are immiscible but blends can be compatibilised by the addition of a natural rubber-MMA graft copolymer. The effect of the added graft copolymer on the blend morphology varies with copolymer concentration, MW, mode of addition and processing conditions [1,2]

Mechanical Properties:
Mechanical Properties General: Tensile strength, tear strength and modulus increase with increasing PMMA content and degree of methacrylate cross-linking whereas elongation at break decreases. Toughness increases with increasing PMMA content up to a max. of approx. 50% PMMA before decreasing. Toughness decreases with increasing methacrylate cross-linking. [3] Blends of graft copolymer prepared by melt mixing rather than soln. casting have lower tensile strength and elongation at break [1,2]

Tensile (Young's) Modulus:

No.	Value	Note
1	10.2–167 MPa	30–70% wt PMMA [2]

Tensile Strength Break:

No.	Value	Note	Elongation
1	2.65–6.5 MPa	30–70% wt. PMMA, ASTM D638 [2]	365–12%
2	4.6–13.8 MPa	soln. cast, 2.5–12.5% graft copolymer, ASTM D638 [1,2]	120%

Mechanical Properties Miscellaneous: Tear strength 6.12–16.5 MPa (30–70% wt. PMMA, ASTM D624-81) [2]; 20.21–23.47 MPa (5–15% graft copolymer, 50:50 blend, ASTM D624-81) [2]
Fracture Mechanical Properties: Blends give fracture surfaces showing an elastic deformation with discontinuous fracture paths. Addition of graft copolymer reduces the size of the dispersed domains and improves interfacial adhesion

Izod Notch:

No.	Value	Notch	Note
1	0.81–1.18 J/m	Y	5–15% graft copolymer, 50:50 blend, ASTM D256 [2]

Applications/Commercial Products:
Processing & Manufacturing Routes: Inter-penetrating networks are synth. by swelling of masticated natural rubber, cross-linked with dicumyl peroxide, in methyl methacrylate. Benzoyl peroxide and divinylbenzene are used to cross-link methyl methacrylate. [3] Blends may be synth. by soln. casting from toluene or melt mixing. In soln. casting a blend of PMMA and natural rubber is cast after 10h immersion at room temp. In melt mixing, melted PMMA is blended with natural rubber then compression moulded [2]

Bibliographic References
[1] Oommen, Z., Gopinathan, Nair, M.R. and Thomas, S., *Polym. Eng. Sci.*, 1996, **36**, 151
[2] Oommen, Z. and Thomas, S., *J. Appl. Polym. Sci.*, 1997, **65**, 1245, (mech props)
[3] Das, B. and Gangopadhyay, T., *Eur. Polym. J.*, 1992, **28**, 867, (morphology, mech props)

NEW-TPI   N-4

Synonyms: *PMDA-3,3'-BAPB. Poly(pyromellitic dianhydride-alt-3,3'-bis(4-aminophenoxy)biphenyldiamine). Poly[3,3'-(4,4'-dioxyphenyl)diphenylene pyromellitimide]. Poly[(5,7-dihydro-1,3,5,7-tetraoxobenzo[1,2-c:4,5-c']dipyrrole-2,6(1H,3H)-diyl)-1,3-phenyleneoxy[1,1'-biphenyl]-4,4'-diyloxy-1,3-phenylene]. Poly(1H,3H-benzo[1,2-c:4,5-c']difuran-1,3,5,7-tetrone-alt-3,3'-bis(4-aminophenoxy)biphenyldiamine)*
Monomers: $1H,3H$-Benzo[1,2-c:4,5-c']difuran-1,3,5,7-tetrone
Material class: Thermoplastic
Polymer Type: polyimide
CAS Number:

CAS Reg. No.	Note
105359-94-2	
105218-97-1	source based

Molecular Formula: $(C_{34}H_{18}N_2O_6)_n$
Fragments: $C_{34}H_{18}N_2O_6$
Morphology: Orthorhombic, a 0.789, b 0.629, c (chain axis) 2.511. Space group P2. Unit cell has 2 chains of 1/1-helical symmetry (zigzag conformn.) [3]. Crystallinity <0.3 depends on crystallisation temp. [4]

Volumetric & Calorimetric Properties:
Density:

No.	Value	Note
1	d 1.32 g/cm^3	non-cryst. [3]
2	d 1.47 g/cm^3	cryst. [3]

Thermal Expansion Coefficient:

No.	Value	Note	Type
1	9.3×10^{-5} K^{-1}	25–275°, unit cell expansion, a axis [2]	L
2	6×10^{-5} K^{-1}	25–275°, unit cell expansion, b axis [2]	L
3	0.000153 K^{-1}	25–275° [2]	V

Latent Heat Crystallization: ΔH_f 6.38 kJ mol^{-1} (116 J g^{-1}, 100% crystalline) [3]
Melting Temperature:

No.	Value	Note
1	388°C	
2	383°C	[1]
3	381–389°C	depends on cryst. temp. [4]

Nitrile rubber

Glass-Transition Temperature:

No.	Value	Note
1	244°C	amorph. [4]
2	249°C	amorph. [4]
3	255°C	semicryst. [3]
4	244–253°C	semicryst., depends on cryst. temp. [1,2,4]

Transition Temperature General: $T°_m$ 406° (equilib. melting temp.) [3]

Transition Temperature:

No.	Value	Note	Type
1	312–319°C	depends on cryst. temp. [4]	Second melting temp.

Mechanical Properties:
Tensile (Young's) Modulus:

No.	Value	Note
1	1950 MPa	23°, 5 mm min^{-1}, injection moulded [6]
2	750 MPa	240°, 5 mm min^{-1}, injection moulded [6]

Complex Moduli: Dynamic tensile modulus 2200 MPa (T <Tg, 10 Hz, semicrystalline) [5], 150 MPa (T >Tg, 10 Hz, semicrystalline) [5], 2000 MPa (T <Tg, 10 Hz, amorph.) [5], 20 MPa (T >Tg, 10 Hz, amorph.) [5]

Electrical Properties:
Dielectric Permittivity (Constant):

No.	Value	Frequency	Note
1	3.21	10 kHz	150°, semicryst. and amorph. [5]
2	3.33	10 kHz	T >Tg, semicryst. [5]
3	3.44	10 kHz	T >Tg, amorph. [5]

Applications/Commercial Products:

Trade name	Supplier
Aurum	Mitsui Toatsu Chemicals
Regulus	Mitsui Toatsu Chemicals

Bibliographic References

[1] Saver, B.B., Hsiao, B.S. and Faron, K.L., *Polymer*, 1996, **37**, 445, (blends)
[2] Brillhart, M.V. and Cebe, P., *J. Polym. Sci., Part B: Polym. Phys.*, 1995, **33**, 927, (thermal expansion)
[3] Hsiao, B.S., Saver, B.B. and Biswas, A., *J. Polym. Sci. Part B: Polym. Phys.*, 1994, **32**, 737, (crystallisation)
[4] Huo, P.P., Friler, J.B. and Cebe, P., *Polymer*, 1993, **34**, 4387, (crystallisation)
[5] Huo, P.P. and Cebe, P., *Polymer*, 1993, **34**, 696, (mech and dielectric relaxation)
[6] Hirade, T., Hama, Y., Sasuga, T. and Seguchi, T., *Polymer*, 1991, **32**, 2499, (radiation resistance)

Nitrile rubber

Synonyms: *NBR. Nitrile butadiene rubber. Acrylonitrile-butadiene copolymer. Poly(acrylonitrile-co-butadiene). Poly(2-propenenitrile-co-1,3-butadiene). Poly(butadiene-co-acrylonitrile). Poly(1-butenylene-co-1-cyanoethylene)*
Related Polymers: Nitrile rubber - general purpose, Nitrile rubber - extrusion unfilled, Nitrile rubber - modifier, Nitrile rubber - injection moulding unfilled, Polyacrylonitrile, Hydrogenated nitrile rubber, Carboxylated nitrile rubber, Nitrile rubber and PVC blends, Butadiene-acrylonitrile copolymer, Hydrogenated nitrile rubbers, Blends of PVC and nitrile rubber, Carbon black filled nitrile rubber
Monomers: 1,3-Butadiene, Acrylonitrile
Material class: Synthetic Elastomers, Copolymers
Polymer Type: acrylonitrile and copolymers, polybutadiene
CAS Number:

CAS Reg. No.	Note
9003-18-3	
106974-60-1	alternating
106974-61-2	block

Additives: Can be vulcanised by sulfur, cadmium and peroxides
General Information: Acrylonitrile content 25–50% (18–22% low, 24–28% medium low, 28–34% medium high, 38–42% high). Different processing routes lead to 'hot' and 'cold' NBR

Applications/Commercial Products:

Trade name	Supplier
Chemigum	Goodyear
Chemigum N	Goodyear
Europrene N	Enichem Elastomers Ltd.
Krynac	Miles
Krynac	Polysar
Nipol DN	Nippon Zeon
Paracril	Uniroyal Chemical
Perrunan N	Bayer Inc.

Nitrile rubber, Carboxylated

Synonyms: *Carboxylated NBR. Carboxylated nitrile rubber. Carboxylated poly(acrylonitrile-co-butadiene). Carboxylated acrylonitrile-butadiene copolymer. Poly(1,3-butadiene-co-acrylonitrile), carboxylated*
Related Polymers: Nitrile rubber, Carbon black filled, Nitrile rubber, Nitrile rubber, Isoprene copolymer, Nitrile rubber, Ethylene copolymer, Nitrile rubber, Hydrogenated
Monomers: Methacrylic acid, Acrylic acid, Acrylonitrile, 1,3-Butadiene
Material class: Synthetic Elastomers, Copolymers
Polymer Type: acrylonitrile and copolymers, polybutadiene
CAS Number:

CAS Reg. No.	Note
123773-26-2	graft, methacrylic acid
9010-81-5	methacrylic acid copolymer
25265-19-4	acrylic acid copolymer
158496-90-3	graft, acrylic acid

Molecular Formula: $[(C_4H_6)_x(C_3H_3N)_y(C_3H_4O_2)_z]_n$ $[(C_4H_6).(C_3H_3N).[C_4H_6O_2)]$
Fragments: C_4H_6 C_3H_3N $C_3H_4O_2$ C_4H_6 C_3H_3N
Molecular Weight: MW 20000-25000
Additives: Zinc oxide, silica, antioxidants
General Information: All data refer to methacrylic acid copolymer unless otherwise stated
Morphology: C-13 nmr spectroscopy and pyrolysis-gc studies used to examine sequence distribution of terpolymer, which varies with mode of addition of methacrylic acid comonomer; mechanism of

polymerisation involves initiation reaction where methacrylic acid is preferentially involved [34]; microstruct. consistent with mech. props. [34]

Miscellaneous: Blend systems with ethylene acrylic rubber and fluoroelastomers have been reviewed [31]

Volumetric & Calorimetric Properties:
Density:

No.	Value	Note
1	d^{25} 0.9382 g/cm^3	[35]

Latent Heat Crystallization: Heat of combustion 41.4–42.3 kJ g^{-1} (9.9–10.1 kcal g^{-1}) (acrylic acid copolymer) [7]

Glass-Transition Temperature:

No.	Value	Note
1	-7--5°C	approx., commercial rubber [8]
2	19–224°C	17.3–48.1 wt% butadiene, 16.5–32.4 wt% acrylonitrile, 66.2–19.5 wt% methacrylic acid [33]

Transition Temperature General: T_g highly dependent on butadiene content of terpolymer [33]. T_g of the acrylic acid copolymer has been reported [6]

Transition Temperature:

No.	Value	Note	Type
1	1°C	[35]	Order-disorder transition
2	24°C	[35]	Order-disorder transition

Surface Properties & Solubility:
Solubility Properties General: Blend systems of carboxylated nitrile rubber (acrylic acid terpolymer) with ethylene acrylic rubber (VAMAC) and fluoroelastomer (Viton E60C) have been reviewed. Mech. props. (modulus, hardness, tensile strength, elongation at break), cure state, and solvent-swelling behaviour (e.g. in 2-butanone and THF), are dependent on blend preparation technique, i.e. preblending (with and without preheating) or master batch process; some phase separation observed in latter technique; fluoroelastomer blends show ageing effects; solvent-swelling studies used to determine phase adhesion between nitrile and additional rubber component; ir spectra of 50/50 blends reported [31]

Cohesive Energy Density Solubility Parameters: Interaction parameter 3.82 (acrylic acid copolymer) [10]

Solvents/Non-solvents: Sl. sol. toluene (acrylic acid copolymer) [11]. Sol. CDCl$_3$ [26]. Swollen by CHCl$_3$; 47–76% wt loss after 72h (various carbon-black mixtures) [9]

Wettability/Surface Energy and Interfacial Tension: Surface tension has been reported [24]. Adhesion to tin has been reported [20]

Transport Properties:
Polymer Melts: Mooney viscosity 57 (100°, commercial rubber, Krynach 7.50) [8]; 50 (100°, butadiene/acrylonitrile/methacrylic acid (66/27/7)) [9]; 60 (butadiene/acrylonitrile/methacrylic acid (67/27/6)) [18]

Permeability of Gases: Gas permeability data have been reported [20]

Water Content: Other water absorption data have been reported [22]

Water Absorption:

No.	Value	Note
1	0.86 %	acrylic acid copolymer [10]

Mechanical Properties:
Mechanical Properties General: Mechanical props. of the acrylic acid copolymer have been reported [1,2]. Other mechanical props. have been reported [13,24,27,28,29]. Other values of tensile strength have been reported [13,14,20,22]. Dynamic mechanical props. of terpolymer moulded in presence of ZnO, and effects of Si filler, have been reported at 28° and 3.5 Hz; variation of storage modulus and loss tangent with temp. reveals occurrence of biphasic struct. (particularly at high ZnO level, e.g. 12 phr); low temp. transition at approx. -5° is due to T_g of terpolymer, while high-temp. transition at 50° appears to result from formation of clusters; latter more prominent in presence of Si filler [8]; comparative studies of terpolymer cross-linked by a sulfur-accelerator system also reported [8]. Other dynamic mechanical investigations on unvulcanised terpolymer (butadiene/acrylonitrile/methacrylic acid (66/28/6)) have examined variation of complex shear compliance with temp. and frequency; frequency variation of compliance reveals existence of broad retardation (or relaxation) dispersion (over temp. range) similar to that found for amorph. polymeric materials; additional sharp resonance features found at temps. above 20° and below 0°; latter believed to be order/disorder transitions. These results, together with microstructural and stress requirements, indicate some type of one-dimensional order (characteristic length) extending over approx. 6μm in terpolymer [35]. Dynamic mechanical studies (-100°-200°, 3.5 Hz) and measurements of failure props. have been used to examine chemical interactions between nitrile rubber and surface-oxidised carbon black; functional groups of oxidised carbon black chemically react with carboxylated rubber when heated at high temps. for prolonged periods; reinforcing ability of surface-oxidised carbon black greater than non-oxidised material [9]; solvent-swelling studies in CHCl$_3$ also reported [9] in complementary studies. Stress-strain behaviour (ASTM D412-87) of various carboxylated rubber carbon-black mixes, moulded at 160° for 60 min., has been reported [9]. Modulus, tensile strength, elongation at break and hardness of blend systems with ethylene acrylic rubbers and fluoroelastomers have been reported [31]. Storage modulus of acrylic acid copolymer has been reported [6]

Tensile Strength Break:

No.	Value	Note	Elongation
1	4.48 MPa	carbon black filled, acrlic acid copolymer [9]	489%
2	0.94–8.11 MPa	various carbon black mixtures [9]	1970–988%

Miscellaneous Moduli:

No.	Value	Note	Type
1	1.26–3.33 MPa	various carbon black mixtures [9]	100% modulus
2	1.6–5.03 MPa	various carbon black mixtures [9]	300% modulus

Viscoelastic Behaviour: Stress-strain curves of carbon black filled, acrylic acid copolymer have been reported [9]

Hardness: Hardness of the acrylic acid copolymer has been reported [2]. Shore A70 (carbon black filled acrylic acid copolymer) [9]. Shore A51–A70 (various carbon black mixtures) [9]

Failure Properties General: Tear strength 39.1 N mm^{-1} (carbon black filled, acrylic acid copolymer) [9]. Other values of tear strength have been reported [14]

Fracture Mechanical Properties: Cut-growth crack props. have been reported [13]

Friction Abrasion and Resistance: Abrasion resistance of the acrylic acid copolymer has been reported [2,3]. Abrasion loss 0.12 cm^3 h^{-1} (BS903 part A9: 1957- method C, carbon black filled acrylic acid copolymer). Other values of tear strength have been reported [14]. Abrasion loss 0.12-0.38 cm^3 h^{-1} (various carbon black mixtures) [9]

Electrical Properties:
Dissipation (Power) Factor:

No.	Value	Frequency	Note
1	0.1–0.35	3.5 Hz	28°, 0–60 phr ZnO [8]
2	0.1–0.2	3.5 Hz	28°, 0–60 phr sulfur [8]

Optical Properties:
Transmission and Spectra: C-13 nmr [4,26,32,34], ir [17,31], mass [25] and H-1 nmr [32,33] spectral data have been reported

Polymer Stability:
Thermal Stability General: Resistance to heat ageing [13] and to high temp. has been reported [19]. Overall thermal stability of terpolymer dependent on thermal stability of each comonomer component, particularly methacrylic acid content; thermal behaviour of terpolymers similar to thermal degradation behaviour of individual homopolymers; evidence that methacrylic acid is thermally the weakest link in terpolymer [33]

Decomposition Details: Thermo-oxidative degradation of the acrylic acid copolymer has been reported [5]. Activation energy for thermal decomposition 52 kJ mol^{-1} (approx., acrylic acid). Rate of decomposition differs with composition. Major products of pyrolysis at 550° and 600° are ethylene, propylene, butadiene, C_6H_6, toluene and vinylcyclohexene (from butadiene part), and acrylonitrile and acetonitrile (from acrylonitrile part); cyclised products also involved, e.g. cyclised PBD, which undergo further thermal decomposition to give various products. The greater the alternating nature of the copolymer the greater the extent of cyclisation reactions [34]. Additional spectroscopic studies of decomposition processes have been reported [12]

Flammability: Scorch control experiments have been reported for terpolymer (butadiene/acrylonitrile/methacrylic acid (67/27/6)); standard compounding recipes contain sulfur and ZnO; however, affinity of latter for carboxyl groups in acid component leads to excessive scorch and short shelf-life; such excessive scorch can be prevented by isolating ZnO from carboxylic acid groups until desired cure temp. is reached; zinc sulfide-coated ZnO, zinc phosphate-coated ZnO, or metallic alkoxide (combined with ZnO) have been used for these purposes; good scorch control achieved without affecting final physical props. [18]

Environmental Stress: Resistance to uv [15] and to humidity has been reported [19]

Chemical Stability: Chemical resistance of kaolin, TiO_2 and lithopane filled material has been reported [20]

Hydrolytic Stability: Water resistance has been reported [14]

Stability Miscellaneous: Resistance to γ-irradiation has been reported [16]

Applications/Commercial Products:
Processing & Manufacturing Routes: Synth. by emulsion polymerisation at 50° using AIBN or potassium persulfate [33] as a catalyst [29,32,34]. Detailed study of polymerisation process, including kinetics and identification of the azeotropic composition, has been reported [33]

Applications: Acrylic acid copolymer is used as a binder fuel in solid boosters of the space shuttle. Other uses in ball mill linings, mud pump pistons, pump impellors, roll coverings, conveyor belting covers, conveyor rollers, blow-out preventers, seals, O-rings

Trade name	Details	Supplier
Chemigum XNBR		Bayer Inc.
Chemigum XNBR		Goodyear
Europrene		Enichem
Hycar 1072	butadiene/acrylonitrile/methacrylic acid (66/28/6)	B.F. Goodrich
Krynack 7.50	butadiene/acrylonitrile/methacrylic acid (66/27/7)	Bayer Polysar
Nipol	various grades	Zeon
Pernuban		Bayer Inc.

Bibliographic References
[1] Seeger, M., *Proc. Int. Rubber Conf. (Milan)*, 1979, **442**; *CA*, **92**, 77782 (mechanical props.)
[2] Weir, R.J., *Rubber India*, 1980, **32**, 13, (uses, mechanical props.)
[3] Starmer, P.H., *Plast. Rubber Process. Appl.*, 1988, **9**, 209, (abrasion resistance)
[4] Kanakavel, M., *Polym. Sci. (Symp. Proc. Polym.)*, 1991, **2**, 588, (C-13 nmr)
[5] Radhakrishnan, T.S., and Rao, M.R., Propellants, *Explos. Pyrotech.*, 1995, **20**, 32, (thermo-oxidative degradation)
[6] Bhagawan, S.S., Prabhakaran, N., Rao, S.S. and Ninan, K.N., *Def. Sci. J.*, 1995, **45**, 17, (transition temp., power factor)
[7] Mastrolia, E.J. and Klager, K., *Adv. Chem. Ser.*, 1969, **88**, 122, (heat of combustion)
[8] Mandal, U.K., Tripathy, D.K. and De, S.K., *J. Appl. Polym. Sci.*, 1995, **55**, 1185, (power factor, transition temp.)
[9] Bandyopadhyay, S, De, P.P., Tripathy, D.K. and De, S.K., *J. Appl. Polym. Sci.*, 1995, **58**, 719, (mechanical props.)
[10] Potts, J.C., *J. Appl. Polym. Sci.*, 1965, **9**, 1841, (water absorption)
[11] Abbott, T.P., James, C. and Otey, F.H., *J. Appl. Polym. Sci.*, 1979, **23**, 1223, (solubility)
[12] Brill, T.B., Arisawa, H., Brush, P.J., Gongwer, P.E. and Williams, G.K., *J. Phys. Chem.*, 1995, **99**, 1384, (use, thermal decomposition)
[13] Dolgoplask, B.A., Reikh, V.N., Tinyakova, E.I., Kalaus, A.E. et al, *Kauch. Rezina*, 1957, **16**, 1, (mechanical props.)
[14] Koloberiin, V.N. and Blokh, G.A., *Khim. Prom. Nauk-Tekhn.-Tekhn. Zb.*, 1962, **1**, 22, (water resistance, mechanical props.)
[15] Kronbergs, V., Eiduss, J. and Borodkin, Y.G., *Vysokomol. Soedin., Ser. A*, 1972, **14**, 2311, (uv resistance)
[16] Saidov, D., Karimov, S.N., Sultanov, A., Bolikekov, U. and Khabibulloev, K., *Dokl. Akad. Nauk SSSR*, 1975, **18**, 20, (γ-irradiation resistance)
[17] Ruzicka, B. and Krotki, E., *Chem. Anal. (Warsaw)*, 1971, **16**, 1207, (ir)
[18] Hallenbeck, V.L., *Rubber Chem. Technol.*, 1973, **46**, 78, (scorch resistance)
[19] Hoang, D. and Belou, B.I., *Tr. Mosk. Inst. Nar. Khoz.*, 1972, **97**, 127, (humidity resistance)
[20] Kulish, M.I., Bespalyi, A.S. and Panchuk, F.D., *Khim. Tekhnol. (Kiev)*, 1975, **1**, 32, (chemical resistance, gas permeability)
[21] Pistrovskii, K.B., Sokolova, I.D. and Pchelintsev, V.V., *Prom.-St. Sint. Kauch.*, 1978, **1**, 4, (oxidative thermodegradation, antioxidants)
[22] Vaskovskaya, A.A., Gnebman, Y.V., Gulimova, D.K. and Shteinberg, S.A., *Kauch. Rezina*, 1982, **1**, 47, (tensile strength, water absorption)
[23] Lyalyushko, D.S., Chebotarevskii, V.V. and Vladimirskii, V.N., *Lakokras. Mater. Ikh Primen.*, 1983, **6**, 12, (miscibility)
[24] Soloveva, T.S., Ivanova, N.N., Afanaseeva, G.V., Goryacheeva, G.B. and Sorokina, T.B., *Izu. Vyssh. Uchebn. Zaved., Tekhnol. Legk. Prom.-Sti.*, 1985, **28**, 48, (surface tension)
[25] Belov, B.I. and Dsipov, N., *Izu. Vyssh. Uchebn. Zaved., Tekhnol. Legk. Prom.-Sti.*, 1987, **30**, 31, (ms, thermal degradation)
[26] Gruzinov, E.V., Panov, V.P., Gusev, V.V., Frosin, V.N. and Rybchinskaya, V.S., *Vysokomol. Soedin., Ser. B*, 1987, **29**, 373, (C-13 nmr)
[27] Ono, Y., Tanigaki, T., Yamaguchi, Y.V., Tanino, K. and Hashizume, G., *Kobunshi Ronbunshu*, 1989, **46**, 389, (mechanical props.)
[28] Marwede, G.W., Oppenheimer-Stix, C., and Magg, H., *Annu. Meet. Proc. Int. Inst. Synth. Rubber Prod.*, 35th, 1994, T523, (mechanical props., uses)
[29] Marinelli, L. and Magg, H., *Muanyag Gumi*, 1996, **33**, 139, (synth., mechanical props.)
[30] *The Materials Selector*, 2nd edn., (eds., N.A., Waterman, and M.F. Ashby), Chapman and Hall, 1997, **3**, 497, (uses)
[31] *Polymeric Materials Encyclopedia*, (ed. J.C. Salamone), CRC Press, 1996, **1**, 709, (miscibility, ir)
[32] Kanakavel, M. and Sebastian, T.V., *Makromol. Chem.*, 1985, **186**, 1313, (H-1 nmr, C-13 nmr, synth.)
[33] Jerman, M.B. and Janovic, Z., *J. Macromol. Sci., Chem.*, 1984, **21**, 887, (H-1 nmr, synth. transition temp.)
[34] Rao, M.R., Sebastian, T.V., Radhakrishnan, T.S. and Ravindran, P.V., *J. Appl. Polym. Sci.*, 1991, **42**, 753, (C-13 nmr, synth., thermal degradation)
[35] Fitzgerald, E.R., *J. Appl. Polym. Sci.*, 1975, **19**, 2015, (transitions, complex shear compliance)

Nitrile rubber, Extrusion unfilled   N-7
Related Polymers: Nitrile rubber
Monomers: 1,3-Butadiene, Acrylonitrile
Material class: Synthetic Elastomers, Copolymers
Polymer Type: acrylonitrile and copolymers, polybutadiene
CAS Number:

CAS Reg. No.
9003-18-3

Applications/Commercial Products:

Trade name	Supplier
Chemigum HR662	Goodyear
Chemigum N	Goodyear
Krynac	Polysar
Nitriflex N	A. Schulman Inc.
NySyn 30-5	Copolymer Rubber
NySyn 30-8	Copolymer Rubber
NySyn 33-3	Copolymer Rubber
Paracril AJLT	Uniroyal Chemical
Paracril AJLT M-60	Uniroyal Chemical
Paracril ALT	Uniroyal Chemical
Paracril BJLT M-30	Uniroyal Chemical
Paracril BJLT M-40	Uniroyal Chemical
Paracril BJLT-HX	Uniroyal Chemical

Nitrile rubber, General purpose   N-8
Related Polymers: Nitrile rubber
Monomers: 1,3-Butadiene, Acrylonitrile
Material class: Synthetic Elastomers, Copolymers
Polymer Type: acrylonitrile and copolymers, polybutadiene
CAS Number:

CAS Reg. No.
9003-18-3

Applications/Commercial Products:

Trade name	Details	Supplier
Chemigum		Goodyear
Humex 3883		Hules Mexicanos
Krynac		Polysar
Nitriflex N-300		A. Schulman Inc.
Nitriflex N-5		A. Schulman Inc.
Nitriflex N-615-B		A. Schulman Inc.
NySyn		Copolymer Rubber
NySyn DE	amine terminated	Copolymer Rubber
NySyn DN	amine terminated	Copolymer Rubber
Nysyn MDN131		Copolymer Rubber
Paracril		Uniroyal Chemical
Paracril 1880		Uniroyal Chemical
Paracril 1880 LM		Uniroyal Chemical
Paracril 4880		Uniroyal Chemical
Paracril AJ		Uniroyal Chemical
Paracril AJLT		Uniroyal Chemical
Paracril B		Uniroyal Chemical
Paracril BJ		Uniroyal Chemical
Paracril BLT		Uniroyal Chemical
Perbunan		Mobay
Perbunan		Bayer Inc.

Nitrile rubber, Hydrogenated   N-9
Synonyms: Poly(1,3-butadiene-co-acrylonitrile), hydrogenated. HNBR. Hydrogenated acrylonitrile-butadiene copolymer. Hydrogenated nitrile rubber. Hydrogenated poly(acrylonitrile-co-butadiene). HSN. Highly saturated nitrile elastomer
Related Polymers: Nitrile rubber, carboxylated, Nitrile rubber, isoprene copolymer, Nitrile rubber, ethylene copolymer, Nitrile rubber, mineral filled, Nitrile rubber, Nitrile rubber, carbon black filled, Nitrile rubber, PVC blends
Monomers: 1,3-Butadiene, Acrylonitrile
Material class: Synthetic Elastomers, Copolymers
Polymer Type: acrylonitrile and copolymers, polybutadiene
CAS Number:

CAS Reg. No.	Note
88254-10-8	
9003-18-3	Nitrile rubber

Molecular Formula: $[(C_4H_6)_x(C_4H_8)_y(C_3H_3N)_z]_n$
Fragments: C_4H_6 C_4H_8 C_3H_3N
Molecular Weight: MW 150000 (100% hydrogenated)
Additives: Curing agents include sulfur, and peroxides with triallyl cyanurate or isocyanurate. Fillers (carbon black); antioxidants; plasticisers; stearic acid; blend components (e.g. PMMA, nylon-6)
General Information: Fully hydrogenated nitrile rubber is a speciality elastomer with excellent physical and mech. props., good resistance to heat, ageing and aggressive environments; has outstanding wear resistance under adverse conditions. Reviews, including physical and mech. props., and applications (in particular those in the oil field industries) have been reported [2,6]. Details of heat- and swell-resistant elastomer containing 34% nitrile component have been reported [1]. Physical props., stability and potential applications of partially hydrogenated nitrile rubber have been reviewed; generally displays similar props. to nitrile rubber, but in addition has improved chemical, environmental and thermal stability [5]
Morphology: Struct. has been reported [20]. Microstructure of hydrogenated nitrile rubbers has been analysed by spectroscopy. Ir spectra indicate preferential hydrogenation reaction of 1,2-butadiene units over 1,4-units; good complementary information about longer sequences and distributions along the chains and mechanisms of hydrogenation reaction are provided by pyrolysis studies and C-13 nmr spectroscopy; C-13 nmr spectra suggest more likelihood of hydrogenation of butadiene units next to acrylonitrile units in polymer chain than of those next to same (butadiene) units [33]
Identification: Degree of unsaturation (iodine value, g $(100g)^{-1}$) 14 (37% acrylonitrile), 28 (37% acrylonitrile), 25 (45% acrylonitrile) (commercial raw elastomers) [2]
Miscellaneous: Blend system with zinc-PMMA has been reported [18]. Studies of blends of hydrogenated rubber with nylon-6 have been reported [37]. General chemical and physical props. of (selectively) hydrogenated nitrile rubber explained by ethylene sequences, which result from 1,4-butadiene sequences, and acrylonitrile content; increased degree of hydrogenation leads to increased crystallinity on account of increased ethylene sequences [38]; in principle, hydrogenated nitrile rubber could be obtained by terpolymerisation of ethene, 1-butene and acrylonitrile [38].

Blends of hydrogenated nitrile rubber and PVC, when processed at high mould temps., undergo cross-linking without aid of any cross-linking agents; known as 'self-cross-linkable polymer blends' [25]

Volumetric & Calorimetric Properties:
Density:

No.	Value	Note
1	d 1.15 g/cm^3	44% acrylonitrile, Zetpol 1020 [34]
2	d 0.95 g/cm^3	fully saturated, double bond content <1%; 34% acrylonitrile [31]
3	d 0.99–1.18 g/cm^3	vulcanisate; dependent on composition
4	d 1.18 g/cm^3	30 phr resin [34,35]
5	d 0.98–1 g/cm^3	commercial raw elastomer; 37–45% acrylonitrile [2]
6	d 1.24 g/cm^3	vulcanisate compound formulation [6]

Glass-Transition Temperature:

No.	Value	Note
1	-40–-15°C	[15]
2	-3°C	Zetpol 1020 [36]
3	-25°C	butadiene:acrylonitrile 2:1 [9]
4	-25°C	vulcanisate compound formulation [6]
5	-26°C	45% acrylonitrile [2]
6	-32°C	37% acrylonitrile [2]
7	-33°C	37% acrylonitrile [2]

Transition Temperature General: Brittleness temp. has been reported [4]. Thermal props., e.g. T_g, affected by degree of crystallinity/hydrogenation [38]. Copolymer shows no melting endotherm at or below 180° indicative of non-crystalline material [9]. Brittleness temp. is approx. 6° lower than that of corresponding non-hydrogenated rubber [2]

Vicat Softening Point:

No.	Value	Note
1	85°C	uncured; disappears after vulcanisation by γ-irradiation [31]

Surface Properties & Solubility:
Solubility Properties General: Effects of compatibilisers or fillers on miscibility and processing of hydrogenated nitrile rubber and nylon-6 have been reported [37]. Forms binary blends with PVC at high moulding temp. without any additional cross-linking agents; known as self-crosslinkable polymer blends [25]

Cohesive Energy Density Solubility Parameters: Solvent interaction parameters with alkanes, cyclohexane, C_6H_6 and CCl_4 have been reported [10]. Interaction parameter 0.61 (Me_2CO) [10]. Cohesive energy has been reported [10]. Relative cohesive energy 7.91 kJ mol^{-1} (calc.) [11]. Solubility parameter δ 6.4–6.43 J$^{1/2}$ cm$^{-3/2}$, (9.81–9.88 cal$^{1/2}$ cm$^{-3/2}$, viscosity method); 6.35–6.37 J$^{1/2}$ cm$^{-3/2}$ (9.64–9.69 cal$^{1/2}$ cm$^{-3/2}$, graphical method); 6.37 J$^{1/2}$ cm$^{-3/2}$ (9.7 cal cal$^{1/2}$ cm$^{-3/2}$, 50–100% hydrogenation) [21]. Huggins constant 0.478 (100% hydrogenation) [21]. Little change observed in value of solubility parameter of nitrile rubber upon hydrogenation, and upon level of hydrogenation [21]

Plasticisers: Di(butoxyethoxyethyl) adipate, trioctyl trimellitate, triisononyl trimellitate, trioctylphosphate [15]

Solvents/Non-solvents: Sol. chlorinated hydrocarbons [21], THF [31]. Insol. MeOH [9]

Wettability/Surface Energy and Interfacial Tension: Interfacial tension 0.24 mN m^{-1} (nylon) [24]. Studies of the adhesion of PVC to hydrogenated nitrile rubber (self-crosslinkable polymer blend system) have been reported; adhesive strength determined by T-peel tests (Instron); peel-fracture energy dependent on presence of stabiliser and plasticiser in PVC phase, in addition to moulding and test conditions; plasticisation improves adhesion; peel strength affected by stabiliser only at high contact temp. and long contact times; high joint strength caused by formation of chemical bonds (cf. self-crosslinkable blend system) [25]. Contact angle measurements have been used to investigate changes in wettability of the hydrogenated rubber surface with ageing time and temp. (75–150°, 0–8h); wettability increases with time and is more rapid when temp. increases [13]

Transport Properties:
Polymer Melts: Mooney viscosity 68 (100°, fully saturated; double-bond content <1%; 34% acrylonitrile) [31], 80 (100°, commercial raw elastomers; 37–46% acrylonitrile) [2]

Permeability of Gases: Permeability coefficients for several gases have been reported; temp. dependence (related to activation energy) found [14]. High-pressure permeation studies of H_2 through hydrogenated nitrile rubber (38% acrylonitrile) containing moderate loadings of filler have been reported [16]

Permeability of Liquids: Permeation rate [MeOH] 0.008 mmol cm^{-2} h^{-1} (23°, 1.1 mm film) [30]

Mechanical Properties:
Mechanical Properties General: Mech. props. of partially hydrogenated nitrile rubbers have been reported, including brittle temp., compression set, tensile strength and abrasion resistance [4]. Temp. dependence of tensile strength of hydrogenated nitrile rubber (filled and unfilled) has been reported over range 20–180° [19]. Mech. props. (e.g. tensile strength) of raw rubber and carbon black filled vulcanisates affected by degree of crystallinity/hydrogenation [38]. Tensile strength of vulcanised (γ-irradiated) rubber increases with dosage rate, while value of elongation at break is reduced; optimum dose to achieve best balance of props. reportedly 14–20 Mrad [31]. Mech. and viscoelastic (dynamic mech.) props. as a function of vulcanisate composition, e.g. carbon black and resin content, cure system and cross-link density have been reported [34,35]; incorporation of resin increases strength props.; 30 phr resin found to be optimum [35]. Comprehensive studies of the mech. props. (tensile strength/elongation, 50% modulus, hardness, elongation at crack) of hydrogenated nitrile rubber vulcanisate compound formulations with varying acrylonitrile contents (e.g. 37% and 45%) after exposure to a severe environment (oil-field type) have been reported [2,5]

Tensile Strength Break:

No.	Value	Note	Elongation
1	30 MPa	approx., carbon black filled	
2	35.7 MPa	44% acrylonitrile, Zetpol 1020 [34,35]	320%
3	17 MPa	100° [35]	
4	11–39 MPa	25° [34,35]	
5	6–30 MPa	75° [34,35]	
6	8–35 MPa	50° [34,35]	
7	4–20 MPa	100° [34,35]	320–600%
8	27–38 MPa	after ageing, 120°, 70h, (vulcanisate; dependent on composition [34,35]	
9	15–35 MPa	various vulcanisate compound formulations [6]	60–500%

Tensile Strength Yield:

No.	Value	Elongation	Note
1	32 MPa		37% acrylonitrile; commercial raw elastomer [2]
2	29.4 MPa	110% extension	37% acrylonitrile; vulcanisate compound formulation [2]
3	28.3 MPa	310% extension	45% acrylonitrile; vulcanisate compound formulation [2]
4	13 MPa		150°; 37% acrylonitrile; vulcanisate compound formulation [2]
5	9 MPa		150°; 45% acrylonitrile; vulcanisate compound formulation [2]

Miscellaneous Moduli:

No.	Value	Note	Type
1	2.5–36.2 MPa	vulcanisate; dependent on composition [34,35]	300% modulus
2	13.9 MPa	37% acrylonitrile; vulcanisate compound formulation [2]	50% modulus
3	3.7 MPa	45% acrylonitrile; vulcanisate compound formulation [2]	
4	26.3 MPa	37% acrylonitrile; vulcanisate compound formulation [2]	100% modulus
5	7.7 MPa	45% acrylonitrile; vulcanisate compound formulation [2]	100% modulus
6	19.5 MPa	45% acrylonitrile; vulcanisate compound formulation [2]	200% modulus
7	3–20 MPa	various vulcanisate compound formulations [6]	100% modulus
8	4.5–26 MPa	various vulcanisate compound formulations [6]	200% modulus
9	6–30 MPa	various vulcanisate compound formulations [6]	300% modulus

Viscoelastic Behaviour: Dynamic mech. props. and relaxation behaviour are affected by degree of crystallinity/hydrogenation [38]. Relaxation behaviour of vulcanised (by γ-irradiation) rubber has been reported. Cross-linking (three-dimensional network) restrains chain motion and therefore material cannot flow (even at higher temps.); behaves like cross-linked elastomer [31]. Dynamic mech. props. and influence of cross-link density, curing system, filler and resin, of hydrogenated nitrile rubber vulcanisates has been reported [34]. Log G′ 6.25–8.1 Pa; unaffected by cross-link density in glassy region, but increases with density in rubber (plateau) zone; G′ for different cure systems (peroxide or sulfur), with same cross-link density, has same values in glassy state; increases with temp. from 90–180°, and reduces above 180°; increases with both resin and carbon-black filler content; also increases with carbon-black surface area; loss tangent (-50–250°) 0.01–1.26; unaffected by cross-link density in glassy region, but decreases with density in rubber (plateau) zone; loss tangent for different cure systems (peroxide or sulfur), with same cross-link density, has same value in glassy state; decreases with temp. from 90–180°, and increases again above 180°; decreases with both resin- and carbon-black filler content; also affected by surface area of filler [34]

Mechanical Properties Miscellaneous: Flexural stiffness measurement at low degrees of deformation has been reported as a method for monitoring ageing behaviour and heat resistance [12] Rebound resilience 30–40% (various vulcanisate compound formulations) [6] Compression set 15% (70h, room temp.), 20% (70h, 150°) (various vulcanisate compound formulations) [6]
Hardness: Shore A79 (44% acrylonitrile, Zetpol 1020) [34]. Shore A51-A85 (vulcanisate; dependent on composition) [34,35]; Shore A70 (commercial raw elastomer; 37% acrylonitrile); A92 (37% acrylonitrile, vulcanisate compound formulation); A84 (45% acrylonitrile, vulcanisate compound formulation) [2]. Shore A50-A90 (various vulcanisate compound formulations) [6]
Failure Properties General: Tear strength 920 N cm^{-1}, (44% acrylonitrile, Zetpol 1020) [34]. Breaking energy (Zetpol 1020) 28.7 kJ m^{-2} (25°), 23.5 kJ m^{-2}, (150°), 1.6 kJ m^{-2} (75°), 11.8 kJ m^{-2} (100°) [35] Tear strength 350–1090 N cm^{-1} (vulcanisate; dependent on composition) [34,35]. Breaking energy 9.2–29.8 kJ m^{-2} (25°), 3.5–26.6 kJ m^{-2} (50°), 1.5–18.8 kJ m^{-2} (75°), 0.6–14.9 kJ m^{-2} (100° vulcanisate; dependent on composition).
Friction Abrasion and Resistance: Displays outstanding abrasion resistance [18]. Comprehensive studies of the influence of compositional variables and testing temp. on the wear of hydrogenated nitrile rubber vulcanisates [35] and investigations of the mechanisms of wear, using optical and scanning electron microscopy [36], have been reported. Abrasion loss v 0.75–5.5 × 10^{-10} m^3 rev^{-1} (Dupont abrader); values increase with decrease in cross-link density (similar values obtained for sulfur and peroxide-cured systems). Incorporation of resin and carbon-black filler decreases abrasion loss, in former case at all temps. (25–100°). [35] μ_{dyn} 1.24–1.96 (Dupont abrader); values increase with carbon-black filler loading, and decrease with cross-link density, temp. and incorporation of resin [35]. Frictional force 44–69.4 N (Dupont abrader); values increase with carbon-black filler loading, and decrease with cross-link density and incorporation of resin (both with temp.) [35]. Abradability 2.2–11.4 × 10^{-15} m^3 J^{-1} m^2 (25–100°); increases linearly with reciprocal of breaking energy [35]. Abraded surfaces at 25° do not show any ridges except ploughing marks along direction of abrasion; ridges observed above 50°, with spacing between adjacent ridges increasing with temp., and decreasing with carbon-black filler loading and incorporation of resin; worn surfaces of swollen vulcanisates show ridges at all normal loads (6–44 kPa); change from abrasive wear mechanism (at 25°) to frictional wear mechanism above 50° [35,36]. Abrasion loss 30–50 mm^3 (various vulcanisate compound formulations) [6]

Electrical Properties:
Dissipation (Power) Factor:

No.	Value	Note
1	0.157	50°, Zetpol 1020 [35]
2	0.118–0.167	50° [34,35]
3	0.11–0.163	75° [34,35]
4	0.108–0.158	100° [34,35]

Optical Properties:
Transmission and Spectra: Ir [9,31,32,33], H-1 nmr [32,33], C-13 nmr [32,33] and mass spectral data have been reported [33]

Polymer Stability:
Polymer Stability General: Degree of crystallinity (cf. hydrogenation) does not have significant effect on ageing behaviour [38]
Thermal Stability General: Has good thermal ageing characteristics (partially hydrogenated nitrile rubber) [4,5,18]; heat resistance has been measured by the use of a flexural stiffness method [12]. Hydrogenated nitrile rubber is thermally more stable than unhydrogenated rubber [13]. Incorporation of resin into vulcanisate compound formulation improves the heat resistance [35]

– Nitrile rubber, Hydrogenated

Upper Use Temperature:

No.	Value	Note
1	140–150°C	[6]

Decomposition Details: Activation energy of degradation 175–220 kJ mol^{-1} [13], 330 kJ mol^{-1} [18], 220kJ mol^{-1} (37% acrylonitrile; partially hydrogenated), 350 kJ mol^{-1} (40% acrylonitrile; highly saturated rubber). Characteristic degradation products at 550°: butadiene, 4-vinylcyclohexane (butadiene dimer), acrylonitrile, hydrocarbons, mononitrile species and dinitrile species; nitriles can be either α-olefinic or saturated species; hydrocarbons represent the methylene chains produced by hydrogenation [33]. Oxidative decomposition behaviour and kinetics of hydrogenated rubber have been studied. Decomposition in air is a two-stage process; in first stage, majority of nitrile rubber (86%) decomposes; onset at 340°, with maximum rate at 495° (for both filled and unfilled material); combustion of carbon-black filled samples and decomposition of remainder of nitrile rubber take place in second stage (onset at approx. 540°, with maximum rate at 580°). Degradation characteristics and kinetics are not influenced by cross-link density, curing system or carbon-black loading [18]. Oxidative degradation initiated by γ-irradiation at 40° has been examined by FT-ir spectroscopy [9]. Studies of degradation at low and high temps. in air and N_2 have been reported [13]

Environmental Stress: Has excellent resistance to ozone [17]. Partially hydrogenated nitrile rubber displays good ozone and weather resistance [5,18]

Chemical Stability: Resistant to aliphatic hydrocarbons and to inorganic chemicals other than oxidising acids [28]. Partially hydrogenated nitrile rubber displays good resistance to lubricating oils and their additives [2,4,5,28], and fuels (in particular, oxidised fuels) [2,4]; ageing behaviour in various corrosive media has been monitored by the use of flexural stiffness measurements; ageing in oil, in presence of air, takes place more rapidly than in oil or air alone [12]; good resistance to 'sour-gas' and petroleum (also in presence of H_2S and amines) [12]. Degradation rates of hydroxylated nitrile rubber are several times slower than those of nitrile rubber and epichlorohydrin- ethylene oxide copolymer (factor of 10) in oxidised fuel [2]; dimensional (elongational) changes of hydroxylated nitrile rubbers in lubricating oil much less than those of polyacrylates [2]. Detailed information on the chemical resistance of hydrogenated rubber to a wide range of reagents has been reported [2]; elastomer resistant to alkali, most acids, water, ASTM standard oils and alcohols [2]; attacked by concentrated sulfuric and nitric acids, toluene, EtOAc and Me_2CO (reduction in tensile strength) [2]

Hydrolytic Stability: Resistance to steam has been reported [22]

Stability Miscellaneous: Service temps. of hydrogenated rubbers (at given service life) are approx. 20° higher than those of corresponding unhydrogenated rubbers [2]. Partially hydrogenated nitrile rubber displays good performance profiles and requires little maintenance [5]

Applications/Commercial Products:

Processing & Manufacturing Routes: Produced by a soln. process that gives selective hydrogenation of the 1,3-butadiene double bond [29]. Also synth. by hydrogenation using Rh or Pd catalysts [32]. Compounding similar to that of standard nitrile rubbers; however, more careful selection of vulcanisation system is required [4]. Hydrogenation carried out under high pressure in presence of rhodium triphenylphosphine catalyst [21]

Applications: Has applications in the oil industry (blow-out preventers, drill pipe protectors, down hole packers, mud pump pistons, valve seals, cable jackets, pressure accumulator bags, fuel diaphragms); in the automotive industry (rotating shaft seals, lip seals, bearing/wheel seals, O-rings, valve cover gaskets, fuel pump diaphragms, fuel pump isolators, fuel hose tubes, timing belts, valve stem seals, water pump seals). General industrial applications in hydraulic hose and packings; as power station cable jacket; in laminating rolls, printing rolls, textile rolls, paper mill rolls; in military components and chemical plant diaphragms. Partially hydrogenated nitrile rubber finds applications on account of good heat and oil resistance, excellent ageing props. and outstanding abrasion resistance (even under adverse conditions) in tank track pads, oil well packers, automotive shaft seals etc. Blend compositions (e.g. with PMMA) also used for similar applications, i.e. under adverse conditions (oil at high temps.) Highly saturated nitrile elastomers have applications on account of excellent resistance to heat, ageing, wear and aggressive environments, and good mech. props., at both low and high temps., in automotive, aircraft, construction and oil field industries

Trade name	Details	Supplier
Therban	various grades	Polysar
Therban	various grades	Bayer Inc.
Tornac	A	Bayer Inc.
Tornac- B3850	38% acrylonitrile	Polysar
Zetpol	2020, vulcanisates, 37% acrylonitrile	Nippon Zeon
Zetpol	HSN, highly saturated nitrile elastomers	Nippon Zeon
Zetpol	1020, partially hydrogenated, 45% acrylonitrile	Nippon Zeon
Zetpol	2010, 37% acrylonitrile	Nippon Zeon

Bibliographic References

[1] Thoermer, J., Marwede, G. and Buding, H., *Kautsch. Gummi Kunstst.*, 1983, **36**, 269, (oil resistance, mech. props.)
[2] Hashimoto, K., Watanabe, N. and Yoshioka, A., *Rubber World*, 1984, **190**, 32, (chemical resistance, mech. props.)
[3] Sugimoto, M. and Kondo, T., *Toyoda Gosei Giho*, 1984, **26**, 51, (oil resistance)
[4] Hashimoto, K., Watanabe, N., Dyama, M. and Todani, Y., *Gummi Fasen. Kunstst.*, 1984, **37**, 602, (brittle temp., mech. props.)
[5] Mirza, J., Schoen, N. and Thoermer, J., *Kautsch. Gummi Kunstst.*, 1986, **39**, 615, (chemical stability, weathering, heat resistance)
[6] Thoenmer, J. Mirza, J. and Schoen, N., *Elastomerics*, 1986, **118**, 28, (chemical resistance)
[7] Cseh, M. Edelenyi, A., Nagy, T.T. and Pandur, J., Proc. Tihany Symp. Radiat. Chem., *6th*, 1987, 639, (radiation resistance)
[8] Skorski, W. and Lipinska, M., *Polimery (Warsaw)*, 1988, **33**, 477, (agrochemical resistance)
[9] Carlsson, D.J., Chmela, S. and Wiles, D.M., *Makromol. Chem., Makromol. Symp.*, 1989, **27**, 139, (oxidative degradation, ir)
[10] Frenkel, R.S. and Kirillova, T.I., *Kauch. Rezina*, 1989, **6**, 44, (cohesive energy, solvent interaction parameter)
[11] Frenkel, R.S., Litinskii, A.D., and Safonov, A.V., *Plaste Kautsch.*, 1990, **37**, 73, (relative cohesive energy)
[12] Dinzburg, B. and Bond, R., *Gummi Fasen. Kunstst.*, 1990, **43**, 54, (flexural stiffness)
[13] Bhattacharjee, S, Bhowmick, A.K. and Avasthi, B.N., *Polym. Degrad. Stab.*, 1991, **31**, 71, (thermal degradation)
[14] Beckmann, W., *Kautsch. Gummi Kunstst.*, 1991, **44**, 323, (gas permeability)
[15] Hayashi, S., Sakakida, H., Dyama, M. and Nakagawa, T., *Rubber Chem. Technol.*, 1991, **64**, 534, (transition temps.)
[16] Campion, R.P. and Morgan, G.J., Plast., *Rubber Compos. Process. Appl.*, 1992, **17**, 51, (gas permeability)
[17] Nakagawa, T., Toya, T. and Dyama, M., *J. Elastomers Plast.*, 1992, **24**, 240, (ozone resistance)
[18] Thauamani, P., Sen, A.K., Khastgir, D. and Bhowmick, A.K., *Thermochim. Acta*, 1993, **219**, 293, (thermal degradation)
[19] Sorokin, G.A., Larionov, V.F. and Siddawa, S.I., *Kauch. Rezina*, 1992, **5**, 23, (tensile strength)
[20] Chen, R., *Xiangjiao Gongye*, 1992, **39**, 628, (struct., mech. props., uses)
[21] Roy, S., Bhattacharjee, S., Bhowmick, A.K., Gupta, B.R. and Kulkami, R.A., *Macromol. Rep.*, 1993, **A30**, 301, (solubility parameter, Huggins constant)
[22] Klinger, R.C., Proc. ADUMAT/91 Int. Symp. Environ. Eff. Adv. Mater., 1991, **31**, 1, (chemical resistance)
[23] Nomura, A, Takano, J, Toyoda, A. and Saito, T., *Nippon Gomu Kyokaishi*, 1993, **66**, 830, (tensile strength)
[24] Bhowmick, A.K. and Inoue, T., *J. Appl. Polym. Sci.: Appl. Polym. Symp.*, 1994, **53**, 77, (interfacial tension)

[25] Manoj, N.R. and De, P.P., *J. Elastomers Plast.*, 1994, **26**, 265, (adhesion)
[26] Ishihara, M., *Nippon Gomu Kyokaishi*, 1995, **68**, 643, (friction, abrasion)
[27] Wada, N., Inoue, A., Kunita, Y., Nakajima, M. *et al*, *Nippon Gomu Kyokaishi*, 1996, **69**, 430, (coefficient of friction)
[28] *The Materials Selector*, 2nd edn. (eds. N.A. Waterman and M.F. Ashby), Chapman and Hall, 1997, **3**, 497, (uses)
[29] *Encyclopaedia of Chemical Technology*, 4th edn., (eds. J.I. Kroschwitz and M. Howe-Grant), John Wiley and Sons, Inc., 1996, **8**, 1019, (synth., uses)
[30] Forte, R. and Leblanc, J.L., *J. Appl. Polym. Sci.*, 1992, **45**, 1473, (liq. permeability)
[31] Zhao, W., Yu, L., Zhong, X., Zhang, Y. and Sun, J., *J. Appl. Polym. Sci.*, 1994, **54**, 1199, (ir, γ-irradiation)
[32] Mohammedi, N.A., and Rempel, G.L., *Macromolecules*, 1987, **20**, 2362, (synth., ir, H-1 nmr, C-13 nmr)
[33] Kondo, A., Dhtani, H., Kosugi, Y., Tsuge, S. *et al*, *Macromolecules*, 1988, **21**, 2918, (spectra, thermal degradation)
[34] Thauamani, P. and Bhowmick, A.K., *J. Mater. Sci.*, 1992, **27**, 3243, (density, mech. props.)
[35] Thauamani, P. and Bhowmick, A.K., *J. Mater. Sci.*, 1993, **28**, 1351, (mech. props., power factor)
[36] Thauamani, P. Khastgir, D. and Bhowmick, A.K., *J. Mater. Sci.*, 1993, **28**, 6318, (wear, transition temp.)
[37] Setua, D.K., Debnath, K.K., Singh, H., Chakrabarti, N. and De, P.P., *Polym. Sci.*, 1994, **2**, 567, (miscibility)
[38] Obrecht, W., Buding, H., Eisele, U., Szentivanyi, Z., and Thoerner, J., *Angew. Makromol. Chem.*, 1986, **145-146**, 161, (transition temp., mech. props.)

Nitrile rubber, Injection moulding, Unfilled   N-10
Related Polymers: Nitrile rubber
Monomers: 1,3-Butadiene, Acrylonitrile
Material class: Synthetic Elastomers, Copolymers
Polymer Type: acrylonitrile and copolymers, polybutadiene
CAS Number:

CAS Reg. No.
9003-18-3

Applications/Commercial Products:

Trade name	Supplier
Chemigum N683B	Goodyear
Krynac 29.60	Polysar
Krynac 34E3S	Polysar
Krynac 822	Polysar
Krynac 825	Polysar
Krynac 826	Polysar
Krynac 827	Polysar
NySyn 35-5	Copolymer Rubber
Paracril 3280B	Uniroyal Chemical

Nitrile rubber modifier   N-11
Related Polymers: Nitrile rubber
Monomers: 1,3-Butadiene, Acrylonitrile
Material class: Synthetic Elastomers, Copolymers
Polymer Type: acrylonitrile and copolymers, polybutadiene
CAS Number:

CAS Reg. No.
9003-18-3

Applications/Commercial Products:

Trade name	Supplier
Chemigum N8	Goodyear
Chemigum N8X1	Goodyear
Krynac 34.140	Polysar
Krynac PXL 34.17	Polysar
Krynac PXL 38.20	Polysar
Krynac XL 29.20	Polysar
Krynac XL 31.25	Polysar
Paracril BJLT M-40	Uniroyal Chemical

Nitrilemethyl silicone rubber, Room temp. vulcanised   N-12
Synonyms: *Methylnitrile silicone rubber, Room temp. vulcanised*
Monomers: Dichlorodimethylsilane, Dichlorocyanopropylmethylsilane
Material class: Synthetic Elastomers, Copolymers
Polymer Type: silicones, dimethylsilicones
CAS Number:

CAS Reg. No.
158465-71-5

Molecular Formula: $[(C_2H_6OSi).(C_5H_9NOSi)]_n$
Fragments: C_2H_6OSi C_5H_9NOSi
Additives: Finely divided silica is usually incorporated as a reinforcer to improve mech. props.
General Information: Silanol groups are present in the pre-cure gum. During the cure they form cross links through condensation. Fully cured elastomer has few free silanol groups

Volumetric & Calorimetric Properties:
Density:

No.	Value	Note
1	d 1.1 g/cm^3	[3]

Thermal Expansion Coefficient:

No.	Value	Note	Type
1	0.0006 K^{-1}	high nitrile content; increases with decrease in nitrile substitution [3]	V

Glass-Transition Temperature:

No.	Value	Note
1	-110°C	5% cyanopropyl, 95% methyl
2	-87°C	20% cyanopropyl, 80% methyl [6]

Transition Temperature:

No.	Value	Note	Type
1	-118°C	-180°F, 10% cyanopropylmethylsilicone [1,2]	S

Surface Properties & Solubility:
Cohesive Energy Density Solubility Parameters: δ 18–19 $J^{1/2}$ cm$^{3/2}$ (8.8–9.3 cal$^{1/2}$ cm$^{3/2}$ (increases with nitrile substitution) [1]

Solvents/Non-solvents: Sol. ethers, esters, ketones. Insol. aliphatic hydrocarbons, aromatic hydrocarbons, chlorinated hydrocarbons, alcohols, H_2O [1]. Vulcanised rubber swollen by solvents rather than dissolved

Surface and Interfacial Properties General: Generally low surface adhesion [5]

Transport Properties:
Permeability of Gases: Gas permeability depends on composition; it decreases as nitrile substitution increases [4]
Gas Permeability:

No.	Gas	Value	Note
1	CO_2	43772–218862 cm^3 mm/(m^2 day atm)	0.05–0.25 cm^2 $(s\ GPa)^{-1}$, 25° [4]
2	O_2	5515–39395 cm^3 mm/(m^2 day atm)	0.0063–0.045 cm^2 $(s\ GPa)^{-1}$, 25° [4]
3	N_2	2188–43772 cm^3 mm/(m^2 day atm)	0.0025–0.05 cm^2 $(s\ GPa)^{-1}$, 25° [4]

Mechanical Properties:
Mechanical Properties General: Requires reinforcing filler to possess acceptable mech. props.
Tensile Strength Break:

No.	Value	Note	Elongation
1	5.6 MPa	825 psi, 40% cyanopropylsilicone, with reinforcing filler [1]	120%

Hardness: Durometer A80–A85 (40% cyanopropyl silicone, with reinforcing filler) [1]

Electrical Properties:
Electrical Properties General: Has low electrical resistance compared to other silicones
Dielectric Permittivity (Constant):

No.	Value	Note
1	4–20	increases with increase in nitrile substitution [3,5]

Optical Properties:
Refractive Index:

No.	Value	Note
1	1.45	increases with increase in nitrile substitution [3]

Polymer Stability:
Polymer Stability General: Solvent resistant material with good low temp. props.
Thermal Stability General: Usable from -120–200° [1,7]
Upper Use Temperature:

No.	Value	Note
1	200°C	
2	200°C	[7]

Environmental Stress: Resistant to oxidative degradation, H_2O, weathering, and ozone. Unaffected by uv radiation [5]
Chemical Stability: Swollen by esters, ethers and ketones. Relatively unaffected by toluene, aliphatic hydrocarbons and CCl_4. Disintegrates in conc. sulfuric acid. Relatively unaffected by weak alkali, weak acids and alcohols [1,2,3,5]
Hydrolytic Stability: Stable in H_2O and aq. solns. at room temp. Damaged by high pressure steam [5]

Biological Stability: Non-biodegradable. If properly vulcanised it is not subject to biological attack [5]

Applications/Commercial Products:
Processing & Manufacturing Routes: Dimethyl and methyl cyanopropyl cyclic polysiloxanes are polymerised together at 150–200° with alkaline catalysis in the presence of small quantities of water. The gum cures at room temp. with alkoxysilane cross-linking agent and metal salt catalyst. Cross-linking requires moisture
Applications: Nitrilemethylsilicone rubbers are solvent resistant and stable over a wide temp. range. They have not found important commercial applications

Bibliographic References
[1] Williams, T.C., Pike, R.A. and Fekete, F., *Ind. Eng. Chem.*, 1959, **51**, 939
[2] Roff, W.J. and Scott, J.R., Fibres, Films, *Plastics and Rubbers*, Butterworths, 1971
[3] Torkelson, A., *Silicone Technology*, (ed. P.F. Bruins), John Wiley and Sons, 1970, 61
[4] Robb, W.L., *Ann. N. Y. Acad. Sci.*, 1968, **146**, 119, (gas permeability)
[5] Noll, W., *Chemistry and Technology of Silicones*, Academic Press, 1968
[6] Borisov, M.F., *Soviet Rubber Technology*, 1966, **25**, 5, (transition temps)
[7] Meals, R.N., *Ann. N. Y. Acad. Sci.*, 1965, **125**, 137
[8] Hardman, B. and Torkelson, A., *Encycl. Polym. Sci. Eng.*, 2nd edn., (ed. J.I. Kroschwitz), John Wiley and Sons, 1985, **15**, 204

para-Nonyl Phenolic resins   N-13

Synonyms: *4-Nonylphenol based resins*
Related Polymers: Phenol-aldehyde resins
Monomers: 4-Nonylphenol, Formaldehyde
Material class: Thermoplastic
Polymer Type: phenolic
CAS Number:

CAS Reg. No.
31605-35-3
134443-11-1

Molecular Formula: $(C_{16}H_{24}O.CH_2O)_n$
Fragments: $C_{16}H_{24}O$ CH_2O
Molecular Weight: M_n 730 MW 470–745
Additives: Additives include kaolin, calcium carbonate, colloidal silica, hydroxymethyl starch, styrene-butadiene rubber latex
General Information: The resins do not crosslink, as the phenol is difunctional (since the *para*-position is blocked)

Volumetric & Calorimetric Properties:
Glass-Transition Temperature:

No.	Value	Note
1	34°C	[3]

Surface Properties & Solubility:
Solvents/Non-solvents: Sol. oils [8]

Transport Properties:
Polymer Solutions Dilute: Apparent shear viscosity 12715–20637 P (uncured novolac, formaldehyde: *p*-nonylphenol ratio 0.25–0.67)

[4]. Resins with M_n 490–1400 show non-Newtonian rheological behaviour of the pseudoplastic type [4,7]

Optical Properties:
Transmission and Spectra: H-1 nmr and ir spectral data have been reported [7]

Polymer Stability:
Decomposition Details: Resins of low MW show greater weight loss than those of higher MW. Activation energy of decomposition 70.3–154 kJ mol^{-1} (16.8–36.8 kcal mol^{-1}) which increases with increasing MW. Below 300°, phenol and formaldehyde are the major decomposition products. At higher temps. aromatic hydrocarbons, substituted phenols and C_1–C_4 gaseous products are formed [4,5]. Char yield 79.83% (200°, uncured novolac) [5], 83.10% (400°, uncured novolac) [5]. Pyrolysis at 770° produces a mixture of branched chain nonenes, p-nonylphenol and methyl-substituted p-nonylphenol [6]
Chemical Stability: Ethoxylation of the phenol hydroxyl groups increases compatibility with oil

Applications/Commercial Products:
Processing & Manufacturing Routes: May be synth. by reaction of p-nonylphenoxymagnesium bromide with paraformaldehyde in dry C_6H_6 for 12h at 80°, followed by acidic quenching [1]. This procedure gives "high-*ortho*" novolaks [10]. Novolak-type resin synth. by reaction of paraformaldehyde and p-nonylphenol with H_2SO_4 catalyst [2,4,7]. Resole-type resin synth., for example, by reaction of paraformaldehyde with p-nonylphenol in the presence of triethylamine catalyst (pH 7–9) at 90° [9]
Applications: Potential application as an antimicrobial agent. Novolak-type used as tackifiers in manufacture of pressure-sensitive adhesives, and in the coated front sheet formulation which receives the image in carbonless paper transfer systems. Resole-type has potential applications in coatings and in adhesives "High-*ortho*" novolaks have potential application as an antioxidant in lard. May be used as a sulfur free alternative curing agent for rubber compounding. The solid resole (5–12%) is compounded with rubbers using activators and acts as a non-sulfur containing vulcanisation agent. For nitrile-butadiene rubber activators are pyromellitic anhydride, phthalic anhydride or fumaric acid. For styrene-butadiene rubber, p-toluenesulfonic acid or chloroacetic acid are used. For neoprene, the activator can be zinc oxide and tin(*II*) chloride. The novolak is compounded with kaolin clay, calcium carbonate, colloidal silica and SBR latex to form a coated front (CF) sheet for carbonless copy paper systems. Ethoxylation of the phenol hydroxyl groups gives a resin compatible with oil, for oil recovery

Trade name	Details	Supplier
Rütaphen KA	adhesive resins	Bakelite
Standard sample		A.C. Hatrick & Co.
Standard sample		CETEC Australia

Bibliographic References
[1] Uchibori, T., Kawada, K., Watanabe, S., Asakura, K. et al, *Bokin Bobai*, 1990, **18**, 215, (synth., use)
[2] Roşu, D., Cascaval, C.N., and Stoleriu, A., *Mater. Plast. (Bucharest)*, 1989, **26**, 147, (synth)
[3] Weill, A., Dechenaux, E. and Paniez, P., *Microelectron. Eng.*, 1986, **4**, 285, (Tg)
[4] Roşu, D., Caşcaval, C.N., and Simionescu, C.I., *Roum. Chem. Q. Rev.*, 1995, **3**, 277, (synth., thermal stability)
[5] Caşcaval, C.N., Agherghinei, I. and Rosu, D., *J. Therm. Anal.*, 1993, **39**, 585, (thermal degradation)
[6] Challinor, J.M., *J. Anal. Appl. Pyrolysis*, 1993, **25**, 349, (thermal degradation)
[7] Caşcaval, C.N., Roşu, D. and Mustaţă, F., *Eur. Polym. J.*, 1994, **30**, 329, (synth., ir, H-1 nmr)
[8] Takahashi, S., Kuriyama, K., and Tobita, S., *Shikizai Kyokaishi*, 1966, **39**, 531, (solubility)
[9] *Rom. Pat.*, 1991, 100883, (synth.)
[10] Abe, Y., Matsumura, S., Asakawa, K. and Kasama, H., *Yukagaku*, 1986, **35**, 751, (synth., use)

Nucleic acid

Monomers: Cytosine, Guanine, Thymine, Adenine, Uracil
Material class: Polynucleotides
General Information: General term describing natural macromolecules found in living cells formed by the polymerisation of mixed nucleotides. There are two main types: Deoxyribonucleic acid which usually has a very high molecular weight, particularly that present in bacterial and animal cells, and Ribonucleic acid which has lower molecular weights.
Morphology: RNA is a polymer of ribonucleotides derived mainly from Adenine, Cytosine, Guanine and Uracil, linked by 3′,5′-phosphodiester bonds, whereas DNA consists of similar polymers, but of deoxyribonucleotides, derived mainly from Adenine, Guanine and Cytosine, as in RNA, but with Thymine in place of Uracil

Bibliographic References
[1] Charagaff, E., *The Nucleic Acids*, Academic Press, 1955
[2] Jordon, D.O., *The Chemistry of the Nucleic Acids*, Butterworths, 1960
[3] *Org. Chem. Nucleic Acids* (Kochetkov, N.K. etal, Ed.), Parts A and B, Plenum Press, 1971
[4] *Basic Princ. Nucleic Acid Chem.* (Ts'O, P.O.P., Ed.), Academic Press, Vols. 1 and 2, 1974
[5] Jankowski, K., *Adv. Heterocycl. Chem.*, 1986, **39**, 79
[6] Taillandier, E., *J. Mol. Struct.*, 1989, **214**, 185 (ir, Raman)
[7] Reid, B.R., *Biochemistry*, 1992, **31**, 3564
[8] Wemmer, D.E., *Biol. Magn. Reson.* Plenum Press, Vol. 10, 1992, 195 (nmr)

Nylon

Applications/Commercial Products:
Processing & Manufacturing Routes: *Commercial Polymerisation of Polyamides*. There are usually at least three stages in a commercial nylon plant and the process is either continuous or semi-continuous. In a typical Nylon 6,6 polym., a purified soln. of salt is charged to thin film evaporator where the water is removed under a constant pressure of 250 psi at 235°. The molten polymer is pumped to a second thin film evaporator where it is further condensed under 15 psi of pressure before passing into a third stage filmtruder[13]. *In the* **DuPont Canada process**, 40% nylon salt soln. is conc. to 50% in an evaporator at 100° after which it is stored in the surge tank prior to condensation. It is next passed to a reactor into which appropriate quantities of additives, (hexamethylene diamine and acetic acid) are fed. The polymer and steam passes through a partial condenser at 265 psi and 210° to minimise diamine losses, before reaching a flasher reactor (230°) to flash off steam. It passes on through two further stages in which temp. is raised up to 280° and the pressure reduced. *In the* **tank continuous process**, purified nylon salt soln. is charged to a stirred autoclave under 18 bar pressure at 275°. It passes onto a second stage pressure reactor equipped with a diamine recirculation system where it is further condensed to a "prepolymer molecular weight" and is still sufficiently mobile to pump to a finishing extruder to be converted to high MW (Plant scale is typically about 13000 tonnes per annum). *In the* **tubular continous process**, purified salt soln. is increased in concentration from 65% to 85% by a stage 1 evaporator operating at 230° and 250 psi. It passes on through a tubular reactor at 280° at low pressure (0.5 psi). In the final step, it is vented under vacuum at 0.05 psi before finishing in a vented extruder. The main problem with this system is the potential for gelling the polymer (Annual throughputs are about 13000 tonnes per annum). In another variant, *a* **column continuous reactor** is fed with 65% soln. of purified Nylon 6,6 salt at 160°. The bottom of the tube reactor is heated to 290° while an upper bank of tubes is kept at 180° by steam heating. The whole reactor is pressured to 8 bar and the vertical upper section is a fractionating column. The salt soln. concentrates to 85% in this three component vertical column tank reactor before passing on through the bottom section as molten polymer to a finishing extruder. The commercial process for the production of *Trogamid T amorphous polyamide* (Huls) relies upon the hydrol. of the initial charge of dimethyl terephthalate by aq. trimethyl hexamethylene diamine at 90° in the first reactor. MeOH is removed under vacuum. Nylon 6(3),T aq.

salt soln. is transferred to a second vessel where the soln. is conc. to 70% at 100°. The conc. soln. is now pumped to an autoclave at 225° and 25 bar for 1.5h in order to ensure sufficient polycondensation and reaction of volatile diamine. The oligomeric Nylon 6(3),T is pumped to a subsequent stirred vessel where it is heated at 270° under 10 bar to form prepolymer with intrinsic viscosity of 0.7–0.8 dl/g. It is discharged to a twin screw extruder and converted to high MW polymer with an intrinsic viscosity in the range 1–1.5 dl/g. Residence time in the extruder is usually less than three mins. The product is removed as strands which are quenched and chopped to pellets. *Nylon 6* is prepared commercially from the polym. of ϵ-caprolactam, in the presence of some water and ϵ-aminocaproic acid catalyst. Caprolactam is hydrolysed under steam pressure. Polym. then proceeds by polycondensation. Pressure reactors release moisture in a similar fashion to Nylon 6,6 reactors. The final polymer contains 10% equilib. concentration of caprolactam which must be removed in an additional leaching stage. Molten polymer is first extruded, quenched and pelletised before passing to a leaching tower to be washed with hot water. The product still contains about 2% caprolactam. After drying this level is reduced to about 0.2%. There is extensive literature on nylon melt polym. processes. Possible variants include a first step solid condensation of salt to prepolymer followed by a second stage through a filmtruder to convert to high MW. A first batch stage to produce prepolymer in a reactor is also feasible. The majority of commercial processes seem to be based on a continuous or semi-continuous process. The commercial preparation of *Aromatic polyamide fibres* from acid chlorides in soln. is also discussed under "Aramids". These materials cannot be prepared by melt condensation processes. *N*-methyl pyrrolidone containing $CaCl_2$ is the preferred solvent for producing high MW aramids

Bibliographic References

[2] *Jpn. Pat.,* 1982, 57-134441A
[3] *U.S. Pat.,* 1964, 3 337 612
[4] *Brit. Pat.,* 1965, 1 034 307, (Nylon 6, 6 salt formation in alcoholic soln)
[5] Dolden, J.G., *Polymer,* 1976, **17**, 876
[6] Dolden, J.G. and Studholme, M., *Eur. Pat. Appl.,* 0 306 165
[7] Khripov, E.G. *et al, Vysokomol. Soedin., Ser. B,* 1976, **18**, 82
[8] Khripov, E.G. *et al, Vysokomol. Soedin., Ser. B,* 1972, **14**, 189
[9] *Encycl. Polym. Sci. Technol.,* 2nd edn., John Wiley and Sons, 1985, **10**, 499
[10] Papaspyrides, C.D. and Kampouris, E.M., *Polymer,* 1984, **25**, 791
[11] Yamazaki, N., Higashi, F. and Kawabata, J., *J. Polym. Sci., Polym. Chem. Ed.,* 1974, **12**, 2149
[12] Yamazaki, N. and Higashi, F., *Tetrahedron,* 1974, **30**, 1323
[14] *Encycl. Polym. Sci. Technol.,* 2nd edn., John Wiley and Sons, 1986, **11**, 400

Nylon 6,I

Synonyms: *Poly(hexamethylene diamine-co-isophthalic acid). Poly(hexamethylene isophthalamide). Poly(iminocarbonyl-1,3-phenylenecarbonylimino-1,6-hexanediyl)*
Related Polymers: Terephthalic/isophthalic acid copolyimides
Monomers: 1,3-Benzenedicarboxylic acid, 1,6-Hexanediamine
Material class: Thermoplastic
Polymer Type: polyamide
CAS Number:

CAS Reg. No.
25668-34-2

Molecular Formula: $(C_{14}H_{18}N_2O_2)$
Fragments: $C_{14}H_{18}N_2O_2$

Molecular Weight: 13000, 30000–35000
Additives: Polymerisation control additives include ammonium hypophosphite or phosphoric acid (thermal stabilisers and catalysts); benzoic acid as MW control agent
General Information: Nylon 6,I is a very transparent essentially amorph. resin with good impact and tensile props. Molecular models show that the chain is too kinked to permit ease of crystallisation [3]. However, crystallisation can be effected by prolonged boiling in water with consequent loss of tensile strength. Crystallisation only occurs with hydrogen bond-forming solvents. This was investigated by FT-IR spectroscopy [7]

Volumetric & Calorimetric Properties:
Density:

No.	Value	Note
1	d 1.185 g/cm³	[1,2]

Thermal Expansion Coefficient:

No.	Value	Note	Type
1	6.6×10^{-5} K^{-1}	[1,2]	L

Melting Temperature:

No.	Value	Note
1	180–210°C	[1,2]

Glass-Transition Temperature:

No.	Value	Note
1	120–130°C	[6]

Vicat Softening Point:

No.	Value	Note
1	143°C	[3]

Transport Properties:
Polymer Solutions Dilute: η_{inh} 0.74 dl g^{-1} (formic acid) [1]; η_{inh} 2.4–2.6 dl g^{-1} (96% H_2SO_4) [2]; η_{inh} 6.1 dl g^{-1} (25°, H_2SO_4) [7]

Mechanical Properties:
Mechanical Properties General: Additives such as terephthalic acid, isophorane diamine and *m*-xylylene diamine are reported to improve impact toughness and flexural modulus [4,5]

Tensile (Young's) Modulus:

No.	Value	Note
1	2800 MPa	[1,2]

Flexural Modulus:

No.	Value	Note
1	3200 MPa	[1,2]

Tensile Strength Break:

No.	Value	Note	Elongation
1	58–65 MPa	[1,2]	20–80%

Flexural Strength at Break:

No.	Value	Note
1	95 MPa	[1,2]

Tensile Strength Yield:

No.	Value	Note
1	86.5 MPa	20°, BS 2782 [3]
2	85–105 MPa	[1,2]

Charpy:

No.	Value	Notch	Note
1	38	Y	20° [1,2]

Applications/Commercial Products:
Processing & Manufacturing Routes: Processed by the melt condensation of the Nylon 6,I salt at 270° stirred under N_2 in the presence of 1% ammonium hypophosphite catalyst and 1% benzoic acid as chain control agent for periods of 3–4h [3]
Applications: Tough, rigid, transparent polyamide similar in props. to Trogamid T; not commercial

Bibliographic References
[1] Gorton, B.S., *J. Appl. Polym. Sci.*, 1969, **9**, 373
[2] Jackson, J.B., *Polymer*, 1969, **10**, 159
[3] Dolden, J.G., *Polymer*, 1976, **17**, 878
[4] *Brit. Pat.*, 1971, 1 228 761, (BP Chemicals)
[5] *Brit. Pat.*, 1971, 1 250 877, (BP Chemicals)
[6] *Encycl. Polym. Sci. Technol.*, 2nd edn., John Wiley and Sons, 1985, **11**, 373
[7] Liu, W., Breault, B. and Brisson J., *J. Polym. Sci., Polym. Phys. Ed.*, 1995, **33**, 619, (crystallisation)

Nylon 3   N-17

Synonyms: *Polypropiolactam. Poly[imino(1-oxo-1,3-propanediyl)]. Poly(3,3'-dimethylpropiolactam). Poly(4,4'-dimethylpropiolactam). N3*
Monomers: Propiolactam, 4,4'-Dimethylpropiolactam, 3,3'-Dimethylpropiolactam, 3-Amino-3-methylbutanoic acid
Material class: Thermoplastic
Polymer Type: polyamide
CAS Number:

CAS Reg. No.
24937-14-2

Molecular Formula: $[(C_3H_5NO)_n(unsubst.).(C_5H_9NO)_n(subst.)]$
Fragments: C_3H_5NO C_5H_9NO
General Information: Unsubstituted polymers decompose on melting. 3,3'-Dimethyl substituted polymer is highly cryst., and gives strong fibres at MW 20000 due to the strong hydrogen bonding. 4,4'-Disubstituted polymer has CO and NH at angles to the fibre axis; MW of 200000 is required for strong fibres

Volumetric & Calorimetric Properties:
Density:

No.	Value	Note
1	d 1.1 g/cm^3	4,4-dimethyl subst.
2	d 1.17 g/cm^3	unsubst., cryst. [3]

Melting Temperature:

No.	Value	Note
1	330°C	cryst., unsubst. [3]
2	296°C	4,4'-dimethyl subst. [2]
3	250°C	4,4'-dimethyl subst. [1]

Glass-Transition Temperature:

No.	Value	Note
1	111°C	cryst., unsubst. [4]

Transport Properties:
Polymer Solutions Dilute: η_{inh} 4.5 dl g^{-1} (1% in H_2SO_4, 4,4'-dimethyl subst.)

Mechanical Properties:
Mechanical Properties General: 4,4'-Dimethyl subst. polymer has tensile strength 50% that of Nylon 6

Polymer Stability:
Environmental Stress: 4,4'-dimethyl subst. polymer fibres are stable to oxidation; 3,3'-dimethyl subst. polymer discolours readily

Applications/Commercial Products:
Processing & Manufacturing Routes: Prepared by low temp. anionic polymerisation of monomer in DMSO at 0–20° with potassium *tert*-butoxide. Initiated by Nylon 6,6 salts, gives MW up to 500000
Applications: Not commercial, though 4,4'-dimethyl polymer can be spun into silk-like fibres. Other fibres and films have been prepared from propiolactams

Bibliographic References
[1] Graf, R., Lohaus, G., Börner, K., Schmidt, E. and Bestian, H., *Angew. Chem., Int. Ed. Engl.*, 1962, **1**, 481, (fibres)
[2] *Encycl. Polym. Sci. Technol.*, 2nd edn., (ed. J.I. Kroschwitz), John Wiley and Sons, 1985
[3] Bestian, H., *Angew. Chem., Int. Ed. Engl.*, 1968, **7**, 278, (T_m)
[4] Wexler, H., *Makromol. Chem.*, 1968, **115**, 262, (T_g)

Nylon 4   N-18

Synonyms: *Poly(4-aminobutyric acid). Poly[imino(1-oxo-1,4-butanediyl)]. Polypyrrolidone. Poly(γ-butyrolactam)*
Monomers: 2-Pyrrolidinone
Material class: Thermoplastic
Polymer Type: polyamide
CAS Number:

CAS Reg. No.
24938-56-5

Molecular Formula: $(C_4H_7NO)_n$
Fragments: C_4H_7NO
Additives: Carbon dioxide and potassium pyrrolidinone to increase thermal stability for melt spinning. Higher MW fibres formed with quaternary ammonium salts
General Information: Has a much higher equilibrium water content than its higher Nylon homologues. While it is not dissimilar to cotton in this respect, it has a lower modulus than cotton. It is inferior to Nylon 6 and Nylon 6,6 in physical property stability (due to high moisture take up). It is difficult to polymerise and less thermally stable than the commercial Nylons
Morphology: Unit cell, a 9.44, b 12.1, c 8.22, β 64°, 8 monomer units per unit cell

Nylon 4,6 (continued)

Volumetric & Calorimetric Properties:
Density:

No.	Value	Note
1	d 1.25 g/cm^3	amorph. [4]
2	d 1.34–1.37 g/cm^3	cryst. [4]

Thermodynamic Properties General: Heat of evaporation proportional to binding force of functional groups being greatest for >1 combined water molecule [5]

Melting Temperature:

No.	Value	Note
1	250–265°C	[4]

Glass-Transition Temperature:

No.	Value	Note
1	81°C	[5]

Transition Temperature General: There is a sharp depression in the T_g with increasing water content up to a critical value at which additional water molecules form clusters around the amide sorption centres [5]

Transition Temperature:

No.	Value	Note	Type
1	-111°C	[6]	T_γ

Surface Properties & Solubility:
Solvents/Non-solvents: Sol. MeOH, cresol, o-chlorophenol, formic acid, sulfuric acid, hot AcOH, hot benzyl alcohol, hot pyrrolidone [13]. Insol. DMF, H_2O [13]

Transport Properties:
Water Content: 9.1% (65% relative humidity), 28% (99% relative humidity) [8,9], 7.3% moisture regain [11,12]

Mechanical Properties:
Mechanical Properties General: The high moisture uptake of Nylon 4 can cause excessive loss of useful props. This seriously limits its use especially as a textile fibre. Tenacity (of fibre) 0.22–0.4 N tex^{-1} [11,12]

Optical Properties:
Transmission and Spectra: C-13 nmr spectral data have been reported [7]

Applications/Commercial Products:
Processing & Manufacturing Routes: Produced by the anionic polymerisation of 2-pyrrolidinone with acid chloride, anhydride or esters. Polymerisation in hydrocarbon dispersion at <60° has prod. high MW polymer in 85% yield. Unreacted monomer and catalyst residues are removed by extraction. Melt spinning from soln. using potassium pyrrolidinone salt and carbon dioxide stabilisation allows spinning at up to 270°
Applications: Not commercially available

Bibliographic References
[1] Dryker, A.B., *Chemiefasern/Textilind.*, 1973, **23**, 527
[2] Sekiguchi, H., Tsourkas, P., Carriere, F. and Audebert, R., *Eur. Polym. J.*, 1974, **10**, 1185
[3] *Encycl. Polym. Sci. Technol.*, 2nd edn., Wiley and Sons, 1985
[4] Van Krevelen, D.W., *Properties of Polymers: Their Correlation with Chemical Structure*, 3rd edn., 1990
[5] Koi, K. and Sugiara, H., *Netsu Sokutei*, 1992, **19**, 145
[6] McKnight, W.J., Bellringer, M.A. and Ng, C.W.A., *Acta Polym.*, 1995, **46**, 361
[7] Kubo, K., Ando, I., Shibashi, T., Yamanobe, T. and Komoto, T., *J. Polym. Sci., Part B: Polym. Phys.*, 1991, **29**, 57
[8] Bacskai, R., *Polym. Prepr. (Am. Chem. Soc., Div. Polym. Chem.)*, 1982, **23**, 162
[9] Backsai, R., *Polym. Bull. (Berlin)*, 1984, **11**, 229
[10] Dachs, K. and Schwartz, E., *Nuova Chim.*, 1972, **48**, 43
[11] Longbottom, R.W., *Mod. Text.*, 1968, **49**, 19
[12] Peters, E.M. and Gervasi, J.A., *Chem. Technol.*, 1972, **2**, 16
[13] Sekiguchi, H., *Bull. Soc. Chim. Fr.*, 1960, 1839

Nylon 4,6

Synonyms: *Poly(imino-1,4-butanediylimino(1,6-dioxo-1,6-hexanediyl)). Poly(tetramethylenediamine-co-adipic acid). Poly(tetramethylene adipamide)*
Related Polymers: Glass fibre reinforced
Monomers: Tetramethylenediamine, Adipic acid
Material class: Thermoplastic
Polymer Type: polyamide
CAS Number:

CAS Reg. No.
50327-22-5

Molecular Formula: $(C_{12}H_{22}N_2O_2)_n$
Fragments: $C_{12}H_{22}N_2O_2$
Molecular Weight: MW 32900, MW 1000000
Additives: Cu and Mn halide salts and halo-organic compounds act as thermal stabilisers. Sb_2O_3 and halopolystyrene used for fire resistant formulations. Glass and carbon fibres give high rigidity, especially at elevated temps. Polyphenylene oxide and EPDM rubber blends used to improve impact resistance
General Information: Nylon 4,6 is the most recently commercialised nylon (1990). It has much higher yield stress and flexural modulus. It melts 30° higher than Nylon 6 at 290° and has outstanding heat and wear, chemical (oil) and fatigue resistance. It withstands 220° under heavy loads and has a continuous use temp. of 150° and is especially suited to applications in the electrical, electronic and automobile industries. High melting fibres of Nylon 4,6 have good friction and abrasion resistance and should compete well with Nylon 6 and 6,6. Recently, it has surprisingly been discovered that Nylon 4,6 is more thermally stable than Nylon 6,6 due to its lower oxygen permeability, resulting from its high crystallinity (70%) [3]. Its market has expanded dramatically from 100 tonnes per annum in 1990 to approx. 20000 tonnes per annum in 1996. Potential property disadvantages which may restrict further usage are high water absorption, which leads to significant dimensional changes after moulding, and rapid crystallisation from the melt, which affects processing and moulding tolerances
Morphology: Nylon 4,6 shows excellent props. compared to other aliphatic nylons in stiffness, dimensional stability, chemical resistance, creep modulus, etc. at higher temps. Because of its high degree of crystallinity (approx. 70%), it is distinctly superior to others in its mech. props., especially in its thermal dimensional stability [10]. These props. are evidently related to a higher degree of crystallinity and a higher melting temp. The determination of the crystallinity of Nylon 4,6 however, presents a problem because of the large discrepancies between the results of different crystallinity determination techniques. These differences and the remarkable thermal behaviour of Nylon 4,6 may largely be explained by imperfections in the cryst. struct. [11]. The struct. and morphology of soln. grown crysts. have been determined and compared with those of other polyamides [12]. They have similar lath-shaped crystallisation products, possessing the characteristics of chain-folded lamellar crysts. with a thickness of approx. 6 nm, and often aggregate to form sheaves. Electron diffraction patterns show that the chains lie perpendicular to the lamellar surfaces and that cryst. twinning occurs. The fold plane corresponds to the plane of the hydrogen bonds, these lying along the long axis of the cryst. in agreement with the electron diffraction results. Sedimented mats of the crysts. have been examined using both wide- and low-angle X-ray diffraction in directions perpendicular to the lamellar normals. The lamellar thickness remains constant up to 125° and then increases with crystallisation temp. Analyses of

Nylon 4,6

the X-ray diffraction patterns confirms that the chain stems in Nylon 4,6 chain-folded crysts. are parallel to the lamellar normal in contrast to previous results from other polyamides, where the chain stems are inclined at substantial angles to the lamellar normal. The consequences of these findings demand an entirely new mechanism for folding in nylons, where an amide group is incorporated in the fold, rather than alkane segments as in previous chain-folded nylons. The fold exhibits similarity with the β-bend in proteins. The lamellar thickness is sufficient to accommodate four structural repeats of the chain. The crystallographic subcell is monoclinic, a 0.96, b 0.826, c 1.47 nm, γ 115°, and contains four chain segments (two sheet segments), which bears similarity with the unit cell of Nylon 6 [12]

Identification: A study of dimensional stability has been reported [34], in which thermal shrinkage, viscoelastic props., wide and small x-ray diffractions, and molecular orientation by fluorescence are all measured on Nylon 4,6 fibre samples. The molecular chains are well oriented along the fibre axis and the onset of thermal movement is shifted to a higher temp. than Nylon 6,6

Volumetric & Calorimetric Properties:

Density:

No.	Value	Note
1	d 1.18 g/cm^3	dry [2,5]

Thermal Expansion Coefficient:

No.	Value	Note	Type
1	8×10^{-5} K^{-1}	23–80°	L
2	0.0001 K^{-1}	80–180° [8]	L

Thermal Conductivity:

No.	Value	Note
1	0.22 W/mK	[8]

Specific Heat Capacity:

No.	Value	Note	Type
1	2.2 kJ/kg.C	[8]	P

Melting Temperature:

No.	Value	Note
1	290°C	[2]
2	319°C	[3,4]
3	295°C	[8]

Glass-Transition Temperature:

No.	Value	Note
1	43°C	dynamic method [1]

Deflection Temperature:

No.	Value	Note
1	160°C	1.8 MPa
2	280°C	0.45 MPa [8]

Transport Properties:

Polymer Solutions Dilute: Only has a narrow temp. window for extruding stock shapes. The rate of degradation of polymer for a characteristic residence time in the extruder can be determined and used as a quality control tool [39]

Water Content: 3.7% (23°, 50% relative humidity) [8]

Water Absorption:

No.	Value	Note
1	3 %	50% relative humidity
2	12 %	100% relative humidity
3	2.3 %	24h, 23°

Mechanical Properties:

Mechanical Properties General: In general, Nylon 4,6 has certain property advantages over the conventional Nylons 6 and 6,6 such as higher melting point, higher tensile and flexural strengths and elastic modulus, which are retained to elevated temps. An adequate level of props. is maintained up to 150° which is considered its maximum long term use temp. Tenacity 0.8 N tex^{-1}

Tensile (Young's) Modulus:

No.	Value	Note
1	3300 MPa	23° [8]
2	800 MPa	120° [8]
3	650 MPa	160° [8]

Flexural Modulus:

No.	Value	Note
1	3000 MPa	23° [8]
2	800 MPa	120° [8]
3	600 MPa	160° [8]

Tensile Strength Break:

No.	Value	Note	Elongation
1	80 MPa	dry [5]	50%
2	65 MPa	50% relative humidity [5]	
3	100 MPa	23° [8]	
4	50 MPa	120° [8]	
5	40 MPa	160° [8]	

Tensile Strength Yield:

No.	Value	Note
1	95 MPa	dry [5]
2	55 MPa	50% relative humidity [5]

Miscellaneous Moduli:

No.	Value	Note	Type
1	1600 MPa	23°, 1000h	Creep modulus
2	700 MPa	120°, 1000h	Creep modulus
3	500 MPa	160°, 1000h	Creep modulus

Hardness: Shore D0 (parallel); D80 (perpendicular)

Izod Area:

No.	Value	Notch	Break	Note
1		Unnotched	Y	-40–23°
2	10–40 kJ/m^2	Notched		23°
3	9–12 kJ/m^2	Notched		-40°

Charpy:

No.	Value	Notch	Break	Note
1	9–12	Y		-40°
2		N	Y	23°
3	12–45	Y		23°

Electrical Properties:

Electrical Properties General: Has less useful insulating props. compared to nylons 6 or 6,6 and is more anisotropic in props. Arc resistance rating 1 (CTI method)

Surface/Volume Resistance:

No.	Value	Note	Type
1	10×10^{15} Ω.cm	parallel [8]	S
2	0.01×10^{15} Ω.cm	perpendicular [8]	S

Dielectric Permittivity (Constant):

No.	Value	Frequency	Note
1	3.9	50 Hz	[8]
2	3.6	1 MHz	[8]

Dielectric Strength:

No.	Value	Note
1	>25 kV.mm^{-1}	min., parallel [8]
2	>15 kV.mm^{-1}	min., perpendicular [8]

Complex Permittivity and Electroactive Polymers: Dielectric props. have been studied as a function of frequency, temp. and moisture content. Above T_g ionic conductivity is observed which gives rise to strong polarisation (due to high crystallinity of nylon). At high temps. conductivity increases to such an extent that it causes electrode polarisation and obscures further dielectric effects. α and β relaxations are significantly enhanced by water uptake, while the corresponding activation energies of the relaxations are reduced [38]

Dissipation (Power) Factor:

No.	Value	Frequency	Note
1	0.007	50 Hz	parallel
2	0.877		perpendicular
3	0.026	1 MHz	parallel
4	0.12	1 MHz	perpendicular

Optical Properties:

Transmission and Spectra: H-1 nmr and C-13 nmr spectral data have been reported [32]

Polymer Stability:

Polymer Stability General: Overall thermal stability and effect of ageing upon physical props. as good as or better than Nylon 6 or Nylon 6,6. However, long residence times in extrusion/moulding have to be avoided because of narrow temp. window between melting and onset of thermal degradation

Thermal Stability General: Better than Nylon 6,6; retains half its tensile strength at 140° for longer than Nylon 6,6. Degradation is a surface phenomenon and is oxygen diffusion limited [6]. Thermal and oxidative stability can be improved by incorporating a copper salt insol. in water with a halogen substituted organic compound [9]

Upper Use Temperature:

No.	Value	Note
1	<163°C	max. 2500h [8]
2	<152°C	max. 5000h [8]
3	141°C	max. 10000h [8]
4	130°C	max. 20000h [8]
5	150°C	[6]

Flammability: Flammability rating V2 (0.9 mm thick). Insulation classification F. Limiting oxygen index 27%. Glow wire time 5 s extinguishing time, (850°, 2 mm) [8]

Chemical Stability: Especially suited for hot oil applications, e.g. engine or transmission oil which runs between 130–165°. Physical props. are not impaired over long testing periods. Otherwise, chemical resistance generally similar to Nylon 6,6

Applications/Commercial Products:

Processing & Manufacturing Routes: Nylon 4,6 is technically more difficult to polymerise than Nylon 6,6 due to its much higher melting point (295°), which makes it more sensitive to degradation and branching. Melt polymerisation and processing also pose problems due to the high volatility of the adipic acid component. Low temp. polymerisation in a solvent [13] or melt polymerisation followed by a solid phase polymerisation has been reported [14]. Continuous process polymerisation [15] may be carried out where aq. salt soln. is first evaporated to form a molten prepolymer, the prepolymer and vapour phase are passed through mass transfer zone to remove the vapour phase prior to discharge of the molten polymer. A method for the synth. of Ultra High MW polymer has been reported [16]. MW exceeding 1000000 has been achieved using tetramethylammonium chloride as the catalyst, in conjunction with a crown ether (18-crown-6) and carbonated potassium pyrrolidonate. Rapid rate of crystallisation makes melt spinning difficult. This problem has been solved by a melt spinning process with controlled temp. and pressure conditions. [33] It possesses high stiffness and low creep at elevated temps., with reduced cycle times and excellent mouldability

Mould Shrinkage (%):

No.	Value	Note
1	1.5–2%	anisotropic shrinkage

Applications: The main property advantages of Nylon 4,6 are heat and wear resistance, chemical (especially oil) and fatigue resistance. It is particularly suited for use in the electric, electronic and automobile industries where a continuous use temp. of at least 150° is required under the bonnet, e.g. vacuum tubing, fuel line retainer, and special grades have been developed for this purpose. Other automotive uses include halogen lamps, Volkswagen's automatic transmission, manual gear change for Range Rover due to excellent high temp. oil and grease resistance. In the electrical and electronic industries, examples of its use include brush carriers for electric motors, circuit breakers for domestic use (where temps. can reach 150°), electromagnetic coils for electric motors, and electronic sensors and connectors

Trade name	Supplier
Stanyl	DSM
	Unitika Ltd.

Bibliographic References

[1] Beaman, R.G. and Cramer, F.B., *J. Polym. Sci.*, 1956, **21**, 223, (T$_g$)
[2] *Encycl. Polym. Sci. Technol.*, 2nd edn., John Wiley and Sons, 1986, **11**, 371
[3] Gayman, R.J. et al, *J. Polym. Sci., Polym. Chem. Ed.*, 1977, **15**, 537
[4] O'Sullivan, D., *Chem. Eng. News*, 1984, **62**, 33
[5] *Stanyl - Preliminary Leaflet*, DSM, 1994, (technical datasheet)
[6] Gijsman, P. et al, *Polym. Degrad. Stab.*, 1995, **49**, 121
[7] U.S. Pat.,252 253
[8] *Stanyl. Typical Properties DSM*, 1995, (technical datasheet)
[9] U.S. Pat., 1992, 5 157 064
[10] Harschnitz, R., Heather, P., Derks, W. and Leeuwendal, R., *Kunststoffe*, 1990, **80**, 1271
[11] Atkins, E.D.T., Hill, M., Hong, S.K., Keller, A. and Organ, S., *Macromolecules*, 1992, **25**, 917
[12] Moonen, W.H.P., Ramaekers, F.J.W., Kooiji, C.J. and Smeets, J., *Integr. Fundam. Polym. Sci. Technol.-5*, April-May, Elsevier Science, 1990, 131
[13] Neth. Pat., 8001, 763/9, (Stamicarbon)
[14] Eur. Pat. Appl., 1983, 77106, (Stamicarbon)
[15] U.S. Pat., 1991, 4 994 550, (BASF)
[16] Bucksai, R., *Polym. Sci. Technol. (Plenum)*, 1984, **24**, 183
[17] Plastics Weekly 7th. Sept, 1992, **51**
[18] Plastics Weekly, 5th Aug., 1991, **1**
[19] New Materials World, July, 1991, **5**
[20] Eur. Pat. Appl., 1992, 0 475 087, (BASF)
[21] Plast. World, 1995, **53**, 58
[22] Autoworld, 1995, **31**, 64
[23] DSM Press Release, 1993
[24] Br. Plast. Rubber, Nov, 1992, **30**
[25] J. Plast. News (Detroit), 23 Mar, 1992, **8**
[26] J. Plast. Ind. News, Apr., 1989, **35**
[27] J. Plastics Technology, Feb., 1991, **37**, 81
[28] J. Plast. Ind. News, Sept., 1991, **37**, 137
[29] Plastics Weekly, 7 Sept., 1992, **2**
[30] Joncken, R., *Conference on New Materials for the Automotive Industry*, Florence, 1-5 June, 1992
[31] de Vries, K., de Linssen, H. and Velden, G., *Macromolecules*, 1989, **22**, 1607
[32] de Vries, K., *Polym. Bull.*, 1991, **26**, 451
[33] Plastics Weekly, 12 May, 1991, **6**
[34] Kudo, K., Mechizuki, M., Kiriyama, S. Watanabe, M. and Hirami, M., *J. Appl. Polym. Sci.*, 1994, **52**, 861
[35] Eur. Pat. Appl., 1988, 0 270 151
[36] Hori, K. and Tsuchikawa, H., *Gosei Gomu*, 1987, **95**
[37] J. New Mater. Japan, 1985, **2**, 11
[38] Steeman, P.A.H. and Maurer, F.H.J., *Polymer*, 1992, **33**, 4236
[39] Shah, P.L., *Shaping the Future. Antec 92*, 1992, **1**, 3

Nylon 5   N-20

$$\mathrm{\left[\!\!-NHCH_2(CH_2)_2CH_2\overset{O}{\underset{\|}{C}}-\!\!\right]_n}$$

Synonyms: *Poly(5-aminovaleric acid). Poly(imino-1-oxopentamethylene). Poly(5-aminopentanoic acid). Poly(δ-valerolactam)*
Monomers: 5-Aminopentanoic acid, 2-Piperidinone
Material class: Thermoplastic
Polymer Type: polyamide
Molecular Formula: $(C_5H_9NO)_n$
Fragments: C_5H_9NO
Morphology: Triclinic a 9.5, b 5.6, c 7.5. Unit cell a 0.965, b 0.978, c 0.726; α 60.4°, β 90.0°, γ 66.7° [3]

Volumetric & Calorimetric Properties:
Density:

No.	Value	Note
1	d 1.3 g/cm^3	[2]

Melting Temperature:

No.	Value	Note
1	258°C	[1]

Bibliographic References

[1] Korshak, V.V. and Frunze, T.M., *Synthetic Hetero-Chain polyamides*, Daniel Darcy, New York, 1964, (temp)
[2] Hasegawa, R.K., Kimoto, Y., Chatani, Y., Tadokoro, H. and Sekiguchi, H., *Disc. Meeting Soc. Polym. Sci. Japan*, 1974, 713, (cryst)
[3] Muñoz-Guerva, S., Prieto, A., Monserrat, J.M. and Sekiguchi, H., *J. Mater. Sci.*, 1992, **27**, 89, (cryst struct)

Nylon 5,6   N-21

$$\mathrm{\left[\!\!-NHCH_2(CH_2)_3CH_2NH\overset{O}{\underset{\|}{C}}CH_2(CH_2)_2CH_2\overset{O}{\underset{\|}{C}}-\!\!\right]_n}$$

Synonyms: *Poly(iminopentamethyleneadipoyl). Poly(pentamethylene diamine-co-adipic acid. Poly[imino-1,5-pentanediylimino(1,6-dioxo-1,6-hexanediyl)]*
Monomers: Hexanedioic acid, 1,5-Pentanediamine
Material class: Thermoplastic
Polymer Type: polyamide
CAS Number:

CAS Reg. No.
41724-56-5

Molecular Formula: $(C_{11}H_{20}N_2O_2)_n$
Fragments: $C_{11}H_{20}N_2O_2$

Volumetric & Calorimetric Properties:
Latent Heat Crystallization: Activation energy for transport 61.1 kJ mol^{-1} [1]
Melting Temperature:

No.	Value	Note
1	254°C	[1]

Glass-Transition Temperature:

No.	Value	Note
1	45°C	heating rate 2° per min. [2,3]

Transport Properties:
Polymer Solutions Dilute: η_{inh} 0.941

Bibliographic References

[1] Van Krevelen, D.W., *Properties of Polymers: Their Correlation with Chemical Structure*, 3rd edn., Elsevier, 1990
[2] Magill, J.H., *J. Polym. Sci., Part A: Polym. Chem.*, 1965, **3**, 1195
[3] Ke, B. and Sisko, A.W., *J. Polym. Sci.*, 1961, **50**, 87

Nylon 6   N-22

$$\mathrm{\left[\!\!-NHCH_2(CH_2)_3CH_2\overset{O}{\underset{\|}{C}}-\!\!\right]_n}$$

Synonyms: *Poly(ε-caprolactam). Poly[imino(1-oxo-1,6-hexanediyl)]. Poly[6-aminohexanoic acid]. Polycaproamide. Polypentamethylenecarbonamide*
Related Polymers: Nylon
Monomers: Caprolactam, Caproamide
Material class: Thermoplastic
Polymer Type: polyamide

Nylon 6

CAS Number:

CAS Reg. No.
25038-54-4

Molecular Formula: $[C_6H_{11}NO]_n$
Fragments: $C_6H_{11}NO$
Additives: Caprolactam monomer remaining from polymerisation acts as plasticiser to increase flexibility and strength. 30–33% Glass fibres blended to improve strength and modulus and reduce mould shrinkage. Mineral filled grades have lower creep. 30–33% Chopped fibreglass (approx. 3 mm) blended and co-extruded. Glass reinforcement increases tensile and flexural props., reduces creep, shrinkage and moisture absorption; toughness is reduced. Mica to improve solvent resistance and strength
General Information: Nylon 6 resembles Nylon 6,6 in chemical and physical props. Nylon 6 is used mainly for fibres and engineering resin moulding applications
Morphology: Nylon 6 develops γ-crystallinity during routine moulding or extrusion but will change to the α-crystalline form by nucleation [43]. Degree of crystallinity 0.4 [1]. α-monoclinic a 9.56, b 8.01, c 17.24, γ 67.5, 8 units in cell [27]; α-monoclinic a 4.81, b 7.61, c 17.1, γ 79.5, 4 units in cell; α-monoclinic a 9.65, b 8.11, c 17.2, γ 66.3; β-monoclinic a 4.9, b 8.0, c 3.6-4.1; γ-monoclinic a 9.14, b 4.84, c 16.68, γ 121, 4 units in cell [18]. Dark spots are seen which react with OsO_4. These are regions of microstructure damage whose number diminish with rising water content/reducing crystallinity. They are low-density areas of cavities formed to relieve stress due to material misfit between domains of lamellae and spherulites during plastic deformation [44]. A collection of all crystallographic data on the α and γ unit cells of Nylon 6 has been reported [45,46]. Crystallisation from dilute soln. has been reported [47]

Volumetric & Calorimetric Properties:
Density:

No.	Value	Note
1	d 1.084 g/cm^3	amorph. [1]
2	d 1.09 g/cm^3	moulded [15]
3	d 1.23 g/cm^3	crystalline [1]
4	d 1.12–1.14 g/cm^3	moulded [1]
5	d 1.13 g/cm^3	[15]

Thermal Expansion Coefficient:

No.	Value	Note	Type
1	8.9×10^{-5} K^{-1}	125°, α-cryst. [31]	L
2	0.0001 K^{-1}	125°, α-cryst. [31]	L
3	9×10^{-5} K^{-1}	-30–30°, ASTM D696 [97]	L

Volumetric Properties General: Molar Volume 100 cm^3 (crystalline rods) [13]; 103 cm^3 (amorph.) [15]. Specific volume 0.917 cm^3 g^{-1} (amorph., 20°); 0.813 cm^3 g^{-1} (α-monoclinic); 1.032 cm^3 g^{-1} (270°, melt) [15]. Other heat capacity data have been reported [1,13]
Thermal Conductivity:

No.	Value	Note
1	0.23 W/mK	23°, mouldings [25]
2	0.43 W/mK	23°, crystalline, moist [25]
3	0.35 W/mK	23°, amorph., moist [25]
4	0.21 W/mK	250°, melt [26]
5	0.23 W/mK	dry, IEC 1006 [30]

Thermal Diffusivity: 1.8×10^{-3} (ambient), 1.4×10^{-3} (100°); 0.65×10^{-3} cm^2 s^{-1} (melt) [26]
Specific Heat Capacity:

No.	Value	Note	Type
1	1.47–1.6 kJ/kg.C	solid state, 23°, 50% relative humidity [1,29]	P
2	2.13–2.47 kJ/kg.C	liquid, 25° [1]	P
3	1.7 kJ/kg.C	dry, IEC 1006 [30]	P

Melting Temperature:

No.	Value	Note
1	260°C	α crystalline form [1,12]
2	231°C	equilibrium [21]
3	218°C	[97]

Glass-Transition Temperature:

No.	Value	Note
1	40°C	[1]
2	60°C	dilatomer, dry [12]
3	53–54°C	27% crystalline sample [12]
4	50–80°C	lower transition [1]
5	125–132°C	[1]

Transition Temperature General: Molar thermal decomposition function $γ_{d½}$ 79 kcal mol^{-1} [10]
Deflection Temperature:

No.	Value	Note
1	190°C	0.45 MPa, ISO 75 [28]
2	95°C	1.8 MPa [28]
3	160°C	0.45 MPa, dry, moulded, ISO R75 [12,28]
4	55–90°C	1.8 MPa [12,28]
5	>160°C	min., 0.45 MPa [30]
6	55–75°C	1.8 MPa [30]
7	180°C	0.45 MPa
8	67°C	1.8 MPa, ASTM D696 [97]

Transition Temperature:

No.	Value	Note	Type
1	40–45°C	[1]	2nd order T_g
2	350°C	[1]	decomposition temp. T_d
3	430°C	[1]	half life temp. of decomposition
4	400°C	[12]	flash ignition temp.
5	440°C	[12]	self ignition temp.

Surface Properties & Solubility:
Cohesive Energy Density Solubility Parameters: δ 22.5 J^{½} cm$^{3/2}$ [1]; d 27.8 J^{½} cm$^{3/2}$ (amorph., 25°) [17]
Plasticisers: Very polar non-volatile liquids e.g. *p*-toluenesulfonamide

Nylon 6

Solvents/Non-solvents: Sol. typical polyamide solvents such as dimethyl acetamide or DMF (dissolution of chloride salts 1–2% assists) and strong acids such as sulfuric (cold), formic or AcOH (hot). Polar solvents will often swell the nylon or dissolve it to a limited extent, e.g. Me_2O, $CHCl_3$, CCl_4, aniline (>40°); cresol, phenol, resorcinol

Surface Tension:

No.	Value	Note
1	40 mN/m	from contact angle [1,24]
2	43 mN/m	[1,24]
3	36 mN/m	[1,24]

Transport Properties:

Transport Properties General: Permeability and diffusion increase significantly with water content, which may vary between 0–10%, and with increasing temp.

Polymer Solutions Dilute: Unperturbed viscosity coefficient $K\theta$ 0.19–0.23 $cm^3\ mol^{1/2}\ g^{-3/2}$ [1].

Polymer Melts: Melt density 0.96 $g\ cm^{-3}$ [12]. Spencer and Gilmore equation constants have been reported [1,4,5]. Coefficient of zero-shear-rate-melt viscosity, K 9.5×10^{-14} (250°) [22]. Melt viscosity 140 Pa s (250°) [30]. Melt viscosity 1000 Pa s (250°, M_n 26000)

Permeability of Gases: Permachor prediction of gas permeability has been reported [8]. Nylon 6 has very low permeability to organic vapours and permanent gases, and is often used for its barrier props. Impermeability to gases may be further enhanced by blending with the commercial amorph. copolyamide of isophthalic acid, terephthalic acid and hexamethylene diamine, Selar AP

Permeability of Liquids: Nylon 6 has the highest permeability of all common thermoplastics to water vapour. $[H_2O]$ 0.13×10^{-13} (30°) $(0.16 \times 10^{-13}\ cm^2\ (s\ Pa)^{-1}$ (0–85% humidity, 23°)

Water Content: Mouldings from Nylon 6 are hygroscopic and gradually take up moisture until equilibrium is reached, but will increase further as humidity rises. The effect upon physical props. is quite profound; toughness and impact strength increase; tensile strength is reduced while elongation is increased; moulding dimensions increase linearly with absorbed water; 0.3% per 1% water uptake. The electrical props. are adversely affected, although generally still OK for domestic purposes. Water absorption is low and it may take several months for a thick section of moulding to equilibrate. Usually mouldings are preconditioned at 60° and 20° prior to despatch to customers, so as to reach equilibrium. As Nylon 6 takes up 10% water, the peeling strength of metal coatings on Nylon 6 is significantly affected by the water absorbed on its surface. Nylon 6 is plated by a wet process to optimise peeling strength, peeling force 1.8 kg cm^{-1} is similar to that of ABS [48]. Structural changes accompanying hydration of Nylon 6 have been reported [49]

Water Absorption:

No.	Value	Note
1	9.5 %	[12]
2	10 %	23°, saturated [28,30]
3	3.2 %	20°, 48h [97]
4	2.6–3.4 %	23°, saturated, 50% relative humidity [30]
5	3.5 %	20° equilibrium., 65% relative humidity
6	3.3–4.5 %	21°, 65% relative humidity
7	7.1–7.8 %	21°, 95% relative humidity

Gas Permeability:

No.	Gas	Value	Note
1	$[N_2]$	26.26 cm^3 mm/(m^2 day atm)	0.03×10^{-13}
2	$[N_2]$	6.26 cm^3 mm/(m^2 day atm)	0.00715×10^{-13} (30°) [1,2,3,4,5,12]
3	$[O_2]$	24.95 cm^3 mm/(m^2 day atm)	0.0285×10^{-13} (30°)
4	$[O_2]$	52.53 cm^3 mm/(m^2 day atm)	0.06×10^{-13} (40% humidity) [30]
5	$[CO_2]$	57.78 cm^3 mm/(m^2 day atm)	0.066×10^{-13} (20°)
6	$[CO_2]$	350.18 cm^3 mm/(m^2 day atm)	0.40×10^{-13} (0% humidity)
7	$[H_2S]$	223.24 cm^3 mm/(m^2 day atm)	0.255×10^{-13} (30°)
8	$[NH_3]$	768.64 cm^3 mm/(m^2 day atm)	$0.878 \times 10^{-13}\ cm^2\ (s\ Pa)^{-1}$ (20°)

Mechanical Properties:

Mechanical Properties General: Mechanical loss factor tan δ (0.1–10 Hz, ISO 537), 0.016 (dry, 20°); 0.048 (dry, 40°); 0.175 (dry, 60°); 0.127 (dry, 80°); 0.064 (dry, 100°) [12]. At moisture content <0.15% mechanical props. are stable, but complex shapes crack [50]. To avoid complications, manufacturers pre-condition granules to 3.5% moisture content. Tensile modulus 2.2–3.1 N tex^{-1} (normal tenacity), 2.9–4.1 N tex^{-1} (65% relative humidity). Tensile stress break 0.4–0.6 N tex^{-1} (normal tenacity), 0.75–0.84 N tex^{-1} (high tenacity)

Tensile (Young's) Modulus:

No.	Value	Note
1	168000 MPa	along chain, oriented fibre [1]
2	3000 MPa	dry, 23° [12,30]
3	1500 MPa	moist, ISO 1110, 23° [12,30]
4	500 MPa	dry, 100° [12]

Flexural Modulus:

No.	Value	Note
1	2200 MPa	dry, 23° [28,29]
2	800–970 MPa	50% relative humidity, 23°, ISO 178 [28,29]
3	4500 MPa	[1,2]
4	5000 MPa	[1]

Tensile Strength Break:

No.	Value	Note	Elongation
1	80 MPa	dry, 23° [12,30]	
2	90 MPa	nucleated [12,30]	
3	50 MPa	moist, 23° [12,30]	
4	22 MPa	moist, saturated in water, 23° [12,30]	
5	30 MPa	dry, 100° [12,30]	
6	70–83 MPa	50% relative humidity, 23° [29]	

– Nylon 6

No.	Value	Note	
7	70000 MPa	ASTM D638 [97]	300%
8	475–550 MPa	draw ratio 4.3:1, 23°, 50% relative humidity, monofilaments	
9	675–775 MPa	draw ratio 5:1, 23°, 50% relative humidity [12,29]	300%
10	60 MPa	dry, 23° [28]	5%
11	45 MPa	50% relative humidity, 23° [28]	25%
12	23–43 MPa	65% relative humidity, normal tenacity, 21°	
13	12–17 MPa	65% relative humidity, high tenacity, 21°	

Flexural Strength at Break:

No.	Value	Note
1	80 MPa	dry, 23° [28]
2	25 MPa	50% relative humidity, 23°, ISO 178 [28]

Compressive Modulus:

No.	Value	Note
1	90 MPa	[1]

Elastic Modulus:

No.	Value	Note
1	157000 MPa	[31]
2	40000 MPa	175° [31]
3	4500 MPa	[1,2]
4	5000 MPa	[1]

Poisson's Ratio:

No.	Value	Note
1	0.33	mouldings, 20° [12]
2	0.46	100° [12]
3	0.5	melt [12]

Tensile Strength Yield:

No.	Value	Elongation	Note
1	44000 MPa		23°, 50% relative humidity [29]
2	55000 MPa	7%	ASTM D638
3	580–635 MPa		normal tenacity fibres
4	855–952 MPa		high tenacity fibres

Flexural Strength Yield:

No.	Value	Note
1	112 MPa	ASTM D790 [99]

Miscellaneous Moduli:

No.	Value	Note	Type
1	5100 MPa	[6]	Bulk modulus
2	4400 MPa	[12]	Bulk modulus
3	1100 MPa	23°, dry [12]	Shear modulus
4	1500 MPa	dry, nucleated, 23° [12]	Shear modulus
5	200 MPa	dry, 100° [12]	Shear modulus
6	800 MPa	dry, 200°, ISO 537 [12]	Shear modulus
7	230 MPa	23°, 50% relative humidity	Apparent creep modulus
8	440 MPa	nucleated	
9	600 MPa	20°, ASTM D1043 [97]	Modulus of rigidity in torsion
10	2100 MPa	20° ASTM D790, [97]	Modulus of elasticity in flexure

Impact Strength: Dispersion of core-shell polymer in caprolactam soln. of telechelic polymerised to form rubber modified Nylon 6 block copolymer with improved impact resistance [96]

Viscoelastic Behaviour: Speed of sound 2700 m s^{-1} (longitudinal), 1120 m s^{-1} (shear) [7]; 1400–2300 m s^{-1} (fibres, 20°) [23]

Hardness: Ball indentation hardness 6.25 [1]; Ball indentation hardness H358/30 (ISO 2039, 150 MPa, dry, 23°). Shore D72 [1]. Rockwell 85 [1], M100 (23°), M70 (moist, ISO 1110, 23°) [1,14]. Rockwell R 118, M 80 [97]

Fracture Mechanical Properties: Fracture energy W_{50} 100 J (dry, 23°); 140 J (moist, 23°); 40 J (dry, -20°, DIN 53443Ti) [30]. Hoop stress 15–12 N m^{-2} (20°, H$_2$O, 100h); 6 N m^{-2} (19000h) [30]. Average toughness 0.07–0.14 N tex^{-1} [1,14,30]

Friction Abrasion and Resistance: Taber abrasion resistance 8 mg (1000 cm^3)$^{-1}$, (ASTM D1044); Taber loss factor 15 mg (DIN 53516) [1]. Coefficient of friction 0.39 [1]; 0.35 (roughness depth 0.1 µm); 0.32 (roughness depth 1 µm, 2µm); 0.43 (roughness depth 6 µm) [14]. Friction and wear coefficients of Nylon 6 pins slid against stainless steel at 20°, -196° and -268.8° have been reported. Although harder at cryogenic temps., non-dimensional wear coefficients are in the range $5 \times 10^{7.5}$ to 10^6. Good correlation between friction and wear coefficients indicates a third to fourth power relationship between them for both unfilled and reinforced nylon. Adhesive wear is the dominant wear mechanism [54]. Mechanism of friction and fatigue limit of Nylon 6 have been reported [94,95]

Izod Notch:

No.	Value	Notch	Note
1	50–100 J/m	N	23°, dry [12]
2	30–50 J/m	N	-40°, dry [12]
3	120 J/m	N	23°, moist, ISO 1110 [12]
4	270 J/m	N	23°, 50% relative humidity, ISO 180/1A [29]

Izod Area:

No.	Value	Notch	Note
1	5–15 kJ/m^2	Notched	dry, 23°, ISO 180/1A [30]
2	5–6 kJ/m^2	Notched	dry, 23°, ISO 180/1A, -40° [30]

– Nylon 6

Charpy:

No.	Value	Notch	Break	Note
1	297	N		304 kg cm cm^{-2}, dry
2	397	Y		dry [30]
3	0.024–0.09	Y		moist [30]
4		N	No break	23°, DIN 53453/ ISO 179 [30]
5	90–380	N		23°, dry, DIN 53453/ ISO 179 [30]

Electrical Properties:
Electrical Properties General: Very good insulating props. and tracking resistance when dry or even conditioned to standard humidity and temp., these deteriorate when nylon is fully saturated due to its high water uptake. Transverse resisitivity 5×10^{13} ohm cm (ASTM D257) [97]

Surface/Volume Resistance:

No.	Value	Note	Type
1	1×10^{15} Ω.cm	dry, [28]	S
2	0.001×10^{15} Ω.cm	50 % relative humidity, 23°, IEC 243 [28]	S
3	0.01×10^{15} Ω.cm	dry [30]	S
4	$1 \times 10^{-5} \times 10^{15}$ Ω.cm	moist [30]	S

Dielectric Permittivity (Constant):

No.	Value	Frequency	Note
1	10.9	100 Hz	50% relative humidity, 23° [16]
2	7.3	1 kHz	50% relative humidity, 23° [16]
3	3.8	1 MHz	50% relative humidity, 23° [16]
4	4.2–4.5		[1]
5	8.3	1 kHz	[29]
6	3.5		dry, 23°
7	4		50% relative humidity, 23° [28]
8	3.5		dry, 23°
9	7	1 kHz	moist, DIN 50014 [30]
10	10.9	100 Hz	23°, 50% relative humidity [16]
11	7.3	1 kHz	23°, 50% relative humidity [16]
12	4.5	100 kHz	23°, 50% relative humidity [16]
13	3.8	1 MHz	23°, 50% relative humidity [16]

Dielectric Strength:

No.	Value	Note
1	100 kV.mm^{-1}	dry [1]
2	40 kV.mm^{-1}	dry [12]
3	60 kV.mm^{-1}	moist, ISO 1110 [17]
4	14 kV.mm^{-1}	dry, 23°
5	10 kV.mm^{-1}	50% relative humidity, IEC 243, 3 mm specimens [28]
6	100 kV.mm^{-1}	dry [31]
7	60 kV.mm^{-1}	moist [31]

Complex Permittivity and Electroactive Polymers: Resistance to tracking CTI 600 (method CTI); CTI-M 600 (method CTI-M) [30]. Comparative tracking index >600 (IEC 112) [28]

Dissipation (Power) Factor:

No.	Value	Frequency	Note
1	0.2	1 kHz	[29]
2	0.023–0.03	1 MHz	moist, DIN 50014, IEC 250 [30]
3	0.06	1 kHz	[97]

Optical Properties:
Refractive Index:

No.	Value	Note
1	1.53	mouldings [12]

Total Internal Reflection: Birefringence 1.53 (perpendicular) [13]; 1.58 (parallel) [1]

Polymer Stability:
Thermal Stability General: Excellent thermal stability to 300°, in absence of air and dry

Upper Use Temperature:

No.	Value	Note
1	120°C	[1]
2	70–85°C	20000h [30]
3	180°C	5–10h

Decomposition Details: Decomposition occurs slowly above 300°, combustible gases given off between 450–500°, plus water, H$_2$O, CO, N$_2$ and other nitrogenous compounds. Above 400° decomposition products are as toxic as burning wood. Surface degradation begins at 10% O$_2$ in inert gas atmosphere [41] above 100° [97]. Graphite, kaolin and other filled grades undergo loss of mass and transformation at 360–500° [39]. Other thermal degradation studies have been reported [40, 42]. Nylon structural residues remain in char after pyrolysis [37]

Flammability: Flammability rating HB (reinforced grades, UL94) [28]; V2 (<1.6 mm, cast sheet, UL94) [30]; V0 (flame retardant, UL94); BH (2–20 mm, IEC 707/VDE 0304T); FH (3–15 mm, IEC 707/ VDE 0304T). Limiting oxygen index 23% [1], 25–27% (dry, ASTM D2863) [12,28]. A range of fire retardants with Nylon 6 have been studied, including brominated polystyrene [59]. Phosphorus compounds [11, 60, 64, 68], phosphine oxide [61], melamine and glass fibre [62]; antimony and zinc compounds [63, 67] and brominated epoxy resins [64]. Polypentabromobenzyl acrylate/reactive butadiene rubber/Nylon 6 flame retardant compounds have been investigated, with improved impact strength, suggesting compatabilisation [36, 65]

Environmental Stress: Unstabilised Nylon 6 mouldings degrade under load in warm wet climate, with creep, loss of fracture energy and stress. Extended exposure causes crystallinity and orientation changes [43]. Embrittlement occurs above 80° on exposure to O$_2$ or Uv radiation. Mechanical props. are also affected by Uv radiation. Stabilisers such as copper halides, amines and esterified phenols, and light stabilisers [56] are used to minimise weathering effects [28]. Nylon 6 undergoes salt induced crazing [32,33]. Above 50°, water and ion sorption increases [55], absorbed salt can be removed by immersion in water. At high temp. T$_g$ and crystallisation changes occur suggesting plasticisation mechanism [34]. Environmental effects on mechanical props. have been reported[52]

Chemical Stability: Nylon 6 is resistant to attack by Me$_2$CO, AcOH, hydrated alumina salts, NH$_3$, amyl acetate, antifreeze, C$_6$H$_6$, borates, butane, butyl alcohol (<40°), citric acid, copper sulfate, decalin, diesel fuel, edible fat, Et$_2$O; fats and fatty esters, fluorine, formaldehyde, formic acid, fuel oils, glucose, glycerol

(<60°), glycol, heptane, hexane, hydraulic fluid, hydrochloric acid, H_2S, H_2O_2, hydroquinone, ink, isooctane, kerosene, lactic acid, 50% $MgCl_2$, Hg, milk, mineral oil, motor oil, naphthalene, nitric acid, nitrobenzene, oleic acid, ozone, perchloroethylene (<40°), petrol, petroleum jelly, phenol, 10% KOH, propane, salicylic acid, sea water, silicone oil, sodium carbonate, 10% NaOH, Na_2SO_4 aq. (<40°), starch, stearates, styrene, sulfur, 10% sulfuric acid, tartaric acid, tetraethyl lead, toluene (<40°), transformer oil, turpentine, urea, uric acid, urine, water, wax, white spirit, wine, xylene (<40°) [28]. Swollen by $CHCl_3$, EtOH. It is attacked by strong acids especially at elevated temps. Flexural and impact strength of glass reinforced Nylon 6 are maintained to at least 80% of their initial values after immersion in engine oil or radiator fluid at 120° for up to 2000h [30]

Hydrolytic Stability: Amorph. regions are attacked first but crystalline areas are not immune to hydrolysis

Applications/Commercial Products:

Processing & Manufacturing Routes: Caprolactam is polymerised to Nylon 6 by heating at 250–270° in an autoclave with water and initiator (aminocaproic acid). AcOH used to control MW. Conversion reaches 90%. Polymer is extruded into water and cut into chips. Polymer is treated with hot water to remove 9% monomer, then dried before processing. Non-woven forms are usually produced by melt spinning and distributing nylon fibres directly onto a mat e.g. on a conveyor belt. Treatments are applied to bond the fibres through mechanical entanglement of the filaments by needle punching. Continuous polymerisation. Caprolactam is continuously metered into steam pressured hydrolyser. Resulting mixture of lactam and polymer are passed through one or more polymerisation reactors designed for moisture release to give satisfactory MW. Molten polymer is metered from final reactor through a strand die and pelletised. Pellets are washed in leaching tower to remove excess lactam. A new invention relates to the polymerisation of caprolactam in which a reactant stream containing caprolactam and a lactam polymerisation initiator are brought into a contact with a second stream containing caprolactam and a lactam magnesium halide. 2-Pyrrolidinone and a pyrrolidinyl capped initiator enhance the rate of polymerisation [101]. During conventional polymerisation of Nylon 6, excess water is evaporated at the degassing stage from the reaction product, which is close to polymerisation equilibrium. Degassing is not always smooth for high viscosity melts. The optimium prepolymer condition at the degassing stage has been determined to be 170–175° [104]. Unique starburst Nylon 6 macromolecules have been synth. which have extraordinary symmetry, extensive branching and a high number of terminal functionality [105]. Nylon 6 two-armed and three-armed systems, prepared from anionic bulk polymerisation, have different characteristics from linear nylon and exhibited lower melting points and higher melt flow than conventional Nylon 6 [106]. Activated anionic polymerisation of caprolactam occurs rapidly reaching equilibrium in a few minutes, compared to conventional polymerisation which takes several hours. This method is used in important industrial applications of Casting and Reaction Injection Moulding (RIM). Polymerisation is conducted below Nylon 6 melting point to permit direct fabrication of parts from monomer [93,107]. Rotational moulding can be used for manufacturing relatively cheap but complex shapes with little built-in mould stress [92]. Nylon rotational moulding was developed for Nylon 11 and 12 powders. Since the 1980s Nylon 6 has entered this market which is growing, and is supplied in pellet form. Powders are employed with particle sizes approx. 500 μm and mould temps. are around 230–280°. Flushing with inert gas during moulding prevents oxidative attack and discolouration of the interior surface. Nylon 6 pellets are processed at oven temps. of 300–360°. Materials of mould construction include cast and sheet aluminium, stainless steel and electroformed nickel [86]. Rotational moulding is a single unique process capable of producing hollow items with very uniform wall thickness [87]. High melt flow powder is placed in a closed, heated mould which is rotated about two perpendicular axes to ensure uniform distribution. The high melt flow index requirement leads to a decrease in toughness and environrental stress cracking resistance which is partly offset by the fact that mouldings are stress free. Where high strength is necessary, then the process can be adopted to coat the inside of a metal tank with a corrosion resistant coating. The technique has been applied successfully to Nylon 6 and HDPE [88]. The props. of Nylon 6 and ethylene-chlorotrifluoroethylene (ECTFE) make them the natural choice for applications requiring better heat resistance, greater strength, modulus and superior environmental stress crack resistance. The chemical resistance of HDPE, Nylon 6 and ECTFE give typical rotomoulded props., tensile creep at 93°, environmental ageing and ambient gasoline permeation compared graphically [89]. Nylon castings are cost effective for making large or heavy-wall parts [70,76,77,78]. Spin casting is a further variation of the process in which molten caprolactam monomer is anionically polymerised in a mould employing a catalyst system comprising typically a metal lactam and a polyacyl lactam. Fillers or fibres may be incorporated into the casting. Spin casting has proved more economical than injection moulding for low volume Nylon 6 materials [73,74]. In order to achieve optimum mechanical strength, stiffness and cold flow, the interfacial bond between the inorganic reinforcing material and the nylon polymer should be made through a chemical reaction. n-((3-Triethoxysilylpropyl)-carbamyl) caprolactam is particularly suitable in anionic cast polymerisation [75]

Applications: Fibres. Nylon 6 and 6,6 dominate the use of fibres in carpets due to excellent wear resistance with retention of appearance and low cost. A full range of colours and lustre are available. Nylon fibres predominated in the carcass of truck tyres, racing car tyres and airplane tyres. Used less in radial tyres and replacement tyres due to temporary flat spots. Nylon continues to be principal fibre in hosiery, women's underwear and stretch fabrics. Nylon lost market share to Polyesters in the apparel market from the mid 80s, but continues to dominate womens underwear. Nylon blends with other fibres, e.g. Lycra, are used by wrapping Lycra filaments with nylon to give a nylon appearance together with high stretch and recovery props. There is a large market in soft-sided luggage due to its strength. Some nylon is also used in upholstery fabrics. Non-woven fabrics are used for artificial leather, component of carpet underlay and for disposable garments in the health care field. Products made from automated moulding operations include many car parts e.g. fluid reservoirs, wheel covers. Electrical applications incude plug sockets, switches. Large amounts of Nylon 6 used in film and wire and cable. Most film is coextruded with polyethylene to utilise oxygen barrier props. of nylon. Nylon 6 also finds use in power tool industry due to excellent pigmentation props. Unreinforced Nylon 6 a) Low viscosity grades used to extrude: monofilaments, bristles, fishing lines, oriented tapes and coatings, machinery parts, housings for hand held tools, thin walled parts, housing/chassis for gasoline pumps, cable sheathing. b) medium viscosity: fittings, car trims, thick walled parts. c) high viscosity: blown film, sheet, thick monofilaments, automotive tubes, blow mouldings. d) anionic mouldings: semi-reinforced prods. and containers. e) impact modified: anchors, rollers. Glass-fibre reinforced applications include automotive mirror housings (high impact), wheels of mountain bikes; also very stiff, dimensionally stable high temp. resistant mouldings e.g. electrically insulating parts. Impact modified glass-fibre reinforced for high notched impact strength applications e.g. BMX bike wheels, steering wheels and power tools. Mineral reinforced grades used for high impact, food dimensional stability and low warpage (e.g. wheel coverings and engine covers, toothbelt guards on engines). Combined glass fibre and mineral reinforced grades used for intermediate stiffness and good dimensional stability and surface finish and low warpage

Trade name	Details	Supplier
2110/2210/2310/2512		Nan Ya Plastics
60 G/6001/6253	various grades	Plastic Materials
AVP	GY 6IL/GY 601/ RY 633	Polymerland Inc.
Adell	various grades	Adell Plastics Inc.

Nylon 6

Akulon	various grades	DSM Engineering
Akuloy	RM J-75/30	DSM Engineering
Amilan	CM series	Toray Industries
Amilon	UTN 121/141	Toray Industries
Ashlene	various grades	Ashley Polymers
Bapolon	760C	Bamberger Polymers
CRI	NX 1170/2011	Custom Resins
CTI	NY range	CTI
Capran Unidraw	101/103/150/200/202	AlliedSignal Corp.
Capron	various grades	AlliedSignal Corp.
Celanese	various grades	Hoechst Celanese
Celstran	N6 range	Polymer Composites
Comalloy	610 series	Comalloy Intl. Corp.
Comco Nylon 6		Commercial Plastic
Comtuf	600 series	Comalloy Intl. Corp.
Dimension	9000 series	AlliedSignal Corp.
Durethan	various grades	Bayer Inc.
EMI-X	various grades	LNP Engineering
Electrafil	various grades	DSM Engineering
Entec	various grades	Entec Polymers
Fiberfil	various grades	DSM Engineering
Fiberstran	G-3/30/40/50	DSM Engineering
Grilon	various grades	EMS
Hiloy	600 series	Comalloy Intl. Corp.
Hylon	500 series	Hale Manufacturing
J-3/30/VO		DSM Engineering
J-3/50		DSM Engineering
J-7/43		DSM Engineering
J-8/30		DSM Engineering
Lubricomp	PFL/PL series	LNP Engineering
Lubrilon	605/613/618	Comalloy Intl. Corp.
M_n 6	FG 10/20/30/40	Modified Plastics
Magnacomp	PL/PM	LNP Engineering
Magnaplas	PA range	Bay Resins
Mon Tor	CM series/UTN	Mon Tor
N	various grades	Thermofil Inc.
NSC Esbrid	NSG series	Thermofil Inc.
NSG	various grades	Thermofil Inc.
Nivionplast	various grades	Enichem America
Novamid	1000 series	Mitsubishi Chem.
Ny-Kon P		LNP Engineering
Nybex	various grades	Ferro Corporation
Nycoa	various grades	NYCOA
Nylamid	various grades	Polymer Service
Nylatron	2000 series	Polymer Corp.
Nylene	various grades	Custom Resins
Nylon	C series	Firestone
Nylon	NST/N60/600 series	Michael Day
Nylon 6 GPL		Albis Corp.
Nyloy	various grades	Nytex Composites
Nypel	various grades	AlliedSignal Corp.
Nyrim	P-1 series	DSM Engineering
Nytron	LNG series	Nytex Composites
PA	200 series	Bay Resins
PA 45/0		Albis Corp.
PA 6-110-EF		MRC Polymers Inc.
PA6	various grades	TP Composites
PRL-Nylon	NY range	Polymer Resources
Plaslube	G-3/J-3	DSM Engineering
Polyamid 6	PA series	Albis Corp.
RTP	200 series	RTP Company
RX14-212/ SC14-2090		Spartech
Schulamid	GB30/GF30/MV5	A.Schulman Inc.
Shinite	C501 range	Shinkong Synthetic
Sniamid	various grades	Snia
Stat-Kon P		LNP Engineering
Stat-Loy P		LNP Engineering
Technyl	C series	Rhone Poulenc Inc.
Texalon	various grades	Texapol Corp.
Texapol	various grades	Texapol Corp.
Thermocomp	PC/PF series	LNP Engineering
UBE	1000 series	UBE Industries Inc.
Ultramid	various grades	BASF
Unitika	A series	Unitika America Co.
Verton	PF-700-10/PF-7007	LNP Engineering
Voloy	612/618/652/658	Comalloy Intl. Corp.
Wellamid	various grades	Wellman Inc.
Zytel	various grades	DuPont

Bibliographic References

[1] Van Krevelen, D.W., *Properties of Polymers: Their Correlation with Chemical Structure*, 3rd edn., 1990
[2] Weyland, H.G., *Text. Res. J.*, 1961, **31**, 629
[3] Wunderlich, B. *et al*, *Encycl. Polym. Sci. Eng.*, Vol. 16, Wiley, 1989, 767
[4] Spencer, R.S. and Gilmore, G.D., *J. Appl. Phys.*, 1950, 523
[5] Sagalev, G.V. *et al*, *B.P. Int. Poly. Sci. Technol.*, 1974, **1**, 76
[6] Warfield, R.W. *et al*, *Angew. Makromol. Chem.*, 1972, **27**, 215
[7] Hartmann, B. *et al*, *J. Polym. Sci., Polym. Phys. Ed.*, 1982, 1269
[8] Salame, M., *Polym. Eng. Sci.*, 1986, **26**, 1543
[9] Hoffmann, J.D., *Ind. Eng. Chem.*, 1966, **2**, 41
[10] Van Krevelen, D.W., *Properties of Polymers: Their Correlation with Chemical Structure*, 1987, 652
[11] Lyons, J.W., *The Chemistry and Uses of Fire Retardants*, Wiley-Interscience, 1970
[12] Sorba, E. and De Chirico, A., *Polymer*, 1976, **17**, 348, (T_g)
[13] Warfield, R.W., Kayser, E.G. and Hartmann, B., *Makromol. Chem.*, 1983, **184**, 1927
[14] Wakelin, J.H., Sutherland, A. and Beck, L.R., *J. Polym. Sci.*, 1960, **42**, 278
[15] Muller, A. and Pfluger, R., *Kunststoffe*, 1960, 203
[16] Bergmann, R., *Konstr. Kunstst.*, (ed. G. Schreyer), Hanser, 1972, 2479
[17] Inoue, M., *J. Polym. Sci., Part A: Polym. Chem.*, 1963, **1**, 2697
[18] Illers, K.H., Haberkorn, H. and Simak, P., *Makromol. Chem.*, 1972, **158**, 285
[19] Starkweather, H.W., *J. Appl. Polym. Sci.*, 1959, **2**, 129
[20] Fukumoto, O., *J. Polym. Sci.*, 1956, **22**, 263
[21] Liberti, F.N. and Wunderlich, B., *J. Polym. Sci., Part A: Polym. Chem.*, 1968, **2**, 833
[22] Laun, H.M., *Rheol. Acta*, 1979, **18**, 478

[23] Jambrich, M., Diavik, I. and Mitterpach, I., *Faserforsch. Textiltech.*, 1972, **23**, 28
[24] Hybart, F., *J. Appl. Polym. Sci.*, 1960, **3**, 118
[25] Hellwege, K., Hoffmann, R. and Knappe, W., *Kolloid Z. Z. Polym.*, 1968, **226**, 109
[26] Dietz, W., *Colloid Polym. Sci.*, 1977, **255**, 755
[27] Holmes, D.R., Bunn, C.W. and Smith, C.J., *J. Polym. Sci.*, 1955, **17**, 159
[28] *BIP*, Nylon 6, technical data booklet, Beetle, 1984, (technical datasheet)
[29] *Encycl. Polym. Sci. Technol.*, 1986, **11**, 366, 426, 461
[30] *Ultramid S Polyamides*, BASF, 1996, (technical datasheet)
[31] Nakamae, K., Nishino, T., Hata, K. and Matsumoto, T., *Kobunshi Ronbunshu*, 1987, **44**, 421-428
[32] Wyzgoski, M.G. and Novak, G.E., *J. Mater. Sci.*, 1987, **22**, 2615
[33] Burford, R.P. and Williams, D.R.G., *J. Mater. Sci. Lett.*, 1988, **7**, 59
[34] Wyzgoski, M.G. and Novak, G.E., *J. Mater. Sci.*, 1987, **22**, 1715
[35] *Mod. Plast.*, 1995, **72**, 69
[36] *Plast. Technol.*, 1993, **39**, 4
[37] Sevecek, P. and Stuzka, V., *Fire Mater.*, 1987, **11**, 89
[38] Seyfarth, H.E., Michels, Ch. and Taeger, E., *Plaste Kautsch.*, 1986, **11**
[39] Bobkov, S.A. and Ushakova, O.B., *Int. Polym. Sci. Technol.*, 1985, **12**, 41, T106
[40] Svoboda, M., Schneider, B. and Stokr, J., *Collect. Czech. Chem. Commun.*, 1991, **56**, 1461
[41] Seyfarth, H.E., Michels, Ch. and Taeger, E., *Plaste Kautsch.*, 1987, **34**, 19
[42] Ionescu, I.V., Gosa, K., Predescu, A., Carp, N. and Topciu, E., *Mater. Plast. (Bucharest)*, 1988, **25**, 139
[43] George, G.A. and O'Shea, M.S., *Mater. Forum*, 1989, **13**, 11
[44] Galeski, A., Argon, A.S. and Cohen, R.E., *Macromolecules*, 1988, **21**, 2761
[45] Salem, D.R. and Weigmann, H.D., *Polym. Commun.*, 1989, **30**, 336-338
[46] Parker, J.P. and Lindenmayer, P.H., *J. Appl. Polym. Sci.*, 1977, **21**, 821
[47] Kiho, H., Miyamoto, Y. and Miyaji, H., *Polymer*, 1986, **27**, 1542
[48] Manabe, K. and Hakayawa, F., *Toyoda Gosei Giho*, 1987, **29**, 8
[49] Murphy, N.S., Stamm, M., Sibilia, J.P. and Krimm, S., *Macromolecules*, 1989, **22**, 1261
[50] Fedotova, M.D., Ivankina, I.V., Pilyaeva, N.F., Ivanofa, G.P. and Kuznetsova, I.G., *Plast. Massy*, 1989, **3**, 30
[51] Stout, R.B., *J. Eng. Mater. Technol.*, 1987, **109**, 259
[52] DeVries, K.L. and Hornberger, L.E., *Polym. Degrad. Stab.*, 1989, **24**, 213
[53] Litak, A., Mazurkiewicz, S. and Targosz, B., *Szklarska Poreba Poland*, 1987, **50**, 235
[54] Michael, P.C., Rabinowicz, E. and Iwasa, Y., *Cryogenics*, 1991, **31**, 695
[55] Wyzgoski, M.G. and Novak, G.E., *J. Mater. Sci.*, 1987, **22**, 1707
[56] Compte, J., *Rev. Plast. Mod.*, 1993, **66**, 510
[57] Kaufmann, W., and Zimmermann, R., *Kunststoffe*, 1989, **79**, 432
[58] Weil, E.D., *Plast. Compd.*, 1987, **10**, 31, 33, 34, 36, 38
[59] Burditt, N.A., *Plast. Compd.*, 1988, **11**, 51
[60] Simeonov, N., Stoilova, M., Moleva, T. and Lefterova, E., *Acta Polym.*, 1988, **39**, 722
[61] Yamamoto, H., *Purasutikku Seikei Gijutsu*, 1988, **5**, 80
[62] *U.S. Pat.*, 1988, 4 789 698
[63] Markezich, R.L., *ANTEC*, 86, Technomic Publishing Co., 1986
[64] *U.S. Pat.*, 1986, 4 584 149
[65] Siegmann, A., Yanai, S. and Dagan, A., Reinforced Plastics Composites, *40th Annual Conference*, Society of Plastic Industry, 1985, 12
[66] Touval, I., Annu. Tech. Conf. - Soc. Plast. Eng., *43rd*, Soc. Plast. Eng., 1985
[67] Ilardo, C.S. and Duffy, J.J., *Plast. Eng. (N.Y.)*, 1985, **41**, 51
[68] Mateva, R.P. and Dencheva, N.V., *J. Appl. Polym. Sci.*, 1993, **47**, 1185
[69] Levchik, S., Camino, G., Costa, L. and Levchik, G., *Fire Mater.*, 1995, **19**, 1
[70] Vass, J.A., *Mach. Des.*, 1994, **66**, 70, 74
[71] *Plast. Rubber Wkly.*, **1063**, 23
[72] Kemeny, *Plast. Technol.*, 1987, **33**, 13, 15, 17
[73] *Plastics News (Detroit)*, 1992, **4**, 7
[74] Raymond, M., *Plastics News (Detroit)*, 1989, **1 (30)**, 1, 20
[75] Krudener, R., *Plastverarbeiter*, 1990, **41 (12)**, 24
[76] Horsky, J. and Skudrna, J., *Plasty Kauc.*, 1989, **26**, 102
[77] *U.S. Pat.*, 1993, 5 179 155
[78] *Engineering (London)*, 1987, **227**, 479
[79] *Eureka*, Polypenco, 1987, **7**, 66
[80] *Anti-Corros. Methods Mater.*, ERTA, 1987, **34 (2)**, 17
[81] *Eur. Plast. News*, Polypenco, 1987, **14 (2)**, 38
[82] *Plast. Rubber Wkly.*, **1226**, 13
[83] *Eur. Plast. News*, 1987, **14**, 37
[84] *Eur. Plast. News*, ERTA, 1985, **12 (10)**, 35
[85] *Miner. Met. Rev.*, 1985, **51 (9)**, 30
[86] 1985, Allied Corp., (technical datasheet)
[87] Rees, R.L., *Engineered Materials Handbook*, 1988, **2**, 360
[88] High Perform. Plast., Nat. Tech. Conf. - Soc. Plast. Eng., *(Prepr.)*, Rotational Moulding Compounds, 1986, **3 (7)**, 1-5
[89] Petrucelli, F., *Managing Corros. Plast.*, 1985, **VI**, 120
[90] *U.S. Pat.*, 1988, 4 729 862
[91] *Plast. Technol.*, 1995, **41**, 84
[92] Crawford, R.J., *Rotational Moulding of Plastics*, Research Studies Press, 1992, 215, (book)
[93] Van Buskirk, B., Akkapaddi, M.K., Polym. Prepr. (Am. Chem. Soc., Div. Polym. Chem.), 1988, **29**, 557
[94] Watanabe, M., Yamaguchi, H., *Wear*, 1986, **110**, 379
[95] Nowak, M. and Zuchowska, D., Pr. Nauk. Politech., Wroclaw., *Konf.*, 1986, **12**, 97
[96] *Eur. Pat. Appl.*, 1987, 0 232 695
[97] *Organamide Nylon 6. General information and properties*, Elf Atochem., 1984, (technical datasheet)
[98] *Plast. Eng. (N.Y.)*, 1995, **51**, 32
[99] Matthew, B.A. and Normandin, M.G.M., *Eng. Plast.*, 1994, **7**, 196
[100] *Br. Plast. Rubber*, 1991, 16
[101] *Eur. Pat. Appl.*, 1986, 0 188 184
[102] *Actual. Chim.*, 1988, **3**, 72
[103] Gupta, S.K., *Encyclopaedia of Engineering Materials, Part A, Polymer Science and Technology, Vol. 1, Synthesis and properties*, 1988, 211
[104] Fujimoto, A., Fujita, T. and Endoh, Y., *Nippon Kagaku Kaishi*, 1988, **3**, 332
[105] Warakomski, J.M., Polym. Prepr. (Am. Chem. Soc., Div. Polym. Chem.), 1989, **39**, 117
[106] Mathias, L.J. and Sikes, A.M., *ACS Symp. Ser.*, 1988, **367**, 66
[107] Van Buskirk, B. and Akkapeddi, M.K., Polym. Prepr. (Am. Chem. Soc., Div. Polym. Chem.), 1988, **29**, 557

Nylon 6,6

$$\left[-NHCH_2(CH_2)_4CH_2NH\overset{O}{\underset{\|}{C}}CH_2(CH_2)_2\overset{O}{\underset{\|}{C}} - \right]_n$$

Synonyms: *Poly(hexamethylenediamine-co-adipic acid). Poly(hexamethylene adipamide). Poly(iminohexamethylene iminoadipoyl). Poly(imino(1,6-dioxo-1,6-hexanediyl)imino-1,6-hexanediyl)*
Related Polymers: Nylon polymerisation
Monomers: 1,6-Hexanediamine, Hexanedioic acid
Material class: Thermoplastic
Polymer Type: polyamide
CAS Number:

CAS Reg. No.
32131-17-2

Molecular Formula: $(C_{12}H_{22}N_2O_2)_n$
Fragments: $C_{12}H_{22}N_2O_2$
Additives: Titanium dioxide is used to reduce lustre. Manganese phosphates and phosphates are thermal stabilisers. Copper salts are used for fibres.
General Information: An important engineering plastic due to its inherent toughness and solvent resistance, it is employed principally in fibre form commercially. Its most important property is its "tenacity" (tensile strength at break). Tensile props. depend upon spin speed, draw ratio and are controlled by the intimate morphology of the fibres, especially the crystalline and amorph. degree of orientation. [3]
Morphology: Max. crystal linear growth rate $20\mu m\ s^{-1}$ [26]. Correlation with density has been reported [37]. Crystalline forms: 40-60% spherulites; triclinic a 4.9, b 5.4, c 17.2. 1 monomer per unit cell [6]; α_{11} triclinic a 4.97, b 5.47, c 17.3 [18]. β triclinic a 4.9, b 8.0, c 17.2 [6]. Above 160° β-pseudo hexagonal form a 5, b 5.9, c 16.2. [19]. Single crystals annealed in glycerol show stepwise increase in lamellar thickness, corresponding to crystal thickness calculated from observing melting point. Two thickening mechanisms are involved, half a monomer unit and $1\frac{1}{2}$ monomer units. [80] Crystallisation studies from quenched melt show defective crystals form as fibrous entities oriented to strain axis; defect migration gives axial density modulation; local softening may relax strain and let crystal relax its orientation to the most stable form [81]

Nylon 6,6

Volumetric & Calorimetric Properties:

Density:

No.	Value	Note
1	d 1.21–1.24 g/cm^3	α, triclinic cryst., [6,7,26,38]
2	d 1.152–1.165 g/cm^3	αII, triclinic cryst. [18]
3	d 1.25 g/cm^3	β, triclinic cryst. [1,6]
4	d 1.1 g/cm^3	triclinic, 170° [19]
5	d 1.13–1.145 g/cm^3	cryst moulding [1]
6	d 1.069 g/cm^3	amorph. [14]
7	d 1.09 g/cm^3	amorph. [1,13]
8	d 0.989 g/cm^3	melt, 270°, 1 bar [1]

Thermal Expansion Coefficient:

No.	Value	Note	Type
1	7×10^{-5}–0.0001 K^{-1}	cryst., 20°	L
2	0.0001–0.00014 K^{-1}	cryst., 100°	L
3	2.1×10^{-5} K^{-1}	αa, triclinic, 20°	L
5	7×10^{-5} K^{-1}	ASTM D696 [54]	L
6	0.000117 K^{-1}	parallel	L
7	0.000114 K^{-1}	normal	L
8	0.00028 K^{-1}	cryst., 20° [32]	V
9	0.00022 K^{-1}	β, triclinic, 20°	L

Latent Heat Crystallization: Energy of activation 54 kJ mol^{-1} [4] Energy of activation of cryst. 64.5 kJ mol^{-1} [26]. Energy of activation of diffusion of water 58 kJ mol^{-1} [1]. Heat of fusion 36.8–68 kJ mol^{-1} (triclinic cryst., α1) [15,16,18,26]. Heat of fusion 196 kJ kg^{-1} [40]. Mandelkern's equation parameters, applicable to the transport process involved in the growth of spherulites in polymer melts, have been reported [24]. Flow activation energy (melt) 60 kJ mol^{-1} [45]. Heat of combustion 31.4 kJ g^{-1} [26]. Enthalpy 85 kJ kg^{-1} (60 Mpa, 20°); 170 kJ kg^{-1} (100 Mpa, 20°); 590 kg kJ (250 MPa 20°) [1]. Enthalpy 250 kJ kg^{-1} (0 Mpa, 150°); 340 kJ kg, (100 Mpa, 150°); 420 kJ kg^{-1} (200 Mpa, 150°) [34]. Stress relaxation activation energy 598 kJ mol^{-1} [71]

Thermal Conductivity:

No.	Value	Note
1	0.24–0.35 W/mK	amorph. [26]
2	0.23 W/mK	mouldings [1,57]
3	0.43 W/mK	cryst., moist, 30°
4	0.36 W/mK	amorph., moist, 30° [53]
5	0.25 W/mK	

Specific Heat Capacity:

No.	Value	Note	Type
1	1.47 kJ/kg.C	[26]	P
2	1.66 kJ/kg.C	cryst. [42]	P
3	1.674 kJ/kg.C	[54]	P
4	1.7 kJ/kg.C	[57]	P

Melting Temperature:

No.	Value	Note
1	263°C	ISO 3146C [56]
2	265–280°C	α1 triclinic [16,17]
3	269.5°C	αII triclinic [16]
4	267–280°C	effective melting point [26]

Glass-Transition Temperature:

No.	Value	Note
1	50°C	[21,22,23]
2	57–58°C	27% cryst. [24,26]
3	80°C	dry [58]
4	26°C	50% relative humidity
5	-17°C	saturated [58]

Deflection Temperature:

No.	Value	Note
1	235°C	0.45 Mpa, ASTM D648
2	235°C	0.5 Mpa, ASTM D648
3	90°C	1.8 Mpa, ASTM D648 [55]
4	200°C	0.45 MPa, dry as moulded
5	100–110°C	1.8 MPa, dry as moulded, ISO-R 75 [1]

Vicat Softening Point:

No.	Value	Note
1	200°C	[58]

Brittleness Temperature:

No.	Value	Note
1	-80°C	ASTM D746 [55]
2	-65°C	[62]

Transition Temperature:

No.	Value	Note
1	390°C	[1] flash ignition temp.

Surface Properties & Solubility:

Cohesive Energy Density Solubility Parameters: 27.8 J$^{1/2}$cm$^{-3/2}$ [26]

Plasticisers: Plasticisation by water and alcohols has been reported. T$_g$ is reduced proportionally to amount of plasticiser per chain unit [68,107]

Solvents/Non-solvents: Sol. strongly polar solvents, formic acid, conc. H$_2$SO$_4$, THF. Absorbs alcohols to high levels without being dissolved; EtOH absorption: 9–12% (mouldings, saturated, 20°); butanol absorption: 4–8% (mouldings, saturated, 20°) glycol absorption: 2–10% (mouldings, saturated, 20°) MeOH absorption: 9–14% (mouldings, saturated, 20°); propanol absorption: 9–12% (mouldings, saturated, 20°) [1]

Surface and Interfacial Properties General: Surface tension 42–46 mN m^{-1}. [26] 40–44 mN m^{-1} (23°) [26,48]. 36 mN m^{-1} (melt) [51,49]

Nylon 6,6

Surface Tension:

No.	Value	Note
1	42–46 mN/m	[26]
2	40–44 mN/m	23°, [26, 48]
3	36 mN/m	melt [51,49]

Transport Properties:

Transport Properties General: Nylon 6,6 melts from cryst. form to a comparatively low viscosity liq. and does not require a high level of power and shear to process. Solid resin has very good resistance to permeation by gases due to its high crystallinity when dry, but is less efficient moist. It absorbs moisture continuously and can take up to 9% by weight. Consequently, Nylon 6,6 film finds applications in packaging where flexibility, strength and barrier props are suited. Poor resistance to water vapour may be overcome by sandwich construction of composite films with a nylon layer between moisture resistant polyethylene layers. Composite film of this nature is ideal for vacuum packaging of perishable foods. Nylon 6,6 film is easily thermoformed to appropriate shapes, and finds many uses in the medical and chemical engineering sectors

Melt Flow Index:

No.	Value	Note
1	50 g/10 min	285°
2	120 g/10 min	melt volume ISO 1133 [57]

Polymer Solutions Dilute: Mark Houwink constants have been reported. [1] [η] 2.6 dl g^{-1} (1g dl^{-1}, H$_2$SO$_4$ M$_n$ 18000); 3.4 dl g^{-1} (1g dl^{-1}, H$_2$SO$_4$ M$_n$ 26000)

Polymer Melts: Melt viscosity 10–1000 Pa s, dependent on temp. and MW. Oxidation increases melt viscosity, leading to loss of mech. props. Melt viscosity 110 Pa s (270° M$_n$ 1400); 70 Pa s (280°) 50 Pa s (200°) [4]. Apparent melt viscosity, shear stress relationship has been reported. [5] Ostwald power law equation parameters have been reported. [27] Other equation of state (melt) information has been reported [26,39,45]. Nylon 6 and Nylon 6,6 have similar rates of degradation and onset of severe degradation as a function of temp. workable melt temp. range for extruding stock shapes without risk of serious polymer breakdown. [77] A qualitative description of the viscoelastic behaviour of Nylon 6,6 was given taking into account temp., MW and MW distribution, branching etc. has been reported [118] Rheological behaviour is related to structural characteristics such as chain stiffness, length, branching and composition [121]

Permeability of Gases: [CO$_2$] 0.51 × 10^{-10} cm^2 (s cmHg)$^{-1}$ (amorph.); Pc = 0.47 × 10^{-10} cm^2 (s cmHg)$^{-1}$ (cryst.) [O$_2$]; Pa = 0.14 × 10^{-10} cm^2 (s cmHg)$^{-1}$ (amorph.); Pc = 0.3 × 10^{-11} cm^2 (s cmHg)$^{-1}$ (dry cryst.) Pc = 0.13 × 10^{-10} cm^2 (s cmHg)$^{-1}$ (amorph.); (saturated cryst); [N$_2$] Pa = 0.3 × 10^{-11} cm^2 (s cmHg)$^{-1}$ (amorph.); Pc = 0.27 × 10^{-11} cm^2 (s cmHg)$^{-1}$ (cryst.)

Permeability of Liquids: Coefficient of diffusion of [H$_2$O] 0.2 × 10^{-8} cm^2 s^{-1} (20°); 3.5 × 10^{-8} cm^2 s^{-1} (60°); 35 × 10^{-8} cm^2 s^{-1} (100°); [52]

Permeability and Diffusion General: The Gruneisen thermodynamic coefficient for several polymers including Nylon 6,6 at room temp. has been reported [113]

Water Content: 8.5% ± 0.5 (saturation 92% water, 25–90°) [ISO 62 [26,55,50]] [26]; 2.9% (50% relative humidity, 100 days). 9.6% (100% water saturation, 20 days). Mouldings from Nylon 6,6 are hygroscopic and gradually take up moisture until equilibrium is reached, about 3% for environment of 50% relative humidity and 23°, but will increase further as the humidity rises. Toughness and impact strength increase; tensile strength is reduced while elongation is increased; moulding dimensions increase linearly with absorbed water: 0.3% per 1% water uptake; the electrical props. are adversely affected, although OK for domestic purposes.

Water absorption is slow and it may take several months for a thick section of moulding to equilibrate. Usually mouldings are preconditioned at 60° and 20° prior to despatch to customers, so as to reach equilibrium at 20°, 65% relative humidity. The effect upon dimensional stability is a maximum increase of 0.9% volume associated with each 1% rise in weight of water absorbed, corresponding to a 0.2–0.3% change in the length of a specimen. This change is limited to <0.1% in the case of glass filled nylon. Max moisture content and diffusion coefficient fit a Fickian diffusion model [112]

Water Absorption:

No.	Value	Note
1	1.2 %	24h, dry moulding, ASTM D570 [54,55]
2	2.8 %	23°, 50% relative humidity [56]

Gas Permeability:

No.	Gas	Value	Note
1	CO$_2$	0.51 cm^3 mm/(m^2 day atm)	amorph.
2		0.47 cm^3 mm/(m^2 day atm)	cryst.
3		0.14 cm^3 mm/(m^2 day atm)	amorph.
4		0.3 cm^3 mm/(m^2 day atm)	dry cryst.
5		0.13 cm^3 mm/(m^2 day atm)	amorph., saturated cryst.
6		0.3 cm^3 mm/(m^2 day atm)	amorph.
7		0.27 cm^3 mm/(m^2 day atm)	cryst.

Mechanical Properties:

Mechanical Properties General: Nylon 6,6 has an excellent combination of medium strength, rigidity, creep strength, high toughness (when moisture absorbed) and excellent tribological props. Nylon 6,6 has the best overall props. of the aliphatic commercial nylon resins, combining greatest hardness, rigidity and resistance to abrasion and heat deformation. It is therefore better suited to fabricate parts which need to withstand high loads, wear and high temps. Nylon resins are extremely tough even when dry. Nylon 6,6 mouldings of housings, panels or specimen boxes survive the impact of a 10kg weight falling from 1.5m in 50% of impacts, even at subzero temps. Impact modified grades are produced for high toughness even in dry conditions. Nylon 6,6 possesses a low coefficient of friction and rate of wear. Moisture regain of Nylon 6,6 fibre is 4.5% at 65% and 7–8% at 90% relative humidity. Tenacity holds up well at 93° but falls dramatically after 300h exposure to 120°. Tensile (Young's) Modulus (includes specific Modulus): 2.2–3.1 N/tex (normal tenacity fibres, conditioned) 2.9–4.1 N/tex (high tenacity fibres, conditioned) [68]. Normal Tenacity (elongation at break %); 0.4–0.6 N/tex (23–43%, conditioned),(wet) ; 0.36–0.51 N/tex (wet); 0.4–0.49 N/tex (loop); 0.35–0.42 N/tex (knot). Average toughness: 0.06–0.14 N/tex (normal tenacity); 0.06–0.14 N/tex (high tenacity). High tenacity (elongation at break %); 0.75–0.84 N/tex (12–17%, conditioned). 0.64–0.7 N/tex (wet); 0.57–0.69 N/tex (loop); 0.52–0.72 N/tex (knot). Maximum tenacity; 1.8 N/tex. [68]

Tensile (Young's) Modulus:

No.	Value	Note
1	2000 MPa	[26]
2	3500 MPa	dry relative humidity
3	1600 MPa	50% relative humidity [58]
4	3000 MPa	dry
5	1500 MPa	50% relative humidity [56]

Nylon 6,6

Flexural Modulus:

No.	Value	Note
1	2300 MPa	[26]
2	2830 MPa	dry ASTM D790
3	1200 MPa	50% relative humidity [55]

Tensile Strength Break:

No.	Value	Note	Elongation
1	80 MPa	[26]	200%
2	90 MPa	dry, 23° [1]	
3	60 MPa	moist, 2% water 23° [1]	
4	35 MPa	moist, saturated, 23° [1]	
5	40 MPa	100° [1]	
6	83 MPa	dry [58]	5%
7	77 MPa	equilibrium	25%
8	103 MPa	-45° ASTM D638	20%
9	83 MPa	21° [57]	80%
10	62 MPa	90° [54]	>300%
11	83 MPa	[55]	60%

Flexural Strength at Break:

No.	Value	Note
1	117 MPa	dry [58]
2	42 MPa	50% relative humidity ASTM D790
3	90 MPa	0.2% H_2O
4	42 MPa	2.5% H_2O [54]

Elastic Modulus:

No.	Value	Note
1	3.3 MPa	23° dry [1]
2	1.7 MPa	moist ISO 1110, 23° [1]
3	0.6 MPa	dry 100° ISO 537 [1]

Poisson's Ratio:

No.	Value	Note
1	0.46	[26]
2	0.3–0.4	mouldings 20°
3	0.1–0.25	extruded rod 0.1*
4	0.3–0.6	extruded rod 0.2*
5	0.4–0.44	extruded rod 0.4*
6	0.43–0.46	extruded rod 0.6* [45]
7	0.44	100° [26]
8	0.5	melt [1]

Tensile Strength Yield:

No.	Value	Elongation	Note
1	57 MPa	25%	[26]
2	83 MPa	4.5% dry	[60]
3	59 MPa	26%	50% relative humidity [56]

Compressive Strength:

No.	Value	Note
1	100 MPa	20° [26]
2	14 MPa	20° 1% strain [1]
3	28 MPa	20° moulded, 2% strain [1]
4	56 MPa	20° moulded, 4% strain [1]
5	70 MPa	20° moulded, 6% strain [1]

Miscellaneous Moduli:

No.	Value	Note	Type
1	8.1 MPa	[26]	
2	3.3 MPa	crystalline rods, dry [55]	Bulk modulus
3	1300 MPa	dry 23° [1]	Shear modulus
4	1700 MPa	dry, nucleated [1]	Shear modulus
5	0.3 MPa	dry, 100°	Shear modulus
6	0.15 MPa	dry, 200° [1]	Shear modulus
7	66.2 MPa	dry Zytel 101 [62]	Shear strength

Impact Strength: Izod 110 J m^{-1} (moist ISO 1110, 23°) [1,26]; no break (unnotched, ASTM D250); 53 J m^{-1} (notched, dry) [54,55]; 112 J m^{-1} (notched 50% relative humidity); 159 J m^{-1} (notched, 25% water) [54]; 5.4 kJ m^{-2} (notched, dry, 23°) [56]; 11.7 kJ m^{-2} (notched, 50% relative humidity, 23°) [56]; 213.7 kJ m^{-2} (notched, dry, 30°); 2.7 kJ m^{-2} (notched, 50% relative humidity). 850 J m (notched super tough Zytel ST801 Nylon 6,6, ex DuPont) [8]; 900–950 J m (notched, Thermotuf V-1000 toughened Nylon 6,6, ex LNP). [8] Charpy no break (unnotched, -30–23°)[51,56]; 6 kJ m^{-2} (notched, dry, 23°); 16 kJ m^{-2} (notched, 50% relative humidity, 23°)[56]; 4 kJ m^{-2} (notched, dry, -30°); 3 kJ m^{-2} (notched, 50% relative humidity, -30°)[50]

Viscoelastic Behaviour: Moisture promotes toughness even at subzero temps. Special impact modified grades may be used when dry. Increasing glass fibre reduces toughness. Fibrous fillings offer better creep resistance than particulates [83]

Mechanical Properties Miscellaneous: Tensile Impact test 504 kJ m^{-2} (long specimen); 167 kJ m^{-2} (short specimen) [55]. Falling Top impact test 140 Nm (min, 50% failure rate, 23°; DIN 53443). Impact modified grades have lower moisture absorption, lower density and better dimensional stability with lower tensile and flexural props. [75]

Hardness: Rockwell Hardness R114; M70 (ASTM D785)[26]; M 105 (dry 23°) ASTM D785; M 95 (moist ISO-1110, 23°) [1] R 121 (dry ASTM D785); R 108 (50% relative humidity) M 85, R 120 (0.2% H_2O); M 60, R 105 (2.5% H_2O) [54] M 79, R 121 [55,56]. Ball indentaion; 72.5 Mpa [26]; Ball Indention Hardness H 358/30: 160 MPa (dry 23°); 170 MPa (dry, nucleated, 23°) 100-110 MPa (moist ISO-1110, 23°); 85 MPa (50% relative humidity) [50]; 160 MPa (H 961/30) (dry) [56]. Shore D: 75 [26]

Failure Properties General: Abrasion loss factor 25 mg [26]. Wear Resistance: 0.28 μm km^{-1} (0.1*); 0.3 μm km^{-1} (0.5*); 0.35 μm km^{-1} (1.0*); 0.48 μm km^{-1} (2.0*); 0.52 μm km^{-1} (4.0*); 0.55 μm km^{-1} (6.0*); * average roughness height μm. Wear resistance; 0.1μm km^{-1} (0.02*); 0.24 μm km^{-1} (0.05*); 0.48 μm km^{-1} (0.10*); 1.2 μm km

(0.25*); 4.8 µm km^{-1} (1.00*); 24 µm km^{-1} (5.0*); 72 µm km^{-1} (15.0*): * average surface pressure (n mm^{-2}). Coefficient of friction, dynamic; 0.15 (dry metal; 0.24 self) 0.5 (0.1 µm); 0.43 (0.5 µm); 0.38 (1 µm); 0.35 (2.0); 0.38 (4.0); 0.46 (6 µm) [30] 0.36 [26]. Creep Modulus: 2500 MPa (dry); 12000 MPa (65% relative humidity) [28]. Apparent Creep modulus; 400 MPa (23°, 50% relative humidity); 450 MPa (nucleated) (ISO 899) [1]

Fracture Mechanical Properties: Fracture energy (housing) W50: 100 J (dry); >140 J (moist) (23°); 30 J (-20°) (IEC 6603/1) [57]. At a given temp., fatigue resistance is greatest for a given water content corresponding to an optimum combination of storage modulus, E', and loss compliance, D. Both quantities influence fatigue crack growth. [86] Upper limit temp. of friction has been reported [116]. Stable crack growth features dimples obtained first by the crack tip blunting, voids initiating, coalescing and extending and the polymer rupturing and contracting to the dimple morphology. [125]. Void coalescence is observed in fracture surface micromorphology of Nylon 6,6, when moisture and impact modifier (IM) levels are low. At higher levels of moisture/IM numerous secondary fissures orient normal to the crack direction to give a rumpled appearance. Trans-spherulitic fracture is observed below T_g [126]. Substantial advances in understanding of the toughening mechanisms have been reported [129]

Friction Abrasion and Resistance: Magnitude of max. friction varies with velocity for continuous sliding, whereas nearly the same value is always obtained for discontinuous sliding between 0.1 and 10 mm s^{-1}. Minimum wear is obtained at maximum friction. [133]. The wear of Nylon 6,6 has been reported under rolling contact conditions [135]. Oil impregnated plastics are used for mechanical parts to reduce friction. A table of props. of four types of oil impregnated plastics (including Nylon 6,6 and glass fibre reinforced Nylon 6,6) have been reported [137]. Wear rate and coeff. of friction were determined for Al, brass, polyethylene and Nylon 6,6 against stainless steel. Nylon had the minimum wear rate and coeff. of friction was 8–12 times less for Nylon compared to aluminium alloy used in bearing bushes. Nylon brushes are now used in energy meters as a direct consequence [138]

Electrical Properties:

Electrical Properties General: Has efficient dielectric props., good volume and surface resistivity and resistance to tracking. Combined with its excellent heat and ageing resistance these props. make it an important material for electrical engineering. Dielectric strength can fall from 30 to 5 kV mm^{-1} when saturated with 8% water at its upper continuous use temp. (80°). Comparative tracking index 600 [56] 550 M (IEC 112/A) [57]. Glass fibre or mineral filled Nylon 6,6 can be injection moulded under special conditions, then electroplated with new activators, followed by a special pre-treatment and plated with commercial nickel or copper baths. Electric attenuation (shielding) 80 dB up to 92 GHz; 60 dB (1 MHz). Excellent at short temp.-time shock tests. No delamination of metal coating when stored 50h at 50° [108].

Surface/Volume Resistance:

No.	Value	Note	Type
8	1 × 10^{15} Ω.cm	50% relative humidity [56]	S
9	0.01 × 10^{15} Ω.cm	dry [57]	S
10	1 × 10^{-5} × 10^{15} Ω.cm	moist [60]	S

Dielectric Permittivity (Constant):

No.	Value	Frequency	Note
1	3.8–4.3		[26]
2	3.1–3.4	100 Hz–1 GHz	-30° [38]
3	3–3.3	100 Hz–1 GHz	0° [38]
4	3–3.6	100 Hz–1 GHz	30° [38]
5	3.1–5	100 Hz–1 GHz	60° [38]
6	10–35	100 Hz–1MHz	90° [38]
7	2–3	1 GHz	90° [38]
8	3.7–7.5	100 Hz–1 GHz	23°, 50% relative humidity [38]
9	3.6	1 Mhz	dry, 8 saturated
10	3.7	100 Hz	0.2% water [54]
11	3.1	1 MHz	0.2% water [50]
12	6	100 Hz	2.5% water
13	3.5	2 Mhz	2.5% water [54]
14	4	100 Hz dry; 1 Mhz, 50% relative humidity	
15	10.9	100 Hz	50% relative humidity [56]

Dielectric Strength:

No.	Value	Note
1	120 kV.mm^{-1}	dry
2	80 kV.mm^{-1}	moist, ISO 1110
3	40 kV.mm^{-1}	dry, 120°
4	31.5 kV.mm^{-1}	IEC 112 [56]
5	24 kV.mm^{-1}	[54]

Dissipation Power Factor: 0.02–0.03 (0.2% H$_2$O, 100 Hzh–1 MHz, 21°); 0.04 (2.5% H$_2$O, 1 kHz, 21°); 0.08 (2.5% H$_2$O, 1 MHz 21°) [54]. 0.01 (dry, 100 Hz); 0.02 (dry 1k Hz–1 MHz) [55]; 0.008 (dry, 100 Hz); 0.026 (dry, 1MHz); 0.58 (100 Hz, 50% relative humidity); 0.07 (1 MHz, 50% relative humidity) [56]

Arc Resistance:

No.	Value	Note
1	116s	dry, ASTM D495 [58]

Magnetic Properties: Magnetic susceptibility χ 0.76 × 10^{-6} cm^3 g^{-1} [26]

Optical Properties:

Optical Properties General: Nylon films tend to lose both clarity and flexibility on cryst. Cast films are clearer and more supple than blown as they are quenched and chilled faster. Finished film can also crystallise in storage at elevated temps. and humidity causing cloud and shrink. A soln. is nucleation, in which fine grain talc is added to provide sites for cryst. Smaller crysts. are generated leading to greater overall transparency. In processing terms, the advantages include better gauge control; broader processing parameters; uniform spherulite size and distribution; improved coefficient of friction; reduction of post cryst.; improved dimensional stability; better roll conformity; easier thermoformity with better side and corner thickness control; improved transparency after thermoforming due to changes of the cryst. composition through heat and orientation. Nucleating agents control the ratios of the crystaline and amorph. phases rate number, type and size of spherulites. [122]. Finely dispersed silicas are added in altering concentrations form 0.1–1.0% by weight [128]. Cryst. Nylon 6,6 is a white translucent material. Dispersion of finely divided titanium dioxide particles (0.1–0.5 microns range) within the nylon fibre delusters to reduce transparency and increase whiteness. 0.3% TiO$_2$ is used for *semi-dull yarns* and 2% TiO$_2$ for *dull* yarns [2]

Nylon 6,6

Refractive Index:

No.	Value	Note
1	1.53	[26]
2	1.475–1.58	[26]
3	1.475	single cryst., α triclinic
4	1.568	single cryst., β triclinic
5	1.58	single cryst., γ
6	1.53	mouldings [6]

Transmission and Spectra: X-ray diffraction [82], ir [43], nmr and other spectral data [131] have been reported

Molar Refraction: Velocity of sound: 2785m s^{-1} (longitudinal); 1120m s^{-1} (transverse) [26]; 2770m s^{-1} (mouldings) [46]

Total Internal Reflection: n: 1.58 (parallel); n: 1.53 (perpendicular) [26]

Polymer Stability:

Polymer Stability General: Mouldings retain shape due to highly crystalline nature. Survives 1000 days in air at 120° losing half tensile strength [58]. Stabilisers include copper halides, secondary amines and ester modified phenols [119]

Thermal Stability General: Exhibits excellent stability up to 300° when dry in absence of air. Heat and oxidation lead to a reduction in MW, discoloration and loss of props. Thermal ageing has been reported [105,106]

Upper Use Temperature:

No.	Value	Note
1	85°C	
2	125°C	mechanical with impact
3	75°C	mechanical
4	200°C	few hours

Decomposition Details: Decomposes slowly above 300° yielding CO_2, NH_3 and amines [103]. Cyclopentanone is also produced [64]. Decomposition onset temp. 350°, 50% decomposition temp. 420° [26]

Flammability: Flammability rating V2 (1 mm–3.2 mm UL94) [54,56,57,62]. Limiting Oxygen Index 23% [26] 28–29% (ASTM D2863) [1] 31% [62]. Flash ignition temp. 390° [1]. Flammability BH2 (10 mm IEC 707); FH2 (20 mm, IEC 707) [57] B1–B2 (1mm DIN 4102). Flame retardants have been reported [94,95,144,145]. These include tin, phosphorus, antimony and halogens [140,141,144]. Non halogenated flame retardants have been reported [151,152]. Flame retardant grades have flammability rating VO

Environmental Stress: Uv radiation causes yellowing and loss of mech. props. Heat and light stabilisers must be used to minimise weathering effects. Cross-linking and loss of crystallinity occurs. Sunlight causes decrease in MW [10,104]. Stress cracking occurs in salt solns. [65] at 50–100° [88,89,120]. Tensile props. decrease significantly. Crazing shows linear growth. Sunlight causes reduction of MW. Cross-linking occurs at elevated temps. in amorph. regions. C-N bond scission may occur. Irradiation causes reduced impact strength and increasing elastic modulus. Ionising radiation produces free radicals leading to discoloration and oxygen permeability

Chemical Stability: Stable to CH_2Cl_2, and 2-butanone. Plasticised and saturated by MeOH. Attacked by oxidising agents at elevated temp., by chlorinated solvs. and metal salt solns. [58,59]. Unfilled grades more susceptible to attack than filled grades [43,60]

Hydrolytic Stability: Slowly affected by boiling water. Stable between pH 5–13. Stable to 10% NaOH at 85° for 16h. Degrades rapidly in 1% H_2SO_4 at 85°. Hydrolyses more slowly then Nylon 6 [61]. Hydrolysis resistant grades are available [111]

Biological Stability: May undergo slight degradation in the body [110]

Recyclability: May be recycled [111]

Applications/Commercial Products:

Processing & Manufacturing Routes: Manufactured from Nylon 6,6 salt of hexamethylene diamine and adipic acid aq. soln. Acid is added to the diamine soln. to obtain a 50% salt concentration, then condensation of the salt soln. in an autoclave at 250–300° under pressures up to 2 MPa or in a continuous reactor [66]. Water of soln. and polymerisation must be removed while minimising loss of diamine to ensure adequate MW in resulting Nylon 6,6. Typically, the salt soln. is concentrated to 60% above 100°, then fed to the autoclave and polymerised at 212° under 1.7 MPa; the temp. is gradually raised to 275°. At this stage MW is 4500. The pressure is slowly reduced over 1.5h and polymerisation is complete in a further hour then extruded to a water bath, cut and dried. Typical MW for the finished polymer is between 12000-20000 [62]

Mould Shrinkage (%):

No.	Value	Note
1	1.6%	60°, ASTM D966 [11]
2	1.5–2%	[54]
3	1.4%	parallel to flow ISO 2577 [56]

Applications: Principal commercial uses are in carpets, apparel, tyre reinforcement and industrial applications. Predominant fibre used in the carcasses of truck, racing and airplane tyres due to excellent strength but not favoured for radial tyres due to flatspotting tendency. Nylon 6,6 and 6 are the principal fibres in hosiery, womens underwear and stretch fabrics. They also find applications in non-woven applications e.g. as a substrate for artificial leather (PVC). Injection moulding applications especially of glass filled grades are the major use of Nylon 6,6 in moulding. Main markets include car industry, electrical and electronics and film and wire coating. Its excellent pigmentation qualities make it especially useful for power tools. Non reinforced grades used in bearings, gear wheels, cable connectors, thin-walled housing, coils formers etc. Extrusion grades are used for semi-finished products, tubes, profiles, sheet and roasting film. Glass fibre reinforced grades are injection moulded less than 30% glass; machinery components, electrical insulating parts, housings such as coil formers and bearing cages, lamp sockets housings, cooling fans, insulating profiles for aluminium window frames, water containers for automotive cooling systems. More than 35% gear wheels, solenoid valves, housings, electric flow heaters, trailing cable attachments. Impact modified glass reinforced: for high stiffness, notch impact and heat resistant, low warpage mouldings, e.g. wheel covers in automobiles and housing components. There has been a dramatic increase in the use of Nylon in the automotive industry in the 1990s, e. g. air intake manifolds in vehicles. Heat stabilised Nylon 6,6 are used for cylinder head cam covers to protect valves, seal the oil and shield against noise. Heat aged grades used for automotive connectors. Flame resistant grades used in electronics and electrical parts such as cable ties, conductors and terminal blocks

Trade name	Details	Supplier
6110/6210/ 6310/6410		Nan Ya Plastics
66/6601/6602	various grades	Plastic Materials
ABS	538	Albis Corp.
AVP	various grades	Polymerland Inc.
Adell	various grades	Adell Plastics Inc.
Akulon	Glass fibre and toughened grades	DSM
Amilan	CM series	Toray Industries
Amilon	UTN 320/325	Toray Industries
Aqualoy	623/624/640	Comalloy Intl. Corp.

Ashlene		Ashley Polymers
Bapolon	766-33G/766L	Bamberger Polymers
CTI	various grades	CTI
Capron		Allied Signal Corporation
Celanese	various grades	Hoechst Celanese
Celstran	N66 range	Hoechst Celanese
Celstran	N66 range	Polymer Composites
Comalloy	various grades	Comalloy Intl. Corp.
Comco Nylon 6/6		Comalloy Intl. Corp.
Comtuf	600 series	Comalloy Intl. Corp.
DC/RX/SC	various grades	Spartech
Dartek	various grades	DuPont
EMI-X	PDX/RC	LNP Engineering
Electrafil	conductive grades	DSM
Entec	N range	Entec Polymers
Fiberfil	various grades	DSM Engineering
Fiberstran	G-1/30/40/50	DSM Engineering
G-1	various grades	DSM Engineering
G-8/40		DSM Engineering
Grilon		EMS
Hiloy	600 series	Comalloy Intl. Corp.
Hylon	500 series	Hale Manufacturing
J-1	various grades	DSM Engineering
J-8/29	various grades	DSM Engineering
J17/29	various grades	DSM Engineering
LCG/LSG		Thermofil Inc.
Leona		Asahi Chem. Co Ltd
Lubricomp	R-1000 series	LNP Engineering
Lubrilon	600 series	Comalloy Intl. Corp.
Lubriloy	R/RF-15/RF-30/RL	LNP Engineering
Lupon	GP/HI/LW/SL	Lucky
M_n	6/6-FG 10/20/30/40	Modified Plastics
Maranyl	A series/TA/XA	ICI Americas Inc.
Minlon	varous grades	DuPont
MonTor	CM series	Mon Tor
N2/N3/N5/N7	various grades	Thermofil Inc.
NSC Esbrid	LSG-440A	Thermofil Inc.
NY-1	TC/30	DSM Engineering
NY-29	MF/40	DSM Engineering
Ny-Kon	R	LNP Engineering
Nybex	20000 series	Ferro Corporation
Nycoa	various grades	NYCOA
Nylalvon	lubricated grades	DSM
Nylamid	various grades	Polymer Service
Nylatron	various grades	Polymer Corp.
Nylon	N range	Michael Day
Nylon 66	various grades	Albis Corp.
Nyloy	M/MC/MG/MS	Nytex Composites
Nytron	LMC/LMG	Nytex Composites
PA	111/113/121/123	Bay Resins
PA	66-110 range	MRC Polymers Inc.
PA	6/6 range	TP Composites
PRL-Nylon	NY	Polymer Resources
Perma Stat	200H	RTP Company
Plaslube	lubricated grades	DSM
Polyamid 66	various grades	Albis Corp.
Polypenco Nylon	101	Polymer Corp.
RTP	various grades	RTP Company
Rimplast	PTA 6601/6602	Huls America
S240-C		DSM Engineering
Schulamid	various grades	A.Schulman Inc.
Sniamid		Technopolimers
Stanyl	46 series	DSM Engineering
Stat-Kon	RC series	LNP Engineering
Technyl		Rhone-Poulenc
Texalon	various grades	Texapol Corp.
Texapol	various grades	Texapol Corp.
Thermocomp	R-1000 series	LNP Engineering
UBE	various grades	UBE Industries Inc.
Ube		Ube Industries Ltd.
Ultramid		BASF
Verton	RF series	LNP Engineering
Voley	600 series	Comalloy Intl. Corp.
Vydyne		Monsanto Chemical
Wellamid	various grades	Wellman Inc.
Xylon	Flame retardant	DSM
Zytel		DuPont

Bibliographic References

[1] Pflüger, R., *Polym. Handb.*, 4th edn., (eds. J. Brandrup, E.H. Immergut and E.A. Grulke), John Wiley and Sons, 1999, **V**, 121, (phys props)
[2] Taylor, G.B., *J. Am. Chem. Soc.*, 1947, **69**, 635
[3] *Polymerisation Processes*, John Wiley and Sons, 1977
[4] Bernhardt, E.L., *Processing of Thermoplastic Materials*, Reinhold, 1959
[5] Perrini, P., Romanini, D., Righi, G.P., *Polymer*, 1976, **17**, 377
[6] Bunn, C.W. and Garner, E.V, *Proc. R. Soc. London A*, 1947, **189**, 39
[7] Starkweather, H.W., Zoller, P and Jones, G.A., *J. Polym. Sci., Polym. Lett. Ed.*, 1984, **22**, 433, (cryst)
[8] Starkweather, H.W. and Jones, G.A., *J. Polym. Sci., Polym. Phys. Ed.*, 1981, **69**, 967
[9] *High Temperature Resistance and Thermal Degradation of Polymers*, Soc. Chem. Ind: London, 1961
[10] Allan, N.S., McKellar, J.F., *J. Polym. Sci., Part D: Macromol. Rev.*, 1978, **13**, 241
[11] Little, K., *Nature (London)*, 1954, **173**, 680
[12] Zimmerman, J., *Text. Manuf.*, 1974, **101**, 19
[13] Starkweather, H., Moore, G., Roder, T., Hansen, J. and Brooks, J., *J. Polym. Sci.*, 1956, **21**, 189
[14] Starkweather, H. and Moynihan, R.E., *J. Polym. Sci.*, 1956, **22**, 363
[15] Schaefgen, J.R., *J. Polym. Sci.*, 1959, **38**, 549
[16] Wunderlich Bond Gaur, U., *Pure Appl. Chem.*, 1980, **52**, 445
[17] Magill, J.H., Girolama, M. and Keller, A., *Polymer*, 1981, **22**, 43
[18] Schlichter, W.P., *J. Polym. Sci.*, 1959, **35**, 77
[19] Haberkorn, H., Illers, K-H and Simak, P., *Polym. Bull. (Berlin)*, 1979, **1**, 485
[20] Colclough, M.L. and Baker, R, *J. Mater. Sci.*, 1978, **13**, 2531

[21] Boyer, R.F., *Rubber Chem. Technol.*, 1963, **63**, 1303
[22] Gordon, G.A., *J. Polym. Sci., Part A: Polym. Chem.*, 1971, **9**, 1693
[23] Temin, S.C., *J. Appl. Polym. Sci.*, 1965, **9**, 471
[24] Kusanagi, H., Takase, M., Chatani, Y. and Tadokoro, H., *J. Polym. Sci., Polym. Phys. Ed.*, 1978, **16**, 131
[26] Van Krevelen, D.W., *Properties of Polymers: Their Correlation with Chemical Structure*, 3rd edn., Elsevier Science, 1990, 596, 704
[27] Ostwald, W., *Kolloid Z. Z. Polym.*, 1925, **36**, 99
[28] Powell, P.C., *Thermoplastics*, (ed. R.M. Ogorkiewicz), John Wiley and Sons, 1974
[29] Rakos, M. *et al*, *J. Phys. B: At., Mol. Phys.*, 1966, **16**, 112, 167
[30] Erhard, G., Strickle, E, *Kunststoffe*, 1972, **62**, 9, 232-234, 282-288
[31] Wakelin, J.H., Sutherland, A., Beck, L.R., *J. Polym. Sci.*, 1960, **42**, 278
[32] Warfield, R.W., Kayser, E.G., Hartmann, B., *Makromol. Chem.*, 1983, **184**, 1927
[33] Tautz, H. and Strobel, L., *Kolloid Z. Z. Polym.*, 1965, **202**, 33
[34] Griskey, R.G., Shou, J.K.P, *Mod. Plast.*, 1968, **45**, 148
[35] Starkweather, H.W., *J. Appl. Polym. Sci.*, 1959, **2**, 363
[36] Starkweather, H.W. and Moynihan, R.E., *J. Polym. Sci.*, 1956, **22**, 363
[37] Muller, A., Pfluger, R., *Kunststoffe*, 1960, **50**, 203
[38] Bergmann, K., *Konstr. Kunstst.*, (ed. G. Schreyer), Hanser, 1972
[39] Hafner, F. and Mietzner, T., *ABT Informabik*, BASF, 1986
[40] Inoune, M., *J. Polym. Sci., Part A: Polym. Chem.*, 1963, **1**, 2697
[41] Gruneisen, E., *Handbuch der Physik*, Springer-Verlag, 1926, **10**
[42] Zosel, A., *Colloid Polym. Sci.*, 1985, **263**, 541
[43] Hummel, D. and Scholl, *Atlas der Kunststoff-Analyse*, 2nd edn. Hanser-Verlag, UCH, 1984
[44] Stamhuis, J.E., Pennings, A.J., *Polymer*, 1977, **18**, 667
[45] Laun, H.M., *Rheol. Acta*, 1979, **18**, 478
[46] Krause, I., Segreto, A.J., Prizirembel, H. and Mach, L., *Mater. Sci. Eng.*, 1966, **1**, 239
[47] Plumer, F., *Kolloid Z. Z. Polym.*, 1961, **2**, 108
[48] Owens, D.K., Nendt, R.C., *J. Appl. Polym. Sci.*, 1969, **13**, 1741
[49] Hybart, F., *J. Appl. Polym. Sci.*, 1960, **3**, 118
[50] Muller, A., Pfluger, R., *Kunststoffe*, 1960, **50**, 203
[51] Hellwege, K.H., Hoffmann, R. and Knappe, W., *Kolloid Z. Z. Polym.*, 1968, **226**, 109
[52] Goldbach, G., *Angew. Makromol. Chem.*, 1973, **32**, 37
[53] Salame, M., *Polym. Eng. Sci.*, 1986, **26**, 1543
[54] *Vydyne Nylon Engineering Thermoplastic Resins*, Monsanto, 1991, (technical datasheet)
[55] *Properties of Zytel Nylon Resins*, E.I. DuPont de Nemours, 1986, (technical datasheet)
[56] *Properties of Zytel Nylon Resins*, DuPont Engineering Resins, 1995, (technical datasheet)
[57] *Ultramid T*, BASF, 1995, (technical datasheet)
[58] *Ultramid Polyamide Product Line, Properties and Processing*, BASF, 1995, (technical datasheet)
[59] *Resistance of Ultramid, Ultraform and Ultradur to Chemicals*, BASF, 1982, (technical datasheet)
[60] *Chem. React. Polym., Monographs on the Chemistry, Physics and Technology of High Polymeric Substances.* no. XIX, John Wiley & Sons, 1964, 561
[61] Haslam, J. and Swift, S.D., *Analyst (London)*, 1954, **79**, 82
[62] *Encycl. Polym. Sci. Technol.*, 1988, **11**, 359
[63] Zimmerman, J., *J. Polym. Sci.*, 1960, **43**, 193
[64] Compte, J., *Rev. Plast. Mod.*, 1993, **66**, 510
[65] Wyzgoskii, M.G., Novak, G.E., *J. Mater. Sci.*, 1987, **22**, 1707
[66] Jacobs, D.B., and Zimmerman, J, *Polymerisation Processes*, (eds. C.E. Schildknecht and I. Skeist), John Wiley and Sons, 1977, 12
[67] *U.S. Pat.*, DuPont, 1963, 3 091 015
[68] *Encycl. Polym. Sci. Technol.*, 1988, **11**, 410
[69] *Br. Plast. Rubber*, 1996, 39, 46
[70] Kawahara, W.A., Brandon, S.L., Korellis, J.S, *Sandia National Laboratories*, 1988, 35
[71] Arridge, R.G.C., Barham, P.J., *J. Phys. D: Appl. Phys.*, 1986, **19**, L89
[72] Shah, P.L., Annu. Tech. Conf. - Soc. Plast. Eng., *50th*, 1992, **1**, 932
[73] Savelev, V.D., Bronnikov, S.V., Vettegren, V.I., Vysokomol. Soedin., *Ser. B*, 1988, **30**, 83
[74] Bucknall, C.B., Heather, P.S. and Lazzeri, A., *J. Mater. Sci.*, 1989, **24**, 2255
[75] Baumer, G., 1987, **19-C**, Cincinnati, Ohio, 6
[77] *Plastics News (Detroit)*, 1994, **6**, 27
[78] Kaufmann, W. and Zimmermann, R., *Kunststoffe*, 1989, 432
[79] Pai, C.C., Jeng, R.J., Grossman, S.J., Huang, J.C., *Adv. Polym. Technol.*, 1989, **9**, 157
[80] Mito, H., *Polymer*, 1988, **29**, 1635
[81] Schultz, J.M., Scattering, *Deformation and Fracture in Polymers*, Materials Research Society, 1986, 215
[82] Murthy, N.S., Curran, S.A., Aharoni, S.M., Minor, H., *Macromolecules*, 1991, **24**, 3215
[83] Newby, G.B., Theberge, J.E., RP/C '84, *39th Annual Conference*, Society of the Plastics Industry, 1984, **16-D**, 11

[84] Huang, D.D., Williams, J.G., *J. Mater. Sci.*, 1987, **22**, 2503
[85] Huang, D.D., *Polym. Prepr. (Am. Chem. Soc., Div. Polym. Chem.)*, 1988, **29**, 159
[86] Hahn, Mt., Hertzberg, R.W., Manson, J.A., Sperling, L.H., *Polymer*, 1986, **27**, 1885
[87] Wyzgoski, M.G., *J. Mater. Sci.*, 1987, **22**, 1707
[88] Wyzgoski, M.G., Novak, G.E, *J. Mater. Sci.*, 1987, **22 (5)**, 1715, 2615
[89] Stolpovskaya, E.M., Elkina, E.V., Bulekina, L.A., Kostrov, V.I., *Plast. Massy*, 1989, **5**, 36
[90] Voss. H., Dolgopolsky, A., Friedrich, K., *Plast. Rubber Process. Appl.*, 1987, **8**, 79
[91] Angelo, R.J., Mirura, H., Gardner, K.H., Chase, D.B., English, A.D., *Macromolecules*, 1989, **22**, 117
[92] Hirschinger, J., Miura, H., English, A.D., Polym. Prepr. (Am. Chem. Soc., *Div. Polym. Chem.)*, 1989, **30**, 312
[93] Miura, H., English, A.D., *Macromolecules*, 1988, **21**, 1543
[94] English, A.D., Miura, H., *Polym. Prepr.*, 1988, **29**, 66
[95] Hotoshi, N., *Porima Daijesuto*, 1988, **40**, 250
[96] Benning, M.A., Flammability Sensitivity Mater. Oxygen-Enriched Atmos., *Symp.*, ASTM, 1986, **2**, 235
[97] Bryan, C.J. and Lowrie, R., Flammability Sensitivity Mater. Oxygen-Enriched Atmos., *Symp.*, ASTM, 1986, **2**, 108
[98] Warton, R.K., Flammability Sensitivity Mater. Oxygen-Enriched Atmos., *Symp.*, ASTM, 1986, **3**, 279
[99] Swindelle, I., Nolan, P.F. and Wharton, R.K., Flammability Sensitivity Mater. Oxygen-Enriched Atmos., *Symp.*, ASTM, 1988, **3**, 206
[100] Allen, N.S., Harrison, M.J., Ledward, M., Follows, G.W., *Polym. Degrad. Stab.*, 1989, **23**, 165
[101] Allen, N.S., Harrison, M.J., Follows, G.W., Matthews, V.J, *Polym. Degrad. Stab.*, 1987, **19**, 77
[102] Allen, N.S., Harrison, M.J., Follows, G.W., *Polym. Degrad. Stab.*, 1988, **21**, 251
[103] Ballisteri, A., Garozzo, D., Giuffrida, M., Montaudo, G., *Macromolecules*, 1987, **20**, 2991
[104] Qayyum, M.M., White, J.R., *J. Mater. Sci.*, 1986, **21**, 2391
[105] Renschler, C.L., *Sandia National Laboratories*, 1985, 4
[106] Renschler, C.L., *J. Mater. Sci. Lett.*, 1985, **4**, 707
[107] Birkinshaw, C., Buggy, M., Daly, S., *Polym. Commun.*, 1987, **28**, 286
[108] Ebneth, H.F., Wolf, G.D., *Materials - Processes: the Intercept Point*,
[109] Curry, J., Farrell, J., Annu. Tech. Conf. - Soc. Plast. Eng., *43rd*, Soc. Plast. Eng., 1985, 522
[110] Smith, R., Williams, D.F., *Polym. Test.*, 1986, 243
[111] *Automot. Eng.*, BIP Chemicals, 1989, **14 (5)**, 63
[112] Valentin, D., Paray, F., Guetta, B., *J. Mater. Sci.*, 1987, **22**, 46
[113] Rodriguez, E.L., Filisko, F.E., *High Temp. - High Pressures*, 1988, **20**, 585
[114] Adams, G.C., Williams, J.G., Annu. Tech. Conf. - Soc. Plast. Eng., *43rd*, Soc. Plast. Eng., 1985
[115] Wu, Y.T., *Mod. Plast.*, 1988, **65**, 89, 93, 96
[116] Ettles, C.M., *J. Tribol.*, 1988, **110**, 678
[117] *Can. Plast.*, 1994, **52**, 7, 12
[118] Pinaud, F., *J. Non-Newtonian Fluid Mech.*, 1986, **23**, 137
[119] Compte, J., *Rev. Plast. Mod.*, 1993, **66**, 510
[120] Wysgoski, M.G., Novak, G.E., *J. Mater. Sci.*, **22**, 1707
[121] Cogswell, F.N., *Polymer Melt Rheology*, George Godwin, 176
[122] *Encycl. Polym. Sci. Technol.*, 2nd edn., 1988, **18**, 493-556
[123] *Encycl. Polym. Sci. Technol.* 2nd edn., Richmond Technology, **11**, 464
[124] *SAMPE J., Sample J.*, 1988, **24**, 97
[125] Zhang, M.J., Zhi, F.X., *Polymer*, 1988, **29**, 2152
[126] Hahn, M.T., Hertzberg, R.W., Manson, J.A., *J. Mater. Sci.*, 1986, **21**, 39
[127] Rolland, L. and Broutman, L.J., Annu. Tech. Conf. - Soc. Plast. Eng., *43rd*, Soc. Plast. Eng., 1985, 634
[128] Wiggins, M., *Expert Systems in Polymer Selection*, Antec '86, Boston, Mass., 1986, 1393
[129] Epstein, B.N., Adams, G.C., *Toughened Plastics*, The Plastics and Rubber Inst., 1985, 10
[130] Vries, K. de Linssen, H., Velden, G. v d, *Macromolecules*, 1989, **22**, 1607
[131] Briber, R.M., Khoury, F., *J. Polym. Sci., Part B: Polym. Phys.*, 1988, **26**, 621
[132] Rymuza, Z., *Wear*, 1991, **142**, 183
[133] Watanabe, M., Yamaguchi, H., *Wear*, 1986, **110**, 379
[134] Ziemanski, K., *Plaste Kautsch.*, 1988, **35**, 98
[135] Lawrence, C.C., Stolarski, T.A., *Wear*, 1989, **132**, 183
[136] Ludema, K.C., *ASM Handbook*, ASM International, 1992, **18**, 236
[137] Imamura, Y.J., *Purasutikku Seikei Gijutsu*, Plast. Moulding Technol., 1989, **6**, 251
[138] Choudhary, T.R., Prashad, H., 1991, Bangalore, India, M43
[139] Lyons, J.W., *The Chemistry and Uses of Fire Retardants*, Wiley-Interscience, 1970
[140] *Mod. Plast.*, 1995, **72**, 69
[141] *Addit. Polym.: Form. Charact.*, 1988, **18**, 3-8

[142] *J. Fire Flammabl. Bull*, 1989, **10 (11)**, 5
[143] Tkac, A., *Dev. Polym. Stab.*, Applied Science, 1982, 153
[144] *Addit. Polym.: Form. Charact.*, 1988, **18**, 5-8
[145] Duffy, J.J., Izv. Akad. Nauk SSSR, *Otd. Khim. Nauk,* Advance Mat. Proc., 1987, 1212
[146] Hotoshi, N., *Porima Daijesuto*, 1988, **40 (11)**, 25
[147] Swindells, I., Nolan, P.F., Wharton, R.K, Flammability Sensitivity Mater. Oxygen-Enriched Atmos., *Symp.*, ASTM, 1987, **3**, 206
[148] Moffett, G.E., Pedley, M.D., Schmidt, N, Flammability Sensitivity Mater. Oxygen-Enriched Atmos., *Symp.*, ASTM, 1987, **3**, 218
[149] Nangrani, K.J., Wenger, R, Daughtery, P.G., *Plast. Compd.*, 1988, **11**, 27, 29
[150] *Plast. Technol.*, 1994, **40**, (13), 48
[151] High Perform. Plast., Nat. Tech. Conf. - Soc. Plast. Eng., *(Prepr.)*, 1989, **6 (3)**, 3
[152] *Plast. Eng. (N.Y.)*, 1989, **45 (5),** 48
[153] *New Materials World*, 1990, 5, 8

Nylon 6,8

Synonyms: *Poly(hexamethylene suberamide). Poly(hexamethylene diamine-co-suberic acid). Poly(iminohexamethyleneiminosuberoyl)*
Monomers: 1,6-Hexanediamine, Suberic acid
Material class: Thermoplastic
Polymer Type: polyamide
CAS Number:

CAS Reg. No.
24936-73-0

Molecular Formula: $(C_{14}H_{26}N_2O_2)_n$
Fragments: $C_{14}H_{26}N_2O_2$
Morphology: Semi-crystalline, triclinic crystals

Volumetric & Calorimetric Properties:
Melting Temperature:

No.	Value	Note
1	232°C	[4]
2	235°C	[8]
3	241°C	[1, 5]

Glass-Transition Temperature:

No.	Value	Note
1	57°C	heating rate 0.5° min^{-1} [1]

Transport Properties:
Polymer Melts: Spencer and Gilmore equation of state parameters have been reported [2,3]

Mechanical Properties:
Tensile (Young's) Modulus:

No.	Value	Note
1	2800 MPa	2.8 × 10^{10} dyn cm^{-2}, 25°C
2	700 MPa	7 × 10^{9} dyn cm^{-2}, 100°C

Bibliographic References
[1] Nishijima, Y., Selki, S. and Kawai, T., *Rep. Prog. Polym. Phys. Jpn.*, 1967, **10**, 473, (T_g)
[2] Spencer, R.S. and Gilmore, G.D., *J. Appl. Phys.*, 1950, **21**, 523

[3] Sagalaev, G.V. *et al*, *Int. Polym. Sci. Technol.*, 1974, **1**, 76
[4] Dachs, K. and Schwartz, E., *Angew. Chem., Int. Ed. Engl.*, 1962, **1**, 430
[5] Scandola, M., Pizzoli, M., Drusiani, A. and Garbuglio, *Eur. Polym. J.*, 1974, **10**, 101

Nylon 6,9

Synonyms: *Poly(hexamethylene azelamide). Poly(hexamethylene-diamine-co-azelaic acid). Poly[iminohexamethyleneiminoazelaoyl]. Poly[imino-1,6-hexanediylimino(1,9-dioxo-1,9-nonanediyl)]*
Monomers: 1,6-Hexanediamine, Nonanedioic acid
Material class: Thermoplastic
Polymer Type: polyamide
CAS Number:

CAS Reg. No.
28757-63-3

Molecular Formula: $(C_{15}H_{28}N_2O_2)_n$
Fragments: $C_{15}H_{28}N_2O_2$
General Information: Nylon 6,9 has lower moisture absorption than Nylon 6,6 and consequently offers better dimensional stability

Volumetric & Calorimetric Properties:
Density:

No.	Value
1	d 1.08 g/cm^3

Thermal Expansion Coefficient:

No.	Value	Type
1	0.000149 K^{-1}	L

Thermal Conductivity:

No.	Value
1	0.216 W/mK

Specific Heat Capacity:

No.	Value	Type
1	1.67 kJ/kg.C	P

Melting Temperature:

No.	Value	Note
1	210°C	DSC

Glass-Transition Temperature:

No.	Value	Note
1	68°C	cryst.

Deflection Temperature:

No.	Value	Note
1	183°C	0.455 MPa, ASTM D648
2	61°C	1.8 MPa, ASTM D648

– **Nylon 6,10**

Surface Properties & Solubility:
Solvents/Non-solvents: Sol. mineral acids, e.g. sulfuric acid, nitric acid; phenols. Insol. aliphatic hydrocarbons, aromatic hydrocarbons

Transport Properties:
Permeability of Liquids: Solubility in and diffusion coefficients of water at different humidities have been reported [1]. Exhibits much lower moisture absorption than Nylon 6,6 and thus has better dimensional stability

Water Absorption:

No.	Value	Note
1	0.48 %	max., 24h, 0.2% moisture content

Mechanical Properties:
Mechanical Properties General: Better dimensional stability due to low moisture absorption gives more stable physical props. than the conventional Nylons 6,6 and Nylon 6

Flexural Modulus:

No.	Value	Note
1	1999.5 MPa	-40°

Flexural Strength at Break:

No.	Value	Note
1	75.8 MPa	21°

Tensile Strength Yield:

No.	Value	Elongation	Note
1	58.6 MPa	10%	21°, ASTM D638

Hardness: Rockwell R111 (21°, ASTM D785)

Izod Notch:

No.	Value	Notch	Note
1	64.16 J/m	Y	½" × ⅛" bar

Electrical Properties:
Electrical Properties General: Well suited to electrical applications e.g. wire jacketing

Dielectric Permittivity (Constant):

No.	Value	Frequency
1	3.7	100 Hz
2	3.6	10 kHz
3	3.2	10 MHz

Dielectric Strength:

No.	Value	Note
1	22.4 kV.mm^{-1}	570 V mil^{-1}, short time
2	19.9 kV.mm^{-1}	507 V mil^{-1}

Dissipation (Power) Factor:

No.	Value	Frequency
1	0.02	100 Hz–1 MHz

Polymer Stability:
Chemical Stability: Attacked by chlorine, bromine and conc. hydrogen peroxide. Temporarily loses stiffness in alcohols
Hydrolytic Stability: Crazed only by very conc. alkali soln.

Applications/Commercial Products:
Processing & Manufacturing Routes: Manufactured by condensation of hexamethylene diamine and azelaic acid. The Nylon 6,9 salt is prepared and then polymerised in an autoclave as for Nylon 6,6. Higher MW grades are made by solid state polymerisation of pellets below melt temp.
Applications: Common uses are electrical wire jacketing, battery cases and cams

Trade name	Details	Supplier
Ashlene	745/746	Ashley Polymers
CRI	826 (CX3395)	Custom Resins
Grilon	CF 62 BSE/XE 3222	EMS
Nylene	826	Custom Resins
Vydyne 69	Nylon Resins	Monsanto Chemical

Bibliographic References
[1] Razumovskij, L.P. and Zaikov, G.E., *Acta Polym.*, 1986, **37**, 146
[2] Hsueh, C.H. and Wendlandt, W.W., *Thermochim. Acta*, 1986, **99**, 37, 42

Nylon 6,10   N-26

Synonyms: Poly(hexamethylene sebacamide). Poly(hexamethylenediamine-co-sebacic acid). Poly[imino-1,6-hexanediylimino(1,10-dioxo-1,10-decanediyl)]
Monomers: 1,6-Hexanediamine, Decanedioic acid
Material class: Thermoplastic
Polymer Type: polyamide
CAS Number:

CAS Reg. No.
9008-66-6

Molecular Formula: $(C_{16}H_{30}N_2O_2)_n$
Fragments: $C_{16}H_{30}N_2O_2$
General Information: Tensile strength and hardness increase with crystallinity, while water absorption and toughness decrease
Morphology: High crystallinity (40-60%) obtained due to regular struct., good chain alignment and high degree of H-bonding. Cryst. forms: α-triclinic, a 4.95, b 5.4, c 22.4, α 49°, β 76.5°, γ 63.5°, 1 monomer per unit cell; β-triclinic, a 4.9, b 8.0, c 22.4, α 90°, β 77°, γ 67.5°, 2 monomers per unit cell

Volumetric & Calorimetric Properties:
Density:

No.	Value	Note
1	d 1.04–1.05 g/cm^3	amorph. [8,9]
2	d 1.16–1.19 g/cm^3	crystalline [8,12]

Thermal Expansion Coefficient:

No.	Value	Note	Type
1	8×10^{-5}–0.0001 K^{-1}	cryst., 20° [4]	L
2	0.00038 K^{-1}	cryst., 20° [3]	V

Equation of State: Spencer and Gilmore equation of state information has been reported [8]

Nylon 6,10

Latent Heat Crystallization: Enthalpy of melting ΔH_m 58.6 kJ mol^{-1} [8]. Entropy of melting ΔS_m 116.6 J mol^{-1} K^{-1} [8]. Heat of fusion 215 kJ kg^{-1} [4]. Enthalpy of activation for transport 61.1 kJ mol^{-1} [8]. Enthalpy of activation of crystallisation 53.6 kJ mol^{-1} [2,8]. Heat of sorption (melt) -58500 [6]. Molar volume 260 cm^3 mol^{-1} (20°, cryst.); 271 cm^3 mol^{-1} (amorph.) [9]. H_f 80 kJ kg^{-1} (60°); 160 kJ kg^{-1} (100°); 400 kJ kg^{-1} (200°); 580 kJ kg^{-1} (250°); 700 kJ kg^{-1} (300°) [4]

Thermal Conductivity:

No.	Value	Note
1	0.23 W/mK	mouldings [4]
2	0.35 W/mK	30°, amorph., moist [5]

Thermal Diffusivity: 1.4×10^{-3} cm^2 s^{-1} [13]

Specific Heat Capacity:

No.	Value	Note	Type
1	1.59 kJ/kg.C	solid, 25° [8]	P
2	2.18 kJ/kg.C	liq. [8]	P
3	502 kJ/kg.C	20°, crystalline [3]	P

Melting Temperature:

No.	Value	Note
1	226–232°C	[8]

Glass-Transition Temperature:

No.	Value	Note
1	45°C	[1,14]

Deflection Temperature:

No.	Value	Note
1	160°C	method B, 0.45 MPa, ISO R75, moulded, dry [4]
2	65–85°C	method A, 1.8 MPa, ISO R75, moulded, dry [4]

Transition Temperature:

No.	Value	Note	Type
1	415°C	ASTM D1929-1977 [4]	Flash ignition temp.

Surface Properties & Solubility:

Cohesive Energy Density Solubility Parameters: δ 26 J$^{1/2}$ cm$^{-3/2}$ [8]

Solvents/Non-solvents: Polymer mouldings absorb alcohols; MeOH 16%; EtOH 8–13%; propanol 10%; butanol 8–12%; glycol 2–4% [4]

Surface Tension:

No.	Value	Note
1	37 mN/m	265°, melt

Transport Properties:

Melt Flow Index:

No.	Value	Note
1	60 g/10 min	MVR 275/5 [7]

Polymer Melts: Melt viscosity flow activation energy 60 kJ mol^{-1}

Volatile Loss and Diffusion of Polymers: Coefficient of diffusion of H$_2$O 0.1×10^{-8} cm^2 s^{-1} (20°); 1.5×10^{-8} cm^2 s^{-1} (60°); 30×10^{-8} cm^2 s^{-1} (100°)

Water Absorption:

No.	Value	Note
1	3.3 %	saturated, in range 20–90°, mouldings [4]
2	1.2–1.6 %	23°, 50% relative humidity [7]

Mechanical Properties:

Tensile (Young's) Modulus:

No.	Value	Note
1	2400 MPa	23°, dry, ISO-R527
2	400 MPa	100°, dry, ISO-R527
3	1500 MPa	23°, moist, ISO-1110

Tensile Strength Break:

No.	Value	Note	Elongation
1	70 MPa	23°, dry	50%
2	50 MPa	1% water, 23°	50%
3	20 MPa	dry, 100°	

Poisson's Ratio:

No.	Value	Note
1	0.3–0.4	mouldings, 20° [4]

Tensile Strength Yield:

No.	Value	Note
1	70 MPa	dry, 23°

Miscellaneous Moduli:

No.	Value	Note	Type
1	800 MPa	23°, dry, ISO 537	Shear modulus
2	120 MPa	100°, dry, ISO 537 [4]	Shear modulus
3	400 MPa	23°, 50% relative humidity, 1000h, ISO 899 [4]	Apparent creep modulus

Impact Strength: Falling weight impact strength W_{50} 140 N m (min., 23°, 50% failure level, 1.5 mm, DIN 53443); W_{50} 110 N m (-20°, 50% failure, 1.5 mm, DIN 53443) [7]

Hardness: Rockwell R75 (dry, 23°, ASTM D785); M60 (moist, 23°, ISO 1110) [3,8]

Izod Notch:

No.	Value	Notch	Note
1	50 J/m	N	23°, dry, ISO 180/1a [4]
2	35 J/m	N	-40°, dry, ISO 180/1a [4]
3	75 J/m	N	23°, moist, ISO 1110 [4]

Electrical Properties:

Electrical Properties General: Tracking resistance CTI 600

– **Nylon 6,12**

Dielectric Permittivity (Constant):

No.	Value	Frequency	Note
1	5.4	1 kHz	23°, 65% relative humidity [10]
2	3.5	1 MHz	23°, 65% relative humidity [10]
3	3	1 GHz	23°, 65% relative humidity [10]

Dielectric Strength:

No.	Value	Note
1	100 kV.mm^{-1}	dry, VDE 0303 part 2, IEC-243, electrode K20/P50
2	60 kV.mm^{-1}	moist, ISO-1110, VDE 0303

Dissipation (Power) Factor:

No.	Value	Frequency	Note
1	0.2	1 kHz	20°, 65% relative humidity [10]
2	0.1	1 MHz	20°, 65% relative humidity [10]
3	0.035	1 GHz	20°, 65% relative humidity [10]

Optical Properties:
Refractive Index:

No.	Value	Note
1	1.565	single cryst.
2	1.53	mouldings [4]

Polymer Stability:
Thermal Stability General: Stable up to 350° for short periods in inert atmosphere
Decomposition Details: Decomposition begins at 310°
Flammability: Flash ignition temp. 415° (ASTM D1929-1977) [4]. Flammability rating FH3 (13 mm min^{-1}); V2 (1.6 mm, UL 94) [7]. Limiting oxygen index 24% (dry, ASTM D2863) [4]
Environmental Stress: Outdoor weathering of up to 10 years can be achieved with uv stabilisers, glass reinforced resins are more vulnerable to erosion, but acceptable mech. props. retained up to 5 years [11]
Chemical Stability: Stable to aliphatic and aromatic hydrocarbons and to many aq. salts and alkalis. Resistant to common wetting agents, oils and alcohols; creep performance not affected [11]. Swollen by some alcohols. Attacked by some chlorinated and polar solvents and aromatic bases at elevated temp. Sensitive to $ZnCl_2$ soln. [11]
Hydrolytic Stability: Hydrolysed by inorganic acid solns. [11]
Biological Stability: Good biological stability
Recyclability: Regrind scrap consisting of solely Nylon 6,10 can be returned to moulding process [11]
Stability Miscellaneous: Electrical props. maintained up to 1000 Mrad of radiation exposure; resin may become discoloured

Applications/Commercial Products:
Processing & Manufacturing Routes: Manufactured by condensation of the Nylon 6,10 salt from conc. aq. soln. in an autoclave at 250–300° and 1.8 MPa
Mould Shrinkage (%):

No.	Value	Note
1	1.2%	60°, ASTM D955 [4]
2	0.8%	longitudinal [7]

Applications: Extrusion grades are manufactured for applications such as semi-finished prods., blown and flat film, monofilaments and bristles. Stabilised injection moulding grades are used for precision parts and highly transparent weather resistant and low temp. impact resistant applications

Trade name	Details	Supplier
610	01/G33/G43	Plastic Materials
Amilan	CM 2001/2006/2402	Toray Industries
Ashlene	990L/990L-30G	Ashley Polymers
CTI	NI range	CTI
EMI-X	PDX/QC	LNP Engineering
Electrafil	J-2/CF/30	DSM Engineering
Fiberfil	J-2/30/40	DSM Engineering
Fiberstran	G-2/30/40	DSM Engineering
G-2/J-2	various grades	DSM Engineering
Lubricomp		LNP Engineering
Magnacomp	QL/QM	LNP Engineering
MonTor	CM 2001/2006/2402	Mon Tor
N15	33 FG-0100	Thermofil Inc.
Ny-Kon	Q	LNP Engineering
RTP	200 series	RTP Company
Stat-Kon	GC-1002	LNP Engineering
Technyl		Rhone-Poulenc
Texapol	GF 1600 A-33/1600 A	Texapol Corp.
Thermocomp	Q/QC/QF	LNP Engineering
Ultramid S		BASF
Verton	QF-700-10/QF 7007	LNP Engineering

Bibliographic References
[1] Gordon, G.A., *J. Polym. Sci.*, Part A-2, 1971, **9**, 1693, (T_g)
[2] Wakelin, J.H., Sutherland, L.R. and Beck. J., *J. Polym. Sci.*, 1960, 278
[3] Warfield, R.W., Kayser, E.G. and Hartmann, B., *Makromol. Chem.*, 1983, **184**, 1927
[4] *Ultramid S Polyamides*, BASF Plastics, 1986
[5] Hellwege, K.H., Hoffmann, R. and Knappe, W., *Kolloid Z. Z. Polym.*, 1968, **226**, 109
[6] Ogata, N., *Makromol. Chem.*, 1960, **42**, 52
[7] *BASF Ultramid Polyamide Range Chart*, BASF, 1996
[8] Van Krevelen, D.W., *Properties of Polymers: Their Correlation with Chemical Structure*, 3rd edn., 1990
[9] Muller, A. and Pfluger, R., *Kunststoffe*, 1960, **50**, 203
[10] Bergmann, K., *Konstr. Kunstst.*, (ed. G. Schreyer), Hanser, 1972
[11] *Ultramid Polyamide Product Line, Properties and Processing*, BASF Plastics, 1996
[12] Bunn, C.W. and Garner, E.V., *Proc. R. Soc. London A*, 1947, **189**, 39
[13] Dietz, W., *Colloid Polym. Sci.*, 1977, **255**, 755
[14] Baird, M.E., *J. Polym. Sci.*, Part A-2, 1970, **8**, 739, (T_g)

Nylon 6,12

Synonyms: Poly(hexamethylenediamine-co-dodecanedioic acid). Poly[imino-1,6-hexanediylimino(1,12-dioxo-1,12-dodecanediyl)]. Poly[iminohexamethyleneiminododecanedioyl]. Poly(hexamethylene dodecanamide)
Monomers: 1,6-Hexanediamine, Dodecanedioic acid
Material class: Thermoplastic
Polymer Type: polyamide
CAS Number:

CAS Reg. No.
24936-74-1

Nylon 6,12

Molecular Formula: $(C_{18}H_{34}N_2O_2)_n$
Fragments: $C_{18}H_{34}N_2O_2$
Additives: Mineral fillers
General Information: A partially crystalline polyamide, with equilibrium water content (1.4%) intermediate between Nylon 6 and Nylon 12. Its props. otherwise closely resemble Nylon 6 but with better dimensional stability and more constant mech. and electrical props. in humid conditions. It is favoured for the production of bristles due to wet strength and rebound resilience

Volumetric & Calorimetric Properties:

Density:

No.	Value
1	d 1.06 g/cm^3

Thermal Expansion Coefficient:

No.	Value	Note	Type
1	9×10^{-5} K^{-1}	[1]	L
2	0.00013 K^{-1}	parallel/normal moulding direction, ASTM E831 [2]	L

Thermal Conductivity:

No.	Value	Note
1	0.22 W/mK	[1]

Specific Heat Capacity:

No.	Value	Note	Type
1	1.67 kJ/kg.C	[1]	P

Melting Temperature:

No.	Value	Note
1	218°C	10 K min^{-1}, ISO 3146C [2]
2	212°C	[1]

Glass-Transition Temperature:

No.	Value	Note
1	46°C	heating rate, 0.5° min.$^{-1}$ [3,4]

Deflection Temperature:

No.	Value	Note
1	180°C	0.45 MPa [1,2]
2	90°C	1.8 MPa [1,2]

Vicat Softening Point:

No.	Value	Note
1	181°C	[2]

Brittleness Temperature:

No.	Value	Note
1	-109°C	[1]

Transport Properties:

Water Content: 3.0% (at saturation)
Water Absorption:

No.	Value	Note
1	1.3 %	23°, 50% relative humidity

Mechanical Properties:

Tensile (Young's) Modulus:

No.	Value	Note
1	2700 MPa	dry [2]
2	1800 MPa	50% relative humidity, 1 mm min^{-1}, ISO 527 parts 1 and 2 [2]

Flexural Modulus:

No.	Value	Note
1	1241 MPa	23°, 50% relative humidity [1]

Tensile Strength Break:

No.	Value	Note	Elongation
1	60.7 MPa	23°, 50% relative humidity [1]	300%

Tensile Strength Yield:

No.	Value	Elongation	Note
1	51 MPa	25%	23°, 50% relative humidity [1]
2	51–61 MPa	7–25%	[2]

Hardness: Rockwell R114 [2]
Fracture Mechanical Properties: Shear strength 55.8 MPa (dry, 23°)

Izod Notch:

No.	Value	Notch	Note
1	75 J/m	N	[1]

Izod Area:

No.	Value	Notch	Note
1	3.5–5 kJ/m^2	Notched	0–50% relative humidity, ISO 180/1A [2]

Charpy:

No.	Value	Notch	Note
1	5–6	Y	0–50% relative humidity [2]

Electrical Properties:

Dielectric Permittivity (Constant):

No.	Value	Frequency	Note
1	5.3	1 kHz	[1]
2	6	100 Hz	[2]
3	4	1 MHz	[2]

Nylon 7

Dissipation (Power) Factor:

No.	Value	Frequency	Note
1	0.15	1 kHz	[1]
2	0.02	1 MHz	dry [2]
3	0.1	1 MHz	50% relative humidity, ISO IEC250 [2]

Polymer Stability:
Flammability: Limiting oxygen index 28% [1]. Flammability rating V2 (UL 94) [2]

Applications/Commercial Products:
Processing & Manufacturing Routes: Prod. by polymerisation of Nylon 6,12 salt from aq. soln., in autoclave at 250–300° and 1.8 MPa, then extruded and pelletised. Higher MW prods. are made by solid state polymerisation of pellets below the soft point

Mould Shrinkage (%):

No.	Value	Note
0	1.1%	DAM test parallel

Applications: Engineering resin, used unfilled or filled. Significant quantities are used in military wire applications due to low moisture resistance. Low to medium viscosity compds. are used to injection mould for monofilaments, sieve fabrics and high quality bristles (toothbrushes). 50% Glass reinforced grades are used for industrial mouldings; 30% carbon fibre reinforced grades are used for injection moulding of medical and sports equipment

Trade name	Details	Supplier
612	01/G23/G43	Plastic Materials
Ashlene		Ashley Polymers
CTI NL	various grades	CTI
Comalloy	640 series	Comalloy Intl. Corp.
Comtuf		ComAlloy International
EMI-X	IC-1008	LNP Engineering
Electrafil		DSM
Fiberfil	J-4 series/TN NY-12	DSM Engineering
Fiberstran	G-4/35/45	DSM Engineering
Grilon	Nylon 6	EMS
Hiloy	640 series/691/693	Comalloy Intl. Corp.
Lubricomp	IFI and IL ranges	LNP Engineering
Lubrilon	643	Comalloy Intl. Corp.
M_n	6/12-FG 10/30/40	Modified Plastics
N6	various grades	Thermofil Inc.
Ny-Kon I		LNP Engineering
Nybex	32001/32008/32009/32012	Ferro Corporation
Nylatron		Polymer Co. Ltd.
Nylon	N6 series	Michael Day
Plaslube	J-4/NY-4	DSM Engineering
RTP	200 series	RTP Company
Texapol	GF 1800 A-33/1800 A	Texapol Corp.
Thermocomp	I/IC/IF ranges	LNP Engineering
UBE	7024B/7125U	UBE Industries Inc.
Vestamid		Hüls AG
Voloy	642 / 648 / 688	Comalloy Intl. Corp.
Zytel 151L		DuPont

Bibliographic References
[1] *Encycl. Polym. Sci. Eng.*, 2nd edn., (ed. J.I. Kroschwitz), John Wiley and Sons, 1985, **11**, 365
[2] DuPont Engineering Polymers Handbook from Concept to Commercialisation DuPont, 1995, 37, (technical datasheet)
[3] Komoto, H., *Rev. Phys. Chem. Jpn.*, 1967, **37**, 105, (T_g)
[4] Saotome, K. and Komoto, H., *J. Polym. Sci., Part A: Polym. Chem.*, 1966, **4**, 1463, (T_g)

Nylon 7

Synonyms: Poly(enantholactam). Poly[imino(1-oxo-1,7-heptanediyl)]. Poly(7-aminoheptanoic acid). Poly(ω-enanthamide)
Monomers: ω-Enantholactam, 7-Aminoheptanoic acid
Material class: Thermoplastic
Polymer Type: polyamide
CAS Number:

CAS Reg. No.
25035-01-2

Molecular Formula: $(C_7H_3NO)_n$
Fragments: C_7H_3NO
General Information: Has similar props. to Nylon 6 but has higher melting point (230-235°) and lower equilibrium monomer content. It has been commercialised in the former Soviet Union
Morphology: Triclinic a 9.8, b 10, c 9.8, α 56°, β 77°, γ 69°, 4 monomers per unit cell (space group C1-1, ρ 1.8 g cm^{-3}) [6]. Struct. and crystallinity of commercial samples have been reported [2]. Melt crystallised samples do not contain γ-form, whereas chemically treated and soln. cryst. samples do have γ-form crystals

Volumetric & Calorimetric Properties:
Density:

No.	Value	Note
1	d < 1.095 g/cm^3	max., amorph.
2	d 1.21 g/cm^3	cryst.

Thermal Expansion Coefficient:

No.	Value	Note	Type
1	0.00035 K^{-1}	T < T_g	L

Specific Heat Capacity:

No.	Value	Note	Type
1	1.67–1.84 kJ/kg.C	solid, 25°	P

Melting Temperature:

No.	Value
1	230–235°C

Glass-Transition Temperature:

No.	Value
1	52–62°C

Transition Temperature General: Annealing increases polymer melt temp. [2]

Surface Properties & Solubility:
Solvents/Non-solvents: Dissolved by strong mineral acids

Optical Properties:
Transmission and Spectra: N-15 nmr spectral data have been reported [2]

Polymer Stability:
Chemical Stability: Possesses similar stability to Nylon 6. Resistant to most aliphatic, aromatic and chlorinated hydrocarbon solvents, dilute acids and alkalis. Strongly attacked by polar solvents

Applications/Commercial Products:
Processing & Manufacturing Routes: Nylon 7 is polymerised from enantholactam in the presence of 4% H_2O in an autoclave for 13h at 200° or 5h at 280°. The polymerisation reaches a specific viscosity of ca. 1.2. It can also be polymerised directly from 7-aminoheptanoic acid at 260° with evolution of H_2O [3]
Applications: Suitable for applications similar to those of Nylon 6; not thought to be available in USA or Europe

Bibliographic References
[1] Van Krevelen, D.W., *Properties of Polymers: Their Correlation with Chemical Structure*, 3rd edn, Elsevier, 1991
[2] Johnson, C.G. and Mathias, L.J., *Polymer*, 1994, **35**, 66, (morphology)
[3] Nesmeyanov, A.N., Strepikheev, A.A., Freidlina, R.Kh., Zakharkin, L.I. et al, *Chem. Tech. (Berlin)*, 1957, **9**, 139, (synth)
[4] Slichter, W.P., *J. Polym. Sci.*, 1959, **36**, 259, (cryst form)
[5] Schmidt, G.F. and Stuart, H.A., *Z. Naturforsch., A: Phys. Sci.*, 1958, **23**, 222

Nylon 7,7

Synonyms: *Poly(heptamethylene pimelamide). Poly[imino(1,7-dioxo-1,7-heptanediyl)imino-1,7-heptanediyl]. Poly(heptamethylenediamine-co-pimelic acid)*
Monomers: 1,7-Heptanediamine, Heptanedioic acid
Material class: Thermoplastic
Polymer Type: polyamide
CAS Number:

CAS Reg. No.
32473-30-6

Molecular Formula: $(C_{14}H_{26}N_2O_2)_n$
Fragments: $C_{14}H_{26}N_2O_2$
Morphology: Crystallises in pseudohexagonal form, a 4.82, b 19.0, c 4.82, β 60°, space group CS-1 [1]

Volumetric & Calorimetric Properties:
Density:

No.	Value	Note
1	d 1.06 g/cm^3	amorph.
2	d 1.108 g/cm^3	cryst.

Melting Temperature:

No.	Value
1	196–214°C

Glass-Transition Temperature:

No.	Value	Note
1	55°C	[2]

Surface Properties & Solubility:
Surface Tension:

No.	Value
1	43 mN/m

Bibliographic References
[1] Kinoshita, Y., *Makromol. Chem.*, 1959, **33**, 1, 21, (cryst struct)
[2] Van Krevelen, D.W., *Properties of Polymers: Their Correlation with Chemical Structure*, 3rd edn., Elsevier, 1990
[3] Nishijima, Y., Selki, T. and Kawai, T., *Rep. Prog. Polym. Phys. Jpn.*, 1967, **10**, 473, (T_g)

Nylon 8

Synonyms: *Poly[imino(1-oxo-1,8-octanediyl)]. Poly(capryllactam). Poly(8-aminooctanoic acid). Poly(8-aminocaprylic acid)*
Monomers: Capryllactam, 8-Aminooctanoic acid
Material class: Thermoplastic
Polymer Type: polyamide
CAS Number:

CAS Reg. No.
25035-02-3

Molecular Formula: $(C_8H_{15}NO)_n$
Fragments: $C_8H_{15}NO$
General Information: Nylon 8 is thermally stable [1]. As its monomer and oligomer content is low, there is no need for the additional monomer extraction step used in the manufacture of Nylon 6. Although fibre and engineering props. are similar to other crystalline aliphatic nylons, its comparatively low melting point (200° compared to 270° for Nylon 6,6) and expensive monomer discourages commercialisation. Much lower water absorption than Nylon 6
Morphology: Monoclinic, a 9.8, b 22.4, c 8.3, β 65°, 8 monomers per unit cell (space group C2-2, ρ 1.14 g cm^{-3})

Volumetric & Calorimetric Properties:
Density:

No.	Value	Note
1	d 1.04 g/cm^3	amorph.
2	d 1.14 g/cm^3	cryst.

Thermal Expansion Coefficient:

No.	Value	Note	Type
1	0.00031 K^{-1}	solid, 25°	L

Melting Temperature:

No.	Value	Note
1	185–209°C	[1]
2	200°C	[2]

Glass-Transition Temperature:

No.	Value	Note
1	50°C	[1,2]
2	51°C	[1,2]

Nylon 8.8 – Nylon 9,6

Surface Properties & Solubility:
Cohesive Energy Density Solubility Parameters: Solubility parameter 26 $J^{1/2}$ $cm^{-3/2}$. Cohesive energy density 91900 J mol^{-1} [2]
Plasticisers: *p*-Toluenesulfonamides
Solvents/Non-solvents: Dissolved by inorganic acids and organic bases such as aniline and pyridine

Polymer Stability:
Chemical Stability: Resistant to most aliphatic and aromatic hydrocarbons and alcohols
Hydrolytic Stability: Resistant to conc. inorganic salt solns. and bases

Applications/Commercial Products:
Processing & Manufacturing Routes: Polymerisation is readily carried out in the melt using a small amount of amino acid initiator, or by anionic initiation. No monomer extraction step is required
Applications: Applications are limited due to low melting point (200°)

Bibliographic References
[1] Dachs, K. and Schwartz, D., *Nuova Chim.*, 1972, **48**, 43
[2] Van Krevelen, D.W., *Properties of Polymers: Their Correlation with Chemical Structure*, 3rd edn., Elsevier, 1990
[3] Vogelsong, D.C., *J. Polym. Sci., Part A: Polym. Chem.*, 1965, **1**, 1055, (T_m)
[4] Schmidt, G.F. and Stuart, H.A., *Z. Naturforsch., A: Phys. Sci.*, 1958, **13**, 222, (T_m)

Nylon 8.8   N-31
Synonyms: *Poly(octamethylene suberamide). Poly(octamethylenediamine-co-suberic acid). Poly[imino(1,8-dioxo-1,8-octanediyl)imino-1,8-octanediyl]*
Monomers: 1,8-Octanediamine, Octanedioic acid
Material class: Thermoplastic
Polymer Type: polyamide
CAS Number:

CAS Reg. No.
33182-53-5

Molecular Formula: $(C_{16}H_{30}N_2O_2)$
Fragments: $C_{16}H_{30}N_2O_2$

Nylon 9   N-32
Synonyms: *Poly(9-aminonanoic acid). Poly(9-aminopelargonic acid). Poly(ω-pelargonamide). Poly[imino(1-oxo-1,9-nonanediyl)]*
Monomers: 9-Aminonanoic acid
Material class: Thermoplastic
Polymer Type: polyamide
CAS Number:

CAS Reg. No.
25035-03-4

Molecular Formula: $(C_9H_{17}NO)_n$
Fragments: $C_9H_{17}NO$
Morphology: Triclinic, a 4.9, b 5.4, c 12.5, α 49°, β 77°, γ 64°, 1 monomer per unit cell (space group C1-1, ρ 1.15 g cm^{-3}) [5]; triclinic, a 9.7, b 9.7, c 12.6, α 64°, β 90°, γ 67°, 4 monomers per unit cell (space group C1-1, ρ 1.07 g cm^{-3}) [6]

Volumetric & Calorimetric Properties:
Density:

No.	Value	Note
1	d 1.052 g/cm^3	amorph. [1]
2	d 1.066 g/cm^3	cryst. [1]
3	d 1.15 g/cm^3	cryst.
4	d 1.07 g/cm^3	cryst.

Thermal Expansion Coefficient:

No.	Value	Note	Type
1	0.00036 K^{-1}	[1]	L

Melting Temperature:

No.	Value
1	194–209°C

Glass-Transition Temperature:

No.	Value	Note
1	51°C	[1]
2	46°C	[3,4]

Applications/Commercial Products:
Processing & Manufacturing Routes: Prod. by polymerisation of ω-aminopelargonic acid by a telomerisation process
Applications: Not commercially available

Bibliographic References
[1] Van Krevelen, D.W., *Properties of Polymers: Their Correlation with Chemical Structure*, 3rd edn., Elsevier Science, 1990
[2] Illers, K.H., *Polymer*, 1977, **18**, 551
[3] Nishijima, Y., Selki, J. and Kawai, T., *Rep. Prog. Polym. Phys. Jpn.*, 1967, **10**, 473, (T_g)
[4] Champetier, G. and Pied, J.P., *Makromol. Chem.*, 1961, **44-46**, 64, (T_g)
[5] Griffin, W.R., *Rubber World*, 1957, **136**, 687, (morphology)
[6] Fishbein, L. and Crowe, B.F., *Makromol. Chem.*, 1961, **48**, 221, (morphology)

Nylon 9,6   N-33

Synonyms: *Poly(imino adipoyl iminononamethylene). Poly[imino(1,6-dioxo-1,6-hexanediyl)imino-1,9-nonanediyl]*
Monomers: 1,9-Nonanediamine, Hexanedioic acid
Material class: Thermoplastic
Polymer Type: polyamide
Molecular Formula: $(C_{15}H_{28}N_2O_2)_n$
Fragments: $C_{15}H_{28}N_2O_2$

Volumetric & Calorimetric Properties:
Latent Heat Crystallization: Activation energy for transport 56.9 kJ mol^{-1}
Glass-Transition Temperature:

No.	Value	Note
1	45°C	[1]
2	42°C	[2]

Bibliographic References
[1] Magill, J.M., *J. Polym. Sci.*, 1965, **3**, 1195, (T_g)
[2] Van Krevelen, D.W., *Properties of Polymers: Their Correlation with Chemical Structure*, 3rd edn., Elsevier, 1990

Nylon 9,9

Synonyms: *Poly(nonamethylene azelaimide). Poly(nonamethylenediamine-co-azelaic acid). Poly(imino(1,9-nonanediyl)imino-1,9-nonanediyl)*
Monomers: 1,9-Nonanediamine, Nonanedioic acid
Material class: Thermoplastic
Polymer Type: polyamide
CAS Number:

CAS Reg. No.
33182-52-4

Molecular Formula: $(C_{18}H_{34}N_2O_2)_n$
Fragments: $C_{18}H_{34}N_2O_2$

Volumetric & Calorimetric Properties:
Density:

No.	Value	Note
1	d <1.043 g/cm^3	max., amorph. [1]

Melting Temperature:

No.	Value	Note
1	177°C	[1]
2	189°C	[2]
3	165°C	[2]

Surface Properties & Solubility:
Surface Tension:

No.	Value
1	36 mN/m

Bibliographic References
[1] Van Krevelen, D.W., *Properties of Polymers: Their Correlation with Chemical Structure*, 3rd edn., Elsevier, 1990
[2] Baker, W.O. and Fuller, C.S., *J. Am. Chem. Soc.*, 1942, **64**, 2399, (cryst)

Nylon 10

Synonyms: *Poly(10-aminodecanoic acid). Poly[imino(1-oxo-1,10-decanediyl)]. Poly(10-aminocapric acid). Poly(azacycloundecan-2-one)*
Monomers: 10-Aminodecanoic acid
Material class: Thermoplastic
Polymer Type: polyamide
CAS Number:

CAS Reg. No.
26970-31-0

Molecular Formula: $(C_{10}H_{19}NO)_n$
Fragments: $C_{10}H_{19}NO$
Morphology: Triclinic, a 9.8, b 5.12, c 27.54, α 54°, β 90°, γ 110°, 4 monomers per unit cell

Volumetric & Calorimetric Properties:
Density:

No.	Value	Note
1	d 1.032 g/cm^3	amorph.
2	d 1.019 g/cm^3	cryst.

Thermal Expansion Coefficient:

No.	Value	Note	Type
1	0.00036 K^{-1}	T > T$_g$	V

Melting Temperature:

No.	Value
1	177–192°C

Glass-Transition Temperature:

No.	Value	Note
1	43°C	[1]
2	42°C	[2]

Bibliographic References
[1] Van Krevelen, D.W., *Properties of Polymers: Their Correlation with Chemical Structure*, Elsevier, 1990
[2] Champetier, G. and Pied, J.P., *Makromol. Chem.*, 1961, **44-46**, 64, (T$_g$)

Nylon 10,9

$$-[NHCH_2(CH_2)_8CH_2NHCCH_2(CH_2)_4CH_2C]_n-$$
(with two C=O groups)

Synonyms: *Poly(decamethylene azelamide). Poly(decamethylene diamine-co-azelaic acid). Poly(imino(1,9-dioxo-1,9-nonanediyl)imino-1,10-decanediyl)*
Monomers: 1,10-Decanediamine, Nonanedioic acid
Material class: Thermoplastic
Polymer Type: polyamide
CAS Number:

CAS Reg. No.
98241-67-9

Molecular Formula: $(C_{19}H_{36}N_2O_2)_n$
Fragments: $C_{19}H_{36}N_2O_2$

Volumetric & Calorimetric Properties:
Density:

No.	Value	Note
1	d <1.044 g/cm^3	max., amorph.

Thermal Expansion Coefficient:

No.	Value	Note	Type
1	0.00066 K^{-1}	linear, T > T$_g$	L

Latent Heat Crystallization: Latent heat of fusion ΔH_m 68.2 kJ mol^{-1} [1]. Entropy of fusion ΔS_m 139.4 J mol K^{-1} [2]

Melting Temperature:

No.	Value
1	216°C

Glass-Transition Temperature:

No.	Value	Note
1	58°C	[3]

Bibliographic References
[1] Wunderlich, B. et al, *Macromolecules,* 1989, **7**, 22
[2] Van Krevelen, D.W., *Properties of Polymers: Their Correlation with Chemical Structure,* 3rd edn., 1990
[3] Schmeider, K. and Wolf, K., *Kolloid Z. Z. Polym.,* 1953, **134**, 149, (T_g)

Nylon 10,10   N-37

$$\left[NHCH_2(CH_2)_8CH_2NH\overset{O}{\overset{\|}{C}}CH_2(CH_2)_6CH_2\overset{O}{\overset{\|}{C}} \right]_n$$

Synonyms: *Poly(decamethylene sebacamide). Poly(decamethylene diamine-co-sebacic acid). Poly[imino(1,10-dioxo-1,10-decanediyl)imino-1,10-decanediyl]*
Monomers: 1,10-Decanediamine, Decanedioic acid
Material class: Thermoplastic
Polymer Type: polyamide
CAS Number:

CAS Reg. No.
28774-87-0

Molecular Formula: $(C_{20}H_{38}N_2O_2)_n$
Fragments: $C_{20}H_{38}N_2O_2$
Morphology: Triclinic, c 25.6, 1 monomer per unit cell [2]

Volumetric & Calorimetric Properties:
Density:

No.	Value	Note
1	d <1.032 g/cm^3	max., amorph. [1]
2	d >1.063	min., cryst. [1]

Thermal Expansion Coefficient:

No.	Value	Note	Type
1	0.00067 K^{-1}	T > T_g	L

Latent Heat Crystallization: ΔH_m 32.7/51.1/72 J (mol K)$^{-1}$ [1,2]; ΔS_m 1.472 J (mol K)$^{-1}$
Melting Temperature:

No.	Value	Note
1	203°C	cryst.
2	198°C	[3]
3	216°C	[4]

Surface Properties & Solubility:
Surface Tension:

No.	Value
1	32 mN/m

Transport Properties:
Water Absorption:

No.	Value	Note
1	2 %	100% relative humidity

Mechanical Properties:
Tensile (Young's) Modulus:

No.	Value	Note
1	130 MPa	1.3 × 10^9 dyn cm^{-2} [2]

Electrical Properties:
Dielectric Permittivity (Constant):

No.	Value
1	3.4–3.8

Applications/Commercial Products:
Processing & Manufacturing Routes: See nylon polymerisation; nylon processing

Bibliographic References
[1] Van Krevelen, D.W., *Properties of Polymers: Their Correlation with Chemical Structure,* 3rd edn., Elsevier, 1990
[2] Baker, W.O. and Fuller, C.S., *J. Am. Chem. Soc.,* 1947, **64**, 2399, (cryst)
[3] Saotome, K. and Komoto, H., *J. Polym. Sci., Part A-1,* 1966, **4**, 1463, (mp)
[4] Dole, M. and Wunderlich, B., *Makromol. Chem.,* 1959, **34**, 29, (mp)

Nylon 11   N-38

$$\left[NHCH_2(CH_2)_8CH_2\overset{O}{\overset{\|}{C}} \right]_n$$

Synonyms: *Poly(11-aminoundecanamide). Poly(ω-aminoundecanoic acid). Poly[imino(1-oxo-1,11-undecanediyl)]*
Related Polymers: Nylon Powder Coatings
Monomers: 11-Aminoundecanoic acid
Material class: Thermoplastic
Polymer Type: polyamide
CAS Number:

CAS Reg. No.
25035-04-5

Molecular Formula: $(C_{11}H_{21}NO)_n$
Fragments: $C_{11}H_{21}NO$
Additives: Glass or carbon fibre to give rigid products for moulding. Fillers are used to control rigidity and shrinkage. Carbon black is employed for uv resistance and antistatic props. Flame retardants such as ammonium polyphosphate are added to improve self-extinguishing props.
General Information: Hydrophobic nylon, easier to prepare than Nylon 6 or Nylon 6,6. The physical props. of Nylon 11 are very similar to Nylon 12, but differ significantly from Nylon 6, and Nylon 66. The long chain confers better water resistance (only 25% of the absorption of Nylon 6 at saturation) resulting in electrical and mech. props. less sensitive to atmospheric variation. Impact props. are high and tensile strength and moduli are reduced compared to Nylon 6/66. The polymer is more flexible and the balance of toughness and flexibility is readily controlled by plasticiser addition. Mouldings are dimensionally stable over a wide range of humidity conditions. It exhibits excellent toughness at low temps. combined with good stress cracking resistance. It is

Nylon 11

less susceptible to strong polar solvents such as formic acid suited to underground cable usage resisting formic acid generated by termites

Morphology: Exists in the stable α triclinic cryst. form at room temp. [10,11,12]. A study of the morphology of melt spun Nylon 11 fibres by wide angle x-ray, DSC and birefringence studies has been reported [13]. Melt spun Nylon 11 possesses the α triclinic struct. Orientation in thin Nylon 11 films has been studied by Trichroic ir spectroscopy to characterise the three dimensional orientation which develops from drawing, annealing or external electric field application [14,15]. Partial dissolution of crystallites and extraction of low MW chains occurs during swelling with formic acid [15]. Reversal of the process leads to amorph. chains recrystallising and the restoration of the original order and morphology. Crystallisation of Polyamide 11 on glass fibre has been studied by chemical etching [16]. Orientation of spherulites in the matrix is related to the mech. process conditions. Nylon 11 polymorphism has been investigated during solid state forging studies [17]. Studies on cross laminates of unidirectionally reinforced Nylon 11 composites show novel aspects of the nature and origin of polymorphism in melt crystallised Nylon 11 prior to and after forging. Two distinct crystalline species are observed; semi-disordered smectic δ and three dimensional α-form. Respective contents depend upon thermal history. The δ-form is kinetically favoured but transforms into the thermodynamically favoured α-form. Thermal analysis shows a characteristic broad transition associated with a cryst.-condis transition. Differences in the mode of hydrogen bonding are given as an explanation of polymorphism. Low temp. forging favours the δ-form, but optimal forging conditions at higher temps. prod. a mixture of crystals. [34] Melt spun Nylon 11 fibres are found to possess an α-triclinic form. The influence of annealing and solid state drawing has been reported [31]. The morphology of EPR/Nylon 11 blends is constructed from a nodular part and a fibrillar part. The stability of the fibrillar part determines whether the blend morphology will evolve into nodules by the Rayleigh mechanism or into phase inversion by coalescence of stable fibres [68]. α-Triclinic cryst. (space group C1-1), a 9.5, b 10, c 15, 60°, 90°, 67°, 4 monomer units per cell [10]; α-triclinic cryst., a 4.78, b 4.13, c 13.1, 90°, 75°, 66°, 1 monomer unit per cell [10]; α-triclinic cryst. (space group C1-1), a 4.9, b 5.4, c 14.9, 49°, 77°, 63°, 1 monomer unit per cell [10]; α-triclinic cryst., a 9.6, b 4.2, c 15, 72°, 90°, 64°, 2 monomer units per cell [10]; α-triclinic cryst., a 4.78, b 4.13, c 14.9, 82°, 75°, 66°, 1 monomer unit per cell [10]

Volumetric & Calorimetric Properties:

Density:

No.	Value	Note
1	d 1.03 g/cm^3	23°, ISO R1183D [1]
2	d 1.01 g/cm^3	amorph. [2]
3	d 1.12–1.23 g/cm^3	cryst. [2]
4	d 1.04 g/cm^3	[6]

Thermal Expansion Coefficient:

No.	Value	Note	Type
1	1×10^{-5} K^{-1}	[1]	L
2	9×10^{-5} K^{-1}	[6]	L

Volumetric Properties General: Other values for heat capacity have been reported [36]
Latent Heat Crystallization: ΔH_{fusion} 41.4 kJ mol^{-1} [2]
Thermal Conductivity:

No.	Value	Note
1	0.19 W/mK	[6]

Specific Heat Capacity:

No.	Value	Note	Type
1	1.26 kJ/kg.C	[6]	P

Melting Temperature:

No.	Value	Note
1	183–187°C	ASTM D789 [1]
2	182–200°C	[2]
3	182–183°C	[10]
4	190°C	[6]

Glass-Transition Temperature:

No.	Value	Note
1	46°C	[2]
2	42°C	DSC, torsion tester [4]
3	43°C	
4	92°C	[5]

Transition Temperature General: Thermal transition behaviour has been reported [35]
Deflection Temperature:

No.	Value	Note
1	149°C	0.46 MPa [6]
2	54°C	1.85 MPa [8]
3	145°C	0.45 MPa [1]
4	50°C	1.85 MPa, ISO 75 [1]

Vicat Softening Point:

No.	Value	Note
1	180°C	10 N [1]
2	160°C	50 N, ISO 306 [1]

Surface Properties & Solubility:

Plasticisers: *p*-Toluenesulfonamide
Solvents/Non-solvents: Sol. *m*-cresol, sulfuric acid, formic acid, alcohol/phenol mixtures. Insol. hydrocarbons, organic acids, Me$_2$CO, gas-oil, glycerine, greases, lactic acid, paraffin, petrol, stearine, turpentine, EtOH [1]
Wettability/Surface Energy and Interfacial Tension: An adhesive composition comprising a blend of polyester and aliphatic polyamides was formed to an adhesive sheet and pressed in a sandwich between Nylon 4,T and Nylon 11 at 250° and found to give a composite with good layer bonding strength. Adhesive strength of Nylon 11 coatings (<300 mm thick) 11 kN m^{-1} (ISSN 0285-3787) [60]
Surface Tension:

No.	Value	Note
1	33–43 mN/m	[2]

Transport Properties:

Polymer Melts: Miscibility, crystallisation and rheological props. of poly-*p*-phenylene terephthalamide/Nylon 11 composites have been studied. The viscosity of Nylon 11 is reduced, unlike Nylon 6

– Nylon 11

where cross hydrogen-bond interactions lead to a magnitude rise in melt viscosity [21]

Permeability and Diffusion General: The diffusion of gases from the wellstream to the annular space in unbonded flexible pipelines has been reported. A thermoplastic Nylon 11 fluid barrier and steel carcass or armour wire is employed for the pipeline. Diffused gas increases the pressure in the annular space and must be vented to prevent the pipeline from bursting [40]

Water Content: 1.9% (saturated, 23°) [1], 2.5% (saturated, 100°) [1], 1.1% (long term) [6]

Water Absorption:

No.	Value	Note
1	0.9 %	23°, 50% relative humidity, moulding [1]
2	0.9–1.1 %	20°, 65% relative humidity, powder [1]
3	1.6–1.9 %	20°, 100% relative humidity, powder [1]
4	2.4–3 %	100°, 100% relative humidity, powder [1]

Gas Permeability:

No.	Gas	Value	Note
1	He	0.959 cm^3 mm/ (m^2 day atm)	1.46×10^{-13} cm^2 (s Pa)$^{-1}$, 10–50° [3]
2	CO$_2$	0.049 cm^3 mm/ (m^2 day atm)	0.754×10^{-13} cm^2 (s Pa)$^{-1}$, 20–60° [3]
3	Ar	0.009 cm^3 mm/ (m^2 day atm)	0.143×10^{-13} cm^2 (s Pa)$^{-1}$, 20–60° [3]
4	Ne	0.017 cm^3 mm/ (m^2 day atm)	0.26×10^{-13} cm^2 (s Pa)$^{-1}$, 20–60° [3]
5	H$_2$	0.088 cm^3 mm/ (m^2 day atm)	1.34×10^{-13} cm^2 (s Pa)$^{-1}$, 20–60° [3]

Mechanical Properties:

Mechanical Properties General: Nylon 11 possesses excellent impact strength and abrasion resistance, has a high resistance to attack by most chemicals, and satisfactory ageing props. It has lower melting point, water absorption and specific gravity compared to Nylons 6, 6,6 or 6,10. Continuous duty at 100° can be achieved by incorporating anti-oxidants. High flexibility and impact even down to low temps. are possible by plasticising the nylon. Good moulding tolerances ($\pm 1.6\%$) and dimensional stability can be achieved by employing a wide range of additives, particularly fillers. Physical props. of nylon injection moulding compounds, including Nylon 11, have been reported. Nylon 11 or 12 may be treated with an alkyl or alkoxyalkyl silane to give a moisture cross-linkable product with improved thermal and mech. props. [69]. Fibre tenacity 0.26–0.44 N tex^{-1} [7]

Tensile (Young's) Modulus:

No.	Value	Note
1	1730 MPa	[6]

Flexural Modulus:

No.	Value	Note
1	1000 MPa	ISO 178 [1]

Tensile Strength Break:

No.	Value	Note	Elongation
1	52 MPa	ISO R257 [1]	300%
2	54 MPa	[6]	330%

Flexural Strength at Break:

No.	Value	Note
1	45 MPa	19 mm max. flexure, ISO 178 [1]

Tensile Strength Yield:

No.	Value	Elongation	Note
1	40 MPa	10%	[1]
2	34 MPa	22%	[6]

Mechanical Properties Miscellaneous: Low acrylonitrile rubbers enhance toughness from 20% levels, but reduce strength and modulus. The two polymers are partially miscible, according to DSC/DMA studies [67]

Hardness: Rockwell R103 (ISO 2039/2) [1]. Shore D70-D72 (ISO 86B) [1]. Surface hardness 820 kg cm^{-2} (10 s under load, DIN 53–456, 20°) [1]

Failure Properties General: Mandrels used for manufacture of hoses comprise 1–20% Rilsan nylon 11 and 7% glass fibre. They show hose pullout time of 2.25 min., compared to 10.25 min. without any glass fibre. [57] Shear strength 35.3–42.2 MPa (3.6–4.3 kg mm^{-2}, ASTM D732) [1]

Fracture Mechanical Properties: The long term resistance of 32 mm Nylon 11 gas pipe to rupture over a range of pressures has been investigated. At any given temp. the relationship between internal pressure and time to rupture is linear on a double log plot. Applying an Arrhenius time-temp. dependence law to pipe stress rupture data gives a high level of confidence for a fifty year lifetime [25]. Mixtures of poly(phenylene terephthalamide) (PPTA) and Nylon 11 were extruded through a capillary die and coagulated in water; 70/30 PPTA/Nylon 11 extrudates were uniaxially oriented. The fracture surfaces of soln. processed extrudates reveal uniformly dispersed microfibrils of PPTA. 5% PPTA compositions have improved tensile modulus and strength [33]

Friction Abrasion and Resistance: Abrasion resistance 20 mg (Elf-Aquitaine test) [1]. The influence of fillers (ZnS, PbS, ZnF$_2$) on transfer film formation was studied by optical and scanning electron microscopy. PbS dissociates during wear and forms a strong adherent film which improves wear resistance. The other fillers increase wear relative to unfilled Nylon 11 [47]. 35 vol% PbS gives the best wear resistance. Wear rate increases considerably upon doubling applied force and speed, and when increasing the surface roughness; coefficient of friction is not affected [44]. With copper fillers (CuS, CuF$_2$, CuO, CuAc), only copper acetate has an adverse effect upon wear rate; the other composites transfer well to the steel counterface to give strongly adherent thin films [48]. Transfer films of CuS and CuF$_2$ decompose under rubbing conditions and produce copper, FeF$_6$ and FeSO$_4$. Highest concentrations are closer to the transfer film-counterface interface. No chemical change is observed with copper acetate [49]. The wear rate of Nylon 11 composites containing CuS filler and short fibres, (CuS has an unusual ability to reduce wear in a number of polymers) is lowest for 20% volume carbon fibre and further reduces in the presence of CuS filler. Synergistic behaviour between the carbon fibre and CuS filler occurs as the filler decomposes which increases adhesion of the composite surface to the steel substrate. A theoretical model for the optimal filler proportion in polymer composites has been reported. [46] Scratch resistance 6 kg (0.4 mm coating on steel) [1]

Charpy:

No.	Value	Notch	Break	Note
1		N		no break -23 - 40°C [1]
2	12	Y		23° [1]
3	6	Y		-40° [1]

Electrical Properties:

Electrical Properties General: Nylon 11 has very good dielectric props. which are much less affected by humidity than Nylons 6 or 6,6 due to its much lower water absorption props. Tracking resistance 600 V (min., NFC 26220)

Surface/Volume Resistance:

No.	Value	Note	Type
1	0.1×10^{15} Ω.cm	ASTM D257 [1]	S

Dielectric Permittivity (Constant):

No.	Value	Frequency	Note
1	3.7		[2]
2	4	1 MHz	[6]

Dielectric Strength:

No.	Value	Note
1	30 kV.mm^{-1}	dry, ASTM D149

Complex Permittivity and Electroactive Polymers: It has been found that the degree of double crystallite orientation resulting from uniaxial stretching below T_g for odd-numbered nylons decreases with increasing amide group content. Unit cells are reoriented without cryst. struct. modification by electric field poling. 90° Reorientation of unit cells occurs for all odd nylons studied (at same moisture level), but Nylon 11 gives only 72° when poled using a dry air purge. The difference may be due to an increase in electric field applied to the cryst. phase when moisture is present in the amorph. regions during poling [66]. Measurements of high frequency dielectric and thickness mode electromechanical props. of ferroelectric Nylon 11 film samples have been reported. Permanent polarisation of the unoriented film 27 mC m^{-2} [69]

Dissipation (Power) Factor:

No.	Value	Frequency	Note
1	0.03	1 MHz	[6]

Optical Properties:

Optical Properties General: Substrate coated with Nylon 11 and gold successively by vacuum deposition gives a recording disc with good lightfastness and storage stability [54]. Matting agents used in plastic coatings are mainly in the form of needles or pellets with average length 1–2000 mm and aspect ratio 10–40:1. Nylon 11 powder coating containing 0.5% terephthalic acid dianilide gives films with 60° gloss 7.2, compared to 44.5 for pure Nylon 11 [64]

Transmission and Spectra: Ir and N-15 nmr spectral data have been reported [14,28,29,41,42]

Volume Properties/Surface Properties: Experimental specific volume data as a function of temp. have been reported [11]

Polymer Stability:

Thermal Stability General: Heat and oxidation lead to a reduction in MW, discoloration and loss of mech. props. Principal stabilisers used are copper halides, secondary amines and esterically modified phenols. Nylon 11 is more readily stabilised against light degradation than Nylons 6 or 6,6

Upper Use Temperature:

No.	Value
1	100°C

Decomposition Details: A photo-oxidation study of N-substituted Nylon 11 in the solid state has been reported [32]. Hydroperoxide and imide groups are characterised as main intermediate products in long wave and at 60°. The N-substituted imide groups are more stable than the unsubstituted and accumulate much faster in substituted Nylon 11. N-Substitution does not promote radical attack on H atoms located in the α position with respect to substituted N atoms

Flammability: Average time of burning 20 s (3.2 mm specimen). Average extent of burning 40 mm (ASTM D635, UL 94) [1]. Flammability rating V2 (3.2 mm); HB (1.6 mm, UL 94) [1]. The pyrolysis of pure Nylon 11 and Nylon 11 containing flame retarding decabromodiphenyl/Sb_2O_3 mixture has been reported [20]. Ammonium polyphosphate (APP) strongly modifies the thermal degradation process of Nylon 11. α and β unsaturated nitriles are the major volatile products of degradation, whereas only α nitriles are given off for the pure Nylon 11. Minor amounts of hydrocarbons, CO, and CO_2 evolved from the pure nylons are suppressed. A charring process occurs which is little affected by APP, but the char intumesces in the presence of APP leading to improved fire retardancy. Formation of intermediate phosphate ester bonds is postulated whose decomposition accounts for the differences observed [34]

Environmental Stress: The susceptibility of nylons to environmental stress cracking in aq. salt soln. has been investigated. Razor cut precracked samples immersed in NaCl, $CaCl_2$ at various initial stress intensities, levels and temps. crack and craze above T_g for Nylon 6 and Nylon 6,6 but Nylon 11 is not susceptible to salt induced cracking [18]

Chemical Stability: Exhibits good resistance to ammonia, $CaCl_2$, $CuSO_4$, gas-oil, glucose, greases, hydrogen, lactic acid, mercury, milk, oils, paraffin, petrol, sea water, NaCl, stearine, turpentine (20–60°), agricultural sprays, fruit juices, sulfur, citric acid, glycerine, oxygen, sodium carbonate (conc.) (20–40°) [1]

Applications/Commercial Products:

Processing & Manufacturing Routes: Manufactured in a continuous process, in an inert gas at atmosphere pressure. Monomer suspended in water, transferred to a converter, melted, water is eliminated and monomer polymerised. Ricin, a protein present in Castor beans from India, Central America and Africa is now being used to make monomer that is polymerised to Nylon 11 [23]. Prepared from Castor oil feedstock, which is treated with MeOH to yield methyl ricinoleate, which upon pyrolysis, hydrol. and HBr addition is converted to 11-aminoundecanoic acid. The acid is polymerised at 200° in the presence of hypophosphoric acid/phosphoric acid to Nylon 11 with continuous removal of water. The polymer is extruded under inert gas and chopped into pellets

Mould Shrinkage (%):

No.	Value
1	1.5%

Applications: Has excellent electrical props. Mainly used in mechanical engineering for moving components e.g. gears, bearings. Injection moulding and extrusion grades are sold with glass or carbon fibre fillers to improve rigidity. Extruded pressure tubing and hoses have important applications in the automobile industry. Precision grades are readily achieved from filled grades. 70% of Nylon 11 finds uses as powder especially for anti-corrosion coatings for metal tubing applications. Plasticised grades are prod. for flexible coatings and ability to withstand subsequent shaping e.g. coated tubes. Applications for Nylon 11 include fuel lines, hydraulic or pneumatic hoses and pumps, bearings, gaskets, housings, gears, flexible teeth on harvesting machinery and the soles of soccer shoes. A new thermoplastic hose for vehicle hydraulics is made by extrusion from improved grade Nylon 11, reinforced with braided synthetic fibre and covered by outer sheaf. It is designed to withstand 70 bar and water/MeOH fluids. Other applications include underfloor heating pipes, shrunk-on hoses and wire insulation. A new development of a Freon gas-escape proof hose, needed for car air conditioners to meet global regulations, has been developed from a laminated construction of Nylon 6/Nylon11/polyolefin for the thin inner wall to satisfy all the requirements of impermeability to Freon gas (reduced to one-tenth of conventional hose) and water with acceptable flexibility and heat resistance. A layered tubing for use in motor vehicles has been patented which incorporates a thick outer layer of Nylon 11 or Nylon 12 (0.2–0.7 mm) with four other internal layers. Nyclad-jacketed ammunition incorporates Nylon 11, the jacketing material introduced with the intention of reducing airborne lead concentrations

Trade name	Details	Supplier
CTI NH	20GF/30GF/40GF	CTI
Ertalon		Chemplast
Lubricomp	H 1000 series	LNP Engineering
MB	3000	Atochem Polymers
N8	30FG-0100	Thermofil Inc.
RTP	200 series	RTP Company
Rilsan		Rilsan Corp.
Rilsan		Elf Atochem (UK) Ltd.
Stat-Kon	H	LNP Engineering
Thermocomp	H/HF/PDX	LNP Engineering

Bibliographic References

[1] *Polyamide II*, Elf-Atochem, 1996, (technical datasheet)
[2] Van Krevelen, D.W., *Properties of Polymers: Their Correlation with Chemical Structure*, 3rd edn., 1990, 798
[3] Ash, R., Barrer, R.M. and Palmer, D.G., *Polymer*, 1970, **11**, 421, (permeability)
[4] Sorta, E. and De Chirico, A., *Polymer*, 1976, **17**, 348
[5] Champetier, G. and Pied, J.P., *Makromol. Chem.*, 1961, **44-46**, 64, (T_g)
[6] *Encycl. Polym. Sci. Technol.*, Vol. 11, 3rd edn., (ed. J.I. Kroschwitz), John Wiley and Sons, 1988, 370
[7] Longbottom, R.W., *Mod. Text.*, 1968, **49**, 19
[8] *J. Mater. Edge.*, 1992, **18(4)**, 11
[9] *K-Plast. Kautsch ZTG*, 1988, **76**, 24
[10] Miller, R.L., *Polym. Handb.*, 4th edn., (eds. J. Brandrup, E.H. Immergut and E.A. Grulke), John Wiley and Sons, 1989, **VI**, 35
[11] Mathias, L.J., Powell, D.G., Autran, J.P. and Porter, R.S., *Mater. Sci. Eng., A*, 1990, **126**, 253
[12] Mathias, L.J., Powell, D.G., Autran, J.P. and Porter, R.S., *Macromolecules*, 1990, **23**, 963
[13] Shi, Y. and White, J.L., *Int. Polym. Process*, 1990, **5**, 25
[14] Fina, L.J. and Yu, H., *Polymeric Materials Science and Engineering Fall meeting*, 1994, **71**, 7
[15] Le Huy, H.M., Huang, X. and Rault, J., *Polymer*, 1993, **34**, 340
[16] Guigon, M., Laporte, P. and Echalier, B., *Ann. Compos.*, 1989, **1-2**, 91
[17] Autran, J.P., *Diss. Abstr. Int. B*, 1991, **51**, 296
[18] Wyzgoski, M.G. and Novak, G.E., *J. Mater. Sci.*, 1987, **22**, 1701
[19] *Rev. Plast. Mod.*, 1993, **66**, 510
[20] Sallet, D., Mailhos-Lefievre, V. and Martel, B., *Polym. Degrad. Stab.*, 1990, **30**, 29
[21] Park, H.S., *Diss. Abstr. Int. B*, 1990, **50**, 264
[22] *Plast. Rubber Wkly.*, **1245**, 7
[23] *Plast. Eng. (N.Y.)*, 1995, **51**, 57
[24] *J. Mater. Edge.*, 1990, **18**, 11
[25] Satyo, H., *Kunststoffe*, 1994, **84**, 1182, 1185
[26] Stenglin, U., *Plastverarbeiter*, 1991, **42**, 56
[27] Kitami, T. and Mito, J., *New Materials and Processes for the Future*, Soc. for Advancement of Material and Process Engineering, 1989, 1403
[28] Mathis, L.J. and Johnson, C.G., *Polym. Prepr. (Am. Chem. Soc., Div. Polym. Chem.)*, 1991, **32**, 80
[29] Svoboda, M., Schneider, B. and Stokr, J., *Collect. Czech. Chem. Commun.*, 1991, **56**, 1461
[30] Rodgers, P.A., *J. Appl. Polym. Sci.*, 1993, **50**, 2075
[31] Shi, Y. and White, J.L., *Int. Polym. Process*, 1990, **5**, 25
[32] Fromageot, D., Lemaire, J. and Sallet, D., *Eur. Polym. J.*, 1990, **26**, 1321
[33] Park, H.S., *Diss. Abstr. Int. B*, 1990, **50**, 264
[34] Levchik, S.V., Costa, L. and Camino, G., *Polym. Degrad. Stab.*, 1992, **36**, 31
[35] Gosa, K., Ionescu, I.V., Predescu, A. and Isbasescu, D., *Mater. Plast. (Bucharest)*, 1991, **28**, 27
[36] Xenopoulos, A. and Wunderlich, B., *Polymer*, 1990, **31**, 1260
[37] Vollmer, F., *Plastverarbeiter*, 1989, **40**, 44, 46, 48, 54, 58
[38] Davies, P., Laporte, P., Echalier, B. and Glemet, M., *Proc. Natl. Symp. Compos., 6th*, Association pour les Materiaux composites, 1988, 301
[39] Chruma, J.L., *Plastics in Automotive Applications: an Overview*, 1983, 71
[40] Korsgaard, J., *Offshore and Arctic Operations*, 1993, 139
[41] *Diss. Abstr. Int. B*, Jan., 1995, **55**, 127
[42] Coleman, M.M. and Painter, P.C., *Encycl. Polym. Sci. Eng.*, 1990, 371
[43] *U.S. Pat.*, 1991, 5 076 329
[44] Kapoor, A. and Bahadur, S., *Tribol. Int.*, 1994, **27**, 323
[45] Bahadur, S., Fu, Q. and Gong, D., *Wear*, 1994, **178**, 123
[46] Bahadur, S. and Gong, D., *Wear Mater.*, 1991, **1**, 177
[47] Bahadur, S. and Gong, D., *Wear*, 1992, **155**, 49
[48] Bahadur, S. and Gong, D., *Wear*, 1992, **154**, 207
[49] Bahadur, S., Gong, D. and Anderegg, J.W., *Wear*, 1993, **165**, 205
[50] Ramasay, A., Wang, Y. and Muzzy, J., *Advanced Materials, Performance Through Technology Insertion*, (eds. V. Bailey, G.C. Janicki and T. Haulik), Anaheim, 1993, **38**, 1882, (application)
[51] Wagner, P. and Colton, J., *Moving Forward with 50 years of Leadership in Advanced Materials*, Soc. for Adv. of Material Process Eng., 1994, **39**, 1536, (application)
[52] Ramai, K., Borgaonkar, H. and Hoyle, C., *Adv. Compos. 10th*, ASM International, 1994, 229, (application)
[53] Poteta, R.M. and Lundblad, W.E., *Materials Challenge: Diversification and the Future*, 1995, **40**, 1058, (application)
[54] *Jpn. Pat.*, 1994, 215 809, (application)
[55] *Jpn. Pat.*, 1995, 203 271, (Tokai Rubber)
[56] *Jpn. Pat.*, 1995, 203 269, (Tokai Rubber)
[57] *Jpn. Pat.*, 1994, 249 995
[58] *Jpn. Pat.*, 1994, 218 735, (Teijin)
[59] Zona, C.A., *Microscope*, 1996, **44**, 11
[60] Handa, T., Noji, F. and Takazawa, H., *Toso Kogaku*, 1996, **31**, 105
[61] *U.S. Pat.*, 1994, 298 076, (application)
[62] *Can. Pat.*, 1996, 2 129 173
[63] *Eur. Pat. Appl.*, 1996, 699 637, (application)
[64] *PCT Int. Appl.*, 1995, 95 33 796, (Elf-Atochem)
[65] *Eur. Pat. Appl.*, 1995, 686 658, (Elf-Atochem)
[66] Mei, B., Scheinbeim, J. and Newman, B., *Ferroelectrics*, 1995, **171**, 177
[67] Mehrabzadeh, M. and Burford, R., *Iran. J. Polym. Sci. Technol. (Engl. Ed.)*, 1995, **4**, 156
[68] Luciani, A. and Jarrin, J., *Polym. Eng. Sci.*, 1996, **36**, 1619
[69] Brown, L.F., Scheinbeim, J.I. and Newman, B.A., *Ferroelectrics*, 1995, **171**, 321

Nylon 12

Synonyms: Poly(ω-dodecanolactam). Poly(12-aminododecanoic acid). Poly[imino(1-oxo-1,12-dodecanediyl)]. PA12
Related Polymers: Nylon
Monomers: ω-Dodecanolactam, 12-Aminododecanoic acid
Material class: Thermoplastic
Polymer Type: polyamide
CAS Number:

CAS Reg. No.
24937-16-4

Molecular Formula: $C_{12}H_{23}NO$
Fragments: $C_{12}H_{23}NO$
Additives: Fillers increase rigidity under load and reduce mould shrinkage. Glass/carbon fibres give rigidity and temp. resistance; conductive fillers (carbon black) added for anti-electrostatic props.; flame retardants and impact modifiers (ABS resins) also used
General Information: Amide groups may be detected by ir or nmr spectroscopy. Nylon 12 may be hydrolysed by acid to the monomer 12-aminododecanoic acid. Complete identification of monomers and ratios present may be carried out by nmr. Nylon 12 differs from Nylon 6 and Nylon 6,6 in that it possesses high dimensional stability over a wide range of humidity conditions. Low water absorption gives rise to substantially constant mech. and electrical props. It exhibits excellent toughness at low temps. combined with good stress crack resistance. These props. make it particularly suitable for tube and hose applications in the automobile industry, and for precision injection mouldings
Morphology: Crystallinity 26% [25]. Hexagonal, a 4.8, b 4.8, c 232.1 [27]. Pseudohexagonal, a 4.79, b 31.9, c 9.58, β 120° [28]. Monoclinic, a 4.90, b 4.67, c 32.1, γ 121.7°. [29] Oriented lamallae form large anisotropic domains on highly oriented substrates of syndiotactic polypropylene [56]

Nylon 12

Miscellaneous: Influence of chain struct. and amount of epoxide silane on morphology and crystallite size of cross-linked Nylon 12 has been reported. Cryst., amorph. and transition regions are observed. Lamellar thickness (5–7 nm) is independent of chain struct. or cross-linking agent. Lamellar length is >100 nm for linear Nylon 12 but reduces to block-like struct. when highly cross-linked. Transition region *ca.* 2 nm [10]. Worldwide market of 40000 ton per annum (1992) [19]

Volumetric & Calorimetric Properties:

Density:

No.	Value	Note
1	d 1.01 g/cm^3	[1,38]
2	d 1.04 g/cm^3	cryst. [36]
3	d 1.034 g/cm^3	monoclinic cryst. [28]
4	d 0.84 g/cm^3	melt density [34]

Thermal Expansion Coefficient:

No.	Value	Note	Type
1	0.00011 K^{-1}	20°, cryst.	L
2	0.0002–0.00023 K^{-1}	100°, cryst. [34]	L
3	0.00029 K^{-1}	20°, cryst.	V
4	9.9 × 10^{-5} K^{-1}	23°, 50% relative humidity [50,51,52]	L
5	0.00015–0.00016 K^{-1}	20–80°, longitudinal	V
6	0.00011–0.00012 K^{-1}	23–80°, transverse [38]	V

Equation of State: Spencer-Gilmore equation of state parameters have been reported [32]
Latent Heat Crystallization: Flow activation energy E_o 60 kJ mol^{-1} [30]. Enthalpy ΔH 80 kJ kg^{-1} (60°); 165 kJ kg^{-1} (100°); 800 kJ kg^{-1} (300°) [31]. ΔH_{fusion} 75 kJ kg^{-1} (amorph., annealed 8h, 50°) [39]. Gruneisen parameters have been reported [33]
Thermal Conductivity:

No.	Value	Note
1	0.25 W/mK	mouldings [34]
2	0.19 W/mK	[50,52]

Specific Heat Capacity:

No.	Value	Note	Type
1	1.26 kJ/kg.C	[50,51,52]	P

Melting Temperature:

No.	Value	Note
1	178–180°C	[1]
2	184°C	γ cryst. [2]
3	177°C	γ-form [2]
4	169°C	α-form [2]
5	175°C	ISO 314, method C [36]
6	187°C	equilibrium [36]

Glass-Transition Temperature:

No.	Value	Note
1	50°C	γ-form
2	45°C	γ-form
3	40°C	α-form [2]
4	43–54°C	MW 4000-12000 [8]
5	40–43°C	[25,26]

Deflection Temperature:

No.	Value	Note
1	80–110°C	0.45 MPa, ISO 75 [38]
2	40–50°C	1.8 MPa, ISO 75 [38]
3	150°C	dry, 0.45 MPa [34]
4	55°C	1.8 MPa, ISO R75 [34]

Vicat Softening Point:

No.	Value	Note
1	130–140°C	50N, ISO 306
2	170°C	10N, ISO 306

Surface Properties & Solubility:

Plasticisers: *p*-Toluenesulfonamide. Plasticised grades of Nylon 12 are available commercially for injection moulding and extrusion applications [18]
Solvents/Non-solvents: Insol. MeOH, EtOH, petrol. Absorbs these solvents, MeOH 8.5% (1000h, 20°) [1], petrol 1.4% (1000h, 20°) [1]
Surface Tension:

No.	Value	Note
1	31 mN/m	[42]
2	25 mN/m	melt [40]

Transport Properties:

Transport Properties General: Nylon 12 is the best moisture barrier and poorest gas barrier of the commercial nylons
Melt Flow Index:

No.	Value	Note
1	4 g/10 min	230°, 0.02% water content [55]
2	11 g/10 min	230°, 0.1% water content [55]
3	20 g/10 min	230°, 0.26% water content [55]

Polymer Melts: Melt viscosity 570 Pa s (270°) [34], 330 Pa s (270°, 0.1% moist) [34]
Permeability of Liquids: [H$_2$O] 9 (0.05 μm, 65% relative humidity); 18 (flat film, 100 μm); 14 cm^3 (m^2 day atm)$^{-1}$ (blown film, 100μm, 38°, 90% relative humidity, BS 1133) [55]
Permeability and Diffusion General: Good aroma and grease impermeability. Nylon 12 has poor fuel diffusion and dimensional stability in fuel lines, which are overcome using suitable barrier materials
Water Content: 0.26%; 0.06% (after 32h drying at 80°) [55]

Nylon 12

Water Absorption:

No.	Value	Note
1	0.85 %	65% relative humidity, 20° [1]
2	1.5 %	1000h, 20° [1]
3	1.6–1.8 %	23°, saturated [38]
4	1.3 %	65% relative humidity [50]
5	2.7 %	99% relative humidity

Gas Permeability:

No.	Gas	Value	Note
1	O_2	0.03 cm^3 mm/(m^2 day atm)	300 cm^3 $(m^2$ day atm$)^{-1}$, 100 µm flat film [1]
2	O_2	0.02 cm^3 mm/(m^2 day atm)	200 cm^3 $(m^2$ day atm$)^{-1}$, 100 µm blown film, ASTM D1434 [1]
3	O_2	15 cm^3 mm/(m^2 day atm)	300 cm^3 $(m^2$ day atm$)^{-1}$, 0.05 mm, 65% relative humidity [1]
4	N_2	0.007 cm^3 mm/(m^2 day atm)	70 cm^3 $(m^2$ day atm$)^{-1}$, 100 µm flat film [55]
5	N_2	0.005 cm^3 mm/(m^2 day atm)	50 cm^3 $(m^2$ day atm$)^{-1}$, 100 µm blown film [55]
6	CO_2	0.13 cm^3 mm/(m^2 day atm)	1300 cm^3 $(m^2$ day atm$)^{-1}$, 100 µm flat film [55]
7	CO_2	0.06 cm^3 mm/(m^2 day atm)	600 cm^3 $(m^2$ day atm$)^{-1}$, 100 µm blown film, ASTM D1434 [55]

Mechanical Properties:

Mechanical Properties General: A general expression to determine the strength props. of polyamides has been derived [14]. Compressive modulus values have been reported [34]. Tenacity (of fibre) 0.22-0.4 N tex^{-1}. [53] Elongation at yield 8% [34]

Tensile (Young's) Modulus:

No.	Value	Note
1	1400 MPa	23°, dry [34]
2	1200 MPa	23°, moist, ISO 1110 [34]
3	200 MPa	100°, dry [34]
4	1350–1450 MPa	ISO 527, 1 mm min^{-1} [38]

Flexural Modulus:

No.	Value	Note
1	1209 MPa	
2	393 MPa	plasticised [18]
3	1100 MPa	[53]

Tensile Strength Break:

No.	Value	Note	Elongation
1	63.7 MPa	[1]	>50%
2	50 MPa	23°, dry [34]	
3	30 MPa	23°, 0.9% water [34]	
4	17 MPa	100°, dry [34]	

Flexural Strength at Break:

No.	Value	Note
1	47 MPa	ISO 178 [53]

Elastic Modulus:

No.	Value	Note
1	1172 MPa	[34]

Poisson's Ratio:

No.	Value	Note
1	0.3–0.4	20°, moulding [34]

Tensile Strength Yield:

No.	Value	Elongation	Note
1	45 MPa	5%	[38]

Flexural Strength Yield:

No.	Value	Note
1	48–64 MPa	20°, 65% relative humidity [55]

Miscellaneous Moduli:

No.	Value	Note	Type
1	4000 MPa	cryst. rods, dry [34]	Bulk modulus
2	500 MPa	dry, 23°	Shear modulus
3	100 MPa	dry, 23°	Shear modulus
4	300 MPa	ISO 899, 23°, 50% relative humidity [34]	Creep modulus
5	450 MPa	DIN 52448-E [38]	Tensile creep modulus

Impact Strength: Impact strength 122.5–205 Nm (notched) [1]
Mechanical Properties Miscellaneous: Mech. loss modulus tan δ 0.01 (20°); 0.048 (40°); 0.16 (60°); 0.08 (80°); 0.048 (100°); (0.1–10 Hz) [34]
Hardness: Ball indentation hardness 63.7 MPa [1], 98 MPa (dry, 23°, ISO 2039) [34]. Rockwell R110 [34]
Friction Abrasion and Resistance: Roughness height 0.1 µm, wear resistance 0.2 µm km^{-1}, coefficient of friction 0.55 (dry). Roughness height 1 µm, wear resistance 0.6 µm km^{-1}, coefficient of friction 0.38 (dry). Roughness height 2 µm, wear resistance 1 µm km^{-1}, coefficient of friction 0.35 (dry). Roughness height 4 µm, wear resistance 2.5 µm km^{-1}, coefficient of friction 0.34 (dry). Roughness height 6 µm, wear resistance 3.9 µm km^{-1}, coefficient of friction 0.4 (dry) [35]

Izod Notch:

No.	Value	Notch	Break	Note
1	200–501 J/m	N		dry, ISO 180/1A [34]
2		N	Y	23°, ISO 180/1C [38]

Izod Area:

No.	Value	Notch	Note
1	7–30 kJ/m^2	Notched	23°, ISO 180/1C [38]
2	6–7 kJ/m^2	unnotched	-30°, ISO 180/1C [38]

Electrical Properties:

Electrical Properties General: Nylon 12 is a good insulator, much better than other commercial semi-crystalline nylons such as Nylon 6 or Nylon 6,6, but a long way short of polyethylene. Dielectric loss factor is half that of Nylon 6 and specific resistivity is two orders of magnitude better - a similar magnitude short of polyethylene. In general all electrical props. lie between Nylon 6 and polyethylene [1]. Nylon 12 conductive filler composites have been studied for effective electromagnetic shielding. Surface treatment of aluminium fibres can be very effective [15]. Electrical props. are unaffected by moisture

Surface/Volume Resistance:

No.	Value	Note	Type
1	0.1×10^{15} Ω.cm	[1]	S
3	0.1×10^{15} Ω.cm	VDE 0303 T30 [38]	S

Dielectric Permittivity (Constant):

No.	Value	Frequency	Note
1	30	100 Hz	-30°
2	3.1	100 Hz	0°
3	4	100 Hz	30°
4	11.5	100 Hz	60°
5	20	100 Hz	90°
6	2.9	100 kHz	-30°
7	3	100 kHz	0°
8	3.1	100 kHz	30°
9	4.5	100 kHz	60°
10	10	100 kHz	90° [37]
11	3.6–3.9	50 Hz	23°
12	2.2–3	1 MHz	23° [38]

Dielectric Strength:

No.	Value	Note
1	90 kV.mm^{-1}	VDE 0303, pt. 2; IEC 243 electrode K20/50, dry [34]
2	25 kV.mm^{-1}	moist [34]
3	27–29 kV.mm^{-1}	VDE 0303, T21; electrode K20/P50 [38]

Complex Permittivity and Electroactive Polymers: Comparative tracking index (ISO 112) has been reported [38]. Tracking resistance CT1 600 [34]

Magnetic Properties: Nylon 12 filled with ferrite magnetic powder produces effective anisotropic magnets, in which orientation is determined by the matrix interacting with the magnetic powder. Magnetic props. are thermally stable [16]

Dissipation (Power) Factor:

No.	Value	Frequency	Note
1	0.05	100 kHz	[1]
2	0.04–0.044	50 Hz	[38]
3	0.025–0.028	1 MHz	[38]

Polymer Stability:

Polymer Stability General: Mouldings have good dimensional stability due to low moisture uptake and, consequently, mech. props. are much more stable. Notched impact is not adversely affected by dryness at low temps., unlike Nylon 6. Like polyethylene, Nylon 12 is a good barrier to moisture but a poor barrier to gas. It has good overall chemical stability, but is still sensitive to hydrol. in the melt. It is a good insulator and its electrical props. are not affected by moisture

Thermal Stability General: Composites with Al or Cu fillers are especially good with a Nylon 12 matrix, exhibiting high thermal stability of the shielding function [4]. Thermal stability is improved by reaction of bisoxasolones with amine terminated Nylon 12 [5]. Embrittlement occurs after 24h at 140° (standard grade), 750h (protected grade)

Upper Use Temperature:

No.	Value	Note
1	45–59°C	DIN 53462 [55]
2	140–160°C	heat and weather resist [55]

Decomposition Details: The main products of thermal degradation are α unsaturated nitriles, plus dodecanolactam, hydrocarbons, CO and CO_2. When ammonium polyphosphate (APP) is employed as a fire retardant then major degradation products are α and β unsaturated nitriles [3]

Flammability: Limiting oxygen index 26% (ASTM D2863, dry). Ammonium polyphosphate is used as a fire retardant, suppression through intumescent char formation

Environmental Stress: Uv exposure causes significant loss of strength in 200h, cracking is observed after ≥1000h [6]. Cast Nylon 12 has superior resistance to chemical, thermal and mech. stresses due to its higher crystallinity and MW [13]

Chemical Stability: Stable to oils, greases, petrol, most hydrocarbons, organic acids including formic acid. Stable to oxidation up to 60° [53]

Hydrolytic Stability: Unaffected by 3% NaCl over 30 d, but some loss of strength occurs after 60 d. Despite low water uptake, it is sensitive to hydrol.. 0.1% Water content is sufficient to reduce melt viscosity from 570–330 Pa s in 20 min. at 270°. Stable to many aq. salt solns., e.g. $CaCl_2$, $CuSO_4$, Na_2CO_3; sea water and dilute mineral acid

Biological Stability: Resistant to formic acid prod. by termites

Stability Miscellaneous: Amorph. Nylon 12 used for sight glasses shows long term resistance to yellowing and uv radiation. Cracks and signs of embrittlement do not occur until after 25 months of continuous outdoor exposure; this compares with only 7 months for other nylons [11]

Applications/Commercial Products:

Processing & Manufacturing Routes: Carothers' original patent [22] first indicated the means of preparation of Nylon 12 and industrial production was first considered after the development of a method for preparing cyclododecatriene [23]. Polymerisation of Nylon 12 is similar to Nylon 6 and is based on high pressure hydrolytic ring opening of laurolactam using acid catalyst and temps. of 300–350° in an autoclave reactor, polymer being extruded into cold water. Unlike Nylon 6, only small levels (<0.5%) of unreacted monomer remain. Alternatively, anionic polymerisation may be carried out in the absence of water using alkali/alkaline earth metals and co-catalysts [1]. The polymerisation of nylons, including 12-aminododecanoic acid [43,44,45,46], can be effected at 270–280° in the presence of a suitable catalyst, such as ammonium or manganese hypophosphite, in a suitable reactor or autoclave to prod. material with relative viscosity 0.7. This low MW prepolymer may be pelletted and fed to a twin screw reactor to be converted to higher MW grades up to relative viscosity 1.4, suitable for injection moulding and extrusion. Lower MW grades may be powdered and used for powder coating applications [48]. The microwave polymerisation of 12-aminododecanoic acid to Nylon 12 has been reported [49]. A new form of Nylon 12 is manufactured by a non-pressurised casting process and can be cast around metal components to provide gears, pinions, propellors etc. It is claimed to have superior creep props. compared to other available nylons, and retains excellent mech. props. up to 120° [12]. Lauramid PA12G is made by this process displaying higher crystallinity and MW than conventional Nylon 12, with advantages in chemical, thermal and mech. props. [13]

Mould Shrinkage (%):

No.	Value	Note
1	1–1.5%	longitudinal
2	0.6–1.4%	transverse

Applications: Injection moulded samples of impact modified conductive Nylon 12 have superior aesthetic qualities, lower warpage and water absorption than Nylon 6,6, and is therefore chosen for electrostatic discharge applications. Plasticised grades are injection moulded or extruded. Applications (extruded) include tubing for the agricultural industry, cable sheathing and pipe for the gas transmission and telephone industries; injection moulded applications include gas filter housings, bearings and gears. Nylon 12 is available in moulding and extrusion grades (granules) and fine powders for powder coating and rotomoulding applications. Major applications are in the automotive, offshore oil, aerospace and most high technology industries. Injection moulded applications include (unfilled): fuel filters, gear wheels, guide rail; (reinforced) gear housings, breathing equipment, centrifugal separators, sports equipment. There are also a wide variety of extruded applications: wire insulation, cable sheathing, fabrics, filters, packaging films, vacuum hoses, window frames, semi-rigid tubing for vacuum, fuel, pneumatics. Nylon 12 is resistant to formic acid, choice material for underground cable coatings, as they resist the action of termites. Fine powder grades are employed in rotomoulding in the temp. range 230–280° and for the formation of anti-corrosion coating of metallic substrates. Coatings are applied by a metal substrate heating process, the most common methods being fluidised bed, electrostatic and flame spraying. Continuous coating methods have been developed to coat welded tubes up to 30 mm diameter with circular, oval or square profiles. Thicknesses of 100 microns can be applied at speeds of 100 m min^{-1}. This type of tubing is employed for central heating installations, garden tools and fencing, etc. Nylon 12 coatings have excellent resistance to corrosion, good impact resistance and electrical insulation, low surface friction, long life and good hygienic props., and no volatile toxic emissions during application. Nylon 12 may be used as part of a multipolymer tubing system, e.g. fuel line hose constructed of five layers with an outer layer of Nylon 12, intermed. and inner layers of Nylon 6 with polyolefin bonding layers in between. Radiation cross-linking technology, long established for the manufacture of PE underfloor heating pipes, is now being applied to moulded reinforced Nylon 12. Also used as an adhesive

Trade name	Details	Supplier
Amilan		Toray
Ashlene 929	Plasticised grades	Ashley Polymers
CTI	AN/NJ	CTI
DC-1015/SC14F-3090		Spartech
Grilamid		EMS-Chemie AG
Lubricomp	SCL/SFL/SL	LNP Engineering
N9-30FG-0100		Thermofil Inc.
Permastat	200F	RTP Company
RTP	205F/207F	RTP Company
Rilsan A		Elf Atochem (UK) Ltd.
Rimplast	PTA 1201/1202	Huls America
Thermocomp	S/SC/SF	LNP Engineering
UBE	various grades	UBE Industries Inc.
Ube		Ube Industries Ltd.
Vestamid		Hüls AG

Bibliographic References

[1] Ruestem, W.G.D., *Ind. Eng. Chem.*, 1970, **62**
[2] Mathias, L.J. and Johnson, C.G., *Macromolecules*, 1991, **24**, 6114
[3] Levch, K., Costa, L. and Camino, G., *Polym. Degrad. Stab.*, 1992, **36**, 31
[4] Osawa, Z. and Kuwabara, S., *Polym. Degrad. Stab.*, 1992, **35**, 33
[5] Acevedo, M. and Fradet, A., *J. Polym. Sci., Part A: Polym. Chem.*, 1993, **31**, 817
[6] Hara, N., Nitra, A., Tsutsumoko, T. and Noji, H., *Bulletin - Ind. Res. Inst. (Djakarta)*, 1993, **36**, 20
[7] Gupta, G. and Smith, B.A.K., *Annu. Tech. Conf. - Soc. Plast. Eng., 50th*, 1992, **1**, 3
[8] Faruque, H.S. and Lacabanne, C., *Indian J. Pure Appl. Phys.*, 1987, **25**, 114
[9] Ogawa, T. and Sakai, M., *J. Polym. Sci., Part A: Polym. Chem.*, 1988, **26**, 3141
[10] Boder, H.G. et al, *J. Mater. Sci.*, 1992, **27**, 4726
[11] Friedman, M., *J. Plast. Des. Forum*, 1987, **12**, 92
[12] *J. Material Manuf.*, Aug., 1986, 15
[13] *J. Plast. Rubber Int.*, 1986, **11**(3), 5
[14] Ivantina, I.V., Kuznetsova, I.G. and Kovriga, V.V., *J. Mech. Compos. Mater. (USSR)*, **24**, 275
[15] Osawa, Z. and Yamanaka, S., *J. Mater. Sci. Lett.*, 1988, **7**, 983
[16] Osawa, Z., Kawouchi, K., Iwata, M. and Haruda, H., *J. Mater. Sci.*, 1988, **23**, 2637
[17] Brunnhofer, E. and Egen, U., *Kunststoffe*, 1988, **78**, 407
[18] *J. Plast. Des.*, Nylon 12 resins, 1994, **19**(8), 47
[19] *J. New Mater. Japan*, 1992, **3**, 14
[20] *J. Plast. Week.*, Feb 3, 1992, 1
[21] *Plastics News (Detroit)*, 1990, **2**(14), 5
[22] U.S. Pat., 1935, 2 071 253, (Carothers WH)
[23] Wilka, G., *Angew. Chem., Int. Ed. Engl.*, 1957, **69**, 397
[24] Endo, R., *Nippon Gomu Kyokaishi*, 1961, **34**, 527, (viscosity)
[25] Sorta, E. and De Chirico, A., *Polymer*, 1976, **17**, 348, (T_g)
[26] Gordon, G.A., *J. Polym. Sci., Part A-2*, 1971, **9**, 1693
[27] Frunze, T.M., Cherdabayev, A.Sh., Schleifman, R.B., Kurashev, V.V. and Tsvankin, D.Ya., *Vysokomol. Soedin., Ser. A*, 1976, **18**, 696
[28] Cojazzi, G., Fichera, A., Garbuglio, C., Malta, V. and Zanetti, R., *Makromol. Chem.*, 1973, **168**, 289
[29] Owen, A.J. and Kollross, P., *Polym. Commun.*, 1983, **24**, 303
[30] Kunze, H., Dissertation T.H. Aachen, 1958
[31] *Polym. Handb.*, 3rd edn., (eds. J. Brandrup and E.H. Immergut), 1989, V/III
[32] Hafner, F. and Mietzner, Th., *BASF Abt. Informatik*, BASF, 1986
[33] Warfield, R.W., Kayser, E.G. and Hartmann, B., *Makromol. Chem.*, 1983, **184**, 1927
[34] Pflüger, R., *Polym. Handb.*, 4th edn., (eds. J. Brandrup, E.H. Immergut and E.A. Grulke), John Wiley and Sons, 1999, V, 121
[35] Erhard, G. and Strickle, E., *Kunststoffe*, 1972, **2**, 232, 282
[36] Goldbach, G., *Angew. Makromol. Chem.*, 1973, **32**, 37
[37] Bergmann, K., *Konstr. Kunstst.*, (ed. G. Schreyer), Hanser, 1972
[38] Engineering for New Ideas Hüls, 1995
[39] Gogolewski, S., Gzerriawska, K. and Gasiorek, M., *Colloid Polym. Sci.*, 1980, **258**, 1130
[40] Hybart, F., *J. Appl. Polym. Sci.*, 1960, **3**, 118
[41] Owens, D.K. and Nendt, R.C., *J. Appl. Polym. Sci.*, 1969, **13**, 1741
[42] Tuzar, Z., Bohdanecky, R. et al, *Eur. Polym. J.*, 1975, **11**, 851
[43] *Brit. Pat.*, 1967, 1 177 154, (BP)
[44] *Brit. Pat.*, 1967, 1 187 419, (BP)
[45] *Brit. Pat.*, 1966, 1 198 423, (BP)
[46] *Brit. Pat.*, 1967, 1 198 422, (BP)
[47] Guide to Properties and Uses of N12 Hüls,(technical datasheet)
[48] *Brit. Pat.*, 1969, 51 294, (BP)
[49] *Brit. Pat.*, 1978, 1 534 151, (BP)
[50] Encycl. Polym. Sci. Technol. John Wiley and Sons, 1986, **11**, 368
[51] *Nylon Plastics*, (ed. M.I. Kohan), John Wiley and Sons, 1973
[52] *Plast. Technol. Manuf. Handbook Buyers Guide*, June, 1983, **29**, 386
[53] Longbottom, R.W., *Mod. Text.*, 1968, **49**, 19
[54] Rilsan A (nylon 12) technical data Elf Atochem U.K., 1996, (technical datasheet)
[55] *Vestamid*, Hüls, May, 1969
[56] Yan, S., Petermann, J. and Yang, D., *Polymer*, 1996, **37**, 2681, (morphology)

Nylon 13

Synonyms: *Poly(imino-1-oxotridecamethylene). Poly(13-aminotridecanoic acid)*
Monomers: 13-Aminotridecanoic acid
Material class: Thermoplastic
Polymer Type: polyamide
Molecular Formula: $(C_{13}H_{25}NO)_n$
Fragments: $C_{13}H_{25}NO$
Morphology: Unit cell a 0.970, b 0.955, c 1.722; α 58.2°, β 90.0°, γ 70.0° [2]

Volumetric & Calorimetric Properties:
Density:

No.	Value	Note
1	d^{39} 1.018 g/cm^3	[1]

Melting Temperature:

No.	Value	Note
1	182–183°C	[1]

Glass-Transition Temperature:

No.	Value	Note
1	41°C	[1]

Transition Temperature:

No.	Value	Note	Type
1	41°C	[1]	Second-order transition temp.

Bibliographic References

[1] Chapetier, G. and Pied, J.P., *Makromol. Chem.*, 1961, **44-46**, 64, (T_g)
[2] Moñoz-Guerra, S., Prieto. A, Montserat, J.M. and Sekiguchi, H., *J. Mater. Sci.*, 1992, **27**, 89, (cryst struct)

Nylon Acrylic Blends N-41

Related Polymers: Nylon 6, ABS, SAN, Poly(ethylene-*co*-acrylic acid), Poly acrylic acid
Material class: Blends
General Information: Nylon-ABS blends. The preparation of true alloys of acrylic amorph. plastics with semi-crystalline nylons produces materials with much improved impact and toughness down to quite low temps. Alloys represent a unique combination of amorph. and cryst. thermoplastic materials with an excellent balance of props., especially toughness, chemical resistance, dimensional stability and mouldability. The alloy has high-impact strength and resistance to chemical cleaning solns. [2], adopted for snowmobile throttles and brake handles and for automotive roof racks [3]. Toughness of nylon/ABS blends measured by J-integral method have been reported. [4] ABS/Nylon 6/SMA (styrene-maleic anhydride copolymer) ternary blends (grafted SMA-Nylon 6 is formed during blending and acts as an effected compatibiliser) are formed when ABS is blended with a 90/10 preblend of SMA/Nylon 6. The graft copolymer formed during the preblending seems to be mainly within the Nylon 6 domain and enhances its modulus and thermal resistance [5]. ABS/Nylon 6 blends can be produced without compatibilisers but props. are inferior to those of the components. The addition of a maleinised polymer acts as compatibiliser by reacting chemically with the Nylon 6 during extrusion to form a graft copolymer which promotes interphase adhesion. The effect of compatibiliser and nylon concentrations on morphology and improved toughness has been reported [6]. The most recent ABS/Nylon alloy to be marketed is Stapron N from DSM, which is a true alloy of ABS and Nylon 6. Within a broad range of blend ratios, very high levels of impact strength are claimed. By using a suitable compatibiliser, a very fine dispersion has been created, which is claimed to impart the high toughness, excellent surface finish and reproduction of mould texture obtained with these alloys. Applications which have been commercialised are those in which the ability to produce a deep matt finish is important. Standard blends are employed for applications such as instrument surrounds, arm rests, air vents, demister grilles, chair seats, lawnmower hoods. Improved grades are used for more demanding applications like car bumpers [7]. Nylon-Acrylic Blends (general). The dispersion of brittle polymer into a ductile polymer matrix followed by chemical grafting is a method frequently employed to obtain toughened plastics. The dispersion of styrene-acrylonitrile (SAN) copolymer into Nylon 6 in the presence of a small amount of styrene-maleic anhydride (SMA) copolymer produces a much finer dispersion of SAN particles, due to the formation of nylon-SMA graft copolymer which acts as a compatibiliser for the two phases. Significant improvements in tensile and impact props. are attributed to the compatibiliser ability to induce a brittle-ductile transition in SAN particles [8]. Nylon 6 blends with poly(ethylene-*co*-acrylic acid) (PEA) containing low levels of acid groups have been prepared by melt processing. Changes in the crystallisation behaviour of the blends indicate that the two polymers are intimately mixed. Fractionation results suggested that the two polymers react at elevated temps. to produce small amounts of copolymer which is believed to act as a compatibilising agent [9]. Blends with polyacrylic acid (PAA) have also been investigated. These are prepared as films cast from formic acid soln. Miscibility and composition of phases have been reported. Crystallinity is observed below 50 wt.% PAA. In the non-crystalline state, the polymers are fully miscible. A max. blend T_g was observed at 25 wt.% PAA [10]. Acrylic core/shell modifiers for reinforced polyamides improve impact resistance. Particular uses envisaged are automotive and tool housings as well as sporting gear [11]. High MW Nylon 6 modified grades using blends with rubber impact modifiers and acrylic-imide viscosity enhancers are suitable for blow moulding.

Mechanical Properties:
Tensile Strength Break:

No.	Value	Note
1	32 MPa	330 kg cm^{-2}, NX-50D
2	38 MPa	nylon/acrylic rubber alloy

Izod Notch:

No.	Value	Notch	Note
1	856 J/m	N	acrylic modified grade
2	53 J/m	N	standard grade

Applications/Commercial Products:

Trade name	Details	Supplier
Elemid	ABS/Nylon 6 alloy	Borg Warner
NX-50D		Monsanto Chemical
Stapron N		DSM
Triax 1120		Monsanto Chemical

Bibliographic References

[1] Owe, D.V. and Kelley, P.D., *ANTEC*, Technomic Publishing Co., Pennsylvania, 1986
[2] *Plast. Rubber Wkly.*, Monsanto, **1254**, 14
[3] *Plast. Rubber Wkly.*, Monsanto, **1354**, 12
[4] Huang, D.D., Adv. Fract. Res., *Proc. Int. Conf. Fract.*, Pergamon, 1989, **4**
[5] Kim, B.K., Lee, Y.M. and Jeong, H.M., *Polymer*, 1993, **34**, 2075

[6] Grmela, V. and Konecny, D., *Int. Polym. Sci. Technol.*, 1995, **22**, T64
[7] *Br. Plast. Rubber*, Sept., 1996, 48
[8] Angola, J.C., Fujita, Y., Sakai, T. and Inoue, T., *J. Polym. Sci., Part B: Polym. Phys.*, April, 1988, **26**, 807
[9] Yoon, K.J., *Diss. Abstr. Int. B*, 1989, **50**, 214
[10] Nishio, Y., Suzuki, H. and Morisaki, K., *Polym. Int.*, 1993, **31**, 15
[11] *Plast. Rubber Wkly.*, **1360**, 1
[12] Gaymans, R.J., Van der Werff, J.W., *Polymer*, 1994, **35**, 3658

Nylon 6,6 Aramid fibre filled   N-42

Synonyms: *Poly(hexamethylenediamine-co-adipic acid). Poly(hexamethylene adipamide). Poly(iminohexamethylene iminoadipoyl). Poly(imino(1,6-dioxo-1,6-hexanediyl)imino-1,6-hexanediyl)*
Related Polymers: Nylon 6,6, Nylon 6,6 Carbon fibre filled, Nylon 6,6, Glass reinforced, Nylon 6,6 Mineral filled, Nylon 6,6 Stainless steel fibre filled
Material class: Thermoplastic
Polymer Type: polyamide
Molecular Formula: $(C_{12}H_{22}N_2O_2)n$
Fragments: $C_{12}H_{22}N_2O_2$
Additives: PTFE, silicone

Volumetric & Calorimetric Properties:
Density:

No.	Value	Note
1	d 1.17–1.19 g/cm^3	(ASTM D792, 15–20% filler) [1]

Deflection Temperature:

No.	Value	Note
1	149°C	ASTM D648, at 1820 kPa, 15–20% filler

Mechanical Properties:
Tensile (Young's) Modulus:

No.	Value	Note
1	4137–4826 MPa	ASTM D638, 15–20% filler [1]

Flexural Modulus:

No.	Value	Note
1	4137 MPa	ASTM D790, 15–20% filler [1]

Tensile Strength Break:

No.	Value	Note	Elongation
1	90–93 MPa	ASTM D638, 15–20% filler [1]	4–8%

Flexural Strength Yield:

No.	Value	Note
1	138 MPa	ASTM D790, 15–20% filler [1]

Izod Notch:

No.	Value	Notch	Note
1	37–43 J/m	Y	ASTM D256, 3.2mm section, 15–20% filler [1]
2	534–561 J/m	N	ASTM D256, 3.2mm section, 15–20% filler [1]

Electrical Properties:
Electrical Properties General: Volume resistivity $<1 \times 10^{14} - <1 \times 10^{16}$ Ω.cm (ASTM D257, 15–20% filler) [1]

Polymer Stability:
Flammability: HB at 1.5mm (ASTM D635) [1]

Applications/Commercial Products:
Processing & Manufacturing Routes: Processing for injection molding [1]:
Injection pressure 69–124Mpa
Melt temp. 277–299°C
Mold temp. 66–107°C
Drying 4 hrs at 79°C
Moisture content 0.2%
Dew point -18°C

Applications/Commercial Products:
Mould Shrinkage (%):

No.	Value	Note
1	1–2%	ASTM D955, 15–20% filler [1]

Trade name	Supplier
Celstran	Ticona
Lubricomp	LNP (GE Plastics)
RTP Series	RTP Co.
	PolyOne Edgetek

Bibliographic References
[1] RTP Co. datasheet http://www.rtpcompany.com, 2006

Nylon Barrier blends   N-43
Related Polymers: Nylon 6, Nylon 12, Nylon 6,12
General Information: The three main barrier polymers are polyvinylidene chloride, ethylene vinyl alcohol copolymers and polyacrylonitrile copolymers. Although Nylon 6, MXD-6 and many amorph. polyamides have been reported to have excellent barrier props. they have found limited opportunites in the barrier and packaging markets due to expense. There have been a number of attempts to blend Nylons with other materials to find a more economic means of utilising their intrinsically useful barrier props. The dispersion of small amounts of a polymer with good permeability resistance to gases, in a matrix of a polymer which is generally quite permeable, can significantly enhance the impermeability of the latter. This can be achieved if the dispersed phase is distributed in effectively parallel and large thin lamellae, rather than the normal homogeneous particle dispersion. The technology for achieving these types of blends has been applied to blow moulding of large containers to enhance their barrier props. Effective barrier enhancement of polyethylene can be achieved with 10–20% Nylon 66/6 or with Selar amorph. polyamide as the dispersed phase [1]
Morphology: V. recent studies of the microstructure and permeability of extruded ribbons of blends of polypropylene (PP/EVOH) and PE/Nylon 6 have shown that the formation of discrete overlapping platelets of the dispersed phases of the barrier

materials can be reproduced in a controllable fashion. Under special processing conditions, laminar morphology was obt. in HDPE/Nylon 6 blends. The toluene permeabilities of extruded ribbons were in the range obt. with conventional multilayer barrier packaging materials [2]. The morphological, thermal, rheological and mechanical props. of blends of Nylons 6, Nylon 12 and Nylon 6/12 with EVOH have been investigated. Vinyl alcohol contents of 62% and 71% were employed and homogeneous phase morphologies were observed in Nylon 6 rich blends with fine phase separations in the EVOH rich regions. The polymers showed compatibility due to enhanced hydrogen bonding. In contrast, 71% EVOH blends with Nylon 12 and Nylon 6/12 gave clean separation with large domains and a decrease in physical props. was noted [3]. The longer Nylon 12 chains appear to inhibit interhydrogen bond links sufficiently to prevent compatibility. In consequence, these blends may be more useful in obtaining platelet morphology of a nylon dispersed phase and could help to improve the moisture barrier resistance of EVOH

Bibliographic References

[1] Subramanian, P.M. and Mehra, V., *ANTEC*, Technometric Publising Co., 1986, 301-305
[2] Karmal, M.R., Garmabi, H., Hozhabr, S. and Arghyris, L., *Polym. Eng. Sci.*, 1995, **35**, 45
[3] Ahn, T.O., Kim, C.K., Kim, B.K., Jeong, H.M. and Huh, J.D., *Polym. Eng. Sci.*, 1990, **30**, 341-349

Nylon 6,6 Carbon fibre filled   N-44

Synonyms: *Poly(hexamethylenediamine-co-adipic acid). Poly(hexamethylene adipamide). Poly(iminohexamethylene iminoadipoyl). Poly(imino(1,6-dioxo-1,6-hexanediyl)imino-1,6-hexanediyl)*
Related Polymers: Nylon 6,6, Nylon 6,6, Glass reinforced, Nylon 6,6 Carbon fibre filled
Monomers: 1,6-Hexanediamine, Hexanedioic acid
Material class: Thermoplastic, Composites
Polymer Type: polyamide
Molecular Formula: $(C_{12}H_{22}N_2O_2)n$
Fragments: $C_{12}H_{22}N_2O_2$
Additives: PTFE, silicone, molybdenum disulfide. Flame retardant, electrically conductive and UV-stabilised grades available [2]

Volumetric & Calorimetric Properties:
Density:

No.	Value	Note
1	d 1.23 g/cm^3	20% carbon fibre [1]
2	d 1.34 g/cm^3	40% carbon fibre [1]
3	d 1.18–1.43 g/cm^3	ASTM D792, 10–60% filler [2]

Thermal Expansion Coefficient:

No.	Value	Note	Type
1	2.5×10^{-5} K^{-1}	ASTM D696, 20% carbon fibre [1]	L
2	1.8×10^{-5} K^{-1}	ASTM D696, 40% carbon fibre [1]	L

Thermal Conductivity:

No.	Value	Note
1	0.5 W/mK	ASTM C177, 20% carbon fibre [1]
2	0.62 W/mK	ASTM C177, 20% carbon fibre [1]

Melting Temperature:

No.	Value	Note
1	260°C	20% carbon fibre

Deflection Temperature:

No.	Value	Note
1	260°C	0.45 MPa, ASTM D648 [1]
2	255°C	1.8 MPa, ASTM D648 [1]
3	252°C	ASTM D648, at 1820 kPa, 60% filler [2]

Transport Properties:
Water Absorption:

No.	Value	Note
1	6.7 %	saturation, ASTM D570, 20% carbon fibre [1]
2	6 %	saturation, ASTM D570, 40% carbon fibre [1]

Mechanical Properties:
Mechanical Properties General: Carbon fibre increases stiffness

Tensile (Young's) Modulus:

No.	Value	Note
1	16000 MPa	[1]
2	8964–41370 MPa	ASTM D638, 10–60% filler, [2]

Flexural Modulus:

No.	Value	Note
1	11000 MPa	ASTM D790, 20% carbon fibre [1]
2	22500 MPa	ASTM D790, 40% carbon fibre [1]
3	6895–36544 MPa	ASTM D790, 10–60% filler, [2]

Tensile Strength Break:

No.	Value	Note	Elongation
1	215 MPa	ASTM D638, 20% carbon fibre [1]	2–3%
2	305 MPa	ASTM D638, 40% carbon fibre [1]	2–2.5%
3	145–221 MPa	ASTM D638, 10–60% filler, [2]	1–3%

Flexural Strength at Break:

No.	Value	Note
1	300 MPa	ASTM D790, 20% carbon fibre [1]
2	395 MPa	ASTM D790, 40% carbon fibre [1]

Flexural Strength Yield:

No.	Value	Note
1	214–365 MPa	ASTM D790, 10–60% filler, [2]

– **Nylon 12 cycloaliphatic/aromatic copolymers**

Hardness: Rockwell M89 (ASTM D785) [1]
Fracture Mechanical Properties: Shear strength 100 MPa (ASTM D732, 20% carbon fibre) [1]; 120 MPa (ASTM D732, 40% carbon fibre) [1]
Friction Abrasion and Resistance: Static coefficient of friction, μ_{stat} 0.16 (20% carbon fibre) [1]; 0.13 (40% carbon fibre) [1]. Dynamic coefficient of friction, μ_{dyn} 0.2 (20% carbon fibre); 0.18 (40% carbon fibre) [1]
Izod Notch:

No.	Value	Notch	Note
1	90 J/m	Y	ASTM D256, 20% carbon fibre [1]
2	110 J/m	Y	ASTM D256, 40% carbon fibre [1]
3	700 J/m	N	ASTM D256, 20% carbon fibre [1]
4	800 J/m	N	ASTM D256, 40% carbon fibre [1]
5	43–69 J/m	Y	ASTM D256, 3.2 mm section, 10–60% filler, [2]
6	427–641 J/m	N	ASTM D256, 3.2 mm section, 10–60% filler, [2]

Electrical Properties:
Electrical Properties General: Carbon fibre filler confers conductive props. Surface resistivity 10–1000 Ω [1]; volume resistivity 10–1000 Ω cm [1]. Volume resistivity $<10-<1 \times 10^{15}$ Ω.cm (ASTM D257, 10–60% filler) [2]

Polymer Stability:
Upper Use Temperature:

No.	Value	Note
1	120°C	40% carbon fibre
2	130°C	20% carbon fibre

Flammability: Flammability rating HB (UL94) [1]. HB at 1.5mm (ASTM D635) [2].
Chemical Stability: Stable to oil, grease and most solvents
Hydrolytic Stability: Limited resistance to acid and alkali

Applications/Commercial Products:
Processing & Manufacturing Routes: Processing for injection molding [2]:
Injection pressure 69–124Mpa
Melt temp. 277–299°C
Mold temp. 66–107°C
Drying 4 hrs at 79°C
Moisture content 0.2%
Dew point -18°C

Applications/Commercial Products:
Mould Shrinkage (%):

No.	Value	Note
1	0.35–0.45%	3 mm, 20% carbon fibre [1]
2	0.2–0.4%	3 mm, 40% carbon fibre [1]
3	0.05–0.4%	ASTM D955, 10-60% filler [2]

Applications: Used where static electricity build-up may cause problems, in printers, copiers, conveyer belts and textile machines

Trade name	Supplier
Electrafil	Techmer Lehvoss Compounds
Frianyl	Frisetta Polymer
Lubricomp	GE Plastics
RTP series	RTP Co.
Reny	Mitsuishi Eng.
Stat-kon	GE Plastics
Thermcomp	GE Plastics
Thermotuf	GE Plastics
Wellamid	CP-Polymer-Technik
	TP Composite
	Albis
	Kolon KOPA
	PolyOne Edgetek

Bibliographic References
[1] *LNP Engineering Plastics Product Information*, LNP, 1996, (technical datasheet)
[2] RTP Co. datasheet http://www.rtpcompany.com, 2006

Nylon 12 cycloaliphatic/aromatic copolymers N-45
Synonyms: *Nylon 12 copolymers with cycloaliphatic and aromatic comonomers. Poly(azacyclotridecan-2-one-co-4,4'-methylenebis(2-methylcyclohexanamine-co-1,3-benzenedicarboxylic acid))*
Related Polymers: Nylon 12
Monomers: 12-Aminododecanoic acid, Azacyclotridecan-2-one, 4,4'-Methylenebis(2-methylcyclohexanamine), 1,3-Benzenedicarboxylic acid
Material class: Thermoplastic, Copolymers
Polymer Type: polyamide
CAS Number:

CAS Reg. No.
79331-75-2

General Information: Commercialised in 1976. Good hydrolytic stability and retains transparency in boiling water. Good performance as medium range engineering polymers [1,2,3,5]
Morphology: Grilamid is amorph. and transparent

Volumetric & Calorimetric Properties:
Density:

No.	Value	Note
1	d 1.06 g/cm^3	TR55 [1]
2	d 1.04–1.06 g/cm^3	Crystamid [2]

Thermal Expansion Coefficient:

No.	Value	Note	Type
1	0.000178 K^{-1}	-40–70°, TR65 [1]	L
2	0.00017–0.00018 K^{-1}	Crystamid 1 [2]	L

Thermal Conductivity:

No.	Value	Note
1	0.23 W/mK	DIN 52612 TR55 [1]
2	0.2 W/mK	DIN 52612 Crystamid [2]

Glass-Transition Temperature:

No.	Value	Note
1	150°C	differential thermal analysis TR55
2	170°C	dry Crystamid
3	140°C	saturated with 3.2% water Crystamid

Nylon 12 cycloaliphatic/aromatic copolymers

Deflection Temperature:

No.	Value	Note
1	145°C	0.46 MPa
2	135°C	1.85 MPa, ISO 75
3	110–150°C	0.46 Mpa, Crystamid
4	95–135°C	1.8 Mpa, Crystamid

Vicat Softening Point:

No.	Value	Note
1	155–157°C	50N DIN 53460 TR55
2	115–160°C	50N DIN 53460 Crystamid

Surface Properties & Solubility:
Solvents/Non-solvents: Sol. formic acid, dimethylacetamide/LiCl, H_2SO_4 (98%). Resistant to a wide range of solvents, especially aliphatic and aromatic hydrocarbons, oils, fats, grease, gasoline. Unaffected by alkalis, detergents or salty solvents, also resistant to attack by alcohols, ketones, ethers, aliphatic and aromatic hydrocarbons, trichloroethylene and most detergents

Transport Properties:
Melt Flow Index:

No.	Value	Note
1	270 g/10 min	25 N DIN 53735

Water Content: [a] 3.1% (saturated, 20°), [b] 2.3–3.2% (23° saturated, Crystamid)

Water Absorption:

No.	Value	Note
1	1.1 %	23° 50% relative humidity
2	0.33 %	23°, 24h, ASTM D570, TR55
3	0.9–1.3 %	23°, 50% relative humidity, Crystamid

Mechanical Properties:
Mechanical Properties General: Mechanical props. after injection welding have been reported [4]

Flexural Modulus:

No.	Value
1	1800–2000 MPa

Tensile Strength Break:

No.	Value	Note	Elongation
1	60 MPa	ASTM D638, Crystamid	50–150%

Elastic Modulus:

No.	Value	Note
1	1800–2000 MPa	ASTM D638, Crystamid

Tensile Strength Yield:

No.	Value	Elongation	Note
1	70–75 MPa	8% elongation	ASTM D638, Crystamid

Hardness: Shore D81-D83 (Crystamid, ISO 868)

Izod Notch:

No.	Value	Notch	Note
1	100–120 J/m	N	ASTM 256, Crystamid

Charpy:

No.	Value	Notch	Break	Note
1	23–40	N	N	ASTM D179, Crystamid
2	8–10	N	N	23°, Crystamid
3	6–8	N	N	-40°, Crystamid

Electrical Properties:
Electrical Properties General: Tracking resistance 600 (min DIN 53480)

Surface/Volume Resistance:

No.	Value	Note	Type
1	0.006×10^{15} Ω.cm	DIN 53482,TR55	S
2	0.001×10^{15} Ω.cm	ASTM D257, Crystamid	S

Dielectric Permittivity (Constant):

No.	Value	Frequency	Note
1	2.9	50 Hz	DIN 53483 TR55
2	2.9	1000 Hz	DIN 53483 TR55
3	3–3.4		ASTM D150, Crystamid

Dielectric Strength:

No.	Value	Note
1	40 kV.mm^{-1}	DIN 53481, TR55
2	50 kV.mm^{-1}	ASTM D1491, Crystamid

Dissipation Power Factor: tan δ 0.005 (50 Hz); 0.007 (1000 Hz, DIN 53483)

Optical Properties:
Refractive Index:

No.	Value	Note
1	1.535	DIN 53491

Transmission and Spectra: Light transmission 85% (3 mm ASTM D1003). 7% Haze (ASTM D1003)

Polymer Stability:
Polymer Stability General: Low moisture absorption provides good dimensional stability at elevated temp. and high electrical resistance even in moist environment

Upper Use Temperature:

No.	Value	Note
1	80°C	[1,2,3]
2	90°C	oil
3	130°C	short term, TR55
4	110–135°C	short term, Crystamid

Decomposition Details: Significant decomposition occurs above 350°

Flammability: Self extinguishing (ASTM D635). Flammability rating V2 (1.6 mm UL94); HB (0.8 mm UL94). Limiting oxygen index 26% (ASTM 1863)
Environmental Stress: Stress cracking may occur with mechanically stressed parts with polar solvents and detergents (JR55). Crack resistant to alcohols, ketones, esters and aromatic solvents Crystamid)
Chemical Stability: Resistant to hydrocarbons, solvents, dilute salts and alkali. Attacked by polar alcohols, AcOH, pyridine
Hydrolytic Stability: Hydrolysed by strong acids. Resistant to hot water and steam

Applications/Commercial Products:
Processing & Manufacturing Routes: Polymerisation route is similar to other amorph. polyamides, e.g. Nylon 6(3), T (see polymerisation of nylons). The amorphous polyamides may be injection moulded or extruded at 100–120° above their softening point (range 240–280°)
Mould Shrinkage (%):

No.	Value	Note
1	0.65%	long
2	0.75%	trans

Applications: In high stress environments, e.g. covers/caps, filter bowls (air, oil and water), pump casings, flow meters, covers for control temps., buttons, fashion jewellery, coffee filters, sanitary fittings, fuse boxes, sight glasses, milking machine covers etc.

Trade name	Supplier
Crystamid	Elf Atochem (UK) Ltd.
Grilamid TR55	EMS-Chemie AG

Bibliographic References
[4] Maskus, P., and Gähwiler, H.U., *Adv. Polym. Technol.*, 1987, **7**, 411, (welding)

Nylon 6/Ethylene ethacrylate copolymer ethylene ethacrylate-maleic anhydride blend N-46
Synonyms: *Nylon 6-EEA/EEA-MA blend*
Related Polymers: Nylon 6
Material class: Thermoplastic, Blends
Polymer Type: polyamide

Transport Properties:
Melt Flow Index:

No.	Value
1	3.8 g/10 min

Mechanical Properties:
Impact Strength: Typical injection grade displays impact strength 20 times that of Nylon 6,6
Izod Notch:

No.	Value	Notch	Note
1	970 J/m	Y	[1]

Applications/Commercial Products:
Processing & Manufacturing Routes: Reactive extrusion product of blending Nylon 6-polyethylene ethacrylate copolymer with polyethylene ethacrylate-maleic anhydride terpolymer

Bibliographic References
[1] *J. Plast. Week.*, 22nd Oct., 1990, 4

Nylon 6-12, Glass and carbon reinforced N-47
Related Polymers: Nylon 6-12
Monomers: 1,6-Hexanediamine, Dodecanedioic acid
Material class: Thermoplastic
Polymer Type: polyamide
Fragments: $C_{18}H_{34}N_2O_2$
Additives: Glass fibre, carbon fibre

Volumetric & Calorimetric Properties:
Density:

No.	Value	Note
1	d 1.31–1.32 g/cm^3	30% glass fibre [1,2]
2	d 1.49 g/cm^3	50% glass fibre [1,2]
3	d 1.2 g/cm^3	30% carbon fibre [1,2]

Thermal Expansion Coefficient:

No.	Value	Note	Type
1	3×10^{-5} K^{-1}	longitudinal, 23–80°, D 53752/B, 30% glass fibre [1]	L
2	2.3×10^{-5} K^{-1}	longitudinal, 23–80°, D 53752/B, 50% glass fibre [2]	L
3	5×10^{-5} K^{-1}	longitudinal, 23–80°, D 53752/B, 30% carbon fibre	L

Melting Temperature:

No.	Value	Note
1	212°C	Fisher John method [2]

Deflection Temperature:

No.	Value	Note
1	200°C	1.8 MPa, 30% glass fibre, DIN 53461 [1]
2	215°C	0.45 MPa, 30% glass fibre, DIN 53461 [1]
3	210°C	91.8 MPa, 30% glass fibre, DIN 53461 [2]
4	205°C	1.8 MPa, 50% glass fibre, DIN 53461 [1]
5	215°C	0.45 MPa, 50% glass fibre, DIN 53461 [1]
6	200°C	1.8 MPa, 30% carbon fibre, DIN 53461 [1]
7	215°C	0.45 MPa, 30% carbon fibre, DIN 53461 [1]

Vicat Softening Point:

No.	Value	Note
1	205°C	50 N, 30% glass fibre
2	205°C	50 N, 50% glass fibre
3	210°C	50 N, 30% carbon fibre

Transport Properties:
Water Content: 0.2% (immersion 24h)
Water Absorption:

No.	Value	Note
1	1.9 %	30% glass fibre
2	1.2 %	50% glass fibre
3	2.2 %	30% carbon fibre

– Nylon 6-12, Glass and carbon reinforced

Mechanical Properties:

Flexural Modulus:

No.	Value	Note
1	8300 MPa	dry, 30% glass fibre [2]
2	5100 MPa	50% relative humidity, 30% glass fibre [2]
3	8270 MPa	23°, 33% glass fibre, ASTM D790
4	6200 MPa	23°, 50% relative humidity, 30% glass fibre, ASTM D790

Tensile Strength Break:

No.	Value	Note	Elongation
1	138 MPa	30% glass fibre [2]	5%
2	130 MPa	5 mm min^{-1}, ISO 527, 30% glass fibre [1]	3%
3	200 MPa	5 mm min^{-1}, ISO 527, 50% glass fibre [1]	3%
4	210 MPa	5 mm min^{-1}, ISO 527, 30% carbon fibre [1]	3%

Elastic Modulus:

No.	Value	Note
1	7600 MPa	30% glass fibre
2	14000 MPa	50% glass fibre
3	15000 MPa	30% carbon fibre, 1 mm min^{-1} in tension, DIN 53447-t

Miscellaneous Moduli:

No.	Value	Note	Type
1	3200 MPa	1000 h, 30 % glass fibre, DIN 53444 [1]	tensile creep modulus

Hardness: Rockwell R118 (33% glass fibre)

Izod Notch:

No.	Value	Notch	Note
1	128 J/m	Y	dry, 30% glass fibre [2]
2	235 J/m	Y	50% relative humidity, 30% glass fibre [2]

Izod Area:

No.	Value	Notch	Note
1	6 kJ/m^2	Notched	-30–23°, 0–50% relative humidity, 30% glass fibre [1]
2	39–46 kJ/m^2	Notched	-30–23°, 0–50% relative humidity, 50% glass fibre [1]
3	15 kJ/m^2	Notched	-30–23°, 0–50% relative humidity, 50% glass fibre [1]
4	70 kJ/m^2	Notched	-30–23°, 0–50% relative humidity, 50% glass fibre [1]
5	9–12 kJ/m^2	Notched	-30–23°, 0–50% relative humidity, 30% carbon fibre [1]
6	60 kJ/m^2	Notched	-30–23°, 0–50% relative humidity, 30% carbon fibre [1]

Electrical Properties:

Electrical Properties General: Surface resistivity 100 Ω (30% carbon fibre) [1, 3]. Volume resistivity 10 Ω cm (30% carbon fibre) [1, 3]

Surface/Volume Resistance:

No.	Value	Note	Type
1	1×10^{15} Ω.cm	30% glass fibre [1, 3]	S
2	1×10^{15} Ω.cm	50% glass fibre [1, 3]	S

Dielectric Permittivity (Constant):

No.	Value	Frequency	Note
1	4.2	50 Hz	30% glass fibre [1,3]
2	3.8	100 Hz	30% glass fibre [1,3]
3	3.4–3.5	1 MHz	30% glass fibre [1,3]
4	4.9	50 Hz	50% glass fibre [1,3]
5	4	1 MHz	50% glass fibre [1,3]

Dielectric Strength:

No.	Value	Note
1	41 kV.mm^{-1}	30% glass fibre [1]
2	33 kV.mm^{-1}	50% glass fibre [1]

Arc Resistance:

No.	Value	Note
1	600s	CTI, VDE 0303 TI, 30% glass fibre [1, 3]
2	600s	CTI, VDE 0303 TI, 50% glass fibre [1, 3]

Dissipation (Power) Factor:

No.	Value	Frequency	Note
1	0.02	50 Hz	[1]
2	0.024	1 MHz	VDE 0303T4 [1]
3	0.0135	100 Hz	[3]
4	0.015	1 MHz	ISO IEC 250, 30% glass fibre [3]
5	0.033	50Hz	
6	0.027	1 MHz	VDE 0303T4, 50% glass fibre [1]

Polymer Stability:

Flammability: Flammability rating HB (all grades, UL94)

Applications/Commercial Products:

Mould Shrinkage (%):

No.	Value	Note
1	0.1–0.15	longitudinal, 30% glass fibre [1]
2	0.15–0.25%	longitudinal, 50% glass fibre [1]
3	0.1–0.15%	longitudinal, 30% carbon fibre [1]
4	1.1–1.35%	transverse, 30% glass fibre [1]
5	0.6–0.9%	transverse, 50% glass fibre [1]
6	1–1.2%	transverse, 30% carbon fibre

– Nylon 4,6 (30% glass fibre reinforced)

Trade name	Details	Supplier
Zytel	77G series	DuPont

Bibliographic References

[1] *Engineering Plastics from New Ideas,* Huls, 1995, 26, (technical datasheet)
[2] *Zytel Nylon Resins,* DuPont, 1985, (technical datasheet)
[3] *From Concept to Commercialisation,* DuPont, 1985, (technical datasheet)
[4] *J. Mater. Eng.,* 1990, **107** (3), 20, (uses)

Nylon 4,6 (30% glass fibre reinforced) N-48

Synonyms: Poly[imino-1,4-butanediylimino(1,6-dioxo-1,6-hexanediyl)]. Poly(tetramethylenediamine-co-adipic acid)
Related Polymers: Nylon 4,6
Monomers: Tetramethylenediamine, Adipic acid
Material class: Thermoplastic
Polymer Type: polyamide
CAS Number:

CAS Reg. No.
50327-22-5

Molecular Weight: MW 32900

Volumetric & Calorimetric Properties:
Density:

No.	Value	Note
1	d 1.41 g/cm^3	dry [1]

Thermal Expansion Coefficient:

No.	Value	Note	Type
1	$2 \times 10^{-5} - 8 \times 10^{-5}$ K^{-1}	23-80° [1]	L
2	$2 \times 10^{-5} - 9 \times 10^{-5}$ K^{-1}	80-180°, due to anisotropy [1]	L

Thermal Conductivity:

No.	Value	Note
1	0.24 W/mK	[1]

Specific Heat Capacity:

No.	Value	Note	Type
1	1.9 kJ/kg.C	[1]	P

Melting Temperature:

No.	Value	Note
1	295°C	[1]

Glass-Transition Temperature:

No.	Value	Note
1	43°C	dynamic method [5]

Deflection Temperature:

No.	Value	Note
1	285°C	1.8 MPa [3]
2	>290°C	min. [1]
3	280°C	50% glass fibre [2]

Transport Properties:
Water Content: 2.6% (23°, 50% relative humidity) [1]
Water Absorption:

No.	Value	Note
1	1.5 %	24h, 23° [1]

Mechanical Properties:
Tensile (Young's) Modulus:

No.	Value	Note
1	6500–10000 MPa	23° [1]
2	5500 MPa	120° [1]
3	5000 MPa	160° [1]

Flexural Modulus:

No.	Value	Note
1	5500–9000 MPa	23°, anisotropic [1]
2	4700 MPa	120° [1]
3	3900 MPa	160° [1]

Tensile Strength Break:

No.	Value	Note	Elongation
1	115–200 MPa	23° [1]	
2	105 MPa	120° [1]	
3	90 MPa	160° [1]	200%

Flexural Strength at Break:

No.	Value	Note
1	180–300 MPa	23° [1]
2	160 MPa	120° [1]
3	130 MPa	160° [1]

Hardness: Shore D90-D95
Failure Properties General: Creep modulus 4500–7000 MPa (23°, 1000h); 4500 MPa (120°, 1000h); 3500 MPa (160°, 1000h)
Izod Area:

No.	Value	Notch	Note
1	60–80 kJ/m^2	Unnotched	-40–23° [1]
2	11–19 kJ/m^2	Notched	23° [1]
3	10 kJ/m^2	Notched	-40° [1]

Nylon 6, glass reinforced

Charpy:

No.	Value	Notch	Note
1	10	Y	-40°
2	11–19	Y	23°

Electrical Properties:
Surface/Volume Resistance:

No.	Value	Note	Type
1	10×10^{15} Ω.cm	parallel [1]	S
2	0.01×10^{15} Ω.cm	perpendicular [1]	S

Dielectric Permittivity (Constant):

No.	Value	Frequency	Note
1	4.3	50 Hz	parallel [1]
2	16	50 Hz	perpendicular [1]
3	4	1 MHz	parallel [1]
4	4.7	1 MHz	perpendicular [1]

Dielectric Strength:

No.	Value	Note
1	>30 kV.mm^{-1}	min., parallel [1]
2	>20 kV.mm^{-1}	min., perpendicular [1]

Dissipation (Power) Factor:

No.	Value	Frequency	Note
1	0.007	50 Hz	parallel [1]
2	0.6	50 Hz	perpendicular [1]
3	0.02	1 MHz	parallel [1]
4	0.1	1 MHz	perpendicular [1]

Polymer Stability:
Thermal Stability General: Polymer may be used in air up to 185° for prolonged periods without serious loss of props. Very high stiffness maintained up to 290°
Upper Use Temperature:

No.	Value	Note
1	>175°C	max. 2500h [1]
2	163°C	max. 5000h [1]
3	151°C	max. 10000h [1]
4	139°C	max. 20000h [1]
5	290°C	short periods
6	185°C	

Flammability: Flammability rating HB (22 mm thick). Limiting Oxygen Index 22%. Glow Wire test extinguishing time 15 s; (temp. 960°; 3 mm thick)
Chemical Stability: Possesses similar chemical stability to Nylon 6,6. Very resistant to greases, oils, aliphatic and aromatic hydrocarbons up to 150° for prolonged periods

Bibliographic References
[1] Stanyl. Typical Properties DSM, 1995, (technical datasheet)
[2] *Plastics News (Detroit)*, 1992, **4**, 8
[3] Wood, A.S., *Mod. Plast.*, 1988, **65(6)**, 34
[4] *Encycl. Polym. Sci. Technol.*, 2nd edn., John Wiley and Sons, 1986, **11**, 371
[5] Beaman, R.G. and Cramer, F.B., *J. Polym. Sci.*, 1956, **21**, 223, (T_g)

Nylon 6, glass reinforced N-49

Synonyms: *Glass reinforced Nylon 6*
Related Polymers: Nylon 6, Nylon 6,6, Nylon 6,6 glass reinforced
Monomers: Caprolactam, Caproamide
Material class: Thermoplastic
Polymer Type: polyamide
Molecular Formula: $(C_6H_{11}NO)_n$
Fragments: $C_6H_{11}NO$
General Information: Long (10mm) fibres improve flexural and creep props with respect to conventional length (5mm) fibre filled materials. Increasing glass fibre content increases stiffness.

Volumetric & Calorimetric Properties:
Density:

No.	Value	Note
1	d 1.22–1.25 g/cm^3	15% glass fibre [2,3,12]
2	d 1.33–1.37 g/cm^3	30% glass fibre [2,3]
3	d 1.55 g/cm^3	50% glass fibre [3]
4	d 1.5 g/cm^3	50% glass spheres [2]

Thermal Expansion Coefficient:

No.	Value	Note	Type
1	1×10^{-5}–1.5×10^{-5} K^{-1}	dry, 50% glass fibre [3]	V
2	5×10^{-5}–6×10^{-5} K^{-1}	moist, 50% glass fibre [3]	V
3	3×10^{-5}–3.5×10^{-5} K^{-1}	dry, 15% glass fibre [3]	V
4	7×10^{-5}–8×10^{-5} K^{-1}	moist, 15% glass fibre [3]	V
5	2×10^{-5}–2.5×10^{-5} K^{-1}	dry, 30% glass fibre [3]	V
6	6×10^{-5}–7×10^{-5} K^{-1}	moist, 30% glass fibre [3]	V

Thermal Conductivity:

No.	Value	Note
1	0.32 W/mK	15–50% glass fibre, DIN 52612 [3]

Specific Heat Capacity:

No.	Value	Note	Type
1	1.3 kJ/kg.C	dry, 50% glass fibre, IEC 1006 [3]	P

Melting Temperature:

No.	Value	Note
1	221°C	15–50% glass fibre [2]
2	220°C	DSC, 15–50% glass fibre [3]

Deflection Temperature:

No.	Value	Note
1	200°C	0.45 MPa, 5% glass fibre [2]
2	190°C	1.8 MPa, 15% glass fibre [2]

— Nylon 6, glass reinforced

3	160°C	1.8 MPa, 15% glass fibre [3]
4	200°C	0.45–1.8 MPa, 30% glass fibre [2]
5	210°C	1.8 MPa, 30% glass fibre [3]
6	215°C	1.8 MPa, 50% glass fibre [3]

Transport Properties:
Transport Properties General: Melt volume rate has been reported [3]
Polymer Solutions Concentrated: Viscosity 135 ml g^{-1} (50% glass fibre, sulfuric acid)
Water Content: 7.2–8.5% (15% glass fibre) 5.97% (30% glass fibre); 4.5–5% (50% glass fibre 23°) [2,3]
Water Absorption:

No.	Value	Note
1	2.1–2.7 %	23°, 25% relative humidity, 15% glass fibre [3]
2	1.8–2.2 %	23°, 50% relative humidity, 30% glass fibre [3]
3	1.3–1.7 %	23°, 50% relative humidity, 50% glass fibre [3]

Mechanical Properties:
Tensile (Young's) Modulus:

No.	Value	Note
1	5200 MPa	dry, 15% glass fibre [3]
2	2800 MPa	moist, 15% glass fibre [3]
3	8400 MPa	dry, 30% glass fibre [3]
4	5200 MPa	moist, 30% glass fibre [3]
5	1600 MPa	dry, 50% glass fibre [2]
6	1200 MPa	moist, 50% glass fibre [3]

Flexural Modulus:

No.	Value	Note
1	5200 MPa	23°, 50% relative humidity, 15% glass fibre [2]
2	2700 MPa	23°, dry, 15% glass fibre [2]
3	7500 MPa	23°, 50% relative humidity, 50% glass fibre [2]
4	4000 MPa	23°, dry, 30% glass fibre [2]
5	6900 MPa	[6,12]

Tensile Strength Break:

No.	Value	Note	Elongation
1	230 MPa	dry, 50% glass fibre [3]	3%
2	160 MPa	moist, 50% glass fibre [3]	3.5%
3	123 MPa	[8,12]	5%
4	159 MPa	[11]	

Flexural Strength at Break:

No.	Value	Note
1	155 MPa	dry, 15% glass fibre [2,3] 8%
2	80 MPa	50% relative humidity, 75% glass fibre [2] 30%
3	55 MPa	moist, 15% glass fibre [3] 17%
4	230 MPa	dry, 30% glass fibre [2] 35%
5	100 MPa	

Elastic Modulus:

No.	Value	Note
1	16000 MPa	-60°, 50% glass fibre [3]
2	10000 MPa	60°, 50% glass fibre [3]
3	7 MPa	100°, 50% glass fibre [3]
4	9000 MPa	-60°, 30% glass fibre [3]
5	5 MPa	50°, 30% glass fibre [3]
6	4000 MPa	100°, 30% glass fibre [3]

Miscellaneous Moduli:

No.	Value	Note	Type
1	500 MPa	150°, 85% glass fibre [4]	shear modulus
2	800 MPa	30% glass fibre [4]	shear modulus
3	1000 MPa	50% glass fibre [4]	shear modulus

Impact Strength: Long fibre composites have increased impact strength [15]. Other impact strength values have been reported. [2,3,5]
Hardness: Ball indentation 150 MPa (dry, 15% glass fibre) [3]; 180 MPa (dry); 110 MPa (moist, 30% glass fibre) [3] 280 MPa (dry); 210 MPa (moist, 50% glass fibre) [3]
Failure Properties General: Failure strength 35–45 MPa (50% glass fibre) [4]
Fracture Mechanical Properties: Fracture energy W_{50} 25 J (dry) 45K (moist, 15% glass fibre); 20 J (dry); 45 J (moist, 30% glass fibre); 3 J (dry); 10 J (moist, 50% glass fibre) [3]
Izod Notch:

No.	Value	Notch	Note
1	954 J/m	n	[7]
2	43 J/m	n	[8]
3	960 J/m	n	[12]
4	183 J/m	y	dry, 22° [3]
5	186 J/m	y	dry [7]
6	400 J/m	y	impact modified [11]

Izod Area:

No.	Value	Notch	Note
1	17 kJ/m^2	Notched	dry, 23°, glass fibre [3]
2	21 kJ/m^2	Notched	moist, 23°, 15% glass fibre [3]
3	22 kJ/m^2	Notched	dry, 23°, 30% glass fibre [3]
4	32 kJ/m^2	Notched	moist, 23°, 30% glass fibre [3]
5	20 kJ/m^2	Notched	dry, 23°, 50% glass fibre [3]
6	24 kJ/m^2	Notched	moist, 23°, 50% glass fibre [3]

Charpy:

No.	Value	Notch	Note
1	8–16	Y	dry, 15% glass fibre [2,3]
2	17–19	Y	dry, 30% glass fibre [2,3]
3	18.5	Y	dry, 50% glass fibre [3]
4	75	N	dry, 15% glass fibre [3]
5	95	N	dry, 15% glass fibre [3]
6	95	N	dry, 50% glass fibre [3]

Nylon 6, glass reinforced

7	30	Y	50% relative humidity, 15% glass fibre [2,3]
8	35	Y	50% relative humidity, 30% glass fibre [2,3]
9	25	Y	moist, 50% glass fibre [3]
10	105	N	moist, 15% glass fibre [3]
11	105	N	moist, 30% glass fibre [3]
12	105	N	moist, 50% glass fibre [3]

Electrical Properties:
Surface/Volume Resistance:

No.	Value	Note	Type
1	1×10^{15} Ω.cm	dry, 15–50% glass fibre, IEC 190 [2]	s
2	0.001×10^{15} Ω.cm	50% relative humidity, IEC 190 [2]	s
3	0.01×10^{15} Ω.cm	dry, 15–50% glass fibre, IEC 93 [3]	s
4	0.001×10^{15} Ω.cm	dry, 30% glass fibre, IEC 93 [3]	s
5	$1 \times 10^{-5} \times 10^{15}$ Ω.cm	moist, 30% glass fibre, IEC 93 [3]	s
6	$1 \times 10^{-5} \times 10^{15}$ Ω.cm	moist, 15–50% glass fibre, IEC 93 [3]	s

Dielectric Permittivity (Constant):

No.	Value	Frequency	Note
1	3.5	1 MHz	dry, 15% glass fibre, IEC 250 [2]
2	4	1 MHz	50% relative humidity, 15% glass fibre, IEC 250 [2]
3	3.8	1 MHz	dry, 30% glass fibre, IEC 250 [2,3]

Dielectric Strength:

No.	Value	Note
1	70 kV.mm^{-1}	moist, 15–30% glass fibre, IEC 243 [3]
2	90 kV.mm^{-1}	dry, 50% glass fibre, IEC 243 [3]
3	80 kV.mm^{-1}	moist, 50% glass fibre, IEC 243 [3]
4	11 kV.mm^{-1}	dry, 15–30% glass fibre, IEC 243 [2]
5	8 kV.mm^{-1}	50% relative humidity, 15–30% glass fibre, IEC 243 [8]
6	80 kV.mm^{-1}	dry, 15–30% glass fibre, IEC 243 [3]

Complex Permittivity and Electroactive Polymers: Comparative tracking index 520 (15–50% glass fibre, IEC 112) [2]; 5550 (15–50% glass fibre) [3]

Dissipation (Power) Factor:

No.	Value	Frequency	Note
1	0.025	1 MHz	dry, 15% glass fibre [3]
2	0.2	1 MHz	moist, 15% glass fibre [3]
3	0.02	1 MHz	dry, 30% glass fibre [3]
4	0.2	1 MHz	moist, 30% glass fibre [3]
5	0.014	1 MHz	dry, 50% glass fibre [3]
6	0.4	1 MHz	moist, 50% glass fibre [3]

Polymer Stability:
Thermal Stability General: Small decreases in tensile and impact strength occur after periods at 70°, resulting from relaxation processes caused by mobility of glass filler. Once ordering has occurred, props remain constant [14]

Upper Use Temperature:

No.	Value	Note
1	180°C	15-30% glass fibre [3]
2	200°C	50% glass fibre [3]

Flammability: Flammability rating UNCL (UL94)[2]; HB (vertical, 1.6–3.2 mm, 15–50% glass fibre, UL94) [3]. BH3 (13 mm, 30% glass fibre); BH2 (70 mm, 50% glass fibre); FH3 (21 mm, 30% glass fibre); FH3 (70 mm). Limiting oxygen index 23% (15% glass fibre); 22% (30% glass fibre, ASTM D863) [2]

Chemical Stability: High resistance to lubricants, engine oils and fuels, hydraulic and radiator fluids, refrigerants, paints, solvents and cleaners, aliphatic and aromatic hydrocarbons. Susceptibility is reduced at high glass content relative to base resin, reducing uptake of polar solvents. [4]

Hydrolytic Stability: Increasing glass fibre content increases susceptibility to alkali, attacked by dil. inorganic acids. [4]

Applications/Commercial Products:
Processing & Manufacturing Routes: Processed by injection moulding or extruding. High viscosity grades used to produce rods and tubes

Mould Shrinkage (%):

No.	Value	Note
1	0.4–0.85%	dry, long, 15% glass fibre [3]
2	0.35–0.8%	dry, long, 30% glass fibre [3]
3	0.2–0.7%	dry, long, 50% glass fibre [3]
4	0.4–0.5%	[6,12]

Applications: Used for materials requiring high stiffness and impact strength such as cycle wheels, tool casings, electrical insulating parts, machine components and housings, and sports and exercise equipment, tool handles, steering wheels, cable sheathing

Trade name	Details	Supplier
Akulon		DSM
Anjamid 6		Janssen
Ashlene	non exclusive	Ashley Polymers
B Taromid		Taro
Beetle MDF	glass fibre filled	BIP
Bergamid B		Bergmann
Capron	glass filled grades	Allied Signal Corporation
Celanese	201,3115,3130,3140	Hoechst
Comalloy 610		ComAlloy International
Denyl 6		Vamp-Technologies S.p.A.
Dorethan B		Bayer Inc.
Fiberfil	non exclusive	DSM
Fiberstran	non exclusive	DSM
Frianyl B		Frisette Polymer
Grilon		EMS
Hylon	504-510	Hale

– Nylon 6,6, Glass reinforced

Jonylon		BIP
M_n 6-FG	10–40	Modified Plastics
Nivionplast		Enichem
Nylon N60G	series	Michael Day
RTP 200A	various grades long fibres	RTP Company
Sriamid		Nyltech
TechnylC		Nyltech
Thermocomp PF	1000 series	LNP Engineering
Ultramid B		BASF
Vampamid		Vamp-Technologies S.p.A.

Bibliographic References

[1] 4th Edn., (Eds. J.I. Kroschwitz and J. Willey), 1988, **11**, 453
[2] 1986, (technical datasheet)
[3] 1993, (technical datasheet)
[4] Ultramid Polyamide Product Line, *Properties and Processing*, 1993, (technical datasheet)
[5] *Plast. World*, 1987, **45**, 75
[6] *Plast. Eng. (N.Y.)*, 1993, **49**, 5
[7] *Plast. Eng. (N.Y.)*, 1986, **42**, 78
[8] *Plastics News (Detroit)*, 1992, **3**, 8
[9] *Plast. Eng. (N.Y.)*, 1991, **47**, 5
[10] *Plast. World*, 1991, **49**, 58
[11] *Plastics Design Forum*, 1991, **16**, 77
[12] *Plast. World*, 1992, **50**, 64
[13] Gennaro, A., Plastics and Rubber Institute, 1986
[14] Afanas'ev, V.G., Yazon, M.G. and Liventseva, D.A., *Plast. Massy*, 1989, 93
[15] Voelker, M.J., *SAMPE J.*, 1989, **16**, 31
[16] *Plastics Design Forum*, Soc. of Plast. Eng., 1991, **16**, 77
[17] Lee, C.S., Jones, E. and Kingsland, R., Annu. Tech. Conf. - Soc. Plast. Eng., *43rd*, 1985, 574

Nylon 6,6, Glass reinforced   N-50

Synonyms: *Poly(hexamethylene adipamide). Poly(hexamethylene-diamine-co-adipic acid). Poly[imino(1,6-dioxo-1,6-hexanediyl)imino-1,6-hexanediyl]*
Related Polymers: Nylon 6,6, Carbon fibre filled Nylon 6,6, Nylon 6, Glass reinforced
Monomers: 1,6-Hexanediamine, Hexanedioic acid
Material class: Thermoplastic
Polymer Type: polyamide
CAS Number:

CAS Reg. No.
32131-17-2

Molecular Formula: $(C_{12}H_{22}N_2O_2)_n$
Fragments: $C_{12}H_{22}N_2O_2$
General Information: Long fibres (20 mm) significantly improve creep resistance and flexural props.

Volumetric & Calorimetric Properties:

Density:

No.	Value	Note
1	d 1.23 g/cm³	15% glass fibre [1]
2	d 1.32 g/cm³	25% glass fibre [1]
3	d 1.36 g/cm³	30% glass fibre [1]
4	d 1.41 g/cm³	35% glass fibre [1]
5	d 1.55 g/cm³	50% glass fibre [1]
6	d 1.4 g/cm³	33% glass fibre [1]
7	d 1.47 g/cm³	[2]
8	d 1.56 g/cm³	50% long glass fibre [5]
9	d 1.36 g/cm³	ASTM D792 [13]

Thermal Expansion Coefficient:

No.	Value	Note	Type
1	$3 \times 10^{-5} – 3.5 \times 10^{-5}$ K⁻¹	long	L
2	$7 \times 10^{-5} – 8 \times 10^{-5}$ K⁻¹	transverse, 15% glass fibre	L
3	$2.5 \times 10^{-5} – 3.5 \times 10^{-5}$ K⁻¹	long	L
4	$6 \times 10^{-5} – 7 \times 10^{-5}$ K⁻¹	transverse, 25% glass fibre	L
5	$1.5 \times 10^{-5} – 2 \times 10^{-5}$ K⁻¹	long, 30% glass fibre	L
6	$6 \times 10^{-5} – 7 \times 10^{-5}$ K⁻¹	transverse, 30% glass fibre	L
7	$1.5 \times 10^{-5} – 2 \times 10^{-5}$ K⁻¹	long, 35% glass fibre	L
8	$6 \times 10^{-5} – 7 \times 10^{-5}$ K⁻¹	transverse, 35% glass fibre	L
9	$5 \times 10^{-5} – 6 \times 10^{-5}$ K⁻¹	transverse, 50% glass fibre, 23–80°, dry, DIN 52612 [1]	L
10	2×10^{-5} K⁻¹	dry, ASTM D696	L
11	7.4×10^{-5} K⁻¹	0.2% moisture [2]	L
12	2.9×10^{-5} K⁻¹	dry [13]	L

Thermal Conductivity:

No.	Value	Note
1	0.23 W/mK	dry, 15–25% glass fibre [1]
2	0.24 W/mK	dry, 30–35% glass fibre [1]
3	0.25 W/mK	dry, 50% glass fibre, DIN 52612 [1]
4	0.22 W/mK	dry, 33% glass fibre, ASTM C177
5	0.35 W/mK	[2]
6	0.47 W/mK	30% glass fibre, ASTM C177 [13]

Specific Heat Capacity:

No.	Value	Note	Type
1	1.8 kJ/kg.C	dry, 15% glass fibre [1]	P
2	1.6 kJ/kg.C	dry, 25% glass fibre	P
3	1.5 kJ/kg.C	dry, 30–35% glass fibre [1]	P
4	1.3 kJ/kg.C	dry, 50% glass fibre, method IEC 1006	P
5	1.38 kJ/kg.C	[2]	P
6	1.633 kJ/kg.C	[13]	P

Melting Temperature:

No.	Value	Note
1	255°C	ASTM D789
2	209°C	DSC [2]
3	253°C	[13]

Nylon 6,6, Glass reinforced

Deflection Temperature:

No.	Value	Note
1	254°C	dry, 0.45 MPa, 33% glass fibre
2	252°C	dry, 1.8 MPa, 33% glass fibre, ASTM D648
3	270°C	moist, 1.8 MPa, 33% glass fibre
4	267°C	1.5% moisture, 1.8 MPa, 33% glass fibre [2]
5	250°C	dry, 1.8 MPa, 30% glass fibre [13]
6	250°C	15% long glass fibre [7]
7	275°C	30% long glass fibre [7]
8	241°C	1.8 MPa, 30% glass fibre [12]
9	249°C	50% glass fibre [6]
10	250°C	0.45-1.8 MPa load in flexure, 15-50% glass fibre, ISO 75 [1]

Vicat Softening Point:

No.	Value	Note
1	200°C	ASTM D1525 [13]

Transport Properties:

Polymer Solutions Dilute: Viscosity number 140 (0.005 g ml^{-1} H_2SO_4 soln.) [1]
Polymer Melts: Melt temp. range 280-300°
Water Content: 6.7-7.3% (15% glass fibre) [1]; 5.7-6.3% (25% glass fibre) [1]; 5.2-5.8% (30% glass fibre) [1]; 4.7-5.3% (35% glass fibre) [1]; 3.7-4.3% (50% glass fibre) [1], 5.0% (24h, 30% glass fibre, ASTM D570) [13]

Water Absorption:

No.	Value	Note
1	1.9-2.5 %	saturation, 23°, 50% relative humidity, 15% glass fibre [1]
2	1.7-2.1 %	saturation, 23°, 50% relative humidity, 25% glass fibre [1]
3	1.5-1.9 %	saturation, 23°, 50% relative humidity, 30% glass fibre [1]
4	1.4-1.8 %	saturation, 23°, 50% relative humidity, 35% glass fibre [1]
5	1-1.4 %	saturation, 23°, 50% relative humidity, 50% glass fibre [1]
6	0.7 %	24h, Vydyne 909 [2]
7	1 %	33% glass fibre
8	1.5 %	30% glass fibre, DIN 53495 [13]

Mechanical Properties:

Mechanical Properties General: Effects of fill time (range 0.8-20 s) on mech. and surface appearance and dimensions of injection moulded specimens have been reported. Peak tensile stress increases by 15% and flexural strength decreases by 10% as fill time increases. Scatter in flexural modulus diminishes with fill time. These variations can be attributed to differences in skin/core orientation of glass fibres. Little effect upon shrinkage is noted for tensile specimens. Dramatic changes in surface appearance observed; at short fill times a dark uniform colour and smooth appearance contrasts with lighter colour and more porous surface at longer fill times, attributed to crystallisation prior to complete pressurisation of mould [11]

Tensile (Young's) Modulus:

No.	Value	Note
1	6000 MPa	dry, 15% glass fibre
2	4500 MPa	moist, 15% glass fibre
3	8500 MPa	dry, 25% glass fibre
4	6000 MPa	moist, 25% glass fibre
5	9500 MPa	dry, 30% glass fibre
6	7500 MPa	moist, 30% glass fibre
7	11000 MPa	dry, 35% glass fibre
8	8500 MPa	moist, 35% glass fibre
9	1600 MPa	dry, 50% glass fibre [1]
10	1300 MPa	moist, 50% glass fibre [1]
11	9700 MPa	dry, 33% glass fibre [2]
12	7500 MPa	moist, 33% glass fibre, ASTM D638
13	10000 MPa	dry, 30% glass fibre, ASTM D638 [13]

Flexural Modulus:

No.	Value	Note
1	6200 MPa	33% glass fibre, ASTM D790
2	7300 MPa	0.2% moisture
3	4840 MPa	1.5% moisture, 33% glass fibre [2]
4	3100 MPa	[3]
5	5200 MPa	30% glass fibre, equilib. 50% relative humidity [4]
6	14000 MPa	50% long glass fibre [5]
7	12000 MPa	50% glass fibre [6]
8	9400 MPa	30% glass fibre [12]

Tensile Strength Break:

No.	Value	Note	Elongation
1	130 MPa	dry, 15% glass fibre [1]	3%
2	80 MPa	moist, 15% glass fibre [1]	6%
3	170 MPa	dry, 25% glass fibre	3%
4	130 MPa	moist, 25% glass fibre [1]	5%
5	210 MPa	dry, 35% glass fibre [1]	3%
6	160 MPa	moist, 35% glass fibre [1]	5%
7	230 MPa	dry, 50% glass fibre	2%
8	180 MPa	moist, 50% glass fibre, DIN 53455 [1]	3%
9	193 MPa	dry, 33% glass fibre	4%
10	152 MPa	moist, 33% glass fibre, ASTM D638	5%
11	114 MPa	0.2% moisture	7%
12	86.5 MPa	1.5% moisture	7%
13	76 MPa	Maranyl AMX9 fibres [3]	
14	200 MPa	Maranyl A-690, 50% glass fibre [6]	
15	127.5 MPa	15% long glass fibre [7]	

– Nylon 6,6, Glass reinforced

16	168 MPa	30% long glass fibre [7]	
17	178 MPa	30% glass fibre, ASTM D638 [13]	3%

Flexural Strength at Break:

No.	Value	Note
1	180 MPa	dry, 15% glass fibre [1]
2	125 MPa	moist, 15% glass fibre [1]
3	220 MPa	dry, 25% glass fibre [1]
4	200 MPa	moist, 25% glass fibre [1]
5	270 MPa	dry, 30% glass fibre [1]
6	220 MPa	moist, 30% glass fibre [1]
7	280 MPa	dry, 35% glass fibre [1]
8	230 MPa	moist, 35% glass fibre [1]
9	320 MPa	dry, 50% glass fibre [1]
10	260 MPa	moist, 50% glass fibre, DIN 53452 [1]
11	200 MPa	33% glass fibre, ASTM D790
12	194 MPa	0.2% moisture [2]
13	138 MPa	1.5% moisture [2]
14	104 MPa	[3]
15	333 MPa	50% long glass fibre [6]
16	321 MPa	50% glass fibre [6]
17	250 MPa	30% glass fibre, ASTM D790 [13]
18	193 MPa	30% glass fibre [18]

Poisson's Ratio:

No.	Value	Note
1	0.375	30% glass fibre [13]

Tensile Strength Yield:

No.	Value	Elongation	Note
1	175 MPa	2.89%	30% glass fibre, ASTM D638 [13]

Miscellaneous Moduli:

No.	Value	Note
1	3190 MPa	30% glass fibre [13]

Impact Strength: Other values of Izod impact strength have been reported. [2,5,7,18] Increasing fibre length confers higher falling weight impact resistance [15,20]. Various impact props. such as notched Izod and Gardner at -40, 0 and 23° of unreinforced and 30% glass reinforced Nylon 6,6, either unmodified, impact modified or ionomer modified have been reported [15]

Mechanical Properties Miscellaneous: Stiffness and strength are both reduced for glass fraction >0.20; the rubber reinforced grades are tougher under impact due to modification in the mode of cracking [16]

Hardness: Rockwell R120 (33% glass fibre, ASTM D785); M89, R120 (0.2% moisture); M70, R110 (1.5% moisture) [2]. Ball indentation hardness H 961/30 200 (dry); 150 (moist, 15% glass fibre), H 961/30 240 (dry); 190 (moist, 25% glass fibre), H 961/30 270 (dry); 200 (moist, 30% glass fibre), H 961/30 280 (dry); 210 (moist, 35% glass fibre), H 961/30 300 (dry); 260 (moist, 50% glass fibre, ISO 2039/1) [1]

Failure Properties General: Fatigue strength for glass reinforced thermoplastics declines linearly with logarithm of load cycles up to 10^6 cycles [8]. Glass fibre improves fatigue strength, unlike glass bead. Fractographic analysis shows that pull out of fibres requires considerable energy and arrests crack growth. Fibre reinforced Nylon 6,6 can withstand higher temps. than fibre reinforced Nylon 6 (based on temp./stress measurements) [9]. Strength of glass composites tested between 10–3000 cycles (min.)$^{-1}$ decreased significantly with increasing loading frequency, particularly in the range 10-600 cycles (min.)$^{-1}$, largely due to softening effect caused by increase of temp. [10]

Fracture Mechanical Properties: Fracture energy (housing) W_{50} 3–4 J (23°, dry); 15 J (23°, moist, 15% glass fibre, ISO 6603/1); 15 J (23°, moist, 25% glass fibre); 15 J (23°, moist, 30% glass fibre); 15 J (23°, moist, 35% glass fibre); 2 J (dry); 10 J (23°, moist, 50% glass fibre). The fracture mechanism of short E-glass fibres/Nylon 6,6 composites has been studied by scanning electron microscopy. The fracture processes, including crack initiation and propagation, were observed *in situ*. There are five stages of fracture: (1) voids located at fibre ends; (2) propagation of interfacial cracks on the fibre sides; (3) occurrence of bands of microcracks and shear bands in the matrix; (4) slow crack propagation in the shear bands with yielding and necking; (5) catastrophic crack propagation [21]

Friction Abrasion and Resistance: Coefficient of friction 0.15 (dry); 0.17 (moist, against metal, 33% glass fibre, ASTM D1894); 0.23 (dry); 0.23 (moist, against self, 33% glass fibre). Wear behaviour of short glass fibre reinforced Nylon 6,6 sliding against smooth steel has been reported [19]

Izod Notch:

No.	Value	Notch	Note
1	170 J/m	N	dry, 30% glass fibre, impact modified [4]
2	370–470 J/m	N	40% long glass fibre [7]

Izod Area:

No.	Value	Notch	Note
1	5.5 kJ/m^2	Unnotched	dry [1]
2	6.5 kJ/m^2	Unnotched	moist, 15% glass fibre
3	11.5 kJ/m^2	Unnotched	dry [1]
4	15.5 kJ/m^2	Unnotched	moist, 35% glass fibre
5	14.5 kJ/m^2	Unnotched	23°, moist, 50% glass fibre [1]

Charpy:

No.	Value	Notch	Note
1	30	N	23°, dry [1]
2	60	N	23°, moist, 15% glass fibre [1]
3	40	N	23°, dry [1]
4	55	N	23°, moist, 25% glass fibre [1]
5	55	N	23°, dry [1]
6	65	N	23°, moist, 50% glass fibre, DIN 53453 [1]
7	25	N	dry, 15% glass fibre [1]
8	45	N	-40°, dry, 50% glass fibre, DIN 53453 [1]
9	8	Y	dry [1]
10	11	Y	moist, 15% glass fibre [1]
11	15	Y	dry [1]
12	21	Y	moist, 50% glass fibre, ISO 179 [1]

– Nylon 6,6, Glass reinforced

Electrical Properties:
Electrical Properties General: Comparative Tracking index 550 (15–50% glass fibre, IEC 112/A) [1]

Surface/Volume Resistance:

No.	Value	Note	Type
1	0.001×10^{15} Ω.cm	dry, 15–50% glass fibre, ISO 93 [1]	S
2	$1 \times 10^{-5} \times 10^{15}$ Ω.cm	moist, 15–50% glass fibre, ISO 93 [1]	S
3	0.1×10^{15} Ω.cm	dry, 30% glass fibre, ASTM D257 [13]	S

Dielectric Permittivity (Constant):

No.	Value	Frequency	Note
1	3.5		dry, 15–25% glass fibre, IEC 250 [1]
2	5.5		moist, 15–25% glass fibre, IEC 250 [1]
3	3.5		dry, 30% glass fibre, IEC 250 [1]
4	5.6		moist, 30% glass fibre, IEC 250 [1]
5	3.5		dry, 35% glass fibre, IEC 250 [1]
6	5.7		moist, 35% glass fibre, IEC 250 [1]
7	3.8		dry, 50% glass fibre, IEC 250 [1]
8	6.6		moist, 50% glass fibre, IEC 250 [1]
9	4	100 Hz	0.2% moisture, 23°, 33% glass fibre
10	3.9	1 kHz	0.2% moisture, 23°, 33% glass fibre
11	3.5	1 MHz	0.2% moisture, 23°, 33% glass fibre
12	5.4	100 Hz	1.5% moisture, 23°, 33% glass fibre
13	5	1 kHz	1.5% moisture, 23°, 33% glass fibre
14	3.6	1 MHz	1.5% moisture, 23°, 33% glass fibre [2]

Dielectric Strength:

No.	Value	Note
1	17 kV.mm^{-1}	33% glass fibre [2]
2	21.6 kV.mm^{-1}	0.2% moist
3	18.4 kV.mm^{-1}	1.5% moist, short time method, 21° [3]
4	19.2 kV.mm^{-1}	0.2% moist
5	14 kV.mm^{-1}	1.5% moist, step by step method, 21° [3]
6	90 kV.mm^{-1}	dry, 15% glass fibre, IEC 243/1 [1]
7	50 kV.mm^{-1}	moist, 15% glass fibre, IEC 243/1 [1]
8	90 kV.mm^{-1}	dry, 25–50% glass fibre, IEC 243/1 [1]
9	80 kV.mm^{-1}	moist, 25–50% glass fibre, IEC 243/1 [1]

Arc Resistance:

No.	Value	Note
1	135s	33% glass fibre, ASTM D495

Dissipation (Power) Factor:

No.	Value	Frequency	Note
1	0.014		dry
2	0.16	1 MHz	moist, 15–30% glass fibre, IEC 250 [1]
3	0.15	1 MHz	moist, 35% glass fibre, IEC 250 [1]
4	0.015	1 MHz	dry, 50% glass fibre, IEC 250 [1]
5	0.13	100 Hz	[2]
6	0.15	1 kHz	[2]
7	0.02	1 MHz	23°, 0.2% moisture, 33% glass fibre [2]
8	0.03	100 Hz	
9	0.3	1 kHz	
10	0.05	1 MHz	23°, 1.5% moisture, 33% glass fibre [2]

Polymer Stability:
Upper Use Temperature:

No.	Value	Note
1	240°C	15–50% glass fibre [1]
2	218°C	Maranyl AMX9 [3]

Flammability: Flammability rating HB (1.6–3.2 mm thick, 15–50% glass fibre, UL 94) [1], 2–50 mm (Method BH, class 707, 25–50% glass fibre), 3–20 mm min.$^{-1}$ (method FH, class 707, 25–50% glass fibre), + rating (FMVSS 302, flammability rating for car interiors, 15–50% glass fibre). Improved fire resistant props. of Nylon 6 or 6,6 containing 3–25 wt.% of melamine, melamine cyanurate or a mixture thereof and 5–45 wt.% unsized glass fibres have been reported [20]

Environmental Stress: Max. dimensional change from 20% relative humidity to 80% relative humidity 0.4% (length), 1–2% (thickness) (30% glass fibre) [24]

Chemical Stability: After 100 days' exposure to aggressive solvents such as EtOH, petrol, detergents, formalin, cyclohexanone etc., the mech. props. of glass filled Nylon 6,6 are affected much less than the unreinforced Nylon [17]. 30% Tensile strength is lost after 88.5h immersion in 50% soln. of ethylene glycol (30% glass fibre). Unreinforced Nylon 6,6 loses 70–80% strength [28]

Applications/Commercial Products:
Processing & Manufacturing Routes: Batch and continuous processes are used. In both, the first step is to form a salt of the diamine and diacid. The salt is heated in a reactor to give polymerisation. Nylon 6,6 must be dried at 80° by passing hot inert gases over it. 70% of reinforced grades are injection moulded compared to 30% extruded. For extrusion L/D ratios of 20:1 to 24:1 are common with screw compression ratios in the range 3.5–4:1. Heating zones must be capable of holding temps. up to 290°. Good pressure and temp. controls are mandatory for safety and quality of product. Extrusion is used typically for pipe and complex shape profiles. Injection moulding is normally carried out with efficient reciprocating screw machines. For reinforced nylons, equipment life can be prolonged by using screws and barrels made of bimetallic alloys or nitrided steel. Nylon moulding grades are very fluid and drool must be controlled by using reverse-taper or positive shut-off nozzles. The mould life may be extended by chromium plating, especially where high gloss finish is required, as nylons precisely reflect the quality of the mould. The mould must be cored for cooling (usually 70–90°) as crystallisation must be critically controlled to ensure dimensional control. For optimum pressure transmission, material should flow from thick sections to thin sections. Nylons are suited to automated moulding operations [25]. Use of surface treated glass fibres as a reinforcement of anisotropic composites results in many improvements in props. such as stiffness. It can also lead to increased fatigue time limit in flow, improved hot shape retention, decreased thermal expansion, slower processing etc. The critical length of the fibre is of importance in processing and in the final props. of the composite [26]

– Nylon 6,6 Mineral filled

Mould Shrinkage (%):

No.	Value	Note
1	2–6%	0.2% moisture, 33% glass fibre [2]

Applications: Fibres (main use) - for carpets, tyres, textiles. Engineering plastics & resins - combines toughness with impacts & abrasion resistance. Fibres, cable sheathing, gears, tubing, connectors, fasteners, powder coating, solvent resistant containers. Reinforced grades of Nylon 6,6 (<30% glass fibre) are used mainly for the injection moulding of machinery parts, electrical insulating parts, housings such as coil formers and bearing cages. Reinforced grades (30–35%) find uses in lamp socket housings, cooling fans, insulating profiles for aluminium window frames; reinforced grades (35–55%) find uses in moulding gear wheels, solenoid valve housings, electric flow heaters, trailing cable attachments. Impact modified glass fibre reinforced grades are employed for applications requiring tough stiffness, notched impact and heat resistance, low warpage, e.g. housing components and wheel covers in automobiles. Other applications include use in high thrust electric trolling motors. Long glass fibre composite (50% glass fibre) for injection moulding of car parts offers improved impact and temp. related stiffness. Externally lubricated 30% glass fibre reinforced Nylon 6,6 modified with PTFE is used in moulded gears, cams, bearings and other moving parts requiring combination of high physical props., wear resistance and dimensional stability. Nylon 6,6 with 30–35% glass fibre is recommended for chains up to 30 m long. Longer chains require 50% glass fibre. Glass reinforced Nylon 6,6 is finding increasing uses in car engine compartments, air intake manifolds, cam covers, rocker covers and other heavy metal parts. Other applications include fan wheels for freezers, dishwasher pump housings, typewriter daisy wheels and rear view mirror housings. Flame retardant grades are used in many electrical applications, e.g. compact electrical contactor systems, multifunctional coffee grinder motor brackets. Nylon 6,6 remains the most widely used material in the electrical/electronics component industry. The thermoplastic air intake manifold for the 1995 GM Cadillac Northstar V-8 engine is made from BASF glass fibre reinforced Ultramid Nylon 6,6. Radiator tanks for European automotive air conditioning and engine cooling systems are now being made from heat and hydrol. resistant Nylon 6,6 containing 30% glass fibre

Trade name	Supplier
Akalon	Akzo
Amilan	Toray
Ashlene	Ashley Polymers
Celanese Nylons	Hoechst Celanese
Comtuf	ComAlloy International
ESD	RTP Company
Fibertil	DSM
Leona	Asahi Chemical
Maranyl	ICI, UK
RTP	RTP Company
Sniamid, Nail on plast	Technopolimeri SpA.
Technyl	Rhone-Poulenc
Ube	Ube Industries Ltd.
Ube	Unitika Ltd.
Ultramid	BASF
Vydyne	Monsanto Chemical
Zytel	DuPont

Bibliographic References

[1] *Ultramid S Polyamides*, BASF Plastics, 1993, (technical datasheet)
[2] Vydyne 909 Glass Reinforced Nylon 6, 6, Monsanto, 1991, (technical datasheet)
[3] *Plastics Design Forum*, 1992, **17**, 66
[4] *J. Mater. Eng.*, 1992, 6
[6] *Plastics Design Forum*, 1993, **18**, 64
[7] High Perform. Plast., Nat. Tech. Conf. - Soc. Plast. Eng., *(Prepr.)*, 1987, **4**, 2
[8] Adkins, D.W. and Kander, R.G., *How to Apply Advanced Composites Technology*, ASM International, 1988, 437
[9] Suzuki, H., Ueki, T. and Kunio, T., Proc. Jpn. Congr. Mater. Res., *30th*, SMSJ, 1986, 257
[10] Buglo, S.T. and Nikonorov, A.S., *Plast. Massy*, 1986, **1**, 59
[11] Cox, H.W. and Mentzer, C.C., *Polym. Eng. Sci.*, 1986, **26**, 488
[12] *Mod. Plast.*, Azco Engineering Plastics, 1989, **66**, 67
[13] *Polyamide 66 PA 140/1 GF30*, Albis, 1994
[14] Johnson, A.E., Moore, D.R., Prediger, R.S., Reed, P.E. and Turner, S., *J. Mater. Sci.*, 1987, **22**, 1724
[15] Nangrani, K.J., 41st Annual Conference, 1986, Atlanta, Georgia Soc. Plastics Ind., 3
[16] Bailey, R.S. and Bader, M.G., Int. Conf. Compos. Mater., ICCM-V, Conf. Proc., 5th Metallurgical Soc., 1985, 947
[17] Stolpovskaya, E.M., Elkina, E.V., Bulekina, L.A. and Kostrov, V.I., *Plast. Massy*, 1989, **5**, 36
[18] Yazbak, G. and Diraddo, R.W., *Plast. Technol.*, 1993, **39**, 61
[19] Voss, H., *Kunststoffe*, 1987, **77**, 272
[20] *U.S. Pat.*, 1988, 4 789 698, (Bayer)
[21] Sato, N., Kurauchi, T., Sato, S. and Kamigaito, O., *ASTM Spec. Tech. Publ.*, 1985, **868**
[22] Weiner, W., *Plaste Kautsch.*, 1988, **35**, 189
[23] From Concept to Commercialisation DuPont Engineering Polymers, 1995
[24] *Br. Plast. Rubber*, 1996, 1
[25] *Encycl. Polym. Sci. Technol.*, 1988, **11**, 453
[26] Miskolci, A., *Plasty Kauc.*, 1988, **25**, 270
[27] Eddy, J., *Plast. Eng. (N.Y.)*, 1994, **50**, 18
[28] Mater. Eng. (Cleveland) Rhone Poulenc, 1990, **107**, 24

Nylon 6,6 Mineral filled

N-51

Synonyms: *Poly(hexamethylenediamine-co-adipic acid). Poly(hexamethylene adipamide). Poly(iminohexamethylene iminoadipoyl). Poly(imino(1,6-dioxo-1,6-hexanediyl)imino-1,6-hexanediyl)*
Related Polymers: Nylon 6,6, Glass reinforced, Nylon 6,6 Carbon fibre filled, Nylon 6,6, Nylon 6,6 Aramid fibre filled, Nylon 6,6 Stainless steel fibre filled
Material class: Thermoplastic
Polymer Type: polyamide
Molecular Formula: $(C_{12}H_{22}N_2O_2)n$
Fragments: $C_{12}H_{22}N_2O_2$
Additives: Molybdenum disulfide. Flame retardant grade available

Volumetric & Calorimetric Properties:

Density:

No.	Value	Note
1	d 1.29–1.5 g/cm^3	(ASTM D792, 20–40% filler) [1]

Deflection Temperature:

No.	Value	Note
1	204°C	ASTM D648, at 1820 kPa, 40% filler
2	238°C	ASTM D648, at 455 kPa, 40% filler

Nylon 6-Nylon 6T copolymer (Semi-Aromatic)

Transport Properties:
Water Absorption:

No.	Value	Note
1	0.5 %	ASTM D570, 24h at 23°C, 40% filler [1]

Mechanical Properties:
Tensile (Young's) Modulus:

No.	Value	Note
1	5516–8274 MPa	ASTM D638, 20–40% filler [1]

Flexural Modulus:

No.	Value	Note
1	4826–6895 MPa	ASTM D790, 20–40% filler [1]

Tensile Strength Break:

No.	Value	Note	Elongation
1	69–76 MPa	ASTM D638, 20–40% filler [1]	2–4%

Flexural Strength Yield:

No.	Value	Note
1	117–138 MPa	ASTM D790, 20–40% filler [1]

Hardness: Rockwell, R 120 (ASTM D785, 40% filler) [1]

Izod Notch:

No.	Value	Notch	Note
1	37–43 J/m	Y	ASTM D256, 3.2 mm section, 20–40% filler [1]
2	267 J/m	N	ASTM D256, 3.2 mm section, 20–40% filler [1]

Electrical Properties:
Electrical Properties General: Volume resistivity $<1\times10^{14}$ $\Omega.cm$ (ASTM D257, 40% filler) [1].

Polymer Stability:
Flammability: HB at 1.5 mm (ASTM D635) [1]

Applications/Commercial Products:
Processing & Manufacturing Routes: Processing for injection moulding [1]:
Injection pressure 69–124Mpa
Melt temp. 277–299°C
Mold temp. 66–107°C
Drying 4 hrs at 79°C
Moisture content 0.2%
Dew point -18°C

Applications/Commercial Products:
Mould Shrinkage (%):

No.	Value	Note
1	0.7–1.3%	ASTM D955, 20–40% filler [1]

Trade name	Supplier
Frianyl	Frisetta Polymer
Nylene	Custom Resins
RTP series	RTP Co.
	PolyOne Edgetek

Bibliographic References
[1] RTP Co. datasheet http://www.rtpcompany.com, 2006

Nylon 6-Nylon 6T copolymer (Semi-Aromatic)

$$\left[\left[-NHCH_2(CH_2)_4\overset{O}{\underset{\|}{C}}-\right]_x-NHCH_2(CH_2)_4CH_2NH\overset{O}{\underset{\|}{C}}-\bigcirc-\overset{O}{\underset{\|}{C}}-\right]_y\Bigg]_n$$

Synonyms: *Ultramid T. Poly(ε-caprolactam-co-hexamethylene diamine-co-terephthalic acid)*
Related Polymers: Nylon 6, Nylon 6T
Monomers: ε-Caprolactam, 1,6-Hexanediamine, 1,4-Benzenedicarboxylic acid
Material class: Thermoplastic, Copolymers
Polymer Type: polyamide
Molecular Formula: $[(C_6H_{14}NO).(C_{14}H_{18}N_2O_2)]_n$
Fragments: $C_6H_{14}NO$ $C_{14}H_{18}N_2O_2$
Additives: Glass fibres

Volumetric & Calorimetric Properties:
Density:

No.	Value	Note
1	d 1.18 g/cm^3	
2	d 1.4 g/cm^3	mineral filled

Thermal Expansion Coefficient:

No.	Value	Note	Type
1	6×10^{-5} K^{-1}	23°, DIN 52612	L
2	8×10^{-5} K^{-1}	80°, DIN 52612	L
3	2.5×10^{-5} K^{-1}	glass fibre filled [1]	L
4	5×10^{-6} K^{-1}	mineral filled [1]	L

Thermal Conductivity:

No.	Value
1	0.23 W/mK

Specific Heat Capacity:

No.	Value	Type
1	1.5 kJ/kg.C	V

Melting Temperature:

No.	Value	Note
1	295°C	ISO 3146

Glass-Transition Temperature:

No.	Value	Note
1	105°C	[1]

Nylon 6-Nylon 6T copolymer (Semi-Aromatic)

Deflection Temperature:

No.	Value	Note
1	100°C	1.8 MPa, DIN
2	120°C	0.45 MPa [1]
3	270°C	1.8 MPa, glass fibre filled
4	280°C	0.45 MPa, glass fibre filled

Vicat Softening Point:

No.	Value	Note
1	280°C	DIN 53460, method B, 50° rate

Transport Properties:

Melt Flow Index:

No.	Value	Note
1	30 g/10 min	325° [1]
2	8 g/10 min	mineral filled [1]

Water Content: 1.6–2% (saturation, 23°, 50% relative humidity) [1]

Water Absorption:

No.	Value	Note
1	6.5–7.5 %	24h, 23°, DIN 53495L [1]
2	4.7 %	mineral filled

Mechanical Properties:

Mechanical Properties General: Mechanical props. stable to 60°

Tensile (Young's) Modulus:

No.	Value	Note
1	3200 MPa	DIN 53457 [1]
2	9000 MPa	glass fibre filled [1]
3	4500 MPa	mineral filled [1]

Tensile Strength Break:

No.	Value	Note	Elongation
1	180 MPa	3.5% glass fibre filled [1]	10–20%
2	85 MPa	12.5% mineral filled [1]	

Tensile Strength Yield:

No.	Value	Note
1	100 MPa	4.5%, DIN 53457 [1]

Miscellaneous Moduli:

No.	Value	Note	Type
1	2300 MPa	100h, 23°, DIN 53444, 0.5% elongation, [1]	tensile creep modulus

Hardness: Ball indentation hardness H961/30 190 MPa (ISO 2039-1) [1]

Failure Properties General: Impact failure energy 100 (23°, DIN 53443); 40 (-40°, DIN 53443) [1]

Izod Notch:

No.	Value	Notch	Note
1	12 J/m	Y	ASTM D256

Charpy:

No.	Value	Notch	Break	Note
1	100	Y	N	23°, 50% relative humidity, DIN 53453
2	90	Y		23°, 50% relative humidity, glass fibre filled [1]
3	200	Y		23°, 50% relative humidity, mineral filled [1]

Electrical Properties:

Electrical Properties General: Electrolytic corrosion AI (DIN 0303 T6) [1]

Surface/Volume Resistance:

No.	Value	Note	Type
1	0.01×10^{15} Ω.cm	DIN 0303-T30 [1]	S

Dielectric Permittivity (Constant):

No.	Value	Frequency	Note
1	4	1 MHz	DIN 0303T4 [1]

Dielectric Strength:

No.	Value	Note
1	100 kV.mm^{-1}	DIN 0303-T21 [1]

Complex Permittivity and Electroactive Polymers: Comparative tracking index CTI 600 (test soln. A, DIN 0303-T1).

Dissipation (Power) Factor:

No.	Value	Frequency	Note
1	0.03	1 MHz	dry, DIN 0303-T4 [1]
2	0.04	1 MHz	saturated, DIN 0303-T4 [1]

Polymer Stability:

Thermal Stability General: Stable in nitrogen to about 310°. 50% loss of tensile strength 20000h (110°, ISO 216-1); 5000h (130°, ISO 216-1)

Upper Use Temperature:

No.	Value
1	250°C

Flammability: Flammability rating HB (1.6 mm, UL94). Limiting oxygen index 24%

Applications/Commercial Products:

Processing & Manufacturing Routes: Prod. by copolymerisation of caprolactam, and Nylon 6,T salt in an autoclave. Processing is similar to that for Nylon 6,6 hence conventional Nylon injection moulding and extruders can be employed

Mould Shrinkage (%):

No.	Value	Note
1	1%	longitudinal
3	1.2%	transverse
4	0.4–0.7%	glass fibre filled [1]

Applications: Automotive engine parts, bearing retainers, mechanical and electrical parts requiring additional heat resistance. It is available unreinforced, in impact modified glass reinforced grades, mineral and carbon-filled grades

Trade name	Details	Supplier
Ultramid T	KR grades include miner.	BASF

Bibliographic References

[1] *Ultramid T*, BASF, 1997, (technical datasheet)

Nylon-Polycarbonate blends   N-53

Synonyms: *Nylon-PC blends*
Related Polymers: Nylon 12, Nylon 6, Nylon 6,6, Polycarbonate
Material class: Thermoplastic, Blends
Polymer Type: polyamide, polycarbonates (miscellaneous)
Additives: Poly(allyl-*co*-maleic anhydride) and poly(ω-caprolactone) added as reactive compatibilisers

Volumetric & Calorimetric Properties:
Glass-Transition Temperature:

No.	Value	Note
1	65°C	75% nylon/25% PC [3]

Transition Temperature General: Miscibility of polymers influences thermal transitions, possibly due to copolymer formation [2]

Surface Properties & Solubility:
Solubility Properties General: Miscibility has been studied as a function of mixing time and composition. The two components show well segregated phases; at high nylon content they become homogeneous after long mixing time, possibly due to block copolymerisation [3]. Phase contrast and glass-transition studies suggest immiscibility; order of homogeneity amorph. Nylon >Nylon 12 >Nylon 6 >Nylon 6,6

Transport Properties:
Polymer Melts: Compatibilisation reduces MW and reduces melt viscosity [3]

Mechanical Properties:
Mechanical Properties General: Useful improvement in phys. props. can be achieved [1]
Impact Strength: No improvement in notched impact strength is observed on blending both polymers [1]

Optical Properties:
Transmission and Spectra: Ir spectral data suggest reactivity between nylon and compatibiliser

Applications/Commercial Products:
Processing & Manufacturing Routes: Difficult to mould from melt blends. Poly(ω-caprolactone) compatibilised ternary blends may be prepared by screw extrusion [5]
Applications: As yet no commercial applications

Bibliographic References

[1] Lapshin, V.V., Andreeva, T.I., Kolerov, A.S., Soloveva, I.I. and Vakhinskaya, T.N., *Plast. Massy*, 1987, **7**, 42
[2] Cortazar, M., Eguiazabal, J.I. and Iruin, J.J., *Br. Polym. J.*, 1989, **21**, 395
[3] Gattiglia, E., La Mantia, F.P., Turturro, A. and Valenza, A., *Polym. Bull. (Berlin)*, 1989, **21**, 47
[4] Sato, M., Akiyama, S. and Honda, S., *Kobunshi Ronbunshu*, 1990, **47**, 287
[5] Kim, W., Park, C. and Burns, C.M., *J. Appl. Polym. Sci.*, 1993, **49**, 1003

Nylon polyester copolymer glass fibre filled   N-54

Synonyms: *Nylon 6 T/Me(5)/PET copolymer. High temperature nylon*
Related Polymers: Nylon 6, PET
Monomers: Ethylene glycol, 1,6-Hexanediamine, Methyl pentamethylene diamine, 1,4-Benzenedicarboxylic acid
Material class: Thermoplastic, Copolymers
Polymer Type: polyamide, polyester

Volumetric & Calorimetric Properties:
Density:

No.	Value	Note
1	d 1.47 g/cm^3	ASTM D792

Thermal Expansion Coefficient:

No.	Value	Note	Type
1	1.8×10^{-5} K^{-1}	-40–23°, ASTM E831, direction of flow	L
2	1.5×10^{-5} K^{-1}	23–125°, ASTM E831, direction of flow	L
3	4×10^{-5} K^{-1}	-47–23°, ASTM E831, transverse	L

Melting Temperature:

No.	Value	Note
1	300°C	ASTM D3418 [1]

Deflection Temperature:

No.	Value	Note
1	260°C	1.8 MPa, 23°, ASTM E831 [1]

Transport Properties:
Water Absorption:

No.	Value	Note
1	0.4 %	24h, immersion, ASTM D570
2	3.5 %	saturation, ASTM D570

Mechanical Properties:
Mechanical Properties General: Mechanical props. decrease with increasing temp.

Tensile (Young's) Modulus:

No.	Value	Note
1	12100 MPa	23°, ASTM D638
2	11200 MPa	100°, ASTM D638
3	7000 MPa	150°, ASTM D638
4	5400 MPa	175°, ASTM D638

Nylon/polyolefin blends

Flexural Modulus:

No.	Value	Note
1	10300 MPa	23°, ASTM D790
2	10000 MPa	100°, ASTM D790
3	5100 MPa	150°, ASTM D790
4	3400 MPa	200°, ASTM D790

Tensile Strength Break:

No.	Value	Note	Elongation
1	230 MPa	-40°, ASTM D638	5.3%
2	214 MPa	23°, ASTM D638	1.75%
3	165 MPa	100°, ASTM D638	
4	70 MPa	175°, ASTM D638	5.3%

Flexural Strength at Break:

No.	Value	Note
1	304 MPa	23°, ASTM D790
2	295 MPa	23°, 50% relative humidity
3	253 MPa	100°, ASTM D790
4	184 MPa	100°, 50% relative humidity
5	86 MPa	200°, ASTM D790
6	78 MPa	200°, 50% relative humidity

Friction Abrasion and Resistance: Taber abrasion resistance 26 mg (1000 cycles)$^{-1}$ (CS-17, 1 kg, ASTM D1044)

Izod Notch:

No.	Value	Notch	Note
1	112 J/m	Y	23°, ASTM D256
2	107 J/m	Y	23°, 50% relative humidity
3	730 J/m	N	23°, ASTM D4812
4	680 J/m	N	23°, 50% relative humidity

Electrical Properties:
Surface/Volume Resistance:

No.	Value	Note	Type
1	0.1×10^{15} Ω.cm	ASTM D257	S

Dielectric Permittivity (Constant):

No.	Value	Frequency	Note
1	4.3	1 kHz	23°
2	4	1 MHz	23°
3	5.1	1 kHz	100°
4	4.7	1 MHz	200°
5	170	1 kHz	200°
6	10	1 MHz	200°

Dielectric Strength:

No.	Value	Note
1	27.1 kV.mm^{-1}	23°, 1.6 mm
2	25.5 kV.mm^{-1}	100°, 1.6 mm
3	19.6 kV.mm^{-1}	175°, 1.6 mm
4	63 kV.mm^{-1}	200°, 1.6 mm, ASTM D149

Dissipation (Power) Factor:

No.	Value	Frequency	Note
1	0.01	1 kHz	23°
2	0.02	1 MHz	23°
3	0.02	1 kHz	100°
4	0.03	1 MHz	100°
5	1.29	1 kHz	200°
6	0.19	1 MHz	200°, ASTM D150

Polymer Stability:
Flammability: Flammability rating HB (0.85 mm, UL 94)

Applications/Commercial Products:
Processing & Manufacturing Routes: Melt condensation. Optimum mould temp. 130–160°

Mould Shrinkage (%):

No.	Value	Note
1	0.3%	flow
2	0.9%	transverse

Applications: Surface mount components, connectors, fittings in distributor transformers. Other electrical applications include coil forms, encapsulated solenoids, lamp reflectors, stock insulators, fuse holders. Automotive industry for underbonnet applications

Trade name	Supplier
Zytel HTN	DuPont

Bibliographic References
[1] *Zytel HTN*, DuPont, 1996, (technical datasheet)

Nylon/polyolefin blends

Synonyms: *Orgalloy RS 6000*
Material class: Thermoplastic, Blends
Polymer Type: polypropylene, polyamide

Volumetric & Calorimetric Properties:
Density:

No.	Value
1	d 1.04 g/cm^3

Thermal Expansion Coefficient:

No.	Value	Note	Type
1	9.3×10^{-5} K^{-1}	longitudinal	L
2	0.00013 K^{-1}	transverse	L

Nylon/polyolefin blends

Melting Temperature:

No.	Value	Note
1	220°C	ISO R1183

Deflection Temperature:

No.	Value	Note
1	130°C	0.46 MPa, ISO R75
2	75°C	1.82 MPa, ISO R75

Vicat Softening Point:

No.	Value	Note
1	195°C	10 N, ISO 306
2	130°C	50 N, ISO 306

Surface Properties & Solubility:
Solvents/Non-solvents: No known solvents

Transport Properties:
Permeability of Gases: Permeability [O_2] 45 cm^3 (m^2 day)$^{-1}$, 23°; [CO_2] 150 cm^3 (m^2 day)$^{-1}$, 23°
Permeability of Liquids: Permeability [H_2O] 30 g (m^2 day)$^{-1}$, 25°, 75% relative humidity; [EtOH] 50 g (m^2 day)$^{-1}$, 20°, 60% relative humidity; [pentane] 0.2 g (m^2 day)$^{-1}$, 20°, 60% relative humidity; [white spirit] 0.2 g (m^2 day)$^{-1}$, 20°, 60% relative humidity

Water Absorption:

No.	Value	Note
1	0.8 %	24h, immersion, 23°, ISO 62
2	1.6 %	equilibrium, 23°, 50% relative humidity

Mechanical Properties:
Flexural Modulus:

No.	Value	Note
1	7000 MPa	glass reinforced, ASTM D638 [1]
2	2200 MPa	dry, moulded, ISO R178 [2]

Tensile Strength Break:

No.	Value	Note	Elongation
1	50 MPa	ASTM D638 [2]	220%

Flexural Strength at Break:

No.	Value	Note
1	73 MPa	ISO R178 [2]

Hardness: Shore D75 (ISO R868) [2]
Izod Notch:

No.	Value	Notch	Note
1	10 J/m	Y	[3]
2	100 J/m	Y	23°, ISO R180 [2]

Charpy:

No.	Value	Notch	Note
1	16	Y	23°, dry, ISO R179 [2]
2	12	Y	-12°, dry, ISO R179 [2]

Electrical Properties:
Surface/Volume Resistance:

No.	Value	Note	Type
1	4.8×10^{15} Ω.cm	ASTM D257	S

Dielectric Strength:

No.	Value	Note
1	19 kV.mm^{-1}	2 mm thick, ASTM D149

Complex Permittivity and Electroactive Polymers: Tracking resistance 600 V (min., NF C26-220)
Dissipation (Power) Factor:

No.	Value	Frequency
1	0.062	100 Hz
2	0.065	1 kHz
3	0.025	1 MHz

Polymer Stability:
Flammability: Flammability rating HB (UL94). Burning rate does not exceed 40 mm min^{-1} greater than 3 mm thick
Environmental Stress: Very good all round chemical resistance
Chemical Stability: Excellent resistance to all hydrocarbons, alcohols
Hydrolytic Stability: Water absorption <20% of Nylon 6 at equilibrium results in marked improvement to resistance to aq. acids and alkalis due to polypropylene content. Resistant to hydrol. by 10% NaOH at 85° but degraded slowly by dilute sulfuric acid at 85°. Resistant to 3% aq. sulfuric acid at 23°

Applications/Commercial Products:
Processing & Manufacturing Routes: Melt blending with grafting reactions in extruder
Mould Shrinkage (%):

No.	Value	Note
1	0.7%	longitudinal
2	1.2%	transverse

Applications: Car under bonnet applications, power tool housings; packaging of hollow vessels, films, moulded compounds, blow moulded engineering components, cable manufacture, tube and profile extrusion, motorcycle cowlings

Trade name	Supplier
Orgalloy	Elf Atochem (UK) Ltd.
Systemer	Showa Denko

Bibliographic References
[1] Gerard, J., *Mater. Tech. (Paris)*, 1980, **778 (11-12)**, 6, (glass reinforced grade)
[2] *Orgalloy Thermoplastic alloy*, Elf Atochem, 1996, (technical datasheet)
[3] Gonzalez-Monteil, A., Keskkula, H. and Paul, D.R., *J. Mater. Eng.*, 1992, **109**, 15, (glass reinforced grade)

[4] Harada, H., *Plast. Age*, 1990, **36**, 145
[5] *J. New Materials World*, 1991, **11**, 5
[6] Gonzalez-Monteil, A., Keskkula, H. and Paul, D.R., *Polym. Mater. Sci. Eng.*, 1993, 194
[7] Lawson, D.F., Hergenrother, W.L. and Matlock, M.C., *Polym. Prepr. (Am. Chem. Soc., Div. Polym. Chem.)*, 1988, **29**, 193
[8] Lohe, P. and Arndt, J., *Kunststoffe*, 1988, **78**, 49
[9] La Mantia, F.P. and Valenza, A., *J. Therm. Anal.*, 1988, **34**, 497
[10] La Mantia, F.P. and Valenza, A., *Eur. Polym. J.*, 1988, **24**, 825
[11] Dagli, S.S., Dey, S., Tupil, R. and Xanthos M., *J. Vinyl Addit. Technol.*, 1995, **1**, 195
[12] Abe, H. and Hosoda, S.I., *New Materials and Processes for the Future*, Conference, Chiba, Japan, 1989
[13] Holsti-Miettinen, R.H., Hietaoja, P.T., Seppala, J.V. and Ikkala, O.T., *J. Appl. Polym. Sci.*, 1994, **54**, 1613
[14] Paul, D.R., Gonzalez-Monteil, A. and Keskkula, H., *Polymer*, 1995, **36**, 4605, 4621
[15] Dagli, S.S., Kamdar, K.M. and Xanthos, M., *Eng. Plast.*, 1995, **8**, 197
[16] Gheluwe, P., Van Favis, B.D. and Chalifoux, J.P., *J. Mater. Sci.*, 1988, **23**, 3910
[17] Hiltner, A., Duvall, J., Selliti, C., Topolkaraev, V. et al, *Polymer*, 1994, **35**, 3948
[18] Sheu, E.Y., Lin, J.S. and Jois, Y.H.R., *J. Appl. Polym. Sci.*, 1995, **55**, 656
[19] Lee, W.C., Wu, J., Kuo, W.F., Kao, H.C., Lee M.S. and Lin J.L., *Adv. Polym. Technol.*, 1995, **14**, 47
[20] Tang, T., Li, H. and Huang, B., *J. Macromol. Sci., Phys.*, 1994, **195**, 2931
[21] Tang, T., Lei, Z., Zhang, X., Chen, H. and Huang, B., *Polymer*, 1995, **36**, 5061

Nylon 12-*block*-poly(tetramethylene glycol)   N-56

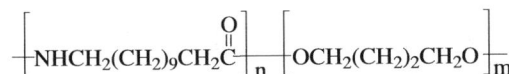

Synonyms: *PA12-block-PTMG*. Polyether Block Amide
Related Polymers: Nylon 12
Monomers: 12-Aminododecanoic acid, Tetramethylene glycol
Material class: Thermoplastic, Copolymers
Polymer Type: polyamide, polyester
Molecular Formula: $[(C_{12}H_{23}NO)_n.(C_4H_8O_2)_m]$
Fragments: $C_{12}H_{23}NO$ $C_4H_8O_2$
Molecular Weight: M_n 2135 or 4200 for the Nylon 12 block. M_n 2032 for the polyether block [1]
General Information: Behaves as a thermoplastic rubber. It is rigid due to crystalline segments at room temp., but rubbery at elevated temps. when hard segments have melted. Pebax is a thermoplastic block copolymer of Nylon 12 and polytetramethylene glycol, with hard nylon and soft ether segments. The polymer has been studied by a thermally stimulated current method [1,2]

Volumetric & Calorimetric Properties:
Density:

No.	Value
1	d 1.01–1.04 g/cm^3

Thermal Expansion Coefficient:

No.	Value	Note	Type
1	0.0002 K^{-1}	5% cryst.	L
2	0.00016 K^{-1}	25% cryst.	L

Melting Temperature:

No.	Value	Note
1	133.5°C	5% cryst. [3]
2	174°C	25% cryst. [3,4]

Deflection Temperature:

No.	Value	Note
1	42°C	0.46 MPa, 5% cryst.
2	99°C	0.46 MPa, 25% cryst. [3]

Vicat Softening Point:

No.	Value	Note
1	165°C	25% cryst., ASTM D1525
2	60°C	5% cryst., ASTM D1525

Transport Properties:
Melt Flow Index:

No.	Value	Note
1	10 g/10 min	10 min., 5% cryst. [3]
2	6 g/10 min	25% cryst. [3]

Polymer Melts: Melt viscosity 2000 Pa s (10 s^{-1}, 170°, 5% cryst.); 100 Pa s (10000 s^{-1}, 170°, 5% cryst.)

Mechanical Properties:
Mechanical Properties General: Very good impact resistance at low temp., good flexibility and dynamic props.

Tensile (Young's) Modulus:

No.	Value	Note
1	10.4 MPa	5% cryst., ASTM D638

Flexural Modulus:

No.	Value	Note
1	15 MPa	5% cryst., ASTM D790
2	390 MPa	25% cryst., ASTM D790

Tensile Strength Break:

No.	Value	Note	Elongation
1	29–33 MPa	ASTM D638	
2	52 MPa	ASTM D638 [3]	420%

Tensile Strength Yield:

No.	Value	Elongation	Note
1	24 MPa	18%	25% cryst., ASTM D638 [3]

Impact Strength: No break (notched, DIN 53453); 50 J m^{-1} (notched, 40°, 25% cryst., ASTM D256) [3]
Hardness: Shore D69 (25% cryst.); D25 (5% cryst.); A75 (5% cryst., ASTM D2240) [3]
Failure Properties General: Tear resistance 58 kN m^{-1} (unnotched, 5% cryst.); 38 kN m^{-1} (notched, 5% cryst., ASTM D1242)
Friction Abrasion and Resistance: Abrasion resistance 94 mg (1000 cycles)$^{-1}$, 5% cryst., ASTM D1242

Electrical Properties:
Surface/Volume Resistance:

No.	Value	Note	Type
3	0.003×10^{15} Ω.cm	5% cryst., ASTM D257 [3]	S
4	0.004×10^{15} Ω.cm	25% cryst., ASTM D257 [3]	S

Polymer Stability:
Polymer Stability General: Has few property variations between -40 and 80°
Chemical Stability: Resistant to most chemical attack

Applications/Commercial Products:
Processing & Manufacturing Routes: May be moulded into a variety of shapes. Moisture levels should be below 0.2% before processing
Mould Shrinkage (%):

No.	Value	Note
1	0.5%	5% cryst.
2	1.1%	25% cryst.

Applications: Suitable for a wide range of applications where good mech. props. are important, casings, fittings, keyboard, sports equipment. May also be extruded into films, sheaths, filaments, tubes and sheets

Trade name	Supplier
Pebax	Elf Atochem (UK) Ltd.

Bibliographic References
[1] Faruque, A.S. and Lacabanne, C., *Polymer*, 1986, **27**, 527, (morphology)
[2] Faruque, A.S. and Lacabanne, C., *J. Phys. D: Appl. Phys.*, 1987, **20**, 939, (constant)
[3] Pebax. Polyether blockamides, *Processing*, Elf Atochem, 1996, (technical datasheet)

Nylon powder coatings   N-57

Related Polymers: Nylon 6, Nylon 6,6, Nylon 11, Nylon 12
General Information: Polyamides find extensive use as anti-corrosion coatings for metallic substrates, but only Nylon 11 and Nylon 12 can readily be prod. in a suitable powder form for powder coating. The most common methods of application are fluidised bed dipping and electrostatic spraying; flame spraying is also used for large articles but can be very wasteful of material. The metal part which is to be covered must be carefully cleaned of oils, greases and oxides, and primered prior to coating. Nylon coatings give low coefficient of friction, high abrasion resistance and other useful props., and act as a barrier to corrosion. Typical film thicknesses are between 0.1 and 0.38 mm. Fluidised bed coating [1] is simple to use and produces thick durable coatings from 0.0008 to 0.25 in. thick on a variety of substrates. It is especially suitable for protective and decorative coatings on metal. The fluidised bed consists of a suitable sized tank to be able to immerse the article to be coated, two thirds filled with nylon powder. Air is blown through a porous base to expand and produce a fluid-like state in the powder into which an article preheated to 250-350° can easily be immersed. The powder melts on contact with the hot metal surface. The coating thickness applied is dependent upon the mass of the part and the time of dipping. Dip coating is readily adapted to production line operations with minimum of labour and very little loss of powder. It is particularly well suited to wire coatings, where the area to be coated is small [2]. Spray methods include electrostatic, airstatic and flock sprays for powders [3]. Electrostatic spraying needs very fine powder with a particle size 60 μm and produces much thinner coatings than those prod. by dip coating, the optimum range being 120-150 μm. The powder is fed at high electrical potentials through a nozzle where it becomes electrically charged, then sprayed onto a metal surface at zero potential and held there by the charge until fused by heating into a continuous coating [4,5]. Flame spraying, a slow operation with high powder losses, is suited to coating very large objects. It can be employed to coat non-metallic flammable substrates such as wood, rubber, masonry and other items, because the flame is used only to melt the plastic powder near the applicator and is not directly involved with the substrate [6]. The advantages of powder coatings over traditional paint systems have been reviewed for Nylon 11 [7]. Flame spraying is singled out as having future potential due to its ease of use on site. Industrially, continuous coating methods have been developed to coat welded tubes up to 30 mm diameter with circular, square or oval profiles, with up to 100 μm thickness applied at rates of 100 m (min)$^{-1}$. The tubing may subsequently be shaped for use in central heating installations, garden tools or fencing. A minicoat system has also been developed for coating small items from a few mm to 10 cm. Elf-Aquitaine, manufacturers of Rylsan Nylon 11 and 12 powders, have been instrumental in the development of these techniques [8]. Powder technology is continuously being developed [9]. Nylon coated aluminium tube components for cars have been prod. [10]; these are 50% lighter than similar stainless-steel tubes. Nylon coated steel has made inroads into the stainless-steel market in certain applications, primarily in the automotive industry, where superior wear resistance or lubricity is needed and where the bright look of stainless steel is not needed, e.g. window trims and wheel opening moulds [11]. Coatings can also be applied to complex geometric shapes to provide increased service life for the chemical processing, petrochemical, pulp and paper, semi-conductor, nuclear electronics and military industries [12]. Techniques have also been developed for prod. Nylon 6 suitable for powder coating.

Bibliographic References
[1] Hornsby, P., *J. Prod. Finish. (London)*, 1987, **40**, 6, 12
[2] Rees, H., *Powder Coatings - Conversion from Wet to dry*, 1985, 7
[3] Brooks, M.F., *Surf. J. Int.*, 1986, **1**, 26
[4] *Encycl. Polym. Sci. Technol.*, 2nd edn., (ed. J.I. Kroschwitz), John Wiley and Sons, 1985
[5] *Finishing*, Nylon powders, 1989, **30**, 39, 41
[6] *Finishing*, Steel Structures Painting Council, 1986, 205
[7] Blackcern, C.E., *Surf. Eng.*, 1987, **3**, 29
[8] *Rylsan nylon 11*, Elf-Aquitaine, (technical datasheet)
[9] Elf Atochem Ltd., *K-Plast. Kautsch ZTG*, 1988, **376**, 24
[10] Ballard, R.L., *J. Alum. Ind.*, 1989, **8**, 12, 14
[11] Balcerek, T., *J. Am. Met. Mark, Metalwork News Ed.*, 1986, **94(52)**, 24
[12] Ogden, S., *Managing Corros. Plast.*, 1987, **8**, 4
[13] Pisev, D., Angelova, N. and Dimov, K., *Plaste Kautsch.*, May, 1989, **36**, 165

Nylon 6,6 Stainless steel fibre filled   N-58

Synonyms: Poly(hexamethylenediamine-co-adipic acid). Poly(hexamethylene adipamide). Poly(iminohexamethylene iminoadipoyl). Poly[imino(1,6-dioxo-1,6-hexanediyl)imino-1,6-hexanediyl]
Related Polymers: Nylon 6,6, Glass reinforced, Nylon 6,6 Carbon fibre filled, Nylon 6,6, Nylon 6,6 Aramid fibre filled, Nylon 6,6 Mineral filled
Material class: Thermoplastic
Polymer Type: polyamide
Molecular Formula: $(C_{12}H_{22}N_2O_2)n$
Fragments: $C_{12}H_{22}N_2O_2$

— Nylon 6,T

Volumetric & Calorimetric Properties:
Density:

No.	Value	Note
1	d 1.18–1.3 g/cm^3	ASTM D792, 5–20% filler [1]

Mechanical Properties:
Tensile (Young's) Modulus:

No.	Value	Note
1	3448–4826 MPa	ASTM D638, 5–20% filler [1]

Flexural Modulus:

No.	Value	Note
1	3103–3448 MPa	ASTM D790, 5–20% filler [1]

Tensile Strength Break:

No.	Value	Note	Elongation
1	69–83 MPa	ASTM D638, 5–20% filler [1]	4–12%

Flexural Strength Yield:

No.	Value	Note
1	110–124 MPa	ASTM D790, 5–20% filler [1]

Izod Notch:

No.	Value	Notch	Note
1	43 J/m	Y	ASTM D256, 3.2mm section, 5–20% filler [1]
2	641–801 J/m	N	ASTM D256, 3.2mm section, 5–20% filler [1]

Electrical Properties:
Electrical Properties General: Surface resistivity <10– <1 Ω.cm (ASTM D257, 5–20% filler) [1]
Volume resistivity <10000– <1x10^6 Ω.cm (ASTM D257, 5–20% filler) [1]

Electrical Properties:
Surface/Volume Resistance:

No.	Value	Note	Type
1	1000–100000 × 10^{15} Ω.cm	(ASTM D257, 5–20% filler) [1]	S

Static Electrification: Static decay: <2s (FTMS101C4046.1, MIL-PRF-81705D, 5kV-50V, 12% RH) [1]

Applications/Commercial Products:
Processing & Manufacturing Routes: Processing for injection molding [1]:
Injection pressure 69–103Mpa
Melt temp. 252–293°C
Mold temp. 79–99°C
Drying 4 hrs at 79°C
Moisture content 0.2%

Applications/Commercial Products:
Mould Shrinkage (%):

No.	Value	Note
1	1–1.8%	ASTM D955, 5–20% filler [1]

Trade name	Supplier
Celstran	Ticona
RTP Series	RTP Co.
	PolyOne Edgetek

Bibliographic References
[1] RTP Co. datasheet http://www.rtpcompany.com, 2006

Nylon 6,T

Synonyms: *Poly(hexamethylene diamine-co-terephthalic acid). Poly(hexamethylene terephthalate). Poly(iminocarbonyl-1,4-phenylenecarbonylimino-1,6-hexanediyl)*
Related Polymers: Terephthalic/Isophthalic acid copolyamides
Monomers: 1,6-Diaminohexane, 1,4-Benzenedicarboxylic acid
Material class: Thermoplastic
Polymer Type: polyamide
CAS Number:

CAS Reg. No.
24938-70-3

Molecular Formula: $(C_{14}H_{18}N_2O_2)_n$
Fragments: $C_{14}H_{18}N_2O_2$
Additives: 0–0.5% Ammonium hypophosphite (usually 0.1%) as polymerisation catalyst and thermal stabiliser
General Information: Highly crystalline, melts at 370° (above its decomposition temp.). Monomers are inexpensive but need solid phase polymerisation to give sufficient MW for spinning. Fibres spun from sulfuric acid have props. on par with Nylon 6,6 fibres and retain strength to much higher temps. [1]. Unit cell parameter c 15.6 [2]

Volumetric & Calorimetric Properties:
Density:

No.	Value	Note
1	d 1.21 g/cm^3	[4]

Melting Temperature:

No.	Value	Note
1	370°C	[4,5]
2	350–371°C	[4,5]

Glass-Transition Temperature:

No.	Value	Note
1	140°C	[3]

Surface Properties & Solubility:
Solvents/Non-solvents: Being very crystalline, it is inert to the majority of solvents. Dissolved by strong acids (sulfuric, formic) and by strong polar solvents such as DMF or *N*-methylpyrrolidone. Solvent action is increased by addition of 1–2% CaCl$_2$ or boric acid. Resistant to alkalis

Nylon 6(3)T

Synonyms: Poly[iminocarbonyl-1,4-phenylenecarbonylimino(trimethyl-1,6-hexanediyl)]. Poly(2,2,4-trimethyl-1,6-hexanediamine-co-2,4,4-trimethyl-1,6-hexanediamine-co-1,4-benzenedicarboxylic acid)
Related Polymers: Nylon
Monomers: 1,4-Benzenedicarboxylic acid, 2,2,4-Trimethylhexanediamine, 2,4,4-Trimethylhexanediamine
Material class: Thermoplastic, Copolymers
Polymer Type: polyamide
CAS Number:

CAS Reg. No.	Note
9071-17-4	Trogamid T
25497-66-9	

Molecular Formula: $(C_{17}H_{24}N_2O_2)_n$, $[(C_8H_4O_2).(C_9H_{20}N_2)]_n$
Fragments: $C_{17}N_{24}N_2O_2$ $C_8H_4O_2$ $C_9H_{20}N_2$
Molecular Weight: M_n 28000. M_n 26700. M_n 20000, MW 44180 (Trogamid T)
Additives: 1% Ammonium hypophosphite (by weight) as antioxidant and catalyst. Glass fibre reinforced up to 35% confers rigidity
General Information: An amorph. transparent copolyamide. It is a rigid and tough material with good mech. props. stable to water absorption and aliphatic and aromatic hydrocarbons. Trogamid T is a commercial engineering resin with good retention of props. up to 120°. It is very hard and has excellent vapour impermeability
Morphology: Irregularities in struct. lead to low crystallinity and transparency

Volumetric & Calorimetric Properties:
Density:

No.	Value	Note
1	d 1.12 g/cm^3	ASTM D792 [1]

Thermal Expansion Coefficient:

No.	Value	Note	Type
1	6×10^{-5} K^{-1}	-30–30°, ASTM D696	L
2	5.1×10^{-5} K^{-1}	-30–30° [1]	L
3	3×10^{-5} K^{-1}	35% glass fibre	L

Thermal Conductivity:

No.	Value	Note
1	0.18 W/mK	[1]
2	0.21 W/mK	[3,8]

Specific Heat Capacity:

No.	Value	Note	Type
1	1.47 kJ/kg.C	[1]	P
2	1.45 kJ/kg.C	[3]	P

Transport Properties:
Polymer Solutions Dilute: η_{inh} 1 dl g^{-1} (96% H$_2$SO$_4$, M$_n$ 15000)
Water Content: 4.5% (70% relative humidity)

Mechanical Properties:
Mechanical Properties General: Tenacity (of fibre) 0.22–0.44 N tex^{-1} (30–45% elongation) [4]

Tensile (Young's) Modulus:

No.	Value
1	4250 MPa

Tensile Strength Break:

No.	Value	Note	Elongation
1	425 MPa	21° [4]	35%
2	100 MPa	185°, 5h [4]	

Failure Properties General: Tensile strength retained after ageing in air, to a certain extent at high temp. [4]

Applications/Commercial Products:
Processing & Manufacturing Routes: May be polymerised in the solid state by heating a suspension of low water sol. salt [11] and 0.2–0.5% ammonium hypophosphite in liq. paraffin or other inert liq. at 200–220° for several hours. Soln. polymerisation may be effected by treating terephthaloyl chloride with hexamethylene diamine in N-methylpyrrolidone containing CaCl$_2$ or boric acid. Soln. polymerisation occurs also from terephthalic acid salt dissolved in N-methylpyrrolidone/dimethylacetamide solvents containing dissolved 1–2% boric acid and refluxing with triphenyl phosphite as dehydrating agent [6,7]. Copolymers containing 20–35% Nylon 6,I which are still crystalline can be prod. by melt polymerisation below 324°. Optimum compositions for fibre fabrication correspond to those containing 35% Nylon 6,I [8]
Applications: Has potential application as a fibre, but due to expensive polymerisation process and need to wet spin from sulfuric acid soln., it has not found commercial acceptance compared with Nylon 6,6 which is much cheaper and easier to produce. Bicomponent fibres comprising a sheath of Nylon 6 (40%) and a core of Nylon 6,T (60%) were manufactured commercially in the 1960s. Bicomponent fabrics have unique props. of shape and helical crimp and were used in hose and tricot fabric. Over 2000 tons were manufactured in 1976, but eventually the bicomponents lost their markets as stretching and texturing of monofibres were improved

Bibliographic References
[1] *Encycl. Polym. Sci. Technol.*, 2nd edn., 1985, **11**, 435, (rev)
[2] Morgan, P.W. and Kwolek, S.L., *Macromolecules*, 1975, **8**, 12, (cryst)
[3] Nielsen, L.E., Buchdahl, R. and Levreault, R., *J. Appl. Phys.*, 1950, **21**, 607, (T$_g$)
[4] Sprague, B.S. and Singleton, R.W., *Text. Res. J.*, 1965, **35**, 999, (physical props)
[5] Van Krevelen, D.W., *Properties of Polymers: Their Correlation with Chemical Structure*, 3rd edn., 1990, (T$_m$)
[6] *Eur. Pat. Appl.*, 1989, 0306165A, (BP Chemicals)
[7] Yamazaki, N., Higashi, F. and Kowabata, J., *J. Polym. Sci., Polym. Chem. Ed.*, 1974, **12**, 2149
[8] Chapman, R.D. et al, *Text. Res. J.*, 1981, **51**, 564
[9] Snider, O.E. and Richardson, R.J., *Encycl. Polym. Sci. Eng.*, 1966, **10**, 46
[10] Tippetts, E.A., *Text. Res. J.*, 1967, **37**, 524
[11] *Brit. Pat.*, 1991, 9 107 687, (BP Chemicals)

– Nylon 6(3)T

Glass-Transition Temperature:

No.	Value	Note
1	150°C	[4]

Deflection Temperature:

No.	Value	Note
1	140°C	0.455 MPa, ASTM D648 [1,3]
2	124°C	1.82 MPa, ASTM D648 [1,3]
3	140°C	1.82 MPa, 35% glass fibre [10]
4	131°C	[5]

Vicat Softening Point:

No.	Value	Note
1	155°C	BS2782 1965 method 102B [4]
2	150°C	DIN 53460, A50

Brittleness Temperature:

No.	Value	Note
1	-12°C	ASTM D746 [1]

Surface Properties & Solubility:
Plasticisers: *p*-Toluenesulfonamide
Solvents/Non-solvents: Sol. MeOH [7], CH_2Cl_2, 2-butanone, phenols, formic acid, conc. H_2SO_4, DMF, *N*-methylpyrrolidinone. Attacked by acids [4,6]

Transport Properties:
Transport Properties General: Has very low gas permeability which makes it of particular interest in packaging applications
Polymer Solutions Dilute: Apparent viscosity 6500 N Sm^{-2} (shear rate 1 s^{-1}, 270°), 4000 N Sm^{-2} (shear rate 100 s^{-1}, 270°). Viscosity 1.36 dl g^{-1} (M_n 26700, 96% sulfuric acid) [1], 1.23 dl g^{-1} (MW 44180, M_n 20000)
Polymer Melts: Melt viscosity 10000 Poise (100% strain, 35% glass fibre). Viscosity number 110 (cresol, 35% glass fibre)
Water Absorption:

No.	Value	Note
1	7.5 %	saturated, 23°, DIN 53495-1L [4]
2	3 %	50 days [1]
3	2 %	4 mm thick, 23°, 300 days, 50% relative humidity [4]
4	5 %	23°, saturated, 35% glass fibre

Gas Permeability:

No.	Gas	Value	Note
1	O_2	3.54 cm^3 mm/(m^2 day atm)	54×10^{-13} cm^2 (s cmHg)$^{-1}$, 0% relative humidity [9]
2	O_2	1.31 cm^3 mm/(m^2 day atm)	20×10^{-13} cm^2 (s cmHg)$^{-1}$, 100% relative humidity [9]
3	O_2	2.62 cm^3 mm/(m^2 day atm)	40×10^{-13} cm^2 (s cmHg)$^{-1}$ [1]
4	CO_2	4.59 cm^3 mm/(m^2 day atm)	70×10^{-13} cm^2 (s cmHg)$^{-1}$ [1]

Mechanical Properties:
Mechanical Properties General: Good range of props. for engineering plastic. Less notch sensitive than polycarbonate. Mech. props. decrease on exposure to boiling water. Does not crack in flexural test, good stress cracking resistance

Tensile (Young's) Modulus:

No.	Value	Note
1	3800 MPa	-40°, ASTM D638 [1,4]
2	2800 MPa	23°, ASTM D638 [1,4]
3	2046 MPa	93°, ASTM D638 [1,4]
4	10000 MPa	1 mm min^{-1}, 35% glass fibre, DIN 53457-1

Flexural Modulus:

No.	Value	Note
1	2659 MPa	23°, ASTM D7960-66 [1]
2	2177 MPa	93°, ASTM D7960-66 [1]
3	2570 MPa	20°, ASTM D790 [2]

Tensile Strength Break:

No.	Value	Note	Elongation
1	67 MPa	23°, ASTM D638 [1]	132%
2	140 MPa	5 mm min^{-1}, 35% glass fibre, DIN 53455	2.1%

Compressive Modulus:

No.	Value	Note
1	2336 MPa	ASTM D695 [1]

Elastic Modulus:

No.	Value	Note
1	3000 MPa	DIN 53457 [5]

Tensile Strength Yield:

No.	Value	Elongation	Note
1	67 MPa	8.8%	ASTM D638 [1]
2	85–90 MPa	9.5%	DIN 53455 [4,5]
3	81 MPa		20°, BS2782 method 301C [2]

Flexural Strength Yield:

No.	Value	Note
1	143.7 MPa	5% elongation, -40°, ASTM D7960-66 [1]
2	91.4 MPa	5% elongation, 23°, ASTM D7960-66

Nylon 6(3)T

No.	Value	Note
3	64.5 MPa	5% elongation, 93°, ASTM D7960-66 [1]
4	190 MPa	35% glass fibre

Compressive Strength:

No.	Value	Note
1	120 MPa	11.5% yield pt, DIN 53454 [5]

Miscellaneous Moduli:

No.	Value	Note	Type
1	6900 MPa	1000h, DIN 53444	tensile creep modulus

Impact Strength: Falling weight impact strength 38 J (2.2 mm thick unnotched disc) [2,7]. Very tough material, impact strength increases with moisture absorption. Charpy impact strength 20 kPa (notched, 23°, ASTM D256) [3]
Hardness: Rockwell M93 (23°, ASTM D785), R120 [1,5]
Failure Properties General: Tear strength 60 kJ m^{-2} (elongation 70–180%, DIN 53455) [5]. Creep strength 50 N mm^{-2} (after 1000h) [3]
Fracture Mechanical Properties: Tensile impact B 1.18 MPa (type 3 mm bar, ASTM D1822) [1]
Friction Abrasion and Resistance: Abrasion resistance 21 mg (1000 cycles)$^{-1}$ (ASTM D1044, Taber CS 17 test)

Izod Notch:

No.	Value	Notch	Break	Note
1		N	N	23° [3,4]
2	177000 J/m	Y	Y	room temp.

Izod Area:

No.	Value	Notch	Note
1	9–15 kJ/m^2	Notched	20°, DIN 53453 [3,4]
2	3–5 kJ/m^2	Notched	-50°, DIN 53453 [3,4]
3	24–25 kJ/m^2	Unnotched	-30–23°, 35% glass fibre
4	6–6.5 kJ/m^2	Notched	23–-30°, 35% glass fibre

Charpy:

No.	Value	Notch	Note
1	67	Y	23°, 3.8 mm notch hp radius [2,7]
2	68	Y	20° [2]

Electrical Properties:

Electrical Properties General: Insulating props. similar to general purpose cryst. nylons. Good tracking resistance. Dielectric props. have slight dependency on temp. and frequency and are not affected by high humidity

Surface/Volume Resistance:

No.	Value	Note	Type
1	>0.01 × 10^{15} Ω.cm	min., dry, DIN 53482 [3]	S
2	1 × 10^{15} Ω.cm	dry, VDE 0303T30 [4]	S

Dielectric Permittivity (Constant):

No.	Value	Frequency	Note
1	3.44	0.3 MHz	-5°, ASTM D15068 [1]
2	4.32	0.3 MHz	38° [1]
3	3.3		-35° [1]
4	3.4	1 MHz	VDE 0303 T4 [4]
5	4.5	50 Hz	
6	3.8	1 MHz	VDE 0303 T4

Dielectric Strength:

No.	Value	Note
1	67 kV.mm^{-1}	ASTM D149 [1]
2	35 kV.mm^{-1}	35% glass fibre, ISO 243 [2]

Arc Resistance:

No.	Value	Note
1	120s	ASTM D495 [1]

Dissipation (Power) Factor:

No.	Value	Frequency	Note
1	0.022	1 kHz	-35° [1]
2	0.036	1 kHz	-5° [1]
3	0.029	1 kHz	38° [1]
4	0.025	1 kHz	52°, ASTM D15068 [1]
5	0.016	50 Hz	
6	0.024	1 MHz	

Optical Properties:

Optical Properties General: A clear transparent material even in relatively thick samples. Light transparency in visible range *ca.* 85%

Refractive Index:

No.	Value	Note
1	1.566	20° [1,9]

Transmission and Spectra: Ir, uv and visible spectral data have been reported [1]

Polymer Stability:

Polymer Stability General: Exhibits good dimensional stability. Low thermal expansion and high heat deflection temp. confer good thermal stability. Changes in temp. and frequency do not affect electrical props. adversely
Thermal Stability General: Amorph. nature has low thermal coefficient of expansion and low mould shrinkage. Good physical props. to within 50° of T_g
Decomposition Details: Stable in N_2 up to 310°. Above this temp. weight loss is observed (gravimetric analysis)
Flammability: Limiting Oxygen Index 26.8%. Flammability rating V2 (1.6 mm specimen, UL 94-1985). Polyamide is self extinguishing
Chemical Stability: Stable to most aliphatic, aromatic and halogenated hydrocarbons. Unaffected by oils, esters and salt solns. Less resistant to polar molecules especially alcohols and nitrogenous bases, leading to stress cracking

Hydrolytic Stability: Susceptible to hot water; prolonged exposure to boiling water leads to loss of transparency [5]. Unaffected by dilute mineral acids and alkalis, but hydrolysed by stronger solns.
Biological Stability: Not biodegradable

Applications/Commercial Products:
Processing & Manufacturing Routes: Commercially prod. from the dimethyl ester (DMT) and a 1:1 mixture of 2,2,4- and 2,4,4-trimethyl-1,6-hexanediamine. Diamine and diacid salt are charged in an autoclave at 270° and 1.8 MPa. Water is slowly removed and benzoic acid or AcOH added to limit MW. Processed at 250-300° using oven dried granules of <0.1% water content. May be compression or injection moulded or extruded
Mould Shrinkage (%):

No.	Value	Note
1	0.5–0.7%	
2	0.05%	longitudinal
3	0.65%	transverse

Applications: Transparent additive-free grades are used for electronic and electrical engineering components, telecommunications, water engineering, machine and apparatus construction, optics and in medical technology applications. Uv stabilised material is used for components with high mechanical strength such as machine elements, electrical switch parts and water engineering applications. 35% Glass fibre reinforced grades are used for rigid, low warpage injection mouldings. Transparent blends with semi-crystalline Nylons are used especially for spectacle frames and orthopaedic products. Advantages of blends are superior toughness and stress cracking resistance. Applications include profiles, lamp covers, inspection windows and light strips

Trade name	Supplier
Trogamid	Dynamit Nobel
Trogamid T	Hüls AG

Bibliographic References
[1] Schneider, J. and Groninger, G.W., High Perform. Plast., Nat. Tech. Conf. - Soc. Plast. Eng., *(Prepr.)*, 1976, 47
[2] Dolden, J., *Polymer*, 1976, **17**, 875
[3] Trogamid T Physical Property Publication Dynamit Nobel Chemicals, 1979, (technical datasheet)
[4] Engineering Plastics from New Ideas Hüls,(technical datasheet)
[5] Schneider, J., *Kunststoffe*, 1974, **64**, 365
[6] Chemical Properties of Trogamid T Dynamit Nobel, 1970, (technical datasheet)
[7] *Eur. Pat. Appl.*, 1991, 411 791, (BP Chemicals)
[8] Van Krevelen, D.W., *Properties of Polymers: Their Correlation with Chemical Structure*, 3rd edn., Elsevier, 1990
[9] Herold, J. and Meyerhoff, G., *Eur. Polym. J.*, 1980, **16**, 1167, (viscosity, MW)
[10] *Trogamid T-GF35*, Hüls America, 1996, (technical datasheet)

para-Octyl Phenolic resins O-1

Synonyms: *4-Octylphenol based resins. p-tert-Octylphenol resins. 4-(1,1,3,3-Tetramethylbutyl)phenol resins*
Related Polymers: Phenol-aldehyde resins, Phenolic resins Rubber adhesives
Monomers: 4-*tert*-Octylphenol, Formaldehyde
Material class: Thermoplastic
Polymer Type: phenolic
CAS Number:

CAS Reg. No.	Note
26678-93-3	*p-tert*-octylphenol-formaldehyde copolymer
124765-80-6	4-*tert*-octylphenol-formaldehyde copolymer

Molecular Formula: $[(C_{14}H_{22}O).(CH_2O)]_n$
Fragments: $C_{14}H_{22}O$ CH_2O
Molecular Weight: M_n 755
Additives: Additives include kaolin, calcium carbonate, colloidal silica, hydroxymethylstarch, styrene-butadiene rubber
General Information: The resins do not cross-link as the phenol is difunctional (the *para*-position is blocked). The *p*-octyl side chain can exist in several isomeric forms. The only isomer of commercial importance is the *p-tert*-octyl form

Volumetric & Calorimetric Properties:
Melting Temperature:

No.	Value	Note
1	72–76°C	[1]

Glass-Transition Temperature:

No.	Value	Note
1	45°C	[3]

Surface Properties & Solubility:
Solvents/Non-solvents: Sol. oils [4]

Optical Properties:
Transmission and Spectra: Mass spectral data have been reported [2]

Polymer Stability:
Thermal Stability General: Details of thermal props. have been reported [6]

Applications/Commercial Products:
Processing & Manufacturing Routes: May be synth. by reaction of *p-tert*-octylphenoxymagnesium bromide with paraformaldehyde in C_6H_6 soln. for 12h at 80° followed by acidic quenching [1]. The effect of catalyst on struct. and props. has been reported [5]
Applications: The solid resole (5–12%) may be compounded with rubbers using activators and acts as a non-sulfur-containing vulcanisation agent. Activators for nitrile-butadiene rubber are pyromellitic anhydride, phthalic anhydride or fumaric acid. For styrene-butadiene rubber (SBR), *p*-toluenesulfonic acid or chloroacetic acid are used. For neoprene, the activator can be zinc oxide and tin (*II*) chloride. The novolak is compounded with kaolin clay, calcium carbonate, colloidal silica and SBR latex to form a coated front (CF) sheet for carbonless copy paper systems. Potential application as an antimicrobial agent. In varnishes and protective coatings. Novolak-type resin used in the production of speciality printing inks. Resoles may be mixed with rubber and alkaline earth hydroxides to make adhesives. May be used as a sulfur free alternative curing agent for rubber compounding. May be used with epoxy resins as coatings for equipment for the chemical industry

Trade name	Details	Supplier
Rütaphen KA	adhesive resins	Bakelite
Standard sample	novolak type	
Sumilite resin R	PR-19900; novolak-type	Sumitoma Chemical, Japan

Bibliographic References
[1] Uchibori, T., Kawada, K., Watanabe, S., Asakura, K. et al, *Bokin Bobai*, 1990, **18**, 215, (synth., use)
[2] Lattimer, R.P., Harris, R.E., Rhee, C.K. and Schulten, H.-R., *Rubber Chem. Technol.*, 1988, **61**, 639, (ms)
[3] Weill, A., Dechenaux, E. and Paniez, P., *Microelectron. Eng.*, 1986, **4**, 285, (Tg)
[4] Takahashi, S., Kuriyama, K. and Tobita, S., *Shikizai Kyokaishi*, 1966, **39**, 531
[5] Yao, F., Wang, X., Meng, J. and Sun, J., *Hecheng Shuzhi Ji Suliao*, 1995, **12**, 21
[6] Pu, Qi, Y., Li, H., Xu, C. et al, *Xiangjiao Gongye*, 1994, **41**, 477, (thermal props.)

Olefin - sulfur dioxide colpolymers O-2

Synonyms: *Poly(olefin sulfone). Poly(alkylene sulfone). Alkene-sulfur dioxide copolymers. Alkyne-sulfur dioxide copolymers*
Material class: Copolymers
Polymer Type: polysulfone
CAS Number:

CAS Reg. No.	Note
64155-66-4	1-heptene/SO_2 copolymer
64155-67-5	2-pentene/SO_2 copolymer
28085-22-5	ethene/SO_2 copolymer
33991-00-3	tetradecene/SO_2 copolymer
108855-21-6	1-butene/SO_2 copolymer, block, graft
	ethene/SO_2 copolymer
	ethene/SO_2 copolymer
51555-48-7	2-methylpentene/SO_2 copolymer
98242-80-9	methylpentene/SO_2 copolymer
76769-82-9	4,4-dimethyl-1-pentene/SO_2 copolymer
110711-58-5	ethene/SO_2 copolymer, alternating
34903-07-6	1-hexene/SO_2 copolymer
34903-08-7	*Z*-2-butene/SO_2 copolymer
122752-93-6	ethene/1-propene/SO_2 terpolymer, alternating
33991-02-5	1-octadecene/SO_2 copolymer
26778-84-7	3-methyl-1-butene/SO_2 copolymer
31975-70-9	2-butene/SO_2 copolymer
34903-09-8	*E*-2-butene/SO_2 copolymer
37382-72-2	butene/SO_2 copolymer
30475-44-6	1-propene/SO_2 copolymer
33990-99-7	1-dodecene/SO_2 copolymer
64155-68-6	2-octadecene/SO_2 copolymer

Olefin - sulfur dioxide copolymers

30795-19-8	1-octene/SO$_2$ copolymer
42255-65-2	1-pentene/SO$_2$ copolymer
33990-98-6	1-decene/SO$_2$ copolymer
29436-19-9	2-methyl-1-propene/SO$_2$ copolymer
51749-80-5	2-heptene/SO$_2$ copolymer
33991-01-4	1-hexadecene/SO$_2$ copolymer
25104-10-3	1-butene/SO$_2$ copolymer

Molecular Formula: $(RSO_2)_n$
Fragments: RSO_2
Molecular Weight: MW of polysulfones is controlled by chain-transfer agents such as $BrCCl_3$, CBr_4 or thiols. [3,4,5]
General Information: Olefin-sulfur dioxide copolymers degrade readily at high temp. preventing commercial application. Their density increases with increasing sulfone to hydrocarbon ratio. Crystallinity has been detected by x-ray diffraction in the poly(ethylene sulfone)s of variable composition, in poly(butadiene sulfone), and in polysulfones with three, four, five, or six main-chain carbon atoms in the repeat unit. No crystallinity was detected in some poly(1-olefin sulfone)s. Alkenes, conjugated and non-conjugated dienes and some alkynes and vinyls combine with SO_2 to form alternating copolymers with a 1:1 SO_2/alkene ratio. Most polysulfones have SO_2 groups separated by two saturated carbon atoms. Separation by more than two carbon atoms can result from the copolymerisation of some dienes and also via polymerisation using a dimercaptan/diolefin or dimercaptan/dihalide reaction with a subsequent oxidation step. Copolymers with lower contents of SO_2 units can be made by polymerisation in the gas phase using γ-irradiation [21]
Tacticity: Are generally atactic [1]. Isotactic and syndiotactic forms of Poly(cyclohexene sulfone) are possible. [6,39]
Morphology: Monoclinic unit cell a 9.88Å, b 9.62Å, c 18.24Å, β 121.7° (hexamethylene); a 9.88Å, b 9.62Å, c 34.00Å, β 121.7° (hexamethylene and pentamethylene); a 9.88Å, b 9.62Å, c 15.68Å, β 121.7° (hexamethylene and tetramethylene); a 9.88Å, b 9.62Å, c 7.76Å, β 121.7° (pentamethylene); a 9.88Å, b 9.62Å, c 28.33Å, β 121.7° (tetramethylene and pentamethylene) [9]. For α-olefin polysulfones the side chain crystallises independently of the backbone when the chain length reaches C_{18} [23]. 1-Hexene and 1-tetradecene polysulfones adopt a helical struct., whereas cyclohexene and other 1,2-disubstituted polysulfones do not [32,33, 34,35]. Alkyne/SO$_2$ copolymers have a *trans* configuration [38]
Identification: Ir [38], H-1 nmr and C-13 nmr spectral data have been reported [6,7,8,14,20,26,34,36,40]

Volumetric & Calorimetric Properties:
Equation of State: Mark-Houwink constants have been reported
Latent Heat Crystallization: The ceiling temp. (T_c) for polysulfone formulation, i.e. where free energy change for the reaction is zero, varies with the parent alkene. For straight chain 1-alkenes (T_c) decreases from above 137° for ethylene to 90° for propylene, and 64° for 1-butene after this it remains reasonably constant. For alkenes T_c decreases from 2-butene (34°) to 2-pentene (8°), and 2-heptene (-38°). Branching in the side chain lowers T_c particularly at the double bond, e.g. T_c 2-for methyl-1-pentene is -34°. Polysulfones of cyclic alkenes have much higher values of T_c than the corresponding acyclic polymer, e.g. T_c of cyclopentene polysulfone is 102°, compared with 63° for 1-pentene [25,29]
Transition Temperature General: T_m of alkyne/SO$_2$ copolymers varies with MW and with crystallinity [38]. Some 1-alkene sulfones (e.g. 1-eicosene, 1-hexadecene) have liq. cryst. phases at temps. between room temp. and 100°. Softening point and clearing temp. for 1-eicosene polysulfone are 54° and 71° respectively; those for 1-hexadecene polysulfone are 65° and 70° respectively [14]

Transport Properties:
Polymer Solutions Dilute: Huggins constants for cyclopentene polysulfone and cyclohexene polysulfone have been reported [22]
Permeability of Gases: Permeabilty of α-olefin polysulfones to O_2 and CO_2 is quite high and reaches a maximum at C_{16} [23]

Mechanical Properties:
Hardness: Brinell hardness 40 propylene polysulfone, 25–30 2-butene polysulfone, 20–23 1-butene polysulfone, 18–19 1-pentene polysulfone [19]
Fracture Mechanical Properties: Films from cycloaliphatic polysulfones tend to crack in solvents although the incorporation of α-olefins in a terpolymer improves the film props. in this respect [28]

Optical Properties:
Molar Refraction: Complex refractive index over the ir range for poly(2-methyl-1-pentene sulfone) has been reported [41]. Some sulfones of, for example, 1-eicosene and 1-hexadecene have liq. cryst. phases between room temp. and 100°. Softening point and clearing temp. for 1-eicosene sulfone are 54° and 71° respectively; those for 1-hexadecene sulfone are 65° and 70° respectively [14]

Polymer Stability:
Thermal Stability General: The polysulfones of norbornene and some of its derivatives are thermally more stable than those of 1-butene and cyclohexene. Stabililty can be enhanced by the use of stabilisers such as thioacetamide or tin(II) oxide [31]
Decomposition Details: Thermal decomposition of 1-alkene polysulfones yields the respective monomers. Cyclic alkene sulfones and 2-alkene sulfones undergo a more complex decomposition yielding a variety of products including aromatics [11]. An ethylene polysulfone is stable up to approx. 225°. Above 275° decomposition proceeds rapidly with 100% weight loss by 350°. A copolymer with only 9 mol% SO_2 decomposes more slowly than the 1:1 copolymer [21]. Weight loss (3h, 190°): cyclohexene polysulfone 16.1%; 1-butene polysulfone 30.5%; 5-norbornene-2-carbonitrile polysulfone 5.4%; methyl 5-norbornene 2-carboxylate polysulfone 9.5%; 5-norbornen-2-yl acetate polysulfone 8.9% [31]
Chemical Stability: Chemical stability of cyclohexene polysulfones may be improved by introduction of aromatic rings into the polymer chain [37]
Stability Miscellaneous: Poly(alkene sulfones) are very sensitive even to relatively low doses of ionising radiation and undergo rapid chain scission alkene. Cleavage occurs predominantly at the C-S bond [29,30]

Applications/Commercial Products:
Processing & Manufacturing Routes: Poly(olefin sulfone)s are obtained by the free-radical reaction of olefins with sulfur-dioxide. They can also be formed from other dienes and alkynes, though not from acetylene. Another method of preparation is by polysulfide oxidation. Poly(olefin sulfone) is formed from an olefin and sulfur dioxide in the gas phase. Following UV or high energy irradiation, copolymerisation has also been carried out in the solid state with isobutene, butadiene and vinyl acetate at low temp. When the polymerisation is homogeneous, the feeds are mixtures of SO_2 with the other monomers and perhaps a solvent. Stainless-steel bombs or sealed tubes are used as reaction vessels at elevated temps. where the pressure may rise to >20 Mpa. For free radical initiation oxygen, ozonides, peroxides, hydroperoxides and hydrogen peroxide-paraldehyde have been used. Initiation is also effected by various forms of ionizing radiation or UV light on monomer feeds. Poly(propylene sulfone) can be moulded between 180 and 200°, and poly(1-butene sulfone) between 125 and 180°. Hot-pressed films have been prepared at 120° from the polysulfones of 1-butene, cyclopentene, and bicyclo-[2.2.1]-2-heptene at high pressures of 140 MN/m^2 which prevented decomposition. Synth. has been reported [6,7,8,10,13] some of the polysulfones may be spun into fibres
Applications: Thermal instability near the moulding temp. has prevented olefin-sulfur dioxide copolymers from being used as bulk thermoplastics, but certain speciality uses have been developed. Gypsum and other minerals have been coated with various polysulfones to impart compatibility with polymeric hydrocarbons, such as polyethylene, to achieve a reinforcing effect. Some polysulfone - based formulations e.g. Stadis 450, are used as antistatic additives for hydrocarbon fuels. Poly(olefin

sulfone)s are considered for use in the membranes of heart-lung machines and other medical purposes due to their biocompatibility and high permeability to oxygen and to CO_2. The most significant use at present exploits sensitivity to high energy radiation. Poly(1-butene sulfone) (PBS) is used as a resist in the electron-beam fabrication of chromium photomasks using electron-beam exposure. It is available as a filtered solution ready for spin coating. Most important application is electron beam resists, photolitho compositions, and in x-ray resists. They also find application in anti-fouling and anti-static formulations and in membranes

Trade name	Details	Supplier
PBS	Poly(1-butene sulfone)	Mead Chemical Co.
PBS		Rolla Mo.
PBS		Chisso Corporation
Stadis 450	polysulfone based formulation	DuPont

Bibliographic References
[1] *Encycl. Polym. Sci. Eng.*, 2nd edn. (eds. Knoschwitz, J.I.), John Wiley & Sons), 1985, 408
[2] *Kirk-Othmer Encycl. Chem. Technol.*, 4th edn. (eds. J.I. Kroschwitz and M. Howe-Grant), Wiley Interscience, 1993, **19**, 945
[3] Matsuda, M., Lino, M., Hirayama, T. and Miyashita, T., *Macromolecules*, 1972, **5**, 240
[4] Stereodyn. Mol. Syst., *Proc. Symp.*, (eds. F.A. Bovey and R.H. Sarma), Pergamon Press, Oxford, UK, 1979, 53
[5] Yamashita, Y., Iwatsuki, S. and Sakai, K., *Adv. Chem. Ser.*, 1971, **99**, 211

Oriented polystyrene sheet and film   O-3

Synonyms: *OPS. Biaxially oriented polystyrene sheet and film. Polystyrene film. Polystyrene, Oriented film*
Related Polymers: Polystyrene, Polystyrene - Foamed, High-Impact polystyrene
Monomers: Styrene
Material class: Fibres and Films, Thermoplastic
Polymer Type: polystyrene
CAS Number:

CAS Reg. No.	Note
9003-53-6	atactic
25086-18-4	isotactic
28325-75-9	syndiotactic

Molecular Formula: $(C_8H_8)_n$
Fragments: C_8H_8
Additives: No plasticisers are usually used. For food packaging applications any additives used must be approved for food contact
General Information: OPS is a large volume, thermoplastic material [1]. It has high optical clarity and high surface gloss, high permeability to water and gases, e.g. oxygen, high flexural modulus, excellent dimensional stability and machinability, low taste and odour transfer and low cost [1,2]. The thickness of OPS film is typically <0.08 mm and for OPS sheet >0.08 mm. Commercial products are available in the range 0.025–0.75 mm
Tacticity: Most commercially produced PS film is atactic. Syndiotactic PS film has been produced commercially for specialist applications [28,29,30,33,34,35,36,37,38,39,40,41]. Isotactic film has been used for research purposes [27] and speciality products [42,43,44]. Syndiotactic film may be white [38] or transparent [36]
Morphology: Most OPS is biaxially oriented. The orientation increases the elongation values and overcomes the problems with brittleness observed for unoriented, cast PS sheet. Uniaxial orientation can also be used to produce fibres and films but this process is not as significant commercially

Volumetric & Calorimetric Properties:
Density:

No.	Value	Note
1	d 1.05–1.06 g/cm^3	[4]

Thermodynamic Properties General: For an oriented sample, the linear coefficients of thermal expansion parallel and perpendicular to the direction of draw are different. The effect of orientation on the linear thermal expansion coefficient for atactic and isotactic films at different draw ratios has been reported [1,45]. α_L perpendicular to the draw is slightly greater than that for unoriented PS. α_L parallel to the draw is much lower than that for unoriented PS

Thermal Conductivity:

No.	Value	Note
1	900–1500 W/mK	0.9–1.5kW m^{-1}, K^{-1} ASTM C177 [4]

Glass-Transition Temperature:

No.	Value	Note
1	105°C	378K [46]

Transition Temperature General: T_g is not affected by orientation or the draw ratio [46]

Surface Properties & Solubility:
Solubility Properties General: Orientation does not affect the solubility or surface props.

Transport Properties:
Transport Properties General: Rheological props. in the melt and the melt flow index are important for the manufacture of PS film and sheet
Permeability of Liquids: The rate of sorption of *n*-pentane increases almost fourfold when the PS is uniaxially oriented [1,53]
Permeability and Diffusion General: Permeability and diffusion are very sensitive to the test conditions and to small changes in the degree of orientation [1,49] and the morphology [47] and vary with film thickness [4,5]. When biaxially oriented PS film is strained uniaxially the gas permeability is affected [1,54,55]. The strain dependence of the permeability and diffusion coefficients for Ar, Kr, Xe, N_2 and CO_2 have been reported [54,55]. Permeability and diffusion coefficients decrease with time at constant rate of strain [54,55]. Other gas permeability values have been reported [1,4,5,49]

Water Absorption:

No.	Value	Note
1	0.04–0.1 %	ASTM D570 [4,5]

Gas Permeability:

No.	Gas	Value	Note
1	O_2	100.6 cm^3 mm/(m^2 day atm)	102000 cm^3 μm m^{-2} day^{-1} bar^{-1}, ASTM D1434/ASTM D3985 [47]
2	O_2	142.5–143.2 cm^3 mm/(m^2 day atm)	2.17–2.18 × 10^{-10} cm^3 cm cm^{-2} (s cm Hg)$^{-1}$ [48]
3	O_2	0.099 cm^3 mm/(m^2 day atm)	2 × 10^{-13} cm^3 cm cm^{-2} (s Pa)$^{-1}$, 25° [50,51]
4	N_2	23.6–26.9 cm^3 mm/(m^2 day atm)	0.36–0.41 × 10^{-10} cm^3 cm cm^{-2} (s cm Hg)$^{-1}$ [48]

5	N_2	0.03 cm^3 mm/(m^2 day atm)	0.59 × 10^{-13} cm^3 cm cm^{-2} (s Pa)$^{-1}$, 25° [50,51]
6	H_2	0.83 cm^3 mm/(m^2 day atm)	17 × 10^{-13} cm^3 cm cm^{-2} (s Pa)$^{-1}$, [50,51]
7	He	0.69 cm^3 mm/(m^2 day atm)	14 × 10^{-13} cm^3 cm cm^{-2} (s Pa)$^{-1}$, 25° [50,51]
8	CO_2	0.39 cm^3 mm/(m^2 day atm)	7.9 × 10^{-13} cm^3 cm cm^{-2} (s Pa)$^{-1}$, 25° [50,51]
9	CO_2	394.76 cm^3 mm/(m^2 day atm)	400000 cm^3, μm m^{-2} day^{-1} bar^{-1}, ASTM D1434/ASTM D3985 [47]
10	H_2O	0.35–0.41 cm^3 mm/(m^2 day atm)	717–840 × 10^{-13} cm^3 cm cm^{-2} (s Pa)$^{-1}$, 25° [50]

Mechanical Properties:

Mechanical Properties General: Biaxially oriented PS has higher values of elongation and greater resistance to brittle fracture compared to unoriented PS [1]. Biaxially oriented PS has a large flexural modulus [1]. Mechanical props. for melt oriented material (uniaxially oriented), at high draw rates and temps. just above T_g, show a fairly good relationship with the birefringence [1]. For example, tensile strength at yield [1,67] and impact strength [1,70] increase with increasing birefringence. Biaxially oriented polymers have increased elongation to break in all directions in the plane of the film [72,73]. The tensile modulus of uniaxially oriented PS increases with birefringence [1,67]

Tensile (Young's) Modulus:

No.	Value	Note
1	2760 MPa	ASTM D638 [4,5]
2	3100 MPa	[1]

Tensile Strength Break:

No.	Value	Note	Elongation %
1	56.2–84.4 MPa	ASTM D882 [4,5]	3–40%
2	62–83 MPa	[1]	5–60%
3	23–33 MPa	ASTM D882, grip speed 12.7 cm min^{-1}, distance 5.1 cm, film thickness 45.5 μm [71]	11–15%

Impact Strength: Pendulum impact strength 1–5 kg cm^{-1} (3.2 mm thick) [4,5]
Mechanical Properties Miscellaneous: Elmendorf tear strength 2 g mm^{-1} (ASTM D1922) [4,5]. Bursting strength 16–35 Mullen points (0.0254 mm thick, ASTM D774) [4,5]. Yield 31.6 m^2 (kg mm)$^{-1}$ (ASTM D2103) [4,5]. Material yield 3.8 m^2 N^{-1} (0.025 mm) [1]
Hardness: Rockwell M85 (ASTM D785) [4,5]
Failure Properties General: Uniaxially oriented PS under normal tensile testing rate, shows brittle to ductile behaviour in the draw direction at a birefringence of -0.002 [1,67]. At this point crazing is suppressed and shear yielding is the predominant mode of deformation. In the cross direction the tensile strength decreases sharply with birefringence [1,68]. Under high speed notched impact conditions the onset of brittle to ductile behaviour occurs at a birefringence of -0.005 [1]. For biaxially oriented PS, brittle to ductile behaviour occurs when the ratio of the birefringence to the intrinsic birefringence is 0.04 at room temp. [1,69]. Biaxial orientation of PS reduces crazing and increases the environmental stress crack resistance [1,70,74,75]. Susceptible to pinhole formation under conditions of repeated flexing [76]

Electrical Properties:

Electrical Properties General: Oriented PS has excellent electrical props.

Dielectric Permittivity (Constant):

No.	Value	Note
1	2.5	22°, ASTM D150 [4,5]

Dielectric Strength:

No.	Value	Note
1	197 kV.mm^{-1}	22°, ASTM D149 [4,5]
2	20 kV.mm^{-1}	60 Hz [2]

Dissipation (Power) Factor:

No.	Value	Note
1	0.0005	22°, ASTM D150 [4,5]

Optical Properties:

Optical Properties General: Optical props. are affected by orientation
Refractive Index:

No.	Value	Note
1	1.6	ASTM D542 [4,5]

Transmission and Spectra: Ir [1,11,12,60,61,62,63,64], uv, Raman, nmr and fluorescence spectral data have been reported [65,66]
Total Internal Reflection: Birefringence may be used to determine the amount of orientation in films [1,56]. Birefringence is proportional to the stress at low stress levels only [1]. Intrinsic birefringence -0.1 [1,57,58]. Stress optical coefficient 4700 × 10^{-12} Pa^{-1} [1,59]. Birefringence may be related to the recoverable strain and heat released just above the T_g [1]
Volume Properties/Surface Properties: Transparency 39% (ASTM D1746) [4,5]. Haze 0.1–3.0% (ASTM D1003) [4,5]

Polymer Stability:

Thermal Stability General: Dimensional shrinkage and polymer disorientation may occur [4]. For uniaxially drawn PS, shrinkage occurs just above T_g [46]
Upper Use Temperature:

No.	Value	Note
1	88 °C	[4,5]
2	-63 °C	min., continuous service [4]
3	88 °C	max., continuous service [4]

Decomposition Details: A method for predicting the temp. for the onset of thermal degradation of films in air has been reported [77]. The activation energy of thermal degradation and the temp. for the onset of thermal degradation are affected significantly by the heat pretreatment of films, e.g. annealing and quenching [77]
Flammability: Flammability 3.8 cm s^{-1} (ASTM D635) [4,5]
Environmental Stress: PS film has fair resistance to uv radiation with tensile strength and heat resistance being affected by irradiation time [4,5]. The stability of PS films to uv radiation has been studied in fresh water, salt-water and air. The tensile strength and elongation at break of the PS films initially increases for all environments with the greatest increase for the uv-air environment and the least for the uv-fresh water environment. This increase may be attributed to the formation of free radicals by photo-induction resulting in cross-linking [71]. Uniaxial orientation enhances craze and environmental stress cracking propagation in the direction parallel to the orientation [1]. Biaxial orientation of PS reduces crazing and increases the resistance to environmental stress cracking [1,70,74]

Oriented polystyrene sheet and film

Chemical Stability: Has good resistance to strong acids and strong alkalis [4,5] with good to poor resistance to greases and organic solvents [4,5]
Hydrolytic Stability: Stable in aq. solns.
Biological Stability: Stable to bacteria and microorganisms. However, bacterial adhesion to the polymer film surface may occur

Applications/Commercial Products:
Processing & Manufacturing Routes: OPS is usually made of a high heat distortion, unplasticised, high MW, general purpose resin. For food packaging applications, resins that are low in residual ethylbenzene and styrene monomer are used. Scrap resin can be added back to the process as either ground sheet or repelletised material. OPS film and sheet may be prepared commercially using a tentering process or a bubble process. The tentering process is most common for OPS sheet and the bubble process for film, both consisting of an extrusion stage where orientation of the film takes place followed by thermoforming. The 'orientation' stretching must be performed at the optimum temp. 95–120°. Biaxially oriented sheet is thermoformed by heating it above T_g (typically 149°) and then shaping in a mould. Vacuum forming is usually used. Biaxial orientation must be maintained during thermoforming. OPS film may be prepared using a blown film (bubble) extrusion process [1,2], where film can be stretched simultaneously in machine and transverse directions [1]. This is followed by rapid cooling to retain biaxial orientation. Although the blown film process is similar to the common bubble process used for making polyethylene, polypropylene and Saran films superficially, there are a number of substantive differences, and the polystyrene process is considerably more difficult to achieve, as the blown film must be stabilised by cooling below T_g. This involves extremely careful control of temp. in the blowing process. Speciality and research applications of polystyrene films may use a solvent casting method [22] or heating latex particles or powder above T_g so that they coalesce followed by cooling to form a film [23,24,25]. Syndiotactic OPS film can be injection moulded [33], laminated and thermoformed [34]

Applications: OPS film is used in window envelopes because of its clarity, stability, its ability to pace through packaging machinery at high speeds, and to be cut and sealed. These same props. have led to the application of OPS film for window cartons. An anti-fog film has been used for the windows of cartons of bacon. This typically has a film thickness of 0.025–0.032 mm. OPS film of thickness 0.06–0.13 mm may be used as inserts in wallets or as sheet protectors. OPS film and sheet is used in food packaging because of low cost and high optical clarity. Transmission characteristics for water and oxygen approximately coincide with the metabolic requirements of the produce being packaged. Hence OPS film and composites containing OPS have been used to wrap fresh fruit, vegetables and flowers and may also be used for packaging bakery goods. For OPS sheet the main use is in meat trays because of its transparency, resistance to moisture and dimensional stability. Rubber modified OPS film may be used for food packaging and for synthetic paper. OPS film is also used as a photographic film base because of its dimensional stability at all humidities and low gel count. OPS film can be used to calibrate ir and FT-IR spectrometers. Film with an adhesive layer may be used as a substrate for pressure sensitive adhesive tape. Transferable xerox indicia may be prepared which contains polystyrene film and a complex may be used to make film based carbon paper. Polystyrene film can be laminated and used for decorative purposes and OPS has been laminated to polystyrene foamed sheet and used for heat-resistant, recyclable foam containers. A composite containing polystyrene film laminated onto paper has been used as antistatic packaging for wrapping electronic devices and liq. cryst. display devices. Laminated PS film has been used as plastic advertising inserts for food packages. PS can be laminated with a gas barrier polymer e.g. EVOH for packaging applications. Coated PS film can be coextruded with EVA and the laminate formed used as push-through lids for paper and plastic containers. Optical compensations for liq. cryst. display devices are made from a composite of polycarbonate film and monoaxial PS film. Luminescent PS film has been used in a fibre optic oxygen sensor.

Other applications of OPS film include transfers for china printing and an artists' substrate. Syndiotactic PS film (biaxially stretched) has been used to form laminated film capacitors. Syndiotactic PS film may also be useful as condenser dielectric, wrap up foil, substrate for adhesive tape and flexible printed circuits; as a base for magnetic tape or recording tape, packaging and photographic plates. Syndiotactic PS film may be injection moulded to form heat-treated mouldings or may be laminated onto foam sheets to give thermoformed food trays. Isotactic PS film has been used in electrical insulation and in heat and chemical resistant mouldings

Trade name	Details	Supplier
Celatron	PS film and sheet	British Celanese Ltd
Cellofoam	PS film and sheet	United States Mineral Products Co.
Dri-Lite	PS film and sheet	Poly Foam Inc
Dylene	PS film and sheet	Sinclair Koppers Co
Kandel	PS film and sheet	Union Carbide Corporation
Polyflex	PS film and sheet	Monsanto Chemical
Santoclear	PS film and sheet	Mitsubishi Monsanto Chem. Co
Styrofilm	PS film	Sekisui Chem. Co Ltd
Styrofilm	PS film	Daicel Chem.
Sumibrite	PS (cast) film	Sumitomo Chem Co
Sumibrite	OPS film (rubber modified)	Idemitsu Petrochemical Ltd
Sumibrite	PS film laminates	Asahi Chem. Co Ltd
Sumibrite	OPS film and sheet	Dainippon Ink and Chemicals
Sumibrite	PS film	Tokyo Cellophane Paper Co Ltd
Sumibrite	syndiotactic PS film	Toyo Boseki
Sumibrite	isotactic PS film	Toray Industries
Sumibrite	syndiotactic PS film	Idemitsu Kosan Co Ltd
Sumibrite	PS sheet laminates	Lustour Corp.
Sumibrite	PS film laminates with EVOH	Kuraray Co
Trycite	PS film and sheet	Dow
Trycite	PS film on PS foam	Oji Kako KK

Bibliographic References
[1] *Encycl. Polym. Sci. Eng.*, 2nd edn., (ed. J.I. Kroschwitz), John Wiley and Sons, 1985, **16**, 210
[2] *Kirk-Othmer Encycl. Chem. Technol.*, 3rd edn., (ed. M. Grayson), Wiley Interscience, 1978, **21**, 830, 840
[3] *Encycl. Polym. Sci. Eng.*, 2nd edn., (ed. J.I. Kroschwitz), John Wiley and Sons, 1985, **10**, 693
[4] *Kirk-Othmer Encycl. Chem. Technol.*, 3rd edn., (ed. M. Grayson), Wiley Interscience, 1978, **6**, 216

[5] Agranoff, J., Guide to Plastics, *Property and Specification Charts*, (ed. W.A. Kaplan), McGraw Hill, 1977
[6] *UK Kompass Register*, Reed Information Services, 1996
[7] *Jpn. Pat.*, 1994, 06 62 728, (use)
[8] *Jpn. Pat.*, 1994, 06 153 785, (use)
[9] *Jpn. Pat.*, 1993, 05 316 942, (use)
[10] *Jpn. Pat.*, 1992, 04 94 641, (use)
[11] Gupta, D., Wang, L., Hanssen, L.M., Hsia, J.J. and Datla, R.U., *NIST Spec. Publ.*, 1995, **260**, 122, (ir)
[12] Barnes, D. and Dent, G., *Spectrosc. Eur.*, 1994, **6**, 8, (ir)
[13] *Jpn. Pat.*, 1994, 06 157 997, (use)
[14] *Jpn. Pat.*, 1995, 07 76 673, (use)
[15] *Brit. Pat.*, 1992, 2 269 342, (use)
[16] *Indian Pat.*, 1992, 171 638, (use)
[17] *Eur. Pat. Appl.*, 1993, 563 812, (use)
[18] *Jpn. Pat.*, 1992, 04 239 638, (use)
[19] *Eur. Pat. Appl.*, 1988, 286 222, (use)
[20] *Jpn. Pat.*, 1991, 03 227 327, (use)
[21] *Jpn. Pat.*, 1991, 03 140 339, (use)
[22] *Eur. Pat. Appl.*, 1990, 390 113, (use)
[23] *Jpn. Pat.*, 1992, 04 140 722, (use)
[24] Papkovskii, D.B., Yaropolov, A.I., Savitskii, A.P., Olah, J. *et al*, *Biomed. Sci. (Tokyo)*, 1991, **2**, 536, (use)
[25] *Encycl. Polym. Sci. Eng.*, 2nd edn., (ed. J.I. Kroschwitz), John Wiley and Sons, 1985, **7**, 98
[26] *Jpn. Pat.*, 1994, 06 29 146
[27] Jandt, K.D., Eng, L.M., Petermann, J. and Fuchs, H., *Polymer*, 1992, **33**, 5331
[28] Vittoria, V. and Gibbs, A.R., *J. Appl. Polym. Sci.*, 1993, **49**, 247
[29] Linne, M.A., Klein, A., Miller, G.A., Sperling, L.H. and Wignall, G.D., *J. Macromol. Sci., Phys.*, 1988, **27**, 217
[30] Winnik, M.A., *Chemtracts: Macromol. Chem.*, 1991, **2**, 35
[31] *Jpn. Pat.*, 1988, 63 151 694, (use)
[32] *U.S. Pat.*, 1987, 4 684 675, (use)
[33] *Ger. Pat.*, 1994, 4 327 293, (use)
[34] *Eur. Pat. Appl.*, 1992, 494 619, (use)
[35] *Jpn. Pat.*, 1991, 03 86 706, (use)
[36] *Jpn. Pat.*, 1994, 06 57 013
[37] *Jpn. Pat.*, 1995, 07 01 643
[38] *Jpn. Pat.*, 1988, 63 98 431
[39] *Eur. Pat. Appl.*, 1993, 563 812
[40] *Eur. Pat. Appl.*, 1989, 318 794
[41] *PCT Int. Appl.*, 1991, 91 04 287
[42] *Jpn. Pat.*, 1987, 62 130 826
[43] *Jpn. Pat.*, 1987, 62 61 202
[44] *Jpn. Pat.*, 1987, 62 134 243
[45] Wang, L.H., Choy, C.L. and Porter, R.S., *J. Polym. Sci., Polym. Phys. Ed.*, 1982, **20**, 633, (thermal props)
[46] Sun, D.C. and Magill, J.H., *Polym. Eng. Sci.*, 1989, **29**, 1503, (thermal props, stability)
[47] Birley, A.W., Haworth, B. and Batchelor, J., Physics of Plastics: Processing, *Properties and Materials Engineering*, Hanser/Oxford University Press, 1991
[48] Burmester, A.F., Manial, T.A., McHattie, J.S. and Wessling, R.A., Polym. Prepr. (Am. Chem. Soc., *Div. Polym. Chem.*), 1986, **27**, 414, (gas permeability)
[49] Salame, M. and Steingiser, S., *Polym.-Plast. Technol. Eng.*, 1977, **8**, 155, (gas permeability)
[50] *Polym. Handb.*, 3rd edn., (eds. J. Brandrup and E.H. Immergut), John Wiley and Sons, 1989
[51] Yasuda, H. and Rosengren, K.J., *J. Appl. Polym. Sci.*, 1970, **14**, 2839, (gas permeability)
[52] Wang, L.H. and Porter, R.S., *J. Polym. Sci.*, 1984, **22**, 1645, (permeability)
[53] Baird, B.R., Hopfenberg, H.B. and Stannett, V.T., *Polym. Eng. Sci.*, 1971, **11**, 274, (gas permeability)
[54] Smith, T.L., Oppermann, W., Chan, A.H. and Levita, G., Polym. Prepr. (Am. Chem. Soc., *Div. Polym. Chem.*), 1983, **24**, 83, (gas permeability)
[55] Levita, G. and Smith T.L., *Polym. Eng. Sci.*, 1981, **21**, 936, (gas permeability)
[56] Hermans, P.H., Contributions to the Physics of Cellular Fibres Elsevier, 1946, (birefringence)
[57] Lefebvre, D., Jasse, B. and Monnerie, L., *Polymer*, 1982, **23**, 706, (birefringence)
[58] Lefebvre, D., Jasse, B. and Monnerie, L., *Polymer*, 1983, **24**, 1240, (birefringence)
[59] Muller, R. and Froelich, D., *Polymer*, 1985, **26**, 1477, (stress optical coefficient)
[60] Krimm, S., *Adv. Polym. Sci.*, 1960, **2**, 51, (ir)
[61] Painter, P. and Koenig, J., *J. Polym. Sci., Polym. Phys. Ed.*, 1977, **15**, 1885, (ir)
[62] *Aldrich Library of Infra Red Spectra*, (ed. C.J. Poucher), Aldrich Chemical Company, 1981, (ir)
[63] Jasse, B. and Koenig, J.L., *J. Polym. Sci., Polym. Phys. Ed.*, 1979, **17**, 799, (ir)
[64] Monnerie, L., *Faraday Symp. Chem. Soc.*, 1983, **18**, 57, (ir)
[65] *Struct. Prop. Oriented Polym.*, (ed. I.M. Ward), Applied Science, 1975
[66] Samuels, R.J., *Structured Polymer Properties*, Wiley-Interscience, 1974
[67] Kanetsuna, H. and Tanabe, Y., *J. Appl. Polym. Sci.*, 1978, **22**, 2707, (mech props)
[68] Jones, T.T., *Pure Appl. Chem.*, 1976, **45**, 39, (mech props)
[69] Matsumoto, K., Fellers, J.F. and White, J.L., *J. Appl. Polym. Sci.*, 1981, **26**, 85, (mech props)
[70] Thomas, L.S. and Cleerman, K.J., *SPE J.*, 1972, **28**, 61
[71] Leonas, K., *J. Appl. Polym. Sci.*, 1993, **47**, 2103, (environmental stability)
[72] *Encycl. Polym. Sci. Eng.*, 2nd edn., (ed. J.I. Kroschwitz), John Wiley and Sons, 1985, **10**, 693, (orientation)
[73] Flood, J.E., White, J.L. and Fellers, J.F., *J. Appl. Polym. Sci.*, 1982, **27**, 2965, (mech props)
[74] *Encycl. Polym. Sci. Eng.*, 2nd edn., (ed. J.I. Kroschwitz), John Wiley and Sons, 1985, **4**, 299, (crazing)
[75] Farrar, N.B. and Kramer, E.J., *Polymer*, 1981, **22**, 691, (environmental stress cracking)
[76] Varughese, J.V. and Gyeszly, S.W., *J. Test. Eval.*, 1993, **21**, 188
[77] Tynysbaev, F.B., Alaniya, M.G., Schlenskii, O.V., Zeelenv, Y.V. *et al*, *Plast. Massy*, 1991, 30, (thermal degradation)
[78] Yartsev, V.P., *Plast. Massy*, 1986, 16, (uv degradation)

Oxetane polymers O-4

$$-[CH_2CH_2CH_2O]_n-$$

(In the general series, any H may be substituted by another group)

Synonyms: *Polyethers (oxetane polymers). Poly(oxetane). Poly(oxy-1,3-propanediyl). Poly(trimethylene oxide). PTO. Poly(3,3-dimethyloxetane). PBCMO. PDEO. Poly(BEMO). Poly[3,3-bis(exothymethyl)oxetane]. Poly(bis(chloromethyl)oxetane). PDMO*
Related Polymers: 1,2-Epoxypropane polymers (polypropylene oxides), Poly[2,2-bis(chloromethyl)trimethylene-3-oxide]
Monomers: Oxetane
Material class: Thermoplastic
Polymer Type: polyetherketones
CAS Number:

CAS Reg. No.	Note
25722-06-9	Poly(oxetane)
25323-58-4	PBCMO
26917-50-0	PBCMO

Molecular Weight: M_n 250000–350000 (for PBCMO)
Morphology: Props. vary greatly with symmetry, bulk and substituents. Oxetane polymers range from completely amorph. liq. to highly cryst., high melting solids [1]. Conformational characteristics have been reported [5]
Miscellaneous: Although oxetane polymers have considerable potential with regard to props., prohibitive costs have meant that only poly(bis(chloromethyl))oxetane (PBCMO) has been commercially prod. and that is now withdrawn

Volumetric & Calorimetric Properties:
Density:

No.	Value	Note
1	d 1.05 g/cm^3	PTO, amorph., 25°
2	d 0.92 g/cm^3	PDMO, amorph., 25°
3	d 0.95 g/cm^3	PDEO, amorph, 25° [3]

Oxetane polymers

Latent Heat Crystallization: ΔH_{fusion} 150.9 J g^{-1} (polyoxetane); 107 J g^{-1} (monoclinic poly(dimethyloxetane)); 86.5 J g^{-1} (orthorhombic poly(dimethyl oxetane)). ΔS_{fusion} 0.468 J g^{-1} K^{-1} (polyoxetane); 0.309 J g^{-1} K^{-1} (monoclinic poly(dimethyl oxetane)); 0.263 J g^{-1} K^{-1} (orthorhombic poly(dimethyl oxetane)) [2]. $\Delta H_{polymerisation}$ 80.9 kJ mol^{-1} (oxetane); 84.6 kJ mol^{-1} (3,3-bis(chloromethyl)oxetane (BMO)); 67.9 kJ mol^{-1} (dimethyloxetane) [1]

Melting Temperature:

No.	Value	Note
1	35°C	unsubstituted polyoxetane
2	47°C	3,3-dimethyl derivative
3	135°C	fluoro-derivative
4	180°C	chloro-derivative
5	220°C	bromo-derivative
6	290°C	iodo-derivative
7	50°C	unsubstituted orthorhombic form [1,2]

Glass-Transition Temperature:

No.	Value	Note
1	-54°C	[10]

Transition Temperature General: Heat capacity and T_g data for various polyoxides including polyoxetane have been reported [12]. T_g variations with alkyl substituents have been reported [8,9]

Surface Properties & Solubility:
Solubility Properties General: Most cryst. oxetane polymers are insol. in common organic solvents. Like most polyethers, oxetane polymers are attacked by strong acids. Amorph./low melting oxetane polymers are sol. in a wide variety of organic solvents [1]
Cohesive Energy Density Solubility Parameters: δ 19.3 J$^{1/2}$ cm$^{-3/2}$ (unsubstituted polyoxetane); 16.2$^{1/2}$ cm$^{-3/2}$ (3,3-dimethyl derivative); 16.2$^{1/2}$ cm$^{-3/2}$ (3,3-diethyl derivative) [1,3]. Experimental values and 3-D solubility parameters have been reported [3]

Transport Properties:
Polymer Solutions Dilute: Theta temp θ 27° (cyclohexane, PTO) [4]. Tabulated intrinsic viscosity data for PTO, PDMO, PDEO as a function of M_n have been reported [3]

Optical Properties:
Refractive Index:

No.	Value	Note
1	1.46	PTO [3]
2	1.44	PDMO [3]
3	1.48	PDEO [3]

Polymer Stability:
Polymer Stability General: Energetic oxetane polymers (and their monomers) require special care in handling because of their explosive nature [1]
Flammability: PBCMO is self-extinguishing due to the chlorine in the side chain [1]
Chemical Stability: PBCMO is resistant to most solvents including ketones, aldehydes, esters, aromatic hydrocarbons
Hydrolytic Stability: PBCMO is resistant to water, ammonia, weak acids, weak and strong alkalis [1]

Applications/Commercial Products:
Processing & Manufacturing Routes: Not currently prod. commercially, but improved methods of polymerisation have been reported [6,7]
Applications: Cost has severely restricted application of oxetane polymers though the potential is great. Newer high tech. applications which may overcome this include solid rocket propellants, explosives, liq. crysts.

Trade name	Details	Supplier
Pentaplast		former USSR
Penton	now withdrawn	Hercules

Bibliographic References

[1] *Kirk-Othmer Encycl. Chem. Technol.*, 4th edn., (eds. J.I. Kroschwitz and M. Howe-Grant), John Wiley and Son, 1996, **19**
[2] Pérez, E., Fatou, J.G. and Bello, A., *Eur. Polym. J.*, 1987, **23**, 469
[3] Pérez, E., Gómez, M.A., Bello, A. and Fatou, J.G., *J. Appl. Polym. Sci.*, 1982, **27**, 3721
[4] Chiu, D.S., Takahashi, Y. and Mark, J.E., *Polymer*, 1976, **17**, 670
[5] Welsh, W.J., Abbassi, Q., Galiatsatos, V. and Taylor, G.K., *Macromolecules*, 1996, **29**, 993
[6] Takeuchi, D., Watanabe, Y., Aida, T. and Inoue, S., *Macromolecules*, 1995, **28**, 651
[7] Takeuchi, D. and Aida, T., *Macromolecules*, 1996, **29**, 8096
[8] Pérez, E., Bello, A. and Perena, J.M., *Polym. Bull. (Berlin)*, 1988, **20**, 291
[9] Bello, A. and Pérez, E., *Thermochim. Acta*, 1988, **134**, 155
[10] Garrido, L., Riande, E. and Guzman, J., *Makromol. Chem., Rapid Commun.*, 1983, **4**, 725
[11] Kawakami, Y., Takahashi, K., Nishiguchi, S. and Toida, K., *Polym. Int.*, 1993, **31**, 35
[12] Gaur, U. and Wunderlich, B., *Polym. Prepr. (Am. Chem. Soc., Div. Polym. Chem.)*, 1979, **20**, 429

PACM-12   P-1

Synonyms: *Poly(bis(4-aminocyclohexyl)methane-co-dodecanoic acid). Poly(bis(4-aminocyclohexyl)methane 1,10-decanedicarboxamide). Poly[imino-1,4-cyclohexanediylmethylene-1,4-cyclohexanediylimino(1,12-dioxo-1,12-dodecanediyl)]. Poly-4,4'-methylenedicyclohexylene dodecanediamide*
Monomers: Bis(4-aminocyclohexyl)methane, Dodecanedioic acid
Material class: Thermoplastic
Polymer Type: polyamide
CAS Number:

CAS Reg. No.
25035-12-5

Molecular Formula: $(C_{27}H_{48}N_2O_2)_n$
Fragments: $C_{27}H_{48}N_2O_2$
General Information: DuPont introduced Qiana in 1968, but it was withdrawn in the 1980s. Fibres were silk-like and contained 70% *trans-trans* PACM. Fibres have good tenacity and dimensional stability and low moisture absorption [3]
Morphology: The *trans-trans* polymer gives rise to orthogonal crystals, a 9.3, b 6.06, c 45.0, 4 monomers per unit cell [1]

Volumetric & Calorimetric Properties:
Density:

No.	Value	Note
1	d 1.034 g/cm^3	amorph. [1, 2]
2	d 1.04 g/cm^3	cryst. [2]

Latent Heat Crystallization: Latent heat of fusion ΔH_m 12 kJ mol^{-1} [2]
Specific Heat Capacity:

No.	Value	Note	Type
1	2.8 kJ/kg.C	[2]	P

Melting Temperature:

No.	Value	Note
1	275–308°C	[2]
2	280°C	fibre [5]

Glass-Transition Temperature:

No.	Value	Note
1	135–147°C	[2]
2	190°C	dry, fibre
3	85°C	wet, fibre [5]

Transport Properties:
Water Content: 2.5% (65% relative humidity) [5]

Mechanical Properties:
Mechanical Properties General: Tenacity 0.78 N tex^{-1} (13% elongation) (fibre) [4]; 0.26–0.29 N tex^{-1} (Qiana fibre) [5]

Applications/Commercial Products:
Processing & Manufacturing Routes: The nylon salt is prepared under pressure above 100° due to its low solubility. The temp. of the conc. soln. is raised in an autoclave with gradual release of H_2O until molten polymer is formed (typically up to 300° and 1.8 MPa). The polymer is extruded to obtain high MW and melt spun above 280° to form fibres
Mould Shrinkage (%):

No.	Value	Note
1	4–10%	Shrinkage in boiling water [5]

Applications: Commercial use of polymer in fibre form (1970–1980) under tradename Qiana. The fibre resembles polyester fibre in props., e.g. for fabric stability and wrinkle resistance. It has a silk-like appearance

Trade name	Details	Supplier
Qiana	Withdrawn in 1980s	DuPont

Bibliographic References
[1] Burton, R., *Bull. Am. Phys. Soc.*, 1987, **32**, 701, (morphology)
[2] Van Krevelen, D.W., *Properties of Polymers: Their Correlation with Chemical Structure*, 3rd edn., Elsevier, 1990
[3] *Encycl. Polym. Sci. Technol.*, Vol. 1, 2nd edn., Wiley Interscience, 1985, 437
[4] Hannell, J.S., *Polym. News*, 1970, **1**, 8
[5] Rumsey, J.S., *Mod. Text.*, 1970, **51 (2)**, 48

PBI   P-2

Synonyms: *Poly([5,5'-bi-1H-benzimidazole]-2,2'-diyl-1,3-phenylene). 1,3-Benzenedicarboxylic acid polymer with [1,1'-biphenyl]-3,3',4,4'-tetramine*
Monomers: 1,3-Benzenedicarboxylic acid, 1,3-Benzenedicarboxylic acid, 3,3',4,4'-Biphenyltetramin
Material class: Thermoplastic
Polymer Type: polybenzimidazole
CAS Number:

CAS Reg. No.	Note
25734-65-0	Poly([5,5'-bi-1H-benzimidazole]-2,2'-diyl-1,3-phenylene)
26101-19-9	1,3-Benzenedicarboxylic acid polymer with [1,1'-biphenyl]-3,3',4,4'-tetramine

Molecular Formula: $(C_{20}H_{12}N_4)_n$
Fragments: $C_{20}H_{12}N_4$
Additives: Polymer can be doped with acids such as phosphoric or sulfuric acid to increase proton conductivity [1].
General Information: High-strength engineering plastic with good thermal stability and flame resistance. Prod. as golden-brown powder.
Morphology: Amorphous polymer with very high glass-transition temperature. No regular crystalline structure according to x-ray diffraction data [2,3]. Acid-doping leads to increase in crystallinity through formation of hydrogen bonds [3].

Volumetric & Calorimetric Properties:

Density:

No.	Value	Note
1	d1.3 g/cm^3	moulded resin [4]

Glass-Transition Temperature:

No.	Value	Note
1	420–427°C	[4,5]

Transition Temperature General: A secondary β-transition reported at 300°C [5]

Deflection Temperature:

No.	Value	Note
1	435°C	[5]

Surface Properties & Solubility:

Solubility Properties General: Dimethylacetamide is the usual solvent for preparation of PBI solns. and fibres.
Solvents/Non-solvents: Sol. dimethylacetamide, dimethylformamide, DMSO [6], formic acid [2], polyphosphoric acid [7], conc. H_2SO_4, strong bases [1]. Insol. most organic solvents.

Transport Properties:

Water Absorption:

No.	Value	Note
1	12–19%	high moisture affinity [1,8]

Mechanical Properties:

Mechanical Properties General: Exhibits a very high shear yield stress, resulting in high compressive and tensile yield stresses [5]. It can be brittle as a result of surface irregularities [5].

Tensile (Young's) Modulus:

No.	Value	Note
1	5900 MPa	[4]

Flexural Modulus:

No.	Value	Note
1	6500 MPa	[4]

Compressive Modulus:

No.	Value	Note
1	5900–6200 MPa	[4]

Poisson's Ratio:

No.	Value	Note
1	0.34	[4]

Tensile Strength Yield:

No.	Value	Elongation	Note
1	160 MPa	3	[4]

Flexural Strength Yield:

No.	Value	Note
1	220 MPa	[4]

Compressive Strength:

No.	Value	Note
1	390–400 MPa	[4,5]

Electrical Properties:

Electrical Properties General: Insulator for electron conduction but exhibits proton conductivity [9]. When doped with acids, the nitrogen atoms are protonated and anions are linked by hydrogen-bonding, while a hopping mechanism for proton transfer occurs [10]. For membrane applications, acid-doped PBI has been tested at high temperatures (150–200°C), with proton conductivities up to 6×10^{-2} S cm^{-1} at different degrees of humidity [11,12]. The conductivity increases markedly on absorption of water by polymer [8]. The electroluminescence of PBI films with light emission at λ_{max} ca. 540nm has been examined in LED devices [13].

Optical Properties:

Optical Properties General: Film exhibits uv absorption centred at ca. 360nm and photoluminescence at ca. 525nm [13].
Transmission and Spectra: pmr [14], ir [2,8,10,14], uv and photoluminescence [13] and x-ray diffraction [2,3] data have been reported.

Polymer Stability:

Thermal Stability General: Water only is evolved from fibre up to ca. 300°C [16]. Polymer starts to decompose at ca. 500°C in inert gases, with rapid decomposition between 550–900°C [2,15].
Flammability: Powder and fibre are nonflammable in air with only non-toxic gases (water and CO_2) evolved up to 560°C [15].

Applications/Commercial Products:

Processing & Manufacturing Routes: Prod. mainly in a two-stage polymerisation of 3,3′,4,4′-tetraaminobiphenyl with diphenyl isophthalate in polyphosphoric acid [2,17]. Isophthalic acid [7,18], hexapropyl orthoisophthalate [19], and isophthalaldehyde bis(bisulfite) [20] have been used in place of diphenyl isophthalate. A one-stage process using sulfolane as solvent has also been reported [21]. Polymer can be processed into moulded products, spun into fibres, or used as solutions in dimethylacetamide.
Applications: Moulded polymer products are used in high heat-resistant applications such as in semiconductors and in the chemical and petrochemical industries. Fibres can be used in the fire service, protective industrial clothing, and also in blended fabrics with aramid fibres. The solutions are used for film castings and coatings. A potential application for acid-doped or modified PBI is as a membrane for fuel cells.

Trade name	Details	Supplier
Celazole	polymer powders, PBI fibres, PBI solutions	PBI Performance Products

Bibliographic References

[1] Li, Q., He, R., Jensen, J.O. and Bjerrum, N.J., *Fuel Cells,* 2004, **4**, 147
[2] Vogel, H. and Marvel, C.S., *J. Polym. Sci.,* 1961, **50**, 511
[3] Staiti, P., Lufrano, F., Arico, A.S., Passalacque, E. and Antonucci, V., *J. Membr. Sci.,* 2001, **188**, 71
[4] Sandor, R.B., *High Performance Polym.,* 1990, **2**, 25
[5] Rose, J., Duckett, R.A. and Ward, I.M., *J. Mater. Sci.,* 1995, **30**, 5328
[6] Dangayach, K.C.B., Karim, K.A. and Bonner, D.C., *J. Appl. Polym. Sci.,* 1981, **26**, 560
[7] Xiao, L., Zhang, H., Scanlon, E., Ramanathan, L.S., Choe, E.-W., Rogers, D., Apple, T. and Benicewicz, B.C., *Chem. Mater.,* 2005, **17**, 5328

[8] Pu, H., Meyer, W.H. and Wegner, G., *J. Polym. Sci., Part B: Polym. Phys.*, 2002, **40**, 663
[9] Hoel, D. and Grunwald, E.J., *J. Phys. Chem.*, 1977, **81**, 2135
[10] Bouchet, R. and Siebert, E., *Solid State Ionics*, 1999, **118**, 287
[11] He, R., Li, Q., Bach, A., Jensen, J.O. and Bjerrum N.J., *J. Membr. Sci.*, 2006, **277**, 38
[12] Ma, Y.-L., Wainwright, J.S., Litt, M.H. and Savinell, R.F., *J. Electrochem. Soc.*, 2004, **151**, A8
[13] Wu, C.-C., Chang, C.-F. and Bai, S.J., *Thin Solid Films*, 2005, **479**, 245
[14] Glipa, X., El Haddad, M., Jones, D.J. and Rozi&fere, J., *Solid State Ionics*, 1997, **97**, 323
[15] Jackson, R.H., *Text. Res. J.*, 1978, **48**, 314
[16] Menczel, J.D., *J. Therm. Anal. Calorim.*, 2000, **59**, 1023
[17] Conciatori, A.B. and Chenevey, E.C., *Macromol. Synth.*, 1968, **3**, 24
[18] Choe, E.-W., *J. Appl. Polym. Sci.*, 1994, **53**, 497
[19] Dudgeon, C.D. and Vogel, O., *J. Polym. Sci., Part A: Polym. Chem.*, 1978, **16**, 1829
[20] Higgins, J. and Marvel, C.S., *J. Polym. Sci., Polym. Chem. Ed.*, 1970, **8**, 171
[21] Hedberg, F.L. and Marvel, C.S., *J. Polym. Sci., Polym. Chem. Ed.*, 1974, **12**, 1823

PBT/ASA blends

Synonyms: *Poly(butylene terephthalate/acrylonitrile-styrene-acrylic) blends. ASA/PBT blends*
Monomers: 1,4-Benzenedicarboxylic acid, 1,4-Butanediol, Methacrylic acid, Styrene, Acrylonitrile
Material class: Blends, Copolymers
Polymer Type: PBT and copolymers, acrylic terpolymers
Additives: Glass fibre (10–30%)
General Information: Produces a good surface finish even with filled grades

Volumetric & Calorimetric Properties:
Density:

No.	Value	Note
1	d 1.32 g/cm^3	10% glass fibre, ISO 1183 [1]
2	d 1.39 g/cm^3	20% glass fibre, ISO 1183 [1]
3	d 1.47 g/cm^3	30% glass fibre, ISO 1183 [1]

Thermal Expansion Coefficient:

No.	Value	Note	Type
1	5.2×10^{-5} K^{-1}	10% glass fibre, DIN 53752 [1]	L
2	3.8×10^{-5} K^{-1}	20% glass fibre, DIN 53752 [1]	L
3	2.7×10^{-5} K^{-1}	30% glass fibre, DIN 53752 [1]	L

Deflection Temperature:

No.	Value	Note
1	168°C	1.8 MPa, 10% glass fibre, ISO 75 [1]
2	194°C	1.8 MPa, 20% glass fibre, ISO 75 [1]
3	204°C	1.8 MPa, 30% glass fibre, ISO 75 [1]

Vicat Softening Point:

No.	Value	Note
1	150°C	50N, 10% glass fibre, ISO 306 [1]
2	157°C	50N, 20% glass fibre, ISO 306 [1]
3	160°C	50N, 30% glass fibre, ISO 306 [1]

Transport Properties:
Melt Flow Index:

No.	Value	Note
1	48 g/10 min	10% glass fibre, ISO 1133 [1]
2	64 g/10 min	20% glass fibre, ISO 1133 [1]
3	73 g/10 min	30% glass fibre, ISO 1133 [1]

Water Absorption:

No.	Value	Note
1	0.67 %	10% glass fibre, ISO 62A [1]
2	0.62 %	20% glass fibre, ISO 62A [1]
3	0.52 %	30% glass fibre, ISO 62A [1]

Mechanical Properties:
Tensile (Young's) Modulus:

No.	Value	Note
1	4600 MPa	10% glass fibre, ISO 527 [1]
2	7500 MPa	20% glass fibre, ISO 527 [1]
3	10500 MPa	30% glass fibre, ISO 527 [1]

Tensile Strength Yield:

No.	Value	Elongation	Note
1	82 MPa	3%	10% glass fibre, ISO 527 [1]
2	110 MPa	2.4%	20% glass fibre, ISO 527 [1]
3	125 MPa	2%	30% glass fibre, ISO 527 [1]

Electrical Properties:
Surface/Volume Resistance:

No.	Value	Note	Type
2	0.1×10^{15} Ω.cm	all grades, ISO 1325 [1]	S

Dielectric Permittivity (Constant):

No.	Value	Frequency	Note
1	3.4	1 MHz	10% glass fibre [1]
2	3.6	1 MHz	20% glass fibre [1]
3	3.7	1 MHz	30% glass fibre [1]
4	3.6	100 Hz	10% glass fibre, IEC 250 [1]
5	3.7	100 Hz	20% glass fibre, IEC 250 [1]
6	3.8	100 Hz	30% glass fibre, IEC 250 [1]

Complex Permittivity and Electroactive Polymers: Comparative tracking index 375 V (10% glass fibre, IEC 112A), 450 V (20% glass fibre, IEC 112A), 500 V (30% glass fibre, IEC 112A) [1]
Dissipation (Power) Factor:

No.	Value	Frequency	Note
1	0.0031	100 Hz	10% glass fibre, ISO 1325 [1]
2	0.003	100 Hz	20% and 30% glass fibre, ISO 1325 [1]
3	0.0205	1 MHz	10% glass fibre [1]

– PBT, Flame retardant

| 4 | 0.019 | 1 MHz | 20% glass fibre [1] |
| 5 | 0.018 | 1 MHz | 30% glass fibre [1] |

Polymer Stability:
Flammability: Flammability rating HB (all grades, UL94) [1]
Recyclability: May be recycled if not thermally degraded or contaminated. Recycled material may alter props. of blend

Applications/Commercial Products:
Processing & Manufacturing Routes: Processed by injection-moulding. Mould temp. 60–100°

Trade name	Supplier
Ultradur S	BASF

Bibliographic Reference
[1] *Ultradur S*, BASF, (technical datasheet)

PBT, Flame retardant P-4
Related Polymers: Polybutylene terephthalate, PBT - Injection Moulding Glass Fibre filled, PBT - High Impact, PBT - Fibres, PBT - Films/Sheets, PBT - Injection Moulding Mineral filled, PBT - Injection Moulding Carbon Fibre filled, PBT - Polyethylene terephthalate blends, PBT - Phenoxy blends
Monomers: Terephthalic acid, 1,4-Butanediol
Material class: Thermoplastic
Polymer Type: PBT and copolymers
CAS Number:

CAS Reg. No.	Note
24968-12-5	repeating unit
26062-94-2	
30965-26-5	from dimethyl terephthalate
59822-52-5	from terephthaloyl chloride

Molecular Formula: $(C_{12}H_{12}O_4)_n$
Fragments: $C_{12}H_{12}O_4$
Additives: Brominated polyoxyphenylene and diglycidyl polymer and antimony trioxide mixed with brominated species are used as flame retardants

Volumetric & Calorimetric Properties:
Density:

No.	Value	Note
1	d 1.47 g/cm^3	Crastin XB 3035, DIN 53479 [2]
2	d 1.66 g/cm^3	Celanex, 30% glass fibre [9]
3	d 1.43 g/cm^3	ASTM D792 [11]

Thermal Expansion Coefficient:

No.	Value	Note	Type
1	7×10^{-5} K^{-1}	Crastin XB 3035, ASTM D696 [2]	L
2	0.00011 K^{-1}	ASTM D696 [11]	V

Melting Temperature:

No.	Value	Note
1	213°C	Crastin XB 3035 [8]

Transition Temperature General: Heat distortion temps. have been reported [8]
Deflection Temperature:

No.	Value	Note
1	208°C	1.82 MPa, Celanex, 30% glass fibre [9]
2	60°C	1.82 MPa, ASTM D645 [11]

Transport Properties:
Melt Flow Index:

No.	Value	Note
1	11.6–19.1 g/10 min	30% glass fibre [10]

Water Absorption:

No.	Value	Note
1	0.1 %	24h, ASTM D570 [11]

Mechanical Properties:
Mechanical Properties General: Mechanical props. of Celanex X-203101 [3] have been reported [8]. Elastic modulus has been reported [2]. Young's modulus 2700 N mm^{-2} (DIN 53457) (Crastin XB 3085) [2]

Tensile (Young's) Modulus:

No.	Value	Note
1	11700 MPa	Celanex, 30% glass fibre

Flexural Modulus:

No.	Value	Note
1	10300 MPa	Celanex, 30% glass fibre [9]
2	2700 MPa	ASTM D790 [11]

Tensile Strength Break:

No.	Value	Note	Elongation
1	60 MPa	Valox 310-SEQ [1]	9.4%
2	100 MPa	Valox 420-SEQ [1]	3.7%
3	135 MPa	Celanex -30% glass fibre [9]	1.5%

Flexural Strength at Break:

No.	Value	Note
1	193 MPa	Celanex, 30% glass fibre [9]

Tensile Strength Yield:

No.	Value	Note
1	50 MPa	ASTM D638 [11]

Impact Strength: Other values of impact resistance have been reported [8]
Hardness: Rockwell M80 (ASTM D785) [1]

– PBT, Flame retardant

Izod Notch:

No.	Value	Notch	Note
1	69.4 J/m	Y	Celanex, 30% glass fibre [9]
2	214 J/m	N	Celanex, 30% glass fibre [9]
3	50 J/m	Y	23°, ASTM D256A [11]

Izod Area:

No.	Value	Notch	Note
1	5 kJ/m²	Notched	Crastin XB 3035, DIN 53453 [2]

Electrical Properties:
Surface/Volume Resistance:

No.	Value	Note	Type
1	0.1×10^{15} Ω.cm	Crastin XB 3035, VDE 0303-3 [2]	S

Dielectric Permittivity (Constant):

No.	Value	Frequency	Note
1	4	50 Hz	Crastin XB 3035, VDE 0303-4 [2]
2	3.5	1 MHz	Crastin XB 3035, VDE 0303-4 [2]
3	3.9		Celanex, 30% glass fibre [9]

Dielectric Strength:

No.	Value	Note
1	150 kV.mm⁻¹	2 mm thick, Crastin XB 3035 [2]
2	19.6 kV.mm⁻¹	Celanex, 30% glass fibre [9]
3	20 kV.mm⁻¹	ASTM D149 [11]

Arc Resistance:

No.	Value	Note
1	17s	ASTM D495
2	104s	20% mica filled [4]

Static Electrification: Tracking strength 250 v (Crastin XB 3035) [2]
Dissipation (Power) Factor:

No.	Value	Frequency	Note
1	0.015	50 Hz	VDE 0303-4, Crastin XB 3035 [2]
2	0.165	1 MHz	VDE 0303-4, Crastin XB 3035 [2]
3	0.03		Celanex X-203101 [3]

Optical Properties:
Transmission and Spectra: Ir spectral data have been reported [6]

Polymer Stability:
Upper Use Temperature:

No.	Value	Note
1	120°C	[11]

Flammability: UL 94 rating V0 (Crastin XB 3035) [2]. Flammability of Celanex X-203101 has been reported [3]. Oxygen index 31% (Crastin XB 3035, ASTM D2863) [2], 26.7–31.2% [10]

Environmental Stress: Resistance to uv radiation has been reported [7]. Valox grade rapidly loses its mechanical props. on ageing at high temps. and humidity [1]. Has good resistance to weathering [9]

Applications/Commercial Products:
Mould Shrinkage (%):

No.	Value	Note
1	2%	ASTM D955 [11]

Trade name	Details	Supplier
Celanex 3112/3210		Hoechst Celanese
Crastine XMB 1069		Ciba-Geigy Corporation
E-9900-0590		Thermofil
Fiberfil	used to refer to other materials	DSM Engineering Plastics
Later	used to refer to other materials	LATI Industria Thermoplastics S.p.A.
Mitsubishi Kasei 5010 N3		Mitsubishi Chemical
Orgater	used to refer to other materials	EY Atochem
Pibiter	used to refer to other materials	Enichem
Pocan KL1-7503		Mobay
RTP 1000 FR		RTP Company
RTP 1000 FRA		RTP Company
RTP 1099 x 29687		RTP Company
Rynite 6900		DuPont
Techster	used to refer to other materials	Rhone-Poulenc
Thermocomp WF-1006 FR		LNP Engineering
Thermofil	used to refer to other materials	Thermofil
Ultradur KR 4015		BASF
Valox	used to refer to other materials	General Electric Plastics
Vandar 8000		Hoechst Celanese
Vestadur	used to refer to other materials	Hüls AG
Voloy	used to refer to other materials	ComAlloy International
Vybex	used to refer to other materials	Ferro Corporation

Bibliographic References
[1] Gardner, R.J. and Marin, J.R., *J. Appl. Polym. Sci.*, 1980, **25**, 2353, (mechanical props., weathering)
[2] Breitenfellner, F., *Kunststoffe*, 1982, **72**, 284, (electrical and mechanical props.)
[3] Baron, A.L., Freed, W.T. and McNally D., Tech. Pap., Reg. Tech. Conf. - Soc. Plast. Eng., 1972, **9**, 1, (mechanical props., flammability)
[4] *Ger. Pat.*, 1979, 2 885 005, (arc resistance)
[5] Burleigh, P.H., Nametz, R.C., Moore, P.D. and Jay, T.A., *J. Fire Retard. Chem.*, 1980, **7**, 47, (flame retardant)
[6] Moehler, H. and Mathias, E., *Experientia, Suppl.*, 1979, **37**, 368, (ir)
[7] Bar-Yaacov, Y., Minke, R., Touval, I., Kourtides, D.A. and Parker, J.A., *J. Fire Retard. Chem.*, 1982, **9**, 181, (weathering)

[8] Green, J. and Chung, J., *J. Fire Sci.*, 1990, **8**, 254, (flammability, mechanical props.)
[9] *Kirk-Othmer Encycl. Chem. Technol.*, 4th edn., (ed J.I. Kroschwitz), Wiley Interscience, 1996, **19**, 630, (mechanical props.)
[10] *Polymeric Materials Encyclopedia*, (ed. J.C. Salamone), CRC Press, 1996, **4**, 2415, (flame retardants)
[11] *The Materials Selector*, 2nd edn., (eds. N.A. Waterman and M.F. Ashby), Chapman and Hall, 1997, **3**, 366, (mechanical and electrical props.)

PBT, High impact, Impact modified P-5

Related Polymers: PBT, Polybutylene terephthalate, PBT - Injection Moulding Glass fibre filled, PBT - fibres, PBT - films/sheets, PBT - Flame Retardant, PBT - Injection Moulding Mineral filled, PBT - Injection Moulding Carbon fibre filled, PBT - Polyethylene terephthalate blends, PBT - Phenoxy blends
Monomers: Terephthalic acid, 1,4-Butanediol
Material class: Composites, Thermoplastic
Polymer Type: PBT and copolymers
CAS Number:

CAS Reg. No.	Note
26062-94-2	
24968-12-5	repeating unit
30965-26-5	from dimethyl terephthalate
59822-52-5	from terephthaloyl chloride

Molecular Formula: $(C_{12}H_{12}O_4)_n$
Fragments: $C_{12}H_{12}O_4$

Volumetric & Calorimetric Properties:
Density:

No.	Value	Note
1	d 1.53 g/cm^3	30% glass fibre, Celanex [2]
2	d 1.25 g/cm^3	ASTM D792 [3]
3	d 1.47 g/cm^3	30% glass fibre, ASTM D792 [3]
4	d 1.21 g/cm^3	Vandar, ISO 1183 [4]

Thermal Expansion Coefficient:

No.	Value	Note	Type
1	0.00014 K^{-1}	Vandar, DIN 53752 [4]	L

Deflection Temperature:

No.	Value	Note
1	190°C	1.82 MPa, 30% glass fibre, ASTM D648 [3]
2	191°C	1.82 MPa, Celanex, 30% glass fibre [2]
3	52°C	1.82 MPa, ASTM D648 [3]
4	50°C	1.8 MPa, Vandar, ISO 75 [4]

Vicat Softening Point:

No.	Value	Note
1	135°C	Vandar, ISO 306 [4]

Transport Properties:
Melt Flow Index:

No.	Value	Note
1	54 g/10 min	Vandar, ISO 1133 [4]

Water Absorption:

No.	Value	Note
1	0.1 %	24h, ASTM D570 [3]
2	0.08 %	24h, 30% glass fibre, ASTM D570 [3]
3	0.6 %	Vandar, DIN 53495 [4]

Mechanical Properties:
Tensile (Young's) Modulus:

No.	Value	Note
1	8300 MPa	Celanex, 30% glass fibre [2]
2	1600 MPa	Vandar, ISO 527 [4]

Flexural Modulus:

No.	Value	Note
1	6900 MPa	30% glass fibre, Celanex [2]
2	1520 MPa	ASTM D790 [3]
3	6380 MPa	30% glass fibre, ASTM D790 [3]

Tensile Strength Break:

No.	Value	Note	Elongation
1	97 MPa	30% glass fibre, Celanex [2]	3.1%
2	40 MPa	ASTM D638 [3]	270%
3	89.7 MPa	ASTM D638 [3]	3.5%

Flexural Strength at Break:

No.	Value	Note
1	152 MPa	30% glass fibre, Celanex [2]
2	48.3 MPa	ASTM D790 [3]
3	134 MPa	30% glass fibre, ASTM D790 [3]

Tensile Strength Yield:

No.	Value	Elongation	Note
1	34 MPa	50%	Vandar, ISO 527 [4]

Izod Notch:

No.	Value	Notch	Break	Note
1	160 J/m	Y		30% glass fibre, Celanex [2]
2	641 J/m	N		30% glass fibre, Celanex [2]
3	960 J/m	Y		ASTM D256 [3]
4	230 J/m	Y		30% glass fibre [3]
5		N	No break	ASTM D256
6	910 J/m	N		30% glass fibre [3]

Izod Area:

No.	Value	Notch	Break	Note
1	kJ/m^2	Unnotched	No break	23°, Vandar, ISO 180/1C [4]

2	kJ/m²	Unnotched	No break	-30°, Vandar, ISO 180/1C [4]
3	70 kJ/m²	Notched		23°, Vandar, ISO 180/1A [4]
4	30 kJ/m²	Notched		-30°, Vandar, ISO 180/1A [4]

Electrical Properties:
Electrical Properties General: Variation of power factor with temp. for Pocan and Ultradur has been reported [1]
Surface/Volume Resistance:

No.	Value	Note	Type
3	1×10^{15} Ω.cm	Vandar, IEC 93 [4]	S

Dielectric Permittivity (Constant):

No.	Value	Frequency	Note
1	4.3		30% glass fibre, Celanex [2]
2	3.7	50 Hz	Vandar, IEC 250 [4]
3	3.3	1 MHz	Vandar, IEC 250 [4]

Dielectric Strength:

No.	Value	Note
1	20 kV.mm^{-1}	30% glass fibre, Celanex [2]
2	24 kV.mm^{-1}	Vandar, IEC 243-1 [4]

Static Electrification: Comparative tracking index 600 V (Vandar, IEC 112) [4]
Dissipation (Power) Factor:

No.	Value	Frequency	Note
1	0.0017	50 Hz	Vandar, ISO 1325 [4]
2	0.014	1 MHz	Vandar, ISO 1325 [4]

Polymer Stability:
Flammability: UL 94 rating HB (0.8 mm thick, Celanex, 30% glass fibre) [2,4]

Applications/Commercial Products:
Mould Shrinkage (%):

No.	Value	Note
1	1.7–1.8%	ASTM D955 [3]
2	1.3–1.4%	30% glass fibre, ASTM D955 [3]

Trade name	Details	Supplier
Celanex		Hoechst
Pocan KU1-7914/1		Bayer Inc.
Pocan KU1-7914/P7		Bayer Inc.
Pocan S1506		Bayer Inc.
Pocan S1517		Bayer Inc.
Ultradur KR 4071		BASF
Vandar	various grades	Hoechst

Bibliographic References
[1] Hourston, D.J., Lane, S. and Zhang, H.X., *Polymer*, 1991, **32**, 2215, (mechanical props.)
[2] *Kirk-Othmer Encycl. Chem. Technol.*, 4th edn., (ed. J.I. Kroschwitz), Wiley Interscience, 1991, **19**, 630
[3] *Encycl. Polym. Sci. Eng.*, 2nd edn., (ed. J.I. Kroschwitz), Wiley Interscience, 1985, **12**, 244, (mechanical props., water absorption)
[4] Vandar Hoechst AG, 1996, (technical datasheet)

PBT, injection moulding glass fibre filled P-6

Synonyms: Polybutylene terephthalate, injection moulding glass fibre filled
Related Polymers: Polybutylene terephthalate, PBT flame retardant, PBT high impact, PBT fibres, PBT films/sheets, PBT mineral filled, PBT carbon fibre filled, PBT Polycarbonate blends, PBT Polyethylene terephthalate blends, PBT phenoxy blends
Monomers: 1,4-Benzenedicarboxylic acid, 1,4-Butanediol
Material class: Composites, Thermoplastic
Polymer Type: PBT and copolymers, polyester
CAS Number:

CAS Reg. No.	Note
24968-12-5	repeating unit
26062-94-2	
30965-26-5	from dimethyl terephthalate
59822-52-5	from terephthaloyl chloride

Molecular Formula: $(C_{12}H_{12}O_4)_n$
Fragments: $C_{12}H_{12}O_4$
General Information: Glass fibre reinforced PBT provides improved mech. and thermal props., and increased fatigue endurance, creep resistance and high temp. resistance

Volumetric & Calorimetric Properties:
Density:

No.	Value	Note
1	d 1.5 g/cm³	30% glass fibre [3,17]
2	d 1.54 g/cm³	30% glass fibre, Celanex [16]
3	d^{23} 1.69 g/cm³	30% glass fibre, Arnite T [17]
4	d 1.53 g/cm³	30% glass fibre, ASTM D792 [18]

Thermal Expansion Coefficient:

No.	Value	Note	Type
1	4×10^{-5} K^{-1}	30% glass fibre [3]	V
2	1.4×10^{-5} K^{-1}	30% glass fibre, ASTM D696 [17]	V
3	2.5×10^{-5} K^{-1}	30% glass fibre, ASTM D696 [17,21]	L
4	4×10^{-5}–5×10^{-5} K^{-1}	flow direction, 30% glass fibre, Arnite T [17]	L
5	7×10^{-5}–8×10^{-5} K^{-1}	cross direction, 30% glass fibre, Arnite T [17]	L

Thermodynamic Properties General: Heat transfer coefficient 13.4 kJ m^{-3} s^{-1} K^{-1}, (30% glass fibre, Novadur 5010) [7]
Thermal Conductivity:

No.	Value	Note
1	12.1 W/mK	30% glass fibre, ASTM C177 [17]
2	0.24 W/mK	30% glass fibre, Arnite T [17]
3	0.21 W/mK	30% glass fibre, ASTM C177 [21]

– PBT, injection moulding glass fibre filled

Specific Heat Capacity:

No.	Value	Note	Type
1	0.46 kJ/kg.C	30% glass fibre [17]	P
2	1.15 kJ/kg.C	40°, 30% glass fibre, Arnite T [17]	P

Melting Temperature:

No.	Value	Note
1	220–225°C	30% glass fibre [3]
2	223°C	30% glass fibre, Arnite T [17]

Deflection Temperature:

No.	Value	Note
1	206°C	1.82 MPa, 30% glass fibre, Celanex [16]
2	213°C	1.82 MPa, 30% glass fibre, ASTM D648 [17]
3	205°C	30% glass fibre, Arnite T [17]
4	210°C	1.82 MPa, 30% glass fibre, ASTM D648 [18]

Vicat Softening Point:

No.	Value	Note
1	205°C	30% glass fibre, DIN 53460 Method B [3]
2	215°C	30% glass fibre, Arnite T, IDN [17]

Transport Properties:
Polymer Melts: Rheological props. have been reported [5]
Permeability of Liquids: Water diffusion coefficient of Duranex 2002 has been reported [8]
Water Content: Water content of Duranex 2002 has been reported [8]. Water absorption of Celanex 917 has been reported [11]
Water Absorption:

No.	Value	Note
1	0.3 %	23°, 30% glass fibre [3]
2	0.06 %	24h, 30% glass fibre, ASTM D570 [17,21]
3	0.26 %	equilibrium, 30% glass fibre, ASTM D570 [18]

Mechanical Properties:
Mechanical Properties General: Mechanical props. of 30% glass fibre filled material have been reported [10,12]. Tensile modulus (Valex 735, mineral filled) has been reported [6]. Dynamic storage modulus (Novadur 5010, 30% glass fibre) has been reported [7]. Other values of tensile strength (Celanex 917) have been reported [11]. Shear modulus of Valex 735 (mineral filled) has been reported [6]

Tensile (Young's) Modulus:

No.	Value	Note
1	8300 MPa	30% glass fibre, ASTM D638 [17]
2	4090 MPa	Duranex 2002 [8]
3	9700 MPa	30% glass fibre, Celanex [16]
4	10200 MPa	30% glass fibre, Arnite T [17]

Flexural Modulus:

No.	Value	Note
1	8300 MPa	30% glass fibre, Celanex [16]
2	8100 MPa	30% glass fibre, ASTM D790 [17]
3	7580 MPa	30% glass fibre, ASTM D790 [18]
4	9900 MPa	30% glass fibre, Arnite T [17]

Tensile Strength Break:

No.	Value	Note	Elongation
1	74.58 MPa	Duranex 2006 [8]	2.26%
2	135 MPa	30% glass fibre, Celanex [16]	2%
3	131 MPa	30% glass fibre, ASTM D638 [17]	4%
4	135 MPa	30% glass fibre, Arnite T [17]	2%
5	117 MPa	30% glass fibre, ASTM D638 [18]	4%

Flexural Strength at Break:

No.	Value	Note
1	193 MPa	30% glass fibre, Celanex [16]
2	193 MPa	30% glass fibre, ASTM D790 [18]

Poisson's Ratio:

No.	Value	Note
1	0.34	Valox 735, mineral filled [6]

Compressive Strength:

No.	Value	Note
1	124 MPa	30% glass fibre, ASTM D695 [17]

Impact Strength: Impact strength 75 J m^{-2} (20°, ASTM D1822) [2].
Viscoelastic Behaviour: Stress-strain curves for Duranex 2002 have been reported [18]. Creep resistance (Celanex 917) has been reported [11]
Hardness: Rockwell R118 (30% glass fibre, ASTM D785) [17,18], M87 (30% glass fibre, Arnite T) [17], M90 (30% glass fibre, ASTM D785) [18]
Failure Properties General: Continuous stress deformation 1.60% (13.8 MPa, 50°, 24h, 30% glass fibre, ASTM D620) [18]
Fracture Mechanical Properties: Creep rupture and fatigue have been reported [1]. Fatigue behaviour (Novadur 5010, 30% glass fibre) has been reported [7]. Shear strength 61 MPa (30% glass fibre, ASTM D732) [28]
Friction Abrasion and Resistance: Abrasion resistance 40 mg (1000 cycles)$^{-1}$ (ASTM D1044) [17]. Taber abrasion 19 mg (1000 cycles)$^{-1}$ (30% glass fibre, ASTM D1044, CS-17) [18]. Coefficient of friction μ 0.15 (against self, ASTM D1894), 0.19 (against metals, 30% glass fibre) [18]
Izod Notch:

No.	Value	Notch	Note
1	136.1 J/m	Y	20°, 30% glass fibre, ASTM D256 [2]
2	979.8 J/m	N	20°, ASTM D256 [2]
3	90.7 J/m	Y	30% glass fibre, Celanex

– PBT, injection moulding glass fibre filled

4	240 J/m	N	[16]
5	96 J/m	Y	30% glass fibre, ASTM D256 [17]
6	96 J/m	Y	23°, 30% glass fibre, ASTM D256
7	800 J/m	N	23°, 30% glass fibre, ASTM D256
8	747 J/m	N	-40°, 30% glass fibre, ASTM D256 [18]
9	147 J/m	Y	-40°, 30% glass fibre, ASTM D256
10	870.9 J/m	N	-40°, ASTM D256 [2]
11	85 J/m	Y	-40°, 30% glass fibre, ASTM D256 [8]

Charpy:

No.	Value	Notch	Note
1	8	Y	30% glass fibre, Arnite T [17]

Electrical Properties:

Electrical Properties General: Electrical props. of 30% glass fibre filled material [10,12] and Celanex 917 have been reported. [11] Volume resistivity (Valox 310) has been reported. [9] Dissipation factor [9] and its variation with temp. [3,7] has been reported

Surface/Volume Resistance:

No.	Value	Note	Type
4	0.03×10^{15} Ω.cm	30% glass fibre, Arnite T [17]	S

Dielectric Permittivity (Constant):

No.	Value	Frequency	Note
1	2.9		30% glass fibre [3]
2	3.7		30% glass fibre, Celanex [16]
3	3.8	60 Hz	30% glass fibre, ASTM D150 [17]
4	3.7	1 kHz	30% glass fibre, Arnite T [17]
5	3.6	1 MHz	30% glass fibre, Arnite T [17]
6	3.8	1 kHz	30% glass fibre, ASTM D150 [18]
7	3.7	1 MHz	30% glass fibre, ASTM D150 [18]

Dielectric Strength:

No.	Value	Note
1	>0.4 kV.mm^{-1}	min., Valox 310 [9]
2	22 kV.mm^{-1}	560 V mil^{-1}, 30% glass fibre, Celanex [16]
3	14 kV.mm^{-1}	357 V mil^{-1}, 30% glass fibre, ASTM D149 [17]
4	47 kV.mm^{-1}	30% glass fibre, Arnite T [19]
5	24.8 kV.mm^{-1}	30% glass fibre, ASTM D149 [21]

Arc Resistance:

No.	Value	Note
1	135s	30% glass fibre, ASTM D495 [17]
2	110–175s	30% glass fibre, ASTM D495 [17]
3	146s	30% glass fibre, ASTM D495 [18]

Strong Field Phenomena General: Breakdown voltage 76 kV (30% glass fibre, Arnite T) [17]
Static Electrification: High voltage track rate 4.65 cm min^{-1} (UL Bulletin 494, flame retardant, mica) [13], 29.46 cm min^{-1} (UL Bulletin 494, flame retardant) [13]

Dissipation (Power) Factor:

No.	Value	Frequency	Note
1	0.01		30% glass fibre [3]
2	0.002	1 kHz	30% glass fibre, Arnite T [17]
3	0.002	1 kHz	30% glass fibre, ASTM D150 [18]
4	0.02	1 MHz	30% glass fibre, ASTM D150 [18]
5	0.011	1 MHz	[17]

Optical Properties:
Transmission and Spectra: Ir spectral data (Novadur 5010, 30% glass fibre) have been reported [7]

Polymer Stability:
Upper Use Temperature:

No.	Value	Note
1	120°C	air, 30% glass fibre
2	65°C	H$_2$O, 30% glass fibre [18]

Flammability: Heat resistance and flame retardancy details have been reported. [12] Flammability rating HB (0.8 mm, Celanex, UL94) [16,17]; HR2 (140°, 30% glass fibre, UL08) [2]; V0 (1.5 mm, Arnite T, UL94). [17] Limiting oxygen index 18.2% (30% glass fibre, ASTM D2863) [18]
Environmental Stress: Weathering of Duranex 2002 has been reported [8]
Chemical Stability: Chemical resistance [12] of Celanex 917 has been reported [11]
Hydrolytic Stability: Hydrolytic stability of 30% glass fibre filled material has been reported [4]
Recyclability: Recycling details have been reported [15]
Stability Miscellaneous: Effect of γ-irradiation has been reported [14]

Applications/Commercial Products:
Mould Shrinkage (%):

No.	Value	Note
1	0.003%	30% glass fibre, ASTM D955 [17]
2	0.2–0.3%	flow direction, 30% glass fibre, Arnite T
3	1.5–1.6%	cross direction, 30% glass fibre, Arnite T [17]

Trade name	Details	Supplier
Arnite	non exclusive	DSM Resins
Beetle PET	non exclusive	BIP
Celanex	1432 Z, 1612 Z, 1632 Z, 1662 Z	Hoechst Celanese
Celanex 3100		Hoechst
Celstran PBTG	30-01-4, 40-01-4, 50-01-4, 60-01-4	Polymer Composites
Comtuf 464	non exclusive	ComAlloy International
Crastine		Ciba-Geigy Corporation
E-30FG-0100		Thermofil
ESD-A-1005		RTP Company
Fiberfil	non exclusive	DSM
Hiloy	non exclusive	ComAlloy International

Later	non exclusive	LATI Industria Thermoplastics S.p.A.
Lubricomp WFL-4036		LNP Engineering
MPBT	non exclusive	Modified Plastics
Mitsubishi Kasei	non exclusive	Mitsubishi Chemical
Orgater	non exclusive	Atochem Polymers
PBT-1100615		Bay Resins
PS-156G/000		Compounding Technology
Petra	non exclusive	Allied Signal Corporation
Pibiter	non exclusive	Enichem
Pocan		Bayer Inc.
Pocan	KLI-7033, KLI-7331, KLI-7391, KUI-7033, KUI-7315	Bayer Inc.
RTP 1001		RTP Company
Rynite 7020/7030		DuPont
Starglass, Star-C PBT	non exclusive	Ferro Corporation
Techster	non exclusive	Rhone-Poulenc
Texapol PBT GF		Texapol
Thermocomp WF		LNP Engineering
Thermofil	non exclusive	Thermofil
Ultradur B4300	G4, G6, G10	BASF
Ultradur B4300 G2		BASF
Valax 414		General Electric Plastics
Vandar 4612R		Hoechst Celanese
Vestodur	non exclusive	Huls America
Voloy	non exclusive	ComAlloy International
Vybex	22003 NA, 22004 NAHS, 22026 NA	Ferro Corporation
Vybex 22001 NA		Ferro Corporation

Bibliographic References

[1] Crawford, R.J. and Benham, P.P., *Polymer*, 1975, **16**, 908, (fracture mechanical props.)
[2] Keuerleher, R.H., *Kunststoffe*, 1980, **70**, 167, (impact strength)
[3] Asmus, K.D., *Kunststoffe*, 1972, **62**, 635, (transition temps., mechanical props.)
[4] Kelleher, P.G., Wentz, R.P., Hellman, M.Y. and Gilbert, E.H., *Polym. Eng. Sci.*, 1983, **23**, 537, (hydrolytic stability)
[5] Guillet, J., Maazous, A. and May, J.F., *Eur. Polym. J.*, 1987, **23**, 1, (rheological props.)
[6] Hedner, G., Selden, R. and Lagererantz, P., *Polym. Eng. Sci.*, 1994, **34**, 513, (mechanical props.)
[7] Takahara, A., Magome, T. and Kajiyama, J., *J. Polym. Sci., Part B: Polym. Phys.*, 1994, **32**, 839, (ir, fatigue, mechanical props.)
[8] Ishak, Z.A.M. and Lim, N.C., *Polym. Eng. Sci.*, 1994, **34**, 1645, (weathering, water diffusion coefficient)
[9] Wambach, A.D. and Kramer, M., Proc. Annu. Conf. - Reinf. Plast./Compos. Inst., *Soc. Plast. Ind.*, 1974, **29**, 24C, (volume resistivity, dielectric strength)
[10] NcNally, D. and Freed, W.T., Soc. Plast. Eng., *Tech. Pap.*, 1974, **20**, 79, (mechanical props, electrical props)
[11] Serle, A.G., Soc. Plast. Eng., *Tech. Pap.*, 1972, **18**, 162, (chemical resistance, electrical props.)
[12] Zimmerman, D.D., *Polym.-Plast. Technol. Eng.*, 1980, **14**, 15, (flame retardancy, mechanical props.)
[13] Ger. Pat., 1978, 2 755 950, (high voltage track rate)
[14] Imai, E., *Gyomu Nenpo-Tochigi-Ken Ken'nan Kogyo Shidosho*, 1986, 31, (γ-irradiation)
[15] Chu, J. and Sullivan, J.L., *Polym. Compos.*, 1996, **17**, 523, (recycling)
[16] *Kirk-Othmer Encycl. Chem. Technol.*, 4th edn., (ed. J.I. Kroschwitz), Wiley Interscience, 1996, **19**, 630, (mechanical props.)
[17] *Encycl. Polym. Sci. Eng.*, 2nd edn., (ed. J.I. Kroschwitz), Wiley Interscience, 1986, **14**, 23, 230, 366, 737, (expansion coefficient, mechanical props., electrical props.)
[18] *The Materials Selector*, 2nd edn., (eds. N.A. Waterman and M.F. Ashby), Chapman and Hall, 1997, **3**, 61, 366, (electrical props., abrasion resistance)

PDMS fluid P-7

Synonyms: *Dimethicones fluid. Dimethylsilicone fluid. Polydimethylsilicone fluid. Poly[oxy(dimethylsilylene)]*
Related Polymers: Polydimethylsiloxane, Polymethyloctylsiloxane fluid, Polydimethyldiphenylsiloxane fluid, Polymethylphenylsiloxane fluid, Polyphenylsiloxane fluid, Fluorosilicone oil, Polymethylhydrogen siloxane, Silly putty, Poly(methylhydrogensiloxane-co-dimethylsiloxane) fluid
Monomers: Dichlorodimethylsilane, Hexamethylcyclotrisiloxane
Material class: Fluids
Polymer Type: dimethylsilicones
CAS Number:

CAS Reg. No.
9006-65-9
9016-00-6
63148-62-9

Molecular Formula: $[C_2H_6OSi]_n$
Fragments: C_2H_6OSi
Molecular Weight: MW 250–25000

Volumetric & Calorimetric Properties:
Density:

No.	Value	Note
1	d 0.96 g/cm^3	n > 50 [1]
2	d 0.8–0.9 g/cm^3	very small n

Thermal Expansion Coefficient:

No.	Value	Note	Type
1	0.001 K^{-1}	n > 10 [1,3]	v
2	0.0015 K^{-1}	very small n	v

Thermal Conductivity:

No.	Value	Note
1	0.16 W/mK	0.00038 cal (cm s °C)$^{-1}$, n > 50 [3,11]
2	0.1 W/mK	very small n

Specific Heat Capacity:

No.	Value	Note	Type
1	1.5 kJ/kg.C	0.36 cal (g °C)$^{-1}$ [3,7,14]	P
2	1.1–1.3 kJ/kg.C	increases with increase in n	V

– PDMS fluid

Melting Temperature:

No.	Value	Note
1	-80--50°C	increases with increase in n [11]

Transition Temperature:

No.	Value	Note	Type
1	-50°C	n >200, [9,11]	Pour point
2	-80°C	low n	Pour point

Surface Properties & Solubility:
Solubility Properties General: Solubility of gases has been reported [5,12]
Cohesive Energy Density Solubility Parameters: δ 15.5 $J^{1/2}$ $cm^{-3/2}$ (7.5–7.6 $cal^{1/2}$ $cm^{-3/2}$); δ_d 15.9; δ_p 0.0; δ_h 4.1 $J^{1/2}$ $cm^{-3/2}$ [4,5,13,15]
Solvents/Non-solvents: Sol. aliphatic hydrocarbons, aromatic hydrocarbons, chlorinated hydrocarbons. Mod. sol. ethers, esters, ketones. Insol. alcohols, H_2O [5,7,8,10,12]
Wettability/Surface Energy and Interfacial Tension: γ_c 20 mJ m^{-2} (20 dyne cm^{-1}, n >10) γ_c 16 mJ m^{-2} (small n). Interfacial tension against water 42 mJ m^{-2} [1,5,9,16,17]

Transport Properties:
Polymer Melts: η 0.0065 St (25°) [1]. Variation of viscosity with MW at 25° has been reported [3]. Other viscosity values have been reported [2,7,10,11]. Viscosity is only moderately affected by pressure. A rise in pressure from 1–2000 atm raises viscosity by a factor of 16
Water Content: Water absorption has been reported [4]

Mechanical Properties:
Mechanical Properties General: Highly compressible
Elastic Modulus:

No.	Value	Note
1	1100 MPa	11000 kg cm^{-2} [8]

Electrical Properties:
Electrical Properties General: Has excellent insulating props.
Dielectric Permittivity (Constant):

No.	Value	Note
1	2.8	25°, n >200 [7,11,18]

Dielectric Strength:

No.	Value	Note
1	10–20 kV.mm^{-1}	[6,18]

Dissipation (Power) Factor:

No.	Value	Frequency	Note
1	0.00015	800 Hz	[7]

Optical Properties:
Refractive Index:

No.	Value	Note
1	1.4	n >10 [1]

Transmission and Spectra: Transparent above 280 nm [5]

Polymer Stability:
Polymer Stability General: Stable over wide temp. range
Thermal Stability General: Usable between -65° and 170°, in presence of air. In absence of air usable up to 300°
Upper Use Temperature:

No.	Value	Note
1	200°C	[9]
2	170°C	

Decomposition Details: Oxidised at temps. >200° [9]
Flammability: Fl. p. 315° (n >100) [7,10]. Flame point 380° (n >100) [11,19]
Environmental Stress: Stable to water and air. Damaged by radiation [7]
Chemical Stability: Chemically inert but decomposed by conc. mineral acids. Very thin films are liable to degradation [4,5,7]
Hydrolytic Stability: Stable to H_2O and in aq. solns. Very thin films spread on water; liable to hydrolytic degradation [4,5,7]
Biological Stability: Non-biodegradable [4,5]
Recyclability: Can be reused. No large scale recycling

Applications/Commercial Products:
Processing & Manufacturing Routes: For low viscosity fluids dimethyldichlorosilane is hydrolysed and the hydrolysate equilibrated under acid catalysis at 180° with small amounts of hexamethyldisiloxane to serve as chain terminator. For high viscosity fluids cyclic dimethylpolysiloxanes are equilibrated together under alkaline catalysis with traces of chain terminator at 150°
Applications: Used as an hydraulic fluid for vibration damping, as antifoamer, in dielectric media, as surfactant, as grease, in heat transfer media, as polish, lubricant, plastic additive and emollient. Particularly important as mould release agent in plastics and rubber industries and in applications where high and low temps. and large temp. ranges are liable to be encountered. PDMS fluid can be copolymerised with polyethers to produce polymers with valuable surfactant props.

Trade name	Details	Supplier
A6,110-100000		Goldschmidt AG
Dow Corning	various grades	Dow Corning STI
Dow Corning 200 Fluid		Dow Corning STI
Masil SF		PPG-Mazer
SF	various grades, not exclusive	General Electric Silicones
Silicone C111		ICI, UK
Silicone fluid		Akron Chemical Company
Union Carbide L-45		Union Carbide Corporation
Viscasil		General Electric Silicones

Bibliographic References
[1] Hardy, D.V.N. and Megson N.J.L., *Q. Rev., Chem. Soc.,* 1948, **2**, 25
[2] Barry, A.J., *J. Appl. Phys.,* 1946, **17**, 1020, (viscosity)
[3] Bates, O.K., *Ind. Eng. Chem.,* 1949, **41**, 1966, (thermal conductivity)
[4] Stark, F.O., Fallender, J.R. and Wright, A.P., *Comprehensive Organometallic Chemistry,* (eds. G. Wilkinson, F.G.A. Stone and E.W. Abel), Pergamon Press, 1982, **2**, 305
[5] Hardman, B. and Torkelson, A., *Encycl. Polym. Sci. Eng.,* 2nd edn., (ed. J.I. Kroschwitz), John Wiley and Sons, 1985, **15**, 204
[6] Roff, W.J. and Scott, J.R., Fibres, Films, *Plastics and Rubbers,* Butterworths, 1971
[7] Noll, W., *Chemistry and Technology of Silicones,* Academic Press, 1968
[8] Meals, R.N. and Lewis, F.M., Silicones Reinhold, 1959
[9] Torkelson, A., *Silicone Technology,* (ed. P.F. Bruins), John Wiley and Sons, 1970, 61

[10] Freeman, G.G., *Silicones* The Plastics Institute, 1962
[11] Garden, W.D., *Silicones,* (ed. S. Fordham), 1960, 145
[12] McGregor, R.R., Silicones and their Uses McGraw-Hill, 1954
[13] Yerrick, K.B. and Beck, H.N., *Rubber Chem. Technol.,* 1964, **37**, 261, (solubility)
[14] Thompson, J.M.C., *Silicones,* (ed. S. Fordham), 1960, 5
[15] Baney, R.H., Voigt, C.E. and Mentele, J.W., Struct.-Solubility Relat. Polym., *(Proc. Symp.),* (eds. F.W. Harris and R.B. Seymour), Academic Press, 1977, 225, (solubility)
[16] Fox, H.W., Taylor, P.W. and Zisman, W.A., *Ind. Eng. Chem.,* 1947, **39**, 1401, (surface tension)
[17] Gaines, G.L., *J. Phys. Chem.,* 1969, **73**, 3143, (surface tension)
[18] Hakim, R.M., Olivier, R.G. and St-Onge, H., *IEEE Trans. Electr. Insul.,* 1977, **12**, 360, (electrical props)
[19] Ames, J., *J. Sci. Instrum.,* 1958, **35**, 1

PDMS-polyether copolymer   P-8

$$\left[\begin{array}{c} Me \\ -O-Si- \\ Me \end{array} \right]_n \left[OCH(CH_3)CH_2 \right]_x \left[OCH_2CH_2 \right]_m$$

Synonyms: *Silicone glycol copolymer. Poly[dimethylsiloxane (oxyethylene cooxypropylene)]*
Related Polymers: PDMS fluid
Monomers: Dichlorodimethylsilane, Ethanediol, Propanediol
Material class: Fluids
Polymer Type: silicones
CAS Number:

CAS Reg. No.
64365-23-7
67762-96-3
68937-54-2
68938-54-5

Molecular Formula: $[(C_2H_6OSi).(C_3H_6O).(C_2H_4O)]_n$
Fragments: C_2H_6OSi C_3H_6O C_2H_4O
Molecular Weight: M_n 600-20000

Volumetric & Calorimetric Properties:
Density:

No.	Value	Note
1	d 1 g/cm^3	[1,2]

Transition Temperature General: Cloud point and pour point increase with increase in fraction of polyoxyethylene in copolymer [3]
Transition Temperature:

No.	Value	Note	Type
1	-48–32°C	-55–90°F [1,2]	Pour point
2	10–64°C		Cloud point

Surface Properties & Solubility:
Solubility Properties General: Solubility in H_2O decreases with increase in temp.; effectively insol. above cloud point [3]
Solvents/Non-solvents: Sol. aromatic hydrocarbons, chlorinated hydrocarbons, ketones, ethers, alcohols, esters. Mod. sol. H_2O. Insol. aliphatic hydrocarbons, polydimethyl siloxanes [1,2,3]
Surface and Interfacial Properties General: Has better lubricating props. than PDMS fluid [6]
Wettability/Surface Energy and Interfacial Tension: γ_c 20.5–30.5 mJ m^{-2} [1,2,3]. Copolymer in aq. soln. reduces γ_c to 21–27 mJ m^{-2} [3,5]

Transport Properties:
Polymer Melts: Viscosity 0.2-20 St. Viscosity is more temp. dependent than for PDMS fluid

Polymer Stability:
Flammability: Fl. p. 80–140° (175–280°F) [1,2]
Hydrolytic Stability: SiC copolymers are resistant to hydrolysis. SiOC copolymers hydrolyse in the presence of catalysts; this limits their use in the chemical industry. Slow hydrolysis occurs, in absence of catalysts, with separation of silicone oil
Recyclability: Non-recyclable

Applications/Commercial Products:
Processing & Manufacturing Routes: Polydimethylsiloxane containing a small number of SiH groups (often only in the terminal positions) is treated with hydroxyl terminal or vinyl terminal polyethers to prod. SiOC or SiC silicone glycol copolymers [3,4,5]
Applications: Surfactant, dispersant and wetting agent. Used as a spreading agent for various applications, a lubricant, slip additive, wetting and gloss agent, defoamer and emollient. Widely used in pharmaceutical and personal care products, mining, agricultural, automotive, textile and plastics applications

Trade name	Supplier
Abil B	Goldschmidt AG
Amercil DMC	Amerchol
Dow Corning 190 series	Dow Corning STI
Masil 280e 1060 series	PPG-Mazer
Silwet	Union Carbide

Bibliographic References
[1] Ash, M. and Ash, I., *Handbook of Plastic Compounds, Elastomers and Resins,* (eds. M.B. Ash and I.A. Ash), Wiley-VCH, 1992
[2] Ash, M. and Ash, I, *Condensed Encyclopedia of Surfactants,* Edward Arnold, 1989
[3] Plumb, J.B. and Atherton, J.H., *Block Copolym.,* (eds. D.C. Allport and W.H. Jones), Applied Science, 1973, 305
[4] Noshay, A. and McGrath, J.E., *Block Copolym.,* Academic Press, 1977
[5] Noll, W., *Chemistry and Technology of Silicones,* Academic Press, 1968
[6] Hardman, B. and Torkelson, A., *Encycl. Polym. Sci. Eng.,* Vol. 15, 2nd edn., (ed. J.I. Kroschwitz), John Wiley and Sons, 1985, 204

PE film   P-9
Synonyms: *Polyethylene film. LLDPE film*
Related Polymers: Polyethylene, Linear low-density PE, Low-density PE, Ultra low-density PE, High-density PE, Medium-density PE, Ultra high molecular weight PE, Copolymer with vinyl acetate
Monomers: 4-methyl-1-pentene, Norbornene, Cyclopentene, Ethylene, 1-Butene, 1-Hexene, 1-Octene
Material class: Thermoplastic
Polymer Type: polyethylene, polyolefins
CAS Number:

CAS Reg. No.	Note
9002-88-4	all density types
25087-34-7	1-butene comonomer
25213-02-9	1-hexene comonomer
26221-73-8	1-octene comonomer
25213-96-1	4-methyl, 1-pentene comonomer
60785-11-7	1-hexene, 1-butene comonomer
74746-95-5	1-butene, 4-methyl-1-pentene comonomer
26007-43-2	norbornene comonomer
72238-82-5	norbornene, 1-butene comonomers
172501-09-6	norbornene, 1-decene comonomer

– PE film

Molecular Formula: $(C_2H_4)_n$
Fragments: C_2H_4
Additives: Antiblock; slip; antioxidants (includes vitamin E for food use); antifogging agents, tackifiers (eg HMW polybutene).
General Information: Film is made from all density types of PE and many copolymers. Film use accounts for around 40% of PE use overall, but about half of LDPE use and about 60% of LLDPE use.
Morphology: Film produced by the 'blown film' method is bi-oriented and data may be segregated into machine direction (MD) and transverse direction (TD). Processing variables and MWD affect morphology. Both stacked lamellae and fibril nuclei have been observed [30]

Surface Properties & Solubility:
Surface and Interfacial Properties General: Due to the inherent inertness of polyethylenes, surface treatment to promote wettability is normally needed prior to printing on PE film.

Transport Properties:
Melt Flow Index:

No.	Value	Note
1	0.15–4.6 g/10 min	LDPE [4]
2	0.5–4.1 g/10 min	LLDPE [2,4,5]
3	0.06–0.08 g/10 min	HDPE [2]

Permeability of Gases: mVLDPE is claimed to have very high oxygen transmission rates [31]
Water Absorption:

No.	Value	Note
1	0.01 %	max, 24 h. ASTM D570 [3]

Gas Permeability:

No.	Gas	Value	Note
1	O_2	96.3–236 cm³ mm/(m² day atm)	$1.1–2.7 \times 10^{-13}$ cm² (s Pa)$^{-1}$, 20–25° LDPE [1,3,4]
2	O_2	49–210 cm³ mm/(m² day atm)	$0.56–2.4 \times 10^{-13}$ cm² (s Pa)$^{-1}$, 20–25°, MDPE [1,3,4]
3	O_2	39.4–78.8 cm³ mm/(m² day atm)	$0.45–0.9 \times 10^{-13}$ cm²s Pa^{-1}, 20–25°, HDPE [1,3,4]
4	CO_2	393–1050 cm³ mm/(m² day atm)	$4.5–12 \times 10^{-13}$ cm²(s Pa)$^{-1}$ 20–25°, LDPE [2,3,4]
5	CO_2	197–980 cm³ mm/(m² day atm)	$2.25–11.2 \times 10^{-13}$ cm²(s Pa)$^{-1}$, 20–25°, MDPE [1,3,4]
6	CO_2	227–275 cm³ mm/(m² day atm)	$2.6–3.15 \times 10^{-13}$ cm²(s Pa)$^{-1}$, 20–25 HDPE [1,3,4]
7	N_2	39.4–78.8 cm³ mm/(m² day atm)	$0.45–0.9 \times 10^{-13}$ cm²(s Pa)$^{-1}$, 20–25°, LDPE [1,3,4]
8	N_2	26.2–1221 cm³ mm/(m² day atm)	$0.3–1.4 \times 10^{-13}$ cm²(s Pa)$^{-1}$, 20–25°, MDPE [1,3,4]
9	N_2	15.8–23.63 cm³ mm/(m² day atm)	$0.18–0.27 \times 10^{-13}$ cm²(s Pa)$^{-1}$, HDPE [1,3,4]
10	H_2	770 cm³ mm/(m² day atm)	8.8×10^{-13} cm²(s Pa)$^{-1}$, 25°, LDPE, MDPE [3]

Mechanical Properties:
Tensile (Young's) Modulus:

No.	Value	Note
1	190–430 MPa	LDPE [4]

Tensile Strength Break:

No.	Value	Note	Elongation
1	17–55 MPa	transverse direction, LLDPE [1,2,3,4]	630–840%
2	14–28 MPa	MDPE [2,3]	50–650%
3	16–42 MPa	HDPE [2,3]	10–650%
4	21–38 MPa	UHMWPE [2,3]	300%
5	10–34 MPa	machine direction, LDPE [1,2,3,4]	200–600%
6	14–21 MPa	transverse direction, LDPE [1,2,3,4]	480–800%
7	30–72 MPa	machine direction, LLDPE [1,2,3,4]	390–700%

Tensile Strength Yield:

No.	Value	Note
1	9–13 MPa	LDPE [4]
2	7–14 MPa	LLDPE [4]

Impact Strength: 27–42 kJ m^{-1} (LDPE); 31–50 kJ m^{-1} (LLDPE); 15–23 kJ m^{-1} (MDPE); 4–12 kJ m^{-1} (HDPE) (ASTA D1822). Dart impact strength 120–520 kg m^{-1} (30–130 g mil^{-1}, LDPE); 320–4200 kg m^{-1} (80–105 g mil^{-1}, LLDPE); 1000–5600 kg m^{-1} (250–1400 g mil^{-1}, metallocene LLDPE)
Failure Properties General: Tear strength 20–115 kN m^{-1} (machine direction, LDPE); 30–125 kN m^{-1} (transverse direction, LDPE) [1,3,4]; 20–50 kN m^{-1} (machine direction); 50–100 kN m^{-1} (transverse direction, conventional LLDPE) [4]; 75–125 kN m^{-1} (machine direction); 145–220 kN m^{-1} (transverse direction, metallocene LLDPE) [4]; 20–118 kN m^{-1} (MDPE, HDPE) [1,3]. Burst strength 10–12 Mullen points (LDPE) [1,3]

Electrical Properties:
Electrical Properties General: Dielectric permittivity increases with increasing polyethylene density
Dielectric Permittivity (Constant):

No.	Value	Frequency	Note
1	2.2–2.3	1 kHz–1 GHz	[1,3]

Dielectric Strength:

No.	Value	Note
1	19 kV.mm^{-1}	475 V mil^{-1}, LDPE [1]
2	20 kV.mm^{-1}	500 V mil^{-1}, MDPE, HDPE [1]
3	51 kV.mm^{-1}	1275 V mil^{-1}, UHMWPE [1]

Dissipation (Power) Factor:

No.	Value	Frequency	Note
1	0.0003	1 kHz–1 GHz	LDPE, MDPE [1,3]
2	0.0005	1 kHz–1 GHz	HDPE [1,3]
3	0.00023	1 kHz–1 GHz	UHMWPE [1,3]

PE film

Optical Properties:
Refractive Index:

No.	Value	Note
1	1.51	LDPE [1]
2	1.52	MDPE [1]
3	1.54	HDPE, UHMWPE [1]

Transmission and Spectra: Raman spectral data have been reported for determination of orientation parameters [3]
Volume Properties/Surface Properties: Transparency 0–47% (LDPE); 30–75% (LLDPE); 10–80% (MDPE); 0.40% (HDPE) [1,4]; Haze 4–50% (LDPE); 4.5–9% (LLDPE); 4–50% (MDPE); 10–50% (HDPE); 2% (mLLDPE) [1,3,4]; Gloss 45–75 (45°), 10–100 (20°, LDPE); 35–97 (45°, LLDPE) [4]. Clarity is improved for MPE [31]

Polymer Stability:
Polymer Stability General: Lower use temp. 57° (LDPE, MDPE); 45° (HDPE, UHMWPE) [1]
Thermal Stability General: Dimensional change at 100° for 30 min. -2% (LLDPE); 0 to -0.7% (MDPE); -0.7–3.0%(HDPE) [1,3]
Upper Use Temperature:

No.	Value	Note
1	88°C	LDPE [1]
2	104°C	MDPE [1]
3	121°C	HDPE, UHMWPE [1]

Biological Stability: Biodegradation of agricultural films, especially with additions of corn starch, pigments and stabilisers [6,7,8,13] have been reported
Recyclability: Unblended film is recyclable and is used in co-extruded film. Industry trends and comments on recycling [36,37,38,39,40] have been reported
Stability Miscellaneous: Vitamin E used as an antioxidant has also been found to eliminate taste and odour from aldehydes, reduce gel formation and improve colour stability [28]

Applications/Commercial Products:
Processing & Manufacturing Routes: Blown film processing: melt temp. 160–210°, BUR 1.5:1 to 3.5:1(LDPE); melt temp. 190–215° (conventional LLDPE), 210–320° (metallocene LLDPE), BUR 1.5:1 to 4:1 ; melt temp. ca. 150°, BUR 3:1 to 4:1 (HDPE) [2,4] LLDPE requires a deeper screw with more gradual transitions than LDPE to overcome the lower degree of viscosity reduction at high shear [9]. There are reports of mLLDPE requiring equipment modification and of mLLDPE needing no equipment modification [10,11,12,14,15,17]. Air ring size and bubble cooling affect throughput [16]. Cast film especially for LLDPE where higher tensile strength and puncture resistance and better low temp. props make it technically competitive to HDPE and LDPE. Flat film extrusion improves transparency, freedom from haze, gloss of LDPE. Typical conditions 150°, 40–50 MPa (HDPE).
Applications: Used in bags, stretch/cling film, industrial sheeting, agricultural mulch, heavy duty films, damp proof membranes, semiconductor polishing, surgical, babywear, food use. Different density grades are more appropriate to different aspects: LLDPE for strength; LDPE for stretch/wrap; HDPE for barrier props, strength, and higher temp; VLDPE for heavy duty film; mPE for food use, clarity, medical applications. PE for film use, especially LLDPE is an area of considerable current development. Blending and/or co-extrusion are commonly used to combine best features.

Trade name	Details	Supplier
Aerowrap	HDPE	BP Chemicals
Affinity		Dow
Aircap	bubblewrap	Plastic Iberica
Alathon	homo-and co-polymers	OxyChem
Alathon	includes food grades	Alathon
Attane	VLDPE	Dow Plastics
Bellox	crosslinkable sheet	Bell Plastics Ltd
Cabelec	used for various products	Cabot Plastics Ltd.
Clysar	LLDPE shrink film	DuPont
Dowlex		Dow
Elite		Dow
Exact	mLLDPE copolymers	Exxon Chemical
Exceed		Exxon Chemical
Flexirene	VLDPE, LLDPE, Clearflex	Enichem
Fortiflex	HDPE homo- and co-polymers	Solvay Polymers
Fric Film		Exxon Chemical
Ipideg	photodegradable	International Plastics Italiana
Lipolen		Northern Industrial Plastics Ltd.
Luflexen	mLLDPE	BASF
Lupolen		BASF
M Stretch	made from 'Dowlex' and 'Affinity'	MMP Packaging
Marlex	homo- and copolymers	Philips Petroleum International
Mattflex	PE and PP	Ace SA
Microflex	other products also	ASP Packaging
Mitsubishi	other products also	Mitsubishi Kasei
Monarflex	thick film/sheeting	Monarflex
Monarfol	thick film/sheeting	Monarflex
Naltene	sheets and rolls	Martyn Industrials
Orbilan	sheet	Danelli Materie Plastiche
Orevac	homo and co-polymer	Atochem Polymers
Pajalen	MDPE	Paja-Kunststoffe Jaecchke
Plastecnics	sheet	Europlastecnics SA
Plastin	HDPE - non-exclusive	4P Folie Forchheim GmbH.
Plastotrans	LDPE	4P Folie Forchheim GmbH.
Polidan	shrink film	Pandoplast
Poliwrap	stretch wrap	British Visqueen
Polystretch	stretch film	Taco Plastics
Portaliner	flame retardent, modified	PC Polythene
Portathene	flame retardent, modified	PC Polythene
Quickwrap	LLDPE stretch film	Centurion Packaging
Reprothene	virgin/recycled coextrusion	Smith Anderson

Saranex	coextrusion non exclusive	MecPac SRL
Satinflex	laminated, coated non exclusive	Ace SA
Schur Finseal	co-extrusion	Schurpack
Sclair film		DuPont
Super Naltene	HDPE sheet and roll	Martyn Industrials
Taffaflex	non exclusive	Ace SA
Trepaphan	copolymer PE, non exclusive	Hoechst
Tuffin		Union Carbide
Ultradur	UHMWPE non exclusive	S.W. Plastics
Ultradur		BASF
Valeron	HDPE	Vauleer Packaging Systems
Velvaflex	non exclusive	Ace SA
Ventoplas	micro perforated	Seevent Plastics
Vistal		SA Plastics Wanters
Vithene		Visual Packaging Ltd.
Weldex		Moore and Buckle (Flexible Packaging)
Xtrathene	LLDPE	Alida Packaging

Bibliographic References

[1] Mackenzie, K.J., *Kirk-Othmer Encycl. Chem. Technol.*, 4th edn (eds. J.I. Kroschwitz and M. Howe-Grant) Wiley Interscience, 1993, **10**, 761
[2] *Kirk-Othmer Encycl. Chem. Technol.*, 4th edn. (eds. J.I. Kroschwitz and M. Howe-Grant), Wiley Interscience, 1996, **17**, 707
[3] Guide to Plastics, *Property and Specification Charts*, McGraw-Hill, 1987, 92
[4] *Polyethylene Films*, (technical datasheet)
[5] Leaversuch, R.D., *Modern Plast. Intern.*, 1997, **27**, 58
[6] Graff, G., *Modern Plast. Intern.*, 1996, **26**, 27
[7] Erlandsson, B., Karlsson, S. and Albertson, A.C., *Polym. Degrad. Stab.*, 1997, **55**, 237
[8] Arias, G., Orona, F., Ruiz, V., Teran, G.E., et al, *Plasticulture*, 1995, **107**, 50
[9] Park, H.C. and Nakmias, A.M., *Encycl. Polym. Sci. Eng.*, 2nd edn., (ed. J.I. Kroschwitz), Wiley Interscience, 1987, 73
[10] Bregar, B., *Plastics News (Detroit)*, 1996, **8**, 1
[11] Bregar, B., *Plastics News (Detroit)*, 1996, **8**, 46
[12] Sukhadia, A.M., *Annu. Tech. Conf. - Soc. Plast. Eng.*, 54th, Indianapolis, 1996, **1**, 1157
[13] Nakashima, T. and Matsuo, M., *J. Macromol. Sci., Phys.*, 1996, **35**, 659
[14] Cardinal, J.C. and Rudolph, G., *Kunststoffe*, 1996, **86**, 25
[15] Vernji, B., *Plastics News (Detroit)*, 1996, **8**, 5
[16] Colvin, R., *Modern Plast. Intern.*, 1996, **26**, 30
[17] Callari, J., *Plast. World*, 1997, **55**, 8
[18] *British Board of Agreement*, Capital Valley Plastics, 1996, 5
[19] *J. Mater. Eng. Perform.*, 1996, **5**, 278
[20] Graff, G., *Modern Plast. Intern.*, 1996, **26**, 27
[21] Pater, H.R., Makwana, M.G. and Pater, B.N., *Plasticulture*, 1995, **107**, 21
[22] Gillet, M., *Eur. Plast. News*, 1997, **24**, 31
[23] Jen-Taut Y., Yu-Lang L. and Chien-Cheng F.-C., *Macromol. Chem. Phys.*, 1996, **197**, 3531, (applications)
[24] Vasselle, J.B., *Plast. Mod. Elastomeres*, 1996, **26**, 112, (applications)
[25] *Modern Plast. Intern.*, 1996, **26**, 112, (applications)
[26] *Packaging Week*, Benn Publications, 1997, **12**, 3
[27] Mueller, C.D., Nazarenko, S., Ebeling, T., Schuman, T.L. et al, *Polym. Eng. Sci.*, 1997, **37**, 355, (applications)
[28] Laermer, S.F., Young, S.S. and Zambetti, P.F., *Plast. '21*, 1995, **46**, 13
[29] Everall, N., Chalmers, J. and Mills, P., *Plast. '21*, 1996, **50**, 1229
[30] Ta-Hua Y. and Wilkes, G.L., *Polymer*, 1996, **37**, 4675
[31] *Eur. Plast. News*, 1996, **23**, 31
[32] *Br. Plast. Rubber*, 1997, 51, (mech props)
[33] *Packaging Week*, 1997, **12**, 2, (mech props)
[34] Stinson, S., *Packaging Week*, Benn Publications, 1997, **75**, 31, (mech props)
[35] *Eur. Chem. News*, 1997, **67**, 24, (mech props)
[36] Smith, S.S., *Eur. Chem. News*, 1996, **8**, 7, (recycling)
[37] Mital, V., *Polym. Recycl.*, 1996, **2**, 173, (recycling)
[38] Khait, K., *J. Vinyl Addit. Technol.*, 1996, **2**, 345, (recycling)
[39] Laguno, O., Collar, E.P. and Martinez, J.M.G., *Rev. Plast. Mod.*, 1995, **69**, 561, (recycling)
[40] Brandrup, J., *Polym. Recycl.*, 1996, **2**, 95, (recycling)
[41] Leaversuch, R.D., *Modern Plast. Intern.*, 1997, **27**, 58, (applications)
[42] Gabriele, M.C., *Modern Plast. Intern.*, 1997, **27**, 24, (applications)

PFA, Carbon filled P-10

Synonyms: *Tetrafluoroethylene-perfluoromethylvinyl ether copolymer, Carbon filled*
Related Polymers: PFA, Mineral and mica filled, PFA, PTFE filled, PFA, PTFE and carbon filled, Filled PFA, PFA, Glass filled
Monomers: Tetrafluoroethylene, Perfluoro(methyl vinyl ether)
Material class: Composites, Copolymers
Polymer Type: PFA, fluorocarbon (polymers)

Volumetric & Calorimetric Properties:
Density:

No.	Value	Note
1	d 2.1 g/cm^3	10% PAN carbon fibre, 15% pitch carbon [1]

Thermal Expansion Coefficient:

No.	Value	Note	Type
1	2.2×10^{-5} K^{-1}	10% PAN carbon fibre [1]	L
2	9×10^{-5} K^{-1}	15% pitch carbon [1]	L
3	8.6×10^{-5} K^{-1}	20% carbon fibre, flow direction, D696 [2]	L

Thermal Conductivity:

No.	Value	Note
1	0.35 W/mK	10% PAN carbon fibre [1]
2	0.55 W/mK	20% carbon fibre, through-plane, C177 [2]

Melting Temperature:

No.	Value	Note
1	307°C	10% carbon fibre [1]

Deflection Temperature:

No.	Value	Note
1	240.56°C	1.8 MPa, 10% PAN carbon fibre [1]
2	85°C	1.8 MPa, 15% pitch carbon [1]
3	149°C	1.82 MPa, 20% carbon fibre, D648 [2]
4	204°C	0.45 MPa, 20% carbon fibre, D648 [2]
5	238°C	1.82 MPa, 30% carbon fibre, D648 [3]

Transport Properties:
Water Absorption:

No.	Value	Note
1	0.041 %	24h, 10% PAN carbon fibre [1]
2	0.01 %	24h, 15% pitch carbon [1]

– PFA, Glass filled

| 3 | 0.04 % | 24h, 23°, 20% carbon fibre, D570 [2] |
| 4 | 0.01 % | 24h, 23°, 30% carbon fibre, D570 [3] |

Mechanical Properties:
Tensile (Young's) Modulus:

No.	Value	Note
1	20670 MPa	30% carbon fibre, D638 [3]

Flexural Modulus:

No.	Value	Note
1	365.2 MPa	10% PAN carbon fibre [1]
2	344.5 MPa	15% pitch carbon [1]
3	7580 MPa	20% carbon fibre, D790 [2]
4	12400 MPa	30% carbon fibre, D790 [3]

Tensile Strength Break:

No.	Value	Note	Elongation
1	34 MPa	20% carbon fibre, D638 [2]	1.5%
2	31 MPa	30% carbon fibre, D638 [3]	0.5%

Tensile Strength Yield:

No.	Value	Elongation	Note
1	34.47 MPa	2.5%	10% PAN carbon fibre [1]
2	33.1 MPa		15% pitch carbon [1]

Flexural Strength Yield:

No.	Value	Note
1	53.08 MPa	10% PAN carbon fibre [1]
2	55 MPa	20% carbon fibre, D790 [2]
3	48 MPa	30% carbon fibre, D790 [3]

Izod Notch:

No.	Value	Notch	Note
1	213.6 J/m	Y	10% PAN carbon fibre [1]
2	160.2 J/m	Y	15% pitch carbon [1]
3	107 J/m	Y	20% carbon fibre, 3.18mm section, D256 [2]
4	91 J/m	Y	30% carbon fibre, 3.18mm section, D256 [3]
5	534 J/m	N	20% carbon fibre, 3.18mm section, D256 [2]
6	267 J/m	N	30% carbon fibre, 3.18mm section, D256 [3]

Electrical Properties:
Electrical Properties General: Volume resistivity 10^4 Ω cm [1]

Polymer Stability:
Upper Use Temperature:

No.	Value	Note
1	260°C	10% PAN carbon fibre [1]

Flammability: Flammability rating V1 (10% PAN carbon fibre, UL 94) [1]

Applications/Commercial Products:
Processing & Manufacturing Routes: Injection Pressure 55–83 MPa [2]
Injection Cylinder Temperature 316–366°C [2]
Mold Temperature 149–232°C [2]

Applications/Commercial Products:
Mould Shrinkage (%):

No.	Value	Note
1	0.9%	10% PAN carbon fibre [1]
2	1%	20% carbon fibre, 3.18mm section, D955 [2]
3	2%	20% carbon fibre, 6.35mm section, D955 [2]
4	0.1%	30% carbon fibre, 3.18mm section, D955 [3]
5	0.2%	30% carbon fibre, 6.35mm section, D955 [3]

Applications: Potential applications due to increased resistance to friction and wear in gears, cams, bearings and other moving parts

Trade name	Details	Supplier
Fluoromelt FP-PC	discontinued material (1002, 1003)	LNP Engineering
RTP 3183	Carbon fibre 20%	RTP Co.
RTP 3185	Carbon fibre 30%	RTP Co.

Bibliographic References
[1] Cen BASE/Materials, (eds. C.E. Nunez and R.A. Nunez), Vol. 1, John Wiley and Sons, Inc., 1990, 242, 243, (technical datasheet)
[2] RTP Co. Winona, *USA*, (technical datasheet)
[3] RTP Co. Winona, *USA*, (technical datasheet)

PFA, Glass filled   P-11
Synonyms: *Tetrafluoroethylene-perfluoro(methyl vinyl) ether copolymer, Glass filled. Poly(tetrafluoroethene-co-trifluoro(trifluoromethoxy)ethene), Glass filled. Poly(tetrafluoroethylene-co-trifluoromethylperfluorovinyl ether), Glass filled*
Related Polymers: PFA, Carbon filled, PFA, Mineral and mica filled, Filled PFA, PFA, PTFE filled, PFA, PTFE and carbon filled
Monomers: Tetrafluoroethylene, Perfluoro(methyl vinyl) ether
Material class: Composites, Copolymers
Polymer Type: PFA, fluorocarbon (polymers)
Additives: Glass (fibre or milled glass)

Volumetric & Calorimetric Properties:
Density:

No.	Value	Note
1	d 2.24 g/cm^3	20% glass fibre or milled glass [2]

Thermal Expansion Coefficient:

No.	Value	Note	Type
1	0.000136 K^{-1}	20% glass fibre [1]	L
2	9.5×10^{-5} K^{-1}	20% glass fibre [2]	L
3	0.000108 K^{-1}	20% milled glass [2]	L

Thermal Conductivity:

No.	Value	Note
1	0.37 W/mK	20% milled glass [2]

– PFA, Mineral and mica filled

Deflection Temperature:

No.	Value	Note
1	260°C	1.8 MPa, 20% glass fibre [2]
2	85°C	1.8 MPa, 20% milled glass [2]
3	82°C	1.82 MPa, 20% glass fibre [1]

Transport Properties:
Water Absorption:

No.	Value	Note
1	0.04 %	24h, 20% glass fibre [1]
2	0.01 %	24h, 20% glass fibre [2]

Mechanical Properties:
Flexural Modulus:

No.	Value	Note
1	114.4 MPa	20% milled glass [2]
2	700 MPa	20% glass fibre [1]
3	365.2 MPa	20% glass fibre [2]

Tensile Strength Break:

No.	Value	Note	Elongation
1	33 MPa	room temp., 20% glass fibre [1]	200%

Tensile Strength Yield:

No.	Value	Elongation	Note
1	42.74 MPa	3.5%	20% glass fibre [2]
2	17.92 MPa	1.7%	20% milled glass [2]

Flexural Strength Yield:

No.	Value	Note
1	72.39 MPa	20% glass fibre [2]
2	25.51 MPa	20% milled glass [2]

Hardness: Shore D68 (20% glass fibre) [1]

Izod Notch:

No.	Value	Notch	Note
1	700 J/m	Y	20% glass fibre [1]
2	299 J/m	Y	20% glass fibre [2]
3	160.2 J/m	Y	20% milled glass [2]

Electrical Properties:
Dielectric Permittivity (Constant):

No.	Value	Frequency	Note
1	2.4	60 Hz	20% glass fibre or milled glass [2]
2	2.9	10 kHz	20% glass fibre [1]
3	3.4	1 MHz	20% glass fibre [2]
4	2.4	1 MHz	20% milled glass [2]

Dielectric Strength:

No.	Value	Note
1	40 kV.mm^{-1}	20% glass fibre [1]
2	22.85 kV.mm^{-1}	20% glass fibre [2]
3	22.85 kV.mm^{-1}	20% milled glass [2]

Arc Resistance:

No.	Value	Note
1	121s	[2]

Dissipation (Power) Factor:

No.	Value	Frequency	Note
1	0.001	10 kHz	dry, 20% glass fibre [1]
2	0.0022	1 MHz	20% glass fibre [2]
3	0.023	60 Hz	20% glass fibre [2]
4	0.0023	1 MHz	20% milled glass [2]
5	0.0023	60 Hz	20% milled glass [2]

Polymer Stability:
Upper Use Temperature:

No.	Value	Note
1	170°C	continuous service, 20% glass fibre [1]
2	260°C	20% milled glass [2]

Flammability: Flammability rating V1 (20% glass fibre or milled glass, UL 94) [2]

Applications/Commercial Products:
Mould Shrinkage (%):

No.	Value	Note
1	0.8%	20% glass fibre [1,2]
2	3.5%	20% milled glass [2]

Trade name	Details	Supplier
Fluoromelt FP-PF	1002, 1004	LNP Engineering
Lubricomp FP-PF	1004 series	LNP Engineering

Bibliographic References
[1] *The Materials Selector*, 2nd edn., (eds. N.A. Waterman and M.F. Ashby), Chapman and Hall, 1997, **3**
[2] *Cen BASE/Materials*, (eds. C.E. Nunez and R.E. Nunez), Vol. 1, John Wiley and Sons, Inc., 1990, 243, (technical datasheet)

PFA, Mineral and mica filled   P-12

Synonyms: *Poly(tetrafluoroethylene-co-trifluoromethylperfluorovinylether), Mineral and mica filled. Tetrafluoroethylene-perfluoromethylvinyl ether copolymer, Mineral and mica filled. Poly(tetrafluoroethene-co-trifluoro(trifluoromethoxy)ethene), Mineral and mica filled*
Related Polymers: PFA, Carbon filled, PFA, Glass filled, PFA, PTFE filled, Tetrafluoroethylene-perfluoromethyl vinyl ether copolymer, PFA, PTFE and carbon filled

— PFA, PTFE and carbon filled

Monomers: Tetrafluoroethylene, Perfluoro(methyl vinyl ether)
Material class: Copolymers, Composites
Polymer Type: PFA, fluorocarbon (polymers)

Volumetric & Calorimetric Properties:
Density:

No.	Value	Note
1	d 2.2 g/cm^3	10% mica [1]

Thermal Expansion Coefficient:

No.	Value	Note	Type
1	0.000103 K^{-1}	10% mica [1]	L

Deflection Temperature:

No.	Value	Note
1	90°C	1.82 MPa, 10% mica [1]

Mechanical Properties:
Flexural Modulus:

No.	Value	Note
1	1382.7 MPa	14100 kg cm^{-2}, 10% mica [1]

Tensile Strength Break:

No.	Value	Note
1	17.26 MPa	176 kg cm^{-2}, 10% mica [1]

Izod Notch:

No.	Value	Notch	Note
1	325.6 J/m	Y	6.1 ft lb in^{-1}, room temp., 10% mica [1]

Applications/Commercial Products:
Processing & Manufacturing Routes: Processed by injection moulding

Trade name	Details	Supplier
Lubricomp FP-PML-3312	discontinued material	LNP Engineering
Thermocomp FP-PML	3312	LNP Engineering

Bibliographic References
[1] Thermocomp FP-PML-3312, LNP Engineering, (technical datasheet)

PFA, PTFE and carbon filled   P-13

Synonyms: *Tetrafluoroethylene perfluoromethyl vinyl ether copolymer, Polytetrafluoroethylene and carbon filled. Poly(tetrafluoroethene-co-trifluoro(trifluoromethoxy)ethene), PTFE and carbon filled. Poly(tetrafluoroethylene-co-trifluoromethylperfluorovinyl ether), PTFE and carbon filled*
Related Polymers: Filled PFA
Monomers: Tetrafluoroethylene, Perfluoro(methyl vinyl ether)
Material class: Copolymers, Composites
Polymer Type: PFA, fluorocarbon (polymers)

Volumetric & Calorimetric Properties:
Density:

No.	Value	Note
1	d 2.05 g/cm^3	10% PTFE, 20% carbon fibre [1,2]

Thermal Expansion Coefficient:

No.	Value	Note	Type
1	8.6 × 10^{-5} K^{-1}	10% PTFE, 20% carbon fibre [1,2]	L

Thermal Conductivity:

No.	Value	Note
1	0.55 W/mK	10% PTFE, 20% carbon fibre [1,2]

Melting Temperature:

No.	Value	Note
1	304°C	10% PTFE, 20% carbon fibre [2]

Deflection Temperature:

No.	Value	Note
1	148.89°C	1.8 MPa, 10% PTFE, 20% carbon fibre [1,2]

Mechanical Properties:
Flexural Modulus:

No.	Value	Note
1	743 MPa	10% PTFE, 20% carbon fibre [1]
2	7443 MPa	75900 kg cm^{-2}, 10% PTFE, 20% carbon fibre [2]

Tensile Strength Break:

No.	Value	Note	Elongation
1	31.67 MPa	323 kg cm^{-2}, 10% PTFE, 20% carbon fibre [2]	1%

Tensile Strength Yield:

No.	Value	Elongation	Note
1	31.71 MPa	1%	10% PTFE, 20% carbon fibre [1]

Flexural Strength Yield:

No.	Value	Note
1	40.68 MPa	10% PTFE, 20% carbon fibre [1]
2	40.59 MPa	414 kg cm^{-2}, 10% PTFE, 20% carbon fibre [2]

Friction Abrasion and Resistance: Wear factor 16 (10% PTFE, 20% carbon fibre) [1]. Coefficient of friction 0.1 (static), 0.09 (dynamic) (10% PTFE, 20% carbon fibre) [1]

Izod Notch:

No.	Value	Notch	Note
1	106.8 J/m	Y	10% PTFE, 20% carbon fibre [1]
2	106.7 J/m	Y	2 ft lb in^{-1}, room temp., 10% PTFE, 20% carbon fibre [2]

PFA, PTFE filled — Phenol Aralkyl resins

Electrical Properties:
Electrical Properties General: Volume resistivity 10^3 Ω cm [1], 100 Ω cm [2]

Polymer Stability:
Upper Use Temperature:

No.	Value	Note
1	260°C	10% PTFE, 20% carbon fibre [1,2]

Flammability: Flammability rating V1 (10% PTFE, 20% carbon fibre, UL 94) [1]

Applications/Commercial Products:
Processing & Manufacturing Routes: Processed by injection moulding

Mould Shrinkage (%):

No.	Value	Note
1	1%	[1,2]

Trade name	Details	Supplier
Fluoromelt FP-PCL 4024	discontinued material	LNP Engineering
Thermocomp FP-PCL 4024		LNP Engineering

Bibliographic References
[1] *Cen BASE/Materials*, (eds. C.E. Nunez and R.A. Nunez), Vol. 1, John Wiley and Sons, Inc., 1990, 243, (technical datasheet)
[2] Thermocomp FP-VCL-4024, LNP Engineering, (technical datasheet)

PFA, PTFE filled   P-14

Synonyms: *Tetrafluoroethylene-perfluoromethyl vinyl ether copolymer, Polytetrafluoroethylene filled. Poly(tetrafluoroethene-co-trifluoro(trifluoromethoxy)ethene), PTFE filled. Poly(tetrafluoroethylene-co-trifluoromethylperfluorovinyl ether), PTFE filled*
Related Polymers: PFA, Mineral and mica filled, Filled PFA, PFA, Glass filled, PFA, PTFE and carbon filled, PFA, Carbon filled
Monomers: Tetrafluoroethylene, Perfluoro(methyl vinyl ether)
Material class: Composites, Copolymers
Polymer Type: PTFE, PFAfluorocarbon (polymers)

Volumetric & Calorimetric Properties:
Density:

No.	Value	Note
1	d 2.2 g/cm^3	5% PTFE [1]

Thermal Expansion Coefficient:

No.	Value	Note	Type
1	0.000103 K^{-1}	5% PTFE [1]	L

Deflection Temperature:

No.	Value	Note
1	90.56°C	1.8 MPa, 5% PTFE [1]

Mechanical Properties:
Flexural Modulus:

No.	Value	Note
1	137.8 MPa	5% PTFE [1]

Tensile Strength Yield:

No.	Value	Note
1	17.24 MPa	5% PTFE [1]

Izod Notch:

No.	Value	Notch	Note
1	325.74 J/m	Y	5% PTFE [1]

Applications/Commercial Products:

Trade name	Supplier
Fluoromelt FP-PL-4020	LNP Engineering
Fluoromelt FP-PML	LNP Engineering
Lubricomp FP-PL	LNP Engineering

Bibliographic References
[1] *Cen BASE/Materials*, (eds. C.E. Nunez and R.A. Nunez), Vol. 1, John Wiley and Sons, Inc., 1990, 242, (technical datasheet)

Phenol Aralkyl resins   P-15

(n = 0 - 6)

Synonyms: *Xylok resins. Phenol-1,4-bis(methoxymethyl)benzene resins. Phenol-α,α'-dimethoxy-p-xylene resins*
Related Polymers: Phenol Aralkyl Resins, Mica filled, Phenol Aralkyl Resins, Glass filled
Monomers: Phenol, 1,4-Bis(methoxymethyl)benzene
Material class: Thermosetting resin
Polymer Type: phenolic
CAS Number:

CAS Reg. No.	Note
26834-02-6	
51937-31-6	copolymer with hexamethylenetetramine

Molecular Weight: MW 600–800 (uncured)
Additives: Fillers include mica, silica, glass fibre, carbon fibre, graphite, asbestos
General Information: Improved temp. resistance, compared with unmodified phenolic resins, owing to reduced oxidative susceptibility of the methylene linkages [8]. Details of molecular struct. and mouldability have been reported [2]

Volumetric & Calorimetric Properties:

Density:

No.	Value	Note
1	d 1.34 g/cm^3	Xylok 225, carbon-fibre filled [6]
2	d 1.68 g/cm^3	Xylok 225, asbestos-filled [6]
3	d 1.87 g/cm^3	Xylok 225, silica-filled [6]

Vicat Softening Point:

No.	Value	Note
1	94°C	uncured [4]

Transport Properties:

Water Absorption:

No.	Value	Note
1	5.5–9.6 %	Xylok 225, filled [6]

Mechanical Properties:

Mechanical Properties General: Xylok 210 graphite-filled composites retain their flexural, tensile and compressive props. at elevated temps. [1]

Tensile Strength Break:

No.	Value	Note
1	42–52 MPa	Xylok 225, silica and asbestos-filled

Flexural Strength at Break:

No.	Value	Note
1	43–58 MPa	250°, Xylok 225, asbestos, silica and carbon fibre filled [6]
2	65–85 MPa	Xylok 225, asbestos, silica and carbon fibre filled [6]

Izod Notch:

No.	Value	Notch	Note
1	15 J/m	Y	Xylok 225, asbestos filled [6]
2	16 J/m	Y	Xylok 225, silica filled [6]
3	20 J/m	Y	Xylok 225, carbon-fibre filled [6]

Electrical Properties:

Electrical Properties General: Lower dielectric props. than phenolic resins and are therefore superior insulating materials

Dielectric Strength:

No.	Value	Note
1	11.8 kV.mm^{-1}	Xylok 225, carbon fibre-filled

Dissipation (Power) Factor:

No.	Value	Frequency	Note
1	0.053–0.076	1 MHz	Xylok 225, asbestos filled [6]

Optical Properties:

Transmission and Spectra: Optical transmission 1.5% (Xylok 210, graphite filled) [1]

Polymer Stability:

Thermal Stability General: The heat stability of Xylok 225 filled resins is superior to that of the corresponding phenolic material. Details of heat resistance have been reported [2]

Upper Use Temperature:

No.	Value	Note
1	150–250°C	Xylok 225 [6]

Decomposition Details: Char yield of Xylok 210 46% whereas that of its graphite composite is 83% (900°, N_2) [1]. Xylok 210 decomposes at 430° with max. rate of weight loss occurring at 525° [1]

Flammability: Limiting oxygen index 46% (Xylok 210, graphite filled, ASTM D2863) [1]. Smoke density reaches a max. after approx. 6 min. (Xylok 210, graphite filled, flaming, NBS-Aminco chamber) [1]

Chemical Stability: The asbestos, silica and carbon-fibre filled mouldings are resistant to most environments, except 98% sulfuric acid and DMF at 90°. Asbestos filled Xylok 210 is resistant to oil, ammonia and fluorocarbons [5]

Bibliographic References

[1] Kourtides, D.A., *Polym. Compos.*, 1984, **5**, 143, (flammability, mechanical props.)
[2] Sugisaki, R. and Morimoto, Y., *Kogyo Zairyo*, 1985, **33**, 91, (struct., heat resistance)
[3] Wang, J. Jiao, Y. and Li, S., *Huagong Xuebao*, 1984, **121**, (synth.)
[4] *Jpn. Pat.*, 1994, 06116369, (synth.)
[5] Harris, C.I., Edwards, A.G. and Coxon, F., *Inf. Chim.*, 1976, **155**, 315
[6] *The Materials Selector*, (eds N.A. Waterman and M.F. Ashby), Elsevier, 1991, **3**, 1903, (props.)

Phenol Aralkyl resins, glass filled P-16

Synonyms: *Xylok resins, Glass filled. Phenol-1,4-bis(methoxymethyl)benzene resins, Glass filled. Phenol-α,α'-dimethoxy-p-xylene resins, Glass filled*

Related Polymers: Phenol Aralkyl Resins, Phenol Aralkyl Resins, Mica filled
Monomers: Phenol, 1,4-Bis(methoxymethyl)benzene
Material class: Thermosetting resin, Composites
Polymer Type: phenolic
Additives: Cycloaliphatic epoxy resin as curing agent
General Information: Props. and uses of glass filled composite have been reported [1,2]

Volumetric & Calorimetric Properties:

Density:

No.	Value	Note
1	d 1.62 g/cm^3	Xylok 225 [4]

Transport Properties:

Water Absorption:

No.	Value	Note
1	0.6 %	168h, 100°, Xylok 210 [3]

Mechanical Properties:

Mechanical Properties General: Xylok 210 has excellent retention of mechanical props. at high temps. with an initial increase in flexural strength at 250° [3]

Tensile Strength Break:

No.	Value	Note
1	81 MPa	Xylok 225 [4]

Flexural Strength at Break:

No.	Value	Note
1	195 MPa	room temp., Xylok 225 [4]
2	124 MPa	250°, Xylok 225 [4]
3	480 MPa	4900 kg cm^{-2}, 250°, Xylok 210 [3]

Izod Notch:

No.	Value	Notch	Note
1	480 J/m	Y	Xylok 225 [4]

Electrical Properties:
Electrical Properties General: Dielectric props. of Xylok 210 are retained after 1400h exposure at 250° [3]. Insulation resistance 400000 (Xylok 210, BS 2782) [3]. Xylok 210 meets the electrical insulation requirements of IEC class C [3]

Dielectric Permittivity (Constant):

No.	Value	Frequency	Note
1	4.77	1 MHz	dry, Xylok 210 [3]
2	4.82	1 MHz	wet, Xylok 210 [3]
3	4.4	1 MHz	dry, Xylok 225 [4]
4	4.4	1 MHz	wet, 24h, Xylok 225 [4]

Dielectric Strength:

No.	Value	Note
1	10.6 kV.mm^{-1}	Xylok 225 [4]
2	28–33 kV.mm^{-1}	Xylok 210 [3]

Dissipation (Power) Factor:

No.	Value	Frequency	Note
1	0.011	1 MHz	dry, Xylok 210 [3]
2	0.013	1 MHz	wet, Xylok 210 [3]
3	0.014	1 MHz	dry, Xylok 225 [4]
4	0.016	1 MHz	wet, Xylok 225 [4]

Polymer Stability:
Polymer Stability General: Diepoxide cured resin is slightly less resistant to high temps. than is hexa cured material [3]
Chemical Stability: Xylok 210 is relatively unaffected by 168h exposure to 30% soln. of antifreeze (90°), motor oil (150°), Skydrol 500 B (100°), transformer oil (100°), toluene (110°), trichloroethylene (85°) [3]. Xylok 210 has good resistance to 10% sodium hydroxide soln., showing only 25% decomposition after 100h immersion at 90°. [3] Xylok 210 shows 5.7% weight loss after 168h exposure to 10% hydrochloric acid at 90° [3]

Applications/Commercial Products:
Mould Shrinkage (%):

No.	Value	Note
1	0.2%	Xylok 225 [4]

Trade name	Supplier
Xylok	Albright and Wilson

Bibliographic References
[1] Harris, G.I. and Edwards, A.G., *Manuf. Qual. Reinf. Plast. Congr.*, 1978, **71**, (props. uses)
[2] Harris, G.I., *Recent Adv. Prop. Appl. Thermosetting Mater. Int. Conf.*, 1979, 6.1-6.6, (use, stability)
[3] Harris, G.I. and Edwards, A.G. and Coxon F., *Inf. Chim.*, 1976, **155**, 315, (struct., props.)
[4] *The Materials Selector*, 2nd edn., (eds. N.A. Waterman and M.F. Ashby), Chapman & Hall, 1991, **3**, 1903, (props)

Phenol Aralkyl resins, Mica filled   P-17

Synonyms: Xylok resins, Mica filled. Phenol-1,4-bis(methoxymethyl)benzene resins, Mica filled. Phenol-α,α'-dimethoxy-p-xylene resins, Mica filled
Related Polymers: Phenol Aralkyl resins, Glass filled, Phenol Aralkyl resins
Monomers: Phenol, 1,4-Bis(methoxymethyl)benzene
Material class: Composites, Thermosetting resin
Polymer Type: phenolic

Volumetric & Calorimetric Properties:
Density:

No.	Value	Note
1	d 1.79 g/cm^3	Xylok 225 [2]

Mechanical Properties:
Tensile Strength Break:

No.	Value	Note
1	30 MPa	Xylok 225 [2]

Flexural Strength at Break:

No.	Value	Note
1	67 MPa	room temp., Xylok 225 [2]
2	43 MPa	250°, Xylok 225 [2]

Izod Notch:

No.	Value	Notch	Note
1	17 J/m	Y	Xylok 225 [2]

Electrical Properties:
Electrical Properties General: Dissipation factor of Xylok 231 increases with increasing temp. [1]
Dielectric Permittivity (Constant):

No.	Value	Frequency	Note
1	4.14	1 MHz	dry, Xylok 225 [2]
2	4.3	1 MHz	24h, wet, Xylok 225 [2]

Dielectric Strength:

No.	Value	Note
1	9.4 kV.mm^{-1}	Xylok 225 [2]

Dissipation (Power) Factor:

No.	Value	Frequency	Note
1	0.017	1 MHz	dry, Xylok 225 [2]
2	0.02	1 MHz	wet, 24h, Xylok 225 [2]

Applications/Commercial Products:
Mould Shrinkage (%):

No.	Value	Note
0	0.3%	Xylok 225 [2]

Trade name	Details	Supplier
Xylok	231	Albright and Wilson

Bibliographic References
[1] Harris, C.I., Edwards, A.G. and Coxon, F., *Inf. Chim.*, 1976, **155**, 315
[2] *The Materials Selector*, 2nd edn., (eds. N.A. Waterman and M.F. Ashby), Chapman & Hall, 1991, **3**, 1903, (props)

Phenol-Butyraldehyde resins P-18

Synonyms: *Phenol-butanal resins*
Monomers: Butyraldehyde, Phenol
Material class: Thermosetting resin
Polymer Type: phenolic
CAS Number:

CAS Reg. No.
27814-12-6

Volumetric & Calorimetric Properties:
Melting Temperature:

No.	Value	Note
1	58–66°C	high-*ortho*, uncured [2]
2	60°C	uncured [3]

Transport Properties:
Polymer Solutions Dilute: Viscosity 280 mPa s (methoxypropanol, 20°, uncured) [3]

Applications/Commercial Products:
Processing & Manufacturing Routes: High-*ortho* resin, with > *ortho-ortho* linkages [1], may be synth. by reaction of anhydrous phenol with butyraldehyde in the presence of a calcium hydroxide catalyst [2]. Novolak resin may be synth. by reaction of phenol and butyraldehyde in the presence of conc. H_2SO_4, in toluene soln. [3]. Butyraldehyde treated with phenol using HCl catalyst produces an oil-soluble resin. When base-catalysed with NaOH, phenol and butyraldehyde produces a moulding powder
Applications: The acid-catalysed phenol-butyraldehyde product can be used as an oil-soluble resin

Bibliographic References
[1] *Jpn. Pat.*, 1977, 77 00 997, (synth.)
[2] *U.S. Pat.*, 1969, 3 425 989, (synth.)
[3] *Ger. Pat.*, 1989, 3 901 930, (synth.)

Phenol-Formaldehyde Novolaks P-19

Uncured novolak

Cured novolak

Synonyms: *PF Novolaks. Novolaks*
Related Polymers: Phenol Formaldehyde Resins, Phenol formaldehyde resoles, Phenolic resins high-ortho, Phenolic Dispersions, Phenolic Resin, Moulding Powders, Mineral filled, Phenolic Resin, Moulding Powders, Graphite filled, Phenolic Resin, Moulding Powders, Mica filled, Phenolic Resin, Moulding Powders, Glass fibre filled, Phenolic Resin, Moulding Powders, Woodflour filled, Phenolic Resin, Moulding Powders, Cellulose filled, Phenolic Resin, Moulding Powders, Cotton filled
Monomers: Phenol, Formaldehyde
Material class: Thermosetting resin
Polymer Type: phenolic
CAS Number:

CAS Reg. No.
9003-35-4

Molecular Formula: $[(C_6H_6O).(CH_2O)]_n$
Fragments: C_6H_6O CH_2O
Molecular Weight: MW 250–900; MW 500–5000. Generally MW < 2000
Additives: Hexamethylene tetramine (curing agent), carbon fibre, asbestos, basalt fibre, graphite fibre, wood flour. Wood flour, cellulose and nutshell flour are added as fillers to novolak moulding compounds to reduce shrinkage during cure, to improve impact strength and provide flow control. Poly(vinyl alcohol) and polyethylene to impove tensile strength
General Information: Novolaks are linear or slightly branched condensation products with methylene bridges of relatively low MW. Uncured resins are thermoplastic and sol. in a variety of solvents. Curing with an amine usual hexamethylene tetramine produces a cross-linked struct. which is hard, insol. and rigid. The cross-link procedure passes through several stages, the resins becoming rubbery (B-stage) and finally hard (C-stage). Props. of cured resin depend on type and amount of curing agent and time and temp. of cure. Props. given in this entry are representative examples of a specific cured resin

Volumetric & Calorimetric Properties:
Thermodynamic Properties General: Calculation of the heat capacity of H41N has been reported [4]
Melting Temperature:

No.	Value	Note
1	80°C	uncured [22]

– Phenol-formaldehyde Novolaks

Glass-Transition Temperature:

No.	Value	Note
1	45–70°C	[9]
2	110–160°C	5–15 phr hexa, moulded resin [15]

Transition Temperature General: Deflection temp. of hexa cured moulded resin increases with increasing amount of curing agent and after heat treatment (210°, 8h) [15]

Deflection Temperature:

No.	Value	Note
1	98–178°C	5–20 phr hexa, moulded resin [15]
2	158–250°C	5–20 phr hexa, moulded resin, heat treated, 210°, 8 h [15]
3	130°C	asbestos filled [16]

Vicat Softening Point:

No.	Value	Note
1	50–60°C	hexa cured [7]
2	100°C	uncured [22]

Surface Properties & Solubility:

Solubility Properties General: Uncured novolak forms miscible blends with polyethylene-*co*-vinyl acetate [1]

Solvents/Non-solvents: Uncured sol. Me_2CO, 2-butanone, EtOAc, EtOH; insol. petroleum, C_6H_6, turpentine, toluene, CCl_4, $CHCl_3$ [22]

Transport Properties:

Polymer Melts: Melt viscosity of hexa cured novolac decreases with increasing temp. with the decrease most marked at temps. up to 150° [7]

Mechanical Properties:

Mechanical Properties General: The mech. props. of hexa cured, carbon fibre laminates are enhanced by increasing the volume of carbon fibre present, and by increasing the curing of the novolac [7]

Tensile (Young's) Modulus:

No.	Value	Note
1	95157 MPa	hexa cured, 15% carbon fibre, DIN 53455 [7]
2	99100 MPa	hexa cured, 30% carbon fibre, DIN 53455 [7]
3	120663 MPa	hexa cured, 45% carbon fibre, DIN 53455 [7]

Flexural Modulus:

No.	Value	Note
1	4964 MPa	720 kpsi, Borden RC-1000 [23]
2	4757 MPa	690 kpsi, BRPA-8152 [23]

Tensile Strength Break:

No.	Value	Note
1	323.7 MPa	hexa cured, 15% carbon fibre, DIN 53455 [7]
2	591.5 MPa	hexa cured, 30% carbon fibre, DIN 53455 [7]
3	20 MPa	asbestos filled [16]
4	663.1 MPa	hexa cured, 45% carbon fibre, DIN 53455 [7]

Flexural Strength at Break:

No.	Value	Note
1	301.2 MPa	hexa cured, 15% carbon fibre, DIN 53452 [7]
2	752.4 MPa	hexa cured, 30% carbon fibre, DIN 53452 [7]
3	1143.8 MPa	hexa cured, 45% carbon fibre, DIN 53452 [7]
4	21.6–110.9 MPa	2–20 phr hexa, moulded resin [15]
5	54.8 MPa	7.95 kpsi, Borden RC-1000 [23]
6	78.1 MPa	11.33 kpsi, BRPA-8152 [23]

Compressive Strength:

No.	Value	Note
1	70–80 MPa	asbestos filled [16]

Impact Strength: Impact strength 35–40 kJ cm^{-2} (asbestos filled) [16]

Failure Properties General: Interlaminar shear strength 14.1 N mm^{-2} (hexa cured, 15% carbon fibre, ASTM Norm D2344-65T), 27.9 N mm^{-2} (hexa cured, 30% carbon fibre, ASTM Norm D2344-65T), 38.9 N $^{-2}$ (hexa cured, 45% carbon fibre, ASTM Norm D2344-65T) [7]

Fracture Mechanical Properties: Fracture strain 1.18% (Borden RC-1000) [23]; 1.81% (BRPA-8152) [23]

Charpy:

No.	Value	Notch	Note
1	34.9	N	hexa cured, 15% carbon fibre, ISO R179-1961 [7]
2	71.4	N	hexa cured, 30% carbon fibre, ISO R179-1961 [7]
3	89.8	N	hexa cured, 45% carbon fibre, ISO R179-1961 [7]

Electrical Properties:

Electrical Properties General: The variation of permittivity and dielectric loss with frequency and temp. has been reported [8]

Optical Properties:

Transmission and Spectra: Ir [7], H-1 nmr [10,11] and C-13 nmr [12,13,14] spectral data have been reported

Volume Properties/Surface Properties: Colour Gardner scale 7 (uncured) [22]

Polymer Stability:

Polymer Stability General: Resistance to heat may be improved by increasing the amount of curing agent used [15]

Thermal Stability General: Cured resins decompose above 250°, with max. rate of decomposition occuring above 300° [9]

Decomposition Details: Char yield at 750° in N_2 59.1% (medium flow novolak, hexa cured) [3]. Char yields increase with increasing hexa content [3]. H41N shows approx. 22% weight loss over the temp. range 300–1100° when heated in an argon atmosphere [4]

Environmental Stress: Weathering in a hot climate for 60 months gives an increase in volume resistivity but decreases dielectric constant and dielectric loss tangent [6]

Chemical Stability: Asbestos filled material is resistant to sulfuric acid, hydrochloric acid and phosphoric acid [16]. Uncured resin is incompatible with linseed oil, tang oil and castor oil at 250°, but is compatible with double boiled linseed oil at the same temp. [22]

Applications/Commercial Products:

Processing & Manufacturing Routes: Novolaks synth. by reaction of phenol and formaldehyde in the presence of a strong acid catalyst. Oxalic acid is the catalyst of choice (ideally less than 0.035%) [18] with sulfuric acid or *p*-toluenesulfonic acid also common. Phosphoric acid and hydrochoric acid are less frequently employed; the use of hydrochloric acid is not recommended owing to the formation of hazardous intermediates. The synth. is either a batch or a continuous process. In the batch process, hot phenol is stirred with formaldehyde and catalyst at 95° until the formaldehyde is consumed. Water is removed by a azeotropic distillation and phenol removed by heating *in vacuo* at 160°. In the continuous process, phenol, formaldehyde and catalyst are mixed and heated before being transferred to a second stage pressurized reactor at 120–180°. In both processes, the product is recovered by flaking onto a cooling belt. The resin may then be cured, typically using hexamethylene tetramine (9–10%). Epoxide resins may also be used to avoid release of gaseous by products. High-*ortho* novolaks are synth. by reaction of a large excess of phenol with formaldehyde in the presence of weak acids (pH 4–6) and a divalent metal catalyst [19,20,21]. High MW (up to 75700) novolaks may be synth. by the reaction of phenol and paraformaldehyde in a variety of organic solvents under acid catalysis at 80° [2]. C-13 nmr and N-15 nmr have been used to elucidate the curing mechanism involving hexamethylenetetramine [5]. Novolaks consisting of an all-*ortho*-substituted struct., may be synth. by repeated reaction of the bromomagnesium salts of phenol and salicyl alcohol [17]

Applications: High-*ortho* novolaks are used in foundry applications and in moulding compounds. Novolaks cured with epoxide resins have uses in commutators and other electrical applications. Powdered novolaks are used as moulding compounds. High-porosity papers impregnated with novolak in solution are used as filters. High MW novolaks in powder form are made into grinding wheels by a cold moulding process. A wide spectrum of phenolic resins are used as binders for friction materials including novolaks, Resoles, novolak/Resole blends, and thermoplastic modified novolaks. A novolak/ (hexamethyltetramine) blend can be used as a tackifier for rubber adhesives. Phenolic novolaks with fine particle hexamethyltetramine can be easily incorporated into rubber blends. On cure the phenolic moiety forms an interpenetrating network providing rubber reinforcement for tyres and shoe soles. Phenolic novolaks in granular form may be used in the hot curing (shell moulding) process for foundry sand binders. novolaks are formulated into a socket putty to bond the glass body to the metal socket in fluorescent lighting

Trade name	Details	Supplier
BRPA-8152	moulding compound	Union Carbide
Borden P	friction binder resins (not exclusive)	Borden
Borden PRL	refractory binder resin	Borden
H41N	60.5% glass and talc. filled	Ametek
Rütaphen DP	socket putty	Bakelite
Rütaphen K	reinforcing resins	Bakelite
Rütaphen SP	grinding wheel powder resin	Bakelite
RC-1000	moulding compound	Borden

Bibliographic References

[1] Mekhilef, N. and Hadjiandreou, P., *Polymer*, 1995, **36**, 2165, (miscibility)
[2] Yamagishi, T.-A., Nomoto, M., Ito, S., Ishida, S. and Nakamoto, Y., *Polym. Bull. (Berlin)*, 1994, **32**, 501, (synth.)
[3] Lemon, P.H.R.B., *Polym. Paint Colour J.*, (suppl.), 1988, 103, (rev.)
[4] Henderson, J.B., and Emmerich, W.D., *Thermochim. Acta*, 1988, **131**, 7, (heat capacity, decomposition)
[5] Hatfield, G.R. and Maciel, G.E., *Macromolecules*, 1987, **20**, 608, (nmr cure)
[6] Melkumov, A.N., Prutkin, V.P. Tavshunskaya, L.I. and Tetenova, S.S., *Plast. Massy*, 1987, **63**, (weathering)
[7] Simitzis, J., *Angew. Makromol. Chem.*, 1989, **165**, 21, (mechanical props.)
[8] Holland, C., Stark, W. and Hinrichsen, G., *Acta Polym.*, 1995, **46**, 64, (dielectric props.)
[9] Kopf, P.W., *Kirk-Othmer Encycl. Chem. Technol.*, 4th edn., (ed. J.I Kroschwitz), 1991, **18**, 603, (rev.)
[10] Woodbrey, J.C., Higginbottom, H.P., Culbertson, H.M., *J. Polym. Sci., Part A-1*, 1965, **3**, 1079, (H-1 nmr)
[11] Yoshikawa, T. and Kumanotani, J., *Makromol. Chem.*, 1970, **131**, 273, (H-1 nmr)
[12] Shaefer, J., *J. Am. Chem. Soc.*, 1976, **98**, 1031, (C-13 nmr)
[13] Fyfe, C., *Macromolecules*, 1983, **16**, 1216, (C-13 nmr)
[14] Bryson, R.L., Hatfield G.R., Early, T.A. Palmer, A.R. and Maciel, G.E., *Macromolecules*, 1983, **16**, 1669
[15] Fukuda, A., *Polymeric Materials Encyclopedia*, (ed. J.C. Salamone), CRC Press, 1996, **7**, 5035, (heat resistance)
[16] Dolezel, B., *Koroze Ochr. Mater.*, 1984, **28**, 77
[17] Casiraghi, G., Cornia, M., Sartori, G., Casnati, et el, *Makromol. Chem.*, 1982, **183**, 2611, (synth.)
[18] Majchrzak, J., gryta, M., Kaledkowski, B. and Studencki, L., *Przemysi Chem.*, 1982, **61**, 325, (synth.)
[19] Peer, H.G., *Recl. Trav. Chim. Pays-Bas*, 1960, **79**, 825, (synth.)
[20] Bender, H.L., *Mod. Plast.*, 1953, **30**, 136, (synth.)
[21] Bender, H.L., Farnham, A.G., Guyer, J.W., Apel, F.N. and Gibb, T.B., *Ind. Eng. Chem.*, 1952, **44**, 1619, (synth.)
[22] Ahisanuddin, Saksena, S.C., Panda, H. and Rakshinda, *Paintindia*, 1982, **32**, 3, (synth., solubility)
[23] Sung, N-H., churchill, G.B. and Suh, N.P., *J. Mater. Sci.*, 1975, **10**, 1741, (flexural props.)

Phenol-formaldehyde resins   P-20

Synonyms: *PF resins. Phenoplasts. Phenolic Resins. Resoles. Novolaks*
Related Polymers: Phenol-formaldehyde Resoles, Phenol-formaldehyde
Monomers: Phenol, Formaldehyde
Material class: Thermosetting resin
Polymer Type: phenolic
Molecular Formula: $[(C_6H_6O).(CH_2O)]_n$
Fragments: C_6H_6O CH_2O

Applications/Commercial Products:

Processing & Manufacturing Routes: Phenol-formaldehyde resins are made by the condensation reaction of molten phenol with formaldehyde, generally as a 37% aq. soln. With a basic catalyst and a molar excess of formaldehyde, a Resole is formed. With an acidic catalyst and excess phenol, a Novolak resin is formed. Resoles are used in soln. generally aq., but may also be used with alcoholic solvents added. They cure by application of heat or acid. Novolaks may be used as solid powdered resin, or in alcoholic solvents. They require a curing agent, generally hexamethylene tetramine. Phenolic resoles are made by treating Phenol with excess formaldehyde (F/p mole ratio above 1.5) with base catalyst, NaOH, Ba(OH)$_2$, Ca(OH)$_2$. A batch process in a resin kettle with good stirring and cooling to become exotherms. Reaction cycles of 60° upwards are chosen to achieve resin viscosities suitable for the application

Applications: See Phenol-aldehyde resins for a general discussion of the uses of phenolic resins

Trade name	Details	Supplier
Aerophen	phenolic resins	Dyno
Borden P	phenolic resins	Borden
Dynosol	phenolic resins	Dyno
Exter	phenolic resins	Neste
Fenorex	phenolic resins	Neste
Pribex	phenolic resins	Neste
Rütaphen	phenolic resins	Bakelite
Vyncolite	engineering thermosets	Vynckier (Perstorp)

Phenol-formaldehyde Resoles   P-21

Synonyms: *PF Resoles. Phenoplasts. Phenolic Resoles. Resoles. A-Stage Resin*
Related Polymers: Phenol-formaldehyde Resins, Phenol-formaldehyde Novolaks, Phenolic Foam Resins
Monomers: Phenol, Formaldehyde
Material class: Thermosetting resin
Polymer Type: phenolic
Molecular Formula: $[(C_6H_6O).(CH_2O)]_n$
Fragments: C_6H_6O CH_2O
General Information: Resoles are produced by treating phenol with formaldehyde in presence of base, producing low MW polyalcohol struct. Cured by heating, which causes cross-linking via the alcohol groups, or cure may occur by adjusting the pH of the initial reaction

Volumetric & Calorimetric Properties:
Specific Heat Capacity:

No.	Value	Note	Type
1	0.0692 kJ/kg.C	20° [12]	P
2	0.418 kJ/kg.C	100° [12]	P
3	2.01 kJ/kg.C	450° [12]	P

Glass-Transition Temperature:

No.	Value	Note
1	35°C	sodium hydroxide catalysed
2	47°C	hexa catalysed

Transport Properties:
Polymer Solutions Dilute: Viscosity 600–1500 mPa s (20°, wood applications)

Optical Properties:
Optical Properties General: Refractive index of uncured resin increases with increasing time of reaction [13]
Refractive Index:

No.	Value	Note
1	1.4844	uncured, reaction time 5 min [13]
2	1.5022	uncured, reaction time 1 h [13]

Transmission and Spectra: C-13 nmr [1, 2, 3, 4], ir [1, 2] and mass [5] spectral data have been reported

Polymer Stability:
Polymer Stability General: Resoles are most stable over prolonged periods of storage at pH approx. 5 [9]
Thermal Stability General: Thermal stability of cured resoles has been reported [8]
Decomposition Details: Complete volatilisation of cured resoles occurs upon heating in air above 300 ° whereas heating in an inert atmosphere causes only 25% weight loss up to 900° owing to formation of carbonaceous char [11]

Applications/Commercial Products:
Processing & Manufacturing Routes: Synth. by reaction of molten phenol with a molar excess of formaldehyde (F/P ratio > 1.5) with base catalyst, generally sodium, calcium or barium hydroxides in a batch process using a resin kettle with good mixing and cooling to overcome exotherms. Reaction cycles of temps. of 60° upwards are chosen to achieve resin viscosities suitable for the application. To increase the solids level, vacuum evaporation at 60° is necessary. Resoles may be cured by application of heat (the most important process) at 130°–200°; by acids [6] typically *p*-toluenesulfonic acid or phenolsulfonic acid; by microwave curing [7]. High-*ortho* liq. resoles may be synth. by reaction of phenol and formaldehyde (1:1.5–1.8 mole ratio) at pH 4–7 in the presence of a divalent metal salt catalyst at 80°–90° for 5–8 h. Synth. of low MW resole has been reported [10] Resoles with a high solids content and low viscosity are best synth. using a formaldehyde: phenol ratio of 1–1.1:1; a short reaction time is also preferable [13]
Applications: Used in wood and joinery applications for bonding woodchip for chipboard, bonding plywood, and decorative laminates. Moulded parts are used in automotive construction, door panels etc. Other uses include sand-based moulds for foundry casting of metals

Trade name	Details	Supplier
Aerodux	Resorcinol resins	Dyno
Aerophen	Phenolic resins	Dyno
Dynosol		Dyno
Exter 4010	PF resin for plywood	Neste
Rutaphen	GA, GC, GH, GK, GN, GS, GW - foundry resins	Bakelite
Rutaphen HW	Wood materials (chipboard, fibreboard, plywood, compreg)	Bakelite
		Borden

Bibliographic References
[1] Grenier-Loustalot, M.-F., Larroque, S. and Grenier, P., *Polymer*, 1996, **37**, 639, (C-13 nmr, ir)
[2] So, S. and Rudin, A., *J. Appl. Polym. Sci.*, 1990, **41**, 205, (C-13 nmr, ir)
[3] Mechin, B., Hanton, D., Le Goff, J. and Tanneur, J.P., *Eur. Polym. J.*, 1984, **20**, 333, (C-13 nmr)
[4] Maciel, G.E., Chuang, I.-S. and Gollob, L., *Macromolecules*, 1984, **17**, 1081, (C-13 nmr)
[5] Prókai, L., *Acta Chim. Hung.*, 1987, **124**, 901, (ms)
[6] Gupta, M.K., Salee, G. and Hoch, D.W., *American Chem. Soc. Div. Polym. Chem. Preprints*, 1986, **27**, 309
[7] Tolmay, A.T. and Focke, W.W., *S. Afr. J. Chem. Eng.*, 1991, **3**, 1, (cure)
[8] Slugin, V.P. and Bolotova, S.V., *Plast. Massy*, 1987, 28, (thermal stability)
[9] Gupta, M.K. and Hindersinn, R.R., *Polym. Eng. Sci.*, 1987, **27**, 976, (stability)
[10] Gryta, M., Majchrzak, J. and Kaledkowski, B., *Przemysi Chem.*, 1982, **61**, 90, (synth.)
[11] Hall, R.W., *Rep. Progr. Appl. Chem.*, 1963, **48**, 223, (rev. degradation)
[12] Chang, S.-S., *American Chem. Soc. Div. Polym. Chem. Preprints*, 1983, **24**, 187, (heat capacity)
[13] Abdel-Mohsen, F.F. and Helday, F.M., *Acta Polym.*, 1987, **38**, 291, (synth., refractive index)

Phenolic dispersions   P-22
Related Polymers: Phenol Aldehyde Resins, Phenolic Resoles, Phenolic Novolaks, Phenolic Resins, Oil-modified, Phenolic Resin Insulation Wool Binders, Phenolic Resins Coatings, Phenolic Resin Abrasives, Phenolic laminating Resins, Phenolic Resin Friction Materials
Monomers: Phenol, Formaldehyde
Material class: Thermosetting resin

Polymer Type: phenolic
Additives: Protective colloids are used to stabilise the resin in water systems

Applications/Commercial Products:
Processing & Manufacturing Routes: Phenolic dispersions are prepared by treating phenol and formaldehyde typically F/P mole ratio 1.5 with hexamethylene tetramine (6–12%) then adding poly(vinyl alcohol) and gum arabic as protective colloids
Applications: Particles of size >1000 microns are used in friction element composition

Phenolic laminating resins   P-23
Related Polymers: Phenol-formaldehyde Resoles, Phenol-formaldehyde Novolaks, Urea-formaldehyde resins, Melamine-formaldehyde resins, Cresol-Formaldehyde Resins, Xylenol-formaldehyde Resins, Phenolic resins, Glass-reinforced plastic
Monomers: Phenol, Urea, Melamine, Formaldehyde
Material class: Thermosetting resin
Polymer Type: phenolic
Additives: Paper (kraft, alpha, or cotton linter), fabric, asbestos felt, glass fabric, nylon, cotton fabric
General Information: Usually based on resoles which contain sufficient methyl groups to allow for cross-linking without addition of formaldehyde donors. Resins are based on phenol or phenol-cresol mixtures when good mech. props are required; electrical grade laminates are based on *m*-cresol. Resins for mech. laminates are made using sodium hydroxide catalyst whereas ammonia catalyst is used for electrical laminates. Props. will depend upon nature of the resin, type of additive and method of synth. [3]. Phenolic resins may also be used for glass reinforced plastics

Volumetric & Calorimetric Properties:
Density:

No.	Value	Note
1	d 1.37–1.4 g/cm^3	paper [3]
2	d 1.36 g/cm^3	fabric [3]
3	d 1.6–1.8 g/cm^3	asbestos felt [3]
4	d 1.4–1.7 g/cm^3	glass fabric [3]
5	d 1.32–1.36 g/cm^3	paper [4]
6	d 1.33–1.36 g/cm^3	cotton fabric [4]
7	d 1.72 g/cm^3	asbestos fabric [4]
8	d 1.65 g/cm^3	glass fabric [4]

Thermal Expansion Coefficient:

No.	Value	Note	Type
1	0.0002 K^{-1}	cotton fabric [4]	L
2	0.00015 K^{-1}	asbestos fabric [4]	L
3	0.00018 K^{-1}	glass fabric [4]	L

Transport Properties:
Polymer Solutions Dilute: Viscosity 0.3–1.5 P (30–150 mPa s 20°, laminate resins)
Water Absorption:

No.	Value	Note
1	0.95–3.3 %	24 h, 3 mm, paper [4]
2	1.3–2.5 %	24 h, 3 mm, cotton fabric [4]
3	2.5 %	24 h, 3 mm, asbestos fabric [4]
4	2 %	24 h, 3mm, glass fabric [4]

Mechanical Properties:
Tensile (Young's) Modulus:

No.	Value	Note
1	6200–7600 MPa	cotton fabric, longitudinal [4]
2	10300 MPa	glass fabric, longitudinal [4]
3	8900–12400 MPa	paper, longitudinal [4]
4	11000 MPa	asbestos fabric, longitudinal [4]

Tensile Strength Break:

No.	Value	Note
1	48–62 MPa	cotton fabric, transverse [4]
2	82 MPa	asbestos fabric, longitudinal [4]
3	70 MPa	asbestos fabric, transverse [4]
4	160 MPa	glass fabric, longitudinal [4]
5	138 MPa	glass fabric, transverse [4]
6	103–138 MPa	paper, longitudinal [4]
7	82–110 MPa	paper, transverse [4]
8	62–90 MPa	cotton fabric, longitudinal [4]
9	69–97 MPa	paper laminates [3]
10	110 MPa	fabric laminate [3]
11	48–104 MPa	asbestos felt [3]
12	83–240 MPa	glass fabric [3]

Flexural Strength at Break:

No.	Value	Note
1	103–107 MPa	min., cotton fabric, longitudinal [4]
2	93–110 MPa	cotton fabric, transverse [4]
3	>125 MPa	min., asbestos fabric, longitudinal [4]
4	93–172 MPa	min., paper, longitudinal [4]
5	81–151 MPa	paper, transverse [4]
6	110 MPa	asbestos fabric, transverse [4]
7	>138 MPa	min., glass fabric, longitudinal [4]
8	125 MPa	glass fabric, transverse [4]

Compressive Strength:

No.	Value	Note
1	162–172 MPa	cotton fabric, edge [4]
2	260 MPa	asbestos fabric, flat [4]
3	145 MPa	asbestos fabric, edge [4]
4	345 MPa	glass fabric, flat [4]
5	120 MPa	glass fabric, edge [4]
6	220–248 MPa	paper, flat [4]
7	130–175 MPa	paper, edge [4]
8	255–269 MPa	cotton fabric, flat [4]

Impact Strength: Impact strength 0.27–0.61 J (paper) [3], 2 J (fabric) [3], 1.35 J (asbestos felt), 13.5 J (glass fabric) [3], 52–210 J m^{-1} (min., paper, flatwise) [4], 95–170 J m^{-1} (min., cotton fabric, flatwise) [4], 190 J m^{-1} (min., asbestos fabric, flatwise) [4], 345 J m^{-1} (min., glass fabric, flatwise) [4]
Failure Properties General: Shear strength 69–83 MPa (paper) [4], 76–83 MPa (cotton fabric) [4], 82 MPa (asbestos fabric) [4], 125 MPa (glass fabric) [4]
Izod Notch:

No.	Value	Notch	Note
1	18–26 J/m	Y	paper, edgewise [4]
2	53–100 J/m	Y	cotton fabric, edgewise [4]
3	160 J/m	Y	asbestos fabric, edgewise [4]
4	290 J/m	Y	glass fabric, edgewise [4]

Electrical Properties:
Dielectric Permittivity (Constant):

No.	Value	Frequency	Note
1	5.8–7	1 MHz	max., cotton fabric [4]
2	<5.5	1 MHz	max., glass fabric [4]
3	5–5.3	1 MHz	max., paper [4]
4	4.6–5.2	1 MHz	paper [3]
5	6.5	1 MHz	fabric [3]
6	6.1	1 MHz	asbestos felt [3]
7	4.5–5.5	1 MHz	glass fabric [3]

Dielectric Strength:

No.	Value	Note
1	18.5–19.6 kV.mm^{-1}	short, 3 mm, paper [4]
2	6–14.5 kV.mm^{-1}	short, 3 mm, cotton fabric [4]
3	2 kV.mm^{-1}	short, 3 mm, asbestos fabric [4]
4	24 kV.mm^{-1}	short, 3 mm, glass fabric [4]
5	15–17.7 kV.mm^{-1}	90°, paper, 0.062 in. thick, normal to laminate [3]
6	1.57 kV.mm^{-1}	90°, fabric, 0.062 in. thick, normal to laminate [3]
7	3.94 kV.mm^{-1}	90°, asbestos felt, 0.062 in. thick, normal to laminate [3]

Dissipation (Power) Factor:

No.	Value	Frequency	Note
1	0.018–0.02	800 Hz	paper [3]
2	0.25	800 Hz	fabric [3]
3	0.01–0.04	800 Hz	glass fabric [3]
4	0.032–0.042	1 MHz	paper [3]
5	0.1	1 MHz	fabric [3]
6	0.11	1 MHz	asbestos [3]
7	0.01–0.02	1 MHz	glass fabric [3]
8	0.04–0.06	1 MHz	max., paper [4]
9	0.05–0.1	1 MHz	max., cotton fabric [4]
10	<0.03	1 MHz	max., glass fabric [4]

Polymer Stability:
Upper Use Temperature:

No.	Value	Note
1	140°C	paper [4]
2	130°C	coton fabric [4]
3	155°C	asbestos fabric [4]
4	180°C	glass fabric [4]

Applications/Commercial Products:
Processing & Manufacturing Routes: Paper and fabric are impregnated with resin, and dried at temps. of 100–190°. Aq. phenolic Resoles and MeOH/toluene solns. of modified cresol and xylenol Resoles are employed for specific laminates. Prepregs of glass filament cloth impregnated with phenolic resins are made into fibreglass reinforced components. The prepreg has a resin content of generally 40–50% and it may be stored at <20° before use. It is cured and compressed at 0.7 Pa and 125° for 1–2h to form FRP laminates having excellent fire-resistant and smoke props.
Applications: Paper- and fabric-based laminates are used as self-supporting insulating materials for electrical applications. Battery separators are made from phenolic resin impregnated cellulose and polyacrylonitrile fibres. The resin level is 25–30% and porosity is achieved by inclusion of surfactants. Industrial filter inserts are made from porous paper and fibrous materials impregnated with phenolic resoles. Fibreglass-reinforced flexible abrasive wheels are made from fibreglass yarn prepregs impregnated with phenolic resole. FRP laminates are used inside aircraft, rail vehicles (especially subway trains), buses and trucks for safety reasons. Cotton fabric laminates used to make gear wheels. Cotton and asbestos laminates used as bearings in steel rolling mills. Other uses in aircraft construction and in chemical plant (but not alkaline environments)

Trade name	Details	Supplier
Phenmat		DSM
Pribex		Neste
Rütaphen I	impregnating resins	Bakelite

Bibliographic References
[1] *Pribex Range*, Neste, (technical datasheet)
[2] Phenmat. DSM,(technical datasheet)
[3] Brydson, J.A., *Plast. Mater.*, 6th edn., Butterworth-Heinemann, 1995, **635**, (synth, props, uses)
[4] *The Materials Selector*, 2nd edn., (eds. N.A. Waterman and M.F. Ashby), Chapman & Hall, 1997, **3**, 722, (props)

Phenolic resin foam
Synonyms: *Phenolic foam. Phenol-formaldehyde foam*
Related Polymers: Phenol-formaldehyde resins, Phenol-formaldehyde Resole
Monomers: Phenol, Formaldehyde
Material class: Thermosetting resin
Polymer Type: phenolic
Molecular Formula: $[(C_6H_6O).(CH_2O)]_n$
Fragments: C_6H_6O CH_2O
Additives: Paper, glass fibre, asbestos, hollow microspheres to improve strength. Boron trioxide, aluminium hydroxide as flame retardants. Calcium carbonate as corrosion inhibitor . Inorganic filler (e.g. perlite) to improve mechanical strength and fire behaviour
General Information: Ratio of open to closed cells is dependent on the density, surfactant used in synth., and on the formulation. Low density forms may contain up to 90% open cells [4]

Phenolic resin foam

Morphology: A crosslinked three dimensional struct. with methylene ether and phenylmethane type 'bridges' within the network

Volumetric & Calorimetric Properties:

Density:

No.	Value	Note
1	d 0.035–0.08 g/cm^3	40% open cells [4]
2	d 0.08 g/cm^3	[3]
3	d 0.015–0.32 g/cm^3	dependent on synth. [1]
4	d 0.04 g/cm^3	Exeltherm [4]
5	d 0.035 g/cm^3	Cellobond [4]
6	d 0.038 g/cm^3	Celotex [4]

Thermal Expansion Coefficient:

No.	Value	Note	Type
1	1.1×10^{-5} K^{-1}	[3]	V
2	3.3×10^{-5} K^{-1}	Exeltherm [4]	L
3	$3 \times 10^{-5} - 4 \times 10^{-5}$ K^{-1}	Cellobond [4]	L

Thermal Conductivity:

No.	Value	Note
1	0.024–0.036 W/mK	
2	0.018 W/mK	90% closed cells
3	0.04 W/mK	0.28 Btu in ft^{-2} h^{-1} °F^{-1}, structural grade, BS874-1956 [1]

Deflection Temperature:

No.	Value	Note
1	90°C	1.8 MPa [3]

Transport Properties:

Polymer Solutions Concentrated: Viscosity 30 P (3000–7000 MPa s) (20°, foam panels and blocks) 150–200 P (15000–20000 MPa s) (20°, closed-cell foams)

Water Content: The amount of water absorbed is dependent upon the percentage of closed cells in the resin. Water absorption is generally low, and any residual water may be removed by drying [4]

Water Absorption:

No.	Value	Note
1	15 %	48h [3]

Mechanical Properties:

Mechanical Properties General: Mechanical props. are dependent upon density; tensile props. and strength increase with increasing density [4]

Tensile (Young's) Modulus:

No.	Value	Note
1	6.4–20.1 MPa	DIN 53423 [4]

Flexural Modulus:

No.	Value	Note
1	20 MPa	[3]

Tensile Strength Break:

No.	Value	Note
1	0.43 MPa	[3]
2	0.19–0.46 MPa	DIN 53571 [4]

Flexural Strength at Break:

No.	Value	Note
1	0.38–0.78 MPa	DIN 53423 [4]

Compressive Strength:

No.	Value	Note
1	0.172 MPa	Exeltherm [4]
2	0.175 MPa	Cellobond [4]
3	0.207 MPa	Celotex [4]
4	0.2–0.55 MPa	DIN 53421 [4]

Viscoelastic Behaviour: Has better creep resistance under load than polystyrene foam [4]

Fracture Mechanical Properties: Shear strength 0.1–0.5 MPa (lengthwise) [3]

Izod Notch:

No.	Value	Notch	Note
1	2 J/m	N	[3]

Electrical Properties:

Dielectric Permittivity (Constant):

No.	Value	Frequency	Note
1	2.8	1 kHz	[3]

Dielectric Strength:

No.	Value	Note
1	7 kV.mm^{-1}	[3]

Polymer Stability:

Thermal Stability General: Heating at 130° causes 1% shrinkage and weight loss [4]

Upper Use Temperature:

No.	Value	Note
1	<120°C	max., [3]
2	-195–130°C	[4]
3	200°C	short time [4]
4	-196–150°C	Exeltherm, Cellobond, Celotex [4]

Flammability: Flammability rating V0 (UL 94) [3]. Has very low smoke emission [3]. Flame spread index 20–25 (ASTM E84) [4]. Self-extinguishing, but slowly char and erode in sustained flame [1,2]. Smoke density 3–15 I M^{-1} (ASTM E84) [4]

Chemical Stability: Resistant to organic solvents [3]. Attacked by concentrated acids and bases

Applications/Commercial Products:

Processing & Manufacturing Routes: A high-solids (75%) resole is mixed with surfactants and a blowing agent, and the mixture stirred rapidly with an acidic curing agent (e.g. *p*-toluenesulfonic acid) then metered into a trough. Foams with a high (>90%) closed-cell content may be synth. by careful selection of the surfactant [1]

Applications: Applications due to chemical durability and 'difficult' flammability characterised by slow surface charring and very low smoke development. Closed-cell foams have good thermal insulation and are used in a variety of building applications; in roofboard, floor insulation, cork composites, cavity board etc. Open cell foams are used in floral displays (as Oasis) where ingress of water is desirable; as an absorbent in chemical industry. Applications may be limited by cost considerations. Other applications in energy-absorbent packaging; in dry cell batteries; in acoustic panels; in sandwich panels

Trade name	Details	Supplier
Cellobond	K	BP Chemicals
Celotex		Celotex Corp.
Ecophen		Recticel
Exeltherm Xtra		Koppers Co. Ltd.
Green stuff	absorbent	Green Stuff Absorbent Products Inc.
K1	roofboard laminates	Kooltherm
K2	roofboard laminates	Kooltherm
K3	floor insulation board	Kooltherm
K4	cork composite roofboard	Kooltherm
K5	roofboard laminates	Kooltherm
K7	sarking board	Kooltherm
K8	cavity board	Kooltherm
Liner board K	foil laminate	Kooltherm
Rütaphen PS	PS range of high solids resoles for PF foams	Bakelite

Bibliographic References

[1] Brydson, J.A., *Plast. Massy,* 6th edn., Butterworth Heinemann, 1995, (rev.)
[2] Mitchell, R.G.B. and Smith, D., *Plastics (London),* 1960, **24**, 224, (flammability)
[3] *The Materials Selector,* 2nd edn., (eds., N.A. Waterman and M.F. Ashby), Chapman & Hall, 1997, **3**, (props.)
[4] Knop, A. and Pilato, L.A., *Phenolic Resins,* Springer-Verlag, 1985, 219, (rev.)

Phenolic resin, Insulation wool binders P-25

Related Polymers: Phenol Formaldehyde Resins
Monomers: Phenol, Formaldehyde, Urea
Material class: Thermosetting resin
Polymer Type: phenolic
Molecular Weight: Generally a low MW resole
Additives: Additives include ammonia, aminosilane (for glass fibre binding), ammonium sulfate and antidusting agent
Miscellaneous: Urea and ammonia are included in the composition to scavenge free formaldehyde during the heat cure of the Resole soln. The amount of urea added also is influenced by economic factors as it is generally a cheap additive. Amino silane is added to improve the binding of the organic phenolic composition to the glass fibre surface. Ammonium sulfate is used as a latent catalyst producing acid during the heat curing step. An aq. emulsion of oil acts as an antidusting agent and is added to improve the handling feel of the cured insulation wool

Applications/Commercial Products:

Processing & Manufacturing Routes: Insulation wool binding Phenolic Resoles are made by treating phenol with a large molar excess of formaldehyde (F/P ratio >2) with base catalyst, is generally sodium, calcium or barium hydroxides. The Resole is mixed with urea and ammonium to scavenge formaldehyde, ammonia sulfate as a latent heat cure catalyst, aminosilane to promote bonding to the inorganic fibre surface and an oil emulsion as an antidusting agent. The composition is sprayed on at about 15% solids to bind insulation wool fibres as they are prod. in a centrifugal melt process. The mat of fibres is cured by conveying it through a curing oven at 200–300°. The resin soln. binds the fibres at crossover points to form a stable insulation product, which is compression packaged for transport

Applications: Insulation wool mats are made from fibreglass by a drum centrifugal blowing process and from rockwool and slagwool fibres by a centrifugal cascade drawing process. They are widely used for thermal and sound insulation in buildings. Insulation wool mats are widely used for domestic loft insulation. Slabs of insulation wool are used for a variety of insulation applications. The product is shaped for thermal insulation of pipework

Trade name	Details	Supplier
Fenorex	insulation wool resin	Neste
Rütaphen M	resins for mineral wool	Bakelite

Phenolic resin, Moulding powders, Cellulose filled P-26

Related Polymers: Phenol Aldehyde Resins, Phenolic Novolaks
Monomers: Phenol, Formaldehyde
Material class: Thermosetting resin
Polymer Type: phenolic
Additives: Moulding powders require the curing agent hexamethylene tetramine
General Information: Detergent resistant grades are available

Volumetric & Calorimetric Properties:

Density:

No.	Value	Note
1	d 1.36–1.38 g/cm^3	[2]
2	d 1.4 g/cm^3	[3]

Deflection Temperature:

No.	Value	Note
1	150–170°C	ISO 75 Method A [1]
2	160°C	1.8 MPa [2]
3	238°C	1.82 MPa [3]

Transport Properties:

Water Absorption:

No.	Value	Note
1	0.8–1.2 %	[2]
2	2.8 %	24h [3]

Mechanical Properties:

Flexural Modulus:

No.	Value	Note
1	6000–9000 MPa	ISO 178 [1]
2	4825 MPa	49200 kg cm^{-2} [3]

Tensile Strength Break:

No.	Value	Note
1	25–35 MPa	ISO R527 [1]
2	45 MPa	[2]
3	51.7 MPa	527 kg cm^{-2} [3]

Flexural Strength at Break:

No.	Value	Note
1	55–65 MPa	ISO 178 [1]
2	62 MPa	[2]

Flexural Strength Yield:

No.	Value	Note
1	86.2 MPa	879 kg cm^{-2} [3]

Compressive Strength:

No.	Value	Note
1	150–180 MPa	ISO 604 [1]
2	152–200 MPa	[2]
3	207 MPa	2110 kg cm^{-2} [3]

Izod Notch:

No.	Value	Notch	Note
1	210–250 J/m	N	[2]
2	32 J/m	Y	0.6 ft lb in^{-1}, room temp. [3]

Charpy:

No.	Value	Notch	Note
1	2.2–4	Y	ISO 178 [1]
2	1.5–5.5	N	ISO 178 [1]

Electrical Properties:

Electrical Properties General: Insulation resistance 10^8–10^{10} Ω (IEC 167) [1]. Exhibits tracking [2]. Tracking resistance 125 V (IEC 112) [1]

Dielectric Strength:

No.	Value	Note
1	4–7 kV.mm^{-1}	90° [1]
2	8–15 kV.mm^{-1}	[2]
3	14.6 kV.mm^{-1}	[3]

Optical Properties:

Volume Properties/Surface Properties: Transparency not transparent [2]

Polymer Stability:

Upper Use Temperature:

No.	Value	Note
1	100°C	[2]

Flammability: Flammability rating V1 (4 mm sample, UL 94) [1]

Applications/Commercial Products:

Processing & Manufacturing Routes: Phenolic Novolak powders with hexamethylenetetramine curing agent containing cellulose fibres are suitable for compression moulding

Mould Shrinkage (%):

No.	Value	Note
1	0.15–0.45%	
2	0.4%	[3]

Applications: Cellulose-filled compositions have medium impact resistance. They are used to make electrical and mechanical housings and bobbins. Electrical uses include solenoid covers, coil formers, contact supports, circuit breaker bases and housings

Trade name	Details	Supplier
Durez	discontinued material	OxyChem
Fiberite	FM series	ICI, UK
G2519		Vyncolite
G5020		Vyncolite
G8320		Vyncolite
RX525		Vyncolite

Bibliographic References

[1] Vincolite Engineering Thermosets Vyncolit NV, 1990, (technical datasheet)
[2] *The Materials Selector*, 2nd edn., (eds. N.A. Waterman and M.F. Ashby), Chapman and Hall, 1997, **3**, 401, (props)
[3] 1996, Resinoid Engineering Corp., (technical datasheet)

Phenolic resin, Moulding powders, Cotton filled P-27

Related Polymers: Phenol Aldehyde Resins, Phenolic Novolaks
Monomers: Phenol, Urea, Melamine, Formaldehyde
Material class: Thermosetting resin
Polymer Type: phenolic
Additives: Moulding powders require the curing agent hexamethylenetetramine
General Information: Cotton as a filler imparts high impact resistance, good mouldability but low resistance to shock [2]

Volumetric & Calorimetric Properties:

Density:

No.	Value	Note
1	d 1.37 g/cm^3	BS2782 509A [2]
2	d 1.37–1.4 g/cm^3	[3]

– Phenolic resin, Moulding powders, Cotton filled

Deflection Temperature:

No.	Value	Note
1	150°C	ISO 75 Method A [1]
2	120–165°C	1.8 MPa [3]

Transport Properties:
Water Content: Absorbs 30–50 mg of water (23°, 24h, BS2782 502F) [2]
Water Absorption:

No.	Value	Note
1	0.9–1 %	48h [3]

Mechanical Properties:
Tensile (Young's) Modulus:

No.	Value	Note
1	6000–10000 MPa	[3]

Flexural Modulus:

No.	Value	Note
1	6000–8500 MPa	ISO 178 [1]

Tensile Strength Break:

No.	Value	Note
1	25–35 MPa	ISO R527 [1]
2	48 MPa	BS2782 301A [2]
3	20–60 MPa	[3]

Flexural Strength at Break:

No.	Value	Note
1	56–65 MPa	ISO I78 [1]
2	62–69 MPa	[3]

Compressive Strength:

No.	Value	Note
1	150 MPa	ISO 604 [1]
2	138–172 MPa	[3]

Impact Strength: Impact strength 0.39 J (BS2782 305A) [2]
Izod Notch:

No.	Value	Notch	Note
1	40–430 J/m	N	[3]

Charpy:

No.	Value	Notch	Note
1	5.5–6	N	[1]
2	5.5–6	Y	[1]
3	1.2	Y	[1]
4	1.2	N	[1]

Electrical Properties:
Electrical Properties General: Insulation resistance 10^8–10^{10} Ω (IEC 167) [1]. Tracking resistance 125V (IEC 112) [1]
Dielectric Permittivity (Constant):

No.	Value	Frequency	Note
1	5.5–5.7	800 Hz	[2]
2	5.2–21	60 Hz	[3]

Dielectric Strength:

No.	Value	Note
1	4–7 kV.mm^{-1}	90° [1]
2	7.8–10.6 kV.mm^{-1}	20° [2]
3	2.9–6.8 kV.mm^{-1}	90°, BS2782 201A [2]
4	8.9–9.8 kV.mm^{-1}	25° [3]

Arc Resistance:

No.	Value	Note
1	28–49s	100° [3]

Dissipation (Power) Factor:

No.	Value	Frequency	Note
1	0.1–0.35	800 Hz	[2]

Optical Properties:
Volume Properties/Surface Properties: Transparency not transparent [3]

Polymer Stability:
Upper Use Temperature:

No.	Value	Note
1	100–121°C	[3]

Flammability: Flammability rating V1 (UL 94). [1] Does not burn [3]

Applications/Commercial Products:
Processing & Manufacturing Routes: Phenolic novolak powders cured with hexamethylenetetramine and containing cotton fibres or fabric are suitable for compression moulding
Mould Shrinkage (%):

No.	Value	Note
1	0.4–0.8%	
2	0.4–0.9%	[3]
3	0.5%	BS2782 106A [2]

Applications: Cotton-filled compositions have high impact resistance. Electrical uses include solenoid covers, coil formers, gear wheels and housings. They are used in the textile industry machinery for shock-resistant parts. Other uses in dynamo fans and in washing-machine agitators

Trade name	Details	Supplier
Fiberite	FM; discontinued material	ICI, UK
G2510		Vyncolite
G8020		Vyncolite

— **Phenolic resin, Moulding powders, Glass fibre filled**

G8120		Vyncolite
RX431		Vyncolite
RX74		Vyncolite
RX84		Vyncolite

Bibliographic References

[1] *Vincolite Engineering Thermosets,* Vyncolit, N.V., 1990, (technical datasheet)
[2] Brydson, J.A., *Plast. Mater.,* 6th edn., Butterworth-Heinemann, 1995, 634, (props)
[3] *The Materials Selector,* 2nd edn., (eds. N.A. Waterman and M.F. Ashby), Chapman and Hall, 1997, **3**, 401, (props)

Phenolic resin, Moulding powders, Glass fibre filled   P-28

Related Polymers: Phenolic Resins, Phenol-Formaldehyde Novolaks
Monomers: Phenol, Urea, Melamine, Formaldehyde
Material class: Thermosetting resin
Polymer Type: phenolic
Additives: Moulding powders require the curing agent hexamethylene tetramine
General Information: Compositions with short or long fibres as fillers may be moulded by compression, transfer or injection flow

Volumetric & Calorimetric Properties:

Density:

No.	Value	Note
1	d 1.7–2 g/cm^3	[2]

Thermal Expansion Coefficient:

No.	Value	Note	Type
1	$1.2 \times 10^{-5} – 1.5 \times 10^{-5}$ K^{-1}	min. [1]	L
2	$3 \times 10^{-5} – 4 \times 10^{-5}$ K^{-1}	max. [1]	L
3	$8 \times 10^{-6} – 2 \times 10^{-5}$ K^{-1}	[2]	L

Thermal Conductivity:

No.	Value	Note
1	0.32–0.88 W/mK	[2]

Deflection Temperature:

No.	Value	Note
1	165–190°C	short fibres, ISO 75 method A
2	>260°C	min., long fibres, ISO 75 method A [1]
3	150–315°C	1.8 MPa [2]

Transport Properties:

Water Absorption:

No.	Value	Note
1	0.03–1.2 %	[2]

Mechanical Properties:

Tensile (Young's) Modulus:

No.	Value	Note
1	20000–27000 MPa	short fibres, ISO R527 [1]
2	20000–23000 MPa	long fibres, ISO R527 [1]
3	13000–23000 MPa	[2]

Flexural Modulus:

No.	Value	Note
1	9–19 MPa	short fibres [1]
2	10–18 MPa	long fibres [1]
3	14000–23000 MPa	[2]

Tensile Strength Break:

No.	Value	Note
1	60–90 MPa	short fibres, ISO R527 [1]
2	45–50 MPa	long fibres, ISO R527 [1]
3	30–120 MPa	[2]

Flexural Strength at Break:

No.	Value	Note
1	70–110 MPa	short fibres, ISO 178 [1]
2	50–75 MPa	long fibres, ISO 178 [1]
3	62–75 MPa	[2]

Compressive Strength:

No.	Value	Note
1	150–300 MPa	short fibres, ISO 604 [1]
2	170–230 MPa	long fibres, ISO 604 [1]
3	260–280 MPa	[2]

Hardness: Ball indentation hardness 280–400 MPa (short fibres, ISO 179) [1]
Charpy:

No.	Value	Notch	Note
1	2.7–5	Y	short fibres, ISO 179 [1]
2	4–9	N	short fibres, ISO 179 [1]
3	4–6.5	Y	long fibres, ISO 179 [1]
4	4–6.5	N	long fibres, ISO 179 [1]

Electrical Properties:

Electrical Properties General: Insulation resistance 10^{11}–10^{12} Ω (IEC 167). Tracking resistance 125–175 V (IEC 112) [1]
Dielectric Permittivity (Constant):

No.	Value	Frequency	Note
1	5–7.1	60 Hz	[2]
2	5–6.9	1 kHz	[2]
3	4.5–6.6	1 MHz	[2]

– Phenolic resin, Moulding powders, Graphite filled

Dielectric Strength:

No.	Value	Note
1	8–12 kV.mm^{-1}	90°, IEC 243 [1]
2	5.6–16 kV.mm^{-1}	[2]

Arc Resistance:

No.	Value	Note
1	<180s	max. [2]

Dissipation (Power) Factor:

No.	Value	Frequency	Note
1	0.02–0.07	1 MHz	IEC 250 [1]

Optical Properties:
Volume Properties/Surface Properties: Not transparent
Polymer Stability:
Upper Use Temperature:

No.	Value	Note
1	175–290°C	[2]

Flammability: Limiting oxygen index 29–53% (short fibres), 57% (long fibres), ISO 4589) [1]. Flammability rating V0 (UL94) [2]
Stability Miscellaneous: Incandescence resistance no visible flame (ISO 181) [1]

Applications/Commercial Products:
Processing & Manufacturing Routes: Phenolic Novolak powders with hexamethylene tetramine curing agent containing glass fibres are suitable for compression, transfer or injection moulding. Compositions with short glass fibres as fillers may be either compression or injection moulded. Compositions with long glass fibre filler can be compression moulded and act as asbestos replacement materials. The compositions have good pourability and plasticity, and are suitable for moulding complicated parts, including hollow components. The compositions are suitable for many complicated precision parts with thin cross sections and narrow tolerances
Mould Shrinkage (%):

No.	Value	Note
1	0.15–0.45%	short fibres [2]
2	0.15–0.3%	long fibres

Applications: Glass-filled compositions are moulded to form components in electric motors, armature insulation, commutators, coil formers, insulators and solenoid covers. Long glass fibre filled compositions are used for high-voltage commutators. Automotive under-bonnet applications include water pump seals, impellers, thermostat housings, pulleys, cam shaft gear wheels, fuel rails, fuel inlet manifolds and starter motor brush holders. Further automotive applications include disc brake pistons, gaskets and automatic gearbox thrust washers, intake manifolds and rocker covers. Glass fibre filled compositions are replacements for asbestos in die-cast automotive metal parts e.g. reactors for automatic gear boxes. The composites are suitable as moulded complex precision parts for video recorders and compact disc players

Trade name	Details	Supplier
Fiberite	various grades	ICI Acrylics
Plenco	discontinued material	Plastics Engr. Co.

RX600	short glass fibre filled	Vyncolite
RX6000	short glass fibre filled	Vyncolite
RX800	long glass fibre filled	Vyncolite

Bibliographic References
[1] Vincolite Engineering Thermosets Vyncolit NV, 1990, (technical datasheet)
[2] *The Materials Selector*, 2nd edn., (eds. N.A. Waterman and M.F. Ashby), Chapman & Hall, 1997, **3**, 401

Phenolic resin, Moulding powders, Graphite filled   P-29

Related Polymers: Phenol Aldehyde Resins, Phenolic Novolaks
Monomers: Phenol, Formaldehyde
Material class: Thermosetting resin
Polymer Type: phenolic
Additives: Moulding powders require curing agent hexamethylenetetramine
General Information: Graphite filled phenolic moulding powders are self-lubricating, have a low coefficient of friction and excellent dimensional stability. They are not suitable for electrical applications

Volumetric & Calorimetric Properties:
Density:

No.	Value	Note
1	d 1.75 g/cm^3	[2]
2	d 1.45 g/cm^3	[3]

Thermal Expansion Coefficient:

No.	Value	Note	Type
1	1.5×10^{-5}–2.5×10^{-5} K^{-1}	[1]	L
2	5.4×10^{-6} K^{-1}	[3]	L

Thermal Conductivity:

No.	Value	Note
1	34.2 W/mK	4.93 Btu in h^{-1} ft^{-2} °F^{-1} [3]

Deflection Temperature:

No.	Value	Note
1	249°C	1.82 MPa [3]

Transport Properties:
Water Absorption:

No.	Value	Note
1	0.04 %	24h [2]
2	0.2 %	24h [3]

Mechanical Properties:
Flexural Modulus:

No.	Value	Note
1	11000–13000 MPa	ISO 178 [1]
2	30989 MPa	316000 kg cm^{-2} [3]

Tensile Strength Break:

No.	Value	Note
1	19.6 MPa	200 kg cm^{-2} [2]

Flexural Strength at Break:

No.	Value	Note
1	60 MPa	ISO 178 [1]

Flexural Strength Yield:

No.	Value	Note
1	37.8 MPa	386 kg cm^{-2} [2]
2	248 MPa	2530 kg cm^{-2} [3]

Compressive Strength:

No.	Value	Note
1	180 MPa	ISO 604 [1]
2	96.5 MPa	984 kg cm^{-2} [2]
3	220 MPa	2250 kg cm^{-2} [3]

Izod Notch:

No.	Value	Notch	Note
1	80 J/m	Y	1.5 ft lb in^{-1}, room temp. [2]
2	534 J/m	Y	10 ft lb in^{-1}, room temp. [3]

Charpy:

No.	Value	Notch	Note
1	1.6	Y	ISO 179 [1]
2	5	N	ISO 179 [1]

Electrical Properties:
Electrical Properties General: Volume resistivity 175 Ω cm [3]

Polymer Stability:
Upper Use Temperature:

No.	Value	Note
1	232°C	continuous service [2]

Applications/Commercial Products:
Processing & Manufacturing Routes: Phenolic Novolak powders with hexamethylene tetramine curing agent containing graphite can be compression or injection moulded

Mould Shrinkage (%):

No.	Value	Note
1	0.2–0.5%	
2	0.25%	[3]

Applications: Graphite-filled materials are used as parts for gas meters, water pump seals, sliding valves and self-lubricating bearings

Trade name	Details	Supplier
3501 CG	graphite filled	Vyncolite
3515 CG	graphite filled	Vyncolite
35260 CG	graphite filled	Vyncolite
Fiberite	various grades; discontinued material	ICI, UK

Bibliographic References
[1] Vincolite Engineering Thermosets Vyncolit NV, 1990, (technical datasheet)
[2] *Polychem,* Budd Company, 1996, (technical datasheet)
[3] *Fiberite,* ICI Fiberite, 1996, (technical datasheet)

Phenolic resin, Moulding powders, Mica filled P-30
Related Polymers: Phenol Aldehyde Resins, Phenolic Novolaks
Monomers: Phenol, Formaldehyde
Material class: Thermosetting resin
Polymer Type: phenolic
Additives: Moulding powders require the curing agent hexamethylenetetramine or an ammonia-free substance
General Information: Mica-filled phenolic moulding powders are heat-resistant, dimensionally stable and have good electrical props. including low dielectric loss. Electrical props. are retained under conditions of high temp. and humidity

Volumetric & Calorimetric Properties:
Density:

No.	Value	Note
1	d 1.7–1.9 g/cm^3	[2]
2	d 1.75 g/cm^3	[3]

Deflection Temperature:

No.	Value	Note
1	160°C	ISO 75, Method A [1]
2	150–180°C	1.8 MPa [2]

Transport Properties:
Water Absorption:

No.	Value	Note
1	0.05 %	48h [2]
2	0.05 %	24h [3]

Mechanical Properties:
Tensile (Young's) Modulus:

No.	Value	Note
1	17000–34000 MPa	[2]
2	17162 MPa	175000 kg cm^{-2} [3]

Flexural Modulus:

No.	Value	Note
1	11000–13000 MPa	ISO 178 [1]
2	17162 MPa	175000 kg cm^{-2} [3]

– Phenolic resin, Moulding powders, Mineral filled

Tensile Strength Break:

No.	Value	Note
1	40 MPa	ISO R527 [1]
2	38–50 MPa	[2]
3	44.7 MPa	456 kg cm^{-2} [3]

Flexural Strength at Break:

No.	Value	Note
1	63 MPa	[2]

Flexural Strength Yield:

No.	Value	Note
1	75.8 MPa	773 kg cm^{-2} [3]

Compressive Strength:

No.	Value	Note
1	200 MPa	ISO 604 [1]
2	175 MPa	[2]
3	199 MPa	2030 kg cm^{-2} [3]

Hardness: Rockwell E88 [3]
Izod Notch:

No.	Value	Notch	Note
1	14–20 J/m	N	[2]
2	16 J/m	Y	0.3 ft lb in^{-1}, room temp. [3]

Charpy:

No.	Value	Notch	Note
1	1.8–2	N	ISO 179 [1]
2	6	N	ISO 179 [1]

Electrical Properties:
Electrical Properties General: Mica-filled products have good electrical insulation props. Insulation resistance 10^{11}–10^{13} Ω (IEC 107) [1]. Tracking resistance 125 V (IEC 112) [1]
Dielectric Permittivity (Constant):

No.	Value	Frequency	Note
1	4.7–6	60 Hz	[2]
2	4.6	1 MHz	[3]

Dielectric Strength:

No.	Value	Note
1	7–14 kV.mm^{-1}	90°, IEC 243 [1]
2	14–16 kV.mm^{-1}	[2]
3	15.7 kV.mm^{-1}	[3]

Dissipation (Power) Factor:

No.	Value	Frequency	Note
1	0.013	1 MHz	[3]

Optical Properties:
Volume Properties/Surface Properties: Transparency not transparent [2]

Polymer Stability:
Upper Use Temperature:

No.	Value	Note
1	120–150°C	[2]
2	176°C	continuous service [3]

Flammability: Flammability rating V1 (4 mm sample, UL 94), HB (1.6 mm sample), SE-0 [2]
Stability Miscellaneous: Incandescence resistance no visible flame (ISO 181) [1]

Applications/Commercial Products:
Processing & Manufacturing Routes: Phenolic Novolak powders with hexamethylene tetramine curing agent containing mica can be compression moulded
Mould Shrinkage (%):

No.	Value	Note
1	0.1–0.2%	
2	0.2–0.6%	[2]

Applications: Mica-filled phenolic moulding powders have exceptional electrical insulation props. They are made into condenser housings, resistor encapsulations, lamp sockets, slip rings, commutators, computer plug boards and volume controls. An ammonia-free version is available which uses a curing agent other than hexamethylenetetramine

Trade name	Details	Supplier
2020 M	mica filled	Vyncolite
3310 M	mica filled	Vyncolite

Bibliographic References
[1] Vincolite Engineering Thermosets Vyncolit NV, 1990, (technical datasheet)
[2] *The Materials Selector*, 2nd edn., (eds. N.A. Waterman and M.F. Ashby), Chapman and Hall, 1997, **3**, 402, (props)
[3] *Polychem*, Budd Company, 1996, (technical datasheet)

Phenolic resin, Moulding powders, Mineral filled P-31

Related Polymers: Phenol Aldehyde Resins, Phenolic Novolaks
Monomers: Phenol, Formaldehyde
Material class: Thermosetting resin
Polymer Type: phenolic
Additives: Moulding powders require curing agents such as hexamethylenetetramine

Volumetric & Calorimetric Properties:
Density:

No.	Value	Note
1	d 1.58 g/cm^3	[2]

– Phenolic resin, Moulding powders, Mineral filled

Thermal Expansion Coefficient:

No.	Value	Note	Type
1	$2 \times 10^{-5} - 3 \times 10^{-5}$ K^{-1}	mineral filled [1]	L
2	$3.5 \times 10^{-5} - 4.5 \times 10^{-5}$ K^{-1}	mineral and organic filled [1]	L
3	$5 \times 10^{-5} - 6 \times 10^{-5}$ K^{-1}	standard grade [1]	L

Deflection Temperature:

No.	Value	Note
1	160°C	1.82 MPa [2]

Transport Properties:
Water Absorption:

No.	Value	Note
1	0.2 %	24h [2]

Mechanical Properties:
Tensile (Young's) Modulus:

No.	Value	Note
1	9000–10000 MPa	ISO 178 [1]
2	14000–16000 MPa	max., ISO 178 [1]

Flexural Modulus:

No.	Value	Note
1	7000–8000 MPa	ISO 178 [1]
2	13000–15000 MPa	ISO 178
3	10297 MPa	105000 kg cm^{-2} [2]

Tensile Strength Break:

No.	Value	Note	Elongation
1		ISO R527 [1]	0.6–0.9%
2	41 MPa	422 kg cm^{-2} [2]	

Flexural Strength at Break:

No.	Value	Note
1	70–85 MPa	most grades, ISO 178 [1]
2	100 MPa	ISO 178

Tensile Strength Yield:

No.	Value	Note
1	1×10^8 MPa	10^{18} N cm^{-2}

Flexural Strength Yield:

No.	Value	Note
1	62 MPa	633 kg cm^{-2} [2]

Compressive Strength:

No.	Value	Note
1	200–240 MPa	ISO 604 [1]
2	166 MPa	1690 kg cm^{-2} [2]

Hardness: Ball indentation method 400–440 MPa (ISO 2039) [1]
Izod Notch:

No.	Value	Notch	Note
1	16 J/m	Y	0.3 ft lb in^{-1}, room temp. [2]

Charpy:

No.	Value	Notch	Note
1	2.1–4	Y	ISO 179 [1]
2	5–9	N	ISO 179 [1]

Electrical Properties:
Electrical Properties General: Insulation resistance $10^8 - 10^{10}$ Ω (mineral and organic filled, IEC 107) [1], $10^{10} - 10^{11}$ Ω (mineral filled, IEC 107) [1]. Tracking resistance 175 V (IEC 112) [1]
Dielectric Strength:

No.	Value	Note
1	4–6 kV.mm^{-1}	mineral and organic filled, IEC 243 [1]
2	7–10 kV.mm^{-1}	mineral filled [1]
3	13.7 kV.mm^{-1}	[2]

Dissipation (Power) Factor:

No.	Value	Frequency	Note
1	0.062	1 MHz	[2]
2	0.03–0.09	1 MHz	[1]

Polymer Stability:
Upper Use Temperature:

No.	Value	Note
1	204°C	continuous service [2]

Stability Miscellaneous: Exposure of graphite-filled material to γ-irradiation (below 10^8 Gy) or to neutrons increases tensile and shear strengths, and flexural modulus

Applications/Commercial Products:
Processing & Manufacturing Routes: Phenolic Novolak powders with hexamethylenetetramine curing agent and mineral powders can be compression or injection moulded. Compositions with both mineral and organic fillers are designed for compression moulding
Mould Shrinkage (%):

No.	Value
1	0.2–0.55%
2	0.4–0.8%

Applications: Used to replace asbestos in fuseholders, brush holders, circuit breakers, thermostat components, automotive ignition and solenoid covers. Other automotive uses include carburettor, fuel pump, oil pump and intake manifold gaskets.

— Phenolic resin, Moulding powders, Woodflour filled

Mineral/organic filled compositions also serve as lamp sockets and lamp holders. Mineral filled compositions are used in household appliances, such as iron heat shields, pan handles etc.

Trade name	Details	Supplier
2023 W	mineral filled	Vyncolite
2930 W	mineral filled	Vyncolite
2932 W	mineral filled	Vyncolite
4623 XB	mineral and organic filled	Vyncolite
Durez	various grades; discontinued material	OxyChem

Bibliographic References

[1] Vincolite Engineering Thermosets Vyncolit NV, 1990, (technical datasheet)
[2] *Plenco*, Plastics Engineering Co., 1996, (technical datasheet)

Phenolic resin, Moulding powders, Woodflour filled   P-32

Related Polymers: Phenolic resin, Moulding powders, Glass fibre filled, Phenolic Resins, Phenol formaldehyde Novolaks, Urea Formaldehyde Resins, Melamine Formaldehyde Resins
Monomers: Phenol, Urea, Melamine, Formaldehyde
Material class: Thermosetting resin
Polymer Type: phenolic
CAS Number:

CAS Reg. No.	Note
9003-35-4	PF resins

Additives: Moulding powders require curing agent hexamethylene tetramine
General Information: Compositions with woodflour as fillers are used as general purpose products by compression or injection moulding. Woodflour reduces exotherm and shrinkage and improves impact strength

Volumetric & Calorimetric Properties:
Density:

No.	Value	Note
1	d 1.3–1.5 g/cm^3	[2]
2	d 1.35 g/cm^3	BS2782 509 A [1]
3	d 1.38 g/cm^3	[3]

Thermal Expansion Coefficient:

No.	Value	Note	Type
1	3×10^{-5}–4.5×10^{-5} K^{-1}	[2]	L
2	5.05×10^{-5} K^{-1}	[3]	L

Thermal Conductivity:

No.	Value	Note
1	0.16–0.3 W/mK	[2]
2	0.187 W/mK	0.027 Btu in ft^{-2} h^{-1}°F^{-1} [3]

Deflection Temperature:

No.	Value	Note
1	130–190°C	1.8 MPa [2]

Transport Properties:
Water Absorption:

No.	Value	Note
1	1 %	48h [2]
2	0.6 %	24h [3]

Mechanical Properties:
Tensile (Young's) Modulus:

No.	Value	Note
1	5000–12000 MPa	[2]
2	9650 MPa	98400 kg cm^{-2} [3]

Flexural Modulus:

No.	Value	Note
1	7000–9000 MPa	[2]
2	7580 MPa	77300 kg cm^{-2} [3]

Tensile Strength Break:

No.	Value	Note
1	51.7 MPa	527 kg cm^{-2} [3]
2	30–60 MPa	[2]
3	55 MPa	BS2782 301A [1]

Flexural Strength at Break:

No.	Value	Note
1	62 MPa	[2]

Flexural Strength Yield:

No.	Value	Note
1	65 MPa	667 kg cm^{-2} [3]

Compressive Strength:

No.	Value	Note
1	165 MPa	[2]
2	185 MPa	1890 kg cm^{-2} [3]

Impact Strength: Impact strength 0.22 J (BS2782 305A) [1]
Hardness: Rockwell M117 [3]
Izod Notch:

No.	Value	Notch	Note
1	10–30 J/m	Y	[2]
2	13.3 J/m	Y	0.25 ft lb in^{-1}, room temp. [3]

Electrical Properties:
Electrical Properties General: Exhibits tracking [2]

Dielectric Permittivity (Constant):

No.	Value	Frequency	Note
1	6–10	800 Hz	[1]
2	4.5–5.5	1 MHz	BS2782 207A [1]
3	4.5	1 MHz	[3]
4	5–13	60 Hz	[2]
5	4–9	1 kHz	[2]
6	4–6	1 MHz	[2]

Dielectric Strength:

No.	Value	Note
1	5.8–11.6 kV.mm^{-1}	20° [1]
2	3.9–9.7 kV.mm^{-1}	90°, BS2782 201A [1]
3	13.3 kV.mm^{-1}	[3]
4	10–16 kV.mm^{-1}	[2]

Dissipation (Power) Factor:

No.	Value	Frequency	Note
1	0.1–0.4	800 Hz	[1]
2	0.03–0.05	1 MHz	BS2782 207A [1]
3	0.033	1 MHz	[3]

Optical Properties:
Volume Properties/Surface Properties: Not transparent [2]

Polymer Stability:
Upper Use Temperature:

No.	Value	Note
1	100–150°C	[2]
2	137°C	continuous service [3]

Flammability: Shows self extinguishing props.

Applications/Commercial Products:
Processing & Manufacturing Routes: General purpose phenolic moulding powders contain phenolic novolak with woodflour filler. They may be compression moulded and injection moulded

Mould Shrinkage (%):

No.	Value	Note
1	0.5–0.8%	
2	0.7–0.9%	
3	0.6%	BS2782 106A [1]
4	0.4–0.9%	[2]

Applications: General purpose applications include screw caps, handles, switch bases, gearwheels and microswitches. May be finely powdered for high gloss finish applications; controlled particle size materials are available for use in accurate automatic machines

Trade name	Details	Supplier
3103-S	injection grade	Vyncolite
3104	compression grade	Vyncolite
3112	compression grade	Vyncolite
3220	compression grade	Vyncolite
Polychem	100, discontinued material	Budd Company

Bibliographic References
[1] Brydson, J.A., *Plast. Mater.*, 6th edn., Butterworth Heinmann, 1995, 634, (props.)
[2] *The Materials Selector*, 2nd edn., (eds N.A. Waterman and M.F.Ashby), Chapman & Hall, 1997, 395, (uses, props.)
[3] Polychem 100, Budd Company, 1996, (technical datasheet)

Phenolic resins

Synonyms: *Phenol-formaldehyde resins. PF resins*
Related Polymers: Cresol-Formaldehyde Resins, Xylenol-Formaldehyde Resins, Resorcinol-Formaldehyde Resins, Cardanol Resins, Phenol-formaldehyde resoles, Phenol-formaldehyde novolaks
Material class: Thermosetting resin
Polymer Type: phenolic
General Information: Phenol has a functionality of 3 by electrophilic substitution in the activated 2, 4 and 6 ring positions. The substituted phenols have a functionality of: 1 when both the 2- and 4- ring positions substituted, 2 when one of the 2- or 4- ring positions substituted otherwise 3. Formaldehyde has a functionality of 2. The resins are generally classified into two broad categories depending on the substrate mole ratios of phenol and formaldehyde, and the nature of the catalyst, whether basic or acidic. The categories are **Resoles** and **Novolaks**.

Resole Resins
When phenol is reacted with excess formaldehyde (F:P ratios 1 to 3) with basic catalyst under controlled conditions to avoid runaway exothermic reactions, the resulting mixture is a **Resole** or A-stage resin. Resoles are inherently thermosetting one-step resins fusible at 150° and soluble in alcohols and ketones.
Basic catalysts used to make resoles include **sodium hydroxide, calcium hydroxide, barium hydroxide and amines**. Condensation of two methylol phenols forms methylene ethers. These species react at 160° through a quinone-methide intermed. to form diphenol methane derivatives with methylene links between phenolic rings. Subsequent reactions form fully crosslinked species with methylene groups between activated ring positions, also known a C-stage resin. This is an infusible and insoluble product. MW of resoles are generally low with one to two phenol rings in the resin species. Resoles are manufactured in batch processes in resin kettles with efficient cooling to eliminate possible exotherms. When amine catalysed, the intermediate products contain nitrogen. On advancement and cure some ammonia is lost and most of the remaining nitrogen content is as dibenzylamine derivatives. Structural studies indicate a preponderance of *ortho* substitution in amine catalysed resoles. The resoles may be used as solns., liqs. and solids. Phenolic resin dispersions are made using a protective colloid in water using an amine catalyst. Colloids include proteinaceous compds., fatty acid amides, poly(vinyl alcohol), cellulosic derivatives or natural gums. In the dispersions much higher MW are achieved, up to as high as 5000. Solid one-step reactive resins are made by recovering and drying the spherical particles from the dispersion process.

Cure of Resoles
During cure of a low MW resole ring-bridging reactions take place building a 3 dimensional network. Resoles are generally heat cured at pH >9 at elevated temps. not exceeding 180°C. In foam and foundry applications acid conditions are used around room temp. to form -CH$_2$- bridges.

Novolak Resins
When phenol reacts with formaldehyde with acid catalyst and in excess the resultant resin product is a Novolak. To achieve increased rate of reaction, formaldehyde is used in 37–50% aq. soln. or as solid paraformaldehyde. The reaction proceeds with the formation of methylene bridges between phenolic rings to give MW <2500. Acid catalysts used in the manufacture of novolaks include sulfuric acid, 4-toluenesulfonic acid, hydrochloric acid,

phosphoric acid and oxalic acid. Oxalic acid has the advantage that it breaks down to volatile gases (CO, CO_2 and H_2O) when the resin is heated during processing. As with Resole resins methylene bridges are formed between the activated ortho- and para- ring positions. A range of highly reactive High-ortho Novolaks may be prepared which contain a high proportion of *ortho*-hydroxymethyl groups. These are made using divalent metallic acetates as catalyst, generally zinc, magnesium or calcium. The metal ion binds to the phenolic OH group directing addition of the formaldehyde to the nearby *ortho*-ring position. Precursors with from 3 to 12 phenolic rings can be made.

Cure of Novolaks
Because they are prepared with a molar deficiency of formaldehyde Novolaks require the addition of a curing agent capable of condensation reactions at the remaining activated phenolic ring positions. Novolaks are generally cured by addition of 8–15% hexamethylenetetramine (hexa). Structural studies indicate that 75% of the nitrogen is chemically bonded mostly as benzylamine derivatives. In some applications phenolic Resoles are used to cure Novolaks. This avoids the incorporation of nitrogen into the product. Other aldehydes may be used, but formaldehyde is by far the most common [12]. Phenolic resins may be prepared and used as dispersions, which can lead to higher MW in some cases.

Transport Properties:
Polymer Solutions Dilute: Viscosity 10 mPa s (room temp., fibre binders), 2000–3000 mPa s (20°, abrasive binders)

Applications/Commercial Products:
Processing & Manufacturing Routes: Resoles for fibre binding are typically made with a large molar excess of formaldehyde (2.5–3 times) which minimises the hazard of free phenol during application and scavenged with urea and ammonia. They may be catalysed with barium hydroxide so that alkali residues can be removed by precipitation to avoid damage to glass fibre during the application. Fleece filled materials are prod. from organic fibres and powdered phenolic resin and curing agent, cured at 160° Resole solns. for plywood bonding application have typically 40–50% solids. They react at F/P ratio 1.2–1.4 with sodium hydroxide catalyst for 3h at 100° then 3h at 65° Solid resoles for moulding materials are made using the stoichiometric F/P mole ratio of 1.5. They are typically cured at 60–70° followed by vaccum reflux at 80–100°. High-solids resoles are made by the reaction of phenol and formaldehyde generally sodium hydroxide catalysed followed by vacuum reflux to evaporate water with controlled/minimal advancement of the resin. Phenolic dispersions are prepared by treating phenol and formaldehyde typically F/P mole ratio 1.5 with hexamethylene tetramine (6–12%), then adding poly(vinyl alcohol) and gum arabic as protective colloids

Applications: Used in plywood, particle board, fibre board, wafer board, oriented strand board and medium-density fibreboard. Phenol-formaldehyde and Resorcinol-formaldehyde water-soluble resoles are the resins of choice when durability in high-humidity environments is required. Otherwise the related classes of Urea-formaldehyde, Melamine-formaldehyde and Melamine-urea-formaldehyde can be used. Used in Fibre Bonding. Glass fibre and rockwool insulation mats used for thermal insulation material are prod by spraying the fibre mat with the resin (Resoles in aq. soln. typically 10–15% solids) and then curing at 200°. fleece filled resins are used as acoustic noise reducing agents in household appliances and cars for bonnet insulation and support Foundry Resins are traditionally used as a binding agent for the consolidation of foundry sand moulds in the shell moulding process and hot-box process. Both phenolic one-step resoles or Novolak/hexamine systems are used. Also used are urea-formaldehyde, furfuraldehyde resins and urethane/phenolic blends. Laminating resins are used for paper, cotton and glass. Industrial uses include electronic circuit boards. Decorative uses for furniture and wall panelling. Abrasive grain material (corundum or silicon carbide), fillers (metal oxides, silicates, chalk) are bonded using low-viscosity resoles as wetting agents for the abrasive grain with the bonding agents powdered high MW novolaks with low free-phenol content and 9% hexamine as curing agent. These are used in grinding wheels, sand papers and pan scourers Brake linings, clutch facings and transmission belts for automotive use are prod from woven fabrics dip impregnated with the resin. Moulding materials are prod from highly-consolidated resin composites with particulate or fibrous fillers and are used as finished parts with a high crosslink density. Bakelite was an early plastic for electrical sockets, switchgears and circuit breakers. Novolaks cured with oxalic acid have also been used. Engineering Thermosets are phenolic resins with reinforcing fibres. Glass-fibre filled materials are used to manufacture many automotive and electrical parts. Other fillers are woodflour, mineral, cellulose, cotton and graphite. Phenolic resins find application as coatings in several systems, as a powder coating for high-temp. wire insulation; for battery separators; and as water dilutable coatings. Phenolics are used in combination with thermoplastic polymers in structural adhesives. The phenolic moiety provides metal/metal and metal/rubber adhesion. Polyvinyl acetate/phenolic resin blend finds uses in the aircraft industry, gas copper adhesives in printed circuit boards, to bind brake linings to brake shoes and in ski manufacture. The nitrile/phenolic and chloroprene/phenolic blends are used to generate bonds with high peel-strength and moisture resistance. Glass-reinforced Phenolics are made from high-solids resins to produce a product with very low smoke emission, suitable for underground railway locomotives. Novolak filaments are post-crosslinked to reduce brittleness. High-solids resoles are made into foams using surface-active agents, blowing agent and an acid catalyst. Phenolic foams have excellent flame-resistance and smoke emission props. in contrast to other foam systems (notably polyurethanes). However these props may be compromised by the phasing out of CFCs as blowing agents and their replacement by flammable alkanes. In the past, phenolic foams suffered by their friability and poor water resistance due to an open cell struct. Use of tailored surfactants has improved these props. Carbon fibre reinforced phenolics have temp. stability up to 230°. Phenolic resins have formed part of the cladding of spacecraft in aerospace application. The material chars on exposure to high temps. and acts as a sacrificial coating during re-entry into the earth's atmosphere. Pheno-formaldehyde resins act as vulcanising agents for rubbers. Particles of size >1000 microns are used in friction element composition. Aq. dispersions used in fibre bonding, paper coating; in laminates and wood bonding. Phenolic dispersions improve the strength of latex contact adhesive applications

Trade name	Details	Supplier
Aerophen	Phenolic resins	Dyno
Borden P	friction applications resins	Borden
Dynosol	Phenolic resins	Dyno
Exter	Phenolic resins	Neste
Fenorex	Phenolic resins	Neste
Pribex	Phenolic resins	Neste
Rütaphen	Phenolic resins	Neste
Rütaphen RL	friction lining aq. resins	Bakelite
Rütaphen RP	friction lining powder resins	Bakelite
Rütaphen RW	friction lining resin solns.	Bakelite
Rütaphen SP	grinding wheel aq. resins	Bakelite
Rütaphen SW	grinding wheel powder resins	Bakelite
Vyncolite	engineering thermosets	Vynckier (Perstorp)

Bibliographic References

[1] Martin, R.W., *Chemistry of Phenolic Resins*, John Wiley and Sons, 1956
[2] Megson, N.J.L., *Phenolic Resins*, Butterworth, 1958
[3] Kopf, P.W., *Kirk-Othmer Encycl. Chem. Technol.*, 4, 18, 603-644
[4] Rolf, W.J. and Scott, J.R., *Fibres, Films, Plastics and Rubbers*, Butterworths, 1971
[5] Prime, R.B., *Therm. Charact. Polym. Mater.*, (ed. E. A. Turi), Academic, 1981, 435
[6] Sandler, S.R. and Karo, W., *Polymer Syntheses*, Vol. 1, Academic Press, New York, 1977
[7] Walker, J.F., *Formaldehyde*, Van Nostrand Reinhold, New York, 1974
[8] Carswell, T.S., *Frontiers in Chemistry*, Interscience, 1947
[9] Knop, A. and Pilato, C., *Carbohydr. Res.*, Springer-Verlag, 1985
[10] Knop, A. and Scheilb, W., *Chemistry and Applications of Phenolic Resins*, Springer-Verlag, 1979
[11] *SPI*, Washington DC, 1994
[12] Brydson, J.A., *Plast. Mater.*, Ch 23, 5th edn., 1989
[13] Whitehouse, A.A.K. and Pritchett, R.G.K., *Carbohydr. Res.*, 2nd edn., 1955

Phenolic resins, Abrasive materials   P-34

Related Polymers: Phenol Aldehyde Resins, Phenol Formaldehyde Resoles, Phenol Formaldehyde Novolaks, Furfuryl Resin
Monomers: Phenol, Furfuryl alcohol, Formaldehyde
Material class: Thermosetting resin
Polymer Type: phenolic
Additives: Additives include corundum, silicon carbide, zirconium corundum, diamond and boron carbide. Fillers are most commonly cryolite, pyrite and antimony sulfide
General Information: These materials comprise bonded or coated abrasive grain wetted with liq. resin and compounded with powdered resin and fillers

Applications/Commercial Products:

Processing & Manufacturing Routes: Bonded or coated alumina or silicon carbide abrasive grain (60–95%) is wetted with liq. phenolic resole or furfuryl resin and compounded with powdered Novolak resin and fillers (0–20%) prior to pressing. Grinding Wheels are made by cold or hot pressing methods. Diamond and some Synthetic Corundum grinding wheels are made by hot pressing. The moistened grains are blended with powdered resin and special fillers and pressed at 15–35MPa at 150–170° for 30–45 min. The cold pressing at 15–25MPa produces "green wheels" which are stripped and cured in ovens with a temp. cycle up to 180° for up to 60h. The composition can also include phenolic resin impregnated glass filament mesh. Coated abrasives are made using binder resins, e.g. animal glues or phenolic resoles with fillers to provide thixotropy, enhance grinding rate and reduce costs. After cure, a second layer of binder (the finishing or sizer coat) is applied to embed the grains. For the binder resin a phenolic resole with viscosity 15000 MPa.s and 75% solids is preferred. For the sizer coat a resole with viscosity 1000 MPa.s and 50% solids is used. Abrasive grains may be blended with phenolic resin with latex as a binder for abrasive mats
Applications: Resin bonded grinding wheels; cold-pressed grinding wheels used for wet grinding; abrasive papers used for dry sanding; vulcan fibre discs and abrasive cloths, and abrasive mats such as those used for pan scourers

Trade name	Details	Supplier
Rütaphen SP	grinding wheel aq. resins	Bakelite
Rütaphen SW	grinding wheel powder resins	Bakelite

Phenolic resins, Carbon materials   P-35

Synonyms: *Glassy carbon. Carbonaceous materials. Carbonaceous char*
Related Polymers: Phenolic Aldehyde Resins
Monomers: Phenol, Formaldehyde
Polymer Type: phenolic
General Information: Carbonaceous char formed by pyrrolysis of cured crosslinked phenolic network. On pyrrolysis at high temps. plastics decompose forming volatile components with varying amounts of solid carbon residue. Phenolic resins have the highest yield of carbon based residue up to 60% of the original weight. The carbon formed retains its bonding property with fillers and the material has a high-temp. resistance in excess of 1000°

Applications/Commercial Products:

Processing & Manufacturing Routes: Pyrolysis of cured phenolic resin in the absence of air can lead to a carbon material having excellent mechanical props. and bonding with fillers. Synthetic carbons are made from coke or graphite powder with phenolic resin as the binding agent, by gradually heating to over 1000° in an inert atmosphere. Carbon fibre reinforcement leads to marked improvement of mechanical strength. Glassy carbon is prod. from phenolic resin by special heat treatment at up to 2000° with air excluded. It is an impermeable, inert, high-strength material which conducts electricity
Applications: Synthetic carbons from graphite are used as special electrodes. Carbon materials form good inert impermeable lining materials for chemical apparatus and nuclear reactors. With carbon-fibre reinforcement they may be used as structural materials for large-scale apparatus in the chemical industry e.g. for heat exchangers. Glassy carbon has found application in several high-technology areas: as compressor wheels for turbo blowers, melting crucibles, vacuum vaporisation boats for metal deposition coatings, spinning nozzles, precision bearings and spot-welding electrodes. Glassy carbon is also used in medical implants: electrodes for heart pacemakers, artificial cardiac valves and bile ducts. Phenolic resins are included in the composition of heat shield laminates for spacecraft. They form a sacrificial ablative coating able to withstand the temp. shock of 14000° during re-entry in the earth's atmosphere

Trade name	Details	Supplier
Rütaphen DP	carbon materials powder resin	Bakelite
Rütaphen DW	carbon materials aq. soln.	Bakelite

Phenolic resins, Cast resin   P-36

Monomers: Phenol, Formaldehyde
Material class: Thermosetting resin
Polymer Type: phenolic
Additives: Graphite to improve friction resistance
General Information: Largely superseded by cheaper, modern thermoplastics

Mechanical Properties:

Mechanical Properties General: Strength of cured resin is dependent upon the nature of the neutralising acid used in the synth. Mineral acids produce weak, brittle resins, whereas lactic acid, phthalic acid etc. produce resins with high strength characteristics. [5]
Compressive Strength:

No.	Value	Note
1	83 MPa	12 kpsi [5]
2	186 MPa	27 kpsi, case-hardened [5]

Polymer Stability:

Chemical Stability: Good chemical resistance; very good resistance to oils and petroleum products

Applications/Commercial Products:

Processing & Manufacturing Routes: Phenolic resole, synth. by reaction of phenol and formaldehyde plus alkali metal hydroxide catalyst [1,2] (1:2.25 mole ratio) for 3h at 70°, in soln. is neutralised then vacuum distilled to remove all H_2O. Product clarity may be improved by addition of a polyol (e.g. glycerol or glycol) [3]. The colour and light transmissibility of the cured resin is dependent on nature of the acid used in neutralisation step [4]. At 55° the warm resin is poured into moulds and cured very slowly (3–10 days) at increasing temp. up to 138° and then slowly cooled taking at least 3h, trapping small droplets of H_2O. Clear castings are produced if the droplets are smaller than the wavelength of visible light. [5]. Two-component resin systems can be used to make unreinforced products using mixing machinery. The system is acid cured. Cast resin is readily machinable by cutting, sawing, slicing etc. Optimum appearance is attained using finishing operations. Mould shrinkage may be reduced by case-hardening of the resin after removal from the mould [5]

Mould Shrinkage (%):

No.	Value	Note
1	0.4–0.6%	[5]

Applications: Applications due to attractive appearance in umbrella handles, knobs, propelling pencil bodies, billiard balls

Trade name	Details	Supplier
Borden LV530M	laminating resin	Borden

Bibliographic References

[1] Brydson, J.A., *Plast. Mater.*, 6th edn., Butterworth-Heinemann, 1995, (rev)
[2] Sandler, S.R. and Karo, W., *Polymer Syntheses*, 2nd edn., Academic Press Inc., 1994, **2**, 65, (synth)
[3] Nelson, J., *Macromol. Synth.*, 1968, **3**, 1, (synth)
[4] Harris, T.G. and Neville, H.A., *J. Polym. Sci.*, 1953, **10**, 19, (synth)
[5] Apley, M., Plast. Inst., *Trans.*, 1952, **20**, 7, (rev)

Phenolic resins, Fibres   P-37

Synonyms: *Novoloid fibres*
Related Polymers: Phenolic Novolaks
Monomers: Phenol, Formaldehyde
Material class: Thermosetting resin, Fibres and Films
Polymer Type: phenolic
Miscellaneous: Light gold in colour, darkening on exposure to heat or light [2]

Volumetric & Calorimetric Properties:
Density:

No.	Value	Note
1	d 1.27 g/cm^3	[4]
2	d 1.25 g/cm^3	[7]

Transport Properties:
Water Content: Moisture regain 6% (20°, 65% relative humidity) [4]

Mechanical Properties:
Mechanical Properties General: Tenacity 0.11–0.15 N tex^{-1} (30–60% elongation) [1]. Elastic recovery (3% elongation) 92–96% [4]. Tensile strength of Kynol decreases with increasing fibre diameter [7]

Tensile (Young's) Modulus:

No.	Value	Note
1	3300–4400 MPa	[4]
2	5900 MPa	[7]

Tensile Strength Break:

No.	Value	Note	Elongation
1	150–200 MPa	[4]	30–60%
2	200 MPa	[7]	30%

Viscoelastic Behaviour: Beyond an elongation of 9% stress required for elongation decreases markedly [7]

Optical Properties:
Volume Properties/Surface Properties: Dyeability is limited by inherent colour [4]

Polymer Stability:
Thermal Stability General: Has good thermal stability at 150° in air and at 200–250° in the absence of air [4]
Flammability: Does not melt when exposed to flame but gradually chars. Has low smoke emission. Limiting oxygen index 30–34% [1,4], 38% (Kynol) [2], 32.6% (Kynol fabric) [6], 33% (Kynol) [7]. Vertical char length 0.1 cm [2], Smoke density 0.002 I m^{-1} (3 min. ignition), 0.2 I m^{-1} (15 min. ignition) [2]
Environmental Stress: Kynol has moderate stability to light and weathering [4]. Exposure to uv radiation causes degradation [5] resulting in loss of strength but only small reduction in limiting oxygen index [6]
Biological Stability: Has good resistance to microbial attack [6]

Applications/Commercial Products:
Processing & Manufacturing Routes: Phenolic Novolak of melt spun high MW (800–1000) is immersed in acidic aq. formaldehyde soln. to form an 85% wt. cross-linked amorph. network. Curing is completed by heat treatment. Fibres are typically 14–33 µm in diameter and 1–100mm long [1]. Other synth. has been reported [8]
Applications: Applications due to flame-resistant props. in textiles, upholstery and papers; as a replacement for asbestos in composites, gaskets and friction materials. Other applications as reinforcing agent for polypropylene and as a precursor to carbon fibre

Trade name	Supplier
Kynol	Nippon Kynol
Kynol	Carborundum Corporation
Kynol	American Kynol

Bibliographic References

[1] Kirk-Othmer Encycl. Chem. Technol. 4th edn., (eds., J.I. Kroschwitz), 1991, **18**, (rev)
[2] Knop, A. and Pilato, L.A., Phenolic Resins Springer-Verlag, 1985, (rev)
[3] Broutman, L.J., *Polym. Eng. Sci.*, 1983, **23**, 776, (use)
[4] Mera, H. and Taketa, T., Ullmanns Encykl. Ind. Chem. 5th edn, (eds., B. Elvers, S. Hawkins, M. Ravenscroft and G. Schulz), VCH, 1989, **A13**, 9, (synth props uses)
[5] Economy, J., *Polym. News*, 1975, **2**, 13, (uv irradiation)
[6] Hamilton, L.E., Gatewood, B.M. and Sherwood, P.M.A., *Text Chem. Color.*, 1994, **26**, 39, (degradation)
[7] Johnson, D.J., *Applied Fibre Science*, (ed. F.Happey), Academic Press, 1979, **3**, 127
[8] Arita, Y. and Iizuka T., *Kino Zairyo*, 1995, **15**, 41, (synth, props, struct, use)

Phenolic resins, Friction materials   P-38

Related Polymers: Phenolic Resoles, Cresol Resoles, Phenolic Novolaks, Cresol Novolaks, Cardanol Resins, Phenolic Resins, Oil-modified, Phenolic Resin Dispersions

– **Phenolic resins, Glass-reinforced plastic**

Monomers: Phenol, Cardanol, Formaldehyde, *o*-Cresol, *p*-Cresol, *m*-Cresol
Material class: Thermosetting resin
Polymer Type: phenolic
General Information: Phenolic and cardanol based resins have applications as friction lining compds.

Applications/Commercial Products:
Processing & Manufacturing Routes: Friction lining compds. contain fibres (up to 60%), fillers (5–30%), metals (5–70%), lubricants (0–5%) and binders (5–25%). Friction Dust (5–15%), derived from Cardanol resins, can be used as an organic filler. The binders comprise solid, liq. and dissolved phenolic resins and rubbers. Woven fabrics or yarns may be dip-impregnated with liq. resin or resin soln., which can also contain added soln. or dispersion of rubber. After drying to remove solvent, the fabric is shaped by calendering or extrusion, fibres, fillers and binders are masticated to give a doughy mass. After shaping, it is dried and cured in tension moulds or by hot pressing. Dry mixes for hot pressing or warm shaping followed by oven curing processes involve pressures of 3–10 MPa at -110–120°. Phenolic dispersions are prepared by treating phenol and formaldehyde with hexamethylene triamine (6–12%), adding poly(vinyl alcohol) and gum arabic as protective colloids. Particles of size >1000 microns are used in friction element composition
Applications: Brake linings are made from impregnated woven fabrics and clutch linings from yarns. Machine and drum brake linings are best made by the wet processing method. Drum brake linings may be made by direct hot pressing to a considerable extent. Disc break linings are prod. by pressing friction lining compds. onto cleaned, adhesive-coated metal base plates

Trade name	Details	Supplier
Borden P	friction applications resins	Borden
Rütaphen RL	friction lining aqueous resins	Bakelite
Rütaphen RP	friction lining powder resins	Bakelite
Rütaphen RW	friction lining resin solutions	Bakelite

Phenolic resins, Glass-reinforced plastic P-39
Related Polymers: Phenolic Resoles
Monomers: Phenol, Formaldehyde
Material class: Thermosetting resin, Composites
Polymer Type: phenolic
Additives: Fillers are chopped strand mat/woven fabric with adhesion promotors such as vinylsilane or Volan (methacrylato chromic chloride). Acid-resistant fillers include wood flour, china clay, and alumina hydrates
Miscellaneous: The use of phenolic resins in hand-layup glass-reinforced plastic or low-pressure/low-temp. compression moulding laminating systems yields a product with inherent fire-retardant props. and low smoke emission. Prepregs may be made from high-solids phenolic resins using volatile solvents to aid prepreg flow props.

Volumetric & Calorimetric Properties:
Density:

No.	Value
1	d 0.00191–0.00195 g/cm^3

Deflection Temperature:

No.	Value
1	250°C

Transport Properties:
Polymer Solutions Dilute: Viscosity 500cP (25°, uncured) [1]

Mechanical Properties:
Flexural Modulus:

No.	Value	Note
1	5000 MPa	[1]
2	12000 MPa	ISO 178, BS2782 335A [2]

Tensile Strength Break:

No.	Value	Note
1	0.459 MPa	[1]
2	55 MPa	ISO R 3268, BS2782 320E [2]

Flexural Strength at Break:

No.	Value	Note
1	183 MPa	[1]
2	170 MPa	ISO 178, BS 2782-335A [2]

Charpy:

No.	Value	Notch	Note
1	65	N	ISO 179 [2]

Electrical Properties:
Surface/Volume Resistance:

No.	Value	Note	Type
1	0.1×10^{15} Ω.cm	IEC 93 [2]	S

Dielectric Permittivity (Constant):

No.	Value	Frequency	Note
1	9.2	1kHz	IEC 250 [2]

Dielectric Strength:

No.	Value	Note
1	9 kV.mm^{-1}	20°, IEC 243 [2]

Arc Resistance:

No.	Value	Note
1	208s	ASTM D 495 [2]

Dissipation (Power) Factor:

No.	Value	Frequency	Note
1	0.15	1 kHz	[2]

Polymer Stability:
Polymer Stability General: Uncured resin (containing 10% Me$_2$CO) has shelf life of 3 months at 20°

Decomposition Details: Burns slowly with very little flame producing CO_2, CO and SO_2 (35% filled)
Flammability: Fire Propagation class 0 (35% filled, BS476 part 6), fire propagation I = 2 i = 0.1 (BS476 part 6). [2] Surface spread of flame class 1 (35% filled, BS476 part 7), class 1 (BS 476 part 7) [2]. Smoke emission test D_s (max.) 11.76 min Time to D_s = 163.83 min; MoD smoke index 1 (NE 5711) [12]. Limiting oxygen index 46–54% (35% filled), 100% (BS5734 part 1 method 4; ISO 4589) [2]
Chemical Stability: 35% Filled material has good resistance to hydrocarbons, aromatic solvents and dilute acids but poor resistance to alkalis, conc. acids and Me_2CO

Applications/Commercial Products:
Processing & Manufacturing Routes: Prepregs are made of glass filament cloth impregnated with phenolic resins. The application viscosity is modified by solvent addition. The prepreg has a resin content of between 20% and 50% and it may be stored at < 20° before use. It is cured and compressed at 0.7 Pa and 90–160° over 1–2h to form laminates having excellent fire-resistant and smoke props. For hand lay-up glass-reinforced plastic or low-pressure/low temp. compression moulding, chopped strand mat is impregnated with a low viscosity acid-cured phenolic resole system in solvent. The inclusion of up to 30% of acid-resistant filler e.g. silica flour or china clay prevents crazing of the resin
Applications: Fibre glass-reinforced flexible abrasive wheels are made from fibre glass yarn prepregs impregnated with phenolic resole. Fibre reinforced laminates are used inside aircraft, rail vehicles (especially for subway trains), buses and trucks for fire safety reasons

Trade name	Details	Supplier
Borden P963/P964	laminating system	Borden
Borden PI105		Borden
Phenmat		DSM Resins
Pribex 7160	impregnating resins	Neste
Rütaphen I	impregnating resins	Bakelite

Bibliographic References
[1] *Borden P963/964*, Borden, (technical datasheet)
[2] *Phenmat.*, DSM Engineering Plastics, 1995, (technical datasheet)

Phenolic resins high-ortho   P-40

Related Polymers: Phenolic Novolaks, Phenolic Resoles, Phenolic Resin, Cast Resin
Monomers: Phenol, Formaldehyde
Material class: Thermosetting resin
Polymer Type: phenolic
Molecular Weight: MW 1800, M_n 550
Additives: A catalyst containing divalent metal ions, e.g. zinc acetate is required
General Information: Divalent metal ions are used to promote *ortho*-substitution
Morphology: Analysis by nmr of a zinc acetate catalysed resin (F/P mole ratio 0.6) shows isomeric distribution: 2,2' - 45%; 2,4' - 45% and 4,4' - 10% indicating a 90% direction to the 2- position for the first condensation reaction

Applications/Commercial Products:
Processing & Manufacturing Routes: Anhydrous high-ortho liq. phenolic Resole is made using F/P mole ratio of between 1.5 and 1, at a pH 4–7 and a divalent salt at 80–90° for 5–8h
Applications: In the Liq. Injection Moulding process, monomers and oligomers are injected into the mould cavity and polymerised *in situ*. Highly-reactive anhydrous high-ortho resins with latent acid catalyst are used which give a peak exotherm at 130–140°

Phenolic Resins, Ion exchangers   P-41

1. strong acid cation exchangers
2. weak base anion exchangers
3. chelation ion exchangers (based on iminodiacetic acid)
4. chelation ion exchangers (based on tyrosine)

Related Polymers: Phenol-formaldehyde resins
Monomers: Phenol, Formaldehyde, Resorcinol, Tyrosine, 4-Hydroxybenzenesulfonic acid, Iminodiacetic acid
Material class: Thermosetting resin
Polymer Type: phenolic
CAS Number:

CAS Reg. No.	Note
29659-43-6	phenol-formaldehyde-tyrosine copolymer
28303-16-4	phenolsulfonic acid-formaldehyde copolymer
98507-54-1	resorcinol-iminodiacetic acid-formaldehyde copolymer

Additives: Carbon fibre to improve thermal stability (strong acid cation exchangers)
General Information: May be categorised into three types [1,2]; strong acid cation exchangers (phenolsulfonic acid-formaldehyde copolymer); weak base anion exchangers (resol-oligomeric amine copolymer); chelation ion exchangers (either resorcinol-iminodiacetic acid-formaldehyde copolymer or phenol-formaldehyde-amino acid copolymer. The amino acid may be tyrosine, typically). The struct. and functionality of strong acid cation exchanger has been reported [7]

Polymer Stability:
Thermal Stability General: Thermal stability of strong acid cation exchangers is improved by addition of carbon fibre [8]
Decomposition Details: Weight loss and rate of thermal degradation of chelation ion exchangers may be reduced by complexation with Cu^{2+} [11]
Hydrolytic Stability: Strong acid cation exchanger (Duolite ARC-359) is stable at high pH [4]. Chelation ion exchanger (Chelex 100) is stable to alkaline soln. [4]

Stability Miscellaneous: Duolite ARC-359 is resistant to radioactivity whereas Chelex 100 is easily damaged [4]

Applications/Commercial Products:
Processing & Manufacturing Routes: Strong acid cation exchangers are synth. by reaction of phenolsulfonic acid with formaldehyde, or by addition of sodium sulfite to phenol-formaldehyde mixture. Chelation ion exchangers may be synth. by reaction of resorcinol with iminodiacetic acid in the presence of 6N NaOH (pH 9) and a template such as $CaCO_3$. Neutralised resin is heat cured at 60° to yield a red-brown solid. The template may be removed by acid hydrolysis. The complexing affinity for ^{137}Cs increases with increase in temp. at which synth. is performed, whereas affinity for ^{85}Sr decreases with synth. temps. above 100°. Nature of the alkaline catalyst also affects affinity; use of NH_4OH produces resins with poor affinity; use of $BaOH$ produces resins with high ^{85}Sr affinity but poor ^{137}Cs affinity [5]. Chelation ion exchangers containing tyrosine may be synth. by reaction of phenol, formaldehyde and tyrosine under basic conditions [9], under reflux at 100° for 15–30 min. followed by heat curing at 120–130° for 24h. [10]
Applications: Strong acid cation exchange resins and chelation ion exchange resins are used to remove ^{137}Cs, Pu and ^{90}Sr from alkaline solns. containing low level radioactive waste. Weak base anion exchangers have applications in the food industry (deacidification of fruit juices, removal of tannins and so on). Chelation ion-exchange resins with tyrosine groups are used to extract uranium ions from sea water. Aslo used to complex divalent metal cations. Aspartic acid as the amino acid component gives the best complexing affinity

Trade name	Details	Supplier
Chelex	100 (iminodiacetic acid-based chelation exchanger)	Bio-Rad Laboratories
Duolite	ARC-359 (strong acid cation exchanger)	Diamond Shamrock
Duolite	A7, ES562/568 (weak base anion exchanger)	Rohm and Haas
Duolite	C-3 (from sodium sulfite), C3R	Rohm and Haas

Bibliographic References
[1] Knop, A. and Pilato, L.A., *Phenolic Resins*, Springer-Verlag, 1985, 169, (rev)
[2] Abrams, I.M. and Benezra, L., *Encycl. Polym. Sci. Technol.*, (eds. H.F. Mark, N.G. Gaylord and N.M. Bikales), Interscience, 1967, **7**, 692, (rev.)
[3] Dickinson, B.N. and Higgins, I.R., *Nucl. Sci. Eng.*, 1967, **27**, 131, (strong acid cation exchanger, use)
[4] Wiley, J.R., *Ind. Eng. Chem. Proc. Des. Dev.*, 1978, **17**, 67, (strong acid cation exchanger, chelation exchanger uses)
[5] Kaczinsky, J.R., Fritz, J.S., Walker, D.D. and Ebra, M.A., *J. Radioanal. Nucl. Chem.*, 1985, **91**, 349, (synth. use)
[6] Cristal, M.J., *Chem. Ind. (London)*, 7 Nov., 1983, 814, (weak base anion exchanger, use)
[7] Schink, J. Belkoura, L., Woermann, D., Yeh F. et al, *Ber. Bunsen-Ges. Phys. Chem.*, 1996, **100**, 1103, (struct.)
[8] Artemenko, S.E., Kardash, M.M. and Svekol'nikova, O. Yu., *Khim. Volokna*, 1992, 29, (strong acid cation exchanger, thermal stability)
[9] Takahashi, K., *Seikatsu Kagaku*, 1971, **9**, 20, (chelation ion exchanger, synth., use)
[10] Tascioglu, S., *J. Macromol. Sci., Pure Appl. Chem.*, 1994, **31**, 367, (chelation ion exchanger, synth., use)
[11] Urushido, K., Tsuchiya, C. and Hojo, N., *Shikizai Kyokaishi*, 1971, **44**, 465, (chelation ion exchanger, thermal degradation)

Phenolic resins, Oil recovery   P-42
Related Polymers: *p*-Nonylphenol Formaldehyde Resins
Monomers: *p*-Nonylphenol, Formaldehyde, Ethanol
Material class: Thermoplastic
Polymer Type: phenolic
General Information: Ethoxylation of the phenolic resin makes the material more hydrophobic and hence compatible with oils

Applications/Commercial Products:
Processing & Manufacturing Routes: The Novolak is partially ethoxylated by reaction with EtOH
Applications: Ethoxylated nonylphenol resin Novolaks of varying MW and degree of ethoxylation are used for enhanced oil recovery operations

Phenolic resin Spheres   P-43
Synonyms: *Phenolic microspheres. Phenolic microballoons*
Related Polymers: Phenolic Resoles
Monomers: Phenol, Formaldehyde
Material class: Thermosetting resin
Polymer Type: phenolic
Additives: Surfactants and blowing agent are required to make spheres

Volumetric & Calorimetric Properties:
Density:

No.	Value	Note
1	d 0.07–0.15 g/cm^3	bulk [2]

Optical Properties:
Volume Properties/Surface Properties: Colour red brown to purple brown

Applications/Commercial Products:
Processing & Manufacturing Routes: Synth. from aq. resole [1] mixed with surfactants and blowing agent (e.g. ammonium carbonate/ammonium nitrate, or dinitrosopentamethylenetetramine). Hollow spheres (5–50 μm diameter) obtained by spray drying
Applications: They act as hollow spherical fillers contributing lightness to composite materials. Phenolic spheres are useful as flotation aids; as a floating cap on petroleum naphtha or oil to retard evaporation. They also have applications in nitroglycerine-based explosives, to allow controlled detonation, and medical applications (artificial skin). May be incorporated into liq. resins (especially epoxy resins) to make syntactic foams which are used as structural core materials in automotive, marine and aerospace applications; in wall panels and ablative heat shields. Other uses as gap filler to deaden sound

Trade name	Supplier
Phenolic Microballoons	Union Carbide

Bibliographic References
[1] Lee, J.D., Bae, M.H., Lyu, S.G. and Sur, G.S., *Pollimo*, 1994, **18**, 1041, (synth.)
[2] *Encycl. Polym. Sci. Eng.*, 2nd edn. (ed. J.I. Kroschwitz), John Wiley & Sons, Inc, 1985, **11**, (uses)
[3] *U.S. Pat.*, 1957, 2 797 201, (use)

Phenolic resins, Rubber adhesives   P-44
Related Polymers: Phenol Aldehyde Resins, *p-tert*-Butyl Phenolic Resin, Phenolic Resoles, Phenolic Novolaks
Monomers: Phenol, 4-*tert*-Butylphenol, Formaldehyde
Material class: Thermosetting resin
Polymer Type: phenolic

Additives: Rubbers, metallic oxides and antioxidants
General Information: Resoles contain 2-Methylol phenols which crosslink by interaction with metallic oxides e.g. MgO

Applications/Commercial Products:
Processing & Manufacturing Routes: Solid *tert*-butyl phenolic Resole (20–40%) is compounded with Neoprene rubber and metallic oxides (e.g. magnesium oxide, zinc oxide) in solvent to make a contact adhesive. Heat-curing phenol Resoles are suitable for compounding with Nitrile rubber to make an adhesive soln.

Phenol Novolak (20–80%) is included in Polyurethane adhesive formulations. Adhesion primers for metallic surfaces to be bonded to rubber are based on elastomers and thermosetting resins. A combination of phenolic Resoles has proved successful as the thermosetting component. Modified phenolic Resoles are used as adhesive primers to copper films in printed circuit boards
Applications: Rubber adhesive soln. based on Neoprene are used in shoe manufacture. Rubber-phenolic adhesives are used widely in furniture, construction and automotive industry applications. Phenolic resoles as adhesion primers are used for electrical applications

Trade name	Details	Supplier
Rütaphen IZ	adhesion primer resin	Bakelite
Rütaphen KA	adhesive resins (Neoprene)	Bakelite
Rütaphen KP	adhesive resins	Bakelite

Phenolic resins, Textile fleece binders   P-45

Synonyms: *Fibre fleece materials*
Related Polymers: Phenolic Novolaks
Monomers: Phenol, Formaldehyde
Material class: Thermosetting resin
Polymer Type: phenolic
Additives: Curing agent usually hexamethylene tetramine at level 4–10%. Anti-dusting agents are required

Applications/Commercial Products:
Processing & Manufacturing Routes: The range of organic fibres used includes wool, cotton, cellulose, flax, jute, polyester and acrylic. Fibres reclaimed from shredded cloth are homogenised and laid on a conveyor as a fluffy mat. The powdered phenolic novolak resin with added curing agent, generally hexamethylene tetramine is spread over the mat by a metering system. The binder is distributed evenly fluidising the mat in a stream of air. The fleece is cured by forcing through hot air at 160–200°. The fleece contains 20–40% of resin. The resin mainly creates point bonds at fibre intersections. The fleece may be shaped after preliminary drying at a lower temp. by increasing the conveyor speed. After shaping and cutting the parts are then finally cured at 160–200°.
Applications: Textile fibre reinforced phenolic resin based materials are useful for acoustic noise reduction in automotive construction. Areas involved include bonnet hood insulation, doors, roofs, sides, footspace, pumps and ventilation areas. The fleece is also used as a self-supporting base for instrument panelling, door panelling, covers for the back rest and reserve panelling. In the roof lining, hat rack, floor damping, mat for the luggage boot and the wheel house insulation, the mat provides both insulation and support. Due to excellent thermal insulation, the fleeces are good climate and humidity regulators for automotive interiors. Fibre fleece materials are also used with laminating materials for direct damping of automobile body vibrations. Flat fleece mats provide excellent acoustic insulation in household appliances, such as washing machines, dishwashers and vacuum cleaners

Trade name	Details	Supplier
Rütaphen TP	textile fleece powder resins	Bakelite

Phenolic Resoles, Coatings   P-46

Related Polymers: Phenol Aldehyde Resins, Phenol Formaldehyde Resins, Phenolic Resoles, Phenolic Novolaks, *p-tert*-Butyl Phenolic Resins
Monomers: Phenol, Formaldehyde
Material class: Thermosetting resin
Polymer Type: phenolic
Additives: Many formulations require resin modifiers to meet specific coating application needs. Additives include adhesion promotors, chemical crosslinkers, and hardening agents

Applications/Commercial Products:
Processing & Manufacturing Routes: Coatings are formulated from substituted and unsubstituted phenols, and may be heat-reactive or non-heat-reactive. Phenolic resoles with high free-formaldehyde content and low MW are cured for 30 min at 150° for use as interior coatings. PF resins prepared using base catalyst is (e.g. benzyl dimethylamine) are used in stoichiometric amounts with epoxy resins and cured at 180° for 2h. Phenolic Resoles from substituted phenols are more compatible with oil and hydrocarbons. Resins from *p-tert*-butyl phenol find application as contact adhesives and as modifiers for baking alkyds, Rosin and estergum. Oleoresinous phenolic varnishes have drying times of 2–4h. The films remain flexible as they are not crosslinked
Applications: Phenol-formaldehyde resoles are applied as 25 micron interior coatings for cans, drum linings and pipes and are used as metal primers. Phenolic baked coatings are used on metals, ceramics and some plastics surfaces. Resoles from substituted phenols are used either alone or as modifiers for alkyds, polyester or epoxy resins for many heat-reactive coating applications. Novolaks are used as a hardener for epoxy resin coatings. Phenolic and epoxy resins in equal stoichiometric amounts give a coating having excellent resistance to heat, chemicals and moisture. Resins from *p-tert*-butyl phenol find application as contact adhesives for electrical insulation varnishes. Non-heat curing resins from substituted phenols are excellent modifiers for oleoresinous varnishes and alkyds

Trade name	Details	Supplier
Rütaphen LA	coating resins for oil use	Bakelite
Rütaphen LE	coating resins for electrodeposition	Bakelite
Rütaphen LG	coating resins for primers	Bakelite
Rütaphen LP	powder coating resins	Bakelite

Phenolic Resoles, Foundry applications   P-47

Related Polymers: Phenolic Resoles, Phenolic Novolaks, Furan Resins
Monomers: Phenol, Furfuryl alcohol, Formaldehyde
Material class: Thermosetting resin
Polymer Type: phenolic
Additives: Additives depend on the foundry moulding process

Applications/Commercial Products:
Processing & Manufacturing Routes: In the shell-moulding (Croning) process quartz sand is hot-coated with a phenolic Novolak (2%), hexamethylene tetramine (0.02–0.04%) and calcium stearate (0.1–0.2%). The sand is then cured in hot pattern plates or core boxes to produce shells or cores. In the hot box process, quartz sand is batch mixed with phenolic Resole or urea-modified furan resin (1.4–2%) and a latent curing agent (e.g. ammonium nitrate 0.3–0.5%). The mix is shaped and precured at 200–260° for up to 2 min. in core boxes. Use of modified furan resins (0.9–1.2%) with curing agents usually sulfonic acids, (0.2–0.4%), enables a warm-box process for binding sand at 140–200°. Phenolic and furan resins with special curing agents may be employed in a vacuum warm box process where the core box is heated at 70–100°. In the thermoshock process, a highly-condensed phenolic or urea resin and acidic salt curing agent is hardened at very high temp. For a no-bake process, sand and phenolic Resole or furan resin (0.7–1.3%) and acidic curing agent (0.2–0.6%) are cured at ambient temp. for 20 min to several hours. In the ester no bake process, phenolic Resole binders are cured with liq. aliphatic esters. In the PUR cold box process, phenolic Resole with ether crosslinks (0.6–1%) is cured with a diisocyanate (0.6–1%) using a tertiary amine curing accelerator (0.02–0.1%). After compacting in the corebox, the material is gassed with an amine-air mixt. In the MF cold box process, aq. phenolic Resoles are gassed with methyl formate to form a solidified Resole
Applications: The processing and manufacturing routes are all used to prepare sand-based moulds for casting of metals in the foundry

Trade name	Details	Supplier
Rütaphen GA	foundry resin-amine cold box	Bakelite
Rütaphen GC	foundry resin-shell moulding	Bakelite
Rütaphen GH	foundry resin-hot box	Bakelite
Rütaphen GK	foundry resin-cold curing	Bakelite
Rütaphen GN	foundry resin-no bake	Bakelite
Rütaphen GS	foundry resin	Bakelite
Rütaphen GW	foundry resin-warm box	Bakelite
	sand mould resin	Neste
Dynosol		Dyno
Exter 4010	PF resin for plywood	Neste
Rütaphen HW	wood materials (chipboard, fibreboard, plywood, compreg)	Bakelite

Phenolic Resoles, Wood applications P-48

Related Polymers: Phenolic Resoles, Phenolic Novolaks, Urea Formaldehyde Resins, Resorcinol Formaldehyde Resins, Wood Applications
Monomers: Phenol, Urea, Melamine, Formaldehyde
Material class: Thermosetting resin
Polymer Type: phenolic
General Information: For superior moisture resistance for outdoor applications refer to melamine urea formaldehyde resins (MUF) and resorcinol formaldehyde resin (RF) or resorcinol-phenol formaldehyde resins (RPF)

Applications/Commercial Products:
Processing & Manufacturing Routes: Chipboard is made from wood chips (82 parts) with liq. phenolic Resole (15 parts), paraffin emulsion and/or wood preservatives. The mix is cured at pressure (2.5MPa 160–220°). Fibreboard is made from wood fibres and phenolic resin binder. The level of binder may vary from 1–2% for standard fibreboards or up to 5–10% for hard-wearing floorbaords with paraffin to limit moisture. The board is prod. by cure (200° and 4–6MPa). Moulded parts (Lignotock process) may be formed from a loose fleece of matted wood fibres with a thermoplastic binder and a dry phenolic resin (5–10%). The final shape is prod. by hot compression at 4MPa and 180°. Phenolic-resin bonded plywood is made from dry weights of wood (81–88%) and phenolic resin (4–10%) with moisture levels at 6–13%. Phenolic Resoles, aq. or in MeOH soln. are used as impregnating resins for compregs. Compregs may be used with decorative wood and paper veneers impregnated with melamine resin. Sodium Kraft paper is generally used for impregnation with phenolic Resole soln. and dried at 130–150° to give a resin level of 30–50%. Several layers are compressed at 7–15MPa and 150–200° for 60–90 min. to make resistant laminates
Applications: Phenolic resins are used to bond woodchips to form chipboard. Hard fibreboard is made from wood fibres from defibrilated wood and scrap. Moulded parts (Lignotock process) are used as light components in automotive construction e.g. dashboard shells (foam backed), hatracks, door and side panels and headliners. Plywood is at least three transverse layers of wood bonded with adhesive with final thickness from 4mm upwards. Reed beech veneer (thickness 0.2–2mm) is generally used to produce compreg - impregnated and compressed wood. Compreg blocks and boards are used for several engineering applications: in machinery, textiles and the electrical industries. Blockboard and chipboard can be covered by adhesive to a decorated laminated sheet or by compression bonding to layers of impregnated paper. Laminates having good resistance to most chemicals are made from layers of impregnated paper. They are used for kitchen surfaces. Structural wood bonding methods have been used to construct loadbearing structs. for both indoor and outdoor applications such as boat building

Trade name	Details	Supplier
Aerodux	resorcinol resins	Dyno
Aerophen	phenolic resins	Dyno

Phenylsilicone rubber, Heat vulcanised P-49

Synonyms: Phenyl based silicone elastomer. Poly(methylphenyl)silicone rubber
Related Polymers: Methylsilicone rubber, Heat vulcanised, Vinyl methyl phenyl silicone rubber, Heat vulcanised
Monomers: Dichloromethylphenylsilane, Dichloroethenylmethylsilane, Cyclic methylphenylsiloxane, Cyclic methylvinylsiloxane
Material class: Synthetic Elastomers, Copolymers
Polymer Type: silicones
CAS Number:

CAS Reg. No.
158865-52-2

Molecular Formula: $[(C_7H_8OSi).(C_3H_6OSi)]_n$
Fragments: C_7H_8OSi C_3H_6OSi
Additives: Finely divided silica is incorporated as a reinforcer to improve mech. props.
General Information: Vinyl groups cause gum to cure more readily with lower activity peroxides

Volumetric & Calorimetric Properties:
Density:

No.	Value	Note
1	d 1.15 g/cm^3	[2]

Thermal Expansion Coefficient:

No.	Value	Note	Type
1	0.00045 K^{-1}	[4]	V

Thermal Conductivity:

No.	Value	Note
1	0.25 W/mK	[18,19]

Specific Heat Capacity:

No.	Value	Note	Type
1	1.3 kJ/kg.C	0.3 Btu lb^{-1}°F^{-1} [2]	P

Melting Temperature:

No.	Value	Note
1	-35°C	[4,16]

Glass-Transition Temperature:

No.	Value	Note
1	-86°C	[4]
2	-30°C	[16]

− Phenylsilicone rubber, Heat vulcanised

Transition Temperature General: There is a conflict in the literature concerning the transition temps. of silicone rubber with high phenyl substitution. As phenyl substitution rises low temp. props. deteriorate eventually becoming worse than those of dimethyl siloxane rubber [4,9,10,11,12,13,14,15,16]
Transition Temperature:

No.	Value	Note
1	−27°C	Stiffening temp. [4]

Surface Properties & Solubility:
Cohesive Energy Density Solubility Parameters: δ 18.4 $J^{1/2}$ $cm^{-3/2}$ (9 $cal^{1/2}$ $cm^{-3/2}$) [3]
Solvents/Non-solvents: Sol. aliphatic hydrocarbons, chlorinated hydrocarbons, aromatic hydrocarbons, ketones, esters, ethers. Insol. alcohols, H_2O [1,3,6]. Vulcanised rubber swollen by solvents rather than dissolved
Surface and Interfacial Properties General: Exhibits generally low surface adhesion [5]

Transport Properties:
Transport Properties General: Viscosity 100000 St (approx. pre-cure gum) [6]
Gas Permeability:

No.	Gas	Value	Note
1	O_2	962 cm^3 mm/(m^2 day atm)	0.0011 cm^2 $(s\ GPa)^{-1}$ [8]
2	N_2	315 cm^3 mm/(m^2 day atm)	0.00036 cm^2 $(s\ GPa)^{-1}$ [8]
3	H_2	2889 cm^3 mm/(m^2 day atm)	0.0033 cm^2 $(s\ GPa)^{-1}$ [8]

Mechanical Properties:
Mechanical Properties General: Requires vulcanisation with a reinforcing filler (usually silica) in order to possess acceptable mech. props.
Tensile Strength Break:

No.	Value	Note	Elongation
1	6 MPa	25°, filled [1]	300–400%

Viscoelastic Behaviour: Compression set 30–40% (150°, 22h compression) [1]

Electrical Properties:
Electrical Properties General: Insulating props. inferior to those of dimethyl silicone rubber
Dielectric Permittivity (Constant):

No.	Value	Note
1	3	[7,17]

Dielectric Strength:

No.	Value	Note
1	20 $kV.mm^{-1}$	[7]

Optical Properties:
Optical Properties General: Refractive index higher than that of dimethyl silicone [20]
Refractive Index:

No.	Value	Note
1	1.5	[2,20]

Transmission and Spectra: No transmission below 290 nm [1,5]

Polymer Stability:
Polymer Stability General: Excellent resistance to very high temps. Poor solvent resistance
Thermal Stability General: Usable from −30–315°
Upper Use Temperature:

No.	Value	Note
1	315°C	[1]

Decomposition Details: Oxidation temp. 375° (dry air) [1]
Environmental Stress: Resistant to oxidative degradation, H_2O, weathering, ozone and corona. Unaffected by uv radiation. Very resistant to high energy radiation (radioactivity)
Chemical Stability: Swollen by toluene, aliphatic hydrocarbons and CCl_4. Disintegrates in conc. H_2SO_4. Relatively unaffected by weak alkali, weak acids and alcohols [5,6]
Hydrolytic Stability: Stable in H_2O and aq. solns. at room temp. Damaged by high pressure steam
Biological Stability: Non-biodegradable. If properly vulcanised it is not subject to biological attack
Recyclability: Can be recycled by depolymerisation using steam or by acid catalysis [5]

Applications/Commercial Products:
Processing & Manufacturing Routes: Methylphenyl- and methyl-vinyl-cyclic siloxanes are polymerised together at 150–200° with an alkaline catalyst. There is a large excess of methylphenyl cyclic polysiloxane producing a linear polymer with less than 1% methylvinylsiloxane. Small quantities of hexamethyldisiloxane are used to terminate the siloxane chains. Readily cured by organic peroxides (typically 1–2 parts per 100 rubber) at temps. >100°. Several hours post-cure treatment required to remove residual peroxide
Applications: Used in laboratory research into silicone rubber chemistry. Has good props. at very high temps. but copolymers with dimethylsiloxane have better electrical props. and solvent resistance

Bibliographic References
[1] Hardman, B. and Torkelson, A., *Encycl. Polym. Sci. Eng.*, 2nd edn. (ed. J.I. Kroshwitz), John Wiley and Sons, 1985, **15**, 204
[2] Ash, M. and Ash, I., *Handbook of Plastic Compounds, Elastomers and Resins*, (eds. M.B. Ash and I.A. Ash), Wiley-VCH, 1992
[3] Yerrick, K.B. and Beck, H.N., *Rubber Chem. Technol.*, 1964, **37**, 261, (solubility)
[4] Polmanteer, K.E. and Hunter, M.J., *J. Appl. Polym. Sci.*, 1959, **1**, 3, (transition temps)
[5] Noll, W., *Chemistry and Technology of Silicones*, Academic Press, 1968
[6] Roff, W.J. and Scott, J.R., *Fibres, Films, Plastics and Rubbers*, Butterworths, 1971
[7] Stark, F.O., Fallender, J.R. and Wright, A.P., *Comprehensive Organometallic Chemistry*, (eds. G. Wilkinson, F.G.A. Stone and E.W. Abel), Pergamon Press, 1982, **2**, 305
[8] Robb, W.L., *Ann. N. Y. Acad. Sci.*, 1968, **146**, 119, (gas permeability)
[9] Borisov, S.N. and Karlin, A.V., *Soviet Rubber Technology*, 1962, **21**, 4, (glass transition)
[10] Clarson, S.J., Dodgson, K. and Semlyen, J.A., *Polymer*, 1985, **26**, 930
[11] Clarson, S.J., Dodgson, K. and Semlyen, J.A., *Polymer*, 1991, **32**, 2823
[12] Ibenisi, J., Gvozdic, N., Kevmin, M., Lynch, M.J. and Maier, D.J., Polym. Prepr. (Am. Chem. Soc., *Div. Polym. Chem.*), 1985, **26**, 18, (glass transition)
[13] Clarson, S.J., McCarthy, D. and Mark, J.E., Polym. Prepr. (Am. Chem. Soc., *Div. Polym. Chem.*), 1989, **30**, 298
[14] Clarson, S.J. and Semlyen, J.A., *Siloxane Polymers*, Prentice Hall, 1993
[15] Clarson, S.J., Mark, J.E. and Dodgson, K., *Polym. Commun.*, 1988, **29**, 208
[16] Borisov, M.F., *Soviet Rubber Technology*, 1966, **25**, 5, (transition temps)
[17] Vincent, G.A., Fearon, F.W.G. and Orbech, T., Annu. Rep., *Conf. Electr. Insul. Dielectr. Phenom.*, 1972, 17, (dielectric constant)
[18] Ames, J., *J. Sci. Instrum.*, 1958, **35**, 1
[19] Jamieson, D.T. and Irving, J.B., *Proc. Int. Conf. Therm. Conduct.*, 1975, 279, (thermal conductivity)
[20] Flaningan, O.L. and Langley, N.R., *Anal. Chem. Silicones*, (ed. A.L. Smith), John Wiley and Sons, 1991, 135

PIPD
P-50

Synonyms: *Poly[(1,4-dihydrodiimidazo[4,5-b:4',5'-e]pyridine-2,6-diyl)(2,5-dihydroxy-1,4-phenylene)]. 2,5-Dihydroxy-1,4-benzenedicarboxylic acid polymer with 2,3,5,6-pyridinetetramine*
Monomers: 2,3,5,6-Pyridinetetramine, 2,5-Dihydroxy-1,4-benzenedicarboxylic acid
Material class: Thermoplastic
CAS Number:

CAS Reg. No.	Note
167304-74-7	Poly[(1,4-dihydrodiimidazo[4,5-b:4',5'-e]pyridine-2,6-diyl)(2,5-dihydroxy-1,4-phenylene)]
169836-78-6	2,5-dihydroxy-1,4-benzenedicarboxylic acid polymer with 2,3,5,6-pyridinetetramine

Molecular Formula: $(C_{13}H_7N_5O_2)_n$
Fragments: $C_{13}H_7N_5O_2$
General Information: Polymer normally produced as high-strength fire-resistant fibres of metallic blue colour.
Morphology: As-spun fibres have crystal hydrate structure, while heat-treated fibres assume a three-dimensional crystal structure with intramolecular and intermolecular hydrogen bonding [1,2]. Monoclinic unit cell, a=13.33Å, b=3.46Å, c=12.16Å [1,2]. The molecular chains deviate slightly from a planar structure [1]. Nematic liquid crystalline polymer solutions are formed in polyphosphoric acid [2,3,4].

Volumetric & Calorimetric Properties:
Density:

No.	Value	Note
1	d −1.7 g/cm³	[3]

Transition Temperature General: No reported glass-transition temperature. Exothermic transitions during polymer degradation in air from 450–650°C [5].

Mechanical Properties:
Mechanical Properties General: Heat-treated fibres have improved mechanical properties compared to as-spun fibres [3,4]. Crystal modulus of *ca.* 440 GPa reported [8].

Tensile (Young's) Modulus:

No.	Value	Note
1	150–360 MPa	[3,4,6,7]

Tensile Strength Break:

No.	Value	Note	Elongation
1	2500–5500 MPa	[3,4,6,7]	1.2–2.7%

Compressive Strength:

No.	Value	Note
1	850–1700 MPa	[3,4,6]

Optical Properties:
Transmission and Spectra: Raman [6,9], cmr [5], x-ray diffraction [2,8] and neutron diffraction [10] data have been reported.

Polymer Stability:
Thermal Stability General: Main polymer degradation in air occurs in range 450–650°C [5].
Flammability: The as-spun fibres with their crystal hydrate structure achieve a fire protection index (FPI) which is higher than that of the heat-treated fibres and other fire-protective polymers [7]. The polymer fibre does not burn even under severe conditions [5]

Applications/Commercial Products:
Processing & Manufacturing Routes: Prod. by polycondensation of 2,3,5,6-pyridinetetramine trihydrochloride with 2,5-dihydroxyterephthalic acid in polyphosphoric acid [3,4]. Fibres are spun from polymer solution in polyphosphoric acid with coagulation in an aqueous medium [3,4].
Applications: Potential applications for fire protection, body armour, ropes and cables, composite materials for the aircraft industry

Trade name	Details	Supplier
M5	AS (as-spun) and HT (heat-treated) fibre grades	Magellan Systems International

Bibliographic References
[1] Takahashi, Y., *Macromolecules*, 2003, **36**, 8652
[2] Klop, E.A. and Lammers, M., *Polymer*, 1998, **39**, 5987
[3] Sikkema, D.J., *Polymer*, 1998, **39**, 5981
[4] Lammers, M., Klop, E.A., Northolt, M.G. and Sikkema, D.J., *Polymer*, 1998, **39**, 5999
[5] Bourgibot, S., Flambard, X., Ferreira, M., Devaux, E. and Poutch, A., *J. Mater. Sci.*, 2003, **38**, 2187
[6] Sirichaisit, J. and Young, R.J., *Polymer*, 1999, **40**, 3421
[7] Northolt, M.G., Sikkema, D.J., Zegers, H.C. and Klop, E.A., *Fire Mater.*, 2002, **26**, 169
[8] Montes-Morán, M.A., Davies, R.J., Riekel, C. and Young, R.J., *Polymer*, 2002, **43**, 5219
[9] Takahashi, Y., *J. Polym. Sci., Part B: Polym. Phys.*, 2001, **39**, 1791
[10] Takahashi, Y., *Macromolecules*, 2002, **35**, 3942

PMDA-4-BDAF
P-51

Synonyms: *Poly(pyromellitic dianhydride-alt-2,2-bis[4-(4-aminophenoxy)phenyl]hexafluoropropane). Poly[(5,7-dihydro-1,3,5,7-tetraoxobenzo[1,2-c:4,5c']dipyrrole-2,6(1H,3H)-diyl)-1,4-phenyleneoxy-1,4-phenylene[2,2,2-trifluoro-1-(trifluoromethyl)ethylidene]-1,4-phenyleneoxy-1,4-phenylene]. Poly[1H,3H-benzo[1,2-c:4,5-c']difuran-1,3,5,7-tetrone-alt-4,4'-[[2,2,2-trifluoro-1-(trifluoromethyl)ethylidene]bis(4,1-phenyleneoxy)bis[benzenamine]]*
Monomers: 1H,3H-Benzo[1,2-c:4,5-c']difuran-1,3,5,7-tetrone, 4,4'-[2,2,2-Trifluoro-1-(trifluoromethyl)ethylidene]bis(4,1-phenyleneoxy)bis[benzenamine]
Material class: Thermoplastic
Polymer Type: polyimide
CAS Number:

CAS Reg. No.	Note
84769-07-3	
84789-95-7	source based

Molecular Formula: $(C_{37}H_{18}F_6N_2O_6)_n$
Fragments: $C_{37}H_{18}F_6N_2O_6$

PMDA-ODA

Volumetric & Calorimetric Properties:
Density:

No.	Value	Note
1	d 1.408 g/cm^3	[4]
2	d^{25} 1.431 g/cm^3	[2,3]

Thermal Expansion Coefficient:

No.	Value	Note	Type
1	4.57×10^{-5} K^{-1}	50–250° [1]	L

Volumetric Properties General: Fractional free volume 0.165 [4]
Melting Temperature:

No.	Value	Note
1	472°C	[5]

Glass-Transition Temperature:

No.	Value	Note
1	295°C	[2]
2	302°C	[4]
3	306°C	[5]
4	310°C	[3]

Surface Properties & Solubility:
Cohesive Energy Density Solubility Parameters: δ 26.9 J$^{1/2}$ cm$^{-3/2}$ [4]
Solvents/Non-solvents: Insol. N-methyl-2-pyrrolidone, N,N-dimethylacetamide, DMSO, pyridine, THF, CHCl$_3$ [2,5]

Transport Properties:
Permeability of Liquids: Water permeability 92.4×10^{-12} g (cm s cmHg)$^{-1}$ [1]
Water Absorption:

No.	Value	Note
1	1.27 %	50h, 25°, 75% relative humidity [1]

Gas Permeability:

No.	Gas	Value	Note
1	CO$_2$	775 cm^3 mm/(m^2 day atm)	11.8×10^{-10} cm^2 (s cmHg)$^{-1}$, 35°, 6.8 atm. [3]
2	CO$_2$	1156 cm^3 mm/(m^2 day atm)	17.6×10^{-10} cm^2 (s cmHg)$^{-1}$, 35°, 10 atm. [4]
3	O$_2$	190.4 cm^3 mm/(m^2 day atm)	2.9×10^{-10} cm^2 (s cmHg)$^{-1}$, 35°, 6.8 atm. [3]
4	O$_2$	327 cm^3 mm/(m^2 day atm)	4.98×10^{-10} cm^2 (s cmHg)$^{-1}$, 35°, 2 atm. [4]
5	CH$_4$	23.6 cm^3 mm/(m^2 day atm)	0.36×10^{-10} cm^2 (s cmHg)$^{-1}$, 35°, 6.8 atm. [3]
6	CH$_4$	41.9 cm^3 mm/(m^2 day atm)	0.638×10^{-10} cm^2 (s cmHg)$^{-1}$, 35°, 10 atm.
7	H$_2$	2252 cm^3 mm/(m^2 day atm)	34.3×10^{-10} cm^2 (s cmHg)$^{-1}$, 35°, 10 atm. [4]
8	CO	97.2 cm^3 mm/(m^2 day atm)	1.48×10^{-10} cm^2 (s cmHg)$^{-1}$, 35°, 10 atm. [4]
9	N$_2$	62 cm^3 mm/(m^2 day atm)	0.943×10^{-10} cm^2 (s cmHg)$^{-1}$ cm^2 (s cmHg)$^{-1}$, 35°, 2 atm. [4]

Mechanical Properties:
Tensile (Young's) Modulus:

No.	Value	Note
1	1400 MPa	[2]

Tensile Strength Break:

No.	Value	Note	Elongation
1	68 MPa	[2]	15%

Hardness: Vickers 20.6 kg mm^{-2} (10 kg load, wear scar) [6]
Friction Abrasion and Resistance: Wear rate 27×10^{-15} m^3 m^{-1} (3.1 m s^{-1}, 9.8 N, 25°, 50% relative humidity) [6]. μ_{dyn} 0.6 (25°), 0.43–0.22 (100–300°) [6]

Electrical Properties:
Surface/Volume Resistance:

No.	Value	Note	Type
1	0.303×10^{15} Ω.cm	conductivity value [2]	S

Dielectric Permittivity (Constant):

No.	Value	Frequency	Note
1	3.4	10 Hz	24° [2]

Polymer Stability:
Decomposition Details: Under N$_2$ atmosphere, single step decomposition occurs, initial decomposition 513°; 10% wt. loss at 525° (heat rate 10° min.$^{-1}$) [2]

Applications/Commercial Products:

Trade name	Details	Supplier
Eymyd	L30, L30N	Ethyl Corporation

Bibliographic References
[1] Numata, S., Fujisaki, K. and Kinja, N., *Polymer*, 1987, **28**, 2282, (thermal expansion)
[2] Negi, Y.S., Suzuki, Y.-I., Kawamura, I., Hagiwara, T. et al, *J. Polym. Sci., Part A: Polym. Chem.*, 1992, **30**, 2281, (general props)
[3] Stern, S.A., Mi, Y., Yamamoto, H. and StClair, A.K., *J. Polym. Sci., Part B: Polym. Phys.*, 1989, **27**, 1887, (gas permeability)
[4] Tanaka, K., Kita, H., Okano, M. and Okamoto, K.-I., *Polymer*, 1992, **33**, 585, (gas permeability)
[5] Rogers, M.E., Brink, M.H., McGrath, J.E. and Brennan, A., *Polymer*, 1993, **34**, 849
[6] Fusaro, R.L. and Hady, W.F., *ASLE Trans.*, 1985, **28**, 542, (friction, wear)

PMDA-ODA

– PMDA-ODA

Synonyms: *Poly(pyromellitic dianhydride-alt-4,4'-oxydianiline). Poly(pyromellitic dianhydride-alt-bis(4-aminophenyl)ether). Poly(4,4'-oxydiphenylene pyromellitimide). Poly[(5,7-dihydro-1,3,5,7-tetraoxobenzo[1,2-c:4,5-c']dipyrrole-2,6(1H,3H)-diyl)-1,4-phenyleneoxy-1,4-phenylene]. Poly(1H,3H-benzo[1,2-c:4,5-c']difuran-1,3,5,7-tetrone-alt-4,4'-oxybisbenzenamine]*
Monomers: 1*H*,3*H*-Benzo[1,2-*c*:4,5-*c'*]difuran-1,3,5,7-tetrone, 4,4'-Oxybis[benzenamine]
Material class: Semithermoplastic
Polymer Type: polyimide
CAS Number:

CAS Reg. No.	Note
25036-53-7	
25038-81-7	source base

Molecular Formula: $(C_{22}H_{10}N_2O_5)_n$
Fragments: $C_{22}H_{10}N_2O_5$
Morphology: Orthorhombic, a 0.631, b 0.402, c (chain axis) 3.258. Space group P2/C [10]

Volumetric & Calorimetric Properties:
Density:

No.	Value	Note
1	d 1.54 g/cm^3	cryst. [10]

Thermal Expansion Coefficient:

No.	Value	Note	Type
1	$1.8 \times 10^{-5} - 4.5 \times 10^{-5}$ K^{-1}	in plane, parallel to optic axis, 50–250°, depends on orientation [5]	L
2	$2.3 \times 10^{-5} - 5.6 \times 10^{-5}$ K^{-1}	in plane, perpendicular to optic axis, 50–250°, depends on orientation [5]	L
3	$7.4 \times 10^{-5} - 0.000126$ K^{-1}	out of plane, 50–250° [5]	V

Volumetric Properties General: Fractional free volume 0.123 [8]
Equation of State: Tait equation of state parameters have been reported [5]

Surface Properties & Solubility:
Cohesive Energy Density Solubility Parameters: CED 918 J cm^{-3} [2]; δ 34.6 J$^{1/2}$ cm$^{-3/2}$ [8]

Transport Properties:
Permeability of Gases: *N*-Methylpyrrolidone D_o 3.3–318 × 10^{-11} cm^2 s^{-1} (30–90°, oriented film, depends on film thickness) [4]. [H$_2$] 5.92 × 10^{-10}; [CO] 0.087 × 10^{-10}; [CO$_2$] 1.18 × 10^{-10}; [CH$_4$] 0.026 × 10^{-10}; [O$_2$] 0.31 × 10^{-10}; [N$_2$] 0.045 × 10^{-10} cm^2 (s cm Hg)$^{-1}$ (35°, H$_2$, CO, CO$_2$, CH$_4$ at 10 atm, O$_2$, N$_2$ at 2 atm) [9]
Permeability of Liquids: Diffusion constant [H$_2$O] 9.8–24.9 × 10^{-10} cm^2 s^{-1}, (0.31 eV, 0–23°, depends on thickness and cure conditions, near Fickian process) [7]. Permeability constant [H$_2$O] 15.1 × 10^{-8} cm^2(s cmHg)$^{-1}$ (50°, H$_2$O activity 0.5) [8]
Water Absorption:

No.	Value	Note
1	2.5 %	50h, 25°, 75% relative humidity [1]

Mechanical Properties:
Tensile (Young's) Modulus:

No.	Value	Note
1	2260 MPa	ASTM D2370 [6]
2	2100 MPa	[13]
3	2700–5000 MPa	depends on orientation [2]
4	4700 MPa	cold drawn film, parallel to drawing direction [13]
5	1700 MPa	cold drawn film, perpendicular to drawing direction [13]

Tensile Strength Break:

No.	Value	Note	Elongation
1	260 MPa	ASTM D2370 [6]	85%
2	130 MPa	[13]	
3	300 MPa	cold drawn film, parallel to drawing direction [13]	
4	80 MPa	cold drawn film, perpendicular to drawing direction [13]	

Poisson's Ratio:

No.	Value	Note
1	0.3	ASTM D2370 [6]

Complex Moduli: Dynamic tensile modulus 2620 MPa (100 rad s^{-1}, ASTM D2370) [6]
Mechanical Properties Miscellaneous: Residual stress 21–31 MPa (room temp., films cast and cured on Si) [12]

Electrical Properties:
Dielectric Permittivity (Constant):

No.	Value	Frequency	Note
1	2.96	10 kHz	room temp., dry [7]

Optical Properties:
Total Internal Reflection: Birefringence Δ_n (η_{TE}-η_{TM}) 0.080–0.065 (isotropic in-plane films, 3–20 μm) [6]. Relative dispersion of birefringence D_b 0.148 (486–656 nm) [11]

Polymer Stability:
Environmental Stress: Good resistance to degradation by atomic oxygen in low earth orbit [13]. Good resistance to γ radiation, low gas evolution with doses up to 25 MGy [15]. Good resistance to ionising radiation for doses over 50 MGy. Mech. props. are degraded by cross-linking. In oxygen, deterioration of mech. props. due to chain scission occurs at 10–20% the rate of deterioration in air [16]
Chemical Stability: Swollen by *N*-methylpyrrolidone [3]

Applications/Commercial Products:

Trade name	Details	Supplier
Apical		Allied Signal Corporation
Kapton		DuPont
Pyralin	PI-2540, PI-2545, PI-5878	DuPont

Pyre ML		DuPont
SPI		Toray
Vespel		DuPont

Bibliographic References

[1] Numata, S., Fujisaki, K. and Kinjo, N., *Polymer*, 1989, **28**, 2282, (thermal expansion)
[2] Numata, S. and Miwa, T., *Polymer*, 1989, **30**, 1170, (thermal expansion)
[3] Ojeda, J.R., Mobley, J. and Martin, D.C., *J. Polym. Sci., Part B: Polym. Phys.*, 1995, **33**, 559, (thermal imidisation)
[4] Chang, Y.-L. and Jou, J.-H, *J. Polym. Sci., Part B: Polym. Phys.*, 1994, **32**, 2143, (NMP diffusion)
[5] Pottiger, M.T., Coburn, J.C. and Edman, J.R., *J. Polym. Sci., Part B: Polym. Phys.*, 1994, **32**, 825, (thermal expansion)
[6] Coburn, J.C., Pottiger, M.T., Noe, S.C. and Senturia, S.D., *J. Polym. Sci., Part B: Polym. Phys.*, 1994, **32**, 1271, (optical props)
[7] Lim, B.S., Nowick, A.S., Lee, K.-W. and Viehbeck, A., *J. Polym. Sci., Part B: Polym. Phys.*, 1993, **31**, 545, (water sorption)
[8] Okamoto, K.-I., Tanihara, N., Watanabe, H., Tanaka, K. et al, *J. Polym. Sci., Part B: Polym. Phys.*, 1992, **30**, 1223, (water sorption)
[9] Tanaka, K., Kita, H., Okano, M. and Okamoto, K.-I., *Polymer*, 1992, **33**, 585, (gas permeability)
[10] Freilich, S.C. and Gardner, K.H., Polyimides: Mater., Chem. Charact., *Proc. Int. Conf. Polyimides*, (eds. C. Feger, M.M. Khojasteh and J.E. McGrath), Elsevier, 1989, 513, (cryst struct)
[11] Cha, C.Y. and Samuels, R.J., *J. Polym. Sci., Part B: Polym. Phys.*, 1995, **33**, 259, (optical props)
[12] Jou, J.-H., Lin, C.-P. and Sheu, W.-H., *J. Polym. Sci., Part B: Polym. Phys.*, 1995, **33**, 1803, (film coating and curing)
[13] Kochi, M., Yokota, R., Iizuka, T. and Mita, M., *J. Polym. Sci., Part B: Polym. Phys.*, 1990, **28**, 2463, (mech props)
[14] Koontz, S. and King, G., *J. Spacecr. Rockets*, 1994, **31**, 475
[15] Hegazy, E.-S.A., Sasuga, T., Nishii, M. and Seguchi, T., *Polymer*, 1992, **33**, 2897, (radiation resistance)
[16] Sasuga, T., *Polymer*, 1988, **29**, 1562, (radiation resistance)

PMR-II (Second generation) resins P-53

Synonyms: *Polymerisation of Monomeric Reactants Second generation resins. 6FDE-pPDA Addition cured (nadic end-groups). α-[4-(1,3,3aα,4α,7α,7aα-Hexahydro-1,3-dioxo-4,7-methano-2H-isoindol-2-yl)phenyl]-ω-(1,3,3aα,4α,7α,7aα-hexahydro-1,3-dioxo-4,7-methano-2H-isoindol-2-yl)poly[(1,3-dihydro-aα,4α,7α,7aα-hexahydro-1,3-dioxo-4,7-methano-2H-isoindol-2-yl)poly[(1,3-dihydroH-isoindole-5,2-diyl)-1,4-phenylene]. PMR-II*
Monomers: Dimethyl bicyclo[2.2.1]hept-5-ene-2,3-dicarboxylate, 1,4-Benzenediamine, ar,ar'-Dimethyl dihydrogen 4,4'-[(2,2,2-trifluoromethyl)ethylidene]bis[1,2-benzenedicarboxylate]
Material class: Thermosetting resin
Polymer Type: polyimide (end capped)
CAS Number:

CAS Reg. No.	Note
70771-19-6	CRU with stereochemically described end-groups

Molecular Formula: $(C_{25}H_{10}F_6N_2O_4)_n$
Fragments: $C_{25}H_{10}F_6N_2O_4$
Molecular Weight: M_n 1829

Volumetric & Calorimetric Properties:

Glass-Transition Temperature:

No.	Value	Note
1	342°C	uncured [2]
2	337°C	uncured [3]
3	362°C	post-cured, air, 343°, 16h [2]
4	366°C	post-cured, air, 341°, 16h [3]

Transport Properties:

Polymer Solutions Dilute: Viscosity increases from 239–270° before decreasing markedly to a minimum at approx. 325°, after which it increases with increasing temp. [2] [η] 0.1 dl g^{-1} (DMF, 25°) [2]

Optical Properties:

Transmission and Spectra: Ir spectral data have been reported [2]

Polymer Stability:

Decomposition Details: Shows an average weight loss of 27.2% after heating in air at 371° for 400h [1]. 5% Weight loss in air at 463° [2]. CO_2 is liberated at 361°; other decomposition products include H_2O, trifluoromethane, CO, and SiF_4 formed by reaction of fluorine radicals with any glass surface [2]

Applications/Commercial Products:

Processing & Manufacturing Routes: Synth. by reaction of 4,4'-(hexafluoroisopropylidene) diphthalic anhydride dimethyl ester diacid (n moles), p-phenylenediamine and nadic acid ester in MeOH in a mole of ratio of n:n+1:2. This yields a viscous gum which is further heated at 210° for 2h [1,2]. Has a processing window of approx. 100° between the onset of melting at 240° and curing at 375° [3]
Applications: Has high temp. applications

Bibliographic References

[1] Sutter, J.K., Jobe, J.M., Crane, E.A., *J. Appl. Polym. Sci.*, 1995, **57**, 1491, (synth., ageing)
[2] Chuang, K.C., Vannucci, R.D., Ansari, I., Cerny, L.L., Scheiman, D.A., *J. Polym. Sci., Part A: Polym. Chem.*, 1994, **32**, 1341, (synth., degradation)
[3] Chuang, K.C., *High Performance Polym.*, 1995, **7**, 81

PMR-15, Unfilled P-54

Synonyms: α-[4-[[4-(1,3,3a,4,7,7a-Hexahydro-1,3-dioxo-4,7-methano-2H-isoindol-2-yl)phenyl]methyl]phenyl]-ω-(1,3,3a,4,7,7a-hexahydro-1,3-dioxo-4,7-methano-2H-isoindol-2-yl)poly[(1,3-dihydro-1,3-dioxo-2H-isoindole-2,5-diyl)carbonyl(1,3-dihydro-1,3-dioxo-2H-isoindole-5,2-diyl)-1,4-phenylenemethylene-1,4-phenylene]
Monomers: Methyl hydrogen bicyclo[2.2.1]hept-5-ene-2,3-dicarboxylate, ar,ar'-Dimethyl dihydrogen 4,4'-carbonylbis[1,2-benzenedicarboxylate], 4,4'-Methylenebis[benzenamine]
Material class: Thermosetting resin
Polymer Type: polyimide (end capped)
CAS Number:

CAS Reg. No.	Note
78392-33-3	CRU with stereochemically described end-groups
68845-78-3	CRU with end-groups
37238-52-1	source based

Molecular Formula: $(C_{30}H_{16}N_2O_5)_n$
Fragments: $C_{30}H_{16}N_2O_5$
Additives: Glass fibre, graphite fibre, carbon fibre
Morphology: The effect of MW, composition and pressure on the crystallinity has been reported [1]

Volumetric & Calorimetric Properties:
Glass-Transition Temperature:

No.	Value	Note
1	370°C	post-cure, in air 371°, 25 h [4]
2	330°C	post-cure, 316°, 10 h [5]

Optical Properties:
Transmission and Spectra: C-13 nmr spectral data have been reported [2]

Polymer Stability:
Upper Use Temperature:

No.	Value	Note
1	300°C	[5]

Decomposition Details: Exhibits approx. 18% weight loss after heating in air at 371° for up to 300 h at both 1 atm and 4 atm pressure [4]. 5% Weight loss at 383°, 10% weight loss at 425°, 20% weight loss at 467°, 30% weight loss at 482°, when heated in air [5]. Activation energy for thermooxidative degradation of uncured resin 174 kJ mol^{-1} [5]

Applications/Commercial Products:
Processing & Manufacturing Routes: A nadic end-capped PMR resin is heated under pressure at high temp. The nadic end-caps undergo addition polymerisation to yield the desired product [3]
Applications: In polymer matrix composites for high temp. applications

Bibliographic References
[1] Fan, Z., Wu, X., Lin, Y. and Li, S., *Huadong Huagong Xueyuan Xuebao*, 1989, **15**, 615
[2] Grenier-Loustalot, M.-F. and Grenier, P., *High Performance Polym.*, 1991, **3**, 113, 263, (C-13 nmr)
[3] Hoyle, N.D., Stewart, N.I., Wilson, D., Baschast, M. et al, *High Performance Polym.*, 1989, **1**, 285, (synth)
[4] Malarik, D.C. and Vannucci, R.D., *SAMPE Q.*, 1992, **23**, 3
[5] Arnold, C.A. and Maskell, R.K., *Int. SAMPE Symp. Exhib.*, 1991, **36**, 1135, (thermal stability)

Poloxamers

P-55

$$HO-[CH_2CH_2O]_x-[CH(CH_3)CH_2O]_y-[CH_2CH_2O]_z-H$$

2 < x < 128   16 < y < 67   2 < z < 128

Synonyms: *Polyoxyethylene-polyoxypropylene block copolymer. Methyl oxirane polymers. Polyethylene-polypropylene glycols. EO/PO block polymer. Poly(oxyethylene-block-oxypropylene). Oxirane polymers with methyl oxirane (block, triblock)*
Related Polymers: Polypropylene oxide, Meroxapol
Monomers: Oxirane, Propylene oxide
Material class: Copolymers
Polymer Type: polyalkylene ether
CAS Number:

CAS Reg. No.	Note
9003-11-6	generic
106392-12-5	block copolymer

Molecular Formula: $[(C_2H_4O).(C_3H_6O).(C_2H_4O)]_n$
Fragments: C_2H_4O C_3H_6O C_2H_4O
Molecular Weight: M_n 1100-14500

Surface Properties & Solubility:
Surface and Interfacial Properties General: Cloud points have been measured with respect to composition [2]. Surface tension vs. concentration, and interfacial tension vs. concentration plots have been reported [2,3]

Transport Properties:
Polymer Solutions Dilute: Bulk viscosity, zero shear viscosity, complex viscosity, storage and loss moduli, and the yield stress value for the gel state have been reported [2,3]

Polymer Stability:
Biological Stability: Mod. toxic by ingestion, intraperitoneal route [1]

Applications/Commercial Products:

Trade name	Supplier
Berol	Berol Kemi AB
Genapol	Hoechst Celanese
Hodag Nonionic	Hodag
Lutrol	BASF
Macol	PPG-Mazer
Pluronic	BASF
Polyglycol	Dow
Synperonic PE	ICI, UK

Bibliographic References
[1] *The Thesaurus of Chemical Products,* 2nd edn., (eds. M. Ash and I. Ash), Edward Arnold, 1992, **1**
[2] Prasad, K.N., Luong, T.T., Florence, A.T., Paris, J. et al, Colloid Interface Sci., *(Proc. Int. Conf.)*, 1979, **69**, 225
[3] Wanka, G., Hoffmann, H. and Ulbricht, W., *Colloid Polym. Sci.*, 1990, **268**, 101

1,2-Poly(1,3-butadiene)

P-56

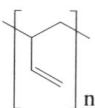

Synonyms: *Vinyl-type butadiene. 1,2-Polybutadiene. 1,2-pb. 1,2-PB*
Monomers: 1,3-Butadiene
Material class: Thermoplastic
Polymer Type: polyolefins
CAS Number:

CAS Reg. No.	Note
9003-17-2	atactic form
29406-96-0	isotactic homopolymer
31567-90-5	syndiotactic
36522-63-1	syndiotactic

Molecular Formula: $(C_4H_6)_n$
Fragments: C_4H_6
Tacticity: Atactic, syndiotactic, isotactic forms known
Morphology: 1,2-Syndiotactic (98%) [2]. Orthorhombic [3], a 10.98, b 6.60, c 5.14; 1,2-isotactic form (99%) [2]; rhombohedral, a 17.3, b 17.3, c 6.5 [2]. Morphology of blends with EPDM has been reported [39]

– 1,2-Poly(1,3-butadiene)

Volumetric & Calorimetric Properties:

Density:

No.	Value	Note
1	d^{20} 0.96 g/cm^3	99% isotactic form [1]
2	d 0.96 g/cm^3	98% syndiotactic [1]
3	d 0.89 g/cm^3	[31,32]
4	d 0.885 g/cm^3	95.3% 1,2-polymer [45]

Equation of State: Equation of state information has been reported [25,27,42]

Melting Temperature:

No.	Value	Note
1	126°C	isotactic form [1]
2	156°C	syndiotactic form [1]
3	206°C	99% 1,2-polymer; 0.8% cis-1,4-polymer [23]

Glass-Transition Temperature:

No.	Value	Note
1	-15°C	tacticity not specified [2]
2	-7°C	[26]
3	-15°C	1,2-polymer [46]
4	40°C	99% 1,2-polymer, 0.8% 1,4-polymer [23]
5	-12°C	95.3% 1,2-polymer [45]

Transition Temperature General: M_n relationship to T_g has been reported. Other values of T_g have been reported [27]

Deflection Temperature:

No.	Value	Note
1	52.5°C	99% 1,2-polymer, 0.8% cis-1,4-polymer [23]
2	56.5°C	99% 1,2-polymer, 0.8% cis-1,4-polymer [23]

Surface Properties & Solubility:

Solubility Properties General: High levels of 1,2-polymer give complete miscibility with cis-1,4-polyisoprene even at high MW [28]

Solvents/Non-solvents: 94% 1,2-Polymerised form sol. toluene, THF [2]. 99% 1,2-Polymer sol. tetralin (140°); insol. boiling toluene, boiling C_6H_6, boiling xylene [23]. Highly stereoregular polymer has very low solubility [23]

Wettability/Surface Energy and Interfacial Tension: Critical surface tension 25 mN m^{-1} [29]

Transport Properties:

Polymer Solutions Dilute: Temp. dependence of rheological props. has been reported [30,31,32]. Rheological props. of 62% 1,2-polymer have been reported [41]

Polymer Melts: Melt viscosity data have been reported [27]

Permeability of Gases: Permeability of He, Ne, N_2 and CO_2 has been reported [34]. Other permeability data have been reported [33]

Mechanical Properties:

Mechanical Properties General: Props. of mixed polymer compositions [35] and blends [49] have been reported. Storage and loss moduli [43] of the syndiotactic [28] and atactic [27] polymers have been reported. Dynamic mech. parameters of blends with EPDM (see ethylene-propylene elastomers), have been reported [39]

Tensile (Young's) Modulus:

No.	Value	Note
1	51 MPa	99% 1,2-polymer, 0.8% 1,4-polymer [23]

Flexural Modulus:

No.	Value	Note
1	51 MPa	99% 1,2-polymer, ASTM D747 [23]
2	78 MPa	99% 1,2-polymer, ASTM D747 [23]

Tensile Strength Break:

No.	Value	Note	Elongation
1	27.5 MPa	99% 1,2-polymer, ASTM D412 [23]	185%

Flexural Strength at Break:

No.	Value	Note
1	24.5 MPa	99% 1,2-polymer, ASTM D747 [23]
2	31 MPa	99% 1,2-polymer, ASTM D747 [23]

Compressive Modulus:

No.	Value	Note
1	320 MPa	99% 1,2-polymer, 0.8% 1,4-polymer [23]
2	420 MPa	99% 1,2-polymer, 0.8% 1,4-polymer [23]

Tensile Strength Yield:

No.	Value	Note
1	36.5 MPa	99% 1,2-polymer

Miscellaneous Moduli:

No.	Value	Note	Type
1	0.58 MPa	[27]	Average plateau modulus
2	0.81 MPa	[31,32]	Plateau modulus

Viscoelastic Behaviour: Temp. dependence of viscoelastic props. described by the Williams-Landel-Ferry equation has been reported [30,44]. Stress-strain behaviour (95.3% 1,2-polymer) has been reported [45]

Mechanical Properties Miscellaneous: Has high uniaxial extension and shear combined with cyclic straining. Flow behaviour related to polymer processing has been reported [38]. Peel strength of syndiotactic polymer has been reported [28]. Peel strength of blends with EPDM has been reported [39]

Hardness: Rockwell R60 (99% 1,2-polymer, 0.8% 1,4-polymer) [23]

Failure Properties General: Critical recoverable deformation ϵ_e 0.44 [36]

Fracture Mechanical Properties: Extension ratio at break has been reported [50]

Izod Area:

No.	Value	Notch	Note
1	5.95 kJ/m^2	Unnotched	6.02 kg cm cm^{-2}, 99% 1,2-polymer, 0.8% 1,4-polymer, ASTM D256-56 [23]
2	19.767 kJ/m^2	Unnotched	20 kg cm cm^{-2}, 99% 1,2-polymer, 0.8% 1,4-polymer, ASTM D256-56 [23]

Optical Properties:
Transmission and Spectra: Ir (syndio, iso forms) [7,10,11,17,28,40], uv [13], esr [19], H-1 nmr [17,18] and C-13 nmr [20,24] spectral data have been reported
Total Internal Reflection: Birefringence measurements have been reported [23]

Polymer Stability:
Thermal Stability General: Thermal stability lies in between that of the *cis* and *trans* 1,4 polymerised forms [21]. Syndiotactic form tends to cross-link at temps. above 150° [23]
Decomposition Details: Cyclisation leads to a loss of unsaturation and some *cis/trans* isomerisation [21]. Heating in nitrogen (20°/min.) causes decomposition by the liberation of hydrocarbon gases (including cyclic ones) [22]. Thermal degradation causes intramolecular cyclisation yielding cyclohexane rings [47]
Flammability: High vinyl polybutadienes are classed as high volatile loss, low-char polymers; they give slightly greater loss of volatiles than does *cis*-1,4-polybutadiene [48]
Chemical Stability: Stable to H_2O, bases, AcOH and HCl. Oxidised by contact with strong acids and acidified dichromate (syndiotactic form) [23]
Stability Miscellaneous: High energy radiation reduces saturation markedly, possibly also leading to intramolecular cyclisation and cross-linking [37]

Applications/Commercial Products:
Processing & Manufacturing Routes: A range of catalysts, mainly chromium and aluminium, [4,5,6,7,8,9,12,15,16] chiefly give the syndiotactic form. The isotactic form is obtainable using chromium-carbonyl complex [7] as the catalyst
Applications: Used for making transparent bottles and films. High permeability and tear resistance makes it useful for packaging of fresh fruit and fish. Also has adhesive applications

Trade name	Details	Supplier
Budene 1255	soln. polymer containing antioxidant	Goodyear
JSRRB	mainly syndiotactic	Japan Synthetic
Ricon		Advanced Resins

Bibliographic References
[1] Natta, G., *Science (Washington, D.C.)*, 1965, **147**, 269, (density, melting point)
[2] *Polym. Handb.*, 3rd edn., (eds. J. Brandrup and E.H. Immergut), John Wiley and Sons, 1989, (density, cryst data, transition temp, solubility, x-ray data)
[3] Natta, G. and Corradini, P., *J. Polym. Sci.*, 1956, **20**, 251, (struct)
[4] *Brit. Pat.*, 1960, 836 189, (synth)
[5] *Brit. Pat.*, 1960, 848 064, (synth)
[6] *Brit. Pat.*, 1960, 832 457, (synth)
[7] *Brit. Pat.*, 1960, 854 615, (synth)
[8] *U.S. Pat.*, 1963, 3 105 828, (synth)
[9] *U.S. Pat.*, 1966, 3 232 920, (synth)
[10] Morero, D., Santambrogio, A., Porri, L. and Ciampelli, F., *Chim. Ind. (Milan)*, 1951, **41**, 758, (ir)
[11] Silas, R.S., Yates, J. and Thornton, V., *Anal. Chem.*, 1959, **31**, 529, (spectra)
[12] Meyer, A.W., Hampton, R.R. and Davison, J.A., *J. Am. Chem. Soc.*, 1952, **74**, 2294, (synth)
[13] Klevens, H.B., *J. Polym. Sci.*, 1953, **10**, 97, (uv spectrum)
[14] *Belg. Pat.*, 1957, 558 148
[15] *Ital. Pat.*, 1959, 599 661
[16] *Ital. Pat.*, 1965, 682 423
[17] Tanaka, Y., Takeuchi, Y., Kobayashi, M. and Tadokoro, H., *J. Polym. Sci., Part A-2*, 1971, **9**, 43, (ir, nmr)
[18] Zymonas, J., Santee, E.R. and Harwood, H.J., *Macromolecules*, 1973, **6**, 129, (H-1 nmr)
[19] Zott, H. and Heusinger, H., *Macromolecules*, 1975, **8**, 182, (epr)
[20] Conti, F., Delfini, M., Segre, A.L., Pini, D. and Porri, L., *Polymer*, 1974, **15**, 816, (C-13 nmr)
[21] Chiantore, O., Luda di Cortemiglia, M.P., Guaita, M. and Renelina, G., *Makromol. Chem.*, 1989, **190**, 3143, (thermal degradation)
[22] Redfern, J.P., *Polym. Int.*, 1991, **26**, 51
[23] Ashitaka, H., Ishikaura, H., Jeno, H. and Nagasaka, A., *J. Polym. Sci., Polym. Chem. Ed.*, 1983, **21**, 1111, 1853, (synth, props)
[24] Kumar, D., Rama Rao, M. and Rao, K.V.C., *J. Polym. Sci.*, 1983, **21**, 365, (C-13 nmr)
[25] Yi, Y.X. and Zoller, P., *J. Polym. Sci., Part B: Polym. Phys.*, 1993, **31**, 779, (equations of state)
[26] Krauss, G., Childers, W. and Gruver, J.T., *J. Appl. Polym. Sci.*, 1967, **11**, 158, (T_g)
[27] Roovers, J. and Toporowski, P.M., *Rubber Chem. Technol.*, 1990, **63**, 734, (T_g)
[28] Roland, C.M., *Rubber Chem. Technol.*, 1988, **61**, 866
[29] Lee, L. and Lee, H., *J. Polym. Sci., Part A-2*, 1967, **5**, 1103, (surface tension)
[30] Roland, C.M. and Ngai, K.L., *Macromolecules*, 1992, **24**, 5315, (rheological props)
[31] Roovers, J. and Toporowski, P.M., *J. Polym. Sci., Polym. Phys. Ed.*, 1988, **26**, 1251, (rheological props)
[32] Roovers, J., *Macromolecules*, 1988, **21**, 1517
[33] Stern, S.A., Krishnakumar, B. and Nadakalti, S.M., *Physical Properties of Polymers Handbook*, (ed. J.E. Mark), AIP Press, 1996, 694, (gas permeability)
[34] Cowling, R. and Park, G.S., *J. Membr. Sci.*, 1979, **5**, 199, (gas permeability)
[35] Short, J.N., Kraus, G., Zelinski, R.P. and Naylor, F.E., Soc. Plast. Eng., *Tech. Pap.*, 1959, **5**, (mech props)
[36] Borisenkova, E.K., Sabsai, O.Yu, Kurbanaliev, M.K., Dreval, V.E. and Vinogradov, G.V., *Polymer*, 1978, **19**, 1473, (deformation)
[37] Bohm, G.G.A. and Tveekram, J.O., *Rubber Chem. Technol.*, 1982, **55**, 575
[38] Vinogradov, G.V., Dreval, V.E. and Yanovsky, Yu. G., *Rheol. Acta*, 1985, **24**, 574
[39] Hamed, G.R., *Rubber Chem. Technol.*, 1982, **55**, 151
[40] Noda, I., Dowrey, A.E. and Marcott, C., *Physical Properties of Polymers Handbook*, (ed. J.E. Mark), AIP Press, 1996, 293
[41] Fetters, L.J., Lohse, D.J. and Colby, R.H., *Physical Properties of Polymers Handbook*, (ed. J.E. Mark), AIP Press, 1996, 338
[42] Carella, J.M. and Graessly, W.W., *Macromolecules*, 1987, **20**, 2226
[43] Yanovsky, Y.G., Vinogradov, G.V. and Ivanova, L.I., *Chem. Eng. Commun.*, 1985, **32**, 219
[44] Zorn, R., McKenna, G.B., Willner, L. and Richter, D., *Macromolecules*, 1995, **28**, 8552
[45] Kramer, O., Carpenter, R.L., Ty, V. and Ferry, J.D., *Macromolecules*, 1974, **7**, 79
[46] Bahary, W.S., Sapper, D.I. and Lane, J.H., *Rubber Chem. Technol.*, 1967, **40**, 1529
[47] Raven, A. and von Heusinger, H., *Angew. Makromol. Chem.*, 1975, **42**, 183
[48] Lawson, D.F., *Rubber Chem. Technol.*, 1986, **59**, 455
[49] Niv, M.M. and Cohen, R.E., *Rubber Chem. Technol.*, 1994, **67**, 342
[50] Ahagon, A., *Rubber Chem. Technol.*, 1986, **59**, 187

Poly(acetaldehyde) P-57

Synonyms: *Poly(ethanal)*
Monomers: Acetaldehyde
Material class: Synthetic Elastomers
Polymer Type: polyalkylene ether
CAS Number:

CAS Reg. No.	Note
9002-91-9	homopolymer
28265-89-6	isotactic

– Poly(acetylene)

Molecular Formula: $(C_2H_4O)_n$
Fragments: C_2H_4O
Additives: May be stabilised by polyamides and aromatic amine antioxidants
Tacticity: Stereoblock [15,16] and isotactic forms are known
Morphology: Isotactic form, tetragonal, a 14.63, c 4.79 [7,9]

Volumetric & Calorimetric Properties:
Density:

No.	Value	Note
1	d 1.14 g/cm^3	[7]

Melting Temperature:

No.	Value	Note
1	-123.3°C	[6]

Glass-Transition Temperature:

No.	Value	Note
1	-30°C	[1,2]

Surface Properties & Solubility:
Solvents/Non-solvents: Sol. EtOH, Et$_2$O, butyl acetate, Me$_2$CO, AcOH, CCl$_4$ [15], 2-butanone [6]. Insol. H$_2$O, petrol, C$_6$H$_6$ [14]

Transport Properties:
Polymer Solutions Dilute: [η] 2.4 dl g^{-1} (27.6°) [6]

Mechanical Properties:
Tensile Strength Break:

No.	Value	Note	Elongation
1	0.17–0.18 MPa	25–27 psi, 23°, compressed film [5]	580%

Optical Properties:
Transmission and Spectra: Ir and nmr spectral data have been reported [5,14,16,18]

Polymer Stability:
Polymer Stability General: Very easily decomposed in the absence of additional stabilisers. Decomposition can be initiated by the presence of a number of different species and proceeds via several different mechanisms [12,19]
Thermal Stability General: Comparatively low thermal stability which may be improved by capping with acetoxy groups [19]
Decomposition Details: Depolymerisation occurs at room temp. (if unstabilised), producing acetaldehyde. Can evaporate completely in a few days or weeks [4]
Chemical Stability: Acidic compds. accelerate the decomposition process [4]
Stability Miscellaneous: Fairly low oxidative stability [5]. May be degraded photolytically in soln. [19]

Applications/Commercial Products:
Processing & Manufacturing Routes: Isotactic polyacetaldehyde is obtained at -75°, using alkali metal alkoxides [10], alkali metals or metal alkyls [11] as initiators. A novel aluminium catalyst has been developed to give a high degree of stereoregularity [17]. Amorph. polymers are obtained when using cationic initiators at temps. below -40° [4]. A number of acidic initiatior are used, as well as alumina [6,8] and silica
Applications: Main uses are in coating and resist applications. Also used as a molluscicide and in photosensitive chemical compositions

Bibliographic References

[1] Williams, G., *Trans. Faraday Soc.*, 1963, **59**, 1397, (glass transition temp)
[2] Read, B.E., *Polymer*, 1962, **3**, 529, (glass transition temp)
[3] *Encycl. Polym. Sci. Eng.*, 2nd edn., Vol. 1, (ed. J.I. Kroschwitz), John Wiley and Sons, 1989, 623, (synth)
[4] *Ullmanns Encykl. Ind. Chem.*, Vol. 1, 5th edn., (ed. W. Gerhartz), VCH, 1985, 42, (synth, stability)
[5] Vogl, O., *J. Polym. Sci.*, *Part A-2*, 1964, **2**, 4591, (spectra, tensile strength)
[6] Furukawa, J., Saegusa, T., Tsuruta, T., Fujii, H. and Tatano, T., *J. Polym. Sci.*, 1959, **36**, 546, (synth, melting point)
[7] Natta, G., Corradini, P. and Bassi, I.W., *J. Polym. Sci.*, 1961, **51**, 505, (struct)
[8] Furukawa, J., Saegusa, T., Tsuruta, T., Fujii, H. et al, *Makromol. Chem.*, 1959, **33**, 32, (isotactic polymerisation)
[9] Natta, G., Mezzanti, G., Corradini, P. and Bassi, I.W., *Makromol. Chem.*, 1960, **37**, 156, (cryst struct)
[10] *Fr. Pat.*, 1960, 1 268 191, (metal alkoxide catalysts)
[11] *Fr. Pat.*, 1960, 1 268 322, (metal alkyl catalysts)
[12] Delzenne, G. and Smets, G., *Makromol. Chem.*, 1955, **18**, 82, (stability)
[13] Novak, A. and Whalley, E., *Can. J. Chem.*, 1959, **37**, 1710, (ir)
[14] Berington, J.C. and Norrish, R.G.W., *Proc. R. Soc. London*, 1949, **A196**, 363, (rev)
[15] Vogl, O., *Polym. Prepr. (Am. Chem. Soc., Div. Polym. Chem.)*, 1966, **7**, 216, (rev)
[16] Fujii, H., Furukawa, J., Saegusa, T. and Kawasaki, A., *Makromol. Chem.*, 1960, **40**, 226, (synth, ir)
[17] Tani, H. and Yasuda, H., *J. Polym. Sci., Part B: Polym. Lett.*, 1969, **7**, 17, (synth)
[18] Goodman, M. and Brandrup, J., *J. Polym. Sci., Part A: Polym. Chem.*, 1965, **3**, 327, (nmr)
[19] Marsh, D.G., *J. Polym. Sci., Polym. Chem. Ed.*, 1976, **14**, 3013, (photolytic degradation)

Poly(acetylene)   P-58

Synonyms: *Poly(ethyne)*
Monomers: Ethyne
Material class: Thermoplastic
Polymer Type: polyacetylene
CAS Number:

CAS Reg. No.	Note
25067-58-7	
73589-68-1	(E)-form
26571-64-2	
74373-36-7	(Z)-form

Molecular Formula: $[(C_2H_2)]_n$
Fragments: C_2H_2
Morphology: Form I *cis*-form, orthorhombic, a 7.74, b 4.32, c 4.47 (a 7.32, b 4.24, c 2.46); form II *trans*-form, orthorhombic, a 5.62, b 4.92, c 2.59; form III, *trans*-form, monoclinic, a 3.73, b 3.73, c 2.44, γ96–100° [1,2]

Volumetric & Calorimetric Properties:
Density:

No.	Value	Note
1	d 1.16 g/cm^3	*cis* [1]
2	d 1.2 g/cm^3	*trans*, orthorhombic [1]
3	d 1.27 g/cm^3	*trans*, monoclinic [1]

Thermal Expansion Coefficient:

No.	Value	Note	Type
1	9 × 10^{-5} K^{-1}	calc. [3]	L

Latent Heat Crystallization: ΔH 7.75 kJ mol^{-1} (-37°, *cis* to *trans* conversion) [9]

Thermal Conductivity:

No.	Value	Note
1	2.1 W/mK	*cis* [5]
2	3.8 W/mK	*trans* [5]

Specific Heat Capacity:

No.	Value	Note	Type
1	60 kJ/kg.C	27°, *cis*-form [4]	P

Surface Properties & Solubility:
Solvents/Non-solvents: Insol. any solvent [6]

Mechanical Properties:
Tensile (Young's) Modulus:

No.	Value	Note
1	100000 MPa	*trans* [7]
2	30000–40000 MPa	*cis* [7]
3	25000–33000 MPa	*trans* [21]
4	384300–590500 MPa	*trans* [8]
5	170100–309000 MPa	*cis* [8]

Tensile Strength Break:

No.	Value	Note
1	900 MPa	*trans* [7]
2	600 MPa	*cis* [7]

Tensile Strength Yield:

No.	Value	Elongation	Note
1	15 MPa	1.2%	*cis* [20]

Electrical Properties:
Electrical Properties General: Conductivity can be increased by seven orders of magnitude by iodine doping [25]

Dielectric Permittivity (Constant):

No.	Value	Note
1	3.4–3.7	*cis* [24]
2	3.9–4.2	*trans* [24]

Complex Permittivity and Electroactive Polymers: Conductivity 4.4×10^{-5} S cm^{-1} (25°, *trans*-form) [11]; other conductivity values have been reported [11,12]. Band gap 1.2–1.7 eV [15]

Dissipation (Power) Factor:

No.	Value	Note
1	0.01–0.025	-100–150° [10]

Optical Properties:
Transmission and Spectra: Ir [17,18], C-13 nmr [16], epr [19], Raman [14] and x-ray [2] spectral data have been reported

Polymer Stability:
Decomposition Details: Pyrolysis proceeds by facile electron-proton exchanges to form various proton-enriched products [23] and C_6H_6

Environmental Stress: Oxidises gradually in air (15 days) with resulting lower conductivity and tensile props. [22]

Applications/Commercial Products:
Processing & Manufacturing Routes: Produced by Ziegler-Natta polymerisation of acetylene in an inert solvent. Electrically conducting films may be produced by halogen doping with Cl, Br or I [25]

Applications: Highly conducting electrical and optical fibre (often doped). Solar cell and battery applications

Bibliographic References
[1] Lieser, G., Wegner, G., Müller, W. and Enkelmann, V., Makromol. Chem., *Rapid Commun.*, 1980, **1**, 627
[2] Shimamura, K., Karasz, F.E., Hirsch, J.A. and Chien, J.C.W., Makromol. Chem., *Rapid Commun.*, 1981, **2**, 473
[3] Ito, T., Shirakawa, H. and Ikeda, S., *J. Polym. Sci.*, 1974, **12**, 11
[4] Domalski, E.S. and Hearing, E.D., *J. Phys. Chem. Ref. Data*, 1990, **19**, 887
[5] Moses, D. and Denenstein, A., *Phys. Rev. B: Condens. Matter*, 1984, **30**, 2090
[6] Chien, J.C.W., Capistran, J., Karasz, F.E., Schen, M. and Fan, J.L., Polym. Prepr. (Am. Chem. Soc., *Div. Polym. Chem.*), 1982, **23**, 76
[7] Akagi, K., Suezaki, M., Shirikana, H., Kyatani, H., et al, *Synth. Met.*, 1989, **28**, D1
[8] Hong, S.Y. and Kertesz, M., *Phys. Rev. B: Condens. Matter*, 1990, **41**, 11368
[9] Simionescu, C.I. and Negulescu, I.I., *J. Macromol. Sci., Chem.*, 1985, **22**, 1001
[10] Chen, S.A. and Li, L.S., Makromol. Chem., *Rapid Commun.*, 1983, **4**, 503
[11] Gau, S.C., Milliken, J., Pron, A., Macdiarmid, A.G. and Heeger, A.J., J.C.S., *Chem. Commun.*, 1979, 662
[12] Leising, G., *Polym. Commun.*, 1984, **25**, 201
[13] Hankin, A.G. and North, A.M., *Trans. Faraday Soc.*, 1967, **63**, 1525
[14] Schügerl, F.B. and Kuzmany, H., *J. Chem. Phys.*, 1981, **74**, 953
[15] Kasowski, R.V., Hsu, W.Y. and Caruthers, E.B., *J. Chem. Phys.*, 1980, **72**, 4896
[16] Gibson, H.W., Pochan, J.M. and Kaplan, S., *J. Am. Chem. Soc.*, 1981, **103**, 4619
[17] Cernia, E., D'ilario, L., Lupoli, M., Mantovani, E., et al, J. Polym. Sci., *Polym. Chem. Ed.*, 1984, **22**, 3393
[18] Tabata, Y., Saito, B., Shibano, H., Sobue, H. and Oshima, K., Makromol. Chem., 1964, **76**, 89
[19] Bernier, P., Rolland, M., Linaya, C., Disi, M., et al, *Polym. J. (Tokyo)*, 1981, **13**, 201
[20] Druy, M.A., Tsang, C.H., Brown, N., Heeger, A.J. and Macdiarmid, A.G., J. Polym. Sci., *Polym. Phys. Ed.*, 1980, **18**, 429
[21] Cao, Y., Smith, P. and Heeger, A.J., *Polymer*, 1991, **32**, 1210
[22] Gibson, H.W. and Pochan, J.M., *Macromolecules*, 1982, **15**, 242
[23] Fan, J.L. and Chien, J.C.W., J. Polym. Sci., *Polym. Chem. Ed.*, 1983, **21**, 3453
[24] Devreux, F., Dory, I., Mihaly, L. Pekker, S., et al, J. Polym. Sci., *Part B: Polym. Phys.*, 1981, **19**, 743
[25] Shirawakawa, H., Lewis, E.J., Macdiarmid, A.G., Chiang, C.K. and Heeger, A.J., J.C.S., *Chem. Commun.*, 1977, 578, (doping)
[26] Cataldo, F., *Polymer*, 1992, **30**, 3073, (polymerisation)

Poly(acetylenes)

Related Polymers: Poly(acetylene), Poly(phenylacetylene), Poly(diphenyldiacetylene)

General Information: These are usually brittle materials that cannot be moulded. Physical props. that are dependent upon soln. characteristics cannot be determined because of insolubility. Variability of props. caused by different laboratory preparation techniques [1,4]. Materials tend to show high electrical conductivity, particularly when doped [3]

Polymer Stability:
Environmental Stress: Poly(acetylenes) are readily oxidised at room temp. Doping decreases their susceptibility to oxygen [1]
Stability Miscellaneous: Polyacetylenes are innocuous and should present no health hazard [1]

Applications/Commercial Products:
Processing & Manufacturing Routes: Acetylenes undergo both additions and condensation polymerisation with homogeneous and heterogeneous Ziegler-Natta catalysts, transition-metal complexes, free-radical initiators, thermal polymerisation and γ-radiation to form linear or cyclic polymeric products [1,2]
Applications: Major uses of poly(acetylenes) have not yet developed due to their intractable and oxygen-sensitive nature. Costs of producing poly(acetylenes) are very high

Bibliographic References
[1] *Encycl. Polym. Sci. Eng.*, (ed. J.I. Kroschwitz), John Wiley and Sons, 1985, 87
[2] Smolin, E.M. and Hoffenberg, D.S., *Encycl. Polym. Sci. Technol.*, Vol. 1, John Wiley and Sons, 1964, 46
[3] *Handbook of Conducting Polymers*, (ed. T.A. Skotheim), Vol. 1, Dekker, 1986, 48
[4] Simionescu, C.I. and Perec, V., *J. Polym. Sci., Polym. Symp.*, 1980, **67**, 43

Polyacrylamide   P-60
Synonyms: *Poly(2-propenamide)*
Related Polymers: Acrylamide copolymer emulsion
Monomers: Acrylamide
Material class: Thermoplastic
Polymer Type: acrylic
CAS Number:

CAS Reg. No.
9003-05-8

Polyacrylamide gels   P-61
Synonyms: *Acrylamide gels*
Related Polymers: Polyacrylamide
Monomers: Acrylamide
Polymer Type: acrylic
General Information: A cross-linked polymer which finds extensive use due especially to its equilibrium swelling props. in water or aq.solns.
Morphology: Polyacrylamide gels, particularly those at lower concentrations, contain heterogeneities believed to be due to cross-link clustering. The gels are transparent and flexible at low degrees of cross-linking but become opaque and fragile at higher degrees of cross-linking. The micromorphologies of polyacrylamide hydrogels have been characterised by multifractal studies [4,5,6,8,19]. Microgels have been characterised by electron microscopy [3,9]

Transport Properties:
Transport Properties General: Perhaps the most important property of polyacrylamide gels is their swelling behaviour. This is very much dependent on the composition of the gel and varies tremendously. Swelling pressure increases with the degree of cross-linking (measured by concentration of cross-linking agent) and the concentration of gel. Ageing gels can dramatically affect the swelling because of hydrol., gel-inhomogeneities and structural changes. [2,4,5]. The dynamic polymer-polymer correlation length is weakly dependent on concentration but is temp-independent. The static screening length is also concentration-dependent. At high cross-linking density the two correlation lengths become approx. equal [12]
Permeability of Liquids: Diffusion coefficient ethylene glycol 4.2×10^{10} ms^{-1}, glycerol 3.2×10^{10} ms^{-1}. Diffusion of carbohydrates in soln. has also been reported. Diffusion coefficient increases with increasing temp. [13]
Permeability and Diffusion General: The diffusion of poly(acrylic acid) and rhodamine dye through polyacrylamide gels has been reported [14]
Water Content: Three types of 'water' have been identified in the hydrogels: ordinary, intermediate and bound (non-freezing). The amount of bound water varies with polymer concentration and the presence of salts [22]

Mechanical Properties:
Mechanical Properties General: Mech. props. depend very much on polymer concentration, degree of swelling and network density [18]. Young's modulus of as-prepared gels increases with degree of cross-linking to a maximum of approx. 78 MPa and then decreases. At equilibrium swelling the modulus behaves similarly but the values are somewhat lower. The strain at break for the same gel falls as the degree of cross-linking increases. [2,4,8] Inhomogeneities in a gel can have significant effects on its mech. props. [6,7] Shear modulus follows scaling law at low concentrations but deviates as the concentration increases. [6,16] The incorporation of a rigid rod linear polyelectrolyte, e.g. poly(4''-disodium 2,5-dimethyl-1,1':4',1''-terphenyl-3',2''-disulfonate) in a polyacrylamide gel can improve the mech. strength of the gel and its water-absorbing props. [15] Yield strength increases with increasing monomer concentration. For a given monomer concentration yield strength increases with increasing cross-link ratio [20]

Poisson's Ratio:

No.	Value	Note
1	0.442–0.467	acrylamide:methylenebisacrylamide 5:0.01, total monomer content 27% wt; equilibrium swelling in water [11]
2	0.318	27°, acrylamide:methylenebisacrylamide 37.5:1, unswollen
3	0.378	73°, acrylamide:methylenebisacrylamide 37.5:1, unswollen [12]

Tensile Strength Yield:

No.	Value	Note
1	0.00271 MPa	monomer concentration 6% [20]
2	0.01013 MPa	monomer concentration 20% [20]

Mechanical Properties Miscellaneous: The effect of added phosphoric and sulfuric acid on the mech.props. of hydrogels has been studied; tensile strength decreases as the acid/polyacrylamide ratio increases [21]

Electrical Properties:
Complex Permittivity and Electroactive Polymers: Proton-conducting hydrogels have ben made by doping with phosphoric and sulfuric acid [21]

Polymer Stability:
Chemical Stability: Microgels are stable in polar solvents and aq. KCl soln. [3]

Applications/Commercial Products:
Processing & Manufacturing Routes: The gels can be made by one of four methods [1]. Three-dimensional polymerisation involves a free-radical polymerisation using an initiator such as persulfate and a cross-linking agent. Perhaps the most widely used cross-linking agent for polyacrylamide gels is N,N'-methylenebisacrylamide (MBAA). The network density can be varied by changing the amount of MBAA used. A second method is cross-linking of water-soluble polymers using multivalent cations such as Cr^{3+}. Gels may also be made by radiation cross-linking or by hydrolysis of polyacrylamide chains to yield acrylamide/acrylic acid copolymers. Gel spheres of particle size 200–1500μm may be synth. by suspension polymerisation of aq. monomer solns. suspended in an organic matrix. [10] The preparation of microgels by precipitation polymerisation has been reported. Particle size (0.06–1.0μm) and distribution controlled by solvent selection [3]
Applications: Used in agriculture (soil improvement, hydroponics); in medicine; in paint and coatings industry, and in adhesives

Bibliographic References

[1] Kazanskii, K.S. and Dubrivskii, S.A., *Adv. Polym. Sci.*, 1992, **104**, 97
[2] Cohen, Y.; Ramon, O.; Kopelman, I.J. and Mizrahi, S., *J. Polym. Sci., Part B: Polym. Lett.*, 1992, **30**, 1055
[3] Kim, K.S.; Cho, S.H. and Shin, J.S., *Polym. J. (Tokyo)*, 1995, **27**, 508, (synth)
[4] Baselga, J.; Hernandez-Fuentes, I.; Masegosa, R.M. and Llorente, M.A., *Polym. J. (Tokyo)*, 1989, **21**, 267
[5] Hsu, T.-P.; Ma, D.S. and Cohen, C., *Polymer*, 1983, **24**, 1273
[6] Mallam, S.; Sorkay, F.; Hecht, A.-M. and Geissler, E., *Macromolecules*, 1989, **22**, 3356
[7] Geissler, E.; Hecht, A.-M.; Horkay, F. and Zrinyi, M., *Macromolecules*, 1988, **21**, 2594, (mech props)
[8] Baselga, J.; Hernandez-Fuentes, I.; Pierola, I.F. and Llorente, M.A., *Macromolecules*, 1987, **20**, 3060, (elastic props)
[9] Hsu, T.-P. and Cohen, C., *Polymer*, 1984, **25**, 1419, (struct)
[10] Patel, S.K.; Rodriguez, F. and Cohen, C., *Polymer*, 1989, **30**, 2198, (mech props)
[11] Takigawa, T.; Morino, Y.; Urayama, K. and Masuda, T., *Polym. Gels Networks*, 1996, **4**, 1, (Poisson'ratio)
[12] Fang, L.; Brown, W. and Konak, C., *Polymer*, 1990, **31**, 1960, (dynamic props)
[13] Brown, W. nad Johnsen, R.M., *Polymer*, 1981, **22**, 185, (diffusion)
[14] Haggerty, L.; Sugarman, J.H. and Prud'hommer, R.K., *Polymer*, 1988, **29**, 1058, (diffusion)
[15] Philippova, O.E.; Rulkens, R.; Kovtunenko, B.I.; Abramchuk, S.S. *et al*, *Macromolecules*, 1998, **31**, 1168
[16] Zhang, X.; Hu, Z. and Li, Y., *Polymer*, 1998, **39**, 2783, (elasticity)
[17] Geissler, E.; Hecht, A.-M. and Torbet, J., *Polymer*, 1986, **27**, 1489, (birefringence)
[18] Opperman, W.; Rose, S.; Rehage, G., *Br. Polym. J.*, 1985, **17**, 175, (elastic behaviour)
[19] Shan, J.; Chen, J.; Liu, Z. and Zhang, M., *Polym. J. (Tokyo)*, 1996, **28**, 886
[20] Goodrich, K.; Yoshimura, A. and Prud'homme, R.K., *J. Rheol. (N.Y.)*, 1989, **33**, 317, (mech props)
[21] Przyluski, J.; Poltartewski, Z. and Wieczorek, W., *Polymer*, 1998, **39**, 4343
[22] Ahad, E., *J. Appl. Polym. Sci.*, 1978, **22**, 1665

Poly(acrylic acid) P-62

Synonyms: Poly(2-propenoic acid). PAA. Poly[1-(carboxy)ethylene]
Related Polymers: Poly(ethylene-*co*-acrylic acid)
Monomers: Acrylic acid
Material class: Polyelectrolyte
Polymer Type: acrylic
CAS Number:

CAS Reg. No.	Note
9003-04-7	sodium salt
25608-12-2	potassium salt
9003-03-6	ammonium salt
9003-01-4	

Molecular Formula: $(C_3H_4O_2)_n$
Fragments: $C_3H_4O_2$
Molecular Weight: The MW varies from 1000 to 5000000
General Information: Poly(acrylic acid) is a clear brittle solid, which cannot be moulded on heating. It dissolves in water and alcohol and behaves as a polyelectrolyte. When prepared in dilute aq. soln. it can be used as prepared, or it can be dried to obtain a powdery white solid. It is advisable not to attempt to dry completely as this will result in cross-linking to yield a product which will be difficult to dissolve. If required freeze-drying methods can be used for complete drying. Salt solns. have high expanded molecular coils and therefore high viscosities. Salts of poly(acrylic acid) can absorb water many times in excess of their weight. Clear brittle solid, soluble in water and alcohol to behave as polyelectrolyte. Salt solns. have high expanded molecular coils and, therefore, high viscosites
Tacticity: Tacticity of the polymer cannot be measured directly. Isotactic samples have been prepared by hydrol. of esters having the appropriate tacticity

Volumetric & Calorimetric Properties:

Equation of State: Equation of state information has been reported
Glass-Transition Temperature:

No.	Value	Note
1	106–150°C	allyl acrylate comonomer

Transition Temperature General: T_g increases with formation of anhydride

Surface Properties & Solubility:

Solvents/Non-solvents: Sol. H_2O, MeOH, EtOH, 2-propanol, dioxane, DMF, ethylene glycol, AcOH (atactic), aq. dioxane (isotactic). Insol. Me_2CO, Et_2O, C_6H_6, aliphatic hydrocarbons (atactic), dioxane (isotactic)

Transport Properties:

Polymer Solutions Dilute: Theta temp. 29° (dioxane). Entropy parameter -0.310. Specific viscosity varies with conc. and follows Huggins equation up to 0.5% concentration. In ionising solvents, viscosity rises with dilution to a max., then decreases at very low concentration
Polymer Solutions Concentrated: Solns. of 3% and more are thixotropic. At low shear rate, viscosity increases with concentration. Viscosity also increases with increasing neutralisation. Theta temp. 30° (dioxane)
Water Content: Linear polymer absorbs water rapidly on exposure to moist air
Water Absorption:

No.	Value	Note
1	8 %	equilibrium, 30°, 50% relative humidity
2	42 %	equilibrium, 30°, 50% relative humidity, Na salt

Optical Properties:

Refractive Index:

No.	Value	Note
1	1.442	20°, 65% in H_2O

Total Internal Reflection: dn/dc 0.137 (25–65°, H_2O); 0.174 (25°, H_2O, dioxane (50:50); 0.09 (25°, dioxane); 0.088 (25°, 0.1 *M* HCl)

Polymer Stability:

Polymer Stability General: Dry polymer is stable but takes up water on exposure to moisture. High MW polymer is stable to degradation by shear, oxidation or conformational changes
Thermal Stability General: Decomposes slowly above 250°
Decomposition Details: On heating loses water above 100°, forming the anhydride. Stable to 200° in vacuum, then above 250° liberates CO_2, and decomposes rapidly above 400°
Chemical Stability: Polymer is a weak acid (pK_a 4.75). Readily neutralised with metal hydroxides

Applications/Commercial Products:

Processing & Manufacturing Routes: Linear polymers can be prepared in acidic 20–30% soln. by free radical polymerisation. Higher concentrations result in uncontrolled exotherms and cross-linked polymers. Because of high heats of polymerisation reaction it is not simple to control polymerisation reaction. Polymerisation of pure acid can be hazardous, as the temp. can rise to values at which the polymer decomposes to generate gaseous product. In general the manufacturer will provide products in a form that does not require further processing
Applications: Thickeners in cosmetics, creams, lotions, hair preparations, pigments, polishing agents and emulsions. Drilling mud additives, ion exchange resins, flocculating agents, paper making, scale inhibitors, detergents, dispersants, sludge conditioners for water treatment, wrap sizes for nylon, adhesives, medical applications, e.g. sustained release tablets for oral

administration. Commercially sodium, potassium, and ammonium salts are used

Trade name	Details	Supplier
Acrysol		Rohm and Haas
Acrysol GS	sodium salt	Rohm and Haas
Acusol		Rohm and Haas
Antiprex	sodium salt	Allied Colloids
Aquakeep		Seistu K K K
Aquatreat	AR 700, sodium salt	Alco Chemicals
Arasorb		Arakawa
Carbopol		B.F. Goodrich
Carbopol 961	ammonium salt	B.F. Goodrich
Drytech		Dow
Favor		Chemische Fabrik
Solsorb		Allied Colloids
		Stockhausen

Bibliographic References

[1] *Ullmanns Encykl. Ind. Chem.*, (ed. W. Gerhartz), VCH, 1992
[2] *Encycl. Polym. Sci. Eng.*, 2nd edn., (ed. J.I. Kroschwitz), John Wiley and Sons, 1985
[3] *Kirk-Othmer Encycl. Chem. Technol.*, 4th edn., 1991
[4] Van Krevelen, D.W., *Properties of Polymers: Their Correlation with Chemical Structure*, Elsevier, 1990
[5] Hughes, L.J.E. and Fordyce, D.B., *J. Polym. Sci.*, 1956, **22**, 509
[6] Pegoraro, M., Szilágye, L., Penati, A. and Alessandrini, G., *Eur. Polym. J.*, 1971, **7**, 1709
[7] Kim, O.-K., Choi, L.S., Long, T., McGrath, K. et al, *Macromolecules*, 1993, **26**, 379
[8] Shefer, A., Grodzinsky, A.J., Prime, K.L. and Busnel, J.-P., *Macromolecules*, 1993, **26**, 5009

Polyacrylonitrile

$$\mathrm{-[CH_2CH(CN)]_n-}$$

Synonyms: *PAN. Poly(2-propenenitrile)*
Related Polymers: SAN, ABS, Nitrile rubber, ASA, Acrylic fibres, Modacrylic fibres, Acrylonitrile-methyl acrylate copolymer, Poly(acrylonitrile-*co*-vinylidene chloride), Acrylonitrile methyl acrylate copolymer, Injection moulding, Polymethacrylonitrile, Acrylonitrile-butadiene-isoprene terpolymer, Butadiene-acrylonitrile copolymers, Poly(vinylidene chloride-*co*-acrylonitrile), Poly(vinyl chloride-*co*-acrylonitrile), Polycarbonate/acrylic-styrene-acrylonitrile alloy, Poly(methyl methacrylate-*co*-acrylonitrile-*co*-butadiene-*co*-styrene)
Monomers: Acrylonitrile
Material class: Thermoplastic
Polymer Type: acrylonitrile and copolymers
CAS Number:

CAS Reg. No.	Note
31631-89-7	syndiotactic
25014-41-9	
29436-44-0	isotactic

Molecular Formula: $(C_3H_3N)_n$
Fragments: C_3H_3N
Molecular Weight: MW 185000; 1100000–1600000
Morphology: The molecular struct. of PAN fibres has been reported [3]. Conformational study has been reported [11]. Orthorhombic, a 10.6 Å, b 11.6 Å, c 5.04 Å (stretched film) [43]. Syndiotactic PAN may assume a *trans-trans* conformn. and a *trans-trans* conformn. including a *gauche-gauche* conformn. [67]. PAN exists mainly in the isotactic form [78] with the molecule itself arranged in an irregular helical conformn. [79,80]. Heterogeneous PAN contains three phases: crystalline, quasi-crystalline and amorph. [81,82]. The relative proportion of each phase depends on the method of synth. PAN is a very poorly crystalline material. However, single crystals may by grown from 105–150° using a film formation crystallisation technique [68]. PAN synth. by radiation-induced bulk polymerisation has, below 50°, a morphology similar to that of polyethylene. Above 70°, PAN synth. by this method comprises large lamellar crystals that contain a large number of growing polymer chains [70]

Volumetric & Calorimetric Properties:

Density:

No.	Value	Note
1	d 0.2–0.4 g/cm³	[5]
2	d 1.1685 g/cm³	flotation method [57]
3	d 1.14–1.17 g/cm³	
4	d^{29} 1.17–1.18 g/cm³	flakes

Thermal Expansion Coefficient:

No.	Value	Type
1	6.6×10^{-5} K^{-1}	L

Equation of State: Mark-Kuhn-Houwink parameters have been reported [12]. Mark-Houwink-Sakurada parameters have been reported [16,33]
Thermodynamic Properties General: Studies on heat capacity have been reported [17]. C_p 63.09 J mol^{-1} K^{-1} (0°) [62]. 86.18 J mol^{-1} K^{-1} (97°) [62].
Latent Heat Crystallization: Enthalpy of reaction 3.05 kJ g^{-1} (Courtelle). [60] ΔH 9.508 kJ mol^{-1} (0°), 16.681 kJ mol^{-1} (97°) [62]. ΔS 67.22 J mol^{-1} K^{-1} (0°), 89.53 J mol^{-1} K^{-1} (97°) [62]
Thermal Conductivity:

No.	Value
1	0.26 W/mK

Specific Heat Capacity:

No.	Value	Note	Type
1	1.285 kJ/kg.C	0.307 kcal g^{-1} °C^{-1} [73]	P

Melting Temperature:

No.	Value	Note
1	320°C	[44]
2	369°C	in N$_2$ [65]
3	317°C	[79,97]
4	326°C	[84]

Glass-Transition Temperature:

No.	Value	Note
1	105°C	[21,98]
2	70.5°C	[58]
3	97°C	[91]

Transition Temperature General: T_m is continually lowered as H_2O content increases, until a critical point when further addition of H_2O causes no further lowering of T_m [23]. T_m in air is obscure at a heating rate of $40°$ min^{-1} or lower [65]. PAN decomposes before its T_m

Transition Temperature:

No.	Value	Note	Type
1	97°C	approx. [7]	First transition temp.
2	167–177°C	[7]	High-temp. transition

Surface Properties & Solubility:

Solubility Properties General: Miscible with poly(*N*-vinyl-2-pyrrolidone) [1]. Compatible with poly(butyl methacrylate) [2], but incompatible with PVC [30,31]. Compatibility with mixtures of poly(vinyl acetate) and cellulose acetate has been reported [36]. Incompatible with poly(methyl methacrylate) in dilute soln. [37]. Partially miscible with poly(styrenesulfonic acid) [77]

Cohesive Energy Density Solubility Parameters: δ_d 9.11; δ_p 7.9; δ_h 3.3 [56]. δ 25.9–31.5 $J^{1/2}$ $cm^{-3/2}$ [64], 42.3 $J^{1/2}$ $cm^{-3/2}$ (11.9 $cal^{1/2}$ $cm^{-3/2}$) [66], 51.2 $J^{1/2}$ $cm^{-3/2}$ (12.2 $cal^{1/2}$ $cm^{-3/2}$)

Solvents/Non-solvents: Sol. DMF, ethylene carbonate [20], di-methylthioformamide, dimethylacetamide, DMSO, propiolactone, *m*-nitrophenol, *p*-nitrophenol, maleic anhydride, conc. sulfuric acid, nitric acid, dimethyl sulfone, conc. aq. solns. of lithium and zinc chloride [92,93]. Amorph. sol. propylene carbonate, formamide (above room temp.). Insol. Me_2CO [19], diethylformamide, aliphatic hydrocarbons, aromatic hydrocarbons, chlorinated hydrocarbons, diethyl sulfone

Wettability/Surface Energy and Interfacial Tension: The relationship between critical surface tension and refractive index has been reported [13]

Transport Properties:

Transport Properties General: The transport of H_2O in glassy PAN follows two modes: simple dual mode sorption at low activity sorption, and clustering at higher pressures [99]

Polymer Solutions Dilute: Intrinsic viscosity 0.727 dl g^{-1} (30°, DMF), 0.704 dl g^{-1} (45°, DMF) [19], 2.01 g cm^{-3} (23.1°, DMF) [69]. Specific viscosity 0.15 dl g^{-1} (25°, DMF) [21]. Van der Waals molar volume 32.6 cm^3 mol^{-1} [66]. Viscosity of PAN in DMF at a constant concentration increases with temp. [74]. Other viscosity studies have been reported [75]

Permeability of Gases: The correlation between temp. parameters and dimension of permeable gas molecules has been reported [25]. Permeability of He, Ne, Ar, Kr, O_2, N_2 and CO_2 has been reported [28]. Permeability of O_2, CO_2 and H_2O has been reported [45]. Has the lowest permeability of any plastic

Permeability of Liquids: Permeability [H_2O] 2.45 g (cm s cmHg)$^{-1}$ (25°) [71]. The diffusion coefficient of sulfuric acid through a PAN rope has been reported [15]

Water Content: Studies on sorbed H_2O at the polymer surface near 0° have been reported [49]

Gas Permeability:

No.	Gas	Value	Note
1	CO_2	0.052 cm^3 mm/ (m^2 day atm)	0.008 × 10^{-11} cm^2 (s cm Hg)$^{-1}$, 25° [71]
2	O_2	0.013 cm^3 mm/ (m^2 day atm)	0.002 × 10^{-11} cm^2 (s cm Hg)$^{-1}$, 25° [71]
3	He	34.8 cm^3 mm/ (m^2 day atm)	53 × 10^{-12} cm^2 (s cm Hg)$^{-1}$, 25°
4	Ne	1.09 cm^3 mm/ (m^2 day atm)	1.67 × 10^{-12} cm^2 (s cm Hg)$^{-1}$, 25°
5	Ar	0.0066 cm^3 mm/ (m^2 day atm)	0.01 × 10^{-12} cm^2 (s cm Hg)$^{-1}$, 25°
6	Kr	0.002 cm^3 mm/ (m^2 day atm)	0.003 × 10^{-12} cm^2 (s cm Hg)$^{-1}$, 25°
7	N_2	0.0105 cm^3 mm/ (m^2 day atm)	0.016 × 10^{-12} cm^2 (s cm Hg)$^{-1}$, 25°
8	O_2	0.0183 cm^3 mm/ (m^2 day atm)	0.028 × 10^{-12} cm^2 (s cm Hg)$^{-1}$, 25°
9	CO_2	0.118 cm^3 mm/ (m^2 day atm)	0.18 × 10^{-12} cm^2 (s cm Hg)$^{-1}$, 25° [94]
10	He	11.16 cm^3 mm/ (m^2 day atm)	0.71 × 10^{-10} cm^2 (s cm Hg)$^{-1}$, 35° [100]
11	N_2	0.002 cm^3 mm/ (m^2 day atm)	2.9 × 10^{-15} cm^2 (s cm Hg)$^{-1}$, 35° [100]
12	O_2	0.035 cm^3 mm/ (m^2 day atm)	5.4 × 10^{-14} cm^2 (s cm Hg)$^{-1}$, 35° [100]
13	Ar	0.0118 cm^3 mm/ (m^2 day atm)	1.8 × 10^{-14} cm^2 (s cm Hg)$^{-1}$, 35° [100]
14	CO_2	0.1838 cm^3 mm/ (m^2 day atm)	2.8 × 10^{-13} cm^2 (s cm Hg)$^{-1}$, 35° [100]

Mechanical Properties:

Mechanical Properties General: The relationship between elastic modulus and the thermal expansion coefficient has been reported [51]. Elastic modulus decreases with increasing temp. [95]

Tensile (Young's) Modulus:

No.	Value	Note
1	7000 MPa	[98]

Elastic Modulus:

No.	Value	Note
1	236000 MPa	syndiotactic [72]

Compressive Strength:

No.	Value	Note
1	76–130 MPa	[98]

Impact Strength: The correlation between notched Izod impact strength and dynamic mech. modulus has been reported; PAN has a low notched Izod impact strength [55]

Failure Properties General: Creep recovery is more complete at lower temps. and recovery is mainly instantaneous. At higher temps. and strains, creep induces permanent structural changes which are not recoverable on heating [50]. Fails by ductile failure at temps. well below T_g [61]

Electrical Properties:

Electrical Properties General: An electropolymerised film has both electrical insulating and conducting props. [26]. Tan δ remains constant with increasing temp. until T_g is reached, then it increases [57,63]

Dielectric Permittivity (Constant):

No.	Value	Frequency	Note
1	5	100 kHz	30° [57]
2	4.2	1 MHz	film
3	5.5	1 kHz	film
4	6.5	60 Hz	film [90]

Complex Permittivity and Electroactive Polymers: Activation energy for electrical conductivity 1.17, 3.41 eV [39]
Magnetic Properties: Paramagnetic data have been reported [40]
Dissipation (Power) Factor:

No.	Value	Frequency	Note
1	<0.27	100 kHz	max. [57]
2	0.033	1 MHz	film [90]
3	0.085	1 kHz	film [90]
4	0.113	60 Hz	film [90]

Optical Properties:
Optical Properties General: Relationship between struct. and refractive index has been reported [46]
Refractive Index:

No.	Value	Note
1	1.514	[56]
2	1.518	25° [91]

Transmission and Spectra: Ir [4,42], Raman [8,47], C-13 nmr [35], reflection [48] and H-1 nmr [53,54] spectral data have been reported

Polymer Stability:
Polymer Stability General: Heat resistance may be increased by heating in boiling DMF for 60–90h. However, similar heating at 20–40h decreases heat resistance [41]
Thermal Stability General: Thermal anal. has been reported [32,34]
Decomposition Details: Initial decomposition temp. 300°.10% Weight loss occurs at 340° and 50% weight loss at 590° [21]. Shows 30% weight loss below 400° [22] but is more stable than other vinyl polymers such as polystyrene, polyacrylic acid, and polyacrylamide. Pyrolysis at 600° prod. a max. of 18.2 g of hydrogen cyanide per 100 g of polymer [27]. The effects of pyrolysis below 300° on the dielectric props. have been reported [38]. Initial decomposition temp. 260°, max. decomposition temp. 345° [57]. Activation energy of thermal degradation 125.6 kJ mol^{-1} (30 kcal mol^{-1}) [57], 71.5 kJ mol^{-1} [59]. Temp. range for decomposition 550–790° with 79% weight loss in this range [59]. Heating PAN in an inert atmosphere up to 800° produces three dimensional cross-linked heterocyclic products which have electrical conductivity [76]. Heating in oxygen at approx. 150° produces an insol. ladder polymer [87]. Subsequent heating up to approx. 3000° prod. carbon fibres. Char oxidation (Courtelle) -1.27 (in air), -7.15 (in O_2) [60]. Activation energy of oxidative thermal degradation 134.4 kJ mol^{-1} (32.1 kcal mol^{-1}) [65]. Char residue at 600° 57.7% [97].
Flammability: Fl. p. 331° [97]. Limiting oxygen index 15.3% (ignition of wick at bottom), 17.8% (ignition of wick at top) [96]. Oxygen index 20.2% [73,97]. 18.2% (Courtelle, ASTM D2863-77) [60]. Ignition temp. 460° [73]. Flash ignition temp. 208–212° (Courtelle) [60]; self-ignition temp. 495–500° (Courtelle, ASTM D1929-68) [60].
Environmental Stress: Has good weather resistance and excellent resistance to photodegradation [86]
Chemical Stability: Resistant to common solvents, oils and chemicals. Has excellent resistance to ozonolysis
Hydrolytic Stability: Attacked by strong bases in soln. whereas weak bases cause only colouration [9]. Decomposition by potassium hydroxide increases with an increase in base concentration, temp. and time, with greatest increase in reaction occurring at 30 min. [18]. May be hydrated by conc. sulfuric acid. Reacts with hydroxylamine [85]
Biological Stability: PAN filaments are degraded by *Cladosporium cladosporioides* [24]. PAN film loses some mechanical props. when immersed for 1 year in a pseudo-extracellular fluid [29]
Recyclability: PAN fibre or textile waste may be recovered by chemical conversion to poly(acrylic acid) [88]

Stability Miscellaneous: Laser or γ-irradiation causes intense chain fragmentation yielding radicals [14]. Annealing at 210° in argon causes changes in physical props. but no weight loss [52]

Applications/Commercial Products:
Processing & Manufacturing Routes: May be synth. by photopolymerisation using a binary catalyst (8-acryloyloxyquinoline polymer and carbon tetrabromide) [6]; by aq. polymerisation at 40–50° using a peroxodisulfate-thiourea redox system [10]. At monomer concentrations of 6 mol l^{-1} in DMF or ethylene carbonate, precipitation may occur during radical polymerisation [20]. Max. bulk polymerisation rate is strongly dependent upon temp. and initiator concentration [33]. Amorph. PAN may be synth. using an organomagnesium catalyst [83, 101]. Synth. commercially by bulk, emulsion, suspension, slurry, or soln. polymerisation. Bulk processes can suffer from polymer precipitation and autocatalytic behaviour. Cannot be melt processed into films or sheeting owing to degradation
Applications: May be thermally converted into carbon fibre used in advanced composites. Applications due to excellent oxygen and CO_2 barrier props. in packaging laminations. Fibres used in blood cell filters. Other uses as catalyst support for Ti and Zr complexes; in preformed trays and in xerography. Its use in bottles for carbonated beverages was prohibited in 1977 by the FDA

Trade name	Details	Supplier
PAN	A-425	Solvay
Solton	S6, S10	Solvay

Bibliographic References

[1] Guo, Q., Huang, J. and Li, X., *Eur. Polym. J.*, 1996, **32**, 423, (miscibility)
[2] Somanathan, N., Senthill, P., Viswanathan, S. and Arumugam, V., *J. Appl. Polym. Sci.*, 1994, **54**, 1537
[3] Lee, W.S., Park, J.S., Cho, S.M. and Cho, S.H., *Han'guk Somyu Konghakhoechi*, 1992, **29**, 835, (struct.)
[4] Minagawa, M., *Jasco Rep.*, 1991, **33**, 6, (ir)
[5] Fritzsche, P., Dautzenberg, H., Noetzel, P. and Rother, G., *Angew. Makromol. Chem.*, 1992, **195**, 171
[6] Zhang, J., Gao, Q., Chen, S. and Li, F., *Gaofenzi Xuebao*, 1991, 250, (synth.)
[7] Tarakanov, B.M. and Andreeva, O.A., *Vysokomol. Soedin., Ser. A*, 1990, **32**, 2105, (transition temps.)
[8] Zimba, C.G. and Rabolt, J.F., *Thin Solid Films*, 1991, **206**, 388, (Raman)
[9] Bashir, Z., Manns, G., Service, D.M., Bott, D.C. et al, *Polymer*, 1991, **32**, 1826
[10] Misra, M., Moharty, A.K. and Singh, B.C., *Angew. Makromol. Chem.*, 1988, **157**, 79, (synth.)
[11] Turska, E., *Chem. Stosow.*, 1986, **30**, 179, (conformn)
[12] Vilenchik, L.Z., Budtov, V.P., Nesterov, V.V., Krasikov, V.D. et al, *Vysokomol. Soedin., Ser. B*, 1989, **31**, 114, (eqn. of state)
[13] Xu, C. and Tong, X., *Hecheng Xiangjiao Gongye*, 1987, **10**, 270
[14] Andreeva, O.A. and Burkova, L.A., *Vysokomol. Soedin., Ser. B*, 1990, **32**, 172, (irradiation)
[15] Alieva, E.R., Kozhevnikov, Yu.P and Serkov, A.T., *Khim. Volokna*, 1990, 23
[16] Kamide, K., Miyazaki, Y. and Kobayashi, H., *Polym. J. (Tokyo)*, 1985, **17**, 607, (eqn. of state)
[17] Bu, H.S., Aycock, W. and Wunderlich, B., *Polymer*, 1987, **28**, 1165, (heat capacity)
[18] Sanli, O., *Eur. Polym. J.*, 1990, **26**, 9, (hydrol.)
[19] Joseph, R., Devi, S. and Rakshit, A.K., *Polym. Int.*, 1991, **26**, 89, (viscosity)
[20] Guyot, A., Makromol. Chem., *Makromol. Symp.*, 1987, **10–11**, 461
[21] Gupta, D.C. and Agrawal, J.P., *J. Appl. Polym. Sci.*, 1989, **38**, 265, (thermal degradation)
[22] Girdhar, H.L. and Peerzada, G.M., *Indian J. Chem., Sect. A*, 1988, **27**, 710, (thermal stability)
[23] Frushour, B.G., *Polym. Bull. (Berlin)*, 1982, **7**, 1
[24] Sato, M., *Sen'i Gakkaishi*, 1984, **40**, T297, (biodegradation)
[25] Teplyakov, V.V. and Durgar'yan, S.G., *Vysokomol. Soedin., Ser. A*, 1984, **26**, 2159
[26] Boiziau, C., Juret, C., Lecayon, G., Nuvolone, R. and Reynaud, C., *Contrib.-Symp. At. Surf. Phys.*, (eds. W. Lindinger, F. Howorka and T.D. Maerk), 1982, 111
[27] Le Moan, G., *Bull. Acad. Natl. Med. (Paris)*, 1982, **166**, 43, (pyrolysis)
[28] Allen, S.M., Fujii, M., Stannett, V., Hopfenberg, H.B. and Williams, J.L., *U.S. NTIS, AD Rep.*, 1976, 21, (gas permeability)

[29] Yamamoto, N., Yamashita, I. and Hayashi, K., *Osaka Kogyo Gijutsu Shikensho Kiho*, 1978, **29**, 203, (biostability)
[30] Schneider, I.A. and Calugaru, E.M., *Eur. Polym. J.*, 1976, **12**, 879, (compatibility)
[31] Feldman, D., Crusos, A. and Ungureanu, C., *J. Macromol. Sci., Phys.*, 1977, **14**, 573, (compatibility)
[32] Ohe, H., Matsuura, K. and Sakai, N., *Fukui Daigaku Kogakubu Sen'i Kogyo Kenkyu Shisetsu Hokoku*, 1977, **15**, 1, (thermal anal.)
[33] Garcia-Rubio, L.H. and Hamielec, A.E., *J. Appl. Polym. Sci.*, 1979, **23**, 1397
[34] Beinoravicius, M. and Bajoras, G., Liet. TSR Aukst. Mokyklu Mokslo Darb., *Chem. Chem. Technol.*, 1978, **19**, 44, (thermal anal.)
[35] Balard, H., Fritz, H. and Meybeck, J., *Makromol. Chem.*, 1977, **178**, 2393, (C-13 nmr)
[36] Vasile, C., Sandru, F., Schneider, I.A. and Asandei, N., *Mater. Plast. (Bucharest)*, 1972, **9**, 473, (compatibility)
[37] Dimov, K. and Dilova, E., *Khim. Volokna*, 1972, **14**, 21, (compatibility)
[38] Ito, S., Ohe, H. and Kajima, K., *Oyo Butsuri*, 1971, **40**, 1079, (pyrolysis)
[39] Neagu, E., Leanca, M. and Neagu, R., An. Stiint. Univ. "Al. I. Cuza" Iasi, *Sect. 1b*, 1974, **20**, 133, (conductivity)
[40] Bikbulatova, L.A., Vetrov, O.D., Pivovarov, S.P. and Messerle, P.E., *Zh. Fiz. Khim.*, 1974, **48**, 818, (magnetic props.)
[41] Salikhova, S.Sh. and Magrupov, M.A., *Sint. Vysokomol. Soedin.*, (ed. M.A. Askarov), Fan, 1972, 55, (heat stability)
[42] Wolfram, L.E., Grasselli, J.G. and Koenig, J.L., *Appl. Polym. Symp.*, 1974, **25**, 27, (ir)
[43] Hinrichsen, G. and Orth, H., *Kolloid Z. Z. Polym.*, 1971, **247**, 844, (struct.)
[44] Hinrichsen, G., *Angew. Makromol. Chem.*, 1971, **20**, 121, (melting point)
[45] Pellino, E., *Mater. Plast. Elastomeri*, 1973, **39**, 623, (gas permeability)
[46] Ayano, S. and Murakawa, T., *Kobunshi Kagaku*, 1972, **29**, 723, (refractive index)
[47] Huang, Y.S. and Koenig, J.L., *Appl. Spectrosc.*, 1971, **25**, 620, (Raman)
[48] Vinokurova, L.N., Cherkasov, Yu.A. and Kisilitsa, P.P., *Opt. Spektrosk.*, 1973, **34**, 805, (reflection spectra)
[49] Jellinek, H.H.G., Luh, M.D. and Nagarajan, V., *Kolloid Z. Z. Polym.*, 1969, **232**, 758
[50] Andrews, R.D. and Okuyama, H., *J. Appl. Phys.*, 1968, **39**, 4909, (creep behaviour)
[51] Perepelkin K.E., *Fiz. Tverd. Tela*, 1969, **11**, 3529
[52] Kasatochkin, V.I., Smutkina, Z.S. and Kazakov, M.E., *Strukt. Khim. Ugleroda Uglei*, (ed. V.I. Kasatochkin), Izd. Nauka, Moscow, 1969, 220
[53] Murano, M., *Rev. Phys. Chem. Jpn.*, 1969, **39**, 93, (H-1 nmr)
[54] Svegliado, G. and Zilio-Grandi, F., *Corsi Semin. Chim.*, 1968, 121, (H-1 nmr)
[55] Wada, Y. and Kasahara, T., *J. Appl. Polym. Sci.*, 1967, **11**, 1661
[56] Koenhen, D.M. and Smolders, C.A., *J. Appl. Polym. Sci.*, 1975, **19**, 1163, (solubility parameters)
[57] Bajaj, P., Gupta, D.C. and Gupta, A.K., *J. Appl. Polym. Sci.*, 1980, **25**, 1673, (props.)
[58] Gupta, A.K., Singhal, R.P. and Agarwal, V.K., *J. Appl. Polym. Sci.*, 1983, **28**, 2745
[59] Joseph, R., Devi, S. and Rakshit, A.K., *J. Appl. Polym. Sci.*, 1993, **50**, 173
[60] Balog, K., Košík, S., Košík, M., Reiser, V. and Simek, I., *Thermochim. Acta*, 1985, **93**, 167, (pyrolysis, flammability)
[61] Aharoni, S.M., *Macromolecules*, 1985, **18**, 2624, (failure characteristics)
[62] Gaur, U., Lau, S., Wunderlich, B.B. and Wunderlich, B., *J. Phys. Chem. Ref. Data*, 1982, **11**, 1065, (thermodynamic data)
[63] Gupta, A.K., Chand, N., Singh, R. and Mansingh, A., *Eur. Polym. J.*, 1979, **15**, 129, (dielectric props.)
[64] Rajulu, A.V., Sab, P.M. and Askadskii, A.A., *Indian J. Chem., Sect. B*, 1994, **33**, 1105, (solubility parameters)
[65] Kotoyori, T., *Thermochim. Acta*, 1972, **5**, 51, (oxidative thermal degradation)
[66] Askadskii, A.A., *Pure Appl. Chem.*, 1976, **46**, 19, (solubility parameter)
[67] Shimanouchi, T., *Pure Appl. Chem.*, 1966, **12**, 287, (conformn.)
[68] Gohil, R.M., Patel, K.C. and Patel, R.D., *Angew. Makromol. Chem.*, 1972, **25**, 83, (morphology)
[69] Osmers, H.R. and Metzner, A.B., *Ind. Eng. Chem. Fundam.*, 1972, **11**, 161
[70] Arai, H., Wada, T. and Kuriyama, I., *J. Polym. Sci., Polym. Phys. Ed.*, 1975, **13**, 2241, (morphology)
[71] Salame, M., *J. Polym. Sci., Polym. Symp.*, 1973, **41**, 1, (gas permeability)
[72] Manley, T.R. and Martin, C.G., *Polymer*, 1973, **14**, 491, (elastic modulus)
[73] Kishore, K. and Mohan Das, K., *Colloid Polym. Sci.*, 1980, **258**, 95, (flammability)
[74] Pfragner, J. and Schurz, J., *Monatsh. Chem.*, 1980, **111**, 1053, (viscosity)
[75] Bercea, M., Ioan, S., Simionescu, B.C. and Simionescu, C.I., *Polym. Bull. (Berlin)*, 1992, **27**, 571, (dilute soln. props.)
[76] Jobst, K., Sawtschenko, L., Schwarzenberg, M. and Wuckel, L., *Synth. Met.*, 1991, **41**, 959
[77] David, M.O., Nguyen, Q.T. and Nél, J., *J. Membr. Sci.*, 1992, **73**, 129
[78] Matsuzaki, K., Uryu, T., Okada, M. and Shiroki, H., *J. Polym. Sci., Polym. Chem. Ed.*, 1968, **6**, 1475
[79] Krigbaum, W.R. and Tokita, N., *J. Polym. Sci.*, 1960, **43**, 467
[80] Henrici-Olivé, G. and Olivé, S., *Adv. Polym. Sci.*, 1979, **32**, 123
[81] Minami, S., *Appl. Polym. Symp.*, 1974, **25**, 145
[82] Kenyon, A.S. and Rayford, McC.J., *J. Appl. Polym. Sci.*, 1979, **23**, 717
[83] Joh, Y., *J. Polym. Sci., Polym. Chem. Ed.*, 1979, **17**, 4051
[84] Dunn, P. and Ennis, B.C., *J. Appl. Polym. Sci.*, 1970, **14**, 1795
[85] Schouteden, F.L., *Chim. Ind. (Paris)*, 1958, **79**, 749
[86] Jellinek, H.H.G. and Bastien, I.J., *Can. J. Chem.*, 1961, **39**, 2056, (photodegradation)
[87] Grassie, N., *Dev. Polym. Degrad.*, 1977, **1**, 137, (pyrolysis)
[88] Weiye, X., *Congr. Proc. - Recycl. World Congr. 3rd*, Basel, 1980, (recycling)
[89] Chiang, R., *J. Polym. Sci., Part A: Polym. Chem.*, 1963, **1**, 2765, (density)
[90] Harris, M., *Handbook of Textile Fibers*, Harris Research Laboratories, 1954
[91] Beevers, R.B., *J. Polym. Sci., Part A: Polym. Chem.*, 1964, **2**, 5257
[92] Houtz, R.C., *Text. Res. J.*, 1956, **20**, 786, (solubility)
[93] Kobayashi, H., *J. Polym. Sci., Part B: Polym. Lett.*, 1963, **1**, 299, (solubility)
[94] Allen, S.M., Fujii, H., Stannett, V., Hopfenberg, H.B. and Williams, J.L., *J. Membr. Sci.*, 1977, **2**, 153, (gas permeability)
[95] Pegoraro, M., Szilágyi, L. and Penati, A., *Rheol. Acta*, 1974, **13**, 557
[96] Meisters, M., *Mod. Plast.*, 1975, **52**, 76
[97] Ohe, H., Matsuura, K. and Sakai, N., *Text. Res. J.*, 1977, **47**, 212
[98] West, A.C., *J. Mater. Sci.*, 1981, **16**, 2025
[99] Haider, M.I., *Diss. Abstr. Int. B*, 1980, **40**, 4396
[100] Berens, A.R., *J. Polym. Sci., Polym. Phys. Ed.*, 1979, **17**, 1757, (gas permeability)
[101] Opitz, G. et al, *Acta Polym.*, 1996, **47**, 67, (synth.)

Poly(L-alanine) fibre   P-64

Synonyms: *PLA fibre*
Related Polymers: Proteins
Monomers: L-Alanine
Material class: Fibres and Films, Proteins and polynucleotides
CAS Number:

CAS Reg. No.	Note
25191-17-7	homopolymer

Molecular Formula: $(C_3H_7NO_2)_n$
Morphology: Undergoes α to β transition at approx. 60°. The greater the β-form content, the more silky the fibre. [1] The α-form content may be increased by using a slow spin speed. The fibre is crystalline only at the surface and contains a large proportion of the β-form. The core of the fibre may contain unoriented amorph. poly (L-alanine) in the α-helical conformn. [7]

Volumetric & Calorimetric Properties:
Density:

No.	Value	Note
1	d 1.26–1.27 g/cm^3	stretched fibre [1]

Transport Properties:
Polymer Solutions Dilute: η 3.84 [1]
Water Absorption:

No.	Value	Note
1	8.7%	20°, 80% relative humidity [1]
2	16%	20°, 100% relative humidity [1]

Mechanical Properties:

Mechanical Properties General: Tenacity 3.5 g den^{-1} (6% elongation) [1], 4.34 g den^{-1} (stretched fibre, dry) (9.3% elongation) [1], 2.75g den^{-1}(stretched fibre, wet) (11.9% elongation), 3.82 g den^{-1} [2], 2.8 g den^{-1} (20.3% elongation) [5]. Tensile modulus 102 g den^{-1} [1]

Polymer Stability:

Decomposition Details: Details of the effects of oxidative and optical degradation of the surface struct. have been reported [4]
Chemical Stability: Effect of solvents have been reported [8]

Applications/Commercial Products:

Processing & Manufacturing Routes: Poly(L-alanine) in dichloroacetic acid or 92% nitric acid at -10° [5] is spun into H$_2$O at 60° [2,3]. The resultant fibre is stretched twice its length in H$_2$O at 80°, then wound through a hot roller at 110°. The fibre is then washed and treated with boiling water to set it into the β-form. [1,7] Other syntheses have been reported [6]
Applications: Synthetic silk

Bibliographic References

[1] Noguchi, J., Tokura, S. and Nishi, N., *Angew. Makromol. Chem.*, 1972, **22**, 107, (synth, props.)
[2] *Jpn. Pat.*, 1967, 20 793, (synth.)
[3] Kirimura, J. and Hiroshi, H., *Nippon Nogei Kagaku Kaishi*, 1962, **36**, 635, (synth.)
[4] Kuwahara, A., Tsutsui, R. and Watanabe, T., *Nippon Sanshigaku Zasshi*, 1970, **39**, 73, (surface degradation)
[5] *Fr. Pat.*, 1972, 2 098 300, (synth.)
[6] Nishi, N. and Noguchi, J., *Zoku Kenshi no Kozo*, 1980, 435, (synth., props.)
[7] Komoto, T., Kawai, T., Bezruk, L.I., Graboshnikova, V.I. and Ebert, G., *Makromol. Chem.*, 1979, **180**, 825, (morphology)
[8] *Schriftenr. Dtsch. Wollforschunginst. (Tech. Hochsch. Aachen)*, 1979, **79**, 26

Poly(alkyl 2-cyanoacrylates)   P-65

Related Polymers: Poly(methyl 2-cyanoacrylate), Poly(ethyl 2-cyanoacrylate), Poly(isobutyl 2-cyanoacrylate), Poly(allyl 2-cyanoacrylate)
Material class: Thermoplastic, Gums and resins
Polymer Type: acrylic, acrylonitrile and copolymers
General Information: Poly(alkyl 2-cyanoacrylates) have been thoroughly reviewed [1,2,3,4]. The interest in these polymers stemmed from a search for acrylic polymers with improved props. and the adhesive props. were discovered serendipitously. Commercial products are unpolymerised monomer; these are low viscosity liquids with good wetting props. which undergo an almost instantaneous polymerisation when in contact with nucleophilic or basic substances via an anionic polymerisation pathway which has no defined end point. The polymers so produced have high MW, which can approximate to living polymerisation, and because of the polar cyano and carbonyl functions have excellent adhesive props., bonding to a range of materials. Kinetics of polymerisation have been reported [5,6] and the polymerisation reaction is initiated by a range of bases such as hydroxide (fast) or amines (slower). Acids terminate and inhibit polymerisation. Cyanoacrylates used as adhesives have several advantages [2] over other adhesives systems. The system requires no mixing of different components; approaches 100% conversion of monomer with a very rapid cure time (in most cases cure is complete within minutes up to 24h). The adhesive bond is strong and many materials can be bonded: metals, other polymers to similar or different materials. The polymers are hard, colourless and amorph. Generally T$_g$ and polarity decrease with increasing alkyl group size [7]. Solubility in organic solvents increases with increasing alkyl group size. The disadvantages of cyanoacrylate adhesives include brittleness and low toughness. Heat resistance is poor, and metal-metal bonds are degraded by moisture. As the polymerisation is catalysed on the material's surface, use as a gap filling adhesive is limited. Commercial materials are susceptible to degradation on storage, and by heat, light, and moisture. Often stabilised with acidic gases. Thickeners such as poly(methyl methacrylate), [9] are used to increase viscosity and to improve bonding. Toughness is improved by addition of copolymers such as ABS or EVA and phthalates are used as plasticisers to improve ageing
Miscellaneous: Bonding props. are improved through proper surface preparation. May also be used to join acidic surfaces such as wood, by surface treatment with base, or addition of curing agents such as crown ethers. [3] Caffeine has been used as a surface treatment to initiate bulk polymerisation [2]

Applications/Commercial Products:

Processing & Manufacturing Routes: Alkyl cyanoacrylates may be prepared from cyanoacetate esters and a halomethyl ester, or more commonly by the base catalysed Knoevenagel condensation of cyanoacetate and formaldehyde. The monomer thus formed spontaneously polymerises, is purified and is heated to 150–300° to induce retropolymerisation. The monomer is then stabilised by addition of acid and radical inhibitors
Applications: Due to the rapid polymerisation and low cure time necessary, adhesives are used on assembly lines to bond materials such as electronic components, electric motors, sports goods, lenses, camera equipment, aircraft interiors, computer housings and medical equipment. Excellent for bonding plastic materials such as polyethylene, though metal-metal bonds tend to degrade. Also suitable for medical applications, as a surgical adhesive for wound repair. Also used in drug delivery systems, where they are used in coating of colloid nanoparticles, where polymerisation occurs at the colloid interface. Also used as photoresists in electronics manufacture

Trade name	Details	Supplier
Accuflo Superglu		Le Pages
Agomet C		Degussa
Cyanolit	various grades	Panacol Elosol
Flash		UHU
Helmitin		Forbo Helmitin
Prisin		Loctite
Sicomet 9000		Henkel Canada
Topfix		CECA

Bibliographic References

[1] Woods, J., *Polymeric Materials Encyclopedia*, (ed. J.C. Salomone), CRC Press, 1996, 1632, (rev)
[2] Millett, G.H., *Structural Adhesives Chemistry and Technology*, (ed. Hartshorn S.R.), Plenum Press, 1986, 249, (rev, props)
[3] Coover, H.W., Dreifus, D.W. and O' Connor, J.T., *Handb. Adhes.*, 3rd edn., (ed. I. Skeist), Van Nostrand Reinhold, 1990, 463, (rev, props)
[4] O' Connor, J.T., *Kirk-Othmer Encycl. Chem. Technol.*, 4th edn., (ed. J.I. Kroschwitz), John Wiley and Sons, 1991, **1**, 344, (rev, props)
[5] Pepper, D.C., Makromol. Chem., *Makromol. Symp.*, 1992, **60**, 267, (kinetics)
[6] Cronin, J.P. and Pepper, D.C., *Makromol. Chem.*, 1988, **189**, 85, (kinetics)
[7] Kukarnu, R.K., Porter, H.S. and Leonard, F., *J. Appl. Polym. Sci.*, 1973, **17**, 3509, (transition temp)
[8] *U.S. Pat.*, (BF Goodrich), 1949, **2 467 927**, (synth)
[9] *U.S. Pat.*, 1979, **4 134 929**, (thickeners)
[10] *Eur. Pat. Appl.*, (3M), 1985, **244 178**, (tougheners)

Polyalkylene dicarboxylates, General   P-66

Related Polymers: Poly(ethylene sebacate), Poly(ethylene succinate), Poly(tetramethylene adipate), Poly(decamethylene seba-

cate), Poly(decamethylene adipate), Poly(ethylene adipate), Poly(ethylene azelate), Polybutylene succinate
Material class: Thermoplastic
Polymer Type: saturated polyester
General Information: Biodegradable polymers, biopolymers.

Volumetric & Calorimetric Properties:
Thermodynamic Properties General: Melting points for a range of polyalkylene dicarboxylates reported [2].

Polymer Stability:
Biological Stability: Biodegradable via enzymatic hydrolysis; biodegrades in soil, seawater, compost and freshwater. Polybutylene succinate adipate degrades faster than polybutylene succinate.

Applications/Commercial Products:
Processing & Manufacturing Routes: Polycondensation data reported [2]. Processing by any conventional method that can be applied to polyolefines but with some modification [2].
Applications: Biodegradable bags (compost, garbage, packaging), rope, tape, packaging film.

Trade name
Bionolle

Bibliographic References
[1] Biopolymers: Polyesters III Wiley-VCH, (Eds. Doi, Y. and Steinbuchel. A.), 2002, **4**, 30
[2] Biopolymers: Polyesters III Wiley-VCH, (Eds. Doi, Y. and Steinbuchel. A.), 2002, **4**, 278

Poly(alkyl ether methacrylate)   P-67

$$\left[CH_2C(CH_3) \atop \underset{O}{C} \underset{OR}{} \right]_n$$

R = CH_2CH_2OMe (2-methoxyethyl)
  = CH_2CH_2OEt (2-ethoxyethyl)
  = $CH_2CH_2OCH_2CH_2CH_3$ (2-propoxyethyl)
  = $CH_2CH_2OCH_2CH_2CH_2CH_3$ (2-butoxyethyl)

Synonyms: *Poly(2-methoxyethyl methacrylate). Poly(2-ethoxyethyl methacrylate). Poly(2-propoxyethyl methacrylate). Poly(2-butoxyethyl methacrylate)*
Related Polymers: Polymethacrylates, General
Monomers: 2-Methoxyethyl methacrylate, 2-Ethoxyethyl methacrylate, 2-Propoxyethyl methacrylate, 2-Butoxyethyl methacrylate
Material class: Thermoplastic
Polymer Type: acrylic
CAS Number:

CAS Reg. No.	Note
36561-33-8	2-ethoxyethyl
26677-80-5	2-butoxyethyl

Volumetric & Calorimetric Properties:
Equation of State: Mark-Houwink constants for the 2-methoxyethyl ester have been reported [3]
Thermodynamic Properties General: Thermal expansivity has been reported [1]

Surface Properties & Solubility:
Cohesive Energy Density Solubility Parameters: Cohesive energy 49.3–60 kJ mol^{-1} (2-ethoxyethyl) [1]. δ 18.4–20.3 J$^{1/2}$ cm$^{-3/2}$ (2-ethoxyethyl) [1]

Wettability/Surface Energy and Interfacial Tension: Contact angle (2-methoxyethyl) 48° (air), 78° (Me_2CO), 74° (dodecane) [2]. Interfacial tension 21.6–24.2 mN m^{-1} (solid/H_2O, 2-methoxyethyl) [2]
Surface Tension:

No.	Value	Note
1	68–74.2 mN/m	γ_d 47.0–52.4; γ_p 21.0–21.8, 2-methoxyethyl [2]
2	71.1 mN/m	20°, 2-methoxyethyl hydrogel, 2% wt H_2O [2]

Transport Properties:
Polymer Solutions Dilute: Molar volumes 80.26 cm^3 mol^{-1} (25°, 2-methoxyethyl, amorph.) [1], 100.72 cm^3 mol^{-1} (25°, 2-propoxyethyl, amorph.) [1]. Theta temps. 64° (butanol, 2-methoxyethyl) [4], 38° (butanol, 2-ethoxyethyl) [4], 17° (butanol, 2-butoxyethyl) [4]

Bibliographic References
[1] Van Krevelen, D.W., *Properties of Polymers: Their Correlation with Chemical Structure*, 3rd edn., Elsevier, 1990
[2] King, R.N., Andrade, J.D., Ma, S.M., Gregonis, D.E. and Brostrom, L.R., *J. Colloid Interface Sci.*, 1985, **103**, 62, (interfacial tension)
[3] Stejskal, J., Janca, J. and Kratochvil, P., *Polym. J. (Tokyo)*, 1976, **8**, 549, (equation of state)
[4] Chawla, R.K., Stejskal, J. and Strkova, D., *Croat. Chem. Acta*, 1987, **60**, 1, (dilute soln. props.)

Poly(allyl cyanoacrylate)   P-68

$$\left[CH_2\underset{\underset{OCH_2CH=CH_2}{\overset{\|}{C}}}{\overset{CN}{C}} \right]_n$$

Synonyms: *Allyl cyanoacrylate polymer*
Related Polymers: Poly(alkyl 2-cyanoacrylates), Poly(methyl cyanoacrylate), Poly(ethyl cyanoacrylate), Poly(isobutyl cyanoacrylate)
Monomers: Allyl 2-cyanoacrylate
Material class: Thermoplastic, Gums and resins
Polymer Type: acrylic, acrylonitrile and copolymers
CAS Number:

CAS Reg. No.
30209-88-2

Molecular Formula: $(C_7H_7NO_2)_n$
Fragments: $C_7H_7NO_2$
Molecular Weight: 182000
General Information: Unsaturated side chain allows potential for crosslinking and somewhat anomalous props. Crosslinking occurs at high temp.
Morphology: Amorph.

Volumetric & Calorimetric Properties:
Thermodynamic Properties General: Thermodynamic parameters have been reported [4]
Specific Heat Capacity:

No.	Value	Note	Type
1	1.216 kJ/kg.C	0.166 KJ mol^{-1}, -23°	p
2	1.476 kJ/kg.C	0.194 KJ mol^{-1}, 25° [4]	P
3	1.623 kJ/kg.C	0.223 KJ mol^{-1}, 77° [3]	p

Poly(1-allyloxy-2,3-epoxypropane)

Glass-Transition Temperature:

No.	Value	Note
1	90°C	dynamic mechanical analysis [1]
2	115°C	[5]
3	170°C	fully cross-linked [5]
4	119.7°C	differential scanning calorimetry

Transition Temperature General: Low T_g may be due to water content [1]
Vicat Softening Point:

No.	Value	Note
1	78°C	[2]

Surface Properties & Solubility:
Solubility Properties General: Incompatible with poly(methyl 2-cyanoacrylate) [3]. Solubility parameter δ 11.5 $J^{1/2}cm^{-3/2}$ [6]

Transport Properties:
Polymer Solutions Dilute: [η] 0.16 dl g^{-1} (1% in C_6H_6)
Water Content: 4.9%
Water Absorption:

No.	Value	Note
1	4.9 %	[1]

Mechanical Properties:
Mechanical Properties General: Increasing temp. increases storage modulus, probably due to temp. induced cross-linking
Flexural Modulus:

No.	Value	Note
1	1751 MPa	254 kpsi, ASTM D790 [2]

Tensile Strength Break:

No.	Elongation
1	10 % [2]

Miscellaneous Moduli:

No.	Value	Note	Type
1	2500 MPa	25° [1]	storage modulus

Impact Strength: 5.9 KJ m^{-2} (23°, 24h, ASTM D950); 3.8 KJ m^{-2} (100°, 24h); 1.6 KJ m^{-2} (150°, 5h) [7]
Mechanical Properties Miscellaneous: Tensile shear strength 12.7 MPa (23°, 24h); 29.1 MPa (100°, 24h); 3.2 Mpa (150°, 5h) [7] 20.82 Mpa (3120 psi, steel-steel, ASTM D1002 [2], 6.55 Mpa (950 psi, Al-Al, ASTM D1002 [2]. Lap shear strength 22 Mpa [2]; 15.8 MPa; 17.2 MPa (100°, 24h); 7.8 MPa (150°, 24h) [5]. Loss factor [1] and dynamic mechanical props. have been reported [5]

Electrical Properties:
Dielectric Permittivity (Constant):

No.	Value	Frequency	Note
1	3.8	1 MHz	ASTM D150 [2]

Dissipation (Power) Factor:

No.	Value	Frequency	Note
1	0.02	1 MHz	ASTM D150 [2]

Polymer Stability:
Thermal Stability General: Better thermal stability than other alkyl cyanoacrylates due to cross-linking [1]
Decomposition Details: Decomp. onset 127° (N_2). Lower than other alkyl cyanoacrylates but retains mechanical props. to high temps.
Environmental Stress: As with other cyanoacrylates, poor resistance to moisture. Unsuitable for use in situations of high heat, high impact, moisture or outdoor use
Chemical Stability: Stable to oils
Hydrolytic Stability: Degrades in aqueous soln. releasing formaldehyde

Applications/Commercial Products:
Processing & Manufacturing Routes: Spontaneous polymerisation of allyl 2-cyanoacrylate monomer catalysed by base. Cure times longer than for lower homologues. see poly(alkyl cyanoacrylates)
Applications: Instant one pot adhesive

Trade name	Supplier
Powerbond	Permabond

Bibliographic References
[1] Cheung, K.H., Guthrie, J., Otterbum, M.S. and Rooney J., *Makromol. Chem.*, 1987, **188**, 3041, (T_g, mechanical props.)
[2] Coover, H.W., Dreifus, D.W. and O'Connor, J.T., *Handb. Adhes.*, 3rd edn. (ed.I. Skiest) Van Nostrand Reinhold, 1990, 463, (rev.)
[3] Agufitel, M. and Dumetrescu, G., *Polym. Prepr.*, 1985, **26**, 226, (compatibility)
[4] Kiparisova, E.G., Bykova, T.A., Lebedev, B.V., Gusera, T.I. etal, Vysokomol. Soedin., *Ser. A*, 1993, **35**, 615, (thermodynamic props)
[5] Kotzev, D.L., Ward, T.C. and Wright, *J. Appl. Polym. Sci.*, 1981, **26**, 1941
[6] Milleth, G.-H., *Structural Adhesives Chemistry and Technology*, (Ed. S.R. Hartshorn), Plenum Press, 1986, 249, (rev)
[7] Denchev, Z.Z. and Kabaivanov, V.S., *J. Appl. Polym. Sci.*, 1993, **47**, 1019, (adhesive props)
[8] Negulescu, I.I., Calugaru, E.M., Vasile, C. and Dumitrescu, G., J. Macromol. Sci., *Pt. A*, 1987, **24**, 75, (thermal degradation)

Poly(1-allyloxy-2,3-epoxypropane)

Synonyms: *Poly[[(2-propenyloxy)methyl]oxirane]*
Monomers: (2-Propenyloxymethyl)oxirane
Material class: Thermosetting resin
Polymer Type: epoxy
CAS Number:

CAS Reg. No.	Note
25639-25-2	homopolymer

Molecular Formula: $(C_6H_{10}O_2)_n$

– Poly(alpha olefin)

Volumetric & Calorimetric Properties:
Density:

No.	Value	Note
1	d 1.048 g/cm^3	[5]

Glass-Transition Temperature:

No.	Value	Note
1	-74°C	by calorimetry [5]
2	-78°C	by dilatometry [5]

Transition Temperature General: Has no T_m [5]
Vicat Softening Point:

No.	Value	Note
1	285°C	polyethylenepolyamine cured [2]

Transport Properties:
Transport Properties General: $[\eta]_{inh}$ 4.6 dl g^{-1} (toluene, 30°). [5] Reduced viscosity 2.6–3.2 dl g^{-1} (cyclohexanone, 50°) [6]

Mechanical Properties:
Tensile Strength Break:

No.	Value	Note
1	68.6 MPa	700 kg cm^{-2}, polyethylenepolyamine cured [2]

Flexural Strength at Break:

No.	Value	Note
1	156.9 MPa	1600 kg cm^{-2}, polyethylenepolyamine cured [2]

Compressive Strength:

No.	Value	Note
1	186.3 MPa	1900 kg cm^{-2}, polyethylenepolyamine cured [2]

Hardness: Brinell hardness 40 kg mm^{-2} (polyethylenepolyamine cured) [2]

Optical Properties:
Refractive Index:

No.	Value	Note
1	1.478	30° [5]

Transmission and Spectra: H-1 nmr and C-13 nmr spectral data have been reported [1]

Applications/Commercial Products:
Processing & Manufacturing Routes: Synth. by cationic polymerisation [7] of allyl glycidyl ether with boron trifluoride etherate at room temp. for 1.5 h. [1] Also synth. by FeCl$_3$ catalysis in the absence of air [3]; by bulk polymerisation using metal caprylates [4]; by soln. polymerisation in C$_6$H$_6$ at 50° using a diethylzinc-sulfur catalyst. [5] Polymerisation with diethyl zinc-nitromethane catalyst yields a rubbery solid [6]
Applications: As a curing agent for acrylates; in adhesives and coatings. Used as a reactive diluent in other epoxy resin compositions

Bibliographic References
[1] Crivello, J.V. and Kim, W.-G., *J. Polym. Sci., Part A: Polym. Chem.*, 1994, **32**, 1639, (synth, H-1 nmr, C-13 nmr)
[2] Mustafaev, R.I., Mamedov, R.I., Ayubov, G.M., Guseinov, F.I. *et al*, *Azerb. Khim. Zh.*, 1973, **79**, (mech props)
[3] Ito, N., Yoshida, H. and Eguchi, K., *Nagoya-Shi Kogyo Kenkyusho Kenkyu Hokoku* 1975, **54**, 13, (synth)
[4] Aoki, S., Yasuzawa, K. and Otsu, T., *Kobunshi Ronbunshu*, 1974, **31**, 262, (synth)
[5] Lal, J. and Trick, G.S., *J. Polym. Sci., Part A-1*, 1970, **8**, 2339
[6] Nakaniwa, M., Kameoka, I., Ozaki, K., Kawabata, N. and Furukawa, J., *Makromol. Chem.*, 1972, **155**, 185, (synth)
[7] Watt, W.R., *ACS Symp. Ser.*, 1978, **114**, 17

Poly(alpha olefin) P-70
Synonyms: *PAO. Poly(α-olefin). Poly(1-decene) oligomeric. Poly-alphaolefin oligomer*
Related Polymers: Poly(1-decene)
Monomers: 1-Decene
Material class: Fluids
Polymer Type: polyolefins
CAS Number:

CAS Reg. No.	Note
68037-01-4	Poly(1-decene) oligomer
25189-70-2	Poly(1-decene)
17438-89-0	1-decene dimer
14638-82-5	1-decene trimer
54545-42-5	1-decene tetramer
75818-30-3	1-decene pentamer
75818-31-4	1-decene hexamer

Molecular Formula: $(C_{10}H_{20})_n$
Fragments: $C_{10}H_{20}$
Molecular Weight: MW 400–600
General Information: Clear liq. with a barely perceptible odour. Oligomers are used as synthetic lubricants

Volumetric & Calorimetric Properties:
Density:

No.	Value	Note
1	d^{16} 0.818 g/cm^3	ASTM D1298 [1]

Surface Properties & Solubility:
Solvents/Non-solvents: Insol. H$_2$O [6]

Transport Properties:
Polymer Melts: Viscosity index 137 (ASTM D2270) [1]. Viscosity 0.039 St (100°, ASTM D445) [1]. Pour point -70° (ASTM D97) [1]
Volatile Loss and Diffusion of Polymers: NDACk volatility 12% (250°, 1h, DIN 51581) [1]
Water Content: 11 ppm (ASTM D1744) [1]

Electrical Properties:
Dielectric Permittivity (Constant):

No.	Value	Frequency	Note
1	2.102	1–1000 kHz	23° [1]

Optical Properties:
Transmission and Spectra: H-1 nmr, C-13 nmr and Raman spectral data have been reported [7]
Volume Properties/Surface Properties: Clear liq., colour 0.5 (max., ASTM D1500) [1]

Polymer Stability:
Environmental Stress: Oxidation occurs above 192° [1] and props. have been reported [8]. The principal products of oxidation are H_2O and CO_2 [9]

Applications/Commercial Products:
Processing & Manufacturing Routes: Linear α-olefins are prod. by catalytic oligomerisation of ethylene with $AlEt_3$ or nickel-based catalysts. α-Olefins are then oligomerised using cationic initiators (BF_3, boron trifluoride/alcohols, $AlCl_3$, $TiCl_4$ and trialkyl aluminium, and $AlBr_3$) at low temps.
Applications: Synthetic lubricant used in lubricant formulation for industrial oils, transformer oils, compressor oils, hydraulic fluids, transmission fluids, gear oils, crankcase fluids, and heat transfer fluids. Also used in multipurpose greases for industrial applications

Trade name	Details	Supplier
Durasyn	164, 166, 168	Amoco Chemical Company
		Collinda Ltd.

Bibliographic References
[1] *Durasyn 164*, Amoco Chemical Belgium S.A., (technical datasheet)
[2] Willermet, P.A., Haakana, C.C. and Sever, A.W., *J. Synth. Lubr.*, 1985, **2**, 22, (dimer)
[3] Brennan, J.A., *Ind. Eng. Chem. Proc. Des. Dev.*, 1980, **19**, 2, (trimer, tetramer)
[4] *U.S. Pat.*, 1986, 4 587 368, (pentamer)
[5] *Eur. Pat. Appl.*, 1988, 288 777, (hexamer)
[6] *Durasyn 164*, Amoco Chemical Belgium S.A., (technical datasheet)
[7] Doskocilova, D., Pecka, J., Dybal, J., Kriz, J. and Mikes, F., *Macromol. Chem. Phys.*, 1994, **195**, 2747, (H-1 nmr, C-13 nmr, Raman)
[8] Zuzi, B., Vercimakova, B. and Bucinska, A., *Ropa Uhlie*, 1991, **33**, 282, (oxidation props)
[9] Koh, C.S. and Butt, J.B., *Ind. Eng. Chem. Res.*, 1995, **34**, 524, (oxidation props)
[10] *Kirk-Othmer Encycl. Chem. Technol.*, 4th edn., (ed. J.I. Kroschwitz), Wiley Interscience, 1996, **17**, 831, 836, 843, (uses)

Poly(anhydrides) P-71

Related Polymers: Poly(oxyisophthaloyl), Poly(oxycarbonyl-1,4-phenylene methylene-1,4-phenylene carbonyl), Poly(oxycarbonyl-1,4-phenylene isopropylidene-1,4-phenylene carbonyl)
General Information: The polyanhydride of an aromatic dicarboxylic acid was first prepared in 1909 [9] and aliphatic polyanhydrides were first synth. in 1932 [10]; they possessed fibre-forming props. but low melting point and hydrolytic stability. It was reported that certain aromatic polyanhydrides had excellent film and fibre-forming props. due to the accumulation of phenylene nuclei along the polymer chain [3]. Research in to this class of polymers was revitalised by researches at MIT in the early 1980's in developing biodegradeable polymeric biomaterials of potential use for drug delivery [1]

Volumetric & Calorimetric Properties:
Transition Temperature General: T_m and T_g values have been reported for a range of poly(anhydrides) [3]

Surface Properties & Solubility:
Cohesive Energy Density Solubility Parameters: Cohesion energies of a series of poly(anhydrides) have been reported [6]
Solvents/Non-solvents: Sol. common organic solvents [1]. Highly aromatic polymers are sol. in polar organic solvents

Polymer Stability:
Decomposition Details: Decomposition rate decreases with increasing aromaticity of the polymer backbone
Hydrolytic Stability: Good hydrolytic stability for aromatic polyanhydrides, aliphatic polyanhydrides are known to decompose on standing [1,7]
Biological Stability: Good biocompatibility [1]

Applications/Commercial Products:
Processing & Manufacturing Routes: Prod. by intramolecular dehydrochlorination of a diacid chloride and a dicarboxylic acid; or melt-polycondensation with a mixed anhydride; or interfacial polycondensation [1,2,4,5,8,13]
Applications: Potential as biodegradeable materials for medical applications, temporary scaffolds and barriers, medical sutures, bottles and bags, microspheres and microcapsules

Bibliographic References
[1] *Concise Encyclopedia of Polymer Science and Engineering*, (eds. H.F. Mark, N.M. Bikales, C.G. Overberger and G. Menges), John Wiley and Sons, 1990, 765
[2] Alger, M.S.M., *Polymer Science Dictionary*, Elsevier, 1989, 327
[3] Conix, A., *Makromol. Chem.*, 1957, **24**, 76, (Tm, Tg)
[4] Yoda, N., *Makromol. Chem.*, 1962, **55**, 174, (synth)
[5] Leong, K.W., Simonte, V. and Langer, R., *Macromolecules*, 1987, **20**, 705, (synth)
[6] Yoda, N., *J. Polym. Sci., Part A: Polym. Chem.*, 1963, **1**, 1323
[7] Domb, A.J. and Langer, R., *Macromolecules*, 1989, **22**, 2117, (solid-state stability)
[8] Yoda, N., *Encycl. Polym. Sci. Technol.*, Vol. 10, John Wiley and Sons, 1969, 630
[9] Bucher, J.E. and Slade, W.C., *J. Am. Chem. Soc.*, 1909, **32**, 1319
[10] Hill, J.W. and Carothers, W.H., *J. Am. Chem. Soc.*, 1932, **54**, 1569

Poly(aniline) P-72

$$\left[\left(\underset{}{\bigcirc}-NH-\underset{}{\bigcirc}-NH\right)_y \left(\underset{}{\bigcirc}-N=\underset{}{\bigcirc}=N\right)_{(1-y)}\right]$$

y = 1 (leucoemeraldine)
y = 0 (Pernigraniline)
y = 0.5 (Emeraldine)

Synonyms: Aniline black. Leucoemeraldine. Pernigraniline. Emeraldine. PANI. Nigraniline. Poly(benzenamine)
Monomers: Aniline
Polymer Type: polyaniline
CAS Number:

CAS Reg. No.
25233-30-1

Molecular Formula: $[(C_{12}H_{10}N_2).(C_{12}H_8N_2)]_n$
Fragments: $C_{12}H_{10}N_2$ $C_{12}H_8N_2$
General Information: Polyaniline may exist in a number of forms [19] and chemical struct. depends upon the method of synth. [1,2,3,4,5,6]. The quinoid-benzenoid-diimine form is an insulator. Redox switching between forms has been studied [20]
Morphology: Orthorhombic, a 7.65, b 5.75 c 10.22 [7]

Surface Properties & Solubility:
Solvents/Non-solvents: Sol. DMF. Spar. sol. CH_2Cl_2, dioxane [8]

Mechanical Properties:
Tensile Strength Break:

No.	Value	Note	Elongation
1	50 MPa	cross-linked [9]	14%
2	10 MPa	cross-linked [9]	4%
3	52 MPa	rigid [9]	33%
4	7 MPa	rigid [9]	20%

Elastic Modulus:

No.	Value	Extension	Note
1	1150 MPa	14%	cross-linked [9]
2	60 MPa	4%	cross-linked [9]
3	1310 MPa	33%	rigid [9]
4	60 MPa	20%	rigid [9]

Electrical Properties:
Electrical Properties General: Conduction in polyaniline can be accounted for in terms of electron flow by the assistance of proton transfer [10,11,13]. Emeraldine is the conducting form [19]. Electrochemical methods allow formation of conducting films at the electrode and doped conductive forms are known [21]

Surface/Volume Resistance:

No.	Value	Note	Type
3	0.00071×10^{15} Ω.cm	250° [14]	S

Complex Permittivity and Electroactive Polymers: Electrorheological fluids with polyaniline particles have been studied. [23] Oriented films show high electrical anisotropy with the highest conductivity in the stretch direction. Conductivities greater than 6000 S cm^{-1} have been reported. [27] The effects of monomer concentration and substrate resistance on redox props. have been reported [28]

Magnetic Properties: Oriented films behave as highly one-dimensional metallic systems with no Curie-type behaviour even at low temps. (below -263°) [27]

Optical Properties:
Transmission and Spectra: Ir [8,15,16], uv [7], H-1 nmr [30] and epr [12,13,30] spectral data have been reported

Applications/Commercial Products:
Processing & Manufacturing Routes: Chemical and electrochemical methods of synth. can be employed [1,8,17,18,19]. Chemical synth. can prod. polymer contaminated with oxidative agents used in synth. Graft copolymerisation on polyacrylamide [24] and conducting soluble copolymers [25] have been made. Conducting composites of polyaniline have been reviewed [26] Conducting films may be synth. by oxidation of aniline with ammonium persulfate at -30° in the presence of lithium chloride. Films are cast from *N*-methylpyrrolidone soln. [27]

Applications: Modified electrodes, corrosion inhibitor for semiconductors in photoelectrochemical assemblies, microwave absorbers, and electromagnetic shields. Oriented films have potential applications in electrical devices (lightweight batteries, sensors, etc.)

Trade name	Supplier
Versicon	Allied Signal Corporation

Bibliographic References
[1] Genies, E.M., Boyle, A., Lapkowski, M. and Tsintaris, C., *Synth. Met.*, 1990, **36**, 139, (rev)
[2] Genies, E.M., Lapkowski, M., Noel, P., Langlois, S. *et al*, *Synth. Met.*, 1991, **43**, 2847, (rev)
[3] Pouget, J.P., Laridjani, M., Jozefowicz, M., Epstein, A.J. *et al*, *Synth. Met.*, 1992, **51**, 95
[4] *Ullmanns Encykl. Ind. Chem.*, (ed. B. Elvers), VCH, 1992, **A21**, 440
[5] Tan, K.L., Tan, B.T.G., Khor, S.H., Neoh, K.G. and Kang, E.T., *J. Phys. Chem. Solids*, 1991, **52**, 673
[6] Cao, Y., Andreatta, A., Heeger, A.J. and Smith, P., *Polymer*, 1989, **30**, 2305
[7] Selvan, S.T., Mari, A., Athinarayanasamy, K., Phani, K.L.N. and Pitchumani, S., *Mater. Res. Bull.*, 1995, **30**, 699
[8] Bingham, A. and Ellis, B., *J. Polym. Sci., Part A-1*, 1969, **7**, 3229
[9] Oka, O., Kiyohara, O., Morita, S. and Yoshino, K., *Synth. Met.*, 1993, **55**, 999
[10] Park, Y.W., Moon, J.S., Bak, M.K. and Jin, J.I., *Synth. Met.*, 1989, **29**, E389
[11] Leclerc, M., D'Aprano, G. and Zotti, G., *Synth. Met.*, 1993, **55**, 1527, (electrochemical props)
[12] Langer, J., *Solid State Commun.*, 1978, **26**, 839
[13] Watanabe, A., Mori, K., Iwabuchi, A., Iwasaki, Y. and Nakamura, Y., *Macromolecules*, 1989, **22**, 3521
[14] Bradley, A. and Hammes, J.P., *J. Electrochem. Soc.*, 1963, **110**, 15
[15] Volkov, A., Tourillon, G., Camille Lacaze, P. and Dubois, J.E., *J. Electroanal. Chem.*, 1980, **115**, 279
[16] Harada, I., Furukawa, Y. and Ueda, F., *Synth. Met.*, 1989, **29**, E303
[17] MacDiarmid, A.G. and Epstein, A.J., *Faraday Discuss. Chem. Soc.*, 1989, **88**, 317, (rev)
[18] Lux, F., *Polymer*, 1994, **35**, 2915
[19] MacDiarmid, A.G., Chiang, J.-C., Halpern, M., Huang, W.-S. *et al*, *Mol. Cryst. Liq. Cryst.*, 1985, **121**, 173, (interconversion)
[20] Gospodinova, N., Mokreva, P. and Terlemezyan, L., *Polym. Int.*, 1996, **41**, 79, (redox switching)
[21] Erden, E., Saçak, M. and Karakişla, M., *Polym. Int.*, 1991, **39**, 153, (oxalic acid doped)
[22] Jinqing, K., Feng, Z., Shalin, M. and Yujun, S., *Sens. Actuators*, 1996, **30**, 7, (polyaniline electrodes)
[23] Xie, H.-G. and Guan, J.-G., *Angew. Makromol. Chem.*, 1996, **235**, 21, (electrorheology)
[24] Xiang, Q. and Xie, H.-Q., *Eur. Polym. J.*, 1996, **32**, 865, (graft copolymers)
[25] Kathirgamanathan, P., Adams, P.N., Quill, K. and Underhill, A.E., *J. Mater. Chem.*, 1991, **1**, 141
[26] Battacharya, A. and Amitabha, D., *Prog. Solid State Chem.*, 1996, **24**, 141, (composites)
[27] Adams, P.N., Laughlin, P.J., Monteman, A.P. and Bernhoeft, N., *Solid State Commun.*, 1994, **91**, 875
[28] Bedekar, A.G., Patil, S.F., Patil, R.C. and Vijayamohanan, K., *Mater. Chem. Phys.*, 1997, **48**, 76
[29] Boara, G. and Sparpaglione, M., *Synth. Met.*, 1995, **72**, 135, (synth)
[30] Pavesi, L. and Tedoldi, F., *Polym. Adv. Technol.*, 1997, **8**, 30, (epr, H-1 nmr)

Poly(arylalkyl methacrylates)

$$\left[\text{CH}_2\text{C}(\text{CH}_3) \atop \underset{\text{OR}}{\overset{\text{O}}{|}} \right]_n$$

R = CH$_2$Ph (benzyl)

   = CHPh$_2$ (diphenylmethyl)

   = CPh$_3$ (triphenylmethyl)

   = CHCH$_3$ (1-phenylethyl)
     |
    Ph

   = CH$_2$CH$_2$Ph (2-phenylethyl)

   = CHCH$_3$–(o-C$_6$H$_4$Cl) (1-(o-chlorophenyl)ethyl)

   = CHCH$_2$Ph (1,2-diphenylethyl)
     |
    Ph

   = CH$_2$CH$_2$CH$_2$CH$_2$Ph (4-phenylbutyl)

   = CH$_2$CH$_2$OCPh$_3$ (triphenylmethoxyethyl)

Synonyms: *Poly(arylalkyl 2-methyl-2-propenoate). Poly(benzyl methacrylate). Poly(diphenylmethyl methacrylate). Poly(triphenylmethyl methacrylate). Poly(1-phenylethyl methacrylate). Poly(1-o-chlorophenylethyl methacrylate). Poly(1,2-diphenylethyl methacrylate). Poly(triphenylmethoxyethyl methacrylate). Poly(2-phenylethyl methacrylate). Poly(4-phenylbutyl methacrylate)*
Related Polymers: Polymethacrylates, General, Poly(aryl methacrylates)
Monomers: Benzyl methacrylate, Diphenylmethyl methacrylate, Triphenylmethyl methacrylate, 1-Phenylethyl methacrylate, 1-(*o*-

– Poly(aryl methacrylates)

Chlorophenylethyl) methacrylate, 2-Phenylethyl methacrylate, 1,2-Diphenylethyl methacrylate, 4-Phenylbutyl methacrylate, Triphenylmethoxyethyl methacrylate
Material class: Thermoplastic
Polymer Type: acrylic
CAS Number:

CAS Reg. No.	Note
25085-83-0	benzyl
27497-74-1	triphenylmethyl

Tacticity: For polymerisations carried out in C_6H_6 using AIBN the isotactic content of the polymer increases in the order: benzyl > diphenylmethyl > triphenylmethyl (<5%, 25% and 40% respectively) [1]

Volumetric & Calorimetric Properties:
Density:

No.	Value	Note
1	d^{20} 1.179 g/cm^3	benzyl, amorph.
2	d^{25} 1.168 g/cm^3	diphenyl, amorph.
3	d^{25} 1.129 g/cm^3	1-phenylethyl, amorph.
4	d^{25} 1.269 g/cm^3	1-(o-chlorophenylethyl)
5	d^{25} 1.147 g/cm^3	1,2-diphenylethyl, amorph. [2]

Thermal Expansion Coefficient:

No.	Value	Note	Type
1	0.0005 K^{-1}	benzyl, M_n 400000, $T > T_g$ [4]	V
2	0.00051 K^{-1}	2-phenylethyl, $T > T_g$ [4]	V
3	0.00017 K^{-1}	benzyl, M_n 400000, $T < T_g$ [4]	V
4	0.00018 K^{-1}	2-phenylethyl, $T < T_g$ [4]	V

Equation of State: Mark-Houwink constants have been reported [1,5,8,9]
Glass-Transition Temperature:

No.	Value	Note
1	54°C	benzyl, M_n 400000 [4]
2	26–56°C	2-phenylethyl [4,5]
3	10°C	4-phenylbutyl [5]

Surface Properties & Solubility:
Cohesive Energy Density Solubility Parameters: Cohesive energy 61100–63500 J mol^{-1} (benzyl) [2]. δ 20.3 J$^{1/2}$ cm$^{-3/2}$ [2], 20.1–20.5 J$^{1/2}$ cm$^{-3/2}$ [2], 38.9 J$^{1/2}$ cm$^{-3/2}$ (9.29 cal$^{1/2}$ cm$^{-3/2}$) [6]
Surface Tension:

No.	Value	Note
1	36 mN/m	benzyl, 20° [7]

Transport Properties:
Polymer Solutions Dilute: Molar volumes (van der Waals, 25°) 98.5 cm^3 mol^{-1} (benzyl, amorph.), 142.7 cm^3 mol^{-1} (diphenylmethyl, amorph.), 108.7 cm^3 mol^{-1} (1-phenylethyl, amorph.), 117.8 cm^3 mol^{-1} (1-o-chlorophenylethyl), 147.7 cm^3 mol^{-1} (1,2-diphenylethyl, amorph.); molar volumes (glassy, 25°) 149.5 cm^3 mol^{-1} (benzyl, amorph.), 216 cm^3 mol^{-1} (diphenylmethyl, amorph.), 285.3 cm^3 mol^{-1} (triphenylmethyl), 168.5 cm^3 mol^{-1} (1-phenylethyl, amorph.), 177.1 cm^3 mol^{-1} (1-o-chlorophenylethyl), amorph.), 232.2 cm^3 mol^{-1} (1,2-diphenylethyl, amorph.) [2,3]. Intrinsic viscosity 0.285 dl g^{-1} (30°, THF, 4-phenylbutyl, MW 143000), 2.98 dl g^{-1} (30°, THF, 4-phenylbutyl, MW 3130000), 0.159 dl g^{-1} (30°, THF, 2-phenylethyl, MW 59000), 1.55 dl g^{-1} (30°, THF, 2-phenylethyl, MW 1180000) [5]. Intrinsic viscosity increases with increasing MW. Theta temps. 73.2° (cyclohexanone, benzyl, VM method) [9], 83.5° (cyclopentanol, benzyl, A method) [10], 45° (3-heptanone, diphenylmethyl) [1], 27.5° (1-chloroheptane, 2-phenylethyl) [5], 23.0° (1-chloroundecane, 4-phenylbutyl) [5], 47° (mesitylene, triphenylmethoxyethyl) [11]

Optical Properties:
Refractive Index:

No.	Value	Note
1	1.568	20°, benzyl [2]

Polymer Stability:
Stability Miscellaneous: Electron beam or γ-irradiation cause scission with rapid degradation of the polymer [12]

Bibliographic References
[1] Mays, J., Hadjichristidis, N. and Lindner, J.S., Polym. Prepr. (Am. Chem. Soc., *Div. Polym. Chem.*), 1991, **32**, 148, (synth.)
[2] Van Krevelen, D.W., *Properties of Polymers: Their Correlation with Chemical Structure*, 3rd edn., Elsevier, 1990
[3] Tricot, M., *Macromolecules*, 1986, **19**, 1268
[4] Krause, S., Gormley, J.J., Roman, N., Shetter, J.A. and Watenabe, W.H., J. Polym. Sci., *Part A: Polym. Chem.*, 1965, **3**, 3573, (glass transition temps.)
[5] Chen, Y., Mays, J.W. and Hadjichristidis, N., J. Polym. Sci., *Polym. Phys. Ed.*, 1994, **32**, 715
[6] Frank, C.W. and Gashgari, M.A., *Macromolecules*, 1979, **12**, 163
[7] Fox, H. and Zisman, W.A., *J. Colloid Sci.*, 1952, **7**, 432
[8] Kurata, M. and Stockmayer, W.H., *Adv. Polym. Sci.*, Springer-Verlag, 1963, **3**, 196
[9] Richards, W.R., *Polymer*, 1977, **18**, 114
[10] Richards, W.R. and Disselhoff, G., *Angew. Makromol. Chem.*, 1978, **66**, 221
[11] Dolezalova, M., Petrus, V., Tuzar, Z. and Bohdanecky, M., *Eur. Polym. J.*, 1976, **12**, 701
[12] Dawes, K. and Glover, L.C., *Physical Properties of Polymers Handbook*, (ed. J.E. Mark), AIP Press, 1996, 557, (degradation)

Poly(aryl methacrylates)

Synonyms: *Poly(aryl 2-methyl-2-propenoate)*. *Poly(phenyl methacrylate)*. *Poly(phenylthiol methacrylate)*. *Poly(2-methylphenyl methacrylate)*. *Poly(2,6-dimethylphenyl methacrylate)*. *Poly(tert-butylphenyl methacrylate)*. *Poly(tetramethylbutylphenyl methacrylate)*. *Poly(5-indanyl methacrylate)*. *Poly(cyclohexylphenyl methacrylate)*. *Poly(4-cyanomethylphenyl methacrylate)*. *Poly(biphenyl methacrylate)*. *Poly(naphthyl methacrylate)*. *Poly(2-chlorophenyl methacrylate)*. *Poly(4-chlorophenyl methacrylate)*. *Poly(2,4,5-trichlorophenyl methacrylate)*. *Poly(pentachlorophenyl methacrylate)*. *Poly(4-cyanophenyl methacrylate)*. *Poly(4-carbomethoxyphenyl methacrylate)*
Related Polymers: Polymethacrylates General, Poly(arylalkyl methacrylates), Poly(methyl methacrylate-co-tert-butylphenyl methacrylate)
Monomers: Phenyl methacrylate, Phenylthiol methacrylate, 2-Methylphenylthiol methacrylate, 2,6-Dimethylphenyl methacrylate, tert-Butylphenyl methacrylate, Tetramethylbutylphenyl methacrylate, 5-Indanyl methacrylate, Cyclohexylphenyl methacrylate, (4-Cyanomethyl)phenyl methacrylate, Biphenyl methacrylate, Naphthyl methacrylate, 2-Chlorophenyl methacrylate, 4-Chlorophenyl methacrylate, 2,4,5-Trichlorophenyl methacrylate, Pentachlorophenyl methacrylate, 4-Cyanophenyl methacrylate, 4-Carbomethoxyphenyl methacrylate
Material class: Thermoplastic
Polymer Type: acrylic

Poly(aryl methacrylates)

CAS Number:

CAS Reg. No.	Note
25189-01-9	phenyl
29696-27-3	*tert*-butylphenyl
76033-34-6	tetramethylbutylphenyl
65930-09-8	[1,1'-biphenyl]-2-yl
40544-65-8	[1,1'-biphenyl]-4-yl
39296-32-7	naphthyl
31547-85-0	1-naphthyl
28702-85-4	2-naphthyl
36876-19-4	2-chlorophenyl
103467-71-6	3-chlorophenyl
34149-11-6	4-chlorophenyl
40921-83-3	2,4,5-trichlorophenyl
36876-23-0	4-cyanophenyl

Volumetric & Calorimetric Properties:
Density:

No.	Value	Note
1	d^{20} 1.21 g/cm^3	Ph ester, amorph. [1,2]
2	d^{25} 1.115 g/cm^3	cyclohexylphenyl ester [2]

Thermal Expansion Coefficient:

No.	Value	Note	Type
1	0.00053 K^{-1}	Ph ester, M_n 816000, $T > T_g$ [3]	V
2	0.00048 K^{-1}	(4-cyanomethyl)phenyl ester, M_n 700000, $T > T_g$ [3]	V
3	0.00041 K^{-1}	4-cyanophenyl ester, M_n 848000, $T > T_g$ [3]	V
4	0.00043–0.00049 K^{-1}	4-carbomethoxyphenyl ester, M_n 129000, $T > T_g$ [3]	V
5	0.00016 K^{-1}	Ph ester, M_n 816000, $T < T_g$ [3]	V
6	0.00018 K^{-1}	(4-cyanomethyl)phenyl ester, M_n 700000, $T < T_g$ [3]	V
7	0.00011 K^{-1}	4-cyanophenyl ester, M_n 848000, $T < T_g$ [3]	V
8	0.00017 K^{-1}	4-carbomethoxyphenyl ester, M_n 129000, $T < T_g$ [3]	V

Equation of State: Mark-Houwink constants have been reported [11,15,20,21,22]

Glass-Transition Temperature:

No.	Value	Note
1	110°C	Ph ester, M_n 816000 [3]
2	112°C	Ph ester [4]
3	98°C	*tert*-butylphenyl ester [5]
4	83°C	5-indanyl ester [6]
5	125–131°C	(4-cyanomethyl)phenyl ester, M_n 700000 [3]
6	75°C	naphthyl ester [7]
7	155°C	4-cyanophenyl ester, M_n 848000 [3]
8	103–109°C	4-carbomethoxyphenyl ester, M_n 129000 [3]

Vicat Softening Point:

No.	Value	Note
1	120°C	Ph ester [1]

Surface Properties & Solubility:
Cohesive Energy Density Solubility Parameters: Cohesive energy 50.5 kJ mol^{-1} (Ph ester) [4]; 38–88 kJ mol^{-1} (naphthyl ester, 9100–21000 cal mol^{-1}) [8,9]. Solubility parameter δ 20.3 J$^{1/2}$ cm$^{-3/2}$ (9.92 cal$^{1/2}$ cm$^{-3/2}$, Ph ester) [10], 19.2 J$^{1/2}$ cm$^{-3/2}$ (9.4 cal$^{1/2}$ cm$^{-3/2}$, *tert*-butylphenyl ester) [5]

Solvents/Non-solvents: *tert*-Butylphenyl ester sol. Me$_2$CO. 5-Indanyl ester sol. ethylbenzene, toluene, C$_6$H$_6$, CCl$_4$, CHCl$_3$, chlorobenzene, dichlorobenzene, CH$_2$Cl$_2$, 2-butanone, isobutyl methyl ketone, dioxane. Insol. hexane, heptane, cyclohexane, Et$_2$O, THF, EtOAc, Me$_2$CO, propanol, isopropanol, DMSO, EtOH, butanol, MeOH, H$_2$O [6]

Surface Tension:

No.	Value	Note
1	35 mN/m	20°, Ph ester [12]

Transport Properties:
Polymer Solutions Dilute: η 3.8 dl g^{-1} (MW 497000), 1.2 dl g^{-1} (MW 80000, 25°, dioxane, Ph ester), 5.15 dl g^{-1} (25°, C$_6$H$_6$, Ph ester), 4.15 dl g^{-1} (25°, 2-butanone, Ph ester), 0.368 dl g^{-1} (25°, THF, phenylthiol ester, MW 252000), 0.312 dl g^{-1} (25°, C$_6$H$_6$, phenylthiol ester, MW 252000), 0.275 dl g^{-1} (25°, toluene, phenylthiol ester, MW 252000), 0.19 dl g^{-1} (25°, 2-butanone, phenylthiol ester, MW 252000), 0.4 dl g^{-1} (25°, THF, methylphenylthiol ester, MW 323000), 0.385 dl g^{-1} (25°, C$_6$H$_6$, methylphenylthiol ester, MW 323000), 0.351 dl g^{-1} (25°, toluene, methylphenylthiol ester, MW 323000), 1.69 dl g^{-1} (C$_6$H$_6$, *tert*-butylphenyl ester, MW 2040000) [5,13,14], 0.39 dl g^{-1} (THF, 5-indanyl ester, MW 170000), 0.37 dl g^{-1} (toluene, 5-indanyl ester, MW 170000), 0.184 dl g^{-1} (2-butanone, 5-indanyl ester, MW 170000) [6], 0.095 dl g^{-1} (25°, C$_6$H$_6$, 4,4'-biphenyl ester, MW 43400), 0.112 dl g^{-1} (25°, CHCl$_3$, 4,4'-biphenyl ester, MW 43400), 0.105 dl g^{-1} (25°, THF, 4,4'-biphenyl ester, MW 43400), 0.1 dl g^{-1} (25°, dioxane, 4,4'-biphenyl ester, MW 43400) [15,16], 0.978 dl g^{-1} (25°, C$_6$H$_6$, 2,2'-biphenyl ester, MW 1420000), 1.11 dl g^{-1} (25°, CHCl$_3$, 2,2'-biphenyl ester, MW 1420000), 0.99 dl g^{-1} (25°, THF, 2,2'-biphenyl ester, MW 1420000), 0.91 dl g^{-1} (25°, dioxane, 2,2'-biphenyl ester, MW 1420000) [17,18]. Molar volumes 58.3 cm^3 mol^{-1} (van der Waals, 25°, amorph., Ph ester) [2], 134 cm^3 mol^{-1} (glassy, 25°, amorph., Ph ester) [2], 194.3 cm^3 mol^{-1} (glassy, 25°, butylphenyl ester) [19], 142.6 cm^3 mol^{-1} (van der Waals, 25°, amorph., cyclohexylphenyl ester) [2], 219.1 cm^3 mol^{-1} (glassy, 25°, amorph., cyclohexylphenyl ester) [2], 272.6 cm^3 mol^{-1} (glassy, 25°, tetramethylbutylphenyl ester) [19], 148 cm^3 mol^{-1} (glassy, 25°, chlorophenyl ester) [19], 178.6 cm^3 mol^{-1} (25°, 3-chlorophenyl ester) [19]. Theta temps. *tert*-butylphenyl ester [23]: 20° (Me$_2$CO), 25° (Me$_2$CO, 4-isomer), 25° (C$_6$H$_6$), 25° (butanone), 18.5° (cyclohexane), 18.4° (cyclohexane), 25° (cyclohexane, 4-isomer), 25° (THF, 4-isomer); biphenyl ester [15]: 10° (C$_6$H$_6$, VM method); naphthyl ester: 20° (tetralin), 20° (tetralin, MW 570000–2620000), 52° (toluene); pentachlorophenyl ester [25,26]: 37° (C$_6$H$_6$), 40° (C$_6$H$_6$), 25° (CHCl$_3$), -126° (o-dichlorobenzene), 25° (ethylbenzene)

Optical Properties:
Refractive Index:

No.	Value	Note
1	1.5706–1.7515	Ph ester [2]
2	1.641	naphthyl ester

Total Internal Reflection: Intrinsic segmental anisotropy [27] -10.5×10^{-25} cm^3 (bromobenzene, Ph ester), -90×10^{-25} cm^3 (bromobenzene, 4-*tert*-butylphenyl ester), 60×10^{-25} cm^3 (tetrabromoethane, 2-naphthyl ester)

Polymer Stability:
Polymer Stability General: Electron beam or γ-irradiation causes scission of the Ph ester, with the polymer degrading rapidly

Bibliographic References
[1] Hoff, E.A.W., Robinson, D.W. and Willbourn, A.H., *Polym. Sci.*, 1955, **18**, 161
[2] Van Krevelen, D.W., *Properties of Polymers: Their Correlation with Chemical Structure*, 3rd edn., Elsevier, 1990
[3] Krause, S., Gormley, J.J., Roman, N., Shetter, J.A. and Watenabe, W.H., *J. Polym. Sci., Part A: Polym. Chem.*, 1965, **3**, 3573
[4] Porter, D., *Group Interaction Modelling of Polymer Properties*, Dekker, 1995, 283
[5] Gargallo, L. and Russo, M., *Makromol. Chem.*, 1975, **176**, 2735
[6] Gargallo, L., Martinez-Pina, F., Leiva, A. and Radic, D., *Eur. Polym. J.*, 1996, **32**, 1303
[7] Lewis, O.G., *Physical Constants of Linear Hydrocarbons*, Springer-Verlag, 1968
[8] Eskin, V.Y. and Nesterov, A.Y., *Polym. Sci. USSR (Engl. Transl.)*, 1966, **8**, 1153
[9] Eskin, V.Y. and Nesterov, A.Y., *J. Polym. Sci., Part C: Polym. Lett.*, 1967, **16**, 1619
[10] Frank, C.W. and Gashgari, M.A., *Macromolecules*, 1979, **12**, 163
[11] Kurata, M. and Stockmayer, W.H., *Adv. Polym. Sci.*, Springer-Verlag, 1963, **3**, 196
[12] Fox, H. and Zisman, W.A., *J. Colloid Sci.*, 1952, **7**, 432
[13] Hadjichristidis, N., Devaleriola, M. and Desreux, V., *Eur. Polym. J.*, 1972, **8**, 1193
[14] Kokkiaris, D., Touloupis, C. and Hadjichristidis, N., *Polymer*, 1981, **22**, 63
[15] Hadjichristidis, N., *Polymer*, 1975, **16**, 848
[16] Alexopoulos, J.B., Hadjichristidis, N. and Vassiliadis, A., *Polymer*, 1975, **16**, 386
[17] Hadjichristidis, N., *Makromol. Chem.*, 1983, **184**, 1043
[18] Alexopoulos, J. and Hadjichristidis, N., *Makromol. Chem.*, 1978, **179**, 549
[19] Tricot, M., *Macromolecules*, 1986, **19**, 1268
[20] Niezette, J., *Polymer*, 1977, **18**, 200
[21] Gargallo, L., Hamidi, N. and Radic, D., *Polymer*, 1990, **31**, 925
[22] Tricot, M. and Desreux, V., *Makromol. Chem.*, 1970, **149**, 185
[23] Tricot, M., Bleus, J.P., Riga, J.P. and Desreux, V., *Makromol. Chem.*, 1974, **175**, 913
[24] Niezette, J., Hadjichristidis, N., and Desreux, V., *Eur. Polym. J.*, 1977, **13**, 41
[25] Radic, D.L., *Polymer*, 1981, **22**, 410
[26] Radic, D. and Gargallo, L., *Makromol. Chem.*, 1979, **180**, 1329
[27] Tsvetkov, V.N., Eskin, V.E. and Frenkel, S.Ya., *Structure of Macromolecules in Solution*, Nauka, 1964

Poly(benzoyl-*p*-phenylene)　　　　　　　　　　　P-75

Synonyms: *Poly(benzoyl-1,4-phenylene)*. 2,5-Dichlorobenzophenone homopolymer. *Poly(2,5-benzophenone)*
Monomers: 2,5-Dichlorobenzophenone
Material class: Thermoplastic
CAS Number:

CAS Reg. No.	Note
150385-13-0	poly(benzoyl-1,4-phenylene)
150347-09-4	2,5-dichlorobenzophenone homopolymer

Molecular Formula: $(C_{13}H_8O)n$
Fragments: $C_{13}H_8O$
General Information: High-strength amorphous polymer with photoluminescent and electroluminescent properties. Prod. as yellow powder.
Tacticity: Polymer produced using bipyridine as co-ligand has either less head-to-tail units in polymer [1] or less head-to-head units restricting conjugation [2] compared to polymer produced without bipyridine.

Volumetric & Calorimetric Properties:
Density:

No.	Value	Note
1	d1.21 g/cm^3	[3]

Glass-Transition Temperature:

No.	Value	Note
1	149–234°C	T_g increases when bipyridine is co-ligand in production process [1,2,4]

Surface Properties & Solubility:
Solvents/Non-solvents: Sol. CHCl$_3$, CH$_2$Cl$_2$, hexane, *sym*-dichloroethane [3,4], dioxan [5], THF (polymer prepared with bipyridine) [4].

Transport Properties:
Transport Properties General: Has high melt viscosity, with chains highly extended in the melt [3].

Mechanical Properties:
Mechanical Properties General: Has very high strength at r.t. but is very brittle at low temperatures [6].

Tensile (Young's) Modulus:

No.	Value	Note
1	9600 MPa	Parmax 1000, 22°C [6]
2	8000 MPa	Parmax 1200, 22°C [6]

Tensile Strength Break:

No.	Value	Note	Elongation
1	188 MPa	Parmax 1000, 22°C [6]	0.9%
2	203 MPa	Parmax 1200, 22°C [6]	4%

Compressive Strength:

No.	Value	Note
1	367 MPa	Parmax 1000, 22°C [6]
2	351 MPa	Parmax 1200, 22°C [6]

Electrical Properties:
Electrical Properties General: For polymers prepared with bipyridine co-ligand, electroluminescence is exhibited in blue region of spectrum with λ_{max} *ca.* 446nm [7,8]. LED devices have been developed with the polymer as emissive layer; turn-on voltage *ca.* 12V. The estimated band gap is 2.74eV [8].

Optical Properties:
Optical Properties General: Polymer (bipyridine ligand process) exhibits uv absorption band centred in the region of 350–380nm [4,5,7,8]. There is a slight red shift on going from solution to film [7]. Photoluminescence for same polymer film occurs in blue region with λ_{max} 433nm [7,8]. For polymer produced without bipyridine, both absorption and photoluminescence occurs at shorter wavelength [4,8].

Transmission and Spectra: Pmr [9], cmr [2, 9], ftir [4], uv and photoluminescence [4,5,7,8] and electroluminescence [7,8] spectral data have been reported.

Polymer Stability:
Thermal Stability General: The polymer is very stable in both air and nitrogen with decomposition (5% weight loss) not occurring on heating until *ca.* 500°C [1,10,11].

Applications/Commercial Products:
Processing & Manufacturing Routes: Prod. by polymerisation of 2,5-dichlorobenzophenone with Ni(0), (generated *in situ* from NiCl$_2$ and Zn) and triphenylphosphine in a suitable solvent such as dimethylformamide or dimethylacetamide [4,8,9,10,11]. The addition of 2,2′-bipyridine as a co-ligand in preparation results in formation of polymer with different physical and optical properties [1,2,4].

Applications: Polymer and composites have possible applications in aircraft and automotive parts, in medical equipment and semiconductor components. Sulfonated derivatives may find application in proton-conducting membranes for fuel cells.

Trade name	Details	Supplier
Parmax 1000	poly(benzoyl-*p*-phenylene) homopolymer	Mississippi Polymer Technologies
Parmax 1200	copolymer with *ca.* 15% 1,3-phenylene units	Mississippi Polymer Technologies

Bibliographic References
[1] Wang, Z.Y., Franklin, J. and Venkatasan, D., *Macromolecules*, 1999, **32**, 1691
[2] Quirk, R.P. and Yu, W., *High Performance Polym.*, 2005, **17**, 349
[3] Vaia, R.A., Krishnamoorti, R., Benner, C. and Trimmer, M., *J. Polym. Sci., Part B: Polym. Phys.*, 1998, **36**, 2449
[4] Wang, Y. and Quirk, R.P., *Macromolecules*, 1995, **28**, 3495
[5] Nakazawa, M., Han, Y.K., Fu, H., Matsuoka, S., Kwei, T.K. and Okamoto, Y., *Macromolecules*, 2001, **34**, 5975
[6] Toplosky, V.J., Walsh, R.P., Tozer, S.W. and Motamedi, F., Adv. Cryog. Eng. Part A, Klewer Academic/Plenum, Balachandran, U.B., Hartwig, K.T., Gubser, D.U. and Bardos, V.A (editors), 2000, **46**, 151
[7] Edwards, A., Blumstengel, S., Sokolik, I., Dorsinville, R., Yun, H., Kwei, T.K. and Okamoto, Y., *Appl. Phys. Lett.*, 1997, **70**, 298
[8] Fu, H., Yun, H., Kwei, T.K., Okamoto, Y., Blumstengel, S., Walser, A. and Dorsinville, R., *Polym. Adv. Technol.*, 1999, **10**, 252
[9] Hagberg, E.C., Olson, D.A. and Sheares, V.V., *Macromolecules*, 2004, **37**, 4748
[10] Phillips, R.W., Sheares, V.V., Samulski, E.T. and DeSimone, J.M., *Macromolecules*, 1994, **27**, 2354
[11] Phillips, R.W. and DeSimone, J.M., *Polym. Prepr. (Am. Chem. Soc., Div. Polym. Chem.)*, 1997, **38**, 484

Poly[bis(4-butylphenyl)silane]   P-76

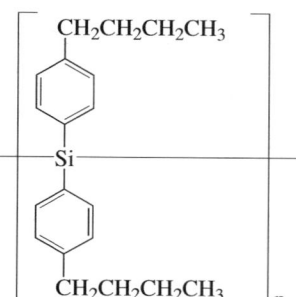

Synonyms: *Poly[bis(4-butylphenyl)silylene]. Bis(4-butylphenyl)-dichlorosilane homopolymer. PBPS*
Material class: Thermoplastic
Polymer Type: polysilanes

CAS Number:

CAS Reg. No.	Note
107999-72-4	poly[bis(4-butylphenyl)silylene]
111939-58-3	bis(4-butylphenyl)dichlorosilane homopolymer

Molecular Formula: $(C_{20}H_{26}Si)n$
Fragments: $C_{20}H_{26}Si$
General Information: Conductive thermochromic polysilane exhibiting photoluminescence and electroluminescence. Prod. as white solid.
Morphology: A disordered conformation is stable at r.t. and normal pressure, which is converted to a *trans*-planar conformation on heating [1,3].

Volumetric & Calorimetric Properties:
Transition Temperature General: At 87–90°C there is an endothermic transition from the disordered form to a *trans*-planar conformation [1,2,3].

Surface Properties & Solubility:
Solvents/Non-solvents: Sol. toluene, THF, hexane, CHCl$_3$ [1,2].

Electrical Properties:
Electrical Properties General: Hole transport predominates, although electron transport has been detected [5]. Stable LED devices with PBPS as the emissive layer have been produced, with electroluminescence in near-UV region with λ_{max} at *ca.* 407nm and with turn-on voltages of *ca.* 18V [1,3,6,9]. Improved LED performance is achieved with lower molecular-weight batches of polymer [7]. Band gap for *trans*-planar conformation (predominates for electroluminescence) is smaller than for disordered conformation [1]. Ionisation potential *ca.* 5.5eV. Hole mobility is increased when the polymer is doped with the electron-acceptor C$_{60}$-fullerene [8].

Optical Properties:
Optical Properties General: In THF solution absorption occurs with λ_{max} at 390nm. Films cast from soln. exhibit with time a decrease in absorption at 390nm and a blue shift with λ_{max} increasing at 315nm. The absorption centred at 390nm increases in magnitude on increasing the temperature. Similar changes occur with the photoluminescence band with a blue shift of about 10nm (410 to 400nm). These changes are attributed to a change in conformation in solution and on heating [1].
Transmission and Spectra: Pmr [2], ir [2], uv and photoluminescence [1,3] and electroluminescence [1,3,6,9] spectral data have been reported.

Applications/Commercial Products:
Processing & Manufacturing Routes: Prod. by Wurtz dehalogenation coupling of bis(4-butylphenyl)dichlorosilane with Na in a solvent such as toluene [2].
Applications: Potential applications for durable polysilane-based LEDs and transistors.

Bibliographic References
[1] Bleyl, I., Ebata, K., Hoshino, S., Furukawa, K. and Suzuki, H., *Synth. Met.*, 1999, **105**, 17
[2] Miller, R.D. and Sooriyakumaran, R., *J. Polym. Sci., Part C: Polym. Lett.*, 1987, **25**, 321
[3] Suzuki, H., Hoshino, S., Furukawa, K., Ebata, K., Yuan, C.-H. and Bleyl, I., *Polym. Adv. Technol.*, 2000, **11**, 460
[4] Yuan, C.-H., Hoshino, S., Toyoda, S., Suzuki, H., Fujiki, M. and Matsumoto, N., *Appl. Phys. Lett.*, 1997, **71**, 3326
[5] Furukawa, K., Yuan, C.-H., Hoshino, S., Suzuki, H. and Matsumoto, N., *Mol. Cryst. Liq. Cryst. Sci. Technol., Sect. A*, 1999, **327**, 181
[6] Suzuki, H., Hoshino, S., Yuan, C.-H., Furukawa, K. and Matsumoto, N., *Polym. Prepr. (Am. Chem. Soc., Div. Polym. Chem.)*, 1998, **39**, 996
[7] Hoshino, S., Furukawa, K., Ebata, K., Breyl, I. and Suzuki, H., *J. Appl. Phys.*, 2000, **88**, 3408
[8] Acharya, A., Seki, S., Koizumi, Y., Saeki, A. and Tagawa, S., *J. Phys. Chem. B*, 2005, **109**, 20174
[9] Suzuki, H., Hoshino, S., Yuan, C.-H., Fujiki, M., Toyoda, S. and Matsumoto, N., *Thin Solid Films*, 1998, **331**, 64

Poly[2,2-bis(chloromethyl)trimethylene-3-oxide]   P-77

Synonyms: *Poly[3,3-bis(chloromethyl)oxacyclobutane]. Poly[3,3-bis(chloromethyl)oxetane]. Poly[oxy(2,2-bis(chloromethyl)-1,3-propanediyl]]. Poly[oxy[2,2-bis(chloromethyl)trimethylene]]*
Related Polymers: Oxetane polymers
Monomers: 3,3-Bis(chloromethyl)oxetane
Material class: Thermoplastic
Polymer Type: chlorinated polyether
CAS Number:

CAS Reg. No.	Note
26917-50-0	
25323-58-4	homopolymer

Molecular Formula: $(C_5H_8Cl_2O)_n$
Fragments: $C_5H_8Cl_2O$
Molecular Weight: M_n 25000–350000
Morphology: Polycrystalline α- and β-forms are known

Applications/Commercial Products:
Mould Shrinkage (%):

No.	Value
1	0.005%

Applications: Lining material for chemical plant equipment (valves, pumps, flow meters etc.). Adhesive and coating applications reported in former USSR.

Trade name	Supplier
Chlorinated Polyether	Dow
Penton	Hercules

Poly(1,2-bis(2,3-epoxyprop-1-oxy)benzene)   P-78

Synonyms: *Poly(2,2'-(1,2-phenylenebis(oxymethylene)bisoxirane))*
Monomers: 1,2-Bis(2,3-epoxypropoxy)benzene
Material class: Thermosetting resin
Polymer Type: epoxy
CAS Number:

CAS Reg. No.	Note
63298-90-8	homopolymer
66485-75-4	copolymer with 1,3-benzenediamine
62570-75-6	copolymer with 4,4'-sulfonylbis(benzeneamine)
84241-79-2	copolymer with aniline
62570-77-8	copolymer with 3,3'-diaminodiphenylsulfone
66485-77-6	copolymer with 2,6-diaminopyridine

Molecular Formula: $(C_{12}H_{14}O_4)_n$
Fragments: $C_{12}H_{14}O_4$
General Information: Reaction of monomer with amines gives structures predominantly containing rings [6,8]. Props. depend on the curing agent and cure time and temp.

Volumetric & Calorimetric Properties:
Density:

No.	Value	Note
1	d 1.299 g/cm^3	*m*-phenylenediamine cured [2]

Glass-Transition Temperature:

No.	Value	Note
1	110°C	*m*-phenylenediamine cured [2]
2	115°C	*m*-phenylenediamine cured, calc. [2,7]
3	88°C	4,4'-diamino-3,3'-dichlorodiphenylmethane cured [3]
4	150°C	4,4'-diaminodiphenylsulfone cured [7]
5	130°C	3,3'-diaminodiphenylsulfone cured [7]
6	120°C	4,4'-diaminodiphenylmethane cured [7]
7	120°C	2,6-diaminopyridine cured [7]

Surface Properties & Solubility:
Solvents/Non-solvents: Sol. common organic solvents [1]

Mechanical Properties:
Miscellaneous Moduli:

No.	Value	Note	Type
1	1600 MPa	25°, glassy, *m*-phenylenediamine cured [2,6]	shear modulus
2	5 MPa	150°, *m*-phenylenediamine cured [2]	shear modulus

Optical Properties:
Transmission and Spectra: Ir spectral data have been reported [1]

Polymer Stability:
Thermal Stability General: Details of thermal stability have been reported [4]

Applications/Commercial Products:
Processing & Manufacturing Routes: May be synth. by cationic polymerisation using boron trifluoride etherate initiator, ideally in CH_2Cl_2 soln. at -27° [1]. Kinetics of cure have been reported [5]

Bibliographic References
[1] Bartulín, J., Parra, M., Ramírez, A. and Zunza, H., *Polym. Bull. (Berlin)*, 1989, **22**, 33, (synth, ir)
[2] Chepel', L.M., Knunyants, M.I., Topolkarayev, V.A., Zelenetskii, A.N. et al, *Vysokomol. Soedin., Ser. A*, 1984, **26**, 362, (dynamic mechanical props.)
[3] Yurechko, N.A., Evtushenko, G.T., Lipskaya, V.A., Shologon, I.M. et al, *Vysokomol. Soedin., Ser. A*, 1978, **20**, 2326
[4] Antal, I. and Csillag, L., *Therm. Anal., Proc. Int. Conf.*, 1974, **3**, 347, (thermal stability)
[5] Galy, J., Pascault, J.P. and Grenier-Loustalot, M.F, Crosslinked Epoxies, *Proc. Discuss. Conf. 9th*, 1986, 169, (cure kinetics)
[6] Chepel', L.M., Topolkarayev, V.A., Zelenetskii, A.N., Prut, E.V. et al, *Vysokomol. Soedin., Ser. A*, 1982, **24**, 1646, (conformn.)
[7] Ponomareva, T.I., Irzhak, V.I. and Rozenberg, B.A., *Vysokomol. Soedin., Ser. A*, 1978, **20**, 597, (T_g)
[8] Topolkarayev, V.A., Berlin, A.A., Oshmyan, V.G. and Prut, E.V., *Makromol. Chem., Makromol. Symp.*, 1986, **4**, 183, (conformn.)

Poly[9,9-bis(2-ethylhexyl)fluorene]   P-79

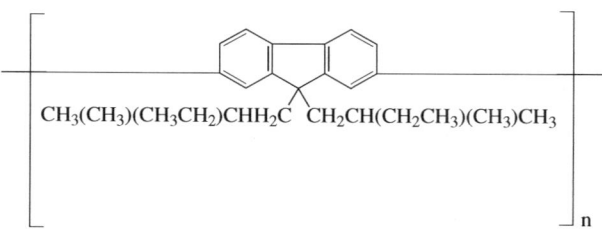

Synonyms: *Poly[9,9-bis(2-ethylhexyl)-9H-fluorene-2,7-diyl]. 2,7-Dibromo-9,9-bis(2-ethylhexyl)-9H-fluorene homopolymer. PF2/6*
Material class: Thermoplastic
Polymer Type: polyfluorene
CAS Number:

CAS Reg. No.	Note
188201-14-1	Poly[9,9-bis(2-ethylhexyl)-9*H*-fluorene-2,7-diyl]
188201-16-3	2,7-Dibromo-9,9-bis(2-ethylhexyl)-9*H*-fluorene homopolymer
460089-72-9	(*R*,*R*)-form

Molecular Formula: $(C_{29}H_{40})_n$
Fragments: $C_{29}H_{40}$
General Information: Stable blue light-emitting photoluminescent and electroluminescent polymer with high hole transport mobility and with liquid crystalline properties. The chiral (*R*,*R*)-form has been prepared [1].
Morphology: Helical conformation of polymer chains with almost hexagonal unit cell after annealing above glass-transition temperature or in liquid crystal phase followed by cooling, a=17.1, b=17.0, c=40.8Å [2,3].

Volumetric & Calorimetric Properties:
Glass-Transition Temperature:

No.	Value	Note
1	80°C	approx. [4,5]

Transition Temperature General: Transition on heating from hexagonal to nematic phase in the region of 140–165°C [4].

Surface Properties & Solubility:
Solvents/Non-solvents: Sol. THF, toluene, $CHCl_3$, CCl_4 [4,13].

Electrical Properties:
Electrical Properties General: Several LED devices with PF2/6 as polarised blue light emitter have been reported, with light turn-on voltages of 4–13V [5,6,7,8,9,10].

Optical Properties:
Optical Properties General: Blue light photoluminescence occurs with λ_{max} 415–420nm Polarised electroluminescence from LED devices with λ_{max} peaks in region 425–475nm and polarisation ratios up to 15 [5,6]. Bands at longer wavelength giving rise to blue-green emission are due to fluorenone defects in polymer [10]. Circularly polarised photoluminescence produced by chiral (*R*,*R*)-form [1]. Refractive indices for aligned and unaligned films have been reported [11].
Transmission and Spectra: Cmr [12], ir [10], uv and photoluminescence [1,5,7,8,9,10,12,13,14], electroluminescence [5,6,7,8,9,10,12] and x-ray [2,3,4,14] spectral data have been reported.

Polymer Stability:
Thermal Stability General: Thermally stable up to 300°C in nitrogen for a range of molecular weights [4].
Decomposition Details: Decomposition (5% weight loss) at 422°C [7].

Applications/Commercial Products:
Processing & Manufacturing Routes: Prod. mainly by Ni(0)-catalysed Yamamoto coupling of 2,7-dibromo-9,9-bis(2-ethylhexyl)fluorene monomer [6,12,15].
Applications: Main potential application is for polarised electroluminescence devices used as backlights in liquid crystal displays.

Details	Supplier
homopolymer with different end-capped groups	American Dye Source

Bibliographic References

[1] Oda, M., Nothofer, H.-G., Scherf, U., Sunjić, V., Richter, D., Regenstein, W. and Neher, D., *Macromolecules*, 2002, **35**, 6792
[2] Tanto, B., Guha, S., Martin, C.M., Scherf, U. and Winokur, M.J., *Macromolecules*, 2004, **25**, 9438
[3] Lieser, G., Oda, M., Miteva, T., Meisel, A., Nothofer, H.-G. and Scherf, U., *Macromolecules*, 2000, **33**, 4490
[4] Knaapila, M., Stepanyan, R., Torkkeli, M., Lyons, B.P., Ikonen, T.P., Almasy, L., Foreman, J.P., Serimaa, R., Güntner, R., Scherf, U. and Monkman, A.P., *Phys. Rev. E*, 2005, **71**, 041802/1
[5] Godbert, N., Burn, P.L., Gilmour, S., Markham, J.P.J. and Samuel, I.D.W., *Appl. Phys. Lett.*, 2003, **83**, 5347
[6] Grell, M., Knoll, W., Lupo, D., Meisel, A., Miteva, T., Neher, D., Nothofer, H.-G., Scherf, U. and Yasuda, A., *Adv. Mater.*, 1999, **11**, 671
[7] Hwang, D.-H., Park, M.-J., Lee, J.-H., Cho, N.-S., Shim, H.-K. and Lee, C., *Synth. Met.*, 2004, **146**, 145
[8] Hwang, D.-H., Park, M.-J. and Lee, J.-H., *Mater. Sci. Eng., C*, 2004, **24**, 201
[9] Pogantsch, A., Trattnig, G., Langer, G., Kern, W., Scherf, U., Tillmann, H., Hörhold, H.-H. and Zojer, E., *Adv. Mater.*, 2002, **14**, 1722
[10] Gong, X., Iyer, P.K., Moses, D., Bazan, G.C., Heeger, A.J. and Xiao, S.S., *Adv. Funct. Mater.*, 2003, **13**, 325
[11] Lyons, B.P. and Monkman A.P., *J. Appl. Phys.*, 2004, **96**, 4735
[12] Nothofer, H.-G., Meisel, A., Miteva, T., Neher, D., Scherf, U., Lupo, D., Yasuda, A. and Knoll, W., *Polym. Prepr. (Am. Chem. Soc., Div. Polym. Chem.)*, 1999, **40**, 1198
[13] Lee, J.-I., Klaerner, G., Davey, M.H. and Miller R.D., *Synth. Met.*, 1999, **102**, 1087
[14] Knaapila, M., Lyons, B.P., Kiske, K., Foreman, J.P., Vainio, A. and Monkman, A.P., *J. Phys. Chem. B*, 2003, **107**, 12425
[15] Jo, J., Chi, C., Höger, S., Wegner, S. and Yoon, D.Y., *Chem. Eur. J.*, 2004, **10**, 2681

Poly[(2,5-bis(2-ethylhexyloxy)-1,4-phenylene ethynylene)-*co*-(2,5-dioctyloxy-1,4-phenylene ethynylene)]   P-80

Synonyms: *Poly[[2,5-bis[(2-ethylhexyl)oxy]-1,4-phenylene]-1,2-ethynediyl[2,5-bis(octyloxy)-1,4-phenylene]-1,2-ethynediyl]. EHO-OPPE*
Material class: Thermoplastic
Polymer Type: poly(arylene vinylene)
CAS Number:

CAS Reg. No.	Note
174592-87-1	poly[[2,5-bis[(2-ethylhexyl)oxy]-1,4-phenylene]-1,2-ethynediyl[2,5-bis(octyloxy)-1,4-phenylene]-1,2-ethynediyl]
173428-83-6	

Poly[2,5-bis(2'-ethylhexyloxy)-1,4-phenylene vinylene]

Molecular Formula: $(C_{48}H_{72}O_4)_n$
Fragments: $C_{48}H_{72}O_4$
General Information: Photoluminescent and electroluminescent polymer with enhanced solubility in organic solvents. Prod. as yellow solid.

Volumetric & Calorimetric Properties:
Glass-Transition Temperature:

No.	Value	Note
1	90–100°C	[1]

Transition Temperature General: Cross-linking transition above 135°C [1].

Surface Properties & Solubility:
Solubility Properties General: Nematic mesophases obtained in 1,2,4-trichlorobenzene solvent [1].
Solvents/Non-solvents: Sol. toluene, xylene, 1,2,4-trichlorobenzene, decahydronaphthalene, $CHCl_3$, THF [1].

Electrical Properties:
Electrical Properties General: Charge transport in EHO-OPPE is ambipolar with high hole and electron mobility at low field at r.t. [2]. LED devices using blends with a polyamine ether have been reported [3].

Optical Properties:
Optical Properties General: Exhibits strong photoluminescence with main emission in region of 470–510nm (in $CHCl_3$) and with a red shift for films to 500–555nm [4]. Polarised emission of yellow-green light occurs in oriented blended films in ultrahigh molecular weight polyethylene [5]. A sensitiser such as 7-diethylamino-4-methylcoumarin enhances the absorption and transfer of light with subsequent polarised emission from polymer [6]. The refractive indices and third-order nonlinear optical susceptibilities have been reported [7].
Refractive Index:

No.	Value	Note
1	1.76	(film, 550nm) [4]

Transmission and Spectra: Pmr [1,4], uv and photoluminescence [1,4] and x-ray diffraction [4] spectral data have been reported.

Applications/Commercial Products:
Processing & Manufacturing Routes: Prod. by Pd-catalysed coupling reaction of 1,4-diethynyl-2,5-dioctyloxybenzene and 1,4-bis(2-ethylhexyloxy)-2,5-diiodobenzene [1,4].
Applications: Potential applications for liquid crystal displays and light-emitting diodes.

Bibliographic References
[1] Steiger, D., Smith, P. and Weder, C., *Macromol. Rapid. Commun.*, 1997, **18**, 643
[2] Kokil, A., Shiyanovskaya, I., Singer, K.D. and Weder, C., *Synth. Met.*, 2003, **138**, 513
[3] Schmitz, C., Pösch, P., Thelakkat, M., Schmidt, H.-W., Montali, A., Feldman, K., Smith, P. and Weder, C., *Adv. Funct. Mater.*, 2001, **11**, 41
[4] Weder, C. and Wrighton, M.S., *Macromolecules*, 1996, **29**, 5157
[5] Montali, A., Bastiaansen, C., Smith, P. and Weder, C., Polym. Prepr. (Am. Chem. Soc., Div. Polym. Chem.), 1998, **39**, 107
[6] Montali, A., Smith, P. and Weder, C., *J. Mater. Sci.: Mater. Electron.*, 2000, **11**, 117
[7] Weder, C., Wrighton, M.S., Spreiter, R., Bosshard, C. and Günter, P., *J. Phys. Chem.*, 1996, **100**, 18931

Poly[2,5-bis(2'-ethylhexyloxy)-1,4-phenylene vinylene]

P-81

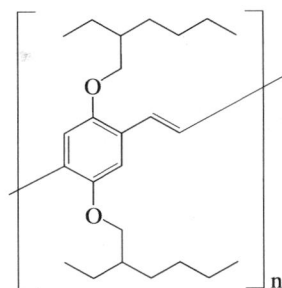

Synonyms: *Poly[[2,5-bis[(2-ethylhexyl)oxy]-1,4-phenylene]-1,2-ethenediyl]. BEH-PPV*
Material class: Thermoplastic
Polymer Type: poly(arylene vinylene)
CAS Number:

CAS Reg. No.	Note
160894-98-4	
191917-70-1	*E*-form

Molecular Formula: $(C_{24}H_{38}O_2)_n$
Fragments: $C_{24}H_{38}O_2$
General Information: Photoluminescent and electroluminescent polymer. Prod. as red solid

Volumetric & Calorimetric Properties:
Glass-Transition Temperature:

No.	Value	Note
1	117–120°C	[2,4]

Transition Temperature General: Thermal transitions in region of 55–80°C; isotropic transition at 200°C [1].

Surface Properties & Solubility:
Solvents/Non-solvents: Sol. toluene, xylene, chlorobenzene, THF, $CHCl_3$, CH_2Cl_2 [3,4].

Electrical Properties:
Electrical Properties General: Conductivity of polymer increased to *ca.* 1 S cm^{-1} by doping with iodine [3]. Photovoltaic devices of blends with electron acceptors such as perylenediimides reported [5].

Optical Properties:
Optical Properties General: Exhibits strong photoluminescence with λ_{max} values in region of 550–620nm. Spectral line narrowing relating to lasing applications has been studied [6,7].
Transmission and Spectra: Pmr [1,8], cmr [4], Raman [7], uv and photoluminescence [7,8,9,10] and electroluminescence [8] spectral data have been reported.

Polymer Stability:
Thermal Stability General: Polymer is unstable on heating in air, with onset of thermooxidative degradation at *ca.* 77°C [2].
Decomposition Details: Decomposition temp. (5% weight loss) in nitrogen at *ca.* 360°C [1,2,4].

Applications/Commercial Products:
Processing & Manufacturing Routes: Several methods for prod. of polymer have been reported, including the Gilch method from 1,4-

bis(halomethyl)-2,5-bis(2'-ethylhexyloxy)benzene and *tert*-BuOK [3,11], the Heck reaction of 1,4-dibromo-2,5-bis(2'-ethylhexyloxy)benzene and ethylene [8], and by Siegrist polycondensation from 2,5-bis(2'-ethylhexyloxy)-4-methylbenzaldehyde [1].

Applications: Polymer has potential applications for lasers and in light-emitting diodes and photovoltaic cells.

Bibliographic References

[1] Van der Veen, M.H., de Boer, B., Stalmach, U., van de Wetering, K.I. and Hadziioannou, G., *Macromolecules,* 2004, **37**, 3673
[2] Schartel, B. and Hennecke, M., *Polym. Degrad. Stab.,* 2000, **67**, 249
[3] *Pat. Coop. Treaty (WIPO),* 1994, 94 20 589
[4] Koch, F. and Heitz, W., *Macromol. Chem. Phys.,* 1997, **198**, 1531
[5] Angadi, M.A., Gosztola, D. and Wasielewski, M.R., *J. Appl. Phys.,* 1998, **83**, 6187
[6] Spiegelberg, Ch., Peyghambarian, N. and Kippelen, B., *Appl. Phys. Lett.,* 1999, **75**, 748
[7] Oliveira, F.A.C., Cury, L.A., Righi, A., Moreira, R.L., Guimarães, P.S.S., Matinaga, F.M., Pimenta, M.A. and Nogueira, R.A., *J. Chem. Phys.,* 2003, **119**, 9777
[8] Klingelhöfer, S., Schellenberg, C., Pommerehne, J., Bässler, H., Greiner, A. and Heitz, W., *Macromol. Chem. Phys.,* 1997, **198**, 1511
[9] Andersson, M.R., Yu, G. and Heeger, A.J., *Synth. Met.,* 1997, **85**, 1275
[10] Gettinger, C.L., Heeger, A.J., Drake, J.M. and Pine, D.J., *J. Chem. Phys.,* 1994, **101**, 1673
[11] Marr, P.C., Crayston, J.A., Halim, M. and Samuel, I.D.W., *Synth. Met.,* 1999, **102**, 1081

Poly[[bis(3-methylphenyl)silylene]methylene]   P-82

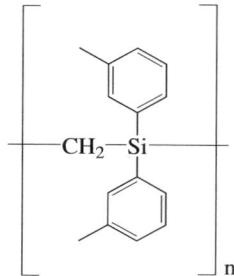

Synonyms: *Poly[(di-m-tolylsilylene)methylene]*. PDmTSM
Related Polymers: Polycarbosilanes, Poly[[bis(4-methylphenyl)silylene]methylene]
Polymer Type: polycarbosilanes
CAS Number:

CAS Reg. No.
170891-62-0

Molecular Formula: $(C_{15}H_{16}Si)n$
Fragments: $C_{15}H_{16}Si$
General Information:
Tacticity: A transition to isotropic state has been suggested [1]
Morphology: Slightly yellow crystalline solid. Wide angle X-ray diffraction analysis exhibited two peaks with strong intensities, suggesting the presence of two types of configuration of tolyl groups in each lattice structure. Observed *d* spacing: PDpTSM (10.5 and 11.2′) > PDmTSM (10.4 and 11.1′) > PDPSM (10.2′), which follows the decreasing steric hindrance of the aryl groups on the backbone [1]
Identification: For $C_{30}H_{32}Si_2$ (monomer): Found: C, 79.03; H, 7.02; Calc: C, 80.30; H, 7.19 [1]

Volumetric & Calorimetric Properties:

Thermodynamic Properties General: See [1]
Melting Temperature:

No.	Value	Note
1	316–321°C	[1,2]

Glass-Transition Temperature:

No.	Value	Note
1	113–148°C	[1,2] A clear glass-transition was not observed, presumably because of its high degree of crystallinity

Surface Properties & Solubility:

Solubility Properties General: Soluble in limited solvents such as diphenyl sulfone at temperatures above 250°C, or slightly soluble in hot diphenyl ether [1]

Transport Properties:

Transport Properties General: A similar tendency was observed for PDmTSM as PDpTSM. The melt viscosity of the polymer indicated that the molecular weight is modifiable by varying the monomer:catalyst ratio when solution polymerisation is employed [1]
Melt Flow Index:

No.	Value	Note
1	24–1300 g/10 min	at 350°C [1,2]

Mechanical Properties:

Mechanical Properties General: Thermally too unstable to be processed in an ambient atmosphere to ascertain its mechanical properties. The degree of crystallinity by X-ray data was found to be $37 \pm 3\% – 50 \pm 3\%$ [2]

Optical Properties:

Transmission and Spectra: Ir [1,2], nmr [1,2], gc [2] and ms [1,2] spectra reported.

Polymer Stability:

Thermal Stability General: Thermostability is comparable to that of PDPSM when conducted under nitrogen, but highly thermally unstable in air compared with PDPSM. This thermal instability may be explained by high sensitivity of the tolyl group toward oxidative decomposition [1,2]
Decomposition Details: Thermogram shows a well-defined double endothermic peak. The polymer exhibited this in the second heating scan also, although the shapes of the peaks were appreciably different from those in the first scan.
Exhibits a gain in weight accompanied with a violent exothermic reaction in the early stage of thermal decomposition in air. This thermal instability of the tolyl-substituted polymer can be explained by high sensitivity of the tolyl groups toward oxidative decomposition. Analysis of pyrolysed products of PDArSMs suggests that these polymers undergo radical cleavage of Si-aryl bonds to provide network materials, which can be precursors for Si-containing ceramics.
Weight percentage of residue after heating to 800°C is 52% in nitrogen and 27% in air.
TG-MS shows that when heated up to 800°C at a heating rate of 10°C/min, the major compound is toluene and the minor ones benzene, xylene and dimethylbenzenes [1,2]

Applications/Commercial Products:

Processing & Manufacturing Routes: Synthesised by ring-opening polymerisation of corresponding 1,1,3,3-tetraaryl-1,3-disilacyclobutanes (yields 38–93%), by bulk polymerisation or precipitation polymerisation [1,2,3,4]
Applications: Provide network materials, which can be precursors for Si-containing ceramics [2]

Bibliographic References

[1] Ogawa, T., Tachikawa, M., Kushibiki, N. and Murakami, M., *J. Polym. Sci., Part A,* 1995, **33**, 2821
[2] Ogawa, T., and Murakami, M., *J. Polym. Sci., Part B: Polym. Phys.,* 1996, **34**, 1317
[3] *Jpn. Pat.,* 1996, 8 109 266
[4] *Jpn. Pat.,* 1998, 10 182 833

Poly[[bis(4-methylphenyl)silylene]methylene]   P-83

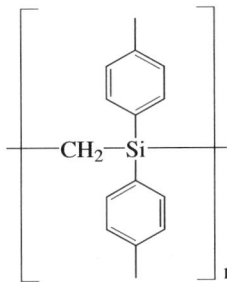

Synonyms: *Poly[(di-p-tolylsilylene)methylene]. PDpTSM*
Related Polymers: Polycarbosilanes, Poly[[bis(3-methylphenyl)silylene]methylene]
Polymer Type: polycarbosilanes
CAS Number:

CAS Reg. No.
170891-63-1

Molecular Formula: $(C_{15}H_{16}Si)n$
Fragments: $C_{15}H_{16}Si$
Morphology: White-slightly yellow crystalline solid [1]. Wide angle X-ray diffraction analysis exhibited two peaks, suggesting the presence of two types of configuration of tolyl groups in each lattice structure. Observed d spacings:
PDpTSM (10.5 and 11.2) > PDmTSM (10.4 and 11.1) > PDPSM (10.2) which follows the decreasing steric bulkiness of the aryl groups on the backbone.
Identification: For $C_{30}H_{32}Si_2$ (monomer): Found: C, 78.54; H, 6.80; Calc: C, 80.30; H, 7.19 [1]

Volumetric & Calorimetric Properties:
Thermodynamic Properties General: See [1]
Melting Temperature:

No.	Value	Note
1	329–330°C	[1,2]

Glass-Transition Temperature:

No.	Value	Note
1	131–160°C	[1,2] Strongly affected by the preparation conditions

Surface Properties & Solubility:
Solubility Properties General: Soluble in limited solvents such as diphenyl sulfone at temperatures above 250°C, or slightly soluble in hot diphenyl ether. This solubility characteristic of poly(diarylsilylmethylene)s is similar to that of poly(oxydiarylsilylene)s. Poly[oxydi(p-tolyl)silylene] is insoluble in common solvents, although poly[oxyphenyl(p-tolyl)silylene] is reasonably soluble [1]

Transport Properties:
Transport Properties General: Melt viscosity was measured to obtain insight into the molecular weight of the polymer, and the results indicated that its molecular weight is modifiable by varying the monomer/catalyst ratio when solution polymerisation is employed [1]
Melt Flow Index:

No.	Value	Note
1	170–5200 g/10 min	at 350°C [1,2]

Mechanical Properties:
Mechanical Properties General: Thermally too unstable to be processed in an ambient atmosphere to ascertain its mechanical properties. The degree of crystallinity by X-ray data was found to be $41 \pm 3\%$

Optical Properties:
Optical Properties General: Wide angle x-ray diffraction analysis exhibited two peaks [1]
Transmission and Spectra: Ir [1,2], nmr [1,2], gc [2], ms [1,2] spectra reported

Polymer Stability:
Thermal Stability General: The thermostability of PDpTSM was comparable to that of PDPhSM when the measurement was conducted under nitrogen, whereas PDpTSM was found to be highly thermally unstable in air compared with PDPhSM. This thermal instability may be explained by high sensitivity of the tolyl group toward oxidative decomposition [1,2]
Decomposition Details: The thermogram of PDpTSM shows one major endothermic peak and a minor one. Decomposition begins around 300°C. Exhibits a gain in weight accompanied by a violent exothermic reaction in the early stage of thermal decomposition in air. This thermal instability of the tolyl-substituted polymer can be explained by high sensitivity of the tolyl groups toward oxidative decomposition. Analysis of pyrolysed products of PDArSMs suggests that these polymers undergo radical cleavage of Si-aryl bonds to provide network materials, which can be precursors for Si-containing ceramics.
Weight percentage of residue after heating to 800°C was 53% in nitrogen and 29% in air.
TG-MS showed that when the PDpTSM is heated up to 800°C at a heating rate of 10°C/min. The major compound is toluene and the minor ones benzene, xylene and dimethylbenzenes [1,2].

Applications/Commercial Products:
Processing & Manufacturing Routes: PDpTSM has been synthesised by ring-opening polymerisation of corresponding 1,1,3,3-tetraaryl-1,3-disilacyclobutanes; also by bulk polymerisation and precipitation polymerisation in 88–93% yields [1,2,3]
Applications: Provide network materials, which can be precursors for Si-containing ceramics [2]

Bibliographic References
[1] Ogawa, T., Tachikawa, M., Kushibiki, N. and Murakami, M., *J. Polym. Sci., Part A*, 1995, **33**, 2821
[2] Ogawa, T. and Murakami, M., *J. Polym. Sci., Part B: Polym. Phys.*, 1996, **34**, 1317
[3] *Jpn. Pat.*, 1996, 08 109 266

Poly[2,5-bis(octyloxy)-1,4-phenylene vinylene]   P-84

Synonyms: *Poly[[2,5-bis(octyloxy)-1,4-phenylene]-1,2-ethenediyl]. DOO-PPV*
Material class: Thermoplastic
Polymer Type: poly(arylene vinylene)
CAS Number:

CAS Reg. No.	Note
133069-19-9	
171757-86-1	*E*-form

Molecular Formula: $(C_{24}H_{38}O_2)_n$
Fragments: $C_{24}H_{38}O_2$
General Information: Photoluminescent and electroluminescent polymer. Prod. as red solid.
Morphology: Formation of a lamellar mesophase in the region of 160–240°C reported [1].

Volumetric & Calorimetric Properties:
Transition Temperature General: Isotropic transition above liquid crystalline phase at *ca.* 300°C [1].

Surface Properties & Solubility:
Solvents/Non-solvents: Sol. THF, $CHCl_3$ [1,2]. Poor solubility in toluene below 50°C [2].

Optical Properties:
Optical Properties General: Exhibits photoluminescence with λ_{max} in the region of 580–630nm. Of the PPV polymers it is particularly suitable for use in laser devices, its films showing high optical gain and low internal optical losses [3,4,5]. The optical gain mechanism is attributed to superradiance [6].
Transmission and Spectra: Pmr [7], cmr [7], ir [7,8,9], Raman [8], uv and photoluminescence [1,2,7,8,9], esr [7] and x-ray diffraction [1] spectral data have been reported.

Applications/Commercial Products:
Processing & Manufacturing Routes: Prod. mainly by Gilch method from 1,4-bis(halomethyl)-2,5-bis(octyloxy)benzene with *tert*-BuOK [7,8,9].
Applications: Main application is in development of lasers.

Bibliographic References
[1] Chen, S.H., Su, A.C., Han, S.R., Chen, S.A. and Lee, Y.Z., *Macromolecules*, 2004, **37**, 181
[2] Hsu, J.-H., Fann, W., Tsao, P.-H., Chuang, K.-R. and Chen, S.-A., *J. Phys. Chem. A*, 1999, **103**, 2375
[3] Frolov, S.V., Ozaki, M., Gellermann, W., Vardeny, Z.V. and Yoshino, K., *Jpn. J. Appl. Phys.*, 1996, **35**, L1371
[4] Frolov, S.V., Fujii, A., Chinn, D., Vardeny, Z.V., Yoshino, K. and Gregory, R.V., *Appl. Phys. Lett.*, 1998, **72**, 2811
[5] Polson, R.C. and Vardeny, Z.V., *Appl. Phys. Lett.*, 2004, **84**, 1893
[6] Frolov, S.V., Gellermann, W., Vardeny, Z.V., Ozaki, M. and Yoshino, K., *Synth. Met.*, 1997, **84**, 471
[7] Dridi, C., Blel, N., Chaieb, A., Majdoub, M., Roudesli, M.S., Davenas, J., Ben Oudas, H. and Maaref, H., *Eur. Polym. J.*, 2001, **37**, 683
[8] Wu, X., Shi, G., Qu, L., Zhang, H. and Chen, F., *J. Polym. Sci., Part A: Polym. Chem.*, 2003, **41**, 449
[9] Barashkov, N.N., Guerrero, D.J., Olivos, H.J. and Ferraris, J.P., *Synth. Met.*, 1995, **75**, 153

Poly[(2,5-bis(3-sulfopropoxy)-1,4-phenylene ethynylene)-(1,4-phenylene ethynylene) disodium salt] P-85

Synonyms: *Poly[[2,5-bis(3-sulfopropoxy)-1,4-phenylene]-1,2-ethynediyl-1,4-phenylene-1,2-ethynediyl disodium salt]. PPE-$SO_3^\ominus$. PPESO3*
Monomers: 1,4-Diethynylbenzene
Material class: Thermoplastic
Polymer Type: poly(arylene ethynylene)

CAS Number:

CAS Reg. No.	Note
439279-76-2	Poly[[2,5-bis(3-sulfopropoxy)-1,4-phenylene]-1,2-ethynediyl-1,4-phenylene-1,2-ethynediyl disodium salt]
433334-80-6	

Molecular Formula: $(C_{22}H_{18}Na_2O_8S_2)_n$
Fragments: $C_{22}H_{18}Na_2O_8S_2$
General Information: Water-soluble polyelectrolyte exhibiting strong fluorescence in soln. Prod. as light yellow fibres.

Surface Properties & Solubility:
Solvents/Non-solvents: Sol. H_2O, MeOH, aq. EtOH [1,2]

Optical Properties:
Optical Properties General: The intense yellow-green fluorescence in water with λ_{max} at *ca.* 530 nm [1,2,3] exhibits a red shift compared to that in MeOH, indicating the presence of monomer in MeOH and aggregation in water [1,2]. There is also a red shift on going from solution to film [1,2,4]. The fluorescence is strongly quenched by cationic cyanine dyes [5], methyl viologen and metal complexes [1,2].
Transmission and Spectra: The UV and photoluminescence spectral data have been reported [1,2,4] as well as pmr and FT-IR data [1].

Applications/Commercial Products:
Processing & Manufacturing Routes: Prod. by Pd-catalysed Sonogashira coupling reaction of 1,4-bis(3-sulfopropoxy)-2,5-diiodobenzene with 1,4-diethynylbenzene [1,6].
Applications: The fluorescence superquenching effect with cationic quenchers can be utilised for analysis of different enzymes, sugars and DNA [3,7,8].

Bibliographic References
[1] Tan, C., Pinto, M.R. and Schanze, K.S., *Chem. Comm.*, 2002, 446
[2] Tan, C., Pinto, M.R. and Schanze, K.S., *Polym. Prepr. (Am. Chem. Soc., Div. Polym. Chem.)*, 2002, **43**, 126
[3] Pinto, M.R. and Schanze, K.S., *Proc. Natl. Acad. Sci. U.S.A.*, 2004, **101**, 7505
[4] Kim, K., Webster, S., Levi, N., Carroll, D.L., Pinto, M.R. and Schanze, K.S., *Langmuir*, 2005, **21**, 5207
[5] Tan, C., Atas, E., Müller, J.G., Pinto, M.R., Kleiman, V.D. and Schanze, K.S., *J. Am. Chem. Soc.*, 2004, **126**, 13685
[6] Pinto, M.R. and Schanze, K.S., *Synthesis*, 2002, 1293
[7] Rininsland, F., Xia, W., Wittenburg, S., Shi, X., Stankewicz, Achyuthan, K., McBranch, D. and Whitten, D., *Proc. Natl. Acad. Sci. U.S.A.*, 2004, **101**, 15295
[8] Kushon, S.A., Bradford, K., Marin, V., Suhrada, C., Armitage, B.A., McBranch, D. and Whitten, D., *Langmuir*, 2003, **19**, 6456

Poly(4-bromostyrene) P-86

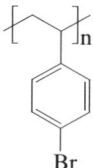

Synonyms: *Poly(p-bromostyrene). Poly(1-bromo-4-ethenylbenzene). Poly(1-(4-bromophenyl)-1,2-ethanediyl)*
Related Polymers: Poly(4-fluorostyrene), Poly(4-chlorostyrene), Poly(4-iodostyrene)
Monomers: 4-Bromostyrene
Material class: Thermoplastic
Polymer Type: styrenes

– Polybutadiene

CAS Number:

CAS Reg. No.
24936-50-3

Molecular Formula: $(C_8H_7Br)_n$
Fragments: C_8H_7Br

Volumetric & Calorimetric Properties:
Density:

No.	Value	Note
1	d 1.011 g/cm^3	[5]

Thermal Expansion Coefficient:

No.	Value	Note	Type
1	0.00016 K^{-1}	$T_g < T < 440K$ [5]	V
2	7.1×10^{-5} K^{-1}	$150K < T < T_g$	V

Volumetric Properties General: ΔC_p 31.9 J K^{-1} mol^{-1}
Glass-Transition Temperature:

No.	Value	Note
1	141°C	[1]
2	137°C	[2]
3	140°C	[3]
4	144°C	[6]

Transport Properties:
Permeability of Gases: [He] 1.64×10^{-10}; [CH$_4$] 0.48×10^{-10} cm^2 (s cmHg)$^{-1}$ [3]

Electrical Properties:
Dielectric Permittivity (Constant):

No.	Value	Frequency	Note
1	3.11		[1]
2	2.476	1 kHz	25° [5]

Complex Permittivity and Electroactive Polymers: μ 1.371 D (25°, C$_6$H$_6$), 1.392 D (50°, C$_6$H$_6$), 1.413 D (25°, 1,4-dioxane), 1.429 D (50°, 1,4-dioxane) [4]

Applications/Commercial Products:
Processing & Manufacturing Routes: Prod. by cationic polymerisation with Lewis acid catalysts

Bibliographic References
[1] Gustafsson, A., Wiberg, G. and Gedde, U.W., *Polym. Eng. Sci.*, 1993, **33**, 549
[2] Judovits, L.H., Bopp, R.C., Gaur, U. and Wunderlich, B., *J. Polym. Sci., Part B: Polym. Lett.*, 1986, **24**, 2725
[3] Puleo, A.C., Muruganandam, N. and Paul, D.R., *J. Polym. Sci., Part B: Polym. Phys.*, 1989, **27**, 2385
[4] Shima, M., Yamaguchi, N. and Sato, M., *Makromol. Chem.*, 1991, **192**, 531
[5] Corrado, L.C., *J. Chem. Phys.*, 1969, **50**, 2261
[6] Dunham, K.R., Faber, J.W.H, Vandenberghe, J. and Fowler, W.F., *J. Appl. Polym. Sci.*, 1963, **7**, 897

Polybutadiene

Synonyms: *Poly(1,3-butadiene)*. Butadiene rubber. Emulsion polymerised polybutadiene
Related Polymers: *cis*-1,4-Polybutadiene, Styrene-butadiene copolymers, Butadiene-acrylonitrile copolymers, Butadiene-pentadiene copolymers, Blends of polybutadiene and styrene-butadiene block copolymers, 1,2-Polybutadiene, Carbon black filled polybutadiene
Monomers: 1,3-Butadiene
Material class: Synthetic Elastomers
Polymer Type: polybutadiene
CAS Number:

CAS Reg. No.
9003-17-2

General Information: 1,2 Polymer may be atactic, isotactic or syndiotactic and is thermoplastic. The elastomeric *trans*-1,4-polymer is semicrystalline, the *cis*-1,4-polymer is most widely used with good low temp. props. Props. also vary with cross-linking and vulcanisation and fillers
Miscellaneous: Polymer struct. can be complex due to the nature of the diene monomer. Polymerisation may take place only at the 1,2 double bond to give the 1,2 polymer with pendant vinyl groups. 1,4 Polymerisation yields a polymer with a *cis* or *trans* double bond in varying compositions. The polymerisation method and conditions influence the composition of the product and hence props.

Volumetric & Calorimetric Properties:
Density:

No.	Value	Note
1	d 0.913 g/cm^3	71% 1,4-*trans*, 10% 1,4-*cis*, 19% 1,2 [4]
2	d 0.893 g/cm^3	54% 1,4-*trans*, 26% 1,4-*cis*, 20% 1,2 [5]

Equation of State: Simha-Somcynsky equation of state parameters have been reported [12]
Thermodynamic Properties General: PVT behaviour of emulsion polymerised material has been reported [9]

Surface Properties & Solubility:
Cohesive Energy Density Solubility Parameters: Solubility parameter 17.2 J$^{1/2}$ cm$^{-3/2}$ [2,3]
Solvents/Non-solvents: Sol. hydrocarbons, THF, higher ketones, higher aliphatic esters. Insol. alcohol, lower ketones and esters, nitromethane, propionitrile, H$_2$O [6,7,8]

Mechanical Properties:
Mechanical Properties General: Mech. property data for a range of microstructs. have been reported [11,12]

Optical Properties:
Refractive Index:

No.	Value	Note
1	1.518	25°, 71% 1,4-*trans*, 10% 1,4-*cis*, 19% 1,2 [5]
2	1.516	25°, 54.5% 1,4-*trans*, 25.4% 1,4-*cis*, 20.1% 1,2 [5]

Polymer Stability:
Polymer Stability General: Oxidative degradation leads to cross-linking and hardening of vulcanisates [14]
Decomposition Details: Thermal analysis has been reported [15,16,17]

Applications/Commercial Products:
Processing & Manufacturing Routes: Emulsion polymerisation yields a high *trans* polymer, where free-radical polymerisation prevents control. Low manufacture cost, mostly made and used for toughening other polymers. Also synth. using a neodymium catalyst system [18] or by a stereoselective polymerisation process using a Ni complex-AlEt$_2$Cl catalyst to give low (MW <11800) MW polymer [19]
Applications: Used in automotive tyres

cis-1,4-Polybutadiene

Synonyms: cis-Polybutadiene
Related Polymers: Polybutadiene, 1,2-Polybutadiene
Monomers: 1,3-Butadiene
Material class: Synthetic Elastomers
Polymer Type: polybutadiene
CAS Number:

CAS Reg. No.
9003-17-2

General Information: Exhibits good heat resilience, low temp. props. and low heat build up
Morphology: Crystallisation induced by strain and temp. has been reported. [49] Crystalline at low temps. but amorph. above -3° [55]

Volumetric & Calorimetric Properties:

Density:

No.	Value	Note
1	d 1.01 g/cm^3	[1]

Equation of State: Tait and Simba-Someynsky equation of state parameters have been reported [31]
Thermodynamic Properties General: Low heat build up in general
Latent Heat Crystallization: Entropy of fusion ΔS_m 33.5 J mol^{-1} K^{-1} (98% cis) [2]. Heat of fusion ΔH_m 2.5 kJ mol^{-1} (98% cis) [5]

Specific Heat Capacity:

No.	Value	Note	Type
1	0.369 kJ/kg.C	-223° [22,23]	P
2	0.897 kJ/kg.C	-123° [22,23]	P
3	1.96 kJ/kg.C	27° [22,23]	P
4	2.21 kJ/kg.C	77° [22,23]	P

Melting Temperature:

No.	Value	Note
1	2°C	98–99% cis/high cis [1,49]

Glass-Transition Temperature:

No.	Value	Note
1	-102°C	high cis [3,22,23]
2	-95°C	98–99% cis [4]
3	-106°C	[5,9]
4	-114°C	[10]

Surface Properties & Solubility:

Cohesive Energy Density Solubility Parameters: Solubility parameter 16.9 J$^{1/2}$ cm$^{-3/2}$ (1,4-cis) [6,7]
Solvents/Non-solvents: Sol. hydrocarbons, THF, higher ketones, higher aliphatic esters. Insol. alcohol, lower ketones and esters, nitromethane, propionitrile, H$_2$O [11,13]. Theta solvents heptane - 1° (97% cis); propyl acetate 35.5° (97% cis) [14]; 3-pentanone 0° (94.6% cis) [15]; diethyl ketone 213° (94% cis); propylene oxide 146° (94% cis) [16]; diethyl ketone 14°, 208° (93% cis) [16]; propylene oxide 35°, 141° (93% cis) [16]; heptane/hexane 5–20° (90% cis) [18]; 5-methyl-2-hexanone 12.6° (90% cis) [19]; 2-pentanone 59.7° (90% cis) [19]. Forms supercritical fluids in 0.27 wt % CO$_2$ (MV 5000, 19.3 MPa)

Trade name	Details	Supplier
Afdene	1,4-cis and -trans	Lehmann and Voss
Bindene	1,4-cis and -trans	Goodyear
Budene	1,2 and 1,4 polymer, various grades	Goodyear
Buna CB	1,4 polymer, various grades	Bayer Inc.
Calprene	1,4 polymer	Repsol Quimica
Cariflex butadiene rubber		Shell Chemicals
Cisdene 1203		American Synthetic Rubber Corporation
Diene, diene 35AC, Diene, 70AC		Firestone
Duragen	1,4-cis and -trans	General Tyre and Rubber Co.
Finaprene	1,4 polymer	Petrafina
Intene	1,4 polymer	Enichem
JSRRB	1,2 polymer	Japan Synthetic
Petcis	1,4-cis and -trans	Petkum Petrokimya
Ricon	1,2 polymer	Advanced Resins
SKB		Soviet Origin
Solprene	1,4-cis and -trans	Negramex
Taktene	1,4-cis and -trans	Polysar
Tufdene	1,4-cis and -trans	Asaki
Unidene	1,4-cis and -trans	International Synthetic Rubber

Bibliographic References

[1] Nelson, R.S., Jessup, R.S. and Roberts, D.E, *J. Res. Natl. Bur. Stand. (U.S.)*, 1952, **48**, 275
[2] Scott, R.L., *J. Chem. Phys.*, 1945, **13**, 178
[3] Grigorovskaya, V.A., Shvarts, A.V. and Bychova, L.P., Vysokomol. Soedin., *Adgez. Polym.*, 1963, 41
[4] Furukawa, J., Yamashita, S., Kotani, T. and Kawashima, M., *J. Appl. Polym. Sci.*, 1969, **13**, 2527
[5] Mandelkern, L., Tyron, M. and Quinn, F.A., *J. Polym. Sci.*, 1956, **19**, 81
[6] Dexheimer, H., Fuchs, O., Nitsche, R. and Wolf, K.A., *Struktur und Physikaleches Verkalten der Kunststoffe*, Vol. 1, Springer Verlag, Berlin-Goettingen-Heidelberg, 1963
[7] Kurata, M. and Stockmeyer, W.H., *Adv. Polym. Sci.*, Springer-Verlag, 1963, **3**, 196
[8] Roff, W.J., Fibres, Films, *Plastics and Rubbers*, Academic Press, 1956
[9] Barlow, J.W., *Polym. Eng. Sci.*, 1978, **18**, 238
[10] Hatmann, B., Simha, R. and Berger, A.E., *J. Appl. Polym. Sci.*, 1989, **27**, 2603
[11] Short, J.N., Kraus, G., Zelinski, R.P. and Naylor, F.E., Soc. Plast. Eng., *Tech. Pap.*, Paper No. 3, 1959, 8
[12] Ahagon, A., *Rubber Chem. Technol.*, 1986, **59**, 187
[13] Bohm, G.G.A. and Tveekrem, J.O., *Rubber Chem. Technol.*, 1982, **55**, 575
[14] Gent, A.N., *Rubber Chem. Technol.*, 1982, **55**, 525
[15] Brazier, D.W., *Rubber Chem. Technol.*, 1980, **53**, 437
[16] Sircar, A.K. and Lamond, T.G., *Rubber Chem. Technol.*, 1972, **45**, 329
[17] Sircar, A.K. and Lamond, T.G., *Thermochim. Acta*, 1973, **7**, 287
[18] Oehme, A., *Angew. Makromol. Chem.*, 1996, **235**, 121, (synth)
[19] Gerbasa, A.E. et al, *Angew. Makromol. Chem.*, 1996, **239**, 33, (synth)

– cis-1,4-Polybutadiene

Surface Tension:

No.	Value	Note
1	32 mN/m	critical surface tension of spreading [8]

Transport Properties:
Polymer Solutions Dilute: Mark-Houwink equation parameters have been reported [17,24]. Williams-Landel-Ferry equation parameters have been reported [32]. Unperturbed dimension values have been reported [24,27,29]

Gas Permeability:

No.	Gas	Value	Note
1	H_2	2766 cm^3 mm/(m^2 day atm)	3.16×10^{-12} cm^2 (s Pa)$^{-1}$, 25° [20]
2	N_2	424 cm^3 mm/(m^2 day atm)	4.84×10^{-13} cm^2 (s Pa)$^{-1}$, 25° [20]
3	O_2	1252 cm^3 mm/(m^2 day atm)	1.43×10^{-13} cm^2 (s Pa)$^{-1}$, 25° [20]
4	CO_2	9104 cm^3 mm/(m^2 day atm)	1.04×10^{-12} cm^2 (s Pa)$^{-1}$, 25° [20]
5	He	2145 cm^3 mm/(m^2 day atm)	2.45×10^{-12} cm^2 (s Pa)$^{-1}$, 25° [21]
6	Ne	1260 cm^3 mm/(m^2 day atm)	1.44×10^{-12} cm^2 (s Pa)$^{-1}$, 25° [21]
7	Ar	2696 cm^3 mm/(m^2 day atm)	3.08×10^{-12} cm^2 (s Pa)$^{-1}$, 25° [21]
8	N_2	1260 cm^3 mm/(m^2 day atm)	1.44×10^{-12} cm^2 (s Pa)$^{-1}$, 25° [21]

Mechanical Properties:
Mechanical Properties General: Mech. props. of a range of polybutadiene structs. have been reported [42,43,44,48]. Possesses low tensile strength and low shear strength
Viscoelastic Behaviour: Viscoelastic behaviour for a range of cis-1,4-polybutadienes has been reported [37,38]
Fracture Mechanical Properties: Fatigue crack propagation has been reported [45]

Electrical Properties:
Complex Permittivity and Electroactive Polymers: Electrical conduction (96–100% cis content) has been reported [55]

Optical Properties:
Transmission and Spectra: Ir spectral data have been reported [25]

Polymer Stability:
Polymer Stability General: Oxidative degradation of polybutadiene leads to cross-linking and to hardening of vulcanisates, unlike polyisoprene, which softens as a result of chain scission [51]
Upper Use Temperature:

No.	Value	Note
1	100°C	[34]

Decomposition Details: Decomposition onset 325°. 50% Wt loss 407° (vacuum) [36]. Low trans contents yields 4-vinyl-1-cyclohexene (dimer), 1,3-butadiene, cyclopentene and 1,3-cyclohexadiene on pyrolysis [39]. Pyrolysis at 450–530° yields monomer in an amount proportional to the temp. Activation energy 37.5 kcal mol^{-1} [40]. Mechanism of decomposition [41] and thermal analysis have been reported [52,53,54]
Flammability: Are high-volatile-loss, low-char polymers with slightly lower volatile loss on pyrolysis than high vinyl polybutadienes [47]

Stability Miscellaneous: High energy radiation increases radical concentration linearly with radiation up to 100 Mrad. Unsaturation decreases markedly. Cis-trans isomerisation occurs and some conjugated double bonds are formed. Yield of cross-linking for cis-1,4 higher than for trans-1,4 but lower than for 1,2 isomer. Stress-strain props. of radiation cross-linked polybutadiene have been reported [50]

Applications/Commercial Products:
Processing & Manufacturing Routes: High cis content produced by Ziegler-Natta catalysis using Ti, Co or Ni catalysts. Processing and props. depend on the catalyst, which affects branching MW and vinyl content
Applications: Used in automotive tyres, usually as blends with natural or SB rubbers. Also used in high-impact polystyrene

Trade name	Details	Supplier
Austrapol	1202, 1220	Australian Syn. Rubber
Budene	1208, 1220	Goodyear
Buna CB		Bayer Inc.
Cariflex BR		Shell Chemicals
Cisdene		American Synthetic Rubber Corporation
Duragen		General Tyre and Rubber Co.
Europrene cis		Enichem
Nipol BR		Nippon Zeon
Solprene	277	Negromex (Brazil)
Unidene	50/50 AC	International Synthetic Rubber

Bibliographic References
[1] Natta, G., *Science (Washington, D.C.)*, 1965, **147**, 269
[2] Natta, G. and Moraglio, G., *Makromol. Chem.*, 1963, **66**, 218
[3] Trick, G.S., *J. Appl. Polym. Sci.*, 1960, **3**, 253
[4] Baccaredda, M. and Butta, E., *Chim. Ind. (Milan)*, 1960, **42**, 978
[5] Bahary, W.S., Sapper, D.I. and Lane, J.H., *Rubber Chem. Technol.*, 1967, **40**, 1529
[6] Scott, R.L., *PhD Thesis*, Princeton University, 1945
[7] Grigorovskaya, V.A., Shvarts, A.V. and Bychova, L.P., Vysokomol. Soedin., Adgez. Polym., 1963, 41
[8] Lee, L. and Lee, H., *J. Polym. Sci., Part A-2*, 1967, **5**, 1103
[9] Takeda, M., Tanaka, K. and Nagao, R., *J. Polym. Sci.*, 1962, **57**, 517
[10] Krauss, G., Childern, W. and Gruver, J.T., *J. Appl. Polym. Sci.*, 1967, **11**, 158
[11] Dexheimer, H. and Fuchs, O., *Struktur und Physikaleches Verkalten der Kunststoffe*, Vol. 1, (eds. R. Nitsche and K.A. Wolf), Springer Verlag, 1963
[12] Kurata, M. and Stockmayer, W.H., *Adv. Polym. Sci.*, Springer-Verlag, 1963, **3**, 196
[13] Roff, W.J., Fibres, Films, *Plastics and Rubbers*, Academic Press, 1956
[14] Moraglio, G., *Eur. Polym. J.*, 1965, **1**, 103
[15] Abe, M. and Fujita, H., *J. Phys. Chem.*, 1965, **69**, 3263
[16] Cowie, J.M.G. and McEwen, J.J., *Polymer*, 1975, **16**, 933
[17] Danusso, F., Moraglio, G. and Gianotti, G., *J. Polym. Sci.*, 1961, **51**, 475
[18] Poddubnyi, I.Ya., Grechanovskii, V.A. and Mosevitskii, M.I., *Vysokomol. Soedin.*, 1963, **5**, 1049
[19] Abe, M. and Fujita, H., *Rep. Prog. Polym. Phys. Jpn.*, 1964, **7**, 42
[20] Amerongen, G.J., *J. Polym. Sci.*, 1950, **5**, 307
[21] Paul, D.R. and DiBenedetto, A.J., *J. Polym. Sci., Part C: Polym. Lett.*, 1965, **10**, 17
[22] Gaur, U., Lau, S.F., Wunderlich, B. et al, *J. Phys. Chem. Ref. Data*, 1983, **12**, 29
[23] Grebowitz, J., Aucock, W. and Wunderlich, B., *Polymer*, 1986, **27**, 575
[24] Poddubnyi, I.Ya., Erenkers, Ye.G. and Yeremina, M.A., *Vysokomol. Soedin.*, 1968, **10**, 1381
[25] Noda, I., Downey, A.E. and Marcott, C., *Physical Properties of Polymers Handbook*, (ed. J.E. Mark), AIP Press, 1996, 293
[26] Fetters, L.J., Lohse, D.J. and Colby, R.H., *Physical Properties of Polymers Handbook*, (ed. J.E. Mark), AIP Press, 1996, 338
[27] Ferry, J.D., *Viscoelastic Properties of Polymers*, 3rd edn., Wiley, 1986
[28] Kurata, M. and Tsunashima, Y., *Polym. Handb.*, 4th edn., (eds. J. Brandrup, E.H. Immergut and E.A. Grulke), John Wiley and Sons, 1999, **VII**, 47

[29] Abe, M., Murakami, Y. and Fujita, H., *J. Appl. Polym. Sci.*, 1965, **9**, 2549
[30] Kurata, M. and Stockmayer, W.H., *Fortschr. Hochpolym.-Forsch.*, 1963, **3**, 196
[31] Jain, R.K. and Simha, R., *Polym. Eng. Sci.*, 1979, **19**, 845
[32] Ngai, K.L. and Plazek, D.J., *Physical Properties of Polymers Handbook*, (ed. J.E. Mark), AIP Press, 1996, 341
[33] Sanders, J.F., *PhD Thesis*, University of Wisconsin, 1968
[34] Billmeyer, F.W., *Textbook of Polymer Science*, Wiley Interscience, 1984, 136, 361
[35] Liggat, J., *Polym. Handb.*, 4th edn., (eds. J. Brandrup, E.H. Immergut and E.A. Grulke), John Wiley and Sons, 1999, **II**, 451
[36] Van Krevelen, D.W. and Hoftyzer, P.J., *Properties of Polymers: Their Correlation with Chemical Structure*, Elsevier Science, 1976, 459
[37] Malkin, A.Ya., Kulichikhin, V.G., Zabugina, M.P. and Vinogradov, G.V., *Vysokomol. Soedin.*, Ser. A, 1970, **12**, 120
[38] Ishikawa, T. and Nagai, K., *J. Polym. Sci.*, Part A-2, 1969, **7**, 1123
[39] Tamura, S. and Gilham, J.K., *J. Appl. Polym. Sci.*, 1978, **22**, 1867
[40] Ericsson, I., *J. Chromatogr. Sci.*, 1978, **16**, 340
[41] Santoso, M., *Diss. Abstr. Int. B*, 1979, **39**, 4383
[42] Short, J.N., Kraus, G., Zelinski, R.P. and Naylor, F.E., Soc. Plast. Eng., Tech. Pap., Paper No. 3, 1959
[43] Natta, G., Crespi, G., Guzzezetta, G., Leghissa, S. and Sabbioni, F., *Rubber Plast. Age*, 1961, **42**, 402
[44] Sharaf, M.A., *Rubber Chem. Technol.*, 1994, **67**, 88
[45] Lee, M.-P. and Moet, A., *Rubber Chem. Technol.*, 1993, **66**, 304
[46] Roland, C.M., *Rubber Chem. Technol.*, 1989, **62**, 456
[47] Lawson, D.F., *Rubber Chem. Technol.*, 1986, **59**, 455
[48] Ahagon, A., *Rubber Chem. Technol.*, 1986, **59**, 187
[49] Goritz, D. and Kiss, M., *Rubber Chem. Technol.*, 1986, **59**, 40
[50] Bohm, G.G.A. and Tveekrem, J.O., *Rubber Chem. Technol.*, 1982, **55**, 575
[51] Gent, A.N., *Rubber Chem. Technol.*, 1982, **55**, 525
[52] Brazier, D.W., *Rubber Chem. Technol.*, 1980, **53**, 437
[53] Sircar, A.K. and Lamond, T.G., *Rubber Chem. Technol.*, 1972, **45**, 329
[54] Sircar, A.K. and Lamond, T.G., *Thermochim. Acta*, 1973, **7**, 287
[55] Tkaczyk, S.W., *Adv. Mater. Opt. Electr.*, 1997, **7**, 87, (electrical props)

Poly(1,3-butadiene-co-acrylonitrile), Carbon black filled P-89

Synonyms: *Nitrile rubber, carbon black filled. NR, carbon black filled. NBR, carbon black filled*
Related Polymers: Nitrile rubber, Isoprene copolymer, Nitrile rubber, Mineral filled, Nitrile rubber, PVC blends, Nitrile rubber, Nitrile rubber, carboxylated, Nitrile rubber, Hydrogenated, Nitrile rubber, Ethylene copolymer
Monomers: 1,3-Butadiene, Acrylonitrile
Material class: Composites, Synthetic Elastomers, Copolymers
Polymer Type: acrylonitrile and copolymers, polybutadiene
CAS Number:

CAS Reg. No.
9003-18-3

Molecular Formula: $[(C_4H_6).(C_3H_3N)]_n$
Fragments: C_4H_6 C_3H_3N
Morphology: Dynamic mech. measurements indicate short-range mobility of nitrile rubber unaffected by addition of carbon black ($\leq 50\%$ loading); no structural changes apparent, although indications that limited number of small crystallites may be formed [26]

Volumetric & Calorimetric Properties:
Density:

No.	Value	Note
1	d 1.06–1.33 g/cm^3	0–96 phr carbon black [25]

Thermodynamic Properties General: Thermal conductivity [1,14] and its variation with temp. [15] have been reported. Heat of combustion has been reported [30]
Thermal Diffusivity: 0.065–0.165 mm^2 s^{-1} [13]
Glass-Transition Temperature:

No.	Value	Note
1	-28°C	33% acrylonitrile, Hycar 1002 [27]

Transition Temperature General: T_g of nitrile rubber unaffected by addition of carbon black reinforcing agent ($\leq 50\%$ filler) [26]

Surface Properties & Solubility:
Wettability/Surface Energy and Interfacial Tension: Surface energy and other surface characteristics, and their effect on nitrile rubber reinforcement, have been reported [21]

Transport Properties:
Transport Properties General: Rheological characterisation, involving effect of filler on nitrile rubber, has been reported. Capillary, compressional and rotational-shearing disk (Mooney) viscometry, and stress-relaxation methods employed; branching and gel content have effect on measurements [22]
Polymer Melts: Capillary rheometry studies have been reported (125°, apparent shear rates 2.9–2900 s^{-1}); steady-shear viscosity, dynamic complex viscosity and viscosity obtained from tensile stress-strain behaviour found to be significantly different from each other (in contrast to unfilled materials); no appreciable effect of pressure on viscosity is observed; data obtained indicate presence of particulate struct., probably consisting of carbon black surrounded with bound rubber [11]

Mechanical Properties:
Mechanical Properties General: Activation energy of deformation has been reported [18]. Poisson's ratio has been reported [8]. Flexural and compressive strength have been reported [14]. Creep modulus has been reported [10]. Loss tangent decreases with increasing surface area of filler [28]. Dynamic mech. measurements have been used to investigate effect of carbon black on chain mobility and struct. of nitrile rubber containing 30–50% acrylonitrile (-170–150°, 2–20 kHz); sound velocity and damping factor measured as function of temp.; mech. spectra do not seem to be significantly affected by addition of carbon black ($\leq 50\%$ loading); thermal coefficients of sound velocity and damping factor, and position of T_g, remain unchanged when reinforcing agent added; indications that short-range mobility of chain elements unaffected by presence of filler particles [26]. Correlations have been reported between dynamic mech. props. and surface props. (surface energy and other characteristics) [21], and with wet-skid behaviour and adhesion [23]. Comprehensive study of tensile and shear behaviour of various carbon black-filled (40phr) nitrile rubbers (with different cross-linked contents and gel contents) has been reported. Results from tensile stress-strain experiments show a strain-time correspondence for filled systems up to yield point, shear curve produced from dynamic measurements (small deformations, frequencies of 0.1 to 100 rad s^{-1}) different from former curve, even when conversion (tensile to shear) factor applied, which can be explained by anisotropic nature of carbon black aggregate density in tensile measurements [17]

Tensile (Young's) Modulus:

No.	Value	Note
1	13.7 MPa	1.4 kgf mm^{-2}, 40% acrylonitrile [24]

Tensile Strength Break:

No.	Value	Note	Elongation
1	0.1–0.31 MPa	1.1–3.2 kgf mm^{-2}, 100°, 40% acrylonitrile [24]	310–700%

Complex Moduli: Dynamic Young's modulus has been reported [28]
Viscoelastic Behaviour: Variation of stress relaxation with temp. has been reported [5]. Viscoelastic props. correlated with surface

props. (wet-skid behaviour/adhesion) have been reported [23]. Stress-relaxation measurements have been reported; sudden jump in relaxation time attributed to formation of rubber-filled network [22]. A comprehensive study of the non-linear viscoelastic behaviour of nitrile rubbers (33% acrylonitrile content) containing 40 phr carbon black has been reported; filled elastomers exhibit strain softening with extrusion, and strain hardening with extension; ultimate props. of filled elastomers are insensitive to bulk viscoelastic behaviour [27]. Model has been reported to describe the linear viscoelastic behaviour of carbon black filled vulcanised nitrile rubber; gives satisfactory representation of experimental results over full range of strain amplitudes, frequencies and temps. studied, and also provides method for describing temp.-dependent deformation at different filler loadings [20]

Mechanical Properties Miscellaneous: Resilience 20% (approx., 40% acrylonitrile) [24]. Compression set 22% (40% compression, 22h, 70°; 30 mins. recovery, room temp., 40% acrylonitrile) [24]

Hardness: 45–78 (IHRD, -30°, 40% acrylonitrile) [24], IHRD hardness 52–72 (surFace area, 20–150 $m^2 g^{-1}$) [28]

Failure Properties General: Tear strength 5.2 kg mm^{-1} (40% acrylonitrile) [24]

Fracture Mechanical Properties: Tearing energy and J-integral measurements have been reported for nitrile rubbers at various carbon-black loadings (0–60 phr). Carbon black has only moderate effect on critical tearing energy due to high T_c of unfilled vulcanisate (4.973 kN m^{-1}); change in crack-tip morphology (crystal formation) has effect on T_c; T_c goes through maximum (6.978 kN m^{-1}) at approx. 40 phr loading; reasonably good correlation between carbon black morphology (particle size, aggregate size and struct.) and T_c at 40 and 60 phr loadings, although compound- and loading-dependent. T_c 4.194–5.777 kN m^{-1} (20 phr); 5.144–6.978 kN m^{-1} (40 phr); 3.816–5.11 kN m^{-1} (60 phr); true modulus 732–1818 kPa (20–60 phr); modulus shows only moderate correlation with carbon black morphology [19]

Friction Abrasion and Resistance: Wear resistance has been reported [3]. Internal friction 87 kP (60 Hz, 40% acrylonitrile) [24]. Wet-skid behaviour (using British pendulum skid-tester) has been investigated μ_A (adhesive) 0.1 (approx.) [23]

Electrical Properties:

Electrical Properties General: Dissipation factor for material containing 20–50% acrylonitrile has been reported [28]. Other values of volume resistivity have been reported [4]

Surface/Volume Resistance:

No.	Value	Note	Type
1	$1 \times 10^{-7} \times 10^{15}$ Ω.cm	approx. [7]	S

Static Electrification: Electrical conductivity has been reported [12,14]. DC electrical conductivity (35–130°) temp. coefficient of resistance 0 [29]. Dielectric constant shows anomalous dispersion with frequency, which increases with carbon content, particularly in low-frequency region; dielectric loss also increases with increase in carbon content and shows maximum at approx. 0.5 MHz, with value of this maximum increasing with increase in carbon concentration [9]. There is no interaction between the NBR rubber and the carbon black [9]. Earlier electrical study of copolymers containing 0–96 phr carbon black had examined dc conductance, dielectric constant and loss tangent; results obtained differ significantly for coarse and fine fillers; variations in dielectric constant indicate likelihood of chain formation; dc conductance decreases (initially) on addition of coarse blacks, while fine blacks show rapid increase in conductance; transients are observed for conductance which are ascribed to presence of particles of carbon black with perfectly conducting graphite core inside semiconducting sheath, the whole being surrounded by ionic atmosphere; maximum ac lossfactor appears at approx. 200 kc. [25] DC conductivity 0.329–1.350 $\times 10^9$ (0–40 phr); DC dielectric loss 5.900–24.5 (100 Hz, 0–40 phr); 0.590–2.45 (1 kHz, 0–40 phr), 0.059–0.245 (10000 Hz, 0–40 phr) [9]. Electrical conduction has been investigated in NBR composites, containing 100 phr carbon (fast extruding furnace) black, subjected to different extents of pre-extension network struct. of rubber does not give particularly good dispersion of filler within NBR matrix and this leads to smaller effect of filler on conductivity upon pre-extension; conductivity 7×10^{-5}–0.01 Ω^{-1} cm^{-1} [29]

Polymer Stability:

Thermal Stability General: Heat resistance has been reported [14]

Flammability: Flammability, oxygen index, ignition temp. and heat flux intensity have been reported [30]. Flammability of vulcanised rubber decreases in presence of fillers; ignition and self-ignition temps. increase by 30–40° with increasing filler content (to 50%) [30]

Chemical Stability: Resistance to organic acids and EtOH has been reported [6] Chemical resistance has been reported [14]

Stability Miscellaneous: Radiation resistance has been reported [2]

Applications/Commercial Products:

Trade name	Supplier
Nysynblak	Copolymer Rubber

Bibliographic References

[1] Bafrnec, M. and Lodes, A., *Chem. Prum.*, 1966, **16**, 171, (thermal conductivity)
[2] Harrington, R., *Rubber Age (N.Y.)*, 1961, **90**, 265, (radiation resistance)
[3] Klitenik, G.S., Ratner, S.B., and Rakova, G.M., *Kauch. Rezina*, 1967, **26**, 32, (wear)
[4] Berezina, N.P., and Kashin, V.A., *Tr. Tomsk Nauch. Issled. Inst. Kabel. Prom.*, 1969, **1**, 274, (volume resistivity)
[5] Takano, Y., Suzuki, U., and Kurihara, C., *Nippon Gomu Kyokaishi*, 1966, **39**, 674, (stress relaxation)
[6] Nedyalkova, K., Taushanski, S. and Mareeva, P., *Khim. Ind. Sofia*, 1970, **10**, 454, (chemical resistance)
[7] Mazin, Y.A., Molodtsova, N.V. and Arsenova, L.I., *Kauch. Rezina*, 1973, **32**, 23, (volume and surface resistivity)
[8] Elektrova, L.M., Melentev, P.V. and Zelenev, Y.U., *Mekh. Polim.*, 1973, **2**, 351, (Poisson's ratio)
[9] Hanna, F.F. and Abou-Bakr, A.F., *Z. Phys. Chem. (Leipzig)*, 1973, **252**, 386, (dielectric constant, dc conductivity)
[11] Nakajima, N. and Collins, E.A., *Rubber Chem. Technol.*, 1975, **48**, 615, (viscosity)
[12] Blinov, A.A., Zhuravlev, U.S., and Komev, A.E., *Kauch. Rezina*, 1975, **10**, 27, (electrical conductivity)
[13] Baldt, V., Kramer, H. and Koopmann, R., *Bayer- Mitt. Gummi. Ind.*, 1978, **50**, 39, (thermal diffusivity)
[14] Chipps, F.R.and Vanderbilt, V.M., *Rev. Plast. Mod.*, 1984, **47**, 520, (compressive strength, flexural strength, conductivity)
[15] Freeman, J.J. and Graig, D., *Adv. Cryog. Eng.*, 1984, **30**, 105, (thermal conductivity)
[16] Liu, J., Zhou, H., and Yu, F., *Yingyong Huaxue*, 1986, **3**, 25, (power factor, tensile modulus)
[17] Nakajima, N., Scobbo, J.J. and Harrell, E.R., *Rubber Chem. Technol.*, 1987, **60**, 761, (storage and loss moduli)
[18] Kovalev, I.M., Vettegren, V.I. and Lazarev, S.D., *Vysokomol. Soedin.*, Ser. B, 1989, **31**, 563
[19] Chung, B., Funt, J.M. and Quyang, G.B., *Rubber World*, 1991, **204**, 46, (mech. props., tearing energy)
[20] Ruddock, N., James, P.W. and Jones, T.E.R., *Rheol. Acta*, 1993, **32**, 286, (complex modulus)
[21] Wolff, S, Wang, M.J. and Tan, E.W., *Kautsch. Gummi Kunstst.*, 1994, **47**, 873, (dynamic modulus, power factor)
[22] Soos, I., *Kautsch. Gummi Kunstst.*, 1994, **47**, 502, (rheological props., stress relaxation)
[23] Gabler, A., Straube, E. and Heinrich, G., *Kautsch. Gummi Kunstst.*, 1993, **46**, 941, (friction coefficient, viscoelastic props., complex elastic modulus)
[24] Roff, W.J. and Scott, J.R., Fibres, Films, *Plastics and Rubbers*, Butterworths Ltd, 1971, 402, (mech. props.)
[25] Gross, B. and Fuoss, R.M., *J. Phys. Chem.*, 1956, **60**, 474, (density, dielectric constant)
[26] Baccaredda, M. and Butta, E., *J. Polym. Sci.*, 1962, **57**, 617, (sound velocity)
[27] Nakajima, N., Bowerman, H.H. and Collins, E.A., *J. Appl. Polym. Sci.*, 1977, **21**, 3063, (transition temp., stress relaxation)
[28] England, W.J., Long, R.F. and Townend, D.J., *Anal. Proc. (London)*, 1981, **18**, 430, (mech. props.)
[29] Amin, M., Nasr, G.M., Khairy, S.A. and Ateia, E., *Angew. Makromol. Chem.*, 1986, **141**, 19, (dc conductivity)

[30] Ushkov, V.A, Lalayan, V.M., Abishev, A.K. and Morozova, N.Y, *Kauch. Rezina*, 1986, 8, (flammability, oxygen index, heat of combustion, ignition temp.)

Poly(1,3-butadiene-*co*-acrylonitrile-*co*-ethylene)   P-90

$$\left[\left[CH_2CH=CHCH_2\right]_x\left[\underset{CN}{CH_2CH}\right]_y\left[CH_2CH_2\right]_z\right]_n$$

Synonyms: *Poly(NBR-co-ethylene). Modified hydrogenated acrylonitrile-butadiene copolymer. Poly(1,3-butadiene-co-propenenitrile-co-ethene). Poly(NR-co-ethylene). Poly(nitrile rubber-co-ethylene*
Related Polymers: NR, Carboxylated, Poly(1,3-butadiene-*co*-acrylonitrile), NR, PVC, NR, Mineral filled, NR, Hydrogenated, NR, Carbon black filled, NR, Isoprene copolymer
Monomers: 1,3 Butadiene, Acrylonitrile, Ethylene
Material class: Copolymers, Synthetic Elastomers
Polymer Type: acrylonitrile and copolymers
CAS Number:

CAS Reg. No.
26835-20-1

Molecular Formula: $[(C_4H_6)_x(C_3H_3N)_y(C_2H_4)_z]_n$
Fragments: C_4H_6 C_3H_3N C_2H_4
Additives: Carbon black
General Information: A modified hydrogenated nitrile rubber developed in order to extend the low-temp. service range of hydrogenated rubber [3]
Miscellaneous: Modifications of nitrile rubber with copolymers containing ethylene and a softening monomer, such as vinyl esters, alkyl (meth)acrylates or acrylonitrile, have been reported [1].: give mixtures with improved workability, providing vulcanisates with good oil and ageing resistance, with reduced costs [1]

Volumetric & Calorimetric Properties:
Glass-Transition Temperature:

No.	Value	Note
1	-47--39°C	modified hydrogenated nitrile rubber [3]

Transition Temperature General: Modified hydrogenated nitrile rubber shows improvement in T_g of up to -10° [3]
Brittleness Temperature:

No.	Value	Note
1	-45–40°C	33% acrylonitrile, ASTM D746 [1]

Mechanical Properties:
Mechanical Properties General: Changes in mechanical props. of vulcanisates containing terpolymers with 33% acrylonitrile have been reported. Tensile strength (and extension), 100% modulus and hardness (Shore A) measured after ageing at room temp. for 70h, (ASTM D471), and after contact with oil (70h 121°, ASTM D573) and MeOH (70h, 64°, ASTM D471) [1]
Tensile Strength Break:

No.	Value	Note	Elongation
1	31.4 MPa	320 kg cm^{-2} room temp.	160%
2	17 MPa	174 kg cm^{-2}, 100° [3]	160%
3	13.6–16.9 MPa	29–41% acrylonitrile [1]	360–470%

Miscellaneous Moduli:

No.	Value	Note	Type
1	3.6–19 MPa	29–41% acrylonitrile [1]	100%

Mechanical Properties Miscellaneous: Significant improvements in cold flexibility (Gehman torsional stiffness ASTM D1053-92) and temp. reduction (ASTM D1329-79)) have been reported for modified hydrogenated nitrile rubber [3]. Dynamic mechanical anal. confirms extension of rubbery plateau modulus by approx. -10° for modified hydrogenated nitrile rubber. This correlates with reduction of T_g by approx.-10° [3]
Hardness: Hardness (JIS A) 90 (room temp., 6% butadiene, 37% acrylonitrile) 85 (100°, 6% butadiene, 37% acrylonitrile) [2]. Shore A57–A76 (29–41% acrylonitrile) [1]
Friction Abrasion and Resistance: Abrasion resistance has been reported [2]

Polymer Stability:
Polymer Stability General: Displays good resistance to ageing [1]
Upper Use Temperature:

No.	Value	Note
1	-40–150°C	[3]

Chemical Stability: Has good resistance to oil [1]. Modified hydrogenated nitrile rubbers show greater resistance to ASTM oils 1, 2 and 3, oil swell, 168h, 150°, than do unmodified elastomers [3]

Applications/Commercial Products:
Processing & Manufacturing Routes: Vulcanisates are prepared from a blend mixture of nitrile rubber with e.g. carbon monoxide-ethylene-vinyl acetate terpolymer, ZnO, sulfur, carbon black, stearic acid and mercaptobenzothiazyl sulfide [1]. Modified hydrogenated nitrile rubber is prepared by emulsion copolymerisation of mixture of acrylonitrile, butadiene and alkyl acrylate, followed by hydrogenation using a selective homogeneous catalyst; raw terpolymer cured using a peroxide formulation (15 min, 170°) [3]
Applications: Potential applications of modified hydrogenated nitrile rubber in automotive and related industries on account of improvements in low-temp. service range

Trade name	Details	Supplier
Zetpol	2000L/2010L	Nippon Zeon

Bibliographic References
[1] *Ger. Pat.*, 1980, 2934931, (synth., mech. props., thermal props., stability)
[2] *Jpn. Pat.*, 1986, 61108645, (hardness, abrasion resistance, tensile strength)
[3] Arsenault, G.J., Brown, T.A. and Jobe, I.R., *Kautsch. Gummi Kunstst.*, 1995, **48**, 418, (synth., uses, mech. props., thermal props., stability)

Poly(1,3-butadiene-*co*-acrylonitrile-*co*-2-methyl-1,3-butadiene)   P-91

$$\left[\left[CH_2CH=CHCH_2\right]_x\left[\underset{CN}{CH_2CH}\right]_y\left[\underset{CH_3}{CH_2C=CHCH_2}\right]_z\right]$$

Synonyms: *Poly(1,3-butadiene-co-propenenitrile-co-2-methyl-1,3-butadiene). Poly(1,3-butadiene-co-acrylonitrile-co-isoprene). Poly(nitrile rubber-co-isoprene). Poly(NR-co-isoprene). Poly(NBR-co-isoprene)*
Related Polymers: NR, PVC blends, NR, Carboxylated, NR, Hydrogenated, NR, Carbon black filled, Poly(1,3-butadiene-*co*-acrylonitrile), NR, Mineral filled, NR, Ethylene copolymer

− Poly(1,3-butadiene-*co*-acrylonitrile), Mineral filled

Monomers: Acrylonitrile, 1,3-Butadiene, 2-Methyl-1,3-butadiene
Material class: Synthetic Elastomers, Copolymers
Polymer Type: acrylonitrile and copolymers
CAS Number:

CAS Reg. No.
25135-90-4

Molecular Formula: $[(C_4H_6)_x.(C_3H_3N)_y(C_5H_8)_z]_n$
Fragments: C_4H_6 C_3H_3N C_5H_8
General Information: Vulcanised material displays improved tensile strength, good low temp. props. and rebound (resilience), good resistance to compression set, and improved retention of props. with ageing, in comparison with acrylonitrile/isoprene or acrylonitrile/butadiene rubbers [4]
Morphology: Degree of crystallinity 15% (27% butadiene), 18% (15% butadiene) [1]
Miscellaneous: Oil-resistant rubber fibres with good strength have been reported [1]

Transport Properties:
Polymer Solutions Dilute: Intrinsic viscosity 0.9 dl g^{-1} (DMF, 30°, 27 mol% butadiene), 0.7 dl g^{-1} (DMF, 30°, 15 mol% butadiene) [1]

Mechanical Properties:
Mechanical Properties General: Blends containing 29% acrylonitrile and 55% butadiene lose 3.1% of their tensile strength after heating at 80° for 72h [2]
Tensile Strength Break:

No.	Value	Note	Elongation
1	22.6 MPa	230 kg cm^{-2}, 29% acrylonitrile, 55% butadiene [2]	680%
2	30.4 MPa	310 kg cm^{-2}, 27% butadiene [1]	510%
3	35.5 MPa	362 kg cm^{-2}, 15 mol% butadiene [1]	515%
4	24.5 MPa	250 kg cm^{-2}, butadiene/isoprene (50/50) [4]	620%
5	22.7 MPa	232 kg cm^{-2}, butadiene/isoprene (75/25) [4]	700%
6	20.1 MPa	205 kg cm^{-2}, butadiene/isoprene (50/50), after 168h, 100° [4]	
7	22.9 MPa	234 kg cm^{-2}, butadiene/isoprene (75/25), after 168h, 100° [4]	

Miscellaneous Moduli:

No.	Value	Note	Type
1	1.5 MPa	15 kg cm^{-2} 27 mol% butadiene [1]	100% modulus
2	1.2 MPa	12 kg cm^{-2}, 15 mol% butadiene [1]	100% modulus
3	3.4 MPa	35 kg cm^{-2}, 27 mol% butadiene [1]	300% modulus
4	3.4 MPa	35 kg cm^{-2}, 15 mol% butadiene [1]	
5	2.9 MPa	30 kg cm^{-2}, 29% acrylonitrile, 55% butadiene [2]	300% modulus

Mechanical Properties Miscellaneous: Tensile set 18% (room temp., 27% mol% butadiene), 190% (room temp., 15 mol% butadiene), 5% (max., 27 mol% butadiene), 28% (50°, 15 mol% butadiene) [1]. Terpolymers containing 15 and 27 mol% butadiene are stable for 35 min at 75° and 350% elongation, or for 120 min. at 80° and 40 min at 100°, both at 200% elongation [1]. Rebound (resilience) 42% (butadiene/isoprene (50/50)), 45.5% (butadiene/isoprene (75/25)) [4]

Hardness: Shore A20 (raw polymer, 27 mol% butadiene), A23 (raw polymer, 15 mol% butadiene); A43 (aged for 7 days at room temp., 27 mol% butadiene); A80 (aged for 7 days at room temp. 15 mol% butadiene) [1]. Spring A61 (butadiene/isoprene (50/50), A60 (butadiene/isoprene(75/25)) [4]

Polymer Stability:
Thermal Stability General: Thermal stability has been reported [3]
Chemical Stability: Resistance to C_6 and oil has been reported [3]. Rubber fibres are oil-resistant [1]

Applications/Commercial Products:
Processing & Manufacturing Routes: Synth., by conventional emulsion polymerisation of mixture of three comonomers at 5° [4]. Also synth. by terpolymerisation in the presence of EtAlCl$_2$/acrylonitrile prereaction mixture and vanadium oxychloride, in hexane at @10° [1]. Vulcanised using a mixture of BaSO$_4$, sulfur, zinc oxide, benzothial disulfide, 2,5-di-*tert*-amylhydroquinone and stearic acid; compounding systems can also include carbon black [4]; vulcanisation is carried out at 145° for 35 min [1], or at 150° for 5 min [4]
Applications: Used in printing rolls, industrial rolls, diaphragms, rubber threads

Trade name	Details	Supplier
Nipol	DN-1201 DN-1201L	Nippon Zeon

Bibliographic References
[1] *Ger. Pat.*, 1973, 2261487, (synth., mech., props., morphology)
[2] *Jpn. Pat.*, 1974, 7413454, (mech. props., thermal stability)
[3] *Fr. Pat.*, 1974, 2195230, (mech., props., synth.)

Poly(1,3-butadiene-*co*-acrylonitrile), Mineral filled

Synonyms: *NBR, mineral filled. NR, mineral filled. Nitrile rubber, mineral filled*
Related Polymers: Nitrile rubber, Nitrile rubber, Isoprene copolymer, Nitrile rubber, Hydrogenated, Nitrile rubber, Ethylene copolymer, Nitrile rubber, Carbon black filled, Nitrile rubber, PVC blends, Nitrile rubber, Carboxylated
Monomers: 1,3-Butadiene, Acrylonitrile
Material class: Synthetic Elastomers, Copolymers, Composites
Polymer Type: acrylonitrile and copolymers, polybutadiene
CAS Number:

CAS Reg. No.
9003-18-3

Molecular Formula: $[(C_4H_6).(C_3H_3N)]_n$
Fragments: C_4H_6 C_3H_3N

Volumetric & Calorimetric Properties:
Density:

No.	Value	Note
1	d 1.81 g/cm^3	glass/fibreglass filled, 43.5% copolymer [16]

Volumetric Properties General: The thermal expansion coefficients of kaolin and chalk filled [2] and asbestos, flax and cotton filled [10,14,30] material have been reported
Thermodynamic Properties General: Thermal conductivity [17] of silica filled [3] material has been reported. Thermally diffusivity has been reported
Latent Heat Crystallization: Heat of combustion of kaolin, chalk, silica gel, magnesium hydroxide and aluminium hydroxide filled material has been reported [19]

Thermal Conductivity:

No.	Value	Note
1	0.628–0.657 W/mK	23–241°; virgin material; glass/fibreglass filled; 43.5% copolymer content [16]
2	0.615–0.649 W/mK	282–492°; 'mixture'; glass/fibreglass filled, 43.5% copolymer content [16]
3	0.582–1.887 W/mK	21–893°; 'char' material (after pyrolysis) glass/fibreglass filled, 43.5% copolymer content [16]

Specific Heat Capacity:

No.	Value	Note	Type
1	1.2–1.35 kJ/kg.C	0.288–0.323 cal $g^{-1\circ}$ C^{-1}, 60–728°; virgin material; glass and fibreglass filled, 43.5% copolymer content [12]	P
2	1.39–1.56 kJ/kg.C	0.333–0.374 cal $g^{-1\circ}$ C^{-1}, 322–728°; virgin material; glass and fibreglass-filled, 43.5% copolymer content [12]	P
3	0.81–1.36 kJ/kg.C	0.195–0.324 cal $g^{-1\circ}$ C^{-1}, 60–728°; 'char' material (after pyrolysis); glass and fibreglass filled, 43.5% co-polymer content [12]	P
4	1.18–1.32 kJ/kg.C	0.315–0.282 cal $g^{-1\circ}$ C^{-1}, 322–728°; 'mixture'; glass and fibreglass filled, 43.5% co-polymer content [12]	P

Transition Temperature General: T_g of kaolin and chalk [2], and MoS_2 and silica-filled material [20] has been reported

Surface Properties & Solubility:

Solubility Properties General: Equilibrium swelling studies of pyrogenic-silica filled material in trichloroethylene and C_6H_6 have been reported. Volume-fraction measurements show an increase in effective cross-links generated by filler; no significant differences observed between type of filler (surface area, surface modification), although type of copolymer (acrylonitrile content) has an effect [28]

Surface and Interfacial Properties General: Soln. absorption studies of nitrile rubber on silicas have been reported for trichloroethylene and C_6H_6. Soln. absorption behaviour is not a reliable guide to the reinforcement props. of the fillers [28,36]

Wettability/Surface Energy and Interfacial Tension: Satisfactory wettability of polymer molecules and hydrated silica (filler) particles indicated by permeability studies of toluene in filled nitrile rubber [21]; permeability decreases with increasing filler loadings [21]. Adhesive props. of silica, calcium carbonate and magnesium carbonate filled material have been reported [26]

Transport Properties:

Transport Properties General: Moisture-absorption characteristics have been investigated for chrysotile-asbestos-reinforced nitrile rubber (70% filler content): results obtained for as- produced samples (calendered sheets), cured using sulfur and peroxide, and sulfur-cured material which had been heat cycled; diffusivity data obtained 0.003–0.026 mm^2 h^{-1} (faces open), 0.12–1.61 mm^2 h^{-1} (long sides open), 0.21–1.64 mm^2 h^{-1} (short sides open) (all for relative humidities 97–33%). Diffusivity is essentially independent of relative humidity and is much higher through the edges of a sample than through the faces (related to fibrous nature). Sulfur cured samples not significantly influenced by heat treatment, and change of curing agent for as-calendered material does not significantly affect diffusivity. Studies important in connection with applications of rubber composites as gaskets, where moisture content has direct relevance to tensile strength and thermal props., such as thermal expansion and thermal conductivity [30].

Polymer Solutions Dilute: Viscosity of glass-filled material has been reported [9]

Permeability of Liquids: Permeation of toluene in nitrile rubber and effects of incorporation of hydrated silica as reinforcing filler have been investigated. Breakthrough time, i.e. elapsed time between initial contact at outside surface and time of detection of chemical at inside surface of material (ASTM F739-85) increases with increasing filler content up to limit of 50 phv, beyond which a plateau appears; such a decrease in permeability indicates good wettability between polymer molecules and filler particles [21]

Gas Permeability:

No.	Gas	Value	Note
1	H_2	288.9 cm^3 mm/ (m^2 day atm)	4.4×10^{-10} cm^2 $(s\ cmHg)^{-1}$, 24h [31]
2	H_2O	198.9 cm^3 mm/ (m^2 day atm)	3.03×10^{-10} cm^2 $(s\ cmHg)^{-1}$, 24h [31]

Mechanical Properties:

Mechanical Properties General: Effects of amount and type (grade) of silica filler on mechanical props. (tensile strength, 300% modulus, Shore hardness, compression set) of heat-resistant and carboxylated nitrile rubbers have been reported [7,23]. Measurements of tensile strength (elongation at break) as function of hydrated silica filler loadings have been reported; improvements observed with increasing filler content (0–60 phr); effects even greater with higher-acrylonitrile content copolymers (e.g. 40%) and carboxylated nitrile rubbers [21]. Prolonged exposure of nitrile rubbers containing various mineral fillers (clay, silica, TiO_2, MoS_2) to γ-irradiation leads to huge increases in tensile strength (around 1500%) and corresponding complete loss of elongation (-100% change) [1]. Effect of silane-based coupling agent on tensile strength of silica-filled nitrile rubber has been reported. Addition of small quantities of γ-mercaptopropyltrimethoxysilane (0.1–1.6 phr silica (5–80 phr)) results in significant improvements in tensile strength; correlations can be made with corresponding improvements in tear strength; comparative studies carried out with styrene-butadiene and natural rubber show improvements in tensile strength greatest for nitrile rubber system; some dependency on method of mixing displayed [34]. Other mechanical props. have been reported [22,24]. Elastic modulus of Kevlar-filled material has been reported [22]. Dynamic endurance of chalk and kaolin filled material has been reported [27]

Tensile Strength Break:

No.	Value	Note	Elongation
1	1.5–23 MPa	nitrile rubber, 30% acrylonitrile; 0–60 phr hydrated silica filler [21]	350–770%
2	8.2–20 MPa	10–35 phr pyrogenic silica filler [25]	660–680%
3	4–22 MPa	0–80 phr silica [34]	

Complex Moduli: Real and imaginary complex shear moduli of silica, kaolin and limestone filled material have been reported [11]

Viscoelastic Behaviour: Effects of silicate filler on dynamic mechanical props./viscoelastic behaviour of nitrile rubbers have been reported; influence of morphology of filled polymers also reported [7]. Stress-relaxation behaviour (in tension) for nitrile rubber vulcanisates containing short jute fibres (10–20 phr) has been reported; effects of strain level, bonding system, fibre orientation and content, temp. and prestraining on rate of stress relaxation have been considered; results indicate existence of relaxation mechanism at very short times (<200 s) [33]. Nature of relaxation process unresolved, although influence of bonding agent indicates involvement of fibre/rubber interface [33]; in general, incorporation of short fibres increases rate of stress relaxation (compared to unfilled systems); orientation has only marginal effect, while prestraining decreases rate considerably- 'stress softening' effect. Generally jute fibres in the presence of a bonding agent impart optimum stress relaxation behaviour to the composites [33]

Mechanical Properties Miscellaneous: Stiffness reinforcement (initial modulus) and stress-strain hysteresis (loading (100% extension)/unloading) measurements have been reported for (pyrogenic) silica-filled vulcanised (γ-irradiated) nitrile rubbers; little differences observed for the reinforcing effects of various fillers (e.g. surface area/surface modification), although acrylonitrile content plays important role in extent of filler reinforcements [28]. Temp. dependence of dynamic mechanical props. of nitrile rubbers (glass-bead filled) has been investigated. Theory developed to explain changes in relative modulus of such composite systems at or around T_g, i.e. jump in modulus near T_g, and subsequent decrease at $T > T_g$; magnitude of jump increases with filler concentration, as does temp. dependence above T_g; reduced (mechanical) damping also increases, at $T > T_g$, with increasing temp.; modulus and damping changes dependent on particle size of filler, with both tending to increase as particle size decreases; such props. dependent on nature of surface of filler particles [29]. Dynamic storage modulus of silica filled material has been reported [25]

Hardness: Shore A48–A60 (10–35 phr pyrogenic silica filler) [25]

Failure Properties General: Addition of small quantities of γ-mercaptopropyltrimethoxysilane (0.1–1.6 phr silica (5–80 phr)) results in enhancement of tear strength (5000–26000 Nm^{-2} (0–80 phr silica); correlations can be made with corresponding improvements in tensile strength. Tear energy measurements on (pyrogenic) silica-filled vulcanised (γ-irradiated) nitrile rubbers have been reported. Tear energy increases with increasing filler content; no significant differences observed with type of filler used; i.e. surface area, surface modification etc [28]. Temp. dependence of the tear strength of silk-fibre-reinforced nitrile rubber composites has been reported. Strength decreases with increasing temp. for both filled and unfilled samples, with filled samples displaying higher strengths at all temps.; decrease in tear strengths of composites at elevated temps. less than that of unfilled vulcanisates; retention of strength at higher temps. is highest for composites with longitudinally oriented rather than transversely orientated fibres [32]. Comparative studies carried out with styrene-butadiene and natural rubber show improvements in (tear) strengths greatest for nitrile rubber systems [34]

Fracture Mechanical Properties: Tensile fatigue resistance of asbestos, flax and cotton filled material has been reported [10]

Electrical Properties:

Electrical Properties General: Dielectric constant of silica-filled material shows anomalous dispersion with frequency. It decreases with increasing silica content. Dielectric loss decreases with increasing silica content in low-frequency region, while displaying little change in high-frequency region. Relaxation times, obtained from theoretical analysis of data, suggest no interactions between filler and rubber [35]. Resistivity of metal-filled material has been reported [6]

Dielectric Permittivity (Constant):

No.	Value	Frequency	Note
1	3–16	500 Hz–10 GHz	10–40 phr silica [35]

Magnetic Properties: Magnetic permeability at room temp. (0.1 MHz-10 GHz) has been reported. Elastic-magnetic materials, filled with powdered nickel/zinc or manganese/zinc ferrites, can be considered as porous magnetic materials in which pores are filled with rubber; polar nature of nitrile rubber gives material with somewhat greater initial magnetic permeability and greater losses in region of dispersion of μ_0) than is case for non-polar rubbers [5]

Optical Properties:

Volume Properties/Surface Properties: Colour brown (silica filled) [28]. Opacity increases with increasing filler content [28]

Polymer Stability:

Thermal Stability General: Oxidative thermal degradation od asbestos filled material has been reported [8]. Studies of the effects of silica fillers on the thermal stability of heat-resistant and carboxylated nitrile rubbers have been reported [23]

Decomposition Details: Heat of decomposition 552 J g^{-1} (132 cal g^{-1}) (glass/fibreglass-filled, 43.5% copolymer content) [12]

Flammability: Flammability, oxygen index, ignition temp. and heat flux intensity of kaolin, chalk, silica gel, magnesium hydroxide and aluminium hydroxide filled material have been reported [19]

Chemical Stability: Resistance to organic acids and EtOH of kaolin filled material has been reported [4]. Studies of the effects of silica fillers on fuel/oil resistance of heat-resistant and carboxylated nitrile rubbers have been reported [23]; chemical resistance of nitrile rubber improves on incorporation of hydrated silica filler [21]; nitrile rubbers with high acrylonitrile content (e.g. 40%) show even greater resistance, as do carboxylated nitrile rubbers [21]

Stability Miscellaneous: Prolonged exposure to γ-irradiation causes hardening and embrittlement in variously filled (clay, silica, TiO$_2$, MoS$_2$) material [1].

Applications/Commercial Products:

Trade name	Details	Supplier
MXBE-350	Glass powder (15.5%) and fibreglass, (41.0%) filled; copolymer (43.5%)	Fiberite Corp.

Bibliographic References

[1] Harrington, R., *Rubber Age (N.Y.)*, 1961, **90**, 265, (radiation resistance)
[2] Vovuodskaya, M.V. and Bartenev, G.M., *Kauch. Rezina*, 1964, **23**, 21, (thermal expansion coefficient)
[3] Bafrnec, M. and Lodes, A., *Chem. Prum.*, 1966, **16**, 171, (thermal conductivity)
[4] Nedyalkova, K., Taushanski, S., and Mareva, P., *Khim. Ind. Sofia*, 1970, **10**, 454, (chemical resistance)
[5] Alekseev, A.G., and Poltinnikov, S.A., *Izv. Akad. Nauk SSSR, Neorg. Mater.*, 1973, **9**, 455, (magnetic permeability)
[6] Zaisnchksuskii, A.D., Avalyan, V.D., Babadzhanova, S.A., Bukankov, E.I. and Kamenetskii, V.D., *Izu. Vyssh. Uchebn. Zaved., Tekhnol. Legk. Prom.-Sti.*, 1974, **4**, 57, (electrical resistance)
[7] Bartenev, G.M. and Zelenev, Y.V., *Plaste Kautsch.*, 1974, **21**, 805, (mech. props.)
[8] Degtyarev, E.V., Koldurovich, G.E. and Malyshev, T.B., *Kauch. Rezina*, 1980, **6**, 46, (friction coefficient, oxidative thermal degradation)
[9] Lutskii, M.S. and Fridman, I.D., *Kauch. Rezina*, 1978, **1**, 10, (viscosity)
[10] Zuev, Y.S., Karpovich, J.I. and Burkhina, M.F., *Kauch. Rezina*, 1978, **6**, 28, (tensile fatigue resistance, expansion coefficient)
[11] Modnov, S.I., Yanovskii, Y.G., Goncharov, G.M. and Bekin, N.G., *Kauch. Rezina*, 1980, **9**, 17, (complex shear modulus)
[12] Henderson, J.B., Wiebelt, J.A., Tant, M.R. and Moore, G.R., *Thermochim. Acta*, 1982, **57**, 161
[13] Alekseev, A.G., Aizikovich, B.V., Klisdt, M.F. and Kiyuts, N.F., *Kauch. Rezina*, 1982, **8**, 33, (magnetic susceptibility)
[14] Zleev, Y.S. and Bukhina, M.F., *Mekh. Kompoz. Mater. (Zinatne)*, 1983, **2**, 359, (expansion coefficient)
[15] Susteric, Z., *Vesti Slov. Kem. Dries*, 1983, **30**, 213, (tearing energy)
[16] Henderson, J.B., Verma, Y.P., Tant, M.R. and Moore, G.R., *Polym. Compos.*, 1983, **4**, 219, (thermal props.)
[17] Kisel, L.D., Kasparov, M.N., Darovskikh, G.T. and Butsevitskaya, D.V., *Deposited Doc.*, 1983, VENITI 78, (thermal props.)
[18] Clark, P.J. and Howard, G.J., *J. Polym. Sci., Polym. Chem. Ed.*, 1973, **11**, 2305, (soln. props., absorption)
[19] Ushkov, V.A., Lalayan, V.M., Abishev, A.K. and Morozova, N.Y., *Kauch. Rezina*, 1986, **3**, 8, (flammability, oxygen index, ignition temp)
[20] Liu, J., Zhou, H, Yan, D., Hou, C. and Yu, F., *Yingyong Huaxue*, 1987, **4**, 53, (transition temp., tensile strength)

[21] Ghouse, A.R., Szamosi, J. and Tobing, S.D., *Int. J. Polym. Mater.*, 1988, **12**, 93, (chemical resistance, toluene permeation)
[22] Ono, Y., Tanigaki, T., Yamaguchi, K., Tanino, K. and Hoshizume, G., *Kobunshi Ronbunshu*, 1989, **46**, 389, (mechanical props.)
[23] Byers, J.T., Hewitt, N.L. and Tultz, J.P., *Gummi Fasen. Kunstst.*, 1989, **42**, 263, (heat resistance, fuel resistance, mech. props.)
[24] Kim, K.J., Kim, J.S., An, B.K., Seu, S.K., et al, *Yongu Pogo-Kungnip Kongop Sihomwon*, 1989, **39**, 163, (mechanical props.)
[25] Pouchelon, A. and Vondracek, P., *Plast., Rubber Compos. Process. Appl.*, 1992, **172**, 109, (electrical conductivity, mechanical props.)
[26] Kiselev, V.Y. and Unukova, V.G., *Kauch. Rezina*, 1996, **3**, 25, (adhesion)
[27] Kim, A.B., Tzoy, C.D., Allayarov, E.S. and Dzhalilov, A.T., *Dokl. Akad. Nauk Resp. Uzb.*, 1995, **5–6**, 42, (dynamic endurance)
[28] Clarke, P.J. and Howard, G.J., *J. Polym. Sci., Polym. Chem. Ed.*, 1973, **11**, 2437, (tear energy, mech. props.)
[29] Lee, B.L. and Nielsen, L.E., *J. Polym. Sci., Polym. Phys. Ed.*, 1977, **15**, 683, (shear modulus)
[30] Loftu P. O'Donnell, J. Wostenholm, G.H., Yates, B. et al, *J. Mater. Sci. Lett.*, 1985, **20**, 1093, (transport props., water absorption)
[31] Kojima, Y., Fukumori, K., Usuki, S., Okada, A. and Kurauchi, T., *J. Mater. Sci. Lett.*, 1993, **12**, 889, (gas permeability)
[32] Setua, D.K., *Polym. Commun.*, 1984, **25**, 345, (tear strength)
[33] Bhagawan, S.S., Tripathy, D.K. and De, S.K., *J. Appl. Polym. Sci.*, 1987, **33**, 1623
[34] Nasir, M., Poh, B.T. and Ng, P.S., *Eur. Polym. J.*, 1988, **24**, 961, (tensile strength, tear strength)
[35] Hanna, F.F. and Abou-Bakr, A.F., *Br. Polym. J.*, 1973, **5**, 49, (dielectric constant, dielectric loss)

Poly(1,3-butadiene-*co*-acrylonitrile)/polyvinyl, Chloride blends   P-93

Synonyms: *Nitrile rubber/polyvinyl chloride blends. NBR/PVC blends. NR/PVC blends. PVC/NBR blends. Nitrile rubber/PVC blends*
Related Polymers: Nitrile rubber, Nitrile rubber, Isoprene copolymer, Nitrile rubber, Carboxylated, Nitrile rubber, Ethylene copolymer, Nitrile rubber, Carbon black filled, Nitrile rubber, Hydrogenated, Nitrile rubber, Mineral filled
Monomers: Vinyl chloride, Acrylonitrile, 1,3-Butadiene
Material class: Blends, Copolymers, Synthetic Elastomers
Polymer Type: PVC, acrylonitrile and copolymerspolybutadiene
Morphology: Morphological struct. depends on composition and nitrile content. PVC may be present in crystalline form [40]. Other morphological studies have been reported [51]

Volumetric & Calorimetric Properties:
Density:

No.	Value	Note
1	d 1.301 g/cm^3	70% PVC, 32% acrylonitrile [32]
2	d 1.26–1.301 g/cm^3	30% wt butadiene-acrylonitrile [52]

Volumetric Properties General: Density increases with increasing PVC content [53]
Thermodynamic Properties General: Relationship between C_p and temp. has been reported [26]. Coefficient of linear thermal expansion has been reported [53]
Glass-Transition Temperature:

No.	Value	Note
1	45°C	75% PVC, 34% acrylonitrile [61]
2	-4°C	25% PVC, 34% acrylonitrile [61]
3	41.1°C	75% PVC, 28% acrylonitrile [61]
4	-28°C	30% acrylonitrile, Hycar VT380 [44]
5	36°C	70% PVC, 32% acrylonitrile [52,56]

Transition Temperature General: T_m [21], T_g [17,48], deflection [14] and softening [14,28] temps., and brittleness temps. [13,34] have been reported
Transition Temperature:

No.	Value	Note	Type
1	-57°C	75% PVC, 31% acrylonitrile [56]	T_β

Surface Properties & Solubility:
Solubility Properties General: Blends of PVC with nitrile rubber containing 31% acrylonitrile are compatible wheras they are compatible when acrylonitrile content is 44% [56]
Cohesive Energy Density Solubility Parameters: Cohesive energy of nitrile rubber components has been reported [30]. δ 17.78 J$^{1/2}$ cm$^{-3/2}$; 8.69 cal$^{1/2}$ cm$^{-3/2}$, 130°, 75% PVC, 34% acrylonitrile [66]. Interaction parameter -4.6 J cm^{-3} (-1.1 cal cm^{-3}, 75% PVC, 34% acrylonitrile) [61]
Wettability/Surface Energy and Interfacial Tension: Surface energy has been reported [30]

Transport Properties:
Transport Properties General: Melt flow index has been reported [60]
Polymer Melts: Melt viscosity has been reported [9]
Permeability of Gases: Water vapour permeability has been reported [36]. Other gas permeability data have been reported [5,16]. Permeability of He, CO_2, O_2 and N_2 decreases with increasing acrylonitrile content [53]
Permeability of Liquids: Permeability of 1,1,2,2-tetrafluoroethane has been reported [36]. Increasing PVC content decreases permeability to ASTM fuel C and to ASTM fuel C/MeOH mixture. Permeability to the latter mixture may also be decreased by increasing acrylonitrile content [29]
Gas Permeability:

No.	Gas	Value	Note
1	[ethylene oxide]	929 cm^3 mm/(m^2 day atm)	1 mg mm in^{-2} min^{-1}, 2.2 mm thick [4]
2	[CO_2]	94559 cm^3 mm/(m^2 day atm)	14.4 × 10^{-8} cm^2] (s cmHg)$^{-1}$, ASTM D1434 [43]
3	[N_2]	1969.9 cm^3 mm/(m^2 day atm)	0.3 × 10^{-8} cm^2 (s cmHg)$^{-1}$, ASTM D1434 [43]
4	[O_2]	19699.8 cm^3 mm/(m^2 day atm)	3.0 × 10^{-8} cm^2 (s cmHg)$^{-1}$, ASTM D1434 [43]

Mechanical Properties:
Mechanical Properties General: Other values of tensile modulus [15], tensile strength [2,5,12,13,22,28,32,34], flexural strength [28] and dynamic elastic modulus [7] have been reported. Shear modulus decreases with increasing temp. [52] Young's modulus reaches a max. at nitrile rubber content of approx. 20% and thereafter decreases [42]. Tensile storage and loss moduli [35] and their variation with temp. [51] have been reported

Tensile (Young's) Modulus:

No.	Value	Note
1	851.3 MPa	75% PVC, 31% acrylonitrile [56]
2	14 MPa	25% PVC, 31% acrylonitrile [56]

Poly(1,3-butadiene-co-acrylonitrile)/polyvinyl, Chloride blends

Tensile Strength Break:

No.	Value	Note	Elongation
1	18–30 MPa	film, ASTM D882 [43]	250–500%
2	28.7 MPa	297 kg cm^{-2}, 10% PVC, hydrogenated [45]	460%
3	7.7 MPa	25% PVC, 31% acrylonitrile [56]	1470%
4	15.7 MPa	160 kg cm^{-2}, 62.5% PVC [16]	372%
5	26.9 MPa	3.9 kpsi, 70% PVC [50]	
6	24.7 MPa	3.59 kpsi, 75% PVC [54]	84%
7	31.1 MPa	75% PVC, 31% acrylonitrile [56]	230%
8	10.7–11.3 MPa	[47]	162–131%
9	18.3–23.7 MPa	hydrogenated [47]	211–240%
10	15.7 MPa	160 kg cm^{-2}, 62.5% PVC [16]	372%

Elastic Modulus:

No.	Value	Note
1	165 MPa	24 kpsi, 80% PVC [50]
2	122.7 MPa	17.8 kpsi, 70% PVC [50]

Miscellaneous Moduli:

No.	Value	Note	Type
1	10.6–11.1 MPa	[47]	100% modulus
2	9.4–10.9 MPa	hydrogenated [47]	100% modulus
3	1580 MPa	229 kpsi, 23°, 75% PVC [54]	torsional modulus
4	3.9 MPa	0.56 kpsi, 120°, 75% PVC [54]	torsional modulus

Impact Strength: Impact strength 20.7 kJ m^{-2} (90% PVC), 8.3 kJ m^{-2} (90% PVC, 25 months ageing) [24]. Other values of impact strength have been reported [2,8,28]. Charpy impact strength shows a maximum at approx. 20% acrylonitrile [62]
Viscoelastic Behaviour: Stress-strain curves [1, 21,50,56] (hydrogenated) [62] have been reported. Creep compliance is increased by a decrease in acrylonitrile content [55]. Stress relaxation data have been reported [31]
Hardness: 66 (JIS, 10% PVC, hydrogenated) [45] Shore D43-D44 [47], A79-A81 (hydrogenated) [47]. Microhardness decreases with increasing nitrile rubber content [42] Other values of hardness have been reported [13,27,28,34,60].
Failure Properties General: Tear resistance 69–73 kN m^{-1} [47], 70 kN m^{-1} (hydrogenated) [47]. Other values of tear strength have been reported [12,13,22,32].
Fracture Mechanical Properties: Rupture energy 105.1 J cm^{-3} (15250 in lb in^{-3}, 70% PVC) conversion (10020 in lb in^3, 80% PVC) [50]. Details of resistance to crack growth [13] at low temp. have been reported [20]. Hydrogenated material containing 10% PVC shows no cracks by bending at a 180° angle after 10 cycles of 72 h of immersion in peroxide containing gasoline and one week of drying in vacuum [45]
Friction Abrasion and Resistance: Coefficient of friction against steel and polyamide has been reported [11]. Abrasion resistance has been reported [13,27]. Internal friction increases with increasing temp. [52]

Electrical Properties:
Electrical Properties General: Dissipation factor has been reported [7,15,21]. The relationship between dielectric permittivity and frequency has been reported [64]. Dielectric loss data have been reported [31]. Volume resistivity has been reported [10,49]
Static Electrification: Other electrical conductivity data have been reported [17]. Activation energy of electrical conduction 0.61 eV (field intensity 100000 V m^{-1}. Activation energy increases above T_g, and with iodine doping [41]

Optical Properties:
Optical Properties General: Optical props. have been reported [54]
Transmission and Spectra: Nmr [19], uv [42], ir [33,40,41,46,47,56,59], H-1 nmr [37], C-13 nmr [44] and thermally stimulated depolarisation current spectral data have been reported [38]. Radiothermoluminesence glow curve has been reported [58]
Volume Properties/Surface Properties: Haze and transparency are dependent upon composition of blend [54]

Polymer Stability:
Polymer Stability General: Resistance to ageing has been reported [1]
Thermal Stability General: Heat resistance has been reported [2]. Thermal stability of hydrogenated polymer is reduced with increasing acrylonitrile content [63]. Resistance to low temps. has been reported [32]. Blends containing hydrogenated nitrile rubber have better resistance to hot air ageing at 100° for 7 days [47]
Decomposition Details: Thermal decomposition details have been reported [18]. Degradation temp. 440° (33.6% PVC) [32]. Thermal oxidative degradation details have been reported [18,33,34]
Flammability: Fire resistance [12] and flammability [25] have been reported
Environmental Stress: Resistance to weathering has been reported [3,6,27]. Resistance to photooxidative ageing [34] and ozone [12,13,32] has been reported. Uv irradiation causes degradation leading to loss of mech. props. [57]
Chemical Stability: Resistance to oil [2], grease [2], acids [2], alkalis [2], oxidising acids [2], aromatic solvents [2], petroleum solvents [3], oil (carboxylated) [27] and fuel [27,29,47] has been reported. Blends of PVC with nitrile rubber and hydrogenated nitrile rubber have similar resistance to ageing (45°, 7 days) with ASTM fuel C. Mech. props. are lost [29,47]
Hydrolytic Stability: Resistance to water [5] and sea water [39] has been reported
Stability Miscellaneous: Corrosion resistance has been reported [23]

Applications/Commercial Products:

Trade name	Details	Supplier
Breon		Aceto
Bur-a-Loy	various grades	Mach-1 Compounding
Chemigum		Goodyear
Europrene		Enichem
Humex	850/870	Hules Mexicanos
JSR		Japan Synthetic Rubber Co.
Krynac	NV 850/870	Polysar
Nipol	various grades	Zeon
Nitriflex	N-7400	A. Schulman Inc.
Nysyn	305V	Copolymer Rubber
PBZ-123/521	hydrogenated	Zeon
Paracril	020	Uniroyal Chemical
Perbunan	N/VC70	Mobay
Perbunan	N/VC70	Bayer Inc.
Rhenoblend	N range	Rhein Chemie
Zetpol		Zeon

Bibliographic References

[1] Pittenger, S.E., and Cohan, G.F., *Mod. Plast.*, 1947, **25**, 81, (viscoelastic props.)
[2] Guzetta, G., *Mater. Plast. (Bucharest)*, 1956, **22**, 203, (chemical resistance, mech. props.)
[3] Sharp, T.J., *Br. Plast.*, 1959, **32**, 431, (weather and solvent resistance)
[4] Dick, M. and Feazel, C.E., *Mod. Plast.*, 1960, **38**, 148, (liq. permeability)
[5] Rezhkov, Y.P. and Frost, A.M., *Lakokras. Mater. Ikh Primen.*, 1965, **4**, 19, (gas permeability, mech. props.)
[6] Ross, J.A., Sharp, T.J. and Pedley, K.A., *Proc. Rubber Technol. Conf.* 4th edn., 1962, 60, (weather resistance)
[7] Zelinger, J. Simunkova, E. and Heidingsfeld, V., *Sb. Vys. Sk. Chem. Technol. Praze. Org. Technol.*, 1966, **9**, 73, (mech. props., power factor)
[8] Zelinger, J. and Zitele, P., *Sb. Vys. Sk. Chem. Technol. Praze. Org. Technol.*, 1966, **9**, 87, (impact strength)
[9] Dimitrou, M., Marko, R. and Pazonyi, J., *Muanyag Gumi*, 1969, **6**, 384, (melt viscosity)
[10] Kevrolova, K.M. and Molodykh, N.E., *Tr. Tomsk Nauch. Issled. Inst. Kabel. Prom.*, 1969, **1**, 267, (volume resistivity)
[11] Luarsabishviti, D.G., *Tekh. Kino Televi.*, 1971, **15**, 21, (coefficient of friction)
[12] Devirts, E.Y. and Izmailova, L.V., *Kauch. Rezina*, 1971, **30**, 13, (ozone and fire resistance)
[13] Huh, D.S., Kim, J.S., Chang, S.B., Kim, S.B. and Choi, C.C., *Kungnip Kangop Yunguso Poge*, 1970, **20**, 99, (brittle point, mech. props.)
[14] Peneva, A. and Nikova, A., *God. Vissh Khimiko Tekhnol. Inst.*, 1969, **6**, 59, (transition temps.)
[15] Aivazov, A.B., Khindanov, M.A., Mindiyarov, K.G., Zelenev, Y.V. and Bartenev, G.M, *Primen Ul'traa Kust. Issled. Veshchestva*, 1971, **25**, 281, (Young's modulus, power factor)
[16] Palimina, I.V. Tikhomirova, T.P., Trepelkova, L.I., Koval, V.G. and Kulesh, L.V., *Plast. Massy*, 1973, 47, (tensile strength, gas permeability)
[17] Belyakova, L.K., Guzeev, V.V. and Malinskii, Y.M., *Plast. Massy*, 1973, 39, (transition temp., electrical conductivity)
[18] Cherenyuk, I.P., Zafranskii, Y.N., Semenova, T.V. and Sorokina, E.A., *Izu. Vyssh. Uchelon. Zaved. Khim.-Khim. Tekhnol.*, 1975, **18**, 122, (thermal degradation)
[19] Mindiyarov, K.G. and Zelenev, Y.V., *Sint. Fiz.-Khim. Polim.*, 1974, **13**, 123, (nmr)
[20] Schwartz, H.F. and edwards, W.S., *Adv. Polym. Sci.*, 1974, **25**, 243, (mech. props.)
[21] Odnolko, V.G. and Mihkin, E.V., *Tr. Mosk. Inst. Khim. Mashinostr.*, 1974, **51**, 152, (melting point, viscoelastic props.)
[22] Pickering, F.G., *Polym. Paint Colour J.*, 1977, **167**, 395, (mech. props.)
[23] Mamedov, S.M., Melikzade, M.M., Malin, V.P., and Mamedov, K.G., *Azerb. Khim. Zh.*, 1978, **5**, 78, (corrosion resistance)
[24] Simuhkova, E., Zelinger, J. and Binko, J., *Plasty Kauc.*, 1979, **16**, 5, (impact strength)
[25] Widenor, W.M., *NASA Tech. Memo.*, N79-12029,479, (flammability)
[26] Godovskii, Y.K. and Bessonova, N.P., *Vysokomol. Soedin., Ser. A*, 1979, **21**, 2293, (heat capacity)
[27] Schwartz, H.F., *Elastomerics*, 1980, **112**, 17, (environmental stress)
[28] Werblinski, W., *Polimery (Warsaw)*, 1985, **30**, 28, (Vicat temp., mech. props.)
[29] Dunn, J.R. and Vara, R.G., *Elastomerics*, 1986, **118**, 29, (fuel resistance, liq. permeability)
[30] Pritykin, L.M., *J. Colloid Interface Sci.*, 1986, **112**, 539, (surface energy, cohesion energy)
[31] Selenev, Y.V. and Aivasov, A.B., *Plaste Kautsch.*, 1984, **31**, 81, (dielectric loss, stress relaxation)
[32] Lysova, L.G, Sorokin, G.A, Sineva, E.V. and Gerasimova, M.F., *Kauch. Rezina*, 1987, **1**, 10, (degradation temp., stability, tear strength)
[33] Ivan, G. and Giurginca, M., *Ind. Usoura*, 1987, **34**, 490, (ir, thermooxidative degradation)
[34] Malac, J., Rektorikova, L., and Sumberova, J., *Plasty Kauc.*, 1989, **26**, 140, (brittleness temp., mech. props.)
[35] Deri, M. and Riczko, A., *Muanyag Gumi*, 1982, **19**, 173, (tensile modulus)
[36] Trexler, H.E., *Rubber Chem. Technol.*, 1983, **56**, 105, (liq. permeability)
[37] Fukumon, K., Sato, N. and Kurauchi, T., *Rubber Chem. Technol.*, 1991, **64**, 522, (H-1 nmr)
[38] Migahed, M.D., Ishra, M., El-khodary, A.and Fahmy, T., *Polym. Test.*, 1993, **12**, 335, (spectra)
[39] Prokopchuk, N.R., Alekseev, A.G., Starostina, T.V. and Barchenko, S.V., *Vesti Akad. Nauk Belanesi Ser. Khim. Nauk*, 1993, **2**, 80, (water resistance)
[40] Migahed, M.D., Ishra, M. and Fahmy, T., *J. Appl. Polym. Sci.: Appl. Polym. Symp.*, 1994, **55**, 207, (morphology, ir)
[41] Ishra, M., *J. Polym. Mater.*, 1993, **10**, 257, (ir, electrical conductivity)
[42] Bakr, N.R. and El-Kady, M., *Polym. Test.*, 1996, **15**, 281, (uv, mech. props.)
[43] *The Materials Selector*, 2nd edn. (eds. N.A. Waterman and M.F. Ashby), Chapman & Hall, 1997, **3**, 497, 665, (uses, gas permeability)
[44] Kwak, S.Y. and Nakajima, N., *Polymer*, 1996, **37**, 195, (C-13 nmr, transition temp.)
[45] *Jpn. Pat.*, 1982, 70135, (mech. props.)
[46] Manoj, N.R., De, S.K. and De, P.P., *Rubber Chem. Technol.*, 1993, **66**, 550, (ir)
[47] Manoj, N.R. and De, P.P., *Polym. Degrad. Stab.*, 1994, **44**, 43, (air and fuel resistance)
[48] Sotiropoulou, D.D., Avramidou, D.E. and Kalfaglou, N.K., *Polymer*, 1993, **34**, 2297, (transition temp.)
[49] Deanin, R.D., Skowronski, B.J. and Wixon, F.R., *Adv. Chem. Ser.*, 1965, **48**, 140, (volume resistivity)
[50] Lee, C.C., Rouatti, W., Skinner, S.M. and Bobalek, E.G, *J. Appl. Polym. Sci.*, 1965, **9**, 2047, (mech. props., viscoelastic props.)
[51] Matsuo, M., Nozaki, C. and Jyo, Y., *Polym. Eng. Sci.*, 1969, **9**, 197, (tensile modulus, morphology)
[52] Zakrzewski, G.A., *Polymer*, 1973, **14**, 347, (density, shear modulus)
[53] Shur, Y.J. and Ranby, B., *J. Appl. Polym. Sci.*, 1975, **19**, 2143, (gas permeability, expansion coefficient)
[54] Jordan, E.F., Artymyshyn, B., Riser, G.R. and Wrigley, A.N., *J. Appl. Polym. Sci.*, 1976, **20**, 2715, (optical props., mech. props.)
[55] Bergman, G., Bertilsson, H. and Shur, Y.J., *J. Appl. Polym. Sci.*, 1977, **21**, 2953, (creep)
[56] Wang, C.B. and Cooper, S.L., *J. Polym. Sci., Polym. Phys. Ed.*, 1983, **21**, 11, (transition temp., morphology)
[57] Skowronski, T., Rabek, J.F. and Ranby, B., *Polym. Eng. Sci.*, 1984, **24**, 278, (photodegradation, uv)
[58] Inoue, T., Hashimoto, T., Tanigami, T. and Miyasaki, K., *Polym. Commun.*, 1984, **25**, 148, (radiothermoluminescence)
[59] Xiaojiang, Z., Pu, H.H., Yanheng, Y. and Junfeng, L., *J. Polym. Sci., Part C: Polym. Lett.*, 1989, **27**, 223, (ir)
[60] Schwarz, H.F., Bley, J.W.F. and Hansmenn, J., *Kunststoffe*, 1987, **77**, 761, (melt flow index, hardness)
[61] Sen, A.K. and Mukherjee, G.S., *Polymer*, 1993, **34**, 2386, (solubility parameter, interaction parameter)
[62] Braun, D., Haufe, A., Leiss, D. and Hellmann, G.P., *Angew. Makromol. Chem.*, 1992, **202–203**, 143, (transitions, viscoelastic props.)
[63] Braun, D., Boehringer, B., Eidam, N, Fischer, M. and Koemmerling, S., *Angew. Makromol. Chem.*, 1994, **216**, 1, (thermal stability)
[64] Nottin, J.P., *J. Chim. Phys.*, 1966, **63**, 1091, (dielectric constant)

Polybutadiene, Carbon black filled P-94

Related Polymers: Polybutadiene
Monomers: 1,3-Butadiene
Material class: Composites
Polymer Type: polybutadiene
Additives: Carbon black
General Information: The incorporation of carbon black into the solid rubber by milling, or at the polymerisation stage, improves physical props. Rubbers with a higher degree of branching have a greater affinity for carbon black [3,8]. Pretreatment (e.g. heat treatment, polymer modification) of the carbon black can have a marked effect on the improvements gained in the filled polymer [5]
Morphology: Addition of carbon black to polybutadiene increases the amorph. portion of the polymer. The three characteristic structs. (two amorph. blocks and one pseudocrystalline) found in an SKD type polybutadiene are present in a filled rubber after the mixing process. There is an increase in both amorph. parts and a decrease in the pseudocrystalline part [4,7]
Miscellaneous: The adsorption props. of polybutadiene on carbon black have been studied using nmr and by swelling and weighing (before and after washing with a 'good' solvent) [9,11]

Surface Properties & Solubility:

Surface and Interfacial Properties General: Hess adhesion index for carbon black 8.4 MPa (86 kg cm^{-2}, untreated), 12.25 MPa (125 kg cm^{-2}, devolatilised), 5.5 MPa (56 kg cm^{-2}, oxidised), 1.86 MPa (19 kg cm^{-2}, graphitised) [5]

Mechanical Properties:

Mechanical Properties General: The relationship between deformation and temp. has been reported [7]. Rubbers filled with carbon black of a smaller particle size have higher moduli, ie. better reinforcement. Tensile stress-strain studies on various unvulcanised polybutadienes with carbon black fillers of different particle sizes have been reported [8]. Emulsion polymerisation in the presence of carbon black gives products which have significantly increased tensile strength and elongation. The contribution of entanglements to the mech. props. of the filled polymer has been reported. [2,3] The surface treatment of carbon black prior to its incorporation in the polymer can have a marked effect on the physical props. of the filled rubber. [5] A linear relationship between the storage modulus of filled and unfilled polymer at temps. above T_g has been reported [6]

Tensile Strength Break:

No.	Value	Note	Elongation
1	17–21 MPa	173–210 kg cm^{-2}, modified carbon black [5]	695–810%

Miscellaneous Moduli:

No.	Value	Note	Type
1	0.82–1.1 MPa	varying rubber type and carbon black particle size [8]	50% modulus
2	1.33–1.86 MPa	varying rubber type and carbon black particle size [8]	100% modulus
3	2.27–3.4 MPa	varying rubber type and carbon black particle size [8]	200% modulus
4	0.98–2.1 MPa	10–21 kg cm^{-2}, modified carbon black [5]	200% modulus

Hardness: Shore A46-A52 (modified carbon black) [5]

Electrical Properties:

Electrical Properties General: Resistivity decreases rapidly after approx. 45 phr carbon black added. The temp. dependence of resistivity depends very much on the loading. At low loadings the resistivity/temp. variation is similar to that of the unfilled rubber; it decreases exponentially with increasing temp. At higher loadings, however, resistivity increases to a maximum then decreases. The temp. at which the max. occurs depends on the loading and on the particle size of the filler (higher temp. for finer filler) [10]. Resistance $29 \times 10^{12} \Omega$ (1 kV, filler particle radius 26.5 nm, loading 50 phr), $67 \times 10^6 \Omega$ (1 kV, filler particle radius 14 nm, loading 50 phr), $50 \times 10^6 \Omega$ (1 kV, filler particle radius 11.3 nm, loading 50 phr), $12 \times 10^{12} \Omega$ (6 kV, filler particle radius 26.5 nm, loading 50 phr), $20 \times 10^6 \Omega$ (6 kV, filler particle radius 14 nm, loading 50 phr), $14 \times 10^6 \Omega$ (6 kV, filler particle radius 11.3 nm; loading 50 phr) [10]

Optical Properties:

Transmission and Spectra: Ir and nmr spectral data have been reported [1,4]

Applications/Commercial Products:

Processing & Manufacturing Routes: Carbon black may be introduced into the rubber by milling, although polymerisation in the presence of carbon black has been reported [3]. The behaviour on milling of various types of polybutadiene with carbon black filler has been reported [12]

Trade name	Supplier
Ameripol synpol E-BR	Ameripol Synpol
Taktene 1359	Polysar

Bibliographic References

[1] Lutz, E.T.G., Luinge, H.J. Van der Maas, J.H., and Van Agen, R., *Appl. Spectrosc.*, 1994, **48**, 1021, (ir)
[2] Heinrich, G. and Vilgis, T.A., *Macromolecules*, 1993, **26**, 1109, (mech. props.)
[3] Maron, S.H., Von Fischer, W., Ellslager, W.M. and Sarvadi, G., *J. Polym. Sci.*, 1956, **16**, 29, (synth.)
[4] Kvlividze, V.I., Klimanov, S.G. and Lezhnev, N.N., *Vysokomol. Soedin., Ser. A*, 1967, **9**, 1924, (nmr)
[5] Donnet, J.B. and Vidal, A., *Adv. Polym. Sci.*, 1986, **76**, 104, (surface props.)
[6] Kraus, G., *Adv. Polym. Sci.*, 1971, **8**, 156
[7] Jurkowska, B., Olkhov, Y.A., Jurkowski, B. and Olkhova, O.M., *J. Appl. Polym. Sci.*, 1999, **71**, 729
[8] Nakajima, N., and Yamaguchi, Y., *J. Appl. Polym. Sci.*, 1997, **66**, 1445, (mech. props.)
[9] Cohen Addad, J.P. and Frebourg, P., *Polymer*, 1996, **37**, 4235, (nmr)
[10] Cashell, E.M., Coey, J.M.D., Wardell, G.E. and McBriety, V.J., *J. Appl. Phys.*, 1981, **52**, 1542, (electrical conduction)
[11] Meissner, B. and Karasek, L., *Polymer*, 1998, **39**, 3083
[12] Nakajima, N., and Yamaguchi, Y., *J. Appl. Polym. Sci.*, 1997, **65**, 1995, (processing)

Polybutadiene/styrene-butadiene block copolymer blends

Synonyms: *Polyblend. Blends of polybutadiene and styrene-butadiene di-block copolymers. B/SB*
Related Polymers: Styrene-butadiene block copolymers, Polybutadiene
Monomers: 1,3-Butadiene, Styrene
Material class: Copolymers, Blends
Polymer Type: polybutadiene, polystyrene
General Information: These blends have been less well studied than the corresponding blends containing polystyrene
Morphology: Morphological studies of SBS triblock/polybutadiene blends have been reported [1,2,3,6]. The effects of blending homopolymer and block copolymers on the morphological struct. have been examined using electron microsopy [1,6] and stress relaxation and dynamic mech. techniques. [1,2,6] Dynamic mech. studies show the presence of a new transition near -40°, intermediate between the T_g values for styrene and butadiene blocks of SBS [1,2,3]. This was originally believed to result from some form of change in the microphase, resulting from entanglement coupling mechanisms [2]. Later work [1], using high and low MW polybutadiene, confirmed intermediate loss peak can be attributed to slippage of untrapped entanglements of polybutadiene chains for low MW homopolymer, whereas in the case of high MW material, transition results from crystallisation and melting of poly(*cis*-)butadiene chains [1]. Complementary electron microscopy investigations show that low MW homopolymer is solubilised into the butadiene domains of the SBS triblocks, whereas high MW polybutadiene is present in a separate phase [1]. Studies using electron microscopy on polybutadiene/(SB)$_x$ radial block copolymer blend systems reveals existence of a polybutadiene continuum containing discrete, spherical polystyrene domains [3,6]
Miscellaneous: Block copolymers are neither compatible nor incompatible with relevant homopolymers. On the molecular scale, they behave like mixtures of homopolymers, even though the mobility of sequences of the latter is restricted by chemical bonds. Each corresponding homopolymer blends into the phase of its own kind (which already exists in the block copolymer), thus altering the size, but not the nature, of this phase. On this basis, there can be no truly compatibilising effect of a block copolymer on a mixture with homopolymers. The effect of the block copolymer is of a colloidal nature, which results in finer and finer dispersions of one phase in the other. [5] Blends containing both polybutadiene and polystyrene with styrene-butadiene block

copolymers have been reported [8,9,10,11]. Addition of copolymers leads to 'compatibility' of otherwise 'incompatible' homopolymers. The emulsifying ability of copolymer leads to polyblends showing microphase rather than macrophase separation; this effect depends on factors such as MW and composition of copolymer, MW of homopolymer(s) and blend composition [1]

Volumetric & Calorimetric Properties:
Thermodynamic Properties General: C_p increases as function of temp. and homopolymer content; deviations found at higher temps. (above approx. 200°) [7]
Specific Heat Capacity:

No.	Value	Note	Type
1	2.67–2.85 kJ/kg.C	127–237°, 22 wt% polybutadiene [7]	P
2	2.695–2.88 kJ/kg.C	127–237° 33 wt% polybutadiene [7]	P

Transition Temperature General: Blending with SBS triblock copolymers results in changes in primary glass-transition temp. of polybutadiene component (phase). This leads to a shift to lower temp. upon incorporation of low MW homopolymer, as consequence of plasticisation by latter. The same transition in blends containing high MW material is split into a doublet on account of presence of homopolymer in separate phase [1,3,6]. T_g of polystyrene enriched domain increases whereas T_g of polybutadiene enriched domain decreases with increasing polybutadiene content [7]
Deflection Temperature:

No.	Value	Note
1	85°C	deflection of 0.254 mm [8,11] or 10 mil [9]
2	98°C	deflection of 1.52 mm [8,11] or 60 mil [9], polystyrene/polybutadiene/SBS (72.5/15/12.5) triblock; S/B in triblock, 9/11-12/8; M_n 225000–250000 [8,9,11]

Surface Properties & Solubility:
Solvents/Non-solvents: Sol. toluene [6], C_6H_6 [1,2]

Mechanical Properties:
Mechanical Properties General: Storage modulus and loss modulus have been reported [1,2,4,6]. Addition of diblock [10] and triblock [8,9,11] copolymers improves compatibility and leads to systems with useful props., e.g. high-impact strength
Flexural Modulus:

No.	Value	Note
1	1765 MPa	256 kpsi, polystyrene/polybutadiene/SB diblock (75/12.5/12.5); S/B in diblock, 2/3-3/2 [10]
2	1916 MPa	278 kpsi, polystyrene/polybutadiene/SBS tri block (72.5/15/12.5); S/B in triblock, 9/11-12/8, M_n 225000–250000 [9]
3	1912 MPa	19500 kg cm^{-2}, polystyrene/polybutadiene/SBS triblock (72.5/15/12.5); S/B in triblock, 9/11-12/8, M_n 225000–250000 [8,11]

Viscoelastic Behaviour: A secondary loss maximum (transition) located between T_gs of polybutadiene and polystyrene blocks is ascribed to motions of untrapped entanglements [3,6]
Fracture Mechanical Properties: Craze plasticity has been reported [12]
Izod Notch:

No.	Value	Notch	Note
1	496 J/m	Y	9.3 ft lb in^{-1} PS/PBD/diblock (75/12.5/12.5); S/B in diblock, 2/3–3/2 [10]
2	402 J/m	Y	7.54 ft lb in^{-1}) [9]

Applications/Commercial Products:
Processing & Manufacturing Routes: Prepared by soln. blending in C_6H_6 or toluene [1,2,6]. Films can be obtained by spin casting [1,2] or pressing of homogeneous blends (after evaporation to dryness) between Teflon-coated plateus [6]. Various methods for preparation and processing of 3-component systems (styrene and butadiene homopolymer plus diblock [10] or triblock copolymers) [8,9,11] have been reported [8,9,10,11]
Applications: Three component polymer blends (polystyrene/polybutadiene/diblock or triblock copolymers) have potential applications as high-impact polymeric materials

Bibliographic References
[1] *Macromolecules*, 1991, **24**, 5324
[2] *Macromolecules*, 1990, **23**, 4305
[3] *Macromolecules*, 1984, **17**, 342
[4] Toy, L., Niinomi, M. and Shen, M., *J. Macromol. Sci., Phys.*, 1975, **11**, 281, (morphology, transport props., viscoelasticity, synth., thermal props.)
[5] Choi, G., Kaya, A. and Shen, M., *Polym. Eng. Sci.*, 1973, **13**, 23, (morphology, transport props., viscoelasticity)
[6] Polym. Blends Vol. 2, (eds., D.R. Paul and S. Newman), Academic, 1978, 243, (morphology, transport props., mech. props., processing, compounding, uses)
[7] Watanake, H. and Kotaka, T., *Macromolecules*, 1983, **16**, 769
[8] Molau, G.E. and Wittbrodt, W.M., *Macromolecules*, 1968, **1**, 260, (compatibility, morphology, microstruct.)
[9] Kraus, G. and Rollmann, K.W., *J. Polym. Sci., Polym. Phys. Ed.*, 1977, **15**, 385, (transport props., viscoelasticity, microstruct., electron microscopy, thermal props.)
[10] Wang, Y.Z., Hseieh, K.H., Chen, L.W. and Tseng, H.C., *J. Appl. Polym. Sci.*, 1993, **49**, 1047, (volumetric props., thermal props.)
[11] *Ger. Pat.*, 1975, 2 342 219, (synth., mech. props.)
[12] *U.S. Pat.*, 1975, 3 906 057, (synth., mech. props.)
[13] *U.S. Pat.*, 1975, 3 907 929, (synth., mech. props.)
[14] *Ger. Pat.*, 1976, 2 432 372
[15] Gebizlioglu, O.S., Argou, A.S. and Cohen, R.E., *Polymer*, 1985, **26**, 529

Poly[1,1-butane bis(4-phenyl)carbonate]   P-96

Synonyms: *Poly[oxycarbonyloxy-1,4-phenylenebutylidene-1,4-phenylene]. Poly(4,4'-dihydroxydiphenyl-1,1-butane-co-carbonic dichloride)*
Monomers: 4,4'-Dihydroxydiphenyl-1,1-butane, Carbonic dichloride
Material class: Thermoplastic
Polymer Type: polycarbonates (miscellaneous)

– Poly[2,2-butane bis(4-phenyl)carbonate]

CAS Number:

CAS Reg. No.
31670-30-1

Molecular Formula: $(C_{17}H_{16}O_3)_n$
Fragments: $C_{17}H_{16}O_3$

Volumetric & Calorimetric Properties:
Density:

No.	Value	Note
1	d 1.17 g/cm^3	25° [1]
2	d 1.17 g/cm^3	20°, amorph. [2]

Melting Temperature:

No.	Value	Note
1	150–170°C	[1]
2	158–170°C	amorph. [2]

Glass-Transition Temperature:

No.	Value	Note
1	123°C	[1]

Mechanical Properties:
Mechanical Properties General: The general shape of the tensile stress-strain curve consists of a viscoelastic region up to the yield point, followed by a stress drop and then necking behaviour (at approx. constant stress); finally the stress increases to the break point [3]

Tensile (Young's) Modulus:

No.	Value	Note
1	1470 MPa	15000 kg cm^{-2}, amorph. [3]

Tensile Strength Break:

No.	Value	Note	Elongation
1	59 MPa	6 kg mm^{-2} [3]	140%

Tensile Strength Yield:

No.	Value	Note
1	44.1 MPa	4.5 kg mm^{-2} [3]

Electrical Properties:
Electrical Properties General: Dielectric constant and dissipation factor do not vary with temp. until approaching T_g
Dielectric Permittivity (Constant):

No.	Value	Frequency	Note
1	3.3	1 kHz	25–100° [1]

Dissipation (Power) Factor:

No.	Value	Frequency	Note
1	0.00052	1 kHz	25–100° [1]

Optical Properties:
Refractive Index:

No.	Value	Note
1	1.5792	25° [1]

Applications/Commercial Products:
Processing & Manufacturing Routes: The main synthetic routes are phosgenation and transesterification. The most widely used phosgenation is interfacial polycondensation in a suitable solvent; soln. polymerisation is also carried out in the presence of pyridine [4]. A slight excess of phosgene is introduced to a soln. or suspension of the aromatic dihydroxy compound in aq. NaOH at 20–30°, in an inert solvent (e.g. MeCl). Catalysts are used to achieve high MWs [1,4]. In transesterification the dihydroxy compound and a slight excess of diphenyl carbonate are mixed in the melt (150–300°) in the absence of O_2, with elimination of phenol. Reaction rates are increased by alkaline catalysts and the use of a medium-high vacuum in the final stages [1,4]

Bibliographic References
[1] Schnell, H., *Ind. Eng. Chem.*, 1959, **51**, 157, (props., synth.)
[2] Kozlov, P.V. and Perepelkin A.N., *Polym. Sci. USSR (Engl. Transl.)*, 1967, **9**, 414, (density, T_m)
[3] Perepelkin, A.N. and Kozlov, P.V., *Polym. Sci. USSR (Engl. Transl.)*, 1968, **10**, 14, (tensile props.)
[4] Schnell, H., *Polym. Rev.*, 1964, **9**, 99, (rev, synth.)

Poly[2,2-butane bis(4-phenyl)carbonate]   P-97

Synonyms: *Poly[oxycarbonyloxy-1,4-phenylene(1-methylpropylidene)-1,4-phenylene]. Poly(4,4'-dihydroxydiphenyl-2,2-butane-co-carbonic dichloride). Bisphenol B polycarbonate*
Monomers: Bisphenol B, Carbonic dichloride
Material class: Thermoplastic
Polymer Type: polycarbonates (miscellaneous)
CAS Number:

CAS Reg. No.
28983-04-2

Molecular Formula: $(C_{17}H_{16}O_3)_n$
Fragments: $C_{17}H_{16}O_3$

Volumetric & Calorimetric Properties:

Density:

No.	Value	Note
1	d 1.18 g/cm^3	25° [1]
2	d 1.18 g/cm^3	20°, amorph. [2]

Melting Temperature:

No.	Value	Note
1	205–222°C	[1]
2	203–218°C	amorph. [2]

Glass-Transition Temperature:

No.	Value	Note
1	134°C	[1,3]
2	132°C	[4]

Transport Properties:

Water Absorption:

No.	Value	Note
1	0.2%	25°, 65% relative humidity [6]

Mechanical Properties:

Mechanical Properties General: The general shape of the tensile stress-strain curve consists of a viscoelastic region up to the yield point, followed by a stress drop and then necking behaviour (at approx. constant stress); finally the stress increases to the break point [5]

Tensile (Young's) Modulus:

No.	Value	Note
1	1570 MPa	16000 kg cm^{-2}, amorph. [5]

Tensile Strength Break:

No.	Value	Note	Elongation
1	71.1 MPa	725 kg cm^{-2} [1]	70%
2	63 MPa	6.4 kg mm^{-2} [5]	103%

Tensile Strength Yield:

No.	Value	Note
1	48.1 MPa	4.9 kg mm^{-2} [5]

Electrical Properties:

Electrical Properties General: Dielectric constant and dissipation factor do not vary with temp. until approaching T_g

Dielectric Permittivity (Constant):

No.	Value	Frequency	Note
1	3.1	1 kHz	25–100° [1]

Dissipation (Power) Factor:

No.	Value	Frequency	Note
1	0.00045	1 kHz	25–100° [1]

Optical Properties:

Refractive Index:

No.	Value	Note
1	1.5827	25° [1]

Transmission and Spectra: Ir spectral data have been reported [6]

Applications/Commercial Products:

Processing & Manufacturing Routes: The main synthetic routes are, phosgenation and transesterification. The most widely used phosgenation approach is interfacial polycondensation in a suitable solvent; soln. polymerisation is also carried out in the presence of pyridine [6]. A slight excess of phosgene is introduced into a soln. or suspension of the aromatic dihydroxy compound in aq. NaOH at 20–30°, in an inert solvent (e.g. MeCl). Catalysts are used to achieve high MWs [1,6]. With transesterification, the dihydroxy compound and a slight excess of diphenyl carbonate are mixed together in the melt (150–300°) in the absence of O$_2$, with elimination of phenol. Reaction rates are increased by alkaline catalysts and the use of a medium-high vacuum in the final stages [1,6]

Bibliographic References

[1] Schnell, H., *Ind. Eng. Chem.*, 1959, **51**, 157, (props, synth)
[2] Kozlov, P.V. and Perepelkin, A.N., *Polym. Sci. USSR (Engl. Transl.)*, 1967, **9**, 414, (density, T_m)
[3] Serini, V., Freitag, D. and Vernaleken, H., *Angew. Makromol. Chem.*, 1976, **55**, 175, (T_g)
[4] Perepelkin, A.N. and Kozlov, P.V., *Polym. Sci. USSR (Engl. Transl.)*, 1966, **8**, 57, (T_g)
[5] Perepelkin, A.N. and Kozlov, P.V., *Polym. Sci. USSR (Engl. Transl.)*, 1968, **10**, 14, (tensile props)
[6] Schnell, H., *Polym. Rev.*, 1964, **9**, 99, (ir, water absorption, synth, rev)

Poly(*trans*-2-butene oxide)

Synonyms: *Poly[oxy(1,2-dimethyl-1,2-ethanediyl)]. Poly(trans-2,3-dimethyloxirane). Poly(trans-2,3-epoxybutane). Poly(trans-epoxybutane). Poly[oxy(1,2-dimethylethylene)]*
Monomers: 2,3-Dimethyloxirane

– Poly(3-butoxypropylene oxide)

Material class: Thermoplastic
Polymer Type: polyalkylene ether
CAS Number:

CAS Reg. No.	Note
26895-26-1	homopolymer
39923-73-4	meso, diisotactic
29796-23-4	threo, diisotactic

Molecular Formula: $(C_4H_8O)_n$
Fragments: C_4H_8O
Molecular Weight: M_w/M_n 2.1–3.4 [1]
General Information: Highly crystalline polymer [1]

Volumetric & Calorimetric Properties:
Density:

No.	Value	Note
1	d^{25} 1.016 g/cm^3	[2]

Melting Temperature:

No.	Value	Note
1	90–100°C	[2]
2	89–105°C	[2]

Surface Properties & Solubility:
Solvents/Non-solvents: Sol. CHCl$_3$ [2]. Insol. heptane, C$_6$H$_6$

Mechanical Properties:
Tensile (Young's) Modulus:

No.	Value	Note
1	0.00033 MPa	0.048 psi [2]

Tensile Strength Break:

No.	Value	Note
1	2.2×10^{-5} MPa	0.0033 psi [2]

Tensile Strength Yield:

No.	Value	Elongation	Note
1	13.79 MPa	14%	2000 psi [2]

Electrical Properties:
Dielectric Permittivity (Constant):

No.	Value	Frequency	Note
1	3.5	1 kHz	[3]

Polymer Stability:
Thermal Stability General: Exhibits low heat resistance [3]

Bibliographic References
[1] Iijima, T., Sashida, N. and Kakiuchi, . H., *J. Polym. Sci., Part A: Polym. Chem.*, 1989, **27**, 3651, (general)
[2] Vandenberg, E.J., *J. Polym. Sci., Part A-1*, 1969, **7**, 525, (density, transition temp, mech props)
[3] Panova, L.M., Cherkanov, S.P., Filatov, I.S., Sazhin, B.I. and Balaev, G.A., *Plast. Massy*, 1974, **4**, 50, (elec props)

Poly(3-butoxypropylene oxide)   P-99

Synonyms: *Poly[oxy[(butoxymethyl)-1,2-ethanediyl]]. Poly[(butoxymethyl)oxirane]. Poly(butylglycidyl ether). Poly(3-butoxy-1,2-epoxypropane). Poly(1-butoxy-2,3-epoxypropane)*
Monomers: (Butoxymethyl)oxirane
Material class: Thermoplastic
Polymer Type: polyalkylene ether
CAS Number:

CAS Reg. No.	Note
133686-41-6	
25610-58-6	homopolymer

Molecular Formula: $(C_7H_{14}O_2)_n$
Fragments: $C_7H_{14}O_2$
Molecular Weight: M_n 5390 (500-5390); M_w/M_n 1.15 (1.10–1.30)

Volumetric & Calorimetric Properties:
Density:

No.	Value	Note
1	d 0.982 g/cm^3	[2]

Melting Temperature:

No.	Value	Note
1	27°C	[2]

Glass-Transition Temperature:

No.	Value	Note
1	-79°C	[2]

Optical Properties:
Refractive Index:

No.	Value	Note
1	1.458	30° [1]

Bibliographic References
[1] Lal, J. and Trick, G.S., *J. Polym. Sci., Part A-1*, 1970, **8**, 2339, (transition temps, density)

Poly(butyl acrylate) P-100

Synonyms: *Poly[1-(2-butoxycarbonyl)ethylene]. Poly(butyl 2-propenoate)*
Monomers: Butyl acrylate
Material class: Natural elastomers
Polymer Type: acrylic
CAS Number:

CAS Reg. No.
9003-49-0
141-32-2

Molecular Formula: $(C_7H_{12}O_2)_n$
Fragments: $C_7H_{12}O_2$
Molecular Weight: The MW can range from 1000–5000000
Additives: Soap-urea, activated thiol, soap-sulfur, lead thiourea, thiourea, trithiocyanuric acid, diamine and carbon black, sodium stearate, ammonium benzoate, ammonium adipate. Antioxidants include Agelite, Stalite S, sodium alkyl sulfates and phosphates
General Information: A tacky, rubbery polymer. Its tackiness decreases with increase in MW. It is a constituent of butyl rubber and a useful impact modifier for polyolefins. It is usually used as copolymer with olefins and vinyl monomers to produce useful commercial products

Volumetric & Calorimetric Properties:
Density:

No.	Value
1	d 1.08 g/cm^3

Thermal Expansion Coefficient:

No.	Value	Note	Type
1	0.00026 K^{-1}	glass	V
2	0.0006 K^{-1}	liq.	V

Specific Heat Capacity:

No.	Value	Note	Type
1	1.64 kJ/kg.C	210 J mol^{-1} K^{-1}, 25°, solid	P
2	1.796 kJ/kg.C	230 J mol^{-1} K^{-1}, 25°, liq.	P

Melting Temperature:

No.	Value
1	47°C

Glass-Transition Temperature:

No.	Value
1	-43°C

Brittleness Temperature:

No.	Value
1	-44°C

Surface Properties & Solubility:
Cohesive Energy Density Solubility Parameters: δ 18 J$^{1/2}$ cm$^{-3/2}$. CED 38.7–41.3 kJ mol^{-1}
Plasticisers: Thiokol TP-759, Plastolein 9720, Admex 760
Solvents/Non-solvents: Sol. aromatic hydrocarbons, chlorinated hydrocarbons, THF, esters, ketones, glycolic esters, phosphorus trichloride, butanol, turpentine. Insol. MeOH, EtOH, cyclohexyl acetate, EtOAc
Surface Tension:

No.	Value
1	28 mN/m

Transport Properties:
Permeability of Gases: Medium permeability to gases
Water Absorption:

No.	Value	Note
1	0.3 %	24h

Mechanical Properties:
Failure Properties General: Fair to good tear resistance
Friction Abrasion and Resistance: Has good abrasion resistance

Electrical Properties:
Electrical Properties General: Has reasonable insulation props.

Optical Properties:
Refractive Index:

No.	Value	Note
1	1.474	25°

Polymer Stability:
Polymer Stability General: Degrades only slowly under extreme conditions. Unlike polymethacrylates it does not depolymerise on heating to high temps.
Thermal Stability General: Stable up to approx. 300°. Good heat ageing props. Decomposes only under extreme heat
Decomposition Details: Decomposition temp. 300–500°. Decomposition products include butene, butanol and CO_2
Flammability: Highly flammable
Environmental Stress: Unaffected by oxygen, ozone and uv light. Stability to uv is improved by addition of uv absorbents
Chemical Stability: Heating in extreme oxygen conditions results in hydroperoxides formed from free radicals
Hydrolytic Stability: Very resistant to hydrol. by acids or alkalis

Applications/Commercial Products:
Processing & Manufacturing Routes: Emulsion polymerisation is the preferred industrial method for the preparation of acrylic polymers. It is usually rapid and gives high MW polymer in a system of low viscosity. The safety hazard and expense of flammable solvents are eliminated. A study of the mechanism of emulsion polymerisation has been reported. [1] Soln. polymerisation is applied to produce sol. polymer, used as prod. The viscosity of the polymer prepared under a constant set of conditions varies with the nature of the solvent due to varying degrees of chain transfer activity in different solvents. Bulk and suspension polymerisation is also used resulting in cross-linked polymer. Stearic acid, TE-80 and Vanfone AP2 are added as processing aids
Applications: Used in architectural paints; coatings and lacquers; radiation-curable systems; applications in the paper, leather and textile industries; adhesives and sealing compounds

Trade name	Supplier
Acronal	BASF
Crilat	Montedison
Luhydran	BASF
Mowilith	Hoechst

– Polybutylene succinate

Plextaol	Rohm and Haas
Primal	Rohm and Haas
Propiofan	BASF
Revacryl	Revertex
Rhodopas	Rhone-Poulenc
Rhoplex	Rohm and Haas
Synthacryl	Synthopol
Uceryl	UCB Films
Uramul	DSM
Vinacryl	Vinyl Products
Vinamul	Vinyl Products
Vinnapas	Wacker
Walpol	Reichhold Chemie

[1] Zirkzee, H.F. et al, *Acta Polym.*, 1996, **47**, 441

Polybutylene succinate   P-101

Synonyms: *Polytetramethylene succinate*
Monomers: Succinic acid, Butane-1,4-diol
Material class: Thermoplastic
Polymer Type: saturated polyester
Molecular Formula: $(C_8H_{12}O_4)_n$
Fragments: $C_8H_{12}O_4$
Additives: The use of bamboo husk as reinforcing agent has been reported [4].
General Information: Biodegradable polymer.
Morphology: Crystalline parameters reported for monofilaments [2].

Volumetric & Calorimetric Properties:
Density:

No.	Value	Note
1	d 1.26 g/cm^3	[3]

Thermodynamic Properties General: Heat of combustion: 5,640 cal g^{-1}, Bionolle #1000 series [3].
Melting Temperature:

No.	Value	Note
1	114–116°C	Bionolle #1000 series [1,3]

Glass-Transition Temperature:

No.	Value	Note
1	-32°C	[1]

Heat Distortion Temperature:

No.	Value	Note
1	97 °C	Bionolle #1000 series [3]

Transition Temperature:

No.	Value	Note
1	75–88°C	Crystallisation temp. T_c, Bionolle #1000 series [3]

Transport Properties:
Melt Flow Index:

No.	Value	Note
1	1.5–25 g/10 min	190°C, 2.16 kg load, Bionolle #1000 series [3]

Mechanical Properties:
Mechanical Properties General: Yarn properties of monofilament fibres reported [2]. Tear strength: 3.6N mm^{-1} (MD, blown film, Bionolle #1000 series) [3]. 11N mm^{-1} (TD, blown film, Bionolle #1000 series) [3].

Tensile (Young's) Modulus:

No.	Value	Note
1	15–25 MPa	Monofilament [1]

Flexural Modulus:

No.	Value	Note
1	657–686 MPa	at 6700–7000 kgf cm^{-2}, Bionolle #1000 series [3]

Tensile Strength Break:

No.	Value	Note	Elongation
1	5.5–6.5 MPa	Monofilament [1]	
2	4.5–5.5 MPa	Multifilament [1]	
3	20.6–56.9 MPa	at 210–580kgf cm^{-2}, Bionolle #1000 series [3]	50–700%

Tensile Strength Yield:

No.	Value	Note
1	32.4–39.2 MPa	at 330–400kgf cm^{-2}, Bionolle #1000 series [3]

Polymer Stability:
Biological Stability: Biodegradable via enzymatic hydrolysis; degradation (in monofilaments) found to proceed preferentially in amorphous regions, details including weight and tensile strength loss on soil burial reported [2]. Biodegradation details reported [3].

Applications/Commercial Products:
Processing & Manufacturing Routes: Good spinnability.
Applications: Mulch film, packaging film, bags, 'flushable' hygiene products, agricultural films, traffic cones, synthetic paper, household goods and industrial trays.

Trade name	Details	Supplier
Bionolle	#1000 series	Showa Highpolymer
SkyGreen BDP		SK Polymers

Bibliographic References

[1] Biopolymers: Polyesters III Wiley-VCH, (Eds. Doi, Y. and Steinbuchel. A.), 2002, **4**, 7
[2] Biopolymers: Polyesters III Wiley-VCH, (Eds. Doi, Y. and Steinbuchel. A.), 2002, **4**, 14
[3] Biopolymers: Polyesters III Wiley-VCH, (Eds. Doi, Y. and Steinbuchel. A.), 2002, **4**, 280
[4] Shih, Y.-F., Lee, W.-C., Jeng, R.-J. and Huang, C.-M., *J. Appl. Polym. Sci.*, 2005, **99**, 188
[5] Lindstrom, A., Albertsson, A.C. and Hakkarainen, M., *Polym. Degrad. Stab.*, 2004, **83**, 487

Polybutylene succinate adipate   P-102

Synonyms: *Polytetramethylene succinate adipate*
Monomers: Butane-1,4-diol, Succinic acid, Hexanedioic acid
Material class: Thermoplastic
Polymer Type: saturated polyester

Volumetric & Calorimetric Properties:
Density:

No.	Value	Note
1	d 1.23 g/cm^3	Bionolle #3000 series [1]

Thermodynamic Properties General: Heat of combustion: 5720 cal g^{-1}, Bionolle #3000 series [1].
Melting Temperature:

No.	Value	Note
1	93–95°C	Bionolle #3000 series [1,2]

Glass-Transition Temperature:

No.	Value	Note
1	-45--41°C	Bionolle #3000 series [1,2]

Heat Distortion Temperature:

No.	Value	Note
1	69 °C	Bionolle #3000 series [1]

Transition Temperature:

No.	Value	Note
1	50–54°C	Crystallisation temp. T_c, Bionolle #3000 series [1,2]

Transport Properties:
Transport Properties General: Data on MFI for Bionolle #3000 series fibres over a range of temperatures, and after different annealing times, has been reported [2].
Melt Flow Index:

No.	Value	Note
1	1.4–25 g/10 min	190°C, 2.16 kg load, Bionolle #3000 series [1]

Permeability and Diffusion General: The measurement of water vapour transmission through films has been reported [3].

Mechanical Properties:
Mechanical Properties General: Tear strength: 4.4N mm^{-1} (MD, blown film, Bionolle #3000 series) [1]; 23N mm^{-1} (TD, blown film, Bionolle #3000 series) [1]. Data on mechanical properties after biodegradation over time at 35°C and 55°C reported [2].
Flexural Modulus:

No.	Value	Note
1	323.6–343.2 MPa	at 3300–3500 kgf cm^{-2}. Bionolle #3000 series [1]

Tensile Strength Break:

No.	Value	Note	Elongation
1	34.3–47.1 MPa	at 350–480 kgf cm^{-2}. Bionolle #3000 series [1]	400–900

Tensile Strength Yield:

No.	Value	Note
1	18.6–19.1 MPa	at 190–195 kgf cm^{-2}. Bionolle #3000 series [1]

Polymer Stability:
Biological Stability: Biodegradation details reported [1,5]. Some data on mechanical property changes with biodegradation reported for Bionolle #3000 series fibres [2].

Applications/Commercial Products:
Processing & Manufacturing Routes:
Applications: Biodegradable bags (compost, refuse, packaging), rope, tape, packaging film.

Trade name	Details	Supplier
Bionolle	#3000 series	Showa Highpolymer
EnPol 4000		Ire Chemical

Bibliographic References

[1] Biopolymers: Polyesters III Wiley-VCH, (Eds. Doi, Y., Steinbuchel. A.), 2002, **4**, 280
[2] Twarowska-Schmidt, K., *Fibres and Textiles in Eastern Europe*, 2004, **12**, 46
[3] Hu, Y., Topolkaraev, V., Hiltner, A. and Baer, E., *J. Appl. Polym. Sci.*, 2001, **81**, 1624

Polybutylene terephthalate   P-103

Synonyms: *PBT. Poly(tetramethylene terephthalate). Poly(1,4-benzenedicarboxylic acid-co-1,4-butanediol). Poly(oxy-1,4-butanediyloxycarbonyl-1,4-phenylenecarbonyl). Poly(oxytetramethyleneoxyterephthaloyl). PTMT. 4GT*
Related Polymers: PBT - High impact-impact modified, PBT - Flame retardant, PBT phenoxy blends, PBT fibres, PBT films/sheets, PBT mineral filled, PBT carbon fibre filled, PBT Polyethylene terephthalate blends, PBT phenoxy blends
Monomers: 1,4-Benzenedicarboxylic acid, 1,4-Butanediol
Material class: Thermoplastic
Polymer Type: PBT and copolymers, polyester
CAS Number:

CAS Reg. No.	Note
26062-94-2	
24968-12-5	
30965-26-5	from dimethyl terephthalate
59822-52-5	from terephthaloyl chloride

– Polybutylene terephthalate

Molecular Formula: $(C_{12}H_{12}O_4)_n$
Fragments: $C_{12}H_{12}O_4$
Molecular Weight: MW 30000–80000
Additives: Glass fibre, carbon fibre, flame retardants, mica, glass beads, talc, kaolin, Wollastonite, barium sulfate. Glass fibre is the principal additive, providing enhanced physical props. proportional to the amount added (30% is the most common, but up to 55% is available). Glass fibre increases strength, modulus, heat distortion temp. and fatigue resistance. Carbon and aramid fibres are also used. Aramid fibre reinforcement does not enhance mech. props., but fibres give excellent wear resistance and near-isotropic props. (characteristics that are not obtained with glass and carbon fibres). Warping is improved by replacing part of the semi-crystalline material with appropriate amorph. polymers, or by using mixed fillers including glass spheres and mica. Addition of PET to PBT enhances gloss and surface appearance. Mineral fillers enhance electrical props., minimise warping and improve modulus under load. Flame retardants include halogenated aromatic compounds, particularly bromo derivatives, in combination with antimony oxide; copolycarbonate of tetrabromobisphenol A and decabromodiphenyl ether. Non-dripping flammability is increased by the addition of PTFE or fumed colloidal silica. Elastomers are added as impact modifiers to increase impact strength. Colourants may be added. Dyes are avoided as they migrate and bleed. Hydroxybenzotriazoles give uv stability. Sterically hindered phenols give thermo-oxidative stability. Bis-epoxides, alkylene bis-fatty acid amides and carbodiimides give chemical and hydrolytic stability. Carbon black and uv stabilisers improve resistance to outdoor ageing

Morphology: α form: triclinic a 0.483, b 0.596, c 1.162, α 99.9°, β 115.2°, γ 111.3° [8]; a 0.483, b 0.605, c 1.145, α 100.5°, β 117°, γ 110.8° [9]; a 0.488, b 0.594, c 1.165, α 98.9°, β 116.6°, γ 110.9° [11]; a 0.483, b 0.594, c 1.159, α 99.7°, β 115.2°, γ 110.8° [12]; a 0.487, b 0.596, c 1.171, α 100.1°, β 116.6°, γ 110.3° [13]; a 0.486, b 0.596, c 1.165, α 99.7°, β 116.0°, γ 110.8° [22]. β Form: triclinic a 0.495, b 0.567, c 1.295, α 101.7°, β 121.8°, γ 99.9° [12]; a 0.496, b 0.580, c 1.30, α 101.9°, β 120.5°, γ 105.0° [22,23]; a 0.463, b 0.588, c 1.306, α 103.3°, β 119.8°, γ 104.4° [46]. Other structural studies have been reported [19,21,57]

Volumetric & Calorimetric Properties:
Density:

No.	Value	Note
1	d 1.41 g/cm³	cryst. [8]
2	d 1.31 g/cm³	[8,10]
3	d 1.3 g/cm³	[12]
4	d 1.283 g/cm³	Ultradur B2550 [50]
5	d 1.256 g/cm³	amorph. [26]
6	d^{23} 1.324 g/cm³	Valox 295 [45]

Thermal Expansion Coefficient:

No.	Value	Note	Type
1	0.0003 K⁻¹	zero pressure, solid [45]	L
2	0.00047 K⁻¹	zero pressure, melt [45]	L

Equation of State: Equation of state relationship (190°–280°, 20.6–103.4 MPa) has been reported [31,83]. Tait equation of state parameters have been reported [45]. Chain of rotators and Bruce Hartmann equation of state information has been reported [61]
Thermodynamic Properties General: ΔC_p 5.9 J mol⁻¹ K⁻¹ (1.4 cal mol⁻¹ °C⁻¹) [14], 0.16 J g⁻¹ K⁻¹ (Crastin LMC 55) [43]. ΔC_p 107 J mol⁻¹ K⁻¹ (-25°), 77 J mol⁻¹ K⁻¹ (47°) [47]. Other values of heat capacity and thermal conductivity have been reported [43,47,53,71]. Linear [36,40,57,89] and volume [83] coefficients of thermal expansion have been reported. Activation energy for conductivity 184 kJ mol⁻¹ (44 kcal mol⁻¹) [54].

Latent Heat Crystallization: ΔH_f 31.8 kJ mol⁻¹ (7.6 kcal mol⁻¹) [2], 9.6 kJ mol⁻¹ (2.3 kcal mol⁻¹) [15], 140 J g⁻¹ [26], 45 J g⁻¹ [55], 28.7 kJ mol⁻¹ [25]; ΔS_f 63.2 J mol⁻¹ K⁻¹ (15.1 cal mol⁻¹ °C⁻¹) [2]; S^{25} 6.03 J mol⁻¹ K⁻¹ (1.44 cal mol⁻¹ °C⁻¹) [24]. Other ΔH, ΔS, ΔG values have been reported [43,53]. S° 19.9 J mol⁻¹ K⁻¹ [53,92]; ΔH 69.891 kJ mol⁻¹ (25°); ΔS 337.1 J mol⁻¹ K⁻¹ (25°); ΔG -30.632 kJ mol⁻¹ (25°) [53]. ΔH_f 142 J g⁻¹ (cryst.) [55]; ΔH_f 32 kJ mol⁻¹. Activation energy for melt flow 47 kJ mol⁻¹ [76], 41.8 kJ mol⁻¹ [93]

Thermal Conductivity:

No.	Value	Note
1	0.17–0.29 W/mK	0.63–1.04 kJ (m h K)⁻¹ [62]

Thermal Diffusivity: Thermal diffusivity has been reported [71]
Specific Heat Capacity:

No.	Value	Note	Type
1	1.38 kJ/kg.C	72.5 cal mol⁻¹ °C⁻¹, 45° [14]	P
2	1.61 kJ/kg.C	25° [53]	P

Melting Temperature:

No.	Value	Note
1	230°C	[1,2]
2	213°C	[14]
3	217°C	[40]
4	232°C	[3,93]
5	220°C	[4,56,77]
6	225°C	[5,8,9,58]
7	224°C	[6,12,31,35]
8	226°C	[15]

Glass-Transition Temperature:

No.	Value	Note
1	80°C	[3,93]
2	20°C	[4]
3	22°C	[5,26,56,93]
4	40°C	[9]
5	45°C	[14]
6	43°C	[31,77]
7	28°C	Crastin LMC 55 [43]
8	69°C	[45]
9	37–52°C	[47]
10	46°C	[58]

Transition Temperature General: Melting behaviour has been reported [87]. Other values of T_m have been reported [55]
Deflection Temperature:

No.	Value	Note
1	48°C	[40]

Vicat Softening Point:

No.	Value	Note
1	231°C	[40]

– Polybutylene terephthalate

Transition Temperature:

No.	Value	Note	Type
1	-60°C	[3,14]	γ-transition temp.
2	-25°C	[56]	γ-transition temp.
3	0°C	[93]	γ-transition temp.
4	236°C	[25]	equilibrium melting temp.
5	245°C	[47,92]	equilibrium melting temp.
6	233°C	[65]	equilibrium melting temp.

Surface Properties & Solubility:
Cohesive Energy Density Solubility Parameters: δ 21.6 $J^{1/2}$ $cm^{-3/2}$ (Valox) [91]
Solvents/Non-solvents: Sol. trifluoroacetic acid [18], warm phenol [63], hexafluoro-2-propanol [93], hexafluoroacetone sesquihydrate [93], phenol/tetrachloroethane (60:40) [93]

Transport Properties:
Transport Properties General: Other melt flow index data have been reported [80]
Melt Flow Index:

No.	Value	Note
1	2.5 g/10 min	ASTM D1238T, PTMT 6P20A [31]

Polymer Solutions Dilute: η_{inh} 0.88 dl g^{-1} (25°, phenol/tetrachloroethane 60:40) [6]; η 0.75 dl g^{-1} (2-chlorophenol) [11]; 0.21 dl g^{-1} (phenol/tetrachloroethane 60:40) [42]; η 1.4 dl g^{-1} (25°, phenol/tetrachloroethane 60:40) [15]; 0.887 dl g^{-1} (25°, phenol/tetrachloroethane 50:50) [35]; 1.04 dl g^{-1} (30°, phenol/tetrachloroethane 60:40) [40]. Other viscosity values have been reported [27,28,55]
Polymer Melts: Melt viscosity data have been reported [28,65,77,80]
Permeability of Gases: Permeability, diffusion and solubility of He, Ne, Ar and CO_2 have been reported [75]. Sorption of CO_2 [50], NO, CO, N_2, CH_4, Ne and He [58,81] has been reported
Permeability of Liquids: Diffusion coefficient 1,2-dichloroethane 0.00078 mm^2 s^{-1} (240°), 1,4-dichlorobutane 0.00063 mm^2 s^{-1} (240°), 1,4-butanediol 0.0024 mm^2 s^{-1} (240°) [38], [H_2O] 0.02–0.06 mm^2 h^{-1} [74]

Mechanical Properties:
Mechanical Properties General: Mechanical props. have been reported [73]. Other elastic modulus values have been reported [77]. Variation of loss and storage moduli with temp. has been reported [14,39]. Other values of Young's modulus [79] for Ultradur B4500 [36] have been reported. Other values of bulk modulus have been reported [79,83]. Electric modulus has been reported [54]. Variation of shear modulus with temp. for Ultradur B4500 has been reported [36]. Stress relaxation modulus has been reported [41]

Tensile (Young's) Modulus:

No.	Value	Note
1	2530 MPa	[10]

Elastic Modulus:

No.	Value	Note
1	13239 MPa	135000 kg cm^{-2}, α-form, parallel to chain axis
2	20594 MPa	210000 kg cm^{-2}, β-form, parallel to chain axis [33]
3	2354 MPa	24000 kg cm^{-2}, 100 plane
4	2157 MPa	22000 kg cm^{-2}, 010 plane [67]
5	4903 MPa	50000 kg cm^{-2}, 135° [78]
6	13238 MPa	135000 kg cm^{-2}, room temp., 104 plane

Poisson's Ratio:

No.	Value	Note
1	0.44	[10]

Miscellaneous Moduli:

No.	Value	Note	Type
1	7390 MPa	[10]	Bulk modulus
2	880 MPa	[10]	Shear modulus

Viscoelastic Behaviour: Stress-strain curves have been reported [11,16,51]. Stress relaxation and creep behaviour have been reported [52,88]
Mechanical Properties Miscellaneous: T-Peel strength 18 kg m^{-1} (1 lb in^{-1}, 23°, ASTM D1876-61T) [6]. Compressibility [31] and bulk and surface compression props. have been reported [49]
Fracture Mechanical Properties: Tensile shear strength 10.3 MPa (1500 psi, 0°, ASTM D1002-64), 13.8 MPa (2000 psi, 23°, ASTM D1002-64), 11 MPa (1600 psi, 70°, ASTM D1002-64), 4.1 MPa (600 psi, 120°, ASTM D1002-64) [6]
Friction Abrasion and Resistance: Coefficient of friction has been reported [69]

Electrical Properties:
Electrical Properties General: Variation of dissipation factor [79] with temp. has been reported [7,14,36,39,68]. Variation of dielectric constant with temp. has been reported [36,56,68]. Volume resistivity has been reported [68]
Dielectric Permittivity (Constant):

No.	Value	Note
1	3.1	[49]

Dielectric Strength:

No.	Value	Note
1	20 kV.mm^{-1}	[68]

Complex Permittivity and Electroactive Polymers: The variation of complex relative permittivity of Valox 315 (-170–175°, 0.1 Hz– 3 MHz) has been reported in graphical form [82]
Static Electrification: Conductivity [54] and its relation to temp. [56] has been reported
Dissipation (Power) Factor:

No.	Value	Note
1	>0.04	min., 60–150° [68]

Optical Properties:
Optical Properties General: Other values of refractive index have been reported [60]. Optical rotation has been reported [60]. Oxyluminescence curve has been reported [37]. Birefringence value for Ultradur B4500 has been reported [36]
Refractive Index:

No.	Value	Note
1	1.411	α-form
2	1.713	α-form
3	1.76	α-form

– Polybutylene terephthalate

No.	Value	Note
4	1.398	β-form [85]
5	1.708	β-form [85]
6	1.768	β-form [85]

Transmission and Spectra: Ir [13,30,64,66,73,84], uv [64], Raman [16,66], fluorescence [70], mass [17,29,34,42,48], XPES [90] and C-13 nmr [18,20,32,35,44] spectral data have been reported
Molar Refraction: dn/dC 0.236 (23°, hexafluoroisopropanol) [27]
Total Internal Reflection: Birefringence 0.183 [11], 0.00145 [38].
Intrinsic birefringence (c axis) 0.153 (α-form), 0.215 (β-form) [85]

Polymer Stability:
Decomposition Details: Decomposition temp. 400° [42]. Thermal degradation details have been reported [17,29,34,42,48,59]
Environmental Stress: Photolysis, photooxidation [64,84] and weathering have been reported [72,93]
Chemical Stability: Solvent resistance has been reported [73]. Resistant to ketones, alcohols, glycols, ethers, aliphatic hydrocarbons, chlorinated aliphatic hydrocarbons, petrol, oil, transmission and brake fluids [93], CCl_4 and detergents [94]
Hydrolytic Stability: Stable to H_2O and dilute acids and alkalis
Recyclability: Details on recyclability have been reported [86]

Applications/Commercial Products:
Processing & Manufacturing Routes: Synth. by transesterification of dimethyl phthalate with 1,4-butanediol using titanium catalysts followed by melt polycondensation using titanium catalysts at 240° under vacuum [5,15,42]. Processed by injection moulding, extrusion and foam moulding (thermoforming with difficulty). Dried at 121° for 4h to optimise props. It is important not to overheat. A melt temp. of 249–271° (480–520°F) should be sufficient. Mould temps. of 37–121° (100–250°F) are commonly used

Mould Shrinkage (%):

No.	Value	Note
1	1.7%	unreinforced, 3 mm sample, parallel flow
2	0.2%	glass filled, 3 mm sample, parallel flow
3	2.3%	unreinforced, 3 mm sample, perpendicular to flow
4	0.6%	glass filled, 3 mm sample, perpendicular to flow

Applications: Used in iron and toaster housings, cooker/fryer handles, hair dryer nozzles, food processor blades. Used for exterior and under bonnet parts of cars. Used in the electrical industry for bobbins, connectors, switches, relays, terminal boards, motor bush holders, TV tuners, fuse cases, ie. carriers and sockets, and belts. Other industrial uses include, pump housings, impellers, valves, brackets, water meter components, tool housings, castings; in drapery hardware, pen barrels, zippers, hair dryers and calculators

Trade name	Details	Supplier
AVP	KVV/RVV	Polymerland Inc.
Adell	HD/HR/HT	Adell Plastics Inc.
Arnite	various grades	DSM Engineering
Ashlene PBT	P123/126/130; 126WO	Ashley Polymers
Azmet	CM range	Azdel Inc.
Beetle		BIP
CTI	PS-15GF/000	CTI
CTXC-301	stainless steel reinforced	Compounding Technology
Celanex		Hoechst Celanese
Celstran PBT	S range	Hoechst Celanese
Celstran PBT	G range	Polymer Composites
Comalloy	410 series	Comalloy Intl. Corp.
Comtuf	415/431/432	Comalloy Intl. Corp.
Comtuf 431	glass and mineral filled	ComAlloy International
Crastine		DuPont
E-30NF-0100	graphite filled	Thermofil Inc.
E-40MI-0100	mica filled	Thermofil Inc.
E-45FM-0319	glass and mineral filled	Thermofil
E-45FM-0393	glass and mica filled	Thermofil Inc.
E/E1	various grades	Thermofil Inc.
EMI-X	PDX/WC	LNP Engineering
Electrafil	G-1854/SS/7 J-1850/CF/30	DSM Engineering
Entec	9000 series	Entec Polymers
FR-PMT		Mitsubishi Chemical
Fiberfil	J range	DSM Engineering
Fiberfil	glass filled, flame retardant	DSM Engineering
Gafite		Celanese Corp.
Gafite		GAF Corp.
Gaftuf		Celanese Corp.
Grilpet XE3060		EMS
Grisuplast U		Premnitz
Hiloy	400 series	Comalloy Intl. Corp.
Hiloy 445/461/462	glass and mineral filled	ComAlloy International
Hostadur		Hoechst Celanese
Kelanex	various grades	Hoechst Celanese
Lubricomp	WCL/WFL/WL	LNP Engineering
Lubricomp	WFL-4036, glass and PTFE filled	LNP Engineering
Lubricomp	WL-4540 PTFE and silicone lubricated	LNP Engineering
Lumax	GP/HF/HR	Lucky
Lupox	various grades	Lucky
MPBT	FG10/20/30/40	Modified Plastics
MPBT-FG10	glass and mineral filled	Modified Plastics
Mitsubishi kasei	various grades	Mitsubishi Chemical
Novadur		Mitsubishi Chemical
Nyloy	BG-0070N-V0	Nytex Composites
Orgatier		Atochem Polymers
PBT		Bay Resins
PBT		Teijin Chem. Ltd.
PBT		Toray Industries
PBT	various grades	TP Composites
PBT	various grades	Nan Ya Plastics

– Polybutylene terephthalate

PBT	500 series	Mitsubishi Chem.
PBT	82G30L/100L/200L	Michael Day
PDX 84369		LNP Engineering
PRL-PBT	TP series	Polymer Resources
PTMT		Dynamit Nobel
Permastat	1000	RTP Company
Petra		Allied Signal Corporation
Pibiter		Montedison
Planac		Dainippon Ink and Chemicals
Pocan	various grades	Bayer Inc.
RPB		Ferro Corporation
RTP	ESD; 1000series	RTP Company
RTP 1001 M20	glass and mineral filled	RTP Company
Rynite	various grades	DuPont
SC21	1090/1230	Spartech
Shinite	D series	Shinkong Synthetic
Shinko-Lac	G/N series	Mitsubishi Rayon
Stat-Kon W		LNP Engineering
Stat-Loy	W	LNP Engineering
Techster		Rhone-Poulenc
Texapol PBT		Texapol
Thermocomp W 1000		LNP Engineering
Thermofil		Thermofil
Toray PBT	various grades	Toray Industries
Tufpet		Mitsubishi Chemical
Ultradur		BASF
Valox 210 HP		General Electric Plastics
Vandar		Hoechst Celanese
Vestadur		Hüls AG
Voloy	400 series	Comalloy Intl. Corp.
Voloy 415	glass and mineral filled	ComAlloy International
Vybex	various grades	Ferro Corporation
Vybex	various grades, flame retardant, glass filled	Ferro Corporation

Bibliographic References

[1] Edgar, O.B. and Hill, R., *J. Polym. Sci.*, 1952, **8**, 1, (melting point)
[2] Corix, A. and Van Kerpel, R., *J. Polym. Sci.*, 1959, **40**, 521, (thermodynamic props.)
[3] Farrow, G., McIntosh, J. and Ward, I.M., *Makromol. Chem.*, 1960, **38**, 147, (transition temps.)
[4] Schulken, R.M., Boy, R.E. and Cox, R.H., *J. Polym. Sci., Part C: Polym. Lett.*, 1963, **6**, 17, (transition temps.)
[5] Smith, J.G., Kibler, C.J. and Sublett, B.J., *J. Polym. Sci., Part A-1*, 1966, **4**, 1851, (transition temps., synth.)
[6] Jackson, W.J., Gray, T.F. and Caldwell, J.R., *J. Appl. Polym. Sci.*, 1970, **14**, 685, (mechanical props, soln. props.)
[7] Nemoz, G., May, J.F. and Vallet, G., *Angew. Makromol. Chem.*, 1976, **49**, 149, (power factor)
[8] Mencik, Z., *J. Polym. Sci., Polym. Phys. Ed.*, 1975, **13**, 2173, (cryst struct)
[9] Alter, U. and Bonart, R., *Colloid Polym. Sci.*, 1976, **254**, 348, (cryst. struct.)
[10] Warfield, R.W. and Barnet, F.R., *Angew. Makromol. Chem.*, 1975, **44**, 181, (mechanical props.)
[11] Jakeways, R., Ward, I.M., Wilding, M.A., Hall, I.H. et al, *J. Polym. Sci., Polym. Phys. Ed.*, 1975, **13**, 799, (cryst struct, birefringence)
[12] Yokouchi, M., Sakakibara, Y., Chatani, Y., Tadokoro, H. et al, *Macromolecules*, 1976, **9**, 266, (cryst. struct.)
[13] Joly, A.M., Nemoz, G., Douillard, A. and Vallet, G., *Makromol. Chem.*, 1975, **176**, 479, (ir, cryst. struct.)
[14] Yip, H.K. and Williams, H.L., *J. Appl. Polym. Sci.*, 1976, **20**, 1217, (thermodynamic props., transition temps.)
[15] Yip, H.K. and Williams, H.L., *J. Appl. Polym. Sci.*, 1976, **20**, 1209, (synth., soln. props.)
[16] Jakeways, R., Smith, T., Ward, I.M. and Wilding, M.A., *J. Polym. Sci., Polym. Lett. Ed.*, 1976, **14**, 41, (Raman, stress-strain curves)
[17] Luderwald, I. and Urrutia, H., *Makromol. Chem.*, 1976, **177**, 2079, (thermal degradation, ms)
[18] Kricheldorf, H.R., *Makromol. Chem.*, 1978, **179**, 2133, (C-13 nmr, solubility)
[19] Alter, U. and Bonart, R., *Colloid Polym. Sci.*, 1980, **258**, 332, (struct.)
[20] Komoraski, R.A., *J. Polym. Sci., Polym. Phys. Ed.*, 1979, **17**, 45, (C-13 nmr)
[21] Stambough, B., Koenig, J.L. and Lando, J.B., *J. Polym. Sci., Polym. Phys. Ed.*, 1979, **17**, 1053, (struct.)
[22] Desborough, I.J. and Hall, I.H., *Polymer*, 1977, **18**, 825, (struct.)
[23] Hall, I.M. and Pass, M.G., *Polymer*, 1976, **17**, 807, (cryst. struct.)
[24] Riande, E., *Eur. Polym. J.*, 1978, **14**, 885, (entropy)
[25] Hasslin, H.W., Droscher, M. and Wegner, G., *Makromol. Chem.*, 1980, **181**, 301, (thermodynamic props.)
[26] Illers, K.H., *Colloid Polym. Sci.*, 1980, **258**, 117, (transition temps.)
[27] Horbach, A., Binsack, R. and Muller, H., *Angew. Makromol. Chem.*, 1981, **98**, 35, (soln. props.)
[28] Borman, W.F.H., *J. Appl. Polym. Sci.*, 1978, **22**, 2119, (melt viscosity, soln. props.)
[29] Lum, R.M., *J. Polym. Sci., Polym. Chem. Ed.*, 1979, **17**, 203, (thermal degradation, ms)
[30] Ouchi, I., Hosoi, M. and Shimotsuma, S., *J. Appl. Polym. Sci.*, 1977, **21**, 3445, (ir)
[31] Wei, K.Y., Cualo, J.A. and Ihm, D.W., *Polym. Phys. Ed.*, 1983, **21**, 1091, (thermodynamic props., flow rate)
[32] Grenier-Loustalot, M.F. and Bocelli, G., *Eur. Polym. J.*, 1984, **20**, 957, (C-13 nmr)
[33] Nakamae, K., Kameyama, M., Yoshikawa, M. and Matsumoto, T., *J. Polym. Sci., Polym. Phys. Ed.*, 1982, **20**, 319, (elastic modulus)
[34] Adams, R.E., *J. Polym. Sci., Polym. Chem. Ed.*, 1982, **20**, 119, (thermal degradation, ms)
[35] Horii, F., Hirai, A., Murayama, K., Kitamura, R. and Suzuki, T., *Macromolecules*, 1983, **16**, 273, (C-13 nmr, soln. props.)
[36] Leung, W.P. and Choy, C.L., *J. Appl. Polym. Sci.*, 1982, **27**, 2693, (birefringence, mechanical props.)
[37] Wendlandt, W.W., *Thermochim. Acta*, 1983, **71**, 129, (oxyluminescence)
[38] Bonatz, E., Rafler, G. and Reinisch, G., *Angew. Makromol. Chem.*, 1983, **119**, 137, (diffusion coefficients)
[39] Rong, S.D. and Williams, H.L., *J. Appl. Polym. Sci.*, 1985, **30**, 2575, (mechanical props)
[40] Ng, T.H. and Williams, H.L., *Makromol. Chem.*, 1981, **182**, 3323, (transition temps., expansion coefficient)
[41] Ng, T.H. and Williams, H.L., *Makromol. Chem.*, 1981, **182**, 3331, (stress relaxation modulus)
[42] Foti, S., Giuffrida, M., Maravigna, P. and Montaudo, G., *J. Polym. Sci., Polym. Chem. Ed.*, 1984, **22**, 1217, (synth., ms, thermal degradation)
[43] Aleman, J.V., *Angew. Makromol. Chem.*, 1985, **133**, 141, (thermodynamic props.)
[44] Gomez, M.A., Cozine, M.H. and Tonelli, A.E., *Macromolecules*, 1988, **21**, 388, (C-13 nmr)
[45] Fakhreddine, Y.A. and Zoller, P., *J. Polym. Sci., Part B: Polym. Phys.*, 1991, **29**, 1141, (thermodynamic props.)
[46] Grasso, R.P., Perry, B.C., Koenig, J.L. and Lando, J.B., *Macromolecules*, 1989, **22**, 1267, (cryst. struct.)
[47] Cheng, S.Z.D., Pan, R. and Wunderlich, B., *Makromol. Chem.*, 1988, **189**, 2443, (heat capacity, transition temps.)
[48] Plage, B. and Schulten, H.R., *Macromolecules*, 1990, **23**, 2642, (thermal degradation, ms)
[49] Aleman, J.V., *Eur. Polym. J.*, 1991, **27**, 221, (dielectric constant)
[50] Thuy, L.P. and Springer, J., *Colloid Polym. Sci.*, 1988, **266**, 614, (CO_2 absorption)
[51] Ng, T.H. and Williams, H.L., *J. Appl. Polym. Sci.*, 1986, **32**, 4883, (stress-strain curves)

[52] Ng, T.H. and Williams, H.L., *J. Appl. Polym. Sci.*, 1987, **33**, 739, (stress relaxation, creep)
[53] Cheng, S.Z.D., Pan, R., Bu, H.S., Cao, M.Y. and Wunderlich, B., *Makromol. Chem.*, 1988, **189**, 1579, (thermodynamic props.)
[54] Starkweather, H.W. and Avakian, P., *J. Polym. Sci., Part B: Polym. Phys.*, 1992, **30**, 637, (conductivity, electric modulus)
[55] Nichols, M.E. and Robertson, R.E., *J. Polym. Sci., Part B: Polym. Phys.*, 1992, **30**, 755, (melting points)
[56] Sandrolini, F., Motori, A. and Saccani, A., *J. Appl. Polym. Sci.*, 1992, **44**, 765, (conductivity, transition temps.)
[57] Huo, P.P., Cebe, P. and Capel, M., *J. Polym. Sci., Part B: Polym. Phys.*, 1992, **30**, 1459, (cryst. struct., expansion coefficient)
[58] Zhou, Z. and Springer, J., *J. Appl. Polym. Sci.*, 1993, **47**, 7, (gas sorption, birefringence)
[59] Goff, L.J., *Polym. Eng. Sci.*, 1993, **33**, 497, (pyrolysis)
[60] Fu, X., Bartus, J. and Vogl, O., *Polym. Int.*, 1993, **31**, 183, (optical rotation, refractive index)
[61] Wohlfarth, C., *J. Appl. Polym. Sci.*, 1993, **48**, 1923, (thermodynamic props.)
[62] Takahara, A., Magome, T. and Kajiyama, J., *J. Polym. Sci., Part B: Polym. Phys.*, 1994, **32**, 839, (coefficient of thermal conductivity)
[63] Dasaradhudu, Y. and Rao, V.V.R.N., *Polym. Int.*, 1994, **35**, 329, (solubility)
[64] Casu, A. and Gardette, J.L., *Polymer*, 1995, **36**, 4005, (photolysis, ir, uv)
[65] Pompe, G., Haussler, L. and Winter, W., *J. Polym. Sci., Part B: Polym. Phys.*, 1996, **34**, 211, (melt viscosity)
[66] Stach, W. and Holland-Moritz, K., *J. Mol. Struct.*, 1980, **60**, 49, (ir, Raman)
[67] Nakamae, K., Kameyama, M., Yoshikawa, M. and Matsumoto, T., *Sen'i Gakkaishi*, 1980, **36**, T33, (elastic modulus)
[68] Lushcheikin, G.A., Kolerov, V.S. and Polevaya-Mansfield, M.K., *Plast. Massy*, 1976, 65, (electrical strength, resistance)
[69] Lapshin, V.V. and Andreeva, T.I., *Plast. Massy*, 1979, 57, (coefficient of friction)
[70] Takai, Y., Mizutani, T. and Ieda, M., *Jpn. J. Appl. Phys.*, 1978, **17**, 651, (fluorescence spectrum)
[71] Piven, A.N., Shimchuk, T.Y. and Chistyakov, V.L., *Prom. Teplotekh.*, 1980, **2**, 48, (thermal conductivity, diffusivity)
[72] Balog, K., *Rev. Plast. Mod.*, 1981, **42**, 197, (weathering)
[73] Kays, A.D. and Hunter, J.D., *ASTM Spec. Tech. Publ.*, 1983, **797**, 119, (ir, solvent resistance)
[74] Aleman, J.V., Gonzalez, D.M. and Munoz, F.C., *Rev. Plast. Mod.*, 1984, **48**, 279, (diffusion coefficient)
[75] Lukashov, A.V., Springer, J. and Phan, T.L., *Vysokomol. Soedin., Ser. B*, 1984, **26**, 875, (gas permeability, solubility, diffusion)
[76] Munari, A., Pilati, F. and Pezzin, G., *Rheol. Acta*, 1985, **24**, 534, (melt flow)
[77] Sanchez-Sancha, M. and Aleman, J.V., *J. Rheol. (N.Y.)*, 1985, **29**, 307, (melt viscosity, transition temps.)
[78] Nishino, T., Yokoyama, F., Nakamae, K. and Matsumoto, T., *Kobunshi Ronbunshu*, 1983, **40**, 357, (elastic modulus)
[79] Galkiewicz, R.K. and Karasz, F.E., *J. Mater. Sci.*, 1983, **18**, 721, (mechanical props.)
[80] Munari, A., Pilati, F. and Pezzin, G., *Rheol. Acta*, 1984, **23**, 14, (melt index, melt viscosity)
[81] Schultze, J.D., Zhou, Z. and Springer, J., *Angew. Makromol. Chem.*, 1991, **185-186**, 265, (gas sorption)
[82] Pratt, G.J. and Smith, M.J.A., *J. Mater. Sci.*, 1990, **25**, 477, (complex relative permittivity)
[83] Fakhreddine, Y.A. and Zoller, P., *Annu. Tech. Conf. - Soc. Plast. Eng., 49th*, 1991, 1642, (thermodynamic props.)
[84] Rivaton, A., *Polym. Degrad. Stab.*, 1993, **41**, 297, (photooxidation, ir)
[85] Ohkoshi, Y. and Nagura, M., *Sen'i Gakkaishi*, 1993, **49**, 601, (refractive index, birefringence)
[86] Mori, H. and Sagawa, M., *Purasuchikkusu*, 1994, **45**, 43, (recycling)
[87] Al-Raheil, I.A. and Qudah, A.M.A., *Polym. Int.*, 1995, **37**, 47, (melting behaviour)
[88] Reif, S.K., Amberge, K.J. and Woodford, D.A., *Mater. Des.*, 1995, **16**, 15, (stress relaxation, creep)
[89] Bikiaris, D.N. and Karayannidis, G.P., *J. Appl. Polym. Sci.*, 1996, **60**, 55, (expansion coefficient)
[90] Teramoto, K., Okajima, T., Baba, F. and Seki, K., *Koen Yoshishu - Nippon Setchaku Gakkai Nenji Takai, 34th*, 1996, 143, (XPES)
[91] Waywood, W.J. and Durning, C.J., *Polym. Eng. Sci.*, 1987, **27**, 1265, (solubility parameter)
[92] *Encycl. Polym. Sci. Eng.*, 2nd edn., Vol. 16, (ed. J.I. Kroschwitz), 1986, 779, (thermodynamic props.)
[93] *Encycl. Polym. Sci. Eng.*, 2nd edn., (ed. J.I. Kroschwitz), John Wiley and Sons, 1986, **12**, 10, 226, (mechanical props., transition temps.)
[94] *The Materials Selector*, 2nd edn., (eds. N.A. Waterman and M.F. Ashby), Chapman and Hall, 1997, **3**, 366, (chemical stability)

Poly(butylene terephthalate-co-butylene ether glycol terephthalate)

P-104

Synonyms: *PBT-poly(oxytetramethylene) copolymer. Poly(butylene terephthalate-co-tetrahydrofuran). Poly(butylene terephthalate-co-butylene oxide). PBT-PTMEG. PBT-PEE. Poly(tetramethylene terephthalate-co-tetramethylene oxide)*

Related Polymers: Polybutylene terephthalate, PBT - Injection moulding, Glass fibre filled, PBT - Flame retardant, PBT - High Impact, Impact modified, PBT - Fibres, PBT - Films/Sheets, PBT - Mineral filled, PBT - Carbon fibre filled, PBT - Structural foam, PBT - PTFE lubricated, PBT - Silicone lubricated, PBT - Uv stabilised, PBT - Phenoxy blends, PBT - Polyethylene terephthalate blends

Monomers: 1,4-Butanediol, Terephthalic acid
Material class: Copolymers
Polymer Type: PBT and copolymers
CAS Number:

CAS Reg. No.	Note
37282-12-5	
9078-71-1	from dimethyl terephthalate
106159-00-6	block

Molecular Formula: $[(C_{12}H_{12}O_4)_x(C_8H_4O_3)_y(C_4H_8O)_z]_n$
Fragments: $C_{12}H_{12}O_4$ $C_8H_4O_3$ C_4H_8O
Molecular Weight: M_n 25000–30000
Additives: Aromatic secondary amines used as stabilisers. Phenolic or amine-containing compounds used as antioxidants. Flame retardants. Carbon black. Polycarbodiimide added as a moisture stabiliser
Morphology: Thin films containing 73% PBT have single cryst. lamellae [30]. Has spherulitic struct. over a wide range of compositions [6,7]. Melt spinning prod. filaments with a two-phase struct. [36]. The morphology of blends with liq. cryst. polymer has been reported [39]. Other morphological studies have been reported [9,10].

Volumetric & Calorimetric Properties:

Density:

No.	Value	Note
1	d 1.15 g/cm^3	Hytrel 4055 [14]
2	d 1.189 g/cm^3	60% PBT [19]
3	d 1.152 g/cm^3	Hytrel 4055 [21]
4	d 1.204 g/cm^3	Hytrel 5556 [32]
5	d 1.13–1.25 g/cm^3	
6	d 1.19 g/cm^3	57% PBT [39]
7	d 1.17 g/cm^3	50% PBT, ASTM D792 [55]

Thermal Expansion Coefficient:

No.	Value	Note	Type
1	0.000172 K^{-1}	57% PBT [39]	L

Thermodynamic Properties General: Melting behaviour has been reported [34]
Latent Heat Crystallization: ΔH_f 10 J g^{-1} (23% PBT) [1], 32.7 J g^{-1} (7.8 cal g^{-1} 50% PBT) [7], 12.31 kJ mol^{-1} (2.94 kcal mol^{-1} 60% PBT) [29]. Activation energy for conduction 77 kJ mol^{-1} (60% PBT) [19].

Entropy of diffusion 17.2 J mol^{-1} K^{-1} (34.5% PBT). Activation energy of diffusion of *p*-nitroaniline 56.38 kJ mol^{-1} (34.5% PBT) [26]

Melting Temperature:

No.	Value	Note
1	198°C	58% PBT [3]
2	176°C	33.1% PBT [4]
3	189°C	50% PBT [7,12]
4	177°C	44% PBT [8]
5	212°C	76% PBT [25]
6	202°C	58% PBT [25]
7	165–227°C	ASTM D3418 [35]
8	203°C	57% PBT, Hytrel 5556 [36]
9	145°C	50% PBT, ASTM D3418 [55]
10	242°C	[27]
11	171–248°C	[56]
12	219°C	10% PTHF [22]
13	240°C	10% PTHF [22]
14	117–200°C	[29]

Glass-Transition Temperature:

No.	Value	Note
1	-78°C	23% PBT [1]
2	-30°C	58% PBT [3]
3	-78°C	33.1% PBT [4]
4	-59°C	50% PBT [7]
5	-65°C	44% PBT [8]
6	-32°C	Hytrel 4055 [11]
7	-48°C	60% PBT [19]
8	-53°C	58% PBT [25]
9	-43°C	57% PBT [39]
10	-57--42°C	[29]
11	-70-20°C	[55]
12	50°C	10% PTHF [22]

Transition Temperature General: Other values of T_g and T_m have been reported [5]

Deflection Temperature:

No.	Value	Note
1	52°C	1.82 MPa, 57% PBT [39]
2	54°C	0.46 MPa, 50% PBT, ASTM D648 [55]

Vicat Softening Point:

No.	Value	Note
1	110–220°C	Shore D32–D63, ASTM D1525 [35]
2	112°C	50% PBT, ASTM D1525 [55]
3	107°C	Shore D35, ASTM D1525 [57]
4	136°C	Shore D40, ASTM D1525
5	181°C	Shore D55, ASTM D1525
6	204°C	Shore D72, ASTM D1525 [57]

Brittleness Temperature:

No.	Value	Note
1	>-70°C	min., 33.1% PBT [4]
2	-60°C	76% PBT, ASTM D746 [5]
3	<-70°C	max., 58% PBT [25]
4	-70°C	58% PBT [55]

Transition Temperature:

No.	Value	Note	Type
1	-110°C	23% PBT [1]	T_β
2	-111°C	Hytrel 4055 [11]	T_β
3	-42°C	34% PBT	T_β
4	27°C	84% PBT [24]	T_β
5	224°C	51% PBT	T_m
6	229°C	65% PBT	T_m
7	233°C	80% PBT [2]	T_m
8	242°C	[27]	T_m
9	246°C	60% PBT [29]	T_m
10	-76°C	33.1% PBT [4]	Softening temp.
11	152°C	86% wt PBT, annealed [18]	T_α
12	42–48°C	86% wt PBT, annealed [18]	T_β
13	-76°C	86% wt PBT, annealed [18]	T_γ

Surface Properties & Solubility:

Solubility Properties General: Miscible with bisphenol A polycarbonate [40]. Immiscible with PBT [2]. Mixing Hytrel 4056 with PVC, gives a dispersion of PVC within a continuous Hytrel phase [16]. Compatible with PVC at Hytrel levels of 25–50%. Incompatible at Hytrel level of 80% [11]. Hytrel 4055 is partially miscible with chlorosulfonated polyethylene [20]. Some mixing between Hytrel 4055 and nitrile rubber occurs [21]

Cohesive Energy Density Solubility Parameters: δ 18.62–22.1 J$^{1/2}$ cm$^{-3/2}$ (9.1–10.8 cal$^{1/2}$ cm$^{-3/2}$) (0–100% PBT) [2]

Solvents/Non-solvents: Sol. phenol, CHCl$_3$, tetrachloroethane, *m*-cresol [5]. Insol. most solvents [5]

Transport Properties:

Transport Properties General: Melt flow index [4,56] for Hytrel [28] has been reported

Melt Flow Index:

No.	Value	Note
1	5.8 g/10 min	58% PBT [25,55]

Polymer Solutions Dilute: $[\eta]_{inh}$ 0.18 dl g^{-1} (30°, *m*-cresol, 50% PBT) [12]. $[\eta]$ 1.52 dl g^{-1} (30°, *m*-cresol, 58% PBT) [25]. $[\eta]_{inh}$ 2.5 dl g^{-1} (25°, *o*-chlorophenol, 57% PBT) [39]

Polymer Melts: Melt viscosity of material containing 58% PBT and its variation with polymerisation time has been reported [4]

Permeability of Gases: Gas permeability data have been reported [45]

Permeability of Liquids: Diffusion coefficient of *p*-nitroaniline 75.9 × 10^{-7}-37.1 × 10^{-6} (67.1–95°, 34.5% PBT) [26]

Water Absorption:

No.	Value	Note
1	0.6 %	24h, 50% PBT, ASTM D570 [55]

Mechanical Properties:

Mechanical Properties General: Variation of dynamic tensile storage and loss moduli with temp. has been reported [7,10,18,29]. Tensile modulus and tenacity of filaments increase with an increase in hard segment content, whereas elongation at break decreases [36]. Variation of tensile modulus with temp. has been reported [32]. Compression set 52% (22h, 70°, 50% PBT, ASTM D395B) [12], 40% (58% PBT, ASTM D395B) [25], 27% (22h, 73°, 9.3 MPa, 50% PBT, ASTM D395A [55]. Other values of flexural modulus have been reported [56]. Other mechanical props. have been reported [9,35,56]. The variation of dynamic storage modulus with temp. for Hytrel 4055 has been reported [11]. The fibre denier and its relation to spinning temp. has been reported [23]. Variation of dynamic tensile storage modulus with temp. has been reported [6]. Compression strain 40% (33.1% PBT, ASTM D395B) [4], compression set 54% (22h, 70°, 50% PBT, ASTM D395-55B) [5]. Tensile creep modulus has been reported [56]. Other values of torsional modulus have been reported [12]. Other values of bulk modulus have been reported [32]

Tensile (Young's) Modulus:

No.	Value	Note
1	1040 MPa	0°, Hytrel 5556 [32]
2	138.3 MPa	1410 kg cm^{-2}, 58% PBT, ASTM D638 [4]
3	84–1250 MPa	depending upon method of synth. [38]

Flexural Modulus:

No.	Value	Note
1	48 MPa	50% PBT, ASTM D790 [55]
2	84.8–2370 MPa	-40°, ASTM D790 [57]
3	210 MPa	57% PBT [39]
4	55–585 MPa	22°, ASTM D790 [55]
5	13.8–2070 MPa	[56]
6	62–1040 MPa	depending upon method of synth. [38]

Tensile Strength Break:

No.	Value	Note	Elongation
1	20 MPa	44% PBT [8]	755%
2	48.4 MPa	50% PBT [12]	755%
3	44.1 MPa	-25°, 58% PBT [25,55]	650%
4	47.6 MPa	25°, 76% PBT	510%
5	24.7 MPa	150°, 76% PBT	530%
6	21.2 MPa	175°, 76% PBT	505%
7	13.7 MPa	200°, 76% PBT [25]	525%
8	18 MPa	25°, Hytrel 40D, ASTM D412-80 [54]	950%
9	39.2 MPa	400 kg cm^{-2}, 25°, 33.1% PBT [4]	810%
10	51 MPa	7.4 kpsi, 50% PBT, ASTM D412 [5]	780%
11	42 MPa	57% PBT [39]	940%
12	30 MPa	50% PBT, ASTM D638 [55]	560%
13	13.2–29.2 MPa	[57]	470–300%
14	18.6 MPa	100°, 4 weeks, H$_2$O [25]	95%
15	15–44 MPa	[35]	

Flexural Strength at Break:

No.	Value	Note
1	11 MPa	57% PBT [39]

Compressive Modulus:

No.	Value	Note
1	86.3 MPa	880 kg cm^{-2}, 58% PBT, ASTM D695 [4]

Poisson's Ratio:

No.	Value	Note
1	0.45	50% PBT [55]

Miscellaneous Moduli:

No.	Value	Note	Type
1	206.9 MPa	2110 kg cm^{-2}, 58% PBT, ASTM D797 [4]	Bending modulus
2	26 MPa	50% PBT, ASTM D1043	Torsional modulus (Clash Berg) T$_{10000}$
3	71 MPa	4°, 58% PBT [25]	Torsional modulus
4	190 MPa	-140°, 58% PBT [25]	Torsional modulus
5	3510 MPa	Hytrel 5556 [32]	Bulk modulus
6	116 MPa	25°, 75% PBT	Torsional modulus
7	310 MPa	-40°, 75% PBT [55]	Torsional modulus

Impact Strength: Drop weight impact failure has been reported [56]
Viscoelastic Behaviour: Longitudinal sonic velocity 0.28 km s^{-1} (20°, Hytrel 4055) [11]. Acoustic impedance 30000 g cm^{-2} s^{-1} (20°, Hytrel 4055) [11]. Stress-strain curves (58% PBT) have been reported [4]; other stress-strain curves have been reported [5,7,8,36]. Tensile creep 8.0% (6.9 MPa, 23°, Hytrel 5556). Compressive creep 0.6%, 1.3% (6.9 MPa, 50°, Hytrel 5556) [56]. Creep resistance may be improved by improving the compression set [44]. Creep behaviour has been reported [55]
Mechanical Properties Miscellaneous: Resilience 62% (50% PBT, Bashore) [55]
Hardness: Shore D55 (58% PBT, ASTM D2240) [55], D35–D72 [55], D50–D80 [38], A92 (Hytrel 40D, ASTM D2240-80) [54], A92 (33.1% PBT) [4], D49 (50% PBT, ASTM D2240) [5], D48 (50% PBT, ASTM D2240) [12], D63 (76% PBT) [25], D55 (58% PBT) [25], D35-D82 [35], D40 (50% PBT, ASTM D2240) [55]
Failure Properties General: Shear strength 24 MPa (50% PBT, ASTM D732) [55]. Tear strength 110 kN m^{-1} (50% PBT, D624 die B), 102.77 kN m^{-1} (Hytrel 40D, ASTM D624-81) [54], 122 kN m^{-1} (die C) [55], 48 KnM^{-1} (50% PBT, ASTM D1938) [12]. Resistance to flex-cut growth Ross, pierced 1000000 cycles (min., 50% PBT, ASTM D1052); unpierced 1000000 (min.); DeMatia, pierced 1000000 cycles (min., ASTM D813) [55], 72000 cycles (22°, ASTM D813), 18000 cycles (121°, ASTM D813) [4]. Resistance to flex-cut growth 300000 (min., 23°, Hytrel, Shore D55, Ross flex, ASTM D1052), 12 (min.) (-40°, Hytrel, Shore D55, Ross flex, ASTM D1052) [56]. Split tear 16 kg cm^{-1} (33.1% PBT, ASTM D1938) [4], 70.6 kN m^{-1} (58% PBT, ASTM D1938) [25]. Tear resistance 290 pli (50% PBT, ASTM D1938) [5]
Fracture Mechanical Properties: Fatigue limit 6.9 MPa (Hytrel 5556, 2.5 million cycles without failure) [56]
Friction Abrasion and Resistance: Abrasion resistance 700% (33.1% PBT, ASTM D1630) [4]. Taber abrasion resistance 90 mg (1000 cycles)$^{-1}$ (58% PBT, ASTM D1044) [25], 5 mg (1000 cycles)$^{-1}$ (CS-17 wheel, ASTM D1044), 64 mg (1000 cycles)$^{-1}$ (H-18 wheel, ASTM D1044) [56]. NBS abrasion index 3540% (ASTM D1630) [56]. Abrasion loss 0.182 cm^3 h^{-1} (Hytrel 40D) [54]

Izod Notch:

No.	Value	Notch	Break	Note
1	1270 J/m	Y		-40°, 58% PBT [25]
2	2040 J/m	N		min., 58% PBT [25]
3		Y	no break	-40°, ASTM D256 [57]

Electrical Properties:
Electrical Properties General: Variation of dissipation factor with temp. has been reported [1,3,6,10,18,35]. Dielectric props. of soln. blends of Hytrel 5526 with PVC have been reported [17]. Other dielectric constants have been reported [3]. The apparent dielectric constant increases with decreasing temp. at temps. below -100° [24].

Dielectric Permittivity (Constant):

No.	Value	Frequency	Note
1	4.6		Hytrel 5556 [56]
2	4.4		25°, 60% PBT [19]
3	4.8		Hytrel 4056 [56]
4	5.146	10 Hz	34% PBT [24]

Dielectric Strength:

No.	Value	Note
1	0.39 kV.mm^{-1}	Hytrel 4056, 0.075" thick [56]
2	0.4 kV.mm^{-1}	Hytrel 5556, ASTM D149 [56]

Dissipation (Power) Factor:

No.	Value	Frequency	Note
1	0.006	100 Hz	Hytrel 5556, ASTM D150 [56]

Optical Properties:
Transmission and Spectra: Ir [7,8,51], C-13 nmr [13,15,16,31,42], H-1 nmr [22,38,42], esr [33], mass [37] and x-ray [37] spectral data have been reported
Total Internal Reflection: Birefringence of filaments increases with an increase in hard segment content [36]
Volume Properties/Surface Properties: Colour off-white [32]

Polymer Stability:
Polymer Stability General: May be used from -55–150° [35,55,56]
Thermal Stability General: Heat resistance has been reported [56]. Details of thermal stability have been reported [25]
Upper Use Temperature:

No.	Value	Note
1	150°C	[35]
2	150°C	[55]
3	148°C	[56]

Decomposition Details: Thermal degradation data have been reported [49]
Flammability: Flammability rating HB (UL94) [56]
Environmental Stress: Weathering data have been reported [48]. Weathering in the Panama jungle rain forest over 5 years causes Hytrel 4056 to lose 105% of its original tensile strength [56]. Undergoes photo-thermal oxidation to yield hydroperoxides [53]
Chemical Stability: Swollen by oil and AcOH [4,25]. Resistant to oil, butyl alcohol, isooctane, isobutyl methyl ketone, xylene, hydraulic fluid, Freon 113, perchloroethylene, trichloroethylene, aniline, dibutyl phthalate, DMF, ethylene glycol, Skydrol 500A [35], acoustic coupling fluids [50]. Resistance to chemicals [25] and fuels [52] has been reported. Has excellent resistance to ASTM oils nos. 1 and 3, ASTM fuels A and C, EtOH, 2-propanol, 2-butanone, mineral oil, Skydrol 500B, soap soln., zinc chloride soln.; has good resistance to Me$_2$CO, formic acid, 10% nitric acid; has poor resistance to 30% nitric acid [57]. Has excellent resistance to oils, aliphatic hydrocarbons, petrol, ethylene glycol, fluorinated hydrocarbons, weak acids, salt solns. Good to moderate resistance to alcohols, ketones, and poor resistance to phenols [56]. Swollen by ASTM oil no. 3 and ASTM fuel B [44]
Hydrolytic Stability: Swollen by H$_2$O [4,55]. Hydrolytic stability in water and seawater of Hytrel 7246 has been reported [50]. Has excellent resistance to seawater [56,57] and boiling H$_2$O [57]
Biological Stability: Biodegradation study has been reported [41]
Recyclability: Details of recycling have been reported [43]
Stability Miscellaneous: Radiation lifetime 150 Cm^{-2} (100 keV, 23°, 73% PBT) [30]. Resistance to radiation has been reported [56]

Applications/Commercial Products:
Processing & Manufacturing Routes: Synth. by a two-stage melt transesterification of dimethyl terephthalate, poly(oxytetramethylene glycol) and 1,4-butanediol at 160–250° with a titanate catalyst to form a prepolymer. The prepolymer is polycondensed at 250° under vacuum. The process is carried out under nitrogen in the presence of an antioxidant [2,4,5,38,55]

Mould Shrinkage (%):

No.	Value	Note
1	0.4–1.8%	Hytrel range, ASTM D955 [56]

Applications: Used in telephone insulation cords; in low pressure tyres, cable jacketing, hose applications, protective coatings, power transmission belts, diaphragms, gaskets, seals and plugs. Other uses in headphones, railcar diaphragms, railcar couplers, medical films, wire clamps and automotive shock absorbers. Blends have uses in automobile fascias, wheel covers, air-bag housings, wire insulation, electrical switches, sealed electrical connectors, electromechanical parts, machine gears, car spoilers, car grilles, car side mouldings, footwear, military holsters, straps, footballs, sports clothing, motorcycle face masks, hoses, tubes, power drive belts, fibre optic coatings, petrol caps, seat belt locking devices, bottle caps, air rifle parts, textile spinning wheels, car door latch covers

Trade name	Details	Supplier
Arnitel		Akzo
Bexloy V	PBT blend	DuPont
Hytrel		DuPont
Kopel		Kolon Ltd.
Lomod		General Electric
Pelprene		Toyo Boseki
Riteflex		Hoechst Celanese
Vandar	PBT blend	Hoechst Celanese

Bibliographic References
[1] Sauer, B.B., Avakian, P. and Cohen, G.M., *Polymer*, 1992, **33**, 2666, (transition temps.)
[2] Gallagher, K.P., Zhang, X., Runt, J.P., Huynh-ba, G. and Lin, J.S., *Macromolecules*, 1993, **26**, 588, (synth., solubility parameter, power factor)
[3] Runt, J., Du, L. and Martynowicz, L.M., *Macromolecules*, 1989, **22**, 3908, (transition temps.)
[4] Von Hoeschele, G.K. and Witsiepe, W.K., *Angew. Makromol. Chem.*, 1973, **29-30**, 267, (mechanical props., melt flow index)
[5] Witsiepe, W.K., *Adv. Chem. Ser.*, 1973, **129**, 39, (solubility, mechanical props.)
[6] Seymour, R.W., Overton, J.R. and Corley, L.S., *Macromolecules*, 1975, **8**, 331, (morphology, power factor)

[7] Lilaonitkul, A., West, J.C. and Cooper, S.L., *J. Macromol. Sci., Phys.,* 1976, **12**, 563, (ir, moduli, transition temps.)
[8] Van Bogart, J.W.C., Lilaonitkul, A., Lerner, L.E. and Cooper, S.L., *J. Macromol. Sci., Phys.,* 1980, **17**, 267, (ir, mechanical props.)
[9] *Thermoplastic Elastomers,* (ed. N.R. Legge, G. Holden and H.E. Schroeder), Hanser, 1987, 163, (morphology, mechanical props.)
[10] Shen, M., Mehra, U., Niinami, M., Koberstein, J.T. and Cooper, S.L., *J. Appl. Phys.,* 1974, **45**, 4182, (mechanical props.)
[11] Hourston, D.J. and Hughes, I.D., *J. Appl. Polym. Sci.,* 1977, **21**, 3099, (sonic velocity, acoustic impedance)
[12] Wolfe, J.R., *Adv. Chem. Ser.,* 1979, **176**, 129, (viscosity)
[13] Jelinski, L.W., *Macromolecules,* 1981, **14**, 1341, (C-13 nmr)
[14] Hourston, D.J. and Hughes, I.D., *J. Appl. Polym. Sci.,* 1981, **26**, 3487, (density)
[15] Jelinski, L.W., Dumais, J.J. and Engel, A.K., *Macromolecules,* 1983, **16**, 403, (C-13 nmr)
[16] Kwak, S.Y. and Nakajima, N., *Macromolecules,* 1996, **29**, 3521, (C-13 nmr, miscibility)
[17] Radhakrishnan, S. and Saini, D.R., *J. Appl. Polym. Sci.,* 1994, **52**, 1577, (conductivity)
[18] Kalfoglou, N.K., *J. Appl. Polym. Sci.,* 1977, **21**, 543, (mechanical props.)
[19] North, A.M., Pethrick, R.A. and Wilson, A.D., *Polymer,* 1978, **19**, 923, (conductivity, dielectric constant)
[20] Hourston, D.J. and Hughes, I.D., *Polymer,* 1980, **21**, 469, (miscibility)
[21] Hourston, D.J. and Hughes, I.D., *Polymer,* 1981, **22**, 127, (miscibility)
[22] Boussias, C.M. and Still, R.H., *J. Appl. Polym. Sci.,* 1980, **25**, 855, (H-1 nmr, transition temps.)
[23] Boussias, C.M., Peters, R.H. and Still, R.H., *J. Appl. Polym. Sci.,* 1980, **25**, 869, (fibre mechanical props.)
[24] Lilaonitkul, A. and Cooper, S.L., *Macromolecules,* 1979, **12**, 1146, (dielectric constant, conductivity)
[25] Hoeschele, G.K., *Angew. Makromol. Chem.,* 1977, **58/59**, 299, (chemical resistance, mechanical props.)
[26] Masuko, T., *Makromol. Chem.,* 1979, **180**, 2183, (diffusion coefficient)
[27] Pedemonte, E., Leva, M., Gattiglia, E. and Turturro, A., *Polymer,* 1985, **26**, 1202, (melting point)
[28] Shenoy, A.V. and Saini, D.R., *Br. Polym. J.,* 1985, **17**, 314, (melt flow index)
[29] Castles, J.L., Vallance, M.A., McKenna, J.M. and Cooper, S.L., *J. Polym. Sci., Polym. Phys. Ed.,* 1985, **23**, 2119, (transition temps., mechanical props.)
[30] Briber, R.M. and Thomas, E.L., *Polymer,* 1986, **27**, 66, (morphology, irradiation)
[31] Belfiore, L.A., *Polymer,* 1986, **27**, 80, (C-13 nmr)
[32] Lagakos, N., Jarzynski, J., Cole, J.H. and Bucaro, J.A., *J. Appl. Phys.,* 1986, **59**, 4017, (mechanical props., appearance)
[33] Lembicz, F. and Ukielski, R., *Makromol. Chem.,* 1985, **186**, 1679, (esr)
[34] Stevenson, J.C. and Cooper, S.L., *J. Polym. Sci., Part B: Polym. Phys.,* 1988, **26**, 953, (melting point)
[35] Hofmann, W. and Koch, R., *Kunststoffe,* 1989, **79**, 606, (transition temp., chemical resistance)
[36] Richeson, G.C. and Spruiell, J.E., *J. Appl. Polym. Sci.,* 1990, **41**, 845, (birefringence, viscoelastic props)
[37] Bhatia, Q.S. and Burrell, M.C., *Polymer,* 1991, **32**, 1948, (x-ray, ms)
[38] Chang, S.J., Chang, F.C. and Tsai, H.B., *Polym. Eng. Sci.,* 1995, **35**, 190, (synth., H-1 nmr)
[39] Jang, S.H. and Kim, B.S., *Polym. Eng. Sci.,* 1995, **35**, 528, (expansion coefficient, viscosity)
[40] Gaztelumendi, M. and Nazabal, J., *J. Polym. Sci., Part B: Polym. Phys.,* 1995, **33**, 603, (miscibility)
[41] Hinrichs, W.L.J., Kuit, J., Feil, H., Wildevuur, C.R.H. and Feijen, J., *Biomaterials,* 1992, **13**, 585, (biodegradation)
[42] Higashiyama, A., Yamamoto, Y., Chujo, R. and Wu, M., *Polym. J. (Tokyo),* 1992, **24**, 1345, (H-1 nmr, C-13 nmr)
[43] Reinhardt, H.G., *Plastverarbeiter,* 1993, **44**, 78, (recycling)
[44] Witsiepe, W.K., Polym. Prepr. (Am. Chem. Soc., Div. Polym. Chem.), 1972, **13**, 588, (solubility, creep resistance)
[45] Jpn. Pat., 1974, 74 73 384, (gas permeability)
[46] Vesperman, W.C. and Wilson, M.K., *West. Electr. Eng.,* 1979, **23**, 10, (additives)
[47] Kane, R.P., *J. Elastomers Plast.,* 1977, **9**, 416, (additives)
[48] Doyle, R.A. and Baker, K.C., Durability Aging Geosynth., *(Pap. Semin.),* 1988, 152, (weathering)
[49] Levantovskaya, I.I., Radetskaya, M.P., Medvedeva, F.M., Pin, L.D. et al, *Plast. Massy,* 1979, **9**, 20, (thermal degradation, stabilisers)
[50] Capps, R.N. and Bush, I.J., *Org. Coat. Appl. Polym. Sci. Proc.,* 1981, **46**, 619, (hydrolytic stability)
[51] Burchell, D.J., Lasch, J.E., Molis, S.E. and Hsu, S.L., Proc. IUPAC, I.U.P.A.C., Macromol. Symp. 28th, 1982, 44, (ir)
[52] Panser, L.M., *J. Elastomers Plast.,* 1983, **15**, 146, (fuel resistance)
[53] Tabankia, M.H., Philippart, J.L. and Gardetta, J.L., *Polym. Degrad. Stab.,* 1985, **12**, 349, (photothermal oxidation)
[54] Thomas, S., Kuriakose, B., Gupta, B.R. and De, S.K., *J. Mater. Sci.,* 1986, **21**, 711, (mechanical failure)
[55] *Encycl. Polym. Sci. Eng.,* 2nd edn., (ed. J.I. Kroschwitz), Wiley-Interscience, 1985, **12**, 75, (synth., uses, mechanical props.)
[56] *Handb. Thermoplast. Elastomers,* 2nd edn., (eds. B.M. Walker and C.P. Rader), Van Nostrand Reinhold, 1988, **181**, 370, (mechanical and electrical props., irradiation, chemical resistance)
[57] *Kirk-Othmer Encycl. Chem. Technol.,* 4th edn., (ed. J.I. Kroschwitz), Wiley-Interscience, 1991, **19**, 633, (chemical resistance, uses, transition temp)

Polybutylene terephthalate, Fibres P-105

Related Polymers: Polybutylene terephthalate, Polybutylene terephthalate, Injection Moulding Glass fibre filled, Polybutylene terephthalate, High Impact, Polybutylene terephthalate, Films/Sheets, Polybutylene terephthalate, Flame Retardant, Polybutylene terephthalate, Injection Moulding Mineral filled, Polybutylene terephthalate, Injection Moulding Carbon fibre filled, Polybutylene terephthalate, Polyethylene terephthalate blends, Polybutylene terephthalate, Phenoxy blends
Monomers: 1,4-Butanediol, Terephthalic acid
Material class: Thermoplastic
Polymer Type: PBT and copolymers
CAS Number:

CAS Reg. No.	Note
26062-94-2	
24968-12-5	repeating unit
30965-26-5	from dimethyl terephthalate
59822-52-5	from terephthaloyl chloride

Molecular Formula: $(C_{12}H_{12}O_4)_n$
Fragments: $C_{12}H_{12}O_4$
Morphology: Cryst. struct. of Celanex 2000 varies with the melt spin speed a 4.785, b 5.905, c 12 (spin speed 1500 m min^{-1}); a 4.745, b 5.878, c 11.84 (4500 m min^{-1}) [4]

Volumetric & Calorimetric Properties:
Volumetric Properties General: Density increases with the melt-spin take up velocity [4]
Glass-Transition Temperature:

No.	Value	Note
1	98°C	[2]

Transition Temperature:

No.	Value	Note	Type
1	214°C	0.2 g den^{-1} loading [1]	Flow point

Mechanical Properties:
Mechanical Properties General: The breaking load of PBT yarn increases with increasing heat-setting temp. up to 140°, when it then decreases [5]. Elongation at break 20–107% depending on draw ratio [2]. Tenacity and elongation of Celanex 2000 to break increase with increase in both take-up velocity and MW [4]. Young's modulus of Celanex 2000 increases with an increase in both take-up velocity and MW [4]
Tensile Strength Break:

No.	Value	Note
1	325 MPa	22°, draw ratio 4
2	278 MPa	90°, draw ratio 4 [3]
3	240–490 MPa	[2]

Miscellaneous Moduli:

No.	Value	Note
1	2200–2600 MPa	2% secant modulus [2]

Viscoelastic Behaviour: Stress-strain curves have been reported [2,3,4]

Optical Properties:
Total Internal Reflection: Birefringence 0.146–0.158 (depending on draw ratio) [2]

Applications/Commercial Products:

Trade name	Supplier
Fortrell II	Fibre Industries Inc.

Bibliographic References

[1] Smith, J.G., Kibler, C.J. and Sublett, B.J., *J. Polym. Sci., Part A-1*, 1966, **4**, 1851, (flow point)
[2] Ward, I.M., Wilding, M.A. and Brady, H., *J. Polym. Sci., Polym. Phys. Ed.*, 1976, **14**, 263, (mech props, transition temp)
[3] Lu, F.M. and Spruiell, J.E., *J. Appl. Polym. Sci.*, 1986, **31**, 1595, (mech props)
[4] Chen, S. and Spruiell, J.E., *J. Appl. Polym. Sci.*, 1987, **33**, 1427, (props)
[5] Mathur, M.R., Skukla, S.R. and Sawant, P.B., *Polym. J. (Tokyo)*, 1996, **28**, 189, (mech props)

Polybutylene terephthalate, Films/Sheets   P-106

Related Polymers: Polybutylene terephthalate, Polybutylene terephthalate, Injection Moulding Glass fibre filled, Polybutylene terephthalate, High Impact, Polybutylene terephthalate, fibres, Polybutylene terephthalate, Flame Retardant, Polybutylene terephthalate, Injection Moulding Mineral filled, Polybutylene terephthalate, Injection Moulding Carbon fibre filled, Polybutylene terephthalate, Polyethylene terephthalate blends, Polybutylene terephthalate, Phenoxy blends
Monomers: 1,4-Butanediol, Terephthalic acid
Material class: Thermoplastic
Polymer Type: PBT and copolymers
CAS Number:

CAS Reg. No.	Note
26062-94-2	
24968-12-5	repeating unit
30965-26-5	from dimethyl terephthalate
59822-52-5	from terephthaloyl chloride

Molecular Formula: $(C_{12}H_{12}O_4)_n$
Fragments: $C_{12}H_{12}O_4$
Molecular Weight: M_n 31000
Morphology: Ultradur a 4.82, b 5.93, c 11.74, α 100.0°, β 115.5°, γ 111.0° (cryst. temp. 220°C) [1]

Volumetric & Calorimetric Properties:
Density:

No.	Value	Note
1	d^{23} 1.289 g/cm^3	dry [2]
2	d 1.283 g/cm^3	Ultradur B2550 [5]
3	d 1.31 g/cm^3	[9]

Volumetric Properties General: The mean square density of Ultradur increases with increasing macroscopic density [1]

Melting Temperature:

No.	Value	Note
1	224.5°C	VFR 4716 [4]
2	224°C	30% glass fibre [4]

Glass-Transition Temperature:

No.	Value	Note
1	44°C	VFR 4716 [4]

Transition Temperature:

No.	Value	Note	Type
1	206°C	air cooled or liq. N_2 quenched	α
2	60°C	[4]	β
3	-85°C	[4]	γ

Transport Properties:
Polymer Solutions Dilute: [η] 0.53 dl g^{-1} (tetrachloroethane/phenol 60/40, Ultradur) [1], [η] 0.8 dl g^{-1} (phenol/tetrachloroethane 60/40, VFR 4716) [4], [η] 2.2 dl g^{-1} (25°, 2-chlorophenol) [3]
Permeability of Gases: The permeability of Ultradur B2550 to He, Ne, Ar and CO_2 increases with increasing temp. [5]
Water Absorption:

No.	Value	Note
1	0.48 %	24h, 20°, 90% relative humidity [2]

Gas Permeability:

No.	Gas	Value	Note
1	CO_2	5.91 cm^3 mm/ (m^2 day atm)	15 cm^3 mil 100 in^{-2} day^{-1} atm^{-1}
2	O_2	1.57 cm^3 mm/ (m^2 day atm)	4 cm^3 mil 100 in^{-2} day^{-1} atm^{-1}
3	H_2O	1.18 cm^3 mm/ (m^2 day atm)	3 cm^3 mil 100 in^{-2} day^{-1} atm^{-1} [9]

Mechanical Properties:
Mechanical Properties General: The dynamic storage and loss moduli decrease with increasing temp. [4]
Tensile (Young's) Modulus:

No.	Value	Note
1	2340 MPa	[9]

Tensile Strength Break:

No.	Value	Note
1	72 MPa	[9]

Viscoelastic Behaviour: Stress-strain curves have been reported [3]
Failure Properties General: Tear strength 1970 g mm^{-1} [9]

Electrical Properties:
Electrical Properties General: The dielectric constant and dielectric loss both increase with an increase in the amount of absorbed water [2]
Static Electrification: Electrical conductivity increases with increasing temp. [6]. Activation energy for electrical conduction 0.74 eV [6]. Field strength 26.6 MV m^{-1} [6]

Optical Properties:
Transmission and Spectra: Ir [7,8], uv [7] and x-ray [4] spectral data have been reported

Polymer Stability:
Upper Use Temperature:

No.	Value	Note
1	165°C	[9]

Environmental Stress: The presence of 30% glass fibre has no influence on the mechanism of photolysis or photooxidation, and the reinforced material yields the same products as PBT [7]

Applications/Commercial Products:

Trade name	Details	Supplier
Duranex	Used to refer to other materials	Polyplastics
Tenite	Used to refer to other materials	Eastman Chemical Company

Bibliographic References
[1] Bomschlegl, E. and Bonart, R., *Colloid Polym. Sci.*, 1980, **258**, 319, (soln props)
[2] Ito, E. and Kobayashi, Y., *J. Appl. Polym. Sci.*, 1980, **25**, 2145, (water absorption, dielectric constant)
[3] Brereton, M.G., Davies, G.R., Jakeways, R., Smith, T. and Ward, I.M., *Polymer*, 1978, **19**, 17, (stress-strain curves)
[4] Chang, E.P. and Slagowski, E.L., *J. Appl. Polym. Sci.*, 1978, **22**, 769, (thermodynamic and mech props)
[5] Thuy, L.P., Lukaschov, A. and Springer, J., *Colloid Polym. Sci.*, 1983, **261**, 973, (gas permeability)
[6] Dasaradhudu, Y. and Rao, V.V.R.N., *Polym. Int.*, 1994, **35**, 329, (electrical conduction)
[7] Casu, A. and Gardette, J.L., *Polymer*, 1995, **36**, 4005, (uv, ir, photooxidation)
[8] Stach, W.W. and Holland-Moritz, K.H., Proc. IUPAC, I.U.P.A.C., *Macromol. Symp.*, 28th, 1982, 66, (ir)
[9] *Kirk-Othmer Encycl. Chem. Technol.*, 3rd edn., (ed. M. Grayson), 1978, **18**, 565, (mech props, gas permeability)

Polybutylene terephthalate, injection moulding carbon fibre filled   P-107

Synonyms: *PBT, injection moulding carbon fibre filled*
Related Polymers: Polybutylene terephthalate, PBT Injection moulding glass fibre filled, PBT fibres, PBT films/sheets, PBT flame retardant, PBT high impact, PBT Injection moulding mineral filled, PBT Polyethylene terephthalate blends, PBT Phenoxy blends
Monomers: 1,4-Butanediol, 1,4-Benzenedicarboxylic acid
Material class: Composites, Thermoplastic
Polymer Type: PBT and copolymers, polyester
CAS Number:

CAS Reg. No.	Note
26062-94-2	
24968-12-5	repeating unit
30965-26-5	from dimethyl terephthalate
59822-52-5	from terephthaloyl chloride

Molecular Formula: $(C_{12}H_{12}O_4)_n$
Fragments: $C_{12}H_{12}O_4$

Volumetric & Calorimetric Properties:
Density:

No.	Value	Note
1	d 1.41 g/cm^3	30% filled, ASTM D792 [7,8]

Thermal Expansion Coefficient:

No.	Value	Note	Type
1	1×10^{-5} K^{-1}	30% filled, ASTM D696 [7]	V

Thermal Conductivity:

No.	Value	Note
1	45.05 W/mK	6.5 Btu (in ft^2 °Fh)$^{-1}$, ASTM C177 [1]

Deflection Temperature:

No.	Value	Note
1	221°C	1.82 MPa, 30% filled, ASTM D645 [7]
2	220°C	1.82 MPa, 30% filled, ASTM D645 [8]

Transport Properties:
Water Absorption:

No.	Value	Note
1	0.04 %	24h, 30% filled, ASTM D570 [1,7,8]

Mechanical Properties:
Mechanical Properties General: Tensile and flexural strengths have been reported [5]. Rigidity and bending strength have been reported [2]
Flexural Modulus:

No.	Value	Note
1	13789 MPa	2000 kpsi, ASTM D790 [1]
2	16000 MPa	30% filled, ASTM D790 [7]
3	15900 MPa	30% filled, ASTM D790 [8]

Tensile Strength Break:

No.	Value	Note	Elongation
1	137.89 MPa	20 kpsi, ASTM D638 [1]	2.3%

Flexural Strength at Break:

No.	Value	Note
1	200 MPa	30% filled, ASTM D790 [8]

Tensile Strength Yield:

No.	Value	Note
1	152 MPa	30% filled, ASTM D638 [7,8]

Flexural Strength Yield:

No.	Value	Note
1	199.95 MPa	29 kpsi, ASTM D790 [1]

Impact Strength: Other values of impact strength have been reported [5]
Viscoelastic Behaviour: Flexural creep resistance has been reported in graphical form [1]
Hardness: Rockwell M90 (30% filled, ASTM D785) [7]
Friction Abrasion and Resistance: Coefficient of friction μ 0.25 (approx., against steel) [1]. Other values of friction coefficient have

been reported [3]. Abrasion resistance has been reported [2]. Wear factor 24×10^{-10} (3 min.) [1]. Other values of wear factor have been reported [3]

Izod Notch:

No.	Value	Notch	Note
1	64 J/m	Y	1.2 ft lb in^{-1}, ASTM D256 [1]
2	240 J/m	N	4.5 ft lb in^{-1}, ASTM D256 [1]
3	60 J/m	Y	23°, 30% filled, ASTM D256A [7]
4	64 J/m	Y	6.25 mm, 30% filled, ASTM D256 [8]
5	35 J/m	N	6.25 mm, 30% filled, ASTM D256 [8]

Electrical Properties:
Electrical Properties General: Specific resistance 6 Ω cm (18% filled) [4]. Intrinsic resistivity has been reported [6]. Volume resistivity 0.01Ω cm (30% filled, ASTM D257) [7]. Surface resistivity 2.4 Ω m (30% filled, LNP #41) [1]
Static Electrification: Electrical conductivity has been reported [6]

Polymer Stability:
Upper Use Temperature:

No.	Value	Note
1	115°C	30% filled, air [7]

Flammability: Flammability rating HB (30% filled, ASTM D635, UL94) [7]

Applications/Commercial Products:
Mould Shrinkage (%):

No.	Value	Note
1	0.15%	30% filled, ASTM D955 [7]
2	0.2%	30% filled, ASTM D955 [8]
3	0.1–0.2%	30% filled, ASTM D955 [1]

Trade name	Supplier
Celanex	Hoechst
ESD-C-1080	RTP Company
Electrafil J1850/CF	Akzo
Lubricomp WC-1006	LNP Engineering
Pibiter	Enichem
Stat-kon WC-1002/1006	LNP Engineering
Thermocomp WC-1006	LNP Engineering
Thermofil	Thermofil
Valox	General Electric Plastics

Bibliographic References
[1] Theberge, J., Arkles, B. and Robinson, R., *Ind. Eng. Chem. Proc. Des. Dev.*, 1976, **15**, 100, (mechanical props., thermal and electrical conductivity)
[2] Keurleber, R., *Kunststoffe*, 1978, **68**, 331, (bending strength, abrasion resistance)
[3] Wolverton, M.P. and Theberge, J.E., *J. Elastomers Plast.*, 1981, **13**, 97, (wear factor, friction coefficient)
[4] *Jpn. Pat.*, 1982, 133 155, (specific resistance)
[5] Suetsugu, K., *Kobunshi Ronbunshu*, 1988, **45**, 555, (mechanical props.)
[6] Sakamoto, Y., Takeuchi, H., Tomonoti, S. and Sawanobori, T., Compos., Proc. Int. Conf. Compos. Mater., *8th*, 1991, **3**, 29P/1, (electrical props.)
[7] *The Materials Selector*, 2nd edn., (eds. N.A. Waterman and M.F. Ashby), Chapman and Hall, 1997, **3**, 366, (mechanical props., electrical props.)
[8] *Encycl. Polym. Sci. Eng.*, 2nd edn., (ed. J.I. Kroschwitz), Wiley Interscience, 1986, **12**, 243, (mechanical props.)

Polybutylene terephthalate, injection moulding mineral filled

Synonyms: *PBT, injection-moulding mineral filled*
Related Polymers: Polybutylene terephthalate, PBT Injection moulding glass fibre filled, PBT fibres, PBT films/sheets, PBT flame retardant, PBT high impact, PBT Injection moulding carbon fibre filled, PBT Polyethylene terephthalate blends, PBT phenoxy blends
Monomers: 1,4-Butanediol, 1,4-Benzenedicarboxylic acid
Material class: Thermoplastic, Composites
Polymer Type: PBT and copolymers, polyester
CAS Number:

CAS Reg. No.	Note
26062-94-2	
24968-12-5	repeating unit
30965-26-5	from dimethyl terephthalate
59822-52-5	from terephthaloyl chloride

Molecular Formula: $(C_{12}H_{12}O_4)_n$
Fragments: $C_{12}H_{12}O_4$

Volumetric & Calorimetric Properties:
Density:

No.	Value	Note
1	d 1.8 g/cm^3	ASTM D792 [5]
2	d 1.66 g/cm^3	25% mica filled, 15% glass filled, ASTM D792 [6]

Thermal Expansion Coefficient:

No.	Value	Note	Type
1	7×10^{-5} K^{-1}	ASTM D696 [5]	V

Deflection Temperature:

No.	Value	Note
1	84°C	30% talc filled [1]
2	190°C	1.82 MPa, ASTM D645 [5]
3	86°C	1.82 MPa, 20% talc filled, ASTM D648
4	215°C	1.82 MPa, 25% mica filled, 15% glass filled, ASTM D648 [6]

Transport Properties:
Water Absorption:

No.	Value	Note
1	0.07 %	24h, ASTM D570 [5]

Mechanical Properties:
Mechanical Properties General: Mechanical props. for mica filled material have been reported [2,3,4]
Flexural Modulus:

No.	Value	Note
1	8000 MPa	ASTM D790 [5]

| 2 | 4410 MPa | 20% talc filled, ASTM D638 [6] |
| 3 | 9660 MPa | 25% mica filled, 15% glass filled, ASTM D790 [6] |

Tensile Strength Break:

No.	Value	Note	Elongation
1	53.1 MPa	7700 psi, 30% talc filled [1]	
2	58 MPa	20% talc filled, ASTM D638 [6]	
3	100 MPa	25% mica filled, 15% glass filled, ASTM D638 [6]	3.1%

Flexural Strength at Break:

No.	Value	Note
1	148 MPa	25% mica filled, 15% glass filled, ASTM D790 [6]

Tensile Strength Yield:

No.	Value	Note
1	85 MPa	ASTM D638 [5]

Miscellaneous Moduli:

No.	Value	Note	Type
1	3600 MPa	530 kpsi, 30% talc filled [1]	bending modulus

Hardness: Rockwell M85 (ASTM D785) [5]
Izod Notch:

No.	Value	Notch	Note
1	40 J/m	Y	23°, ASTM D256A [5]
2	27 J/m	Y	6.25 mm, 20% talc filled, ASTM D256 [6]
3	43 J/m	Y	6.25 mm, 25% mica filled, 15% glass filled, ASTM D256 [6]
4	390 J/m	N	6.25 mm, 20% talc filled, ASTM D256 [6]
5	590 J/m	N	6.25 mm, 25% mica filled, 15% glass filled, ASTM D256 [6]

Electrical Properties:
Dielectric Strength:

No.	Value	Note
1	17 kV.mm^{-1}	ASTM D149 [5]

Dissipation (Power) Factor:

No.	Value	Frequency
1	0.002	100 Hz
2	0.01	1 MHz

Polymer Stability:
Upper Use Temperature:

No.	Value	Note
1	120°C	air [5]

Flammability: Flammability rating V0 (ASTM D635, UL94) [5]
Stability Miscellaneous: Stability of the mica filled material to γ-radiation has been reported [3]

Applications/Commercial Products:
Mould Shrinkage (%):

No.	Value	Note
1	0.7%	ASTM D955 [5]
2	0.65%	25% mica filled, 15% glass filled, ASTM D955 [6]

Trade name	Details	Supplier
Celanex	non exclusive	Hoechst Celanese
Comtuf 431		ComAlloy International
Crastine XMB 1055/1059		Ciba-Geigy Corporation
E-45FM-0319/ 40MI-0100		Thermofil
Hiloy	non exclusive	ComAlloy International
Later	non exclusive	LATI Industria Thermoplastics S.p.A.
MPBT-F610		Modified Plastics
Pibiter	non exclusive	Enichem
Pocan B7375/ KU1-7341		Bayer Inc.
RTP 1001 M20		RTP Company
Techster	non exclusive	Rhone-Poulenc
Thermofil	non exclusive	Thermofil
Ultradur	KR 4000 series	BASF
Upstodur	non exclusive	Hüls AG
Valox	700 series	General Electric Plastics
Vandar 2122		Hoechst Celanese
Voloy 415		ComAlloy International

Bibliographic References
[1] *Ger. Pat.*, 1971, 2 051 233, (mechanical props.)
[2] Xanthos, M., Hawley, G.C. and Antonacci, J., *Soc. Plast. Eng., Tech. Pap.*, 1977, **23**, 352, (mechanical props.)
[3] Yamaoka, H., Miyata, K., Nakayama, Y. and Yoshida, H., *Proc. Int. Cryog. Mater. Conf.*, 1982, 282, (mechanical props.)
[4] Naik, S., Fenton, M. and Carmel, M., *Mod. Plast.*, 1985, **62**, 179, (mechanical props.)
[5] *The Materials Selector*, 2nd edn., (eds. N.A. Waterman and M.F. Ashby), Chapman and Hall, 1997, **3**, 366, (mechanical props., electrical props.)
[6] *Encycl. Polym. Sci. Eng.*, 2nd edn., (ed. J.I. Kroschwitz), Wiley Interscience, 1986, **12**, 243, (mechanical props.)

Polybutylene terephthalate, Phenoxy blends   P-109
Synonyms: *Polybutylene terephthalate poly(hydroxy ether) of bisphenol A blends. PBT - PHE*
Related Polymers: Polybutylene terephthalate, Polybutylene terephthalate, Injection Moulding Glass fibre filled, Polybutylene terephthalate, Flame Retardant, Polybutylene terephthalate, High Impact, Impact Modified, Polybutylene terephthalate, fibres, Polybutylene terephthalate, films/sheets, Polybutylene terephthalate, Mineral filled, Polybutylene terephthalate, Carbon fibre filled, Polybutylene terephthalate, Structural foam, Polybutylene terephthalate, PTFE lubricated, Polybutylene terephthalate, Silicone lubricated, Polybutylene terephthalate, Uv stabilised, Polybutylene terephthalate, Polyethylene terephthalate blends, Poly(butylene terephthalate-*co*-butylene ether glycol terephthalate)

— Polybutylene terephthalate-...

Monomers: 1,4-Butanediol, Terephthalic acid, Epichlorohydrin, Bisphenol A
Material class: Blends
Polymer Type: PBT and copolymers

Volumetric & Calorimetric Properties:
Density:

No.	Value	Note
1	d 1.24 g/cm^3	50% PBT, annealed [1]

Volumetric Properties General: Other density values have been reported [7]
Latent Heat Crystallization: ΔH_f 24.3 J g^{-1} (5.8 cal g^{-1}, 50% PBT) [1]
Melting Temperature:

No.	Value	Note
1	220°C	50% PBT [1]

Glass-Transition Temperature:

No.	Value	Note
1	51°C	50% PBT [2]
2	59°C	50% PBT [1]

Transition Temperature General: Other values of T_m have been reported [5]. T_m of blends is lower than that of pure PBT [3]. T_g is dependent upon blend composition [3]
Transition Temperature:

No.	Value	Note
1	-140°C	50% PBT
2	-80°C	50% PBT [1]

Surface Properties & Solubility:
Solubility Properties General: Interaction energy -11.7 J cm^{-3} (-2.8 cal cm^{-3}) [5]
Wettability/Surface Energy and Interfacial Tension: Surface free energy data have been reported [5]

Transport Properties:
Polymer Melts: The variation of melt viscosity with time at 250° shows an initial fall owing to melt plasticisation. After a time melt viscosity increases due to occurrence of transesterification reactions. Ultimately, melt failure occurs with subsequent decrease in melt viscosity [1]

Mechanical Properties:
Mechanical Properties General: Mechanical props. have been reported [7]. The variation in secant and shear moduli with temp. has been reported. Secant modulus decreases with increasing temp. [1]
Flexural Modulus:

No.	Value	Note
1	2549 MPa	26000 kg cm^{-2}, 97% PBT [4]

Tensile Strength Break:

No.	Value	Note
1	54.6 MPa	557 kg cm^{-2}, 97% PBT [4]

Flexural Strength at Break:

No.	Value	Note
1	82.8 MPa	844 kg cm^{-2}, 97% PBT [4]

Bibliographic References
[1] Robeson, L.M. and Furtek, A.B., *J. Appl. Polym. Sci.*, 1979, **23**, 645, (transition temps, mechanical props.)
[2] Seymour, R.W. and Zehner, B.E., *J. Polym. Sci., Polym. Phys. Ed.*, 1980, **18**, 2299, (transition temp)
[3] Equiazabal, J.I., Cortazar, M., Iruin, J.J. and Guzman, G.M., *J. Macromol. Sci., Phys.*, 1988, **27**, 19, (transition temp)
[4] *Jpn. Pat.*, 1989, 01 203 433, (mech props)
[5] Jo, W.H. and Lee, H.S., *Korea Polym. J.*, 1993, **1**, 123, (surface free energy, interaction energy, melting point)
[6] *Encycl. Polym. Sci. Eng.*, 2nd edn., (ed. J.I. Kroschwitz), Wiley Interscience, 1985, **12**, 449, (uses)
[7] Martinez, J.M., Equiazabal, J.L. and Nazabal, J., *J. Macromol. Sci., Phys.*, 1991, **30**, 345, (density, mech props)

Polybutylene terephthalate-polyethylene terephthalate blends

Synonyms: *PBT/PET blends*
Related Polymers: Polybutylene terephthalate, Poly(ethylene terephthalate), PBT-phenoxy blends, Poly(butylene terephthalate-*co*-butylene ether glycol terephthalate)
Material class: Blends
Polymer Type: PBT and copolymers, PET and copolymers
Additives: Tetrabromophthalic anhydride, triphenyl phosphite, antimony trioxide (flame retardant)
Morphology: Blending of the individual components disorders their spherulitic struct., which may be lost as the amount of either component in the blend is increased [3]. Other morphological studies have been reported [8]

Volumetric & Calorimetric Properties:
Volumetric Properties General: The variation of density with time has been reported [3]
Latent Heat Crystallization: ΔH 0.04 J (0.0096 cal) [1]. ΔH_f 87 J g^{-1} (PBT, 50% PBT), 54 J g^{-1} (PET, 50% PBT) [5]. Other enthalpies of formation have been reported [2,18]. Enthalpy of mixing has been reported [1,8]. ΔH_{cryst} 54.85 J g^{-1} (13.1 cal g^{-1}) (fibres, 97% PET) [30]. ΔH_{fusion} 61.97 J g^{-1} (14.8 cal g^{-1}) (fibres, 97% PET) [30]
Melting Temperature:

No.	Value	Note
1	263–264°C	10% PBT [2]
2	220°C	PBT, 50% PBT [5]
3	254°C	PET, 50% PBT [5]
4	266°C	fibres, 97% PBT [30]

Glass-Transition Temperature:

No.	Value	Note
1	36°C	50% PET [3]
2	47°C	50% PBT [5]
3	69°C	fibres, 97% PET [30]

Transition Temperature General: Other values for T_m [17,18] and other transition temps. [14] have been reported. T_g increases with increasing PET content [3]
Deflection Temperature:

No.	Value	Note
1	66°C	1.8 MPa, 18.5 kg cm^{-2}, 50% PBT [19]

Transition Temperature:

No.	Value	Note	Type
1	115°C	fibres, 97% PET [30]	crystallisation temp.

Surface Properties & Solubility:

Cohesive Energy Density Solubility Parameters: Interaction parameter is at a minimum when the weight fraction of PBT is 0.2–0.8 [1]. Interaction parameter -0.064 (220°, Scott equation) [11]. Other interaction and solubility parameter data have been reported [8]

Solvents/Non-solvents: Sol. trifluoroacetic acid [33]

Transport Properties:

Transport Properties General: Melt flow index has been reported [10]

Polymer Melts: [η] 0.59 dl g^{-1} [3]. Other [η] values have been reported [14,17]

Mechanical Properties:

Mechanical Properties General: Tensile modulus of fibres decreases with increasing PBT content [6]. Sonic modulus 126 g den^{-1} (fibres, 97% PET) [30]. Elongation at break of the fibres decreases with increasing PBT content [6]. Tenacity 2.5 g den^{-1} (fibres, 97% PET) [30]. Elongation at break 48% (fibres, 97% PET) [30], 400% (50% PBT) [5]

Tensile (Young's) Modulus:

No.	Value	Note
1	8400 MPa	film, 50% PBT [4]

Flexural Modulus:

No.	Value	Note
1	2912 MPa	29700 kg cm^{-2}, 23°, 50% PBT [20]
2	2020 MPa	20600 kg cm^{-2}, 80° [20]

Tensile Strength Break:

No.	Value	Note	Elongation
1	244 MPa	film, 50% PBT [4]	26%
2	50.2 MPa	573 kg cm^{-2}, 50% PBT [19]	5.5%
3	245 MPa	25 kg mm^{-2}, 12 μm film, 25% PBT [22]	130%
4	220–248 MPa	50% PBT [31]	5–20%

Flexural Strength at Break:

No.	Value	Note
1	94.9 MPa	968 kg cm^{-2}, 50% PBT [19]

Elastic Modulus:

No.	Value	Note
1	1540 MPa	50% PBT [5]
2	4000–5540 MPa	50% PBT [31]
3	1800–2400 MPa	fibres, various moulding conditions [7,16]

Viscoelastic Behaviour: Viscoelastic props. have been reported [8,10]

Mechanical Properties Miscellaneous: Dynamic modulus values have been reported [14]. Warp 1 mm (max.) initially, 20 mm (max.) (177°, 3 min., 60% PBT) [25]

Fracture Mechanical Properties: The toughness is governed by competition between shear yielding and crazing in the matrix [13]

Friction Abrasion and Resistance: Static friction coefficient 0.33 (12 μm film, 25% PBT) [22]. Dynamic friction coefficient 0.3 (12 μm film, 25% PBT) [22]

Izod Notch:

No.	Value	Notch	Note
1	27 J/m	Y	60% PBT [25]
2	600 J/m	N	60% PBT [25]
3	30.3 J/m	Y	0.03 kg m cm^{-1}, 50% PBT [19]
4	123–240 J/m	Y	fibres, various moulding conditions, ASTM D256 [7,16]

Electrical Properties:

Arc Resistance:

No.	Value	Note
1	182s	6 d, H20, 23°, ASTM D495, 29.6% PBT, 10% PET, 20% glass fibre, 30% talc [26]

Optical Properties:

Transmission and Spectra: Ir [3,9,24,28], H-1 nmr [9], C-13 nmr [9,27,33] and photoacoustic [28] spectral data have been reported

Volume Properties/Surface Properties: Haze 3.2% (12 μm film, 25% PBT) [22]

Polymer Stability:

Decomposition Details: Thermal degradation [9,12] studies have been reported. Max. decomposition temp. 522° (fibres, 97% PET). Thermal stability decreases with increasing PBT content [30]

Flammability: Flammability rating V0 (44.17% PBT, 44.17% PET, 0.02% antimony trioxide, 11.67% halogenated bisphenol A - type epoxy resin, UL 94) [29]. Limiting oxygen index 32% [21]

Environmental Stress: Films with densities less than 1.39 g cm^{-3} have good weathering resistance [23]

Applications/Commercial Products:

Processing & Manufacturing Routes: Polymers are coextruded at 271° [3], 280° [4] and 310° [5]

Mould Shrinkage (%):

No.	Value	Note
1	1.65%	80% PBT [20]

Applications: Applications include electrical and electronic usage and as brake and fuel lines

Bibliographic References

[1] Mishra, S.P. and Despura, B.L., *Polym. Commun.*, 1985, **26**, 5, (thermodynamic props.)
[2] Mishra, S.P. and Despura, B.L., *Makromol. Chem.*, 1985, **186**, 641, (mp, enthalpy of fusion)
[3] Escala, A. and Stein, R.S., *Adv. Chem. Ser.*, 1979, **176**, 455, (ir, transition temps., density)
[4] Eustatiev, M. and Fakirov, S., *Polymer*, 1992, **33**, 877, (mechanical props.)
[5] Auramova, N., *Polymer*, 1995, **36**, 801, (transition temps.)
[6] Gutmann, R. and Hertinger, H., *J. Appl. Polym. Sci.: Appl. Polym. Symp.*, 1991, **47**, 199, (mechanical props.)
[7] Vu-Khanh, T., Denault, J., Habib, P. and Low, A., *Compos. Sci. Technol.*, 1991, **40**, 423, (mechanical props.)
[8] Tao, J., Jin, H. and Sun, T., *Plast., Rubber Compos. Process. Appl.*, 1991, **16**, 49, (morphological study, viscoelastic props, solubility paramater)
[9] Sui, W. and Zhang, J., *Zhongguo Fangzhi Daxue Xuebao*, 1991, **17**, 70, (ir, H-1 nmr, C-13 nmr, thermal degradation)
[10] Jin, H. and Tao, J., *Plast., Rubber Compos. Process. Appl.*, 1991, **16**, 45, (melt flow, viscoelastic props.)
[11] Liu, J., Xu, W., Wan, G., Deng, Z. and Chen, Y., *Gaofenzi Cailiao Kexue Yu Gongcheng*, 1991, **7**, 45, (interaction parameter)
[12] Kim, J.H. and Ha, W.S., *Han'guk Somyu Konghakhoechi*, 1991, **28**, 801, (thermal degradation)
[13] Vu-Khanh, T. and Denault, J., *J. Compos. Mater.*, 1992, **26**, 2262, (toughness)

[14] Liu, Y., Zhang, F., Tong, Y. and Yang, S., *Gaofenzi Cailiao Kexue Yu Gongcheng,* 1994, **10**, 75, (dynamic modulus, transition temps., viscosity)
[15] Schmack, G., Hofmann, H. and Schoene, A., *Chemiefasern/Textilind.,* 1993, **43**, 898, (mechanical props.)
[16] Vu-Khanh, T. and Denault, J., *Advances in Polymer Blends and Alloys Technology,* (ed. K. Finlayson), Technomic, 1994, **5**, 74, (mechanical props.)
[17] Liu, Y., Tong, Y., Zhang, F. and Yu, R., *Gaofenzi Cailiao Kexue Yu Gongcheng,* 1996, **12**, 68, (mp, viscosity)
[18] Liu, Y., Tong, Y., Yang, Z. and Li, Z., *Gongneng Gaofenzi Xuebao,* 1996, **9**, 47, (mp, enthalpy of fusion)
[19] *Ger. Pat.,* 1973, 2 255 654, (heat distortion temp., mechanical props.)
[20] *Jpn. Pat.,* 1973, 38 343, (mould shrinkage, flexural modulus)
[21] *Jpn. Pat.,* 1975, 126 920, (fire resistance)
[22] *Jpn. Pat.,* 1976, 05 357, (mechanical props.)
[23] *Jpn. Pat.,* 1979, 87 776, (weather resistance)
[24] Escala, A., Balizer, E. and Stein, R.S., *Polym. Prepr. (Am. Chem. Soc., Div. Polym. Chem.),* 1978, **19**, 152, (ir, transition temp.)
[25] *Neth. Pat.,* 1980, 79 03 369, (mechanical props.)
[26] *Jpn. Pat.,* 1983, 179 261, (arcing resistance)
[27] Li, H.M. and Wong, A.H., *MMI Press Symp. Ser.,* 1982, **2**, 395, (C-13 nmr)
[28] Balizer, E., Talaat, H. and Bucara, J.A., *Ultrason. Symp. Proc.,* 1982, **2**, 571, (ir, photoacoustic spectroscopy)
[29] *Jpn. Pat.,* 1983, 34 854, (flame retardants)
[30] Viswanath, C.S., Despura, B.L. and Mishra, S.P., *Indian J. Text. Res.,* 1988, **13**, 23, (mech props, thermal stability)
[31] *Eur. Pat. Appl.,* 1989, 300 836, (mech props)
[32] *Kirk-Othmer Encycl. Chem. Technol.,* Vol. 19, 4th edn., (ed. J.I. Kroschwitz), Wiley Interscience, 1996, 869, (uses)
[33] Jacques, B., Devaux, J., Legras, R. and Nield, E., *J. Polym. Sci., Part A: Polym. Chem.,* 1996, **34**, 1189, (C-13 nmr, solubility)
[34] *Encycl. Polym. Sci. Eng.,* Vol. 12, 2nd edn., (ed. J.I. Kroschwitz), Wiley Interscience, 1988, 449, (use)
[35] Auramov, I. and Auramova, N., *J. Macromol. Sci., Phys.,* 1991, **30**, 335, (transition temp.)

Polybutylene terephthalate, PTFE lubricated P-111

Related Polymers: Polybutylene terephthalate, Polybutylene terephthalate, Injection moulding, glass fibre filled, Polybutylene terephthalate, Flame retardant, Polybutylene terephthalate, High impact, Polybutylene terephthalate, Fibres, Polybutylene terephthalate, Films/Sheets, Polybutylene terephthalate, Injection moulding, Mineral filled, Polybutylene terephthalate, Injection moulding, Carbon fibre filled, Polybutylene terephthalate, Polyethylene terephthalate blends, Polybutylene terephthalate, Phenoxy blends, Polybutylene terephthalate, Structural foam, Polybutylene terephthalate, Silicone lubricated, Polybutylene terephthalate, Uv stabilised, Poly(butylene terephthalate-*co*-butylene ether glycol terephthalate)
Monomers: 1,4-Butanediol, Terephthalic acid
Material class: Thermoplastic
Polymer Type: PBT and copolymers
CAS Number:

CAS Reg. No.	Note
26062-94-2	
24968-12-5	repeating unit
30965-26-5	from dimethyl terephthalate
59822-52-5	from terephthaloyl chloride

Molecular Formula: $(C_{12}H_{12}O_4)_n$
Fragments: $C_{12}H_{12}O_4$

Volumetric & Calorimetric Properties:
Density:

No.	Value	Note
1	d 1.42 g/cm^3	ASTM D792 [1]
2	d 1.43 g/cm^3	20% PTFE, ASTM D792 [2]

Thermal Expansion Coefficient:

No.	Value	Note	Type
1	0.0001 K^{-1}	ASTM D696 [1]	V
2	9×10^{-5} K^{-1}	20% PTFE, ASTM D696 [2]	L

Thermal Conductivity:

No.	Value	Note
1	0.23 W/mK	20% PTFE, ASTM C177 [2]

Deflection Temperature:

No.	Value	Note
1	190°C	0.45 MPa, ASTM D645
2	80°C	1.8 MPa, ASTM D645 [1]
3	60°C	1.81 MPa, 20% PTFE, ASTM D648 [2]

Transport Properties:
Water Absorption:

No.	Value	Note
1	0.05 %	24h, ASTM D570 [1]
2	0.08 %	24h, 20% PTFE, ASTM D570 [2]
3	0.45 %	saturation, 20% PTFE, ASTM D570 [2]

Mechanical Properties:
Flexural Modulus:

No.	Value	Note
1	1700 MPa	ASTM D790 [1]
2	2300 MPa	20% PTFE, ASTM D790 [2]

Tensile Strength Break:

No.	Value	Note	Elongation
1	40 MPa	20% PTFE, ASTM D638 [2]	10–15%

Flexural Strength at Break:

No.	Value	Note
1	70 MPa	20% PTFE, ASTM D790

Tensile Strength Yield:

No.	Value	Note
1	42 MPa	ASTM D638 [1]

Compressive Strength:

No.	Value	Note
1	45 MPa	20% PTFE, ASTM D695 [2]

Hardness: Rockwell M70 (ASTM D785) [1], M80 (ASTM D785) [2]
Failure Properties General: Shear strength 20 MPa (20% PTFE, ASTM D732) [2]
Friction Abrasion and Resistance: Wear factor 15 (20% PTFE, LNP SOP) [2]. Coefficient of friction 0.09 (static), 0.17 (dynamic) (20% PTFE, LNP SOP) [2]

– Polybutylene terephthalate, Silicone lubricated

Izod Notch:

No.	Value	Notch	Note
1	50 J/m	Y	23°, ASTM D256 A [1]
2	40 J/m	Y	20% PTFE, ASTM D256 [2]
3	300 J/m	N	20% PTFE, ASTM D256 [2]

Electrical Properties:
Surface/Volume Resistance:

No.	Value	Note	Type
3	0.1×10^{15} Ω.cm	20% PTFE, ASTM D257 [2]	S

Dielectric Permittivity (Constant):

No.	Value	Frequency	Note
1	3	60 Hz	20% PTFE, ASTM D150 [2]
2	2.8	1 MHz	20% PTFE, ASTM D150 [2]

Dielectric Strength:

No.	Value	Note
1	20 kV.mm^{-1}	ASTM D149 [1,2]

Static Electrification: Tracking resistance KC 600 (DIN IEC 112) [2]
Dissipation (Power) Factor:

No.	Value	Frequency	Note
1	0.002	60 Hz	20% PTFE, ASTM D150 [2]
2	0.018	1 MHz	20% PTFE, ASTM D150 [2]

Polymer Stability:
Upper Use Temperature:

No.	Value	Note
1	120°C	[1]
2	140°C	continuous service, 20% PTFE, UL 746 B [2]

Flammability: Flammability rating HB (3.2 mm thick, 20% PTFE, UL 94) [1,2]

Applications/Commercial Products:
Mould Shrinkage (%):

No.	Value	Note
1	2%	ASTM D955 [1]
2	1.2–1.8%	20% PTFE, 3 mm section, ASTM D955 [2]

Trade name	Details	Supplier
Celanex	non-exclusive	Hoechst
Later	non-exclusive	LATI Industria Thermoplastics S.p.A.
Lubricomp	non-exclusive	LNP Engineering
Pibiter	non-exclusive	Enichem
Thermofil	non-exclusive	Thermofil
Valox	non-exclusive	General Electric Plastics

Bibliographic References
[1] *The Materials Selector,* 2nd edn., (eds. N.A. Waterman and M.F. Ashby), Chapman and Hall, 1997, **3**, 372, (mechanical, electrical and thermodynamic props.)
[2] *LNP Engineering Plastics Product Information,* LNP Engineering Plastics, 1996, (technical datasheet)

Polybutylene terephthalate, Silicone lubricated

P-112

Related Polymers: Polybutylene terephthalate, Polybutylene terephthalate, Injection moulding, Glass fibre filled, Polybutylene terephthalate, Flame retardant, Polybutylene terephthalate, High impact, Polybutylene terephthalate, Fibres, Polybutylene terephthalate, Films/Sheets, Polybutylene terephthalate, Injection moulding, Mineral filled, Polybutylene terephthalate, Injection moulding, Carbon fibre filled, Polybutylene terephthalate, Polyethylene terephthalate blends, Polybutylene terephthalate, Phenoxy blends, Polybutylene terephthalate, Structural foam, Polybutylene terephthalate, PTFE lubricated, Polybutylene terephthalate, Uv stabilised, Poly(butylene terephthalate-*co*-butylene ether glycol terephthalate)
Monomers: 1,4-Butanediol, Terephthalic acid
Material class: Thermoplastic
Polymer Type: PBT and copolymers
CAS Number:

CAS Reg. No.	Note
26062-94-2	
24968-12-5	repeating unit
30965-26-5	from dimethyl terephthalate
59822-52-5	from terephthaloyl chloride

Molecular Formula: $(C_{12}H_{12}O_4)_n$
Fragments: $C_{12}H_{12}O_4$

Volumetric & Calorimetric Properties:
Density:

No.	Value	Note
1	d 1.29 g/cm^3	ASTM D792 [1]

Thermal Expansion Coefficient:

No.	Value	Note	Type
1	0.0001 K^{-1}	ASTM D696 [1]	V

Deflection Temperature:

No.	Value	Note
1	180°C	0.45 MPa, ASTM D645
2	80°C	1.8 MPa, ASTM D645 [1]

Transport Properties:
Water Absorption:

No.	Value	Note
1	0.07 %	24h, ASTM D570 [1]

Mechanical Properties:
Flexural Modulus:

No.	Value	Note
1	2300 MPa	ASTM D790 [1]

— Polybutylene terephthalate, Structural foam

Tensile Strength Yield:

No.	Value	Note
1	55 MPa	ASTM D638 [1]

Hardness: Rockwell M70 (ASTM D785) [1]
Izod Notch:

No.	Value	Notch	Note
1	50 J/m	Y	23°, ASTM D256A [1]

Electrical Properties:
Dielectric Strength:

No.	Value	Note
1	20 kV.mm^{-1}	ASTM D149 [1]

Polymer Stability:
Upper Use Temperature:

No.	Value	Note
1	120°C	[1]

Flammability: Flammability rating HB (UL 94, ASTM D635) [1]

Applications/Commercial Products:
Mould Shrinkage (%):

No.	Value	Note
1	2%	ASTM D955 [1]

Trade name	Details	Supplier
Celanex	non-exclusive	Hoechst
Pibiter	non-exclusive	Enichem
Thermocomp	non-exclusive	LNP Engineering
Thermofil	non-exclusive	Thermofil
Valox	non-exclusive	General Electric Plastics

[1] *The Materials Selector,* 2nd edn., (eds. N.A. Waterman and M.F. Ashby), Chapman and Hall, 1997, **3**, 372, (mechanical, electrical and thermodynamic props.)

Polybutylene terephthalate, Structural foam P-113

Related Polymers: Polybutylene terephthalate, Polybutylene terephthalate, Injection moulding, Glass fibre filled, Polybutylene terephthalate, Flame retardant, Polybutylene terephthalate, High impact, Polybutylene terephthalate, Fibres, Polybutylene terephthalate, Films/Sheets, Polybutylene terephthalate, Injection moulding, Mineral filled, Polybutylene terephthalate, Injection moulding, Carbon fibre filled, Polybutylene terephthalate, Polyethylene terephthalate blends, Polybutylene terephthalate, Phenoxy blends, Polybutylene terephthalate, PTFE lubricated, Polybutylene terephthalate, Silicone lubricated, Polybutylene terephthalate, Uv stabilised, Poly(butylene terephthalate-*co*-butylene ether glycol terephthalate)
Monomers: 1,4-Butanediol, Terephthalic acid
Material class: Thermoplastic
Polymer Type: PBT and copolymers

CAS Number:

CAS Reg. No.	Note
26062-94-2	
24968-12-5	repeating unit
30965-26-5	from dimethyl terephthalate
59822-52-5	from terephthaloyl chloride

Molecular Formula: $(C_{12}H_{12}O_4)_n$
Fragments: $C_{12}H_{12}O_4$

Volumetric & Calorimetric Properties:
Density:

No.	Value	Note
1	d 1.29 g/cm^3	ASTM D792 [1]

Thermal Expansion Coefficient:

No.	Value	Note	Type
1	9×10^{-5} K^{-1}	ASTM D696 [1]	V

Deflection Temperature:

No.	Value	Note
1	140°C	0.45 MPa, ASTM D645
2	80°C	1.8 MPa, ASTM D645 [1]

Transport Properties:
Water Absorption:

No.	Value	Note
1	0.07 %	24h, ASTM D570 [1]

Mechanical Properties:
Flexural Modulus:

No.	Value	Note
1	2200 MPa	ASTM D790 [1]

Tensile Strength Break:

No.	Note	Elongation
1	ASTM D638 [1]	4%

Tensile Strength Yield:

No.	Value	Note
1	55 MPa	ASTM D638 [1]

Hardness: Rockwell M50 (ASTM D785) [1]
Izod Notch:

No.	Value	Notch	Note
1	50 J/m	Y	23°, ASTM D256A [1]

– Polybutylene terephthalate, Uv stabilised

Electrical Properties:
Dielectric Strength:

No.	Value	Note
1	18 kV.mm^{-1}	ASTM D149 [1]

Polymer Stability:
Upper Use Temperature:

No.	Value	Note
1	120°C	[1]

Flammability: Flammability rating HB (UL 94, ASTM D635) [1]

Applications/Commercial Products:
Mould Shrinkage (%):

No.	Value	Note
1	1.5%	ASTM D955 [1]

Trade name	Details	Supplier
Celanex	non-exclusive	Hoechst
Pibiter	non-exclusive	Enichem
Thermofil	non-exclusive	Thermofil
Valox	non-exclusive	General Electric Plastics

[1] *The Materials Selector,* 2nd edn., (eds. N.A. Waterman and M.F. Ashby), Chapman and Hall, 1997, **3**, 372, (mechanical, electrical and thermodynamic props.)

Polybutylene terephthalate, Uv stabilised   P-114

Related Polymers: Polybutylene terephthalate, Polybutylene terephthalate, Injection moulding, Glass fibre filled, Polybutylene terephthalate, Flame retardant, Polybutylene terephthalate, High impact, Polybutylene terephthalate, Fibres, Polybutylene terephthalate, Films/Sheets, Polybutylene terephthalate, Injection moulding, Mineral filled, Polybutylene terephthalate, Injection moulding, Carbon fibre filled, Polybutylene terephthalate, Phenoxy blends, Polybutylene terephthalate, Structural foam, Polybutylene terephthalate, PTFE lubricated, Poly(butylene terephthalate-*co*-butylene ether glycol terephthalate, Polybutylene terephthalate, Polyethylene terephthalate blends, Polybutylene terephthalate, Silicone lubricated
Monomers: 1,4-Butanediol, Terephthalic acid
Material class: Thermoplastic
Polymer Type: PBT and copolymers
CAS Number:

CAS Reg. No.	Note
26062-94-2	
24968-12-5	repeating unit
30965-26-5	from dimethyl terephthalate
59822-52-5	from terephthaloyl chloride

Molecular Formula: $(C_{12}H_{12}O_4)_n$
Fragments: $C_{12}H_{12}O_4$
Additives: Carbon black improves uv stability

Volumetric & Calorimetric Properties:
Density:

No.	Value	Note
1	d 1.3 g/cm^3	ASTM D792 [1]

Thermal Expansion Coefficient:

No.	Value	Note	Type
1	0.00012 K^{-1}	ASTM D696 [1]	V

Deflection Temperature:

No.	Value	Note
1	150°C	0.45 MPa, ASTM D645
2	60°C	1.8 MPa, ASTM D645 [1]

Transport Properties:
Water Absorption:

No.	Value	Note
1	0.13 %	24h, ASTM D570 [1]

Mechanical Properties:
Flexural Modulus:

No.	Value	Note
1	2100 MPa	ASTM D790 [1]

Tensile Strength Break:

No.	Note	Elongation
1	ASTM D638 [1]	200%

Tensile Strength Yield:

No.	Value	Note
1	50 MPa	ASTM D638 [1]

Hardness: Rockwell M70 (ASTM D785) [1]
Izod Notch:

No.	Value	Notch	Note
1	50 J/m	Y	23°, ASTM D256A [1]

Electrical Properties:
Dielectric Strength:

No.	Value	Note
1	20 kV.mm^{-1}	ASTM D149 [1]

Polymer Stability:
Upper Use Temperature:

No.	Value	Note
1	120°C	[1]

Flammability: Flammability rating HB (UL94) [1]
Applications/Commercial Products:
Mould Shrinkage (%):

No.	Value	Note
1	2%	ASTM D955 [1]

Trade name	Details	Supplier
Celanex	non-exclusive	Hoechst
Gastine	non-exclusive	DuPont
Later	non-exclusive	LATI Industria Thermoplastics S.p.A.
Orgater	non-exclusive	EY Atochem
Pibiter	non-exclusive	Enichem
Pocan	non-exclusive	Bayer Inc.
Techster	non-exclusive	Rhone-Poulenc
Thermofil	non-exclusive	Thermofil
Ultradur	non-exclusive	BASF
Valox	non-exclusive	General Electric Plastics
Vestodur	non-exclusive	Hüls AG

[1] *The Materials Selector*, 2nd edn., (eds. N.A. Waterman and M.F. Ashby), Chapman and Hall, 1997, **3**, 372, (mechanical, electrical and thermodynamic props.)

Poly(butylethylsilane) P-115

$$\left[\begin{array}{c} CH_2CH_2CH_2CH_3 \\ | \\ Si \\ | \\ CH_2CH_3 \end{array} \right]_n$$

Synonyms: *Poly(butylethylsilylene). Poly(ethyl-n-butylsilylene). PBES. PEBS*
Polymer Type: polysilanes
CAS Number:

CAS Reg. No.
132230-58-1

Molecular Formula: $(C_6H_{14}Si)n$
Fragments: $C_6H_{14}Si$
Molecular Weight: $2.0 \times 10^4 - 1.4 \times 10^6$ [2]
Morphology: Adopts a hexagonal columnar liquid crystalline structure at room temperature and below. Crystallinity is less than that of the polylsilylenes having symmetrical side-chains. Conformations and packings have been proposed on the basis of experimental and theoretical X-ray diffraction patterns. The Si backbone of the polymer has a helical 7/3 conformation, whereas the backbones of poly(PrMesilane) and poly(iso-PrMesilane) have *all-trans* conformations.
Unit cell parameters a=1.0865, b=1.0865, c=1.3880nm and γ=60° [1,2,5,6]

Volumetric & Calorimetric Properties:
Melting Temperature:

No.	Value	Note
1	106°C	low MW [2]
2	185°C	high MW [2]

Transport Properties:
Transport Properties General: Flow properties have been investigated as a function of molecular weight [2].

Optical Properties:
Optical Properties General: Two successive thermochromic transitions observed in the UV, one near the first-order exothermic transition and one near the -20°C transition [2].
Transmission and Spectra: Uv-vis [2] spectra reported.

Polymer Stability:
Thermal Stability General: Undergoes a weak endothermic transition at -20°C and a first-order phase transition to a nematic liquid crystalline form at 90°C for the low and 170°C for the high MW fraction [2]

Applications/Commercial Products:
Processing & Manufacturing Routes: Synthesised by Na coupling of ethyl-*n*-butyldichorosilane, into two fractions with differing molecular weights [2].
Applications: Electrophotographic photoreceptor having polysilane-containing excellent charge-transport layer and gives a low residual potential [3,4].

Bibliographic References
[1] Furukawa, S. and Koga, T., *J. Phys.: Condens. Matter*, 1997, **9**, L99
[2] Asuke, T. and West, R., *J. Inorg. Organomet. Polym.*, 1994, **4**, 45
[3] *Jpn. Pat.*, 1992, 04 151 669
[4] *Jpn. Pat.*, 1990, 02 150 853
[5] Furukawa, S., *J. Organomet. Chem.*, 2000, **611**, 36
[6] Furukawa, S., Takeuchi, K. and Shimana, M., *J. Phys.: Condens. Matter*, 1994, **6**, 11007

Poly(butylhexylsilane) P-116

$$\left[\begin{array}{c} CH_2(CH_2)_4CH_3 \\ | \\ Si \\ | \\ CH_2CH_2CH_2CH_3 \end{array} \right]_n$$

Synonyms: *Poly(butylhexylsilylene). n-Butylhexyldichlorosilane homopolymer. PBHS*
Polymer Type: polysilanes
CAS Number:

CAS Reg. No.
125121-32-6

Molecular Formula: $(C_{10}H_{22}Si)n$
Fragments: $C_{10}H_{22}Si$
Molecular Weight: 2×10^6 [1,2,8]
General Information: Atactic. Crystallises with difficulty [2]. Properties widely different from corresponding dibutyl or dihexyl polymers [1].
Morphology: Rubbery solid at room temperature. Exists at high temperature as a hexagonal columnar mesophase (hcm) and undergoes a complicated first-order phase transition on cooling. The process of the phase transformation begins at *ca.* -23°C and continues down to *ca.* -63°C. Depending on the mode of cooling in the temperature interval from -23°C to -33°C, this transformation can proceed differently, but it necessarily involves a transformation of the hcm into a 7/3 helical form with simultaneous appearance of a phase with the planar zigzag *all-trans* backbone conformation. The *all-trans* form does not appear directly from the hcm. The process of nucleation and growth of the form within the helical form is kinetically limited and does not take place to completion but ceases at the point of a quasi-second-completion order phase transition of a glassification type. Even at -246°C, the polymer exists as a frozen mixture of at least two forms [1,2].
Identification: Calculated: C, 70.6; H, 12.9; Si, 16.5. Found: C, 69.9; H, 12.4: Si. 16.6.

— Poly(n-butyl methacrylate)

Volumetric & Calorimetric Properties:
Volumetric Properties General: Oxidation Potential: Eonset 1.00V. Epeak 1.14V. λ_{max} approx. 317nm
Glass-Transition Temperature:

No.	Value	Note
1	-40°C	[1,2]

Optical Properties:
Optical Properties General: Thin films show a broad UV absorption band at 320 nm above 25°C. When cooled, λ_{max} shifts hypsochromically to 312nm at approx. -25°C, and at still lower temperatures (approx. -40°C), a new band is seen at 345nm, suggesting two different phase transitions occur below 0°C [1,2].
Transmission and Spectra: Ir [2,6], Raman [2,6], uv [1,2,6,7] and nmr [1] spectra reported.

Polymer Stability:
Thermal Stability General: DSC shows a second-order discontinuity due to a probable glass-transition at -40°C and an endotherm with ΔH=4.6cal/g from a first-order melting-type transition at -22°C. A weak tan delta peak accompanied by a sharp decrease in loss modulus is observed near -45°C, and a rapid increase in tan delta and further decrease in modulus are found near -20°C, at the first-order transition. The phase transition at -20°C appears to represent a partial crystallisation, perhaps due to locking-in of motion of the side chains, as has been observed for PDHS. The second-order transition at -45°C may be associated with loss of motion of the main chain [1].

Applications/Commercial Products:
Processing & Manufacturing Routes: Prepared from n-butyl-n-hexyldichlorosilylene by condensation with Na metal in toluene at 110°C, by the usual Wurtz-type reaction [1].
Applications: Optical functional device [4]. Electrophotographic photoreceptor [6].

Bibliographic References
[1] Asuke, T. and West, R., *Macromolecules*, 1991, **24**, 343
[2] Bukalov, S, S., Leites, L.A., West, R. and Asuke, T., *Macromolecules*, 1996, **29**, 907
[3] Diaz, A.F., Baler, M., Wallraff, G.M. and Miller, R.D., *J. Electrochem. Soc.*, 1991, **138**, 742
[4] *Jpn. Pat.*, 1995, 07 239 487
[5] Bukalov, S.S., Leites, L.A., West, R. and Asuke, T., *Mendeleev Commun.*, 1994, **6**, 205
[6] *Jpn. Pat.*, 1990, 02 150 853
[7] Wallraff, G.M., Baier, M., Miller, R.D., Rabolt, J.F., Hallmark, V., Cotts, P. and Shukla, P., *Polym. Prepr. (Am. Chem. Soc., Div. Polym. Chem.)*, 1989, **30**, 245
[8] Embs, F.W., Wegner, G., Neher, D., Albouy, P., Miller, R.D., Willson, C.G. and Schrepp, W., *Macromolecules*, 1991, **24**, 6068

Poly(n-butyl methacrylate) P-117

$$\mathrm{-[CH_2C(CH)_3\; |\; O{=}C{-}OCH_2CH_2CH_2CH_3]-}_n$$

Synonyms: PBMA. Poly(butyl 2-methyl-2-propenoate). Poly[1-(butoxycarbonyl)-1-methylethylene]. Acrylic
Related Polymers: Polymethacrylates-General
Monomers: Butyl methacrylate
Material class: Thermoplastic
Polymer Type: acrylic
CAS Number:

CAS Reg. No.	Note
9003-63-8	
27836-17-5	syndiotactic

Molecular Formula: $(C_8H_{14}O_2)_n$
Fragments: $C_8H_{14}O_2$
Tacticity: Isotactic, syndiotactic and atactic forms known

Volumetric & Calorimetric Properties:
Density:

No.	Value	Note
1	d 1.06 g/cm³	[1,2,3]
2	d 1.05 g/cm³	Uv polymerisation 25°, T_g
3	d 0.99 g/cm³	Uv polymerisation, 120°

Thermal Expansion Coefficient:

No.	Value	Note	Type
1	0.00061 K⁻¹	Uv polymerisation; T >T_g [4]	V
2	0.000605 K⁻¹	free radical polymerisation; T >T_g [5]	V
3	0.00038 K⁻¹	Uv polymerisation; TTg [4]	V
4	0.000436 K⁻¹	Uv polymerisation; TTg [5]	V

Volumetric Properties General: Molar volume 86.8 cm³ mol⁻¹ (amorph. polymer, 25°); 135 cm³ mol⁻¹ (rubbery polymer)
Equation of State: Mark-Houwink constants have been reported. [28,29] Tait, Simka-Somcynsky, Bruce-Hartmann and Sanchez and Lacombe equation of state information has been reported [5,6,7,8]
Specific Heat Capacity:

No.	Value	Note	Type
1	1.68 kJ/kg.C	solid, 25° [9]	P
2	1.86 kJ/kg.C	liq., 25° [9]	P
3	263.4 kJ/kg.C	melt, 27° [10]	P

Melting Temperature:

No.	Value	Note
1	17°C	refractometric method [1]

Glass-Transition Temperature:

No.	Value	Note
1	20°C	Uv polymerisation; dilatometric method [4,5]
2	20°C	atactic [12]
3	81°C	syndiotactic [9,13]
4	88°C	syndiotactic [12,14]
5	-24°C	isotactic [9,12,13,14,15]

Vicat Softening Point:

No.	Value	Note
1	30°C	[1]

Brittleness Temperature:

No.	Value	Note
1	18°C	refractometric method [11]

Transition Temperature:

No.	Value	Note	Type
1	-158 – -140°C	[16]	Tγ

Surface Properties & Solubility:

Cohesive Energy Density Solubility Parameters: δ 30.9–46.5 $J^{1/2}$ $cm^{-3/2}$ (7.4–11.1 $cal^{1/2}$ $cm^{-3/2}$, poor H-bonding solvents) [20], 30.9–41.4 $J^{1/2}$ $cm^{-3/2}$ (7.4–9.9 $cal^{1/2}$ $cm^{-3/2}$, moderate H-bonding solvents) [20], 39.8–47.7 $J^{1/2}$ $cm^{-3/2}$ (9.5–11.4 $cal^{1/2}$ $cm^{-3/2}$, strong H-bonding solvents) [20], 17.8–18.4 $J^{1/2}$ $cm^{-3/2}$ [9], 36 $J^{1/2}$ $cm^{-3/2}$ (8.6 $cal^{1/2}$ $cm^{-3/2}$) [21], 30.1 $J^{1/2}$ $cm^{-3/2}$ (7.18 $cal^{1/2}$ $cm^{-3/2}$) [22], 36.6–37.9 $J^{1/2}$ $cm^{-3/2}$ (8.75–9.05 $cal^{1/2}$ $cm^{-3/2}$) [23], cohesive energy 43.5–46.5 kJ mol^{-1} [9], 38–45.043 kJ mol^{-1} (9100–10750 cal mol^{-1}; varying methods) [17,18], 41.5 kJ mol^{-1} [12], 339 J cm^{-3} (35°; 81 cal cm^{-3}) [19], γ_d 20.3, (140°), γ_p 3.8 (140°) [27] γ_d 31.3, γ_p 2.0

Solvents/Non-solvents: Sol. hexane, cyclohexane, gasoline, Nujol (hot), turpentine, castor oil (hot), linseed oil (hot), CCl_4, EtOH (hot), isopropanol (>23.7°), Et_2O, Me_2CO. Insol. formic acid, EtOH (cold) [24]

Surface Tension:

No.	Value	Note
1	31.2 mN/m	20°, MV 37000
2	32 mN/m	critical surface tension [27]

Transport Properties:

Polymer Solutions Dilute: Theta temps. [29,30,31,32], 10.7° (isobutanol), 84.8° (n-decane), 23.6° (DMF), 101° (n-dodecane), 119.8° (hexadecane), 87.9° (methyl carbitol), 59.6° (methyl cellosolve), 68.9° (n-octane), 20.9° (isopropanol), 21.5° (isopropanol), 23.7° (isopropanol), 25° (isopropanol). Polymer-liq. interaction parameters [33], 0.78 (hexane), 0.62 (2-butanone), 1.4 (acetonitrile), 0.1 ($CHCl_3$). 1.01 (n-propanol), 0.93 (Me_2CO), 0.67 (EtOAc)

Permeability of Liquids: Diffusion coefficients [EtOAc] 0.145–0.174 × 10^{-7} cm^2 s^{-1} (20°, MW 254000); [THF] 0.181–0.191 × 10^{-7} (MW 340000–380000); [H_2O] 11.4 × 10^{-7} (37°) [26,34]

Water Absorption:

No.	Value	Note
1	0.9%	max. [26]

Mechanical Properties:

Tensile Strength Break:

No.	Value	Note	Elongation
1	6.9 MPa	[35,36]	230%
2	3.4 MPa	23° [2]	300%

Viscoelastic Behaviour: Williams-Landel-Ferry constants for temp. superposition [16,37]; and Vogel parameters [37] have been reported. Other viscoelastic props. have been reported [41]

Hardness: Knoop number

Electrical Properties:

Dielectric Permittivity (Constant):

No.	Value	Note
1	2.5–3.1	[9]

Optical Properties:

Refractive Index:

No.	Value	Note
1	1.483	[9,38]

Total Internal Reflection: Intrinsic segmental anisotropy, -14 × 10^{-25} cm^3 (atactic, C_6H_6), -2 × 10^{-25} cm^3 (isotactic, C_6H_6)

Polymer Stability:

Polymer Stability General: Large doses of ionising radiation cause brittleness, strength loss and general loss of desirable props. Polymethacrylates exhibit predominantly main-chain fracture during irradiation and the deterioration of props is due to lowering of MW

Decomposition Details: Decomposition at 250° yields 40% monomer and traces of 1-butene; at 170° with 253.7 nm radiation decomposition yields 100% monomer [40]

Stability Miscellaneous: Undergoes shear-induced degradation in soln. in poor solvents [41]

Applications/Commercial Products:

Trade name	Details	Supplier
Elvacite	2044	ICI, UK
Neocryl		Zeneca Netherlands
Plexigum	P range	Rohm and Haas
Standard Samples	MW 200000–5000000	Aldrich
Standard Samples		Polysciences Inc
Standard Samples		Scientific Polymer Products Inc.

Bibliographic References

[1] Hoff, E.A.W., Robinson, D.W. and Willbourn, A.H., *Polym. Sci.*, 1955, **18**, 161
[2] *Elvacite*, ICI, 1996, (technical datasheet)
[3] Chee, K.K., *J. Appl. Polym. Sci.*, 1987, **33**, 1067
[4] Rogers, S.S. and Mandelkern, L., *J. Phys. Chem.*, Elsevier, 1957, **61**, 985
[5] Beret, S. and Prausnitz, J.M., *Macromolecules*, 1975, **8**, 206
[6] Olabisi, O. and Simha, R., *Macromolecules*, 1975, **8**, 211, (eqn of state)
[7] Hartmann, B., Simha, R. and Berger, A.E., *J. Appl. Polym. Sci.*, 1991, **43**, 983, (eqn of state)
[8] Sanchez, I.C. and Lacombe, R.H., *J. Polym. Sci., Polym. Lett. Ed.*, 1977, **15**, 71, (eqn of state)
[9] Van Krevelen, D.W., *Properties of Polymers: Their Correlation with Chemical Structure*, Elsevier, 1976, 79
[10] Gaur, U., Lau, S.F., Wunderlich, B.B. and Wunderlich, B., *J. Phys. Chem. Ref. Data*, 1982, **11**, 1065, (heat capacity)
[11] Wiley, R.H. and Brauer, G.M., *J. Polym. Sci.*, 1948, **3**, 647
[12] Porter, D., *Group Interaction Modelling of Polymer Properties*, Dekker, 1995, 283
[13] Crawford, J.W.C., *J. Soc. Chem. Ind., London*, 1949, **68**, 201
[14] Karey, F.E. and MacKnight, W.J., *Macromolecules*, 1968, **1**, 537
[15] Shetter, J.A., *J. Polym. Sci., Part B: Polym. Phys.*, 1962, **1**, 209
[16] Plazek, D.J. and Ngai, K.L., *Physical Properties of Polymers Handbook*, (ed. J.E. Mark), AIP Press, 1996, **139**, 341
[17] Eskin, V.Y. and Nesterov, A.Y., *Polym. Sci. USSR (Engl. Transl.)*, 1966, **8**, 1153, (cohesive energy)
[18] Eskin, V.Y. and Nesterov, A.Y., *J. Polym. Sci.*, 1967, **16**, 1619
[19] Lee, W.A. and Sewell, J.H., *J. Appl. Polym. Sci.*, 1968, **12**, 1397
[20] Mangaraj, D., Patra, S. and Rashid, S., *Makromol. Chem.*, 1963, **65**, 39, (solubility parameters)
[21] Allen, G., Sims, D. and Wilson, G.J, *Polymer*, 1961, **2**, 375
[22] DiPaola-Baranyi, G., *Macromolecules*, 1982, **15**, 622, (solubility parameter)
[23] Mangaraj, D., Patra, S. and Roy, P.C., *Makromol. Chem.*, 1965, **81**, 173
[24] Strain, D.E., Kennelly, R.G. and Dittmar, H.R., *Ind. Eng. Chem.*, 1939, **31**, 382, (solubility)
[25] Wu, S., *J. Polym. Sci., Part C: Polym. Lett.*, 1971, **34**, 19, (interfacial tension)
[26] Kalachandra, S. and Kusy, R.P., *Polymer*, 1991, **32**, 2428
[27] Wu, S., *J. Phys. Chem.*, Springer-Verlag, 1968, **72**, 3332
[28] Kurata, M. and Stockmayer, W.H., *Adv. Polym. Sci.*, Springer-Verlag, 1963, **3**, 196, (eqn of state)
[29] Chinai, S.N. and Guzzi, R.A., *J. Polym. Sci.*, 1956, **21**, 417
[30] Chinai, S.N. and Valles, R.J., *J. Polym. Sci.*, Springer-Verlag, 1959, **39**, 363
[31] Lath, D. and Bohdanecky, M, *J. Polym. Sci., Polym. Lett. Ed.*, 1977, **15**, 555
[32] Nakajima, A. and Okazaki, K, *Kobunshi Kagaku*, 1965, **22**, 791, (theta temps)
[33] Walsh, D.J. and McKeown, J.G., *Polymer*, 1980, **21**, 1335

[34] Kunz, D., Thurn, A. and Burchard, W., *Colloid Polym. Sci.*, 1980, **21**, 1335
[35] Brearley, W.H., *J. Paint Varnish Products*, 1983, **63**, 23
[36] Cramer, A.S., *Kunststoffe*, 1940, **30**, 337
[37] Child, W.C. and Ferry, J.D., *J. Colloid Sci.*, 1957, **12**, 327, (dynamic mech props)
[38] Lewis, O.G., *Physical Constants of Linear Hydrocarbons*, Springer-Verlag, 1968
[39] Tsvetkov, V.N., Ljubina, S. Ja, *Vysokomol. Soedin.*, 1968, **1**, 577
[40] Grassie, N. and MacCallum, J.R., *J. Polym. Sci., Part A: Polym. Chem.*, 1964, **27**, 59, (shear degradation)
[41] Herold, F.K., Schultz, G.V. and Wolf, B.A., *Polym. Commun.*, 1986, **2**, 983, (degradation)
[42] Aharoni, S.M., *Macromolecules*, 1983, **16**, 1722

Poly(*sec*-butyl methacrylate)   P-118

Synonyms: *Poly(2-methylbutyl methacrylate). Poly(sec-butyl 2-methyl-2-propenoate)*
Related Polymers: Polymethacrylates-General
Monomers: 1-Methylpropyl methacrylate
Material class: Thermoplastic
Polymer Type: acrylic
CAS Number:

CAS Reg. No.
29356-88-5

Molecular Formula: $(C_8H_{14}O_2)_n$
Fragments: $C_8H_{14}O_2$

Volumetric & Calorimetric Properties:
Density:

No.	Value	Note
1	d 1.052 g/cm^3	[1,2]

Thermal Expansion Coefficient:

No.	Value	Note	Type
1	0.00066 K^{-1}	M_n 160000; T > T_g [3]	V
2	0.00035 K^{-1}	M_n 160000; T g [3]	V

Volumetric Properties General: Molar volume 86.8 cm^3 mol^{-1} (van der Waals, amorph., 25°) [2]; 135.2 cm^3 mol^{-1} (glassy, amorph. 25°) [2]

Glass-Transition Temperature:

No.	Value	Note
1	60°C	atactic [2,4]
2	58–62°C	free-radical bulk polymerisation, 60°; M_n 160000 [3]

Vicat Softening Point:

No.	Value	Note
1	60°C	[1]

Surface Properties & Solubility:
Cohesive Energy Density Solubility Parameters: δ 36.5 J$^{1/2}$ cm$^{3/2}$ (8.72 cal$^{1/2}$ cm$^{3/2}$) [5], 30.1 J$^{1/2}$ cm$^{3/2}$ (7.2 cal$^{1/2}$ cm$^{3/2}$) [6]

Mechanical Properties:
Hardness: Vickers 10 (5 kg load) [4]

Optical Properties:
Refractive Index:

No.	Value	Note
1	1.48	20° [4]

Polymer Stability:
Decomposition Details: Decompostion at 250° yields monomer and small amounts of olefin by side-chain cracking [4]

Bibliographic References
[1] Hoff, E.A.W., Robinson, D.W. and Willbourn, A.H., *Polym. Sci.*, 1955, **18**, 161
[2] Van Krevelen, D.W., *Properties of Polymers: Their Correlation with Chemical Structure*, Elsevier, 1976, 79
[3] Krause, S., Gormley, J.J., Roman, N., Shetter, J.A. and Watenabe, W., *J. Polym. Sci., Part A: Polym. Chem.*, 1965, **3**, 3573, (glass transition temps)
[4] Crawford, J.W.C., *J. Soc. Chem. Ind., London*, 1949, **68**, 201
[5] Frank, C.W. and Gashgari, M.A., *Macromolecules*, 1979, **12**, 163
[6] DiPaola-Baranyi, G., *Macromolecules*, 1982, **15**, 622

Poly(*tert*-butyl methacrylate)   P-119

Synonyms: *Poly(1,1-dimethylethyl methacrylate). Poly(tert-butyl 2-methyl-2-propenoate)*
Related Polymers: Polymethacrylates-General
Monomers: *tert*-Butyl methacrylate
Material class: Thermoplastic
Polymer Type: acrylic
CAS Number:

CAS Reg. No.
25189-00-8

Molecular Formula: $(C_8H_{14}O_2)_n$
Fragments: $C_8H_{14}O_2$
Tacticity: Syndiotactic, atactic and isotactic forms known

Volumetric & Calorimetric Properties:
Density:

No.	Value	Note
1	d 1.022 g/cm^3	[1,2]
2	d^{20} 1.021 g/cm^3	[3]

Thermal Expansion Coefficient:

No.	Value	Note	Type
1	0.00072 K^{-1}	M_n 250000; T > T_g [4]	V
2	0.00028 K^{-1}	M_n 250000; T Tg [4]	V

Volumetric Properties General: Molar volumes 86.8 cm^3 mol^{-1} (amorph. 25° Van der Waals) [2], 139.1 cm^3 mol^{-1}, (amorph. 25°, glassy) [2]
Equation of State: Mark-Houwink constants have been reported [14,17]

Melting Temperature:

No.	Value	Note
1	165°C	approx., syndiotactic [5]
2	104°C	approx., isotactic [5]

Glass-Transition Temperature:

No.	Value	Note
1	107°C	M_n 250000 [4]
2	122°C	[6,7]
3	118°C	atactic [4]
4	114°C	syndiotactic [4]
5	84°C	52% syndiotactic, 46% atactic, 2% isotactic [8]
6	47°C	99% isotactic, 1% atactic [8]
7	7°C	isotactic [4]

Vicat Softening Point:

No.	Value	Note
1	104°C	[1,9]

Surface Properties & Solubility:

Cohesive Energy Density Solubility Parameters: Cohesive energy 40100 J mol^{-1} [2]; 17 J$^{1/2}$ cm$^{-3/2}$ [2,10] 39.3 J$^{1/2}$ cm$^{-3/2}$ (9.4 cal$^{1/2}$ cm$^{-3/2}$) [11]; 36.2 J$^{1/2}$ cm$^{-3/2}$ (8.65 cal$^{1/2}$ cm$^{-3/2}$) [12]

Wettability/Surface Energy and Interfacial Tension: dγ/dT) 0.059 mN m^{-1} K^{-1} [14]. Contact angle 72° (advancing, H$_2$O) [15]. Interfacial tension 5.9 mN m^{-1} (polyethylene, branched; 20°) [14]; 3 mN m^{-1} (PMMA; 20°) [14]; 3.6 mN m^{-1} (polyoxydimethylsiloxane; 20°) [14]

Surface Tension:

No.	Value	Note
1	30.5 mN/m	20°, MV 6000 [14]

Transport Properties:

Polymer Solutions Dilute: Theta temps 64° (n-heptane, high syndiotactic content) [16]. 5° (cyclohexane). [14] Intrinsic viscosity varies with MW [16] η 0.13 dl g^{-1} (30°, THF), 0.124 dl g^{-1} (25°, butanone), 0.106 dl g^{-1} (10°, cyclohexane), 0.088 dl g^{-1} (64°, n-heptane)

Permeability of Liquids: Diffusion coefficient [H$_2$O] 10.3 × 10^{-8} cm^2 s^{-1} (37°) [15]

Water Absorption:

No.	Value	Note
1	0.5%	max. [15]

Mechanical Properties:
Hardness: Vickers 15 (5 kg load) [9]

Optical Properties:
Refractive Index:

No.	Value	Note
1	1.4638	20° [2]

Total Internal Reflection: Intrinsic segmental anisotropy 2.1 × 10^{-25} cm^3 (atactic; C$_6$H$_6$) [18]; 19.3 × 10^{-25} cm^3 (isotactic; C$_6$H$_6$) [18]

Polymer Stability:
Polymer Stability General: Large doses of ionising radiation cause brittleness, loss of strength and general loss of desirable props. Irradiation causes main-chain fracture, deterioration of props. is due to lowering of MW.
Thermal Stability General: Changes to the anhydride via the acid on heating [5]
Decomposition Details: Decomposition at 180–200° yields high levels of isobutylene and water, low levels of monomer, traces of methacrylic acid and a residue of polymethacrylic anhydride [19,20]. Decomposition below 180° with irradiation at 253.7 nm yields 100% monomer [9,19]

Bibliographic References
[1] Hoff, E.A.W., Robinson, D.W. and Willbourn, A.H., *Polym. Sci.*, 1955, **18**, 161
[2] Van Krevelen, D.W., *Properties of Polymers: Their Correlation with Chemical Structure*, Elsevier, 1976, 79
[3] Struik, L.C.E, *Polymer*, 1987, **28**, 1869
[4] Krause, S., Gormley, J.J., Roman, N., Shetter, J.A. and Watenabe, W.H., *J. Polym. Sci., Part A: Polym. Chem.*, 1965, **3**, 3573, (glass transition temps)
[5] Matsuzaki, K., Okamoto, T., Ishida, A. and Sobue, H., *J. Polym. Sci., Part A: Polym. Chem.*, 1964, **2**, 1105
[6] Lewis, O.G., *Physical Constants of Linear Hydrocarbons*, Springer-Verlag, 1968
[7] *Unpublished Data*, Rohm and Haas,
[8] Walstrom, A.M., Subramanian, R., Long, T.E., McGrath, J.E. and Ward, T.C., *Polym. Prepr. (Am. Chem. Soc., Div. Polym. Chem.)*, 1986, **32**, 135
[9] Crawford, J.W.C., *J. Soc. Chem. Ind., London, Trans. Commun.*, 1949, **68**, 201
[10] Mangaraj, D., Patra, S. and Roy, P.C., *Makromol. Chem.*, 1965, **81**, 173
[11] Gargallo, L. and Russo, M., *Makromol. Chem.*, 1975, **176**, 2735
[12] Frank, C.W. and Gashgari, M.A., *Macromolecules*, 1979, **12**, 163
[13] Barton, A.F.M., *CRC Handbook of Polymer-Liquid Interaction Parameters and Solubility Parameters*, CRC Press, 1990, 265
[14] Wu, S., *J. Polym. Sci., Part C: Polym. Lett.*, 1971, **34**, 19, (interfacial tension)
[15] Kalachandra, S. and Kusy, R.P., *Polymer*, 1991, **32**, 2428
[16] Karndinos, A., Nan, S. and Mays, J.W., *Macromolecules*, 1991, **24**, 2007, (soln props)
[17] Tricot, M., Bleus, J., Riga, J.-P. and Desreux, V., *Makromol. Chem.*, 1974, **175**, 913, (eqn of state)
[18] Tsvetkov, V.N., Boitsova, N.N. and Vitovskaja, M.G., *Vysokomol. Soedin.*, 1964, **6**, 297
[19] Grassie, N. and MacCallum, J.R., *J. Polym. Sci., Part A: Polym. Chem.*, 1964, **2**, 983, (thermal degradation)
[20] Grant, D.H. and Grassie, N., *Polymer*, 1960, **1**, 445, (thermal degradation)

Poly(butyl methacrylate-*co*-vinyl chloride)

$$\left[\left[CH_2C(CH_3) \right]_x \left[\begin{array}{c} CH_2CH \\ | \\ Cl \end{array} \right]_y \right]_n$$
$$ \| $$
$$ O OCH_2(CH_2)_2CH_3$$

Related Polymers: Polymethacrylates, General, Poly(butyl methacrylate), Poly(vinyl chloride)
Monomers: Butyl methacrylate, Vinyl chloride
Material class: Thermoplastic
Polymer Type: acrylic copolymers
Molecular Formula: $[(C_8H_{14}O_2)_x (C_2H_3Cl)_y]_n$
Fragments: $C_8H_{14}O_2$ C_2H_3Cl

Volumetric & Calorimetric Properties:
Glass-Transition Temperature:

No.	Value	Note
1	18–100°C	[1]

Transition Temperature General: T_g varies with the percentage content of methacrylate [1]

Transport Properties:
Polymer Solutions Dilute: Intrinsic viscosity varies with percentage content of methacrylate. η 0.34–0.59 dl g^{-1} [1]

[1] Johnstone, N.W., Polym. Prepr. (Am. Chem. Soc., Div. Polym. Chem.), 1969, **10**, 608

Poly[(butylmethylsilane)methylene]   P-121

$$\left[-CH_2-\underset{\underset{CH_3}{|}}{\overset{\overset{CH_2CH_2CH_2CH_3}{|}}{Si}}-\right]_n$$

Synonyms: *Poly[(butylmethylsilylene)methylene]. PMBSM*
Related Polymers: Polycarbosilanes
Polymer Type: polycarbosilanes
CAS Number:

CAS Reg. No.
198066-89-6

Molecular Formula: $(C_6H_{14}Si)_n$
Fragments: $C_6H_{14}Si$
Molecular Weight: 73,600 [1]
General Information: Atactic configuration due to polymerisation with racemic monomer [1].
Morphology: Amorphous solid [1].
Identification: Found: C, 63.47%; H, 12.68%; Calculated: C, 63.71%; H, 12.39%.

Volumetric & Calorimetric Properties:
Glass-Transition Temperature:

No.	Value	Note
1	-63°C	[1]

Optical Properties:
Transmission and Spectra: Nmr [7] spectra reported.

Applications/Commercial Products:
Processing & Manufacturing Routes: Prepared by ring-opening polymerisation of 1,3-dibutyl-1,3-dimethyl-1,3-disilacyclobutane [1].
Applications:

[1] Shen, Q.H. and Interrante, L.V., J. Polym. Sci., *Part A*, 1997, **35**, 3193

Poly(butylpentylsilane)   P-122

$$\left[-\underset{\underset{CH_2CH_2CH_2CH_3}{|}}{\overset{\overset{CH_2(CH_2)_3CH_3}{|}}{Si}}-\right]_n$$

Synonyms: *Poly(butylpentylsilylene). Dichlorobutylpentylsilane homopolymer, sru. PBPS. C4–C5*
Polymer Type: polysilanes
CAS Number:

CAS Reg. No.
132230-79-6

Molecular Formula: $(C_9H_{20}Si)_n$
Fragments: $C_9H_{20}Si$
Molecular Weight: 1×10^6 [1,8]
General Information: Strongest of the poly(*n*-pentyl-*n*-alkylsilanes); Only member of the series with an order-disorder transition temperature >25°C [2]
Tacticity: Forms a strong, tough, but flexible film that feels similar to low-density polyethylene.
The film is not brittle and shows superior fracture toughness when compared with symmetric poly(di-n-alkylsilanes) [1].
Morphology: Has a 7/3 helical backbone conformation at 25°C for film or normal spherulites, as has been observed for poly(dibutylsilane) and poly(dipentylsilane). The X-ray diffraction pattern of a uniaxially oriented film of PBPS may be indexed with an orthorhombic unit cell containing hexagonally packed chains with a = 1.323 nm., b = 2.292 nm, and c = 1.388 nm. At 45°C a sharp first order phase transition to a hexagonal, columnar structure with lattice parameter a=b=13.4Å is observed [1,9].
Miscellaneous: Large polysilane spherulites have been observed. The spherulites were prepared by controlling the removal rate of solvents from PBPS under a xylene atmosphere. The diameter of the spherulites was greater than 0.1mm [8].

Volumetric & Calorimetric Properties:
Glass-Transition Temperature:

No.	Value	Note
1	45–50°C	[1,2,8,9]

Transport Properties:
Transport Properties General: Anomalous transition observed [5].

Mechanical Properties:
Mechanical Properties General: Films can be stretched to moderately high draw ratios. A film with a draw ratio of 4 was obtained for X-ray diffraction [1].

Tensile (Young's) Modulus:

No.	Value	Note
1	296 MPa	at 23°C [2]
2	50 MPa	at 100°C [2]

Miscellaneous Moduli:

No.	Value	Note
1	3.07 MPa	Yield stress [1]

Failure Properties General: The average crosshead displacement at which fracture initiated was 0.91±0.08 mm. Little or no dependence of this value on the initial crack length was found. Using an Instron tensile testing machine and specimens in the single edge notch (SEN) geometry yielded plane stress J1c (critical value for J for fracture initiation) of 1745±305J/m^2 [2].
Fracture Mechanical Properties: Average fracture strain of 85% for defect-free specimens at 23°C [2].

Optical Properties:
Optical Properties General: At 25°C the sample gives a relatively sharp absorption maximum at 314nm, which lies in the range (305–326nm) usually characteristic of a helical or disordered conformation of the silicon backbone. When cooled to -45°C, a very small shoulder appears at 338nm, but the intensity of the 314nm band does not diminish as the temperature is lowered further. Upon warming, the shoulder disappears [1,8].
Transmission and Spectra: Ir [4], Raman[3], uv [1,4,8] and nmr [3] spectra reported.

Polymer Stability:
Thermal Stability General: A second-order transition is observed at -44°C, followed by a sharp endotherm at 45°C possessing a heat of transition of 2.5cal g^{-1}. This sharp endotherm corresponds to a solid-state disordering transition seen in poly(dialkylsilanes) with side chains longer than three carbon atoms. It is seen that the

asymmetry of PBPS lowers the temperature of the disordering transition relative to those of other alkylpentylsilylenes [1,9].

Applications/Commercial Products:
Processing & Manufacturing Routes: Prepared via the dichlorosilane and then by the usual Wurtz condensation reaction with Na in toluene at 110°C [1,2,8].
Applications: Evaporated film for electroluminescent devices [6]. An electrophotographic photoreceptor; a binder resin for a single photosensitive layer in which a charge-generating substance and a charge-transporting substance are mixed [7].

Bibliographic References
[1] Klemann, B.M., West, R. and Koutsky, J.A., *Macromolecules*, 1993, **26**, 1042
[2] Klemann, B.M., DeVilbiss, T. and Koutsky, J.A., *Polym. Eng. Sci.*, 1996, **36**, 135
[3] Miller, R.D., Farmer, B.L., Fleming, W., Sooriyakumaran, R. and Rabolt, J., *J. Am. Chem. Soc.*, 1987, **109**, 2509
[4] Kigook, S., Miller, R.D., Wallraff, G.M. and Rabolt, J.F., *Macromolecules*, 1991, **24**, 4084
[5] Majima, Y., Nishizawa, H., Hiraoka, T., Nakano, Y. and Hayase, S., Jpn. J. Appl. Phys., Part 1, 1995, **34**, 3820
[6] Furukawa, S., *Thin Solid Films*, 1998, **331**, 222
[7] *Jpn. Pat.*, 1990, 02 150 853
[8] Majima, Y., Kawata, Y., Nakano, Y. and Hayase, S., J. Polym. Sci., Part A, 1997, **35**, 427
[9] Klemann, B.M., West, R. and Koutsky, J.A., *Macromolecules*, 1996, **29**, 198

Poly(3-butylthiophene)   P-123

Synonyms: *P3BT. Poly(3-butyl-2,5-thiophenediyl)*
Monomers: 3-Butylthiophene
Material class: Thermoplastic
Polymer Type: polythiophene
CAS Number:

CAS Reg. No.	Note
98837-51-5	
189102-88-3	Poly(3-butyl-2,5-thiophenediyl)

Molecular Formula: $(C_8H_{10}S)_n$
Fragments: $C_8H_{10}S$
Additives: Dopants such as hexafluorophosphate [1], perchlorate [2], I_2 and $FeCl_3$ are added to increase conductivity.
General Information: Conductive polymer which exhibits thermochromism [3] and has greater solubility and processability than polythiophene.
Morphology: Orthorhombic unit cell, a=13.18, b=7.52, c=7.77Å [4]. Polymorphic with two crystal phases and a nematic mesophase [5].

Volumetric & Calorimetric Properties:
Melting Temperature:

No.	Value	Note
1	222–243°C	[5]

Glass-Transition Temperature:

No.	Value	Note
1	58.7–75.4°C	Glass-transition temperature increases with addition of $FeCl_3$ dopant [6,7]

Surface Properties & Solubility:
Solvents/Non-solvents: Sol. $CHCl_3$, toluene [8,9], limited sol. THF [3].

Electrical Properties:
Electrical Properties General: Conductivity is dependent on polymer structure, nature and concentration of dopant and temperature, with typical values for polymer film from electrochemical synth. of 40 S cm^{-1} (perchlorate dopant) [2] and 110 S cm^{-1} (hexafluorophosphate dopant) [1].

Optical Properties:
Transmission and Spectra: Pmr [3,10], cmr [3], ir [3], uv [2,3] and x-ray [4] spectra have been reported.
Volume Properties/Surface Properties: Polymer has red or brown colour [3] with thermochromic changes on heating.

Polymer Stability:
Polymer Stability General: Photochemical degradation of polymer has been reported [9].
Decomposition Details: Decomposition temperature of polymer produced by chemical oxidation process 357.8°C [8].

Applications/Commercial Products:
Processing & Manufacturing Routes: Produced by chemical or electrochemical oxidative polymerisation of monomer with different catalysts [2,3,8,9] to produce mostly regiorandom (head-to-head/head-to-tail) polymers or by catalytic polymerisation of 2-bromo-5-(bromomagnesio)-3-dodecylthiophene [11,12] or 2-bromo-5-(bromozincio)-3-dodecylthiophene [3] to produce regioregular (head-to-tail) polymers.
Applications: Potential applications in photovoltaic devices, light-emitting diodes, rechargeable batteries, field-effect transistors and chemical and optical sensors [13].

Trade name	Details	Supplier
Rieke	regioregular and regiorandom polymers	Rieke Metals

Bibliographic References
[1] Sato, M., Tanaka, S. and Kaeriyama, K., *Polymer*, 1987, **28**, 1071
[2] Hotta, S., Rughooputh, S., Heeger, A.J. and Wudl, F., *Macromolecules*, 1987, **20**, 212
[3] Chen, T.-A., Wu, X. and Rieke, R.D., *J. Am. Chem. Soc.*, 1995, **117**, 233
[4] Kawai, T., Nakazono, M., Sugimoto, R. and Yashino K., *J. Phys. Soc. Jpn.*, 1992, **61**, 3400
[5] Causin, V., Marega, C., Marigo, A., Valenti, L. and Kenny, J.M., *Macromolecules*, 2005, **38**, 409
[6] Chen, S.-A. and Ni, J.M., *Polym. Bull. (Berlin)*, 1991, **26**, 673
[7] Chen, S.-A. and Ni, J.M., *Macromolecules*, 1992, **25**, 6081
[8] Liu, Y., Xu, Y. and Zhu, D., *Macromol. Chem. Phys.*, 2001, **202**, 1010
[9] Caronna, T., Forte, M., Catellani, M. and Meille, S.V., *Chem. Mater.*, 1997, **9**, 991
[10] Krische, B., Zagorska, M. and Hellberg, J., *Synth. Met.*, 1993, **58**, 295
[11] McCullough, R.D., Tristram-Nagle, S., Williams, S.P., Lowe, R.D. and Jayaraman, M., *J. Am. Chem. Soc.*, 1993, **115**, 4910
[12] Buvat, P. and Hourquebie, P., *Macromolecules*, 1997, **30**, 2685
[13] *Handbook of Conducting Polymers*, 2nd Edn., (Eds. Skotheim, T.A., Elsenbaumer, R.L. and Reynolds, J.R.) Marcel Dekker, 1998, 225

Poly(butyl vinyl ether)   P-124

Synonyms: *PVnBE. Poly(vinyl butyl ether). Poly(butoxyethene). Poly(1-(ethenyloxy)butane)*
Related Polymers: Poly(vinyl ethers)
Monomers: Butyl vinyl ether
Material class: Thermoplastic, Synthetic Elastomers
Polymer Type: polyvinyls
CAS Number:

CAS Reg. No.	Note
25232-87-5	homopolymer
28214-13-3	isotactic

– Poly(*sec*-butyl vinyl) ether

Molecular Formula: $(C_6H_{12}O)_n$
Fragments: $C_6H_{12}O$

Volumetric & Calorimetric Properties:
Density:

No.	Value	Note
1	d^{30} 0.926 g/cm^3	[6]

Thermal Expansion Coefficient:

No.	Value	Note	Type
1	2.42×10^{-5} K^{-1}	T > -55° [8]	L
2	1.3×10^{-5} K^{-1}	-105° < T < -55°	L
3	1.1×10^{-5} K^{-1}	T < -105°	L

Volumetric Properties General: Molar volume 108 cm^3 mol^{-1} [15]
Thermodynamic Properties General: Thermodynamic props. have been reported [13,14]
Glass-Transition Temperature:

No.	Value	Note
1	-55°C	[6]
2	-54°C	[13]

Vicat Softening Point:

No.	Value	Note
1	54–62°C	[3]

Transition Temperature:

No.	Value	Note	Type
1	-21°C	Inflection temp. [6]	
2	-105°C	Tgg (1) (168K) [8]	
3	-41°C	[6]	Inflection temp.
4	-32°C	[6]	Low temp. stiffening

Surface Properties & Solubility:
Cohesive Energy Density Solubility Parameters: δ 16.06 J$^{1/2}$ cm$^{3/2}$ (7.85 cal$^{1/2}$ cm$^{-3/2}$) [6], 18.1 J$^{1/2}$ cm$^{-3/2}$ (8.85 cal$^{1/2}$ cm$^{-3/2}$). Other solubility parameters have been reported [18]
Solvents/Non-solvents: Sol. cyclohexanone, C_6H_6, butanol, 2-butanone, toluene, cyclohexane, Et_2O. Insol EtOH, 2-ethoxyethanol, MeOH, Me_2CO, heptane [6,7,9]

Transport Properties:
Polymer Solutions Dilute: [η] 4.3 dl g^{-1} (C_6H_6, 30°) [6]. η$_{sp}$ 0.11–0.17 dl g^{-1} (C_6H_6, 30°) [16], 0.5 dl g^{-1} [17]. Huggins constant k$_H$ 2.15–8.61 [17]

Mechanical Properties:
Hardness: Shore A0 [3]

Optical Properties:
Refractive Index:

No.	Value	Note
1	1.4563	30° [6]
2	1.401	20° [3]

Molar Refraction: Molar refraction 29.5 [6]

Applications/Commercial Products:
Processing & Manufacturing Routes: May be polymerised by cationic initiation [1,2,3] or by the use of zeolites [10]. The isotactic polymer is prepared in toluene [9], the syndiotactic polymer in nitroethane or CH_2Cl_2 [9]. Most procedures use aluminium or iron-H_2SO_4 catalysts [11,12]

Bibliographic References
[1] Yagci, Y. and Schabel, W., Makromol. Chem., *Rapid Commun.*, 1987, **8**, 209, (polymerisation)
[2] Cho, C.G. and McGrath, J.E., Polym. Prepr. (Am. Chem. Soc., *Div. Polym. Chem.*), 1987, **28**, 455, (polymerisation)
[3] Fishbein, L. and Crowe, B.F., *Makromol. Chem.*, 1961, **48**, 221, (cationic polymerisation, props)
[4] Lal, J. and Scott, K.W., J. Polym. Sci., *Part C: Polym. Lett.*, 1965, **9**, 113, (props, struct)
[5] Lal, J. and McGrath, J.E., J. Polym. Sci., *Part A: Polym. Chem.*, 1964, **2**, 3369
[6] Lal, J. and Trick, G.S., J. Polym. Sci., *Part A-2*, 1964, **2**, 4559, (props)
[7] Vandenberg, E.J., Heck, R.F. and Breslow, D.S., *J. Polym. Sci.*, 1959, **41**, 519, (melting point, solubility)
[8] Haldon, R.A., Schell, W.J. and Sinha, R., J. Macromol. Sci., Phys. Part B, 1967, **1**, 759, (transition temp)
[9] Lal, J., McGrath, J.E. and Trick, G.S., J. Polym. Sci., *Part A-1*, 1967, **5**, 795, (polymerisation)
[10] Barrer, R.M. and Oei, A.T.T., *J. Catal.*, 1973, **30**, 460, (polymerisation)
[11] *U.S. Pat.*, 1968, 3 377 328
[12] *U.S. Pat.*, 1968, 3 386 980
[13] Aharoni, S.M., Polym. Prepr. (Am. Chem. Soc., *Div. Polym. Chem.*), 1973, **14**, 334
[14] Haldon, R.A., Schell, W.F. and Simha, R., J. Macromol. Sci., *Phys.*, 1967, 759
[15] Van Krevelen, D.W., *Properties of Polymers: Their Correlation with Chemical Structure*, Elsevier, 1972, 44
[16] Aoki, S. and Stille, J.K., *Macromolecules*, 1970, **3**, 473
[17] Manson, J.A. and Arquette, G.L., *Makromol. Chem.*, 1960, **37**, 187
[18] Ahmad, H. and Yaseen, M., *Polym. Eng. Sci.*, 1979, **19**, 858, (solubility parameters)

Poly(*sec*-butyl vinyl) ether

Synonyms: *PVsBE. PVSBE. Poly(2-methylpropyl vinyl) ether. Poly(vinyl sec-butyl ether). Poly(2-(ethenyloxy)butane). Poly(vinyl 2-butyl ether). Poly(sec-butoxy ethene)*
Related Polymers: Poly(vinyl ethers)
Monomers: *sec*-Butyl vinyl ether
Material class: Synthetic Elastomers
Polymer Type: polyvinyls
CAS Number:

CAS Reg. No.	Note
9003-44-5	homopolymer, atactic
29298-86-0	(±)-form
28214-12-2	isotactic homopolymer
28574-30-3	isotactic S-(-)-homopolymer

Molecular Formula: $(C_6H_{12}O)_n$
Fragments: $C_6H_{12}O$
Tacticity: Isotactic and atactic forms known
Morphology: Isotactic form tetragonal a = b 18.25A, c 35.5A [4] 17/5 helix [3,4]

Volumetric & Calorimetric Properties:
Density:

No.	Value	Note
1	d^{30} 0.924 g/cm^3	[1]
2	d^{30} 0.916 g/cm^3	[1]
3	d^{25} 0.924 g/cm^3	[7]

– Poly(*tert*-butyl vinyl) ether

Thermal Expansion Coefficient:

No.	Value	Note	Type
1	0.000212 K^{-1}	T > -32° [2]	L
2	0.000122 K^{-1}	-32° <T <-69°	L
3	0.000107 K^{-1}	-96° <T <-69° [2]	L
4	9.6 × 10^{-5} K^{-1}	T <-96°	L

Volumetric Properties General: Molar volume V_r 108.3 cm^3 mol^{-1} [7]
Thermodynamic Properties General: Thermodynamic props. have been reported [8]
Melting Temperature:

No.	Value	Note
1	170°C	[7]

Glass-Transition Temperature:

No.	Value	Note
1	-32°C	[2]
2	-20°C	[1]

Transition Temperature:

No.	Value	Note	Type
1	-69°C	[2]	T_{gg} (1)
2	-96°C	[2]	T_{gg} (2)
3	-5°C	[1]	Inflection temp.

Surface Properties & Solubility:
Solvents/Non-solvents: Very sol. Et$_2$O, pentane, heptane, C$_6$H$_6$ [6]. Prac. insol. 2-propanol [6]. Insol. MeOH, Me$_2$CO [6]. (*S*)-form sol. Me$_2$CO; insol. Et$_2$O

Transport Properties:
Polymer Solutions Dilute: η_{inh} 0.8 dl g^{-1} (C$_6$H$_6$, 30°) [1]

Mechanical Properties:
Hardness: Shore A10 (6 d) [9]

Optical Properties:
Refractive Index:

No.	Value	Note
1	1.474	30° [1]

Total Internal Reflection: [M]$_D^{25}$ +200 (Me$_2$CO, *S*-form), +272 (Me$_2$CO, Et$_2$O) [10], +312 (Et$_2$O) [10]

Applications/Commercial Products:
Processing & Manufacturing Routes: Polymerised using a tri-isopropyl aluminium(*III*) catalyst in H$_2$SO$_4$ [3]. May also be polymerised at low temp. [5]

Bibliographic References
[1] Lal, J. and Trick, G.S., *J. Polym. Sci., Part A: Polym. Chem.*, 1964, **2**, 4559, (physical props)
[2] Haldon, R.A., Schell, W.J. and Simha, R., *J. Macromol. Sci., Phys.*, 1967, **1**, 759, (transition temps, thermal expansion)
[3] Natta, G., Allegra, G., Bassi, I.W., Carlini, C., et al, *Macromolecules*, 1969, **2**, 311, (isomorphism)
[4] Natta, G., Bassi, I.W. and Allegra, G., *Makromol. Chem.*, 1965, **89**, 81, (crystalline isotactic form)
[5] Hallensleben, M.L. and Möller, K., *Polym. Bull. (Berlin)*, 1984, **11**, 7, (polymerisation)
[6] Dall'asta, G. and Oddo, N., *Chim. Ind. (Milan)*, 1960, **42**, 1234, (synth, solubility)
[7] Van Krevelen, D.W., *Properties of Polymers: Their Correlation with Chemical Structure*, Elsevier, 1972, 44, 387
[8] Aharoni, S.M., *Polym. Prepr. (Am. Chem. Soc., Div. Polym. Chem.)*, 1973, **14**, 334
[9] Schildknecht, C.E., *Polym. Prepr. (Am. Chem. Soc., Div. Polym. Chem.)*, 1973, **13**, 1071
[10] Pino, P., Lorenzi, G.P. and Chiellini, E., *J. Polym. Sci., Part C: Polym. Lett.*, 1965, **16**, 3279

Poly(*tert*-butyl vinyl) ether P-126

Synonyms: *PtBVE. PTBVE. PtBuVE. Poly(2,2-dimethylpropyl vinyl) ether. Poly(tert-butoxyethene). Poly(tert-butyl vinyl ether). Poly(vinyl 2,2-dimethylpropyl ether)*
Related Polymers: Poly(vinyl ethers)
Monomers: *tert*-Butyl vinyl ether
Material class: Synthetic Elastomers
Polymer Type: polyvinyls
CAS Number:

CAS Reg. No.	Note
25655-00-9	homopolymer
52079-39-7	syndiotactic homopolymer
25654-99-3	isotactic homopolymer

Molecular Formula: (C$_6$H$_{12}$O)$_n$
Fragments: C$_6$H$_{12}$O
Tacticity: Atactic, isotactic and syndiotactic forms known
Morphology: Isotactic form, tetragonal I4$_1$/a. a = b 18.84Å, c 7.65Å with 4/1 helix [3]

Volumetric & Calorimetric Properties:
Density:

No.	Value	Note
1	d 0.94 g/cm^3	[3]

Thermodynamic Properties General: Thermodynamic props. have been reported [6]
Melting Temperature:

No.	Value	Note
1	240–260°C	[1]
2	160°C	isotactic form [3]

Glass-Transition Temperature:

No.	Value	Note
1	88°C	[14]

Transition Temperature General: B. p. 77–80° [3]
Vicat Softening Point:

No.	Value	Note
1	68–79°C	[2]

Surface Properties & Solubility:
Solvents/Non-solvents: Very sol. *n*-heptane, C$_6$H$_6$ [3]. Mod. sol. Et$_2$O (hot); sol. Me$_2$CO, 2-butanone [15]. Isotactic form prac. insol. MeOH, Me$_2$CO, 2-propanol, Et$_2$O (cold). Insol. heptane, C$_6$H$_6$ [1]

Transport Properties:
Polymer Solutions Dilute: η_{inh} 0.37 dl g^{-1} [2]. η_{sp} 0.08 dl g^{-1} (c, 0.5 g dl^{-1}, C$_6$H$_6$, 30°) [16]

Poly(caprolactam-co-laurolactam)

Synonyms: *Nylon 6.12 copolyamide*
Related Polymers: Nylon 6, Nylon 12, Nylon 6.12
Monomers: ε Caprolactam, Laurolactam
Material class: Thermoplastic, Copolymers
Polymer Type: polyamide
Molecular Formula: $[(C_6H_{11}NO)_x(C_{12}H_{23}NO)_y]_n$
Fragments: $C_6H_{11}NO$ $C_{12}H_{23}NO$
General Information: Degree of crystallinity is a minimum for copolymers with 70:30–40:60 caprolactam:laurolactam ratio [1]. Composite crystallinity has been reported [2]

Volumetric & Calorimetric Properties:
Density:

No.	Value	Note
1	d 1.0098 g/cm^3	25% caprolactam, 75% laurolactam [2]
2	d 1.0414 g/cm^3	50% caprolactam, 50% laurolactam [2]

Thermodynamic Properties General: Heat of fusion 22.95 J g^{-1} (50% caprolactam, 50% laurolactam), 25.54 J g^{-1} (40% caprolactam, 60% laurolactam) [1]. Coefficient of linear thermal expansion is highest for 70:30 and 50:50 mole ratios of caprolactam:laurolactam [1]. Heat of melting 47.8 J g^{-1} (25% caprolactam, 75% laurolactam), 34.3 J g^{-1} (50% caprolactam, 50% laurolactam) [2]

Melting Temperature:

No.	Value	Note
1	132–138°C	50% caprolactam, 50% laurolactam [1]
2	146.2°C	[2]
3	158.3°C	25% caprolactam, 75% laurolactam [2]

Glass-Transition Temperature:

No.	Value	Note
1	26–30°C	70% caprolactam, 30% laurolactam, cryst.
2	16–20°C	[1]
3	23.4°C	50% caprolactam, 50% laurolactam, cryst. [2]
4	20–24°C	40% caprolactam, 60% laurolactam, cryst. [1]
5	31.8°C	25% caprolactam, 75% laurolactam [2]

Transition Temperature General: Min. T_m and T_g occur at 1:1 molar ratio of comonomers [1]

Transport Properties:
Water Content: Water absorption decreases with increasing content of laurolactam [1]

Mechanical Properties:
Tensile (Young's) Modulus:

No.	Value	Note
1	630 MPa	70% caprolactam, 30% laurolactam, ASTM D638-82
2	500 MPa	50% caprolactam, 50% laurolactam, ASTM D638-82 [1,2]
3	550 MPa	40% caprolactam, 60% laurolactam, ASTM D638-82 [1]
4	800 MPa	25% caprolactam, 75% laurolactam [2]

Tensile Strength Break:

No.	Value	Note	Elongation
1	51 MPa	70% caprolactam, 30% laurolactam, ASTM D638-82	580%

Mechanical Properties:
Tensile Strength Break:

No.	Value	Note
1	5.52 MPa	800 psi [2]

Elastic Modulus:

No.	Value	Note
1	4.02×10^{-5} MPa	[17]

Hardness: Shore A95 [2]. Rockwell 95 [14]

Optical Properties:
Refractive Index:

No.	Value	Note
1	1.3969	20° [2]

Transmission and Spectra: Critical ir bands 736, 722 cm^{-1} [4,9]. C-13 nmr [8], ir [9] and H-1 nmr [10] spectral data have been reported

Polymer Stability:
Polymer Stability General: Very susceptible to degradation by γ-radiation, much more so than the other lower polyvinyl ethers [6]

Applications/Commercial Products:
Processing & Manufacturing Routes: High degree of polymerisation using Friedel-Crafts catalysts (BF$_3$, AlCl$_3$, SnCl$_4$ or their complexes) [3], especially for isotactic form [12,13]. May also be prepared cationically [4,5], by radiation - induced polymerisation [7] or by the use of fluorides [11]
Applications: Uses include adhesives, lubricants, greases and surface coatings

Bibliographic References
[1] Vandenberg, E.J., Heck, R.F. and Breslow, D.S., *J. Polym. Sci.*, 1959, **41**, 519, (solubility, melting point)
[2] Fishbein, L. and Crowe, B.F., *Makromol. Chem.*, 1961, **48**, 221, (mechanical props)
[3] Bassi, I.W., Dall'Asta, G., Campigli, U. and Strepparola, E., *Makromol. Chem.*, 1963, **60**, 202, (prep, cryst data)
[4] Higashimura, T., Suzuoki, K. and Okamura, S., *Makromol. Chem.*, 1965, **86**, 259, (stereospecific polymerisation)
[5] Kunitake, T. and Takerabe, K., *Makromol. Chem.*, 1981, **182**, 817, (stereospecific polymerisation)
[6] Suzuki, Y., Rooney, J.M. and Stannett, V., *J. Macromol. Sci., Chem.*, 1978, **12**, 1055, (radiation-induced degradation)
[7] Goineau, A.M., Kohler, J. and Stannett, V., *J. Macromol. Sci., Chem.*, 1977, **11**, 99, (radiation-induced polymerisation)
[8] Matsuzaki, K., Ito, H., Kawamura, T. and Urgu, T., *J. Polym. Sci.*, 1973, **11**, 971, (C-13 nmr)
[9] Matsuzaki, K., Hamada, M., and Arita, K., *J. Polym. Sci., Part A-1*, 1967, **5**, 1233, (stereoregularity, ir)
[10] Ramey, K.C., Field, N.D. and Borchert, A.E., *J. Polym. Sci., Part A: Polym. Chem.*, 1965, **3**, 2885, (H-1 nmr)
[11] U.S. Pat., 1966, 3 252 953, (alkaline earth fluoride catalysts)
[12] Ital. Pat., 1964, 675 072, (Friedel-Crafts catalysts)
[13] Belg. Pat., 1962, 619 277, (Friedel-Crafts catalysts)
[14] Van Krevelen, D.W., *Properties of Polymers: Their Correlation with Chemical Structure*, Elsevier, 1972, 387
[15] Kern, R.J., Hawkins, J.J. and Calfee, J.D., *Makromol. Chem.*, 1963, **66**, 126
[16] Aoki, S. and Stille, J.K., *Macromolecules*, 1970, **3**, 473
[17] Sakurada, I. and Kaji, K., *J. Polym. Sci., Part C: Polym. Lett.*, 1970, **31**, 57

Poly(caprolactam-*co*-laurolactam)

$$\left[(NHCH_2(CH_2)_3CH_2\overset{O}{\underset{\|}{C}})_x - (NHCH_2(CH_2)_9CH_2\overset{O}{\underset{\|}{C}})_y \right]_n$$

– Poly(caprolactone)

| 2 | 50 MPa | 50% caprolactam, 50% laurolactam, ASTM D638-82 | 560% |
| 3 | 43 MPa | 40% caprolactam, 60% laurolactam, ASTM D638-82 [1] | 500% |

Tensile Strength Yield:

No.	Value	Note
1	13.2 MPa	25% caprolactam, 75% laurolactam [2]
2	8.9 MPa	50% caprolactam, 50% laurolactam [2]

Compressive Strength:

No.	Value	Note
1	47.5 MPa	25% caprolactam, 75% laurolactam [2]
2	30 MPa	50% caprolactam, 50% laurolactam [2]

Izod Notch:

No.	Value	Notch	Break	Note
1	700 J/m	Y		70% caprolactam, 30% laurolactam, ASTM D256-81, method A [1]
2		N	Nobreak	50% caprolactam, 50% laurolactam, ASTM D256-81, method A [1]
3		N	Nobreak	40% caprolactam, 60% laurolactam, ASTM D256-81, method A [1]

Applications/Commercial Products:
Processing & Manufacturing Routes: Synth. from caprolactam and laurolactam at 145° [1]. Addition of 1–2 mol % laurolactam decreases the time taken to reach max. reaction rate [3]
Applications: Studied as means of reducing time required for reaction and crystallisation to occur in reaction injection moulding [3]

Bibliographic References
[1] Garner, D.P. and Fasulo, P.D., *J. Appl. Polym. Sci.*, 1988, **36**, 495, (cryst., mechanical props.)
[2] Arvanitoyannis, I. and Psomiadou, E., *J. Appl. Polym. Sci.*, 1994, **51**, 1883, (mech props)
[3] Garner, D.P. and Iobst, S.A., *Polym. Mater. Sci. Eng.*, 1988, **58**, 937

Poly(caprolactone)   P-128

Synonyms: *Poly(ϵ-caprolactone)*. *PCL*
Material class: Thermoplastic
Polymer Type: polyester
CAS Number:

CAS Reg. No.
24980-41-4

Molecular Formula: $(C_6H_{10}O_2)n$
Fragments: $C_6H_{10}O_2$
Molecular Weight: 2000–80000

Additives: Available with mineral fillers.
General Information: A biodegradable, linear polyester available in liquid, powder, granular and fibre forms. Capable of producing hard crystalline polymers with low melt low temperatures (58–60°C) and very good hot melt adhesive characteristics.
Morphology: Semicrystalline, degree of crystallinity around 50% [11]. Crystal structure reported [13].

Volumetric & Calorimetric Properties:
Thermodynamic Properties General: Some thermal parameters only slightly influenced by cross-linking [4].
Melting Temperature:

No.	Value	Note
1	58–60°C	MW 50,000 to 80,000 [1]
2	51.9°C	MW 2,000 [10]

Glass-Transition Temperature:

No.	Value	Note
1	-60°C	[11,12]

Surface Properties & Solubility:
Solubility Properties General: Data including specific retention volume and Flory-Huggins parameters reported for various solvents at a range of temperatures [2].
Cohesive Energy Density Solubility Parameters: Solubility parameter: 9.34–9.43 $(cal/cm^3)^{1/2}$ [1].
Surface and Interfacial Properties General: Surface energy parameters determined at 160 and 180°C [5]. Contact angle, surface morphology of cast films reported with information on adhesion and proliferation of fibroblasts by direct contact [8].

Transport Properties:
Melt Flow Index:

No.	Value	Note
1	3–7 g/10 min	2.16 kg, 1″ PVC die, 160°C MW 50,000–80,000 [1]

Water Content: <1% [1]
Mechanical Properties:
Mechanical Properties General: The PCL chain is flexible with high elongation at break and low modulus [11]. Transient shear and elongational properties have been reported [14].

Tensile (Young's) Modulus:

No.	Value	Note
1	90–150 MPa	Electrospun fibre [9]

Tensile Strength Break:

No.	Value	Note	Elongation
1	30–50 MPa	Electrospun fibre [9]	100–300%

Tensile Strength Yield:

No.	Value	Elongation	Note
1	6–20 MPa	10–30%	Electrospun fibre [9]

Mechanical Properties Miscellaneous: The effect of crystallinity on dynamic mechanical relaxation behaviour (flexural storage modulus and loss tangent) has been studied [7].

Optical Properties:
Transmission and Spectra: Ir and pmr spectra reported with images [10].

Polymer Stability:
Polymer Stability General: Crosslinking can diminish ability to biodegrade [4].
Biological Stability: Biodegradation has been investigated with caprolactone, 6-hydroxyhexanoic acid, cyclic dimer and cyclic trimer identified as products. Degradation begins in amorphous regions, crystalline regions degrading later [6]. Details of biotic degradation of PCL reported [11,12].

Applications/Commercial Products:
Processing & Manufacturing Routes: PCL is manufactured by ring-opening polymerisation of ε-caprolactone. The use of mixtures of CO_2 and N_2 as blowing agents in the foaming of PCL has been reported [3]. PCL can be processed without significant molecular weight reduction, e.g. filmblowing, injection moulding and sheet extrusion [11].
Applications: Hot melt and laminate adhesives, blown films, moulding materials freely transformable using hot air or hot water, medical materials, plastic modifiers, ceramic binders, metamould binders, biodegradable plastics, heat-transfer inks, fluid modifiers for resin moulding, refractive-index regulators and polyurethane production.

Trade name	Supplier
CAPA	Solvay
PLACCEL range	Daicel Chemical Industries
TONE	Dow

Bibliographic References
[1] CASA datasheets www.solvaychemicals.com, 2005
[2] Sarac, A., Sakar, D., Cankurtaran, O. and Karaman, F.Y., *Polym. Bull.,* 2005, **53**, 349
[3] Di Maio, E., Mensitieri, G., Iannace, S., Nicolais, L., Li . W. and Flumerfelt, R.W., *Polym. Eng. Sci.,* 2005, **45**, 432
[4] El-Rehim, H.A.A., *Nucl. Instrum. Methods Phys. Res., Sect. B,* 2005, **229**, 293
[5] Biresaw, G. and Carriere, C.J., *J. Adhes. Sci. Technol.,* 2004, **18**, 1675
[6] Hakkarainen, M. and Albertsson, A.C., *Macromol. Chem. Phys.,* 2002, **203**, 1357
[7] Harrison, K.L. and Jenkins, M.J., *Polym. Int.,* 2004, **53**, 1298
[8] Tang, Z.G., Black, R.A., Curran, J.M., Hunt, J.A., Rhodes, N.P. and Williams, D.F., *Biomaterials,* 2004, **25**, 4741
[9] Tan, E. P.S., Ng, S.Y. and Lim, C.T., *Biomaterials,* 2005, **26**, 1453
[10] Kweon, H.Y., Yoo, M.K., Park, I.K., Kim, T.H., Lee, H.C., Lee, H.S., Oh, J.S., Akaike, T. and Cho, C.S., *Biomaterials,* 2003, **24**, 801
[11] Hakkarainen, M., *Adv. Polym. Sci.,* 2002, **157**, 115
[12] Mochizuki, M., *Biopolymers,* Wiley-VCH, 2002, **4**, 1
[13] Iwata, T. and Doi, Y., *Biopolymers,* Wiley-VCH, 2002, **3b**, 215
[14] Kapoor, B. and Bhattacharya, M., *Polym. Eng. Sci.,* 1999, **39**, 676

Polycarbazole P-129

Synonyms: *9H-Carbazole homopolymer*
Monomers: Carbazole
Material class: Thermoplastic
Polymer Type: polycarbazole
CAS Number:

CAS Reg. No.
51555-21-6

Molecular Formula: $(C_{12}H_7N)n$
Fragments: $C_{12}H_7N$
Additives: Polymer is usually prepared in doped form, containing perchlorate or tetrafluoroborate ions.
General Information: Conductive electrochromic polymer, usually prod. as film or powder with yellow colour in reduced form. Carbazole units in the polymer are connected via 3- and 6- positions [1].

Surface Properties & Solubility:
Solvents/Non-solvents: Sol. DMSO; poorly sol. CH_2Cl_2, THF [2].

Electrical Properties:
Electrical Properties General: Polymer in reduced (de-doped) form has low conductivity, while with dopant, when polymer is partly present in form of radical cations, conductivity can reach 7.5×10^{-3} S cm^{-1} (with boron trifluoride diethyl etherate) [2], 10^{-3} to 10^{-4} S cm^{-1} (with tetrafluoroborate) [3], or 10^{-4} S cm^{-1} (with perchlorate) [4]. Post-doping of polymer with iodine gave a maximum conductivity of 1.5×10^{-4} S cm^{-1} [5]. Electrochemical oxidation of monomer at voltages above the oxidation potential of carbazole results in a loss of conductivity [6]. The band gap in reduced form is ca. 3.5eV [7,8]. The de-doped polymer exhibits Schottky diode behaviour [8]. Electroluminescence has been reported for the polymer as the emissive layer in an OLED device [9].

Optical Properties:
Optical Properties General: Films in reduced form exhibit uv absorption at shorter wavelength, while oxidised forms exhibit absorption bands at longer wavelength (λ_{max} ca. 385, 715–735nm) [2,4,7,10]. Photoluminescence bands in the blue region (420–440nm) have been reported for polymer as solid and in soln. [2,9]. The polymer exhibits electrochromic behaviour, having a yellow colour in the reduced form and changing to green and blue in oxidised forms [4,10,11].
Transmission and Spectra: Ir [1,2,3,4,12,13,17], esr [5,6], uv [1,2,4,9,10], photoluminescence [2,9], electroluminescence [9] and XPS [5,7] spectral data have been reported.

Polymer Stability:
Thermal Stability General: Perchlorate-doped polymer exhibits weight loss due to dopant (up to 35%) between 70 and 400°C on heating in air. Decomposition of polymer backbone starts at 450°C [15,16]. The de-doped polymer in air loses 10% weight below 450°C, then undergoes rapid decomposition with 50% weight loss at 660°C (747°C in nitrogen) [16].

Applications/Commercial Products:
Processing & Manufacturing Routes: Prod. electrochemically by anodic oxidation of carbazole in non-aqueous solvents such as boron trifluoride etherate, acetonitrile, CH_2Cl_2 or DMF [2,3,13,17,18] with a supporting electrolyte such as tetraalkylammonium perchlorate or in an aqueous methanolic soln. of perchloric acid [1,4,10]. Films have been reported to be of better quality in aqueous media [4]. Carbazole monomer can also be deposited by vacuum evaporation or coating on to a suitable anode and then oxidised electrochemically [11,14]. Doped films can be de-doped in water or aqueous KCl [8,17]
Applications: Polymer films have been tested for potential application in an ammonia gas sensor [19]

Bibliographic References
[1] Mengoli, G., Musiani, M.M., Schreck, B. and Zecchin, S., *J. Electroanal. Chem.,* 1988, **246**, 73
[2] Wei, Z., Xu, J., Nie, G., Du, Y. and Pu, S., *J. Electroanal. Chem.,* 2006, **589**, 112
[3] Macit, H., Sen, S. and Saçak, M., *J. Appl. Polym. Sci.,* 2005, **96**, 894
[4] Verghese, M.M., Sunderasan, N.S., Basu, T. and Malhotra, B.D., *J. Mater. Sci. Lett.,* 1995, **14**, 401
[5] Bern&fede, J.C., Taoudi, H., Bonnet, A., Molinie, P., Morsli, M., Del Valle, M.A. and Diaz, F., *J. Appl. Polym. Sci.,* 1999, **71**, 115
[6] Taoudi, H., Bern&fede, J.C., Del Valle, M.A., Bonnet, A. and Morsli, M., *J. Mater. Sci.,* 2001, **36**, 631
[7] Kessel, R. and Schultze, J.W., *Surf. Interface Anal.,* 1990, **16**, 401
[8] Clergereaux, R., Séguy, I., Jolinat, P., Farenc, J. and Destruel, P., *J. Phys. D: Appl. Phys.,* 2000, **33**, 1947
[9] Abe, S.Y., Bern&fede, J.C., Del Valle, M.A., Tregouet, Y., Ragot, F., Diaz, F.R. and Lefrant, S., *Synth. Met.,* 2002, **126**, 1

[10] Verghese, M.M., Ram., M.K., Vardhan, H., Malhotra, B.D. and Ashraf, S.M., *Polymer*, 1997, **38**, 1625
[11] Inzelt, G., *J. Solid State Electrochem.*, 2003, **7**, 503
[12] Hayashida, S. and Sukegawa, K., *Synth. Met.*, 1990, **35**, 253
[13] Desbene-Monvernay, A., Lacaze, P.C. and Dubois, J.E., *J. Electroanal. Chem.*, 1981, **129**, 229
[14] Taoudi, H., Bern&fede, J.C., Bonnet, A., Morsli, M. and Godoy, A., *Thin Solid Films*, 1997, **304**, 48
[15] Abthagir, P.S., Dhalanakshmi, K. and Saraswathi, R., *Synth. Met.*, 1998, **93**, 1
[16] Abthagir, P.S. and Saraswathi, R., *Thermochim. Acta*, 2004, **424**, 25
[17] Pandey, P.C., Prakash, R., Singh, G., Tiwari, I. and Tripathi, V.S., *J. Appl. Polym. Sci.*, 2000, **75**, 1749
[18] O'Brien, R.N., Sundaresan, N.S. and Santhanam., K.S.V., *J. Electrochem. Soc.*, 1984, **131**, 2028
[19] Saxena, V., Choudhury, S., Gadkari, S.C., Gupta, S.K. and Yakhmi, J.V., *Sens. Actuators*, 2005, **107**, 277

Polycarbonate/acrylic-styrene-acrylonitrile alloy P-130

Synonyms: *Polycarbonate/ASA alloy. PC/ASA blend. ASA/PC blend*
Related Polymers: Poly[2,2-propanebis(4-phenyl)carbonate], Acrylic-styrene-acrylonitrile
Monomers: Bisphenol A, Acrylic acid, Styrene, Acrylonitrile
Material class: Blends, Copolymers
Polymer Type: ASA, polycarbonate alloy blendsacrylic terpolymersbisphenol A polycarbonate

Volumetric & Calorimetric Properties:
Density:

No.	Value	Note
1	d 1.15 g/cm^3	ISO 1183 [1]
2	d 1.14 g/cm^3	[2]
3	d 1.11 g/cm^3	[1]

Thermal Expansion Coefficient:

No.	Value	Note	Type
1	8×10^{-6} K^{-1}	DIN 53752 [1]	L
2	0.000149 K^{-1}	[1]	L

Deflection Temperature:

No.	Value	Note
1	109°C	1.8 MPa, ISO 75 [1]
2	103°C	0.45 MPa, 0.25 in. thick [2]
3	90°C	1.8 MPa, 0.25 in. thick [2]
4	108°C	1.82 MPa [1]
5	127°C	0.45 MPa [1]
6	96–99°C	1.82 MPa [1]

Vicat Softening Point:

No.	Value	Note
1	130°C	50 N, ISO 306 [1]
2	128°C	[1]

Transport Properties:
Melt Flow Index:

No.	Value	Note
1	5 g/10 min	220°, 10 kg, ISO 1133 [1]
2	1.6 g/10 min	230°, 3.8 kg [2]
3	4 g/10 min	[1]
4	1 g/10 min	[1]

Water Absorption:

No.	Value	Note
1	1 %	ISO 62A [1]
2	0.24 %	24h [2]
3	0.35 %	24h [1]
4	0.29 %	[1]

Mechanical Properties:
Mechanical Properties General: Elongation at break 25% [2]

Tensile (Young's) Modulus:

No.	Value	Note
1	2500 MPa	ISO 527

Flexural Modulus:

No.	Value	Note
1	2588.96 MPa	26400 kg cm^{-2} [2]
2	2206.5 MPa	22500 kg cm^{-2} [1]

Tensile Strength Break:

No.	Value	Note	Elongation
1	51.68 MPa	527 kg cm^{-2} [1]	100%
2	45 MPa	[1]	70%

Flexural Strength at Break:

No.	Value	Note
1	50–55 MPa	3.2 mm [1]

Tensile Strength Yield:

No.	Value	Elongation	Note
1	62 MPa	>50%	ISO 527 [1]
2	59.3 MPa		605 kg cm^{-2} [2]
3	51.68 MPa	5%	527 kg cm^{-2} [1]

Flexural Strength Yield:

No.	Value	Note
1	88.26 MPa	900 kg cm^{-2} [2]

Hardness: Rockwell R99 [1]
Izod Notch:

No.	Value	Notch	Note
1	170.8 J/m	Y	3.2 ft lb in^{-1}, room temp. [2]
2	597 J/m	Y	11.2 ft lb in^{-1}, room temp., $\frac{1}{8}$ in. thick [1]
3	690 J/m	Y	3.2 mm [1]

Polycarbonate, Carbon fibre reinforced, PTFE lubricated
P-131

Related Polymers: Polycarbonate, carbon fibre reinforced, Bisphenol A polycarbonate, PTFE lubricated, Bisphenol A polycarbonate
Material class: Thermoplastic, Composites
Polymer Type: polycarbonates (miscellaneous)
Molecular Formula: $(C_{16}H_{14}O_3)_n$
Fragments: $C_{16}H_{14}O_3$
Additives: Contains carbon fibre (15–30%) and PTFE (15–20%)

Volumetric & Calorimetric Properties:
Density:

No.	Value	Note
1	d 1.36 g/cm^3	15% PTFE, 15% carbon fibre [1]
2	d 1.37 g/cm^3	15% PTFE, 20% carbon fibre [1]
3	d 1.465 g/cm^3	20% PTFE, 30% carbon fibre [1]
4	d 1.41 g/cm^3	15% PTFE, 25% carbon fibre, ASTM D792 [2]
5	d 1.35 g/cm^3	

Thermal Expansion Coefficient:

No.	Value	Note	Type
1	1.85×10^{-5} K^{-1}	15% PTFE, 25% carbon fibre, ASTM D696 [2]	L

Thermal Conductivity:

No.	Value	Note
1	0.62 W/mK	15% PTFE, 25% carbon fibre, ASTM C177 [2]

Deflection Temperature:

No.	Value	Note
1	150°C	1.81 MPa, 15% PTFE, 25% carbon fibre, ASTM D648 [2]
2	155°C	0.45 MPa, 15% PTFE, 25% carbon fibre, ASTM D648 [2]
3	143°C	290°F, 1.8 MPa, 15% PTFE, 15% carbon fibre [1]
4	146°C	295°F, 1.8 MPa, 15% PTFE, 20% carbon fibre [1]

Transport Properties:
Water Absorption:

No.	Value	Note
1	0.08 %	24h, 15% PTFE, 25% carbon fibre, ASTM D570 [2]
2	0.2 %	24h, saturation, 15% PTFE, 25% carbon fibre, ASTM D570 [2]
3	0.09–0.2 %	15% PTFE, 15% carbon fibre [1]
4	0.15 %	15% PTFE, 20% carbon fibre [1]
5	0.2 %	20% PTFE, 30% carbon fibre [1]

Mechanical Properties:
Flexural Modulus:

No.	Value	Note
1	12000 MPa	15% PTFE, 25% carbon fibre, ASTM D790 [2]

Charpy:

No.	Value	Notch	Note
1	70	Y	23°, ISO179/1A [1]
2	15	Y	-30° [1]

Electrical Properties:
Surface/Volume Resistance:

No.	Value	Note	Type
2	0.1×10^{15} Ω.cm	ISO 1325 [1]	S

Dielectric Permittivity (Constant):

No.	Value	Frequency	Note
1	3.1	100 Hz	IEC 250 [1]
2	3	1 MHz	[1]
3	3.3		[1]

Dielectric Strength:

No.	Value	Note
1	3.74 kV.mm^{-1}	[1]
2	21 kV.mm^{-1}	3.2 mm [1]

Complex Permittivity and Electroactive Polymers: Comparative tracking index 225 V (IEC 112A) [1]
Dissipation (Power) Factor:

No.	Value	Frequency	Note
1	0.002	100 Hz	ISO 1325 [1]
2	0.01	1 MHz	[1]

Polymer Stability:
Flammability: Flammability rating HB (0.8 mm thick, UL94) [1]
Environmental Stress: Has excellent weatherability [1]

Applications/Commercial Products:
Mould Shrinkage (%):

No.	Value	Note
1	0.6%	[2]
2	0.45%	injection moulded [1]

Trade name	Details	Supplier
Geloy XP 4001 HG		General Electric Plastics
Terblend S		BASF
Terblend S	KR 2861/1; discontinued material	BASF

Bibliographic References
[1] Ash, M. and Ash, I., Handbook of Plastic Compounds, *Elastomers and Resins*, VCH, 1992, 218, (range of props.)
[2] *Geloy Range*, Geloy, GE Plastics, (technical datasheet)

Tensile Strength Break:

No.	Value	Note	Elongation
1	100–117 MPa	14.5–17 kpsi, 15% PTFE, 15% carbon fibre [1]	
2	134 MPa	19.5 kpsi, 15% PTFE, 20% carbon fibre [1]	
3	117 MPa	17 kpsi, 20% PTFE, 30% carbon fibre [1]	
4	130 MPa	15% PTFE, 25% carbon fibre, ASTM D638 [2]	1–2%

Flexural Strength at Break:

No.	Value	Note
1	110–186 MPa	16–27 kpsi, 15% PTFE, 15% carbon fibre [1]
2	121 MPa	17.5 kpsi, 15% PTFE, 20% carbon fibre [1]
3	159 MPa	23 kpsi, 20% PTFE, 30% carbon fibre [1]
4	180 MPa	15% PTFE, 25% carbon fibre, ASTM D790 [2]

Compressive Strength:

No.	Value	Note
1	140 MPa	15% PTFE, 25% carbon fibre, ASTM D695 [2]

Impact Strength: Izod 96–133 J m^{-1} (1.8–2.5 ft lb in^{-1}, ¼" thick, 15% PTFE, 15% carbon fibre) [1]; 75 J m^{-1} (1.4 ft lb in^{-1}, 15% PTFE, 15% carbon fibre) [1]: 96 J m^{-1} (1.8 ft lb in^{-1}, ¼" thick, 15% PTFE, 20% carbon fibre) [1]; 107–133 J m^{-1} (2.0–2.5 ft lb in^{-1}, 20% PTFE, 30% carbon fibre) [1]; 95 J m^{-1} (notched, 25% carbon fibre, 15% PTFE, ASTM D256) [2]; 300 J m^{-1} (unnotched, 25% carbon fibre, 15% PTFE, ASTM D256) [2]
Hardness: Rockwell M95 (15% PTFE, 25% carbon fibre, ASTM D785) [2]
Fracture Mechanical Properties: Shear strength 90 MPa (15% PTFE, 25% carbon fibre, ASTM D732) [2]
Friction Abrasion and Resistance: Wear factor 15 (15% PTFE, 25% carbon fibre, LNP SOP) [2]. Coefficient of friction 0.2 (static), 0.19 (dynamic) (15% PTFE, 25% carbon fibre, LNP SOP) [2]

Electrical Properties:
Electrical Properties General: Surface resistivity 10–1000 Ω sq^{-1} (15% PTFE, 25% carbon fibre, ASTM D257) [2]. Volume resistivity 10–1000 Ω cm (15% PTFE, 25% carbon fibre, ASTM D257) [2]; 1000 Ω cm (15% PTFE, 15% carbon fibre) [1]

Polymer Stability:
Upper Use Temperature:

No.	Value	Note
1	130°C	continuous service, 15% PTFE, 25% carbon fibre, UL 746 B [2]

Flammability: Flammability rating V1 (15% PTFE, 15% carbon fibre, UL94) [1]; V0/V1 (15% PTFE, 20% carbon fibre, UL94) [1]; V0/V1 (20% PTFE, 30% carbon fibre, UL94) [1]; V1 (3.2 mm thick, 25% carbon fibre, 15% PTFE, UL94) [2]

Applications/Commercial Products:
Mould Shrinkage (%):

No.	Value	Note
1	0.1–0.2%	⅛" thick [1]
2	0.1–0.3%	3 mm, ASTM D955 [2]

Trade name	Details	Supplier
PC	15CF/15T FCR, 20CF/15T, 30 CF/20T	Compounding Technology
Stat-Kon	DCL-4033	LNP Engineering

Bibliographic References
[1] Ash, M. and Ash, I., Handbook of Plastic Compounds, *Elastomers and Resins*, (eds. M.B. Ash and I.A. Ash), Wiley-VCH, 1992, 325, (props)
[2] *LNP Engineering Plastics Product Information*, LNP Engineering Plastics, (technical datasheet)

Polycarbonate foam, Glass fibre reinforced P-132
Related Polymers: Glass fibre reinforced polycarbonate, Polycarbonate structural foam
Monomers: Bisphenol A, Carbonic dichloride
Material class: Thermoplastic
Polymer Type: bisphenol A polycarbonate
Molecular Formula: $(C_{16}H_{14}O_3)_n$
Fragments: $C_{16}H_{14}O_3$
Additives: 5–30% Glass fibre reinforced. Flame retardants and blowing agents required

Applications/Commercial Products:
Mould Shrinkage (%):

No.	Value
1	0.3–0.7%

Trade name	Details	Supplier
Makrolon SF	various grades	Bayer Inc.
Makrolon SF-600	5%, opaque	Mobay
Makrolon SF-800	5%, opaque	Mobay
Makrolon SF-810	10%, opaque	Mobay
Makrolon SF-820	20%, opaque	Mobay
Makrolon SF-830	30%	Mobay

Polycarbonate, Glass fibre and carbon fibre reinforced P-133
Related Polymers: Glass fibre reinforced polycarbonate, Carbon fibre reinforced polycarbonate
Monomers: Bisphenol A, Carbonic dichloride
Material class: Thermoplastic
Polymer Type: bisphenol A polycarbonate
CAS Number:

CAS Reg. No.
24936-68-3

Molecular Formula: $(C_{16}H_{14}O_3)_n$
Fragments: $C_{16}H_{14}O_3$
Additives: 20% glass fibre and 10% carbon fibre reinforced (typically)

Volumetric & Calorimetric Properties:
Density:

No.	Value	Note
1	d 1.39 g/cm^3	10% carbon fibre, 20% glass fibre [1]

– **Polycarbonate, Glass fibre reinforced, PTFE lubricated**

Deflection Temperature:

No.	Value	Note
1	149°C	300°F, 1.82 MPa, 10% carbon fibre, 20% glass fibre [1]

Transport Properties:
Water Absorption:

No.	Value	Note
1	0.07 %	10% carbon fibre, 20% glass fibre [1]

Mechanical Properties:
Tensile Strength Break:

No.	Value	Note	Elongation
1	145 MPa	21 kpsi, 10% carbon fibre, 20% glass fibre [1]	4.0%

Flexural Strength at Break:

No.	Value	Note
1	214 MPa	31 kpsi, 10% carbon fibre, 20% glass fibre [1]

Izod Notch:

No.	Value	Notch	Note
1	133 J/m	Y	2.5 ft lb in^{-1}, 10% carbon fibre, 20% glass fibre [1]

Electrical Properties:
Electrical Properties General: Volume resistivity 100–1000000 ohm-cm (10% carbon fibre, 20% glass fibre [1]

Polymer Stability:
Flammability: Flammability rating VI (10% carbon fibre, 20% glass fibre) [1]

Applications/Commercial Products:
Mould Shrinkage (%):

No.	Value
1	0.15%

Trade name	Supplier
Stat Kon DCF-1006	LNP Engineering

Bibliographic References

[1] Ash, M. and Ash, I., Handbook of Plastic Compounds, *Elastomers and Resins*, VCH, 1992, 218, (props.)

Polycarbonate, Glass fibre reinforced, PTFE lubricated   P-134

Related Polymers: Polycarbonate, glass fibre reinforced, Bisphenol A polycarbonate, PTFE lubricated, Bisphenol A polycarbonate
Monomers: Bisphenol A, Carbonic dichloride
Material class: Thermoplastic, Composites
Polymer Type: bisphenol A polycarbonate

CAS Number:

CAS Reg. No.
24936-68-3

Molecular Formula: $(C_{16}H_{14}O_3)_n$
Fragments: $C_{16}H_{14}O_3$
Additives: 20–40% Glass fibre; 10–15% PTFE added as lubricant
General Information: Flame retardant versions available

Volumetric & Calorimetric Properties:
Density:

No.	Value	Note
1	d 1.45 g/cm^3	15% PTFE, 20% glass fibre [1,2]
2	d 1.55 g/cm^3	15% PTFE, 30% glass fibre [1,2]
3	d 1.64 g/cm^3	15% PTFE, 40% glass fibre [1]

Thermal Expansion Coefficient:

No.	Value	Note	Type
1	2.8×10^{-5} K^{-1}	15% PTFE, 20% glass fibre, ASTM D696 [2]	L
2	2.4×10^{-5} K^{-1}	15% PTFE, 30% glass fibre, ASTM D696 [2]	L

Thermal Conductivity:

No.	Value	Note
1	0.23 W/mK	15% PTFE, 20% glass fibre, ASTM C177 [2]
2	0.25 W/mK	15% PTFE, 30% glass fibre, ASTM C177 [2]

Deflection Temperature:

No.	Value	Note
1	145°C	1.81 MPa, 15% PTFE, 20–30% glass fibre, ASTM D648 [2]
2	150°C	0.45 MPa, 15% PTFE, 20–30% glass fibre, ASTM D648 [2]
3	141–143°C	285–290°F, 1.8 MPa, 15% PTFE, 20% glass fibre [1]
4	141–146°C	285–295°F, 1.8 MPa, 15% PTFE, 30% glass fibre [1]
5	138–143°C	280–290°F, 1.8 MPa, 15% PTFE, 40% glass fibre [1]

Vicat Softening Point:

No.	Value	Note
1	152–157°C	305–315°F, 15% PTFE, 20% glass fibre [1]
2	152–157°C	305–315°F, 15% PTFE, 30% glass fibre [1]
3	152–157°C	305–315°F, 15% PTFE, 40% glass fibre [1]

Transport Properties:
Water Absorption:

No.	Value	Note
1	0.08–0.09 %	15% PTFE, 20% glass fibre [1]
2	0.06–0.08 %	15% PTFE, 30% glass fibre [1,2]
3	0.05–0.07 %	15% PTFE, 40% glass fibre [1]

– Polycarbonate, Glass fibre reinforced, PTFE lubricated

| 4 | 0.22 % | saturation, 15% PTFE, 30% glass fibre [2] |
| 5 | 0.25 % | saturation, 15% PTFE, 20% glass fibre [2] |

Mechanical Properties:

Mechanical Properties General: The addition of glass fibre to PTFE lubricated polycarbonate improves tensile and flexural strengths, while maintaining good frictional and wear characteristics

Flexural Modulus:

No.	Value	Note
1	8200 MPa	15% PTFE, 30% glass fibre, ASTM D790 [2]
2	6100 MPa	15% PTFE, 20% glass fibre, ASTM D790 [2]

Tensile Strength Break:

No.	Value	Note	Elongation
1	100–103 MPa	14.5–15 kpsi, 15% PTFE, 20% glass fibre [1]	
2	103–124 MPa	15.0–18 kpsi, 15% PTFE, 30% glass fibre [1]	
3	134–138 MPa	19.5–20 kpsi, 15% PTFE, 40% glass fibre [1]	
4	100 MPa	15% PTFE, 20% glass fibre, ASTM D638 [2]	3–4%
5	115 MPa	15% PTFE, 30% glass fibre, ASTM D638 [2]	2–3%

Flexural Strength at Break:

No.	Value	Note
1	152 MPa	22 kpsi, 15% PTFE, 20% glass fibre [1]
2	152–165 MPa	22–24 kpsi, 15% PTFE, 30% glass fibre [1]
3	186 MPa	27 kpsi, 15% PTFE, 40% glass fibre [1]
4	135 MPa	15% PTFE, 20% glass fibre, ASTM D790 [2]
5	155 MPa	15% PTFE, 30% glass fibre, ASTM D790 [2]

Compressive Strength:

No.	Value	Note
1	105 MPa	15% PTFE, 20% glass fibre, ASTM D695 [2]
2	130 MPa	15% PTFE, 30% glass fibre, ASTM D695 [2]

Hardness: Rockwell M93; R115-R118 (15% PTFE, 30% glass fibre) [1,2]; M91 (15% PTFE, 20% glass fibre, ASTM D785) [2]
Fracture Mechanical Properties: Shear strength 65 MPa (15% PTFE, 20% glass fibre, ASTM D732); 70 MPa (15% PTFE, 30% glass fibre, ASTM D732) [2]
Friction Abrasion and Resistance: Has low frictional props. and good wear resistance [1]. Wear factor 40 (15% PTFE, 20% glass fibre, LNP SOP) [2]; 30 (15% PTFE, 30% glass fibre, LNP SOP) [2]. Static coefficient of friction 0.19 (15% PTFE, 20% glass fibre, LNP SOP) ; 0.18 (15% PTFE, 30% glass fibre, LNP SOP) [2]. Dynamic coefficient of friction 0.22 (15% PTFE, 20% glass fibre, LNP SOP); 0.2 (15% PTFE, 30% glass fibre, LNP SOP) [2]

Izod Notch:

No.	Value	Notch	Note
1	96–101 J/m	Y	1.8–1.9 ft lb in^{-1}, $\frac{1}{4}$" thick, 15% PTFE, 20% glass fibre [1]
2	96–107 J/m	Y	1.8–2 ft lb in^{-1}, $\frac{1}{4}$" thick, 15% PTFE, 30% glass fibre [1]
3	107–112 J/m	Y	2.0–2.1 ft lb in^{-1}, $\frac{1}{4}$" thick, 15% PTFE, 40% glass fibre [1]
4	130 J/m	Y	15% PTFE, 20 and 30% glass fibre, ASTM D256 [2]
5	550 J/m	N	15% PTFE, 20% glass fibre, ASTM D256 [2]
6	400 J/m	N	15% PTFE, 30% glass fibre, ASTM D256 [2]

Electrical Properties:

Surface/Volume Resistance:

No.	Value	Note	Type
5	1×10^{15} Ω.cm	15% PTFE, 20–30% glass fibre, ASTM D257 [2]	S

Dielectric Permittivity (Constant):

No.	Value	Frequency	Note
1	3.2	60 Hz-1 MHz	15% PTFE, 20% glass fibre, ASTM D150 [2]
2	3.3	60 Hz-1 MHz	15% PTFE, 30% glass fibre, ASTM D150 [2]

Dielectric Strength:

No.	Value	Note
1	17.7–18.9 kV.mm^{-1}	450–480 V mil^{-1}, 15% PTFE, 30% glass fibre [1]
2	20 kV.mm^{-1}	15% PTFE, 20–30% glass fibre, ASTM D149 [2]

Arc Resistance:

No.	Value	Note
1	120s	15% PTFE, 20–30% glass fibre, ASTM D495 [2]

Complex Permittivity and Electroactive Polymers: Tracking resistance 175 (15% PTFE, 20–30% glass fibre, DIN IEC 112) [2]

Dissipation (Power) Factor:

No.	Value	Frequency	Note
1	0.0009	60 Hz	15% PTFE, 20% glass fibre, ASTM D150 [2]
2	0.001	60 Hz	15% PTFE, 30% glass fibre, ASTM D150 [2]
3	0.007	1 MHz	15% PTFE, 20–30% glass fibre, ASTM D150 [2]

Polymer Stability:

Upper Use Temperature:

No.	Value	Note
1	125°C	continuous service, 15% PTFE, 20% glass fibre, UL 746 B [2]
2	130°C	continuous service, 15% PTFE, 30% glass fibre, UL 746 B [2]

Flammability: Flammability rating V0/V1 (15% PTFE, 20% glass fibre, UL94) [1,2]; V0/V1 (15% PTFE, 30% glass fibre, UL94) [1,2]; V0/V1 (15% PTFE, 40% glass fibre, UL94) [1]

Applications/Commercial Products:
Mould Shrinkage (%):

No.	Value	Note
1	0.1–0.15%	
2	0.2–0.4%	3 mm section, 15% PTFE, 20% glass fibre [2]
3	0.15–0.35%	3 mm section, 15% PTFE, 30% glass fibre [2]

Trade name	Details	Supplier
Lubricomp	DFL-4036 (30% glass fibre, 15% PTFE)	LNP Engineering
PC-20GF/15T	series 20% glass fibre; 15% PTFE, flame retardant	Compounding Technology
Plaslube J-50/30/TF/15	30% glass fibre, 15% PTFE	Akzo
RTP 303TFE10	20% glass fibre, 10% PTFE	RTP Company
RTP 305TFE15	30% glass fibre, 15% PTFE	RTP Company
Thermocomp DFL-4000	series 15% PTFE, 20-40% glass fibre	LNP Engineering

Bibliographic References
[1] Ash, M. and Ash, I., Handbook of Plastic Compounds, *Elastomers and Resins*, (eds. M.B. Ash and I.A. Ash), Wiley-VCH, 1992, 325, (props)
[2] *LNP Engineering Plastics Product Information*, LNP Engineering Plastics, 1996, (technical datasheet)

Polycarbonate/polybutylene terephthalate alloy P-135

Synonyms: *Polycarbonate/polybutylene terephthalate blend. PC/PBT blend*
Related Polymers: Polybutylene terephthalate, Poly[2,2-propane-bis(4-phenyl)carbonate], Polycarbonate/polyester alloys, Glass fibre reinforced polycarbonate/polybutylene terephthalate alloy
Monomers: Bisphenol A, Carbonic dichloride, 1,4-Benzenedicarboxylic acid, 1,4-Butanediol
Material class: Blends
Polymer Type: PBT and copolymers, polycarbonate alloy blends-bisphenol A polyesters
Additives: Commercial blends contain low levels of stabiliser such as organic phosphites. MMA-styrene (3:2) copolymer as an impact modifier. Impact modifiers (3–30%)
General Information: Impact modified version available. Structural foam version available. PC/PBT blends form complicated systems, the props. of which depend not only on the composition and any additives used, but also on the preparation, processing and subsequent exploitation. [1,2,3,4,5,6]. Transesterification, phase separation, crystallisation. and degradation processes can all occur as a result of thermal treatment, and the relative degrees of these processes depend on various factors including: the melt processing temp. and the duration of extrusion [1]; the number of times the material is extruded (i.e. reprocessed) [7]; the moulding process [1,8]; annealing of the material after processing, which facilitates crystallisation and phase separation [1,2,8]. Depending on conditions, transesterification reactions may also occur at high temps. (270°) [9]. Commercial blends generally contain PC and PBT in approx. equal proportions, with added impact modifier (3–30%) [5,7,10,11,12,13,14]

Morphology: Solid PC/PBT blends show partial miscibility and comprise two amorph. phases (one PC-rich, the other PBT-rich) and a crystalline PBT phase. [2,4,5,6,9,15,16] The mass fraction of PBT in the PC-rich phase is higher than the mass fraction of PC in the PBT-rich phase (suggesting greater solubility of PBT in the PC-rich phase). This is the case for both melt-extruded and soln. cast blends. [6,16] For blends with less than 60% PC, the PBT-rich phase forms the continuous phase and the PC-rich phase the dispersed phase [5,14]. As the proportion of PC increases, the dispersed phase becomes highly interconnected, and, for blends containing 60% PC, an interpenetrating network is formed [5,14]. Above 60% PC the PC-rich phase forms the continuous phase. [5,16] When a core-shell impact modifier is included, it becomes located exclusively in the PC-rich phase at all PC concentrations. [5,14,17]. The morphology of the solid is affected by composition; for blends melt extruded at 250°, phase separation is more pronounced for 60 or 70% PC than for 10, 20 or 90% PC [6]; the method of blend preparation; blends formed by casting from soln. at 25° show distinct phases for all compositions, and compatibility is less than for extruded blends [6,16,27]. Blends formed in the melt which are subsequently dissolved in solvent and cast from soln. show phase separation characteristic of the solvent cast blends, rather than the melt extruded blends. [16,27] In the melt, blends may form one or two phases depending on temp. and composition. [2,4,6,15] Depending on the cooling method, some of the morphology of the melt is preserved in the solid state [8,18]. This is maximised for samples which are rapidly cooled (quenched) from the melt. PBT forms a crystalline phase in PC/PBT blends. However, its rate of crystallisation in the blend is slower than in the pure component. [15,18] The crystallinity of the blend depends on post-processing conditions, such as the rate of cooling from the melt and any annealing. Crystallisation may also occur during thermal measurements [2]. Annealing (e.g. at 130–160°) and slow cooling give rise to phase separation [8,14,18], which leads to an increase in the crystallinity of the blend [2,5,8,18], and to changes in the amorph. phases [8,18]. The crystalline phase increases in volume on heating, using PBT crystallised from both amorph. phases [2,3,15] As the PC content increases from 10–90% the degree of crystallinity decreses from 18.4–0.7% - both as a percentage of the total blend and of the PBT component alone [2,3]. At a given PC content, annealed material is more crystalline than the unannealed material [2,3]

Volumetric & Calorimetric Properties:
Density:

No.	Value	Note
1	d 1.263 g/cm^3	10% PC, unannealed [2]
2	d 1.237 g/cm^3	10% PC, amorph, calc [2]
3	d 1.273 g/cm^3	10% PC, unannealed [3]
4	d 1.289 g/cm^3	10% PC, annealed, 100 min, 50° [3]
5	d 1.255 g/cm^3	20% PC, unannealed [2]
6	d 1.234 g/cm^3	20% PC, amorph., calc [2]
7	d 1.256 g/cm^3	30% PC, unannealed [3]
8	d 1.264 g/cm^3	30% PC, annealed, 100 min, 50° [3]
9	d 1.241 g/cm^3	40% PC, unannealed [2]
10	d 1.224 g/cm^3	40% PC, amorph., calc. [2]
11	d 1.233 g/cm^3	50% PC, unannealed [2]
12	d 1.216 g/cm^3	50% PC, amorph., calc [2]
13	d 1.24 g/cm^3	50% PC, unannealed [3]
14	d 1.244 g/cm^3	50% PC, annealed, 100 min, 50° [3]
15	d 1.226 g/cm^3	60% PC, unannealed [2]
16	d 1.214 g/cm^3	60% PC, amorph., calc [2]
17	d 1.21 g/cm^3	80% PC, unannealed [2]
18	d 1.204 g/cm^3	80% PC, amorph., calc. [2]

– Polycarbonate/polybutylene terephthalate alloy

No.	Value	Note
19	d 1.202 g/cm^3	90% PC, unannealed [2]
20	d 1.22 g/cm^3	Xenoy CL-100 [10]
21	d 1.222 g/cm^3	Xenoy CL-100 [18]
22	d 1.222 g/cm^3	Xenoy CL-100, ASTM D782 [8]
23	d 1.17–1.22 g/cm^3	Xenoy, various grades, all impact modified [22]
24	d 1.21–1.24 g/cm^3	Makroblend PR 52, DIN 53 478 [1,21]

Thermal Expansion Coefficient:

No.	Value	Note	Type
1	7×10^{-5}–7.2×10^{-5} K^{-1}	Makroblend PR 52, ASTM D696	L

Volumetric Properties General: The density depends on post-processing conditions (e.g. annealing) [8,18]
Latent Heat Crystallization: ΔH_m 24 J g^{-1} (50% PC) [10]; 40.2 J g^{-1} (9.6 cal g^{-1}) (50% PC) [6]; 45.2 J g^{-1} (10.8 cal g^{-1}) (60% PC) [6]; 45.2 J g^{-1} (10.8 cal g^{-1}) (60% PC) [6]; 46.9 J g^{-1} (11.2 cal g^{-1}) (80% PC) [6]; 18 J g^{-1} (Xenoy CL-100) [18]; 18 J g^{-1} (Xenoy CL-100, injection moulded, 260°) [8]; 18.6 J g^{-1} (Xenoy CL-100, injection moulded, 280°) [8]; 21.7 J g^{-1} (Xenoy Cl-100, injection moulded, 300°) [8]

Melting Temperature:

No.	Value	Note
1	222°C	independent of composition [2]
2	231–233°C	0–70% PC [15]
3	223°C	50% PC [23]
4	221.5°C	50% PC [24]
5	218.8–219.8°C	50–80% PC [6]
6	215–220°C	Xenoy CL-100 [9]
7	215°C	Xenoy CL-100 [7]
8	223°C	Xenoy CL-100 [18]
9	222.8–223.3°C	Xenoy CL-100 [8]

Glass-Transition Temperature:

No.	Value	Note
1	69–128°C	10% PC, unannealed, melt mixed, 250° [15]
2	52–109°C	20% PC, unannealed, melt mixed, 250° [15]
3	51–150°C	20% PC, quenched, melt mixed, 260° [4]
4	95–151°C	20% PC, DMTA scanned, melt mixed, 280° [2]
5	47.3–134.2°C	30% PC, unannealed, soln. cast, 25° [6]
6	51.8–122.6°C	30% PC, unannealed, melt mixed, 250° [6]
7	46.5–134.4°C	40% PC, unannealed, soln. cast, 25° [6]
8	49.6–123.7°C	40% PC, unannealed, melt mixed, 250° [6]
9	65–128°C	40% PC, unannealed, melt mixed, 280° [2]
10	91–152°C	40% PC, DMTA scanned, melt mixed, 280° [2]
11	44.6–136.8°C	50% PC, unannealed, soln. cast, 25° [6]
12	48.8–125°C	50% PC, unannealed, melt mixed, 250° [6]
13	45–125°C	50% PC, unannealed, melt mixed, 250° [15]
14	43.6–141.2°C	70% PC, unannealed, soln. cast, 250° [6]
15	47.8–128.4°C	70% PC, unannealed, melt mixed, 250° [6]
16	50–125°C	75% PC, unannealed, melt mixed, 250° [15]
17	55–135°C	80% PC, unannealed, melt mixed, 250° [2]
18	47–101°C	Xenoy CL-100, unannealed [7]
19	46–47°C	Xenoy CL-100, unannealed [9]
20	100–101°C	Xenoy CL-100, unannealed [9]
21	47–100°C	Xenoy CL-100, conditioned 250°, 20 min [9]
22	46–47°C	Xenoy CL-100, conditioned, 270°, 5–15 min [9]
23	100–101°C	Xenoy CL-100, conditioned, 270°, 5–15 min [9]
24	99°C	Xenoy CL-100, conditioned, 270°, 20 min [9]

Transition Temperature General: T_g varies with prior thermal history, including the method of blending (melt mixing or solution blending), the temp. and duration of melt mixing, any annealing, even the methor of measurement [4,6,8]. It is therefore difficult to discuss trends. However, for a given series, the upper T_g increases with percentage PC content, while the lower T_g decreases

Deflection Temperature:

No.	Value	Note
1	80°C	1.81 MPa, DIN 53 461, Makroblend PR 52 [21]
2	108°C	0.45 MPa, DIN 53 461, Makroblend PR 52 [21,25]
3	98°C	50% PC, 1.82 MPa, ASTM D648 [23]
4	96–121°C	1.82 MPa, Xenoy, various grades, all impact modified [22]

Vicat Softening Point:

No.	Value	Note
1	150°C	DIN 53 460, A, Makroblend PR 52 [21]
2	120°C	DIN 53 460, B, Makroblend PR 52 [21]

Surface Properties & Solubility:
Cohesive Energy Density Solubility Parameters: Flory-Huggins interaction parameter 0.033–0.042 (calc., melt extruded, 250°) [6]; 0.039–0.054 (calc., soln. cast, 25°) [6]; $_{12}$ (critical) 0.026 [6]
Solvents/Non-solvents: Sol. CCl$_4$ [9], boiling tetrachloroethane [19], 1,1,1,3,3,3-hexafluoro-2-propanol [6], CHCl$_3$/CF$_3$COOH (90:10) [5]
Surface and Interfacial Properties General: PC/PBT blends (around 50:50 composition) show strong bonding between the phases (-30°–-25°) [10,20]

Transport Properties:
Transport Properties General: Melt flow index increases with increasing temp., which is indicative of partial thermal degradation [1] and with increase in the number of extrusions [7]
Melt Flow Index:

No.	Value	Note
1	18 g/10 min	Xenoy CL-100, 260°, 5 kg, DIN 53 735 [1]
2	1.57–2.41 g/10 min	Xenoy CL-100, 270°, 2.16 kg, ISO 1133 [7]

– Polycarbonate/polybutylene terephthalate alloy

3	20 g/10 min	Makroblend PR 52, 260°, 5 kg, DIN 53 735 [1]
4	10–16 g/10 min	Makroblend PR 52, 49.05 N, DIN 53 735 [21]
5	22 g/10 min	260°, 5 kg, Xenoy CL-100, DIN 53 735 [1]
6	29 g/10 min	280°, 5 kg, Xenoy CL-100, DIN 53 735 [1]

Permeability of Liquids: Diffusivity [H_2O] 6.64×10^{-8} cm$_2$ s^{-1} (21°) [24]; 3.84×10^{-8} cm$_2$ s^{-1} [24]. Activation energy for diffusivity of H_2O 46.9 kJ mol^{-1} (11.2 kcal mol^{-1}, 21–98°) [24]

Water Absorption:

No.	Value	Note
1	0.12–0.14 %	Xenoy, various grades, all impact modified [22]

Mechanical Properties:

Mechanical Properties General: Poisson's ratio of a 50% PC blend increases after 10 h annealing and decreases after 100 h. Stress transfer across the phases is one of the critical factors determining the load bearing capacity and the impact behaviour of the material [10]. The degree of crystallisation has a great influence on the phase behaviour, which in turn affects the mechanical props. [2] The degree of crystallinity is affected by processing and post-processing conditions. PC/PBT blends combine the toughness of polycarbonate with the chemical and stress cracking resistance of PBT. Tensile strength at break has a minimum at 30% PC content. Elongation at break decreases with increasing PC content [2]. Tensile strength at yield increases with increasing PC content [2]. Young's modulus shows a maximum at 80% PC content, indicating synergism in the blend. [2] The variation of loss tangent with temp. for blends containing 50% [24] and 80% PC [2] and for commercial blends (Xenoy CL-100) [7,9] has been reported. The variation of tensile storage and tensile loss moduli with temp. for blends containing 60% [15] and 80% PC [15] and for commercial blends (Xenoy CL-100) [7,9] has been reported. Flexural strength decreases on exposure to H_2O at elevated temps. [24]

Tensile (Young's) Modulus:

No.	Value	Note
1	1970–2070 MPa	20% PC, unannealed [2]
2	2200–2360 MPa	50% PC, unannealed [2]
3	2160 MPa	22000 kg cm^{-2}, 50% PC [23]
4	2460–2500 MPa	80% PC, unannealed [2]
5	2360–2460 MPa	90% PC, unannealed [2]

Tensile Strength Break:

No.	Value	Note	Elongation
1	55–57.4 MPa	10% PC, unannealed [2]	299–363%
2	52.3–56.4 MPa	20% PC, unannealed [2]	228–257%
3	54.3–59.2 MPa	50% PC, unannealed [2]	116–138%
4	63.9–67.2 MPa	80% PC, unannealed [2]	77–96%
5	43.79–44.26 MPa	Xenoy CL-100 [7]	29.9–48.5%

Flexural Strength at Break:

No.	Value	Note
1	130 MPa	50% PC [24]

Poisson's Ratio:

No.	Value	Note
1	0.4	50% PC [10]
2	0.42	50% PC, annealed 10h [10]
3	0.39	50% PC, annealed 100 h [10]

Tensile Strength Yield:

No.	Value	Elongation	Note
1	34.7–38.7 MPa		10% PC, unannealed [2]
2	40.6–43.9 MPa		20% PC, unannealed [2]
3	46.5–50.3 MPa		50% PC, unannealed [2]
4	54.9 MPa		560 kg cm^{-2}, 50% PC, unannealed, ASTM D638 [23]
5	54.3 MPa		80% PC, unannealed [2]
6	57.9 MPa		90% PC, unannealed [2]
7	55.2–55.7 MPa	14.2%	Xenoy CL-100 [7]

Flexural Strength Yield:

No.	Value	Note
1	94.1 MPa	960 kg cm^{-2}, 50% PC, ISO R178 [23]

Impact Strength: Low temp. impact strength decreases with increasing number of extrusion cycles. [7] Tensile impact strength 245 kJ m^{-2} (250 kg cm cm^{-2}, 50% PC, ASTM D1822). [23] Commercial blends usually have impact modifiers added; the impact strength of the blend depends on their nature and level. Impact energies decrease drastically on exposure to H_2O at elevated temps., and are found to vary with sample thickness [24], temp. and time of H_2O immersion. For example, immersion at 80° for 12 d gives brittle fracture (ductile before immersion, flexural impact testing) [24]. Impact energies/unit volume 4.4 J cm^{-3} (unnotched, as moulded) [24]; 4.1 J cm^{-3} (unnotched, 21°, 90 d) [24]; 3.9 J cm^{-3} (unnotched, 55°, 7 d) [24]; 0.93 J cm^{-3} (unnotched, 80°, 7 d) [24]; 0.05 J cm^{-3} (unnotched, 98°, 7 d) [24]; 0.5 J cm^{-3} (unnotched, 98°, 3 d) [24]; 3.9 J cm^{-3} (unnotched, 98°, 1 d) [24]

Viscoelastic Behaviour: The load-bearing props. of a 50% PC blend (as illustrated by creep) are affected very little after 10 h annealing at 160°. [10] Volumetric strain-creep time curves for stress levels of 20–50 MPa for blends containing 50% PC have been reported. The blend shows cooperative behaviour, with props. of PBT dominating at low stresses and those of PC dominating at higher stresses [10]

Hardness: Ball indentation hardness H30 (100 MPa, Makroblend PR 52, DIN 53 456) [21,25]; Rockwell R112–R115 (Xenoy, various grades, all impact modified) [22]

Failure Properties General: Failure stress 53–55 MPa (1000s creep time). [10] Activation energy of embrittlement 92 kJ mol^{-1} (22 kcal mol$_{-1}$, H_2O immersion, 55–98°) [24]

Fracture Mechanical Properties: Annealing results in significant embrittlement [1,8,24], with a reduction in fracture energy. Fracture energy, G_{Ic} annealed/G_{Ic} unannealed 0.9 (annealed 80°, 100h, Xenoy CL-100), 0.55 (annealed, 130°, 100 h, Xenoy CL-100) [1]; G_{Ic} annealed/G_{Ic} unannealed 0.64 (130°, 10 h, Makroblend PR 52); 0.42 (130°, 100 h). [1] In notched impact tests at -30°–25°, 55% PC/45% PBT blends show brittle failure [20]; 45% PC/40% PBT/15% core shell impact modified blend shows ductile failure [20]. Lower levels of impact modifier and PC give temp. dependent failure props. For example, 15% PC/70% PBT/ 15% impact modified blends show semi-brittle failure (-30°–-20°) [20], and ductile failure (-20°–25°). [20] At low temps. (-70°) Xenoy CL-100 displays brittle failure [7], whereas at 25° the specimens are completely ductile (Xenoy 1102, impact modified). [28] The fracture energy, G_{Ic}, increases smoothly with increasing testing temp. (i.e. no sharp ductile-brittle transition) [1]. G_{Ic} 4.1 kJ m^{-2} (-27°, Makroblend PR 52) [1]; 5.3 kJ m^{-2} (10′ Makroblend PR 52). [1] At temps. above 10°, the samples (Makroblend PR 52) do not break.[1] Impact fracture energy of compression-moulded samples is lower than that of injection-moulded samples [1,8]. Processing affects fracture energy, which is significantly reduced at higher temps: G_{Ic} 3.9 kJ m^{-2} (260°, Xenoy CL-100), 4.5 kJ m^{-2} (280°, Xenoy CL-100), 3.1 kJ m^{-2} (300°, Xenoy CL-100) [1]

Izod Notch:

No.	Value	Notch	Note
1	80–100 J/m	Y	-50°, 3.2 mm, ASTM D256 [21,25]
2	640–800 J/m	Y	Xenoy, various grades [22]
3	98 J/m	Y	10 kg cm cm^{-1}, 50% PC, ISO R180 [23]
4	1060 J/m	Y	108 kg cm cm^{-1}, 23°, impact modified [12]
5	680 J/m	Y	69 kg cm cm^{-1}, -30°, impact modified [12]
6	670 J/m	Y	68 kg cm cm^{-1}, 23°, impact modified [13]
7	610 J/m	Y	62 kg cm cm^{-1}, 23°, impact modified [13]
8	700–800 J/m	Y	23°, 3.2 mm, Makroblend PR 52, ASTM D256 [21,25]

Charpy:

No.	Value	Notch	Note
1	40–45	Y	room temp., Makroblend PR 52, DIN 53 453 [21]
2	12–15	Y	-20°, Makroblend PR 52, DIN 53 453 [21]
3	6–8	Y	-40°, Makroblend PR 52, DIN 53 453 [21]

Electrical Properties:

Electrical Properties General: Contour maps of ϵ' and dissipation factor (-180°–200°, 0.1 Hz–3MHz) for PC:PBT (5:4) have been reported

Dielectric Permittivity (Constant):

No.	Value	Frequency	Note
1	2.93–3.6	100 Hz	Xenoy, various grades [22]

Dielectric Strength:

No.	Value	Note
1	28–29 kV.mm^{-1}	3.2 mm, Xenoy, various grades [22]

Optical Properties:

Transmission and Spectra: Ir spectral data have been reported [4,9]
Volume Properties/Surface Properties: Films containing more than 75% PC are transparent or sometimes hazy [9,15], whereas blends with less than 75% PC are not fully transparent/opaque because of the crystallinity of PBT [2,9,15]. Films cast from soln. are slightly opaque. [6] The appearance of the melt at 250° varies from transparent (25% and 90% PC), through a satiny, opaque state (50% and 80% PC) to a light brown, translucent opaque state (60% and 70% PC) [15]. At 280° the melt is transparent at all compositions [2]

Polymer Stability:

Polymer Stability General: PC/PBT blends have excellent chemical stability. However, they are hydrolysed at elevated temps., and heating at moderate temps. (130–175°, >10 h depending on temp.) leads to embrittlement

Thermal Stability General: The props. of PC/PBT blends are affected by processing conditions and subsequent treatment. [4,5,6,9,10,18,23] Parts manufactured from PC/PBT alloys are known to undergo changes in their props., especially during service at elevated temps. These changes include a considerable embrittlement during long term use, which is significantly worse than found in the pure components. [8] Changes can be attributed to ester interchange reactions, degradation of the polymers, phase separation and crystallisation equilibria. [4,8,18] There is some debate concerning the existence of ester interchange reactions at high (>260°) temps., but its relative likelihood depends on the conditions. [2,4,6,9,18,23] Where observed, the extent of transesterification depends on the temp. and the composition [4,26], e.g. at higher PC content the reaction proceeds much more slowly than at lower PC content. [4,26] Over short time intervals (about 10–60 min., depending on temp.) a block copolyester is formed, whose composition becomes statistical at longer times (>40 min., depending on temp.) [4,9,19,23] It is thought that the exchange reaction proceeds when PBT degrades [4], and that residual titanium in the PBT catalyses the reaction [23,26]; low levels of organic phosphorus stabilisers (e.g. di- or triphosphites) can be added to complex with these residues thus limiting transesterification. [3,5,14,23] During melt processing the high temps. (in excess of 240°) cause degradation, transesterification and miscibility changes to occur, which affect the physical props. Such changes depend on the temp. and duration of processing [1,4]

Decomposition Details: Usually unstable above 230° [23]. Blends containing 50% PC evolve CO_2 at temps. above 260° [23]

Flammability: Flammability rating HB (0.081–1.57 mm, Xenoy, various impact-modified grades) [22]

Chemical Stability: Xenoy has excellent chemical resistance [22]. Elongation to break is unaffected by immersion in automotive fuels for 5 min. under load [21]. Waxing and dewaxing also have no effect [21]

Hydrolytic Stability: Immersion of Xenoy CL-100 samples in warm (40°) deionised water increases melt flow index, significantly decreases low temp. impact strength but has little effect on tensile props. [7] Immersion of a 50% PC blend in H_2O at elevated temps. (98°, 10 d) results in very fragile specimens, with pitted and dull surfaces [24]. It is thought that hydrolytic chain scission at the ester linkage occurs on immersion in H_2O [7,24] resulting in a decrease in MW [24]

Recyclability: Reprocessing (the number of times the material is extruded) increases melt flow index and tensile strength (yield and break), and decreases the low temp. impact strength [7]

Applications/Commercial Products:

Processing & Manufacturing Routes: Drying prior to processing (at 100–110°) is required. Melt temps. should not exceed 270–290°. Prepared by melt mixing or by combining two solns. and then removing the solvent. In melt mixing, the two components are first dried (mixed or separately) in an air circulating oven or under vacuum to prevent hydrol. (120°, 4–48 h) [2,5,7,9,10,23]; (80°, vacuum, 48 h). [6] The polymers are then combined in the melt by extrusion [5,6,10,18] or by mechanical mixing. [2,5,15,19,23] The temp. should be kept above 232°, i.e. melting point of PBT. [15] The extrudate can be ground in a pelletiser; moisture absorption is prevented by keeping the material in a closed container. [10] In soln. blending each of the two polymers is dissolved in solvent (e.g. hot tetrachloroethane), mixed and then the solvent is evaporated. The blend is then dried at 120° for 12 h [9]

Mould Shrinkage (%):

No.	Value
1	0.8–1%

Applications: Automotive (e.g. bumpers), electrical, mechanical and sport, medical products, fluid handling equipment, lightweight lawnmower casings

Trade name	Details	Supplier
AVP	GLV8U/GLV80	Polymerland Inc.
Azloy	AM70350	Azdel Inc.
Carbotex	K-1000	Kotetsu
Iupilon	MB4302/4402	Mitsubishi Gas
Makroblend	PR52 (impact modified)	Mobay
Makroblend		Bayer Inc.
Nyloy	E range	Nytex Composites
PC/PBT	5502	Nan Ya Plastics
Shinblend	A274G20/A724	Shinkong Synthetic
Ultrablend	2230 (impact modified)	BASF
Xenoy	2230, 5220, 5720	General Electric Plastics

Bibliographic References

[1] Bertilsson, H., Franzén, B. and Kubát, J., *Plast. Rubber Process. Appl.*, 1988, **10**, 137, (props)
[2] Sánchez, P., Remiro, P.M. and Nazábal, J., *J. Appl. Polym. Sci.*, 1993, **50**, 995, (props, synth)
[3] Halder, R.S., Joshi, M. and Misra, A., *J. Appl. Polym. Sci.*, 1990, **39**, 1251, (crystallinity, density, stability, synth)
[4] Konyukhova, Y.V., Bessonova, N.P., Belousov, S.I., Fel'dman, V.I. and Godovskii, Y.K., *Polym. Sci.*, 1991, **33**, 2262, (morphology, T_g, ir, stability)
[5] Delimoy, D., Bailly, C., Devaux, J. and Legras, R., *Polym. Eng. Sci.*, 1988, **28**, 104, (props, synth)
[6] Kim, W.N and Burns, C.M., *Makromol. Chem.*, 1989, **190**, 661, (props, synth)
[7] Birley, A.W. and Chen, X.-Y., *Br. Polym. J.*, 1985, **17**, 297, (props, synth)
[8] Bertilsson, H., Franzén, B. and Kubát, J., *Plast. Rubber Process. Appl.*, 1988, **10**, 145, (props)
[9] Birley, A.W. and Chen, X.-Y., *Br. Polym. J.*, 1984, **16**, 77, (props, synth)
[10] Bertilsson, H., Franzén, B. and Kubát, J., *Plast. Rubber Process. Appl.*, 1989, **11**, 167, (props, synth)
[11] Stokes, V.K., *Polymer*, 1992, **33**, 1237, (composition)
[12] *Eur. Pat. Appl.*, 1989, 299 469, (composition, impact strength)
[13] *Eur. Pat. Appl.*, 1989, 299 468, (composition, impact strength)
[14] Hobbs, S.Y., Dekkers, M.E.J. and Watkins, V.H., *J. Mater. Sci.*, 1988, **23**, 1219, (morphology, stability)
[15] Wahrmund, D.C., Paul, D.R. and Barlow, J.W., *J. Appl. Polym. Sci.*, 1978, **22**, 2155, (props, synth)
[16] Pratt, G.J. and Smith, M.J.A., *Polymer*, 1989, **30**, 1113, (morphology, complex permittivity)
[17] Dekkers, M.E.J., Hobbs, S.Y., Bruker, I. and Watkins, V.H., *Polym. Eng. Sci.*, 1990, **30**, 1628, (morphology)
[18] Bertilsson, H., Franzén, B. and Kubát, J., *Makromol. Chem., Makromol. Symp.*, 1990, **38**, 115, (props, synth)
[19] Devaux, J., Godard, P., Mercier, J.P., Touillaux, R. and Dereppe, J.M., *J. Polym. Sci., Polym. Phys. Ed.*, 1982, **20**, 1882, (stability, solvents, synth)
[20] Dekkers, M.E.J., Hobbs, S.Y. and Watkins, V.H., *J. Mater. Sci.*, 1988, **23**, 1225, (cohesion, fracture props)
[21] Nouvertne, W., Peters, H. and Beicher, H., *Br. Plast. Rubber*, 1983, **17**, (props)
[22] Ash, M. and Ash, I., Handbook of Plastic Compounds, Elastomers and Resins VCH, 1992, 218, (props)
[23] Devaux, J., Godard, P. and Mercier, J.P., *Polym. Eng. Sci.*, 1982, **22**, 229, (props, synth)
[24] Golovoy, A., Cheung, M.F. and Van Oene, H., *Polym. Eng. Sci.*, 1988, **28**, 200, (props)
[25] *Encycl. Polym. Sci. Eng.*, Wiley-Interscience, (ed. J.I. Kroschwitz), 1985, **11**, 692, (props)
[26] Devaux, J., Godard, P. and Mercier, J.P, *J. Polym. Sci., Polym. Phys. Ed.*, 1982, **20**, 1901, (stability)
[27] Hobbs, S.Y., Groshans, V.L., Dekkers, M.E.J. and Shultz, A.R., *Polym. Bull. (Berlin)*, 1987, **17**, 335, (morphology)
[28] Dekkers, M.E.J. and Hobbs, S.Y., *Polym. Eng. Sci.*, 1987, **27**, 1164, (fracture)

Polycarbonate/polybutylene terephthalate alloy, Glass reinforced P-136

Related Polymers: Polycarbonate/polybutylene terephthalate blend
Monomers: Bisphenol A, Carbonic dichloride, 1,4-Benzenedicarboxylic acid, 1,4-Butanediol
Material class: Blends
Polymer Type: PBT and copolymers, polycarbonate alloy blends
Additives: Impact modifiers are added

Applications/Commercial Products:
Mould Shrinkage (%):

No.	Value
1	0.4–0.9%

Trade name	Details	Supplier
Xenoy 6240	10%, impact modified	General Electric Plastics
Xenoy 6370	30%, impact modified	General Electric Plastics
Xenoy 6380	30%, impact modified	General Electric Plastics

Polycarbonate/polyester alloys P-137

Synonyms: *Polycarbonate/polyester blend*
Related Polymers: Poly[2,2-propanebis(4-phenyl)carbonate], Polycarbonate/polybutylene terephthalate alloy, Polycarbonate/polyethylene terephthalate alloy
Material class: Blends
General Information: High impact and flame retardant versions available

Applications/Commercial Products:
Mould Shrinkage (%):

No.	Value	Note
1	0.5–1.1%	depending on composition

Trade name	Details	Supplier
Makroblend	PR 52, UT400, UT1018	Mobay
PC/Polyester	1400	Engineered Polymer
Sabre	1628, 1647, 1664	Dow
Ultrablend	2230	BASF
Vybex	40004 NAFC, 40001 NA	Ferro Corporation
Xenoy		General Electric Plastics
		Bayer Inc.

Polycarbonate/polyethylene alloy P-138

Synonyms: *Polycarbonate/polyethylene blend*
Related Polymers: Polyethylene, Poly[2,2-propanebis(4-phenyl)carbonate]
Monomers: Bisphenol A, Ethylene
Material class: Blends
Polymer Type: polyethylene, polycarbonate alloy blends
Molecular Formula: $[(C_6H_{14}O_3).(C_2H_4)]_n$
Fragments: $C_6H_{14}O_3$ C_2H_4
General Information: Polycarbonate/polyethylene blends are tougher than polycarbonate alone. They exhibit less stress-cracking andd have a higher resistance to some chemicals [1]

– Polycarbonate/polyethylene terephthalate alloy

Volumetric & Calorimetric Properties:

Density:

No.	Value	Note
1	d 1.19 g/cm^3	ASTM D792 [1]

Deflection Temperature:

No.	Value	Note
1	132°C	1.82 MPa, ASTM D648 [1]

Vicat Softening Point:

No.	Value	Note
1	150°C	ASTM D1525, A [1]

Transport Properties:

Water Absorption:

No.	Value	Note
1	0.12 %	24h, ASTM D750 [1]
2	0.34 %	long term, ASTM D570 [1]

Mechanical Properties:

Tensile (Young's) Modulus:

No.	Value	Note
1	2300 MPa	ASTM D638 [1]

Flexural Modulus:

No.	Value	Note
1	2200 MPa	ASTM D790 [1]

Tensile Strength Break:

No.	Value	Note	Elongation
1	59 MPa	ASTM D638 [1]	115%

Tensile Strength Yield:

No.	Value	Elongation	Note
1	59 MPa	8%	ASTM D638 [1]

Flexural Strength Yield:

No.	Value	Note
1	86 MPa	ASTM D790 [1]

Compressive Strength:

No.	Value	Note
1	69 MPa	yield, strain 9%, ASTM D695 [1]

Impact Strength: 640–800 J m^{-1} (3.2 mm thick, ASTM D256) [1], 480–640 J m^{-1} (6.4 mm thick, ASTM D256) [1]
Hardness: Rockwell A115-A118 (ASTM D785) [1]

Electrical Properties:

Dielectric Strength:

No.	Value	Note
1	>16 kV.mm^{-1}	min., 3.2 mm thick, ASTM D149 [1]

Arc Resistance:

No.	Value	Note
1	10–12s	stainless-steel electrodes, ASTM D495 [1]
2	30–70s	tungsten electrodes, ASTM D495 [1]

Optical Properties:

Volume Properties/Surface Properties: Clarity opaque [1]

Polymer Stability:

Upper Use Temperature:

No.	Value	Note
1	121°C	intermittent use [1]
2	121°C	continuous use, no load [1]

Applications/Commercial Products:

Processing & Manufacturing Routes: Processed at the temp. usually employed for Bisphenol A polycarbonate
Mould Shrinkage (%):

No.	Value	Note
1	0.6–0.8%	ASTM D955 [1]

Applications: Protective headgear, sporting goods, bobbins for the textile industry, automobile bumpers

Trade name	Supplier
Makrolan T7700	Bayer Inc.

Bibliographic References

[1] *Encycl. Polym. Sci. Eng.*, (ed., J.I. Kroschwitz), Wiley-Interscience, 1987, **11**, 692, (props.)

Polycarbonate/polyethylene terephthalate alloy P-139

Synonyms: *PET/PC blends. PC/PET blends. Polycarbonate/polyethylene terephthalate blend*
Related Polymers: Poly[2,2-propanebis(4-phenyl)carbonate], Polycarbonate/polyester alloys, PET, Glass fibre reinforced polycarbonate/polyethylene terephthalate alloy
Monomers: Bisphenol A, Carbonic dichloride, 1,4-Benzenedicarboxylic acid, 1,2-Ethanediol
Material class: Blends
Polymer Type: PET and copolymers, polycarbonate alloy blends
General Information: PC/PET blends form complicated systems, the props. of which depend not only on the composition and any additives used, but also on the synth., processing and subsequent exploitation. [1,2,3,4,5,6,7,8] Transesterification, phase separation, crystallisation and degradation processes can all occur as a result of thermal treatment during processing or annealing at lower temps. (125–200°) during post-processing. There is some debate over whether the blends form one or two phases, especially at lower PC levels, and this feature is influenced by the above processes
Morphology: The phase morphology of PC/PET blends depends on various factors, including the composition, the method of mixing, the temp. and duration of mixing and the presence of additives. [4,5] Above 50% PC two phases are formed, one PC-rich, the other PET-rich [7]. The mass fraction of PET in the PC-

rich phase is higher than the mass fraction of PC in the PET-rich phase (suggesting PET dissolves more in the PC-rich phase than PC does in the PET-rich phase). This is the case for both melt-extruded and soln. cast blends. [7] Below 50% PC the components are less compatible [7], and phase separation is more pronounced at higher PC levels (60 and 80%) than at lower levels (10 and 30%) [7]. There is some dispute over the number of phases present. Some reports have found more than one phase present (as indicated by T_g measurement and phase-contrast microscopy) [4,6], others only one [2,5]. Another report concludes the blends are compatible but not microscopically miscible [7]. An additional complication is that the method of measurement of T_g may influence the conclusions [5,6]. This may be a consequence of different thermal treatments during measurement, [6] e.g. DSC measurement of a 20% PC blend shows a single T_g while DMTA shows two. [6] One proposed explanation for the differences in morphology observed is transesterification [4], although this, too, is disputed [5]. Transesterification initially gives rise to block copolymer formation, which homogenises the separate phases, eventually leading to a single phase [4]; blends with 20% or less PC appear to homogenise more quickly. [4] Soln. cast blends form distinct phases at all compositions. [4,7] PET can crystallise in the blend, and the extent of this is increased by annealing [1]

Volumetric & Calorimetric Properties:
Density:

No.	Value	Note
1	d^{20} 1.2737 g/cm^3	50% PC, unannealed, ASTM D792 [1]
2	d^{20} 1.2782 g/cm^3	50% PC, annealed, 130°, ASTM D792 [1]

Latent Heat Crystallization: Latent heat of fusion, ΔH_m 33–38 J g^{-1} (8.0–9.2 cal g^{-1}, 10% PC) [2]; 6.6 kJ mol^{-1} (10% PC, annealed, 200°, 1 h) [4], 6.5 kJ mol^{-1} (20% PC, annealed, 200°, 1 h) [4]; 27–31 J g^{-1} (6.5–7.4 cal g^{-1}, 25% PC) [2]; 27 J g^{-1} (6.3 cal g^{-1}, 40% PC) [2]; 18–22 J g^{-1} (4.2–5.3 cal g^{-1}, 50% PC) [2]; 4.9 kJ mol^{-1} (50% PC, annealed, 200°, 1 h) [4]; 11–12 J g^{-1} (2.7–2.9 cal g^{-1}, 65% PC) [2]; 6.3–6.7 J g^{-1} (1.5–1.6 cal g^{-1}, 75% PC) [2]; 4.9 kJ mol^{-1} (80% PC, annealed, 200°, 1h) [4]; 4.2 kJ mol^{-1} (90% PC, annealed, 200°, 1h). [4] Latent heat of crystallisation, ΔH_c, -26 J g^{-1} (-6.3 cal g^{-1}, 10% PC) [2]; -20 J g^{-1} (-4.7 cal g^{-1}, 25% PC) [2]; -13 J g^{-1} (-3.2 cal g^{-1}, 40% PC) [2]; -10 J g^{-1} (-2.4 cal g^{-1}, 50% PC) [2]; -3.3. J g^{-1} (-0.8 cal g^{-1}, 65%PC) [2]; -1.3 J g^{-1} (-0.3 cal g^{-1}, 75% PC) [2]

Melting Temperature:

No.	Value	Note
1	260°C	10% PC [2]
2	261°C	10% PC, quenched [5]
3	267°C	10% PC, annealed, 180°, 2 h [5]
4	257°C	10% PC, annealed, 200°, 1 h [4]
5	238°C	20% PC [6]
6	257°C	20% PC, annealed, 200°, 1 h [4]
7	259°C	25% PC [2]
8	262°C	30% PC, quenched [5]
9	268°C	30% PC, annealed, 180°, 2 h [5]
10	257°C	40% PC [2]
11	238°C	40% PC [6]
12	260°C	50% PC [1]
13	256°C	50% PC [2]
14	261°C	50% PC, quenched [5]
15	263°C	50% PC, annealed, 180°, 2 h [5]
16	256°C	50% PC, annealed, 200°, 1 h [4]
17	254°C	65% PC [2]
18	260°C	70% PC [1]
19	254°C	70% PC, quenched [5]
20	264°C	70% PC, annealed, 180°, 2 h [5]
21	252°C	80% PC [2]
22	238°C	80% PC [6]
23	261°C	80% PC, annealed, 200°, 1 h [4]
24	266°C	Vybex 40001 NA [11]

Glass-Transition Temperature:

No.	Value	Note
1	90°C	10% PC, melt processed, 290–300°, 8–10 min [5]
2	80°C	10% PC, melt processed, 280°, 4–5 min [2]
3	81–147°C	10% PC, melt processed, 270° [4]
4	84°C	10% PC, melt processed, 270°, 40 min [4]
5	95°C	20% PC, melt processed, 290–300°, 8–10 min [5]
6	85°C	20% PC, melt processed, 280°, 4–5 min [2]
7	85–140°C	20% PC, melt processed, 270° [6]
8	80–150°C	20% PC, melt processed, 270° [4]
9	84°C	20% PC, melt processed 270°, 40 min [4]
10	88°C	25% PC, melt processed, 280°, 4–5 min [2]
11	95–131°C	40% PC, melt processed, 290–300°, 8–10 min [5]
12	79–136°C	40% PC, melt processed 290° [6]
13	76–144°C	40% PC, melt processed, 280°, 4–5 min [2]
14	84.9–134°C	40% PC, melt processed, 275° [7]
15	83.5–136.5°C	40% PC, soln. cast [7]
16	95–131°C	50% PC, melt processed, 290–300°, 8–10 min [5]
17	148°C	50% PC, melt processed, 280° [1]
18	81–145°C	50% PC, melt processed, 280°, 4–5 min [2]
19	83.7–135.3°C	50% PC, melt processed, 275° [7]
20	82–148°C	50% PC, melt processed, 270° [4]
21	81.5–136.9°C	50% PC, soln. cast [7]
22	91–147°C	50% PC, soln. cast [4]
23	75–141°C	60% PC, melt processed, 290° [6]
24	81.7–138.9°C	60% PC, melt processed, 275° [7]
25	80.9–138.4°C	60% PC, soln. cast [7]
26	76–145°C	65% PC, melt processed, 280°, 4–5 min [2]
27	92–137°C	70% PC, melt processed, 290–300°, 8–10 min [5]
28	80.3–140°C	70% PC, melt processed, 275° [7]
29	79.8–139.9°C	70% PC, soln. cast [7]
30	90–144°C	80% PC, melt processed, 290–300°, 8–10 min [5]
31	75–143°C	80% PC, melt processed, 280° [6]
32	73–144°C	80% PC, melt processed, 280°, 4–5 min [2]
33	84–152°C	80% PC, melt processed, 270° [4]

– Polycarbonate/polyethylene terephthalate alloy

| 34 | 88–136°C | 80% PC, soln. cast [4] |
| 35 | 95–144°C | 90% PC, melt processed, 290–300°, 8–10 min [5] |

Transition Temperature General: The melting point of the PET component varies with composition and also with prior thermal history. [2,5] T_g varies with prior thermal history, e.g. processing time and temp., and, to a lesser extent, with any post-processing such as annealing. It can even depend on the method of measurement [5,6]

Deflection Temperature:

No.	Value	Note
1	120°C	1.82 MPa, Makroblend UT 400 [11]
2	88°C	1.82 MPa, Makroblend UT 1018 [11]
3	74°C	1.82 MPa, Vybex 40001 NAFC[11]
4	76°C	1.82 MPa, Vybex 40004 NAFC[11]
5	114°C	1.82 MPa, Makroblend UT 620 G, 1% glass fibre [11]
6	122°C	1.82 MPa, Makroblend UT 640 G, 20% glass fibre[11]

Heat Distortion Temperature:

No.	Value	Note
1	120 °C	1.82 MPa, Makroblend UT 400 [11]
2	88 °C	1.82 MPa, Makroblend UT 1018 [11]
3	74 °C	1.82 MPa, Vybex 4001 NA [11]
4	76 °C	1.82 MPa, Vybex 4004 NAFC [11]
5	114 °C	1.82 MPa, Makroblend UT 620 G, 10% glass fibre [11]
6	122 °C	1.82 MPa, Makroblend UT 640 G, 20% glass fibre [11]

Surface Properties & Solubility:

Solubility Properties General: The 50:50 blend is miscible [1]. All blends are initially immiscible, but homogenise on melt processing [4]. The two polymers have interaction parameters that suggest immiscibility over the whole composition range [9]

Cohesive Energy Density Solubility Parameters: The Flory Huggins interaction parameter, χ_{12}, increases with percentage PC content. 0.037–0.04 (30% PC, extruder blended) [7]; 0.040–0.043 (30% PC, soln. cast) [7]; 0.050–0.054 (70% PC, extruder blended) [7]; 0.054–0.058 (70% PC, soln. cast) [7]

Solvents/Non-solvents: Sol. 1,1,1,3,3,3-hexafluoro-2-propanol [4,7], tetrachloroethane [6], 2-chlorophenol, trifluoroacetic acid [8]; insol. EtOH [10]

Transport Properties:
Melt Flow Index:

No.	Value	Note
1	12.6 g/10 min	20% PC, 270°, 2.16 kg [6]
2	14.6 g/10 min	40% PC, 270°, 2.16 kg [6]
3	15.4 g/10 min	60% PC, 270°, 2.16 kg [6]
4	17.3 g/10 min	80% PC, 270°, 2.16 kg [6]
5	8–16 g/10 min	Makroblend UT 1018 [11]

Water Absorption:

No.	Value	Note
1	0.16 %	Makroblend UT 1018 [11]

Mechanical Properties:

Mechanical Properties General: In polymer blends, stress transfer across the phases is one of the critical factors determining the load bearing capacity and the impact behaviour of the material [13]. PC/PET blends are highly ductile and show large elongations to break [2]; in the necked region whitening occurs due to stress-induced crystallisation. [2] Elongation at break 119% (10% PC), [2] 200% (min. 20–40% PC) [2], 156% (50% PC) [2], 139% (65% PC) [2], 116% (75% PC) [2], 104% (90% PC) [2], 95% (Vybex 40001 NA) [11], 283 % (Vybex 40004 NAFC) [11]. Annealing (125–200°, 1–24 h) gives a significant increase in tensile strength and modulus, whilst the elongation at break decreases [1,6]

Tensile (Young's) Modulus:

No.	Value	Note
1	2080 MPa	50% PC, ASTM D638 [1]

Flexural Modulus:

No.	Value	Note
1	1240 MPa	50% PC, ASTM D790 [1]

Tensile Strength Break:

No.	Value	Note	Elongation
1	33.59 MPa	20% PC, unannealed, BS 2782 [6]	150%
2	43.99 MPa	20% PC, annealed 125°, 18h, BS 2782 [6]	40%
3	62.8 MPa	ASTM D638 [1]	125%
4	77.3 MPa	50% PC, annealed, 130°, 24h, air [1]	19%
5	42.48 MPa	60% PC, unannealed, BS 2782 [6]	110%
6	50.34 MPa	60% PC, annealed 125°, 18h, BS 2782 [6]	100%
7	52–55 MPa	7600 psi, Makroblend UT 400/UT 1018 [11]	151%

Flexural Strength at Break:

No.	Value	Note
1	75–88 MPa	10900 psi, Makroblend UT 400/UT 1018 [11]
2	68 MPa	Vybex 40001 NA [11]
3	89 MPa	Vybex 40004 NAFC [11]
4	134 MPa	19400 psi, Makroblend UT 602 G, 10% glass fibre [11]
5	165 MPa	24 kpsi, Makroblend UT 640G, 20% glass fibre [11]

Tensile Strength Yield:

No.	Value	Elongation	Note
1	56.89 MPa	11.0%	20% PC, unannealed, BS 2782 [6]
2	78.9 MPa	14.2%	20% PC, annealed 125°, 18h, BS 2782 [6]
3	59.71 MPa	12.6%	60% PC, unannealed, BS 2782 [6]
4	71.55 MPa	13.9%	60% PC, annealed 125°, 18h, BS 2782 [6]

— Polycarbonate/polyethylene terephthalate alloy

| 5 | 47 MPa | | Vybex 40001 NA [11] |
| 6 | 54 MPa | | Vybex 40004 NAFC [11] |

Flexural Strength Yield:

No.	Value	Note
1	95.9 MPa	50% PC, ASTM D790 [1]

Complex Moduli: The variation of loss tangent and tensile storage modulus with temp. (-110°–160°, 1 Hz) has been reported [6]

Impact Strength: The impact strength for PC/PET blends depends on several factors, including the notch size and any annealing. The impact strength of a PC/PET/impact modified blend increases with increasing MW of the PET component; little change is found when the MW of the PC component is increased. [14] PC/PET blends remain ductile after annealing, but the notched impact strength can deteriorate on heating [1,6]

Hardness: Rockwell R114 (Makroblend UT 1018) [11]; R118 (Vybex 40004 NAFC) [11]

Izod Notch:

No.	Value	Notch	Note
1	70 J/m	Y	7.1 kg cm cm^{-1}, 50% PC, ASTM D256 [1]
2	620 J/m	Y	63 kg cm cm^{-1}, room temp. M_n 8800 (PET), ASTM D256 [14]
3	750 J/m	Y	76 kg cm cm^{-1}, room temp., M_n 12700 (PET), ASTM D256 [14]
4	730 J/m	Y	74 kg cm cm^{-1}, room temp., M_n 5100 (PC), ASTM D256 [14]
5	720 J/m	Y	73 kg cm cm^{-1}, room temp., M_n 8300 (PC), ASTM D256 [14]
6	66 J/m	Y	6.7 kg cm cm^{-1}, 50% PC, annealed, 130°, 24 h, air [1]
7	800–960 J/m	Y	18 ft lb in^{-1}, Makroblend UT 400/UT 1018 [11]
8	101.6 J/m	Y	Vybex 40004 NAFC [11]
9	139 J/m	Y	2.6 ft lb in^{-1}, Makroblend UT 620G, 10% glass fibre [11]
10	128 J/m	Y	2.4 ft lb in^{-1}, Makroblend UT 640G, 20% glass fibre [11]
11		N	Vybex 40001 NA [11]

Charpy:

No.	Value	Notch	Note
1	0.000477	Y	20% PC, unannealed, BS 4618 [6]
2	0.000357	Y	20% PC, annealed, BS 4618 [6]
3	0.00135	Y	20% PC, unannealed, BS 4618 [6]
4	0.000272	Y	20% PC, annealed, BS 4618 [6]
5	0.00054	Y	60% PC, unannealed, BS 4618 [6]
6	0.00303	Y	60% PC, annealed, BS 4618 [6]
7	0.01205	Y	60% PC, unannealed, BS 4618 [6]
8	0.00206	Y	60% PC, annealed, BS 4618 [6]

Electrical Properties:
Electrical Properties General: Volume resistivity 4.82×10^8 Ω cm (Vybex 40004 NAFC/Vybex 40001 NA) [11]

Dielectric Permittivity (Constant):

No.	Value	Frequency	Note
1	2.96	60 Hz	Makroblend UT 1018 [11]

Optical Properties:
Transmission and Spectra: Ir spectral data have been reported [4,6,8]

Polymer Stability:
Thermal Stability General: PC/PET blends are less stable in a N_2 atmosphere than PC or PET. They have lower initial decomposition temps. [10], and do not form the same decomposition products as are generated in the pyrolysis of PC or PET alone. [10] This is because of intermolecular exchange reactions occurring between 250° and 300°, which give compounds that are less stable than the starting materials [4,10]

Decomposition Details: PC/PET blends are found to have a broad decomposition range in a N_2 atmosphere, with several maxima in the decomposition rate between 380 and 500°. These maxima occur at 380° and 450° (25% PC) [10]; 410°, 460° and 500° (50% PC) [10]; 42°, 460° and 500° (75% PC). [10] PC/PET blends (0–75% PC) heated at 290° for at least 20 mins become amber to dark brown, indicating some thermal degradation. [2] For blends containing 50% PC the decomposition occurs in three stages; intermolecular exchange reactions; [4,10] evolution of pyrolysis products [10]; formation of a cyclic oligomer [10]

Flammability: Flammability rating HB (Makroblend UT 400/UT 1018, $1/16$" thick); [11] HB (Makroblend UT 620 G, 10% glass fibre, $1/16$" in thick) [11]; HB (Makroblend UT 640 G, 20% glass fibre, $1/16$" in thick) [11]

Chemical Stability: Makroblend UT 1018 is resistant to gasoline and a variety of chemicals, including motor oils, antifreeze, trichloroethylene, xylene and Me_2CO [12]. It has limited resistance to 40% sulfuric acid [12]

Applications/Commercial Products:
Processing & Manufacturing Routes: Drying (at 100–110°) prior to processing is required. Melt temps. should not exceed 270–290°. Blends are prepared by melt mixing or by combining two solns. and then removing the solvent. In melt mixing, the two dry components [2,5] are combined in the melt by extrusion [1,2,6,14] or by mechanical mixing. [5,8] The temp. should be kept above the melting point of PET, i.e. 260°. [1] In soln. mixing, the polymers are dissolved and mixed in a suitable solvent, e.g. 1,1,1,3,3,3-hexafluoro-2-propanol at room temp. [4,7], or tetrachloroethane [6]. Films are then cast from soln., and subsequently dried at 60° for 1 d, then at room temp. for a few weeks [4,6,7]. In soln. precipitation the polymers are dissolved in solvent, e.g. phenol:1,1,2,2-tetrachloroethane (60:40) at 120° [10] and the hot soln. added dropwise to EtOH, from which the solid precipitates [10]

Mould Shrinkage (%):

No.	Value
1	0.4–0.9%

Applications: Automotive, electrical-electronics, mechanical and sport applications, including vacuum cleaner housings and nozzles, protective face guards, tractor parts, instrument housings and speaker grilles

Trade name	Details	Supplier
Iupilon	GMB/MB	Mitsubishi Gas
Makroblend UT	400, 1018	Mobay
SC7A21	2063	Spartech
Vybex 40001 NA	high impact	Ferro Corporation
Vybex 40004 NAFC		Ferro Corporation
		Bayer Inc.

Bibliographic References

[1] Halder, R.S., Deopura, B.L. and Misra, A., *Polym. Eng. Sci.*, 1989, **29**, 1766, (props, synth)
[2] Murff, S.R., Barlow, J.W. and Paul, D.R., *J. Appl. Polym. Sci.*, 1984, **29**, 3231, (props, synth)
[3] Pilati, F., Marianucci, E. and Berti, C., *J. Appl. Polym. Sci.*, 1985, **30**, 1267, (stability)
[4] Suzuki, T., Tanaka, H. and Nishi, T., *Polymer*, 1989, **30**, 1287, (props, synth)
[5] Nassar, T.R., Paul, D.R. and Barlow, J.W., *J. Appl. Polym. Sci.*, 1979, **23**, 85, (morphology, stability, T_m, T_g, synth)
[6] Chen, X.-Y. and Birley, A.W., *Br. Polym. J.*, 1985, **17**, 347, (props, synth)
[7] Kim, W.N. and Burns, C.M., *J. Polym. Sci., Polym. Phys. Ed.*, 1990, **28**, 1409, (morphology, interaction parameter, solvent, T_g, synth)
[8] Godard, P., Dekoninck, J.M., Devlesaver, V. and Devaux, J., *J. Polym. Sci., Polym. Chem. Ed.*, 1986, **24**, 3301, (solvent, ir, stability, synth)
[9] Kugo, K., Kitaura, T. and Nishino, J., *Chem. Express*, 1993, **8**, 357, (morphology)
[10] Montaudo, G., Puglisi, C. and Samperi, F., *Polym. Degrad. Stab.*, 1991, **31**, 291, (solvent, thermal stability, synth)
[11] Ash, M. and Ash, I., Handbook of Plastic Compounds, Elastomers and Resins VCH, 1992, 218, (props)
[12] Encycl. Polym. Sci. Eng. 2nd edn., Wiley-Interscience, (ed. J.I. Kroschwitz), 1987, **11**, 692, (chemical stability)
[13] Bertilsson, H., Franzén, B. and Kubát, J., *Plast. Rubber Process. Appl.*, 1989, **11**, 167, (mech props.)
[14] Kanai, H., Sullivan, V. and Auerbach, A., *J. Appl. Polym. Sci.*, 1994, **53**, 527, (impact strength, synth)

Polycarbonate/polyethylene terephthalate alloy, Glass fibre reinforced P-140

Related Polymers: Polycarbonate/polyethylene terephthalate alloy
Monomers: Bisphenol A, Carbonic dichloride, 1,4-Benzenedicarboxylic acid, 1,2-Ethanediol
Material class: Blends, Composites
Polymer Type: PET and copolymers, polycarbonate alloy blends
Additives: 10–20% Glass fibre reinforced

Applications/Commercial Products:
Mould Shrinkage (%):

No.	Value
1	0.1–0.6%

Trade name	Details	Supplier
Makroblend UT 600	600 series (6204 - 10%, 6404 - 20%)	Mobay

Polycarbonate/polyurethane alloy P-141

Synonyms: *Polycarbonate/polyurethane blend*
Related Polymers: Poly[2,2-propanebis(4-phenyl)carbonate], Polyurethanes
Monomers: Bisphenol A, Carbonic dichloride
Material class: Blends
Polymer Type: polycarbonate alloy blends, polyurethane

Volumetric & Calorimetric Properties:
Density:

No.	Value	Note
1	d 1.21–1.22 g/cm^3	[1]

Vicat Softening Point:

No.	Value	Note
1	141–150°C	[1]

Mechanical Properties:
Tensile Strength Break:

No.	Value	Note	Elongation
1	31–41 MPa	4500–6000 psi [1]	150–340%

Hardness: Shore D58-D75 [1]

Applications/Commercial Products:

Trade name	Details	Supplier
Texin 3200	3203, 3215	Mobay
Texin 4200	4203, 4206, 4210, 4215	Mobay

[1] Ash, M. and Ash, I., Handbook of Plastic Compounds, *Elastomers and Resins*, VCH, 1992, 218, (props.)

Polycarbonate/styrene-maleic anhydride alloy P-142

Synonyms: *Polycarbonate/SMA alloy*
Related Polymers: Poly[2,2-propanebis(4-phenyl)carbonate]
Monomers: Bisphenol A, Carbonic dichloride, Styrene, Maleic anhydride
Material class: Blends
Polymer Type: styrene-maleic anhydride (SMA), bisphenol A polycarbonate

Volumetric & Calorimetric Properties:
Density:

No.	Value	Note
1	d 1.13–1.14 g/cm^3	opaque blend [1]
2	d 1.13 g/cm^3	clear blend [1]

Deflection Temperature:

No.	Value	Note
1	108–117°C	1.82 MPa, opaque blend [1]
2	103°C	1.82 MPa clear blend [1]

Vicat Softening Point:

No.	Value	Note
1	138–143°C	opaque blend [1]
2	131°C	clear blend [1]

Transport Properties:
Melt Flow Index:

No.	Value	Note
1	0.7–1.3 g/10 min	opaque blend [1]
2	3.2 g/10 min	clear blend [1]

Mechanical Properties:
Tensile Strength Break:

No.	Value	Note	Elongation
1	45–55 MPa	opaque blend [1]	30–80%
2	63 MPa	clear blend [1]	

Flexural Strength at Break:

No.	Value	Note
1	83–103 MPa	opaque blend [1]
2	116 MPa	clear blend [1]

Izod Notch:

No.	Value	Notch	Note
1	530–640 J/m	Y	opaque blend [1]
2	80 J/m	Y	clear blend [1]

Optical Properties:
Volume Properties/Surface Properties: Clarity opaque or clear (commercial blends) [1]

Applications/Commercial Products:

Trade name	Details	Supplier
Arloy 1000 series	1000, 1100, 1200 (blend of PC with Dylark styrene copolymers, opaque)	ARCO Chemical Company
Arloy 2000	transparent	ARCO Chemical Company

[1] Ash, M. and Ash, I., Handbook of Plastic Compounds, *Elastomers and Resins*, VCH, 1992, 218, (props.)

Polycarbosilanes   P-143

$$\left[\begin{array}{c} R_2 \\ | \\ -SiR_1- \\ | \\ R_3 \end{array} \right]_n$$

R_1, R_2 and R_3 = alkyl or aryl groups

Related Polymers: Polymethylphenylsilylenemethylene, Poly[(phenylsilylene)methylene]
Polymer Type: polycarbosilanes

Polymer Stability:
Polymer Stability General: Stable under extreme conditions providing air and oxidising agents are excluded [1,2]
Hydrolytic Stability: Extremely resistant to hydrol. [2]

Applications/Commercial Products:
Processing & Manufacturing Routes: Synth. varies widely with precise chemical composition of polymer. See specific polycarbosilanes for more detailed information. Main methods include Wurtz coupling, hydrosilation and ring opening polymerisations [1]
Applications: Thermolysis at temps. of >800° produces silicon carbide ceramic fibres. May have value as resistant elastomer at high temp. in the absence of oxygen

Bibliographic References
[1] Trefonas, P., *Encycl. Polym. Sci. Eng.*, Vol. 13, 2nd edn., (ed. J.I. Kroschwitz), John Wiley and Sons, 1985, 162
[2] Koopmann, F., Bargath, A., Knsschka, R., Leukel, J. and Frey, H., *Acta Polym.*, 1996, **47**, 377

Poly(5-carboxyindole)   P-144

Synonyms: *1H-Indole-5-carboxylic acid homopolymer*
Monomers: 1H-Indole-5-carboxylic acid
Material class: Thermoplastic
Polymer Type: polyindole
CAS Number:

CAS Reg. No.
91201-83-1

Molecular Formula: $(C_9H_5NO_2)_n$
Fragments: $C_9H_5NO_2$
Additives: Film can contain dopant ions such as perchlorate or tetrafluoroborate after electrolysis
General Information: Conductive electrochromic polymer. According to the most widely proposed model, the polymer exists as indole units linked via the 2,3-positions [1,2]. In another model, the polymer consists of linked indole trimer units [3].

Surface Properties & Solubility:
Solvents/Non-solvents: Sol. DMSO; insol. DMF (indole trimer model) [3].

Electrical Properties:
Electrical Properties General: Conductivity of 3 to 6×10^{-3} S cm^{-1} reported for 'self-doped' polymer [4,5]. The polymer exists in a reduced insulating form, a half-oxidised conducting form with a delocalised benzenoid-quinoid structure and a fully oxidised insulating form [1,2,7].

Optical Properties:
Optical Properties General: Polymer films oxidised electrochemically change colour from yellow (reduced form) via green to black (oxidised form) [1]. In the reduced state, absorption occurs in the UV region with λ_{max} *ca.* 350nm, while on oxidation a band centred at 774nm increases and then disappears in the fully oxidised form [1]. Fluorescence with a band centred at *ca.* 540nm has been reported for the polymer based on the indole trimer model [9].
Transmission and Spectra: FT-IR [2,3,5,6,7], Raman [1] and uv [1,3,5,9] spectra reported.

Polymer Stability:
Thermal Stability General: Initial weight loss above 117°C is due to loss of carbon dioxide in side chain. Thermal decomposition of backbone starts at *ca.* 400°C, with 25% weight loss at 424°C [8]

Applications/Commercial Products:
Processing & Manufacturing Routes: Prod. mainly electrolytically by anodic oxidation of 5-carboxyindole in acetonitrile with lithium perchlorate or tetraalkylammonium tetrafluoroborate as supporting electrolyte [1,3,5,6,7,10]

Bibliographic References
[1] Talbi, H., Billaud, D., Louarn, G. and Pron, A., *Spectrochim. Acta, Part A*, 2001, **57**, 423
[2] Zotti, G., Zecchia, S., Schiavon, G., Seraglia, R., Berlin, A. and Canavesi, A., *Chem. Mater.*, 1994, **6**, 1742

[3] Mackintosh, J.G., Redpath, C.R., Jones, A.C., Langridge-Smith, P.R.R., Reed, D. and Mount, A.R., *J. Electroanal. Chem.*, 1994, **375**, 163
[4] Abthagir, P.S. and Saraswathi, R., *Organic Electronics*, 2004, **5**, 299
[5] Sivakkumar, S.R., Angulakshmi, N. and Saraswathi, R., *J. Appl. Polym. Sci.*, 2005, **98**, 917
[6] Billaud, D., Humbert, B., Thevenot, L., Thomas, P. and Talbi, H., *Spectrochim. Acta, Part A*, 2003, **59**, 163
[7] Bartlett, P.N., Dawson, D.H. and Farrington, J., *J. Chem. Soc., Faraday Trans.*, 1992, **88**, 2685
[8] Abthagir, P.S. and Saraswathi, R., *Thermochim. Acta*, 2004, **424**, 25
[9] Jennings, P., Jones, A.C. and Mount, A.R., *J. Chem. Soc., Faraday Trans.*, 1998, **94**, 3619
[10] Bartlett, P.N. and Farrington, J., *Bull. Electrochem.*, 1992, **8**, 208

Poly(2-chloroethyl vinyl ether) P-145

$\mathrm{-\!\!\!\!-[CH_2CH]_n\!\!\!\!-}$
        |
        OCH_2CH_2Cl

Synonyms: *Poly(2-chloroethoxyethene). Poly(vinyl 2-chloroethyl ether)*
Related Polymers: Poly(vinyl ethers)
Monomers: 2-Chloroethyl vinyl ether
Material class: Thermoplastic
Polymer Type: polyvinyls
CAS Number:

CAS Reg. No.
29160-08-5

Molecular Formula: $(C_4H_7ClO)_n$
Fragments: C_4H_7ClO

Volumetric & Calorimetric Properties:
Melting Temperature:

No.	Value	Note
1	150°C	[1]

Transition Temperature General: Boiling point 108° [3]

Surface Properties & Solubility:
Solvents/Non-solvents: Insol. Me_2CO [1]

Transport Properties:
Polymer Solutions Dilute: [η] 0.17 dl g^{-1} (C_6H_6, 30°) [3]

Optical Properties:
Refractive Index:

No.	Value	Note
1	1.4377	20° [3]

Polymer Stability:
Chemical Stability: High sensitivity (180 mJ cm^{-2}) to X-ray irradiation [2]

Applications/Commercial Products:
Processing & Manufacturing Routes: Prod. by cationic polymerisation with various catalysts

Bibliographic References
[1] Vandenberg, E.J., Heck, R.F. and Breslow, D.S., *J. Polym. Sci.*, 1959, **41**, 519
[2] Imamura, S., Sugawara, S. and Murase, K., *J. Electrochem. Soc.*, 1977, **124**, 1139
[3] Fishbein, L. and Crowe, B.F., *Makromol. Chem.*, 1961, **48**, 221

Poly(2-chloroethyl vinyl ether-co-isobutylene-co-butyl acrylate) P-146

$\mathrm{-\!\!\!-[CH_2CH]_x[CH_2C(CH_3)_2]_y[CH_2CH]_z\!\!-_n}$
        | |
    OCH_2CH_2Cl $O{=}C{-}OH$

Related Polymers: General information
Monomers: Vinyl 2-chloroethyl ether, Isobutylene, Butyl acrylate
Material class: Thermoplastic, Copolymers
Polymer Type: polyolefins, acrylicpolyvinyls
CAS Number:

CAS Reg. No.
31049-61-3

Molecular Formula: $[(C_4H_7ClO).(C_4H_8).(C_7H_{12}O_2)]_n$
Fragments: C_4H_7ClO C_4H_8 $C_7H_{12}O_2$
General Information: Is a white amorph. solid. Introduction of 2-chloroethyl vinyl ether into the polymer allows the copolymer to be cured using polyamides to give greater mechanical strength [1]

Transport Properties:
Polymer Solutions Dilute: Intrinsic viscosity 3.55 dl g^{-1} (C_6H_6, 30°) [2]

Mechanical Properties:
Tensile Strength Break:

No.	Value	Note	Elongation
1	10 MPa	102 kg cm,p1>-2, 43.2 mol% isobutylene, 50.9 mol% butyl acrylate, 5.9 mol% 2-chloroethyl vinyl ether, polyamine cured [1]	370%

Miscellaneous Moduli:

No.	Value	Note	Type
1	6.6 MPa	67 kg cm^{-2}, 43.2 mol% isobutylene, 50.9 mol% butyl acrylate, 5.9 mol% 2-chloroethyl vinyl ether, polyamine cured [1]	300% modulus

Hardness: Shore A72 (43.2 mol% isobutylene, 50.9 mol% butyl acrylate, 5.9 mol% 2-chloroethyl vinyl ether, polyamine-cured) [1]

Applications/Commercial Products:
Processing & Manufacturing Routes: Prod. by free radical initiated copolymerisation

Bibliographic References
[1] Mashita, K., Imai, S., Yasui, S and Hirooka, M., *Polymer*, 1995, **36**, 3943
[2] *U.S. Pat.*, 1973, 3752788

Poly(2-chloroethyl vinyl ether-co-tetracyanoquinodimethane) P-147

$\mathrm{-\!\!\!\!-[CH_2CH]_n\!\!\!\!-}$
        |
        OCH_2CH_2Cl

Synonyms: *Poly[(2-chloroethoxy)ethene-co-2,2'-(2,5-cyclohexadiene-1,4-diylidene)bispropanedinitrile]*
Related Polymers: General information
Monomers: 2-Chloroethyl vinyl ether, Tetracyanoquinodimethane
Material class: Thermoplastic, Copolymers

Polymer Type: polyvinyls
CAS Number:

CAS Reg. No.
78747-51-0

Molecular Formula: [(C$_4$H$_7$ClO).(C$_9$H$_4$N$_4$)]$_n$
Fragments: C$_4$H$_7$ClO C$_9$H$_4$N$_4$
General Information: Is a pale brown powder [1]

Surface Properties & Solubility:
Solvents/Non-solvents: Sol. MeCN, Me$_2$CO, DMF, DMSO [1]

Transport Properties:
Polymer Solutions Dilute: Viscosity 0.449 dl g^{-1} (MeCN, 30°, approx. 50/50 copolymer) [1]

Optical Properties:
Transmission and Spectra: Ir and H-1 nmr spectral data have been reported [1]

Applications/Commercial Products:
Processing & Manufacturing Routes: An alternating copolymer may be synth. by free-radical copolymerisation (50–60°, MeCN) using azobis(isobutyronitrile) as initiator. Polymerisation may be carried out without AIBN but yields are lower. The use of alternative solvents has been reported [1]

Bibliographic References
[1] Iwatsuki, S. and Itoh, T., *Macromolecules*, 1979, **12**, 208, (synth., props.)

Poly(2-chloroethyl vinyl ether-co-N-vinylpyrrolidinone) P-148

Synonyms: *Poly[(2-chloroethoxy)ethene-co-1-ethenyl-2-pyrrolidinone]*
Related Polymers: Poly(vinyl ethers)
Monomers: 2-Chloroethyl vinyl ether, N-Vinylpyrrolidinone
Material class: Thermoplastic, Copolymers
Polymer Type: polyvinyls
CAS Number:

CAS Reg. No.
54803-58-6

Molecular Formula: [(C$_4$H$_7$ClO).(C$_6$H$_9$NO)]$_n$
Fragments: C$_4$H$_7$ClO C$_6$H$_9$NO

Volumetric & Calorimetric Properties:
Melting Temperature:

No.	Value	Note
1	170°C	[1]

Deflection Temperature:

No.	Value	Note
1	140°C	[1]

Applications/Commercial Products:
Processing & Manufacturing Routes: Prod. by copolymerisation of 2-chloroethyl vinyl ether and N-vinylpyrrolidinone with AIBN at 60° for 24h [3]
Applications: Used in photochemical cross-linking of polymers to immobilise enzymes

Bibliographic References
[1] Barnes, J., Esslemont, G. and Holt, P., *Makromol. Chem.*, 1975, **176**, 275
[2] Ichimura, K. and Watanabe, S., *J. Polym. Sci., Polym. Chem. Ed.*, 1980, **18**, 891, (use)
[3] Chubinskaya, S.G., Fedorova, I.P., Veksler, I.G., Berdova, A. *et al*, *Khim.-Farm. Zh.*, 1990, **24**, 121, (synth)

Polychloroprene P-149

Synonyms: *Unvulcanised polychloroprene. Polychloroprene, Unvulcanised. Chloroprene rubber. CR. Poly(2-chloro-1,3-butadiene)*
Related Polymers: Polychloroprene, Vulcanised
Monomers: Chloroprene
Material class: Synthetic Elastomers
Polymer Type: polybutadiene
CAS Number:

CAS Reg. No.
9010-98-4

Molecular Formula: (C$_4$H$_5$Cl)$_n$
Fragments: C$_4$H$_5$Cl
Molecular Weight: High monomer reactivity leads to high MW
General Information: First introduced in 1932 as Du Prene by DuPont; first commercial successful synthetic elastomer. Chlorine substituent in monomer increases reactivity, restricts incorporation of other monomers. Gel formation results from the formation of branching and cross-linking, controlled by temp. and addition of free-radical inhibitors. Commercial polymer is mainly 1,4-*trans*; regularity decreases with increasing temp. of polymerisation and crystallinity is highest when formed at low temp. Vulcanisation increases mechanical props.; see polychloroprene, vulcanised. Available in dry and latex forms
Morphology: Consists of 1,4-*trans*, 1,2-, 1,4-*cis* and other isomerised variations in varying proportions. Free-radical emulsion polymerisation yields mainly 1,4-*trans* polymer, composition increases at low polymerisation temp. and with the use of catalysts. Highly ordered polymer is crystalline [117]

Volumetric & Calorimetric Properties:
Density:

No.	Value	Note
1	d 1.23 g/cm^3	amorph. [28,29,30]
2	d 1.35 g/cm^3	cryst. [28,29,30]
3	d 1.23 g/cm^3	[27]
4	d 1.25 g/cm^3	[58]
5	d 1.2–1.25 g/cm^3	[87]
6	d 1.23 g/cm^3	[2,7,10]

Thermal Expansion Coefficient:

No.	Value	Note	Type
1	0.0006 K^{-1}	[2,3]	V
2	0.000196–0.000206 K^{-1}	[35]	L

– Polychloroprene

Volumetric Properties General: Heat of fusion for crystalline phase 22.7 ± 2 cal g^{-1} [53]. Heat of fusion of crystals, ΔH_c 8.41 kJ mol^{-1} [6]. Heat of fusion of crystalline phase, ΔH_c 94.6 kJ kg^{-1} [6]
Thermodynamic Properties General: Thermodynamic props. of crystallisation have been reported [28,29,30]
Thermal Conductivity:

No.	Value	Note
1	0.192 W/mK	[2]

Specific Heat Capacity:

No.	Value	Note	Type
1	1.38 kJ/kg.C	[99]	P

Melting Temperature:

No.	Value	Note
1	90°C	polymerisation temp. -130° [54,55]
2	30°C	polymerisation temp. 100° [54,55]
3	70–80°C	polymerisation temp. -60° [53]
4	10–20°C	polymerisation temp. 110° [53]
5	107°C	infinite chain length [121]

Glass-Transition Temperature:

No.	Value	Note
1	-50°C	polymerisation temp. -130° [54,55]
2	-24°C	polymerisation temp. 100° [54,55]
3	-40°C	[56]
4	-41°C	[4,99]
5	49°C	*trans* [120]

Transition Temperature General: Other melting temps. have been reported [5,6,12]. T_m decreases with increasing polymerisation temp. [31]

Surface Properties & Solubility:
Cohesive Energy Density Solubility Parameters: Solubility parameter 18.9 J$^{1/2}$ cm$^{-3/2}$ [35]. 18.42 J$^{1/2}$ cm$^{-3/2}$ [84], 19.19 J$^{1/2}$ cm$^{-3/2}$ (calc.) [85], 17.6 J$^{1/2}$ cm$^{-3/2}$ (swelling) [86]
Solvents/Non-solvents: The dispersibility of polychloroprene in solvents depends on polymerisation temp. At 10–40° dispersion is complete; at 50° dispersion is partial and at 100° swelling occurs [47]. Sol. C_6H_6, EtOAc, pyridine, chlorinated hydrocarbons, toluene. Insol. alcohols, aliphatic hydrocarbons
Surface Tension:

No.	Value	Note
1	40.5 mN/m	liq., 50° [25]
2	28 mN/m	liq., 200° [25]
3	43.6 mN/m	liq./vapour, 20° [91]

Transport Properties:
Polymer Solutions Dilute: Mark-Houwink-Sakurada constants have been reported for linear monodisperse and branched polychloroprenes in theta solvents [74]. Mark-Houwink constants have been reported [17,18,19,20]. Polymer/solvent interaction parameter χ has been reported for a wide range of solvents. Values range from 0.06 (THF), 2.27 (EtOH), 0.1 (chlorobenzene), 0.14 (toluene) to 1.61 (butanol) [80]. Theta temp. θ 25° (2-butanone) [81,82]; 45.5 (cyclohexane) [81]; 56.3 (cyclopentane) [81]

Permeability of Gases: Other permeability data for polychloroprene at 17–50° have been reported for H_2, He, N_2, O_2, CO_2 and CH_4 [92]
Gas Permeability:

No.	Gas	Value	Note
1	H_2	889 cm^3 mm/(m^2 day atm)	1.03×10^{-8} cm^2 (s atm)$^{-1}$, 25° [35]
2	H_2	2462 cm^3 mm/(m^2 day atm)	2.85×10^{-7} cm^2 (s atm)$^{-1}$, 50° [35]
3	N_2	77.7 cm^3 mm/(m^2 day atm)	9×10^{-9} cm^2 (s atm)$^{-1}$, 25° [35]
4	N_2	311 cm^3 mm/(m^2 day atm)	3.6×10^{-8} cm^2 (s atm)$^{-1}$, 50° [35]
5	O_2	259 cm^3 mm/(m^2 day atm)	3×10^{-8} cm^2 (s atm)$^{-1}$, 25° [35]
6	O_2	872 cm^3 mm/(m^2 day atm)	1.01×10^{-7} cm^2 (s atm)$^{-1}$, 50° [35]
7	CO_2	1685 cm^3 mm/(m^2 day atm)	1.95×10^{-7} cm^2 (s atm)$^{-1}$, 25° [35]
8	CO_2	4881 cm^3 mm/(m^2 day atm)	5.65×10^{-7} cm^2 (s atm)$^{-1}$, 50° [35]

Mechanical Properties:
Mechanical Properties General: Tensile strength decreases with increasing polymerisation temp. [47]. Compressibility has been reported [8,9]
Tensile Strength Break:

No.	Value	Note	Elongation
1	493 MPa	polymerisation temp. 10° [47]	530%
2	278 MPa	polymerisation temp. 40° [47]	510%
3	232 MPa	polymerisation temp. 100° [47]	430%

Mechanical Properties Miscellaneous: Capillary rheometer data for Neoprene W have been reported as shear stress vs shear rate plots [33]
Hardness: Polymerised at 100° remains soft and flexible indefinitely, at 55° requires months for hardening, 40° takes days and at 10° hardening occurs in hours [47]

Optical Properties:
Refractive Index:

No.	Value	Note
1	1.558	[7]

Transmission and Spectra: Raman [24], ir [102], nmr [116] and epr [112] spectral data have been reported

Polymer Stability:
Decomposition Details: The initial stages of the thermal degradation of polychloroprene have been reported [79]. Pyrolysis of polychloroprene yields 2% of monomer, 70% of the chlorine is released as HCl [97]. Initial decomposition temp. 170° [90]. The rate of heat and smoke release from polychloroprene has been reported [105]. Weight loss occurs slowly 230° (volatiles and CO), faster at 275–375° (rapid loss of HCl) and gradually lessens at 700° (combustion of residual carbon). Degradation characteristics have been reported for commercial Neoprenes. Initial weight loss (4%) 265–280° with fastest weight loss at 380°. Carbonaceous residue at 550° 21–23% [111]
Flammability: Flame retardation and smoke suppression of polychloroprene by tin additives have been reported [75]
Environmental Stress: The ageing and degradation of polychloroprene and its blends with ethylene/propylene rubber have been

reported [70]. The effect of water absorption, with and without salts present, on resistance to tearing has been reported [113]. The presence of $FeCl_3$ accelerates tearing

Stability Miscellaneous: Ozone degradation of polychloroprene has been found to give acid chloride groups, partially ozonised polychloroprene is less stable than partially ozonised 1,4-cis-polybutadiene, [69]. Blends of polychloroprene (10–40%) have been reported with respect to thermal stability, light resistance, storage modulus and flammability [72]

Applications/Commercial Products:

Processing & Manufacturing Routes: Produced by free-radical emulsion polymerisation; 1,4 addition is thermodynamically favoured. Increase in temp decreases regularity of polymer. Soft enough to be processed without plasticisers

Applications: Mainly used as vulcanised form. Applications include use in mechanical and automotive components, belts, hoses, mouldings, footwear and cables

Trade name	Details	Supplier
Aquastik	adhesive	DuPont
Baypren	various grades	Bayer Inc.
Baypren		Miles
Butachlor		Rhone-Poulenc
Daubond		Daubert
Neoprene	various grades	DuPont
Skyprene		Tosoh

Bibliographic References

[1] Wood, L.A., *Polym. Handb.*, 3rd edn., (eds. J. Brandrup and E.H. Immergut), Wiley Interscience, 1989, **V**, 7
[2] Cotton, N.L., *The Neoprenes*, Rubber Chemicals Division, 1953
[3] Payne, A.R. and Scott, J.R., *Engineering Design with Rubber*, Interscience, 1960
[4] Kell, R.M., Bennett, B. and Stickney, P.B., *J. Appl. Polym. Sci.*, 1959, **2**, 8
[5] Krigbaum, W.R., Dawkins, J.V., Via, G.H. and Balta, Y.I., J. Polym. Sci., Part A-2, 1972, **4**, 475
[6] Maynard, J.T. and Mochel, W.E., *J. Polym. Sci.*, 1954, **13**, 235
[7] Wood, L.A., *Rubber Chem. Technol.*, 1940, **13**, 861
[8] Cramer, W.S. and Silver, I., 1951, U.S. Naval Ordnance Lab., U.S. Naval Ordnance Lab., Whiteoak, MD
[9] Naunton, W.H.S. et al, *Rubber Engineering*, Chemical Publishing Co., 1946, 30
[10] Wood, L.A., Bekkedahl, N. and Roth, F.L., *Rubber Chem. Technol.*, 1943, **16**, 244
[11] Yin, T.P. and Pariser, R., *J. Appl. Polym. Sci.*, 1963, **7**, 667
[12] Gent, A.N., *J. Polym. Sci.*, 1965, **3**, 3787
[13] McPherson, A.T., *Rubber Chem. Technol.*, 1963, **36**, 1230
[14] Ball, J.M. and Maassen, G.C., ASTM, *Symp. Appl. Synth. Rubbers*, 1944, 27
[15] Boomstra, B.B.S.T., Elastomers. Their Chemistry, *Physics and Technology*, (ed. R. Houwink), Elsevier, 1948, **3**
[16] Koros, W.J. and Hellums, M.W., *Encycl. Polym. Sci. Eng.*, 2nd edn., (ed. J.I. Kroschwitz), John Wiley and Sons, 1989, **Supplement**, 742
[17] Mochel, W.E. and Nichols, J.B., *J. Am. Chem. Soc.*, 1949, **71**, 3425
[18] Mochel, W.E., Nichols, J.B. and Mighton, C.J., *J. Am. Chem. Soc.*, 1948, **70**, 2185
[19] Mochel, W.E., Nichols, J.B. and Mighton, C.J., *Ind. Eng. Chem.*, 1951, **43**, 154
[20] Coleman, M.M. and Fuller, R.E., *J. Macromol. Sci., Phys.*, 1975, **11**, 419
[21] Cabasso, I., *Encycl. Polym. Sci. Eng.*, 2nd edn., (ed. J.I. Kroschwitz), John Wiley and Sons, 1987, **9**, 562
[22] Clough, R., *Encycl. Polym. Sci. Eng.*, 2nd edn., (ed. J.I. Kroschwitz), John Wiley and Sons, 1988, **13**, 675
[23] Lancaster, J.K., *Encycl. Polym. Sci. Eng.*, 2nd edn., (ed. J.I. Kroschwitz), John Wiley and Sons, 1985, **1**, 18
[24] Petcavich, R.J. and Coleman, M.M., *J. Macromol. Sci., Phys.*, 1980, **18**, 47
[25] Koberstein, J.T., *Encycl. Polym. Sci. Eng.*, 2nd edn., (ed. J.I. Kroschwitz), John Wiley and Sons, 1985, **8**, 242
[26] Struik, L.C.E., *Polym. Eng. Sci.*, 1979, **19**, 223
[27] Witsiepe, W.K., *Kirk-Othmer Encycl. Chem. Technol.*, 4th edn., (ed. J.I. Kroschwitz), John Wiley and Sons, 1993, **8**, 1045
[28] Coleman, M.M., Petcavich, R.J. and Painter P.C., *Polymer*, 1978, **19**, 1253
[29] Tsereteli, G.I., Sochava, I.V. and Buka, A., Vestn. Leningr. Univ., Ser. 4: Fiz., Khim., 1975, 67
[30] Teitelbaum, B.Y. and Aroshina, N.P., *Vysokomol. Soedin.*, 1965, **7**, 978
[31] Maynard, J.T. and Mochel, W.E., *J. Polym. Sci.*, 1954, **13**, 242
[32] Murray, R.M. and Detenber, J.D., *Rubber Chem. Technol.*, 1961, **34**, 668
[33] McCabe, C.C., *Reinforcement of Elastomers*, (ed. G. Kraus), Interscience, 1965, 231
[34] Greensmith, H.W., Mullins, L. and Thomas, A.G., *The Chemistry and Physics of Rubber-Like Substances*, (ed. L. Bateman), The Garden City Press, 1963, 261
[35] Fletcher, W.P., *Rubber Technology and Manufacture*, 2nd edn., (eds. C.M. Blow and C. Hepburn), Butterworth Scientific, 1982, 118
[36] Weir, C.E., *J. Res. Natl. Bur. Stand. (U.S.)*, 1951, **46**, 207
[37] Cooper, D.H., *Ind. Eng. Chem.*, 1951, **43**, 365
[38] Kean, W.N., *Trans. ASME*, 1946, **68**, 237
[39] Kean, W.N., *India Rubber World*, 1953, **128**, 351
[40] Torii, C., Hoshii, K. and Isshiki, S., J. Soc. Rubber Ind., *Jpn*, 1956, **29**, 38
[41] Hoshii, K. and Isshiki, S., J. Soc. Rubber Ind., *Jpn*, 1958, **29**, 151
[42] Carter, W.C., Magat, M., Schneider, W.C. and Smyth, C.P., *Trans. Faraday Soc.*, 1946, **42A**, 213
[43] Hillier, K.W., *Trans., Inst. Rubber Ind.*, 1950, **26**, 64
[44] Nolle, A.W., *J. Polym. Sci.*, 1950, **5**, 1
[45] Fenner, J.V., *Rubber Age (London)*, 1956, **79**, 445
[46] James, R.R. and Guyfon, C.W., *Rubber Age (London)*, 1956, **78**, 725
[47] Walker, H.W. and Mochel, W.E., *Rubber Chem. Technol.*, 1950, **23**, 652
[48] Boomstra, B.S.T.T., *Rubber Chem. Technol.*, 1950, **23**, 338
[49] Tager, A and Sanatina, V., *Rubber Chem. Technol.*, 1951, **24**, 773
[50] Pisarenko, A.P. and Rebinder, P.A., *Rubber Chem. Technol.*, 1951, **24**, 569
[51] Buist, J.M., *Rubber Chem. Technol.*, 1953, **26**, 935
[52] Thompson, D.C., Baker, R.H. and Browlow, R.W., *Rubber Chem. Technol.*, 1952, **25**, 928
[53] Maynard, J.T. and Mochel, W.E., *Rubber Chem. Technol.*, 1954, **27**, 634
[54] Garett, R.R., Hargreaves, C.A. and Robinson, D.N., *J. Macromol. Sci., Chem.*, 1970, **48**, 1679
[55] Aufdermarsh, C.A. and Pariser, R., *J. Polym. Sci.*, 1964, **2**, 4727
[56] Musch, R. and Magg, H., *Polymeric Materials Encyclopedia*, (ed. J.C. Salamone), CRC Press, 1996, 1238
[57] Schweitzer, P.A., *Corrosion Resistance of Elastomers*, Marcel Dekker, 1990, 23
[58] Hertz, D.L., Handbook of Elastomers, *New Developments and Technology*, (eds. A.K. Bhowmick and H.L. Stephens), Marcel Dekker, 1988, 464
[59] Delor, F., Lacoste, J., Lemair, J., Barroisoudin, N. and Cardinet, C., *Polym. Degrad. Stab.*, 1996, **53**, 361
[60] Kundu, P.P., Banerjee, S. and Tripathy, D.K., *Int. J. Polym. Mater.*, 1996, **32**, 125
[61] Kundu, P.P. and Tripathy, D.K., *Kautsch. Gummi Kunstst.*, 1996, **49**, 268
[62] Kundu, P.P., Tripathy, D.K. and Banerjee, S., *Polymer*, 1996, **37**, 2423
[63] Bandyopadhyay, S., De, P.P., Tripathy, D.K. and De, S.K., *Polymer*, 1995, **36**, 1979
[64] *Polym.-Plast. Technol. Eng.*, 1995, **34**, 143
[65] Ghosh, P. and Ray, P, Plast., *Rubber Compos. Process. Appl.*, 1994, **22**, 51
[66] Xue, Y.P., Chen, Z.F. and Frisch, H.C., *J. Appl. Polym. Sci.*, 1996, **52**, 1833
[67] Jana, P.B., Plast., *Rubber Compos. Process. Appl.*, 1993, **20**, 107
[68] Lawandy, S.N., Botros, S.H. and Mounir, A., *Polym.-Plast. Technol. Eng.*, 1993, **32**, 255
[69] Anachkov, M.P., Rakovsky, S.K., Sefanova, R.V. and Stoyanova, A.K., *Polym. Degrad. Stab.*, 1993, **41**, 185
[70] Kalidaha, A.K., De, P.P. and Sen, A.K., *Polym. Degrad. Stab.*, 1993, **39**, 179
[71] Pilai, V.B. and Das, J.N., Plast., *Rubber Compos. Process. Appl.*, 1992, **18**, 155
[72] Rang, D.I., Ha, C.S. and Cho, W.J., *Eur. Polym. J.*, 1992, **28**, 565
[73] Gjurova, K. and Karaivanova, M., *J. Therm. Anal.*, 1991, **37**, 2593
[74] Itoyama, K., Shimizu, N. and Natsuzawa, S., *Polym. J. (Tokyo)*, 1991, **23**, 1139
[75] Hornsby, P.R., Mitchell, P.A. and Cusack, P.A., *Polym. Degrad. Stab.*, 1991, **32**, 299
[76] Budrugeae, P. and Ciutacu, S., *Polym. Degrad. Stab.*, 1991, **33**, 377
[77] Nagode, J.B. and Roland C.M., *Polymer*, 1991, **32**, 505
[78] Lawandy, S.N. and Wassef, M.S., *J. Appl. Polym. Sci.*, 1990, **40**, 323

[79] Miyata, V. and Atsumi, M., *J. Polym. Sci., Part A: Polym. Chem.,* 1988, **26**, 2561
[80] Munk, P., Hattam, P., Du, Q. et al, *J. Appl. Polym. Sci.: Appl. Polym. Symp.,* 1990, **45**, 289
[81] Hanafus, K., Teramoto, A. and Fujita, H., *J. Phys. Chem.,* 1966, **70**, 4004
[82] Kawahara, R., Norisuye, T. and Fujita, H., *J. Chem. Phys.,* 1968, 4330
[83] Norisuya, T., Kawahara, K., Teramoto, A. et al, *J. Chem. Phys.,* 1968, **49**, 4330
[84] Shvarts, A.G., *Kolloidn. Zh.,* 1956, **18**, 755
[85] Small, P.A., *J. Appl. Chem.,* 1953, **3**, 71
[86] Gee, G., Trans., *Inst. Rubber Ind.,* 1943, **18**, 266
[87] Brostow, W., Kubát, J. and Kubát, M.M., *Physical Properties of Polymers Handbook,* (ed. J.E. Mark), AIP Press, 1996, 333
[88] Tewarson, A., *Physical Properties of Polymers Handbook,* (ed. J.E. Mark), AIP Press, 1996, 587
[89] Billmeyer, F.W., *Textbook of Polymer Science,* John Wiley and Sons, 1984, 136
[90] Liggat, J., *Polym. Handb.,* 4th edn., (eds. J. Brandrup, E.H. Immergut and E.A. Grulke), John Wiley and Sons, 1999, **II**, 451
[91] Wu, S., *J. Polym. Sci., Part C: Polym. Lett.,* 1971, **34**, 19
[92] van Amerongen, G.J., *J. Appl. Phys.,* 1946, **17**, 972
[93] Hartmann, B. and Jarzynski, J., *J. Acoust. Soc. Am.,* 1974, **56**, 1469
[94] Hartmann, A. and Glander, F., *Rubber Chem. Technol.,* 1956, **29**, 166
[95] Murray, R.M., *Rubber Chem. Technol.,* 1959, **32**, 1117
[96] Murray, R.M. and Detenber, J.D., *Rubber Chem. Technol.,* 1961, **34**, 668
[97] Well, L.A., *J. Res. Natl. Bur. Stand. (U.S.),* 1948, **41**, 315
[98] Robbins, R.F., Ohori, Y. and Weitzel, D.H., *Rubber Chem. Technol.,* 1964, **37**, 154
[99] Wood, L.A., *Rubber Chem. Technol.,* 1976, **49**, 189
[100] Dillon, J.H., Prettyman, I.B. and Hall, G.L., *J. Appl. Phys.,* 1944, **15**, 309
[101] Boomstra, B.B.S.T., *Rubber Chem. Technol.,* 1951, **24**, 199
[102] Ferguson, R.C., *Rubber Chem. Technol.,* 1965, **38**, 532
[103] Becker, R.O., *Rubber Chem. Technol.,* 1964, **37**, 76
[104] Hennig, J., *Rubber Chem. Technol.,* 1966, **39**, 678
[105] Stewart, C.W., Dawson, R.L. and Johnson, P.R., *Rubber Chem. Technol.,* 1975, **48**, 132
[106] Holownia, B.P., *Rubber Chem. Technol.,* 1975, **48**, 246
[107] Johnson, P.R., *Rubber Chem. Technol.,* 1976, **49**, 158
[108] Fabris, H.J. and Sommer, J.G., *Rubber Chem. Technol.,* 1977, **50**, 523
[109] Sircar, A.K. and Lamond, T.G., *Rubber Chem. Technol.,* 1978, **51**, 126
[110] Brazier, D.W. and Nickel, G.H., *Rubber Chem. Technol.,* 1979, **52**, 735
[111] Brazier, D.W., *Rubber Chem. Technol.,* 1980, **53**, 437
[112] Mean, W.T., *Rubber Chem. Technol.,* 1980, **53**, 245
[113] Bhowmick, A.K. and Gent, A.N., *Rubber Chem. Technol.,* 1983, **56**, 845
[114] Bhowmick, A.K., Gent, A.N. and Pulford, C.T.R., *Rubber Chem. Technol.,* 1983, **56**, 226
[115] Sircar, A.K., *Rubber Chem. Technol.,* 1992, **65**, 503
[116] Stewart, C.A., Takeshita, T. and Coleman, M.L., *Encycl. Polym. Sci. Eng.,* 2nd edn., (ed. J.I. Kroschwitz), John Wiley and Sons, 1985, **3**, 4412, (nmr)
[117] Gent, A.N., *J. Polym. Sci., Part A-2,* 1966, **4**, 447
[118] Gent, A.N., *Trans. Faraday Soc.,* 1977, **50**, 55
[119] Kanaya, T., Kawaguchi, T and Kaji, K., *J. Chem. Phys.,* 1996, **105**, 4342
[120] Aufdermarsh, C.A. and Pariser, R., *J. Polym. Sci., Part A-2,* 1964, 4727
[121] Krigbaum, W.R. and O'Mara, J.H., *J. Polym. Sci., Polym. Phys. Ed.,* 1970, **8**, 1011
[122] Garret, R.R., Hargreaves, C.A. and Robinson, D.N., *J. Macromol. Sci., Chem.,* 1970, **8**, 1679
[123] Hill, D.J.T., O'Donnell, J.H., Perera, M.C.S. and Pomery, P.J., *ACS Symp. Ser.,* 1993, **527**, 74
[124] Gent, A.N., *Rubber Chem. Technol.,* 1967, **40**, 1071
[125] Krigbaum, W.R., Dawkins, J.V., Via, G.H. and Balta, Y.I., *Rubber Chem. Technol.,* 1967, **40**, 788

Polychloroprene, Vulcanised   P-150

Synonyms: *Poly(2-chloro-1,3-butadiene), Vulcanised. Chloroprene rubber, Vulcanised. Vulcanised polychloroprene*
Related Polymers: Polychloroprene
Monomers: Chloroprene
Material class: Synthetic Elastomers
Polymer Type: polybutadiene
Additives: Vulcanised by zinc oxide, thiourea, magnesium oxides, sulfur. Antimony trioxide, zinc borate and aluminium trihydrate as flame retardants
General Information: Vulcanisation increases tear strength and adhesion to fabrics, and gives high tensile strength and low gas permeability

Volumetric & Calorimetric Properties:

Density:

No.	Value	Note
1	d 1.32 g/cm^3	unfilled [99]
2	d 1.42 g/cm^3	50 phv carbon black [99]
3	d 1.32 g/cm^3	[1]

Thermal Expansion Coefficient:

No.	Value	Note	Type
1	0.0144 K^{-1}	-203–24° [98]	L
2	0.00013 K^{-1}	rubbery [98]	V
3	0.00061–0.00072 K^{-1}	[1,2,11,57]	V
4	0.00035 K^{-1}	glass [98]	V

Thermodynamic Properties General: The effect of stretching on thermal conductivity has been reported [104]
Latent Heat Crystallization: Rates of crystallisation have been reported for different degrees of vulcanisation, carbon black loadings and plasticiser levels for temps. ranging from 0–-20° [96]

Thermal Conductivity:

No.	Value	Note
1	0.21 W/mK	50 phv carbon black [2,99]
2	0.19 W/mK	[26]
3	0.192 W/mK	unfilled [3,99]

Specific Heat Capacity:

No.	Value	Note	Type
1	2.1–2.2 kJ/kg.C	unfilled [3]	P
2	2.175 kJ/kg.C	[26]	P
3	1.7 kJ/kg.C	[57]	P
4	1.7–1.8 kJ/kg.C	50 phv carbon black [3,99]	P

Glass-Transition Temperature:

No.	Value	Note
1	-41°C	[4,12,99]
2	-40°C	[4,58]
3	-50°C	[35]

Brittleness Temperature:

No.	Value	Note
1	-40°C	[57]
2	-43°C	0 phv butyl oleate [35]
3	-68°C	35 phv butyl oleate [35]

Transport Properties:

Permeability and Diffusion General: The penetration of oils into polychloroprene rubber has been reported [78]

Mechanical Properties:

Mechanical Properties General: Effect of heat ageing on mechanical props. has been reported [45,103]. Cross-linking increases hardness and tensile strength [22]. The effect of carbon black on Poisson's ratio has been reported [106]. The ratio decreases from 0.4997 at 0 phv carbon to 0.4945 at 120 phv carbon black. Tensile strength decreases with temp. [48]. Effects of filling with carbon black have been reported [7]. Other values for tensile strength have been reported [45]. Complex dynamic Young's moduli for neoprene have been reported over the temp. range -60–100° [44]. Young's modulus has been measured for various oil and carbon black loadings over a range of temps. [46]. Compressibility measurements have been reported [8,9,36]. Ultimate elongation 800–1000% [14,15]

Tensile (Young's) Modulus:

No.	Value	Note
1	1–3 MPa	unfilled, vulcanised [87]
2	3–5 MPa	50 pph carbon black filled, vulcanised [87]

Tensile Strength Break:

No.	Value	Note	Elongation
1	22–170 MPa	[57]	200–600%
2	13–22 MPa	unfilled, vulcanised [87]	800–1000%
3	23–25 MPa	50 pph carbon black filled, vulcanised [87]	200–450%
4	48 MPa	0° [48]	750%
5	25 MPa	80° [34,48]	750%
6	33 MPa	0°, 30 pph carbon black [48]	350%
7	6 MPa	80°, 30 pph carbon black [48]	
8	60 MPa	0° [34]	
9	7–17 MPa	[35]	200–600%
10	21 MPa	[27]	
11	22 MPa	[58]	600%
12	14.5 MPa	27° [50]	
13	2 MPa	127° [50]	

Miscellaneous Moduli:

No.	Value	Note	Type
1	1.2 MPa	20° [35]	Torsional rigidity modulus
2	3.5 MPa	-30° [35]	Torsional rigidity modulus
3	70 MPa	-40° [5]	Torsional rigidity modulus
4	2270 MPa	[8]	Isothermal bulk modulus

Complex Moduli: Storage modulus 0.65 MPa (unfilled), 2.8 (50 phv carbon black). Loss modulus 0.11 MPa (unfilled), 0.5 (50 phv carbon black)

Viscoelastic Behaviour: Longitudinal speed of sound 1730 m s^{-1} (carbon black filled) [46,93]. Compression set 15–35% [57]. The dependence of stiffness and creep on proportion of carbon black has been reported [37]. Creep measurements have been made in shear deformation as a function of degree of vulcanisation, preconditioning and composition for up to 4 years [38,39]. Resistance 60–65% (unfilled) [15,101], 48% (50 phv carbon black) [10]

Mechanical Properties Miscellaneous: Thermo-mechanical analysis of vulcanised polychloroprene has been reported, showing crystalline transition at 42° and 47° [110]. Filaments have been subjected to oscillations in the range 0.5–4 kHz at temps. in the range 0–40° and at a strain of 300% to give the dependence of dynamic modulus on strain and to compare with static moduli [43]

Hardness: Shore A40–A95 [27,57]. Other values of hardness have been reported [45]. Typically hardness for compounded polymer ranges from 40–95 IRHD [35]

Failure Properties General: The effect of the rate of tearing and of temp. on the tear strength and tensile strength of neoprene compounds has been reported [113]. Fracture energy and tensile strength were measured from 0–150°. The tear strength of Neoprene WRT has been reported for threshold conditions which minimise viscoelastic and other dissipative processes [114] for different kinds of cross-linking. Flex-cracking and crack growth for Neoprene GN has been (36 pph carbon black) compared with other rubbers from 4–100°. The rate of crack growth becomes faster as the temp. rises from 40–100° [51]. Compression set measurements have been reported for different forms of polychloroprene under compressive stress at -10° (1–300h) [32]

Electrical Properties:
Electrical Properties General: Dielectric constant values have been reported
Dielectric Permittivity (Constant):

No.	Value	Frequency	Note
1	6.5–8.1	1 kHz	[13]

Complex Permittivity and Electroactive Polymers: The electrical conduction of carbon fibre-filled polychloroprene has been reported [67]. The dependence of electrical conductivity on carbon black particle size distribution has been investigated. Specific conductivity 0.0001 S cm^{-1} (20 phv carbon black); 0.01 S cm^{-1} (20 phv carbon black) [109]

Dissipation (Power) Factor:

No.	Value	Frequency	Note
1	0.031–0.086	1 kHz	[13]

Optical Properties:
Refractive Index:

No.	Value	Note
1	1.56	[57]

Polymer Stability:
Upper Use Temperature:

No.	Value	Note
1	82–93°C	[57]
2	80°C	VDE 0304 [56]
3	100°C	[89]
4	120°C	[27]

Decomposition Details: Thermal degradation studies show three stages of volatilisation [115]. HCl released by neoprene on heating is absorbed by carbon black. Oxygen accelerates decomposition [40]. Decomposes at high temp. yielding CO_2, H_2O, HCl [56]

Flammability: The effect of antipyrene additives on the thermal behaviour and flammability of polychloroprene compositions has been investigated [73]. The chlorine content in polychloroprene (40%) and a tendency to form char results in more resistance to burning than for hydrocarbon rubbers. The main weight loss occurs between 290–400°, all of the chlorine is lost as HCl. Limiting oxygen index 26.3–40 [88]; 26–31 (unfilled) [108]; 41 [107]. Polychloroprene is 56% combustible; heat of combustion 22.6 kJ g^{-1} [107]. Inherently flame retardant

Environmental Stress: The effect of carbon black and cross-linking on the photo and thermal ageing of polychloroprene has been reported [59]. Accelerated thermal ageing has been reported by measuring residual deformation under constant deformation in air

[76]. The effect of fillers and other additives (including accelerators, plasticisers, softeners, extenders, tackifiers and antioxidants) on the ozone resistance of neoprene GN, W, FR for different cure times at constant strain and at constant load has been reported [52]. The effect of ozone on crack propagation [94] and flex resistance [95] for different additives has been reported. Rain erosion rate 2 μm min^{-1} (whirling arm, 25 mm h^{-1} rain) [23]

Chemical Stability: Medium resistance to oils; good resistance to paraffinic/naphthalenic oil and hydraulic fluids. Limited resistance to aromatic fuels [56]. Resistance to H_2O and oil has been reported [41]. Effect of carbon black on swelling and resistance to hydrocarbon oil has been reported [64,68]

Biological Stability: Contact with soil for long periods results in attack by soil bacteria and fungi; resistance to microorganisms in soil is higher than for hydrocarbon rubbers [56]

Stability Miscellaneous: Exposure of lightly cross-linked polychloroprene to high energy radiation (γ rays) has been found to increase hardness (Shore D17-D70), increase tensile strength and reduce elongation at break as a result of cross-linking [22]

Applications/Commercial Products:

Processing & Manufacturing Routes: Vulcanisation is carried out using metal oxides (e.g. zinc or magnesium oxide) and with sulfur and mercaptan transfer agents - the latter are produced in highest quantity

Applications: Used in mechanical and automotive components such as belts, hoses, gaskets; footwear, electrical cable covering. Higher crystallinity grades used in adhesives

Bibliographic References

[1] Wood, L.A., *Polym. Handb.*, 3rd edn., (eds. J. Brandrup and E.H. Immergut), Wiley Interscience, 1989, **V**, 7
[2] Catton, N.L., *The Neoprenes*, Rubber Chemicals Division, 1953
[3] Payne, A.R. and Scott, J.R., *Engineering Design with Rubber*, Interscience, 1960
[4] Kell, R.M., Bennett, B. and Stickney, P.B., *J. Appl. Polym. Sci.*, 1959, **2**, 8
[5] Krigbaum, W.R., Dawkins, J.V., Via, G.H. and Balta, Y.I., *J. Polym. Sci., Part A-2*, 1972, **4**, 475
[6] Maynard, J.T. and Mochel, W.E., *J. Polym. Sci.*, 1954, **13**, 235
[7] Wood, L.A., Synthetic Rubbers: A Review of their Compositions, Properties and Uses, Natl. Bur. Std. Circ. C427, 1940
[8] Cramer, W.S. and Silver, I., 1951, U.S. Naval Ordnance Lab., Whiteoak, MD
[9] Naunton, W.H.S. *et al*, *Rubber Engineering*, Chemical Publishing Co., 1946, 30
[10] Wood, L.A., Bekkedahl, N. and Roth, F.L., *Rubber Chem. Technol.*, 1943, **16**, 244
[11] Yin, T.P. and Pariser, R., *J. Appl. Polym. Sci.*, 1963, **7**, 667
[12] Gent, A.N., *J. Polym. Sci., Part A-1*, 1965, **3**, 3787
[13] McPherson, A.T., *Rubber Chem. Technol.*, 1963, **36**, 1230
[14] Ball, J.M. and Maassen, G.C., ASTM, Symp. Appl. Synth. Rubbers, 1944, 27
[15] Boomstra, B.B.S.T., Elastomers. Their Chemistry, *Physics and Technology*, (ed. R. Houwink), Elsevier, 1948, **3**
[16] Koros, W.J. and Hellums, M.W., *Encycl. Polym. Sci. Eng.*, 2nd edn., (eds. H.F. Mark, N.M. Bikales, C.G. Overberger, G. Menges and J.I. Kroschwitz), John Wiley and Sons, 1989, **Supplement**, 742
[17] Mochel, W.E. and Nichols, J.B., *J. Am. Chem. Soc.*, 1949, **71**, 3425
[18] Mochel, W.E., Nichols, J.B. and Mighton, C.J., *J. Am. Chem. Soc.*, 1948, **70**, 2185
[19] Mochel, W.E., Nichols, J.B. and Mighton, C.J., *Ind. Eng. Chem.*, 1951, **43**, 154
[20] Coleman, M.M. and Fuller, R.E., *J. Macromol. Sci., Phys.*, 1975, **11**, 419
[21] Cabasso, I., *Encycl. Polym. Sci. Eng.*, 2nd edn., Vol. 9 (eds. H.F. Mark, N.M., Bikales, C.G. Overberger, G. Menges and J.I. Kroschwitz), John Wiley and Sons, 1987, 562
[22] Clough, R., *Encycl. Polym. Sci. Eng.*, 2nd edn., Vol. 13 (eds H.F. Mark, N.M. Bikales, C.G. Overberger, G. Menges and J.I. Kroschwitz), John Wiley and Sons, 1988, 675
[23] Lancaster, J.K., *Encycl. Polym. Sci. Eng.*, 2nd edn., Vol. 1 (eds. H.F. Mark, N.M. Bikales, C.G. Overberger, G. Menges and J.I. Kroschwitz), John Wiley and Sons, 1985, 18
[24] Petcavich, R.J. and Coleman, M.M., *J. Macromol. Sci., Phys.*, 1980, **18**, 47
[25] Koberstein, J.T., *Encycl. Polym. Sci. Eng.*, 2nd edn., Vol. 8 (eds H.F. Mark, N.M. Bikales, C.G. Overberger, G. Menges and J.I. Kroschwitz), John Wiley and Sons, 1985, 242
[26] Struik, L.C.E., *Polym. Eng. Sci.*, 1979, **19**, 223
[27] Witsiepe, W.K., *Kirk-Othmer Encycl. Chem. Technol.*, 4th edn., Vol.8 (eds. J.I. Kroschwitz and M. Howe-Grant), John Wiley and Sons, 1993, 1045
[28] Coleman, M.M., Petcavich, R.J. and Painter, P.C., *Polymer*, 1978, **19**, 1253
[29] Tsereteli, G.I., Sochava, I.V. and Buka, A., Vestn. Leningr. Univ., Ser. 4: Fiz., Khim., 1975, 67
[30] Teitelbaum, B.Y. and Aroshina, N.P., *Vysokomol. Soedin.*, 1965, **7**, 978
[31] Maynard, J.T. and Mochel, W.E., *J. Polym. Sci.*, 1954, **13**, 242
[32] Murray, R.M. and Detenber, J.D., *Rubber Chem. Technol.*, 1961, **34**, 668
[33] McCabe, C.C., *Reinforcement of Elastomers*, (ed. G. Kraus), Interscience, 1965, 231
[34] Greensmith, H.W., Mullins, L. and Thomas, A.G., *The Chemistry and Physics of Rubber-Like Substances*, (ed. L. Bateman), The Garden City Press, 1963, 261
[35] Fletcher, W.P., *Rubber Technology and Manufacture*, 2nd edn., (eds. C.M. Blow and C. Hepburn), Butterworth Scientific, 1982, 118
[36] Weir, C.E., *J. Res. Natl. Bur. Stand. (U.S.)*, 1951, **46**, 207
[37] Cooper, D.H., *Ind. Eng. Chem.*, 1951, **43**, 365
[38] Kean, W.N., *Trans. ASME*, 1946, **68**, 237
[39] Kean, W.N., *India Rubber World*, 1953, **128**, 351
[40] Torii, C., Hoshii, K. and Isshiki, S., J. Soc. Rubber Ind., Jpn, 1956, **29**, 38
[41] Hoshii, K. and Isshiki, S., J. Soc. Rubber Ind., Jpn, 1958, **29**, 151
[42] Carter, W.C., Magat, M., Schneider, W.C. and Smyth, C.P., *Trans. Faraday Soc.*, 1946, **42**, 213
[43] Hillier, K.W., Trans., *Inst. Rubber Ind.*, 1950, **26**, 64
[44] Nolle, A.W., *J. Polym. Sci.*, 1950, **5**, 1
[45] Fenner, J.V., *Rubber Age (London)*, 1956, **79**, 445
[46] James, R.R. and Guyton, C.W., *Rubber Age (London)*, 1956, **78**, 725
[47] Walker, H.W. and Mochel, W.E., *Rubber Chem. Technol.*, 1950, **23**, 652
[48] Boomstra, B.S.T.T., *Rubber Chem. Technol.*, 1950, **23**, 338
[49] Tager, A. and Sanatina, V., *Rubber Chem. Technol.*, 1951, **24**, 773
[50] Pisarenko, A.P. and Rebinder, P.A., *Rubber Chem. Technol.*, 1951, **24**, 569
[51] Buist, J.M., *Rubber Chem. Technol.*, 1953, **26**, 935
[52] Thompson, D.C., Baker, R.H. and Browlow, R.W., *Rubber Chem. Technol.*, 1952, **25**, 928
[53] Maynard, J.T. and Mochel, W.E., *Rubber Chem. Technol.*, 1954, **27**, 634
[54] Garett, R.R., Hargreaves, C.A. and Robinson, D.N., *J. Macromol. Sci., Chem.*, 1970, **48**, 1679
[55] Auftermarsh, C.A. and Pariser, R., *J. Polym. Sci., Part A: Polym. Chem.*, 1964, **2**, 4727
[56] Musch, R. and Magg, H., *Polymeric Materials Encyclopedia*, (ed. J.C. Salamone), CRC Press, 1996, 1238
[57] Schweitzer, P.A., *Corrosion Resistance of Elastomers*, Marcel Dekker, 1990, 23
[58] Hertz, D.L., Handbook of Elastomers, *New Developments and Technology*, (eds. A.K. Bhowmick and H.L. Stephens), Marcel Dekker, 1988, 464
[59] Delor, F., Lacoste, J., Lemair, J., Barroisoudin, N. and Cardinet, C., *Polym. Degrad. Stab.*, 1996, **53**, 361
[60] Kundu, P.P., Banerjee, S. and Tripathy, D.K., *Int. J. Polym. Mater.*, 1996, **32**, 125
[61] Kundu, P.P. and Tripathy, D.K., *Kautsch. Gummi Kunstst.*, 1996, **49**, 268
[62] Kundu, P.P., Tripathy, D.K. and Banerjee, S., *Polymer*, 1996, **37**, 2423
[63] Bandyopadhyay, S., De, P.P., Tripathy, D.K. and De, S.K., *Polymer*, 1995, **36**, 1979
[64] *Polym.-Plast. Technol. Eng.*, 1995, **34**, 143
[65] Ghosh, P. and Ray, P., Plast., *Rubber Compos. Process. Appl.*, 1994, **22**, 51
[66] Xue, Y.P., Chen, Z.F. and Frisch, H.C., *J. Appl. Polym. Sci.*, 1994, **52**, 1833
[67] Jana, P.B., Plast., *Rubber Compos. Process. Appl.*, 1993, **20**, 107
[68] Lawandy, S.N., Botros, S.H. and Mouniv, A., *Polym.-Plast. Technol. Eng.*, 1993, **32**, 255
[69] Anachkov, M.P., Rakovsky, S.K., Sefanova, R.V. and Stoyanova, A.K., *Polym. Degrad. Stab.*, 1993, **41**, 185
[70] Kalidaha, A.K., De, P.P. and Sen, A.K., *Polym. Degrad. Stab.*, 1993, **39**, 179
[71] Pilai, V.B. and Das, J.N., Plast., *Rubber Compos. Process. Appl.*, 1992, **18**, 155
[72] Rang, D.I., Ha, C.S. and Cho, W.J., *Eur. Polym. J.*, 1992, **28**, 565
[73] Gjurova, K. and Karaivanova, M., *J. Therm. Anal.*, 1991, **37**, 2593
[74] Itoyama, R., Shimizu, N. and Natsuzawa, S., *Polym. J. (Tokyo)*, 1991, **23**, 1139
[75] Hornsby, P.R., Mitchell, P.A. and Cusack, P.A., *Polym. Degrad. Stab.*, 1991, **32**, 299
[76] Budrugeae, P. and Ciutacu, S., *Polym. Degrad. Stab.*, 1991, **33**, 377

[77] Nagode, J.B. and Roland, C.M., *Polymer*, 1991, **32**, 505
[78] Lawandy, S.N. and Wassef, M.S., *J. Appl. Polym. Sci.*, 1990, **40**, 323
[79] Miyata, V. and Atsumi, M., *J. Polym. Sci., Part A: Polym. Chem.*, 1988, **26**, 2561
[80] Munk, P., Hattam, P., Du, Q. et al, *J. Appl. Polym. Sci.: Appl. Polym. Symp.*, 1990, **45**, 289
[81] Hanafusu, R., Teramoto, A. and Fujita, H., *J. Phys. Chem.*, 1966, **70**, 4004
[82] Kawahara, R., Norisuye, T. and Fujita, H., *J. Chem. Phys.*, 1968, 4330
[83] Norisuya, T., Kawahara, K., Teramoto, A. et al, *J. Chem. Phys.*, 1968, **49**, 4330
[84] Shvarts, A.G., *Kolloidn. Zh.*, 1956, **18**, 755
[85] Small, P.A., *J. Appl. Chem.*, 1953, **3**, 71
[86] Gee, G., *Trans. Inst. Rubber Ind.*, 1943, **18**, 266
[87] Brostow, W., Kubát, J. and Kubát, M.M., *Physical Properties of Polymers Handbook*, (ed. J.E. Mark), AIP Press, 1996, 333
[88] Tewarson, A., *Physical Properties of Polymers Handbook*, (ed. J.E. Mark), AIP Press, 1996, 587
[89] Billmeyer, F.W., *Textbook of Polymer Science*, John Wiley and Sons, 1984, 136
[90] Liggat, J., *Polym. Handb.*, 4th edn., (eds. J. Brandrup, E.H. Immergut and E.A. Grulke), John Wiley and Sons, 1999, **II**, 451
[91] Wu, S., *J. Polym. Sci., Part C: Polym. Lett.*, 1971, **34**, 19
[92] van Amerongen, G.J., *J. Appl. Phys.*, 1946, **17**, 972
[93] Hartmann, B. and Jarzynski, J., *J. Acoust. Soc. Am.*, 1974, **56**, 1469
[94] Hartmann, A. and Glander, F., *Rubber Chem. Technol.*, 1956, **29**, 166
[95] Murray, R.M., *Rubber Chem. Technol.*, 1959, **32**, 1117
[96] Murray, R.M. and Detenber, J.D., *Rubber Chem. Technol.*, 1961, **34**, 668
[97] Well, L.A., *J. Res. Natl. Bur. Stand. (U.S.)*, 1948, **41**, 315
[98] Robbins, R.F., Ohori, Y. and Weitzel, D.H., *Rubber Chem. Technol.*, 1964, **37**, 154
[99] Wood, L.A., *Rubber Chem. Technol.*, 1976, **49**, 189
[100] Dillon, J.H., Prettyman, I.B. and Hall, G.L., *J. Appl. Phys.*, 1944, **15**, 309
[101] Boomstra, B.B.S.T., *Rubber Chem. Technol.*, 1951, **24**, 199
[102] Ferguson, R.C., *Rubber Chem. Technol.*, 1965, **38**, 532
[103] Becker, R.O., *Rubber Chem. Technol.*, 1964, **37**, 76
[104] Hennig, J., *Rubber Chem. Technol.*, 1966, **39**, 678
[105] Stewart, C.W., Dawson, R.L. and Johnson, P.R., *Rubber Chem. Technol.*, 1975, **48**, 132
[106] Holownia, B.P., *Rubber Chem. Technol.*, 1975, **48**, 246
[107] Johnson, P.R., *Rubber Chem. Technol.*, 1976, **49**, 158
[108] Fabris, H.J. and Sommer, J.G., *Rubber Chem. Technol.*, 1977, **50**, 523
[109] Sitcar, A.K. and Lamond, T.G., *Rubber Chem. Technol.*, 1978, **51**, 126
[110] Brazier, D.W. and Nickel, G.H., *Rubber Chem. Technol.*, 1979, **52**, 735
[111] Brazier, D.W., *Rubber Chem. Technol.*, 1980, **53**, 437
[112] Mean, W.T., *Rubber Chem. Technol.*, 1980, **53**, 245
[113] Bhowmick, A.K. and Gent, A.N., *Rubber Chem. Technol.*, 1983, **56**, 845
[114] Bhowmick, A.K., Gent, A.N. and Pulford, C.T.R., *Rubber Chem. Technol.*, 1983, **56**, 226
[115] Sircar, A.K., *Rubber Chem. Technol.*, 1992, **65**, 503

Poly(2-chlorostyrene)   P-151

Synonyms: *Poly(o-chlorostyrene)*. *Poly(1-chloro-2-ethenylbenzene)*. *Poly(1-(2-chlorophenyl)-1,2-ethanediyl)*
Related Polymers: Poly(3-chlorostyrene), Poly(4-chlorostyrene)
Monomers: 2-Chlorostyrene
Material class: Thermoplastic
Polymer Type: styrenes
CAS Number:

CAS Reg. No.
26125-41-7

Molecular Formula: $(C_8H_7Cl)_n$
Fragments: C_8H_7Cl

Volumetric & Calorimetric Properties:
Density:

No.	Value	Note
1	d^{25} 1.25 g/cm^3	[1]

Glass-Transition Temperature:

No.	Value	Note
1	119°C	[6]

Surface Properties & Solubility:
Cohesive Energy Density Solubility Parameters: δ 9.22 (CHCl$_3$) [4], δ 9.201 (CH$_2$Cl$_2$) [4], $δ_p$ 8.9 [5]
Solvents/Non-solvents: Sol. CCl$_4$, toluene, *o*-xylene, C$_6$H$_6$, 2-butanone, chlorobenzene, CH$_2$Cl$_2$, methyl isopropyl ketone, *o*-dichlorobenzene, cyclopentanone, pyridine, *o*-toluidine [5]. Insol. Freon-215, isopentane, isooctane, pentane, hexane, octane, methylcyclohexane, cyclohexane, Me$_2$CO, amyl alcohol, pentyl alcohol, EtOH, MeOH

Transport Properties:
Polymer Solutions Dilute: [η] 0.445–1.418 dl g^{-1} (30°, toluene) [3]. Theta temp. Θ 23–27° (butanone) [1]

Optical Properties:
Molar Refraction: dη/dC 0.204 (25°, butanone) [1], 0.118 (35°, toluene), 0.211 (30°, butanone) [2]

Applications/Commercial Products:
Processing & Manufacturing Routes: Prod. by cationic polymerisation with Lewis acid catalysts

Bibliographic References
[1] Nozaki, M., Shimada, K. and Okamoto, S., *Jpn. J. Appl. Phys.*, 1971, **10**, 179
[2] Matsumura, K., *Bull. Chem. Soc. Jpn.*, 1969, **42**, 1874
[3] Matsumura, K., *Makromol. Chem.*, 1969, **124**, 204
[4] Suh, K.W. and Clarke, D.H., *J. Polym. Sci., Part A: Polym. Chem.*, 1967, **5**, 1671
[5] Suh, K.W. and Corbett, J.M., *J. Appl. Polym. Sci.*, 1968, **12**, 2359
[6] Dunham, K.R., Faber, J.W.H., Vandenberghe, J. and Fowler, W.F., *J. Appl. Polym. Sci.*, 1963, **7**, 897

Poly(3-chlorostyrene)   P-152

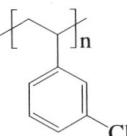

Synonyms: *Poly(m-chlorostyrene)*. *Poly(1-chloro-3-ethenylbenzene)*. *Poly(1-(3-chlorophenyl)-1,2-ethanediyl)*. *Poly (1-chloro-3-vinylbenzene)*
Related Polymers: Poly(2-chlorostyrene), Poly(4-chlorostyrene)
Monomers: 3-Chlorostyrene
Material class: Thermoplastic
Polymer Type: styrenes
CAS Number:

CAS Reg. No.	Note
26100-04-9	
116002-24-5	isotactic
107830-48-8	syndiotactic

Molecular Formula: $(C_8H_7Cl)_n$
Fragments: C_8H_7Cl

Volumetric & Calorimetric Properties:
Density:

No.	Value	Note
1	d 1.241 g/cm^3	[1]

Glass-Transition Temperature:

No.	Value	Note
1	90°C	[2]

Bibliographic References
[1] Nozaki, M., Shimada, K. and Okamoto, S., *Jpn. J. Appl. Phys.*, 1971, **10**, 179
[2] Dunham, K.R., Faber, J.W.H., Vandenberghe, J. and Fowler, W.F., *J. Appl. Polym. Sci.*, 1963, **7**, 897

Poly(4-chlorostyrene)   P-153

Synonyms: *Poly(p-chlorostyrene). Poly(1-chloro-4-vinylbenzene). Poly(1-(4-chlorophenyl)-1,2-ethanediyl)*
Related Polymers: Poly(2-chlorostyrene), Poly(3-chlorostyrene)
Monomers: 4-Chlorostyrene
Material class: Thermoplastic
Polymer Type: styrenes
CAS Number:

CAS Reg. No.	Note
24991-47-7	
29297-96-9	isotactic
62319-29-3	syndiotactic

Molecular Formula: $(C_8H_7Cl)_n$
Fragments: C_8H_7Cl

Volumetric & Calorimetric Properties:
Density:

No.	Value	Note
1	d 1.229 g/cm^3	25° [3]
2	d 1.246 g/cm^3	[7]

Thermodynamic Properties General: ΔC_p 31.1 J K^{-1} mol^{-1} [2]
Glass-Transition Temperature:

No.	Value	Note
1	129°C	[1]
2	133°C	[2]
3	125°C	[5]
4	126°C	[7]

Transition Temperature General: Boils at 56–57° (3 mmHg)

Surface Properties & Solubility:
Cohesive Energy Density Solubility Parameters: Solubility and solubility parameters for polymer in 97 solvents have been reported [9]

Transport Properties:
Polymer Solutions Dilute: Mark-Houwink constants have been reported [8]. Theta temp. θ -14.7° (ethylbenzene), 59.0° (isopropylbenzene), 50.7° (CCl$_4$), 44.4° (tetrachloroethylene), 64.6° (methyl chloroacetate), -1.8° (ethyl chloroacetate) [9]. Other values have been reported [9]
Permeability of Gases: [He] 16.4 × 10^{-10}; [CH$_4$] 0.26 × 10^{-10} cm^2 (s cmHg)$^{-1}$ [7]

Electrical Properties:
Dielectric Permittivity (Constant):

No.	Value	Note
1	2.95	[1]

Optical Properties:
Refractive Index:

No.	Value	Note
1	1.5662	[6]

Transmission and Spectra: Ir spectral data have been reported [4]

Polymer Stability:
Decomposition Details: In the decomposition of the polymer, random scissions reduce the chain length of the polymer and depolymerisation reactions cause weight loss [5]

Applications/Commercial Products:
Processing & Manufacturing Routes: Prod. by cationic polymerisation with Lewis acid catalysts

Bibliographic References
[1] Gustafsson, A, Wiberg, G. and Gedde, U.W., *Polym. Eng. Sci.*, 1993, **33**, 549
[2] Judovits, L.H., Bopp, R.C., Gaur, V. and Wunderlich, B., *J. Polym. Sci., Part B: Polym. Lett.*, 1986, **24**, 2725
[3] Nozaki, M., Shimada, K. and Okamoto, S., *Jpn. J. Appl. Phys.*, 1971, **10**, 179
[4] Kobayashi, M., Tsumura, K. and Tadokoro, H., *J. Polym. Sci.*, 1968, **6**, 1493
[5] Malhotra, S.L., Lessard, P. and Blanchard, L.P., *J. Macromol. Sci., Chem.*, 1981, **15**, 279
[6] Noguchi, Y., Aoki, A., Tanaka, G. and Yamakawa, H., *J. Chem. Phys.*, 1970, **52**, 2651
[7] Puleo, A.C., Muruganandam, N. and Paul, D.R., *J. Polym. Sci., Part B: Polym. Phys.*, 1989, **27**, 2385
[8] Kuwahara, N., Ogino, K., Kasai, A., Ueno, S. and Kaneko, M., *J. Polym. Sci., Part A: Polym. Chem.*, 1965, **3**, 985
[9] Izumi, Y. and Miyake, Y., *Polym. J. (Tokyo)*, 1972, **3**, 647

Polychlorotrifluoroethylene   P-154

$+CF_2CFCl+_n$

Synonyms: *PCTFE. Polychlorotrifluoroethene. CTFE*
Related Polymers: Poly(chlorotrifluoroethylene-*co*-ethylene)
Monomers: Chlorotrifluoroethylene
Material class: Thermoplastic
Polymer Type: PCTFE, polyhaloolefins
CAS Number:

CAS Reg. No.
9002-83-9

Polychlorotrifluoroethylene

Molecular Formula: $(C_2ClF_3)_n$
Fragments: C_2ClF_3
Molecular Weight: M_n 70000–400000
Morphology: Polymer chain is helical, with an average of 16.8 monomer units in one turn of the helix. Crystallinity ranges from 40–80% depending on MW and thermal history

Volumetric & Calorimetric Properties:

Density:

No.	Value	Note
1	d 2.14 g/cm^3	Kel-F81 [2]

Thermal Expansion Coefficient:

No.	Value	Note	Type
1	8×10^{-5} K^{-1}	[3]	L
2	7×10^{-5} K^{-1}	30–-30° [4]	L
3	5.1×10^{-5} K^{-1}	-30–-100° [4]	L
4	3.6×10^{-5} K^{-1}	-100–-190° [4]	L

Thermodynamic Properties General: Heat of formation ΔH_f 656.8 kJ mol^{-1}. Heat of combustion ΔH_c 130.6 kJ mol^{-1} [4]

Thermal Conductivity:

No.	Value	Note
1	1.195 W/mK	ASTM C177 [4]

Specific Heat Capacity:

No.	Value	Note	Type
1	0.77 kJ/kg.C	91.8 J mol^{-1} K^{-1}, -3° [6]	P
2	0.73 kJ/kg.C	87.5 J mol^{-1} K^{-1}, -3° [6,7]	V

Melting Temperature:

No.	Value	Note
1	212°C	Kel-F81 [2]
2	211–216°C	[1]

Glass-Transition Temperature:

No.	Value	Note
1	50°C	[2]
2	71–99°C	[1]

Deflection Temperature:

No.	Value	Note
1	126°C	0.46 MPa, ASTM D648 [4]
2	70°C	1.82 MPa, ASTM D648 [4]

Brittleness Temperature:

No.	Value	Note
1	-240°C	[1]

Transition Temperature:

No.	Value	Note	Type
1	150°C	1 Hz [4]	α transition (relaxation)
2	90°C	1 Hz [4]	β transition (relaxation)
3	-37°C	1 Hz [4]	γ transition (relaxation)

Surface Properties & Solubility:

Solvents/Non-solvents: Highly fluorinated or chlorinated solvents swell the polymer but do not degrade it [4]. Sol. 2,5-dichlorobenzotrifluoride (135°) [4]
Wettability/Surface Energy and Interfacial Tension: Critical surface tension 31 mN m^{-1} [4]

Transport Properties:

Transport Properties General: PCTFE has the lowest water-vapour permeability of all known gases [4]
Polymer Melts: Melt viscosity 114000 Pa s [4]
Permeability of Gases: [O$_2$] 0.47 m^3 mm m^{-2} h^{-1} MPa^{-1} (1 mm thick film); [N$_2$] 0.17 m^3 mm m^{-2} h^{-1} MPa^{-1} (1 mm thick film); [CO$_2$] 1.1 m^3 mm m^{-2} h^{-1} MPa^{-1} (1 mm thick film) [4]
Water Content: 0%
Water Absorption:

No.	Value	Note
1	0 %	24h [8]

Mechanical Properties:

Mechanical Properties General: Has good resistance to creep and cold flow; has good rigidity and good low temp. toughness

Tensile (Young's) Modulus:

No.	Value	Note
1	1180–1450 MPa	ASTM D638 [4]

Flexural Modulus:

No.	Value	Note
1	1340–1820 MPa	ASTM D790 [4]

Tensile Strength Break:

No.	Value	Note	Elongation
1	40 MPa	ASTM D638, KEL-F 81 6050 [4]	100–400%
2	150 MPa	-129°	9%
3	200 MPa	-252° [1]	5%

Compressive Modulus:

No.	Value	Note
1	1350 MPa	KEL-F 81 6050

Tensile Strength Yield:

No.	Value	Note
1	32–39 MPa	ASTM D638 [4]

Flexural Strength Yield:

No.	Value	Note
1	65–71 MPa	ASTM D790 [4]

– Polychlorotrifluoroethylene

Compressive Strength:

No.	Value	Note
1	36–43 MPa	0.2% offset, ASTM D695 [4]
2	11–13 MPa	1% strain, ASTM D695 [4]
3	380 MPa	[1]

Miscellaneous Moduli:

No.	Value	Note
1	1200–1600 MPa	compressive modulus of elasticity, ASTM D695 [4]

Viscoelastic Behaviour: Deformation under load 0.2% (7 MPa, 24h, 25°, ASTM D621); 1.4–1.9% (7 MPa, 24h, 80°, ASTM D621); 5.0–9.0% (7 MPa, 24h, 100°, ASTM D621) [4]
Hardness: Shore D75-D85 (ASTM D1706) [4]
Friction Abrasion and Resistance: Abrasion resistance 5 mg (1000 cycles)$^{-1}$ (low crystallinity, Taber abrader, CS-10 wheel); 12 mg (1000 cycles)$^{-1}$ (high crystallinity Taber abrader, CS-10 wheel)
Izod Notch:

No.	Value	Notch	Note
1	133–187 J/m	Y	ASTM D256 [4]

Electrical Properties:
Surface/Volume Resistance:

No.	Value	Note	Type
2	1×10^{15} Ω.cm	ASTM D257 [4]	S

Dielectric Permittivity (Constant):

No.	Value	Frequency	Note
1	2.3–2.7	1 kHz	ASTM D150 [4]
2	2.3–2.5	1 MHz	ASTM D150 [4]

Dielectric Strength:

No.	Value	Note
1	25.4 kV.mm^{-1}	1.5 mm, short time, ASTM D149 [4]
2	25.6 kV.mm^{-1}	1.5 mm, step by step, ASTM D149 [4]
3	120 kV.mm^{-1}	100 μm, short time, ASTM D149 [4]

Arc Resistance:

No.	Value	Note
1	360s	ASTM D495 [4]

Dissipation (Power) Factor:

No.	Value	Frequency	Note
1	0.025	10 kHz	[4]
2	0.01	10 MHz	[4]

Optical Properties:
Refractive Index:

No.	Value	Note
1	1.425	20° [4]

Transmission and Spectra: Light transmittance 90% (ASTM D1003, KEL-F 81 6050)
Volume Properties/Surface Properties: Clarity transparent to opaque. Haze 4.0% (ASTM D1003, KEL-F 81 6050) [8]

Polymer Stability:
Polymer Stability General: Has good thermal and chemical stability
Thermal Stability General: Thermally stable up to 250° [1]
Upper Use Temperature:

No.	Value	Note
1	205°C	[1]

Decomposition Details: Thermal decomposition at 350–370° yields 26% chlorotrifluoroethylene monomer. The remainder is a mixture of halocarbon waxes of M_n 904 [4]
Flammability: Non-flammable. Limiting oxygen index 95% (ASTM D2863). Vertical burn rating VEO (UL 94) [4]
Environmental Stress: Has good uv resistance. Vacuum degassing at 130–0.013 mPa under uv radiation up to 30 kJ cm^{-3} shows no significant changes in MW, or tensile props. [4]
Chemical Stability: Resistant to attack by most industrial chemicals, including strong acids and alkalis, over a wide range of temps. and concentrations [4]. Attacked by alkali metal complexes and certain organic amines [4]
Stability Miscellaneous: Has good radiation resistance (better than PTFE). The G-value for scission with γ radiation is 0.67 [4]

Applications/Commercial Products:
Processing & Manufacturing Routes: Processed using compression moulding, injection moulding, extrusion, fluidised bed coating techniques. Manufactured by free radical bulk, suspension or aq. emulsion techniques
Mould Shrinkage (%):

No.	Value	Note
1	0.01–0.015%	[4]

Applications: Cryogenic applications, high performance packaging, seals, gaskets, valve parts, pipe, laboratory ware and medical equipment

Trade name	Supplier
Daiflon	Daikin Kogyo
Halon	Allied Chemical
Hostaflon C2	Hoechst Celanese
Kel-F81	3M Industrial Chemical Products
Voltalef	Atochem Polymers

Bibliographic References
[1] *Kirk-Othmer Encycl. Chem. Technol.*, Vol. 11, 4th edn., (ed. J.I. Kroschwitz), Wiley Interscience, 1980, 49
[2] Samara, G.A. and Fritz, I.J., *J. Polym. Sci., Polym. Lett. Ed.*, 1975, **13**, 93
[3] *Encyclopedia of Advanced Materials*, (eds. D. Bloor, R.J. Brook, M.C. Flemings, S. Mahajan and R.W. Cahn), Pergamon Press, 1994, **3**, 863
[4] Chandrasekan, S., *Encycl. Polym. Sci. Eng.*, 2nd edn., (ed. J.I. Kroschwitz), Wiley Interscience, 1989, **3**, 467
[5] Kerbow, D.L. and Sperati, C.A., *Polym. Handb.*, 4th edn., (eds. J. Brandrup, E.H. Immergut and E.A. Grulke), John Wiley and Sons, 1999, **V**, 55
[6] Lee, W.K., Lau, P.C. and Choy, C.L., *Polymer*, 1974, **15**, 487
[7] Choy, C.L., *J. Polym. Sci., Polym. Phys. Ed.*, 1975, **13**, 1263
[8] Kerbow, D.L. and Sperati, C.A., *Polym. Handb.*, 4th edn., (eds. J. Brandrup, E.H. Immergut and E.A. Grulke), John Wiley and Sons, 1999, **V**, 57

Poly(cycloaliphatic methacrylates) P-155

$$\left[CH_2C(CH_3) \right]_n$$
$$|$$
$$C{-}OR$$
$$\|$$
$$O$$

R = cyclo(CH₂)ₙ structures:
- n = 1 (cyclobutyl)
- n = 2 (cyclopentyl)
- n = 3 (cyclohexyl)
- n = 5 (cyclooctyl)
- n = 7 (cyclodecyl)
- n = 9 (cyclododecyl)
- n = 14 (cycloheptadecyl)

R = t-butylcyclohexyl

Polymer Stability:
Decomposition Details: The cyclohexyl derivative has an initial decomposition temp. of 200° followed by three maxima at 284°, 356° and 446° [12]

Bibliographic References
[1] Hoff, E.A.W., Robinson, D.W. and Willbourn, A.H., *Polym. Sci.*, 1955, **18**, 161
[3] Struik, L.C.E., *Polymer*, 1987, **28**, 1869
[4] Lewis, O.G., *Physical Constants of Linear Hydrocarbons*, Springer-Verlag, 1968
[5] Olabisi, O. and Simba, R., *Macromolecules*, 1975, **8**, 206
[6] Matsumoto, A., Mizuta, K. and Otsu, T., *Macromolecules*, 1993, **26**, 1659

Poly[1,1-cyclohexane bis[4-(2,6-dichlorophenyl)]-carbonate] P-156

Synonyms: *Poly[oxycarbonyloxy(2,6-dichloro-1,4-phenylene)cyclohexylidene(3,5-dichloro-1,4-phenylene)]. Poly(3,3',5,5'-tetrachloro-4,4'-dihydroxydiphenyl-1,1-cyclohexane-co-carbonic dichloride)*
Related Polymers: Poly[1,1-cyclohexanebis(4-phenyl)carbonate]
Monomers: 3,3',5,5'-Tetrachloro-4,4'-dihydroxydiphenyl-1,1-cyclohexane, Carbonic dichloride
Material class: Thermoplastic
Polymer Type: polycarbonates (miscellaneous)
CAS Number:

CAS Reg. No.
104934-65-8

Molecular Formula: $(C_{19}H_{14}Cl_4O_3)_n$
Fragments: $C_{19}H_{14}Cl_4O_3$

Volumetric & Calorimetric Properties:
Density:

No.	Value	Note
1	d 1.38 g/cm³	25° [1]

Melting Temperature:

No.	Value	Note
1	260–270°C	[1]

Electrical Properties:
Electrical Properties General: Dielectric strength does not vary with temp. until T_g. Dissipation factor increases with temp.
Dielectric Permittivity (Constant):

No.	Value	Frequency	Note
1	2.6	1 kHz	25–100° [1]

Dissipation (Power) Factor:

No.	Value	Frequency	Note
1	0.0005	1 kHz	25° [1]
2	0.004	1 kHz	100° [1]

Optical Properties:
Refractive Index:

No.	Value	Note
1	1.5858	25° [1]

Transmission and Spectra: Ir spectral data have been reported [2]
Volume Properties/Surface Properties: Transparent [3]

Applications/Commercial Products:
Processing & Manufacturing Routes: The main synthetic routes are, phosgenation and transesterification. The most widely used approach is interfacial polycondensation in a suitable solvent, but soln. polymerisation is also carried out in the presence of pyridine [2]. A slight excess of phosgene is introduced into a soln. or suspension of the aromatic dihydroxy compound in aq. NaOH at 20–30°, in an inert solvent (e.g. MeCl). Catalysts are used to achieve high MWs [1,2]. With transesterification, the dihydroxy compound and a slight excess of diphenyl carbonate are mixed together in the melt (150–300°) in the absence of O_2, with elimination of phenol. Reaction rates are increased by alkaline catalysts and the use of a medium-high vacuum in the final stages [1,2]

Bibliographic References
[1] Schnell, H., *Ind. Eng. Chem.*, 1959, **51**, 157, (density, T_m, dielectric props, refractive index, synth)
[2] Schnell, H., *Polym. Rev.*, 1964, **9**, 99, (ir, synth, rev)
[3] Smirnova, O.V., Korovina, Y.V., Kolesnikov, G.S., Lipkin, A.M. and Kuzina, S.I., *Polym. Sci. USSR (Engl. Transl.)*, 1966, **8**, 776, (volume props)

Poly[1,1-cyclohexane bis(4-phenyl)carbonate] P-157

Synonyms: *Poly[oxycarbonyloxy-1,4-phenylenecyclohexylidene-1,4-phenylene]. Poly(4,4'-dihydroxydiphenyl-1,1-cyclohexane-co-carbonic dichloride). Polycarbonate Z. Bisphenol Z polycarbonate*
Related Polymers: Poly[1,1-cyclohexane bis(4-(2,6-dichlorophenyl)carbonate), Copolymers of Bisphenol Z and Bisphenol A
Monomers: Bisphenol Z, Carbonic dichloride
Material class: Thermoplastic
Polymer Type: polycarbonates (miscellaneous)
CAS Number:

CAS Reg. No.
25135-52-8

Molecular Formula: $(C_{19}H_{18}O_3)_n$
Fragments: $C_{19}H_{18}O_3$

Volumetric & Calorimetric Properties:

Density:

No.	Value	Note
1	d 1.2 g/cm^3	25° [4]
2	d 1.2 g/cm^3	20°, amorph. [5]
3	d 1.206 g/cm^3	[2]
4	d 1.2 g/cm^3	30° [6]
5	d 1.201 g/cm^3	[7]

Equation of State: Mark-Houwink constant has been reported [13]

Melting Temperature:

No.	Value	Note
1	250–260°C	[4]
2	248–259°C	amorph. [5]

Glass-Transition Temperature:

No.	Value	Note
1	145°C	[8]
2	186°C	[9]
3	177–181°C	dependent on MW [2]
4	185°C	[6]
5	182°C	[7]
6	171°C	[10]
7	166°C	[11]
8	176°C	[1]

Vicat Softening Point:

No.	Value	Note
1	250–260°C	[12]

Transition Temperature:

No.	Value	Note	Type
1	-58°C	[9]	Gamma relaxation temp.

Surface Properties & Solubility:

Solubility Properties General: Forms miscible blends with Bisphenol A polycarbonate and tetramethylpolycarbonate (clear blends are formed at the melt mixing temp., 250°, but phase separation occurs before decomposition) [1]; immiscible with polymethyl methacrylate and polystyrene [1]

Cohesive Energy Density Solubility Parameters: Flory-Huggins polymer-solvent interaction parameters have been reported: 0.47 (chloromethane, 250°) [2]; 0.31 (CHCl$_3$, 250°) [2]; 1.33 (decane, 250°) [2]; 0.06 (cyclohexanol, 250°) [2]

Solvents/Non-solvents: Sol. chloromethane, *m*-cresol, cyclohexanone, THF, C$_6$H$_6$, pyridine, DMF; swollen by EtOAc, Me$_2$CO [3]

Transport Properties:

Water Absorption:

No.	Value	Note
1	0.2%	25°, 65% relative humidity [3]

Gas Permeability:

No.	Gas	Value	Note
1	CO$_2$	144 cm^3 mm/(m^2 day atm)	1.65 × 10^{-13} cm^2 (s Pa)$^{-1}$, 35°, 10 atm [6]
2	CH$_4$	6.04 cm^3 mm/(m^2 day atm)	6.9 × 10^{-15} cm^2 (s Pa)$^{-1}$, 35°, 10 atm, calc. [6]
3	O$_2$	37.4 cm^3 mm/(m^2 day atm)	4.28 × 10^{-14} cm^2 (s Pa)$^{-1}$, 35° [8]
4	N$_2$	6.92 cm^3 mm/(m^2 day atm)	7.9 × 10^{-15} cm^2 (s Pa)$^{-1}$, 35°, 10 atm, calc. [6]
5	He	656 cm^3 mm/(m^2 day atm)	7.5 × 10^{-13} cm^2 (s Pa)$^{-1}$, 35°, 10 atm [6]
6	H$_2$	595 cm^3 mm/(m^2 day atm)	6.8 × 10^{-13} cm^2 (s Pa)$^{-1}$, 35°, 10 atm, calc. [6]

Mechanical Properties:

Mechanical Properties General: The general shape of the tensile stress-strain curve consists of a viscoelastic region up to the yield point, followed by a stress drop and then some necking behaviour (at approx. constant stress); finally the stress increases to the break point [14]

Tensile (Young's) Modulus:

No.	Value	Note
1	2350 MPa	24000 kg cm^{-2}, amorph. [14]

Tensile Strength Break:

No.	Value	Note	Elongation
1	79.8 MPa	814 kg cm^{-2} [4]	56%
2	82 MPa	8.4 kg mm^{-2} [14]	62%

Tensile Strength Yield:

No.	Value	Note
1	71.6 MPa	7.3 kg mm^{-2} [14]

Complex Moduli: The variations of E′, E″ and tan δ with temp. (-150–200°, 110 Hz) have been reported [6]. Tan δ is practically independent of frequency at normal temps. [16]

Electrical Properties:

Electrical Properties General: Dielectric constant and dissipation factor do not vary with temp. until T_g [4]

Dielectric Permittivity (Constant):

No.	Value	Frequency	Note
1	3	1 kHz	25–100° [4]

Dissipation (Power) Factor:

No.	Value	Frequency	Note
1	0.00112	1 kHz	25–100° [4]

Optical Properties:

Refractive Index:

No.	Value	Note
1	1.59	25° [4]
2	1.59	calc. [1]

Transmission and Spectra: Ir spectral data have been reported [3]

Polymer Stability:
Thermal Stability General: Stable under N_2 atmosphere up to 400° [2]. Evolution of gases starts at 320° in a vacuum [16]
Decomposition Details: Thermooxidation onset 260–270° (O_2); at 280° and 30 cmHg, O_2, CO_2, CO, MeOH, H_2O and other gases are evolved. Activation energy 73.7 kJ mol^{-1} (O_2, 260–320°) [16]

Applications/Commercial Products:
Processing & Manufacturing Routes: The main synthetic routes are, phosgenation and transesterification. The most widely used approach is interfacial polycondensation in a suitable solvent, but soln. polymerisation is also carried out in the presence of pyridine [3]. A slight excess of phosgene is introduced into a soln. or suspension of the aromatic dihydroxy compound in aq. NaOH at 20–30°, in an inert solvent (e.g. MeCl). Catalysts are used to achieve high MWs [3,4]. With transesterification, the dihydroxy compound and a slight excess of diphenyl carbonate are mixed together in the melt (150–300°) in the absence of O_2, with the elimination of phenol. Reaction rates are increased by alkaline catalysts and the use of a medium-high vacuum in the final stages [3,4]

Bibliographic References

[1] Kim, C.K. and Paul, D.R., *Macromolecules*, 1992, **25**, 3097, (polymer-polymer compatibility, T_g, refractive index)
[2] Dipaola-Baranyi, G., Hsiao, C.K., Spes, M., Odell, P.G. and Burt, R.A., *J. Appl. Polym. Sci.: Appl. Polym. Symp.*, 1992, **51**, 195, (polymer-solvent interaction parameters, density, T_g, thermal stability)
[3] Schnell, H., *Polym. Rev.*, 1964, **9**, 99, (solvents, water absorption, ir, synth, rev)
[4] Schnell, H., *Ind. Eng. Chem.*, 1959, **51**, 157, (props, synth)
[5] Kozlov, P.V. and Perepelkin, A.N., *Polym. Sci. USSR (Engl. Transl.)*, 1967, **9**, 414, (density, T_m)
[6] McHattie, J.S., Koros, W.J. and Paul, D.R., *J. Polym. Sci., Polym. Phys. Ed.*, 1991, **29**, 731, (density, T_g, gas permeability, complex moduli)
[7] Cais, R.E., Nozomi, M., Kawai, M. and Miyake, A., *Macromolecules*, 1992, **18**, 4588, (density, T_g)
[8] Schmidhauser, J.C. and Longley, K.L., *J. Appl. Polym. Sci.*, 1990, **39**, 2083, (T_g, gas permeability)
[9] Sommer, K., Batoulis, J., Jilge, W., Morbitzer, L. et al, *Adv. Mater.*, 1991, **3**, 590, (transition temps.)
[10] Serini, V., Freitag, D. and Vernaleken, H., *Angew. Makromol. Chem.*, 1976, **55**, 175, (T_g)
[11] Perepelkin, A.N. and Kozlov, P.V., *Polym. Sci. USSR (Engl. Transl.)*, 1966, **8**, 57, (T_g)
[12] Morgan, P.W., *Macromolecules*, 1970, **3**, 536, (softening temp.)
[13] Garmonova, Y.I., Vikovskaya, M.G., Lavrenko, P.N., Tsvetkov, V.N. and Korovina, Y.V., *Polym. Sci. USSR (Engl. Transl.)*, 1971, **13**, 996, (dilute soln. props.)
[14] Perepelkin, A.N. and Kozlov, P.V., *Polym. Sci. USSR (Engl. Transl.)*, 1968, **10**, 14, (tensile props.)
[15] Kolesnikov, G.S., Kotrelev, V.N., Kostryukova, T.D., Lyamkina, Z.V. et al, *Plast. Massy*, 1966, 20, (complex moduli)
[16] Kovarskaya, B.M., Kolesnikov, G.S., Levantovskaya, I.I., Smirnova, O.V. et al, *Plast. Massy*, 1966, 40, (thermal stability, oxidation)

Poly[[1,1-cyclohexane bis(4-phenyl)carbonate]-*co*-[2,2-propane bis(4-phenyl)carbonate]]   P-158

Synonyms: *Bisphenol Z-Bisphenol A copolycarbonates. Poly[oxycarbonyloxy-1,4-phenylenecyclohexylidene-1,4-phenyleneoxycarbonyloxy-1,4-phenylene(1-methylethylidene)-1,4-phenylene]. Poly[4,4'-(cyclohexylidene)bisphenol-co-4,4'-(1-methylethylidene)-co-diphenyl carbonate]*
Related Polymers: Poly[1,1-cyclohexanebis(4-phenyl)carbonate], Poly[2,2-propanebis(4-phenyl)carbonate]
Monomers: Bisphenol A, Bisphenol Z, Carbonic dichloride, Diphenyl carbonate
Material class: Thermoplastic, Copolymers
Polymer Type: polycarbonates (miscellaneous), bisphenol A polycarbonate
CAS Number:

CAS Reg. No.	Note
109759-89-9	
31193-48-3	source: diphenylcarbonate

Molecular Formula: $[(C_{19}H_{18}O_3).(C_{16}H_{14}O_3)]_n$
Fragments: $C_{19}H_{18}O_3$ $C_{16}H_{14}O_3$
Applications/Commercial Products:

Trade name	Details	Supplier
Apec HT KU 1-9360	Source: diphenylcarbonate	Bayer Inc.

Poly(1,4-cyclohexane dimethylene terephthalate)   P-159

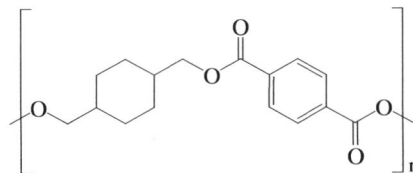

Synonyms: *PCHDMT. PCT. PDCT. Poly(cyclohexyl dimethylene terephthalate). PCHDT. Poly(cyclohexylene dimethylene terephthalate). Poly(oxymethylene-1,4-cyclohexylenemethyleneoxyterephthaloyl). Poly(oxyterephthaloyl oxymethylene-1,4-cyclohexylenemethylene). Poly(oxycarbonyl-1,4-phenylenecarbonyloxymethylene-1,4-cyclohexanediylmethylene)*
Monomers: 1,4-Cyclohexanedimethanol, 1,4-Benzenedicarboxylic acid
Material class: Thermoplastic
Polymer Type: unsaturated polyester
CAS Number:

CAS Reg. No.	Note
100218-60-8	*cis/trans* diol
25037-99-4	
31289-45-9	*cis* diol
28064-00-8	*trans* diol
24936-69-4	
28182-62-9	*trans*
25135-20-0	from dimethyl terephthalate
51176-34-2	from dimethyl terephthalate, *cis* and *trans*
31546-29-9	repeating unit, *cis*

Molecular Formula: $(C_{16}H_{18}O_4)_n$
Fragments: $C_{16}H_{18}O_4$
Additives: Glass fibre. Triphenyl phosphate and tris(2,3-dibromopropyl)phosphate are added as flame retardants
Morphology: *Trans*-form triclinic a 0.637, b 0.663, c 1.42, α 89.35°, β 47.11°, γ 114.36° [1]; triclinic a 0.646, b 0.665, c 1.42, α 89.4°, β 47.0°, γ 114.9° [9]. *Cis*-form triclinic a 0.602, b 0.601, c 1.37, α 89.14°, β 53.08°, γ 112.49° [1]

Volumetric & Calorimetric Properties:
Density:

No.	Value	Note
1	d 1.23 g/cm^3	*trans*-form [9]
2	d 1.222 g/cm^3	[18]
3	d 1.283 g/cm^3	cryst. [17]
4	d 1.195 g/cm^3	amorph. [18]
5	d 1.196 g/cm^3	[19]
6	d 1.226 g/cm^3	film [24]
7	d 1.43 g/cm^3	30% glass fibre [35]

– Poly(1,4-cyclohexane dimethylene terephthalate)

Thermal Expansion Coefficient:

No.	Value	Note	Type
1	0.0001 K^{-1}	[10]	V

Volumetric Properties General: Heat of combustion -27.197 kJ g^{-1} (-6.496 kcal g^{-1}) [12]. Other values of α_v [10] and α_l [29] and their variation with temp. have been reported. Density increases with melt-spin speed [16]. Other values of density have been reported [21]
Equation of State: Einstein and Debye equation of state constants have been reported [10]
Latent Heat Crystallization: ΔH_f 124.8 J g^{-1} (29.8 cal g^{-1}) [17]. Heat of combustion -27.197 kJ g^{-1} (-6.496 kcal g^{-1}) [12].

Melting Temperature:

No.	Value	Note
1	260°C	*cis*-form [1]
2	251–256°C	*cis*-form [2]
3	250°C	*cis*-form [6]
4	245°C	*cis*-form [8]
5	320°C	*trans*-form [1]
6	312–318°C	*trans*-form [2]
7	305°C	*trans*-form [6]
8	340°C	*trans*-form [8]
9	312–318°C	[4]
10	290°C	fibre [13]
11	294–295°C	[15]
12	286°C	[22]
13	260–290°C	film [24]

Glass-Transition Temperature:

No.	Value	Note
1	85°C	[3]
2	85°C	*trans*-form [6]
3	65°C	*cis*-form [6]
4	91°C	[15]
5	95°C	[19]
6	92°C	[22]

Transition Temperature General: T_m increases with increasing amounts of the *trans*-glycol [6]

Deflection Temperature:

No.	Value	Note
1	194°C	1.82 MPa, 30% glass fibre, ASTM D648 [20]
2	165°C	0.34 MPa, film, 2% distortion [24]
3	263°C	1.82 MPa, 30% glass fibre [35]

Brittleness Temperature:

No.	Value	Note
1	49°C	[36]

Transition Temperature:

No.	Value	Note	Type
1	-98°C	1 Hz, *trans*-form	T_γ
2	-73°C	1 Hz, *cis*-form [5]	T_γ
3	-60°C	110 Hz [33]	T_γ
4	-228°C	*trans*-form	T_δ
5	-225°C	*cis*-form [5]	T_δ
6	-90°C	11 Hz [33]	T_δ
7	-31°C	[15]	T_β
8	-35°C	[19]	T_β

Surface Properties & Solubility:

Solubility Properties General: Miscible with polycarbonate [15] but immiscible with poly(vinylphenol) [23]. Polymer blends with amorph., crystalline and liq. cryst. polymers give improvements in processability in extrusion processes [21]
Cohesive Energy Density Solubility Parameters: Cohesive energy density 356–440 J cm^{-3} (85–105 cal cm^{-3}) [34]
Solvents/Non-solvents: Sol. trifluoroacetic acid [20], conc. sulfuric acid [23]
Wettability/Surface Energy and Interfacial Tension: Contact angle 27.0° (aq. EtOH, advancing), 30° (aq. EtOH, receding) [27]. Surface energy 32.25 N m^{-1} (formamide/2-ethoxyethanol) [33]

Surface Tension:

No.	Value	Note
1	44.6 mN/m	[27]

Transport Properties:

Polymer Solutions Dilute: $[\eta]_{inh}$ 0.85 dl g^{-1} (25°, phenol/tetrachloroethane (60:40), *cis*-form) [2], 0.58 dl g^{-1} (25°, phenol/tetrachloroethane (60:40), *trans*-form) [2], 0.83 dl g^{-1} (25°, phenol/tetrachloroethane (60:40)) [3], 0.33 dl g^{-1} (phenol/tetrachloroethane (60:40)) [4], 0.297 dl g^{-1} (25°, phenol/tetrachloroethane (60:40), *cis*-form) [8], 0.203 dl g^{-1} (25°, phenol/tetrachloroethane (60:40), *trans*-form) [8]. $[\eta]$ 0.89 dl g^{-1} [21]. Other viscosity values have been reported [8]
Permeability of Gases: Other gas permeability data have been reported [31]

Water Absorption:

No.	Value	Note
1	0.3%	24h, room temp., film [24]

Gas Permeability:

No.	Gas	Value	Note
1	O_2	15.94 cm^3 mm/(m^2 day atm)	40.5 cm^3 mil 100 in^{-2} day^{-1} atm^{-1} 2.5 mm film [19]

Mechanical Properties:

Mechanical Properties General: Mechanical props have been reported [33]. The variation of storage and loss moduli with temp. has been reported [15]. Variation of mechanical props. with temp. and melt-spin speed has been reported [16]. Other values of tensile strength and modulus have been reported [21]

Tensile (Young's) Modulus:

No.	Value	Note
1	7720 MPa	30% glass fibre, ASTM D638 [20]

– Poly(1,4-cyclohexane dimethylene terephthalate)

| 2 | 2800 MPa | 400 kpsi, film [24] |
| 3 | 1600 MPa | [34] |

Flexural Modulus:

No.	Value	Note
1	7240 MPa	30% glass fibre, ASTM D648 [20]
2	8300 MPa	30% glass fibre [35]

Tensile Strength Break:

No.	Value	Note	Elongation
1	101 MPa	30% glass fibre, ASTM D638 [20]	2.1%
2	117 MPa	17 kpsi, film [4]	45%
3	130 MPa	30% glass fibre [35]	2.6%

Flexural Strength at Break:

No.	Value	Note
1	138 MPa	2.2% elongation, 30% glass fibre, ASTM D648 [20]
2	192 MPa	30% glass fibre [35]

Elastic Modulus:

No.	Value	Note
1	1000 MPa	[34]
2	2200 MPa	[34]

Tensile Strength Yield:

No.	Value	Note
1	69 MPa	10 kpsi, film [24]

Mechanical Properties Miscellaneous: T-Peel strength 18 kg m^{-1} (1 lb in^{-1}) (23°, ASTM D1876-61T) [3]
Failure Properties General: Elmendorf tear strength 6 g mm^{-1} (film) [24]. MIT folding endurance 60000 cycles (min., film) [24]. Mullen burst strength 0.19 MPa (28 psi, film) [24]
Fracture Mechanical Properties: Tensile shear strength 13.8 MPa (2000 psi, 23°), 6.2 MPa (900 psi, 70°, ASTM D1002-64) [3]
Izod Notch:

No.	Value	Notch	Note
1	37 J/m	Y	30% glass fibre, ASTM D256 [20]
2	74.7 J/m	Y	30% glass fibre [35]

Electrical Properties:
Electrical Properties General: The dielectric props. are affected greatly by humid conditions [25]
Dielectric Permittivity (Constant):

No.	Value	Note
1	3.4	140°, film [24]

Dissipation (Power) Factor:

No.	Value	Note
1	1.6	140°, film [24]

Optical Properties:
Optical Properties General: Oxyluminescence curves have been reported [18]
Refractive Index:

No.	Value	Note
1	1.709	parallel [21]
2	1.538	perpendicular [21]

Transmission and Spectra: Ir [1,7,16,22], Raman [28], electron [14], H-1 nmr [20,33] and C-13 nm r[33] spectral data have been reported
Total Internal Reflection: Birefringence has been reported [17]. Intrinsic birefringence 0.172 [21]

Polymer Stability:
Decomposition Details: 10% Weight loss occurs at 414° and 50% weight loss at 430° [4]. Heating to 340° in oxygen causes 50% weight loss in 11 min., compared with 39 min. when heated in a N_2 atmosphere [4]
Chemical Stability: Chemical resistance of films has been reported [24]
Biological Stability: Biodegradation details have been reported [30]
Stability Miscellaneous: Stability to γ-irradiation has been reported [26]

Applications/Commercial Products:
Processing & Manufacturing Routes: Synth. by condensation polymerisation of dimethyl terephthalate and 1,4-cyclohexane dimethanol at 300–310° under vacuum using tetraisopropyl titanate as a catalyst [2,8]. Can undergo rapid hydrol. during melt extrusion if sample is not sufficiently dry [11]
Applications: Used as a fibre in carpets, blankets and in synthetic furs. Also used as a moulding resin in the electronic and automotive industries

Trade name	Details	Supplier
Kodar	film	Eastman Chemical Company
Kodel	fibre	Eastman Chemical Company
Thermx		Eastman Chemical Company
Vestan	fibre	Eastman Chemical Company

Bibliographic References
[1] Boye, C.A., *J. Polym. Sci.*, 1961, **55**, 263, 275, (struct., ir, melting points)
[2] Kibler, C.J., Bell, A. and Smith, J.G., *J. Polym. Sci., Part A: Polym. Chem.*, 1964, **2**, 2115, (synth, soln props)
[3] Jackson, W.T., Gray, T.F. and Caldwell, J.R., *J. Appl. Polym. Sci.*, 1970, **14**, 685, (mech props)
[4] Tamir, L. and Smith, J.G., *J. Polym. Sci., Part A-1*, 1971, **9**, 1203, (thermal degradation)
[5] Hiltner, A. and Baer, E., *J. Macromol. Sci., Phys.*, 1972, **6**, 545, (transition temps)
[6] Schulken, R.M., Boy, R.E. and Cox, R.H., *J. Polym. Sci., Part C: Polym. Lett.*, 1964, **6**, 17, (transition temps, thermal degradation)
[7] Manley, T.R. and Williams, D.A., *Polymer*, 1971, **12**, 2, (ir)
[8] Lenz, R.W. and Go, S., *J. Polym. Sci., Polym. Chem. Ed.*, 1973, **11**, 2927, (synth, soln props)
[9] Remillard, B. and Brisse, F., *Polymer*, 1982, **23**, 1960, (struct)
[10] Simha, R., Roe, J.M. and Nanda, V.S., *J. Appl. Phys.*, 1972, **43**, 4312, (thermodynamic props, expansion coefficient)
[11] Wampler, F.C. and Gregory, D.R., *J. Appl. Polym. Sci.*, 1972, **16**, 3253, (thermal and hydrolytic degradation)
[12] Birky, M.M. and Yeh, K.N., *J. Appl. Polym. Sci.*, 1973, **17**, 239, (heat of combustion)
[13] Bostic, J.E., Yeh, K.N. and Barker, R.H., *J. Appl. Polym. Sci.*, 1973, **17**, 471, (flame retardants)
[14] Clark, D.T. and Thomas, H.R., *J. Polym. Sci., Polym. Chem. Ed.*, 1978, **16**, 791, (electon spectroscopy)
[15] Mohn, R.N., Paul, D.R., Barlow, J.W. and Cruz, C.A., *J. Appl. Polym. Sci.*, 1979, **23**, 575, (mech props, miscibility)
[16] Falkai, B., Giessler, W., Schultze-Gebhardt, F. and Spilgies, G., *Angew. Makromol. Chem.*, 1982, **108**, 9, (ir, birefringence, mech props)

[17] Barnum, R.S., Barlow, J.W. and Paul, D.R., *J. Appl. Polym. Sci.*, 1982, **27**, 4065, (density, thermodynamic props)
[18] Wendlandt, W.W., *Thermochim. Acta*, 1983, **71**, 129, (oxyluminescence)
[19] Light, R.R. and Seymour, R.W., *Polym. Eng. Sci.*, 1982, **22**, 857, (gas permeability)
[20] Auerbach, A.B. and Sell, J.W., *Polym. Eng. Sci.*, 1990, **30**, 1041, (mech props)
[21] Kikutani, T., Morohoshi, K., Yoo, H.Y., Umemoto, S. and Okui, N., *Polym. Eng. Sci.*, 1995, **35**, 942, (refractive index, birefringence)
[22] Landry, C.J.T., Massa, D.J., Teegarden, D.M., Landry M.R. et al, *Macromolecules*, 1993, **26**, 6299, (ir, miscibility)
[23] Mifune, A., *Kogyo Kagaku Zasshi*, 1962, **65**, 276, (solubility)
[24] Watson, M.T., *SPE J.*, 1961, **17**, 1083, (mechanical props)
[25] Scott, A.H. and Kinard, J.R., *J. Res. Natl. Bur. Stand. (U.S.)*, 1967, **71**, 119, (dielectric props)
[26] Price, H.L., *NASA Tech. Memo.*, 1967, (γ-irradiation)
[27] Dann, J.R., *J. Colloid Interface Sci.*, 1970, **32**, 302, (surface tension)
[28] McGraw, G.E., Am. Chem. Soc., Div. Org. Coat. Plast. Chem., *Pap*, 1970, **30**, 20, (Raman)
[29] Roe, J.M. and Simha, R., *Int. J. Polym. Mater.*, 1974, **3**, 193, (expansion coefficient)
[30] Potts, J.E., Clendinning, R.A., Ackart, W.B. and Niegisch, W.D., Tech. Pap., Reg. Tech. Conf. - Soc. Plast. Eng., 1972, **63**, (biodegradation)
[31] Mou, L., *Hsin Hsien Wei*, 1983, **25**, 33, (gas permeability)
[32] Lee, M.C.H., *Polym. Sci. Technol. (Plenum)*, 1984, **29**, 95, (surface energy)
[33] Benavente, R. and Perena, J.M., *Rev. Plast. Mod.*, 1989, **57**, 585, (H-1 nmr, C-13 nmr, transition temps)
[34] O'Reilly, J.M. and Perchak, D., *Polym. Mater. Sci. Eng.*, 1993, **69**, 94, (cohesive energy density, elastic moduli)
[35] *Kirk-Othmer Encycl. Chem. Technol.*, 4th edn., (ed. J.I. Kroschwitz), Wiley Interscience, 1991, **19**, 628, (mech props)
[36] *Ullmanns Encykl. Ind. Chem.*, (ed. B. Elvers), VCH, 1993, **A10**, 603, (brittleness temp)

Poly(4-cyclohexyl-1-butene)   P-160

Synonyms: *Poly(1-(2-cyclohexylethyl)-1,2-ethanediyl)*
Monomers: 4-Cyclohexyl-1-butene
Material class: Thermoplastic
Polymer Type: polyolefins
CAS Number:

CAS Reg. No.
31325-02-7

Molecular Formula: $(C_{10}H_{18})_n$
Fragments: $C_{10}H_{18}$

Volumetric & Calorimetric Properties:
Melting Temperature:

No.	Value	Note
1	170°C	[2,3]

Glass-Transition Temperature:

No.	Value	Note
1	40°C	[3]

Transition Temperature:

No.	Value	Note	Type
1	40°C	[2]	Softening point
2	145–147°C	[1]	Softening point

Transport Properties:
Polymer Solutions Dilute: $[\eta]_{inh}$ 2.03 dl g^{-1} (tetrahydronaphthalene, 145°) [1]

Applications/Commercial Products:
Processing & Manufacturing Routes: Synth. using Ziegler-Natta catalysts such as VCl$_3$-AlEt$_3$ at 50° [2], or MoCl$_5$ and butyllithium at 150° [1]

Bibliographic References
[1] *Brit. Pat.*, 1962, 899 946, (soln props, softening point)
[2] Dunham, K.R., Vandenberghe, J., Faber, J.W.H. and Contois, L.E., *J. Polym. Sci., Part A: Polym. Chem.*, 1963, **1**, 751, (synth, melting point)
[3] *High Polymers*, (eds. R.A.V. Raff and K.W. Doak), Interscience, 1965, **20**, (transition temps)

Poly(1,4-cyclohexylidene dimethylterephthalate)   P-162

Synonyms: *Poly(oxycarbonyl-1,4-phenylenecarbonyloxymethylene-1,4-cyclohexanediyl methylene). Poly(cyclohexylenedimethylene terephthalate). Poly(1,4-benzenedicarboxylic acid-co-1,4-cyclohexane dimethanol)*
Monomers: Dimethyl terephthalate, 1,4-Cyclohexanedimethanol
Material class: Thermoplastic
Polymer Type: saturated polyester
CAS Number:

CAS Reg. No.
100218-60-8
28182-62-9
25037-99-4
24936-69-4

Molecular Formula: $(C_{16}H_{16}O_4)_n$
Additives: Reinforcing agents, flame retardants, heat stabilisers, uv stabilisers
General Information: Is heat-resistant and crystalline. It is available commercially in a variety of forms; reinforced, flame-retardant, chemically resistant etc., and also as a copolymer with ethylene terephthalate. *Cis/trans* isomers are possible owing to the presence of the cyclohexane ring
Morphology: *Trans*-form triclinic a 6.46Å, b 6.65Å, c 14.2Å; α 89.4°, β 47.0°, γ 114.9°, space group P1[2], triclinic, a 6.37Å, b 6.63Å, c 14.2Å; α 89.35°, β 47.11°, γ 114.36°; unit cell volume 359Å [3,10]. *Cis*-form probably triclinic, a 6.02Å, b 6.01Å, c13.7Å α 89.14°, β 53.08°, γ 112.49°; unit cell volume 349.9Å [3]. Conformational anal. has been reported [8]

Volumetric & Calorimetric Properties:
Density:

No.	Value	Note
1	d 1.265 g/cm^3	cryst., *trans*-form [7,13]
2	d 1.19 g/cm^3	amorph., *trans*-form [7,13]
3	d 1.209 g/cm^3	amorph., *cis*-form [13]
4	d 1.303 g/cm^3	cryst., *cis*-form [13]

– Poly(1,4-cyclohexylidene dimethylterephthalate)

Thermal Expansion Coefficient:

No.	Value	Note	Type
1	$1.4 \times 10^{-5} – 2.5 \times 10^{-5}$ K^{-1}	20°, in parallel direction, varying levels of glass fibre [12]	L
2	$5 \times 10^{-5} – 0.000103$ K^{-1}	20°, transverse to flow, varying levels of glass fibre [12]	L

Melting Temperature:

No.	Value	Note
1	290°C	approx., [1]
2	294°C	*trans* [2]
3	312–318°C	*trans*-form [4,13]
4	251–257°C	*cis*-form [13]
5	283–287°C	varying levels of glass fibre [12]

Glass-Transition Temperature:

No.	Value	Note
1	130°C	approx. [1]
2	65°C	approx., *cis*-form
3	85°C	approx., *cis*-form
4	69–71°C	varying levels of glass fibre [12]
5	92°C	*trans*-form [13]

Transition Temperature General: T_m increases with increasing *cis* content. T_g also increases with increasing *cis* content but to lesser extent [4,5]

Deflection Temperature:

No.	Value	Note
1	260°C	1.8 MPa 30% glass fibre ASTM D648 [3]
2	227–264°C	1.8 MPa, varying levels of glass fibre [12]
3	268–280°C	0.46 MPa, varying levels of glass fibre [12]

Transport Properties:
Polymer Solutions Dilute: Intrinsic viscosity 0.89 dl g^{-1} (approx., 68% *trans*-form) [9]
Gas Permeability:

No.	Gas	Value	Note
1	[O$_2$]	16.63 cm^3 mm/ (m^2 day atm)	0.19×10^{-13} cm^2 (s Pa)$^{-1}$, 30° [11]

Mechanical Properties:
Mechanical Properties General: The variation of mechanical props. with take-up velocity of as-spun fibres has been reported [9]. Elongation at break 20–22% (approx., as-spun fibre, varying take-up velocities; approx. 68% *trans*-form) [9]

Tensile (Young's) Modulus:

No.	Value	Note
1	1000–4000 MPa	as-spun fibre, varying take-up velocities, approx. 68% *trans*-form) [9]
2	1100–11500 MPa	in tension, 1mm min^{-1}, varying levels of glass fibre [12]

Tensile Strength Break:

No.	Value	Note	Elongation
1	96.5 MPa	14 kpsi, 30% glass fibre, ASTM D638 [3]	
2	95–129 MPa	5 mm/min., varying levels of glass fibre [12]	1.2–2.2%

Flexural Strength at Break:

No.	Value	Note
1	137.8 MPa	20 kpsi, 30% glass fibre, ASTM D790 [3]

Izod Notch:

No.	Value	Notch	Note
1	80 J/m	Y	1.5 ft lb in^{-1}, 30% glass fibre, ASTM D256 [3]
2	261 J/m	N	4.9 ft lb in^{-1}, 30% glass fibre, ASTM D256 [3]

Charpy:

No.	Value	Notch	Note
1	5.6–11.7	Y	varying levels of glass fibre [12]
2	6.6–10.9	Y	-30°, varying levels of glass fibre [12]
3	30.5–63.1	N	varying levels of glass fibre [12]
4	25–64.7	N	-30°, varying levels of glass fibre [12]

Electrical Properties:
Dielectric Strength:

No.	Value	Note
1	16.9–17.5 kV.mm^{-1}	varying levels of glass fibre [12]

Complex Permittivity and Electroactive Polymers: Comparative tracking index 390–600 V (varying levels of glass fibre) [12]
Dissipation (Power) Factor:

No.	Value	Frequency	Note
1	0.005–0.019	1 MHz	varying levels of glass fibre [12]

Optical Properties:
Optical Properties General: The variation of optical props. with take-up velocity for as-spun fibres has been reported [9]. Molar polarisability 332.7×10^{-31} m^3 (parallel to chain axis, as-spun fibre, approx. 68% *trans*-form), 266.6×10^{-31} m^3 (perpendicular to chain axis, as-spun fibre, approx. 68% *trans*-form [9]

Poly[1,1-cyclopentane bis(4-phenyl)carbonate]

Refractive Index:

No.	Value	Note
1	1.709	parallel to chain axis, as-spun fibre, approx. 68% *trans* form [9]
2	1.538	perpendicular to chain axis, as-spun fibre, approx. 68% *trans*-form [9]

Transmission and Spectra: Absorption and emission spectral data have been reported [8]. Ir, H-1 nmr and C-13 nmr spectral data have been reported [6,10]
Total Internal Reflection: Intrinsic birefringence 0.172 (as-spun fibre, approx., 68% *trans*-form [9]

Polymer Stability:
Flammability: Flammability rating V0 (1.6 mm, UL94) [3,12]
Environmental Stress: Is more resistant to weathering than poly(ethylene terephthalate) [1]
Hydrolytic Stability: Is more resistant to water than poly(ethylene terephthalate) [1]

Applications/Commercial Products:
Processing & Manufacturing Routes: May be processed at 302° [12]
Mould Shrinkage (%):

No.	Value	Note
1	2–3%	parallel to flow, linear, glass/mineral or glass fibre reinforced, ASTM D255 [12]

Applications: Used in electrical and electronic applications (e.g. electrical connectors)

Trade name	Details	Supplier
Ektar		Eastman Chemical Company
Kodar	film	Eastman Chemical Company
Kodel	fibre, discontinued	Eastman Chemical Company
Thermx		Eastman Chemical Company
Thermx		Sun Koung Industry

Bibliographic References
[1] Alger, M.S.M., *Polymer Science Dictionary*, Elsevier, 1989
[2] Remillard, B. and Brisse, F., *Polymer*, 1982, **23**, 1960
[3] *U.S. Pat.*, 1989, 4837254
[4] Kibler, C.J., Bell, A. and Smith, J.G., *J. Polym. Sci. Part A*, 1964, **2**, 2115
[5] Schulken, R.M., Boy, R.E. and Cox, R.H., *J. Polym. Sci. Part C*, 1964, **6**, 17, (thermal anal.)
[6] Yoshie, N., Inoue, Y., Yoo, H.Y. and Okui, *Polymer*, 1994, **35**, 1931, (nmr)
[7] Yoo, H.Y., Umemoto, S., Kikutani, and Okui, N., *Polymer*, 1994, **35**, 117
[8] Mendicuti, F. and Mattice, W.L., *Polymer*, 1992, **33**, 4180
[9] Kikutani, T., Morohoshi, K., Yoo, H.Y., Umemoto, S. and Okui, N., *Polym. Eng. Sci.*, 1995, **35**, 942, (fibres)
[10] Boye, C.A., *J. Polym. Sci.*, 1961, **55**, 263, 275, (ir, xrd)
[11] Light, R., and Seymour, R., *Soc. Plast. Eng., Tech. Pap.*, 1983, **29**, 417
[12] *Thermx*, Eastman Chemical Co, (technical datasheet)
[13] *Polymer Yearbook 6*, (ed. R.A. Pethrick), Harwood Academic Press, 1990

Poly[1,1-cyclopentane bis(4-phenyl)carbonate] P-163

Synonyms: *Poly[oxycarbonyloxy-1,4-phenylenecyclopentylidene-1,4-phenylene]*. *Poly(4,4'-dihydroxydiphenyl-1,1-cyclopentane-co-carbonic dichloride)*
Monomers: 4,4'-Dihydroxydiphenyl-1,1-cyclopentane, Carbonic dichloride
Material class: Thermoplastic
Polymer Type: polycarbonates (miscellaneous)
CAS Number:

CAS Reg. No.
31694-76-5

Molecular Formula: $(C_{18}H_{16}O_3)_n$
Fragments: $C_{18}H_{16}O_3$

Volumetric & Calorimetric Properties:
Density:

No.	Value	Note
1	d 1.21 g/cm^3	25° [2]
2	d 1.21 g/cm^3	20°, amorph. [3]

Melting Temperature:

No.	Value	Note
1	240–250°C	[2]
2	239–248°C	amorph. [3]
3	205–215°C	[1]

Glass-Transition Temperature:

No.	Value	Note
1	155°C	[4]
2	163°C	[5]
3	166°C	[1]

Surface Properties & Solubility:
Solvents/Non-solvents: Very sol. chlorinated hydrocarbons [1]

Transport Properties:
Water Absorption:

No.	Value	Note
1	0.2 %	25°, 65% relative humidity [6]
2	0.2 %	[1]

Gas Permeability:

No.	Gas	Value	Note
1	O_2	88.4 cm^3 mm/(m^2 day atm)	1.01×10^{-13} cm^2 (s Pa)$^{-1}$, 25°, 1 atm [4]

Mechanical Properties:
Mechanical Properties General: The general shape of the tensile stress-strain curve consists of a viscoelastic region up to the yield point, followed by a stress drop and then necking behaviour (at approx. constant stress); finally the stress increases to the break point [7]

– Poly(decamethylene adipate)

Tensile (Young's) Modulus:

No.	Value	Note
1	2350 MPa	24000 kg cm^{-2}, amorph. [7]

Tensile Strength Break:

No.	Value	Note	Elongation
1	69.4 MPa	708 kg cm^{-2} [2]	119%
2	71 MPa	7.2 kg mm^{-2} [7]	100%
3	73.5 MPa	750 kg cm^{-2} [1]	40%

Tensile Strength Yield:

No.	Value	Note
1	64.7 MPa	6.6 kg mm^{-2} [7]

Electrical Properties:
Electrical Properties General: Dielectric constant and dissipation factor do not vary with temp. until T_g [1,2]
Dielectric Permittivity (Constant):

No.	Value	Frequency	Note
1	2.9	1 kHz	25–100° [2]

Dissipation (Power) Factor:

No.	Value	Frequency	Note
1	0.0005	1 kHz	25–100° [2]
2	0.004		0–180° [1]

Optical Properties:
Refractive Index:

No.	Value	Note
1	1.5993	25° [2]

Polymer Stability:
Thermal Stability General: Substantial weight loss occurs above 400°
Decomposition Details: Thermo-oxidation starts at 129°. Weight loss 7% (300°); 10% (350°); 23% (400°); 40% (450°) [1]

Applications/Commercial Products:
Processing & Manufacturing Routes: The main synthetic routes are phosgenation and transesterification. The most widely used phosgenation approach is interfacial polycondensation in a suitable solvent, but soln. polymerisation is also carried out in the presence of pyridine [6]. A slight excess of phosgene is introduced into a soln. or suspension of the aromatic dihydroxy compound in aq. NaOH at 20–30° in an inert solvent (e.g. MeCl). Catalysts (quaternary ammonium salts and tertiary amines, especially triethylbenzylammonium chloride) are used to achieve high MWs [1,2,6]. With transesterification, the dihydroxy compound and a slight excess of diphenyl carbonate are mixed together in the melt (150–300°) in the absence of O_2 with elimination of phenol. Reaction rates are increased by alkaline catalysts and the use of a medium-high vacuum in the final stages [2,6]

Bibliographic References
[1] Smirnova, O.V., Korovina, Y.V., Kolesnikov, G.S., Lipkin, A.M. and Kuzina, S.I., *Polym. Sci. USSR (Engl. Transl.)*, 1966, **8**, 776, (props., synth.)
[2] Schnell, H., *Ind. Eng. Chem.*, 1959, **51**, 157, (props., synth.)
[3] Kozlov, P.V. and Perepelkin, A.N., *Polym. Sci. USSR (Engl. Transl.)*, 1967, **9**, 414, (density, T_m)
[4] Schmidhauser, J.C. and Longley, K.L., *J. Appl. Polym. Sci.*, 1990, **39**, 2083, (T_g, gas permeability)
[5] Perepelkin, A.N. and Kozlov, P.V., *Polym. Sci. USSR (Engl. Transl.)*, 1966, **8**, 57, (T_g)
[6] Schnell, H., *Polym. Rev.*, 1964, **9**, 99, (water absorption, synth, rev)
[7] Perepelkin, A.N. and Kozlov, P.V., *Polym. Sci. USSR (Engl. Transl.)*, 1968, **10**, 14, (tensile props.)

Poly(decamethylene adipate)   P-164

Synonyms: *PDMA. Poly(oxyadipoyloxydecamethylene). Poly(oxydecamethylene oxycarbonyl tetramethylene carbonyl). Poly(1,10-decanediol-co-hexanedioic acid). Poly[oxy(1,6-dioxo-1,6-hexanediyl)oxy-1,10-decanediyl]*
Monomers: 1,10-Decanediol, Hexanedioic acid
Material class: Thermoplastic
Polymer Type: saturated polyester
CAS Number:

CAS Reg. No.
28552-31-0
25212-79-7

Molecular Formula: $(C_{16}H_{28}O_4)_n$
Fragments: $C_{16}H_{28}O_4$
Morphology: Monoclinic, a 0.5, b 0.74, c 2.21 [6]; a 0.511, b 0.743 [14]

Volumetric & Calorimetric Properties:
Density:

No.	Value	Note
1	d^{109} 0.9732 g/cm^3	[7]

Thermal Expansion Coefficient:

No.	Value	Note	Type
1	0.000715 K^{-1}	[7]	V
2	0.00072 K^{-1}	[12]	V

Latent Heat Crystallization: ΔH_f 40.9 kJ mol^{-1} [1] ΔH_f 15.9 kJ mol^{-1} [4]. ΔH_f 42.5 kJ mol^{-1} [10]. ΔH_f 44 kJ mol^{-1} [14]. ΔS_f 127 J mol^{-1} K^{-1} [14]. ΔH_f 42.6 kJ mol^{-1} [5]. Activation energy of polymerisation 49 kJ mol^{-1} (11.7 kcal mol^{-1}) [19]. Activation energy of polymerisation 50.2 kJ mol^{-1} (12 kcal mol^{-1}) [16]. Melt flow activation energy 34.46 kJ mol^{-1} (8.23 kcal mol^{-1}) [7]. Entropy of activation of polymerisation -150.7 J mol^{-1} K^{-1} (-36 cal mol^{-1} K^{-1}) [19]

Melting Temperature:

No.	Value	Note
1	77°C	[1]
2	79.5–82°C	[4]
3	74°C	[6,14]
4	79.5°C	[10]

5	74.5°C	[11]
6	79°C	[15]
7	70°C	343 K [21]

Glass-Transition Temperature:

No.	Value	Note
1	-98°C	175K [12]
2	-56°C	[23]

Surface Properties & Solubility:
Solubility Properties General: Miscibility with other polymers has been reported [15]
Internal Pressure Heat Solution and Miscellaneous: Phase diagrams with toluene [18], naphthalene [18,21], methyldioxane[18], dichloroethane [18], 1,4-dioxane [18,20,21], 1,4-dichlorobenzene [18,21], tetrachlorobenzene [21], 1,4-dimethylbenzene [21], chloronitrobenzene [21], phenanthrene [21], 1,2,4,5-tetrachlorobenzene have been reported [21]

Transport Properties:
Polymer Solutions Dilute: [η] 0.246 dl g^{-1} (chlorobenzene, 25°) [1]. [η] 0.0222 dl g^{-1} (diethyl succinate, 79°) [13]. η$_{inh}$ 0.96 dl g^{-1} (40:60 phenol/tetrachloroethane, 30°) [14]. Soln. props. have been reported [2,9]
Polymer Solutions Concentrated: Conc. soln. props. have been reported [3]
Polymer Melts: Melt viscosities have been reported [7,17]

Mechanical Properties:
Mechanical Properties General: Tensile strength yield has been reported [22]

Electrical Properties:
Electrical Properties General: Volume resistivity has been reported in graphical form [11]

Optical Properties:
Transmission and Spectra: Ir and Raman spectral data have been reported [8]

Applications/Commercial Products:
Processing & Manufacturing Routes: Synth. by condensation polymerisation of 1,10-decanediol and adipoyl chloride at 200° [4,14]

Bibliographic References

[1] Howard, G.J. and Knutton, S., *Polymer*, 1968, **9**, 527, (soln props, thermodynamic props)
[2] Van Krevelen, D.W. and Hoftyzer, P.J., *J. Appl. Polym. Sci.*, 1967, **11**, 1409, (soln props)
[3] Chee, K., *J. Macromol. Sci., Phys.*, 1981, **19**, 257, (conc soln props)
[4] Evans, R.D., Mighton, H.R. and Flory, P.J., *J. Am. Chem. Soc.*, 1950, **72**, 2018, (synth)
[5] Chickos, J.S. and Stemberg, M.J.E., *Thermochim. Acta*, 1995, **264**, 13, (thermodynamic props)
[6] Fuller, C.S. and Frosch, C.J., *J. Am. Chem. Soc.*, 1939, **61**, 2575, (cryst struct)
[7] Flory, F.J., *J. Am. Chem. Soc.*, 1940, **62**, 1057, (melt viscosity, density)
[8] Holland-Moritz, K., Modric, I., Heinen, K.U. and Hummel, D.O., *Kolloid Z. Z. Polym.*, 1973, **251**, 913, (ir, Raman)
[9] Flory, P.J. and Stickney, P.B., *J. Am. Chem. Soc.*, 1940, **62**, 3032, (soln props)
[10] Mandelkern, L., Garrett, R. and Flory, P.J., *J. Am. Chem. Soc.*, 1952, **74**, 3949, (thermodynamic props)
[11] Warfield, R.W., *Makromol. Chem.*, 1965, **89**, 269, (volume resistivity)
[12] Pezzin, G., *J. Appl. Polym. Sci.*, 1966, **10**, 21, (transition temp)
[13] Rink, M., Pavan, A., Roccasalvo, S. and Anic, S., *Polym. Eng. Sci.*, 1978, **18**, 755, (soln props)
[14] Maglio, G., Marchetta, C., Botta, A., Palumbo, R. and Pracella, M., *Eur. Polym. J.*, 1979, **15**, 695, (cryst struct, synth)
[15] Braun, D., Leiss, D., Bergmann, M.J. and Hellmann, G.P., *Eur. Polym. J.*, 1993, **29**, 225, (polymer miscibility)
[16] Rafikov, S.R. and Korshak, V.V., *Dokl. Akad. Nauk SSSR*, 1949, **64**, 211, (thermodynamic props)
[17] Takayanagi, M., *Nippon Kagaku Kaishi*, 1956, **59**, 49, (melt viscosity)
[18] Ponge, C., Rosso, J.C. and Carbonnel, L., *J. Therm. Anal.*, 1979, **15**, 101, (phase diagrams)
[19] Ivanoff, N., *Bull. Soc. Chim. Fr.*, 1950, 347, (thermodynamic props)
[20] Carbonnel, L., Rosso, J.C. and Ponge, C., *Bull. Soc. Chim. Fr.*, 1972, 941, (phase diagram)
[21] Ponge, C., Rosso, J.C. and Carbonnel, L., *Bull. Soc. Chim. Fr.*, 1981, 273, (phase diagrams)
[22] Cowan, J.C., Wheeler, D.H., Teeter, H.M. and Pashke, R.F. et al, *Ind. Eng. Chem.*, 1949, **41**, 1647, (tensile strength)
[23] Grieveson, B.M., *Polymer*, 1960, **1**, 499, (transition temp)

Poly(decamethylene sebacate)

Synonyms: *Poly(oxydecamethyleneoxysebacoyl). Poly(oxydecamethyleneoxycarbonyloctamethylenecarbonyl). PDS. Poly(sebacic acid-co-decanediol). Poly[oxy-1,10-decanediyloxy(1,10-dioxo-1,10-decanediyl)]*
Monomers: 1,10-Decanediol, Decanedioic acid
Material class: Thermoplastic
Polymer Type: saturated polyester
CAS Number:

CAS Reg. No.	Note
27514-86-9	Poly(decamethylene sebacate)
25482-94-4	repeating unit

Molecular Formula: $(C_{20}H_{36}O_4)_n$
Fragments: $C_{20}H_{36}O_4$
Morphology: Monoclinic, a 0.50, b 0.74, c 2.71 [9]. Crystal struct. has been studied [17]

Volumetric & Calorimetric Properties:
Density:

No.	Value	Note
1	d^{109} 0.9378 g/cm^3	[10]
2	d^{25} 1.086 g/cm^3	[11]
3	d 1.15 g/cm^3	[13,14]

Thermal Expansion Coefficient:

No.	Value	Note	Type
1	0.0007 K^{-1}	[10]	V

Latent Heat Crystallization: ΔS_f 151.2 J K^{-1} mol^{-1} [7]. ΔH_f 150.7 J g^{-1} (36 cal g^{-1}) [1]. ΔH_f 57.8 J g^{-1} (13.8 cal g^{-1}) [6]. ΔH_f 167.1 J g^{-1} (39.9 cal g^{-1}) [13,16]. ΔH_f 50.3 kJ mol^{-1} [7]. Activation energy for viscous flow 33.7 kJ mol^{-1} (8.05 kcal mol^{-1}) [10]

Melting Temperature:

No.	Value	Note
1	77°C	[1]
2	80°C	353 K [4]
3	80–84.5°C	[6]
4	71°C	344 K [8]
5	85°C	358 K [8]
6	73°C	[9]
7	77.2°C	[13]

| 8 | 75°C | [14] |
| 9 | 73.7°C | [16] |

Glass-Transition Temperature:

No.	Value	Note
1	-53°C	220 K [4]
2	-58°C	[13,16]
3	-75°C	[14]

Transition Temperature:

No.	Value	Note	Type
1	88°C	361K [4]	Tm

Surface Properties & Solubility:
Solubility Properties General: Miscible with polyepichlorohydrin [13], styrene-allyl alcohol copolymers [14], polyvinyl chloride [15] and tetramethyl bisphenol-A polycarbonate [16]
Surface and Interfacial Properties General: Interfacial free energy 0.89 m J cm^{-2} (8.9 erg cm^{-2}, parallel) [4]. Interfacial free energy 2.04 m J cm^{-2} (20.4 erg cm^{-2}, perpendicular) [4]

Transport Properties:
Polymer Solutions Dilute: [η] 0.6 dl.g^{-1} (toluene, 25°) [4]. [η] 0.28 dl.g^{-1} (CHCl$_3$, 25°) [14]. Soln. props. in C$_6$H$_6$ [18], 1,4-dioxane [18], CHCl$_3$ [18], cyclohexanone [18], THF [20,21], DMF [18], toluene [20,21] and 2-butanone have been reported [20,21]. Soln. props. (a prediction of inherent viscosity), have been reported [2]. Theta temp. has been reported [21]
Polymer Melts: Studies on melt viscosity variation with M_n have been reported [10]

Mechanical Properties:
Tensile Strength Yield:

No.	Value	Note
1	6.13 MPa	890 lb in^{-2} [22]

Electrical Properties:
Dielectric Permittivity (Constant):

No.	Value	Note
1	3.32	[11]

Optical Properties:
Transmission and Spectra: Raman [3,12] and ir [12] spectral data have been reported

Polymer Stability:
Biological Stability: Biodegradation studies have been reported [19]

Applications/Commercial Products:
Processing & Manufacturing Routes: Synth. by condensation polymerisation of 1,10-decanediol and sebacoyl dichloride at 200° [5,6]

Bibliographic References
[1] Howard, G.J. and Knietton, S., *Polymer*, 1968, **9**, 527, (thermodynamic props)
[2] Van Krevelen, D.W. and Hoftyzer, P.J., *J. Appl. Polym. Sci.*, 1967, **11**, 1409, (soln props)
[3] Folkes, M.J., Keller, A., Stejny, A., Goggin, P.L. et al, *Colloid Polym. Sci.*, 1975, **253**, 354, (Raman)
[4] Godovskii, Y.K. and Slonimskii, G.L., *J. Polym. Sci., Polym. Phys. Ed.*, 1974, **12**, 1053, (interfacial free energy, transition temps)
[5] Girolamo, M., Keller, A. and Stejny, J., *Makromol. Chem.*, 1975, **176**, 1489, (synth)
[6] Evans, R.D., Mighton, H.R. and Flory, P.J., *J. Am. Chem. Soc.*, 1950, **72**, 2018, (thermodynamic props, synth)
[7] Chickos, J.S. and Sternberg, M.J.E., *Thermochim. Acta*, 1995, **264**, 13, (thermodynamic props)
[8] Koehler, M.G. and Hopfinger, A.J., *Polymer*, 1989, **30**, 116, (melting points)
[9] Fuller, C.S. and Fresch, C.J., *J. Am. Chem. Soc.*, 1939, **61**, 2575, (cryst struct)
[10] Flory, P.J., *J. Am. Chem. Soc.*, 1940, **62**, 1057, (melt viscosity, expansion coefficient)
[11] Yager, W.A. and Baker, W.D., *J. Am. Chem. Soc.*, 1942, **64**, 2164, (dielectric constant, density)
[12] Holland-Moritz, K., Modric, I., Heinen, K.U. and Hummel, D.O., *Kolloid Z. Z. Polym.*, 1973, **251**, 913, (ir, Raman)
[13] Fernandes, A.C., Barlow, J.W. and Paul, D.R., *J. Appl. Polym. Sci.*, 1984, **29**, 1971, (transition temps)
[14] Woo, E.M., Barlow, J.W. and Paul, D.R., *J. Appl. Polym. Sci.*, 1984, **29**, 3837, (polymer miscibility, transition temps)
[15] Woo, E.M., Barlow, J.W. and Paul, D.R., *Polymer*, 1985, **26**, 763, (polymer miscibility)
[16] Fernandes, A.C., Barlow, J.W. and Paul D.R., *Polymer*, 1986, **27**, 1799, (polymer miscibility, transition temps)
[17] Mnyukh, Y.V., *Izv. Akad. Nauk SSSR, Otd. Khim. Nauk*, 1958, 1128, (cryst struct)
[18] Orszagh, A. and Fejgin, J., *Vysokomol. Soedin.*, 1963, **5**, 1861, (soln props)
[19] Diamond, M.J., Freedman, B. and Garibaldi, J.A., *Int. Biodeterior. Bull.*, 1975, **11**, 127, (biodegradation)
[20] Tarasova, G.I., Pavlova, S.I. and Korshak, V.V., *Vysokomol. Soedin., Ser. B*, 1973, **15**, 219, (soln props)
[21] Tarasova, G.I., Pavlova, S.I. and Korshak, V.V., *Vysokomol. Soedin., Ser. A*, 1975, **17**, 712, (soln props)
[22] Cowan, J.C., Wheeler, D.H., Teeter, H.M., Pashke, R.F. et al, *Ind. Eng. Chem.*, 1949, **41**, 1647, (tensile strength)

Poly(decamethylene terephthalate)

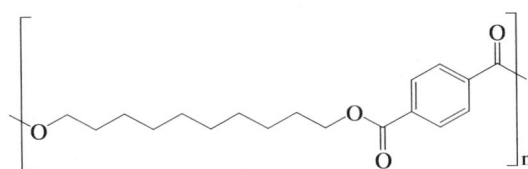

Synonyms: *Poly(oxyterephthaloyloxydecamethylene). Poly(1,10-decanediol-co-1,4-benzenedicarboxylic acid)*
Monomers: 1,10-Decanediol, 1,4-Benzenedicarboxylic acid
Material class: Thermoplastic
Polymer Type: saturated polyester
CAS Number:

CAS Reg. No.
27043-73-8
27055-32-9

Molecular Formula: (C$_{18}$H$_{24}$O$_4$)$_n$
Fragments: C$_{18}$H$_{24}$O$_4$
Morphology: Morphological studies [20,22] have been reported. Triclinic, a 0.455, b 0.628, c 2.010, α 107.5°, β 96°, γ 113.5° (90K) [5]; triclinic, a 0.462, b 0.630, c 2.010, α 107.5°, β 96°, γ 113.5° (291K) [5]; triclinic, a 0.482, b 0.645, c 2.010, α 107.5°, β 96°, γ 113.5° (390K) [5]. Two polymorphic forms [17] known, form I, c 1.629 [18]

Volumetric & Calorimetric Properties:
Density:

No.	Value	Note
1	d 1.046 g/cm^3	[4]

Poly(decamethylene terephthalate)

2	d^{150} 1.0324 g/cm^3		[6]
3	d 1.159 g/cm^3		[17]

Thermal Expansion Coefficient:

No.	Value	Note	Type
1	0.00055 K^{-1}	138° [4]	V
2	0.0002 K^{-1}	80° [16]	L

Volumetric Properties General: C_p values have been reported [10]
Latent Heat Crystallization: ΔH_f 46.1 kJ mol^{-1} (11 kcal mol^{-1}), ΔS_f 113 J mol^{-1} K^{-1} (27 cal mol^{-1} K^{-1}) [4], ΔS_f 107.2 J mol^{-1} K^{-1} (25.6 cal mol^{-1} K^{-1}) [12], ΔH_f 18 kJ mol^{-1} (4.3 kcal mol^{-1}) [11], ΔH_f 46.1 kJ mol^{-1} [9], ΔH_f 43.5 kJ mol^{-1} (10.4 kcal mol^{-1}), 48.6 kJ mol^{-1} (11.6 kcal mol^{-1}) [12]. Enthalpies of melting have been reported. [8] Other entropy values have been reported [13]

Melting Temperature:

No.	Value	Note
1	123°C	396K [2]
2	138°C	411K [2,4,12]
3	129°C	[3,5]
4	137.5°C	[6]
5	131°C	[7]
6	125°C	398K [8]
7	110°C	[10]
8	123–127°C	[11,13]
9	134°C	[15]
10	120°C	[16]

Glass-Transition Temperature:

No.	Value	Note
1	-5°C	268K [2,7,16]
2	25°C	298K [2,5]
3	-9°C	[10]
4	-8°C	[16]

Transition Temperature General: Other T_g values have been reported [19]

Deflection Temperature:

No.	Value	Note
1	-14°C	Heat deflection temp. [16]

Vicat Softening Point:

No.	Value	Note
1	133°C	[16]

Transition Temperature:

No.	Value	Note	Type
1	-125°C	[5]	T_γ
2	20°C	[10]	$T_{\beta max}$
3	-97°C	[10]	$T_{\gamma max}$
4	95°C	[17]	T_{cryst}

Surface Properties & Solubility:
Solvents/Non-solvents: Sol. hot diphenyl ether [7], phenol/tetrachloroethane (50:50) [9]. Insol. EtOH [9], MeOH [15]
Wettability/Surface Energy and Interfacial Tension: Lateral surface energy 5.7 erg cm^{-2} [25]

Transport Properties:
Polymer Solutions Dilute: η_{inh} 0.85 dl g^{-1} (20°, phenol/tetrachloroethane (60:40)) [7]; [η] 0.91 dl g^{-1} (25°, CHCl$_2$) [11]; [η] 1.222 dl g^{-1} (25°, phenol/tetrachloroethane (50:50)) [15]; [η] 0.71 dl g^{-1} (30°, phenol/tetrachloroethane (60:40)) [16]
Polymer Melts: Melt viscosity 740 poise (225°) [6]

Mechanical Properties:
Mechanical Properties General: Tensile storage and loss moduli, in graphical form, have been reported [10]

Tensile (Young's) Modulus:

No.	Value	Note
1	0.4 MPa	[19]

Viscoelastic Behaviour: Stress relaxation modulus 200 MPa (16 s) [16]. Viscoelastic props., stress relaxation and creep have been reported [18]

Electrical Properties:
Electrical Properties General: Dielectric props. have been reported [19]
Complex Permittivity and Electroactive Polymers: Complex permittivity has been reported [24]

Optical Properties:
Transmission and Spectra: Nmr [5], esr [14] and C-13 nmr [15] spectral data have been reported

Polymer Stability:
Decomposition Details: Thermal degradation has been reported [21,23]
Stability Miscellaneous: Prolonged exposure to γ-radiation in air leads to a decrease in T_m possibly due to crosslinking and degradation [8]

Applications/Commercial Products:
Processing & Manufacturing Routes: Synth. by transesterification of dimethyl terephthalate with 1,10-decanediol at 280° under vacuum using tetraisopropyl titanate as catalyst [7,4,9,11,16,17]. Also synth. by bulk polymerisation of 1,10-decanediol and dimethyl terephthalate in the presence of tetraisopropyl titanate at 180° for 2h, followed by heating to 260° at 1 mmHg for 3h [16]

Bibliographic References
[1] Chickos, J.S. and Sternberg, M.J.E., *Thermochim. Acta*, 1995, **264**, 13, (thermodynamic props)
[2] Koehler, M.G. and Hopfinger, A.J., *Polymer*, 1989, **30**, 116, (transition temps)
[3] Izard, E.F., *J. Polym. Sci.*, 1952, **8**, 503, (melting point)
[4] Flory, P.J., Bedon, H.D. and Keefer, E.H., *J. Polym. Sci.*, 1958, **28**, 151, (thermodynamic props, synth)
[5] Farrow, G., McIntosh, J. and Ward, I.M., *Makromol. Chem.*, 1960, **38**, 147, (cryst struct, nmr, transition temps)
[6] Sharples, A. and Swinton, F.L., *Polymer*, 1963, **4**, 119, (melt viscosity)
[7] Smith, J.G., Kibler, C.J. and Sublett, B.J., *J. Polym. Sci., Part A-1*, 1966, **4**, 1851, (synth, soln props)
[8] Pietrzak, M. and Kroh, J., *Eur. Polym. J.*, 1972, **8**, 237, (γ-radiation study)
[9] Wlochowicz, A., Pietrzak, M. and Kroh, J., *Eur. Polym. J.*, 1972, **8**, 313, (synth, solubility)
[10] Yip, H.K. and Williams, H.L., *J. Appl. Polym. Sci.*, 1976, **20**, 1217, (thermodynamic props, mech props)
[11] Yip, H.K. and Williams, H.L., *J. Appl. Polym. Sci.*, 1976, **20**, 1209, (synth, soln props)
[12] Kirshenbaum, I., *J. Polym. Sci., Part A: Polym. Chem.*, 1965, **3**, 1869, (thermodynamic props)
[13] Riande, E., *Eur. Polym. J.*, 1978, **14**, 885, (entropy)
[14] Sundholm, F., Wasserman, A.M., Barashkova, I.I., Timofeev, V.P. and Buchachenko, A.L., *Eur. Polym. J.*, 1984, **20**, 733, (esr)

- Poly(1-decene)

[15] Horii, F., Hirai, A., Murayama, K., Kitamura, R. and Suzuki, T., *Macromolecules*, 1983, **16**, 273, (C-13 nmr)
[16] Ng, T.H. and Williams, H.L., *Makromol. Chem.*, 1981, **182**, 3323, 3331, (viscoelastic behaviour)
[17] Daniewska, I. and Wasiak, A., *Makromol. Chem., Rapid Commun.*, 1982, **3**, 897, (cryst struct)
[18] Ng, T.H. and Williams, H.L., *J. Appl. Polym. Sci.*, 1986, **32**, 4883, (viscoelastic props)
[19] Maklakov, A.I. and Pimenov, G.G., *Vysokomol. Soedin., Ser. A*, 1973, **15**, 107, (dielectric props)
[20] Daniewska, I., *Polym. J. (Tokyo)*, 1980, **12**, 417, (morphology)
[21] Rafter, G., Blaesche, J., Moeller, B. and Stromeyer, M., *Acta Polym.*, 1981, **32**, 608, (thermal degradation)
[22] Daniewska, I., Hashimoto, T. and Nakai, A., *Polym. J. (Tokyo)*, 1984, **16**, 49, (morphology)
[23] McNeill, I.C. and Bounekhel, M., *Polym. Degrad. Stab.*, 1991, **34**, 187, (thermal degradation)
[24] Tatsumi, T. and Ito, E., *Kobunshi Ronbunshu*, 1992, **49**, 229, (complex permittivity)
[25] Dobreeva, A., Alonso, M., Gonzalez, M., Gonzalez, A. and deSaja, J.A., *Thermochim. Acta*, 1995, **258**, 197, (surface energy)

Poly(1-decene)   P-167

$$-[CH_2CH]_n-$$
$$\quad\ \ |$$
$$\ \ CH_2(CH_2)_6CH_3$$

Synonyms: *Poly(1-octylethylene). Poly(1-octyl-1,2-ethanediyl)*
Related Polymers: Poly(alpha olefin)
Monomers: 1-Decene
Material class: Thermoplastic
Polymer Type: polyolefins
CAS Number:

CAS Reg. No.	Note
25189-70-2	
26746-85-0	isotactic

Molecular Formula: $(C_{10}H_{20})_n$
Fragments: $C_{10}H_{20}$
General Information: Rubbery appearance [1]
Morphology: Struct.: c 1.32 [2]. Three forms: form I monoclinic, a 0.59, b 4.77, c 1.34, β 96° [3]

Volumetric & Calorimetric Properties:
Density:

No.	Value	Note
1	d^{30} 0.847 g/cm^3	[4]

Latent Heat Crystallization: ΔH_m 10-24 J g^{-1} [7]. ΔH_m 35 J g^{-1} (Form I) [3]. ΔH_m 5 J g^{-1} (Form III) [3]
Melting Temperature:

No.	Value	Note
1	22–27°C	[6,7,8]
2	29°C	[8]
3	31°C	[3]
4	32°C	[9]
5	34°C	[1,2]
6	35°C	[5]
7	40°C	[10]
8	-48°C	Form III [3]

Glass-Transition Temperature:

No.	Value	Note
1	-33–-26°C	[6]
2	-35°C	[1]
3	-41°C	[11]

Transition Temperature:

No.	Value	Note	Type
1	3.3–9.5°C	[6]	First T_m
2	5–7°C	[7]	First T_m
3	-160°C	590 Hz [1]	β
4	-60°C	[12]	Transition temp.
5	-170°C	[12]	Transition temp.
6	-190°C	[12]	Transition temp.

Surface Properties & Solubility:
Solvents/Non-solvents: Sol. CO_2 [13]

Transport Properties:
Polymer Solutions Dilute: [η] 1.6 dl g^{-1} (135°, tetrahydronaphthalene) [11]. [η] 2.76 dl g^{-1} (25°, toluene) [8]. Other viscosity values have been reported [6,14]. Theta temp. has been reported [5]
Polymer Solutions Concentrated: Conc. soln. props. have been reported [4,15]
Polymer Melts: Melt flow activation energy 58.6 kJ mol^{-1} (14 kcal mol^{-1}) [16]

Mechanical Properties:
Mechanical Properties General: Dynamic tensile storage and loss moduli have been reported in graphical form [17]

Optical Properties:
Refractive Index:

No.	Value	Note
1	1.473	[18]

Transmission and Spectra: Ir [19], Raman [20,21,22], C-13 nmr [7,22] and H-1 nmr [22] spectral data have been reported
Molar Refraction: dn/dC 0.110–0.152 (heptane), dn/dC 0.1689 (heptane) [6]
Total Internal Reflection: Stress optical coefficient -3370 [8]

Applications/Commercial Products:
Processing & Manufacturing Routes: Synth. using Ziegler-Natta catalysts, such as $MgCl_2$ supported $TiCl_4$-$AlEt_3$ in heptane at 70° [7]

Bibliographic References
[1] Clark, K.J., Turner-Jones, A. and Sandiford D.J.H., *Chem. Ind. (London)*, 1962, 2010, (transition temps)
[2] Turner-Jones, A., *Makromol. Chem.*, 1964, **71**, 1, (struct, melting point)
[3] Trafara, G., *Makromol. Chem., Rapid Commun.*, 1980, **1**, 319, (struct)
[4] Tait, P.J.T. and Livesey, P.J., *Polymer*, 1970, **11**, 359, (conc soln props, density)
[5] Wang, J.S., Porter, R.S. and Knox, J.R., *Polym. J. (Tokyo)*, 1978, **10**, 619, (expansion coefficient, soln props)
[6] Chien, J.C.W. and Ang, T., *J. Polym. Sci., Part A: Polym. Chem.*, 1986, **24**, 2217, (refractive index, transition temp)
[7] Pena, B., Delgado, J.A., Bello, A. and Perez, E., *Makromol. Chem., Rapid Commun.*, 1991, **12**, 353, (C-13 nmr)
[8] Philippoff, W. and Tomquist, E., *J. Polym. Sci., Part C: Polym. Lett.*, 1966, **23**, 881, (stress optical coefficient)
[9] Beck, D.L., Knox, J.R. and Price, J.A., *Prepr. - Am. Chem. Soc., Div. Pet. Chem.*, 1963, (melting point)
[10] Turner-Jones, A., *Polymer*, 1965, **6**, 249, (melting point)

[11] Natta, G., Danusso, F. and Moraglio, G., *CA*, 1958, **52**, 17789e, (transition temp)
[12] Manabe, S., Nakamura, H., Uemura, S. and Takayanagi, M., *Kogyo Kagaku Zasshi*, 1970, **73**, 1587, (transition temps)
[13] Heller, J.P., Dandge, D.K., Card, R.J. and Donanema, L.G., *Int. Symp. Oilfield Geotherm. Chem.*, 1983, (solubility props)
[14] Dandge, D.K., Heller, J.P., Lien, C.L. and Wilson, K.V., *Polym. Prepr. (Am. Chem. Soc., Div. Polym. Chem.)*, 1986, **27**, 273, (rheological props)
[15] Daniels, C.A., Maron, S.H. and Livesey, P.J., *J. Macromol. Sci., Phys.*, 1970, **4**, 47, (rheological props)
[16] Wang, J.S., Porter, R.S. and Knox, J.R., *J. Polym. Sci., Part B: Polym. Lett.*, 1970, **8**, 671, (melt flow activation energy)
[17] Takayanagi, M., *Pure Appl. Chem.*, 1970, **23**, 151, (tensile modulus)
[18] *Polym. Handb.*, 3rd edn., (eds. J. Brandrup and E.H. Immergut), Wiley Interscience, 1989, 458, (refractive index)
[19] McRae, M.A. and Maddams, W.F., *Makromol. Chem.*, 1976, **177**, 449, (ir)
[20] Fraser, G.V., Hendra, P.J., Chalmers, J.M., Cudby, M.E.A. and Willis, H.A., *Makromol. Chem.*, 1973, **173**, 195, (ir, Raman)
[21] Holland-Moritz, K., Djudovic, P. and Hummel, D.D., *Prog. Colloid Polym. Sci.*, 1975, **57**, 206, (ir, Raman)
[22] Doskocilova, D., Pecka, J., Dybal, J., Kriz, J. and Mikes, F., *Macromol. Chem. Phys.*, 1994, **195**, 2747, (H-1 nmr, C-13 nmr)

Poly(3-decylthiophene) P-168

Synonyms: *Poly(3-decyl-2,5-thiophenediyl)*. P3DT
Monomers: 3-Decylthiophene
Material class: Thermoplastic
Polymer Type: polythiophene
CAS Number:

CAS Reg. No.	Note
110851-65-5	
216000-50-9	Poly(3-decyl-2,5-thiophenediyl)

Molecular Formula: $(C_{14}H_{22}S)_n$
Fragments: $C_{14}H_{22}S$
Additives: Dopants such as hexafluorophosphate [1], $FeCl_3$, perchlorate and I_2 are added to increase conductivity.
General Information: Thermochromic conductive polymer which exhibits electroluminescence [2] and has greater solubility and processability than polythiophene.
Morphology: Orthorhombic unit cell, a=26.43, b=7.73, c=7.77 Å [3]. Polymorphic with two crystal phases and one mesophase [4].

Volumetric & Calorimetric Properties:
Density:

No.	Value	Note
1	d^{25} 1.02 g/cm^3	crystal [4]

Melting Temperature:

No.	Value	Note
1	85–180°C	actual max. temp. 170–180°C [4,5]

Surface Properties & Solubility:
Solvents/Non-solvents: Sol. $CHCl_3$ [6], THF [5].

Electrical Properties:
Electrical Properties General: Conductivity is dependent on polymer structure, nature and concentration of dopant and temperature, with typical values for regiorandom polymer films of 5 S cm^{-1} (hexafluorophosphate dopant) and 25 S cm^{-1} ($FeCl_3$ dopant) [1].

Optical Properties:
Transmission and Spectra: Pmr [1,5,6,7,8], cmr [5,6], ir [5,6], uv [5,6,8] and x-ray [3,4] spectra have been reported.
Volume Properties/Surface Properties: Polymer has red or brown colour [5] with thermochromic changes on heating.

Applications/Commercial Products:
Processing & Manufacturing Routes: Produced by chemical or electrochemical oxidative polymerisation of monomer with different catalysts [1,6,7,9] to give mainly regiorandom (head-to-head/head-to-tail) polymers, or by polymerisation of 2-halo-5-(halomagnesio)-3-decylthiophene [4] or 2-bromo-5-(bromozincio)-3-decylthiophene [5] to give mainly regioregular (head-to-tail) polymers.
Applications: Potential applications in photovoltaic devices, light-emitting diodes, rechargeable batteries, field-effect transistors and chemical and optical sensors [10].

Trade name	Details	Supplier
Rieke	regiorandom and regioregular polymers	Rieke Metals

Bibliographic References
[1] Leclerc, M., Martinez Diaz, F. and Wegner, G., *Makromol. Chem.*, 1989, **190**, 3105
[2] Barta, P., Cacialli, F., Friend, R.H. and Zagorska, M., *J. Appl. Phys.*, 1998, **84**, 6279
[3] Kawai, T., Nakazono, M., Sugimoto, R. and Yoshino, K., *J. Phys. Soc. Jpn.*, 1992, **61**, 3400
[4] Bolognesi, A., Porzio, W. and Provasoli, F., *Makromol. Chem.*, 1993, **194**, 817
[5] Chen, T.-A., Wu, X. and Rieke, R.D., *J. Am. Chem. Soc.*, 1995, **117**, 233
[6] Lanzi, M., Della-Casa, C., Costa-Bizzarri, P. and Bertinelli, F., *Macromol. Chem. Phys.*, 2001, **202**, 1917
[7] Daoud, W.A. and Xin, J.H., *J. Appl. Polym. Sci.*, 2004, **93**, 2131
[8] Botta, C., Stein, P.C., Bolognesi, A. and Catellani, M., *J. Phys. Chem.*, 1995, **99**, 3331
[9] Oztemiz, S., Beaucage, G., Ceylan, O. and Mark, H.B., *J. Solid State Electrochem.*, 2004, **8**, 928
[10] *Handbook of Conducting Polymers* 2nd Edn., (Eds. Skotheim, T.A., Elsenbaumer, R.L. and Reynolds, J.R.) Marcel Dekker, 1998, 225

Poly(1,4-diaminobenzene) P-169

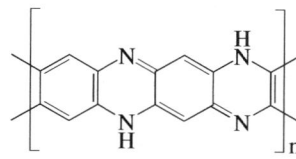

Synonyms: *Poly(p-phenylenediamine). Polyaniline (emeraldine salt). Poly(p-diaminophenylene). Poly(p-aminoaniline). PPD. PPDA. p-Phenylenediamine homopolymer*
Related Polymers: Poly(aniline)
Monomers: 1,4-Benzenediamine
Polymer Type: polyaniline
CAS Number:

CAS Reg. No.
25168-37-0

Molecular Formula: $(C_6H_8N_2)_n$
Fragments: $C_6H_8N_2$
Molecular Weight: When treated with ammonium peroxodisulfate, brown polymers of Mn 10,000 soluble in DMSO are obtained [2,9].

General Information: Blue-black solid. Polymerises under oxidation with potassium persulfate to yield a polyquinoxaline, with ladder structure and in a fully oxidised state somewhat similar to that shown by pernigraniline [2].
Morphology: Structure based on the comparison of the electronic spectra of pernigraniline with that of PPD. This shows a strict analogy, consisting of alternating units of benzenoid and quinoid moieties linked together by -NH- and =N- bridges [2].
Identification: Elemental composition and product yield vary upon method of preparation [14]

Volumetric & Calorimetric Properties:
Density:

No.	Value	Note
1	d 1.34 g/cm³	[12]

Thermodynamic Properties General: *Ab initio* study of the polymerisation mechanism of poly(*p*-phenylenediamine) with thermodynamic properties has been reported [7].

Surface Properties & Solubility:
Solubility Properties General: Displays limited solubility in organic solvents [12].

Electrical Properties:
Electrical Properties General: A compressed pellet of the polymer shows a conductivity of 6.3×10^{-6} S cm^{-1}, consistent with a low conductivity semiconductor, and with band gap of 2.37eV. The joint chemical oxidation of aniline with phenylenediamines yields products with d.c. conductivity spanning more than eleven orders of magnitude. The dependence of the conductivity on the composition of the reaction mixture has been investigated for all three co-monomers, *o*-, *m*- and *p*-phenylenediamine and the electrical properties compared [2,9,12,14]

Optical Properties:
Transmission and Spectra: FT-IR and the pmr spectra support the proposed ladder structure composed of alternating sequence of benzenoid and quinoniimine moieties linked together by -NH- and =N- bridges [2,7,14]. The uv-vis spectrum is similar to the pernigraniline spectrum with bands at 275, 344, 404 and 543 (*vs.* pernigraniline bands in DMF at 273, 340, 364 and 500 nm) [2,8,12,14].

Polymer Stability:
Thermal Stability General: The thermal stability of poly(*p*-phenylenediamine-terephthalamide) (PPT) and poly(*p*-phenylenebenzobisoxazole) (PBO) fibres has been studied [15].
Biological Stability: PPD is a regulated substance [1]. OSHA and NIOSH give a recommended exposure limit (REL) of 0.1 mg m^{-3} TWA (skin). Known health effects include skin and respiratory sensitisation and bronchial asthma.

Applications/Commercial Products:
Processing & Manufacturing Routes: Obtained from *p*-Phenylenediamine hydrochloride dissolved in HCl then treated with potassium persulfate [2,8].
Applications: Free radicals of Wurster's salts [8]. Lead determination by anodic stripping voltammetry using a *p*-phenylenediamine modified carbon paste electrode [11]. Determination of lead in wastewater samples [11]. Biosensors prepared from electrochemically-synthesised conducting polymers [13].

Bibliographic References
[1] 2002, www.epa.gov/ttn/atw/188polls.html
[2] Cataldo, F., *Eur. Polym. J.*, 1996, **32(1)**, 43
[3] *ACD/Labs*, calcd. using ACD/Labs v8.14 for Solaris, 1994
[4] National Institute for Occupational Safety and Health International Chemical Safety Cards, 2006
[5] National Library of Medicine (US) Hazardous Substances Data Bank, http://toxnet.nlm.nih.gov/,
[6] Syracuse Research Corporation of Syracuse, New York (US) PhysProp, www.syrres.com/esc/physdemo.htm, 2006
[7] Herlem, G., Lakard, S. and Fahys, B., *THEOCHEM*, 2003, **638**, 177
[8] Michaelis, L., Schubert, M. P. and Granick, S., *J. Am. Chem. Soc.*, 1939, 1981
[9] Prokes, J., Stejskalb, J., Kfivka, I. and Tobolkova, E., *Synth. Met.*, 1999, **102**, 1205
[10] Chandrasekhar, P. and Gumbs, R.W., *J. Electrochem. Soc.*, 1991, **138**, 1337
[11] Adraoui, I., El Rhazi, M., Amine, A., Idrissi, L., Curulli, A. and Palleschi, G., *Electroanalysis (N.Y.)*, 2005, **17**, 685
[12] Sulimenko, T., Jaroslav, S. and Proke, J., *J. Colloid Interface Sci.*, 2001, **236**, 328
[13] Deshpande, M.V. and Amalnerkar, D.P., *Prog. Polym. Sci.*, 1993, **18**, 623
[14] Neob, K.G., Kang, E.T. and Tan, K.L., *J. Phys. Chem.*, 1992, **96**, 6711
[15] Bourbigot, S., Flambard, X. and Duquesne, S., *Polym. Int.*, 2001, **50**, 157

Poly(2,6-dibromophenylene oxide)

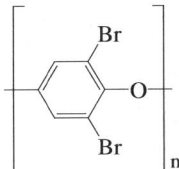

Synonyms: *Poly[oxy(dibromophenylene)]*
Monomers: 2,4,6-Tribromophenol
Polymer Type: polyphenylene oxide
CAS Number:

CAS Reg. No.
59779-25-8
26023-27-8
74082-93-2

Molecular Formula: (C$_6$H$_2$BrO)$_n$
Fragments: C$_6$H$_2$BrO

Volumetric & Calorimetric Properties:
Glass-Transition Temperature:

No.	Value	Note
1	110–150°C	[1]

Vicat Softening Point:

No.	Value	Note
1	240–245°C	[2]

Surface Properties & Solubility:
Solvents/Non-solvents: Sol. in a range of solvents [2]

Optical Properties:
Transmission and Spectra: C-13 nmr and ir spectral data have been reported [1]

Polymer Stability:
Thermal Stability General: 50% weight loss occurs in 2h at 445° [2]

Applications/Commercial Products:
Processing & Manufacturing Routes: Prod. by oxidative coupling of 2,4,6-tribromophenol with alkali
Applications: Non-migrating fire retardant

Trade name	Supplier
Firemaster TSA	Unitika Ltd.

Poly(2,5-dibutoxy-1,4-phenylene ethynylene) P-171

Synonyms: *Poly[(2,5-dibutoxy-1,4-phenylene)-1,2-ethynediyl]. 1,4-Dibutoxy-2,5-diethynylbenzene homopolymer. PDBOPA*
Material class: Thermoplastic
Polymer Type: poly(arylene ethynylene)
CAS Number:

CAS Reg. No.	Note
145130-55-8	Poly[(2,5-dibutoxy-1,4-phenylene)-1,2-ethynediyl]
128834-30-0	1,4-Dibutoxy-2,5-diethynylbenzene homopolymer

Molecular Formula: $(C_{16}H_{20}O_2)_n$
Fragments: $C_{16}H_{20}O_2$
General Information: Conductive polymer with greater solubility than poly(1,4-phenylene ethynylene). Prod. as red-brown solid.

Volumetric & Calorimetric Properties:
Transition Temperature General: Cross-linking transition at 347°C [1].

Surface Properties & Solubility:
Solvents/Non-solvents: Sol. THF, $CHCl_3$, chlorobenzene, toluene [1,2,3].

Electrical Properties:
Electrical Properties General: Increase in conductivity when polymer is doped with iodine or $FeCl_3$ [1,3] or when exposed to humidity or MeOH vapour [4]. Photoconductivity 40 times higher than dark conductivity [1]. Electroluminescence has been reported [2].

Optical Properties:
Optical Properties General: Exhibits photoluminescence with relatively large Stokes shift and band gap of *ca.* 2.7 eV [5]. There is a considerable red shift in both absorption and photoluminescence on changing from solution ($CHCl_3$) to film, with luminescence λ_{max} at *ca.* 480nm ($CHCl_3$) and *ca.* 540nm (film) [6].
Transmission and Spectra: Pmr [3,7], cmr [1,7], ir [7], uv and photoluminescence [5,6] spectral data have been reported.

Polymer Stability:
Polymer Stability General: When irradiated with light at wavelength close to absorption maximum polymer is stable towards photooxidation [8].

Applications/Commercial Products:
Processing & Manufacturing Routes: Prod. by coupling reaction between 1,4-dibutoxy-2,5-dihalobenzene and 1,4-dibutoxy-2,5-diethynylbenzene in the presence of Pd catalyst [2,3], by reaction of 1,4-diiodo-2,5-dibutoxybenzene and tributylethynyltin [6,7], or by oxidative coupling of 1,4-dibutoxy-2,5-diethynylbenzene [1].
Applications: Potential application in gas and vapour sensors [4,9].

Bibliographic References
[1] Okawa, H. and Uryu, T., *Polym. J. (Tokyo)*, 1990, **22**, 539
[2] *U.S. Pat.*, 1994, 5 334 539
[3] Pelter, A. and Jones, D.E., *J. Chem. Soc., Perkin Trans. 1*, 2000, 2289
[4] Bearzotti, A., Fratoddi, I., Palummo, L., Petrocco, S., Furlani, A., Lo Sterzo, C. and Russo, M.V., *Sens. Actuators*, 2001, **76**, 316
[5] Ni, Q.-X., Swanson, L.S., Lane, P.A., Shinar, J., Ding, Y.W., Ijadi-Maghsoodi, S. and Barton, T.J., *Synth. Met.*, 1992, **49/50**, 447
[6] Pizzoferrato, R., Berliocchi, D., Di Carlo, A., Lugli, P., Venanzi, M., Micozzi, A., Ricci, A. and Lo Sterzo, C., *Macromolecules*, 2003, **36**, 2215
[7] Giardina, G., Rosi, P., Ricci, A. and Lo Sterzo, C., *J. Polym. Sci., Part A: Polym. Chem.*, 2000, **38**, 2603
[8] Cumpston, B.H. and Jensen, K.F., *J. Appl. Polym. Sci.*, 1998, **69**, 2451
[9] Penza, M., Cassano, G., Sergi, A., Lo Sterzo, C. and Russo, M.V., *Sens. Actuators*, 2001, **81**, 88

Poly(dibutylsilane) P-172

Synonyms: *Poly(dibutylsilylene). Dibutyldichlorosilane homopolymer. PDBS*
Monomers: Dibutyldichlorosilane
Material class: Thermoplastic
Polymer Type: polysilanes
CAS Number:

CAS Reg. No.	Note
95999-72-7	poly(dibutylsilylene)
97036-65-2	dibutyldichlorosilane homopolymer

Molecular Formula: $(C_8H_{18}Si)_n$
Fragments: $C_8H_{18}Si$
Additives: Polymer can be doped with I_2, $FeCl_3$, SbF_5, etc. to increase conductivity [1].
General Information: Conductive thermochromic polysilane exhibiting photoluminescence and electroluminescence in near-UV and visible region. Prod. as white elastomeric solid or powder.
Morphology: Exhibits polymorphism, with a 7/3 helical conformation of the polymer backbone stable at r.t. and normal pressure [2,3]. At low temperatures [3,4], by applying high pressure [4,5,6], or by controlling precipitation or solvent evaporation [7,8], a near planar *trans*-conformation is obtained.

Volumetric & Calorimetric Properties:
Glass-Transition Temperature:

No.	Value	Note
1	-40°C	[2]

Transition Temperature General: A weak transition at 36°C [2] and a transition to a disordered mesophase at *ca.* 85°C have been reported [2,9].

Surface Properties & Solubility:
Solvents/Non-solvents: Sol. hexane, toluene, $CHCl_3$, THF [10]. Hexane is theta solvent at 19°C [11]

Electrical Properties:
Electrical Properties General: Hole transport predominates in films [12]. Polymer exhibits electroluminescence in UV and in visible region (a broad band in the region of 400–450nm), and has been used as the emissive layer in LED devices with turn-on voltage of 16 V [13,14]. Ionisation potential 5.83 eV [1].

Optical Properties:
Optical Properties General: Exhibits thermochromism with a red shift in absorption on reducing the temperature in solution [2,3,4], or by increasing the pressure [4,5] on films with the absorption maximum at 315nm shifting to 355nm. This partly reversible effect is attributed to a change in backbone conformation [3,4,5]. Films on heating to 70°C exhibit a broadening of the absorption centred at *ca.* 315nm with very little shift of λ_{max} [2].

Photoluminescence is exhibited with a principal band for films in the UV region (λ_{max} 340nm) and a much weaker band in the visible region [3,13,15].
Transmission and Spectra: Pmr [10], cmr [2,10], Si-29 nmr [2,7], Raman [5], uv and photoluminescence [2,3,5,7,13,15], electroluminescence [13,14] and x-ray diffraction [2,7,8,15] spectral data have been reported.

Polymer Stability:
Polymer Stability General: Photochemical degradation occurs on prolonged exposure to UV radiation by scission of σ-bonds. The degradation and trapped products have been identified [10,16].

Applications/Commercial Products:
Processing & Manufacturing Routes: Prod. by Wurtz dehalogenation coupling of dibutyldichlorosilane monomer with Na in a solvent such as toluene [8,10,11].
Applications:

Bibliographic References
[1] Fukushima, M., Hamada, Y., Tabei, E., Aramata, M., Mori, S. and Yamamoto, Y., *Synth. Met.*, 1998, **94**, 299
[2] Schilling, F.C., Lovinger, A.J., Zeigler, J.M., Davis, D.D. and Bovey, F.A., *Macromolecules*, 1989, **22**, 3055
[3] Itoh, T. and Mita, I., *Macromolecules*, 1992, **25**, 479
[4] Walsh, C.A., Schilling, F.C., Macgregor, R.B., Lovinger, A.J., Davis, D.D., Bovey, F.A. and Zeigler, J.M., *Synth. Met.*, 1989, **28**, C559
[5] Song, K., Miller, R.D., Wallraff, G.M. and Rabolt, J.F., *Macromolecules*, 1991, **24**, 4084
[6] Furukawa, S., Takeuchi, K. and Shimana, M., *J. Phys.: Condens. Matter*, 1994, **6**, 11007
[7] Walsh, C.A., Schilling, F.C., Lovinger, A.J., Davis, D.D., Bovey, F.A. and Zeigler J.M., *Macromolecules*, 1990, **23**, 1742
[8] Hu, Z., Zhang, F., Huang, X., Zhang, M. and He, T., *Macromolecules*, 2004, **37**, 3310
[9] Hu, Z., Du, B., Zhang, F., Xie, F. and He, T., *Polymer*, 2002, **43**, 6005
[10] Trefonas, P., West, R. and Miller, R.D., *J. Am. Chem. Soc.*, 1985, **107**, 2737
[11] Kato, H., Sasanuma, Y., Kaito, A., Tanigaki, N., Tanabe, Y. and Kinugasa, S., *Macromolecules*, 2001, **34**, 262
[12] Abkowitz, M. and Stolka, M., *Solid State Commun.*, 1991, **78**, 269
[13] Suzuki, H., Hoshino, S., Yuan, C.-H., Fujiki, M., Toyoda, S. and Matsumoto, N., *Thin Solid Films*, 1998, **331**, 64
[14] Hoshino, S., Suzuki, H., Fujiki, M., Morita, M. and Matsumoto, N., *Synth. Met.*, 1997, **89**, 221
[15] Winokur, M.J. and West, R., *Macromolecules*, 2003, **36**, 7338
[16] Karatsu, T., Miller, R.D., Sooriyakumaran, R. and Michl, J., *J. Am. Chem. Soc.*, 1989, **111**, 1141

Poly[1,1-dichloroethylene bis(4-phenyl) carbonate]   P-173

Synonyms: *Poly[oxycarbonyloxy-1,4-phenylene(dichloroethenylidene)-1,4-phenylene]. Poly(4,4′-dihydroxydiphenyl-1,1-dichloroethylene-co-carbonic dichloride). Chloral polycarbonate*
Related Polymers: Copolymers of Bis-phenol dichloroethylene with Bisphenol A
Monomers: 4,4′-Dihydroxydiphenyl-1,1-dichloroethylene, Carbonic dichloride
Material class: Thermoplastic
Polymer Type: polycarbonates (miscellaneous)
CAS Number:

CAS Reg. No.
31546-39-1

Molecular Formula: $(C_{15}H_8Cl_2O_3)_n$
Fragments: $C_{15}H_8Cl_2O_3$

Volumetric & Calorimetric Properties:
Density:

No.	Value	Note
1	d^{30} 1.39 g/cm^3	[3,5]
2	d^{23} 1.414 g/cm^3	ASTM D1505-79 [6]
3	d 1.4 g/cm^3	[7]

Equation of State: A generalised Flory/Van der Waals equation of state for the polymer melt has been reported [8]. Tait [2], Sanchez-Lacombe [2] and Mark-Houwink [12,13] equation of state data have been reported
Glass-Transition Temperature:

No.	Value	Note
1	166°C	[9]
2	164.5°C	[10]
3	164°C	[11]
4	165°C	[1,4]
5	163°C	[2]
6	168°C	[7]

Deflection Temperature:

No.	Value	Note
1	144°C	1.8 MPa, (264 psi), ASTM D648 [7]

Surface Properties & Solubility:
Solubility Properties General: Forms miscible blends with Bisphenol A Polycarbonate [1], tetramethylpolycarbonate [1] and poly(methyl) methacrylate), which shows a lower critical soln. temp at higher temps. [1,2] Cloud points for blends with poly(methyl methacrylate) have been reported [3] Immiscible with polystyrene [1] and hexafluoropolycarbonate [2]
Cohesive Energy Density Solubility Parameters: Solubility parameter 23.04 J$^{1/2}$ cm$^{-3/2}$ (11.24 cal$^{1/2}$ cm$^{-3/2}$, calc) [2] Cohesive energy density 427 J m^{-3} (calc) [4]

Transport Properties:
Melt Flow Index:

No.	Value	Note
1	0.8–0.9 g/10 min	ASTM D1238, procedure A, condition O [7]

Permeability of Gases: Conditioning at 20 atm for 300h makes little difference to the permeability of CO_2 [14]. Diffusivity constants CO_2 13 × 10^{-9} cm^2 s^{-1}, 35°, 1–2 atm [3], 14 × 10^{-9} cm^2 s^{-1}, 35°, 1–2 atm [14]; CH_4 30 × 10^{-10} cm^2 s^{-1}, 35°, 1–2 atm [3,14]; O_2 40 × 10^{-9} cm^2 s^{-1}, 35°, 2 atm [3,16]; N_2 12 × 10^{-9} cm^2 s^{-1}, 35°, 2 atm [3,16]; Ar 16 × 10^{-9} cm^2 s^{-1}, 35°, 1–2 atm [3]
Permeability of Liquids: [H_2O] 7.5 g (100 in^2 day)$^{-1}$, 38°, 90% relative humidity [7]
Water Absorption:

No.	Value	Note
1	0.26%	25° [7]

Gas Permeability:

No.	Gas	Value	Note
1	[CO_2]	490.25 cm^3 mm/ (m^2 day atm)	5.6 × 10^{-13} cm^2 (s Pa)$^{-1}$, 35°, 1–2 atm [3]
2	[CO_2]	455.23 cm^3 mm/ (m^2 day atm)	5.2 × 10^{-13} cm^2 (s Pa)$^{-1}$, 35°, 1–2 atm [14]
3	[CO_2]	402.91 cm^3 mm/ (m^2 day atm)	4.6 × 10^{-13} cm^2 (s Pa)$^{-1}$, 35°, 5 atm [15]

– Poly[1,1-dichloroethylene bis(4-phenyl) carbonate]

4	[CO_2]	323.96 cm^3 mm/(m^2 day atm)	3.7×10^{-13} cm^2 (s Pa)$^{-1}$, 35°, 20 atm [15]
5	[CH_4]	176.5 cm^3 mm/(m^2 day atm)	20.16×10^{-14} cm^2 (s Pa)$^{-1}$, 35°, 1–2 atm [3]
6	[CH_4]	18.38 cm^3 mm/(m^2 day atm)	2.1×10^{-14} cm^2 (s Pa)$^{-1}$, 35°, 1–2 atm [14]
7	[CH_4]	15.76 cm^3 mm/(m^2 day atm)	1.8×10^{-14} cm^2 (s Pa)$^{-1}$, 35°, 5 atm [15]
8	[CH_4]	14.88 cm^3 mm/(m^2 day atm)	1.7×10^{-14} cm^2 (s Pa)$^{-1}$, 35°, 20 atm [15]
9	[O_2]	96.3 cm^3 mm/(m^2 day atm)	1.10×10^{-13} cm^2 (s Pa)$^{-1}$, 25°, 1 atm [9]
10	[O_2]	63.9 cm^3 mm/(m^2 day atm)	7.3×10^{-14} cm^2 (s Pa)$^{-1}$, 25° [7]
11	[O_2]	96.3 cm^3 mm/(m^2 day atm)	1.10×10^{-13} cm^2 (s Pa)$^{-1}$, 35°, 2 atm [3,16]
12	[N_2]	13.1 cm^3 mm/(m^2 day atm)	1.5×10^{-14} cm^2 (s Pa)$^{-1}$, 25° [7]
13	[N_2]	18.4 cm^3 mm/(m^2 day atm)	2.1×10^{-14} cm^2 (s Pa)$^{-1}$, 35°, 2 atm [16]
14	[N_2]	16.63 cm^3 mm/(m^2 day atm)	1.9×10^{-14} cm^2 (s Pa)$^{-1}$, 35°, 1–2 atm [3]
15	[Ar]	41.14 cm^3 mm/(m^2 day atm)	4.7×10^{-14} cm^2 (s Pa)$^{-1}$, 35°, 1–2 atm [3]
16	[He]	822.92 cm^3 mm/(m^2 day atm)	9.4×10^{-13} cm^2 (s Pa)$^{-1}$, 35°, 1–2 atm [3]
17	[H_2]	857.94 cm^3 mm/(m^2 day atm)	9.8×10^{-13} cm^2 (s Pa)$^{-1}$, 35°, 1–2 atm [3]
18	[H_2O] vapour	15670 cm^3 mm/(m^2 day atm)	1.79×10^{-11} cm^2 (s Pa)$^{-1}$, 25° [7]
19	[Freon 22]	39395 cm^3 mm/(m^2 day atm)	4.5×10^{-11} cm^2 (s Pa)$^{-1}$, 25° [7]
20	[Freon 13]	39395 cm^3 mm/(m^2 day atm)	4.5×10^{-11} cm^2 (s Pa)$^{-1}$, 25° [7]

Mechanical Properties:

Mechanical Properties General: The variation of tensile storage and loss moduli and loss tangent with temp. (-150°–200°, 110 Hz) has been reported [5]. The variation of loss tangent and storage modulus with temp. (-150°–200°, 1 Hz) has been reported [11]. The variation of loss tangent with temp. (-160°–160°, 3 Hz) has been reported [17]

Tensile (Young's) Modulus:

No.	Value	Note
1	2570 MPa	[4]

Flexural Modulus:

No.	Value	Note
1	2590 MPa	376 kpsi, ASTM D790 [7]

Tensile Strength Break:

No.	Value	Note	Elongation
1	79.3 MPa	11.5 kpsi, ASTM D638 [7]	70–113%

Flexural Strength at Break:

No.	Value	Note
1	111.7 MPa	16.2 kpsi, ASTM D790 [7]

Tensile Strength Yield:

No.	Value	Note
1	69 MPa	[4]
2	76.5 MPa	11.1 kpsi, ASTM D638 [7]

Impact Strength: Gardner impact strength 30 J (min., 320 in Ib^{-1}) [7]
Failure Properties General: Critical crazing strain in air 2.0%, (23°, 20h) [4]
Izod Notch:

No.	Value	Notch	Note
1	750–850 J/m	Y	14–16 ft Ib in^{-1}, ASTM D256 [7]

Optical Properties:
Refractive Index:

No.	Value	Note
1	1.61	25° [7]
2	1.61	calc. [1]

Polymer Stability:

Decomposition Details: Decomposition is a two step process: heating below 550° produces volatile tars and fuel gases and primary char; heating this primary char above 550° produces a carbonaceous residue, best described as a conglomerate of loosely linked small graphitic regions [18]. Thermal oxidation is a two stage process, depending on temp. At temp. up to 450°, thermo-oxidative pyrolysis occurs, with a predominance of exothermic scission reactions. At higher temps., exothermic oxidation occurs [10]. The activation energies for the two processes are 158 kJ mol^{-1} (up to 450°), 72 kJ mol^{-1} (above 450°) [10]. Maximum rate of decomposition (pyrolysis) occurs at 438° and for oxidation at 612° [10]. Decomposition in N_2 or He is a single step process [10]
Flammability: Has excellent flame resistance. Oxygen index 56 (ASTM D2863) [7]. Flammability rating 5V (UL94) [7]. Radiant panel flame spread index 0 (ASTM E162, 1/8") [7]. Smoke test NBS chamber: 67 (flaming, 1/8", NFPA No. 258), 1 (non flaming, 1/8", NFPA No.258) [7]. Arapahoe smoke chamber 0.8% smoke (based on amount burned) [7]
Chemical Stability: Tertiary amines e.g. triethylamine, catalyse conversion into trimeric cyclic carbonate [19]
Recyclability: Is relatively easily degraded by activated sewage sludge (after hydrol. of the polymer). The rate of biodegradation is greatly increased by the addition of nutrients [7]

Applications/Commercial Products:

Processing & Manufacturing Routes: Synth. by interfacial polycondensation: phosgene in $CH_2 Cl_2$ is introduced into a soln. of Bisphenol C in aq. sodium hydroxide at 20–30°. A catalyst, e.g. triethylamine, is used to achieve high MW's [7]

Bibliographic References

[1] Kim, C.K. and Paul, D.R., *Macromolecules*, 1992, **25**, 3097, (miscibility, T_g, refractive index)
[2] Kim, C.K. and Paul, D.R., *Polymer*, 1992, **33**, 4929, 4941, (miscibility, solubility parameter, equation of state, T_g)
[3] Chiou, J.S. and Paul, D.R., *J. Appl. Polym. Sci.*, 1987, **33**, 2935, (miscibility, density, gas permeability)

[4] Kambour, R.P., *Polym. Commun.*, 1983, **24**, 292, (cohesive energy density, T_g, tensile props, mechanical failure)
[5] McHattie, J.S., Koros, W.J. and Paul, D.R., *J. Polym. Sci., Polym. Phys. Ed.*, 1991, **29**, 731, (density, complex moduli)
[6] Dobkowski, Z. and Grzelak, D., *Eur. Polym. J.*, 1984, **20**, 1045, (density)
[7] Factor, A. and Orlando, C.M., *J. Polym. Sci., Polym. Chem. Ed.*, 1980, **18**, 579, (props., synth.)
[8] Brannock, G.R. and Sanchez, I.C., *Macromolecules*, 1993, **26**, 4970, (equation of state)
[9] Schmidhauser, J.C. and Longley, K.L., *J. Appl. Polym. Sci.*, 1990, **39**, 2083, (T_g, gas permeability)
[10] Dobkowski, Z. and Rudnik, E., *J. Therm. Anal.*, 1992, **38**, 2211, (T_g, thermal stability)
[11] Yee, A.F. and Smith, S.A., *Macromolecules*, 1981, **14**, 54, (T_g, complex moduli)
[12] Horbach, A., Muller, H. and Bottenbruch, L., *Makromol. Chem.*, 1981, **82**, 2873, (soln. props.)
[13] Dobkowski, Z., *J. Appl. Polym. Sci.*, 1984, **29**, 2683, (soln. props.)
[14] Raymond, P.C. and Paul, D.R., *J. Polym. Sci., Polym. Phys. Ed.*, 1990, **28**, 2103, 2213, (gas permeability)
[15] Raymond, P.C., Koros, W.J. and Paul, D.R., *J. Membr. Sci.*, 1993, **77**, 49, (gas permeability)
[16] Koros, W.J., Coleman, M.R. and Walker, D.R.B., *Annu. Rev. Mater. Sci.*, 1992, **22**, 47, (gas permeability)
[17] de los Santos Jones, H., Liu, Y., Inglefield, P.T., Jones, A.A. et al, *Polymer*, 1994, **35**, 57, (complex moduli)
[18] Factor, A., *ACS Symp. Ser.*, 1990, **425**, 274, (thermal stability)
[19] Hallgren, J.E. and Matthews, R.O., *J. Polym. Sci., Polym. Chem. Ed.*, 1979, **17**, 3781, (chemical stability)

Poly[[1,1-dichloroethylene bis(4-phenyl)carbonate]-*co*-[2,2-propane bis(4-phenyl)carbonate] P-174

Synonyms: *Bisphenol-dichloroethylene-Bisphenol A copolycarbonates. Poly[oxycarbonyloxy-1,4-phenylene(dichloroethenylidene)-1,4-phenyleneoxycarbonyloxy-1,4-phenylene(1-methylethylidene)-1,4-phenylene]*
Related Polymers: Poly[1,1-dichloroethylenebis(4-phenyl)carbonate], Poly[2,2-propanebis(4-phenyl)carbonate]
Monomers: Bisphenol A, 4,4′-Dihydroxydiphenyl-1,1-dichloroethylene, Carbonic dichloride
Material class: Thermoplastic, Copolymers
Polymer Type: polycarbonates (miscellaneous), bisphenol A polycarbonate
CAS Number:

CAS Reg. No.
109759-93-5

Molecular Formula: $[(C_{15}H_8Cl_2O_3).(C_{16}H_{14}O_3)]_n$
Fragments: $C_{15}H_8Cl_2O_3$ $C_{16}H_{14}O_3$

Volumetric & Calorimetric Properties:
Density:

No.	Value	Note
1	d 1.39 g/cm^3	30° [6]
2	d 1.414 g/cm^3	23°, ASTM D1505-79 [7]
3	d 1.4 g/cm^3	[8]

Equation of State: Flory/Van der Waals data [9]. Tait and Sanchez-Lacombe equation
Glass-Transition Temperature:

No.	Value	Note
1	166°C	[10]
2	165°C	[1]
3	168°C	[8]
4	164.5°C	[11]
5	165°C	[5]
6	164°C	[12]
7	163°C	[2]

Deflection Temperature:

No.	Value	Note
1	144°C	1.8 MPa, 264 psi, ASTM D648 [8]
2	121°C	5% chloral BPA [8]
3	130°C	10% chloral BPA [8]
4	120°C	25% chloral BPA [8]
5	127°C	50% chloral BPA [8]
6	141°C	85% chloral BPA [8]

Surface Properties & Solubility:
Solubility Properties General: Compatibilities with other polymers: Chloral polycarbonate forms miscible blends with Bisphenol A polycarbonate and tetramethylpolycarbonate [1]. Blends with polymethylmethacrylate, show a lower critical soln. temp. at higher temps. [1,2]. Cloud points: 223° (10 wt % chloral polycarbonate/90% polymethylmethacrylate) [3]; 217° (35 wt % chloral polycarbonate/65% polymethylmethacrylate) [3]; 258° (90 wt % chloral polycarbonate/10% polymethylmethacrylate) [3]. Forms immiscible blends with polystyrene [1,2], and hexafluoropolycarbonate [2]
Cohesive Energy Density Solubility Parameters: Solubility parameter 23.04 J$^{1/2}$ cm$^{-3/2}$ (11.24 cal$^{1/2}$ cm$^{-3/2}$, calc.) [2,4]. Cohesive energy density 427 J m^{-3} (calc.) [5]
Solvents/Non-solvents: Sol. CH_2Cl_2, $CHCl_3$, THF [8]; insol. H_2O [8]

Transport Properties:
Melt Flow Index:

No.	Value	Note
1	0.8–0.9 g/10 min	ASTM D1238, procedure A, condition O [8]
2	1.8 g/10 min	85% chloral BPA, ASTM D1238, procedure A, condition O [8]

Polymer Solutions Dilute: Mark-Houwink constants have been reported. [13,14] Intrinsic viscosity for a solution of the 85% chloral BPA copolymer in THF is given by the Staudinger-Kuhn equation [8]
Permeability of Gases: Conditioning at 20 atm for 300h makes little difference to the permeability of CO_2 [17]. Gas selectivity [15,16] and diffusivity [3,15,18] have been reported
Water Absorption:

No.	Value	Note
1	0.26%	25° [8]

Gas Permeability:

No.	Gas	Value	Note
1	CO_2	490 cm^3 mm/ (m^2 day atm)	5.6 × 10^{-13} cm^2 (s Pa)$^{-1}$, 35°, 1–2 atm [3,15]
2	CO_2	402 cm^3 mm/ (m^2 day atm)	4.6 × 10^{-13} cm^2 (s Pa)$^{-1}$, 35°, 5 atm [16]

3	CO_2	324 cm³ mm/ (m² day atm)	3.7×10^{-13} cm² (s Pa)$^{-1}$, 35°, 20 atm [16]
4	CH_4	20.1 cm³ mm/ (m² day atm)	2.3×10^{-14} cm² (s Pa)$^{-1}$, 35°, 1–2 atm [5,15]
5	CH_4	15.75 cm³ mm/ (m² day atm)	1.8×10^{-14} cm² (s Pa)$^{-1}$, 35°, 1–2 atm [16]
6	O_2	96.3 cm³ mm/ (m² day atm)	1.1×10^{-13} cm² (s Pa)$^{-1}$, 35°, 1–2 atm [3,10,18]
7	O_2	63.9 cm³ mm/ (m² day atm)	7.3×10^{-14} cm² (s Pa)$^{-1}$, 25° [8]
8	N_2	13.1 cm³ mm/ (m² day atm)	1.5×10^{-14} cm² (s Pa)$^{-1}$, 25° [8]
9	N_2	18.4 cm³ mm/ (m² day atm)	8.1×10^{-14} cm² (s Pa)$^{-1}$, 35°, 1–2 atm [3,18]
10	Freon	3.94 cm³ mm/ (m² day atm)	4.5×10^{-15} cm² (s Pa)$^{-1}$, 25° [8]
11	Ar	41.1 cm³ mm/ (m² day atm)	4.7×10^{-14} cm² (s Pa)$^{-1}$, 35°, 1–2 atm [3]
12	He	823 cm³ mm/ (m² day atm)	9.4×10^{-13} cm² (s Pa)$^{-1}$, 35°, 1–2 atm [3]
13	H_2	856 cm³ mm/ (m² day atm)	9.8×10^{-13} cm² (s Pa)$^{-1}$, 35°, 1–2 atm [3]

Mechanical Properties:

Tensile (Young's) Modulus:

No.	Value	Note
1	2570 MPa	[5]

Flexural Modulus:

No.	Value	Note
1	2590 MPa	376 kpsi, ASTM D790 [8]
2	2300 MPa	340 kpsi, 5% Bisphenol C [8]
3	2400 MPa	350 kpsi, 10% Bisphenol C [8]
4	2400 MPa	350 kpsi, 25% Bisphenol C [8]
5	2500 MPa	360 kpsi, 50% Bisphenol C [8]
6	2500 MPa	360 kpsi, 85% Bisphenol C [8]

Tensile Strength Break:

No.	Value	Note	Elongation
1	79.3 MPa	11.5 kpsi, ASTM D638 [8]	70–113%
2	68.7 MPa	9971 psi, 5% Bisphenol C [8]	95%
3	80.3 MPa	11644 psi, 10% Bisphenol C [8]	115%
4	71.4 MPa	10364 psi, 25% Bisphenol C [8]	87%
5	59.5 MPa	8625 psi, 50% Bisphenol C [8]	37%
6	77.9 MPa	11.3 kpsi, 85% Bisphenol C [8]	61%

Flexural Strength at Break:

No.	Value	Note
1	111.7 MPa	16.2 kpsi, ASTM D790 [8]
2	103 MPa	15 kpsi, 5–10% Bisphenol C [8]
3	110 MPa	16 kpsi, 25–85% Bisphenol C [8]

Tensile Strength Yield:

No.	Value	Elongation	Note
1	69 MPa		[5]
2	76.5 MPa		11.1 kpsi, ASTM D638 [8]
3	63.4 MPa	10.2%	9200 psi, 5% Bisphenol C [8]
4	62.5 MPa	10.7%	9067 psi, 10% Bisphenol C [8]
5	63.9 MPa	10.7%	9265 psi, 25% Bisphenol C [8]
6	65.9 MPa	11.4%	9559 psi, 50% Bisphenol C [8]
7	69.6 MPa	11.0%	10.1 kpsi, 85% Bisphenol C [8]

Complex Moduli: Variations of E′, E″ [6], tan δ [12,19] and G′ [19] with temp. (-150–200°) have been reported
Impact Strength: Gardner impact strength 30 J (min. 320 in lb) [8]
Failure Properties General: Critical crazing strain 2.0% (air, 23°, 20h) [5]
Friction Abrasion and Resistance: Taber abrasion resistance 29.2% haze (100 cycles)$^{-1}$ (90% chloral BPA) [8]

Izod Notch:

No.	Value	Notch	Note
1	750–850 J/m	Y	14–16 ft lb in^{-1}, ASTM D256 [8]
2	160 J/m	Y	3 ft lb in^{-1}, 5% Bisphenol C [8]
3	1010 J/m	Y	19 ft lb in^{-1}, 5% Bisphenol C [8]
4	960 J/m	Y	18 ft lb in^{-1}, 10% Bisphenol C [8]
5	53 J/m	Y	2 ft lb in^{-1}, 25% Bisphenol C [8]
6	910 J/m	Y	17 ft lb in^{-1}, 25% Bisphenol C [8]
7	910 J/m	Y	17 ft lb in^{-1}, 50% Bisphenol C [8]
8	850 J/m	Y	16 ft lb in^{-1}, 85% Bisphenol C [8]

Optical Properties:
Refractive Index:

No.	Value	Note
1	1.61	25° [1,8]

Polymer Stability:
Decomposition Details: Decomposition by two step process; heating below 550° produces volatile tars and fuel gases and a primary char. Further heating above 550° produces a carbonaceous residue, a conglomerate of loosely linked small graphitic regions [20]
Flammability: Excellent flame resistance. Oxygen index 56 (ASTM D2963); 50 (85% Bisphenol C, ASTM D2863) [8]. Flammability rating 5V (85–90% Bisphenol C, UL 94) [8]. Radiant panel flame spread index (Is) 0 (ASTM E162, ⅛ in); 11 (85% Bisphenol C, ⅛ in, ASTM E162) [8]. Smoke density 67 (Flaming, ⅛ in, NFPA No. 258, NBS chamber) [8]; 1 (Nonflaming, ⅛ in, NFPA No. 258, NBS chamber); 19 (90% Bisphenol C, flaming, ⅛ in, NFPA No. 258, NBS chamber); 31 (85% Bisphenol C, flaming, ⅛ in, NFPA No. 258, NBS chamber); 10 (85% Bisphenol C, non flaming, ⅛ in, NFPA No. 258, NBS chamber) [8]. Arapahoe smoke chamber 0.8% smoke; 1.7% smoke (85% Bisphenol C, based on amount burned) [8] Smoke density rating 4.5 (85% Bisphenol C, XP-2 chamber, ASTM D2843). [8] Flame spread rating 20 (85% Bisphenol C, ¼ in, ASTM E84, 25ft tunnel test) [8]; smoke rating 200 (85% Bisphenol C, ¼ in, ASTM E84) [8]; fuel contribution value 0 (85% Bisphenol C, ¼ in, ASTM E84) [8]
Chemical Stability: Thermal oxidation is a two stage process. At temps. up to 450°, thermo-oxidative pyrolysis occurs, with a predominance of exothermic scission reactions. At higher temps. exothermic oxidation takes place [11]. Max. rate of decomposition: 438° (pyrolysis, heating rate 10° min^{-1}) [11]; 612° (oxidation,

broad peak, heating rate 10° min^{-1} [11]. With N_2 or He only one step decomposition is found [11]. Tertiary amines catalyse conversion into the trimeric cyclic carbonate e.g. triethylamine precipitates the cyclic trimer from a soln. in CH_2Cl_2 in 24h (room tem., N_2) [21]
Recyclability: Relatively easily degraded by activated sewage sludge (after hydrolysis of the polymer). The rate of biodegradation is greatly increased by the addition of nutrients [8]
Stability Miscellaneous: Not toxic oral route of inhalation

Applications/Commercial Products:
Processing & Manufacturing Routes: Synth. by interfacial polycondensation. Phosgene is introduced into a soln. of Bisphenol C and Bisphenol A in aq. NaOH at 20–30°, in an inert solvent (CH_2Cl_2). Phenol is used as a chain stopper; triethylamine catalyst is used to achieve high MW. Isol. by steam precipitation [8]

Bibliographic References
[1] Kim, C.K. and Paul. D.R., *Macromolecules*, 1992, **25**, 3097, (polymer-polymer compatibilities, T_g, refractive index)
[2] Kim, C.K. and Paul, D.R., *Polymer*, 1992, **33**, 4929, (polymer-polymer compatibilities, solubility parameter, equation of state, T_g)
[3] Chiou, J.S. and Paul, D.R., *J. Appl. Polym. Sci.*, 1987, **33**, 2935, (polymer-polymer compatibilities, density, gas permeability)
[4] Kim, C.K. and Paul, D.R., *Polymer*, 1992, **33**, 4941, (solubility parameter)
[5] Kambour, R.P., *Polym. Commun.*, 1983, **24**, 292, (cohesive energy density, T_g, tensile props, mech failure)
[6] McHattie, J.S., Koros, W.J. and Paul, D.R., *J. Polym. Sci., Polym. Phys. Ed.*, 1991, **29**, 731, (density, complex moduli)
[7] Dobkowski, A. and Grzelak, D., *Eur. Polym. J.*, 1984, **20**, 1045, (density)
[8] Factor, A. and Orlando, C.M., *J. Polym. Sci., Polym. Chem. Ed.*, 1980, **18**, 579, (props, synth)
[9] Brannock, G.R. and Sanchez, I.C., *Macromolecules*, 1993, **26**, 4970, (equation of state)
[10] Schmidhauser, J.C. and Longley, K.L., *J. Appl. Polym. Sci.*, 1990, **39**, 2083, (T_g, gas permeability)
[11] Dobkowski, Z. and Rudnik, E., *J. Therm. Anal.*, 1992, **38**, 2211, (T_g, thermal stability)
[12] Yee, A.F. and Smith, S.A., *Macromolecules*, 1981, **14**, 54, (T_g, complex moduli)
[13] Horbach, A., Muller, H. and Bottenbruch, L., *Makromol. Chem.*, 1981, **82**, 2873, (dilute soln props)
[14] Dobowski, Z., *J. Appl. Polym. Sci.*, 1984, **29**, 2683, (dilute soln props)
[15] Raymond, P.C. and Paul, D.R., *J. Polym. Sci., Polym. Phys. Ed.*, 1990, **28**, 2103, (gas permeability)
[16] Raymond, P.C., Koros, W.J. and Paul, D.R., *J. Membr. Sci.*, 1993, **77**, 49, (gas permeability)
[17] Raymond, P.C. and Paul, D.R., *J. Polym. Sci., Polym. Phys. Ed.*, 1990, **28**, 2213, (gas permeability)
[18] Koros, W.J., Coleman, M.R. and Walker, D.R.B., *Annu. Rev. Mater. Sci.*, 1992, **22**, 47, (gas permeability)
[19] de los Santos Jones, H., Liu, Y., Inglefield, P.T., Jones, A.A., Kim, C.K. and Paul, D.R., *Polymer*, 1994, **35**, 57, (complex moduli)
[20] Factor, A., *ACS Symp. Ser.*, 1990, **425**, 274, (thermal stability)
[21] Hallgren, J.E. and Matthews, R.O., *J. Polym. Sci., Polym. Chem. Ed.*, 1979, **17**, 3781, (acid/base stability)

Poly(didecylsilane)

$$\left[\begin{array}{c} CH_2(CH_2)_8CH_3 \\ | \\ Si \\ | \\ CH_2(CH_2)_8CH_3 \end{array} \right]_n$$

Synonyms: *Poly(didecylsilylene). Poly(di-n-decylsilane). PdDecSi. PD10S. PDDS*
Related Polymers: Polycarbosilanes, Poly(methyl phenyl silylene methylene), Dichloropoly(methylphenylsilane), Poly(methylphenylsilane)
Polymer Type: polysilanes

CAS Number:

CAS Reg. No.
107999-69-9

Molecular Formula: $(C_{20}H_{42}Si)n$
Fragments: $C_{20}H_{42}Si$
Molecular Weight: 412,000–610,000 [2,3,7]
General Information: Five distinct ordered forms (crystalline or semicrystalline), which vary in their respective chain structure, packing and optical properties. Of these only one appears to be thermodynamically stable. One of the metastable polymorphic structures exhibits an unprecedented tripling of the basic unit cell structure [2]
Morphology: Some disagreement exists in the literature for data on PDDS [5].
It has been suggested that the Si backbone is dominated by T56 and D conformers with a TnD-Tn'D+ construction [2].
Unit cell with dimensions a=21.4Å; b=24.9Å; c=7.8Å and γ=90° [1]
Miscellaneous: There is also strong evidence indicative of a crossover in the intrachain and interchain ordering of the alkyl side chains when progressing from a hexyl to a decyl moiety. The actual thermodynamically stable form should be the low-temperature type II phase [2,4,5]

Volumetric & Calorimetric Properties:
Density:

No.	Value	Note
1	d 0.96 g/cm^3	PdDecSi I [2]
2	d 0.98 g/cm^3	PdDecSi II* [2]
3	d 0.97 g/cm^3	PdDecSi II [2]
4	d 1.02 g/cm^3	PdDecSi III [2]
5	d 0.91 g/cm^3	PdDecSi QM [2]

Thermodynamic Properties General: See [3]
Melting Temperature:

No.	Value	Note
1	7°C	Freezing temp. [7]

Electrical Properties:
Electrical Properties General: KAT 1950 × 10^{-9}S m^2/J; KTG 210 × 10^{-9}S m^2/J; M 28 × 10^{-9}S m^2/J
The highest mobility of 4x10^{-5}m^2/(V s) is found for the *all-trans* conformation of the polysilane backbone [5]

Optical Properties:
Optical Properties General: At least two different crystalline phases with λ_{max} of 350nm and 378nm have been observed; the solid state composition has been reported to vary with molecular weight and thermal history. Further heating into the mesophase leads to a uv absorption with λ_{max} of 318nm [3,5]
Transmission and Spectra: Uv [2,4,5,7], Raman [4], nmr [3,6,7], esr [6] and gc-ms [6] spectra reported.

Polymer Stability:
Thermal Stability General: Can be thermally cycled or solvent cast to form at least five nominally crystalline polymorphs. Referred to as type II*, I, II, III and QM (for quenched mesophase). Phases III and I appear to be monotropic while the type II structure is clearly enantiotropic. The QM form may be only representative of a kinetically limited structure and so has not been assigned. The QM phase can be warmed to sequentially give the type III > I > II structures [2,4]
Decomposition Details: The first is a broad endotherm centred near -22°C, and the second is an exotherm near -2°C. The first corresponds to the QM to III transition while the second is

suggestive of a cold crystallisation process in the transformation from the type III to I phase [2,4]

Applications/Commercial Products:
Processing & Manufacturing Routes: Prepared from dichlorodecylsilane by dehalogenation coupling with Na under standard conditions for Wurtz polysilane synthesis [2,3]
Applications:

Bibliographic References

[1] KariKari, E.K., Greso, A.J., Farmer, B.L., Miller, R.D. and Rabolt, R.F., *Macromolecules*, 1993, **26**, 3937
[2] Chunwachirasiri, W., West, R. and Winokur, M.J., *Macromolecules*, 2000, **33**, 9720
[3] Mueller, C., Frey, H. and Schmidt, C., *Monatsh. Chem.*, 1999, **130**, 175
[4] Bukalov, S.S., Teplitsky, M.V., Gordeev, Y.Y., Leites, L.A. and West, R., *Russ. Chem. Bull. (Engl. Transl.)*, 2003, **52**, 1066
[5] van der Laan, G.P, de Haas, M.P., Hummel, A., Frey, H. and Moller, M., *J. Chem. Phys.*, 1996, **100**, 5470
[6] McKinley, A.J., Karatsu, T., Wallraff, G.M., Miller, R.D., and Soorlyakumaran, R, *Organometallics*, 1988, **7**, 2567
[7] Seki, S., Koizumi, Y., Kawaguchi, T., Habara, H. and Tagawa, S., *J. Am. Chem. Soc.*, 2004, **126**, 3521

Poly(diethylsilane)   P-176

$$\left[\begin{array}{c}CH_2CH_3 \\ | \\ Si \\ | \\ CH_2CH_3\end{array}\right]_n$$

Synonyms: *Poly(diethylsilylene). Diethylsilane homopolymer, sru. PDES*
Related Polymers: Polycarbosilanes, Poly(methyl phenyl silylene methylene), Dichloropoly(methylphenylsilane), Poly(methylphenylsilane), Poly(dipropylsilane), Poly[[bis(3-methylphenyl)silylene]methylene], Poly[[bis(4-methylphenyl)silylene]methylene], Poly(dihexylsilane)
Polymer Type: polysilanes
CAS Number:

CAS Reg. No.
125457-34-3

Molecular Formula: $(C_4H_{10}Si)n$
Fragments: $C_4H_{10}Si$
Morphology: The Si backbone conformation of the crystal is *all-trans*, and the sidechain conformation is mainly determined by intramolecular steric hindrance. A monoclinic unit cell with a=1.108nm, b=1.210nm, c=0.399nm and γ=84.8° [3,4,5]

Surface Properties & Solubility:
Surface and Interfacial Properties General: From contact angle measurements using water and methylene iodide, surface tensions for solid films of poly(silylene) polymers were obtained using the geometric-mean method; value depends strongly on nature of substituents [6]

Optical Properties:
Optical Properties General: Soft X-ray emission and absorption spectra in the Si L region of a number of substituted polysilanes, $(SiR_2)n$, were obtained using synchrotron radiation to study their electronic structures. Studied polysilanes were substituted with Me (R=CH_3), Et, Pr, Bu, n-pentyl and Ph groups. Although similar spectral features in both X-ray emission and absorption are observed among alkyl-substituted polysilanes, slight differences are distinguished between alkyl- and phenyl-substituted ones. These spectral features are qualitatively reproduced by summing calculated d-of-state spectra for Si-3s- and Si-3d-orbitals. Thus, spectral features are explained through the hybridisation of electronic orbitals in both backbone Si atoms and substituent C atoms [2]. The length of alkyl substituents has little effect on the electronic structure of the Si backbone [6]
Transmission and Spectra: nmr [3] and uv-vis [1,3] spectra reported.

Polymer Stability:
Decomposition Details: Photolysis proceeds via silylene extrusion competing with homolytic cleavage [8]

Applications/Commercial Products:
Processing & Manufacturing Routes: PDES on smooth quartz plates has been prepared by the friction deposition technique [1]. Alsosynth. by Wurtz-type reductive coupling of corresponding dichlorosilane [5]
Applications: Used in fabricated electroluminescent device [5]

Bibliographic References

[1] Hajihedidari, D., Iran Nat. Chem. Eng. Congress., *8th*, 2003, 480/1
[2] Muramatsu, Y., Fujino, M., Yamamoto, T., Gullikson, E.M. and Perera, R.C.C., *Nucl. Instrum. Methods Phys. Res.*, *Sect. B*, 2003, **199**, 260
[3] Lovinger, A.J., Davis, D.D., Schilling, F., Bovey, F. A., Frederic, C. and Zeigler, J.M., *Polym. Commun.*, 1989, **30**, 356
[4] Furukawa, S., Takeuchit, K., Mizoguchit, M, Shimanat, M., and Tamurat, M., *J. Phys.: Condens. Matter*, 1993, **5**, L461
[5] Furukawa, S., *Thin Solid Films*, 1988, **331**, 222
[6] Crespo, R., Carmen, M.P. and Tomas, F., *J. Chem. Phys.*, 1994, **100**, 6953
[7] Fujisaka, T. and West, R., *J. Organomet. Chem.*, 1993, **449**, 105
[8] McKinley A.J., Karatsu T., Wallraff, G.M., Miller, R.D., Soorlyakumaran, R. and Michl, J., *Organometallics*, 1988, **7**, 2567

Poly(diheptylsilane)   P-177

$$\left[\begin{array}{c}CH_2(CH_2)_5CH_3 \\ | \\ Si \\ | \\ CH_2(CH_2)_5CH_3\end{array}\right]_n$$

Synonyms: *Poly(diheptylsilylene). PDHepS. PDHepSi. PdnHS*
Related Polymers: Polycarbosilanes, Poly(methyl phenyl silylene methylene), Dichloropoly(methylphenylsilane), Poly(methylphenylsilane)
Polymer Type: polysilanes
CAS Number:

CAS Reg. No.	Note
100044-91-5	homopolymer
100044-92-6	sru

Molecular Formula: $(C_{14}H_{30}Si)n$
Fragments: $C_{14}H_{30}Si$
Molecular Weight: 1.9×10^6 [6]
Morphology: Planar zigzag structure together with a small amount of a disordered component.
The *d*-spacing, 4.54Å, is characteristic of the packing of *trans*-planar alkyl side chains into a triclinic unit cell. It has a chain repeat distance of 13.8Å, corresponding to a 7/3 helical conformation.
Orthorhombic unit cell having dimensions a=14.7Å, b=27.4Å and c=4.0Å [1,2,3]

Volumetric & Calorimetric Properties:
Density:

No.	Value	Note
1	d 0.93 g/cm^3	[3]

Thermodynamic Properties General: The smaller enthalpy associated with the transition probably due to the conformation and packing [3]

– Poly(9,9-dihexylfluorene)

Melting Temperature:

No.	Value	Note
1	38–39°C	[1]

Transition Temperature:

No.	Value	Note	Type
1	142°C	at 3 Torr [6]	Bp

Surface Properties & Solubility:
Surface Tension:

No.	Value	Note
1	25–52.8 mN/m	dependent on the substituent groups [7]

Optical Properties:
Optical Properties General: Exhibits similar piezochromic behaviour to that found with PDBS and PDPS. At 42°C, the polymer undergoes a reversible structural transition, which has been observed by ir, Raman and X-ray diffraction [1,2,3]
Transmission and Spectra: Ir [1,2,3], Raman [1,3], uv [1,2,3,6,8] and nmr [3,6] spectra reported.

Polymer Stability:
Thermal Stability General: The long wavelength UV absorption of PDHepS rapidly disappears upon heating above 45°C and the original spectrum is restored. This thermal effect is reversible, as observed with PDHS [1]

Applications/Commercial Products:
Processing & Manufacturing Routes: n-Heptyltrichlorosilane was rapidly added to n-heptylmagnesium bromide in Et_2O at 0°C, allowed to warm to room temperature and stirred for 12h. The solution was refluxed for another 2h, and after filtration, the solution was concentrated and distilled three times [6]
Applications: Electrophotographic photoreceptor [4]. Effective mid-uv resists [5]

Bibliographic References
[1] Rabolt, J.F., Hofer, D., Miller, R.D. and Fickes, G.N., *Macromolecules,* 1986, **19**, 611
[2] Song, K., Miller, R.D., Wallraff, G.M and Rabolt, J.F., *Macromolecules,* 1991, **24**, 4084
[3] KariKari, E.K., Greso, A.J. and Farmer, B.L., *Macromolecules,* 1993, **26**, 3937
[4] *Jpn. Pat.,* 1990, 02 150 853
[5] *Res. Discl.,* 1987, **273**, 26
[6] Seki, S., Koizumi, Y., Kawaguchi, T., Habara, H. and Tagawa, S., *J. Am. Chem. Soc.,* 2004, **126**, 3521
[7] Fujisaka, T. and West, R., *J. Organomet. Chem.,* 1993, **449**, 105
[8] Schilling, F.C., Lovinger, A.J., Davis, D.D., Bovey, F.A. and Zeigler, J.M., *J. Inorg. Organomet. Polym.,* 1992, **2**, 47

Poly(9,9-dihexylfluorene) P-178

Synonyms: *Poly(9,9-dihexyl-9H-fluorene-2,7-diyl). PDHF. F6*
Material class: Thermoplastic
Polymer Type: polyfluorene

CAS Number:

CAS Reg. No.	Note
123863-98-9	Poly(9,9-dihexylfluorene)
133019-09-7	Poly(9,9-dihexyl-9H-fluorene-2,7-diyl)
201807-75-2	2,7-Dibromo-9,9-dihexyl-9H-fluorene homopolymer

Molecular Formula: $(C_{25}H_{32})_n$
Fragments: $C_{25}H_{32}$
General Information: Stable blue-light-emitting electroluminescent and photoluminescent polymer with liquid crystalline properties. Polymer is produced as yellow powder or flakes.
Morphology: Hexagonal unit cell with vectors $a_1=a_2=12.4$Å [1]. Liquid crystalline phase occurs above ca. 160°C [1,2].

Volumetric & Calorimetric Properties:
Glass-Transition Temperature:

No.	Value	Note
1	55–103°C	dependent on molecular weight [2,3,4]

Transition Temperature General: Transition to nematic phase at about 160–200°C. Liquid becomes isotropic from about 300°C [1,2].

Surface Properties & Solubility:
Solvents/Non-solvents: Sol. $CHCl_3$, toluene, CH_2Cl_2, THF, xylene, chlorobenzene [2,3,5]; insol. alcohols, Me_2CO, H_2O [6].

Mechanical Properties:
Elastic Modulus:

No.	Value	Note
1	16600–23200 MPa	thin film [7]

Hardness: 0.37 ± 0.07GPa (thin film) [7].

Electrical Properties:
Electrical Properties General: Several LED devices with PDHF as blue light-emitter have been reported, with light turn-on voltages up to 17V [8,9,10]. Very low conductivity for undoped polymer; conductivity of about 0.001S cm^{-1} for samples doped with tetrafluoroborate [6].

Optical Properties:
Optical Properties General: Blue-light photoluminescence occurs with λ_{max} in region of 410–420 nm and band gap of approx. 2.9eV [4,6]. High photoluminescent quantum efficiency of up to 90% in $CHCl_3$ solution [11]. Electroluminescence from LED devices with λ_{max} in region of 470–488nm [8,10].
Transmission and Spectra: pmr [3,4,9], cmr [3,4,9], ir [3,10,11], uv and photoluminescence [2,3,4,5,9,10,11,12,13], electroluminescence [8,9,10] and x-ray [1] spectral data have been reported.

Polymer Stability:
Polymer Stability General: Good stability in presence of oxygen, moisture and light [2]. The electrical and photoinduced degradation of films have been reported [12].
Thermal Stability General: Good thermal stability with 5% weight loss only above 400°C [5,9].

Applications/Commercial Products:
Processing & Manufacturing Routes: Prod. by oxidation of 9,9-dihexylfluorene monomer with $FeCl_3$ [3], Ni(0)-catalysed polymerisation of 2,7-dibromo-9,9-dihexylfluorene [9,11,14], or by Suzuki coupling of the dibromo monomer with diboronate derivative to give higher molecular-weight polymers [4,10].

Applications: Main potential application is as a blue-light emitter for light-emitting diodes, also for thin film transistors, lasers and light-emitting electrochemical cells.

Bibliographic References

[1] Kawana, S., Durrell, M., Lu, J., Macdonald, J.E., Grell, M., Bradley, D.D.C., Jukes, P.C., Jones, R.A.L. and Bennett, S.L., *Polymer,* 2002, **43**, 1907
[2] Teetsov, J. and Fox, M.A., *J. Mater. Chem.,* 1999, **9**, 2117
[3] Fukuda, M., Sawada, K. and Yoshino, K., *J. Polym. Sci., Part A: Polym. Chem.,* 1993, **31**, 2465
[4] Liu, B., Yu, W.-L., Lai, Y.-H. and Huang, W., *Chem. Mater.,* 2001, **13**, 1984
[5] Kreyenschmidt, M. Klaerner, G., Fuhrer, G., Ashenhurst, J. and Miller R.D., *Macromolecules,* 1998, **31**, 1099
[6] Fukuda, M., Sawada, K. and Yoshino, K., *Jpn. J. Appl. Phys.,* 1989, **28**, L1433
[7] Zeng, K., Chen, Z.-K., Shen, K. and Liu B., *Thin Solid Films,* 2005, **477**, 111
[8] Yoshida, M., Fujii, A., Ohmori, Y. and Yoshino, K., *Appl. Phys. Lett.,* 1996, **69**, 734
[9] Klärner, G., Lee, J.-I., Lee, V.Y., Chan, E., Chen, J.-P., Nelson, D. and Miller R.D., *Chem. Mater.,* 1999, **11**, 1800
[10] Assaka, A.M., Rodrigues, P.C., de Oliveira, A.R.M., Ding, L., Hu, B., Karasz, F.E. and Akcelrud, L., *Polymer,* 2004, **45**, 7071
[11] Stéphan, O. Tran-Van, F. and Chevrot, C., *Synth. Met.,* 2002, **131**, 31
[12] Bliznyuk, V.N., Carter, S.A., Scott, J.C., Klärner, G., Miller, R.D. and Miller D.C., *Macromolecules,* 1999, **32**, 361
[13] Cho, H.N., Kim, D.Y., Kim, J.K. and Kim, C.Y., *Synth. Met.,* 1997, **91**, 293
[14] Carter, K.R., *Macromolecules,* 2002, **35**, 6757

Poly(dihexylsilane)   P-179

$$\left[\begin{array}{c} CH_2(CH_2)_4CH_3 \\ -Si- \\ CH_2(CH_2)_4CH_3 \end{array}\right]_n$$

Synonyms: *Poly(dihexylsilylene).* Dichlorodihexylsilane homopolymer. *PDHS*
Monomers: Dichlorodihexylsilane
Material class: Thermoplastic
Polymer Type: polysilanes
CAS Number:

CAS Reg. No.	Note
94904-85-5	poly(dihexylsilylene)
97036-67-4	dichlorodihexylsilane homopolymer

Molecular Formula: $(C_{12}H_{26}Si)n$
Fragments: $C_{12}H_{26}Si$
Additives: Dopants such as I_2 and $FeCl_3$ have been added to increase conductivity.
General Information: Thermochromic polymer exhibiting photoluminescence and electroluminescence. Prod. as white solid.
Morphology: At r.t. in polymer films an *all-trans* planar conformation of the backbone is predominant [1,2,3] with ordered sidechains and with orthorhombic unit cell, a = 13.76, b = 23.86, c = 3.99 Å [4]. Above *ca.* 42°C the planar conformation is completely converted to a columnar mesophase conformation with disordered sidechains [3,4]. The effect of pressure is to distort the *all-trans* conformation at lower temperatures [5]. In very thin films (thickness <2000Å) crystallisation is hindered and the backbone and sidechains become disordered [6,7].

Volumetric & Calorimetric Properties:

Glass-Transition Temperature:

No.	Value	Note
1	-52.5°C	heat capacity data [9]
2	-33°C	hole transport data [10]

Transition Temperature General: The strong endothermic transition at 42°C corresponds to the change to a disordered conformation [1,3,8,9].

Surface Properties & Solubility:

Solvents/Non-solvents: Sol. THF, hexane [11], octane, toluene.

Mechanical Properties:

Mechanical Properties General: Polymer generally has low ductility with films breaking when under stress at the transition point (41°C) [12].
Elastic Modulus:

No.	Value	Note
1	126 MPa	[12]

Fracture Mechanical Properties: Fracture strain 2.7% [12].

Electrical Properties:

Electrical Properties General: Conductivity increased by dopants such as I_2 or $FeCl_3$ [13]. Hole transport occurs by chain hopping [10]. Electroluminescence observed only at low temperatures, with quantum yield decreasing to zero above 230K [14]. LED devices used with PDHS as emissive layer had a turn-on voltage of 26V [15]. Ionisation potential increases above transition temperature: 5.78eV (20°C), 5.94eV (50°C) [28].

Optical Properties:

Optical Properties General: For polymer films at r.t. there is a strong UV absorption band with λ_{max} *ca.* 370nm and a weak band with λ_{max} *ca.* 316nm. On heating to 42°C the longer wavelength band disappears and the higher energy band increases in intensity [1,2,3,8,16]. When films are subjected to pressure, the longer wavelength band can be restored above 42°C and the shorter wavelength band decreases in intensity [1,5]. Oriented films exhibit a reduction in absorption anisotropy for polarised UV light on going from the planar to disordered conformations [17]. When the thickness of films is decreased the longer wavelength band decreases relative to the shorter wavelength band [6]. Polymer solutions at r.t. exhibit only the shorter wavelength band (*ca.* 316nm) [2], while on cooling there is a transition to a longer wavelength band (λ_{max} *ca.* 254 nm) [18,19]. A third band has been detected on heating solutions from low temperatures [20]. A photoluminescence band occurs in UV region with λ_{max} *ca.* 375nm for films; the quantum yield increases at lower temperatures [14,15].

Refractive Index:

No.	Value	Note
1	1.55–1.62	Increase in refractive index on going from planar to disordered state on heating [21]

Transmission and Spectra: Pmr [22,23], cmr [22,23], Si-29 nmr [25], ir [2,5,8,23,24], Raman [3,5], uv [2,3,8,14,16,17,20], photoluminescence [14,15,16,24], electroluminescence [14,15] and x-ray diffraction [1,4,26,27] data have been reported.

Polymer Stability:

Polymer Stability General: Photoscission of chains occurs on exposure to uv radiation.
Thermal Stability General: Decomposes (5% weight loss) at 288°C [9].
Chemical Stability: Undergoes slow chemical degradation by alkali metal cations and K metal in THF solution [29].

Applications/Commercial Products:

Processing & Manufacturing Routes: Prod. mainly by Wurtz dehalogenation coupling of dichlorodihexylsilane with Na in a solvent such as toluene, THF or octane [22,23]. The addition of a crown ether increases polymer yield [30,31]. Polymer also produced by electrolytic methods [32].
Applications: Potential applications as optical modulating component and in photoaligned films.

Bibliographic References

[1] Schilling, F.C., Bovey, F.A., Davis, D.D., Lovinger, A.J., Macgregor, R.B., Walsh, C.A. and Zeigler, J.M., *Macromolecules*, 1989, **22**, 4645
[2] Rabolt, J.F., Hofer, D., Miller, R.D and Fickes, G.N., *Macromolecules*, 1986, **19**, 611
[3] Bukalov, S.S., Leites, L.A., Magdanurov, G.I. and West, R., *J. Organomet. Chem.*, 2003, **685**, 51
[4] Patnaik, S.S. and Farmer, B.L., *Polymer*, 1992, **33**, 4443
[5] Song, K., Kuzmany, H., Wallraff, G.M., Miller, R.D. and Rabolt, J.F., *Macromolecules*, 1990, **23**, 3870
[6] Despotopoulou, M.M., Frank, C.W., Miller, R.D. and Rabolt, J.F., *Macromolecules*, 1995, **28**, 6687
[7] Despotopoulou, M.M., Miller, R.D., Rabolt, J.F. and Frank, C.W., *J. Polym. Sci., Part B: Polym. Phys.*, 1996, **34**, 2335
[8] Miller, R.D., Hofer, D., Rabolt, J. and Fickes, G.N., *J. Am. Chem. Soc.*, 1985, **107**, 2172
[9] Varma-Nair, M., Cheng, J., Jin, Y. and Wunderlich, B., *Macromolecules*, 1991, **24**, 5442
[10] Stolka, M., Alkowitz, M.A., Knier, F.E., Weagley, R.J. and McGrane, K.M., *Synth. Met.*, 1990, **37**, 295
[11] Cotts, P.M., *Polym. Mater. Sci. Eng.*, 1985, **53**, 336
[12] Klemann, B.M., DeVilbliss, T. and Koutsky, J.A., *Polym. Eng. Sci.*, 1996, **36**, 135
[13] Fukushima, M., Hamada, Y., Tabei, E., Aramata, M., Mori, S. and Yamamoto, Y., *Synth. Met.*, 1998, **94**, 299
[14] Ebihara, K., Matsushita, S., Koshihara, S., Minami, F., Miyazawa, T., Obata, K. and Kira, M., *J. Lumin.*, 1997, **72/74**, 43
[15] Hoshino, S., Suzuki, H., Fujiki, M., Morita, M. and Matsumoto, N., *Synth. Met.*, 1997, **89**, 221
[16] Ostapenko, N., Zaika, V. and Suto, S., *Mol. Cryst. Liq. Cryst.*, 2002, **384**, 93
[17] Tachibana, H., Matsumoto, M. and Tokura, Y., *Macromolecules*, 1993, **26**, 2520
[18] Sanji, T., Sakamoto, K. and Sakurai, H., *Bull. Chem. Soc. Jpn.*, 1995, **68**, 1052
[19] Trefonas, P., Damewood, J.R., West, R. and Miller, R.D., *Organometallics*, 1985, **4**, 1318
[20] Bukalov, S.S., Leites, L.A. and West, R., *Macromolecules*, 2001, **34**, 6003
[21] Sato, T., Nagayama, N. and Yokoyama, M., *J. Mater. Chem.*, 2004, **14**, 287
[22] Miller, R.D., Thompson, D., Sooriyakumaran, R. and Fickes, G.N., *J. Polym. Sci., Part A: Polym. Chem.*, 1991, **29**, 813
[23] Holder, S.J., Achilleos, M. and Jones, R.G., *Macromolecules*, 2005, **38**, 1633
[24] Radhakrishnan, J., Kaito, A., Tanigaki, N. and Tanabe, Y., *Polymer*, 1999, **40**, 6199
[25] Schilling, F.C. and Bovey, F.A., Polym. Prepr. (Am. Chem. Soc., *Div. Polym. Chem.)*, 1998, **29**, 72
[26] Lovinger, A.J., Schilling, F.C., Bovey, F.A. and Zeigler, J.M., *Macromolecules*, 1986, **19**, 2657
[27] Furukawa, S., *Thin Solid Films*, 1998, **331**, 222
[28] Yokoyama, K.. and Yokoyama, M., *Solid State Commun.*, 1989, **70**, 241
[29] Kim, H.K. and Matyjaszewski, K., *J. Polym. Sci., Part A: Polym. Chem.*, 1993, **31**, 299
[30] Fujino, M. and Isaka, H., *Chem. Comm.*, 1989, 466
[31] Cragg, R.H., Jones, R.G., Swain, A.C. and Webb, S.J., *Chem. Comm.*, 1990, 1147
[32] Martins, L., Aeiyach, S., Jouini, M., Lacaze, P.C., Satgé, J. and Martins, J.P., *Appl. Organomet. Chem.*, 2002, **16**, 76

Poly(*p*-2,5-dimethoxyphenylene vinylene) P-180

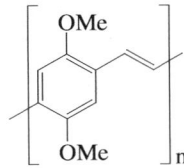

Synonyms: *PDMV. MO-PPV. Poly(2,5-dimethoxy-1,4-phenylene-1,2-ethenediyl)*
Related Polymers: Poly(*p*-phenylene vinylene)
Monomers: 2,5-Dimethoxy-1,4-benzenedicarboxaldehyde
Material class: Thermoplastic
Polymer Type: poly(arylene vinylene)

CAS Number:

CAS Reg. No.	Note
25190-54-9	
144225-26-3	(*E*)-form

Molecular Formula: $(C_{10}H_{10}O_2)_n$
Fragments: $C_{10}H_{10}O_2$
Morphology: Bond lengths and angles of polymer repeat unit have been calculated [1]

Volumetric & Calorimetric Properties:
Density:

No.	Value	Note
1	d 1.25 g/cm^3	[6]

Electrical Properties:
Electrical Properties General: Chemical doping increases conductivity [2,3,4,5]

Optical Properties:
Refractive Index:

No.	Value	Note
1	1.7–1.99	700–2500 nm [8]

Transmission and Spectra: C-13 nmr [9], H-1 nmr [10], ir [4,6,10,11], x-ray [13] and uv spectral data [12] have been reported

Applications/Commercial Products:
Processing & Manufacturing Routes: Prod. by base or thermal elimination of dialkyl sulfonium salt [2,10,4,14]; Wittig reaction of 2,5-dimethoxyterephthaldehyde [7]; or dehydrochlorination of 2,5-dimethoxy-*p*-xylylene dichloride [7]
Applications: Of potential application as electroluminescent diodes

Bibliographic References

[1] Eckhardt, H., Shacklette, L.W., Jen, K.Y. and Elsenbaumer, R.L., *J. Chem. Phys.*, 1989, **91**, 1303
[2] Antoun, S., Karasz, F.E. and Lenz, R.W., *J. Polym. Sci., Part A: Polym. Chem.*, 1988, **26**, 1809
[3] Gregorius, R.M., Lahti, P.M. and Karasz, F.E., *Macromolecules*, 1992, **25**, 6664
[4] Murase, I., Ohnishi, T., Noguchi, T. and Hirooka, M., *Polym. Commun.*, 1985, **26**, 362
[5] Schlenoff, J.B., Obrzut, J. and Karasz, F.E., *Phys. Rev. B: Condens. Matter*, 1989, **40**, 11822
[6] Tokito, S., Momii, T., Murata, H., Tsutsui, T. and Saito, S., *Polymer*, 1990, **31**, 1137
[7] Hörhold, H.H. and Opfermann, J., *Makromol. Chem.*, 1970, **131**, 105
[8] Yang, C.J. and Jenekhe, S.A., *Chem. Mater.*, 1995, **7**, 1276
[9] Schröter, B., Hörhold, H.H. and Raabe, D., *Makromol. Chem.*, 1981, **182**, 3185
[10] Burn, P.L., Bradley, D.D.C., Friend, R.H., Halliday, D.A. et al, *J. Chem. Soc., Perkin Trans. 1*, 1992, 3225
[11] Voss, K.F., Foster, C.M., Smilowitz, L., Mihailovic, D. et al, *Phys. Rev. B: Condens. Matter*, 1991, **43**, 5109
[12] Jin, J.I., Kim, J.H., Lee, Y.H., Lee, G.H. and Park, Y.W., *Synth. Met.*, 1993, **57**, 3742
[13] Martens, J.H.F., Marseglia, E.A., Bradley, D.D.C., Friend, R.H. et al, *Synth. Met.*, 1993, **55**, 449
[14] Halliday, D.A., Bradley, D.D.C., Burn, P.L., Friend, R.H. and Holmes, A.B., *Synth. Met.*, 1991, **41**, 931
[15] Era, M., Kamiyama, K., Yoshiura, K., Momii, T. et al, *Thin Solid Films*, 1989, **1**, 179

Poly(*N*,*N*-dimethylacrylamide) P-181

Synonyms: *Poly(N,N-dimethyl-2-propenamide)*. *PDMA*
Related Polymers: Polyacrylamide, Poly(*N*-isopropylacrylamide)
Monomers: 2-Propenoic acid, dimethylamide
Polymer Type: polyamide
CAS Number:

CAS Reg. No.	Note
26793-34-0	
37165-17-6	isotactic
89616-11-5	syndiotactic

Molecular Formula: $(C_4H_9NO)_n$
Fragments: C_4H_9NO
Molecular Weight: MW 34000
Morphology: Amorph. at and below 19% isotactic content; partially crystalline at 23% isotactic content. Crystallinity increases with increasing isotacticity [22]
Miscellaneous: A review of props. and uses has been reported [14]

Volumetric & Calorimetric Properties:
Equation of State: Mark-Houwink parameters have been reported [17,26]
Melting Temperature:

No.	Value	Note
1	309°C	28% isotactic [22]
2	326°C	41% isotactic [22]

Glass-Transition Temperature:

No.	Value	Note
1	119°C	14% isotactic [22]
2	112°C	41% isotactic [22]
3	89°C	[23]
4	110°C	[23]
5	119°C	[25]

Transition Temperature General: Other values of T_g have been reported [11]. T_g decreases noticeably above an isotactic content of 39% [22]. T_m only observed in isotactic polymers at an isotactic content above 23% [22]

Surface Properties & Solubility:
Solubility Properties General: Miscible with poly(vinylidene fluoride), provided the wt. fraction of PVDF is less than 0.7 [1]; with poly(styrene-*co*-acrylonitrile), when the acrylonitrile content is 68% [2]; with poly(chloromethyl methacrylate), poly(2-chloroethyl methacrylate), poly(3-chloropropyl methacrylate), poly(2-bromoethyl methacrylate) and poly(2-iodoethyl methacrylate) [7]. Also miscible with poly(acrylonitrile-*co*-methyl acrylate) [20], poly(vinylphenol) [21,27] and poly(styrene-*co*-vinylphenol) [27]. Immiscible with polystyrene, poly(styrene-*co*-styrene-4-carboxylic acid) and poly(styrene-*co*-styrene-4-sulfonic acid) [27]
Solvents/Non-solvents: Sol. H_2O [24,26,28], MeOH, 2-methoxyethanol, 2-butanone, Et_2O, toluene, $CHCl_3$, dioxane [28]

Transport Properties:
Polymer Solutions Dilute: $[\eta]_{inh}$ 0.5 dl g^{-1} (MeOH) [21], $[\eta]$ 1.33 dl g^{-1} (CH_2Cl_2, 14% isotactic), 1.91 dl g^{-1} (CH_2Cl_2, 41% isotactic) [22]. Shear viscosity shows a gradual decrease with time [24]

Water Content: An increase in tacticity results in reduced water uptake [22]. The rate of water vapour sorption decreases with increasing tacticity [22]

Mechanical Properties:
Mechanical Properties General: Modulus and yield stress of isotactic polymer are strongly dependent upon the degree of tacticity and crystallinity. For polymers of low tacticity, water absorption significantly affects mech. props. [22]

Optical Properties:
Transmission and Spectra: H-1 nmr [3,18], fluorescence [6] and C-13 nmr [9,13] spectral data have been reported

Polymer Stability:
Decomposition Details: Decomposition temp. 347° (min., 14% isotactic) [22], 338° (41% isotactic) [22]
Chemical Stability: Alkaline hydrol. at 100° is more complete for the isotactic form than for the atactic form (56% and 4% respectively) [15]

Applications/Commercial Products:
Processing & Manufacturing Routes: Synth. of the block copolymer has been reported [4]. May be synth. by anionic polymerisation [12,22] in THF at -78° using lithium and caesium alkyl catalysts [5]. Synth. by electropolymerisation in aq. sulfuric acid [8]; by plasa polymerisation [10] and by radical polymerisation [16]. May also be synth. by "group-transfer" polymerisation using ketene silyl acetals as initiator and zinc halide catalyst [19]
Applications: Has applications in coatings and adhesives

Bibliographic References
[1] Galin, M., *Makromol. Chem.*, 1987, **188**, 1391, (miscibility)
[2] Yeo, Y.T., Goh, S.H. and Lee, S.Y., *Polym. Polym. Compos.*, 1996, **4**, 235, (miscibility)
[3] Bulai, A., Jimeno, M.L., Alencar de Queiroz, A.-A., Gallardo, A. and Roman, J.S., *Macromolecules*, 1996, **29**, 3240, (H-1 nmr)
[4] Xie, X., *Diss. Abstr. Int. B*, 1995, **56**, 4912, (synth.)
[5] Xie, X. and Hogen-Esch, T.E., *Macromolecules*, 1996, **29**, 1746, (synth.)
[6] Soutar, I., Swanson, L., Thorpe, F.G. and Zhu, C., *Macromolecules*, 1996, **29**, 918, (fluorescence spectrum)
[7] Low, S.M., Lee, S.Y. and Goh, S.H., *Eur. Polym. J.*, 1994, **30**, 139, (miscibility)
[8] Iroh, J.O., Bell, J.P. and Scola, D.A., *J. Appl. Polym. Sci.*, 1993, **49**, 583, (synth.)
[9] Huang, S.S. and McGrath, J.E., Polym. Prepr. (Am. Chem. Soc., Div. Polym. Chem.), 1983, **24**, 138, (C-13 nmr)
[10] Hirotsu, T., *Kenkyu Hokoku - Sen'i Kobunshi Zairyo Kenkyusho*, 1982, 49, (synth.)
[11] Camino, G., Casorati, E., Chiantore, O., Costa, L. *et al*, Conv. Ital. Sci. Macromol., *(Atti)*, 1977, 134, (T_g)
[12] Okamoto, Y., Adachi, M., Shohi, H. and Yuki, H., *Polym. J. (Tokyo)*, 1981, **13**, 175, (synth.)
[13] Huynh, B.G. and McGrath, J.E., *Polym. Bull. (Berlin)*, 1980, **2**, 837, (nmr)
[14] Ogino, K., *Yuki Gosei Kagaku Kyokaishi*, 1975, **33**, 504, (props., uses)
[15] Sakaguchi, Y., Hamada, T., Tamaki, K. and Nishino, J., *Kobunshi Kagaku*, 1972, **29**, 562, (hydrolysis)
[16] Sakaguchi, Y., Mayama, H., Tamaki, K. and Nishino, J., *Kobunshi Kagaku*, 1973, **30**, 541, (synth.)
[17] Van Krevelen, D.W. and Hoftyzer, P.J., *J. Appl. Polym. Sci.*, 1967, **11**, 1409, (equation of state)
[18] Miron, Y. and Morawetz, H., *Macromolecules*, 1969, **2**, 162, (H-1 nmr)
[19] Sogah, D.Y., Hertler, W.R., Webster, O.W. and Cohen, G.M., *Macromolecules*, 1987, **20**, 1473, (synth.)
[20] Percec, S. and Melamud, L., *J. Appl. Polym. Sci.*, 1990, **41**, 1853, (miscibility)
[21] Landry, M.R., Massa, D.J., Landry, C.J.T., Teegarden, D.M. *et al*, *J. Appl. Polym. Sci.*, 1994, **54**, 991, (miscibility)
[22] Mohajer, Y., Wilkes, G.L. and McGrath, J.E., *J. Appl. Polym. Sci.*, 1981, **26**, 2827, (props.)
[23] Krause, S., Gormley, J.J., Roman, N., Shetter, J.A. and Watanabe, W.H., *J. Polym. Sci., Part A: Polym. Chem.*, 1965, **3**, 3573, (glass transition temps.)
[24] Chmelir, M., Künschner, A. and Barthell, A., *Angew. Makromol. Chem.*, 1980, **89**, 145, (shear viscosity)
[25] Chiantore, O., Costa, L. and Guaita, M., *Makromol. Chem., Rapid Commun.*, 1982, **3**, 303, (glass transition temp.)

[26] Chiantore, O., Costa, L. and Guaita, M., *Makromol. Chem.*, 1982, **183**, 2257, (conformn)
[27] Landry, C.J.T. and Teegarden, D.M., *Macromolecules*, 1991, **24**, 4310, (miscibility)
[28] Buchholz, F.L., *Ullmanns Encykl. Ind. Chem.*, 5th edn., (eds. B. Elvers, S. Hawkins and G. Schulz), VCH, 1992, **A21**, 150

Polydimethylbutadiene P-182

Synonyms: *Methyl rubber. Poly(2,3-dimethyl-1,3-butadiene)*
Monomers: 2,3-Dimethyl-1,3-butadiene
Material class: Synthetic Elastomers
Polymer Type: polybutadiene
CAS Number:

CAS Reg. No.
25034-65-5

General Information: *trans*-1,4-Polydimethylbutadiene. *cis*-1,4-Polydimethylbutadiene

Volumetric & Calorimetric Properties:
Melting Temperature:

No.	Value	Note
1	253–259°C	*trans*-1,4; 2,3-dimethyl-1,3-butadiene [1]
2	189–192°C	*cis*-1,4 [1]
3	267–272°C	*trans*-1,4 highly stereoregular [1]

[1] Henderson, N., *Encycl. Polym. Sci. Eng.*, 2nd edn., (ed. J.I., Kroschwitz), John Wiley & Sons, 1985, **2**

Polydimethyldiphenylsiloxane fluid P-183

Synonyms: *Polydiphenyldimethylsiloxane fluid*
Related Polymers: Polyphenylsiloxane fluid, PDMS fluid
Monomers: Dichlorodiphenylsilane, Dichlorodimethylsilane, Hexamethylcyclotrisiloxane
Material class: Fluids, Copolymers
Polymer Type: silicones, dimethylsilicones
CAS Number:

CAS Reg. No.
156048-34-9

Molecular Formula: $[(C_{12}H_{10}OSi).(C_2H_6OSi)]_n$
Fragments: $C_{12}H_{10}OSi$ C_2H_6OSi
Identification: Silicone fluids containing phenyl groups have higher values of density than dimethylsiloxane

Volumetric & Calorimetric Properties:
Density:

No.	Value	Note
1	d 0.98 g/cm^3	low phenyl
2	d 1.05 g/cm^3	high phenyl [3]

Thermal Expansion Coefficient:

No.	Value	Note	Type
1	0.0009 K^{-1}	low phenyl [6]	V
2	0.00055 K^{-1}	high phenyl	V

Thermal Conductivity:

No.	Value	Note
1	0.13–0.15 W/mK	0.00032–0.00036 cal (cm s °C)$^{-1}$, falls as phenyl substitution rises [12,13]

Specific Heat Capacity:

No.	Value	Note	Type
1	1.4 kJ/kg.C	0.34 Btu lb^{-1}°F^{-1} [3]	p

Transition Temperature General: Low phenyl subsitution lowers pour point; high phenyl silicone raises pour point above that of dimethyl silicone [1,8]
Transition Temperature:

No.	Value	Note
1	-73°C	Pour point (-100°F, low phenyl) [3]
2	-40°C	Pour point (-40°F, high phenyl)

Surface Properties & Solubility:
Cohesive Energy Density Solubility Parameters: δ 16–18.4 J$^{1/2}$ cm$^{-3/2}$ (7.8–9 cal$^{1/2}$ cm$^{-3/2}$) [5,9]
Solvents/Non-solvents: Sol. aliphatic hydrocarbons, aromatic hydrocarbons, chlorinated hydrocarbons, ethers, esters, ketones. Insol. alcohols, H_2O. (Low MW methyl phenyl silicones are sol. alcohols)
Surface and Interfacial Properties General: The high phenyl substituted form is a better lubricant than methyl silicone oils [17]
Wettability/Surface Energy and Interfacial Tension: γ_c 21 mJ m^{-2} (21 dyne cm^{-1}, low phenyl). γ_c 25 mJ m^{-2} (25 dyne cm^{-1}, high phenyl) [3,6]

Transport Properties:
Polymer Melts: Viscosity temp. coefficient 0.65–0.85 (rises with rise in phenyl subsitution) (see [poly phenyl siloxane fluid] [3,10,14,15]

Mechanical Properties:
Miscellaneous Moduli:

No.	Value	Note	Type
1	1650–2000 MPa	16700–20000 kg cm^{-2} [10,11]	Bulk modulus

Mechanical Properties Miscellaneous: Compressibility decreases with increasing phenyl substitution [10,11]

Electrical Properties:
Electrical Properties General: Has excellent insulating props. but these are inferior to those of dimethylsilicone fluids
Dielectric Permittivity (Constant):

No.	Value	Note
1	2.9–3	[4,16]

Dielectric Strength:

No.	Value	Note
1	20 kV.mm^{-1}	[4]

Optical Properties:
Optical Properties General: Refractive index higher than that of dimethyl silicone [7]

Refractive Index:

No.	Value	Note
1	1.422	low phenyl [3]
2	1.498	high phenyl [3]

Transmission and Spectra: Transparent above 280 nm [5]

Polymer Stability:
Polymer Stability General: Usable over wide temp. range; good low temp. props.
Thermal Stability General: Usable between -75° and 225° in presence of air. In absence of air usable up to 300° (low phenyl for low temps. and high phenyl for high temps.) [3]
Upper Use Temperature:

No.	Value
1	250°C
2	225°C

Decomposition Details: Oxidised at temps. of over 250° [2]
Flammability: Fl. p. 230° (440°F) [3]
Environmental Stress: Stable to H_2O, air, oxygen and other materials. The high phenyl form is resistant to radiation [1]
Chemical Stability: Chemically inert but decomposed by conc. mineral acids. Very thin films liable to degradation [1,4,5]
Hydrolytic Stability: Stable to H_2O and aq. solns. Very thin films spread on H_2O; liable to hydrolytic degradation [1,4,5]
Biological Stability: Non-biodegradable [4,5]
Recyclability: Can be reused; no large scale recycling

Applications/Commercial Products:
Processing & Manufacturing Routes: For low viscosity fluids dimethyldichlorosilane and diphenyldichlorosilane are hydrolysed and the hydrolysate equilibrated under acid catalysis at 180° with small amounts of hexamethyldisiloxane to serve as chain terminator. For high viscosity fluids dimethyldiphenyl and dimethyl cyclic polysiloxanes are polymerised together under alkaline catalysis with traces of chain terminator at 150°
Applications: Used for low temp. damping for aerospace instruments and electronic equipment, heat transfer media and solder baths. Used as dielectric coolant and coupler and in greases for ball bearing lubrication. Other applications in chromatography columns

Trade name	Supplier
SF1150 series	General Electric Silicones

Bibliographic References
[1] Noll, W., *Chemistry and Technology of Silicones,* Academic Press, 1968
[2] Murphy, C.M., Saunders, C.E. and Smith, D.C., *Ind. Eng. Chem.,* 1950, **42**, 2462, (thermal stability)
[3] Ash, M. and Ash, I., Handbook of Plastic Compounds, *Elastomers and Resins,* (eds. M.B. Ash and I.A. Ash), Wiley-VCH, 1992
[4] Stark, F.O., Fallender, J.R. and Wright, A.P., *Comprehensive Organometallic Chemistry,* (eds. G. Wilkinson, F.G.A. Stone and E.W. Abel), Pergamon Press, 1982, **2**, 305
[5] Hardman, B. and Torkelson, A., *Encycl. Polym. Sci. Eng.,* 2nd edn. (ed. J.I. Kroshwitz), John Wiley and Sons, 1985, **15**, 204
[6] Fox, H.W., Taylor, P.W. and Zisman, W.A., *Ind. Eng. Chem.,* 1947, **39**, 1401
[7] Flaningan, O.L. and Langley, N.R., *Anal. Chem. Silicones,* (ed. A.L. Smith), John Wiley and Sons, 1991, 135
[8] Warrick, E.L., Hunter, M.J. and Barry, A.J., *Ind. Eng. Chem.,* 1952, **44**, 2196
[9] Yerrick, K.B. and Beck, H.N., *Rubber Chem. Technol.,* 1964, **37**, 261, (solubility)
[10] Meals, R.N. and Lewis, F.M., *Silicones,* Reinhold, 1959
[11] Thompson, J.M.C., *Silicones,* (ed. S. Fordham), 1960, 5
[12] Bates, O.K., *Ind. Eng. Chem.,* 1949, **41**, 1966, (thermal conductivity)
[13] Jamieson, O.T. and Irving, J.B., *Proc. Int. Conf. Therm. Conduct.,* 1975, 279, (thermal conductivity)
[14] Buch, R.R., Klimisch, H.M. and Johannson, O.K., *J. Polym. Sci., Part A-2,* 1970, **8**, 541, (viscosity)
[15] Lee, C.L. and Haberland, C.G., *J. Polym. Sci., Part B: Polym. Lett.,* 1965, **3**, 883, (viscosity)
[16] Vincent, G.A., Feuron, F.W.G. and Orbech, T., *Annu. Rep., Conf. Electr. Insul. Dielectr. Phenom.,* 1972, 17, (dielectric constant)
[17] Clarson, S.J. and Semlyen, J.A., *Siloxane Polymers,* Prentice Hall, 1993

Poly(4,4-dimethyl-1-hexene)   P-184

$$-[CH_2CH]_n-$$
$$\quad\; CH_2C(CH_3)_2CH_2CH_2CH_3$$

Synonyms: *Poly(1-(2,2-dimethylbutyl)-1,2-ethanediyl)*
Monomers: 4,4-Dimethyl-1-hexene
Material class: Thermoplastic
Polymer Type: polyolefins
CAS Number:

CAS Reg. No.
35064-73-4

Molecular Formula: $(C_8H_{16})_n$
Fragments: C_8H_{16}
General Information: Hard, crystalline polymer [1]
Morphology: Three isomers are known [3]

Volumetric & Calorimetric Properties:
Melting Temperature:

No.	Value	Note
1	>350°C	min. [1,2]

Transition Temperature:

No.	Value	Note	Type
1	314°C	[1]	Polymer melt

Transport Properties:
Polymer Solutions Dilute: η_{inh} 1.12 (decahydronaphthalene, 130°) [1,2]

Applications/Commercial Products:
Processing & Manufacturing Routes: Synth. using Ziegler-Natta catalysts such as $TiCl_4$-$AlEt_3$ [1,3]. Catalysts may be prepared by treating $TiCl_4$ with a lithium aluminium tetraalkyl in an inert atmosphere and an inert organic solvent [2]

Bibliographic References
[1] Campbell, T.W. and Haven, A.C., *J. Appl. Polym. Sci.,* 1959, **1**, 73, (melting point, synth, soln props)
[2] *U.S. Pat.,* 1966, 3 257 367, (synth)
[3] Buniyat-Zade, A.A., Mamedov, E.L., Avakyan, L.G., Pokatilov, V.D. et al, *Plast. Massy,* 1972, **3**, 12, (synth)

Poly[2-(dimethyloctylsilyl)-1,4-phenylene vinylene]   P-186

Synonyms: *Poly[[(dimethyloctylsilyl)-1,4-phenylene]-1,2-ethenediyl]. DMOS-PPV*
Material class: Thermoplastic
Polymer Type: poly(arylene vinylene)
CAS Number:

CAS Reg. No.	Note
189084-69-3	poly[[(dimethyloctylsilyl)-1,4-phenylene]-1,2-ethenediyl]
184687-87-4	

Molecular Formula: $(C_{18}H_{28}Si)_n$
Fragments: $C_{18}H_{28}Si$
General Information: Silyl-substituted green light emitting photoluminescent and electroluminescent polymer. Prod. as yellow solid.

Surface Properties & Solubility:
Solvents/Non-solvents: Sol. $CHCl_3$, THF, toluene [1].

Electrical Properties:
Electrical Properties General: Polymer used in LED devices as green light emitter [1,2,3,4]. The high turn-on voltage of 15V can be reduced with the use of fluorinated derivatives [3] or copolymers with MEH-PPV [4].

Optical Properties:
Optical Properties General: Exhibits photoluminescence with high quantum efficiency in green region of spectrum (λ_{max} in region of 520-540nm). Band gap of polymer 2.34eV [3].
Transmission and Spectra: Uv and photoluminescence [1,3,4,5], electroluminescence [1,2,3,4] and ftir [1] spectral data have been reported.

Applications/Commercial Products:
Processing & Manufacturing Routes: Prod. by Gilch method from 1,4-bis(halomethyl)-2-(dimethyloctylsilyl)benzene and *tert*-BuOK [1,6].
Applications: Can be used as emissive layer in light-emitting diodes as homopolymer or in form of copolymers.

Bibliographic References
[1] Hwang, D.-H., Kim, S.T., Shim, H.-K., Holmes, A.B., Moratti, S.C. and Friend R.H., *Synth. Met.*, 1997, **84**, 615
[2] Kim, S.T., Hwang, D.-H., Li, X.C., Grüner, J., Friend, R.H., Holmes, A.B. and Shim, H.K., *Adv. Mater.*, 1996, **8**, 979
[3] Jin, Y., Kim, J., Lee, S., Kim, J.Y., Park, S.H., Lee, K. and Suh, H., *Macromolecules*, 2004, **37**, 6711
[4] Hwang, D.-H., Chuah, B.S., Li, X.-C., Kim, S.T., Moratti, S.C., Holmes, A.B., De Mello, J.C. and Friend R.H. Makromol. Chem., *Makromol. Symp.*, 1997, **125**, 111
[5] Grey, J.K., Kim, D.Y., Lee, Y.J., Gutierrez, J.J., Luong, N., Ferraris, J.P. and Barbara, P.L., Angew. Chem., *Int. Ed. Engl.*, 2005, **44**, 6207
[6] Geneste, F., Fischmeister, C., Martin, R.E. and Holmes, A.B., *Synth. Met.*, 2001, **121**, 1709

Poly(4,4-dimethyl-1-pentene)   P-187

$$\left[CH_2-CH(CH_2C(CH_3)_3) \right]_n$$

Synonyms: *Poly(1-(2-methylpropyl)-1,2-ethanediyl)*
Monomers: 4,4-Dimethyl-1-pentene
Material class: Thermoplastic
Polymer Type: polyolefins
CAS Number:

CAS Reg. No.	Note
29252-76-4	atactic
30847-58-6	isotactic

Molecular Formula: $(C_7H_{14})_n$
Fragments: C_7H_{14}
General Information: Hard, crystalline polymer [1]
Morphology: Tetragonal, a 2.035, b 2.035, c 0.701, helix 4_1 [2]; or tetragonal, a 2.03, b 2.03, c 1.38, helix 7_2 [3]

Volumetric & Calorimetric Properties:
Density:

No.	Value	Note
1	d 0.9 g/cm³	cryst. [2]

Thermodynamic Properties General: ΔH_f -156.2 kJmol⁻¹ (-37.3 kcal mol⁻¹, calc.) [7]
Melting Temperature:

No.	Value	Note
1	>350°C	min. [1,2]
2	>320°C	min. [4]
3	>380°C	min. [5]
4	>400°C	min. [3]

Glass-Transition Temperature:

No.	Value	Note
1	190°C	approx., film [1]

Transition Temperature:

No.	Value	Note
1	59°C	softening temp. [6]
2	330°C	polymer melt temp. [1]

Surface Properties & Solubility:
Solvents/Non-solvents: Insol. organic solvents [3]

Applications/Commercial Products:
Processing & Manufacturing Routes: Synth. using Ziegler-Natta catalysts, such as VCl_3AlEt_3 at 50° [6]

Bibliographic References
[1] Campbell, T.W. and Haven, A.C., *J. Appl. Polym. Sci.*, 1959, **1**, 73, (transition temps)
[2] Tumer-Jones, A., *Polymer*, 1966, **7**, 23, (struct)
[3] Noether, H.D., *J. Polym. Sci., Part C: Polym. Lett.*, 1967, **16**, 725, (struct)
[4] Reding, F.P., *J. Polym. Sci.*, 1956, **21**, 547, (melting point)
[5] *High Polymers*, (eds. R.A.V. Raff and K.W. Doak), Interscience, 1965, **20**, 691, (melting point)
[6] Dunham, K.R., Vandenberghe, J., Faber, J.W.H. and Contois, L.E., *J. Polym. Sci., Part A: Polym. Chem.*, 1963, **1**, 751, (transition temps)
[7] Joshi, R.M., *J. Macromol. Sci., Chem.*, 1970, **4**, 1819, (heats of formation)

Poly(2,6-dimethyl-1,4-phenylene oxide)   P-188

Poly(2,6-dimethyl-1,4-phenylene oxide)

Synonyms: *DMPPO. Modified PPO. PPO-M. PPE. Poly(2,6-dimethylphenol). Polyphenylene oxide. Polyphenylene ether. Poly(2,6-dimethyl-1,4-phenylene ether). Poly[oxy(2,6-dimethyl-1,4-phenylene)]*
Related Polymers: Poly(phenylene oxide)
Monomers: 2,6-Dimethylphenol
Material class: Thermoplastic
Polymer Type: polyphenylene oxide
CAS Number:

CAS Reg. No.	Note
25134-01-4	homopolymer
24938-67-8	

Molecular Formula: $(C_8H_8O)_n$
Fragments: C_8H_8O
Molecular Weight: MW 40000, M_n 18000
Additives: Blended with polystyrene to improve flow props. Blends are often referred to as PPO
General Information: The most important polyphenylene oxide polymer, this material is often confusingly called Polyphenylene oxide. An attractive, tough, stiff material with a wide temp. range of use with its principal applications as a blend. Linear crystalline polymer of good thermal stability but limited flow props.

Volumetric & Calorimetric Properties:
Density:

No.	Value	Note
1	d 1.21 g/cm^3	[4]
2	d^{23} 1.06 g/cm^3	[26]
3	d 0.958 g/cm^3	[26]

Thermal Expansion Coefficient:

No.	Value	Note	Type
1	0.00053 K^{-1}	[3]	L
2	0.00021 K^{-1}	[3]	V
3	5.2×10^{-5} K^{-1}	ASTM D696 [26,27,29]	L
4	0.000209 K^{-1}	[9]	V

Equation of State: Einstein/Debye low temp. equation constants have been reported [13]. Hartmann-Haque equation of state constant values have been reported [14]
Latent Heat Crystallization: S 247 J K^{-1} mol^{-1} (210°), 295.1 J K^{-1} mol^{-1} (307°) [1]. ΔC_p 32.2 J K^{-1} mol^{-1} [1]. ΔC_p 28.8 J K^{-1} mol^{-1} (0.24 J g^{-1} K^{-1}) [3]. ΔH_f 1.97 kJ mol^{-1} (16.4 J g^{-1}) [8]. ΔH_f 5.0–5.9 kJ mol^{-1} [28]. ΔS_f 7.7–11.3 J K^{-1} mol^{-1} [10]. ΔH_c 4200 kJ mol^{-1} (35 kJ g^{-1}) [11]. ΔH_f 5.4 kJ mol^{-1} (1290 cal mol^{-1}) [12]. Other specific heat capacity values have been reported [29]
Thermal Conductivity:

No.	Value	Note
1	0.22 W/mK	ASTM C177 [29]

Specific Heat Capacity:

No.	Value	Note	Type
1	251.8 kJ/kg.C	210° [1]	P
2	274 kJ/kg.C	307° [1]	P
3	48.3 kJ/kg.C	-193° [7]	P
4	202 kJ/kg.C	147°[7]	P
5	53.4 kJ/kg.C	0.445 J K^{-1} g^{-1}, 81.839K crystalline [10]	P
6	265.6 kJ/kg.C	2.213 J K^{-1} g^{-1}, 544.167K crystalline [10]	P
7	163.3 kJ/kg.C	1.361 J K^{-1} g^{-1}, 334.235K amorph. [10]	P
8	268.2 kJ/kg.C	2.235 J K^{-1} g^{-1}, 567.33K amorph.	P
9	53.08 kJ/kg.C	-193°	P
10	136.5 kJ/kg.C	-3° [28]	P

Melting Temperature:

No.	Value	Note
1	307°C	[1]
2	257°C	[4]
3	262–267°C	[26,29]

Glass-Transition Temperature:

No.	Value	Note
1	210°C	[1,2,4]
2	207°C	[3,5]
3	203–219°C	[6,26,29]

Deflection Temperature:

No.	Value	Note
1	145°C	182 kN m^{-2}, ASTM D648 [4,26,29]
2	100°C	1.82 MPa [26,29]
3	191°C	19 kg cm^{-2}, ASTM D648 [26,29]

Surface Properties & Solubility:
Cohesive Energy Density Solubility Parameters: δ_d 9.41, δ_p 1.3, δ_h 2.4 [16]. Other solubility parameter values have been reported [30]
Solvents/Non-solvents: Very sol. C_6H_6, carbon disulfide, CCl_4, chlorobenzene, $CHCl_3$, *o*-dichlorobenzene, *cis*-dichloroethylene, 1,1,2,2-tetrachlorotrichloroethylene. Spar. sol. anisole, benzonitrile, cyclohexanone, *trans*-dichloroethylene, 1,4-dioxane, diphenyl ether, CH_3Cl, nitrobenzene, pyridine, *m*-xylene, *m*-cresol, DMF, Me_2CO. Prac. insol. AcOH, Me_2CO, Et_2O, DMSO, EtOAc, *n*-hexane, MeOH, 2-butanone, trifluoroacetic acid [15]. Will dissolve in CH_2Cl_2 but forms insoluble complex at room temp.

Transport Properties:
Permeability of Gases: Permeability of He, O_2, N_2, CO_2 and CH_4 has been reported [17]
Water Absorption:

No.	Value	Note
1	0.06%	24h, 23° [26]
2	0.3%	equilibrium, 100° [26]

– Poly(2,6-dimethyl-1,4-phenylene oxide)

Gas Permeability:

No.	Gas	Value	Note
1	N_2	229.8 cm³ mm/ (m² day atm)	3.5×10^{-10} cm² (s cmHg)$^{-1}$ [25]
2	O_2	958.7 cm³ mm/ (m² day atm)	14.6×10^{-10} cm² (s cmHg)$^{-1}$ [25]
3	CO_2	4301.1 cm³ mm/ (m² day atm)	65.5×10^{-10} (s cmHg)$^{-1}$ [25]

Mechanical Properties:
Tensile (Young's) Modulus:

No.	Value	Note
1	1200–1360 MPa	[20]
2	2690 MPa	23° [26,27]
3	2480 MPa	93°, ASTM D638 [26,27]

Flexural Modulus:

No.	Value	Note
1	5100 MPa	23°, ASTM D790 [4]
2	2650 MPa	-17° [26,27]
3	2590 MPa	23° [26,27]
4	2480 MPa	93°, ASTM D638 [26,27]

Tensile Strength Break:

No.	Value	Note	Elongation
1	56.2–62 MPa	[20]	21–41%
2	98–99 MPa	23°, ASTM D1708 [4]	4–6%
3	80 MPa	23° [26,27]	20–40% extension
4	55 MPa	93°, ASTM D638 [26,27]	30–70% extension

Elastic Modulus:

No.	Value	Note
1	1000000 MPa	100°, ASTM D256-56 [18]
2	2255 MPa	230 kgf mm^{-2} [26,27]

Tensile Strength Yield:

No.	Value	Note
1	67.8–75.2 MPa	[20]

Flexural Strength Yield:

No.	Value	Note
1	14 MPa	23°, ASTM D790 [4]
2	134 MPa	-17° [26,27]
3	114 MPa	23° [26,27]
4	87 MPa	93°, ASTM D790 [26,27]

Compressive Strength:

No.	Value	Note
1	12140 MPa	23°, ASTM D1621-64 [4]
2	83 MPa	[26,27]

Impact Strength: Impact strength 15 kJ m^{-2} (150 kgf cm cm^{-2}, 100°, DIN 53453) [18]. Puncture impact 50 J (23°, ASTM D3763)
Hardness: Rockwell M78 (ASTM D785)
Friction Abrasion and Resistance: μ_{dyn} (polymer/self) 0.18-0.23. Abrasion resistance 17 mg (1000 cycles)$^{-1}$ (Taber CS-17 test)
Izod Notch:

No.	Value	Notch	Note
1	85.4 J/m	N	23°, ASTM D256 [4]
2	53 J/m	N	-40° [26,27]
3	91 J/m	Y	93°, ASTM D256 [26,27]
4	2000 J/m	N	min., ASTM D256 [26,27]

Charpy:

No.	Value	Notch	Note
1	4	Y	-20–100° [19]

Electrical Properties:
Electrical Properties General: Dielectric constant and dissipation factor remain low and constant over broad ranges of temp., frequency and humidity [26,27]. Exhibits high volume resistivity and dielectric strength [24]
Dielectric Permittivity (Constant):

No.	Value	Frequency	Note
1	2.58	60 Hz	23°, ASTM D150 [26,27]

Dielectric Strength:

No.	Value	Note
1	20 kV.mm^{-1}	3.18 mm thick, ASTM D149 [26,27]

Arc Resistance:

No.	Value	Note
1	75s	ASTM D195 [26,27]

Complex Permittivity and Electroactive Polymers: μ 0.88 D [16]
Dissipation (Power) Factor:

No.	Value	Frequency	Note
1	0.00035	60 Hz	23° [26,27]
2	0.0009	1 MHz	23°, ASTM D150 [26,27]
3	0.01	1 Hz	approx., 100° [18]

Optical Properties:
Refractive Index:

No.	Value	Note
1	1.567	[16]

Transmission and Spectra: Mass spectral data have been reported [21]

Volume Properties/Surface Properties: Colour beige

Polymer Stability:
Polymer Stability General: Sensitive to photochemical and thermal oxidation
Thermal Stability General: Exhibits high heat deflection temp. [24]
Upper Use Temperature:

No.	Value	Note
1	105°C	[24]
2	80–105°C	

Decomposition Details: Thermal degradation occurs in two steps; a rapid exothermic process between 430° and 500° producing phenolic products, H_2O and a cross-linked residue followed by a slower, char-forming process above 500° [22]. Char yield 17% (800°, N_2), 17% (800°, air) [4]. Rapidly degrades above 250° in oxygen atmosphere. Decomposition temp. 450° (10° min^{-1}, N_2) [25]
Flammability: Limiting oxygen index 32% (ASTM D2863) [4]. Smoke evolution D_m 775.35 [4]
Environmental Stress: Shows low radiation resistivity [20]. Prone to environmental stress cracking. Prolonged outdoor exposure causes discoloration, crazing and loss of props. [24]
Chemical Stability: Resistant to strong acids and alkalis. Attacked by many organic solvents [24]
Biological Stability: Methyl groups in the aromatic ring hinder biodegradation [23]
Stability Miscellaneous: Toxicity T_i (time to incapacitation) 8.65 min. T_d (time to death) 19.96 min. [4]

Applications/Commercial Products:
Processing & Manufacturing Routes: Prod. by oxidative coupling with a homogeneous oxidation catalyst (e.g. oxygen, Cu(*I*) salt and pyridine catalyst in a chlorinated hydrocarbon solvent). Most often used as a blend with polystyrene and also nylons
Mould Shrinkage (%):

No.	Value	Note
1	0.005–0.007%	ASTM D955 [29]

Applications: Automotive industry, electronic appliances, office equipment, replacement for zinc die-castings, water distribution and environmental appliances. Mouldings have good dimensional stability. Blended with polystyrene to give Noryl resins, a major engineering thermoplastic. PPO/PA alloys (Vestoblend and Ultranyl) are also prod. commercially

Trade name	Details	Supplier
Aryloxa		former USSR
Biapen		Polish
Biapen		Polish
Laril		Freeman
Luranyl		BASF
Noryl		General Electric
PPO		General Electric
Prevex		Borg Warner
Thermocomp		LNP Engineering
Ultranyl		BASF
Vestoblend		former USSR
Vestoran	blends with polystyrene and related polymers	Hüls AG

Bibliographic References
[1] Cheng, S.Z.D., Pan, R., Bu, H.S., Cao, M.Y. and Wunderlich, B., *Makromol. Chem.*, 1988, **189**, 1579
[2] Rigby, S.J. and Dew-Hughes, D., *Polymer*, 1974, **15**, 639
[3] O'Reilly, J.M., *J. Appl. Phys.*, 1977, **48**, 4043
[4] Kourtides, D.A. and Parker, J.A., *Polym. Eng. Sci.*, 1978, **18**, 855
[5] Sharma, S.C., Mandelkern, L. and Stehling, F.C., *J. Polym. Sci., Polym. Lett. Ed.*, 1972, **10**, 345
[6] Savolavnen, A.V., *Makromol. Chem.*, 1973, **172**, 213
[7] Cheng, S.Z.D., Lim, S., Judovits, L.H. and Wunderlich, B., *Polymer*, 1987, **28**, 10
[8] Karasz, F.E. and O'Reilly, J.M., *J. Polym. Sci., Polym. Lett. Ed.*, 1965, **3**, 561
[9] Matheson, R.R., *Macromolecules*, 1987, **20**, 1847
[10] Karasz, F.E., Bair, H.E. and O'Reilly, J.M., *J. Polym. Sci., Part A-2*, 1968, **6**, 1141
[11] Ng, S.C. and Chee, K.K., *Polymer*, 1993, **34**, 3871
[12] Karasz, F.E. and Mangaraj, D., *Polym. Prepr. (Am. Chem. Soc., Div. Polym. Chem.)*, 1971, **12**, 317
[13] Simha, R., Roe, J.M. and Nanda, V.S., *J. Appl. Phys.*, 1972, **43**, 4312
[14] Hartmann, B. and Haque, M.A., *J. Appl. Polym. Sci.*, 1985, **30**, 1553
[15] White, D.M. and Klopfer, H.J., *J. Polym. Sci., Part A: Polym. Chem.*, 1972, **10**, 1565
[16] Koenhen, D.M. and Smolders, C.A., *J. Appl. Polym. Sci.*, 1975, **19**, 1163
[17] Maeda, Y. and Paul, D.R., *J. Polym. Sci., Part B: Polym. Phys.*, 1987, **25**, 981
[18] Heijboer, J., *J. Polym. Sci., Part C: Polym. Lett.*, 1968, **16**, 3755
[19] Vincent, P.I., *Polymer*, 1974, **15**, 111
[20] Sasuga, T., Hayakawa, N., Yoshida, K. and Hagiwara, M., *Polymer*, 1985, **26**, 1039
[21] Wiley, R.H., *J. Polym. Sci.*, 1971, **9**, 129
[22] Factor, A., *J. Polym. Sci., Part A: Polym. Chem.*, 1969, **7**, 363
[23] Li, L.X., Grulke, E.A. and Oriel, P.J., *J. Appl. Polym. Sci.*, 1993, **48**, 1081
[24] *The Materials Selector*, (eds. N.A. Waterman and M.F. Ashby), Elsevier, 1991, 1816
[25] Aguilar-Vega, M. and Paul, D.R., *J. Polym. Sci., Part B: Polym. Phys.*, 1993, **31**, 1577, (decomposition)
[26] Aycock, D., Abolius, V. and White, D.M., *Encycl. Polym. Sci. Eng.*, Vol. 13, (ed. J.I. Kroschwitz), John Wiley and Sons, 1988, 1, (props)
[27] Rosato, D., *Encycl. Polym. Sci. Eng.*, Vol. 14, (ed. J.I. Kroschwitz), John Wiley and Sons, 1988, 366, (props)
[28] Gaur, U. and Wunderlich, B., *J. Phys. Chem. Ref. Data*, 1981, **10**, 1001, (thermodynamic props)
[29] *ASM Engineered Materials Reference Book*, (ed. M. Bauccio), ASM International, 1994, 428, (thermodynamic props)
[30] *CRC Handbook of Polymer-Liquid Interaction Parameters and Solubility Parameters*, (ed. A.F.M. Barton), CRC Press, 1990, 120, (solubility parameters)

Poly(2,6-dimethylphenylene oxide)/polystyrene blends   P-189

Related Polymers: Poly(2,6-dimethylphenylene oxide), Polystyrene
Monomers: 2,6-Dimethylphenol, Styrene
Material class: Blends
Polymer Type: polyphenylene oxide, polystyrene
Molecular Formula: $[(C_8H_8O).(C_8H_8)]_n$
Fragments: C_8H_8O C_8H_8
Additives: Organic phosphates added to increase flame retardancy. Glass fillers must then be added to offset adverse electrical props.
General Information: Poly(2,6-dimethylphenylene oxide) and polystyrene are miscible in all proportions. Props. of the blends can be varied over a wide range by changing the ratio of components and the amount and type of additives [1,2]

Volumetric & Calorimetric Properties:
Density:

No.	Value	Note
1	d 1.06 g/cm^3	[2]
2	d 1.21 g/cm^3	[2]

Poly(2,6-dimethylphenylene oxide)/polystyrene blends

Thermal Expansion Coefficient:

No.	Value	Note	Type
1	6×10^{-5} K^{-1}	unreinforced, ASTM D696 [2]	V
2	4×10^{-5} K^{-1}	20% glass reinforced, ASTM D696 [2]	V

Thermodynamic Properties General: High heat deflection temp. Low coefficient of thermal expansion [3]

Thermal Conductivity:

No.	Value	Note
1	0.22 W/mK	unreinforced, ASTM C177 [2]
2	0.24 W/mK	20% glass reinforced, ASTM C177 [2]

Deflection Temperature:

No.	Value	Note
1	130°C	1.8 MPa, unreinforced [2]
2	132°C	1.8 MPa, 20% glass reinforced [2]

Vicat Softening Point:

No.	Value	Note
1	148°C	unreinforced [2]
2	145°C	20% glass reinforced [2]

Transport Properties:
Water Content: Very low water absorption [2]
Water Absorption:

No.	Value	Note
1	0.07 %	24h, ASTM D570 [2]
2	0.14 %	equilib., ASTM D570 [2]

Mechanical Properties:
Mechanical Properties General: Relatively constant mech. props. over a wide temp. range (-40–105°) [2,3]

Tensile (Young's) Modulus:

No.	Value	Note
1	2500 MPa	unreinforced, ASTM D638 [2]
2	6500 MPa	20% glass reinforced, ASTM D638 [2]

Flexural Modulus:

No.	Value
1	2500 MPa

Tensile Strength Break:

No.	Value	Note	Elongation
1	90 MPa	20% glass reinforced, ASTM D638 [2]	3%
2	50 MPa	unreinforced, ASTM D638 [2]	50%

Flexural Strength Yield:

No.	Value	Note
1	98 MPa	unreinforced, ASTM D790 [2]

Impact Strength: High-impact strength which is retained at -40° [1,3]
Hardness: Rockwell R115 (ASTM D785) [2], L106 (20% glass reinforced, ASTM D785) [2]

Izod Notch:

No.	Value	Notch	Note
1	80 J/m	Y	20% glass reinforced, ASTM D256 [2]
2	200 J/m	Y	ASTM D256 [2]

Charpy:

No.	Value	Notch	Note
1	0.005	Y	unreinforced [2]

Electrical Properties:
Electrical Properties General: High volume resistivity and dielectric strength unaffected by changes in humidity [3]
Dielectric Permittivity (Constant):

No.	Value	Frequency	Note
1	2.7	50 Hz	unreinforced, ASTM D150 [2]
2	2.9	50 Hz	glass reinforced, ASTM D150 [2]

Dielectric Strength:

No.	Value	Note
1	22 kV.mm^{-1}	3.2 mm, unreinforced, ASTM D149
2	17 kV.mm^{-1}	3.2 mm, glass reinforced, ASTM D149 [2]

Dissipation (Power) Factor:

No.	Value	Frequency	Note
1	0.0004	50 Hz	unreinforced, ASTM D150 [2]
2	8×10^{-5}	50 Hz	glass reinforced, ASTM D150 [2]

Polymer Stability:
Flammability: Variation of polymer components will produce different blends with different flame retardant props. Flame retardancy of blends is intermed. between high flame retardancy of PPE and high flammability of polystyrene. Limiting oxygen index 24% (ASTM D2863) [1]
Chemical Stability: Attacked by many organic solvents, particularly aromatic and chlorinated aliphatics [3]. Swollen by aliphatic compds.
Hydrolytic Stability: Highly resistant to hydrol. under acidic and basic conditions [2] and very stable to water

Applications/Commercial Products:
Processing & Manufacturing Routes: Can be extruded or injection moulded and fabricated by solvent, ultrasonic or heat melding [3]
Applications: Water distribution including plumbing fixtures, pumps and filters, electrical-electronic appliances; cable covers, junction boxes, motor housings, business machines, and automobiles. Instrument panels, electrical connectors and filler panels. Some grades suitable for use with foodstuffs

Trade name	Supplier
Laril	Freeman
Luranyl	BASF
Noryl	General Electric
Prevex	Borg Warner
Thermocomp	LNP Engineering
Ultranyl	BASF

Bibliographic References
[1] *Kirk-Othmer Encycl. Chem. Technol.*, Vol. 19, 4th edn., (eds. J.I. Kroschwitz and M. Howe-Grant), Wiley Interscience, 1993, 688
[2] *Ullmanns Encykl. Ind. Chem.*, (ed. V.B. Elvers), VCH, 1992, **A21**, 605
[3] *The Materials Selector*, (eds. N.A. Waterman and M.F. Ashby), Chapman and Hall, 1997, 1816
[4] Kramer, M., *Appl. Polym. Symp.*, 1971, **15**, 227, (additives)
[5] Cooper, G.D., Lee, G.F., Katchmann, A. and Shank, C.D., *Mater. Technol.*, 1981, 12, (blend composition)

Poly(α,α-dimethyl-β-propiolactone)   P-190

Synonyms: *Poly(pivalolactone). PVL. Poly(β-pivalolactone). Poly(3,3-dimethyl-2-oxetanone). Poly(oxycarbonyl(1,1-dimethylethylene)). Poly(oxy(2,2-dimethyl-1-oxo-1,3-propanediyl)). Poly(hydroxypivalic acid). Poly(2,2-dimethylhydracrylic acid). Poly(3-hydroxy-2,2-dimethylpropanoic acid)*
Monomers: 3-Hydroxy-2,2-dimethylpropanoic acid
Material class: Thermoplastic
Polymer Type: saturated polyester
CAS Number:

CAS Reg. No.	Note
30622-81-2	Poly(3-hydroxy-2,2-dimethyl propanoic acid)
24969-13-9	Poly(3,3-dimethyl-2-oxetanone)

Molecular Formula: $(C_5H_8O_2)_n$
Fragments: $C_5H_8O_2$
Morphology: Displays three crystalline polymorphs α, β and γ. Cryst. struct. has been reported [5,32,39]. Three polymorphic forms have been reported [4,17,26]. Morphological study has been reported [18]. α form monoclinic a 0.905, b 1.158, c 0.603, β 121.5°, helix 2_1 [6]; α form monoclinic a 0.902, b 1.164, c 0.602, β 121° 29", helix 2_1 [7]; β form c 0.476 [18]; γ form orthorhombic a 0.823, b 1.127, c 0.604, helix 2_1 [18]. γ form orthorhombic a 0.823, b 1.128, c 0.602, helix 2_1 [21]

Volumetric & Calorimetric Properties:
Density:

No.	Value	Note
1	d^{20} 1.223 g/cm^3	cryst. [4]
2	d^{20} 1.097 g/cm^3	amorph. [4]
3	d^{25} 1.1 g/cm^3	amorph. [9]
4	d 1.19 g/cm^3	[35]
5	d^{25} 1.23 g/cm^3	cryst. [9]
6	d^{235} 0.94 g/cm^3	amorph. [9]
7	d 1.18 g/cm^3	[10]
8	d^{23} 1.185 g/cm^3	[15]

Thermal Expansion Coefficient:

No.	Value	Note	Type
1	0.00065 K^{-1}	200-250° [9]	V

Thermodynamic Properties General: C_p values have been reported [25,38]
Latent Heat Crystallization: Heat of polymerisation -84.2 kJ mol^{-1} (-20.1 kcal mol^{-1}) [3]. $\Delta H°_f$ 14.86 kJ mol^{-1} (3.55 kcal mol^{-1}); $\Delta°S_f$ 28.97 J mol^{-1} K^{-1} (6.92 cal mol^{-1}°C^{-1}) [9]; ΔH_f 118 Jg^{-1} [13]; ΔH_f 14.5 kJ mol^{-1} [15]; ΔH_f 120 Jg^{-1} [35]. Heat of polymerisation 77 kJ mol^{-1} [33]. Other enthalpy, entropy and Gibbs functions have been reported [38]

Melting Temperature:

No.	Value	Note
1	245°C	[4]
2	228°C	β-form [4]
3	240°C	[6,10,15,24,33]
4	235°C	508K, α-form [8]
5	242°C	[12]
6	234°C	[14]
7	230–240°C	α-form [18]
8	239°C	512K, β-form [8]
9	210–220°C	γ-form [18]

Glass-Transition Temperature:

No.	Value	Note
1	-10°C	[4,10]
2	96°C	369K [13]
3	100°C	373K [15,35]
4	6°C	[23]
5	137°C	410K [24,37]
6	-25°C	[25]
7	-3°C	270K [34]
8	67°C	340K [34]
9	-13°C	260K [38]

Deflection Temperature:

No.	Value	Note
1	180–200°C	264 lb f in^{-2} [4]

Vicat Softening Point:

No.	Value	Note
1	220°C	[4]

Transition Temperature:

No.	Value	Note	Type
1	238–240°C	[9]	$T°_m$
2	265–271°C	[16]	$T°_m$

3	269°C	[20]	$T°_m$
4	272°C	[22]	$T°_m$
5	240°C	513K [35]	$T°_m$
6	27°C	300K [13]	T_γ
7	2°C	275K [24,37]	T_γ

Surface Properties & Solubility:
Solubility Properties General: Immiscible with bisphenol A polycarbonate [12] and polyvinyl chloride [15]
Solvents/Non-solvents: Sol. CH_2Cl_2 [9], hot phenol, hot 2-chlorophenol, hot cresols, hot trichloroacetic acid [27]. Insol. common organic solvents [23]
Wettability/Surface Energy and Interfacial Tension: Interfacial surface free energy 6.1×10^{-6} J cm^{-2} (61 erg cm^{-2}, perpendicular) [20]; 4.3×10^{-6} J cm^{-2} (43 erg cm^{-2}, perpendicular) [16]; 2.9×10^{-6} J cm^{-2} (29 erg cm^{-2}, parallel) [20]; 1.3×10^{-6} J cm^{-2} (13 erg cm^{-2}, parallel) [16]

Transport Properties:
Polymer Solutions Dilute: η_{inh} 0.3 dl.g^{-1} (trifluoroacetic acid) [2]; η_{inh} 1.48 dl.g^{-1} (phenol/tetrachloroethane (60:40), 30°) [19]. Other viscosity values have been reported [23]
Polymer Melts: Melt viscosity has been reported [4]
Water Content: Other water absorption values have been reported [31]
Water Absorption:

No.	Value	Note
1	0.2 %	24h [4]
2	0.14–0.22 %	11 days [23]

Mechanical Properties:
Mechanical Properties General: Fibre-mechanical props. have been reported [10,33]. Tensile strength break values have been reported. [4,23,30] Yield stress and strain have been reported in graphical form [23]. Storage and loss moduli have been reported in graphical form [13,23]

Tensile (Young's) Modulus:

No.	Value	Note
1	1130 MPa	[35]

Elastic Modulus:

No.	Value	Note
1	6276 MPa	64000 kg cm^{-2} [1]

Tensile Strength Yield:

No.	Value	Elongation	Note
1	38 MPa	4-6%	[4]

Impact Strength: Tensile impact strength 100 kJ m^{-2} [4]. Impact resistance has been reported [30]
Viscoelastic Behaviour: Creep resistance 0.8% length increase (10000h, 20°, 20.6 MPa stress) [4]. Stress-strain [8,23] and stress-relaxation curves have been reported [13]. Viscoelastic props. have been reported [13]
Hardness: Shore D85 [4]
Izod Area:

No.	Value	Notch	Note
1	4 kJ/m^2	Unnotched	[4]

Charpy:

No.	Value	Notch	Note
1	3.4	Y	[4]
2	4.6	N	[4]

Electrical Properties:
Electrical Properties General: Dielectric relaxation data has been reported. [37] Dissipation factor data have been reported in graphical form [13,14,23,24]

Optical Properties:
Optical Properties General: Optical props. have been reported [30]
Transmission and Spectra: Mass [2,11,19], H-1 nmr [14], C-13 nmr [17] and x-ray PES spectral data have been reported [36]

Polymer Stability:
Decomposition Details: Stable for 1000h at 200° (fibres) [33]. Thermal decomposition causes depolymerisation by unzipping [4]. Thermal degradation details have been reported [11,19]. Activation energy of thermal degradation 98.4 kJ mol^{-1} (23.5 kcal mol^{-1}, 400–600°) [28]. Temp. of max. decomposition rate 410° [19]
Environmental Stress: Fibres have high resistance to weathering [4] and uv light [3]
Chemical Stability: Resistant to aliphatic and aromatic hydrocarbons, alcohols, ketones, and AcOH [4]. Fibres are resistant to acids, alkalis and solvents [4,33]
Biological Stability: Biodegradation study has been reported [29]

Applications/Commercial Products:
Processing & Manufacturing Routes: Prod. by nucleophilic or electrophilic ring-opening polymerisation of 2,2-dimethyl-β-propiolactone. Also synth. by ring opening polymerisation of pivalolactone using a weakly basic substance such as a tertiary amine or a phosphine as a catalyst, e.g. triphenylphosphine at 75° [3,9,39]
Applications: Fibre used in modification of rubbers. Matrix for fibre composites

Bibliographic References
[1] Sakurada, I. and Kaji, K., *J. Polym. Sci., Part C: Polym. Lett.*, 1970, **31**, 57, (mech props)
[2] Wiley, R.H., *J. Macromol. Sci., Chem.*, 1970, **4**, 1797, (ms)
[3] Hall, H.K., *Macromolecules*, 1969, **2**, 488, (synth)
[4] Oosterhof, H.A., *Polymer*, 1974, **15**, 49, (transition temps, mech props)
[5] Cornibert, J. and Marchessault, R.H., *Macromolecules*, 1975, **8**, 296, (cryst struct)
[6] Perego, G., Melis, A. and Cesari, M., *Makromol. Chem.*, 1972, **157**, 269, (cryst struct)
[7] Carazzolo, G., *Chem. Ind. (London)*, 1964, **46**, 525, (cryst struct)
[8] Prud'homme, R.E. and Marchessault, R.H., *Macromolecules*, 1974, **7**, 541, (melting points, mech props)
[9] Boni, C., Bruckner, S., Crescenzi, V., Della Fortuna, G. et al, *Eur. Polym. J.*, 1971, **7**, 1515, (synth, thermodynamic props, density)
[10] Ubesternenk, J.M., *Angew. Makromol. Chem.*, 1978, **71**, 117, (transition temps)
[11] Kricheldorf, H.R. and Luderwald, I., *Makromol. Chem.*, 1978, **179**, 421, (thermal degradation, ms)
[12] Cruz, C.A., Barlow, J.W. and Paul, D.R., *J. Appl. Polym. Sci.*, 1979, **24**, 2399, (polymer miscibility)
[13] Noah, J. and Prud'homme, R.E., *Macromolecules*, 1979, **12**, 300, (mech props, viscoelastic props)
[14] Allegrezza, A.E., Lenz, R.W., Cornibert, J. and Marchessault, R.H., *J. Polym. Sci., Polym. Chem. Ed.*, 1978, **16**, 2617, (H-1 nmr, dissipation factor)
[15] Aubin, M. and Prud'homme, R.E., *Macromolecules*, 1980, **13**, 365, (transition temps, polymer miscibility)
[16] Noah, J. and Prud'homme, R.E., *Eur. Polym. J.*, 1981, **17**, 353, (surface free energy)
[17] Veregin, R.P., Fyfe, C.A. and Marchessault, R.H., *Macromolecules*, 1986, **19**, 2379, (C-13 nmr)
[18] Meille, S.V., Konishi, T. and Geil, P.H., *Polymer*, 1984, **25**, 773, (cryst struct, morphological study)
[19] Garozzo, D., Giuffrido, M. and Montaudo, G., *Macromolecules*, 1986, **19**, 1643, (thermal degradation, ms)
[20] Morand, H. and Hoffman, J.D., *Macromolecules*, 1990, **23**, 3682, (surface free energy)

[21] Meille, S.V., Bruckner, S. and Lando, J.B., *Polymer,* 1989, **30**, 786, (cryst struct)
[22] Khanna, Y.P. and Kumar, R., *J. Polym. Sci., Part B: Polym. Phys.,* 1989, **27**, 369, (melting point)
[23] Paul, C.W., *J. Appl. Polym. Sci.,* 1988, **36**, 675, (mech props)
[24] Plesu, R., Malik, T.M. and Prud'homme, R.E., *Polymer,* 1992, **33**, 4463, (transition temps, dissipation factor)
[25] Beshouri, S.M., Grebowicz, J.S. and Chuah, H.H., *Polym. Eng. Sci.,* 1994, **34**, 69, (thermodynamic props)
[26] Meille, S.V., *Polymer,* 1994, **35**, 2607, (polymorphic forms)
[27] Yamashita, Y., Ishikawa, Y. and Tsuda, T., *Kogyo Kagaku Zasshi,* 1964, **67**, 252, (solubility)
[28] Nikolaeva, N.A., Belenkii, B.G., Glukhov, N.A., Zhuraulev, Y.V. and Sazanov, Y.N., *Vysokomol. Soedin., Ser. A,* 1970, **12**, 2625, (thermal degradation)
[29] Potts, J.E., Clendinning, R.A. and Ackart, W.B., *Degrad. Polym. Plast., (Prepr.) Conf.,* 1973, 12, (biodegradation)
[30] Ames, W.A., *High Perform. Plast., Nat. Tech. Conf. - Soc. Plast. Eng., (Prepr.),* 1976, 152, (optical and mechanical props)
[31] Fuzek, J.F., *Ind. Eng. Chem. Proc. Des. Dev.,* 1985, **24**, 140, (water absorption)
[32] Hiroshi, K., *Kobunshi Ronbunshu,* 1996, **53**, 111, (cryst struct)
[33] Mayne, N.R., *Adv. Chem. Ser.,* 1973, **129**, 175, (fibre mechanical props)
[34] Pratt, C.F. and Geil, P.H., *J. Macromol. Sci., Phys.,* 1982, **21**, 617, (transition temps)
[35] Duchesne, D. and Prud'homme, R.E., *Polymer,* 1979, **20**, 1199, (thermodynamic props)
[36] Briggs, D. and Bearnson, G., *Anal. Chem.,* 1992, **64**, 1729, (XPES)
[37] Malik, T.M. and Prud'homme, R.E., *J. Macromol. Sci., Phys.,* 1984, **23**, 323, (dielectric relaxation data)
[38] Grebowicz, J., Varma-Nair, M. and Wunderlich, B., *Polym. Adv. Technol.,* 1992, **3**, 51, (specific heat capacity, thermodynamic data)
[39] Tijsma, E.J., van der Does, L., Bantjers, A. and Vulic, I., *J. Macromol. Sci., Rev. Macromol. Chem. Phys.,* 1994, **34**, 515, (synth, struct)

Poly(dimethylsilane)

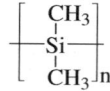

Synonyms: *Poly(dimethylsilylene).* PDMS. PDMSi
Related Polymers: Polycarbosilanes, Poly(silane)
Polymer Type: polysilanes
CAS Number:

CAS Reg. No.
28883-63-8

Molecular Formula: $(C_2H_6Si)_n$
Fragments: C_2H_6Si
Molecular Weight: 2,175 [17]
Morphology: Off-white powder [7,10] with a Si-Si backbone chain conformation of *all-trans* configuration, similar to that found with analogues PDES and PDPrS. The room temperature d-spacing can be indexed by a monoclinic unit cell with dimensions a=0.745nm (12.2Å); b=0.724nm (8.0Å); c=0.389nm (3.88Å) and $\gamma=90°$ [1,3,5,20]
Miscellaneous: Side-chain interactions are minimal, yet the Si backbone is exceptionally rigid.
Chain orientation is strongly affected by vacuum pressure and also by the substrate temperature during deposition. The substrate temperature also affects the orientation of the polymer. An ordered phase is observed at low temperature, followed by disordered and ordered phases with an increase in temperature. As temperature is further increased, a transformation to a more disordered phase (gauche-rich conformation) occurs slowly [2,3,5,7]

Volumetric & Calorimetric Properties:
Thermodynamic Properties General: T_{CM} 162°C; ΔH_{CM} 0.6kJ mol^{-1}; T_{ML} 226°C; ΔH_{ML} 0.2kJ/mol [7,13]
The kinetic parameters of thermal degradation were evaluated by different integral methods using TG data. The activation energy of decomposition (E) for the polymer is 122kJ mol^{-1}, and the corresponding value of order of reaction is 1 [16]

Surface Properties & Solubility:
Solubility Properties General: Insoluble in organic solvents at room temperature [3,12]

Mechanical Properties:
Mechanical Properties General: See [20]

Electrical Properties:
Electrical Properties General: Electrical conductivity along the c axis is expected to be greater than that perpendicular to the c axis [3]. It has been found that electrical conductivity increases with increasing irradiation time, and found from alternating current measurements that both the real and the imaginary parts of the complex dielectric constant increase with increasing irradiation time [8]

Optical Properties:
Optical Properties General: Heating the polymer to 160°C does not lead to any significant change of the UV spectrum, pointing to the absence of thermochromic effect, as found with other poly(dialkylsilyene)s. Only a slight shift of the band to 334nm and its broadening are observed, the picture being not reversible. However, ir spectra samples after heating in air to 200°C show the presence of SiOSi bonds due to the partial oxidation of the polymer [8,12,20]
Transmission and Spectra: Ir [7,8,9,12,21], Raman [12], uv-vis [3,8,9,10], nmr [2,7,20] and pes [11] spectra reported.

Polymer Stability:
Thermal Stability General: Thermal degradation has been investigated by pyrolysis-gas chromatography and thermogravimetry. Decomposition produces linear and cyclic oligomeric products [16]. The polymer was found to be much less stable than the styrene copolymer.
Decomposition Details: The DSC behavior of the polymer is complicated and shows a number of weak endotherms. Second and subsequent heating scans show only two well defined peaks at 162°C and 226°C [6,7]

Applications/Commercial Products:
Processing & Manufacturing Routes: Synthesised by the reaction of dimethyldichlorosilane with Na in toluene [12]. Due to its insolubility, thin films have been prepared by a new vacuum evaporation technique [3,4]. Epitaxial growth of evaporated films on poly(tetrafluoroethylene) layer [15]
Applications: Electro luminescent device [4]. Good ceramic material precursors for making meltable and soluble polycarbosilanes [6]. Preparation of polycarbosilane in the presence of zeolite as a catalyst [14]. Correction of pattern defects of colour filters for liquid crystal displays [18]. Rare earth magnets with surface protection layers [19]. Method of synthesising the polycarbosilane at normal pressure by adding several wt. % of polyborodiphenylsiloxane to PDMS [20]

Trade name	Supplier
788	Scientific Polymer Products, Inc.
93-1497	Strem Chemicals, Inc.
Dimethylpolysilane	ChemPur GmbH
PSS-1M01	ABCR GmbH KG
S 81433	
S93-1497	ABCR GmbH KG
SF 18-350	
SF 96-350	Nippon Soda Co.
	GELEST Inc.

Bibliographic References

[1] KariKari, E.K., Greso, A.J., Farmer, B.L., Miller, R.D. and Rabolt, R.F., *Macromolecules*, 1993, **26**, 3937
[2] Takayama, T., *J. Mol. Struct.*, 1998, **441**, 101
[3] Furukawa, S. and Ohta, H., *Thin Solid Films*, 2003, **438-439**, 48
[4] Furukawa, S., *Thin Solid Films*, 1998, **33**, 222
[5] Crespo, R., Piqueras, M.C. and Tomas, F., *J. Chem. Phys.*, 1994, **100**, 6953
[6] Shukla, S.K., Tiwari, R.K., Ranjan, A., Saxena, A.K. and Mathur, G.N., *Thermochim. Acta*, 2004, **424**, 209
[7] Nair, M.V., Cheng, J., Jin, Y. and Wunderlich, B., *Macromolecules*, 1991, **24**, 5442
[8] Ohta, H., Takamoto, T., Yasuda, T. and Furukawa, S., *Thin Solid Films*, 2006, **499**, 129
[9] Shimomura, M., Okumoto, H., Kaito, A., Ueno, K., Shen, J. and Ito, K., *Macromolecules*, 1998, **31**, 7483
[10] Muramatsu, Y., Fujino, M., Yamamoto, T., Gullikson, E.M. and Perera, R.C.C., *Nucl. Instrum. Methods Phys. Res.*, *Sect. B*, 2003, **199**, 260
[11] Nath, K.G., Shimoyama, I., Sekiguchi, T. and Baba, Y., *J. Electron. Spectrosc. Relat. Phenom.*, 2005, **144-147**, 323
[12] Leites, L.A., Bukalov, S., Yadritzeva, T.S., Mokhov, M.K., Antipova, B.A., Frunzc, T.M. and Dement'ev, V.V., *Macromolecules*, 1992, **25**, 2991
[13] Allegra, G. and Meille, S.V., *Macromolecules*, 2004, **37**, 3487
[14] Kim, Y.H., Jung, S.J., Kim, H.R., Kim, H.D., Shin, D.G. and Riu, D.H., *Adv. Tech. Mat. Mat. Proc.*, 2004, **6**, 192
[15] Hattori, R., Aoki, Y., Sugano, T., Shirafuji, J. and Fujiki, T., Jpn. J. Appl. Phys., *Part 1*, 1997, **36**, 819
[16] Radhakrishnan, T.S., *J. Appl. Polym. Sci.*, 2006, **99**, 2679
[17] Kulandainathan, M.A., Elangovan, M., Kulangiappar, K., Raju, T., Muthukumaran, A. and Sanjeeviraja, C., Solid State Ionics Proc. 8th Asian Conf., 2002, 303
[18] *Jpn. Pat.*, 2006, 06 154 398
[19] *Jpn. Pat.*, 2005, 05 322 810
[20] Lovinger, A.J., Davis, D.D., Schilling, F.C., Padden, F.J. and Bovey, F.A., *Macromolecules*, 1991, **24**, 132
[21] Yajima, S., Hasegawa, Y., Okamura, K. and Matsuzawa, T., *Nature (London)*, 1978, **273**, 525

Polydimethylsiloxane   P-192

Synonyms: *PDMS. Dimethicone. Dimethylsilicone. Dimethyl polysiloxane. Polydimethylsilicone. Poly[oxy(dimethylsilylene)]. Methylsilicone*
Related Polymers: Polysiloxanes, PDMS Fluids, PDMS Gum, Dimethiconol
Monomers: Dichlorodimethylsilane, Hexamethylcyclotrisiloxane
Polymer Type: dimethylsilicones
CAS Number:

CAS Reg. No.
9006-65-9
9016-00-6
63148-62-9

Molecular Formula: $[C_2H_6OSi]_n$
Fragments: C_2H_6OSi

Volumetric & Calorimetric Properties:
Glass-Transition Temperature:

No.	Value	Note
1	-123°C	[3]

Surface Properties & Solubility:
Cohesive Energy Density Solubility Parameters: δ 15.5 $J^{1/2}$ $cm^{3/2}$ (7.5-7.6 $cal^{1/2}$ $cm^{3/2}$). δ_d 15.9; δ_p 0.0; δ_h 4.1 [1,5,6]
Solvents/Non-solvents: Sol. chlorinated hydrocarbons, aromatic hydrocarbons, aliphatic hydrocarbons. Mod. sol. ketones, esters, ethers. Insol. alcohols, H_2O [1,4,5,6]

Electrical Properties:
Electrical Properties General: Has excellent insulating props.

Optical Properties:
Refractive Index:

No.	Value	Note
1	1.404	[4]

Polymer Stability:
Polymer Stability General: Stable material usable over wide temp. range
Flammability: Low flammability [2,7]
Environmental Stress: Resistant to oxidative degradation, water, and corona discharge [7]
Chemical Stability: Chemically inert [4,7]
Hydrolytic Stability: Stable in H_2O and aq. solns. at room temp. [1,2,7]
Biological Stability: Non-biodegradable

Applications/Commercial Products:
Processing & Manufacturing Routes: Cyclic dimethylsiloxanes are polymerised at 150-200° with an alkaline catalyst. Small quantities of hexamethyldisiloxane are added to terminate the siloxane chains
Applications: Low MW PDMS is the basis of silicone fluids while high MW PDMS is the basis of PDMS gum which is cured to produce rubbers. Liq. prepolymer used in manufacture of optical channel waveguides

Trade name	Details	Supplier
A6,110-100000	fluid	Goldschmidt AG
C-156	gum	Wacker Silicones
COHR lastic 9655U	gum	CHR Industries
Dow Corning 200 Fluid	fluid	Dow Corning STI
Masil SF	fluid	PPG-Mazer
SF	various grades, not exclusive	General Electric Silicones
Silicone 0111	fluid	ICI, UK
Silicone Fluid	fluid	Akrochem
Viscasil	fluid	General Electric Silicones

Bibliographic References

[1] Stark, F.O., Fallender, J.R. and Wright, A.P., *Comprehensive Organometallic Chemistry*, (eds. G. Wilkinson, F.G.A. Stone and E.W. Abel), Pergamon Press, 1982, **2**, 305
[2] Hardman, B. and Torkelson, A., *Encycl. Polym. Sci. Technol.*, Vol. 15, 2nd edn., (ed. J.I. Kroschwitz), John Wiley and Sons, 1985, 204
[3] Polmanteer, K.E. and Hunter, M.J., *J. Appl. Polym. Sci.*, 1959, **1**, 3, (transition temps)
[4] Roff, W.J. and Scott, J.R., Fibres, Films, *Plastics and Rubbers*, Butterworths, 1971
[5] Baney, R.H., Voigt, C.E. and Mentele, J.W., Struct.-Solubility Relat. Polym., (*Proc. Symp.*), (eds. F.W. Harris and R.B. Seymour), Academic Press, 1977, 225, (solubility)
[6] Yerrick, K.B. and Beck, H.N., *Rubber Chem. Technol.*, 1964, **37**, 261, (solubility)
[7] Noll, W., *Chemistry and Technology of Silicones*, Academic Press, 1968
[8] Kim, E. *et al*, *Adv. Mater.*, 1996, **8**, 139, (use)

Polydimethylsiloxane gum   P-192a

Synonyms: *Dimethylsilicone gum. Poly[oxy(dimethylsilylene)]. Methylsilicone gum. PDMS gum*
Related Polymers: More general information, Methylsilicone rubber, Fibre glass reinforced elastomer
Monomers: Dichlorodimethylsilane, Hexamethylcyclotrisiloxane
Material class: Synthetic Elastomers, Gums and resins
Polymer Type: dimethylsilicones

CAS Number:

CAS Reg. No.	Note
9006-65-9	
9016-00-6	
63148-62-9	
63394-02-5	elastomer
31900-57-9	PDMS

Molecular Formula: $(C_2H_6OSi)_n$
Fragments: C_2H_6OSi
Molecular Weight: MW 10000–1000000

Volumetric & Calorimetric Properties:
Density:

No.	Value	Note
1	d 0.97 g/cm^3	[1]

Specific Heat Capacity:

No.	Value	Note	Type
1	1.3 kJ/kg.C	0.3 Btu lb^{-1}°F^{-1}) [2]	P

Melting Temperature:

No.	Value	Note
1	-54°C	[3]

Glass-Transition Temperature:

No.	Value	Note
1	-123°C	[3]

Surface Properties & Solubility:
Cohesive Energy Density Solubility Parameters: δ 15.5 J$^{1/2}$ cm$^{-3/2}$ (7.5–7.6 cal$^{1/2}$ cm$^{-3/2}$). δ_d 15.9, δ_p 0.0, δ_h 4.1 J$^{1/2}$cm$^{-3/2}$ [4,5,6]
Solvents/Non-solvents: Sol. chlorinated hydrocarbons, aromatic hydrocarbons, aliphatic hydrocarbons. Mod. sol. ketones, esters, ethers. Insol. alcohols, H$_2$O [1,4,5,6]

Transport Properties:
Polymer Melts: Kinematic viscosity 100000 St (approx.) [1]

Optical Properties:
Refractive Index:

No.	Value	Note
1	1.404	[1]

Applications/Commercial Products:
Processing & Manufacturing Routes: Cyclic dimethylsiloxanes are polymerised at 150–200° with an alkaline catalyst. Small quantities of hexamethyldisiloxane are added to terminate the siloxane chains. To reduce mould shrinkage low MW siloxanes can be removed by solvent extraction or thermal treatment. In solvent extraction volatile siloxanes are removed by washing in MeOH. In thermal treatment the gum is steam or air stripped at 250°. These processes reduce the volatile fraction of the gum from around 10% by weight to less than 2%
Applications: Gum precursor for rubber manufacture. See methylsilicone rubber

Trade name	Details	Supplier
B-2	elastomer	Wacker Silicones
B-210	elastomer	Wacker Silicones
C-156	gum	Wacker Silicones
COHR lastic	various grades, not exclusive	CHR Industries
COHR lastic 9655U	gum	CHR Industries
Rhodorsil	RS4U series, elastomer	Rhone-Poulenc
Rhodorsil RS50	elastomer	Rhone-Poulenc
SE	various grades, not exclusive	General Electric Silicones
Silastic GP	elastomer	Dow Corning STI
Silopren HV	elastomer	Bayer Inc.

Bibliographic References
[1] Roff, W.J. and Scott, J.R., Fibres, Films, *Plastics and Rubbers*, Butterworths, 1971
[2] Ash, M. and Ash, I., Handbook of Plastic Compounds, *Elastomers and Resins*, (eds. M.B. Ash and I.A. Ash), Wiley-VCH, 1992
[3] Polmanteer, K.E. and Hunter, M.J., *J. Appl. Polym. Sci.*, 1959, **1**, 3, (transition temps)
[4] Stark, F.O., Fallender, J.R. and Wright, A.P., *Comprehensive Organometallic Chemistry*, (eds. G. Wilkinson, F.G.A. Stone and E.W. Abel), Pergamon Press, 1982, **2**, 305
[5] Barney, R.H., Voigt, C.E. and Mentele, J.W., Struct.-Solubility Relat. Polym., *(Proc. Symp.)*, (eds. F.W. Harris and R.B. Seymour), Academic Press, 1977, 225, (solubility)
[6] Yerrick, K.B. and Beck, H.N., *Rubber Chem. Technol.*, 1964, **37**, 261, (solublity)
[7] Hardman, B. and Torkelson, A., *Encycl. Polym. Sci. Technol.*, 2nd edn., (ed. J.I. Kroschwitz), John Wiley and Sons, 1985, **15**, 204
[8] Drake, J., Peters, D.B. and McGuire, S., *Silicones*, (ed. S. Fordham), 1960, 117
[9] *Brit. Pat.*, 1957, 773 324, (low shrinkage gum)

Poly(dimethylsiloxanes) with liquid crystal sidechains

Synonyms: *Side chain liquid crystalline polydimethylsiloxane copolymers*
Related Polymers: Poly(hydrogenmethyl siloxanes) with liquid crystal sidechains, Poly(methylhydrogensiloxane-*co*-dimethylsiloxane) fluid
Monomers: Dichlorodimethylsilane
Material class: Polymer liquid crystals, Copolymers
Polymer Type: dimethylsilicones
Molecular Formula: $[(C_2H_6OSi)(C_{24}H_{36}OSi)]_n$
Fragments: C_2H_6OSi $C_{24}H_{36}OSi$
Morphology: Shows liq. cryst. phase below isotropisation temp. [4]

Volumetric & Calorimetric Properties:
Volumetric Properties General: For copolymers with alkyl ester side chains the thickness of films falls by over 10% with rise in temp. above the liq. crystalline 'mesophonic' temp. range [3]
Melting Temperature:

No.	Value	Note
1	51°C	alkyl ether side chain [2]

Glass-Transition Temperature:

No.	Value	Note
1	-121°C	alkyl ether side chain [2]
2	15°C	alkyl ether side chain [2]

– Poly[(dimethylsilylene)methylene]

Transition Temperature:

No.	Value	Note	Type
1	222°C	[2]	Isotropisation temp

Surface Properties & Solubility:
Solvents/Non-solvents: Sol. aromatic hydrocarbons, chlorinated hydrocarbons. Insol. alcohols, H_2O [1]

Polymer Stability:
Thermal Stability General: Stable over wide temp. range, primarily above T_g
Chemical Stability: Stability depends on props. of the side chain
Hydrolytic Stability: Stable to H_2O, and aq. solns. Depolymerised by high pressure steam [5]
Recyclability: Not recyclable

Applications/Commercial Products:
Processing & Manufacturing Routes: Substituted alkene and polymethyldimethylsiloxane fluid are mixed together in dry toluene under nitrogen with platinum catalyst at 100° for about 48h. If there is a small excess of the substituted alkene this results in complete conversion of all SiH bonds [1,6]

Bibliographic References
[1] Kozlovsky, M.V., Bustamante, E.A.S. and Haase, W., *Liq. Cryst.*, 1996, **20**, 35
[2] Castelvetro, V. and Ciardelli, F., *Polym. Int.*, 1996, **39**, 37
[3] Kozlovsky, M.V. and Haase, W., *Acta Polym.*, 1996, **47**, 361
[4] Finkelmann, H. and Rehage, G., *Makromol. Chem., Rapid Commun.*, 1980, **1**, 31
[5] Noll, W., *Chemistry and Technology of Silicones,* Academic Press, 1968
[6] Clarson, S.J. and Semlyen, J.A., *Siloxane Polymers,* Prentice Hall, 1993

Poly[(dimethylsilylene)methylene]   P-194

$$-\left[CH_2-\underset{\underset{Me}{|}}{\overset{\overset{Me}{|}}{Si}}\right]_n-$$

Synonyms: *Poly(1,1,3,3-tetramethyl-1,3-disilacyclobutane),* SRU. PDMSM
Monomers: 1,1,3,3-Tetramethyl-1,3-disilacyclobutane
Polymer Type: polycarbosilanes
CAS Number:

CAS Reg. No.	Note
25722-25-2	
98261-12-2	α-Et, ω-Br
98241-69-1	α-Me, ω-Cl

Molecular Formula: $(C_3H_8Si)_n$
Fragments: C_3H_8Si
Molecular Weight: MW 357,000–721,000 [11,12,13,14]; M_n 120,000 and 170,000 [5,6,7]
General Information: Has significant inorganic character. A candidate for analysis with regard to the problem of the effect of structure on the statistical properties of a chain molecule. Like PDMSO, it is an amorphous, non-glassy material at room temperature, has excellent thermal stability, and is soluble in a variety of solvents [3]. Also available end-capped α=Et, ω=Br or α=Me, ω=Cl [2].
Tacticity: Relatively small characteristic ratio [4]. The melt is much more strongly structured, and the individual chains have mean square dimensions 40% higher than the ones predicted by the RIS model for the single chain. Intermolecular packing interactions in the structured PDMSM melt are accompanied by expansion of individual chains [5]

Morphology: An amorphous, non-glassy material at room temperature. The packing of poly(di-*n*-alkylsilylenemethylene) (PDASM) chains has been studied using X-ray electron diffraction. The PDASM with the shorter substituent showed a lack of ability to interlock its side chains due to the short length of the alkyl groups. It has been found that the length of the alkyl side chains could change the packing arrangement from monoclinic to orthorhombic to hexagonal with only short-range order as the alkyl side chain length decreases at room temperature [1,15]

Volumetric & Calorimetric Properties:
Density:

No.	Value	Note
1	d 0.906 g/cm^3	[12]

Thermodynamic Properties General: Enthalpy of sorption of *n*-alkanes has been calculated and shown to be in good agreement simulation with available experimental [15]
Glass-Transition Temperature:

No.	Value	Note
1	-95--88°C	Low temperature is a direct result of the larger Si-C bond length compared to the C-C bond length [11,14,16]

Surface Properties & Solubility:
Solubility Properties General: Soluble in a variety of solvents [3,12,15,16]. Viscosity and osmotic pressure of dilute solutions have been measured [3]. High molar mass PDMSM is miscible only with PIB oligomers with MW=1500 g mol^{-1} [11].
Wettability/Surface Energy and Interfacial Tension: Cohesive energy density 32–34cal cm^{-1} at 373K. Surface energy 30–31erg cm^{-1} at 373K [5]

Transport Properties:
Transport Properties General: PDMSM has the highest gas permeability coefficient, but poor film-forming properties of this amorphous polymer complicate its application as a membrane material. PDMSTM, due to its crystallinity and higher glass-transition temperature, shows much lower permeability coefficients [12]
Polymer Melts: PDMSM melt densities in the temperature range 296–308K and 0.1MPa [15]
Permeability and Diffusion General: Experimental and simulated gas permeability of polysilylmethylenes has been reported [12,14]

Mechanical Properties:
Mechanical Properties General: Elasticity and thermoelastic measurements have been recorded [3]. Poor mechanical characteristics, such as inability to form stable films, may be solved by direct controlled curing of the polymer (radiation induced or chemical, in particular, peroxide-induced curing), or by introduction into the polymer of a necessary amount of reactive functional groups (for example, unsaturated) by copolymerisation of the monomer with comonomers containing these groups. This would enable cross-linking following film formation. However, both these approaches have proven difficult to accomplish, because of the high chemical stability of the polymer, and the labour-consuming preparation of the soluble functionalised silylmethylene polymer [14]

Electrical Properties:
Electrical Properties General: Dielectric constants and refractive indices have been measured [3]

Optical Properties:
Transmission and Spectra: Ir [8,10,13], uv [13], nmr [7,13] and X-ray diffraction pattern has been studied [1,13].

Polymer Stability:
Thermal Stability General: Orders of thermal stability are as follows: In He, polydimethylsilylmethylene > polydimethylsiloxane > polyisobutylene; in air, polydimethylsiloxane > polyisobutylene [7]
Decomposition Details: A fractionally pptd. PDMSM had a weight loss of 5% at 266°C in air [11], and less than 5% at 585–600°C in He [7,11]

Applications/Commercial Products:
Processing & Manufacturing Routes: Prepared by the Grignard synthesis of the monomer, 1,1,3,3-tetramethyl-1,3-disilacyclobutane, followed by its ring-opening polymerisation using a suitable platinum catalyst [2,3,6,11]. Photocatalysed ring-opening polymerisation of 1,1,3,3-tetramethyl-1,3-disilacyclobutane [1,2,12]
Applications: PDMSM gums have been converted into a new type of rubber [6]. Polycarbosilanes have been of significant interest owing to their potential applications as silicon carbide ceramic precursors [12]. Most promising polymer for use as a material separating hydrocarbon gases [14]. Membrane material [15]. Secondary battery [17]

Bibliographic References
[1] Park, S.Y., *Macromol. Rapid. Commun.*, 2003, **24**, 793
[2] *Jpn. Pat.*, 1985, 60 084 330
[3] Ko, J.H. and Mark, J.E., *Macromol. Rapid. Commun.*, 1975, **8**, 869
[4] Ko, J.H. and Mark, J.E., *Macromol. Rapid. Commun.*, 1975, **8**, 874
[5] Chen, D. and Mattice, W.L., *Polymer*, 2004, **45**, 3877
[6] Bamford, W.R., Lovie, J.C. and Watt, J.A.C., *J.C.S.(C)*, 1966, **13**, 1137
[7] Levin, G. and Carmichael, J.B., *J. Polym. Sci.*, Part A: Polym. Chem., 1968, **6**, 1
[8] Grigoriev, V.P. and Ushakov, N.V., *Vysokomol. Soedin., Ser. B*, 1975, **17**, 342
[9] Tsyba, V.T. and Egorov, Y.P., *Spektrosk. At. Mol.*, 1969, 454
[10] Tsyba, V.T., *Vysokomol. Soedin., Ser. B*, 1969, **11**, 351
[11] Maier, R.D., Kopf, M., Mader, D., Koopmann, F., Frey H. and Kressler, J., *Acta Polym.*, 1998, **49**, 356
[12] Alentiev A., Economou I.G., Finkelshtein E., Petrou J., Raptis V.E., Sanopoulou M., Soloviev, S., Ushakov, N. and Yampolskii, Y., *Polymer*, 2004, **45**, 6933
[13] Wu, X. and Neckers, D.C., *Macromol. Rapid. Commun.*, 1999, **32**, 6003
[14] Finkelshtein, E.S., Ushakov, N.V., Krasheninnikov, E.G. and Yampolskii, Y.P., *Russ. Chem. Bull. (Engl. Transl.)*, 2004, **53**, 2604
[15] Raptis, V.E., Economou, I.G., Theodorou, D.N., Petrou, J. and Petropoulos, J.H., *Macromol. Rapid. Commun.*, 2004, **37**, 1102
[16] Soloviev, S., Yampolskii, Y. and Economou, I.G., *Polym. Mater. Sci. Eng.*, 2001, **85**, 341
[17] *Jpn. Pat.*, 1998, 10 208 747
[18] *Jpn. Pat.*, 1988, 63 152 673

Poly(dinonylsilane)   P-195

$$\left[\begin{array}{c}CH_2(CH_2)_7CH_3\\|\\Si\\|\\CH_2(CH_2)_7CH_3\end{array}\right]_n$$

Synonyms: *Poly(dinonylsilylene)*. *PDNS*
Related Polymers: Polycarbosilanes, Poly(methyl phenyl silylene methylene), Dichloropoly(methylphenylsilane), Poly(methylphenylsilane)
Polymer Type: polysilanes
CAS Number:

CAS Reg. No.
148160-57-0

Molecular Formula: $(C_{18}H_{38}Si)_n$
Fragments: $C_{18}H_{38}Si$
Morphology: Si-Si backbone conformation *trans*-gauche-*trans*-gauche′ (TGTG′) has been observed.

The room temperature d-spacings can be indexed by an orthorhombic unit cell with dimensions a=20.5Å; b=25.1Å; c=7.8Å and γ=90° [1,2]

Electrical Properties:
Electrical Properties General: A value of $2.6 \times 10^{-7} S\ m^2/J$ is found for phase KTG. Above the KTG to M transition at 23°C, a value of $2.3 \times 10^{-8} S\ m^2/J$ in phase M is obtained close to the values obtained for the other polymers in phase M [2]

Optical Properties:
Optical Properties General: At high temperatures, a single absorption at 318nm is observed for phase M. On cooling, a small absorption centered at 375nm appeared at 9°C, while at -9°C a hypsochromic shift to 358nm is observed with a shoulder at 350nm. On heating, this transforms into a single absorption centered at 350nm characteristic for phase KTG. This change is related to the cold crystallisation. At 25°C, the normal thermochromism at the transition is observed, resulting in a λ_{max} of 318nm.
The crystallisation to the ktg phase is a slow process so that part of the polymer is immobilised below -15°C in a phase that is less crystalline but shows a somewhat larger λ_{max} of 358nm. At -5°C, the cold crystallisation completes the formation of the ktg phase. Further heating results in the normal ktg to M transition at 23°C

Optical Properties:
Transmission and Spectra: uv [2]

Polymer Stability:
Thermal Stability General: The DSC scan obtained for PDNS shows an exothermic cold crystallisation peak at -5°C with an enthalpy change of *ca.* 4kJ mol^{-1} on heating followed at 23°C by the endothermic peak with a ΔH of 12kJ mol^{-1} associated with the ktg to M transition. On cooling, a single exothermic crystallisation peak is observed at -9°C with a ΔH of 8kJ mol^{-1} [2]

Bibliographic References
[1] KariKari, E.K., Greso, A.J., Farmer, B.L., Miller, R.D. and Rabolt, R.F., *Macromolecules*, 1993, **26**, 3937
[2] van der Laan, G.P., de Haas, M.P., Hummel, A., Frey, H. and Moller, M., *J. Chem. Phys.*, 1996, **100**, 5470

Poly(9,9-dioctylfluorene)   P-196

Synonyms: *Poly(9,9-dioctyl-9H-fluorene-2,7-diyl)*. *PDOF*. *PFO*. *F8*
Material class: Thermoplastic
Polymer Type: polyfluorene
CAS Number:

CAS Reg. No.	Note
123864-00-6	Poly(9,9-dioctylfluorene)
195456-48-5	Poly(9,9-dioctyl-9*H*-fluorene-2,7-diyl)

Molecular Formula: $(C_{29}H_{40})_n$
Fragments: $C_{29}H_{40}$
General Information: Stable blue light-emitting electroluminescent and photoluminescent polymer with liquid crystalline properties. Polymer is produced as yellow powder or flakes.
Morphology: Morphology of crystal phases reported in detail [1,2,3]. Heat treatment of liquid crystalline polymer followed by quenching results in orientation as glassy monodomain on a suitable polymer layer for use in LED devices [4].

Volumetric & Calorimetric Properties:
Glass-Transition Temperature:

No.	Value	Note
1	47–80°C	dependent on molecular weight [1,5,6,7,8]

Transition Temperature General: Transition to nematic phase at about 160° [1]. Liquid becomes isotropic at about 270–280° [1,20].

Surface Properties & Solubility:
Solvents/Non-solvents: Sol. $CHCl_3$, toluene, CH_2Cl_2, THF [5,6].

Electrical Properties:
Electrical Properties General: Several LED devices with PDOF as blue light-emitter have been reported, with light turn-on voltages of 4–7.5V [4,7,9,10]. Ionisation potential of PDOF film 5.6–5.8eV [11,12].

Optical Properties:
Optical Properties General: Blue-light photoluminescence occurs with λ_{max} in region of 410–420nm and band gap of 3.1–3.6eV [7,12]. Electroluminescence from LED devices with λ_{max} in region of 435–456nm [7,9,10], including polarised electroluminescence [4].

Refractive Index:

No.	Value	Note
1	1.614–1.781	min. at 900nm; max. at 450nm [13]

Transmission and Spectra: Pmr [14,15], cmr [14,15], Raman [16,17], uv and photoluminescence [3,5,7,8,9,16,18], electroluminescence [4,7,9,10,19] and x-ray [1,2,3] spectral data have been reported.

Polymer Stability:
Polymer Stability General: Good stability in presence of oxygen, moisture and light [5].
Thermal Stability General: Thermal degradation above 348°C [7,20].

Applications/Commercial Products:
Processing & Manufacturing Routes: Prod. originally by oxidation of 9,9-dioctylfluorene monomer with $FeCl_3$ [6], and later by Ni(0)-catalysed polymerisation of 2,7-dibromo-9,9-dioctylfluorene, or Suzuki coupling of the dibromo monomer with diboronate derivative to give higher molecular-weight polymers [14,15].
Applications: Main potential application is as a blue light-emitter for light-emitting diodes, also for thin film transistors, lasers and light-emitting electrochemical cells.

Details	Supplier
homopolymers with different end-capped groups	American Dye Source

Bibliographic References
[1] Kawana, S., Durrell, M., Lu, J., Macdonald, J.E., Grell, M., Bradley, D.D.C., Jukes, P.C., Jones, R.A.L. and Bennett, S.L., *Polymer*, 2002, **43**, 1907
[2] Grell, M., Bradley, D.D.C., Ungar. G., Hill, J. and Whitehead, K.S., *Macromolecules*, 1999, **32**, 5810
[3] Grell, M., Bradley, D.D.C., Inebaskaran, M., Ungar, G., Whitehead, K.S. and Woo, E.P., *Synth. Met.*, 2000, **111/2**, 579
[4] Whitehead, K.S., Grell, M., Bradley, D.D.C., Jandke, M. and Strohriegl, P., *Appl. Phys. Lett.*, 2000, **76**, 2946
[5] Teetsov, J. and Fox, M.A., *J. Mater. Chem.*, 1999, **9**, 2117
[6] Fukuda, M., Sawada, K. and Yoshino K., *J. Polym. Sci., Part A: Polym. Chem.*, 1993, **31**, 2465
[7] Yang, W., Hou, Q., Liu, C., Niu, Y., Huang, J., Yang, R. and Cao, Y., *J. Mater. Chem.*, 2003, **13**, 1351
[8] Blondin, P., Bouchard, J., Beaupré, S., Belletete, M., Durocher, G. and Leclerc, M., *Macromolecules*, 2000, **33**, 5874
[9] Grice, A.W., Bradley, D.D.C., Bernius, M.T., Inebaskaran, M., Wu, W.W. and Woo, E.P., *Appl. Phys. Lett.*, 1998, **73**, 629
[10] Cheun, H. and Winokur, M.J., *Synth. Met.*, 2005, **154**, 137
[11] Janietz, S., Bradley, D.D.C., Grell, M., Giebeler, C., Inebaskaran, M. and Woo, E.P., *Appl. Phys. Lett.*, 1998, **73**, 2453
[12] Liao, L.S., Fung, M.K., Lee, C.S., Lee, S.T., Inebaskaran, M., Woo, E.P. and Wu, W.W., *Appl. Phys. Lett.*, 2000, **76**, 3582
[13] Wang, X.H., Grell, M., Lane, P.A. and Bradley D.D.C., *Synth. Met.*, 2001, **119**, 535
[14] Ranger, M., Rondeau, D. and Leclerc, M., *Macromolecules*, 1997, **30**, 7686
[15] Kameshima, H., Nemoto, N. and Endo, T., *J. Polym. Sci., Part A: Polym. Chem.*, 2001, **39**, 3143
[16] Ariu, M., Lidzey, D.G. and Bradley, D.D.C., *Synth. Met.*, 2000, **111/2**, 607
[17] Ariu, M., Lidzey, D.G., Lavrentiev, M., Bradley, D.D.C., Jandke, M. and Strohriegl, P., *Synth. Met.*, 2001, **116**, 217
[18] Cadby, A.J., Lane, P.A., Wohlgenannt, An, C., Vardeny, Z.V. and Bradley D.D.C., *Synth. Met.*, 2000, **111/2**, 515
[19] Fujishima, D., Mori, T., Mizutani, T., Yamamoto, T. and Kitamura, N., *Jpn. J. Appl. Phys.*, 2005, **44**, 546
[20] Grell, M., Bradley, D.D.C., Inebaskaran, M. and Woo, E.P., *Adv. Mater.*, 1997, **9**, 798

Poly(9,9-dioctylfluorene-*co*-2,1,3-benzothiadiazole)

Synonyms: Poly[2,1,3-benzothiadiazole-4,7-diyl(9,9-dioctyl-9H-fluorene-2,7-diyl)]. 9,9-Dioctylfluorene 2,1,3-benzothiadiazole copolymer. F8BT
Related Polymers: Poly(9,9-dioctylfluorene)
Monomers: 2,1,3-Benzothiadiazole
Material class: Thermoplastic
Polymer Type: polyfluorene
CAS Number:

CAS Reg. No.	Note
210347-52-7	Poly[2,1,3-benzothiadiazole-4,7-diyl (9,9-dioctyl-9H-fluorene-2,7-diyl)]
316825-94-2	9,9-Dioctylfluorene-2,1,3-benzothiadiazole copolymer

Molecular Formula: $(C_{35}H_{42}N_2S)_n$
Fragments: $C_{35}H_{42}N_2S$
General Information: Stable green-yellow light-emitting photoluminescent and electroluminescent polymer with liquid crystalline properties and with good electron transport mobility. Polymer is prod. as yellow fibrous solid.
Morphology: Morphology of films is dependent on molecular weight and annealing temperature [1] as well as solvent used for casting [2].

Volumetric & Calorimetric Properties:
Glass-Transition Temperature:

No.	Value	Note
1	125–135°C	[3]

Transition Temperature General: Transition to liquid crystalline phase in the region of 220–280°C depending on molecular weight [3,16]. High molecular-weight samples still exhibited no transition to isotropic liquid above 300°C [3].

Surface Properties & Solubility:
Cohesive Energy Density Solubility Parameters: Solubility parameter 19.4–21.6 $(J\ cm^{-3})^{1/2}$ [20].
Solvents/Non-solvents: Tetrachloroethane, p-xylene, o-xylene, toluene, $CHCl_3$, isodurene [2].

Electrical Properties:
Electrical Properties General: Electron mobility has been recorded; ionisation potential 5.9eV, estimated electron affinity 3.2eV [4]. Photocurrent in photovoltaic devices is greatly enhanced by blending with a hole-transporting fluorene copolymer [5,6,7,8]. LED devices with F8BT as green-yellow light emitter have been reported [9,10,11].

Optical Properties:
Optical Properties General: Green-yellow light photoluminescence and electroluminescence with λ_{max} in the region of 550nm reported, with dichroic ratios up to 9 from aligned films [11]. Refractive indices for aligned and unaligned films at different wavelengths have been reported [12,13].
Transmission and Spectra: Pmr [14], Raman [1,15], uv and photoluminescence [1,6,7,8,9,10,11,16], electroluminescence [9,10,11] and x-ray [1] spectral data have been reported.

Applications/Commercial Products:
Processing & Manufacturing Routes: Produced mainly by Suzuki coupling between 4,7-dibromo-2,1,3-benzothiadiazole and the diboronate derivative obtained from 2,7-dibromo-9,9-dioctylfluorene [14,17,18].
Applications: Main potential applications as polymer blend in photovoltaic devices, for light-emitting diodes and as gain medium for lasers.

Details	Supplier
copolymer endcapped with 3,5-dimethylphenyl groups	American Dye Source

Bibliographic References
[1] Donley, C.L., Zaumseil, J., Andreasen, J.W., Nielsen, M.M., Sirringhaus, H., Friend, R.H. and Kim, J.-S., *J. Am. Chem. Soc.*, 2005, **127**, 12890
[2] Banach, M.J., Friend, R.H. and Sirringhaus, H., *Macromolecules*, 2004, **37**, 6079
[3] Banach, M.J., Friend, R.H. and Sirringhaus, H., *Macromolecules*, 2003, **36**, 2838
[4] Campbell, A.J., Bradley, D.D.C. and Antoniadis, H., *Appl. Phys. Lett.*, 2001, **79**, 2133
[5] Halls, J.J.M., Arias, A.C., MacKenzie, J.D., Wu, W., Inbasekaran, M., Woo, E.P. and Friend, R.H., *Adv. Mater.*, 2000, **12**, 498
[6] Kietzke, T., Neher, D., Kumske, M., Montenegro, R., Landfester, K. and Scherf, U., *Macromolecules*, 2004, **37**, 4882
[7] Arias, A.C., MacKenzie, J.D., Stevenson, R., Halls, J.J.M., Inbasekaran, M., Woo, E.P., Richards, D. and Friend, R.H., *Macromolecules*, 2001, **34**, 6005
[8] Pacios, R. and Bradley D.D.C., *Synth. Met.*, 2002, **127**, 261
[9] Morgado, J., Moons, E., Friend, R.H. and Cacialli, F., *Synth. Met.*, 2001, **124**, 63
[10] He, Y., Gong, S., Hattori, R. and Kanicki, J., *Appl. Phys. Lett.*, 1999, **74**, 2265
[11] Whitehead, K.S., Grell, M., Bradley, D.D.C., Inbasekaran, M. and Woo, E.P., *Synth. Met.*, 2000, **111/2**, 181
[12] Xia, R., Campoy-Quiles, M., Heliotis, G., Stavrinou, P., Whitehead, K.S. and Bradley, D.D.C., *Synth. Met.*, 2005, **155**, 274
[13] Ramsdale, C.M. and Greenham N.C., *J. Phys. D: Appl. Phys.*, 2003, **36**, L29
[14] Yan, H., Lee, P., Armstrong, N.R., Graham, A., Evmenenko, G.A., Dutta, P. and Marks, T.J., *J. Am. Chem. Soc.*, 2005, **127**, 3172
[15] Stevenson, R., Arias, A.C., Ramsdale, C., MacKenzie, J.D. and Richards, R., *Appl. Phys. Lett.*, 2001, **79**, 2178
[16] Grell, M., Redecker, M., Whitehead, K.S., Bradley, D.D.C., Inbasekaran, M., Woo, E.P. and Wu, W., *Liq. Cryst.*, 1999, **26**, 1403
[17] U.S. Pat., Dow Chemical, 1998, 5 777 070
[18] Pat. Coop. Treaty (WIPO), 2000, 00 53 656
[19] Moons, E., *J. Phys.: Condens. Matter*, 2002, **14**, 12235
[20] Kim, J.-S., Ho, P.K.H., Murphy, C.E. and Friend, R.H., *Macromolecules*, 2004, **37**, 2861

Poly(9,9-dioctylfluorene-*co*-bis-*N,N'*-(4-butylphenyl)-*N,N'*-diphenyl-1,4-benzenediamine)   P-198

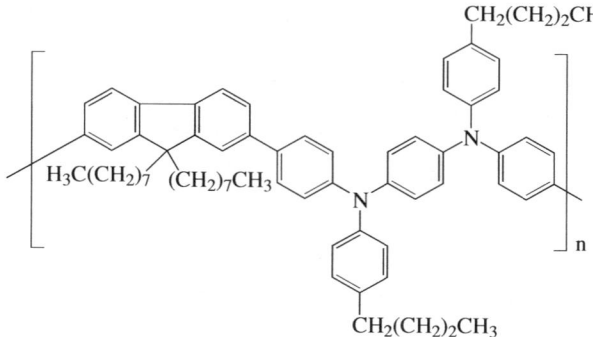

Synonyms: PFB
Material class: Thermoplastic
Polymer Type: polyfluorene
CAS Number:

CAS Reg. No.
223569-28-6

Molecular Formula: $(C_{67}H_{78})_n$
Fragments: $C_{67}H_{78}$
General Information: Stable blue light-emitting photoluminescent and electroluminescent polymer with high hole-transporting mobility.

Surface Properties & Solubility:
Solvents/Non-solvents: Sol. $CHCl_3$, xylene [1,2].

Electrical Properties:
Electrical Properties General: Ionisation potential 5.09eV [3]. The hole mobility of PFB films has been studied [3,4]. In photovoltaic devices the photocurrent from PFB alone is much less than from blends with F8BT [5].

Optical Properties:
Optical Properties General: Blue-light photoluminescence and electroluminescence with λ_{max} values in the region of 450–460nm reported [1,6]. A graph of refractive indices vs. wavelength in films has been reported [7].
Transmission and Spectra: UV and photoluminescence [1,2,6], electroluminescence [6], and Raman [8] spectral data have been reported.

Applications/Commercial Products:
Processing & Manufacturing Routes: Prod. mostly by Suzuki coupling of the diboronate derivative of 2,7-dibromo-9,9-dioctylfluorene with the dibromo derivative of the benzenediamine monomer [3,9].

Applications: Mostly used as a polymer blend with dioctylfluorene-benzothiadiazole copolymer for development of photovoltaic devices.

Details	Supplier
copolymer with different end-capped groups	American Dye Source

Bibliographic References

[1] Kietzke, T., Neher, D., Kumske, M., Montenegro, R., Landfester, K. and Scherf, U., *Macromolecules*, 2004, **37**, 4882
[2] Arias, A.C., MacKenzie, J.D., Stevenson, R., Halls, J.J.M., Inbasekaran, M., Woo, E.P., Richards, D. and Friend, R.H., *Macromolecules*, 2001, **34**, 6005
[3] Redecker, M., Bradley, D.D.C., Inbasekaran, M., Wu, W.W. and Woo E.P., *Adv. Mater.*, 1999, **11**, 241
[4] Poplavskyy, D., Nelson, J. and Bradley, D.D.C., *Makromol. Chem., Makromol. Symp.*, 2004, **212**, 415
[5] Halls, J.J.M., Arias, A.C., MacKenzie, J.D., Wu, W., Inbasekaran, M., Woo, E.P. and Friend, R.H., *Adv. Mater.*, 2000, **12**, 498
[6] Cirpan, A., Ding, L. and Karasz, F.E., *Synth. Met.*, 2005, **150**, 195
[7] Ramsdale, C.M. and Greenham, N.C., *J. Phys. D: Appl. Phys.*, 2003, **36**, 129
[8] Stevenson, R., Arias, A.C., Ramsdale, C., MacKenzie, J.D. and Richards, D., *Appl. Phys. Lett.*, 2001, **79**, 2178
[9] *Pat. Coop. Treaty (WIPO)*, 2005, 05 49 689

Poly(9,9-dioctylfluorene-*co*-2,2′-bithiophene) P-199

Synonyms: *Poly[[2,2′-bithiophene]-5,5′-diyl(9,9-dioctyl-9H-fluorene-2,7-diyl)]*. 9,9-Dioctylfluorene-2,2′-bithiophene copolymer. F8T2. PBTF
Related Polymers: Poly(9,9-dioctylfluorene), Poly(thiophene)
Material class: Thermoplastic
Polymer Type: polyfluorene
CAS Number:

CAS Reg. No.	Note
210347-56-1	Poly[[2,2′-bithiophene]-5,5′-diyl(9,9-dioctyl-9H-fluorene-2,7-diyl)]
289625-34-9	9,9-Dioctylfluorene-2,2′-bithiophene copolymer

Molecular Formula: $(C_{37}H_{44}S_2)n$
Fragments: $C_{37}H_{44}S_2$
General Information: Stable green light-emitting photoluminescent and electroluminescent polymer with liquid crystalline properties and with good hole mobility. Polymer is prod. as yellow fibrous solid.
Morphology: Morphology has been described, with nematic phase usually occurring above 265°C [1,2,3].

Volumetric & Calorimetric Properties:
Glass-Transition Temperature:

No.	Value	Note
1	73–132°C	[3,4]

Transition Temperature General: Melting to nematic phase in the region of 200–265°C [1,3]. Recrystallisation occurs at 195°C on slow cooling [3].

Surface Properties & Solubility:
Solvents/Non-solvents: Sol. xylenes, THF, $CHCl_3$ [2,5]

Electrical Properties:
Electrical Properties General: Hole-transporting polymer with p-channel characteristics in thin-film transistors [2,3]. Aligned films show mobility anisotropies of current flow in transistor [6]. LED devices with F8T2 as green light emitter have been reported [4,7,8].

Optical Properties:
Optical Properties General: Green light photoluminescence and electroluminescence with λ_{max} in the region of 550–590nm; a red shift is observed on changing from $CHCl_3$ soln. to film [4,5,7,9].
Transmission and Spectra: Pmr [5,9,10], cmr [9,10], ir [10], uv and photoluminescence [1,2,4,5,6,7,9,10] and electroluminescence [4,7,8] spectral data have been reported.

Polymer Stability:
Decomposition Details: Decomposition temp. >400°C [2].

Applications/Commercial Products:
Processing & Manufacturing Routes: Prod. mainly by Suzuki coupling between 5,5′-dibromo-2,2′-bithiophene and the diboronate derivative obtained from 2,7-dibromo-9,9-dioctylfluorene [5,8,11]; also by Stille-type coupling with distannylated bithiophene derivative [9].
Applications: Potential applications in thin-film transistors, light-emitting diodes and photovoltaic devices.

Bibliographic References

[1] Grell, M., Redecker, M., Whitehead, K.S., Bradley, D.D.C., Inbasekaran, M., Woo, E.P. and Wu, W., *Liq. Cryst.*, 1999, **26**, 1403
[2] Kinder, L., Kanicki, J. and Petroff, P., *Synth. Met.*, 2004, **146**, 181
[3] Lim, E., Jung, B.-J., Lee, J., Shim, H.-K., Lee, J.-I., Yang, Y.S. and Do, L.-M., *Macromolecules*, 2005, **38**, 4531
[4] Donat-Bouillud, A., Lévesque, I., Tao, Y., D'Iorio, M., Beaupré, S., Blondin, P., Ranger, M., Bouchard, J. and Leclerc, M., *Chem. Mater.*, 2000, **12**, 1931
[5] Ranger, M. and Leclerc, M., *Can. J. Chem.*, 1998, **76**, 1571
[6] Sirringhaus, H., Wilson, R.J., Friend, R.H., Inbasekaran, M., Wu, W., Woo, E.P., Grell, M. and Bradley, D.D.C., *Appl. Phys. Lett.*, 2000, **77**, 406
[7] Lévesque, I., Donat-Bouillud, A., Tao, Y., D'Iorio, M., Beaupré, S., Blondin, P., Ranger, M., Bouchard, J. and Leclerc, M., *Synth. Met.*, 2001, **122**, 79
[8] Lim, E., Jung, B-J. and Shim, H.-K., *Macromolecules*, 2003, **36**, 4288
[9] Asawapirom, U., Güntner, R., Forster, M., Farrell, T. and Scherf, U., *Synthesis*, 2002, 1136
[10] D'Amore, F., Osmond, J., Destri, S., Pasini, M., Rossi, V. and Porzio, A., *Synth. Met.*, 2005, **149**, 123
[11] *U.S. Pat.*, Dow Chemical, 1998, 5 777 070

Poly(9,9-dioctylfluorene-co-N-(4-(1-methylpropyl)phenyl)diphenylamine) P-200

Synonyms: Poly[[4-(1-methylpropyl)phenyl]imino]-1,4-phenylene(9,9-dioctyl-9H-fluorene-2,7-diyl)-1,4-phenylene]. TFB
Related Polymers: Poly(9,9-dioctylfluorene)
Material class: Thermoplastic
Polymer Type: polyfluorene
CAS Number:

CAS Reg. No.
220797-16-0

Molecular Formula: $(C_{51}H_{61}N)_n$
Fragments: $C_{51}H_{61}N$
General Information: Stable photoluminescent and electroluminescent polymer with good hole transporting properties. Polymer is prod. as pale yellow solid.

Volumetric & Calorimetric Properties:
Glass-Transition Temperature:

No.	Value	Note
1	148°C	reported for sample with M_n=56000 [1]

Surface Properties & Solubility:
Cohesive Energy Density Solubility Parameters: Solubility parameter 17.3–19.3 $(J\ cm^{-3})^{1/2}$ [1]
Solvents/Non-solvents: Sol. p-xylene [1], toluene [2]

Electrical Properties:
Electrical Properties General: Polymer has good hole transport mobility, with ionisation potential of 5.30eV and band gap of 3eV [1,3]. LED devices of TFB blended with F8BT with strong electroluminescence and light turn-on voltages as low as 2V have been reported [1,2,4,5,6,9].

Optical Properties:
Optical Properties General: Polymer exhibits blue-light photoluminescence with λ_{max} in the region of 440nm [7]; yellow-green light electroluminescence from LED devices with TFB:F8BT blends.
Transmission and Spectra: Pmr [2], Raman [1], uv and photoluminescence [7,8] and photoelectron [8] spectral data have been reported.

Applications/Commercial Products:
Processing & Manufacturing Routes: Prod. by Suzuki coupling between N,N-bis(4-bromophenyl)-4-(1-methylpropyl)aniline and the diboronate compd. derived from 2,7-dibromo-9,9-dioctylfluorene [2,3].
Applications: Used mainly as blend with F8BT for development of light-emitting diodes.

Supplier
H. W. Sands

Bibliographic References
[1] Kim, J.-S., Ho, P.K.H., Murphy, C.E. and Friend, R.H., *Macromolecules*, 2004, **37**, 2861
[2] Yan, H., Lee, P., Armstrong, N.R., Graham, A., Evmenenko, G.A., Dutta, P. and Marks, T.J., *J. Am. Chem. Soc.*, 2005, **127**, 3172
[3] Redecker, M., Bradley, D.D.C., Inbasekaran, M., Wu, W.W. and Woo, E.P., *Adv. Mater.*, 1999, **11**, 241
[4] Corcoran, N., Arias, A.C., Kim, J.S., MacKenzie, J.D. and Friend, R.H., *Appl. Phys. Lett.*, 2003, **82**, 299
[5] Kim, J.S., Friend, R.H., Grizzi, I. and Burroughes, J.H., *Appl. Phys. Lett.*, 2005, **87**, 023506-1
[6] Moons, E., *J. Phys.: Condens. Matter*, 2002, **14**, 12235
[7] Hou, Y., Koelberg, M. and Bradley, D.D.C., *Synth. Met.*, 2003, **139**, 859
[8] Sancho-Garcia, J.C., Foden, C.L., Grizzi, I., Greczynski, G., de Jong, M.P., Salaneck, W.R., Brédas, J.L. and Cornil, J., *J. Phys. Chem. B*, 2004, **108**, 5594
[9] Xia, Y. and Friend, R.H., *Macromolecules*, 2005, **38**, 6466

Poly(2,5-dioctyloxy-1,4-phenylene ethynylene) P-201

Synonyms: Poly[[2,5-bis(octyloxy)-1,4-phenylene]-1,2-ethynediyl]. O-OPPE
Material class: Thermoplastic
Polymer Type: poly(arylene ethynylene)
CAS Number:

CAS Reg. No.	Note
153033-25-1	poly[[2,5-bis(octyloxy)-1,4-phenylene]-1,2-ethynediyl]
153033-33-1	

Molecular Formula: $(C_{24}H_{36}O_2)n$
Fragments: $C_{24}H_{36}O_2$
General Information: Photoluminescent and electroluminescent polymer with greater solubility than poly(1,4-phenylene ethynylene). Prod. as orange solid.
Morphology: The annealed polymer exhibits a layered structure with main chains coplanar and side chains extending obliquely from main chains and not interleaved with adjacent side chains [1].

Surface Properties & Solubility:
Solvents/Non-solvents: Sol. $CHCl_3$, toluene, CH_2Cl_2 [1]

Electrical Properties:
Electrical Properties General: Exhibits electroluminescence in LED devices in yellow-green region of spectrum with turn-on voltage of 11v [2].

Optical Properties:
Optical Properties General: Exhibits photoluminescence in green region of spectrum (λ_{max} 470–480 nm in $CHCl_3$) [1,3,4] with a considerable red shift on going from soln. to film (λ_{max} in region of 540–580 nm) [1,5]. Refractive indices and third-order nonlinear optical susceptibilities have been reported [6].
Refractive Index:

No.	Value	Note
1	1.78	(550nm, film) [1]

Transmission and Spectra: Pmr [3,4], cmr [4], ir [4], uv and photoluminescence [1,3,4,5], electroluminescence [2] and x-ray diffraction [1] spectral data have been reported

Applications/Commercial Products:

Processing & Manufacturing Routes: Prod. by catalytic coupling reaction between 1,4-dioctyloxy-2,5-diiodobenzene and 1,4-diethynyl-2,5-dioctyloxybenzene [1], or Pd-catalysed reaction of 1,4-dioctyloxy-2,5-diiodobenzene with tributylethynyltin [4,5].
Applications: Potential applications in chemical sensors [7] and for light-emitting diodes [2].

Bibliographic References

[1] Weder, C. and Wrighton M.S., *Macromolecules*, 1996, **29**, 5157
[2] Montali, A., Smith, P. and Weder, C., *Synth. Met.*, 1998, **97**, 123
[3] Beck, J.B., Kokil, A., Ray, D., Rowan, S.J. and Weder, C., *Macromolecules*, 2002, **35**, 590
[4] Giardina, G., Rosi, P., Ricci, A. and Lo Sterzo, C., *J. Polym. Sci., Part A: Polym. Chem.*, 2000, **38**, 2603
[5] Pizzoferrato, R., Berliocchi, A., Di Carlo, A., Lugli, P., Venanzi, M., Micozzi, A., Ricci, A. and Lo Sterzo, C., *Macromolecules*, 2003, **36**, 2215
[6] Weder, C., Wrighton, M.S., Spreiter, R., Bosshard, C. and Günter, P., *J. Phys. Chem.*, 1996, **100**, 18931
[7] Penza, M., Cassano, G., Sergi, A., Lo Sterzo, C. and Russo, M.V., *Sens. Actuators*, 2001, **81**, 88

Poly(dioctylsilane) P-202

$$\left[\begin{array}{c} CH_2(CH_2)_6CH_3 \\ | \\ Si \\ | \\ CH_2(CH_2)_6CH_3 \end{array} \right]_n$$

Synonyms: *Poly(dioctylsilylene)*. PDOctSi. PDOS. PDOctS
Related Polymers: Polycarbosilanes, Poly(methyl phenyl silylene methylene), Dichloropoly(methylphenylsilane), Poly(methylphenylsilane)
Polymer Type: polysilanes
CAS Number:

CAS Reg. No.	Note
97036-68-5	PDOS homopolymer
98865-30-6	PDOS SRU

Molecular Formula: $(C_{16}H_{34}Si)n$
Fragments: $C_{16}H_{34}Si$
Molecular Weight: $32,000-10^6$. The two samples were separated from a reaction product which had a bimodal molecular weight distribution. The higher molecular weight material adopts the *all-trans* conformation [2,5,6]
Tacticity: Exhibits the hexagonal, columnar mesophase both above and below its first-order transition temp. of -32°C. This type of behaviour is characteristic of poly(*n*-alkyl-*n*-alkylsilanes) with a side chain asymmetry of 2 or 3 C atoms.
Morphology: Soft white waxy solid.
3 types of Si-Si backbone conformations, i.e., helical, planar zigzag, and *trans-gauche-trans-gauche′* (TGTG′), have been observed.
Unit cells parameters a=16.4Å; b=26.4Å; c=4.0Å and γ=90° [1,2,3,4,5,7,10,11]
PDOS has been found to yield at least three structures, and these are referred to as QM, ordered I, and II. All three appear to have qualitatively similar analogues in the literature. PDOS samples evidence two additional structures which are, from their X-ray spectra, reminiscent of the higher temperature thermotropic mesophase, type M, and so are designated M′ and M″. This M′ phase can best be distinguished by other characterisation methods. Isolating the claimed M″ was especially problematic because it is transient and only appeared as a minority phase in PDOS during the type I to II transition. Assuming solely a simple two-phase superposition of the type I and II phases does not fit the indicated M″ profile.
The type II structure is monotropic and appears only after warming from either the QM or type I phases above 10°C. In the latter instance, this process occurs extremely slowly. The PDOS type I phase itself forms only on cooling from the mesophase within a narrow temperature window spanning from 5 to 10°C. The precise conditions for formation of the M′ and M″ phases are not fully clarified. In this study the M′ phase was only observed on heating, but there is evidence that this phase may also form on cooling at sufficiently slow cooling rates.
Miscellaneous: Exhibits up to five distinct structural forms at temperatures below that of a 25°C transition to the thermotropic mesophase. Three of these are clearly ordered while the other two are mesophases with features suggestive of a less ordered state [5]

Volumetric & Calorimetric Properties:
Density:

No.	Value	Note
1	d 0.91–1.01 g/cm^3	[5]

Thermodynamic Properties General: See [7,10]

Surface Properties & Solubility:
Solvents/Non-solvents: Quite hydrophobic and impervious to water, but soluble in toluene and other nonpolar organic solvents [1]

Electrical Properties:
Electrical Properties General: In the crystalline phase, the polymer displays a considerably higher conductivity than PDBS and PDHS in phase radiation-induced KHE. A similar value of 7×10^{-7} Sm^2/J has been observed [7]. Resistivity properties have been reported [13].

Optical Properties:
Optical Properties General: Exhibits a dramatic change in UV absorption above the phase-transition temperature. A shift of the 315nm absorption band to 370nm and a continuous blue shift of this band with pressure have been observed [4,10]

Optical Properties:
Transmission and Spectra: Ir [3,4]; Raman [3]; uv-vis [1,3,4,5,6,7,9,10,11]; nmr [9]; epr [12] spectra reported.

Polymer Stability:
Thermal Stability General: The type II and M′ structures both appear thermally stable up to 25°C. Thus, the reported 10°C crystal-HCM transition represents alternative structural pathways (on heating either QM to M′ or I to M″) [5,10]
Heating: *all-trans* phase to mesophase change at 11°C
Cooling: mesophase to *all-trans* phase change at -3°C [7]

Polymer Stability:
Decomposition Details: The DSC reveals a large cooling peak at 282K and two very small peaks at 252K and 316K. From this it has been observed that the polymer has a three-structure below 282K but a dimensional-ordered hcm with melted side chains and a disordered structure backbone above 316K. In addition, the small DSC peak at 316K should be attributed to the remaining *all-trans* component disappearance [10]

Applications/Commercial Products:
Processing & Manufacturing Routes: The dichlorodioctylsilane was polymerised by a Wurtz condensation reaction with Na in toluene at 110°C (36% yield) [1,5,10]
Applications: Used in electrically conductive polymer compositions [13,14]. Electrophotographic photoreceptor [15]. Effective mid-UV resists [16]

Bibliographic References

[1] Klemann, B.M., West, R. and Koutsky, J.A., *Macromolecules*, 1996, **29**, 198
[2] KariKari, E.K., Greso, A.J., Farmer, B.L., Miller, R.D. and Rabolt, R.F., *Macromolecules*, 1993, **26**, 3937
[3] Rabolt, J.F., Hofer, D., Miller, R.D. and Fickes, G.N., *Macromolecules*, 1986, **19**, 611
[4] Song, K., Miller, R.D., Wallraff, G.M., and Rabolt, J.F., *Macromolecules*, 1991, **24**, 4084

[5] Chunwachirasiri, W., West, R. and Winokur, M.J., *Macromolecules*, 2000, **33**, 9720
[6] French, R.H., Meth, J.S., Thorne, J.R.G, Hochstrasser, R.M. and Miller, R.D., *Synth. Met.*, 1992, **50**, 499
[7] van der Laan, G.P, de Haas, M.P., Hummel, A., Frey, H. and Moller, M., *J. Chem. Phys.*, 1996, **100**, 5470
[8] Fujisaka, T. and West, R., *J. Organomet. Chem.*, 1993, **449**, 105
[9] Seki, S., Koizumi, Y., Kawaguchi, T., Habara, H., and Tagawa, S., *J. Am. Chem. Soc.*, 2004, **126**, 3521
[10] Kanai, T., Ishibashi, H., Hayashi, Y., Oka, K., Dohmaru, T., Ogawa, T., Furukawa, S., and West, R., *J. Polym. Sci.*, Part B: Polym. Lett., 2001, **39**, 1085
[11] Chunwachirasiri, W., Kanaglekar, I., Lee, G.H., West, R. and Winokur, M.J., *Synth. Met.*, 2001, **119**, 31
[12] Gregory, M., Thompson, D.P., Miller, R.D. and Michl, J., *J. Am. Chem. Soc.*, 1991, **113**, 2003
[13] *Jpn. Pat.*, 1996, 08 069 709
[14] *Jpn. Pat.*, 1995, 07 254 307
[15] *Jpn. Pat.*, 1990, 02 150 853
[16] *Res. Discl.*, 1987, **273**, 26

Poly(dipentene dioxide) P-203

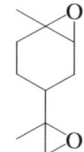

Synonyms: *Poly[1-methyl-4-(2-methyloxiranyl)-7-oxabicyclo[4.1.0]heptane]. Poly(limonene dioxide)*
Material class: Thermosetting resin
Polymer Type: epoxy
CAS Number:

CAS Reg. No.
29616-43-1

Molecular Formula: $(C_{10}H_{16}O_2)_n$

Applications/Commercial Products:
Processing & Manufacturing Routes: May be polymerised by photoinduction in the presence of an aryl substituted sulfonium salt [1]
Applications: Used in coatings. Uncured resin used as a reactive diluent in epoxy resin compositions

Bibliographic References
[1] Akhtar, S.R., Crivello, J.V., Lee, J.L. and Schmitt, M.L., *Chem. Mater.*, 1990, **2**, 732, (synth)

Poly(dipentylsilane) P-204

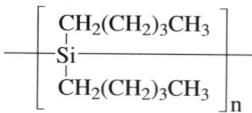

Synonyms: *Poly(dipentylsilylene). Dichlorodipentylsilane homopolymer. PDPS*
Material class: Thermoplastic
Polymer Type: polysilanes
CAS Number:

CAS Reg. No.	Note
96228-24-9	poly(dipentylsilylene)
97036-66-3	dichlorodipentylsilane homopolymer

Molecular Formula: $(C_{10}H_{22}Si)n$
Fragments: $C_{10}H_{22}Si$
General Information: Thermochromic polymer exhibiting photoluminescence and electroluminescence in uv region. Prod. as white solid.
Morphology: Most stable conformation of backbone at r.t. is a 7/3 helical structure [1,2,3,4]. At low temperatures a *trans*-conformation is stable [2,5], which can also exist at r.t. [1]. On increasing the temperature to 35°C only the 7/3 helical conformation remains [1] and between 56 and 70°C there is a transition to a disordered columnar mesophase [1,2,6]. The *trans*-conformation can be formed at normal temperatures by application of high pressure [7].

Volumetric & Calorimetric Properties:
Glass-Transition Temperature:

No.	Value	Note
1	-45.6°C	derived from heat capacity data [8]

Transition Temperature General: Endothermic transition at 35°C [1] and a weaker transition between 56 and 70°C [1,2,8], which is dependent on molecular weight of the sample [2]. These correspond to changes in conformation [1,2].

Surface Properties & Solubility:
Solvents/Non-solvents: Sol. toluene, $CHCl_3$, hexane, cyclohexane, THF; insol. alcohols [9].

Electrical Properties:
Electrical Properties General: Hole transport predominates by interchain hopping. There is little difference in hole transport for oriented and unoriented films [5]. LED devices using the polymer as an emissive layer have been produced, with electroluminescence in UV region (λ_{max} 355nm) [10]

Optical Properties:
Optical Properties General: For polymer solutions the UV absorption spectrum exhibits a red shift with a reduction in temperature (λ_{max} 320nm at r.t. to 353nm at -36°C) [3]. Polymer films on heating exhibit a broadening of absorption with little change in position (λ_{max} *ca.* 315nm) [3,4]. An increase from normal to high pressure applied to polymer films results in a considerable red shift (λ_{max} 315 to 360nm), which is not completely reversible on reducing the pressure [11,12]. Photoluminescence peaks occur in UV region (λ_{max} 340, 370nm) [10]
Transmission and Spectra: Si-29 nmr [3,6,13], cmr [3], Raman [1,2,4,12], uv [1,3,4,11,12], electroluminescence [10] and x-ray diffraction data [1,3,4,7] have been reported

Polymer Stability:
Polymer Stability General: Photochemical degradation occurs on prolonged exposure to UV radiation by scission of σ-bonds
Decomposition Details: Decomposes (5% weight loss) at 272°C [8]

Applications/Commercial Products:
Processing & Manufacturing Routes: Prod. by Wurtz dehalogenation coupling of dichlorodipentylsilane with Na in toluene [9,13]

Bibliographic References
[1] KariKari, E.K., Greso, A.J., Farmer, B.L., Miller, R.D. and Rabolt, J.F., *Macromolecules*, 1993, **26**, 3937
[2] Bukalov, S.S., Leites, L.A., Magdanurov, G.I. and West, R., *J. Organomet. Chem.*, 2003, **685**, 51
[3] Schilling, F.C., Lovinger, A.J., Zeigler, J.M., Davis, D.D. and Bovey, F.A., *Macromolecules*, 1989, **22**, 3055
[4] Miller, R.D., Farmer, B.L., Fleming, W., Sooriyakumaran, R. and Rabolt, J., *J. Am. Chem. Soc.*, 1987, **109**, 2509
[5] Nakayama, Y., Hirooka, K., Oka, K. and West, R., *Solid State Commun.*, 1999, **109**, 45
[6] Schilling, F.C. and Bovey, F.A., *Polym. Prepr. (Am. Chem. Soc., Div. Polym. Chem.)*, 1998, **29**, 72
[7] Furukawa, S., Takeuchi, K.-I. and Shimana, M., *J. Phys.: Condens. Matter*, 1994, **6**, 11007

[8] Varma-Nair, M., Cheng, J., Jin, Y. and Wunderlich, B., *Macromolecules*, 1991, **24**, 5442
[9] Schwegler, L.A., Meyer-Pundsack, C. and Möller, M., *J. Polym. Sci., Part A: Polym. Chem.*, 2000, **38**, 2306
[10] Suzuki, H., Hoshino, S., Yuan, C.-H., Fujiki, M., Toyoda, S. and Matsumoto, N., *Thin Solid Films*, 1998, **331**, 64
[11] Rabolt, J.F., Song, K., Kuzmany, H., Sooriyakumaran, R., Fickes, G. and Miller, R.D., *Polym. Prepr. (Am. Chem. Soc., Div. Polym. Chem.)*, 1990, **31**, 262
[12] Song, K., Miller, R.D., Wallraff, G.M. and Rabolt, J.F., *Macromolecules*, 1991, **24**, 4084
[13] Frey, H., Matyjaszewski, K., Möller, M. and Oelfin, D., *Colloid Polym. Sci.*, 1991, **269**, 442

Poly(*N,N'*-diphenylbenzidine) P-205

Synonyms: *Poly(DPBz). NNPheBz. Poly(DPB)*
Monomers: *N,N'*-Diphenylbenzidine
Polymer Type: polyaniline
CAS Number:

CAS Reg. No.
108443-85-2

Molecular Formula: $(C_{24}H_{18}N_2)_n$
Fragments: $C_{24}H_{18}N_2$
Molecular Weight: Initial GPC studies indicated sharp MW distributions for all polymers [3]
General Information: Can be reversibly doped [2]
Morphology: Green-dark brown [3]. *sec*-Amine hydrogen signatures observed in ir and nmr and accompanying strong five adjacent aromatic H signatures in the ir indicate alternating N- and *p*- linkages for the polymer [2,3]
Identification: C, 6; H, 4.6; N, 0.49 [3,5,6]

Surface Properties & Solubility:
Solubility Properties General: Can be recast from films from saturated solutions in a number of solvents, including $CHCl_3$ and DMF, of which the latter was found to be the most solubilising. Insoluble in H_2O [3]

Electrical Properties:
Electrical Properties General: $\sigma = 2 \times 10^{-5}$ to 9.7×10^{-7} S cm^{-1} [2,3,6]

Optical Properties:
Optical Properties General: Forms a green to brown (dependent on doping state), environmentally stable, electroactive film, with electrochromic changes from a glass-clear transparency at extreme cathodic (reducing) potential to dark blue at large oxidative potential [3]
Transmission and Spectra: Ir [2,3,4,5,6], uv-vis [3,5], xps [2,5] and nmr [3] data reported

Polymer Stability:
Thermal Stability General: The thermogram of Poly(DPBz) indicates that the onset of the first major weight loss step occurs at about the same temperature as that of the monomer. The weight loss at this temperature constitutes approx. 20–40% of the initial weight, depending on the sample, and is attributed to the unreacted monomer and also possibly short-chain oligomers. The subsequent weight loss occurs either gradually or as a second weight loss step at approx. 500°C [5]
Biological Stability: Toxic [5]

Applications/Commercial Products:
Processing & Manufacturing Routes: Electrolytically polymerised in MeCN in the presence of $[Bu_4N]ClO_4$ and 2,6-lutidine [1,5]. The polymer has been synthesised from DPBz using ammonium persulfate as oxidising agent [5,6]
Applications: Corrosion protection of aluminium alloys by coating [7]. Organic electroluminescent devices [8]. Secondary batteries with improved cathodes [9]

Bibliographic References
[1] *Jpn. Pat.*, (Ricoh), 1987, 62 053 328
[2] Hagiwara, T., Demura, T. and Iwata, K., *Synth. Met.*, 1987, **18**, 317
[3] Chandrasekhar, P. and Gumbs, R.W., *J. Electrochem. Soc.*, 1991, **138**, 1337
[4] Santana, H. and Dias, F.C., *Mater. Chem. Phys.*, 2003, **82**, 882
[5] Neob, K.G., Kang, E.T. and Tan, K.L., *J. Phys. Chem.*, 1992, **96**, 6711
[6] Hagiwara, T., Yamaura, M. and Iwata, K., *Nippon Kagaku Kaishi*, 1989, 1791
[7] Perucki, M. and Chandrasekhar, P., *Synth. Met.*, 2001, **119**, 385
[8] *Jpn. Pat.*, (Ricoh), 2002, 02 151 256
[9] *Jpn. Pat.*, (Ricoh), 1994, 06 052 859

Poly(diphenyldiacetylene) P-206

Synonyms: *Poly(1,1-(1,3-butadiyne-1,4-diyl)bisbenzene)*
Monomers: Diphenyldiacetylene
Material class: Thermoplastic
Polymer Type: polyacetylene
CAS Number:

CAS Reg. No.
25135-09-5

Molecular Formula: $(C_{32}H_{20})_n$
Fragments: $C_{32}H_{20}$

Volumetric & Calorimetric Properties:
Melting Temperature:

No.	Value	Note
1	270–300°C	[1]
2	160–230°C	[4]

Vicat Softening Point:

No.	Value	Note
1	215–260°C	[2]

Surface Properties & Solubility:
Solvents/Non-solvents: Sol. $CHCl_3$, C_6H_6, carbon disulfide, dioxane [2]

Optical Properties:
Transmission and Spectra: Ir [3,4], uv [2], ms [1] and epr [2] spectral data have been reported

Polymer Stability:
Thermal Stability General: Weight loss 3% (approx., 6h, 250°, in Ar atmosphere) [2]

Applications/Commercial Products:
Processing & Manufacturing Routes: Prod. by thermal polymerisation of 1,4-diphenylbuta-1,3-dyne

Bibliographic References

[1] Wiley, R.H. and Lee, J.Y., *J. Macromol. Sci., Chem.*, 1971, **5**, 513
[2] Teyssie, P.H. and Korn-Girard, A.C., *J. Polym. Sci., Part A: Polym. Chem.*, 1964, **2**, 2849
[3] Kojima, Y., Tsuji, M., Matsuoka, T. and Takahashi, H., *J. Polym. Sci., Part A: Polym. Chem.*, 1994, **32**, 1371
[4] Farafonov, V., Grovu, M. and Simionescu, C., *J. Polym. Sci., Part A: Polym. Chem.*, 1977, **15**, 2041

Poly[diphenylmethane bis(4-phenyl)carbonate] P-207

Synonyms: *Poly[oxycarbonyloxy-1,4-phenylene(diphenylmethylene)-1,4-phenylene]. Poly(4,4'-dihydroxydiphenyldiphenylmethane-co-carbonic dichloride)*
Monomers: 4,4'-Dihydroxydiphenylmethane, Carbonic dichloride
Material class: Thermoplastic
Polymer Type: polycarbonates (miscellaneous)
CAS Number:

CAS Reg. No.
28934-44-3

Molecular Formula: $(C_{26}H_{18}O_3)_n$
Fragments: $C_{26}H_{18}O_3$

Volumetric & Calorimetric Properties:
Density:

No.	Value	Note
1	d 1.27 g/cm^3	25° [2]
2	d 1.219 g/cm^3	[1]

Melting Temperature:

No.	Value	Note
1	210–230°C	[2]
2	260°C	[3]

Glass-Transition Temperature:

No.	Value	Note
1	204°C	[1]
2	213°C	[3]

Surface Properties & Solubility:
Cohesive Energy Density Solubility Parameters: Flory-Huggins polymer-solvent interaction parameters have been reported: 0.56 (chloromethane, 250°) [1]; 0.28 (CHCl$_3$, 250°) [1]; 1.25 (decane, 250°) [1]; -0.36 (cyclohexanol, 250°) [1]

Optical Properties:
Refractive Index:

No.	Value	Note
1	1.6539	[2]

Transmission and Spectra: Ir spectral data have been reported [3]

Polymer Stability:
Thermal Stability General: Stable under N$_2$ atmosphere up to 400° [1]

Decomposition Details: Polymer decomposition temp. 423° [3]. Max. rate of weight loss occurs at 450° [3]

Applications/Commercial Products:
Processing & Manufacturing Routes: The main synthetic routes are phosgenation and transesterification. The most widely used approach is interfacial polycondensation in a suitable solvent, but soln. polymerisation is also carried out in the presence of pyridine [4]. A slight excess of phosgene is introduced into a soln. or suspension of the aromatic dihydroxy compound in aq. NaOH at 20–30°, in an inert solvent (e.g. chloromethane). Catalysts (quaternary ammonium salts) are used to achieve high MWs [2,3,4]. In transesterification, the dihydroxy compound and a slight excess of diphenyl carbonate are mixed together in the melt (150–300°) in the absence of O$_2$ with elimination of phenol. Reaction rates are increased by alkaline catalysts and the use of a medium-high vacuum in the final stages [2,4]

Bibliographic References

[1] Dipaola-Baranyi, G., Hsiao, C.K., Spes, M., Odell, P.G. and Burt, R.A., *J. Appl. Polym. Sci.: Appl. Polym. Symp.*, 1992, **51**, 195, (polymer-solvent interaction parameters, density, T$_g$, thermal stability)
[2] Schnell, H., *Ind. Eng. Chem.*, 1959, **51**, 157, (density, T$_m$, refractive index, synth.)
[3] Mikroyannidis, J.A., *Eur. Polym. J.*, 1985, **21**, 895, (T$_m$, T$_g$, ir, thermal stability, synth.)
[4] Schnell, H., *Polym. Rev.*, 1964, **9**, 99, (synth., rev)

Poly(2,6-diphenylphenylene oxide) P-208

Synonyms: *Poly(oxy-2,6-diphenyl-1,4-phenylene)*
Monomers: 2,6-Diphenylphenol
Material class: Fibres and Films
Polymer Type: polyphenylene oxide
CAS Number:

CAS Reg. No.
24938-68-9

Molecular Formula: $(C_{18}H_{12}O)_n$
Fragments: $C_{18}H_{12}O$

Volumetric & Calorimetric Properties:
Density:

No.	Value	Note
1	d 1.214 g/cm^3	[3]

Thermal Expansion Coefficient:

No.	Value	Note	Type
1	4.55×10^{-5} K^{-1}	T < Tg [2]	V
2	0.000102 K^{-1}	T > Tg [2]	V

Latent Heat Crystallization: ΔH_f 11.72–12.72 kJ mol^{-1} (2796–3036 cal mol^{-1}) [2], ΔH_f 12.2 kJ mol^{-1} [3], ΔS_f 15.29–16.97 J K^{-1} mol^{-1} (3.65–4.05 cal°C^{-1} mol^{-1}), ΔC_p 66.96–71.15 J K^{-1} mol^{-1} (15.98–16.98 cal °C^{-1} mol^{-1}, Tg) [2]

– Poly(2,6-diphenyl-1,4-phenylene oxide)

Specific Heat Capacity:

No.	Value	Note	Type
1	180.6 kJ/kg.C	-93°	P
2	395 kJ/kg.C	120°	P

Melting Temperature:

No.	Value	Note
1	484°C	[2]
2	480°C	[3]

Glass-Transition Temperature:

No.	Value	Note
1	225°C	[1]
2	220°C	[3]

Optical Properties:
Optical Properties General: Mass spectral data have been reported [4]

Applications/Commercial Products:
Processing & Manufacturing Routes: Prod. by oxidative coupling of 2,6-diphenylphenol
Applications: High voltage cable insulation. Adsorbent for gas chromatography

Trade name	Supplier
Tenax	Akzo

Bibliographic References
[1] Weyland, H.G., Hoftyzer, P.J. and Van Krevelen, D.W., *Polymer*, 1970, **11**, 79
[2] Wrasidlo, W., *Macromolecules*, 1971, **4**, 642
[3] Gaur, U. and Wunderlich, B., *J. Phys. Chem. Ref. Data*, 1981, **10**, 1001
[4] Hummel, D.O., Dussel, H.J., Rosen, H. and Rubenacker, K., *Makromol. Chem.*, Suppl. 1, 1975, 471

Poly(2,6-diphenyl-1,4-phenylene oxide) P-209

Synonyms: *Poly(2,6-diphenyl-1,4-phenylene ether). Poly(oxy[1,1′:3′,1′′-terphenyl]-2′,5′-diyl). Poly(oxy[2,6-diphenyl-1,4-phenylene]). Poly(oxy-m-terphenyl-2′,5′-ylene)*
Monomers: 2,6-Diphenylphenol
Material class: Thermoplastic
Polymer Type: polyphenylene oxide
CAS Number:

CAS Reg. No.
24938-68-9

Molecular Formula: $(C_{18}H_{12}O)_n$
Fragments: $C_{18}H_{12}O$

Molecular Weight: M_n 42000–280000; MW 55000–500000; MV 143000
Morphology: Amorph. but becomes highly crystalline when heated above Tg [1,5]

Volumetric & Calorimetric Properties:
Density:

No.	Value
1	d 1.146 g/cm^3

Latent Heat Crystallization: ΔH_f 12.2 kJ mol^{-1} [2]
Specific Heat Capacity:

No.	Value	Note	Type
1	253.8 kJ/kg.C	0° [2]	P
2	272.8 kJ/kg.C	25° [2]	P
3	395.1 kJ/kg.C		P
4	471.7 kJ/kg.C	220°, Tg [2]	P
5	650.7 kJ/kg.C	480°, T$_m$ [2]	P

Melting Temperature:

No.	Value	Note
1	475–500°C	[1]

Glass-Transition Temperature:

No.	Value	Note
1	210–240°C	[1]

Transition Temperature:

No.	Value	Note	Type
1	258°C	[3]	Crystallisation Temp.
2	270°C	[4]	Crystallisation Temp

Surface Properties & Solubility:
Solvents/Non-solvents: Amorph. form sol. wide range of solvents. Crystallised form sol. halogenated solvents

Transport Properties:
Permeability of Gases: Gas permeability at 1.5 atm, 35° [3] [He] 32.7 × 10^{-10} cm^2 (s cmHg)$^{-1}$, [O$_2$] 7.7 × 10^{-10}, [N$_2$] 1.5 × 10^{-10}, [CH$_4$] 2.7 × 10^{-10}, [CO$_2$] 39.9 × 10^{-10}

Water Absorption:

No.	Value
1	0.5 %

Mechanical Properties:
Tensile Strength Break:

No.	Value	Note
1	61 MPa	8900 psi, 0.59 nm film, room temp. [1]
2	13 MPa	1850 psi, 1.1 mm film, 200° [1]

Polymer Stability:
Thermal Stability General: Exhibits excellent thermal stability. In contrast to PPO has very high thermooxidative stability

Poly[(diphenylsilylene)(methylene)]

Upper Use Temperature:

No.	Value
1	375°C

Decomposition Details: Decompostion temp. 564° [4]

Applications/Commercial Products:

Trade name	Supplier
Tenax	General Electric
Tenax GC	General Electric
Tenax TA	General Electric

Bibliographic References

[2] Gaur, U. and Wunderlich, B., *J. Phys. Chem. Ref. Data*, 1981, **10**, 1001, (ΔH^f, Cp values)
[3] Aguilar-Vega, M. and Paul, D.R., *J. Polym. Sci., Part B: Polym. Phys.*, 1993, **31**, 1577, (Tc)
[4] Yang, H. and Hay, A.S., *J. Polym. Sci., Part A: Polym. Chem.*, 1993, **31**, 1261, (Tc)
[5] *U.S. Pat.*, 1969, 3 432 466, (synth, MW)

Poly[(diphenylsilylene)(methylene)] P-210

$$-[CH_2-Si(Ph)(Ph)]_n-$$

Synonyms: *Poly(diphenylcarbosilane). Poly(diphenylsilylmethylene). 1,1,3,3-Tetraphenyl-1,3-disilacyclobutane homopolymer, sru. PDPhSM*
Related Polymers: Polycarbosilanes
Monomers: 1,1,3,3-Tetraphenyl-1,3-disilacyclobutane
Polymer Type: polycarbosilanes
CAS Number:

CAS Reg. No.
86490-33-7

Molecular Formula: $(C_{13}H_{12}Si)n$
Fragments: $C_{13}H_{12}Si$
Molecular Weight: 90,000–2,640,000 solvent dependent [3,5]
General Information: PDPhSM is of particular interest because of its high thermal stability (>670K), high rigidity, low toughness and long material lifetime visible light emission induced by irradiation with an ultraviolet laser light [1,2,6]
The thermostability of PDArSMs with tolyl groups was found to be comparable to that of PDPhSM when carried out in nitrogen, whereas the polymers were highly thermally unstable in air compared with PDPhSM [6,7,10,14]
Morphology: Yellow crystalline solid [14]. The surface morphology and crystallinity of the films were greatly affected by the metal species used. The surface roughness of the films ranged from 2 to 400nm, depending on the metal, but could be adjusted within the range by use of a combination of several metals. The films were transparent and smooth, like glass, with Pt or Pt/Pd, but opaque and very rough with Cu and Au [2,9]

Volumetric & Calorimetric Properties:
Thermodynamic Properties General: See [14]
Melting Temperature:

No.	Value	Note
1	350–352°C	[5,10,14]. Thin films are difficult to synthesise due to high melting point [1]

Glass-Transition Temperature:

No.	Value	Note
1	130°C	[3]
2	140–141°C	[5,10,14]

Surface Properties & Solubility:
Solubility Properties General: In addition to its high melting point, lack of solubility in most organic solvents means thin films are difficult to synthesise [1,5,14]. Slightly soluble in hot diphenyl ether and appreciable solubility in hot diphenyl sulfone [14]. Solubility has been improved appreciably by incorporation of an MP unit into the polymer backbone (PMPSM) [3]

Transport Properties:
Transport Properties General: The melt viscosity of PDPhSM was found to be higher than that of PMPSM at 360°C. The melt viscosities of the polymer blends consisting of PDPhSM and PMPSM decreased with increasing PMPSM content. [4,14]
Melt Flow Index:

No.	Value	Note
1	30–77 g/10 min	[14]

Mechanical Properties:
Mechanical Properties General: Polymerisation induces mechanical stress at the substrate-film interface, probably caused by change of the film thickness during the polymerisation. These mechanical stresses could lead to the cracking of coatings; the higher the thickness change, the easier the cracking. This cracking was not observed in films made with Pt or Pt/Pd. However, it appeared a little in films made with Cu and more extensively when Au was used to activate the polymerisation [2]. To improve its mechanical properties, binary polymer blends consisting of PDPSM as a hard component and an amorphous PSM as a soft component have been studied [4]
Tensile Strength Break:

No.	Value	Note	Elongation
1	6 MPa	[4,6]	500% at break [4]

Hardness: PDPhSM has a high hardness [4]

Optical Properties:
Optical Properties General: The polymer films have been shown to exhibit the same long lifetime visible light emission upon UV laser irradiation as those reported for chemically synthesised polymers [2].
When PDPhSM films were irradiated with an excimer laser (KrF, 248nm), light emission, whose spectra were broad ranging from 300–500nm and exhibiting a large peak at 340nm, was observed. Blue-green light emission was observed for several seconds after the laser light was turned off. Furthermore, the emission spectra were found to change with a number of laser shots irradiated. The emission peak at 340nm decreased rapidly, while the emission in the range from 400–600nm increased gradually as the sample was repeatedly irradiated with the laser [7]

Optical Properties:
Transmission and Spectra: Ir [1,9,14]; uv [1,9]; nmr [1,3,5,9,12,14]; xps [1]; ms [6,9].

Polymer Stability:
Polymer Stability General: Thickness of PDPhSM thin films was found to be dependent upon the nature of sputtered metal [1]
Thermal Stability General: The thermogravimetric behaviour of PDPhSM films fabricated at 550K for 10min. in air and initially subjected to 4min. of different metal sputtering in air, together with that of chemically synthesised PDPhSM has been studied. The temperature at which the weight loss reached 5% (Td5) was

considered to be one criterion of the thermal stability. The Td5 was 610K for Pt/Pd, 650K for Cu, 735K for Au and 695K for chemically synthesised PDPhSM. These films exhibited a thermal stability similar to that of chemically synthesised film, depending on the kind of metal, and they exhibited different crystallinities. The film prepared with Pt/Pd appeared to be the most amorphous, while those with Cu, Au, and chemically synthesised PDPhSM were more crystalline. Logically, the thermal stability tended to increase with the crystallinity of PDPhSM [1,2,3]. Thermostability of PDArSMs with tolyl groups were comparable to that of PDPhSM when the measurement was conducted under nitrogen, whereas the polymers having tolyl substituents were highly thermally unstable in air compared with PDPhSM [6,7,10,14]

Decomposition Details: Started to decompose at around 400°C and is complete at about 600°C under nitrogen. Exhibits a two-step weight loss in the temperature range ca. 400 to 700°C in air. The first decomposition step is assignable to elimination of phenyl groups, while the second one is presumably due to oxidative decomposition, which is evident from a large exothermic peak in the DTA trace [6,7,10]

Applications/Commercial Products:
Processing & Manufacturing Routes: New and unique method for fabricating PDPhSM thin films, consisting of four-step process: monomer deposition; preparation of polymerisation catalysts; thermal polymerisation activated by adding metal nanosized particles, and removal of unreacted monomer [1,2,9]. Also by Cu-catalysed polymerisation of 1,1,3,3-tetraphenyl-1,3-disilacyclobutane (TPDC) [3]

Applications: Polymer-based optical devices [1]. Porous insulation films for semiconductor devices [11]. A SiC/SiC composite [13]

Bibliographic References
[1] Rossignol, F., Nakata, Y., Nagai, H., Konno, S., Okutani, T. and Suzuki, M., *Chem. Mater.*, 1998, **10**, 2047
[2] Rossignol, R., Konno, S., Nakata, Y., Nagai, H., Okutani, T. and Suzuki, M., *Chem. Mater.*, 1999, **11**, 367
[3] Ogawa, T., *Polymer*, 1998, **39**, 2715
[4] Ogawa, T. and Murakami, M., *J. Polym. Sci.*, *Part A*, 2003, **41**, 257
[5] Ogawa, T. and Murakami, M., *J. Polym. Sci.*, *Part A*, 1997, **35**, 399
[6] Ogawa, T. and Murakami, M., *J. Polym. Sci.*, *Part B: Polym. Phys.*, 1996, **34**, 1317
[7] Ogawa, T. and Murakami, M., *J. Polym. Sci.*, *Part A*, 1997, **35**, 1431
[8] Suzuki, M., Nakata, Y., Nagai, H., Okutani, T., Kushibiki, N. and Murakami, M., *Mater. Sci. Eng.*, *B*, 1997, **49**, 172
[9] Rossignol, R., Konno, S., Nakata, Y., Nagai, H., Okutani, T. and Suzuki, M., *Chem. Mater.*, 1999, **11**, 358
[10] Ogawa, T., Suzuki, T. and Murakami, M., *J. Polym. Sci.*, *Part B: Polym. Phys.*, 1998, **36**, 755
[11] *Eur. Pat.*, 2003, 1 308 476
[12] Kuroki, S., Nanba, J., Ando, I., Ogawa, T. and Murakami, M., *Polym. J. (Tokyo)*, 1999, **31**, 369
[13] Tanoglu, M. and Parvizi-Majidi, A., *Sci. Eng. Compos. Mater.*, 1998, **7**, 239
[14] Ogawa T., Tachikawa M., Kushibiki, N. and Murakami, M., *J. Polym. Sci.*, *Part A*, 1995, **33**, 2821

Poly(dipropylsilane)   P-211

$$\left[\begin{array}{c} CH_2CH_2CH_3 \\ | \\ Si \\ | \\ CH_2CH_2CH_3 \end{array} \right]_n$$

Synonyms: *Poly(dipropylsilylene)*. PDPrS
Related Polymers: Polycarbosilanes, Poly(methyl phenyl silylene methylene), Dichloropoly(methylphenylsilane), Poly(methylphenylsilane), Poly(dihexylsilane)
Polymer Type: polysilanes

CAS Number:

CAS Reg. No.
96228-25-0

Molecular Formula: $(C_6H_{14}Si)n$
Fragments: $C_6H_{14}Si$
Tacticity: Propyl groups exist in a single conformation [7]
Morphology: The conformation of the Si backbone is *all-trans*. The two propyl groups stretch asymmetrically due to the intramolecular steric hindrance of the second CH_2 group in the side chains bonded to the same Si atom. A tetragonal unit cell with a = b = 0.980nm; c = 0.399nm [3,5,7]

Surface Properties & Solubility:
Solubility Properties General: Insoluble [7].
Surface Tension:

No.	Value	Note
1	28.6 mN/m	Due almost entirely to depression forces. Obtained using geometric-mean method [3,10]

Optical Properties:
Optical Properties General: The polymer displays reversible thermochromic behaviour in solution with a bathochromic shift of 16–25nm occuring from 92 to -67°C. The temperature dependence of the UV absorption maxima is thought to be due to conformational changes occurring along the polymer backbone [3].
Transmission and Spectra: Ir and Raman [1,7], uv-vis [2,5,7,10] and nmr [5,7,10] spectra reported

Polymer Stability:
Thermal Stability General: Two clear transitions, a weak one at -17.0°C (onset at -29°C) with ΔH=0.66kJ/mol and a sharp intense peak at 235°C (onset at 222°C), which corresponds to a first order transition with ΔH=7.3 kJ/mol (0.064kJ/g, 15.2cal/g) [7]. When the sample was cooled to -40°C, no significant changes were observed, which shows that no drastic structural changes are involved below the weak phase transition at -29°C
Decomposition Details: First order DSC transition near 222°C. At 250°C, weight loss was 3% and decomposition ensued rapidly above 338°C [7]

Applications/Commercial Products:
Processing & Manufacturing Routes: Synth. by Wurtz-type reductive coupling of corresponding dichlorosilane [3,5]
Applications: Composing and placing security markings on items which need protection against forgery or counterfeiting [1]. In fabrication of electroluminescent devices [4]. Uv absorbers for cosmetics [9]

Bibliographic References
[1] *U.S. Pat.*, 2002, 02 025 490
[2] Muramatsu, Y., Fujino, M., Yamamoto, T., Gullikson, E.M. and Perera, R.C.C., Nucl. Instrum. Methods Phys. Res., *Sect. B*, 2003, **199**, 260
[3] Crespo, R., Carmen, M.P. and Tomas, F., *J. Chem. Phys.*, 1994, **100**, 6953
[4] Furukawa, S., *Thin Solid Films*, 1998, **331**, 222
[5] Lovinger, A.J., Davis, D.D., Schilling, F., Bovey, F.A., Frederic, C. and Zeigler, J.M., *Polym. Commun.*, 1989, **30**, 356
[6] Menescal, R., West, R. and Murray, C., *Macromolecules*, 1991, **24**, 329
[7] Menescal, R., Eveland, J., West, R., Leites, L.L., Bukalov, S.S., Yadritseva, T.D. and Blazso, M., *Macromolecules*, 1994, **27**, 5885
[8] Furukawa, S., Takeuchit, K., Mizoguchit, M, Shimanat M. and Tamurat, M., *J. Phys.: Condens. Matter*, 1993, **5**, L461
[9] *Jpn. Pat.*, 1990, 02 204 410
[10] Fujisaka, T. and West, R., *J. Organomet. Chem.*, 1993, **449**, 105

Poly[(dipropylsilylene)methylene]   P-212

$$\left[-CH_2 - \underset{\underset{CH_2CH_2CH_3}{|}}{\overset{\overset{CH_2CH_2CH_3}{|}}{Si}} - \right]_n$$

Synonyms: *Poly(di-n-propylsilylenemethylene)*. PDPrSM. PDPSM
Related Polymers: Polycarbosilanes, Poly(methyl phenyl silylene methylene), Dichloropoly(methylphenylsilane), Poly(methylphenylsilane)
Polymer Type: polycarbosilanes
CAS Number:

CAS Reg. No.
163934-33-6

Molecular Formula: $(C_7H_{16}Si)_n$
Fragments: $C_7H_{16}Si$
Molecular Weight: 166,500–181,000 [3,4]
Tacticity: The fibre repeat, 4.86Å, indicates an all-gauche conformation in the main chain, giving a 4₁ helical structure. *Ab initio* and semi-empirical energy calculations show that with attachment of *n*-alkyl side chains an all-gauche conformation for the main chain gives the lowest energy. This is in contrast to the unsubstituted polymer, poly(silaethylene), which has a minimum energy for the *all-trans* conformation. The all-gauche conformation of the main chain of PDPrSM relieves the steric hindrance between side chains found in the *all-trans* conformation [1,2,3,4]
Morphology: Amorphous [3]. The alkyl side chains are interlocked with each other like cross-shaped gears in the two-dimensional monoclinic *ab* projections in the unit cell [1]. Unit cell parameters a = 10:52Å; b = 8:66Å; c = 4:86Å; α = 78:4°; β = 100:0° and γ = 98:2° [2]

Volumetric & Calorimetric Properties:
Density:

No.	Value	Note
1	d 0.995 g/cm³	[1]

Melting Temperature:

No.	Value	Note
1	-77.6°C	[3]

Glass-Transition Temperature:

No.	Value	Note
1	-41.2°C	Does not show any first-order transitions [5]

Transition Temperature General: Exhibits lower backbone flexibility than that observed for the analogous poly(dipropylsiloxane) [4]

Optical Properties:
Transmission and Spectra: Nmr spectrum was consistent with α (anti-Markovnikov) addition of the olefins, with no sign of a β (Markovnikov) addition product in either the pmr or cmr spectra [3,4]

Polymer Stability:
Thermal Stability General: Part-transformed into a mobile state below isotropisation, which may possibly be classified as a conformationally disordered mesophase.

Decomposition Details: 5% decomposition temp=349°C; 10% decomposition temp = 380°C [3,5].
Chemical Stability: Oxidative stability [3,5]

Applications/Commercial Products:
Processing & Manufacturing Routes: Obt. by hydrosilylation of poly[hydridomethylsilylene(methylene)]; an excess of olefin was added followed by Karstedt's catalyst in xylene to yield PDPrSM [3]
Reaction of trichloro(chloromethyl)silane with respective Grignard reagent and subsequent cyclisation, 1,1,3,3-tetrapropyldisilacyclobutane, followed by catalytic polymerisation with H_2PtCl_6 gave PDPrSM, with strictly alternating SiR_2/CH_2 backbone structure [4]

Applications/Commercial Products:

Bibliographic References
[1] Park, S.Y., *Macromol. Rapid. Commun.*, 2003, **24**, 793
[2] Park, S.Y., Interrante, L.V. and Farmer, B.L., *Polymer*, 2001, **42**, 4253
[3] Rushkin, I.L. and Interrante, L.V., *Macromolecules*, 1996, **29**, 5784
[4] Koopmann F. and Holger, F., *Macromol. Rapid. Commun.*, 1995, **16**, 363
[5] Koopmann F. and Holger, F., *Macromolecules*, 1996, **29**, 3701

Poly(ditetradecylsilane)   P-213

$$\left[-\underset{\underset{CH_2(CH_2)_{12}CH_3}{|}}{\overset{\overset{CH_2(CH_2)_{12}CH_3}{|}}{Si}} - \right]_n$$

Synonyms: *Poly(ditetradecylsilylene)*. PDTS. PTDSi. PdnTDS
Related Polymers: Polycarbosilanes
Polymer Type: polysilanes
CAS Number:

CAS Reg. No.	Note
117652-56-9	homopolymer
107999-70-2	sru

Molecular Formula: $(C_{28}H_{58}Si)_n$
Fragments: $C_{28}H_{58}Si$
Molecular Weight: 3.071×10^6 [7]
Morphology: White, semi-crystalline solid. Adopts a TGTG′ backbone conformation and orthorhombic unit cell, with dimensions a=21.5Å, b=29.5Å, c=7.9Å and γ=90°. Probably assumes a columnar hexagonal packing mode (HPM) with disordered backbone conformation[1,2,3]
Identification: Exhibits a continuous blue shift of the uv absorption band at 345nm. Does not show a red shift of the uv band with pressure, implying that pressure cannot induce a transition to the planar zigzag conformation.

Volumetric & Calorimetric Properties:
Density:

No.	Value	Note
1	d 0.925	calc. 1.01 [1]

Thermodynamic Properties General: An endotherm leads to a value of ΔS_d 119J/(K mol) and accounts for about half of the maximum possible conformations participating in disordering [4]
Transition Temperature General: No glass-transition was observed, but has been estimated at -23°C from heat capacity data [4,8].

Optical Properties:
Optical Properties General: Thermochromic transition at 54°C. At room temperature, the polymer absorbs at 350nm, a value intermediate to those observed for PDHS (370nm) and PDPS

(313nm). Above its transition it exhibits a blue shift; PDTS absorbs at 313nm, as do the other poly(di-*n*-alkylsilanes) above their respective transition temperatures [1,2]
Transmission and Spectra: Ir [1,2], Raman [2], uv-vis [1,2,6], epr [5] and nmr [5,6,8] spectra reported.

Polymer Stability:
Thermal Stability General: The backbone becomes disordered above the transition temperature. Undergoes an order-disorder transition at 55°C [1]
Observation under a polarising microscope shows the edge of birefringent objects become isotropic and start flowing at 210°C. Beyond this temperature a complete transition to the isotropic state is observed.

Polymer Stability:
Decomposition Details: A broad endotherm appears as a broad doublet with some shoulders in the first run, changes in the subsequent scans, leaving a single peak with a broad onset. The peak temperature from an average of three measurements is at 55.9°C with a heat of transition of 39.2 ± 0.35 kJ mol^{-1} for all samples, irrespective of the different thermal histories of the samples. No further transition was observed when the samples were heated to just below the decomposition temperature [8]

Applications/Commercial Products:
Processing & Manufacturing Routes: Prepared from dichlorotetrasilane by dehalogenation coupling with Na under standard conditions for Wurtz polysilane synthesis (60% yield) [7].
Applications: Fabry Perot miniature resonator optical switch [10]

Bibliographic References
[1] KariKari, E.K., Greso, A.J. and Farmer, B.L., *Macromolecules*, 1993, **26**, 3937
[2] Song, K., Miller, R.D., Wallraff, G.M. and Rabolt, J.F., *Macromolecules*, 1991, **24**, 4084
[3] French, R.H., Meth, J.S., Thorne, J.R.G, Hochstrasser, R.M. and Miller, R.D., *Synth. Met.*, 1992, **50**, 499
[4] Varma-Nair, M., Cheng, J., Jin, Y. and Wunderlich, B., *Macromolecules*, 1991, **24**, 5442
[5] McKinley, A.J., Karatsu, T., Wallraff, G.M., Miller, R.D. and Soorlyakumaran, R., *Organometallics*, 1988, **7**, 2567
[6] Seki, S., Koizumi, Y., Kawaguchi, T., Habara, H. and Tagawa, S., *J. Am. Chem. Soc.*, 2004, **126**, 3521
[7] Miller, R.D. and Michl, J., *Chem. Rev.*, 1989, **89**, 1359
[8] Lovinger, A.J., Davis, D.D., Schilling, F.C., Padden, F.J. and Bovey, F.A., *Macromolecules*, 1991, **24**, 132
[9] *Jpn. Pat.*, 1999, 11 084 437

Poly(divinyl-1,4-butanediyl ether-*co*-maleic anhydride)   P-214
Related Polymers: General information
Monomers: Divinyl-1,4-butanediyl ether, Maleic anhydride
Material class: Thermoplastic
CAS Number:

CAS Reg. No.
33520-84-2

Molecular Formula: $[(C_8H_{14}O_2).(C_4H_2O_3)]_n$
Fragments: $C_8H_{14}O_2$ $C_4H_2O_3$

Applications/Commercial Products:
Processing & Manufacturing Routes: Prod. by free-radical initiated copolymerisation
Applications: Enzyme carrier, fibrinolytic active polymers to prevent blood coagulation [1,2]

Bibliographic References
[1] *Jpn. Pat.*, 1977, 77 142 772
[2] *Ger. Pat.*, 1971, 2 008 990

Poly(divinyl ether)   P-215

Related Polymers: Poly(vinyl ethers)
Monomers: Divinyl ether
Material class: Synthetic Elastomers
Polymer Type: polyvinyls
CAS Number:

CAS Reg. No.
9003-19-4

Molecular Formula: $(C_{16}H_{23}O_4)_n$
Fragments: $C_{16}H_{23}O_4$

Volumetric & Calorimetric Properties:
Latent Heat Crystallization: Thermodynamic props. have been reported [1]

Surface Properties & Solubility:
Cohesive Energy Density Solubility Parameters: δ_{small} 33.48 J$^{1/2}$ cm$^{-3/2}$ (7.99 cal$^{1/2}$ cm$^{-3/2}$) [2]. Other solubility parameters have been reported [2]

Optical Properties:
Transmission and Spectra: C-13 nmr spectral data have been reported [3]

Applications/Commercial Products:
Processing & Manufacturing Routes: Free-radical initiated polymerisation with stereoselective cyclisation
Applications: Applications include use in hair conditioners, electric conductive films, microcapsules, photosensitive films, adhesives and coatings

Bibliographic References
[1] Guaita, M., *Makromol. Chem.*, 1972, **157**, 111
[2] Ahmad, H. and Yaseen, M., *Polym. Eng. Sci.*, 1979, **19**, 858, (solubility parameters)
[3] Tsukino, M. and Kunitake, T., *Polym. J. (Tokyo)*, 1981, **13**, 657
[4] *Czech. Pat.*, 1984, 215 635
[5] *Ger. Pat.*, 1986, 3 512 565
[6] *Ger. Pat.*, 1987, 3 609 137
[7] *Jpn. Pat.*, 1989, 01 297 091

Poly(divinyl ether-maleic anhydride)   P-216

Synonyms: *DIVEMA*. Pyran polymer. *Poly[2,5-furandione-co-1,1'-oxybisethene]*
Related Polymers: Poly(vinyl ethers)
Monomers: Divinyl ether, Maleic anhydride
Material class: Synthetic Elastomers
CAS Number:

CAS Reg. No.
27100-68-1

Molecular Formula: $(C_{12}H_{10}O_7)_n$
Fragments: $C_{12}H_{10}O_7$
General Information: An alternating copolymer

Volumetric & Calorimetric Properties:
Latent Heat Crystallization: ΔE_{act} 112.7 kJ mol^{-1} (26.9 kcal mol^{-1}, CHCl$_3$); 111.9 kJ mol^{-1} (26.7 kcal mol^{-1}, DMF) [2]

Optical Properties:
Transmission and Spectra: H-1 nmr, C-13 nmr and uv spectral data have been reported [3,4]

Applications/Commercial Products:
Processing & Manufacturing Routes: Prod. by free radical initiated copolymerisation of divinyl ether and maleic anhydride with stereoselective cyclisation
Applications: Modulates a variety of biological responses related to bacteria, fungi and viruses. A neoplasm inhibitor, antitumour and antiviral agent

Bibliographic References
[1] Samuels, R.J., *Polymer*, 1977, **18**, 452
[2] Butler, G.B. and Fujimori, K., *J. Macromol. Sci., Chem.*, 1972, **6**, 1533
[3] Butler, G.B. and Chu, Y.C., *J. Polym. Sci.*, 1979, **17**, 858
[4] Tsukino, M. and Kunitake, T., *Polym. J. (Tokyo)*, 1981, **13**, 671
[5] *Ger. Pat.*, 1973, 2 262 449, (use)
[6] *U.S. Pat.*, 1975, 3 859 433, (use)
[7] *Fr. Pat.*, 1978, 2 368 222, (use)
[8] *U.S. Pat.*, 1976, 3 996 347, (use)

Poly(1-docosene) P-217

Synonyms: *Poly(1-eicosylethylene). Poly(1-eicosyl-1,2-ethanediyl)*
Monomers: 1-Docosene
Material class: Thermoplastic
Polymer Type: polyolefins
CAS Number:

CAS Reg. No.	Note
31441-79-9	atactic
59284-12-7	isotactic

Molecular Formula: $(C_{22}H_{44})_n$
Fragments: $C_{22}H_{44}$
Morphology: Three polymorphic forms: form I orthorhombic, a 0.75, b 8.8, c 0.67, helix 4$_1$ [4]; form II monoclinic, a 0.53, b 10.8, c 0.77, β 94°, helix 4$_1$ [4]

Volumetric & Calorimetric Properties:
Thermodynamic Properties General: ΔH_f 145 J g^{-1} (form I); 120 J g^{-1} (form II); 40 J g^{-1} (form III) [3]
Melting Temperature:

No.	Value	Note
1	91°C	364K [1]
2	92°C	365K [2]
3	90°C	363K, form I [3]
4	83°C	356K, form II [3]
5	75°C	348K, form III [3]
6	90°C	363K [4]

Transition Temperature:

No.	Value	Note	Type
1	69°C	342K	T_1
2	72°C	345K [2]	T_2

Optical Properties:
Transmission and Spectra: Ir and Raman [1,3] spectral data have been reported

Applications/Commercial Products:
Processing & Manufacturing Routes: Synth. using Ziegler-Natta catalysts, such as TiCl$_3$-AlEt$_2$Cl at room temp. [1]

Bibliographic References
[1] Modric, I., Holland-Moritz, K. and Hummel, D.O., *Colloid Polym. Sci.*, 1976, **254**, 342, 976, (synth, ir, Raman)
[2] Trafara, G., Koch, R., Blum, K. and Hummel, D., *Makromol. Chem.*, 1976, **177**, 1089, (transition temps)
[3] Trafara, G., Koch, R. and Sausen, E., *Makromol. Chem.*, 1978, **179**, 1837, (struct, Raman)
[4] Blum, K. and Trafara, G., *Makromol. Chem.*, 1980, **181**, 1097, (melting point)

Poly(1-dodecene) P-218

$$-[CH_2CH]_n-$$
$$\quad\quad |$$
$$CH_2(CH_2)_8CH_3$$

Synonyms: *Poly(1-decylethylene). Poly(1-decyl-1,2-ethanediyl)*
Monomers: 1-Dodecene
Material class: Thermoplastic
Polymer Type: polyolefins
CAS Number:

CAS Reg. No.	Note
25067-08-7	
26746-86-1	isotactic

Molecular Formula: $(C_{12}H_{24})_n$
Fragments: $C_{12}H_{24}$
General Information: Rubbery appearance
Morphology: Three forms: form I monoclinic, a 0.59, b 5.60, c 1.34 (1.32), β 96° [7]

Volumetric & Calorimetric Properties:
Density:

No.	Value	Note
1	d^{30} 0.856 g/cm^3	[15]

Thermodynamic Properties General: ΔH_f 4.5–4.8 kJ mol^{-1} [1]. ΔS_f 14.4–15.4 J mol^{-1} K^{-1}. [1] Values of expansion coefficients have been reported [10]
Latent Heat Crystallization: ΔH_m 45 J g^{-1} (T_m 46°, Form I) [7]. ΔH_m 12 J g^{-1} (T_m -25°, Form III) [7]
Melting Temperature:

No.	Value	Note
1	38°C	311K [1]
2	45°C	[2,18]
3	49°C	322K [4,13,14,17]
4	46°C	[16]
5	46°C	319K, Form I
6	32°C	305K, Form II
7	-25°C	248K, Form III [7]

Glass-Transition Temperature:

No.	Value	Note
1	-25°C	[2]
2	-6°C	[4]
3	-36°C	[21]

Transition Temperature:

No.	Value	Note	Type
1	45–50°C	[3]	Softening point
2	-45°C		α
3	-165°C	[5]	β
4	-145°C	520 Hz [4]	β

Surface Properties & Solubility:
Cohesive Energy Density Solubility Parameters: Energy of cohesion has been reported [8]
Wettability/Surface Energy and Interfacial Tension: Critical surface tension has been reported [8]

Transport Properties:
Melt Flow Index:

No.	Value	Note
1	50 g/10 min	approx., ASTM D1238-15T η_{inh} 1 dl g^{-1} (tetrahydronaphthalene, 100° or 145°) [11]

Polymer Solutions Dilute: [η] 0.96–1.45 dl g^{-1} (25°, CHCl$_3$) [1]. η_{inh} 2.43–2.98 (25°, C$_6$H$_6$) [3]. [η] 2.4 dl g^{-1} (25°, toluene) [16]. Theta temp. and dilute soln. props. have been reported [10]
Polymer Solutions Concentrated: Conc. soln. props. have been reported [15, 20]
Polymer Melts: Melt flow activation energy 29.3 kJ mol^{-1} (7 kcal mol^{-1}) [11]

Mechanical Properties:
Mechanical Properties General: Dynamic tensile storage and loss moduli in relation to struct. have been reported [19]
Viscoelastic Behaviour: Viscoelastic props. have been reported [6]

Optical Properties:
Transmission and Spectra: Ir, Raman [9,12,13,14] and H-1 nmr [18] spectral data have been reported
Total Internal Reflection: Stress optical coefficient -4900–-5070 [16]

Applications/Commercial Products:
Processing & Manufacturing Routes: Synth. using Ziegler-Natta catalysts such as TiCl$_3$-AlEt$_2$Cl

Bibliographic References

[1] Magagnini, P.L., Lupinacci, D., Cotrazzi, F. and Andruzzi, F., *Makromol. Chem., Rapid Commun.*, 1980, **1**, 557, (thermodynamic props)
[2] Reding, F.P., *J. Polym. Sci.*, 1956, **21**, 547, (transition temps)
[3] Marvel, C.S. and Rogers, J.R., *J. Polym. Sci.*, 1961, **49**, 335, (transition temps)
[4] Clark, K.J., Turner-Jones, A. and Sandiford, D.J.H., *Chem. Ind. (London)*, 1962, 2010, (transition temps)
[5] Manabe, S., Nakamura, H., Uemura, S. and Takayanagi, M., *Kogyo Kagaku Zasshi*, 1970, **73**, 1587, (transition temps)
[6] Manabe, S. and Takayanagi, M., *Kogyo Kagaku Zasshi*, 1970, **73**, 1595, (viscoelastic props)
[7] Trafara, G., *Makromol. Chem., Rapid Commun.*, 1980, **1**, 319, (cryst struct)
[8] Sewell, J.H., *Rev. Plast. Mod.*, 1971, **22**, 1367, (energy of cohesion)
[9] Holland-Moritz, K., Modric, I., Heinen, K.U. and Hummel, D.O., *Kolloid Z. Z. Polym.*, 1973, **251**, 913, (ir, Raman)
[10] Wang, J.S., Porter, R.S. and Knox, J.R., *Polym. J. (Tokyo)*, 1978, **10**, 619, (dilute soln props)
[11] Combs, R.L., Slonaker, D.F. and Coover, H.W., *J. Appl. Polym. Sci.*, 1969, **13**, 519, (melt props)
[12] McRae, M.A. and Maddams, W.F., *Makromol. Chem.*, 1976, **177**, 449, (ir)
[13] Holland-Moritz, K., *Colloid Polym. Sci.*, 1975, **253**, 922, (ir, Raman)
[14] Modric, I., Holland-Moritz, K. and Hummel, D.O., *Colloid Polym. Sci.*, 1976, **254**, 342, (ir, Raman)
[15] Tait, P.J.T. and Livesey, P.J., *Polymer*, 1970, **11**, 359, (conc soln props, density)
[16] Philippoff, W. and Tomquist, E., *J. Polym. Sci., Part C: Polym. Lett.*, 1966, **23**, 881, (stress optical coefficient)
[17] Turner-Jones, A., *Makromol. Chem.*, 1964, **71**, 1, (struct, melting point)
[18] Slichter, W.P. and Davis, D.D., *J. Appl. Phys.*, 1964, **35**, 10, (nmr)
[19] Takayanagi, M., *Pure Appl. Chem.*, 1970, **23**, 151, (mech props)
[20] Daniels, C.A, Maron, S.H. and Livesey, P.J., *J. Macromol. Sci., Phys.*, 1970, **4**, 47, (conc soln props)
[21] Natta, G., Danusso, F. and Moraglio, G., *CA*, 1958, **52**, 17789e, (transition temp)

Poly(*n*-dodecylmethylsilane) P-219

$$\left[\begin{array}{c} CH_2(CH_2)_{10}CH_3 \\ | \\ Si \\ | \\ CH_3 \end{array} \right]_n$$

Synonyms: *Poly(dodecylmethylsilylene). (DodecMeSi)n. ME12*
Polymer Type: polysilanes
CAS Number:

CAS Reg. No.
88018-84-2

Molecular Formula: (C$_{13}$H$_{28}$Si)n
Fragments: C$_{13}$H$_{28}$Si
Molecular Weight: 0.87×10^5–10^6 [2,3,4,6].
Morphology: Forms an ordered structure on cooling [5]

Volumetric & Calorimetric Properties:
Transition Temperature General: Melts without decomposition [11].

Surface Properties & Solubility:
Solubility Properties General: Readily soluble in common organic solvents [11].

Optical Properties:
Optical Properties General: Displays an abrupt thermochromic transition (302nm to 328nm, after red shift is complete), with an onset temperature of 26°C [2,6].
Cooled through the transition temperature, it adopts a more extended conformation with a greater number of *all-trans* segments along the polymer chain. Support for this interpretation is provided by the variable-temperature Si-29 CP MAS NMR spectra. As the polymer goes through the thermochromic transition, a new downfield resonance appears and grows in intensity [2].
The photoefficiency for both scissioning and cross-linking has been studied, which is reported only to fragment upon irradiation [9,10].

Optical Properties:
Transmission and Spectra: Ir [6], uv-vis [2,3,5,6], pes [3] and nmr [4,6,7] spectra reported.

Applications/Commercial Products:
Processing & Manufacturing Routes: Synthesised from dichloromethyl-*n*-dodecylsilane by dehalogenative coupling reaction with Na under standard conditions for polysilane synthesis by Wurtz coupling [1,2,3,4,11].
Polymerisation of dodecylmethyldichlorosilane was carried out by the Kipping method with melted Na metal in toluene under vigorous stirring [6].

Applications/Commercial Products:
Applications: Small light source for optical recording disks [8].

Bibliographic References

[1] Miller, R.D., Thompson, D., Sooriyakumaran, R. and Fickes, G.N., *J. Macromol. Sci, Part A*, 1991, **29**, 813
[2] Yuan, C.H. and West, R., *Macromolecules*, 1994, **27**, 629
[3] Loubreil, G., and Zeigler, J., *Phys. Rev. B*, 1986, **33**, 4203
[4] Ishino, Y., Hirao, A. and Nakahama, S., *Macromolecules*, 1986, **19**, 2309
[5] Chunwachirasiri, W., Kanaglekar, I., Winokur, M.J., Koe, J.C. and West, R., *Macromolecules*, 2001, **34**, 6719
[6] Seki, S., Koizumi, Y., Kawaguchi, T., Habara, H. and Tagawa, S., *J. Am. Chem. Soc.*, 2004, **126**, 3521
[7] Wolff, A.R., Maxka, J. and West, R., *J. Polym. Sci., Part A*, 1988, **26**, 713
[8] *Jpn. Pat.*, 1997, 09 202 878
[9] Miller, R.D., Hofer, D., McKean, D.R., Willson, C.G., West, R. and Trefonas, P.T., *ACS Symp. Ser. Mater. Microlithogr.*, 1984, **266**, 293
[10] Miller, R.D., Hofer, D., Willson, C.G., Trefonas, P. and West, R., *Polym. Prepr. (Am. Chem. Soc., Div. Polym. Chem.)*, 1984, **25**, 307
[11] Trefonas, P., Djurovich, P.I., Zhang, X.H., West, R., Miller, R.D. and Hofer, D., *J. Polym. Sci., Polym. Lett. Ed.*, 1983, **21**, 819

Poly(dodecylpentylsilane)   P-220

$$\left[\begin{array}{c} CH_2(CH_2)_{10}CH_3 \\ | \\ Si \\ | \\ CH_2(CH_2)_3CH_3 \end{array} \right]_n$$

Synonyms: *Poly(dodecylpentylsilylene). C12-C5*
Polymer Type: polysilanes
CAS Number:

CAS Reg. No.
171673-35-1

Molecular Formula: $(C_{17}H_{36}Si)_n$
Fragments: $C_{17}H_{36}Si$
General Information: Films are the weakest in the series of poly(*n*-alkylpentylsilylenes) [1].
Morphology: The X-ray diffraction pattern indicates little order, probably a manifestation of a partial alignment of the molecules as in a nematic phase, but the degree of order is significantly diminished with respect to the hexagonal, columnar liquid crystalline structure seen in so many poly(*n*-alkyl-*n*-alkylsilylene)s at high temperatures. It is thought that two or more different conformations are present at low temperatures [1].
Miscellaneous: The X-ray diffraction pattern changes on cooling to around -20°C, corresponding to the higher-temperature peak seen in the DSC cooling scan. The onset of this DSC transition is at -19°C with the peak at -23°C.
Below -25°C, very little change is seen in the pattern, as evidenced by the similarity diffraction of the patterns at -25 and -50°C. At -50°C, the broad, sharp-tipped peak observed is split into two reflection peaks. There also may be one or more weaker peaks present between these maxima, as there is some structure to this region. This corroborates what is suggested by the low-temperature uv spectra that there must be several backbone conformations present in C5-C12 at low temperatures. The presence of a shoulder on the high-angle side of the amorphous halo may indicate multiple preferred conformations of the alkyl side chains as well, because the d-spacing of the amorphous halo for alkyl polysilanes is approximately that of the interchain distance commonly seen for alkyl chains [1].

Optical Properties:
Optical Properties General: The variable temperature UV spectra of a film upon slow cooling shows one broad transition. At 25°C a relatively broad peak is seen with a maximum at 320nm, which is characteristic of a helical backbone conformation. Upon cooling, a shoulder on the 320nm absorption band develops between -25 and -40°C. Below -40°C it is apparent that there are at least two overlapping bands. It is thought that two or more different conformations are present at low temperatures [1].
Transmission and Spectra: Uv [1] spectra reported.

Polymer Stability:
Thermal Stability General: Single equilibrium phase transition observed via DSC. Possible multiple phases or conformations at low temperature [1]

[1] Klemann, B.M., West, R. and Koutsky, J.A., *Macromolecules*, 1996, **29**, 198

Poly(dodecylsilane)   P-221

$$\left[\begin{array}{c} SiH \\ | \\ CH_2(CH_2)_{10}CH_3 \end{array} \right]_n$$

Synonyms: *Poly(dodecylsilylene). PD12S*
Related Polymers: Polycarbosilanes, Poly(methyl phenyl silylene methylene), Dichloropoly(methylphenylsilane), Poly(methylphenylsilane)
Polymer Type: polysilanes
CAS Number:

CAS Reg. No.
187536-66-9

Molecular Formula: $(C_{12}H_{26}Si)_n$
Fragments: $C_{12}H_{26}Si$
Molecular Weight: 1.4×10^6 [4]
Morphology: White oil [2], with a Si-Si backbone chain conformation of TGTG'. The room temperature d-spacing can be indexed by an orthorhombic unit cell with dimensions a=21.6Å; b=25.0Å; c=7.8Å and γ=90° [1]

Optical Properties:
Transmission and Spectra: Uv [4]; nmr [2,4] spectra reported.

Applications/Commercial Products:
Processing & Manufacturing Routes: Prepared from dichlorododecylsilane by dehalogenation coupling with Na under standard conditions for Wurtz polysilane synthesis [2,3,4]
Applications: Process stabiliser for an organic material (e.g. polypropylene fibres) against degradation induced by light, heat or oxidation [2,3]

Bibliographic References

[1] KariKari, E.K., Greso, A.J., Farmer, B.L., Miller, R.D. and Rabolt, R.F., *Macromolecules*, 1993, **26**, 3937
[2] *U.S. Pat.*, 1997, 6 005 036
[3] *U.S. Pat.*, 2003, 6 538 055
[4] Seki, S., Koizumi, Y., Kawaguchi, T., Habara, H., and Tagawa, S., *J. Am. Chem. Soc.*, 2004, **126**, 3521
[5] Miller, R.D. and Michl, J., *Chem. Rev.*, 1989, **89**, 1359

Poly(3-dodecylthiophene)   P-222

Synonyms: *Poly(3-dodecyl-2,5-thiophenediyl). P3DT. P3DDT*
Monomers: 3-Dodecylthiophene
Material class: Thermoplastic
Polymer Type: polythiophene

– Poly(1-eicosene)

CAS Number:

CAS Reg. No.	Note
104934-53-4	
137191-59-4	poly(3-dodecyl-2,5-thiophenediyl)

Molecular Formula: $(C_{16}H_{26}S)_n$
Fragments: $C_{16}H_{26}S$
Additives: Dopants such as I_2 [1], hexafluorophosphate [2], $FeCl_3$ [3] and perchlorate are added to increase conductivity.
General Information: Conductive thermochromic polymer [4] which exhibits electroluminescence [5] and has greater solubility and processability than polythiophene.
Morphology: Orthorhombic unit cell, a=34.75Å, b=7.52Å, c=7.77Å [6]. Polymorphic with two crystal phases [7].

Volumetric & Calorimetric Properties:
Density:

No.	Value	Note
1	d 1.07 g/cm^3	fibre [4]

Melting Temperature:

No.	Value	Note
1	81–156°C	[3,7,8]

Glass-Transition Temperature:

No.	Value	Note
1	-20–5.6°C	Glass-transition temperature increases with addition of FeCl$_3$ dopant [3,8]

Surface Properties & Solubility:
Solvents/Non-solvents: Sol. $CHCl_3$, C_6H_6, tetrahydronaphthalene [9], THF [10]

Mechanical Properties:
Mechanical Properties General: Mechanical properties have been reported [11]. Strength and stiffness of drawn polymer are much lower than for polythiophene.

Electrical Properties:
Electrical Properties General: Conductivity depends on polymer structure, nature and concentration of dopant, and temperature, with maximum values of about 1000 S cm^{-1} reported for I_2-doped samples [1].

Optical Properties:
Transmission and Spectra: Pmr [2,10,12], cmr [2], ir [4,9,10], uv [10,12,13] and x-ray [4,6,14] spectra have been reported.
Volume Properties/Surface Properties: Thermochromic polymer changes from red colour at 30°C to yellow at 80°C [4].

Polymer Stability:
Decomposition Details: Samples that are undoped and doped with I_2 decompose at about 350°C [15]. Dedoping occurs on heating.

Applications/Commercial Products:
Processing & Manufacturing Routes: Produced by chemical or electrochemical oxidative polymerisation of monomer with different catalysts [2,4,9,12,16] to produce mostly regiorandom (head-to-head/head-to-tail) polymers, or by catalytic polymerisation of 2-bromo-5-(bromomagnesio)-3-dodecylthiophene [1,13] or 2-bromo-5-(bromozincio)-3-dodecylthiophene [10] to produce regioregular (head-to-tail) polymers.
Applications: Potential applications in photovoltaic devices, light-emitting diodes, rechargeable batteries, field-effect transistors and chemical and optical sensors [5].

Trade name	Details	Supplier
Rieke	regioregular and regiorandom polymers	Rieke Metals

Bibliographic References
[1] McCullough, R.D., Tristram-Nagle, S., Williams, S.P., Lowe, R.D. and Jayaraman, N., *J. Am. Chem. Soc.*, 1993, **115**, 1719
[2] Leclerc, M., Martinez Diaz, F. and Wegner, G., *Makromol. Chem.*, 1989, **190**, 3105
[3] Chen, S.-A. and Ni, J.-M., *Polym. Bull. (Berlin)*, 1991, **26**, 673
[4] Tashiro, K., Ono, K., Minagawa, Y., Kobayashi, M., Kawai, T. and Yoshino, K., *J. Polym. Sci., Part B: Polym. Phys.*, 1991, **29**, 1223
[5] *Handbook of Conducting Polymers* 2nd edn., (Eds. Skotheim, T.A., Elsenbaumer, R.L. and Reynolds, J.R.) Marcel Dekker, 1998, 225
[6] Kawai, T., Nakazono, M., Sugimoto, R. and Yashino, K., *J. Phys. Soc. Jpn.*, 1992, **61**, 3400
[7] Causin, V., Marega, C., Marigo, A., Valenti, L. and Kenny, J.M., *Macromolecules*, 2005, **38**, 409
[8] Chen, S.-A. and Ni, J.-M., *Macromolecules*, 1992, **25**, 6081
[9] Sato, M.-A., Tanaka, S. and Kaeriyama, K., *Makromol. Chem.*, 1987, **188**, 1763
[10] Chen, T.-A., Wu, X. and Rieke, R.D., *J. Am. Chem. Soc.*, 1995, **117**, 233
[11] Moulton, J. and Smith, P., *Polymer*, 1992, **33**, 2340
[12] Daoud, W.A. and Xin J.H., *J. Appl. Polym. Sci.*, 2004, **93**, 2131
[13] Buvat, P. and Hourquebie, P., *Macromolecules*, 1997, **30**, 2685
[14] Chen, S.-A. and Lee, S.-J., *Polymer*, 1995, **36**, 1719
[15] Wang, J., *Polym. Degrad. Stab.*, 2005, **89**, 15
[16] Oztemiz, S., Beaucage, G., Ceylan, O. and Mark, H.B., *J. Solid State Electrochem.*, 2004, **8**, 928

Poly(1-eicosene) P-223

Synonyms: *Poly(1-octadecylethylene). Poly(1-octadecyl-1,2-ethanediyl)*
Monomers: 1-Eicosene
Material class: Thermoplastic
Polymer Type: polyolefins
CAS Number:

CAS Reg. No.	Note
27323-11-1	atactic
72108-27-1	isotactic

Molecular Formula: $(C_{20}H_{40})_n$
Fragments: $C_{20}H_{40}$
Morphology: Three polymorphic forms [2]: form II orthorhombic [1]; isotactic orthorhombic [7]; atactic hexagonal (room temp.) and orthorhombic (low temp.)

Volumetric & Calorimetric Properties:
Density:

No.	Value	Note
1	d 0.912 g/cm^3	[5]

Thermodynamic Properties General: ΔH_f 27.46 kJ mol^{-1}, ΔS_f 82.96 J mol$_{-1}$ K^{-1} [2]
Melting Temperature:

No.	Value	Note
1	58°C	[1,2]
2	54°C	327 K, actactic
3	80°C	353 K, isotactic [3]
4	81.8°C	[5]

Transition Temperature General: Behaviour of T_m of the isotactic form has been reported [4]

Transition Temperature:

No.	Value	Note	Type
1	40.5°C	[5]	First T_m

Surface Properties & Solubility:
Solvents/Non-solvents: Sol. $CHCl_3$ [3]

Transport Properties:
Polymer Solutions Dilute: [η] 1.73 dl g^{-1} (25°, toluene) [1]; [η] 0.56 dl g^{-1} (25°, $CHCl_3$) [2]

Optical Properties:
Transmission and Spectra: C-13 nmr [3] and ir [6,7] spectral data have been reported

Applications/Commercial Products:
Processing & Manufacturing Routes: Synth. using Ziegler-Natta catalysts, such as $TiCl_3$-$AlEt_3$ at 40° in heptane [1,5]

Bibliographic References
[1] Magagnini, P.L., Andruzzi, F. and Benetti, G.F., *Macromolecules*, 1980, **13**, 12, (synth, struct)
[2] Magagnini, P.L., Lupinacci, D., Cotrozzi, F. and Andruzzi, F., Makromol. Chem., *Rapid Commun.*, 1980, **1**, 557, (struct, props)
[3] Segre, A.L., Andruzzi, F., Lupinacci, D. and Magagnini, P.L., *Macromolecules*, 1981, **14**, 1845, (C-13 nmr, melting point)
[4] Lupinacci, D., Andruzzi, F., Paci, M. and Magagnini, P.L., *Polymer*, 1982, **23**, 277, (transition temps)
[5] Soga, K., Lee, D.H., Shiono, T. and Kashiwa, N., *Makromol. Chem.*, 1989, **190**, 2683, (synth, melting points)
[6] Benedetti, E., Vergamini, P., Andruzzi, F. and Magagnini, P.L., *Polym. Bull. (Berlin)*, 1980, **2**, 241, (ir)
[7] Benedetti, E., Magagnini, P.L., Andruzzi, F. and Vergamini, P., *CA*, 1979, **93**, 72494, (ir, struct)

Poly(2-(2,3-epoxycyclopentyl)phenyl glycidyl ether)   P-224

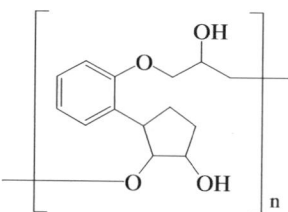

Synonyms: *Poly[2-[o-(2,3-epoxypropoxy)phenyl]-6-oxabicyclo[3.1.0]hexane]*
Monomers: 2-(2,3-Epoxycyclopentyl)phenyl glycidyl ether
Material class: Thermosetting resin
Polymer Type: epoxy
Molecular Formula: $(C_{14}H_{16}O_3)_n$
Fragments: $C_{14}H_{16}O_3$

Volumetric & Calorimetric Properties:
Deflection Temperature:

No.	Value	Note
1	157°C	*m*-phenylenediamine cured, ASTM D648-56 [1]

Transport Properties:
Polymer Solutions Dilute: Viscosity 247 cSt (uncured resin) [1]

Mechanical Properties:
Mechanical Properties General: Other mechanical props. have been reported [2,3]

Compressive Modulus:

No.	Value	Note
1	5260 MPa	763 kpsi, *m*-phenylenediamine cured, ASTM D695-61 [1]

Compressive Strength:

No.	Value	Note
1	224.7 MPa	32.6 kpsi, *m*-phenylenediamine cured, ASTM D695-61 [1]

Applications/Commercial Products:

Trade name	Details	Supplier
ERNA	0386	Union Carbide

Bibliographic References
[1] *U.S. Pat.*, Union Carbide, 1967, 3 332 908, (mechanical props.)
[2] Burhans, A.S., Pitt, C.F., Sellers, R.F. and Smith, S.G., Proc. Ann. Tech. Conf. SPI Reinf. Plast. Div., 21st, *Chicago*, 1966, **10**, (mechanical props.)
[3] Madden, J.J., Norris, D.P., Sellers, R.F. and Smith, S.G., *Proc. Ann. Tech. Management Conf. Reinforced Plastics*, 1964, **19**, 22

Poly(3,4-epoxy-6-methylcyclohexylmethyl-3,4-epoxy-6-methylcyclohexane carboxylate)   P-225

Synonyms: *Poly[(4-methyl-7-oxabicyclo[4.1.0]hept-3-yl)methyl 4-methyl-7-oxabicyclo[4.1.0]heptane-3-carboxylate]*
Monomers: 3,4-Epoxy-6-methylcyclohexylmethyl-3,4-epoxy-6-methylcyclohexane carboxylate
Material class: Thermosetting resin
Polymer Type: epoxy
CAS Number:

CAS Reg. No.
25086-23-1

Molecular Formula: $(C_{16}H_{24}O_4)_n$

Volumetric & Calorimetric Properties:
Density:

No.	Value	Note
1	d 1.2203 g/cm^3	Unox 201 [1]
2	d 1.121 g/cm^3	Unox 201 [2]

Volumetric Properties General: Density of Unox 201 decreases with increasing temp. [2]
Transition Temperature:

No.	Value	Note	Type
1	215°C	Unox 201 [2]	Boiling point 5 mmHg

Surface Properties & Solubility:
Wettability/Surface Energy and Interfacial Tension: Contact angle 22–38° (20–100°, static, Unox 201) [2]. Surface tension of Unox 201 decreases linearly with increasing temp. [2] Other surface tension data have been reported [4]

Transport Properties:
Polymer Solutions Dilute: Viscosity 1520 cP (25°, Unox 201). [1] Viscosity decreases with increasing temp. [2]

Electrical Properties:
Dielectric Strength:

No.	Value	Note
1	650 kV.mm^{-1}	hexahydrophthalic anhydride cured, 17h, 100° [3]
2	630 kV.mm^{-1}	hexahydrophthalic anhydride cured, 45h, 100° [3]
3	690 kV.mm^{-1}	hexachloroendomethylenetetrahydrophthalic anhydride cured, 23h, 100° [3]

Optical Properties:
Refractive Index:

No.	Value	Note
1	1.494	20°, Unox 201 [1]

Applications/Commercial Products:

Trade name	Details	Supplier
Unox	201	Union Carbide

Bibliographic References
[1] Pilný, M. and Mleziva, J., *Kunststoffe*, 1977, **67**, 783
[2] Zvonar, V., *Kolloid Z. Z. Polym.*, 1966, **212**, 113
[3] Beard, J.H. and Orman, S., *Proc. Inst. Electr. Eng.*, 1967, **114**, 989, (dielectric strength)
[4] Herczog, A., Ronay, G.S. and Simpson, W.C., Aerosp. Adhes. Elastomers, Nat. SAMPE (Soc. Aerosp. Mater. Process Eng.) Tech. Conf., Proc., 2nd, 1970, 221, (surface tension)

Poly(1,2-epoxy-3-phenoxypropane)   P-226

Synonyms: *Poly[(phenoxymethyl)oxirane]. Poly(phenyl glycidyl ether)*
Monomers: (Phenoxymethyl)oxirane
Material class: Thermosetting resin
Polymer Type: epoxy
CAS Number:

CAS Reg. No.
25265-27-4

Volumetric & Calorimetric Properties:
Melting Temperature:

No.	Value	Note
1	203°C	isotactic [6]

Glass-Transition Temperature:

No.	Value	Note
1	18°C	approx., isotactic [4]

Surface Properties & Solubility:
Solvents/Non-solvents: Isotactic insol. most solvents [4], Me$_2$CO [5], Et$_2$O, C$_6$H$_6$ [6]

Transport Properties:
Polymer Solutions Dilute: [η]$_{inh}$ 1.61 dl g^{-1} (tetrachloroethane, 30°, isotactic) [4], 2.2 dl g^{-1} (chloronaphthalene, 135°, isotactic) [6]

Applications/Commercial Products:
Processing & Manufacturing Routes: Anionic polymerisation initiated by sodium caprolactam yields oligo(hydroxy ethers) of various structures depending on concentration of the initiator and reaction temp. [1] Phenyl glycidyl ether can be cationically polymerised to yield cyclic dimers when catalysed by 'onium' salts [2]; or thermally polymerised using (4-octyloxyphenyl)iodonium hexafluorophosphate. [3] Isotactic polymer can be synth. using aluminium isopropoxide in combination with a zinc salt [4] or diethylzinc [5] as initiator. Isotactic polymer in 64% yield can be obtained with the aluminium isopropoxide-diethyl zinc system. [5] Crystalline isotactic polymer may be synth. using an AlEt$_3$-H$_2$O catalyst [6]
Applications: Uncured resin used as a reactive diluent in epoxy resin compositions

Bibliographic References
[1] Tänzer, W., Büttner, K. and Ludwig, I., *Polymer*, 1996, **37**, 997
[2] Kushch, P.P., Karateyev, A.M., Kuzayev, A.I., Dzhavadyan, E.A., and Rozenberg, V.A., *Vysokomol. Soedin.*, Ser. A, 1987, **29**, 1832
[3] Crivello, J.V. and Lee, J.L., *J. Polym. Sci., Part A: Polym. Chem.*, 1989, **27**, 3951
[4] Ronda, J.C., Serra, A., Mantecón, A. and Cádiz, V., *Acta Polym.*, 1996, **47**, 269, (synth)
[5] Bero, M., *J. Polym. Sci., Polym. Chem. Ed.*, 1982, **19**, 191, (synth)
[6] Vandenberg, E.J., *J. Polym. Sci., Part A-1*, 1969, **7**, 525, (synth)

Polyester carbonates   P-227

Related Polymers: Poly[2,2-propanebis(4-phenyl)carbonate], Bisphenol A-terephthalic acid polyester carbonate, Bisphenol A-isophthalic acid polyester carbonate, Bisphenol A-terephthalic acid-isophthalic acid polyester carbonate
Material class: Thermoplastic
Molecular Formula: [(C$_{16}$H$_{14}$O$_3$).(C$_{23}$H$_{18}$O$_4$)]$_n$
Fragments: C$_{16}$H$_{14}$O$_3$ C$_{23}$H$_{18}$O$_4$
General Information: Glass fibre reinforced versions are available

Applications/Commercial Products:
Processing & Manufacturing Routes: Prod. by interfacial polymerisation at low temps. (0–30°) of the sodium salt of Bisphenol A (aq. phase) with aromatic acid chlorides and phosgene in an organic phase. The catalysts are tertiary amines, quaternary ammonium or phosphonium salts. Phenols, phenylchlorocarbonates and aromatic carbonyl halides are used as chain terminators. The polyester carbonates are isol. from the organic phase
Mould Shrinkage (%):

No.	Value
1	0.7–1%

Applications: Lighting engineering, electronics and electrical engineering applications

Trade name	Details	Supplier
Apec	KL 1-9308	Bayer Inc.
Copec		Mitsubishi Chemical
Lexan	4501, 4701, -PPC, KL-9306, 3250	General Electric Plastics
XP 73		Dow
	KU 1-9318	Bayer Inc.
	KU 1-9312 (glass fibre reinforced)	Bayer Inc.

Poly[1,1-ethane bis(4-phenyl)carbonate] P-228

Synonyms: *Poly[oxycarbonyloxy-1,4-phenyleneethylidene-1,4-phenylene]. Poly(4,4′-dihydroxydiphenyl-1,1-ethane-co-carbonic dichloride)*
Monomers: 4,4′-Dihydroxydiphenyl-1,1-ethane, Carbonic dichloride
Material class: Thermoplastic
Polymer Type: polycarbonates (miscellaneous)
CAS Number:

CAS Reg. No.
28774-91-6

Molecular Formula: $(C_{15}H_{12}O_3)_n$
Fragments: $C_{15}H_{12}O_3$

Volumetric & Calorimetric Properties:
Density:

No.	Value	Note
1	d 1.22 g/cm^3	25° [3]

Melting Temperature:

No.	Value	Note
1	185–195°C	[3]

Glass-Transition Temperature:

No.	Value	Note
1	130°C	[1,3]

Surface Properties & Solubility:
Solubility Properties General: Forms miscible blends with Bisphenol A polycarbonate and tetramethylpolycarbonate [1]; it forms immiscible blends with polymethyl methacrylate and polystyrene [1]
Solvents/Non-solvents: Sol. MeCl, *m*-cresol, cyclohexanone, THF, C_6H_6, pyridine, DMF; swollen by EtOAc, Me_2CO [2]

Mechanical Properties:
Tensile Strength Break:

No.	Value	Note	Elongation
1	74.4 MPa	759 kg cm^{-2} [3]	167%

Electrical Properties:
Electrical Properties General: Dielectric strength and dissipation factor do not vary with temp. until T_g
Dielectric Permittivity (Constant):

No.	Value	Frequency	Note
1	2.9	1 KHz	25–100° [3]

Dissipation (Power) Factor:

No.	Value	Frequency	Note
1	0.00049	1 KHz	25–100° [3]

Optical Properties:
Refractive Index:

No.	Value	Note
1	1.5937	25° [3]
2	1.594	calc. [1]

Transmission and Spectra: Ir spectral data have been reported [2]

Applications/Commercial Products:
Processing & Manufacturing Routes: Synth. by two routes; phosgenation or ester interchange. Phosgenation is accomplished by interfacial polycondensation (most widely used) involving base-catalysed reaction of phosgene with an aromatic dihydroxy compound in CH_2Cl_2, with a catalyst to give high MW products [2,3], or by soln. polymerisation in pyridine [2] Ester exchange involves melt (150–300°) polymerisation of the aromatic dihydroxy compound with excess diphenyl carbonate in the absence of air. Reaction rate may be increased by use of alkaline catalysts or a medium-high vacuum in the final stages [2,3]

Bibliographic References
[1] Kim, C.K. and Paul, D.R., *Macromolecules*, 1992, **25**, 3097, (miscibility, T_g, refractive index)
[2] Schnell, H., *Polym. Rev.*, 1964, **9**, 99, (solvents, ir, rev)
[3] Schnell, H., *Ind. Eng. Chem.*, 1959, **51**, 157, (props.)

Poly(1-ethenyl-2-pyrrolidinone-*co*-ethenyl acetate-*co*-ethenyl propanoate) P-229

Synonyms: *Poly(1-vinyl-2-pyrrolidinone-co-vinyl acetateco-vinyl propanoate)*
Material class: Copolymers
Polymer Type: acrylic, polyvinyls
CAS Number:

CAS Reg. No.
26124-25-4

Molecular Formula: $[(C_6H_9NO).(C_4H_6O_2).(C_5H_8O_2)]_n$
Fragments: C_6H_9NO $C_4H_6O_2$ $C_5H_8O_2$

Surface Properties & Solubility:
Solubility Properties General: Solns. of the terpolymer are compatible with propane and butane. Cloud point data for various solvent mixtures have been reported [1]
Solvents/Non-solvents: Sol. EtOH, 2-propanol, CH_2Cl_2 [1]

Applications/Commercial Products:

Trade name	Supplier
Luviskol VAP343	BASF

Bibliographic References
[1] *Luviskol*, BASF, 1998, (technical datasheet)

Poly(1-ethenyl-2-pyrrolidinone-co-[ethyl 2-methyl-2-propenoate]-co-2-methyl-2-propenoic acid)   P-230

Synonyms: *PVP-ethyl methacrylate-methacrylic acid copolymer. Acrylates-PVP copolymer. Poly(1-vinyl-2-pyrrolidinone-co-ethyl methacrylate-co-methacrylic acid)*
Related Polymers: Poly N-vinylpyrrolidone, Methacrylate/vinyl copolymers, Hydroxyalkyl methacrylate copolymers
Monomers: Methacrylic acid, Ethyl methacrylate, N-Vinylpyrrolidinone
Material class: Copolymers
Polymer Type: acrylic, polyvinyls
CAS Number:

CAS Reg. No.
26589-26-4

Molecular Formula: $[(C_4H_6O_2).(C_6H_{10}O_2).(C_6H_9NO)]_n$
Fragments: $C_4H_6O_2$ $C_6H_{10}O_2$ C_6H_9NO

Applications/Commercial Products:

Trade name	Details	Supplier
Luviflex VBM	discontinued	BASF
Luviflex VBM 35		BASF
Stepanhold		Stepan
Stepanhold Extra		Stepan

Polyetheretherketone   P-231

Synonyms: *PEEK. Polyaryletheretherketone. Poly(oxy-1,4-phenyleneoxy-1,4-phenylenecarbonyl-1,4-phenylene)*
Related Polymers: Polyetherketone, Polyetherketoneketone, Polyetherketoneetherketoneketone, Polyetheretherketone 30% Glass fibre filled, Polyetheretherketone 30% Carbon fibre filled, Polyetheretherketoneketone
Monomers: 4,4′-Difluorobenzophenone, Hydroquinone
Material class: Thermoplastic
Polymer Type: polyetherketones
CAS Number:

CAS Reg. No.
31694-16-3

Molecular Formula: $(C_{19}H_{12}O_3)_n$
Fragments: $C_{19}H_{12}O_3$
Additives: Glass fibre, carbon fibre, PTFE, graphite
Miscellaneous: Commercial material is approx. 35% crystalline

Applications/Commercial Products:
Processing & Manufacturing Routes: Prod. by polycondensation under anhydrous conditions of an aromatic diol and an aromatic dihalide in which the halogen atoms are activated by carbonyl groups. Processed by injection moulding, extrusion, compression and transfer moulding

Mould Shrinkage (%):

No.	Value
1	1%

Applications: Pipelines, pump and valve parts, cable insulation, composites, film, circuit boards, switchgear, bearings and sensor coatings. Generally used for applications in 'aggressive environments'

Trade name	Details	Supplier
Arlon	various grades	Greene, Tweed and Co.
CTI PK	filled grades	CTI
EGC	XC-2	EGC Corporation
Electrafil	J-1105/CF/30	DSM Engineering
Fiberfil	J-1105/30	DSM Engineering
Fluorotemp	103	Furon
K2	varous grades	Thermofil Inc.
Lubricomp	various grades	LNP Engineering
RTP	various grades	LNP Engineering
Tetra-Temp	PEEK 1000	Tetrafluor Inc.
Victrex		Victrex Ltd.
Xytrex	450 series	EGC Corporation

Polyetheretherketone, 30% Carbon fibre filled   P-232

Synonyms: *Polyaryletheretherketone, 30% Carbon fibre filled. PEEK, 30% Carbon fibre filled*
Related Polymers: Polyetheretherketone, Polyetheretherketone, 30% Glass fibre filled
Monomers: 4,4′-Difluorobenzophenone
Material class: Thermoplastic, Composites
Polymer Type: polyetherketones
CAS Number:

CAS Reg. No.
31694-16-3

Molecular Formula: $(C_{19}H_{12}O_3)_n$
Fragments: $C_{19}H_{12}O_3$
Additives: Carbon fibre
General Information: Both standard and "easy-flow" grades are available in a black granular form
Morphology: Commercial products are typically 35% crystalline [1]
Miscellaneous: Addition of carbon fibre improves tensile strength and heat distortion temp.

Volumetric & Calorimetric Properties:
Density:

No.	Value	Note
1	d 1.44 g/cm^3	cryst., ISO R1183 [1]

Thermal Expansion Coefficient:

No.	Value	Note	Type
1	1.5×10^{-5} K^{-1}	$T < T_g$, ASTM D696 [1]	L

– Polyetheretherketone, 30% Carbon fibre filled

Thermal Conductivity:

No.	Value	Note
1	0.92 W/mK	ASTM C177 [1]

Melting Temperature:

No.	Value	Note
1	340–343°C	[1]
2	326–330°C	short fibre reinforced [25]

Glass-Transition Temperature:

No.	Value	Note
1	143°C	[1]

Transition Temperature General: The presence of short fibres has a comparatively small effect on T_m [25]

Deflection Temperature:

No.	Value	Note
1	300–315°C	1.82 MPa, ISO R75 [1]

Heat Distortion Temperature:

No.	Value	Note
1	300–315 °C	1.82 MPa, ISO R 74 [1]

Transport Properties:

Permeability and Diffusion General: Diffusion coefficient [H_2] 3.93×10^9 cm^2 s^{-1} (60°, 75% relative humidity), 5.91×10^9 cm^2 s^{-1} (70°, 75% relative humidity), 8.43×10^9 cm^2 s^{-1} (80°, 75% relative humidity) [24]

Water Absorption:

No.	Value	Note
1	0.12 %	24h, 23°, ASTM D570 [10]
2	0.3 %	saturation, air, 23° [8]
3	0.14 %	50% relative humidity, saturation, 23° [8]
4	0.06 %	24h, 23°, ISO R62A [1]

Mechanical Properties:

Mechanical Properties General: The effect on processing parameters on the ultimate flexural strength, flexural modulus and tensile strength have been reported. The use of higher forming temps., 400–440°, gives an improvement in mechanical props. [7,21]. Shows good mechanical props. at elevated temps. and high humidity, e.g. 95% retention of flexural strength and 70% retention of tensile strength after 1400h at 80° and 75% relative humidity [24]

Tensile (Young's) Modulus:

No.	Value	Note
1	24100 MPa	3500 kpsi, ASTM D638 [10]
2	12400 MPa	short fibre [25]

Flexural Modulus:

No.	Value	Note
1	19200–20200 MPa	23°, ISO R178 [1]
2	18600 MPa	120°, ISO R178 [1]
3	5100 MPa	250°, ISO R178 [1]

Tensile Strength Break:

No.	Value	Note	Elongation
1	228 MPa	33 kpsi, ASTM D638	1.3%
2	215.8 MPa	short fibre [25]	2.5%

Flexural Strength at Break:

No.	Value	Note
1	355 MPa	23°, ISO R178 [1]
2	260 MPa	120°, ISO R178 [1]
3	105 MPa	250°, ISO R178 [1]

Compressive Strength:

No.	Value	Note
1	240 MPa	23°, with flow, ASTM D695 [1]
2	153 MPa	23°, across flow, ASTM D695 [1]

Viscoelastic Behaviour: Elastic costants over the range 20–200° have been reported [5]

Hardness: Rockwell R124 (ASTM D785) [1], M107 (ASTM D785) [1]

Failure Properties General: The energy absorption capability of tubes has been investigated; tubes with fibres orientated at 15° to the tube axis give the highest value 225 kJ kg^{-1}. Crush zone morphology has been reported [18]. The testing of compressive strength of unidirectional and cross-ply laminates has been reported. Failure in unidirectional laminates occurs through in-plane fibre buckling. In the cross ply laminate out-of-plane fibre buckling is the mode of failure and the failure of longitudinal plies occurs at higher compressional stress than in the unidirectional case [16]. The effect of varying the forming temp. on mode I and mode II delamination resistance has been reported [21]. The application of heat (330–380°) at a pressure of 1.4 MPa allows recovery of mode I delamination resistance to varying degrees [23]

Fracture Mechanical Properties: Ultimate shear strength 97 MPa (23°, ASTM D3846) [1]. The effects of MW, volume fraction of fibre and strength of fibre/matix bond on fracture behaviour have been investigated. Crack stability and temp. variation have been reported [3,4]. Shows excellent fracture toughness at temps. up to 100° above which creep causes a decrease in toughness. Two types of crack propagation have been described in high temp. fatigue: time dependent and cycle dependent

Friction Abrasion and Resistance: Wear behaviour as a function of counterface (steel) roughness and temp. has been reported [2]. Coefficient of friction 0.16 (20°, 160 kg load), (200°, 13 kg load) [1]. Wear rate 88 μm h^{-1} (20°, 160 kg load), 200 μm h^{-1} (200°, 120 kg load), 225 μm h$^-$ (20°, 22 kg load) [1]

Izod Notch:

No.	Value	Notch	Note
1	9.1 J/m	Y	23°, 0.25 mm notch, ISO R180/1A [1]
2	42.1 J/m	N	23°, ISO R180/1A [1]

– Polyetheretherketone, 30% Glass fibre filled

Electrical Properties:
Electrical Properties General: Volume resistivity 140000 Ω cm (IEC 93) [1]

Optical Properties:
Optical Properties General: Specific optical density, D_s 5 (3.2 mm thick, flaming mode, BS6401/ASTM E662), 2 (3.2 mm thick, non-flaming mode, BS6401/ASTM E662) [1]. Specific optical density is zero after 1.5 min. [1]. Time to 90% D_s 19 min. (3.2 mm thick, flaming mode, BS6401/ASTM E662) [1]

Polymer Stability:
Upper Use Temperature:

No.	Value	Note
1	240–260°C	UL 746B [1]

Decomposition Details: Products of decomposition include CO and CO_2 [1]
Flammability: Flammability rating V0 (1.45 mm thick, UL94) [1]

Applications/Commercial Products:
Processing & Manufacturing Routes: Processed by conventional injection moulding and extrusion techniques. Injection moulding conditions are generally the same as those used for the non-filled grade although temps. are approx. 20° lower for the latter. Machining conditions such as turning, milling and drilling generally require lower speeds than the unreinforced grade [1]

Mould Shrinkage (%):

No.	Value	Note
1	0.3%	across flow [1]
2	0.03%	with flow [1]

Trade name	Supplier
APC-2	Fiberite Corp.
APC-2	ICI Acrylics
Erta PEEK	Erta
Franplas	Franklin
Victrex	Victrex Ltd.

Bibliographic References
[1] *Victrex Peek,* Victrex, 1997, (technical datasheet)
[2] Friedrich, K., Karger-Kocsis, J. and Lu, Z., *Wear,* 1991, **235**, (friction, wear)
[3] Hine, P.J., Brew, B., Duckett, R.A. and Ward, I.M, *Compos. Sci. Technol.,* 1991, **40**, 47, (failure props)
[4] Hine, P.J., Brew, B., Duckett, R.A. and Ward, I.M, *Compos. Sci. Technol.,* 1989, **35**, 31, (failure props)
[5] Uematsu, Y., Kitamura, T. and Ohtari, R, *Compos. Sci. Technol.,* 1995, **53**, 333
[6] Davies, P., Cartwell, W.J., Bourban, P.-E., Zysman, V. and Kausch, H.H., *Composites,* 1991, **22**, 425
[7] Astrom, B.T., Larsson, P.H., Hepola, P.J. and Pipes, R.B., *Compos. Sci. Technol.,* 1994, **25**, 814, (flexural props)
[8] *Erta,* Internet: http://www.erta.be/e/sp/products/erta peek.htm, (technical datasheet)
[9] *Advanced Composite Materials Products and Manufacturers,* (ed. D.J. De Ronzo), 1988, **NDC**, 16, (fiberite products)
[10] *Advanced Composite Materials Products and Manufacturers,* (ed. D.J. De Ronzo), 1988, **NDC**, 115, (RTP products)
[11] Friedrich, K. and Jacobs, O., *Compos. Sci. Technol.,* (ed. D.J. De Ronzo), 1992, **43**, 71
[12] Buggy, M. and Carew, A., *J. Mater. Sci.,* 1994, **29**, 1925
[13] Buggy, M. and Carew, A., *J. Mater. Sci.,* 1994, **29**, 2255
[14] Ramakrishna, S., Tan, W.K., Teoh, S.H. and Lai, M.O., *Key Eng. Mat.,* 1998, **137**, 1, (recycling)
[15] Sarasua, J.R. and Pouyet, J., *J. Mater. Sci.,* 1997, **32**, 533, (recycling)
[16] Kominar, V., Narkis, M., Sigemann, A. and Vaxman, A, *J. Mater. Sci.,* 1995, **30**, 2620
[17] Choy, C.L., Kwok, K.W., Leung, W.P. and Lau, F.P., *J. Polym. Sci., Part B: Polym. Lett.,* 1994, **32**, 1389, (thermal conductivity)
[18] Hamada, H., Ramakrishna, S. and Sato, H, *J. Compos. Mater.,* 1996, **30**, 947
[19] Barnes, J.A., *J. Mater. Sci.,* 1993, **28**, 4974
[20] Phillips, R., Glauser, T., Manson, J.-A., *Polym. Compos.,* 1997, **18**, 500, (thermal stability)
[21] Jar, P.-Y. B., Mulone, R., Davies, P., Kausch, H.-H., *Compos. Sci. Technol.,* 1993, **46**, 7, (mech. props)
[22] Wang, W., Qi, Z. and Jeronimidis, G., *J. Mater. Sci.,* 1991, **26**, 5915
[23] Davies, P., Cartwell, W., Kausch, H.H., *J. Mater. Sci. Lett.,* 1989, **8**, 1247
[24] Ma, C.-C.M., Yur, S.-W., *Polym. Eng. Sci.,* 1991, **31**, 34
[25] Sarasua, J.R., Remiro, P.M., Pouyet, J., *J. Mater. Sci.,* 1995, **30**, 3501

Polyetheretherketone, 30% Glass fibre filled P-233
Synonyms: *Polyaryletheretherketone, 30% Glass fibre filled. PEEK, 30% Glass fibre filled*
Related Polymers: Polyetheretherketone, Polyetheretherketone, 30% Carbon fibre filled
Monomers: 4,4′-Difluorobenzophenone, Hydroquinone
Material class: Thermoplastic, Composites
Polymer Type: polyetherketones
CAS Number:

CAS Reg. No.
31694-16-3

Molecular Formula: $(C_{19}H_{12}O_3)_n$
Fragments: $C_{19}H_{12}O_3$
Additives: Glass fibre
General Information: Both standard and "easy-flow" grades are available in a brown granular form
Morphology: Commercial products are typically 35% crystalline [1]
Miscellaneous: Addition of glass fibre improves tensile strength and heat distortion temperature

Volumetric & Calorimetric Properties:
Density:

No.	Value	Note
1	d 1.49 g/cm^3	cryst., ISO R1183 [1]

Thermal Expansion Coefficient:

No.	Value	Note	Type
1	$2.2 \times 10^{-5} – 2.5 \times 10^{-5}$ K^{-1}	$T < T_g$, ASTM D696 [1,2]	L

Thermal Conductivity:

No.	Value	Note
1	0.43 W/mK	ASTM C177 [1]
2	0.43 W/mK	2.98 Btu in (h ft$^{2\circ}$F)$^{-1}$, ASTM D696 [2]

Melting Temperature:

No.	Value	Note
1	340–343°C	[1,2]
2	329.5°C	short fibre reinforced

Glass-Transition Temperature:

No.	Value	Note
1	142–143°C	[1,2]

Polyetheretherketone, 30% Glass fibre filled

Transition Temperature General: The incorporation of short glass fibres has only a comparatively small effect on T_m [4]

Deflection Temperature:

No.	Value	Note
1	300–315°C	1.82 MPa, ISO R75 [1]

Heat Distortion Temperature:

No.	Value	Note
1	300–315 °C	1.82 Mpa, ISO R75 [1]

Transport Properties:
Water Absorption:

No.	Value	Note
1	0.11 %	24h, 23°, ISO R62A [1]

Mechanical Properties:
Tensile (Young's) Modulus:

No.	Value	Note
1	17900 MPa	2600 kpsi, ASTM D638 [5]
2	7400 MPa	short fibre [4]

Flexural Modulus:

No.	Value	Note
1	9700–10000 MPa	23°, ISO R178 [1]
2	7600 MPa	1100 kpsi, 23°, ASTM D790 [2]
3	9200 MPa	120°, ISO R178 [1]
4	3000 MPa	250°, ISO R178 [1]

Tensile Strength Break:

No.	Value	Note	Elongation
1	196.5 MPa	28500 psi, ASTM D638	2–3%
2	128.7 MPa	short fibre [4]	2.4%

Flexural Strength at Break:

No.	Value	Note
1	233 MPa	23°, ISO R178 [1]
2	186 MPa	27 kpsi, 23°, ASTM D790 [2]
3	175 MPa	120°, ISO R178 [1]
4	70 MPa	250°, ISO R178 [1]

Poisson's Ratio:

No.	Value	Note
1	0.45	23°, across flow, ASTM D638 [1]

Compressive Strength:

No.	Value	Note
1	215 MPa	23°, with flow, ASTM D695 [1,2]
2	149 MPa	23°, across flow, ASTM D695 [1,2]

Miscellaneous Moduli:

No.	Value	Note	Type
1	2400 MPa	23° [1]	shear modulus

Hardness: Rockwell R124 (ASTM D785) [1,2], M103 (ASTM D785) [1]
Fracture Mechanical Properties: Ultimate shear strength 97 Mpa (23°, ASTM D3846 [1], 100 Mpa 14065 psi, 23°, ASTM D3846 [2]
Friction Abrasion and Resistance: Wear behaviour as function of counterface (steel) roughness and temp. has been reported [3]

Izod Notch:

No.	Value	Notch	Note
1	102 J/m	Y	23°, 0.25 mm notch, ISO R180/1A [1]
2	96 J/m	Y	23°, 0.25 mm notch, ASTM D256 [2]
3	409 J/m	N	23°, ISO R180/1A [1]
4	726 J/m	N	23°, ASTM D256 [2]

Electrical Properties:
Surface/Volume Resistance:

No.	Value	Note	Type
1	10×10^{15} Ω.cm	ASTM D257 [2]	V

Dielectric Permittivity (Constant):

No.	Value	Frequency	Note
1	3.7	50–100 Hz	1–150°, ASTM D150 [2]

Dielectric Strength:

No.	Value	Note
1	17.5 kV.mm^{-1}	50μm film, ASTM D149 [2]

Strong Field Phenomena General: Comparative tracking index 175 V (23°, IEC112) [1]

Dissipation (Power) Factor:

No.	Value	Frequency	Note
1	0.004	1 MHz	23°, ASTM D150 [2]

Polymer Stability:
Upper Use Temperature:

No.	Value	Note
1	240–260°C	UL 746B [1]

Flammability: Flammability rating V0 (1.45 mm thick, UL94) [1]
Hydrolytic Stability: Exposure to water at 200° and under pressure of 14 bar causes loss of tensile and flexural strength and flexural modulus. Elongation to weak and falling weight impact strength are affected [1]

Applications/Commercial Products:
Processing & Manufacturing Routes: Processed by conventional injection moulding and extrusion techniques. Injection moulding conditions are generally the same as those used for the non-filled grade although temps. are approx. 20° lower for the latter. Machining conditions such as turning, milling and drilling generally require lower speeds than the reinforced grade [1]

Mould Shrinkage (%):

No.	Value	Note
1	1.1%	across flow [1]
2	0.2%	with flow [1]

Trade name	Supplier
Victrex	Victrex Ltd.

Bibliographic References

[1] *Victrex Peek,* Victrex, (technical datasheet)
[2] *Arlon,* 1999, (technical datasheet)
[3] Friedrich, K., Karger-Kocsis, and Lu, Z., *Wear,* 1991, 235, (friction, wear)
[4] Sarasua, J.R., Remiro, P.M., Pouyet, J.J., *J. Mater. Sci.,* 1995, **39**, 3501, (mechanical props.)

Polyetheretherketoneketone P-234

Synonyms: *Polyaryletheretherketoneketone. PEEKK. Poly(oxy-1,4-phenyleneoxy-1,4-phenylenecarbonyl-1,4-phenylenecarbonyl-1,4-phenylene)*
Related Polymers: Polyetheretherketone, Polyetherketone, Polyetherketoneketone, Polyetherketoneetherketoneketone
Material class: Thermoplastic
Polymer Type: polyetherketones
CAS Number:

CAS Reg. No.
60015-03-4

Molecular Formula: $(C_{32}H_{20}O_4)_n$
Fragments: $C_{32}H_{20}O_4$

Applications/Commercial Products:

Trade name	Supplier
Hostatec	Hoechst Celanese

Polyetherketone P-235

Synonyms: *PEK. mPEK. Poly(oxy-1,3-phenylenecarbonyl-1,4-phenylene). Poly(oxy-1,4-phenylenecarbonyl-1,4-phenylene).*
Related Polymers: Polyetheretherketone, Polyetherketoneketone, Polyetherketoneetherketoneketone, Polyetheretherketoneketone
Monomers: 4,4′-Dihydroxybenzophenone
Material class: Thermoplastic
Polymer Type: polyetherketones
CAS Number:

CAS Reg. No.	Note
27380-27-4	1,4-isomer
110418-45-6	1,3-isomer
127288-82-8	
54991-68-3	
104381-45-5	

Molecular Formula: $(C_{13}H_8O_2)_n$
Fragments: $C_{13}H_8O_2$
Additives: Glass fibre, carbon fibre, PTFE
General Information: Is a semi-crystalline engineering polymer with very good thermal and chemical stability. The normal polymer has *para*-substitution throughout (the 1,4-isomer) but a form with *meta*-substituted units (the 1,3-isomer) is known (mPEK). PEK/mPEK copolymers have been synth. [10]
Morphology: Can be crystallised over a wide temp. range or quenched to an amorph. glass. Two crystalline polymorphs have been reported [5]. Orthorhombic a 0.776, b 0.600, c 1.001; volume 466.0Å^3; density 1.398 g cm^{-3} [1]; a 0.762, b 0.587, c 1.003 [12]. Crystallinity 31%; crystallite size a 6.9, b 5.0, c 7.7 [12]. mPEK is amorph. when isolated but is capable of solvent-induced crystallisation to give a mixture of polymorphs. PEK/mPEK copolymers can be thermally crystallised [10]. Differences in the morphology of polymeric and oligomeric PEK are attributable to chain folding in the polymeric form [13]

Volumetric & Calorimetric Properties:
Density:

No.	Value	Note
1	d 1.317 g/cm^3	[12]
2	d 1.43 g/cm^3	cryst. [12]
3	d 1.272 g/cm^3	amorph. [12]
4	d 1.29–1.32 g/cm^3	mPEK, amorph. [16]

Thermal Expansion Coefficient:

No.	Value	Note	Type
1	3.2×10^{-5} K^{-1}	21–149°, ASTM D696 [18]	L
2	3.8×10^{-5} K^{-1}	149–199°, ASTM D696 [18]	L
3	4.4×10^{-5} K^{-1}	199–249°, ASTM D696 [18]	L

Thermal Conductivity:

No.	Value	Note
1	0.245 W/mK	1.7 Btu in h^{-1} ft^{-2}°F^{-1}), ASTM D695 [18]

Melting Temperature:

No.	Value	Note
1	365°C	[1]
2	365–373°C	[7]
3	373°C	[18]
4	375°C	[12]
5	180°C	mPEK [10]
6	264–307°C	PEK/mPEK copolymer, varying composition [10]

Glass-Transition Temperature:

No.	Value	Note
1	153°C	[4]
2	162°C	[18]
3	131–132°C	mPEK [10,16]
4	142–150°C	PEK/mPEK copolymer, varying composition [10]

Vicat Softening Point:

No.	Value	Note
1	340°C	[2]

– Polyetherketone

Transport Properties:
Polymer Solutions Dilute: Intrinsic viscosity 5.1 dl g^{-1} (conc. H_2SO_4, MW 11500), 11.2 dl g^{-1} (conc. H_2SO_4, MW 34900), 15.2 dl g^{-1} (conc. H_2SO_4, MW 49600) [17]

Mechanical Properties:
Mechanical Properties General: Tensile stress at 200° increases with increasing crystallisation temp., particularly when T_c is significantly higher than T_g. The total crystallinity of the polymers, however, is essentially the same despite the increasing crystallisation temp. The variation of tensile stress is not observed when measurements are carried out at room temp. [14]. The elastic moduli in the crystalline regions, both parallel and perpendicular to the chain axis have been reported [12]

Tensile (Young's) Modulus:

No.	Value	Note
1	1800 MPa	[2]
2	4500 MPa	650 kpsi, ASTM D638 [18]
3	5400 MPa	[12]
4	57000 MPa	parallel to chain axis [12]
5	5400–5700 MPa	perpendicular to chain axis [12]
6	4100–4200 MPa	mPEK [16]

Flexural Modulus:

No.	Value	Note
1	4900 MPa	500 kg mm^{-2} [7]
2	4500 MPa	650 kpsi, ASTM D790 [18]

Tensile Strength Break:

No.	Value	Note	Elongation
1	96.5 MPa	14 kpsi, ASTM D638 [18]	14%
2	128 MPa	13 kg mm^{-2} [7]	
3	323 MPa	[12]	30.2%
4	56–77 MPa	mPEK [16]	1–2%

Flexural Strength at Break:

No.	Value	Note
1	196 MPa	28500 psi, ASTM D790 [18]
2	206 MPa	21 kg mm^{-2} [7]

Tensile Strength Yield:

No.	Value	Note
1	117 MPa	17 kpsi, ASTM D638 [18]

Compressive Strength:

No.	Value	Note
1	75 MPa	[2]

Viscoelastic Behaviour: Stress at constant strain rate decreases with increasing temp. At room temp. stress at yield increases with increasing strain rate [15]
Hardness: Rockwell R126 (ASTM D785) [18]
Failure Properties General: Bending strength 59 MPa [2]
Fracture Mechanical Properties: Crack velocities during catastrophic failure have been measured at approx. 200 m s^{-1}. The possible decomposition of polymer in the crack zone of a running crack has been reported [11]

Izod Notch:

No.	Value	Notch	Note
1	69.4 J/m	N	23°, 1.3 ft Ib in^{-1}, ASTM D256 [18]

Electrical Properties:
Dielectric Permittivity (Constant):

No.	Value	Frequency	Note
1	3.4–3.5	50–100 kHz	0–150°, ASTM D150 [18]

Dielectric Strength:

No.	Value	Note
1	19 kV.mm^{-1}	50μm film, ASTM D149 [18]

Dissipation (Power) Factor:

No.	Value	Frequency	Note
1	0.005	1 MHz	23°, ASTM D150 [18]

Polymer Stability:
Upper Use Temperature:

No.	Value	Note
1	260°C	continuous, ASTM D695 [18]

Decomposition Details: 5% Weight loss in N_2 occurs at 520° [7]. Onset of degradation 420° (mPEK) [16]. The main decomposition products at 500° are CO, CO_2 and low MW hydrocarbons [11]

Applications/Commercial Products:
Processing & Manufacturing Routes: Produced by condensation reaction between dihalides and diphenoxides or self condensation of acid chlorides with Lewis acid catalysts. Processed using conventional injection moulding and extrusion equipment. High MW PEK reacts with 4,4′-difluorobenzophenone in the presence of potassium carbonate to give low MW PEK. The reaction via oligomerisation of the monomer and trans-etherification between oligomers and PEK have been reported [9]. High MW PEK can be synth. by nucleophilic polycondensation of 4,4′-dihydroxybenzophenone or the dihalo analogue in the presence of sodium carbonate and a silicon/copper catalyst. Can also be made by electrophilic polycondensation of 4-phenoxybenzoyl chloride in the presence of a Friedel-Crafts catalyst [6,7]. Synth. via a ketimine route has been reported [8]. The preparation of mPEK and PEK/mPEK copolymers has been reported [10,16]

Trade name	Supplier
Arlon	Raychem Inc.
Hostatec	Hoechst Celanese
Stilan	
Ultrapek	BASF
Victrex PEK	ICI Americas Inc.

Bibliographic References
[1] Blundell, D. J., and Newton, A. B., *Polymer*, 1991, **32**, 309
[2] Askadskii, A. A., Salazkin, S. N., Bychko, K. A., Gileva, N. G. *et al*, *Polym. Sci. USSR (Engl. Transl.)*, 1989, **31**, 2930, (struct., mechanical props.)

[3] Zolotukhin, M. G., Gileva, N. G., Salazkin, S. N., Sangalov, Y. A., et al, *Polym. Sci. USSR (Engl. Transl.)*, 1988, **31**, 2748, (synth.)
[4] Goodwin, A. A. Hay, J. N., *J. Polym. Sci. Part B*, 1998, **36**, 851, (dielectric props.)
[5] Gardner, K. H., Hsiao, B. S. and Faron, K. L., *Polymer*, 1994, **35**, 2290, (morphology)
[6] Fukawa, I., Tanabe, T., Hachiya, H., *Polym. J. (Tokyo)*, 1992, **24**, 173, (synth.)
[7] Fukawa, I., and Tanabe, T., *Macromolecules*, 1991, **24**, 3838, (synth.)
[8] Bourgeois, Y., Devaux, J., Legras, R., and Parsons, I. W., *Polymer*, 1996, **37**, 3171, (synth.)
[9] Fukawa, I., Tanabe, T., *J. Polym. Sci. Part A*, 1993, **31**, 535, (synth.)
[10] Teasley, M. F., and Hsiao, B. S., *Macromolecules*, 1996, **29**, 6432, (synth., props.)
[11] Swallowe, G. M., Dawson, P. C., Tang, T. B. and Xu, Q. L., *J. Mater. Sci.*, 1995, **30**, 3853
[12] Nishimo, T., Tada, K., and Nakamae, K., *Polymer*, 1992, **33**, 736, (mech. props.)
[13] Waddon, A. J., Keller, A. and Blundell, D. J., *Polymer*, 1992, **33**, 27, (morphology)
[14] Tregub, A., Karger-Kocsis, J., Konnecke, K., and Zimmerman, H. J., *Macromolecules*, 1995, **28**, 3890
[15] Hamad, S., Swallowe, G. M., *J. Mater. Sci.*, 1996, **31**, 1415, (mech. props)
[16] Bai, S. J., Dotrong, M., Soloski, E. J. and Evers, R. C., *J. Polym. Sci. Part B*, 1991, **29**, 119, (synth., props.)
[17] Wei-Berk, C., and Berry, G. C., *J. Polym. Sci. Part B*, 1990, **28**, 1873, (soln. props.)
[18] *Arlon*, 1999, (technical datasheet)

Polyetherketoneetherketoneketone   P-236

Synonyms: *Poly(oxy-1,4-phenylenecarbonyl-1,4-phenyleneoxy-1,4-phenylenecarbonyl-1,4-phenylenecarbonyl-1,4-phenylene). Poly(-oxy-1,4-phenylenecarbonyl-1,4-phenyleneoxy-1,4-phenylenecarbonyl-1,3-phenylenecarbonyl-1,4-phenylene). PEKEKK*
Related Polymers: Polyetheretherketone, Polyetherketone, Polyetherketoneketone, Polyetheretherketoneketone
Material class: Thermoplastic
Polymer Type: polyetherketones
CAS Number:

CAS Reg. No.
132210-87-8
88030-82-4
60015-05-6
104381-49-9

Molecular Formula: $(C_{26}H_{16}O_4)_n$
Fragments: $C_{26}H_{16}O_4$
Additives: Is a semi-crystalline polymer with good thermal stability
Morphology: Two crystalline polymorphs exist, the relative proportions of which depend on the mode of crystallisation. As prepared the polymer is mainly in one crystalline form but after annealing the second form predominates. Reprecipitation alters the crystalline form [3,5,6,7]. Orthorhombic a 0.774 b 0.604, 1.005; cubic volume 469.9 Å [1]. The crystallisation in miscible and immiscible blends does not affect the unit cell dimensions [4]

Volumetric & Calorimetric Properties:
Density:

No.	Value	Note
1	d 1.403 g/cm³	[1]

Melting Temperature:

No.	Value	Note
1	370°C	[1]
2	388°C	approx. [4]
3	342–386°C	varying MW and cryst. structrure [5,7]
4	392°C	[6]

Glass-Transition Temperature:

No.	Value	Note
1	158°C	[2]
2	172.5°C	[4]
3	162–172°C	varying MW [5]

Transition Temperature General: T_g and T_m vary with MW and also with the crystalline form of the polymer. Shows double melting behaviour when crystallised above T_g. The presence of isomeric defect struct. can also modify transition temps. [5,6,7]
Transition Temperature:

No.	Value	Note	Type
1	344°C	approx. [4]	cryst. temp.
2	199–212°C	[5]	cryst. temp.

Surface Properties & Solubility:
Surface and Interfacial Properties General: The effect of time and cooling rate on bonding with titanium has been reported [9]

Transport Properties:
Polymer Solutions Dilute: Intrinsic viscosity 9.6 dl g^{-1} (conc. H_2SO_4, MW 23600), 11.6 dl g^{-1} (conc. H_2SO_4, MW 33700) [11]

Mechanical Properties:
Mechanical Properties General: Mechanical props. are dependent on morphology and therefore can vary with the processing conditions

Tensile (Young's) Modulus:

No.	Value	Note
1	4000 MPa	[9]

Tensile Strength Break:

No.	Value	Note
1	118 MPa	[9]

Flexural Strength at Break:

No.	Value	Note
1	130 MPa	[9]

Optical Properties:
Transmission and Spectra: C-13 nmr spectral data have been reported

Polymer Stability:
Decomposition Details: Undergoes a trans-etherification reaction when heated at 340° in the presence of potassium carbonate. The resulting polymer contains etherketone and etherketoneketone units in addition to the original etherketoneetherketoneketone units [8]

Polyetherketoneketone

Applications/Commercial Products:

Processing & Manufacturing Routes: Can be synth. by nucleophilic polycondensation in the presence of sodium carbonate or potassium carbonate or by electrophilic polycondensation in the presence of a Friedel-Crafts catalyst. If the nucleophilic polycondensation reaction is carried out at higher temps. in the presence of a larger amount of potassium carbonate then transetherification can occur [8,10]

Trade name	Supplier
Ultrapek	BASF

Bibliographic References

[1] Blundell, D.J. and Newton, A.B., *Polymer*, 1991, **32**, 309, (struct.)
[2] Goodwin, A.A., and Hay, J.N., *J. Polym. Sci.*, 1998, **36**, 851
[3] Gardner, K.H., Hsiao, B.S. and Faron, K.L., *Polymer*, 1994, **35**, 2290, (morphology)
[4] Androsch, R., Radusch, H.J., Zahradnik, F. and Munstedt, M., *Polymer*, 1997, **38**, 397
[5] Rueda, D.R., Zolotukhin, M.G. and Cagiao, M.E., Calleja, F.J.B. et al, *Macromolecules*, 1996, **29**, 7016, (transition temps.)
[6] Kruger, K.N. and Zachmann, H.G., *Macromolecules*, 1993, **26**, 5202
[7] Rueda, D.R., Gutierrez, M.C.G., Ania, F., Zolotukhin, M.G. and Calleja, F.J.B., *Macromolecules*, 1998, **31**, 8201, (morphology)
[8] Fukawa, I., Tanabe, T. and Hachiya, H., *Polym. J. (Tokyo)*, 1992, **24**, 173, (synth.)
[9] Ramani, K., Tagle, J., Devanathan, D., Nazre, A. et al, *Polym. Eng. Sci.*, 1995, **35**, 1972
[10] Fukawa, I. and Tanabe, T., *Macromolecules*, 1991, **24**, 3838, (synth.)
[11] Wei-Berk, C. and Berry, G.C., J. Polym. Sci., Part B: Polym. Lett., 1990, **28**, 1873, (soln. props.)

Polyetherketoneketone

Synonyms: *Polyaryletherketoneketone. PEKK. Poly(oxy-1,4-phenylenecarbonyl-1,4-phenylenecarbonyl-1,4-phenylene)*
Related Polymers: Polyetheretherketone, Polyetherketone, Polyetherketoneetherketoneketone, Polyetheretherketoneketone, Polyetherketoneketone, glass-filled, Polyetherketoneketone, carbon-filled
Material class: Thermoplastic
Polymer Type: polyetherketones
CAS Number:

CAS Reg. No.
54991-67-2

Molecular Formula: $(C_{18}H_{12}O_3)_n$
Fragments: $C_{18}H_{12}O_3$
Additives: Carbon- and glass-filled products available
General Information: Available in amorphous and crystalline grades
Morphology: Two cryst. struct. have been reported; 2 chain orthorhombic unit cell a 0.786, b 0.575, c 1.016; 1 chain (metrically) orthorhombic unit cell a 0.786, b 0.575, c 1.016 [1]. An orthorhombic struct. with different dimensions has been reported; a 0.417, b 1.134, c 1.008 cubic volume 476.4 Å; density 1.395 g cm^{-3} [1]

Volumetric & Calorimetric Properties:

Density:

No.	Value	Note
1	d 1.28 g/cm^3	(amorphous, ASTM D792) [2]
2	d 1.31 g/cm^3	(crystalline grade, ASTM D792) [2]

Thermal Expansion Coefficient:

No.	Value	Note
1	$3.8 \times 10^{-5} - 7.7 \times 10^{-5}$ K^{-1}	ASTM D696, $< T_g$ [2]

Thermal Conductivity:

No.	Value	Note
1	1.75 W/mK	ASTM C1777 [2]

Melting Temperature:

No.	Value	Note
1	307–360°C	585–680°F, DSC [2]

Glass-Transition Temperature:

No.	Value	Note
1	154–163°C	310–325°F, DSC [2]

Heat Distortion Temperature:

No.	Value	Note
1	141–175 °C	286–347°F, at 264 psi, ASTM D648 [2]

Transport Properties:

Melt Flow Index:

No.	Value	Note
1	20–120 g/10 min	8.4 kg at 380°C [2]

Water Absorption:

No.	Value	Note
1	0.2–0.3 %	(min. and max. both less than) 24 hr, ASTM D570 [2]

Mechanical Properties:

Tensile (Young's) Modulus:

No.	Value	Note
1	3447–4413 MPa	0.50–0.64 Mpsi, ASTM D368 [2]

Flexural Modulus:

No.	Value	Note
1	3379–4551 MPa	0.49–0.66 Mpsi, ASTM D570 [2]

– Polyetherketoneketone, Carbon-filled

Tensile Strength Break:

No.	Value	Note	Elongation
1	89.6–110.3 MPa	13–16 Kpsi, ASTM D638 [2]	12 > 80

Flexural Strength Yield:

No.	Value	Note
1	137.9–193.1 MPa	20–28 Kpsi, ASTM D570 [2]

Compressive Strength:

No.	Value	Note
1	103.4–206.9 MPa	15–30 Kpsi, ASTM D695 [2]

Failure Properties General: MIT fold endurance 5–13 kcycle [2]
Friction Abrasion and Resistance: Coefficient of friction 0.173–0.186 (ASTM D1894) [2]. Coefficient of friction, static 0.262–0.285 (ASTM D1894) [2].

Izod Notch:

No.	Value	Note
1	48.1–69.4 J/m	0.9–1.3 ftlb/in, ASTM D256 [2]

Electrical Properties:
Electrical Properties General: Electrical resistivity 1.00×10^{16} Ω cm (ASTM D257) [2]

Surface/Volume Resistance:

No.	Value	Note	Type
1	2×10^{15} Ω.cm	ASTM D150 [2]	S

Dielectric Permittivity (Constant):

No.	Value	Frequency	Note
1	3.3	1	ASTM D150 [2]

Dielectric Strength:

No.	Value	Note
1	0.6 kV.mm^{-1}	ASTM D149 [2]

Dissipation (Power) Factor:

No.	Value	Frequency	Note
1	0.004	1	[2]

Polymer Stability:
Upper Use Temperature:

No.	Value	Note
1	260°C	[2]

Flammability: Flammability rating V-0 (UL94) [2]. Limiting oxygen index 40% (ASTM D2863) [2]. NBS smoke density <10 [2]

Applications/Commercial Products:
Mould Shrinkage (%):

No.	Value	Note
1	0.5–1.4%	(min. value less than) <0.005–0.014 in/in, [2]

Trade name	Details	Supplier
OXPEKK-C	Crystalline injection grade	Oxford Performance Materials, Inc.
OXPEKK-E	Crystalline extrusion grade	Oxford Performance Materials, Inc.
OXPEKK-SP	Superior process grade	Oxford Performance Materials, Inc.

Bibliographic References
[1] Blundell, D.J. and Newton, A.B., *Polymer*, 1991, **32**, 309, (cryst. struct.)
[2] Oxford Performance Materials (web) http://www.oxfordpm.com/polymer~xproducts.htm, 2006

Polyetherketoneketone, Carbon-filled   P-238

Synonyms: PEKK, carbon-filled
Related Polymers: Polyetherketoneketone, Polyetherketoneketone, glass-filled
Material class: Thermoplastic
Polymer Type: polyetherketones
Additives: Unfilled and glass-filled products available

Volumetric & Calorimetric Properties:
Density:

No.	Value	Note
1	d1.36–1.36 g/cm³	(crystalline, 30% carbon, ASTM D792) [1]
2	d1.45–1.45 g/cm³	(crystalline, 40% carbon, ASTM D792) [1]

Thermal Expansion Coefficient:

No.	Value	Note
1	6×10^{-6}–1×10^{-5} K^{-1}	ASTM D696, 30–40% carbon, $<T_g$ [1]

Melting Temperature:

No.	Value	Note
1	360°C	680°F, DSC, 30–40% carbon [1]

Glass-Transition Temperature:

No.	Value	Note
1	163°C	325°F, DSC, crystalline, 30–40% carbon [1]

– Polyetherketoneketone, Glass-filled

Heat Distortion Temperature:

No.	Value	Note
1	321–327 °C	610–621°F, ASTM D648, crystalline, 30–40% carbon, at 264 psi [1]

Transport Properties:
Melt Flow Index:

No.	Value	Note
1	14–18 g/10 min	30% carbon, 8.4 kg at 380°C [1]
2	10–12 g/10 min	40% carbon, 8.4 kg at 380°C [1]

Water Absorption:

No.	Value	Note
1	0.1 %	(greater than) at 24 hr, ASTM D570, 30% carbon [1]
2	0.08 %	(greater than) at 24 hr, ASTM D570, 40% carbon [1]

Mechanical Properties:
Tensile (Young's) Modulus:

No.	Value	Note
1	27940 MPa	4 Mpsi, ASTM D638, 30% carbon [1]
2	48895 MPa	7 Mpsi, ASTM D638, 40% carbon [1]

Flexural Modulus:

No.	Value	Note
1	22064 MPa	3.2 Mpsi, ASTM D570, 30% carbon [1]
2	31027 MPa	4.5 Mpsi, ASTM D570, 40% carbon [1]

Tensile Strength Break:

No.	Value	Note	Elongation
1	248 MPa	36 Kpsi, ASTM D638, 30% carbon [1]	1.2
2	324 MPa	47 Kpsi, ASTM D638, 40% carbon [1]	1.3

Flexural Strength Yield:

No.	Value	Note
1	386 MPa	56 Kpsi, ASTM D570, 30% carbon [1]
2	448 MPa	65 Kpsi, ASTM D570, 40% carbon [1]

Friction Abrasion and Resistance: Coefficient of friction 0.160–0.162 (30–40% carbon, ASTM D1894) [1].
Coefficient of friction, static 0.220–0.218 (30–40% carbon, ASTM D1894) [1].

Mechanical Properties:
Izod Notch:

No.	Value	Note
1	56–96 J/m	1.05–1.8 ftlb/in, 30–40% carbon-filled, ASTM D256 [1]

Polymer Stability:
Upper Use Temperature:

No.	Value	Note
1	260°C	500°F, 30–40% carbon [1]

Flammability: Flammability rating V-0 (UL94, 30–40% carbon) [1]

Applications/Commercial Products:
Mould Shrinkage (%):

No.	Value	Note
1	0.05–0.1%	0.0005–0.001 in/in, 30–40% carbon [1]

[1] Oxford Performance Materials (web) http://www.oxfordpm.com/polymer~xproducts.htm, 2006

Polyetherketoneketone, Glass-filled

Synonyms: *PEKK, glass-filled*
Related Polymers: Polyetherketoneketone, Polyetherketoneketone, carbon-filled
Material class: Thermoplastic
Polymer Type: polyetherketones
Additives: Unfilled and carbon-filled products available.

Volumetric & Calorimetric Properties:
Density:

No.	Value	Note
1	d 1.5 g/cm^3	(amorphous, 30% glass, ASTM D792) [1]
2	d 1.51 g/cm^3	(crystalline, 30% glass, ASTM D792) [1]
3	d 1.62 g/cm^3	(crystalline, 40% glass, ASTM D792) [1]

Thermal Expansion Coefficient:

No.	Value	Note
1	1.1×10^{-5}–1.6×10^{-5} K^{-1}	ASTM D696, 30–40% glass, $<T_g$ [1]

Melting Temperature:

No.	Value	Note
1	360°C	680°F, DSC, 30–40% glass [1]

Glass-Transition Temperature:

No.	Value	Note
1	154°C	310°F, DSC, amorphous, 30% glass [1]
2	163°C	325°F, DSC, crystalline, 30–40% glass [1]

– Polyethersulfone

Heat Distortion Temperature:

No.	Value	Note
1	321–326 °C	610–619°F, crystalline, 30–40% glass, at 264 psi [1]

Transport Properties:
Water Absorption:

No.	Value	Note
1	0.1 %	at 24 hr, ASTM D570, amorphous, 30% carbon [1]
2	0.1 %	(greater than) at 24 hr, ASTM D570, crystalline, 30% carbon [1]
3	0.08 %	(greater than) at 24 hr, ASTM D570, crystalline, 40% carbon [1]

Mechanical Properties:
Tensile (Young's) Modulus:

No.	Value	Note
1	11032 MPa	1.6 Mpsi, ASTM D638, amorphous, 30% glass [1]
2	12411 MPa	1.8 Mpsi, ASTM D638, crystalline, 30% carbon [1]
3	19995 MPa	2.9 Mpsi, ASTM D638, crystalline, 40% carbon [1]

Flexural Modulus:

No.	Value	Note
1	11032 MPa	1.6 Mpsi, ASTM D570, amorphous, 30% glass [1]
2	11032 MPa	1.6 Mpsi, ASTM D570, crystalline, 30% carbon [1]
3	14479 MPa	2.1 Mpsi, ASTM D570, crystalline, 40% carbon [1]

Tensile Strength Break:

No.	Value	Note	Elongation
1	179 MPa	26 Kpsi, ASTM D638, amorphous, 30% glass [1]	2.3%
2	186 MPa	27 Kpsi, ASTM D638, crystalline, 30% carbon [1]	1.8%
3	200 MPa	29 Kpsi, ASTM D638, crystalline, 40% carbon[1]	1.6%

Flexural Strength Yield:

No.	Value	Note
1	241 MPa	35 Kpsi, ASTM D570, amorphous, 30% glass [1]
2	255 MPa	37 Kpsi, ASTM D570, crystalline, 30% carbon [1]
3	290 MPa	42 Kpsi, ASTM D570, crystalline, 40% carbon [1]

Izod Notch:

No.	Value	Note
1	85.4–106.8 J/m	1.6–2 ftlb/in, 30–40% glass filled, ASTM D256 [1]

Polymer Stability:
Upper Use Temperature:

No.	Value	Note
1	260°C	500°F, 30–40% glass [1]

Flammability: Flammability rating V-0 (UL94, 30–40% glass) [1]

Applications/Commercial Products:
Mould Shrinkage (%):

No.	Value	Note
1	0.2–0.3%	0.002–0.003 in/in, 30–40% glass [1]

[1] Oxford Performance Materials (web) http://www.oxfordpm.com/polymer~xproducts.htm, 2006

Polyethersulfone   P-240

Synonyms: *PES. Poly(oxy-1,4-phenylenesulfonyl-1,4-phenylene)*
Related Polymers: Polyethersulfone, 20% Glass fibre reinforced, Polyethersulfone, 30% Glass fibre reinforced, Polyethersulfone, 30% Carbon fibre reinforced
Monomers: 4,4′-Dichlorodiphenyl sulfone, 4,4′-Dihydroxydiphenyl sulfone, Bis(4-chlorosulfonylphenyl) ether, Diphenyl ester
Material class: Thermoplastic
Polymer Type: polysulfone
CAS Number:

CAS Reg. No.	Note
25667-42-9	Victrex, Ultrason E

Molecular Formula: $[(C_{24}H_{16}O_6S_2).(C_{12}H_8O_3S)]_n$
Fragments: $C_{24}H_{16}O_6S_2$ $C_{12}H_8O_3S$
Additives: Fillers include carbon fibre and glass
General Information: Polyethersulfone is an amorph., engineering thermoplastic which at room temp. is tough, strong and rigid. It is distinguished by excellent high temp. performance, impressive creep resistance, hydrolytic stability, low flammability, and smoke emission. Long-term thermal stability is excellent. A lifetime of 50000h can be expected at 190°. Props. show little change up to 200°. Dimensional stability is also excellent up to this temp. A rapid fall in props. occurs beyond 200° with very little strength or rigidity being maintained over 230°. Compared with other plastics PES provides a greater margin of performance in such areas as toughness. Unreinforced polyethersulfone is available in transparent and opaque forms in both injection moulding and extrusion grades. Light amber colour [6]

Volumetric & Calorimetric Properties:
Density:

No.	Value	Note
1	d 1.37 g/cm^3	ASTM D792 [1]

– Polyethersulfone

Thermal Expansion Coefficient:

No.	Value	Note	Type
1	5.5×10^{-5} K^{-1}	ASTM D696 [2]	L

Volumetric Properties General: Has higher coefficient of linear thermal expansion compared to reinforced grades

Thermal Conductivity:

No.	Value	Note
1	0.18 W/mK	ASTM C177 [4]

Specific Heat Capacity:

No.	Value	Note	Type
1	1.12 kJ/kg.C	23° [6]	P

Melting Temperature:

No.	Value	Note
1	345–390°C	[1]

Glass-Transition Temperature:

No.	Value	Note
1	220°C	[6]

Transition Temperature General: Tg of PES is 35° higher than that of PSU

Deflection Temperature:

No.	Value	Note
1	204°C	1.8 MPa, ASTM D648 [1]
2	208°C	0.45 MPa, ISO/IEC 75 [5]

Vicat Softening Point:

No.	Value	Note
1	220°C	ISO/IEC 306, VST/A/50 [5]

Transition Temperature:

No.	Value	Note	Type
1	-100°C	approx. [6]	Second order (β)

Surface Properties & Solubility:
Solubility Properties General: Undergoes stress-cracking or dissolves in polar compds. such as ketones and some halogenated hydrocarbons [6]
Cohesive Energy Density Solubility Parameters: δ 22 J$^{1/2}$ cm$^{-3/2}$ (min.) [4]
Solvents/Non-solvents: Sol. polar solvents, DMF, N-methylpyrrolidinone. Sol. CH$_2$Cl$_2$ (may crystallise when dissolved, resulting in an unstable soln.) [4]

Transport Properties:
Transport Properties General: PES has higher water absorption than PSU [2]

Melt Flow Index:

No.	Value	Note
1	30 g/10 min	0.3 MPa, 400°, ASTM D1238 [1]

Water Absorption:

No.	Value	Note
1	0.61 %	24h [6]
2	2.1 %	equilib., ASTM D570 [6]

Mechanical Properties:
Mechanical Properties General: Tough and rigid thermoplastic. Provides a greater margin of performance in such areas as toughness in comparison with other plastics. Creep resistance at room temp. is particularly impressive, after 10 years under a constant tensile stress of 7250 psi, samples have shown only 3% strain. Impact behaviour is similar to that of nylon, both being tough materials but notch-sensitive. Mech. props. show low sensitivity to temp. changes [1,3,7]

Tensile (Young's) Modulus:

No.	Value	Note
1	2660 MPa	ASTM D638 [1]

Flexural Modulus:

No.	Value	Note
1	2900 MPa	ASTM D790 [1]

Tensile Strength Break:

1	ASTM D638 [6]	40%

Compressive Modulus:

No.	Value	Note
1	2680 MPa	ASTM D695 [6]

Poisson's Ratio:

No.	Value	Note
1	0.4	0.5% strain [4]

Tensile Strength Yield:

No.	Value	Note
1	83 MPa	ASTM D638 [1]

Flexural Strength Yield:

No.	Value	Note
1	111 MPa	ASTM D790 [6]

Compressive Strength:

No.	Value	Note
1	100 MPa	ASTM D695 [6]

Complex Moduli: Tensile creep modulus 2700 MPa (1000h, 23°, Dehnung ≤0.5%, ISO/IEC 899) [5]
Impact Strength: Tensile impact strength 337 kJ m^{-2} (ASTM D1822) [4]
Hardness: Rockwell M88 (ASTM D785) [4]
Failure Properties General: Shear strength (yield) 50 MPa (ASTM D732) [6]

— Polyethersulfone

Friction Abrasion and Resistance: Abrasion resistance 19 mg (1000 cycles)$^{-1}$ (ASTM D1040, Taber CS-17 test) [6]
Izod Notch:

No.	Value	Notch	Break	Note
1	83 J/m	Y		3.1 mm, ASTM D256 [4]
2		N	Y	[4]

Electrical Properties:

Electrical Properties General: PES offers excellent insulating and other electrical props. It has low dielectric constants and dissipation factors even in the GHz (microwave) frequency range. This performance is retained over a wide temp. range
Surface/Volume Resistance:

No.	Value	Note	Type
2	$>0.1 \times 10^{15}$ Ω.cm	min. ISO/IEC 93 [5]	S

Dielectric Permittivity (Constant):

No.	Value	Frequency	Note
1	3.65	60 Hz	ASTM D150 [6]
2	3.65	1 kHz	ASTM D150 [6]
3	3.52	1 MHz	ASTM D150 [6]

Dielectric Strength:

No.	Value	Note
1	15.5 kV.mm^{-1}	3.2 mm, ASTM D149 [6]

Dissipation (Power) Factor:

No.	Value	Frequency	Note
1	0.0019	60 Hz	ASTM D150 [6]
2	0.0023	1 kHz	ASTM D150 [6]
3	0.0048	1 MHz	ASTM D150 [6]

Optical Properties:

Optical Properties General: Exhibits optical transparency
Refractive Index:

No.	Value	Note
1	1.65	ASTM D1505 [6]

Transmission and Spectra: Light transmittance 70% [6]
Volume Properties/Surface Properties: Haze <7% (3.1 mm thick, ASTM D1004) [6]

Polymer Stability:

Polymer Stability General: Long-term performance at high temps. in severe service environments. Withstands exposure to elevated temp. in air and water for prolonged periods [1]
Thermal Stability General: Excellent thermal stability and resistance to thermal oxidation
Upper Use Temperature:

No.	Value	Note
1	180°C	UL 746 [1]
2	180°C	UL 746B [1]

Flammability: Flammability rating V0 (0.8 mm thick, UL 94) [6]. Limiting oxygen index 38% (ASTM D286) [6]. Exhibits excellent resistance to burning. Has outstanding flame retardancy and very low smoke release characteristics [6]. Smoke density 35 (6.2 mm, ASTM E662) [6]
Environmental Stress: High degree of oxidative stability. Poor resistance to uv light. Absorbance in the uv region causes discoloration and loss of mechanical props. due to polymer degradation. For outdoor application protective lacquers or grades of the material containing carbon black are recommended [6]. Has good resistance to microwave, visible and infrared radiation as well as α, β and γ radiation [6,9]
Chemical Stability: Highly resistant to mineral acids, alkali and salt solns. Resistant to aliphatic hydrocarbons and most alcohols. Is crazed, swollen or dissolved by chlorinated hydrocarbons, esters and ketones [6]
Hydrolytic Stability: Superb resistance to hydrolysis by hot water and steam [6]
Stability Miscellaneous: Good dimensional stability [6]

Applications/Commercial Products:

Processing & Manufacturing Routes: Commercial synth. of polyethersulfone is obtained mainly via the nucleophilic substitution polycondensation route. This synth. is based on reaction of essentially equimolar quantities of 4,4′-dichlorodiphenylsulfone with Bisphenol S in dipolar aprotic solvent in the presence of an alkali base. Diphenyl sulfone, sulfolane and N-methyl-2-pyrrolidinone are examples of suitable solvents. Potassium carbonate is used as base. Polyethersulfone can be synthesised by the electrophilic Friedel-Crafts reaction of bis(4-chlorosulfonylphenyl) ether with diphenyl ether. The same reaction can be carried out using 4-chlorosulfonyldiphenyl ether as the single source monomer. Polyethersulfone can be compression moulded, injection moulded and extruded using conventional equipment. Processing times are comparable with those of conventional engineering thermoplastics. Prior to processing it requires drying for 3.5h at 150°. The recommended mould temp. is 140–165° and melt temp. 345–390°. The resin can be extruded into film, sheet and profiles and it can be thermoformed. Mould shrinkage is low and it therefore is suitable for applications requiring close tolerances and little dimensional change over a wide temp. range. PES can be plated by an electrodeless nickel or copper process [1,7,9,10,11]
Mould Shrinkage (%):

No.	Value	Note
1	0.6%	ASTM D955 [1]

Applications: Mixtures of polyethersulfone with continuous carbon or other high performance fibres provide high strength, high temp. composites which can be shaped by thermoforming. A range of grades is available to meet special requirements. Easy flow unreinforced and glass reinforced grades are available for difficult mouldings. Typical applications are lighting fixtures, electrical/electronic, medical devices, chemical process equipment and automotive applications. Electrical applications include high temp. electrical multipin connectors, coil bobbins, integrated circuit sockets, edge and round multipin connectors, terminal blocks, printed circuit boards and DIP switches. The material mechanical strength accounts for its use in radomes, pump housings, bearing cages, and power saw manifolds. Small appliance manufacturers use PES for hair dryer outlets, hot combs, and projector lamp grilles. Transparency has led to its use in indicator lights and sight glasses. Exceptional fire-safety characteristics make it suitable for use in commercial aircraft interiors. Due to its amorph. character it can be dissolved and spray coated into metals. It meets U.S. Food and Drugs Administration (FDA) requirements for direct food contact

Trade name	Details	Supplier
CTI ES	various grades	CTI
EMI-X	PDX-J-90315/91021	LNP Engineering

Electrafil	G-1100/SS/10, J-1100/CF/30	DSM Engineering
Fiberfil	J-1100/20/40	DSM Engineering
J-1100/30		DSM Engineering
K	various grades	Thermofil Inc.
Lubricomp	various grades	LNP Engineering
RTP	various grades	RTP Company
Radel A	standard moulding grade	Amoco Performance Products
Stat-Kon	various grades	LNP Engineering
Sumikaexcel		Sumitomo Chem Co
Thermocomp	JF-1002/1004/1006/1008	LNP Engineering
Ultrason E	various grades available	BASF
Victex	various grades	Victrex

Bibliographic References

[1] *Engineering Plastics for Performance and Value*, Amoco Performance Products, Inc., (technical datasheet)
[2] The Plastics Compendium, *Key Properties and Sources*, (ed. R. Dolbey), Rapra Technology Ltd., 1995, **1**
[3] Attwood, T.E., Cinderey, M.B. and Rose, J.B., *Polymer*, 1993, **34**, 1322
[4] Harris, J.E. and Johnson, R.N., *Encycl. Polym. Sci. Eng.*, Vol. 13, 2nd edn., (ed. J.I. Kroschwitz), John Wiley and Sons, 1985, 197, (rev)
[5] Ultrason (Polyethersulfone/Polysulfone PES/PSU) Range Chart, Features Applications, *Typical Values*, BASF, (technical datasheet)
[6] *Kirk-Othmer Encycl. Chem. Technol.*, 4th edn., (eds. J.I. Kroschwitz and M. Howe-Grant), Wiley Interscience, 1993, **9**, 945
[7] Guide to Plastics, *Property and Specification Charts*, McGraw-Hill, 1991
[8] Allen, G. and McAinsh, J., *Eur. Polym. J.*, 1970, **6**, 1635, (soln props)
[9] Searle, O.B. and Pfeiffer, R.H., *Polym. Eng. Sci.*, 1985, **25**, 8, 474, (props)
[10] Jennings, B.E., Jones, M.E.B. and Rose, J.B., *J. Polym. Sci., Part C: Polym. Lett.*, 1967, **16**, 715, (synth)
[11] Rose, J.B., *Polymer*, 1974, **15**, 456, (synth, struct, props)

Polyethersulfone, 30% Carbon fibre reinforced

P-241

Synonyms: PES, 30% Carbon fibre reinforced
Related Polymers: Polyethersulfone
Monomers: 4,4′-Dichlorodiphenylsulfone, 4,4′-Dihydroxydiphenyl sulfone, Bis(4-chlorosulfonylphenyl)ether, Diphenyl ether
Material class: Thermoplastic
Polymer Type: polysulfone
CAS Number:

CAS Reg. No.	Note
25667-42-9	Victrex, Ultrason E

Molecular Formula: $[(C_{24}H_{16}O_6S_2).(C_{12}H_8O_3S)]_n$
Fragments: $C_{24}H_{16}O_6S_2$ $C_{12}H_8O_3S$
Additives: Carbon fibre
General Information: Extremely strong and rigid thermoplastic which offers a weight saving over metals. Of interest to the aerospace industry and in general transportation. Due to the incorporation of carbon fibre it is conductive and offers full resistance to uv but at high expense

Volumetric & Calorimetric Properties:
Density:

No.	Value	Note
1	d 1.48 g/cm^3	ASTM D792 [1]

Thermal Expansion Coefficient:

No.	Value	Note	Type
1	1.9×10^{-5} K^{-1}	[1]	L

Volumetric Properties General: Has lower coefficient of linear thermal expansion compared to unreinforced PES
Thermal Conductivity:

No.	Value	Note
1	0.45 W/mK	ASTM C177 [1]

Melting Temperature:

No.	Value	Note
1	360–380°C	[3]

Glass-Transition Temperature:

No.	Value	Note
1	225°C	[2]

Deflection Temperature:

No.	Value	Note
1	220°C	1.8 MPa, ASTM D648 [1]
2	220°C	0.45 MPa, ASTM D648 [1]

Transition Temperature:

No.	Value	Note	Type
1	-100°C	[4]	Second order (β)

Surface Properties & Solubility:
Solubility Properties General: Undergoes stress-cracking, or dissolves, in polar compounds such as ketones and some halogenated hydrocarbons [4]
Solvents/Non-solvents: Sol. polar solvents, DMF, N-methylpyrrolidinone. Sol. CH_2Cl_2, may crystallise when dissolved, resulting in an unstable soln. [5]

Transport Properties:
Water Absorption:

No.	Value	Note
1	0.5 %	24h, ASTM D570 [1]
2	1.6 %	saturation, ASTM D570 [1]

Mechanical Properties:
Mechanical Properties General: Compared with unreinforced PES it has superior tensile and flexural props., but reduced notched Izod impact strength and lower elongation at break

Tensile (Young's) Modulus:

No.	Value	Note
1	14615–19993 MPa	2120–2900 kpsi, ASTM D638 [2]

Flexural Modulus:

No.	Value	Note
1	17000 MPa	ASTM D790 [1]

– Polyethersulfone, 20% Glass fibre reinforced

Tensile Strength Break:

No.	Note	Elongation
1	ASTM D638 [1]	1–2%

Tensile Strength Yield:

No.	Value	Note
1	180 MPa	ASTM D638 [1]

Flexural Strength Yield:

No.	Value	Note
1	255 MPa	ASTM D790 [1]

Compressive Strength:

No.	Value	Note
1	150 MPa	ASTM D695 [1]

Hardness: Rockwell M99 (ASTM D785) [1]
Failure Properties General: Shear strength 90 MPa (ASTM D732) [1]
Friction Abrasion and Resistance: Coefficient of friction (static) 0.15. Coefficient of friction (dynamic) 0.18 (LNP SOP) [1]
Izod Notch:

No.	Value	Notch	Note
1	90 J/m	Y	ASTM D256 [1]
2	300 J/m	N	ASTM D256 [1]

Electrical Properties:
Electrical Properties General: Due to the incorporation of carbon fibre this material is conductive. Volume and surface resistivities have been reported, 10–100 Ω cm [1]
Polymer Stability:
Polymer Stability General: Stable in the long-term at high temps. in severe service environments
Thermal Stability General: Excellent thermal stability and resistance to thermal oxidation
Upper Use Temperature:

No.	Value	Note
1	180°C	UL 746B [1]

Flammability: Flammability rating V0 (3.2 mm, UL 94) [1]. Limiting oxygen index 42% [3]. Has outstanding flame retardancy and very low smoke release characteristics [4]
Environmental Stress: High degree of oxidative stability. Resistant to uv light [4]
Chemical Stability: Highly resistant to mineral acids, alkali and salt solns. Resists aliphatic hydrocarbons and most alcohols. Is crazed, swollen, or dissolved by chlorinated hydrocarbons, esters and ketones [4]
Hydrolytic Stability: Superb resistance to hydrol. by hot water and steam [4]

Applications/Commercial Products:
Processing & Manufacturing Routes: Mould temp. of 90–160° and melt temp. of 360–380° are recommended for injection moulding. Prior to melt processing the resin is dried for 3h at 150° to prevent bubbling, surface streaks, and splash marks in moulded parts. Injection moulded components can have anisotropic props. [3]

Mould Shrinkage (%):

No.	Value	Note
1	0.2%	ASTM D955 [3]

Applications: Used for electrical and electronic components and in aerospace applications (nose cones and air ducting)

Trade name	Supplier
CTI PES	Compounding Technology
RTP	Vigilant Plastics
Thermocomp JC-1006	LNP Engineering
Thermofil PES	Thermofil

Bibliographic References
[1] LNP Engineering Plastics, *Product Data Book,*
[2] Guide to Plastics, *Property and Specification Charts,* McGraw-Hill, 1993
[3] The Plastics Compendium, *Key Properties and Sources,* Vol. 1, (ed. R. Dolbey), Rapra Technology Ltd., 1995
[4] *Kirk-Othmer Encycl. Chem. Technol.,* Vol. 19, 4th edn., (eds. J.I. Kroschwitz and M. Howe-Grant), Wiley Interscience, 1993, 945
[5] Harris, J.E. and Johnson, R.H., *Encycl. Polym. Sci. Eng.,* Vol. 13, 2nd edn., (ed. J.I. Kroschwitz), John Wiley and Sons, 1985, 196
[6] Searle, O.B. and Pfeiffer, R.H., *Polym. Eng. Sci.,* 1985, 25, 8, 474, (props)

Polyethersulfone, 20% Glass fibre reinforced P-242

Synonyms: *PES, 20% Glass fibre reinforced*
Related Polymers: Polyethersulfone
Monomers: 4,4′-Dichlorodiphenyl sulfone, 4,4′-Dihydroxydiphenyl sulfone, Bis(4-chlorosulfonylphenyl)ether, Diphenyl ether
Material class: Thermoplastic
Polymer Type: polysulfone
CAS Number:

CAS Reg. No.	Note
25667-42-9	Victrex, Ultrason E

Molecular Formula: $[(C_{24}H_{16}O_6S_2).(C_{12}H_8O_3S)]_n$
Fragments: $C_{24}H_{16}O_6S_2$ $C_{12}H_8O_3S$
Additives: Glass fibre
General Information: Tough and rigid engineering thermoplastic with excellent thermal stability. Tensile and flexural props. as well as resistance to cracking in chemical environments are substantially enhanced compared with unreinforced grades and it is used in severe chemical environments for enhanced resistance and long service life

Volumetric & Calorimetric Properties:
Density:

No.	Value	Note
1	d 1.51 g/cm^3	ASTM D792 [1]

Thermal Expansion Coefficient:

No.	Value	Note	Type
1	3.1×10^{-5} K^{-1}	ASTM D696 [1]	L

Volumetric Properties General: Has lower coefficient of linear thermal expansion compared to unreinforced PES
Thermal Conductivity:

No.	Value	Note
1	0.22 W/mK	ASTM C177 [2]

Polyethersulfone, 20% Glass fibre reinforced

Melting Temperature:

No.	Value	Note
1	345–390°C	[1]

Glass-Transition Temperature:

No.	Value	Note
1	220–225°C	[6]

Deflection Temperature:

No.	Value	Note
1	215°C	1.8 MPa, ASTM D648 [2]
2	220°C	0.45 MPa, ASTM D648 [2]

Vicat Softening Point:

No.	Value	Note
1	225°C	ISO/IEC 306 VST/A/50 [3]

Transition Temperature:

No.	Value	Note	Type
1	-100°C	[4]	Second order (β)

Surface Properties & Solubility:

Solubility Properties General: Undergoes stress-cracking, or dissolves, in polar compds. such as ketones and some halogenated hydrocarbons

Cohesive Energy Density Solubility Parameters: δ 22 $J^{1/2}$ $cm^{-3/2}$ (min.) [5]

Solvents/Non-solvents: Sol. polar solvents, DMF and N-methylpyrrolidinone. Sol. CH_2Cl_2 (may crystallise when dissolved, resulting in an unstable soln.) [5]

Transport Properties:

Melt Flow Index:

No.	Value	Note
1	6 g/10 min	0.3 MPa, 343°, ASTM D1238 [1]

Water Absorption:

No.	Value	Note
1	0.4 %	24h, 23°, 3.2 mm, ASTM D570 [1]

Mechanical Properties:

Mechanical Properties General: Improved tensile strength and flexural modulus, significant redn. in elongation at break and increased tensile creep modulus compared with unreinforced PES

Tensile (Young's) Modulus:

No.	Value	Note
1	5690 MPa	ASTM D638 [1]

Flexural Modulus:

No.	Value	Note
1	5190 MPa	ASTM D790 [1]

Tensile Strength Break:

No.	Value	Note	Elongation
1	105 MPa	ASTM D638 [1,4]	3.2%

Tensile Strength Yield:

No.	Value	Note
1	105 MPa	ASTM D638 [1]

Flexural Strength Yield:

No.	Value	Note
1	162 MPa	ASTM D790 [1]

Compressive Strength:

No.	Value	Note
1	110 MPa	ASTM D695 [2]

Complex Moduli: Tensile creep modulus 5600 MPa (1000h, 23°, Dehnung $\leq$0.5%, ISO/IEC 899) [3]

Impact Strength: Tensile impact strength 65 kJ m^{-2} (ASTM D1822) [1]

Hardness: Rockwell M99 (ASTM D785) [2]

Failure Properties General: Shear strength 70 MPa (ASTM D732) [2]

Friction Abrasion and Resistance: Coefficient of friction (static) 0.23 (LNP SOP). Coefficient of friction (dynamic) 0.21 (LNP SOP) [2]

Izod Notch:

No.	Value	Notch	Note
1	85 J/m	Y	[2]
2	250 J/m	N	ASTM D256 [2]

Electrical Properties:

Electrical Properties General: Has excellent insulating and other electrical props. Dielectric constants and dissipation factors are low even in the GHz (microwave) frequency range. This performance is retained over a wide temp. range

Surface/Volume Resistance:

No.	Value	Note	Type
2	0.1×10^{15} Ω.cm	ISO/IEC 93 [3]	S

Dielectric Permittivity (Constant):

No.	Value	Frequency	Note
1	3.84	60 Hz	ASTM D150 [1]
2	3.88	1 MHz	ASTM D150 [1]
3	3.84	1 kHz	ASTM D150 [1]

Dielectric Strength:

No.	Value	Note
1	17 kV.mm^{-1}	3.2 mm, ASTM D149 [1]

– Polyethersulfone, 30% Glass fibre reinforced

Dissipation (Power) Factor:

No.	Value	Frequency	Note
1	0.0015	60 Hz	ASTM D150 [1]
2	0.0018	1 kHz	ASTM D150 [1]
3	0.0081	1 MHz	ASTM D150 [1]

Optical Properties:
Volume Properties/Surface Properties: Opaque. Amber coloured [8]

Polymer Stability:
Polymer Stability General: Long-term performance at high temps. in severe service environments. Withstands exposure at elevated temps. in air and water for prolonged periods [1]
Thermal Stability General: Excellent thermal stability and resistance to thermal oxidation
Upper Use Temperature:

No.	Value	Note
1	190°C	UL 746B [1]

Flammability: Flammability rating V0 (0.8 mm thick, UL 94) [1]. Limiting oxygen index 41% [7]. Has outstanding flame retardancy and very low smoke release characteristics [4]
Environmental Stress: High degree of oxidative stability. Poor resistance to uv light. Resistant to α, β and γ radiation [8]. Stress cracking resistance is substantially enhanced compared with unreinforced PES [4]
Chemical Stability: Highly resistant to mineral acids, alkali and salt solns. Resists aliphatic hydrocarbons and most alcohols. Is crazed, swollen or dissolved by chlorinated hydrocarbons, esters and ketones [4]
Hydrolytic Stability: Superb resistance to hydrol. by hot water and steam [4]
Stability Miscellaneous: Good dimensional stability [4]

Applications/Commercial Products:
Processing & Manufacturing Routes: Moulded in modern injection-moulding equipment. Mould temp. of 140–165° and melt temp. of 345–390° are recommended. Prior to melt processing the resin is dried for 3.5 h at 150° to prevent bubbling, surface streaks, and splash marks in moulded parts [1]. Incorporation of glass fibre can cause anisotropy in moulded parts [7]
Mould Shrinkage (%):

No.	Value	Note
1	0.4%	ASTM D955 [1]

Applications: Applications in the automotive industry, under the bonnet or in the gear box area, (carburettor parts, bearing cages). Aerospace applications and electrical components such as printed circuit boards and connectors

Trade name	Details	Supplier
CTI PES		Compounding Technology
RTP		Vigilant Plastics
Radel A		Amoco Performance Products
Thermocomp IF-1004		LNP Engineering
Thermofil PES		Thermofil
Ultrason E	several grades	BASF

Bibliographic References
[1] *Engineering Plastics for Performance and Value*, Amoco Performance Products, Inc.,
[2] LNP Engineering Plastics, *Product Data Book*,(technical datasheet)
[3] Ultrason (Polyethersulfone/Polysulfone PES/PSU) Range Chart, Features Applications, *Typical Values*, BASF, (technical datasheet)
[4] *Kirk-Othmer Encycl. Chem. Technol.*, Vol. 19, 4th edn., (eds. J.I. Kroschwitz and M. Howe-Grant), Wiley Interscience, 1993, 945
[5] Harris, J.E. and Johnson, R.H., *Encycl. Polym. Sci. Eng.*, Vol. 13, 2nd edn., (ed. J.I. Kroschwitz), John Wiley and Sons, 1985, 196
[6] Guide to Plastics, *Property and Specification Charts*, McGraw-Hill, 1993
[7] The Plastics Compendium, *Key Properties and Sources*, Vol. 1, (ed. R. Dolbey), Rapra Technology Ltd., 1995
[8] Searle, O.B. and Pfeiffer, R.H., *Polym. Eng. Sci.*, 1985, 25, 474, (props)

Polyethersulfone, 30% Glass fibre reinforced P-243
Synonyms: *PES, 30% Glass fibre reinforced*
Related Polymers: Polyethersulfone
Monomers: 4,4′-Dichlorodiphenylsulfone, 4,4′-Dihydroxydiphenyl sulfone, Bis(4-chlorosulfonylphenyl)ether, Diphenyl ether
Material class: Thermoplastic
Polymer Type: polysulfone
CAS Number:

CAS Reg. No.	Note
25667-42-9	Victrex, Ultrason E

Molecular Formula: $[(C_{24}H_{16}O_6S_2).(C_{12}H_8O_3S)]_n$
Fragments: $C_{24}H_{16}O_6S_2$ $C_{12}H_8O_3S$
Additives: Glass fibre
General Information: Tough and rigid engineering thermoplastic with excellent thermal stability. Tensile and flexural props. as well as resistance to cracking in chemical environments are substantially enhanced compared with unreinforced grades and with 20% glass fibre reinforced grades. It is used in harsh chemical environments for enhanced resistance and long service life

Volumetric & Calorimetric Properties:
Density:

No.	Value	Note
1	d 1.58 g/cm^3	ASTM D792 [2]

Thermal Expansion Coefficient:

No.	Value	Note	Type
1	3.1×10^{-5} K^{-1}	ASTM D696 [2]	L

Volumetric Properties General: Has lower coefficient of linear thermal expansion compared to unreinforced PES
Thermal Conductivity:

No.	Value	Note
1	0.26 W/mK	ASTM C177 [1]

Melting Temperature:

No.	Value	Note
1	345–390°C	[2]

Deflection Temperature:

No.	Value	Note
1	215°C	1.8 MPa [1]
2	220°C	0.45 MPa, ASTM D648 [1]

– Polyethersulfone, 30% Glass fibre reinforced

Vicat Softening Point:

No.	Value	Note
1	225°C	ISO/IEC 306, VST/A/50 [3]

Transition Temperature:

No.	Value	Note	Type
1	-100°C	[4]	Second order (β)

Surface Properties & Solubility:
Solubility Properties General: Undergoes stress-cracking, or dissolves, in polar compds. such as ketones and some halogenated hydrocarbonsJ
Cohesive Energy Density Solubility Parameters: δ 22 $J^{1/2}$ $cm^{-3/2}$ (min.) [5]
Solvents/Non-solvents: Sol. polar solvents, DMF, N-methylpyrrolidinone. Sol. CH_2Cl_2 (may crystallise when dissolved, resulting in an unstable soln.) [5]

Transport Properties:
Melt Flow Index:

No.	Value	Note
1	4.5 g/10 min	0.3 MPa, 343 °, ASTM D1238 [2]

Water Absorption:

No.	Value	Note
1	0.39 %	24h, 23°, 3.2 mm, ASTM D570 [2]

Mechanical Properties:
Mechanical Properties General: Excellent stiffness and tensile strength. Significantly higher flexural modulus, increased tensile creep modulus and reduced elongation at break compared with unreinforced PES

Tensile (Young's) Modulus:

No.	Value	Note
1	8630 MPa	ASTM D638 [2]

Flexural Modulus:

No.	Value	Note
1	10100 MPa	ASTM D990 [2]

Tensile Strength Break:

No.	Value	Note	Elongation
1	126 MPa	ASTM D638 [2,4]	1.9%

Tensile Strength Yield:

No.	Value	Note
1	130 MPa	ASTM D638 [2]

Flexural Strength Yield:

No.	Value	Note
1	180 MPa	ASTM D790 [2]

Compressive Strength:

No.	Value	Note
1	120 MPa	ASTM D695 [1]

Complex Moduli: Tensile creep modulus 8300 MPa (1000h, 23°, Dehnung $\leq$0.5%, ISO/IEC 899) [3]
Impact Strength: Tensile impact strength 72 kJ m^{-2} (ASTM D1822) [2]
Hardness: Rockwell M99 (ASTM D785) [1]
Failure Properties General: Shear strength 70 MPa (ASTM D732) [1]
Friction Abrasion and Resistance: Coefficient of friction (static) 0.23. Coefficient of friction (dynamic) 0.21 (LNP SOP) [1]

Izod Notch:

No.	Value	Notch	Note
1	85 J/m	Y	[1]
2	250 J/m	N	ASTM D256 [1]

Electrical Properties:
Electrical Properties General: Has excellent electrical props. which are maintained over a wide temp. range. Dielectric constants and dissipation factors are low, even in GHz (microwave) frequency range

Surface/Volume Resistance:

No.	Value	Note	Type
2	0.1×10^{15} Ω.cm	ISO/IEC [3]	S

Dielectric Permittivity (Constant):

No.	Value	Frequency	Note
1	4.11	60 Hz	[3]
2	4.13	1 kHz	[2]
3	4.17	1 MHz	ASTM D150 [2]

Dielectric Strength:

No.	Value	Note
1	17 $kV.mm^{-1}$	3.2 mm, ASTM D149 [2]

Dissipation (Power) Factor:

No.	Value	Frequency	Note
1	0.0019	60 Hz	ASTM D150 [2]
2	0.0018	1 kHz	ASTM D150 [2]
3	0.0094	1 MHz	ASTM D150 [2]

Optical Properties:
Volume Properties/Surface Properties: Opaque [2]

Polymer Stability:
Polymer Stability General: Long-term performance at high temps. in severe service environments. Withstands exposure at elevated temps. to air and water for prolonged periods [2]
Thermal Stability General: Excellent thermal stability and resistance to thermal oxidation

Upper Use Temperature:

No.	Value	Note
1	190°C	UL 746B [2]

– **Poly(ethyl acrylate)**

Flammability: Flammability rating V0 (0.8 mm thick, UL 94) [2]. Limiting oxygen index 41% [6]. Has outstanding flame retardancy and very low smoke release characteristics [4]
Environmental Stress: High degree of oxidative stability. Poor resistance to uv light. Resistant to α, β and γ radiation [7]. Stress-cracking resistance is substantially enhanced compared with unreinforced PES [4]
Chemical Stability: Highly resistant to mineral acids, alkali and salt solns. Resists aliphatic hydrocarbons and most alcohols. Is crazed, swollen or dissolved by chlorinated hydrocarbons, esters and ketones [4]
Hydrolytic Stability: Superb resistance to hydrol. by hot water and steam [4]
Stability Miscellaneous: Good dimensional stability [4]

Applications/Commercial Products:
Processing & Manufacturing Routes: Mould temp. of 140–165° and melt temp. of 345–390° are recommended for injection moulding. Prior to melt processing the resin is dried for 3.5h at 150° to prevent bubbling, surface streaks, and splash marks in moulded parts [2]. Injection moulded components can have anisotropic props. [6]
Mould Shrinkage (%):

No.	Value	Note
1	0.3%	ASTM D955 [2]

Applications: Used for electrical components, such as printed circuit boards and connectors, in automotive applications especially under the bonnet or in the gearbox area, (fittings and connectors, nose cones, air ducting, car heater fans)

Trade name	Supplier
CTI PES	Compounding Technology
RTP	Vigilant Plastics
Radel AG-330	Amoco Performance Products
Thermocomp IF-1006	LNP Engineering
Thermofil PES	Thermofil
Ultrason E-101096	BASF

Bibliographic References
[1] LNP Engineering Plastics, *Product Data Book,*
[2] *Engineering Plastics for Performance and Value,* Amoco Performance Products, Inc., (technical datasheet)
[3] Ultrason (Polyethersulfone/Polysulfone PES/PSU) Range Chart, Features Applications, *Typical Values,* BASF, (technical datasheet)
[4] *Kirk-Othmer Encycl. Chem. Technol.,* Vol. 19, 4th edn., (eds. J.I. Kroschwitz and M. Howe-Grant), Wiley Interscience, 1993, 945
[5] Harris, J.E. and Johnson, R.H., *Encycl. Polym. Sci. Eng.,* Vol. 13, 2nd edn., (ed. J.I. Kroschwitz), John Wiley and Sons, 1985, 196
[6] The Plastics Compendium, *Key Properties and Sources,* (ed. R. Dolbey), Rapra Technology Ltd., 1995, **1**
[7] Searle, O.B. and Pfeiffer, R.H., *Polym. Eng. Sci.,* 1985, **25**, 8, 474, (props)

Poly(ethyl acrylate)   P-244
Synonyms: *PEA. Poly(ethyl 2-propenoate). Poly[1-(ethoxycarbonyl)ethylene]*
Monomers: Ethyl acrylate
Material class: Natural elastomers
Polymer Type: acrylic
CAS Number:

CAS Reg. No.
9003-32-1

Molecular Formula: $(C_5H_8O_2)_n$
Fragments: $C_5H_8O_2$
Molecular Weight: The MW ranges from 10000–5000000
Additives: Soap-urea, activated thiol, soap-sulfur, diamines, lead thiourea, carbon black, trithiocyanuric acid and Agerite stalites
General Information: Poly(ethyl acrylate) is a considerably soft, rubbery and slightly tacky low polymer with low T_g. It is resistant to swelling in nonpolar oils and liqs. and has good thermal stability. There is no significant change in T_g with changes in tacticity. It forms the basis of acrylate rubber. Copolymers with a variety of monomers (e.g. polyolefins, vinyl acetate, vinyl chloride) give a wide range of products. Cross-linking elevates and extends the rubbery plateau with little effect on T_g until extensive cross-linking has been introduced. Cross-linking decreases thermo-plasticity and solubility and increases resilience.

Volumetric & Calorimetric Properties:
Density:

No.	Value
1	d 1.12 g/cm^3

Thermal Expansion Coefficient:

No.	Value	Note	Type
1	0.00028 K^{-1}	glass	V
2	0.00061 K^{-1}	liq.	V

Volumetric Properties General: Van der Waals volume 56.11 cm^3 mol^{-1}. Molar volume 86.6 cm^3 mol^{-1}
Specific Heat Capacity:

No.	Value	Note	Type
1	145 kJ/kg.C	145 J mol^{-1} K^{-1}, 25°, solid	P
2	182 kJ/kg.C	182 J mol^{-1}, K^{-1}, 25°, liq.	P

Glass-Transition Temperature:

No.	Value	Note
1	-24°C	conventional
2	-24°C	syndiotactic
3	-25°C	isotactic

Brittleness Temperature:

No.	Value
1	-24°C

Surface Properties & Solubility:
Cohesive Energy Density Solubility Parameters: δ 18.8 J$^{1/2}$ cm$^{-3/2}$. Molar cohesive energy 30.6–31.9 kJ mol^{-1}. CED 353.34 J cm^{-3}
Plasticisers: Thiokol TP-759, Plastolein 9720, Admex 760
Solvents/Non-solvents: Sol. THF, DMF, Me$_2$CO, butanone, EtOAc, CHCl$_3$, toluene, C$_6$H$_6$. Insol. H$_2$O, EtOH, cyclohexanol, Et$_2$O, aliphatic hydrocarbons. Generally sol. in polar solvents, insol. in non polar solvents. No phase separation or precipitation occurs on solvent evaporation
Surface Tension:

No.	Value
1	35–42 mN/m

Mechanical Properties:
Tensile Strength Break:

No.	Value	Elongation
1	0.2 MPa	1800%

Optical Properties:
Optical Properties General: Specific refractive index increment 0.109 (Me_2CO, 4360°); 0.106 (Me_2CO, 5460°)
Refractive Index:

No.	Value	Note
1	1.464	25°

Polymer Stability:
Polymer Stability General: Retains original props. under normal use conditions
Thermal Stability General: Decomposes slowly unless subjected to extreme heat
Decomposition Details: When heated to 300–500° decomposes to ethylene, EtOH and CO_2
Flammability: Burns readily in air and oxygen
Environmental Stress: Unaffected by normal degradation; primary uv absorption occurs below the solar spectrum of 290 nm. Uv absorbers improve stability
Chemical Stability: When heated in presence of free radical initiators, undergoes cross-linking and some degradation. Resistant to oxygen
Hydrolytic Stability: Reasonably stable to both acid and alkaline hydrol.

Applications/Commercial Products:
Processing & Manufacturing Routes: Commercial poly(ethyl acrylate) is usually prepared by free radical soln. and emulsion polymerisation. It is processed by injection moulding, blow moulding, extrusion and thermoforming
Applications: Acrylics are widely used in paints, coatings and lacquers, in paper industry, as radiation curable systems, as adhesives and sealing compds., in gaskets for automotive industry and in textile and leather industries

Trade name	Supplier
Acryloid	Rohm and Haas
Cyanacryl	American Cyanamide
Deglan	Degussa
Elvacite	DuPont
Hycar	B.F. Goodrich
Joncryl	Johnson
Lucite	DuPont
Synthacryl	Synthopol
Vinnapas	Wacker
Vultacryl	General Latex

Bibliographic References
[1] *Ullmanns Encykl. Ind. Chem.*, 5th edn., (ed. W. Gerhartz), VCH, 1992
[2] *Encycl. Polym. Sci. Eng.*, 2nd edn., (ed. J.I. Kroschwitz), 1985, **1**
[3] *Kirk-Othmer Encycl. Chem. Technol.*, 4th edn., (ed. J.I. Kroschwitz), 1991, **1**
[4] Hofmann, W., *Rubber Technology Handbook*, Hanser, 1989
[5] Gunawan, L. and Haken, J.K., *J. Polym. Sci., Polym. Chem. Ed.*, 1985, **23**, 2539
[6] Srinivasan, K.S.V. and Sentappa, M., *Polymer*, 1973, **14**, 5
[7] Yokota, K., Miwa, M., Hirabayashi, T. and Inai, Y., *Macromolecules*, 1992, **25**, 5821
[8] Davis, R.F.B. and Reynolds, G.E.J., *J. Appl. Polym. Sci.*, 1968, **12**, 47
[9] Nucessle, A.C. and Kine, B.B., *Ind. Eng. Chem.*, 1953, **45**, 1287
[10] De-Marco, R.D., *Rubber Chem. Technol.*, 1979, **52**, 173
[11] Rehberg, C.F. and Fisher, C.H., *J. Am. Chem. Soc.*, 1944, **66**, 1203
[12] Riddle, E.H., *Monomeric Acrylic Esters*, Reinhold, 1954

Polyethylbutadiene   P-245
Synonyms: *Poly(2-ethyl-1,3-butadiene)*
Monomers: 2-Ethyl-1,3-butadiene
Material class: Synthetic Elastomers
Polymer Type: polydienes

Applications/Commercial Products:
Processing & Manufacturing Routes: Polymerisation in an emulsion system has been reported [3]
Applications: Potential application as protective coating on optical products

Bibliographic References
[1] Marconi, W, Mazzei, A, Cucinella, S. and Cseari, M., *J. Polym. Sci., Part A-1*, 1964, **2**, 4261
[2] Overberger, C.G., Arond, L.H., Wiley, R.H., and Garrett, R.R, *J. Polym. Sci.*, 1951, **7**, 431, (props.)
[3] Marvel, C.S., Williams, J.L.R. and Baumgarten, H.E., *J. Polym. Sci.*, 1949, **4**, 583, (synth.)
[4] Hattam, P., Gauntlett, S, Mays, J.W., Hadjichristidis, N. *et al*, *Macromolecules*, 1991, **24**, 6199, (conformn.)

Poly(ethyl cyanoacrylate)   P-246

$$\left[-CH_2-\underset{\underset{O}{|}}{\overset{\overset{CN}{|}}{C}}- \right]_n$$
$$ OCH_2CH_3$$

Synonyms: *Poly(ethyl 2-cyano-2-propenoate). PECA. PEtCA. Ethyl cyanoacrylate polymer*
Related Polymers: Poly(alkyl 2-cyanoacrylates), Poly(methyl 2-cyanoacrylate), Poly(isobutyl 2-cyanoacrylate), Poly(allyl 2-cyanoacrylate)
Monomers: Ethyl 2-cyanoacrylate
Material class: Gums and resins
Polymer Type: acrylic, acrylonitrile and copolymers
CAS Number:

CAS Reg. No.
66547-34-0
25067-30-5

Molecular Formula: $(C_5H_7NO_2)_n$
Molecular Weight: 108000
Additives: Additives may increase mech. props
General Information: Most widely used commercial cyanoacrylate polymer. Is a clear, hard resin with good adhesive props. though brittle with low peel and impact strength
Morphology: Amorph.

Volumetric & Calorimetric Properties:
Density:

No.	Value	Note
1	d 1.2 g/cm^3	[10]
2	d 1.248 g/cm^3	20° [2]

Thermodynamic Properties General: Thermodynamic props. have been reported [19]. Heat capacity increases with increasing temp.

– Poly(ethyl cyanoacrylate)

Specific Heat Capacity:

No.	Value	Note	Type
1	1.044 kJ/kg.C	0.118 kJ mol^{-1}, -73° [19]	P
2	1.37 kJ/kg.C	0.155 kJ mol^{-1}, 25° [19]	P
3	1.65 kJ/kg.C	0.186 kJ mol^{-1}, 127° [19]	P

Glass-Transition Temperature:

No.	Value	Note
1	116°C	dynamic mechanical analysis [1]
2	140°C	differential scanning calorimetry [2]
3	115°C	dilatometry [3]
4	133°C	differential scanning calorimetry [18]

Vicat Softening Point:

No.	Value	Note
1	126°C	[4]

Surface Properties & Solubility:
Cohesive Energy Density Solubility Parameters: δ 11.2–11.4 J$^{1/2}$ cm$^{-3/2}$ (calc.) [8,10]
Solvents/Non-solvents: Sol. THF, CH$_2$Cl$_2$, Me$_2$CO, acetonitrile, DMF, nitromethane, N-methylpyrrolidone, nitrobenzene. Insol. EtOAc, MeOH, toluene, hexane, Et$_2$O, C$_6$H$_6$, CHCl$_3$, H$_2$O [5]

Transport Properties:
Polymer Solutions Dilute: Mark-Houwink parameters have been reported [5]. η 0.51 (30°, THF) [9]; 0.58 (30°, 28 d, THF) [9]; η 0.32 dl g^{-1} (20°, 1% in nitromethane) [19]. Commercial materials available in low, medium and high viscosity
Water Content: 0.7%

Mechanical Properties:
Mechanical Properties General: Increasing polymer viscosity increases adhesive bond strength, which is also increased by additives. Storage modulus decreases with increasing temp.; breaking strength decreases on heating [8] or on exposure to moisture [9]

Tensile (Young's) Modulus:

No.	Value	Note
1	2200 MPa	bulk polymer [1]

Flexural Modulus:

No.	Value	Note
1	2068 MPa	300 kpsi, ASTM D790

Tensile Strength Break:

No.	Value	Note	Elongation
1	28 MPa	steel-steel, ASTM D638 [4]	2%
2	46 MPa	Bulk polymer [10]	4.3%
3	35 MPa	cold rolled steel	
4	5.7 MPa	bone	

Miscellaneous Moduli:

No.	Value	Note	Type
1	8800 MPa	25°[1]	storage modulus

Impact Strength: Shear impact strength 267–533 J m^{-1} (5–10 ft lb in^{-1}, ASTM D950) [4] 4.8 kJ m^{-2} (23°, 24h) [18]; 2.2 kJ m^{-2} (100°, 24h) [18]; 1.2 kJ m^{-2} (150°, 5h) [18]
Mechanical Properties Miscellaneous: Tensile shear strength 14 MPa (cold rolled steel); 17.2 MPa (2500 psi, steel) [4]; 13.7 MPa (2000 psi, Al) [4]. Overlap shear strength 14.5 MPa (low viscosity, cold rolled steel); 1.8 MPa (high viscosity, cold rolled steel); 10 MPa (low viscosity, brass); 13 MPa (high viscosity, brass) [10>; >22.8 MPa (cold rolled steel, added trihydroxy benzoic acid). T peel strength 0.09 kg cm^{-1}, 0.11 kg cm^{-1} [16], 2.5 kg cm^{-1} [17] (cold rolled steel). Lap shear break strength 17 MPa. Mechanical failure of bonded polymers usually occurs before bond failure. Dynamic mechanical props. and loss factor values have been reported [1,10]
Hardness: Rockwell M58
Izod Notch:

No.	Value	Notch	Note
1	40 J/m	N	cold rolled steel

Izod Area:

No.	Value	Notch	Note
1	0.3 kJ/m^2	Notched	3 kg cm cm^{-2} [14]
2	0.65 kJ/m^2	Notched	6.5 kg cm cm^{-2} [15]

Electrical Properties:
Dielectric Permittivity (Constant):

No.	Value	Frequency	Note
1	3.98	1 MHz	[4]

Optical Properties:
Refractive Index:

No.	Value	Note
1	1.45	20° [4,10]

Polymer Stability:
Thermal Stability General: Loss of bond strength occurs above 90°
Upper Use Temperature:

No.	Value	Note
1	80°C	approx.

Decomposition Details: Depolymerisation begins as low as 90° by chain fragmentation without MW loss [8]. Decomposition onset 168° (N$_2$), 153° (air) [8]. Depolymerisation is virtually complete above 260°
Environmental Stress: Has poor resistance to moisture [9] when adhered to metals. Unsuitable for use in situations of high heat, high impact, moisture or outdoor conditions
Chemical Stability: is stable to motor oils and alcohols
Hydrolytic Stability: Degraded rapidly by alkali soln., releasing formaldehyde [9,12]

Applications/Commercial Products:
Processing & Manufacturing Routes: Spontaneous polymerisation of ethyl cyanoacrylate catalysed by base see poly(alkyl cyanoacrylates)
Applications: Used for binding small parts in many industries where an instant bond is required

Trade name	Details	Supplier
7431		Bostik

Agomet C	various adhesive grades	Degussa
CA	various grades	Delo
Cyanolit	200 series	Panacol Elosol
Cyanolit Cristal	various grades	Panacol Elosol
Helmatrin	various grades	Forbo Helmitin
Maxibond SAS	various grades	Dunlop
Plus C	various grades	Permabond
Prism	400 series	Loctite
Pronto CA	various adhesive grades	3M
Sicomet	various grades	Henkel Canada
Superglue	various adhesive grades	Hylomar Bostik
TB	various grades	Threebond

Bibliographic References

[1] Cheung, K.H., Guthner, J., Otterburn, M.S. and Rooney, J.M., *Makromol. Chem.*, 1987, **188**, 3041, (Tg, mechanical props.)
[2] Tseng, Y-C., Hyon, S.H. and Ikada, Y., *Biomaterials*, 1990, **11**, 73, (synth, mechanical props., degradation)
[3] Kukarni, R.K., Porter, H. and Leonard F., *J. Appl. Polym. Sci.*, 1973, **17**, 3509, (Tg)
[4] Coover, H.W., Dreifus, D.W. and O' Connor, J.T., *Handb. Adhes.*, 3rd edn., (ed., I. Skiest), Van Nostrand Reinhold, 1990, **463**, (rev)
[5] Donnelly, E.F., and Pepper D.C., Makromol. Chem., *Rapid Commun.*, 1981, **2**, 439, (solubility)
[6] Negulescu, I.I., Calugaru, E.M., Vasile, C and Dumitrescu, G., J. Macromol. Sci., *Pt. A*, 1987, **24**, 75, (thermal degradation)
[7] Guthrie, J., Otterburn, M.S., Rooney, J.M. and Tsang, C.N., *J. Appl. Polym. Sci.*, 1985, **30**, 2863, (thermal degradation)
[8] Drain, K.F., Guthmer, J., Martin, F.R. and Otterburn, M.S., *J. Adhes.*, 1984, **17**, 71, (steel adhesion)
[9] Millet, G.H., *Structural Adhesives Chemistry and Technology*, (ed. S.R. Hartshorn), Plenum Press, 1996, 1632, (rev)
[10] Woods, J., *Polymeric Materials Encyclopedia*, (ed. J.C. Salamone), CRC Press, 1996, **2**, 1632, (rev)
[11] Leonard, F., Kulkarni, R.K., Brandes, G., Nelson, J. and Cameron, J.J., *J. Appl. Polym. Sci.*, 1966, **10**, 259, (hydrolytic stability)
[12] Jpn. Pat., 1972, 51807, (composition)
[13] U.S. Pat., (Loctite), 1978, 4102945, (tougheners)
[14] PCT Int. Appl., (Loctite), 1983, 02 450, (tougheners)
[15] Denchev, Z.Z. and Kabaivanov, V.S., *J. Appl. Polym. Sci.*, 1993, **47**, 1019, (thermal props.)
[16] Bykova, T.A., Kiparisova, Y.G., Lebedev, B.V., Mayer, K.A. and Gololobov, Y.G., Vysokomol. Soedin., *Ser. A*, 1991, **33**, 614, (thermodynamic props.)
[17] Papatheofanis, F.J., *Biomaterials*, 1989, **10**, 185, (adhesive strength)
[18] Hussey, B., and Wilson, J., Structural Adhesives, *Directory and Databook*, Chapman & Hall, 1996, 118, (technical datasheets)

Polyethylene

P-247

$$\{CH_2CH_2\}_n$$

Synonyms: *Polyethene. Polythene. PE*
Related Polymers: High-density polyethylene, Low-density polyethylene, Linear low-density polyethylene, Ultra high MW polyethylene, Medium-density polyethylene, Ultra (or very) low-density polyethylene, Polyethylene films, Polyethylene fibres, Polyethylene blends
Monomers: Ethylene, Butene, Hexene, Octene
Material class: Thermoplastic
Polymer Type: polyethylene, polyolefins
CAS Number:

CAS Reg. No.
9002-88-4

Molecular Formula: $(C_2H_4)_n$
Fragments: C_2H_4

Molecular Weight: M_n 20000–40000 (typical commercial value). MW 300–600 (for waxes). M_n 3000000 (min., ultra high MW)
Additives: Antioxidants. Carbon black for electroconductive composites
General Information: Polyethylene is the generic name for a large family of olefin homopolymers and copolymers. The method of production, co-monomer and consequent MW, degree of polymerisation, crystallinity and density give rise to a broad range of props. divided into groups related to density. High-density polyethylene (HDPE), density 0.941 g cm^{-3}; ultra high MW polyethylene (UHMWPE), density 0.93; medium-density polyethylene (MDPE), density 0.93–0.94; linear low-density polyethylene (LLDPE), density 0.915–0.925; low-density polyethylene (LDPE), density 0.91–0.94; very low-density polyethylene (VLDPE, ULDPE), density 0.88–0.91. This classification is based on 2 parameters readily measured in 1950's. It cannot easily describe important distinctions between struct. and props. of resin brands [2]
Morphology: The smallest crystalline units are planar lamellae comprised of chains perpendicular to the plane, folding back and forth. A few chains interconnecting adjacent lamellae through an amorph. region lead to the formation of the larger spherulites. The high nucleation rate gives rise to small spherulites - which are always present even after rapid quenching. The spherulite size may greatly affect props. [3]
Identification: Cl, S, P absent; N may be present in antioxidant. No simple chemical tests for identification. Low density, lack of chemical reactivity, limited solubility are indicative [4]. May be distinguished from polypropylene by sinking in isophorone, polypropylene floats
Miscellaneous: The repeating unit $(CH_2CH_2)_n$ is strictly polymethylene $(CH_2)_m$. Originally prepd. by the polymerisation of diazomethane, it predated PE to which it is nominally identical but much more linear. It is now basically only of historical interest [1]. Ethylene polymers in practice can contain short and long-chain branched sections [2]

Volumetric & Calorimetric Properties:
Density:

No.	Value	Note
1	d 1 g/cm^3	25° cryst. phase,
2	d 0.852–0.862 g/cm^3	25° amorph. phase, [1,2,3]

Melting Temperature:

No.	Value	Note
1	135°C	crystallinity 70–90%, density 0.96–0.97
2	105–115°C	crystallinity 40–50%, density 0.910–0.915 [3]
3	142.6°C	theoretical 100% crystallinity [4]

Transition Temperature General: Controversy whether β-transition temp. or γ-transition temp. is T_g [1]. Transition temps. are dependent on crystallinity
Transition Temperature:

No.	Value	Note	Type
1	50°C	approx., cryst. phase motion	α transition temp.
2	-20°C	approx., branch point motion	β transition temp.
3	-120°C	approx., CH_2 gp motion	γ transition temp.

Surface Properties & Solubility:
Plasticisers: Plasticisers are little used since flexibility can be adjusted by polymerisation conditions and blending, without deterioration of phys. props. They include vinyl plasticisers (at ca. 0–2 wt.%), hydrocarbons (up to ca. 20 wt.%), rubbers and chlorinated paraffins in conjunction with antimony oxide for flame retardance [4, 5]
Solvents/Non-solvents: Sol. aliphatic hydrocarbons, aromatic hydrocarbons, halogenated hydrocarbons, elevated temp. above

– Polyethylene

50° for low density types. Solubility decreases with increasing density and crystallinity. Temps. up to 170° may be required for highly crystalline PE. Insol. all liquids at room temp., polar liquids at elevated temp. [4, 6]

Transport Properties:
Melt Flow Index:

No.	Value	Note
1	0.1–17 g/10 min	load 2.16 kg, 190° [20]

Polymer Melts: Viscosity is very dependent on shear rate [20]
Permeability of Gases: $[N_2]$ 12.25–131.3 0.14–1.5 $\times 10^{-13}$ cm^3 $(cm\ s\ Pa)^{-1}$. 11.82–131.3 (0.18–2 $\times 10^{-10}$ cm^2 $(s\ cmHg)^{-1}$, 25°); 113.8–595.3 1.3–6.8 $\times 10^{-13}$ cm^3 $(cm\ s\ Pa)^{-1}$. (1.75–9.1 $\times 10^{-10}$ cm^2 $(s\ cmHg)^{-1}$, 50°); $[O_2]$ 0.53–4.2 $\times 10^{-13}$ cm^3 $(cm\ s\ Pa)^{-1}$. (0.7–5.6 $\times 10^{-10}$ cm^2 $(s\ cmHg)^{-1}$, 25°); 1.6–5.3 $\times 10^{-13}$ cm^3 $(cm\ s\ Pa)^{-1}$. (2.1–7.1 $\times 10^{-10}$ cm^2 $(s\ cmHg)^{-1}$, 50°); $[CO_2]$ 2.3–26.3 $\times 10^{-13}$ cm^3 $(cm\ s\ Pa)^{-1}$. (3–35 $\times 10^{-10}$ cm^2 $(s\ cmHg)^{-1}$, 25°); 8.9–53 $\times 10^{-13}$ $(cm\ s\ Pa)^{-1}$ (11.8–71 $\times 10^{-10}$ cm^2 $(s\ cmHg)^{-1}$, 50°). Transmission decreases with increasing crystallinity [4]

Water Absorption:

No.	Value	Note
1	0.2 %	1 year, 20° [4]

Mechanical Properties:
Mechanical Properties General: It is difficult to correlate stress-strain props. of polyethylenes with density, crystallinity and melt flow index since test conditions and specimen preparation are significant. In general, props. involving small deformation depend on crystallinity and density while those involving large deformations depend on MW and chain branching. Typically LDPE is soft and flexible while HDPE is harder and stiffer [1,4]

Electrical Properties:
Electrical Properties General: Being completely non-polar, polyethylenes have very high electrical resistivity and exceptionally low loss, making them ideal for high frequency insulation. Occluded air or voids, oxidation and mech. stress are deleterious [1,4]
Surface/Volume Resistance:

No.	Value	Note	Type
2	>0.1 $\times 10^{15}$ $\Omega.cm$	min.	S

Dielectric Strength:

No.	Value	Note
1	20–160 $kV.mm^{-1}$	greater for thin films, increases with density and MW, but more affected by thickness and temp. [4]

Dissipation Power Factor: tan δ <0.0002 (but increased by process residues, modification, etc.) [4]
Dissipation (Power) Factor:

No.	Value	Note
1	<0.0002	Max. but increased by process residues, modification, etc. [4]

Optical Properties:
Refractive Index:

No.	Value	Note
1	1.51–1.52	increases with density
2	1.49	amorph.
3	1.52	cryst. α, β
4	1.582	cryst. γ [4]

Total Internal Reflection: Birefringence 0.001–0.004 (varies irregularly with stress in the specimen) [4]
Volume Properties/Surface Properties: Opaque, translucent in thin film (due to light scattering by spherulites)

Polymer Stability:
Thermal Stability General: Stable in absence of air to 280–300°. Melts and becomes transparent on pyrolysis. Does not melt if cross-linked
Decomposition Details: Above 300° decomposes to yield low MW hydrocarbons
Flammability: Burns with a yellow tipped blue flame with little smoke; continues to burn and may drip when removed from flame. On extinguishing smells similar to hot paraffin wax
Environmental Stress: Susceptible to environmental stress cracking, especially if subject to strain and exposure to polar liquids or vapours [1,4]
Chemical Stability: Highly inert to swelling by all solvents. Resistant to polar solvents, vegetable oils, H_2O, alkalis, most conc. acids (inc. HF) at room temp. Decomposed by strong oxidising agents e.g. fuming HNO_3, H_2SO_4. Slowly attacked by halogens and chlorinating agents [1,4]

Applications/Commercial Products:
Processing & Manufacturing Routes: There are 4 types of reaction system operated under 3 synthesis technologies for the polymerisation of ethylene of commercial importance; reaction systems are 1. High pressure (60–360 MPa) free radical process using oxygen, peroxide or other oxidiser as initiator at a temp. up to 350° to produce LDPE by a liquid phase reaction; (a recent variant, uses low pressure (0.7–2.1 MPa) and low temp. (<100°) but requires ultra high purity ethylene to produce LDPE by a gas phase reaction); 2. Low pressure (0.1–20 MPa) using a heterogeneous catalyst (supported molybdenum or chromium oxides) at a temp. of 50–300° to produce HDPE (Philipps Process); 3. A low pressure process using ionic catalysts (titanium halides and aluminium alkyls - Ziegler catalysts) and 4. Low pressure process using Ziegler catalysts supported on inorganic carriers. Synthesis technologies are supercritical ethylene at a temp. above the melting temp. of the PE being produced and high pressure, the mixture of supercritical ethylene and molten PE serves as the polymerisation medium. As well as reaction systems 1, 3, 4 above, metallocene (Kaminsky) catalysts are accomodated; B. Solution or Slurry - A variety of catalysts may be used with a hydrocarbon solvent at 120–150°. This versatile method was the first to use Philipps and Ziegler catalysts. Solvent strip stage required; C. Gas Phase - Many catalysts can be adapted to use in a mechanically stirred, or fluidised, bed of PE particles in a gas phase reactor. Absence of solvent recirculation gives an economical process [2, 7]. In the early 1930's, Perrin at ICI discovered the high pressure reaction system [8]. Standard Oil Co. developed a process using molybdenum oxide supported on alumina as a catalyst in the early 1950's [9]. This process is not now commercially used. Also in the early 1950's, Ziegler discovered the titanium halide + alkyl aluminium compound catalyst [10, 11, 12] and the potential of chromium oxide supported on silica (Philipps catalyst) was discovered [13, 14]. A DuPont process in the mid 1950's gave linear PE but using a pressure of 707 MPa was not commercially viable [15]. More recent developments are the metallocene (Kaminsky) catalysts [16, 17] and the Union Carbide, Unipol process [18], which both allow a high degree of product control
Applications: Due to its great versatility, PE is the largest quantity manufactured polymer in the world (ca. 4.5 $\times 10^7$ tonnes total in 1994). PE is primarily used for commodity plastics but also for some speciality polymers. Applications include various types of films; coatings for paper, metal, wire and glass; containers for both household and industrial use; toys; pipe and tubing; and fibres

– Polyethylene

Trade name	Details	Supplier
A-C	Also includes	Allied Signal Corporation
ACtone		Allied Signal Corporation
ACumist	Also includes	Allied Signal Corporation
Akrowax (PE)	LMWPE based process aid	Akrochem
Amoco	Also includes various polyolefins	Amoco
Amsoft	PE emulsion	American Emulsion Co
Arpak	4322	ARCO Chemical Co.
Bellox	cross-linkable PE sheet	Bell Plastics Ltd
Betadyr		Lunds of Bingley
Bocolene E	PE compound	Boliden Compound
Carlona	Also includes	Shell Chemicals
Cestidur		DSM (UK) Ltd.
Clysar	ECL/EH	DuPont
Compound 403, 401	Silane cross-linkable, flame retardant	AEI Compounds
Daplen	Also includes	PCD Polymere GmbH.
Dow	Many other names	Dow
DuPont	Many other names	DuPont
Ekill		Eki Eeknoff Kunstoffe
Elite	metallocene catalysed	Dow Plastics
Elite	metallocene catalysed	Dow Plastics
Epolene	Includes polypropylene and other polyolefins	Eastman Chemical Company
Escomer	polyethylene wax	Exxon Chemical
Escomer		Hays Colours
Escomer		Ashland Plastics
Esi-Cryl	PE wax emulsions	Emulsion Systems Inc
Estralen		Estra Kunstoffe Gmbh
Evolue	metallocene catalysed	Mitsui Petrochemicals
Evolue	metallocene catalysed	Mitsui Petrochemicals
Exceed	metallocene catalysed	Exxon Chemical
Exceed	metallocene catalysed	Exxon Chemical
Forbest	PE compound for paints	Lucas Meyer
Graf		Otto Graf
Hakorit		Kors Plastics
Hartolon	Also includes esters	Hart Chemicals
Hoechst Wax	PE waxes and oxidised PE waxes	Hoechst
Industrial Wavingas		Wavin Industrial Products
Isoplas P		Micropol
Ispaleen H		Stokvis Plastics
Microcoat	PE powder	Micropol
Microlink	Cross-linkable PE powder	Micropol
Micropol	PE powder	Micropol
Micropoly		Presperse
Microstat	Semiconductive PE powder	Micropol
Minicel	various grades	Voltek
Mobil	Also used for other materials	W. Boesch
Monolywe		Monoplas Industries Ltd.
Moplen	Used for ethylene, propylene polymers and copolymers	Enimont Iberica SA
Moplen	Used for ethylene, propylene polymers and copolymers	Himont
Moplen		Montecatini Edison SpA.
Moplen		Novamont
Multibase	Masterbatches and compounds also PP, polymers and copolymers	Omya
Nortuff		Quantum Chemical
Norvan		Artilabo
Novatec	Also includes other polyolefin adhesives	British Traders and Shippers Ltd
Novatec		Mitsubishi Chemical
Novex	homo- and copolymers	BP Chemicals
Novex		KD Thermoplastics
Orbilan		Danelli Materie Plastiche
Paxon	Includes mica filled, talc filled and other products	Paxon Polymer
Pionier	PE compound with paraffin oil	Hansen and Rosenthal
Plasbak	PE compound	Cabot Plastics Ltd.
Plasdeg	PE compound	Cabot Plastics Ltd.
Plasgrey	PE compound	Cabot Plastics Ltd.
Plasinter	PE coating powder	Plascoat Systems Ltd
Plaspol	PE powders	Ashland Plastics
Plastecnics	Also used for other materials	Europlastecnics SA
Plasthen	Hot melt PE	Elf Atochem (UK) Ltd.
Plasthene		Plastribution Ltd.
Plaswite	PE compound	Cabot Plastics Ltd.
Plexar		Quantum Chemical
Poligen PE	Aq. dispersions and wax emulsions - PE denotes use of the name for polyethylene	BASF
Polysoft	PE emulsion	Sybran
Polywax	Also used for other products	Petrolite
Portaliner	Flame retardant modified PE	PC Polythene
Portathene	Flame retardant modified PE	PC Polythene

Profax	Includes PE but mostly PP and copolymers	Hercules
Purflo	Also includes polyesters	Purflo DTL
Retain	Contains recycled material also includes other material	Dow Plastics
Sclair		DM and C Watering
Sclairlisk	8000	DuPont, Canada
Senden	Also includes PP	Senova Kunststoffe GmbH.
Siltek		Petrolite
Struktol	LMW homopolymer wax	Struktol
Sustain	recycled; formerly RE-NU	Novacor Chem. Int.
Sustain	recycled PE-formerly RE-NU	Nova Chemicals
TTC		Ten Castle Mouldings
Tecafine PE		Ensinger Ltd.
Tekalen		Terbaack Kunstoffe
Tenite	Also includes PP and copolymers and cellulosics	Eastman Chemical Company
Thermocomp	F-series. Glass filled and blended	LNP Engineering
Trovidur	Also includes PP and PVC	VT Plastics
Trovidur		Building and Engineering Plastics
Typopack		Ten Castle Mouldings
Vapladur	PE sheet	Vapla
Vaplasol	PE sheet	Vapla
Vaplatec	PE sheet	Vapla
Vestowax	Compound PE and paraffin and hydrocarbon waxes	Astor Wax
Visqueen		MV Distributors
Vitrathene	Sheet, rod and block	Stanley Smith and Company Plastics Ltd.
Volara	various grades	Voltek
Wavin		Wavin Industrial Products
Weldex		Moore and Buckle (Flexible Packaging)
Zemid		DuPont

Bibliographic References

[1] Alger, M., *Polymer Science Dictionary*, 2nd edn., Chapman and Hall, 1997
[2] Kissin, Y.V., *Kirk-Othmer Encycl. Chem. Technol.*, Vol. 17, 4th edn., (eds. J.I. Kroschwitz, Wiley Interscience, 1996, 702
[3] Doak, K.W., *Encycl. Polym. Sci. Eng.*, 2nd edn., (ed. J.I. Kroschwitz), Wiley Interscience, 1986, **6**, 383
[4] Roff, W.J. and Scott, J.R., Fibres, Films, *Plastics and Rubbers*, Butterworths, 1971
[5] Sears, J.K. and Touchette, N.W., *Encycl. Polym. Sci. Eng.*, 2nd edn., (eds. J.I. Kroschwitz), Wiley Interscience, 1989, 626, (Supplement Volume)
[6] Orwell, R.A., *Encycl. Polym. Sci. Eng.*, Vol. 15, 2nd edn., (ed. J.I. Kroschwitz), Wiley Interscience, 1989
[7] Sundaram, K.M., Shreehan, M.M. and Olszewski, E.F., *Kirk-Othmer Encycl. Chem. Technol.*, 4th edn., (eds. J.I. Kroschwitz and M. Howe-Grant), Wiley Interscience, 1994, **9**, 878
[8] *Brit. Pat.*, 1937, 471 590
[9] *U.S. Pat.*, 1952, 2 692 257
[10] Ziegler, K., *Kunststoffe*, 1955, **45**, 506
[11] *Ger. Pat.*, 1960, 973 626
[12] *Belg. Pat.*, 1955, 533 362
[13] *Belg. Pat.*, 1955, 530 617
[14] *U.S. Pat.*, 1958, 2 825 721
[15] *U.S. Pat.*, 1957, 2 816 883
[16] Andresen, A., Cordes, H.-G., Herwig, J., Kaminsky, W. *et al*, *Angew. Chem., Int. Ed. Engl.*, 1976, **15**, 630
[17] Sinn, H. and Kaminsky, W., *Adv. Organomet. Chem.*, 1980, **18**, 99, (rev)
[18] *Chem. Eng. News*, 1977, **55**, Nov. 21, 21
[19] *Ger. Pat.*, 1976, 2 609 889
[20] Paschke, E., *Kirk-Othmer Encycl. Chem. Technol.*, Vol. 16, 3rd edn., (eds. M. Grayson and D. Eckroth), Wiley Interscience, 1981, 440

Poly(ethylene-*co*-acrylic acid) P-248

Synonyms: *EAA copolymer*
Related Polymers: Poly(acrylic acid), Poly(ethylene)
Monomers: Ethylene, Acrylic acid
Material class: Copolymers
Polymer Type: polyethylene, acrylic
CAS Number:

CAS Reg. No.	Note
9003-06-9	
25750-82-7	sodium salt
27515-34-0	potassium salt

Molecular Formula: $[(C_2H_4).(C_4H_4O_2)]_n$
Fragments: C_2H_4 $C_4H_4O_2$
Molecular Weight: The MW can range from 1000–100000
Additives: Carbon black, flame retardants, and titanium dioxide
General Information: Can be amorph. or partially crystalline. It is one phase material if amorph. and two phase if crystalline. It has excellent adhesion to most metallic and non-metallic substrates and good toughness, tear strength and stiffness. The hydrogen bonds between neighbouring carboxylic acid groups induce a cross-linking effect resulting in good toughness. Usually prod. as a neutralised dispersion with alkali and metal ions which forms a clear glassy material

Volumetric & Calorimetric Properties:
Density:

No.	Value	Note
1	d 0.96 g/cm^3	20% acrylic acid

Volumetric Properties General: Density of the polymer is dependent on the content of acrylic acid and it reaches a maximum at 20% of acrylic acid content
Glass-Transition Temperature:

No.	Value
1	0°C

Vicat Softening Point:

No.	Value	Note
1	50°C	XO [3]
2	86°C	EAA [2]

Poly(ethylene-co-acrylic acid)

3	85°C	[2]
4	82°C	[2]

Brittleness Temperature:

No.	Value	Note
1	-20°C	XO [3]
2	-73°C	EAA [2]

Surface Properties & Solubility:
Solvents/Non-solvents: Sol. H_2O. Swollen by organic solvents at elevated temps. Ionic bonding decreases solubility in organic solvents

Transport Properties:
Melt Flow Index:

No.	Value	Note
1	300 g/10 min	XO grades [3]
2	11 g/10 min	EAA 435 [2]
3	5.5 g/10 min	EAA 449 [2]
4	9 g/10 min	EAA 459 [2]
5	67 g/10 min	14% acrylic acid
6	12.2 g/10 min	88% Na salt
7	0.3 g/10 min	33% Na salt

Polymer Melts: Melt viscosity increases with acid content and also increases with neutralisation. Can be controlled by adjusting copolymer MW. Melt strength and elasticity values are similar to those of polyethylene
Permeability and Diffusion General: Permeability of liquids and gases is high for low crystallinity polymer and low for maximum crystallinity. Low permeability of natural oils
Water Content: 1% (approx.)
Water Absorption:

No.	Value
1	1 %

Mechanical Properties:
Mechanical Properties General: Tensile strength decreases with increasing acid unit content, to a minimum at 12–15%, then increases. Fracture strength exhibits a max. at same acid content. Increasing neutralisation increases tensile strength, reducing elongation at break
Tensile Strength Break:

No.	Value	Note	Elongation
1	11 MPa	XO grades [3]	300%
2	14.1–19 MPa	EAA grades [2]	650%
3	14.8 MPa	14% acrylic acid	470%
4	21.7 MPa	88% Na salt	420%
5	33 MPa	33% Na salt	280%

Tensile Strength Yield:

No.	Value	Note
1	6.2 MPa	XO grades [3]
2	9.3–9.7 MPa	EAA grades [2]

Hardness: Shore D54 (XO grades) [3], Shore D47-D49 (EAA grades) [2]
Friction Abrasion and Resistance: Has good abrasion resistance

Electrical Properties:
Dielectric Permittivity (Constant):

No.	Value	Frequency
1	2.4–2.5	60 Hz

Dielectric Strength:

No.	Value	Note
1	39.37 $kV.mm^{-1}$	1 kV mil^{-1}, $\frac{1}{8}$ in thick

Dissipation (Power) Factor:

No.	Value	Frequency
1	0.006–0.008	60 Hz

Applications/Commercial Products:
Processing & Manufacturing Routes: Ethylene can be polymerised with acrylic acid under high pressure using free radical initiators to produce a highly branched polymer. Copolymer can be processed like LDPE. The carboxylic acid groups present in the copolymer cause corrosion. When the acid content is up to 8% and melt temp. less than 177° temper-hardened steel is good enough but at higher acid contents nickel or chrome plated surfaces are required.
Applications: Coatings, laminates and bonding with paper, metal foils, nylon, polyethylene and polyester. Processing aid in sheet extrusion and injection moulding. It is also recommended as a flush vehicle

Trade name	Details	Supplier
A-C copolymer 540		Dow
EAA	several grades	Dow
XD-60899		Dow
XO-2375.33		Dow

Bibliographic References
[1] Young, M.C., *Tech. Pap., Reg. Tech. Conf. - Soc. Plast. Eng.*, 1975, 62
[2] Dow EAA Copolymers Dow Chemical Co., 1976, (technical datasheet)
[3] 1979, (technical datasheet)
[4] *Modern Plast. Intern.*, 1978, **8/12 Dec.,** 54
[5] 1979, (technical datasheet)
[6] *Modern Plast. Intern.*, Aug., 1982, 52
[7] Bonotto, S. and Boner, E.F., *Macromolecules*, 1968, **1,** 510
[8] Hirasawa, E., Yamamoto, Y., Tadano, K. and Yano, S., *J. Appl. Polym. Sci.*, 1991, **42,** 351
[9] Elias, H.-G.and Vohwinkel, F., *New Commercial Polymers,* 2nd edn., Gordon and Breach, 1986

Poly(ethylene-co-acrylic acid) P-249

$$\left[-(CH_2CH_2)_x-(CH_2CH)_y-\underset{\underset{OH}{\overset{\|}{C}}}{\underset{O}{}}\right]_n$$

copolymer with acrylic acid

$$\left[-(CH_2CH_2)_x-(CH_2C(CH_3))_y-\underset{\underset{OH}{\overset{\|}{C}}}{\underset{O}{}}\right]_n$$

copolymer with methacrylic acid

Polymer Stability:
Thermal Stability General: Props. maintained at very low temps. [1]
Environmental Stress: Environmental stress crack resistance increases with increasing carboxyl content [1]
Chemical Stability: Polymers tend to be corrosive due to presence of acid groups [1]
Stability Miscellaneous: FDA approved grades available

Poly(ethylene adipate) P-250

Synonyms: *Poly(oxyethyleneoxyadipoyl). PEA. Poly(ethylene glycol adipate). Adipic acid-ethylene glycol copolymer. Poly[oxy-1,2-ethanediyloxy(1,6-dioxo-1,6-hexanediyl)]*
Monomers: 1,2-Ethanediol, Hexanedioic acid
Material class: Thermoplastic
Polymer Type: saturated polyester
CAS Number:

CAS Reg. No.	Note
24938-37-2	Poly(ethylene adipate)
24937-05-1	

Molecular Formula: $(C_8H_{12}O_4)_n$
Fragments: $C_8H_{12}O_4$
Morphology: Monoclinic a 0.547, b 0.723, c 1.172, β 113.5° [12]; (c 1.191) [3], monoclinic a 0.547, b 0.724, c 1.155, β 113.5° [13]; monoclinic a 0.540, b 0.726, c 1.085, β 113.3° [14]; monoclinic a 0.50, b 0.74, c 1.171 [15]; monoclinic a 2.57, b 3.07, c 1.171, β 103.8° [2]. Morphology has been studied [11,47]. There are two crystalline modifications [27,50,55,56]

Volumetric & Calorimetric Properties:
Density:

No.	Value	Note
1	d^{25} 1.25 g/cm^3	[2,5]
2	d 1.221 g/cm^3	[17]
3	d 1.263 g/cm^3	[25]
4	d 1.21 g/cm^3	[35]

Thermal Expansion Coefficient:

No.	Value	Note	Type
1	0.00065 K^{-1}	amorph. [17]	v
2	0.00064 K^{-1}	crystalline [17]	v
3	0.00071 K^{-1}	[19]	v
4	0.00051 K^{-1}	[24]	v

Volumetric Properties General: Compressibility 0.00065 cm^3 g^{-1} atm^{-1} [17], -5.3 × 10^{-5} cm^3 g^{-1} bar^{-1} [24]

Equation of State: Equation of state information has been reported [19]
Thermodynamic Properties General: ΔH_f 15.91 kJ mol^{-1} (3800 cal mol^{-1}), S_f 50 J mol^{-1} (12 cal mol^{-1}) [8]; ΔH_m 12.1 kJ mol^{-1} (70.4 J g^{-1}), ΔS_m 6.7 J mol^{-1} K^{-1} [24]; ΔH_f 10.73 kJ mol^{-1} (62.4 J g^{-1}) (14.9 cal g^{-1}) [17]; $\Delta H_{cryst.}$ 147 J cm^{-3} (35 cal cm^{-3}) [18]. Other ΔH values have been reported [50]. C_p values have been reported [52]. ΔH_f 2.1 kJ mol^{-1} [66]; ΔS_f 7.8 J mol^{-1} K^{-1} [66]
Thermal Conductivity:

No.	Value	Note
1	0.169 W/mK	Coefficient of thermal conductivity [58]

Melting Temperature:

No.	Value	Note
1	50°C	[1,8]
2	52–54°C	[2,19,34]
3	47°C	[9,18]
4	65°C	[17]
5	53°C	[24]
6	49.7°C	[35]
7	47–52°C	[38]
8	53–55°C	[65]

Glass-Transition Temperature:

No.	Value	Note
1	-70°C	203K [8]
2	-68°C	[10]
3	-50°C	223K [18]
4	-26°C	110 Hz [23]
5	-30°C	[25]
6	-46°C	[34]
7	-47°C	[35]
8	-60°C	β-form [56]

Transition Temperature General: Other transition temps. have been reported [30,44]
Transition Temperature:

No.	Value	Note	Type
1	55°C	328K [18]	$T°_m$
2	-95°C	110 Hz [23]	T_γ
3	70°C	343K [51]	$T°_m$

Surface Properties & Solubility:
Solubility Properties General: Phase diagrams with organic solvents have been reported [22]
Cohesive Energy Density Solubility Parameters: δ_o 19.8 J$^{1/2}$ cm$^{-3/2}$; δ_d 16.3 J$^{1/2}$ cm$^{-3/2}$; δ_p 6.1 J$^{1/2}$ cm$^{-3/2}$; δ_h 9.4 J$^{1/2}$ cm$^{-3/2}$ [33]. Energy of cohesion has been reported [48]. Molar cohesive energy 54.18 kJ mol^{-1} (12.94 kcal mol^{-1}) [63]. Cohesive energy density 5.19 kJ mol^{-1} [66]
Internal Pressure Heat Solution and Miscellaneous: Polymer - solvent interaction parameters for C_6H_6, $CHCl_3$ and THF have been reported [32, 64]
Solvents/Non-solvents: Sol. C_6H_6, Me_2CO, EtOAc, AcOH [1], DMSO [28], $CHCl_3$ [4,31], trichloroethene [4], 1,1,2,2-tetrachloroethane [4], trifluoroacetic acid [21]. Slightly sol. CCl_4,

Poly(ethylene adipate)

tetrachloroethene [4]. Insol. H_2O [1], EtOH [1], petrol [1], Et_2O [1,31]

Surface and Interfacial Properties General: Interfacial free energy 0.89 mJ cm^{-2} (8.9 erg cm^{-2}, parallel) [18]. Interfacial free energy 2.27 mJ cm^{-2} (22.7 erg cm^{-2}, perpendicular) [18]

Transport Properties:
Polymer Solutions Dilute: [η] 0.15 dl g^{-1} (25°, C_6H_6) [19]; $η_{inh}$ 0.16 dl g^{-1} (30°, $CHCl_3$) [29]. Other viscosity values have been reported [37]. Theta temp. 114.5° (cyclohexanol) [62]
Polymer Melts: Melt viscosity has been reported [39]
Water Content: Water absorption has been reported [43]

Mechanical Properties:
Complex Moduli: Real and imaginary parts of complex shear modulus have been reported [25]

Electrical Properties:
Electrical Properties General: Other dielectric constants and loss factors have been reported in graphical form [23]. Volume resistivity has been reported [59]. Dissipation factor has been reported [31,57,59]
Dielectric Permittivity (Constant):

No.	Value	Note
1	5.16	[5]

Static Electrification: Effective dipole moment $μ_e$ 0.8 [23]. Electrical conductivity has been reported [57]

Optical Properties:
Transmission and Spectra: Ir [6,27,41,50,53,54], H-1 nmr [45,61], mass [20,29,60], Raman [54], C-13 nmr [21,28,61], esr [26,46], XPES [36] and uv [40] spectral data have been reported

Polymer Stability:
Decomposition Details: Thermooxidative decomposition occurs without formation of hydroperoxides. [16] Max. decomposition occurs at 380° [29]. Energy of activation for decomposition 177.5–188 kJ mol^{-1} (42.4–45 kcal mol^{-1}) [42]. Decomposition has been studied by mass spectroscopy [20]. Activation energy of oxidation 80–92 kJ mol^{-1} (19–22 kcal mol^{-1}) [16]
Biological Stability: Biodegradability has been reported [49]

Applications/Commercial Products:
Processing & Manufacturing Routes: Synthesised by condensation polymerisation of ethylene glycol and adipic acid at 200–250° [1,7,19,21]
Applications: Used as plasticiser for PVC. Used in polyurethane industry for manufacturing of pre-polymers, thermoplastic elastomers, coatings and adhesives, dispersing agents, and cast elastomers

Trade name	Supplier
Formrez 22	Witco
Formrez 22	Baxenden Chemicals Ltd
Paraplex	C.P. Hall
Plasthall	C.P. Hall
Ultramoll	Bayer Inc.

Bibliographic References

[1] Carothers, W.H. and Arvin, J.A., *J. Am. Chem. Soc.*, 1929, **51**, 2560, (synth, solubilities)
[2] Fuller, C.S. and Erickson, C.L., *J. Am. Chem. Soc.*, 1937, **59**, 344, (cryst struct)
[3] Storks, K.H., *J. Am. Chem. Soc.*, 1938, **60**, 1753, (cryst struct)
[4] Marvel, C.S., Frederick, C.D. and Copley, M.J., *J. Am. Chem. Soc.*, 1940, **62**, 2273, (solubilities)
[5] Yager, W.A. and Baker, W.O., *J. Am. Chem. Soc.*, 1942, **64**, 2164, (dielectric constant)
[6] Davison, W.T. and Corish, P.J., *J. Chem. Soc.*, 1995, 2428, (ir)
[7] Youngson, G.W. and Melville, H.W., *J. Chem. Soc.*, 1950, 1613, (synth)
[8] Edgar, D.B. and Hill, R., *J. Polym. Sci.*, 1952, **8**, 1, (thermodynamic props)
[9] Edgar, D.B. and Ellery, E., *J. Chem. Soc.*, 1952, 2633, (melting point)
[10] Edgar, D.B., *J. Chem. Soc.*, 1952, 2638, (transition temp)
[11] Takayanagi, M. and Yamashita, T., *J. Polym. Sci.*, 1956, **22**, 552, (morphology)
[12] Turner-Jones, A. and Bunn, C.W., *Acta Cryst.*, 1962, **15**, 105, (cryst struct)
[13] Hobbs, S.Y. and Billmeyer, F.W., *J. Polym. Sci., Part A-2*, 1969, **7**, 1119, (cryst struct)
[14] Point, J.J., *Bull. Cl. Sci., Acad. R. Belg.*, 1953, **39**, 435, (cryst struct)
[15] Fuller, C.S. and Frosch, C.J., *J. Phys. Chem.*, 1939, **43**, 323, (cryst struct)
[16] Gomory, I., Stimel, J., Reya, R. and Gornoryova, A., *J. Polym. Sci., Part C: Polym. Lett.*, 1967, **16**, 451, (thermooxidative decomposition)
[17] Hobbs, S.Y. and Billmeyer, F.W., *J. Polym. Sci., Part A-2*, 1970, **8**, 1387, (expansion coefficient, compressibility)
[18] Godovskii, Y.K. and Slonimskii, G.L., *J. Polym. Sci., Polym. Phys. Ed.*, 1974, **12**, 1053, (interfacial free energy, transition temps)
[19] Manzini, G. and Crescenzi, V., *Macromolecules*, 1975, **8**, 195, (equation of state, synth)
[20] Luederwald, I. and Urrutia, H., *Makromol. Chem.*, 1976, **177**, 2093, (thermal degradation, ms)
[21] Kricheldorf, H.R., *Makromol. Chem.*, 1978, **179**, 2133, (C-13 nmr, synth)
[22] Guieu, R., Ponge, C., Rosso, J.C. and Carbonnel, L., *Bull. Soc. Chim. Fr.*, 1973, **9**, 2776, (phase diagrams)
[23] Ito, M., Nakatini, S., Gokan, A. and Tanaka, K., *J. Polym. Sci., Polym. Phys. Ed.*, 1977, **15**, 605, (dielectric constants, transition temps)
[24] Ueberreiter, K., Karl, V.H. and Altmeyer, A., *Eur. Polym. J.*, 1978, **14**, 1045, (thermodynamic props)
[25] Ito, M., Kubo, M., Tsunita, A. and Tanaka, K., *J. Polym. Sci., Polym. Phys. Ed.*, 1978, **16**, 1435, (complex shear modulus)
[26] Lembicz, F. and Kuriata, J., *Makromol. Chem.*, 1986, **187**, 1551, (esr)
[27] Teitelbaum, B.Y., Palikhov, N.A., Maklakov, L.I., Anoshina, N.P. et al, *Vysokomol. Soedin., Ser. A*, 1967, **9**, 1672, (cryst struct, ir)
[28] Delides, C., Pettirick, R.A., Cunliffe, M. and Montaudo, G., *Macromolecules*, 1986, **19**, 1643, (C-13 nmr)
[30] Miller, G.W. and Saunders, J.W., *J. Polym. Sci., Part A-1*, 1970, **8**, 1923, (thermal degradation, ms)
[31] Pelmore, D.R. and Simons, E.L., *Proc. R. Soc. London London*, 1940, **175**, 468, (dissipation factor)
[32] Simek, L. and Bohdanecky, M., *Eur. Polym. J.*, 1987, **23**, 343, (soln props)
[33] Mieczkowski, R., *Eur. Polym. J.*, 1991, **27**, 377, (solubility parameters)
[34] Ellis, T.S., *Macromolecules*, 1995, **28**, 1882, (transition temps)
[35] Fernandes, A.C., Barlow, J.W. and Paul, D.R., *J. Appl. Polym. Sci.*, 1984, **29**, 1971, (transition temps)
[36] Briggs, D. and Beamson, G., *Anal. Chem.*, 1992, **64**, 1729, (XPES)
[37] Batzer, H., *Makromol. Chem.*, 1950, **5**, 5, (soln props)
[38] Korshak, V.V., Vinogradova, S.V. and Ulasova, E.S., *Izv. Akad. Nauk SSSR, Otd. Khim. Nauk*, 1954, 1089, (melting point, solubilities)
[39] Takayanagi, M. and Kuriyama, S., *Kogyo Kagaku Zasshi*, 1954, **57**, 876, (melt viscosity)
[40] Voight, J., *Makromol. Chem.*, 1958, **27**, 80, (uv)
[41] Giger, A., Henniker, J. and Jacque, L., *Chim. Ind. (Paris)*, 1958, **79**, 757, (ir)
[42] Farre-Rius, F. and Guiochori, G., *Bull. Soc. Chim. Fr.*, 1965, 455, (thermal decomposition)
[43] Willmott, P.W.T. and Billmeyer, F.W., *CA*, 1963, **60**, 679b, (water absorption, diffusion coefficient)
[44] Ishido, Y., Yamafuji, K. and Shimada, K., *Kolloid Z. Z. Polym.*, 1964, **200**, 49, (transition temp)
[45] Urman, Y.G., Khramova, T.S., Gorbunova, V.G., Barshtein, R.S. and Slonim, I.Y., *Vysokomol. Soedin., Ser. A*, 1970, **12**, 160, (H-1 nmr)
[46] Nozowa, Y., *Bull. Chem. Soc. Jpn.*, 1970, **43**, 657, (esr)
[47] Teitelbaum, B.Y. and Palikhov, N.A., *Vysokomol. Soedin., Ser. A*, 1968, **10**, 1468, (morphology)
[48] Sewell, J.H., *Rev. Plast. Mod.*, 1971, **22**, 1367, (energy of cohesion)
[49] Potts, J.E., Clendinning, R.A., Ackart, W.B. and Niegisch, W.D., *Polym. Sci. Technol. (Plenum)*, 1973, **3**, 61, (biodegradability)
[50] Tarasov, A.I., Sinitsyna, T.A., Serebryakova, V.M. and Surovtsev, A.G., *Vysokomol. Soedin., Ser. B*, 1975, **17**, 176, (ir, thermodynamic props)
[51] Privalko, V.P., *Sint. Fiz.-Khim. Polim.*, 1974, **14**, 95, (transition temps)
[52] Privalko, V.P. and Lipatov, Y.S., *Vysokomol. Soedin., Ser. A*, 1972, **14**, 1420, (heat capacity)
[53] Gribov, L.A., Zhitlova, I.V. and Krondratov, O.I., *Opt. Spektrosk.*, 1974, **36**, 911, (ir)
[54] Holland-Moritz, K., *Kolloid Z. Z. Polym.*, 1973, **251**, 906, (ir, Raman)
[55] Schuster, P., *Thermochim. Acta*, 1972, **3**, 485, (two polymorphic forms)

[56] Privalko, V.P., Vysokomol. Soedin., *Ser. A*, 1972, **14**, 1235, (two polymorphic forms)
[57] Sandrolini, F., Manaresi, P. and Zappia, G., *Mater. Chem.*, 1979, **4**, 201, (electrical conductivity, dissipation factor)
[58] Grishchenko, G.V., Lepilin, V.N. and Medvedev, N.N., *CA*, 1975, **87**, 68780, (coefficient of thermal conductivity)
[59] Yang, S. and Pan, J., *CA*, 1981, **97**, 24476, (dissipation factor, volume resistivity)
[60] Doerr, M., Luederwald, I. and Schulten, H.R., *J. Anal. Appl. Pyrolysis*, 1985, **8**, 109, (ms)
[61] Yu, H., Chen, X. and Ren, L., *CA*, 1990, **114**, 208417, (H-1 nmr, C-13 nmr)
[62] Zavaglia, E.A. and Billmeyer, F.W., *CA*, 1964, **60**, 15996h, (soln props)
[63] Hayes, R.A., *J. Appl. Polym. Sci.*, 1961, **5**, 318, (molar cohesive energy)
[64] Riedl, B. and Prud'homme, R.E., *J. Polym. Sci., Polym. Phys. Ed.*, 1986, **24**, 2565, (polymer-solvent interaction parameters)
[65] Iwakura, Y., Taneda, Y. and Vichida, S., *J. Appl. Polym. Sci.*, 1961, **5**, 108, (mp)
[66] Wunderlich, B. and Gaur, U., *Pure Appl. Chem.*, 1980, **52**, 445, (thermodynamic props)

Poly(ethylene azelate)   P-251

Synonyms: *Poly(oxyethyleneoxyazelaoyl). Poly[oxy-1,2-ethanediyloxy(1,9-dioxo-1,9-nonanediyl)]. PE7. Poly(1,2-ethanediol-cononanedioic acid)*
Monomers: 1,2-Ethanediol, Nonanedioic acid
Material class: Thermoplastic
Polymer Type: saturated polyester
CAS Number:

CAS Reg. No.
26760-99-6
26762-07-2

Molecular Formula: $(C_{11}H_{18}O_4)_n$
Fragments: $C_{11}H_{18}O_4$
Morphology: Cryst. struct. has been studied [3]. Orthorhombic, a 0.745, b 0.497, c 3.12 [17]; monoclinic, a 2.57, b 3.07, c 3.12, β 103.8° [13]

Volumetric & Calorimetric Properties:
Density:

No.	Value	Note
1	d^{25} 1.172 g/cm^3	[13]
2	d 1.163 g/cm^3	[15]

Thermal Expansion Coefficient:

No.	Value	Note	Type
1	0.00062 K^{-1}	[15]	V

Volumetric Properties General: Compressibility 6.8×10^{-5} cm^3 g^{-1} bar^{-1} [15]
Latent Heat Crystallization: ΔH_m 83.6 J g^{-1}, ΔS_m 10.5 J mol^{-1} K^{-1} [15]
Melting Temperature:

No.	Value	Note
1	44–47°C	[1,13,18]
2	34–36°C	[7]
3	47°C	[15]
4	37°C	[16]

Glass-Transition Temperature:

No.	Value	Note
1	-54°C	[16]

Transition Temperature:

No.	Value	Note	Type
1	45°C	[2]	Softening temp.
2	47°C	[2]	Thread formation temp.

Surface Properties & Solubility:
Cohesive Energy Density Solubility Parameters: Cohesive energy density 309 J cm^{-3} (74 cal cm^{-3}) [6]
Solvents/Non-solvents: Sol. C_6H_6, boiling EtOAc, boiling cyclohexanone [1,2]. Slightly sol. EtOH [1]. Insol. EtOAc, cyclohexanone [1]

Transport Properties:
Polymer Solutions Dilute: [η] values have been reported. Viscosity 0.28 [2]

Electrical Properties:
Electrical Properties General: Dielectric constant and loss factor have been reported [14]
Dielectric Permittivity (Constant):

No.	Value	Note
1	3.96	[14]

Optical Properties:
Optical Properties General: Refractive index has been reported [14]
Transmission and Spectra: Esr [8], ir and Raman [9] spectral data have been reported

Polymer Stability:
Biological Stability: Microbial degradation [10] and biodegradation [11] have been reported

Applications/Commercial Products:
Processing & Manufacturing Routes: Synth. by condensation polymerisation of ethylene glycol and azelaic acid at 250° under vacuum [1,5,6,7,12,15]

Bibliographic References
[1] Korshak, V.V., Vinogradova, S.V. and Vlasova, E.S., *Izv. Akad. Nauk SSSR, Otd. Khim. Nauk*, 1954, 1089, (synth, solubility)
[2] Korshak, V.V., Golubev, V.V. and Karpova, G.V., *Izv. Akad. Nauk SSSR*, 1958, 88, (solubility)
[3] Mnyukh, Y.V., *Izv. Akad. Nauk SSSR, Otd. Khim. Nauk*, 1958, 1128, (cryst struct)
[4] Tanaka, H., *CA*, 1959, **54**, 9350c, (refractive index)
[5] Orszagh, A., Czarnodola, H., Gorska, J. and Zurakowska-Orszagh, J., *CA*, 1962, **58**, 9780d, (synth, soln props)
[6] Nakatsuka, R., *Bull. Chem. Soc. Jpn.*, 1964, **37**, 403, (synth, cohesive energy density)
[7] Karpov, O.N., Fedosyuk, L.G. and Kovanko, Y.U., *Zh. Prikl. Khim. (Leningrad)*, 1969, **42**, 952, (synth, melting point)
[8] Nozawa, Y., *Bull. Chem. Soc. Jpn.*, 1970, **43**, 657, (esr)
[9] Holland-Moritz, K., *Kolloid Z. Z. Polym.*, 1973, **251**, 906, (ir, Raman)
[10] Tokiwa, Y., Ando, T. and Suzuki, T, *Hakko Kogaku Zasshi*, 1976, **54**, 603, (microbial degradation)
[11] Diamond, M.J., Freedman, B. and Garibaldi, J.A., *Int. Biodeterior. Bull.*, 1975, **11**, 127, (biodegradation)
[12] Baile, M., Chou, Y.J. and Saam, J.C., *Polym. Bull. (Berlin)*, 1990, **23**, 251, (synth)
[13] Fuller, C.S. and Erickson, C.L., *J. Am. Chem. Soc.*, 1937, **59**, 344, (cryst struct, density)
[14] Yager, W.A. and Baker, W.O., *J. Am. Chem. Soc.*, 1942, **64**, 2164, (dielectric constant)
[15] Ueberreiter, K., Karl, V.H. and Altmeyer, A., *Eur. Polym. J.*, 1978, **14**, 1045, (thermodynamic props, expansion coefficient)

[16] Ellis, T.S., *Macromolecules,* 1995, **28**, 1882, (transition temp)
[17] Fuller, C.S. and Frosch, C.J., *J. Phys. Chem.,* 1939, **43**, 323, (cryst struct)
[18] Iwakura, Y., Taneda, Y. and Uchida, S., *J. Appl. Polym. Sci.,* 1961, **5**, 108, (melting point)

Polyethylene blends   P-252

Synonyms: *Polyethylene alloys. Blended polyethylenes*
Related Polymers: PE blends with styremics, Linear low density (low pressure process) PE, Medium-density PE, Ultrahigh molecular weight PE, High-density (low pressure process) PE, General information on polyethylenes, Low-density (highpressure process) PE, PE blends with rubbers, Very low or ultralow density PE
Material class: Blends, Thermoplastic
Polymer Type: polyethylene, polyolefins
General Information: Due to the wide variation in props. between polyethylene classes and within each class there is enormous scope for further modification of props. by blending; classes within a single class (e.g. low MW HDPE with high MW HDPE); class to class (e.g. LDPE with LLDPE); and with other polymers/ elastomers e.g. with polyamides and rubbers. Immiscibility is common and compatibilisation and strength improvement may be achieved by using copolymers, grafted polymers, functionalised polyethylene, rubbers, CaO (coated/uncoated) and glass fibres. Common PE/PE blends include use of low MW HDPE waxes to improve melt flow characteristics, hardness, abrasion resistance, grease resistance of other PE grades; LLDPE blended with LDPE to give LDPE processability with LLDPE dart impact strength, puncture strength and tear strength; addition of mLLDPE to commodity LLDPE to improve sealability; addition of mLLDPE to various PE grades to improve oxygen barrier properties, toughness, clarity [1]. Rapidly growing important area [2,3,4,5,6,7,9]. With proper resin selection it is possible to pick and choose the desired specific physical props. of the contributing resin selection in order to achieve a finished (film) with very unique end use props. [9]
Morphology: Viscosity ratio, processing conditions and compatibiliser level affect size and number of EVOH layers in LDPE/ EVOH blend [10]. Partial co-crystallisation can occur between branched and linear PE [11]. Morphology of LDPE or LLDPE blended with gelatinised starch plasticised with glycerol is controlled by the starch content [12]. Moulding and storage conditions affect the morphology of starch blended with maleated HDPE, maleated LLDPE, or maleated EVA [14]. Phase morphology in HDPE/PP/PS and HDPE/PS/PMMA blends can be predicted and manipulated [13]. Morphology and its effect on props. of octene based LLDPE blended with ethylene-propylene-butylene terpolymer is given [15]. Phase structures in binary blends of any two of HDPE, LDPE, LLDPE are discussed as a basis for understanding of more complex variations [16]
Miscellaneous: TEM can detect phase separation at just below 1 mol% co-monomer (butene or octene) and DSC can detect phase separation at approx. 1.3 mol% (butene) or 1.5 mol% (octene) in linear blends with lightly branched copolymer [17]. Small angle neutron scattering has been used to investigate liquid/liquid phase separation in linear PE/branched PE blends [18]. In LLDPE blended with a given weight of LDPE, FTIR to identify the α-olefin followed by DSC to quantify the amount of α-olefin form a rapid anal. method [3]. Blends of EVA with LDPE or LLDPE and blends of Dow Affinity PE with LDPE or LLDPE are compared [6]. In blends of butene LLDPE or octene LLDPE with linear PE's, for lightly branched copolymers melt phase separation is more affected by the number of branches than their length [19]. The effect of long-chain branching in EPM in blends including EPM/PP/HDPE has been reported [20]

Volumetric & Calorimetric Properties:
Equation of State: An equation of state model based on generalised Flory theory is given which accounts for effects of MW and degree of branching on miscibility in polyolefin blends (including PE/PP, HDPE/LLDPE, HDPE/iPP) [4]
Thermodynamic Properties General: Thermal and other props. of HDPE blended with homogenous branched PE have been reported [21] also of octene based LLDPE with ethylene-propylene-butylene terpolymer [15] and UHMWPE with low MW short chain branched PE [22]

Surface Properties & Solubility:
Surface and Interfacial Properties General: Interfacial adhesion in PE/SBH blends is improved by adding the product from extended reactive blending of functionalised PE (containing free carboxylic groups) with SBH [23]. The contact angle between LLDPE blended with LLDPE-graft- maleic anhydride and water or DMF decreases with increasing grafting [7]

Transport Properties:
Polymer Melts: Effects of blending 2 miscible metallocene catalysed HDPE'S of very different MW on dynamic viscoelasticity, steady state viscosity and dimensional stability are reported [24]. Rheological props. of octene based LLDPE blended with ethylene-propylene-butylene terpolymer are given [15]. The blend components are immiscible but processability is improved if there is less than 20 wt% terpolymer. Flow of flame retardant HDPE is improved by blending with mLLDPE [25]. Viscosity of UHMWPE/low MW short chain branched PE is given [22]. Rheological props. of HDPE virgin/recycled blends are examined [26]
Permeability and Diffusion General: Permeability, esp. of oxygen, can be reduced by large amounts (740 × for 6 phr compatibiliser) by increasing the size and number of EVOH layers in LDPE/ EVOH blends [10]

Mechanical Properties:
Mechanical Properties General: Mechanical and other props. of HDPE/homogenous branched PE blends are described. [21] Blends of LDPE or LLDPE with gelatinised starch plasticised with glycerol have high elongation at high starch loading even without an interfacial modifier [12]. Blending an octene based LLDPE with less than 20 wt% of an ethylene-propylene-butylene terpolymer shows improved mechanical props. [15]. 70/30 LDPE/ m(LLD)PE gives better film props. (for heavy duty packaging) than 70/30 LDPE/LLDPE- even when downgauged [27]. mPE's use as a performance enhancing blend component with other polyolefins is also discussed elsewhere [9,28]. Orientation of UHMWPE in blends with low MW (short chain branched) PE increases tensile strength and modulus [2]. The effect of multiple extrusion cycles on blends of virgin HDPE with recycled HDPE has been analysed [26]
Mechanical Properties Miscellaneous: In maleated PE (HDPE or LLDPE) blends with starch, tensile props. increase during the first 5 weeks of storage (at 50° or -10°), tensile strength increasing more at the higher storage temp., but flexural strength remains constant [14]
Failure Properties General: Slow crack growth resistance in a blend of pipe grade branched (0.6 mol% butyl branches) PE with a linear PE was worse than for the pure branched PE despite partial co-crystallisation [11]
Fracture Mechanical Properties: The presence and distribution of the chain molecules is important in determining the fracture behaviour of binary low MW linear PE/high MW linear or branched PE [29]

Optical Properties:
Transmission and Spectra: FTIR has been used in the identification of α-olefin in LLDPE blends [3]

Polymer Stability:
Polymer Stability General: Blending of a starch grafted methyl-vinyl ketone into PE gives a photobiodegradable polymer [8] yet LDPE/corn starch blends appear less biodegradable than LDPE (under thermoxidative conditions- not uv irradiation) [31]
Thermal Stability General: LLDPE blended with mLLDPE increases heat resistance (relative to mLLDPE) [5]
Recyclability: Blending of recycled/virgin material of the same classification is often a primary method of recycling PE's [26,30]

Applications/Commercial Products:
Processing & Manufacturing Routes: Produced by extrusion mixing, solid state shear extrusion pulverisation [30]. Processing

props. may be blend dependent [12,21,26]. Processing additives may be used e.g. fluoropolymers [2]
Applications: As for 'parent class' of PE but with higher specfication, e.g. film with heat resistance comparable to LLDPE and sealability comparable to mLLDPE [5]. Reduction of defects e.g. shark skin, slipstick [24]. But note that on occasion, a 'new class' may outperform or replace existing common blend [32]

Trade name	Details	Supplier
Hiplen		HIP Petrochemija DP
Kompact	PU/PE	Ubbink Nederland BV
Microthene	various PE	Quantum Chemical
Pemuplasten	also other polyolefins	Plastenterprises Pemú Ltd.
Pionier	PE/paraffin oil compounds	Hansen and Rosenthal
Thermocomp	filled and blends	LNP Engineering
Vestowax	PE compounds and other products	Astor Wax

Bibliographic References

[1] *Kirk-Othmer Encycl. Chem. Technol.*, 4th edn., (eds. Kroschwitz, J.I. and Howe-Grant, M), Wiley Interscience, 1996, **17**, 702
[2] Herne, J., *SPE Antec 97 Conf. Proc.*, Toronto, 1997, **I**, 64, (general description, processing)
[3] Prasad, A., Mowery, D., *SPE Antec 97 Conf. Proc.*, **II**, 2310, (general description, misc., spectra)
[4] Mehta, S.D., Honnell, K.G., *SPE Antec 97 Conf. Proc.*, **II**, 2648, (general description, equation of state)
[5] Tomatsuri, T., Sekine, N., Furusawa, N., *SPE Antec 97 Conf. Proc.*, **III**, 2670, (general description, stability)
[6] Patel, R.M., Saavedra, P., deGroot, J., Hinton, C., Guerra, R., *SPE Antec 97 Conf. Proc.*, (general description, misc.)
[7] Sanchez-Valdez, S., Guerrero-Salazar, C., Rames de Valle, L.P., Lopez-Quintanilla, M., Yanez-Hores, I., Orona-Villarreal, F., Ramirez-Vargas, R., *SPE Antec 97 Conf. Proc.*, **II**, 2526, (general description, surface props.)
[8] Hwu, H.-D., Lin, H.-M., Jiang, S, -F., Chen, J.-H., *SPE Antec 97 Conf. Proc.*, **II**, 2026, (general description, stability)
[9] Brandenburg, J.S., Davis, M., *SPE Antec 97 Conf. Proc.*, Houston, (Corporate eds. Schotland Business Research Inc.), 1996, 213, (general description, mechanical props.)
[10] Lee, S.Y., Kim, S.C., *Polym. Eng. Sci.*, 1997, **37**, 463, (morphology, permeability)
[11] Trankner, T., Hedenqvist, M., Gedde, U.W., *Polym. Eng. Sci.*, 1997, **37**, 346, (morphology, failure props.)
[12] St. Pierre, N, Favis, B.D., Ramsay, B.A., Ramsay, J.A., Verhoogt, H., *Polymer*, 1997, **38**, 647, (morphology, mechanical props.)
[13] Guo, H.F., Packirisamy, S., Gvozdic, N.V., Meier, D.J., *Polymer*, 1997, **38**, 785, (morphology)
[14] Ramkumar, D.H.S., Bhattacharya, M., *J. Mater. Sci.*, 1997, **32**, 2565, (morphology, mechanical props.)
[15] Cho, K., Ahn, T.K., Lee, B.H., Choe, S., *J. Appl. Polym. Sci.*, 1997, **63**, 1265, (morphology, volumetric props., rheological props., mechanical props.)
[16] Mandelkern, L., Alamo, R.G., Wignall, G.D., Stehling, F.C., *Trends Polym. Sci.*, 1996, **4**, 377, (morphology)
[17] Hill, M.J., Morgan, R.L., Barham, P.J, *Polymer*, 1997, **38**, 3003
[18] Crist, B., *Polymer*, 1997, **38**, 3145
[19] Morgan, R.L., Hill, M.J., Barham, P.J., Frye, C.J., *Polymer*, 1997, **38**, 1903
[20] Brignac, S.D., Young, H.W., 1996, **22**, 11, (misc.)
[21] Schellenberg, J., *Adv. Polym. Technol.*, 1997, **16**, 135, (volumetric props., mechanical props., processing)
[22] Jen-Taut Yeh, Yu-Lang Lin, Chien-Cheng Fan-Chiang, *Macromol. Chem. Phys.*, 1996, **197**, 3531, (volumetric props., rheological props., mechanical props.)
[23] Minkova, L.I., Miteva, T., Sek, D., Kaczmarczyk, B., Magagnini, P.L., Paci, M., La Mantia, F.P., Scaffaro, R., *J. Appl. Polym. Sci.*, 1996, **62**, 1613, (interfacial props.)
[24] Munoz-Escalona, A., Lafuente, P., Vega, J.F., Munoz, M.E., *Polymer*, 1997, **38**, 589, (rheological props., processing)
[25] Huggard, M., *J. Fire Sci.*, 1996, **14**, 393, (rheological props., mechanical props.)
[26] Zahavich, A.T.P., Latto, B., Takacs, E., Vlachopoulos, J., *Adv. Polym. Technol.*, 1997, **16**, 11, (rheological props., mechanical props., recyclability, processing)
[27] *Packaging Week*, 1997, **12**, 3, (mechanical props.)
[28] Hartzmann, G., Plesske, P., *SPO 96 Conf. Proc.*, Houston, (Corporate eds. Schotland Business Research Inc.), 1996, 495, (mechanical props.)
[29] Hedenqvist, M., Conde Brana, M.T., Gedole, U.W., Martinez-Salazar, J., *Polymer*, 1996, **37**, 5123, (fracture props.)
[30] Nesarikar, A.R., Carr, S.H., Khait, K., Mirabella, F.M., *J. Appl. Polym. Sci.*, 1997, **63**, 1179, (recyclability, processing)
[31] Erlandsson, B., Karlsson, S., Albertson, A.C., *Polym. Degrad. Stab.*, 1997, **55**, 237, (stability)
[32] Lee, M., *Eur. Plast. News*, 1997, **24**, 23, (processing)

Polyethylene blends with polystyrenes

Synonyms: *PE/PS blends. Polyethylene blends with styrenics*
Related Polymers: Low-density polyethylene, Polystyrene, HDPE
Material class: Blends
Polymer Type: polyethylene, polystyrene
Additives: Compatibilisers (especially ethylene-styrene block copolymers and interpolymers)
Morphology: Phase struct. and miscibility depend greatly on the mixing equipment [2] and method of mixing [4]; the compatibiliser and its nature [1,3,5,6]

Transport Properties:

Polymer Melts: Relative viscosity of low-density polyethylene/ polystyrene blends shows marked negative deviation from unity [5]

Mechanical Properties:

Mechanical Properties General: Tensile props. depend strongly on matrix orientation and dispersed phase morphology and orientation. [4] Tensile modulus shows a negative deviation from additivity rule which is greatest if SEBS compatibiliser is present [4]

Tensile (Young's) Modulus:

No.	Value	Note
1	1000–4000 MPa	approx., 100% polystyrene [4]

Flexural Modulus:

No.	Value	Note
1	250 MPa	approx., 50–100% LDPE [5]
2	1800 MPa	approx., 100% polystyrene [5]

Tensile Strength Break:

No.	Value	Note
1	20 MPa	approx., 100% HDPE-50% HDPE/ 50% polystyrene [4]
2	30–45 MPa	100% polystyrene [4]

Bibliographic References

[1] Champagne, M.F. and Dumoulin, M.M., *SPE Antec 97 Conf. Proc.*, 1997, **2**, 2562, (morphology)
[2] Yu, D.-W., Esseghir, M. and Curry, J., *SPE Antec 97 Conf. Proc.*, 1997, **2**, 181, (morphology)
[3] Park, C.P. and Clingerman, G.P., *Plast. Eng. (N.Y.)*, 1997, **53**, 97, (morphology)
[4] Bureau, M.N., El Kadi, H., Denault, J. and Dickson, J.I., *Polym. Eng. Sci.*, 1997, **37**, 377, (morphology, mech. props)
[5] Caze, C. and Pak, S.H., *J. Mater. Sci. Lett.*, 1997, **16**, 53, (morphology, rheology, mech props)
[6] McCoy, M., *Chem. Mark. Reporter*, 1996, **250**, 5, (morphology)

Polyethylene blends with rubbers P-254

Related Polymers: Polyethylene blends
Material class: Blends, Synthetic Elastomers

Transport Properties:
Polymer Melts: Melt viscosity increases with rubber content [1,2]. At rubber concentrations of 10% and 30% blends are compatible. The type of rubber and the blend ratio may affect props. [1]. For HDPE blended with recycled tyre rubber, viscosity decreases with increasing shear rate and increases with rubber content [2]

Mechanical Properties:
Mechanical Properties General: Mechanical props are improved if the rubber is partially cross-linked [1]. For recycled tyre rubber, tensile and tear strengths are higher if the components are co-pulverised than if the powdered components are mixed [3]. In a tri-blend of LDPE/LLDPE/EPDM a double yield phenomenon is observed [4]

Bibliographic References
[1] Rodriguez-Fernandez, O.S., Sanchez-Valdez, S., Ramirez-Vargas, R., Yanes-Hores, I.G. and Lopez-Quintanilla, M.L., *SPE Antec 97 Conf. Proc.*, 1997, **2**, 2636, (rheology, mechanical props.)
[2] Li, Y., and Lin, P., *SPE Antec 97 Conf. Proc.*, 1997, **1**, 1082, (rheology)
[3] Khait, K., *ACS Division of Polymer Chemistry Meeting L.A. J. Polym. Prepr.*, 1996, 9, (mechanical props.)
[4] Plaza, A.R., Ramos, E., Manzur, A., Olayo, R. and Escobar, A., *J. Mater. Sci.*, 1997, **32**, 549

Poly(ethylene-*co*-carbon monoxide) P-255

$$\left[\left[CH_2CH_2\right]_x \left[\begin{array}{c}O\\\|\\C\end{array}\right]_y\right]_n$$

Synonyms: ECO. *Ethylene copolymer with carbon monoxide*
Monomers: Ethylene, Carbon monoxide
Material class: Copolymers
Polymer Type: polyethylene, polyolefinspolyketones
CAS Number:

CAS Reg. No.	Note
25052-62-4	
111190-67-1	alternating

Molecular Formula: $[(C_2H_4)_x(CO)_y]_n$
General Information: While moderately extensive work has been done [1] on ECO particularly because of its biodegradability there are, as yet, few major applications. [2] Polymers containing low levels of CO have props. similar to those of LDPE, but are highly sensitive to photodegradation. 1:1 Copolymers have higher melting points and improved mech. props.
Morphology: Good uniformity is obtained in autoclave reactors. At approx. 50 mol. % CO, the composition is repeating C_2H_4CO units leading to a crystal lattice and high melting point, higher than that of polyethylene. Crystal repeat unit 0.25 nm until ethylene:carbon monoxide ratio reaches approx. 1.1:1 when repeat unit changes to approx. 0.75 nm corresponding to 2 CH_2CH_2CO units in a linear zigzag form [1]

Volumetric & Calorimetric Properties:
Latent Heat Crystallization: ΔH_{fusion} 2.5 kJ mol^{-1}; ΔS_{fusion} 4.8 J mol^{-1} K^{-1}. Values are lower than corresponding values for polyethylene
Melting Temperature:

No.	Value	Note
1	125°C	approx., 38–40% CO
2	242°C	49.8% CO [1]
3	111°C	27.5% CO
4	122°C	35.1% CO
5	126°C	38.7% CO
6	147°C	40.6% CO
7	181°C	43.9% CO
8	203°C	46.3% CO
9	235°C	49.0% CO [4]

Transition Temperature General: T_m rises linearly with CO content

Polymer Stability:
Polymer Stability General: ECO polymers are photodegradable because light absorption by ketone groups causes chain scission [1]
Decomposition Details: Decomposes via chain scission to yield alkene and ketone fragments [4]
Biological Stability: Is biodegradable

Applications/Commercial Products:
Processing & Manufacturing Routes: Single stage autoclave can be used to give polymer with up to approx. 20 mol. % CO with good uniformity. Max. temp. is lower than for a tubular reactor and degradation is minimised. Polymers with up to 12 mol. % CO have been produced in tubular reactors. Terpolymers with vinyl acetate or alkyl acrylates have been produced
Applications: Used for photodegradable requirements e.g. drink can carriers. Terpolymers used for foam, and as PVC plasticisers

Bibliographic References
[1] Doak, K.W., *Encycl. Polym. Sci. Eng.*, 2nd edn., Vol. 6, (ed. J.I. Kroschwitz.), Wiley Interscience, 1986, 386, (rev)
[2] Pebsworth, L.W., *Kirk-Othmer Encycl. Chem. Technol.*, 4th edn., Vol. 17, (eds. J.I. Kroschwitz and M. Howe-Grant), Wiley Interscience, 1996, 707, (rev)
[3] Boysen, R.L., *Kirk-Othmer Encycl. Chem. Technol.*, 3rd edn., Vol. 16 (eds. M. Grayson and D. Eckroth), Wiley Interscience, 1981, 402, (rev)
[4] Starkweather, H., *Encycl. Polym. Sci. Eng.*, 2nd edn., Vol. 10 (ed. J.I. Kroschwitz), Wiley Interscience, 1987, 369, (thermal props.)

Poly(3,4-ethylenedioxypyrrole) P-256

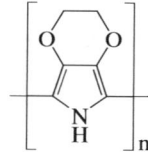

Synonyms: *2,3-Dihydro-6H-1,4-dioxino[2,3-c]pyrrole homopolymer*. PEDOP. PEDP
Monomers: 3,4-Ethylenedioxypyrrole
Material class: Thermoplastic
Polymer Type: polypyrrole
CAS Number:

CAS Reg. No.
259737-85-4

Molecular Formula: $(C_6H_5NO_2)n$
Fragments: $C_6H_5NO_2$
Additives: Various dopants such as sulfonate, phosphate, perchlorate or iodine can be added or incorporated at the polymerisation stage.
General Information: Stable conductive polymer with electrochromic properties.

Electrical Properties:
Electrical Properties General: Polymer has a very low half-wave potential for oxidation (p-doping) and band gap of *ca*. 2 eV [1].

Typical conductivity values in oxidised form of 40 S cm^{-1} (CF$_3$SO$_3$ dopant) [2] and 15.5 S cm^{-1} (perchlorate dopant) [3] have been reported.

Optical Properties:
Optical Properties General: In neutral state polymer is red in colour and changes in oxidised state to transmissive blue/grey colour [1,5]. UV absorption with λ_{max} in the region of 505nm recorded [3], while in oxidised form absorption is at longer wavelength. Electrochromic effects have been recorded at different voltages [1].

Polymer Stability:
Polymer Stability General: Polymer in oxidised form generally has high stability and is not susceptible to biological reducing agents such as glutathione and Cleland's reagent [4].

Applications/Commercial Products:
Processing & Manufacturing Routes: Prod. from the monomer by electrochemical oxidation in a suitable solvent such as acetonitrile or propylene carbonate [1,3,5], or by chemical polymerisation with oxidising agents such as Fe^{3+}, Cu^{2+} or even air [6].
Applications: Polymer can be used for printer toners [7] or in electrochromic devices [8]. In a blend with poly(styrene sulfonate) it can be used as an electrically conducting antistatic layer for photographic films [9].

Bibliographic References
[1] Gaupp, C.L., Zong, K., Schottland, P., Thompson, B.C., Thomas, C.A. and Reynolds, J.R., *Macromolecules*, 2000, **33**, 1132
[2] Giurgiu, L., Zong, K., Reynolds, J.R., Lee, W.-P., Brenneman, K.R., Saprigin, A.V., Epstein, A.J., Hwang, J. and Tanner D.B., *Synth. Met.*, 2001, **119**, 405
[3] Zotti, G., Zecchin, S., Schiavon, G. and Groenendaal, L., *Chem. Mater.*, 2000, **12**, 2996
[4] Thomas, C.A., Zong, K., Schottland, P. and Reynolds, J.R., *Adv. Mater.*, 2000, **12**, 222
[5] Schottland, P., Zong, K., Gaupp, C.L., Thompson, B.C., Thomas, C.A., Giurgiu, L., Hickman, R., Abboud, K.A. and Reynolds, J.R., *Macromolecules*, 2000, **33**, 7051
[6] Sönmez, G., Schottland, P., Zong, K. and Reynolds, J.R., *J. Mater. Chem.*, 2001, **11**, 289
[7] *U.S. Pat.*, 2002, 6 387 581
[8] *Pat. Coop. Treaty (WIPO)*, 2003, 03 46 106
[9] *U.S. Pat.*, 1997, 5 665 498

Poly(3,4-ethylenedioxythiophene)   P-257

Synonyms: *PEDOT. PEDT*
Monomers: 3,4-Ethylenedioxythiophene
Polymer Type: polythiophene
CAS Number:

CAS Reg. No.
163359-60-2

Molecular Formula: $(C_6H_4O_2S)_n$
Fragments: $C_6H_4O_2S$
General Information: A low band gap polymer which is black or dark blue when neutral but light blue, almost transparent, after doping [1,3]
Morphology: Electrochemically produced films are smooth and homogeneous [6]. The doped polymer is amorph. [1]. Doping is accompanied by a rapid contraction followed by a gradual expansion. This is attributed to structural changes associated with the doping process [6]

Electrical Properties:
Electrical Properties General: Possesses exceptionally high electrical stability. Can be *n*-doped resulting in conductivity 200 S cm^{-1}. Surface resistance is not affected by atmospheric humidity [9]. Conductivity of the polymer varies with the oxidising agent used in its preparation: conductivity 5–10 S cm^{-1} (FeCl$_3$, 20°, MeCN), 31 S cm^{-1} (FeCl$_3$, 180°, benzonitrile), 200 S cm^{-1} (anodic oxidation, MeCN), 40 S cm^{-1} (anodic oxidation, H$_2$O) [10]. Surface resistivity 200–1000000 Ω [9]. Volume resistivity 0.005–0.2 (conductivity value) [10]
Magnetic Properties: Doped polymers exhibit Curie-like susceptibility below -198° and Pauli-like susceptibility above this temp. [4]

Optical Properties:
Optical Properties General: Low band gap; exhibits significant electrochromic props. Optical density 0.01 (max.) [9]. Optical band gap 16 eV (approx., 760–780 nm). Onset of optical absorption approx. 1.5 eV, maximum optical absorption 2.2 eV [7]
Transmission and Spectra: XANES and XPS spectral data [3] have been reported. Has a strong absorption in the visible region but after doping is quite transparent in the uv/visible region and shows strong absorption in near infrared range [1,6]

Polymer Stability:
Polymer Stability General: Superior thermal stability compared to other alkylthiophene polymers
Thermal Stability General: Has good thermal stability [6]; the resistance varies little on standing at 110° or for short periods (less than 10 min.) at 200°. After 70 min. the resistance starts to increase. Doped polymer is stable up to 150° [1,4]
Decomposition Details: Thermal degradation leads to loss of the thiophene struct. within the chain; conversion into sulfates takes place (accompanied by pairing with a suitable counter-ion) [3]. Onset of degradation of the doped polymer at approx. 150°. Major decomposition occurs between 390° and 450° [4]
Chemical Stability: Has good chemical stability [6]
Hydrolytic Stability: Is more resistant than polypyrrole to hydrol. [6,9]

Applications/Commercial Products:
Processing & Manufacturing Routes: Prepared by electro or FeCl$_3$-catalysed polymerisation of the monomer [1], which may be obtained indirectly from diethyl oxalate [2]. The blue form is prepared as a tetrachloroferrate which stabilises the cationic nature of the chain [3]. Films containing doping agents are prepared by anodic oxidation [4]. Electropolymerisation in aq. media using anionic surfactants to increase monomer solubility has been reported [5]. Doping is a three stage process [8]. Processable colloidal solns. can be produced by the chemical polymerisation of 3,4-ethylenedioxythiophene in aq. solns. of polystyrenesulfonic acids [9,10]
Applications: Has been mainly used as light-emitting diodes to date. Also has various potential uses in optical and electrooptical devices. Used in antistatic transparent films, and in electroplating for metallisation of insulators. Other uses as electrodes in solid electrolyte capacitors and in electrochromic devices

Trade name	Supplier
Baytron	Bayer Inc.

Bibliographic References
[1] Pei, Q., Zuccarello, G., Ahlskog, M. and Inganas, O., *Polymer*, 1994, **35**, 1347, (synth)
[2] Heywang, G. and Jonas, F., *Adv. Mater.*, 1992, **4**, 116, (synth)
[3] Winter, I., Reese, C., Hormes, J., Heywang, G. and Jonas, F., *Chem. Phys. Lett.*, 1995, **194**, 207, (XANES, XPS)
[4] Kiebooms, R., Aleshin, A., Hutchison, K. and Wudl, F., *J. Phys. Chem. B*, 1997, **101**, 11037, (thermal props., electromagnetic props.)
[5] Sakmeche, N., Aaron, J.J., Fall, M., Aeiyach, S. *et al*, *Chem. Comm.*, 1996, 2723, (synth.)
[6] Carlberg, C., Chen, X. and Inganas, O., *Solid State Ionics*, 1996, **85**, 73
[7] Gustafsson, J. C., Liedberg, B. and Inganas, G., *Solid State Ionics*, 1994, **69**, 145, (spectra)
[8] Chen, X. and Inganas, O., *J. Phys. Chem.*, 1996, **100**, 15202

[9] Lerch, K, Jonas, F. and Linke, M., *J. Chim. Phys.*, 1998, **95**, 1506, (props., uses)
[10] Jonas, F. and Heywang, G., *Electrochim. Acta*, 1994, **39**, 1345, (uses)

Poly(3,4-ethylenedioxythiophene)/poly(4-styrenesulfonate) blend
P-258

Synonyms: *Ethenylbenzenesulfonic acid homopolymer compound with 2,3-dihydrothieno[3,4-b]-1,4-dioxin homopolymer.* PEDOT-PSS. PEDOT:PSS. PEDOT/PSS. PEDT-PSS
Monomers: 2,3-dihydrothieno[3,4-*b*]-1,4-dioxin
Material class: Thermoplastic
Polymer Type: polythiophene, polystyrene
CAS Number:

CAS Reg. No.
155090-83-8

Molecular Formula: $(C_6H_4O_2S)n.(C_8H_8O_3S)n$
Fragments: $C_6H_4O_2S$ $C_8H_8O_3S$
General Information: A polymer blend of poly(3,4-ethylenedioxythiophene) (PEDOT) and polystyrenesulfonic acid (PSS), which can be regarded as a doped form of positively charged PEDOT with PSS as the counterion (usually in excess). The polymer has good hole conducting properties, with PEDOT acting as the charge carrier. Commercially produced as aqueous colloidal dispersions that can be cast into films. The aqueous dispersions are dark blue in colour, while the films are transparent or very pale blue.
Morphology: In general, the morphology has a considerable effect on conductivity and is also influenced by addition of solvent to the dispersions. The films cast from colloidal dispersions consist of grains of conjugated polymer random coils, PEDOT chains being attached ionically to PSS chains with some neutral species such as PSSH between the grains [1]. PSS predominates on the surface [2]. There is an interaction between the chains on addition to the dispersions of a solvent such as sorbitol [3] or glycerol [4] leading to a rearrangement of the polymer chains and enhanced clustering into PEDOT domains. Addition of ethylene glycol may change the polymer chains from a random coil conformation to a linear or expanded coil structure with PEDOT changing from a benzenoid to quinoid structure [5]. Another possible result from the addition of DMSO is a screening effect and reduction of coulombic attraction between PSS counterion and PEDOT charge carrier [6]. With an increase in content of PSS in the blend, the more water is taken up and the films become more hygroscopic [7].

Surface Properties & Solubility:
Solubility Properties General: Polymer blend is in the form of colloidal dispersions in water; the films can be redispersed in water at 100°C [5]. Insoluble in common organic solvents [8]. The addition of ethylene glycol to dispersions makes the films insoluble in water [5]

Electrical Properties:
Electrical Properties General: Films cast from Baytron P dispersions have conductivities in the region of $0.05-10$ S cm^{-1}, with minimum surface resistivity of 300Ω/square and PEDOT work function of about 5.2 eV [9]. An increased proportion of PEDOT in the blend results in greater conductivity [1,15]. The effect on conductivity of the addition of solvents to the dispersions has been widely reported. Addition of ethylene glycol increased conductivity in films from 0.4 to $160-200$ S cm^{-1} [5,10], while with DMSO conductivity increased from 0.8 to 80 S cm^{-1} [6]. Glycerol [4] and sorbitol [1,3] also increased conductivity. A water rinse of films removes surface PSS and increases conductivity by 50% [11]. Addition of NaOH or CsOH to the dispersions reduced the work function of PEDOT from about 5.1 to 4 eV [12]. The effect of heat [7] and hydrochloric acid [13] on conductivity and sheet resistance of films has been reported.
Magnetic Properties: Magnetic susceptibility dependence on temperature has been studied [6].

Optical Properties:
Optical Properties General: The films have good transparency with no significant absorption up to 900nm. Electrochromic behaviour from electrochemical doping is observed with an absorption maximum (dark blue colour) at about 600nm [14].
Transmission and Spectra: XPS spectroscopy has been widely used to study morphology [1,2,6,13,15,16]. Raman [5,10,13,17], ftir [4,13,17], esr [5] and uv/visible [5,14] spectral data also reported.

Polymer Stability:
Thermal Stability General: The films are generally stable on heating in air up to 1000h with no change in conductivity [7]. No significant changes in composition and conductivity were observed on heating to 250°C [13].
Chemical Stability: Treatment of films with hydrochloric acid followed by heating leads to loss of polystyrenesulfonic acid [13]. Any PSS present as Na salt is acidified to PSSH with subsequent loss of free acid [2,16].
Hydrolytic Stability: The films are redispersed by hot water at 100°C [5]. Water treatment reduces the surface content of PSS [11].
Stability Miscellaneous: UV light induces photooxidation of PEDT with a loss of conductivity [15]. The stability of polymer films deposited on indium-tin oxide (ITO) anodes has also been studied, with indium ions detected in the polymer structure [13,15,21].

Applications/Commercial Products:
Processing & Manufacturing Routes: Baytron P dispersions are produced by oxidation of ethylenedioxythiophene monomer in aqueous polystyrenesulfonic acid using $Na_2S_2O_8$ as oxidising agent [8,9,18]. Powders have been produced by Fe(*III*) oxidation of monomer and PSS in MeCN/H_2O mixtures [19]. Electrochemical oxidation of monomer and PSS has also been reported [20]
Applications: Of increasing importance is use of the polymer blend in light-emitting diodes as a hole-injection layer in the form of a thin film between the anode, usually indium-tin oxide (ITO), and an emissive polymer [15,17,21,22]. Other applications include antistatic layers for photographic films [8], photovoltaic devices [4,23], electrochromic windows [14], capacitors and high-power lithium batteries [15].

Trade name	Details	Supplier
Baytron P	various grades of aqueous dispersions	H.C. Starck (Bayer)

Bibliographic References
[1] Jönsson, S.K.M., Birgerson, Crispin, X., Greczynski, G., Osikowicz, W., Denier van der Gon, A.W., Salaneck, W.R. and Fahlman, M., *Synth. Met.*, 2003, **139**, 1

[2] Greczynski, G., Kugler, Th. and Salaneck, W.R., *Thin Solid Films,* 1999, **354,** 129
[3] Timpanaro, S., Kemerink, M., Touwslager, F.J., De Kok, M.M. and Schrader, S., *Chem. Phys. Lett.,* 2004, **394,** 339
[4] Snaith, H.J., Kenrick, H., Chiesa, M. and Friend, R.H., *Polymer,* 2005, **46,** 2573
[5] Ouyang, J., Xu, Q., Chu, C.-W., Yang, Y., Li, G. and Shinar, J., *Polymer,* 2004, **45,** 8443
[6] Kim, J.Y., Jung, J.H., Lee, D.E. and Joo, J., *Synth. Met.,* 2002, **126,** 311
[7] Huang, J., Miller, P.F., de Mello, J.C., de Mello, A.J. and Bradley D.D.C., *Synth. Met.,* 2003, **139,** 569
[8] Groenendaal, L., Jonas, F., Freitag, D., Pielartzik, H. and Reynolds, J.R., *Adv. Mater.,* 2000, **77,** 2255
[9] Starck, H.C., Baytron - Conducting Chemistry http://www.baytron.com/index,
[10] Ouyang, J., Chu, C.-W., Chen, F.-C., Xu, Q. and Yang, Y., *Adv. Funct. Mater.,* 2005, **15,** 203
[11] DeLongchamp, D.M., Vogt, B.D., Brooks, C.M., Kano, K., Obrzut, J., Richter, C.A., Kirillov, O.A. and Lin, E.K., *Langmuir,* 2005, **21,** 11480
[12] Weijtens, C.H.L., van Elsbergen, V., de Kok, M.M. and de Winter, S.H.P.M., *Organic Electronics,* 2005, **6,** 97
[13] Nguyen, T.P., Le Rendu, P., Long, P.D. and De Vos, S.A., *Surf. Coat. Technol.,* 2004, **180/1,** 646
[14] Heuer, H.W., Wehrmann, R. and Kirchmeyer, S., *Adv. Funct. Mater.,* 2002, **12,** 89
[15] Crispin, X., Marciniak, S., Osikowicz, W., Zoffi, G., Denier van der Gon, A.W., Louwet, F., Fahlman, M., Groenendaal, L., de Schryver, F. and Salaneck, W.R., *J. Polym. Sci., Part B: Polym. Phys.,* 2003, **41,** 2561
[16] Greczynski, G., Kugler, Th., Keil, M., Osikowicz, W., Fahlman, M. and Salaneck W.R., *J. Electron. Spectrosc. Relat. Phenom.,* 2001, **121,** 1
[17] Ouyang, J., Chu, C.-W., Chen, F.-C., Xu, Q. and Yang, Y., *J. Macromol. Sci., Part A: Pure Appl. Chem.,* 2004, **41,** 1497
[18] Louwet, F., Groenendaal, L., Dhaen, J., Manca, J., Van Luppen, J., Verdonck, E. and Leenders, L., *Synth. Met.,* 2003, **135/6,** 115
[19] Lefebvre, M., Qi, Z., Rana, D. and Pickup, P.G., *Chem. Mater.,* 1999, **11,** 262
[20] Li, G. and Pickup, P.G., *Phys. Chem. Chem. Phys.,* 2000, **2,** 1249
[21] de Jong, M.P., van Ijzendoorn, L.J. and de Voigt, M.J.A., *Appl. Phys. Lett.,* 2000, **77,** 2255
[22] Cao, Y., Yu, G., Zhang, C., Menon, R. and Hoeger, A.J., *Synth. Met.,* 1997, **87,** 171
[23] Zhang, F., Johansson, M., Andersson, M.R., Hummelen, J.C. and Inganäs, O., *Adv. Mater.,* 2002, **14,** 662

Poly(ethylene-*co*-ethyl acrylate)   P-259

Related Polymers: Polyethylene, Poly(ethyl acrylate)
Monomers: Ethylene, Ethyl acrylate
Material class: Natural elastomers, Copolymers
Polymer Type: polyethylene, acrylic
CAS Number:

CAS Reg. No.
9010-86-0

Molecular Formula: $[(C_2H_4).(C_5H_8O_2)]_n$
Fragments: C_2H_4 $C_5H_8O_2$
Molecular Weight: The MW can range from 1000–200000
Additives: Diphenylguanidine, methylenedianiline and N774
General Information: Rubbery, tough and flexible. The polymers are low melting resins, suitable for hot melt adhesives to polyethylene-like products. Ethyl acrylate is about 15% to 30% of the copolymer composition and it contributes flexibility and polarity to the polymer. The hot melt adhesives have a unique combination of props., including high shear fail temps., toughness at low temps. and excellent adhesion to non-polar substrates. When compounded with other polymers it helps to improve the low temp. toughness and stress cracking props. When compounded with carbon black it is used for microchip packaging and a variety of hospital uses where static electricity presents a potential hazard. This polymer with 8% ethyl acrylate is approved by the FDA for food contact applications

Volumetric & Calorimetric Properties:
Density:

No.	Value	Note
1	d 0.93 g/cm³	18% ethyl acrylate

Melting Temperature:

No.	Value
1	152°C

Vicat Softening Point:

No.	Value
1	116°C

Transport Properties:
Melt Flow Index:

No.	Value	Note
1	6 g/10 min	18% ethyl acrylate

Polymer Solutions Dilute: η_{inh} 0.81 (18% ethyl acrylate)

Applications/Commercial Products:
Processing & Manufacturing Routes: Resins are produced by free radical polymerisation in specially modified high pressure polyethylene reactors. With the increase in ethyl acrylate content the copolymer becomes more flexible, tougher and more resilient. It is usually sold as an unmodified pelletised resin, but it can also be the base resin for specially highly filled compds., such as flame retardant compds. and semiconductive materials for wire coating. The copolymers can be processed on standard extrusion, blow moulding or injection moulding equipment used for LDPE of similar melt index. Typical extrusion melt temp. is 178° (320°F). The range of injection moulding temp. varies between 155° (280°F) and 222° (400°F) depending on melt index and the ethyl acrylate content. These copolymers usually do not require the use of additives for processing, whether it is extrusion, injection moulding or blow moulding
Applications: Used as a polymer modifier, in hot-melt adhesives and sealants, flexible hose and tubing, laminate sheets, multilayer films, injection moulded and extruded parts and wire and cable applications. It is compatible with almost all other olefin polymers. It is compounded with polyolefins, polyamides and polyesters to produce specifically required high modulus polymers while retaining much of its original props.

Bibliographic References
[1] *Plastics Reference Handbook,* McGraw-Hill, 1994, 38
[2] Bailor, F.V., *Rubber World,* 1979, **180,** 42

Polyethylene fibres   P-260

Synonyms: *PE fibres*
Related Polymers: Polyethylene, LDPE, HDPE, Polypropylene fibre, UHMWPE
Monomers: Ethylene
Material class: Thermoplastic
Polymer Type: polyethylene, polyolefins
CAS Number:

CAS Reg. No.	Note
9002-88-4	all density types

Molecular Formula: $(C_2H_4)_n$
Fragments: C_2H_4

Molecular Weight: MW 20000–1000000; M_w/M_n 2–15
Additives: Hindered amine light stabilisers; substituted hindered phenol thermal stabilisers
General Information: Although fibres can be formed from several polyolefins, polyethylene fibres are very much of secondary importance to polypropylene fibres, and no others are of practical importance. Early fibres of LDPE, lacked dimensional stability, abrasion resistance, resilience, light stability. Improvements were obtained when MDPE, became available. Current interest centres on performance fibres mainly of UHMPE, [4,5,7] and blends. [8] Structure contains fewer short chain branches
Morphology: Drawing increases orientation. In the case of gel spun UHMWPE, fibre, this can be virtually complete. [4] Linear density is also reduced during drawing. High temp. annealing of spun fibre under tension gives a hard elastic fibre of strength comparable to drawn fibre but with substantially increased elastic recovery. The hard elastic fibres are thought to deform through opening of the stacked lamellae structure producing voids. [1] In high modulus UHMWPE fibre, longitudinal tension enhances orthorhombic to pseudohexagonal transformation while axial tension inhibits or prevents the transformation. [4,5]

Volumetric & Calorimetric Properties:
Volumetric Properties General: Density values have been reported .[6] Thermal expansion coefficient is higher in a direction than b direction for ultra high modulus UHMWPE fibres, enhancing transformation from orthorhombic to psuedohexagonal struct. [4]

Mechanical Properties:
Mechanical Properties General: Props tend to be far inferior to theoretical values. Super drawing, high pressure extrusion, spinning of liquid crystalline polymers, gel spinning and hot drawing have been used to obtain improvement in mech. props. [1]
Tensile Strength Break:

No.	Value	Note	Elongation
1	76 MPa	commercial fibre	
2	2600 MPa	high strength commercial fibre, diameter 0.038 mm [1,2]	3.5%

Elastic Modulus:

No.	Value	Note
1	117000 MPa	high strength commercial fibre, diameter 0.038 mm [1,2]

Mechanical Properties Miscellaneous: Resilience 70% (min., 40°); 90% (approx., -40°, 120°) [1,3]

Polymer Stability:
Polymer Stability General: Stability is generally as for the type of polyethylene used in the fibre
Flammability: May be reduced by application of a fabric finish.

Applications/Commercial Products:
Processing & Manufacturing Routes: Processed by melt spinning, melt blowing, spun bonding, spurting, [1] gel spinning [4,9,10,11,12,13,14,15,16,17]
Applications: Applications include high strength, low weight packaging (spunbond PE fibre), high modulus fibre (UHMWPE), bicomponent fibres (eg with PP, for melt bonding), meltblown fabrics, felted fabric and paper making (spurted fibre). All olefin fibres are of most use where low moisture absorption (hence good stain resistance and almost equal wet/dry props) and/or low density outweigh the relatively poor oxidation stability and light stability.

Trade name	Details	Supplier
Dyneema	ultra high modulus	Toyo Boseki

Bibliographic References
[1] Wust, C.J. and Landoll, L.M., *Kirk-Othmer Encycl. Chem. Technol.*, 4th edn. (eds. J.I. Kroschwitz and M. Howe-Grant) Wiley Interscience, 1993, **10**, 639
[2] Wills, A.F., Capaccio, G. and Ward, I.M., *J. Polym. Sci., Polym. Phys. Ed.*, 1980, **18**, 493
[3] Bryant, G.M., *Text. Res. J.*, 1967, **37**, 552
[4] You-Lo Hsieh, Xiao-Ping Hu, *J. Polym. Sci., Polym. Phys. Ed.*, 1997, **35**, 623
[5] Tashiro, K., Sasaki, S. and Kobayashi, M., *Macromolecules*, 1996, **29**, 7460
[6] Shambaugh, R.L and Bansal, V., Annu. Tech. Conf. - Soc. Plast. Eng., *55th*, Toronto, 1997, 1677
[7] Mead, J., Gabriel, R., Murray, T., Foley, G. and McMorrow, J., Annu. Tech. Conf. - Soc. Plast. Eng., *55th*, Toronto, 1997, 1722
[8] Chen, J.C. and Harrison, I.R., Annu. Tech. Conf. - Soc. Plast. Eng., *55th*, Toronto, 1997, 2621
[9] Jen-Taut Y., Yu-Lang L., Chien-Cheng F.-C., *Macromol. Chem. Phys.*, 1996, **197**, 3531
[10] *U.S. Pat.*, 1987, 4 643 865
[11] Ohta, T., Okada, F., Hayashi, M. and Mihoichi, M., *Polymer*, 1989, **30**, 2170
[12] Kanamoto, T., Ooki, T., Tanaka, K. and Takeda, M., *Technol. Rep. Osaka Univ.*, 1983, **32**, 741
[13] Smith, P., Chanzy, H.D. and Rotzinger, B.P., *Polym. Commun.*, 1985, **26**, 258
[14] *PCT Int. Appl.*, 1987, 8 703 288
[15] Smith, P., Chanzy, H.D. and Rotzinger, B.P., *J. Mater. Sci.*, 1987, **22**, 523

Polyethylene, High density P-261
Synonyms: High-density polyethylene. PE-HD. Low pressure polyethylene. HDPE
Related Polymers: Polyethylene, HDPE film, HDPE blends, LLDPE
Monomers: Ethylene, 1-Butene, 1-Hexene
Material class: Thermoplastic
Polymer Type: polyethylene, polyolefins
CAS Number:

CAS Reg. No.	Note
9002-88-4	all density types

Molecular Formula: $(C_2H_4)_n$
Fragments: C_2H_4
Molecular Weight: M_n 10000–1000000 (typical commercial values; waxes may be lower and high/ultra high MW polymers higher)
Additives: Stabilisers, uv stabilisers, flame retardants, antistatic additives (especially carbon black)
General Information: The most molecularly regular polyethylene grade. Approximates to structural unit with five Me groups per 1000 carbon atoms. Metallocene catalysts are used to improve chain regularity. Virtually no branches or a small number of branches introduced by copolymerisation with another α-olefin. The number (or degree) of branches varies from 0.5–10 branches per 1000 carbon atoms. One chain end is a Me group and the other is either a Me or a vinyl group. The degree of branching, MW and the MW distribution influence most physical and mechanical props. [2,3]
Morphology: Is semicryst. with crystallinity between 40 and 80%. Principal cryst. form orthorhombic, a 0.740, b 0.493, c 0.2534 nm. The pseudo monoclinic form is also known, which is stable at temps. below 50°; annealing at 80–100° restores the orthorhombic form (monoclinic, a 0.405, b 0.485, c 0.254 nm, $\alpha = \beta = 90°$, γ 105°). Chains have a flat zigzag configuration folded every 5–15 nm. These form planar lamellae, the plane of which they are perpendicular to. Interlamellar tie molecules give mech. strength and rigidity, forming the lamellae into fibrils. Fibrils spread outwards from central points to form spherulites of 1–10 μm diameter. Amorph. regions between lamellae give flexibility and high-impact strength [2,3]. Most HDPE articles exhibit a degree of molecular and crystalline orientation [30]. In uniaxial orientation the c-axis alignment with the stretch direction can approach 100%. In the second orientation mode the a-axis aligns to the film

— Polyethylene, High density

machine direction [31]. Effect of processing variables and MW distribution on extruded tubular HDPE film morphology has been reported [52,53]. Deformation of spherulites has been reported [43,57]

Miscellaneous: Due to the customisable nature of polyethylenes, and differences between manufacturing processes, data given here are overall ranges. Specific manufacturer's grades normally have closer tolerances; with certain processes, they may even be available outside the ranges quoted [4]

Volumetric & Calorimetric Properties:
Density:

No.	Value	Note
1	d^{25} 0.94–0.97 g/cm^3	
2	d 0.94–0.96 g/cm^3	Ziegler-Natta process
3	d 0.96–0.97 g/cm^3	Philipps process [1,2,3,5]
4	d 0.964 g/cm^3	[8]
5	d 1 g/cm^3	orthorhombic
6	d 0.965 g/cm^3	monoclinic

Thermal Expansion Coefficient:

No.	Value	Note	Type
1	5.9×10^{-5}–0.00011 K^{-1}	homo and copolymers, ASTM D696,	L
2	4.8×10^{-5} K^{-1}	30% glass fibre reinforced, ASTM D696 [5]	L
3	0.00012–0.00018 K^{-1}	23–80°, DIN 53752 [14]	L

Equation of State: Mark-Houwink equation parameters have been reported [3,38,39,41,42,43,44,45,46,47,48,49,50]
Thermodynamic Properties General: Other thermodynamic props. have been reported [11,35,36,37]. Variation of thermal expansion coefficient with temp. has been reported [11]
Latent Heat Crystallization: ΔH°_{fusion} 280–300 J g^{-1} (100% crystalline) [2]
Thermal Conductivity:

No.	Value	Note
1	0.46–0.5 W/mK	0.00155 cal (cm s °C)$^{-1}$, homopolymer
2	0.42 W/mK	0.001 cal (cm s °C)$^{-1}$
3	0.36–0.46 W/mK	0.0008–0.0011 cal (cm s °C)$^{-1}$, 30% glass fibre reinforced, ASTM C177 [5]
4	0.35–0.43 W/mK	DIN 52612 [14]

Specific Heat Capacity:

No.	Value	Note	Type
1	1.7–1.9 kJ/kg.C	ISO 1006 [14]	P

Melting Temperature:

No.	Value	Note
1	130–137°C	homopolymer [5]
2	122–127°C	rubber modified [5]
3	125–135°C	copolymers
4	120–140°C	glass fibre reinforced [5]
5	135°C	crystallinity 70–90%, unbranched PE [5]
6	146–147°C	extrapolated equilibrium melting point for orthorhombic HDPE [2,26]

Glass-Transition Temperature:

No.	Value	Note
1	-140–-100°C	[2,27,28,29]

Transition Temperature General: T_g not clearly defined due to high crystallinity; usually associated with γ transition. Brittleness temp. is related. T_m for commercial resins decreases linearly with branching; 1° per branch per 1000 carbon atoms [2]
Deflection Temperature:

No.	Value	Note
1	79–91°C	175–196°F, 0.455 MPa (66 psi), homopolymer [5]
2	65–80°C	149–176°F, 0.455 MPa (66 psi), copolymers [5]
3	127–129°C	260–265°F, 0.455 MPa (66 psi) [5]
4	121°C	250°F, 1.84 MPa (264 psi), 30% glass fibre reinforced [5]
5	121–127°C	250–260°F, 0.455 MPa (66 psi) [5]
6	115–121°C	240–250°F, 1.84 MPa (264 psi), 20–30% glass fibre reinforced, ASTM D648 [5]
7	41–49°C	1.8 MPa, ISO 75 [14]

Vicat Softening Point:

No.	Value	Note
1	124–133°C	VST/A/50
2	72–80°C	VST/B/50, DIN/ISO 306 [14]

Brittleness Temperature:

No.	Value	Note
1	-140–-100°C	[2]

Surface Properties & Solubility:
Solvents/Non-solvents: Sol. (above 80°) aromatic hydrocarbons, aliphatic hydrocarbons, halogenated hydrocarbons, xylenes, tetralin, decalin, 1,2,4-trimethylbenzene, o-dichlorobenzene, 1,2,4-trichlorobenzene [2]; binary systems comprising carbon disulfide with xylene, cyclohexane, C_6H_6, toluene, CCl_4, THF or $CHCl_3$ [56]. Insol. all known single solvents at room temp.

Transport Properties:
Transport Properties General: Difficulties and apparent contradictions arise because of MW and molecular struct. effects. [9,10]. Means of overcoming these effects have been reported [59]
Melt Flow Index:

No.	Value	Note
1	0.1–1.8 g/10 min	ISO 1133–1991, 2.16 kg, medium MW, blow moulding and extrusion grades [14]
2	1.4–26 g/10 min	ISO 1133–1991, 2.16 kg, injection and rotational moulding grades [14]
3	2–10 g/10 min	ISO 1133–1991, 21.6 kg (condition 7), high MW blow moulding grades [14]
4	0.001–500 g/10 min	overall range, melt flow ratio 25–150 [2]

Polymer Solutions Dilute: Viscosity number 135–540 ml g^{-1} (0.001 g ml^{-1} decalin, ISO 1628) [14]

Polymer Melts: Rheological props. of conventional and metallocene catalysed HDPE have been compared [9,10]. New melt flow information showing a sharp minimum in flow resistance with a narrow temp. window has been reported [56]. The die orifice, die and barrel cleanliness, temp. and sample mass are the more crucial parameters affecting melt index variability. [60] Wall slip effects have been studied using sliding plate rheometer and capillary rheometer [61], or rectangular conduit [62]. A superfluid-like stick-slip transition occurs above a critical stress at temps. of 180–260°. The amount of slip is a material property which decreases with decreasing MW [63]. Viscosity before transition and the stick-slip transition are MW dependent [63]. If MW exceeds 1000000 a flow curve discontinuity occurs at 148–152°. Flow is a maximum, surface roughness and periodic bulk distortions are absent but melt fracture is always present [55,58,64,65,66]. Elsewhere between T_m and 160° (normal lower limit temp. for processing) flow induced crystallisation to a highly oriented solid occurs [55,67]. The nature of chain branching strongly affects melt strength and flow props. [9,10,68,69]. Effect of viscous heating in capillary rheometry [70] and of non-linear flow on melt mixing [71] has been reported. Low MW mHDPE acts as a processing aid for higher MW mHDPE, the $G'G''$ correlation corresp. to complete miscibility. Dynamic viscoelasticity and steady state viscosity are also reported for this blend [72]. Effects of fillers on rheology are reported for cross-linked HDPE granules [73] and pulverised recycled tyre rubber [74].

Permeability of Gases: Permeabilities of a range of gases have been reported [8,12]

Water Absorption:

No.	Value	Note
1	0.01 %	max., 24h, ⅛ in thick, homopolymer and copolymers
2	0.02–0.06 %	24h, ⅛ in thick, glass fibre reinforced, ASTM D570 [5]
3	0.01 %	max., saturation, 23°, DIN 53495/1L [14]

Gas Permeability:

No.	Gas	Value	Note
1	He	75.3 cm³ mm/(m² day atm)	8.6×10^{-14} cm² (s Pa)⁻¹, 25° [8,12]
2	O_2	26.3 cm³ mm/(m² day atm)	3×10^{-14} cm² (s Pa)⁻¹, 25° [8,12]
3	Ar	113.8 cm³ mm/(m² day atm)	1.3×10^{-13} cm² (s Pa)⁻¹, 25° [8,12]
4	CO_2	23.6 cm³ mm/(m² day atm)	2.7×10^{-14} cm² (s Pa)⁻¹, 25° [8,12]
5	CO	13.13 cm³ mm/(m² day atm)	1.5×10^{-14} cm² (s Pa)⁻¹, 25° [8,12]
6	N_2	9.63 cm³ mm/(m² day atm)	1.1×10^{-14} cm² (s Pa)⁻¹, 25° [8,12]
7	CH_4	25.4 cm³ mm/(m² day atm)	2.9×10^{-14} cm² (s Pa)⁻¹, 25° [8,12]
8	NH_3	700 cm³ mm/(m² day atm)	8×10^{-13} cm² (s Pa)⁻¹, 20° [8,13]
9	H_2O	788 cm³ mm/(m² day atm)	9×10^{-13} cm² (s Pa)⁻¹, 25° [8,32]
10	propane	76.2 cm³ mm/(m² day atm)	8.7×10^{-14} cm² (s Pa)⁻¹, 25° [8,12]

Mechanical Properties:

Mechanical Properties General: Mech. props. (e.g. bending modulus) [25] are affected by the thermal history of the sample and its corresponding morphology. Mech. props. are improved by biaxial orientation [4]

Tensile (Young's) Modulus:

No.	Value	Note
1	1070–1090 MPa	155–158 kpsi, homopolymer, ASTM D638 [5]
2	620–895 MPa	90–130 kpsi, low-medium MW copolymers, ASTM D638 [5]
3	940 MPa	136 kpsi, high MW copolymers, ASTM D638 [5]
4	4825–6205 MPa	700–900 kpsi, 30% glass fibre reinforced, ASTM D638 [5]

Flexural Modulus:

No.	Value	Note
1	1000–1550 MPa	145–225 kpsi, 23°, homopolymer, ASTM D790 [5]
2	830–1240 MPa	120–180 kpsi, 23°, copolymer, ASTM D790 [5]
3	4830–5515 MPa	700–800 kpsi, 23°, 30% glass fibre reinforced, ASTM D790 [5]

Tensile Strength Break:

No.	Value	Note	Elongation
1	22–31 MPa	3200–4500 psi, homopolymer [5]	10–1200%
2	16–20 MPa	2300–2900 psi, rubber modified [5]	600–700%
3	21–45 MPa	3000–6500 psi, low-medium MW copolymer [5]	10–1300%
4	17–30 MPa	2500–4300 psi, high MW copolymer [5]	170–800%
5	52–62 MPa	7500–9000 psi, 30% glass fibre reinforced [5]	1.5–2.5%
6	48–59 MPa	7000–8500 psi, 20–30% long glass fibre reinforced, ASTM D638 [5]	2.0–2.5%

Flexural Strength at Break:

No.	Value	Note
1	76–83 MPa	11000–12000 psi, 30% glass fibre reinforced [5]
2	55–65 MPa	8000–9500 psi, 20–30% long glass fibre reinforced, ASTM D790 [5]

Tensile Strength Yield:

No.	Value	Elongation	Note
1	26–33 MPa		3800–4800 psi, homopolymer [5]
2	10–18 MPa		1400–2600 psi, rubber modified [5]
3	18–29 MPa		2600–4200 psi, low-medium MW copolymer [5]
4	19–27 MPa	8–15% elongation	2800–3900 psi, high MW copolymer, ASTM D638 [5]

– Polyethylene, High density

Compressive Strength:

No.	Value	Note
1	19–25 MPa	2700–3600 psi, homopolymer, low-medium MW copolymer [5]
2	41–48 MPa	6000–7000 psi, 30% glass fibre reinforced [5]
3	34–41 MPa	5000–6000 psi, 20–30% long glass fibre reinforced, ASTM D695 [5]

Miscellaneous Moduli:

No.	Value	Note	Type
1	250–400 MPa	1000h, elongation ≤0.5%, 23°, ISO 899 [14]	Tensile Creep modulus
2	550–1000 MPa	ISO 6721-2 [14]	Shear modulus

Impact Strength: Tensile impact resistance 30–700 kJ m^{-2} (23°, ASTM D1822, ISO 8256/1B). Charpy impact values over a range of temps. and annealing conditions have been reported [6]
Hardness: Shore D66-D73 (homopolymer); D55–D60 (rubber modified) ; D58–D70 (low-medium MW copolymers); D63–D65 (high MW copolymers, ASTM D2240) [5]. Ball indentation hardness H 132/30 35–65 MPa (DIN/ISO 2039-1) [14]
Failure Properties General: Bottle test-time to failure 10–4000h (300 ml/22 g test bottle, internal pressure 2 bar, 5% aq. Lutensol FSA 10 soln., 50°) [14]
Fracture Mechanical Properties: Non-linear-elastic thermal decohesion model for impact fracture correlates well with notched Charpy specimens [15,21]
Izod Notch:

No.	Value	Notch	Note
1	21–210 J/m	Y	0.4–4 ft lb in^{-1}, 1/8 in, homopolymer [5]
2	19–320 J/m	Y	0.35–6 ft lb in^{-1}, 1/8 in, copolymers [8]
3	170–240 J/m	Y	3.2–4.5 ft lb in^{-1}, 1/8 in, high MW copolymers [5]

Electrical Properties:
Electrical Properties General: Data given for electrically unmodified polymers. Up to 50% aluminium powder [22] or 40% carbon black [23] are added for antistatic/conductivity props. Changes in mech. props. also occur. Value of dissipation factor depends on temp. and struct. [11]
Surface/Volume Resistance:

No.	Value	Note	Type
2	>0.1 × 10^{15} Ω.cm	min., ISO 93, DIN 0303-30 [5,14]	S

Dielectric Permittivity (Constant):

No.	Value	Frequency	Note
1	2.3–2.6	100 Hz–1 GHz	ASTM D150, ISO 250, DIN 0303-4 [5,14]

Dielectric Strength:

No.	Value	Note
1	118 kV.mm^{-1}	3000 V mil^{-1}, 5 mil thick [5]
2	>100 kV.mm^{-1}	min., 3800 V mil^{-1}, ISO 243/1 [14]
3	17.71–19.68 kV.mm^{-1}	450–500 V mil^{-1}, 1/8 in, short time, ASTM D149 [5]
4	1968 kV.mm^{-1}	5000 V mil^{-1}, 1 mil thick [5]

Complex Permittivity and Electroactive Polymers: Comparative Tracking Index CTI 600 (test soln. A); CTI 600M (test soln. B) (ISO 112/A, DIN 0303-1) [14]
Dissipation (Power) Factor:

No.	Value	Frequency	Note
1	0.0005	1 kHz–1 GHz	ASTM D150 [5]
2	<0.0002	100 Hz–1 MHz	max., ISO 250, DIN 0303-4 [14]

Optical Properties:
Refractive Index:

No.	Value	Note
1	1.53–1.54	25°, decreases with increasing branching [3]

Transmission and Spectra: Films are opaque if thick, translucent if thin [2,3]. Uv transmission limit approx. 230 nm
Total Internal Reflection: Specific birefringence 0.059 (crystalline phase); 0.28 (amorph. phase) [3]

Polymer Stability:
Polymer Stability General: Is non-polar, and largely chemically and environmentally inert. Unless used in multi-material products, HDPE is readily recyclable or useable as a fuel [16]
Thermal Stability General: (-0.7)–(-3.0)% change in linear dimensions at 100° for 30 min. [5]
Upper Use Temperature:

No.	Value	Note
1	90°C	50% loss of tensile strength after 20000h, ISO 216-1 [14]

Decomposition Details: At 225° small quantities of fumes evolved. Decomposition and oxidative pyrolysis occur at 300°. Heat of reaction may cause rapid temp. rise and accelerate decomposition. CO, H_2O, CO_2, aldehydes and hydrocarbons are evolved [2,14]. In inert atmosphere, pyrolysis yields waxes and hydrocarbons above 500° [2]
Flammability: Above 300° decomposition products may ignite. Burning may be accompanied by the release of flaming molten droplets. Flammability rating HB (0.4 mm, 1–6 mm thick, ISO UL94) [14]
Environmental Stress: Light of wavelength <400 nm initiates photooxidative degradation. Initial products accelerate the process. Final products are similar to those of thermal oxidation. Environmental stress cracking resistance has been reported [14]
Chemical Stability: May be swollen by non-polar reagents.; is less susceptible than lower density types. Can be nitrated at room temp. or decomposed by nitric acid/sulfuric acid mixtures at temps of 100–150° [2]
Hydrolytic Stability: Has good resistance to acids and alkalis [14,17,18]. Reacts only slowly with sulfuric acid at elevated temp.
Recyclability: Readily recyclable [20] either by itself or in blends with other polyethylenes. Some applications (*e.g.* bottle crates) employ closed loop recycling [16]. Some PE tradenames refer exclusively to recycled material. Oil-containing vessels may be recycled [19]

Applications/Commercial Products:
Processing & Manufacturing Routes: An immense variety of catalysts can be used for low pressure PE production but only the following are of current commercial interest. Philips catalyst, based on Cr^{VI} compounds supported on silica or similar. Temp. 85–110°. activity approx. 300–1000 kg HDPE per g Cr. MW and

MW distribution are controlled by temp. and catalyst composition/preparation. M_w/M_n approx. 6–12, MFR 90–120. Organo-chromium catalyst activity 800–1000 kg HDPE per g Cr Ziegler catalysts - Ti or V compound with organoaluminium cocatalyst, may be supported or not. Temp. 70–110°, 0.1–2 MPa, using inert liq. (e.g hexane, isobutane) or gas phase. MW typically high. Hydrogen used as chain transfer agent to reduce MW; metallocene catalysts [33] most recent commercial catalyst based on complex of Zr, Ti or Hf (usually containing 2 cyclopentadienyl rings) plus an organoaluminium or perfluorinated boron aromatic cocatalyst. MW of polymer very sensitive to hydrogen. Polymer is very regular; bimodal catalysts (combinations of catalyst) may be used to obtain broad MW distribution. Slurry processing is the oldest method and gives full range of HDPEs from low to ultra-high MW; approx. 60% world production is from the slurry process. Uses loop, continuous stirred or liq. pool reactors. Temp. 70–110°, 3–4.5 MPa, 0.5–1.5h, conversion rate up to 98%; Gas phase process (Unipol) can accommodate a wide variety of solid and supported catalysts using a fluidised bed of dry PE particles. Temp. 70–80°, 2 MPa; Solution process, in heavy solvent or molten PE (PE sol. saturated hydrocarbons (C_6-C_9) above 120°). Used mostly for LMWPE as soln. viscosity increases rapidly with MW of PE. Temp. 150–200°, 5–10 MPa, 5–10 min. (solvent) or 170–350°, 30–200 MPa, ≤ 1 min. [34] (molten PE). The solution process was the first to use metallocene catalysts. Bimodal processes using series reactors may be used for broad MW distribution. Typical processing temp. range 150–250° at which most HDPE behaviour non-Newtonian; effective viscosity is reduced by increasing melt flow speed. Most conventional techniques may be used. Injection moulding (temp. 200–260°, 70–140 MPa, typical cycle time 10–30s) [53]; Blow moulding is largest volume technique for HDPE; uses high MW, high melt viscosity resins; Rotational moulding uses a speed of rotation 10–40 rpm [51]; extrusion (temp. approx. 150°, 40–50 MPa) [58]. Blown film: continuous film 0.007–0.125 mm thick prod. by extrusion through a circular die (diameter 40–100 cm, gap approx. 1 mm) usually on 'high stalk' film lines. Blow-up ratios up to 4:1 used. Post consumer HDPE may be reprocessed by the same methods. Can be electron beam cross-linked [24]

Mould Shrinkage (%):

No.	Value	Note
1	1.6–2.3%	longitudinal
2	1.2–2.4%	transverse, free
3	1.3–1.4%	transverse, restricted [14]
4	1.5–4%	0.015–0.04 in in^{-1}, homopolymers
5	1.2–4%	0.012–0.04 in in^{-1}, copolymers, ASTM D955 [5]
6	1.2–1.3%	longitudinal

Applications: Main applications are in hollow articles (blown bottles, containers) and in film and pipe (growth rate for these three approx. 2.5–4% per annum). Also used in injection moulding (growth rate approx. 1% per annum)

Trade name	Details	Supplier
Alathon	homo and copolymers also includes EVA, EPR	OxyChem
Alathon		Alathon
Alathon		Transmare BV
Alcudia	also used for other densities	Repsol Quimica
Alcudia		Petroplas
Atochem		Paklink Ltd.
Cabelec	modified/carbon black filled	Cabot Plastics Ltd.
Conductomer HDC-22H	22% carbon black-conductive	Synthetic Rubber Technologies
Ecothene		Quantum Chemical
Electrafil PE	carbon black filled - conductive	Akzo
Eltex	also PP and ethene-propene copolymers	Solvay
Empee	also LDPE, flame retardant	Monmouth Plastics
Eraclene		Enichem
Eraclene		Ashland
Eucalene	also LDPE	Van Raaijen Kunststoffen
Finathene	also LDPE	Fina Chemicals (Division of Petrofina SA)
Fortiflex	includes HMW HDPE	Solvay
HDPE ***	copolymer, includes HMW HDPE and glass filled	Dow
HDPEX	auto cross-linkable	Borealis Compounds AB
HiD	homo and copolymer	Chevron Chemical Co.
Hilex		Royalite Plastics Ltd.
Hiplex		HIP Petrochemija DP
Hobolene		Riddle and Hobbs Ltd.
Hostalen	name covers HD, MD, HMW, UHMW, as well as PP and EVA copolymers	Hoechst (UK) Ltd
Hostalen		Tinto Plastics
Lacqtene	also includes LDPE, LLDPE	Elf Atochem (UK) Ltd.
Lacqtene		Ashland
Lupolen	includes most types of PE as well as several other polymers	BASF
Lupolen		Northern Industrial Plastics Ltd.
Luwax	includes homo and copolymer, also LDPE and oxidised PE	BASF
Marlex	includes LDPE, MDPE and copolymers	Philips Petroleum International
Mirathen		Anilac NV/SA
Mitsubishi	includes other types and other products	Mitsubishi Chemical
Novapol	includes LLDPE, LDPE, linear MDPE and copolymers	Novacor
Octowax	wax emulsion	Tiarco
Okulen		Christian Berner Oy
PEX-C	electron beam cross-linked	Mercury Plastics, Ohio
Pemuplasten		Plastenterprises Pemú Ltd.

— Polyethylene, High density

Petrothene	covers most grades of PE homo and copolymers as well as EVA, PP, EPR	Quantum Chemical
Plastin	film	4P Folie Forchheim GmbH.
Polybleu	drinking water grade	Meridionale des Plastiques SA
Polybond	also includes various copolymers	Uniroyal
Polycere		BP Chemicals
RTP	10% glass filled, also used for other products	RTP Company
Repsol	also LDPE and PP	Commercial Quimica Insular SA
Rigidex	also MDPE and various ethylene copolymers	BP Chemicals
Rigidex		Hellyar Plastics Ltd.
Sabic	also LDPE	NV Chemach SA
Scolefin		Poole and Davidson Plastics
Scolefin		Anilac NV/SA
Stamylan	also LDPE, LLDPE, UHMWPE copolymers and PP	DSM
Stamylan		Macrodan A/S
Stamylan		Ashland
Stat-Kon	electrically conducting composite, also other products	LNP Engineering
Stat-Kon	electrically conducting composite, also other products	ICI Chemicals and Polymers Ltd.
Statoil	includes LDPE, PP copolymers and other products	Northern Industrial Plastics Ltd.
Sumiplene	also LDPE	Sumica SA
Suntec	also LDPE and EVA	Asahi Chem. Co Ltd
Super Naltene	PE sheet and rolls	Martyn Industrials
Tecnodure	also includes PP and blends, PC and blends, PS, ABS	Campos 1925 SA
Tipelin	also hexene-ethene copolymer, iPP	Tiszai Vegyi Kombinát
Tipplen	also PP and ethene-propene copolymers	Tiszai Vegyi Kombinát
Tipplen	also PP and ethene-propene copolymers	Alfordshire Ltd.
Valeron	film	Vauleer Packaging Systems
Vestolen	narrow MW distribution, specials also includes PB, PP, EPR	Hüls AG
Zemid		DuPont

Bibliographic References

[1] Alger, M., *Polymer Science Dictionary*, 2nd edn., Chapman and Hall, 1997
[2] Kissin, Y.V., *Kirk-Othmer Encycl. Chem. Technol.*, Vol. 17, 4th edn., (eds. J.I. Kroschwitz and M. Howe-Grant), Wiley Interscience, 1996, 724, (rev)
[3] Beach, D.L. and Kissin, Y.V., *Encycl. Polym. Sci. Eng.*, 2nd edn., (ed. J.I. Kroschwitz), Wiley Interscience, 1986, **6**, 454, (rev)
[4] Prins, A.J., Kortschof, M.T. and Woodhams, R.T., *Polym. Eng. Sci.*, 1997, **37**, 261, (mech props)
[5] Guide to Plastics, *Property and Specification Charts*, McGraw-Hill, 1987, 73, 93, 150, (thermal props, mech props, electrical props)
[6] Pauly, S., *Polym. Handb.*, 3rd edn., (eds. J. Brandrup and E.H. Immergut), John Wiley and Sons, 1989, **VI**, 435
[7] Vega, J.F., Muñoz-Escalona, A., Santamaria, A., Muñoz, M.E. and Lafuente, P., *Macromolecules*, 1996, **29**, 960
[8] Carella, J.M., *Macromolecules*, 1996, **29**, 8280
[9] Zhu, L., Chiu, F.C., Fu, Q., Quirk, R.P. and Cheng, S.Z.D., *Polym. Handb.*, 4th edn., (eds. J. Brandrup, E.H. Immergut and E.A. Grulke), John Wiley and Sons, 1999, **V**, 9
[10] Michaels, A.S. and Bixler, H.J., *J. Polym. Sci.*, 1961, **50**, 413
[11] Braunisch, V.L. and Lenhart, H., *Kolloid Z. Z. Polym.*, 1961, **177**, 24
[12] Lupolen High Density Polyethylene, BASF, (technical datasheet)
[13] Leevers, P.S. and Morgan, R.E., *Eng. Fract. Mech.*, 1995, **52**, 999
[14] Kaps, R., Lecht, R. and Schulte, U., *Kunststoffe*, 1996, **86**, 12
[15] Marshall, D., *Eur. Plast. News*, 1996, **23**, 33
[16] *Plast. South. Afr.*, 1997, **26**, 18
[17] *Plastics Recycling Update*, 1997, **10**, 3
[18] Zahavich, A.T.P., Latto, B., Takacs, E. and Vlachopoulos, J., *Adv. Polym. Technol.*, 1997, **16**, 11
[19] Leevers, P.S., *Polym. Eng. Sci.*, 1996, **36**, 2296
[20] Tavman, I.H., *J. Appl. Polym. Sci.*, 1996, **62**, 216
[21] Mather, P.J. and Thomas, K.M., *J. Mater. Sci.*, 1997, **32**, 401
[22] Schut, J.H., *Plast. World*, 1997, **55**, 12
[23] Maxwell, A.S., Unwin, A.P., Ward, I.M., Abo el Maaty, M.I. et al, *J. Mater. Sci.*, 1997, **32**, 567
[24] Hay, J.N., *Polymer*, 1981, **22**, 718
[25] Gaur, U. and Wunderlich, B., *Macromolecules*, 1980, **13**, 445
[26] Lam, R. and Geil, P.H., *J. Macromol. Sci.*, *Phys.*, 1981, **20**, 37
[27] Dechter, J.J., Axelson, D.E., Dekmezian, A., Glotin, M., and Mandelkern, L., *J. Polym. Sci., Polym. Phys. Ed.*, 1982, **20**, 641
[28] *Int. J. Polym. Mater.*, 1993, **20**, 1
[29] Keller, A. and Sanderman, I., *J. Polym. Sci.*, 1955, **15**, 31, 133, 137
[30] Myers, A.W., Meyer, J.A., Rogers, C.E., Stannett, V. and Szwarc, M., *Tappi*, 1961, **44**, 58
[31] Sinn, H. and Kaminsky, W., *Adv. Organomet. Chem.*, 1980, **18**, 99
[32] Choi, K.-Y. and Ray, W.H., *J. Macromol. Sci.*, *Rev. Macromol. Chem. Phys.*, 1985, **25**, 57
[33] Gaur, U. and Wunderlich, B., *J. Phys. Chem. Ref. Data*, 1981, **10**, 119
[34] Pyda, M. and Wunderlich, B., *Polym. Handb.*, 4th edn., (eds. J. Brandrup, E.H. Immergut and E.A. Grulke), John Wiley and Sons, 1999, **VI**, 483
[35] Wunderlich, B., Macromolecular Physics, Vol. 3, *Crystal Melting*, Academic Press, 1980
[36] Francis, P.S., Cooke, R. and Elliott, J.M., *J. Polym. Sci.*, 1958, **31**, 453
[37] Henry, P.M., *J. Polym. Sci.*, 1959, **36**, 3
[38] Chiang, R., *J. Phys. Chem.*, 1965, **69**, 1945
[39] Keii, T., *Kinetics of Ziegler-Natta Polymerisation*, Kodansha, 1972
[40] Tung, L.M., *J. Polym. Sci.*, 1957, **24**, 333
[41] De La Cuesta, M.O. and Billmeyer, J.W., *J. Polym. Sci., Part A: Polym. Chem.*, 1963, **1**, 1721
[42] Wilson, T.P. and Hurley, C.F., *J. Polym. Sci., Polym. Chem. Ed.*, 1963, **1**, 281
[43] Duch, E. and Kuechler, L., *Z. Elektrochem.*, 1956, **60**, 218
[44] Wesslau, H., *Makromol. Chem.*, 1958, **26**, 96
[45] Kaufmann, H.S. and Walsh, E.K., *J. Polym. Sci.*, 1957, **26**, 124
[46] Stacy, C.J. and Arnett, R.L., *J. Polym. Sci., Part A: Polym. Chem.*, 1964, **2**, 167
[47] Yau, W.W., Kirkland, J.J. and Bly, D.D., *Modern Size-Exclusion Liquid Chromatography*, John Wiley and Sons, 1979
[48] Rodionov, A.G., Domareva, H.M., Baulin, A.A., Ponomareva, E.L. and Ivanchev, C.C., *Vysokomol. Soedin., Ser. A*, 1981, **23**, 1560
[49] Yu, T.-H. and Wilkes, G.L., *Polymer*, 1996, **37**, 4675
[50] Yu, T.-H. and Wilkes, G.L., *J. Rheol. (N.Y.)*, 1996, **40**, 1079
[51] Butcher, J., *Plast. World*, 1997, **55**, 28
[52] Wang, H., Cao, B., Jen, C.K., Nguyen, K.T., and Viens, M., *Polym. Eng. Sci.*, 1997, **37**, 363
[53] Kolnaar, J.W.H. and Keller, A., *J. Non-Newtonian Fluid Mech.*, 1996, **67**, 213
[54] Vidyarthi, N., and Palit, S.R., *J. Polym. Sci., Polym. Chem. Ed.*, 1980, **18**, 3315
[55] Matsuo, M., and Xu, C., *Polymer*, 1997, **38**, 4311
[56] Kolnaar, J.W.H., and Keller, A., *Polymer*, 1997, **38**, 1817

Poly(ethylene isophthalate) — P-262

Synonyms: *Poly(oxyethyleneoxyisophthaloyl). Poly(oxy-1,2-etha-nediyloxycarbonyl-1,3-phenylenecarbonyl)*
Monomers: Isophthalic acid, Ethylene glycol
Material class: Thermoplastic
Polymer Type: unsaturated polyester
CAS Number:

CAS Reg. No.	Note
26948-62-9	repeating unit
26810-06-0	from isophthalic acid
68103-50-4	from dimethyl isophthalate

Molecular Formula: $(C_{10}H_8O_4)_n$
Fragments: $C_{10}H_8O_4$
Morphology: Two polymorphic forms known [3,21] α-form c 1,48, β-form c 2.1 [3]. Struct. studies have been reported [18]. α-form triclinic a 0.520, b 0.708, c 1.48, α 109°, β 136°, γ 96° [21]

Volumetric & Calorimetric Properties:
Density:

No.	Value	Note
1	d^{25} 1.346 g/cm^3	amorph. [2,11]
2	d^{25} 1.358 g/cm^3	cryst. [2]
3	d^0 1.3596 g/cm^3	[22]

Thermal Expansion Coefficient:

No.	Value	Note	Type
1	0.00013 K^{-1}	T < 51° [22]	V
2	0.000524 K^{-1}	T > 51° [22]	V

Latent Heat Crystallization: ΔH_β 54 kJ mol^{-1} (13 kcal mol^{-1}) [4]. E_{act} (T_g) 283 kJ mol^{-1} (70 kcal mol^{-1}) [10]
Melting Temperature:

No.	Value	Note
1	240°C	[2]
2	143°C	[3]
3	102–108°C	[15]

Glass-Transition Temperature:

No.	Value	Note
1	61°C	334K [10]
2	68°C	[11]
3	51°C	[22]

Transition Temperature General: Other transition temps. have been reported [19]
Transition Temperature:

No.	Value	Note	Type
1	110°C	[2]	Softening point
2	-50°C	[11]	β-transition temp.

Surface Properties & Solubility:
Solvents/Non-solvents: Sol. tetrachloroethane [3], 5% trichloroacetic acid [5], CH_2Cl_2 [7], DMF, DMSO, 1-methyl-2-pyrrolidinone, *m*-cresol [12,15], H_2SO_4 [15]. Insol. Me_2CO, 2-butanone, C_6H_6, toluene, THF, chlorobenzene, nitrobenzene, 1,4-dioxane, cyclohexanone [12], *n*-heptane [3], MeOH [7], H_2O [8], CH_2Cl_2 [12], $CHCl_3$ [12], *n*-hexane [12]

Transport Properties:
Polymer Solutions Dilute: [η] 0.33 dl g^{-1} (phenol/tetrachloroethane (60:40)) [2]; [η]$_{inh}$ 0.07 dl g^{-1} (1-methyl-2-pyrrolidinone, 30°) [8]; [η]$_{inh}$ 0.09 dl g^{-1} (1-methyl-2-pyrrolidinone, 30°) [12]; [η] 0.75 dl g^{-1} (phenol/tetrachloroethane, (60:40), 25°) [22]. Other η values have been reported [3]
Permeability of Gases: Diffusion coefficient [O_2] 2.7×10^{-7} mm^2 s^{-1} (1.2 mm film) [11]
Gas Permeability:

No.	Gas	Value	Note
1	O_2	0.98 cm^3 mm/(m^2 day atm)	2.5 cm^3 mil 100 in^{-2} day atm^{-1} [11]
2	CO_2	2.52 cm^3 mm/(m^2 day atm)	0.0384×10^{-10} cm^2 (s cmHg)$^{-1}$, 1.2 mm film [11]

Mechanical Properties:
Complex Moduli: Complex dynamic modulus from -170–200° has been reported [23]

Electrical Properties:
Electrical Properties General: Volume resistivity [20] and electrical conductivity [16] values have been reported. Other dissipation power values have been reported. [1,11,20] Dielectric constant (82–94°) has been reported [4]
Dissipation (Power) Factor:

No.	Value	Note
1	0.03	-30° [1]

Optical Properties:
Transmission and Spectra: Ir [3,12], fluorescence [9], H-1 nmr [5,12,13], C-13 nmr [7,14] and mass spectral [8] data have been reported

Polymer Stability:
Decomposition Details: Thermal degradation produces cyclic oligomers [8]. Polymer decomposition temp. 445° [8]
Biological Stability: Biodegradation studies have been reported [17]

Applications/Commercial Products:
Processing & Manufacturing Routes: Synth. by transesterification of dimethyl isophthalate with ethylene glycol at 240° followed by condensation polymerisation at 240° under vacuum [2,3,6,7,8]

Bibliographic References
[1] Kawaguchi, T., *J. Polym. Sci.*, 1958, **32**, 417, (power factor)
[2] Conix, A. and Vankerpel, R., *J. Polym. Sci.*, 1959, **40**, 521, (synth, mp, density)
[3] Yamadera, R. and Sonoda, C., *J. Polym. Sci., Part B: Polym. Lett.*, 1965, **3**, 411, (ir, cryst struct)
[4] Yamafuji, K. and Ishida, Y., *Kolloid Z. Z. Polym.*, 1962, **183**, 15, (dielectric constant)
[5] Murano, M., Kancishi, Y. and Yamadera, R., *J. Polym. Sci., Part A: Polym. Chem.*, 1965, **3**, 2698, (H-1 nmr)
[6] Ogata, N. and Okamoto, S., *J. Polym. Sci., Polym. Chem. Ed.*, 1973, **11**, 2537, (synth)
[7] Kricheldorf, H.R., *Makromol. Chem.*, 1978, **179**, 2133, (C-13 nmr, synth)
[8] Foti, S., Giuffrida, M., Maravigna, P. and Montaudo, G., *J. Polym. Sci., Polym. Chem. Ed.*, 1984, **22**, 1217, (thermal degradation, ms, soln props)
[9] Mendicuti, F., Patel, B., Waldeck, D.H. and Mattice, W.L., *Polymer*, 1989, **30**, 1680, (fluorescence spectrum)
[10] Aharoni, S.M., *Macromolecules*, 1985, **18**, 2624, (transition temp)
[11] Light, R.R. and Seymour, R.W., *Polym. Eng. Sci.*, 1982, **22**, 857, (gas permeability, power factor)

- Poly(ethylene-co-methyl acrylate)

[12] Shit, S.C., Mahato, B.M., Maiti, M.M. and Maiti, S., *J. Appl. Polym. Sci.*, 1986, **31**, 55, (solubility, synth, ir, H-1 nmr)
[13] Spera, S., Po, R., and Abis, L., *Polymer*, 1993, **34**, 3380, (H-1 nmr)
[14] Spear, S., Po, R. and Abis, L., *Polymer*, 1996, **37**, 729, (C-13 nmr)
[15] Korshak, V.V. and Soboleva, T.A., *Izv. Akad. Nauk SSSR, Otd. Khim. Nauk*, 1952, 526, (mp, solubility)
[16] Ambrorski, L.E., N. A. S.-N. R. C., *Publ.*, 1961, 11, (electrical conductivity)
[17] Tokiwa, Y., Ando, T. and Suzuki, T., *Hakko Kogaku Zasshi*, 1976, **54**, 603, (biodegradation)
[18] Vancso-Szmercsanyi, I. and Bodor, G., *Vorabdruck-Oeff. Jahrestag. Arbeitsgem. Verstaerkte Kunstst.*, 1977, **14**, 42, (struct)
[19] Budin, J. and Vanicek, J., *Thermochim. Acta*, 1979, **28**, 15, (transition temps)
[21] Kobayashi, S. and Hachiboshi, M., *Rep. Prog. Polym. Phys. Jpn.*, 1970, **13**, 161, (cryst struct)
[22] Kolb, H.J. and Izard, E.F., *J. Appl. Phys.*, 1949, **20**, 564, (expansion coefficient, transition temp)
[23] Tajiri, K., Fujii, Y., Aida, M. and Kawai, H., *J. Macromol. Sci., Phys.*, 1970, **4**, 1, (complex dynamic modulus)

Poly(ethylene-co-methyl acrylate) P-263

Synonyms: *Ethylene-acrylic elastomer. Poly(ethylene-co-methyl 2-propenoate)*
Related Polymers: Polyethylene, Poly(methyl acrylate)
Monomers: Methyl acrylate, Ethylene
Material class: Natural elastomers, Copolymers
Polymer Type: polyethylene, acrylic
Molecular Formula: $[(C_2H_4).(C_4H_6O_2)]_n$
Fragments: C_2H_4 $C_4H_6O_2$
Molecular Weight: The MW can range from 1000–100000
Additives: Octadecylamine, diphenylamine, hexamethylenediamine carbamate, diphenylguanidine, stearic acid, organophosphate (Gafac RC210), pentaerythritoltetrastearate, alumina, FEF carbon black (N-550), and SRF carbon black (N-774)
General Information: One of the most thermally stable of all the α-olefin copolymers. It is compatible with all the polyolefin resins and can be filled to more than 50% without loss of its elastomeric props. Four different base polymers are commercially available Vamac G, Vamac LS, Vamac D and Vamac DLS. Vamac G polymer is a terpolymer of 55% methyl acrylate, ethylene and a cure site monomer. Vamac LS is a higher methyl acrylate content terpolymer the composition of which was chosen to increase the oil resistance while minimising loss in low temp. flexibility. Vamac D and Vamac DLS are peroxide cure copolymers. The removal of carboxylic acid cure site in these polymers has improved the stability towards amine based additives. Its vulcanisates remain flexible at low temp. and possess high resistance to degradation, unique vibration damping props., and resistance to compression set and flex fatigue

Volumetric & Calorimetric Properties:
Brittleness Temperature:

No.	Value
1	-39°C

Surface Properties & Solubility:
Plasticisers: Santicizer 409 (polyester plasticiser), dioctyl sebacate
Solvents/Non-solvents: Sol. aromatic hydrocarbons, polar solvents. Insol. aliphatic hydrocarbons, H_2O

Mechanical Properties:
Tensile Strength Break:

No.	Value	Note	Elongation
1	11.6 MPa		460%
2	11.6 MPa	oven-aged, 7 d, 177°	300%
3	11.8 MPa	oven-aged, 3 d, 200°	250%

Miscellaneous Moduli:

No.	Value	Note	Type
1	3.2 MPa		100% Modulus
2	5.6 MPa	oven-aged, 3 d, 200°	100% Modulus
3	6.8 MPa		200% Modulus
4	5 MPa	oven-aged, 7 d, 177°	100% Modulus

Hardness: Durometer A85 (oven-aged, 3 d, 200°). Durometer A82 (oven-aged, 7 d, 177°). Durometer A71

Polymer Stability:
Thermal Stability General: Excellent thermal stability. At elevated temps., ageing occurs by an oxidative cross-linking process which causes brittleness
Upper Use Temperature:

No.	Value
1	121°C
2	200°C

Decomposition Details: Heat resistance profile shows that useful life exceeds 24 months at 121° but falls to 7 d at 200°
Flammability: Flammable. Rate of burning may be retarded by compounding
Environmental Stress: More resistant than polyethylene
Chemical Stability: Good stability but attacked by conc. sulfuric acid, aromatic hydrocarbons and polar solvents
Hydrolytic Stability: Stable to dilute acids and bases. Immersion in boiling H_2O for 70h causes only 5% change in volume

Applications/Commercial Products:
Processing & Manufacturing Routes: It is produced in a conventional high pressure reactor where methyl acrylate monomer is injected with the ethylene gas resulting in a random copolymer. Commercial polymers contain 18% to 24% by weight of methyl acrylate
Mould Shrinkage (%):

No.	Value	Note
1	2%	unplasticised
2	4.5%	compounded

Applications: In the automotive industry (static and dynamic seals), cable and wire jacketing, pipe seals, hydraulic system seals, dampers for machinery and high speed printers. Also used as peroxide cure copolymers

Trade name	Details	Supplier
Acrythene	EM 802-009	Quantum Chemical
Bynel	various grades	DuPont
EMAC	SP range	Chevron Chemical
Nucrel	various grades	DuPont
Optema	TC range; XS-12.04	Exxon Chemical
Vamac	Several grades	E.I. DuPont de Nemours & Co., Inc.

Bibliographic References
[1] *Encycl. Polym. Sci. Eng.*, 2nd edn., (ed. J.I. Kroschwitz), 1985, **1**
[2] *Mod. Plast.*, 1994
[3] *Kirk-Othmer Encycl. Chem. Technol.*, 4th edn., (ed. J.I. Kroschwitz), Wiley Interscience, 1991
[4] Crary, J.W., *Vamac Ethylene-Acrylic Elastomer-Balanced properties for Demanding Exposures*, Contribution 372, E.I. DuPont de Nemours and Co., Inc., Wilmington, Del., May, 1980

Poly(ethylene-1,5-naphthalate) P-264

Synonyms: *Poly(ethylene naphthalene-1,5-dicarboxylate). Poly(-oxyethyleneoxycarbonyl-1,5-naphthalenecarbonyl). Poly(oxy-1,2-ethanediyloxycarbonyl-1,5-naphthalenediylcarbonyl)*
Monomers: 1,2-Ethanediol, 1,5-Naphthalenedicarboxylic acid
Material class: Thermoplastic
Polymer Type: unsaturated polyester
CAS Number:

CAS Reg. No.	Note
51778-83-7	repeating unit
51778-87-1	from 1,5-naphthalene dicarboxylic acid
52973-02-1	from 1,5-dimethylnaphthalate

Molecular Formula: $(C_{14}H_{10}O_4)_n$
Fragments: $C_{14}H_{10}O_4$

Volumetric & Calorimetric Properties:
Density:

No.	Value	Note
1	d 1.37 g/cm^3	amorph. [4]
2	d 1.39 g/cm^3	cryst. [4]

Thermal Expansion Coefficient:

No.	Value	Note	Type
1	0.000185 K^{-1}	$T < T_g$, cryst. [4]	V
2	0.000409 K^{-1}	$T > T_g$, cryst. [4]	V
3	0.000214 K^{-1}	$T < T_g$, amorph. [4]	V
4	0.00047 K^{-1}	$T > T_g$, cryst. [4]	V

Melting Temperature:

No.	Value	Note
1	226–230°C	[1,3]
2	259°C	[2]

Glass-Transition Temperature:

No.	Value	Note
1	71°C	amorph.
2	78°C	cryst. [4]

Transport Properties:
Polymer Solutions Dilute: [η] 0.48 dl g^{-1} (phenol/tetrachloroethane, 60/40, 25°) [4]. η$_{inh}$ 0.77 dl g^{-1} [2]

Applications/Commercial Products:
Processing & Manufacturing Routes: Synth. by condensation polymerisation of ethylene glycol with 1,5-naphthalene dicarboxylic acid or dimethyl-1,5-naphthalate [1,2]

Bibliographic References
[1] *Brit. Pat.*, 1948, 604 073, (synth, melting point)
[2] *Res. Discl.*, 1988, **294**, 807, (synth, transition temps)
[3] Hill, R. and Walker, E.E., *J. Polym. Sci.*, 1948, **3**, 609, (melting point)
[4] Kolb, H.J. and Izard, E.F., *J. Appl. Phys.*, 1949, **20**, 564, (density, expansion coefficient, transition temp., intrinsic viscosity)
[5] Hagman, J.F., *Bulletin EA-220*, E.I. DuPont de Nemours and Co., Inc. Wilmington, Del., Feb., 1983, (technical datasheet)
[6] Bailor, F.V., *Rubber World*, 1979, **180**, 42

Poly(ethylene-2,6-naphthalate) films P-265

Synonyms: *Poly(ethylene-2,6-naphthalenedicarboxylate). PEN. Poly(oxyethyleneoxycarbonyl-2,6-naphthalenecarbonyl). PEN-2,6. Poly(oxy-1,2-ethanediyloxycarbonyl-2,6-naphthalenediylcarbonyl)*
Related Polymers: Poly(ethylene-2,6-naphthalate) general purpose
Monomers: Ethylene glycol, 2,6-Naphthalenedicarboxylic acid
Material class: Thermoplastic
Polymer Type: unsaturated polyester
CAS Number:

CAS Reg. No.	Note
24968-11-4	repeating unit
25230-87-9	
25853-85-4	

Molecular Formula: $(C_{14}H_{10}O_4)_n$
Fragments: $C_{14}H_{10}O_4$

Volumetric & Calorimetric Properties:
Density:

No.	Value	Note
1	d^{23} 1.33 g/cm^3	[3,4]
2	d 1.325 g/cm^3	amorph. [5]
3	d 1.407 g/cm^3	cryst. [5]
4	d^{30} 1.362 g/cm^3	α-form [6]
5	d^{30} 1.366 g/cm^3	[10]

Thermal Expansion Coefficient:

No.	Value	Note	Type
1	1.56×10^{-5} K^{-1}	22–228°, 4110 plane [10]	L

Latent Heat Crystallization: ΔH_f 56.5 J g^{-1} (α-form) [6]; ΔH_f 69.9 J g^{-1} [10]

Melting Temperature:

No.	Value	Note
1	266°C	α-form [6]
2	265°C	[7]
3	245–275°C	[9]
4	272°C	[10]
5	273°C	[11]

Glass-Transition Temperature:

No.	Value	Note
1	124°C	[4]
2	113°C	[5,11]

Poly(ethylene-2,6-naphthalate) films

3	116°C	[7]
4	117°C	[9]

Transition Temperature:

No.	Value	Note	Type
1	100°C		$T\beta_1$
2	-20°C	[4]	$T\beta_2$
3	60°C	[5]	T_β
4	270.9°C	[12]	Softening point
5	270.9°C		Equilibrium softening point
6	155°C	[7]	Endurance temp.

Transport Properties:
Permeability of Gases: Other permeability data have been reported [4]

Water Absorption:

No.	Value	Note
1	0.84 %	24h, 90% relative humidity, 20° [3]
2	0.3 %	[11]

Gas Permeability:

No.	Gas	Value	Note
1	O_2	5.25 cm^3 mm/(m^2 day atm)	0.08 × 10^{-10} cm^2 (s cmHg)$^{-1}$ [11]
2	CO_2	2.43 cm^3 mm/(m^2 day atm)	0.037 × 10^{-10} cm^2 (s cmHg)$^{-1}$ [11]

Mechanical Properties:
Mechanical Properties General: Variation of Young's Modulus and tensile strength with temp. has been reported in graphical form [10]. Tensile modulus and tensile strength break and yield have been reported [9]

Tensile (Young's) Modulus:

No.	Value	Note
1	5540–6256 MPa	56500–63800 kg cm^{-2} [12]
2	6080 MPa	620 kg mm^{-2} [11]

Tensile Strength Break:

No.	Value	Note	Elongation
1	588 MPa	60 kg mm^{-2} [7]	
2	274 MPa	[11]	90%
3	138–142 MPa	[12]	28–33%

Elastic Modulus:

No.	Value	Note
1	145000 MPa	[6,10]
2	17651 MPa	1800 kg mm^{-2} [7]

Viscoelastic Behaviour: Stress-strain curves [6,9], necking behaviour [9] and creep behaviour [9] have been reported.

Electrical Properties:
Electrical Properties General: Variation of dielectric constant and dielectric loss with temp. of both wet and dry samples has been reported in graphical form [3]. Other values of dissipation factor have been reported [10,13]. Other values of dielectric strength have been reported [13]

Surface/Volume Resistance:

No.	Value	Note	Type
1	200 × 10^{15} Ω.cm	25° [11]	S

Dielectric Permittivity (Constant):

No.	Value	Frequency	Note
1	2.9	1 kHz	[11]

Dielectric Strength:

No.	Value	Note
1	287 kV.mm^{-1}	[12]

Static Electrification: Dependence of electrical conductivity on temp. has been reported in graphical form [5]. Breakdown voltage 400 V μm^{-1} [7]

Dissipation (Power) Factor:

No.	Value	Frequency	Note
1	0.005	1 kHz	[11]

Optical Properties:
Optical Properties General: Refractive index has been reported [9]
Transmission and Spectra: Fluorescence [1], esr [8], Raman [14], XPS [8], C-13 nmr [15] and ir [16] spectral data have been reported. Film blocks light below 380 nm [11]

Polymer Stability:
Thermal Stability General: Heat resistance has been reported [13]
Upper Use Temperature:

No.	Value	Note
1	180°C	Electrical UL 746B E52857 (m) [11]
2	160°C	Mechanical UL 746B E52857 (m) [11]

Environmental Stress: Degrades photochemically with scission of the naphthyl-carbon bond [2]
Chemical Stability: Stable towards most chemicals except conc. sulfuric acid, conc. nitric acid and conc. hydrochloric acid [11]
Hydrolytic Stability: Stable to immersion in water for 200h (maintains 60% elongation under water at 130°) [7]
Stability Miscellaneous: Radiation resistance 11 MGY (to reduce elongation by 50%) [7]

Applications/Commercial Products:
Processing & Manufacturing Routes: Synth. by the transesterification of dimethyl naphthalate with ethylene glycol at 185° followed by polycondensation at 285° under vacuum using a titanium catalyst [7,17]
Applications: Used for audio and videotape. Has electronics applications

Trade name	Supplier
Q-Film	Teijin Chem. Ltd.
Q-Polymer	Teijin Chem. Ltd.
Teonex	Teijin Chem. Ltd.

Poly(ethylene-2,6-naphthalate) general purpose

Bibliographic References

[1] Ouchi, I., Hosoi, M. and Matsumoto, F., *J. Appl. Polym. Sci.*, 1976, **20**, 1983, (fluorescence spectrum)
[2] Allen, N.S. and McKellar, J.F., *J. Appl. Polym. Sci.*, 1978, **22**, 2085, (photodegradation)
[3] Ito, E. and Kobayashi, M., *J. Appl. Polym. Sci.*, 1980, **25**, 2145, (dielectric constant, water absorption)
[4] Light, R.R. and Seymour, R.W., *Polym. Eng. Sci.*, 1982, **22**, 857, (transition temps., gas permeability)
[5] Ishiharada, M., Hayashi, S. and Saito, S., *Polymer*, 1986, **27**, 349, (electrical conductivity, transition temps.)
[6] Nakamae, K., Nishino, T., Tada, K., Kanamoto, T. and Ito, M., *Polymer*, 1993, **34**, 3322, (mechanical props.)
[7] Wang, C.S. and Sun, Y.M., *J. Polym. Sci., Part A: Polym. Chem.*, 1994, **32**, 1295, (stability, synth., mechanical props.)
[8] Niino, H., Imura, T., Ohana, T., Nagai, C. and Yabe, A., *Chem. Lett.*, 1994, **7**, 1341, (esr, XPS)
[9] Cakmak, M., Wang, Y.D. and Simhambhatla, M., *Polym. Eng. Sci.*, 1990, **30**, 721, (mechanical props., refractive index)
[10] Nakamae, K., Nishino, T. and Gotoh, Y., *Polymer*, 1995, **36**, 1401, (mechanical props., expansion coefficient)
[11] *Polymeric Materials Encyclopedia*, (ed. J.C. Salamone), CRC Press, 1996, **7**, 5355, (mechanical props., electrical props., stability)
[12] *Ger. Pat.*, 1974, 2 337 815, (dielectric strength, mechanical props., softening point)
[13] Masui, M., Tsunita, K. and Nagasaki, H., *Ibaraki Daigaku Kogakubu Kenkyu Shuho*, 1974, **21**, 103, (electrical props., power factor, heat resistance)
[14] Huang, T. and Wang, P., *Sichuan Daxue Xuebao, Ziran Kexueban*, 1990, **27**, 423, (Raman)
[15] Yamanobe, T., Maksuda, H., Imai, K., Hirota, A., et al, *Polym. J. (Tokyo)*, 1996, **28**, 177, (C-13 nmr)
[16] Wang, S., Shen, D. and Qian, R., *J. Appl. Polym. Sci.*, 1996, **60**, 1385, (ir)
[17] Wang, C.S. and Sun, Y.M., *J. Polym. Sci., Part A: Polym. Chem.*, 1994, **32**, 1305, (synth.)

Poly(ethylene-2,6-naphthalate) general purpose P-266

Synonyms: *Poly(ethylene-2,6-naphthalenedicarboxylate)*. PEN. *Poly(oxyethyleneoxycarbonyl-2,6-naphthalenecarbonyl)*. PEN-2,6. *Poly(oxy-1,2-ethanediyloxycarbonyl-2,6-naphthalenediylcarbonyl)*
Related Polymers: Poly(ethylene-2,6-naphthalate) films
Monomers: 1,2-Ethanediol, 2,6-Naphthalenedicarboxylic acid
Material class: Thermoplastic
Polymer Type: unsaturated polyester
CAS Number:

CAS Reg. No.	Note
24968-11-4	repeating unit
25230-87-9	from dicarboxylic acid
25853-85-4	from 2,6-dimethyl naphthalate

Molecular Formula: $(C_{14}H_{10}O_4)_n$
Fragments: $C_{14}H_{10}O_4$
Morphology: Two crystalline forms are known. α-Form triclinic a 0.651, b 0.575, c 1.32, α 81.33°, β 144°, γ 100° [7], β-form triclinic a 0.926, b 1.559, c 1.273, α 121.6°, β 95.57°, γ 122.52° [9]

Volumetric & Calorimetric Properties:
Density:

No.	Value	Note
1	d^{23} 1.407 g/cm^3	cryst. α-form [7]
2	d 1.439 g/cm^3	cryst. β-form [9]
3	d 1.34 g/cm^3	amorph. [9]
4	d^{23} 1.3654 g/cm^3	fibres [23]
5	d^0 1.33 g/cm^3	amorph. [25]
6	d^0 1.345 g/cm^3	cryst. [25]

Thermal Expansion Coefficient:

No.	Value	Note	Type
1	4.3×10^{-5} K^{-1}	30–70° [16]	L
2	0.000187 K^{-1}	$T < T_g$ [25]	V
3	0.000646 K^{-1}	$T > T_g$ [25]	V

Volumetric Properties General: Variation of C_p with temp. has been reported in graphical form [8]
Latent Heat Crystallization: S_o 3.04 J mol^{-1} K^{-1} [6]; Enthalpy, entropy and Gibbs [6] values have been reported; ΔH_f 25 kJ mol^{-1}; ΔS_f 41 J mol^{-1} K^{-1} [8]; ΔH_f 190 J g^{-1} (calc.) [9]; ΔH_f 38 J g^{-1} [16]; ΔC_p 81.3 J mol^{-1} K^{-1} [6]
Specific Heat Capacity:

No.	Value	Note	Type
1	5.19 kJ/kg.C	-263°	P
2	99.01 kJ/kg.C	-173°	P

Melting Temperature:

No.	Value	Note
1	254°C	[6]
2	282°C	[9]
3	261.9°C	[13]
4	265°C	[16]
5	274°C	fibres [23]

Glass-Transition Temperature:

No.	Value	Note
1	130°C	[1]
2	117°C	390K [6]
3	120.4°C	[9]
4	128°C	[10]
5	123.1°C	[13]
6	121°C	[16]
7	125°C	fibres [20]
8	113°C	[25]

Transition Temperature General: Other T_g values have been reported [22]
Deflection Temperature:

No.	Value	Note
1	91°C	1.82 MPa, ASTM D648 [5]
2	95°C	1.82 MPa, ASTM D648 [16]

Transition Temperature:

No.	Value	Note	Type
1	337°C	[6]	$T_m°$
2	300°C	[9]	$T_m°$
3	62°C		$T_{\beta 1}$ - transition temp.
4	-42°C	[10]	$T_{\beta 2}$ - transition temp.

Surface Properties & Solubility:
Solubility Properties General: Miscibility with poly(ethylene terephthalate) [12] and poly(tetramethylene terephthalate) [22] has been reported
Wettability/Surface Energy and Interfacial Tension: Surface free energy 0.06 J m^{-2} (calc.) [9]

Transport Properties:
Polymer Solutions Dilute: [η]$_{inh}$ 0.8 dl g^{-1} (phenol/p-chlorophenol/tetrachloroethane (25:40:35), 25°) [5]; [η] 0.593 dl g^{-1} (phenol/o-dichlorobenzene, 25°) [8]; [η] 0.480–0.565 dl g^{-1} (hexafluoroisopropanol, 30°) [9]; [η] 0.57 dl g^{-1} (phenol/tetrachloroethane, (60:40), 25°) [25]
Polymer Melts: Melt viscosity [16] and rheological props [16] have been reported

Mechanical Properties:
Mechanical Properties General: Fibre-mechanical props have been reported [24]. Storage moduli and loss modulus relationships with temp. have been reported [16,23]. Shear storage and loss moduli (-140–270°) have been reported [10]
Flexural Modulus:

No.	Value	Note
1	2550 MPa	370 kpsi, ASTM D790 [5]

Tensile Strength Break:

No.	Value	Note	Elongation
1	83 MPa	12 kpsi, ASTM D1708 [5]	16%

Miscellaneous Moduli:

No.	Value	Note	Type
1	40200 MPa	[23]	Sonic modulus

Viscoelastic Behaviour: Stress-strain curves (fibres) have been reported [20,23]
Izod Notch:

No.	Value	Notch	Note
1	37 J/m	Y	0.7 ft lb in^{-1}, ASTM D256 method A [5]

Electrical Properties:
Electrical Properties General: Variation of tan δ with temp. has been reported in graphical form [10,16]

Optical Properties:
Optical Properties General: Thermal luminescence [1,17] has been reported. Photoluminescence and photoconductivity have been reported
Refractive Index:

No.	Value	Note
1	1.5293	[13]

Transmission and Spectra: Ir [2,13,19], Raman [14], esr [3], optical absorption [17], uv [4], C-13 nmr [12,15,22] and H-1 nmr [15] spectral data have been reported
Molar Refraction: dn/dC 0.115 (phenol/tetrachloroethane, (60:40)) [13]
Total Internal Reflection: Intrinsic birefringence 0.487 [14,23]. Birefringence of fibres has been reported [24]

Polymer Stability:
Decomposition Details: Thermal degradation studies have been reported [21]
Stability Miscellaneous: γ-Radiation causes cross-linking [3]

Applications/Commercial Products:
Processing & Manufacturing Routes: Synth. by the transesterification of dimethyl naphthalate with ethylene glycol at 185° followed by polycondensation at 285° under vacuum using a titanium catalyst [9,13,18,23]
Mould Shrinkage (%):

No.	Value	Note
1	1%	ASTM D955 [5]

Bibliographic References
[1] Hendricks, H.D., *Mol. Phys.*, 1971, **20**, 189, (thermal luminescence)
[2] Ouchi, I., Hosoi, M. and Shimotsuma, S., *J. Appl. Polym. Sci.*, 1977, **21**, 3445, (ir)
[3] Rogowski, R.S. and Pezdirtz, G.F., *J. Polym. Sci., Part A-2*, 1971, **9**, 2111, (esr, radiation resistance)
[4] Richards, R.R. and Rogowski, R.S., *J. Polym. Sci., Polym. Phys. Ed.*, 1974, **12**, 89, (uv)
[5] Jackson, W.J., *Macromolecules*, 1983, **16**, 1027, (mech props)
[6] Cheng, S.Z.D., Pan, R., Bu, H.S., Cao, M.Y. and Wunderlich, B., *Makromol. Chem.*, 1988, **189**, 1579, (thermodynamic props)
[7] Mencik, Z., *Chem. Prum.*, 1967, **17**, 78, (cryst struct)
[8] Cheng, S.Z.D. and Wunderlich, B., *Macromolecules*, 1988, **21**, 789, (thermodynamic props)
[9] Buchner, S., Wiswe, D. and Zachmann, H.G., *Polymer*, 1989, **30**, 480, (cryst struct, synth, surface tension)
[10] Chen, D. and Zachmann, H.G., *Polymer*, 1991, **32**, 1612, (transition temps, mech props)
[11] Darsey, J.A., Noid, D.W., Wunderlich, B. and Tsoukalas, L., *Makromol. Chem., Rapid Commun.*, 1991, **12**, 325, (heat capacity)
[12] Guo, M. and Zachmann, H.G., *Polymer*, 1993, **34**, 2503, (C-13 nmr)
[13] Wang, C.S. and Sun, Y.M., *J. Polym. Sci., Part A: Polym. Chem.*, 1994, **32**, 1305, (synth, ir, transition temps)
[14] Huijts, R.A. and Peters, S.M., *Polymer*, 1994, **35**, 3119, (Raman, birefringence)
[15] Abis, L., Merlo, E. and Po, R., *J. Polym. Sci., Part B: Polym. Phys.*, 1995, **33**, 691, (H-1 nmr, C-13 nmr)
[16] Kim, B.S. and Jang, S.H., *Polym. Eng. Sci.*, 1995, **35**, 1421, (rheological props, mech props)
[17] Ieda, M., Takai, Y. and Mizutani, T., *Mem. Fac. Eng., Nagoya Univ.*, 1977, **29**, 1, (optical props, electrical props)
[18] *Res. Discl.*, 1988, **294**, 816, (synth)
[19] Chen, D. and Xie, H., *Guangpuxue Yu Guangpu Fenxi*, 1987, **7**, 24, (ir)
[20] Huijts, R.A. and DeVries, A.J., *Int. J. Polym. Mater.*, 1993, **22**, 231, (viscoelastic props)
[21] Park, S.S., Kim, D.K. and Im, S.S., *Han'guk Somyu Konghakhoechi*, 1994, **31**, 225, (thermal degradation)
[22] Guo, M. and Zachmann, H.G., *Polym. Prepr. (Am. Chem. Soc.), Div. Polym. Chem.)*, 1996, **37**, 829, (C-13 nmr, miscibility)
[23] Jager, J., Juijn, J.A., Van den Heuvel, C.J.M. and Huijts, R.A., *J. Appl. Polym. Sci.*, 1995, **57**, 1429, (synth, mech props)
[24] Carr, P.L., Zhang, H. and Ward, I.M., *Polym. Adv. Technol.*, 1996, **7**, 39, (birefringence, mech props)
[25] Kolb, H.J. and Izard, E.F., *J. Appl. Phys.*, 1949, **20**, 564, (expansion coefficient, densities, soln props)

Poly(ethylene oxide)   P-267

Synonyms: *Polyoxydin. PEG. Poly(ethylene glycol). Polyoxyethylene glycol. Poly(α-hydroxy-ω-hydroxy oxy-1,2-ethanediyl). PEO*
Related Polymers: Poly(propylene oxide-co-ethylene oxide)
Monomers: Ethylene oxide
Polymer Type: polyalkylene ether, polyethylene oxide
CAS Number:

CAS Reg. No.
25322-68-3

Molecular Formula: $(C_2H_4O)_n$
Fragments: C_2H_4O
Molecular Weight: MW 200–600 (clear, colourless, viscous liqs.), 600–3000 (low melting, waxy solids), 800 (approx., partly crystalline), 100000–1000000 (highly crystalline, thermoplastic solids)
Additives: Metal salts for polyelectrolytes, 2,6-Di-*tert*-butyl-4-methylphenol, substituted phenols, cresols, quinones, mixtures of

– Poly(ethylene oxide)

aliphatic amines and dihydric phenols, carbamates and other compounds as stabilisers
General Information: Available in a wide range of MWs up to several million. Can be linear or branched, amorph., or crystalline.
Morphology: Spherulitic; monoclinic; a 0.796 nm, b 1.311 nm, c 1.939 nm; β 124.8°; 4 molecular chains per unit cell [1]. Crystal morphology varies with the crystallisation process and crystallisation temp. T_c. Crystal thickness increases as T_c increases. Can crystallise with integral-folded (IFC) or non-integral folded chains (NIFC) depending on the conditions during crystallisation. Raman longitudinal acoustic mode studies on the isothermal crystallisation of PEO (MW 3000) have shown that the initial lamellae are made up of non-integer-fold chains but integer-fold chains eventually develop in a solid-state conversion. SAXS studies have also been reported on NIFC and IFC formation in PEO complexes with some disubstituted aromatic compounds [8,36,37]. Crystallisation and melting have been studied and the lamellar growth rate measured in real time by a combination of optical and atomic force microscopy [34]
Miscellaneous: Viscometry evidence suggests that PEO prepared using calcium or zinc catalysts retains metal atoms in the polymer in a bonded form. Consequently the polymer carries a positive charge and will behave as a polyelectrolyte [31,37]. Can form association complexes with various electron acceptors e.g. poly(methacrylic acid), phenolic compounds, urea, polyacrylic acid. Complexes with electrolytes, particularly lithium salts, are of special interest for their potential uses, e.g. in batteries and electrochromic sensors/displays. The synth. and struct. of PEO-based polyelectrolytes has been widely reported. [2,46,47,57]. The name PEG generally refers to polymer of lower MW. The uses of PEO have been reviewed [2]

Volumetric & Calorimetric Properties:
Density:

No.	Value	Note
1	d^{20} 1.127 g/cm^3	amorph., [1]
2	d^{20} 1.21 g/cm^3	cryst. [1]

Specific Heat Capacity:

No.	Value	Note	Type
1	5.65 kJ/kg.C	1.35 cal (g °C)$^{-1}$, 25°, MW 750 [3]	P
2	1.54 kJ/kg.C	0.367 cal (g °C)$^{-1}$, 25°, MW 4000 [3]	P
3	1.49 kJ/kg.C	0.358 cal (g °C)$^{-1}$, 25°, MW 20000 [3]	P

Melting Temperature:

No.	Value	Note
1	-15–-8°C	PEG, MW approx., 300 [3]
2	4–8°C	PEG, MW approx., 400 [3]
3	20–25°C	PEG, MW approx., 600 [3]
4	37–45°C	PEG, MW approx., 1000 [3]
5	43–46°C	PEG, MW approx., 1540 [3]
6	53–56°C	PEG, MW approx., 4000 [3]
7	60–63°C	PEG, MW approx., 6000 [3]
8	50–55°C	PEG, MW approx., 20000 [3]
9	37°C	M_n 1000, crystalinity 0.78 [10]
10	54°C	M_n 2000, crystalinity 0.85 [10]
11	58°C	M_n 35000, crystalinity 0.87 [10]
12	62°C	M_n 6700, crystalinity 0.91 [10]
13	62.5°C	M_n 13000, crystallinity 0.88 [10]
14	63°C	M_n 13000, crystallinity 0.91 [10]
15	66.2°C	M_n 13000, crystallinity 0.99 [10]
16	60°C	M_n 17000, crystallinity 0.89 [10]
17	65°C	high MW [1]
18	69°C	equilibrium, MW 10000 [28]
19	76°C	equilibrium [3]

Glass-Transition Temperature:

No.	Value	Note
1	-52°C	high MW, partly crystalline [1]
2	-17°C	MW 6000; 95% crystalline [1]
3	-95–-15°C	varying MW [2]

Transition Temperature General: T_m varies with MW and crystallinity/crystallisation temp. [3,10,28]. Increasing the draw ratio increases the melting temp. of PEO filaments (MW 300000–4000000) from approx. 65° to approx. 72° at draw ratio 30 [14]. The glass-transition temp. varies with increasing MW, rising from -95° at an approximate MW of 100 to between -15° and -20° at MW values around 10000. After this point it falls, levelling off at approx. 60° when MW is approx. 1000000. This variation has been attributed to the effect of changing crystallinity. Quenched polymers tend to exhibit T_gs rising in a monotonic manner with increasing MW [2]

Surface Properties & Solubility:
Solubility Properties General: Shows inverse solubility in water, becoming less sol. as the temp. rises. When a certain temp. (the cloud point), is reached the polymer precipitates and remains insol. if the temp. rises further. The cloud point depends on concentration and MW, falling as concentration increases. The presence of salts usually brings about a decrease in the cloud point; the magnitude of the change depends on the concentration, valence and size of the ionic species present. Phase separation of aq. solns. of PEO as a result of spinodal degradation is caused by increasing pressure. The effect is independent of concentration but varies with MW [52]. At concentrations of <1% aq. solns. of high MW PEO become stringy. At concentrations of approx. 20% the solns. are gel-like and above this they resemble hard, water plasticised polymers [1,2]. PEO was believed to aggregate in water and other solvents. However, more recent work has shown that evidence of aggregation is not found providing that the solvent is adequately purified. Anomalous results can occur with solvent which is not pretreated, leading to erroneous conclusions. [15,32]. The swelling behaviour of a PEO hydrogel in a wide variety of solvents has been reported. Water and formamide show anomalous swelling/temp. relationships. [40,42,45]
Cohesive Energy Density Solubility Parameters: Solubility parameters 10.3 J$^{1/2}$ cm$^{-3/2}$ (70°, MW 10700), 6.3 J$^{1/2}$ cm$^{-3/2}$ (100°, MW 10700), 6.07 J$^{1/2}$ cm$^{-3/2}$ (120°, MW 10700), 5.9 J$^{1/2}$ cm$^{-3/2}$ (150°, MW 10700) [30]
Solvents/Non-solvents: Sol. H$_2$O, chlorinated hydrocarbons, 2-butanone, 2-ethoxyethyl acetate, butyl acetate, cyclohexanone, esters, DMF, C$_6$H$_6$ (elevated temp.), toluene (elevated temp.) [1]. Insol. ethers, aliphatic hydrocarbons [1]
Surface and Interfacial Properties General: The surface activity of PEO in water increases with MW up to a value of approx. 10000 above which the interfacial absorption characteristics remain essentially constant. An anomalous additional reduction in surface tension is observed in aq. PEO solns. at high polymer concentrations. The phenomenon is attributed to the formation of dense layers of PEO. Although the reduction in surface tension at low concentrations is independent of MW, the additional surface tension reduction is strongly MW dependent [3,51,53,55]. Surface free energy of low MW PEO decreases with increasing MW [10]

Poly(ethylene oxide)

Surface Tension:

No.	Value	Note
1	33.8 mN/m	[23]

Transport Properties:

Transport Properties General: The solubility of many alkali metal salts in PEO soln. decreases as the temp. increases. With one exception (NaNO$_3$), it is found that salts which form no complex with PEO display this negative temp. dependence in their solubility [49]. At low concentrations of aq. PEO the solubilities of gases are little different to those in pure water, but at polymer concentrations greater than approx. 70% they increase appreciably. Solubilities in neat polymer are higher for PEO of greater MW [54]. Melt flow index has been reported [2]

Polymer Solutions Dilute: At approx. 50% PEO (actual value MW dependent) a PEO/water system behaves as a binary eutectic mixture [35]. The effects of temp., pressure and the presence of salts on the rheological props. of aq. PEO soln. have been studied. Aq. solns. of high MW PEO exhibit shear thinning behaviour. Increasing temp. and the addition of salts cause a decrease in shear thinning and in the apparent viscosity. However, an increase is observed on the addition of EtOH, a poor solvent for PEO [8]. The viscosity of aq. PEO solns. increases with concentration, and MW but decreases with increasing temp. and shear rate [2]. Under normal conditions higher MW PEO exhibits LCST behaviour in aq. soln. With increasing pressure the LCST falls and the polymer eventually exhibits UCST beaviour [21]. Studies on PEO in aq. THF have shown that as temp. falls to near the LCST the polymer contracts, the inherent viscosity decreasing to approx. 33% of its previously virtually constant value [33]. A detailed study of local dynamics in toluene soln. has been reported [20,24,25]. Intrinsic viscosity [16,18] 3.7 dl g^{-1} (MeOH, 25°, MW 40000), 26.5 dl g^{-1} (MeOH, 25°, MW 1200000), 0.221 dl g^{-1} (H$_2$O, 25°, MW 62), 10.2 dl g^{-1} (H$_2$O, 25°, MW 86000), 257 dl g^{-1} (H$_2$O, 25°, MW 11000000), 0.28 dl g^{-1} (C$_6$H$_6$, 25°, MW 400), 22.6 dl g^{-1} (C$_6$H$_6$, 25°, MW 348000), 158 dl g^{-1} (C$_6$H$_6$, 25°, MW 5200000), 0.24 dl g^{-1} (aq. 0.45M K$_2$SO$_4$, 34.5°, MW 106), 1.19 dl g^{-1} (aq. 0.45M K$_2$SO$_4$, 34.5°, MW 5190). Huggins constants [24] and Mark-Houwink constants [1,16, 19, 24, 25, 26] have been reported. Theta temp. 17° (approx., MeOH), 35° (aq. 0.45M K$_2$SO$_4$), 45° (aq. 0.45M K$_2$SO$_4$) [16]. Second virial coefficients have been reported [9,15,16]. Flory-Huggins interaction parameter has been reported [9,19]

Polymer Melts: Melt behaviour is pseudoplastic, melt viscosity decreases significantly with an increase in shear rate. The effect is more pronounced for higher MW polymers

Permeability of Gases: [O$_2$] 85.8 µmol (m s GPa)$^{-1}$ (film) [2]

Water Content: PEG hydrogels form various hydrates depending on the MW. Lower MWs form a tetrahydrate and higher ones a trihydrate possibly with a hexahydrate. The transition point for tetrahydrate to trihydrate lies between M_n 429 and 949 [6]

Gas Permeability:

No.	Gas	Note
1	[O$_2$]	85.8 µmol (m s GPa)$^{-1}$ (film) [2]

Mechanical Properties:

Mechanical Properties General: The effects of draw ratio on the tensile strength and modulus of PEO filaments, prepared by solid-state extrusion, have been reported. Both strength and modulus increase with the draw ratio, the former levelling off to a constant value. The increase in tensile strength is more pronounced for higher MW polymer [14]

Tensile (Young's) Modulus:

No.	Value	Note
1	5000–9000 MPa	approx., MW 300000–4000000 [14]

Tensile Strength Break:

No.	Value	Note	Elongation
1	16 MPa	film, machine direction [2]	550%
2	13 MPa	film, transverse direction [2]	650%
3	10 MPa	approx., filament MW 300000–4000000 [14]	
4	82.7 MPa	12000 lb in^{-2} [3]	700%

Tensile Strength Yield:

No.	Value	Elongation	Note
1	11.7 MPa	10%	1700 lb in^{-2} [3]

Miscellaneous Moduli:

No.	Value	Note	Type
1	290 MPa	film, machine direction [2]	Secant modulus
2	480 MPa	film, transverse direction [2]	Secant modulus

Viscoelastic Behaviour: The viscoelastic parameters of PEO in compression creep have been reported and the effects of temp. and pressure investigated. PEO chains exhibit volume deformation increasing with rising temp. [50]. The ultrasonic velocity and absorption depend on MW and concentration [56]

Failure Properties General: Tear strength 100 kN m^{-1}(film, machine direction), 240 kN m^{-1}(film, transverse direction) [2]

Fracture Mechanical Properties: Is resistant to cold crack at -46 [2]

Friction Abrasion and Resistance: PEO is effective as a friction reducer in aq. solns. at high dilution [3]

Electrical Properties:

Electrical Properties General: Has a much higher conductivity than expected, particularly so in the liq. state. Conductivity increases as the temp. rises to 65–70° and then tends to level off. At temps. below T_m conductivity rises with increasing frequency. At higher temps. the conductivity is virtually the same for frequencies of 1 kHz and above [3,57]. The dielectric constant and loss are strongly dependent on temp., pressure and frequency [39]

Complex Permittivity and Electroactive Polymers: Polymer electrolytes based on PEO have been studied extensively. Complexes with alkali, alkaline earth and transition metal salts are possible all with a variety of counter ions. The relationship between struct. and conductivity has been investigated as has the variation of conductivity with temp. and composition. In the case of PEO-LiBF$_4$ based electrolytes the relationship between Li salt concentration and morphological changes in the polymers complex. The highest conductivity is obtained for PEO:LiBF$_4$ ration of 8:1.Permittivity data on PEO-LiBF4 mixtures agrees with the findings on conductivity [46,47,57]

Optical Properties:

Refractive Index:

No.	Value	Note
1	1.46	PEG, MW approx. 200 [3]
2	1.463	PEG, MW approx. 300 [3]
3	1.467	PEG, MW approx. 400 [3]
4	1.469	PEG, MW approx. 600 [3]

Transmission and Spectra: Ir, Raman and nmr spectral data have been reported [3,11,12].

Total Internal Reflection: Increasing the draw ratio increases the birefringence of PEO filaments [14]

– Poly(ethylene oxide)

Polymer Stability:
Polymer Stability General: PEO is susceptible to oxidation. It can degrade by autodegradation via the formation and cleavage of intrachain hydroperoxide bonds. Uv light or the presence of heavy metal ions or strong acids can promote hydroperoxide formation. Degradation is also caused by mechanical means such as high shear [2]. The stability of aq. solns. can be improved by the addition of isopropanol, EtOH, ethylene glycol, propylene glycol or manganese ion. [1,2]
Thermal Stability General: After 100h at 80° in air a free-flowing powder form of the polymer (MW 6000) is transformed into a waxy, lower melting point material. Thermal treatment *in vacuo* at 80° for 1000h, however, has little effect on either form or melting point. The addition of an antioxidant such as 2,2′-methylenebis(4-methyl-6-*tert*-butylphenol) stabilises the polymer against thermo-oxidative degradation [17]. In an inert atmosphere the presence of salts reduces the thermal stability of PEO but in air the polymer is stabilised against thermooxidation [11]
Decomposition Details: Pyrolysis studies on PEO (MW 900000) show that degradation, by C-O and C-C bond scission, occurs in the range 235–255°. The products include CO_2, formaldehyde, CH_3CHO and $(CH_3CH_2)_2O$. The thermal and thermooxidative degradation of the homopolymer, with or without added salts, has been reported [5,11]
Flammability: Fl. p. 171–182° (MW 200, Cleveland open cup), 254–265° (MW 1000, Cleveland open cup), 271° (MW 4000, Cleveland open cup), 148° (min., MW 1000, Cleveland closed cup), 177° (min., MW 200, Cleveland closed cup), 177° (min., MW 4000, Cleveland open cup)[1]
Chemical Stability: Is hygroscopic. Its aq. solns. are corrosive to carbon steel but not to stainless steel. For prolonged storage, especially in hot environments, PEO should be stored under an inert atmosphere such as nitrogen to avoid degradation [1]
Biological Stability: Is biodegradable
Stability Miscellaneous: Exposure to ionising radiation can lead to either cross-linking or degradation. In the presence of oxygen, exposure results in degradation. In an inert atmosphere, however, cross-linking occurs, producing polymer networks which exhibit hydrogel props. Uv-induced cross-linking of PEO in the presence of a variety of initiators, and the gel fraction and swelling behaviour of the resultant products have been reported. Network density can be controlled by varying the irradiation temp. [2,3,4,41]. Ultrasonic irradiation (500 kHz) of an aq. soln. of PEO causes a gradual decrease in the MW from (narrow range) to 1000–10000 range centred approx. 1500) over a two hour period [48]

Applications/Commercial Products:
Processing & Manufacturing Routes: Ethylene oxide can be polymerised by three methods [1]; acid or cationic initiation using Bronsted or Lewis acids; base or anionic initiation using metal alkoxides; ionic coordinate initiation usually using alkaline earth or transition metal complexes. Star-shaped PEO of various types (three-arm, multi-arm and comb-shaped) has been prepared by the anionic polymerisation of oxirane using a plurifunctional organometallic initiator [43]. The efficient separation of linear and star polmers by supercritical fluids and the fractional precipitation of star molecules have been reported. [22,44]. Can be cross-linked to form gels by reaction with isocyanates [40,42];γ-irradiation of degassed aq. PEO soln. [41]; electron beam irradiation in aq. soln. [45]. High MW PEO can be moulded, extruded and calendered. Sheets and films are heat-sealable [2]
Applications: Low MW PEO (MW 800) is used as an intermediate in chemical manufacture, e.g. of surfactants and thickeners. Slightly higher MW (1000–2000) polymers are used in pharmaceutical applications (e.g. ointments, suppositories) and in cosmetics (creams and lotions). Medium MW polymers are used as adhesives, binders, plasticisers, lubricants, moulding compounds and preservatives for items e.g. archaeological findings which have been subjected to long submersion under water. High MW PEO is used as textile sizes, hydrodynamic friction reducing agents, as a water sol. coating for seeds and also in cosmetics and toiletries, packaging, adhesives, chromatography and dentistry. Association complexes of PEO are used in medical applications (controlled release formulations, microencapsulation and artificial kidneys). Complexes with electrolytes, particularly lithium salts, are of special interest for their potential uses in batteries and electrochromic sensors/displays.

Trade name	Supplier
Carbowax	
Polyox	Union Carbide

Bibliographic References

[1] Bailey, F.E., Koleske, J.V., *Ullmanns Encykl. Ind. Chem.*, 1992, **A21**, 579
[2] Back, D.M., Clark, E.M., Ramachandran, R., *Kirk-Othmer Encycl. Chem. Technol.*, 4th edn., 1996, **19**, 701
[3] Bailey, F.E., Koleske, J.V., *Polyethylene Oxide*, Academic Press, 1976
[4] Doytcheva, M; Dotcheva, D; Stamenova, R. Orahovats. A. et al, *J. Appl. Polym. Sci.*, 1997, **64**, 2299
[5] Fares, M.M., Hacalogu, J. Suzer, S., *Eur. Polym. J.*, 1994, **30**, 845, (degradation)
[6] Graham, N.B., Chen, C.F., *Eur. Polym. J.*, 1993, **29**, 149
[7] Pracella, M., *Eur. Polym. J.*, 1985, **21**, 551, (morphology)
[8] Briscoe, B., Luckham, P., Zhu, S., *J. Appl. Polym. Sci.*, 1998, **70**, 419, (rheological props.)
[9] Venohr, H., Fraaije, V., Strunk, H., Borchard, W., *Eur. Polym. J.*, 1998, **34**, 723
[10] Balta Calleja, F.J., Santa Cruz, C., *Acta Polym.*, 1996, **47**, 303, (microstuct.)
[11] Costa, L., Gad, A.M., Camino, G., *Macromolecules*, 1992, **25**, 5512, (degradation)
[12] Dechter, J.J., *J. Polym. Sci. Part C*, 1985, **23**, 261, (H-1 nmr, C-13 nmr)
[13] Allen, R.C., Mandelkern, L., *J. Polym. Sci. Part B*, 1982, **20**, 1465, (struct.)
[14] Kim, B.S., Porter, R.S., *Macromolecules*, 1985, **18**, 1214
[15] Kinugasa, S, Nakahara, H, Kawahara, J.-I., Koga, Y., Takaya, H., *J. Polym. Sci. Part B*, 1996, **34**, 583
[16] Zhou, P., Brown, W., *Macromolecules*, 1990, **23**, 1131
[17] Han, S., Kim, C., Kwon, D., *Polymer*, 1997, **38**, 317, (degradation)
[18] Kawaguchi, S., Imai, G., Suzuki, J., Miyahara, A. et al, *Polymer*, 1997, **38**, 2885, (soln., props.)
[19] Gregory, P., Huglin, M.B., *Makromol. Chem.*, 1986, **187**, 1745, (viscosity)
[20] Fuson, *M.M.*, *Macromolecules*, Hauser, K.H., Ediger, M.D., 1997, **30**, 5704, 5714
[21] Sun, T., King, H.E., *Macromolecules*, 1998, **31**, 6383, (soln., props.)
[22] Cansell, F., Botella, P., Six, J.-L., Garrabos, Y. et al, *Polym. J. (Tokyo)*, 1997, **29**, 910
[23] Koberstein, J.T., *Encycl. Polym. Sci. Eng.*, 1987, **8**, 237, (interfacial props.)
[24] Thomas, D.K., Charlesby, A., *J. Polym. Sci.*, 1960, **42**, 195, (viscosity)
[25] Allen, G., Booth, C., Hurst, S.J., Jones, M.N., Price, C., *Polymer*, 1967, **8**, 391, (intrinsic viscosity)
[26] Amu, T.C., *Polymer*, 1982, **23**, 1775, (theta temps.)
[27] Kambe, Y., Honda, C., *Polym. Commun.*, 1984, **25**, 154
[28] Murphy, C.J., Henderson, G.V.S., Murphy, E.A., Sperling, L.H., *Polym. Eng. Sci.*, 1987, **27**, 781
[29] Kugler, J., Fischer, E.W., Peuscher, M., Eisenbach, C.D., *Makromol. Chem.*, 1983, **184**, 2325
[30] Galin, M., *Polymer*, 1983, **24**, 865, (solubility parameters)
[31] Zhivkova, I., Zhivkov, A., Stoychev, D., *J. Mater. Sci. Lett.*, 1996, **15**, 2115, (viscosity)
[32] Kinugasa, S., Nakahara, H., Fudagawa, N., Koga, Y., *Macromolecules*, 1994, **27**, 6889
[33] Park, I.H., Kim, M.J., *Macromolecules*, 1997, **30**, 3879
[34] Pearce, R., Vancso, G.J., *Macromolecules*, 1997, **30**, 5843
[35] Bogdanov, B., Mihailov, M., *J. Polym. Sci. Part B*, 1985, **23**, 2149
[36] Kim, I., Krimm, S., *Macromolecules*, 1996, **29**, 7186
[37] Paternostre, L., Damman, P., Dosiere, M., *Macromolecules*, 1997, **30**, 3946
[38] Zhivkova, I., Zhivkov, A., Stoychev, D., *Eur. Polym. J.*, 1998, **34**, 531, (electrostatic behaviour)
[39] Fanggao, C., Saunders, G.A., Lambson, E.F., Hampton, R.N. et al, *J. Polym. Sci. Part B*, 1996, **34**, 425, (dielectric props)
[40] Graham, N.B., Nwachuku, N.E., Walsh, D., *Polymer*, 1982, **23**, 1345
[41] Minkova, L., Stamenova, R., Tsvetanova, C., Nedkov, E., *J. Polym. Sci. Part B*, 1989, **27**, 621, (struct.)
[42] Gnanou, Y., Hild, G., Rempp, P., *Macromolecules*, 1987, **20**, 1662
[43] Gnanou, Y., Lutz, P., Rempp, P., *Makromol. Chem.*, 1988, **189**, 2885, (synth.)

[44] Yen, D.R., Raghavan, S., Merrill, E.W., *Macromolecules*, 1996, **29**, 8977
[45] Cima, L.G., Lopina, S.T., *Macromolecules*, 1995, **28**, 6787
[46] Chandrasekhar, V., *Adv. Polym. Sci.*, 1998, **135**, 140
[47] Magistris, A., Singh, K.E., *Polym. Int.*, 1992, **28**, 277
[48] Koda, S., Mori, H., Matsumoto, K., Nomura, H., *Polymer*, 1994, **35**, 30, (degradation)
[49] Ohno, H., Ito, K., *Polymer*, 1993, **35**, 4171
[50] Aleman, J.V., *Polymer*, 1991, **32**, 2555, (viscoelastic props.)
[51] Kim, M.W., Cao, B.H., *Europhys. Lett.*, 1993, **24**, 229, (surface tension)
[52] Cook, R.L., King, H.E., Peiffer, D.G., *Phys. Rev. Lett.*, 1992, **69**, 3072
[53] Cao, B.H., Kim, M.W., *Faraday Discuss. Chem. Soc.*, 1994, **98**, 245, (surface tension)
[54] King, A.D., *J. Colloid Interface Sci.*, 1991, **144**, 579, (gas solubility)
[55] Kim, M.W., *Colloids Surf.*, 1997, **128**, 283, (surface activity)
[56] Bass, H.E., Tan, J., *J. Acoust. Soc. Am.*, 1993, **93**, 283, (ultrasonic props.)
[57] Sircar, A.K., Weissman, P.T., Kumar, B., Marsh, R.A., *Thermochim. Acta*, 1993, **226**, 281

Poly(ethylene oxybenzoate)   P-268

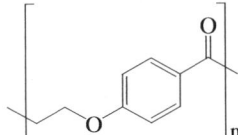

Synonyms: *PEB. PEOB. Poly(ethylene benzoate). Poly(ethylene-p-hydroxybenzoate). Poly(4-(2-hydroxyethoxy)benzoic acid). Poly(oxyethyleneoxy-1,4-phenylenecarbonyl). Poly(p-(ethyleneoxy)benzoate). Poly(oxy-1,2-ethanediyloxy-1,4-phenylenecarbonyl). Poly(methyl 4-(2-hydroxyethoxy)benzoate)*
Monomers: Methyl 4-(2-hydroxyethoxy)benzoate
Material class: Thermoplastic
Polymer Type: unsaturated polyester
CAS Number:

CAS Reg. No.	Note
25248-22-0	repeating unit
25302-46-9	Poly(ethylene-*p*-hydroxy benzoate)
25585-19-7	Poly(4-(2-hydroxyethoxy)benzoic acid
25232-94-4	Poly(methyl 4-(2-hydroxyethoxy)benzoate)

Molecular Formula: $(C_9H_8O_3)_n$
Fragments: $C_9H_8O_3$
Morphology: α form - orthorhombic a 1.046, b 0.476, c 1.560, helix 2_1 [2]; α form - orthorhombic a 1.049, b 0.475, c 1.560, helix 2_1 [6]; α form - orthorhombic a 1.052, b 0.475, c 1.565, helix 2_1 [17]; β form - monoclinic a 0.819, b 1.107, c 1.905, β 114.8°, helix 2_1 [8]. Other forms have been reported [3]

Volumetric & Calorimetric Properties:
Density:

No.	Value	Note
1	d 1.3 g/cm^3	amorph. [1]
2	d 1.39 g/cm^3	cryst. [1]
3	d 1.34 g/cm^3	fibre [5]
4	d 1.312 g/cm^3	amorph. [5]
5	d^{25} 1.34 g/cm^3	α-form [8]
6	d^{25} 1.31 g/cm^3	β-form [8]
7	d^{30} 1.337 g/cm^3	[9]

Volumetric Properties General: Bulk compressibility 13.8×10^{-6} atm^{-1} [9]. Linear thermal expansion coefficient has been reported [16]

Latent Heat Crystallization: Enthalpy and entropy values have been reported [18]
Melting Temperature:

No.	Value	Note
1	223°C	[5]
2	225°C	[7]
3	205°C	478 K [11]
4	229°C	[17]
5	208°C	[19]
6	220°C	[20]

Glass-Transition Temperature:

No.	Value	Note
1	65°C	amorph. [5]
2	84°C	cryst. [5,7]
3	71°C	344K [11]
4	80°C	[17]

Transition Temperature General: Other T_g values have been reported [18,19]
Transition Temperature:

No.	Value	Note	Type
1	197–201°C	fibre [5]	Softening point
2	182–185°C	fibre [13]	Softening point

Surface Properties & Solubility:
Solvents/Non-solvents: Sol. 2-chlorophenol [5], 4-chlorophenol [11], phenol/tetrachloroethane (50:50) [20]. Insol. most organic solvents [5]
Transport Properties:
Polymer Solutions Dilute: [η] 0.62 dl g^{-1} (2-chlorophenol, film, 25°) [5]; [η]$_{inh}$ 0.53 dl g^{-1} (4-chlorophenol, 50°) [11]; [η] 0.61 dl g^{-1} [19]; [η] 0.72 dl g^{-1} (tetrachloroethane/phenol (50:50), 25°) [20]
Permeability of Gases: Gas permeability (25 μm film) [N$_2$] 0.1 ml m^{-2} h^{-1} atm^{-1}; [O$_2$] 3.49 ml m^{-2} h^{-1} atm^{-1}; [CO$_2$] 7.43 ml m^{-2} h^{-1} atm^{-1} [5]
Water Absorption:

No.	Value	Note
1	0.4–0.5 %	20°, 65% relative humidity, fibre [5]

Mechanical Properties:
Tensile (Young's) Modulus:

No.	Value	Note
1	4707 MPa	480 kg mm^{-2}, fibre [12]
2	4903–5884 MPa	500–600 kg mm^{-2}, fibre

Tensile Strength Break:

No.	Value	Note	Elongation
1	4–5.3 MPa	fibre [5]	15–30%
2	5.3 MPa	fibre [7]	25%
3	176 MPa	18 kg mm^{-2}, 25 μm film, 200 mm min^{-1}	127%

Elastic Modulus:

No.	Value	Note
1	6000 MPa	α-form, crystallite [2,16]
2	8900 MPa	macroscopic [7]
3	5900 MPa	crystallite [7]
4	2000 MPa	α-form, crystallite, 175° [16]

Poisson's Ratio:

No.	Value	Note
1	0.49	[9]

Impact Strength: Missile impact strength 31 kg cm (25 μm film) [5]
Viscoelastic Behaviour: Stress-strain curves for fibre have been reported [1,5,9,16]

Electrical Properties:
Electrical Properties General: Dielectric constant has been reported [10,15]

Optical Properties:
Transmission and Spectra: Esr [1,4], and ir [14,20] spectral data have been reported. Transmittance 92% (25 μm film, 560 μm) [5]
Total Internal Reflection: Birefringence 0.12 [12]

Polymer Stability:
Upper Use Temperature:

No.	Value	Note
1	150°C	fibre [5]

Environmental Stress: The fibre has good weathering and uv resistance [5]
Chemical Stability: The fibre has good resistance to sulfuric acid, nitric acid, hydrochloric acid, formic acid, AcOH, sodium hydroxide and phenol [5]
Hydrolytic Stability: The fibre has good resistance to hydrol. [5]

Applications/Commercial Products:
Processing & Manufacturing Routes: Synth. by treating 4-hydroxybenzoic acid with ethylene oxide to form 4-(2-hydroxyethoxy) benzoic acid, which is then treated with MeOH to form methyl 4-(2-hydroxyethoxy) benzoate. This monomer then undergoes condensation polymerisation at 250–260° under vacuum [5,11]
Applications: Used as a filament yarn in the silk clothing industry

Trade name	Supplier
A-Tell	Unitika Ltd.

Bibliographic References
[1] Nagamura, T., Fukitani, K. and Takayanagi, M., *J. Polym. Sci., Polym. Phys. Ed.*, 1975, **13**, 1515, (esr, viscoelastic props)
[2] Sakurada, I. and Kaji, K., *J. Polym. Sci., Part C: Polym. Lett.*, 1970, **31**, 57, (cryst struct, mech props)
[3] Korematsu, M. and Kuriyama, S., *Nippon Kagaku Zasshi*, 1960, **81**, 852, (cryst struct)
[4] Nagamura, T. and Takayanagi, M., *J. Polym. Sci., Polym. Phys. Ed.*, 1974, **12**, 2019, (esr)
[5] Mihara, K., *Angew. Makromol. Chem.*, 1974, **40–41**, 41, (synth, use, mech props)
[6] Kusanagi, H., Tadokoro, H., Chatani, Y. and Suehira, K., *Macromolecules*, 1977, **10**, 405, (cryst struct)
[7] Tashiro, K., Kobayashi, M. and Tadokoro, H., *Macromolecules*, 1977, **10**, 413, (transition temps, mech props)
[8] Takahashi, Y., Kurumizawa, T., Kusanagi, H. and Tadokoro, H., *J. Polym. Sci., Polym. Phys. Ed.*, 1978, **16**, 1989, (cryst struct, densities)
[9] Ito, T., *Polymer*, 1982, **23**, 1412, (mech props)
[10] Ishida, Y., Yamafuji, K., Ito, H. and Takayanagi, M., *Kolloid Z. Z. Polym.*, 1962, **184**, 97, (dielectric constant)
[11] Kumar, A. and Ramakrishnan, S., *Macromolecules*, 1996, **29**, 2524, (synth, transition temps)
[12] U.S. Pat., 1966, 3 291 778, (birefringence, mech props)
[13] Tsuji, W., *Senshoku Kogyo*, 1969, **17**, 129, (softening point)
[14] Yamamoto, Y., Sangen, O. and Kishimoto, Y., *CA*, 1972, **81**, 64125, (ir)
[15] Stetsovskii, A.P. and Tarasova, L.V., *Vysokomol. Soedin., Ser. A*, 1978, **20**, 1116, (dielectric constant)
[16] Nakamae, K., Nishino, T., Hata, K. and Matsumoto, T., *Kobunshi Ronbunshu*, 1985, **42**, 361, (expansion coefficient, mech props)
[17] Takayanagi, M., *J. Macromol. Sci., Phys.*, 1974, **9**, 391, (cryst struct, transition temps)
[18] Korematsu, M., Masuda, H. and Kuriyama, S., *Kogyo Kagaku Zasshi*, 1960, **63**, 884, (thermodynamic props)
[19] Horio, M., Imamura, R., Kiyotsukuri, T. and Takegoshi, T., *CA*, 1962, **62**, 16427d, (soln props, transition temps)
[20] Ishibashi, M., *Polymer*, 1964, **5**, 305, (ir, melting point, soln props)

Poly(ethylene sebacate)

Synonyms: *Poly(oxyethyleneoxysebacoyl). PES. Poly[oxy-1,2-ethanediyloxy(1,10-dioxo-1,10-decanediyl)]. Poly(ethylene sebacinate). Poly(1,2-ethanediol-co-decanedioic acid)*
Monomers: 1,2-Ethanediol, Decanedioic acid
Material class: Thermoplastic
Polymer Type: saturated polyester
CAS Number:

CAS Reg. No.
25037-32-5
25034-96-2

Molecular Formula: $(C_{12}H_{20}O_4)_n$
Fragments: $C_{12}H_{20}O_4$
Morphology: Two polymorphic forms [20,36] have been studied [7,24,30], form I, monoclinic, a 0.552, b 0.730, c 1.665, β 115.0° [21]; monoclinic, a 0.552, b 0.740, c 1.690, β 65° [22]; monoclinic, a 0.55, b 1.50, c 1.69, β 65° [23]; monoclinic, a 0.558, b 0.731, c 1.676, β 115.5° [24]; form II, triclinic, a 0.539, b 0.760, c 1.676, α 104.8°, β 111.7°, γ 72.4° [36]

Volumetric & Calorimetric Properties:
Density:

No.	Value	Note
1	d^{25} 1.167 g/cm^3	[17]
2	d^{25} 1.113 g/cm^3	amorph. [17]
3	d 1.211 g/cm^3	crystalline [17]
4	d^{25} 1.148 g/cm^3	[18,19]
5	d 1.244 g/cm^3	crystalline [21]
6	d 1.083 g/cm^3	amorph. [26]
7	d 1.173 g/cm^3	[24]
8	d 1.101 g/cm^3	amorph. [33]
9	d 1.156 g/cm^3	[35]
10	d 1.136 g/cm^3	[35]

Poly(ethylene sebacate)

Thermal Expansion Coefficient:

No.	Value	Note	Type
1	0.00075 K^{-1}	amorph. [26]	V
2	0.00053 K^{-1}	crystalline [26]	V
3	0.00074 K^{-1}	25° [32]	V
4	0.000196 K^{-1}	T < Tg [37]	V
5	0.000358 K^{-1}	T > Tg [37]	V

Volumetric Properties General: Compressibility 0.0001 cm^3 g^{-1} atm^{-1} [26]
Equation of State: Equation of state information has been reported [31]
Latent Heat Crystallization: ΔH_f 80.8 kJ mol^{-1} (19.3 kcal mol^{-1}, film), 88 kJ mol^{-1} (21 kcal mol^{-1}, block) [9]; ΔH_f 13.8 kJ mol^{-1} (3300 cal mol^{-1}), S_f 40.2 J mol^{-1} K^{-1} (9.6 cal mol^{-1} °C^{-1}) [14]; ΔH_f 29.1 kJ mol^{-1} (30.5 cal g^{-1}), S_f 55.3 J mol^{-1} K^{-1} (13.2 cal mol^{-1} °C^{-1}, 277°) [16]; S_f 83.7 J mol^{-1} K^{-1} (20 cal mol^{-1} °C^{-1}) [17]; ΔH 35.12 kJ mol^{-1} (36.8 cal g^{-1}) [35]. Other ΔH values have been reported [4,26,34]

Specific Heat Capacity:

No.	Value	Note	Type
1	0.12 kJ/kg.C	6.78 cal mol^{-1} K^{-1}, 0° [13]	P

Melting Temperature:

No.	Value	Note
1	72°C	[15]
2	76°C	[16,29]
3	69°C	[17]
4	68–74°C	[18]
5	69.5°C	[20]
6	80–83.8°C	[25]
7	83°C	[26,38]
8	81°C	[24]
9	77°C	[31]
10	72.6°C	[35]
11	74–75°C	[37]

Glass-Transition Temperature:

No.	Value	Note
1	<-80°C	max. [14]
2	-53°C	220K [29]
3	-44°C	[35]
4	-30°C	[37]

Transition Temperature:

No.	Value	Note	Type
1	88°C	361K [29]	T°$_m$
2	-23°C	[33]	T$_\beta$
3	-70°C	[34]	T$_\beta$
4	-130°C	[34]	T$_\gamma$

Surface Properties & Solubility:
Solubility Properties General: Miscibility with polyepichlorohydrin has been reported [35]
Cohesive Energy Density Solubility Parameters: Cohesion energy 52.3 kJ mol^{-1} (12.5 kcal mol^{-1}) [41]
Internal Pressure Heat Solution and Miscellaneous: Absorption of EtOAc has been reported [8]. Toluene phase diagram has been reported [20]
Solvents/Non-solvents: Sol. warm isoamyl acetate, warm hexanol [24], 2-butanone [36]. Insol. MeOH [36]. Calculated solvent interaction parameters predict sol. C_6H_6, $CHCl_3$, THF [39]
Surface and Interfacial Properties General: Interfacial free energy 0.89 mJ cm^{-2} (8.9 erg cm^{-2}, parallel) [29]. Interfacial free energy 2.24 mJ cm^{-2} (22.4 erg cm^{-2}, perpendicular) [29]

Transport Properties:
Polymer Solutions Dilute: Soln. props. in DMF and cyclohexane have been reported [6]. [η] 0.15 dl.g^{-1} (C_6H_6, 25°) [31]. [η] 0.55 dl.g^{-1} (m-cresol, 30°) [37]. Other viscosity values have been reported [6]
Permeability and Diffusion General: Diffusion coefficient of water has been reported [5]
Water Content: Water sorption has been reported [5]

Mechanical Properties:
Complex Moduli: Real and imaginary complex shear moduli have been reported [34]
Viscoelastic Behaviour: Viscoelastic props. have been reported [42]

Electrical Properties:
Electrical Properties General: Volume resistivity has been reported [10]. Dielectric constant and loss factor have been reported. [19,33] Dissipation factor value has been reported [10]
Magnetic Properties: Diamagnetic susceptibility has been reported [41]

Optical Properties:
Transmission and Spectra: Esr [1], ir [2,3], Raman [2], H-1 nmr [27] and C-13 nmr [32] spectral data have been reported

Polymer Stability:
Decomposition Details: Thermal [11] and oxidative thermal degradation have been reported [11,28]. Activation energy for oxidation 80–84 kJ mol^{-1} (19–20 kcal mol^{-1}) [28]
Biological Stability: Biodegradation studies have been reported [12,40]

Applications/Commercial Products:
Processing & Manufacturing Routes: Synth. by transesterification of dimethyl sebacate with ethylene glycol under vacuum at 170–225° [18,31]

Bibliographic References
[1] Nozawa, Y., *Bull. Chem. Soc. Jpn.*, 1970, **43**, 657, (esr)
[2] Holland-Moritz, K., *Kolloid Z. Z. Polym.*, 1973, **251**, 906, (ir, Raman)
[3] Giger, A., Henniker, J. and Jacque, L., *Chim. Ind. (Paris)*, 1958, **79**, 757, (ir)
[4] Kozlov, P.V., Iouleva, M.M. and Shinjaeva, L.L., *Vysokomol. Soedin.*, 1959, **1**, 1106, (thermodynamic props)
[5] Willmott, P.W.T. and Billmeyer, F.W., Off. Dig., *Fed. Soc. Paint Technol.*, 1963, **35**, 847, (diffusion coefficient, water sorption)
[6] Orszagh, A. and Fejgin, J., *Polimery (Warsaw)*, 1963, **8**, 233, (soln props)
[7] Vasilevskaya, L.P., Bakeev, N.F., Kozlov, P.V. and Kargin, V.A., *Dokl. Akad. Nauk SSSR*, 1966, **168**, 846, (morphology)
[8] Bogaevskaya, T.A., Gatovskaya, T.V. and Kargin, V.A., *Vysokomol. Soedin., Ser. A*, 1970, **12**, 243, (EtOAc adsorption)
[9] Malinskii, Y.M. and Titova, N.M., *Vysokomol. Soedin., Ser. B*, 1976, **18**, 259, (thermodynamic props)
[10] Yang, S. and Pan, J., *CA*, 1981, **97**, 24476, (power factor, volume resistivity)
[11] Miroshnichenko, A.A., Platitsa, M.S., Volkov, V.I. and Doubnya, V.A., *Vysokomol. Soedin., Ser. A*, 1988, **30**, 2516, (thermal degradation)
[12] Doi, Y., Kasuya, K., Abe, H., Koyama, N. et al, *Polym. Degrad. Stab.*, 1996, **51**, 281, (biodegradation)
[13] Dole, M., *Kolloid Z. Z. Polym.*, 1959, **165**, 40, (specific heat capacity)
[14] Edgar, O.B., *J. Chem. Soc.*, 1952, 2633, 2638, (thermodynamic props, transition temp)

[15] Izard, E.F., *J. Polym. Sci.*, 1952, **8**, 503, (melting point)
[16] Dole, M. and Wunderlich, B., *Makromol. Chem.*, 1959, **34**, 29, (thermodynamic props)
[17] Wunderlich, B. and Dole, M., *J. Polym. Sci.*, 1958, **32**, 125, (density, specific heat capacity)
[18] Fuller, C.S. and Erickson, C.L., *J. Am. Chem. Soc.*, 1937, **59**, 344, (synth)
[19] Yager, W.A. and Baker, W.O., *J. Am. Chem. Soc.*, 1942, **64**, 2164, (dielectric constant)
[20] Carbonnel, L., Guieu, R. and Rossa, J.C., *Bull. Soc. Chim. Fr.*, 1970, 2855, (phase diagram)
[21] Hobbs, S.Y. and Billmeyer, F.W., *J. Polym. Sci., Part A-2*, 1969, **7**, 1119, (cryst struct)
[22] Fuller, C.S., *Chem. Rev.*, 1940, **26**, 143, (cryst struct)
[23] Esipova, N.G., Pan-Tun, L.N.S., Andreeva, N.S. and Kozlov, P.V., *Vysokomol. Soedin.*, 1960, **2**, 1109, (cryst struct)
[24] Kanamoto, T., Tanaka, K. and Nagai, H., *J. Polym. Sci., Part A-2*, 1971, **9**, 2043, (cryst struct)
[25] Theil, M. and Mandelkern, L., *J. Polym. Sci., Part A-2*, 1970, **8**, 957, (melting point)
[26] Hobbs, S.Y. and Billmeyer, F.W., *J. Polym. Sci., Part A-2*, 1970, **8**, 1387, (thermal expansion coefficient)
[27] Yamadera, R. and Murano, M., *J. Polym. Sci., Part A-1*, 1967, **5**, 2259, (H-1 nmr)
[28] Gomory, I., Stimd, J., Reya, R. and Gomoryova, A., *J. Polym. Sci., Part C: Polym. Lett.*, 1966, **16**, 451, (thermooxidative degradation)
[29] Godovsky, Y.K. and Slonimsky, Y.L., *J. Polym. Sci., Polym. Phys. Ed.*, 1974, **12**, 1053, (interfacial free energies)
[30] Kanamoto, T., *J. Polym. Sci., Polym. Phys. Ed.*, 1974, **12**, 2535, (morphology)
[31] Manzini, G. and Crescenzi, V., *Macromolecules*, 1975, **8**, 195, (equation of state)
[32] Kricheldorf, H.R., *Makromol. Chem.*, 1978, **179**, 2133, (C-13 nmr)
[33] Ito, M., Nakatani, S., Gokan, A. and Tanaka, K., *J. Polym. Sci., Polym. Phys. Ed.*, 1977, **15**, 605, (dielectric constant, transition temp)
[34] Ito, M., Kubo, M., Tsuruta, T. and Tanaka, K., *J. Polym. Sci., Polym. Phys. Ed.*, 1978, **16**, 1435, (complex shear moduli, transition temps)
[35] Fernandes, A.C., Barlow, J.W. and Paul, D.R., *J. Appl. Polym. Sci.*, 1984, **29**, 1971, (transition temp)
[36] Keith, H.D., *Macromolecules*, 1982, **15**, 114, 122, (morphology, cryst struct)
[37] Iwakura, Y., Taneda, Y. and Uchida, S., *J. Appl. Polym. Sci.*, 1961, **5**, 108, (thermal expansion coefficients, soln props)
[38] Lim, S. and Wunderlich, B., *Polymer*, 1987, **28**, 777, (specific heat capacity)
[39] Simek, L. and Bohdanecky, M., *Eur. Polym. J.*, 1987, **23**, 343, (polymer-solvent interaction parameters)
[40] Tokiwa, Y. and Suzuki, T., *J. Appl. Polym. Sci.*, 1981, **26**, 441, (biodegradation)
[41] Balta-Calleja, F.J. and Barrales-Rienda, J.M., *J. Macromol. Sci., Phys.*, 1972, **6**, 387, (cohesion energy, diamagnetic susceptibility)
[42] Tajiri, K., Fujii, Y., Aida, M. and Kawai, H., *J. Macromol. Sci., Phys.*, 1970, **4**, 1, (viscoelastic props)

Poly(ethylene succinate)   P-270

Synonyms: Polyethylene glycol succinate. Poly(oxyethyleneoxysuccinoyl). Poly(oxyethyleneoxysuccinyl). Ethylene glycol succinate polyester. Poly[oxy-1,2-ethanediyloxy(1,4-dioxo-1,4-butanediyl)]
Monomers: 1,2-Ethanediol, Succinic acid
Material class: Thermoplastic
Polymer Type: saturated polyester
CAS Number:

CAS Reg. No.	Note
25667-11-2	repeating unit
25569-53-3	Poly(ethylene succinate)

Molecular Formula: $(C_6H_8O_4)_n$
Fragments: $C_6H_8O_4$
General Information: Hard, brittle and white [30]
Morphology: Structural [45] and morphological [27,53,54] studies have been made and two polymorphic forms are known [40,58]. Orthorhombic, a 0.760, b 1.075, c 0.833 [11]; monoclinic, a 0.905, b 1.109, c 0.832, β 102.8° [32]. Other cryst. measurements have been reported (c 0.83, 0.842, 0.832 [33,36,58]).

Volumetric & Calorimetric Properties:
Density:

No.	Value	Note
1	d^{25} 1.358 g/cm^3	[32,34]
2	d 1.32 g/cm^3	[48]

Latent Heat Crystallization: Activation energy of polymerisation 54–59 kJ mol^{-1} (13–14 kcal mol^{-1}) [10]. Other ΔH_f values have been reported [7]. ΔH_f 13.8 kJ mol^{-1} (22.9 cal g^{-1}) [38]. Activation energy of melting 47.1 kJ mol^{-1} (11.25 kcal mol^{-1}) [38]. Conformational energy of fusion 335 J mol^{-1} (80 cal mol^{-1}) [39]. Conformational entropy of fusion 1.62 J K^{-1} mol^{-1} [39]

Melting Temperature:

No.	Value	Note
1	102–103°C	[1,2]
2	104°C	[5]
3	105–107°C	[8]
4	102°C	[29,31]
5	104–106°C	[32]
6	98–99°C	[35]
7	112°C	[39]
8	95–96°C	[56]
9	108°C	[42]
10	101°C	[47]
11	98°C	[48]
12	91°C	[49]

Glass-Transition Temperature:

No.	Value	Note
1	-12°C	[47]
2	-15°C	[48]
3	-1°C	272K [50]
4	-14°C	[40]
5	-23°C	[57]

Transition Temperature General: Other T_g, T_m°, T_m values have been reported [7,29]

Surface Properties & Solubility:
Solvents/Non-solvents: Sol. hot EtOAc, AcOH, Ac$_2$O, Me$_2$CO, ethyl succinate [30], CHCl$_3$ [42,46], o-chlorophenol [42], DMSO [46], 2,2,2-trifluoroethanol [46], CHCl$_3$/trifluoroacetic acid (50:50) [46]. Slightly sol. C$_6$H$_6$ [2]. Insol. petrol [8,30], MeOH [9,42], H$_2$O, Et$_2$O, EtOH [2,30]
Wettability/Surface Energy and Interfacial Tension: Surface free energy 2×10^{-5} J mm^{-2} (2 erg cm^{-2}) [38]. Other surface free energy values have been reported [16]

Transport Properties:
Polymer Solutions Dilute: Soln. props. have been reported [18]. [η] 0.367 dl g^{-1} (CHCl$_3$, 25°) [42], η$_{inh}$ 0.12 dl g^{-1} CHCl$_3$, 30°) [43]
Polymer Melts: Melt viscosity values have been reported [3,28,38]

Mechanical Properties:
Mechanical Properties General: Tensile strength of rods has been reported [54]

Electrical Properties:
Electrical Properties General: Volume resistivity and dissipation factor have been reported [19]

Poly(ethylene terephthalate)

Dielectric Permittivity (Constant):

No.	Value	Note
1	5.6	[34]

Optical Properties:
Refractive Index:

No.	Value	Note
1	1.4744	25° [52]

Transmission and Spectra: Ir [4,13,37,55], esr [9,50], Raman [13,55], mass [20,23,25,41,43,51] and C-13 [21,42,46,47] nmr spectral data have been reported

Polymer Stability:
Thermal Stability General: Thermal stability studies have been reported [14,20,22,23,25,41,43]. Decomposition temp. 350° [30]. Temp. of max. decomposition rate 410° [43]
Chemical Stability: Depolymerisation occurs in $CHCl_3$ and increases with increasing temp. [44]
Biological Stability: Biodegradability studies have been reported [12,15]

Applications/Commercial Products:
Processing & Manufacturing Routes: Synth. by condensation polymerisation of ethylene glycol and succinic acid at 200–250° under vacuum [8,9,25,26,30,31]

Bibliographic References

[1] Korshak, V.V., Vinogradova, S.V. and Vlasova, E.S., *Dokl. Akad. Nauk SSSR*, 1954, **94**, 61, (melting point)
[2] Korshak, V.V., Vinogradova, S.V. and Vlasova, E.S., Izv. Akad. Nauk SSSR, *Otd. Khim. Nauk*, 1954, 1089, (soln property, melting point)
[3] Takayanagi, M., *Nippon Kagaku Kaishi*, 1956, **59**, 49, (melt viscosity)
[4] Giger, A., Henniker, J. and Jacque, L., *Chim. Ind. (Paris)*, 1958, **79**, 757, (ir)
[5] Ueberreiter, K. and Steiner, K., *Makromol. Chem.*, 1966, **91**, 175, (melting point)
[6] Korshak, V.V., Golubev, V.V. and Karpova, G.V., Izv. Akad. Nauk SSSR, *Otd. Khim. Nauk*, 1958, 88, (solubilities)
[7] Hirai, N, *Nippon Kagaku Zasshi*, 1959, **80**, 1396, (thermodynamic props)
[8] Karpov, O.N., Fedosyuk, L.G. and Kovanko, Y.U., *Zh. Prikl. Khim. (Leningrad)*, 1969, **42**, 952, (synth)
[9] Nozawa, Y., *Bull. Chem. Soc. Jpn.*, 1970, **43**, 657, (esr, synth)
[10] Vancso-Szmercsany, I., Makay-Bodi, E., Szabo-Rethy, E. and Hirschberg, P., Kinet. Mech. Polyreactions, Int. Symp. Macromol. Chem., *Prepr.*, 1969, **1**, 103, (thermodynamic props)
[11] Uedo, A.S., Chatani, Y. and Tadokoro, H., *Polym. J. (Tokyo)*, 1971, **2**, 387, (cryst struct)
[12] Potts, J.E., Clendinning, R.A., Ackart, W.B. and Niegisch, W.D., *Polym. Sci. Technol. (Plenum)*, 1973, **3**, 61, (biodegradation)
[13] Holland-Moritz, K. and Hummel, D.O., *J. Mol. Struct.*, 1973, **19**, 289, (ir, Raman)
[14] Styskin, E.L., Taradai, E.P., Chikishev, Y.G. and Lukoyanova, L.V., *Zavod. Lab.*, 1975, **41**, 667, (thermal stability)
[15] Diamond, M.J., Freedman, B. and Garibaldi, J.A., *Int. Biodeterior. Bull.*, 1975, **11**, 127, (biodegradation)
[16] Privalko, V.P., *Polym. J. (Tokyo)*, 1978, **10**, 607, (surface free energy)
[17] Ponge, C., Rosso, J.C. and Carbonnel, L., *J. Therm. Anal.*, 1979, **15**, 101, (phase diagrams)
[18] Asmussen, F., *Ber. Bunsen-Ges. Phys. Chem.*, 1979, **83**, 309, (soln props)
[19] Yang, S. and Pan, J., *CA*, 1981, **97**, 24476, (volume resistivity, dissipation factor)
[20] Aguilera, C. and Luderwald, I., *Bol. Soc. Chil. Quim.*, 1983, **28**, 31, (ms, thermal degradation)
[21] Belfiore, L.A., Qin, C., Pires, A.T.N. and Ueda, E., Polym. Prepr. (Am. Chem. Soc., Div. Polym. Chem.), 1990, **31**, 170, (C-13 nmr)
[22] Mirashnichenko, A.A., Platitsa, M.S., Volkov, V.I. and Dovbyna, V.A., Vysokomol. Soedin., *Ser. A*, 1988, **30**, 2516, (thermal degradation)
[23] Plage, B. and Schulten, H.-R., *J. Anal. Appl. Pyrolysis*, 1988, **15**, 197, (ms, thermal degradation)
[24] Lele, A.K. and Shine, A.D., Polym. Prepr. (Am. Chem. Soc., Div. Polym. Chem.), 1990, **31**, 677, (solubility)
[25] Huang, F., Wang, X. and Li, S., *J. Anal. Appl. Pyrolysis*, 1991, **22**, 139, (synth, ms)
[26] *Jpn. Pat.*, 1993, 05 70 572, (synth)
[27] Al-Raheil, I.A. and Qudah, A.M.A., *Polym. Int.*, 1995, **37**, 249, (morphology)
[28] Yoshida, Y., Nakatsuka, R. and Oshima, K., *Kagaku Kogyo*, 1957, **31**, 74, (melt viscosity)
[29] Korenkova, O.P., *Khim. Nauka Promst.*, 1958, **3**, 824, (melting point)
[30] Carothers, W.H. and Arvin, J.A., *J. Am. Chem. Soc.*, 1929, **51**, 2560, (synth, solubility)
[31] Carothers, W.H. and Dorough, G.L., *J. Am. Chem. Soc.*, 1930, **52**, 711, (synth)
[32] Fuller, C.S. and Erickson, C.L., *J. Am. Chem. Soc.*, 1937, **59**, 344, (cryst struct, density)
[33] Storks, K.H., *J. Am. Chem. Soc.*, 1938, **60**, 1753, (struct)
[34] Yager, W.A. and Baker, W.O., *J. Am. Chem. Soc.*, 1942, **64**, 2164, (dielectric constant)
[35] Baker, W.D. and Fuller, C.S., *Ind. Eng. Chem.*, 1946, **38**, 272, (melting point)
[36] Fuller, C.S. and Baker, W.O., *J. Chem. Educ.*, 1943, **20**, 3, (struct)
[37] Davison, W.H.T. and Corish, P.J., *J. Chem. Soc.*, 1955, 2428, (ir)
[38] Ueberreiter, K. and Lucas, K.J., *Makromol. Chem.*, 1970, **140**, 65, (surface free energy, thermodynamic props)
[39] Tonelli, A.E., *J. Chem. Phys.*, 1972, **56**, 5533, (thermodynamic props)
[40] Schuster, P., *Thermochim. Acta*, 1972, **3**, 485, (transition temp)
[41] Luederwald, I. and Urrutia, H., *Makromol. Chem.*, 1976, **177**, 2093, (ms, thermal degradation)
[42] Horii, F., Hirai, A., Murayama, K., Kitamura, R. and Suzuki, T., *Macromolecules*, 1983, **16**, 273, (C-13 nmr)
[43] Garozzo, D., Giuffrida, M. and Montaudo, G., *Macromolecules*, 1986, **19**, 1643, (ms, thermal degradation)
[44] Domb, A.J. and Langer, R., *Macromolecules*, 1989, **22**, 2117, (soln stability)
[45] Liau, W.B. and Boyd, R.H., *Macromolecules*, 1990, **23**, 1531, (struct)
[46] Kricheldorf, H.R., *Makromol. Chem.*, 1978, **179**, 2133, (C-13 nmr)
[47] Belfiore, L.A., Qin, C., Ueda, E. and Pires, A.T.N., *J. Polym. Sci., Part B: Polym. Phys.*, 1993, **31**, 409, (C-13 nmr, transition temp)
[48] Fernandes, A.C., Barlow, J.W. and Paul, D.R., *J. Appl. Polym. Sci.*, 1984, **29**, 1971, (transition temp)
[49] Braun, D., Leiss, D., Bergmann, M.J. and Hellmann, G.P., *Eur. Polym. J.*, 1993, **29**, 225, (melting point)
[50] Lee, S.H. and Gelerinter, E., *Polymer*, 1994, **35**, 2725, (esr)
[51] Kim, Y.L. and Hercules, D.M., *Macromolecules*, 1994, **27**, 7855, (ms)
[52] *Polym. Handb.*, 3rd edn., (eds. J. Brandrup and E.H. Immergut), John Wiley and Sons, 1989, (refractive index)
[53] Steiner, K., Lucas, K.J. and Ueberreiter, K., *Kolloid Z. Z. Polym.*, 1966, **214**, 23, (morphology)
[54] Steiner, K., Engelbart, W., Asmussen, F. and Ueberreiter, K., *Kolloid Z. Z. Polym.*, 1969, **233**, 849, (morphology, tensile strength)
[55] Holland-Moritz, K., *Kolloid Z. Z. Polym.*, 1973, **251**, 906, (ir, Raman)
[56] Mayrhofer, R.C. and Sell, P.J., *Angew. Makromol. Chem.*, 1971, **20**, 153, (melting point)
[57] Rosso, J.C., Guieu, R. and Carbonnel, L., *Bull. Soc. Chim. Fr.*, 1972, **3**, 934, (transition temp)
[58] Fuller, C.S. and Frosch, C.J., *J. Phys. Chem.*, 1939, **43**, 323, (struct)

Poly(ethylene terephthalate) P-271

$$\left[\overset{O}{\underset{\|}{C}} - \underset{}{\bigcirc} - \overset{O}{\underset{\|}{C}} OCH_2CH_2O \right]_n$$

Synonyms: *PET. Poly(oxy-1,2-ethanediyloxycarbonyl-1,4-phenylenecarbonyl). Poly(oxyethylene oxyterephthaloyl)*
Monomers: Dimethyl terephthalate, 1,2-Ethanediol
Material class: Thermoplastic
Polymer Type: polyester
CAS Number:

CAS Reg. No.
25038-59-9

Molecular Formula: $(C_{10}H_8O_4)_n$
Fragments: $C_{10}H_8O_4$
Additives: Antioxidants (Irganox B-215 and 1098); compatibilisers (polypropylene-*graft*-maleic anhydride, polypropylene-*graft*-acrylic acid)

Poly(ethylene terephthalate)

Morphology: May be crystallised from the glassy state at 120° and 200°. Struct. is not a simple biphasic system and may be mesomorphic [13]

Volumetric & Calorimetric Properties:

Density:

No.	Value	Note
1	d 1.2 g/cm^3	melt, 300° [4]
2	d 1.333 g/cm^3	amorph. [4]
3	d 1.365 g/cm^3	biaxially oriented [4]
4	d 1.515 g/cm^3	cryst., calc. [6]

Thermal Expansion Coefficient:

No.	Value	Note	Type
1	0.00016 K^{-1}	30–60° [4]	L
2	0.00037 K^{-1}	90–190° [4]	L
3	0.00065 K^{-1}	melt [5]	L

Latent Heat Crystallization: ΔH_{fusion} 37.7–67 J g^{-1} (9–16 cal g^{-1})

Thermal Conductivity:

No.	Value	Note
1	0.147 W/mK	33°, film [12]
2	0.14 W/mK	0.000336 cal (cm s K)$^{-1}$, ASTM C177 [4]

Thermal Diffusivity: 0.000929 cm^2 s^{-1} (33°, film) [12]

Melting Temperature:

No.	Value	Note
1	245–258°C	[4]
2	280°C	[6]

Glass-Transition Temperature:

No.	Value	Note
1	73°C	[4]
2	67°C	amorph. [8]
3	81°C	cryst. [8]
4	125°C	cryst., oriented [9]

Deflection Temperature:

No.	Value	Note
1	65°C	150°F [4]

Surface Properties & Solubility:

Solvents/Non-solvents: Sol. phenol, chlorophenol, chlorobenzene, dichloroethane. Insol. hexane, C_6H_6, EtOH, cyclohexane

Surface Tension:

No.	Value	Note
1	39.5 mN/m	25°, solid/liq. [10]
2	27 mN/m	290°, melt [10]

Transport Properties:

Polymer Solutions Dilute: $[\eta]_{inh}$ 0.72–0.8 dl g^{-1} [4]

Gas Permeability:

No.	Gas	Value	Note
1	CO_2	7.87 cm^3 mm/(m^2 day atm)	20 cm^3 mil (100 in^2 day atm)$^{-1}$, 25°, unoriented [4]
2	CO_2	4.72 cm^3 mm/(m^2 day atm)	12 cm^3 mil (100 in^2 day atm)$^{-1}$, 25°, oriented [4]
3	O_2	3.94 cm^3 mm/(m^2 day atm)	10 cm^3 mil (100 in^2 day atm)$^{-1}$, 25°, unoriented [4]
4	O_2	1.96 cm^3 mm/(m^2 day atm)	5 cm^3 mil (100 in^2 day atm)$^{-1}$, 25°, oriented [4]
5	N_2	0.18 cm^3 mm/(m^2 day atm)	0.46 g mil (100 in^2 day atm)$^{-1}$, film, oriented
6	N_2	0.28–0.39 cm^3 mm/(m^2 day atm)	25°, oriented film

Mechanical Properties:

Tensile (Young's) Modulus:

No.	Value	Note
1	2757 MPa	400 kpsi, ASTM D638, amorph. [4]
2	4136 MPa	600 kpsi, ASTM D638, oriented film [4]
3	14100 MPa	oriented film
4	63 MPa	oriented film, transverse [11]

Tensile Strength Break:

No.	Value	Note	Elongation
1	48 MPa	7000 psi, amorph. ASTM D638 [4]	250–500%
2	248 MPa	36 kpsi, longitudinal [4]	60%
3	69 MPa	10 kpsi, transverse [4]	150%
4	172 MPa	25 kpsi, biaxially oriented, ASTM D638 [4]	100%
5	172 MPa	film [7]	

Poisson's Ratio:

No.	Value	Note
1	0.44	extension [11]
2	0.37	transverse [11]

Impact Strength: Impact strength 240 J m^{-1} (film) [7]

Electrical Properties:

Surface/Volume Resistance:

No.	Value	Note	Type
1	2×10^{15} Ω.cm	23°, 30% relative humidity [7]	S
2	0.0002×10^{15} Ω.cm	23°, 80% relative humidity [7]	S

Dielectric Permittivity (Constant):

No.	Value	Frequency	Note
1	3.3	60 Hz	23° [7]
2	3.25	1 kHz	23° [7]
3	3	1 MHz	23° [4]
4	3.7	60 Hz	150° [7]

Poly(ethylene terephthalate)

Dielectric Strength:

No.	Value	Note
1	295 kV.mm^{-1}	60 Hz, 23° [7]
2	275 kV.mm^{-1}	60 Hz, 150° [7]

Dissipation (Power) Factor:

No.	Value	Frequency	Note
1	0.0025	60 Hz	23° [7]
2	0.005	1 kHz	23° [7]
3	0.0016	1 MHz	23° [7]
4	0.0004	60 Hz	150° [7]

Polymer Stability:
Upper Use Temperature:

No.	Value
1	150°C

Flammability: Is completely combustible, leaving little residue. Burns more cleanly than coal and other fossil fuels
Chemical Stability: Good resistance to oils and fats, organic alcohols and hydrocarbons. Poor resistance to phenols, ketones, esters and chlorinated organic solvents
Hydrolytic Stability: Good resistance to dilute acids and alkalis, poor resistance to conc. bases
Recyclability: Can be recycled through depolymerisation

Applications/Commercial Products:
Processing & Manufacturing Routes: Synth. by two routes; reaction of dimethyl terephthalate with 1,4-ethanediol with catalyst (usually acetate salts of zinc, calcium, manganese, magnesium and cobalt) at 150–210°. MeOH and excess diol removed via condenser. The intermediate is then polymerised at 270–280° at 0.5–1 mmHg with an antimony (III) oxide catalyst. Stabilisers or colour improvers, e.g. triaryl phosphites or phosphates, are added to prevent precipitation of insol. terephthalate salts. Higher MW products can be obtained by a solid-state post-polymerisation process or by chain extension reactions. Also prepd. by reaction of terephthalic acid and aq. ethylene glycol at 250° and 276 kPa for 3h. It is then further treated at 245° and atmospheric pressure for 2h. Common processing techniques used include injection moulding, extrusion and thermoforming. Polymer must be thoroughly dried otherwise loss of MW and deterioration of phys. props. occurs

Mould Shrinkage (%):

No.	Value	Note
1	0.003–0.005%	3 mm thickness, in direction of flow
2	0.003–0.005%	perpendicular to direction of flow

Applications: Clothing, films, bottles, food containers and engineering plastics for precision-moulded parts. A wide range of applications is possible because of the excellent balance of props. PET possesses and because the degree of crystallinity and the level of orientation in the finished part can be controlled. Can be made into articles with excellent clarity, and outstanding dimensional stability can be achieved. For packaging PET combines optimum processing, mech. and barrier props. Barrier props. are adequate for retention of carbon dioxide for packaging carbonated beverages and good oxygen barrier props. allow for its use in many types of food containers. However, barrier props. not sufficient for beer and foods that are highly sensitive to oxygen. Glass reinforced PET used for electrical connectors, auto body parts. Copolymer used for clear mouldings, e.g. medical devices, toys, cosmetic jars

Trade name	Details	Supplier
Arnite	Injection-mouldable with nucleating agents	Akzo
Arnite	AO/AV/DO	DSM Engineering
Bexloy	AP V-170/171/172 K 550	DuPont
Celstran PET	G30/40/50/60	Polymer Composites
Cleartuf		Shell Chemicals
Comtuf	461/462/463/464	Comalloy Intl. Corp.
E2	various grades	Thermofil Inc.
Eastalloy	EA001	Eastman Perform
Eastapak PET	various grades	Eastman Chem. Prod.
Electrafil	J-1800/CF/30	DSM Engineering
Hiloy	400 range	Comalloy Intl. Corp.
Hostalen		Hoechst
Impet	various grades	Hoechst Celanese
J-1800	various grades	DSM Engineering
Kodapak		Eastman Chemical Company
Kodar PETG copolyester	Copolymer	Eastman Chemical Company
Lumirror	F6G	Toray Industries
Melinar		ICI, UK
Mylar	various grades	DuPont
PET	200G30L/200G45L	Michael Day
PET	4410/G3/G6/G8	Nan Ya Plastics
Petlon		Mobay
Petra		Allied Chemical
Plenco	50030/50045/50125/ 50130/50240/50335	Plastics Engr. Co.
RTP	1100 series	RTP Company
Repete	80310C/80320C	Shell Chemical Co.
Ropet		Rohm and Haas
Rynite	Fast crystallising PET moulding compounds	E.I. DuPont de Nemours & Co., Inc.
Selar		DuPont
Shinite	T series	Shinkong Synthetic
Spectar	PETG 14471	Eastman Chem. Prod.
Stanuloy	ST series	MRC Polymers Inc.
Tenite		Eastman Chemical Company
Traytuf	various grades	Shell Chemical Co.
Valox	various grades	General Electric
Voloy	440/441/460/461/462	Comalloy Intl. Corp.

Bibliographic References
[1] *Handbook of Plastic Materials and Technology,* (ed. I.I. Rubin), Wiley Interscience, 1990
[2] *Encycl. Polym. Sci. Eng.* 2nd edn., (ed. J.I. Kroschwitz), John Wiley and Sons, 1988
[3] *Concise Encyclopedia of Polymer Science and Engineering,* (eds. H.F. Mark, N.M. Bikales, C.G. Overberger and G. Menges), John Wiley and Sons, 1990

[4] Cleartuf® PET Packaging Resins Shell Chemical, 1996, (technical datasheet)
[5] Starkweather, H.W., Zoller, P. and Jones, G.A., *J. Polym. Sci., Polym. Phys. Ed.*, 1983, **21**, 295, (thermal expansion)
[6] Fakirov, S., Fisher, E.W. and Schmidt, G.H., *Makromol. Chem.*, 1975, **176**, 2459, (density)
[7] Heffelfinger, C.L. and Knox, K.L., *Sci. Technol. Polym. Films*, (ed. O.J. Sweeting), Wiley, 1971, 587, (electric props)
[8] Edgar, O.B. and Hill, R, *J. Polym. Sci.*, 1952, **8**, 1, (T_g)
[9] Woods, D.W., *Nature (London)*, 1954, **174**, 173
[10] Kaelble, D.H., *J. Adhes.*, 1970, **2**, 66, (surface tension)
[11] Ward, I.M., *J. Macromol. Sci., Phys.*, 1967, **1**, 667
[12] Steere, R.C., *J. Appl. Phys.*, 1966, **37**, 3338
[13] Bove, L. et al, *Polym. Adv. Technol.*, 1996, **7**, 858

Poly(ethylene-*co*-vinylacetate)   P-272

$$\left[-(CH_2CH_2)_x-(CH_2CH)_y- \atop { \quad\quad\quad\quad\quad | \atop \quad\quad\quad\quad\quad O-\underset{\underset{O}{\|}}{C}-CH_3 }\right]_n$$

Polymer Stability:
Environmental Stress: Stress crack resistance is significantly greater than for LDPE. Cross-linking gives further improvement [1]

Poly(ethylene-*co*-vinyl alcohol)   P-273

Synonyms: *EVOH. ETVA. EVAL. EVA. Hydrolysed ethylene-vinyl alcohol copolymer. Poly(ethenol-co-ethene)*
Related Polymers: Poly(vinyl alcohol), Poly(ethylene), Ethylene-vinyl acetate copolymer, Ethylene-vinyl alcohol copolymer films
Monomers: Ethylene, Vinyl acetate
Material class: Gums and resins, Copolymers
Polymer Type: polyethylene, polyvinyl alcohol
CAS Number:

CAS Reg. No.
25067-34-9

Molecular Formula: $[(C_2H_4)_x(C_2H_4O)_y]_n$
Fragments: C_2H_4 C_2H_4O
General Information: Degree of hydrolysis is usually greater than 98%. Ethylene-vinyl alcohol copolymer can be prod. in any ratio. Polymer composition affects physical props. and applications. When ethylene content is 82–90 mol% it is used as an adhesive not as a barrier polymer; when vinyl alcohol content is 60–75 mol% it is used as barrier polymer and for processing into films [2]. The copolymer is random and amphoteric consisting of hydrophobic and hydrophilic segments
Morphology: EVOH copolymers are co-crystallisable over the entire range of copolymer compositions and can form isomorphic crysts.

Volumetric & Calorimetric Properties:
Density:

No.	Value	Note
1	d 1.14–1.19 g/cm³	ASTM D792 [1,4,7,36]
2	d 1.162 g/cm³	62 mol % vinyl alcohol [3]
3	d 1.2 g/cm³	73 mol % vinyl alcohol [3]
4	d 1.14 g/cm³	44 mol % ethylene [6,7]

Thermal Expansion Coefficient:

No.	Value	Note	Type
1	0.0006264 K^{-1}	melt. 59 mol % ethylene, degree of polymerisation 680 [9]	L

Volumetric Properties General: Density increases with increasing vinyl alcohol content [3,5]. The increase in density with increasing vinyl alcohol content may be related to the sample crystallinity [5]. Density may also depend on the melt index and hence MW [5]
Equation of State: Equation of state information has been reported [8]
Latent Heat Crystallization: The enthalpy of melting is affected by sample processing and increases with increasing vinyl alcohol content. ΔH_m 74–86 J g^{-1} (56 mol % vinyl alcohol), 89–101 J g^{-1} (68 mol % vinyl alcohol) [10]. Heat of combustion 27.3–40.8 kJ g^{-1} [17,18]. Heat of fusion depends on the vinyl alcohol content of the polymer. At low vinyl alcohol content, the heat of fusion decreases with increasing vinyl alcohol content, reaching a min. value of 2.35 J g^{-1} (9.84 cal g^{-1}) at 40 mol % vinyl alcohol. The heat of fusion then increases with vinyl alcohol content attaining a max. value of 4.73 J g^{-1} (19.8 cal g^{-1}) at 82 mol % vinyl alcohol [3].
Thermal Conductivity:

No.	Value	Note
1	0.2 W/mK	[8]

Specific Heat Capacity:

No.	Value	Type
1	1.9 kJ/kg.C	P

Melting Temperature:

No.	Value	Note
1	183°C	ethylene copolymer ratio 0.32, DSC method [4]
2	158°C	ethylene copolymer ratio 0.44, DSC method [4]
3	191°C	ethylene copolymer ratio 0.27, DSC method [4]
4	159°C	59 mol % ethylene [9]
5	165–187.5°C	56–71 mol % vinyl alcohol [10]
6	107–205°C	13.8–82 mol % vinyl alcohol [3]
7	165°C	44 mol % ethylene [6]
8	142–181°C	[36]

Glass-Transition Temperature:

No.	Value	Note
1	52°C	29% ethylene, DSC method [11]
2	65°C	29% ethylene, DMA method [11]
3	31°C	59 mol % ethylene, DSC method [9]
4	58°C	dynamic mechanical analysis [12]
5	68°C	dry, ethylene copolymer ratio 0.32, DMA method [4]
6	69°C	[1]
7	55–69°C	[36]

Transition Temperature General: T_g is reduced by H_2O absorption [12,13] and by the sorption of alcohols [3,11], with MeOH and EtOH giving the lowest T_g values due to interactions with the polymer [11]. The effect on T_g of MeOH absorption is similar to that observed for poly(vinyl alcohol) [4]. A variation of approx. 15° was noted for the T_g of different EVOH samples [4]. T_m increases with increasing vinyl alcohol content and decreases with increasing ethylene content. [3,10] T_m is similar to that of LDPE samples at low concentrations of vinyl alcohol ($\leq$ 20 mol %) [5].

The Vicat softening point is higher than that of the corresponding ethylene-vinyl acetate copolymer and depends slightly on the vinyl alcohol content [5]. Samples with high melt index have a lower softening point [5]. T_m is lowest at 13.8 mol % vinyl alcohol [3]. T_m is affected by a sample's thermal history, with the effect more pronounced at higher vinyl alcohol content [10]. Brittleness temp. may be lower for samples with a high melt index [5]

Vicat Softening Point:

No.	Value	Note
1	90–110°C	≤ 20 mol% vinyl alcohol, ASTM D1525-76 [5]

Brittleness Temperature:

No.	Value	Note
1	-76°C	≤ 20 mol % vinyl alcohol [5]

Transition Temperature:

No.	Value	Note
1	-75–2°C	low temp. relaxation
2	124°C	high temp. relaxation [12]

Surface Properties & Solubility:
Solubility Properties General: EVOH (56 mol% vinyl alcohol) is compatible with poly(4-vinylpyridine) when the poly(4-vinylpyridine) content is less than 60% (w/w) [16]. Blends of EVOH (59 mol%) ethylene) and nylon 6–12 (60 mol% nylon) are miscible in the molten state in the overall blend composition range at 150° [9]. The Flory interaction parameter for this blend is -0.58 [9]
Cohesive Energy Density Solubility Parameters: Cohesive energy density 41.8 J cm^{-3} (175 cal ml^{-1}) [11]
Plasticisers: Water [13], MeOH [4,11,14], EtOH [11], glycerol, glycerine [6]
Solvents/Non-solvents: Sol. DMSO. Insol. C_6H_6, Me_2CO, MeOH, EtOH [15]

Transport Properties:
Transport Properties General: Melt flow index varies widely and depends on the MW and MW distribution [5]

Melt Flow Index:

No.	Value	Note
1	1.3 g/10 min	32 mol % ethylene [7]
2	5.5 g/10 min	44 mol % ethylene [7]
3	12 g/10 min	210°, 44 mol % ethylene [6]
4	1.9–22 g/10 min	ASTM D1238-73 [5]

Polymer Solutions Dilute: $[\eta]_{inh}$ 0.89 dl g^{-1} (30°, DMSO) [15]
Polymer Melts: The temp. dependence of the melt rheology has been reported [7]
Permeability and Diffusion General: Permeability increases with ethylene fraction (but moisture sensitivity decreases). 60–75 mol.% ethylene copolymer is a compromise chosen to allow good barrier properties with acceptably low moisture sensitivity
Water Content: Equilib. H_2O content at 23° after 7 d 0.5% (10% relative humidity); 0.8% (20% relative humidity); 1.7% (50% relative humidity); 3.5% (80% relative humidity); 4.4% (90% relative humidity). Equilib. H_2O content increases dramatically upon plasticisation with 15% glycerol [6]. WVTR 0.35 nmol m^{-1} s^{-1} (44 mol% ethylene); 0.95 nmol m^{-1} s^{-1} (32 mol% ethylene) [37]

Water Absorption:

No.	Value	Note
1	6.7–8.6 %	24h 1/8 inch thick specimen, ASTM D570 [36]

Mechanical Properties:
Mechanical Properties General: Mechanical props. are affected by H_2O absorption, the presence of plasticisers [6], the thermal history, preparation conditions [5,10,35] and the copolymer composition [3,5]. The temp. dependence of elasticity and tensile modulus is found to be consistent with a stepwise increase in the molecular mobility [26]. The change in yield strain with annealing depends on the vinyl alcohol content. The yield stress and the Young's modulus of all samples increases with annealing below T_g [35]. Tensile strength at break and yield, and the elastic modulus of compression-moulded samples increase with increasing vinyl alcohol content. Tensile strength at yield is at a max. when vinyl alcohol content is 15 mol %. Elongation at break decreases with increasing vinyl alcohol content [5]

Tensile (Young's) Modulus:

No.	Value	Note
1	345 MPa	50 kpsi, approx., injection moulding, moisture content 6.5% [6]
2	2000–2700 MPa	300–385 kpsi, ASTM D638 [36]

Tensile Strength Break:

No.	Value	Note	Elongation
1	22.8 MPa	3.3 kpsi, injection moulding, moisture content 6.5%	330%
2	10–15 MPa	compression moulding [5]	150–300%
3	59–80 MPa	8520–11600 psi, ASTM D638 [36]	180–280%
4	43 MPa	film, 44 mol% ethylene	4%
5	66.2 MPa	film, 32 mol% ethylene [38]	5%

Flexural Strength at Break:

No.	Value	Note
1	1.2–2 MPa	230–285 psi, ASTM D790 [36]

Tensile Strength Yield:

No.	Value	Note
1	10–20 MPa	compression moulding [5]
2	51–71 MPa	7385–10365 psi, ASTM D638 [36]

Impact Strength: Izod 53–62 J m^{-1} (1–1.7 ft lb/in, 1/8 inch thick, ASTM D256A [36]
Viscoelastic Behaviour: Horizontal shift activation energy 3.69 kJ mol^{-1} (15.45 kcal mol^{-1}) (32 mol% ethylene), 3.29 kJ mol^{-1} (13.78 kcal mol^{-1}) (44 mol% ethylene) [7]. The vertical shift activation energy is zero [7]. Data on viscoelastic behaviour have been reported [22]
Mechanical Properties Miscellaneous: The torsional stiffness of EVOH polymers is higher than that of LDPE and ethylene-vinyl acetate [5]. Torsional stiffness decreases with increasing temp. and the same temp. effect is observed for increasing vinyl alcohol concentration [5]
Hardness: Shore A and D values have been measured under standard conditions (ASTM D2240-75). [5] Hardness increases as the vinyl alcohol content is increased [5]. Microhardness increases with vinyl alcohol content but is also affected by the samples' thermal history [10]. Quenched samples have a lower value of microhardness than slow cooled samples [10]. Microhardness increases with increasing annealing temp. for quenched samples with the opposite effect for slow cooled samples [10]. Microhardness is in the range 125–200 MPa depending on the sample [10]

Failure Properties General: Stress-strain curves show that EVOH is a soft, tough material [6]

Electrical Properties:
Electrical Properties General: Dielectric props. are affected by moisture absorption [12,13] and temp. [12,13,26]

Optical Properties:
Transmission and Spectra: Ir [27], Raman [28], C-13 nmr [29,30,31,32,33] and H-1 nmr spectral data [30,34] have been reported
Volume Properties/Surface Properties: Haze 2.3% (film, 44 mol% ethylene); 1.8% (film, 32mol% ethylene) [38]

Polymer Stability:
Thermal Stability General: Weight loss is very low (approx. 2%) at 300° for a sample containing 32 mol % ethylene [15]. Thermal stability depends on the copolymer composition and is related to the thermal stabilities of poly(ethylene) and poly(vinyl alcohol). The processing conditions affect the decomposition and MW degradation. Degradation by melt shearing can occur [24]
Flammability: Limiting oxygen index 18.0–19.6 (12–74 mol % vinyl alcohol) and increases as the vinyl alcohol content increases [18]
Environmental Stress: Environmental stress cracking (ASTM D1693-70) by aq. detergent and pure detergent is reduced by increasing the melt index and the detergent concentration [5]. Samples with the lowest vinyl alcohol content have the lowest resistance to environmental stress cracking in dil. detergent [5]
Chemical Stability: The vinyl alcohol groups in EVOH copolymer can react with formaldehyde to prod. formals [19,20,21] and with acetaldehyde [22] to form acetals. EVOH copolymers undergo a base-catalysed Michael-type addition reaction with vinyl sulf-oxides [15]. Reaction with phosphoric acid in suspension or aq. soln. forms phosphorylated copolymers [17,18]. Swollen by H_2O and alcohols [4,6,11,14]
Hydrolytic Stability: EVOH samples prepared from ethylene-vinyl acetate copolymer may undergo further hydrol. with the addition of strong acid or base in a manner analogous to that of poly (vinyl alcohol), which increases the number of vinyl alcohol groups

Applications/Commercial Products:
Processing & Manufacturing Routes: EVOH copolymers are prepared by hydrolysis of the corresponding ethylene-vinyl acetate copolymer. Methanolic KOH in C_6H_6 has been used. Injection or extrusion moulding 185–250° [36]. Manufactured by hydrolysis of poly(ethylene-*co*-vinyl acetate)
Applications: The main use of EVOH copolymers is in the formation of film and sheeting material. High ethylene content copolymers may be used as adhesives (ethylene content 82–90 mol%) and as coating materials for treatment of poly(vinyl alcohol) fibres. The copolymer may be used as an additive for rubber materials. Barrier applications (typically as inner layer of laminate to protect from moisture- 25–45 mol% ethylene); adhesive (10–12 mol% ethylene [38]

Trade name	Details	Supplier
Clarene	various grades	Solvay & Cie
Clarene		Solvay
Eval		Kuraray Co
Eval		Nippon Synthetic Chemical Industry Co
NCPE 0210		Neste Chemicals
Novatec	adhesive	Mitsubishi Chemical
Selar OH		DuPont
Soarnol		Nippon Gohsei
Soarnol		Mitsubishi Chemical
Vinese	various grades	Air Products

Bibliographic References
[1] *Encycl. Polym. Sci. Eng.,* 2nd edn., (ed. J.I. Kroschwitz), John Wiley and Sons, 1989, 176
[2] Iwanami, T. and Hirai, V., *Tappi,* 1992, 339
[3] Nishino, T., Takano, K. and Nakamae, K., *Polymer,* 1995, **36**, 959, (props.)
[4] Samus, M.A. and Rossi, G., *Macromolecules,* 1996, **29**, 2275, (T_g, MeOH absorption)
[5] Koopmans, R.J., van der Linden, R. and Vansant, E., *Polym. Eng. Sci.,* 1983, **23**, 306, (mechanical and thermal props.)
[6] George, E.R., Sullivan, T.M. and Park, E.H., *Polym. Eng. Sci.,* 1994, **34**, 17, (water absorption)
[7] Mavridis, H. and Shroff, R.H., *Polym. Eng. Sci.,* 1992, **32**, 1778, (melt rheology)
[8] Puissant, S., Vergnes, B., Demay, Y. and Agassant, J.F., *Polym. Eng. Sci.,* 1992, **32**, 213, (rheological props.)
[9] Akiba, I. and Akiyama, S., *Polym. J. (Tokyo),* 1994, **26**, 873
[10] Fonsenca, C., Perena, J.M., Benavente, R., Cerrada, M.L. *et al, Polymer,* 1995, **36**, 1887, (thermal props., hardness)
[11] Gavara, R., Catala, R., Aucejo, S., Cabedo, D. and Hernandez, R., J. Polym. Sci., *Part B: Polym. Phys.,* 1996, **34**, 1907, (sorption of alcohols)
[12] Starkweather, H.W., Avakian, P., Fontanella, J.J. and Wintersgill, M.C., Plast., *Rubber Compos. Process. Appl.,* 1991, **16**, 255, (dielectric props., T_g)
[13] Bizet, A., Nakamura, N., Teramoto, Y. and Hatakeyama, T., *Thermochim. Acta,* 1994, **241**, 191, (dielectric props., T_g)
[14] Samus, M.A. and Rossi, G., *ACS Symp. Ser.,* 1995, **598**, 535, (liq. sorption)
[15] Imai, K., Shiomi, T., Tezuka, Y. and Inokami, K., *Polym. Int.,* 1992, **27**, 309, (solubility)
[16] Isasi, J.R., Cesteros, L.C. and Katime, I., Makromol. Chem., *Rapid Commun.,* 1994, **15**, 903, (miscibility)
[17] Banks, M., Ebdon, J.R. and Johnson, M., Flame Retard. '94, (Proc. Conf.), 6th, 6th, 1994, 183, (flammability)
[18] Banks, M., Ebdon, J.R. and Johnson, M., *Polymer,* 1993, **34**, 4547, (flammability)
[19] Yonezu, K., *Kobunshi Ronbunshu,* 1993, **50**, 637, (props.)
[20] Yonezu, K., *Kobunshi Ronbunshu,* 1995, **52**, 698, (props.)
[21] Yonezu, K., *Kobunshi Ronbunshu,* 1992, **49**, 993, (props.)
[22] Akiyama, S. and Akiba, I., *Adv. Sci. Technol.,* 1995, **4**, 273, (viscoelastic props)
[23] Fedtke, M. and Henning, K., Makromol. Chem., *Makromol. Symp.,* 1989, **28**, 203, (props.)
[24] Park, W.R.R., Therm. Anal., Proc. Int. Conf. 7th, 1982, **2**, 1057, (thermal anal, stability)
[25] La Mantia, F.P., and Spadaro, G., *Mater. Chem. Phys.,* 1983, **8**, 315, (dielectric props.)
[26] Kabisch, O., Lorenz, P., Kemp, M., Bauer, K. and Juergens, G., *Plaste Kautsch.,* 1983, **30**, 675, (dielectric and mechanical props.)
[27] Koopmans, R.J., van den Linden, R. and Vansant, E.F., *Spectrosc.: Int. J,* 1982, **1**, 352, (ir)
[28] Cooney, T.F., Wang, L., Sharma, S.K., Gauldie, R.W. and Montana, A.J., J. Polym. Sci., *Part B: Polym. Phys.,* 1994, **32**, 1163, (Raman)
[29] van der Hart, D.L., Simmons, S. and Gilman, J.W., *Polymer,* 1995, **36**, 4223, (C-13 nmr)
[30] Bruch, M.D., *Macromolecules,* 1988, **21**, 2707, (C-13 nmr)
[31] Montani, T. and Iwasaki, H., *Macromolecules,* 1978, **11**, 1251, (C-13 nmr)
[32] Cheng, H.N. and Bennett, M.A., *Anal. Chem.,* 1984, **56**, 2320, (C-13 nmr)
[33] Ketels, H., de Haan, J., Aerdts, A. and van der Velden, G., *Polymer,* 1990, **31**, 1419, (C-13 nmr)
[34] Cheng, H.N. and Lee, G.H., *Macromolecules,* 1988, **21**, 3164, (H-1 nmr)
[35] Yoshida, H., Kanbara, H. and Takemura, N., *Sen'i Gakkaishi,* 1983, **39**, T512, (mechanical props.)
[36] *Modern Plastics Encyclopedia,* Guide to plastics- Property and Specification Charts McGraw-Hill, 1987, 73, (mechanical, thermal, physical and processing data)
[37] De Lassus, P., *Kirk-Othmer Encycl. Chem. Technol.,* 4th edn. (edn. Kroschwitz, J.I. and Howe-Grant, M., Wiley Interscience, 1992, **3**, 940, (barrier polymers-section starts p.31, permeability and diffusion, composition general info)
[38] Combellick, W.A., *Encycl. Polym. Sci. Eng.,* 2nd ed. (ed. J.I. Kroscharitz), Wiley-Interscience, 1985, **2**, 188, (general inforrmation)

Poly(ethylene-*co*-vinyl alcohol), Films

Synonyms: *EVOH films. EVAL films. Hydrolysed ethylene-vinyl acetate films*
Related Polymers: Poly(vinyl alcohol) films, Poly(ethylene) films, Ehtylene-vinyl alcohol copolymers, general information

– Poly(ethylene-*co*-vinyl alcohol), Films

Monomers: Ethylene, Vinyl acetate
Material class: Copolymers, Fibres and Films
Polymer Type: polyethylene, polyvinyl alcohol
CAS Number:

CAS Reg. No.
25067-34-9

Molecular Formula: $[(C_2H_4).(C_2H_4O)]_n$
Fragments: C_2H_4 C_2H_4O
General Information: The vinyl alcohol content of the copolymer used to form films is 60–75 mol%
Miscellaneous: Multilayer sheets and films consist of base film, adhesive, EVOH, adhesive, heat sealant. The following commercial coextruded films are prod.: Nylon 6/EVOH/adhesive/Nylon 6 and poly(propylene)/adhesive/EVOH/adhesive/poly(propylene). Polyethylene and ethylene-propylene copolymer layers may also be used. Soarnol is an ACBCA film where A, ethylene-propylene copolymer; B, EVOH and C, blend of ethylene-propylene copolymer and maleated ethylene-propylene copolymer. Poly-ethylene-nylon blends may also be used

Volumetric & Calorimetric Properties:
Density:

No.	Value	Note
1	d 1.19 g/cm^3	commercial film, ethylene content 32 mol %, ASTM D638 [3,9,10]
2	d 1.14 g/cm^3	commercial film, ethylene content 44 mol %, ASTM D638 [3,9,10]
3	d 1.15 g/cm^3	commercial film [11]

Volumetric Properties General: Density increases with increasing vinyl alcohol content [12] and may be affected by water absorption [13,14]
Thermodynamic Properties General: Thermal conductivity has been measured by ir [15]
Transition Temperature General: T_g is significantly increased by γ-irradiation. T_m is only slightly affected [32,33]

Surface Properties & Solubility:
Solvents/Non-solvents: Sol. hot H_2O [6]. Swollen by H_2O, with swelling increasing with increasing vinyl alcohol content [21]
Wettability/Surface Energy and Interfacial Tension: Contact angle 53–61° (H_2O, 25°) [16], 60–68° (H_2O, 20°). The contact angle measurements in H_2O are affected by immersion in H_2O and differences between the advancing and receding contact angles are observed [20]. Contact angle 55–59° (glycerol, 20°), 48–52° (formamide, 20°), 34–37° (methylene iodide, 20°), 31–33° (ethylene glycol, 20°), 50.2–84.9° (hexane/ethylene glycol) [18], 54.2–93.6° (cyclohexane/ethylene glycol) [18]. Contact angle decreases with increasing vinyl alcohol content [17,21]. Contact angle 62.4–92.6° (heptane/ethylene glycol) [18], 53.6–86.8° (octane/ethylene glycol) [18], 49.4–83.8° (isooctane/ethylene glycol) [18]. Contact angles are affected by the copolymer composition and the degree of polymerisation [18]. γ_d 35.9–36.0, γ_p 9.5–13.9, γ_c 38.0–39 [17], γ_d 27.5–30.3, γ_p 12.0–15.5 [17]; γ_d 35.9–36.7; γ_p 9.7–11.2 [17]. γ_d increases slightly with increasing vinyl alcohol content whereas γ_p increases dramatically [17]. γ_d 28–30; γ_p 2.7–9.1 [18]. The dispersive component of the surface free energy is virtually independent of the ethylene content [18]. The other components increase with increasing ethylene content but are also affected by the degree of polymerisation [18]. The effect of surface roughness on the surface free energy has been reported [19]. Adhesion props. (hot tack strength) have been measured. Higher hot tack strength is obtained for films with lower ethylene content [22]

Transport Properties:
Permeability of Gases: Oxygen permeability [38,39,41,43] is affected by humidity [40,42,44] and temp. [42] Oxygen transmission decreases slightly as the relative humidity increases from 0–50% and increases significantly above 50% relative humidity [44]. The diffusion rates of O_2 and N_2 through EVOH films are lower than those of polyethylene owing to the presence of hydrogen bonds [45]. The permeability of N_2 and CO_2 is sensitive to moisture [40]. The permeation of EtOAc vapour has been reported [46]
Permeability of Liquids: The diffusion of MeOH [9] and MeOH/toluene mixtures [47] has been reported. The diffusion of alcohols is decreased by high humidity whereas diffusion of aldehydes is increased [48]. The absorption of aq. sodium chloride has been reported [14]
Water Content: Water absorption of EVOH films (vinyl alcohol content 70 mol %) at 25° has been reported in a graphical form [14]. Equilib. moisture content after 7d at 23° at levels of relative humidity shown in brackets: 0.5% (10%), 0.8% (20%), 1.7% (50%), 3.5% (80%), 4.4% (90%) [13]
Gas Permeability:

No.	Gas	Value	Note
1	O_2	0.089 cm^3 mm/(m^2 day atm)	0.46×10^{-18} mol (m s Pa)$^{-1}$, 44 mol % ethylene, dry [3]
2	O_2	0.01 cm^3 mm/(m^2 day atm)	0.052×10^{-18} mol (m s Pa)$^{-1}$, 32 mol % ethylene, dry [3]
3	O_2	0.008 cm^3 mm/(m^2 day atm)	0.04 n mol (m s GPa)$^{-1}$, dry [1]
4	N_2	0.066 cm^3 mm/(m^2 day atm)	0.001×10^{-10} cm^3 cm (cm^2 s cmHg)$^{-1}$, max. [38,39]
5	SO_2	6.5 cm^3 mm/(m^2 day atm)	0.1×10^{-10} cm^3 cm (cm^2 s cmHg)$^{-1}$, max. [38]

Mechanical Properties:
Mechanical Properties General: Mechanical props. depend on the copolymer composition [3,12]. Tensile modulus depends on the vinyl alcohol content and has a minimum value at 40 mol % vinyl alcohol [12]. Young's modulus and yield stress increase with the degree of microhardness [35]. Tensile strength also varies with composition with a min. value for vinyl alcohol content of 40 mol % and a max. value at vinyl alcohol content of 62 mol % [12]. Elongation at break varies widely with vinyl alcohol content with no obvious trend in the data

Tensile (Young's) Modulus:

No.	Value	Note
1	1300–8000 MPa	25°, depending on copolymer composition [12]
2	2300 MPa	25°, 45% ethylene content [33]
3	2820 MPa	25°, 25% ethylene content [33]

Tensile Strength Break:

No.	Value	Note	Elongation
1	43 MPa	44 mol % ethylene, ASTM D638 [3]	4%
2	66.2 MPa	32 mol % ethylene, ASTM D638 [3]	5%
3	58–483 MPa	25°, depending on vinyl alcohol content [12]	6.96–24.1%
4	57.5 MPa	25°, 45% ethylene [33]	41.5%
5	78.5 MPa	25°, 25% ethylene [33]	21%

Poly(ethylene-co-vinyl alcohol), Films

Tensile Strength Yield:

No.	Value	Elongation	Note
1	71.5 MPa	7%	25°, 45% ethylene [33]
2	98.5 MPa	8.5%	25°, 25% ethylene [33]

Viscoelastic Behaviour: Variation of complex modulus and loss factor with temp. has been reported [33]
Mechanical Properties Miscellaneous: Tear strength 120 kg cm^{-1} (machine direction, 50 μm thick), 180 kg cm^{-1} (transverse direction, 50 μm thick) [36]. Heat sealing strength 3.5 kg mm^{-1} [36]. Heat sealing strength increases at lower initiation temps. for films with higher ethylene content [22]
Hardness: Microhardness increases with increasing vinyl alcohol content and with the degree of crystallinity [35].
Failure Properties General: Stress-strain curves have been reported [12,33]. Uniaxial orientation is unsuccessful due to splitting and cracking of films upon drawing [37]

Electrical Properties:
Static Electrification: EVOH film is free of static [3]

Optical Properties:
Transmission and Spectra: Raman [23] and ir [25] spectral data have been reported. Uv permeability (λ 200–400 μm) 75% (33 mol % vinyl alcohol units, film thickness 50 μm) [24]. Ir permeability rate (wave number 600–1700 cm^{-1}) 0% (33 mol % vinyl alcohol units, film thickness 50 μm) [24]
Volume Properties/Surface Properties: Haze 1.8% (32 mol % ethylene), 2.3% (44 mol % ethylene) [3]. Transparency excellent [3,5,26]

Polymer Stability:
Environmental Stress: Has good weatherability [3]. Undergoes oxidation on exposure to ozone with formation of peroxides [27]. Oxidative degradation has been studied and the oxidation products identified [28]. Hydroperoxide and carbonyl species are prod. and extensive preferential attack at the hydroxyl sites gives backbone ketone groups in films with less than 20% w/w vinyl alcohol [27]
Chemical Stability: Has high resistance to solvents and oils [3,26]. Resistance to olive oil at 70° is excellent [11]
Biological Stability: Some commercial films are biodegradable and show a weight reduction of 68.9% after a two week incubation with bacteria [6]. The biocompatibility of EVOH and other polymer films has been reported [16]
Recyclability: May be recycled [30]
Stability Miscellaneous: γ-Irradiation causes both degradation (chain scission) and cross-linking, [32,33] with subsequent reduction of tensile modulus, strength and elongation and elastic modulus. Brittleness is increased and degree of crystallinity reduced [33]

Applications/Commercial Products:
Processing & Manufacturing Routes: Multilayer films are prepared commercially by coextrusion in a one-step solvent-free process. Blow or cast coextrusion methods can be used. Films and sheeting with five or more layers can be prod. by coextrusion. Coextruded bottles may also be manufactured using blow-moulding technology. Processing of multilayer sheets, tubes and bottles is the same as for films
Applications: Films containing EVOH may be used in food packaging, e,.g. for cheese. The barrier props. of EVOH have led to its use in multilayer and laminated packaging products. It may be used as an inner lining for canisters and tanks used to transport chemicals and agrochemicals. In the building industry it is used as an additive for sound insulation materials and as a protective surface for building materials. EVOH has been used for medical applications e.g. membrane for blood purification by dialysis. Laminated bottles containing an EVOH layer are recyclable and have been used for food packaging

Trade name	Details	Supplier
Eval		Kuraray-Mitsui Toatsu
Eval		Okura Industrial Co Ltd
Eval		Nippon Petrochemicals Co Ltd
Eval		Nippon Synthetic Chemical Industry Co
Eval		W R Grace
Gamma	laminated bottles	Showa Denko
Lamicon	laminated bottles	DuPont
Soarnol		Mitsubishi Chemical
Soarnol		Idemitsu Chemicals

Bibliographic References

[1] *Encycl. Polym. Sci. Eng.*, 2nd edn., Vol.7, (ed. J.I. Kroschwitz), John Wiley and Sons, 1989, **7**, 73
[2] *Encycl. Polym. Sci. Eng.*, 2nd edn., Vol. 17, (ed. J.I. Kroschwitz), John Wiley and Sons, 1989, 167
[3] *Encycl. Polym. Sci. Eng.*, 2nd edn., Vol. 2, (ed. J.I. Kroschwitz), John Wiley and Sons, 1989, **2**, 176
[4] Iwanami, T. and Hirai, V., *Tappi J.*, 1992, 339
[5] *Jpn. Pat.*, 1994, 06 155 665
[6] *Jpn. Pat.*, 1993, 05 86 209, (biodegradability)
[7] *Eur. Pat. Appl.*, 1995, 641 825
[8] *U.S. Pat.*, 1993, 44 669
[9] Samus, M.A. and Rossi, G., *Macromolecules*, 1996, **29**, 2275, (T_g, liq. permeability)
[10] Mavridis, H. and Shroff, R.H., *Polym. Eng. Sci.*, 1992, **32**, 1778, (melt rheology)
[11] Marjanski, W., Jarvela, P. and Penttinen, T., *J. Mater. Sci. Lett.*, 1995, **14**, 1171, (grease resistance)
[12] Nishino, T., Takano, K. and Nakamae, K., *Polymer*, 1995, **36**, 959, (props.)
[13] George, E.R., Sullivan, T.M. and Park, E.H., *Polym. Eng. Sci.*, 1994, **34**, 17, (water absorption)
[14] Apicella, A. and Hopfenberg, H., *J. Appl. Polym. Sci.*, 1982, **27**, 1139, (water absoprtion)
[15] Rantala, J., Hartikainen, J. and Jaarinen, J., *Rev. Prog. Quant. Nondestr. Eval.*, 1991, **10B**, 1735, (thermal conductivity)
[16] Tamada, Y. and Ikada, Y., *J. Colloid Interface Sci.*, 1993, **155**, 334, (biocompatibility, surface props.)
[17] Erbil, H.Y., *J. Appl. Polym. Sci.*, 1987, **33**, 1397, (surface props.)
[18] Shiomi, T., Nishioka, S., Tezuka, Y. and Imai, K., *Polymer*, 1985, **26**, 429, (surface props.)
[19] Bischof, C., Shulze, R.D., Possart, W. and Kamusewitz, H., *Adhesion (London)*, 1988, **12**, 1, (surface props.)
[20] Yasuda, T. and Miyama, M., *Langmuir*, 1994, **10**, 583, (surface props.)
[21] Ou, S.H., Ishida, H. and Lando, J.B., *J. Polym. Sci., Part B: Polym. Phys.*, 1991, **29**, 67, (surface props.)
[22] Kiang, W.W., *Adesione*, 1993, **4**, 17, (adhesive props.)
[23] Cooney, T.F., Wang, L., Sharma, S.K., Gauldie, R.W. and Montana, A.J., *J. Polym. Sci., Part B: Polym. Phys.*, 1994, **32**, 1163, (Raman)
[24] Moroi, H., *Br. Polym. J.*, 1988, **20**, 335, (props.)
[25] Koopmans, R.J., van den Linden, R. and Vansant, E.F., *Spectrosc.: Int. J.*, 1982, **1**, 352, (ir)
[26] Iwanami, T. and Hirai, Y., *Tappi J.*, 1983, **66**, 85
[27] Fujimoto, K., Takebayashi, Y., Inoue, H. and Ikada, Y., *J. Polym. Sci., Part A: Polym. Chem.*, 1993, **31**, 1035, (ozone oxidation)
[28] Carlsson, D.J., Chmela, S. and Wiles, D.M., *Polym. Degrad. Stab.*, 1991, **31**, 255, (oxidation)
[29] Arutyunyan, Zh.S. and Tavakalyan, N.B., *Plast. Massy*, 1991, 57
[30] Kitchel, F.T., *Tappi J.*, 1991, **74**, 183, (recycling)
[31] Djordjevic, D., *J. Plast. Film Sheeting*, 1988, **4**, 276, (recycling)
[32] Acierno, D., Calderaro, E., Napoli, L. and Spadero, G., *Colloid Polym. Sci.*, 1983, **261**, 121, (γ-irradiation)
[33] Spadero, G., Calderaro, E. and Acierno, D., *Colloid Polym. Sci.*, 1983, **261**, 231, (γ-irradiation)
[34] Spadero, G., Piccarolo, S., Canto, E., Calderaro, E. and Acierno, D., *Polym. Bull.*, 1982, **6**, 611, (γ-irradiation)
[35] Fonseca, C., Perena, J.M. and Fatou, J.G., *J. Mater. Sci. Lett.*, 1991, **10**, 739, (mechanical props.)
[36] *Jpn. Pat.*, 1989, 01 153 733, (mechanical props.)

[37] Shastri, R. and Dollinger, S.E., *Polym. Prepr. (Am. Chem. Soc., Div. Polym. Chem.)*, 1989, **30**, 539, (barrier props.)
[38] Imai, K., Shiomi, T., Tezuka, Y. and Inokami, K., *Polym. Int.*, 1992, **27**, 309, (gas permeability)
[39] Iwasaki, H. and Hoashi, K., *Kobunshi Ronbunshu*, 1977, **34**, 785, (gas permeability)
[40] Granger, V.M. and Tigami, M.A., *Tappi J.*, 1993, **76**, 98, (gas permeability)
[41] Axelson-Larsson, L., *Packag. Technol.*, Hillsdale, N.J., 1992, **5**, 297, (gas permeability)
[42] Axelson-Larsson, L., *Spec. Publ. - R. Soc. Chem.*, 1995, **162**, 129, (gas permeability)
[43] Foster, R.H., *Polym. News*, 1986, **11**, 264, (gas permeability)
[44] Kollen, W. and Gray, D., *J. Plast. Film Sheeting*, 1991, **7**, 103, (gas permeability)
[45] Koshihara, S., Aoyagi, T., Yamanaka, S. and Komatsu, T., *Chem. Express*, 1992, **7**, 537, (gas permeability)
[46] Sajiki, T., and Giacin, J.R., *J. Plast. Film Sheeting*, 1993, **9**, 97, (diffusion props.)
[47] Samus, M.A. and Rossi, G., *ACS Symp. Ser.*, 1995, **598**, 535, (diffusion props.)
[48] Johansson, F. and Leufven, A., *J. Food Sci.*, 1994, **59**, 1328, (diffusion props.)

Poly(ethyl methacrylate)

P-275

$$-[CH_2C(CH_3)(COOEt)]_n-$$

Synonyms: *Poly(ethyl-2-methyl-2-propenoate. Poly[1-(ethoxycarbonyl)-1-methyl ethylene]. PEMA*
Related Polymers: Polymethacrylates - General
Monomers: Ethyl methacrylate
Material class: Thermoplastic
Polymer Type: acrylic
CAS Number:

CAS Reg. No.	Note
9003-42-3	(general)
26814-01-7	(isotactic)
26814-02-8	(syndiotactic)

Molecular Formula: $(C_6H_{10}O_2)_n$
Fragments: $C_6H_{10}O_2$
Tacticity: Isotactic and syndiotactic forms known

Volumetric & Calorimetric Properties:
Density:

No.	Value	Note
1	d^{20} 1.12 g/cm^3	amorph. [1]
2	d^{25} 1.119 g/cm^3	amorph. [2]
3	d^{25} 1.19 g/cm^3	cryst. [2]

Thermal Expansion Coefficient:

No.	Value	Note	Type
1	0.00054 K^{-1}	Uv polymerisation, >T_g [3]	V
2	0.000275 K^{-1}	Uv polymerisation, >T_g [3]	V

Volumetric Properties General: Molar volumes: Van der Waal's, V_w 66.3 cm^3 mol^{-1} (amorph. polymer; 25°)^{254}, glassy polymer, V_g 102 cm^3 mol^{-1} (amorph. polymer; 25°)^{254}, cryst. polymer, V_c 96 cm^3 mol^{-1} (cryst. polymer; 25°)^{254}
Equation of State: Mark-Houwink constants have been reported [22]

Thermodynamic Properties General: Molar heat capacity at constant pressure 166 J mol^{-1} K^{-1} solid, 25° [2]
Specific Heat Capacity:

No.	Value	Note	Type
1	1.45 kJ/kg.C	J kg^{-1} K^{-1} solid; 25° [2]	P
2	1.67 kJ/kg.C	J kg K 27° [4]	P

Melting Temperature:

No.	Value	Note
1	47°C	refractometric method [7]

Glass-Transition Temperature:

No.	Value	Note
1	65°C	Uv polymerisation, dilatometric method [3]
2	63–65°C	atactic [8,9]
3	66°C	free-radical polymerisation at 44–60°, dilatometric method [10]
4	120°C	syndiotactic [9]
5	86°C	77% syndiotactic, 2% isotactic, 21% atactic [11]
6	8°C	isotactic [9]
7	8–12°C	anionic polymerisation/non-polar solvent, dilatometric method [10]

Transition Temperature General: In some composites (nylon fibre) T_g decreases as the fibre content increases. [5] T_g can be lowered by sorption of CO_2 [6]
Vicat Softening Point:

No.	Value	Note
1	81°C	[1]

Brittleness Temperature:

No.	Value	Note
1	50°C	refractometric method [7]

Transition Temperature:

No.	Value	Note	Type
1	37°C	[12]	β-transition temp.
2	-223°C	[12]	δ-transition temp.

Surface Properties & Solubility:
Cohesive Energy Density Solubility Parameters: Cohesive energy density 351 J cm^{-3} (35°, 83.7 cal cc^{-1}) -1 [2], 32500 J mol^{-1} [9]. 17.4–22.7 J$^{1/2}$ cm$^{3/2}$ (poor H-bonding solvents) [14], 16.0–27.2 J$^{1/2}$ cm$^{3/2}$ (moderate H-bonding solvents) [14], 19.4–23.3 J$^{1/2}$ cm$^{3/2}$ (strong H-bonding solvents) [14], 18.3 J$^{1/2}$ cm$^{3/2}$ [2], 18.2–18.7 J$^{1/2}$ cm$^{3/2}$ [2], 22.1 J$^{1/2}$ cm$^{3/2}$ [2], 37.6 J$^{1/2}$ cm$^{3/2}$ (8.99 cal$^{1/2}$ cm$^{3/2}$) [16]. 37.8–38.3 J$^{1/2}$ cm$^{3/2}$ (8.95–9.15 cal$^{1/2}$ cm$^{3/2}$); [15]γ_d 18.8 γ_p 10.8, γ_{hh} 43 [2]
Plasticisers: Argon and CO_2 plasticise PEMA (at temps. above T_g) [17]
Solvents/Non-solvents: Sol. tetralin, turpentine (hot), CCl_4, EtOH (hot), isopropanol (>37°), Et_2O, formic acid. Insol. linseed oil, isopropyl ether [18]
Surface and Interfacial Properties General: 60° (air) [20], 81° (Me_2CO) [20], 82° (dodecane) [20]. 33 mN m^{-1} [2], 35.9 mN m^{-1}

(γ_d 26.9 mN m^{-1}, g$_p$ 9 mN m^{-1}) [2], 49.2–52.2 mN m^{-1} (g$_d$ 29.9–31.2 mN m^{-1} 19.3–21 mN m^{-1}) [20]

Wettability/Surface Energy and Interfacial Tension: Contact angle 62° (advancing, H$_2$O) [19]. Interfacial tension 13.9–15.2 mN m^{-1} (solid/H$_2$O) [20]

Transport Properties:
Polymer Solutions Dilute: Molar volumes 66.3 cm^3 mol^{-1} (van der Waal's, 25°, amorph.) [2], 102 cm^3 mol^{-1} (glassy, amorph. 25°) [2], 96 cm^3 mol^{-1} (cryst., 25°) [2]. Intrinisic viscosity [21] 1.45 dl g^{-1} (35°, C$_6$H$_6$), 4.7 dl g^{-1} (45°, n-butanol), 1.33 dl g^{m1} (35°, EtOAc), 0.88 dl g^{-1} (36.9°, isopropanol), 1.57 dl g^{-1} (23°, 2-butanone), 1.31 dl g^{-1} (47°, m-xylene). Theta temps. 44° (n-butanol, VM method), 23° (butanone), 0° (n-butyl bromide, VM method) -13° (n-butyl chloride, VM method), -1° (methyl propyl ketone, VM method), 36.9° (isopropanol, MW 220000–130000), -3° (m-xylene, VM method) [21]. Polymer-liq. interaction parameters [23], 1.75 (hexane), 0.6 (-2-butanone), 0.98 (acetonitrile), -0.25 (CHCl$_3$), 0.98 (n-propanol), 0.72 (Me$_2$CO) 0.61 (EtOAc), 62.8 (isopropanol)

Permeability of Gases: [He] 0.34 × 10^{-4}, [Ne] 0.323 × 10^{-4}, [O$_2$] 2.8 × 10^{-4}, [Ar] 2.7 × 10^{-4}, [N$_2$] 2.9 × 10^{-4}, [CO$_2$] 0.58 × 10^{-4}, [Kr] 13.7 × 10^{-4}, [H$_2$S] 30.38 × 10^{-4}, [SF$_6$] 0.000118 × 10^{-4}, [H$_2$O] 0.0074 × 10^{-4} (below T$_g$), [H$_2$O] 2.1 × 10^{-4} cm^2 (s cmHg)$^{-1}$ (above T$_g$) [24]

Permeability of Liquids: Diffusion coefficients [Me$_2$CO] 17.4–0.85 cm^2 s^{-1} (20°, MW 34100-9350000), 20–0.95 × 10^{-7} cm^2 s^{-1} (20°, MW 24600-7440000); [25,26] [H$_2$O] 1.39 × 10^{-7} cm^2 s^{-1} (37°) [19]

Water Content: 0.35–0.5% (max.) [8]

Water Absorption:

No.	Value	Note
1	1 %	[19]

Mechanical Properties:
Tensile Strength Break:

No.	Value	Note	Elongation
1	34 MPa	[27,28]	7%
2	7–37 MPa	23° [8]	0.6–25%

Viscoelastic Behaviour: Williams-Landel-Ferry constants for temp. superposition and Vogel parameters have been reported [30]. Longitudinal sound speed 2700 m s^{-1} (room temp. 1 MHz) [31], 2400 m s^{-1} [33]. Longitudinal sound absorption 3.9 dB cm^{-1} (room temp. 2 MHz) [31]

Hardness: Knoop number 11 (Tukon) [28]. Vickers 11 (5 kg load) [29]

Electrical Properties:
Dielectric Permittivity (Constant):

No.	Value	Note
1	2.7–3.4	[2]

Optical Properties:
Refractive Index:

No.	Value	Note
1	1.485	20° [2,32]

Polymer Stability:
Polymer Stability General: Large doses of ionising radiation can cause brittleness, strength loss and general loss of desirable props. Irradiation causes main chain fracture; deterioration of props. is due to lowering of MW

Decomposition Details: Decomposes to monomer at 250° [29]

Applications/Commercial Products:

Trade name	Details	Supplier
Elvacite	2042, 2043	ICI, UK
Standard Samples	MW 200000–500000	Aldrich
Standard Samples		Polysciences Inc

Bibliographic References

[1] Hoff, E.A.W, Robinson, D.W. and Willbourn, A.H., *Polym. Sci.*, 1955, **18**, 161, (density)
[2] Van Krevelen, D.W., Properties of Polymers: Their Correlation with Chemical Structure Elsevier, 1976, 79, (density, heat capacity)
[3] Rogers, S.S. and Mandelkern, L, *J. Phys. Chem.*, 1957, **61**, 985
[4] Gaur, U., Lau, S.F., Wunderlich, B.B. and Wunderlich, B., *J. Phys. Chem. Ref. Data*, 1982, **11**, 1065, (heat capacity)
[5] Estelles, J.M., Ribelles, J.L.G., and Pradas, M.M., *Polymer*, 1993, **34**, 3837, (transition temp)
[6] Shine, A.D., *Physical Properties of Polymers Handbook*, (ed. J.E. Mark), AIP Press, 1996, 249
[7] Wiley, R.H. and Brauer, G.M., *J. Polym. Sci.*, 1948, **3**, 647, (transition temp)
[8] *Datasheet on Elvacite*, ICI, 1996, (technical datasheet)
[9] Porter, D., *Group Interaction Modelling of Polymer Properties*, Dekker, 1995, 283
[10] Shetter, J.A., *J. Polym. Sci., Part B: Polym. Lett.*, 1962, **1**, 209
[11] Walstrom, A.M., Subramanian, R., Long, T.E., McGrath, J.E and Ward, T.C., Polym. Prepr. (Am. Chem. Soc., Div. Polym. Chem.),
[12] Plazek, D.J. and Ngai, K.L., *Physical Properties of Polymers Handbook*, (ed. J.E. Mark), AIP Press, 1996, 139
[13] Lee, W.A. and Sewell, J.H., *J. Appl. Polym. Sci.*, 1968, **12**, 1397
[14] Mangaraj, D., Patra, S. and Rashid, S., *Makromol. Chem.*, 1963, **65**, 39, (solubility parameters)
[15] Mangaraj, D., Patra, S. and Roy, P.C., *Makromol. Chem.*, 1965, **81**, 173
[16] Frank, C.W. and Gashgari, M.A., *Macromolecules*, 1979, **12**, 163
[17] Kamiya, Y., Mizoguchi, K. and Naito, Y., *J. Polym. Sci., Part B: Polym. Phys.*, 1992, **30**, 1183
[18] Strain, D.E., Kennelly, R.G. and Dittmar, H.R., *Ind. Eng. Chem.*, 1939, **31**, 382
[19] Kalachandra, S. and Kusy, R.P., *Polymer*, 1991, **32**, 2428
[20] King, R.N., Andrake, J.D., Ma, S.M., Gregonis, D.E. and Brostrom, L.R., *J. Colloid Interface Sci.*, 1985, **103**, 62, (interfacial tension)
[21] Vasudevan, P. and Santappa, M., *Makromol. Chem.*, 1970, **137**, 262, (viscosity)
[22] Kurata, M. and Stockmayer, W.H., *Adv. Polym. Sci.*, Springer-Verlag, 1963, **3**, 196, (eqn of state)
[23] Walsh, D.J. and McKeown, J.G., *Polymer*, 1980, **21**, 1335
[24] Stannett, V. and Williams, J.L., *J. Polym. Sci., Part C: Polym. Lett.*, 1966, **10**, 45, (gas permeability)
[25] Meyerhoff, G., *Makromol. Chem.*, 1955, **15**, 68, (liq permeability)
[26] Schulz, G.V. and Meyerhoff, G.Z., *Z. Elektrochem.*, 1952, **56**, 545, (liq permeability)
[27] Brearley, W.H., *J. Paint Varnish Products*, 1973, **63**, 23
[28] Cramer, A.S., *Kunststoffe*, 1940, **30**, 337
[29] Crawford, J.W.C., *J. Soc. Chem. Ind.*, London, 1949, **68**, 201
[30] Ferry, J.D., Child, W.C., Zard, R., Stern, D.M., et al, *J. Colloid Sci.*, 1957, **12**, 53, (viscoelastic props)
[31] Hartman, B., *Physical Properties of Polymers Handbook*, (ed. J.E. Mark), AIP Press, 1996, 677, (acoustic props)
[32] Lewis, O.G., *Physical Constants of Linear Hydrocarbons*, Springer-Verlag, 1968, (refractive index)
[33] North, A.M., Prethrick, R.A. and Phillips, D.W., *Polymer*, 1977, **18**, 324

Poly[(ethylmethylsilane)(methylene)]

P-276

$$-\left[CH_2-\underset{\underset{CH_3}{|}}{\overset{\overset{CH_2CH_3}{|}}{Si}}\right]_n-$$

Synonyms: *Poly[(ethylmethylsilylene)(methylene)]*
Related Polymers: Polycarbosilanes
Polymer Type: polysilanes

CAS Number:

CAS Reg. No.
124479-34-1

Molecular Formula: $(C_4H_{10}Si)n$
Fragments: $C_4H_{10}Si$
Molecular Weight: 91,600 [1]
Identification: Found: C, 56.89%; H, 12.53%; Calculated: C, 60.61%; H, 12.12%

Volumetric & Calorimetric Properties:
Glass-Transition Temperature:

No.	Value	Note
1	-78.2°C	[1]

Optical Properties:
Transmission and Spectra: Nmr [1] spectra reported.

Applications/Commercial Products:
Processing & Manufacturing Routes: Prepared by ring-opening polymerisation of 1,3-diethyl-1,3-dimethyl-1,3-disilacyclobutane [1].
Applications: Polydialkylsilalkylene polymers are useful in enhanced oil recovery using carbon dioxide flooding [2].

Bibliographic References
[1] Shen, Q.H. and Interrante, L.V., *J. Polym. Sci., Part A,* 1997, **35**, 3193
[2] U.S. Pat. (Chevron Research), 1989, 4 852 651

Poly(ethylpentylsilane)   P-277

$$\left[\begin{array}{c} CH_2(CH_2)_3CH_3 \\ -Si- \\ CH_2CH_3 \end{array} \right]_n$$

Synonyms: *Poly(ethylpentylsilylene). PEPS. C2-C5*
Polymer Type: polysilanes
CAS Number:

CAS Reg. No.
132230-59-2

Molecular Formula: $(C_7H_{16}Si)n$
Fragments: $C_7H_{16}Si$
General Information: Soft, rubbery texture. Weakest in the poly(*n*-alkylpentylsilylene) series; films tear most easily [1,2].
Morphology: The diffraction pattern of the polymer at 25°C shows only a strong refection at 1.09nm. Probably a hexagonal, columnar liquid crystal [2,4]. Poorly ordered with a second order phase transition at -35°C [2].
Miscellaneous: Thin films cast from toluene are sticky, with properties similar to typical pressure sensitive adhesives. Because of the large asymmetry between the two sidechains and the small size of the ethyl group, this polymer is poorly ordered. Thus, no order-disorder transition is seen [1].

Volumetric & Calorimetric Properties:
Glass-Transition Temperature:

No.	Value	Note
1	-35--33°C	[1]

Mechanical Properties:
Mechanical Properties General: The polymer film is less cohesive and tends to retract rather than retain its elongation when the tension on it is removed [2].
The modulus drops four orders of magnitude at -35°C so the polymer is quite soft at ambient conditions [1].

Mechanical Properties:
Tensile (Young's) Modulus:

No.	Value	Note
1	77.1 MPa	at 23°C [1]
2	0.7 MPa	at 100°C [1]

Optical Properties:
Optical Properties General: Variable-temperature uv spectrum at ambient conditions shows a somewhat broad absorption maximum is centred at 306nm. A shoulder at 330nm becomes perceptible upon cooling to -16°C, which grows as the temperature is lowered, reaching an isosbestic point at -42°C. At -50°C, the 330nm band is just a little more intense than the original absorption peak (which has shifted slightly to 310nm). The ratio of these two peak intensities is unchanged upon further cooling to -80°C. Evidently, the backbone conformations are frozen due to lack of mobility at temperatures below -60°C. The behaviour described above is fully reversible. Upon heating, the isosbestic point changed only 1°C and the second band disappeared at -16°C [2].
Transmission and Spectra: Uv-vis [2,4] spectra reported.

Polymer Stability:
Thermal Stability General: A second-order transition at -35°C is observed in the DSC [2,4].
The tan gamma shows two peaks and suggests that another transition may be present *ca.* 10°C [2].

Applications/Commercial Products:
Processing & Manufacturing Routes: The monomer is prepared by Grignard reaction between *n*-pentylmagnesium bromide and ethyltrichlorosilanes in THF under Ar. After isolation by two fractional distillations, the resulting dichlorosilane was polymerised by the usual Wurtz condensation reaction with Na in toluene at 110°C [1].
Applications: Electrophotographic photoreceptor; binder resin for a single photosensitive layer in which a charge-generating and a charge-transporting substance are mixed [3].

Bibliographic References
[1] Klemann, B.M., DeVilbiss, T. and Koutsky, J.A., *Polym. Eng. Sci.,* 1996, **36**, 135
[2] Klemann, B.M., West, R. and Koutsky, J.A., *Macromolecules,* 1993, **26**, 1042
[3] *Jpn. Pat.,* 1990, 02 150 853
[4] Klemann, B.M., West, R. and Koutsky, J.A., *Macromolecules,* 1996, **29**, 198

Poly(ethyl vinyl ether-*co*-maleic anhydride)   P-278

Synonyms: *Poly(ethoxyethene-co-2,5-furandione). Poly(2,5-furandione-co-ethoxyethane)*
Related Polymers: Poly(vinyl ethers)
Monomers: Ethyl vinyl ether, Maleic anhydride
Material class: Synthetic Elastomers, Copolymers
Polymer Type: polyvinyls
CAS Number:

CAS Reg. No.	Note
26711-22-8	
62601-51-8	ethyl ester
36350-24-0	sodium salt
106209-34-1	alternating

Molecular Formula: $[(C_4H_8O).(C_4H_2O_3)]_n$
Fragments: C_4H_8O $C_4H_2O_3$

Volumetric & Calorimetric Properties:

Glass-Transition Temperature:

No.	Value	Note
1	68–69°C	[1]
2	146°C	[2]

Surface Properties & Solubility:

Solvents/Non-solvents: Sol. Me_2CO, 2-butanone, THF, acetophenone, EtOAc, MeOH, EtOH. Insol. decane, octane, *iso*-octane, hexane, cyclohexane, C_6H_6, dioxane, $CHCl_3$ [3]

Transport Properties:

Polymer Solutions Dilute: [η] 0.392 dl g^{-1} (THF, 30°), 0.215 dl g^{-1} (Me_2CO, 30°) [3]. $η_{inh}$ 1.92 dl g^{-1} (0.5% in DMF, 30°) [2]

Optical Properties:

Refractive Index:

No.	Value	Note
1	1.406	25° [3]

Transmission and Spectra: H-1 and C-13 nmr spectral data have been reported [2,4]
Total Internal Reflection: dη/dC 0.0883 ml g^{-1} (THF, 25°) [3]

Applications/Commercial Products:

Processing & Manufacturing Routes: Prod. by free radical initiated copolymerisation with AIBN
Applications: Stabilised pretreating liquid for metallic surfaces, conversions coatings on metals, surface soiling inhibitor with surfactant, resin drug carrier, denture fixative composition [5,6,7,8]

Bibliographic References

[1] Dhal, P.K., Ramakrishna, M.S., Srinivasan, G. and Chaudhari, S.N.K., *J. Polym. Sci., Polym. Chem. Ed.,* 1985, **23**, 2679
[2] Fu, X., Bassett, W. and Vogl, O., *J. Polym. Sci., Polym. Chem. Ed.,* 1983, **21**, 891
[3] Shimizu, T., Minakata, A. and Tomiyama, T., *Polymer,* 1980, **21**, 1427
[4] Reddy, B.S.R., Arshady, R. and George, M.H., *Polymer,* 1984, **25**, 115
[5] *Ger. Pat.,* 1971, 2 125 963, (use)
[6] *Austrian Pat.,* 1975, 463 181, (use)
[7] *Jpn. Pat.,* 1983, 58 83 041, (use)
[8] *U.S. Pat.,* 1985, 4 521 551, (use)

Poly(fluorene) P-279

Synonyms: *Poly(9H-fluorene-2,7-diyl). Poly(9H-fluorene)*
Monomers: Fluorene
Material class: Thermoplastic
Polymer Type: polyfluorene
CAS Number:

CAS Reg. No.	Note
95270-88-5	
99627-56-2	Poly(9*H*-fluorene-2,7-diyl)

Molecular Formula: $(C_{13}H_8)_n$
Fragments: $C_{13}H_8$
General Information: Parent polymer structure of a group of conductive and electroluminescent polymers usually substituted at the 9-position.

Electrical Properties:

Electrical Properties General: Films are conductive in electrochemically oxidised or doped state and nonconductive in neutral state [1]. Conductivity of film doped with tetrafluoroborate 0.0001 S cm^{-1} [2].

Optical Properties:

Transmission and Spectra: Spectral data for ir [3], uv [4,5] and epr [6] have been reported.
Volume Properties/Surface Properties: Films are green in neutral state and red in oxidised state [1].

Applications/Commercial Products:

Processing & Manufacturing Routes: Prod. by electrochemical anodic oxidation of monomer in a suitable solvent such as acetonitrile [1,3,4,5,6].
Applications:

Bibliographic References

[1] Janiszewska, L. and Osteryoung, R.A., *J. Electrochem. Soc.,* 1988, **135**, 116
[2] Simonet, J. and Rault-Berthelot, J., *Prog. Solid State Chem.,* 1991, **21**, 1
[3] Rault-Berthelot, J. and Simonet J., *Nouv. J. Chim.,* 1986, **10**, 169
[4] Hapiot, P., Lagrost, C., Le Floch, F., Raoult, E. and Rault-Berthelot, J., *Chem. Mater.,* 2005, **17**, 2003
[5] Sharma, H.S. and Park, S.-M., *J. Electrochem. Soc.,* 2004, **151**, E61
[6] Oudard, J.F., Allendoerfer, R.D. and Osteryoung, R.A., *Synth. Met.,* 1988, **22**, 407

Poly(4-fluorostyrene) P-280

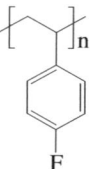

Synonyms: *Poly(p-fluorostyrene). Poly(4-fluoro-1-vinyl-benzene). Poly(1-(4-fluorophenyl)-1,2-ethanediyl)*
Monomers: 4-Fluorostyrene
Material class: Thermoplastic
Polymer Type: styrenes
CAS Number:

CAS Reg. No.	Note
24936-47-8	
53745-75-8	isotactic
54193-27-0	syndiotactic

Molecular Formula: $(C_8H_7F)_n$
Fragments: C_8H_7F

Volumetric & Calorimetric Properties:

Density:

No.	Value	Note
1	d 1.179 g/cm^3	[3]
2	d 1.022 g/cm^3	[4]

Thermodynamic Properties General: ΔC_p 33.3 J K^{-1} mol^{-1} [2]
Melting Temperature:

No.	Value	Note
1	265°C	[5]

Glass-Transition Temperature:

No.	Value	Note
1	108°C	[1]
2	111°C	[2]
3	80°C	[5]
4	104°C	[6]

Transition Temperature General: Boils at 29–30° (4 mmHg) [4]

Transport Properties:
Permeability of Gases: [He] 34.4×10^{-10}; [CH_4] 1.06×10^{-10} cm^2 $(s\ cmHg)^{-1}$ [6]

Electrical Properties:
Dielectric Permittivity (Constant):

No.	Value	Note
1	2.96	[1]

Complex Permittivity and Electroactive Polymers: μ 1.82 D (25°, C_6H_6) [9]

Optical Properties:
Refractive Index:

No.	Value	Note
1	1.514–1.5158	20° [4]

Transmission and Spectra: Raman [7] and ir [10] spectral data have been reported

Polymer Stability:
Environmental Stress: Produces cross-linking products with liberation of H_2 and HF via radical mechanism on uv degradation [8]

Applications/Commercial Products:
Processing & Manufacturing Routes: Prod. by free radical polymerisation with catalyst

Bibliographic References
[1] Gustafsson, A., Wiberg, G. and Gedde, U.W., *Polym. Eng. Sci.*, 1993, **33**, 549
[2] Judovits, L.H., Bopp, R.C., Gaur, U. and Wunderlich, B., *J. Polym. Sci., Part B: Polym. Lett.*, 1986, **24**, 2725
[3] Nozaki, M., Shimada, K. and Okamoto, S., *Jpn. J. Appl. Phys.*, 1971, **10**, 179
[4] Matsuo, K. and Stockmayer, W.H., *J. Polym. Sci.*, 1973, **11**, 43
[5] Danusso, F. and Polizzotti, G., *Makromol. Chem.*, 1963, **61**, 157
[6] Puleo, A.C., Muruganandam, N. and Paul, D.R., *J. Polym. Sci., Part B: Polym. Phys.*, 1989, **27**, 2385
[7] Spells, S.J., Shepherd, I.W. and Wright, C.J., *Polymer*, 1977, **18**, 905
[8] Weir, N.A. and Milkie, T.H., *J. Polym. Sci., Polym. Chem. Ed.*, 1979, **17**, 3735
[9] Baysal, B.M. and Aras, L., *Macromolecules*, 1985, **18**, 1693
[10] Kobayashi, M., Nagai, K. and Nagai, E, *Bull. Chem. Soc. Jpn.*, 1960, **33**, 1421

Poly(2-furanmethanol)   P-281

Synonyms: *Furan resins. Poly(furfuryl alcohol-co-formaldehyde). Poly(furfuryl alcohol)*
Related Polymers: Furan Resins, Foundry Applications
Monomers: Furfuryl alcohol, Formaldehyde, Phenol, Melamine, Urea
Material class: Thermosetting resin
Polymer Type: furan

CAS Number:

CAS Reg. No.	Note
25989-02-0	copolymer with formaldehyde
25212-86-6	2-furan-methanol homopolymer

Molecular Formula: $(C_5H_4O)_n$
Fragments: C_5H_4O
Additives: The curing agent is typically *p*-sulfamic acid. Glass fibre. Hexabromo-2-butene and 2,3-dibromopropanol as flame retardants for the furfuryl alcohol-formaldehyde copolymer. Asbestos to improve corrosion resistance (furfuryl alcohol-formaldehyde copolymer). Phenol and urea as modifiers (furfuryl alcohol-formaldehyde copolymer)
General Information: Furfuryl alcohol undergoes self condensation reactions to produce a linear polymeric struct. with furfuryl rings linked by methylene groups. Condensation reactions with formaldehyde, phenol, melamine and urea produce a range of useful products derived at least partly from renewable raw materials. Furan resins, or furan polymers as they are also known, is a general term for polymers of derivatives of furan. The two most important of these are poly(furfuryl alcohol) and its copolymer with formaldehyde. Unless otherwise indicated, all data refers to poly(furfuryl alcohol)

Volumetric & Calorimetric Properties:
Density:

No.	Value	Note
1	d 1.75 g/cm^3	asbestos filled
2	d 1.4–1.6 g/cm^3	glass filled laminate
3	d 1.48 g/cm^3	31% glass fibre filled [5]
4	d 1.21 g/cm^3	formaldehyde copolymer, oxalic acid catalysed [7]
5	d 1.25 g/cm^3	formaldehyde copolymer, sulfuric acid catalysed [7]

Thermal Expansion Coefficient:

No.	Value	Note	Type
1	1.1×10^{-5}–3.2×10^{-5} K^{-1}	carbon filled [2]	L
2	6.1×10^{-5} K^{-1}	Alkor [4]	L
3	2.4×10^{-5} K^{-1}	31% glass filled [5]	L

Volumetric Properties General: Density of furfuryl alcohol-formaldehyde copolymer is dependent upon the nature of the acid catalyst used in its synth.; higher values are obtained with inorganic acids. Density also dependent to a lesser degree on mole ratio of the comonomers [7]
Thermal Conductivity:

No.	Value	Note
1	0.12 W/mK	0.203 Btu ft^{-2} h^{-1} ft, asbestos filled [2]

Surface Properties & Solubility:
Solvents/Non-solvents: Low MW resin sol. Me_2CO, alcohols, esters, aromatic hydrocarbons [3]

Transport Properties:
Polymer Solutions Dilute: Viscosity 139 mPa s (formaldehyde copolymer, uncured, oxalic acid catalysed), 1264 mPa s (formaldehyde copolymer, uncured, sulfuric acid catalysed) [7]. Use of inorganic acids as catalysts in the synth. of formaldehyde copolymer gives a large increase in the viscosity of the uncured resin [7]

— Poly(2-furanmethanol)

Water Absorption:

No.	Value	Note
1	0.01–0.2 %	
2	0.5 %	max., carbon filled [2]
3	0.2 %	Alkor, ASTM C143 [4]
4	2.5 %	31% glass fibre filled [5]

Mechanical Properties:
Tensile (Young's) Modulus:

No.	Value	Note
1	10000–12000 MPa	asbestos filled
2	5500 MPa	32% glass-filled
3	10892 MPa	1580 kpsi, 21°, asbestos filled [2]
4	7159 MPa	73000 kg cm^{-2}, 31% glass fibre filled [5]

Tensile Strength Break:

No.	Value	Note	Elongation
1	20–32 MPa	asbestos filled	
2	82 MPa	32% glass-filled laminate	
3	10.3 MPa	1500 psi, carbon filled [2]	
4	27 MPa	4000 psi, 21°, asbestos filled [2]	
5	80.9 MPa	825 kg cm^{-2}, 31% glass fibre filled [5]	2.25%

Flexural Strength at Break:

No.	Value	Note
1	4–65 MPa	asbestos filled
2	138 MPa	32% glass filled
3	10.3 MPa	1500 psi, carbon filled [2]
4	51.7 MPa	7500 psi, 21°, asbestos filled [2]
5	6.9 MPa	1000 psi, Alkor, 7 days, 24°, ASTM C306 [4]
6	145 MPa	1475 kg cm^{-2}, 31% glass fibre filled [5]

Compressive Strength:

No.	Value	Note
1	70–100 MPa	asbestos filled
2	80 MPa	32% glass filled
3	82.7 MPa	12 kpsi, carbon filled [2]
4	79.3 MPa	11.5 kpsi, 21°, asbestos filled [2]
5	62.7 MPa	9100 psi, Alkor, 7 days, 24°, ASTM C306 [4]
6	102.5 MPa	1045 kg cm^{-2}, 31% glass fibre filled [5]

Miscellaneous Moduli:

No.	Value	Note	Type
1	26.2 MPa	3800 psi, Alkor, 24° [4]	Rupture modulus
2	6914 MPa	70500 kg cm^{-2}, 31% glass fibre filled [5]	Bending modulus

Impact Strength: Impact strength 12.8 J m^{-1} (135 kg cm cm^{-2}, 31% glass fibre filled) [5]
Hardness: Rockwell R110 (asbestos filled), Rockwell R106 (24°, asbestos filled) [2], R99 (66°, asbestos filled) [2], R78 (107°, asbestos filled) [2]. Ball indentation hardness (900 kg cm^{-2}, (31% glass fibre filled) [5]
Failure Properties General: Shear strength 24 MPa (3500 psi, 21°, asbestos filled) [2]

Electrical Properties:
Dielectric Strength:

No.	Value	Note
1	4.7 kV.mm^{-1}	31% glass fibre filled [5]

Dissipation Power Factor: 0.61 (31% glass fibre filled) [5]
Dissipation (Power) Factor:

No.	Value	Note
1	0.61	31% glass fibre filled [5]

Optical Properties:
Transmission and Spectra: Ir spectral data have been reported [12]

Polymer Stability:
Upper Use Temperature:

No.	Value	Note
1	125–165°C	asbestos filled
2	148°C	glass-filled laminate
3	149–193°C	carbon filled [2]
4	193°C	Alkor [4]
5	149°C	filled furfuryl alcohol-formaldehyde copolymer [8]

Decomposition Details: Furfuryl alcohol resin (cured with alcoholic ZnCl$_2$ soln.) decomposes in Ar atmosphere at 340–540° with CO$_2$ and CO as the main products. Activation energy of decomposition 92.9 kJ mol^{-1} (22.2 kcal mol^{-1}); char yield 60% (800°) [1]. Char yield 31% (31% glass fibre filled) [5]. Filled furfuryl alcohol-formaldehyde copolymer is completely decomposed after 0.5h at 538° [8]
Flammability: Has good flame resistance and low smoke emission [5]. Flame spread 11.8 (formaldehyde copolymer, 11.6 phr hexabromo-2-butene, Monsanto Two-Foot tunnel) [6], 12 (formaldehyde copolymer, 15 phr 2,3-dibromopropanol, Monsanto Two-Foot tunnel) [6]. Max. smoke density 24 (formaldehyde copolymer, 11.6 phr hexabromo-2-butene, NBS chamber) [6], 22 (formaldehyde copolymer, 15 phr 2,3-dibromopropanol, NBS chamber) [6]
Chemical Stability: Asbestos-filled resin attacked by conc. nitric acid, conc. sulfuric acid, conc. chromic acid, sodium hypochlorite, bromine, iodine, aniline, pyridine and fluorine compounds [2]. Fully cured resin resistant to strong acids and organic solvents, but attacked by strong oxidising agents [3]. Filled furfuryl alcohol-formaldehyde copolymer has excellent resistance to sulfuric acid, sodium hydroxide, C$_6$H$_6$, CCl$_4$, sodium chloride, and MeOH; good resistance to hydrochloric acid; fair resistance to Me$_2$CO but poor resistance to nitric acid and wet chlorine gas [8]. Glass filled resins resistant to weak acids and bases, aliphatic hydrocarbons and aromatic hydrocarbons, esters, alcohols, ketones and chlorinated hydrocarbons; mod. resistant to strong acid and bases; attacked by halogens and oxidising agents [5]
Hydrolytic Stability: Glass filled resins stable to H$_2$O and salt solns. [5]

Applications/Commercial Products:
Processing & Manufacturing Routes: Synth. [2,9,11] by polymerisation of furfuryl alcohol under acid catalysis (PH 3–5) at 80–

100° with the reaction rate moderated by external cooling. If a liq. resin is desired [10], the mixture is neutralised. Otherwise, the resin is cured in situ using acids such as p-toluenesulfonic acid to give dark solids. Processed by transfer and compression press moulding; extrusion is possible but is not reasonable for many requirements [8]. Furfuryl alcohol-formaldehyde copolymer may be synth. by reaction of 1 mole furfuryl alcohol with 0.5–0.75 mole aq. formaldehyde in the presence of oxalic acid catalyst. Alternatively, paraformaldehyde may be employed and an organic medium used [7]. Latent cure agents used for heat cure, sulfonic acids used for cure at lower temp. A furan resin two-component system can be mixed with a range of fillers to produce chemically-resistant compositions. Furan resins with latent heat curing catalyst are used as chemically-resistant coatings for chemical process equipment. A Furan resin two-component system can be used for refractory mixes by a cold-mixing process. In the foundry, for the hot box process, quartz sand is batch mixed with urea-modified furan resin (1.4–2%) and latent curing agent e.g. ammonium nitrate (0.3–0.5%). The mix is shaped and precured at 200–260° for up to 2 min in core boxes. Use of modified furan resins (0.9–1.2%) with curing agents (0.2–0.4%, usually sulfonic acids) enables a warm-box process for binding sand at 140–200°. Furan resins with special curing agents may be employed in a vacuum warm box process where the corebox is heated at 70–100°. For a no-bake process, sand and phenolic resole or furan resin (0.7–1.3%) and an acidic curing agent (0.2–0.6%) are cured at ambient temps. for 20 mins to several hours. Binding of foundry sand is also done using furan resins (0.8–2%) with peroxide hardeners (0.3–0.9%) in the Hardox process.

Applications: Chemically-resistant putties using furan resins have exceptional solvent and alkali resistance. They are used as bonding and grouting materials for industrial floors and walls. Industrail concrete and screed-topped floors are protected by laminated coatings e.g. furan resin and fibrous filler. Chemical processing equipment may be coated with filled furan resin systems to enhance chemical resistance. Shaped refractory products can be made using furan resins by a cold-mixing technique. Furan resins (homopolymers of 2-furanmethanol) may be carbonised to produce glassy carbon parts; active carbon materials e.g. molecular sieve, catalysts; and graphitised films. Furfuryl alcohol resins used mainly in the foundry industry as sand binders for casting moulds and cores. Applications due to good corrosion resistance in mortar, cement, waste disposal systems and sewers. As rapid repair materials in oilfield applications; in adhesives and binders in wood technology. Also used to make impregnating solns. to improve the imperviousness of materials. Glass-reinforced grades used in piping. Furfuryl alcohol-formaldehyde copolymer used in adhesives and as a binder

Trade name	Details	Supplier
Alkor		
Fenofur	Furan/Phenolic	Haveg Corp.
Haveg 60	cast moulded, asbestos filled	Atlas Minerals and Chemicals
Quacorr RP 100A	glass reinforced	Quaker Oats Co.
Rütaphen AF	acid-proof coating resin	Bakelite
Rütaphen FL	refractory mix resin	Bakelite
Rütaphen GH	foundry resins hot box	Bakelite
Rütaphen GK	foundry resins cold-cure	Bakelite
Rütaphen GN	foundry resins no-bake	Bakelite

Bibliographic References

[1] Lapina, N.A., Ostrovskii, V.S. and Kamenskii, I.V., *Vysokomol. Soedin., Ser. A*, 1969, **11**, 2073, (decomposition)
[2] Dunlop, A.P. and Peters, F.N., *The Furans*, 1953, **220**, 783, (synth, uses, mechanical props.)
[3] Laszlo-Hedvig, Z. and Szesztay, M., *Polymeric Materials Encyclopedia*, (ed. J.C. Salamone), CRC Press, 1996, 2691, (rev)
[4] Siegfried, K.J., *Encycl. Polym. Sci. Technol.*, 1967, **7**, 432
[5] Radcliffe, A.T. and Lens, T.J., *Kunststoffe*, 1973, **63**, 854, (props)
[6] Leitheiser, R.H., Londrigan, Fire Retard., *Proc. Eur. Conf. Flammability Fire Retard.*, 1978, 126, (flammability)
[7] Zmihorska-Gottfryd, A. and Szlezyngier, W., *Plaste Kautsch.*, 1984, **31**, 53, (synth, viscosity, density)
[8] Barton, H.D., *Mater. Prot. Perform.*, 1973, **12**, 16, (stability)
[9] Gandini, A., *Adv. Polym. Sci.*, 1997, **25**, 47, (synth)
[10] Delgado, R., Primelles, E., Rodriguez, V.I. and Ulloa, H., *Cent. Azucar*, 1985, 63, (synth)
[11] Nowicki, J. Maciejewski, Z., Majnusz, J. and Rudnicka, I., *Przemysl Chem.*, 1996, **75**, 219, (synyh, props,)
[12] Stepen, R.A., Peryshkina, G.I. and Solovev, L.S., *Spektrosk. At. Mol.*, 1969, 408, (ir)

Polyglycidyl ether of Tetraphenolethane   P-282

Synonyms: *1,1,2,2-Tetrakis[4-(2,3-epoxypropoxy)phenyl]ethane. 2,2',2'',2'''-[1,2-Ethanediylidenetetrakis(phenyleneoxymethylene)]tetrakisoxirane. Tetrakis (4-glycidyloxyphenyl)ethane*
Related Polymers: Epoxy resins
Monomers: Epichlorohydrin, Phenol, Glyoxal
Material class: Thermosetting resin
Polymer Type: epoxy
CAS Number:

CAS Reg. No.	Note
27043-37-4	
53228-95-8	homopolymer

Molecular Formula: $C_{38}H_{38}O_8$
Fragments: $C_{38}H_{38}O_8$
Additives: Glass fibre
Miscellaneous: All props. are dependent upon type of curing agent and/or the method of cure used

Volumetric & Calorimetric Properties:
Melting Temperature:

No.	Value	Note
1	80°C	approx. [2]
2	96°C	Epon 1031 [9]

Transition Temperature General: Vicat softening temp. is higher for resin cured with maleic anhydride than for resin cured with a BF_3/ethylamine complex. Vicat temp. increases with increased time of ageing at 230°

Transport Properties:
Polymer Solutions Dilute: Viscosity 39 P (137°, Epon 1031) [9]
Polymer Melts: Melt viscosity 15 P (150°, Epon 1031) [9]

Mechanical Properties:
Flexural Modulus:

No.	Value	Note
1	20682 MPa	3000 kpsi, 23°, reinforced grade, nadic methyl anhydride and benzyldimethylamine cured [5]
2	13099 MPa	1900 kpsi 260°, reinforced grade, nadic methyl anhydride and benzyldimethylamine cured [5]

| 3 | 19993 MPa | 2900 kpsi, 23°, reinforced grade, diaminophenylsulfone/BF$_3$.ethylamine cured [5] |
| 4 | 16546 MPa | 2400 kpsi, 260°, reinforced grade, diaminophenylsulfone/BF$_3$.ethylamine cured [5] |

Flexural Strength at Break:

No.	Value	Note
1	489 MPa	71 kpsi, 23°, reinforced grade, nadic methyl anhydride and benzyldimethylamine cured [5]
2	86 MPa	12.5 kpsi, 260°, reinforced grade, nadic methyl anhydride and benzyldimethylamine cured [5]
3	420 MPa	61 kpsi, 23°, reinforced grade, diaminophenylsulfone/BF$_3$.ethylamine cured [5]
4	172 MPa	25 kpsi, 260°, reinforced grade, diaminophenylsulfone/BF$_3$.ethylamine cured [5]

Optical Properties:
Transmission and Spectra: Ir [3] spectral data have been reported
Volume Properties/Surface Properties: Colour tan; Gardner 18 (min., 80% soln. in 2-butanone, Epon 1031) [9]

Polymer Stability:
Thermal Stability General: Weight loss of resin aged at 230° is greater for that cured with BF$_3$/ethylamine complex than that cured with maleic anhydride [7]. Resin cured with either maleic auhydride or *m*-phenylenediamine shows two exothermic peaks when subjected to differential thermal anal. under N$_2$, with these peaks occurring at 100–150° and 300–400°. The latter peaks may be due to isomerisation of the epoxy groups to carbonyl groups [8]
Decomposition Details: Activation energy of thermal degradation 21.4 kJmol^{-1} (5.1 kcal mol^{-1}) (cured with BF$_3$.ethylamine complex) [4]. Up to 500° the major pyrolysis product is CO$_2$; above this temp. the major product is CO [4]. Resin cured with 2-nitroresorcinol is stable to 300° when heated *in vacuo*, with 50% weight loss occurring at approx. 500° [6]

Applications/Commercial Products:
Processing & Manufacturing Routes: May be synth. by the reaction of a tetraphenol-substituted ethane with epichlorohydrin in aq. 2-propanol under basic conditions [1] or the acid-catalysed reaction of glyoxal with phenol followed by treatment with epichlorohydrin [2]. Curing with pyromellitic dianhydride results in premature gelation [7]
Applications: As an additive to moulding compounds; in adhesives; in electrical laminates; in high performance composites for the aerospace industry. Not recommended for food contact use according to FDA regulations

Trade name	Details	Supplier
Araldite	0163	Ciba-Geigy Corp.
E-1031 S		Yuka-Shell
Epon	1310/1031	Shell Chemical Co.

Bibliographic References
[1] Lee, M.-C., T.-H., and Wang, C.-S., *J. Appl. Polym. Sci.*, 1996, **62**, 217, (synth.)
[2] Gannon, J., *Kirk-Othmer Encycl. Chem. Technol.*, 4th edn., Vol.9 (ed. J.I.Kroschwitz), 1991, 735
[3] Serboli, G., *Kunststoffe*, 1966, **13**, 106, (ir)
[4] Modorsky, S.L. and Straus, S., *Mod. Plast.*, 1961, **38**, 134, (thermal degradation)
[5] Delmonte, J., May, C.A., Hoggatt, J.T., *Epoxy Resins: Chemistry and Technology*, 2nd edn., (ed. C.A. May), Marcel Dekker, 1988, 885
[6] Fleming, G.J., *J. Appl. Polym. Sci.*, 1969, **13**, 2579, (thermal anal.)
[7] Ehlers, G.F.L., *Polymer*, 1960, **1**, 304, (thermal stability)
[8] Anderson, H.C., *Anal. Chem.*, 1960, **32**, 1592, (thermal anal.)
[9] 1997, sheet no 235-89. 131R, Shell Chemical Company, (technical datasheet)

Poly(glycolic acid)   P-283

Synonyms: *Polyoxyacetyl. Poly(hydroxyacetic acid). PGA. Poly(hydroxyacetic ester). Polyglycolide. Polyglycollide. Poly(glycollic acid). Poly(glycollic ester). Poly(methylene carboxylate). Poly(oxy-1-oxoethylene). Poly[oxy(1-oxo-1,2-ethanediyl)]. 1,4-Dioxane-2,5-dione homopolymer. PGLY. PHAc*
Monomers: Glycolic acid, Glycolide
Material class: Thermoplastic
Polymer Type: saturated polyester
CAS Number:

CAS Reg. No.	Note
26124-68-5	Poly(glycolic acid)
26009-03-0	repeating unit
26202-08-4	polyglycolide

Molecular Formula: $(C_2H_2O_2)_n$
Fragments: $C_2H_2O_2$
Molecular Weight: MW 20000–145000 (fibre). MW 60000–90000
Additives: Polyamides as stabilisers against irradiation
Morphology: Orthorhombic, a 0.522, b 0.619, c 0.702, helix 2$_1$ [11]; orthorhombic, a 0.636, b 0.513, c 0.704, helix 2$_1$ [12]; orthorhombic, a 0.5201, b 0.6168, c 0.6948 [26]; fibre, c 0.703 [2,41]. Crystallises as spherulites, hedrites and as a hedritic rosette [44]

Volumetric & Calorimetric Properties:
Density:

No.	Value	Note
1	d 1.6 g/cm^3	[2]
2	d 1.575 g/cm^3	[11]
3	d 1.605 g/cm^3	[12]
4	d 1.58 g/cm^3	[13]
5	d 1.7 g/cm^3	[37]
6	d^{25} 1.584 g/cm^3	fibre-Dexon 2-0 [21]
7	d 1.69 g/cm^3	[44]

Thermodynamic Properties General: ΔC_p 40 J mol^{-1} K^{-1} [28]ΔC_p 31.8 J mol^{-1} K^{-1} [31]. Variation of C_p with temp. has been reported [38]
Latent Heat Crystallization: ΔH_f 191.3 J g^{-1} (45.7 cal g^{-1}) [4]; ΔH_f 206.6 J g^{-1} (49.34 cal g^{-1}) [6]; ΔH_f 71 Jg^{-1} [82]. Heat of polymerisation 26.4 kJ mol^{-1} (6.3 kcal mol^{-1}) [9]; ΔH_f 11.1 kJ mol^{-1} [35]. ΔS_f 22 J mol^{-1} K^{-1} [35]; $\Delta H_m°$ 15.7 kJ mol^{-1}; $\Delta S_m°$ 47 J mol^{-1} K^{-1} [24,28]; S 150.9 J mol^{-1} K^{-1} (25°) [24]; ΔH_f 9.74 kJ mol^{-1} [31]. Other ΔH_f values have been reported [4,6,56,61]. ΔG and ΔH values have been reported [32]. S_{250}^c 75.67 J mol^{-1} K^{-1}, S_{250}^a 83.22 J mol^{-1} K^{-1}, S_{Tg}^a 87.66 J mol^{-1} K^{-1} [31], S_{Tm}^a 137.89 J mol^{-1} K^{-1}, S_o^a 7.55 J mol^{-1} K^{-1}, $H_o^a - H_o^c$ 5.03 kJ mol^{-1} [31], $H_{25}^a - H_o^c$ 16.287 kJ mol^{-1}, $H_{Tg}^a - H_o^c$ 17.652 kJ mol^{-1}, $H_{Tm}^a - H_o^c$ 37.999 kJ mol^{-1} [31]. Heat of combustion -1435 kJ mol^{-1} [74], -1411.4 kJ mol^{-1} (cryst.) [74]. Heat of formation -734.4 kJ mol^{-1} (cryst.). Heat of polymerisation -34 kJ mol^{-1} (cryst) [74]; -10.8 kJ mol^{-1} (glassy) [74]. Heats of crystallisation have been reported [79]
Specific Heat Capacity:

No.	Value	Note	Type
1	130.3 kJ/kg.C	25° [24, 28]	P

Poly(glycolic acid)

Melting Temperature:

No.	Value	Note
1	187–222°C	MW dependent [18]
2	231–232°C	[2]
3	224–227°C	[4]
4	227–230°C	[10]
5	233°C	[11,35]
6	220°C	[13,26,32]
7	223°C	[25]
8	228°C	501K [24,28,31]
9	216–218°C	[30]
10	222°C	[34]
11	219°C	[37]
12	210°C	[82]
13	225–230°C	[40]
14	214–219°C	[83]

Glass-Transition Temperature:

No.	Value	Note
1	36°C	[8,17]
2	45°C	318K [24,28,31]
3	36.5°C	[29]
4	37°C	[32]
5	44°C	[34]
6	35°C	[82]

Transition Temperature General: Other glass-transition temps. have been reported [64]

Transition Temperature:

No.	Value	Note	Type
1	59°C	332K, 1 kHz	T_α
2	-18°C	255K, 1 kHz [34]	T_β

Surface Properties & Solubility:

Cohesive Energy Density Solubility Parameters: Cohesive energy 5.86 kJ mol^{-1} [35]. CED has been reported [61]

Solvents/Non-solvents: Sol. hot DMF, hot DMSO [2], hot phenol/trichlorophenol (60:40) [5], hot glycolide [9], hot nitromethane, hot nitrobenzene, hot 1-chloro-1-nitropropane, 10% sodium hydroxide soln., hexafluoroacetone sesquihydrate [30], 1,1,1,3,3,3-hexafluoro-2-propanol [36]. Slightly sol. CHCl$_3$/trifluoroacetic acid mixture [63]. Insol. common organic solvents [2,8]

Transport Properties:

Melt Flow Index:

No.	Value	Note
1	<2 g/10 min	max., 230° [50]
2	2.2 g/10 min	[78]

Polymer Solutions Dilute: [η] 0.4 dl g^{-1} (DMSO, 90°) [2]; η_{inh} 0.35 dl g^{-1} (trichlorophenol/phenol (60:40), 30°) [5]; η_{inh} 0.51 dl g^{-1} (phenol/tetrachloroethane (60:40), 30°) [15]; η_{inh} 1.42 dl g^{-1} (hexafluoroacetone sesquihydrate, 30°) [49]; [η] 1.14 dl g^{-1} (phenol/trichlorophenol (10:7)) [55]

Polymer Melts: Melt viscosity 500 Pa s (5000 poise, 245°) [48]

Permeability of Liquids: Water diffusion coefficient 5×10^{-9} cm^2 s^{-1} (37°). Other water diffusion coefficients have been reported [7,75]

Mechanical Properties:

Mechanical Properties General: Dynamic loss [2], storage [2], flexural and Young's moduli [64] have been reported. Tensile strength break has been reported [6,16]. Stiffness 1.84 MPa (fibre-Dexon 3-0) [20]. Elastic stiffness 41.9 kg (fibre-Dexon 2-0) [21]. Tensile strength at break 4.9 gpd (35% elongation, fibre- Dexon 2.0) [23], 9 gpd (20% elongation, fibre) [49], 10 gpd (fibre) [55], 6.5 gpd (fibre) [78]

Tensile (Young's) Modulus:

No.	Value	Note
1	6500 MPa	[80]

Flexural Modulus:

No.	Value	Note
1	10000–15000 MPa	rods [65]

Tensile Strength Break:

No.	Value	Note	Elongation
1	647 MPa	66 kg mm^{-2}, fibre-Dexon 2 [21]	33.1%
2	338–440 MPa	34.5–44.7 kg mm^{-2}, fibre-Dexon Plus [22]	
3	57 MPa	17.5×10^{-8} kg mm^{-2}, fibre-Dexon O [29,80]	34%
4	375 MPa	38.24 kg mm^{-2}, Dexon fibre [46]	20%
5	304 MPa	31 kg mm^{-2}, fibre [60]	

Flexural Strength at Break:

No.	Value	Note
1	355–395 MPa	rods [65]

Elastic Modulus:

No.	Value	Note
1	6760 MPa	690 kg mm^{-2}, fibre-Dexon 2-0 [21]

Miscellaneous Moduli:

No.	Value	Note	Type
1	6800 MPa	[2]	Modulus of glassy state

Viscoelastic Behaviour: Viscoelastic props. for Dexon fibre [58] and stress relaxation studies have been reported [67,81]

Failure Properties General: Break load 3873 N (fibre-Dexon 3-0) [20]. Shear strength 240 MPa (rods) [65]

Fracture Mechanical Properties: Toughness per unit length 0.806 kg cm cm^{-1} [21]

Friction Abrasion and Resistance: Coefficient of friction 0.191 (fibre-Dexon 3-0) [20]

Electrical Properties:

Electrical Properties General: Variation of tan δ with temp. has been reported in graphical form [34]

Dissipation Power Factor: Dielectric loss factor has been reported [2]

Optical Properties:

Transmission and Spectra: Ir [1,2,3,13,30,83], uv [1], Raman [27], XPES [33,39,62], SIMS [62], H-1 nmr [30,38,63], C-13 nmr [63,83] and mass [15,25] spectral data have been reported

Polymer Stability:

Decomposition Details: $T_{dec.}$ 240° (approx.) [10]; $T_{dec.}$ 254° [82] 50% weight loss at 346° [10,32]. Eact. 136.5 kJ mol^{-1} (32.6 kcal mol^{-1}, degradation, N$_2$) [14]. Eact. 154.1 kJ mol^{-1} (36.8 kcal mol^{-1}, degradation, O$_2$) [14]. Max. decomposition occurs at 340° [15], 390° (approx.) [83]. Thermal degradation curves have been reported [8,71]. Details of thermal degradation [14] and other decomposition temp. values [64] have been reported

Biological Stability: Fibre-biodegradability studies have been reported [6,7,16,17,29,51,52,53,57,67]

Stability Miscellaneous: γ-Radiation degradation studies have been reported [5,8,16]

Applications/Commercial Products:

Processing & Manufacturing Routes: Prod. by esterification of glycolic acid. Synth. by cationic ring opening polymerisation of the cyclic diester glycolide, e.g.: using stannous octoate as the catalyst and lauryl alcohol as the chain control agent at 220° [6,8,9,30,36,40,41,42,43,54,55,59,60,66,68,69,70,78,83]

Applications: Used in surgical sutures, dentistry, orthopaedics, and drug delivery systems. Dexon is patented

Trade name	Supplier
Dexon	American Cyanamid Company
Dexon Plus	Davis and Beck

Bibliographic References

[1] Barker, S.A., Grant, P.M., Stacey, M. and Ward, R.B., *J. Chem. Soc.*, 1959, 2648, (ir, uv)
[2] Ishida, Y., Ito, H. and Takayanagi, M., *J. Polym. Sci., Part B: Polym. Lett.*, 1965, **3**, 87, (mech props, solubility, ir)
[3] Nichols, J.B., *J. Appl. Phys.*, 1954, **25**, 840, (ir)
[4] Chu, C.C., *Polymer*, 1980, **21**, 1480, (melting point)
[5] Pittman, C.U., Iqbal, M., Chen, C.Y. and Helbert, A.N., *J. Polym. Sci., Polym. Chem. Ed.*, 1978, **16**, 2721, (soln props, radiation degradation)
[6] Chu, C.C., *J. Appl. Polym. Sci.*, 1981, **26**, 1727, (synth, biodegradation)
[7] Moiseev, Y.V., Daurova, T.T., Voronkova, O.S., Gumargalieva, K.Z. and Privalova, L.G., *J. Polym. Sci., Polym. Symp.*, 1979, **66**, 269, (water diffusion coefficient, biodegradation)
[8] Gilding, D.K. and Reed, A.M., *Polymer*, 1979, **20**, 1459, (synth, transition temp)
[9] Chujo, K., Kobayashi, H., Suzuki, J., Tokuhara, S. and Tanabe, M., *Makromol. Chem.*, 1967, **100**, 262, (synth, thermodynamic props)
[10] Chujo, K., Kobayashi, H., Suzuki, J. and Tokuhara, S., *Makromol. Chem.*, 1967, **100**, 267, (thermal degradation, density)
[11] Chatani, Y., Suehiro, K., Okita, Y., Tadokoro, H. and Chujo, K., *Makromol. Chem.*, 1968, **113**, 215, (struct)
[12] Hirono, K., Wasai, K., Saegusa, T. and Furukawa, J., *Kogyo Kagaku Zasshi*, 1964, **67**, 604, (struct)
[13] Tadokoro, H., Kobayashi, M., Yoshidome, H., Tai, K. and Makino, D., *J. Chem. Phys.*, 1968, **49**, 3359, (ir)
[14] Cooper, D.R., Sutton, G.J. and Tighe, B.J., *J. Polym. Sci., Polym. Chem. Ed.*, 1973, **11**, 2045, (thermal degradation)
[15] Garozzo, D., Giuffrida, M. and Montaudo, G., *Macromolecules*, 1986, **19**, 1643, (ms, thermal degradation)
[16] Chu, C.C., *Polymer*, 1985, **26**, 591, (biodegradation)
[17] Pourdeyhimi, B., *J. Mater. Sci. Lett.*, 1987, **6**, 1039, (biodegradation, transition temp)
[18] Cohn, D., Younes, H. and Marom, G., *Polymer*, 1987, **28**, 2018, (transition temps)
[19] Eilert, J.B., Binder, P., McKinney, P.W., Beal, J.M. and Conn, J., *Am. J. Surg.*, 1971, **121**, 561, (use)
[20] Rodeheaver, G.T., Thacker, J.G., Owen, J., Strauss, M. *et al*, *J. Surg. Res.*, 1983, **35**, 525, (mech props)
[21] Pavan, A., Bozio, M. and Longo, T., *J. Biomed. Mater. Res.*, 1979, **13**, 477, (mech props)
[22] Von Fraunhofer, J.A., Storey, R.S., Stone, I.K. and Masterson, B.J., *J. Biomed. Mater. Res.*, 1985, **19**, 595, (mech props)
[23] Chu, C.C., *J. Biomed. Mater. Res.*, 1981, **15**, 19, (mechanical props)
[24] Domalski, E.S. and Hearing, E.D., *J. Phys. Chem. Ref. Data*, 1990, **19**, 881, (thermodynamic props)
[25] Short, R.D. and Davies, M.C., *Int. J. Mass Spectrom. Ion Processes*, 1989, **89**, 149, (ms)
[26] Marega, C., Marigo, A., Zanetti, R. and Paganetto, G., *Eur. Polym. J.*, 1992, **28**, 1485, (struct)
[27] Cassanas, G., Kister, G., Fabregue, E., Morssli, M. and Bardet, L., *Spectrochim. Acta, Part A*, 1993, **49**, 271, (Raman)
[28] Lebedev, B. and Eustropov, A., *Makromol. Chem.*, 1984, **185**, 1235, (thermodynamic props)
[29] Reed, A.M. and Gilding, D.K., *Polymer*, 1981, **22**, 494, (biodegradation, mech props)
[30] Pinkus, A.G. and Subramanyam, R., *J. Polym. Sci., Polym. Chem. Ed.*, 1984, **22**, 1131, (ir, H-1 nmr, synth, solubilities)
[31] Varma-Nair, M., Pan, R. and Wunderlich, B., *J. Polym. Sci., Part B: Polym. Phys.*, 1991, **29**, 1107, (thermodynamic props)
[32] Jin, X., Carfagna, C., Nicolais, L. and Lanzetta, R., *J. Polym. Sci., Part A: Polym. Chem.*, 1994, **32**, 3115, (solubilities, transition temps)
[33] Briggs, D. and Beamson, G., *Anal. Chem.*, 1993, **65**, 1517, (XPES)
[34] Starkweather, H.W., Avakian, P., Fontanella, J.J. and Wintersgill, M.C., *Macromolecules*, 1993, **26**, 5084, (dissipation factor, transition temps)
[35] Wunderlich, B. and Gaur, U., *Pure Appl. Chem.*, 1980, **52**, 445, (thermodynamic props)
[36] Braun, D. and Kohl, P.R., *Angew. Makromol. Chem.*, 1986, **139**, 191, (synth)
[37] Kricheldorf, H.R., Mang, T. and Jonte, J.M., *Macromolecules*, 1984, **17**, 2173, (melting point)
[38] Lim, S. and Wunderlich, B., *Polymer*, 1987, **28**, 777, (heat capicity)
[39] Briggs, D. and Beamson, G., *Anal. Chem.*, 1992, **64**, 1729, (XPES)
[40] Asahara, T., Yamashita, K. and Katayama, S., *Kogyo Kagaku Zasshi*, 1963, **66**, 485, (synth, melting point)
[41] Asahara, T. and Katayama, S., *Kogyo Kagaku Zasshi*, 1964, **67**, 362, (synth, struct)
[42] Andreas, F., Sowada, R. and Scholz, J., *Prakt. Chem.*, 1962, **18**, 141, (synth)
[43] *Belg. Pat.*, 1965, 654 236, (synth)
[44] Grabar, D.G., *Microscopy*, 1970, **18**, 203, (morphological props)
[45] *U.S. Pat.*, 1969, 3 468 853, (synth)
[46] *Fr. Pat.*, 1969, 1 563 261, (synth)
[47] *Jpn. Pat.*, 1968, 3017, (stabilisers)
[48] *U.S. Pat.*, 1967, 3 297 033, (synth, melt viscosity)
[49] *U.S. Pat.*, 1975, 3 890 283, (soln props, mech props)
[50] *Ger. Pat.*, 1971, 2 025 468, (melt flow velocity)
[51] Razumova, L.L., Veretennikova, A.A. and Zaikov, G.E., *Dokl. Akad. Nauk SSSR*, 1979, **245**, 654, (biodegradation)
[52] Privalova, L.G., Daurova, T.T., Voronkova, O.S., Gumargalieva, K.Z. *et al*, *Vysokomol. Soedin., Ser. A*, 1980, **22**, 1891, (biodegradation)
[53] Razumova, L.L., Daurova, T.T., Veretennikova, A.A., Privalova, L.G. *et al*, *Polim. Med.*, 1979, **9**, 119, (biodegradation)
[54] *U.S.S.R. Pat.*, 1976, 525 714, (synth)
[55] *Jpn. Pat.*, 1985, 60 144 325, (synth)
[56] Chu, C.C., *Adv. Biomater.*, 1982, **3**, 781, (heat of fusion)
[57] Browning, A. and Chu, C.C., *Polym. Mater. Sci. Eng.*, 1985, **53**, 510, (biodegradation)
[58] Hasegawa, M., Yamagu, K., Takenaka, I., Dobashi, T. and Sakanishi, I., *CA*, 1985, **104**, 116031, (viscoelastic props)
[59] *Brit. Pat.*, 1983, 2 169 609, (synth)
[60] *Jpn. Pat.*, 1983, 58 98 329, (synth)
[61] Schultze-Gebhardt, F., *Acta Polym.*, 1987, **38**, 36, (cohesive energy density, thermodynamic props)
[62] Davies, M.C., Short, R.D., Khan, M.A. and Watts, J.F. *et al*, *CA*, 1989, **111**, 160136, (ms, XPS)
[63] Hariharan, R. and Pinkus, A.G., *Polym. Bull. (Berlin)*, 1993, **30**, 91, (H-1 nmr, C-13 nmr)
[64] Engelberg, I. and Kohn, J., *Isr. Mater. Eng. Conf.*, 5th, 1991, 277, (mechanical props)
[65] Zhang, X., Goosen, M.F.A., Wyss, U.P. and Pichora, D., *J. Macromol. Sci., Rev. Macromol. Chem. Phys.*, 1993, **33**, 81, (mech props)
[66] *Jpn. Pat.*, 1993, 05 287 056, (synth)
[67] Hayes, M.J. and Lauren, M.D., *J. Appl. Biomater.*, 1994, **5**, 215, (stress relaxation, biodegradation)
[68] *S. African Pat.*, 1968, 68 01 143, (synth)
[69] Frazza, E.J. and Schmitt, E.E., *Biomed. Mater. Symp.*, 1970, **1**, 43, (synth)
[70] Sanina, G.S., Fornina, M.V., Khomyakov, A.K. and Livshits, V.S. *et al*, *Vysokomol. Soedin., Ser. A*, 1975, **17**, 2726, (synth)
[71] Dundurs, J., Alksnis, A., Surna, J. and Gruzins, I., *CA*, 1975, **83**, 164787, (thermal degradation)
[72] Lebedev, B.V., Eustropov, A.A., Kiparisova, E.G., Lyudvig, E.B. and Sanina, G.S., *Dokl. Akad. Nauk SSSR*, 1977, **236**, 669, (thermodynamic props)

- Poly(glycol methacrylates)

[73] Lebedev, B.V., Eustropov, A.A., Kiparisova, E.G. and Belov, V.I., Vysokomol. Soedin., Ser. A, 1978, **20**, 29, (thermodynamic props)
[74] Kiparisova, E.G., Bykova, T.A. and Lebedev, B.V., CA, 1977, **92**, 22892, (thermodynamic props)
[75] Livshits, V.S. and Selezneva, V.E., Mekh. Polim., 1978, **2**, 367, (diffusion coefficient)
[76] Lebedev, B.V. and Rabinovich, I.B., Dokl. Akad. Nauk SSSR, 1977, **237**, 641, (thermodynamic props)
[77] Rabinovich, I.B. and Lebedev, B.V., Vysokomol. Soedin., Ser. A, 1979, **21**, 2025, (thermodynamic props)
[78] Jpn. Pat., 1988, 63 17 929, (synth, melt flow index)
[79] Wang, Y., Sui, W., Tu, T. and Shao, Y., CA, 1991, **117**, 27615, (heat of crystallisation)
[80] Winter, D.G., Gibbons, D.F. and Plench, J., Adv. Biomater., 1982, **3**, 271, (mech props)
[81] Metz, S.A., Von Fraunhofer, J.A. and Masterson, B.J., Biomaterials, 1990, **11**, 197, (stress relaxation)
[82] Engelberg, I. and Kohn, J., Biomaterials, 1990, **12**, 292, (transition temps)
[83] Mathias, L.J., Kusefoglu, S.A. and Kress, A.O., J. Appl. Polym. Sci., 1989, **38**, 1037, (synth, H-1 nmr, C-13 nmr, ir)

Poly(glycol methacrylates)   P-284

$$\mathrm{+CH_2C(CH_3)+_n}$$
$$\mathrm{O\!=\!\!C\!-\!O(CH_2CH_2O)_xH}$$

where x = 1 (ethyleneglycol methacrylate)
x = 2 (diethyleneglycol methacrylate)

$$\mathrm{+CH_2C(CH_3)\quad (CH_3)CCH_2+_n}$$
$$\mathrm{O\!=\!\!C\!-\!O(CH_2CH_2)_yO\!-\!C\!=\!\!O}$$

where x = 1 (ethyleneglycol dimethacrylate)
x = 2 (diethyleneglycol dimethacrylate)

Synonyms: Poly(ethylene oxide) polymethacrylate. Poly(1,2-ethanediylbis(oxy-2,1-ethanediyl) methacrylate). Poly(ethylene glycol methacrylate). Poly(ethylene glycol dimethacrylate). Poly(ethylene glycol trimethacrylate). Poly(triethylene glycol methacrylate). Poly(tetraethylene glycol methacrylate). Poly(triethylene glycol dimethacrylate). Poly(tetraethylene glycol dimethacrylate)
Related Polymers: Polymethacrylates, General
Monomers: Glycol dimethacrylate
Material class: Thermoplastic
Polymer Type: acrylic
CAS Number:

CAS Reg. No.	Note
25101-31-9	triethylene glycol dimethacrylate
25721-76-0	ethylene glycol methacrylate

Molecular Formula: $(C_{10}H_{14}O_4)_n$
Fragments: $C_{10}H_{14}O_4$
General Information: The physical appearance of the polymers depends on the length of the glycol chain and the degree of polymerisation. For example the poly(ethylene oxide) polymethacrylates vary from crystalline solids (long chain) to viscous liqs. (short chain). As the MW increases the polymer tends towards greater solid character. Highly cross-linked polymers of poly(mono, di, tri and tetraethylene glycol methacrylates) can be prod. with 2–3 repeat units between cross-links [1]
Identification: Contains only C, H and O

Volumetric & Calorimetric Properties:
Density:

No.	Value	Note
1	d 1.229 g/cm^3	ethylene glycol methacrylate [1]
2	d 1.26 g/cm^3	ethylene glycol dimethacrylate [2]
3	d 1.221 g/cm^3	diethylene glycol methacrylate [1]
4	d 1.211 g/cm^3	triethylene glycol methacrylate [1]
5	d 1.175 g/cm^3	tetraethylene glycol methacrylate [1]

Specific Heat Capacity:

No.	Value	Note	Type
1	0.0919–0.2047 kJ/kg.C	tetraethylene glycol methacrylate, MW 20000–23400 [3]	P

Melting Temperature:

No.	Value	Note
1	261–313.8°C	8–23 ethylene oxide units in chain, various MW [3]

Glass-Transition Temperature:

No.	Value	Note
1	204.6–240.3°C	4–23 ethylene oxide units in chain, various MW [3]
2	90°C	ethylene glycol methacrylate [4]
3	105°C	ethylene glycol methacrylate, preheated to 220° [4]
4	125°C	ethylene glycol, twice preheated to 220° [4]
5	271°C	ethylene glycol dimethacrylate [1]
6	250°C	diethylene glycol dimethacrylate [1]
7	135–248°C	triethylene glycol dimethacrylate [1,5]
8	27°C	tetraethylene glycol methacrylate [5]
9	115–279°C	tetraethylene glycol dimethacrylate [1,5]
10	-5°C	ethylene glycol 400 dimethacrylate [5]
11	-35°C	ethylene glycol 600 dimethacrylate [5]
12	-53°C	ethylene glycol 1000 dimethacrylate [5]

Vicat Softening Point:

No.	Value	Note
1	200°C	glycol dimethacrylate [2]

Surface Properties & Solubility:
Solubility Properties General: The swelling characteristics of ethylene, diethylene, triethylene and tetraethylene glycol methacrylates in xylene, MeOH, 2-butanone and isobutanol have been reported [1]
Cohesive Energy Density Solubility Parameters: δ 72.4 J$^{1/2}$ cm$^{-3/2}$ (17.3 cal$^{1/2}$ cm$^{-3/2}$) (diethylene glycol methacrylate), 74.1 J$^{1/2}$ cm$^{-3/2}$ (17.7 cal$^{1/2}$ cm$^{-3/2}$) (triethylene glycol methacrylate), 71.6 J$^{1/2}$ cm$^{-3/2}$ (17.1 cal$^{1/2}$ cm$^{-3/2}$) (ethylene glycol 600 methacrylate) [6]
Wettability/Surface Energy and Interfacial Tension: The contact angle of poly(methoxypolyethylene glycol methacrylates) decreases as the number of ethylene glycol units (n) increases. Contact angle 70° (approx., n = 1), 10° (max., n = 3). Contact angle remains essentially constant with further increase in n [7]

Surface Tension:

No.	Value	Note
1	33 mN/m	33 dyne cm^{-1}, glycol dimethacrylate [8]

Transport Properties:
Water Content: The water content of poly(methoxypolyethylene glycol methacrylates) increases rapidly as the number of ethylene glycol units (n) increases reaching a value of 90% (approx., n = 3) with a moderate rate of increase to a value of approx. 120% (n = 9) [7]

Water Absorption:

No.	Value	Note
1	450 %	approx., methoxypolyethylene glycol methacrylate, 3 ethylene glycol units [7]
2	600 %	approx., methoxypolyethylene glycol methacrylate, 9 ethylene glycol units [7]

Mechanical Properties:
Mechanical Properties General: The tensile strength of poly(methoxypolyethylene glycol methacrylates) decreases as the number of ethylene glycol units (n) increases. Swelling the polymer in H_2O reduces the tensile strength slightly; the effect is more pronounced at lower values of n. Elongation at break 50% (approx., n = 1 and n = 9)-300% (min., n = 2), 110% (approx., n = 4) [7]

Tensile Strength Break:

No.	Value	Note
1	9.8 MPa	100 kg cm^{-2}, min., methoxypolyethylene glycol methacrylate, 1-ethylene glycol unit [7]
2	0.98 MPa	10 kg cm^{-2}, max., methoxypolyethylene glycol methacrylate, 3 ethylene glycol units [7]
3	0.098 MPa	1 kg cm^{-2}, max., methoxypolyethylene glycol methacrylate, 4 ethylene glycol units [7]

Poisson's Ratio:

No.	Value	Note
1	0.2	triethylene glycol dimethacrylate
2	0.25	tetraethylene glycol dimethacrylate
3	0.5	tetraethylene glycol methacrylate
4	0.41	ethylene glycol 400 dimethacrylate
5	0.3	ethylene glycol 600 dimethacrylate [5]

Electrical Properties:
Electrical Properties General: Solid polymer electrolyte complexes can be formed between certain salts and poly(methacrylates) with pendant ethylene oxide chains. The crystalline complex so formed results in a polymer with conducting props.

Optical Properties:
Total Internal Reflection: Intrinsic segmental anisotropy of ethylene glycol methacrylate [4] 1×10^{-25} cm^3 (DMF), -6×10^{-25} cm^3 (EtOH), -6×10^{-25} cm^3 (H_2O)

Polymer Stability:
Thermal Stability General: On heating pendant glycol groups can undergo a disproportionation reaction causing cross-linking of the chains [4]

Applications/Commercial Products:
Processing & Manufacturing Routes: May be synth. by free-radical polymerisation, using 2,2-dimethoxy-2-phenylacetophenone as a photoinitiator, in air or under N_2 at 22–27° [1]
Applications: Mono, di, tri and tetraethylene glycol methacrylates are used in laser video discs. Ethylene glycol 1000 methacrylates are used in solid state battery electrolytes. Methoxypolyethylene glycol methacrylates have biomedical applications

Bibliographic References
[1] Bowman, C.N., Carver, A.L., Kennet, S.N., Williams, M.M. and Peppas, N.A., *Polymer*, 1990, **31**, 135
[2] Hoff, E.A.W., Robinson, D.W. and Willbourn, A.H., *Polym. Sci.*, 1955, **18**, 161
[3] Yan, F., Dejardin, P., Frere, Y. and Gramain, P., *Makromol. Chem.*, 1990, **191**, 1197, 1209, (heat capacity, transition temps)
[4] Grishchenko, A.Y. and Yezrielev, R.I., *Polym. Sci. USSR (Engl. Transl.)*, 1972, **A14**, 582
[5] Allen, P.E.M., Simon, G.P., Williams, D.R.G. and Williams, E.H., *Macromolecules*, 1989, **22**, 809, (dynamic mech props)
[6] Ore, S., Kulvic, E., Ohr, W. and Zapffe, J.W., *Makromol. Chem.*, 1971, **148**, 211
[7] Kumakuru, M. and Kaetsu, I., *J. Mater. Sci.*, 1983, **18**, 2430
[8] Lee, L.H., *Coat. Plast. Prepr. Pap. Meet. (Am. Chem. Soc., Div. Org. Coat. Plast. Chem.)*, 154th, 1967, 61

Poly(haloalkyl methacrylate)   P-285

Synonyms: *Poly(2-bromoethyl methacrylate). Poly(1-chloroethyl methacrylate). Poly(2-chloroethyl methacrylate). Poly(2,3-dibromopropyl methacrylate). Poly(2,2,2-trichloroethyl methacrylate). Poly (2,2,2-trifluoroethyl methacrylate). Poly(tetrafluoropropyl methacrylate). Poly(1,1,1-trifluoroisopropyl methacrylate). Poly(1H-tetrafluoroisopropyl methacrylate). Poly(1H-hexafluoroisopropyl methacrylate). Poly(heptafluoroisopropyl methacrylate). Poly(chlorohexafluoroisopropyl methacrylate). Poly(heptafluorobutyl methacrylate). Poly(tert-nonafluorobutyl methacrylate). Poly(octafluoropentyl methacrylate). Poly(dodecafluoroheptyl methacrylate). Poly(1H,1H-pentadecafluorooctyl methacrylate). Poly(heptadecafluorooctyl methacrylate). Poly(hexadecafluorononyl methacrylate). Poly(2-cyanoethyl methacrylate)*

Related Polymers: Polymethacrylates-General, Polymethacrylate esters of substituted alkyls, Poly(methyl methacrylate), Poly(ethyl methacrylate), Poly(propyl methacrylate), Poly(isopropyl methacrylate), Poly(butyl methacrylate), Poly(*tert*-butyl methacrylate), Poly(pentyl methacrylate), Poly(octyl methacrylate), Poly(nonyl methacrylate)

Monomers: 2-Bromoethyl methacrylate, 1-Chloroethyl methacrylate, 2-Chloroethyl methacrylate, 2-Cyanoethyl methacrylate, 2,3-Dibromopropyl methacrylate, 2,2,2-Trichloroethyl methacrylate, 2,2,2-Trifluoroethyl methacrylate, Tetrafluoropropyl methacrylate, 1,1,1-Trifluoroisopropyl methacrylate, 1*H*-Tetrafluoroisopropyl methacrylate, 1*H*-Hexafluoroisopropyl methacrylate, Heptafluoroisopropyl methacrylate, Chlorohexafluoroisopropyl methacrylate, Heptafluorobutyl methacrylate, *tert*-Nonafluorobutyl methacrylate, Octafluoropentyl methacrylate, Dodecafluoroheptyl methacrylate, 1*H*,1*H*-Pentadecafluorooctyl methacrylate, Heptadecafluorooctyl methacrylate, Hexadecafluorononyl methacrylate

Material class: Thermoplastic
Polymer Type: acrylic
CAS Number:

CAS Reg. No.	Note
107194-75-2	1-chloroethyl
26937-47-3	2-chloroethyl
40715-87-5	2,3-dibromopropyl

Identification: All contain C, H, O and halogen

Volumetric & Calorimetric Properties:

Density:

No.	Value	Note
1	d^{25} 1.32 g/cm^3	amorph., 1-chloroethyl [1]
2	d^{20} 1.4934 g/cm^3	tetrafluoropropyl [2]
3	d^2 1.34 g/cm^3	1,1,1-trifluoroisopropyl [1,3]

Thermal Expansion Coefficient:

No.	Value	Note	Type
1	0.00032 K^{-1}	2-bromoethyl, bulk polymerisation, M$_n$ 431000, T > T$_g$ [4]	V
2	0.00031 K^{-1}	2-cyanoethyl, bulk polymerisation, M$_n$ 985000, T > T$_g$ [4]	V
3	0.00099 K^{-1}	2-bromoethyl, bulk polymerisation, M$_n$ 431000, T < T$_g$ [4]	V
4	0.0001 K^{-1}	2-cyanoethyl, bulk polymerisation, M$_n$ 985000, T < T$_g$ [4]	V

Glass-Transition Temperature:

No.	Value	Note
1	52°C	2-bromoethyl, bulk polymerisation, M$_n$ 431000 [4]
2	92°C	1-chloroethyl [5]
3	91°C	2-cyanoethyl, bulk polymerisation, M$_n$ 985000 [4]
4	132°C	2,2,2-trichloroethyl [6]
5	72°C	2,2,2-trifluoroethyl [6]
6	64°C	tetrafluoropropyl [2]
7	81°C	1,1,1-trifluoroisopropyl [3]
8	67°C	heptafluorobutyl [7]
9	156°C	*tert*-nonafluorobutyl [8]
10	36°C	octafluoropentyl [9]
11	13°C	dodecafluoroheptyl [9]
12	-15°C	hexadecafluorononyl [9]

Vicat Softening Point:

No.	Value	Note
1	118°C	1-chloroethyl [3]
2	81°C	1,1,1-trifluoroisopropyl [3]

Surface Properties & Solubility:

Solvents/Non-solvents: 2,2,2-Trichloroethyl: sol. CHCl$_3$, acetonitrile, THF [6]. 2,2,2-Trifluoroethyl: sol. Me$_2$CO, acetonitrile, CHCl$_3$, THF [6].

Surface and Interfacial Properties General: Critical surface tension: 14.8–15.4 mN m^{-1} (20°, 1*H*-hexafluoroisopropyl) [10], 15 mN m^{-1} (20°, heptafluoroisopropyl) [10], 19.1 mN m^{-1} (chlorohexafluoroisopropyl) [10], 14.7 mN m^{-1} (20°, *tert*-nonafluorobutyl) [8], 10.6 mN m^{-1} (20°, 1*H*,1*H*-pentadecafluorooctyl [10,11,13], 10.6 mN m^{-1} (heptadecafluorooctyl) [10]

Surface Tension:

No.	Value	Note
1	19 mN/m	tetrafluoropropyl, 20° [9]
2	15 mN/m	tetrafluoroisopropyl, 20° [10]
3	10.5 mN/m	pentadecafluorooctyl, γ_d 10.0, γ_s 0.5 [11]
4	15.3 mN/m	heptadecafluorooctyl, 20° [12]

Optical Properties:

Refractive Index:

No.	Value	Note
1	1.4215	tetrafluoropropyl [2]

Polymer Stability:

Decomposition Details: Decomposition of the 2-bromoethyl ester below 500° yields monomer (>95%) as the major product with <1% of both CO$_2$ and vinyl bromide [14]. Decomposition of the 2,2,2-trifluoroethyl ester at room temp. under λ-irradiation yields many fluorinated products plus methane, carbon monoxide, MeOH and CO$_2$ [6]

Applications/Commercial Products:

Processing & Manufacturing Routes: 2-Bromoethyl ester polymer may be synth. by bulk polymerisation at 60°. The 2-cyanoethyl ester polymer may be synth. by free-radical polymerisation at 60° in acetonitrile using AIBN initiator [15]

Applications: Used in low surface energy or non-wettable coatings, and in electron lithography

Bibliographic References

[1] Van Krevelen, D.W., *Properties of Polymers: Their Correlation with Chemical Structure*, 3rd edn., Elsevier, 1990
[2] *Unpublished Data*, Rohm and Haas,
[3] Hoff, E.A.W., Robinson, D.W. and Willbourn, A.H., *Polym. Sci.*, 1955, **18**, 161
[4] Krause, S., Gormley, J.J., Roman, N., Shetter, J.A. and Watenabe, W.H., J. Polym. Sci., *Part A: Polym. Chem.*, 1965, **3**, 3573, (glass transition temps.)
[5] Heijboer, J., *Br. Polym. J.*, 1969, **1**, 3
[6] Babu, G.N., Lu, P.H., Hsu, S.L. and Chien, J.C.W., J. Polym. Sci., *Polym. Chem. Ed.*, 1984, **22**, 195
[7] Lee, W.M., *Diss. Abstr. Int. B*, 1967, **28**, 1897
[8] Roitman, J.N. and Pittman, A.G., J. Polym. Sci., *Polym. Lett. Ed.*, 1972, **10**, 499
[9] Tamaribuchi, K., Polym. Prepr. (Am. Chem. Soc., *Div. Polym. Chem.*), 1967, **8**, 631
[10] Pittman, A.G., Sharp, D.L. and Ludwig, B.A, J. Polym. Sci., *Part A-1*, 1968, **6**, 1729, (wetting props.)
[11] Kaelble, D.H., *Physical Chemistry of Adhesion*, Wiley-Interscience, 1971
[12] Gott, V.L. and Baier, R.E., *Public Document PH43-68-84-4-2 Nat. Tech. Info. Service*, 1973
[13] Bernett. M.K. and Zisman, W.A., *J. Phys. Chem.*, 1962, **66**, 1207
[14] Grassie, N., Johnstone, A. and Scotney, A., *Eur. Polym. J.*, 1981, **17**, 589, (thermal degradation)
[15] Sandler, S.R. and Karo, W., *Polymer Syntheses*, 2nd edn., Academic Press, 1992, **1**, (synth.).

Poly[4,4-heptane bis(4-phenyl)carbonate]   P-286

Synonyms: *Poly[oxycarbonyloxy-1,4-phenylene(1-propylbutylidene)-1,4-phenylene]. Poly(4,4'-dihydroxydiphenyl-4,4-heptane-co-carbonic dichloride)*
Monomers: 4,4'-Dihydroxydiphenyl-4,4-heptane, Carbonic dichloride
Material class: Thermoplastic
Polymer Type: polycarbonates (miscellaneous)

P-287 Poly(1-heptene)

CAS Number:

CAS Reg. No.
32243-07-5

Molecular Formula: $(C_{20}H_{22}O_3)_n$
Fragments: $C_{20}H_{22}O_3$

Volumetric & Calorimetric Properties:
Density:

No.	Value	Note
1	d 1.16 g/cm^3	25° [2]

Melting Temperature:

No.	Value	Note
1	190–200°C	[2]

Glass-Transition Temperature:

No.	Value	Note
1	130°C	[3]
2	148°C	[2]

Surface Properties & Solubility:
Solvents/Non-solvents: Sol. CH_2Cl_2, *m*-cresol, EtOAc, THF, C_6H_6, pyridine, DMF; swollen by cyclohexanone, Me_2CO [1]

Transport Properties:
Gas Permeability:

No.	Gas	Value	Note
1	O_2	154 cm^3 mm/(m^2 day atm)	1.76 × 10^{-13} cm^2 (s Pa)$^{-1}$, 25°, 1 atm [3]

Optical Properties:
Refractive Index:

No.	Value	Note
1	1.5602	25° [2]

Applications/Commercial Products:
Processing & Manufacturing Routes: The main synthetic routes are phosgenation and transesterification. The most widely used phosgenation approach is interfacial polycondensation in a suitable solvent, but soln. polymerisation is also carried out in the presence of pyridine [1]. A slight excess of phosgene is introduced into a soln. or suspension of the aromatic dihydroxy compound in aq. NaOH at 20–30°, in an inert solvent (e.g. CH_2Cl_2). Catalysts are used to achieve high MWs [1,2]. With transesterification, the dihydroxy compound and a slight excess of diphenyl carbonate are mixed together in the melt (150–300°) in the absence of O_2 with elimination of phenol. Reaction rates are increased by alkaline catalysts and the use of a medium-high vacuum in the final stages [1,2]

Bibliographic References
[1] Schnell, H., *Polym. Rev.*, 1964, **9**, 99, (solubility, synth., rev)
[2] Schnell, H., *Ind. Eng. Chem.*, 1959, **51**, 157, (props., synth.)
[3] Schmidhauser, J.C. and Longley, K.L., *J. Appl. Polym. Sci.*, 1990, **39**, 2083, (T_g, gas permeability)

Poly(1-heptene) P-287

$-[CH_2CH]_n-$
           $|$
       $CH_2(CH_2)_3CH_3$

Synonyms: *PHEP*. Poly(1-pentylethylene). Poly(1-pentyl-1,2-ethanediyl)
Monomers: 1-Heptene
Material class: Thermoplastic
Polymer Type: polyolefins
CAS Number:

CAS Reg. No.	Note
25511-64-2	atactic
26746-83-8	isotactic

Molecular Formula: $(C_7H_{14})_n$
Fragments: C_7H_{14}
General Information: Sticky rubber [1]
Morphology: Two forms, c 0.645, helix 3_1 (probable) [1]

Volumetric & Calorimetric Properties:
Density:

No.	Value	Note
1	d^{30} 0.863 g/cm^3	[2]

Thermal Expansion Coefficient:

No.	Value	Note	Type
1	0.0006 K^{-1}	[3]	V

Melting Temperature:

No.	Value	Note
1	17°C	[1,7]

Glass-Transition Temperature:

No.	Value	Note
1	-31°C	[7]
2	-45°C	[8]

Transition Temperature:

No.	Value	Note	Type
1	-148°C	320 Hz [7]	β

Transport Properties:
Polymer Solutions Dilute: [η] 0.51 dl g^{-1} (30°, C_6H_6) [11]. [η] 2.49 dl g^{-1} (25°, toluene) [12]. Other viscosity values have been reported [9]. Soln. props. have been reported [10]. Theta temp. has been reported [4]
Polymer Melts: Melt flow activation energy 71.2 kJ mol^{-1} (17 kcal mol^{-1}) [3,5]. Zero shear viscosity 4500 (80°) [13]. Melt props. have been reported [13]

Optical Properties:
Optical Properties General: Refractive index has been reported [9]
Transmission and Spectra: Ir [11], Raman [14,15], H-1 nmr [16] and C-13 nmr [17,18] spectral data have been reported
Total Internal Reflection: Stress optical coefficient -1000 [12]

Polymer Stability:
Environmental Stress: Undergoes oxidation [19]

Applications/Commercial Products:
Processing & Manufacturing Routes: Synth. using Ziegler-Natta catalysts, such as $TiCl_4Al(^iBu)_3$ in toluene at room temp. [17]

Bibliographic References
[1] Turner-Jones, A., *Makromol. Chem.*, 1964, **71**, 1, (struct, melting point)
[2] Tait, P.J.T. and Livesey, P.J., *Polymer*, 1970, **11**, 359, (soln props)
[3] Porter, R.S., Chuah, H.H. and Wang, J.S., *J. Rheol. (N.Y.)*, 1995, **39**, 649, (melt props)
[4] Wang, J.S., Porter, R.S. and Knox, J.R., *Polym. J. (Tokyo)*, 1978, **10**, 619, (expansion coefficient, soln props)
[5] Wang, J.S. and Porter, R.S., *Rheol. Acta*, 1995, **34**, 496, (melt props)
[6] Bu, H.S., Aycock, W. and Wunderlich, B., *Polymer*, 1987, **28**, 1165, (specific heat capacities)
[7] Clark, K.J., Turner-Jones, A. and Sandiford, D.J.H., *Chem. Ind. (London)*, 1962, **47**, 2010, (transition temps)
[8] Natta, G., Danusso, F. and Moraglio, G., *CA*, 1958, **52**, 17789e, (transition temp)
[9] Shalaeva, L.F., *Vysokomol. Soedin., Ser. B*, 1968, **10**, 449, (soln props)
[10] Daniels, C.A., Maron, S.H. and Livesey, P.J., *J. Macromol. Sci., Phys.*, 1970, **4**, 47, (soln props)
[11] Otsu, T., Nagahama, H. and Endo, K., *J. Polym. Sci., Polym. Lett. Ed.*, 1972, **10**, 601, (ir)
[12] Philippoff, W. and Tomquist, E., *J. Polym. Sci., Part C: Polym. Lett.*, 1966, **23**, 881, (optical props)
[13] Wang, J.S., Knox, J.R. and Porter, R.S., *J. Polym. Sci., Polym. Phys. Ed.*, 1978, **16**, 1709, (mech props)
[14] Holland-Moritz, K., Modric, I., Heinen, K.U. and Hummel, D.D., *Kolloid Z. Z. Polym.*, 1973, **251**, 913, (ir, Raman)
[15] Modric, I., Holland-Moritz, K. and Hummel, D.D., *Colloid Polym. Sci.*, 1976, **254**, 342, (ir, Raman)
[16] Slichter, W.P. and Davis, D.D., *J. Appl. Phys.*, 1964, **35**, 10, (H-1 nmr)
[17] Delfini, M., DiCocco, M.E., Paci, M., Aglietto, M. et al, *Polymer*, 1985, **26**, 1459, (C-13 nmr)
[18] Asakura, T., Demura, M. and Nishiyama, Y., *Macromolecules*, 1991, **24**, 2334, (C-13 nmr)
[19] Emanuel, N.M., Marin, A.P., Kinyushkin, S.G., Moiseev, Y.V. and Shlyapnikov, Y.A, *Dokl. Akad. Nauk SSSR*, 1985, **280**, 416, (polymer oxidation)

Poly(heptylhexylsilane)   P-288

$$\left[\begin{array}{c} CH_2(CH_2)_5CH_3 \\ | \\ Si \\ | \\ CH_2(CH_2)_4CH_3 \end{array} \right]_n$$

Synonyms: *Poly(heptylhexylsilylene)*. *Dichloroheptylhexylsilane homopolymer*. *PHHS*
Polymer Type: polysilanes
CAS Number:

CAS Reg. No.
120534-41-0

Molecular Formula: $(C_{13}H_{28}Si)_n$
Fragments: $C_{13}H_{28}Si$
Molecular Weight: $0.5 \times 10^6 - 1 \times 10^6$ [1,3,5].
General Information: Similar to PDHS as both sets of chains are transplanar, as are the Si backbones [1].
Morphology: Two forms have been observed [1]. Forms a polymer with an *all-trans* arrangement of the Si backbone and hexagonal unit cell dimensions $a = b = 1.53$nm [2].

Volumetric & Calorimetric Properties:
Glass-Transition Temperature:

No.	Value	Note
1	-5°C	form 1 [1]
2	25°C	form 2 [1]

Surface Properties & Solubility:
Solubility Properties General: Soluble in common organic solvents such as toluene, THF and isooctane [1]

Transport Properties:
Transport Properties General: The spreading behavior of polysilanes at the air-H_2O interface of a Langmuir trough has been investigated and stable monomolecular layers have not been obtained [5].

Optical Properties:
Optical Properties General: Exhibits a thermochromic transition, two forms being observed, having absorbances at 350 and 370nm depending on the thermal history.
Raman band intensities bear a marked resemblance to those of PDHS. Both exhibit UV absorbances near 370nm. The lowering of the transition temperature for the 370nm form from 42°C for PDHS to 25°C for PHHS is understandable in light of the disorder introduced by the mismatch of side chain length. PHHS also exhibited an isolated absorption at 350nm when quenched to low temperatures [1,3].

Optical Properties:
Transmission and Spectra: Ir [1,2], Raman [1] and uv [1,3,4,6] spectra reported.

Applications/Commercial Products:
Processing & Manufacturing Routes: Prepared from *n*-heptyl-*n*-hexyldichlorosilylene by condensation with Na metal in toluene at 110°C by the usual Wurtz-type reaction [1,5].
Applications:

Bibliographic References
[1] Hallmark, V.M., Sooriyakumaran, R., Miller, R.D. and Rabolt, J.F., *J. Chem. Phys.*, 1989, **90**, 2486
[2] KariKari, E.K., Greso, A.J. and Farmer, B.L., *Macromolecules*, 1993, **26**, 3937
[3] Miller, R.D., Wallraff, G.M., Baier, M., Cotts, P.M., Shukla, P., Russell, T.P., De Schryver, F.C. and Declercq, D., *J. Inorg. Organomet. Polym.*, 1991, **1**, 505
[4] Song, K., Miller, R.D., Wallraff, G.M. and Rabolt, J.F., *Macromolecules*, 1992, **25**, 3629
[5] Embs, F.W., Wegner, G., Neher, D., Albouy, P., Miller, R.D., Willson, C.G. and Schrepps, W., *Macromolecules*, 1991, **24**, 5068
[6] Wallraff, G.M., Baier, M., Miller, R.D., Rabolt, J.F., Hallmark, V., Cotts, P. and Shukla, P., *Polym. Prepr. (Am. Chem. Soc., Div. Polym. Chem.)*, 1989, **30**, 245

Poly(1-hexadecene)   P-289

Synonyms: *Poly(1-tetradecylethylene)*. *Poly(1-tetradecyl-1,2-ethanediyl)*
Monomers: 1-Hexadecene
Material class: Thermoplastic
Polymer Type: polyolefins
CAS Number:

CAS Reg. No.	Note
25655-58-7	atactic
26746-88-3	isotactic
133578-05-9	syndiotactic

Molecular Formula: $(C_{16}H_{32})_n$
Fragments: $C_{16}H_{32}$
General Information: Stiff rubber [2]
Morphology: Three polymorphic forms [9]; form I orthorhombic, a 0.75, b 6.3, c 0.67, helix 4_1 [5,6,10]; form II orthorhombic, 0.75, b 6.32, c 0.67 [25]

Volumetric & Calorimetric Properties:
Density:

No.	Value	Note
1	d 0.894 g/cm^3	[16]

– Polyhexafluoropropene

Latent Heat Crystallization: ΔH_f 17.16 kJ mol^{-1}; ΔS_f 50.48 J mol^{-1} K^{-1} [7]. ΔH_m 94–107 J g^{-1} [17]. Other enthalpy values have been reported [9]

Melting Temperature:

No.	Value	Note
1	67.5°C	[2,25]
2	67°C	[3,4,7]
3	66°C	[10]
4	67.6°C	[16]
5	60–65°C	[5,22]
6	58–63°C	[17,18,24]

Glass-Transition Temperature:

No.	Value	Note
1	40°C	[2]
2	36°C	[12]

Transition Temperature:

No.	Value	Note	Type
1	-125°C	200Hz [2]	β
2	28°C		1
3	31°C	[5]	2
4	-60–-10°C	[11]	α
5	35°C	[16]	T_m
6	22°C	[17]	T_m
7	10°C	110Hz	α
8	-10°C	110Hz	α
9	-30°C	110Hz	α
10	-60°C	110Hz	α
11	-150°C	110Hz [22]	β

Surface Properties & Solubility:
Solvents/Non-solvents: Sol. heptane, insol. Me$_2$CO [21]

Transport Properties:
Polymer Solutions Dilute: [η] 1.83 dl g^{-1} (CHCl$_3$, 25°) [7]. [η] 1.08–1.55 dl g^{-1} [24]. Other viscosity values have been reported [13,20]
Polymer Melts: Zero shear viscosity 320000 (80°) [8]. Flow activation energy 31.8 kJ mol^{-1} (7.6 kcal mol^{-1}) [26]

Mechanical Properties:
Mechanical Properties General: Variation of dynamic tensile storage modulus and loss modulus with temp. has been reported [1]. Shear modulus increases with increasing shear rate at 100° [8]
Viscoelastic Behaviour: Viscoelastic props. have been reported [23]

Optical Properties:
Transmission and Spectra: Ir, Raman and C-13 nmr spectral data have been reported [3,4,12,15,16,17,19]
Total Internal Reflection: Stress optical coefficient -8370–-8680 [24]

Applications/Commercial Products:
Processing & Manufacturing Routes: Synth. using Ziegler-Natta catalysts such as TiCl$_3$-AlEt$_3$ at 40° in heptane [3,4,11,13,16,17,21,24]

Trade name	Supplier
Solef X8ZY	Solvay

Bibliographic References

[1] Takayanagi, M., *Pure Appl. Chem.*, 1970, **23**, 151, (mech props)
[2] Clark, K.J., Jones, A.T. and Sandiford, D.J.H., *Chem. Ind. (London)*, 1962, 2010, (transition temps)
[3] Holland-Moritz, K., *Colloid Polym. Sci.*, 1975, **253**, 922, (ir, Raman, synth)
[4] Modric, I., Holland-Moritz, K. and Hummel, D.D., *Colloid Polym. Sci.*, 1976, **254**, 342, (ir, Raman, synth)
[5] Trafara, G., Koch, R., Blum, K. and Hummel, D., *Makromol. Chem.*, 1976, **177**, 1089, (transition temps)
[6] Privalko, U.P., *Macromolecules*, 1980, **13**, 370, (struct)
[7] Magagnini, P.L., Lupinacci, D., Cotrozzi, F. and Andriezzi, F., *Makromol. Chem., Rapid Commun.*, 1980, **1**, 557, (thermodynamic props)
[8] Wang, J.S., Porter, R.S. and Knox, J.R., *J. Polym. Sci., Polym. Phys. Ed.*, 1978, **16**, 1709, (mech props)
[9] Trafara, G., *Makromol. Chem.*, 1980, **181**, 969, (struct)
[10] Blum, K. and Trafara, G., *Makromol. Chem.*, 1980, **181**, 1097, (struct)
[11] Mishra, A., *J. Appl. Polym. Sci.*, 1982, **27**, 381, (synth)
[12] Graener, H., Ye, T.Q. and Lauburau, A., *Chem. Phys. Lett.*, 1989, **164**, 12, (ir)
[13] Jeong, Y.T. and Lee, D.H., *Pollimo*, 1990, **14**, 401, (C-13 nmr)
[14] *Eur. Pat. Appl.*, 1990, 403 866, (syndiotactic synth)
[15] Graener, H., Ye, T.Q. and Lauburau, A., *J. Phys. Chem.*, 1989, **93**, 7044
[16] Soga, K., Lee, D.H., Shiono, T. and Kashiwa, N., *Makromol. Chem.*, 1989, **190**, 2683, (synth, ir, C-13 nmr)
[17] Pena, B., Delgado, J.A., Perez, E. and Bello, A., *Makromol. Chem., Rapid Commun.*, 1992, **13**, 447, (synth, C-13 nmr, thermodynamic props)
[18] Pena, B., Delgado, J.A., Bello, A. and Perez, E., *Polymer*, 1994, **35**, 3039, (melting point)
[19] Holland-Moritz, K., Modric, I., Heinen, K.U. and Hummel, D.D., *Kolloid Z. Z. Polym.*, 1973, **251**, 913, (ir, Raman)
[20] Wang, J.S., Porter, R.S. and Knox, J.R., *Polym. J. (Tokyo)*, 1978, **10**, 619, (expansion coefficient, soln props)
[21] Natta, G., Danusso, F. and Moraglio, G., *Atti Accad. Naz. Lincei, Rend., Cl. Sci. Fis., Mat. Nat., Rend.*, 1958, **24**, 254, (synth)
[22] Manabe, S. and Takayanagi, M., *Kogyo Kagaku Zasshi*, 1970, **73**, 1581, (transition temps)
[23] Manabe, S. and Takayanagi, M., *Kogyo Kagaku Zasshi*, 1970, **73**, 1595, (viscoelastic props)
[24] Philippoff, W. and Tomquist, E., *J. Polym. Sci., Part C: Polym. Lett.*, 1966, **23**, 881, (stress optical coefficient)
[25] Turner-Jones, A., *Makromol. Chem.*, 1964, **71**, 1, (struct)
[26] Wang, J.S., Porter, R.S. and Knox, J.R., *J. Polym. Sci., Part B: Polym. Lett.*, 1970, **8**, 671, (melt props)

Polyhexafluoropropene

$$-\left[CF_2-\underset{\underset{F}{|}}{\overset{\overset{CF_3}{|}}{C}}\right]_n-$$

Synonyms: *Polyhexafluoropropylene. Hexafluoropropylene polymer. Poly(1,1,2,3,3,3-hexafluoro-1-propene). Polyperfluoropropylene. PHFP*
Related Polymers: Polyhexafluoropropylene copolymers
Monomers: Hexafluoropropene
Material class: Thermoplastic
Polymer Type: fluorocarbon (polymers)
CAS Number:

CAS Reg. No.	Note
25120-07-4	
27458-51-1	isotactic

Molecular Formula: (C$_3$F$_6$)$_n$
Molecular Weight: MW 3000 (approx.: plasma polymerised films); M$_n$ 5700000 (max.)
General Information: Plasma polymerisation produces thin, well-defined polymer films with potential applications for coatings, dielectrics, sensors etc. Such materials have been extensively reported [1,2,3,4,5,7,11,14,15,16,17,18,20,21,22,30,37,43] The quality of films/surfaces can be modified by variations in RF duty

cycles when using pulsed RF plasmas [24]. A comprehensive study of the mechanism and kinetics of the polymerisation process has been reported. [18] Such films have a low coefficient of friction, low surface tension, thermal stability, biocompatability and chemical resistance. Their greatest commercial importance is as copolymers with other fluorine-containing comonomers, e.g. vinylidene fluoride (Viton rubber), or as terpolymers, with e.g. vinylidene fluoride and tetrafluoroethylene [19]. Can be white (chemically polymerised) [27] or yellow (plasma polymerised) [16,17] amorph. solid or transparent highly-adhering films (on various substrates) [5]

Morphology: Struct. has not been determined but claims of isotactic struct. with four-fold helix conformn. have been reported [19,28]. Theoretical studies involving calculations of conformational energies of isotactic form confirm such predictions. The most stable conformn. is helix with CCC valence angle close to 113°, with number of monomer units per turn of the helix being 3.83–3.91 [36] ESCA, F-19 nmr and ir studies of plasma polymerised films reveal high degree of branching: approx. 57% of tertiary C atoms in polymer are bonded with CF_3 groups [7,23]. Evidence for cross-linking has also been reported [11,16]

Miscellaneous: High quality (flawless) thin films, suitable as dielectrics, can be produced by low-pressure glow-discharge (approx. 10 kHz) plasma polymerisation [14]. Quality can be further improved by annealing in vacuum (e.g. 15 min at 250°) immediately afer deposition [17]. Chemical struct. of amorph. cross-linked material obtained by plasma polymerisation consists largely of similar amounts of C^*-CF, CF, CF_2 and CF_3 units [3,5,16,30]

Volumetric & Calorimetric Properties:
Density:

No.	Value	Note
1	d 1.4–1.9 g/cm^3	plasma-polymerised films; determined optically [11]
2	d 2.2 g/cm^3	plasma-polymerised films; determined from deposition data [5,16]

Melting Temperature:

No.	Value	Note
1	110–120°C	crude polymer, obtained by Ziegler-Natta catalysis [28]

Glass-Transition Temperature:

No.	Value	Note
1	152°C	[39]

Transition Temperature General: Plasma-polymerised films can be sublimed at 350° [17]

Vicat Softening Point:

No.	Value	Note
1	225–250°C	chemically polymerised [27]

Transition Temperature:

No.	Value	Note
1	322°C	half-decomposition temp., plasma-polymerised films [21]

Surface Properties & Solubility:
Cohesive Energy Density Solubility Parameters: Solubility parameters 5.2–5.6 $J^{1/2}$ cm$^{3/2}$, (6.5–7.5 cal$^{1/2}$ cm$^{3/2}$, calc) [42]

Solvents/Non-solvents: Plasma-polymerised films insol. common organic solvents. [17] γ-ray polymerised material sol. fluorocarbons. [24] Ziegler-Natta polymerised material sol. C_6H_6 (boiling), CCl_4 (boiling); slightly sol. Et_2O, Me_2CO, heptane, cyclohexanone, DMF; insol. MeOH, petrol (boiling), CH_2Cl_2 (boiling) [28]

Wettability/Surface Energy and Interfacial Tension: Work of adhesion between polymer and water, formamide, methylene iodide and hexadecane has been reported. [25] Contact angles for acrylic fibres coated with plasma-polymerised layers 116° (H_2O, advancing); 34° (H_2O, receding) [1]. Contact angles for plasma-polymerised polymer films deposited on glass slides, for both continuous and pulsed radiofrequency discharges 107° (continuous), 104° (pulsed). [18] Surface energy of plasma-polymerised films 18.5 mN m^{-1}, γ_p 0.02; γ_d 18.5; 21.2 mN m^{-1}, γ_p 0.07; γ_d 21.1 (polymerised in H_2); 20.8 mN m^{-1}, γ_p 1.3; γ_d 19.5 (polymerised in N_2) [16]; 18.6 mN m^{-1}, γ_p 0.05; γ_d 18.5 (polymerised in Ar) [30]; 18.5 mN m^{-1}, γ_p 0.04; γ_d 18.5 (polymerised in O_2) [30]. Basis of glueing in relation to welding has been reported, including the role of forces involved in wetting and adhesion [41]

Surface Tension:

No.	Value	Note
1	20 mN/m	plasma-polymerised film [17]

Transport Properties:
Polymer Solutions Dilute: Intrinsic viscosity 0.10-0.13 dl g^{-1} (C_6H_6, 30°, Ziegler-Natta catalysis) [28]; 2 dl g^{-1} (γ-irradiated polymer) [39]

Permeability of Gases: Migration of plasma gas (e.g. N_2) into polymer films during plasma polymerisation process has been reported [21]

Permeability of Liquids: Plasma-polymerised thin films (250 nm thick) are impermeable to K_2SO_4/H_2SO_4 electrolytes [14]

Mechanical Properties:
Mechanical Properties Miscellaneous: Chemically polymerised material retains 82.5% of its stiffness when heated from room temp. to 90° [27]

Fracture Mechanical Properties: Pressed films are often brittle [38]

Electrical Properties:
Electrical Properties General: Plasma-polymerised films withstand electric fields of 300 V μm^{-1} [17]. Plasma-polymerised material has lower dielectric constant than chemically polymerised material, attributed to presence of oxygen impurities and corresponding CO orientation effects

Dielectric Permittivity (Constant):

No.	Value	Note
1	1.5–1.9	plasma-polymerised [11]

Strong Field Phenomena General: Plasma-polymerised films have an electrical breakdown strength of 61 MV m^{-1}, which increases to 72 MV m^{-1} when polymerised in H_2, but decreases to 23 MV m^{-1} when polymerised in N_2 [16]. Electrical breakdown strength 68 MV m^{-1} (polymerised in Ar), 40 MV m^{-1} (polymerised in O_2) [30]

Dissipation (Power) Factor:

No.	Value	Frequency	Note
1	0.002	100 kHz	plasma-polymerised [17]

Optical Properties:
Optical Properties General: Yellow colour of plasma-polymerised films (compared to white colour of chemically polymerised material) evidenced by strong absorption in uv and near-uv regions. This is explained by presence of unsaturation and fluorine-deficient sites [11]

– Polyhexafluoropropene

Refractive Index:

No.	Value	Note
1	1.39	plasma-polymerised [11]

Transmission and Spectra: Uv [11] and XPS spectral data [1,2,4,11,15] for plasma-polymerised films have been reported [2,4,5,7,15,16,23,30]. Ir, F-19 nmr [7,23] and esca [5,7,16,23,30,38] spectral data for plasma polymerised material have been reported. Esr spectral data for plasma polymerised films have been reported [7,18,20,22]. Ir spectral data for polymer obtained by Ziegler-Natta catalysis have been reported. [28] Reflection and Raman spectral data of plasma-polymerised films have been reported [37]

Polymer Stability:
Thermal Stability General: High-temp. ageing in air at 225° for 87h causes little change. Corresponding experiments *in vacuo* show light cross-linking occurs [40]
Environmental Stress: Plasma-polymerised material oxidised in air [11]. Chemically polymerised material slowly oxidised in air [24]. Commercially available films display excellent resistance to ozone [6]
Chemical Stability: Reacts with dilute fluorine (5% in He at 25°, 1 atm) [9,33]
Stability Miscellaneous: Can be electrochemically carbonised (reductive splitting of CF bonds) to give so-called electrochemical carbon [11,12]. Partial cleavage of CF bonds occurs during plasma-polymerisation process [1]. Plasma-polymerised films show good electrochemical stability during electrolysis in liq. solns. of both inert and redox electrolytes [37]

Applications/Commercial Products:
Processing & Manufacturing Routes: Synth. by either chemical or plasma polymerisation. A typical chemical process involves free-radical polymerisation of monomer in inert perfluorinated solvents. (e.g. perfluoroheptane or perfluorocyclobutane) at temps. above 200° and pressures ideally 2000–100000 atm. The absence of water and air is essential. Suitable free-radical initiators include $(F_3CS)_2$, lead tetrakis(perfluorocarboxylates) or mercury bis(perfluoroalkylmercaptide) [27]. Chemically polymerised material can be injection or compression moulded, or extruded above 250°. [27] Liq. and solid phase polymerisation using γ-irradiation at temps. between -196° and 30° gives low MW, high-boiling-point liq. polymers [19,34,35]; polymer with MW approx. 5000000 can be produced using γ-irradiation at high temps. and high pressures [24,32,39]; detailed kinetic studies (effects of temp., pressure and MW) have been reported. [24] Typical conditions for plasma polymerisation: AC glow- (450V, 10μA, 20 kHz) of 1:1 (by volume) mixture of monomer and Ar (pressure 40 Pa, flow rate 10 ml min^{-1}. [11] Use of Ziegler-Natta catalysts (CH_2Cl_2, 40°) produces highly crystalline, isotactic material [19,28]
Applications: Used as matrix for obtaining absorption spectra of molecular ions; in laminates, electrothermographic recording materials, optical recording media, sensors and secondary batteries; as phosphor compositions for lamps, electrical insulators, dielectrics (as plasma-polymerised thin films), chromatography column packings (in reduced form); in mouldings, electrochemical applications, in electrical barrier coatings; in dry lubricants and release coatings and agents; in antiwear and non-stick coatings, biomedical applications (prosthetics and dental compositions), cation-exchange membranes; electrodes for fuel cells; electrocatalysis; fabric treatment (e.g. hydrophobic coating on acrylic fibres)

Details	Supplier
Polymer films (thickness, 0.2 mm)	Mitsui Fluorochemical

Bibliographic References

[1] Occhiello, E., *Angew. Makromol. Chem.*, 1994, **222**, 189, (uses, surface props, xps)
[2] Savage, C.R. and Timmons, R.B., *Polym. Mater. Sci. Eng.*, 1991, **64**, 95, (plasma polymerisation, ir, xps)
[3] Silverstein, M.S. and Chen, R., *Polym. Mater. Sci. Eng.*, 1995, **73**, 237, (plasma polymerisation, ir, surface props)
[4] Panchalingam, V. Chen, X., Savage, C.R. and Timmons, R.B., *Polym. Prepr. (Am. Chem. Soc., Div. Polym. Chem.)*, 1993, **34**, 681, (plasma polymerisation, xps)
[5] Chen, R., Gorelik, V. and Silverstein, M.S., *J. Appl. Polym. Sci.*, 1995, **56**, 615, (plasma polymerisation, optical microscopy, morphology, sem, esca, ir)
[6] Fujimoto, K., Takebayashi, Y., Inoue, H. and Ikada, Y., *J. Polym. Sci., Polym. Chem. Ed.*, 1993, **31**, 1035, (environmental stability)
[7] Wang, D. and Chen, J., *J. Appl. Polym. Sci.*, 1991, **42**, 233, (plasma polymerisation, F-19 nmr, esca, ir, esr, morphology, ordering)
[8] Ikematie, S., Kodera, Y. and Hikida, T., *J. Photochem. Photobiol. A*, 1990, **52**, 193, (uses)
[9] Florin, R.E., *J. Fluorine Chem.*, 1979, **14**, 253, (chemical stability, esr)
[10] Bernett, M.K., *Macromolecules*, 1974, **7**, 917, (surface props)
[11] Kavan, L., Dousek, F.P. and Doblhofer, K., *J. Fluorine Chem.*, 1991, **55**, 37, (synth, electrical props, optical props, uv, ir, xps, electrochem)
[12] Kavan, L., *Chem. Phys. Carbon*, 1991, **23**, 69, (electrochem, stability)
[13] Khanna, R., Jameel, A.T. and Sharma, A., *Ind. Eng. Chem. Res.*, 1996, **35**, 3081, (surface props)
[14] Doblhofer, K., *Makromol. Chem., Makromol. Symp.*, 1987, **8**, 323, (plasma polymerisation, electrochem props, electron microscopy, transport props)
[15] Savage, C.R.; Timmons, R.B. and Lin, J., *Adv. Chem. Ser.*, 1993, **236**, 745, (plasma polymerisation, xps, ir)
[16] Silverstein, M.S. and Chen, R., *ACS Symp. Ser.*, 1996, **648**, 451, (plasma polymerisation, esca, ir, sem, surface props, morphology, microstruct, electrical props)
[17] Ristow, D., *J. Mater. Sci.*, 1977, **12**, 1411, (plasma polymerisation, electrical props, surface props)
[18] Yasuda, H. and Hsu, T.S., *J. Polym. Sci., Polym. Chem. Ed.*, 1977, **15**, 2411, (plasma polymerisation, kinetics, esr, surface props)
[19] *Fluoropolymers*, (ed. L.A. Wall), Wiley-Interscience, 1972, **17**, 143, (synth, copolymerisation, MW)
[20] Liu, X., Wang, T., Pan, Z. and Liu, G., *Huaxue Xuebao*, 1987, **45**, 276, (esr)
[21] Wang, D. and Chen, J., *Yingyong Huaxue*, 1988, **5**, 62, (plasma polymerisation, thermal props, esca)
[22] Wang, D. and Chen, J., *Yingyong Huaxue*, 1989, **6**, 89, (esr)
[23] Wang, D., Chen, J. and Chen, C., *Fushe Yanjiu Yu Fushe Gongyi Xuebao*, 1988, **6**, 6, (struct, ordering, esca, F-19 nmr, ir)
[24] Lowry, R.E., Brown, D.W. and Wall, L.A., *J. Polym. Sci., Part A-1*, 1966, **4**, 2229, (synth, MW, environmental stability)
[25] Kaelble, D.H., *J. Adhes.*, 1970, **2**, 66, (surface props, surface tension, wettability, adhesion)
[26] Volkova, E.V. and Skobina, A.I., *Vysokomol. Soedin.*, 1964, **6**, 964, (synth)
[27] *U.S. Pat.*, 1960, 2958685, (synth, thermal props, mech props)
[28] Sianesi, D. and Caporiccio, G., *Makromol. Chem.*, 1963, **60**, 213, (synth, ir, soln props, solubility, thermal props, transport props, morphology, microstruct)
[29] Bernett, M.K. and Zisman, W.A., *J. Phys. Chem.*, 1961, **65**, 2266, (surface props)
[30] Chen, R. and Silverstein, M.S., *J. Polym. Sci., Polym. Chem. Ed.*, 1996, **34**, 207, (plasma polymerisation, spectra, surface props, electrical props)
[31] Kodera, Y. and Hikida, T., *Spectrosc. Lett.*, 1989, **22**, 1229, (uses)
[32] Wall, L.A., *Polym. Prepr. (Am. Chem. Soc., Div. Polym. Chem.)*, 1966, **7**, 1112, (synth)
[33] Brown, D.W., Florin, R.F. and Wall, L.A., *Appl. Polym. Symp.*, 1973, **22**, 169, (chemical stability)
[34] Skobina, A.I. and Volkova, E.V., *Radiats. Khim. Polim., Mater. Simp.*, Moscow, 1964, 126, (synth)
[35] Volkova, E.V., *Radiats. Khim. Polim., Mater. Simp.*, Moscow, 1964, 113, (synth)
[36] Bates, T.W., *Trans. Faraday Soc.*, 1968, **64**, 3180, (theory, ordering)
[37] Doblhofer, K., Nölte, D. and Ulstrup, J., *Ber. Bunsen-Ges. Phys. Chem.*, 1978, **82**, 403, (plasma polymerisation, uses, electrical props, optical props, electrochem)
[38] Clark, D.T., Feast, W.J., Kilcast, D. and Musgrave, W.K.R., *J. Polym. Sci., Polym. Chem. Ed.*, 1973, **11**, 389, (esca, theory)
[39] Brown, D.W. and Wall, L.A., *J. Polym. Sci., Part A-2*, 1969, **7**, 601, (synth, thermal props, transport props)
[40] Brown, D.W. and Wall, L.A., *J. Polym. Sci., Polym. Chem. Ed.*, 1972, **10**, 2967, (thermal stability)
[41] Dubois, P. and Henni, J., *J. Chim. Ind., Genie Chim.*, 1969, **101**, 1143, (surface props, adhesion)
[42] Ahmed, H. and Yaseon, M., *Paint Manufact.*, 1978, **48**, 28, (solubility parameters)
[43] Kieser, J. and Neusch, M., *Thin Solid Films*, 1984, **118**, 203, (plasma polymerisation)

Poly(hexafluoropropylene oxide)   P-292

$$CF_3CF_2\!-\!\!\left[CF_2OCF\atop CF_3\right]_n\!\!-\!\!\overset{O}{\underset{}{C}}F$$

Oligomer, $1 < n < 100$, monofunctional

$$F\overset{O}{C}\!-\!\!\left[CFOCF_2\atop CF_3\right]_m\!\!-\!R_f\!-\!\!\left[CF_2OCF\atop CF_3\right]_n\!\!-\!C\overset{O}{F}$$

R_f is a completely fluorinated group (e.g. $CF(CF_3)OCF_2CF_2OCF(CF_3)$) $n \sim 100$, functional

Synonyms: *Poly(hexafluoropropene oxide). Poly(trifluoro(trifluoromethyl)oxirane)*
Related Polymers: Propylene oxide elastomers, Poly(propylene oxide)
Monomers: trifluoro(trifluoromethyl)oxirane
Material class: Fluids, Synthetic Elastomers
Polymer Type: polyurethane, polyether-imide
CAS Number:

CAS Reg. No.
25038-02-2

Molecular Formula: $(C_3F_6O)_n$
Fragments: C_3F_6O
Molecular Weight: M_n 74000 (max.). M_w/M_n 1.02–1.40
General Information: Lower MW polymers are fluids. Cross-linking gives elastomeric networks, in which the type of link plays a significant part. At room temp., triazine-linked networks are invariably soft rubbers but urethane-linked networks can be tough rubbers. [2] Struct. obtained depends on catalyst system and reaction conditions. May be mono- or di-functional [2,3]

Volumetric & Calorimetric Properties:
Volumetric Properties General: Thermal props., plotted against temp., pressure and MW for Krytox (100% hexafluoro-oxypropylene chain) [4] and Fomblin Y (95 mol% hexafluoro-oxypropylene chain), have been reported [5]
Glass-Transition Temperature:

No.	Value	Note
1	-60°C	Fomblin Y, extrapolated to infinite molar mass [2]

Transport Properties:
Polymer Melts: Rheological props. of Krytox (100% hexafluoro-oxypropylene chain) [4] and Fomblin Y (95 mol% hexafluoro-oxypropylene chain) have been reported [5]
Water Absorption:

No.	Value	Note
1	0.1 %	triazine-linked networks [2]
2	1 %	urethane-linked networks [2]

Mechanical Properties:
Mechanical Properties General: Mech. props. for cross-linked poly(hexafluoropropylene oxide) networks have been reported [2]

Tensile Strength Break:

No.	Value	Note	Elongation
1	0.21–0.25 MPa	triazine-linked networks [2]	160–550%
2	0.2–5.8 MPa	20°, urethane-linked networks [2]	160–300%

Miscellaneous Moduli:

No.	Value	Note	Type
1	0.045–0.12 MPa	[2]	Shear modulus
2	0.04–0.06 MPa	triazine-linked networks [2]	Storage modulus
3	0.044–4.1 MPa	20°	Shear modulus
4	0.053–1.08 MPa	70°, urethane-linked networks [2]	Shear modulus

Polymer Stability:
Polymer Stability General: The large size of the fluorine atoms combined with the high strength of the CO bonds gives polymers which are generally thermally and chemically stable
Upper Use Temperature:

No.	Value	Note
1	300°C	triazine-linked networks, short term [2]
2	150°C	urethane-linked networks [2]
3	270°C	triazine cross-linked networks

Chemical Stability: Both triazine- and urethane-linked networks are highly resistant to paraffin and silicone oils. Triazine-linked networks are resistant to a wide range of organic liquids whereas urethane-linked networks are attacked by polar organic solvents. Both types are significantly attacked by fluorinated species [2]
Hydrolytic Stability: Water-resistant

Applications/Commercial Products:
Processing & Manufacturing Routes: Synth. by one of two routes: anionic ring opening polymerisation of hexafluoropropylene oxide, by addition of fluoride ion to perfluorinated mono- or di-acid fluorides; or photo oxidation of hexafluoropropylene, by introduction of oxygen gas into liq. hexafluoropropylene at -40° and irradiating with uv light (below 300 nm) [2]
Applications: Thermally and chemically stable oils, greases and elastomers

Trade name	Details	Supplier
Fomblin	Various grades. Fomblin Y is homopolymer	Montefluoro
Krytox		DuPont

Bibliographic References
[1] Millauer, H., Schwertfeger, W. and Siegemund, G., *Angew. Chem., Int. Ed. Engl.*, 1985, **24**, 161
[2] Mobbs, R.H., Heatley, F., Price, C. and Booth, C., *Prog. Rubber Plast. Technol.*, 1995, **11**, 94
[3] Hill, J.T., *J. Macromol. Sci., Chem.*, 1974, **8**, 499
[4] Alper, T., Barlow, A.J., Gray, R.W., Kim, M.G. et al, *J. Chem. Soc., Faraday Trans.*, 1980, **76**, 205
[5] Marchionni, G., Ajroldi, G. and Pezzin, G., *Eur. Polym. J.*, 1988, **24**, 1211

Poly(hexamethylenediamine-*co*-isophorone diamine-*co*-isophthalic acid-*co*-terephthalic acid) P-293

Synonyms: *6/IPD/I/T copolymer*
Related Polymers: Nylon 6
Monomers: Hexamethylene diamine, Terephthalic acid, Isophthalic acid, Isophorone diamine
Material class: Thermoplastic, Copolymers
Polymer Type: polyamide
Molecular Formula: $[(C_6H_{14}N_2).(C_{10}H_{20}N_2).(C_8H_4O_2)]_n$
Fragments: $C_6H_{14}N_2$ $C_{10}H_{20}N_2$ $C_8H_4O_2$
Molecular Weight: 15000–40000
General Information: Is amorph.

Volumetric & Calorimetric Properties:
Density:

No.	Value	Note
1	d 1.18 g/cm^3	7.5–33% isophoronediamine [1]

Glass-Transition Temperature:

No.	Value	Note
1	150°C	7.5–33% isophoronediamine

Deflection Temperature:

No.	Value	Note
1	130°C	7.5–33% isophoronediamine

Vicat Softening Point:

No.	Value	Note
1	150–217°C	7.5–33% isophoronediamine [2]

Surface Properties & Solubility:
Plasticisers: *p*-Toluenesulfonamide
Solvents/Non-solvents: Sol. sulfuric acid, formic acid, dimethylacetamide, *N*-methylpyrrolidone (containing 1–2% boric acid)

Transport Properties:
Melt Flow Index:

No.	Value	Note
1	11 g/10 min	285°, 7.5–33% isophoronediamine, isophthalic acid:terephthalic acid (3:2)

Water Absorption:

No.	Value	Note
1	0.67 %	24h, 20°, 7.5–33% isophthalic acid:terephthalic acid (3:2)

Gas Permeability:

No.	Gas	Value	Note
1	[O$_2$]	0.66–1.77 cm^3 mm/(m^2 day atm)	10–27×10^{-13} cm^3 cm cm^{-2} (s cm Hg)$^{-1}$; 7.5–33% isophoronediamine, isophthalic acid:terephthalic acid (3:2) [3]

Mechanical Properties:
Flexural Modulus:

No.	Value	Note
1	2500–3000 MPa	20°, 7.5–33% isophoronediamine [2]
2	2000 MPa	120°, 7.5–33% isophoronediamine [2]

Tensile Strength Break:

No.	Value	Note
1	95–102 MPa	20°, 7.5–33% isophoronediamine [2]
2	35 MPa	120°, 7.5–33% isophoronediamine [2]

Impact Strength: Falling weight impact strength 40–50 J (2mm disc, 0% failure rate, 7.5–33% isophoronediamine
Hardness: Rockwell R126 (7.5–33% isophoronediamine)
Friction Abrasion and Resistance: Taber abrasion resistance 0.29g (1000 cycles)$^{-1}$ (7.5–33% isophoronediamine)
Izod Notch:

No.	Value	Notch	Note
1	150 J/m	N	7.5–33% isophoronediamine, injection moulded [1]
2	32 J/m	Y	7.5–33% isophoronediamine, injection [1]

Electrical Properties:
Dielectric Strength:

No.	Value	Note
1	67 kV.mm^{-1}	7.5–33% isophoronediamine

Arc Resistance:

No.	Value	Note
1	120s	7.5–33% isophoronediamine

Optical Properties:
Transmission and Spectra: Transmits approx. 85% of visible light
Volume Properties/Surface Properties: Transparency transparent

Polymer Stability:
Thermal Stability General: Up to 36% of the polymer is oxidised at 270° (0.2 mm thick dry film)
Upper Use Temperature:

No.	Value	Note
1	80°C	7.5–33% isophoronediamine

Decomposition Details: Volatilisation of isophoronediamine occurs above 300°
Flammability: Limiting oxygen index 25% (approx.)
Environmental Stress: Ageing in air above 80° causes yellowing
Chemical Stability: Is resistant to most aliphatic and aromatic hydrocarbons. Injection moulded polymer is resistant to C_6H_6,

paraffin, oil, CCl$_4$; slightly cracked by petroleum ether, aq. alkali, hydrochloric acid, sulfuric acid and salt soln. Severe cracking caused by Me$_2$CO, lower alcohols, THF and brake fluid.
Hydrolytic Stability: Treatment with boiling H$_2$O causes crystallisation after 24h, and loss of tensile strength

Applications/Commercial Products:
Processing & Manufacturing Routes: Polymer containing up to 80 mol% terephthalic acid is synth. by aq. melt polymerisation of nylon 6, isophoronediamine, isophthalic acid and terephthalic acid salts at up to 275° for 3h [3]
Applications: Potential applications due to impact toughness and general performance. Uses have not been investigated commercially

Bibliographic References
[1] *Brit. Pat.*, 1969, 52194, (synth., use)
[2] Dolden, J.G., *Polymer*, 1976, **17**, 875, (rev.)
[3] *Encycl. Polym. Sci. Eng.*, Vol. 11, 2nd edn., (ed. J.I. Kroschwitz), John Wiley and Sons, 1985, 372, (rev.)

Poly(hexamethylenediamine-*co*-terephthalic acid-*co*-isophthalic acid) P-294

Synonyms: *Nylon-6, T/nylon-6, I copolymer. 6//T/I copolymer. Poly(caproamide-co-terephthalic acid-co-isophthalic acid)*
Monomers: Hexamethylenediamine, Isophthalic acid, Terephthalic acid
Material class: Thermoplastic, Copolymers
Polymer Type: polyamide
Molecular Formula: [(C$_6$H$_{14}$N$_2$).(C$_8$H$_4$O$_2$)]$_n$
Fragments: C$_6$H$_{14}$N$_2$ C$_8$H$_4$O$_2$
Molecular Weight: 15000–40000
General Information: Amorph. when the terephthalic acid content is less than 70 mol%

Volumetric & Calorimetric Properties:
Density:

No.	Value	Note
1	d 1.23 g/cm^3	isophthalic acid, terephthalic acid 2:1 [1]

Melting Temperature:

No.	Value	Note
1	310°C	isophthalic acid : terephthalic acid 2:1 [2]

Glass-Transition Temperature:

No.	Value	Note
1	165°C	isophthalic acid : terephthalic acid 2:1

Surface Properties & Solubility:
Plasticisers: *p*-Toluenesulfonamide
Solvents/Non-solvents: Sol. H$_2$SO$_4$, formic acid, dimethylacetamide, *N*-methylpyrrolidone (containing 1–2% boric acid)

Transport Properties:
Water Content: 3.5% (21°, 65% relative humidity, terephthalic acid isophthalic acid 2:1) [1]

Gas Permeability:

No.	Gas	Value	Note
1	[O$_2$]	0.47–0.98 cm^3 mm/(m^2 day atm)	7.2–15 × 10^{-13} cm^3 cm cm^{-2} (s cm Hg)$^{-1}$, terephthalic acid : isophthalic acid 2:1 [1]

Mechanical Properties:
Mechanical Properties General: Tenacity 0.22–0.44 N tex^{-1} (8–40% elongation) terephthalic acid : isophthalic acid 2:1 [2]
Tensile Strength Break:

No.	Value	Note
1	440 MPa	terephthalic acid : isophthalic acid 2:1, oriented fibre [3]

Electrical Properties:
Electrical Properties General: Has good electrical insulation props., particularly when dry

Polymer Stability:
Flammability: Limiting oxygen index 25% (approx.)
Environmental Stress: Ageing in air above 80° causes yellowing
Chemical Stability: Boiling H$_2$O causes crystallisation after 24h. Injection moulding polymer is resistant to C$_6$H$_6$, paraffin, oil, CCl$_4$; slight cracking caused by petroleum ether, aq. alkali, hydrochloric acid, sulfuric acid and salt soln. Severe cracking caused by Me$_2$CO, lower alcohols, THF and brake fluid. Is resistant to most aliphatic and aromatic hydrocarbons
Hydrolytic Stability: Boiling H$_2$O causes crystallisation after 24h

Applications/Commercial Products:
Processing & Manufacturing Routes: Polymer containing up to 80 mol% terephthalic acid may be synth. by aq. melt polymerisation of nylon, isophthalic acid and terephthalic acid salts at up to 275° for 3h [1]
Applications: Applications due to excellent barrier props. Not exploited commercially

Bibliographic References
[1] *Encycl. Polym. Sci. Eng.*, 2nd edn., eds., John Wiley & Sons, 1985, **11**, 372, (rev.)
[2] Chapman, R.D., Holmer, D.A., Pickett, O.A., Lea, K.R. and Saunders, J.H., *Text. Res. J.*, 1981, **51**, 564, (rev.)
[3] Dolden, J.G., *Polymer*, 1976, **17**, 875, (rev.)

Poly(hexamethylenediamine-*co*-*m*-xylylenediamine-*co*-isophthalic acid-*co*-terephthalic acid) P-295

Synonyms: *6/MXD//I/T copolymer*
Related Polymers: Nylon 6
Monomers: Terephthalic acid, Hexamethylenediamine, Isophthalic acid, *m*-Xylylenediamine
Material class: Thermoplastic, Copolymers
Polymer Type: polyamide
Molecular Formula: [(C$_6$H$_{14}$N$_2$).(C$_8$H$_{10}$N$_2$).(C$_8$H$_4$O$_2$)]$_n$
Fragments: C$_6$H$_{14}$N$_2$ C$_8$H$_{10}$N$_2$ C$_8$H$_4$O$_2$

Poly(hexamethylene terephthalate)

Molecular Weight: 15000–4000
General Information: Amorph. when the terephthalic acid content is less than 70 mol%

Volumetric & Calorimetric Properties:
Density:

No.	Value	Note
1	d 1.2 g/cm³	7.5–10% *m*-xylylenediamine, isophthalic acid:terephthalic acid (3:2) [1]

Glass-Transition Temperature:

No.	Value	Note
1	178°C	35% *m*-xylylenediamine, isophthalic:terephthalic acid (1:1)
2	166°C	25% *m*-xylylenediamine, isophthalic:terephthalic acid (1:1)

Vicat Softening Point:

No.	Value	Note
1	140–145°C	7.5–10% *m*-xylylenediamine, isophthalic:terephthalic acid (3:2) [2]

Surface Properties & Solubility:
Plasticisers: *p*-Toluenesulfonamide
Solvents/Non-solvents: Sol. H_2SO_4, formic acid, dimethylacetamide, *N*-methylpyrrolidone (containing 1–2% boric acid)

Transport Properties:
Water Content: Has low water adsorption compared to aliphatic nylons
Gas Permeability:

No.	Gas	Value	Note
1	[O_2]	0.36–1.12 cm³ mm/(m² day atm)	5.517×10^{-13} cm² (s cm Hg)⁻¹, 25% *m*-xylylenediamine, isophthalic:terephthalic acid (1:1) [3]
2	[O_2]	0.16–0.2 cm³ mm/(m² day atm)	$2.5-3.1 \times 10^{-13}$ cm² (s cm Hg)⁻¹, 35% *m*-xylylenediamine, isophthalic:terephthalic acid (1:1) [3]

Mechanical Properties:
Flexural Modulus:

No.	Value	Note
1	2600 MPa	20°, 7.5–10% *m*-xylylenediamine, isophthalic acid:terephthalic acid (3:2) [2]

Tensile Strength Break:

No.	Value	Note
1	106 MPa	7.5–10% *m*-xylylenediamine, isophthalic acid:terephthalic acid (3:2) [2]

Charpy:

No.	Value	Notch	Note
1	38–50	Y	7.5–10% *m*-xylylenediamine, isophthalic acid:terephthalic acid (3:2)

Electrical Properties:
Electrical Properties General: Has good electrical insulation props., particularly when dry

Polymer Stability:
Decomposition Details: Volatilisation of *m*-xylylenediamine occurs above 300°
Flammability: Limiting oxygen index 25% (approx.)
Environmental Stress: Ageing in air above 80° causes yellowing
Chemical Stability: Is resistant to most aliphatic and aromatic hydrocarbons. Injection moulded polymer is resistant to C_6H_6, paraffin, oil, CCl_4; slight cracking caused by petroleum ether, aq., hydrochloric acid, sulfuric acid and salt soln. Me_2CO, lower alcohols, THF and brake fluid cause severe cracking
Hydrolytic Stability: Boiling H_2O may induce crystallisation after 24h

Applications/Commercial Products:
Processing & Manufacturing Routes: Polymer containing up to 80 mol% terephthalic acid synth. by aq. melt polymerisation of nylon, isophthalic acid and terephthalic acid salts at up to 275° for 3h [3]
Applications: Potential applications due to low gas permeability have been restricted by cost

Bibliographic References
[1] Chapman, R.D., Holmer, D.A., Pickett, O.A., Lea, K.R. and Saunders, J.H., *Text. Res. J.*, 1981, **51**, 564, (rev.)
[2] Dolden, J.G., *Polymer*, 1976, **17**, 875, (rev.)
[3] Encycl. Polym. Sci. Technol. vol II, 2nd edn. (ed. J.I. Kroschwitz) John Wiley and Sons, 1986, 372, (rev.)

Poly(hexamethylene terephthalate)

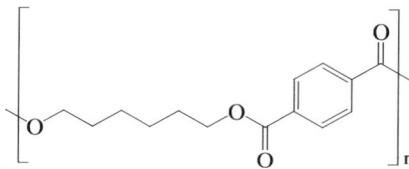

Synonyms: Poly(oxycarbonyl-1,4-phenylenecarbonyloxy-1,6-hexanediyl). Poly(1,6-hexanediyl terephthalate). Poly(hexane terephthalate). Poly(oxyterephthaloyloxyhexamethylene). PHT. PHMT
Monomers: Terephthalic acid, 1,6-Hexanediol
Material class: Thermoplastic
Polymer Type: unsaturated polyester
CAS Number:

CAS Reg. No.	Note
58517-23-0	
26637-42-3	repeating unit
26618-59-7	Poly(hexamethylene terephthalate)

Molecular Formula: $(C_{14}H_{16}O_4)_n$
Fragments: $C_{14}H_{16}O_4$
Morphology: Three polymorphic forms:- α form - monoclinic a 0.9100, b 1.7560, c 1.574, α 127.8°; β form - triclinic a 0.5217, b 0.5284, c 1.5783, α 129.4°, β 97.6°, γ 95.6°; β¹ form - triclinic a 0.5217, b 1.0568, c 1.5738, α 129.4°, β 97.6°, γ 95.6° [19]. Cryst. struct. has been reported [6,11,30]. Cryst. struct. at -183°, 18° and 125° has been reported [3]. Three polymorphic forms:- α form - monoclinic a 0.91, b 1.72, c 1.55, α 127.3°; β form - triclinic a 0.48, b 0.57, c 1.57, α 104.4°, β 116.0°, γ 107.8°; γ form - triclinic a 0.53, b 1.39, α 123.6°, β 129.6°, γ 88.0° [17]; δ form - c 1,275 [21]. β form - triclinic a 0.484, b 0.574, c 1.553, α 104.1°, β 116.3°, γ 106.2°; β¹ form - triclinic a 0.468, b 1.557, c 1.553, α 105.2°, β 114.1°, γ 109.0° [27]. Morphological study has been reported [34]

– Poly(hexamethylene terephthalate)

Volumetric & Calorimetric Properties:
Density:

No.	Value	Note
1	d^{180} 1.166 g/cm^3	[2]
2	d 1.242 g/cm^3	α-form [17]
3	d 1.251 g/cm^3	β-form [18]
4	d 1.256 g/cm^3	γ-form [17]

Thermal Expansion Coefficient:

No.	Value	Note	Type
1	0.000103 K^{-1}	20° [22]	L

Latent Heat Crystallization: ΔH_f 35.59 kJ mol^{-1} (8.5 kcal mol^{-1}), ΔS_f 81.6 J mol^{-1} K^{-1} (19.5 cal mol^{-1} °C^{-1}) [2]; ΔH_f 34.8 kJ mol^{-1} (8.3 kcal mol^{-1}); ΔS_f 77 J mol^{-1} K^{-1} (18.4 cal mol^{-1} K^{-1}) [7]; ΔH_f 14.24 kJ mol^{-1} (3.4 kcal mol^{-1}) [8]; ΔH_f 143.4 J g^{-1} [11]

Specific Heat Capacity:

No.	Value	Note	Type
1	322 kJ/kg.C	77 cal mol^{-1} °C^{-1}, 8° [9]	P

Melting Temperature:

No.	Value	Note
1	152°C	[1,8,35]
2	160.5°C	[2]
3	154°C	[3]
4	148°C	[4]
5	157°C	[4,20]
6	161°C	[7]
7	147°C	[8,9]
8	156°C	[10,11]
9	143°C	[22]
10	148–154°C	[28]

Glass-Transition Temperature:

No.	Value	Note
1	45°C	[3]
2	-9°C	[4]
3	8°C	[9]
4	-8°C	[10]
5	9°C	[15]
6	4°C	[22]
7	6°C	[22]
8	23°C	[35]

Deflection Temperature:

No.	Value	Note
1	5°C	[22]

Vicat Softening Point:

No.	Value	Note
1	151°C	[22]

Transition Temperature:

No.	Value	Note	Type
1	-110°C	[3]	T_γ
2	134°C	Under loading, fibre [4]	Flow point
3	43°C	[9]	$T_{\beta max}$
4	-95°C	[9]	$T_{\gamma max}$

Surface Properties & Solubility:
Solubility Properties General: Miscibility with the polyhydroxyether of bisphenol A has been reported [15]
Solvents/Non-solvents: Sol. CH_2Cl_2 [9], $CHCl_3$ [12], hexafluoroisopropanol [14]. Insol. MeOH [12]. Solubilities in EtOH and C_6H_6 have been reported [28]

Transport Properties:
Polymer Solutions Dilute: η_{inh} 0.77 dl g^{-1} (phenol/tetrachloroethane (60:40), 20°) [4]; [η] 1 dl g^{-1} (CHCl$_3$, 25°) [8]; [η] 0.404 dl g^{-1} (phenol/tetrachloroethane (50:50), 25°) [20]; [η] 1.05 dl g^{-1} (phenol/tetrachloroethane (60:40), 30°) [22]; η_{inh} 0.08 dl g^{-1} (1-methyl-2-pyrrolidinone, 30°) [24]; [η] 1.01 dl g^{-1} (2-chlorophenol, 35°) [32]. Other viscosity values have been reported [16]
Permeability of Liquids: D 7.7×10^{-6} cm^2 s^{-1} (1,2-dichloroethane, 240°), 5.3×10^{-6} cm^2 s^{-1} (1,4-dichlorobutane, 240°), 1.5×10^{-5} cm^2 s^{-1} (1,6-hexanediol, 240°) [18]

Mechanical Properties:
Mechanical Properties General: Shear modulus values have been reported [35]. Dynamic loss modulus and storage modulus have been reported [2]

Tensile (Young's) Modulus:

No.	Value	Note
1	0.75 MPa	[25]

Tensile Strength Break:

No.	Value	Note	Elongation
1	3 MPa	fibre [4]	9%
2	3.3 MPa	fibre [33]	37%
3	0.514 MPa	fibre [35]	854%

Miscellaneous Moduli:

No.	Value	Note	Type
1	43 MPa	fibre [4]	Modulus
2	5.135 MPa	fibre [35]	Modulus

Complex Moduli: Complex dynamic modulus has been reported in graphical form [36]
Impact Strength: Izod impact strength has been reported [32]
Viscoelastic Behaviour: Stress relaxation modulus has been reported in a graphical form [23]. Stress-strain curves, viscoelastic props. [25,35], stress relaxation and creep have been reported [26].

Electrical Properties:
Electrical Properties General: Dissipation factor values have been reported in graphical form [9]
Complex Permittivity and Electroactive Polymers: Complex permittivity has been reported [31]

Optical Properties:
Transmission and Spectra: Ir [6,19], C-13 nmr [13,14,20] and mass [24,29] spectral data have been reported
Molar Refraction: dn/dC 0.231 (hexafluoroisopropanol, 23°, 633 nm) [16]

Polymer Stability:
Decomposition Details: Thermal degradation details have been reported [24,29,37]. Max. rate of polymer decomposition at 405° [24]
Stability Miscellaneous: Effects of γ-irradiation have been reported [5,30]

Applications/Commercial Products:
Processing & Manufacturing Routes: Synth. by condensation polymerisation of dimethyl terephthalate and 1,6-hexanediol using tetraisopropyl titanate as a catalyst at 280° under vacuum [2,4,5,8,11,13,16,21,22,24,28]

Bibliographic References
[1] Izard, E.F., *J. Polym. Sci.*, 1952, **8**, 503, (melting point)
[2] Flory, P.J., Bedon, H.D. and Keefer, E.H., *J. Polym. Sci.*, 1958, **28**, 151, (synth, density, thermodynamic props)
[3] Farrow, G., McIntosh, J. and Ward, I.M., *Makromol. Chem.*, 1960, **38**, 147, (cryst struct, transition temps)
[4] Smith, J.G., Kibler, C.J. and Sublett, B.J., *J. Polym. Sci., Part A-1*, 1966, **4**, 1851, (synth, transition temps)
[5] Wlochowicz, A., Pietrzak, M. and Kroh, J., *Eur. Polym. J.*, 1972, **8**, 313, (synth, radiation)
[6] Joly, A.M., Nemoz, G., Douillard, A. and Vallet, G., *Makromol. Chem.*, 1975, **176**, 479, (cryst struct, ir)
[7] Kirshenbaum, I., *J. Polym. Sci., Part A-1*, 1965, **3**, 1869, (thermodynamic props)
[8] Yip, H.K. and Williams, H.L., *J. Appl. Polym. Sci.*, 1976, **20**, 1209, (synth, melting point)
[9] Yip, H.K. and Williams, H.L., *J. Appl. Polym. Sci.*, 1976, **20**, 1217, (heat capacity, mech props)
[10] Gilbert, M. and Hybart, F.J., *Polymer*, 1974, **15**, 407, (transition temps)
[11] Gilbert, M. and Hybart, F.J., *Polymer*, 1972, **13**, 327, (synth, cryst struct)
[12] Hall, I.H. and Ibrahim, B.A., *J. Polym. Sci., Polym. Lett. Ed.*, 1980, **18**, 183, (solubility)
[13] Kricheldorf, H.R., *Makromol. Chem.*, 1978, **179**, 2133, (C-13 nmr, synth)
[14] Komoroski, R.A., *J. Polym. Sci., Polym. Phys. Ed.*, 1979, **17**, 45, (C-13 nmr)
[15] Seymour, R.W. and Zehner, B.E., *J. Polym. Sci., Polym. Phys. Ed.*, 1980, **18**, 2299, (miscibility)
[16] Horbach, A., Binsack, R., Muller, H. and Grunewald, H., *Angew. Makromol. Chem.*, 1981, **98**, 35, (soln props, synth)
[17] Hall, I.H. and Ibrahim, B.A., *Polymer*, 1982, **23**, 805, (cryst struct)
[18] Bonatz, E., Rafler, G. and Reinisch, G., *Angew. Makromol. Chem.*, 1983, **119**, 137, (diffusion coefficients)
[19] Palmer, A., Poulin-Dandurand, S., Revol, J.F. and Brisse, F., *Eur. Polym. J.*, 1984, **20**, 783, (cryst struct)
[20] Horii, F., Hirai, A., Murayama, K., Kitamaru, R. and Suzuki, T., *Macromolecules*, 1983, **16**, 273, (C-13 nmr, soln props)
[21] Daniewska, I. and Wasiak, A., *Makromol. Chem., Rapid Commun.*, 1982, **3**, 897, (cryst struct, synth)
[22] Ng, T.H. and Williams, H.L., *Makromol. Chem.*, 1981, **182**, 3323, (transition temps, synth)
[23] Ng, T.H. and Williams, H.L., *Makromol. Chem.*, 1981, **182**, 3331, (stress relaxation modulus)
[24] Foti, S., Giuffrida, M., Maravigna, P. and Montaudo, G., *J. Polym. Sci., Polym. Chem. Ed.*, 1984, **22**, 1217, (thermal degradation, ms)
[25] Ng, T.H. and Williams, H.L., *J. Appl. Polym. Sci.*, 1986, **32**, 4883, (Young's modulus, viscoelastic props)
[26] Ng, T.H. and Williams, H.L., *J. Appl. Polym. Sci.*, 1987, **33**, 739, (stress relaxation, creep)
[27] Inomata, K. and Sasaki, S., *J. Polym. Sci., Part B: Polym. Phys.*, 1996, **34**, 83, (cryst struct)
[28] Korshak, V.V., Vinogradova, S.V. and Belyakov, B.M., *Izv. Akad. Nauk SSSR, Otd. Khim. Nauk*, 1957, 730, (solubility, synth)
[29] Sugimura, Y. and Tsuge, S., *J. Chromatogr. Sci.*, 1979, **17**, 269, (thermal degradation, ms)
[30] Zhang, Y., Ji, D., Li, Z., Chen, X. and Tang, A., *Chem. Res. Chin. Univ.*, 1993, **9**, 70, (cryst struct, γ-irradiation)
[31] Tatsumi, T. and Ito, E., *Kobunshi Ronbunshu*, 1992, **49**, 229, (complex permittivity)
[32] *Jpn. Pat.*, 1976, 76 61 554, (impact strength)
[33] *Jpn. Pat.*, 1978, 78 147 815, (mech props)
[34] Daniewska, I., *Polym. J. (Tokyo)*, 1980, **12**, 417, (morphology)
[35] Ghaffar, A., Goodman, I. and Peters, R.H., *Br. Polym. J.*, 1978, **10**, 115, (mech props)
[36] Tajiri, K., Fujii, Y., Aida, M. and Kawai, H., *J. Macromol. Sci., Phys.*, 1970, **4**, 1, (complex dynamic modulus)
[37] Ohtani, H., Kimura, T. and Tsuge, S., *Anal. Sci.*, 1986, **2**, 179, (thermal degradation)

Poly(1-hexene) P-297

$$-[CH_2CH]_n-$$
$$\ \ \ \ \ \ |$$
$$CH_2(CH_2)_2CH_3$$

Synonyms: PH. PHE. PHEX. Poly(1-hexenylene). Poly(1-butylethylene). Poly(1-butyl-1,2-ethanediyl)
Monomers: 1-Hexene
Material class: Thermoplastic
Polymer Type: polyolefins
CAS Number:

CAS Reg. No.	Note
25067-06-5	atactic
26746-82-7	isotactic
133578-03-7	syndiotactic

Molecular Formula: $(C_6H_{12})_n$
Fragments: C_6H_{12}
Morphology: Orthorhombic, a 1.17, b 2.69, c 1.37, helix 7_2 (isotactic). Monoclinic, a 2.22, b 0.889, c 1.37, γ 94.5°, helix 7_2 (isotactic) [2]

Volumetric & Calorimetric Properties:
Density:

No.	Value	Note
1	d^{20} 0.854 g/cm^3	[4]
2	d^{20} 0.852 g/cm^3	[5]
3	d 0.86 g/cm^3	amorph. [2]
4	d 0.73 g/cm^3	monoclinic [2]
5	d 0.91 g/cm^3	orthorhombic [2]

Thermal Expansion Coefficient:

No.	Value	Note	Type
1	0.00016 K^{-1}	$T < T_g$ [6]	L
2	0.00062 K^{-1}	[7]	V

Volumetric Properties General: ΔC_p 25.1 J mol^{-1} K^{-1} [8]. Molar volume 97.9 cm^3 mol^{-1}. [3] Ratio of heat capacities has been reported [4,12,13]
Latent Heat Crystallization: H_{T_g} 14.188 kJ mol^{-1} [8]. S_{T_g} 121.3 J mol^{-1} K^{-1} [8]; 171.3 J mol^{-1} K^{-1} (25°) [9,10]. ΔS, ΔH, ΔG values have been reported [10,11]. ΔH_T 26.93 kJ mol^{-1} (6.433 kcal mol^{-1}). C_p, C_v values have been reported [4,12,13]
Specific Heat Capacity:

No.	Value	Note	Type
1	70.02 kJ/kg.C	-173.63° [4]	P
2	68.7 kJ/kg.C	-173.63° [4]	V

Poly(1-hexene)

Glass-Transition Temperature:

No.	Value	Note
1	-55°C	218K [14]
2	-57.5°C	[4,9]
3	-50°C	[8,33]

Transition Temperature:

No.	Value	Note	Type
1	-110°C	Low temp. discontinuity [6]	γ
2	30°C	Equilib. melting temp. [13]	m
3	-79°C	Kauzmann isoentropic temp. [14]	z
4	-23°C	170 Hz	α
5	-130°C	500 Hz [15]	β
6	134°C	[16]	Polymer melt temp.
7	-36°C	Softening point [17]	Soft

Surface Properties & Solubility:
Cohesive Energy Density Solubility Parameters: CED 262.9 J cm^{-3} (62.8 cal cm^{-3}) [32]
Solvents/Non-solvents: Sol. heptane, toluene, cyclohexane, CCl$_4$, decahydronaphthalene [19], THF [18]. Slightly sol. phenetole [18], CO$_2$ [20]

Transport Properties:
Melt Flow Index:

No.	Value	Note
1	12 g/10 min	approx., ASTM D1238-15T
2	20 g/10 min	approx., ASTM D1238-15T [21]

Polymer Solutions Dilute: η_{inh} 3.86 (decahydronaphthalene, 130°) [16]. [η] 3.29 dl g^{-1} (isotactic, toluene, 55°) [22]. [η] 3.22 dl g^{-1} (amorph., toluene, 55°) [22]. [η] 2.56 dl g^{-1} (toluene, 25°) [23], θ 61.3° (phenetole) [18]
Polymer Melts: Melt flow activation energy E* 71.2 kJ mol^{-1} [7], 75.4 kJ mol^{-1} [24], 50.7 kJ mol^{-1} [5], 46.1 kJ mol^{-1} [21]. Zero shear viscosity η_0 4400 (80°) [25]. Critical shear rate values have been reported [21]. Dynamic viscosity, dynamic rigidity and apparent viscosity results at various temps. have been reported [5]

Mechanical Properties:
Complex Moduli: Complex shear modulus [26] and complex tensile modulus [27] have been reported

Electrical Properties:
Complex Permittivity and Electroactive Polymers: Complex permittivity 2.5 [23]
Dissipation (Power) Factor:

No.	Value	Frequency	Note
1	0.002–0.04	1–5 MHz	20–31° [23,27]

Optical Properties:
Optical Properties General: Light scattering results have been reported [18]
Transmission and Spectra: C-13 nmr [1,28,31] and ir [29,30] spectral data have been reported
Molar Refraction: dη/dC 0.12 (heptane, 20°) [23]
Total Internal Reflection: Stress optical coefficient -264–492 [22]

Applications/Commercial Products:
Processing & Manufacturing Routes: Synth. using Ziegler-Natta catalysts, such as TiCl$_4$AlEt$_3$ at 100° [20]

Bibliographic References
[1] Asanuma, T., Nishimori, Y., Ito, M., Uchikawa, N. and Shiomura, T., *Polym. Bull. (Berlin)*, 1991, **25**, 567, (synth, syndiotactic, C-13 nmr)
[2] Turner-Jones, A., *Makromol. Chem.*, 1964, **71**, 1, (struct, density)
[3] Van Krevelen, D.W., *Properties of Polymers: Their Correlation with Chemical Structure*, 2nd edn., Elsevier, 1976, 55, (props)
[4] Bourdariat, J., Isnard, R. and Odin, J., *J. Polym. Sci., Polym. Phys. Ed.*, 1973, **11**, 1817, (specific heat capacities)
[5] Shirayama, K., Matsuda, T. and Shinichiro, K., *Makromol. Chem.*, 1971, **147**, 155, (rheological props)
[6] Damnis, M.L., *J. Appl. Polym. Sci.*, 1959, **1**, 121, (expansion coefficient)
[7] Porter, R.S., Chuah, H.H. and Wang, J.-S., *J. Rheol. (N.Y.)*, 1995, **39**, 649, (rheological props)
[8] Gaur, U., *J. Phys. Chem. Ref. Data*, 1983, **12**, 29, (thermodynamic props)
[9] Domalski, E.S. and Hearing, E.D., *J. Phys. Chem. Ref. Data*, 1990, **19**, 881, (thermodynamic props)
[10] Lebedev, B.V. and Lebedev, N.K., *Zh. Fiz. Khim.*, 1974, **48**, 2899, (thermodynamic props)
[11] Aycock, W., PhD Thesis Rennsselaer Polytechnic Institute, Troy, N.Y., 1989, (thermodynamic props)
[12] Bourdariat, J., Berton, A., Chaussy, J., Isnard, A. and Odin, J., *Polymer*, 1973, **14**, 167, (specific heat capacities)
[13] Bu, H.S., Aycock, W. and Wunderlich, B., *Polymer*, 1987, **28**, 1165, (specific heat capacities)
[14] Dubcy, K.S., Ramachandrarao, P. and Lele, S., *Polymer*, 1987, **28**, 1341, (Kauzmann temp)
[15] Clark, K.J., Turner-Jones, A. and Sandiford, D.J.H., *Chem. Ind. (London)*, 1962, **47**, 2010, (transition temps)
[16] Campbell, T.W. and Haven, A.C., *J. Appl. Polym. Sci.*, 1959, **1**, 73, (transition temps)
[17] Dunham, K.R., Vandenberghe, J., Faber, J.W.H. and Contois, L.E., *J. Polym. Sci., Part A: Polym. Chem.*, 1963, **1**, 751, (transition temps)
[18] Lin, F.C., Stiuala, S.S. and Biesenberger, J.A., *J. Appl. Polym. Sci.*, 1973, **17**, 1073, (dilute soln props)
[19] Mamedov, T.I., Ibragimova, L.S., Mirzakhanov, I.S. and Sadykhzade, S.I., *Azerb. Khim. Zh.*, 1965, 34, (synth, solubilities)
[20] Dandge, D.K., Heller, J.P. and Wilson, K.V., *J. Appl. Polym. Sci.*, 1986, **32**, 3775, (synth, solubility)
[21] Combs, R.L., Slonaker, D.F. and Hoover, H.W., *J. Polym. Sci.*, 1969, **13**, 519, (rheological props)
[22] Philippoff, W. and Tomquist, E., *J. Polym. Sci., Part C: Polym. Lett.*, 1966, **23**, 881, (optical props)
[23] Grandaud, J.L., May, J.F., Berticat, P. and Vallet, G., *C.R. Seances Acad. Sci., Ser. C*, 1971, **273**, 500, (synth, dielectric props)
[24] Wang, J.-S., Porter, R.S. and Knox, J.R., *J. Polym. Sci., Part B: Polym. Lett.*, 1970, **8**, 671, (melt flow activation energy)
[25] Wang, J.-S., Knox, J.R. and Porter, R.S., *J. Polym. Sci., Polym. Phys. Ed.*, 1978, **16**, 1709, (rheological props)
[26] Sheldon, F.K., Passaglia, E. and Pariser, R., *J. Appl. Phys.*, 1957, **28**, 499, (dynamic mech props)
[27] Piloz, A., Decroix, J.Y. and May, J.F., *Angew. Makromol. Chem.*, 1975, **44**, 77, (dynamic mech props)
[28] Asakura, T., Hiraro, K., Demura, M. and Kato, K., *Makromol. Chem., Rapid Commun.*, 1991, **12**, 215, (C-13 nmr)
[29] Ciampelli, F. and Tosi, C., *Spectrochim. Acta*, 1968, **24**, 2157, (ir)
[30] Harvey, M.C. and Ketley, A.D., *J. Appl. Polym. Sci.*, 1961, **5**, 247, (ir)
[31] Asakura, T., Demura, M. and Nishiyama, Y., *Macromolecules*, 1991, **24**, 2334, (C-13 nmr)
[32] He, T., *J. Appl. Polym. Sci.*, 1985, **30**, 4319, (solubility props)
[33] Natta, G., Danusso, F. and Moraglio, G., *CA*, 1957, **52**, 17789e, (transition temp)

Poly(hexene oxide) P-298

(structure: [-O-CH(-(CH$_2$)$_3$CH$_3$)-CH$_2$-]$_n$)

Synonyms: *Poly[oxy(butyl-1,2-ethanediyl)]. Poly(butyloxirane). Poly(1,2-epoxyhexane). Poly(1,2-hexane oxide)*
Monomers: Butyl oxirane
Material class: Thermoplastic
Polymer Type: polyalkylene ether
CAS Number:

CAS Reg. No.	Note
26402-36-8	Homopolymer

Molecular Formula: $(C_6H_{12}O)_n$
Fragments: $C_6H_{12}O$
General Information: Semitransparent elastomer [1]

Volumetric & Calorimetric Properties:
Density:

No.	Value	Note
1	d 0.925–0.97 g/cm^3	[2]

Melting Temperature:

No.	Value	Note
1	71–72°C	[2]

Glass-Transition Temperature:

No.	Value	Note
1	-70 – -68°C	[2]

Optical Properties:
Transmission and Spectra: C-13 nmr spectral data have been reported [1]

Bibliographic References
[1] Bansleben, D.A., Hersman, M.J. and Vogl, O., J. Polym. Sci., *Polym. Chem. Ed.*, 1984, **22**, 2489, (C-13 nmr, ir)
[2] Lai, J. and Trick, G.S., J. Polym. Sci., *Part A-1*, 1970, **8**, 2339, (density, transition temp)

Poly(hexyl methacrylates) P-299

(structure: [-CH$_2$C(CH$_3$)(C(=O)OR)-]$_n$)

R = C$_6$H$_{13}$ (hexyl)

R = CH(CH$_3$)CH$_2$CH$_2$CH$_2$CH$_3$ (1-methylpentyl)

R = CH$_2$CH(CH$_2$CH$_3$)CH$_2$CH$_3$ (2-ethylbutyl)

R = CH(CH$_3$)CH$_2$CH(CH$_3$)CH$_3$ (1,3-dimethylbutyl)

R = CH(CH$_3$)C(CH$_3$)$_2$CH$_3$ (1,2,2-trimethylbutyl)

R = CH$_2$CH$_2$C(CH$_3$)$_2$ (1,2,2-trimethylbutyl)

Synonyms: *Poly(hexyl 2-methyl-2-propenoate). Poly(1-methylpentyl methacrylate). Poly(2-ethylbutyl methacrylate). Poly(1,3-dimethylbutyl methacrylate). Poly(1,2,2-trimethylpropyl methacrylate). Poly(3,3-dimethylbutyl methacrylate). Poly(n-hexyl methacrylate)*
Related Polymers: Polymethacrylates-General
Monomers: n-Hexyl methacrylate, 1,2,2-Trimethylpropyl methacrylate, 1,3-Dimethylbutyl methacrylate, 1-Methylpentyl methacrylate, 3,3-Dimethylbutyl methacrylate, Ethylbutyl methacrylate
Material class: Thermoplastic
Polymer Type: acrylic
CAS Number:

CAS Reg. No.	Note
25087-17-6	n-hexyl
25087-19-8	ethylbutyl

Molecular Formula: $(C_{10}H_{18}O_2)_n$
Fragments: $C_{10}H_{18}O_2$
General Information: There are several isomeric hexyl methacrylates. Data refer to the n-hexyl isomer, unless specified otherwise

Volumetric & Calorimetric Properties:
Density:

No.	Value	Note
1	d^{25} 1.008 g/cm^3	25° [1]
2	d^{100} 0.959 g/cm^3	100° [1]
3	d^{25} 0.991 g/cm^3	25° 1,2,2-trimethylpropyl [2]
4	d^{25} 1.005 g/cm^3	25° 1,3-dimethylbutyl [2]
5	d^{25} 1.013 g/cm^3	25° 1-methylpentyl [2]
6	d^{25} 1.001 g/cm^3	25° 3,3-dimethylbuty [2]
7	d 1.04 g/cm^3	2-ethylbutyl [2]
8	d^{100} 0.995 g/cm^3	100° 2-ethylbutyl [3]
9	d^{20} 0.99 g/cm^3	3,3-dimethylbutyl [7,8]

Thermal Expansion Coefficient:

No.	Value	Note	Type
1	0.00066–0.0007 K^{-1}	uv polymerised, T >T$_g$	L
2	0.00042–0.00046 K^{-1}	uv polymerised, T Tg [29]	V

Equation of State: Mark-Houwink constants have been reported [12,15]
Specific Heat Capacity:

No.	Value	Note	Type
1	324.83 kJ/kg.C	27°, melt [5]	P
2	310 kJ/kg.C	-3° [5]	P
3	381 kJ/kg.C	147°	P

Glass-Transition Temperature:

No.	Value	Note
1	-5°C	[4,6]
2	108°C	3,3-dimethylbutyl [7]
3	11°C	2-ethylbutyl [2]

– Poly(hexyl methacrylates)

Vicat Softening Point:

No.	Value	Note
1	55°C	1,3-dimethylbutyl [8]
2	23°C	1-methylpentyl [8]
3	123°C	3,3-dimethylbutyl [7]

Surface Properties & Solubility:
Cohesive Energy Density Solubility Parameters: Cohesive energy 50.5 kJ mol^{-1} [6]. Solubility parameters δ 36 J$^{1/2}$ cm$^{-3/2}$ (8.6 cal$^{1/2}$ cm$^{-3/2}$) [9], 35.1 J$^{1/2}$ cm$^{-3/2}$ (8.4 cal$^{1/2}$ cm$^{-3/2}$) [10], 31.4–33.5 J$^{1/2}$ cm$^{-3/2}$ (7.5–8 cal$^{1/2}$ cm$^{-3/2}$) [11].
Solvents/Non-solvents: Sol. isopropanol (>33°), 2-butanone [12]
Wettability/Surface Energy and Interfacial Tension: (dγ/dT) 0.062 mN m^{-1} K^{-1} (MV 52000) [2,13]. Contact angle 70° (H$_2$O, advancing) [14]

Surface Tension:

No.	Value	Note
1	30 mN/m	20°, MV 52000) [2,13]
2	22 mN/m	150°, MV 52000) [13]
3	18.9 mN/m	200°, MV 52000) [13]

Transport Properties:
Polymer Solutions Dilute: Molar volumes of amorph. forms, 25° van der Waal's 107.3 : cm^3 mol^{-1}, 107.2 (1,2,2-trimethylpropyl, 1,3-dimethylbutyl, 1-methylpentyl, 3,3-dimethylbutyl, 2-ethylbutyl). Glassy: 169.1 cm^3 mol^{-1}, 171.9 (1,2,2-trimethylpropyl), 169.5 (1,3-dimethylbutyl), 168.1 (1-methylpentyl), 170.1 (3,3-dimethylbutyl), 163.8 (2-ethylbutyl) [2]. Theta temps. 30° (butanone), 32.6° (isopropanol), 37° (propanol) [15,16,17], 25° (butanone; 2-ethylbutyl), 27.4° (isopropanol; 2-ethylbutyl) [15,18]. Poylmer-liquid interaction parameters, 0.91 (hexane), 0.81 (2-butanone), 1.47 (acetonitrile), 0.22 (CHCl$_3$), 1.36 (n-propanol), 1.2 (Me$_2$CO), 0.9 (EtOAc)
Permeability of Liquids: Diffusion coefficient [H$_2$O] 519 × 10^{-8} cm^2 s^{-1} (37°) [14]

Water Absorption:

No.	Value	Note
1	6.9 %	max. [14]

Mechanical Properties:
Mechanical Properties General: Plateau modulus 0.087 dyne cm^{-2}; 0.14 dyne cm^{-2} (2-ethylbutyl) [3]
Viscoelastic Behaviour: Williams-Landel-Ferry constants for temp. superposition and Vogel parameters have been reported [1,20,21]. Other viscoelastic props. have been reported [24]
Hardness: Vickers (5 kg load) [8] 10 (1,3-dimethylbutyl), too rubbery (1-methylpentyl), 22 (3,3-dimethylbutyl)

Optical Properties:
Refractive Index:

No.	Value	Note
1	1.481	[22]
2	1.474	1,3-dimethylbutyl [8]
3	1.478	1-methylpentyl [8]
4	1.468	20°, 3,3-dimethylbutyl [8]

Total Internal Reflection: Intrinsic segmental anisotropy -40 × 10^{-25} cm^3 (C$_6$H$_6$), -9.7 × 10^{-25} cm^3 (CCl$_4$) [23]

Polymer Stability:
Decomposition Details: Decomposition at 250° yields monomer and a small amount of olefin by side-chain cracking [8]

Bibliographic References
[1] Abralkhodja, S., Nigmankhodjaev, R., Bi, L.K., Wong, C.P. et al, *J. Polym. Sci., Part A-2*, 1970, **8**, 1927, (dynamic mech props)
[2] Van Krevelen., D.W., *Properties of Polymers: Their Correlation with Chemical Structure*, 1976, 79
[3] Grassely, W.W. and Edwards, S.F., *Polymer*, 1981, **22**, 1329
[4] Rogers, S.S. and Mandelkern, L., *J. Phys. Chem.*, 1957, **61**, 985
[5] Gaur, U., Lau, S.F., Wunderlich, B.B. and Wunderlich, B., *J. Phys. Chem. Ref. Data*, 1982, **11**, 1065, (thermodynamic props)
[6] Porter, D., *Group Interaction Modelling of Polymer Properties*, Dekker, 1995, 283
[7] Hoff, E.A.W., Robinson, D.W. and Willbourn, A.H., *Polym. Sci.*, 1955, **18**, 161
[8] Crawford, J.W.C., *J. Soc. Chem. Ind., London, Trans. Commun.*, 1949, **68**, 201
[9] Hughes, L.J. and Britt. G.E., *J. Appl. Polym. Sci.*, 1961, **5**, 337
[10] Van Krevelen, D.W. and Hoftyzer, P.J., *J. Appl. Polym. Sci.*, 1967, **11**, 2189
[11] Ahmed, H. and Yaseen, M., *Polym. Eng. Sci.*, 1979, **19**, 858, (solubility parameters)
[12] Kurata, M. and Stockmayer, W.H., *Adv. Polym. Sci.*, Springer-Verlag, 1963, **3**, 196, (solubility data)
[13] Wu, S., *Polymer Interface and Adhesion*, Marcel Dekker, 1982
[14] Kalachandra, S. and Kusy, R.P., *Polymer*, 1991, **32**, 2428
[15] Chinai, S.N., *J. Polym. Sci.*, 1957, **25**, 413, (eqn of state)
[16] Bohdanecky, M. and Tuzar, Z., *Collect. Czech. Chem. Commun.*, 1969, **34**, 3318
[17] Burchard, W., *Z. Phys. Chem. (Munich)*, 1964, **43**, 265
[18] Didot, E.F., Chinai, S.N. and Levi, D.W., *J. Polym. Sci.*, 1960, **43**, 557, (dilute soln props)
[19] Walsh, D.J. and McKeown, J.G., *Polymer*, 1980, **21**, 1335
[20] Child, W.C. and Ferry, J.D., *J. Colloid Sci.*, 1957, **12**, 389, (viscoelastic props)
[21] Ngai, K.L. and Plazek, D.J., *Physical Properties of Polymers Handbook*, (ed. J.E. Mark), AIP Press, 1996, 341, (viscoelastic props)
[22] Lewis, O.G., *Physical Constants of Linear Hydrocarbons*, Springer-Verlag, 1968
[23] Grishchenko, A.E., Vitovskaja, M.G., Tsvetkov, V.N. and Andreeva, L.N., *Vysokomol. Soedin.*, 1966, **8**, 800
[24] Aharoni, S.M., *Macromolecules*, 1983, **16**, 1722, (viscoelastic props.)

Poly(hexylmethylsilane)   P-300

$$\left[\begin{array}{c} CH_2(CH_2)_4CH_3 \\ | \\ -Si- \\ | \\ CH_3 \end{array} \right]_n$$

Synonyms: *Poly(hexylmethylsilylene)*. *Hexylmethylsilane homopolymer, sru. PHMS. P61S. (HexMeSi)n. ME6*
Polymer Type: polysilanes
CAS Number:

CAS Reg. No.
88003-15-0

Molecular Formula: $(C_7H_{16}Si)n$
Fragments: $C_7H_{16}Si$
Molecular Weight: $5,200-2.8 \times 10^5$ [1,2,11,12]
Morphology: Amorphous white solid [3,12]. The Si-backbone is dominated by the presence of an intermediate-; mixed transoid (T), deviant (D) and possibly ortho (O) conformations with a dihedral angle *ca.* $\pm 90°$. There is no evidence for a significant fraction of *all-anti* (A, i.e., *trans*-planar), *all-trans*, or *anti*-gauche conformers. The polymer incorporates a D/T-based Si backbone, and this, locally, approximates a D+T+D-T- motif. Displays a number of low-temperature polymorphic structures, one of which exhibits Bragg-like scattering features. Temperature-dependent uv absorption measurements of the polymer resolve 2 distinguishable isosbestic points and are indicative of mesomorphic behaviour [6]

Surface Properties & Solubility:
Solubility Properties General: Readily soluble in common org. solvents [9]

Electrical Properties:
Electrical Properties General: The charge carrier mobility is 2×10^{-7} $m^2/V \; s$ [3]

Optical Properties:
Optical Properties General: Uv-vis spectroscopic analysis shows a broad absorption with λ_{max} 305nm, consistent with an uninterrupted Si chain conjugation length greater than 30 Si atoms [1]. Exists as phase A at all temperatures. A small reversible thermochromic shift (from 325nm to 298nm) with an onset at *ca.* -23°C on heating has been reported, which is broadened over *ca.* 20°C. The presence of an isosbestic point indicates that the thermochromism is related to a structural transition and not a gradual shift to shorter wavelengths. The transition seems to take place in the polymer backbone only, since the material as a whole remains amorphous, only limited conformational changes of the silicon backbone are possible [3].
The dispersion interaction between the polarisable side chains and polysilane backbone appears to be important in this series. Thus the greater polarisability induced by the longer hexyl substituent in PHMS, increases V_D/ϵ to the point where a discontinuous transition is triggered (strong coupling). The polarisability of the side chain increases with the number of carbon atoms, leading to larger values of V_D/ϵ and a higher onset temperature of 10°C [7]

Optical Properties:
Transmission and Spectra: Ir [1,8,11], uv-vis [1,2,3,6,7,11] and nmr [1,4,8,11,13] spectra reported. The proton assignments in the PHMS spectrum have been assigned by means of 2D COSY spectroscopy, which clearly revealed the scalar coupling relations. From these assignments the 13C resonances were assigned by a 2D heterocorrelated spectrum [4].

Polymer Stability:
Thermal Stability General: The DSC of PHMS displays three endothermic features at temperatures near 0, -20, and -40°C. These have been assigned as follows: the 0°C endotherm to the first conformational change and a partial transition to the M phase. The second endotherm with formation of the K2 phase; since there are no overt signatures of a backbone conformational change, the underlying impetus for this partial crystallisation must arise from the influence of the alkyl side chains. The lowest temperature at -40°C therefore correlates with the second chain conformational change [6].
Heating: amorphous low temperature to amorphous high temperature $\Delta H = 0 kJ \; mol^{-1}$ [3]
Cooling: amorphous high temperature to amorphous low temperature $\Delta H = -20 kJ \; mol^{-1}$

Applications/Commercial Products:
Processing & Manufacturing Routes: The synthesis of PHMS was accomplished with high yields of up to 83% by performing Wurtz-type reductive coupling polymerisation of dichloromethyl-*n*-hexylsilane in THF at room temperature [1,4,8,14].
Polymerisation of *n*-hexylmethyldichlorosilane was carried out by the Kipping method with melted Na metal in toluene under vigorous stirring [2].

Applications/Commercial Products:
Applications: Pyrolysis to silicon carbides [8]. Efficient photoinitiators for free radical polymerisation of Methyl methacrylate and styrene [10].

Bibliographic References
[1] Holder, S.J., Achilleos, M. and Jones, R.G., *Macromolecules*, 2005, **38**, 1633
[2] Kawaguchi, T., Seki, S., Okamoto, K., Saeki, A., Yoshida, Y. and Tagawa, S., *Chem. Phys. Lett.*, 2003, **374**, 353
[3] van der Laan, G.P., de Haas, M.P., Hummel, A., Frey, H. and Moller, M., *J. Chem. Phys.*, 1996, **100**, 5470
[4] Ishino, Y., Hirao, A. and Nakahama, S., *Macromolecules*, 1986, **19**, 2309
[5] Skryshevskii, Y.A., *J. Appl. Spectrosc. (Engl. Transl.)*, 2003, **70**, 855
[6] Chunwachirasiri, W., Kanaglekar, I., Winokur, M.J., Koe, J.C. and West, R., *Macromolecules*, 2001, **34**, 6719
[7] Yuan, C.H. and West, R., *Macromolecules*, 1994, **27**, 629
[8] Abu-Eid, M.A., King, R.B. and Kotliar, A.M., *Eur. Polym. J.*, 1992, **28**, 315
[9] Trefonas, P., Djurovich, P.I., Zhang, X.H., West, R., Miller, R.D. and Hofer, D., J. Polym. Sci., *Polym. Lett. Ed.*, 1983, **21**, 819
[10] Peinado, C., Alonso, A., Catalina, F. and Schnabel, W., *Macromol. Chem. Phys.*, 2000, **201**, 1156
[11] Seki, S., Koizumi, Y., Kawaguchi, T., Habara, H. and Tagawa, S., *J. Am. Chem. Soc.*, 2004, **126**, 3521
[12] Sakurai, H., Yoshida, M. and Sakamoto, K., *J. Organomet. Chem.*, 1996, **521**, 287
[13] Wolff, A.R., Maxka, J. and West, R., J. Polym. Sci., *Part A*, 1988, **26**, 713
[14] Miller, R.D., Thompson, D., Sooriyakumaran, R. and Fickes, G.N., J. Polym. Sci., *Part A*, 1991, **29**, 813

Poly[(hexylmethylsilane)(methylene)]   P-301

$$\left[-CH_2-\begin{array}{c} CH_2(CH_2)_4CH_3 \\ | \\ Si \\ | \\ CH_3 \end{array}- \right]_n$$

Synonyms: *Poly[(hexylmethylsilylene)(methylene)]*. *1,3-Dihexyl-1,3-dimethyl-1,3-disilacyclobutane homopolymer. PMHSM*
Related Polymers: Polycarbosilanes
Polymer Type: polycarbosilanes
CAS Number:

CAS Reg. No.
160053-09-8

Molecular Formula: $(C_8H_{18}Si)n$
Fragments: $C_8H_{18}Si$
Molecular Weight: 95,600 [1]
Identification: Found: C, 65.23%; H, 13.05%; Calculated: C, 65.6%; H, 12.5%

Volumetric & Calorimetric Properties:
Glass Transition Temperature:

No.	Value	Note
1	-71.1°C	[1]

Applications/Commercial Products:
Processing & Manufacturing Routes: Prepared by ring-opening polymerisation of 1,3-dichloro-1,3-dihexyl-1,3-disilacyclobutane [1,2].

Bibliographic References
[1] Shen, Q.H. and Interrante, L.V., *J. Polym. Sci., Part A*, 1997, **35**, 3193
[2] Shen, Q.H., Interrante, L.V. and Wu, H.J., *Polym. Prepr. (Am. Chem. Soc., Div. Polym. Chem.)*, 1994, **35**, 395

Poly(hexyloctylsilane) P-302

$$\left[\begin{array}{c} CH_2(CH_2)_6CH_3 \\ | \\ -Si- \\ | \\ CH_2(CH_2)_4CH_3 \end{array} \right]_n$$

Synonyms: *Poly(hexyloctylsilylene). Poly(n-octyl-n-hexylsilane). Dichlorohexyloctylsilane homopolymer. PHOS. POHS*
Polymer Type: polysilanes
CAS Number:

CAS Reg. No.
120517-01-3

Molecular Formula: $(C_{14}H_{30}Si)_n$
Fragments: $C_{14}H_{30}Si$
Molecular Weight: $> 5 \times 10^5$ [2,3].
General Information: Disruption of conformation due to the mismatch of the side-chain length [3].
Tacticity: Atactic sequences have been observed [3].
Morphology: Side chains are mainly planar, with higher *trans*-content in the backbone [3].

Volumetric & Calorimetric Properties:
Melting Temperature:

No.	Value	Note
1	-5°C	[3]

Optical Properties:
Optical Properties General: Exhibits a thermochromic transition with two forms being observed, having absorbances at 350 and 320nm depending on the thermal history. Transition temperature is higher than for PHHS, which suggests more order in the low temperature phase; this is borne out by the presence of a more intense IR band at 689cm^{-1}, indicating a higher (35%±5%) *trans* content in the backbone [3].
Transmission and Spectra: Ir [3] and uv [1,2,3,4] spectra reported.

Applications/Commercial Products:
Processing & Manufacturing Routes: Prepared from *n*-hexyl-*n*-octyldichlorosilylene by condensation with Na metal in toluene at 110°C (Wurtz reaction) [3].

Bibliographic References
[1] Song, K., Miller, R.D., Wallraff, G.M. and Rabolt, J.F., *Macromolecules*, 1992, **25**, 3629
[2] Miller, R.D., Wallraff, G.M., Baier, M., Cotts, P.M., Shukla, P., Russell, T.P., De Schryver, F.C. and Declercq, D., *J. Inorg. Organomet. Polym.*, 1991, **1**, 505
[3] Hallmark, V.M., Sooriyakumaran, R., Miller, R.D. and Rabolt, J.F., *J. Chem. Phys.*, 1989, **90**, 2486
[4] Wallraff, G.M., Baier, M., Miller, R.D., Rabolt, J.F., Hallmark, V., Cotts, P. and Shukla, P., *Polym. Prepr. (Am. Chem. Soc., Div. Polym. Chem.)*, 1989, **30**, 245

Poly(3-hexyloxypropylene oxide) P-303

Synonyms: *Poly[(hexyloxymethyl)oxirane]. Poly(1,2-epoxy-3-hexyloxypropane). Poly[oxy[(hexoxymethyl)-1,2-ethanediyl]]. Poly(hexylglycidyl ether). Poly(3-n-hexoxy-1,2-epoxypropane)*
Monomers: (Hexyloxymethyl)oxirane
Polymer Type: epoxy, polyalkylene ether
CAS Number:

CAS Reg. No.
28476-69-9

Molecular Formula: $(C_9H_{18}O_2)_n$
Fragments: $C_9H_{18}O_2$
General Information: Obtained as a tacky rubber

Volumetric & Calorimetric Properties:
Density:

No.	Value	Note
1	d 0.966 g/cm^3	[1]

Melting Temperature:

No.	Value	Note
1	44°C	small endotherm [1]

Glass Transition Temperature:

No.	Value	Note
1	-83°C	[1]
2	-88°C	[1]

Surface Properties & Solubility:
Solvents/Non-solvents: Sol. Et$_2$O [2], C$_6$H$_6$ [1]. Slightly sol. toluene [1], CHCl$_3$ [2]. Insol. MeOH [1]

Transport Properties:
Polymer Solutions Dilute: η_{inh} 2.8 dl g^{-1} (30°, 0.1% soln. in toluene) [1]. Reduced specific viscosity 0.69 dl g^{-1} (25°, 0.1% soln. in CHCl$_3$) [2]

Optical Properties:
Refractive Index:

No.	Value	Note
1	1.459	30° [1]

Bibliographic References
[1] Lal, J. and Trick, G.S., *J. Polym. Sci., Part A-1*, 1970, **8**, 2339, (synth, glass transition temp)
[2] *U.S. Pat.*, 1965, 3 214 390

Poly(hexylpentylsilane) P-304

$$\left[\begin{array}{c} CH_2(CH_2)_4CH_3 \\ | \\ -Si- \\ | \\ CH_2(CH_2)_3CH_3 \end{array} \right]_n$$

– Poly(3-hexylthiophene)

Synonyms: *Poly(hexylpentylsilylene). Poly(pentylhexylsilane).
PPHS. C5–C6*
Polymer Type: polysilanes
CAS Number:

CAS Reg. No.
120517-00-2

Molecular Formula: $(C_{11}H_{24}Si)_n$
Fragments: $C_{11}H_{24}Si$
Molecular Weight: $0.92 \times 10^6 - 2.6 \times 10^6$ [1,6].
Tacticity: A large conformational change occurs from the disordered conformation to an *all-trans* conformation [2].
Morphology: Hard, waxy and brittle. The diffraction pattern shows a sharp peak followed by a weak peak at 25°C. The ratio of these d-spacings is 1.76, slightly greater than 3x. Consequently, the structure is that which is expected at high temperatures: hexagonal, columnar with $a = b = 1.51$nm. X-ray diffraction pattern at -50°C is consistent with the monoclinic unit cell and *trans* backbone conformation [1,2,4,6].
Miscellaneous: The polymer chains pack in a hexagonal lattice, which is oriented such that the chains lie parallel to the surface. In addition, a remarkable degree of orientation order was found with the planes containing the neighbouring molecules lying in the surface plane. Parallel to the film surface, the average crystallite size is 600Å, while perpendicular to the surface, there is nearly perfect interchain stacking of the polymers throughout the thickness of the film [6].

Volumetric & Calorimetric Properties:
Glass Transition Temperature:

No.	Value	Note
1	-20 – -13°C	[3,5]

Mechanical Properties:
Mechanical Properties General: The ability of the polymer to crystallise into a structure with an ordered (*all-trans*) backbone conformation does not influence its mechanical properties at ambient conditions.

Tensile (Young's) Modulus:

No.	Value	Note
1	4.33 MPa	at 23°C
2	4.3 MPa	at 100 °C

Optical Properties:
Optical Properties General: A discontinuous red shift in which the 314nm absorption band present at room temperature is replaced by a peak at 344nm is observed at -25°C upon cooling. This corresponds to the lower transition found for other polyalkylsilylenes [2].
Transmission and Spectra: Ir [3], Raman [3] and uv [1,2,3] spectra reported.

Polymer Stability:
Thermal Stability General: DSC reveals a broad single first-order thermal transition. The transition enthalpy is 6.9 cal g^{-1} (5.3kJ mol^{-1}). This indicates that a greater change in structure occurs for the first order transition for C5-C6 than for C5-C8. A transition onset temperature of -13°C (260K) and a peak at -6°C (267K) have been observed [1,2,5].

Applications/Commercial Products:
Processing & Manufacturing Routes: The monomer was formed by the reaction of *n*-pentylmagnesium bromide Grignard with hexyltrichlorosilane in THF under Ar. After isolation by two fractional distillations, the resulting dichlorosilane was polymerised by the usual Wurtz condensation reaction with Na in toluene at 110°C to better than 98% purity [2,3,4].
Applications: Electrophotographic photoreceptor having polysilane-containing excellent charge-transport layer and gives a low residual potential [7].

Bibliographic References
[1] Shukla, P., Cotts, P.M., Miller, R.D., Russell, T.P., Smith, B.A., Wallraff, G.M. and Baier, M., *Macromolecules*, 1991, **24**, 5606
[2] Klemann, B.M., West, R. and Koutsky, J.A., *Macromolecules*, 1996, **29**, 198
[3] Hallmark, V.M., Sooriyakumaran, R., Miller, R.D. and Rabolt, J.F., *J. Chem. Phys.*, 1989, **90**, 2486
[4] Embs, F.W., Wegner, T.G., Neher, D., Albouy, P., Miller, R.D., Willson, C.G. and Schrepps, W., *Macromolecules*, 1991, **24**, 5068
[5] Klemann, B.M., DeVilbiss, T. and Koutsky, J.A., *Polym. Eng. Sci.*, 1996, **36**, 135
[6] Factor, B.J., Russell, T.P., Toney, M.F. and Miller, R.D., *Acta Polym.*, 1995, **46**, 60
[7] *Jpn. Pat.*, 1990, 02 150 853

Poly(3-hexylthiophene)   P-305

Synonyms: *P 3-HT*
Related Polymers: Poly(thiophene)
Monomers: 3-Hexylthiophene
Polymer Type: polythiophene
CAS Number:

CAS Reg. No.
104934-50-1

Molecular Formula: $(C_{10}H_{14}S)_n$
Fragments: $C_{10}H_{14}S$
Molecular Weight: Regiorandom P3HT [14,18,24] MW = 24400; M_n = 5650; PD = 4.32
Regioregular P3HT [14,18,24] MW = 37 680; M_n = 25500; PD = 1.48
Additives: Organically modified montmorillonite (om-MMT) clay. [18]
General Information: The chemistry of polythiophenes has been reviewed. [23]
P 3-HT is an important polymer because of its excellent electrical conducting conductivity, electroluminescence, and nonlinear optical properties. However, its physical, thermal, and properties are not very good and finds difficulty in many applications. Recently, polymer-clay composites (PNCs) have received significant research nano-composites attention for the large-scale improvement in the mechanical and physical properties. [18]
Morphology: Crystal form, a 17.4, b 3.9, c 7.8 [5]. Regioregular P 3-HT is a purple film, with a comb-like polymer having a pendent hexyl group with interchain lamella structure and a 98% head tail [18,21,22,24]. Both regioregular and regio random isomers have been isolated. [24]
Identification: Regioregular P 3-HT: Calc: C, 72.22; H, 8.49; S, 19.29; Found C, 71.91; H, 8.48; S, 18.73
Regiorandom P 3-HT: Calc: C, 72.22; H, 8.49; S, 19.29; Found C, 71.50; H, 8.28; S, 19.77

Volumetric & Calorimetric Properties:
Density:

No.	Value	Note
1	d 1.02–1.5 g/cm^3	[15,16]

Thermodynamic Properties General: Enthalpy of fusion=4.5 J/g [18]
The thermodynamics of cocrystallization for the P 3-HT(R) and (P 3-HT-1) system have been observed [26]

Volumetric & Calorimetric Properties:
Melting Temperature:

No.	Value	Note
1	161°C	[17]
2	231°C	[18]
3	222°C	[22]

Glass Transition Temperature:

No.	Value	Note
1	33°C	[1]
2	10–30°C	[2,3,4]
3	20.3°C	[22]

Surface Properties & Solubility:
Solubility Properties General: The regiorandom polymer shows a rather higher solubility than the regioregular one [14]
Solvents/Non-solvents: Sol. $CHCl_3$ [6]

Transport Properties:
Polymer Solutions Dilute: [η] 0.91 dl g^{-1} (25°, $CHCl_3$) [5]

Mechanical Properties:
Mechanical Properties General: The storage modulus value indicates that there is a large mechanical reinforcement by the clay particles for all the PNCs versus P 3-HT. [18]

Electrical Properties:
Electrical Properties General: Electrical conductivity depends on method of synth. [1,7,8]. Solvent-cast films from chloroform show different behavior than that of the melt-cooled samples. The solvent cast films show a sharp increase followed by a decrease, whereas the melt-cooled films show only a sigmoidal increase [9]

Optical Properties:
Transmission and Spectra: H-1 nmr [1,12,24], ir [11,18,24], x-ray, [8,11] epr [13] uv, [7,8,9,10,18,20,23,24] and C-13 nmr spectral data [1,12,24] have been reported

Polymer Stability:
Polymer Stability General: Degradation temperature 428.6°C [18]
Thermal Stability General: In the thermogram there are two breaks - one at lower temperature (approx. -100°C) and the other at higher temperature (approx. 20°C), indicating two different types of transitions in the P 3-HT chains [18]

Applications/Commercial Products:
Processing & Manufacturing Routes: Prod. by polymerisation of the Grignard reagent of 2-halo-3-hexylthiophenes [12], or chemical or electrochemical oxidation polymerisation with various catalysts. Oxidative coupling polymerisation of P 3-HT with $FeCl_3$ to give regioregularity P 3-HT. As a result, both the polymerisation temperature and monomer concentration were effective to the regioregularity. The lower temperature and the lower concentration produced the polymer having the higher head-tail content [14,24]
2-bromo-5-(bromomagnesio)-3-alkylthiophene is polymerised with catalytic amounts of Ni(dppp)Cl_2 leading to 98% headtail couplings [21]

Applications/Commercial Products:
Applications: White LEDs using conjugated polymer blends [19]. Flexible photovoltaic devices [20]. Polymer thin film field effect transistors by friction-transfer technique [25]

Bibliographic References
[1] Hotta, S., Soga, M. and Sonoda, N., *Synth. Met.*, 1988, **26**, 267
[2] Van de Leur, R.H.M., de Ruiter, B. and Breen, J., *Synth. Met.*, 1993, **54**, 203
[3] Jianguo, M. and Kalle, L., Polym. Prepr. (Am. Chem. Soc., *Div. Polym. Chem.*), 1993, **34**, 727
[4] Zhao, Y., Yuan, G., Roche, P. and Leclerc, M., *Polymer*, 1995, **36**, 2211
[5] Moulton, J. and Smith, P., *Polymer*, 1992, **33**, 2340
[6] Assadi, A., Svensson, C., Willander, M. and Inganas, O., *Synth. Met.*, 1989, **28**, C8, 63
[7] Hotta, S., *Synth. Met.*, 1987, **22**, 103
[8] Mao, H., Xu, B. and Holderoft, S., *Macromolecules*, 1993, **26**, 1163
[9] Yoshino, K., *Synth. Met.*, 1989, **28**, C669
[10] McCullough, R.D. and Lowe, R.D., Polym. Prepr. (Am. Chem. Soc., *Div. Polym. Chem.*), 1992, **33**, 195
[11] Winokur, M.J., Spiegel, D., Kim, Y., Hotta, S. and Heeger, A.J., *Synth. Met.*, 1989, **28**, C419
[12] McCullough, R.D., Lowe, R.D., Jayaraman, M., Ewbank, P.C. and Anderson, D.L., *Synth. Met.*, 1993, **55**, 1198
[13] Souto Maior, R.M., Hin-Kelmann, K., Eckert, H. and Wudl, F., *Macromolecules*, 1990, **23**, 1268
[14] Amou S., Haba O., Shirato K., Hayakawa T., Ueda M., Takeuchi K., and Asai M., *J. Polym. Sci., Part A*, 1999, **37**, 1943
[15] Saeki, A., *J. Phys. Chem. B*, 2005, **109**, 10015
[16] Chang C.C., *Thin Solid Films*, 2005, **479**, 254
[17] Sudip M. and Nandi, A.K., *J. Polym. Sci., Part B: Polym. Phys.*, 2002, **40**, 2073
[18] Kuila B.P., *Macromolecules*, 2004, **37**, 8577
[19] Hwang D.H., Park M.J. and Lee C., *Synth. Met.*, 2005, **152**, 205
[20] Ibrahim M.A., Roth H.K., Schroedner M., Konkin A., Zhokhavets U., Gobsch G., Schar P., and Sensfuss S., *Organic Electronics*, 2005, **6**, 65
[21] McCullough R.D., Tristram-Nagle S., Williams S.P., Lowe R.D., and Jayaraman M., *J. Am. Chem. Soc.*, 1993, **115**, 4910
[22] Pal S., Roy S., Nandi A.K., *J. Phys. Chem. B*, 2005, **109**, 18332
[23] McCullough R.D., *Adv. Mater.* (Weinheim, Ger.), 1998, **10**
[24] Chen T.A., Wu X., and Rieke R.D., *J. Am. Chem. Soc.*, 1995, **117**, 233
[25] Nagamatsu S., Tannigaki N., Yoshida Y., Takashima W., Yase K., and Kaneto K., *Synth. Met.*, 2003, **137**, 923
[26] Pal S., Roy S., Nandi A.K., *J. Phys. Chem. B*, 2005, **109**, 18332

Poly(3-hexylthiophene)/PCBM blend

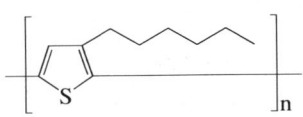

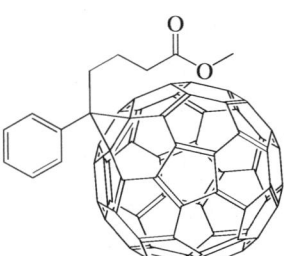

Synonyms: *Poly(3-hexylthiophene)/[6,6]-phenyl-C_{61}-butyric acid methyl ester blend.* P3HT:PCBM
Material class: Thermoplastic
Polymer Type: polythiophene
CAS Number:

CAS Reg. No.	Note
104934-50-1	3-hexylthiophene homopolymer
110134-47-9	poly(3-hexyl-2,5-thiophenediyl)
160848-22-6	PCBM

Molecular Formula: $(C_{10}H_{14}S)n.(C_{72}H_{14}O_2)n$
Fragments: $C_{10}H_{14}S$ $C_{72}H_{14}O_2$
General Information: A blend of the conducting polymer poly(3-hexylthiophene) in regioregular (head-to-tail) form with the

soluble fullerene PCBM in varying proportions, usually about 1:1 by weight. Films are cast from the solid mixture using organic solvents. The polymer acts as an electron donor and the PCBM as electron acceptor for heterojunction solar cell devices.
Morphology: Films of P3HT:PCBM (usually about 1:1 proportion by weight) in photovoltaic devices are usually subjected to thermal treatment (annealing) in the region of 90–150°C prior to use. Generally the morphology has a marked effect on performance, with annealing leading to increased photocurrent. After annealing the amorphous matrix is converted to one containing domains of P3HT crystallites [1,2,3,4,5], with the main chains parallel and the side chains perpendicular to the substrate in photovoltaic devices [1]. After annealing the PCBM is reported to form clusters or aggregates [1,3,6] or needle-like crystals [7] in the matrix. The annealed matrix has formed percolating pathways for charge transfer, where holes and electrons cannot easily recombine [6,8].

Surface Properties & Solubility:
Solvents/Non-solvents: Sol. 1,2-dichlorobenzene [9,10], xylene [3], chlorobenzene and chloroform [4,8] for film casting.

Electrical Properties:
Electrical Properties General: The blends are normally incorporated as films into solar cell devices, which would typically consist of an indium-tin oxide (ITO) anode, a layer of the hole transporter PEDOT-PSS, a thin film of the blend and a cathode such as aluminium. Annealing of the devices (90–150°C) increases in particular the hole mobility of the P3HT [5,10,11]. Application of an external voltage prior to use also increases photocurrent [12]. An increase in molecular weight of P3HT results in greater efficiency [3]. Power conversion efficiencies of 4–5% have been reported [2,10,13], with fill factors up to 67% [10], short-circuit current densities of 10–15 mA cm^{-2} and open-circuit voltages of about 0.6V.

Optical Properties:
Optical Properties General: Application of an external light source to the blend results in charge transfer from P3HT to PCBM. External quantum efficiencies of 60–70% have been reported [10,12]. Band gap of P3HT in the blend 2.14 eV [14]. Optical absorption (with λ_{max} in the region of 520nm) is increased by annealing [4,5,6,8,11] with some red shift of absorption maximum. Differences in absorption using chloroform and chlorobenzene as solvents have been reported [4,8]. The photoluminescence of P3HT alone is quenched on addition of PCBM [14,15], while annealing of the blend increases photoluminescence [11].
Transmission and Spectra: UV [6,8,9,10,15], photoluminescence [11,14,15], FT-IR [9], XPS [9] and x-ray diffraction [1,2] spectral data have been reported.

Polymer Stability:
Thermal Stability General: Blends are thermally stable in solar cell devices annealed at 150°C in nitrogen [2].

Applications/Commercial Products:
Processing & Manufacturing Routes: Regioregular P3HT is blended with PCBM usually in a weight ratio of about 1:1 and formed into films, normally by spin-casting in a suitable organic solvent, and incorporated into solar cell devices as described.
Applications: Blends are used in development of heterojunction solar cells. Other possible applications include field-effect transistors.

Trade name	Details	Supplier
Rieke	regioregular poly(3-hexylthiophene)	Rieke Metals
	regioregular poly(3-hexylthiophene)	American Dye Source
	PCBM	American Dye Source

Bibliographic References
[1] Erb, T., Zhokhavets, U., Gobsch, G., Raleva, S., Stühn, B., Schilinsky, P., Waldauf, C. and Brabec, C.J., *Adv. Funct. Mater.*, 2005, **15**, 1193
[2] Ma, W., Yang, C., Gong, X., Lee, K. and Heeger, A.J., *Adv. Funct. Mater.*, 2005, **15**, 1617
[3] Schilinsky, P., Asawapirom, U., Scherf, U., Biele, M. and Brabec, C.J., *Chem. Mater.*, 2005, **17**, 2175
[4] Zhokhavets, U., Erb, T., Gobsch, G., Al-Ibrahim, M. and Ambacher, O., *Chem. Phys. Lett.*, 2006, **418**, 347
[5] Mihailetchi, V.D., Xie, H., de Boer, B., Koster, J.A. and Blom, P.W.M., *Adv. Funct. Mater.*, 2006, **16**, 699
[6] Chirvase, D., Parisi, J., Hummelen, J.C. and Dyakonov, V., *Nanotechnology*, 2004, **15**, 1317
[7] Swinnen, A., Haeldermans, van de Ven, M., D'Haen, J., Vanhoyland, G., Aresu, S., D'Olieslaeger, M. and Manca, J., *Adv. Funct. Mater.*, 2006, **16**, 760
[8] Al-Ibrahim, M., Ambacher, O., Sensfuss, S. and Gobsch, G., *Appl. Phys. Lett.*, 2005, **86**, 201120/1
[9] Shrotiya, V., Ouyang, J., Tseng, R.J., Li, G.and Yang, Y., *Chem. Phys. Lett.*, 2005, **411**, 138
[10] Li, G., Shrotiya, V., Huang, J., Yao, Y., Moriarty, T., Emery, K. and Yang, Y., *Nature Materials*, 2005, **4**, 864
[11] Zhokhavets, U., Erb, T., Hoppe, H., Golsch, G. and Sariciftci, N.S., *Thin Solid Films*, 2006, **496**, 679
[12] Padinger, F., Rittberger, R.S. and Sariciftci, N.S., *Adv. Funct. Mater.*, 2003, **13**, 85
[13] Reyes-Reyes, M., Kim, K. and Carroll, D.L., *Appl. Phys. Lett.*, 2005, **87**, 083506/1
[14] Chirvase, D., Chiguvare, Z., Knipper, M., Parisi, J., Dyakonov, V. and Hummelen, J.C., *Synth. Met.*, 2003, **138**, 299
[15] Kim, Y., Choulis, S.A., Nelson, A., Bradley, D.D.C., Cook, S. and Durrant, J.R., *J. Mater. Sci.*, 2005, **40**, 1371

Poly(hydrogenmethylsiloxanes) with liquid crystal sidechains P-307

$$Me_3Si-\left[O-\underset{Me}{\underset{|}{Si}}-\overset{H}{\underset{|}{}}\right]_x\left[O-\underset{Me}{\underset{|}{Si}}-\overset{CH_2R}{\underset{|}{}}\right]_y\Bigg]_n O-SiMe_3$$

$$R = -CH_2(CH_2)_6CH_2CH_2O-\!\!\left\langle\;\;\right\rangle\!\!-Ph$$

Synonyms: Sidechain liquid crystalline polymethylhydrogensiloxane copolymers
Related Polymers: Copolymers with PDMS (sidechain polysiloxane polydimethylsiloxane copolymer), Poly(methylhydrogensiloxane)
Material class: Polymer liquid crystals, Copolymers
Polymer Type: silicones
Molecular Formula: $(CH_4OSi)(C_{24}H_{36}OSi)_n$
Fragments: CH_4OSi $C_{24}H_{36}OSi$
General Information: Data refer principally to polymers substituted with a long-chain alkyl group and two aromatic groups, unless specified as referring to the copolymer
Morphology: Shows liq. crystal phase below isotropisation temp. [5]

Volumetric & Calorimetric Properties:
Melting Temperature:

No.	Value	Note
1	59°C	alkyl ether side chain

Glass Transition Temperature:

No.	Value	Note
1	-120 – 10°C	alkyl ether side chain [4]
2	90°C	ester side chain copolymer [2]

| 3 | 28°C | ester side chain | |
| 4 | 19.5°C | ester side chain copolymer [2] | |

Transition Temperature:

No.	Value	Note	Type
1	221°C	alkyl ether side chain [4]	Isotropisation temp.
2	45°C	ester side chain copolymer	Nematic-isotropic temp.
3	89°C	ester side chain	Nematic-isotropic temp.
4	68°C	ester side chain copolymer [2]	Nematic-isotropic temp.

Surface Properties & Solubility:
Solvents/Non-solvents: Sol. aromatic hydrocarbons, THF, chlorinated hydrocarbons. Insol. alcohols, H_2O [1,2,3]

Transport Properties:
Transport Properties General: Gas permeability rises by a factor of about 50 as the polymer changes from glass state to isotropic state via a liq. cryst. phase. The rise in permeability is particularly marked as the polymer changes from glass state to liq. cryst. phase [3]
Permeability of Gases: Permeability of gases $[O_2]$ 2.9×10^{-5}; $[CO_2]$ 0.00016; $[N_2]$ 9.5×10^{-5} cm^2 $(s\ GPa)^{-1}$ [3]
Permeability of Liquids: For liq. cryst. membrane below T_g membrane has low permeability to both H_2O and EtOH but is preferentially permeable to H_2O by a factor of about 10 because the rigid, disordered polymer has small volume. The membrane at temps. above T_g but below T_{NI} has higher permeability and is preferentially permeable to H_2O by a factor which decreases with increasing temp. At the isotropic temp. H_2O and EtOH have equal permeabilities. Above T_{NI} the flexible disordered polymer membrane is highly permeable and is preferentially permeable to EtOH by a factor of about 1.5 because of its higher solubility in a hydrophobic material. Between T_g and T_{NI} the relative permeability of the liq. cryst. membrane depends upon a balance between the effects of orientation of the mesogenic groups and the flexibility of the siloxane chain [2]. Permeability of liqs. has been reported [2]
Permeability and Diffusion General: Permeability of water 3×10^{-9} (10°); 4×10^{-8} (30°); 7×10^{-7} $kg\ (m\ h)^{-1}$ (80°). EtOH 2×10^{-11} (10°); 2×10^{-9} (30°); 1.2×10^{-7} $kg\ (m\ h)^{-1}$ (80°)

Polymer Stability:
Polymer Stability General: For high stability all SiH groups must be replaced, see Poly(dimethylsiloxanes) with liquid crystal sidechains, P-193
Thermal Stability General: Usable primarily at room temp.
Chemical Stability: If SiH groups are present then they can be oxidised quite readily causing crosslinking with elimination of hydrogen [6]

Applications/Commercial Products:
Processing & Manufacturing Routes: Side-chain substituted alkene and polymethylhydrogensiloxane are mixed together in dry toluene under nitrogen with platinum catalyst at 100° for about 48h. With excess alkene all SiH bonds may be substituted else 'copolymers' are formed [1,2,3,7]

Bibliographic References
[1] Kozlovsky, M.V., Bustamante, E.A.S. and Haase, W., *Liq. Cryst.*, 1996, **20**, 35
[2] Inui, K., Miyata, T. and Uragami, T., *Angew. Makromol. Chem.*, 1996, **240**, 241
[3] Chen, O.S. and Hsiue, G.H., *Makromol. Chem.*, 1991, **192**, 2021
[4] Castelvetro, V. and Ciardelli, F., *Polym. Int.*, 1996, **39**, 37
[5] Finkelmann, H. and Rehage, G., Makromol. Chem., *Rapid Commun.*, 1980, **1**, 31
[6] Noll, W., *Chemistry and Technology of Silicones*, Academic Press, 1968
[7] Clarson, S.J. and Semlyen, J.A., *Siloxane Polymers*, Prentice Hall, 1993
[8] Kazlovsky, M.V. and Haase, W., *Acta Polym.*, 1996, **47**, 361

Poly(hydroxyalkanoate) P-308
Synonyms: *PHA*
Related Polymers: Poly(3-hydroxybutyrate-*co*-3-hydroxyvalerate), Poly(3-hydroxybutyrate)
Material class: Thermoplastic
Polymer Type: polyester
General Information: A group of biopolymers, the most commonly found example being poly(3-hydroxybutyrate). They are microbially derived materials, occurring as granules within the cytoplasm of cells. The physical properties of this material may be improved by increasing MW and by incorporating other hydroxyalkanoate units, e.g. 3-hydroxyvalerate. Mechanical properties can be modified to resemble elastic rubber or crystalline plastic.
Morphology: Data on variation of crystallinity with level and type of second PHA component in poly(3-hydroxybutyrate) copolymers is reported [2].

Volumetric & Calorimetric Properties:
Density:

No.	Value	Note
1	d 1.24	poly(hydroxybutyrate) [3]
2	d 1.2	poly(hydroxyvalerate) [3]
3	d 1.02	poly(hydroxyoctanoate) [3]
4	d 1.03	poly(hydroxynonanoate) [3]
5	d 1.03	poly(hydroxydecanoate) [3]

Thermodynamic Properties General: Data on variation of T_m and T_g with level and type of second PHA component in poly(3-hydroxybutyrate) copolymers is reported [2].
Melting Temperature:

No.	Value	Note
1	180°C	poly(hydroxybutyrate) [3]
2	105–108°C	poly(hydroxyvalerate) [3]
3	61°C	poly(hydroxyoctanoate) [3]
4	54°C	poly(hydroxynonanoate) [3]
5	54°C	poly(hydroxydecanoate) [3]

Glass Transition Temperature:

No.	Value	Note
1	5°C	poly(hydroxybutyrate) [3]
2	-11°C	poly(hydroxyvalerate) [3]
3	-36°C	poly(hydroxyoctanoate) [3]
4	-39°C	poly(hydroxynonanoate) [3]
5	-40°C	poly(hydroxydecanoate) [3]

Mechanical Properties:
Mechanical Properties General: Data on variation of some mechanical properties with level and type of second PHA component in poly(3-hydroxybutyrate) copolymers have been reported [2,4].

Polymer Stability:
Decomposition Details: PHAs biodegrade in microbially active environments. In aerobic environments final products are water and carbon dioxide, in anaerobic environments methane is also produced.
Biological Stability: PHAs are biodegradable in natural environments. Decomposition occurs via enzymes secreted by microorganisms. The effects of structure and solid state on biodegradability have been reported [1].

Stability Miscellaneous: Data on deformation and morphology have been reported [4].

Applications/Commercial Products:
Processing & Manufacturing Routes: Production is performed via fermentation, and processing by conventional moulding and extrusion techniques is possible. The biosynthesis can be tailored to give polymers with specific properties and recent work has been directed towards the modification of the enzymes involved in PHA synthesis by genetic engineering [5]. Data on deformation and morphology are reported [4]. Recent work on PHA metabolic engineering has been reviewed [6].

Applications: PHAs have many potential uses: in medicine (controlled drug release, sutures, bone plates, wound care), personal hygiene products and packaging. Initial pricing was high compared to non-biopolymers but falling costs have meant increasing usage.

Trade name	Supplier
Biomer	Biomer
Nodax	Procter and Gamble
	Metabolix Inc.

Bibliographic References
[1] Doi, Y. and Abe, H., *Biopolymers: Polyesters II* Wiley-VCH, (Eds. Doi, Y. and Steinbuchel, A.), 2002, **3b**, 116
[2] Doi, Y. and Abe, H., *Biopolymers: Polyesters II* Wiley-VCH, (Eds. Doi, Y. and Steinbuchel, A.), 2002, **3b**, 109
[3] Marchessault, R.H. and Yu, G., *Biopolymers: Polyesters II* Wiley-VCH, (Eds. Doi, Y. and Steinbuchel, A.), 2002, **3b**, 176
[4] Satkowski, M.M., Melik, D.H., Autran, J.-P., Green, P.R., Noda, I. and Schechtman, L.A., *Biopolymers: Polyesters II* Wiley-VCH, (Eds. Doi, Y. and Steinbuchel, A.), 2002, **3b**, 231
[5] Poillon, F., *Biopolymers: Making Materials Nature's Way*, 1993, 29
[6] Aldor, I.S. and Keaslingy, J.D., *Curr. Opin. Biotechnol.*, 2003, **14**, 475

Poly(hydroxyalkyl methacrylate)   P-309

Synonyms: Poly(2-hydroxyethyl methacrylate). Poly(2-hydroxypropyl methacrylate)
Related Polymers: Polymethacrylates General, Hydroxyalkyl methacrylate copolymers
Monomers: 2-Hydroxyethyl methacrylate, 2-Hydroxypropyl methacrylate
Material class: Thermoplastic
Polymer Type: acrylic
CAS Number:

CAS Reg. No.	Note
25249-16-5	2-hydroxyethyl
70614-47-0	2-hydroxyethyl, isotactic
70559-14-7	2-hydroxyethyl, syndiotactic
25703-79-1	2-hydroxypropyl

Molecular Formula: $(C_6H_{10}O_3)_n$ $(C_7H_{12}O_3)_n$
Fragments: $C_{10}H_{10}O_3$ $C_7H_{12}O_3$
General Information: Data refer to poly(2-hydroxyethyl methacrylate), unless stated otherwise
Tacticity: Isotactic and syndiotactic forms known
Miscellaneous: PHEMA when cross-linked (e.g. with ethylene diacrylate) forms pliable, soft hydrogels with a low network density and poor mechanical strength. The latter may be improved by incorporation of hydroxypropyl methacrylate into the network. The hydrophilicity of methacrylic hydrogels may be improved by the incorporation of a second hydroxyl group e.g. using 2,3-dihydroxypropyl methacrylate as a comonomer

Volumetric & Calorimetric Properties:
Density:

No.	Value	Note
1	d 1.274 g/cm^3	[1]

Thermal Expansion Coefficient:

No.	Value	Note	Type
1	0.00026 K^{-1}	free-radical bulk polymerisation at 60°, T > T$_g$ [2]	V
2	0.000102 K^{-1}	free-radical bulk polymerisation at 60°, T < T$_g$ [2]	V

Equation of State: Equation of state information has been reported [16]
Glass Transition Temperature:

No.	Value	Note
1	55–100°C	[4,5,6]
2	25°C	isotactic [4]
3	109°C	syndiotactic [4]

Transition Temperature General: The presence of water or other plasticising additives, e.g. polyethylene glycol, lowers the T$_g$ [3,4]
Transition Temperature:

No.	Value	Note	Type
1	50°C	[7]	T$_\delta$

Surface Properties & Solubility:
Cohesive Energy Density Solubility Parameters: Cohesive energy density 39.1 J cm^{-3} [8], 39.5 J cm^{-3} (hydroxypropyl) [8]
Solvents/Non-solvents: Isotactic sol. H$_2$O (when not cross-linked). Syndiotactic, atactic insol. H$_2$O. Syndiotactic and atactic forms swollen by H$_2$O to form gels [1]
Wettability/Surface Energy and Interfacial Tension: Surface tension 37 mN m^{-1} (20°, dry) [9], 69 mN m^{-1} (40% wt. H$_2$O); γ_d 20.2; γ_p 48.8 [10]; 51.6 mN m^{-1}; γ_d 31.4; γ_p 20.2 [8]; 75.59 mN m^{-1} (10°, isotactic); γ_d 25.28; γ_p 50.31 [11]; 74.54 mN m^{-1} (10°, syndiotactic); γ_d 24.52; γ_p 50.02 [11]; 71.39 mN m^{-1} (20°, isotactic; γ_d 23.10; γ_p 48.29 [11]; 71.58 mN m^{-1} (20°, syndiotactic); γ_d 23.57; γ_p 48.01 [11]; 68.97 mN m^{-1} (40°, isotactic); γ_d 22.06; γ_p 46.91 [11]; 71.08 mN m^{-1} (40°, syndiotactic); γ_d 24.64; γ_p 46.44 [11]; 62.88 mN m^{-1} (60°, isotactic); γ_d 18.13; g$_p$ 44.75 [11]; 66.83 mN m^{-1} (60°, syndiotactic); γ_d 24.92; γ_p 43.91 [11]; 49.7 mN m^{-1} (hydroxypropyl); γ_d 33.1; γ_p 16.6 [8]. Contact angle 15° (air, 40% H$_2$O) [10], 16° (Me$_2$CO, 40% H$_2$O) [10], 20° (dodecane, 40% H$_2$O) [10]. Hamilton contact angle 148 [12]. Surface energy 51.6 mN m^{-2}; γ_p 20.2; γ_d 31.4 [12]. Interfacial tension 0.06–0.09 mN m^{-1} (solid/H$_2$O, 40% H$_2$O) [10]

Transport Properties:
Polymer Solutions Dilute: η_{inh} 1.48 dl g^{-1} (2-methoxyethanol, 25°, MW 34000), 3.26 dl g^{-1} (2-methoxyethanol, 25°, MW 529000) [13], 1.935 dl g^{-1} (DMF, isotactic, MW 816000), 0.356 dl g^{-1} (H$_2$O, isotactic, MW 816000), 1.011 dl g^{-1} (MeOH, isotactic, MW 816000), 0.652 dl g^{-1} (EtOH, isotactic, MW 816000), 0.52 dl g^{-1} (propanol, isotactic, MW 816000), 1.31 dl g^{-1} (*tert*-butanol, isotactic, MW 816000) [14]. Viscosity increases with increasing MW [13] and with electrolyte [15]. Theta temps. 15.3° (H$_2$O), 15.8° (EtOH), 32.1° (propanol), 14° (2-propanol), 3.7° (2-butanol) [14]. The polymer-solvent interaction parameter has been related to the level of cross-linking (in diethylene glycol at 25°) [17]

Permeability of Gases: Permeability coefficient [O_2] 29.5×10^{-6}–0.00295 mol (m s Pa)$^{-1}$ (10^{-11}–10^{-9} cm^2 (s cmHg)$^{-1}$) [19]; [CO_2] 29.5×10^{-6}–0.00295 mol (m s Pa)$^{-1}$ (10^{-11}–10^{-9} cm^2 (s cmHg)$^{-1}$) [18,19]

Water Content: 44% (10°, isotactic), 41.2% (20°, isotactic); 37.5% (10°, syndiotactic), 32.4% (20°, syndiotactic); 37.5% (40°, isotactic), 36.7% (40°, syndiotactic); 35.5% (60°, isotactic), 38.7% (60°, syndiotactic) [11]. Water content decreases with increasing level of cross-linking agent [1]

Water Absorption:

No.	Value	Note
1	36.4 %	MW 1500000 [20]
2	20–70 %	[19]
3	21.3 %	hydroxypropyl [20]

Mechanical Properties:
Tensile (Young's) Modulus:

No.	Value	Note
1	500–2600 MPa	dehydrated [8]
2	1.9–7.6 MPa	37.6% water [8]

Tensile Strength Break:

No.	Value	Note	Elongation
1	0.49–2.94 MPa	5–30 kg cm^{-2} [19]	100–300%

Miscellaneous Moduli:

No.	Value	Note	Type
1	200–1000 MPa	dehydrated [8]	rigidity modulus
2	0.5–2.4 MPa	37.6% water [8]	rigidity modulus

Optical Properties:
Refractive Index:

No.	Value	Note
1	1.5119	[21]

Polymer Stability:
Decomposition Details: Decomposition at 375–500° yields 20% monomer (at 500°), 6% ethylene dimethacrylate (at 500°) as major products. CO_2 and carbon monoxide are liberated [22]

Applications/Commercial Products:
Processing & Manufacturing Routes: Synth. by radical polymerisation with a cross-linking agent (e.g., ethylene dimethacrylate) present [19]

Applications: Applications increasing in medicine; particularly hydrogels. Uses include soft contact lenses, implant matrices, burn wound dressings and drug delivery systems

Trade name	Supplier
Hydron	National Patent Development Corp.
Softlens	Bausch and Lomb

Bibliographic References
[1] Perera, D.I. and Shanks, R.A., *Polym. Int.*, 1995, **36**, 303
[2] Krause, S., Gormley, J.J., Roman, N., Shetter, J.A. and Watenabe, W.H., *J. Polym. Sci., Part A: Polym. Chem.*, 1965, **3**, 3573
[3] Verhoeven, J., Schaeffer, R., Bouwstra, J.A. and Junginger, H.E., *Polymer*, 1989, **30**, 1946
[4] Sung, Y.K., Gregonis, D.E., Russell, D.E. and Andrade, J.D., *Polymer*, 1978, **19**, 1362
[5] Wiley, R.H. and Brauer, G.M., *J. Polym. Sci.*, 1948, **3**, 647
[6] Kyu, T. and Saldanha, J.M., *Macromolecules*, 1988, **21**, 1021
[7] Plazek, D.J. and Ngai, K.L., *Physical Properties of Polymers Handbook*, (ed. J.E. Mark), AIP Press, 1996, 139
[8] Barnes, A. Corkhill, P.H. and Tighe, B.J., *Polymer*, 1988, **29**, 2191, (surface props)
[9] Wu, S., *Polymer Interface and Adhesion*, Marcel Dekker, 1982
[10] King, R.N., Andrade, J.D., Ma, S.M., Gregonis, D.E. and Brostrom, L.R., *J. Colloid Interface Sci.*, 1985, **103**, 62, (interfacial tension)
[11] Yuk, S.H. and Jhon, M.S., *J. Colloid Interface Sci.*, 1987, **116**, 25
[12] Baker, D.A., Corkhill, P.H., Ng, C.O., Skelly, P.J. and Tighe, B.J., *Polymer*, 1988, **29**, 691
[13] Carenza, M., Palma, G., Lora, S. and Flira, F., *Eur. Polym. J.*, 1987, **23**, 741
[14] Oh, S.H. and Jhon, M.S., *J. Polym. Sci.*, 1989, **27**, 1731
[15] Kim, W.G. and Jhon, M.S., *J. Polym. Sci., Part A: Polym. Chem.*, 1988, **26**, 859
[16] Fort, R.J. and Polyzoidis, T.M., *Eur. Polym. J.*, 1976, **12**, 685, (eqn of state)
[17] Bahar, I., Erbit, H.Y., Baysal, B.M. and Erman, B., *Macromolecules*, 1987, **20**, 1353
[18] Yang, W.H., Smolen, V.F. and Peppas, N.A., *J. Membr. Sci.*, 1981, **9**, 53, (gas permeability)
[19] Liu, Y. and Yang, Y., *Fushe Yanjiu Yu Fushe Gongyi Xuebao*, 1988, **6**, 44
[20] Stevenson, W.T.K. and Sefton, M.V., *J. Appl. Polym. Sci.*, 1988, **36**, 1541, (water content)
[21] Lewis, O.G., *Physical Constants of Linear Hydrocarbons*, Springer-Verlag, 1968
[22] Razga, J. and Petranek, J., *Eur. Polym. J.*, 1975, **11**, 805, (thermal degradation)

Poly(*p*-hydroxybenzoate)   P-310

Synonyms: *Poly(poxybenzoyl)*. PHBA. *Poly(4-hydroxybenzoic acid)*. *Poly(oxy-1,4-phenylenecarbonyl)*. PDB. *Poly(4-hybe)*. *Poly(p-oxybenzoate)*

Monomers: 4-Hydroxybenzoic acid
CAS Number:

CAS Reg. No.	Note
26099-71-8	repeating unit
30729-36-3	Poly(*p*-hydroxybenzoate)

Molecular Weight: M_n 8000–12000 (Ekonol)

Morphology: Two polymorphs at room temp. Form I orthorhombic a 0.752, b 0.570, c 1.249, helix 2_i [6]; a 0.747, b 0.567, c 1.255, helix 2_i [15]; a 0.742, b 0.570, c 1.245, helix 2_i [23]. Form II orthorhombic a 0.377, b 1.106, c 1.289, helix 2_i [6]; a 0.383, b 1.116, c 1.256, helix 2_i [23]. Cryst. structs. have been reported [13,24,39]. Form III pseudo-hexagonal a 0.924, b 0.528, c 1.25 [15]. Morphological study has been reported [8]; orthorhombic a 0.92, b 0.53, c 1.24, helix 2_i [7]. Form IV hexagonal a 0.935, b 0.54, c 1.245 [24]. Forms I and II become form III at $T\alpha$ [24]

Volumetric & Calorimetric Properties:
Density:

No.	Value	Note
1	d 1.419 g/cm^3	[1]
2	d 1.43 g/cm^3	[9]
3	d 1.49 g/cm^3	Form I
4	d 1.48 g/cm^3	Form II [6]
5	d 1.32 g/cm^3	Form III [7]
6	d 1.44 g/cm^3	[33]

– Poly(p-hydroxybenzoate)

Thermal Expansion Coefficient:

No.	Value	Note	Type
1	$1.6 \times 10^{-5} - 2.8 \times 10^{-5}$ K^{-1}	[9]	L
2	0.000104 K^{-1}	0° [13]	L
3	0.00026 K^{-1}	300° [13]	L
4	5.6×10^{-5} K^{-1}	ASTM D696 [32]	L

Latent Heat Crystallization: ΔH_f 4.6 kJ mol^{-1} [1]. ΔH_f 3.8 kJ mol^{-1} [10]. ΔH_f 4.3-6.6 kJ mol^{-1} [5]. Other ΔH, ΔS and ΔG values have been reported [36,38]. ΔC_p [5,10] and C_p values [10,36] have been reported

Thermal Conductivity:

No.	Value	Note
1	0.75 W/mK	0.0018 cal (cm s °C)$^{-1}$ [9]
2	0.72 W/mK	ASTM C177 [32]

Specific Heat Capacity:

No.	Value	Note	Type
1	34 kJ/kg.C	[10]	p

Melting Temperature:

No.	Value	Note
1	510–530°C	[17]
2	510°C	783K [22]

Glass Transition Temperature:

No.	Value	Note
1	180°C	[1]
2	158–165°C	431-438 K [5]
3	120°C	[20]
4	161°C	434 K [36]

Transition Temperature General: Other transition temps. have been reported [29,36]. Softening temp. has been reported [29]

Deflection Temperature:

No.	Value	Note
1	>550°C	min. [31]

Transition Temperature:

No.	Value	Note	Type
1	325–360°C	cryst. form I/II to form III transition [1]	Tα
2	350°C	[34]	Tα
3	345°C	618 K [5]	Tα
4	343.5°C	616.5 K [10]	Tα
5	340°C	[16,18]	Tα
6	347°C	620 K [22]	Tα
7	330°C	[38]	Tα
8	510°C		Tγ
9	528–532°C	[17]	Tδ
10	430°C	cryst. form III to form IV transition [15,18,39]	Tβ
11	445°C	[16]	Tβ
12	500°C	773 K [22]	Tβ

Surface Properties & Solubility:
Solvents/Non-solvents: Insol. all common solvents [4]

Transport Properties:
Water Content: Other water absorption values have been reported [26]

Water Absorption:

No.	Value	Note
1	0.02 %	ASTM D570 [32]
2	0.02 %	24h, room temp. [33]
3	0.4 %	100h, 100° [33]

Mechanical Properties:
Mechanical Properties General: Mech. props. have been reported [37]. Other flexural moduli values have been reported [9]

Tensile (Young's) Modulus:

No.	Value	Note
1	157 MPa	[19]
2	4000 MPa	ASTM D638 [32]

Flexural Modulus:

No.	Value	Note
1	6900 MPa	1000 kpsi [9]
2	7100 MPa	[31]
3	7000 MPa	20°, ASTM D790 [32]

Tensile Strength Break:

No.	Value	Note	Elongation
1	18 MPa	20°, ASTM D638 [32]	6.5%

Flexural Strength at Break:

No.	Value	Note
1	73.8 MPa	10700 psi [9]
2	74 MPa	[31]
3	35–40 MPa	ASTM D790 [32]

Elastic Modulus:

No.	Value	Note
1	7100 MPa	1030 kpsi [3]

Compressive Strength:

No.	Value	Note
1	110 MPa	ASTM D695 [32]
2	265 MPa	38500 psi [33]

– Poly(*p*-hydroxybenzoate)

Viscoelastic Behaviour: Stress-strain curves have been reported [35]
Friction Abrasion and Resistance: Coefficient of friction 0.10–0.16 [9]. Other coefficient of friction values have been reported [27]. Specific wear rate 0.12 mm^3 kg^{-1} km^{-1} [27]

Electrical Properties:
Electrical Properties General: Other dielectric constants have been reported [15,18,32]
Dielectric Permittivity (Constant):

No.	Value	Frequency	Note
1	3.8		[33]
2	2.958	100 Hz	25°
3	2.982	100 Hz	100° [1]

Dielectric Strength:

No.	Value	Note
1	0.66 kV.mm^{-1}	0.125" thick [33]

Strong Field Phenomena General: Breakdown temp. 360° (110V, 60 Hz) [25]
Dissipation (Power) Factor:

No.	Value	Note
1	0.000198	[33]

Optical Properties:
Transmission and Spectra: Ir [2,17,19,21,22], C-13 nmr [3,30], H-1 nmr [4,36] and mass spectral data [12,22] have been reported

Polymer Stability:
Upper Use Temperature:

No.	Value	Note
1	315°C	[32]
2	270–300°C	[32]

Decomposition Details: Thermal degradation has been studied [2,10,12,22,29]. Max. decomposition temp. 542° (815K) [22]. 1% Weight loss at 425° (698K) [10], 260° (2000h) [33]. 5% Weight loss at 520° (793K) [10]. 10% Weight loss at 545° (818K) [10], 482° (in air), [12] 454° (in N$_2$) [12]
Flammability: Flammability rating V0 (UL94) [32]. Limiting oxygen index 36% [34]
Chemical Stability: Attacked by trifluoromethanesulfonic acid [14]. Resistant to most aromatic and chlorinated hydrocarbons [32], dilute acids [33], dilute alkalis [33], conc. phosphoric acid [33] and trifluoroacetic acid [33]. Attacked by hot, conc. sulfuric acid or sodium hydroxide soln. [33]
Biological Stability: Biodegradation has been reported [28]
Stability Miscellaneous: Resistant to radiation (100 Grad) [32]

Applications/Commercial Products:
Processing & Manufacturing Routes: Prod. by ionic polymerisation of the phenyl ester of *p*-hydroxybenzoic acid or by polymerisation of *p*-acetoxybenzoic acid at 250° [1,11,21,23]
Mould Shrinkage (%):

No.	Value	Note
1	0.7%	ASTM D955 [32]

Applications: Metal coatings, plasma coatings, liquid crystals, antifriction materials. Used in plasma-sprayed coatings in bearings, abradable seals, rotors or vanes in processing pumps

Trade name	Supplier
Ekonol	Carborundum Corporation
	Sumitomo Chem Co

Bibliographic References
[1] Economy, J., Storm, R.S., Matkovic, V.I., Cottis, S.G. and Nowak, B.E., *J. Polym. Sci., Polym. Chem. Ed.*, 1976, **14**, 2207, (synth, dielectric constant)
[2] Jellinek, H.H.G. and Fujwara, H., *J. Polym. Sci., Part A-1*, 1972, **10**, 1719, (ir, thermal degradation)
[3] Fyfe, C.A., Lyerla, J.A., Volksen, W. and Yannoni, C.S., *Macromolecules*, 1979, **12**, 757, (C-13 nmr)
[4] Kricheldorf, H.R. and Schwarz, G., *Makromol. Chem.*, 1983, **184**, 475, (H-1 nmr, solubility)
[5] Meesin, W., Menczel, J., Gaur, U. and Wunderlich, B., *J. Polym. Sci., Polym. Phys. Ed.*, 1982, **20**, 719, (heat capacity, transition temps)
[6] Lieser, G., *J. Polym. Sci., Polym. Phys. Ed.*, 1983, **21**, 1611, (cryst struct)
[7] Blackwell, J., Lieser, G. and Gutierrez, G.A., *Macromolecules*, 1983, **16**, 1418, (cryst struct)
[8] Lieser, G., Schwarz, G. and Kricheldorf, H.R., *J. Polym. Sci., Polym. Phys. Ed.*, 1983, **21**, 1599, (morphology)
[9] Economy, J., *J. Macromol. Sci., Chem.*, 1984, **21**, 1705, (mech props, thermal conductivity)
[10] Cao, M.Y. and Wunderlich, B., *J. Polym. Sci., Polym. Phys. Ed.*, 1985, **23**, 521, (heat capacity, thermal degradation)
[11] Kricheldorf, H.R. and Schwarz, G., *Polymer*, 1984, **25**, 520, (synth)
[12] Crossland, B., Knight, G.J. and Wright, W.W., *Br. Polym. J.*, 1986, **18**, 371, (ms, thermal degradation)
[13] Hanna, S. and Windle, A.H., *Polym. Commun.*, 1988, **29**, 236, (expansion coefficient)
[14] Johnson, R.D., Niessner, N., Muehlebach, A., Economy, J. and Lyerla, J., *Polym. Commun.*, 1990, **31**, 383, (chemical stability)
[15] Yoon, D.Y., Masciocchi, N., Depero, L.E., Viney, C. and Parrish, W., *Macromolecules*, 1990, **23**, 1793, (cryst struct, dielectric constant)
[16] Economy, J., Volksen, W., Viney, C. and Geiss, R. *et al*, *Macromolecules*, 1988, **21**, 2777, (transition temps)
[17] Kricheldorf, H.R. and Schwarz, G., *Polymer*, 1990, **31**, 481, (ir, melting point, transition temps)
[18] Kalika, D.S. and Yoon, D.Y., *Macromolecules*, 1991, **24**, 3404, (transition temps, dielectric constant)
[19] Tashiro, K. and Kobayashi, M., *Polymer*, 1991, **32**, 454, (ir, Young's modulus)
[20] Chen, D. and Zachmann, H.G., *Polymer*, 1991, **32**, 1612, (transition temp)
[21] Mathew, J., Bahulekar, R.V., Ghadage, R.S. and Rajan, C.R. *et al*, *Macromolecules*, 1992, **25**, 7338, (synth, ir)
[22] Hummel, D.O., Neuhoff, U., Bretz, A. and Dussel, H.J., *Makromol. Chem.*, 1993, **194**, 1545, (ir, thermal degradation)
[23] Iannelli, P. and Yoon, D.Y., *J. Polym. Sci., Part B: Polym. Phys.*, 1995, **33**, 977, (synth, cryst struct)
[24] Lakasheva, N.V., Mosell, T., Sariban, A. and Brickmann, J., *Macromolecules*, 1996, **29**, 1286, (cryst struct)
[25] *Ger. Pat.*, 1970, 2 025 516, (strong field phenomenon)
[26] Kitamura, A. and Shibamoto, A., *Nippon Sanshigaku Zasshi*, 1975, **44**, 307, (water absorption)
[27] Sekiguchi, I., Yamaguchi, Y., Souma, I. and Sakamoto, S. *et al*, *CA*, 1980, **94**, 104987, (friction coefficients, wear rate)
[28] Tokiwa, Y., Ando, T., Suzuki, T. and Takeda, K., *ACS Symp. Ser.*, 1990, **433**, 136, (biodegradation)
[29] Karis, T., Siemens, R., Volksen, W. and Economy, J., *Mol. Cryst. Liq. Cryst.*, 1988, **157**, 567, (softening temp, thermal degradation)
[30] Lyerla, J.A., Economy, J., Maresch, G.G., Muehlebach, A. *et al*, *ACS Symp. Ser.*, 1990, **435**, 359, (H-1 nmr, C-13 nmr)
[31] Dyson, R.W., *Speciality Polymers*, Blackie, 1987, 56, (heat distortion temp, mech props)
[33] *Encycl. Polym. Sci. Technol.*, (eds. H.F. Mark, N.G. Gaylord and N.M. Bikales), Interscience, 1964, **15**, 292, (volume resistivity, dissipation factor, MW)
[34] *Ullmanns Encykl. Ind. Chem.*, (ed. B. Elvers), VCH, 1993, **A10**, 464, (flammability)
[35] Sauer, J.A., Pae, K.D. and Bhateja, S.K., *J. Macromol. Sci., Phys.*, 1973, **8**, 649, (viscoelastic props)
[36] Cao, M.Y., Varma-Nair, M. and Wunderlich, B., *Polym. Adv. Technol.*, 1990, **1**, 151, (thermodynamic props, transition temps)
[37] Economy, J., *Mol. Cryst. Liq. Cryst.*, 1989, **169**, 1, (mech props)
[38] Hsiue, L.T., Ma, C.C.M. and Tsai, H.B., *J. Appl. Polym. Sci.*, 1995, **56**, 471, (transition temp)
[39] Liu, J., Rybnikar, F. and Geil, P.H., *J. Macromol. Sci., Phys.*, 1993, **32**, 395, (struct, transition temps)

Poly(4-hydroxybenzoate-co-6-hydroxy-2-naphthalenecarboxylate) P-311

Synonyms: *Poly(4-hydroxybenzoate-co-2-oxy-6-naphthoyl). HBA/HNA*
Monomers: 4-Hydroxybenzoic acid, 6-Hydroxy-2-naphthalenedicarboxylic acid
Material class: Copolymers
Polymer Type: unsaturated polyester
CAS Number:

CAS Reg. No.	Note
81843-52-9	
106004-41-5	block copolymer

Molecular Formula: $[(C_7H_4O_2).(C_{11}H_6O_2)]_n$
Fragments: $C_7H_4O_2$ $C_{11}H_6O_2$
Molecular Weight: 25000-45000
Additives: Wollastonite and glass fibre
Morphology: Electron microscopy studies have been reported [1]. Polymer containing 75% HBA has two polymorphic forms: orthorhombic a 0.76, b 0.57 and orthorhombic (pseudohexagonal) a 0.92, b 0.52, c 1.26 [2]. Polymer containing 70% HBA orthorhombic a 0.796, b 0.549, c 1.354 [35]. Cryst. struct. has been reported [27,36]. Morphological studies have been reported [19,25]. Hexagonal crystal form has been reported. [3] The polymer chains self-align in the direction of processing to create a self-reinforcing network

Volumetric & Calorimetric Properties:
Density:

No.	Value	Note
1	d 1.393–1.395 g/cm^3	58% HBA [9]
2	d 1.4 g/cm^3	73% HBA [12,23,41]
3	d 1.43 g/cm^3	70% HBA [35]
4	d 1.14 g/cm^3	fibre [43]

Thermal Expansion Coefficient:

No.	Value	Note	Type
1	0.00023 K^{-1}	20-95°, 58% HBA	V
2	0.00043 K^{-1}	95-205°, 58% HBA [9]	V
3	0.0008 K^{-1}	>205°, 58% HBA [9]	V
4	7×10^{-6} K^{-1}	-173° [7]	V
5	2.7×10^{-5} K^{-1}	73% HBA, flow [42]	V
6	2.7×10^{-5}–5.4×10^{-5} K^{-1}	73% HBA, transverse [42]	V
7	1×10^{-5} K^{-1}	[44]	L

Thermodynamic Properties General: Other thermodynamic props. have been reported [10]. Variation of thermal conductivity with temp. has been reported [7]
Latent Heat Crystallization: ΔH_f 1.82 kJ mol^{-1}, ΔS_f 3.19 J mol^{-1} K^{-1} (58% HBA) [9]. Other ΔH_f values (fibres) have been reported [16]. ΔH_f 1.57 J g^{-1} (73% HBA) [23]. Other values of C_p, ΔH, ΔS and ΔG have been reported [8,37]. ΔC_p 32 J mol^{-1} K^{-1} (75% HBA) [8]. Values of C_v over temp. range of -271°–-173° have been reported [7]
Melting Temperature:

No.	Value	Note
1	285°C	73% HBA [15]
2	279°C	[16]
3	291°C	[16]
4	280°C	film [21]
5	278°C	73% HBA [23]
6	295°C	[33]
7	330°C	fibre [43]

Glass Transition Temperature:

No.	Value	Note
1	110°C	[5]
2	152°C	75% HBA [8]
3	100°C	58% HBA [9]
4	110°C	film [21]
5	103°C	73% HBA [23]
6	160°C	dynamic mech. anal.

Transition Temperature General: Other transition temps. have been reported. [7,36] Forms a nematic thermotropic phase above 280°
Deflection Temperature:

No.	Value	Note
1	180°C	1.82 MPa, 73% HBA [41]
2	240–250°C	1.82 MPa [45]
3	222°C	0.46 MPa, A950 grade, ASTM D648

Transition Temperature:

No.	Value	Note	Type
1	50°C	[5]	β temp.
2	75°C	75% HBA [6]	β temp.
3	29°C	73% HBA [23]	β temp.
4	-40°C	[5]	γ temp.
5	-25°C	75% HBA [6]	γ temp.
6	280°C	75% HBA [6]	High temp. transition
7	278°C	73% HBA [14]	High temp. transition
8	282°C	75% HBA [8]	Disordering transition temp.

Surface Properties & Solubility:
Solvents/Non-solvents: Sol. pentafluorophenol, 3,5-bis(trifluoromethyl)phenol [28], 4-chlorophenol, pentafluorophenol, trifluoroacetic acid at high temp. Insoluble in and inert to many solvents owing to high crystallinity and polyester struct.

Transport Properties:
Transport Properties General: Rheological props. have been reported [4,32]
Polymer Solutions Dilute: $[\eta]_{inh}$ 5.37 dl g^{-1} (pentafluorophenol/4-chlorophenol 60/40) [25]. $[\eta]$ 5.3 dl g^{-1} (pentafluorophenol, 60°) [33]. $[\eta]_{inh}$ 7.57 dl g^{-1} (73% HBA, film), 5.7 dl g^{-1} (73% HBA, fibre) [42]
Polymer Melts: Melt viscosity has been reported in graphical form. [24] Melt viscosity 600 P (300°, 1000 s^{-1}) [46]
Permeability of Liquids: Water diffusion coefficient has been reported [39]
Water Absorption:

No.	Value	Note
1	0.02–0.03 %	24h, ASTM D570 [43,44]

Gas Permeability:

No.	Gas	Value	Note
1	O_2	0.03 cm³ mm/(m² day atm)	47×10^{-15} cm² (s cmHg)$^{-1}$, 35°, 73% HBA [23]
2	N_2	0.197 cm³ mm/(m² day atm)	3×10^{-13} cm² (s cmHg)$^{-1}$, 35°, 73% HBA [23]
3	CO_2	0.046 cm³ mm/(m² day atm)	70×10^{-15} cm² (s cmHg)$^{-1}$, 35°, 73% HBA [23]
4	He	10.8 cm³ mm/(m² day atm)	16400×10^{-15} cm² (s cmHg)$^{-1}$, 35°, 73% HBA [23]
5	H_2	3.01 cm³ mm/(m² day atm)	4590×10^{-15} cm² (s cmHg)$^{-1}$, 35°, 73% HBA [23]
6	Ar	0.007 cm³ mm/(m² day atm)	9.9×10^{-15} cm² (s cmHg)$^{-1}$, 35°, 73% HBA [23]

Mechanical Properties:

Mechanical Properties General: General mech. props. have been reported in graphical form [4]. Other mech. props. have been reported [36]. Self-alignment of the polymer chains under shear in the melt creates exceptionally good mech. props., with stiffness and notched impact strength resistance much greater than those of conventional glass-fibre reinforced engineering resins. It has a fibrous nature but its props. are very anisotropic. May be glass reinforced to obtain adequate strength in the transverse direction. Self-reinforcement nature and high orientation are achieved during draw-down of fibres giving extremely high mech. stiffness and strength. Sonic modulus as a function of draw-down ratio has been reported in graphical form [14]. Variation of dynamic tensile modulus [13,25,26,31] and shear modulus [31] with temp. has been reported in graphical form. Elastic moduli have been reported. [20] Tensile modulus decreases with increaing sample thickness [11]. Elongation at break 1.8% (A950 grade, ASTM D882). [5] Tensile storage and loss moduli decrease with increasing temp. [47,53,54]

Tensile (Young's) Modulus:

No.	Value	Note
1	72900 MPa	fibre [43]
2	60000 MPa	fibres [16]
3	100000 MPa	-80° [36]
4	25000 MPa	130° [36]
5	9600–38000 MPa	73% HBA [42]
8	9700–10500 MPa	23°, direction of orientation, ASTM D638 [44,45,55]
9	11100 MPa	0°, direction of orientation, ASTM D638 [44,45,55]
10	14600 MPa	-140°, direction of orientation, ASTM D638 [44,45,55]
11	19300 MPa	Vectra B950

Flexural Modulus:

No.	Value	Note
1	8970 MPa	73% HBA [41]
2	9000 MPa	23°, A950 grade, ASTM D790 [44]
3	15200 MPa	23°, B950 grade, ASTM D790 [44]

Tensile Strength Break:

No.	Value	Note	Elongation
1	1100 MPa	fibres [16]	2%
2	3200 MPa	fibre [43]	3.8%
3	207 MPa	30% glass fibre [42]	
4	138–241 MPa	73% HBA [42]	
5	160–165 MPa	23°, A950 grade, ASTM D638 [45]	
6	180 MPa	0°, A950 grade, ASTM D638 [45]	
7	165 MPa	-140°, A950 grade, ASTM D638 [45]	
8	165 MPa	-200°, A950 grade, ASTM D638 [45]	
9	188 MPa	B950 grade	1.3%

Flexural Strength at Break:

No.	Value	Note
1	151–296 MPa	73% HBA [42]
2	169 MPa	23°, A950 grade, ASTM D790 [43]
3	118 MPa	50°, A950 grade
4	55 MPa	100°, A950 grade
5	31 MPa	150°, A950 grade
6	245 MPa	23°, B950 grade, ASTM D790 [43]

Compressive Modulus:

No.	Value	Note
1	6300 MPa	A950 grade, ASTM D695
2	2620 MPa	B950 grade, ASTM D695

Flexural Strength Yield:

No.	Value	Note
1	140 MPa	3% elongation, axial stress [55]

Compressive Strength:

No.	Value	Note
1	70 MPa	A950 grade, ASTM D695 [44]
2	84 MPa	B950 grade, ASTM D695 [44]

Miscellaneous Moduli:

No.	Value	Note
1	5000 MPa	bulk modulus [7]
2	200–1600 MPa	-80°–-130°, shear modulus [37]

Viscoelastic Behaviour: Creep props. (73% HBA) have been reported. [26] Flexural creep 8200 MPa (10h, 23°, 50 MPa stress), 6100 MPa (1000h, 23°, 50 MPa stress) [48]

Hardness: Rockwell M60 (A950 grade, ASTM D785), M100 (B950 grade, ASTM D785)

Failure Properties General: Shear strength 105 MPa (A950 grade, ASTM D732), 129 MPa (B950 grade, ASTM D732) [43]
Fracture Mechanical Properties: Fracture toughness 3 NM m$^{-3/2}$ [26]
Friction Abrasion and Resistance: Coefficient of friction 0.12 (dynamic, metal), 0.14 (static, metal) [43]. Tabor abrasion 56 mg (1000 cycles)$^{-1}$ (ASTM D1044) [43]
Izod Notch:

No.	Value	Notch	Note
1	690 J/m	Y	23°, A950 grade [45]
2	505 J/m	Y	-200°, A950 grade [45]
3	520 J/m	N	A950 grade [44]
4	415 J/m	N	B950 grade [44]
5	50 J/m	N	[55]
6	520 J/m	Y	ASTM D256 [26]
7	534 J/m	Y	73% HBA [41]
8	53–534 J/m	Y	73% HBA [42]

Electrical Properties:
Electrical Properties General: Variation of tan δ with temp. has been reported in graphical form [5,12,13,25,26]. Variation of dielectric constant with temp. has been reported [5,6,12]. Variation of dielectric constant with temp. has been reported in graphical form [15]. Has excellent dielectric strength. Electrical props. may be enhanced using fillers
Dielectric Permittivity (Constant):

No.	Value	Frequency	Note
1	2.6	1 kHz	ASTM D150 [46]
2	3.3	1 kHz	ASTM D150 [43]

Dielectric Strength:

No.	Value	Note
1	39 kV.mm^{-1}	23°, 50% relative humidity, 1.5 mm thick, short term, A950 grade [43,46]
2	32 kV.mm^{-1}	23°, 50% relative humidity, 1.5 mm thick, short term, B950 grade [43,46]

Arc Resistance:

No.	Value	Note
1	137s	A950 grade, ASTM D495 [43,46]
2	74s	B950 grade, ASTM D495 [43,46]

Complex Permittivity and Electroactive Polymers: Comparative tracking index 175 V (A950 grade, ASTM D3636), 150 V (B950 grade, ASTM D3636) [43,46]
Static Electrification: Impedance behaviour has been reported [38]
Dissipation (Power) Factor:

No.	Value	Frequency	Note
1	0.004	1 kHz	ASTM D150
2	0.004	1 GHz	ASTM D150

Optical Properties:
Transmission and Spectra: H-1 nmr [3,18,22,40], C-13 nmr [11,12], ir [14,17,29], esr [30] and mass spectral [29] data have been reported
Total Internal Reflection: Birefringence has been reported [34]

Polymer Stability:
Polymer Stability General: Has excellent stability; chemical, mech. and electrical props. are retained. Has good dimensional stability to temp. and moisture
Thermal Stability General: Has excellent thermal stability
Upper Use Temperature:

No.	Value
1	325°C

Decomposition Details: Decomposition temp. 450° [8]; max. decomposition temp. 522° [29]
Flammability: Limiting oxygen index 28% (fibre) [43], 35% (A950 grade), 50% (B950 grade)[43]. Flammability rating V0 (0.8 mm thick, UL94) [43]. Smoke generation conforms to proposed FAA standards [49]
Environmental Stress: Has excellent weathering resistance even after 2000h exposure (ASTM D2565) [46]
Chemical Stability: Polymer containing 73% HBA is inert for 30 days to 80% formic acid (216°), 37% HCl (190°), 10% NaOH soln. (190°), Me$_2$CO (133°), CH$_2$Cl$_2$ (148°), EtOAc (131°), gasoline (250°), Skydrol (160°) [41]. Is resistant to strong acids and bases, alcohols, hydrocarbons, aromatics, esters and ketones. Is not stress-crazed by solvents even at elevated temps. [49]
Hydrolytic Stability: Stable after 200h exposure at 110° [46]
Recyclability: Vectra is claimed to be 100% recyclable. Up to 25% regrind can be incorporated into virgin feed for processing [49]
Stability Miscellaneous: Stable to γ-irradiation of up to 500 Mrad [46] and microwave radiation of 2.45 kHz

Applications/Commercial Products:
Processing & Manufacturing Routes: Synth. by melt polycondensation of the acetoxy derivatives of 4-hydroxybenzoic acid and 6-hydroxy-2-naphthalenecarboxylic acid at 340° [11,33] Vectra may be processed by thermoforming, ultrasonic forming, pultrusion, metallising (sputter coating) or injection moulding [49,56]. Care must be taken in injection moulding to avoid weld lines. Vectra moulds easily in thin-walled moulds, but in thick-walled moulds, fibres align only in the surface layer and not in the core. Moulding and weld line problems may be overcome using multi-live feed injection moulding [50]. Vectra is processed by extrusion into sheets, tubes, films, extrusion coatings, rods or strands.
Mould Shrinkage (%):

No.	Value	Note
1	0.006%	73% HBA [42]
2	0.1%	[44]
3	0.1%	A950 grade, flow direction [43]
4	0.6%	A950 grade, transverse direction [43]
5	0.2%	B950 grade, flow direction [43]
6	0.4%	B950 grade, transverse direction [43]

Applications: Applications include injection moulding resin used in electronics, fibre optics, automotive industry, aircraft/aerospace industry, in consumer products, chemical processing and manufacturing fields. Applications due to exceptionally low ion content as an encapsulant for integrated circuits, transistors and hybrid electronic components. Other uses include medical components, safety equipment, leisure goods, watch components, chemical analysis equipment

Trade name	Details	Supplier
SRP2	discontinued material	ICI, UK
Vectra	Wollastonite and glass fibre filled grades available	Hoechst Celanese
Vectra	A950/B950	Hoechst
Vectran	Wollastonite and glass fibre filled grades available	Kuraray Co

Bibliographic References

[1] Donald, A.M. and Windle, A.H., *J. Mater. Sci.*, 1983, **18**, 1143, (electron microscopy)
[2] Sun, Z., Cheng, H.M. and Blackwell, J., *Macromolecules*, 1991, **24**, 4162, (struct)
[3] Clements, J., Humphreys, J. and Ward, I.M., *J. Polym. Sci., Part B: Polym. Phys.*, 1986, **24**, 2293, (H-1 nmr)
[4] Siegmann, A., Dagan, A. and Kenig, S., *Polymer*, 1985, **26**, 1325, (rheological props)
[5] Blundell, D.J. and Buckingham, K.A., *Polymer*, 1985, **26**, 1623, (transition temps, power factor)
[6] Takase, Y., Mitchell, G.R. and Odajima, A., *Polym. Commun.*, 1986, **27**, 76, (transition temps, dielectric constant)
[7] Crispin, A.J. and Greig, D., *Polym. Commun.*, 1986, **27**, 246, 264, (thermal conductivity, heat capacity, bulk modulus)
[8] Cao, M.Y. and Wunderlich, B., *J. Polym. Sci., Polym. Phys. Ed.*, 1985, **23**, 521, (thermodynamic props)
[9] Butzbach, G.D., Wendorff, J.H. and Zimmermann, H.J., *Makromol. Chem., Rapid Commun.*, 1985, **6**, 821, (thermodynamic props, expansion coefficients)
[10] Cheng, S.Z.D., *Macromolecules*, 1988, **21**, 2475, (thermodynamic props)
[11] Muhlebach, A., Johnson, R.D., Lyerla, J. and Economy, J., *Macromolecules*, 1988, **21**, 3115, (synth, C-13 nmr)
[12] Alhaj-Mohammed, M.H., Davies, G.R., Jawad, S.A. and Ward, I.M., *J. Polym. Sci., Part B: Polym. Phys.*, 1988, **26**, 1751, (dielectric constant)
[13] Troughton, M.J., Davies, G.R. and Ward, I.M., *Polymer*, 1989, **30**, 58, (mech props)
[14] Kaito, A., Kyotani, M. and Nakayama, K., *Macromolecules*, 1991, **24**, 3244, (ir, sonic modulus)
[15] Kalika, D.S. and Yoon, D.Y., *Macromolecules*, 1991, **24**, 3404, (dielectric constant)
[16] Sarlin, J. and Tormala, P., *J. Polym. Sci., Part B: Polym. Phys.*, 1991, **29**, 395, (mech props)
[17] Jansen, J.A.J., Van der Maas, J.H. and Posthuma de Boer, A., *Macromolecules*, 1991, **24**, 4278, (ir)
[18] Allen, R.A. and Ward, I.M., *Polymer*, 1991, **32**, 202, (H-1 nmr)
[19] Hanna, S, Lemmon, T.J., Spontak, R.J. and Windle, A.H., *Polymer*, 1992, **33**, 3, (morphological study)
[20] Choy, C.L., Leung, W.P. and Yee, A.F., *Polymer*, 1992, **33**, 1788, (elastic modulus)
[21] Yonetake, K., Sagiya, T., Koyama, K. and Masuko, T., *Macromolecules*, 1992, **25**, 1009, (transition temps)
[22] Fyfe, C.A., Fahie, B.J., Lyerla, J.R., Economy, J., *et al*, *Macromolecules*, 1992, **25**, 1623, (H-1 nmr, C-13 nmr)
[23] Weinkauf, D.H. and Paul, D.R., *J. Polym. Sci., Part B: Polym. Phys.*, 1992, **30**, 837, (gas transport coefficients, transition temps)
[24] Heino, M.T. and Seppala, J.V., *J. Appl. Polym. Sci.*, 1992, **44**, 2185, (melt viscosity)
[25] Kyotani, M., Kaito, A. and Nakayama, K., *J. Appl. Polym. Sci.*, 1993, **47**, 2053, (mech props, soln props)
[26] Plummer, C.J.G., Wu, Y., Davies, P., Zulle, B., *et al*, *J. Appl. Polym. Sci.*, 1993, **48**, 731, (creep props, mech props)
[27] Wilson, D.J., Vonk, C.G. and Windle, A.H., *Polymer*, 1993, **34**, 227, (struct)
[28] Schoffeleers, H.M., Tacx, J.C.J.F., Kingma, J.A. and Vulic, I., *Polymer*, 1993, **34**, 557, (solubility)
[29] Hummel, D.D., Neuhoff, U., Bretz, A. and Dussel, H.J., *Makromol. Chem.*, 1993, **194**, 1545, (ms, ir, thermal degradation)
[30] Hubrich, M., Maresch, G.G. and Spiess, H.W., *Chem. Phys. Lett.*, 1994, **218**, 81, (esr)
[31] Green, D.I., Zhang, H., Davies, G.R. and Ward, I.M., *Polymer*, 1993, **34**, 4803, (shear modulus, tensile modulus)
[32] Izu, P., Munoz, M.E., Pena, J.J. and Santamaria, A., *Polym. Eng. Sci.*, 1996, **36**, 721, (rheological props)
[33] *U.S. Pat.*, 1983, 4 393 191, (synth, soln props)
[34] Alderman, N.J. and Mackley, M.R., *Faraday Discuss. Chem. Soc.*, 1985, **79**, 149, (birefringence)
[35] Fa, X.C. and Takahashi, T., *Sen'i Gakkaishi*, 1990, **46**, 49, (struct)
[36] Davies, G.R., *Makromol. Chem., Makromol. Symp.*, 1988, **20/21**, 293, (mech props, struct)
[37] Cao, M.Y., Varma-Nair, M. and Wunderlich, B., *Polym. Adv. Technol.*, 1990, **1**, 151, (thermodynamic props, transition temps)
[38] Abdul-Jawad, S. and Ahmad, M.S., *Mater. Lett.*, 1993, **17**, 91, (impedance behaviour)
[39] Urtis, J.K. and Farns, R.J., *J. Appl. Polym. Sci.*, 1996, **59**, 1849, (water diffusion coefficient)
[40] Gentzler, M., Reimer, J.A. and Denn, M.M., Polym. Prepr. (Am. Chem. Soc., Div. Polym. Chem.), 1996, **37**, 764, (H-1 nmr)
[41] *Kirk-Othmer Encycl. Chem. Technol.*, 4th edn., (ed. J.I. Kroschwitz), Wiley Interscience, 1996, **19**, 645, (mech props, chemical stability)
[42] *Polymeric Materials Encyclopedia*, (ed. J.C. Salamone), CRC Press, 1996, **5**, 3652, 3692, 3703, 3709, (mech props, mould shrinkage, flammability, heat distortion temp)
[43] *Vectra LCP*, sheet V2, Hoechst Celanese, 1987, (technical datasheet)
[44] *Vectra LCP* A900, Hoechst Celanese, 1987, (technical datasheet)
[45] *Vectra LCP*, sheet no. V26, Hoechst Celanese, 1986, (technical datasheet)
[46] *Vectra LCP*, sheet no A900, Hoechst Celanese, 1987, (technical datasheet)
[47] *Vectra LCP*, sheet no V45/10, Hoechst Celanese, 1989, (technical datasheet)
[48] *Vectra LCP*, sheet no. V21, Hoechst Celanese, 1986, (technical datasheet)
[49] *Vectra LCP*, Publ. 107, Hoechst Celanese, (technical datasheet)
[51] *Vectra LCP*, sheet no. V1, Hoechst Celanese, 1988, (technical datasheet)
[52] Williams, G., *Vectra LCP*, Vectra LCP, Hoechst Celanese,
[53] *Vectra LCP*, sheet no. V45/15, Hoechst Celanese, 1989, (technical datasheet)
[54] *Vectra LCP*, sheet no. V45/19, Hoechst Celanese, 1989, (technical datasheet)
[55] Van Krevelen, D.W., *Properties of Polymers: Their Correlation with Chemical Structure*, 3rd edn., 1990, 804

Poly(4-hydroxybenzoate-*co*-isophthalic acid-*co*-4,4′-dihydroxydiphenylether) P-312

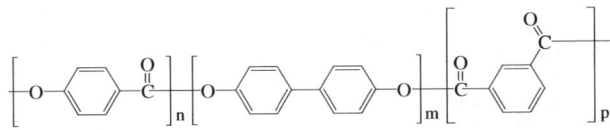

Synonyms: *Poly(1,3-benzenedicarboxylic acid-co-1,1′-biphenyl-4,4′-diolco-4-hydroxybenzoic acid)*
Monomers: 4-hydroxybenzoic acid, isophthalic acid, 4,4′-dihydroxydiphenyl ether
Material class: Copolymers
Polymer Type: unsaturated polyester
CAS Number:

CAS Reg. No.	Note
52724-09-1	
113214-02-1	block copolymer

Molecular Formula: $[(C_7H_4O_2)(C_{12}H_8O_2)(C_8H_4O_2)]_n$
Fragments: $C_7H_4O_2$ $C_{12}H_8O_2$ $C_8H_4O_2$

Transport Properties:
Water Content: Water absorption values have been reported

Mechanical Properties:
Mechanical Properties General: Mechanical props. have been reported

Optical Properties:
Transmission and Spectra: Mass spectral [1] data has been reported

Polymer Stability:
Thermal Stability General: Resistance to heat has been reported [3]
Decomposition Details: Thermal degradation and degradation activation energies [1]. 10% weight loss temp., 486° (air), 504° (N_2 [1], 50% weight loss temp., 523° (air), 575° (N_2 [1], have been reported
Recyclability: α Irradiation resistance has been reported

Applications/Commercial Products:
Processing & Manufacturing Routes: Synth. by slurry polymerisation of acetylated isophthalic acid 4,4′-biphenol and 4-hydroxybenzoic acid at 300–350°
Applications: Used as an injection moulding resin for electrical, aerospace, automotive and lighting applications

Trade name	Supplier
Ekkcel 1000	Sumitomo Chem. Co.

Bibliographic References

[1] Crossland, B., Knight, G.J. and Wright, W.W, *Br. Polym. J.,* 1986, **18**, 371, (thermal degradation, mass spectrum)
[2] Cottis, S.G., *Mod. Plast.,* 1975, **52**, 62, (mechanical props.)
[3] Cook, R.E., High Perform. Plast., Nat. Tech. Conf. - Soc. Plast. Eng., *(Prepr.),* 1976, 229, (mechanical props., water absorption, heat and α irradiation resistance)

Poly(4-hydroxybenzoate-*co*-terephthalic acid-*co*-4,4′-dihydroxydiphenyl ether)   P-313

Synonyms: *Poly(1,4-benzenedicarboxylic acid-co-1,1′-biphenyl-4,4′-diol-co-4-hydroxybenzoic acid)*
Monomers: 4-Hydroxybenzoic acid, Terephthalic acid, 4,4′-Dihydroxydiphenyl ether
Material class: Thermoplastic, Copolymers
Polymer Type: unsaturated polyester
CAS Number:

CAS Reg. No.	Note
31072-56-7	
96218-35-8	from terephthaloyl chloride

Molecular Formula: $[(C_7H_4O_2).(C_{12}H_8O_2).(C_8H_4O_2)]_n$
Fragments: $C_7H_4O_2$ $C_{12}H_8O_2$ $C_8H_4O_2$
Additives: Glass fibre
General Information: Stiffness, thermal stability and melting point increase with increasing Bisphenol content
Morphology: X-ray studies [5,7,8,16,19] have been reported. Two polymorphic forms exist both pseudo-orthorhombic [6]. Orthorhombic a 0.788, b 0.557, c 1.255 [8]. Has liq. crystal phase

Volumetric & Calorimetric Properties:
Density:

No.	Value	Note
1	d 1.4 g/cm^3	Ekkcel C2000 [10]
2	d 1.35 g/cm^3	Xydar SRT-300 [17]
3	d 1.6 g/cm^3	Xydar G-930, ASTM D792 [23]
4	d 1.35 g/cm^3	Xydar [24]

Thermal Expansion Coefficient:

No.	Value	Note	Type
1	2.88×10^{-5} K^{-1}	1.6×10^{-5} °F^{-1} [10]	L
2	$3 \times 10^{-6} - 7 \times 10^{-6}$ K^{-1}	Xydar G-630, ASTM D696 [23]	L

Volumetric Properties General: Other density values have been reported [21]
Latent Heat Crystallization: ΔH_f 10.5 J g^{-1} [6]
Melting Temperature:

No.	Value	Note
1	380°C	[2]
2	427°C	[6]
3	430°C	[7]
4	421°C	[8]
5	425°C	[10]
6	420°C	[11,23]

Glass Transition Temperature:

No.	Value	Note
1	115°C	DSC, KR4002
2	200°C	HX2000

Deflection Temperature:

No.	Value	Note
1	265°C	1.82 MPa, ASTM D648 [2]
2	293°C	Ekkcel C2000 [3]
3	355°C	1.82 MPa, ASTM D648 [17]
4	240–260°C	1.82 MPa, Xydar - glass fibre reinforced [24]
5	271°C	1.82 MPa, ASTM D648, Xydar G-830 [23]

Vicat Softening Point:

No.	Value	Note
1	369°C	Xydar SRT-300 [17]
2	358°C	Xydar [24]

Transition Temperature:

No.	Value	Note	Type
1	100°C	[6]	cryst. transition
2	472°C	[6]	Nematic-isotropic transition
3	0°C	[8]	T_β

Surface Properties & Solubility:
Solubility Properties General: Blend with polyether ether ketone has been reported [11]
Solvents/Non-solvents: Insol. all organic solvents [10]. Xydar SRT-300 sol. 3,5-bis(trifluoromethyl)phenol (25°), pentafluorophenol (40°) [12]

Transport Properties:
Polymer Solutions Dilute: [η] 16.8 dl g^{-1} (3,5-bis(trifluoromethyl)phenol, 25°, Xydar SRT-300) [12]; [η] 14.5 dl g^{-1} (pentafluorophenol, 40°, Xydar SRT-300) [12]
Polymer Melts: Rheological props. have been reported [11]
Water Absorption:

No.	Value	Note
1	0.1 %	max., 24h, ASTM D570, Xydar G-930 [23]

Mechanical Properties:
Mechanical Properties General: Other mech. props. have been reported [18,21]. Dynamic mech. modulus has been reported [19]

Tensile (Young's) Modulus:

No.	Value	Note
1	2460 MPa	350 kpsi, Ekkcel C2000 [10]
2	8300 MPa	Xydar [24]
3	15800 MPa	2300 kpsi, ASTM D638, Xydar G-830 [23]

— Poly(4-hydroxybenzoate-co-terephthalic...

Flexural Modulus:

No.	Value	Note
1	10500 MPa	1520 kpsi, ASTM D790 [2]
2	4800 MPa	700 kpsi, 23°, Ekkcel C2000 [3]
3	7000 MPa	Xydar SRT-300 [11]
4	16200 MPa	235 kpsi, 260°, Ekkcel C2000 [10]
5	13400 MPa	1950 kpsi, ASTM D790, Xydar G-930 [23]

Tensile Strength Break:

No.	Value	Note	Elongation
1	156.5 MPa	22700 psi, ASTM D1708 [2]	9%
2	96.5 MPa	14000 psi, Ekkcel C2000 [3]	8%
3	135 MPa	Xydar SRT-300 [11]	
4	138 MPa	20000 psi, Xydar SRT-300 [17]	4.9%
5	135 MPa	ASTM D638, Xydar G-930 [23]	1.6%

Flexural Strength at Break:

No.	Value	Note
1	117 MPa	17000 psi, 23°, Ekkcel C2000 [10]
2	28 MPa	4000 psi, 260°, Ekkcel C2000 [10]
3	131 MPa	19000 psi, Xydar SRT-300 [17,24]
4	172 MPa	ASTM D790, Xydar G-930 [23]

Impact Strength: Impact strength 200 J m^{-1} (Xydar SRT-300) [11]. Other values of impact strength have been reported [21]
Viscoelastic Behaviour: Stress-strain curves for Xydar SRT-300 have been reported [11]. Viscoelastic props. have been reported [13]
Izod Notch:

No.	Value	Notch	Note
1	27 J/m	Y	0.5 ft lb in^{-1}, ASTM D256 method A [2]
2	53 J/m	N	1 ft lb in^{-1}, Ekkcel C2000 [3]
3	130 J/m	N	2.4 ft lb in^{-1}, Xydar SRT-300 [17]
4	454 J/m	N	Xydar [24]
5	587 J/m	Y	Xydar G-930 [23]
6	208 J/m	Y	Xydar [24]
7	95 J/m	Y	Xydar G-930, ASTM D256 [23]
8	425 J/m	N	Xydar G-930, ASTM D256 [23]

Electrical Properties:
Electrical Properties General: Dielectric props. have been reported [13]. Variation of dielectric loss with temp. has been reported in graphical form [8]. Dissipation factor for Xydar SRT-300 has been reported [11]. Dielectric constant and strength have been reported [22]. Other resistivity values have been reported [21]
Dielectric Permittivity (Constant):

No.	Value	Frequency	Note
1	4.2	60 Hz	ASTM D150, Xydar G-930 [23]
2	3.9	1 MHz	ASTM D150, Xydar G-930 [25]

Dielectric Strength:

No.	Value	Note
1	0.6 kV.mm^{-1}	Xydar SRT-300 [17]
2	39 kV.mm^{-1}	1000 V mil^{-1}, 1.6 mm, ASTM D149, Xydar G-930 [23]

Arc Resistance:

No.	Value	Note
1	>300s	min., UL-746A, Xydar G-930 [23]

Strong Field Phenomena General: Electrical breakdown characteristics have been reported [15]
Static Electrification: Electrical conduction of Xydar SRT-300 has been reported [14]
Optical Properties:
Transmission and Spectra: C-13 nmr [1,7], mass [4,9], nmr [20] and ir [20] spectral data have been reported
Polymer Stability:
Thermal Stability General: Heat resistance has been reported [18]
Upper Use Temperature:

No.	Value
1	330°C

Decomposition Details: Activation energy of thermal degradation has been reported. [4] 10% Weight loss at 475° (air), 482° (N$_2$) [4]; 50% weight loss at 514° (air), 514° (N$_2$) [4]. Main products of pyrolysis include C_6H_6 and phenol. [9] Other decomposition details have been reported [20,22]
Flammability: Flame retardance has been reported. [22] Flammability rating V0 (Xydar SRT-300, UL 94) [17], V0 (Xydar G-930, UL 94) [23]
Environmental Stress: Possesses good weathering resistance [23]. Stable to high levels of uv and microwave radiation [23]
Chemical Stability: Exhibits resistance to most chemicals, acids, solvents, hydrocarbons. [23]
Hydrolytic Stability: Resistant to boiling water [23]
Applications/Commercial Products:
Processing & Manufacturing Routes: Synth. by the slurry polymerisation of acetylated terephthalic acid, 4,4′-biphenol, and 4-hydroxybenzoic acid at 300–350° [6]. Xydar G-930 resin should be dried before use at 150° for 8h. Processed by injection moulding (melt temp. 280–360°, mould temp. 65–95°). Excellent flow characteristics enable them to fill long, thin moulds
Mould Shrinkage (%):

No.	Value	Note
1	0%	ASTM D955 [2]
2	1.2%	Ekkcel C2000 [3]

Applications: Used as an injection moulding resin for electrical, aerospace, automotive and lighting applications. Applications include usage in electrical and electronic connectors, surface interconnection devices, sockets, memory card frames. Other applications involve the aerospace industry, automotive industry and in machinery and equipment

Trade name	Details	Supplier
Ekkcel		Sumitomo Chem Co
Ekkcel		Carborundum Corporation

HX2000		DuPont
Ultrox	KR4000	BASF
Xydar	G-930, G-960	Amoco Chemical Company

Bibliographic References

[1] Fyfe, C.A., Lyerla, J.A., Volksen W. and Yannoni, C.S., *Macromolecules*, 1979, **12**, 757, (C-13 nmr)
[2] Jackson, W.J., *Br. Polym. J.*, 1980, **12**, 154, (mech props)
[3] Ekkcel C2000 Product Specifications, The Carborundum Company, Niagara Falls, N.Y., (mech props)
[4] Crossland, B., Knight, G.J. and Wright, W.W., *Br. Polym. J.*, 1986, **18**, 371, (thermal degradation, ms)
[5] Blackwell, J., Cheng, H.M. and Biswas, A., *Macromolecules*, 1988, **21**, 39, (x-ray study)
[6] Field, N.D., Baldwin, R., Layton, R., Frayer, P. and Scardiglia, F., *Macromolecules*, 1988, **21**, 2155, (synth)
[7] Volksen, W., Lyerla, J.R., Economy, J. and Dawson, B., *J. Polym. Sci., Polym. Chem. Ed.*, 1983, **21**, 2249, (C-13 nmr, x-ray study)
[8] Kalika, D.S., Yoon, D.Y., Ianelli, P. and Parrish, W., *Macromolecules*, 1991, **24**, 3413, (cryst struct, dielectric constant)
[9] Sueoka, K., Nagata, M., Ohtani, H., Nagai, N. and Tsuge, S., *J. Polym. Sci., Part A: Polym. Chem.*, 1991, **29**, 1903, (thermal degradation, ms)
[10] Economy, J., *J. Macromol. Sci., Chem.*, 1984, **A21**, 1705, (solubility, mech props)
[11] Isayev, A.I. and Subramanian, P.R., *Polym. Eng. Sci.*, 1992, **32**, 85, (viscoelastic props, rheological props, mech props)
[12] Schoffeleers, H.M., Tacx, J.C.J.F., Kingma, J.A. and Vulic, I., *Polymer*, 1993, **34**, 557, (soln props)
[13] Satoh, N. and Nakanishi, A., *Toso Kenkyu Hokoku*, 1992, **36**, 87, (viscoelastic props, dielectric props)
[14] Ohki, Y., Hirai, T., Uda, S., Tanaka, Y. and Ikeda, M., Proc. Int. Conf. Conduct. Breakdown Solid Dielectr., *4th*, 1992, 22, (electrical conduction)
[15] Akiyama, K., Ohki, Y., Umeshima, Y. and Ikeda, M., Proc. Int. Conf. Conduct. Breakdown Solid Dielectr., *4th*, 1994, **1**, 79, (electrical breakdown)
[16] Antipov, E., Godovsky, Y., Stamm, M. and Fischer, E.W., *Polym. Mater. Sci. Eng.*, 1995, **72**, 435, (struct)
[17] Ash, M. and Ash, I., *Chemical Products Desk Reference*, Edward Arnold, 1990, 973, (mech props, flammability)
[18] Cook, R.E., High Perform. Plast., Nat. Tech. Conf. - Soc. Plast. Eng. *(Prepr.)*, 1976, 229, (heat resistance, water absorption)
[19] Frayer, P.D., *Polym. Compos.*, 1987, **8**, 379, (struct, dynamic mech modulus)
[20] Sueoka, K., Nagai, Y., Nagata, M. and Segawa, S., *Anal. Sci.*, 1990, **6**, 371, (nmr, ir, thermal degradation)
[21] Reilly, J.J., Thoman, S.J. and Lin, W.W., *SAMPE J.*, 1987, **23**, 22, (electrical resistivity, mech props)
[22] Fein, M.M., Natl. SAMPE Symp. Exhib., *(Proc.)*, 1985, **30**, 556, (chemical resistance, uv resistance, dielectric props)
[23] *Xydar Liquid Crystal Polymer*, Amoco Chemical Co, Alpharetta, Ga, 1996, (technical datasheet)
[24] *Polymeric Materials Encyclopedia*, (ed. J.C. Salamone), CRC Press, 1996, **5**, 3652, 3709, (mech props, transition temps)

Poly(4-hydroxybenzoic acid-*co*-6-hydroxy-2-naphthoic acid), Carbon fibre filled   P-314

Synonyms: *Poly(4-hydroxybenzoic acid-co-2-oxy-6-naphthoic acid), carbon fibre filled*
Related Polymers: Poly(4-hydroxybenzoic acid-*co*-6-hydroxy-2-naphthoic acid), Poly(4-hydroxybenzoic acid-*co*-6-hydroxy-2-naphthoic acid), glass fibre filled, Poly(4-hydroxybenzoic acid-*co*-6-hydroxy-2-naphthoic acid), mineral filled
Monomers: 4-Hydroxybenzoic acid, 6-Hydroxy-2-naphthoic acid
Material class: Copolymers, Polymer liquid crystals, Composites
Polymer Type: polyester
CAS Number:

CAS Reg. No.
81843-52-9

Additives: Carbon fibre (30%)

Volumetric & Calorimetric Properties:
Density:

No.	Value	Note
1	d 1.5 g/cm^3	A230 and B230 grades [1]

Thermal Expansion Coefficient:

No.	Value	Note	Type
1	2×10^{-6} K^{-1}	A230 grade, flow direction [1]	L
2	3×10^{-6} K^{-1}	B230 grade, flow direction [1]	L
3	6.5×10^{-5} K^{-1}	A230 grade, transverse to flow [1]	L
4	4.5×10^{-5} K^{-1}	B230 grade, transverse to flow [1]	L

Melting Temperature:

No.	Value	Note
1	260°C	A230 grade [1]
2	280°C	B230 grade [1]

Deflection Temperature:

No.	Value	Note
1	221°C	1.82 MPa, A230 grade, ASTM D648 [1]
2	225°C	1.82 MPa, B230 grade, ASTM D648 [1]

Transport Properties:
Water Absorption:

No.	Value	Note
1	0.03 %	ASTM D570 [1]

Mechanical Properties:
Mechanical Properties General: Specific moduli are similar to those of aluminium. Has greater stiffness than for example, Nylon 6 and Nylon 6,6. Flexural modulus and flexural strength at break decrease with increasing temp. [2] and sample thickness [3]. Tensile storage modulus decreases with increasing temp. [4]

Tensile (Young's) Modulus:

No.	Value	Note
1	31000 MPa	A230 grade, ASTM D638 [1]
2	37000 MPa	B230 grade, ASTM D638 [1]

Flexural Modulus:

No.	Value	Note
1	24000 MPa	A230 grade [1]
2	32000 MPa	B230 grade [1]

Tensile Strength Break:

No.	Value	Note	Elongation
1	165 MPa	A230 grade [1]	1.2%
2	161 MPa	B230 grade [1]	1.1%

Flexural Strength at Break:

No.	Value	Note
1	255 MPa	A230 grade [1]
2	317 MPa	B230 grade [1]

Compressive Modulus:

No.	Value	Note
1	17200 MPa	A230 grade [1]
2	33100 MPa	B230 grade [1]

Compressive Strength:

No.	Value	Note
1	124 MPa	A230 grade [1]
2	238 MPa	B230 grade [1]

Miscellaneous Moduli:

No.	Value	Note	Type
1	<800 MPa	max, 80° [1]	tensile loss modulus

Fracture Mechanical Properties: Shear strength 121 MPa (A230 grade, ASTM D732) [1]

Electrical Properties:
Electrical Properties General: Has poor dielectric and insulating props. owing to the presence of carbon fibre

Polymer Stability:
Flammability: Has low smoke emission and low flammability [1]
Chemical Stability: Has good resistance to chemicals [1]

Applications/Commercial Products:
Processing & Manufacturing Routes: Suitable for injection moulding

Mould Shrinkage (%):

No.	Value	Note
1	0%	flow direction [1]
2	0.15%	transverse direction [1]

Applications: Applications due to low weight in injection moulding controlled surfaces; as replacement for aluminium and zinc parts in automotive and aerospace industries

Trade name	Details	Supplier
Vectra LCP	A230/B230 (30% filled; B215 (15% filled)	Hoechst Celanese

Bibliographic References
[1] *Vectra LCP*, Sheet no. V2, Hoechst Celanese, (technical datasheet)
[2] *Flexural Props. of Selected LCP Formulations*, Hoechst Celanese, (technical datasheet)
[3] *Vectra LCP*, Sheet no. V19, Hoechst Celanese, (technical datasheet)
[4] *Vectra LCP*, Sheet no. V45/3, Hoechst Celanese, (technical datasheet)

Poly(4-hydroxybenzoic acid-*co*-6-hydroxy-2-naphthoic acid), Glass fibre filled P-315

Synonyms: *Poly(4-hydroxybenzoic acid-co-2-oxy-6-naphthoyl), glass fibre filled*
Related Polymers: Poly(4-hydroxybenzoic acid-*co*-6-hydroxy-2-naphthoic acid), Poly(4-hydroxybenzoic acid-*co*-6-hydroxy-2-naphthoic acid), carbon fibre filled, Poly(4-hydroxybenzoic acid-*co*-6-hydroxy-2-naphthoic acid), mineral filled

Monomers: 4-Hydroxybenzoic acid, 6-Hydroxy-2-naphthoic acid
Material class: Composites, Polymer liquid crystals, Copolymers
Polymer Type: polyester
CAS Number:

CAS Reg. No.
81843-52-9

Volumetric & Calorimetric Properties:
Density:

No.	Value	Note
1	d 1.6 g/cm^3	30% filled [1]
2	d 1.8 g/cm^3	50% filled [1]

Thermal Expansion Coefficient:

No.	Value	Note	Type
1	5×10^{-6} K^{-1}	flow [1]	L
2	$6 \times 10^{-5} - 6.5 \times 10^{-5}$ K^{-1}	transverse, A130 and A150 grades [1]	L
3	2×10^{-6} K^{-1}	flow [1]	L
4	5×10^{-5} K^{-1}	transverse, B130 grade [1]	L
5	6×10^{-6} K^{-1}	flow, C130 and C150 grades [1]	L
6	7×10^{-5} K^{-1}	transverse, C130 grade [1]	L
7	6×10^{-5} K^{-1}	C150 grade [1]	L

Melting Temperature:

No.	Value	Note
1	280°C	A and B resins [1]
2	327°C	C resin [1]
3	355°C	E resin [1]

Deflection Temperature:

No.	Value	Note
1	251–254°C	A resin [1]
2	284°C	0.46 MPa, C resin [1]
3	230–232°C	A and B resin [1]
4	243°C	C resin [1]
5	270°C	E resin [1]
6	230°C	1.8 MPa, 30% filled

Vicat Softening Point:

No.	Value	Note
1	140°C	A resin

Transport Properties:
Water Absorption:

No.	Value	Note
1	0.02 %	A and C resin
2	0.01 %	24h, B resin, ASTM D792

Mechanical Properties:

Mechanical Properties General: Tensile storage moduli of all grades decrease with increasing temp. when measured in a N_2 atmosphere [6,7,8,9,10,11]. Flexural modulus and flexural strength all grades decrease with increasing sample thickness [4] and temp.

Tensile (Young's) Modulus:

No.	Value	Note
1	16600 MPa	A130 grade, ASTM D790 [1]
2	22000 MPa	A150 grade, ASTM D638 [1]
3	24100 MPa	B150 grade, ASTM D638 [1]
4	14500 MPa	C130 grade, ASTM D638 [1]
5	21000 MPa	C150 grade, ASTM D638 [1]
6	13800 MPa	E130 grade, ASTM D638 [1]

Flexural Modulus:

No.	Value	Note
1	12750–23100 MPa	0.79–6.35 mm thick, A130 grade [4]
2	18430–27360 MPa	0.79–6.35 mm thick, B130 grade [4]
3	10500–16600 MPa	0.79–6.35 mm thick, C130 grade [4]
4	3700–19450 MPa	23–220°, B130 grade
5	4400–21400 MPa	23–220°, B150 grade
6	3900–13200 MPa	23–220°, C130 grade
7	15000 MPa	A130 grade, ASTM D790 [1]
8	20000 MPa	A150 grade, ASTM D790 [1]
9	19000 MPa	B130 grade and C150 grade, ASTM D790 [1]
10	13100 MPa	C130 grade, ASTM D790 [1]
11	13800 MPa	E130 grade, ASTM D790 [1]

Tensile Strength Break:

No.	Value	Note
1	161 MPa	C130 grade, ASTM D638 [1]
2	165 MPa	C150 grade, ASTM D638 [1]
3	124 MPa	E130 grade, ASTM D638 [1]
4	207 MPa	A130 grade, ASTM D638 [1]
5	180 MPa	A150 grade, ASTM D638 [1]
6	221 MPa	B130 grade, ASTM D638 [1]

Flexural Strength at Break:

No.	Value	Note
1	254 MPa	A130 grade, ASTM D790 [1]
2	248 MPa	A150 grade, ASTM D790 [1]
3	300 MPa	B130 grade, ASTM D790 [1]
4	212 MPa	C130 grade, ASTM D790 [1]
5	288–396 MPa	0.79–6.35 mm thick, B130 grade [4]
6	165–273 MPa	0.79–6.35 mm thick, C130 grade [4]
7	24.8–276 MPa	50–200°, B130 and B150 grades, ASTM D790 [5]
8	17.3–324 MPa	23–220°, B130 grade, ASTM D790
9	15.9–290 MPa	23–220°, B150 and C150 grades, ASTM D790
10	220 MPa	C150 grade, ASTM D790 [1]
11	172 MPa	E130 grade, ASTM D790 [1]
12	208–335 MPa	0.79–6.35 mm thick, A130 grade [4]

Compressive Modulus:

No.	Value	Note
1	12000 MPa	A130 grade [5]
2	15400 MPa	A150 grade [5]
3	20000 MPa	B130 grade [5]
4	20600 MPa	B150 grade [5]
5	12000 MPa	A130 grade, ASTM D790 [1]
6	15000 MPa	A150 grade, ASTM D790 [1]
7	20000 MPa	B130 and C130 grades, ASTM D790 [1]

Compressive Strength:

No.	Value	Note
1	195 MPa	B130 grade [5]
2	198 MPa	B150 grade [5]
3	124 MPa	C130 and C150 grades, ASTM D790 [1]
4	142 MPa	A130 grade [5]
5	156 MPa	A150 grade [5]
6	140 MPa	A130 grade, ASTM D790 [1]
7	155 MPa	A150 grade, ASTM D790 [1]
8	192 MPa	B130 grade, ASTM D790 [1]

Miscellaneous Moduli:

No.	Value	Note	Type
1	<750 MPa	max., 1060°, B150 grade [9]	tensile loss modulus
2	<500 MPa	max., 80°, C130 grade [10]	tensile loss modulus
3	<450 MPa	max., 80°, C150 grade [11]	tensile loss modulus
4	<600 MPa	max., 60°, A130 grade [6]	tensile loss modulus
5	<600 MPa	max., 80°, A150 grade [7]	tensile loss modulus
6	<900 MPa	max., 95°, B130 grade [8]	tensile loss modulus

Hardness: Rockwell M80 (A resins, ASTM D785), M98 (B resins, ASTM D785), M75 (C resins, ASTM D785) [1]
Fracture Mechanical Properties: Shear strength 123 MPa (A130 grade, ASTM D695) [1], 132 MPa (B130 grade, ASTM D695) [1]
Friction Abrasion and Resistance: Tabor abrasion resistance 7.4 mg (1000 cycles)$^{-1}$ (A130 grade, ASTM D1044) [1]. Dynamic coefficient of friction 0.14 (against metal, A130 grade), 0.19 (against metal, A150 grade); static coefficient of friction 0.19 (against metal, A130 grade), 0.16 (against metal, A150 grade)

Izod Notch:

No.	Value	Notch	Note
1	150 J/m	Y	A130 grade, ASTM D256 [1,2]

Poly(4-hydroxybenzoic acid-co-6-hydroxy-2-naphthoic acid), Mineral filled

No.	Value		Note
2	90 J/m	Y	A150 and B130 grades, ASTM D256 [1,2]
3	125 J/m	Y	C130 grade, ASTM D256 [1,2]
4	85 J/m	Y	C150 grade, ASTM D256 [1,2]
5	107 J/m	Y	E150 grade, ASTM D256 [1,2]

Electrical Properties:
Dielectric Permittivity (Constant):

No.	Value	Frequency	Note
1	4.1–4.3	1 kHz	
2	3.7–3.9	1 MHz-1 GHz	all grades, ASTM D150 [1]
3	3.7	1 kHz	E130 grade, ASTM D150 [2]

Dielectric Strength:

No.	Value	Note
1	42–43 kV.mm^{-1}	23°, 50% relative humidity, short-term, 1.5 mm thick, 30% filled, ASTM D149

Arc Resistance:

No.	Value	Note
1	137–140s	30% filled, ASTM D495 [1,2]
2	181s	50% filled, ASTM D495 [1,2]

Dissipation (Power) Factor:

No.	Value	Note
1	0.006	A and E resins
2	0.004	C resin

Polymer Stability:
Flammability: Flammability rating V0 (0.8 mm thick, UL94) V0 (0.4 mm thick, B resin, UL94) [1,2]. Limiting oxygen index 37% (ASTM D2863)

Applications/Commercial Products:

Trade name	Details	Supplier
Vectra LCP	A130/B130/B150/C130/E130	Hoechst Celanese

Bibliographic References
[1] Vectra LCP V2, Typical Properties, Hoechst Celanese, 1988, (technical datasheet)
[2] Vectra E130, Hoechst Celanese, 1990, (technical datasheet)
[3] Vectra LCP V21, Hoechst Celanese, 1986, (technical datasheet)
[4] Vectra E130, sheet no. V19, Hoechst Celanese, 1988, (technical datasheet)
[5] Vectra LCP, Celanese Corp., 1986, (technical datasheet)
[6] Vectra LCP, sheet no. V45/1, Hoechst Celanese, 1989, (technical datasheet)
[7] Vectra LCP, sheet no. V45/13, Hoechst Celanese, 1989, (technical datasheet)
[8] Vectra LCP, sheet no. V45/12, Hoechst Celanese, 1989, (technical datasheet)
[9] Vectra LCP, sheet no. V45/13, Hoechst Celanese, 1989, (technical datasheet)
[10] Vectra LCP, sheet no. V45/16, Hoechst Celanese, 1989, (technical datasheet)
[11] Vectra LCP, sheet no. V45/17, Hoechst Celanese, 1989, (technical datasheet)

Poly(4-hydroxybenzoic acid-co-6-hydroxy-2-naphthoic acid), Mineral filled

Synonyms: *Poly(4-hydroxybenzoic acid-co-2-oxy-6-naphthoic acid), mineral filled*
Related Polymers: Poly(4-hydroxybenzoic acid-co-6-hydroxy-2-naphthoic acid), Poly(4-hydroxybenzoic acid-co-6-hydroxy-2-naphthoic acid), glass fibre filled, Poly(4-hydroxybenzoic acid-co-6-hydroxy-2-naphthoic acid), carbon fibre filled
Monomers: 4-Hydroxybenzoic acid, 6-Hydroxy-2-naphthoic acid
Material class: Composites, Polymer liquid crystals, Copolymers
Polymer Type: polyester
CAS Number:

CAS Reg. No.
81843-52-9

Additives: Wollastonite
Miscellaneous: Special bearing and wear grades are available. These are filled with various materials and give improved wear resistance and load-bearing props.

Volumetric & Calorimetric Properties:
Density:

No.	Value	Note
1	d 1.76 g/cm^3	A540 grade
2	d 1.89 g/cm^3	C550 grade

Thermal Expansion Coefficient:

No.	Value	Note	Type
1	9×10^{-6} K^{-1}	A540 grade	L
2	1.2×10^{-5} K^{-1}	C550 grade	L

Melting Temperature:

No.	Value	Note
1	280°C	A540 grade
2	327°C	C550 grade

Deflection Temperature:

No.	Value	Note
1	204°C	1.8 MPa, A540 grade
2	221°C	1.8 MPa, C550 grade

Transport Properties:
Water Absorption:

No.	Value	Note
1	0.02 %	24h, ASTM D570

Mechanical Properties:
Mechanical Properties General: Tensile storage modulus of the C550 [3], A540 [4] and A515 [5] grades decreases with increasing temp.

Tensile (Young's) Modulus:

No.	Value	Note
1	12000 MPa	A515 grade, ASTM D638 [1]

Tensile Modulus (continued):

No.	Value	Note
2	17200 MPa	A540 grade, ASTM D638 [1]
3	13800 MPa	C550 grade, ASTM D638 [1]

Flexural Modulus:

No.	Value	Note
1	10000 MPa	A515 grade, ASTM D790 [1]
2	15000 MPa	A540 grade, ASTM D790 [1]
3	17200 MPa	C550 grade, ASTM D790 [1]

Tensile Strength Break:

No.	Value	Note	Elongation
1	185 MPa	A515 grade, ASTM D638 [1]	93.9%
2	145 MPa	A540 grade, ASTM D638 [1]	3.5%
3	115 MPa	C550 grade, ASTM D638 [1]	2%

Flexural Strength at Break:

No.	Value	Note
1	103 MPa	50°, A540 grade, ASTM D790 [7]
2	51.7 MPa	100°, A540 grade, ASTM D790 [7]
3	26.2 MPa	150°, A540 grade, ASTM D790 [7]
4	11 MPa	200°, A540 grade, ASTM D790 [7]
5	161 MPa	A515 grade [1]
6	181 MPa	A540 grade [1]
7	163 MPa	C550 grade [1]

Compressive Modulus:

No.	Value	Note
1	9600 MPa	A515 grade, ASTM D695 [1,6]
2	14000 MPa	A540 grade, ASTM D695 [1,6]

Compressive Strength:

No.	Value	Note
1	101 MPa	A540 grade [6]
2	79 MPa	A515 grade, ASTM D695 [1]
3	80 MPa	A515 grade [6]
4	100 MPa	A540 grade, ASTM D695 [1]

Miscellaneous Moduli:

No.	Value	Note	Type
1	<850 MPa	max., 80° [3,4]	tensile loss modulus
2	<750 MPa	max., 65° [5]	tensile loss modulus

Impact Strength: Tensile impact strength 76 kJ m^{-2} (A515 grade, ASTM D1822), 42 kJ m^{-2} (A540 grade, ASTM D1822), 53 kJ m^{-2} (C550 grade, ASTM D1822) [1]
Viscoelastic Behaviour: Flexural creep 10.7 MPa (10h, 50 MPa stress, A540 grade), 8.6 MPa (100h, 50 MPa stress, A540 grade), 7.1 MPa (1000h, 50 MPa stress, A540 grade)
Hardness: Rockwell M72 (A540 grade, ASTM D785) [1]

Izod Notch:

No.	Value	Notch	Note
1	370 J/m	Y	A515 grade, ASTM D256 [1]
2	70 J/m	Y	A540 grade, ASTM D256 [1]
3	55 J/m	Y	C550 grade, ASTM D256 [1]

Electrical Properties:
Dielectric Permittivity (Constant):

No.	Value	Frequency	Note
1	4.2	1 kHz	A515 grade, ASTM D150 [1]
2	3.8	1 MHz	A515 grade, ASTM D150 [1]
3	3.9	1 GHz	A515 grade, ASTM D150 [1]

Dielectric Strength:

No.	Value	Note
1	35 kV.mm^{-1}	23°, 50% relative humidity, short term, 1.5 mm thick, A515 grade

Complex Permittivity and Electroactive Polymers: Comparative tracking index 175 V (A515 grade, ASTM D3636) [1]
Dissipation (Power) Factor:

No.	Value	Frequency	Note
1	0.004	1 GHz	A515 grade, ASTM D150 [1]

Polymer Stability:
Upper Use Temperature:

No.	Value
1	250°C

Flammability: Flammability rating V0 (3 mm thick, A515 and C550 grades, UL94), V0 (1.5 mm thick, A540 grade, UL94) [1]. Limiting oxygen index 33% (A515 grade, ASTM D2863), 37% (A540 and C550 grades, ASTM D2863) [1]
Recyclability: 100% Regrind retains 80–100% of its strength after five mouldings. Recycled filled grades lose some impact strength

Applications/Commercial Products:
Mould Shrinkage (%):

No.	Value	Note
1	0.1%	A515 grade, flow direction, ASTM D955
2	0.4%	A515 grade, transverse direction, ASTM D955
3	0.2%	A540 and C550 grades, flow direction, ASTM D955
4	0.4%	A540 and C550 grades, transverse direction, ASTM D955

Trade name	Details	Supplier
SRP 2	discontinued material	ICI Chemicals and Polymers Ltd.
Vectra	A515/A540/C550	Hoechst Celanese

Bibliographic References

[1] *Vectra LCP,* Sheet no. V2, Hoechst Celanese, 1987, (technical datasheet)
[2] *Vectra LCP,* Sheet no. V21, Hoechst Celanese, 1986, (technical datasheet)
[3] *Vectra LCP,* Sheet no. V45/18, Hoechst Celanese, 1989, (technical datasheet)
[4] *Vectra LCP,* Sheet no. V45/8, Hoechst Celanese, 1989, (technical datasheet)
[5] *Vectra LCP,* Sheet no. V45/7, Hoechst Celanese, 1989, (technical datasheet)
[6] Vectra LCP, *Compression Props.,* 1986, (technical datasheet)
[7] Vectra LCP, *Flexural Strength,* Hoechst Celanese, (technical datasheet)
[8] *Performance Chemicals,* Nov., 1987, 46

Poly(3-hydroxybutyrate)   P-317

Synonyms: *PHB. Poly(α-hydroxybutyrate). Poly(3-hydroxybutanoate)*
Related Polymers: Poly(3-hydroxybutyrate-*co*-3-hydroxyvalerate)
Monomers: 3-Hydroxybutanoic acid
Material class: Thermoplastic
Polymer Type: polyester
CAS Number:

CAS Reg. No.	Note
26063-00-3	homopolymer

Molecular Formula: $(C_4H_6O_2)_n$
Fragments: $C_4H_6O_2$
Molecular Weight: MW up to 500000
Additives: Glass fibre acts as a strengthening agent, increasing flexural modulus and softening point and decreasing tensile strength
Tacticity: Biosynth. material is isotactic, with the β-carbon in the chain having an *R*-configuration
Morphology: Prevalent in aerobic and anaerobic bacteria in a range of structs., used as a storage material, in membranes and in physiological processes [5,6]. Biosynthetic pathway involves stereoselective reduction and polymerase enzymes [7]. Cryst. struct. orthorhombic, unit cell a 5.76, b 13.2, c 5.96 [9]. Exists in a helical conformation and forms lamellar crysts. Crystallation kinetics have been reported [10]. Partial melting and crystallisation (self seeding) gives rise to smaller spherulites and better mech. props. [24]

Volumetric & Calorimetric Properties:
Density:

No.	Value	Note
1	d 1.177 g/cm^3	amorph. [24]
2	d 1.243 g/cm^3	81% cryst. [23]
3	d 1.25 g/cm^3	80% cryst. [4]

Thermal Expansion Coefficient:

No.	Value	Note	Type
1	6×10^{-5} K^{-1}	[23]	L

Thermodynamic Properties General: Thermodynamic props. have been reported [24]

Melting Temperature:

No.	Value	Note
1	174–177°C	optically active [2]
2	167–169°C	racemic [2]
3	179°C	injection moulded [24]
4	175°C	80% cryst. [4]

Glass Transition Temperature:

No.	Value	Note
1	9°C	DSC study [23]
2	15°C	80% cryst. [29]

Deflection Temperature:

No.	Value	Note
1	157°C	injection moulded [24]

Vicat Softening Point:

No.	Value	Note
1	160°C	[23]

Surface Properties & Solubility:
Plasticisers: Polar compounds can act as plasticisers. Only the amorph. phase of the polymer is affected
Solvents/Non-solvents: Sol. CHCl$_3$, trifluoroethanol, dichloroethane, dichloroacetic acid, CH$_2$Cl$_2$, pyridine. Insol. MeOH, hexane, Et$_2$O

Transport Properties:
Polymer Solutions Dilute: Mark-Houwink equation parameters have been reported for a range of solvents [24,25]
Polymer Melts: Unstable in melt. Melt viscosity 15000 NS m^{-2} (calc.) [24]
Permeability of Gases: Less gas permeable than poly(ethylene terephthalate) by a factor of five [2]
Water Absorption:

No.	Value	Note
1	0.13 %	23°, 1 month [24]

Gas Permeability:

No.	Gas	Value	Note
1	O$_2$	45 cm^3 mm/(m^2 day atm)	film [24]

Mechanical Properties:
Mechanical Properties General: Young's modulus decreases with increasing temp. [23]. Processing methods affect mech. props., spherulites and cracks result in brittle polymers, these defects can be removed by rolling and melt processing

Tensile (Young's) Modulus:

No.	Value	Note
1	5500 MPa	-150° [23]
2	4500 MPa	10° [23]

Poly(3-hydroxybutyrate)

No.	Value	Note
3	1500 MPa	100° [23]
4	3500 MPa	unnucleated moulding [24]
5	3200 MPa	self seeded moulding [24]
6	4000 MPa	film [24]

Flexural Modulus:

No.	Value	Note
1	9000 MPa	30% glass filled [24]
2	4000 MPa	80% cryst. [4]

Tensile Strength Break:

No.	Value	Note	Elongation
1	28 MPa	[24]	7%
2	30 MPa	45% cold rolled [24]	180%
3	40 MPa	injection moulded [4,24]	6%
4	10 MPa	0°, amorph. [24]	1140%
5	0.2 MPa	23°, amorph. [24]	130%
6	700 MPa	film [24]	75%

Tensile Strength Yield:

No.	Value	Elongation	Note
1	20 MPa	5%	amorph. [24]
2	11.4 MPa		7°, amorph. [24]

Mechanical Properties Miscellaneous: Other dynamic mech. props. have been reported [23]

Izod Notch:

No.	Value	Notch	Note
1	20 J/m	Y	unnucleated moulded [24]
2	65 J/m	Y	self seeded moulded [24]
3	50 J/m	Y	injection moulded [24]

Electrical Properties:

Dielectric Permittivity (Constant):

No.	Value	Note
1	3	room temp. [24]

Complex Permittivity and Electroactive Polymers: Piezoelectric; produces a voltage when deformed, or deforms when an external voltage is applied [22,24]

Optical Properties:

Transmission and Spectra: Nmr and ir [24] spectral data have been reported [26,27]
Total Internal Reflection: Optical rotation varies with wavelength, max. in uv range
Volume Properties/Surface Properties: Films are transparent with good clarity

Polymer Stability:

Decomposition Details: Decomposes to crotonic acid and oligomers (dimers, trimers) up to 340°. Above this CO_2 and other volatile products are liberated. Degradation occurs via random scission
Environmental Stress: Has good resistance to uv radiation
Chemical Stability: Has good stability to motor oil, but poor stability to chlorinated solvents [24]. Has moderate stability to alcohols
Hydrolytic Stability: Stable for several months in water but surface modification occurs resulting in hydrolysis and degradation of films by crystallisation [19]. Unstable to base and hot conc. acid [24]
Biological Stability: Undergoes biological attack within the cell as a consequence of normal biological processes. Processed polymer is also attacked by bacteria, fungi and algae [20]. Degradation occurs rapidly in moist soil, where bacteria dissolve the polymer before metabolising it. Degradation is negligible in air, anaerobic conditions dramatically increase degradation rate. Degrades *in vivo* to the monomer which is a natural metabolite
Recyclability: Biodegradability outweighs usefulness of recycling, though may possibly be recycled with other polymers [21]

Applications/Commercial Products:

Processing & Manufacturing Routes: Isol. from a range of bacterial species [28] including *Alcaligenes eutrophus*, *Pseudomonas oleovorans* [1], *Bacillus megaterium* [13], *Pseudomonas cepacia* [14] among others, fed with a suitable carbon source *e.g.* fructose, glucose, MeOH, AcOH, CO_2 [11,12]. The biosynth. polymer is removed from the bacteria in one of three ways. Solvent extraction [15] involves dissolving the polymer from the cells using $CHCl_3$, CH_2Cl_2 or other chlorinated hydrocarbons, followed by filtration to remove cell material, then precipitation. High MW polymers are obtained but large solvent quantities are involved. Sodium hypochlorite [16] is also used to digest bacteria, degrading cell walls and non-polymeric material. Washing the intact polymer with an organic solvent such as Et_2O leaves the pure polymer. The alkalinity of the hypochlorite can cause degradation of the polymer and lower MW. Biochemical degradation of the bacteria using enzymes [17] followed by drying or solvent extraction also is used to purify the polymer. MW can be controlled. May be melt processed at temp. <190° for short time. Can be blow-moulded and injection-moulded
Applications: Applications include degradable packaging, bottles and films. Useable in surgery for implants and sutures due to low toxicity, slowly degraded *in vivo*. Piezoelectric props. may also be useful in bone fracture healing. May be used as a chiral organic building block or as a stationary phase in chromatography

Bibliographic References

[1] Gross, R.A., De Mello, C., Lenz, R.W., Brandl, H. and Fuller, R.C., *Macromolecules*, 1989, **22**, 1106, (biosynth)
[2] Yokouchi, M., Chatani, Y., Tadokoro, H., Teranishi, K. and Tani, H., *Polymer*, 1973, **14**, 267, (cryst struct)
[3] Marchessault, R.H., Monasterios, C.J., Morin, F.G. and Sundarajan, P.R., *Int. J. Biol. Macromol.*, 1990, **12**, 158, (cryst struct, C-13 nmr)
[4] Howells, E.R., *Chem. Ind. (London)*, 1982, 508, (rev)
[5] Ellar, D., Lundgren, D.G., Okamura, K. and Marchessault, R.H., *J. Mol. Biol.*, 1968, **35**, 439
[6] Reusch, R.N., Sparrow, A.W. and Gardiner, J., *Biochim. Biophys. Acta*, **1123**, 33, (occur)
[7] Senior, P.J. and Dawes, E.A., *Biochem. J.*, 1973, **134**, 225, (biosynth)
[8] Haywood, G.W., Anderson, A.J., Chu, L. and Dawes, E.A., *Biochem. Soc. Trans.*, 1988, **16**, 1046, (biosynth)
[9] Okamura, K. and Marchessault, R.H., *Conformation of Biopolymers*, (ed. G.M. Ramachandran), Academic Press, 1967, **2**, 709
[10] Organ, S.J. and Barham, P.J., *J. Mater. Sci.*, 1991, **28**, 1368, (cryst struct)
[11] U.S. Pat., 1962, 3 036 959, (isol)
[12] U.S. Pat., 1962, 3 044 942, (isol)
[13] Findlay, R.H. and White, P.C., *Appl. Environ. Microbiol.*, 1983, 45, 71
[14] Ramsay, B.A., Ramsay, R.A. and Copper, D.G., *Appl. Environ. Microbiol.*, 1989, **55**, 584
[15] Lundgren, D.G., Alper, R., Schnaitmann, C. and Marchessault, R.H., *J. Bacteriol.*, 1965, **89**, 245
[16] Williamson, D.H. and Williamson, J.F., *J. Gen. Microbiol.*, 1958, **19**, 198
[17] Merrick, J.M. and Doudoroff, M., *J. Bacteriol.*, 1964, **88**, 60
[18] Grassie, N., Murray, E.J. and Holmes, P.A., *Polym. Degrad. Stab.*, 1984, **6**, 47, 95, (thermal degradation)
[19] Doi, Y., Kanesawa, Y., Kawaguchi, Y. and Kunioka, M., *Makromol. Chem., Rapid Commun.*, 1989, **10**, 227, (hydrolysis)
[20] Holmes, P.A., *Phys. Technol.*, 1985, **16**, 32, (rev, degradation)
[21] Hocking, P.J. and Marchessault, R.H., *Chemistry and Technology of Biodegradable Polymers*, (ed. G.L.J. Griffin), Blackie Academic and Professional, 1994, 48, (rev)

[22] Basta, N., *High Technol.*, 1984, **4(2)**, 67, (rev)
[23] Mitomo, H., Barham, P.J. and Keller, A., *Polym. Commun.*, 1988, **29**, 112, (dynamic mech props)
[24] Holmes, P.A., *Dev. Cryst. Polym.*, (ed. D.C. Bassett), Elsevier Applied Science, 1988, **2**, 1, (rev)
[25] Hiroyse, T., Einaga, Y. and Fujita, H., *Polymer*, 1979, **11**, 819, (soln props)
[26] Bloembergen, S., Holden, D.A., Bluhm, T.L., Hamer, G.K. and Marchessault, R.H., *Macromolecules*, 1987, **20**, 3086, (nmr)
[27] Bluhm, T.L., Hamer, G.K., Marchessault, R.H., Fyfe, C.A.N. and Veregin, R.P., *Macromolecules*, 1986, **19**, 2871, (nmr)
[28] Byron, D., *Trends Biotechnol.*, 1987, **5**, 246, (occur, biosynth)
[29] King, P.P., *J. Chem. Technol. Biotechnol.*, 1982, **32**, 2, (rev)

Poly(3-hydroxybutyrate-*co*-3-hydroxyvalerate)   P-318

Synonyms: PHBV. Poly(α-hydroxybutyrate-co-α-hydroxyvalerate). Poly(3-hydroxybutanoate-co-3-hydroxypentanoate)
Related Polymers: Poly(3-hydroxybutyrate)
Monomers: 3-Hydroxybutanoic acid, 3-Hydroxypentanoic acid
Material class: Thermoplastic, Copolymers
Polymer Type: polyester
CAS Number:

CAS Reg. No.
80181-31-3

Molecular Formula: $[(C_4H_6O_2)_x(C_5H_8O_2)_y]_n$
Fragments: $C_4H_6O_2$ $C_5H_8O_2$
General Information: Increasing amount of hydroxyvalerate decreases transition temps., improves flexibility and increases toughness [7]
Tacticity: Biosynth. confers inherent tacticity, every β-carbon has R-configuration [5]
Morphology: Displays isodimorphism, where the comonomer components crystallise in the same crystal lattice, the crystal lattice for either hydroxybutyrate or hydroxyvalerate [5]. This co-crystallisation enhances toughness, and increases melting point due to higher lattice energies and an increase in the *a* cell parameter [2,3,9]

Volumetric & Calorimetric Properties:
Density:

No.	Value	Note
1	d 1.231 g/cm^3	11% hydroxyvalerate [21]
2	d 1.2045 g/cm^3	28% hydroxyvalerate [21]

Thermal Expansion Coefficient:

No.	Value	Note	Type
1	8×10^{-5} K^{-1}	28% hydroxyvalerate [14,21]	L

Thermodynamic Properties General: Thermal expansion coefficient increases with increasing hydroxyvalerate content [21]
Melting Temperature:

No.	Value	Note
1	160°C	10% hydroxyvalerate [2]
2	85°C	32% hydroxyvalerate [2]
3	105°C	86% hydroxyvalerate [21]
4	112–127°C	18% hydroxyvalerate [9]
5	68–90°C	55% hydroxyvalerate [9]

Glass Transition Temperature:

No.	Value	Note
1	2°C	11% hydroxyvalerate
2	-8°C	28% hydroxyvalerate [14,21]

Transition Temperature General: T_m decreases with increasing hydroxyvalerate content to a min. at 30 mol %, then increases with increasing hydroxyvalerate content [2]. T_m also depends on cryst. temp. Racemic polymer has lower T_m. T_g decreases with increasing hydroxyvalerate content [21]
Deflection Temperature:

No.	Value	Note
1	140°C	3% hydroxyvalerate [7]
2	112°C	14% hydroxyvalerate [7]
3	92°C	25% hydroxyvalerate [7]

Vicat Softening Point:

No.	Value	Note
1	90°C	28% hydroxyvalerate [7]

Surface Properties & Solubility:
Plasticisers: Polar compounds may be used as plasticisers [7]. 3-Hydroxyvalerate acts as an internal plasticiser for polyhydroxybutyrate
Solvents/Non-solvents: Sol. $CHCl_3$, CH_2Cl_2, dichloroethane, pyridine, trifluoroethanol. Insol. MeOH, Et$_2$O, hexane
Surface Tension:

No.	Value	Note
1	48 mN/m	37°, 12% hydroxyvalerate [15]

Transport Properties:
Polymer Solutions Dilute: [η] 1.03 (4% hydroxyvalerate, MV 113000) ; 3.42 (17% hydroxyvalerate, MV 526000); 1.49 (32% hydroxyvalerate, MV 181000, synthetic) [2]
Polymer Melts: Unstable in the melt
Gas Permeability:

No.	Gas	Value	Note
1	O$_2$	60 cm^3 mm/(m^2 day atm)	film, 17% hydroxyvalerate [7]

Mechanical Properties:
Mechanical Properties General: Tensile strength and Young's modulus decrease with hydrolytic degradation [16] and with increasing temp. [21]. Increasing hydroxyvalerate content decreases tensile props. and increases impact strength

Tensile (Young's) Modulus:

No.	Value	Note
1	1300 MPa	14% hydroxyvalerate [16]
2	3500 MPa	-150°, 28% hydroxyvalerate [21]

| 3 | 1500 MPa | 10°, 28% hydroxyvalerate [21] |
| 4 | 700 MPa | 25% hydroxyvalerate [7] |

Tensile Strength Break:

No.	Value	Note	Elongation
1	113 MPa	14% hydroxyvalerate [16]	42%
2	38 MPa	3% hydroxyvalerate [7]	
3	30 MPa	25% hydroxyvalerate [7]	

Mechanical Properties Miscellaneous: Dynamic mech. props. have been reported [19,20,21]. Dynamic modulus decreases with increasing temp. [20]. Copolymer fibres have been reported [7]
Izod Notch:

No.	Value	Notch	Note
1	60 J/m	Y	3% hydroxyvalerate [7]
2	120 J/m	Y	14% hydroxyvalerate [7]
3	400 J/m	Y	25% hydroxyvalerate [7]

Electrical Properties:
Complex Permittivity and Electroactive Polymers: Piezoelectric: produces a voltage when deformed, and is deformed by an applied voltage [10]

Optical Properties:
Transmission and Spectra: H-1 nmr, C-13 nmr [2,5], ir [5] and uv [5] spectral data have been reported

Polymer Stability:
Thermal Stability General: Is susceptible to thermal degradation
Decomposition Details: Pyrolysis yields crotonic acid and pentenoic acid derivatives (250°) [14]
Hydrolytic Stability: Increasing hydroxyvalerate content and increasing temp. decrease hydrolytic stability [13]. Random chain scission occurs decreasing MW and thus mech. props. [16]
Biological Stability: Degraded via surface attack of bacteria, algae and fungi [10], which dissolve and metabolise the polymer to CO_2 and H_2O. Decompositon rates are increased in moist anaerobic conditions such as in soil, compost, sewage and landfill sites. Degradation is negligible in air
Recyclability: Degradability outweighs any merits of recycling

Applications/Commercial Products:
Processing & Manufacturing Routes: Isol. from *Alcalgeus entrophus* fed with acetate and propionate sources, such as fructose, glucose and propionic acid [4,5,6]. This produces a random copolymer, the composition of the two components is controlled by the amount of feedstock materials. The polymer is then extracted from the bacterial broth using one of three methods. Solvent extraction [11] involves dissolving the polymer from the bacterial cells using chlorinated solvents, followed by filtration to remove cell material, then precipitation. This yields high MW polymers, but large quantities of solvent are required. Sodium hypochlorite [12] digests bacteria, degrading all walls and non-polymeric material. The polymer remains intact and is washed with an organic solvent. This process may degrade the polymer due to the alkalinity of the hypochlorite, leading to lower MW. Biochemical degradation of the bacteria using enzymes [13], followed by drying or solvent extraction is also used. MW may be controlled by the enzymes and conditions used. May be produced synthetically from butyrolactone and valerolactone polymerisation with a stereoselective aluminoxane catalyst. Melt processing must be carried out below 165°
Applications: Used in biodegradable packaging such as bags, films and bottles. Used also for items which are difficult to recycle such as sanitary and food products. Used as an impermeable coating for paper, to increase biodegradability for use in liquid carriers. Possible uses include biodegradable capsules for pharmaceuticals and insecticides, as biocompatible medical and veterinary materials, sutures, dressings and fibres. Applications due to piezoelectricity such as bone fracture healing and in electronics

Trade name	Supplier
Biopol	Monsanto Chemical

Bibliographic References
[1] King, P.P., *J. Chem. Technol. Biotechnol.,* 1982, **32,** 2, (rev)
[2] Bloembergen, S., Holden, D.A., Bluhm, T.L., Hamer, G.K. and Marchessault, R.H., *Macromolecules,* 1987, **20,** 3086, (nmr, synth)
[3] Kunioka, M., Tamaki, A. and Doi, Y., *Macromolecules,* 1989, **22,** 694, (cryst struct)
[4] Doi, Y., Kunioka, M., Nakamura, Y. and Soga, K., *Macromolecules,* 1987, **20,** 2988, (biosynth)
[5] Bluhm, T.L., Hamer, G.K., Marchessault, R.H., Fyfe, C.A.N. and Veregin, R.P., *Macromolecules,* 1986, **19,** 2871
[6] Byrom, D., *Trends Biotechnol.,* 1987, **5,** 246, (occur, synth)
[7] Holmes, P.A., *Dev. Cryst. Polym.,* (ed. D.C. Bassett), Elsevier Applied Science, 1988, **2,** 1, (rev)
[8] Hocking, P.J. and Marchessault, R.H., *Chemistry and Technology of Biodegradable Polymers,* (ed. G.L.J. Griffin), Blackie Academic and Professional, 1994, 48, (rev)
[9] Yoshie, N., Sakurai, M., Inoue, Y. and Chujo, R., *Macromolecules,* 1992, **25,** 2046
[10] Holmes, P.P., *Phys. Technol.,* 1985, **16,** 32, (rev)
[11] Lundgren, D.G., Alper, R., Schnaitmann, C. and Marchessault, R.H., *J. Bacteriol.,* 1965, **89,** 245
[12] Williamson, D.H. and Williamson, J.F., *J. Gen. Microbiol.,* 1958, **19,** 198
[13] Merrick, J.M. and Doudoroff, M., *J. Bacteriol.,* 1964, **88,** 60
[14] Morikawa, H. and Marchessault, R.H., *Can. J. Chem.,* 1981, **59,** 2306, (thermal degradation)
[15] Holland, S.J., Jolly, A.M., Yasin, M. and Tighe, B.J., *Biomaterials,* 1987, **8,** 289, (hydrolytic stability)
[16] Kanesawa, Y. and Doi, Y., Makromol. Chem., *Rapid Commun.,* 1990, **11,** 679
[17] Matavulju, M., Moss, S.T. and Molitoris, H.P., *FEMS Microbiol. Rev.,* 1992, **103,** 465, (biodegradation)
[18] Püchner, P. and Müller, W.R., *FEMS Microbiol. Rev.,* 1992, **103,** 469, (biodegradation)
[19] Bauer, H. and Owen, A.J., *Helv. Phys. Acta,* 1985, **58,** 941, (dynamic mech props)
[20] Owen, A.J., *Colloid Polym. Sci.,* 1985, **263,** 799, (dynamic mech props)
[21] Mitomo, H., Barham, P.J. and Keller, A., *Polym. Commun.,* 1988, **29,** 112, (dynamic mech props)
[22] Marchessault, R.H., Monasterios, C.J., Morin, F.G. and Sundarajan, P.R., *Int. J. Biol. Macromol.,* 1990, **12,** 158, (cryst struct, C-13 nmr)

Polyindole

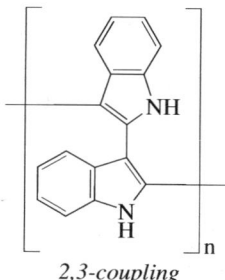

2,3-coupling

Synonyms: *1H-Indole homopolymer*
Monomers: Indole
Material class: Thermoplastic
Polymer Type: polyindole
CAS Number:

CAS Reg. No.
82451-55-6

Molecular Formula: $(C_8H_5N)_n$
Fragments: C_8H_5N
Additives: Dopant ions such as tetrafluoroborate, perchlorate, hexafluorophosphate and chloride are present in the oxidised form obtained electrochemically or chemically.
General Information: Conductive polymer with electrochromic properties. In oxidised (doped) form the polymer obtained as film or powder is usually dark green in colour, while in reduced (de-doped) form it is golden-yellow.
Tacticity: 2,3-Coupling has been established as the most likely form of unit coupling in the polymer [1,2], with regular, statistical or random 2,2-, 2,3-, or 3,3-links possible.
Morphology: Morphology of films is dependent on experimental conditions [3].

Surface Properties & Solubility:
Solubility Properties General: Polymer is usually insoluble in common organic solvents, although electrochemically prepared samples using some supporting electrolytes such as boron trifluoride diethyl etherate exhibit limited solubility in MeCN, THF, Me_2CO, and DMSO [4].

Electrical Properties:
Electrical Properties General: Undoped polymer is an insulator but its conductivity increases considerably when doped, for example with $PF_6^{\ominus}$ (0.6 S cm^{-1}) [5], boron trifluoride diethyl etherate (up to 0.1 S cm^{-1}) [4], $ClO_4^{\ominus}$ (0.03 S cm^{-1}) [6] or $BF_4^{\ominus}$ (0.01 S cm^{-1}) [7]. Samples prepared electrochemically have higher levels of dopant and are more conductive than those prepared by chemical oxidation [3].

Optical Properties:
Optical Properties General: Polymer films exhibit an absorption band in UV region with λ_{max} ca. 400nm; fluorescence bands at longer wavelength have also been reported [4].
Transmission and Spectra: Pmr [2], FT-IR [1,2,4,8,9,10,11], Raman [1,11], uv [4,8,9,10] and esr [8,9,10] spectral data of doped and de-doped polymer have been reported.

Polymer Stability:
Polymer Stability General: Doped polymers are stable at r.t. [3].
Thermal Stability General: Polymer is stable up to ca. 400°C with small weight loss due to dopant only. Decomposition with rupture of polymer backbone occurs between 400°C and 650°C in air [6,12]. Dedoped polymer undergoes 50% weight loss at 544°C in air and 681°C in nitrogen [6].

Applications/Commercial Products:
Processing & Manufacturing Routes: Prod. mainly by electrochemical anodic oxidation of indole in a suitable solvent such as acetonitrile [1,3,11] or dichloromethane [5] with a supporting electrolyte such as lithium perchlorate [1,3,11], boron trifluoride etherate [4] or tetraethylammonium tetrafluoroborate [10]. The polymer obtained is in the doped form and can be dedoped by treatment with, for example, sodium hydroxide. A chemical method of producing polyindole has been reported [3,13], where indole is treated with a suitable oxidising agent such as $FeCl_3$, $CuCl_2$ or KIO_3 in a solvent such as acetonitrile or an alcohol (with $FeCl_3$).
Applications: Potential applications as anode material in batteries and for corrosion protection coatings of metals.

Bibliographic References
[1] Talbi, H., Ghanbaja, J., Billaud, D. and Humbert, B., *Polymer*, 1997, **38**, 2099
[2] Xu, J., Hou, J., Zhou, W., Nie, G., Pu, S. and Zhang, S., Spectrochim. Acta, Part A, 2006, **63**, 723
[3] Maarouf, E.B., Billaud, D. and Hannecart, E., *Mater. Res. Bull.*, 1994, **29**, 1239
[4] Xu, J., Nie, G., Zhang, S., Han, X., Hou, J. and Pu, S., *J. Polym. Sci., Part A: Polym. Chem.*, 2005, **43**, 1444
[5] Pandey, P.C. and Prakash, R., *J. Electrochem. Soc.*, 1998, **145**, 999
[6] Abthagir, P.S. and Saraswathi, R., *Thermochim. Acta*, 2004, **424**, 25
[7] Abthagir, P.S. and Saraswathi, R., *Organic Electronics*, 2004, **5**, 299
[8] Choi, K.M., Jang, J.H. and Kim, K.H., *Mol. Cryst. Liq. Cryst.*, 1992, **220**, 201
[9] Choi, K.M., Jang, J.H., Rhee, H.-W. and Kim, K.H., *J. Appl. Polym. Sci.*, 1992, **46**, 1695
[10] Choi, K.M., Kim, C.Y. and Kim, K.H., *J. Phys. Chem.*, 1992, **96**, 3782
[11] Jackowska, K. and Bukowska, J., *Pol. J. Chem. (Rocz. Chem.)*, 1992, **66**, 1477
[12] Abthagir, P.S., Dhanalakshmi, K. and Saraswathi, R., *Synth. Met.*, 1998, **93**, 1
[13] Billaud, D., Maarouf, E.B. and Hannecart, E., *Synth. Met.*, 1995, **69**, 571

Poly(4-iodostyrene) P-320

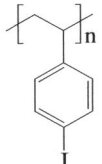

Synonyms: *Poly(p-iodostyrene). Poly(4-iodo-1-vinyl-styrene). Poly(1-(4-iodophenyl)-1,2-ethanediyl)*
Monomers: 4-Iodostyrene
Polymer Type: styrenes
CAS Number:

CAS Reg. No.	Note
24936-53-6	
66572-70-1	isotactic

Volumetric & Calorimetric Properties:
Thermodynamic Properties General: C_p 37.9 J K^{-1} mol^{-1}
Glass Transition Temperature:

No.	Value	Note
1	151°C	[1]

Applications/Commercial Products:
Processing & Manufacturing Routes: Prod. by cationic polymerisation with Lewis acid catalysts

Bibliographic References
[1] Judovits, L.H., Bopp, R.C., Gaur, U. and Wunderlich, B., *J. Polym. Sci., Part B: Polym. Lett.*, 1986, **24**, 2725

Poly(isobutyl cyanoacrylate) P-321

$$\left[CH_2\underset{\underset{\underset{OCH_2CH(CH_3)_2}{|}}{C=O}}{\overset{\overset{CN}{|}}{C}} \right]_n$$

Synonyms: *Isobutyl cyanoacrylate polymer*
Related Polymers: Poly(alkyl cyanoacrylate), Poly(methyl cyanoacrylate), Poly(ethyl cyanoacrylate), Poly(allyl cyanoacrylate)
Monomers: Isobutyl cyanoacrylate
Material class: Thermoplastic, Gums and resins
Polymer Type: acrylic, acrylonitrile and copolymers
CAS Number:

CAS Reg. No.
26809-38-1

Molecular Formula: $(C_8H_{11}NO_2)_n$
Morphology: Amorph.

– Poly(isobutyl methacrylate)

Volumetric & Calorimetric Properties:
Density:

No.	Value	Note
1	d 1.125 g/cm^3	20° [2]

Melting Temperature:

No.	Value	Note
1	192°C	[5]

Glass Transition Temperature:

No.	Value	Note
1	130°C	differential scanning calorimetry [2]
2	51°C	dilatometry [4]

Transition Temperature General: T_g varies with composition of polymer and impurities

Vicat Softening Point:

No.	Value	Note
1	107°C	[5]

Surface Properties & Solubility:
Cohesive Energy Density Solubility Parameters: δ 11 J$^{1/2}$ cm$^{-3/2}$
Wettability/Surface Energy and Interfacial Tension: Contact angle 55° [3]

Mechanical Properties:
Mechanical Properties General: Is a good adhesive for bone fractures; tensile strength is max. with addition of 10% hydroxyapatite [2]

Tensile Strength Break:

No.	Value	Note	Elongation
1	20.4 MPa	2960 psi, steel, ASTM D638 [5]	2%
2	8.3 MPa	bone [3]	

Mechanical Properties Miscellaneous: Tensile shear strength 9.79 MPa (1420 psi, steel-steel, Al-Al, ASTM D1002) [7]

Electrical Properties:
Dielectric Permittivity (Constant):

No.	Value
1	4.2

Optical Properties:
Refractive Index:

No.	Value	Note
1	1.26	20°

Polymer Stability:
Polymer Stability General: Is unstable at temps. above 150° (approx.)
Decomposition Details: Depolymerisation occurs at increased temp.
Environmental Stress: Unsuitable for use in situations of high heat, high impact, moisture or outdoor conditions
Hydrolytic Stability: Degraded by alkali; unstable to moisture

Applications/Commercial Products:
Processing & Manufacturing Routes: Prepared by polymerisation of isobutyl cyanoacrylate monomer when exposed to base. Forms nanoparticles if polymerised in dilute soln. in presence of surfactant [1]. Particle size up to 50 nm
Applications: Instant adhesive, may be used for bone fractures. Nanoparticles may be used for drug transport if polymerisation occurs in a dilute soln. of pharmacological agent, which can be used for targetted therapy

Bibliographic References
[1] Seijo, B., Fattal, E., Roblot-Treupel, L., and Couvreur, P., *Int. J. Pharm.*, 1990, **62**, 1, (nanoparticles)
[2] Tseng, Y.C., Hyon, S.H. and Ikadu, Y., *Biomaterials*, 1990, **11**, 73, (synth., mechanical props., degradation)
[3] Papatheofanis, F.J., *Biomaterials*, 1989, **10**, 185, (mechanical props.)
[4] Kulkarni, R.K., Porter, H.S. and Leonard, F.J., *J. Appl. Polym. Sci.*, 1973, **17**, 3509, (Tg)
[5] Coover, H.W., Dreifus, D.W. and O' Connor, J.T., *Handb. Adhes.*, 3rd edn., (ed. I. Skeist), Van Nostrand, 1990, 463, (rev)
[6] Millet, G.H., *Structural Adhesives Chemistry and Technology*, (ed, S.R. Hartshorn), Plenum Press, 1986, 249, (rev)

Poly(isobutyl methacrylate)   P-322

$$\left[\begin{array}{c} CH_2C(CH_3) \\ | \\ O\!\!=\!\!C\!-\!OCH_2CH(CH_3)_2 \end{array} \right]_n$$

Synonyms: *Poly(2-methylpropyl methacrylate). Poly(isobutyl 2-methyl-2-propenoate)*
Related Polymers: Polymethacrylates-General
Monomers: Isobutyl methacrylate
Material class: Thermoplastic
Polymer Type: acrylic
CAS Number:

CAS Reg. No.	Note
9011-15-8	general
26814-04-0	isotactic
26814-03-9	syndiotactic

Molecular Formula: $(C_8H_{14}O_2)_n$
Fragments: $C_8H_{14}O_2$
Tacticity: Isotactic, syndiotactic and atactic forms known

Volumetric & Calorimetric Properties:
Density:

No.	Value	Note
1	d^{20} 1.045 g/cm^3	[1]
2	d^{20} 1.088 g/cm^3	[2]

Thermal Expansion Coefficient:

No.	Value	Note	Type
1	0.0006 K^{-1}	M_n 912000, T > T_g [3]	V
2	0.00026 K^{-1}	M_n 912000, T g [3]	V

Equation of State: Mark-Houwink constants have been reported [18,19]
Specific Heat Capacity:

No.	Value	Note	Type
1	223.4 kJ/kg.C	27°, melt [4]	P

Poly(isobutyl methacrylate)

2	109 kJ/kg.C	-43°	P
3	137.7 kJ/kg.C	27°	P
4	207 kJ/kg.C	127°	P

Glass Transition Temperature:

No.	Value	Note
1	48°C	bulk polymerisation, 60°; M_n 912000 [3]
2	53°C	atactic [5]
3	53°C	free-radical polymerisation, 44–60° [6]
4	120°C	syndiotactic [5,7]
5	8°C	isotactic [5,7]
6	8°C	anionic polymerisation/non-polar solvent [6]
7	55°C	[2]

Vicat Softening Point:

No.	Value	Note
1	67°C	[1]

Transition Temperature:

No.	Value	Note	Type
1	-148°C	[8]	γ-transition

Surface Properties & Solubility:
Cohesive Energy Density Solubility Parameters: Cohesive energy 38.3–62.7 kJ mol-l [9], 41500 J mol^{-1} [5]. Cohesive energy density 313 J cm^{-3} (74.8 cal cm^{-3}) (35°) [10]. δ 35.6–46.5 J$^{1/2}$ cm$^{-3/2}$ (8.5–11.1 cal$^{1/2}$ cm$^{-3/2}$, poor H-bonding solvents) [11], 35.6–41.4 J$^{1/2}$ cm$^{-3/2}$ (8.5–9.9 cal$^{1/2}$ cm$^{-3/2}$) (moderate H-bonding solvents) [11], 39.8–47.7 J$^{1/2}$ cm$^{-3/2}$ (9.5–11.4 cal$^{1/2}$ cm$^{-3/2}$) (strong H-bonding solvents) [11], 16.8–21.5 J$^{1/2}$ cm$^{-3/2}$ [9], 35.7 J$^{1/2}$ cm$^{-3/2}$ (8.53 cal$^{1/2}$ cm$^{-3/2}$) [12] 29.9 J$^{1/2}$ cm$^{-3/2}$ (7.15 cal$^{1/2}$ cm$^{-3/2}$) [13], 36.2 J$^{1/2}$ cm$^{-3/2}$ (8.65 cal$^{1/2}$ cm$^{-3/2}$) [14]; γ_d 20.4, γ_p 3.3

Solvents/Non-solvents: Sol. hexane, cyclohexane, gasoline, Nujol (hot), turpentine, castor oil (hot), linseed oil (hot), CCl_4, EtOH (hot), isopropanol (>23.7°), Et_2, Me_2CO. Insol. formic acid, EtOH (cold) [15].

Wettability/Surface Energy and Interfacial Tension: Contact angle 73° (advancing, H_2O) [16]; (dγ/dT) 0.06 mN m^{-1}k^{-1} [17]. Interfacial tension 5.5 mN m^{-1} (polyethylene, branched; 20° [17].

Surface Tension:

No.	Value	Note
1	30.9 mN/m	20°; MV 35000 [17]
2	23.7 mN/m	140°; MV 35000 [17]
3	21.3 mN/m	180°; MV 35000 [17]

Transport Properties:
Polymer Solutions Dilute: Theta temps. 48.1° (DMF isotactic), 115.6° (methyl carbitol), 74.9° (methyl cellosolve) [19,20]

Permeability of Liquids: Diffusion coefficient [H_2O] 24.3 × 10^{-8} cm^2 s^{-1} (37°) [16]

Water Content: 0.4% (max.) [2]

Water Absorption:

No.	Value	Note
1	0.4 %	max. [16]

Mechanical Properties:
Tensile Strength Break:

No.	Value	Note	Elongation
1	25 MPa	23° [2]	1%

Miscellaneous Moduli:

No.	Value	Note	Type
1	0.29 MPa	[21]	Bulk modulus

Viscoelastic Behaviour: Longitudinal sound speed 2140 m s^{-1} (room temp., 1 MHz) [22], 2350 m s^{-1} [23]. Shear sound speed 945 m s^{-1} (room temp., 1 MHz) [22]

Hardness: Knoop number 8 (Tukon) [2]. Vickers 9 (5 kg load) [1]

Optical Properties:
Refractive Index:

No.	Value	Note
1	1.477	20° [9]

Polymer Stability:
Decomposition Details: Decompostion at 250° yields monomer [1]

Applications/Commercial Products:

Trade name	Details	Supplier
Acryloid (Paraloid)	B67	Rohm and Haas
Elvacite	2045	ICI, UK
Standard Samples	MW 200000–500000	Aldrich
Standard Samples		Scientific Polymer Products Inc.

Bibliographic References
[1] Crawford, J.W.C., *J. Soc. Chem. Ind.*, London, *Trans. Commun.*, 1949, **68**, 201
[2] *Elvacite*, ICI, 1996, (technical datasheet)
[3] Krause, S., Gormley, J.J., Roman, N., Shetter, J.A. and Watenabe, W.H., *J. Polym. Sci., Part A: Polym. Chem.*, 1965, **3**, 3573
[4] Gaur, U., Lau, S.F., Wunderlich, B.B. and Wunderlich, B., *J. Phys. Chem. Ref. Data*, 1982, **11**, 1065, (heat capacity)
[5] Porter, D., *Group Interaction Modelling of Polymer Properties*, Dekker, 1995, 283
[6] Shetter, J.A., *J. Polym. Sci., Part B: Polym. Lett.*, 1962, **1**, 209
[7] Karey, F.E. and MacKnight, W.J., *Macromolecules*, 1968, **1**, 537
[8] Plazek, D.J. and Ngai, K.L., *Physical Properties of Polymers Handbook*, (ed. J.E. Mark), AIP Press, 1996, 139
[9] Van Krevelen, D.W., *Properties of Polymers: Their Correlation with Chemical Structure*, Elsevier, 1976, 79
[10] Lee, W.A. and Sewell, J.H., *J. Appl. Polym. Sci.*, 1968, **12**, 1397
[11] Mangaraj, D., Patra, S. and Rashid, S., *Makromol. Chem.*, 1963, **65**, 39, (solubility parameters)
[12] Frank, C.W. and Gashgari, M.A., *Macromolecules*, 1979, **12**, 163
[13] DiPaola-Baranyi, G., *Macromolecules*, 1982, **15**, 622, (solubility parameters)
[14] Mangaraj, D., Patra, S. and Roy, P.C., *Makromol. Chem.*, 1965, **81**, 173
[15] Strain, D.E., Kennelly, R.G. and DiHmar, H.R., *Ind. Eng. Chem.*, 1939, **31**, 382, (solubility)
[16] Kalachandra, S. and Kusy, R.P., *Polymer*, 1991, **32**, 2428
[17] Wu, S., *J. Polym. Sci., Part C: Polym. Lett.*, 1971, **34**, 19, (interfacial tension)
[18] Valles, R.J., *J. Polym. Sci., Part A: Polym. Chem.*, 1965, **3**, 3853, (eqn of state)
[19] Zafar, M.M., Mahmood, R. and Wadood, S.A., *Makromol. Chem.*, 1972, **160**, 313
[20] Nakajima, A. and Okazaki, K, *Kobunshi Kagaku*, 1965, **12**, 791, (theta temps)
[21] Warfield, R.W., Cueves, J.E. and Barnet, F.R., *J. Appl. Polym. Sci.*, (ed. J.E. Mark), AIP Press, 1968, **12**, 1147, (acoustic props)

[22] Hartman, B., *Physical Properties of Polymers Handbook*, (ed. J.E. Mark), AIP Press, 1996, **18**, 1147, (acoustic props)
[23] Mangaraj, D., Patra, S. and Rashid, S, *Polymer*, 1963, **65**, 39, ((Solubility parameters))

Poly(isobutyl vinyl ether) P-323

Synonyms: *Poly(2-methylpropyl) ether. PIBVE. PiBVE. Poly(2-methylpropoxyethene). Poly(isobutoxyethene). Poly(vinyl isobutyl ether). Poly(vinyl 2-methylpropyl ether)*
Related Polymers: Poly(vinyl ethers), Poly(vinyl isobutyl ether-*co*-maleic anhydride
Monomers: Isobutyl vinyl ether
Material class: Synthetic Elastomers
Polymer Type: polyvinyls
CAS Number:

CAS Reg. No.	Note
28853-85-2	homopolymer
28299-00-5	*S*-(-)-homopolymer
9003-44-5	

Molecular Formula: $(C_6H_{12}O)_n$
Fragments: $C_6H_{12}O$
General Information: Polymerisation can give rise to (*S*)-(-)-form and an (*R,S*)-form
Tacticity: Isotactic and isotactic (*S*)-form have been reported
Morphology: (*S*)-(-)-form crystallises with 17/5 helix [13]

Volumetric & Calorimetric Properties:
Density:

No.	Value	Note
1	d^{30} 0.916 g/cm^3	[2,9]
2	d^{25} 0.93 g/cm^3	rubbery [40]

Thermal Expansion Coefficient:

No.	Value	Note	Type
1	2.26×10^{-5} K^{-1}	T >-22° [1]	L
2	1.26×10^{-5} K^{-1}	-22° >T >-63°	L
3	1.14×10^{-5} K^{-1}	-63° >T >-121°	L
4	1.03×10^{-5} K^{-1}	T <-121° [1]	L
5	0.0007922 K^{-1}	20° [17]	V

Volumetric Properties General: Molar volume V_r 107.6 cm^3 mol^{-1} [40]
Equation of State: Tyagi-Deshpande equation of state information has been reported [17]
Thermodynamic Properties General: Thermodynamic props. have been reported [41]
Specific Heat Capacity:

No.	Value	Note	Type
1	1.63 kJ/kg.C	3° [42]	P
2	1.68 kJ/kg.C	23° [42]	P

Melting Temperature:

No.	Value	Note
1	165°C	[4]
2	120–130°C	[43]

Glass Transition Temperature:

No.	Value	Note
1	-19°C	[2]
2	-24°C	[5]

Vicat Softening Point:

No.	Value	Note
1	65–76°C	[5]

Transition Temperature:

No.	Value	Note	Type
1	-63°C	[1]	T_{gg}
2	-121°C	[1]	T_{gg}
3	-7°C	[2]	Inflection temp.
4	<-10°C	max. [2]	Low temp. stiffening

Surface Properties & Solubility:
Cohesive Energy Density Solubility Parameters: Solubility parameter 31.5 J$^{1/2}$ cm$^{-3/2}$ (7.5 cal$^{1/2}$ cc$^{1/2}$) [2]. δ 38 J$^{1/2}$ cm$^{-3/2}$ (9.07 cal$^{1/2}$ cm$^{-3/2}$) [26]
Solvents/Non-solvents: Mod. sol. 2-butanone, heptane, C_6H_6, THF [3]. Insol. MeOH, Me$_2$CO [3]. Isotactic form insol. heptane, C_6H_6 [4]; isotactic (*S*)-form insol. methyl isobutyl ketone, 2-butanone, nitrobenzene [9]. Amorph. sol. C_6H_6, halogenated hydrocarbons, butanol, 2-butanone, cyclohexanone, cyclohexane, heptane, Et$_2$O [27], insol. EtOH, 2-ethoxyethanol. Crystalline sol. CHCl$_3$ (50°), insol. heptane, hot 2-propanol, 2-butanone
Surface Tension:

No.	Value	Note
1	29 mN/m	23° [12]

Transport Properties:
Polymer Solutions Dilute: [η] 0.11–0.12 dl g^{-1} (C_6H_6, 30°) [35], 0.197 dl g^{-1} (C_6H_6, 30°, isotactic) [36], 0.193 dl g^{-1} (C_6H_6, 30°, atactic) [36], 0.0067 dl g^{-1} (6.7 ml g^{-1}, atactic) [17], 0.0089 dl g^{-1} (8.9 ml g^{-1}, isotactic) [17]. η$_{inh}$ 4 dl g^{-1} [1], 0.28 dl g^{-1} (CHCl$_3$, 30°) [30]. η$_{sp}$ 0.19–0.28 dl g^{-1} (C_6H_6, 30°) [31]. Other viscosity values have been reported [4,5,32,33,34]

Mechanical Properties:
Tensile Strength Break:

No.	Value	Note
1	2.96 MPa	430 psi [5]

Elastic Modulus:

No.	Value	Note
1	3.51–4.56 MPa	509–661 psi [5]

Miscellaneous Moduli:

No.	Value	Note	Type
1	0.94–1.08 MPa	137–157 psi [5]	100% modulus

Hardness: Shore A55 [5], A35 (24h) [29]. Rockwell 55 [5]

Electrical Properties:
Dielectric Permittivity (Constant):

No.	Value	Note
1	2.3214–2.3362	25°, isotactic [26]
2	2.3115–2.3422	25°, atactic [26]
3	2.2683–2.2822	50°, isotactic [36]
4	2.2625–2.2919	50°, atactic [36]

Magnetic Properties: 1.16D (25°, isotactic), 1.07 (25°, atactic) [36], 1.21D (50°, isotactic), 1.11 (50°, atactic) [36], 0.976–0.983D (atactic), 0.983–0.987D (isotactic) [36]. Average dipole moment of polymer per monomeric unit has been reported [33]

Dissipation (Power) Factor:

No.	Value	Note
1	0.00016	[37]

Optical Properties:
Optical Properties General: Forms an optically active homopolymer

Refractive Index:

No.	Value	Note
1	1.4507	30° [2]
2	1.3965	20° [4]

Transmission and Spectra: C-13 nmr [8] and H-1 nmr spectral data of poly (R,S)-form have been reported [22]. Broad-banded uv spectrum between 185–200 nm for (S)-(-)-form [22]has been reported. Ir [25,38,39] and vibrational circular dichroism spectral data have been reported [23,25]. H-1 nmr spectral data have been reported [38,39]

Molar Refraction: Molar refraction 29.41 [2]

Total Internal Reflection: $[\alpha]_D^{25}$ +0.425 (plate, low crystallinity) [9]; $[\alpha]_D^{14}$ +209 ((S)-form, p-xylene) [15]

Polymer Stability:
Polymer Stability General: Oxidatively degraded by air. Also degraded by γ-radiation [14]

Decomposition Details: Decomposes by oxidative chain degradation [14]

Applications/Commercial Products:
Processing & Manufacturing Routes: Polymerised mainly by the use of aluminium catalysts [3,21,24,25], or by using calcinated iron(II) sulfate [7,11,12]. Stereospecific polymerisation achieved with vanadium pentoxide [6]. Has also been prepared by using zeolites [16,19] and Ziegler-Natta catalysts [10]. May also be prepared by catalysis using boron trifluoride [18,20]

Applications: Used in adhesives, lubricants, greases and surface coatings

Trade name	Details	Supplier
Gantrez B		GAF
Lutonal I		BASF
Oppanol C	solid	BASF

Bibliographic References
[1] Haldon, R.A., Schell, W.J. and Simha, R., *J. Macromol. Sci., Phys.*, 1967, **1**, 759, (transition temp, thermal expansion)
[2] Lal, J. and Trick, G.S., *J. Polym. Sci., Part A: Polym. Chem.*, 1964, **2**, 4559, (glass transition temp)
[3] Lal, J. and McGrath, J.E., *J. Polym. Sci., Part A: Polym. Chem.*, 1964, **2**, 3369, (polymerisation)
[4] Vandenberg, E.J., Heck, R.F. and Breslow, D.S., *J. Polym. Sci.*, 1959, **41**, 519, (isotactic form)
[5] Fishbein, L. and Crowe, B.F., *Makromol. Chem.*, 1961, **48**, 221, (mech props)
[6] Biswas, M. and Ahsanul Kabir, G.M., *Polymer*, 1978, **19**, 357, (stereospecific polymerisation)
[7] Hino, M. and Arata, K., *J. Polym. Sci., Polym. Lett. Ed.*, 1978, **16**, 529, (nonstereoselective polymerisation)
[8] Matsuzaki, H., Ito, H., Kawamura, T. and Uryu, T., *J. Polym. Sci., Polym. Chem. Ed.*, 1973, **11**, 971, (C-13 nmr)
[9] Bonsignori, O. and Lorenzi, G.P., *J. Polym. Sci., Part A-2*, 1970, **8**, 1639, (optical activity)
[10] Mishra, A., *Can. J. Chem.*, 1981, **59**, 1425, (Ziegler-Natta catalysis)
[11] Arata, K. and Hino, M., *Bull. Chem. Soc. Jpn.*, 1980, **53**, 535, (polymerisation)
[12] Hino, M. and Arata, K., *J. Polym. Sci., Polym. Chem. Ed.*, 1980, **18**, 235, (polymerisation)
[13] Lorenzi, G.P., Benedetti, E. and Chiellini, E., *Chim. Ind. (Milan)*, 1964, **46**, 1474, (optical activity)
[14] Suzuki, Y., Rooney, J.M. and Stannett, V., *J. Macromol. Sci., Chem.*, 1978, **12**, 1055, (radiation-induced degradation)
[15] Pino, P., Lorenzi, G.P. and Chiellini, E., *J. Polym. Sci., Part C: Polym. Lett.*, 1968, **16**, 3279, (optical rotation)
[16] Barrer, R.M. and Oei, A.T.T., *J. Catal.*, 1974, **34**, 19, (zeolites polymerisation)
[17] Tyagi, O.S. and Deshpande, D.D., *J. Appl. Polym. Sci.*, 1989, **37**, 2041, (thermal expansion, density)
[18] U.S. Pat., 1936, 2 061 934, (boron trifluoride catalysis)
[19] U.S. Pat., 1966, 3 228 923, (zeolite catalysis)
[20] Brit. Pat., 1936, 441 064, (boron trifluoride catalysis)
[21] Ger. Pat., 1960, 1 084 918, (aluminium catalysis)
[22] Pino, P., Salvadori, P., Lorenzi, G.P., Chiellini, E. et al, *Chim. Ind. (Milan)*, 1973, **55**, 182, (H-1 nmr)
[23] Nafie, L.A., Kiederling, T.A. and Stephens, P.J., *J. Am. Chem. Soc.*, 1976, **98**, 2715, (vibrational circular dichroism)
[24] Saegusa, T., Ueshima, T. and Kato, K., *Bull. Chem. Soc. Jpn.*, 1968, **41**, 1694, (aluminium silanolates catalysis)
[25] Saegusa, T., Imai, H. and Furukawa, J., *Makromol. Chem.*, 1963, **64**, 224, (triethyl aluminium catalysis)
[26] Ahmad, H. and Yaseen, M., *Polym. Eng. Sci.*, 1979, **19**, 858, (calculated solubility parameters)
[27] Bello, A., Barrales-Rienda, J.M. and Guzmán, G.M., *Polym. Handb.*, 4th edn., (eds. J. Brandrup, E.H. Immergut and E.A. Grulke), John Wiley and Sons, 1999, **VII**, 353
[28] Bates, R., *J. Appl. Polym. Sci.*, 1976, **20**, 2941
[29] Schildknecht, C.E., Polym. Prepr. (Am. Chem. Soc., Div. Polym. Chem.), 1973, **13**, 1071
[30] Gong, M.S. and Hall, H.K., *Macromolecules*, 1986, **19**, 3011
[31] Aoki, S. and Stille, J.K., *Macromolecules*, 1970, **3**, 473
[32] Imai, H., Saegusa, T., Ohsugi, S. and Furakawa, J., *Makromol. Chem.*, 1965, **81**, 119
[33] Luisi, P.L., Chiellini, E., Franchini, P.F. and Orienti, M., *Makromol. Chem.*, 1968, **112**, 197
[34] Matsumato, T., Furukawa, J. and Morimura, H., *J. Polym. Sci., Part A: Polym. Chem.*, 1971, **9**, 875
[35] Aoki, S., Otsu, T. and Imato, M., *Makromol. Chem.*, 1966, **99**, 133
[36] Takeda, M., Inamura, Y., Okamura, S. and Higashimura, J., *J. Chem. Phys.*, 1960, **33**, 631
[37] Amrhein, E. and Mueller, F.H., *Mater. Res. Bull.*, 1968, **3**, 169
[38] Hodnett, E.M., Amirmoazzami, J. and Tai, J.T.H., *J. Med. Chem.*, 1978, **21**, 652
[39] Nakano, S., Iwasaki, K. and Fukutoni, H., *J. Polym. Sci., Part A: Polym. Chem.*, 1963, **1**, 3277
[40] Van Krevelen, D.W., *Properties of Polymers: Their Correlation with Chemical Structure*, Elsevier, 1972, 44

Poly(isobutyl vinyl ether-co-maleic anhydride) P-324

Synonyms: *Poly(vinyl 2-methylpropyl ether-co-maleic anhydride). Poly[1-(ethenyloxy)-2-methylpropane-co-2,5-furandione)*

Related Polymers: Poly(vinyl ethers)

Monomers: Isobutyl vinyl ether, Maleic anhydride
Material class: Synthetic Elastomers, Copolymers
Polymer Type: polyvinyls
CAS Number:

CAS Reg. No.	Note
26298-64-6	
107247-04-1	alternating

Molecular Formula: $[(C_6H_{12}O).(C_4H_2O_3)]_n$
Fragments: $C_6H_{12}O$ $C_4H_2O_3$

Volumetric & Calorimetric Properties:
Glass Transition Temperature:

No.	Value	Note
1	145°C	[1]

Transport Properties:
Polymer Solutions Dilute: η_{inh} 0.62 dl g^{-1} (0.5% in DMF, 30°) [1]

Optical Properties:
Transmission and Spectra: H-1 nmr spectral data have been reported [1]

Applications/Commercial Products:
Processing & Manufacturing Routes: Prod. by free radical initiated copolymerisation with AIBN
Applications: Cellulose fleece compound, photographic film supports, thickener for printing pastes [2,3,4]

Bibliographic References
[1] Fu, X., Bassett, W. and Vogl, O., *J. Polym. Sci., Polym. Chem. Ed.*, 1983, **21**, 891
[2] *Ger. Pat.*, 1972, 2 049 832, (use)
[3] *Fr. Pat.*, 1974, 2 221 750, (use)
[4] *Ger. Pat.*, 1977, 2 526 688, (use)

1,2-Polyisoprene   P-325

Synonyms: *Polyisoprene. Vinyl-polyisoprene. 1,2-Poly(2-methyl-1,3-butadiene)*
Monomers: Isoprene
Material class: Synthetic Elastomers
Polymer Type: polybutadiene
CAS Number:

CAS Reg. No.
9003-31-0

cis-1,4-Polyisoprene   P-326

Synonyms: *cis-Polyisoprene. cis-1,4-Poly(2-methyl-1,3-butadiene). IR*
Related Polymers: Natural rubber
Monomers: Isoprene
Material class: Synthetic Elastomers
Polymer Type: polybutadiene
CAS Number:

CAS Reg. No.
9003-31-0

General Information: Polyisoprene with > 90% *cis*-1,4 microstruct. is rubbery. Inferior props. to natural rubber, poor retention at high temp. and may be difficult to compound. Easier to process, and generally has more consistent props. than natural rubber
Morphology: The less *cis* content in the polymer, the less crystalline it is, and the less like natural rubber

Volumetric & Calorimetric Properties:
Density:

No.	Value	Note
1	d 0.906 g/cm^3	[23]
2	d 0.91 g/cm^3	98% *cis*-1,4, 1.4% *trans*-1,4, 0.3% 3,4 [2]

Thermal Conductivity:

No.	Value	Note
1	0.13 W/mK	[22]

Specific Heat Capacity:

No.	Value	Note	Type
1	1.885 kJ/kg.C	[22]	P

Glass Transition Temperature:

No.	Value	Note
1	-40°C	37% *cis*, 30% *trans*, 33% 3,4 [1]
2	-34°C	28% *cis*, 25% *trans*, 45% 3,4 [1]
3	-70--64°C	high *cis* content, low MW [32]
4	-72°C	98.3% *cis*-1,4, 1.4% *trans*-1,4, 0.3% 3,4 [2]
5	-71°C	100% *cis* [1]
6	-68°C	70% *cis*, 23% *trans*, 7% 3,4 [1]
7	-52°C	50% *cis*, 27% *trans*, 24% 3,4 [1]

Transition Temperature General: T_g varies with composition, increasing with increasing 3,4-polymer content [32]. Other T_g values have been reported [3]

Surface Properties & Solubility:
Solubility Properties General: *Cis*-1,4-polyisoprene gives the closest approximation to ideal mixing with poly-1,2-butadiene rubber [26]
Cohesive Energy Density Solubility Parameters: Solubility parameter δ 16.9 J$^{1/2}$ cm$^{-3/2}$ [21]
Solvents/Non-solvents: Sol. THF, hydrocarbons; insol. alcohols, esters, nitromethane
Surface Tension:

No.	Value	Note
1	32 mN/m	20° [3]

Transport Properties:
Polymer Solutions Dilute: θ temp. 31.2° (dioxane, 96% *cis*) [4]; 16.5° (methyl isobutyl ketone, 94% *cis*) [5]; 12–34° (dioxane, branched butanones) [6,7,8]. Mark-Houwink constants have been reported [24,45,46]

Mechanical Properties:
Mechanical Properties General: Mech. props. compared to other rubbers have been reported [24]
Tensile Strength Break:

No.	Value	Note	Elongation
1	26 MPa	Natsyn [24]	480%

Tensile Strength Yield:

No.	Value	Elongation	Note
1	2.35 MPa	630%	Natsyn [24]

– trans-1,4-Polyisoprene

Viscoelastic Behaviour: Williams-Landel-Ferry equation parameters for viscoelastic response have been reported [11,12,13]. Other viscoelastic props. have been reported [28,29]. Zero shear viscosity 2.55 MPa s (Natsyn) [33], 0.68 MPa s (Cariflex) [33]. Longitudinal speed of sound 1550 m s^{-1} (MHZ, cis) [17]
Mechanical Properties Miscellaneous: Tear strength 72.3 kN m^{-1} (Natsyn) [24]. Compression set 17% (Natsyn, 22h, 70°) [24]
Friction Abrasion and Resistance: MST abrasion resistance 168 (Natsyn) [24]

Optical Properties:
Refractive Index:

No.	Value	Note
1	1.519	25° [23]

Transmission and Spectra: Ir spectral data have been reported [10]

Polymer Stability:
Polymer Stability General: Has similar resistance to natural rubber, and Guayule rubber, though lacks the natural antioxidants of *Hevea* rubber [25]
Thermal Stability General: Degrades at moderate temp.
Upper Use Temperature:

No.	Value	Note
1	60–80°C	[15]

Decomposition Details: Thermal degradation yields volatile, low MW hydrocarbons [19]; may also produce hydroxyl, ether and carbonyl functionalised compounds and polynuclear aromatic compounds [25]. 50% Volatilisation occurs at 325° [19]. Depolymerisation activation energy 234–264 kJ mol^{-1} [20].
Flammability: Limiting oxygen index 18.5% [14]. High volatile loss, low char polymer. Aluminium trihydrate acts as flame and smoke retardant. Decomposition volatiles are flammable [19,27]
Biological Stability: Stability to soil cultures has been reported [16]
Stability Miscellaneous: γ-Radiation reduces level of unsaturation, producing free radicals, and also produces hydrogen gas release [30]

Applications/Commercial Products:
Processing & Manufacturing Routes: Prod. by Ziegler-Natta or alkyl lithium catalysis of isoprene. Has good moulding props. and often used as processing aid
Applications: More expensive alternative to natural rubber. Has better moulding props. but poorer mech. and green strength props. Used in similar applications, *e.g.* footwear, hoses, flexible mounts

Trade name	Supplier
Cariflex IR	Shell Chemicals
Isolene	Hardman
Kuraray IR	Kuraray Co
Natsyn	Goodyear
Nipol IR	Nippon Zeon

Bibliographic References

[1] Kow, C., Morton, M., Fetters, L.J. and Hadjichristidis, N., *Rubber Chem. Technol.*, 1981, **54**, 940
[2] Senyck, M., *Kirk-Othmer Encycl. Chem. Technol.*, Vol. 19, 4th edn., (eds. J.I. Kroschwitz and M. Howe-Grant), John Wiley and Sons, 1994, 2
[3] Lee, L.H., *J. Polym. Sci., Part A-2*, 1967, **5**, 1103
[4] Ansorena, F.J., Revuetta, L.M. and Guzman, G.M., *Eur. Polym. J.*, 1982, **18**, 19
[5] Candau, F., Strazielle, C. and Benoit, H., *Makromol. Chem.*, 1973, **170**, 165
[6] Hadjichristidis, N., Xu, Z. and Fetters, L.J., *J. Polym. Sci., Polym. Phys. Ed.*, 1982, **20**, 743
[7] Hadjichristidis, N. and Roovers, J.E.L., *J. Polym. Sci., Polym. Phys. Ed.*, 1974, **12**, 2521
[8] Bauer, B.J., Hadjichristidis, N., Fetters, L.J. et al, *J. Am. Chem. Soc.*, 1980, **102**, 2410
[9] Balsara, N.P., *Physical Properties of Polymers Handbook*, (ed. J.E. Mark), AIP Press, 1996, 266
[10] Noda, I., Dowrey, A.E. and Marcott, C., *Physical Properties of Polymers Handbook*, (ed. J.E. Mark), AIP Press, 1996, 396
[11] Boesse, D., Kremer, F. and Fetters, L.J., *Macromolecules*, 1990, **23**, 1826
[12] Boesse, D. and Kremer, F., *Macromolecules*, 1990, **23**, 829
[13] Nemoto, N., Moriwaki, M., Odani, H. and Onogi, S., *Macromolecules*, 1971, **4**, 215
[14] Tewarson, A. and Macaione, D., *J. Fire Sci.*, 1993, **11**, 421
[15] Billmeyer, F.W., *Textbook of Polymer Science*, Wiley Interscience, 1984, 136, 361
[16] Tsuchii, A., Suzuki, T. and Takahara, Y., *Agric. Biol. Chem.*, 1979, **43**, 2441
[17] Hartmann, B. and Jarzynski, J., *J. Acoust. Soc. Am.*, 1974, **56**, 1469
[18] Keller, R.W., *Handbook of Polymer Science and Technology*, (ed. N.P. Cheremisinoff), Marcel Dekker, 1989, **2**, 156
[19] Lawson, D.F., *Handbook of Polymer Science and Technology*, (ed. N.P. Cheremisinoff), Marcel Dekker, 1989, **2**, 206
[20] Sawada, H., *Encycl. Polym. Sci. Eng.*, (eds. H.F. Mark, N.M. Bikales, C.G. Overberger, G. Menges and J.I. Kroschwitz), Wiley Interscience, 1987, **4**, 735
[21] Cowie, J.M.G., *Encycl. Polym. Sci. Eng.*, Supplement, (eds. H.F. Mark, N.M. Bikales, C.G. Overberger, G. Menges and J.I. Kroschwitz), Wiley Interscience, 1987, 466
[22] Thompson, E.V., *Encycl. Polym. Sci. Eng.*, (eds. H.F. Mark, N.M. Bikales, C.G. Overberger, G. Menges and J.I. Kroschwitz), Wiley Interscience, 1987, **16**, 737
[23] Mills, N.J., *Encycl. Polym. Sci. Eng.*, (eds. H.F. Mark, N.M. Bikales, C.G. Overberger, G. Menges and J.I. Kroschwitz), Wiley Interscience, 1987, **10**, 497
[24] *Natsyn Polyisoprene Rubber Bulletin*, Goodyear Chemical Division, The Goodyear Tyre and Rubber Co., (technical datasheet)
[25] Lin, S.-S., *Rubber Chem. Technol.*, 1989, **62**, 315
[26] Roland, C.M. and Trask, C.A., *Rubber Chem. Technol.*, 1989, **62**, 896
[27] Lawson, D.F., *Rubber Chem. Technol.*, 1986, **59**, 455
[28] Vinogradov, G.V. et al, *J. Polym. Sci., Polym. Phys. Ed.*, 1975, **13**, 1721
[29] Kurata et al, *Macromolecules*, 1971, **4**, 215
[30] Böhm, G.A. and Tveekrem, J.O., *Rubber Chem. Technol.*, 1982, **55**, 612
[31] Kow, C., Morton, M. and Fetters, L.J., *Rubber Chem. Technol.*, 1982, **55**, 245
[32] Widmaier, J.M. and Meyer, G.C., *Rubber Chem. Technol.*, 1981, **54**, 940
[33] Senyak, M., *Encycl. Polym. Sci. Eng.*, (eds. H.F. Mark, N.M. Bikales, C.G. Overberger, G. Menges and J.I. Kroschwitz), Wiley Interscience, 1987, **8**, 507
[34] Fabris, H.J. and Sommer, J.G., *Flame Retard. Polym. Mater.*, (eds. W.C. Kuryla and A.J. Papa), Marcel Dekker, 1973, 2

trans-1,4-Polyisoprene

Synonyms: *trans*-Polyisoprene. *trans*-1,4-Poly(2-methyl-1,3-butadiene)
Related Polymers: Natural balata, *cis*-1,4-Polyisoprene
Monomers: Isoprene
Material class: Synthetic Elastomers, Thermoplastic
Polymer Type: polybutadiene
CAS Number:

CAS Reg. No.
9003-31-0

General Information: >90% *trans*-1,4 struct. is crystalline

Volumetric & Calorimetric Properties:
Density:

No.	Value	Note
1	d 0.96 g/cm^3	99% *trans*-1,4, 0.2% 3,4 [1]

3,4-Polyisoprene

Melting Temperature:

No.	Value	Note
1	67°C	99% trans-1,4, 0.2% 3,4 [1]

Surface Properties & Solubility:
Surface Tension:

No.	Value	Note
1	31 mN/m	20° [2]

Transport Properties:
Polymer Solutions Dilute: Theta temps. θ 47.7° (dioxane, 96%trans) [3,4], 250.5° (toluene, propanol) [3,4]. Mark Houwink constants have been reported [3,4]

Permeability of Gases: Gas permeability increases markedly at approx. 45° [6].
$[H_2]$ (11×10^8 cm^2 $(s\ atm)^{-1}$, 25°, vulcanised) [6]
$[O_2]$ (4.7×10^8 cm^2 $(s\ atm)^{-1}$, 25°, vulcanised) [6]
$[N_2]$ 1.65×10^8 cm^2 $(s\ atm)^{-1}$, 25°, vulcanised) [6]
$[CO_2]$ (27×10^8 cm^2 $(s\ atm)^{-1}$, 25°, vulcanised) [6]

Transport Properties:
Gas Permeability:

No.	Gas	Note
1	$[H_2]$	25°, vulcanised [6]
2	$[O_2]$	25°, vulcanised [6]
3	$[N_2]$	25°, vulcanised [6]
4	$[CO_2]$	25°, vulcanised [6]

Mechanical Properties:
Friction Abrasion and Resistance: Good resistance to abrasion, scuffing and cutting

Polymer Stability:
Thermal Stability General: Thermal degradation occurs at moderate temp. [7]
Decomposition Details: Decomposition yields volatile MW hydrocarbons [7]; 50% volatilisation temp. 325° [7]
Flammability: Volatile degradation productss are flammable [7]
Hydrolytic Stability: Softens in hot water

Bibliographic References
[1] Senyek, M., *Kirk-Othmer Encycl. Chem. Technol.*, 4th edn., vol. 9 (eds. J.I. Kroschwitz and M. Howe-Grant), John Wiley & Sons, 1994, 2
[2] Lee, H., *J. Polym. Sci.*, Part A-2, 1967, **5**, 1103
[3] Cooper, W., Eaves, D.E., Vaughan, G., *J. Polym. Sci.*, 1962, **59**, 241
[4] Chatievedi, P.N., Patel, C.K., *J. Polym. Sci., Polym. Phys. Ed.*, 1985, **23**, 1255
[5] Balsara, N.P., *Physical Properties of Polymers Handbook*, (ed. J.E. Mark), A.I.P. Press, 1966, 266
[6] Van Amerongen, G.J., *J. Polym. Sci.*, 1947, **2**, 381

3,4-Polyisoprene

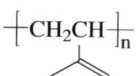

Synonyms: *Polyisoprene. 3,4-Poly(2-methyl-1,3-butadiene)*
Monomers: Isoprene
Material class: Synthetic Elastomers
Polymer Type: polybutadiene

CAS Number:

CAS Reg. No.
9003-31-0

General Information: Amorph.
Tacticity: Syndiotactic and isotactic forms

Polyisoprene star polymers

Synonyms: *Poly(2-methyl-1,3-butadiene) star polymers. Star-branched polymer of polyisoprene*
Monomers: 2-Methyl-1,3-butadiene
Material class: Synthetic Elastomers
Polymer Type: polydienes, polybutadiene
CAS Number:

CAS Reg. No.	Note
9003-31-0	Polyisoprene

Volumetric & Calorimetric Properties:
Glass Transition Temperature:

No.	Value	Note
1	-60.8 - -59.5°C	4-arm star, varying MW [9]
2	-60.5 - -59.9°C	18-arm star, varying MW [9]

Surface Properties & Solubility:
Solvents/Non-solvents: Sol. dioxane, toluene, cyclohexane

Transport Properties:
Polymer Solutions Dilute: Dilute soln. props. and chain dimensions in good solvents and in theta solvents have been reported [7]. Intrinsic viscosity [4,7,11] 0.354 dl g^{-1} (toluene, 3-branch star, MW 35200), 0.958 dl g^{-1} (toluene, 3-branch star, MW 137000), 1.85 dl g^{-1} (toluene, 3-branch star, MW 322000), 0.193 dl g^{-1} (dioxane, 4-branch star, MW 37800), 0.476 dl g^{-1} (dioxane, 4-branch star, MW 215000), 1.112 dl g^{-1} (toluene, 4-branch star, MW 215000), 1.47 dl g^{-1} (dioxane, 4-branch star, MW 1950000), 5.7 dl g^{-1} (toluene, 4-branch star, MW 1950000). Values for higher branching have been reported [4,7,11]. Huggins coefficients and second virial coefficients have been reported [7]. Theta temps. [1,3,7] 34.1° (dioxane, linear star), 33.4° (4-arm star), 33.5° (6-arm star), 32.8° (8-arm star, MW (arm) 39000, MW (star) 276000), 33.5° (8-arm star, MW (arm) 760000, MW (star) 5900000), 28.7° (12-arm star, MW (arm) 21800, MW (star) 250000), 33.0° (12-arm star, MW (arm) 445000, MW (star) 5260000), 24.5° (18-arm star, MW (star) 64000), 33.2° (18-arm star, MW (star) 6800000). The conformn. of star polymers with up to 128 arms in a good solvent has been reported [2,7].

Polymer Melts: The dynamic mech. spectra of polyisoprenes show that the terminal relaxation times of the star polymers are more sensitive to temp. changes than are the linear polymers [8]. Terminal relaxation times of star molecules are determined by the length not the number of arms [9]

Permeability and Diffusion General: Diffusion coefficients Cyclohexane 2.4×10^{-7} cm^2 s^{-1} (23°, 3-branch star, MW 137000), 1.87×10^{-7} cm^2 s^{-1} (23°, 4-branch star, MW 173000), 0.52×10^{-7} cm^2 s^{-1} (23°, 4-branch star, MW 1760000), 0.89×10^{-7} cm^2 s^{-1} (23°, 12-branch star, MW 810000), 0.305×10^{-7} $cm^2 s^{-1}$ (23°, 12-branch star, MW 5280000), 1.88×10^{-7} cm^2 s^{-1} (23°, 18-branch star, MW 384000), 0.455×10^{-7} cm^2 s^{-1} (23°, 18-branch star, MW 3570000) [7]

Optical Properties:
Transmission and Spectra: H-1 nmr and C-13 nmr spectral data have been reported [12]

Total Internal Reflection: Refractive index increments for a series of polymers in cyclohexane with degree of branching 3.9–18.1 and varying MW have been reported. Dn/dc appears independent of branching above the critical weight with the exception of 18-arm stars where dn/dc increases slightly [5]

Applications/Commercial Products:
Processing & Manufacturing Routes: May be synth. by anionic methods using coupling agents such as divinyl benzene and chlorosilane. For example, 1,2-bis(dichloromethylsilyl)ethane and 1,2-bis(trichlorosilyl)ethane are used to make 4-arm and 6-arm stars. Tetra(methyldichlorosilylethane)silane and tetra(trichlorosilylethane)silane are used to synth. 8-arm and 12-arm stars [4,6]

Bibliographic References
[1] Elias, H.-G., *Polym. Handb.*, 4th edn., (eds. J. Brandrup, E.H. Immergut and E.A. Grulke), John Wiley and Sons, 1999, **VII**, 300
[2] Willner, L, Jucknischke, O., Richter, D., Roovers, J. et al, *Macromolecules*, 1994, **27**, 3821, (struct)
[3] Bauer, B.J.; Hadjichristidis, N., Fetters, L.J. and Roovers, J.E.L., *J. Am. Chem. Soc.*, 1980, **102**, 2410, (theta temps)
[4] Hadjichristidis, N. and Roovers, J.E.L., *J. Polym. Sci., Part B: Polym. Lett.*, 1974, **12**, 2521, (synth, soln props)
[5] Hadjichristidis, N. and Fetters, L.J., *J. Polym. Sci., Part B: Polym. Lett.*, 1982, **20**, 2163, (refractive index)
[6] Hadjichristidis, N., Guyot, A. and Fetters, L.J., *Macromolecules*, 1978, **11**, 668, (synth)
[7] Bauer, B.J., Fetters, L.J., Graessley, W.W., Hadjichristidis, N. and Quack, G.F., *Macromolecules*, 1989, **22**, 2337, (soln props)
[8] Bero, C.A. and Roland, C.M., *Macromolecules*, 1996, **29**, 1562
[9] Pakula, T., Geyler, S., Edling, T. and Boese, D., *Rheol. Acta*, 1996, **35**, 631, (viscoelastic props)
[10] Richter, D.; Stuhn, B., Ewen, B. and Nerger, D., *Phys. Rev. Lett.*, 1987, **58**, 2462
[11] Adam, M.; Fetters, L.J.; Graessley, W.W. and Witten, T.A., *Macromolecules*, 1991, **24**, 2434
[12] Pitsikalis, M. and Hadjichristidis, N., *Macromol. Chem. Phys.*, 1995, **196**, 2767, (nmr)

Poly(*N*-isopropylacrylamide)   P-330

Synonyms: *PNIPAM. Poly(N-(1-methylethyl)-2-propenamide)*
Monomers: *N*-Isopropylacrylamide
Polymer Type: polyamide
CAS Number:

CAS Reg. No.
25189-55-3

Molecular Formula: $(C_6H_{11}NO)_n$
Fragments: $C_6H_{11}NO$
General Information: Much of the interest in this polymer centres on its unusual behaviour of its aq. soln. during heating. Can be made in a number of forms including macroscopic gels, microgels, latexes, films, coatings, and fibres [1]. Is a fast-response, temp.-sensitive hydrogel
Morphology: Can exist in both cryst. and amorph. forms. The gel has an opaque microporous struct., with a fibrous network consisting of fine fibres of diameter less than 5μm

Volumetric & Calorimetric Properties:
Density:

No.	Value	Note
1	d 1.07 g/cm^3	amorph. [5,22]
2	d 1.118 g/cm^3	cryst. [22]

Equation of State: Mark-Houwink-Sakurada equation of state parameters have been reported [8]
Melting Temperature:

No.	Value	Note
1	170–200°C	cryst. [22]
2	100–125°C	amorph. [22]

Glass Transition Temperature:

No.	Value	Note
1	134°C	M_n 190000–882000 [2]

Transition Temperature:

No.	Value	Note
1	33°C	LCST

Surface Properties & Solubility:
Solubility Properties General: Exhibits inverse solubility on heating [1]. The polymer can be solubilised in water at temps. above its cloud point by the addition of surfactants [7,30]. Mechanism of solubilisation for poly(*N*-isopropylacrylamide)/sodium alkyl sulfafte has been studied. [23,24] Microporous gel swells by factor of approx. 1.7 (by wt) in H_2O at room temp. It swells below and shrinks above its LCST (33°). A complete contraction (in less than 2 min) is observed when gel is immersed in hot H_2O (50°). Its original size/shape is recovered when shrunk gel is immersed in cold H_2O (10°). Rapid/reversible volume change observed; thermal volume change repeatable between 10° and 50°. MeOH/H_2O mixtures (less than 64.4% (v/v) MeOH) are non-solvents at certain temps. Aq. solns. of other org. solvents e.g. Me_2CO, THF, dioxane, DMSO and EtOH may also be non-solvents. Phase diagram and interaction parameters for MeOH/H_2O system have been reported. Hydrogen bond breaking agents such as urea or LiCl cause precipitation from aq. soln. [6,7,25,26]. Forms a strong interpolymer complex with polyacrylic acid) with a poly(*N*-isopropylacrylamide):polyacrylic acid) ratio of 1.5:1 [27]
Cohesive Energy Density Solubility Parameters: Cohesive energy density 393 J cm^{-3} [2]
Solvents/Non-solvents: Sol H_2O, MeOH, THF, Me_2Co, $CHCl_3$ [6,7,13]
Surface and Interfacial Properties General: Solns. of poly(*N*-isopropylacrylamide) give steady-state surface tensions independent of concentration but varying with temp [15,16]
Wettability/Surface Energy and Interfacial Tension: The contact angle of water on water-swollen poly(*N*-isopropylacrylamide) gels is very temp.-sensitive. Contact angle increases by approx. 60° between 30° and 40° [17]

Transport Properties:
Polymer Solutions Dilute: The effect of anionic surfactant on the coil-to-globule transition has been reported. Coil-to-globule and globule-to-coil transitions for a single chain have been studied and the radius of gyration through both transitions has been reported [18,20]. Spinodal decomposition of aq. solns. occurs at temps just above the cloud point. Kinetics of spinodal decomposition have been reported [4,10]. The plateau adsorbed amount of films at the air-water interface is lower at 16° than at 31.3° but the thickness of the adsorbed layer is larger at the lower temp. At the interface the polymer does not undergo a phase change between 25 and 40° [13,14,15,16]. Adsorption onto silica has been studied as a function of temp.; there is a steep increase in the amount adsorbed at approx. 30° [12]. Aq. solns. are shear-thinning: the viscosity curve changing with temp. [21]. Drag reduction props. of the polymer in aq. soln have been reported [26]. The presence of fairly low levels of surfactant can cause a large increase in the intrinsic viscosity [7,21]. Intrinsic viscosity 4.4 dl g^{-1} (H_2O, 20°, MW 138000); 0.16 dl g^{-1} (H_2O, 20°, MW 652000); 16.7 dl g^{-1} (H_2O, 20°, MW 1630000); 3.8 dl g^{-1} (MeOH, 25°, MW 72300); 8.4 dl g^{-1} (MeOH, 25°, MW 225000); 24.2 dl g^{-1} (MeOH, 25°, MW 1250000) [8,9,28]
Polymer Solutions Concentrated: Conc. solns. display spinnability and Weissenberg effect; no negative thixotropy. Above approx. 28° conc. solns. form a stiff gel [7]
Water Content: 7.8% (film prepared by redox catalysis) [5]

Mechanical Properties:
Mechanical Properties Miscellaneous: The gel is very elastic and flexible in the swollen state. Shrunk gel can be deformed by

mechanical stress; the original state is not recovered, even after removal of stress (plastic deformation)

Optical Properties:
Transmission and Spectra: H-1 nmr and C-13 nmr spectral data have been reported [3,28]

Polymer Stability:
Decomposition Details: Decomposition is a single-stage process with the main weight loss occurring at approx. 420°. Dimers or trimers are liberated and possibly, low MW oligomers [11]

Applications/Commercial Products:
Processing & Manufacturing Routes: The gel is prepared by soln. polymerisation of aq. N-isopropylacrylamide using α-irradiation for 15.3h). May be synth. by free-radical initiation in organic solvent; by redox initiation in aq. soln. by ionic polymerisation or radiation polymerisation [1]
Applications: Used in oil recovery; as thickener (food and coatings applications); in paper finishing; as a viscosity control agent; in immunoassay technology, and in drug-delivery systems

Bibliographic References

[1] Schild, H.G., *Prog. Polym. Sci.*, 1992, **17**, 418, (synth.)
[2] Chinatore, O., Costa, L. and Guaita, M., *Makromol. Chem.*, 1982, **3**, 303, (T_g)
[3] Tokuhiro, T., Amiya, T., Mamada, A. and Tanaka, T., *Macromolecules*, 1991, **24**, 2936, (nmr)
[4] Inomata, H., Yagi, Y., Otake, K., Konno, M. and Saito, S., *Macromolecules*, 1989, **22**, 3494, (spinodal decomposition)
[5] Chiklis, C.K. and Grasshoff, J.M., *J. Polym. Sci., Part A-2*, 1970, **8**, 1617, (soln. props.)
[6] Winnik, F.M., Ringsdorf, H. and Venzmer, J., *Macromolecules*, 1990, **23**, 2415, (solubility)
[7] Eliassaf, J., *J. Appl. Polym. Sci.*, 1978, **22**, 873, (soln. props.)
[8] Chinatore, O., Guaita, M. and Trossarelli, L., *Makromol. Chem.*, 1979, **180**, 969, (soln. props.)
[9] Kubota, K., Fujishige, S. and Ando, I., *Polym. J. (Tokyo)*, 1990, **22**, 15, (soln. props.)
[10] Wang, G. and Hu, Z., *Macromolecules*, 1995, **28**, 4194
[11] Schild, H.G., *J. Polym. Sci., Part A-1*, 1994, **34**, 2259, (thermal decomposition)
[12] Tanahashi, T., Kawaguchi, M., Honda, T. and Takahashi, A., *Macromolecules*, 1996, **27**, 606, (adsorption)
[13] Kawaguchi, M., Saito, W. and Kato, T., *Macromolecules*, 1994, **27**, 5882, (interfacial props.)
[14] Saito, W., Kawaguchi, M., Kato, T. and Imae, T., *Langmuir*, 1996, **12**, 5947, (interfacial props.)
[15] Zhang, J. and Pelton, R., *Langmuir*, 1996, **12**, 2611, (interfacial props.)
[16] Kawaguchi, M., Hirose, Y. and Kato, T., *Langmuir*, 1996, **12**, 3523, (surface tension)
[17] Zhang, J., Pelton, R. and Deng, Y., *Langmuir*, 1995, **11**, 2301, (contact angles)
[18] Wang, X., Qiu, X. and Wu, C., *Macromolecules*, 1998, **31**, 2972
[19] Chiantore, O., Trossarelli, L. and Guaiata, M., *Makromol. Chem.*, 1982, **183**, 2257, (conformn.)
[20] Zhu, P.W. and Napper, D.H., *Langmuir*, 1996, **12**, 5992
[21] Tam, K.C., Wu, X.Y. and Pelton, R.H., *J. Polym. Sci., Part A-1*, 1993, **31**, 963, (viscosity)
[22] Shields, D.J. and Coover, H.W., *J. Polym. Sci.*, 1959, **39**, 532
[23] Lee, L.-T. and Cabane, B., *Macromolecules*, 1997, **30**, 6559
[24] Schild, H.G. and Tirrell, D.A., *Langmuir*, 1991, **7**, 665
[25] Schild, H.G., Muthukumar, M. and Tirrell, D.A., *Macromolecules*, 1991, **24**, 948, (solubility)
[26] Mumick, P.S. and McCormick, C.L., *Polym. Eng. Sci.*, 1994, **34**, 1419, 1429
[27] Staikos, G., Bokias, G. and Karayanni, K., *Polym. Int.*, 1996, **41**, 345
[28] Zhou, S., Fan, S., Au-Yeung, S.C.F. and Wu, C., *Polymer*, 1995, **36**, 1341
[29] Schild, H.G. and Tirrell, D.A., *J. Phys. Chem.*, 1990, **94**, 4352
[30] Kishi, R., et al, *Polym. Gels Networks*, 1997, **5**, 145, (synth., morphology, mechanical props.)

Poly(isopropyl acrylate)   P-331

Synonyms: *Poly[1-(2-propoxycarbonyl)ethylene]. Poly(1-methylethyl 2-propenoate)*
Monomers: Isopropyl acrylate
Material class: Natural elastomers
Polymer Type: acrylic

CAS Number:

CAS Reg. No.
26124-32-3

Molecular Formula: $(C_6H_{10}O_2)_n$
Fragments: $C_6H_{10}O_2$
Molecular Weight: The MW can range from 10000–5000000
Additives: Soap-urea, activated thiol, soap sulfur, lead thiourea, thiourea, diamine, carbon black, trithiocyanuric acid, etc.
General Information: A soft, tacky polymer. Only used as a copolymer

Volumetric & Calorimetric Properties:
Density:

No.	Value
1	d^{25} 1.08 g/cm^3

Thermal Expansion Coefficient:

No.	Value	Note	Type
1	0.00022–0.00026 K^{-1}	glass	V
2	0.00061–0.00063 K^{-1}	liq.	V

Volumetric Properties General: Molar volume 19.3 cm^3 mol^{-1}. Van der Waals volume 66.33 cm^3 mol^{-1}
Glass Transition Temperature:

No.	Value	Note
1	-6 – -3°C	conventional
2	-2 – 7°C	syndiotactic
3	-11°C	isotactic

Transition Temperature General: There is very little change in T_g with changes in tacticity

Surface Properties & Solubility:
Solubility Properties General: Solubility of acrylate ester polymers is dependent on the alcohol derived side group. Short chain groups dissolve in polar solvents, whereas longer chains are sol. in non-polar solvents
Cohesive Energy Density Solubility Parameters: δ 17.8–22.5 J$^{1/2}$ cm$^{-3/2}$ CED 37.8–60.4 kJ mol^{-1}
Plasticisers: Thiokol TP-759, Plastolein 9720, Admex 760
Solvents/Non-solvents: Sol. aromatic hydrocarbons, chlorinated hydrocarbons, THF, esters, ketones, DMF. Insol. aliphatic hydrocarbons, hydrogenated naphthalenes, Et$_2$O, alcohols

Transport Properties:
Permeability of Gases: Has moderate permeability to gases

Mechanical Properties:
Mechanical Properties General: Mech. props. improve with increasing MW up to MW 100000–200000
Impact Strength: Has poor impact resistance
Fracture Mechanical Properties: Has fair flexural cracking resistance and fair to good tear resistance
Friction Abrasion and Resistance: Has good abrasion resistance

Optical Properties:
Molar Refraction: dn/dc -0.103 (bromobenzene, 436 nm), -0.087 (bromobenzene, 546 nm), 0.124 (TFP, 546 nm, isotactic, atactic)

Polymer Stability:
Thermal Stability General: Decomposes in oxygen rich atmospheres under extreme heat
Decomposition Details: Depolymerises much less readily than other polymethacrylates. Between 300 and 500° decomposes to give isoprorene, isopropanol and CO_2

Flammability: Highly flammable
Environmental Stress: Very stable to weathering and resistant to oxygen and ozone. Fair resistance to radiation and synthetic lubricants
Chemical Stability: When heated in presence of free radical initiators, may undergo cross-linking and degradation. Under extreme conditions it can undergo chemical reaction on the ester group
Hydrolytic Stability: Resistant to acid and alkaline hydrol.

Applications/Commercial Products:
Processing & Manufacturing Routes: Commercially it is manufactured by free radical emulsion polymerisation or free radical soln. polymerisation. It is used to produce paints, lacquers and adhesives. It is usually processed by injection moulding, blow moulding, extrusion and thermoforming. Stearic acid, TE-80 and Vanfone APL are added as processing aids
Applications: Acrylics are widely used in paints, coatings and lacquers; in paper industry; as radiation curable systems; adhesives and sealing compds.; in gaskets for the automotive industry and in the textile and leather industries

Bibliographic References
[1] *Ullmanns Encykl. Ind. Chem.*, 5th edn., (ed. W. Gerhartz), VCH, 1992
[2] *Encycl. Polym. Sci. Eng.*, 2nd edn., (eds. H.F. Mark, N.M. Bikales, C.G. Overberger and G. Menges), John Wiley and Sons, 1985
[3] *Kirk-Othmer Encycl. Chem. Technol.*, 4th edn., (ed. J.I. Kroschwitz), 1991, **1**
[4] Krause, S., Gormley, J.J., Roman, N., Shetter, J.A. and Watenabe, W.H., *J. Polym. Sci.*, 1965, **3**, 3573
[5] Warson, H., *Makromol. Chem., Suppl.*, 1985, **10-11**, 265
[6] Goode, W.E., Fellmann, R.P. and Owens, F.H., *Macromol. Synth.*, 1970, **1**, 25
[7] Riddle, E.H., *Monomeric Acrylic Esters*, Reinhold, 1954

Polyisopropylbutadiene P-332
Synonyms: *Poly(2-isopropyl-1,3-butadiene)*
Monomers: 2-Isopropyl-1,3-butadiene
Material class: Synthetic Elastomers
Polymer Type: polydienes

Applications/Commercial Products:
Processing & Manufacturing Routes: Polymerisation in an emulsion system has been reported [4]. Stereoregular polymers are prepared by inclusion polymerisation in deoxycholic acid and apocholic canals [5]

Bibliographic References
[1] Marconi, W, Mazzei, A, Cucinella, S. and Cseari, M., *J. Polym. Sci., Part A-1*, 1964, **2**, 4261
[2] Overberger, C.G., Arond, L.H., Wiley, R.H., and Garrett, R.R, *J. Polym. Sci., Part A-1*, 1951, **7**, 431, (props.)
[3] Seferis, J.C., *Polym. Handb.*, 4th edn., (eds. J. Brandrup, E.H. Immergut and E.A. Grulke), John Wiley and Sons, 1999, **VI**, 575
[4] Marvel, C.S., Williams, J.L.R. and Baumgarten, H.E., *J. Polym. Sci.*, 1949, **4**, 583, (synth.)
[5] Miyata, M., Tsuzuki, T. and Takemoto, K., *Makromol. Chem., Rapid Commun.*, 1987, **8**, 501, (synth.)

Poly(isopropyl methacrylate) P-333

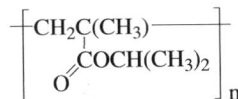

Synonyms: *Poly(2-propyl methacrylate)*
Related Polymers: Polymethacrylates - General
Monomers: Isopropyl methacrylate
Material class: Thermoplastic
Polymer Type: acrylic

CAS Number:

CAS Reg. No.
26655-94-7

Molecular Formula: $(C_7H_{12}O_2)_n$
Tacticity: Syndiotactic, isotactic and atactic form known

Volumetric & Calorimetric Properties:
Density:

No.	Value	Note
1	d^{20} 1.033 g/cm^3	[1,2]

Thermal Expansion Coefficient:

No.	Value	Note	Type
1	0.00067 K^{-1}	free radical polymerisation, 44°, in methyl propionate, M_n 153000, >T_g [3]	V
2	0.00021 K^{-1}	free radical polymerisation, 44°, in methyl propionate, M_n 1530, T_g [3]	V

Glass Transition Temperature:

No.	Value	Note
1	78°C	free radical polymerisation, 44° in methyl propionate [3]
2	81°C	atactic [4]
3	81°C	free radical polymerisation, 44–60°; dilatometric method [5]
4	85°C	free radical polymerisation,
5	139°C	syndiotactic [4]
6	86°C	65% syndiotactic, 29% atactic, 6% isotactic [6]
7	34°C	83% isotactic, 17% atactic [6]
8	27°C	isotactic [4]
9	27°C	anionic polymerisation/non-polar solvent, dilatometric method [5]

Vicat Softening Point:

No.	Value	Note
1	88°C	[7]

Transition Temperature:

No.	Value	Note	Type
1	-223°C	[8]	γ-transition temp.

Surface Properties & Solubility:
Cohesive Energy Density Solubility Parameters: Cohesive energy 37 J mol^{-1} [4], 35.6–37.7 J$^{1/2}$ cm$^{3/2}$ (8.5–9 cal$^{1/2}$ cm$^{-3/2}$) [9,10]
Wettability/Surface Energy and Interfacial Tension: Contact angle 75° (advancing, H_2O) [11]

Transport Properties:
Polymer Solutions Dilute: Molar volumes 76.6 cm^3 mol^{-1} (van der Waal's, amorph., 25°) [2], 124.1 cm^3 mol^{-1} (glassy, amorph., 25°) [2]

Permeability of Liquids: Diffusion coefficient [H_2O] 3.9×10^{-8} cm^2 s^{-1} (37°) [11]
Water Absorption:

No.	Note
1	max. [11]

Mechanical Properties:
Hardness: Vickers 13 (5 kg load) [1]

Electrical Properties:
Dielectric Permittivity (Constant):

No.	Value	Note
1	3	[2]

Optical Properties:
Refractive Index:

No.	Value	Note
1	1.472	20° [1]

Polymer Stability:
Decomposition Details: Decomposes at 250° to yield monomer [1]

Bibliographic References
[1] Crawford, J.W.C., *J. Soc. Chem. Ind.*, London, 1949, **68**, 201
[2] Van Krevelen, D.W., *Properties of Polymers: Their Correlation with Chemical Structure*, Elsevier, 1976, 79
[3] Krause, S., Gormley, J., Roman, N., Shetter, J.A. and Watenabe, W.H., *J. Polym. Sci., Part A: Polym. Chem.*, 1965, **3**, 3573
[5] Shettet, J.A., *J. Polym. Sci., Part A: Polym. Chem.*, 1962, **1**, 209
[6] Walstrom, A.M., Subramanian, R., Long, T.E., McGrath, J.E. and Ward, T.C., Polym. Prepr. (Am. Chem. Soc., *Div. Polym. Chem.*), 1986, **32**, 135
[8] Plazek, D.J. and Ngai, K.L., *Physical Properties of Polymers Handbook*, (ed. J.E. Mark), AIP Press, 1996, 139
[9] Ahmed, H. and Yaseen, M., *Polym. Eng. Sci.*, 1979, **19**, 858
[10] Frank, C.W. and Gashgari, M.A., *Macromolecules*, 1979, **12**, 163
[11] Kalachandra, S. and Kusy, R.P., *Polymer*, 1991, **32**, 2428

Poly(*N*-isopropyl-*N'*-phenyl-*p*-phenylenediamine)   P-334

Synonyms: *N-(1-Methylethyl)-N'-phenyl-1,4-benzenediamine homopolymer. Poly(PrPPD)*
Monomers: *N*-Isopropyl-*N'*-phenyl-*p*-phenylenediamine
Polymer Type: polyaniline
CAS Number:

CAS Reg. No.
108443-86-3

Molecular Formula: $(C_{15}H_{18}N_2)_n$
Fragments: $C_{15}H_{18}N_2$
Morphology: Structure substantially identical to polyaniline. In the formation of polyaniline, the radical cations of semiquinone type were assumed to play an important role [2].
Identification: Chloride derivative: $C_{6.6}H_{5.7}NCl_{0.67}$ [2].

Electrical Properties:
Electrical Properties General: $\sigma = 0.13$ S cm^{-1} [3]

Optical Properties:
Transmission and Spectra: Ir [1,2], esr [2] spectra reported

Applications/Commercial Products:
Processing & Manufacturing Routes: Synthesised from PrPPD using ammonium persulfate as oxidising agent [1,2,3]
Applications: Secondary batteries, which maintain a high charging and discharging efficiency [4,5]

Bibliographic References
[1] Hagiwara, T., Demura, T. and Iwata, K., *Synth. Met.*, 1987, **18**, 317
[2] Hagiwara, T., Yamaura, M. and Iwata, K., *Nippon Kagaku Kaishi*, 1989, 1791
[3] *Jpn. Pat.*, 1987, 62 119 231
[4] *Eur. Pat.* (Sanyo Electric; Mitsubishi Chemical), 1988, 261 837
[5] *Jpn. Pat.* (Bridgestone), 1985, 60 262 355

Poly(isopropyl vinyl ether)   P-335

Synonyms: *PIPVE. Poly(2-propyl vinyl ether). Poly(vinyl isopropyl ether). Poly(2-propoxyethene). Poly(vinyl 2-propyl ether)*
Related Polymers: Poly(vinyl ethers)
Monomers: Isopropyl vinyl ether
Material class: Synthetic Elastomers
Polymer Type: polyvinyls
CAS Number:

CAS Reg. No.	Note
25585-49-3	homopolymer
25620-40-0	isotactic homopolymer

Molecular Formula: $(C_5H_{10}O)_n$
Fragments: $C_5H_{10}O$
Additives: May be stabilised to prevent discoloration
Tacticity: Isotactic and atactic forms known
Morphology: Isotactic form tetragonal with a = b 17.2Å and 17/5 helix [7]

Volumetric & Calorimetric Properties:
Density:

No.	Value	Note
1	d^{30} 0.924 g/cm^3	[4]
2	d^{20} 0.926 g/cm^3	[6]

Thermal Expansion Coefficient:

No.	Value	Note	Type
1	2.23×10^{-5} K^{-1}	T > -12° [5]	L
2	1.02×10^{-5} K^{-1}	-12° > T > -70°	L
3	9.5×10^{-6} K^{-1}	T < -70°	L

Volumetric Properties General: Molar volume V_r 93.2 cm^3 mol^{-1} [23]
Thermodynamic Properties General: Thermodynamic props. have been reported [8,22]
Melting Temperature:

No.	Value	Note
1	190°C	[2]

Glass Transition Temperature:

No.	Value	Note
1	-3°C	[4]
2	-12°C	[22]

– Poly(lactide)

Vicat Softening Point:

No.	Value	Note
1	70–82°C	Softening point [1]

Transition Temperature:

No.	Value	Note	Type
1	6°C	[4]	Inflection temp.
2	-70°C	[5]	T_g

Surface Properties & Solubility:
Cohesive Energy Density Solubility Parameters: δ 18.86 $J^{1/2}$ $cm^{-3/2}$ (9.22 $cal^{1/2}$ $cm^{-3/2}$) [25]. Other solubility parameters have been reported [25]
Solvents/Non-solvents: Sol. C_6H_6, Me_2CO, $CHCl_3$, 2-propanol [6], Et_2O, pentane [9]. Mod. sol. diethyl ketone [3]. Isotactic form sol. hexane [12], insol. MeOH, Me_2CO [2]

Transport Properties:
Polymer Solutions Dilute: η_{inh} 4.5 dl g^{-1} [1], 3.6 dl g^{-1} (toluene, 30°), 0.37–0.51 dl g^{-1} [26]. η_{sp} 0.05–0.07 dl g^{-1} (C_6H_6, 30°) [27]. Huggins constant k_H 5.36–8.94 [26]

Mechanical Properties:
Hardness: Shore A0 [1], A20 (6 d) [28]

Optical Properties:
Refractive Index:

No.	Value	Note
1	1.3849	20° [1]
2	1.3608	25° [24]

Transmission and Spectra: C-13 nmr [10,11] and H-1 nmr [13] spectral data have been reported

Polymer Stability:
Stability Miscellaneous: Stability to γ-irradiation has been reported, 0.86 G scissions $(100\ eV)^{-1}$ [8]

Applications/Commercial Products:
Processing & Manufacturing Routes: Polymerised using an aluminium or iron hydrosulfate catalyst. [3,21] Friedel-Crafts type catalysts [14,17,19,20] and bimetallic catalysts have been reported [18]
Applications: Uses include adhesives, adhesive tapes and antifoaming agents

Bibliographic References
[1] Fishbein, L. and Crowe, B.F., *Makromol. Chem.*, 1961, **48**, 221, (physical props)
[2] Vandenberg, E.J., Heck, R.F. and Breslow, D.S., *J. Polym. Sci.*, 1959, **41**, 519, (melting points, solubility)
[3] Lal, J. and McGrath J.E., *J. Polym. Sci., Part A: Polym. Chem.*, 1964, **2**, 3369, (aluminium hydrosulfate catalyst, polymerisation)
[4] Lal, J. and Trick, G.S., *J. Polym. Sci., Part A: Polym. Chem.*, 1964, **2**, 4559, (transition temp)
[5] Haldan, R.A., Schell, W.J. and Siniha, R., *J. Macromol. Sci., Phys.*, 1967, **1**, 759, (transition temp, thermal expansion)
[6] Shostakovsky, M.F., Mikhantyev, B.I. and Orchinnikova, N.N., *Bull. Acad. Sci. USSR, Div. Chem. Sci. (Engl. Transl.)*, 1953, 959, (density, solubility)
[7] Natta, G., Allegra, G., Bassi, I.W., Carlini, C., *et al*, *Macromolecules*, 1969, **2**, 311, (cryst data)
[8] Suzuki, Y., Rooney, J.M. and Stannett, V., *J. Macromol. Sci., Chem.*, 1978, **12**, 1055, (radiation-induced degradation)
[9] Dall'Asta, G. and Oddo, N., *Chim. Ind. (Milan)*, 1960, **42**, 1234, (solubility, cationic polymerisation)
[10] Deffieux, A., Subira, F. and Stannett, V.T., *Polymer*, 1984, **25**, 1131, (C-13 nmr, radiation-induced polymerisation)
[11] Matsuzaki, K., Ito, H., Kawamura, T. and Uryu, T., *J. Polym. Sci., Polym. Chem. Ed.*, 1973, **11**, 971, (C-13 nmr)
[12] Okamura, S., Hishigamura, T. and Yamamoto, H., *J. Polym. Sci.*, 1958, **33**, 510, (isotactic polymerisation)
[13] Ramey, K.C., Field, N.D. and Borchert, A.E., *J. Polym. Sci., Part A: Polym. Chem.*, 1965, **3**, 2885, (H-1 nmr)
[14] *U.S. Pat.*, 1951, 2 555 179, (boron trifluoride-etherate catalyst)
[15] *U.S. Pat.*, 1960, 2 962 476, (stabilisation)
[16] *U.S. Pat.*, 1965, 3 201 366, (stabilisation)
[17] *U.S. Pat.*, 1963, 3 080 352, (boron trifluoride-etherate catalyst)
[18] *Ger. Pat.*, 1960, 1 084 918, (bimetallic catalysts)
[19] *Ger. Pat.*, 1958, 1 091 754, (Friedel-Crafts type catalysts)
[20] *Brit. Pat.*, 1948, 610 203, (boron trifluoride-etherate system)
[21] *Brit. Pat.*, 1960, 846 690, (iron hydrosulfate catalyst)
[22] Aharoni, S.M., *Polym. Prepr. (Am. Chem. Soc., Div. Polym. Chem.)*, 1973, **14**, 334
[23] Van Krevelen, D.W., *Properties of Polymers: Their Correlation with Chemical Structure*, 1972, 44
[24] Luisi, P.L., Chellini, E., Franchini, P.F. and Orienti, M., *Makromol. Chem.*, 1968, **112**, 197
[25] Ahmad, H. and Yaseen, M., *Polym. Eng. Sci.*, 1979, **19**, 858, (solubility parameters)
[26] Manson, J.A. and Arquette, G.J., *Makromol. Chem.*, 1960, **37**, 187
[27] Aoki, S., Otsu, T. and Imato, M., *Makromol. Chem.*, 1966, **99**, 133
[28] Schildknecht, C.E., *Polym. Prepr. (Am. Chem. Soc., Div. Polym. Chem.)*, 1973, **13**, 1071

Poly(lactide)

Synonyms: *PLA. Poly(lactic acid). Poly(sarcolatic acid). Poly(2-hydroxypropanoic acid). Poly(lactilic acid). Poly(dimethylglycolic acid). Poly(oxycarbonylethylidene)*
Related Polymers: Poly(lactide-*co*-glycolide)
Monomers: Lactide
Material class: Thermoplastic
Polymer Type: saturated polyester
CAS Number:

CAS Reg. No.	Note
26023-30-3	repeat unit
26161-42-2	repeat D (+) unit
26917-25-9	repeat L (-) unit
51063-13-9	repeat (±) unit
70504-40-4	S-*co* (±)
80531-02-8	(3*R*-*cis*)-*co*-(3*S*-*cis*)
135796-12-2	block (3*R*-*cis*)-*co*-(3*S*-*cis*)
85114-66-5	syndiotactic
85066-50-8	isotactic
85075-50-9	*trans* syndiotactic

Molecular Formula: $(C_3H_4O_2)_n$
Fragments: $C_3H_4O_2$
Additives: CMPA fibre, Ca/P-based fibres, carbon fibre
General Information: Can exist as a dimer
Morphology: Struct. has been investigated [30]. Two polymorphic forms exist [14]. α-form - pseudo-orthorhombic a 1.07, b 0.645, c 2.78 (S) [4]; pseudo-orthorhombic a 1.034, b 0.597, c 2.78 (L) [12]; pseudo-orthorhombic a 1.06, b 0.61, c 2.88 (L, helix 10_3) [25]; pseudo-orthorhombic a 1.0700, b 0.6126, c 2.8939 (L) [28]. Other morphology studies have been reported [19]. β-form - orthorhombic a 1.031, b 1.821, c 0.9 (L, helix 3_1) [25]. Mixt. (L) and (D) - triclinic a 0.916, b 0.916, c 0.870, α 109.2°, β 109.2°, γ 109.8° (helix 3_1) [35]

Volumetric & Calorimetric Properties:
Density:

No.	Value	Note
1	d 1.29 g/cm^3	cryst., L-form [14,39]
2	d 1.248 g/cm^3	amorph., L-form [14,39]
3	d^{25} 1.248 g/cm^3	DL-form [23]

Poly(lactide)

Thermodynamic Properties General: Other thermodynamic props. have been reported [37]
Latent Heat Crystallization: ΔH_f 50.7 J g^{-1} (L-form) [12]; ΔH_f 93.7 J g^{-1} (L-form) [19]; ΔH_f 55 J g^{-1} [20]; ΔH_f 14.7 kJ mol^{-1} (3.5 kcal mol^{-1}) [22]; ΔH_f 59 J g^{-1} (L-form) [29]; ΔH_f 76 J g^{-1} [39]; ΔH_f 52.2 J g^{-1} (L-form) [50]. ΔC_p 0.54 J g^{-1} K^{-1} (L-form) [29]

Melting Temperature:

No.	Value	Note
1	60°C	DL-form [1]
2	170°C	L(+)-form [1]
3	160–170°C	L-form [3,7]
4	181°C	[13]
5	148°C	[16,17]
6	170°C	[20]
7	184°C	L-form [22]
8	185°C	α-form, L-form
9	175°C	β-form, L-form [25]
10	186°C	L-form [29]
11	192.1°C	[39]
12	174.8°C	L-form [50]

Glass Transition Temperature:

No.	Value	Note
1	110–115°C	L(+)-form [1]
2	57°C	[11]
3	55°C	L-form [12]
4	56°C	[20]
5	58°C	L-form [22]
6	57°C	(±)-form [22]
7	64°C	L-form [29]
8	59.2°C	[39]
9	55.5°C	L-form [50]
10	49.1°C	DL-form [50]

Transition Temperature:

No.	Value	Note	Type
1	188°C	461K [9]	$T°_m$
2	215°C	L-form [12]	$T°_m$

Surface Properties & Solubility:

Solubility Properties General: Miscible with poly(ethylene glycol) [21,43], poly(ε-caprolactone) [43]. Immiscible with poly(vinyl chloride) [23] and poly(β-hydroxybutyrate) [44]
Cohesive Energy Density Solubility Parameters: Solvent interaction parameters (120°) pentane 1.6, hexane 2.0, heptane 2.0, EtOAc 0.46, Me$_2$CO 0.56, 2-butanone 0.53, CCl$_4$ (DL-form) 0.89, CHCl$_3$ 0.32, CH$_2$Cl$_2$ 0.99, C$_6$H$_6$ 0.52 [23]
Solvents/Non-solvents: Sol. CHCl$_3$ [2,6,34], trichloroethane [2,6], CH$_2$Cl$_2$ [2,6,34], m-cresol [2], dichloroacetic acid [2,6], conc. H$_2$SO$_4$ [2], dioxane [6,34], acetonitrile [6], glycol sulfite [6], hexafluoroacetone sesquihydrate [6], hexafluoroisopropanol [6], DMF [34], pyridine [34]. Insol. petrol [7], MeOH [13,34], hexane [34]
Wettability/Surface Energy and Interfacial Tension: Fold surface free energy 2450 J mol^{-1} (585 cal mol^{-1}) [9]. Surface free energy 0.0265 J m^{-2} (26.5 erg cm^{-2}) [9]. Surface free energy 0.075 J m^{-2} (L-form) [12]

Transport Properties:

Polymer Solutions Dilute: [η] 0.295 dl g^{-1} (bromobenzene, 85°, L-form) [3]; [η] 0.016 dl g^{-1} (C$_6$H$_6$, 30°) [16]; [η]$_{inh}$ 1.34 dl g^{-1} (dioxane, 30°) [7]; [η]$_{inh}$ 1.23 dl g^{-1} (CHCl$_3$, 30°) [10]. Other viscosity values have been reported [13,36,41]

Water Absorption:

No.	Value	Note
1	1.2 %	30° [18]
2	0.5 %	24h, 23°, L-form [50]
3	0.4 %	24h, 23°, DL-form [50]

Mechanical Properties:

Mechanical Properties General: Variation of tensile storage modulus, [29] loss modulus and tensile strength [19,40] with temp. has been reported in graphical form [22]. Flexural storage and Young's moduli of the L and DL-forms have been reported [40]

Tensile (Young's) Modulus:

No.	Value	Note
1	6000–10000 MPa	L-form [14]
2	1720 MPa	L-form [50]
7	1930 MPa	DL-form [50]

Flexural Modulus:

No.	Value	Note
1	6160 MPa	plates, L-form [51]

Tensile Strength Break:

No.	Value	Note	Elongation
1	59.5 MPa	[39]	
2	260–1000 MPa	L-form [14]	12–26%

Tensile Strength Yield:

No.	Value	Note
1	42–51 MPa	DL-form [51]

Flexural Strength Yield:

No.	Value	Note
1	109 MPa	L-form, plates [51]
2	180 MPa	L-form, rods [51]

Impact Strength: Dynstat impact strength 12.7–13.5 kJ m^{-2} (unnotched) [39]
Viscoelastic Behaviour: Stress-strain curves (L-form) have been reported [14,39,50]. Stress relaxation [38] has been reported
Hardness: Shore D77 [50].
Fracture Mechanical Properties: Shear strength 110 MPa (L-form, rods) [51]
Izod Notch:

No.	Value	Notch	Note
1	40 J/m	Y	[39]

– Poly(lactide)

Electrical Properties:
Electrical Properties General: Variation of tan δ with temp. has been reported in graphical form. [29] Dielectric constant of the L-form has been reported [42]
Static Electrification: Piezoelectrical props. (L-form) have been reported [42]

Optical Properties:
Optical Properties General: Optical props. have been reported [30]
Refractive Index:

No.	Value	Note
1	1.47	(R)-form and (S)-form, carbon disulfide-n-butanol [46]

Transmission and Spectra: Ir [2,5,8,17,21,26,36,45,47], nmr [36], esr [49], H-1 nmr [2,5,31,45], C-13 nmr [8,15,31,45,50], Raman [26,47], mass spectrum [10,45] and XPES [27,48] spectral data have been reported
Molar Refraction: dn/dC -0.06 (bromobenzene, 85°) [3]
Total Internal Reflection: $[\alpha]_D$ -150° (CHCl$_3$, 25°, L-form) [3]. $[\alpha]_D^{26}$ -144° (CHCl$_3$, L-form) [7]; $[\alpha]_D$ -162.8° (CH$_2$Cl$_2$, 25°, S-form) [13]; $[\alpha]_D^{25}$ -155° (CHCl$_3$, L-form) [24]; $[\alpha]_D^{25}$ +157° (CHCl$_3$, D-form) [24]. Other [α] values have been reported. [2,5,14,45] Circular dichroism has been reported [46]

Polymer Stability:
Decomposition Details: Thermal degradation details have been reported [10,11,17]. Max. decomposition temp. 365° [10]. Thermal degradation activation energy 92–105 kJ mol^{-1} (22 –25 kcal mol^{-1}) [17]. Decomposition temp. 190° [22], 127° [37]
Hydrolytic Stability: Hydrolytic degradation details have been reported [1,33,39]
Biological Stability: Biodegrades to give lactic acid, CO_2 and H_2O. Polymer has a half-life of 3 months. [32] Self-reinforced fibres degrade faster *in vivo* than *in vitro* [52]
Stability Miscellaneous: γ-Irradiation in air at room temp. causes simultaneous chain scission and cross-linking; crystallinity is decreased [16]

Applications/Commercial Products:
Processing & Manufacturing Routes: Synth. by ring opening polymerisation of lactide using tin catalysts at 130° [1,3,7,11,13,14,24,34,36,39,45]
Applications: Used as surgical sutures and internal bone fracture fixation devices

Trade name	Details	Supplier
PLA	Also reinforced grades (CMPA fibre, Ca/P-based fibres and carbon fibre)	Purac

Bibliographic References

[1] Kulkami, R.K., Moore, E.G., Hegyeli, E.F. and Leonard, F., *J. Biomed. Mater. Res.*, 1971, **5**, 169, (synth, transition temps)
[2] Schutz, R.C. and Schwaab, J., *Makromol. Chem.*, 1965, **87**, 90, (H-1 nmr, solubility)
[3] Tonelli, A.E. and Flory, P.J., *Macromolecules*, 1969, **2**, 225, (synth, soln props)
[4] DeSantis, P. and Kovacs, A., *Biopolymers*, 1968, **6**, 299, (struct)
[5] Goodman, M. and D'Alagni, M., *J. Polym. Sci., Part B: Polym. Lett.*, 1967, **5**, 515, (ir, H-1 nmr)
[6] Schulz, R.C. and Guthmann, A., *J. Polym. Sci., Part B: Polym. Lett.*, 1967, **5**, 1099, (solubility)
[7] Matsuo, S. and Iwakura, Y., *Makromol. Chem.*, 1972, **152**, 203, (optical rotation, synth)
[8] Lillie, E. and Schulz, R.C., *Makromol. Chem.*, 1975, **176**, 1901, (H-1 nmr, C-13 nmr)
[9] Sanchez, I.C. and Eby, R.K., *Macromolecules*, 1975, **8**, 638, (surface free energy)
[10] Garozzo, D., Giuffrida, M. and Montaudo, G., *Macromolecules*, 1986, **19**, 1643, (thermal degradation, ms)
[11] Gilding, D.K. and Reed, A.M., *Polymer*, 1979, **20**, 1459, (synth, thermal degradation)
[12] Kalb, B. and Pennings, A.J., *Polymer*, 1980, **21**, 607, (surface free energy, transition temps)
[13] Schindler, A. and Harper, D., *J. Polym. Sci., Polym. Chem. Ed.*, 1979, **17**, 2593, (soln props)
[14] Eling, B., Gogolewski, S. and Pennings, A.J., *Polymer*, 1982, **23**, 1587, (viscoelastic props, mech props)
[15] Chabot, F., Vert, M., Chapelle, S. and Granger, P., *Polymer*, 1983, **24**, 53, (C-13 nmr)
[16] Gupta, M.C. and Deshmukh, V.G., *Polymer*, 1983, **24**, 827, (radiation stability)
[17] Gupta, M. and Deshmukh, V.G., *Colloid Polym. Sci.*, 1982, **260**, 308, (ir, thermal degradation)
[18] Siemann, U., *Thermochim. Acta*, 1985, **85**, 513, (water absorption)
[19] Gogolewski, S. and Pennings, A.J., *J. Appl. Polym. Sci.*, 1983, **28**, 1045, (morphological study, mech props)
[20] Cohn, D., Younes, H. and Marom, G., *Polymer*, 1987, **28**, 2018, (transition temps)
[21] Younes, H. and Cohn, D., *Eur. Polym. J.*, 1988, **24**, 765, (ir, miscibility)
[22] Jamshidi, K., Hyon, S.H. and Ikada, Y., *Polymer*, 1988, **29**, 2229, (mech props, transition temps)
[23] Riedl, B. and Prud'homme, R.E., *J. Polym. Sci., Part B: Polym. Phys.*, 1986, **24**, 2565, (density, polymer-solvent parameters)
[24] Yui, N., Dijkstra, P.J. and Feijin, J., *Makromol. Chem.*, 1990, **191**, 481, (synth, optical rotation)
[25] Hoogsteen, W., Postema, A.R., Pennings, A.J. and tenBrinke, G., *Macromolecules*, 1990, **23**, 634, (struct, mp)
[26] Kister, G., Cassanas, G., Fabreque, E. and Bardet, L., *Eur. Polym. J.*, 1992, **28**, 1273, (ir, Raman)
[27] Briggs, D. and Beamson, G., *Anal. Chem.*, 1992, **64**, 1729, (XPES)
[28] Marega, C., Marigo, A., DiNoto, V., Zannetti, R., et al, *Makromol. Chem.*, 1992, **193**, 1599, (struct)
[29] Celli, A. and Scandola, M., *Polymer*, 1992, **33**, 2699, (thermodynamic props, mech props)
[30] Kobayashi, J., Asahi, T., Ichiki, M., Oikawa, A., et al, *J. Appl. Phys.*, 1995, **77**, 2957, (struct, optical props)
[31] Espartero, J.L., Rashkov, I., Li, S.M., Manolova, N. and Vert, M., *Macromolecules*, 1996, **29**, 3535, (H-1 nmr, C-13 nmr)
[32] Sinclair, R.G., *Environ. Sci. Technol.*, 1973, **7**, 955, (biodegradation)
[33] Mason, N.S., Miles, C.S. and Sparkes, R.E., *Polym. Sci. Technol. (Plenum)*, 1981, **14**, 279, (hydrolytic degradation)
[34] Kricheldorf, H.R. and Serra, A., *Polym. Bull. (Berlin)*, 1985, **14**, 497, (solubility)
[35] Okihara, T., Tsuji, M., Kawaguchi, A., Katayama, K., et al, *J. Macromol. Sci., Phys.*, 1991, **30**, 119, (struct)
[36] Rak, J., Ford, J.L., Rostron, C. and Walters, V., *Pharm. Acta Helv.*, 1985, **60**, 162, (synth, ir, nmr, soln props)
[37] Kulagina, T.G., Lebedev, B.V., Kiparisova, E.G., Lyudvig, E.B. and Barskaya, I.G., *Vysokomol. Soedin., Ser. A*, 1982, **24**, 1496, (thermodynamic props)
[38] Hasegawa, M., Yamaga, K., Takenaka, I., Dobashi, T. and Sokanishi, A., *Nihon Reoroji Gakkaishi*, 1985, **13**, 131, (stress relaxation)
[39] Grijpma, D.W. and Pennings, A.J., *Macromol. Chem. Phys.*, 1994, **195**, 1633, 1649, (thermodynamic props, mech props)
[40] Engelberg, I. and Kohn, J., *Biomaterials*, 1991, **12**, 292, (mech props)
[41] Rafler, G., Dahlmann, J. and Ruhnau, I., *Acta Polym.*, 1991, **42**, 408, (soln props)
[42] Fukada, E., Proc.-Int. Symp. Electrets, 7th, 1991, 695, (piezoelectrical props)
[43] Zhang, L., Xiong, C. and Deng, X., *Gaofenzi Cailiao Kexue Yu Gongcheng*, 1993, **9**, 70, (miscibility)
[44] Deng, X. and Zhang, L., *Chin. Chem. Lett.*, 1993, **4**, 269, (miscibility)
[45] Hariharan, R. and Pinkus, A.G., Polym. Prepr. (Am. Chem. Soc., Div. Polym. Chem.), 1993, **34**, 526, (ir, nmr, ms, synth)
[46] Bartus, J., Weng, D. and Vogl, O., *Polym. Int.*, 1994, **34**, 433, (refractive index, optical rotation)
[47] Kister, G., Gassanas, G., Vert, M., Pauvert, B. and Terol, A., *J. Raman Spectrosc.*, 1995, **26**, 307, (ir, Raman)
[48] Shard, A.G., Volland, C., Kissel, T. and Davies, M.C., Polym. Prepr. (Am. Chem. Soc., Div. Polym. Chem.), 1995, **36**, 74, (XPES)
[49] Babanalbandi, A., Hill, D.J.T., O'Donnell, J.H., Pomery, P.J. and Whittaker, A., *Polym. Degrad. Stab.*, 1995, **50**, 297, (esr, radiation stability)
[50] Karjalainen, T., Hiljanen-Vainio, M., Malin, M. and Seppala, J., *J. Appl. Polym. Sci.*, 1996, **59**, 1281, (water absorption, thermodynamic props, mech props)
[51] *Polymeric Materials Encyclopedia*, (ed. J.C. Salamone), CRC Press, 1996, **1**, 596, (mech props)
[52] Pohjonen, T. et al, *J. Mater. Sci.: Mater. Med.*, 1997, **8**, 311, (use)

Poly(lactide-co-glycolide) P-337

Synonyms: *PLA-PGA. PGLA. Poly(lactic acid-co-glycolic acid)*
Related Polymers: Poly(lactide)
Monomers: Lactide, Glycolide
Material class: Copolymers
Polymer Type: saturated polyester
CAS Number:

CAS Reg. No.	Note
26780-50-7	Poly(lactide-*co*-glycolide)
34346-01-5	Poly(lactic acid-*co*-glycolic acid)
30846-39-0	L(-) lactide
31213-75-9	DL ($\pm$) lactide
57579-59-6	*cis*-lactide
107131-73-7	*cis*, block
120682-19-1	*trans*
153439-97-5	alternating lactic acid-glycolic acid

Molecular Formula: $[(C_3H_4O_2).(C_2H_2O_2)]_n$
Fragments: $C_3H_4O_2$ $C_2H_2O_2$
Morphology: Orthorhombic, a 0.525, b 0.624, c 0.702 [2]. Other details of struct. have been reported [14,19]
Miscellaneous: Unless otherwise specified, data refer to polymer containing 92% glycolic acid, 8% lactic acid

Volumetric & Calorimetric Properties:
Density:

No.	Value	Note
1	d 1.548 g/cm^3	[1]

Volumetric Properties General: Other density values have been reported [14]
Latent Heat Crystallization: Thermodynamic props (-273–57°) have been reported [15]. ΔH_f 20 J g^{-1} (75% LA, 25% GA). [21] Other enthalpy values have been reported [9]
Melting Temperature:

No.	Value	Note
1	215–217°C	[1,2]
2	210°C	[6]
3	178.5°C	5.7% GA
4	147°C	25% GA, 75% LA [21]
5	147°C	26.6% GA [9]

Glass Transition Temperature:

No.	Value	Note
1	37°C	90% GA, 10% LA [4]
2	43°C	[6]
3	58.5°C	5.7% GA
4	45.5°C	26.6% GA [9]

Transition Temperature General: Other T_m values have been reported [14]
Transition Temperature:

No.	Value	Note	Type
1	65°C	[1]	Softening temp.

Surface Properties & Solubility:
Solubility Properties General: Miscible with poly(3-hydroxybutyric acid-*co*-3-hydroxyvaleric acid), poly(3-hydroxybutyric acid), poly-caprolactone, poly(mandelic acid), poly(propylene fumarate) [11]
Solvents/Non-solvents: Sol. hot γ-butyrolactone [1], CHCl$_3$ [9], hexafluoro-2-propanol [9]

Transport Properties:
Polymer Solutions Dilute: [η] 0.87 dl g^{-1} (trichlorophenol/phenol (7:10), 30°) [1]; viscosity values for 5.7–53.7% GA have been reported [9]; [η] 0.65 dl g^{-1} (hexafluoroisopropanol, 30°, 35% GA) [11]. Rheological props. have been reported [19]. Calculated values for viscosity have been reported [18]
Water Absorption:

No.	Value	Note
1	1.5 %	60h, 37°, 26.6% GA [9]

Mechanical Properties:
Mechanical Properties General: The decrease in tensile strength after immersion of the polymer in phosphate buffer soln. has been reported in graphical form [8]. Other tensile strength values have been reported [13]
Flexural Modulus:

No.	Value	Note
1	2070 MPa	300 kpsi, 30°, ASTM 1043-51 [1]

Tensile Strength Break:

No.	Value	Note	Elongation
1	87.6 MPa	12700 psi [1]	8.4%

Tensile Strength Yield:

No.	Value	Elongation	Note
1	116.5 MPa	4.6%	16900 psi [1]

Viscoelastic Behaviour: Stress-strain curves have been reported [13]

Optical Properties:
Transmission and Spectra: Mass [3,12,20], H-1 nmr [4,5,7], nmr [17,19], C-13 nmr [7,9], XPES [20], esr [10] and ir [14,19] spectral data have been reported
Total Internal Reflection: Birefringence has been reported [6]. Optical rotations for polymer synth. from varying mole ratios and catalysts have been reported [7]
Volume Properties/Surface Properties: Slightly yellow and opaque [1]

Polymer Stability:
Decomposition Details: Decomposition temp. 240° [1]. Cyclic oligomers formed [3]. 50% Weight loss at 360° [1]
Hydrolytic Stability: Hydrol. of the ester bonds occurs causing rapid loss of mech. props. and a colour change from amber to white [9]
Biological Stability: Biodegradation details have been reported [19]
Stability Miscellaneous: γ-Irradiation causes degradation by unzipping, with the decrease in M_n more rapid than that of MW [4]

Applications/Commercial Products:
Processing & Manufacturing Routes: Synth. by ring opening copolymerisation of lactide and glycolide at 110° using tin octoate as a catalyst [1,4,7,9,14,17,18,19,21]

– Polymethacrylamide

Applications: Used as surgical suture material

Trade name	Details	Supplier
Polyglactin 910	92% GA, 8% LA	Ethicon Inc.
Vicryl	92% GA, 8% LA	Ethicon Inc.

Bibliographic References

- [1] Chujo, K., Kobayashi, H., Suzuki, J. and Tokuhara, S., *Makromol. Chem.*, 1968, **113**, 267, (synth, mech props)
- [2] Chatani, Y., Suehiro, K., Okita, Y., Todokoro, H. and Chujo, K., *Makromol. Chem.*, 1968, **113**, 215, (struct)
- [3] Jacobi, E., Luderwald, I. and Schulz, R.C., *Makromol. Chem.*, 1978, **179**, 429, (thermal degradation, ms)
- [4] Gilding, D.K. and Reed, A.M., *Polymer*, 1979, **20**, 1459, (transition temp, stability)
- [5] Aydin, O., Jacobi, E. and Schulz, R.C., *Angew. Makromol. Chem.*, 1980, **86**, 193, (H-1 nmr)
- [6] Mohajer, Y., Wilkes, G.L. and Orler, B., *Polym. Eng. Sci.*, 1984, **24**, 319, (birefringence, transition temps)
- [7] Kricheldorf, H.R., Jonte, J.M. and Berl, M., Makromol. Chem., *Suppl.*, 1985, **12**, 25, (H-1 nmr, C-13 nmr, optical rotation)
- [8] Chu, C.C., *Polymer*, 1985, **26**, 591, (biodegradation)
- [9] Grijpma, D.W., Nijenhuis, A.J. and Pennings, A.J., *Polymer*, 1990, **31**, 2201, (synth, C-13 nmr, water absorption)
- [10] Pitt, C.G, Wang, J., Shah, S.S., Sik, R. and Chignell, C.F., *Macromolecules*, 1993, **26**, 2159, (esr)
- [11] Domb, A.J., *J. Polym. Sci., Part A: Polym. Chem.*, 1993, **31**, 1973, (miscibility, soln prop)
- [12] Shard, A.G., Volland, C., Davies, M.C. and Kissel, T., *Macromolecules*, 1996, **29**, 748, (ms)
- [13] Chu, C.C. and Moncrief, G., *Ann. Surg.*, 1983, **198**, 223, (use, mech props)
- [14] Asahara, T. and Katayama, S., *Kogyo Kagaku Zasshi*, 1965, **68**, 983, (ir, density, struct, synth)
- [15] Kulagina, T.G. and Lebedev, B.V., *Fiz.-Khim. Osn. Sint. Pererab. Polim.*, 1980, 109, (thermodynamic props)
- [16] Athanasiou, K.A., Niederauer, G.G. and Agrawal, C.M., *Biomaterials*, 1996, **17**, 93, (uses)
- [17] Khomyakov, A.K., Ulasova, T.V. and Lyudvig, E.B., Vysokomol. Soedin., *Ser. A*, 1986, **28**, 2217, (synth, nmr)
- [18] Rafler, G., Dahlmann, J. and Ruhnau, T., *Acta Polym.*, 1990, **41**, 628, (soln props, synth)
- [19] Zhu, J., Shao, Y., Zhang, S. and Zheng, Q., *Zhongguo Fangzhi Daxue Xuebao*, 1991, **17**, 24, (ir, nmr, struct, rheological props, biodegradation)
- [20] Shard, A.G., Volland, C., Kissel, T. and Davies, M.C., Polym. Prepr. (Am. Chem. Soc., *Div. Polym. Chem.*), 1995, **36**, 74, (XPES, ms)
- [21] Grijpma, D.W. and Pennings, A.J., *Macromol. Chem. Phys.*, 1994, **195**, 1633, (synth, hydrolytic degradation, thermodynamic props)

Polymethacrylamide   P-338

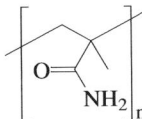

Synonyms: *Poly(2-methyl-2-propenamide). PMAM. PMAAm*
Related Polymers: Poly(methyl methacrylate)
Monomers: Methacrylamide
Material class: Thermoplastic
Polymer Type: acrylic
CAS Number:

CAS Reg. No.	Note
25014-12-4	
28157-21-3	(−)-form
28157-24-6	(+)-form

Molecular Formula: $(C_4H_7NO)_n$
Fragments: C_4H_7NO

Molecular Weight: 133000
Morphology: Forms micelles in aq. soln. [5]

Volumetric & Calorimetric Properties:
Density:

No.	Value	Note
1	d 0.34–1.86 g/cm^3	Rohacell [8]

Thermodynamic Properties General: Other values of C_p have been reported [18]. Heat capacity parameters have been reported [20]
Specific Heat Capacity:

No.	Value	Note	Type
1	0.0025–1.395 kJ/kg.C	0.2127–118.7 J mol^{-1} K^{-1}, -263–27°, glass [28]	P

Melting Temperature:

No.	Value	Note
1	317°C	[20]

Surface Properties & Solubility:
Cohesive Energy Density Solubility Parameters: Solubility parameter data have been reported [12]
Solvents/Non-solvents: Sol. MeOH, H_2O, ethylene, glycol, Me_2CO aq. [26], AcOH (above 50°), formamide (above 50°), formic acid, 2M urea, 0.4M magnesium perchlorate [29]. Insol. tetrafluoropropanol, DMSO, DMF, cresol [29]

Transport Properties:
Polymer Solutions Dilute: Viscosity of aq. solns. may be reduced by the addition of formic acid, MeOH and formaldehyde in amounts of approx. 10% [16]. [η] 1.214 dl g^{-1} (8M urea, MW 950000), 0.61 dl g^{-1} (8M urea, MW 380000), 0.78 dl g^{-1} (0.4M magnesium perchlorate, MW 1380000) [29], 1.74 dl g^{-1} (25°, H_2O) [31]

Mechanical Properties:
Mechanical Properties General: Mech. props. of foams as a function of density have been reported [8]. Temp. dependence of the loss modulus of polymer swollen by H_2O has been reported [25]

Tensile (Young's) Modulus:

No.	Value	Note
1	3600 MPa	Rohacell [8]

Tensile Strength Yield:

No.	Value	Note
1	360 MPa	Rohacell [8]

Viscoelastic Behaviour: Ultrasonic velocity increases linearly with increasing polymer concentration in aq. soln. whereas adiabatic compressibility decreases [31]. Compressive stress of rigid foams is constant for applied strains below approx. 0.75 [8]

Electrical Properties:
Electrical Properties General: Studies on dielectric relaxation in aq. soln. have been reported [32]
Magnetic Properties: Studies on paramagnetism have been reported [14]

Optical Properties:
Transmission and Spectra: XPS [7], H-1 nmr [13], C-13 nmr [13] and esr [17] spectral data have been reported

Polymer Stability:

Decomposition Details: Thermal degradation above 340° [30] yields cyclic imides [21]. Details of char yield have been reported [22]. Thermal degradation is a two-step process; evolution of gaseous material at low temps., with extensive chain scission between 360° and 450° [24]. The only volatile products formed below 340° are NH_3 and H_2O [30]

Chemical Stability: The rate of production of NH_3 upon hydrol. with sodium hydroxide decreases markedly at approx. 66–67% of the theoretical NH_3 content [23]

Stability Miscellaneous: γ-Irradiation causes cleavage of the amide group followed by main chain cleavage [27]

Applications/Commercial Products:

Processing & Manufacturing Routes: May be synth. by electro-polymerisation using sodium acetate as a supporting electrolyte [1]; by aq. polymerisation [2,4,6]. Polymerisation is retarded by the presence of VCl_3 or $MoCl_3$ [3]. Also synth. by vibratory milling [9], plasma-initiated polymerisation [10], high pressure extrusion [11] and ultrasound irradiation in aq. soln. [19]. Optically active material may be synth. using barium and calcium salts of amyl alcohol [15]

Applications: In beads used in hplc columns for chiral resolution and as a flocculant. Possible use in biomedical applications, and as a positive working resist in lithographic applications

Trade name	Supplier
"Standard Samples"	Scientific Polymer Products Inc.
Rohacell	Rohm

Bibliographic References

[1] Davies, M., Venkataraman, B. and Krishnamoorthy, S., *J. Electrochem. Soc. India*, 1990, **39**, 107, (synth.)
[2] Behari, K., Raja, G.D. and Lal, M., *Angew. Makromol. Chem.*, 1987, **150**, 151, (synth.)
[3] Moszner, N., Hartmann, M. and Beil, D., *Acta Polym.*, 1989, **40**, 420
[4] Behari, K., Gupta, K.C., Raja, G.D. and Das, P., *Acta Polym.*, 1991, **42**, 206, (synth.)
[5] Makarewicz, E., *Polimery (Warsaw)*, 1994, **39**, 692, (soln. props.)
[6] Behari, K., Agrawal, U. and Das, R., *Polymer*, 1993, **34**, 4557, (synth.)
[7] Briggs, D. and Beamson, G., *Anal. Chem.*, 1993, **65**, 1517, (XPS)
[8] Maiti, S.K., Gibson, L.J. and Ashby, M.F., *Acta Metall.*, 1984, **32**, 1963, (mechanical props.)
[9] Simionescu, C.I. and Oprea, C.V., *J. Polym. Sci., Polym. Chem. Ed.*, 1985, **23**, 501, (synth.)
[10] Kuzuya, M., Kawaguchi, T., Yanagihara, Y., Nakai, S. et al, *J. Polym. Sci., Part A: Polym. Chem.*, 1986, **24**, 707, (synth.)
[11] Bogdanov, A.Yu., Zharov, A.A. and Zhulin, V.M., *Izv. Akad. Nauk SSSR, Ser. Khim.*, 1986, 250, (synth.)
[12] Ahmad, H., *J. Colour Soc.*, 1981, **20**, 108, (solubility parameters)
[13] Hatada, K., Kitayama, T. and Ute, K., *Polym. Bull. (Berlin)*, 1983, **9**, 241, (nmr)
[14] Zhorin, V.A. and Mel'nikov, V.P., *Vysokomol. Soedin., Ser. B*, 1980, **22**, 389, (EPR)
[15] Yamaguchi, K. and Minoura, Y., *J. Polym. Sci., Part A-1*, 1972, **10**, 1217, (synth.)
[16] Sakaguchi, Y. and Tamaki, K., *Kobunshi Ronbunshu*, 1974, **31**, 208, (viscosity)
[17] Faucitano, A., *Chim. Ind. (Milan)*, 1970, **52**, 427, (esr)
[18] Lebedev, B.V. and Rabinovich, I.B., *Tr. Khim. Khim. Tekhnol.*, 1967, 30, 36, (heat capacity)
[19] Tazuke, S., Tsukamoto, K., Hayashi, K. and Okamura, S., *Kobunshi Kagaku*, 1967, **24**, 302
[20] Xenopoulos, A. and Wunderlich, B., *Polymer*, 1990, **31**, 1260, (heat capacity)
[21] Zurakowska-Orszagh, J., Busz, W. and Soerjosoeharto, K., *Bull. Acad. Pol. Sci., Ser. Sci. Chem.*, 1977, **25**, 845, (thermal degradation)
[22] Hilado, C.J. and Machado, A.M., *J. Fire Flammability*, 1978, **9**, 367
[23] Noma, K. and Niwa, M., *Doshisha Daigaku Rikogaku Kenkyu Hokoku*, 1975, **16**, 53
[24] McNeill, I.C. and Zulfiqar, M., *J. Polym. Sci., Polym. Chem. Ed.*, 1978, **16**, 2465, (thermal degradation)
[25] Kolařík, J. and Dušek, K., *J. Macromol. Sci., Phys.*, 1974, **10**, 157, (relaxation behaviour)
[26] Buchholz, F.L., *Ullmanns Encykl. Ind. Chem.*, (eds. B. Elvers, S. Hawkins and G. Schulz), Vol. A21, VCH, 1992, 143
[27] O'Connor, D., Yang, N.-L. and Woodward, A.E., *Polym. Prepr. (Am. Chem. Soc., Div. Polym. Chem.)*, 1983, **24**, 172, (γ-irradiation)
[28] Gaur, U., Lau, S.-F., Wunderlich, B.B. and Wunderlich, B., *J. Phys. Chem. Ref. Data*, 1982, **11**, 1065, (heat capacity)
[29] Titkova, L.V., Prokopová, E., Sedláček, B., Petrus, V. et al, *Eur. Polym. J.*, 1978, **14**, 145, (solubility, viscosity)
[30] Grassie, N., McNeill, I.C. and Samson, J.N.R., *Eur. Polym. J.*, 1978, **14**, 931, (thermal degradation)
[31] Roy-Chowdhury, P., *Indian J. Chem.*, 1969, **7**, 692, (acoustic props.)
[32] Kaatze, U., *Prog. Colloid Polym. Sci.*, 1978, **65**, 214, (dielectric relaxation)

Poly(methacrylate-co-butadiene-co-styrene) P-339

$$\left[\left[\begin{array}{c} CH_2C(CH_3) \\ | \\ COMe \\ | \\ O \end{array} \right]_x \left[\begin{array}{c} CH_2CH \\ | \\ CH=CH_2 \end{array} \right]_y \left[\begin{array}{c} CH_2CH \\ | \\ Ph \end{array} \right]_z \right]_n$$

Related Polymers: Polymethacrylates General, Poly(methyl methacrylate) General, Poly(methyl methacrylate-co-acrylonitrile-co-butadiene-co-styrene), Polystyrene, Polyacrylonitrile, Polybutadiene

Monomers: Methyl methacrylate, Butadiene, Styrene
Material class: Thermoplastic
Polymer Type: acrylic copolymers
CAS Number:

CAS Reg. No.	Note
25053-09-2	
105935-35-1	block
114180-16-4	block, graft
107080-92-2	graft

Molecular Formula: $(C_5H_8O_2)_m(C_4H_6)_n(C_8H_8)_o$
Fragments: $C_5H_8O_2$ C_4H_6 C_8H_8
Additives: Polyorganosiloxanes to increase toughness
General Information: Is available in a range of colours and transparencies, with good clarity, heat resistance and ageing props. but poor toughness. The terpolymer is two-phase as the components are only partly compatible

Volumetric & Calorimetric Properties:

Density:

No.	Value	Note
1	d 1.07 g/cm^3	ASTM D792 [2]
2	d 1.11 g/cm^3	R1183 [2]

Thermal Expansion Coefficient:

No.	Value	Note	Type
1	$6 \times 10^{-5} - 0.0001$ K^{-1}	ASTM D696 [1]	L
2	9×10^{-5} K^{-1}	ASTM D696 [2]	L

Volumetric Properties General: Specific volume 0.000975 m^3 kg^{-1}, 27 (in^3 lb^{-1}, ASTM D792) [1]

Thermal Conductivity:

No.	Value	Note
1	0.19–0.34 W/mK	0.00045–0.0008 cal s^{-1} cm^{-1} °C^{-1}, ASTM C177 [1]

– Poly(methacrylate-co-...

Specific Heat Capacity:

No.	Value	Note	Type
1	1.257–1.676 kJ/kg.C	0.3–0.4 cal g^{-1} °C^{-1} [1]	P

Deflection Temperature:

No.	Value	Note
1	102–107°C	1.82 MPa, annealed, ASTM D648 [1]
2	107–110°C	0.455 MPa, annealed, ASTM D648 [1]
3	85°C	1.82 MPa [2]

Vicat Softening Point:

No.	Value	Note
1	95°C	[2]
2	96–99°C	[3]

Transport Properties:
Melt Flow Index:

No.	Value	Note
1	0.3–0.6 g/10 min	ASTM 1110 [2]

Water Absorption:

No.	Value	Note
1	0.5 %	24h, ASTM D570 [1]

Mechanical Properties:
Tensile (Young's) Modulus:

No.	Value	Note
1	2280–2550 MPa	330–370 kpsi, ASTM D638 [1]

Flexural Modulus:

No.	Value	Note
1	2000–2200 MPa	[2]

Tensile Strength Break:

No.	Value	Note	Elongation
1	41.4–55.2 MPa	6000–8000 psi, ASTM D638 [1]	20–60%
2	42–55 MPa	[3]	15–18%

Compressive Modulus:

No.	Value	Note
1	1034–1724 MPa	150–250 kpsi, ASTM D695 [1]

Elastic Modulus:

No.	Value	Note
1	2275–2551 MPa	330–370 kpsi, ASTM D790 [1]

Tensile Strength Yield:

No.	Value	Note
1	41–45 MPa	R527 [2]

Flexural Strength Yield:

No.	Value	Note
1	68.9–82.7 MPa	10–12 kpsi, ASTM D790 [1]
2	66–70 MPa	[2]

Compressive Strength:

No.	Value	Note
1	65.5–148.2 MPa	9.5–21.5 kpsi, ASTM D695 [1]

Impact Strength: Izod 320–427 J m^{-1} (6.0–8 ft lb in^{-1}) (notched, ½" × ⅛" bar, ASTM D256 [1]. Charpy 60 kJ m^{-2} (unnotched, 23°) [2]; 34 kJ m^{-2} (unnotched, -40°) [2]; 6.5 kJ m^{-2} (notched, 23°) [2]; 4.9 kJ m^{-2} (notched, -20°) [2]
Mechanical Properties Miscellaneous: Loss factor 0.029 [3]
Hardness: Rockwell R102-108 (ASTM D785) [1]. Ball indentation 105 (2039-1) [2]

Electrical Properties:
Electrical Properties General: Specific resistivity 2.7 × 10" Ω cm
Dielectric Permittivity (Constant):

No.	Value	Frequency	Note
1	3–4	60 Hz	ASTM D150 [1]
2	3–3.8	1 kHz	ASTM D150 [1]
3	3–3.5	1 MHz	ASTM D150 [1]

Dielectric Strength:

No.	Value	Note
1	13.7–19.7 kV.mm^{-1}	350–500 V mil^{-1}, short time, ⅛" thick, ASTM D149 [1]
2	14.2–15.5 kV.mm^{-1}	360–400 V mil^{-1}, step-by-step, ⅛" thick, ASTM D149 [1]

Dissipation Power Factor: tan δ 0.02–0.05 (60 Hz, ASTM D150) [1]; 0.02–0.04 (1 kHz, ASTM D150) [1]; 0.02–0.03 (1 MHz, ASTM D150) [1]
Static Electrification: Arc resistance no track (ASTM D495) [1]

Optical Properties:
Refractive Index:

No.	Value	Note
1	1.538	25°, ASTM D542 [1]
2	1.535	25° [3]

Transmission and Spectra: Transmission 85% (DIN 5036) [2]; 89% [3]

Polymer Stability:
Upper Use Temperature:

No.	Value	Note
1	71–93°C	continuous [1]

Environmental Stress: Shows good resistance to prolonged exposure to sunlight
Chemical Stability: Has good chemical resistance especially to oils, fats and petrol
Stability Miscellaneous: Has good resistance to uv and gamma irradiation

Applications/Commercial Products:
Processing & Manufacturing Routes: Processed by compression and injection moulding
Applications: Refrigerator/freezer drawers, writing instruments, blister-packaging for medical supplies, disposable medical instruments, containers for medicines, cosmetics etc. Large scale use as an additive to increase toughness in PVC (5–15% by weight of the copolymer is added to PVC to improve the clarity and impact strength for use in the manufacture of films and bottles)

Trade name	Supplier
Acryloid	Rohm and Haas
Cyrolite G20	Rohm and Haas
Zylar	Novacor

Bibliographic References
[1] Guide to Plastics, *Property and Specification Charts*, McGraw-Hill,
[2] *Handbook of Industrial Materials,* 2nd edn., Elsevier Advanced Technology, 1992, 393
[3] Svec, P., Rosik, L., Horak, Z. and Vecerka, F., *Styrene-based Plastics and their Modification,* Horwood, 1989
[4] Miura, Y., Sato, T. and Hosoda, A., *Abstr. Pap.-Am. Chem. Soc.,* 1987, **194**, 163, (synth, props)

Polymethacrylate esters of substituted alkyls P-340

$$\left[\begin{array}{c} CH_2C(CH_3) \\ | \\ O=C-OR \end{array} \right]_n$$

where R = $CH_2CH_2NMe_2$ (dimethylaminoethyl)
      = $CH_2CH_2NEt_2$ (diethylaminoethyl)
      = $CH_2CH_2NC(CH_3)_3$ (*tert*-butylaminoethyl)
      = CH_2CH_2SEt (ethylthioethyl)
      = $CH_2CH_2ONO_2$ (2-nitroethyl)
      = $SiMe_3$ (trimethylsilyl)

Synonyms: Poly(dimethylaminoethyl methacrylate). Poly(diethylaminoethyl methacrylate). Poly(tert-butylaminoethyl methacrylate). Poly((ethylthio)ethyl methacrylate). Poly(2-nitroethyl methacrylate). Poly(trimethylsilyl methacrylate)
Related Polymers: Polymethacrylates - General, Polymethacrylate esters of halogenated alkyls, Poly(hydroxyalkyl methacrylates), Poly(ethyl methacrylate)
Monomers: Dimethylaminoethyl methacrylate, Diethylaminoethyl methacrylate, *tert*-Butylaminoethyl methacrylate, (Ethylthio)ethyl methacrylate, 2-Nitroethyl methacrylate, Trimethylsilyl methacrylate
Material class: Thermoplastic
Polymer Type: acrylic
CAS Number:

CAS Reg. No.	Note
25154-86-3	Dimethylaminoethyl ester
94943-81-4	isotactic
25119-82-8	Diethylaminoethyl ester
27273-87-6	(ethylthio)ethyl ester

Volumetric & Calorimetric Properties:
Thermal Expansion Coefficient:

No.	Value	Note	Type
1	0.0006 K^{-1}	dimethylaminoethyl ester, free-radical polymerisation at -20°, $T > T_g$ [1]	L
2	0.00058 K^{-1}	dimethylaminoethyl ester, free-radical polymerisation at 44°, $T > T_g$ [1]	L
3	0.00027 K^{-1}	dimethylaminoethyl ester, free-radical polymerisation at -20°, T g [1]	L
4	0.00037 K^{-1}	dimethylaminoethyl ester, free-radical polymerisation at 44°, T g [1]	L
5	0.00057 K^{-1}	*tert*-butylaminoethyl ester, free-radical polymerisation at 60°, $T > T_g$ [1]	L
6	0.0002 K^{-1}	*tert*-butylaminoethyl ester, free-radical polymerisation at 60°, T g [1]	L
7	0.00046 K^{-1}	(ethylthio)ethyl ester, free-radical polymerisation at 60°, $T > T_g$ [1]	L
8	0.00015 K^{-1}	(ethylthio)ethyl ester, free-radical polymerisation at 60°, T g [1]	L

Melting Temperature:

No.	Value	Note
1	127°C	trimethylsilyl ester, isotactic [2]
2	224°C	trimethylsilyl ester, syndiotactic [2]

Glass Transition Temperature:

No.	Value	Note
1	19°C	dimethylaminoethyl ester, free-radical polymerisation at -20° [1]
2	17°C	dimethylaminoethyl ester, free-radical polymerisation at 44°, M_n 242000 [1]
3	18°C	dimethylaminoethyl ester [3]
4	33°C	*tert*-butylaminoethyl ester, free-radical polymerisation at 60° [1]
5	25°C	(ethylthio)ethyl ester, free-radical polymerisation at 60°, M_n 54000 [1]
6	-20°C	(ethylthio)ethyl ester [4]
7	45°C	2-nitroethyl ester [5]
8	68°C	trimethylsilyl ester, isotactic [2]
9	27°C	trimethylsilyl ester, syndiotactic [2]

Surface Properties & Solubility:
Solubility Properties General: The syndiotactic trimethylsilyl ester has very poor solubility in most solvents
Cohesive Energy Density Solubility Parameters: 39.31 $J^{1/2}$ $cm^{-3/2}$ (9.39 $cal^{1/2}$ $cm^{-3/2}$) (diethylaminoethyl ester) [6]
Surface Tension:

No.	Value	Note
1	36 mN/m	20°, dimethylaminoethyl ester [7]
2	34 mN/m	20°, *tert*-butylaminoethyl ester [7]

Optical Properties:
Refractive Index:

No.	Value	Note
1	1.53	(ethylthio)ethyl ester [4]

Transmission and Spectra: Nmr spectral data have been reported [2]

Applications/Commercial Products:
Processing & Manufacturing Routes: The trimethylsilyl ester is synth. using butyl lithium in toluene for the isotactic form and in THF for the syndiotactic [2]
Applications: Used in the manufacture of dialysis membranes and as fuel additives

Bibliographic References
[1] Krause, S., Gormley, J.J., Roman, N., Shetter, J.A. and Watenabe, W.H., *J. Polym. Sci., Part A: Polym. Chem.*, 1965, **3**, 3573, (glass transition temps)
[2] Aylward, N.N., *J. Polym. Sci., Part A-1*, 1970, **8**, 319
[3] Hopfinger, A.J., Koehler, M.G., Pearlstein, R.A. and Tripathy, S.K., *J. Polym. Sci., Part B: Polym. Phys.*, 1988, **26**, 2007, (glass transition temps)
[4] *Kirk-Othmer Encycl. Chem. Technol.*, 3rd edn., (ed. M. Grayson), John Wiley and Sons, 1978
[5] Borisova, T.I., *Polym. Sci. USSR (Engl. Transl.)*, 1970, **12**, 1060
[6] Golender, B.A., Larin, P.P. and Tashmukhamedov, S., *Polym. Sci. USSR (Engl. Transl.)*, 1976, **18**, 1522
[7] Wu, S., *Polymer Interface and Adhesion*, Marcel Dekker, 1982

Poly(methacrylate ionomers)   P-341

Related Polymers: Polymethacrylates-General, Methacrylic copolymers, Hydroxyalkyl methacrylate copolymers, Methacrylic acid copolymers, Poly(methacrylic acid salt-*co*-ethylene)
Monomers: Methacrylic acid, Methyl methacrylate, Ethyl methacrylate, Butyl methacrylate, Ethyl acrylate
Material class: Thermoplastic
Polymer Type: acrylic
CAS Number:

CAS Reg. No.	Note
130097-34-6	(Methacrylic acid-*co*-acrylic acid, ammonium salt)
25930-98-7	(Butyl methacrylate-*co*-acrylic acid)
173390-78-8	(Methacrylic acid-*co*-butyl acrylate-*co*-acrylic acid, Zn salt)
173390-77-7	(Methacrylic acid-*co*-butyl acrylate-*co*-acrylic acid-*co*-ethyl hexyl methacrylate, Zn salt)
122161-53-9	1,4-butanediyl methacrylate-*co*-methacrylic acid
122161-54-0	1,4-butanediyl methacrylate-*co*-methacrylic acid, Na salt
25035-69-2	(Methyl methacrylate-*co*-butyl acrylate-*co*-methacrylic acid)
26300-51-6	(Methyl methacrylate-*co*-butyl acrylate-*co*-acrylic acid)
28262-63-7	(Methyl methacrylate-*co*-*n*-butyl methacrylate-*co*-methacrylic acid)
130405-07-1	(block copolymer)
121415-00-7	(graft copolymer)
68460-16-2	(Methyl methacrylate-*co*-methacrylic acid-*co*-methacrylic anhydride)
25133-90-8	1,2-ethandiyl methacrylate-*co*-methacrylic acid
26284-14-0	(Butyl methacrylate-*co*-methacrylic acid)
37624-88-7	(Butyl methacrylate-*co*-methacrylic acid, ammonium salt)

Morphology: Poly(methyl methacrylate-*co*-methacrylic acid) is an amorph., glassy polymer. Morphology has been studied by dynamic mechanical themal analysis and small-angle x-ray scatttering [1]
Identification: Contains only C, H, O

Volumetric & Calorimetric Properties:
Density:

No.	Value	Note
1	d 1.01 g/cm^3	methyl methacrylate-*co*-methacrylic acid, 70% methacrylic acid, MV 30500 [2]
2	d 1.13 g/cm^3	methyl methacrylate-*co*-methacrylic acid, 85% methacrylic acid, MV 28530 [2]

Glass Transition Temperature:

No.	Value	Note
1	113°C	methyl methacrylate-*co*-methacrylic acid [1]
2	163°C	methyl methacrylate-*co*-methacrylic acid, 6 mol% Na salt [1]
3	244°C	methyl methacrylate-*co*-methacrylic acid, 6 mol% Na salt [1]
4	123–143°C	methyl methacrylate-*co*-methacrylic acid, 15–30% methacrylic acid, MW 430000–474000 [3]
5	111–115°C	methacrylic acid-*co*-ethyl acrylate, 20–25% methacrylic acid [4]

Transition Temperature General: Ionomers with an ion content between certain values can show two glass-transition temps. corresponding to ion-rich clusters and ion-poor matrix [1]. T_g is affected by composition and MW [3]
Transition Temperature:

No.	Value	Note	Type
1	37°C	methyl methacrylate-*co*-methacrylic acid, 6 mol % Na salt	First order transition temp.
2	50°C	methyl methacrylate-*co*-methacrylic acid, 12.4 mol % Na salt	First order transition temp.
3	85°C	methyl methacrylate-*co*-methacrylic acid, 6 mol % Na salt	Second order transition temp.
4	112°C	methyl methacrylate-*co*-methacrylic acid, 12.4 mol % Na salt [1]	Second order transition temp.

Transport Properties:
Melt Flow Index:

No.	Value	Note
1	5 g/10 min	177°, methacrylic acid-*co*-ethyl acrylate, 25% methacrylic acid
2	5.5 g/10 min	177°, methacrylic acid-*co*-ethyl acrylate, 22.2% methacrylic acid
3	8 g/10 min	177°, methacrylic acid-*co*-ethyl acrylate, 20% methacrylic acid [4]

Poly(methacrylate ionomers)

Polymer Solutions Dilute: For butyl methacrylate, methacrylic acid-co-methacrylic acid in DMF soln. a polyelectrolyte expansion effect is seen at low concentrations. Unravelling of chains tends to increase with the degree of neutralisation [5]. Huggins constants of butyl methacrylate-co-methacrylic acid tend to increase with the level of neutralisation. Only the acid form and ionomers with a low level of ionic groups give "normal" Huggins constants, i.e., 0.3–0.5. Others are around unity or higher [5]

Permeability and Diffusion General: Methyl methacrylate-co-methacrylic acid membranes are permeable to sodium chloride, urea and creatinine. Permeability was not affected by the addition of $Zn^{2\oplus}$ ions [6]

Water Content: Water absorption of methyl methacrylate-methacrylic acid copolymer decreases with increasing methacrylic acid content [2]

Water Absorption:

No.	Value	Note
1	1.96 %	48h, 52% relative humidity, methyl methacrylate-co-methacrylic acid, 70% methacrylic acid
2	9.88 %	624h, 52% relative humidity, methyl methacrylate-co-methacrylic acid, 70% methacrylic acid [2]
3	0.54 %	48h, 52% relative humidity, methyl methacrylate-co-methacrylic acid, 85% methacrylic acid
4	4.35 %	624h, 52% relative humidity, methyl methacrylate-co-methacrylic acid, 85% methacrylic acid [2]

Gas Permeability:

No.	Gas	Value	Note
1	O_2	846 cm^3 mm/ (m^2 day atm)	methacrylic acid-co-ethyl acrylate, 25% methacrylic acid [4]
2	O_2	931 cm^3 mm/ (m^2 day atm)	methacrylic acid-co-ethyl acrylate, 20% methacrylic acid [4]

Mechanical Properties:

Mechanical Properties General: The tensile strength of the methyl methacrylate/methacrylic acid copolymers increases on annealing, rising with annealing time to a plateau value

Tensile (Young's) Modulus:

No.	Value	Note
1	758.3 MPa	110 kpsi, methacrylic acid-co-ethyl acrylate, 25% methacrylic acid
2	213.7 MPa	31 kpsi, methacrylic acid-co-ethyl acrylate, 20% methacrylic acid [4]

Tensile Strength Break:

No.	Value	Note	Elongation
1	24 MPa	3480 psi, methyl methacrylate-co-ethyl acrylate, 25% methylacrylic acid [4]	200%
2	20.9 MPa	3040 psi, methyl methacrylate-co-ethyl acrylate, 22.2% methacrylic acid [4]	250%
3	9.1 MPa	1320 psi, methyl methacrylate-co-ethyl acrylate, 20% methacrylic acid [4]	290%
4	269.3 MPa	methyl methacrylate-co-methacrylic acid, 70% methacrylic acid, MV 28530 [2]	133%
5	40 MPa	methyl methacrylate-co-methacrylic acid, 0.85% methacrylic acid, unannealed [7]	approx. 0.2–0.4%
6	60 MPa	methyl methacrylate-co-methacrylic acid, 0.85% methacrylic acid, annealed [7]	approx. 0.2–0.4 %

Failure Properties General: Heat-treated methyl methacrylate-co-methacrylic acid, Na salt ionomers with 6–12 mol % Na salt deform by crazing with shear. Ionomers with >12 mol % Na salt content deform by shear only. This trend towards shear deformation makes the polymers increasingly fracture resistant [1]

Electrical Properties:

Electrical Properties General: Zeta potential -43.4–-38.8 mV (methyl methacrylate-co-methacrylic acid 85:15), -36–-32.2 mV (methyl methacrylate-co-methacrylic acid 97:3)

Optical Properties:

Refractive Index:

No.	Value	Note
1	0.12	methyl methacrylate-co-methacrylic acid, 70% methacrylic acid, MV 30500 [2]
2	0.1	methyl methacrylate-co-methacrylic acid, 85% methacrylic acid, MV 28530 [2]

Polymer Stability:

Polymer Stability General: The stability of methacrylic acid-co-ethyl acrylate depends greatly on the carboxylate content

Decomposition Details: Methacrylic acid/tert-butyl methacrylate copolymers decompose at 250° to give mainly polymethylacrylic anhydride with the anhydride bonds being intra rather than intermolecular. At 200° the copolymer is converted into primarily methacrylic acid/tert-butyl methacrylate/methacrylic anhydride terpolymers [8]. The copolymers undergo an intramolecular cyclisation at moderate temps. (130–230°) to give terpolymers with anhydride moieties [9]

Environmental Stress: Methacrylic acid/ethyl acrylate copolymers with high acid group content are broken down readily in the soil by rainwater. Resins with high acid group content are swollen upon outdoor exposure and eventually disintegrate completely [4]

Chemical Stability: Methacrylic acid/ethyl acrylate copolymers with high acid group content dissolve readily in brine [4]

Applications/Commercial Products:

Processing & Manufacturing Routes: May be synth. by partial hydrol. in conc. sulfuric acid [3]. Poly(butyl methacrylate-co-methacrylic acid) is synth. by partial hydrol. of poly(butyl methacrylate) followed by refluxing the product with hydrogen chloride in THF, then hydrol. with base in isopropanol [5]. Methyl methacrylate/methacrylic acid copolymer salt ionomers can be synth. by hydrol. of poly(methyl methacrylate) with sodium hydroxide in THF/H_2O (50:50) followed by precipitation into MeOH. Ionomers with higher salt content can be synth. by neutralisation of poly(methyl methacrylate)/methacrylic acid random copolymers [1]

Applications: Used in heat-resistant pressure-sensitive adhesives, coatings, photoresists; as a binder in inks; in dental cements, drug-delivery systems and in microlithographic resists. Applications due to biodegradability include short-term packaging for foodstuffs; as disposable labels, and in agricultural applications. Other uses include, specialised moulding applications where solubility of the mould is required

Bibliographic References

[1] Ma, X., Sauer, J.A. and Hara, M., *Macromolecules*, 1995, **28**, 3953, 5526
[2] Sanghavi, N.M. and Fruitwala, M., *J. Appl. Polym. Sci.*, 1994, **51**, 1673, (synth.)
[3] Smith, P. and Goulet, L., *J. Polym. Sci., Part B: Polym. Phys.*, 1993, **31**, 327
[4] Wielgolinski, L.J., *Polym. Prepr. (Am. Chem. Soc., Div. Polym. Chem.)*, 1991, **32**, 135
[5] Niezette, J., Vanderschueren, J. and Aras, L., *J. Polym. Sci., Polym. Phys. Ed.*, 1984, **22**, 1845, (dilute soln. props.)
[6] Sanli, O. and Aras, L., *Br. Polym. J.*, 1990, **22**, 155, (transport props.)
[7] Yoo, S., Harelle, L., Daniels, E.S., El-Asser, M.S. and Klein, A., *J. Appl. Polym. Sci.*, 1995, **58**, 367
[8] Lai, J.H., *Macromolecules*, 1984, **17**, 1010, (thermal degradation)
[9] Reinhardt, M., Pfeiffer, K. and Lorkowski, H.J., *J. Appl. Polym. Sci.*, 1994, **51**, 297

Polymethacrylate/polybutadiene blends P-342

Synonyms: *Polybutadiene/PMMA blends. PMMA/polybutadiene interpenetrating networks. Poly(butadiene)/poly(methyl methacrylate) blends. Poly(methyl methacrylate)/ABS blends. ABS/PMMA blends. PMMA/ABS blends*
Related Polymers: Poly(methyl methacrylate), Polybutadiene, Polystyrene, Polyacrylonitrile
Monomers: Methyl methacrylate, 1,3-Butadiene, Acrylonitrile, Styrene
Material class: Thermoplastic, Blends
Polymer Type: polydienes, acrylic

Volumetric & Calorimetric Properties:
Density:

No.	Value	Note
1	d 1.056–1.075 g/cm^3	poly(butadiene)/poly(methyl methacrylate) 65:35 [1]
2	d 1.065–1.081 g/cm^3	poly(butadiene)/poly(methyl methacrylate) 50:50 [1]
3	d 1.075–1.103 g/cm^3	poly(butadiene)/poly(methyl methacrylate) 35:65 [1]

Mechanical Properties:
Mechanical Properties General: Fully interpenetrating networks of poly(methyl methacrylate)/poly(butadiene) show better tensile strength, modulus and tear strength than do semi-interpenetrating networks. The latter show improved toughness and elongation at break [1]. Tensile strength, flexural modulus, flexural strength and impact strength of poly(methyl methacrylate)/ABS blends vary with composition [2]

Flexural Modulus:

No.	Value	Note
1	2334 MPa	23800 kg cm^{-2}, ABS/poly(methyl methacrylate) 90:10 [2]
2	3192 MPa	32545 kg cm^{-2}, ABS/poly(methyl methacrylate) 35:75 [2]
3	2477 MPa	25255 kg cm^{-2}, ABS/poly(methyl methacrylate) 90:10 [2]
4	3082 MPa	31425 kg cm^{-2}, ABS/poly(methyl methacrylate) 35:75 [2]

Tensile Strength Break:

No.	Value	Note	Elongation
1	46.8 MPa	477 kg cm^{-2}, ABS/poly(methyl methacrylate) 90:10, PMMA MW 66000 [2]	
2	63.5 MPa	648 kg cm^{-2}, ABS/poly(methyl methacrylate) 35:75 [2]	
3	45.6 MPa	465 kg cm^{-2}, ABS/poly(methyl methacrylate) 90:10 [2]	
4	56.5 MPa	576 kg cm^{-2}, ABS/poly(methyl methacrylate) 35:75 [2]	
5	6.6–8.2 MPa	poly(butadiene)/poly(methyl methacrylate) 65:35 [1]	425–270%
6	11.5–13.1 MPa	poly(butadiene)/poly(methyl methacrylate) 50:50 [1]	310–185%
7	14–15.7 MPa	poly(butadiene)/poly(methyl methacrylate) 35:65 [1]	210–72%

Flexural Strength at Break:

No.	Value	Note
1	65.2 MPa	665 kg cm^{-2}, ABS/poly(methyl methacrylate) 90:10 [2]
2	94.2 MPa	961 kg cm^{-2}, ABS/poly(methyl methacrylate) 35:75 [2]
3	64.5 MPa	658 kg cm^{-2}, ABS/poly(methyl methacrylate) 90:10 [2]
4	87.1 MPa	888 kg cm^{-2}, ABS/poly(methyl methacrylate) 35:75 [2]

Miscellaneous Moduli:

No.	Value	Note	Type
1	0.82–0.98 MPa	poly(butadiene)/poly(methyl methacrylate) 65:35 [1]	50% modulus
2	3.5–5.1 MPa	poly(butadiene)/poly(methyl methacrylate) 50:50 [1]	50% modulus
3	12–14.5 MPa	poly(butadiene)/poly(methyl methacrylate) 35:65 [1]	50% modulus

Impact Strength: 150.2 J m^{-1} (15.3 kg cm cm^{-1}, ABS/poly(methyl methacrylate) 90:10 [2], 31.4 J m^{-1} (3.2 kg cm cm^{-1}, ABS/poly(methyl methacrylate) 35:75 [2], 148.9 J m^{-1} (15.21 kg cm cm^{-1}, ABS/poly(methyl methacrylate) 90:10 [2], 28.7 J m^{-1} (2.92 kg cm cm^{-1}, ABS/poly(methyl methacrylate) 35:75 [2]
Hardness: Rockwell 20.7 (ABS/poly(methyl methacrylate) 90:10 [2], 74.4 (ABS/poly(methyl methacrylate) 35:75 [2], 17.7 (ABS/poly(methyl methacrylate) 90:10 [2], 64.8 (ABS/poly(methyl methacrylate) 35:75 [2]
Fracture Mechanical Properties: Tear strength 10.1–12.3 N mm^{-1} (poly(butadiene)/poly(methyl methacrylate) 65:35) [1], 16.0–20.7 N mm^{-1} (poly(butadiene)/poly(methyl methacrylate) 50:50) [1], 22.8–26.1 N mm^{-1} (poly(butadiene)/poly(methyl methacrylate) 35:65) [1]. Toughness 0.7924–0.982 MPa (poly(butadiene)/poly(methyl methacrylate) 65:35) [1], 1.080–1.925 MPa (poly(butadiene)/poly(methyl methacrylate) 50:50) [1], 1.080–1.925 MPa (poly(butadiene)/poly(methyl methacrylate) 35:65) [1]

Applications/Commercial Products:
Processing & Manufacturing Routes: Polybutadiene rubber is masticated and cured with varying levels of dicumyl peroxide. The resulting polymer sheet is swollen with methyl methacrylate; benzoyl peroxide is added together with varying amounts of divinylbenzene. The methyl methacrylate is then allowed to polymerise slowly

Bibliographic References
[1] Das, B., Gangopadhyay, T. and Sinha, S., *J. Appl. Polym. Sci.*, 1994, **54**, 367
[2] Kim, B.K., Shin, G.S., Kim, Y.J. and Park, T.S., *J. Appl. Polym. Sci.*, 1993, **47**, 1581, (mechanical props., surface props.)

Polymethacrylate/polycarbonate blends   P-343

Synonyms: *Poly(methyl methacrylate)/polycarbonate blends. Poly(phenyl methacrylate)/polycarbonate blends. Polycarbonate/ acrylic blend. Polycarbonate/acrylic alloy. PMMA/PC blends. PC/ PMMA blends*
Related Polymers: Polymethacrylates, General, Poly(methyl methacrylate), Poly(phenyl methacrylate), Poly(2,2-propanebis(4-phenyl)carbonate)
Monomers: Methacrylic acid, Bisphenol A, Carbonic dichloride, Acrylic acid
Material class: Thermoplastic, Blends
Polymer Type: acrylic/polycarbonate
General Information: High-impact grades are available
Morphology: The blends are single phase but phase separation occurs on heating above the cloud point temp. [1]

Volumetric & Calorimetric Properties:
Density:

No.	Value	Note
1	d 1.188 g/cm^3	poly(methyl methacrylate)/polycarbonate 3:7, as blended
2	d 1.161	poly(methyl methacrylate)/polycarbonate 3:7, heated at 230° for 30 min.

Volumetric Properties General: Density of a single phase blend decreases on heating above its cloud point [1]
Glass Transition Temperature:

No.	Value	Note
1	147°C	90% poly(phenyl methacrylate) [5]
2	134°C	50% poly(phenyl methacrylate) [5]
3	125°C	10% poly(phenyl methacrylate) [5]
4	375°C	50% PMMA
5	369°C	90% PMMA [4]
6	416°C	90% PMMA [4]
7	376°C	10% PMMA [4]
8	422°C	10% PMMA [4]
9	419°C	50% PMMA

Transition Temperature General: Although most blends of poly(methyl methacrylate) and polycarbonate show two glass-transition points the two polymers have been found to be miscible [2,3]. T_g varies with composition [4]

Mechanical Properties:
Mechanical Properties General: Tensile props. of a single phase blend deteriorate due to phase separation on heating [1]

Tensile (Young's) Modulus:

No.	Value	Note
1	2610 MPa	50% PMMA [4]
2	2420 MPa	10% PMMA [4]
3	2800 MPa	90% PMMA [4]

Tensile Strength Break:

No.	Value	Note	Elongation
1	62.5 MPa	90% PMMA [4]	13.8%
2	52.4 MPa	50% PMMA [4]	19.9%
3	59.8 MPa	10% PMMA [4]	63.2%

Tensile Strength Yield:

No.	Value	Elongation	Note
1	68.6 MPa	11.5%	90% PMMA [4]
2	66 MPa	11.8%	50% PMMA [4]
3	66.8 MPa	11.8%	10% PMMA [4]

Miscellaneous Moduli:

No.	Value	Note	Type
1	969–1321 MPa	30°, 10–90% PMMA	Storage shear modulus
2	4.5–589.4 MPa	125°, 10–90% PMMA	Storage shear modulus

Optical Properties:
Transmission and Spectra: Blends can have glass-like transparency [5]
Total Internal Reflection: Birefringence-free alloy is obtained at a ratio of poly(phenyl methacrylate) to polycarbonate of 12:88 [5]

Applications/Commercial Products:
Processing & Manufacturing Routes: Blends can be made by soln. (e.g. in THF), melt mixing at approx. 250° or by precipitation methods (using THF and *n*-hexane) [1,5]
Mould Shrinkage (%):

No.	Value
1	0.4–0.6%

Applications: Has applications in optics

Trade name	Details	Supplier
Acrolex 40001 NA	PMMA/PC alloy	Ferro Corporation
SD-9100	9101, 9104 (modified acrylic/PC alloy)	Polysar

Bibliographic References
[1] Kodama, M., *Polym. Eng. Sci.*, 1993, **33**, 640
[2] Kyu, T. and Saldanha, J.M., *Macromolecules*, 1988, **21**, 1021
[3] Garlund, Z., *Abstr. Pap.-Am. Chem. Soc.*, 1983, **186**, 10, (transition temps.)
[4] Kolarik, J., Lednicky, F., Pukanszky, B. and Pegoraro, M., *Polym. Eng. Sci.*, 1992, **32**, 886, (mechanical props.)
[5] Kyu, T., Park, D. and Cho, W., *J. Appl. Polym. Sci.*, 1992, **44**, 2233

Polymethacrylate/polyethylene oxide blends   P-344
Synonyms: *PMMA/PEO blends*
Related Polymers: Polymethacrylates, General, Poly(methyl methacrylate), Poly(ethylene oxide)
Monomers: Methyl methacrylate, Ethylene oxide
Material class: Thermoplastic, Blends
Polymer Type: acrylic, polyalkylene ether
General Information: Blending is preferred with isotactic polymethacrylate rather than with syndiotactic polymethacrylate as polyethylene oxide is more compatible and therefore more stable with the former [1]
Morphology: Polyethylene oxide adopts a planar zig-zag struct. [1]. Blends with a polyethylene oxide content below 20% are amorph. and miscible, showing only one glass-transition temp. [3]

Volumetric & Calorimetric Properties:
Melting Temperature:

No.	Value	Note
1	39.2–58.1°C	[4]

– Polymethacrylate/polyvinyl blends

Glass Transition Temperature:

No.	Value	Note
1	96–98°C	5% polyethylene oxide, MW 600–600000 [3]
2	84–91°C	10% polyethylene oxide, MW 600–600000 [3]
3	78°C	15% polyethylene oxide, MW 20000 [3]
4	70°C	20% polyethylene oxide, MW 400–50000 [3]

Transition Temperature General: The melting temp. and glass-transitions of polyethylene oxide in interpenetrating networks are affected by the presence of cross-linker, the polymerisation temp. and the MW of the polyethylene oxide [4]

Surface Properties & Solubility:
Solubility Properties General: Compatibility in the melt state has been reported [2]

Mechanical Properties:
Viscoelastic Behaviour: Williams-Landel-Ferry parameters have been reported [3]

Optical Properties:
Transmission and Spectra: Ir [1,3] and nmr [2] spectral data have been reported

Applications/Commercial Products:
Processing & Manufacturing Routes: Semi-interpenetrating polymer networks can be made by polymerising a soln. of polyethylene oxide in methyl methacrylate [4]

Bibliographic References
[1] Rao, G.R., Castiglioni, C., Gussoni, M. and Zerbi, G., *Polymer*, 1985, **26**, 811
[2] Martuscelli, E., Demma, G, Rossi, E. and Segre, A.L., *Polym. Commun.*, 1983, **24**, 266, (compatibility)
[3] Zhao, Y., Jasse, B. and Monnerie, L., *Polymer*, 1989, **30**, 1643
[4] Ahn, S.-H., An, J.H., Lee, D.S. and Kim, S.C., *J. Polym. Sci., Part B: Polym. Phys.*, 1993, **31**, 1627

Polymethacrylate/polyvinyl blends   P-345

Synonyms: Poly(methyl methacrylate)/poly(vinyl chloride). Poly(methyl methacrylate-co-methyl acrylate)/poly(vinyl chloride). Poly(methyl methacrylate-co-ethyl acrylate)/poly(vinyl chloride). Poly(methyl methacrylate-co-butyl acrylate)/poly(vinyl chloride). Poly(methyl methacrylate-co-ethylhexyl acrylate)/poly(vinyl chloride). Poly(methyl methacrylate)/poly(vinyl chloride-co-vinyl acetate). Poly(ethyl methacrylate)/poly(vinyl chloride-co-vinyl acetate). Poly(butyl methacrylate)/poly(vinyl chloride-co-vinyl acetate). Poly(methyl methacrylate)/nitrile rubber/poly(vinyl chloride). Poly(ethyl methacrylate)/nitrile rubber/poly(vinyl chloride). Poly(butyl methacrylate)/nitrile rubber/poly(vinyl chloride). PMMA/PVC blends. PVC/PMMA blends. PMMA/NBR/PVC blends. NBR/PMMA/PVC blends. PVC/NBR/PMMA blends
Related Polymers: Polymethacrylate, General, Poly(methyl methacrylate), Poly(ethyl methacrylate), Poly(butyl methacrylate), Poly(ethylhexyl methacrylate), Poly(vinyl chloride), Poly(vinyl acetate), Poly(acrylonitrile)
Monomers: Methyl methacrylate, Ethyl methacrylate, Butyl methacrylate, Ethylhexyl methacrylate, Vinyl chloride, Vinyl acetate, Acrylonitrile
Material class: Thermoplastic, Blends
Polymer Type: acrylic/PVC

Volumetric & Calorimetric Properties:
Density:

No.	Value	Note
1	d 1.3 g/cm^3	poly(methyl methacrylate)/PVC, moulding grade, ASTM D792 [4]
2	d 1.35 g/cm^3	poly(methyl methacrylate)/PVC, sheet [4,5]

Thermal Expansion Coefficient:

No.	Value	Note	Type
1	6.3×10^{-5} K^{-1}	poly(methyl methacrylate)/PVC, sheet, ASTM D696 [4,5]	L

Thermal Conductivity:

No.	Value	Note
1	6.7 W/mK	0.98 Btu (h ft^2 in)$^{-1}$, PMMA/PVC moulding [4]
2	7 W/mK	1.01 Btu (h ft^2 in)$^{-1}$, PMMA/PVC sheet [4]

Specific Heat Capacity:

No.	Value	Note	Type
1	1.23 kJ/kg.C	0.293 Btu lb^{-1} °F^{-1}, poly(methyl methacrylate)/PVC, extruded sheet [4]	P

Glass Transition Temperature:

No.	Value	Note
1	90°C	50–70% PMMA/poly(vinyl chloride-co-vinyl acetate) [1]
2	52–70°C	50–70% poly(ethyl methacrylate)/poly(vinyl chloride-co-vinyl acetate)
3	40–50°C	50–70% poly(butyl methacrylate)/poly(vinyl chloride-co-vinyl acetate
4	70–90°C	poly(methyl methacrylate)/PVC, <60% poly(methyl methacrylate) [6]
5	70–78°C	poly(methyl methacrylate-co-methyl acrylate)/PVC, 80% PVC [8]
6	62–78°C	poly(methyl methacrylate-co-ethyl acrylate)/PVC, 80% PVC [8]
7	60–76°C	poly(methyl methacrylate-co-butyl acrylate)/PVC, 80% PVC [8]
8	58–68°C	poly(methyl methacrylate-co-ethylhexyl acrylate)/PVC, 80% PVC [8]

Transition Temperature General: T_g of poly(methyl methacrylate)/PVC blends is higher than that of PVC alone. T_g of syndiotactic poly(methyl methacrylate)/PVC blends decreases as the level of poly(methyl methacrylate) increases. Two T_gs are observed when the amount of poly(methyl methacrylate) exceeds 60% [6]. Blends of poly(vinyl alcohol) and poly(ethyl methacrylate) become more brittle as the amount of poly(ethyl methacrylate) increases [3]. Poly(methyl methacrylate)/PVC blends with up to 50% by weight poly(methyl methacrylate) have improved heat distortion temps.

Deflection Temperature:

No.	Value	Note
1	82°C	0.45 MPa, poly(methyl methacrylate)/PVC, moulding grade, ASTM D648 [4]
2	77°C	1.81 MPa, poly(methyl methacrylate)/PVC, moulding grade, ASTM D648 [4]
3	80.5°C	0.45 MPa, poly(methyl methacrylate)/PVC, extruded sheet, ASTM D648 [4,5]
4	71°C	1.81 MPa, poly(methyl methacrylate)/PVC, extruded sheet, ASTM D648 [4,5]

Surface Properties & Solubility:
Solubility Properties General: Syndiotactic poly(methyl methacrylate) is miscible with PVC at up to 60% poly(methyl

methacrylate). Isotactic poly(methyl methacrylate) is immiscible with PVC over the whole range of compositions; the blends are opaque and show two glass-transition temps. Blends of a vinyl chloride/vinyl acetate copolymer with simple polymethacrylates are homogeneous and transparent [1]. Blends of poly(methyl methacrylate) and poly(vinyl alcohol) prepared from $CHCl_3$ and cyclohexane soln. are miscible over the range of compositions. Blends prepared from THF soln. are opaque and show two glass-transition temps. which indicates phase separation [2]. Poly(ethyl methacrylate) and poly(vinyl alcohol) form compatible blends over a wide range of compositions although there is a tendency for phase separation where the MW of poly(vinyl alcohol) is very high [3]. Blends of poly(vinyl acetate) and poly(methyl methacrylate) prepared from $CHCl_3$ and cyclohexane solns. show lower critical soln. temp. behaviour on heating [2]. Literature reports suggest syndiotactic poly(methyl methacrylate) and PVC may only be miscible up to 10% poly(methyl methacrylate) and that MW is an important factor [7]

Transport Properties:
Transport Properties General: The melt-flow behaviour of PVC is improved by addition of a poly(methyl methacrylate-*co*-acrylate) copolymer [8]

Water Absorption:

No.	Value	Note
1	0.13 %	24h, poly(methyl methacrylate)/PVC, moulding grade, ASTM D570 [4]
2	0.06 %	24h, poly(methyl methacrylate)/PVC, extruded sheet, ASTM D570 [4]
3	0.1 %	24h, poly(methyl methacrylate)/PVC, sheet, ASTM D570 [5]

Mechanical Properties:
Mechanical Properties General: The ductility of poly(methyl methacrylate-*co*-acrylate)/PVC blends is improved relative to that of PVC [8]. Elongation at break 150% (poly(methyl methacrylate)/PVC, moulding grade, ASTM D638), 100% (min.) (poly(methyl methacrylate)/PVC, extruded sheet, ASTM D638) [4]. Blends of poly(vinyl acetate) and poly(methyl methacrylate) have high-impact strength with the poly(vinyl acetate) content and its MW being important factors in determining the impact strength of the blend. The optimum level of poly(vinyl acetate) is 20% [2]. The tensile strength has a max. value at approx. 20% poly(ethyl methacrylate). It decreases as the poly(ethyl methacrylate) content increases before increasing from approx. 50% poly(ethyl methacrylate). The elongation at break follows a similar pattern with a max. value of approx. 340% at 20% poly(ethyl methacrylate) decreasing to approx. 100% at 80–100% poly(ethyl methacrylate). Tensile strength increases with increasing MW of poly(ethyl methacrylate) but decreases as poly(vinyl acetate) MW increases [3]

Tensile (Young's) Modulus:

No.	Value	Note
1	1895 MPa	275 kpsi, poly(methyl methacrylate)/PVC, moulding grade, ASTM D638 [4]
2	2308 MPa	335 kpsi, poly(methyl methacrylate)/PVC, extruded sheet, ASTM D638 [4]
3	2300 MPa	poly(methyl methacrylate)/PVC, sheet, ASTM D638 [5]

Flexural Modulus:

No.	Value	Note
1	2067 MPa	300 kpsi, poly(methyl methacrylate)/PVC, moulding grade, ASTM D790, [4]
2	2756 MPa	400 kpsi, poly(methyl methacrylate)/PVC, extruded sheet [4]
3	2760 MPa	poly(methyl methacrylate)/PVC, sheet, ASTM D790 [5]

Tensile Strength Break:

No.	Value	Note	Elongation
1	39.52 MPa	403 kg cm^{-2}, 50% poly(methyl methacrylate)/PVC-*co*-vinyl acetate [1]	3%
2	33.15 MPa	338 kg cm^{-2}, PVC-*co*-vinyl acetate, 50% poly(ethyl methacrylate) [1]	4%
3	29.71 MPa	303 kg cm^{-2}, PVC-*co*-vinyl acetate, 50% poly(butyl methacrylate) [1]	5%
4	29.11 MPa	PVC/nitrile rubber, 58% poly(methyl methacrylate) [1]	50%
5	14.41 MPa	PVC/nitrile rubber, 55% poly(ethyl methacrylate) [1]	110%
6	4.12 MPa	PVC/nitrile rubber, 58% poly(butyl methacrylate) [1]	206%
7	44.8 MPa	6500 lb in^{-2}, poly(methyl methacrylate)/PVC, sheet, ASTM D638 [9]	100%

Flexural Strength at Break:

No.	Value	Note
1	59.9 MPa	8.7 kpsi, poly(methyl methacrylate)/PVC, moulding grade, ASTM D790 [4]
2	73.7 MPa	10.7 kpsi, poly(methyl methacrylate)/PVC, extruded sheet, ASTM D790 [4,5]

Compressive Modulus:

No.	Value	Note
1	27560 MPa	4000000 lb in^{-2}, poly(methyl methacrylate)/PVC, sheet, ASTM D695 [9]

Tensile Strength Yield:

No.	Value	Note
1	37.9 MPa	5.5 kpsi, poly(methyl methacrylate)/PVC, moulding grade, ASTM D638 [4]
2	44.8 MPa	6.5 kpsi, poly(methyl methacrylate)/PVC, extruded sheet, ASTM D638 [4,5]

Flexural Strength Yield:

No.	Value	Note
1	73.7 MPa	10700 lb in^{-2}, poly(methyl methacrylate)/PVC, sheet, ASTM D790 [9]

Compressive Strength:

No.	Value	Note
1	42.7 MPa	6.2 kpsi, poly(methyl methacrylate)/PVC, moulding grade, ASTM D695 [4]
2	57.9 MPa	8.4 kpsi, poly(methyl methacrylate)/PVC, extruded sheet, ASTM D695 [4]

— Polymethacrylate/polyvinyl blends

Impact Strength: Poly(methyl methacrylate-*co*-acrylate)/PVC blends have slightly improved impact strength compared to PVC [8]
Mechanical Properties Miscellaneous: Storage modulus varies with blend composition [1]
Hardness: R104 (poly(methyl methacrylate)/PVC, moulding grade, ASTM D785) [4], R105 (poly(methyl methacrylate)/PVC, extruded sheet, ASTM D785) [4], R105 (poly(methyl methacrylate)/PVC, sheet, ASTM D789) [5]
Friction Abrasion and Resistance: Taber abrasion resistance 0.0058 mg g^{-1} loss (1000 cycles)$^{-1}$ (CS-10 wheel, poly(methyl methacrylate)/PVC, moulding grade, ASTM D1044) [4], 0.073 mg g^{-1} loss (1000 cycles)$^{-1}$ (CS-10 wheel, poly(methyl methacrylate)/PVC, extruded sheet, ASTM D1044) [4]
Izod Notch:

No.	Value	Notch	Note
1	800 J/m	Y	15 ft lb in^{-1}, poly(methyl methacrylate)/PVC, moulding grade and extruded sheet, ASTM D638 [4,5]

Electrical Properties:
Dielectric Permittivity (Constant):

No.	Value	Frequency	Note
1	4	60 Hz	poly(methyl methacrylate)/PVC, moulding grade [4]
2	3.4	1 MHz	poly(methyl methacrylate)/PVC, moulding grade [4]
3	3.9	60 Hz	poly(methyl methacrylate)/PVC, sheet, ASTM D150 [4,5]
4	3.4	1 MHz	poly(methyl methacrylate)/PVC, sheet, ASTM D150 [4,5]

Dielectric Strength:

No.	Value	Note
1	0.4 kV.mm^{-1}	poly(methyl methacrylate)/PVC, moulding grade, short time, ASTM D149 [4]
2	>0.429 kV.mm^{-1}	min., poly(methyl methacrylate)/PVC, extruded sheet, short time, ASTM D149 [4]
3	17 kV.mm^{-1}	poly(methyl methacrylate)/PVC, sheet, ASTM D149 [5]

Arc Resistance:

No.	Value	Note
1	25s	poly(methyl methacrylate)/PVC, moulding grade, ASTM D495 [4]
2	80s	poly(methyl methacrylate)/PVC, extruded sheet, ASTM D495 [4]

Dissipation (Power) Factor:

No.	Value	Frequency	Note
1	0.037	60 Hz	poly(methyl methacrylate)/PVC, moulding grade [4]
2	0.031	1 MHz	poly(methyl methacrylate)/PVC, moulding grade [4]
3	0.076	60 Hz	poly(methyl methacrylate)/PVC, extruded sheet [4]
4	0.094	1 MHz	poly(methyl methacrylate)/PVC, extruded sheet [4]
5	0.002	60 Hz	poly(methyl methacrylate)/PVC, sheet, ASTM D150 [5]
6	0.03	1 MHz	poly(methyl methacrylate)/PVC, sheet, ASTM D150 [5]

Optical Properties:
Volume Properties/Surface Properties: Transparency opaque (poly(methyl methacrylate)/PVC, sheet, ASTM D542), (poly(methyl methacrylate)/PVC, 3 mm sheet, ASTM D1003-61) [4,5]. The optical clarity of poly(methyl methacrylate-*co*-acrylate)/PVC blends is reduced on blending with the acrylic homopolymer [8]. Blends of poly(vinyl acetate) and poly(methyl methacrylate) have good optical transparency [2]

Polymer Stability:
Upper Use Temperature:

No.	Value	Note
1	71°C	160°F, poly(methyl methacrylate)/PVC, extruded sheet [4]

Flammability: Poly(methyl methacrylate)/PVC does not burn (ASTM D635) [4]
Environmental Stress: Poly(methyl methacrylate)/PVC sheet is unaffected by sunlight [9]
Chemical Stability: Poly(methyl methacrylate)/PVC has excellent resistance to acids, bases and oils but is attacked by esters, ketones, aromatic hydrocarbons and chlorinated hydrocarbons [9]

Applications/Commercial Products:
Processing & Manufacturing Routes: Poly(methyl methacrylate)/PVC blends may be processed by injection-moulding at 171–193° [4]
Mould Shrinkage (%):

No.	Value	Note
1	0.7%	[4]

Applications: Poly(methyl methacrylate)/PVC blends have applications in machine housings, safety helmets, boxes/trays, corrosion-resistant ducting, luggage, power tool housing, sports goods, leather-finishing. Poly(methyl methacrylate-*co*-acrylate)/PVC blends containing 1–4% wt of the copolymer are used in the manufacture of bottles

Trade name	Supplier
Hostalit H	Hoechst
Kydene	Rohm and Haas
Vinidur	BASF

Bibliographic References
[1] Mathew, A. and Deb., P.C., *J. Appl. Polym. Sci.*, 1994, **53**, 1103, 1107
[2] Qipeng, G., *Polym. Commun.*, 1990, **31**, 217
[3] Olayemi, J.Y. and Ibiyeye, M.K., *J. Appl. Polym. Sci.*, 1986, **31**, 237
[4] *Handbook of Materials Science*, (ed. C.T. Lynch), CRC Press, 1975, **3**
[5] *The Materials Selector*, 2nd edn., (eds. N.A. Waterman and M.F. Ashby), Chapman and Hall, 1997
[6] Schurer, J.W., De Boer, A. and Challa, G., *Polymer*, 1975, **16**, 201
[7] Vanderschueren, J., Janssens, A., Ladang, M. and Niezette, J., *Polymer*, 1982, **23**, 395
[8] Saroop, U.K., Sharma, K.K., Jain, K.K., Misra, A. and Maiti, S.N., *J. Appl. Polym. Sci.*, 1989, **38**, 1401, 1421
[9] *Handb. Plast. Elastomers*, (ed. C.A. Harper), McGraw-Hill, 1975

Polymethacrylates with cyclic sidechains containing heteroatoms
P-346

Synonyms: *Poly(furfuryl methacrylate). Poly(2-(9H-carbazol-9-yl)pentyl methacrylate). Poly(tetrahydrofurfuryl methacrylate). Poly(2-(9H-carbazol-9-yl)hexyl methacrylate). Poly(tetrahydro-4H-2-pyranyl methacrylate). Poly(2-(9H-carbazol-9-yl)undecyl methacrylate). Poly(2-(9H-carbazol-9-yl)ethyl methacrylate). Poly(2-ferrocenylmethyl methacrylate). Poly(2-(9H-carbazol-9-yl)propyl methacrylate). Poly(2-ferrocenylethyl methacrylate)*
Related Polymers: Polymethacrylates General
Monomers: Furfuryl methacrylate, Tetrahydrofurfuryl methacrylate, Tetrahydro-4H-2-pyranyl methacrylate, 2-(9H-Carbazol-9-yl)ethyl methacrylate, 2-(9H-Carbazol-9-yl)propyl methacrylate, 2-(9H-Carbazol-9-yl)pentyl methacrylate, 2-(9H-Carbazol-9-yl)hexyl methacrylate, 2-(9H-Carbazol-9-yl)undecyl methacrylate, 2-Ferrocenylmethyl methacrylate, 2-Ferrocenylethyl methacrylate
Material class: Thermoplastic
Polymer Type: acrylic
CAS Number:

CAS Reg. No.	Note
29320-19-2	furfuryl
25035-85-2	tetrahydrofurfuryl
29692-07-7	2-(9H-carbazol-9-yl)ethyl
110017-52-2	2-(9H-carbazol-9-yl)ethyl, isotactic

Molecular Weight: MW 11000–84000, M_n 1600–28000

Volumetric & Calorimetric Properties:
Density:

No.	Value	Note
1	d 1.2603 g/cm^3	furfuryl [1,2]

Equation of State: Mark-Houwink constants have been reported [4]

Glass Transition Temperature:

No.	Value	Note
1	60°C	furfuryl [1,2]
2	60°C	tetrahydrofurfuryl [1]
3	148°C	2-(9H-carbazol-9-yl)ethyl, MW 83000 [3]
4	113°C	2-(9H-carbazol-9-yl)propyl, MW 24000 [3]
5	91°C	2-(9H-carbazol-9-yl)pentyl, MW 49000 [3]
6	73°C	2-(9H-carbazol-9-yl)hexyl, MW 84000 [3]
7	31°C	2-(9H-carbazol-9-yl)undecyl, MW 11000 [3]
8	185–195°C	2-ferrocenylmethyl [4]

Transport Properties:
Polymer Solutions Dilute: Theta temps. 31.2° (2-hydroxymethyl-tetrahydrofuran, tetrahydrofurfuryl), 30.4° (isobutanol, tetrahydropyranyl, MW 40000–850000), 25° (C_6H_6, 2-ferrocenylmethyl, MW 6000–36000) [5,6]

Optical Properties:
Refractive Index:

No.	Value	Note
1	1.5381	20°, furfuryl, [1]
2	1.5096	20°, tetrahydrofurfuryl, [1]

Transmission and Spectra: Ir spectral data have been reported [3]

Applications/Commercial Products:
Processing & Manufacturing Routes: Synth. by free-radical polymerisation. The monomers are mixed in dry toluene with AIBN as initiator, and polymerised at 60° for 48h [3]. May also be synth. by mixing monomers in dry THF in the presence of BuLi and prepolymer such as polymethylacryloylchloride and polymerising at room temp. for 3h [3]
Applications: The carbazole polymers (and their copolymers with alkyl methacrylates) are used as photoconducting polymers in the reprographic industry

Bibliographic References
[1] Lewis, O.G., *Physical Constants of Linear Hydrocarbons*, Springer-Verlag, 1968
[2] Unpublished Data Rohm and Haas,
[3] Strohriegl, P., *Mol. Cryst. Liq. Cryst.*, 1990, **183**, 261, (synth.)
[4] Pittman, C.U., Lai, J.C., Vanderpool, D.P., Good, M. and Prado, R., *Macromolecules*, 1970, **3**, 746
[5] Zafar, M.M. and Mahmood, R., *Makromol. Chem.*, 1974, **175**, 903, (theta temps.)
[6] Tsvetkov, V.N., Khardi, D., Shtennikova, I.N., Korneyeva, Ye.V. *et al*, Vysokomol. Soedin., Ser. A, 1969, **11**, 349, (theta temps)

Polymethacrylates with fused-ring side chains
P-347

where R = (adamantyl)

(bornyl)

(*iso*-bornyl)

(norbornyl)

(decahydronaphthyl)

Synonyms: *Poly(norbornyl methacrylate). Poly(exo-1,7,7-trimethylbicyclo[2.2.1]hept-2-yl methacrylate). Poly(tricyclo[3.3.3.1]dec-1-yl methacrylate). Poly(bicyclo[2.2.1]hept-2-yl methacrylate). Poly(endo-1,7,7-trimethylbicyclo[2.2.1]hept-2-yl methacrylate). Poly(isobornyl methacrylate). Poly(bornyl methacrylate). Poly(adamantyl methacrylate)*
Related Polymers: Polymethacrylates-General
Monomers: 2-decahydronaphthyl methacrylate, Adamantyl methacrylate, Norbornyl methacrylate, Bornyl methacrylate, Isobornyl methacrylate, 3,5-dimethyladamantyl methacrylate
Material class: Thermoplastic
Polymer Type: acrylic

CAS Number:

CAS Reg. No.	Note
28854-39-9	isobornyl

Molecular Formula: $(C_{14}H_{20}O_2)_n$, $(C_{16}H_{24}O_2)_n$, $(C_{14}H_{22}O_2)_n$,
Fragments: $C_{14}H_{20}O_2$ $C_{16}H_{24}O_2$ $C_{14}H_{22}O_2$

Volumetric & Calorimetric Properties:
Density:

No.	Value	Note
1	d 1.06 g/cm^3	d^{20}; isobornyl [1,2]

Equation of State: Mark-Houwink constants for the isobornyl ester have been reported [8]

Glass Transition Temperature:

No.	Value	Note
1	141°C	adamantyl
2	183°C	adamantyl, isotactic [5]
3	196°C	3,5-dimethyladamantyl
4	194°C	bornyl [6]
5	110°C	isobornyl [1,2,7]
6	111°C	isobornyl, syndiotactic [7]
7	140°C	isobornyl [8]
8	150–206°C	isobornyl, varying tacticity and MW [3,4]
9	145°C	decahydronaphthyl [6]

Transition Temperature General: T_g of the isobornyl ester varies with tacticity [3] and increases with increasing MW [4]. T_g of the (predominantly syndiotactic) adamantyl ester has not been detected below decomp. temp. [5]

Surface Properties & Solubility:
Cohesive Energy Density Solubility Parameters: 16.6 J$^{1/2}$ cm$^{-3/2}$ (8.1 cal$^{1/2}$ cm$^{-3/2}$) (isobornyl) [9], 16.8 J$^{1/2}$ cm$^{-3/2}$ (8.2 cal$^{1/2}$ cm$^{-3/2}$) (isobornyl) [8]
Solvents/Non-solvents: Adamantyl sol.[5] C_6H_6, CHCl$_3$, CCl$_4$, THF, cyclohexanone. Insol. DMF, Me$_2$CO, 2-butanone, hexane, MeOH. Bornyl sol. [6,10] C_6H_6, cyclohexane, CCl$_4$, toluene, THF, CHCl$_3$, cyclohexanone. Insol. MeOH, Me$_2$CO, DMF. Isobornyl sol.[6] hexane, cyclohexane, CCl$_4$, toluene, THF, C_6H_6, CHCl$_3$, cyclohexanone. Insol. Me$_2$CO, DMF, MeOH. Decahydronaphthyl sol. [6] cyclohexane, CCl$_4$ toluene, THF, C_6H_6, CHCl$_3$, cyclohexanone. Insol. hexane, Me$_2$CO, DMF, MeOH

Transport Properties:
Polymer Solutions Dilute: Intrinsic viscosity 0.74–3.1 dl g^{-1} (C_6H_6, bornyl, MW 146017–1920184) [10], 6.2–29 dl g^{-1} (C_6H_6, isobornyl, MW 106219–250956) [10], 2.53 dl g^{-1} (30°, THF, isobornyl), 13.1 dl g^{-1} (39.6°, 1-octanol, isobornyl) [8]. Theta temp. 39.6 (1-octanol, isobornyl, MW 90000–1200000) [8]

Water Absorption:

No.	Value	Note
1	0.2 %	norbornyl

Optical Properties:
Optical Properties General: norbornyl [11]
Refractive Index:

No.	Value	Note
1	1.5059	bornyl 20° [1]
2	1.5	isobornyl 20° [1,2]
3	1.512	norbornyl [11]

Transmission and Spectra: Transmittance 92% (norbornyl) [11]. Nmr spectral data have been reported [5,6]

Polymer Stability:
Decomposition Details: Initial decomposition temp. of the adamantyl ester 254°; max. decomposition occurs at 304° [5]. Initial decomposition temp. of bornyl ester 252°, with three maxima at 307°, 382° and 434°. Decomposition yields monomer. Initial decomposition temp. of isobornyl ester 264°, with two maxima at 314° and 436°. At 264° camphene is produced; at 290° both camphene and monomer are produced. Initial decomposition temp. of decahydronaphthyl ester 212°, with three maxima at 296°, 350°, and 442°[6]

Applications/Commercial Products:
Processing & Manufacturing Routes: Synth. by free-radical polymerisation using AIBN initiator in C_6H_6 at 60°. Polymers isol. by precipitation from MeOH [5,6]

Bibliographic References

[1] Lewis, O.G., *Physical Constants of Linear Hydrocarbons*, Springer-Verlag, 1968
[2] *Kirk-Othmer Encycl. Chem. Technol.*, Vol.15, 3rd edn., (ed. M.Grayson), Wiley Interscience, 1978, 377
[3] Yu, J.M., Dubois, P. and Jerome, R., *Macromolecules*, 1996, **29**, 7316
[4] Zhang, X.Q. and Wang, C.H., J. Polym. Sci. Part B, 1994, **32**, 1951
[5] Matsumoto, A., Tanaka, S. and Otsu, T., *Macromolecules*, 1991, **24**, 4017, (synth.)
[6] Matsumoto, A., Mizuta, K. and Otsu, T., J. Polym. Sci., *Part A-1*, 1993, **31**, 2531, (synth., thermal props.)
[7] Shetter, J.A., J. Polym. Sci. Part B, 1962, **1**, 209
[8] Hadjichristidis, N., Mays, J., Ferry, W. and Fetters, L.J., J. Polym. Sci., Part B: Polym. Phys., 1984, **22**, 1745
[9] Frank, C.W. and Gashgari, M.A., *Macromolecules*, 1979, **12**, 163
[10] Imoto, M., Otsu, T., Tsuda, K. and Ito, T., J. Polym. Sci., *Part A-1*, 1964, **2**, 1407
[11] *Jpn. Pat.*, 1986, 61152708

Polymethacrylate/vinylidene halide blends   P-348

Synonyms: *PMMA/PVDF blends. PVDF/PMMA blends. PVF$_2$/PMMA blends. PMMA/PVF$_2$ blends*
Related Polymers: Polymethacrylates, General, Poly(methyl methacrylate), Poly(ethyl methacrylate), Poly(vinylidene fluoride)
Monomers: Methacrylic acid, 1,1-Difluoroethylene
Material class: Thermoplastic, Blends
Polymer Type: PVDF, acrylic
Morphology: Blends of poly(vinylidene fluoride) with atactic, isotactic and syndiotactic poly(methyl methacrylate) are completely compatible. The interactions of poly(vinylidene fluoride) segments are stronger for blends with isotactic poly(methyl methacrylate) than for those with syndiotactic poly(methyl methacrylate) [1]. Poly(ethyl methacrylate)/poly(vinylidene fluoride) blends are amorph. and miscible at poly(ethyl methacrylate) levels of 80% or more. Crystallisation of poly(vinylidene fluoride) occurs when the amount of poly(ethyl methacrylate) is low. The microstruct. of the blends depends on preparation and treatment [2]. Poly(methyl methacrylate)/poly(vinylidene fluoride) blends are amorph. when amount of poly(vinylidene fluoride) is lower than 40% [4]

Volumetric & Calorimetric Properties:
Thermodynamic Properties General: Heat capacity decreases with increasing poly(vinylidene fluoride) content [3]
Melting Temperature:

No.	Value	Note
1	157°C	40% PVDF/PMMA [3]
2	173°C	90% PVDF/PMMA [3]

– Poly(methacrylic acid)

Glass Transition Temperature:

No.	Value	Note
1	55°C	60% PVDF/PMMA [3]
2	80°C	20% PVDF/PMMA [3]
3	55–105°C	PMMA/PVDF, 60–0% PVDF [3]

Transition Temperature General: T_m increases with increasing poly(vinylidene fluoride) content [4]. Multiple melting occurs in poly(ethyl methacrylate)/poly(vinylidene fluoride) blends with up to four transitions observed in some blends [2,5]. T_g increases with decreasing poly(vinylidene fluoride) content [3]

Transition Temperature:

No.	Value	Note	Type
1	-26–-25°C	PMMA/PVDF, 33–100% [6], 33–100% poly(vinylidene fluoride)	

Volumetric & Calorimetric Properties:
Transition Temperature Miscellaneous:

No.	Value	Note	T_β	°C

Surface Properties & Solubility:
Solubility Properties General: Upper and lower critical soln. temp. behaviour has been reported for blends of the two polymers [7]

Transport Properties:
Permeability of Gases: Diffusion coefficient [CO_2] 8.53×10^{-9} cm^2 s^{-1} (35°, 90% poly(vinylidene fluoride)), 3.17×10^{-9} cm^2 s^{-1} (35°, 20% poly(vinylidene fluoride)) [3]

Mechanical Properties:
Impact Strength: An abrupt increase in Izod impact strength of poly(methyl methacrylate)/poly(vinylidene fluoride) blends is found when the amount of poly(methyl methacrylate) is 20%; for greater amounts relatively little change occurs. Annealing increases the impact strength slightly [4]

Viscoelastic Behaviour: Longitudinal sonic velocity increases with decreasing poly(vinylidene fluoride) content [6]

Optical Properties:
Transmission and Spectra: Nmr spectral data have been reported [8]

Bibliographic References
[1] Boscher, F., Ten Brinke, G., Eshuis, A. and Challa, G., *Macromolecules*, 1982, **15**, 1364
[2] Kwei, T.K. and Wang, T.T., *Macromolecules*, 1976, **9**, 780
[3] Chiou, J.S. and Paul, D.R., *J. Appl. Polym. Sci.*, 1986, **32**, 2897
[4] Mijovic, J., Lus, H.L. and Han, C.D., *Polym. Eng. Sci.*, 1982, **22**, 234
[5] Eshuis, A., Roerdink, E. and Challa, G., *Polymer*, 1982, **23**, 735
[6] Hourston, D.J. and Hughes, I.D., *Polymer*, 1977, **18**, 1175
[7] Saito, H., Fujita, Y. and Inoue, T., *Polym. J. (Tokyo)*, 1987, **19**, 405, (soln. props.)
[8] Douglass, D.C. and McBriety, V.J., *Macromolecules*, 1978, **11**, 766, (nmr)

Poly(methacrylic acid)   P-349

$$\mathrm{-\!\!\!-\!\!\!\left[CH_2C(CH_3)(COOH)\right]\!\!-\!\!\!\!\!-_n}$$

Synonyms: *Poly(2-methyl-2-propenoic acid)*. *PMA*
Related Polymers: Polymethacrylates General, Poly(methacrylic acid), salts
Monomers: 2-Methyl-2-propenoic acid
Material class: Thermoplastic

Polymer Type: acrylic
CAS Number:

CAS Reg. No.	Note
25087-26-7	
25750-36-1	(syndiotactic)
25068-55-7	(isotactic)

Molecular Formula: $(C_4H_6O_2)_n$
Fragments: $C_4H_6O_2$
General Information: PMAA is a weak acid
Miscellaneous: Forms complexes with basic polymers, e.g. poly(4-vinylpyridine) [21]

Volumetric & Calorimetric Properties:
Specific Heat Capacity:

No.	Value	Note	Type
1	112.5 kJ/kg.C	27° [1]	P
2	1.23 kJ/kg.C	-263° [1]	P
3	94.16 kJ/kg.C	-33° [1]	P
4	40.91 kJ/kg.C	-183° [1]	P

Glass Transition Temperature:

No.	Value	Note
1	228°C	[3]
2	149.7°C	[2]

Transition Temperature General: The presence of water or other plasticising additives lowers the T_g [2]

Surface Properties & Solubility:
Solubility Properties General: δ 0 (poor H-bonding solvents); δ 20.3 J$^{1/2}$ cm$^{-3/2}$ (moderate H-bonding solvents); δ 26.0–29.7 J$^{1/2}$ cm$^{-3/2}$ 26.0–29.7 (strong H-bonding solvents) [4]
Solvents/Non-solvents: Sol. alcohols, H$_2$O, aq. HCl (0.002M, 30°), dil. aq. NaOH. Insol. hydrocarbons, carboxylic acids, esters [5,6,7]
Wettability/Surface Energy and Interfacial Tension: Surface energy 57 mN m^{-2}; γ_p 23.6; γ_d 33.4 [22] Hamilton contact angle 151° [22]

Transport Properties:
Melt Flow Index:

No.	Value	Note
1	5.8 g/10 min	[8]

Polymer Solutions Dilute: The intrinsic viscosity and Huggins coefficients in acidified EtOH show unusually large variations with two minima and two maxima seen in the plot of viscosity vs. mole fraction of EtOH. A similar variation was found in the behaviour of mixtures of other acidified aliphatic alcohols. Propanol showed the greatest variation and ethanediol the least. The dependence of viscosity on MW varied with alcohol concentration [9]. Mark-Houwink constants have been reported [7,10]. Theta temps. 30° (0.002 *M* HCl) [11], 25° (aq. NaCl) [12], 36.9° (0.5 *M* NaCl) [12], 68° (0.05 *M* NaCl) [12]. Intrinsic viscosity constant K$_\theta$ 0.066 cm$^{1/2}$ mol$^{1/2}$ g$^{1/2}$. Second virial coefficients have been reported [13]

Polymer Solutions Concentrated: Aq. solns. at concentrations in the region of 8–15% form stiff gels at room temp. At lower concentrations they form solns. which are converted into gels on heating, reverting to a soln. of the original consistency on cooling. Such solns. exhibit negative thixotropy i.e., they can show a large increase in consistency on stirring or shaking and can set to a firm gel [14]

– Poly(methacrylic acid salt-...

Water Absorption:

No.	Value	Note
1	73.5 %	equilib. [22]

Mechanical Properties:
Tensile Strength Break:

No.	Value	Note	Elongation
1	23.4 MPa	[8]	553%

Tensile Strength Yield:

No.	Value	Note
1	6.1 MPa	[8]

Miscellaneous Moduli:

No.	Value	Type
1	0.00024 MPa	Rigidity modulus

Viscoelastic Behaviour: Longitudinal sound speed 3350 m s^{-1}, room temp., 1 MHz

Optical Properties:
Total Internal Reflection: Birefringence. Intrinsic segmental anisotropy 50×10^{-25} cm^3 (MeOH), 150×10^{-25} cm^3 (0.002 M HCl) [16]

Volume Properties/Surface Properties: Clarity hazy

Polymer Stability:
Decomposition Details: Initial decomposition temp. 200°. At 200° decomposition yields, almost quantitatively, water with traces of monomer and a residue of polymethacrylic anhydride [18]. At low temps. monomer is released with CO_2 and H_2O and the anhydride is formed. At higher temps. the decomposition is more complex [19]. Thermal degradation is a two stage process: decomposition to the anhydride (activation energy 40.5 kcal mol^{-1}) and fragmentation of the anhydride (activation energy 37.4 kcal mol^{-1}) [20]

Applications/Commercial Products:

Trade name	Supplier
Daxad 34	WR Grace
Tamol	Rohm and Haas

Bibliographic References

[1] Gaur, U., Lau, S.F., Wunderlich, B.B. and Wunderlich, B., *J. Phys. Chem. Ref. Data*, 1982, **11**, 1065, (heat capacity)
[2] Verhoeven, J., Schaeffer, R., Bouwstra, J.A. and Junginger, H.E., *Polymer*, 1989, **30**, 1946, (glass transition temp)
[3] Van Krevelen, D.W., *Properties of Polymers: Their Correlation with Chemical Structure*, Elsevier, 1976, 79
[4] Grulke, E.A., *Polym. Handb.*, 4th edn., (eds. J. Brandrup, E.H. Immergut and E.A. Grulke), John Wiley and Sons, 1999, **VII**, 675
[5] Fuchs, O., *Polym. Handb.*, 4th edn., (eds. J. Brandrup, E.H. Immergut and E.A. Grulke), John Wiley and Sons, 1999, **VII**, 479, (solubility)
[6] Dexheimer, H., Fuchs, O., Nitsche, R. and Wolf, K.A., *Struktur und Physikaliches Verkalten der Kunststoffe*, Springer-Verlag, 1961, 1
[7] Kurata, M. and Stockmayer, W.H., *Adv. Polym. Sci.*, Springer-Verlag, 1963, **3**, 196
[8] Nielsen, L.E., *Encycl. Polym. Sci. Eng.*, **8**
[9] Priel, Z. and Silberberg, A., *J. Polym. Sci., Part A-2*, 1970, **8**, 689, 705, (conformn)
[10] Kurata, M. and Tsumashima, Y., *Polym. Handb.*, 4th edn., (eds. J. Brandrup, E.H. Immergut and E.A. Grulke), John Wiley and Sons, 1999, **VII**, 1
[11] Katchalsky, A. and Eisenberg, H., *J. Polym. Sci.*, 1951, **6**, 145, (theta temp)
[12] Kanevskaya, Ye.A., Zubov, P.I., Ivanova, L.V. and Lipatov, Yu.S., *Polym. Sci. USSR (Engl. Transl.)*, 1964, **6**, 1080, (theta temps)
[13] Bimedina, L.A., Roganov, V.V. and Bekturov, E.A., *J. Polym. Sci., Polym. Symp.*, 1974, **44**, 65, (virial coefficients)
[14] Eliassaf, J., Silberberg, A. and Katchalsky, A., *Nature (London)*, 1955, **176**, 1119
[15] Hartman, B., *Physical Properties of Polymers Handbook*, (ed. J.E. Mark), AIP Press, 1996, 677, (acoustic props)
[16] Tsvetkov, V.N., Ljubina, S.Y. and Bolevskii, K.L., *Vysokomol. Soedin., Ser. A*, 1963, **4**, 33, 36, (anisotropy)
[17] Welsh, W.J., *Physical Properties of Polymers Handbook*, (ed. J.E. Mark), AIP Press, 1996, 605, (stability)
[18] Grant, D.H. and Grassie, N., *Polymer*, 1960, **1**, 125, (degradation)
[19] Schild, H.G., *J. Polym. Sci., Polym. Chem. Ed.*, 1993, **31**, 2403, (degradation)
[20] Mead, G. and Willing, R.I., *Polym. J. (Tokyo)*, 1991, **23**, 1401
[21] Inal, Y., Kato, S-I., Hirabayashi, T. and Yokota, K., *J. Polym. Sci., Part A: Polym. Chem.*, 1996, **34**, 2341
[22] Baker, D.A., Corkhill, P.H., Ng, C.O., Skelly, P.J. and Tighe, B.J., *Polymer*, 1988, **29**, 691

Poly(methacrylic acid salt-*co*-ethylene) P-350

$$\left[\begin{array}{c} -\text{CH}_2\text{C}(\text{CH}_3)- \\ | \\ \text{O}=\text{C} \\ | \\ \text{O}^- \text{M}^+ \end{array} \right]_x \left[-\text{CH}_2\text{CH}_2- \right]_y \Bigg]_n$$

Synonyms: *Ionomer*
Related Polymers: Polymethacrylates General, Poly(methacrylic acid), Poly(methacrylic acid salts), Polyethylene, Poly(methacrylic acid-*co*-styrene) and methacrylate salts
Monomers: Methacrylic acid salt, Ethylene
Material class: Thermoplastic
Polymer Type: acrylic copolymers
CAS Number:

CAS Reg. No.	Note
25608-26-8	
108644-30-0	salt with complex cation Zn-BAC

Molecular Formula: $(C_4H_5MO_2)_m(C_2H_4)_n$
Fragments: $C_4H_5MO_2$ C_2H_4

General Information: Ethylene-methacrylic ionomers are usually supplied in pellet form, typically contain 3–10% methacrylic acid and are neutralised to a level between 20% and 80%. The polymers described above are tough, transparent, solvent resistant materials which possess characteristics of both thermoplastics and thermosets. They have very good resilience, flexibility at low temps., high elongation, impact strength and melt viscosities plus good adhesion and abrasion props. and outstanding hot tack. Of great interest for practical applications is the reduction in haze which occurs on neatralisation. Generally, there are strong similarities between the physical props. of polymers with inorganic cations and those with organic cations. The soln. and melt props., however, can be completely different. Aliphatic amines will react with ethylene-methacrylic acid copolymers to form salts but most aromatic amines are too weak for successful reaction [1]

Morphology: The ethylene chain has pendant anionic carboxy groups with metal cations. Chains are linked by electrostatic forces arising because of the metal ions. When temp. increases these forces are overcome allowing the polymer to be physically deformed using conventional techniques. The ionic groups in ionomers neutralised with metal cations tend to separate from the polymer matrix to form ionic clusters (ordered assemblies of ionic groups) which have a profound effect on the physical props. With amine cations replacing some of the metal ions, however, there can be changes to the polymer behaviour and props. The orderliness of the ionic clusters depends on the cation; those with a coordination number higher than the second shell have a greater degree of order [2,3,12,13]

– Poly(methacrylic acid salt-...

Identification: Contains only C. H, O and metal. Characterised by ir and x-ray scattering [1,14,15]

Volumetric & Calorimetric Properties:
Density:

No.	Value	Note
1	d 0.93–0.94 g/cm^3	[4]
2	d 0.93–0.96 g/cm^3	[5]
3	d 0.94–0.96 g/cm^3	Zn salt [16]
4	d 0.94–0.96 g/cm^3	Na salt [16]
5	d 0.93–0.96 g/cm^3	moulding grade [21]

Thermal Expansion Coefficient:

No.	Value	Note	Type
1	0.00058–0.00078 K^{-1}	0–60% neutralisation	L
2	0.00047–0.000506 K^{-1}	0–60% neutralisation, 0–97% BAC	L
3	0.0001–0.00012 K^{-1}	[6]	L
4	0.00014 K^{-1}	Na salt	L
5	0.00016 K^{-1}	Zn salt	L

Volumetric Properties General: Coefficient of thermal expansion for the Zn salt varies with the degree of neutralisation and the amount of added 1,3-bis(aminomethyl)cyclohexane (BAC). Specific volume for the Zn salt ranges from 1.008–1.0368 cm^3 g^{-1} (degree of neutralisation 0–60, no added BAC) and 1.0632–1.037 cm^3 g^{-1} (degree of neutralisation 0–60, 0.97% BAC) at 25°
Thermodynamic Properties General: Ionomers have good heat-sealing props. and very good hot tack
Thermal Conductivity:

No.	Value	Note
1	0.24 W/mK	DIN 52612 [7]
2	0.22–0.26 W/mK	0.00053–0.00062 cal cm^{-1} s^{-1} °C^{-1} [6]
3	0.24 W/mK	[5]

Specific Heat Capacity:

No.	Value	Note	Type
1	2303 kJ/kg.C	0.55 cal °C^{-1} g^{-1} [8]	P
2	2554 kJ/kg.C	0.61 cal °C^{-1} g^{-1} [6]	P

Melting Temperature:

No.	Value	Note
1	94°C	Na salt
2	81–96°C	Na salt [16]
3	90–95°C	Zn salt [12]
4	81–98°C	Zn salt [16]
5	89–91°C	Zn-BAC salt [12]
6	99–100°C	Na salt, 4.1 mol% MAA [17]

Transition Temperature General: Mechanical relaxations for acid and Na salts (4.1 mol% MAA) [17]

Deflection Temperature:

No.	Value	Note
1	0.45°C	40–47° [7]
2	0.455°C	43.3°, ASTM D648 [8]
3	0.455°C	40°, Na salt
4	0.455°C	41°, Zn salt
5	1.82°C	37.8–48.9°, ASTM D648 [8]
6	0.45°C	41°, moulding grade, ASTM D648 [21]

Vicat Softening Point:

No.	Value	Note
1	40–47°C	1525/306 [7]
2	70–90°C	[6]

Brittleness Temperature:

No.	Value	Note
1	-112–-50°C	ASTM D746 [7]
2	>-100°C	min. [6]
3	-140–-100°C	ASTM D746 [14]
4	-107°C	moulding grade, ASTM D746-57T [21]

Transition Temperature:

No.	Value	Note	Type
1	-120°C	approx., virtually independent of level of ionisation	γ
2	-10°C	approx., absent from acid copolymre and increases with increasing ionisastion	β
3	23°C	approx., decreases with increasing ionisation	β
4	29°C	3.61% acid: 100% ethylene	β relaxation temp.
5	54°C	6.80% acid: 100% ethylene	β relaxation temp.
6	-120°C	3.61% acid: 100% ethylene	γ relaxation temp.
7	-112°C	6.80% acid: 100% ethylene [11]	γ relaxation temp.

Surface Properties & Solubility:
Solvents/Non-solvents: Solubility in organic solvents decreases as ionic bonding with metal ions increases. Ionomers can usually be swollen by certain solvents e.g., aromatic hydrocarbons at elevated temps. The resistance to solvent surface etching is normally high

Transport Properties:
Transport Properties General: Melt flow index for samples containing 10% wt. methacrylate increases with ionic charge on the metal ion [5] and decreases with increasing degree of neutralisation [14,18]. Melt viscosity is unaffected by neutralisation with diamines [1]
Melt Flow Index:

No.	Value	Note
1	0.7–14 g/10 min	10 min. [7]
2	1.3 g/10 min	10 min., Na salt

3	0.9–2.8 g/10 min	10 min., Na salt [16]
4	5.5 g/10 min	10 min., Zn salt
5	0.7–14 g/10 min	10 min., Zn salt [16]
6	2–10 g/10 min	moulding grade, ASTM D1238

Polymer Solutions Dilute: Dispersions of ionomer in water can range from water-like to paste-like depending on the amount of solid and the acid level. The viscosity of Na ionomers, but not Zn ionomers is reduced by free acid groups [19]. Na and Zn ionomers exhibit Newtonian behaviour at low shear rates [19]. Zero shear viscosity increases with increasing neutralisation for the Na and Zn salts

Permeability of Gases: $[O_2]$ 5×10^{-10} cm^2 (s cmHg)$^{-1}$ (25°) [6], 130–229 $\times 10^{-10}$ cm^3 mm m^2 day atm. (0.051 mm film, Na salt) [16], 134–233 $\times 10^{-10}$ cm^3 mm m^2 day atm. (0.051 mm film, Zn salt) [16]; $[N_2]$ 0.9 t 10^{-10} cm^2 (s cmHg)$^{-1}$ (25°); $[CO_2]$ 6×10^{-10} cm^2 s^{-1} (cmHg)$^{-1}$ (25°); $[H_2O]$ 0.3×10^{-10} cm^2 s^{-1} (cmHg)$^{-1}$ (1 mm film, 25°, 90% relative humidity), 0.51–0.95 $\times 10^{-10}$ g m m^2 day (0.051 mm film, Na salt) [16], 0.47–0.79 g mm m^2 day (o.051 mm film, Zn salt) [16]

Permeability of Liquids: Acid polymers are more resistant to oil permeation than the ethylene homopolymer and the difference increases greatly when the ionomer is neutralised

Water Content: 0.04–3.47% (increases with increasing relative humidity, and after immersion). Increase in water content reduces the stiffness of the ionomer; this is particularly marked for Na and K salts [18]

Water Absorption:

No.	Value	Note
1	0.1–1.4 %	24h [5]
2	0.1–0.4 %	24h, moulding grade, ASTM D570

Mechanical Properties:

Mechanical Properties General: Ionomer films remain flexible down to approx. -100° [3] The moulding grade remains flexible down to -105°. Both stiffness and yield point increase, irrespective of the nature of the cation, as the polymer is progressively neutralised with a diamine, but the ultimate tensile strength shows little change [1]. Neutralisation with metal ions has a significant effect on the ultimate tensile strength however

Tensile (Young's) Modulus:

No.	Value	Note
1	192–276 MPa	28–40 kpsi [9]
2	138–413 MPa	20–60 kpsi, ASTM D638 [8]
3	193–517 MPa	19.7–52.7 kg f mm^{-2} [6]

Flexural Modulus:

No.	Value	Note
1	70–380 MPa	[7]
2	220 MPa	Na salt
3	130 MPa	Zn salt
4	180 MPa	moulding grade, ASTM D790

Tensile Strength Break:

No.	Value	Note	Elongation
1	35.8 MPa	Na salt, 4.8% wt. cross-linking agent	330%
2	33.9 MPa	Li salt, 2.8% wt. cross-linking agent	317%
3	33.9 MPa	Ba salt, 9.6% wt. cross-linking agent	370%
4	40.4 MPa	Mg salt, 8.4% wt. cross-linking agent	326%
5	29.7 MPa	Zn salt, 12.8% wt. cross-linking agent	313%
6	22 MPa	Al salt, 14% wt. cross-linking agent	347%
7	21–35 MPa	R 527 [7]	280–529%
8	24.1–37.9 MPa	[4]	
9	24.1–37.9 MPa	3.5–5.5 kpsi [9]	300–450%
10	24.1–38 MPa	2.46–3.87 kg f mm^{-2} [6]	200–600%
11	24.1–34.4 MPa	[5]	350–450%
12	29.6–40 MPa	Na salt, ASTM D882 [16]	300–500%
13	24.1–35.8 MPa	Zn salt, ASTM D882 [16]	200–450%
14	28 MPa	moulding grade, ASTM D638	450%

Tensile Strength Yield:

No.	Value	Note
1	13.2 MPa	Na salt, 4.8% wt. cross-linking agent
2	13.1 MPa	Li salt, 2.8% wt. cross-linking agent
3	13.4 MPa	Ba salt, 9.6% wt. cross-linking agent
4	15 MPa	Mg salt, 8.4% wt. cross-linking agent
5	13.2 MPa	Zn salt, 12.8% wt. cross-linking agent
6	7.1 MPa	Al salt, 14% wt. cross-linking agent [5]
7	9–29 MPa	R527 [7]
8	13.8–17.2 MPa	[4]
9	7.6–19 MPa	Na salt, ASTM D882 [16]
10	9–25.5 MPa	Zn salt, ASTM D882 [16]

Miscellaneous Moduli:

No.	Value	Note	Type
1	69–372 MPa	Na salt, ASTM D882 [16]	Secant modulus
2	90–400 MPa	Zn salt, ASTM D882 [16]	Secant modulus

Impact Strength: Tensile impact strength 7.6–12.8 $\times 10^{22}$ kJ m^{-2} (23°, 180) [7], 560–1180 kJ m^{-2} (-40°, 180) [7]. Izod 304–694 J m^{-1} (notched, 5.7–13 ft lb in^{-1}, ½″ bar) [9]; 320–800 J m^{-1} (notched) [5]; 610 J m^{-1} (notched, Na salt); no break (notched, Zn salt). Spencer 0.12–0.22 J m^{-1} (Na salt, ASTM D3420) [16]; 0.08–0.27 J m^{-1} (Zn salt, ASTM D3420) [16]

Mechanical Properties Miscellaneous: Stiffness 190.3 MPa (Na salt, 4.8% wt. cross-linking agent), 206.8 MPa (Li salt, 2.8% wt. cross-linking agent), 223.4 MPa (Ba salt, 9.6% wt. cross-linking agent), 164.1 MPa (Mg salt, 8.4% wt. cross-linking agent), 208 MPa (Zn salt, 12.8% wt. cross-linking agent), 103.4 MPa (Al salt, 14% wt. cross-linking agent) [5]. Elmendorf tear resistance 630–6457 g mm^{-1} (Na salt, ASTM D1922) [16]; 315–5118 g mm^{-1},(Zn salt, ASTM D1922) [16]. Graves tear strength 400–760 g (Na salt, ASTM D1004) [16]; 500–800 g (Zn salt, ASTM D1004) [16]

Hardness: Shore D56-68 (DIN 53505) [7], D60-65 [6], D60 (Na salt), D54 (Zn salt), D56-68 (moulding grade)

Friction Abrasion and Resistance: Ionomers have good abrasion resistance

Izod Notch:

No.	Value	Notch	Note
1	450–700 J/m	Y	moulding grade, ASTM D256

Electrical Properties:
Electrical Properties General: The majority of ionomers have good dielectric props. over a broad range of frequencies. Despite the presence of ionic groups there is not a great difference between the electrical props. of an ionomer and its parent polyethylene [20]. Surface resistivity generally increases with increasing degree of neutralisation

Surface/Volume Resistance:

No.	Value	Note	Type
2	0.0001×10^{15} Ω.cm	70% neutralisation	S

Dielectric Permittivity (Constant):

No.	Value	Frequency	Note
1	2.4	60 Hz	[10]
2	2.5		Zn salt [14]
3	2.4–2.5	60 Hz	moulding grade

Dielectric Strength:

No.	Value	Note
1	1 kV.mm^{-1}	1 kV mm^{-1} [10]
2	402 kV.mm^{-1}	kV mm^{-3}, 40.2 V mm^{-1}, Zn salt [14]
3	25.5 kV.mm^{-1}	kV mm^{-1}, moulding grade, ASTM D149

Dissipation Power Factor: tan δ 0.0018 (Zn salt) [14]

Static Electrification: High static charges can be generated on some ionomer surfaces but those highly neutralised with alkali-metal cations exhibit anti-electrostaticity. The effect is thought to be due to the higher equilibrium water content of these ionomers [20]

Dissipation (Power) Factor:

No.	Value	Frequency	Note
1	0.001–0.005	60 Hz	moulding grade

Optical Properties:
Optical Properties General: Formation of salts (e.g., with diamines) results in increased transparency. The level of change is related to both carboxylate content and the degree of neutralisation - at 3.5 mol% methacrylic acid the ionomer must be completely neutralised for transparency to be achieved whereas at 5.5 mol% only 70% neutralisation is necessary [1]

Refractive Index:

No.	Value	Note
1	1.51	moulding grade, ASTM D542

Transmission and Spectra: Transmittance 75–85% (ASTM D1003) [5]
Volume Properties/Surface Properties: Haze 3–17% [8]; 1–7% (Na salt, ASTM D1003-61) [16]; 1–9% (Zn salt, ASTM D1003-61) [16]. Clarity 10–80% (Na salt, ASTM D1746) [16]; 20–70% (Zn salt, ASTM D1746) [16]. Transparent (moulding grade). Gloss 20–90 (Na salt, 20°, ASTM D5236) [16]; 20–90 (Na salt, 45°, ASTM D5236) [16]; 10–90 (Zn salt, 20°, ASTM D5236) [16]; 20–90 (Zn salt, 45°, ASTM D5236) [16]

Polymer Stability:
Thermal Stability General: An amide cross-linked polymer can be formed from a diamine salt by heating to 275° under nitrogen followed by a period under vacuum. The product is clear and colourless but has a much lower tensile strength and elongation than the free acid precursor; stiffness and yield are little changed. Diamine salts are stable at 190° but cross-linking oxidation and yellowing occur at higher temps. [1]

Upper Use Temperature:

No.	Value	Note
1	60°C	140°F, continuous [9]
2	71–82°C	160–180°F, no load [10]
3	70°C	moulding grade [21]

Decomposition Details: Combustion products are CO_2 and H_2O. At low oxygen levels combustion can yield CO, unsaturated hydrocarbons and trace amounts of acrolein [5]
Flammability: Burning rate 2.54 cm min^{-1} (ASTM D635) [8], 23–28 mm min^{-1} (moulding grade, ASTM D635)
Environmental Stress: Exposure to uv light and weather can cause deterioration of both optical and mechanical props. Incorporation of a stabiliser can give an outdoor lifetime of up to one year with no deterioration; addition of carbon black can provide uv resistance
Chemical Stability: Ionomers are resistant to fats, oils, (Na salt better than Zn), alkalis, alcohols and ketones; many other organic solvents can cause swelling. Resistance to acids is poor and the polymer is degraded by oxidising agents such as bromine or nitric acid [7]

Applications/Commercial Products:
Processing & Manufacturing Routes: The copolymer is usually manufactured in an autoclave with comonomers and initiator supplied by continuous feed. Temps. of 150–280° and pressures above 100 MPa are used. Due to reactivity of methacrylic acid rapid end-over-end stirring is best employed. After polymerisation the acid copolymer is neutralised using an appropriate metal salt. Neutralisation using the melt process has proved useful in obtaining purified ethylene ionomers [15]. Processed by thermoforming, injection moulding, blow-moulding or extrusion. Ionomers have unusually high resistance to tear-off in high-speed extrusion operations. The very high melt strength permits the use of ionomers in situations where the molten polymer comes into contact with a sharp metal object, e.g., in the packaging of fishhooks etc. Ionomers also have superior performance where good hot tack strength is required

Mould Shrinkage (%):

No.	Value
1	0.003–0.02%

Applications: Major use in the manufacture of films and coatings where formability, visual appearance and toughness are important. other applications include golf ball covers and other sporting uses, automobile bumper guards, bottles, tubes, hoses, heat-sealing applications, food packaging, blister wrap and metal-coated articles

Trade name	Supplier
Aclyn	Allied Signal Corporation
Escor	Exxon Chemical
Hi-Milan	Mitsui Petrochemicals
Lucalen	BASF
Surlyn	DuPont

Bibliographic References

[1] Rees, R.W., *Polym. Prepr. (Am. Chem. Soc., Div. Polym. Chem.)*, 1973, **14**, 796
[2] Yano, S., Yamamoto, H., Tadano, K. and Hirasawa, E., *J. Polym. Sci., Part B: Polym. Phys.*, 1989, **27**, 2647, (dielectric props)
[3] *Paraloid Range*, Rohm and Haas, 1996, (technical datasheet)
[4] Tirrell, D.A., *Encycl. Polym. Sci. Eng.*, (ed. J.I. Kroschwitz), Wiley Interscience, 1988, **3**, 194, (copolymerisation)
[5] *Encycl. Polym. Sci. Eng.*, (ed. J.I. Kroschwitz), Wiley Interscience, 1986, **8**
[6] Roff, W.J. and Scott, J.R., *Fibres, Films, Plastics and Rubbers*, Butterworths, 1971, 332
[7] *Handbook of Industrial Materials*, 2nd edn., Elsevier Advanced Technology, 1992, 393
[8] *Guide to Plastics*, *Property and Specification Charts*, McGraw-Hill,
[9] *Plast. World*, 1973
[10] *Plastics Engineering Handbook*, 5th edn., (ed. M.L. Berins), Van Nostrand Reinhold, 1991
[11] MacKnight, W.J. and Emerson, F.A., *Polym. Prepr. (Am. Chem. Soc., Div. Polym. Chem.)*, 1971, **12**, 149, (relaxation temps)
[12] Tadano, K., Hirasawa, E., Yamamoto, H. and Yano, S., *Macromolecules*, 1989, **22**, 226, 2776
[13] Hirasawa, E., Yamamoto, Y., Tadano, K. and Yano, S., *J. Appl. Polym. Sci.*, 1991, **42**, 351
[14] *Kirk-Othmer Encycl. Chem. Technol.*, Vol. 14, 4th edn., (ed. J.I. Kroschwitz), John Wiley and Sons, 1993
[15] Kutsumitzu, S., Ikeno, T., Osada, S., Hara, H. *et al*, *Polym. J. (Tokyo)*, 1996, **28**, 299
[16] *Permeability and other Film Properties of Plastics and Elastomers*, Plastic Design Library, 1996, 63
[17] MacKnight, W.J., McKenna, L.W. and Read, B.E., *J. Appl. Phys.*, 1967, **18**, 4208
[18] Tachino, H., Hisaaki, H., Hirasawa, E., Kutsumizu, S. and Yano, S., *J. Appl. Polym. Sci.*, 1995, **55**, 131
[19] Vanhoorne, P. and Register, R.A., *Macromolecules*, 1996, **29**, 598, (melt viscosity)
[20] Tachino, H., Hara, H., Hirasawa, E., Kutsumizu, S. adn Yano, S., *Polym. J. (Tokyo)*, 1994, **36**, 1170
[21] *The Materials Selector*, 2nd edn., (eds. N.A. Waterman and M.F. Ashby), Chapman & Hall, 1997

Poly(methacrylic acid), Salts   P-351

$$\left[\text{CH}_2\text{C}(\text{CH}_3) \atop \text{O}^{\ominus}\text{C}\text{O}\ \text{M}^{\oplus} \right]_n$$ where M = metal cation

Synonyms: *Poly(2-methyl-2-propenoic acid), salts*
Related Polymers: Polymethacrylates General, Polymethacrylic acid
Monomers: Methacrylic acid salt
Material class: Thermoplastic
Polymer Type: acrylic
CAS Number:

CAS Reg. No.	Note
25086-62-8	Na salt
67536-64-5	Al salt
26715-36-6	Ca salt
142224-56-4	Co salt
69420-35-5	Li salt
136262-01-6	Ni salt
27360-06-1	K salt
116357-22-3	Cu salt
82334-36-9	Pb salt
50862-17-4	Mg salt
60088-51-9	Zn salt
54193-36-1	Na salt
29297-91-4	Li salt
29297-93-6	K salt

Molecular Formula: $(C_4H_5O_2M)_n$
Fragments: $C_4H_5O_2M$

Volumetric & Calorimetric Properties:
Glass Transition Temperature:

No.	Value	Note
1	310°C	Na salt [1]
2	490°C	Mg salt [1]

Transport Properties:
Polymer Solutions Dilute: Viscosity in aq. soln. 2.561 cm s^{-1} (c, 0.1 g dl^{-1}), 1.38 cm s^{-1} (c, 0.01 g dl^{-1}), 1.051 cm s^{-1} (c, 0.001 g dl^{-1}) [2]. The hydrogen-bonding of water molecules to polymer becomes weaker as temp. decreases [3]
Permeability of Liquids: Diffusion coefficients of aq. sodium chloride at 20° 1.15 × 10^7 cm^2 s^{-1} (c, Na ions/PMA 4 mol dm^{-3}, MW 237000), 0.602 × 10^7 cm^2 s^{-1} (c, Na ions/PMA 59 mol dm^{-3}, MW 1014000) [4]

Electrical Properties:
Electrical Properties General: Suspensions of PMAA salts can show high electrorheological activity; for example, yield stress increases with electric field strength, average particle diameter or acid content of salt. Electrorheological activity of suspensions increases in the order: Li salt > Na salt > K salt > NH$_4$ salt [5]

Optical Properties:
Total Internal Reflection: Intrinsic segmental anisotropy 150 × 10^{-25} cm^3 (Na salt, 0.012M NaCl, pH 7) [6]; 400 × 10^{-25} cm^3 (Na salt, 0.0012M NaCl, pH 7) [6]; 56–300 × 10^{-25} cm^3 (Na salt, H$_2$O) [7]

Polymer Stability:
Polymer Stability General: In dil. aq. soln. sodium polymethacrylate can undergo a two-stage degradation process during viscosity measurement. The first stage is a mechanochemical degradation due to shear stress and the action of oxygen, and the second is due to oxidation and photolysis [2]
Decomposition Details: The Na, Li, K and Cs salts decompose at approx. 350°, whereas the Mg and other alkaline earth metal salts decompose at 500°. Similar products including ethylene, Me$_2$CO, 1-butene and various ketones are produced [8,9]. The NH$_4$ salt decomposes via cyclisation followed by fragmentation at higher temps. At 320° products include NH$_3$ and H$_2$O; at 500° they include isocyanic acid and HCN [8]

Applications/Commercial Products:

Trade name	Supplier
Aquatreat	Alco Chemicals
Darvan	R.T. Vanderbilt
Daxad 30	WR Grace
Tamol (Orotan)	Rohm and Haas
Vulcastab	Akzo
Xenacryl	Baxenden Chemicals Ltd

Bibliographic References

[1] Otocka, E.P. and Kwei, T.K., *Macromolecules*, 1968, **1**, 401
[2] Saita, T., *Jpn. J. Appl. Phys.*, 1984, **23**, 87
[3] Kim, H.G. and Jhon, M.S., J. Polym. Sci., *Polym. Chem. Ed.*, 1995, **33**, 63

- Poly(methacrylimide)

[4] Howard, G.J. and Jordan, D.O., *J. Polym. Sci.*, 1954, **12**, 209, (liq permeability)
[5] Hong-Quen, X., Jian-Guo, G. and Junshi, G., *J. Appl. Polym. Sci.*, 1995, **58**, 951, (electrorheological props)
[6] Tsvetkov, V.N., Ljubina, S.Ya. and Bolevskii, K.L., Vysokomol. Soedin., *Karbotsepnye Vysokomol. Soedin.*, 1963, **4**, 33
[7] Kuhn, W., Kuenzle, O. and Katchalsky, A., *Helv. Chim. Acta*, 1948, **31**, 1994
[8] McNeill, L.C. and Zulfiquar, M., J. Polym. Sci., *Polym. Chem. Ed.*, 1978, **16**, 2465, 3208, (thermal degradation)
[9] McNeill, L.C. and Zulfiquar, M., *Polym. Degrad. Stab.*, 1979, **1**, 89, (thermal degradation)

Poly(methacrylimide)   P-352

Synonyms: *Rohacell, PMI*
Related Polymers: Poly(methyl methacrylate), Polymethacrylonitrile, Polymethacrylic acid
Monomers: Methacrylic acid, Methacrylonitrile
Material class: Thermoplastic
Polymer Type: polyimide
CAS Number:

CAS Reg. No.	Note
53112-45-1	
56273-40-6	Rohacell 51
56273-41-7	Rohacell 71
89468-92-8	Rohacell 110
110070-73-0	Rohacell WF
89468-95-1	Rohacell 110WF
56273-39-3	Rohacell 31
89468-93-9	Rohacell 51 WF
89468-94-0	Rohacell 71 WF
70535-56-7	Rohacell 41S

Molecular Formula: $(C_7H_9NO_2)_n$
Fragments: $C_7H_9NO_2$
General Information: A closed-cell foam with average cell size 200–500 μm [1]

Volumetric & Calorimetric Properties:
Density:

No.	Value	Note
1	d 0.056 g/cm^3	51 WF [1]
2	d 0.11 g/cm^3	110 WF [1,8]
3	d 0.341 g/cm^3	300 WF [1]
4	d 0.07 g/cm^3	71 WF [1]
5	d 1.2 g/cm^3	solid [1]
6	d 0.032 g/cm^3	Rohacell 31, DIN 53420 [8]
7	d 0.052 g/cm^3	Rohacell 51, DIN 53420 [8]
8	d 0.075 g/cm^3	Rohacell 75, DIN 53420 [8]
9	d 0.11 g/cm^3	Rohacell 110, DIN 53420 [8]
10	d 0.052 g/cm^3	51 WF, DIN 53420 [8]
11	d 0.075 g/cm^3	71 WF, DIN 53420 [8]
12	d 0.205 g/cm^3	200 WF, DIN 53420 [8]

Thermal Conductivity:

No.	Value	Note
1	0.03–0.034 W/mK	[6]
2	0.031 W/mK	Rohacell 31, DIN 52612 [8]
3	0.029 W/mK	Rohacell 51, DIN 52612 [8]
4	0.03 W/mK	Rohacell 71, DIN 52612 [8]

Glass Transition Temperature:

No.	Value	Note
1	140°C	approx.

Deflection Temperature:

No.	Value	Note
1	180°C	[6]
2	205°C	annealed [6]

Transition Temperature:

No.	Value	Note	Type
1	360°C	[7]	T_α
2	100°C	[7]	T_β
3	-120°C	[7]	T_γ

Transport Properties:
Water Absorption:

No.	Value	Note
1	18 %	50 days, Rohacell 31, DIN 53428
2	14 %	50 days, Rohacell 51 and 71, DIN 53428 [8]
3	0.59 %	saturated, 20°, 98% relative humidity, Rohacell 31
4	0.88 %	saturated, 20°, 98% relative humidity, Rohacell 51
5	1.1 %	saturated, 20°, 98% relative humidity, Rohacell 71 [8]

Mechanical Properties:
Mechanical Properties General: Foams show no Cosserat elasticity [1]. Other mech. props. have been reported [3,4]

Tensile (Young's) Modulus:

No.	Value	Note
1	213 MPa	WF 110 [1]
2	36 MPa	Rohacell 31, DIN 53457 [8]
3	70 MPa	Rohacell 51, DIN 53457 [8]
4	92 MPa	Rohacell 71, DIN 53457 [8]
5	160 MPa	Rohacell 110, DIN 53457 [8]
6	75 MPa	51 WF, DIN 53457 [8]

Poly(methacrylimide)

7	105 MPa	71 WF, DIN 53457 [8]
8	180 MPa	110 WF, DIN 53457 [8]
9	260 MPa	200 WF, DIN 53457 [8]

Tensile Strength Break:

No.	Value	Note
1	1 MPa	Rohacell 31, DIN 53455 [8]
2	1.9 MPa	Rohacell 51, DIN 53455 [8]
3	2.8 MPa	Rohacell 71, DIN 53455 [8]
4	3.5 MPa	Rohacell 110, DIN 53455 [8]
5	1.6 MPa	51 WF, DIN 53455 [8]
6	2.2 MPa	71 WF, DIN 53455 [8]
7	3.7 MPa	110 WF, DIN 53455 [8]
8	6.8 MPa	200 WF, DIN 53455 [8]

Flexural Strength at Break:

No.	Value	Note
1	5.2 MPa	110 WF, DIN 53423 [8]
2	12 MPa	200 WF, DIN 53423 [8]
3	4.5 MPa	Rohacell 110, DIN 53423 [8]
4	1.6 MPa	51 WF, DIN 53423 [8]
5	2.9 MPa	71 WF, DIN 53423 [8]
6	0.8 MPa	Rohacell 31, DIN 53423 [8]
7	1.6 MPa	Rohacell 51, DIN 53423 [8]
8	2.5 MPa	Rohacell 71, DIN 53423 [8]

Elastic Modulus:

No.	Value	Note
1	4400 MPa	solid [1]
2	5200 MPa	solid, heat treated 178°, 20h [1]
3	76–87 MPa	51 WF [1]
4	107–120 MPa	71 WF [1]
5	196–212 MPa	110 WF [1]
6	563–743 MPa	300 WF [1]

Compressive Strength:

No.	Value	Note
1	15.4–21.7 MPa	300 WF, ASTM 1621 [1]
2	0.4 MPa	Rohacell 31, DIN 53421 [8]
3	0.9 MPa	Rohacell 51, DIN 53421 [8]
4	1.12–1.51 MPa	51 WF, ASTM 1621 [1]
5	1.73–1.92 MPa	71 WF, ASTM 1621 [1]
6	3.77–4.08 MPa	110 WF, ASTM 1621 [1]
7	1.5 MPa	Rohacell 71, DIN 53421 [8]
8	3 MPa	Rohacell 110, DIN 53421 [8]
9	0.8 MPa	51 WF, DIN 53421 [8]
10	1.7 MPa	71 WF, DIN 53421 [8]
11	3.6 MPa	110 WF, DIN 53421 [8]
12	9 MPa	200 WF, DIN 53421 [8]

Miscellaneous Moduli:

No.	Value	Note	Type
1	150 MPa	200 WF, DIN 53294 [8]	Shear modulus
2	40 MPa	51 WF	Shear modulus
3	75 MPa	110 WF	Shear modulus
4	360 MPa	300 WF [1]	Shear modulus
5	14 MPa	Rohacell 31, DIN 53445 [8]	Shear modulus
6	21 MPa	Rohacell 51, DIN 53455	Shear modulus
7	30 MPa	Rohacell 71, DIN 53455	Shear modulus
8	58 MPa	Rohacell 110, DIN 53455 [8]	Shear modulus
9	13 MPa	Rohacell 31, DIN 53294	Shear modulus
10	19 MPa	Rohacell 51, DIN 53294	Shear modulus
11	29 MPa	Rohacell 71, DIN 53294	Shear modulus
12	50 MPa	Rohacell 110, DIN 53294 [8]	Shear modulus
13	19 MPa	51 WF, DIN 53294	Shear modulus
14	29 MPa	71 WF, DIN 53294	Shear modulus
15	50 MPa	110 WF, DIN 53294	Shear modulus

Fracture Mechanical Properties: Undergoes surface cracking on exposure to liq. hydrogen temps. with the amount of cracking increasing with increasing thermal cycling [7]

Electrical Properties:
Surface/Volume Resistance:

No.	Value	Note	Type
1	0.002×10^{15} Ω.cm	Rohacell 31	S
2	0.0009×10^{15} Ω.cm	Rohacell 51	S
3	0.00055×10^{15} Ω.cm	Rohacell 71, 23°, 50% relative humidity [8]	S

Dielectric Permittivity (Constant):

No.	Value	Frequency	Note
1	1.04	2.8 GHz	20°, Rohacell 31 [8]
2	1.07	2.8 GHz	20°, Rohacell 51 [8]
3	1.1	2.8 GHz	20°, Rohacell 71 [8]

Dissipation (Power) Factor:

No.	Value	Frequency	Note
1	0.0006	2.8 GHz	20°, Rohacell 31 [8]
2	0.0008	2.8 GHz	20°, Rohacell 51 [8]
3	0.001	2.8 GHz	20°, Rohacell 71 [8]

Polymer Stability:
Upper Use Temperature:

No.	Value
1	160°C

Decomposition Details: Decomposes at approx. 240°
Flammability: Details of flammability have been reported [5]. Burns, but emits little smoke [6]. Oxygen index 20 [6]
Chemical Stability: Is resistant to organic solvents but attacked by alkalis [6]. Resistant at 20° (DIN standard 53428) to Me$_2$CO, Et$_2$O, C$_6$H$_6$, diesel oil, EtOAc, 2-propanol, 2-butanone, petrol, 10% sulfuric acid, styrene, supergrade gasoline, CCl$_4$, toluene, trichloroethene. Has moderate resistance to dibutyl phthalate. Attacked by MeOH, AcOH, THF and 5% NaOH [8]

Applications/Commercial Products:
Processing & Manufacturing Routes: Synth. from cast sheets of methacrylic acid-*co*-methacrylonitrile at approx. 200° [2]
Applications: Applications in aircraft industry (landing flaps, landing gear doors, engine cowling, helicopter rotor blades). Other uses in core materials for x-ray tables, mammography cassettes and in cores of golf clubs, skis and tennis rackets. Applications due to excellent dielectric props. in antennae and radar domes

Trade name	Details	Supplier
Rohacell	various grades	Rohm and Haas

Bibliographic References
[1] Anderson, W.B., Chen, C.P. and Lakes, R.S., *Cell. Polym.*, 1994, **13**, 1, 16, (mech props)
[2] Pip, W., *Int. SAMPE Tech. Conf.*, 1996, **28**, 293, (uses)
[3] Pip, W., *Ind.-Anz.*, 1983, **105**, 12, (props, uses)
[4] Lachmann, Z., Konf. Lepeni Kovov "Intermetalbond" (Pr.), *5th*, 5th, 1972, 201, (mech props)
[5] Hilado, C.J., Cumming, H.J. and Casey, C.J., *J. Cell. Plast.*, 1979, **15**, 205, (flammability)
[6] Weber, H., De Grave, I. and Röhrl, E., *Ullmanns Encykl. Ind. Chem.*, 5th edn., (eds. B. Elvers, J.F. Rounsaville and G. Schultz), VCH, 1988, **A11**, 435
[7] Sharpe, E.L. and Helenbrook, R.G., Nonmet. Mater. Compos. Low Temp., *(Proc. ICMC Symp.)*, 1978, 207
[8] Pip, W., *Kunststoffe*, 1974, **64**, 23
[9] Pip, W., *Kunststoffe*, 1988, **78**, 201, (props)

Polymethacrylonitrile P-353
Synonyms: *Poly(2-methyl-2-propenenitrile)*. PMAN. PMCN
Related Polymers: Polyacrylonitrile
Monomers: Methacrylonitrile
Material class: Thermoplastic
Polymer Type: acrylic
CAS Number:

CAS Reg. No.
25067-61-2

Molecular Formula: (C$_4$H$_5$N)$_n$
Fragments: C$_4$H$_5$N
Molecular Weight: MW 4000000, M_n 25000, M_w/M_n 2.2 (radical polymerisation). M_n 16000–49000, M_w/M_n 1.71–1.78 (radical polymerisation). M_n 500000 (from membrane osmometry), MW 1330000 (from gpc), MW 13000 (degree of polymerisation 190, clathrate polymerised), MW 7000 (degree of polymerisation 100, clathrate polymerised)
General Information: Has good mech. strength and high resistance to solvents, acids and alkalis. It softens at a lower temp. than its analogue polyacrylonitrile, but discolours at temps. required for moulding
Tacticity: 2-Dimensional spectral techniques have proved usesful in determining tacticity [7,9]. An enhanced isotactic-heterotactic struct. is possessed by polymer produced by γ-irradiation of induced complex of methacrylonitrile with cyclotriphosphazene (clathrate polymerisation system [5,8])
Miscellaneous: Can be soln. blended (using DMSO) with polyvinyl alcohol [34]. Grafting of polymer to metallic surfaces (e.g. Ni, Pt) has been achieved by electrochemical polymerisation (cathodic polarisation) [2,11,13,20,22]; various spectroscopic techniques suggest strong chemical bonds betweeen carbon and metal atoms [20,22]. Studies of nitrogen levels (bulk and surface) using neutron activation analysis and proton track counting have been reported. [32] Highly monodisperse polymer beads of varying size and cross-link density have been reported [18]

Volumetric & Calorimetric Properties:
Melting Temperature:

No.	Value	Note
1	>200°C	min. [36]

Glass Transition Temperature:

No.	Value	Note
1	118°C	approx. [40,45]
2	120°C	[41]
3	104°C	[19]

Transition Temperature General: T_g and its correlation with cohesive energy density and chain-stiffness parameter (Flory ratio) has been reported [19]
Vicat Softening Point:

No.	Value	Note
1	115°C	approx. [36]

Surface Properties & Solubility:
Cohesive Energy Density Solubility Parameters: Cohesive energy density 440 J cm^{-3} (105 cal cm^{-3}) [19]. Solubility parameters 7.2 J$^{1/2}$ cm$^{-3/2}$ 12.4 cal$^{1/2}$ cm$^{-3/2}$), 6.6–7.3 J$^{1/2}$ cm$^{-3/2}$ (10.4–12.8 cal$^{1/2}$ cm$^{-3/2}$) [26]
Solvents/Non-solvents: Sol. DMSO [17], Me$_2$CO [17,40], cyclohexanone [30,40], THF [25], acetonitrile [11], trifluoroacetic acid [1], CH$_2$Cl$_2$ [5,8], C$_6$H$_6$ (hot) [5], xylenes (hot) [8]. Sl. sol. diethylketone [41], methyl propyl ketone [41], DMF [47]. γ-Irradiated (bulk) material insol. THF (non-clathrate material) [8]. Insol. MeOH [17,40], hexane [5,8], nitromethane [43], hexane [5,8]. Sl. sol. diethylketone [40], methyl propyl ketone [40], DMF [46].The solubility and swelling behaviour of homopolymer in more than 50 liqs. have been reported [26,27]

Transport Properties:
Permeability and Diffusion General: Diffusion coefficient [O$_2$] 1.1×10^{-8} cm^2 s^{-1} (after γ-irradiation and exposure to air). Diffusion coefficient shows an inverse dependence on dose [23]

Mechanical Properties:
Mechanical Properties General: Has high mech. strength

Optical Properties:
Transmission and Spectra: H-1 nmr [9], C-13 nmr [9] and xps spectral data have been reported. Xps, ups [2,20,22] and electrochemical impedance spectral data for electrochemically polymerised material have been reported. C-13 nmr spectral data for γ-irradiated, clathrate-polymerised material (inclusion complex with cyclotriphosphazene) have been reported [5,8]. Finely resolved C-13 nmr spectral data for atactic and syndiotactic-rich polymer have been reported [7,21]

Polymer Stability:
Thermal Stability General: Thermal degradation behaviour has been reported [1,12,21,31,33,37,38]. Depending on experimental conditions, polymer can either undergo a low-temp. coloration reaction or undergo almost qualitative depolymerisation to monomer [12]. The coloration reaction can be eliminated if polymer is prepared *in vacuo* using carefully purified monomer and AIBN initiator [12,37,38]
Decomposition Details: Thermal degradation occurs by depolymerisation process to give greater than 90% monomer; nitrile groups appear to prevent intramolecular H-transfer reactions [31].

Degradation to monomer occurs above 200° and is complete at 270° [33]. Anal. of pure polymer (prepared using AIBN or benzoyl peroxide) shows almost quantitative degradation to monomer, with trace quantities of isobutene and HCN; 2-stage decomposition process with onset of decomposition above 250° [12]. Other studies show two initiation processes for polymer degradation in N_2, namely chain-end and random-scission initiation; internal cyclisation reaction leads to formation of thermally stable residue having conjugated polyimine struct. [1,16]. Recent studies indicate the formation of at least two additional cyclic forms

Hydrolytic Stability: Alkaline hydrol. [21] (aq. NaOH soln.) of polymer in DMSO leads to formation of cyclic sequences [14,35,40]. Hydrol. of nitrile groups in aq. alkali solns. leads to formation of carboxylates and acid groups (water sol.) [14]

Stability Miscellaneous: Irradiation using high-energy radiation leads to main-chain scission without simultaneous cross-linking [30]. Sensitivity to X-rays has been reported [40]. The effects of radiation in the absence of O_2 at room temp. have been reported [24]. The predominant reaction is main-chain scission [24,44,45]. γ-Irradiation of thin polymer sheets and exposure to air has been reported [23]. Radicals produced by γ-irradiation are primarily the result of main-chain scission, hydrogen atom addition and abstraction [4]

Applications/Commercial Products:

Processing & Manufacturing Routes: Generally produced by free-radical polymerisation of methacrylonitrile in a similar way to polyacrylonitrile [36]. However, methacrylonitrile polymerises less readily so polymerisation is performed in emulsion. Anionic polymerisation using Grignard reagents gives stereoregular polymer [17]. Using BuLi as catalyst at 0° gives atactic product [1,46] whereas Et_2Mg at 80° gives isotactic rich product [1,7] free-radical initiators such as benzoyl peroxide, potassium persulfate and AIBN can also be used. Living polymerisation process [6] using aluminium porphyrin initiators has been reported. [3] Inclusion polymerisation, e.g. within cyclophosphazene clathrates, using γ-irradiation has been reported [5,8]; can be used to control stereochemistry of polymer [8]. Free-radical polymerisation using the *tert*-butylthiyl radical as initiator has also been reported [10]. Electrochemical polymerisation on metal surfaces has been reported [2,11,13,20,22]. Preparative laser photolysis, using dianisoyl as initiator, gives narrow-banded atactic oligomeric products (degree of polymerisation 3–26) plus insol. glassy polymer [25]. Methacrylonitrile can also be polymerised by group transfer polymerisation using (2-trimethylsilyl)isobutyronitrile as initiator, either with [24] or without [28] coinitiators (e.g. tetraethylammonium cyanide, or tetrabutylammonium fluoride). This is a living polymerisation reaction giving material of almost random tacticity, [28] which is of limited practical importance on account of severe side reactions which reduce conversions and MW [29]. Highly monodisperse polymethacrylonitrile beads have been prepared by emulsion copolymerisation of monomer and allyl methacrylate in the absence of emulsifier using potassium persulfate initiator [18]. Polymers wth narrow MW distribution are obtained using BulLi as initiator [46]

Applications: Uses include electrode compositions, fuel cells, optical recording media, electrophotography, magnetorhelogical compositions, electromagnetic materials, sensors, (lubricating) oils, films/fibres, paints/coatings, adhesives, resists (photoresists in x-ray lithography) integrated circuits, insulating compositions, conducting composites, thermal printing, lithography, photographic compositions, mouldings, biomedical applications. Blends with poly(vinyl alcohol) have potential applications in porous, permselective membranes; copolymers with fluorinated comonomers have potential applications as photo(x-ray and electron-beam) resists, as well as copolymers containing methyl methacrylate. Thin films, produced by electrochemical polymerisation techniques, have potential applications in electronics, catalysis, corrosion inhibition, etc.

Bibliographic References

[1] Nagata, A., Ohta, K., and Iwamoto, R., *Macromol. Chem. Phys.*, 1996, **197**, 1959, (synth., thermal stability,)
[2] Bureau, C., Deniau, G., Valin, F., Guittet, M.-J., *et al*, *Surf. Sci.*, 1996, **355**, 177, (synth., electrochem., theory, surface studies, spectra)
[3] Sugimoto, H., Saika, M., Hosokawa, Y., Aida, T. and Inoue, S., *Macromolecules*, 1996, **355**, 177, (synth., electrochem., theory, surface studies, spectra)
[4] Dong, L. Hill, D.J.T., O' Donnell, J.H. and Pomeny, P.J, *Polymer*, 1995, **36**, 2873, (radiation stability, degradation, esr)
[5] Allcock, H.R., Silverberg, E.N., Dudley, G.K. and Pucher, S.R., *Macromolecules*, 1994, **27**, 7550, (synth., C-13 nmr)
[6] Otsu, T., Matsunaga, T., Doi, T. and Matsumoto, A., *Eur. Polym. J.*, 1995, **31**, 67, (synth.)
[7] Kawamura, T., Toshima, N. and Matsuzaki, K., *Macromol. Chem. Phys.*, 1995, **195**, 3343, (synth., H-1 nmr, C-13 nmr)
[8] Allcock, H.R., Silverberg, E.N., and Dudley, G.K., *Macromolecules*, 1994, **27**, 1033, (synth., C-13 nmr)
[9] Dong, L., Hill, D.J.T., O'Donnell, J.H. and Whittaker, A.K., *Macromolecules*, 1994, **27**, 1830, (tacticity, H-1 nmr, C-13 nmr)
[10] Busfield, W.K., Heiland, K. and Jenkins, I.D., *Tetrahedron Lett.*, 1994, **35**, 6541, (synth.)
[11] Tanguy, J., Viel, P., Deniau, G. and Lécayon, G., *Electrochim. Acta*, 1993, **38**, 1501, (synth., electrochem, surface studies, impedance spectroscopy)
[12] McNeill, I.C. and Mahmood, J., *Polym. Degrad. Stab.*, 1994, **45**, 285, (thermal stability, thermal degradation, T_g)
[13] Tanguy, J., Deniau, G. Augé, C., Zalczer, G. and Lécayon, G., *J. Electroanal. Chem.*, 1994, **377**, 115, (synth., electrochem., surface studies)
[14] Andreeva, O.A. and Burkova, L.A., *Int. J. Polym. Mater.*, 1994, **26**, 177, (chemical stability, uv)
[15] Ogawa, T., Ikegami, M., and Kitamura, A., *Bunseki Kagaku*, 1996, **45**, 77, (xps, theory)
[16] Hill, D.J.T., Dong, L., O'Donnell, J.H., George, G. and Pomeny, P., *Polym. Degrad. Stab.*, 1993, **40**, 143, (thermal stability, thermal degradation)
[17] Ono, H., Hisatani, K. and Kamide, K., *Polym. J. (Tokyo)*, 1993, **25**, 245, (synth., C-13 nmr)
[18] Zou, D., Aklonis, J.J. and Salovey, R., *J. Polym. Sci., Polym. Chem. Ed.*, 1992, **30**, 2443, (synth., sem)
[19] Boyer, R.F, *Macromolecules*, 1992, **25**, 5326, (thermal props., solubility)
[20] Deniau, G., Viel, P., Lécayon, G. and Delhalle, J., *Surf. Interface Anal.*, 1992, **18**, 443, (electrochem, ir, surface studies, synth., spectra)
[21] Bashir, Z., Packer, E.J., Herbert, I.R. and Price, D.M., *Polymer*, 1992, **33**, 373, (synth., tacticity, C-13 nmr, thermal studies, chemical stability)
[22] Deniau, G., Lécayon, G., Viel, P., Hennico, G. and Delhallé, J., *Langmuir*, 1992, **8**, 267, (electrochem, synth, surface studies, spectra, theory)
[23] Katircioglu, Y., Kaptan, Y. and Guven, O., Proc. Tihany Symp. Radiat. Chem., *7th*, 1991, 377, (γ-radiation stability, transport props., esr)
[24] Schnabel, W., Zhu, Q.Q. and Klaumuenzer, S., *ACS Symp. Ser.*, 1991, **475**, 44, (gel formation, γ-irradiation, stability)
[25] Kaupp, G., Sauerland, O., Marquardt, T. and Plagmann, M., *J. Photochem. Photobiol. A*, 1991, **56**, 381, (synth.)
[26] Ho, B.-C., Chin, W.-K. and Lee, Y.-D., *J. Appl. Polym. Sci.*, 1991, **42**, 99, (solubility)
[27] Schlegel, L. and Schnabel, W., *J. Appl. Polym. Sci.*, 1990, **41**, 1797, (solubility props., chemical stability, surface props.)
[28] Sogah, D.Y., Hertler, W.R., Webster, O.W. and Cohen, G.M., *Macromolecules*, 1987, **20**, 1473, (synth.)
[29] Bandermann, F. and Witkowski, R., *Makromol. Chem.*, 1986, **187**, 2691, (synth.)
[30] Schnabel, W. and Klaumünzer, S., *Radiat. Phys. Chem.*, 1989, **33**, 323, (irradiation stability)
[31] Buniyat-Zade, A.A., Androsova, V.M., Pokatilova, S.D. and Aliev, A.M., *Plast. Massy*, 1990, 71, (thermal stability, thermal degradation)
[32] Mitchell, J.W., Yegnasubramanian, S. and Shepherd, L., *J. Radioanal. Nucl. Chem.*, 1987, **112**, 425, (anal.)
[33] Hodder, A.N., Holland, K.A. and Rae, I.D., *Aust. J. Chem.*, 1986, **39**, 1883, (thermal stability, ms, thermal degradation)
[34] Hu, C.-M. and Chiang, W.-Y., *Angew. Makromol. Chem.*, 1990, **179**, 157, (blends, uses)
[35] Rumynskaya, I.G., Romanova, E.P., Agranova, S.A. and Frenkel, S.Ya., *Acta Polym.*, 1991, **42**, 250, (stability, uv)
[36] Kern, W. and Fernow, H., *J. Prakt. Chem.*, 1942, **160**, 296, (synth., transport props., thermal props.)
[37] Grassie, N. and McNeill, I.C., *J. Chem. Soc.*, 1956, 3929, (thermal stability, degradation, ir)
[38] Grassie, N. and McNeill, I.C., *J. Polym. Sci.*, 1959, **27**,30, 39, 37, 207, 211, (thermal stability, degradation, ir)
[39] Batty, N.S. and Guthrie, J.T., *Makromol. Chem.*, 1981, **182**, 71, (chemical stability, hydrolytic stability)
[40] Schlegel, L. and Schnabel, W., J. Vac. Sci. Technol., *B*, 1986, **6**, 82, (synth., uses, stability)

[41] Pittman, C.U., *J. Electrochem. Soc.*, 1981, **128**, 1758, (copolymers, radiation stability, uses)
[42] Acar, M.H. and Yagci, Y., *Macromol. Rep.*, 1991, **A28 (suppl. 2)**, 177, (copolymers, radiation stability, uses)
[43] Helbert, J.N., Cook, C.F., Chen, C.-Y. and Pittman, C.U., *J. Electrochem. Soc.*, 1979, **126**, 694, (uses, stability)
[44] Helbert, J.N., Poindexter, E.H., Stahl, G.A., Chen, C.Y. and Pittman, C.U., *J. Polym. Sci., Polym. Chem. Ed.*, 1979, **17**, 49, (uses, stability)
[45] Helbert, J.N., Iafrate, G.J., Pittman, C.U. and Lai, J.H., *Polym. Eng. Sci.*, 1980, **20**, 1077, (uses, copolymers, stability)
[46] Feit, B.-A., Heller, E. and Zilkha, A., *J. Polym. Sci., Part A-1*, 1966, **4**, 1151, (synth.)

Poly[methane bis(4-phenyl)carbonate]   P-354

Synonyms: *Poly[oxycarbonyloxy-1,4-phenylenemethylene-1,4-phenylene]. Poly(4,4'-dihydroxydiphenylmethane-co-carbonic dichloride)*
Monomers: 4,4'-Dihydroxydiphenylmethane, Carbonic dichloride
Material class: Thermoplastic
Polymer Type: polycarbonates (miscellaneous)
CAS Number:

CAS Reg. No.
9070-69-3
28935-53-7

Molecular Formula: $(C_{14}H_{10}O_3)_n$
Fragments: $C_{14}H_{10}O_3$

Volumetric & Calorimetric Properties:
Density:

No.	Value	Note
1	d 1.22 g/cm^3	20°, amorph. [3]
2	d 1.28 g/cm^3	cryst. [3]

Melting Temperature:

No.	Value	Note
1	223–225°C	[2]
2	236–253°C	amorph. [3]
3	275–278°C	cryst. [3]

Glass Transition Temperature:

No.	Value	Note
1	147°C	[2]
2	102°C	[1]

Surface Properties & Solubility:
Solubility Properties General: Forms miscible blends with Bisphenol A polycarbonate, tetramethylpolycarbonate, and polymethylacrylate [1]; it forms immiscible blends with polystyrene, [1]
Solvents/Non-solvents: Swollen by MeCl, *m*-cresol, pyridine, DMF; insol. EtOAc, Me$_2$CO, cyclohexanone, THF, C$_6$H$_6$ [2]

Mechanical Properties:
Mechanical Properties General: Tensile stress-strain curve shows a viscoelastic region up to the yield point, followed by a stress drop and then necking behaviour (at approx. constant stress); finally the stress increases to the break point [4]

Tensile (Young's) Modulus:

No.	Value	Note
1	1770 MPa	18000 kg cm^{-2}, amorph. [4]
2	1960 MPa	20000 kg cm^{-2}, cryst. [4]

Tensile Strength Break:

No.	Value	Note	Elongation
1	68 MPa	6.9 kg mm^{-2}, amorph. [4]	180%
2	79 MPa	8 kg mm^{-2}, cryst. [4]	176%

Tensile Strength Yield:

No.	Value	Note
1	54.9 MPa	5.6 kg mm^{-2} [4]

Optical Properties:
Refractive Index:

No.	Value	Note
1	1.608	calc. [1]

Transmission and Spectra: Ir spectral data have been reported [2]

Bibliographic References
[1] Kim, C.K. and Paul, D.R., *Macromolecules*, 1992, **25**, 3097, (miscibility, T$_g$, refractive index)
[2] Schnell, H., *Polym. Rev.*, 1964, **9**, 99, (props.)
[3] Kozlov, P.V. and Perepelkin, A.N., *Polym. Sci. USSR (Engl. Transl.)*, 1967, **9**, 414, (density, T$_m$)
[4] Perepelkin, A.N. and Kozlov, P.V., *Polym. Sci. USSR (Engl. Transl.)*, 1968, **10**, 14, (tensile props)

Poly(2-methoxyaniline-5-sulfonic acid)   P-355

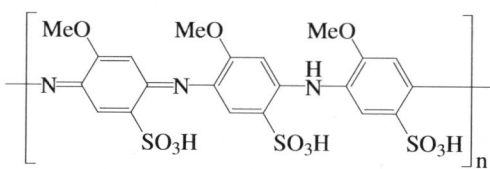

Synonyms: *3-Amino-4-methoxybenzenesulfonic acid homopolymer. PMAS*
Material class: Thermoplastic
Polymer Type: polyaniline
CAS Number:

CAS Reg. No.
167860-86-8

Molecular Formula: $(C_7H_7NO_4S)n$
Fragments: $C_7H_7NO_4S$
General Information: Water-soluble conductive polymer with self-doped properties, in which the backbone has a polyemeraldine structure analogous to polyaniline [1,2]. Prod. as yellow-green solid or aqueous soln.

Surface Properties & Solubility:
Solvents/Non-solvents: Sol. water; sparingly sol. EtOH; insol. MeCN [2]

Electrical Properties:
Electrical Properties General: Conductivity of self-doped polymer up to 0.04 S cm^{-1} reported [1]. The conductivity remains fairly constant even in $2M$ NaOH [3].

Optical Properties:
Optical Properties General: Polymer has characteristic absorptions in uv-vis region with λ_{max} ca. 330 and 475nm [4,5,6]. An optically active form of the polymer has been prepared with UV absorption band centred at ca. 483nm [6]. The yellow-brown aqueous soln. of the ammonium salt changes to a blue colour on addition of NaOH, with the appearance of absorption bands with λ_{max} ca. 350 and 750nm [3].
Transmission and Spectra: Uv [4,5,6] and ftir [4] spectral data have been reported.

Polymer Stability:
Thermal Stability General: Polymer is stable up to 215°C [2].

Applications/Commercial Products:
Processing & Manufacturing Routes: Prod. mainly by oxidation of 2-methoxyaniline-5-sulfonic acid with ammonium peroxydisulfate in basic media such as aqueous pyridine [1,5]. It has also been prepared electrochemically from the same monomer in ammonium hydroxide soln. [4]. Optically active polymer has been prepared chemically using an optically active peroxydisulfate salt [6].
Applications: PMAS has been evaluated in electrochemical biosensors for hydrogen peroxide using horseradish peroxidase [7,8].

Bibliographic References
[1] Shimizu, S., Saitoh, T., Uzawa, M., Yuasa, M., Yano, K., Maruyama, T. and Watanabe, K., *Synth. Met.*, 1997, **85**, 1337
[2] Tallman, D.E. and Wallace, G.G., *Synth. Met.*, 1997, **90**, 13
[3] Strounina, E.V., Shepherd, R., Kane-Maguire, L.A.P. and Wallace, G.G., *Synth. Met.*, 2003, **135/6**, 289
[4] Guo, R., Barisci, J.N., Innis, P.C., Too, C.O., Wallace, G.G. and Zhou, D., *Synth. Met.*, 2000, **114**, 267
[5] Masdarolomoor, F., Innis, P.C., Ashraf, S. and Wallace, G.G., *Synth. Met.*, 2005, **153**, 181
[6] Strounina, E.V., Kane-Maguire, L.A.P. and Wallace, G.G., *Synth. Met.*, 1999, **106**, 129
[7] Tatsuma, T., Ogawa, T., Sato, R. and Oyama, N., *J. Electroanal. Chem.*, 2001, **501**, 180
[8] Ngamna, O., Morrin, A., Moulton, S.E., Killard, A.J., Smyth, M.R. and Wallace, G.G., *Synth. Met.*, 2005, **153**, 181

Poly[2-methoxy-5-(3′,7′-dimethyloctyloxy)-1,4-phenylene vinylene]

P-356

Synonyms: *Poly[[[2-(3,7-dimethyloctyl)oxy]-5-methoxy-1,4-phenylene]-1,2-ethenediyl]. Poly[[[(3,7-dimethyloctyl)oxy]methoxy-1,4-phenylene]-1,2-ethenediyl]. MDMO-PPV. OC$_1$C$_{10}$-PPV*
Material class: Thermoplastic
Polymer Type: poly(arylene vinylene)
CAS Number:

CAS Reg. No.	Note
177716-59-5	poly[[[2-(3,7-dimethyloctyl)oxy]-5-methoxy-1,4-phenylene]-1,2-ethenediyl]
250164-11-5	poly[[[(3,7-dimethyloctyl)oxy]methoxy-1,4-phenylene]-1,2-ethenediyl]

Molecular Formula: $(C_{19}H_{28}O_2)n$
Fragments: $C_{19}H_{28}O_2$
Additives: The polymer is blended with the soluble methanofullerene PCBM, usually in the ratio of about 1:3, for photovoltaic applications [1,2,3].
General Information: Photoluminescent and electroluminescent polymer with hole conducting properties. Prod. as red or orange solid.
Morphology: The morphology has been reported. Circular features are indicative of aggregation in films [12].

Volumetric & Calorimetric Properties:
Glass Transition Temperature:

No.	Value	Note
1	48–50°C	[4]

Surface Properties & Solubility:
Solubility Properties General: Polymer films cast with chlorobenzene as solvent give best results for photovoltaic cells and field-effect transistors [5,6].
Solvents/Non-solvents: Sol. THF, toluene, xylene, chlorobenzene [4,5,6].

Electrical Properties:
Electrical Properties General: Polymer can act as electron donor (ionisation potential 5.33eV [7]) in combination with the electron acceptor PCBM in heterojunction photovoltaic cells [1,2,3]. The hole mobility of pure polymer is increased in the MDMO-PPV/PCBM blend [1,2]. MDMO-PPV can also be used as emissive layer in LEDs with turn-on voltages of 2.5–3.5V [4].

Optical Properties:
Optical Properties General: Polymer exhibits photoluminescence and electroluminescence in orange-red region of spectrum with λ_{max} at about 600nm, and band gap of about 2.4 eV [3,4,7]. Birefringence exhibited in region of 600 nm with in-plane refractive index much higher than out-of-plane index [8].
Transmission and Spectra: pmr [4,9], cmr [4], ir [4,9], uv and photoluminescence [3,4,7,10,12], electroluminescence [4] and esr [11] spectral data have been reported.

Polymer Stability:
Thermal Stability General: Initial loss of weight above ca. 315°C. Longer side-chain degradation from 330–450°C [4].

Applications/Commercial Products:
Processing & Manufacturing Routes: Polymer prod. mainly by Gilch method from 1,4-bis(halomethyl)-2-methoxy-5-(3′,7′-dimethyloctyloxy)benzene with *tert*-BuOK, or by sulfinyl route from 1-(chloromethyl)-2-(3′,7′-dimethyloctyloxy)-5-methoxy-4-(alkylsulfoxymethyl)benzene [4,9,10]. Polymer prod. by the sulfinyl route is reported to have greater efficiency in blends for photovoltaic cells [13]. Polymer produced by the Gilch route has up to 2% tolane-bisbenzyl defects [14].
Applications: Mainly used as blends with the methanofullerene PCBM for development of solar cells. Also possible applications in light-emitting diodes and field-effect transistors.

Details	Supplier
polymer end-capped with different groups (solid and in toluene soln.)	American Dye Source

Bibliographic References
[1] Tuladhar, S.M., Poplavskyy, D., Choulis, S.A., Durrant, J.R., Bradley, D.D.C. and Nelson, T., *Adv. Funct. Mater.*, 2005, **15**, 1171
[2] Mihailetchi, V.D., Koster, L.J.A., Blom, P.W.M., Metzer, C., de Boer, B., van Duren, J.K.J. and Janssen, R.A.J., *Adv. Funct. Mater.*, 2005, **15**, 795
[3] Al-Ibrahim, M., Konkin, A., Roth, H.-S., Egbe, D.A.M., Klemm, E., Zhokhavets, U., Gobsch, G. and Senssfuss, S., *Thin Solid Films*, 2005, **474**, 201

[4] Lutsen, L., Adriaensens, P., Becker, H., Van Breemen, A.J., Vanderzande, D. and Gelan, J., *Macromolecules*, 1999, **32**, 6517
[5] Rispens, M.T., Meetsma, A., Riffberger, R., Brabec, C.J., Sariciftci, N.S. and Hummelen, J.C., *Chem. Comm.*, 2003, 2116
[6] Geens, W., Shaheen, S.E., Wessling, B., Brabec, C.J., Poortmans, J. and Sariciftci, N.S., *Organic Electronics*, 2005, **3**, 105
[7] Schubert, U.S. and Wienk, M.M., *Chem. Mater.*, 2004, **16**, 2503
[8] Ramsdale, C.M. and Greenham, N.C., *Adv. Mater.*, 2002, **14**, 212
[9] Hontis, L., Vrindts, V., Vanderzandre, D. and Lutsen, L., *Macromolecules*, 2003, **36**, 3035
[10] Mozer, A.J., Denk, P., Scharber, M.C., Neugebauer, H., Sariciftci, N.S., Wagner, P., Lutsen, L. and Vanderzande, D., *J. Phys. Chem. B*, 2004, **108**, 5235
[11] Parisi, J., Dyakonov, V., Pientka, M., Riedel, I., Deibel, C., Brabec, C.J., Sariftci, N.S. and Hummelen, J.C., *Z. Naturforsch., A: Phys. Sci.*, 2002, **57**, 995
[12] Kemerink, M., van Duren, J.K.J., van Breemen, A.J.J., Wildeman, J., Wienk, M.M., Blom., P.W.M., Schoo, H.F.M. and Janssen R.A.J., *Macromolecules*, 2005, **38**, 7784
[13] Munters, T., Martens, T., Goris, L., Vrindts, V., Manca, J. and Brabec, C.J., *Thin Solid Films*, 2005, **474**, 201
[14] Becker, H., Spreitzer, H., Ibrom, K. and Kreuder, W., *Macromolecules*, 1999, **32**, 4925

Poly[2-methoxy-5-(2′-ethylhexyloxy)-1,4-phenylene vinylene]   P-357

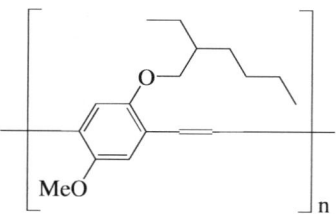

Synonyms: *Poly[[2-[(2-ethylhexyl)oxy]-5-methoxy-1,4-phenylene]-1,2-ethenediyl]. Poly[[[(2-ethylhexyl)oxy]methoxy-1,4-phenylene]-1,2-ethenediyl]. MEH-PPV*
Material class: Thermoplastic
Polymer Type: poly(arylene vinylene)
CAS Number:

CAS Reg. No.	Note
133030-00-9	poly[[2-[(2-ethylhexyl)oxy]-5-methoxy-1,4-phenylene]-1,2-ethenediyl]
138184-36-8	poly[[[(2-ethylhexyl)oxy]methoxy-1,4-phenylene]-1,2-ethenediyl]

Molecular Formula: $(C_{17}H_{24}O_2)_n$
Fragments: $C_{17}H_{24}O_2$
Additives: Dopants such as I_2 have been added to increase conductivity. The polymer has been blended with fullerenes for use in photovoltaic devices.
General Information: Stable conductive photoluminescent and electroluminescent polymer with greater solubility in organic solvents compared to poly(*p*-phenylene vinylene). The polymer is prod. as an orange-red solid.
Morphology: Orthorhombic unit cell, a = 7.12, b = 20.89, c = 6.47 Å. Chain orientation is anisotropic, with the polymer backbones mainly parallel to film planes [1]. Nematic-like phases in the region of 200–260°C detected [2]

Volumetric & Calorimetric Properties:
Glass Transition Temperature:

No.	Value	Note
1	57–67°C	[2,3]

Transition Temperature General: Nematic region 200–260°C with isotropic transition at about 290°C [2].

Surface Properties & Solubility:
Solubility Properties General: High molecular weight fractions and gelated forms obtained during production are insoluble in solvents such as THF and toluene [4,5,6].
Solvents/Non-solvents: Sol. chlorobenzene, toluene, xylene, 1,1,2,2-tetrachloroethane, $CHCl_3$, CH_2Cl_2, THF, 1,4-dioxan [5,7].

Mechanical Properties:
Mechanical Properties General: Mechanical properties of fibres spun from THF and drawn fibres have been reported [8].

Electrical Properties:
Electrical Properties General: Polymer is used as emissive layer in LEDs [9,10,11,12,26] with low turn-on voltages and also has photovoltaic properties. Electron mobility in polymer is comparable to hole mobility [13].

Optical Properties:
Optical Properties General: Polymer has band gap (2.2–2.3eV) [25] lower than that of poly(*p*-phenylene vinylene). Films exhibit photoluminescence and electroluminescence in the red-orange region of the spectrum with λ_{max} values in the region of 590–630nm [5,12,14]. The longer-wavelength maximum is dependent on spinning speed for film casting [14]. Efficiency of photoluminescence is markedly reduced on increase of pressure [15]. Birefringence is exhibited in region of 550–700nm, with in-plane refractive index higher than out-of-plane index [16].
Transmission and Spectra: Pmr [7,24], cmr [7], ir [7,12,17,18,22,24], uv and photoluminescence [7,10,12,17,18,19], electroluminescence [9,10,12,14], esr [20], UPS and XPS [21,22] and x-ray diffraction [1,2] spectral data have been reported.

Polymer Stability:
Polymer Stability General: Polymer undergoes oxidative degradation on exposure to UV radiation and with heat treatment [3,22].
Thermal Stability General: Polymer is much more unstable on heating in air compared with nitrogen. For polymer films, lifetime to 5% weight loss at 200°C is 2.32min. (air), 71h (nitrogen) [3].

Applications/Commercial Products:
Processing & Manufacturing Routes: Prod. mainly by modified Gilch reaction of 1,4-bis(halomethyl)-2-methoxy-5-(2-ethylhexyl)benzene with a base such as *tert*-BuOK. The formation of insoluble gel form can be avoided by the addition of 4-*tert*-butylbenzyl chloride [4,24], tetrabutylammonium bromide [12], 4-methoxyphenol [23] or with DMF as a solvent [6]. Also prod. by condensation of 1,4-bis(diethoxyphosphonomethyl)-2-methoxy-5-(2-ethylhexyloxy)benzene with the corresponding 1,4-dialdehyde in a Horner-type reaction [7].
Applications: The principal application is as an emissive polymer in light-emitting diodes, also potential applications for lasers and photovoltaic cells.

Details	Supplier
polymer end-capped with different groups (solid and in toluene soln.)	American Dye Source

Bibliographic References
[1] Yang, C.Y., Hide, F., Diaz-Garcia, M.A., Heeger, A.J. and Cao, Y., *Polymer*, 1998, **39**, 2299
[2] Chen, S.-H., Su, A.-C., Huang, Y.-F., Su, C.-H., Peng, G.-Y. and Chen, S.-A., *Macromolecules*, 2002, **35**, 4229
[3] Wang, H., Tao, X. and Newton, E., *Polym. Int.*, 2004, **53**, 20
[4] Hsieh, B.R., Yu, Y., VanLaeken, A.C. and Lee, H., *Macromolecules*, 1997, **30**, 8094
[5] Kraft, A., Grimsdale, A.C., and Holmes, A.B., *Angew. Chem., Int. Ed. Engl.*, 1998, **37**, 402
[6] Parekh, B.P., Tangonan, A.A., Newaz, S.S., Sanduja, S.K., Ashraf, A.Q., Krishnamoorti, R. and Lee, T.R., *Macromolecules*, 2004, **37**, 8883
[7] Pfeiffer, S. and Hörhold, H.-H., *Macromol. Chem. Phys.*, 1999, **200**, 1870
[8] Motamedi, F., Ihn, K.J., Ni, Z., Srdanov, G., Wudl, F. and Smith, P., *Polymer*, 1992, **32**, 1102
[9] Braun, D. and Heeger, A.J., *Appl. Phys. Lett.*, 1991, **58**, 1982

[10] Burn, P.L., Grice, A.W., Tajbakhsh, A., Bradley, D.D.C. and Thomas, A.C., *Adv. Mater.*, 1997, **9**, 1171
[11] Yang, Y. and Bharathan, J., *Polym. Prepr. (Am. Chem. Soc., Div. Polym. Chem.)*, 1998, **39**, 98
[12] Wu, X., Shi, G., Chen, F., Han, S. and Peng, J., *J. Polym. Sci., Part A: Polym. Chem.*, 2004, **42**, 3049
[13] Scott, J.C., Brock, P.J., Salem, J.R., Ramos, S., Malliaras, G.G., Carter, S.A. and Bozano, L., *Synth. Met.*, 2000, **111/2**, 289
[14] Shi, Y. and Yang, Y., *Polym. Prepr. (Am. Chem. Soc., Div. Polym. Chem.)*, 2000, **41**, 802
[15] Tikhoplav, R.K. and Hess, B.C., *Synth. Met.*, 1999, **101**, 236
[16] Boudrioua, A., Hobson, P.A., Matterson, B., Samuel, I.D.W. and Barnes W.L., *Synth. Met.*, 2000, **111/2**, 545
[17] McCallien, D.W.J., Thomas, A.C. and Burn, P.L., *J. Mater. Chem.*, 1999, **9**, 847
[18] Wu, X., Shi, G., Qu, L., Zhang, J. and Chen, F., *J. Polym. Sci., Part A: Polym. Chem.*, 2003, **41**, 449
[19] Barashkov, N.N., Guerrero, D.J., Olivos, H.J. and Ferraris, J.P., *Synth. Met.*, 1995, **75**, 153
[20] Kuroda, S., Marumoto, K., Greenham, N.C., Friend, R.H., Shimoi, Y. and Abe, S., *Synth. Met.*, 2001, **119**, 655
[21] Burn, P.L., Holmes, A.B., Kaeriyama, K., Sonoda, Y., Llost. O., Meyers, F. and Bredas, J.L., *Macromolecules*, 1995, **28**, 1959
[22] Atreya, M., Li, S., Kang, E.T., Neoh, K.G., Ma, Z.H., Tan, K.L. and Huang, W., *Polym. Degrad. Stab.*, 1999, **65**, 287
[23] Neef, C.J. and Ferraris, J.P., *Macromolecules*, 2000, **33**, 2311
[24] Ram, M.K., Sarkar, N., Bertoncello, P., Sarkar, A., Narizzano, R. and Nicolini, C., *Synth. Met.*, 2001, **122**, 369
[25] Yang, Y., Pei, Q. and Heeger, A.J., *Synth. Met.*, 1996, **78**, 263
[26] Parker, I.D., *J. Appl. Phys.*, 1994, **75**, 1656

Poly(3-methoxypropylene oxide)   P-358

Synonyms: *Poly[oxy(methoxymethyl)-1,2-ethanediyl]. Oxirane, (methoxymethyl), homopolymer, 9CI. Poly[(methoxymethyl)oxirane]. Poly(methylglycidyl ether). Poly(1,2-epoxy-3-methoxypropane). Poly(3-methoxy-1,2-epoxypropane)*
Monomers: (Methoxymethyl)oxirane
Material class: Thermoplastic
Polymer Type: polyalkylene ether
CAS Number:

CAS Reg. No.	Note
137597-55-8	
28325-89-5	homopolymer

Molecular Formula: $(C_4H_8O_2)_n$
Fragments: $C_4H_8O_2$
Molecular Weight: M_n 610; M_w/M_n 1.07 and 1.8 [1,2]

Volumetric & Calorimetric Properties:
Density:

No.	Value	Note
1	d 1.095 g/cm^3	[3]

Melting Temperature:

No.	Value	Note
1	57°C	[2,3]

Glass Transition Temperature:

No.	Value	Note
1	-56°C	[2,3]
2	-62°C	[2,3]

Optical Properties:
Refractive Index:

No.	Value	Note
1	1.467	30° [3]
2	1.459	30° [3]

Bibliographic References
[2] Dumont, M., Boils, D., Harvey, P.E. and Prud'homme, J., *Macromolecules*, 1991, **24**, 1791
[3] Lal, J. and Trick, G.S., *J. Polym. Sci., Part A-1*, 1970, **8**, 2339, (transition temps, density)

Poly(4-methoxystyrene)   P-359

Synonyms: *Poly(p-methoxystyrene). Poly(4-methoxy-1-ethenylbenzene). Poly(1-(4-methoxyphenyl)-1,2-ethanediyl)*
Monomers: 4-Methoxystyrene
Material class: Thermoplastic
Polymer Type: styrenes
CAS Number:

CAS Reg. No.	Note
24936-44-5	
54190-50-0	isotactic
107911-23-9	syndiotactic

Molecular Formula: $(C_9H_{10}O)_n$
Fragments: $C_9H_{10}O$

Volumetric & Calorimetric Properties:
Density:

No.	Value	Note
1	d 1.118 g/cm^3	[6]

Equation of State: Hartmann-Haque equation of state information has been reported [3]
Glass Transition Temperature:

No.	Value	Note
1	109°C	[1]
2	105°C	[6]

Transport Properties:
Polymer Solutions Dilute: Theta temps. 52.5° (*tert*-butylbenzene), 75.0° (isoamyl acetate), 92.6° (dichlorodecane), 23.4° (methyl isobutyl ketone) [7]
Permeability of Gases: [He] 15×10^{-10}; [CH_4] 0.91×10^{-10} cm^2 (s cmHg)$^{-1}$ [6]

Electrical Properties:
Dielectric Permittivity (Constant):

No.	Value	Note
1	3.73	[1]
2	2.2741	C_6H_6, 25° [2]
3	2.2446	C_6H_6, 40° [2]

Complex Permittivity and Electroactive Polymers: μ 1.204–1.213D (C_6H_6, 25°). μ 1.209–1.221D (C_6H_6, 40°) [2,4]. μ 1.23 (C_6H_6, 50°) [5]

Optical Properties:
Refractive Index:

No.	Value	Note
1	1.4979	C_6H_6, 25° [2]
2	1.4883	C_6H_6, 40° [2]

Applications/Commercial Products:
Processing & Manufacturing Routes: Prod. by cationic polymerisation with Lewis acid catalysts

Bibliographic References
[1] Gustafsson, A., Wiberg, G. and Gedde, U.W., *Polym. Eng. Sci.*, 1993, **33**, 549
[2] Yamaguchi, N., Sato, M. and Shima, M., *Polym. J. (Tokyo)*, 1988, **20**, 97
[3] Hartmann, B., Simha, R. and Berger, A.E., *J. Appl. Polym. Sci.*, 1989, **37**, 2603
[4] Baysal, B.M. and Aras, L., *Macromolecules*, 1985, **18**, 1693
[5] Shima, M., Yamaguchi, N. and Sato, M., *Makromol. Chem.*, 1991, **192**, 531
[6] Puleo, A.C., Muruganandam, N. and Paul, D.R., *J. Polym. Sci., Part B: Polym. Phys.*, 1989, **27**, 2385
[7] Pizzoli, M., Stea, G., Ceccorulli, G. and Gechele, G.B., *Eur. Polym. J.*, 1970, **6**, 1219

Poly[2-methoxy-5-(3-sulfopropoxy)-1,4-phenylene vinylene] P-360

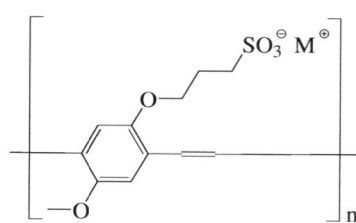

M=Na, Li or NH_4

Synonyms: *Poly[[2-methoxy-5-(3-sulfopropoxy)-1,4-phenylene]-1,2-ethenediyl]*. *MPS-PPV*
Material class: Thermoplastic
Polymer Type: poly(arylene vinylene)
CAS Number:

CAS Reg. No.	Note
125714-87-6	Na salt
258531-40-7	Li salt
125714-86-5	

Molecular Formula: $(C_{12}H_{13}O_5S.M)_n$
Fragments: $C_{12}H_{13}O_5S.M$

General Information: Water-soluble polyelectrolyte exhibiting strong fluorescence in soln. Forms red aqueous solns. and black films [1].

Electrical Properties:
Electrical Properties General: Conductivity of ammonium salt reaches 10^{-2} S cm^{-1} [1].

Optical Properties:
Optical Properties General: Exhibits strong fluorescence in aqueous soln. which is quenched by cationic reagents such as methyl viologen [2]. Broad fluorescence band in region of 500-550nm exhibits red shift and an increase in photoluminescence quantum efficiency when dendrimer assemblies are added to aqueous solns. [3]. Quantum efficiency is also enhanced in different polyelectrolytes, particularly anionic ones [4]. Band gap for aqueous polymer *ca.* 2 eV [1].
Transmission and Spectra: Ir [1], uv and photoluminescence [1,2,3,4] spectral data have been reported.

Applications/Commercial Products:
Processing & Manufacturing Routes: Prod. as different salts *via* precursor polymer derived from 5-methoxy-2-(3-chlorosulfonyl)-propoxy-1,4-xylene-α,α'-bis(tetrahydrothiophenium chloride) [1].
Applications: The amplified quenching effect with cationic reagents is utilised for assay of proteins such as avidin, employing a biotin-methyl viologen ligand [2,5,6].

Bibliographic References
[1] Shi, S. and Wudl, F., *Macromolecules*, 1990, **23**, 2119
[2] Chen, L., McBranch, D.W., Wang, H.-L., Helgeson, R., Wudl, F. and Whitten D.G., *Proc. Natl. Acad. Sci. U.S.A.*, 1999, **96**, 12287
[3] Montaño, G.A., Dattelbaum, A.M., Wang, H.-L. and Shreve, A.P., *Chem. Comm.*, 2004, 2490
[4] Abe, S. and Chen, L., *J. Polym. Sci., Part B: Polym. Lett.*, 2003, **41**, 1676
[5] Jones, R.M., Bergstedt, T.S., McBranch, D.W. and Whitten, D.G., *J. Am. Chem. Soc.*, 2001, **123**, 6726
[6] Dwight, S.J., Gaylord, B.S., Hong, J.W. and Bazan, G.C., *J. Am. Chem. Soc.*, 2004, **126**, 16850

Poly(methyl acrylate) P-361

Synonyms: *PMA. Poly(methyl 2-propenoate). Poly[1-(methoxy carbonyl)ethylene]. Methylacrylate polymer*
Related Polymers: Poly(ethylene-*co*-methylacrylate) (EMA)
Monomers: Methyl acrylate
Material class: Natural elastomers
Polymer Type: acrylic
CAS Number:

CAS Reg. No.
9003-21-8

Molecular Formula: $(C_4H_6O_2)_n$
Fragments: $C_4H_6O_2$
Molecular Weight: MW 1000–5000000
Additives: Soap-urea, activated thiol, lead thiourea, diamines, soap sulfur, carbon black, trithiocyanuric acid, ethylene thiourea. Plasticisers include: Thiokol TP-759, Plastolein 9720 and Admex 760. Process aids include: TE-80 and Vanfre AP-2. Antioxidants include sodium alkyl sulfate, inorganic and organic phosphates, diphenylamine, Agerite stalites
General Information: Poly(methyl acrylate) is an elastomer belonging to a family called acrylic elastomers, acrylic ester polymers or simply polyacrylates (ACM's), known since early 1940's. When compounded with additives they offer some unique props. which make them useful where other elastomers (rubbers)

Poly(methyl acrylate)

fail. Among speciality elastomers this ranks above all except silicon and fluorinated hydrocarbon elastomers. It is a tough, rubbery polymer with low T_g. It is water-sensitive but resistant to swelling in oils and non-polar liqs., with good thermal stability. T_g changes with tacticity. It is inherently resistant to uv radiation, oxygen, ozone and heat. In the absence of oxygen thermal degradation does not occur until approx. 300°. A tough, rubbery polymer with low T_g. It is water-sensitive but resistant to swelling in oils and non-polar liqs., and has good thermal stability. Its T_g changes with tacticity. It has a saturated backbone struct. which makes it inherently resistant to uv radiation, oxygen, ozone and heat. In the absence of oxygen, thermal degradation does not occur until 300° approx. Specialised automotive needs have been the key to most of its applications. Copolymers with suitable co-monomers and small amounts of proper cure site monomers constitute the building blocks of current acrylic elastomers. This greatly increases the range and value of the physical props. for a variety of applications. These polymers are further compounded with typical additives and ingredients to produce a commercial product. A general recipe includes polymer, stearic acid, processing aid, antioxidant, filler, plasticiser amd cure system. This recipe depends upon the cure site present and the application requirements

Volumetric & Calorimetric Properties:
Density:

No.	Value	Note
1	d 1.22 g/cm³	25°

Thermodynamic Properties General: ΔC_p 42.3 kJ kg^{-1} K^{-1}
Latent Heat Crystallization: ΔG_1^α 8.0–8.8 kJ mol^{-1}; ΔH_1^α 9.8–11 kJ mol^{-1}
Specific Heat Capacity:

No.	Value	Note	Type
1	1340 kJ/kg.C	115 J mol^{-1} K^{-1}, 25°, solid	P
2	1800 kJ/kg.C	155 J mol^{-1} K^{-1}, 25°, liq.	P

Glass Transition Temperature:

No.	Value	Note
1	10–11°C	
2	5°C	syndiotactic, head-to-tail
3	31°C	isotactic, head-to-tail

Brittleness Temperature:

No.	Value
1	0–6°C

Surface Properties & Solubility:
Cohesive Energy Density Solubility Parameters: E_{coh} 27800–31800 J mol^{-1}; CED 396.57 J cm^{-3}; δ 19.9–21.3 J$^{1/2}$ cm$^{-3/2}$
Solvents/Non-solvents: Sol. aromatic hydrocarbons, chlorinated hydrocarbons, THF, esters, ketones, glycolic ester ethers, phosphorus trichloride. Insol. aliphatic hydrocarbons, hydrogenated naphthalenes, Et$_2$O, alcohols, CCl$_4$
Wettability/Surface Energy and Interfacial Tension: γ_s 45 mN m^{-1} (calculated)

Transport Properties:
Polymer Melts: Critical Molar Mass M_{cr} 24000. Critical value of the number of chain atoms in the polymer molecule Z_{cr} 560. Unperturbed viscosity coefficient K_θ 0.065 cm^3 mol$^{1/2}$ g$^{-3/2}$ $K_\theta M_{cr}^{1/2}$ 10 cm^3 g^{-1}

Mechanical Properties:
Mechanical Properties General: Exhibits acceptable compressive strength

Tensile (Young's) Modulus:

No.	Value
1	6.9 MPa

Mechanical Properties Miscellaneous: Elongation 350%. Elongation at break 750%
Hardness: Shore A40-A95 (Durometer)
Friction Abrasion and Resistance: Exhibits good abrasion resistance

Electrical Properties:
Dielectric Strength:

No.	Value	Note
1	31.5 kV.mm^{-1}	800 V mil^{-1}

Complex Permittivity and Electroactive Polymers: Dielectric relaxation frequency 1.8 GHz (25°, toluene). Electrical temp. 15° (1s)
Dissipation (Power) Factor:

No.	Value
1	0.2

Optical Properties:
Refractive Index:

No.	Value	Note
2	1.479	25°

Polymer Stability:
Polymer Stability General: Has excellent shelf life when stored under normal dry conditions. Suitably formulated and cured polymer systems are resistant to oils and greases, ozone, aliphatic hydrocarbons at low and elevated temps., and sunlight discoloration and weathering. Shelf life of compounded polymer depends on the activity of the cure system employed. Generally, aged compounds should be mill-freshened prior to processing
Thermal Stability General: Temp. for half life of 30 min 330°
Upper Use Temperature:

No.	Value	Note
1	<80°C	max., homopolymer
2	190–204°C	70h max., compounded
3	163–177°C	1000h max.

Decomposition Details: $T_{1/2}$ 328° (601K), 332° (605K)
Flammability: Ignites and burns quite readily and completely to leave no residue
Environmental Stress: Acrylic polymers are not sensitive to normal uv degradation since the primary uv absorption of acrylics occurs below the solar spectrum of 290 nm
Chemical Stability: Resistant to oxygen and only decomposes slowly under extreme conditions, e.g. high temp. in an oxygen atmosphere. Does not readily depolymerise on heating. Decomposes at 292–399° to produce mainly MeOH and carbon dioxide
Hydrolytic Stability: Resistant to both acidic and alkaline hydrol.

Applications/Commercial Products:

Processing & Manufacturing Routes: Commercially poly(methyl acrylate) is prod. by soln. and emulsion polymerisation. It is processed by injection moulding, blow moulding, extrusion and thermoforming. Co-polymers with suitable co-monomers and small amounts of proper cure site monomers constitute the building blocks of the current acrylic elastomers, greatly increasing the range and value of the physical props. for a variety of applications. These polymers are further compounded with typical additives and ingredients to produce a commercial product: typically this includes polymer, stearic acid, processing aid, antioxidant, filler, plasticiser and cure system. This recipe depends upon the cure site present and the application requirements
Applications: The chief market for acrylic elastomers is automotive industry. Main applications are as seals, gaskets, actuating diaphragms, valve seats, adhesives, coating industry caulks, sealants, textile back coatings, hoses, paper saturation and leather finishing. Actuating disphragms, valve seats etc.

Trade name	Supplier
Cyanacryl	American Cyanamide
Emulsion E1614	Rohm and Haas
Fulacryl	H.B. Fuller
Hycar	B.F. Goodrich
Nipol AR	Zeon
Nobestos	Lydall
Ucar Latex	Union Carbide
Vultacryl	General Latex

Bibliographic References

[1] *Encycl. Polym. Sci. Eng.*, 2nd edn., (ed. J.I. Kroschwitz), 1985, **1**
[2] *Kirk-Othmer Encycl. Chem. Technol.*, 4th edn., (ed. J.I. Kroschwitz), 1991, **1**
[3] Hofmann, W., *Rubber Technology Handbook*, Hanser, 1989
[4] *Ullmanns Encykl. Ind. Chem.*, 5th edn., (ed. W. Gerhartz), VCH, 1992
[5] Van Krevelen, D.W., *Properties of Polymers: Their Correlation with Chemical Structure*, Elsevier, 1990
[6] DeMarco, R.D., *Rubber Chem. Technol.*, 1979, **52**, 173
[7] Brunacci, A., Pedemonte, E. and Turturro, A., *Polymer*, 1992, **33**, 4429
[8] Smith, G.D. and Boyd, R.H., *Macromolecules*, 1992, **25**, 1326
[9] Suchopárck, M., Spváček, J. and Masař, B., *Polymer*, 1994, **35**, 3389
[10] *U.S. Pat.*, 1970, 3 544 535
[11] Nucessle, A.C. and Kine, B.B., *Ind. Eng. Chem.*, 1953, **45**, 1287
[12] Busfield, W.K. and Methven, J.M., *Polymer*, 1973, **14**, 137
[13] Riddle, E.H., *Monomeric Acrylic Esters*, Reinhold, 1954

Poly(3-methyl-1-butene)

$$-[CH_2CH]_n-\ |\ CH(CH_3)_2$$

Synonyms: *Polyisopropylethylene. Poly(2-vinylpropane). Poly(1-(1-methylethyl)-1,2-ethanediyl)*
Monomers: 3-Methyl-1-butene
Material class: Thermoplastic
Polymer Type: polyolefins
CAS Number:

CAS Reg. No.	Note
25085-05-6	atactic
26703-14-0	isotactic
138875-96-4	syndiotactic

Molecular Formula: $(C_5H_{10})_n$
Fragments: C_5H_{10}

Morphology: Monoclinic, a 0.955, b 1.708, c 0.684, γ 116°30′ [14], helix 2_1. Tetragonal, a 3.432, b 3.432, c 0.684 [7], helix 4_1. Other variations have been reported [18,32,33,34,35,36]

Volumetric & Calorimetric Properties:
Density:

No.	Value	Note
1	d 0.91 g/cm^3	[7,9,14]
2	d 0.9 g/cm^3	[15,16,17,18,19]
3	d^{25} 0.906 g/cm^3	[10]

Thermal Expansion Coefficient:

No.	Value	Note	Type
1	0.000375 K^{-1}	T <87.5° [8]	V
2	0.000609 K^{-1}	T >87.5° [8]	V

Volumetric Properties General: Molar volume 78.12 cm^3 mol^{-1} [31]
Thermodynamic Properties General: ΔH_f 17.3 kJ mol^{-1} [37]
Melting Temperature:

No.	Value	Note
1	300°C	[1,2,3,4]
2	310°C	[5,25]
3	305°C	[6]
4	305–310°C	[7,8,9]

Glass Transition Temperature:

No.	Value	Note
1	50°C	[2,4,10,11,12,28]
2	66–74°C	[8]
3	60°C	[13]

Deflection Temperature:

No.	Value	Note
1	>185°C	min. [6]

Transition Temperature:

No.	Value	Note	Type
1	262°C	[5]	Polymer melt temp.
2	100°C	approx. [6]	Transition temp.
3	53°C	[3]	Softening point
4	87.5°C	[8]	Second order transition temp.

Surface Properties & Solubility:
Cohesive Energy Density Solubility Parameters: CED 282.6 J cm^{-3} (67.5 cal cm^{-3}) [20]. δ 33.0–36.3 J$^{1/2}$ cm$^{-3/2}$ (7.88–8.68 cal$^{1/2}$ cm$^{-3/2}$) [31]
Solvents/Non-solvents: Slightly sol. o-dichlorobenzene [23], C_6H_6 [19]. Insol. Et_2O, heptane [6,19]

Transport Properties:
Melt Flow Index:

No.	Value	Note
1	0.8 g/10 min	320°, 2.16 kg load [43]

Poly(3-methyl-1-butene)

Polymer Solutions Dilute: [η] 3 dl g^{-1} (135°, tetrahydronaphthalene) [15]

Gas Permeability:

No.	Gas	Value	Note
1	N$_2$	656660 cm^3 mm/ (m^2 day atm)	1 × 10^{-6} cm^2 (s cmHg)$^{-1}$ [44]

Mechanical Properties:

Mechanical Properties General: Mech. props. have been reported [8,40,44]. Dynamic [40], shear [11] and dynamic elastic moduli [10] have been reported in graphical form. Dynamic tensile storage modulus and loss modulus have been reported [45]

Flexural Modulus:

No.	Value	Note
1	3380 MPa	490 kpsi [6]

Tensile Strength Break:

No.	Value	Note	Elongation
1	12.4 MPa	1800 psi, 100° [6]	
2	34.5 MPa	5000 psi, 25° [6]	3% extension

Flexural Strength Yield:

No.	Value	Note
1	47.6 MPa	6900 psi [6]

Viscoelastic Behaviour: Stress relaxation has been reported [8]

Izod Notch:

No.	Value	Notch	Note
1	16 J/m	Y	0.3 ft lb in^{-1}, 25° [6]
2	74.7 J/m	N	1.4 ft lb in^{-1}, 25° [6]

Optical Properties:

Optical Properties General: Thermoluminescence has been reported [42]

Transmission and Spectra: Ir [13,22,46], Raman [29,30], esr [4], H-1 nmr [21,22,23] and C-13 nmr [24,25,26,27] spectral data have been reported

Polymer Stability:

Thermal Stability General: Undergoes rapid oxidative degradation above 100° [8,39]. Thermal degradation has been reported [39]

Decomposition Details: Initial decomposition 340–400° [41]. E_{act} 84.6 kJ mol^{-1} (20.2 kcal mol^{-1}, cryst., decomposes) [39]. E_{act} 63.6 kJ mol^{-1} (15.2 kcal mol^{-1}, amorph., decomposes) [39]. E_{act} 193 kJ mol^{-1} (46 kcal mol^{-1}, decomposes) [41]

Flammability: Pyrolysis produces isopentane as the main product, caused by 'unzipping' [38]

Applications/Commercial Products:

Processing & Manufacturing Routes: Synth. using Ziegler-Natta catalysts, such as TiCl$_3$AlEt$_2$Cl in *n*-heptane at 70°

Mould Shrinkage (%):

No.	Value	Note
1	5%	approx. [6]

Bibliographic References

[1] Reding, F.P., *J. Polym. Sci.*, 1956, **21**, 547, (transition temp)
[2] Boyer, R.F., *Rubber Chem. Technol.*, 1963, **36**, 1303, (transition temp)
[3] Dunham, K.R., Vandenberghe, J., Faber, J.W.H. and Contois, L.E., *J. Polym. Sci., Part A: Polym. Chem.*, 1963, **1**, 751, (transition temp)
[4] Kusumoto, N., Suhiro, K. and Takayanagi, N., *J. Polym. Sci., Polym. Lett. Ed.*, 1972, **10**, 81, (esr)
[5] Campbell, T.W. and Haven, A.C., *J. Appl. Polym. Sci.*, 1959, **1**, 73, (transition temp)
[6] Kirshenbaum, I., Feist, W.C. and Isaacson, R.B., *J. Appl. Polym. Sci.*, 1965, **9**, 3023, (mech props)
[7] Huguet, M.G., *Makromol. Chem.*, 1966, **94**, 205, (struct)
[8] Quynn, R.G. and Sprague, B.S., *J. Polym. Sci., Part A-2*, 1970, **8**, 1971, (mech props)
[9] Bacskai, R., Goodrich, J.E. and Wilkes, J.B., *J. Polym. Sci., Part A-1*, 1972, **10**, 1529, (transition temp)
[10] Woodward, A.E., Sauer, J.A. and Wall, R.A., *J. Polym. Sci.*, 1961, **50**, 117, (mech props)
[11] Kirshenbaum, I., Isaacson, R.B. and Druin, M., *J. Polym. Sci., Part B: Polym. Lett.*, 1965, **3**, 525, (mech props)
[12] Kawasaki, N. and Takayanagi, M., *Rep. Prog. Polym. Phys. Jpn.*, 1967, **10**, 337, (transition temp)
[13] Bevza, T.I., Pokatilo, N.A. and Topchiev, A.V., *Neftekhimiya*, 1964, **4**, 727, (transition temp, ir)
[14] Corradini, P., Ganis, P. and Petraccone, V., *Eur. Polym. J.*, 1970, **6**, 281, (struct)
[15] *Brit. Pat.*, 1960, 835 759, (viscosity)
[16] Natta, G., *Chim. Ind. (Milan)*, 1956, **38**, 751, (transition temp)
[17] Natta, G., *Angew. Chem.*, 1956, **68**, 393, (transition temp)
[18] Noether, H.D., *J. Polym. Sci., Part C: Polym. Lett.*, 1967, **16**, 725, (struct)
[19] Natta, G., Pino, P., Mazzanti, G., Corradini, P. and Giannini, U., *CA*, 1955, **50**, 16256c, (transition temp)
[20] He, T., *J. Appl. Polym. Sci.*, 1985, **30**, 4319, (solubility)
[21] Woodward, A.E., Odajima, A. and Sauer, J.A., *J. Phys. Chem.*, 1961, **65**, 1384, (H-1 nmr)
[22] Kennedy, J.P., Minckler, L.S., Wanless, G. and Thomas, R.M., *J. Polym. Sci., Part A: Polym. Chem.*, 1964, **2**, 2093, (H-1 nmr, ir)
[23] Suzuki, T., Koshiro, S. and Takegami, Y., *J. Polym. Sci., Polym. Lett. Ed.*, 1972, **10**, 829, (H-1 nmr)
[24] Tanaka, Y. and Sato, H., *J. Polym. Sci., Polym. Lett. Ed.*, 1976, **14**, 335, (C-13 nmr)
[25] Borriello, A., Busico, V., DeRosa, C. and Schulze, D., *Macromolecules*, 1995, **28**, 5679, (C-13 nmr)
[26] Asakura, T. and Nakayama, N., *Polym. Commun.*, 1991, **32**, 213, (C-13 nmr)
[27] *Jpn. Pat.*, 1991, 03 200 812, (syndiotactic, C-13 nmr)
[28] Reding, F.P., Faucher, J.A. and Whitman, R.D., *J. Polym. Sci.*, 1962, **57**, 483, (transition temp)
[29] Deroualt, J., Hendra, P.J., Cudby, M.E.A., Fraser, G. *et al*, *Adv. Raman Spectrosc.*, 1972, **1**, 277, (Raman)
[30] Ferraro, J.R., Ziomek, J.S. and Mack, G., *Spectrochim. Acta*, 1961, **17**, 802, (Raman)
[31] Ahmad, H. and Yaseen, M., *Polym. Eng. Sci.*, 1979, **19**, 858, (solubility)
[32] Natta, G., Corradini, P. and Bassi, I.W., *CA*, 1955, **50**, 16256g, (struct)
[33] Utsunomiya, H., Kawasaki, N., Niinoni, N. and Takayanagi, M., *J. Polym. Sci., Part B: Polym. Lett.*, 1967, **5**, 907, (struct)
[34] Turner-Jones, A. and Aizlewood, J.M., *J. Polym. Sci., Part B: Polym. Lett.*, 1963, **1**, 471, (struct)
[35] Sakaguchi, F., Kitamura, R. and Tsuji, W., *CA*, 1966, **65**, 20230a, (struct)
[36] Nechitailo, N.A., Sanin, P.I., Bevza, T.I. and Pokatilo, N.A., *Plast. Massy*, 1964, 3, (struct)
[37] Schaefgen, J.R, *J. Polym. Sci.*, 1959, **38**, 549, (thermodynamic props)
[38] van Schooten, J. and Evenhuis, K., *Polymer*, 1965, **6**, 343, (pyrolysis)
[39] Minsker, K.S., Liakumovich, A.G., Sangalov, Y.A., Svirskaya, D.D. *et al*, *Vysokomol. Soedin., Ser. A*, 1974, **16**, 2751, (degradation)
[40] Tsuji, W., Kitamura, R. and Sakaguchi, F., *CA*, 1967, **68**, 78707, (mech props)
[41] Konovalov, V.V., Blyumenfeld, A.B., Kerber, M.L. and Akutin, M.S., *CA*, 1974, **85**, 63578, (degradation)
[42] Korobeinikova, V.N., Kazakov, V.P., Minsker, K.S. and Soldaeva, N.P., *CA*, 1972, **77**, 140704, (thermoluminescence)
[43] *Jpn. Pat.*, 1986, 61 181 844, (transport props)
[44] *Jpn. Pat.*, 1985, 60 255 107, (gas permeability)
[45] Takayanagi, M., *Pure Appl. Chem.*, 1970, **23**, 151, (mech props)
[46] Kennedy, J.P., Minckler, L.S., Wanless, G.G. and Thomas, R.M., *J. Polym. Sci., Part A: Polym. Chem.*, 1964, **2**, 1441, (ir)

Poly(methyl-*n*-butylsilane)

P-363

$$\left[\begin{array}{c} \text{CH}_2\text{CH}_2\text{CH}_2\text{CH}_3 \\ | \\ -\text{Si}- \\ | \\ \text{CH}_3 \end{array}\right]_n$$

Synonyms: *Poly(butylmethylsilylene)*. *pMeBuSi*. *ME4*
Polymer Type: polysilanes
CAS Number:

CAS Reg. No.
88003-14-9

Molecular Formula: $(\text{C}_5\text{H}_{12}\text{Si})_n$
Fragments: $\text{C}_5\text{H}_{12}\text{Si}$
Molecular Weight: $5.4 \times 10^3 – 1.0 \times 10^5$ [3,4].
General Information: The polymer is an asymmetrical substituted methyl-substituted polysilylene.
Morphology: Amorphous white solid [1,3]. Si backbone dominated by the presence of mixed transoid (T), deviant (D) and, possibly, ortho (O) conformations with dihedral angles approximating 155°, corroborating the formation of a T+D+T-D- Si backbone [1].
There is no evidence for a significant fraction of *all-anti* (A, i.e., *trans*-planar), *all-trans*, or *anti*-gauche conformers [1].

Volumetric & Calorimetric Properties:
Transition Temperature General: Melts without decomposition [8]

Surface Properties & Solubility:
Solubility Properties General: Readily soluble in common organic solvents [8]

Optical Properties:
Optical Properties General: Modest 25nm blue shift on warming with no evidence of an isosbestic point [1]. The temperature-dependent uv spectra of the polymer showed only a continuous bathochromic shift (302nm to 313nm). The short side chain interacts weakly with the polysilane chromophore, resulting in a gradual bathochromic shift with decreasing temperature [2].
Transmission and Spectra: Ir [4], uv [1,2,4], and nmr [4,5,6] spectra reported.

Polymer Stability:
Polymer Stability General: Does not form an ordered phase [1]
Thermal Stability General: The thermogram shows three endotherms at 0, -20, and -40°C: The first endotherm at 0°C has been assigned to a first conformational change and to a partial transition to the M phase (the most thermodynamically stable). The second endotherm with formation of the metastable K2 phase, since there are no signs of change in the backbone conformation. The third endotherm at -40°C correlates with a second chain conformational change [1,2].

Applications/Commercial Products:
Processing & Manufacturing Routes: Synthesised from dichloromethyl-*n*-butylsilane by dehalogenative coupling reaction with Na under standard conditions for Wurtz polysilane synthesis [1,2,7,8].
Synthesised in yields of 51–60% from 1,2-dichloro-1,1-isobutyl-2,2-dimethyldisilane or dichloro-1,2-dimethyl-1,2-dibutyldisilane using 3,3′,5,5′-tetramethylbiphenyl anion radical as a reducing agent, which shows a remarkable chemoselectivity in the polymerisation reaction [3]

Applications/Commercial Products:
Applications: X-ray lithography [7].

Bibliographic References
[1] Chunwachirasiri, W., Kanaglekar, I., Winokur, M.J., Koe, J.C. and West, R., *Macromolecules*, 2001, **34**, 6719
[2] Yuan, C.H. and West, R., *Macromolecules*, 1994, **27**, 629
[3] Sakurai, H., Yoshida, M. and Sakamoto, K., *J. Organomet. Chem.*, 1996, **521**, 287
[4] Seki, S., Koizumi, Y., Kawaguchi, T., Habara, H. and Tagawa, S., *J. Am. Chem. Soc.*, 2004, **126**, 3521
[5] Abu-Eid, M.A., King, R.B. and Kotliar, A.M., *Eur. Polym. J.*, 1992, **28**, 315
[6] Wolff, A.R., Maxka, J. and West, R., *J. Polym. Sci.*, *Part A*, 1988, **26**, 713
[7] Miller, R.D., Thompson, D., Sooriyakumaran, R. and Fickes, G.N., *J. Polym. Sci.*, *Part A*, 1991, **29**, 813
[8] Trefonas, P., Djurovich, P.I., Zhang, X. H., West, R., Miller, R.D. and Hofer, D., *J. Polym. Sci.*, *Polym. Lett. Ed.*, 1983, **21**, 819

Poly[(2-methylbutyl)undecylsilylene]

P-364

$$\left[\begin{array}{c} \text{CH}_2(\text{CH}_2)_9\text{CH}_3 \\ | \\ -\text{Si}- \\ | \\ \text{CH}_2\text{CH}(\text{CH}_3)\text{CH}_2\text{CH}_3 \end{array}\right]_n$$

Polymer Type: polysilanes
CAS Number:

CAS Reg. No.
162844-16-8

Molecular Formula: $(\text{C}_{16}\text{H}_{34}\text{Si})_n$
Fragments: $\text{C}_{16}\text{H}_{34}\text{Si}$
Molecular Weight: 45,600–51,000 [1,2]

Volumetric & Calorimetric Properties:
Transition Temperature General: Bp 135-136°C (at 0.4 mmHg) [2]

Applications/Commercial Products:
Processing & Manufacturing Routes: Polymerisation of methyl[(S)-2-methylbutyl]undecylsilane dichloride in Et₂O in the presence of Na and 18-crown-6 [1,2].
Applications: Useful as enantio-recognitive material in chromatography and as standard materials in 1-dimensional semiconductor-quantum wire structures [1].

Bibliographic References
[1] *Eur. Pat.*, 1994, 612 756
[2] *U.S. Pat.*, 1998, 5 710 301

Poly(methyl cyanoacrylate)

P-365

$$\left[\begin{array}{c} \text{CN} \\ | \\ -\text{CH}_2\text{C}- \\ | \\ \text{COOMe} \end{array}\right]_n$$

Synonyms: *Methyl cyanoacrylate polymer*
Related Polymers: Poly(alkyl cyanoacrylates), Poly(ethyl cyanoacrylate), Poly(isobutyl cyanoacrylate), Poly(allyl cyanoacrylate)
Monomers: Methyl 2-cyanoacrylate
Material class: Thermoplastic, Gums and resins
Polymer Type: acrylic, acrylonitrile and copolymers
CAS Number:

CAS Reg. No.
25067-29-2

Molecular Formula: $(\text{C}_5\text{H}_5\text{NO}_2)_n$
Fragments: $\text{C}_5\text{H}_5\text{NO}_2$
Molecular Weight: Up to 75000
Additives: Adhesive strength and toughness can be improved by addition of additives
General Information: Amorph.

Poly(methyl cyanoacrylate)

Morphology: Clear, hard resin with good adhesive props., though with low peel and impact strength. Most polar alkyl cyanoacrylate polymer

Volumetric & Calorimetric Properties:
Density:

No.	Value	Note
1	d 1.25 g/cm^3	[14]
2	d^{20} 1.283 g/cm^3	[2]

Melting Temperature:

No.	Value
1	205°C

Glass Transition Temperature:

No.	Value	Note
1	96°C	dynamic mechanical analysis [1]
2	-36°C	differential scanning calorimetry [2]
3	100°C	dilatometry [3]

Transition Temperature General: T_g may vary with polymerisation conditions and presence of contaminants

Deflection Temperature:

No.	Value	Note
1	157–163°C	free-radical polymerisation [13]

Vicat Softening Point:

No.	Value	Note
1	165°C	[4]

Surface Properties & Solubility:
Solubility Properties General: Incompatible with poly(allyl-2-cyanoacrylate) [8] δ 11.8 J$^{1/2}$cm$^{-3/2}$ (calc) [14]

Solvents/Non-solvents: Sol. acetonitrile, DMSO, DMF, nitromethane, N-methylpyrrolidone, pyridine. Insol. hexane, toluene, MeOH, EtOAc, THF, Me$_2$CO, CH$_2$Cl$_2$ [9], C$_6$H$_6$, CHCl$_3$, H$_2$O, Et$_2$O

Transport Properties:
Water Absorption:

No.	Value	Note
1	5.2 %	[1]

Mechanical Properties:
Mechanical Properties General: Strengtheners and tougheners have been reported [10]. Storage modulus decreases with increasing temp.

Flexural Modulus:

No.	Value	Note
1	3654–3840 MPa	530–557 kpsi [13]
2	3399 MPa	493 kpsi, ASTM D790

Tensile Strength Break:

No.	Value	Note	Elongation
1	43 MPa	bone-bone [13]	
2	31 MPa	4500 psi, ASTM D638 [4]	2%
3	13.2 MPa	135 kg cm^{-2} [2]	

Flexural Strength at Break:

No.	Value	Note
1	104–122 MPa	15.1–17.7 kpsi, ASTM D790 [13]

Miscellaneous Moduli:

No.	Value	Note	Type
1	15000 MPa	25° [1]	Storage modulus

Impact Strength: Impact shear strength 213–533 J m^{-1} (4–10 ft lb m^{-1}, ASTM D950) [4]

Mechanical Properties Miscellaneous: Tensile shear strength 22.1 MPa (3200 psi, steel-steel) [4]; 17.24 MPa (2500 psi, Al-Al) [4]. Overlap shear strength 21.5 MPa (cold rolled steel); 16 MPa (Al); 25 MPa (copper) [14]. Peel strength 0.11 kg cm^{-1} [14]. Loss factor values have been reported [4]. For polymer-polymer bonds, substrate failure occurs before bond failure [4]

Hardness: Rockwell M65

Electrical Properties:
Dielectric Permittivity (Constant):

No.	Value	Frequency	Note
1	3.34	1 MHz	[4]
2	3.98		[14]

Optical Properties:
Refractive Index:

No.	Value	Note
1	1.45	20° [4]
2	1.49	[14]

Polymer Stability:
Polymer Stability General: Poor heat resistance due to degradation at substrate/adhesive interface, leading to loss of adhesion [4]

Upper Use Temperature:

No.	Value
1	80°C

Decomposition Details: Above 260° depolymerisation is virtually complete. Decomposition onset at 160° (N$_2$), 148° (air) [8]

Environmental Stress: Unsuitable for use in situations of high temp, high impact, long-term moisture and for outdoor use

Chemical Stability: Shows generally good resistance to solvents, oils, etc [4]. Degraded by Me$_2$CO

Hydrolytic Stability: Degrades more rapidly in water than other cyanoacrylate polymers [12]

Stability Miscellaneous: Degraded by radiation [13]

Applications/Commercial Products:
Processing & Manufacturing Routes: Synth. by spontaneous polymerisation of methyl cyanoacrylate monomer catalysed by base see poly(alkyl cyanoacrylates)

Applications: Instant one pot adhesive, used for a range of applications including production line product assembly

Trade name	Details	Supplier
Cyanolib	100 series	Panacol Elosol

Plus C910	100 series	Permabond
Pronto CA	7	3M
Sicomet	85,99,7000	Sichel Henhel
Tb	1701,1702,1703	Threebond

Bibliographic References
[1] Cheung, K.H., Guthrie, J., Otterburn, M.S. and Rooney, J.R., *Makromol. Chem.*, 1987, **188**, 3041, (Tg, mech props)
[2] Tseng, Y.C., Hyon, S.H and Ikada, Y., *Biomaterials*, 1990, **11**, 73, (synth, mech props, degradation)
[3] Kukarni, R.K., Porter, H.S. and Leonard, F., *J. Appl. Polym. Sci.*, 1973, **17**, 3509, (Tg)
[4] Coover, H.W., Dreifus, D.W. and O'Connor, J.T., Handb. Adhes. 3rd edn., Van Nostrand Reinhold, 1990, 463, (rev)
[5] Woods, J., *Polymeric Materials Encyclopedia*, (ed. J.C. Salamone), CRC Press, 1996, 1632, (rev)
[6] Hussey, B. and Wilson J., Adhesives Handbook Directory & Databook, Chapman and Hall, 1996, 118
[7] Negulescu, I.I., Calugaru, E.M., Vasile, C., Agafitel, M. and Dumitrescu, G., *Polym. Bull. (Berlin)*, 1986, **15**, 43, (compatibility)
[8] Negulescu, I.I., Calugaru, E.M., Vasile, C. and Dumitrescu, G., *J. Macromol. Sci., Chem.*, 1987, **24**, 75, (thermal degradation)
[9] Donnely, E.F. and Pepper, D.C., Makromol. Chem., *Rapid Commun.*, 1981, **2**, 439, (solubility)
[10] U.S. Pat., (Loctite), 1978, **4** 102 945, (peel strength)
[11] Papatheofainis, F.J., *Biomaterials*, 1989, **10**, 185, (adhesive strength)
[12] Leonard, F., Kulkerni, R.K., Brandes, G., Nelson, J, and Cameron J.J., *J. Appl. Polym. Sci.*, 1966, **10**, 259, (synth, degradation)
[13] Canale, A.J., Goode, W.E., Kinsinger, J.B., Panchak, J.R. et al, *J. Appl. Polym. Sci.*, 1960, **4**, 231, (synth, props)
[14] Millet, G.H., *Structural Adhesives Chemistry and Technology*, (ed. S.R. Hartshorn), Plenum Press, 1986, 249

Poly(γ-methyl-L-glutamate) fibre   P-366

Synonyms: *Poly(4-methyl-L-glutamic acid) fibre*
Related Polymers: Proteins
Monomers: 4-Methyl-L-glutamic acid
Material class: Proteins and polynucleotides, Fibres and Films
CAS Number:

CAS Reg. No.	Note
29967-97-3	homopolymer

Molecular Formula: $C_6H_{11}NO_4$
Morphology: Contains predominantly the β-form [1]

Volumetric & Calorimetric Properties:
Density:

No.	Value	Note
1	d 1.29–1.33 g/cm^3	stretched fibre [1]

Surface Properties & Solubility:
Solvents/Non-solvents: Insol. CH_2Cl_2 (spun thread) [2]

Transport Properties:
Water Absorption:

No.	Value	Note
1	2.4 %	20°, 80% relative humidity [1]
2	3.5 %	20°, 100% relative humidity [1]

Mechanical Properties:
Mechanical Properties General: Tenacity 3.1 g den^{-1} (11.8% elongation) 2.23 g den^{-1} (12.8% elongation, stretched fibre, dry), 2.02 g den^{-1} (16.8% elongation, stretched fibre, wet) [1], 2.7 g den^{-1} (15% elongation, wet). [2] Tensile modulus 41.2 g den^{-1} [1]

Optical Properties:
Transmission and Spectra: Ir and X-ray spectral data have been reported [1]
Volume Properties/Surface Properties: Dyeability can be improved by treatment with hydrazine or copolymerisation with L-methionine [1]

Polymer Stability:
Hydrolytic Stability: Unstable to alkali

Applications/Commercial Products:
Processing & Manufacturing Routes: Poly (γ-methyl-L-glutamate) in CH_2Cl_2/dioxane (80:20 v/v) soln. is spun into Me_2CO. The resultant fibre is rolled, then stretched to 1.7 times its length in a water-bath at room temp., before being rolled again at 110° [1]
Applications: Potential application as synthetic silk. Its use in the manufacture of fabrics has been retarded owing to the instability to alkali of the resultant fabric

Bibliographic References
[1] Noguchi, J., Tokura, S. and Nishi, N., *Angew. Makromol. Chem.*, 1972, **22**, 107, (synth, props)
[2] Moncrieff, R.W., *Man-Made Fibers*, 6th edn., Newnes-Butterworth, 1975, 686, (rev)

Poly(methylglycidyl ether)   P-367

$$-[OCH_2CH]_n-$$
$$\quad\quad | $$
$$\;\;CH_2OMe$$

Synonyms: *Poly(3-methoxypropylene oxide). Poly[(methoxymethyl)oxirane]. Poly(1,2-epoxy-3-methoxypropane)*
Related Polymers: Epoxy Resins
Monomers: (Methoxymethyl)oxirane
Polymer Type: epoxy, polyalkylene ether
CAS Number:

CAS Reg. No.	Note
144569-35-7	dimer
144569-36-8	trimer
28325-89-5	polymer
64491-71-0	(R)-form
64491-69-6	(-)-form
64491-71-0	(+)-form

Molecular Formula: $(C_4H_8O_2)_n$
Fragments: $C_4H_8O_2$

Volumetric & Calorimetric Properties:
Density:

No.	Value	Note
1	1.095 g/cm^3	[1]

Melting Temperature:

No.	Value	Note
1	57°C	calorimetric value [1]

Glass Transition Temperature:

No.	Value	Note
1	-62°C	calorimetric value [1]

Surface Properties & Solubility:
Solvents/Non-solvents: Sol. decahydronaphthalene [3], trichloroethylene, *o*-dichlorobenzene [3]

Transport Properties:
Polymer Solutions Dilute: η_{inh} 1.5 dl g^{-1} (30°, 0.1% soln. in toluene) [1]

Optical Properties:
Refractive Index:

No.	Value	Note
1	1.459	rapidly quenched sample [1]
2	1.467	30° [1]

Transmission and Spectra: C-13 nmr spectral data have been reported [4]

Bibliographic References
[1] Lal, J. and Trick, G.S., *J. Polym. Sci., Part A-1*, 1970, **8**, 2339, (synth)
[2] Vincens, V., Le Borgne, A. and Spassky, N., *ACS Symp. Ser.*, 1992, **496**, 205, (synth)
[3] Kozlova, I.K., Zhomov, A.K. and Sapunov, V.N., *CA*, 1970, **73**, 131411q, (mechanisms)
[4] Spassky, N., Pourdjavadi, A. and Sigwalt, P., *Eur. Polym. J.*, 1977, **13**, 467, (synth)

Poly(5-methyl-1-heptene)   P-368

Synonyms: *Poly[(3-methylpentyl)ethylene]. Poly(1-(3-methylpentyl)-1,2-ethanediyl)*
Monomers: 5-Methyl-1-heptene
Material class: Thermoplastic
Polymer Type: polyolefins
CAS Number:

CAS Reg. No.	Note
31783-74-1	(*S*)-form
50897-39-7	(*R*)-form
28574-32-5	isotactic, (*S*)-form
29407-51-0	(+)-form
25189-90-6	(*S*)-form, (+)-form
9047-33-0	isotactic, (*R-co-S*)-form
28962-84-7	isotactic, (*RS*)-form

Molecular Formula: $(C_8H_{16})_n$
Fragments: C_8H_{16}
Tacticity: Isotactic forms of known absolute stereochemistry are known
Morphology: Isotactic, (*S*)-form, pseudo-orthorhombic, a 1.840, b 1.062, c 0.636, helix 3_1. Isotactic, (*RS*)-form, tetragonal, a 2.000, b 2.000, c 3.876, helix 19_6 [4]

Volumetric & Calorimetric Properties:
Density:

No.	Value	Note
1	d^{25} 0.86 g/cm^3	Isotactic (*S*)-form [4,7]
2	d 0.88 g/cm^3	Isotactic (*RS*)-form [4]

Melting Temperature:

No.	Value	Note
1	60–61°C	isotactic (*S*)-form [4]
2	48–52°C	(*S*)(+)-form [8]
3	59–60°C	isotactic (*RS*)-form [4]
4	52–56°C	isotactic (*S*)-form [10]

Surface Properties & Solubility:
Solvents/Non-solvents: Isotactic (*SR*)-form sol. CHCl$_3$, insol. MeOH, insol. 2-butanone. [4] (*R*)-form, (*S*)-form insol. boiling MeOH, sol. boiling Et$_2$O. [1] (*S*), (+)-form insol. Me$_2$CO. [8] (+)-form sol. CCl$_4$ [2]

Transport Properties:
Polymer Solutions Dilute: [η] 3.8 dl g^{-1} (tetrahydronaphthalene, 120°, isotactic (*S*)-form) [4]. [η] 1.9 dl g^{-1} (tetrahydronaphthalene, 120° (*S*)-form) [1]. [η] 1.85 dl g^{-1} (tetrahydronaphthalene, 120°, (*R*)-form) [1]. [η] 0.4 dl g^{-1} (tetrahydronaphthalene, 120°, isotactic (*S*)-form) [9,11]. Other viscosity values have been reported [2]

Optical Properties:
Transmission and Spectra: Ir and C-13 nmr spectral data have been reported [9,11,12,13]
Total Internal Reflection: $[\alpha]_D^{25}$ +52.3 (C$_6$H$_6$, (*S*)-form) [1]. $[\alpha]_D^{25}$ +55.7 (CHCl$_3$, isotactic (*S*)-form) [11]. $[M]_D^{25}$ +68.1 (C$_6$H$_6$, isotactic (*S*)-form) [4]. $[M]_D^{25}$ +69 (CHCl$_3$, isotactic (*S*)-form) [9]. $[M]_D^{25}$ +63 (CHCl$_3$, isotactic (*S*)-form) [11]. $[M]_D^{25}$ +68.1 (C$_6$H$_6$, (*S*)-form) [10]. $[\alpha]_D^{18.5}$ +56.5 (*S*), (+)-form, decahydronaphthalene) [8]. $[\alpha]_D^{25}$ -49.6 (C$_6$H$_6$, (*R*)-form) [1]. Other values have been reported [2,4,7,14]

Applications/Commercial Products:
Processing & Manufacturing Routes: Synth. using Ziegler-Natta catalysts such as TiCl$_3$Al(Bu^i)$_3$ in pentane at -55° [4]

Bibliographic References
[1] Chiellini, E. and Marchetti, M., *Makromol. Chem.*, 1973, **169**, 59, (solubility, optical rotation)
[2] Bacskai, R., *J. Polym. Sci., Part A-1*, 1967, **5**, 619, (optical rotation)
[3] Pino, P., Lorenzi, G.P. and Bonsignori, O., *Corsi Semin. Chim.*, 1968, **8**, 183, (optical rotation)
[4] Corradini, P., Martiscelli, E., Montagnoli, G. and Petraccone, V., *Eur. Polym. J.*, 1970, **6**, 1201, (struct, solubility)
[5] Luis, P.L. and Bonsignori, O., *J. Chem. Phys.*, 1972, **56**, 4298, (optical rotation)
[6] Pino, P., Ciardelli, F., Montagnoli, G. and Pieroni, O., *J. Polym. Sci., Part B: Polym. Lett.*, 1967, **5**, 307, (optical rotation)
[7] Bonsignori, O. and Lorenzi, G.P., *J. Polym. Sci., Part A-2*, 1970, **8**, 1639, (optical rotation)
[8] Pino, P. and Lorenzi, G.P., *J. Am. Chem. Soc.*, 1960, **82**, 4745, (optical rotation)
[9] Benedetti, E. and Chiellini, E., *J. Polym. Sci., Polym. Phys. Ed.*, 1977, **15**, 1251, (ir)
[10] Pino, P., Ciardelli, F., Lorenzi, G.P. and Montagnoli, G., *Makromol. Chem.*, 1963, **61**, 207, (optical rotation)
[11] Delfini, M., DiCocco, M.E., Paci, M., Aglietto, M. *et al*, *Polymer*, 1985, **26**, 1459, (C-13 nmr)
[12] Benedetti, E., Martino, P., Vergamini, P. and Chiellini, E., *Chim. Ind. (Milan)*, 1976, **58**, 218, (ir)
[13] Benedetti, E., Ciardelli, F., Chiellini, E. and Pino, P., *Conv. Ital. Sci. Macromol.*, (*Atti*), 1977, 168, (ir)
[14] Pino, P. and Luisi, P.L., *J. Chim. Phys. Phys. - Chim. Biol.*, 1968, **65**, 130, (optical rotation)

Poly(6-methyl-1-heptene)   P-369

$$\left[CH_2CH\right]_n$$
$$|$$
$$CH_2CH_2CH_2CH(CH_3)_2$$

Synonyms: *Poly(1-(4-methylpentyl)-1,2-ethanediyl)*
Monomers: 6-Methyl-1-heptene
Material class: Thermoplastic
Polymer Type: polyolefins
CAS Number:

CAS Reg. No.
77136-88-0
95831-09-7

Molecular Formula: $(C_8H_{16})_n$
Fragments: C_8H_{16}
General Information: Soft tacky rubber [1]

Volumetric & Calorimetric Properties:
Transition Temperature:

No.	Value	Note	Type
1	180°C	[1]	Polymer melt temp.
2	-34°C	[2]	Softening point

Transport Properties:
Polymer Solutions Dilute: $[\eta]_{inh}$ 2.37 (decahydronaphthalene, 130°) [1]

Applications/Commercial Products:
Processing & Manufacturing Routes: Synth. using Ziegler-Natta catalysts such as VCl_3-$AlEt_3$ at 50° [2]

Bibliographic References
[1] Campbell, T.W. and Haven, A.C., *J. Appl. Polym. Sci.*, 1959, **1**, 73, (synth, soln props)
[2] Dunham, K.R., Vandenberghe, J., Faber, J.W.H. and Contois, L.E., *J. Polym. Sci., Part A: Polym. Chem.*, 1963, **1**, 751, (synth, softening point)

Poly(3-methyl-1-hexene) P-370

$$\text{+CH}_2\text{CH+}_n$$
$$\text{CH(CH}_3)\text{CH}_2\text{CH}_2\text{CH}_3$$

Synonyms: *P3MH*. Poly(1-(1-methylbutyl)-1,2-ethanediyl)
Monomers: 3-Methyl-1-hexene
Material class: Thermoplastic
Polymer Type: polyolefins
CAS Number:

CAS Reg. No.	Note
112265-74-4	
25721-71-5	(*S*)-form

Molecular Formula: $(C_7H_{14})_n$
Fragments: C_7H_{14}
General Information: Powder [2]

Volumetric & Calorimetric Properties:
Melting Temperature:

No.	Value	Note
1	>350°C	min. [3]
2	285°C	[5]

Transition Temperature:

No.	Value	Note	Type
1	210–215°C	[2]	Softening point

Surface Properties & Solubility:
Solvents/Non-solvents: Insol. MeOH [2], cyclohexane [1]

Transport Properties:
Polymer Solutions Dilute: $[\eta]_{inh}$ 1.05 dl g^{-1} [2]

Optical Properties:
Total Internal Reflection: $[\alpha]_D^{25}$ -3.7--16 ((*R*)-form) [1]

Applications/Commercial Products:
Processing & Manufacturing Routes: Synth. using Ziegler-Natta catalysts such as VCl_3-$AlEt_3$ in C_6H_6 at 75° [1,2,5]

Bibliographic References
[1] Vizzini, J., Ciardelli, F. and Chien, J.C.W., *Macromolecules*, 1992, **25**, 108, (synth, optical props)
[2] *Brit. Pat.*, 1965, 999 727, (synth, soln props)
[3] *Neth. Pat.*, 1965, 6 506 740, (ICI, melting point)
[4] Vizzini, J. and Chien, J.C.W., *Polym. Mater. Sci. Eng.*, 1991, **64**, 111, (synth)
[5] *High Polymers*, (eds. R.A.V. Raff and K.W. Doak), Interscience, 1965, **20**, (melting point)

Poly(4-methyl-1-hexene) P-371

Synonyms: *PMHE*. Poly(1-(2-methylbutyl)-1,2-ethanediyl)
Monomers: 4-Methyl-1-hexene
Material class: Thermoplastic
Polymer Type: polyolefins
CAS Number:

CAS Reg. No.	Note
26101-60-0	
25620-38-6	isotactic
131724-40-8	syndiotactic
26680-37-5	isotactic, (*S*)-form
56781-40-9	isotactic, (*R*)-form
29036-76-8	syndiotactic, (*S*)-form
25189-88-2	(*S*), (+)-form
29504-00-5	(*S*), (-)-form
25585-24-4	(*R*), (-)-form
29407-50-9	(+)-form
30775-40-7	(±)-form
33409-74-4	isotactic, (±)-form

Molecular Formula: $(C_7H_{14})_n$
Fragments: C_7H_{14}
General Information: Hard, non-tacky, non-rubbery. Optically active variants known
Tacticity: Isotactic and syndiotactic forms known
Morphology: Isotactic, (*S*)-form: form I tetragonal, a 1.985, b 1.985, c 1.350, helix 7_2; form II tetragonal a 1.964, b 1.964, c 1.400, helix 7_2

Volumetric & Calorimetric Properties:
Density:

No.	Value	Note
1	d 0.86 g/cm^3	[16,17]
2	d 0.83 g/cm^3	[18]
3	d 0.851 g/cm^3	(*S*)-form [19]

Melting Temperature:

No.	Value	Note
1	157–160°C	[2]
2	171–174°C	[20]
3	188°C	[17]
4	200°C	[21]
5	234°C	[22]
6	238–240°C	[18]

– Poly(5-methyl-1-hexene)

7	210–215°C	S-form [8]
8	228–238°C	S-form [23]
9	190–195°C	(±)-form [24]
10	147°C	syndiotactic [11]

Transition Temperature General: Other T_m measurements have been reported [6,9,25]
Transition Temperature:

No.	Value	Note	Type
1	196°C	[2]	Polymer melt temp.

Surface Properties & Solubility:
Solvents/Non-solvents: Sol. boiling C_6H_6, heptane. Mod. sol. C_6H_6. Slightly sol. Et_2O [16,24]. Insol. boiling Me_2CO (S-form, (±)-form), EtOAc (S-form, (±)-form) [25]

Transport Properties:
Polymer Solutions Dilute: Theta temp. θ 165° (o-chloronaphthalene, S-isotactic), θ 133° (o-dichlorobenzene, S-isotactic) [23]. [η] 4.97 dl g^{-1} (135°, decahydronaphthalene) [18]; [η] 3.18 dl g^{-1} (120°, decahydronaphthalene) [8]. Other viscosity values dependent upon method of synth. have been reported [26]. Dilute soln. props. have been reported [23,27]
Polymer Melts: Melt viscosity 3100 [22]

Optical Properties:
Optical Properties General: Optical props. have been reported [12,14,28,29,30]. Vibrational circular dichroism has been reported [31,32,33]
Transmission and Spectra: Ir [6,14,34] and C-13 nmr [1,7,26,35,36,37,38] spectral data have been reported
Molar Refraction: dn/dC 0.0714 (80°, methylcyclohexane, (±)-isotactic) [23]; dn/dC 0.0643 (80°, methylcyclohexane, (S)-isotactic), 0.0672 (60°, 2-methyltetrahydrofuran, (S[h0])-isotactic), 0.0684 (50°, 2,5-dimethyltetrahydrofuran, (S)-isotactic), 0.0911 (70°, 2,5-dimethyltetrahydrofuran, (S)-isotactic) [23]
Total Internal Reflection: $[M]_D^{21}$ 249° (S-form) [20], $[M]_D^{25}$ 260° (S-form) [13], $[M]_D^{25}$ 278° (S-form) [24,39], $[M]_D^{25}$ 279–288° (S-form) [19], $[M]_D^{25}$ 286° (S-form) [23,39], $[M]_D^{25}$ 288 (S-form) [8,10], $[M]_D^{25}$ -40° ((S)-isotactic) [3]. Other molar optical rotation values have been reported [3,15,40]. Variation in optical rotation with different methods of synth. has been reported [26]

Applications/Commercial Products:
Processing & Manufacturing Routes: Synth. using Ziegler-Natta cataysts, such as $TiCl_4Al(^iBu)_3$ at 60° [1]

Bibliographic References
[1] Zambelli, A., Grassi, A., Galimberti, M. and Perego, G., *Makromol. Chem., Rapid Commun.*, 1992, **13**, 269, (syndiotactic, C-13 nmr)
[2] Campbell, T.W. and Haven, A.C., *J. Appl. Polym. Sci.*, 1959, **1**, 73, (melting point)
[3] Bassi, I.W., Bonsignori, O., Lorenzi, G.P., Pino, P. et al, *J. Polym. Sci., Part A-2*, 1971, **9**, 193, (struct)
[4] Natta, G., *Makromol. Chem.*, 1960, **35**, 94, (struct)
[5] Natta, G., Allegra, G., Bassi, I.W., Carlini, C. et al, *Macromolecules*, 1969, **2**, 311, (struct)
[6] Benedetti, E., Bonsignori, O., Chiellini, E. and Pino, P., *CA*, 1978, **89**, 180514, (ir)
[7] Acquiviva, L., Conti, F., Paci, M., Delfini, M. et al, *Proc. Eur. Conf. NMR Macromol.*, (ed. F. Conti), Lerici, 1978, 215, (C-13 nmr)
[8] Luisi, P.L. and Pino, P., *J. Phys. Chem.*, 1968, **72**, 2400, (optical props)
[9] Bonsignori, O., Pino, P., Manzini, G. and Crescenzi, V., *Makromol. Chem., Suppl.*, 1975, 317, (melting points)
[10] Abe, A., *J. Am. Chem. Soc.*, 1968, **90**, 2205, (syndiotactic)
[11] *Eur. Pat. Appl.*, 1990, 387 609, (syndiotactic)
[12] Montagnoli, G., Pini, D., Lucherini, A., Ciardelli, F. and Pino, P., *Macromolecules*, 1969, **2**, 684, (optical props)
[13] Pino, P., Ciardelli, F., Montagnoli, G. and Pieroni, O., *J. Polym. Sci., Part B: Polym. Lett.*, 1967, **5**, 307, (optical props)
[14] Bacskai, R., *J. Polym. Sci., Part A-1*, 1967, **5**, 619, (optical props, ir)
[15] Ciardelli, F., Montagnoli, G., Pini, D., Pieroni, O. et al, *Makromol. Chem.*, 1971, **147**, 53, (optical props)
[16] Natta, G., Pino, P., Mazzanti, G., Corradini, P. and Giannini, U., *CA*, 1955, **50**, 16256e, (solubility)
[17] Natta, G., *Chim. Ind. (Milan)*, 1956, **38**, 751, (melting point)
[18] Bacskai, R., Goodrich, J.E. and Wilkes, J.B., *J. Polym. Sci., Part A-1*, 1972, **10**, 1529, (soln props)
[19] Bonsignori, O. and Lorenzi, G.P., *J. Polym. Sci., Part A-2*, 1970, **8**, 1639, (optical props)
[20] Nozakura, S., Takeuchi, S., Yuki, H. and Murahashi, S., *Bull. Chem. Soc. Jpn.*, 1961, **34**, 1673, (optical props)
[21] Natta, G., Corradini, P. and Bassi, I.W., *CA*, 1955, **50**, 16256g, (melting point)
[22] *Brit. Pat.*, 1964, 968 471, (melt props)
[23] Neuenschwander, P. and Pino, P., *Makromol. Chem.*, 1980, **181**, 737, (soln props)
[24] Pino, P., Lorenzi, G.P. and Lardicci, L., *Chim. Ind. (Milan)*, 1960, **42**, 712, (optical props)
[25] Luisi, P.L. and Pezzana, F., *Eur. Polym. J.*, 1970, **6**, 259, (solubility)
[26] Conti, F., Acquaviva, L., Chiellini, E., Ciardelli, F. et al, *Polymer*, 1976, **17**, 901, (optical props, C-13 nmr)
[27] Bonner, F.J. and Luisi, P.L., *Ark. Kemi*, 1967, **27**, 129, (soln props)
[28] Pino, P., Ciardelli, F. and Lorenzi, G.P., *J. Polym. Sci., Part C: Polym. Lett.*, 1964, **4**, 21, (optical props)
[29] Pino, P., Montagnoli, G., Ciardelli, F. and Benedetti, E., *Makromol. Chem.*, 1966, **93**, 158, (optical props)
[30] Pino, P. and Luisi, P.L., *J. Chim. Phys. Phys. - Chim. Biol.*, 1968, **65**, 130, (optical props)
[31] Nafie, L.A., Keiderling, T.A. and Stephens, P.J., *J. Am. Chem. Soc.*, 1976, **98**, 2715, (optical props)
[32] Bertucci, C., Pini, D., Rosini, C. and Salvadori, P., *CA*, 1983, **100**, 210777, (optical props)
[33] Bertucci, C., Salvadori, P., Ciardelli, F. and Fatti, G., *Polym. Bull. (Berlin)*, 1985, **13**, 469, (optical props)
[34] Benedetti, E., Ciardelli, F. and Chiellini, E., *Chim. Ind. (Milan)*, 1977, **59**, 654, (ir)
[35] Neuenschwander, P., *Makromol. Chem.*, 1976, **177**, 1231, (C-13 nmr)
[36] Acquaviva, L., Conti, F., Paci, M., Delfini, M. et al, *Polymer*, 1978, **19**, 1453, (C-13 nmr)
[37] Segre, A.L. and Solaro, R., *Polym. Commun.*, 1986, **27**, 216, (C-13 nmr)
[38] Zambelli, A., Grassi, A., Galimberti, M. and Perego, G., *Makromol. Chem., Rapid Commun.*, 1992, **13**, 467, (syndiotactic, C-13 nmr)
[39] Pino, P. and Lorenzi, G.P., *J. Am. Chem. Soc.*, 1960, **82**, 4745, (optical props)
[40] Pino, P., Ciardelli, F., Lorenzi, G.P. and Natta, G., *J. Am. Chem. Soc.*, 1962, **84**, 1487, (optical props)

Poly(5-methyl-1-hexene)

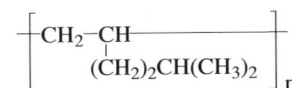

Synonyms: *Poly(1-(3-methylbutyl)-1,2-ethanediyl)*
Monomers: 5-Methyl-1-hexene
Material class: Thermoplastic
Polymer Type: polyolefins
CAS Number:

CAS Reg. No.	Note
25302-96-9	atactic
28962-82-5	isotactic

Molecular Formula: $(C_7H_{14})_n$
Fragments: C_7H_{14}
Morphology: Somewhat rubbery, but crystalline [1]. Isotactic, monoclinic, a 1.762, b 1.017, c 0.633, helix 3_1 [2]

Volumetric & Calorimetric Properties:
Density:

No.	Value	Note
1	d^{25} 0.85 g/cm^3	[3]
2	d 0.84 g/cm^3	[2]

Polymethylhydrogen siloxane

Melting Temperature:

No.	Value	Note
1	110°C	[4]
2	130°C	isotactic [5]

Glass Transition Temperature:

No.	Value	Note
1	-15°C	isotactic [6]

Transition Temperature:

No.	Value	Note
1	138°C	polymer melt temp. [1]
2	-14°C	softening point [4]

Surface Properties & Solubility:
Solvents/Non-solvents: V. sol. boiling C_6H_6 [3]. Sol. C_6H_6, $CHCl_3$, boiling heptane. Slightly sol. boiling Et_2O. Insol. Me_2CO [2]

Transport Properties:
Polymer Solutions Dilute: $[\eta]$ 6.5 dl g^{-1} (tetrahydronapthalene, 120°) [2]. η_{inh} 2.62 (decahydronapthalene, 130°) [1]

Optical Properties:
Transmission and Spectra: Ir (CH_2 rocking frequency) [7] has been reported

Applications/Commercial Products:
Processing & Manufacturing Routes: Synth. using Ziegler-Natta catalysts, such as $TiCl_3Al(Bu^i)_3$ in pentane at -20° [2]

Bibliographic References
[1] Campbell, T.W. and Haven, A.C., *J. Appl. Polym. Sci.*, 1959, **1**, 73, (physical props)
[2] Corradini, P., Martuscelli, E., Montagnoti, G. and Petraccone, V., *Eur. Polym. J.*, 1970, **6**, 1201, (struct)
[3] Natta, G., Pino, P., Mazzanti, G., Corradini, P. and Giannini, U., *Atti Accad. Naz. Lincei, Rend., Cl. Sci. Fis., Mat. Nat., Rend.*, 1955, **19**, 397, (synth, props)
[4] Dunham, K.R., Vandenberghe, J., Faber, J.W.H. and Contois, L.E., *J. Polym. Sci., Part A: Polym. Chem.*, 1963, **1**, 751, (transition temps)
[5] Natta, G., *Angew. Chem.*, 1956, **68**, 393, (transition temps)
[6] Natta, G., Mazzanti, G., Crespi, G. and Valvassori, A., *XXXVII Congr. Intern. Quim. Ind. Madrid*, 1967, (transition temps)
[7] Ciampelli, F. and Tasi, C., *Spectrochim. Acta, Part A*, 1968, **24**, 2157, (ir)

Polymethylhydrogen siloxane   P-373

Synonyms: *Polymethylsiloxane. Poly[oxy(methylsilylene)]*
Related Polymers: PDMS fluid, Poly(methylhydrogensiloxane-*co*-dimethylsiloxane) fluid, Poly(hydrogenmethylsiloxanes) with liquid crystal sidechains
Monomers: Dichloromethylsilane, Cyclic methylhydrogensiloxane
Material class: Fluids
Polymer Type: silicones
CAS Number:

CAS Reg. No.
49718-23-2

Molecular Formula: $[CH_4OSi]_n$
Fragments: CH_4OSi
Molecular Weight: MW 1000–10000

Volumetric & Calorimetric Properties:
Density:

No.	Value	Note
1	d 0.98 g/cm^3	[1]

Glass Transition Temperature:

No.	Value	Note
1	-138°C	large values of *n* [4,6]

Surface Properties & Solubility:
Surface and Interfacial Properties General: Strongly water repelling [3]
Wettability/Surface Energy and Interfacial Tension: γ_c 20 mJ m^{-2} (20 dyne cm^{-1}) [1]

Transport Properties:
Polymer Melts: Viscosity 0.25 St (25°) [1,4]. Viscosity temp. coefficient 0.5

Electrical Properties:
Dielectric Permittivity (Constant):

No.	Value	Note
1	3.2	[2,5]

Dielectric Strength:

No.	Value	Note
1	15 kV.mm^{-1}	[2]

Dissipation (Power) Factor:

No.	Value	Note
1	1×10^{-5}	[2]

Optical Properties:
Refractive Index:

No.	Value	Note
1	1.397	[1]

Polymer Stability:
Polymer Stability General: Unstable at elevated temps. and in the presence of water
Decomposition Details: Cross-linking of chains occurs with elimination of hydrogen [3]
Flammability: Fl. p. 125° (260°F) [1]
Chemical Stability: Oxidation causes cross-linking and elimination of hydrogen
Hydrolytic Stability: In aq. soln. cross-linking of chains occurs with elimination of hydrogen [3]. Mild acid conditions increase stability in water [6]
Recyclability: Not recyclable

Applications/Commercial Products:
Processing & Manufacturing Routes: For low viscosity fluids methylhydrogendichlorosilane is hydrolysed under slightly acid conditions with small quantities of trimethylchlorosilane. For high viscosity fluids cyclic methylhydrogen polysiloxanes are equilibrated together under acid catalysis with traces of hexamethyl disiloxane

Applications: Used to provide a water repellent coating and used in textile processing. Polymethyl hydrogen siloxane is used as a cross-linking agent with low viscosity vinyl methyl gums to prod. rubber by the liq. silicone process see (vinylmethyl silicone rubber (heat vulcanised)). Can be used in the addition of mesogens by hydrosilation

Bibliographic References
[1] Meals, R.N. and Lewis, F.M., *Silicones*, Reinhold, 1959
[2] Stark, F.O., Fallender, J.R. and Wright, A.P., *Comprehensive Organometallic Chemistry*, (eds. G. Wilkinson, F.G.A. Stone and E.W. Abel), Pergamon Press, 1982, **2**, 305
[3] Noll, W., *Chemistry and Technology of Silicones*, Academic Press, 1968
[4] Lee, C.L. and Haberland, C.G., *J. Polym. Sci., Part B: Polym. Lett.*, 1965, **3**, 883, (viscosity)
[5] Vincent, G.A., Fearon, F.W.G. and Orbech, T., Annu. Rep., *Conf. Electr. Insul. Dielectr. Phenom.*, 1972, 17, (dielectric constant)
[6] Clarson, S.J. and Semlyen, J.A., *Siloxane Polymers*, Prentice Hall, 1993

Poly(methylhydrogensiloxane-co-dimethylsiloxane) fluid   P-374

Synonyms: *Methylhydrogen dimethylsiloxane copolymer*
Related Polymers: Methylsiloxane homopolymer (Polymethyl hydrogen siloxane), Dimethylsiloxane homopolymer (PDMS fluid), With merogenic sidechain (side chain polysiloxane polydimethylsiloxane copolymer), Functional siloxane (poly 1-propenyl ether functional siloxane)
Monomers: Dichlorodimethylsilane, Cyclic methylhydrogensiloxane
Material class: Fluids, Copolymers
Polymer Type: silicones, dimethylsilicones
CAS Number:

CAS Reg. No.
115118-35-3

Molecular Formula: $[(C_2H_6OSi).(CH_4OSi)]_n$
Fragments: C_2H_6OSi CH_4Si
Molecular Weight: MW 1000–10000

Volumetric & Calorimetric Properties:
Density:

No.	Value	Note
1	d 0.97 g/cm^3	[1]

Surface Properties & Solubility:
Surface and Interfacial Properties General: Strongly water repelling [3]
Wettability/Surface Energy and Interfacial Tension: τ_c 20 m.J.m^{-2} (20 dyne.cm^{-1}) [1]

Transport Properties:
Polymer Melts: Viscosity temp. coefficient 0.55 [1,4]

Electrical Properties:
Dielectric Permittivity (Constant):

No.	Value	Note
1	3	[2,5]

Dielectric Strength:

No.	Value	Note
1	15 kV.mm^{-1}	[2]

Optical Properties:
Refractive Index:

No.	Value	Note
1	1.4	[1]

Polymer Stability:
Polymer Stability General: Unstable at elevated temps. and in the presence of water
Decomposition Details: Crosslinking of chains occurs with elimination of hydrogen [3]
Chemical Stability: Oxidation causes crosslinking, with elimination of hydrogen [3]
Hydrolytic Stability: In aq. soln. crosslinking of chains occurs with elimination of hydrogen. Mild acid conditions increase stability in water [3,6]
Recyclability: Not recyclable

Applications/Commercial Products:
Processing & Manufacturing Routes: Cyclic dimethyl and methyl hydrogen polysiloxanes are polymerised together under acid catalysis with traces of hexamethyldisiloxane to serve as chain terminator [1,2,6]
Applications: Can be used for addition of mesogens by hydrosilation and in the synth. of functional siloxanes

Bibliographic References
[1] Meals, R.N. and Lewis, F.M., *Silicones*, Reinhold, 1959
[2] Stark, F.O., Fallender, J.R. and Wright, A.P., *Comprehensive Organometallic Chemistry*, (eds. G. Wilkinson, F.G.A. Stone and E.W. Abel), Pergamon Press, 1982, **2**, 305
[3] Noll, W., *Chemistry and Technology of Silicones*, Academic Press, 1968
[4] Lee, C.L. and Haberland, C.G., *Polym. Int.*, 1965, **3**, 883
[5] Vincent, G.A., Fearon, F.W.G. and Orbech, T., Annu. Rep., *Conf. Electr. Insul. Dielectr. Phenom.*, 1972, 17
[6] Clarson, S.J. and Semlyen, J.A., *Siloxane Polymers*, Prentice Hall, 1993

Poly(methyl isopropenyl ketone)   P-375

$$\left[\begin{array}{c} CH_2C(CH_3) \\ | \\ O=C-CH_3 \end{array} \right]_n$$

Synonyms: *Poly(3-methyl-3-buten-2-one)*. *PMIPK*. *Poly(methyl-1-methylvinyl ketone)*. *Poly(methylene ethyl methyl ketone)*
Monomers: 3-Methyl-3-buten-2-one
Polymer Type: polyolefins
CAS Number:

CAS Reg. No.	Note
54789-11-6	dimer
25988-32-3	
29691-29-0	isotactic
92755-59-4	homopolymer radical ion

Molecular Formula: $(C_5H_8O)_n$
Fragments: C_5H_8O
General Information: Light sensitive. Used as a positive photoresist in place of poly(methyl methacrylate). Varying polymerisation conditions leads to different polymer forms [8,9]

Volumetric & Calorimetric Properties:
Density:

No.	Value	Note
1	d 1.11–1.15 g/cm^3	[5]
2	d^{20} 1.116 g/cm^3	[7]

Melting Temperature:

No.	Value	Note
1	240°C	[8]
2	160°C	anionic polymerised [9]
3	200–220°C	crystalline-form [9]

Deflection Temperature:

No.	Value	Note
1	95°C	approx., 1.82 MPa, (264 psi) [10]

Transition Temperature:

No.	Value	Note	Type
1	60–80°C	[5,8]	Softening range
2	114°C	[8]	Softening point

Surface Properties & Solubility:
Solvents/Non-solvents: Sol. Me_2CO, dioxane, esters, $CHCl_3$, nitrobenzene. Slightly sol. dioxane, pyridine. Insol. EtOH, C_6H_6, cyclohexane, petrol, octane [5,7,8,9,10]. Solubility depends on method of polymerisation

Transport Properties:
Polymer Solutions Dilute: Intrinsic viscosity values derived from Staudinge-Kuhn relation have been reported [6]

Water Absorption:

No.	Value	Note
1	0.3 %	24h, 20° [5]
2	0.67 %	7 d, 20° [5]

Mechanical Properties:
Tensile Strength Break:

No.	Value	Note
1	68.94 MPa	approx., 10 kpsi [10]

Flexural Strength at Break:

No.	Value	Note
1	117.19 MPa	approx., 17 kpsi [10]

Compressive Strength:

No.	Value	Note
1	130.99 MPa	approx., 19 kpsi [10]

Hardness: Brinell 20 (approx.) [5]. Rockwell M98 (approx.) [10]

Izod Notch:

No.	Value	Notch	Note
1	18.11 J/m	Y	0.34 ft lb in^{-1} [10]

Optical Properties:
Refractive Index:

No.	Value
1	1.5212

Transmission and Spectra: Ir [1,9] and x-ray [8,9] spectral data have been reported

Polymer Stability:
Polymer Stability General: Exhibits poor thermal and photochemical stability [10]
Decomposition Details: The polymer cyclises via an Aldol condensation eliminating water easily at 250° [1]. Above 170° the polymer changes colour [1] and the ir spectrum begins to alter. The crystalline form decomposes faster than the amorph. form
Environmental Stress: Compression moulded disks become dark and brittle after 22h exposure in a weatherometer [10]. Undergoes photolytic decomposition. Stability can be improved by reducing carbonyl groups with lithium aluminium hydride; disks of this modified polymer can withstand 200h in weatherometer [10]
Chemical Stability: Rapidly cyclised by hot alkali [3]. Decomposed by acids [9,10]

Bibliographic References
[1] Matsuzaki, K. and Lay, T.C., *Makromol. Chem.*, 1967, **110**, 185, (synth)
[2] Wissbrun, K.F., *J. Am. Chem. Soc.*, 1959, **81**, 58, (photolysis)
[3] Cooper, W. and Catterall, E., *Chem. Ind. (London)*, 1954, 1514
[4] Kador, U. and Mehnert, P., *Makromol. Chem.*, 1971, **144**, 37, (thermal degradation)
[5] Morgan, G.T., Megson, N.J.L. and Pepper, K.W., *Chem. Ind. (London)*, 1938, **16**, 885, (props)
[6] Chaudhuri, A.K. and Basu, S., *Makromol. Chem.*, 1959, **29**, 48, (kinetics)
[7] Schildknecht, C.E., Vinyl and Related Polymers Chapman and Hall, 1952, (polymerisation)
[8] Watanabe, H., Koyama, R., Nagai, H. and Nishioka, A., *J. Polym. Sci.*, 1962, **62**, S74, (synth)
[9] Tsuruta, T., Fujio, R. and Furukawa, J., *Makromol. Chem.*, 1964, **80**, 172, (synth)
[10] Brown, F., Berardinelli, F., Kray, R.J. and Rosen, L.J., *Ind. Eng. Chem.*, 1959, **51**, 79, (polymerisation, props)

Poly(methyl methacrylate-*co*-acrylonitrile)   P-376

$$\left[\left[CH_2C(CH_3) \right]_x \left[CH_2CH \right]_y \right]_n$$
$$\quad\quad |\quad\quad\quad\quad\quad\quad |$$
$$\quad\quad C=O\quad\quad\quad\quad CN$$
$$\quad\quad |$$
$$\quad\quad OMe$$

Related Polymers: Polymethacrylates, General, Poly(methyl methacrylate), Poly(acrylonitrile)
Monomers: Methyl methacrylate, Acrylonitrile
Material class: Thermoplastic
Polymer Type: acrylic copolymers
Molecular Formula: $[(C_5H_8O_2)_x (C_3H_3N)_y]_n$
Fragments: $C_5H_8O_2$ C_3H_3N

Volumetric & Calorimetric Properties:
Equation of State: Mark-Houwink constants have been reported [2,3]

Glass Transition Temperature:

No.	Value	Note
1	87–99°C	[1]

Transition Temperature General: T_g generally increases with decreasing acrylonitrile content [1]

Transport Properties:
Polymer Solutions Dilute: Polymer-solvent interaction parameters 0.48 (30°, EtOAc, 23.6% acrylonitrile), 0.486 (30°, C_6H_6, 23.6% acrylonitrile), 0.497 (30°, DMF, 28.9% acrylonitrile), 0.4695 (30°, DMSO, 50% acrylonitrile), 0.492 (30°, DMSO, 74% acrylonitrile) [2,3]. η 0.90–6.26 dl g^{-1} (15.9–96.5% acrylonitrile) [1,4,5]

Optical Properties:
Refractive Index:

No.	Value	Note
1	1.4938–1.5161	25° [4]

Bibliographic References

[1] Johnstone, N.W., Polym. Prepr. (Am. Chem. Soc., Div. Polym. Chem.), 1969, **10**, 608
[2] Mangalam, P.V. and Kalpagam, V., *Polymer*, 1982, **23**, 991
[3] Kashyap, A.K., Kalpagam, V. and Rami Reddy, C., *Polymer*, 1977, **18**, 878
[4] Beevers, R.B. and White, E.F.T., *Trans. Faraday Soc.*, 1960, **56**, 1529
[5] Asaduzzaman, A.K.M., Rabshit, A.K. and Devi, S., *J. Appl. Polym. Sci.*, 1993, **47**, 1813, (soln props)

Poly(methyl methacrylate-*co*-acrylonitrile-*co*-butadiene-*co*-styrene)   P-377

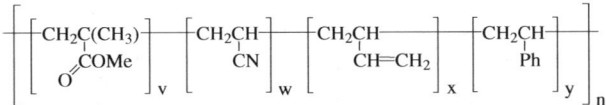

Synonyms: *MABS*
Related Polymers: Polymethacrylates - General, Poly(methyl methacrylate) General, Poly(methyl methacrylate-*co*-butadiene-*co*-styrene), Polystyrene, Polyacrylonitrile, Polybutadiene
Monomers: Methyl methacrylate, Acrylonitrile, Butadiene, Styrene
Material class: Thermoplastic
Polymer Type: acrylic copolymers
CAS Number:

CAS Reg. No.	Note
9010-94-0	
125638-62-2	(block)
109216-34-4	(block, graft)
107592-06-3	(graft)

Molecular Formula: $[(C_5H_8O_2)_w(C_3H_3N)_x(C_4H_6)_y(C_8H_8)_z]_n$
Fragments: $C_5H_8O_2$ C_3H_3N C_4H_6 C_8H_8
General Information: A tough, transparent, thermally formable polymer with good impact resistance and chemical resistance, available in pellet form and in a wide range of colours
Morphology: The copolymer is amorph. and shows no sign of crystalline zones. It has a butadiene rubber phase submicroscopically embedded in an S/A/M matrix
Identification: Contains only C, H, N and O

Volumetric & Calorimetric Properties:
Density:

No.	Value	Note
1	d 1.08 g/cm^3	DIN 53479; ISO 1043 [1]

Thermal Expansion Coefficient:

No.	Value	Note	Type
1	8×10^{-5}–0.00011 K^{-1}	linear, DIN 53752 [1]	L
2	6×10^{-5}–9×10^{-5} K^{-1}	linear, CI77 [2]	L

Thermodynamic Properties General: The substitution of α-methyl styrene gives copolymers with higher heat distortion temps. [2]
Thermal Conductivity:

No.	Value	Note
1	0.17 W/mK	W^{-1}K^{-1}, DIN 52612 [1]

Transition Temperature General: The highest second order transitions occur around 93° and the polymers become fluid above 149° [2]

Deflection Temperature:

No.	Value	Note
1	77–91°C	15.6 MPa, ASTM D648 [2]
2	87–90°C	1.8 MPa, ISO 75 [1]
3	93–94°C	0.45 MPa, ISO 75 [1]

Vicat Softening Point:

No.	Value	Note
1	105°C	VST/A/50, ISO 306 [1]
2	87–91°C	VST/B/50, ISO 306 [1]

Surface Properties & Solubility:
Solubility Properties General: Incompatible with other thermoplastics; opalescence may occur
Transport Properties:
Permeability of Gases: High levels of acrylonitrile improve resistance of the copolymer to gas penetration [2]
Water Absorption:

No.	Value	Note
1	0.35 %	24h, saturated, DIN 53495/11, DIN 53495/1 [1]

Mechanical Properties:
Tensile (Young's) Modulus:

No.	Value	Note
1	1900–2000 MPa	DIN 53457, ISO 527 [1]
2	1998–2963 MPa	290–430 kpsi, ASTM D638 [2]

Tensile Strength Break:

No.	Value	Note	Elongation
1	42–48 MPa	DIN 53455, ISO 527 [1]	4%
2	38–55 MPa	5600–8000 psi, ASTM D638 [2]	

Flexural Strength at Break:

No.	Value	Note
1	60–70 MPa	DIN 53452, ISO 178 [1]
2	65.5–89.6 MPa	9500–13000 psi, ASTM D790 [2]

Miscellaneous Moduli:

No.	Value	Note	Type
1	800 MPa	DIN 53445; ISO 537 [1]	shear modulus
2	207–276 MPa	30–40 kpsi, ASTM D790 [27]	flexural modullus
3	1250 MPa	DIN 53444, ISO 899 [1]	tensile creep modullus

Impact Strength: Izod 60 kJ m^{-2} (23°, ISO 180/C) [1], (-30°, ISO 180/C) [1], 7 kJ m^2 (notched, 23°, ISO 180/1A) [1], 2 kJ m^2 (notched, -30°, ISO 180/1A) [1], 213.6–267 J m^{-1} (notched, high-impact grade, ASTM D256) [2], 80.1–160.2 J m^{-1} (notched, medium impact grade, ASTM D256) [2], 37.4–53.4 J m^{-1}

(notched, high modulus grade, ASTM D256) [2]. Charpy 50–65 kJ m^2 (23°, DIN 53453) [1], 20–30 kJ m^2 (-20°, DIN 53453) [1], 3 kJ m^{-2} (23°, DIN 53453) [1], 1 kJ m^{-2} (-40°, DIN 53453) [1]
Hardness: Ball indentation 70–75 MPa (ISO 2039-1) [1]. Rockwell R98–119 (ASTM D785)

Electrical Properties:
Electrical Properties General: Comparative tracking index CTI 600 (ISO 112/A; DIN 0303T1) [1]; CTI 600M (ISO 112A, DIN 0303 T1) [1]
Surface/Volume Resistance:

No.	Value	Note	Type
1	$0.01–0.1 \times 10^{15}$ Ω.cm	(ISO 93, DIN 303-T30) [1]	S

Dielectric Permittivity (Constant):

No.	Value	Frequency	Note
1	2.8–3	100 Hz-1 MHz	ISO 250, DIN 0303-T4 [1]
2	2.8–3.2	1 MHz	ASTM D150 [2]

Dielectric Strength:

No.	Value	Note
1	77–85 kV.mm^{-1}	kV mm^{-1} (ISO 243/1; DIN 0303-T21) [1]

Dissipation Power Factor: tan δ 0.013–0.016 (100 Hz–1 MHz, ISO 250, DIN 0303-T4) [1]; 0.015–0.02 (1 MHz), ASTM D150) [2]
Static Electrification: MABS can be supplied with an antistatic finish

Optical Properties:
Optical Properties General: The reflective indices of matrix and elastomer phases differ at high temps. causing the polymer to become cloudy. Clarity is restored on cooling
Refractive Index:

No.	Value	Note
1	1.51–1.54	(ISO 489, ASTM D542) [1,2]

Transmission and Spectra: Transmission 80–87% (D1003) [2]. Haze 6–12% (D1003) [2]

Polymer Stability:
Polymer Stability General: Hot water causes irreversible hazing
Thermal Stability General: Is thermally stable up to 260° [2]
Upper Use Temperature:

No.	Value	Note
1	75°C	few hours [1]

Flammability: Flammability rating HB (UL94) [1]; BH3 rating (DIN VDE 0304/3) [1]
Environmental Stress: Intense uv radiation, atmospheric oxygen at elevated temps. or prolonged outdoor exposure can cause yellowing and lead to a reduction in impact resistance [1,2]
Chemical Stability: Resistant to saturated hydrocarbons, vegetable and animal fats and oils, H_2O, dilute acids and alkalis and aq. solns. of salts. It is attacked by aromatic hydrocarbons, chlorinated hydrocarbons, esters, ethers and ketones [1]
Recyclability: Can be recycled providing that it is not contaminated. Production waste can be reground and reused, although preferably not for high quality uses
Stability Miscellaneous: γ-Irradiation at the levels required for sterilisation purposes (e.g. 2.5–3.5 Mrad) has very little effect on mechanical props. but can cause slight yellowing [1]

Applications/Commercial Products:
Processing & Manufacturing Routes: Two general routes; bulk or bulk suspension polymerisation. Rubber is dissolved or dispersed in the monomer mix before polymerisation. This gives a linear copolymer (continuous phase) and a graft copolymer (disperse phase) or styrene/butadiene or butadiene latexes can be emulsion polymerised together with styrene, methyl methacrylate and acrylonitrile. Again part of the monomers graft onto the rubber to give a dispersed phase. May be processed by injection moulding or extrusion. The polymer is hygroscopic and may require drying before use in extrusion or moulding applications [1]
Applications: Used in medical applications (equipment housings and parts, inhaler mouthpieces, containers), food (containers, lids, packaging), cosmetics (containers, toothbrushes), writing accessories, watches, toys, household items (trays/baskets for freezers/refrigerators, containers, video cassette housing), office materials (machinery covers, containers). 5–15% by weight of the copolymer is added to PVC to improve the clarity and impact strength for use in the manufacture of films and bottles

Trade name	Supplier
Cryolite XT	CYRO Industries
Terlux	BASF
Zylar	Novacor

Bibliographic References
[1] *Terlux Range,* BASF, 1997, (technical datasheet)
[2] *Encycl. Polym. Sci. Technol.,* Suppl. 1, 1976, 307

Poly(methyl methacrylate-*co*-acrylonitrile-*co*-α-methylstyrene)

$$\left[\left[CH_2C(CH_3) \right]_x \left[CH_2CH \right]_y \left[CH_2C(CH_3) \right]_z \right]_n$$
$$\qquad\;\; O\;\;\;OMe \qquad\;\; CN \qquad\qquad Ph$$

Related Polymers: Polymethacrylates, General, Poly(methyl methacrylate), Poly(acrylonitrile)
Monomers: Methyl methacrylate, Acrylonitrile, α-Methylstyrene
Material class: Thermoplastic, Copolymers
Polymer Type: acrylic copolymers
Molecular Formula: $[(C_5H_8O_2)_x (C_3H_3N)_y (C_9H_{10})_z]_n$
Fragments: $C_5H_8O_2$ C_3H_3N C_9H_{10}

Volumetric & Calorimetric Properties:
Density:

No.	Value	Note
1	d 1.09 g/cm^3	ASTM D792 [2]

Thermal Expansion Coefficient:

No.	Value	Note	Type
1	$6 \times 10^{-5} – 8 \times 10^{-5}$ K^{-1}	ASTM D696 [2]	L

Equation of State: Mark-Houwink constants have been reported [1]
Thermal Conductivity:

No.	Value	Note
1	0.189 W/mK	0.0004–0.0005 cal (cm s K)$^{-1}$, ASTM C177 [5]

Specific Heat Capacity:

No.	Value	Note	Type
1	1.42 kJ/kg.C	0.34 cal (g °C)$^{-1}$ [2]	p

Glass Transition Temperature:

No.	Value	Note
1	130–148°C	[3]

Deflection Temperature:

No.	Value	Note
1	96–100°C	205–212 °F, 1.82 MPa, ASTM D648 [2]

Transport Properties:
Polymer Solutions Dilute: Theta temps. 93° (1-chlorohexane), 84° (3-heptanone), 62° (2-methylcyclohexanol), 94° (2-octanone) [4]. η 0.4–1.3 dl g^{-1} [3]

Water Absorption:

No.	Value	Note
1	0.15 %	24h, 1/8" thick, ASTM D570 [2]

Mechanical Properties:
Tensile (Young's) Modulus:

No.	Value	Note
1	2960 MPa	430 kpsi, ASTM D638 [2]

Flexural Modulus:

No.	Value	Note
1	1790–2620 MPa	260–380 kpsi, ASTM D790 [2]

Tensile Strength Break:

No.	Value	Note	Elongation
1	68.9 MPa	10 kpsi, ASTM D638 [2]	3%

Compressive Modulus:

No.	Value	Note
1	1650–2550 MPa	240–370 kpsi, ASTM D695 [2]

Flexural Strength Yield:

No.	Value	Note
1	110–131 MPa	16–19 kpsi, ASTM D790 [2]

Compressive Strength:

No.	Value	Note
1	75.8–103.3 MPa	11–15 kpsi, ASTM D695 [2]

Hardness: Rockwell M75 [2]

Izod Notch:

No.	Value	Notch	Note
1	16 J/m	Y	0.3 ft lb in^{-1}, ASTM D256 [2]

Optical Properties:
Refractive Index:

No.	Value	Note
1	1.567	ASTM D542 [2]

Transmission and Spectra: Transmittance 90% (ASTM D1003) [2]

Polymer Stability:
Upper Use Temperature:

No.	Value	Note
1	82–93°C	180–200 °F, continuous [2]

Flammability: Burning rate 1.3 in min^{-1} (ASTM D635) [2]
Environmental Stress: Unaffected by sunlight [2]
Chemical Stability: Stable to weak acids, but attacked by strong oxidising acids. Resistant to alkalis. Attacked by ketones, esters, aromatic hydrocarbons and chlorinated hydrocarbons [2]

Applications/Commercial Products:
Processing & Manufacturing Routes: Processed by compression or injection moulding [2]

Mould Shrinkage (%):

No.	Value	Note
1	0.2–0.6%	[2]

Bibliographic References
[1] Niezette, J., Vanderschueren, J. and Aras, L., *J. Polym. Sci.*, *Part B: Polym. Phys.*, 1984, **22**, 1845
[2] *Modern Plastics Encyclopedia*, (ed. W.A. Kaplan), McGraw-Hill,
[3] Johnstone, N.W., Polym. Prepr. (Am. Chem. Soc., *Div. Polym. Chem.*), 1969, **10**, 608
[4] Kotaka, T., Ohnuma, H. and Inagaki, Polym. Prepr. (Am. Chem. Soc., *Div. Polym. Chem.*), 1970, **11**, 660, (theta temps.)

Poly(methyl methacrylate), cast sheet, rods, and tubes

Related Polymers: Methacrylates General, Poly(methyl methacrylate) General
Monomers: Methyl methacrylate
Material class: Thermoplastic
Polymer Type: acrylic
Molecular Formula: $(C_5H_8O_2)_n$
Fragments: $C_5H_8O_2$
Molecular Weight: M_n 300000–6000000
General Information: Cast sheet is available in a wide range of colours (transparent, translucent and opaque) and sizes (up to approx. 1.2 × 2.4 m for cell cast and reels and 183 × 2.75 m for continuous cast). Available thicknesses 0.75–250 mm (cell cast); 9.4 mm (continuous cast). A range of special surface finishes including matt, mirror-finish, satin, rippled, stippled and abrasion resistant are available. Special grades including heat resistant, uv impermeable, flame-retardant and improved chemically resistant grades are available. ASTM D4802 specifies three categories of cast PMMA sheet; A1 (cell cast sheet), A2 (continuous cast sheet), B1 (continuous manufactured sheet). ASTM D5436 specifies four categories for cast PMMA rods, tubes and shapes; type UVA (uv absorbing), type UVT (uv transmitting), finish 1 (smooth) and finish 2 (rough or unfinished)

Volumetric & Calorimetric Properties:
Density:

No.	Value	Note
1	d 1.18 g/cm^3	DIN 53479 [1]
2	d 1.17–1.2 g/cm^3	ASTM D792 [2]

– Poly(methyl methacrylate), cast sheet, rods, and tubes

3	d 1.18–1.19 g/cm³	ISO 1183 [3]
4	d 1.2 g/cm³	ISO 1183 [4]
5	d²³ 1.19 g/cm³	50% relative humidity, ISO 1183 [5]
6	d 1.17–1.19 g/cm³	general purpose type I [6]
7	d 1.18–1.2 g/cm³	general purpose type II [6]

Thermal Expansion Coefficient:

No.	Value	Note	Type
1	7.7×10^{-5} K^{-1}	ASTM D696 [3,4]	L
2	4.5×10^{-5} K^{-1}	general purpose types I and II [6]	L
3	7×10^{-5} K^{-1}	0–50°, DIN 53752-A [5]	L

Thermal Conductivity:

No.	Value	Note
1	0.17–0.19 W/mK	DIN 52612 [7]
2	0.17 W/mK	clear, NBN B62-202 [4]
3	0.21 W/mK	1.44 Btu (h ft² °F in)$^{-1}$, types I and II [6]
4	0.19 W/mK	DIN 4701 [5]

Thermal Diffusivity: U value 5.8 W m^{-2} (1 mm thick, DIN 4701) [5]; 5.6 W m^{-2} (3 mm thick, DIN 4701) [5]; 5.3 W m^{-2} (5 mm thick, DIN 4701) [5]; 4.4 W m^{-2} (10 mm thick, DIN 4701) [5]

Specific Heat Capacity:

No.	Value	Note	Type
1	1.467 kJ/kg.C	0.35 cal (g °C)$^{-1}$ [4,5]	P

Melting Temperature:

No.	Note
1	no melt, ASTM D3418 [2]

Glass Transition Temperature:

No.	Value	Note
1	90–105°C	ASTM D3418 [2]

Deflection Temperature:

No.	Value	Note
1	71–102°C	1.8 MPa, ASTM D648 [2]
2	66–82°C	150–180°F, 1.8 MPa, general purpose type I [6]
3	88–107°C	190–220°F, 1.8 MPa, general purpose type II [5,6]
4	102°C	ISO 75A [4]
5	74–113°C	0.45 MPa, ASTM D648 [2]
6	113–115°C	0.45 MPa, ISO 75 [5]
7	100°C	1.82 MPa, ASTM D648 [8]

Vicat Softening Point:

No.	Value	Note
1	110°C	min., ISO 306A [3]
2	115°C	ISO 306/B [5]
3	95–110°C	ASTM D1525 [2]
4	111–129°C	5 kg, DIN 53460 [7]
5	119°C	5 kg, impact modified, DIN 53460 [7]

Transport Properties:
Permeability of Gases: [H$_2$O] 2.3×10^{-10} g cm cm^{-2} h^{-1} Pa^{-1} [5]; [O$_2$] 2×10^{-14} [5]; [N$_2$] 4.5×10^{-15} [5]; [CO$_2$] 1.1×10^{-13} [5]; [air] 8.3×10^{-15} [5]

Water Absorption:

No.	Value	Note
1	0.2–0.4 %	24h, ASTM D570, ISO 62 [3]
2	0.3–0.4 %	24h, type I [6]
3	0.2–0.4 %	24h, type II [6]

Mechanical Properties:
Tensile (Young's) Modulus:

No.	Value	Note
1	2900 MPa	5 mm thick [9]
2	3300 MPa	short term, ISO 527 [5]
3	3200 MPa	long term, ISO 527 [5]
4	2413–3102 MPa	350–450 kpsi, general purpose type I, ASTM D638 [2,6,10]
5	2758–3447 MPa	400–500 kpsi, general purpose type II, ASTM D638 [6,10]

Flexural Modulus:

No.	Value	Note
1	2960–3210 MPa	ISO 178 [3]
2	2758 MPa	400 kpsi, general purpose type I [6]
3	3102 MPa	450 kpsi, general purpose type II [6]
4	2413–3102 MPa	350–450 kpsi, general purpose type I [10]
5	2758–3447 MPa	400–500 kpsi, general purpose type II [2,10]

Tensile Strength Break:

No.	Value	Note	Elongation
1	62–75 MPa	ISO 527/1 [3]	4%
2	110 MPa	-40°, ISO 527 [5]	5.5%
3	80 MPa	23°, ISO 527 [5]	5.5%
4	40 MPa	70°, ISO 527 [5]	5.5%
5	41–62 MPa	general purpose type I, ASTM D638 [10]	2–7%
6	55–69 MPa	general purpose type II, ASTM D638 [2,10]	2–7%

Flexural Strength at Break:

No.	Value	Note
1	105–116 MPa	ISO 178 [3]
2	129 MPa	2 mm min^{-1}, ISO 178 [4]
3	115 MPa	ISO 178 [5]
4	112 MPa	ASTM D695 [7]

– Poly(methyl methacrylate), cast sheet, rods, and tubes

Elastic Modulus:

No.	Value	Note
1	1700 MPa	10 Hz, cast and extruded grades, ISO 527 [5]

Tensile Strength Yield:

No.	Value	Elongation	Note
1	41.4–62 MPa	2–7%	6–9 kpsi, type I [6]
2	55.2–68.9 MPa	2–7%	8–10 kpsi, type II [6]

Flexural Strength Yield:

No.	Value	Note
1	82.7–96.5 MPa	12–14 kpsi, general purpose type I, ASTM D790 [6]
2	103.4–117.2 MPa	15–17 kpsi, general purpose type II, ASTM D790 [6]

Compressive Strength:

No.	Value	Note
1	82.7–96.5 MPa	12–14 kpsi, 2% offset, general purpose type I, ASTM D695 [6]
2	96.5–124.1 MPa	14–18 kpsi, 2% offset, general purpose type II, ASTM D695 [6]
3	110 MPa	ISO 604 [5]
4	126 MPa	ASTM D695 [7]

Hardness: Rockwell M98.5-M102 (ISO 2039-2) [3]; M80-M90 (general purpose type I, ASTM D785) [6]; M96–M102 (general purpose type II, ASTM D785) [6]. Ball indentation 200 MPa (ISO 2039-1) [1]; 175 MPa (ISO 2039-1) [5]. Barcol 87. Shore 90 (durometer) [11]
Fracture Mechanical Properties: Fatigue strength 40 MPa (unnotched, 1000000 cycles, cast), 20 MPa (notched, 1000000 cycles, cast) [5]
Friction Abrasion and Resistance: Friction coefficient 0.8 (plastic/plastic), 0.45 (steel/plastic) [7]. Abrasion resistance 0.025 N (Martens method) [5]

Izod Notch:

No.	Value	Notch	Note
1	21.4 J/m	Y	0.4 ft lb in^{-1}, general purpose types I and II, ASTM D256 [6]
2	16.2–32.4 J/m	Y	room temp., ASTM D256 [2]

Izod Area:

No.	Value	Notch	Note
1	1.2 kJ/m^2	Notched	IS grade, ISO 179/2D [3]
2	1.6 kJ/m^2	Notched	ISO 180/1A [5]
3	1.2–2.17 kJ/m^2	Notched	ISO 179/2D [3]

Charpy:

No.	Value	Notch	Note
1	11–19	N	ISO 179 [4]
2	15	N	ISO 179/1fu [5]

Electrical Properties:
Surface/Volume Resistance:

No.	Value	Note	Type
3	0.1×10^{15} Ω.cm	IEC 93 [3]	S
4	0.05×10^{15} Ω.cm	DIN VDE 0303, part 3 [5]	S

Dielectric Permittivity (Constant):

No.	Value	Frequency	Note
1	3.6	50 Hz	23°, cast, DIN VDE 0303, part 4 [5]
2	3.4–4.5	60 Hz	general purpose type I, ASTM D150 [6]
3	3.5–4.5	60 Hz	general purpose type II, ASTM D150 [6]
4	2.7	100 kHz	23°, cast, DIN VDE 0303, part 4 [5]
5	2.7–3.2	1 MHz	general purpose types I and II, ASTM D150 [6]

Dielectric Strength:

No.	Value	Note
1	0.45–0.53 kV.mm^{-1}	short time, general purpose type I, ASTM D149 [6]
2	0.45–0.5 kV.mm^{-1}	short time, general purpose type II, ASTM D149 [6]
3	0.45–0.5 kV.mm^{-1}	short time, 1/8" thick, ASTM D149 [12]
4	0.431 kV.mm^{-1}	step-by-step, 1/8" thick, ASTM D149 [12]
5	15 kV.mm^{-1}	IEC 243 [3]
6	30 kV.mm^{-1}	1 mm thick, DIN VDE 0303, part 2 [5]

Arc Resistance:

No.	Note
1	no track, general purpose grades I and II, ASTM D495 [6,10]

Static Electrification: Tracking resistance KC 600 (min., DIN VDE 0303, part 1) [5]
Dissipation (Power) Factor:

No.	Value	Frequency	Note
1	0.06	50 Hz	23°, DIN VDE 0303, part 4 [5]
2	0.05–0.06	60 Hz	general purpose types I and II, ASTM D150 [6]
3	0.02	100 kHz	23°, DIN VDE 0303, part 4 [5]
4	0.02–0.03	1 MHz	general purpose types I and II, ASTM D150 [6]

Optical Properties:
Refractive Index:

No.	Value	Note
1	1.49	ISO 489/A [3,5]
2	1.491	20°, DIN 53491

– Poly(methyl methacrylate), Extruded sheet

Transmission and Spectra: Transmission > 92% (ASTM D1003) [3]
Volume Properties/Surface Properties: Haze 20–30% [5]

Polymer Stability:
Upper Use Temperature:

No.	Value	Note
1	80°C	continuous use [3,5]
2	60–71°C	140–160°F, general purpose type I, no load [6]
3	80–85°C	flat sheet [4]
4	82–93°C	180–200°F, general purpose type II, no load [6]
5	80°C	continuous use [3,5]
6	75–80°C	thermoformings [4]
7	80–85°C	flat sheet [4]
8	75–80°C	thermoformings [4]

Flammability: Flammability rating B2 (DIN 4102) [3,5]; B1 (flame retardant) [5]; HB (UL94) [3]; Class 3 (BS 466, part 7) [3]. Ignition temp. 425° (DIN 51794) [5]
Environmental Stress: Outdoor weathering causes very slight yellowing after 1015 days (ASTM D1435) [2]

Applications/Commercial Products:
Processing & Manufacturing Routes: Manufactured by bulk polymerisation, which involves an initial partial polymerisation to produce a syrup containing 10–30% polymerised material. Plasticiser, colouring agents, uv absorbers etc., may be added to the syrup prior to its use. The syrup may be refrigerated if not immediately required. A typical syrup preparation might involve heating monomer (with about 0.5% initiator, e.g., benzoyl peroxide) at 90° for about 8 mins. with agitation and then cooling to room temp. An alternative, simpler route for the manufacture of a syrup is to dissolve polymer in monomer and add peroxide to the soln. The syrup is then used in the casting process, its viscosity making it easier to contain. The partial polymerisation reduces both the reaction time and the amount of shrinkage in the cell. Polymerisation is carried out at 40° for 4 h in an inert gas atmosphere or in a full cell to avoid oxygen inhibition followed by reaction at 100° for 30 mins. The reaction rate and product are affected by the presence of oxygen because of methacrylate peroxide side reactions. The production of rods requires the use of special techniques to overcome the problem of shrinkage. Vertical aluminium tubes are filled with syrup and slowly lowered into a bath held at 40°. Thus the syrup is polymerised from the bottom of the tube up allowing gradual contraction to take place without creating voids or distortions in the finished product. Tubes are manufactured by rotating an aluminium tube containing a set amount of syrup at a constant rate with heating. Polymerisation takes place on the walls of the tube. Processed by thermoforming
Applications: Cast sheet is used mainly in building applications and glazing; in illuminated signs and shopfittings; in technical components, functional models; in furniture; in bathroom suites and fittings; in medical products (incubators, X-ray tubes, spectacles etc.) and in specialist glazing applications (aircraft, museums/art galleries)

Trade name	Details	Supplier
Perspex	VA (uv absorbing), IM (impact resistant), AS (acoustic screening), AG, CQ, SW, ME, SW (speciality grades)	ICI, UK
Plexiglas	GS209 (uv impermeable), GS237 (flame retardant)	Rohm and Haas

Bibliographic References
[1] Oberbach, K, *Kunststoffe*, 1989, **79**, 713
[2] Harrington, M., *Handbook of Plastic Materials and Technology*, (ed. I.I. Rubin), Wiley Interscience, 1990, 355
[3] *Perspex Cast Sheet*, ICI, 1996, (technical datasheet)
[4] *Perspex SW Cast Sheet*, ICI, 1996, (technical datasheet)
[5] *Plexiglas*, Rohm, 1996, (technical datasheet)
[6] *The Materials Selector*, 2nd edn., (eds. N.A. Waterman and M.F. Ashby), Chapman and Hall, 1997
[7] *Handbook of Industrial Materials*, 2nd edn., Elsevier Advanced Technology, 1992, 393
[8] Brydson, J.A., *Plast. Mater.*, Butterworths, 1982, 371
[9] Aly El-Sayed, A., Takeda, N. and Takahashi, K., *J. Phys. D: Appl. Phys.*, 1989, **22**, 687
[10] *CRC Materials Science and Engineering Handbook*, 2nd edn., (eds. J.F. Shackleford, W. Alexander and J.S. Park), CRC Press, 1994, 515
[11] *Test. Polym.*, (ed. J.V. Schmitz), Interscience, 1965, **2**
[12] *Guide to Plastics, Property and Specification Charts*, (ed. W.A. Kaplan), McGraw-Hill,

Poly(methyl methacrylate), Extruded sheet   P-380
Related Polymers: Polymethacrylates General, Poly(methyl methacrylate) General
Monomers: Methyl methacrylate
Material class: Thermoplastic
Polymer Type: acrylic
Molecular Formula: $(C_5H_8O_2)_n$
Fragments: $C_5H_8O_2$
Molecular Weight: M_n 100000–200000
General Information: Extruded sheet is available in a wide range of colours (transparent, translucent and opaque) and thicknesses of 1.5–18 mm. Mullet-skin sheets are available if thicker grades are required. Special grades are available for specific requirements, e.g. heat resistant, uv-impermeable, flame-retardant, improved chemical resistance, etc. ASTM D4802 specifies the category B2 for extruded PMMA sheet and details the property requirements

Volumetric & Calorimetric Properties:
Density:

No.	Value	Note
1	d^{23} 1.18–1.19 g/cm^3	50% relative humidity, ISO 1183 [1,2]

Thermal Expansion Coefficient:

No.	Value	Note	Type
1	7×10^{-5} K^{-1}	DIN 53752-A [1]	L
2	7×10^{-5}–7.8×10^{-5} K^{-1}	ISO 53752A [2]	L

Thermal Conductivity:

No.	Value	Note
1	0.19–0.22 W/mK	DIN 52612 [2]
2	0.19 W/mK	DIN 4701 [1]

Thermal Diffusivity: 5.8 W m^{-2} (1 mm thick, DIN 4701) [1]; 5.6 W m^{-2} (3 mm thick, DIN 4701) [1]; 5.3 W m^{-2} (5 mm thick, DIN 4701) [1]; 4.4 W m^{-2} (10 mm thick, DIN 4701) [1]
Specific Heat Capacity:

No.	Value	Note	Type
1	1.47 kJ/kg.C	[1]	P

Deflection Temperature:

No.	Value	Note
1	90°C	1.8 MPa, ISO 75 [1]
2	95°C	0.45 MPa, ISO 75 [1]

– Poly(methyl methacrylate), Extruded sheet

Vicat Softening Point:

No.	Value	Note
1	102°C	ISO 306-B50 [1]
2	102–106.5°C	53460 [2]

Transport Properties:
Gas Permeability:

No.	Gas	Note
1	[H_2O]	2.3×10^{-10} g cm cm^{-2} h^{-1} Pa^{-1} [1]
2	[O_2]	2×10^{-14} [1]
3	[N_2]	4.5×10^{-15} [1]
4	[CO_2]	1.1×10^{-13} [1]
5	[air]	8.3×10^{-15} [1]

Mechanical Properties:
Tensile (Young's) Modulus:

No.	Value	Note
1	3210–3300 MPa	short term, ISO 527 [2]
2	3300 MPa	short term, ISO 527 [1]
3	3200 MPa	long term, ISO 527 [1]

Tensile Strength Break:

No.	Value	Note	Elongation
1	100 MPa	-40°, ISO 527 [1]	4.5%
2	72 MPa	23°, ISO 527 [1]	4.5%
3	35 MPa	70°, ISO 527 [1]	4.5%

Flexural Strength at Break:

No.	Value	Note
1	105 MPa	ISO 178 [1]

Elastic Modulus:

No.	Value	Note
1	1700 MPa	10 Hz, ISO 527 [1]

Poisson's Ratio:

No.	Value	Note
1	0.45	[1]

Compressive Strength:

No.	Value	Note
1	103 MPa	extruded, ISO 604 [1]

Hardness: Ball indentation 175 MPa (ISO 2039-1) [1]
Fracture Mechanical Properties: Fatigue strength 30 MPa (unnotched, 10000000 cycles) [1]; 10 MPa (notched, 10000000 cycles) [1]

Friction Abrasion and Resistance: Friction coefficient 0.8 (plastic/plastic) [2]; 0.45 (steel/plastic) [2]. Abrasion resistance 0.025 N (Martens method) [1]
Izod Area:

No.	Value	Notch	Note
1	1.6 kJ/m^2	Unnotched	ISO 180/1 A [1]

Charpy:

No.	Value	Notch	Note
1	15	N	ISO 179/1 fu [1]

Electrical Properties:
Surface/Volume Resistance:

No.	Value	Note	Type
1	0.05×10^{15} Ω.cm	DIN VDE 0303, part 3 [1]	S

Dielectric Permittivity (Constant):

No.	Value	Frequency	Note
1	3.7	50 Hz	23°, DIN VDE 0303, part 4 [1]
2	2.8	100 kHz	23°, DIN VDE 0303, part 4 [1]

Dielectric Strength:

No.	Value	Note
1	30 kV.mm^{-1}	1 mm thick, DIN VDE 0303, part 2 [1]

Static Electrification: Tracking resistance KC 600 (min., DIN VDE 0303, part 1) [1]
Dissipation (Power) Factor:

No.	Value	Frequency	Note
1	0.06	50 Hz	23°, DIN VDE 0303, part 4 [1]
2	0.03	100 kHz	23°, DIN VDE 0303, part 4 [1]

Optical Properties:
Transmission and Spectra: Total energy transmittance 85% (DIN 67507) [1]
Volume Properties/Surface Properties: Haze 20–30% [1]

Applications/Commercial Products:

Trade name	Details	Supplier
Perspex XT		ICI Acrylics
Plexiglas XT		Rohm and Haas
Plexiglas XT soundstop	acoustic barrier	Rohm and Haas
Plexiglas XTS	impact modified	Rohm and Haas

Bibliographic References
[1] *Plexiglas Datasheet*, Rohm, 1996, (technical datasheet)
[2] *Handbook of Industrial Materials*, 2nd edn., Elsevier Advanced Technology, 1992, 393

Poly(methyl methacrylate), General   P-381

$$\mathrm{-[CH_2C(CH_3)(COOMe)]}_n\mathrm{-}$$

Synonyms: Poly(methyl 2-methyl-2-propenoate). Poly[1-(methoxycarbonyl)-1-methyl ethylene]. Acrylic. PMMA
Related Polymers: Polymethacrylates-General, Moulding and Extrusion Compounds, Cast Sheet (rods and tubes), Extruded Sheet, Butyl methacrylate copolymer
Monomers: Methyl methacrylate
Material class: Thermoplastic
Polymer Type: acrylic
CAS Number:

CAS Reg. No.	Note
9011-14-7	general
25188-98-1	isotactic
25188-97-0	syndiotactic

Molecular Formula: $(C_5H_8O_2)_n$
Fragments: $C_5H_8O_2$
Molecular Weight: M_n 300000–6000000 (cast material); 100000–200000 (moulded or extruded material); 252000 (GPC determined); MW 481000 (GPC determined); MZ 766000 (GPC determined); MZ +1 1196000 (GPC determined)
Additives: PMMA materials may contain residual monomer or processing additives. Typical additives used in the manufacture of PMMA by suspension polymerisation (usually for moulding and extrusion compounds) include suspending agents, buffering agents, chain transfer agents (to control MW), lubricants and emulsifier. A number of additives may be blended with PMMA, perhaps the most important being dyes and pigments. Listings of commonly used colorants for PMMA have been reported. [1] Sucrose-based additives improve heat stability. Plasticisers are sometimes used with PMMA, e.g. in moulding compositions to enhance the melt flow. Levels vary from 1–50%; 5% is typical. Lubricants reduce heat dissipation during machining or polishing. Some commercial grades contain uv absorbers (290–350 nm). This both screens the user from sunburn, e.g. in glazing applications, and protects the polymer against long term degradation from light. Typical levels are in the range 0.001–2%. Alumina hydrate is used as a flame-retardant filler in some forms of PMMA used for bathroom fittings, kitchen worktops and 'synthetic marble'. Fillers and reinforcements are not often used although chalk is sometimes incorporated, e.g. for opaque illuminated panels in advertising displays. PMMA can be reinforced, e.g. with glass fibre, polyethylene, Kevlar or carbon to ensure high mechanical strength at temps. around 100°. Polystyrene (0.2–5%) may be added as an opalescent. Rubber particles increase toughness
General Information: PMMA may be fractionated by means of supercritical fluid chromatography. [3] Commercially the most important member of the acrylate/methacrylate group of polymers, PMMA was first introduced in 1936 and is available as cast and extruded sheet, tubes, rods, powders, pellets, solns, dispersions and emulsions. In addition to its use in sheet form, PMMA is also suitable for many extrusion and moulding applications, coating and surface treatments and has many industrial uses. It is non-toxic, odourless and tasteless. Nothing is extracted from PMMA by oil or water. PMMA may be coloured to give transparent, translucent or opaque forms. It is possible to texture the surface of PMMA products during moulding or by later embossing and products may be further decorated by spray-painting, vacuum-metallising, hot stamping etc. It is a clear, colourless, hard, brittle, fairly rigid material which can be drilled, carved or sawn and has exceptional optical clarity and resistance to degradation by uv light. The hardest and most heat resistant PMMA materials are the cast sheet products, followed by continuous cast materials, then the extruded or moulded products.

Its total internal reflection permits a wide light beam to be transmitted through long lengths of PMMA and around corners with little loss provided that the radius of curvature of the sheet or rod is three times its thickness. Has good electrical insulating props. at low frequencies. Above its glass-transition temp. PMMA is tough, pliable, extensible and easily bent or formed into complex shapes; at high temps., however, it is very susceptible to depolymerisation. It is a polar material and will absorb some moisture but is resistant to alkali, detergent, oils and dilute acids. Although PMMA has comparatively limited impact resistance this can be improved by copolymerisation, e.g. Butyl acrylate/MMA or acrylonitrile/MMA, or by blending e.g. with Poly(n-butyl acrylate) or a butadiene copolymer. PMMA has low flammability, a high melt viscosity, low resistance to creep at temps. only a little above room temp., poor solvent resistance and a comparative lack of abrasion resistance
Tacticity: Although nominally atactic, conventionally produced PMMA is more syndiotactic than atactic. Differences in the adsorption behaviour of atactic and tactic polymers have been reported: isotactic PMMA is absorbed onto silica (from acetonitrile soln.) whereas atactic PMMA is not. [4] Stereoregular polymers may be prod. in soln. by use of anionic catalysts, e.g. organolithium compounds or Grignard reagents. Isotactic polymers tend to be prod. in non-polar solvents and syndiotactic in polar solvents. The change in composition of the product polymer in a series of reactions carried out in various toluene/dimethoxyethane mixtures using n-butyl lithium as initiator at -30° has been reported. [5] Stereoregular PMMA can be made by living polymerisation in toluene using tert-butyl MgBr (for isotactic) or tert-butyl lithium aluminium alkyl (for syndiotactic). Living polymerisation also allows the preparation of star and comb-like polymers with stereoregular main and side chains, e.g. isotactic main chain and syndiotactic side chain [13,14]
Morphology: Commercial grades are normally amorph. Isotactic: orthorhombic a 41.96Å, b 24.34Å, c 10.50Å, α 92.9°, β 88.2°, γ 90.6° containing eight double-strand 10/1 helices per cell. [6,7,8] Repeat unit volume 87-89 cm^3 mol^{-1} [9]; 89.3 cm^3 mol^{-1}. [10] The head-to-head polymer has been reported [11]
Identification: Contains only C, H and O. The surface of any acrylic plastic is immediately fogged or crazed by Me_2CO

Volumetric & Calorimetric Properties:
Density:

No.	Value	Note
1	d^{30} 1.19 g/cm^3	conventional [18,19,20]
2	d^{30} 1.19 g/cm^3	syndiotactic [18]
3	d^{30} 1.21–1.22 g/cm^3	isotactic [6,18]
4	d^{23} 1.19 g/cm^3	cast and extruded grades, 50% relative humidity, ISO 1183 [20]
5	d^{25} 1.23 g/cm^3	cryst. [2]
6	d^{25} 1.17 g/cm^3	amorph. [2,21]
7	d^{25} 1.15–1.23 g/cm^3	[2,6,18,19,22,23]

Thermal Expansion Coefficient:

No.	Value	Note	Type
1	0.00053–0.000575 K^{-1}	$T > T_g$, 1 atm, free radical polymerisation	V
2	0.000225–0.000295 K^{-1}	$T < T_g$, 1 atm, free radical polymerisation [26]	V
3	$4.8 \times 10^{-5} – 8 \times 10^{-5}$ K^{-1}	impact modified, ASTM D696 [24]	L
4	7×10^{-5} K^{-1}	cast and extruded grades, 0–50°, DIN 53752-A [20]	L
5	$5 \times 10^{-5} – 6 \times 10^{-5}$ K^{-1}	general purpose, cast sheet, ASTM D696 [24]	L

Poly(methyl methacrylate), General

Volumetric Properties General: Molar volumes at 25° [2]: 56.1 cm^3 mol^{-1} van der Waals, 86.5 cm^3 mol^{-1} glassy, 81.8 cm^3 mol^{-1} cryst.
Equation of State: Tait, Bruce-Hartmann, Sanchez and Lacombe, Spencer-Gilmore and Smka-Somcynsky equation of state information has been reported [27,28,29,30,31,32,33]
Thermodynamic Properties General: Thermal conductivity varies with both temp. and pressure although at ordinary pressure the variation is quite small. Control over thermomechanical props. can be improved by the addition of low MW poly(butyl methacrylate) or poly(butyl-co-methyl methacrylate). [35]. Molar heat capacity data for amorph. PMMA has been reported [42]
Latent Heat Crystallization: Enthalpy of melting 9.6 kJ mol^{-1}. [2] Entropy data for a range of temps. and pressures have been reported. [28] Entropy, enthalpy and Gibbs free energy data for temps. -263–13° have been reported in a graphical form [36]

Thermal Conductivity:

No.	Value	Note
1	0.15–0.2 W/mK	[2]
2	0.15–0.161 W/mK	-50° [37]
3	0.197 W/mK	0.00047 cal (s cm °C)$^{-1}$, 35° [38,39]
4	0.25 W/mK	100° [40]
5	0.17 W/mK	clear cast sheet, NBN B62-202 [41]
6	0.19 W/mK	cast and extruded grades, DIN 4701 [20]

Specific Heat Capacity:

No.	Value	Note	Type
1	1.38 kJ/kg.C	solid, 25° [2]	P
2	1.8 kJ/kg.C	liq. 25° [2]	P
3	1.47 kJ/kg.C	cast and extruded grades [20]	P

Melting Temperature:

No.	Value	Note
1	72°	emulsion polymerisation, refractometric method [43]
2	>200°C	min., syndiotactic [18]
3	160°C	isotactic [18]

Glass Transition Temperature:

No.	Value	Note
1	85–105°C	ASTM D3418 [24]
2	80–100°C	impact-modified, ASTM D3418 [24]
3	90–105°C	cast sheet, ASTM D3418 [24]
4	105°C	conventional [26]
5	115°C	syndiotactic [18]
6	160°C	syndiotactic [44]
7	43–45°C	isotactic [18,44,45]
8	41.5°C	isotactic 95%, atactic 5% [46]
9	54.3°C	isotactic 73%, atactic 16%, syndiotactic 11% [45]
10	61.6°C	isotactic 62%, atactic 20%, syndiotactic 18% [45]
11	104°C	isotactic 6%, atactic 30%, syndiotactic 56% [46]
12	114.2°C	isotactic 10%, atactic 31%, syndiotactic 59% [45]
13	119.5°C	isotactic 10%, atactic 20%, syndiotactic 70% [46]
14	125.6°C	isotactic 9%, atactic 36%, syndiotactic 64% [45]

Deflection Temperature:

No.	Value	Note
1	74–105°C	1.82 MPa, ASTM D648 [24]
2	79–107°C	0.455 MPa, ASTM D648 [24]
3	88–102°C	ISO 75A [23]
4	91–106°C	ISO 75B [23]
5	71–102°C	1.82 MPa, cast sheet, ASTM D648 [54]
6	74–113°C	0.455 MPa, cast sheet ASTM D648 [54]
7	102°C	cast, ISO 75A [41]
8	105–107°C	1.8 MPa, cast, ISO 75 [20]
9	113–115°C	0.45 MPa, cast, ISO 75 [20]
10	90°C	1.8 MPa, extruded, ISO 75 [20]
11	95°C	0.45 MPa, extruded, ISO 75 [20]
12	74–99°C	1.82 MPa, impact-modified, ASTM D648 [54]
13	82–102°C	0.455 MPa, impact-modified, ASTM D648 [54]
14	100°C	1.82 MPa, sheet, ASTM D648 [46]
15	85–95°C	1.82 MPa, moulding compound, ASTM D648 [46]
16	76–106°C	1.82 MPa, MEC, ASTM D648, DIN 53461, ISO R-75 [22,23,46]
17	108°C	1.82 MPa, 75% glass fibre, pultrusion [47]
18	110°C	1.82 MPa, 75% glass fibre, pultrusion, post formed [47]
19	118°C	1.82 MPa, 58% carbon fibre, pultrusion [47]
20	113°C	carbon fibre, pultruded [48]
21	116°C	carbon fibre, pultruded and postformed [48]
22	117°C	Kevlar fibre, pultruded [48]
23	119°C	Kevlar fibre, pultruded and postformed [48]

Vicat Softening Point:

No.	Value	Note
1	109–112°C	moulding compound [46]
2	119°C	[19]
3	80–118°C	MEC, ASTM D1525, DIN 53460, ISO R-306 [22,23,46]
4	>110°C	min., cast sheet, ISO 306A [25]
5	115°C	cast, ISO 306-B50 [20]
6	102°C	extruded, ISO 306-B50 [20]
7	82–110°C	ASTM D1525 [24]
8	85–110°C	impact-modified, ASTM D1525 [24]
9	95–110°C	cast sheet, ASTM D1525 [24]

– Poly(methyl methacrylate), General

Brittleness Temperature:

No.	Value	Note
1	90°C	emulsion polymerisation; refractometric method [43]
2	80°C	Charpy-type impact, velocity 244 cm s^{-1}, 3 mm thick [49]

Transition Temperature:

No.	Value	Note
1	95°C	cast, DIN 53458 [20]
2	85°C	extruded, DIN 53458 [20]
3	56°C	MEC, ASTM grade 5, DIN grade 525 [51]
4	64°C	MEC, ASTM grade 6, DIN grade 526 [51]
5	72°C	MEC, DIN grade 527 [51]
6	80°C	MEC, ASTM grade 8, DIN grade 528 [51]
7	27°C	T_β [2]
8	8–26°C	T_β [50]
9	24°C	atactic T_β [50]
10	27°C	syndiotactic [50]
11	12°C	isotactic [50]
12	38°C	isotactic, crystalline [50]

Surface Properties & Solubility:

Solubility Properties General: The solubility of PMMA can vary with tacticity, e.g. highly syndiotactic polymer is insol. in C_6H_6 whereas the atactic material dissolves [52]. In some solvents, e.g. 2-butanone, butyl acetate, 2-ethoxyethanol, syndiotactic PMMA will only dissolve molecularly above 70°. Below this temp. aggregation occurs and, eventually, precipitation. Isotactic PMMA also precipitates but at lower temps. which vary significantly with solvent. Atactic PMMA does not precipitate from soln. even when cooled to -70° [52]. When solns. of syndiotactic and isotactic PMMA are mixed gelation results, probably as a result of stereocomplex formation. the solvents used exert a major influence on the formation of the stereocomplexes and the resultant effects on viscosity and associated props. [52]. Crystalline syndiotactic PMMA forms complexes with chloroacetone and other solvents. The solvent is necessary to maintain crystallinity [53]

Cohesive Energy Density Solubility Parameters: 29.90–59.4 kJ mol^{-1} [2]; 29.058–31.471 kJ mol^{-1} (25–35°) [54]; 347 J cm^{-3} [55]; 346–374 J cm^{-3} (25–35°) [54]. Solubility parameter, δ 18.2–26 J$^{1/2}$ cm$^{-3/2}$ (poor H-bonding solvents), 17.4–27.2 J$^{1/2}$ cm$^{-3/2}$ (moderate H-bonding solvents), 0 (strong H-bonding solvents), [56] 38.7 J$^{1/2}$ cm$^{-3/2}$, [57] 23.1 J$^{1/2}$ cm$^{-3/2}$, [2] 18.6–26.3 J$^{1/2}$ cm$^{-3/2}$, 18.0–23.1 J$^{1/2}$ cm$^{-3/2}$ [9], δ_d 15.6–18.6 [9], 18.8 [2]. δ_p 1.4–10.6 [9], 10.2, [2] δ_h 1–7.8, [9] 8.6. [2]

Internal Pressure Heat Solution and Miscellaneous: π 266 MPa (42.9°, M_n 15000), 400 MPa (114.1°, M_n 15000) [58]

Plasticisers: Dimethyl phthalate, diethyl phthalate, dibutyl phthalate, benzyl butyl phthalate, dibutyl sebacate, diethylhexyl sebacate, phosphates are used as plasticisers. Antiplasticisation can occur. [59,60] The presence of plasticisers affects uptake of water [83]

Solvents/Non-solvents: Sol. C_6H_6 (atactic) [52], xylene, CH_2Cl_2, $CHCl_3$, chlorobenzene, isobutanol (hot), cyclohexanol (hot), ethoxyethanol, dioxane, 2-butanone, cyclohexanone, AcOH, isobutyric acid, EtOAc, cyclohexyl acetate, EtOH aq, EtOH/CCl_4, isopropanol/2-butanone (1:1, >25°), formic acid, nitroethane, Me_2CO, isopropyl methyl ketone, o-dichlorobenzene, cyclopentanone, pyridine, o-toluidene. Insol. hexane, cyclohexane, gasoline, Nujol, castor oil, MeOH, ethylene glycol, glycerol, formamide, turpentine, linseed oil, hydrogenated naphthalenes, CCl_4, EtOH, butylene glycol, Et_2O, higher esters, m-cresol, Freon-215, iso-pentane, isooctane, n-pentane, n-octane, methyl cyclohexane, n-propanol, n-amyl alcohol, C_6H_6 (highly syndiotactic) [52,61,62]

Surface and Interfacial Properties General: The surface props. of PMMA can be altered by modification of the surface using graft copolymerisation [62] or by the incorporation of a surfactant into the polymer matrix. [63] PMMA forms an extremely viscoelastic condensed monolayer. At a surface conc. of 5×10^{-5} mg cm^{-2} PMMA enters a biphasic state when small patches of condensed material develop [64]. Monolayer and surface viscosity parameters have been reported. [64] The nature of spread films at an air-water interface determined by neutron reflectometry has been reported [155,156,157]

Wettability/Surface Energy and Interfacial Tension: Contact angle, θ 80° (H_2O, advancing, 20°) [65,66]; 58° (H_2O, advancing) [84]; 41° (methylene iodide, 20°) [65]; 59° (air, 3% H_2O present) [69]; 89° (Me_2CO, 3% H_2O present) [69]. Surface energy Parachlor 224 [67], MacLeod's 4.2 (MV 3000) [68]. Polarity 0.281 (MV 3000) [68]. -$d\gamma/dT$ 0.076 mN m^{-1} K^{-1} (MV 3000) [68]. Critical surface tension 40 mN m^{-1} [65], 39 mN m^{-1} [66]. Interfacial tension 11.9 mN m^{-1} (polyethylene, linear, 20°) [68], 3.2 mN m^{-1} (polystyrene, 20°) [68], 3.4 mN m^{-1} (poly n-butyl methacrylate, 20°) [68], 3 mN m^{-1} (poly tert-butyl methacrylate, 20°) [71], 23.6–24.1 mN m^{-1} (H_2O, 3% H_2O present) [69]

Surface Tension:

No.	Value	Note
1	27–44 mN/m	[2]
2	60.1–61.8 mN/m	3% H_2O present [69]
3	62 mN/m	after contact with H_2O [70]

Transport Properties:

Melt Flow Index:

No.	Value	Note
1	0.5–24 g/10 min	230°, 3.8 kg, MEC, ASTM D1238, DIN 53735, ISO 1133 [22,23,72]

Polymer Solutions Dilute: Intrinsic viscosity [73]. η_{inh} 850 dl g^{-1} (25°, C_6H_6), 190 dl g^{-1} (25°, $CHCl_3$), 560 dl g^{-1} (25°, DMF), 530 dl g^{-1} (25°, EtOAc), 90–4140 dl g^{-1} (25°, toluene). Viscosity is not greatly affected by tacticity but varies with MW [74]. Theta temps. -126° (Me_2CO), -223° (C_6H_6), 27° (CCl_4), -273° ($CHCl_3$), -98° (EtOAc), 5–85° (propanol), -65° (toluene) [58,75]

Polymer Melts: The melt viscosity of MEC compounds is usually higher than that of most other polymers. Newtonian viscosity 50 kN s m^{-2} (200°). PMMA has a small consistency index and so the viscosity is very sensitive to temp. changes. Control over the viscosity of PMMA can be improved by the addition of low MW poly(butyl methacrylate) or poly(butyl-co-methyl methacrylate) [35]. Because of its small power law index [76] PMMA is very sensitive to shear rate [77]

Permeability of Gases: Permeability [20] [H_2O] 2.3×10^{-10} g cm cm^{-2} h^{-1} Pa^{-1} (cast and extruded grades); [O_2] 2×10^{-14} (cast and extruded grades); [N_2] 4.5×10^{-15} (cast and extruded grades); [CO_2] 1.1×10^{-13} (cast and extruded grades); [air] 8.3×10^{-15} (cast and extruded grades). Permeability coefficient, P [78,79]; [O_2] $0.032–0.105 \times 10^{-10}$ cm^3 cm^{-2} s^{-1} (cmHg)$^{-1}$, 35°, 1–2 atm; [He] $3.75–9.57 \times 10^{-10}$; [$H_2$] $1.29–4.7 \times 10^{-10}$; [N_2] $0.0014–0.013 \times 10^{-10}$; [Ar] $0.0047–0.0308 \times 10^{-10}$; [$CH_4$] $0.00098–0.0064 \times 10^{-10}$; [$CO_2$] $0.050–0.415 \times 10^{-10}$

Permeability of Liquids: Diffusion coefficients [Me_2CO] $2.2–6.8 \times 10^{-7}$ cm^2 s^{-1} (20°, MW 201000–6350000); [n-butyl chloride] $14.9–22.4 \times 10^{-7}$ (34.5°, MW 29000–16000), $1.15–7.18 \times 10^{-7}$ (35.6°, 6550000–200000); [EtOAc] $2.34–7.21 \times 10^{-7}$ (20°, 931000–79000); [methyl methacrylate] $1.51–1.95 \times 10^{-7}$ (MW 454000-370000); [toluene] 5.8×10^{-7} (20°, MW 115000); [H_2O] 51×10^{-7} (37°), 1.3×10^{-7} (50°) [84,85]; [H_2O] 5.2 g cm cm^{-2} h^{-1} (mmHg)$^{-1}$ (25°) [82]

Permeability and Diffusion General: Isotactic PMMA is less permeable to gases than is the syndiotactic form [78]

– Poly(methyl methacrylate), General

Water Content: Water absorption decreases with increasing amount of plasticiser. 0.3–0.36% (MEC, ASTM D570, DIN 53495, ISO R-62) [22,23]

Water Absorption:

No.	Value	Note
1	0.2–0.4 %	ISO 62 [25]
2	0.1–0.4 %	24h, ASTM D570 [24]
3	0.2–0.6 %	24h, impact modified, ASTM D570 [24]
4	0.2–0.4 %	24h, cast sheet, ASTM D570 [24]
5	2.3 %	max. [84]

Mechanical Properties:

Mechanical Properties General: The deformation of PMMA is temp. dependent. Increasing temp. leads to greater elongation but reduced strength. The temp. dependence of the viscoelastic props. is due to the temp. dependence of the α and β relaxation times which are in turn controlled by the rate of molecular movement of the side group (β process) and the backbone (α process). The temp. at which these movements start corresponds to a change in failure mode from brittle to ductile. The storage modulus is affected by the presence of plasticiser [87]. The tensile strength of PMMA can be increased when the polymer is drawn into fibres. A sixfold increase compared to the bulk resin has been found at high draw ratios, with the Young's modulus of the polymer increasing tenfold [88]. Fibre reinforcement can greatly improve the tensile strength at elevated temps.

Tensile (Young's) Modulus:

No.	Value	Note
1	2241–3240 MPa	ASTM D638 [24]
2	2415–3100 MPa	impact modified, ASTM D638 [24]
3	2413–3103 MPa	cast, ASTM D638 [24]
4	3200 MPa	[20,21,89,92]
5	3180 MPa	23°, MW 1600000 [90]
6	2900 MPa	cast, 5 mm thick [91]
7	3800 MPa	[93]
8	3300 MPa	cast and extruded grades, short term, ISO 527 [20]
9	2700 MPa	H$_2$O, 37°, MW 1600000 [90]
10	318 MPa	in air, 23° [90]
11	1300 MPa	75°
12	12000 MPa	35% glass fibre

Flexural Modulus:

No.	Value	Note
1	2900–3400 MPa	MEC, ASTM D790, DIN 53452, ISO R-178 [22,23]
2	2241–3240 MPa	ASTM D790 [24]
3	2760–3450 MPa	impact modified, ASTM D790 [24]
4	2758–3448 MPa	cast sheet, ASTM D790 [24]
5	3000 MPa	[2]
6	2960–3210 MPa	cast, ISO 178 [25]

Tensile Strength Break:

No.	Value	Note	Elongation
1	80 MPa	[1]	5%
2	65 MPa	[2]	10%
3	38–85 MPa	MEC, ASTM D638, DIN 53455, ISO 527 [22,23]	4–40%
4	62–75 MPa	ISO 527/1 [25]	4%
5	48–76 MPa	ASTM D638 [24]	
6	55–76 MPa	cast sheet, ASTM D638 [24]	
7	110 MPa	-40°, cast, ISO 527 [20]	5.5%
8	80 MPa	23°, cast, ISO 527 [20]	5.5%
9	40 MPa	70°, cast, ISO 527 [20]	5.5%
10	100 MPa	-40°, extruded, ISO 527 [20]	4.5%
11	72 MPa	23°, extruded, ISO 527 [20]	4.5%
12	35 MPa	70°, extruded, ISO 527 [20]	4.5%
13	55–76 MPa	impact modified, ASTM D638 [24]	2–40%
14	120 MPa	35% glass fibre [1]	
15	848 MPa	75% glass fibre, pultruded [47]	
16	1035 MPa	75% glass fibre, pultrusion/ postformed [47]	
17	1503 MPa	58% carbon fibre, pultruded [47]	
18	1435 MPa	carbon fibre, pultruded [48]	
19	1662 MPa	carbon fibre, pultruded, postformed [48]	
20	1345 MPa	Kevlar fibre, pultruded [48]	
21	1621 MPa	Kevlar fibre, pultruded, postformed [48]	
22	83.3 MPa	25° [86]	
23	65.5 MPa	50° [86]	
24	38.8 MPa	75° [86]	
25	0 MPa	100° [86]	
26	153 MPa	25°, 10% glass fibre [86]	
27	112 MPa	100°, 10% glass fibre [86]	

Flexural Strength at Break:

No.	Value	Note
1	110 MPa	[2]
2	105–116 MPa	ISO 178 [25]
3	129 MPa	2 mm min^{-1}, ISO 178 [41]
4	115 MPa	cast, ISO 178 [20]
5	105 MPa	extruded, ISO 178 [20]
6	112 MPa	cast, ASTM D695 [51]
7	120 MPa	moulding powder, ASTM D695 [51]
8	579 MPa	75% glass fibre, pultruded [47]
9	862 MPa	75% glass fibre, pultrusion, postformed [47]
10	434 MPa	58% carbon fibre, pultruded [47]
11	366 MPa	carbon fibre, pultruded [48]

– Poly(methyl methacrylate), General

12	524 MPa	carbon fibre, pultruded, postformed [48]
13	248 MPa	Kevlar fibre, pultruded [48]
14	345 MPa	Kevlar fibre, pultruded, postformed [48]

Elastic Modulus:

No.	Value	Note
1	2400–9130 MPa	Plexiglas, no applied pressure [94]
2	3300 MPa	25° [86]
3	3100 MPa	50° [86]
4	1300 MPa	75° [86]
5	0 MPa	100°, [86]

Poisson's Ratio:

No.	Value	Note
1	0.4	[24]
2	0.32	[93]
3	0.327	Plexiglas, no applied pressure [94]
4	0.36	[95]
5	0.34	cast, -25–50° [91]
6	0.45	cast and extruded grades [20]

Tensile Strength Yield:

No.	Value	Elongation	Note
1	41.4–62 MPa	2–7%	6–9 kpsi, cast, type I [96]
2	55.2–68.9 MPa	2–7%	8–10 kpsi, cast, type II [96]

Flexural Strength Yield:

No.	Value	Note
1	82.73–131 MPa	12–19 kpsi, ASTM D790 [97]
2	82.7–96.5 MPa	12–14 kpsi, cast, type I [96]
3	103.4–117.2 MPa	15–17 kpsi, cast, type II [96]

Compressive Strength:

No.	Value	Note
1	75–131 MPa	11–19 kpsi, ASTM D695 [97]
2	41–124 MPa	MEC, ASTM D695, DIN 53454, ISO R-604 [22,97]
3	110 MPa	cast, ISO 604 [20]
4	103 MPa	extruded, ISO 604 [20]
5	82.7–96.5 MPa	12–14 kpsi, 2% offset, cast, type I [96]
6	96.5–124.1 MPa	14–18 kpsi, 2% offset, cast, type II [96]

Miscellaneous Moduli:

No.	Value	Note	Type
1	1000–1500 MPa	$1.0–1.5 \times 10^{-9}$ N m^{-2} [28]	shear modulus
2	1700 MPa	cast and extruded grades, 10 Hz, ISO 527 [20]	shear modulus
3	6000 MPa	approx. 6 GN m^{-2} [98]	bulk modulus
4	5100 MPa	0.51 dynes cm^{-2} [92]	bulk modulus
5	5930 MPa	Plexiglas, no pressure [94]	bulk modulus

Complex Moduli: Compressive modulus $25.51–32.75 \times 10^{-8}$ MPa ($3.70–4.75 \times 10^{-5}$ psi, ASTM D695)[97]

Impact Strength: Instrumented falling weight impact strength 0.3–0.6 J (MEC, ISO R-6603/2) [23]. There is some loss of impact strength at low temps. but it remains high enough for use at -40°

Viscoelastic Behaviour: Creep compliance data for the atactic form have been reported [100]. Compressibility at a given pressure increases with increasing temp. [105]. Other viscoelastic props. have been reported [154]

Mechanical Properties Miscellaneous: Williams-Landel-Ferry constants for temp. superposition [100,101,102], Vogel parameters [101], entanglement MW [2,98,103] and relative entanglement moduli [104] have been reported. Failure strength (alternating bending test) 1000000 cycles [20], 40 MPa (cast, unnotched), 30 MPa (extruded, unnotched) 20 MPa (cast, notched), 10 MPa (extruded, notched)

Hardness: Rockwell M38-M102 [2,106,107], R125 [2]. Ball indentation hardness 172 MPa (17.2×10^7 N m^{-2}) [2]; 200 MPa (ISO 2039-1) [108]; 175 MPa (cast and extruded grades, ISO 2039-1) [20]. Moh hardness 2–3 [106,107]. Brinell hardness 20 [106,107]. Vickers hardness 5 [106], 17.1 (0.500" thick), 17.9 (0.125" thick), 19 (0.060" thick) [107]. Knoop hardness 16 [106], 17.7 (MW 45000), 18.6 (MW 64000), 20.4 (MW 123000), 21.3 (MW 393000), 21.8 (MW 1615000) [107]. Barcol hardness 80 [106], 87 (sheet), 85 (MEC grade 8), 80 (MEC grade 6), 77 (MEC grade 5), 56 (high impact) [107]. Shore 90 [106], 85 [2], 90 (sheet), 90 (MEC grade 8), 88 (MEC grade 6), 87 (MEC grade 5), 77 (high impact) [107]. Schleroscope 99 [106,107]. Sward rocker 46 [107]. Kohinoor pencil 9H [107]. Bierbaum scratch 16.4 [107]

Failure Properties General: Craze resistance 30–3600 s (min.) to break (MEC, ICI cantilever test, isopropanol) [23]. When the temp. is below the brittle-ductile transition point, PMMA fails by brittle fracture. Above this temp. it undergoes general yielding [109]. The presence of sorbed water at levels below 1% reduces the fatigue resistance of both cast and extruded PMMA. At levels above 1% there is little further loss of resistance. [109] For both high and low MW PMMA the strain to fracture increases to a max. and then drops to low values as the water content rises [109]. In low MW PMMA craze-dominated fracture occurs at all water content levels whereas in high MW resins this is only true for water-saturated material. [109] PMMA mouldings usually show considerable molecular orientation and it has been noted that a moulding with a high degree of 'frozen-in' orientation is stronger and tougher in the direction parallel to the orientation than in the transverse direction. When stress is applied parallel to the orientation direction the resistance to crazing is high; when stress is applied at 90° to this direction resistance to crazing is low. Large numbers of short, thin crazes are prod. in the former case and a comparatively small number of long, thick crazes in the latter [110]

Fracture Mechanical Properties: Fracture surface energy per unit area, γ, 300 J m^{-2} (3×10^5 erg cm^{-2}, M_n 3000000) [111], 200–400 J m^{-2} [112], 120 J m^{-2} (1.2×10^5 erg cm^{-2}) [113]. Fracture surface energy increases with decreasing temp. and varies with MW. The value of γ is lowered significantly by cross-linking. 80 J m^{-2} (lightly cross-linked, 80000 erg cm^{-2}) [113]; 46 J m^{-2} (cross-linked-PMMA/EGDMA, 46500 erg cm^{-2}) [114]; 23 J m^{-2} (densely cross-linked, 23000 erg cm^{-2}) [113]. Critical stress intensity factor (fracture toughness), K_c 1.1 M_n m$^{-3/2}$, 1.21 MPa m$^{1/2}$ (MW 1600000, 23°) [90], 0.7–1.6 M_n m$^{-3/2}$ [98]; 1.76 MPa m$^{1/2}$ (in water, 37°, MW 1600000), 1.21 MPa m$^{1/2}$ (air value) [90]. Because of the dependence of its K_c value on crack velocity PMMA is prone to time-dependent failure during periods of prolonged loading

Friction Abrasion and Resistance: Coefficients of friction 0.4–0.8 (self), 0.5 (steel) [1,20,115], 0.45 (steel) [1,20]. Abrasion resistance

0.025 N (Martens method, cast and extruded grades) [20]. Data for falling abrasive methods have been reported [107]

Izod Notch:

No.	Value	Notch	Note
1	15–64 J/m	Y	MEC, ASTM D256, ISO R180-4a [22]
2	16–26.7 J/m	Y	0.3–0.5 lbf in^{-1}, standard [99]
3	53.4–122.8 J/m	Y	1.0–2.3 lbf in^{-1}, high impact [99]
4	21.4 J/m	Y	0.4 ft lb in^{-1}, cast, types I and II [96]
5	16.2–32.4 J/m	Y	general purpose, cast sheet, ASTM D256 [24]
6	35–135 J/m	Y	impact modified, ASTM D256 [24]
7	2000 J/m	Y	75% glass fibre, pultruded [47]
8	2200 J/m	Y	75% glass fibre, pultruded, postformed [47]
9	1200 J/m	Y	58% carbon fibre, pultruded [47]
10	1400 J/m	Y	carbon fibre, pultruded [48]
11	1700 J/m	Y	carbon fibre, pultruded, postformed [48]
12	2400 J/m	Y	Kevlar fibre, pultruded [48]
13	2800 J/m	Y	Kevlar fibre, pultruded, postformed [48]

Izod Area:

No.	Value	Notch	Note
1	1200 kJ/m^2	Unnotched	cast sheet, ISO 179/2D
2	18 kJ/m^2	Unnotched	DIN 53453 [28]
3	16 kJ/m^2	Unnotched	cast and extruded grades, ISO 180/1A [20]
4	2000 kJ/m^2	Notched	DIN 53453 [28]
5	1200–2170 kJ/m^2	Notched	cast sheet, ISO 179/2D

Charpy:

No.	Value	Notch	Note
1	11–60	N	MEC, DIN 53453, ISO R179-2D [22]
2	11–19	N	cast, ISO R179 [41]
3	15	N	cast and extruded grades, ISO 179/1fu [20]
4	5.9	Y	0.11 ft lb in^{-1} [99]
5	2–7	Y	MEC, DIN 53453, ISO R179-2C [22]

Electrical Properties:

Electrical Properties General: PMMA has a low electrical conductivity, good arc resistance and high dielectric strength. The electrical props. are hardly affected by the absorption of water or the presence of plasticiser. PMMA is about midway in the triboelectric series for polymers [115]. The dielectric constant and dissipation factor of PMMA vary with both temp. and frequency. The effects of ion-implantation on the props. of PMMA can be quite extensive [116].

Surface/Volume Resistance:

No.	Value	Note	Type
1	1×10^{15} Ω.cm	DIN 53482 [1]	S
2	0.1×10^{15} Ω.cm	MEC, ASTM D257, DIN 53482 [22]	S
3	0.1×10^{15} Ω.cm	cast sheet, IEC 93 [25]	S
4	0.05×10^{15} Ω.cm	cast and extruded grades, DIN VDE 0303, part 3 [20]	S
5	600×10^{15} Ω.cm	cast, ASTM D257 [51]	S

Dielectric Permittivity (Constant):

No.	Value	Frequency	Note
1	2.2–3.7		ASTM D150 [2]
2	3.6	50 Hz	25° [118]
3	3.6	50 Hz	23°, cast, DIN VDE 0303, part 4 [20]
4	3.7	50 Hz	23°, extruded, DIN VDE 0303, part 4 [20]
5	3.7–3.9	50 Hz	MEC, ASTM D150, DIN 53483 [22]
6	3	1 kHz	25° [118]
7	2.7	100 kHz	23°, cast, DIN VDE 0303, part 4 [20]
8	2.8	100 kHz	23°, extruded, DIN VDE 0303, part 4 [20]
9	2.6	1 MHz	25° [118]
10	2.57	30 GHz	25° [119,120]
11	2.59	138 GHz	25° [120]

Dielectric Strength:

No.	Value	Note
1	15–20 kV.mm^{-1}	ASTM D149, DIN 53581, IEC 243 [22,25]
2	16–20 kV.mm^{-1}	ASTM D149 [24]
3	15–20 kV.mm^{-1}	impact modified, ASTM D149 [24]
4	18–22 kV.mm^{-1}	cast sheet, ASTM D149 [24]
5	30 kV.mm^{-1}	cast and extruded grades, 1 mm thick, DIN VDE 0303, part 2 [20]

Arc Resistance:

No.	Note
1	no track, ASTM D495

Strong Field Phenomena General: The dielectric strength of PMMA is high at low temps. but falls rapidly at the critical temp., T_c (approx. 20°) [121]

Static Electrification: The high surface resistance can give rise to a surface static charge but this can be prevented by the application of an anti-static surface coating [51]

Magnetic Properties: Magnetic susceptibility, c 0.59×10^6 cm^3 g^{-1} [122]. The effect of ion-implantation on the props. of PMMA can be quite extensive [116]

Poly(methyl methacrylate), General

Dissipation (Power) Factor:

No.	Value	Frequency	Note
1	0.062	50 Hz	25° [118]
2	0.06	50 Hz	23°, cast and extruded grades, DIN VDE 0303, part 4 [20]
3	0.055	1 kHz	25° [118]
4	0.02	100 kHz	23°, cast, DIN VDE 0303, part 4 [20]
5	0.03	100 kHz	23°, extruded, DIN VDE 0303, part 4 [20]
6	0.014	1 MHz	25° [118]
7	0.04	1 MHz	MEC, ASTM D150, DIN 53483 [22]
8	0.0056	30 GHz	25° [119,120]
9	0.01	138 GHz	25° [120]

Optical Properties:

Optical Properties General: Temp. coefficient of refractive index -0.001°. Extinction modulus 55000 mm^{-1} (647 nm) [124]; 50000 mm^{-1} (633 nm) [117]; 19000 mm^{-1} (514 nm) [124]; 23000 mm^{-1} (488 nm) [124]; 10000 mm^{-1} (400 nm) [117]. PMMA exhibits the Kerr effect with a Kerr constant of 0.2×10^{-14} mV^2 (25°, commercial cast grade) [125]. Stress optical coefficient -160×10^{12} Pa^{-1} [126]. Large heterogeneities in PMMA give rise to light-scattering loss. Light-scattering loss 13 dB km^{-1} (633 nm; polymer with no heterogeneities prepd. by radical polymerisation above T_g) [127,128]

Refractive Index:

No.	Value	Note
1	1.49	589 nm [2,20,22,23,129]
2	1.4831	633 nm, syndiotactic [130]
3	1.4832	633 nm, isotactic [130]
4	1.488	656 nm, n_c [129]
5	1.496	486 nm, n_f [129]

Transmission and Spectra: Nmr [12,13,14], Raman [15], ir [16], esr [123] and time-of-flight negative ion mass [17] spectral data have been reported. PMMA transmits light in the range 360–1000 nm; at a thickness of 2.54 cm or less it absorbs virtually no visible light; above 2800 nm essentially all ir radiation is absorbed. λ_{max} 213 nm (syndiotactic, extinction coefficient 170–173 M^{-1} cm^{-1}, 16.5–65°) [131]; λ_{max} 207 nm, with shoulders at 211.5 and 216.5 nm (isotactic, 17.5°) [131]. Deuteration of the monomer prior to polymerisation can extend the transmission window and improve the loss of a resin, the degree of improvement depending on the level of deuteration [132]. Deuterated PMMA has a lower optical absorption coefficient than PMMA in both the visible and the near infrared regions [133]. Water in PMMA causes some loss of transmission around 950 nm but otherwise has little effect on the absorption props. Water in deuterated PMMA, however, has a much greater effect [133]. The transparency of PMMA to X-rays is similar to that of human flesh or water. Finely-divided inorganic salts have been added to PMMA to render it opaque to X-rays but there is a detrimental effect on the mechanical props. Bismuth and barium-containing glasses have been similarly used but again the mechanical props. suffer with the polymer increasing in weight and becoming brittle. PMMA can be made radio-opaque by the incorporation of heavy metal salts during polymerisation. The presence of the salt can sometimes raise the T_g of the polymer [134]. Sheets are opaque to α-particles and for thicknesses above 6.35 nm they are essentially opaque to β-radiation. PMMA sheet may be used as a neutron stopper. Most colourless sheet formulations have high transmittance to standard broadcast and TV waves and most radar bands. Total energy transmittance 85% (cast and extruded grades, DIN 67507) [20]

Molar Refraction: Dispersion 57.8 [129]. The complex refractive index of syndiotactic and isotactic PMMA over the ir region has been calculated [130]

Total Internal Reflection: Critical angle 42° (air, 5893Å) [46,51]. Intrinsic segmental anisotropy, 2×10^{25} cm^3 (C_6H_6, atactic), 25×10^{25} cm^3 (C_6H_6, isotactic) [135]

Volume Properties/Surface Properties: Haze 20–30% (Taber method, ISO 9352, cast and extruded grades). Clarity 90–92% transmission (D1003, DIN 5036) [20,22,23]. Haze 0.4–2.5% (max., D1003, DIN 5036) [22,23]

Polymer Stability:

Upper Use Temperature:

No.	Value	Note
1	80°C	continuous use, cast sheet [25]
2	80–85°C	flat sheet [41]
3	85–100°C	[1]
4	80°C	cast [20]
5	75–80°C	thermoformings [41]
6	70°C	extruded [20]
7	65–90°C	[1]

Decomposition Details: On heating the polymer above 300° an unzipping process occurs giving monomer at >95% yield. The main prods. of combustion in a well-ventilated atm. are CO_2 and CO [136,137,138,139,140]. k_{350} 5.2% min^{-1} [46]. Heat of combustion 16600 kJ kg^{-1} [140]. Char yield 0% [98]

Flammability: Burns slowly with a blue, crackling flame and little smoke; liable to drip. Burns to completion if not extinguished. Flame-retardant grades are available. Limiting oxygen index 17.3% [2,141]. Flammability rating HB (UL 94) [25]. Ignition temp. 425° [1], 425° (cast, DIN 51794) [20], 430° (extruded, DIN 51794) [20], 350° (moulding powder) [51]

Environmental Stress: Can last up to 20 years with little appreciable loss of props. After a period of 17 years' exposure in New Mexico a sample of sheet was found to have reduced flexural strength (-51%) and strain at rupture (-66%). The material had a slightly reduced glass-transition temp. and was also more brittle, although the brittleness was attributed partly to crack initiation by surface roughness. The sheet was still transparent and the loss in transmission less than 10%. The original degree of transmission could be almost completely restored by polishing, showing that the loss was due to surface roughness rather than degradation of the polymer [142]. PMMA is more stable to photolysis in air than in a vacuum [143]. Outdoor weathering causes very slight yellowing after 1095 days (general purpose/cast sheet, ASTM D1435) and embrittlement after 730 days (ASTM D1435) [24]. Rain drops may cause ring cracks [144]. Resistance to environmental stress cracking may be improved by annealing

Chemical Stability: Resistant to oxidation at ambient temps. May be crazed by solvents and organic compounds [98]. Cast sheet may be rendered opaque by MeOH [145]. Unaffected by weak acids and alkalis. Attacked by strong acids [24]. Ozonolysis causes degradation [146]

Hydrolytic Stability: There is a marked difference between crystallisable and non-crystallisable forms. Conventional (i.e., free-radical bulk polymerised) and syndiotactic polymers hydrolyse more slowly than isotactic [147]. The config. of the polymer is unchanged by hydrol.

Biological Stability: PMMA with a small amount of N-benzyl-4-vinylpyridinium chloride in the main chain shows a reduction in MW after time in the aeration tank of a sewage works

Recyclability: May be recycled via pyrolysis at 720° when depolymerisation yields monomer at 97% or better

Stability Miscellaneous: Uv irradiation of thin films at 25° can cause a rapid decrease in MW with small amounts of volatile prods. formed [148]. The carbonyl group absorbs at 215 nm leading to chain scission and consequent degradation. Large doses of ionising radiation can cause brittleness, loss of strength and

– Poly(methyl methacrylate), General

general reduction in props. Main chain fracture occurs during irradiation; deterioration of props. is due to lowering of MW. Non-heat resistant grades distorted by boiling H$_2$O. High energy radiation causes irreversible yellowing but no change in mechanical or electrical props. γ-Irradiation causes chain rupture [149] leading to decrease in T$_g$ and intrinsic viscosity [150]

Applications/Commercial Products:
Processing & Manufacturing Routes: Manufactured by bulk (sheet, rod or tube), suspension, soln. or dispersion [152] polymerisation. Free-radical polymerisation can be initiated by AIBN or a peroxide at 100°, by bacteria and yeast [151], or by heat or microwave-inducement [158]. Stereoregular material is made by living polymerisation using metal alkyl catalysts [153]. Processed by manipulation (sheet, rod, tube) or by extrusion and injection moulding

Mould Shrinkage (%):

No.	Value	Note
1	0.4–0.7%	[23]
2	0.2–0.6%	ASTM D955 [25]

Applications: Applications in consumer goods; in medicine as dental cements/resins, hard contact lenses and in bone reconstruction. Also used in the manufacture of membranes for dialysis and ultrafiltration. Used as a positive photoresist both with uv (at 215 nm) and with an electron beam, although its sensitivity to the latter is low. Also used as a low-profile additive to eliminate polymerisation shrinkage in sheet moulding compounds and bulk moulding formulations e.g., in PVC

Trade name	Details	Supplier
Acrifix		Rohm and Haas
Acrigel DH/MP		Proquigel
Acry Sirop		Mitsubishi Chemical
Acrybase		Rohm and Haas
Acryex	CM 200 range	Chi Mei Industrial
Acrylite		CYRO Industries
Acryloid		Rohm and Haas
Acrylt	WD/WH/WM	Sumitomo Chem Co
Acrylub		Rohm and Haas
Acrypanel		Mitsubishi Chemical
Acrypet		Mitsubishi Chemical
Altuglas and Altucite		Altulor Groupe CdF, France
Aravite	CW 8730/HW 8730/CY 8740	Ciba-Geigy Corp.
Asterite		ICI Acrylics
Barex	200 range; extrusion and injection moulded	BP Chemicals
Biodrak		A. Drakopoulos, Greece
Casoglas and Casorcryl		Casolith BK, Netherlands
Conap	CE-1170	Conap Inc.
Cryolite	G-20-100/G-20-300	CYRO Industries
Degalan		Degussa
Degaplast		Degussa
Deglas		Degussa
Delpet and Delpet		Asahi Chem. Co Ltd
Dewoglas		Degussa
Diakon		ICI Acrylics
Elecroglas		Glasfix Corp., USA
Elvacite		ICI Acrylics
Eska		Mitsubishi Chemical
Exolite		CYRO Industries
Gardlite		Southern Plastics Co., USA
Kamax	T series	AtoHaas N.A.
Lacrilex		LATI Industria Thermoplastics S.p.A.
Lucite		ICI Acrylics
Lucky		Lucky Ltd.
MS	100 series; 300/600	Network Polymers
Modar	various grades	ICI Acrylics Inc.
NAS/P-605	various grades	Novacor Chemicals
NSC	A-101/A-126	Thermofil Inc.
Neocryl		Zeneca
Optix	PL range	Plaskolite Inc.
Oroglas		Rohm and Haas
PMMA	various grades	Lucky
Palapet		Kyowa Gas Chem., Japan
Paraglas		Degussa
Paraloid		Rohm and Haas
Perma Stat	1800	RTP Company
Perspex		ICI Acrylics
Piacryl		VEB Piesteritz, Germany
Plexidur		Rohm and Haas
Plexiglas		Rohm and Haas
RX9-901		Spartech
Resartglas		Resart-Ihm, Germany
Resartit		Resart-Ihm, Germany
Shinkolilthe		Mitsubishi Chemical
Sumipex		Sumitoma Chemical, Japan
Swedcast		Swedlow Inc., USA
Umaplex		Synthesia, Czechslovakia
Unilock		British Vita Co., UK
Vedril		Vedril SpA, Italy
Vestiform		Hüls AG
Vetredil		Vetril, Italy
Zylar	ST range/93 series	Novacor Chemicals

Bibliographic References
[1] Domininghaus, H., *Plastics for Engineers,* Hanser, 1993, 276, (colourings)
[2] Van Krevelen, D.W., *Properties of Polymers: Their Correlation with Chemical Structure,* Elsevier, 1976, 79, (MW)
[3] De Boer, A., Van Ekenstein, G.O.R.A. and Challa, G., *Polymer,* 1975, **16,** 930, (chromatography)
[4] Hamori, E., Forsman, W.E. and Hughes, R.E., *Macromolecules,* 1971, **4,** 193

[5] Cowie, J.M.G., *Polymers: Chemistry and Physics of Modern Materials*, 2nd edn., Blackie, 1991, 406, (tacticity)
[6] Boscher, F., Ten Brinke, G., Eshuis, A. and Challa, G., *Macromolecules*, 1982, **15**, 1364, (morphology)
[7] Kusanagi, Chatani, Y. and Takadoro, H., *Macromolecules*, 1976, **9**, 531, (morphology)
[8] Hadjichristidis, N., *Makromol. Chem.*, 1977, **178**, 1463, (morphology)
[9] Barton, A.F.M., *CRC Handbook of Polymer-Liquid Interaction Parameters and Solubility Parameters*, CRC Press, 1990, 265
[10] Ahmed, H. and Yaseen, M., *Polym. Eng. Sci.*, 1979, **19**, 858
[11] Vogel, O. and Grossman, S., *Encycl. Polym. Sci. Eng.*, 2nd edn., (ed. J.I. Kroschwitz), Wiley Interscience, 1988, **7**, 632
[12] Konishi, T., Tamai, Y., Fuji, M., Einaga, Y. and Yamakawa, H., *Polym. J. (Tokyo)*, 1989, **21**, 329, (C-13 nmr, H-1 nmr)
[13] Kitayama, T., Ute, K. and Hatada, K., *Br. Polym. J.*, 1990, **23**, 5, (synth., nmr)
[14] Kitayama, T., Nakagawa, O. and Hatada, K., *Polym. J. (Tokyo)*, 1995, **27**, 1180, (synth.)
[15] Neppel, A. and Butler, I.S., *J. Raman Spectrosc.*, 1984, **15**, 247, (Raman)
[16] Miyamoto, T. and Inagaki, H., *Polym. J. (Tokyo)*, 1970, **1**, 46, (ir)
[17] Lub, J., Van Vroonhoven, F.C.B.M., Van Leyen, D. and Benninghoven, A., *J. Polym. Sci., Part B: Polym. Phys.*, 1989, **27**, 2071, (ms)
[18] Shetter, J.A., *J. Polym. Sci., Part B: Polym. Lett.*, 1962, **1**, 209, (density)
[19] Hoff, E.A.W., Robinson, D.W. and Willbourn, A.H., *Polym. Sci.*, 1955, **18**, 161, (density)
[20] *Plexiglas Datasheet*, Rohm and Haas, 1996, (technical datasheet)
[21] Chee, K.K., *J. Appl. Polym. Sci.*, 1987, **33**, 1067, (density)
[22] *Oroglas Granules*, ICI, 1996, (technical datasheet)
[23] *Diakon Range*, ICI, 1996, (technical datasheet)
[24] Harrington, M., *Handbook of Plastic Materials and Technology*, (ed. I.I. Rubin), Wiley Interscience, 1990, 355, (thermal expansion)
[25] *Perspex Cast Extruded Sheet*, ICI, 1996, (technical datasheet)
[26] Olabisi, O. and Simba, R., *Macromolecules*, 1975, **8**, 206
[27] Spencer, R.S. and Gilmore, G.D., *J. Appl. Phys.*, 1950, **21**, 523, (eqn. of state)
[28] Progelhof, R.C. and Throne, J.L., *Polymer Engineering Principles*, Hanser, 1993, 254, (eqn. of state)
[29] Simha, R., Wilson, P.S. and Olabisi, O., *Kolloid Z. Z. Polym.*, 1973, **251**, 402, (eqn. of state)
[30] Beret, S. and Prausnitz, J.M., *Macromolecules*, 1975, **8**, 206, (eqn. of state)
[31] Quach, A., Wilson, P.S. and Simha, R., *J. Macromol. Sci., Phys.*, 1974, **9**, 533, (eqn. of state)
[32] Olabasi, O. and Simha, R., *Macromolecules*, 1975, **8**, 211, (eqn of state)
[33] Hartmann, B., Simha, R. and Berger, A.E., *J. Appl. Polym. Sci.*, 1991, **43**, 983, (eqn. of state)
[34] Sanchez, I.C. and Lacombe, R.H., *J. Polym. Sci., Polym. Lett. Ed.*, 1977, **15**, 71, (eqn. of state)
[35] Emel'yanov, D.N. and Myachev, V.A., *Plast. Massy*, 1988, **5**, 10, (thermodynamic props.)
[36] Warfield, R.W. and Petree, M.C., *J. Polym. Sci., Part A: Polym. Chem.*, 1963, **1**, 1701
[37] Barber, R.E., Chen, R.Y.S. and Frost, R.S., *J. Polym. Sci., Polym. Phys. Ed.*, 1977, **15**, 1199, (thermal conductivity)
[38] Calvert, E., Bros, J.P. and Pinelle, H., *C.R. Hebd. Seances Acad. Sci.*, 1965, **260**, 1164, (thermal conductivity)
[39] Eiermann, K., *Kolloid Z. Z. Polym.*, 1964, **198**, 5, (thermal conductivity)
[40] Baschirow, A.B., Selenew, J.W. and Aschundow, S.K., *Plaste Kautsch.*, 1976, **23**, 351, (thermal conductivity)
[41] *Perspex SW Cast Sheet*, ICI, 1996, (technical datasheet)
[42] Gaur, U., Lau, S.F., Wunderlich, B.B. and Wunderlich, B., *J. Phys. Chem. Ref. Data*, 1982, **11**, 1065, (heat capacity)
[43] Wiley, R.H. and Brauer, G.M., *J. Polym. Sci.*, 1948, **3**, 647
[44] Karey, F.E. and MacKnight, W.J., *Macromolecules*, 1968, **1**, 537
[45] Thompson, E.V., *J. Polym. Sci., Part A-2*, 1966, **4**, 199
[46] Brydson, J.A., *Plast. Mater.*, Butterworth, 1982, 371, 383
[47] Chen-Chi, M.M.A. and Chin-Hsing, C., *Polym. Eng. Sci.*, 1991, **31**, 1094, (mechanical and thermal props.)
[48] Chen-Chi, M.M.A. and Chin-Hsing, C., *J. Appl. Polym. Sci.*, 1992, **44**, 819, (mechanical props.)
[49] Lal, J. and Trick, G.S., *J. Polym. Sci., Part A: Polym. Chem.*, 1964, **2**, 4559
[50] Plazek, D.J. and Ngai, K.L., *Physical Properties of Polymers Handbook*, (ed. J.E. Mark), AIP Press, 1996, 139
[51] *Handbook of Industrial Materials*, Trade and Technical Press, 1978, 343
[52] Belnikevitch, N.G., Mrkvickova, L. and Quadrat, O., *Polymer*, 1983, **24**, 700, 713, 719, (solubility)
[53] Kusuyama, H., Miyamoto, N., Chatani, Y. and Tadokoro, H., *Polym. Commun.*, 1983, **24**, 119, 1256
[54] Lee, W.A. and Sewell, J.H., *J. Appl. Polym. Sci.*, 1968, **12**, 1397
[55] Billmeyer, F.W., *Textbook of Polymer Science*, Wiley Interscience, 1962
[56] Bloch, D.R., *Polym. Handb.*, 4th edn., (eds. J. Brandrup, E.H. Immergut and E.A. Grulke), John Wiley and Sons, 1999, **VII**, 497, (solubility parameters)
[57] Frank, C.W. and Gashgari, M.A., *Macromolecules*, 1979, **12**, 163
[58] Allen, G., Sims, D. and Wilson, G.J., *Polymer*, 1961, **2**, 375
[59] *Mod. Plast.*, 1958, **36**, 135
[60] Soc. Plast. Eng., *Tech. Pap.*, 1963, **19**, 623
[61] Bloch, D.R., *Polym. Handb.*, 4th edn., (eds. J. Brandrup, E.H. Immergut and E.A. Grulke), John Wiley and Sons, 1999, **VII**, 497
[62] Ichijima, H., Okada, T., Uyama, Y. and Ikada, Y., *Makromol. Chem.*, 1991, **192**, 1213
[63] Torstensson, M., Randby, B. and Hult, A., *Macromolecules*, 1990, **23**, 126
[64] Kawaguchi, M., Sauer, B.B. and Yu, H., *Macromolecules*, 1989, **22**, 1735
[65] Zisman, W.A., *Adv. Chem. Ser.*, 1964, **43**, 1
[66] Owen, M.J., *Physical Properties of Polymers Handbook*, (ed. J.E. Mark), AIP Press, 1996, 669
[67] Sewell, J.H., *J. Appl. Polym. Sci.*, 1973, **17**, 1741
[68] Wu, S., *J. Phys. Chem.*, 1970, **74**, 632
[69] King, R.N., Andrade, J.D., Ma, S.M., Gregonis, D.E. and Brostrom, L.R., *J. Colloid Interface Sci.*, 1985, **103**, 62, (interfacial tension)
[70] Ayme, J.C., Emery, J., Lavielle, G. and Schultz, J., *J. Mater. Sci.: Mater. Med.*, 1992, **1**, 387
[71] Wu, S., *J. Polym. Sci., Part C: Polym. Lett.*, 1971, **34**, 19
[72] *Degalan Range*, Degussa, 1997, (technical datasheet)
[73] Schoff, C.K., *Polym. Handb.*, 4th edn., (eds. J. Brandrup, E.H. Immergut and E.A. Grulke), John Wiley and Sons, 1999, **VII**, 265, (viscosity)
[74] Sawatari, N., Konishi, T., Yoshizaki, T. and Yamakawa, H., *Macromolecules*, 1995, **28**, 1089, 1095, (viscosity)
[75] Elias, H.-G., *Polym. Handb.*, 4th edn., (eds. J. Brandrup, E.H. Immergut and E.A. Grulke), John Wiley and Sons, 1989, **VII**, 291, (theta temps.)
[76] Tadmor, Z. and Gogos, C.G., *Principles of Polymer Processing*, Wiley, 1979
[77] *Thermoplastics*, Wiley, 1974
[78] Min, K.E. and Paul, D.R., *J. Polym. Sci., Polym. Phys. Ed.*, 1988, **26**, 1021, (gas permeability)
[79] Chiou, J.S. and Paul, D.R., *J. Appl. Polym. Sci.*, 1986, **32**, 2897, (gas permeability)
[80] Barker, R.E., *J. Polym. Sci.*, 1962, **58**, 553
[81] MacCullum, R.J. and Rudkin, A.L., *Eur. Polym. J.*, 1978, **14**, 655, (diffusion coefficients)
[82] Bueche, F., *J. Polym. Sci.*, 1954, **14**, 414, (liq. permeability)
[83] Kalachandra, S. and Turner, D.T., *J. Polym. Sci., Part B: Polym. Phys.*, 1987, **25**, 697
[84] Kalachandra, S. and Kusy, R.P., *Polymer*, 1991, **32**, 2428, (liq. permeability)
[85] Lechner, M.D. Nordmeier, L. and Steineiger, D.G., *Polym. Handb.*, 4th edn., (eds. J. Brandrup, E.H. Immergut and E.A. Grulke), John Wiley and Sons, 1999, **VII**, 85, (liq. permeability)
[86] Lin, H., Day, D.E. and Stoffer, J.O., *Polym. Eng. Sci.*, 1992, **32**, 5, (mechanical props.)
[87] Kalachandra, S., Kusy, R.P., Wilson, T.W., Shin, I.D. and Stejskal, E.O., *J. Mater. Sci.: Mater. Med.*, 1993, **4**, 509
[88] Buckley, C.A., Lautenschlager, E.P. and Gilbert, J.L., *J. Appl. Polym. Sci.*, 1992, **44**, 1321
[89] Froelich, D, *PhD Thesis*, University of Strasbourg, 1966
[90] Johnson, J.A. and Jones, D.W., *J. Mater. Sci.*, 1994, **29**, 870
[91] Aly El-Sayed, A.-E.-E., Takeda, N. and Takahashi, K., *J. Phys. D: Appl. Phys.*, 1989, **22**, 687
[92] Warfield, R.W., Cueves, J.E. and Barnet, F.R., *J. Appl. Polym. Sci.*, 1968, **12**, 1147
[93] Kusy, R.P. and Turner, D.T., *Polymer*, 1974, **15**, 394
[94] Weischaupt, K., Krebecek, H. and Pietalle, M., *Polymer*, 1995, **36**, 3267
[95] Sandor, B.I., *Strength Mater. (Engl. Transl.)*, Prentice-Hall, 1978, 412
[96] *The Materials Selector*, 2nd edn., (eds. N.A. Waterman and M.F. Ashby), Chapman and Hall, 1997
[97] *Guide to Plastics, Property and Specification Charts*, (ed. W.A. Kaplan), McGraw-Hill,
[98] Mills, N.J., *Plastics Microstructure and Engineering Applications*, 2nd edn., Edward Arnold, 1993, 38
[99] *Encycl. Polym. Sci. Technol.*, (eds. N.F. Mark, N.G. Gaylord and N.M. Bikales), Wiley Interscience, 1971, 589, 607, (impact strength)
[100] Ferry, J.D., *Viscoelastic Properties of Polymers*, 3rd edn., Wiley, 1980, 330
[101] Masuda, T., Toda, N., Aota, Y. and Onogi, S., *Polym. J. (Tokyo)*, 1972, **3**, 315, (viscoelastic props.)
[102] Ngai, K.L. and Plazek, D.J., *Physical Properties of Polymers Handbook*, (ed. J.E. Mark), AIP Press, 1996, 341, (viscoelastic props.)
[103] Wu, S. and Beckerbauer, *Polym. J. (Tokyo)*, 1992, **24**, 1437, (viscoelastic props.)

[104] Osaki, K., Takatori, E., Watenabe, H. and Kotaka, T., *Rheol. Acta,* 1993, **32**, 132, (viscoelastic props.)
[105] Shtarkmann, B.P., Monich, J.M. and Arzhakov, S.A., *Polym. Sci. USSR (Engl. Transl.),* 1976, **18**, 1206, (compressibility)
[106] Crawford, R.J., *Plast. Eng. (N.Y.),* Pergamon Press, 1981
[107] *Test. Polym.,* (ed. J.V. Schmitz), Wiley Interscience, 1965, **2**
[108] Oberbach, K., *Kunststoffe,* 1989, **79**, 713
[109] Chen, C.C., Shen, J. and Sauer, J.A., *Polymer,* 1985, **26**, 89, 511
[110] Beardmore, P. and Rabinowitz, S., *J. Mater. Sci.,* 1975, **10**, 1763
[111] Berry, J.P., *J. Polym. Sci.,* 1961, **50**, 107
[112] Young, R.J. and Lovell, P.A., *Introduction to Polymers,* 2nd edn., Chapman and Hall, 1991, 400
[113] Broughton, L.J. and McGarry, F.J., *J. Appl. Polym. Sci.,* 1965, **9**, 609
[114] Berry, J.P., *J. Polym. Sci., Part A-1,* 1963, **1**, 993
[115] Van Krevelen, D.W. and Hoftyzer, P.J., *Properties of Polymers: Their Correlation with Chemical Structure,* 1972
[116] Jenekhe, S.A., *Encycl. Polym. Sci. Eng.,* Wiley Interscience, 1988, 352
[117] Schreyer, G., *Konstr. Kunstst.,* (ed. G. Schreyer), Hanser, 1972
[118] Schreyer, G., *Kunststoffe,* 1965, **55**, 771
[119] Amrhein, G., *Kolloid Z. Z. Polym.,* 1967, **216–217**, 38
[120] Zeil, W., Sistig, R., Frank, W. and Hoffman, V., *Ber. Bunsen-Ges. Phys. Chem.,* 1970, **74**, 883
[121] Oakes, W.G., *Proc. Inst. Electr. Eng.,* 1949, **96**, 37
[122] Bedwell, M.E., *J. Chem. Soc.,* 1947, 1350
[123] Tsay, F-D. and Gupta, A., *J. Polym. Sci., Part B: Polym. Phys.,* 1987, **25**, 855, (esr)
[124] Crist, B., Marhic, M.E., Raviv, G. and Epstein, M., *J. Appl. Phys.,* 1980, **51**, 1160, (extinction modulus)
[125] Kim, K.S., Cheng, T.C. and Cooper, D.E., *J. Appl. Phys.,* 1983, **54**, 449, (Kerr effect)
[126] Dibbs, M.G., *Encycl. Polym. Sci. Eng.,* 1988, **16**, 151
[127] Tanio, N., Koike, Y. and Ohtsuka, Y., *Polym. J. (Tokyo),* 1989, **21**, 119, (light-scattering)
[128] Koike, Y., Tanio, N. and Ohtsuka, Y, *Macromolecules,* 1989, **22**, 1367, (light-scattering)
[129] Almand, P. and Byrd, R., *Mater. Eng. (Cleveland),* 1972, **76**, 42, (refractive index)
[130] Pacansky, J., England, C. and Waltman, R.J., *J. Polym. Sci., Part B: Polym. Phys.,* 1987, **25**, 901, (refractive index)
[131] D'Alagri, M., De Santis, P. Liquori, A.M. and Savino, M., *J. Polym. Sci., Polym. Lett. Ed.,* 1964, **2**, 925, (uv)
[132] Kaius, T., Jinguji, K. and Nara, S., *Appl. Phys. Lett.,* 1982, **41**, 802
[133] Avakian, P., Hsu, W.Y., Meakin, P. and Snyder, H.L., *J. Polym. Sci., Polym. Phys. Ed.,* 1984, **22**, 1607
[134] Cabasso, I., Smid, J. and Sahni, S.K., *J. Appl. Polym. Sci.,* 1989, **38**, 1653
[135] Tsetkov, V.N. and Boitzkova, N., *Vysokomol. Soedin., Ser. A,* 1960, **3**, 1176
[136] Zemany, P.D., *Nature (London),* 1953, **171**, 391, (thermal degradation)
[137] Straus, S. and Madorsky, S.L., *J. Res. Natl. Bur. Stand., Sect. A,* 1953, **50**, 165, (pyrolysis)
[138] Cowley, P.R.E.J. and Melville, H.W., *Proc. R. Soc. London,* 1952, **A210**, 461, (photodegradation)
[139] Grassie, N., Scotney, A. and Makinson, L., *J. Polym. Sci., Polym. Chem. Ed.,* 1977, **15**, 251, (photodegradation)
[140] Tewarson, A., *Physical Properties of Polymers Handbook,* (ed. J.E. Mark), AIP Press, 1996, 577, (flammability)
[141] Vaccari, J.A., *Mater. Eng. (Cleveland),* 1977, **85**, 31, (flammability)
[142] Rainhart, L.G. and Schimmel, J., *Effect of Ageing on Acrylic Sheet,* Inter-Solar Energy, Soc. Paper, 1974
[143] Doležel, B., *Die Beständigkeit von Kunststoffen und Gummi,* Carl Hanser, 1978, 209
[144] *Ind. Photogr.,* 1960, **9**, 23
[145] Lin, C.B., Liu, K.S. and Lee, S., *J. Polym. Sci., Part B: Polym. Phys.,* 1991, **29**, 1457, (stability)
[146] Barnard, D., *J. Polym. Sci.,* 1956, **22**, 213, (ozonolysis)
[147] *U.S. Pat.,* 1962, 3 029 228
[148] Fox, R.B., Isaacs, L.G. and Stokes, S., *J. Polym. Sci., Part A: Polym. Chem.,* 1963, **1**, 1079, (uv stability)
[149] Schultz, A.R., *J. Polym. Sci.,* 1959, **35**, 369, (irradiation)
[150] Lin, C.B. and Lee, S., *J. Appl. Polym. Sci.,* 1992, **44**, 2213, (irradiation)
[151] Imoto, M., Ouchi, T., Inaba, M., Tokuyama, T. el al, *Polym. J. (Tokyo),* 1981, **13**, 105, (synth)
[152] Maury, E.E. and DeSimone, J.M., *Abstr. Pap.-Am. Chem. Soc.,* Poly 490, (synth.)
[153] Kityama, T., Ute, K. and Hatada, K., *Br. Polym. J.,* 1990, **23**, 5, (synth.)
[154] Aharoni, S.M., *Macromolecules,* 1983, **16**, 1722, (viscoelastic props.)
[155] Henderson, J.A., Richards, R.W., Penfold, J., Shackleton, C. and Thomas, R.K., *Polymer,* 1991, **32**, 3284
[156] Henderson, J.A., Richards, R.W., Penfold, J. and Thomas, R.K., *Macromolecules,* 1993, **26**, 65
[157] Henderson, J.A., Richards, R.W., Penfold, J. and Thomas, R.K., *Acta Polym.,* 1993, **44**, 184
[158] Albert, P. *et al, Acta Polym.,* 1996, **47**, 74
[159] Albert, P., *et al, Acta Polym.,* 1996, **47**, 74
[160] Henderson, J.A., *et al, Polymer,* 1991, **32**, 3284
[161] Henderson, J.A., *et al, Macromolecules,* 1993, **26**, 65
[162] Henderson, J.A., *et al, Macromolecules,* 1993, **44**, 184

Poly(methyl methacrylate-co-itaconic acid) — P-382

Related Polymers: Polymethacrylates, General, Poly(methyl methacrylate)
Monomers: Methyl methacrylate, Itaconic acid
Material class: Thermoplastic
Polymer Type: acrylic copolymers
Molecular Formula: $[(C_5H_8O_2)_x \, (C_5H_7O_4)_y]_n$
Fragments: $C_5H_8O_2$ $C_5H_7O_4$

Surface Properties & Solubility:
Solvents/Non-solvents: Sol. MeOH, EtOH, THF; insol. H_2O, cyclohexanone, 2-butanone, $CHCl_3$, CCl_4, octanol [1]

Polymer Stability:
Stability Miscellaneous: Irradiation causes decrease in MW [1]

Bibliographic References
[1] Parsonage, E.E. and Peppas, N.A., *Br. Polym. J.,* 1987, **19**, 469

Poly(methyl methacrylate-co-itaconic anhydride) — P-383

Related Polymers: Polymethacrylates, General, Poly(methyl methacrylate)
Monomers: Methyl methacrylate, Itaconic anhydride
Material class: Thermoplastic
Polymer Type: acrylic copolymers
Molecular Formula: $[(C_5H_8O_2)_x \, (C_5H_4O_3)_y]_n$
Fragments: $C_5H_8O_2$ $C_5H_4O_3$

Volumetric & Calorimetric Properties:
Glass Transition Temperature:

No.	Value	Note
1	<170°C	max. [1]

Transition Temperature General: T_g increases with increasing itaconic anhydride content [1]

Polymer Stability:
Thermal Stability General: Thermal stability increases with decreasing itaconic anhydride content [1]. Heating at 170° causes cross-linking [1]

Applications/Commercial Products:
Processing & Manufacturing Routes: Synth. by free-radical polymerisation [1]
Applications: Applications in electron beam resists

Bibliographic References
[1] Miles, A.F. and Cowie, J.M.G., *Eur. Polym. J.,* 1991, **27**, 165

Poly(methyl methacrylate-co-methyl itaconate)

P-384

$$\left[\left[CH_2C(CH_3) \right]_x \left[CH_2C \right]_y \right]_n$$
$$\quad\quad\quad\quad\; O\quad OMe \quad\quad CH_2COOMe$$
$$\quad\quad\quad\quad\quad\quad\quad\quad\quad\quad COOH$$

Related Polymers: Polymethacrylates, General, Poly(methyl methacrylate)
Monomers: Methyl methacrylate, Methyl itaconate
Material class: Thermoplastic
Polymer Type: acrylic copolymers
Molecular Formula: $[(C_5H_8O_2)_x\,(C_6H_8O_4)_y]_n$
Fragments: $C_5H_8O_2$ $C_6H_8O_4$

Volumetric & Calorimetric Properties:
Transition Temperature General: T_g increases with increasing methyl itaconate content. Precise measurement is not possible owing to cross-linking [1]

Polymer Stability:
Thermal Stability General: Thermal stability decreases as the methyl itaconate content increases [1]

Applications/Commercial Products:
Processing & Manufacturing Routes: Synth. by free-radical polymerisation [1]
Applications: Applications in electron beam resists

Bibliographic References
[1] Miles, A.F. and Cowie, J.M.G., *Eur. Polym. J.*, 1991, **27**, 165

Poly(methyl methacrylate), Moulding and extrusion compounds

P-385

Related Polymers: Methacrylates General, Poly(methyl methacrylate) General
Monomers: Methyl methacrylate
Material class: Thermoplastic
Polymer Type: acrylic
Molecular Formula: $(C_5H_8O_2)_n$
Fragments: $C_5H_8O_2$
Molecular Weight: M_n 100000–200000
General Information: Compounds are classified according to ASTM standard D788. This comprises three groups (unmodified, impact modified and heat resistance modified). Each group is subdivided into a number of classes, which in turn are divided into grades (grade 1: general purpose; grade 2: uv transmitting; grade 3: uv stabilised; grade 4: impact modified). Low levels of acrylate comonomers may be incorporated to modify the melt flow and thermal stability props. Available in granule or bead (600 μm) form with a full range of transparent, translucent and opaque colours

Volumetric & Calorimetric Properties:
Density:

No.	Value	Note
1	d 1.15–1.19 g/cm^3	ASTM D792, DIN 53479, ISO R1183 [1,2,3]
2	d 1.18–1.19 g/cm^3	ASTM D788 [4]
3	d 1.12–1.16 g/cm^3	high impact grade [3,4]
4	d 1.02–1.1 g/cm^3	modified [4]
5	d^{23} 0.67 g/cm^3	DIN 53466 [3]

Thermal Expansion Coefficient:

No.	Value	Note	Type
1	6.5×10^{-5}–0.0001 K^{-1}	ASTM D696, DIN 53752A [1]	L
2	3×10^{-5}–4×10^{-5} K^{-1}	ASTM D788 [4]	L
3	4×10^{-5}–6×10^{-5} K^{-1}	high impact grade [4]	L
4	4.4×10^{-5}–4.5×10^{-5} K^{-1}	modified [4]	L
5	7×10^{-5} K^{-1}	0–50°, ASTM D788, DIN 52612, VDE 0304 [3,5]	L
6	9.5×10^{-5}–0.00011 K^{-1}	impact modified, DIN 52612 [3]	L

Thermal Conductivity:

No.	Value	Note
1	0.21 W/mK	1.44 Btu (h ft^2 °F in)$^{-1}$, high impact, ASTM D788 [4]
2	0.23 W/mK	1.56 Btu (h ft^2 °F in)$^{-1}$, modified [4]
3	0.19 W/mK	ASTM D788, VDE 0304/1 [5]
4	0.18 W/mK	[3]

Specific Heat Capacity:

No.	Value	Note	Type
1	1.5 kJ/kg.C	ISO 1006 [3]	P

Deflection Temperature:

No.	Value	Note
1	74–121°C	166–250°F, 1.8 MPa, general purpose [4]
2	74°C	1.82 MPa, general purpose, ASTM D648 [6]
3	75–98°C	1.8 MPa, ISO 75 [3]
4	80–103°C	0.45 MPa, ISO 75 [3]
5	80°C	1.82 MPa, general purpose, ASTM D648 [6]
6	90–102°C	1.82 MPa, general purpose, ASTM D648 [6]
7	87–114°C	0.46 MPa, ISO 75 [7]
8	82–109°C	1.82 MPa, ISO 75 [7]
9	88–102°C	ISO 75A [2]
10	91–106°C	ISO 75B [2]
11	91°C	195°F, 1.8 MPa, modified [4]
12	75–93°C	1.8 MPa, impact modified, ISO 75 [3]
13	80–98°C	0.45 MPa, impact modified, ISO 75 [3]
14	76–96°C	169–205°F, 1.8 MPa, high impact grade [4]
15	68–92°C	1.82 MPa, high impact grade, ISO 75 [7]
16	76–106°C	1.82 MPa [1,2,8]
17	74°C	1.85 MPa, general purpose, ASTM D788, DIN 525, DIN 53461 [5]
18	82°C	1.85 MPa, general purpose, ASTM D788, DIN 526, DIN 53461 [5]
19	90°C	1.85 MPa, DIN 527, DIN 53461 [5]
20	98°C	1.85 MPa, general purpose, ASTM D788, DIN 528, DIN 53461 [5]

— Poly(methyl methacrylate), Moulding and extrusion compounds

21	79°C	0.46 MPa, general purpose, ASTM D788, DIN 525, DIN 53461 [5]
22	87°C	0.46 MPa, general purpose, ASTM D788, DIN 526, DIN 53461 [5]
23	95°C	0.46 MPa, DIN 527, DIN 53461 [5]
24	103°C	0.46 MPa, general purpose, ASTM D788, DIN 528, DIN 53461 [5]
25	69–79°C	1.82 MPa, ASTM D788
26	80–86°C	1.82 MPa, ASTM D788
27	87°C	1.82 MPa, ASTM D788

Vicat Softening Point:

No.	Value	Note
1	80–118°C	ASTM D1525, DIN 53460, ISO R306 [1,2,8]
2	109–112°C	[8]
3	86°C	general purpose, ASTM D788, DIN 525, DIN 53460 [5]
4	94°C	general purpose, ASTM D788, DIN 526, DIN 53460 [5]
5	102°C	DIN 527, DIN 53460 [5]
6	110°C	general purpose, DIN 528, DIN 53460 [5]
7	75–98°C	ISO 306 [3]
8	76–106°C	high impact grade, DIN 53460 [7]
9	98–106°C	impact modified, ISO 306 [3]

Surface Properties & Solubility:
Plasticisers: Plasticisers are sometimes used in moulding compositions, e.g. to enhance the melt flow. Levels vary from 1–50% although 5% is most common. Phthalates (e.g. dimethyl, diethyl, dibutyl, butyl benzyl), sebacates (e.g. dibutyl, diethylhexyl) and phosphates are some of the more common plasticisers

Transport Properties:
Melt Flow Index:

No.	Value	Note
1	0.9–3 g/10 min	standard grades, 230°, 3.8 kg, DIN 53735 [7]
2	24 g/10 min	grade 5, ASTM D1238 [6]
3	17 g/10 min	grade 6, ASTM D1238 [6]
4	2–8 g/10 min	grade 8, ASTM D1238 [6]
5	6 g/10 min	low level acrylic-base impact modification [6]
6	3.2–9 g/10 min	medium level acrylic-base impact modification [6]
7	1–3.5 g/10 min	high level acrylic-base impact modification [6]
8	0.1–2.6 g/10 min	high impact grade, 230°, 3.8 kg, DIN 53735 [7]

Permeability of Gases: [H_2O] 0.8–1 g m^{-2} d^{-1} (DIN 53122) [3], 0.8–1.1 (impact modified, DIN 53122) [3]
Water Content: 0.3–0.36% (ASTM D570, DIN 53495, ISO R62) [1,2]
Water Absorption:

No.	Value	Note
1	0.3–0.4 %	24h, general purpose grades, ASTM D788 [4]
2	0.2–0.3 %	24h, high impact grade [4]
3	0.3 %	24h, modified [4]

Mechanical Properties:
Tensile (Young's) Modulus:

No.	Value	Note
1	2930 MPa	425 kpsi, general purpose grades, ASTM D788 [4]
2	1750–2200 MPa	280 kpsi, high impact grade [3,4]
3	2758 MPa	400 kpsi, modified [4]

Flexural Modulus:

No.	Value	Note
1	2900–3400 MPa	ASTM D790, DIN 53452, ISO R178 [1,2]
2	2930 MPa	425 kpsi, general purpose grades, ASTM D788 [4]
3	2137 MPa	310 kpsi, high impact grade [4]
4	2622 MPa	low level acrylic-base impact modification, ASTM D790 [6]
5	2484 MPa	medium level acrylic-base impact modification, ASTM D790 [6]
6	1863 MPa	high level acrylic-base impact modification, ASTM D790 [6]
7	2250 MPa	375 kpsi, modified [4]

Tensile Strength Break:

No.	Value	Note	Elongation
1	38–85 MPa	ASTM D638, DIN 53455, ISO R257 [1,2]	4–40%
2	64 MPa	ASTM D788, DIN 525, DIN 53455 [5]	2.5%
3	68 MPa	ASTM D788, DIN 526, DIN 53455 [5]	3–4%
4	72 MPa	DIN 527, DIN 53455 [5]	3.5–4.5%
5	75 MPa	ASTM D788, DIN 528, DIN 53455 [5]	3.5–4.5%
6	45–60 MPa	high impact grade, ASTM D638 [3,10]	
7	59 MPa	low level acrylic-base impact modification, ASTM D638 [6]	
8	47 MPa	medium level acrylic-base impact modification, ASTM D638 [6]	
9	38 MPa	high level acrylic-base impact modification, ASTM D638 [6]	

Flexural Strength at Break:

No.	Value	Note
1	120 MPa	ASTM D695 [5]

Compressive Modulus:

No.	Value	Note
1	2551–3171 MPa	370–460 kpsi, ASTM D695 [9]

– Poly(methyl methacrylate), Moulding and extrusion compounds

Elastic Modulus:

No.	Value	Note
1	1700 MPa	1700 N mm^{-2}, approx. 10 c/s, general purpose grades, ASTM D788, DIN 525-8, DIN 53445 [5]

Tensile Strength Yield:

No.	Value	Elongation	Note
1	65.5–72.4 MPa	3–5%	9.5–10.5 kpsi, general purpose grades, ASTM D788 [4]
2	37.9–55.2 MPa	>25%	5.5–8 kpsi, high impact grade [4]
3	48.3–55.2 MPa	12–30%	7–8 kpsi, modified [4]

Flexural Strength Yield:

No.	Value	Note
1	62–131 MPa	ASTM D790, DIN 53452, ISO R178 [1,2]
2	103.4–110.3 MPa	15–16 kpsi, general purpose grades, ASTM D788 [4]
3	60–92 MPa	8.7–12 kpsi, high impact grade [3,4]
4	75.8–89.6 MPa	11–13 kpsi, modified [4]
5	100 MPa	ASTM D788, DIN 525, DIN 53452 [5]
6	110 MPa	ASTM D788, DIN 526, DIN 53452 [5]
7	120 MPa	DIN 527, DIN 53452 [5]
8	130 MPa	ASTM D788, DIN 528, DIN 53452 [5]

Compressive Strength:

No.	Value	Note
1	41–124 MPa	ASTM D695, DIN 53454, ISO R604 [1,9]
2	100–117.2 MPa	14.5–17 psi, 2% offset, general purpose, ASTM D788 [4]
3	50.3–82.7 MPa	7.3–12 psi, 2% offset, high impact grade [4]
4	65.5–79.3 MPa	9.5–11.5 psi, 2% offset, modified [4]
5	97 MPa	ASTM D788, DIN 525, DIN 53454 [5]
6	100 MPa	ASTM D788, DIN 526, DIN 53454 [5]
7	103 MPa	DIN 527, DIN 53454 [5]
8	107 MPa	ASTM D788, DIN 528, DIN 53454 [5]

Miscellaneous Moduli:

No.	Value	Note	Type
1	1400 MPa	ISO 899, DIN 53444 [3]	tensile creep modulus

Impact Strength: 2 kJ m^{-2} (notched, 23°, ISO 180/1A) [3]; 2.8–5.5 kJ m^{-2} (23°, impact modified, ISO 180/1A) [3] Falling weight impact strength 0.3–0.6 J (ISO R6603/2) [2]
Hardness: Rockwell M38-M99 (ASTM D785, ISO R2039-2) [1,2], M80-M103 (ASTM D788) [4], M84 (ASTM D788, ASTM D785) [6], M89 (ASTM D788, ASTM D785) [6], M97 (ASTM D788, ASTM D785) [6], M81 (low level acrylic-base impact modification, ASTM D785) [6], M65-M68 (medium level acrylic-base impact modification, ASTM D785) [6], M38–M45 (high level acrylic-base impact modification, ASTM D785) [6], M45–M68 (modified) [4], L60-L94 (high impact grade) [4], R121, S118, L109, M92, P68 (general purpose grade, ASTM D788) [11], R120, S112, L99, M76, P46 (general purpose grade, ASTM D788) [11]. Ball indentation 75–190 MPa (ISO R2039-2, H961/30) [2], 170–197 MPa (23°, ISO 2039-1) [3], 83–125 MPa (23°, impact modified, ISO 2039-1) [3]. Barcol (GYZJ935) 77–85 (general purpose grades, ASTM D788) [11]. Shore 87–90 (durometer, general purpose grades, ASTM D788) [11]
Failure Properties General: Craze resistance 30–3600 s (min) to break (isopropanol, ICI cantilever test) [2]
Friction Abrasion and Resistance: Martens scratch resistance 0.025 N (standard grade) [7]

Izod Notch:

No.	Value	Notch	Note
1	15–64 J/m	Y	ASTM D256, ISO R180-4a [1]
2	10.7–21.4 J/m	Y	0.2–0.4 ft lb in^{-1}, ASTM D788 [4]
3	42.8–123.1 J/m	Y	0.8–2.3 ft lb in^{-1}, high impact grade [4]
4	21.3 J/m	Y	low level acrylic-base impact modification, ASTM D256 [6]
5	32 J/m	Y	medium level acrylic-base impact modification, ASTM D256 [6]
6	53.4 J/m	Y	high level acrylic-base impact modification, ASTM D256 [6]
7	53.5–107 J/m	Y	1–2 ft lb in^{-1}, modified [4]

Izod Area:

No.	Value	Notch	Note
1	15 kJ/m^2	Y	23°C, ISO 180/1C [3]

Charpy:

No.	Value	Notch	Note
1	11	N	23°, ASTM D788 [5]
2	18	N	30°, ASTM D788 [5]
3	20	N	23°, ISO 179 [3]
4	2–3.6	N	23°, ASTM D788 [3,5,7]

Electrical Properties:
Surface/Volume Resistance:

No.	Value	Note	Type
6	>0.1 × 10^{15} Ω.cm	min., ASTM D257, DIN 53482 [1]	S
7	0.1 × 10^{15} Ω.cm	ISO 93, DIN 0303-30 [3]	S

Dielectric Permittivity (Constant):

No.	Value	Frequency	Note
1	3.7–3.9	50 Hz	ASTM D150, DIN 53483 [1]
2	3.5–3.9	60 Hz	general purpose and high impact grades, ASTM D788 [4]
3	2.7–2.9	1 MHz	general purpose grades, ASTM D788 [4]
4	2.5–3	1 MHz	high impact grade [4]
5	2.78–2.86	1 MHz	modified [4]
6	2.9–3.2	100 Hz–1 MHz	ISO 250, DIN 0303-4 [3]

− Poly(methyl methacrylate), Moulding and extrusion compounds

Dielectric Strength:

No.	Value	Note
1	15–19.7 kV.mm^{-1}	15–19.7 MV m^{-1}, ASTM D149, DIN 53581 [1]
2	0.4 kV.mm^{-1}	general purpose grades, ASTM D788 [4]
3	0.4–0.5 kV.mm^{-1}	high impact grade [4]
4	0.4–0.5 kV.mm^{-1}	short time, 1/8" thick, ASTM D149 [9]
5	0.35–0.4 kV.mm^{-1}	step-by-step, 1/8" thick, ASTM D149 [9]
6	60 kV.mm^{-1}	ISO 243/1, DIN 0303-21 [3]

Arc Resistance:

No.	Note
1	no track [4]

Complex Permittivity and Electroactive Polymers: Comparative tracking index 600-0 (ISO 112/A, DIN 0303-1) [3], 600 M-0.1 (ISO 112/A, DIN 0303-1) [3]

Dissipation (Power) Factor:

No.	Value	Frequency	Note
1	0.04	1 MHz	ASTM D150, DIN 53483 [1]
2	0.04–0.06	60 Hz	general purpose grades, ASTM D788 [4]
3	0.03–0.04	60 Hz	high impact grade [4]
4	0.026–0.029	60 Hz	modified [4]
5	0.02–0.03	1 MHz	general purpose grades, ASTM D788 [4]
6	0.01–0.02	1 MHz	high impact grade [4]
7	0.022–0.025	1 MHz	modified [4]
8	0.03–0.04	100 Hz–1 MHz	ISO 250, DIN 0303-4 [3]

Optical Properties:
Refractive Index:

No.	Value	Note
1	1.492	ASTM D542, DIN 53491, ISO R489 [1,2,3]

Transmission and Spectra: Transmittance 90–92% (ASTM D1003, DIN 5036) [1,2,3]
Volume Properties/Surface Properties: Haze 0.4–2.5% (max., ASTM D1003, DIN 5036) [1,2,3], 1 (max., ISO D1003) [3], 2 (max., impact modified, ISO D1003) [3]

Polymer Stability:
Upper Use Temperature:

No.	Value	Note
1	68–88°C	155–190°F, no load, ASTM D788 [4]
2	71°C	160°F, no load, modified [3]
3	74°C	no load, general purpose, ASTM D788, DIN 525 [5]
4	82°C	no load, general purpose, ASTM D788, DIN 526 [5]
5	90°C	no load, DIN 527 [5]
6	98°C	no load, general purpose, ASTM D788, DIN 528 [5]
7	75–98°C	[3]
8	75–93°C	impact modified [3]

Flammability: Flammability rating HB (UL94) [1,2,3]; BH 3–30 mm min^{-1} (ISO 707) [3], FH 3–35 mm min^{-1} (ISO 707) [3]. Glow wire test 650° (IEC 695-2-1) [2]. Ignition point 350° (moulding powder) [7]

Applications/Commercial Products:
Processing & Manufacturing Routes: The MW of most bulk-polymerised PMMA is too high to allow satisfactory flow props. for it to be used in injection moulding, extrusion, blow moulding and coating applications. Although it is possible to greatly reduce the MW by mechanical scission by rolling on a two-roll mill, moulding/extrusion compounds are usually manufactured by suspension polymerisation. Suspension polymerisation is usually carried out in a comparatively short time (1h max.) as there is no serious exotherm problem. A catalyst (e.g., peroxide) is dissolved in the monomer and the soln. dispersed in approx. two parts water along with various additives. Typical additives include suspending agents (aluminium oxide, talc and magnesium carbonate); protective colloids e.g., PVA and sodium polymethacrylate; buffering agents e.g., sodium hydrogen phosphate, chain transfer agents (to control MW) (lauryl mercaptan, trichloroethylene); lubricants (stearic acid); emulsifier (sodium lauryl sulfate). The size and shape of the beads formed during suspension polymer-isation can be difficult to control. The size of the beads (usually between 50 and 1000 µm) is determined by the size and shape of the reactor, the type and rate of agitation and the nature of the additives. Processed by standard injection or compression moulding, extrusion and thermoforming. Regrind can be used at high ratios in most applications but must be dried if used immediately. If a properly designed, vented barrel is used PMMA can generally be injection-moulded or extruded without drying. Good temp., pressure and speed control are necessary for consistent high quality products. Moisture levels should be below 0.3% before extruding and below 0.1% before moulding. Products may require annealing to improve stress crack resistance if they are to come into contact with solvents or are to be load-bearing

Mould Shrinkage (%):

No.	Value	Note
1	0.3–0.8%	[3]
2	0.5–0.8%	impact modified [3]
3	0.2–0.8%	ASTM D955 [1,2,8]
4	0.1–0.4%	[9]

Applications: Has applications in building components (e.g., panels, frames etc.); glazing and display fittings; automotive industry (lights, number plates, spoilers and instrumentation); road lighting, domestic lighting; domestic equipment (e.g., bowls, glasses, furniture etc); bathroom fittings; medical products (e.g. diagnostic test-tubes, optical lenses etc); engineering components (e.g. photocopier parts); fashion items (e.g. shoe heels, costume jewellery)

Trade name	Details	Supplier
Diakon	various grades	ICI, UK
Lucryl		BASF
Oroglas	various grades	Atohaas

Bibliographic References
[1] *Oroglas Granules*, ICI, 1996, (technical datasheet)
[2] *Diakon Range*, ICI, 1996, (technical datasheet)
[3] *Lucryl Range*, BASF, 1997, (technical datasheet)
[4] *The Materials Selector*, 2nd edn., (eds. N.A. Waterman and M.F. Ashby), Chapman and Hall, 1997
[5] *Handbook of Industrial Materials,* Trade and Technical Press, 1978, 343

[6] Cassidy, R.T., *Engineered Materials Handbook,* ASM International, 1988, **2**, 103
[7] *Handbook of Industrial Materials,* 2nd edn., Elsevier Advanced Technology, 1992, 393
[8] Brydson, J.A., *Plast. Mater.,* Butterworth, 1982, 371
[9] Guide to Plastics, *Property and Specification Charts,* (ed. W.A. Kaplan), McGraw-Hill,
[10] *CRC Materials Science and Engineering Handbook,* 2nd edn., (eds. J.F. Shackelford, W. Alexander and J.S. Park), CRC Press, 1994, **515**
[11] *Test. Polym.,* (ed. J.V. Schmitz), Wiley Interscience, 1965, **2**
[12] Quach, A., Wilson, P.S. and Simha, R., *J. Macromol. Sci., Phys.,* 1974, **9**, 533

Poly(methyl methacrylate-co-vinyl alcohol) P-386

$$-[CH_2C(CH_3)]_x-[CH_2CH]_y-_n$$
$$\quad\quad |\quad\quad\quad\quad\quad\quad |$$
$$\quad\; C(=O)OMe\quad\quad OH$$

Related Polymers: Polymethacrylates, General, Poly(methyl methacrylate), Poly(vinyl alcohol)
Monomers: Methyl methacrylate, Vinyl alcohol
Material class: Thermoplastic
Polymer Type: acrylic copolymers
Molecular Formula: $[(C_5H_8O_2)_x (C_2H_4O)_y]_n$
Fragments: $C_5H_8O_2$ C_2H_4O
Morphology: Graft copolymers exhibit lamellar microphase morphology. [1] Various methods of synth. all produce copolymers with the same struct. [2]
Identification: Poly(methyl methacrylate) branches may be isol. from graft copolymers by oxidative degradation with nitric acid [1,3]

Volumetric & Calorimetric Properties:
Glass Transition Temperature:

No.	Value	Note
1	105–110°C	[1]

Transition Temperature General: T_m increases with increasing methacrylate content [1]. T_g varies with the percentage content of methacrylate [1]

Applications/Commercial Products:
Processing & Manufacturing Routes: Synth. by free-radical polymerisation using potassium persulfate or ceric ion initiators [2,4]
Applications: Used as alkali-resistant sizing agents for textiles

Bibliographic References
[1] Yao, Y., Liu, L., Li, H., Fang, T. and Zhou, E., *Polymer,* 1994, **35**, 3122
[2] Sakurada, I., Ikada, Y., Uehara, H., Nishizaki, Y. and Horii, F., *Makromol. Chem.,* 1970, **139**, 183, (struct.)
[3] Nagaoka, S., *Polym. J. (Singapore),* 1989, **21**, 847
[4] Iwakura, Y. and Imai, Y., *Makromol. Chem.,* 1966, **98**, 1

Poly(methyl methacrylate-co-vinyl chloride) P-387

$$-[CH_2C(CH_3)]_x-[CH_2CH]_y-_n$$
$$\quad\quad |\quad\quad\quad\quad\quad\quad |$$
$$\quad\; C(=O)OMe\quad\quad Cl$$

Related Polymers: Polymethacrylates, General, Poly(methyl methacrylate), Poly(vinyl chloride)
Monomers: Methyl methacrylate, Vinyl chloride
Material class: Thermoplastic
Polymer Type: acrylic copolymers
Molecular Formula: $[(C_5H_8O_2)_x (C_2H_3Cl)_y]_n$
Fragments: $C_5H_8O_2$ C_2H_3Cl

Volumetric & Calorimetric Properties:
Glass Transition Temperature:

No.	Value	Note
1	65–105°C	[13]

Transition Temperature General: T_g varies with the percentage content of methacrylate [1]

Transport Properties:
Polymer Solutions Dilute: Intrinsic viscosity varies with composition. η 0.87–1.02 dl g^{-1} (0–33% MMa) [3]

Mechanical Properties:
Mechanical Properties General: Yield stress and initial modulus of the graft copolymer increase with increase in methacrylate grafting, whereas yield strain and elongation decrease [3]

Polymer Stability:
Decomposition Details: Pyrolysis at 150–200° liberates methyl chloride and causes intramolecular cyclisation yielding terpolymers [2,3]

Bibliographic References
[1] Johnstone, N.W., *Rev. Macromol. Chem.,* (ed. G.B. Butler), Dekker, 1976, **14B**, 215
[2] Zutty, N.L. and Welch, F.J., *J. Polym. Sci., Part A-1,* 1963, **1**, 2289
[3] Saroop, U.K., Sharma, K.K., Jain, K.K., Misra, A. and Maiti, S.N., *Eur. Polym. J.,* 1988, **24**, 693

Poly(methyl methacrylate-co-N-vinylpyrrolidone) P-388

$$-[CH_2C(CH_3)]_x-[CH_2CH]_y-_n$$
$$\quad\quad |\quad\quad\quad\quad\quad\quad |$$
$$\quad\; C(=O)OMe\quad\quad N\text{-pyrrolidone}$$

Related Polymers: Polymethacrylates, General, Poly(methyl methacrylate)
Monomers: Methyl methacrylate, *N*-Vinylpyrrolidone
Material class: Thermoplastic
Polymer Type: acrylic copolymers
Molecular Formula: $[(C_5H_8O_2)_x (C_6H_9NO)_y]_n$
Fragments: $C_5H_8O_2$ C_6H_9NO
Morphology: Forms hydrogels known as "snake-cage" polymers; linear poly(methyl methacrylate) "snakes" held in a cage of cross-linked *N*-vinylpyrrolidone [1]

Mechanical Properties:
Tensile Strength Break:

No.	Value	Note	Elongation
1	157 MPa	16 kg mm^{-2}, approx., 25% MMA, 74% water content [1]	450%
2	98 MPa	10 kg mm^{-2}, 20% MMA, 76% water content, approx. [1]	400%
3	19.6 MPa	2 kg mm^{-2}, 10% MMA, 85% water content, approx. [1]	200%

Applications/Commercial Products:
Processing & Manufacturing Routes: Synth. by dissolution of *N*-vinylpyrrolidone in poly(methyl methacrylate) in the presence of cross-linking agent (diethylene glycol dimethacrylate) then polymerising by photoinitiation (24h, 10°). The resultant polymer is then heated at 90–100° for a further 24h [1]
Applications: Applications in medicine (membranes, drug delivery systems)

Poly(methyl-*n*-octylsilane) P-389

$$\left[\begin{array}{c} CH_2(CH_2)_6CH_3 \\ | \\ -Si- \\ | \\ CH_3 \end{array} \right]_n$$

Synonyms: *Poly(methyloctylsilylene)*. PMOS. ME8
Polymer Type: polysilanes
CAS Number:

CAS Reg. No.
114528-74-4

Molecular Formula: $(C_9H_{20}Si)_n$
Fragments: $C_9H_{20}Si$
Molecular Weight: 2.9×10^5 [1,2,3,9]
General Information: Hydrophobic [2].
Morphology: Shows a highly elastic character. Conformational studies carried out using uv-vis of films on a hydrophobic surface [2,7].

Surface Properties & Solubility:
Solubility Properties General: See [7].

Transport Properties:
Transport Properties General: Reported to have the viscosity of rubber cement [4].

Optical Properties:
Optical Properties General: Broad absorption band around 308nm, indicating a noncrystalline disordered state including gauche conformations of the Si backbone [2]. Displays an abrupt thermochromic transition (301nm to 330nm, after red shift) [3,9]. The anisotropic polarisability and its isotropic fluid structure allow the possibility of the Kerr effect [4].
Transmission and Spectra: Ir [1,7,9], uv-vis [1,2,3,8,9,10], nmr [1,7,9,11] and fluorescence [5,9] data reported.

Polymer Stability:
Thermal Stability General: Unlike lower analogues, a first-order transition occurs sumultaneously, thermochromism being observed at an onset temperature of -1°C, with an increased ΔH [3].

Applications/Commercial Products:
Processing & Manufacturing Routes: Synth. by Wurtz-type reductive-coupling polymerisation of dichloromethyl-*n*-octylsilane in THF at room temperature with yields of 54–58% (highest yield achieved to-date) [1,3,7,12]. There are no other reported syntheses in the literature [1].
Applications: Organic semiconductor [4]. Membrane plate module [6]. Pyrolysis to silicon carbides [7].

Bibliographic References
[1] Holder, S.J., Achilleos, M. and Jones, R.G., *Macromolecules*, 2005, **38**, 1633
[2] Nagano, S. and Seki, T., *J. Am. Chem. Soc.*, 2002, **124**, 2074
[3] Yuan, C.H. and West, R., *Macromolecules*, 1994, **27**, 629
[4] McGraw, D.J., Siegman, A.E., Wallraff, G.M. and Miller, R.D., *Appl. Phys. Lett.*, 1989, **54**, 1713
[5] Rauscher, U., Bässler, H. and Taylor, R., *Chem. Phys. Lett.*, 1989, **162**, 127
[6] *PCT Int. Appl.*, 2004, 04 082 810
[7] Abu-Eid, M.A., King, R.B. and Kotliar, A.M., *Eur. Polym. J.*, 1992, **28**, 315
[8] Johnson, G.E. and McGrane, K.M., ACS Symp. Ser. Photophys. Polym., 1987, **358**, 499
[9] Seki, S., Koizumi, Y., Kawaguchi, T., Habara, H. and Tagawa, S., *J. Am. Chem. Soc.*, 2004, **126**, 3521
[10] Chunwachirasiri, W., Kanaglekar, I., Winokur, M.J., Koe, J.C. and West, R., *Macromolecules*, 2001, **34**, 6719
[11] Wolff, A.R., Maxka, J. and West, R., *J. Polym. Sci., Part A*, 1988, **26**, 713
[12] Miller, R.D., Thompson, D., Sooriyakumaran, R. and Fickes, G.N., *J. Polym. Sci., Part A*, 1991, **29**, 813

Polymethyloctylsiloxane fluid P-390

Synonyms: *Polymethylalkylsiloxane fluid. Poly[oxy(methyloctylsilylene)]*
Related Polymers: PDMS fluid
Monomers: Dichloromethyloctylsilane, Cyclic methyloctylsiloxane
Material class: Fluids
Polymer Type: silicones
CAS Number:

CAS Reg. No.
108644-23-1
158865-53-3

Molecular Formula: $[C_9H_{20}OSi]_n$
Fragments: $C_9H_{20}OSi$
Molecular Weight: MW 1500–10000
Miscellaneous: Methylalkylsiloxanes with alkyl chains of more than eight carbon atoms have similar props. to polymethyloctylsiloxane fluid. Commercial methylalkylsiloxanes may contain mixts.

Volumetric & Calorimetric Properties:
Density:

No.	Value	Note
1	d 0.91 g/cm³	[1,5]

Glass Transition Temperature:

No.	Value	Note
1	-92°C	large values of *n* [10,11]

Transition Temperature:

No.	Value	Note
1	-50°C	Pour point (-58°F) [1,5]

Surface Properties & Solubility:
Solvents/Non-solvents: Sol. aliphatic hydrocarbons, chlorinated hydrocarbons, aromatic hydrocarbons. Mod. sol. ketones, esters, ethers, higher alcohols. Insol. H_2O, lower alcohols [3,4]
Surface and Interfacial Properties General: Coefficient of friction 0.08. Has much better lubricating props. than does dimethyl silicone fluid [5]
Wettability/Surface Energy and Interfacial Tension: γ_c 30.4 mJ m^{-2} (30.4 dyne cm^{-1}) [1,5]

Transport Properties:
Polymer Melts: Viscosity 1.5 St (38°). Viscosity increases sharply with increasing pressure. Viscosity temp. coefficient 0.75 [5,6,7]

Electrical Properties:
Electrical Properties General: Electrical props. are similar to hydrocarbon fluids
Dielectric Permittivity (Constant):

No.	Value	Note
1	2.4	[2,9]

Poly(2-methylpentadiene)

Dissipation (Power) Factor:

No.	Value	Note
1	1×10^{-5}	[2]

Optical Properties:
Refractive Index:

No.	Value	Note
1	1.445	[1,5]

Polymer Stability:
Polymer Stability General: Less stable than dimethyl silicone fluids
Thermal Stability General: Usable between -50–150° [5]
Upper Use Temperature:

No.	Value
1	150°C

Decomposition Details: Oxidised above 150° [5]
Flammability: Fl. p. 315° [5]
Environmental Stress: Stable to H_2O, air, oxygen and other materials. Props. deteriorate due to cross-linking when exposed to radiation [8]
Chemical Stability: Chemically inert but decomposed by conc. mineral acids. Very thin films liable to degradation [4,5,7]
Hydrolytic Stability: Stable to H_2O and aq. solns. Very thin films spread on water; liable to hydrolytic degradation [2,7,8]
Biological Stability: Non-biodegradable [2,7]
Recyclability: Can be reused; no large scale recycling

Applications/Commercial Products:
Processing & Manufacturing Routes: For low viscosity fluids methyloctyldichlorosilane is hydrolysed and the hydrolysate equilibrated under acid catalysis at 180° with small amounts of hexamethyldisiloxane. For high viscosity fluids cyclic methyloctylpolysiloxanes are equilibrated together under alkaline catalysis with traces of chain terminator at 150°
Applications: Used as a lubricant, mold release agent and hydraulic fluid

Trade name	Supplier
Masil 263	PPG-Mazer
Masil 264	PPG-Mazer
SF 1080	General Electric Silicones
SF 1147	General Electric Silicones

Bibliographic References

[1] Torkelson, A., *Silicone Technology*, (ed. P.F. Bruins), John Wiley and Sons, 1970, 61
[2] Stark, F.O., Fallender, J.R. and Wright, A.P., *Comprehensive Organometallic Chemistry*, (eds. G. Wilkinson, F.G.A. Stone and E.W. Abel), Pergamon Press, 1982, **2**, 305
[3] Ash, M. and Ash, I., *Handbook of Plastic Compounds, Elastomers and Resins*, (eds. M.B. Ash and I.A. Ash), Wiley-VCH, 1992
[4] Roff, W.J. and Scott, J.R., *Fibres, Films, Plastics and Rubbers*, Butterworths, 1971
[5] Brown, E.P., *ASLE Trans.*, 1966, **9**, 31
[6] Barry, A.J. and Beck, H.N., *Inorg. Polym.*, (eds. F.G.A. Stone and W.A.G. Graham), Academic Press, 1962, 189
[7] Hardman, B. and Torkelson, A., *Encycl. Polym. Sci. Technol.*, 2nd edn. (ed. J.I. Kroshwitz), John Wiley and Sons, 1985, **15**, 204
[8] Noll, W., *Chemistry and Technology of Silicones*, Academic Press, 1968
[9] Vincent, G.A., Fearon, F.W.G. and Orbech, T., Annu. Rep., *Conf. Electr. Insul. Dielectr. Phenom.*, 1972, 17, (dielectric constant)
[10] Clarson, S.J. and Semlyen, J.A., *Siloxane Polymers*, Prentice Hall, 1993
[11] Stem, S.A., Shah, V.M. and Hardy, B.J., *J. Polym. Sci., Polym. Phys. Ed.*, 1987, **25**, 1263

Poly(2-methylpentadiene) P-391

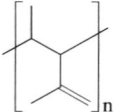

Synonyms: *Poly(1,3-dimethyl-1,3-butadiene). PMPD. Poly(2-methyl-1,3-pentadiene). PDMB. Poly(2-methyl-1,4-pentadiene)*
Monomers: 2-Methyl-1,4-pentadiene, 2-Methyl-1,3-pentadiene
Polymer Type: polyolefins, polydienes
CAS Number:

CAS Reg. No.	Note
105639-53-0	Poly((*E*)-2-methyl-1,3-pentadiene), isotactic
95763-37-4	Poly(2-methyl-1,4-pentadiene), isotactic
25931-14-0	Poly(2-methyl-1,3-pentadiene)
83174-63-4	Poly(2-methyl-1,3-pentadiene), isotactic
26714-20-5	Poly((*E*)-2-methyl-1,3-pentadiene)

Molecular Formula: $(C_6H_{10})_n$
Fragments: C_6H_{10}
Morphology: There are two cryst. forms of (*E*)-2-methyl-1,3-pentadiene (greater than 98% *cis*-content). α-form: orthorhombic a 10.74Å, b 13.04Å, c 7.87Å; space group Pbca; Z=8 (isotactic); β-form a 9.3Å, b 7.73Å, c 7.90Å; space group $P2_12_12_1$; Z=4 [2]. *trans*-1,4-poly(2-methylpentadiene) has a monoclinic cryst. struct: a 4.82Å, b 9.19Å, c 12.89Å; space group $P2_1/c$; β 93.5° (25°); a 4.82Å, b 9.08Å, c 12.78Å; space group $P2_1/c$; β 93.6° (-120°) [8]

Volumetric & Calorimetric Properties:
Density:

No.	Value	Note
1	d 0.85 g/cm³	1,4-form [4,9]

Thermodynamic Properties General: Variation of C_p with temp. has been reported [4]
Melting Temperature:

No.	Value	Note
1	165°C	*cis*-form; β cryst form [2]
2	175°C	*cis*-form; α cryst form [2]
3	200°C	*trans*-form [2]

Glass Transition Temperature:

No.	Value	Note
1	3.7–4.7°C	varying MW [4]
2	1.1–2.6°C	*cis/trans* (60/40), varying MW [9]
3	-13°C	high *cis*-1,4 content [10]

Transport Properties:
Polymer Solutions Dilute: The effects of temp. and MW on rheological props. have been reported. [9] Intrinsic viscosity 0.47 dl g^{-1} (THF, 30°, MW 445 000), 0.71 dl g^{-1} (THF, 30°, MW 763000), 2.18 dl g^{-1} (THF, 30°, MW 3770000); 0.23 dl g^{-1} (2-octanol, 28.9°, MW 445000), 0.3 dl g^{-1} (2-octanol, 28.9°, MW 763000), 0.65 dl g^{-1} (2-octanol, 28.9°, MW 3770000) [4]. Huggins

– Poly(3-methyl-1-pentene)

coefficients 0.77 (MW 445000), 0.88 (MW 763000), 1.02 (MW 3770000) [4]. Theta temp. 28.9° (2-octanol, cis/trans (64/36)) [11]

Mechanical Properties:
Miscellaneous Moduli:

No.	Value	Note	Type
1	0.42 MPa	[4,9]	Plateau modulus
2	0.3 MPa	calc., 1,4-form [4,9]	Plateau modulus

Optical Properties:
Transmission and Spectra: Ir and C-13 nmr spectral data have been reported [2,4,5,6,7]
Total Internal Reflection: Birefringence dn/dc 0.0989 cm^3 g^{-1} (633 nm, 30°, cyclohexane) [4]

Applications/Commercial Products:
Processing & Manufacturing Routes: Polymerised using vanadium, titanium or nickel-based catalysts; the product has varying isomeric compositions. A cis-1,4-polymer has been obtained using the catalyst system AlR$_3$-TiCl$_4$ and also a ternary catalyst system: AlEt$_2$Cl-Nd(OCOC$_7$H$_{15}$)$_3$-triisobutyl aluminium. A trans-1,4-polymer has been produced using Ni(AcAc)$_2$-MAO. [1,2,7] Anionic polymerisation has also been reported [4]. Optically active polymer can be obtained by asymmetric inclusion polymerisation, e.g. in apocholic acid canals [3]
Applications: Used as intermediates in synth. of polypropylene

Bibliographic References
[1] Ricci, G., Panagia, A. and Porri, L., *Polymer*, 1996, **37**, 363, (synth)
[2] Bruckner, S, Meille, S.V., Porzio, W. and Ricci, G., *Makromol. Chem.*, 1988, **189**, 2135, 2145, (synth, struct)
[3] Mijata, M., Kitahara, Y. and Takemoto, K., *Makromol. Chem.*, 1983, **184**, 1771, (synth)
[4] Zhongde, X., Mays, J., Xuexin, C., Hadjichristidis, N. *et al*, *Macromolecules*, 1985, **18**, 2560, (synth, anal)
[5] Cuzin, D., Chauvin, Y. and Lefebvre, G., *Eur. Polym. J.*, 1987, **3**, 581, (synth)
[6] Sozzani, P., Silvestro, G., Grassi, M. and Farina, M., *Macromolecules*, 1984, **17**, 2532, (C-13 nmr)
[7] Oliva, L., Longo, P., Grassi, A., Ammendola, P. and Pellecchia, C., *Makromol. Chem., Rapid Commun.*, 1990, **11**, 519, (synth)
[8] Bruckner, S., Luzzati, S., Porzio, W. and Sozzani, P., *Macromolecules*, 1987, **20**, 585, (cryst struct)
[9] Pearson, D. S., Fetters, L.J., Younghouse, L.B. and Mays, J.W., *Macromolecules*, 1988, **21**, 478, (rheology)
[10] Henderson, J.N. and Throckmorton, M.C., Encycl. Polym. Sci. Eng. 2nd edn., (ed. J.I. Kroschwitz), 1985, **10**, 811
[11] Xu, Z., Mays, J.W. and Xaexin, C., *Macromolecules*, 1991, **24**, 6199

Poly(3-methyl-1-pentene) P-392

$$-[CH_2CH]_n-$$
$$\quad\quad |$$
$$\quad CH(CH_3)CH_2CH_3$$

Synonyms: *P3M1P. Poly(2-vinylbutane). Poly(1-(1-methylpropyl)-1,2-ethanediyl)*
Monomers: 3-Methyl-1-pentene
Material class: Thermoplastic
Polymer Type: polyolefins
CAS Number:

CAS Reg. No.	Note
25266-25-5	
30919-81-4	(+)-form
30920-01-5	(+)-form
28575-86-2	isotactic, (S)-form
34590-92-6	(S)-form
25189-87-1	(S)-form, (+)-form
66906-10-3	(±)-form
88611-18-1	isotactic, (±)-form
29036-72-4	syndiotactic

Molecular Formula: $(C_6H_{12})_n$
Fragments: C_6H_{12}
Tacticity: Optically active isotactic and atactic forms known
Morphology: Isotactic, (S)-form, tetragonal, a 1.395, b 1.335, c 0.68 [9], helix 4$_4$ (or possibly tetragonal, a 1.92, b 1.92 [15])

Volumetric & Calorimetric Properties:
Density:

No.	Value	Note
1	d 0.89 g/cm^3	[9]
2	d 0.88 g/cm^3	[14]
3	d 0.831 g/cm^3	[11]
4	d 0.832 g/cm^3	[11]

Melting Temperature:

No.	Value	Note
1	271–273°C	(S)(+)-form [10,13]
2	260–264°C	(S)(+)-form [11]
3	271–278°C	(S)(+)-form [12]
4	229–237°C	[12]
5	200–205°C	[11]
6	360–362°C	[14]
7	266–268°C	[13]

Surface Properties & Solubility:
Solvents/Non-solvents: Sol. 1,2,4-trichlorobenzene, boiling 1,1-ditolylethane, 1,1-diphenylethane [6]. Slightly sol. dibutyl phthalate, dioctyl phthalate (above 264°) [11]. Insol. boiling xylene [12]

Transport Properties:
Polymer Solutions Dilute: [η] 0.20–0.34 dl g^{-1} (30°, CCl$_4$) [2]. [η] 0.8 dl g^{-1} [13]

Optical Properties:
Transmission and Spectra: Ir [16] and C-13 nmr [3,6,7,17] spectral data have been reported
Total Internal Reflection: $[\alpha]_D^{25}$ +257 ((S)(+)-form) [12]. $[\alpha]_D^{24}$ +255 ((S)(+)-form) [11]. Other optical rotation values have been reported [10,19]. Temp. coefficients of optical rotation have been reported [5,20]. $[M]_D^{25}$ +154 ((S)(+)-form) [21,22]. $[M]_D^{20}$ +163 ((S)(+)-form) [13]. $[M]_D$ +180 ((S)(+)-form) [8]. Temp. effects on birefringence intensity have been reported [15]

Polymer Stability:
Environmental Stress: Undergoes oxidation by ozonolysis [18]

Applications/Commercial Products:
Processing & Manufacturing Routes: Polymerised using Ziegler-Natta catalysts, such as TiCl$_3$AlMe$_3$-ZnMe$_2$ in toluene at 50° [3]

Bibliographic References
[1] Bacskai, R., Polym. Prepr. (Am. Chem. Soc., *Div. Polym. Chem.*), 1965, **6**, 687, (synth)
[2] Bacskai, R., *J. Polym. Sci., Part A-1*, 1967, **5**, 619, (viscosity)
[3] Sacchi, M.C., Locatelli, P., Zetta, L. and Zambelli, A., *Macromolecules*, 1984, **17**, 483, (C-13 nmr)
[4] Corradini, P. and Petraccone, V., *CA*, 1972, **79**, 5795, (struct)
[5] Pino, P., Lorenzi, G.P. and Bonsignori, O., *CA*, 1968, **72**, 3901, (optical props)
[6] Zambelli, A., Ammendola, P., Locatelli, P. and Sacchi, M.C., *Macromolecules*, 1984, **17**, 977, (C-13 nmr)

- [7] Ferro, D.R. and Ragazzi, M., *Macromolecules*, 1984, **17**, 485, (C-13 nmr)
- [8] Abe, A., *J. Am. Chem. Soc.*, 1968, **90**, 2205, (optical props)
- [9] Petraccone, V., Ganis, P., Corradini, P. and Montagnoli, G., *Eur. Polym. J.*, 1972, **8**, 99, (struct)
- [10] Pino, P. and Lorenzi, G.P., *J. Am. Chem. Soc.*, 1960, **82**, 4745, (optical props)
- [11] Nozakura, S., Takeuchi, S., Yuki, H. and Murahashi, S., *Bull. Chem. Soc. Jpn.*, 1961, **34**, 1673, (optical props)
- [12] Bailey, W.J. and Yates, E.T., *J. Org. Chem.*, 1960, **25**, 1800, (optical props)
- [13] Pino, P., Lorenzi, G.P. and Lardicci, L., *Chim. Ind. (Milan)*, 1960, **42**, 712, (optical props)
- [14] Bacskai, R., Goodrich, J.E. and Wilkes, J.B., *J. Polym. Sci., Part A-1*, 1972, **10**, 1529, (melting point)
- [15] Gallegos, E.J., *J. Polym. Sci., Part A: Polym. Chem.*, 1965, **3**, 3982, (struct)
- [16] Benedetti, E. and Ciandelli, F., *Chim. Ind. (Milan)*, 1977, **59**, 588, (ir)
- [17] Sacchi, M.C., Locatelli, P., Zetta, L., Zambelli, A. *et al*, *CA*, 1983, **101**, 73317, (C-13 nmr)
- [18] Tarasevich, B.N. and Atyaksheva, L.F., *CA*, 1990, **113**, 60001, (ozonolysis)
- [19] Pino, P., Ciardelli, F., Lorenzi, G.P. and Montagnoli, G., *Makromol. Chem.*, 1963, **61**, 207, (optical props)
- [20] Pino, P. and Luisi, P.L., *J. Chim. Phys. Phys. - Chim. Biol.*, 1968, **65**, 130, (optical props)
- [21] Bonsignori, O. and Lorenzi, G.P., *J. Polym. Sci., Part A-2*, 1970, **8**, 1639, (optical props)
- [22] Pino, P., Bartus, J. and Vogl, O., *Polym. Prepr. (Am. Chem. Soc., Div. Polym. Chem.)*, 1988, **29**, 254, (optical props)

Poly(4-methyl-1-pentene) P-393

Synonyms: *P4MP*
Related Polymers: Injection moulding Glass filled, Injection moulding Mineral filled, Glass reinforced, Moulded, Extrusion unfilled
Monomers: 4-Methyl-1-pentene
Material class: Thermoplastic
Polymer Type: polyolefins
CAS Number:

CAS Reg. No.	Note
24979-98-4	isotactic
25068-26-2	
131724-39-5	syndiotactic
101239-82-1	
113834-30-3	
26680-37-5	isotactic *S*-form
33409-74-4	isotactic ($\pm$)-form

Molecular Formula: $(C_6H_{12})_n$
Fragments: C_6H_{12}
Additives: Antioxidants
General Information: Refers to commercial product, which is a copolymer with 3–5% of an α-olefin having a linear alkyl chain C_6-C_{16}.
Tacticity: Predominantly isotactic, head-to-tail struct.
Morphology: Form III tetragonal a 19.46, b 19.46Å, c 7.022Å, α 90° β 90° μ 90° 4, helix [63,64]. Isotactic, tetragonal a 18.70Å, c 13.68Å [49]. Tetragonal a = b 18.35Å, c 13.52Å (lower draw); orthrohombic a 17.70Å, b 8.85Å, c 12.33Å (higher draw) [54]. Can crystallise in five forms depending on crystallisation conditions [57]. Other morphological studies have been reported [59]. Syndiotactic tetragonal a 18.03Å [64]

Volumetric & Calorimetric Properties:
Density:

No.	Value	Note
1	d 0.833 g/cm^3	[44]
2	d^{23} 0.829 g/cm^3	[48]
3	d 0.83 g/cm^3	isotactic [49]
4	d 0.83 g/cm^3	[52]
5	d 0.838 g/cm^3	amorph. [53]
6	d 0.828 g/cm^3	cryst. [53]

Thermal Expansion Coefficient:

No.	Value	Note	Type
1	0.00069 K^{-1}	T <-185° [61]	l
2	0.00089 K^{-1}	T >-185° [61]	L

Equation of State: Equation of state information has been reported [17,33]
Thermodynamic Properties General: Thermodynamic props. over the temp. range -266–397° have been studied [2]. Heat of combustion has been reported [25]. Variation of Cp with temp. has been reported [65]. Thermal conductivity increases with increasing temp. [53]
Latent Heat Crystallization: Heat of fusion 5.2-5.8 kJ mol^{-1} (isotactic) [15], 61.96 J g^{-1} (14.8 cal g^{-1}) [16]. ΔH 92 J g^{-1} [43], 102 J g^{-1} [43]. Entropy of fusion 10.3 JK^{-1} mol^{-1} (2.46 cal K^{-1} mol^{-1}). [16] Heat of combustion has been reported [25]. Variation of Cp with temp. has been reported [65] ΔH (melt) 9.93 kJ mol^{-1} [45]. ΔS (melt) 19.5 J K^{-1} mol^{-1} [45]. Heat of polymerisation -95 kJ mol^{-1} [45]. Entropy of polymerisation -131 J mol^{-1} K^{-1} [45]. ΔG$_{polym}$ -56 kJ mol^{-1} [45]. Heat of fusion 10.04–13.8 kJ mol^{-1} (2400–3300 cal mol^{-1}) [52].
Thermal Conductivity:

No.	Value	Note
1	0.11 W/mK	melt [44]
2	0.2 W/mK	cryst. [44]

Specific Heat Capacity:

No.	Value	Note	Type
1	0.56 kJ/kg.C	-194°	P
2	2.57 kJ/kg.C	147° [55]	p

Melting Temperature:

No.	Value	Note
1	245°C	[43]
2	230°C	[44]
3	235°C	[45]
4	238°C	[47]
5	240°C	[52,57]

Glass Transition Temperature:

No.	Value	Note
1	17°C	[44]
2	38°C	[46]
3	42°C	[47]
4	34°C	[55]
5	23–27°C	amorph. [75]

Transition Temperature General: T_m of crystalline polymer increases with pressure up to a max. pressure of 3 k bar [3]

– Poly(4-methyl-1-pentene)

Deflection Temperature:

No.	Value	Note
1	80–90°C	0.46 MPa, ASTM D686 [67]
2	48–50°C	1.82 Mpa, ASTM D686 [67]
3	58°C	ASTM D648 [52]

Vicat Softening Point:

No.	Value	Note
1	173–180°C	ASTM D1525 [67]

Transition Temperature:

No.	Value	Note
1	172°C	fusion temp. [55]

Surface Properties & Solubility:
Cohesive Energy Density Solubility Parameters: δ 30.98–33.5 $J^{1/2}cm^{-3/2}$ (7.4–8 $cal^{1/2}\,cm^{-3/2}$) [51]
Solvents/Non-solvents: Sol. isopropylcyclohexane (154°), tert-butylcyclohexane (154°), cyclohexylbenzene (212°), butyl stearate (220°), n-octyl acetate (210°), decalin (190°), dodecane (197°), mineral oil (248°) [51]. Insol. Me_2CO, MeOH, EtOH, 2-propanol [51]
Surface and Interfacial Properties General: Critical surface tension has been reported. [31]
Wettability/Surface Energy and Interfacial Tension: Wettability data have been reported. [68] Contact angle 95° (glycerol/polymer melt-vapour interface), 96° (glycerol/gold-polymer melt interface) [74]

Transport Properties:
Polymer Solutions Dilute: η 6.5 dl g^{-1} (65°, cyclohexane) [28], 2.8 dl g^{-1} (135°, decalin) [28], 2.5 dl g^{-1} (135°, decalin) [52]. Viscosity determined from flow curves has been reported [37]. η 5.5 dl g^{-1} (150°, decalin) [46]. Viscosity is dependent upon method of synth. [46]
Permeability of Gases: The diffusion coefficient of CO_2 increases with increasing pressure (2–6.3 MPa) whereas those of N_2 and CH_4 do not change with increasing pressure below Tg [18]. The diffusion coefficients of H_2, He, N_2 and Ar have been determined graphically. [24] Gas flux [O_2] 0.61 × 10^{-6} $cm^3\,cm^{-2}\,s^{-1}$ (cm Hg)$^{-1}$ [62]
Permeability of Liquids: Permeability coefficients of C_1–C_4 alkanes have been reported [29].
Water Absorption:

No.	Value	Note
1	0.01 %	saturation, ASTM D570 [67]

Gas Permeability:

No.	Gas	Value	Note
1	O_2	32.3 cm^3 mm/(m^2 day atm)	
2	CO_2	92.6 cm^3 mm/(m^2 day atm)	
3	N_2	8.4 cm^3 mm/(m^2 day atm)	
4	H_2	136 cm^3 mm/(m^2 day atm)	
5	He	101 cm^3 mm/(m^2 day atm)	
6	CH_4	10.79 cm^3 mm/(m^2 day atm)	[57]

Mechanical Properties:
Mechanical Properties General: Variation of mechanical loss modulus with temp. has been reported. [48]. Other mechanical props. have been reported [70,71].

Tensile (Young's) Modulus:

No.	Value	Note
1	2230 MPa	[54]
2	800–1200 MPa	[57]
3	1590 MPa	[58]

Tensile Strength Break:

No.	Value	Note	Elongation
1	31 MPa	4500 psi, ASTM D638 [52]	5%
2	17–20 MPa	[57]	10–25%

Flexural Strength at Break:

No.	Value	Note
1	1172 MPa	170 kpsi, ASTM D747 [52]

Compressive Modulus:

No.	Value	Note
1	800–1200 MPa	ASTM D695 [67]

Elastic Modulus:

No.	Value	Note
1	343.2 MPa	3500 kg cm^{-2} [30]
2	2900 MPa	isotactic, room temp. [40]
3	2620 MPa	isotactic, room temp., chain axial direction [40]

Poisson's Ratio:

No.	Value	Note
1	0.33	isotactic [40]
2	0.43	[58]

Tensile Strength Yield:

No.	Value	Note
1	9.81 MPa	100 kg cm^{-2} [30]
2	23–28 MPa	[57]

Miscellaneous Moduli:

No.	Value	Note
1	1544 MPa	224 kpsi secant modulus [52]
2	1654 MPa	240 kpsi tangent modulus ASTM D638 [52]
3	4030 MPa	

Viscoelastic Behaviour: Thermoacoustic parameters have been reported [5,19]. The acoustic impedance at 2MHz and room temp. is similar to that of human skin [7] thereby making it suitable for medical applications.
Hardness: Rockwell R80 (ASTM D785) [52], Shore D80 (ASTM D1706) [52]

Friction Abrasion and Resistance: Wear data have been reported [69]

Izod Notch:

No.	Value	Notch	Note
1	26.68 J/m	Y	0.5 ft lb in^{-1}, ASTM D256
2	100–200 J/m	Y	[57]

Electrical Properties:

Electrical Properties General: A continuous spectrum of complex dielectric permittivity and loss tangent has been reported. [11] Dielectric props. have been reported [23]. Power factor is unaffected by ageing at 150° for 1500h or at 200° for 21h [32]. Loss tangent increases with increasing frequency and decreasing temp. [42]. Other electrical props. have been reported [72]

Dielectric Permittivity (Constant):

No.	Value	Frequency	Note
1	2.1		approx. [32]
2	2.1	30–100 Hz, 25° [57]	

Dielectric Strength:

No.	Value	Note
1	63–65 kV.mm^{-1}	ASTM D149 [67]

Magnetic Properties: Magnetic susceptibility and anisotropy have been reported [20]

Dissipation (Power) Factor:

No.	Value	Frequency	Note
1	8×10^{-5}	100–1000 kHz	min., [32]
2	<0.00045	10000 MHz	max., [32]
3	<0.266		max. [46]

Optical Properties:

Optical Properties General: Study of complex refractive index has been reported. [9,10,11]. A study of far-ir refractive index has been reported [12]. The temp. variation of optical constants of TPX has been reported [13]

Refractive Index:

No.	Value	Note
1	1.46	17° [57]

Transmission and Spectra: C-13 nmr [4,8,27], FT-IR [6], ir [22,34,39,76], Raman [36], esr [41] and H-1 nmr [77] spectral data have been reported. Useful transmittance in the medium and long-wave ir regions [26]

Volume Properties/Surface Properties: Clarity Highly transparent due to limited spherulite growth [52]. Haze 1.2–1.5% [67]

Polymer Stability:

Thermal Stability General: Degrades above 280°

Decomposition Details: Thermal degradation proceeds via intramolecular hydrogen transfer. Pyrolysis at 700° gives short-chain (C_1-C_4) hydrocarbons, butadiene and isopentane as the main products [14,21]. Oxidative thermal degradation at 130–170° gives Me_2CO and H_2O plus short-chain hydrocarbons [35]. Activation energy of degradation 96.29 kJ mol^{-1} [35]. Pyrolysis at 380–400° yields 1-alkene oligomers formed by free-radical processes, backbiting and β-scission [56]

Flammability: Self-ignition temp. has been reported [38]

Environmental Stress: Degrades readily in the presence of oxygen and light [67]. Other weathering data have been reported [73]

Chemical Stability: Has high stability to inorganic substances, acids and alkalis. Attacked by conc. solns. of chromic acid, chlorine, bromine and fuming nitric acid [67]

Biological Stability: Biodegradation study has been reported [1]

Stability Miscellaneous: Electron beam irradiation significantly decreases elongation at break, T_m and heat of fusion [50]

Applications/Commercial Products:

Processing & Manufacturing Routes: Polymerisation performed in liq. monomer at 50–70° using Ziegler-Natta catalysts. Reaction product is a granular powder which is stabilised and extruded into pellet form. Isotactic, head-to-tail polymer is synth. using Ziegler-Natta catalysts containing chromium, zirconium, vanadium or titanium. [57] Also synth. by cationic polymerisation [60] at low temps. (-130–20°) using aluminium halide initiators [57]. Syndiotactic polymer can by synth. using a complex zirconium dichloride or methylaluminoxane catalyst at 30° [66]

Mould Shrinkage (%):

No.	Value
1	1.2–3.4%

Applications: Medical equipment: syringes, transfusion equipment, pacemaker parts and respiration equipment. Laboratory equipment: spectroscopic cells, laboratory ware, animal cages. Food contact products: microwave oven trays (major use), food packaging. Electrical and electronic applications including release film for printed circuit boards, cables and connectors. Automotive applications, and in wire and cable coating. Uses due to microwave transparency (microwave oven cookware, food packing, service trays). Potential use in contact lenses

Trade name	Details	Supplier
Crystalor		Phillips 66
FR-TPX	various grades	Mitsui Petrochem
J-98	10/20/30/40	DSM Engineering
MP-98	various grades	DSM Engineering
Opulent		Mitsui Petrochemicals
TPX		Mitsui Petrochemicals

Bibliographic References

[1] Mlinac-Misak, M., *Polimeri (Zagreb)*, 1994, **15**, 214, (biodegradation)
[2] Lebedev, B.V., Smirnova, N.N., Vasil'ev, V.G., Kiparisova, E.G. and Kleiner, V.I., *Vysokomol. Soedin., Ser. B*, 1994, **36**, 1413, (thermodynamic props)
[3] Rastogi, S., Newman, M., Keller, A. and Hikosaka, M, NATO ASI Ser., Ser. C, 1993, **405**, 135, ($T^{\wedge}_m$, phase behaviour)
[4] Ma, S., Chen, M.C. and Wang, S.P., *J. Appl. Polym. Sci.*, 1994, **51**, 1861, (C-13 nmr)
[5] Kumar, M.R., Reddy, R.R., Rao, T.V.R and Sharma, B.K., *J. Appl. Polym. Sci.*, 1994, **51**, 1805, (thermoacoustic props)
[6] Birch, J.R., *Infrared Phys.*, 1993, **34**, 89, (ir)
[7] Konda, T., Kishimoto, S. and Sato, Y, *Jpn. J. Appl. Phys., Part 1*, 1992, **31**, 163, (acoustic props)
[8] Mizuno, A. and Kawachi, H, *Polymer*, 1992, **33**, 57, (C-13 nmr)
[9] Stead, M. and Simonis, G.J., *Proc. SPIE-Int. Soc. Opt. Eng.*, **1039**, 306, (refractive index)
[10] Nakashima, S., Hattori, T., Hangyo, M., Yamamoto, A., et, Oyo Butsuri, *Oyo Butsuri*, 1990, **59**, 1093, (refractive index)
[11] Afsar, M.N., *IEEE Trans. Instrum. Meas.*, 1987, **36**, 530, (dielectric props, refractive index)
[12] Stuetzel, P., Tegtmeier, H.D. and Tacke, M., *Infrared Phys.*, 1988, **28**, 67, (refractive index)
[13] Meny, C. Leotin, J and Birch, J.R., *Infrared Phys.*, 1991, **31**, 211, (optical props)
[14] Buniyat-Zade, A.A., Androsova, V.M., Pokatilova, S.D. and Aliev, A.M, *Plast. Massy*, 1990, 71, (degradation)
[15] Charlet, G. and Delmas, G., *J. Polym. Sci., Polym. Phys. Ed.*, 1988, **26**, 1111, (heat of fusion)
[16] Zoller, P., Starkweather, H.W. and Jones, G.A., s*J. Polym. Sci Polym. Phys. Ed.*, 1986, **24**, 1451, (heat of fusion)
[17] Hartmann, B. and Haque, M.A., *J. Appl. Phys.*, 1985, **58**, 2831, (eqn of state)

[18] Tigina, O.N. and Golubev, I.F, *Teor. Osn. Khim. Tekhnol.*, 1983, **17**, 687, (diffusion coefficients)
[19] Sharma, B.K., *J. Phys. D: Appl. Phys.*, 1982, **15**, 1273, (thermoacoustic props.)
[20] Kestel'man, V.N., Stadnik, A.D. and Fischer, A.D., *Plaste Kautsch.*, 1983, **30**, 75, (magnetic props)
[21] Yasufuku, S. and Ishioka, Y, *Kobunshi Ronbunshu*, 1985, **42**, 121, (degradation)
[22] Gribov, L.A., Demukhamedova, S.D. and Zubkova, O.B., *Zh. Prikl. Spektrosk.*, 1983, **38**, 230, (ir)
[23] Romanovskaya, O.S., Eidel'nant, M.P., Shuvaev, V.P., Lobanov, A.M. et al, *Plast. Massy*, 1978, 36, (dielectric props)
[24] Kapanin, V.V. and Reitlinger, S.A, *Vysokomol. Soedin., Ser. B*, 1976, **18**, 770, (gas permeability)
[25] Dobkowski, Z., *Polimery (Warsaw)*, 1977, **22**, 229, (heat of combustion)
[26] Lytle, J.D., Wilkerson, G.W. and Jaramillo, J.G., *Appl. Opt.*, 1979, **18**, 1842, (transmittance)
[27] Ferraris, G., Corno, C., Priola, A. and Cesca, S., *Macromolecules*, 1977, **10**, 188, (C-13 nmr, cationic polymerisation)
[28] Aharoni, S.M., Charlet, G., Delmas, G., *Macromolecules*, 1981, **14**, 1390, (viscosity)
[29] Yampol'skii, Yu.P., Durgar'yan, S.G. and Nametkin, N.S., *Vysokomol. Soedin., Ser. B*, 1979, **21**, 616, (liq permeability)
[30] Volodin, V.P., Andreeva, I.N., Vavilova, I.I., Mashkova, L.V et al, *Plast. Massy*, 1978, 46
[31] Sewell, J.H., *Rev. Plast. Mod.*, 1971, **22**, 1367, (critical surface tension)
[32] Eidel'nant, M.P., Lobanov, A.M., Luk'yanova, N.V., Demidova, V.M. et al, *Plast. Massy*, 1974, 51, ((dielectric props.))
[33] Sagalaev, G.V., Ismailov, T.M., Ragimov, A.M., Makhmudov, A.A. and Svyatodukhov, B.P., *Plast. Massy*, 1974, 76, (eqn of state)
[34] Gabbay, S.M. and Stivala, S.S., *Polymer*, 1976, **17**, 121, (ir)
[35] Yasina, L.L., Silina, A.G. and Pudov, V.S., *Vysokomol. Soedin., Ser. B*, 1975, **17**, 698, (oxidative degradation)
[36] Kim, J.J., McLeish, J., Hyde, A.J. and Bailey, R.T., *Chem. Phys. Lett.*, 1973, **22**, 503, (Raman)
[37] Bukhgalter, V.I., Meshcherova, F.F., Mulin, Yu.A. and Shervud, M.A., *Plast. Massy*, 1976, 76, ((rheological props.))
[38] Baillet, C. and Delfosse, L., *Nehorlavost Plast. Hmot, Dreva Text.*, 1974, 288
[39] Chantry, G.W., Fleming, J.W., Smith, P.M., Cudby, M. and Willis, H.A., *Chem. Phys. Lett.*, 1971, **10**, 473, (ir)
[40] Kaji, K., Sakurada, I., Nakamae, K., Shintaku, T., and Shikata, E., *Bull. Inst. Chem. Res., Kyoto Univ.*, 1974, **52**, 308, (elastic moduli)
[41] Goodhead, D.T., *J. Polym. Sci., Part A-2*, 1971, **9**, 999, (esr)
[42] Allan, R.N. and Kuffel, E., *Proc. Inst. Electr. Eng.*, 1968, **115**, 432, (dielectric loss)
[43] Phong-Nguyen, H., Charlet, G. and Delmas, G., *J. Therm. Anal.*, 1996, **46**, 809, (heat of fusion)
[44] Privalko, V.P. and Rekhteta, N.A., *J. Therm. Anal.*, 1992, **38**, 1083, (thermal conductivity)
[45] Lebedev, B.V., Tsvetkova, L.Ya. and Smirnova, N.N., *J. Therm. Anal.*, 1994, **41**, 1371, (thermodynamic props)
[46] Hewett, W.A. and Weir, F.E., *J. Polym. Sci., Part A: Polym. Chem.*, 1963, **1**, 1239, (dynamic props)
[47] Kirshenbaum, I., Isaacson, R.B. and Druin, M, *J. Polym. Sci., Part B: Polym. Phys.*, 1965, **3**, 525
[48] Crissman, J.M., Sauer, J.A, and Woodward, A.E., *J. Polym. Sci., Part A: Polym. Chem.*, 1964, **2**, 5075
[49] Kusanagi, H., Minoru, T., Chatani, Y. and Tadokoro, H., *J. Polym. Sci., Polym. Phys. Ed.*, 1978, **16**, 131, (cryst struct)
[50] El-Naggar, A.M., Lopez, L.C. and Wilkes, G.L., *J. Appl. Polym. Sci.*, 1990, **39**, 427, (irradiation)
[51] Beck, H.N., *J. Appl. Polym. Sci.*, 1993, **50**, 897, (solubility)
[52] Isaacson, R.B., Kirshenbaum, I. and Feist, W.C., *J. Appl. Polym. Sci.*, 1964, **8**, 2789, (mechanical props)
[53] Choy, C.L., Ong, E.L. and Chen, F.C, *J. Appl. Polym. Sci.*, 1981, **26**, 2325, (thermal conductivity)
[54] He, T. and Porter, R.S., *Polymer*, 1987, **28**, 1321, (cryst struct)
[55] Karasz, F.E., Bair, H.E. and O'Reilly, J.M., *Polymer*, 1967, **8**, 547, (thermodynamic props)
[56] Lattimer, R.P., *J. Anal. Appl. Pyrolysis*, 1995, **31**, 203, (pyrolysis)
[57] Lopez, C.C., Wilkes, G.L., Stricklen, P.M. and White, S.A., *J. Macromol. Sci., Rev. Macromol. Chem. Phys.*, 1992, **32**, 301, (synth, struct, props)
[58] Warfield, R.W. and Barnet, F.R., *Angew. Makromol. Chem.*, 1972, **27**, 215
[59] Patel, D. and Bassett, D.C., *Proc. R. Soc. London A*, 1994, **445**, 577, (morphology)
[60] Bogomolnij, V., Shelekhova, O. and Skorokhodov, S., *Makromol. Chem.*, 1978, **179**, 1847, (cationic polymerisation)
[61] Haldon, R.A. and Simha, R, *Macromolecules*, 1968, **1**, 340, (thermal expansivity)
[62] Lin, F-C, Wang, D-M and Lai, J-Y, *J. Membr. Sci.*, 1996, **110**, 25
[63] DeRosa, C., Auriemma, F., Borriello, A. and Corradini, P., *Polymer*, 1995, **36**, 4723, (cryst struct)
[64] DeRosa, C., Borriello, A., Venditto, V. and Corradini, P., *Macromolecules*, 1992, **25**, 6938, (cryst struct)
[65] Chang, S-S, Polym. Prepr. (Am. Chem. Soc., Div. Polym. Chem.), 1987, **28**, 244, (heat capacity)
[66] Asanuma, T., Nishimori, Y., Ito, M., Uchikawa, N. and Shiomura, T., *Polym. Bull. (Berlin)*, 1991, **25**, 567, (synth)
[67] Kissin, Y.V., *Encycl. Polym. Sci. Eng.*, (ed. J.I. Kroschwitz), 2nd edn., 1985, **9**, 707
[68] Lieng-Huang, L., Proc. Annu. Tech. Conf., SPI Reinf. Plast. Div., 22nd, Washington, D.C., Reinf. Plast. Div. 22nd. Washington DC, 1967, **13-C**, 12, (wettability)
[69] Lagally, P. and Nagy, R., *ASLE Trans.*, 1971, **14**, 12, (wear)
[70] Visser, P.J., *Ned. Chim. Ind.*, 1968, **10**, 501, (synth, mechanical props)
[71] Parducci, M, *Mater. Plast. Elastomeri*, 1966, **32**, 94, (mechanical props)
[72] Mortillaro, L., *Mater. Plast. Elastomeri*, 1967, **33**, 1259, (electrical props)
[73] Winslow, F. and Hawkins, W.L., *Mod. Plast.*, 1967, **44**, 41, (weathering)
[74] Schonhorn, H., *J. Polym. Sci., Part B: Polym. Phys.*, 1967, **5**, 919
[75] Ranby, B.G., Chan, K.S. and Brumberger, J., *J. Polym. Sci.*, 1962, **58**, 545
[76] Ciampelli, F. and Tosi, C., *Spectrochim. Acta, Part A*, 1968, **24**, 2157, (ir)
[77] Bacskai, R., Lindeman, L.P., Ransley, D.L. and Sweeney, W.A., *J. Polym. Sci., Part A-1*, 1969, **7**, 247, (H-1 nmr)

Poly(4-methyl-1-pentene), Extrusion, Unfilled

Related Polymers: Poly(4-methyl-1-pentene)
Monomers: 4-Methyl-1-pentene
Material class: Thermoplastic
Polymer Type: polyolefins
CAS Number:

CAS Reg. No.
24979-98-4
25068-26-2

Molecular Formula: $(C_6H_{12})_n$
Fragments: C_6H_{12}
General Information: FDA compliant

Volumetric & Calorimetric Properties:
Density:

No.	Value	Note
1	d 0.84 g/cm^3	ISO 1183 [1]

Vicat Softening Point:

No.	Value	Note
1	130°C	Mandrel SX, ASTM D1525
2	145°C	Mandrel HX, Mandrel HXO, ASTM D1525
3	160°C	Mandrel HXP, ASTM D1525 [1]

Applications/Commercial Products:
Processing & Manufacturing Routes: Extrusion and thermoforming
Applications: Release film for printed circuit boards. Food packaging: microwave trays and covers, food containers, paper coatings. Separators in manuf. of laminates. Automotive component housings. FDA compliance

Trade name	Details	Supplier
Crystalor		Phillips 66

– Poly(4-methyl-1-pentene) films

Mandrel	SX, HX grades	Mitsui Sekka
TPX		Mitsui Petrochemicals

Bibliographic References

[1] Bashford, D., *Thermoplastics Directory and Databook*, Chapman and Hall, 1997, 186

Poly(4-methyl-1-pentene) films   P-395

Related Polymers: Poly(4-methyl-1-pentene)
Monomers: 4-Methyl-1-pentene
Material class: Thermoplastic, Fibres and Films
Polymer Type: polyolefins
CAS Number:

CAS Reg. No.
24979-98-4
25068-26-2

Molecular Formula: $(C_6H_{12})_n$
Fragments: C_6H_{12}
Morphology: Form I isotactic tetragonal a 18.70Å, c 13.68Å 7_2 helix [5]
Miscellaneous: Studies on struct. of isotactic film have been reported [1]

Volumetric & Calorimetric Properties:
Density:

No.	Value	Note
1	d^{30} 0.83 g/cm^3	isotactic [5]
2	d 0.835 g/cm^3	Mandrel X-22, X-88, ISO 1183 [8]
3	d 0.834 g/cm^3	Mandrel X-44, ISO 1183 [8]

Thermal Expansion Coefficient:

No.	Value	Type
1	0.00017 K^{-1}	L

Thermal Conductivity:

No.	Value	Note
1	0.04 W/mK	0.0004 cal (cm s °C)$^{-1}$

Specific Heat Capacity

No.	Value	Note	Type
1	1.96 kJ/kg.C	0.47 cal (g °C)$^{-1}$	P

Melting Temperature:

No.	Value	Note
1	243°C	isotactic [5]

Vicat Softening Point:

No.	Value	Note
1	145°C	Mandrel X-22, ASTM D1525
2	160°C	Mandrel X-44, ASTM D1525
3	185°C	Mandrel X-88 [8]

Transport Properties:
Permeability of Gases: Permeation and diffusion coefficients of He, Ne, Ar, CH$_4$, propane, isobutane and neopentane in isotactic films increase with increasing temp. (30–80°) but are independent of film thickness and gas pressure [3]. Permeability constants *in vacuo*; sample thickness 0.01076 cm [4] [N$_2$] 7.83 × 10^{-10} cm^3 cm cm^{-2} s^{-1} (cm Hg)$^{-1}$, [O$_2$] 32.3 × 10^{-10}, [CO$_2$] 92.6 × 10^{-10}, [He] 101 × 10^{-10} [H$_2$] 136 × 10^{10}. Diffusion constants *in vacuo*, sample thickness 10760 cm^3 [4] [N$_2$] 5.5 × 10^7 cm^2 s^{-1} [O$_2$] 10.1 × 10^7 cm^2 s^{-1}. Electron-beam irradiation (1–80 M rad in N$_2$) has little effect on the permeability coefficient [6]

Mechanical Properties:
Tensile Strength Break:

No.	Value	Note	Elongation
1	360 MPa	18° isotactic [5]	30%
2	37 MPa	150° isotactic [5]	67%

Elastic Modulus:

No.	Value	Note
1	687 MPa	23°, 0.05 mm, TPX-22
2	1079 MPa	23°, 0.05 mm, TPX X-44
3	1472 MPa	23°, 0.05 mm, TPX X-88

Tensile Strength Yield:

No.	Value	Note
1	16.7 MPa	23°, 0.05 mm, TPX-22
2	23.1 MPa	23°, 0.05 mm, TPX X-44
3	28.5 MPa	23°, 0.05 mm, TPX X-88

Mechanical Properties Miscellaneous: Stress-strain behaviour of ultra-drawn isotactic films has been reported [7]
Failure Properties General: Tear resistance 3500 g mm^{-1} (25°, 0.05 mm, TPX X-22); 1000 g mm^{-1} (23°, 0.05 mm, TPX X-44); 500 g mm^{-1} (23°, 0.05 mm, TPX X-88)

Electrical Properties:
Surface/Volume Resistance:

No.	Value	Type
1	10 × 10^{15} Ω.cm	S

Dielectric Permittivity (Constant):

No.	Value
1	2.12

Dielectric Strength:

No.	Value
1	63–65 kV.mm^{-1}

Optical Properties:
Transmission and Spectra: 92–94% transmission
Volume Properties/Surface Properties: Haze 1.5% (23°, TPX X-22); 20% (23° TPX X-88)

– Poly(4-methyl-1-pentene), Glass reinforced

Polymer Stability:
Decomposition Details: Oxidative thermal stability and oxidation induction time both increase with increasing degree of orientation of the film [2]

Applications/Commercial Products:

Trade name	Details	Supplier
Mandrel	X-22, X44, X-88	Mitsui Sekka
Opulent		Mitsui Petrochemicals
TPX		Mitsui Petrochemicals

Bibliographic References
[1] Tsuji, M., Tosaka, M., Kawaguchi, A., Katayama, K. and Iwatsuki, M., *Sen'i Gakkaishi*, 1992, **48**, 384, (struct)
[2] Rapoport, N.Ya, Shibryaeva, L.S. and Miller, V.B, *Vysokomol. Soedin., Ser. A*, 1983, **25**, 831, (degradation)
[3] Tschamler, H., Pesta, O. and Rudorfer, D. D, *Mitt. Chem. Forschungsinst. Wirtsch. Oesterr. Oesterr. Kunststoffinst.*, 1974, **28**, 19, (gas permeability)
[4] Yasuda, H. and Rosengren, K.J., *J. Appl. Polym. Sci.*, 1970, **14**, 2839, (gas permeability)
[5] Nakamae, K, Nishino, T. and Takagi, S., *J. Macromol. Sci., Phys.*, 1991, **30**, 47, (elastic modulus)
[6] Kita, H., Muraoka, M., Tanaka, K. and Okamoto, K., *Polym. J. (Tokyo)*, 1988, **20**, 485, (gas permeability)
[7] Kanamoto, T. and Ohtsu, O., *Polym. J. (Tokyo)*, 1988, **20**, 179, (stress-strain behaviour)
[8] Bashford, D, *Thermoplastics Directory and Databook*, Chapman and Hall, 1997, 186

Poly(4-methyl-1-pentene), Glass reinforced P-396
Related Polymers: Poly(4-methyl-1-pentene)
Monomers: 4-Methyl-1-pentene
Material class: Thermoplastic, Composites
Polymer Type: polyolefins
Molecular Formula: $(C_6H_{12})_n$
Fragments: C_6H_{12}
Additives: 10–40% Short Glass fibre reinforced
General Information: Glass reinforcement increases mechanical props. relative to unfilled grades

Volumetric & Calorimetric Properties:
Density:

No.	Value	Note
1	d 0.91 g/cm³	10% filled [1]
2	d 0.99 g/cm³	20% filled [1]
3	d 1.04 g/cm³	30% filled [1]
4	d 1.1 g/cm³	40% filled [1]

Deflection Temperature:

No.	Value	Note
1	75°C	1.8 Mpa, 10% filled [1]
2	87°C	1.8 Mpa, 20% filled [1]
3	116°C	1.8 Mpa, 30% filled [1]
4	118°C	1.8 Mpa, 40% filled [1]
5	135°C	0.45 Mpa, 10% filled [1]
6	150°C	0.45 Mpa, 20% filled [1]
7	161°C	0.45 Mpa, 30% filled [1]
8	164°C	0.45 Mpa, 40% filled [1]

Mechanical Properties:
Tensile (Young's) Modulus:

No.	Value	Note
1	1569 MPa	16000 kg cm^{-2}, 10% filled [1]
2	2559 MPa	26100 kg cm^{-2}, 20% filled [1]
3	3383 MPa	34500 kg cm^{-2}, 30% filled [1]
4	4158 MPa	42400 kg cm^{-2}, 40% filled [1]

Flexural Modulus:

No.	Value	Note
1	1235.6 MPa	12600 kg cm^{-2}, 10% filled [1]
2	2157.5 MPa	22000 kg cm^{-2}, 20% filled [1]
3	2745.9 MPa	28000 kg cm^{-2}, 30% filled [1]
4	4099.1 MPa	41800 kg cm^{-2}, 40% filled [1]

Tensile Strength Break:

No.	Value	Note	Elongation
1	41.28 MPa	421 kg cm^{-2}, 40% filled [1]	
2	21.3 MPa	217 kg cm^{-2}, 10% filled [1]	6%
3	29.6 MPa	302 kg cm^{-2}, 20% filled [1]	3%
4	38.5 MPa	393 kg cm^{-2}, 30% filled [1]	2%

Izod Notch:

No.	Value	Notch	Note
1	36.8 J/m	Y	0.69 ft lb in^{-1}, room temp. 10% filled [1]
2	48.04 J/m	Y	0.9 ft lb in^{-1}, room temp. 20% filled [1]
3	50.1 J/m	Y	0.94 ft lb in^{-1}, room temp. 30% filled [1]
4	53.4 J/m	Y	1 ft lb in^{-1}, room temp. 40% filled [1]

Applications/Commercial Products:
Processing & Manufacturing Routes: Melt temp. can range from 274° to 316° - optimum 293–313°. Mould temp. is 66°. Low pressures are desirable for optimum props. 2.4–3.4 MPa (350–500 psi). Hold pressure 25–50% less than injection pressure. Back pressure 0–0.7 MPa (0–100 psi). Screw speed 50–100 rpm

Mould Shrinkage (%):

No.	Value	Note
1	0.1–1%	
2	1%	10% filled [1]
3	0.5%	20% filled [1]
4	0.2%	30% and 40% filled [1]

Applications: Automotive and electrical applications. Flame retardant version available at 30% fibre

Trade name	Details	Supplier
Crystalor		Phillips 66
J-98	discontinued	DSM Engineering Plastics

Methafil J-98	Wilson-Fiberfil
TPX	Mitsui Petrochemicals

Bibliographic References
[1] *TPX*, Mitsui Petrochemicals, (technical datasheet)

Poly(4-methyl-1-pentene), Injection moulding, Glass filled
P-397

Related Polymers: Poly(4-methyl-1-pentene)
Monomers: 4-Methyl-1-pentene
Material class: Thermoplastic, Composites
Polymer Type: polyolefins
Molecular Formula: $(C_6H_{12})_n$
Fragments: C_6H_{12}
Additives: 30–40% Glass filled

Volumetric & Calorimetric Properties:
Density:

No.	Value	Note
1	d 0.93 g/cm^3	15% filled, ASTM D792 [1]
2	d 1.03 g/cm^3	30% filled, ASTM D792 [1]
3	d 1.43 g/cm^3	30% filled, flame retardant, ASTM D792 [1]

Thermal Expansion Coefficient:

No.	Value	Note	Type
1	6×10^{-5} K^{-1}	15% filled, ASTM D696 [1]	L
2	3.9×10^{-5} K^{-1}	30% filled, ASTM D696 [1]	L
3	3×10^{-5} K^{-1}	30% filled, flame retardant, ASTM D696 [1]	L

Deflection Temperature:

No.	Value	Note
1	175°C	1.8 MPa, 15% filled, ASTM D648 [1]
2	190°C	1.8 MPa, 30% filled, ASTM D648 [1]
3	165°C	1.8 MPa, 30% filled, flame retardant, ASTM D648 [1]

Transport Properties:
Water Absorption:

No.	Value	Note
1	0.01 %	max., 24h, 15% and 30% filled, ASTM D570 [1]
2	0.01 %	24h, 30% filled, flame retardant, ASTM D570 [1]

Mechanical Properties:
Flexural Modulus:

No.	Value	Note
1	3000 MPa	15% filled, ASTM D790 [1]
2	5500 MPa	30% filled, ASTM D790 [1]
3	7100 MPa	30% filled, fire retardant, ASTM D790 [1]

Flexural Strength at Break:

No.	Value	Note
1	75 MPa	15% filled, ASTM D790 [1]
2	93 MPa	30% filled, ASTM D790 [1]
3	89 MPa	30% filled fire retardant, ASTM D790 [1]

Tensile Strength Yield:

No.	Value	Elongation	Note
1	47 MPa	2%	15% filled, ASTM D638 [1]
2	60 MPa	2%	30% filled, ASTM D638 [1]
3	52 MPa	2%	30% filled, fire retardant, ASTM D638

Hardness: Rockwell R105 (15% filled, ASTM D785) [1], R107 (30% filled, ASTM D785) [1], R95 (30% filled, flame retardant, ASTM D785) [1]

Izod Area:

No.	Value	Notch	Note
1	8 kJ/m^2	Notched	23°, 15% filled, ASTM D256 [1]
2	7 kJ/m^2	Notched	23°, 30% filled, ASTM D256 [1]
3	4 kJ/m^2	Notched	23°, 30% filled, fire retardant, ASTM D256 [1]

Electrical Properties:
Dielectric Permittivity (Constant):

No.	Value	Frequency	Note
1	2.3	1 MHz	15% filled, ASTM D150 [1]
2	2.5	1 MHz	30% filled, ASTM D150 [1]
3	2.9	1 MHz	30% filled, fire retardant, ASTM D150 [1]

Dielectric Strength:

No.	Value	Note
1	30 kV.mm^{-1}	15% filled, ASTM D149 [1]
2	27 kV.mm^{-1}	30% filled, ASTM D149 [1]
3	24 kV.mm^{-1}	30% filled, fire retardant, ASTM D149 [1]

Complex Permittivity and Electroactive Polymers: Comparative tracking index 600 V (min) (all grades, IEC 112A) [1]
Dissipation (Power) Factor:

No.	Value	Frequency	Note
1	0.0002	1 MHz	15% filled, ASTM D150 [1]
2	0.0007	1 MHz	30% filled, ASTM D150 [1]
3	0.0012	1 MHz	30% filled, fire retardant, ASTM D150 [1]

Polymer Stability:
Flammability: Flammability rating HB (15% and 30% filled, UL94) [1]; V0 (30% filled, flame retardant UL94) [1]

Applications/Commercial Products:
Mould Shrinkage (%):

No.	Value	Note
1	0.3–0.7%	
2	0.6–0.8%	15% filled, ASTM D955 [1]
3	0.4–0.6%	30% filled, ASTM D955 [1]
4	0.3–0.5%	30% filled, flame retardant, ASTM D955 [1]

Trade name	Supplier
Crystalor HBG-30	Phillips 66
FR-TPX	Mitsui Petrochemicals
	Wilson-Fiberfil

Bibliographic References
[1] Bashford, D., *Thermoplastics Directory and Databook*, Chapman and Hall, 1997, 186

Poly(4-methyl-1-pentene), Injection moulding, Mineral filled   P-398

Related Polymers: Poly(4-methyl-1-pentene)
Monomers: 4-Methyl-1-pentene
Material class: Thermoplastic, Composites
Polymer Type: polyolefins
Molecular Formula: $(C_6H_{12})_n$
Fragments: C_6H_{12}
Additives: 15–35% Mica filler or mica copolymer

Volumetric & Calorimetric Properties:
Density:

No.	Value	Note
1	d 0.91 g/cm^3	15% filled [1]
2	d 1.07 g/cm^3	mica copolymer [1]

Mechanical Properties:
Tensile (Young's) Modulus:

No.	Value	Note
1	2343.8 MPa	23900 kg cm^{-2}, 15% filled [1]
2	3373.5 MPa	34400 kg cm^{-2}, mica copolymer [1]

Flexural Modulus:

No.	Value	Note
1	2961.6 MPa	30200 kg cm^{-2}, 15% filled [1]
2	4962.2 MPa	50600 kg cm^{-2}, mica copolymer [1]

Tensile Strength Break:

No.	Value	Note	Elongation
1	28.9 MPa	295 kg cm^{-2}, 15% filled [1]	2%
2	32.3 MPa	330 kg cm^{-2}, mica copolymer [1]	1%

Izod Area:

No.	Value	Notch	Note
1	26.7 kJ/m^2	notched	0.5 ft lb in^{-1}, room temp. [1]

Applications/Commercial Products:
Mould Shrinkage (%):

No.	Value	Note
1	1.2–3.5%	
2	1.8%	15% filled [1]
3	1.2%	mica copolymer [1]

Trade name	Details	Supplier
MP-98/MB	(discontinued)	DSM Engineering Plastics
Methanfil MP-98/MB		Wilson-Fiberfil

Bibliographic References
[1] *DSM Engineering Plastics Product Information*, DSM Engineering, (technical datasheet)

Poly(4-methyl-1-pentene), Moulded   P-399

Related Polymers: Poly(4-methyl-1-pentene)
Monomers: 4-Methyl-1-pentene
Material class: Thermoplastic
Polymer Type: polyolefins
CAS Number:

CAS Reg. No.
113834-30-3
24979-98-4
25068-26-2

Molecular Formula: $(C_6H_{12})_n$
Fragments: C_6H_{12}
Additives: May be 'white filled' (opaque). Easily coloured
General Information: Includes blow moulded and injection moulded materials

Volumetric & Calorimetric Properties:
Density:

No.	Value	Note
1	d 0.833 g/cm^3	RT18, RT18XB, DX820, ASTM D792 [3]
2	d 0.835 g/cm^3	MX002, ASTM D792 [3]
3	d 0.834 g/cm^3	MX004, ASTM D792 [3]
4	d 1.08 g/cm^3	MBZ230; ASTM D792

Thermal Expansion Coefficient:

No.	Value	Note	Type
1	0.000117 K^{-1}	ASTM D696 [3]	L

Poly(4-methyl-1-pentene), Moulded

Melting Temperature:

No.	Value	Note
1	236°C	injection moulded [2]

Glass Transition Temperature:

No.	Value	Note
1	40–44°C	injection moulded [2]

Deflection Temperature:

No.	Value	Note
1	90°C	0.45 MPa, RT18, RT18XB; ASTM D648 [3]
2	80°C	0.45 MPa, MX002; ASTM D648 [3]
3	85°C	0.45 MPa, MX004; ASTM D648 [3]
4	110°C	0.45 MPa, MBZ230; ASTM D648 [3]

Vicat Softening Point:

No.	Value	Note
1	173°C	9.8N, RT18, RT18XB, DX820; ASTM D1525 [3]
2	145°C	9.8N, MX002; ASTM D1525 [3]
3	160°C	9.8N, MX004; ASTM D1525 [3]
4	175°C	9.8N, MBZ230; ASTM D1525 [3]

Transport Properties:
Melt Flow Index:

No.	Value	Note
1	26 g/10 min	TPX RT18, TPX RT18XB, TPX MX004; 260°, ASTM D1238 [3]
2	180 g/10 min	TPX DX820; 260°, ASTM D1238 [3]
3	22 g/10 min	TPX MX002, ASTM D1238 [3]
4	30 g/10 min	TPX MBZ230, ASTM D1238 [3]

Water Absorption:

No.	Value	Note
1	0.01 %	24h, ASTM D570 [3]
2	0.11 %	24h, TPX MBZ230, ASTM D570 [3]

Mechanical Properties:
Tensile Strength Yield:

No.	Value	Elongation	Note
1	2.35 MPa	25%	TPX RT18, RT18XB; ASTM D638 [3]
2	2.3 MPa	20%	TPX DX820; ASTM D638 [3]
3	1.5 MPa	120%	TPX MX002; ASTM D638 [3]
4	2 MPa	85%	TPX MX004; ASTM D638 [3]
5	2.4 MPa	25%	TPX MBZ230; ASTM D638 [3]

Fracture Mechanical Properties: Fatigue fracture behaviour of an injection-moulded sampling is not very sensitive to post-moulding conditioning. Room temp. ageing has little effect on stress-relaxation behaviour [1]

Electrical Properties:
Dielectric Permittivity (Constant):

No.	Value	Frequency	Note
1	2.12	100Hz, ASTM D150	[3]

Dielectric Strength:

No.	Value	Note
1	65 kV.mm^{-1}	ASTM D149 [3]
2	63 kV.mm^{-1}	TPX MX002; ASTM D149 [3]

Optical Properties:
Volume Properties/Surface Properties: Transparency 91–92% [3]

Applications/Commercial Products:
Processing & Manufacturing Routes: Blow moulding, preferred temp. at lip part of the core 280–300°. Injection moulding at 260–330°

Mould Shrinkage (%):

No.	Value	Note
1	1.2–3.4%	
2	1.7–2.1%	TPX RT18 [3]
3	1.7–1.9%	TPX MX004 [3]
4	1.4–1.7%	TPX MBZ230 [3]

Applications: Bottles for medical use. Cultivating vessels. Laboratory ware

Trade name	Details	Supplier
Crystalor		Phillips 66
TPX	RT and MX series injection or extrusion moulded grades	Mitsui Petrochemicals

Bibliographic References
[1] Sandilands, G.J. and White, J.R., *J. Appl. Polym. Sci.*, 1985, **30**, 4771, (ageing)
[2] Krüger, J., Peetz, L. and Pietralla, M., *Polymer*, 1978, **19**, 1397, (surface props)
[3] Bashford, D., *Thermoplastics Directory and Databook*, Chapman and Hall, 1997, 186

Poly(4-methyl-1-pentene) rubber   P-400

Related Polymers: Poly(4-methyl-1-pentene)
Monomers: 4-Methyl-1-pentene
Material class: Synthetic Elastomers
Polymer Type: polyolefins
Molecular Formula: $(C_6H_{12})_n$
Fragments: C_6H_{12}
Miscellaneous: Displays rubber props.

Volumetric & Calorimetric Properties:
Density:

No.	Value	Note
1	d 0.84 g/cm^3	all grades [1]

Melting Temperature:

No.	Value	Note
1	235–240°C	HX grades [1]
2	220°C	SX [1]

Deflection Temperature:

No.	Value	Note
1	80°C	0.45 MPa, HX [1]
2	75°C	0.45 MPa, SX [1]

Vicat Softening Point:

No.	Value	Note
1	145°C	HX [1]
2	130°C	SX [1]

Mechanical Properties:
Flexural Modulus:

No.	Value	Note
1	470.7 MPa	4800–5000 kg cm^{-2}, HX [1]
2	245.2 MPa	2500 kg cm^{-2}, SX [1]

Tensile Strength Break:

No.	Value	Note	Elongation
1	15.7 MPa	160 kg cm^{-2}, HX [1]	120%
2	9.8 MPa	100 kg cm^{-2}, SX [1]	160%

Tensile Strength Yield:

No.	Value	Note
1	14.7 MPa	150 kg cm^{-2} [1]
2	8.8 MPa	90 kg cm^{-2}, SX [1]

Hardness: Shore D63 (HX) [1]; D48 (SX) [1]
Izod Notch:

No.	Value	Notch	Note
1	39.2 J/m	Y	0.735 ft lb in^{-1}, room temp., HX [1]
2	490.5 J/m	Y	9.19 ft lb in^{-1}, room temp., SX [1]

Applications/Commercial Products:

Trade name	Details	Supplier
TPX	HX and SX grades	Mitsui Petrochemicals

Bibliographic References
[1] *TPX*, Mitsui Petrochemicals, 1996, (technical datasheet)

Poly(methylphenylsilane)   P-401

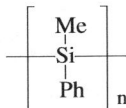

Synonyms: *Poly(phenylmethylsilane)*. *Poly(methylphenylsilylene)*
Related Polymers: Dichloropoly(methylphenylsilane), Polysilanes
Monomers: Dichloromethylphenylsilane
Material class: Thermoplastic
Polymer Type: polysilanes
CAS Number:

CAS Reg. No.
76188-55-1

Molecular Formula: $(C_7H_8Si)_n$
Fragments: C_7H_8Si
Molecular Weight: MW 20000–1000000

Volumetric & Calorimetric Properties:
Melting Temperature:

No.	Value	Note
1	220°C	[5]

Surface Properties & Solubility:
Solvents/Non-solvents: Sol. aromatic hydrocarbons, chlorinated hydrocarbons, THF. Mod. sol. ethers, aliphatic hydrocarbons. Slightly sol. acetonitrile. Insol. alcohols, H_2O [1,2,3,8]

Electrical Properties:
Electrical Properties General: Volume resistivity for polymer doped with AsF_5 1–10 Ω cm[5]
Complex Permittivity and Electroactive Polymers: Holemobility 2×10^{-4} cm^2 V^{-1} s^{-1} (2×10^5 V cm^{-1}, room temp.) [7,10]

Optical Properties:
Transmission and Spectra: Strong absorption 338–341 nm. Fluorescence 360 nm [4,6,9]

Polymer Stability:
Upper Use Temperature:

No.	Value
1	220°C

Environmental Stress: Decomposed by uv below 350 nm. Cross-linked by uv of 350 nm or above [6,9], radical cations are generated
Chemical Stability: Inert to air and atmospheric moisture. Gases break down SiSi bond with evolution of hydrogen. Oxidised by strong oxidising agents [3]
Hydrolytic Stability: Stable to H_2O and acid solns. Basic solns. decompose SiSi bonds; hydrogen is evolved [3]

Applications/Commercial Products:

Processing & Manufacturing Routes: Prepd. by Wurtz coupling of [1,2,8] dichloromethylphenylsilane in organic solvents at around 110°. The reaction is quenched by addition of water. Other synth. has been reported [11]

Applications: Potentially useful electronic (light-emitting diodes) and non-linear optical props. Can be used in microlithography due to uv sensitivity

Bibliographic References

[1] Jones, R.G., Benfield, R.E., Cragg, R.H., Swain, A.C. and Webb, S.J., *Macromolecules*, 1993, **26**, 4878
[2] Demoustier-Champagne, S., Marchand-Brynaert, J. and Devaux, J., *Eur. Polym. J.*, 1996, **32**, 1037
[3] West, R. and Maxka, J., *ACS Symp. Ser.*, 1988, **360**, 6
[4] Yu-Ling, H., Banovetz, J.P. and Waymouth, R.M., *ACS Symp. Ser.*, 1994, 55
[5] West, R., *Comprehensive Organometallic Chemistry*, (eds. G. Wilkinson, F.G.A. Stone and E.W. Abel), Pergamon Press, 1982, **2**, 365
[6] Trefonas, P., *Encycl. Polym. Sci. Eng.*, Vol. 13, 2nd edn., (ed. J.I. Kroschwitz), John Wiley and Sons, 1985, 162
[7] Stolka, M., Yuh, H.-J., McGrane, K. and Pai, D.M., *J. Polym. Sci., Part A: Polym. Chem.*, 1987, **25**, 823
[8] Trefonas, P., Djurovich, P.I., Zhang, X.-H., West, R., Miller, R.D. and Hofer, D., *J. Polym. Sci., Polym. Lett. Ed.*, 1983, **21**, 819
[9] Trefonas, P., West, R., Miller, R.D. and Hofer, D., *J. Polym. Sci., Polym. Lett. Ed.*, 1983, **21**, 823
[10] Eckhardt, A. and Schnabel, W., *J. Inorg. Organomet. Polym.*, 1996, **6**, 95
[12] Suzuki, H., *Adv. Mater.*, 1993, **5**, 743, (use)

Polymethylphenylsiloxane fluid P-402

Synonyms: *Poly((dimethyl)(methylphenyl)siloxane) fluid*
Related Polymers: PDMS fluid, Polyphenylsiloxane fluid
Monomers: Dichlorodimethylsilane, Dichloromethylphenylsilane, Hexamethylcyclotrisiloxane, Cyclic methylphenylsiloxane
Material class: Fluids, Copolymers
Polymer Type: silicones, dimethylsilicones
CAS Number:

CAS Reg. No.
156048-35-0

Molecular Formula: $[(C_7H_8OSi).(C_2H_6OSi)]_n$
Fragments: C_7H_8OSi C_2H_6OSi
Identification: Silicone fluids containing phenyl groups have higher values of density than dimethylsiloxane

Volumetric & Calorimetric Properties:

Density:

No.	Value	Note
1	d 1.05 g/cm^3	[3]

Thermal Expansion Coefficient:

No.	Value	Note	Type
1	0.0009 K^{-1}	[7]	V

Thermal Conductivity:

No.	Value	Note
1	0.15 W/mK	0.087 Btu ft^{-1} h^{-1} °F^{-1} [3,11]

Specific Heat Capacity:

No.	Value	Note	Type
1	1.4 kJ/kg.C	0.34 Btu lb^{-1} °F^{-1} [3]	P

Transition Temperature General: Low levels of phenyl substitution lower the pour point [1,10]
Transition Temperature:

No.	Value	Note
1	-73°C	Pour point (-100°F) [3]

Surface Properties & Solubility:

Cohesive Energy Density Solubility Parameters: δ 16 J$^{1/2}$ cm$^{-3/2}$ (7.8–7.9 cal$^{1/2}$ cm$^{-3/2}$) [5,9]
Solvents/Non-solvents: Sol. aliphatic hydrocarbons, aromatic hydrocarbons, chlorinated hydrocarbons, ethers, esters, ketones. Insol. alcohols, H_2O [1,4,5]. (Low MW methylphenyl silicones are sol. alcohols)
Wettability/Surface Energy and Interfacial Tension: $γ_c$ 21 mJ m^{-2} (21 dyne cm^{-1}) [3,7]. $γ_c$ Increases with increase in phenyl substitution

Transport Properties:

Polymer Melts: Viscosity 0.7 St. Viscosity temp. coefficient 0.65 see [poly phenyl siloxane fluid] [3,4,12,13]

Mechanical Properties:

Miscellaneous Moduli:

No.	Value	Note
1	1650 MPa	16700 kg cm^{-2} [4]

Mechanical Properties Miscellaneous: Compressibility decreases with increasing phenyl substitution [4]

Electrical Properties:

Electrical Properties General: Has excellent insulating props.
Dielectric Permittivity (Constant):

No.	Value	Note
1	2.9	[3,6,14]

Dielectric Strength:

No.	Value	Note
1	29 kV.mm^{-1}	[3]

Dissipation (Power) Factor:

No.	Value	Note
1	0.0013	[3]

Optical Properties:

Optical Properties General: Refractive index higher than that of dimethyl silicone [8]
Refractive Index:

No.	Value	Note
1	1.428	[3]

Transmission and Spectra: Transparent above 280 nm [5]

Polymer Stability:

Polymer Stability General: Stable over wide temp. range
Thermal Stability General: Usable between -75–205° in the presence of air. In the absence of air usable up to 300°

Upper Use Temperature:

No.	Value
1	225°C
2	205°C

Decomposition Details: Oxidised above 225° [2]
Flammability: Fl. p. 220° (430°F) [3]
Environmental Stress: Stable to H_2O, air, oxygen and other materials. Not as resistant to radiation as [polyphenylsiloxane fluid] [1]
Chemical Stability: Chemically inert but decomposed by conc. mineral acids. Very thin films are liable to degradation [1,5,6]
Hydrolytic Stability: Stable to H_2O and aq. solns. Very thin films spread on water; liable to hydrolytic degradation [1,5,6]
Biological Stability: Non-biodegradable [5,6]
Recyclability: Can be reused; no large scale recycling

Applications/Commercial Products:

Processing & Manufacturing Routes: For low viscosity fluids dichlorodimethylsilane and dichloromethylphenylsilane are hydrolysed and equilibrated under acid catalysis at 180° with small amounts of hexamethyldisiloxane. For high viscosity fluids dimethyl- and methylphenyl-cyclic polysiloxanes are polymerised together under alkaline catalysis with traces of chain terminator at 150°

Applications: Used as a hydraulic fluid and for lubrication between -75° and 205°. Possess good electrical props.

Trade name	Supplier
F50	General Electric Silicones

Bibliographic References

[1] Noll, W., *Chemistry and Technology of Silicones,* Academic Press, 1968
[2] Murphy, C.M., Saunders, C.E. and Smith, D.C., *Ind. Eng. Chem.,* 1950, **42**, 2462, (thermal stability)
[3] Ash, M. and Ash, I., *Handbook of Plastic Compounds, Elastomers and Resins,* (eds. M.B. Ash and I.A. Ash), Wiley-VCH, 1992
[4] Meals, R.N. and Lewis, F.M., *Silicones,* Reinhold, 1959
[5] Hardman, B. and Torkelson, A., *Encycl. Polym. Sci. Eng.,* 2nd edn. (ed. J.I. Kroshwitz), John Wiley and Sons, 1985, **15**, 204
[6] Stark, F.O., Fallender, J.R. and Wright, A.P., *Comprehensive Organometallic Chemistry,* (eds. G. Wilkinson, F.G.A. Stone and E.W. Abel), Pergamon Press, 1982, **2**, 305
[7] Fox, H.W., Taylor, P.W. and Zisman, W.A., *Ind. Eng. Chem.,* 1947, **39**, 1401
[8] Flaningan, O.L. and Langley, N.R., *Anal. Chem. Silicones,* (ed. A.L. Smith), John Wiley and Sons, 1991, 135
[9] Yerrick, K.B. and Beck, H.N., *Rubber Chem. Technol.,* 1964, **37**, 261, (solubility)
[10] Warrick, E.L., Hunter, M.J. and Barry, A.J., *Ind. Eng. Chem.,* 1952, **44**, 2196
[11] Jamieson, D.T. and Irving, J.B., *Proc. Int. Conf. Therm. Conduct.,* 1975, 279, (thermal conductivity)
[12] Buch, R.R., Klimisch, H.M. and Johannson, O.K., *J. Polym. Sci., Part A-2,* 1970, **8**, 541, (viscosity)
[13] Lee, C.L. and Haberland, C.G., *J. Polym. Sci., Part B: Polym. Lett.,* 1965, **3**, 883, (viscosity)
[14] Vincent, G.A., Fearon, F.W.G. and Orbech, T., *Annu. Rep., Conf. Electr. Insul. Dielectr. Phenom.,* 1972, 17, (dielectric constant)

Poly[(methylphenylsilylene)(methylene)] P-403

$$\left[-CH_2-\underset{\underset{Ph}{|}}{\overset{\overset{CH_3}{|}}{Si}}- \right]_n$$

Synonyms: *1,3-Dimethyl-1,3-diphenyl-1,3-disilacyclobutane homopolymer, SRU. Poly(methylphenylcarbosilane).* PMPSM
Related Polymers: Polycarbosilanes

Polymer Type: polycarbosilanes
CAS Number:

CAS Reg. No.	Note
32202-09-8	Poly[(methylphenylsilylene)(methylene)]

Molecular Formula: $(C_8H_{10}Si)_n$
Fragments: $C_8H_{10}Si$
Molecular Weight: MW 82,500 [2], 200,000 [1]. There is a molecular weight dependence on the preparative method employed [9]. The molecular weight decrease caused by main chain scission and insolubilisation, which is attributable to cross-linking, were observed when the reaction proceeded heterogeneously [9]. The highest MW polymer was synthesised when a pure trans isomer was used as monomer [12].
General Information: Slightly yellow solid. Suitable starting polymer for the preparation of numerous functional substituted poly(silylenemethylenes).
Tacticity: Atactic [2]. Rh- and Pt-catalysed polymerisation products show slightly different tacticities [12].
Morphology: Elastic thermalisation lifetime analysis shows a slow relaxation process, leading to an equilibrium structure with small radii of the cavities [5]. Scanning electron microscopy shows the polymer to have a relatively fine-grained surface [8].
Identification: Elemental analysis: %C calc. 71.57 found 70.98; %H calc. 7.51 found 7.24 [2].

Volumetric & Calorimetric Properties:
Melting Temperature:

No.	Value	Note
1	125–140°C	[2]

Glass-Transition Temperature:

No.	Value	Note
1	22–25°C	[2,6,10] Measured for a series of PMPSMs and tacticities by DSC [12].

Surface Properties & Solubility:
Solvents/Non-solvents: Soluble in $CHCl_3$ and THF. Insoluble in alcohols and aliphatic hydrocarbons [2].

Transport Properties:
Volatile Loss and Diffusion of Polymers: 10% wt. loss temp. 476°C [1,12]

Mechanical Properties:
Tensile (Young's) Modulus:

No.	Value	Note
1	100–850 MPa	[12]

Tensile Strength Break:

No.	Value	Note	Elongation
1	4–16 MPa	[12]	Show great elongation above 200%: 200%–850%

Optical Properties:
Transmission and Spectra: pmr ($CDCl_3$) = 0.35 ($SiCH_2$), 0.56 ($SiCH_3$), 6.9–7.8 (Ph) [2,12].
cmr ($CDCl_3$) = 1.2 ($SiCH_3$), 5.5 ($SiCH_2$), 127.0–140.5 (Ph) [2,12]. Si-29 ($CDCl_3$) = -4.7 [2].
A theoretical model (ETLA) is tested by means of amorphous poly(methylphenylsilenmethylene). Estimated values of the localisation time constants are 137ps [7].

IR: Characteristic absorption at 1055 cm^{-1} for the Si-C-Si backbone, Si-H at 2115cm^{-1} [10,12].

Polymer Stability:
Polymer Stability General: Exhibits very good thermal stability in the absence of oxidising agents [9,12,13,]. Weight percentage of material obtained after heating the polymer to 800°C are tabulated [12]
Thermal Stability General: Weight remains unchanged below about 300°C. Weight loss occurs at 467°C in air, and around 600°C in nitrogen. Showed a two step weight loss in the temperature range 400°C and 700°C in an air atmosphere [10,12]. Other studies claim stable up to 500°C in inert atmosphere; easily oxidised in air above 220°C [15,16]
Decomposition Details: The residue after heating to 800°C was an off-white solid forming a silica-like material. The major products of this pyrolysis were benzene, toluene, xylene and biphenyl [10,12]. A spectroscopic analysis of the pyrolysed products are tabulated [12]

Applications/Commercial Products:
Processing & Manufacturing Routes: Radical polymerisation of 1,3-disilacyclobutanes having one or more aryl groups in the presence of radical generators and 1,3-disilacyclobutanes having no radially reactive groups, then ring-opening and cross-linking of them under heating [1,12]. Bulk polymerisation by Rh, with or without catalysed catalyst polymerisation [4,12]. Pt catalysed [13,15,16,17]
Applications: Thermolysis >800°C produces silicon carbide ceramic fibre [1]. Can be used for copolymer with polystyrene. May have value as resistant elastomer at high temp. in the absence of oxygen. A chemical sensor having a transducer element and a layer of composite material including a polymer matrix and a solid particulate filler disposed in the polymer matrix provides chemical sensors exhibiting improved properties [11].

Bibliographic References
[1] Ogawa T., Murakami M. and Lee S.D., *Jpn. Pat.*, (Dow Corning asia Ltd), 2000, 200336170
[2] Uhlig, W., *Polym. Adv. Technol.*, 1999, **10**, 513
[3] Rushkin, I.L. and Interrante, L.V., *Macromolecules*, 1996, **29**, 3123
[4] Ogawa, T., and Murakami, M., *Chem. Mater.*, 1996, **8**, 1260
[5] Kansy, J., Suzuki, T., Ogawa, T., and Murakami, M., *Radiat. Phys. Chem.*, 2000, **58**, 545
[6] Ogawa, T., *Polymer*, 1998, **39(13)**, 2715
[7] Kansy, J., and Suzuki, T., *Radiat. Phys. Chem.*, 2003, **68**, 497
[8] Ogawa, T., and Murakami, M., J. Polym. Sci., Part A: Polym. Chem., 1997, **35**, 399
[9] Ogawa, T., and Murakami, M., J. Polym. Sci., Part A: Polym. Chem., 1997, **35**, 1431
[10] Ogawa, T., Dolee, S., and Murakami, M., J. Polym. Sci., Part A: Polym. Chem., 2002, **40**, 416
[11] Hartmann-Thompson, C., *U.S. Pat.*, 2005, 20050090015
[12] Ogawa, T., and Murakami, M., *Chem. Mater.*, 1996, **8**, 1260
[13] Koopmann, F., Burgath, A., Knischka, J.L and Frey, H., *Acta Polym.*, 1996, **47**, 377
[14] Seyferth, D., *ACS Symp. Ser.*, 1988, **360**, 21
[15] Krinev, W.A., J. Polym. Sci., Part A-1, 1966, **4**, 444
[16] Trefonas, P., *Encycl. Polym. Sci. Eng.*, Vol. 13, 2nd edn., (ed. J.I. Kroschwitz), John Wiley and Sons, 1985, 162
[17] Shen, Q., Interrante, L.V. and Wu, H.-J., Polym. Prepr. (Am. Chem. Soc., Div. Polym. Chem.), 1994, **35**, 395

Poly[(methylphenylsilylene)methylene]   P-404

$$\left[\text{CH}_2 - \underset{\underset{\text{Ph}}{|}}{\overset{\overset{\text{Me}}{|}}{\text{Si}}} \right]_n$$

Polymer Type: carbosilane polymers
CAS Number:

CAS Reg. No.
98241-68-0

Molecular Formula: $(C_8H_{10}Si)n$
Fragments: $C_8H_{10}Si$
Molecular Weight: 2,000–3,000 [1]
General Information: End capped α-ethyl-ω-bromo-

Applications/Commercial Products:
Processing & Manufacturing Routes: Prepared by treating Grignard reagents and silanes [1]

Bibliographic References
[1] Jpn. Pat. (Shin-Etsu Chemical Industry), *CA*,124130, 1985, 60 084 330

Poly[1,1-(2-methylpropane) bis(4-phenyl) carbonate]   P-405

$$\left[-O-\!\!\!\bigcirc\!\!\!-\underset{\underset{\text{CH(CH}_3)_2}{|}}{\text{CH}}-\!\!\!\bigcirc\!\!\!-O-\overset{\overset{O}{\|}}{C} \right]_n$$

Synonyms: *Poly[oxycarbonyloxy-1,4-phenylene(2-methylpropylidene)-1,4-phenylene]. Poly(4,4'-dihydroxydiphenyl-1,1-isobutane-co-carbonic dichloride)*
Monomers: 4,4'-Dihydroxydiphenyl-1,1-isobutane, Carbonic dichloride
Material class: Thermoplastic
Polymer Type: polycarbonates (miscellaneous)
CAS Number:

CAS Reg. No.
32242-75-4

Molecular Formula: $(C_{17}H_{16}O_3)_n$
Fragments: $C_{17}H_{16}O_3$

Volumetric & Calorimetric Properties:
Density:

No.	Value	Note
1	d 1.18 g/cm^3	25° [2]

Melting Temperature:

No.	Value	Note
1	170–180°C	[2]

Glass-Transition Temperature:

No.	Value	Note
1	149°C	[2]

Surface Properties & Solubility:
Solvents/Non-solvents: Sol. CH_2Cl_2, *m*-cresol, EtOAc, cyclohexanone, THF, pyridine, DMF; swollen by C_6H_6, Me_2CO [1]

Mechanical Properties:
Tensile Strength Break:

No.	Value	Note	Elongation
1	76 MPa	775 kg cm^{-2} [2]	147%

Electrical Properties:
Electrical Properties General: Dielectric constant and dissipation factor do not vary with temp. until approaching T_g

– Poly(methylpropylsilane)

Dielectric Permittivity (Constant):

No.	Value	Frequency	Note
1	2.4	1 kHz	25–100° [2]

Dissipation (Power) Factor:

No.	Value	Frequency	Note
1	0.00052	1 kHz	25–100° [2]

Optical Properties:
Refractive Index:

No.	Value	Note
1	1.5702	25° [2]

Transmission and Spectra: Ir spectral data have been reported [1]

Applications/Commercial Products:
Processing & Manufacturing Routes: The main synthetic routes are phosgenation and transesterification. The most widely used phosgenation approach is interfacial polycondensation in a suitable solvent, but soln. polymerisation is also carried out in the presence of pyridine [1]. A slight excess of phosgene is introduced into a soln. or suspension of the aromatic dihydroxy compound in aq. NaOH at 20-30°, in an inert solvent (e.g. CH_2Cl_2). Catalysts are used to achieve high MWs [12]. With transesterification, the dihydroxy compound and a slight excess of diphenyl carbonate are mixed together in the melt (150-300°) in the absence of O_2 with elimination of phenol. Reaction rates are increased by alkaline catalysts and the use of a medium-high vacuum in the final stages [1,2]

Bibliographic References
[1] Schnell, H., *Polym. Rev.*, 1964, **9**, 99, (solvents, ir, synth., rev)
[2] Schnell, H., *Ind. Eng. Chem.*, 1959, **51**, 157, (props., synth.)

Poly(methylpropylsilane) P-406

$$\left[\begin{array}{c} CH_2CH_2CH_3 \\ | \\ Si \\ | \\ CH_3 \end{array} \right]$$

Synonyms: *Poly(methylpropylsilylene). Dichloromethylpropylsilane homopolymer*
Monomers: Dichloromethylpropylsilane
Material class: Thermoplastic
Polymer Type: polysilanes
CAS Number:

CAS Reg. No.	Note
88003-13-8	poly(methylpropylsilylene)
88002-81-7	dichloromethylpropylsilane homopolymer

Molecular Formula: $(C_4H_{10}Si)n$
Fragments: $C_4H_{10}Si$
General Information: Conductive polymer which exhibits photoluminescence and thermochromism. Prod. as white solid.
Morphology: At r.t. adopts hexagonal columnar unit cell, with a=11.50, b=9.36, c=3.87Å [1]. Exist in an all-trans planar conformation of the backbone [1,2,3], which changes to a less ordered or isotropic state above 40°C [2,3]. The chains pack with little interpenetration, and the crystals may be considered as bundles of long, closely packed prisms. The restricted interlocking of neighboring chains results, in turn, in a poor register of the chains along the c-axis. Transmission electron microscopy reveals that the crystallised polymer adopts a lamellar microstructure, with parallel lamellae tending to form tight bundles. Crystallises with a nucleation-controlled type of kinetics [1,2]

Volumetric & Calorimetric Properties:
Melting Temperature:

No.	Value	Note
1	48.4–51.8°C	[2,4]

Glass-Transition Temperature:

No.	Value	Note
1	-35°C	[7]

Transition Temperature General: Small endothermic transition at 38–39°C and larger melting transition at 45–50°C [2,7].

Surface Properties & Solubility:
Solvents/Non-solvents: Sol. C_6H_6, toluene, THF, chloroform, alkanes [5,7].

Electrical Properties:
Electrical Properties General: Photoconductive behaviour indicates hole transport predominates. Band gap 3.6eV [6]

Optical Properties:
Optical Properties General: Thermochromism exhibited with blue shift on increasing the temperature of films, λ_{max} at 325nm (20°C) changes to 304nm (55°C) [3]. Photoluminescence spectrum with λ_{max} at ca. 350nm is quenched by electron acceptors such as C_{60} fullerene [8]. Birefringence is evident at 25°C, while at temperatures above 40°C polymer films are isotropic [7].
Transmission and Spectra: pmr [5,7,9,10,11], cmr [5,7,9,10], Si-29 nmr [9,10], ir [11], uv [3,18], Raman [19], photoluminescence [8] and x-ray diffraction [1,7,12,13] spectral data have been reported.

Polymer Stability:
Polymer Stability General: Prolonged exposure to UV light results in photoscission of σ-bonds, which can be reduced with the addition of electron acceptors such as fullerenes [8].
Thermal Stability General: Thermal degradation at 400°C has been studied by pyrolysis mass spectrometry [14].
Decomposition Details: Onset of thermal decomposition at 260°C, complete decomposition at 540°C [15].

Applications/Commercial Products:
Processing & Manufacturing Routes: Prod. by Wurtz dehalogenation coupling of dichloromethylpropylsilane with sodium in toluene [5,7,10]. The addition of a crown ether [16] or ethyl acetate [17] or the use of THF as solvent [11] to increase the rate of polymerisation has been reported.

Bibliographic References
[1] Furukawa S. and Ohta H., *Jpn. J. Appl. Phys., Part 1*, 2005, **44**, 495
[2] Jambe B., Jonas A., and Devaux J., *J. Polym. Sci., Part B: Polym. Lett.*, 1997, **35**, 1533
[3] Yokoyama, K. and Yokoyama, M., *Solid State Commun.*, 1989, **70**, 241
[4] Radhakrishnan, J., Tanigaki, N. and Kaito, A., *Polymer*, 1999, **40**, 1381
[5] Trefonas, P., Djurovich, P.I., Zhang, X.-H., West, R., Miller, R.D. and Hofer, D., *J. Polym. Sci., Polym. Lett. Ed.*, 1983, **21**, 819
[6] Fujino, M., *Chem. Phys. Lett.*, 1987, **136**, 451
[7] Bukalov S.S., and Leites L.A., *Proc. SPIE- Int. Soc. Opt. Eng.*, **4069**, 2
[8] Ninomiya, S., Ashihara, Y., Nakayama, Y., Oka, K. and West, R., *J. Appl. Phys.*, 1998, **83**, 3652
[9] Schilling, F.C., Bovey, F.A. and Zeigler, J.M., *Macromolecules*, 1986, **19**, 2309
[10] Saxena A., Okoshi K., Fujiki M., Naito M., Guo G., Hagihara T., and Ishikawa M., *Macromolecules*, 2004, **37**, 367
[11] Holder S.J., Achilleos M., and Jones R.G., *Macromolecules*, 2005, **38**, 1633
[12] KariKari, E.K., Greso, A.J., Farmer, B.L., Miller, R.D. and Rabolt, J.F., *Macromolecules*, 1993, **26**, 3937

[13] Winokur, M.J., Koe, J. and West, R., *Polym. Prepr. (Am. Chem. Soc., Div. Polym. Chem.)*, 1997, **38**, 57
[14] Dave, P., Israel, S.C. and Sawan, S.P., *Polym. Prepr. (Am. Chem. Soc., Div. Polym. Chem.)*, 1990, **31**, 566
[15] Abu-Eid, M.A., King, R.B. and Kotliar, A.M., *Eur. Polym. J.*, 1992, **28**, 315
[16] Fujino, M. and Isaka, H., *Chem. Comm.*, 1989, 466
[17] Miller R.D. and Jenkner, P.K., *Macromolecules*, 1994, **27**, 5921
[18] Kawaguchi T., *Chem. Phys. Lett.*, 2003, **374**, 353
[19] Bukalov S.S., and Leites L.A., *Proc. SPIE- Int. Soc. Opt. Eng.*, **4069**, 2

Poly(*N*-methylpyrrole) P-407

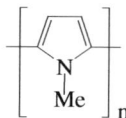

Synonyms: *PNMP*
Related Polymers: Poly(pyrrole)
Monomers: *N*-Methylpyrrole
Polymer Type: polypyrrole
CAS Number:

CAS Reg. No.
72945-66-5

Molecular Formula: $(C_5H_5N)_n$
Fragments: C_5H_5N

Volumetric & Calorimetric Properties:
Density:

No.	Value	Note
1	d 1.33–1.46 g/cm^3	[1]

Electrical Properties:
Dielectric Permittivity (Constant):

No.	Value	Frequency	Note
1	3.8–30.8	10 kHz	[2]

Complex Permittivity and Electroactive Polymers: Conductivity values have been reported, 1×10^{-3} [3]; 10^{-3}–10^{-6} [1]; 6×10^{-12}–5.2×10^{-4} Ω^{-1} cm^{-1} [2]
Dissipation (Power) Factor:

No.	Value	Frequency	Note
1	0.045–10	10 kHz	min. [2]

Optical Properties:
Transmission and Spectra: X-ray photoelectron spectral data have been reported [4]

Applications/Commercial Products:
Processing & Manufacturing Routes: Prod. by chemical or electrochemical oxidative polymerisation of *N*-methylpyrrole [1,5]

Bibliographic References
[1] Diaz, A.F., Castillo, J., Kanazawa, K.K. and Logan, J.A., *J. Electroanal. Chem.*, 1982, **133**, 233
[2] Binder, M., Mammone, R.J. and Schlotter, N.E., *Synth. Met.*, 1990, **39**, 215, (electrical props, spectroscopic props)
[3] Kanazawa, K.K., Diaz, A.F., Geiss, R.H., Gill, W.D. *et al*, J.C.S., *Chem. Commun.*, 1979, 854
[4] Tourillan, G. and Jugnet, J., *J. Chem. Phys.*, 1988, **89**, 1905
[5] Wei, Y., Tian, J. and Yang, D., *Makromol. Chem., Rapid Commun.*, 1991, **12**, 617

Poly[(methylsilylene)(methylene)] P-408

$$\left[\left(\begin{array}{c}Si-CH_2\\|\\CH_3\end{array}\right)_x\left(\begin{array}{c}SiH-CH_2\\|\\CH_3\end{array}\right)_y\left(\begin{array}{c}SiH_2\\|\\CH_3\end{array}\right)_z\right]_n$$

Synonyms: *Poly(1-methyl-1-silylenemethylene)*. *Polycarbomethylsilane*. *Polysilapropylene*. *PMSM*. *OMSM*. *Oligo[(methylsilylene)(methylene)]*
Related Polymers: Polycarbosilanes
Polymer Type: polycarbosilanes
CAS Number:

CAS Reg. No.
62306-27-8

Molecular Formula: $(C_2H_6Si)_n$
Fragments: C_2H_6Si
Molecular Weight: For OMSM: M_n 442; M_w/M_n 1.4 (60%); M_n 28,900, M_w/M_n 2.4 (40%) [3]
General Information: Refers to a series of oligo[(methylsilylene)(methylenes)] (OMSM) and poly[(methylsilylene)(methylenes)] (PMSM). An important feature of this polymer is the presence of a large proportion of dihydridosilicon groups [3]
Tacticity: From the methylene pmr spectrum, syndiotactic and isotactic sequences are both involved in PMSM [2,4]
Morphology: Yellow oil [2]
Identification: For -(SiH(CH$_3$)CH$_2$)n: Calc: C, 41.34; H, 10.33; Cl, 0; Si, 48.33
For -(CH$_3$)$_2$HSiCH$_2$-SiH(CH$_3$)-(CH$_2$)$_{23}$-SiH(CH$_3$)$_2$: Calc: C 41.67; H 10.56; Cl, 0; Found C, 41.6; H, 10.35, Cl, 0.14; Si not given

Surface Properties & Solubility:
Solubility Properties General: Insol. pentane [2]

Optical Properties:
Transmission and Spectra: Ir [2,3], Si-29 nmr and C-13 mas nmr [1,2,3] and gc-ms [2] spectra reported

Polymer Stability:
Decomposition Details: Heating PMSM at 350–450°C (476°C see [4]) in a carefully purged atmosphere (dry argon) gave products identified by gc-ms as: H$_2$, CH$_4$, C$_2$H$_6$, MeSiH$_3$, Me$_2$SiH$_2$, Me$_3$SiH and Me$_4$Si. Small amounts of CH$_2$=CH$_2$ and SiH$_4$ also were detected [2,3]
Pyrolysis of PMSM at 1000°C under an argon flow resulted only in poor ceramic yield (5%). This was due to the linear structure of the polymer. The cleavage of Si-CH$_2$ bonds gave volatile products, which were eliminated since a condenser could not trap them, thermolysis being at atmospheric pressure in an open vessel [2]
Polytitanocarbosilane containing an excess amount of titanium alkoxide was synthesised by the mild reaction of PMSM with titanium(*IV*) tetra-*n*-butoxide at 220°C under nitrogen atmosphere. The obtained precursor polymer was melt-spun at 150°C continuously using melt-spinning equipment with a winding drum. The spun fibre, which contained excess amount of nonreacted titanium alkoxide, was pre-heated to 100°C and subsequently fired up to 1200°C heat-treated in air to obtain continuous lucid fibre [5]

Applications/Commercial Products:
Processing & Manufacturing Routes: Synthesised like other polycarbosilanes by ring-opening polymerisation of disilacyclobutanes using a Pt catalyst (66% yield) [4]. Can also be produced from polydimethylsilane by Kumada rearrangement at 470°C [4]. OMSM was prepared by a Grignard coupling reaction followed by reduction [4]. The chlorinated polycarbosilane was readily converted into the corresponding hydrido or deuterio polymer after refluxing the reaction mixture for 24h in an inert atmosphere (96.5% yield) [2]. Visible light-absorbing platinum(*II*) bis(β-diketonate) complexes for photocatalytic hydrosilylation

cross-linking of OMSM with tetravinylsilane also used [3]
Applications: Coated silicon wafers [4]. Pyrolysis to produce semiconductor-grade SiC layers [3,4]. Strong photocatalytic fibre (TiO$_2$-covered SiO$_2$ fibre) [5]. Coliform-sterilisation ability of TiO$_2$-covered SiO$_2$ fibre [5]. Powder metallurgical compositions containing organometallic lubricants [6]. Antireflective hard mask compositions [7].

Trade name	Supplier
Polycarbomethylsilane	Aldrich

Bibliographic References

[1] Yong, Y. and You, Z., *Bopuxue Zazhi*, 1996, **13**, 567
[2] Bacque, E., Pillot, J.P, Birot, M. and Dunogugs, J., *Macromolecules*, 1998, **21**, 34
[3] Guo, A., Fry, B.E. and Neckers, D.C., *Chem. Mater.*, 1998, **10**, 531
[4] Fry, B.E, Guo, A. and Neckers, D.C., *J. Organomet. Chem.*, 1997, **538**, 151
[5] Ishikawa, T., *Int. J. Appl. Ceram. Technol.*, 2004, **1**, 49
[6] *U.S. Pat. Appl. Publ.*, 2006, 06 034 723
[7] *U.S. Pat. Appl. Publ.*, 2005, 05 042 538

Poly(2-methylstyrene)   P-409

Synonyms: *Poly(o-methylstyrene). Poly(2-methyl-1-vinyl-benzene). Poly(1-(1-methylphenyl)-1,2-ethanediyl)*
Related Polymers: Poly(4-methylstyrene)
Monomers: 2-Methylstyrene
Material class: Thermoplastic
Polymer Type: styrenes
CAS Number:

CAS Reg. No.	Note
25087-21-2	
54190-45-3	isotactic
54193-22-5	syndiotactic

Molecular Formula: (C$_9$H$_{10}$)$_n$
Fragments: C$_9$H$_{10}$

Volumetric & Calorimetric Properties:

Equation of State: Equations of state have been reported [2,3,4]. Tait, hale and cell theory parameters calculated for polymer at different temps. have been reported [7]
Melting Temperature:

No.	Value	Note
1	>360°C	min. [1]

Glass-Transition Temperature:

No.	Value	Note
1	96°C	[1]
2	131°C	[2]
3	135.5°C	[6]

Deflection Temperature:

No.	Value	Note
1	119°C	[6]

Surface Properties & Solubility:
Cohesive Energy Density Solubility Parameters: δ 37.71–39.6 J$^{1/2}$ cm$^{-3/2}$ (9.00–9.45 cal$^{1/2}$ cm$^{-3/2}$) [5]

Transport Properties:
Water Content: 0.01% [6]

Mechanical Properties:
Tensile (Young's) Modulus:

No.	Value	Note
1	40200 MPa	41000 kg cm^{-2} [6]

Tensile Strength Break:

No.	Value	Note	Elongation
1	69.1 MPa	705 kg cm^{-2} [6]	3%

Applications/Commercial Products:
Processing & Manufacturing Routes: Prod. by cationic polymerisation with Lewis acid catalysts

Bibliographic References

[1] Danusso, F. and Polizzotti, G., *Makromol. Chem.*, 1963, **61**, 157
[2] Olabisi, O. and Simha, R., *Macromolecules*, 1975, **8**, 211
[3] Beret, S. and Prausnitz, J.M., *Macromolecules*, 1975, **8**, 878
[4] Roszkowski, Z., *Makromol. Chem.*, 1982, **183**, 669
[5] Ahmad, H. and Yaseen, M., *Polym. Eng. Sci.*, 1979, **19**, 858
[6] Dunham, K.R., Faber, J.W.H., Vandenberghe, J. and Fowler, W.F., *J. Appl. Polym. Sci.*, 1963, **7**, 897
[7] Quach, A. and Simha, R., *J. Appl. Phys.*, 1971, **42**, 4592

Poly(4-methylstyrene)   P-410

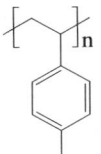

Synonyms: *Poly(p-methylstyrene). Pp-MS. Poly(1-(4-methylphenyl)-1,2-ethanediyl). Poly(4-methyl-1-vinyl-benzene)*
Related Polymers: Poly(2-methylstyrene)
Monomers: 4-Methylstyrene
Material class: Thermoplastic
Polymer Type: styrenes
CAS Number:

CAS Reg. No.	Note
24936-41-2	
54190-46-4	isotactic
54193-24-7	syndiotactic

Volumetric & Calorimetric Properties:
Density:

No.	Value	Note
1	d 1.019 g/cm^3	25° [3]

– Poly(α-methylstyrene)

Thermodynamic Properties General: ΔC_p 34.6 J K^{-1} mol^{-1}
Glass-Transition Temperature:

No.	Value	Note
1	108°C	[1]
2	107°C	[2]
3	111°C	[6]

Transport Properties:
Polymer Solutions Dilute: Mark-Houwink constants have been reported [7]
Permeability of Gases: [He] 37.1 × 10^{-10}; [CH$_4$] 2.2 × 10^{-10} [1]; [CO$_2$] 9 × 10^{-10} cm^2 (s cmHg)$^{-1}$ [5]

Polymer Stability:
Decomposition Details: Decomposition of the polymer occurs with random scissions [6] which reduce chain length; depolymerisation reactions account for volatilisation
Environmental Stress: γ-Radiation causes cross-linking and scission; H$_2$ is liberated [4]

Applications/Commercial Products:
Processing & Manufacturing Routes: Prod. by cationic polymerisation with Lewis acid catalyst

Bibliographic References
[1] Puleo, A.C., Muruganandam, N. and Paul, D.R., *J. Polym. Sci., Part B: Polym. Phys.*, 1989, **27**, 2385
[2] Judovits, L.H., Bopp, R.C., Gaur, U. and Wunderlich, B., *J. Polym. Sci., Part B: Polym. Lett.*, 1986, **24**, 2725
[3] Nozaki, M., Shimada, K. and Okamoto, S., *Jpn. J. Appl. Phys.*, 1971, **10**, 179
[4] Burlant, W., Neeman, J. and Serment, V., *J. Polym. Sci.*, 1962, **58**, 491
[5] Greenwood, R. and Weir, N., *Makromol. Chem.*, 1975, **176**, 2041
[6] Malhatra, S.L., Lessard, P., Minh, L. and Blanchard, L.P., *J. Macromol. Sci., Chem.*, 1980, **14**, 517
[7] Kuwahara, N., Ogino, K., Kasai, A., Ueno, S. and Kaneko, M., *J. Polym. Sci., Part A: Polym. Chem.*, 1965, **3**, 985

Poly(α-methylstyrene)   P-411

Synonyms: Pα-MS. Poly[(1-methylethenyl)benzene]. Poly(1-methyl-1-phenyl-1,2-ethanediyl)
Monomers: α-Methylstyrene
Material class: Thermoplastic
Polymer Type: styrenes
CAS Number:

CAS Reg. No.	Note
25014-31-7	
30939-97-0	isotactic
25086-17-3	syndiotactic

Molecular Formula: (C$_9$H$_{10}$)$_n$
Fragments: C$_9$H$_{10}$
General Information: The high MW polymer is a hard clear material. Low MW material is liq.

Volumetric & Calorimetric Properties:
Density:

No.	Value	Note
1	d 1.04 g/cm^3	[1]
2	d 1.065 g/cm^3	[2]

Equation of State: Equation of state information has been reported [8]
Thermodynamic Properties General: $\Delta\alpha$ 0.000398 K^{-1}. ΔC_p 37.58 J K^{-1} mol^{-1} (0.076 cal g^{-1} K^{-1}) [16]
Latent Heat Crystallization: ΔH_f 73.3 kJ mol^{-1} (17.5 kcal mol^{-1}) [3]. -ΔH_c 5010.9 kJ mol^{-1} (1195.91 kcal mol^{-1}) [3]. -ΔH_c 4991.4 kJ mol^{-1} (42.3 kJ g^{-1}) [4]. S 134.7 J K^{-1} mol^{-1} (27°) [5]. ΔC_p 26.3 J K^{-1} mol^{-1} [6]
Specific Heat Capacity:

No.	Value	Note	Type
1	1.492 kJ/kg.C	27° [5]	P
2	1.555 kJ/kg.C	37° [7]	P
3	2.757 kJ/kg.C	217° [7]	P

Glass-Transition Temperature:

No.	Value	Note
1	165°C	[2]
2	168°C	[6]

Surface Properties & Solubility:
Cohesive Energy Density Solubility Parameters: δ 8.6 [14]. δ_p 38.55–38.97 J$^{1/2}$ cm$^{-3/2}$ (9.20–9.3 cal$^{1/2}$ cm$^{-3/2}$) [9]

Transport Properties:
Polymer Solutions Dilute: Theta temps. 36.5–37.5° (cyclohexane) [11], 31.5–33.5° (cyclohexane) [14], 92.0–96.0° (methylcyclohexane) [14]
Permeability of Gases: [He] 14.5 × 10^{-10}; [CH$_4$] 0.14 × 10^{-10} cm^2 (s cmHg)$^{-1}$ [2]
Permeability of Liquids: Liq. permeability values have been reported [10]

Mechanical Properties:
Elastic Modulus:

No.	Value	Note
1	0.32 MPa	200° [1]

Optical Properties:
Transmission and Spectra: Raman spectral data have been reported [12]

Polymer Stability:
Flammability: Limiting oxygen index 18% [15]
Environmental Stress: Exposure to uv radiation *in vacuo* causes random chain scission, with subsequent depolymerisation of the radicals formed to yield small amounts of the monomer [13]

Applications/Commercial Products:
Processing & Manufacturing Routes: Prod. by low temp. free radical polymerisation (below 0°, due to exceptionally low ceiling temp.)
Applications: Plasticiser in paints, waxes and adhesives. Plasticisers are generally of low MW and are liq.

Trade name	Supplier
Resin 18	Dow

Bibliographic References
[1] Fetters, L.J., Lohse, D.J., Richter, D., Witten, T.A. and Zirkel, A., *Macromolecules*, 1994, **27**, 4639
[2] Puleo, A.C., Muruganandam, N. and Paul, D.R., *J. Polym. Sci., Part B: Polym. Phys.*, 1989, **27**, 2385
[3] Joshi, R.M. and Zwolinski, B.J., *Macromolecules*, 1968, **1**, 25
[4] Ng, S.C. and Chee, K.K., *Polymer*, 1993, **34**, 3870
[5] Domalski, E.S. and Hearing, E.D., *J. Phys. Chem. Ref. Data*, 1990, **19**, 881

[6] Judovits, L.H., Bopp, R.C., Gaur, U. and Wunderlich, B., *J. Polym. Sci., Part B: Polym. Lett.*, 1986, **24**, 2725
[7] Gaur, U. and Wunderlich, B., *Macromolecules*, 1980, **13**, 1618
[8] Simha, R., Rae, J.M. and Nanda, V.S., *J. Appl. Phys.*, 1972, **43**, 4312
[9] Cowie, J.M.G., *Polymer*, 1969, **10**, 708
[10] Qian, J.W. and Rudin, A., *J. Appl. Polym. Sci.*, 1989, **37**, 2007
[11] Cowie, J.M.G., Bywater, S. and Worsfold, D.J., *Polymer*, 1967, **8**, 105
[12] Spells, S.J., Shepherd, I.W. and Wright, C.J., *Polymer*, 1977, **18**, 905
[13] Stokes, S. and Fox, R.B., *J. Polym. Sci.*, 1962, **56**, 507
[14] Goldnasser, D.J. and Williams, D.J., Polym. Prepr. (Am. Chem. Soc., Div. Polym. Chem.), 1971, **13**, 539
[15] Pummer, W.J. and Wall, L.A., Polym. Prepr. (Am. Chem. Soc., Div. Polym. Chem.), 1972, **13**, 1046
[16] Ichihara, S., Komatsu, A. and Hata, J., *Polym. J. (Tokyo)*, 1971, **2**, 650

[5] Sato, M., Shimizu, T. and Yamauchi, A., *Makromol. Chem.*, 1990, **191**, 313
[6] Ochmanska, J. and Pickup, P.G., *J. Electroanal. Chem.*, 1991, **297**, 211
[7] Roncali, J., Garreau, R., Yassar, A., Marque, P. et al, *J. Phys. Chem.*, 1987, **91**, 6706, (electrical props)
[8] Ritter, S.K. and Noftle, R.E., *Chem. Mater.*, 1992, **4**, 872
[9] Amer, A., Zimmer, H., Mulligan, K.J. and Mark, H.B., *J. Polym. Sci., Polym. Lett. Ed.*, 1984, **22**, 77
[10] Krische, B. and Zagorska, M., *Synth. Met.*, 1989, **28**, C263
[11] Hotta, S., *Synth. Met.*, 1987, **22**, 103
[12] Tourillan, G. and Garnier, F., *J. Phys. Chem.*, 1983, **87**, 2289
[13] Stöckert, D., Kessel, R. and Schutze, J.W., *Synth. Met.*, 1991, **41**, 1295
[14] Choi, K.M., Kim, K.H. and Choi, J.S., *J. Phys. Chem.*, 1989, **93**, 4659
[15] Krische, B., Zagorska, M. and Hellberg, J., *Synth. Met.*, 1993, **58**, 295

Poly(3-methylthiophene)   P-412

Related Polymers: Poly(thiophene)
Monomers: 3-Methylthiophene
Polymer Type: polythiophene
CAS Number:

CAS Reg. No.
84928-92-7

Molecular Formula: $(C_5H_4S)_n$
Fragments: C_5H_4S

Volumetric & Calorimetric Properties:
Glass-Transition Temperature:

No.	Value	Note
1	145°C	[1]

Surface Properties & Solubility:
Solvents/Non-solvents: Sol. THF, nitropropane, toluene, xylene, $CHCl_3$, anisole, nitrobenzene, benzonitrile [1]

Electrical Properties:
Electrical Properties General: Electrical conductivity depends on method of synth. [2,6,7,8,9,14]. Doping increases electrical conductivity [10]
Complex Permittivity and Electroactive Polymers: Conductivity values have been reported. σ 100–120 Ω^{-1} cm^{-1} [2], 750 Ω^{-1} cm^{-1} [11]

Optical Properties:
Transmission and Spectra: Ir [3,4,5], uv [9,12,13], H-1 nmr [9] and x-ray photoelectron [12] spectral data have been reported

Applications/Commercial Products:
Processing & Manufacturing Routes: Prod. by chemical or electrochemical oxidn. polymerisation of 3-methylthiophene with various catalysts [14], or oxidn. polymerisation of bithiophenes [15]
Applications: Potential applications in liq. crystal films and diodes

Bibliographic References
[1] Jen, K.Y., Miller, G.G. and Elsenbaumer, R.L., *J.C.S., Chem. Commun.*, 1986, 1346
[2] Sato, M., Tanaka, S. and Kaeriyama, K., *J.C.S., Chem. Commun.*, 1985, 713
[3] Neugebauer, H., Nauer, G., Neckel, A., Tourillon, G. et al, *J. Phys. Chem.*, 1984, **88**, 652
[4] Hotta, S., Rughooputh, S.D.D.V., Heeger, A.J. and Wudl, F., *Macromolecules*, 1987, **20**, 212

Poly(*N*-methyl-*N*-vinylacetamide)   P-413

Synonyms: *Poly(N-ethenyl-N-methylacetamide). PMVAc. Poly(N-vinyl-N-methylacetamide)*
Monomers: *N*-Methyl-*N*-vinylacetamide
Polymer Type: polyvinyls
CAS Number:

CAS Reg. No.
26616-03-5

Molecular Formula: $(C_5H_9NO)_n$
Fragments: C_5H_9NO
Morphology: Exists predominantly in the *trans*-configuration [6]

Volumetric & Calorimetric Properties:
Vicat Softening Point:

No.	Value	Note
1	202°C	[3]

Surface Properties & Solubility:
Solubility Properties General: Miscible with poly(vinylidene fluoride) provided the wt fraction of PVF is less than 0.7 [11]; with poly(styrene-*co*-acrylonitrile), when the acrylonitrile content is 65% [2]
Surface and Interfacial Properties General: Films have good adhesion to glass and fair adhesion to aluminium [3]
Surface Tension:

No.	Value	Note
1	65.3 mN/m	1% soln. [4]
2	67.1 mN/m	[4]

Transport Properties:
Polymer Solutions Dilute: η 2.08 dl g^{-1} (20°, H_2O), 0.48 dl g^{-1} (20°, Me_2CO) [3], 4.43 dl g^{-1} (30°, H_2O) [8]

Optical Properties:
Transmission and Spectra: Ir spectral data have been reported [9]

Polymer Stability:
Thermal Stability General: Becomes rubbery above 260° and darkens at 290° [3]
Chemical Stability: May be completely hydrolysed by heating with 3*M* hydrochloric acid at 101° for approx. 150h [5]
Biological Stability: The ability of an aq. soln. to transport the pharmaceutical agent Paraoxon through the skin has been reported [8]

Applications/Commercial Products:
Processing & Manufacturing Routes: May be synth. by free-radical polymerisation [7,8] in alkaline soln. under N_2 at 50° for 18h using AIBN initiator [3] or in heptane under N_2 at 95° [4]

Bibliographic References
[1] Galin, M., *Makromol. Chem.*, 1987, **188**, 1391, (miscibility)
[2] Yeo, Y.T., Goh, S.H. and Lee, S.Y., *Polym. Polym. Compos.*, 1996, **4**, 235, (miscibility)
[3] *Brit. Pat.*, 1967, 1 082 016, (synth.)
[4] *U.S. Pat.*, 1972, 3 696 085, (synth., surface tension)
[5] Buys, H.C.W.M., Vercauteren, F.F., van Elven, A. and Tinnemans, A.H.A., *Recl. Trav. Chim. Pays-Bas*, 1989, **108**, 123, (hydrolysis)
[6] Kirsh, Yu.E., Berestova, S.S., Aksenov, A.I. and Karaputadze, T.M., *Russ. J. Phys. Chem. (Engl. Transl.)*, 1990, **64**, 1013, (conformn)
[7] Kirsh, Yu.E., Aksenov, A.I., Berestova, S.S. and Karaputadze, T.M., *Vysokomol. Soedin., Ser. B*, 1988, **30**, 323, (synth.)
[8] Hofmann, V., Ringsdorf, H. and Muacevic, G., *Makromol. Chem.*, 1975, **176**, 1929, (synth.)
[9] Wood, F., Ramsden, D.K. and King, G., *Nature (London)*, 1966, **212**, 606, (ir)

Poly(2-methylvinylcyclohexane)   P-414

Synonyms: *Poly(1-ethenyl-2-methylcyclohexane). Poly(1-(2-methylcyclohexyl)-1,2-ethanediyl)*
Monomers: 2-Methyl-1-vinylcyclohexane
Material class: Thermoplastic
Polymer Type: polyolefins
Molecular Formula: $(C_9H_{16})_n$
Fragments: C_9H_{16}

Surface Properties & Solubility:
Solvents/Non-solvents: Slightly sol. boiling tetrahydronaphthalene, boiling decahydronaphthalene. Insol. C_6H_6 [1]

Applications/Commercial Products:
Processing & Manufacturing Routes: Synth. using Ziegler-Natta catalysts such as $TiCl_4$-$Al(^iBu)_3$ in heptane at 75–85° [1]

Bibliographic References
[1] Overberger, C.G. and Mulvaney, J.E., *J. Am. Chem. Soc.*, 1959, **81**, 4697, (synth, solubility)

Poly(3-methylvinylcyclohexane)   P-415

Synonyms: *Cyclohexane-1-ethenyl-3-methyl homopolymer. Poly(1-ethenyl-3-methylcyclohexane). Poly(1-(3-methylcyclohexyl)-1,2-ethanediyl)*
Monomers: 3-Methyl-1-vinylcyclohexane
Material class: Thermoplastic
Polymer Type: polyolefins
CAS Number:

CAS Reg. No.
126895-43-0

Molecular Formula: $(C_9H_{16})_n$
Fragments: C_9H_{16}
General Information: Solid [1]

Volumetric & Calorimetric Properties:
Melting Temperature:

No.	Value	Note
1	276–355°C	[1]

Surface Properties & Solubility:
Solvents/Non-solvents: Sol. C_6H_6, tetrahydronaphthalene [1]. Mod. sol. Et_2O [1]. Insol. MeOH [1]

Transport Properties:
Polymer Solutions Dilute: $[\eta]$ 0.74 dl g^{-1} (C_6H_6, 29.2°) [1]

Applications/Commercial Products:
Processing & Manufacturing Routes: Synth. using Ziegler-Natta catalysts such as $TiCl_4$-$Al(^iBu)_3$ in heptane at 60–85° [1]

Bibliographic References
[1] Overberger, C.G. and Mulvaney, J.E., *J. Am. Chem. Soc.*, 1959, **81**, 4697, (synth, melting point, solubility)

Poly(4-methylvinylcyclohexane)   P-416

Synonyms: *Poly(1-ethenyl-4-methylcyclohexane). Poly(1-(4-methylcyclohexyl)-1,2-ethanediyl)*
Monomers: 4-Methyl-1-vinylcyclohexane
Material class: Thermoplastic
Polymer Type: polyolefins
Molecular Formula: $(C_9H_{16})_n$
Fragments: C_9H_{16}

Volumetric & Calorimetric Properties:
Melting Temperature:

No.	Value	Note
1	225–250°C	[1]

Surface Properties & Solubility:
Solvents/Non-solvents: Sol. Et_2O, C_6H_6 [1]. Slightly sol. boiling 1-butanol [1]. Insol. MeOH [1]

Transport Properties:
Polymer Solutions Dilute: $[\eta]$ 0.45 dl g^{-1} (C_6H_6, 29.2°) [1]

Applications/Commercial Products:
Processing & Manufacturing Routes: Synth. using Ziegler-Natta catalysts such as $TiCl_4$-$Al(^iBu)_3$ in heptane at 75–85° [1]

Bibliographic References
[1] Overberger, C.G. and Mulvaney, J.E., *J. Am. Chem. Soc.*, 1959, **81**, 4697, (synth, melting point, solubility)

Poly(methyl vinyl ketone)   P-417

Synonyms: *Poly(acetyl ethylene). Poly(3-buten-2-one). POLY MVK. PMVK*
Monomers: 3-Buten-2-one
Material class: Gums and resins
Polymer Type: polyvinyls

– Poly(1-nonadecene)

CAS Number:

CAS Reg. No.	Note
78-94-4	3-buten-2-one polymers
25038-87-3	homopolymer
31175-23-2	isotactic
86992-29-2	syndiotactic
50940-42-6	dimer

Molecular Formula: $(C_4H_6O)_n$
Fragments: C_4H_6O
General Information: Isotactic, syndiotactic and atactic forms known [17]
Morphology: Crystal Lattice Constants a 14.56Å, b 14.56Å, c 14.10Å [2]

Volumetric & Calorimetric Properties:
Density:

No.	Value	Note
1	d 1.12 g/cm^3	[15]
2	d 1.17 g/cm^3	[2]
3	d 1.216 g/cm^3	[2]

Volumetric Properties General: Molar volume 66.59 cm^3 (25°, calc.) [14]
Melting Temperature:

No.	Value	Note
1	30–40°C	[2]
2	165–170°C	[2]
3	140–160°C	[13]
4	170–195°C	[13]
5	110–125°C	[13]

Glass-Transition Temperature:

No.	Value	Note
1	35°C	[1]

Surface Properties & Solubility:
Solubility Properties General: Solubility parameters have been reported [14]. Molar heat of sorption has been reported [2]
Solvents/Non-solvents: Very sol. CHCl$_3$. Sol. phenylacetylene, thiophenol, nitromethane, 2-methyl-3-butyn-2-ol, pyrrole, trichloroethylene, 1,1,2,2-tetrachloroethane, Me$_2$CO, EtOH, EtOAc, CH$_2$Cl$_2$. Insol. hept-1-yne, N-methylacetamide, CCl$_4$, tetrachloroethane, H$_2$O, cyclohexane [9,10,13,15]
Wettability/Surface Energy and Interfacial Tension: Dispersive surface energy γ 18–29 mJ m^{-2} (temp. dependent) [1]

Transport Properties:
Polymer Solutions Dilute: Reduced viscosity η_{sp} 0.197 (Me$_2$CO, 25°), 0.234 (Me$_2$CO, 25°), 0.242 (Me$_2$CO, 25°) [13]

Electrical Properties:
Electrical Properties General: Equilibrium dipole moment 0.35 D [11]. Relaxation dipole moment 1.02 D (50 MHz), 1.25 D (1 GHz)
Complex Permittivity and Electroactive Polymers: Complex permittivity calculation for polymer in dioxane soln. has been reported [11]

Dissipation (Power) Factor:

No.	Value	Note
1	0.012–0.03	[16]

Optical Properties:
Refractive Index:

No.	Value	Note
1	1.5	[15]

Transmission and Spectra: Ir [2,8,13], C-13 nmr [3], H-1 nmr [13] and ion spectral data [4] have been reported

Polymer Stability:
Decomposition Details: Decomposition occurs at 200° (approx.) causing aldol condensations with the elimination of water. The mechanism of decomposition has been reported [8,13]
Environmental Stress: Radiation decomposition leads to free radical dissociation giving CO, CH$_4$ and acetaldehyde [12]
Chemical Stability: Degraded and coloured by alkali [8,13]. Attacked by acid

Applications/Commercial Products:

Trade name	Supplier
Poly(methyl vinyl ketone)	Aldrich
Poly(methyl vinyl ketone)	Merck

Bibliographic References

[1] Al-Saigh, Z.Y., *Polym. Int.*, 1996, **40**, 25, (thermodynamic and surface props)
[2] Tsuruta, T., Fujio, R. and Furukawa, J., *Makromol. Chem.*, 1964, **80**, 172, (solute interaction values)
[3] Merle-Aubry, L. and Merle, Y., *Eur. Polym. J.*, 1980, **16**, 227, (C-13 nmr)
[4] Chilikoti, A., Ratner, B.D. and Briggs, D., SIA *Surf. Interface Anal.*, 1992, **18**, 604, (ion spectra)
[5] Esumi, K., Schwarz, A.M. and Zettlemoyer, A.C., *J. Colloid Interface Sci.*, 1983, **95**, 102
[6] Aida, M., Kaneda, Y., Kobayashi, N., Endo, K. and Chong, D.P., *Bull. Chem. Soc. Jpn.*, 1994, **67**, 2972, (x-ray spectra)
[7] Fyfe, C.A. and McKinnon, M.S., *Can. J. Chem.*, 1985, **63**, 232, (nmr, uv, ir)
[8] Grassie, N. and Hay, J.N., *Makromol. Chem.*, 1963, **64**, 82
[9] Marvel, C.S., Harkema, J. and Copley, M.J., *J. Am. Chem. Soc.*, 1941, **63**, 1609, (solubility)
[10] Marvel, C.S., Dietz, F.C. and Copley, M.J., *J. Am. Chem. Soc.*, 1940, **62**, 2273, (solubility)
[11] Mashimo, S., Winsor, P., Cole, R.H., Matsuo, K. and Stockmayer, W.H., *Macromolecules*, 1983, **16**, 965, (dielectric constant, nmr)
[12] Wissbrun, K.F., *J. Am. Chem. Soc.*, 1959, **81**, 58, (photolysis)
[13] Nasrallah, E. and Baylouzian, S., *Polymer*, 1977, **18**, 1173
[14] Ahmad, H. and Yaseen, M., *Polym. Eng. Sci.*, 1979, **19**, 858
[15] Schildknecht, C.E., Vinyl and Related Polymers Chapman and Hall, 1952
[16] Kobeko, P.P., Mikhailov, G.P. and Novikova, Z.I., *Zh. Tekh. Fiz.*, 1944, **14**, 24

Poly(1-nonadecene)   P-418

Synonyms: *Poly(1-heptadecylethylene). Poly(1-heptadecyl-1,2-ethanediyl)*
Monomers: 1-Nonadecene
Material class: Thermoplastic
Polymer Type: polyolefins
CAS Number:

CAS Reg. No.	Note
62132-73-4	atactic

Molecular Formula: $(C_{19}H_{38})_n$
Fragments: $C_{19}H_{38}$

Applications/Commercial Products:
Processing & Manufacturing Routes: Synth. using $Al(^iBu)_2H$ as catalyst to form a dimer, which is then polymerised using a strong acid catalyst in an inert solvent at 0–50° for up to 5h [1]

Bibliographic References
[1] *U.S. Pat.*, 1976, 3 999 960, (synth)

Poly(1-nonene) P-419

$-[(CH_2CH)]_n-$
            |
            $CH_2(CH_2)_5CH_3$

Synonyms: *Poly(1-heptylethylene). Poly(1-heptyl-1,2-ethanediyl)*
Monomers: 1-Nonene
Material class: Thermoplastic
Polymer Type: polyolefins
CAS Number:

CAS Reg. No.	Note
29254-71-5	atactic
27458-69-1	isotactic

Molecular Formula: $(C_9H_{18})_n$
Fragments: C_9H_{18}
General Information: Gum
Morphology: Monoclinic, a 0.59, b 4.3, c 1.34, β 96° [8]

Volumetric & Calorimetric Properties:
Thermal Expansion Coefficient:

No.	Value	Note	Type
1	0.00078 K^{-1}	[10]	V

Latent Heat Crystallization: ΔH_m 33 J g^{-1} [8]
Melting Temperature:

No.	Value	Note
1	19°C	[5]
2	17–18°C	[6]
3	22°C	[7]
4	14°C	[8]

Glass-Transition Temperature:

No.	Value	Note
1	-47°C	[5]

Transition Temperature:

No.	Value	Note	Type
1	-160°C	310 Hz [5]	β

Transport Properties:
Polymer Solutions Dilute: Theta temp. has been reported [2]. Dilute soln. props. and viscosity values have been reported [3]

Mechanical Properties:
Mechanical Properties General: Shear modulus values have been reported [11]

Optical Properties:
Transmission and Spectra: Raman [7], C-13 nmr [9] and ir [12] spectral data have been reported

Polymer Stability:
Thermal Stability General: Oxidative thermal stability [1] has been reported

Applications/Commercial Products:
Processing & Manufacturing Routes: Synth. using Ziegler-Natta catalysts, such as $TiCl_4$-$AlEt_3$, at -15° to 70° [4,9]

Bibliographic References
[1] Emanuel, N.M., Marin, A.P., Kiryushkin, S.G., Moiseev, Y.V. and Shlyapriikov, Y.A., *Dokl. Akad. Nauk SSSR*, 1985, **280**, 416, (oxidation props)
[2] Wang, J.S., Porter, R.S. and Knox, J.R., *Polym. J. (Tokyo)*, 1978, **10**, 619, (expansion coefficient, dilute soln props)
[3] Shalaeva, L.F., *Vysokomol. Soedin.*, *Ser. B*, 1968, **10**, 449, (dilute soln props)
[4] Sadykhzade, S.I., Mamedov, T.I. and Mirzakhanov, I.S., *Dokl. Akad. Nauk. Az. SSR*, 1968, **24**, 15, (synth)
[5] Clark, K.J., Jones, A.T. and Sandiford, D.J.H., *Chem. Ind. (London)*, 1962, 2010, (transition temps)
[6] Turner-Jones, A., *Makromol. Chem.*, 1964, **71**, 1, (melting point)
[7] Modric, I., Holland-Moritz, K. and Hummel, D.O., *Colloid Polym. Sci.*, 1976, **254**, 342, (ir, Raman)
[8] Trafara, G., Makromol. Chem., *Rapid Commun.*, 1980, **1**, 319, (struct)
[9] Asakura, T., Demura, M. and Nishiyama, Y., *Macromolecules*, 1991, **24**, 2334, (C-13 nmr)
[10] Porter, R.S., Chuah, H.H. and Wang, J.S., *J. Rheol. (N.Y.)*, 1995, **39**, 649, (expansion coefficient)
[11] Wang, J.S., Knox, J.R. and Porter, R.S., *J. Polym. Sci., Polym. Phys. Ed.*, 1978, **16**, 1709, (shear modulus)
[12] Lecomte, J., *Bull. Soc. Chim. Fr.*, 1949, 923, (ir)

Poly(1-octadecene) P-420

$-[(CH_2CH)]_n-$
            |
            $CH_2(CH_2)_{14}CH_3$

Monomers: 1-Octadecene
Material class: Thermoplastic
Polymer Type: polyolefins
CAS Number:

CAS Reg. No.	Note
25511-67-5	atactic
26746-89-4	isotactic
133578-06-0	syndiotactic

Molecular Formula: $(C_{18}H_{36})_n$
Fragments: $C_{18}H_{36}$
General Information: White fibrous powder [2]
Morphology: Isotactic poly(1-octadecene) has two polymorphic forms. Form I orthorhombic, form II orthorhombic, a 0.75, b 7.04, c 0.67, α 130°36′ [3]

Volumetric & Calorimetric Properties:
Density:

No.	Value	Note
1	d^{20} 0.92 g/cm^3	[4]
2	d 0.905 g/cm^3	isotactic, form I [3]
3	d 0.91–0.92 g/cm^3	isotactic, form II [3]
4	d^{20} 0.95 g/cm^3	cryst. [5]
5	d^{20} 0.85 g/cm^3	amorph. [5]

– Poly(1-octene)

Thermal Expansion Coefficient:

No.	Value	Note	Type
1	0.00076 K^{-1}	T > T$_m$ [6]	V

Melting Temperature:

No.	Value	Note
1	71°C	[9]
2	68–70°C	isotactic [4]
3	71°C	isotactic [7]
4	68°C	[8]
5	58°C	calc [10]

Glass-Transition Temperature:

No.	Value	Note
1	-30°C	[18]
2	41.5°C	atactic [4]
3	39°C	calc. [10]

Transition Temperature:

No.	Value	Note	Type
1	55°C	25 Hz	α
2	-110°C	170 Hz [2]	β
3	43°C	isotactic	first order
4	46°C	isotactic [9]	second order
5	-140°C	[18]	Low temp.

Surface Properties & Solubility:
Solvents/Non-solvents: Mod. sol. hexane [11]

Transport Properties:
Polymer Solutions Dilute: [η] 1.53 dl g^{-1} (tetrahydronaphthalene 135°) [12]. [η] 2.26 dl g^{-1} (toluene, 55°) [7]. η$_{spec}$ 70.4 (toluene, 55°) [7]
Polymer Melts: Dynamic viscosity, dynamic rigidity, apparent viscosity results have been reported at various temps. [12]. Flow activation energy 31.4–35.2 kJ mol^{-1} (7.5–8.4 kcal mol^{-1}) [13]. Zero shear viscosity η$_0$ 2600 (80°) [13]

Mechanical Properties:
Mechanical Properties General: Dynamic tensile storage modulus and loss modulus values have been reported (110 Hz, -160–70°) [19]
Miscellaneous Moduli:

No.	Value	Note	Type
1	0.0004 MPa	0.1 Hz, 100° [13]	Shear

Optical Properties:
Refractive Index:

No.	Value	Note
1	1.495	29° [4]

Transmission and Spectra: Ir and Raman spectral data have been reported [14,15,16]
Total Internal Reflection: Stress optical coefficient -10500 (25°) [7]

Polymer Stability:
Stability Miscellaneous: Stability to uv light in presence of stabilisers has been reported [17]

Applications/Commercial Products:
Processing & Manufacturing Routes: Synth. using Ziegler-Natta catalysts such as TiCl$_4$AlEt$_3$ [11]

Bibliographic References

[1] *Eur. Pat. Appl.*, 1990, 403 866, (synth, syndiotactic)
[2] Clark, K.J., Turner-Jones, A. and Sandiford, D.J.H., *Chem. Ind. (London)*, 1962, **47**, 2010, (transition temps)
[3] Turner-Jones, A., *Makromol. Chem.*, 1964, **71**, 1, (struct)
[4] Aubrey, D.W. and Barnatt, A., *J. Polym. Sci., Part A-2*, 1968, **6**, 241, (transition temps)
[5] Tait, P.J.T. and Livesey, P.J., *Polymer*, 1970, **11**, 359, (thermodynamic props)
[6] Porter, R.S., Chuah, H.H. and Wang, J.S., *J. Rheol. (N.Y.)*, 1995, **39**, 649, (thermal expansion coefficient, melt flow activation energy)
[7] Philippoff, W. and Tomquist, E., *J. Polym. Sci., Part C: Polym. Lett.*, 1968, **23**, 881, (optical props)
[8] Blum, K. and Trafara, G., *Makromol. Chem.*, 1980, **181**, 1097, (transition temps)
[9] Trafara, G., Koch, R., Blum, K. and Hummel, D., *Makromol. Chem.*, 1976, **177**, 1089, (transition temps)
[10] Van Krevelen, D.W., *Properties of Polymers: Their Correlation with Chemical Structure*, 3rd edn., Elsevier, 1990, 144, 164, (transition temps)
[11] Aubrey, D.W. and Barnatt, A., *J. Polym. Sci., Part A-1*, 1966, **4**, 1709, (synth)
[12] Shirayama, K., Matsuda, T. and Tadashi, K.S., *Makromol. Chem.*, 1971, **147**, 155, (rheological props)
[13] Wang, J.-S., Knox, J.R. and Porter, R.S., *J. Polym. Sci., Polym. Phys. Ed.*, 1978, **16**, 1709, (rheological props)
[14] Modric, I., Holland-Moritz, K. and Hummel, D.O., *Colloid Polym. Sci.*, 1976, **254**, 342, (Raman, ir)
[15] Holland-Moritz, K., *Colloid Polym. Sci.*, 1975, **253**, 922, (Raman, ir)
[16] Fraser, G.V., Hendra, P.J., Chalmers, J.M., Cudby, M.E.A. and Willis, H.A., *Makromol. Chem.*, 1973, **173**, 195, (Raman, ir)
[17] *U.S. Pat.*, 1968, 3 391 106, (uv stabilisers)
[18] *High Polymers*, (eds. R.A.V. Raff and K.W. Doak), Interscience, 1965, **20**, 706, (transition temps)
[19] Takayanagi, M., *Pure Appl. Chem.*, 1970, **23**, 151, (mech props)

Poly(1-octene)

$$-[(CH_2CH)]_n-$$
$$CH_2(CH_2)_4CH_3$$

Synonyms: *Poly(caprylene). Poly(1-hexylethylene). Poly(1-hexyl-1,2-ethanediyl)*
Monomers: 1-Octene
Material class: Thermoplastic
Polymer Type: polyolefins
CAS Number:

CAS Reg. No.	Note
25068-25-1	atactic
26746-84-9	isotactic
133578-04-8	syndiotactic

Molecular Formula: (C$_8$H$_{16}$)$_n$
Fragments: C$_8$H$_{16}$
General Information: Viscous, tacky, yellow, rubbery gum [3,4]
Morphology: Isotactic, probably monoclinic a 0.56, b 3.8, c 0.76, β 97° helix 4$_1$ [2]

Volumetric & Calorimetric Properties:
Thermal Expansion Coefficient:

No.	Value	Note	Type
1	0.00023 K^{-1}	[5]	L

– Poly(octene oxide)

Thermodynamic Properties General: ΔH_f 30 J g^{-1} [2]. Other thermophysical props. have been reported [26]. Specific heat capacity values have been reported [7]

Melting Temperature:

No.	Value	Note
1	-38°C	[8,10]
2	5°C	[2]
3	10°C	[9]
4	21°C	calc. [10]

Glass-Transition Temperature:

No.	Value	Note
1	-65°C	[5,10,12]
2	-63°C	[11]
3	-53°C	[28]
4	-45°C	[12,32]
5	-52°C	calc. [11]

Transition Temperature:

No.	Value	Note	Type
1	-42°C		α
2	-160°C	[3]	β
3	-105°C	low temp. discontinuity [5]	γ
4	-70°C		α
5	-170°C	[24]	β

Surface Properties & Solubility:
Solvents/Non-solvents: Sol. boiling MeOH [13]

Transport Properties:
Melt Flow Index:

No.	Value	Note
1	30 g/10 min	ASTM D1238-15, η_{inh} 1 dl g^{-1} (tetrahydronaphthalene, 100° or 145°) [14]

Polymer Solutions Dilute: 50.4° (phenetole) [4]. [η] 2.66 dl g^{-1} (toluene, 25°) [19]. Other values for viscosity and θ values have been reported [1,4,15,16,18,23,25]
Polymer Melts: Melt flow activation energy 35.6 kJ mol^{-1} (8.5 k cal mol^{-1}) [14]; 77.5 kJ mol^{-1} (18.5 k cal mol^{-1}) [20]; 41.9 kJ mol^{-1} (10 k cal mol^{-1}) [16]. Zero shear viscosity 4500 (80°) [21]. Dynamic viscosity, dynamic rigidity, apparent viscosity, and critical shear rates at various temps. have been reported [14,16]

Mechanical Properties:
Mechanical Properties General: Dynamic tensile storage modulus and loss modulus values have been reported [31]
Miscellaneous Moduli:

No.	Value	Note	Type
1	0.003 MPa	0.1 Hz, 100° [21]	Shear modulus

Optical Properties:
Transmission and Spectra: C-13 nmr [17,27,29], H-1 nmr [29], ir [6,22,28,29,30] and Raman [6,22,30] spectral data have been reported
Molar Refraction: $\frac{dn}{dC}$ -0.105 (bromobenzene, 30°) [4]
Total Internal Reflection: Stress optical coefficient C_T -1590 (25°) [19]

Applications/Commercial Products:
Processing & Manufacturing Routes: Synth. using Ziegler-Natta catalysts such as MgCl$_2$ supported TiCl$_4$-AlR$_3$ [29]

Bibliographic References
[1] Asanuma, T., Nishimori, Y., Ito, M., Uchikawa, N. and Shiomura, T., *Polym. Bull. (Berlin)*, 1991, **25**, 567, (syndiotactic, C-13 nmr)
[2] Trafara, G., *J. Polym. Sci., Polym. Chem. Ed.*, 1980, **18**, 321, (struct)
[3] Clark, K.J., Jones, A.T. and Sandiford, D.H., *Chem. Ind. (London)*, 1962, **47**, 2010, (transition temps)
[4] Kinsinger, J.B. and Ballard, L.E., *J. Polym. Sci., Part A: Polym. Chem.*, 1965, **3**, 3963, (soln props)
[5] Dannis, M.L., *J. Appl. Polym. Sci.*, 1959, **1**, 121, (thermal expansion)
[6] Fraser, G.V., Hendra, P.J., Chalmers, J.M., Cudby, M.E.A. and Willis, H.A., *Makromol. Chem.*, 1973, **173**, 195, (ir, Raman)
[7] Bu, H.S., Aycock, W. and Wunderlich, B., *Polymer*, 1987, **28**, 1165, (specific heat capacity)
[8] Reding, F.P., *J. Polym. Sci.*, 1956, **21**, 547, (transition temp)
[9] Modric, I., Holland-Moritz, K. and Hummel, D.O., *Colloid Polym. Sci.*, 1976, **254**, 342, (ir, Raman)
[10] Koehler, M.G. and Hopfinger, A.J., *Polymer*, 1989, **30**, 116, (transition temps)
[11] Morley, D.C.W., *J. Mater. Sci.*, 1974, **9**, 619, (transition temps)
[12] Hopfinger, A.J., Koehler, M.G., Pearlstein, R.A. and Tripathy, S.K., *J. Polym. Sci., Polym. Phys. Ed.*, 1988, **26**, 2007, (transition temp)
[13] Bailey, W.J. and Yates, E.T., *J. Org. Chem.*, 1960, **25**, 1800, (solubility)
[14] Combs, R.L., Slonaker, D.F. and Coover, H.W., *J. Appl. Polym. Sci.*, 1969, **13**, 519, (rheological props)
[15] Kinsinger, J.B. and Ballard, L.E., *J. Polym. Sci., Part B: Polym. Lett.*, 1964, **2**, 879, (soln props)
[16] Shirayama, K., Matsuda, T. and Kita, S., *Makromol. Chem.*, 1971, **147**, 155, (rheological props)
[17] Asakura, T., Demura, M. and Nishiyama, Y., *Macromolecules*, 1991, **24**, 2334, (C-13 nmr)
[18] IUPAC Commission on Polymer Characterisation and Properties, *Pure Appl. Chem.*, 1985, **57**, 823, (soln props)
[19] Philippoff, W. and Tomquist, E., *J. Polym. Sci., Part C: Polym. Lett.*, 1966, **23**, 881, (optical props)
[20] Wang, J.-S., Porter, R.S. and Knox, J.R., *J. Polym. Sci., Part B: Polym. Lett.*, 1970, **8**, 671, (flow activation energy)
[21] Wang, J.-S., Knox, J.R. and Porter, R.S., *J. Polym. Sci., Polym. Phys. Ed.*, 1978, **16**, 1709, (rheological props)
[22] Holland-Moritz, K., Djudovic, P. and Hummel, D.O., *Prog. Colloid Polym. Sci.*, 1975, **57**, 206, (ir, Raman)
[23] Shalaeva, L.F., *Vysokomol. Soedin., Ser. B*, 1968, **10**, 449, (soln props)
[24] Manabe, S., Nakamura, H., Uemura, S. and Takayanagi, M., *Kogyo Kagaku Zasshi*, 1970, **73**, 1587, (transition temps)
[25] Feng, L., Fan, Z., Xu, X., Bo, Y. and Yang, S., *Gaofenzi Xuebao*, 1989, **3**, 349, (soln props)
[26] Privalko, V.P., Severova, N.N. and Shmorgun, A.V., *Prom. Teplotekh.*, 1986, **8**, 83, (thermophysical props)
[27] McLaughlin, K.W. and Vanderwal, R.P., Polym. Prepr. (Am. Chem. Soc., *Div. Polym. Chem.*), 1987, **28**, 137, (C-13 nmr)
[28] Graener, H., Ye, T.Q. and Laubereau, A., *Chem. Phys. Lett.*, 1989, **164**, 12, (ir)
[29] Kothandaraman, K. and Devi, M.S., *J. Polym. Sci., Part A: Polym. Chem.*, 1994, **32**, 1283, (ir, nmr)
[30] Holland-Moritz, K., Djudovic, P. and Hummel, D.O., *Strukt. Polym.-Syst., Vortr. Diskuss. Hauptversamml. Kolloid-Ges.*, 26th, 1973, 206, (ir, Raman)
[31] Takayanagi, M., *Pure Appl. Chem.*, 1970, **23**, 151, (mech props)
[32] Natta, G., Danusso, F. and Moraglio, G., *Atti Accad. Naz. Lincei, Rend., Cl. Sci. Fis., Mat. Nat., Rend.*, 1958, **24**, 254, (transition temp)

Poly(octene oxide) P-422

Synonyms: *Poly[oxy(hexyl-1,2-ethanediyl)]. Poly(hexyloxirane). Poly(1,2-epoxyoctane). Poly(1,2-octane oxide)*
Monomers: 1-Octene oxide (*n*-hexyloxirane)
Material class: Thermoplastic
Polymer Type: polyalkylene ether
CAS Number:

CAS Reg. No.	Note
25639-83-2	homopolymer
71343-28-7	α-hydro-ω-hydroxyhomopolymer

Molecular Formula: $(C_8H_{16}O)_n$
Fragments: C_8H_{16}
General Information: More details available in patents [1,3]

Volumetric & Calorimetric Properties:
Density:

No.	Value	Note
1	d 0.944 g/cm^3	[2]
2	d 0.974 g/cm^3	[2]

Melting Temperature:

No.	Value	Note
1	86–87°C	[2]

Glass-Transition Temperature:

No.	Value	Note
1	-67°C	[2]
2	-74°C	[2]

Bibliographic References
[1] *U.S. Pat.,* 1968, 3 409 565
[2] Lal, J. and Trick, G.S., *J. Polym. Sci., Part A-1,* 1970, **8**, 2339, (density, transition temps)
[3] *U.S. Pat.,* 1970, 3 509 068

Poly(octyl methacrylate) — P-423

$R = CH_2(CH_2)_6CH_3$ (octyl)
$= CH(CH_3)(CH_2)_6CH_3$ (1-methylheptyl)
$= CH_2CH(CH_2)_3CH_3$
$\quad\ \ |$
$\quad CH_2CH_3$ (2-ethylhexyl)
$= CH_2(CH_2)_4CH(CH_3)_2$ (iso-octyl)

Structure: $-[CH_2C(CH_3)]_n-$ with $-C(=O)OR$ side group

Synonyms: *Poly(octyl 2-methyl-2-propenoate)*
Related Polymers: Polymethacrylates General
Monomers: Octyl methacrylate, 1-Methylheptyl methacrylate, 2-Ethylhexyl methacrylate, Isooctyl methacrylate
Material class: Thermoplastic
Polymer Type: acrylic
CAS Number:

CAS Reg. No.	Note
25087-18-7	n-octyl
25719-51-1	2-ethylhexyl

Molecular Formula: $(C_{12}H_{22}O_2)_n$
Fragments: $C_{12}H_{22}O_2$
General Information: There are several isomeric octyl methacrylates. Unless specified otherwise, data refer to the *n*-octyl isomer

Volumetric & Calorimetric Properties:
Density:

No.	Value	Note
1	d^{25} 0.971 g/cm^3	[1,2]
2	d^{100} 0.927 g/cm^3	[3]
3	d^{20} 0.988 g/cm^3	1-methylheptyl [4]

Thermal Expansion Coefficient:

No.	Value	Note	Type
1	0.00058–0.00062 K^{-1}	uv polymerised, T > T$_g$ [5]	V

Equation of State: Mark-Houwink constants have been reported [12,14]
Melting Temperature:

No.	Value	Note
1	< -55°C	max. [6]

Glass-Transition Temperature:

No.	Value	Note
1	-20°C	uv polymerisation, dilatometric method [5]
2	-70°C	[6,7]
3	-10°C	2-ethylhexyl [8]

Brittleness Temperature:

No.	Value	Note
1	-15°C	refractometric method [6]

Surface Properties & Solubility:
Cohesive Energy Density Solubility Parameters: δ 12.13 J$^{1/2}$ cm$^{-3/2}$ (8.4 cal$^{1/2}$ cm$^{-3/2}$) [9]; 12.05 J$^{1/2}$ cm$^{-3/2}$ (8.3 cal$^{1/2}$ cm$^{-3/2}$) [10]; 13.02 J$^{1/2}$ cm$^{-3/2}$ (9.7 cal$^{1/2}$ cm$^{-3/2}$, 2-ethylhexyl) [11]
Solvents/Non-solvents: Sol. *n*-butanol, 2-butanone [12]
Wettability/Surface Energy and Interfacial Tension: -(dγ/dT) 0.062 mN m^{-1} K^{-1} [13]
Surface Tension:

No.	Value	Note
1	28.8 mN/m	20°, 2-ethylhexyl, MV 64000 [13]
2	20.8 mN/m	150°, 2-ethylhexyl, MV 64000 [13]
3	17.7 mN/m	200°, 2-ethylhexyl, MV 64000 [13]

Transport Properties:
Polymer Solutions Dilute: Molar volumes 127.7 cm^3 mol^{-1} (van der Waals, amorph., 25°) [2]; 204.2 (rubber, amorph., 25°) [2]. Theta temps. 16.8° (*n*-butanol, MW 330000–12500000) [14]; 20° (*n*-butanol); 20° (butanone); 25° (Me$_2$CO/*n*-heptane, isooctyl) [15]

Mechanical Properties:
Mechanical Properties General: Plateau modulus 0.033 dyne cm^{-2} [3]
Viscoelastic Behaviour: Williams-Landel-Ferry constants for temp. superposition and Vogel parameters have been reported [16,17]. Other viscoelastic props. have been reported [19]

Optical Properties:
Refractive Index:

No.	Value	Note
1	1.478	20°, 1-methylheptyl [4]

Total Internal Reflection: Intrinsic segmental anisotropy -47 × 10^{-25} cm^3 (C$_6$H$_6$) [18]; -12.5 × 10^{-25} cm^3 (CCl$_4$) [18]

Bibliographic References
[1] Van Krevelen, D.W., *Properties of Polymers: Their Correlation with Chemical Structure,* Elsevier, 1976, 79
[2] Chee, K.K., *J. Appl. Polym. Sci.,* 1987, **33**, 1067

[3] Grassely, W.W. and Edwards, S.F., *Polymer*, 1981, **22**, 1329
[4] Crawford, J.W.C., *J. Soc. Chem. Ind., London*, 1949, **68**, 201
[5] Rogers, S.S. and Mandelkern, L., *J. Phys. Chem.*, 1957, **61**, 985
[6] Wiley, R.H. and Brauer, G.M., *J. Polym. Sci.*, 1948, **3**, 647
[7] Lal, J. and Trick, G.S., *J. Polym. Sci., Part A: Polym. Chem.*, 1964, **2**, 4559, (glass transition temp)
[8] Lewis, O.G., *Physical Constants of Linear Hydrocarbons*, Springer-Verlag, 1968
[9] Hughes, L.J. and Britt, G.E., *J. Appl. Polym. Sci.*, 1961, **5**, 337
[10] Van Krevelen, D.W. and Hoftyzer, P.J., *J. Appl. Polym. Sci.*, 1967, **11**, 2189
[11] Takemura, A., Tomita, B.I. and Migumachi, H., *J. Appl. Polym. Sci.*, 1986, **32**, 3489
[12] Kurata, M. and Stockmayer, W.H., *Adv. Polym. Sci.*, Springer-Verlag, 1963, **3**, 196
[13] Wu, S., *Polymer Interface and Adhesion*, Marcel Dekker, 1982
[14] Chinai, S.N., Resnick, A.L. and Lee, H.T., *J. Polym. Sci.*, 1958, **33**, 471, (eqn of state)
[15] Kalfus, M. and Mitus, J., *J. Polym. Sci., Part A-1*, 1966, **4**, 953
[16] Dannhauser, W., Child, W.C. and Ferry, J.D., *J. Colloid Sci.*, 1958, **13**, 103, (dynamic mech props)
[17] Yin, T.P. and Ferry, J.D., *J. Colloid Sci.*, 1961, **16**, 166, (dynamic mech props)
[18] Grishchenko, A.E., Vitovskaja, M.G., Tsvetkov, V.N. and Andreeva, L.N., *Vysokomol. Soedin.*, 1966, **8**, 800
[19] Aharoni, S.M., *Macromolecules*, 1983, **16**, 1722, (viscoelastic props)

Poly(octylpentylsilylene)   P-424

Synonyms: *Dichlorooctylpentylsilane homopolymer. C5-C8*
Polymer Type: polysilanes
CAS Number:

CAS Reg. No.
171673-30-6

Molecular Formula: $(C_{13}H_{28}Si)_n$
Fragments: $C_{13}H_{28}Si$
General Information: Produces a non-brittle film. Shows superior fracture toughness when compared with the symmetric poly(alkylsilanes) [3].
Morphology: Structure is very similar to that of C3–C5 (hexagonal, columnar packing both above and below the order-disorder transition). Exhibits a large degree of order in the hexagonal, columnar mesophase above and below its first-order temperature [1,3].
Miscellaneous: Stronger than C2–C5 but it retains a relatively large degree of flexibility and extensibility. Ordering of the polymer occurs at low temperatures, as DSC and X-ray diffraction show an order-disorder transition [1].

Volumetric & Calorimetric Properties:
Glass-Transition Temperature:

No.	Value	Note
1	36–37°C	[1]

Mechanical Properties:
Mechanical Properties General: The large degree of order in the hexagonal, columnar mesophase is confined by the relatively high modulus of 2×10^7 Pa. Hexagonal packing of the condis phase inhibits liquid-like flow under pressure. The transition merely reduces the elastic modulus by one to two orders of magnitude relative to the glassy state [1].

Tensile (Young's) Modulus:

No.	Value	Note
1	21.4 MPa	at 23°C [1]
2	16 MPa	at 100°C [1]

Optical Properties:
Optical Properties General: The existence of first-order transitions and the observation of UV thermochromic red shifts upon cooling for these polymers which do not crystallise are most likely the result of polarisation interactions between the σ-bonded Si backbone and the alkyl side chains, which is consistent with the predictions of the model of Schweizer for conjugated polymers [2,3].
Transmission and Spectra: Uv [2,3] spectra reported.

Polymer Stability:
Thermal Stability General: The disordering transition occurs at -36°C in the DMA of C5–C8 at 100rad sec^{-1} (16Hz) [1]. The transition enthalpy of 3.9cal g^{-1} (3.5kJ mol^{-1}) for C5-C8 is relatively small, indicating that a greater change in structure occurs at the first order transition [3].

Bibliographic References
[1] Klemann, B.M., DeVilbiss, T. and Koutsky, J.A., *Polym. Eng. Sci.*, 1996, **36**, 135
[2] Klemann, B.M., West, R. and Koutsky, J.A., *Macromolecules*, 1993, **26**, 1042
[3] Klemann, B.M., West, R. and Koutsky, J.A., *Macromolecules*, 1996, **29**, 198

Poly(3-octylthiophene)   P-425

Monomers: 3-Octylthiophene
Polymer Type: polythiophene
CAS Number:

CAS Reg. No.
104934-51-2

Molecular Formula: $(C_{12}H_{18}S)_n$
Fragments: $C_{12}H_{18}S$
Morphology: Crystal form, a 21.1, b 4.0, c 7.8 [3]

Volumetric & Calorimetric Properties:
Melting Temperature:

No.	Value	Note
1	150°C	approx. [1]

Transition Temperature General: T_g values have been reported [2]

Surface Properties & Solubility:
Solvents/Non-solvents: Sol. THF, decalin, $CHCl_3$ [4]

Transport Properties:
Polymer Solutions Dilute: [η] 1.05 dl g^{-1} (25°, $CHCl_3$) [3]

Electrical Properties:
Complex Permittivity and Electroactive Polymers: Conductivity 200 Ω^{-1} cm^{-1} [1]

Optical Properties:
Transmission and Spectra: Ir [5,7] and uv spectral data [1,6] have been reported

Applications/Commercial Products:
Processing & Manufacturing Routes: Prod. by oxidation polymerisation of bithiophenes, [8] or chemical or electrochemical oxidation polymerisation with various catalysts
Applications: Potential applications as liq. crystal films, diodes and solid electrolytes

Bibliographic References
[1] McCullough, R.D., Lowe, R.D., Jayaraman, M., Ewbank, P.C. and Anderson, D.L., *Synth. Met.*, 1993, 1198
[2] Jianguo, M. and Kalle, L., *Polym. Prepr. (Am. Chem. Soc., Div. Polym. Chem.)*, 1993, **34**, 727
[3] Moulton, J. and Smith, P., *Polymer*, 1992, **33**, 2340
[4] Taka, T., Nyholm, P., Laakso, J., Lopanen, M.T. and Osterholm, J.E., *Synth. Met.*, 1991, **41**, 899
[5] Louarn, G., Mevellec, J.Y., Buisson, J.P. and Lefrant, S., *Synth. Met.*, 1993, **55**, 587
[6] Yoshino, K., *Synth. Met.*, 1989, **28**, 69
[7] Gustafsson, G., Inganäs, O., Stafström, S., Osterholm, H. and Laakso, J., *Synth. Met.*, 1991, **41**, 593
[8] Krische, B., Zagorska, M. and Hellberg, J., *Synth. Met.*, 1993, **58**, 295

Poly(1,3,4-oxadiazole-2,5-diyl-1,3-phenylene) P-426

Synonyms: *Poly(m-phenylene-1,3,4-oxadiazole)*
Related Polymers: Poly(oxadiazoles), Poly(1,3,4-oxadiazole-2,5-diyl-1,4-phenylene)
Monomers: 1,3-Benzenedicarboxylic acid, Hydrazine sulfate
Material class: Thermoplastic
CAS Number:

CAS Reg. No.
26100-80-1

Molecular Formula: $(C_8H_4N_2O)_n$
Fragments: $C_8H_4N_2O$

Volumetric & Calorimetric Properties:
Melting Temperature:

No.	Value	Note
1	>400°C	min. [1]

Surface Properties & Solubility:
Solvents/Non-solvents: Sol. conc. H_2SO_4 [6]. Insol. DMF

Transport Properties:
Polymer Solutions Dilute: η_{inh} 0.134–0.172 (0.025–0.2% H_2SO_4) [3], 0.24 (H_2SO_4) [4], 0.73 dl g^{-1} (c, 0.2 g dl^{-1} in H_2SO_4, 30°) [5], 0.6 (H_2SO_4) [1]

Optical Properties:
Transmission and Spectra: Uv spectral data have been reported [1,3]

Polymer Stability:
Thermal Stability General: Rapid weight loss occurs at 400° and 500–560° then slow weight loss up to 700° [2]

Applications/Commercial Products:
Processing & Manufacturing Routes: Isophthalic acid is treated with hydrazine sulfate in the presence of a strong dehydrating agent

Bibliographic References
[1] Frazer, A.H., Sweeny, W. and Wallenberger, F.T., *J. Polym. Sci., Part A: Polym. Chem.*, 1964, **2**, 1157
[2] Frazer, A.H. and Sarasohn, I.M., *J. Polym. Sci., Part A-1*, 1966, **4**, 1649
[3] Abshire, C.J. and Marvel, C.S., *Makromol. Chem.*, 1961, **44-46**, 388
[4] Korshak, V.V., Krongauz, E.S. and Rusanov, A.L., *J. Polym. Sci., Part C: Polym. Lett.*, 1967, **16**, 2635
[5] Ueda, M. and Sugita, H., *J. Polym. Sci., Part A: Polym. Chem.*, 1988, **26**, 159
[6] Frazer, A.H. and Wallenberger, F.T., *J. Polym. Sci., Part A: Polym. Chem.*, 1964, **2**, 1171

Poly(1,3,4-oxadiazole-2,5-diyl-1,4-phenylene) P-427

Synonyms: *Poly(p-phenylene-1,3,4-oxadiazole)*. POD. PODZ
Related Polymers: Poly(oxadiazoles)
Monomers: 1,4-Benzenedicarboxylic acid, Hydrazine sulfate
Material class: Thermoplastic
CAS Number:

CAS Reg. No.
26023-46-1

Molecular Formula: $(C_8H_4N_2O)_n$
Fragments: $C_8H_4N_2O$

Volumetric & Calorimetric Properties:
Thermal Conductivity:

No.	Value	Note
1	0.11 W/mK	[7]

Specific Heat Capacity:

No.	Value	Note	Type
1	0.47 kJ/kg.C	[7]	P

Melting Temperature:

No.	Value	Note
1	>400°C	min. [1]

Surface Properties & Solubility:
Solvents/Non-solvents: Sol. H_2SO_4, trifluoroacetic acid. Insol. m-cresol, DMF, $CHCl_3$, nitrobenzene, formic acid, tetrafluoropropanol [1]

Transport Properties:
Polymer Solutions Dilute: η_{inh} 0.042–0.129 (0.025–0.2% H_2SO_4) [3]; 0.5 (H_2SO_4) [1]; 0.24 (H_2SO_4) [4]; 1.1dl g^{-1} (c, 0.2 g dl^{-1} in H_2SO_4, 30°) [5]

Mechanical Properties:
Tensile Strength Break:

No.	Value	Note	Elongation
1	64.7 MPa	66 kg mm^{-2} [11,12]	5.3%

Electrical Properties:
Complex Permittivity and Electroactive Polymers: Conductivity values have been reported [6]

Optical Properties:
Optical Properties General: Ir [2,6], uv [1], C-13 nmr [6] and x-ray photoelectron [6] spectral data have been reported

Polymer Stability:
Thermal Stability General: Rapid weight loss occurs from 450° to 560° then slow weight loss to 700° [8]. Degrades thermally by nitrogen elimination and also by formation of benzonitrile, terephthalonitrile and/or isophthalonitrile [9,10]

Applications/Commercial Products:
Processing & Manufacturing Routes: Terephthalic acid is treated with hydrazine sulfate in the presence of a strong dehydrating agent
Applications: Potential application as a high temp. resistant fibre

Bibliographic References
[1] Frazer, A.H., Sweeny, W. and Wallenberger, F.T., *J. Polym. Sci., Part A: Polym. Chem.*, 1964, **2**, 1157
[2] Iwakura, Y., Uno, K. and Hara, S., *J. Polym. Sci., Part A: Polym. Chem.*, 1965, **3**, 45
[3] Abshire, C.J. and Marvel, C.S., *Makromol. Chem.*, 1961, **44-46**, 388
[4] Korshak, V.V., Krongauz, E.S. and Rusanov, A.L., *J. Polym. Sci., Part C: Polym. Lett.*, 1967, **16**, 2635
[5] Ueda, M. and Sugita, H., *J. Polym. Sci., Part A: Polym. Chem.*, 1988, **26**, 159
[6] Murakami, M., Mizogami, S., Yasujima, H., Naitoh, S. and Yoshimura, S., *J. Polym. Sci., Part A: Polym. Chem.*, 1990, **28**, 1483
[7] Yasujima, H., Murakami, M. and Yoshimura, S., *Appl. Phys. Lett.*, 1986, **49**, 499
[8] Frazer, A.H. and Sarasohn, I.M, *J. Polym. Sci., Part A-1*, 1966, **4**, 1649
[9] Cotter, J.L., Knight, G.J. and Wright, W.W., *J. Gas Chromatogr.*, 1967, **5**, 86
[10] Varma, I.K., Sambandam, R.M. and Varma, D.S., *Makromol. Chem.*, 1973, **170**, 117
[11] Slutsker, L.I., Utevskii, L.E., Chereiskii, Z.Y. and Perepelkin, K.E., *J. Polym. Sci., Polym. Symp.*, 1977, **58**, 339
[12] Jones, R.S. and Soehngen, J.W., *J. Appl. Polym. Sci.*, 1980, **25**, 315

Poly(oxadiazoles)   P-428

Related Polymers: Poly(1,3,4-oxadiazole-2,5-diyl-1,3-phenylene), Poly(1,3,4-oxadiazole-2,5-diyl-1,4-phenylene)
General Information: Polymers containing oxadiazole units have excellent thermal and hydrolytic stability and can form fibres, films and membranes [1,4,5]

Polymer Stability:
Thermal Stability General: Aromatic polymers decompose between 450–500°; aliphatic polymers decompose between 400–450°. Fibres have excellent thermal stability [2]

Applications/Commercial Products:
Processing & Manufacturing Routes: Prod. by dehydration of a polyhydrazide or dihydrazide; reaction of a bistetrazole with a diacid chloride; or heating a polyacylamidazone with acid [1]
Applications: The oxadiazole-containing polymers have potential applications as reverse-osmosis membranes, fibres, films coatings, felts, laminates and moulded objects, gas and liq. filters

Bibliographic References
[1] *Encycl. Polym. Sci. Eng.*, Vol. 12, (ed. H.F. Mark), John Wiley and Sons, 1988, 332
[2] Frazer, A.H., Sweeny, W. and Wallenberger, F.T., J. Polym. Sci., Part A: Polym. Chem., 1964, **2**, 1157, 1171
[4] Hasegawa, M. and Unishi, T., J. Polym. Sci., *Part B: Polym. Lett.*, 1964, **2**, 237
[5] Ehlers, G.F.L. and Fisch, K.R., *Appl. Polym. Symp.*, 1969, **8**, 171, (props)

Poly(oxycarbonyl-1,4-phenyleneisopropylidene-1,4-phenylenecarbonyl)   P-429

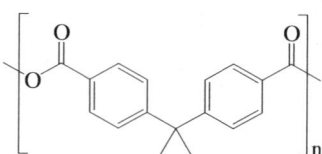

Synonyms: *Poly(4,4'-(1-methylethylidene)bisbenzoic acid)*
Related Polymers: General information
Monomers: 4,4'-(1-Methylethylidene)bisbenzoic acid
Material class: Thermoplastic
Polymer Type: polyanhydride
Molecular Formula: $(C_{17}H_{14}O_3)_n$
Fragments: $C_{17}H_{14}O_3$

Volumetric & Calorimetric Properties:
Melting Temperature:

No.	Value	Note
1	230–235°C	[1]
2	238–240°C	[2]

Glass-Transition Temperature:

No.	Value	Note
1	60°C	[1]
2	140°C	[2]

Applications/Commercial Products:
Processing & Manufacturing Routes: The dibasic acid is converted to mixed anhydride with acetic acid. Mixed anhydride polymerises on heating under vacuum

Bibliographic References
[1] Conix, A., *J. Polym. Sci.*, 1958, **29**, 343
[2] Yoda, N., *Makromol. Chem.*, 1959, **32**, 1

Poly(oxycarbonyl-1,4-phenylenemethylene-1,4-phenylene carbonyl)   P-430

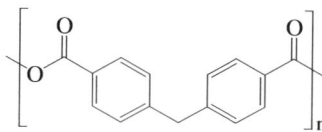

Synonyms: *Poly(4,4'-methylenebisbenzoic acid). Poly(diphenylmethane-4,4'-dicarboxylic acid)*
Related Polymers: Poly(anhydrides)
Monomers: 4,4'-Methylenebisbenzoic acid
Material class: Thermoplastic
Polymer Type: polyanhydride
Molecular Formula: $(C_{15}H_{10}O_3)_n$
Fragments: $C_{15}H_{10}O_3$
General Information: Crystallises to prod. weak fibres [1]

Volumetric & Calorimetric Properties:
Melting Temperature:

No.	Value	Note
1	332°C	[2]

Glass-Transition Temperature:

No.	Value	Note
1	122°C	[2]

Applications/Commercial Products:
Processing & Manufacturing Routes: The dibasic acid is converted to the mixed anhydride with AcOH. The mixed anhydride polymerises on heating under vacuum

Bibliographic References
[1] Conix, A., *J. Polym. Sci.*, 1958, **29**, 343
[2] Yoda, N., *Makromol. Chem.*, 1959, **32**, 1

Poly(oxyisophthaloyl) — P-431

Synonyms: *Poly(isophthalic anhydride). Poly(oxycarbonyl-1,3-phenylene carbonyl)*
Related Polymers: Poly(anhydrides)
Monomers: 1,3-Benzenedicarboxylic acid
Material class: Thermoplastic
Polymer Type: polyanhydride
CAS Number:

CAS Reg. No.
26913-46-2
25950-44-1

Molecular Formula: $(C_8H_4O_3)_n$
Fragments: $C_8H_4O_3$

Volumetric & Calorimetric Properties:
Glass-Transition Temperature:

No.	Value	Note
1	130°C	[1]

Applications/Commercial Products:
Processing & Manufacturing Routes: Prod. by reaction of 1,4-bis(trichloromethyl)benzene and isophthalic acid with various Lewis Acid catalysts via acid chloride
Applications: Biodegradable medical applications

Bibliographic References
[1] Ward, I.M., *Text. Res. J.*, 1961, **13**, 650

Polyoxymethylene — P-432

Synonyms: *Polyacetal. Acetal. Polyformaldehyde. POM*
Monomers: Formaldehyde, Trioxane
Material class: Thermoplastic
Polymer Type: acetal, polyalkylene ether
CAS Number:

CAS Reg. No.
9002-81-7

Molecular Formula: $(CH_2O)_n$
Fragments: CH_2O
Molecular Weight: MW 20000–90000; M_w/M_n 2

Additives: Antioxidants, uv stabilisers, pigments, glass fibre, lubricants, PTFE
Morphology: Two cryst. forms. Trigonal a 0.447 nm, b 0.447 nm, c 1.73 nm; orthorhombic a 0.476 nm, b 0.766 nm, c 0.356 nm [6]
Miscellaneous: Crystalline polymers with approx. 60–80% crystallinity. End groups are usually modified to prevent unzipping. The main chain may also be modified to improve stability by the inclusion of copolymers e.g. ethylene oxide. Many commercial grades are co- and terpolymers

Volumetric & Calorimetric Properties:
Density:

No.	Value	Note
1	d 1.25 g/cm^3	amorph., trig. [6]
2	d 1.49 g/cm^3	cryst. [6]
3	d 1.533 g/cm^3	cryst., orthorhombic [6]
4	d 1.39–1.43 g/cm^3	ASTM D742 [1]

Thermal Expansion Coefficient:

No.	Value	Note	Type
0	9.2×10^{-5} K^{-1}	[2]	l
1	8.1×10^{-5} K^{-1}	[1]	l

Thermal Conductivity:

No.	Value	Note
1	0.224 W/mK	[1]

Specific Heat Capacity:

No.	Value	Note	Type
1	1.47 kJ/kg.C	[1,2]	p

Melting Temperature:

No.	Value	Note
1	166°C	[2]
2	175°C	[3]
3	177°C	[5]
4	186–189°C	[7]

Glass-Transition Temperature:

No.	Value	Note
1	-13°C	[3]
2	-82°C	[6]

Transition Temperature General: -73° [3]
Deflection Temperature:

No.	Value	Note
1	170°C	0.45 MPa ASTM D684 [1]
2	121–124°C	1.8 MPa, ASTM D684 [1]

Vicat Softening Point:

No.	Value	Note
1	185°C	ASTM D569

Polyoxymethylene

Surface Properties & Solubility:
Solvents/Non-solvents: No solvents at room temp. but liquids with a similar solubility parameter will cause swelling. Sol phenol, aniline, benzyl alcohol (above 110°)

Transport Properties:
Transport Properties General: Low gas and vapour permeability. Little efffect on props. from humidity

Melt Flow Index:

No.	Value	Note
1	9 g/10 min	general purpose [3]
2	2.5 g/10 min	[2]
3	6 g/10 min	[5]

Polymer Melts: Viscosity, MW relationships have been reported in a variety of solvents [8,9]. Huggins coefficient 0.31–0.41 [9,10]

Water Absorption:

No.	Value	Note
1	0.25 %	after 24h immersion [1]
2	0.16 %	50% relative humidity [3]
3	0.2 %	equilibrium, 50% relative humidity

Gas Permeability:

No.	Gas	Value	Note
1	CO_2	14.6–19.7 cm^3 mm/(m^2 day atm)	23°, 50% relative humidity, Delrin [5]
2	O_2	4.7–6.7 cm^3 mm/(m^2 day atm)	23°, 50% relative humidity, Delrin [5]

Mechanical Properties:
Mechanical Properties General: Acetals have excellent fatigue resistance and resistance to creep. They also have a low coefficient of friction and excellent abrasion resistance

Tensile (Young's) Modulus:

No.	Value	Note
1	3500–5900 MPa	ASTM D638 [1,2,4,5]

Flexural Modulus:

No.	Value	Note
1	2390–3020 MPa	ASTM D790 [1]

Tensile Strength Break:

No.	Value	Note	Elongation
1	62–70 MPa	ASTM D638 [1,2]	25–65% elongation, break; 15–31% elongation, yield

Flexural Strength at Break:

No.	Value	Note
1	88–100 MPa	ASTM D790 [1,2]

Poisson's Ratio:

No.	Value	Note
1	0.35	[2]

Flexural Strength Yield:

No.	Value	Note
1	98.5 MPa	[5]

Compressive Strength:

No.	Value	Note
1	112.31 MPa	ASTM D695; Delrin 500P [5]

Miscellaneous Moduli:

No.	Value	Note	Type
1	2190 MPa	[2]	shear modulus

Mechanical Properties Miscellaneous: Shear strength 54.84 MPa (ASTM D732) [2]. Stress-strain relationships have been reported
Hardness: Rockwell M78-M94 [1,2] Rockwell R120 (Delrin 500P) [5]
Fracture Mechanical Properties: Has better fatigue resistance than Nylon 66
Friction Abrasion and Resistance: Coefficient of friction (static) μ_{stat} 0.1–0.3 (dry, steel) [1,3]. Abrasion resistance Taber CS-17 14–20 mg (1000 cycles)$^{-1}$ (ASTM D1044). [1] Coefficient of friction 0.3 (self) [4]

Izod Notch:

No.	Value	Notch	Note
1	62–72 J/m	Y	ASTM D256 [2]
2	70–120 J/m	Y	ASTM D256 [1]

Electrical Properties:
Electrical Properties General: Typical electrical props. for thermoplastics used in non-critical applications. Not suitable for application at high frequencies (1GHz) or applications involving high electric stress above 70°

Surface/Volume Resistance:

No.	Value	Note	Type
2	10×10^{15} Ω.cm	0.2% water, ASTM D257 [3]	S

Dielectric Permittivity (Constant):

No.	Value	Frequency	Note
1	3–7	100Hz–1MHz	ASTM D150 [1,3]

Dielectric Strength:

No.	Value	Note
1	19.7 kV.mm^{-1}	short time [3]

Arc Resistance:

No.	Value	Note
1	129s	ASTM D495 [1]

Dissipation (Power) Factor:

No.	Value	Frequency	Note
1	0.004–0.007	100Hz–1MHz	[2,8]

Optical Properties:
Optical Properties General: Opaque

Polymer Stability:
Thermal Stability General: Generally good resistance to thermal and oxidative degradation
Upper Use Temperature:

No.	Value	Note
1	90°C	[1]
2	82°C	air [3]
3	65°C	water [3]

Decomposition Details: Thermal decomposition leads to the production of formaldehyde gas; this can lead to the chain unzipping if unchecked. Modified end groups are used to prevent this
Flammability: Flammability rating HB (UL94) [5]
Environmental Stress: Prolonged exposure to uv light will induce chalking and gradual embrittlement. Additives be can used to improve this
Chemical Stability: Good resistance to common chemicals but contact with strong oxidising agents is not recommended. Good resistance to aromatics, alcohols, oils, grease, chlorinated hydrocarbons; fair resistance to ketones and detergents
Hydrolytic Stability: Has good resistance to hot water. Poor resistance to mineral acids

Applications/Commercial Products:
Processing & Manufacturing Routes: Polymerised using ionic initiation. Anionic initiators are thought to be used to produce commercial polymers. Carried out in an inert hydrocarbon solvent. Processed by injection moulding, extrusion and rotational moulding. Further fabrication can be carried out using conventional metal-working techniques
Applications: Consumer goods (kettles). Plumbing (fittings and valves). Automotive components (fuel caps or pump housings). Gears, bearings and air flow valves, filter housings

Trade name	Supplier
Celcon	Celanese Plastics
Delrin	DuPont
Fulton	LNP Engineering
Tenac	Asahi Chemical

Bibliographic References
[1] The Materials Selector Elsevier Applied Science, 1991, 1644
[2] *Ashlene 180*, Ashlene 180, Ashley Polymers, 1994, (technical datasheet)
[3] Brydson, J.A., *Plast. Mater.*, 5th edn., Butterworth-Heinemann, 1989
[4] *Mach. Des.*, 1995, **67 (3),** 79
[5] *Delrin 500P*, DuPont, 1994, (technical datasheet)
[6] Gaur, U. and Wunderlich, B., *J. Phys. Chem. Ref. Data*, 1981, **10,** 1001, (cryst. struct.)
[7] Salaris, F., Turturro, A., Bianchi, U. and Martuscelli, *Polymer*, 1978, **19,** 1163, (melt temp.)
[8] Wagner, M.L. and Wissbrun, K.F., *Makromol. Chem.*, 1965, **81,** 14, (viscosity)
[9] Hoehr, L., Jaacks, V., Cherdron, H., Iwabuchi, S. and Kern, W., *Makromol. Chem.*, 1967, **103,** 279, (MW)
[10] Grassie, N., Roche, R.S., *J. Polym. Sci., Part C: Polym. Lett.*, 1968, **16,** 4207, (Huggins coefficient)
[11] Linton, W.H. and Goodman, H.K., *J. Appl. Polym. Sci.*, 1959, **1,** 179, (stress-strain)

Poly(oxysiliconphthalocyanin) P-433

Synonyms: *Phthalocyaninatopoly(siloxane). Poly(phthalocyaninato silicone oxo)*
Related Polymers: Polysiloxanes
Monomers: Dichlorosilicon phthalocyanine, Dihydroxysilicon phthalocyanine
Material class: Polymer liquid crystals
Polymer Type: silicones
CAS Number:

CAS Reg. No.
39114-20-0

Molecular Formula: $[C_{68}H_{88}N_8O_9Si]_n$
Fragments: $C_{68}H_{88}N_8O_9Si$
Molecular Weight: MW 40000–100000
General Information: Is dark purple, highly crystalline and axially-linked. Is of interest owing to its electrical props. [6]
Morphology: Hexagonal packing of cylinders, inter rod distance 2.57 nm repeat distance of monomers along chain 0.333 nm [1]. Crystallises naturally into an orthorhombic struct. a 13.80Å, b 27.59Å, C 6.66Å, space group Ibam. A tetrahedral struct. is produced during oxidation of the polymer and this is retained after reduction [6,10,11]. Doping with halogen and $NO^{\oplus} X^{\ominus}$ (where X is $BF_4^{\ominus}$, $PF_6^{\ominus}$ or $SbF_6^{\ominus}$) gives a tetrahedral struct. a 13.97Å, c 6.60Å; space group P4/mcc; Si:I or Br 1:1.12. Doping is heterogeneous [11,12,13]

Volumetric & Calorimetric Properties:
Density:

No.	Value	Note
1	d 1.432 g/cm^3	[6]
2	d 1.802 g/cm^3	iodine-doped [11]

Transition Temperature General: Decomposes before melting [8]

Surface Properties & Solubility:
Solvents/Non-solvents: Sol. conc. mineral acids, chlorinated hydrocarbons, aromatic hydrocarbons, ethers [2,4,5,6,7]. Insol. H_2O, alcohols
Surface and Interfacial Properties General: Forms stable film on air-water interface readily transferred onto hydrophobic surface to form Langmuir-Blodgett film with liq. crystalline props. [1,7]
Wettability/Surface Energy and Interfacial Tension: Collapse pressure of film on water 23 mN m^{-1} (65°); 39 mN m^{-1} (6°) [1,5]

Transport Properties:
Transport Properties General: Readily transferred from film on water air interface onto hydrophobic solid surface to form Langmuir-Blodgett film [1,5]

Electrical Properties:
Electrical Properties General: Resistance falls sharply on doping with iodine to 100 Ω cm [3,8].
Complex Permittivity and Electroactive Polymers: Electrochemical doping produces a wide range of conductive polymers with tuneable electrical, magnetic and optical props. Although oxidative doping is usual, reductive doping has also been reported [10,11]. The effects of doping on conductivity have been reported [11,12]. Doping with electron-acceptors such as 2-chloroaniline, results in a decrease in conductivity [9]
Magnetic Properties: Iodine-doped polymer exhibits Curie-tailing at low temps. [11]

Optical Properties:
Optical Properties General: Optical reflectance data for the doped polymer have been reported [11]
Transmission and Spectra: Absorption max. 545 nm (polymer). Absorption max. 680 nm (monomer) [5]. Ir, Raman, esr, C-13 nmr and uv spectral data have been reported [6,11,13]
Total Internal Reflection: Chromophore dipole transition moment is perpendicular to polymer backbone [5]. Dichroic ratio of Langmuir-Blodgett film increases from 2.7–5.8 on annealing at 140° for 30 mins. (Dichroic ratio absorption of light polarised perpendicular to direction of dipping/light polarised parallel to dipping) [5,7]

Polymer Stability:
Polymer Stability General: The bulk of phthalocyanine ring protects the siloxane backbone from some reagents [6,8]
Upper Use Temperature:

No.	Value	Note
1	280°C	[7]

Decomposition Details: Onset of decomposition (without melting) occurs at 550° [8]
Environmental Stress: Not degraded by oxygen or moisture [3,7]. Reduced in electrolytic cell at -1.5 V, oxidised at 0.25 V
Chemical Stability: Stable in conc. acids. Sol. conc. acids and most organic solvents [5,6]
Hydrolytic Stability: Insol. in water; not degraded by water or moisture [2,3]
Recyclability: Not recyclable

Applications/Commercial Products:
Processing & Manufacturing Routes: The dichloro silicon phthalocyanine monomer (or the dihydroxy silicon phthalocyanine monomer obtained from it by hydrolysis or ion exchange) is condensed in organic solvent at around 200° in the presence of metal catalyst [2,4,8]. Catalyst necessary in order for polymerisation to occur at temps. low enough to avoid degradation of alkoxy substituted phthalocyanines [2]. Fibres may be spun from solns. in strong acid [11]. Doping is carried out by stirring powdered polymer with halogen in dry, deoxygenated C_6H_6 for 48h. Iodine doping can also be performed in conc. sulfuric acid soln. [13]
Applications: Main commercial interest is in the optical props. of Langmuir-Blodgett films formed from this material. These films are stable and of particularly high optical quality

Bibliographic References
[1] Albouy, P.A., Schaub, M. and Wegner, G., *Acta Polym.*, 1994, **45**, 210
[2] Orthmann, E. and Wegner, G., Makromol. Chem., *Rapid Commun.*, 1986, **7**, 243
[3] Dirk, C.W., Mintz, E.A., Schoch, K.F. and Marks, T.J., J. Macromol. Sci., *Chem.*, 1981, **16**, 275
[4] Caseri, W., Sauer, T. and Wegner, G., Makromol. Chem., *Rapid Commun.*, 1988, **9**, 651
[5] Schwiegk, S., Vahlenkamp, T., Yuanze, X. and Wegner, G., *Macromolecules*, 1992, **25**, 2513
[6] Dirk, C.W., Inabe, T., Schoch, K.F. and Marks, T.J., *J. Am. Chem. Soc.*, 1983, **105**, 1539
[7] Orthmann, E. and Wegner, G., Angew. Chem., *Int. Ed. Engl.*, 1986, **25**, 1105
[8] Snow, A.W. and Griffith, J.R., *Encycl. Polym. Sci. Eng.*, 2nd edn., (ed. J.I. Kroschwitz), John Wiley and Sons, 1985, **11**, 212
[9] Wohnle, D., *Adv. Polym. Sci.*, 1983, **50**, 45
[10] Marks, T.J., Gaudiello, J.G., Kellogg, G.E. and Tetrick, S.M., *ACS Symp. Ser.*, 1980, **360**, 224
[11] Dirk C.W., Inabe, T., Lyding, J.W., Schoch, K.F. et al, J. Polym. Sci., *Polym. Symp.*, 1983, **70**, 1, (conductivity)
[12] Inabe, T., Gaudiello, J.G., Moguel, M.K. Lyding, J.W. et al, *J. Am. Chem. Soc.*, 1986, **108**, 7595
[13] Diel, B.N., Inabe, T., Lyding, J.W., Schoch, K.F. et al, *J. Am. Chem. Soc.*, 1983, **105**, 1551

Poly(1-pentadecene)   P-434
Synonyms: *Poly(1-tridecylethylene). Poly(1-tridecyl-1,2-ethanediyl)*
Monomers: 1-Pentadecene
Material class: Thermoplastic
Polymer Type: polyolefins
CAS Number:

CAS Reg. No.	Note
59284-11-6	Isotactic

Molecular Formula: $(C_{15}H_{30})_n$
Fragments: $C_{15}H_{30}$
Morphology: Three forms known [2]

Volumetric & Calorimetric Properties:
Melting Temperature:

No.	Value	Note
1	54°C	[1]
2	60°C	[3]
3	59°C	[2]

Transition Temperature:

No.	Value	Note
1	13°C	Isotactic first order transitions [2]
2	20°C	

Optical Properties:
Transmission and Spectra: Ir and Raman spectral data have been reported [1]

Applications/Commercial Products:
Processing & Manufacturing Routes: Synth. using Ziegler-Natta catalysts, such as $TiCl_3$-$AlEt_2Cl$ [1]

Bibliographic References
[1] Modric, I., Holland-Moritz, K. and Hummel, D.O., *Colloid Polym. Sci.*, 1976, **254**, 342, (ir, Raman)
[2] Trafara, G., Koch, R., Blum, K. and Hummel, D.O., *Makromol. Chem.*, 1976, **177**, 1089, (melting points)
[3] Blum, K. and Trafara, G., *Makromol. Chem.*, 1980, **181**, 1097, (melting point)

Poly[2,2-pentane bis(4-phenyl)carbonate]   P-435

$$-\left[O-\underset{}{\underset{}{C_6H_4}}-\underset{CH_2CH_2CH_3}{\overset{CH_3}{C}}-\underset{}{C_6H_4}-O-\overset{O}{\underset{}{C}}\right]_n$$

Synonyms: *Poly[oxycarbonyloxy-1,4-phenylene(1-methylbutylidene)-1,4-phenylene]. Poly(4,4'-dihydroxydiphenyl-2,2-pentane-co-carbonic dichloride)*
Monomers: 4,4'-Dihydroxydiphenyl-2,2-pentane, Carbonic dichloride
Material class: Thermoplastic
Polymer Type: polycarbonates (miscellaneous)
CAS Number:

CAS Reg. No.
32243-05-3

Molecular Formula: $(C_{18}H_{18}O_3)_n$
Fragments: $C_{18}H_{18}O_3$

Volumetric & Calorimetric Properties:
Density:

No.	Value	Note
1	d 1.13 g/cm^3	25° [1]

Melting Temperature:

No.	Value	Note
1	200–220°C	[1]

Poly(1-pentene)

Synonyms: *PP1. PPe. PPE. PPEN. Poly(1-propylethylene). Poly(1-propyl-1,2-ethanediyl)*
Monomers: 1-Pentene
Material class: Thermoplastic
Polymer Type: polyolefins
CAS Number:

CAS Reg. No.	Note
25587-79-5	atactic
25587-78-4	isotactic
133578-02-6	syndiotactic

Molecular Formula: $(C_5H_{10})_n$
Fragments: C_5H_{10}
Morphology: Form Ia monoclinic, a 1.135, b 2.085, c 0.649, β 99.6°, helix 3_1 [2]; or monoclinic, a 2.115, b 1.120, c 0.649, β 99.6°, helix 3_1 [2]; or monoclinic a 2.24, b 2.12, c 0.649, β 91°, helix 3_1 [3]. Form Ib monoclinic, a 2.43, b 1.93, c 0.650, β 96°, helix 3_1 [3]. Form IIa monoclinic, a 1.930, b 1.690, c 0.708, γ 116°, helix 4_1 [2]. Form IIb monoclinic, a 1.960, b 1.675, c 0.708, γ 115° 20', helix 4_1 [2]. Form II monoclinic, a 3.565, b 2.02, c 0.713, β 91°, helix 5_1, 24_5, 26_5 [4]. Form III orthorhombic, a 2.120, b 1.148, c 1.439, helix 7_2 [3]. Cryst. struct. has been reported [5,6]

Volumetric & Calorimetric Properties:
Density:

No.	Value	Note
1	d 0.896 g/cm³	[3]
2	d 0.916 g/cm³	[3]
3	d 0.85 g/cm³	amorph. [8]
4	d 0.87 g/cm³	[8,9,10]
5	d^{25} 0.858 g/cm³	[11]
6	d^{25} 0.89 g/cm³	[12]

Thermal Expansion Coefficient:

No.	Value	Note	Type
1	0.00015 K⁻¹	T < T_g [13]	L
2	0.00092 K⁻¹	[8]	V
3	0.000806 K⁻¹	[11]	V
4	0.00079 K⁻¹	[14]	V

Thermodynamic Properties General: Molar volume 82.5 cm³ mol⁻¹. C_p [18,19,20,22,25,27] and C_v [24] values have been reported
Latent Heat Crystallization: ΔH, ΔS and ΔG values have been reported [18,19,20,21,22,23,24]
Specific Heat Capacity:

No.	Value	Note	Type
1	87.9 kJ/kg.C	200K	P
2	97.67 kJ/kg.C	233K	P

Melting Temperature:

No.	Value	Note
1	101°C	[3]
2	111°C	[22]
3	130°C	approx. [30]
4	80°C	[31,32]

Glass-Transition Temperature:

No.	Value	Note
1	137°C	[1,2]

Transport Properties:
Water Absorption:

No.	Value	Note
1	0.2 %	25°, 65% relative humidity [3]

Gas Permeability:

No.	Gas	Value	Note
1	O_2	97.17 cm³ mm/(m² day atm)	1.11 × 10⁻¹³ cm² (s Pa)⁻¹, 25 atm, 1 atm [2]

Mechanical Properties:
Tensile Strength Break:

No.	Value	Note	Elongation
1	65.2 MPa	665 kg cm⁻² [1]	66%

Electrical Properties:
Electrical Properties General: Dielectric constant and dissipation factor do not vary with temp. until approaching T_g
Dielectric Permittivity (Constant):

No.	Value	Frequency	Note
1	2.3	1 kHz	25–100° [1]

Dissipation (Power) Factor:

No.	Value	Frequency	Note
1	0.00042	1 kHz	25–100° [1]

Optical Properties:
Refractive Index:

No.	Value	Note
1	1.5745	25° [1]

Transmission and Spectra: Ir spectral data have been reported [3]

Applications/Commercial Products:
Processing & Manufacturing Routes: The main synthetic routes are phosgenation and transesterification. The most widely used phosgenation approach is interfacial polycondensation in a suitable solvent, but soln. polymerisation is also carried out in the presence of pyridine [3]. A slight excess of phosgene is introduced into a soln. or suspension of the aromatic dihydroxy compound in aq. NaOH at 20–30°, in an inert solvent (e.g. CH_2Cl_2). Catalysts are used to achieve high MWs [1,3]. With transesterification, the dihydroxy compound and a slight excess of diphenyl carbonate are mixed together in the melt (150-300°) in the absence of O_2, with elimination of phenol. Reaction rates are increased by alkaline catalysts and the use of a medium-high vacuum in the final stages [1,3]

Bibliographic References
[1] Schnell, H., *Ind. Eng. Chem.*, 1959, **51**, 157, (props., synth.)
[2] Schmidhauser, J.C. and Longley, K.L., *J. Appl. Polym. Sci.*, 1990, **39**, 2083, (T_g, gas permeability)
[3] Schnell, H., *Polym. Rev.*, 1964, **9**, 99, (water absorption, ir, synth., rev)

Poly(1-pentene)

5	79°C	[22]
6	78°C	[33,34]
7	77°C	[35]
8	76°C	[3,4]
9	75°C	[13,36]
10	75–80°C	Form II [8,9,10]
11	73°C	Form II [37,38]
12	71°C	Form II [39]
13	70°C	[40]
14	70°C	Form III [3]

Glass-Transition Temperature:

No.	Value	Note
1	-40°C	[5,25]
2	-38°C	[29]
3	-37°C	[39]
4	-29°C	[22]
5	-24°C	[40]
6	-20°C	[41]
7	-53°C	[74]
8	-29°C	calc. [29]
9	-60°C	calc. [42]

Transition Temperature:

No.	Value	Note	Type
1	-29°C	[33]	$T_{softening\ point}$
2	-90°C	[13]	T_γ
3	-13°C		T_α
4	-150°C	[43]	T_β
5	-3°C	80 Hz	T_α
6	-130°C	300 Hz [31]	T_β
7	-140°C	[44]	$T_{transition}$
8	-25°C	[44]	$T_{transition}$
9	-52°C	[8]	$T_{transition}$
10	-10°C	[8]	$T_{transition}$
11	13°C	[8]	$T_{transition}$
12	225–230°C	[45]	$T_{transition}$
13	130°C	equilibrium melting temp. [25]	T°_m

Surface Properties & Solubility:
Solubility Properties General: Solubility props. and polymer solvent interaction parameter values have been reported [46,47]
Cohesive Energy Density Solubility Parameters: Cohesive energy density parameters have been reported [48]
Internal Pressure Heat Solution and Miscellaneous: Heats of mixing at infinite dilution in various solvents have been reported at 25° [11]
Solvents/Non-solvents: Sol. boiling heptane, toluene, C_6H_6 [12], Et_2O, EtOAc. Insol. MeOH [8,10]. Other solubility data have been reported [52]
Wettability/Surface Energy and Interfacial Tension: Surface free energy [53] and critical surface tension have been reported [48]

Transport Properties:
Polymer Solutions Dilute: θ 56° (isotactic, phenetole), 48.5° (amorph., phenetole) [49]; θ 31–32° (isoamyl acetate) [50]; θ 62.4° (2-pentanol) [12]; θ 32.5° (isobutyl acetate), 64° (phenetole), 85° (anisole), 121° (diphenylmethane), 149° (diphenyl ether) [51]. [η] 1.695 dl g^{-1} (30°, toluene, isotactic) [50]; [η] 3.48 dl g^{-1} (30°, toulene, isotactic) [51]; [η] 0.215 dl g^{-1} (21°, heptane) [3]; [η] 1.68 dl g^{-1} (25°, C_6H_6) [54]; [η] 2–3 dl g^{-1} (100°, tetrahydronaphthalene) [8]; [η] 3.29 dl g^{-1} (55°, toluene) [37]; [η] 4.7 dl g^{-1} (135°, tetrahydronaphthalene) [34]. Other viscosity values have been reported [10,55,56,57,58]

Mechanical Properties:
Mechanical Properties General: Mech. props. [5,59,60], flexibility [57] and rigidity [61] have been reported. Tensile, dynamic tensile and loss moduli have been reported [44,62]. Variation of dynamic modulus with temp. has been reported in graphical form [41]
Tensile Strength Break:

No.	Value	Note
1	58 MPa	-60° [63]

Impact Strength: Impact strength values have been reported [59]
Viscoelastic Behaviour: Acoustic [44] and thermoacoustic props. [65] have been reported
Hardness: Hardness has been reported [52]

Electrical Properties:
Dissipation (Power) Factor:

No.	Value	Frequency	Note
1	0.015–0.07	40–600 Hz	-180–-30° [41]

Optical Properties:
Optical Properties General: Refractive index has been reported [57]
Transmission and Spectra: H-1 nmr [45], C-13 nmr [1,66,67], esr [68], ir [32,34,69] and Raman [38,70,71] spectral data have been reported
Total Internal Reflection: Birefringence has been reported [5]. Stress optical coefficient 325–378 [37]

Polymer Stability:
Decomposition Details: The isotactic form decomposes at 115° to give CO_2, AcOH and propanoic acid as the major products [34,73]. Other decomposition details have been reported [72]

Applications/Commercial Products:
Processing & Manufacturing Routes: Synth. using Ziegler-Natta catalysts such as $TiCl_3AlEt_2Cl$ [32,38]. Moulding behaviour has been studied

Bibliographic References
[1] Asanuma, T., Nishimori, Y., Ito, M., Uchikawa, N. and Shiomura, T., *Polym. Bull. (Berlin)*, 1991, **25**, 567, (syndiotactic)
[2] Turner-Jones, A. and Aizlewood, J.M., *J. Polym. Sci., Part B: Polym. Lett.*, 1963, **1**, 471, (struct)
[3] Moser, M. and Boudeulle, M., *J. Polym. Sci., Polym. Phys. Ed.*, 1976, **14**, 1161, (struct)
[4] Decroix, J.Y., Moser, M. and Boudeulle, M., *Eur. Polym. J.*, 1975, **11**, 357, (struct)
[5] Powers, J., Hoshino, S. and Stein, R.S., *CA*, 1961, **58**, 14123f, (struct, birefringence)
[6] Finberg, A.D., *Diss. Abstr. Int. B*, 1978, **39**, 2328, (struct)
[7] Quinn, F.A. and Powers, J., *J. Polym. Sci., Part B: Polym. Lett.*, 1963, **1**, 341, (crystallisation rate)
[8] Danusso, F. and Gianotto, G., *Makromol. Chem.*, 1963, **61**, 164, (transition temps, solubility)
[9] Natta, G., *Chim. Ind. (Milan)*, 1955, **37**, 888, (melting point)
[10] Natta, G., Pino, P. and Mazzanti, G., *Chim. Ind. (Milan)*, 1955, **37**, 927, (rheological props)
[11] Phuong-Nguyen, H. and Delmas, G., *Macromolecules*, 1979, **12**, 740, (soln props)
[12] Mark, J.E. and Flory, P.J., *J. Am. Chem. Soc.*, 1965, **87**, 1423, (solubility props)

[13] Dannis, M.L., *J. Appl. Polym. Sci.*, 1959, **1**, 121, (transition temps)
[14] Lewis, D.G., *Physical Properties of Linear Homopolymers*, Springer-Verlag, 1968, (expansion coefficient)
[15] Boyer, R.F., *J. Macromol. Sci., Phys.*, 1973, **7**, 487, (calorimetric props)
[16] Wang, J.S., Porter, R.S. and Knox, J.R., *Polym. J. (Tokyo)*, 1978, **10**, 619, (solubility props)
[17] Sewell, J.H., *J. Appl. Polym. Sci.*, 1973, **17**, 1741, (molar volume)
[18] Lebedev, B.V., Smirnova, N.N., Kiparisova, E.G., Faminskii, G.G. et al, Vysokomol. Soedin., *Ser. A*, 1993, **35**, 1951, (calorimetric props)
[19] Lebedev, B.V., Tsvetkova, L.Y. and Smirnova, N.N., *J. Therm. Anal.*, 1994, **41**, 1371, (calorimetric props)
[20] Lebedev, B.V., Tsvetkova, L.Y. and Smirnova, N.N., *Calorim. Anal. Therm.*, 1993, **24**, 245, (calorimetric props)
[21] Aycock, W., *PhD Thesis*, Rensselauer Polytechnic Institute, Troy, N.Y., 1989, (calorimetric props)
[22] Gianotti, G. and Capizzi, A., *Eur. Polym. J.*, 1968, **4**, 677, (calorimetric props)
[23] Throne, J.L. and Griskey, R.G., *Mod. Plast.*, 1972, **49**, 96, (calorimetric props)
[24] Aharoni, S.M., Polym. Prepr. (Am. Chem. Soc., *Div. Polym. Chem.*), 1973, **14**, 334, (calorimetric props)
[25] Bu, H.S., Aycock, W. and Wunderlich, B., *Polymer*, 1987, **28**, 1165, (calorimetric props)
[26] Gaur, U., Cao, M., Pan, R. and Wunderlich, B., *J. Therm. Anal.*, 1986, **31**, 421, (heat capacity)
[27] Chang, S.S., Polym. Prepr. (Am. Chem. Soc., *Div. Polym. Chem.*), 1987, **28**, 244, (heat capacity)
[28] Gaur, U., Wunderlich, B.B. and Wunderlich, B., *J. Phys. Chem. Ref. Data*, 1983, **12**, 29, (calorimetric props)
[29] Morley, D.C.W., *J. Mater. Sci.*, 1974, **9**, 619, (transition temps)
[30] Danusso, F. and Gianotti, G., *Makromol. Chem.*, 1964, **80**, 1, (melting point)
[31] Clark, K.J., Turner-Jones, A. and Sandiford, D.J.H., *Chem. Ind. (London)*, 1962, 2010, (transition temps)
[32] Rubin, I.D., *J. Polym. Sci., Part A-2*, 1967, **5**, 1323, (ir)
[33] Dunham, K.R., Vandenberghe, J., Faber, J.W.H. and Contois, L.E., *J. Polym. Sci., Part A: Polym. Chem.*, 1963, **1**, 751, (transition temps)
[34] Gabbay, S.M. and Stiuala, S.S., *Polymer*, 1976, **17**, 61, (ir, degradation)
[35] Slichter, W.P. and Davis, D.D., *J. Appl. Phys.*, 1964, **35**, 10, (melting point)
[36] Campbell, T.W. and Haven, A.C., *J. Appl. Polym. Sci.*, 1959, **1**, 73, (melting point)
[37] Philippoff, W. and Tomquist, E., *J. Polym. Sci., Part C: Polym. Lett.*, 1968, **23**, 881, (optical props)
[38] Modric, I., Holland-Moritz, K. and Hummel, D.D., *Colloid Polym. Sci.*, 1976, **254**, 342, (ir, Raman)
[39] Sauer, J.A., Woodward, A.E. and Fuschillo, N., *J. Appl. Phys.*, 1959, **30**, 1488, (transition temps)
[40] Reding, F.P., *J. Polym. Sci.*, 1956, **21**, 547, (transition temps)
[41] Willbourn, A.H., *Trans. Faraday Soc.*, 1958, **54**, 717, (mech props)
[42] Hopfinger, A.J., Koehler, M.G., Pearlstein, R.A. and Tripathy, S.K., *J. Polym. Sci., Part B: Polym. Phys.*, 1988, **26**, 2007, (transition temps)
[43] Manabe, S., Nakamura, H., Uemura, S. and Takayanagi, M., *Kogyo Kagaku Zasshi*, 1970, **73**, 1587, (transition temps)
[44] Baccaredda, M. and Butta, E., *Chim. Ind. (Milan)*, 1962, **44**, 1288, (mech props)
[45] Woodward, A.E., Odajima, A. and Sauer, J.A., *J. Phys. Chem.*, 1961, **65**, 1384, (H-1 nmr)
[46] Filiatrault, D. and Delmas, G., *J. Polym. Sci., Polym. Phys. Ed.*, 1981, **19**, 773, (solubility props)
[47] Charlet, G., Ducasse, R. and Delmas, G., *Polymer*, 1981, **22**, 1190, (solubility props)
[48] Sewell, J.H., *Rev. Plast. Mod.*, 1971, **22**, 1367, (surface tension, solubility props)
[49] Krigbaum, W.R. and Woods, J.D. *J. Polym. Sci., Part A: Polym. Chem.*, 1964, **2**, 3075, (solubility props)
[50] Moraglio, G. and Brzezinski, J., *J. Polym. Sci., Part B: Polym. Lett.*, 1964, **2**, 1105, (solubility props)
[51] Moraglio, G. and Gianotti, G., *Eur. Polym. J.*, 1969, **5**, 781, (rheological props)
[52] Schildknecht, C.E., Tannebring, J.N. and Williams, R.F., Polym. Prepr. (Am. Chem. Soc., *Div. Polym. Chem.*), 1974, **15**, 421, (moulding behaviour, hardness)
[53] Miller, R.L. and Boyer, R.F., *Polym. News*, 1978, **4**, 255, (surface free energy)
[54] Cooper, G.D. and Gilbert, A.R., *J. Polym. Sci.*, 1959, **38**, 275, (rheological props)
[55] Gianotti, G. and Moraglio, G., *Chim. Ind. (Milan)*, 1967, **49**, 927, 944, (rheological props)
[56] Moraglio, G. and Oddo, N., *Chim. Ind. (Milan)*, 1966, **48**, 224, (rheological props)
[57] Shalaeva, L.F., Vysokomol. Soedin., *Ser. B*, 1968, **10**, 449, (rheological props, flexibility)
[58] Filiatrault, D., Phuong-Nguyen, H. and Delmas, G., *Macromolecules*, 1979, **12**, 763, (rheological props)
[59] Sakaguchi, F. and Tsuji, W., *CA*, 1967, **67**, 117454, (mech props)
[60] Tsuji, W., Kitamura, R. and Sakaguchi, F., *CA*, 1967, **68**, 78707, (mech props)
[61] Privalko, V.P., *CA*, 1974, **81**, 170033, (rigidity)
[62] Takayanagi, M., *Pure Appl. Chem.*, 1970, **23**, 151, (mech props)
[63] Vincent, P.I., *Polymer*, 1972, **13**, 558, (mech props)
[64] Woodward, A.E., Sauer, J.A. and Wall, R.A., *J. Polym. Sci.*, 1961, **50**, 117, (mech props)
[65] Sharma, B.K., *J. Polym. Mater.*, 1984, **1**, 193, (thermoacoustic props)
[66] Asakura, T., Demura, M. and Nishiyama, Y., *Macromolecules*, 1991, **24**, 2334, (C-13 nmr)
[67] Asakura, T, Hirano, K., Demura, M. and Kato, K., *Makromol. Chem., Rapid Commun.*, 1991, **12**, 215, (C-13 nmr)
[68] Kusomoto, N., *Chem. Lett.*, 1976, **2**, 141, (esr)
[69] Gabbay, S.M. and Shivala, S.S., *Polymer*, 1976, **17**, 121, (ir)
[70] Holland-Moritz, K., Modric, I., Heinen, K.U. and Hummel, D.O., *Kolloid Z. Z. Polym.*, 1973, **251**, 913, (ir, Raman)
[71] Holland-Moritz, K., Sausen, E., Djudovic, P., Coleman, M.M. and Painter, P.C., J. Polym. Sci., *Polym. Phys. Ed.*, 1979, **17**, 25, (ir, Raman)
[72] Gabbay, S.M., *Diss. Abstr. Int. B*, 1976, **36**, 4517, (thermal oxidation)
[73] Gabbay, S.M., Stivala, S.S. and Reed, P.R., *Anal. Chim. Acta*, 1975, **78**, 359, (thermal oxidation)
[74] Natta, G., Danusso, F. and Moraglio, G., *CA*, 1958, **52**, 17789e, (transition temp)

Poly(pentyl methacrylate) P-437

$-[CH_2C(CH_3)]_n-$
$\quad\quad |$
$\quad\quad O=C-OR$

$R = C_5H_{11}$ (pentyl)
$= CH_2CH_2CH(CH_3)_2$ (*iso*-pentyl)
$= CH_2C(CH_3)_3$ (*neo*-pentyl)
$= CH(CH_3)CH_2CH_2CH_3$ (1-methylbutyl)
$= CH(CH_3)CH(CH_3)$ (1,2-dimethylpropyl)

Synonyms: *Poly(amyl methacrylate)*. *Poly(pentyl 2-methyl-2-propenoate)*
Related Polymers: Polymethacrylates General
Monomers: *n*-Pentyl methacrylate, Isopentyl methacrylate, Neopentyl methacrylate, 1-Methylbutyl methacrylate, 1,2-Dimethylpropyl methacrylate
Material class: Thermoplastic
Polymer Type: acrylic
CAS Number:

CAS Reg. No.	Note
34903-87-2	neopentyl

Molecular Formula: $(C_9H_{16}O_2)_n$
Fragments: $C_9H_{16}O_2$
General Information: There are several isomeric pentyl methacrylates. Where no isomer is specified, the data refer to the *n*-pentyl isomer

Volumetric & Calorimetric Properties:
Density:

No.	Value	Note
1	d^{20} 1.032 g/cm^3	
2	d 1.03 g/cm^3	1-methylbutyl [1]
3	d^{25} 0.993 g/cm^3	neopentyl [2]
4	d 1.032 g/cm^3	isopentyl [2]

Glass-Transition Temperature:

No.	Value	Note
1	15°C	[3]
2	26–39°C	neopentyl [4]

Vicat Softening Point:

No.	Value	Note
1	115°C	neopentyl
2	46°C	isopentyl
3	63°C	1-methylbutyl [1]

Surface Properties & Solubility:
Cohesive Energy Density Solubility Parameters: δ 5.9 $J^{1/2}$ $cm^{-3/2}$ (8.4 $cal^{1/2}$ $cm^{-3/2}$) (neopentyl) [4]
Wettability/Surface Energy and Interfacial Tension: Polymer liq. interaction parameters 0.62 (hexane), 0.79 (2-butanone), 1.41 (acetonitrile), 0.11 ($CHCl_3$), 1.19 (n-propanol), 1.13 (Me_2CO), 0.79 (EtOAc) [5]

Transport Properties:
Polymer Solutions Dilute: Intrinsic viscosity 0.98 dl g^{-1} (C_6H_6, neopentyl, MW 1930000), 0.32 dl g^{-1} (C_6H_6, neopentyl, MW 150000) [4]. Molar volumes 97 cm^3 mol^{-1} (van der Waals, amorph., 25°, isopentyl and neopentyl), 151.4 cm^3 mol^{-1} (glassy, amorph., 25°, isopentyl), 157.3 cm^3 mol^{-1} (glassy, amorph., 25°, neopentyl) [2]

Mechanical Properties:
Hardness: Vickers (5 kg load) 17 (neopentyl), 4 (isopentyl), 12 (1-methylbutyl) [1]

Optical Properties:
Refractive Index:

No.	Value	Note
1	1.481	20°
2	1.461	20°, neopentyl
3	1.48	20°, isopentyl
4	1.477	20°, 1-methylbutyl [1]

Polymer Stability:
Decomposition Details: Decomposition at 250° yields monomer for each isomer with a small amount of olefin by side-chain cracking for some isomers (e.g., 1,2-dimethylpropyl) [1]

Bibliographic References
[1] Crawford, J.W.C., *J. Soc. Chem. Ind., London, Trans. Commun.*, 1949, **68**, 201
[2] Van Krevelen, D.W., *Properties of Polymers: Their Correlation with Chemical Structure*, Elsevier, 1976, 79
[3] Rehberg, C.E. and Fisher, C.H., *Ind. Eng. Chem.*, 1948, **40**, 1429
[4] Gargallo, L. and Russo, M., *Makromol. Chem.*, 1975, **176**, 2735
[5] Walsh, D.J. and McKeown, J.G., *Polymer*, 1980, **21**, 1335

Poly(3-phenoxypropylene oxide) — P-438

Synonyms: *Poly[oxy(phenoxymethyl)-1,2-ethanediyl]. Poly(phenylglycidyl ether). Poly[(phenoxymethyl)oxirane]. Poly(1,2-epoxy-3-phenoxypropane)*
Monomers: (Phenoxymethyl)oxirane
Material class: Thermoplastic
Polymer Type: polyalkylene ether
CAS Number:

CAS Reg. No.	Note
100629-24-1	
25265-27-4	homopolymer
26895-29-4	homopolymer, isotactic

Molecular Formula: $(C_9H_{10}O_2)_n$
Fragments: $C_9H_{10}O_2$

Molecular Weight: > M_n 3500–5700; M_w/M_n 1.42–2.07
General Information: Highly crystalline polymer

Bibliographic References
[1] Hamazu, F., Akashi, S., Koizumi, T., Takata, T., Endo, T., *J. Polym. Sci., Part A: Polym. Chem.*, 1991, **29**, 1845

Poly(phenylacetylene) — P-439

Synonyms: *Poly(ethynylbenzene)*
Related Polymers: Poly(acetylenes)
Monomers: Phenylethyne
Material class: Thermoplastic
Polymer Type: polyacetylene
CAS Number:

CAS Reg. No.	Note
25038-69-1	
120851-94-7	(Z)-form
27306-15-6	(E)-form
26970-23-0	

Molecular Formula: $(C_8H_6)_n$
Fragments: C_8H_6
General Information: Poly(phenylacetylene) is air stable and insulating when undoped, becoming semiconducting when doped with both donors or acceptors. [14] Formation of high MW materials has been studied [15]

Volumetric & Calorimetric Properties:
Melting Temperature:

No.	Value	Note
1	225°C	[1]
2	200–210°C	[2]

Vicat Softening Point:

No.	Value	Note
1	214–227°C	[3]

Surface Properties & Solubility:
Solvents/Non-solvents: Sol. aromatic and halogenated hydrocarbons, cyclic ethers [7], Me_2CO, C_6H_6 [8]

Transport Properties:
Polymer Solutions Dilute: [η] 0.16 dl g^{-1} (toluene, 30°) [12]

Optical Properties:
Refractive Index:

No.	Value	Note
1	1.7–1.72	1064 nm [13]
2	1.67–1.71	354 nm [13]

Transmission and Spectra: Uv [3], ir [4], H-1 nmr [5] and C-13 nmr [9] spectral data have been reported

Polymer Stability:
Thermal Stability General: In soln. *cis*-transoid struct. isomerises above 80° to produce a *trans*-cisoidal struct. At temps. >120° *cis-trans* isomerisation is accompanied by cyclisation and scission

Decomposition Details: Decomposes above 325° with hydrogen migration accompanied by crosslinking reactions; at 420° thermal decomposition leads to C_6H_6 [11]
Environmental Stress: No changes observed when stored at -20° in the dark for 6 months. MW decreases by half on standing at room temp. in light for 6 months [3,10]

Applications/Commercial Products:
Processing & Manufacturing Routes: Prod. by addition polymerisation with Ziegler-Natta catalysts, transition-metal complexes and free radical initiators. Can also be prod. by an Aldol-type polycondensation from phenylacetaldehyde [14] and by an anionic polymerisation mechanism [16]
Applications: Potential applications as selective gas permeable membranes, photoconductors and in non-linear optics

Bibliographic References
[1] Woon, P.S. and Farona, M.F, J. Polym. Sci., *Polym. Chem. Ed.*, 1974, **12**, 1749
[2] Chiang, A.C., Waters, P.F., and Aldridge, M.H., J. Polym. Sci., *Polym. Chem. Ed.*, 1982, **20**, 1807
[3] Masuda, T., Sasaki, N. and Higashimura, T., *Macromolecules*, 1975, **8**, 717
[4] Kambara, V.S. and Noguchi, H., *Makromol. Chem.*, 1964, **73**, 244
[5] Simionescu, C.I. and Perec, V., J. Polym. Sci., *Polym. Symp.*, 1980, **67**, 43
[6] Simionescu, C.I., Perec, V. and Dumitrescu, S., J. Polym. Sci., *Polym. Chem. Ed.*, 1977, **15**, 2497
[7] Masuda, T., Takahashi, T., Yamamoto, K. and Higashimura, T., J. Polym. Sci., *Polym. Chem. Ed.*, 1982, **20**, 2603
[8] Berlin, A.A., *J. Polym. Sci.*, 1961, **55**, 621
[9] Furlani, A., Napoletano, C., Russo, M.V., Marisch, N. and Camus, A., J. Polym. Sci., *Part A: Polym. Chem.*, 1989, **27**, 75
[10] Masuda, T., Tang, B.Z., Higashimura, T. and Yamaoka, H., *Macromolecules*, 1985, **18**, 2369
[11] Ito, T., Shirakana, H. and Ikeda, S., *J. Polym. Sci.*, 1975, **13**, 1943
[12] Pohl, H.A. and Chartoff, R.P., *J. Polym. Sci.*, 1964, **2**, 2787
[13] Neher, D., Kalbeitzel, A., Wolf, A., Bubeck, C. and Wegner, G., *J. Phys. D: Appl. Phys.*, 1991, **24**, 1193
[14] Cataldo, F., *Polym. Int.*, 1996, **39**, 91
[15] Masuda, T., *Macromolecules*, 1974, **7**, 728, (high MW polymerisation)
[16] Kobryanskii, V.M., J. Polym. Sci., *Polym. Chem. Ed.*, 1992, **30**, 1935

Poly(*N*-phenylaniline) P-440

Synonyms: *Poly(diphenylamine). N-phenylbenzenamine homopolymer. Diphenylamine homopolymer. PDPA*
Monomers: Diphenylamine
Polymer Type: polyaniline
CAS Number:

CAS Reg. No.
25656-57-9

Molecular Formula: $(C_{12}H_9N)n$
Fragments: $C_{12}H_9N$
Molecular Weight: 57,458 [3]
General Information: Polymerisation proceeds via 4,4′ coupling as opposed to the head-tail polymerisation observed in polyaniline and derivatives.
Morphology: Yellow-green-blue [7]. The morphology of the films change from fibrillar to granular with a change in monomer concentration of 0.05–0.4 M. TEM revealed that PDPA nanofibrils have a uniform and well-aligned array structure [5]
Miscellaneous: Two kinds of electrochemically synthesised film have been characterised by *ex situ* resonance Raman, FT-IR and UV-vis-NIR Reflectance spectroscopies. The *ex situ* spectra of the films showed differences that were rationalised assuming the predominance of the diphenosemiquinone aminoimine structure in one film, while in the other the diphenoquinone diimine segments were predominant [8].

Surface Properties & Solubility:
Solubility Properties General: Shows ca. 8% wt/wt (0.3M) or higher solubility in doped and undoped states in organic media [7].

Electrical Properties:
Electrical Properties General: Electrical conductivity 0.02S cm^{-1} to 1×10^{-8} S cm^{-1} [1,7]. Films of PDPA-BF_4^- anion show an increase in conductivity [9]. Conducting blends of PDPA doped with H_2SO_4 or *p*-toluenesulfonic acid (*p*-TSA) and PVDF showed an increase in conductivity from about 10^{-9} to 10^{-5} S cm^{-1}, with a percolation threshold of 5% of PDPA-H_2SO_4/*p*-TSA [13]

Optical Properties:
Transmission and Spectra: Ir [2,3,5,6,7,8], pes [2], uv-vis [2,3,5,7,8], uv-vis for copolymerisation of aniline with diphenylamine-4-sulfonic acid [4] and nmr [7] spectra reported.

Polymer Stability:
Decomposition Details: The polymer decomposes in three steps corresponding to an initial loss of H_2O molecules, volatilisation of ClO_4 along with low molecular weight fragments and complete decomposition and degradation of the polymer matrix. The decreasing trend in the % weight loss of the second step is observed with increasing monomer concentration due to presence of a non-ionic insulating phase [5,8]

Applications/Commercial Products:
Processing & Manufacturing Routes: Synthesised electrochemically under galvanostatic conditions in non-aqueous MeCN media at different monomer concentrations [5]. 4-Aminobiphenyl and diphenylamine were electropolymerised in acidic and organic media to produce poly(4-aminobiphenyl) and poly(diphenylamine) [6]. Synthesised by chemical oxidative method using salt of dodecyl benzene sulfonic acid, in conc. HCl and diphenylamine in HCl followed by pre-cooling and addition of freshly prepared ammonium persulfate [3]
Applications: A corrosion inhibitor for Fe [3]. Fuel cells containing conductive polymer electrode catalysts [10]. Application of a single electrode, modified with PDPA and dodecyl sulfate, for the simultaneous amperometric determination of electro-inactive anions and cations in ion chromatography [11]. Sensitive and selective method and device for the detection of trace amounts of a substance [12]

Bibliographic References
[1] Matnishian, H.A. and Beylerian, N.M., *Oxid. Commun.*, 2005, **28**, 67
[2] Yanchun, Z., Miao, C., Xiang, L., Tao X. and Weimin, L., *Mater. Chem. Phys.*, 2005, **91**, 518
[3] Jeyaprabha, C., Sathiyanarayanan, S., Phani, K.L.N. and Venkatachari, G., *J. Electroanal. Chem.*, 2005, **585**, 250
[4] Wen, T.C., Sivakumar, C. and Gopalan, A., *Electrochim. Acta*, 2001, **46**, 1071
[5] Athawale, A.A., Deore, B.A. and Chabukswar, V.V., *Mater. Chem. Phys.*, 1999, **58**, 94
[6] Guay, J. and Dao, L.H., *J. Electroanal. Chem.*, 1989, **274**, 135
[7] Chandrasekhar, P. and Gumbs, R.W., *J. Electrochem. Soc.*, 1991, **138**, 1337
[8] Santana, H., Matos, J.R. and Temperini, M.L.A., *Polym. J. (Tokyo)*, 1998, **30**, 315
[9] Santana, H. and Dias, F.C., *Mater. Chem. Phys.*, 2003, **82**, 882
[10] *Jpn. Pat.*, Nitto Denko Corp., Japan, 2004, 04 146 358
[11] Qun, X., Chun, X., Qingjiang, W., Kazuhiko, T., Hiroshi, T., Wen, Z. and Litong, J., *J. Chromatogr., A*, 2003, **997**, 65
[12] *U.S. Pat.*, M.S. Tech Ltd., Israel, 2002, 02 103 340
[13] Ten-Chin W., Sivakumar C., and Gopalan A., *Mater. Lett.*, 2002, **54**, 430

Poly(4-phenyl-1-butene) P-441

Synonyms: *Poly(1-(2-phenylethyl)-1,2-ethanediyl)*
Monomers: 4-Phenyl-1-butene

Material class: Thermoplastic
Polymer Type: polyolefins
CAS Number:

CAS Reg. No.	Note
30137-21-4	
28827-17-0	isotactic

Molecular Formula: $(C_{10}H_{12})_n$
Fragments: $C_{10}H_{12}$
Additives: Tetra-substituted phenols as stabilisers
General Information: Slightly rubbery solid [6]
Morphology: Monoclinic, a 1.04, b 1.80, c 0.661, helix 3_1 [9]; c 0.655, helix 3_1 [10]

Volumetric & Calorimetric Properties:
Density:

No.	Value	Note
1	d 1.042 g/cm^3	[7]
2	d 1.045 g/cm^3	isotactic [9]
3	d 0.962 g/cm^3	atactic [9]
4	d 1.04 g/cm^3	[13]

Latent Heat Crystallization: Heat of fusion has been reported [9]. ΔH_f 4.36 kJ mol^{-1}, 4.6 kJ mol^{-1} [12]
Melting Temperature:

No.	Value	Note
1	159°C	[4]
2	158–160°C	[6]
3	162–168°C	[7]
4	158°C	[9]

Glass-Transition Temperature:

No.	Value	Note
1	10°C	[7,9]
2	40°C	[11]

Transition Temperature:

No.	Value	Note	Type
1	130–160°C	[1]	Softening temp.
2	77–85°C	[3]	Softening temp.
3	40°C	[4]	Softening temp.
4	168°C	[6]	Polymer melt temp.

Surface Properties & Solubility:
Solvents/Non-solvents: Sol. heptane [3,7], warm C_6H_6 [3]. Insol. Me$_2$CO [3], isopropanol [7], EtOH [9]

Transport Properties:
Polymer Solutions Dilute: [η] 1.19 dl g^{-1} [4]; η$_{inh}$ 3.65 dl g^{-1} (decahydronaphthalene, 130°) [6]; [η] 0.63 dl g^{-1} (tetrahydronaphthalene, 80°) [9]; [η] 1.6–2.9 dl g^{-1} (tetrahydronaphthalene, 135°) [13]. Other viscosity values have been reported [1]

Optical Properties:
Transmission and Spectra: Ir spectral data have been reported [3]

Polymer Stability:
Environmental Stress: A study into the effects of oxidation by atomic oxygen has been reported. Bulk props. are unaffected as only the polymer surface is affected [8]

Applications/Commercial Products:
Processing & Manufacturing Routes: Synth. using Ziegler-Natta catalysts such as TiCl$_4$-AlEt$_3$ in heptane at 55° [1,3,4,5,6,7,9,13]

Bibliographic References
[1] Topchiev, A.V., Chernyi, G.I. and Andronov, V.N., *Dokl. Akad. Nauk SSSR*, 1962, **146**, 833, (synth)
[2] *Brit. Pat.*, 1963, 917 100, (stabilisers)
[3] Topchiev, A.V., Andronov, V.N. and Chernyi, G.I., *Neftekhimiya*, 1963, **3**, 725, (synth, ir, solubilities)
[4] Dunham, K.R., Vandenberghe, J., Faber, J.W.H. and Contois, L.E., *J. Polym. Sci., Part A: Polym. Chem.*, 1963, **1**, 751, (synth, melting point)
[5] Topchiev, A.V., *J. Polym. Sci.*, 1961, **53**, 195, (synth)
[6] Campbell, T.W. and Haven, A.C., *J. Appl. Polym. Sci.*, 1959, **1**, 73, (synth, soln props)
[7] Price, J.A., Lytton, M.R. and Ranby, B.G., *J. Polym. Sci.*, 1961, **51**, 541, (synth, transition temps)
[8] Hansen, R.H., Pascale, J.V., DeBenedictus, T. and Rentzepis, P.M., *J. Polym. Sci., Part A: Polym. Chem.*, 1965, **3**, 2205, (oxidation)
[9] Golemba, F.J., Guillet, J.E. and Nyburg, S.C., *J. Polym. Sci., Part A-1*, 1968, **6**, 1341, (struct, synth, heat of fusion)
[10] Natta, G., *Makromol. Chem.*, 1960, **35**, 94, (struct)
[11] *High Polymers*, (eds. R.A.V. Raff and K.W. Doak), Interscience, 1965, **20**
[12] *Polym. Handb.*, 3rd edn., (eds. J. Brandrup and E.H. Immergut), John Wiley and Sons, 1989, 10, (heats of fusion)
[13] Pregagtia, G. and Binagtii, M., *Gazz. Chim. Ital.*, 1960, **90**, 1554, (synth, density, soln props)

Poly(*p*-phenylene) P-442

Synonyms: *Poly(paraphenylene)*. *P.P.P.* *p-Polyphenyl*
Monomers: Biphenyl
CAS Number:

CAS Reg. No.	Note
26008-28-6	homopolymer

Molecular Formula: $(C_{12}H_8)_n$
Fragments: $C_{12}H_8$

Electrical Properties:
Electrical Properties General: Conductivity 3.3 × 10^{-13} S cm^{-1} (non-doped); 2.2 × 10^{-10} S cm^{-1} (iodine atmosphere) [1]

Optical Properties:
Transmission and Spectra: Ir and uv spectral data have been reported [1,2,4,5]

Polymer Stability:
Polymer Stability General: Possesses excellent oxidative and thermal stability [3,4]
Thermal Stability General: Thermal stability is dependent upon thermal treatment during the aromatisation process [3]
Decomposition Details: Decomposes at 600–800° forming oligomers, but with no evidence of C_6H_6 or biphenyl being produced [3]
Chemical Stability: Has excellent oxidative stability in air, withstanding temps. up to 350° [3]

Applications/Commercial Products:
Processing & Manufacturing Routes: Main methods of preparation are the Kovacic synthesis [6] and the Yamamoto synthesis [7,10]. May also be polymerised electrochemically [4,9]
Applications: Has applications in electroactive materials, including use in batteries, cathodes, LEDs. Applications due to heat resistance include electrical insulators, fabrics for heat-resistance, and rocket nozzles

Bibliographic References

[1] Miyashita, K. and Kaneko, M., *Synth. Met.*, 1995, **68**, 161, (ir, uv, conductivity)
[2] McKean, D.R. and Stille, J.K., *Macromolecules*, 1987, **20**, 1787, (ir, uv)
[3] Ballard, D.G.H., Courtis, A., Shirley, I.M. and Taylor, S.C., *Macromolecules*, 1988, **21**, 294, (stability)
[4] Kovacic, P. and Jones, M.B., *Chem. Rev.*, 1987, **87**, 357, (rev)
[5] Toshina, N. and Asakura, T., *Bull. Chem. Soc. Jpn.*, 1993, **66**, 948, (synth)
[6] Kovacic, P. and Kyriakis, A., *J. Am. Chem. Soc.*, 1963, **85**, 454, (synth)
[7] Yamamoto, T. and Yamamoto, A., *Chem. Lett.*, 1977, 353, (synth)
[8] *Ger. Pat.*, 1986, 3 617 777, (electroactive material)
[9] *Ger. Pat.*, 1987, 3 627 242, (synth)
[10] Yamamoto, T., Hayashi, Y. and Yamamoto, A., *Bull. Chem. Soc. Jpn.*, 1978, **51**, 2091, (synth)

Poly(*p*-phenylenebenzobisthiazole) P-443

Synonyms: *Poly(benzo[1,2-d:4,5-d′]bisthiazole-2,6-diyl-1,4-phenylene)*. PBZT. PBT
Monomers: 1,4-Benzenedicarboxylic acid, 2,5-Diamino-1,4-benzenedithiol
Material class: Thermoplastic
Polymer Type: polybenzothiazole
CAS Number:

CAS Reg. No.
69794-31-6

Molecular Formula: $(C_{14}H_6N_2S_2)_n$
Fragments: $C_{14}H_6N_2S_2$
General Information: High-strength crystalline rigid-rod polymer with good thermal stability. Prod. as yellow or brown powder.
Morphology: Monoclinic unit cell, a=11.60Å, b=3.59Å, c=12.51Å [1,2]. Heat treatment increases crystallinity [2]. Concentrated solutions of polymer in acids such as sulfuric acid [3] or polyphosphoric acid [4] used for producing fibres are liquid crystalline, with crystal solvates formed in conc. aq. solns [5].

Volumetric & Calorimetric Properties:
Density:

No.	Value	Note
1	d 1.5 g/cm^3	fibre [6]

Thermal Expansion Coefficient:

No.	Value	Note
1	2.9×10^{-6} K^{-1}	axial chain value (20-445°C) [7]

Transition Temperature General: No glass-transition temperature or melting point recorded up to decomposition temperature (*ca.* 700°C in nitrogen) [8,9]

Surface Properties & Solubility:
Solubility Properties General: Polymer sol. only in strong acids but will dissolve in aprotic solvents such as nitromethane in the presence of Lewis acids such as BCl$_3$, AlCl$_3$ and GaCl$_3$ [8,9]
Solvents/Non-solvents: Sol. sulfuric acid, polyphosphoric acid, methanesulfonic acid, chlorosulfonic acid, trifluoroacetic acid [9,10]. Insol. most organic solvents, H$_2$O

Mechanical Properties:
Tensile (Young's) Modulus:

No.	Value	Note
1	168000–269000 MPa	fibre [11,13]
2	132000–238000 MPa	film [12]

Tensile Strength Break:

No.	Value	Note	Elongation
1	1000–2900 MPa	fibre [11,13]	0.5–1.7% [13]
2	1230 MPa	film [12]	2.5%

Elastic Modulus:

No.	Value	Note
1	372000 MPa	[7]

Electrical Properties:
Electrical Properties General: After electrochemical oxidation in THF with tetrabutylammonium tetrafluoroborate, doped polymer had conductivity of 20 S cm^{-1} [14]. Polymer is reported to exhibit both hole [15] and electron [16] transport properties. Electroluminescence has been exhibited in LED devices [15,17,18], with a blue shift observed (λ_{max} 570nm to 496nm) on increasing bias voltage [18]

Optical Properties:
Optical Properties General: UV absorption (λ_{max} 437nm) and photoluminescence (λ_{max} 587nm) bands for polymer film have been recorded [17,18]. Optical nonlinear third order susceptibilities as high as 10^{-11} to 10^{-10} esu have been reported [10].
Transmission and Spectra: Pmr [9], ir [8,19,20,21], uv [8,9,15,17,18], photoluminescence [15,17,18] and x-ray diffraction [1,2,7] and neutron diffraction [24] data have been reported.

Polymer Stability:
Thermal Stability General: Rapid thermal degradation occurs above 600°C in air and at *ca.* 700°C in nitrogen [10]. After 200h in air there is no weight loss at 316°C, only 10% at 343°C and 30% at 370°C [10]

Applications/Commercial Products:
Processing & Manufacturing Routes: Prod. mainly by reaction of 2,5-diamino-1,4-benzenedithiol dihydrochloride with terephthalic acid in polyphosphoric acid [22]. The diphosphate instead of the dihydrochloride has been used [23], and modified dithiol compounds with terephthaloyl dichloride as monomers have been reported [19,20]. Fibres are obtained from liquid crystalline solutions of polymer in acid by extrusion and coagulation in water.
Applications: Potential applications for high-strength fibres, in electronics and for fuel cell membranes.

Bibliographic References

[1] Takahashi, Y. and Sul, H., *J. Polym. Sci., Part B: Polym. Phys.*, 2000, **38**, 376
[2] Odell, J.A., Keller, A., Atkins, E.D.T. and Miles M.J., *J. Mater. Sci.*, 1981, **16**, 3309
[3] Wang, C.-S., Price, G.E., Lee, J.-W. and Dean, D.R., *Polym. Prepr. (Am. Chem. Soc., Div. Polym. Chem.)*, 1998, **39**, 404
[4] Wolfe, J.F., Loo, B.H. and Sevilla, E.R., *Polym. Prepr. (Am. Chem. Soc., Div. Polym. Chem.)*, 1981, **22**, 60
[5] Cohen, Y., Sariyama, Y. and Thomas, E.L., *Macromolecules*, 1991, **24**, 1161
[6] Allen, S.R., Filippov, A.G., Farris, R.J., Thomas, E.L., Wong, C.-P., Berry, G.C. and Chenevy, E.C., *Macromolecules*, 1981, **14**, 1135

[7] Nakamae, K., Nishino, T., Gotoh, Y., Matsui, R. and Nagura, M., *Polymer*, 1999, **40**, 4629
[8] Jenekhe, S.A., Johnson, P.O. and Agrawal, A.K., *Macromolecules*, 1989, **22**, 3216
[9] Roberts, M.F. and Jenekhe, S.A., *Chem. Mater.*, 1993, **5**, 1744
[10] Hu, X.-D., Jenkins, S.E., Min, B.G., Polk, M.B. and Kumar, S., *Macromol. Mater. Eng.*, 2003, **288**, 823
[11] So, Y.-H., Zaleski, J.M., Murlick, C. and Ellaboudy, A., *Macromolecules*, 1996, **29**, 2783
[12] Feldman, L., Farris, R.J., and Thomas, E.L., *J. Mater. Sci.*, 1985, **20**, 2719
[13] Mehta, V.R. and Kumar, S., *J. Appl. Polym. Sci.*, 1999, **73**, 305
[14] DePra, P.A., Gaudiello, J.G. and Marks, T.J., *Macromolecules*, 1988, **21**, 2297
[15] Wu, C.C., Cheng, H.-Y., Wang, W. and Bai, S.J., *Thin Solid Films*, 2005, **477**, 174
[16] Alam, M.M. and Jenekhe, S.A., *Chem. Mater.*, 2002, **14**, 4775
[17] Wu, C.C., Chang, C.-F. and Bai, S.J., *Thin Solid Films*, 2005, **479**, 245
[18] Bai, S.-J., Wu, C.C., Tu, L.W. and Lee, K.H., *J. Polym. Sci., Part B: Polym. Phys.*, 2002, **40**, 1760
[19] Hattori, T., Akita, H., Kakimoto, M. and Imai, Y., *Macromolecules*, 1992, **25**, 3351
[20] Hattori, T., Kagawa, K., Kakimoto, M. and Imai, Y., *Macromolecules*, 1993, **26**, 4089
[21] Chang, C. and Hsu, S.L., *J. Polym. Sci., Polym. Phys. Ed.*, 1985, **23**, 2307
[22] Wolfe, J.F., Loo, B.H. and Arnold, F.E., *Macromolecules*, 1981, **14**, 915
[23] Odnoralova, V.N., Shchel'tsyn, V.K., Ermolova, N.P., Kiya-Oglu, V.N., Platonov, V.A. and Budnitskii, G.A., *Fibre Chem. (Engl. Transl.)*, 1998, **30**, 392
[24] Takahashi, Y., *Macromolecules*, 2001, **34**, 2012

Poly(phenylenediselenocarbonate) P-444

Synonyms: *Polyphenylenediselenothiocarbonate*
Monomers: 1,4-Dibromobenzene
Molecular Formula: $(C_7H_4OSe_2)_n$; $(C_7H_4SSe_2)_n$
Fragments: $C_7H_4OSe_2$ $C_7H_4SSe_2$
Additives: Doping agents include $AlCl_3$, $FeCl_3$, SbF_5 and I_2

Surface Properties & Solubility:
Solvents/Non-solvents: Insol. all common solvents, conc. H_2SO_4, $1M$ NaOH [1]

Applications/Commercial Products:
Processing & Manufacturing Routes: 1,4-Dibromobenzene is dissolved in DMF and refluxed at 120° with Na_2Se_2 for several days. The resultant phenyldiselenide polymer is suspended in THF and excess sodium borohydride is added followed by thiophosgene or phosgene [1]
Applications: Potential limited use as semiconductor when doped with iodine

Bibliographic References
[1] Diaz, F.R., Godoy, A., Tagle, L.H., Valdebenito, N. and Bernede, J.C., *Eur. Polym. J.*, 1996, **32**, 1155

Poly(1,4-phenylene ethynylene) P-445

Synonyms: *Poly(1,4-phenylene-1,2-ethynediyl). Poly(p-phenyleneethynylene). Poly(p-phenylene xylidine)*
Monomers: 1,4-Diethynylbenzene, 1,4-Diiodobenzene, 1-Ethynyl-4-iodobenzene
Material class: Thermoplastic
CAS Number:

CAS Reg. No.	Note
26520-99-0	poly(1,4-phenylene-1,2-ethynediyl)
90216-60-7	

Molecular Formula: $(C_8H_4)n$
Fragments: C_8H_4
General Information: Parent polymer of group of conductive polymers. Polymer is prod. as stable yellow-brown solid with very limited solubility in organic solvents. Improved solubility obtained by substitution in benzene ring
Morphology: Rigid-rod structure with p2gg two-dimensional symmetry, a=7.74, b=4.98 Å [1]

Surface Properties & Solubility:
Solubility Properties General: Virtually insol. most organic solvents, incl. THF, benzene, toluene, CH_2Cl_2, [2,3]

Electrical Properties:
Electrical Properties General: Properties of insulator when undoped [4]. Increase in conductivity on doping with iodine or $FeCl_3$ [3,5]

Polymer Stability:
Polymer Stability General: Stable up to 360° with no melting point [3]
Thermal Stability General: Degradation occurs at *ca.* 380° in N_2 [4]

Applications/Commercial Products:
Processing & Manufacturing Routes: Prod. by Pd-catalysed coupling of 1,4-dihalobenzene and 1,4-diethynylbenzene [3,4], from 4-iodophenylacetylene with $CuSO_4$ [6], by Pd-catalysed reaction of 1,4-diiodobenzene with acetylene [2], or by electrolysis of α,α,α,α′,α′,α′-hexachloroxylene in propylene carbonate [5]

Bibliographic References
[1] Yamamoto, T., Muramatsu, Y., Shimizu, T. and Yamada, W., *Macromol. Rapid. Commun.*, 1998, **19**, 293
[2] Li, C-J., Slaven, W.T., John, V.T. and Banerjee, S., *Chem. Comm.*, 1997
[3] Pelter, A. and Jones, D.E., *J. Chem. Soc., Perkin Trans. 1*, 2000, 1569
[4] Sanechika, K., Yamamoto, T. and Yamamoto, A., *Bull. Chem. Soc. Jpn.*, 1984, **57**, 2289
[5] Tateishi, M., Nishihara, H. and Aramaki, K., *Chem. Lett.*, 1987, 1727
[6] Lakshmikantham, M.V., Vartikar, J., Jen, K-Y., Cava, M.P., Huang, W.S. and MacDiarmid, A.G., *Polym. Prepr. (Am. Chem. Soc., Div. Polym. Chem.)*, 1983, **24**, 75

Poly(*m*-phenylene isophthalamide) P-446

Synonyms: *Nylon MPD-I. Nylon HT. Poly(imino-1,3-phenyleneiminocarbonyl-1,3-phenylenecarbonyl). Poly(m-phenylenediamine-co-isophthalic acid)*
Related Polymers: Aramids, Nylon polyester copolymer, Glass fibre filled
Monomers: 1,3-Benzenediamine, 1,3-Benzenedicarboxylic acid
Material class: Thermoplastic
Polymer Type: aromatic polyamides
CAS Number:

CAS Reg. No.
24938-60-1

Molecular Formula: $(C_{14}H_{10}N_2O_2)_n$
Fragments: $C_{14}H_{10}N_2O_2$
Additives: Common uv stabilisers are not effective
Morphology: Orthogonal, a 6.7, b 4.71, c 11, 1 monomer per unit cell (ρ 1.14 g cm^{-3}); triclinic, a 5.27, b 5.25, c 11.3, 1 monomer per unit cell (space group Cl-1, ρ 1.147 g cm^{-3})

– Poly(phenylene oxide)

Volumetric & Calorimetric Properties:
Density:

No.	Value	Note
1	d < 1.33 g/cm^3	max., amorph.
2	d > 1.36	min., cryst. [1]
3	d 1.14 g/cm^3	cryst., orthogonal
4	d 1.147 g/cm^3	cryst., triclinic

Thermal Expansion Coefficient:

No.	Value	Note	Type
1	4.5 × 10^{-5} K^{-1}	T > T$_g$ [1]	L

Thermodynamic Properties General: Heat of combustion 28.7 kJ g^{-1}
Specific Heat Capacity:

No.	Value	Note	Type
1	1.42 kJ/kg.C	solid, 25° [1]	P
2	1.214 kJ/kg.C	[5]	V

Melting Temperature:

No.	Value	Note
1	387–427°C	[1]

Glass-Transition Temperature:

No.	Value	Note
1	272°C	[1]

Transport Properties:
Polymer Solutions Dilute: η_{inh} 1.86 dl g^{-1} (sulfuric acid, 30°, 0.5 g fibre (100 ml)$^{-1}$

Mechanical Properties:
Mechanical Properties General: Tensile modulus 10.4 N tex^{-1} (20°), 7.9 N tex^{-1} (300°) [3]
Mechanical Properties Miscellaneous: Tenacity 0.49 N tex^{-1} (28% elongation, 20°); 0.27 N tex^{-1} (22% elongation, 250°); 0.11 N tex^{-1} (23% elongation, 330°); 0 N tex^{-1} (440°) [3]

Polymer Stability:
Polymer Stability General: Has excellent chemical, radiation and hydrolytic stability
Thermal Stability General: Prolonged exposure to air at 300° reduces tenacity remarkably [3]. Nomex paper has useful life 1400h at 250°, 40h at 300°. Retains 80% strength at 177° for several thousand hours [5]
Upper Use Temperature:

No.	Value
1	180–200°C

Flammability: Burns with difficulty forming black insulating char. Limiting oxygen index 26% (top of sample). Surface chlorine treatment increases limiting oxygen index to 36% [2,4]. Limiting oxygen index 28.5% [1]
Environmental Stress: Uv light discolours surface layer which then protects material from further degradation
Chemical Stability: More resistant than Nylon 6,6 to acids, less resistant than polyester. Stable to phosphoric acid, formic acid, Me$_2$CO, formaldehyde, gasoline, dimethylacetamide, fluorocarbons and trichloroethylene. Degraded by conc. mineral acids and NaOH at ambient temp. [5]

Hydrolytic Stability: Good base and hydrolytic stability compared with polyesters. Withstands washing cycles at 95°. Resistant to steam, but undergoes hydrol. on exposure to SO$_2$
Stability Miscellaneous: Fadometer test, 50% strength loss (25h, AATCC method, 16 A-1960)

Applications/Commercial Products:
Processing & Manufacturing Routes: Low temp. soln. polymerisation or interfacial polymerisation using diamine and diacid chloride. Fibres prod. by soln. spinning
Applications: High performance fibres used in radiator hoses and steam hoses

Trade name	Supplier
Apyeil	Unitika Ltd.
Conex	Teijin Chem. Ltd.
Durette	Monsanto Chemical
KM-21	Kuraray-Mitsui Toatsu
Nomex	DuPont
Phenylon	Soviet Origin
Teijinconex	Teijin Chem. Ltd.

Bibliographic References
[1] Van Krevelen, D.W., *Properties of Polymers: Their Correlation with Chemical Structure,* 3rd edn., Elsevier, 1990
[2] *Encycl. Polym. Sci. Technol.,* 2nd edn., Wiley Interscience, 1985
[3] *U.S. Pat.,* 1962, 3 049 518, (DuPont)
[4] Hathaway, C.E. and Early, C.L., *Appl. Polym. Symp.,* 1973, **21,** 101
[5] Nomex Meta-aramid Fibre Technical Data and Processing Guidelines DuPont Engineering Fibres,(technical datasheet)
[6] Herlinger, H., Horner, H.P., Druschke, F., Knöll, H. and Haiber, F., *Angew. Makromol. Chem.,* 1973, **29/30,** 229, (cryst)
[7] Kakida, H., Chatani, Y. and Tadokoro, H., *J. Polym. Sci., Polym. Phys. Ed.,* 1976, **14,** 427, (cryst)
[8] Sakaoku, K., Itoh, T and Kitamura, K., *Rep. Prog. Polym. Phys. Jpn.,* 1981, **24,** 139, (cryst)

Poly(phenylene oxide) P-447

Synonyms: *PPO. Poly(oxyphenylene)*
Monomers: Phenol
Material class: Thermoplastic
Polymer Type: polyphenylene oxide
CAS Number:

CAS Reg. No.
25667-40-7
9041-80-9

Molecular Formula: (C$_6$H$_4$O)$_n$
Fragments: C$_6$H$_4$O
General Information: Linear crystalline polymer of good thermal stability but limited flow props. The name also refers confusingly to Poly(2,6-dimethyl-1,4-phenylene oxide), of greater commercial use

Volumetric & Calorimetric Properties:
Density:

No.	Value	Note
1	d 1.32 g/cm^3	amorph. [1]
2	d 1.43 g/cm^3	cryst. [1]

– Poly(phenylene sulfide)

Latent Heat Crystallization: S 138.4 J mol^{-1} K^{-1} (T_g 358 K) [2], 203.4 J mol^{-1} K^{-1} (T_m 535 K) [2]. ΔC_p 25.7 J K^{-1} mol^{-1} [2]. ΔH_c 2.56 kJ mol^{-1} (27.8 J g^{-1}), ΔH_m 3.46 kJ mol^{-1} (37.6 J g^{-1}) [3], ΔH_f 5.23 kJ mol^{-1} [9]

Specific Heat Capacity:

No.	Value	Note	Type
1	150 kJ/kg.C	85° [2]	P
2	175.2 kJ/kg.C	285° [2]	P
3	109.1 kJ/kg.C	27° [9]	P
4	128.6 kJ/kg.C	85° [9]	P

Melting Temperature:

No.	Value	Note
1	285°C	[1]
2	262°C	[2]
3	268°C	[3]

Glass-Transition Temperature:

No.	Value	Note
1	85°C	[1,2]
2	86°C	[3]

Mechanical Properties:
Tensile (Young's) Modulus:

No.	Value	Note
1	2440 MPa	24400 M dyne cm^{-2} [4]

Tensile Strength Break:

No.	Value	Note	Elongation
1	47.7 MPa	477 M dyne cm^{-2} [4]	79.2%

Tensile Strength Yield:

No.	Value	Elongation	Note
1	56.2 MPa	6.5%	562 M dyne cm^{-2} [4]

Optical Properties:
Transmission and Spectra: Ir, C-13 nmr and x-ray spectral data have been reported [5,6,7].

Polymer Stability:
Decomposition Details: Ether linkages decompose without rearrangement though cyclisation to benzofuran systems is observed (25–600°) [8]. Decomposes to give H_2 and methane between 350 and 620° [5]

Applications/Commercial Products:
Processing & Manufacturing Routes: Prod. by oxidative coupling with a homogeneous oxidation catalyst or via halogen displacement (Ullmann coupling)

Trade name	Details	Supplier
Ashlene PPO	various grades	Ashley Polymers
CTI	MP-25NCF/000/MP-30GF/000	CTI
EMI-X	ZC-1008	LNP Engineering
Electrafil	various grades	DSM Engineering
Fiberfil	J-1700/20/30/40	DSM Engineering
L-10FG	0100	Thermofil Inc.
L-20FG	0100	Thermofil Inc.
L-30FG	0100	Thermofil Inc.
L1-30FG	0580	Thermofil Inc.
Lubricomp	various grades	LNP Engineering
Noryl	various grades	General Electric
Noryl Plus	various grades	General Electric
PPO	730 Nat	Multibase
PRL-PPX	PPX	Polymer Resources
RTP	1701/1703/1705/1707	RTP Company
RX8	2230	Spartech
SC8F	2088	Spartech
Stat-Kon	ZC-1003/1006	LNP Engineering
Styvex	72001/72002	Ferro Corporation
Thermocomp	Z/ZF	LNP Engineering

Bibliographic References
[1] Tabor, B.J., Magré, E.P. and Boon, J., *Eur. Polym. J.*, 1971, **7**, 1127
[2] Cheng, S.Z.D., Pan, R., Bu, H.S., Cao, M.Y. and Wunderlich, B., *Makromol. Chem.*, 1988, **189**, 1579
[3] Könnecke, K., *J. Macromol. Sci.*, *Phys.*, 1994, **33**, 37
[4] Takano, M. and Nielsen, L.E., *J. Appl. Polym. Sci.*, 1976, **20**, 2193, (mech props)
[5] Ehlers, G.F.L., Fisch, K.R. and Powell, W.R., *J. Polym. Sci., Part A: Polym. Chem.*, 1969, **7**, 2955
[6] Brown, C.E., Khoury, I., Bezaari, M.D. and Kovacic, P., *J. Polym. Sci., Part A: Polym. Chem.*, 1982, **20**, 1697
[7] Montando, G., Bruno, G., Maravigna, P. and Bottino, F., *J. Polym. Sci., Part A: Polym. Chem.*, 1974, **12**, 2881
[8] Jachowicz, J., Kryszewski, M. and Sobol, A., *Polymer*, 1979, **20**, 995
[9] Wrasidlo, W., *J. Polym. Sci., Part A-2*, 1972, **10**, 1719

Poly(phenylene sulfide)

Synonyms: *PPS. Poly(thiophenylene)*
Related Polymers: Poly(phenylene oxide)
Monomers: 1,4-Dichlorobenzene, 4-Bromobenzenethiol
Material class: Thermoplastic
Polymer Type: polyphenylene sulfide
CAS Number:

CAS Reg. No.
25212-74-2
9016-75-5

Molecular Formula: $(C_6H_4S)_n$
Fragments: C_6H_4S
Additives: Glass fibre, mineral fillers, carbon fibre, lubricants (PTFE, silicones)
General Information: Linear polymers with approx. 60–65% crystallinity. One of the most important properties is the material's tendency to change when heated in air. The change is indicative of an increase in molecular weight due to chain extension or cross-linking. This process is often referred to as curing. Filled grades have better impact strength, stiffness and strength

Poly(phenylene sulfide)

Volumetric & Calorimetric Properties:

Density:

No.	Value	Note
1	d 1.32 g/cm^3	amorph.
2	d 1.43 g/cm^3	crystalline [2]
3	d 1.348 g/cm^3	[9]
4	d 1.36 g/cm^3	[6]

Thermal Expansion Coefficient:

No.	Value	Note	Type
1	4.9 × 10^{-5} K^{-1}	[14]	v

Latent Heat Crystallization: S 142.7 J mol^{-1} K^{-1} (363K). 230.4 J mol^{-1} K^{-1} (593K) [1]. ΔC_p 33 J K^{-1} mol^{-1} [1]. ΔH_c 2.85 kJ mol^{-1} (26.4 J g^{-1}), ΔH_m 4.45 kJ mol^{-1} (41.2 J g^{-1}) [4]. ΔH_f 4.98 kJ mol^{-1} (11 cal g^{-1}) [7], 3.15 kJ mol^{-1} (6.97 cal g^{-1}) [9]. Other thermodynamic props. have been reported [5,7]

Specific Heat Capacity:

No.	Value	Note	Type
1	165.4 kJ/kg.C	363K [1]	P
2	194.3 kJ/kg.C	593K [1]	P

Melting Temperature:

No.	Value	Note
1	320°C	[1]
2	295°C	[2]
3	281°C	[4]

Glass-Transition Temperature:

No.	Value	Note
1	90°C	[1,4]
2	92°C	[2,3]
3	83°C	[5]

Deflection Temperature:

No.	Value	Note
1	136.7°C	0.18 MPa, ASTM D648 [6]
2	100°C	unannealed [8]
3	128°C	annealed [8]

Transport Properties:
Permeability of Gases: [He] 5.15 × 10^{10}, [O$_2$] 0.19 × 10^{10}, [N$_2$] 0.046 × 10^{10}, [CH$_4$] 0.066 × 10^{10}, [CO$_2$] 1.6 × 10^{10} cm^2 (s cmHg)$^{-1}$ [10]
Water Content: 0.02% (max., 24h, ASTM D570) [14]
Water Absorption:

No.	Value	Note
1	0.07 %	saturation, ASTM D955

Mechanical Properties:
Tensile (Young's) Modulus:

No.	Value	Note
1	8100 MPa	Specimen modulus [9]
2	2100 MPa	amorph. [13]
3	2700 MPa	crystalline [13]

Flexural Modulus:

No.	Value	Note
1	4400 MPa	ASTM D790, 23° [6]
2	3800 MPa	ASTM D790 [14]

Tensile Strength Break:

No.	Value	Note	Elongation
1	74.47 MPa	ASTM D1708, 23° [6]	3%
2	609 MPa	[9]	16.6%
3	45 MPa	amorph. [13]	23%
4	70 MPa	crystalline [13]	18%

Flexural Strength at Break:

No.	Value	Note
1	103 MPa	15000 psi, unannealed [8]
2	89.6 MPa	13000 psi, annealed [8]
3	140 MPa	ASTM D790 [14]

Elastic Modulus:

No.	Value	Note
1	14000–28000 MPa	[9]
2	3300 MPa	ASTM D638 [14]

Tensile Strength Yield:

No.	Value	Elongation	Note
1	75 MPa	1.6%	ASTM D638 [14]
2	59 MPa	3%	amorph. [13]
3	85 MPa	4.9%	crystalline [13]

Hardness: Rockwell R124 (ASTM D785) [14]
Friction Abrasion and Resistance: Coefficient of friction 0.24 (dynamic)
Izod Notch:

No.	Value	Notch	Note
1	16.2 J/m	N	23°, ASTM D256 [6]
2	20 J/m	N	ASTM D256 [14]

Electrical Properties:
Dielectric Permittivity (Constant):

No.	Value	Frequency	Note
1	3	1 kHz	20° [3]

Poly(phenylene sulfide), 40% Glass fibre filled

Dielectric Strength:

No.	Value	Note
1	150 kV.mm^{-1}	3.2 mm, ASTM D149 [14]

Arc Resistance:

No.	Value
1	120s

Dissipation (Power) Factor:

No.	Value	Frequency	Note
1	0.0005	10 kHz	ASTM D150 [14]

Optical Properties:
Transmission and Spectra: Ir spectral data have been reported [11]

Polymer Stability:
Thermal Stability General: Stable after continuous use at 240°. Non-burning [14]

Upper Use Temperature:

No.	Value
1	200°C

Decomposition Details: Between 350–450° decomposition produces hydrogen sulfide, dimers and trimers via radicals [11]. Above 450° evolution of hydrogen becomes predominant producing radicals which undergo cross-linking. Char yield 66–72% (N_2, 800°) [6], 42–47% (air, 800°)

Flammability: Limiting oxygen index 48–50% (ASTM D2863) [6]. Non-burning. Non drip. Flammability V0 (UL 94). Smoke evolution D_m 124.88 [6]

Environmental Stress: Polymer shows considerable resistance to degradation from electron beam radiation [13]

Chemical Stability: Resistant to weak acids, strong alkalis and organic solvents. Attacked slowly by strong oxidising acids. Not resistant to halogens or halogenated organics at higher temps. [14]. Rarely used unfilled

Stability Miscellaneous: Toxicity: Ti (time to incapacitation) 9.59–10.84 min. Td (time to death) 10.57–12.4 min. [6]

Applications/Commercial Products:
Processing & Manufacturing Routes: Prepared by reaction of 1,4-dichlorobenzene with sodium sulfide (Phillips Process), or self condensation of a metal salt of a 4-halothiophenol. May be compression moulded, extruded or injection moulded

Mould Shrinkage (%):

No.	Value	Note
1	1%	[14]

Applications: Chemical and high temp. resistant coating and moulding material for electrical connectors, bearing material with low friction. Electrical conducting polymer when doped

Trade name	Details	Supplier
CTI	SF series	CTI
Celstran	PPS	Hoechst Celanese
Celstran PPS	various grades	Polymer Composites
Comalloy	710 series	Comalloy Intl. Corp.
Craston		Ciba-Geigy Corporation
EMI-X	OC/PDX	LNP Engineering
Electrafil	various grades	DSM Engineering
Fiberfil	J	DSM Engineering
Fortron		Hoechst Celanese
Hyvex	various grades	Ferro Corporation
IPC	various grades	IPC
Lubricomp	various grades	LNP Engineering
Lusep	various grades	Lucky
Plaslube	1305/TF/20Nat	DSM Engineering
RTP	various grades	RTP Company
Ryton	R series	Philips Petroleum International
Stat-Kon	OC/OCL	LNP Engineering
Sulfil		Wilson-Fiberfil
Supec		General Electric Plastics
Susteel		Tosoh
T	various grades	Thermofil Inc.
Tedur		Bayer Inc.
Tedur	L	Albis Corp.
Thermocomp	O series	LNP Engineering
Tonen PPS		Tohpren
Torelina		Toray
Verton	OF-700-10	LNP Engineering
Xtyrex	641/642/645/655	EGC Corporation

Bibliographic References
[1] Cheng, S.Z.D., Pan, R., Bu, H.S., Cao, M.Y. and Wunderlich, B., *Makromol. Chem.*, 1988, **189**, 1579
[2] Tabor, B.J., Magré, E.P. and Boon, J., *Eur. Polym. J.*, 1971, **7**, 1127
[3] Rigby, S.J. and Dew-Hughes, D., *Polymer*, 1974, **15**, 639
[4] Könnecke, K., *J. Macromol. Sci., Phys.*, 1994, **33**, 37
[5] Seo, K.H., Park, L.S., Baek, J.B. and Brostow, W., *Polymer*, 1993, **34**, 2524
[6] Kourtides, D.A. and Parker, J.A., *Polym. Eng. Sci.*, 1978, **18**, 855
[7] Huo, P. and Cebe, P., *Colloid Polym. Sci.*, 1992, **270**, 840
[8] Brady, D.G., *J. Appl. Polym. Sci.*, 1976, **20**, 2541
[9] Nishino, T., Tada, K. and Nakamae, K., *Polymer*, 1992, **33**, 736
[10] Aguilar-Vega, M. and Paul D.R. *J. Polym. Sci., Part B: Polym. Phys.*, 1993, **31**, 1577
[11] Ehlers, G.F.L., Fisch, K.R. and Powell, W.R., *J. Polym. Sci., Part A: Polym. Chem.*, 1969, **7**, 2955
[12] Yoshino, K., Yun, M.S., Ozaki, M., Kim, S.H., *et al*, *Jpn. J. Appl. Phys.*, 1983, **22**, 1510
[13] El-Naggar, A.M., Kim, H.C., Lopez, L.C. and Wilkes, G.L., *J. Appl. Polym. Sci.*, 1989, **81**, 1655
[14] *The Materials Selector*, 2nd edn., (eds. N.A. Waterman and M.F. Ashby), Chapman and Hall, 1997, **3**, 336

Poly(phenylene sulfide), 40% Glass fibre filled P-449
Monomers: 1,4-Dichlorobenzene
Material class: Composites
Polymer Type: polyphenylene sulfide
CAS Number:

CAS Reg. No.
9016-75-5
25212-74-2

Volumetric & Calorimetric Properties:
Density:

No.	Value	Note
1	d 1.6 g/cm^3	ASTM D1505 [1]

– Poly(*p*-phenylene terephthalamide)

Thermal Expansion Coefficient:

No.	Value	Note	Type
1	2.2×10^{-5} K^{-1}	[1]	L

Deflection Temperature:

No.	Value	Note
1	243°C	ASTM D648 [1]
2	>260°C	min., ASTM D648, annealed [1]

Transport Properties:
Water Absorption:

No.	Value	Note
1	0.05 %	max., 24h

Mechanical Properties:
Mechanical Properties General: Possesses good dimensional stability, with low creep and good fatigue resistance
Flexural Modulus:

No.	Value	Note
1	11700 MPa	ASTM D790 [1]

Tensile Strength Yield:

No.	Value	Elongation	Note
1	135 MPa	1.3%	ASTM D638 [1]

Flexural Strength Yield:

No.	Value	Note
1	200 MPa	ASTM D790 [1]

Compressive Strength:

No.	Value	Note
1	145 MPa	ASTM D695 [1]

Hardness: Rockwell R123 (ASTM D785) [1]
Izod Area:

No.	Value	Notch	Note
1	0.075 kJ/m^2	Notched	75 N m^{-1}, ASTM D256 [1]

Electrical Properties:
Surface/Volume Resistance:

No.	Value	Note	Type
1	46×10^{15} Ω.cm	ASTM D257 [1]	S

Dielectric Permittivity (Constant):

No.	Value	Frequency	Note
1	3.9	1 kHz	ASTM D150 [1]
2	3.8	1 MHz	ASTM D150 [1]

Dielectric Strength:

No.	Value	Note
1	177 kV.mm^{-1}	32 mm, ASTM D149 [1]

Arc Resistance:

No.	Value	Note
1	34s	ASTM D495 [1]

Dissipation (Power) Factor:

No.	Value	Frequency	Note
1	0.001	1 kHz	ASTM D150 [1]
2	0.0013	1 MHz	ASTM D150 [1]

Polymer Stability:
Upper Use Temperature:

No.	Value
1	170–200°C

Flammability: Flammability rating V0, 5 V (UL 94)
Environmental Stress: Resistant to radiation
Chemical Stability: Has good resistance up to 200°; resistant to organic solvents, weak acids and strong alkalis

Applications/Commercial Products:
Mould Shrinkage (%):

No.	Value	Note
1	0.2–0.6%	[1]

Trade name	Supplier
Larton	LATI Industria Thermoplastics S.p.A.
Promef	Solvay
Ryton R-4	Philips Petroleum International
Supec	General Electric Plastics
Tedor	Bayer Inc.

Bibliographic References
[1] *The Materials Selector*, 2nd edn., (eds. N.A. Waterman and M.F. Ashby), Chapman and Hall, 1997, **3**, 337
[2] Ryton R-4 Phillips,(technical datasheet)

Poly(*p*-phenylene terephthalamide)   P-450

Synonyms: *Poly(p-phenylene diamine-co-terephthalic acid). Nylon PPD-T. Poly(imino-1,4-phenyleneiminocarbonyl-1,4-phenylenecarbonyl)*
Related Polymers: Aramids, Nylon polyester copolymer, Glass fibre filled
Monomers: 1,4-Benzenediamine, 1,4-Benzenedicarboxylic acid
Material class: Thermoplastic, Polymer liquid crystals
Polymer Type: aromatic polyamides
CAS Number:

CAS Reg. No.
24938-64-5

Poly(p-phenylene terephthalamide)

Molecular Formula: $(C_{14}H_{10}N_2O_2)_n$
Fragments: $C_{14}H_{10}N_2O_2$
Additives: 1% Phosphorous to reduce flammability
General Information: Exists in soln. as rigid rodlike liq. cryst. structs. When spun from soln. to fibre random coil structs. do not form, but rigid chains pack in quasi parallel bundles and shearing forces orient bundles to give a product with high modulus and tenacity, due to highly orientated fully extended chains which crystallise easily
Morphology: Monoclinic, a 7.728, b 5.184, c 12.81, γ 90°, 2 monomers per unit cell

Volumetric & Calorimetric Properties:

Density:

No.	Value	Note
1	d 1.48 g/cm^3	cryst.
2	d 1.542 g/cm^3	cryst.

Melting Temperature:

No.	Value	Note
1	497–600°C	[1]

Glass-Transition Temperature:

No.	Value	Note
1	307–347°C	[1]

Surface Properties & Solubility:
Solvents/Non-solvents: Sol. sulfuric acid

Transport Properties:
Polymer Solutions Dilute: Inherent viscosity 4.8 dl g^{-1} (sulfuric acid, 30°, MW 60000, 0.5 g fibre (100 ml)$^{-1}$)

Mechanical Properties:
Tensile (Young's) Modulus:

No.	Value	Note
1	60000–160000 MPa	[5]

Flexural Modulus:

No.	Value	Note
1	100000 MPa	ASTM D1044 [1]

Tensile Strength Break:

No.	Value	Note	Elongation
1	2900–3320 MPa	[5]	1.5–3.6%
2	2100 MPa	aged 500h, 180° [5]	
3	1200 MPa	aged 500h, 200° [5]	

Compressive Strength:

No.	Value
1	250 MPa

Complex Moduli: Specific modulus 16000–23000 N tex^{-1}
Mechanical Properties Miscellaneous: Tenacity 20500 N tex^{-1} (23°, fibre) [5]; 8800 N tex^{-1} (300°, fibre) [1]
Failure Properties General: Has excellent wear resistance
Friction Abrasion and Resistance: Abrasion resistance 250 cycles (perforation, Martindale test, P180 sandpaper) [6]

Electrical Properties:
Electrical Properties General: Has very high resistivity. Electrical props. are retained at high temps.
Complex Permittivity and Electroactive Polymers: Breakdown voltage -76 V mm^{-1} (180°) [7]

Polymer Stability:
Thermal Stability General: Short term property retention at 300°. Heat ageing results in loss of elongation and tenacity, but ultimate failure due to loss of tenacity [7]. High level of strength up to 250°

Upper Use Temperature:

No.	Value
1	250°C

Decomposition Details: Decomposition temp. 460° (onset of wt loss); 530° (50% wt loss) [9,10]. Decomposes releasing mainly CO_2; CO, N_2O and HCN are also released [5]
Flammability: Fibres burn with difficulty and do not melt producing a thick insulating char layer. Limiting oxygen index 28.5–29% (top of sample); 24.5% (top); 16.5% (bottom) [2,3]. Values can be increased to 40–42% by cross-linking with incorporation of 1% phosphorous
Environmental Stress: Susceptible to uv radiation but stable to γ-irradiation and x-rays
Chemical Stability: Stable to long term exposure to phosphoric acid, Me_2CO, formaldehyde, gasoline, fluorocarbons
Hydrolytic Stability: Withstands steam and boiling water. Resistant to hot alkali, but degraded by hot dil. mineral acids and conc. acids at ambient temp.
Biological Stability: Has good biological stability

Applications/Commercial Products:
Processing & Manufacturing Routes: Diamine and acid chloride are mixed in N-methylpyrrolidone containing dissolved lithium chloride or calcium chloride. Optimum polymer concentration is approx. 6–7% for optimum MW. Commercially H_2O is used to wash and slurry the polymer. The polymer is redissolved in H_2SO_4 to give a soln. for spinning into fibres. Wet spinning gives the highest MW and tensile props. Dry yarn is hot drawn, draw ratio depending on spin speed affecting tenacity and elongation of the fibre. The polymer forms an anisotropic soln. which assists molecular alignment, so hot drawing is less important to attain high strength fibres

Trade name	Supplier
Arenka	Enka
Kevlar	DuPont
Technora	Teijin Chem. Ltd.
Twaron	Akzo
Vniivlon	Soviet Origin

Bibliographic References

[1] Van Krevelen, D.W., *Properties of Polymers: Their Correlation with Chemical Structure*, 3rd edn., Elsevier, 1990
[2] Preston, J., *Polym. Eng. Sci.*, 1976, **16**, 298
[3] Hathaway, C.E. and Early, C.I., *Appl. Polym. Symp.*, 1973, **21**, 101
[4] Preston, J. and Hofferbert, W.J., *Text. Res. J.*, 1979, **44**, 283
[5] Kevlar, Visible Performance, Invisible Strength DuPont Engineering Fibres, 1993, (technical datasheet)
[6] Kevlar, Hand Protection DuPont Engineering Fibres, 1994
[7] *Encycl. Polym. Sci. Technol.*, 2nd edn., 1986, **11**, 388
[8] Northolt, M.G., *Eur. Polym. J.*, 1974, **10**, 799, (cryst)
[9] Korshak, U., *Chemical Structure and Thermal Characteristics of Polymers. Israel Program for Science Translations*, 1971
[10] Arnold, C., *J. Polym. Sci., Part D: Macromol. Rev.*, 1979, **14**, 265, (stability)
[11] Tashiro, K., Kobayashi, M. and Tadokoro, H., *Macromolecules*, 1977, **10**, 413
[12] Northolt, M.G. and Stuut, H.A., *J. Polym. Sci., Polym. Phys. Ed.*, 1978, **16**, 939, (cryst)

Poly(*p*-phenylene vinylene)   P-451

Synonyms: *Poly(p-xylylidene). Poly(1,4-phenylene-1,2-ethenediyl). PPV*
Related Polymers: Poly(*p*-2,5-dimethoxyphenylene vinylene)
Monomers: 1,4-Benzenedicarboxaldehyde, 4-Bromostyrene
Material class: Thermoplastic
Polymer Type: poly(arylene vinylene)
CAS Number:

CAS Reg. No.	Note
26009-24-5	
33379-71-4	(*E*)-form

Molecular Formula: $(C_8H_6)_n$
Fragments: C_8H_6
General Information: Conducting polymer. Yellow powder in appearance [2,3,4]
Morphology: Monoclinic; a 0.790, b 0.605, c 0.658, α 123° [1] bond lengths and angles of polymer repeat unit have been calculated [5,6,7]

Volumetric & Calorimetric Properties:
Density:

No.	Value	Note
1	d 1.283 g/cm^3	[1]

Thermodynamic Properties General: ΔC_p 0.14 kJ kg^{-1} K^{-1} [11]. ΔH_f 2.36 kJ mol^{-1} (23.1 J g^{-1}) [11]
Melting Temperature:

No.	Value	Note
1	>340°C	min. [8,9,10]

Glass-Transition Temperature:

No.	Value	Note
1	8°C	[11]

Surface Properties & Solubility:
Solvents/Non-solvents: Insol. all common solvents [11]

Transport Properties:
Polymer Solutions Dilute: η 1.5 (0.012*M* Na$_2$SO$_4$, 25°) [12]

Mechanical Properties:
Mechanical Properties General: Mech. props. are anisotropic and highly dependent upon the degree of molecular orientation. Chemical doping reduces the tensile modulus and tensile strength [13]

Tensile (Young's) Modulus:

No.	Value	Note
1	2300–37000 MPa	[13]

Tensile Strength Break:

No.	Value	Note	Elongation
1	48.2–271 MPa	[13]	2.1–38%

Tensile Strength Yield:

No.	Value	Elongation	Note
1	45.2–152 MPa	1.4–3.3 %	[13]

Electrical Properties:
Electrical Properties General: Electrical props. depend upon method of synth. [14,15,16]. Chemical doping increases conductivity [17,18,19,20,21,22,23]
Complex Permittivity and Electroactive Polymers: The redox behaviour has been studied [43]

Optical Properties:
Refractive Index:

No.	Value	Note
1	1.95–2.28	700–2500 nm [35]

Transmission and Spectra: Ir [2,24,25,26,27,28,29,30], Raman [31,32,33,34], uv [36], C-13 nmr [38] and mass [37] spectral data have been reported

Polymer Stability:
Decomposition Details: Major decomposition in nitrogen occurs above 400°; 45% weight loss occurs at 900° [39]

Applications/Commercial Products:
Processing & Manufacturing Routes: Prod. in a variety of ways including Wittig reaction of terephthaldehyde [2,8]; base elimination of xylylene bisdialkyl sulfonium salts [40]; Heck reaction of 4-bromostyrene; [9] electrochemical polymerisation of tetrabromo-*p*-xylene [41] and by a precursor route [44]. Easily processed and fabricated
Applications: Many potential applications in optical and electronic areas for use in light emitting diodes, electroluminescent devices

Bibliographic References
[1] Granier, T., Thomas, E.L., Gagnon, D.R., Karasz, F.E. and Lenz, R.W., *J. Polym. Sci., Polym. Phys. Ed.*, 1984, **24**, 2793
[2] Kossmehl, V.G., Hörtel, M. and Manecke, G., *Makromol. Chem.*, 1970, **131**, 37
[3] Hörhold, V.H.H., *Z. Chem.*, 1972, **12**, 41, (rev)
[4] Bradley, D.D.C., *J. Phys. D: Appl. Phys.*, 1987, **20**, 1389, (rev)
[5] Eckhardt, H., Shacklette, C.W., Jen, K.Y. and Elsenbaumer, R.L., *J. Chem. Phys.*, 1989, **91**, 1303
[6] Obrzut, J. and Karasz, F.E., *J. Chem. Phys.*, 1987, **87**, 2349
[7] Darsey, J.A. and Sumpter, B.G., *Polym. Prepr. (Am. Chem. Soc., Div. Polym. Chem.)*, 1989, **30**, 3
[8] Saikachi, H. and Muto, H., *Chem. Pharm. Bull.*, 1971, **19**, 959
[9] Heitz, W., Brügging, W., Freund, L., Gailberger, M. *et al*, *Makromol. Chem.*, 1988, **189**, 119
[10] McDonald, R.N. and Campbell, T.W., *J. Am. Chem. Soc.*, 1960, **82**, 4669
[11] Leung, L.M. and Chik, G.L., *Polymer*, 1993, **34**, 5174
[12] Jin, J.I., Kim, J.H., Lee, Y.H., Lee, G.H. and Park, Y.W., *Synth. Met.*, 1993, **57**, 3742
[13] Machado, J.M., Masse, M.A. and Karasz, F.E., *Polymer*, 1989, **30**, 1992
[14] Hörhold, V.H.H. and Opfermann, J., *Makromol. Chem.*, 1970, **131**, 105
[15] Pellegrin, E., Fink, J., Martens, J.H.F., Bradley, D.D.C. *et al*, *Synth. Met.*, 1991, **41**, 1353
[16] Tokito, S., Tsutsui, T., Saito, S. and Tanaka, R., *Polym. Commun.*, 1986, **27**, 333
[17] *Handbook of Conducting Polymers*, (ed. T.A. Skotheim), Marcel Dekker, 1986, **1**, 365
[18] Brédas, J.L., Beljonne, D., Shuai, Z. and Toussaint, J.M., *Synth. Met.*, 1991, **41**, 3743
[19] Tanaka, S. and Reynolds, J.R., *J. Macromol. Sci., Chem.*, 1995, **32**, 1049
[20] Yoshino, K., Takiguchi, T., Hayashi, S., Park, D.H. and Sugimoto, R., *Jpn. J. Appl. Phys., Part 1*, 1986, **25**, 881
[21] Massardier, V., Guyat, A. and Tran, V.H., *Polymer*, 1994, **35**, 1561
[22] Wnek, G.E., Chien, J.C.W., Karasz, F.E. and Lillya, C.P., *Polymer*, 1979, **20**, 1441
[23] Gagnon, D.R., Capistran, J.D., Karasz, F.E. and Lenz, R.W., *Polym. Prepr. (Am. Chem. Soc., Div. Polym. Chem.)*, 1984, **25**, 284

- [24] Kossmehl, G. and Yaridjanian, A., *Makromol. Chem.*, 1981, **182**, 3419
- [25] Rakovic, D., Kostic, R., Davidova, I.E. and Gribov, L.A., *Synth. Met.*, 1993, **55**, 541
- [26] Rakovic, D., Kostic, R., Gribov, L.A., Stepanyan, S.A. and Davidova, I.E., *Synth. Met.*, 1991, **41**, 275
- [27] Furukawa, Y., Sakamoto, A. and Tasumi, M., *J. Phys. Chem.*, 1989, **93**, 5354
- [28] Sakamoto, A., Furukawa, Y. and Tasumi, M., *J. Phys. Chem.*, 1992, **96**, 1490
- [29] Bradley, D.D.C., Friend, R.H., Lindenberger, H. and Roth, S., *Polymer*, 1986, **27**, 1709
- [30] Rakovic, D., Kostic, R., Gribov, L.A. and Davidova, I.E., *Phys. Rev. B: Condens. Matter*, 1990, **41**, 10744
- [31] Buisson, J.P., Mevellec, J.Y., Zeraoui, S. and Lefrant, S., *Synth. Met.*, 1987, **41**, 287
- [32] Lefrant, S., Buisson, J.P. and Eckhardt, H., *Synth. Met.*, 1990, **37**, 91
- [33] Eckhardt, H., Baughman, R.H., Buisson, J.P., Lefrant, S. et al, *Synth. Met.*, 1991, **43**, 3413
- [34] Buisson, J.P., Lefrant, S., Louarn, G., Mevellec, J.Y. et al, *Synth. Met.*, 1992, **49**, 305
- [35] Yang, C.J. and Jenekhe, S.A., *Chem. Mater.*, 1995, **7**, 1276
- [36] Burn, P.L., Bradley, D.D.C., Friend, R.H., Halliday, D.A. et al, *J. Chem. Soc., Perkin Trans. 1*, 1992, 3225
- [37] Holzmann, G. and Kossmehl, G., *Org. Mass Spectrom.*, 1980, **15**, 336
- [38] Schröter, B., Hörhold, H.H. and Raabe, D., *Makromol. Chem.*, 1982, **181**, 3185
- [39] Hoeg, D.F., Lusk, D.I. and Goldberg, E.P., *J. Polym. Sci., Polym. Lett. Ed.*, 1964, **2**, 697
- [40] Wessling, R.A., *J. Polym. Sci., Polym. Symp.*, 1985, **72**, 55
- [41] Nishihawa, H., Tateishi, M., Aramaki, K., Ohsawa, T. and Kimura, O., *Chem. Lett.*, 1987, **3**, 539
- [42] Burroughs, J.H., Bradley, D.D.C., Brown, A.R., Marks, R.N. et al, *Nature (London)*, 1990, **347**, 539
- [43] Helbig, M. and Hörhold, H.-H., *Makromol. Chem.*, 1993, **194**, 1607
- [44] Schmid, W., Dankesreiter, R., Gmeiner, J., Vogtmann, Th. and Schwoerer, M., *Acta Polym.*, 1993, **44**, 208, (synth)

Poly[1,1-(1-phenylethane) bis(4-phenyl) carbonate]

P-452

Synonyms: *Poly[oxycarbonyloxy-1,4-phenylene(1-phenylethylidene)-1,4-phenylene]. Poly(4,4'-dihydroxydiphenylphenylmethylmethane-co-carbonic dichloride). Bisphenol ACP polycarbonate*
Monomers: Bisphenol ACP, Carbonic dichloride
Material class: Thermoplastic
Polymer Type: polycarbonates (miscellaneous)
CAS Number:

CAS Reg. No.
26985-42-2

Molecular Formula: $(C_{21}H_{16}O_3)_n$
Fragments: $C_{21}H_{16}O_3$

Volumetric & Calorimetric Properties:
Density:

No.	Value	Note
1	d 1.21 g/cm^3	25° [4]
2	d 1.084 g/cm^3	[2]
3	d 1.207 g/cm^3	30° [5]

Melting Temperature:

No.	Value	Note
1	210–230°C	[4]

Glass-Transition Temperature:

No.	Value	Note
1	176°C	[4]
2	179°C	[5,6]
3	183°C	[2]
4	174°C	[1]

Surface Properties & Solubility:
Solubility Properties General: Forms miscible blends with tetramethylpolycarbonate [1]; and immiscible blends with bisphenol A polycarbonate, polymethyl methacrylate, and polystyrene [1]
Cohesive Energy Density Solubility Parameters: Flory-Huggins polymer-solvent interaction parameters have been reported: 0.55 (CH_2Cl_2, 250°) [2]; 0.32 ($CHCl_3$, 250°) [2]; 1.4 (decane, 250°) [2]; 0.09 (cyclohexanol, 250°) [2]
Solvents/Non-solvents: Sol. CH_2Cl_2, *m*-cresol, cyclohexanone, THF, C_6H_6, pyridine, DMF; swollen by EtOAc, Me_2CO [3]

Transport Properties:
Permeability of Gases: The permeability of CO_2 decreases with increasing pressure (2-20 atm) [5]
Water Absorption:

No.	Value	Note
1	0.2 %	25°, 65% relative humidity [3]

Gas Permeability:

No.	Gas	Value	Note
1	CO_2	622 cm^3 mm/(m^2 day atm)	7.11 × 10^{-13} cm^2 (s Pa)$^{-1}$, 35°, 2 atm [5]
2	CH_4	27.49 cm^3 mm/(m^2 day atm)	3.14 × 10^{-14} cm^2 (s Pa)$^{-1}$, 35°, 2 atm, calc. [5]
3	O_2	133 cm^3 mm/(m^2 day atm)	1.52 × 10^{-13} cm^2 (s Pa)$^{-1}$, 25°, 1 atm [6]
4	N_2	23.72 cm^3 mm/(m^2 day atm)	2.71 × 10^{-14} cm^2 (s Pa)$^{-1}$, 35°, 2 atm, calc. [5]
5	He	910 cm^3 mm/(m^2 day atm)	1.04 × 10^{-12} cm^2 (s Pa)$^{-1}$, 35°, 2 atm [5]
6	H_2	910 cm^3 mm/(m^2 day atm)	1.05 × 10^{-12} cm^2 (s Pa)$^{-1}$, 35°, 2 atm [5]

Mechanical Properties:
Tensile Strength Break:

No.	Value	Note	Elongation
1	80.8 MPa	824 kg cm^{-2} [4]	55%

Complex Moduli: The variation of tan δ with temp. (-150–240°) has been reported [5,7]

Electrical Properties:
Electrical Properties General: Dielectric constant and dissipation factor do not vary with temp. until approaching T_g
Dielectric Permittivity (Constant):

No.	Value	Frequency	Note
1	3.3	1 kHz	25–100° [4]

Dissipation (Power) Factor:

No.	Value	Frequency	Note
1	0.0005	1 kHz	25–100° [4]

Optical Properties:
Refractive Index:

No.	Value	Note
1	1.613	25° [4]
2	1.616	calc. [1]

Transmission and Spectra: Ir spectral data have been reported [8]
Molar Refraction: dn/dc 0.1288 (c →0, 25°, C_6H_6) [9]

Polymer Stability:
Thermal Stability General: Stable under N_2 atmosphere up to 400° [2]
Decomposition Details: Decomposition temp. 427° (N_2) [5]. Thermo-oxidation is slow below 280° with weight loss onset at 300° [8]. Weight loss on heating in air 7.04% (350°, 30 min.) [5]. High temp. oxidation proceeds through scission of carbonate bridges, removal of phenyl substituents at the central carbon atom and formation of aldehyde, ketone and quinenoid carbonyl groups [8]

Applications/Commercial Products:
Processing & Manufacturing Routes: The main synthetic routes, phosgenation and transesterification. The most widely used approach is interfacial polycondensation in a suitable solvent, but soln. polymerisation is also carried out in the presence of pyridine [3]. A slight excess of phosgene is introduced into a soln. or suspension of the aromatic dihydroxy compound in aq. NaOH at 20–30°, in an inert solvent (e.g. CH_2Cl_2). Catalysts (quaternary ammonium and phosphonium salts) are used to achieve high MWs [3,4,10]. With transesterification, the dihydroxy compound and a slight excess of diphenyl carbonate are mixed together in the melt (150–300°) in the absence of O_2 with the elimination of phenol. Reaction rates are increased by alkaline catalysts and the use of a medium-high vacuum in the final stages [3,4]

Bibliographic References
[1] Kim, C.K. and Paul, D.R., *Macromolecules,* 1992, **25**, 3097, (polymer-polymer compatibility, T_g, refractive index)
[2] Dipaola-Baranyi, G., Hsiao, C.K., Spes, M., Odell, P.G. and Burt, R.A., *J. Appl. Polym. Sci.: Appl. Polym. Symp.,* 1992, **51**, 195, (polymer-solvent interaction parameters, density, T_g, thermal stability)
[3] Schnell, H., *Polym. Rev.,* 1964, **9**, 99, (solvents, water absorption, synth., rev)
[4] Schnell, H., *Ind. Eng. Chem.,* 1959, **51**, 157, (props., synth.)
[5] Aguilar-Vega, M. and Paul, D.R., *J. Polym. Sci., Polym. Phys. Ed.,* 1993, **31**, 1599, (props.)
[6] Schmidhauser, J.C. and Longley, K.L., *J. Appl. Polym. Sci.,* 1990, **39**, 2083, (T_g, gas permeability)
[7] Reding, F.P., Faucher, J.A. and Whitman, R.D., *J. Polym. Sci.,* 1961, **54**, S56, (complex moduli)
[8] Levantovskaya, I.I., Dralyuk, G.V., Pshenitsyna, V.P., Smirnova, O.V. et al, *Polym. Sci. USSR (Engl. Transl.),* 1968, **10**, 1892, (ir, thermo-oxidative stability)
[9] Fabre, M.J., Tagle, L.H., Gargallo, L., Radic, D. and Hernandez-Fuentes, I., *Eur. Polym. J.,* 1989, **25**, 1315
[10] Tagle, L.H. and Diaz, F.R., *Eur. Polym. J.,* 1987, **23**, 109, (synth)

Poly(phenylglycidyl ether)   P-453

$$-[OCH_2CH]_n-$$
$$\quad\quad CH_2OPh$$

Synonyms: *Poly[(phenoxymethyl)oxirane]. Poly(3-phenoxypropylene oxide). Poly(1,2-epoxy-3-phenoxypropane). PGE*
Monomers: (Phenoxymethyl)oxirane
Polymer Type: epoxy, polyalkylene ether
CAS Number:

CAS Reg. No.	Note
25265-27-4	
26895-29-4	isotactic

Molecular Formula: $(C_9H_{10}O_2)_n$
Fragments: $C_9H_{10}O_2$
General Information: Hard, tough, opaque solid [4]

Volumetric & Calorimetric Properties:
Melting Temperature:

No.	Value	Note
1	203°C	[2,4]
2	195°C	isotactic [4]
3	72–120°C	[3]
4	186–189°C	isotactic [5]

Glass-Transition Temperature:

No.	Value	Note
1	9°C	C_6H_6 extracted polymer, xylene recrystallised polymer

Surface Properties & Solubility:
Solvents/Non-solvents: Sol. hot cyclohexanone, hot *o*-dichlorobenzene, hot DMF, hot phenylglycidyl ether [5]. Slightly sol. C_6H_6 [2], chloronaphthalene [2]. Insol. Et_2O [2]

Transport Properties:
Polymer Solutions Dilute: η_{inh} 2.2 dl g^{-1} (0.1% in chloronaphthalene, 135°) [2]. η 2 dl g^{-1} (0.1% in chloronaphthalene, 135°, crude polymer) [4]. η 1.9 dl g^{-1} (0.1% in chloronaphthalene, 135°, extracted with C_6H_6) [4]. η 1.6 dl g^{-1} (0.1% in chloronaphthalene, 135°, xylene recrystallised polymer) [4]. $[\eta]$ 0.467 dl g^{-1} (*o*-dichlorobenzene, 121°) [5]

Optical Properties:
Optical Properties General: Electron micrographs have been reported [6]
Transmission and Spectra: Ir [5], x-ray [5,6], H-1 nmr [7] and mass spectral data [3] have been reported

Bibliographic References
[1] Chabanne, P., Tighzert, L. and Pascault, J.-P., *J. Appl. Polym. Sci.,* 1994, **53**, 769, (polymerisation)
[2] Vandenberg, E.J., *J. Polym. Sci., Part A-1,* 1969, **7**, 525, (isotactic)
[3] Tanzer, W., Buttner, K. and Ludwig, I., *Polymer,* 1996, **37**, 997, (anionic polymerisation)
[4] Vandenberg, E.J. and Senyek, M.L., *Macromol. Synth.,* 1972, **4**, 55, (isotactic)
[5] Noshay, A. and Price, C.C., *J. Polym. Sci.,* 1959, **34**, 165, (isotactic)
[6] Takada, Y., *J. Macromol. Sci., Chem.,* 1967, **1**, 1369, (x-ray)
[7] Nakano, S. and Endo, T., *J. Polym. Sci., Part A-1,* 1995, **33**, 505, (H-1 nmr)

Poly(*N*-phenyl-*p*-phenylenediamine)   P-454

Synonyms: *1,4-Benzenediamine, N-phenyl-, homopolymer. Poly(4-aminodiphenylamine). PPPD*
Monomers: 4-Aminodiphenylamine
Polymer Type: polyaniline
CAS Number:

CAS Reg. No.
89230-95-5

Molecular Formula: $(C_{12}H_{12}N_2)n$
Fragments: $C_{12}H_{12}N_2$
Additives: PANi

– Poly(phenylsilane)

General Information: Methods of preparation result in head-tail polymerisation. The base polymers show some dissimilarities in the thermal decomposition behaviour, solubility, and intrinsic oxidation state [5].
Morphology: Structure is similar to polyaniline [5]
Identification: Chloride derivative: $C_{7.0}H_{5.9}NCl_{0.58}$ [2,5]

Surface Properties & Solubility:
Solubility Properties General: The monomer does not dissolve in the aqueous acid [5]

Electrical Properties:
Electrical Properties General: $\sigma = 1.8 \times 10^{-2}$ to 2×10^{-4} S cm^{-1} [1,2,3,5]

Optical Properties:
Transmission and Spectra: Ir [1,5]; uv-vis [1,5] and xps [5] spectra reported.

Polymer Stability:
Polymer Stability General: Polymer stability depends upon synthesis [5]
Thermal Stability General: The polymer shows weight loss behaviour entirely different from the monomer. Retains a significantly higher proportion of weight for temperatures higher than 500°C. This difference in degradation behaviour is not as evident in the TG scans of polyaniline base obtained by the different methods. The onset of the major weight loss step of the DPAI base polymer is approx. 100°C higher than those of the polyaniline bases and the weight retained at 700°C is also slightly higher. In the case of PPDI base polymer, the weight loss is slow and gradual with no distinct steps, and the weight retained at 700°C [5]

Applications/Commercial Products:
Processing & Manufacturing Routes: The aniline dimer, N-(4-aminophenyl)aniline, has been polymerised cleanly under mild conditions to obtain an emeraldine base form of polyaniline using [MeB(3-(Mes)Pz)$_3$]CuCl as the catalyst and H$_2$O$_2$ as the oxidant, while the subsequent acidification of the emeraldine base gives the conducting emeraldine salt form of polyaniline [1]
The polymer has been synthesised from PrPPD using ammonium persulfate in dilute acids or by Cu(ClO$_4$)$_2$.6H$_2$O in MeCN [2,4,8]
Electrically conducting thin films of aniline derivatives have been deposited by the Langmuir-Blodgett (LB) technique and vacuum evaporation [3]

Applications/Commercial Products:
Applications: Secondary batteries, which maintain a high charging and discharging efficiency [6,7]. Photopolymerisation of aniline derivatives by photoinduced electron transfer for application to image formation [9]. Used in the manufacture of electric cable [10]

Bibliographic References
[1] Rasika, D.H.V., Wang, X., Rajapakse, R.M.G. and Elsenbaumer, R.L., *Chem. Comm.*, 2006, 976
[2] Hagiwara, T., Demura, T. and Iwata, K., *Synth. Met.*, 1987, **18**, 317
[3] Punkka, E., Laakso, K., Stubb, H., Levon, K. and Zheng, W.Y., *Thin Solid Films*, 1994, **243**, 515
[4] Neob, K.G., Kang, E.T. and Tan, K.L., *J. Phys. Chem.*, 1992, **96**, 6711
[5] Hagiwara, T., Michio, Y. and Kaoru, I., *Nippon Kagaku Kaishi*, 1989, **10**, 1791
[6] *Eur. Pat.*, 1988, 261 837
[7] *Jpn. Pat.*, 1985, 60 262 355
[8] Chandrasekhar, P. and Gumbs, R.W., *J. Electrochem. Soc.*, 1991, **138**, 1337
[9] Sei, U., Takayuki, N. and Norihisa, K., *J. Mater. Chem.*, 2001, **11**, 1585
[10] *Eur. Pat.*, 1997, 782 151
[11] Feng, J., Zhang, W., Macdiarmid, A.G. and Epstein, A.J., *Annu. Tech. Conf. - Soc. Plast. Eng.*, *55th*, 1997, 1373

Poly(phenylsilane)

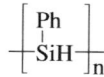

Synonyms: *Poly(phenylsilylene)*. *Phenylpolysilane*. *PPSi*
Polymer Type: polysilanes
CAS Number:

CAS Reg. No.
95584-36-4

Molecular Formula: $(C_6H_6Si)n$
Fragments: C_6H_6Si
Molecular Weight: 1,900–14,000 depending upon metallocene catalyst used. Ti-based catalysts gave samples with a lower average degree of polymerisation compared with Zr analogues [1,2,3,5,11,12]
General Information: Polysilanes (RSR′) have significant potential as preceramics and in the fabrication of optical devices.
Tacticity: Atactic (a-PPSi), isotactic (i-PPSi) and syndiotactic stereoarrangements are known for PPSi. Mostly found as atactic microstructure with a small syndiotactic bias. Syndioselective probability (1-Pm) of 0.55 during propagation of the pseudochiral chain. The d.p. was found to be dependent on the metal and the degree of Cp-ring substitution of the metallocene precatalyst [1,3,5]
Morphology: Off-white gum/crystal [1,3]. Amorphous structure, where the arrangement of Ph rings is comparable to the statistical disorder of benzene molecules in the liquid state [10]. An atactic structure has been suggested for the polymer morphology based on the large number of ^{29}Si environments indicated to be present [5,12]

Volumetric & Calorimetric Properties:
Thermodynamic Properties General: See [11]

Surface Properties & Solubility:
Solubility Properties General: Si-H moiety on which a Ph group is substituted may be responsible for the good solubility in basic aqueous solutions [6]

Optical Properties:
Transmission and Spectra: Ir [2,8], uv-vis [9], nmr [1,2,5,8,9,12] and ms [8] spectra reported

Applications/Commercial Products:
Processing & Manufacturing Routes: Phenylsilane can be converted catalytically to polyphenylsilane using substituted achiral group IV metallocene dichloride precatalysts (CpR)$_2$MCl$_2$, where M=Ti, Zr and CpR=C$_5$H$_4$C(Me)$_2$CHMe$_2$ or Cp(CpR)MCl$_2$ where M=Ti, Zr and Cp=C$_5$H$_5$ in combination with 2 equiv. of n-BuLi [1,5,9].
Synth. by Wurtz-type coupling of prochiral organochlorosilanes [1].
Presence of B(C$_6$F$_5$)$_3$ in reactions catalysed by the Cp$_2$ZrCl$_2$/BuLi gives (PhSiH)n. [2,3]
Ni-catalysed polymerisation of PhSiH$_3$ [2]
PPSi can be developed in basic aqueous solution when exposed to 254 nm light. [6]

Applications/Commercial Products:
Applications: Co-catalyst for Si-Si bond formation reactions [2]. Polysilane and epoxy resin used as primer coat with good adhesion [4]. Battery electrodes, solar battery and electromagnetic shields [6]

Trade name
Phenylpolysilane 10

Bibliographic References
[1] Grimmond, B.J., Rath, N.P. and Corey, J.Y., *Organometallics*, 2000, **19**, 2975
[2] Georges Fontaine, F. and Zargarian, D., *J. Am. Chem. Soc.*, 2004, **126**, 8786
[3] Hubmann, J.L., Corey, J.Y., and Rath, N.P., *J. Organomet. Chem.*, 1997, **533**, 61
[4] *Jpn. Pat.*, 2000, 2000 265 084

[5] Grimmond, B. J.; Corey, J.Y. and Joyce, Y., *Organometallics*, 1999, **18**, 2223
[6] *Jpn. Pat.*, 1998, 10 120 907
[7] Rikako, K., Yoshihiko, N. and Shuzi, H., *Polym. Mater. Sci. Eng.*, 1996, **75**, 431
[8] Aitken C. and Harrod, J.F., *Can. J. Chem.*, 1987, **65**, 1804
[9] Goldslager, B.A. and Clarson, S.J., *J. Inorg. Organomet. Polym.*, 1999, **9**, 123
[10] Resel, R., Leising, G., Lunzer, F., and Marschner, C., *Polymer*, 1998, **39**, 5257
[11] Georges Fontaine, F. and Zargarian, D., *Organometallics*, 2002, **21**, 401
[12] Dioumaev, V.K., Rahimian, K., Gauvin F. and Harrod, J.F., *Organometallics*, 1999, **18**, 2249

Polyphenylsiloxane fluid   P-456

Synonyms: *Polymethylphenylsiloxane fluid. Poly[oxy(methylphenylsilylene)]*
Related Polymers: More general information, Dimethyl substituted
Monomers: Dichloromethylphenylsilane, Cyclic methylphenylsiloxane
Material class: Fluids
Polymer Type: silicones
CAS Number:

CAS Reg. No.
9005-12-3
31230-04-3

Molecular Formula: $[C_7H_8COSi]_n$
Fragments: C_7H_8OSi
Molecular Weight: MW 1000–30000
General Information: Silicone fluids containing phenyl groups have higher values of density than dimethylsiloxane

Volumetric & Calorimetric Properties:
Density:

No.	Value	Note
1	d 1.1 g/cm^3	[3]

Thermal Expansion Coefficient:

No.	Value	Note	Type
1	0.00055 K^{-1}	[7]	V

Thermal Conductivity:

No.	Value	Note
1	0.13 W/mK	(0.00032 cal.S^{-1}.cm^{-1}.°C^{-1}) [13,14]

Specific Heat Capacity:

No.	Value	Note	Type
1	1.4 kJ/kg.C	(0.34 Btu.16^{-1}°F^{-1}) [3]	P

Transition Temperature General: High level of phenyl substitution raises pour point above that for dimethyl silicone [9]
Transition Temperature:

No.	Value	Note
1	-30°C	Pour point [1]

Surface Properties & Solubility:
Cohesive Energy Density Solubility Parameters: δ 18.4 J$^{1/2}$cm$^{-3/2}$ (9 cal$^{1/2}$cm$^{-3/2}$) [5,10]
Solvents/Non-solvents: Sol. aliphatic hydrocarbons, aromatic hydrocarbons, chlorinated hydrocarbons, esters, ethers, ketones. Insol. alcohols, H_2O [1,5,11]
Surface and Interfacial Properties General: Better lubricating props. than dimethyl silicone fluid [6]
Wettability/Surface Energy and Interfacial Tension: τ_c 25 mJm^{-2} (25 dyne.cm^{-1}). τ_c increases with increasing phenyl substitution [3,7]

Transport Properties:
Polymer Melts: Viscosity has been reported [1,2,3,11,15,16]. Viscosity temp. coefficient 0.85. The viscosity of methyl phenyl silicone oils drops more rapidly with temp. than that of methyl silicone oils, but at room. temp. methyl phenyl silicones have higher viscosity than methyl silicones of the same MW

Mechanical Properties:
Mechanical Properties General: Much less compressible than [PDMS fluid]
Elastic Modulus:

No.	Value	Note
1	2000 MPa	Bulk modulus (approx.) (200000 kg.cm^{-2}, (approx.)) [11,12]

Electrical Properties:
Electrical Properties General: Excellent insulating props. but inferior to those of dimethyl silicone fluids
Dielectric Permittivity (Constant):

No.	Value	Note
1	3	[4,17]

Dielectric Strength:

No.	Value	Note
1	20 kV.mm^{-1}	[4]

Optical Properties:
Optical Properties General: Refractive index higher than that of dimethyl silicone [8]
Refractive Index:

No.	Value	Note
1	1.525	[3,8]

Transmission and Spectra: Transparent above 280 nm [5]

Polymer Stability:
Polymer Stability General: The homopolymer is particularly stable at high temps.
Thermal Stability General: Usable between -40° and 250° in the presence of air. Usable up to 300° in the absence of air [3]
Upper Use Temperature:

No.	Value
1	250°C
2	275°C

Decomposition Details: Oxidised above 275° [2]
Flammability: Fl. p. 315° (600°F) [3]
Environmental Stress: Stable to H_2O, air, oxygen, metals and other materials. Highly resistant to radiation [1]
Chemical Stability: Chemically inert but decomposed by conc. mineral acids. V. thin films liable to degradation [1,4,5]
Hydrolytic Stability: Stable to H_2O and aq. solns. V. thin films spread on water; liable to hydrolytic degradation [1,4,5]

Biological Stability: Non-biodegradable [4,5]
Recyclability: Can be reused; no large scale recycling

Applications/Commercial Products:
Processing & Manufacturing Routes: For low viscosity fluids dichloromethylphenylsilane is hydrolysed and equilibrated under acid catalysis at 180° with small amounts of hexamethyldisiloxane. For high viscosity fluids cyclic methylphenylpolysiloxanes are equilibrated under alkaline catalysis with traces of chain terminator at 150°
Applications: Used as plastics additive and for high temp. heat transfer. Used in high temp. greases and in solder baths

Trade name	Supplier
SF 1265	General Electric Silicones

Bibliographic References
[1] Noll, W., *Chemistry and Technology of Silicones*, Academic Press, 1968
[2] Murphy, C.M., Saunders, C.E. and Smith, D.C., *Ind. Eng. Chem.*, 1950, **42**, 2462, (thermal stability)
[3] Ash, M. and Ash, I., Handbook of Plastic Compounds, *Elastomers and Resins*, (eds. M.B. Ash and I.A. Ash), Wiley-VCH, 1992
[4] Stark, F.O., Fallender, J.R. and Wright, A.P., *Comprehensive Organometallic Chemistry*, (eds. G. Wilkinson, F.G.A. Stone and E.W. Abel), Pergamon Press, 1982, **2**, 305
[5] Hardman, B. and Torkelson, A., *Encycl. Polym. Sci. Eng.*, 2nd edn. (ed. J.I. Kroshwitz), John Wiley and Sons, 1985, **15**, 204
[6] Clarson, S.J. and Semlyen, J.A., *Siloxane Polymers*, Prentice Hall, 1993
[7] Fox, H.W., Taylor, P.W. and Zisman, W.A., *Ind. Eng. Chem.*, 1947, **39**, 1401
[8] Flaningan, O.L. and Langley, N.R., *Anal. Chem. Silicones*, (ed. A.L. Smith), John Wiley and Sons, 1991, 135
[9] Worrick, E.L., Hunter, M.J. and Barry, A.J., *Ind. Eng. Chem.*, 1952, **44**, 2196
[10] Yerrick, K.B. and Beck, H.N., *Rubber Chem. Technol.*, 1964, **37**, 261
[11] Meals, R.N. and Lewis, F.M., *Silicones*, Reinhold, 1959
[12] Thompson, J.M.C., (ed. S. Fordham), *Silicones*, 1960, 5
[13] Bates, O.K., *Ind. Eng. Chem.*, 1949, **41**, 1966, (thermal conductivity)
[14] Jamieson, D.T. and Irving, J.B., *Proc. Int. Conf. Therm. Conduct.*, 1975, 279, (thermal conductivity)
[15] Buch, R.R., Klimisch, H.M. and Johannson, O.K., *J. Polym. Sci., Part A-2*, 1970, **8**, 541, (viscosity)
[16] Lee, C.L. and Haberland, C.G., *Polym. Int.*, 1965, **3**, 883, (viscosity)
[17] Vincent, G.A., Feuron, F.W.G. and Orbech, T., *Annu. Rep., Conf. Electr. Insul. Dielectr. Phenom.*, 1972, 17, (dielectric constant)

Poly[(phenylsilylene)methylene]   P-457

$$\left[\mathrm{CH_2} - \underset{\mathrm{H}}{\overset{\mathrm{Ph}}{\mathrm{Si}}} \right]_n$$

Synonyms: *PPSM*
Related Polymers: Polycarbosilanes
Polymer Type: polycarbosilanes
CAS Number:

CAS Reg. No.
321667-67-8

Molecular Formula: $(C_7H_8Si)_n$
Fragments: C_7H_8Si
Molecular Weight: MW 11,200; M_w/M_n 2.5. All asymmetrically substituted polymers were obtained in high molecular weight form [2]
General Information: Polymerisation was conducted in toluene since the monomer is a solid which sublimes very easily during heating. Even at a higher temperature and for longer reaction times, the resultant polymer was still obtained in a relatively low molecular weight compared to other PSM's. The difficulty of this polymerisation could originate from the strongly electronegative nature of the chloro and phenyl groups. PPSM is a monosilicon analogue of polystyrene [2]

Morphology: Polymer found to be atactic [2]
Identification: Found: C, 65.23%; H, 13.05%; calc: C, 65.6%; H, 12.50% [2]

Volumetric & Calorimetric Properties:
Glass-Transition Temperature:

No.	Value	Note
1	-37.6°C	[2]

Optical Properties:
Transmission and Spectra: Pmr, cmr, Si-29 nmr spectra reported [2]

Polymer Stability:
Thermal Stability General: The phenyl attached directly to the polymer backbone imparts increased thermal stability, relative to alkyl and hydrogen substituents. In addition, the polymer begins to lose weight at a lower temperature than PMPSM resulting in a higher char yield [2]

Applications/Commercial Products:
Processing & Manufacturing Routes: Improved synthesis routes to the DSCB monomers were developed which proceed through Grignard ring closure reactions on alkoxy-substituted chlorocarbosilanes [2]
Applications: Insulating film for semiconductor device [1]

Bibliographic References
[1] *Jpn. Pat.*, 2001, 01 022 089
[2] Shen, Q.H. and Interrante, L.V., *J. Polym. Sci., Part A*, 1997, **35**, 3193

Poly(4-phenylstyrene)   P-458

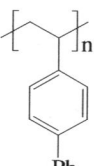

Synonyms: *Poly(4-ethenyl-1,1'-biphenyl). Poly(1-(4-[1,1'-biphenyl])-1,2-ethanediyl)*
Monomers: 4-Phenylstyrene
Material class: Thermoplastic
Polymer Type: styrenes
CAS Number:

CAS Reg. No.	Note
25232-08-0	
29793-42-8	isotactic
135268-47-2	syndiotactic

Molecular Formula: $(C_{14}H_{12})_n$
Fragments: $C_{14}H_{12}$

Volumetric & Calorimetric Properties:
Glass-Transition Temperature:

No.	Value	Note
1	161°C	[4]

Transport Properties:
Polymer Solutions Dilute: [η] 1.22-21.2 dl g^{-1} (c, 0.869–20.925 g dl^{-1}) [1]. [η] 0.007 dl g^{-1} (tetralin, 100°) [2]

Optical Properties:
Transmission and Spectra: Uv spectral data have been reported [3]

Polyphenylsulfone, Unreinforced

Bibliographic References
[1] Rudin, A., *J. Appl. Polym. Sci.*, 1975, **19**, 619
[2] Notta, G., Danusso, F. and Sianesi, D., *Makromol. Chem.*, 1958, **28**, 253
[3] Kern, V.W., Heitz, W., Jager, M., Pfitzner, K. and Wirth, H.O., *Makromol. Chem.*, 1969, **126**, 73
[4] Utracki, L.A. and Simha, R., *Makromol. Chem.*, 1968, **117**, 94

Polyphenylsulfone, Unreinforced

P-459

Synonyms: *PPSF. Poly[oxy(1,1'-biphenyl)-4,4'-diyloxy-1,4-phenylenesulfonyl-1,4-phenylene]*
Monomers: 4,4'-Dichlorodiphenyl sulfone, 4,4'-Biphenyldiol
Material class: Thermoplastic
Polymer Type: polysulfone
CAS Number:

CAS Reg. No.	Note
25608-64-4	Radel R

Molecular Formula: $(C_{24}H_{16}O_4S)_n$
Fragments: $C_{24}H_{16}O_4S$
General Information: Polyphenylsulfone is a transparent, tough, rigid, amorph. thermoplastic. It offers outstanding thermal stability, superb resistance to hydrol. in hot water and steam environments and excellent inherent burning resistance. The impact performance of polyphenylsulfone is very high when compared with the other sulfone resins. It continues to exhibit useful mech. props. at temps. up to 180° under prolonged or repeated thermal exposure. Light amber colour

Volumetric & Calorimetric Properties:
Density:

No.	Value	Note
1	d 1.29 g/cm³	ASTM D792 [2]

Thermal Expansion Coefficient:

No.	Value	Note	Type
1	5.6×10^{-5} K^{-1}	ASTM D696 [2]	L

Thermal Conductivity:

No.	Value	Note
1	0.35 W/mK	ASTM C177 [1]

Specific Heat Capacity:

No.	Value	Note	Type
1	1.17 kJ/kg.C	23° [1]	P

Melting Temperature:

No.	Value	Note
1	360–390°C	[2]

Glass-Transition Temperature:

No.	Value	Note
1	220°C	[1]

Transition Temperature General: T_g of PPSF is 30° higher than that of PSU
Deflection Temperature:

No.	Value	Note
1	207°C	1.8 MPa, ASTM D648 [1]

Transition Temperature:

No.	Value	Note	Type
1	-100°C	approx. [1]	Second order (β)

Surface Properties & Solubility:
Cohesive Energy Density Solubility Parameters: δ 22 J$^{1/2}$ cm$^{-3/2}$ (min.) [3]
Solvents/Non-solvents: Sol. polar solvents such as *N*-methylpyrrolidinone, dimethylacetamide, pyridine and aniline [1]

Transport Properties:
Melt Flow Index:

No.	Value	Note
1	18 g/10 min	0.3 MPa, 400°, ASTM D1238 [2]

Water Absorption:

No.	Value	Note
1	0.37 %	24h
2	1.1 %	equilibrium, ASTM D570 [1]

Mechanical Properties:
Mechanical Properties General: Tough and rigid. Has exceptional resistance to creep. Very high notched impact strength and has the ability to retain a high degree of ductility after prolonged heat exposure. Impact and toughness props. are improved over those of PES and PSU [1,4,5]
Tensile (Young's) Modulus:

No.	Value	Note
1	2350 MPa	ASTM D638 [2]

Flexural Modulus:

No.	Value	Note
1	2410 MPa	ASTM D790 [2]

Tensile Strength Break:

No.	Note	Elongation
1	ASTM D638 [1]	90%

Compressive Modulus:

No.	Value	Note
1	1730 MPa	ASTM D695 [1]

Poisson's Ratio:

No.	Value	Note
1	0.42	0.5% strain [1]

– Polyphenylsulfone, Unreinforced

Tensile Strength Yield:

No.	Value	Elongation	Note
1	70 MPa	7.2%	ASTM D638 [2]

Flexural Strength Yield:

No.	Value	Note
1	91 MPa	ASTM D790 [2]

Compressive Strength:

No.	Value	Note
1	99 MPa	ASTM D695 [1]

Impact Strength: Tensile impact strength 400 kJ m^{-2} (ASTM D1822) [2]
Hardness: Rockwell M86 (ASTM D785) [1]
Failure Properties General: Shear strength (yield) 62 MPa (ASTM D732) [1]
Friction Abrasion and Resistance: Abrasion resistance 20 mg (1000 cycles)$^{-1}$ (ASTM D1040, Taber CS-17 test) [1]
Izod Notch:

No.	Value	Notch	Break	Note
1	693 J/m	Y		ASTM D256 [2]
2		N	Y	ISO 180/IC [2]

Electrical Properties:
Electrical Properties General: Offers excellent insulating and other electrical props. The dielectric constants and dissipation factors are low even in the GHz (microwave) frequency range. This performance is retained over a wide temp. range
Dielectric Permittivity (Constant):

No.	Value	Frequency	Note
1	3.44	60 Hz	ASTM D150 [1]
2	3.45	1 kHz	ASTM D150
3	3.45	1 MHz	ASTM D150 [1]

Dielectric Strength:

No.	Value	Note
1	14.6 kV.mm^{-1}	3.2 mm thick, ASTM D149 [1]

Dissipation (Power) Factor:

No.	Value	Frequency	Note
1	0.0006	60 Hz	ASTM D150 [1]
2	0.0076	1 MHz	ASTM D150 [1]

Optical Properties:
Optical Properties General: Exhibits optical transparency
Refractive Index:

No.	Value	Note
1	1.67	ASTM D1505 [1]

Transmission and Spectra: Light transmittance 70% [1]
Volume Properties/Surface Properties: Haze 7% (max., 3.1 mm thick, ASTM D1004) [1]

Polymer Stability:
Polymer Stability General: Withstands exposure to elevated temps. in air and water for prolonged periods
Thermal Stability General: Excellent thermal stability and resistance to thermal oxidation
Upper Use Temperature:

No.	Value	Note
1	180°C	UL 746 [1]

Flammability: Flammability rating V0 (0.8 mm thick, UL94) [1]. Limiting oxygen index 38% (ASTM D286) [1]. Smoke density 30 (6.2 mm, ASTM E662) [1]. Exhibits excellent resistance to burning. Has outstanding flame retardancy and very low smoke release characteristics [1]
Environmental Stress: High degree of oxidative stability. Poor resistance to uv light. Absorbance in the uv region occurs with attendant discoloration and losses in mech. props. due to polymer degradation [1]. Good resistance to many forms of ionising radiation [1,6,7,8]
Chemical Stability: Highly resistant to mineral acids, alkali and salt solns. Resists aliphatic hydrocarbons and most alcohols. Crazed, swollen or dissolved by chlorinated hydrocarbons, esters and ketones [1]
Hydrolytic Stability: Superb resistance to hydrol. by hot water and steam [1]
Stability Miscellaneous: Good dimensional stability [1]

Applications/Commercial Products:
Processing & Manufacturing Routes: Polyphenylsulfone is obtained via the nucleophilic substitution polycondensation route. This synth. is based on reaction of essentially equimolar quantities of 4,4'-dichlorodiphenylsulfone with the biphenol in a dipolar aprotic solvent in the presence of alkali base. Diphenyl sulfone, sulfolane, and N-methyl-2-pyrrolidinone are examples of suitable solvents. Potassium carbonate is used as base. High injection pressure is usually required for injection moulding. Melt temps. are in the range of 360–400° and mould temp. range is 120–190°. Prior to melt processing the resin must be dried for 2.5h at 150° to avoid structural and appearance defects in the fabricated parts due to bubbling and foaming of the trapped moisture. It is easily processed by other thermoplastic fabrication techniques including extrusion, thermoforming, and blow moulding. Extrusion into film, sheet, tubing or profile can be accomplished on conventional extrusion equipment. Stock temps. during extrusion are in the range of 315–375°. Once formed can be annealed to reduce moulded-in stress [1,2,3]
Mould Shrinkage (%):

No.	Value	Note
1	0.7%	ASTM D955 [2]

Applications: Typical applications include aircraft cabinet interior components, sterilisable medical devices, electrical/electronic components, chemical process equipment and automotive parts. It complies with U.S. Pharmacopeia Class VI requirements for use in medical device components

Trade name	Details	Supplier
Radel R-5000	medical applications; injection moulding grade	Amoco Performance Products
Radel R-7000	commercial aircraft interior	Amoco Performance Products

Bibliographic References

[1] *Kirk-Othmer Encycl. Chem. Technol.*, 4th edn., (eds. J.I. Kroschwitz and M. Howe-Grant), Wiley Interscience, 1993, **19**, 945
[2] *Engineering Plastics for Performance and Value*, Amoco Performance Products, Inc., (technical datasheet)

[3] Harris, J.E. and Johnson, R.H., *Encycl. Polym. Sci. Eng.*, 2nd edn., (ed. J.I. Kroschwitz), John Wiley and Sons, 1985, **13**, 196
[4] Brydson, J.A., *Plast. Mater.*, 5th edn., Butterworths, 1988
[5] Attwood, T.E., Cinderey, M.B. and Rose, J.B., *Polymer*, 1993, **34**, 1322
[6] Davis, A., Gleaves, M.H., Golden, J.H. and Huglin, M.B., *Makromol. Chem.*, 1969, **129**, 63
[7] Brown, J.R. and O'Donnell, J.H., *J. Polym. Sci., Part B: Polym. Lett.*, 1970, **8**, 121
[8] Lyons, A.R., Symons, M.C.R. and Yandel, J.K., *Makromol. Chem.*, 1972, **157**, 103
[9] Rose, J.B., *Polymer*, 1974, **15**, 456, (synth)

Polyphosphazenes   P-460

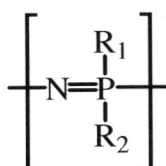

$R_1 = R_2$ = alkoxy groups (OR), where R = F, etc.

Related Polymers: Fluoroalkoxyphosphazene
Material class: Synthetic Elastomers
Polymer Type: polyphosphazenes

Volumetric & Calorimetric Properties:
Transition Temperature General: Low T_g (typically below -40°) causes polymers to have elastomeric props. at room temp. [1]

Mechanical Properties:
Mechanical Properties General: Good elastomeric props. [1]

Polymer Stability:
Thermal Stability General: Good low temp. props. [1]
Flammability: Non-flammable [1]
Environmental Stress: Resistant to oxidation [1]

Applications/Commercial Products:
Processing & Manufacturing Routes: Synth. varies widely with precise chemical composition of polymer. See specific polyphosphazenes for more detailed information [1]

Bibliographic References
[1] Allcock, H.R., *Encycl. Polym. Sci. Eng.*, Vol. 13, 2nd edn., (ed. J.I. Kroschwitz), John Wiley and Sons, 1985, 31

Polyphthalamide   P-461

Synonyms: *PPA*
Related Polymers: Aramids
Material class: Thermoplastic
Polymer Type: aromatic polyamides
Molecular Formula: $C_8H_6N_2O_2$
Fragments: $C_8H_6N_2O_2$
General Information: Copolyamide of isophthalic acid and terephthalic acid. It possibly contains some phthalic acid and is polymerised with aliphatic and substituted diamines
Miscellaneous: Water absorption is 30% that of Nylon 6,6. Chemical resistance on par with Nylon 6,6

Volumetric & Calorimetric Properties:
Density:

No.	Value	Note
1	d 1.46 g/cm^3	33% glass fibre reinforced
2	d 1.15 g/cm^3	impact modified
3	d 1.18 g/cm^3	unfilled

Thermal Expansion Coefficient:

No.	Value	Note	Type
1	2.4×10^{-5} K^{-1}	33% glass fibre reinforced, ASTM D696	L
2	8×10^{-5} K^{-1}	unfilled, ASTM D696	L

Specific Heat Capacity:

No.	Value	Note	Type
1	1.5 kJ/kg.C	50°	V
2	3.8 kJ/kg.C	300°	V

Melting Temperature:

No.	Value
1	310°C

Glass-Transition Temperature:

No.	Value
1	121–138°C

Deflection Temperature:

No.	Value	Note
1	285°C	1.8 MPa, 33% glass fibre reinforced, ASTM D648
2	297°C	0.455 MPa, 33% glass fibre reinforced, ASTM D648
3	180°C	0.45 MPa, unfilled, ASTM D648
4	120°C	1.8 MPa, impact modified, ASTM D648

Transition Temperature:

No.	Value	Type
1	321–338°C	Injection moulding temp.

Transport Properties:
Water Absorption:

No.	Value	Note
1	0.21 %	24h, 33% glass fibre filled
2	0.12 %	24h, 45% glass fibre filled
3	0.1 %	24h, 65% glass fibre filled
4	0.7 %	24h, unfilled
5	2.5 %	saturation, unfilled

Mechanical Properties:
Tensile (Young's) Modulus:

No.	Value	Note
1	13100 MPa	33% glass fibre, ASTM D638
2	2400 MPa	impact modified

Flexural Modulus:

No.	Value	Note
1	11400 MPa	33% glass fibre

Polyphthalamide

No.	Value	Note
2	2067 MPa	unfilled
3	2600 MPa	impact modified
4	3500 MPa	unfilled, ASTM D790

Tensile Strength Break:

No.	Value	Note	Elongation
1	221 MPa	33% glass fibre, ASTM D638	2.5%
2	76 MPa	impact modified, ASTM D638	30%
3	62.7 MPa		
4	90 MPa	unfilled, ASTM D638	5%

Flexural Strength at Break:

No.	Value	Note
1	317 MPa	33% glass fibre
2	128 MPa	impact modified, ASTM D790
3	105 MPa	unfilled, ASTM D790

Poisson's Ratio:

No.	Value
1	0.41

Compressive Strength:

No.	Value	Note
1	276 MPa	dry
2	247 MPa	50% relative humidity, DIN 53454, ISO R604
3	70 MPa	unfilled, ASTM D695

Hardness: Rockwell M80 (unfilled), M89 (30% glass fibre), 125 (33% glass fibre, ASTM D-785, ISO 2039/2A)
Failure Properties General: Shear strength 101 MPa (dry, 33% glass fibre, ASTM D732); 89 MPa (50% relative humidity, 33% glass fibre, ASTM D732)
Friction Abrasion and Resistance: Coefficient of friction 0.22 (static, LND SOP); 0.28 (dynamic)

Izod Notch:

No.	Value	Notch	Break	Note
1	112 J/m	Y		3.2 mm, dry, 33% glass fibre, ASTM D256
2	101 J/m	Y		3.2 mm, 50% relative humidity, 33% glass fibre, ASTM D256
3	600 J/m	N		unfilled, ASTM D785
4		N	No break	impact modified
5	960 J/m	Y		impact modified, ASTM D256

Izod Area:

No.	Value	Notch	Note
1	11.7 kJ/m^2	Notched	3.2 mm, dry, 33% glass fibre, ISO
2	7.8 kJ/m^2	Notched	3.2 mm, 50% relative humidity, 33% glass fibre, ISO
3	62.5 kJ/m^2	Unnotched	dry, 33% glass fibre, ISO
4	38.5 kJ/m^2	Notched	50% relative humidity, 33% glass fibre
5	59 kJ/m^2	Notched	impact modified, ISO 180/1A

Charpy:

No.	Value	Notch	Break	Note
1	15.2	Y		Impact modified, ISO 179/1A
2		N	No break	

Electrical Properties:

Surface/Volume Resistance:

No.	Value	Note	Type
1	10×10^{15} Ω.cm	33% glass fibre, UDE 0303/1	S

Dielectric Permittivity (Constant):

No.	Value	Frequency	Note
1	4.4	60 Hz	33% glass fibre, ASTM D150
2	4.2	1 MHz	33% glass fibre, ASTM D150

Dielectric Strength:

No.	Value	Note
1	21.6 kV.mm^{-1}	3.2 mm, 33% glass fibre, ASTM D149

Arc Resistance:

No.	Value	Note
1	140s	33% glass fibre, ASTM D495
2	125s	30% glass fibre, ASTM D495

Complex Permittivity and Electroactive Polymers: Comparative tracking index 550 V (UDE 0303/1)
Dissipation (Power) Factor:

No.	Value	Frequency	Note
1	0.005	60 Hz	33% glass fibre, ASTM D150
2	0.017	1 MHz	33% glass fibre, ASTM D150

Polymer Stability:

Polymer Stability General: The polymer is stable in air up to 115° and up to 349° in nitrogen
Thermal Stability General: Thermogravimetric analysis in air at a heating rate of 10° min^{-1} shows that Amodel PPA are stable to 349°. Rapid thermal degradation occurs above 340°
Upper Use Temperature:

No.	Value	Note
1	185°C	5000h, 33% glass fibre reinforced, ASTM D3045
2	35°C	unfilled
3	115°C	tensile impact
4	165°C	20000h, 33% glass fibre reinforced, ASTM D3045

Flammability: Glow wire test: Polyphthalamide (PPA) passes 30 second contact test at 960°. Flammability rating HB (0.8 mm, 40 mm min^{-1}, 33% glass fibre, UL94); V0 (1.6 mm, 50% glass fibre, UL94)

Environmental Stress: Weight loss after 30 days immersion at 23°; 0.8% in 10% NH$_4$OH, 2.1% in 10% aq. NaCl, 4.8% in 50% ZnCl$_2$, 1.8% in 36% H$_2$SO$_4$, 1.6% in 10% NaOH, 1.4% in 5% NaOCl soln.. In Arc Model x-w weatherometer test (alternate wet/dry cycles), PPA losses 50% tensile strength after 6000h; Uv modified grades lost <5% after 6000h
Chemical Stability: Commercial grades contain Uv stabiliser. Chemically resistant to aliphatic and aromatic hydrocarbons, oils, greases, chlorinated hydrocarbons, ketones, alkalis. Acceptable resistance to MeOH and strong acids and to ethylene glycol soln. Attacked by phenols. Dissolves in H$_2$SO$_4$
Hydrolytic Stability: Resistant to steam at 132°. Measured on tensile bars in 30 min. cycles from room temp. to 132°, about 30% loss of strength after 50 cycles. When tensile bars subject to exposure of water containing 30 ppm chlorine at 82°, effect measured as weight change. 3% loss of weight occurs after 90 days. 3.4% weight loss after 30 days in deionised water at 93°

Applications/Commercial Products:
Processing & Manufacturing Routes: Minimum mould temp. of 135–150° is required. PPA is hydroscopic so must be pre-dried
Mould Shrinkage (%):

No.	Value	Note
1	0.4–0.8%	33% glass fibre, ASTM D955
2	1.3–1.8%	impact modified, ASTM D955
3	1.1–1.5%	unfilled, ASTM D955

Applications: Used in automobile under the bonnet applications; fuel component systems. Also used in electronic and electrical systems for connectors, switch components and housings

Trade name	Details	Supplier
Amodel		Amoco
Lubricomp UCL	carbon fibre/PTFE filled grades	LNP Engineering
Permastat	4000/4005	RTP Company
RTP	various grades	RTP Company
Stat-Kon		LNP Engineering
Thermocomp UF-1000	includes filled products series	LNP Engineering
Verton UF-700-10	structural polyphthalamide	LNP Engineering

Bibliographic References
[1] Naitove, M.H., *Plast. Technol.*, 1991, **37**, 45, (thermal props)
[2] Wood, A.S., *Mod. Plast.*, 1991, **68**, 24
[3] *Mod. Plast.*, 1995, **72 (9)**, 156, (mech props)
[4] *Amodel Polyphthalamide*, Amoco Corp., (technical datasheet)
[5] *Structural phthalamide UF 700-10*, LNP Engineering Plastics, (technical datasheet)

Poly(piperazine sebacamide) P-462

Synonyms: *Poly(piperazine-co-sebacic acid). Poly[1,4-piperazine-diyl(1,10-dioxo-1,10-decanediyl)]*
Monomers: Piperazine, Decanedioic acid
Material class: Thermoplastic
Polymer Type: polyamide

CAS Number:

CAS Reg. No.
26967-92-0

Molecular Formula: (C$_{14}$H$_{24}$N$_2$O$_2$)$_n$
Fragments: C$_{14}$H$_{24}$N$_2$O$_2$

Volumetric & Calorimetric Properties:
Latent Heat Crystallization: ΔH_m 26.0–26.4 kJ mol^{-1} [1,2]
Melting Temperature:

No.	Value	Note
1	268–300°C	[1]
2	180°C	[2]

Glass-Transition Temperature:

No.	Value
1	115°C

Bibliographic References
[1] Van Krevelen, D.W., *Properties of Polymers: Their Correlation with Chemical Structure*, 3rd edn., Elsevier, 1990
[2] Miller, R.L., *Polym. Handb.*, 4th edn., (eds. J. Brandrup, E.H. Immergut and E.A. Grulke), John Wiley and Sons, 1999, **VI**, 39, (mp)

Poly[2,2-propane bis[4-(2-chlorophenyl)] carbonate] P-463

Synonyms: *Poly[oxycarbonyloxy(2-chloro-1,4-phenylene)(1-methylethylidene)(3-chloro-1,4-phenylene)]. Poly(4,4'-dihydroxy-3,3'-dichlorodiphenyl-2,2-propane-co-carbonic dichloride)*
Monomers: 3,3'-Dichloro-4,4'-dihydroxydiphenyl-2,2-propane, Carbonic dichloride
Material class: Thermoplastic
Polymer Type: polycarbonates (miscellaneous)
CAS Number:

CAS Reg. No.
31884-94-3

Molecular Formula: (C$_{16}$H$_{12}$Cl$_2$O$_3$)$_n$
Fragments: C$_{16}$H$_{12}$Cl$_2$O$_3$

Volumetric & Calorimetric Properties:
Density:

No.	Value	Note
1	d 1.32 g/cm^3	[1]
2	d 1.22 g/cm^3	20°, amorph. [2]
3	d 1.27 g/cm^3	cryst. [2]

Melting Temperature:

No.	Value	Note
1	190–210°C	[1]

| 2 | 192–214°C | amorph. [2] |
| 3 | 224–230°C | cryst. [2] |

Glass-Transition Temperature:

No.	Value	Note
1	146°C	[3]
2	142°C	[4]

Mechanical Properties:

Mechanical Properties General: Tensile stress-strain curve shows a viscoelastic region up to the yield point, followed by a stress drop and then a short necking region (at approx. constant stress) until the stress increases to the break point. [5] Tensile strength decreases with temp., but the polymer is still relatively strong to approx. 35-40% of original value at 180° [5]

Tensile (Young's) Modulus:

No.	Value	Note
1	2650 MPa	27000 kg cm^{-2}, amorph. [5]
2	2750 MPa	28000 kg cm^{-2}, cryst. [5]

Tensile Strength Break:

No.	Value	Note	Elongation
1	97.6 MPa	995 kg cm^{-2}	19%
2	94 MPa	9.6 kg mm^{-2} [5]	34%
3	100 MPa	10.2 kg mm^{-2}, cryst. [5]	36%

Tensile Strength Yield:

No.	Value	Note
1	90.2 MPa	9.2 kg mm^{-2} [5]

Complex Moduli: Variation of power dissipation factor (tan δ) with temp. has been reported

Electrical Properties:

Electrical Properties General: Dielectric constant and power dissipation factor increase with increasing temp. [1]

Dielectric Permittivity (Constant):

No.	Value	Frequency	Note
1	3.3	1 kHz	25° [1]
2	3.6	1 kHz	100° [1]

Dissipation (Power) Factor:

No.	Value	Frequency	Note
1	0.00115	1 kHz	25° [1]
2	0.0016	1 kHz	100° [1]

Optical Properties:

Refractive Index:

No.	Value	Note
1	1.59	25° [1]

Transmission and Spectra: Ir spectral data [6] and absorbance [7] have been reported

Applications/Commercial Products:

Processing & Manufacturing Routes: Synth. by two routes; phosgenation or ester interchange. Phosgenation is accomplished by interfacial polycondensation (most widely used) involving base-catalysed reaction of phosgene with an aromatic dihydroxy compound in CH_2Cl_2, with a catalyst to give high MW products [1,6,7], or by soln. polymerisation in pyridine [6] Ester exchange involves melt (150-300°) polymerisation of the aromatic dihydroxy compound with excess diphenyl carbonate in the absence of air. Reaction rate may be increased by use of alkaline catalysts or a medium-high vacuum in the final stages [1,6]

Bibliographic References

[1] Schnell, H., *Ind. Eng. Chem.*, 1959, **51**, 157, (props.)
[2] Kozlov, P.V. and Perepelkin, A.N., *Polym. Sci. USSR (Engl. Transl.)*, 1967, **9**, 414, (density, T_m)
[3] Yee, A.F. and Smith, S.A., *Macromolecules*, 1981, **14**, 54, (T_g, complex moduli)
[4] Perepelkin, A.N. and Kozlov, P.V., *Polym. Sci. USSR (Engl. Transl.)*, 1966, **8**, 57, (T_g)
[5] Perepelkin, A.N. and Kozlov, P.V., *Polym. Sci. USSR (Engl. Transl.)*, 1968, **10**, 14, (tensile props.)
[6] Schnell, H., *Polym. Rev.*, 1964, **9**, 99, (ir, rev)
[7] Tagle, L.H., Diaz, F.R. and Fuenzalida, R., *J. Macromol. Sci., Pure Appl. Chem.*, 1994, **31**, 283, (ir)

Poly[2,2-propane bis[4-(2,6-dibromophenyl)]carbonate]

P-464

Synonyms: *Poly[oxycarbonyloxy(2,6-dibromo-1,4-phenylene)(1-methylethylidene)(3,5-dibromo-1,4-phenylene)]. Poly(4,4'-dihydroxy-3,3',5,5'-tetrabromodiphenyl-2,2-propane-cocarbonic dichloride).* Tetrabromobisphenol A polycarbonate

Related Polymers: Copolymers of Tetrabromobisphenol A and Bisphenol A
Monomers: Tetrabromobisphenol A, Carbonic dichloride
Material class: Thermoplastic
Polymer Type: polycarbonates (miscellaneous)
CAS Number:

CAS Reg. No.
28774-93-8
50641-20-8

Molecular Formula: $(C_{16}H_{10}Br_4O_3)_n$
Fragments: $C_{16}H_{10}Br_4O_3$

Volumetric & Calorimetric Properties:

Density:

No.	Value	Note
1	d 1.91 g/cm^3	[5]
2	d 1.25 g/cm^3	20°, amorph. [6]
3	d 1.27 g/cm^3	cryst. [6]
4	d 1.953 g/cm^3	[3,7]

Volumetric Properties General: Volume expansion coefficient 0.000329 cm^3 (g °C)$^{-1}$ (melt) [3]; 63.6 × 10^{-6} cm^3 (g °C)$^{-1}$ (glass) [3]

Melting Temperature:

No.	Value	Note
1	240–260°C	[5]
2	238–256°C	amorph. [6]
3	263–267°C	cryst. [6]

Glass-Transition Temperature:

No.	Value	Note
1	263°C	[3,7]
2	248°C	[1]

Deflection Temperature:

No.	Value	Note
1	243°C	1.8 MPa, 264 psi, ASTM D1708-59T [10]

Surface Properties & Solubility:
Solubility Properties General: Forms immiscible blends with Bisphenol A polycarbonate; tetramethylpolycarbonate, polymethyl methacrylate and polystyrene [1]
Cohesive Energy Density Solubility Parameters: Solubility parameter 21 $J^{1/2}$ $cm^{-3/2}$ (calc.) [2]; 20.7 $J^{1/2}$ $cm^{-3/2}$ (10.1 $cal^{1/2}$ $cm^{-3/2}$) (calc.) [3]
Solvents/Non-solvents: Sol. (room temp.) 1-methyl-2-pyrrolidinone, methyl phenyl ketone, cyclohexanone, bis(2-methoxyethyl)ether, triethyleneglycol dimethyl ether, tetraethyleneglycol dimethyl ether, CH_2Cl_2 [4]; insol. hydrocarbons, alcohols, diethylene glycol, triethylene glycol, 2-ethoxyethanol, H_2O. [4] A lower critical soln. temp. is found for bis(2-ethoxyethyl) ether at 25° for polymer concns. of less than 10% wt [4]

Transport Properties:
Gas Permeability:

No.	Gas	Value	Note
1	CO_2	0.208 cm^3 mm/(m^2 day atm)	3.17×10^{-13} cm^2 $(s\ Pa)^{-1}$, 35°, 20 atm [3,7]
2	CH_4	0.0062 cm^3 mm/(m^2 day atm)	9.45×10^{-15} cm^2 $(s\ Pa)^{-1}$, 35°, 20 atm [3,7]
3	O_2	0.0669 cm^3 mm/(m^2 day atm)	1.02×10^{-13} cm^2 $(s\ Pa)^{-1}$, 35°, 1-2 atm [3,7,8]
4	N_2	0.0089 cm^3 mm/(m^2 day atm)	1.37×10^{-14} cm^2 $(s\ Pa)^{-1}$, 35°, 1-2 atm [3,7,8]
5	He	0.867 cm^3 mm/(m^2 day atm)	1.32×10^{-12} cm^2 $(s\ Pa)^{-1}$, 35°, 20 atm [3,7]

Mechanical Properties:
Mechanical Properties General: Tensile stress-strain curve shows a viscoelastic region up to the yield point followed by an increase in stress to the break point; little or no neck formation is observed. The tensile strength (yield) is high compared to that of Bisphenol A polycarbonate [9]

Tensile (Young's) Modulus:

No.	Value	Note
1	3140 MPa	32000 kg cm^{-2}, amorph. [9]
2	3240 MPa	33000 kg cm^{-2}, cryst. [9]
3	3200 MPa	470 kpsi, ASTM D1708-59T [10]

Tensile Strength Break:

No.	Value	Note	Elongation
1	109 MPa	1112 kg cm^{-2} [5]	8%
2	107 MPa	10.9 kg mm^{-2}, amorph. [9]	10%
3	110 MPa	11.2 kg mm^{-2}, cryst. [9]	8%
4	108 MPa	15.7 kpsi, ASTM D1708-59T [10]	7%

Tensile Strength Yield:

No.	Value	Note
1	98.1 MPa	10 kg mm^{-2} [9]
2	109 MPa	15.8 kpsi, ASTM D1708-59T [10]

Electrical Properties:
Electrical Properties General: Dielectric constant varies only with temp. at T_g. Dissipation factor increases with increasing temp. [5]
Dielectric Permittivity (Constant):

No.	Value	Frequency
1	2.7	1 kHz, 25–100° [5]

Dissipation (Power) Factor:

No.	Value	Frequency	Note
1	0.002	1 kHz	25° [5]
2	0.0038	1 kHz	100° [5]

Optical Properties:
Refractive Index:

No.	Value	Note
1	1.6147	25° [6]
2	1.615	calc. [1]

Applications/Commercial Products:
Processing & Manufacturing Routes: Synth. by two routes; phosgenation or ester interchange. Phosgenation is accomplished by interfacial polycondensation (most widely used) involving base-catalysed reaction of phosgene with an aromatic dihydroxy compound in CH_2Cl_2, with a catalyst to give high MW products [5,11], or by soln. polymerisation in pyridine [11] Ester exchange involves melt (150-300°) polymerisation of the aromatic dihydroxy compound with excess diphenyl carbonate in the absence of air. Reaction rate may be increased by use of alkaline catalysts or a medium-high vacuum in the final stages [5,11]
Applications: Flame retardant applications

Trade name	Supplier
Lexan DL 616	General Electric Plastics

Bibliographic References
[1] Kim, C.K. and Paul, D.R., *Macromolecules*, 1992, **25**, 3097, (miscibility, T_g, refractive index)
[2] Sommer, K., Batoulis, J., Jilge, W., Morbitzer, L., et al, *Adv. Mater.*, 1991, **3**, 590, (solubility parameter)
[3] Muruganandam, N., Koros, W.J. and Paul, D.R., *J. Polym. Sci., Polym. Phys. Ed.*, 1987, **25**, 1999, (props.)
[4] Beck, H.N., *J. Appl. Polym. Sci.*, 1993, **48**, 13, 21, (solvents)
[5] Schnell, H., *Ind. Eng. Chem.*, 1959, **51**, 157, (props.)
[6] Kozlov, P.V. and Perepelkin, A.N., *Polym. Sci. USSR (Engl. Transl.)*, 1967, **9**, 414, (density, T_m)

[7] Muruganandam, N. and Paul, D.R., *J. Membr. Sci.*, 1987, **34**, 185, (density, T_g, gas permeability)
[8] Koros, W.J., Coleman, M.R. and Walker, D.R.B., *Annu. Rev. Mater. Sci.*, 1992, **22**, 47, (gas permeability)
[9] Perepelkin, A.N. and Kozlov, P.V., *Polym. Sci. USSR (Engl. Transl.)*, 1968, **10**, 14, (tensile props.)
[10] Jackson, W.J. and Caldwell, J.R., *J. Appl. Polym. Sci.*, 1967, **11**, 227, (heat distortion temp.)
[11] Schnell, H., *Polym. Rev.*, 1964, **9**, 99, (synth., rev)

Poly[(2,2-propane bis[4-(2,6-dibromophenyl)]carbonate]-*co*-(2,2-propane bis(4-phenyl)carbonate)]   P-465

Synonyms: *Bisphenol A-Tetrabromobisphenol A copolycarbonates. Poly[oxycarbonyloxy(2,6-dibromo-1,4-phenylene)(1-methylethylidene)(3,5-dibromo-1,4-phenylene)oxycarbonyloxy-1,4-phenylene(1-methylethylidene)-1,4-phenylene]. Poly(4,4'-(1-methylethylidene)bis(2,6-dibromophenol)co-4,4'-(1-methylethylidene)bisphenol-co-carbonic acid)*
Monomers: Carbonic acid
CAS Number:

CAS Reg. No.	Note
61361-57-7	
30583-65-4	source: carbonic acid
32844-27-2	source: carbonic dichloride

Volumetric & Calorimetric Properties:
Density:

No.	Value	Note
1	d 1.24 g/cm^3	Novarex 7025NB [2]

Thermal Expansion Coefficient:

No.	Value	Note	Type
1	5.6×10^{-5} K^{-1}	Novarex 7025NB	L

Glass-Transition Temperature:

No.	Value	Note
1	185°C	50 mol. 50 % tetrabromobisphenol A MW 28263 [1]
2	207°C	50 mol. 50 % tetrabromobisphenol A MW 44568 [1]
3	209–215°C	50 5 polycarbonate, BuT4 block copolymer

Deflection Temperature:

No.	Value
1	142°C

Surface Properties & Solubility:
Solvents/Non-solvents: Block copolymer Sol. CH$_2$Cl$_2$ [1,3] Insol. hot H$_2$O [1,3]

Transport Properties:
Water Absorption:

No.	Value	Note
1	0.14 %	Novarex 7025NB [2]

Mechanical Properties:
Tensile Strength Break:

No.	Value	Note	Elongation
1	72 MPa	730 Kg cm^{-2} Novarex 7025NB [2]	90–110%

Flexural Strength at Break:

No.	Value	Note
1	94 MPa	960 Kg cm^{-2} Novarex 7025NB [2]

Complex Moduli: The variation of tan δ with temp. (-150 to 180°, 1Hz) for B4T4, block copolymer has been reported [1]
Hardness: Rockwell M85 (Novarex 702NB) [2]

Electrical Properties:
Dielectric Permittivity (Constant):

No.	Value	Frequency	Note
1	2.83	1 KHz	(Novarex 7025NB [2]

Dielectric Strength:

No.	Value	Note
1	15 kV.mm^{-1}	1/8 Novarex 7025NB [2]

Optical Properties:
Refractive Index:

No.	Value	Note
1	1.585	Novarex 7025NB [2]

Transmission and Spectra: Transmittance 90.5% (Novarex 7025NB) [2] Data for 50 mol. % TBBPA copolymer has been reported [1]
Volume Properties/Surface Properties: Haze 1.0% (Novarex 7025NB) [2]

Polymer Stability:
Thermal Stability General: Heating the block copolymer at 300° for 30 min results in some loss in MW decrease from 41205 to 36274 (sol. polymerisation method [3] decrease from 44899 to 41027 (interfacial method) [3] some discoloration is found in both cases [3]
Decomposition Details: Block copolymer decomposition temp. Td 454° N2, 50% polycarbonate sol. polymerisation synthesis, total MW 41205 [3] 446° (N2 50 % polycarbonate interfacial polymerisation synthesis, total MW 44899 [3]
Flammability: The copolymer is self-extinguishing (Novarex 7025NB) [2] Flammability rating V0 (1/16 in, Novarex 7025NB) [2]
Chemical Stability: Block decomposition temp. Td 450° (air, 50% polycarbonate soln. polymerisation synthesis total MW 41205) [3] 440 ° (air, 50% polycarbonate, interfacial polymerisation synthesis, total MW 44889 [3]

Applications/Commercial Products:
Mould Shrinkage (%):

No.	Value
1	0.5–0.7%

Trade name	Supplier
Lexan RL 1624	General Electric Plastics
Novarex 7025 NB	Mitsubishi Chemical

Poly[2,2-propane bis[4-(2,6-dichlorophenyl)] carbonate]

P-466

Synonyms: *Poly[oxycarbonyloxy(2,6-dichloro-1,4-phenylene)(1-methylethylidene)(3,5-dichloro-1,4-phenylene)]. Poly(4,4'-dihydroxy-3,3',5,5'-tetrachlorodiphenyl-2,2-propane-co-carbonic dichloride). Tetrachloro Bisphenol A polycarbonate*
Monomers: 3,3',5,5'-Tetrachloro-4,4'-dihydroxydiphenyl-2,2-propane, Carbonic dichloride
Material class: Thermoplastic
Polymer Type: polycarbonates (miscellaneous)
CAS Number:

CAS Reg. No.
26913-25-7
50923-69-8

Molecular Formula: $(C_{16}H_{10}Cl_4O_3)_n$
Fragments: $C_{16}H_{10}Cl_4O_3$

Volumetric & Calorimetric Properties:
Density:

No.	Value	Note
1	d 1.42 g/cm^3	25° [4]
2	d 1.39 g/cm^3	20°, amorph. [5]
3	d 1.41 g/cm^3	cryst. [5]
4	d 1.415 g/cm^3	[2,6]

Volumetric Properties General: Volume expansion coefficient 0.000533 cm^3 (g °C)$^{-1}$ (melt) [2]; 0.00014 cm^3 (g °C)$^{-1}$ (glass) [2]
Equation of State: Mark-Houwink parameters have been reported [10]
Melting Temperature:

No.	Value	Note
1	250–260°C	[4]
2	253–264°C	amorph. [5]
3	270–275°C	cryst. [5]

Glass-Transition Temperature:

No.	Value	Note
1	230°C	[2,6]
2	231°C	[7]
3	217°C	[8]
4	230°C	[1]

Deflection Temperature:

No.	Value	Note
1	225°C	1.8 MPa, 264 psi, ASTM D1708 [9]

Surface Properties & Solubility:
Solubility Properties General: Forms immiscible blends with Bisphenol A polycarbonate, tetramethylpolycarbonate, polymethyl methacrylate and polystyrene [1]
Cohesive Energy Density Solubility Parameters: Solubility parameter 20.4 J$^{1/2}$ cm$^{-3/2}$ (9.97 cal$^{1/2}$ cm$^{-3/2}$, calc) [2]
Solvents/Non-solvents: Sol. MeCl, EtOAc, cyclohexanone, THF, C_6H_6, pyridine, DMF; insol. *m*-cresol swollen by Me_2CO

Transport Properties:
Gas Permeability:

No.	Gas	Value	Note
1	CO_2	0.328 cm^3 mm/(m^2 day atm)	5 × 10^{-13} cm^2 (s Pa)$^{-1}$, 35°, 20 atm [2,6]
2	CH_4	0.011 cm^3 mm/(m^2 day atm)	1.68 × 10^{-14} cm^2 (s Pa)$^{-1}$, 35°, 20 atm [2,6]
3	O_2	0.11 cm^3 mm/(m^2 day atm)	1.71 × 10^{-13} cm^2 (s Pa)$^{-1}$, 35°, [2,6,11]
4	N_2	0.0177 cm^3 mm/(m^2 day atm)	2.7 × 10^{-14} cm^2 (s Pa)$^{-1}$, 35° [2,6,11]
5	He	1.35 cm^3 mm/(m^2 day atm)	2.6 × 10^{-12} cm^2 (s Pa)$^{-1}$, 35°, 20 atm [2,6]

Mechanical Properties:
Mechanical Properties General: Tensile stress-strain curve shows a viscoelastic region up to the yield point, followed by an increase in stress to the break point; little or no neck formation is observed. Tensile strength (yield) is relatively high compared to Bisphenol A polycarbonate [12]

Tensile (Young's) Modulus:

No.	Value	Note
1	3140 MPa	32000 kg cm^{-2}, amorph. [12]
2	3240 MPa	33000 kg cm^{-2}, cryst. [12]
3	2800 MPa	410 kpsi, ASTM D1708-59T [9]

Tensile Strength Break:

No.	Value	Note	Elongation
1	113.2 MPa	1154 kg cm^{-2} [4]	10%
2	113 MPa	11.5 kg mm^{-2}, amorph. [12]	12%
3	115 MPa	11.7 kg mm^{-2}, cryst. [12]	10%
4	77.9 MPa	11.3 kpsi, ASTM D1708-59T [9]	27%

Tensile Strength Yield:

No.	Value	Note
1	103 MPa	10.5 kg mm^{-2} [12]
2	92.4 MPa	13.4 kpsi, ASTM D1708-59T [9]

Complex Moduli: The variation of power dissipation factor with temp. (-150–240°) has been reported [13]
Impact Strength: The impact strength is greatly reduced compared to Bisphenol A polycarbonate. Dart impact strength 24.3–29.7 mJ (0.4 mm thick) [14]

Electrical Properties:
Electrical Properties General: Dielectric constant varies with temp. only at T_g; power dissipation factor increases with increasing temp.

Dielectric Permittivity (Constant):

No.	Value	Frequency	Note
1	3	1 kHz	25–100° [4]

Dissipation (Power) Factor:

No.	Value	Frequency	Note
1	0.001	1 kHz	25° [4]
2	0.0037	1 kHz	100° [4]

Optical Properties:
Refractive Index:

No.	Value	Note
1	1.6056	25° [4]
2	1.606	calc. [1]

Transmission and Spectra: Ir spectral data have been reported [3]

Polymer Stability:
Thermal Stability General: Possesses a high degree of thermal stability
Decomposition Details: Principal degradation mechanism involves chain scission; HCl, CO_2, CO and CH_4 are evolved. Residual polymer is insol. even at early stages of degradation. [15] Activation energy for rate of CO_2 evolution 141 kJ mol^{-1} (33.6 k cal mol^{-1}, vacuum, 360–420°) [15]
Chemical Stability: Resistant to MeOH [4]
Hydrolytic Stability: Very resistant to basic hydrolysis; withstands 20% NaOH at room temp. [4]; hydrolysis rate at 100° in 10% NaOH is 3% that of Bisphenol A polycarbonate [16]

Applications/Commercial Products:
Processing & Manufacturing Routes: Synth. by two routes; phosgenation or ester interchange. Phosgenation is accomplished by interfacial polycondensation (most widely used) involving base-catalysed reaction of phosgene with an aromatic dihydroxy compound in CH_2Cl_2, with a catalyst to give high MW products [3,4], or by soln. polymerisation in pyridine [3] Ester exchange involves melt (150-300°) polymerisation of the aromatic dihydroxy compound with excess diphenyl carbonate in the absence of air. Reaction rate may be increased by use of alkaline catalysts or a medium-high vacuum in the final stages [3,4]

Trade name	Supplier
Lexan DL 616	General Electric Plastics

Bibliographic References
[1] Kim, C.K. and Paul, D.R., *Macromolecules*, 1992, **25**, 3097, (miscibility, T_g, refractive index)
[2] Muruganandam, N., Koros, W.J. and Paul, D.R., *J. Polym. Sci., Polym. Phys. Ed.*, 1987, **25**, 1999, (solubility parameter, density, T_g, thermal expansion coefficient, gas permeability)
[3] Schnell, H., *Polym. Rev.*, 1964, **9**, 99, (solvents, ir, rev)
[4] Schnell, H., *Ind. Eng. Chem.*, 1959, **51**, 157, (props.)
[5] Kozlov, P.V. and Perepelkin, A.N., *Polym. Sci. USSR (Engl. Transl.)*, 1967, **9**, 414, (density, T_g)
[6] Muruganandam, N. and Paul, D.R., *J. Membr. Sci.*, 1987, **34**, 185, (density, T_g, gas permeability)
[7] Garfield, L.J., *J. Polym. Sci., Polym. Symp.*, 1970, **30**, 551, (T_g)
[8] Perepelkin, A.N. and Kozlov, P.V., *Polym. Sci. USSR (Engl. Transl.)*, 1966, **8**, 57, (T_g)
[9] Jackson, W.J. and Caldwell, J.R., *J. Appl. Polym. Sci.*, 1967, **11**, 227, (heat distortion temp., tensile props.)
[10] Horbach, A., Muller, H. and Bottenbruch, L., *Makromol. Chem.*, 1981, **82**, 2873, (dilute soln.)
[11] Koros, W.J., Coleman, M.R. and Walker, D.R.B., *Annu. Rev. Mater. Sci.*, 1992, **22**, 47, (gas permeability)
[12] Perepelkin, A.N. and Kozlov, P.V., *Polym. Sci. USSR (Engl. Transl.)*, 1968, **10**, 14, (tensile props.)
[13] Reding, F.P., Faucher, J.A. and Whitman, R.D., *J. Polym. Sci.*, 1961, **54**, 56, (complex moduli)
[14] Steger, T.R., Schaefer, J., Stejskal, E.O. and McKay, R.A., *Macromolecules*, 1980, **13**, 1127, (impact strength)
[15] Selkine, Y., Ikeda, K. and Taketani, H., *Kogyo Kagaku Zasshi*, 1969, **72**, 791, (thermal stability)
[16] Boranowska, Z. and Wielgosz, Z., *Polimery (Warsaw)*, 1970, **15**, 12, (chemical stability)

Poly[2,2-propane bis[4-[2,6-dimethylphenyl]] carbonate]

Synonyms: Poly[oxycarbonyloxy(2,6-dimethyl-1,4-phenylene)(1-methylethylidene)(3,5-dimethyl-1,4-phenylene)]. Poly(4,4'-dihydroxy-3,3',5,5'-tetramethyldiphenyl-2,2-propane-co-carbonic dichloride). Tetramethyl Bisphenol A polycarbonate
Related Polymers: Blends with high impact polystyrene
Monomers: 4,4'-Dihydroxy-3,3',5,5'-tetramethyldiphenyl-2,2-propane, Carbonic dichloride
Material class: Thermoplastic
Polymer Type: polycarbonates (miscellaneous)
CAS Number:

CAS Reg. No.
38797-88-5
39399-36-5

Molecular Formula: $(C_{20}H_{22}O_3)_n$
Fragments: $C_{20}H_{22}O_3$

Volumetric & Calorimetric Properties:
Density:

No.	Value	Note
1	d 1.083 g/cm^3	[14,18,19,20]
2	d 1.08 g/cm^3	DIN 53479 [21]
3	d 1.0824 g/cm^3	[9]
4	d^{30} 1.083 g/cm^3	[11]
5	d^{23} 1.083 g/cm^3	[22]
6	d^{30} 1.084 g/cm^3	[16,23]
7	d^{25} 1.09 g/cm^3	[24]

Thermal Expansion Coefficient:

No.	Value	Note	Type
1	0.000187 K^{-1}	glass [14]	V
2	0.00021 K^{-1}	glass [25]	V
3	0.000128 K^{-1}	200–230°, glass [22]	V
4	0.00013 K^{-1}	glass, 23–47° [26]	V
5	0.000659 K^{-1}	melt [14]	V
6	0.00072 K^{-1}	melt [25]	V

Equation of State: Flory/Van der Waals equation of state information for the polymer melt (0.1–200 MPa) has been reported [27]. Sanchez-Balazs [28], Tait [7] and Sanchez-Lacombe [7] equation of state information have been reported [29]

Latent Heat Crystallization: Enthalpy of fusion 14–20 J g^{-1} (3.4–4.7 cal g^{-1}) [9]

Melting Temperature:

No.	Value	Note
1	280°C	[9]

Glass-Transition Temperature:

No.	Value	Note
1	191°C	[37]
2	203°C	[13]
3	181.8°C	[42]
4	193°C	[14,18]
5	202°C	[19,20]
6	207°C	[21]
7	200°C	[17]
8	193.5°C	[9]
9	192°C	[13]
10	197–199°C	[1,22,25,29]
11	201°C	[30]

Vicat Softening Point:

No.	Value	Note
1	197°C	rate B [8]
2	198°C	rate B [31]

Transition Temperature:

No.	Value	Note	Type
1	45°C	[32]	γ relaxation temp.

Surface Properties & Solubility:

Solubility Properties General: Forms miscible blends in all proportions with Bisphenol A polycarbonate [1,2,3,4], from room temp. -280°; solvent induced phase separation may also occur [3]. Also miscible in all proportions with polystyrene [1,4,5,6,7,8], phase separation occurs on heating (lower critical soln. temp. behaviour) [1,6,7,9,10]. Cloud points 241–255° (30 wt. % polystyrene) [6]; 249–254° (30 wt. % polystyrene) [7]; 237–244° (50 wt. % polystyrene) [6]; 242–246° (50 wt. % polystyrene) [7]; 250° (57% polystyrene) [10]; 257° (68% polystyrene/32% TMPC) [10]; 250–256° (70 wt. % polystyrene) [7]; 265–295° (70 wt. % polystyrene) [6]. Forms miscible blends at room temp. with styrene-acrylonitrile copolymers (18 wt. % acrylonitrile) [11]; styrene-maleic anhydride (10 wt. % maleic anhydride) [9]; styrene-methyl methacrylate (20–30 wt. % methylmethacrylate) [11]. Above these compositions immiscible blends are observed. Compatibility with copolymers of a given monomer ratio may be influenced by blend composition, preparation method and temp. [9,11]. Miscible with linear aliphatic polyesters having 5–10 aliphatic carbons per ester group in the repeat struct. With poly(butylene adipate) and poly(1,4-cyclohexane dimethylene succinate), partial miscibility is found; blends with poly(ethylene adipate) and poly(2,2-dimethyl-1,3-propylene succinate), are completely phase separated [12]. Tetramethylpolycarbonate forms immiscible blends with poly methyl methacrylate [1]

Cohesive Energy Density Solubility Parameters: Solubility parameter 18.5 J$^{1/2}$ cm$^{-3/2}$ (calc.) [13]; 19.2 J$^{1/2}$ cm$^{-3/2}$ (9.39 cal$^{1/2}$ cm$^{-3/2}$, calc.) [14]; 21.4 J$^{1/2}$ cm$^{-3/2}$ (10.46 cal$^{1/2}$ cm$^{-3/2}$, calc.) [15,16]. Cohesive energy density 352 J m^{-3} (calc.) [17]

Transport Properties:

Polymer Solutions Dilute: Mark-Houwink constants have been reported [21,33]

Permeability of Gases: The permeability of CO_2 decreases with pressure up to 300 psi (20 atm) [19,34,35,36] after which plasticisation occurs and permeability increases; the permeability-pressure curve shows hysteresis [35]. Conditioning at 550 psi (37 atm) for 7 days leads to an increase in permeability [35]. Temp. dependence [14,24,26] and gas selectivity [14,18,19,23,34] have been reported

Gas Permeability:

No.	Gas	Value	Note
1	CO_2	1610 cm^3 mm/(m^2 day atm)	1.84 × 10^{-12} cm^2 (s Pa)$^{-1}$, 35°, 1 atm [36]
2	CO_2	1426 cm^3 mm/(m^2 day atm)	1.63 × 10^{-12} cm^2 (s Pa)$^{-1}$, 35°, 5 atm [36]
3	CO_2	1225 cm^3 mm/(m^2 day atm)	1.4 × 10^{-12} cm^2 (s Pa)$^{-1}$, 35°, 10 atm [34,36]
4	CO_2	1138 cm^3 mm/(m^2 day atm)	1.3 × 10^{-12} cm^2 (s Pa)$^{-1}$, 35°, 20 atm [14,18,34,36]
5	O_2	367 cm^3 mm/(m^2 day atm)	4.2 × 10^{-13} cm^2 (s Pa)$^{-1}$, 35°, 2 atm [14,18,23,34]
6	O_2	262 cm^3 mm/(m^2 day atm)	3 × 10^{-13} cm^2 (s Pa)$^{-1}$, 25°, 1-2 atm [24]
7	N_2	46.7 cm^3 mm/(m^2 day atm)	5.34 × 10^{-14} cm^2 (s Pa)$^{-1}$, 25°, 1-2 atm [24]
8	N_2	71.8 cm^3 mm/(m^2 day atm)	8.2 × 10^{-14} cm^2 (s Pa)$^{-1}$, 35°, 1-2 atm [14,18,34]
9	Ar	107.7 cm^3 mm/(m^2 day atm)	1.23 × 10^{-13} cm^2 (s Pa)$^{-1}$, 25°, 1-2 atm [24]
10	Ar	148 cm^3 mm/(m^2 day atm)	1.7 × 10^{-13} cm^2 (s Pa)$^{-1}$, 35°, 2 atm [23]
11	H_2	3589 cm^3 mm/(m^2 day atm)	4.1 × 10^{-12} cm^2 (s Pa)$^{-1}$, 35°, 2 atm [23]
12	He	3064 cm^3 mm/(m^2 day atm)	3.5 × 10^{-12} cm^2 (s Pa)$^{-1}$, 35°, 2-20 atm [14,18,23,34]
13	CH_4	52.5 cm^3 mm/(m^2 day atm)	6 × 10^{-14} cm^2 (s Pa)$^{-1}$, 35°, 20 atm [14,18,19]
14	CH_4	66.3 cm^3 mm/(m^2 day atm)	7.58 × 10^{-14} cm^2 (s Pa)$^{-1}$, 35°, 10 atm [36]
15	CH_4	75.5 cm^3 mm/(m^2 day atm)	8.6 × 10^{-14} cm^2 (s Pa)$^{-1}$, 35°, 1 atm [36]
16	CH_4	59 cm^3 mm/(m^2 day atm)	6.8 × 10^{-14} cm^2 (s Pa)$^{-1}$, 35°, 10 atm, calc. [34]

Mechanical Properties:

Tensile (Young's) Modulus:

No.	Value	Note
1	2400 MPa	DIN 53457 [8,21]
2	2180 MPa	[17]

Tensile Strength Break:

No.	Value	Note	Elongation
1	62 MPa	DIN 53457 [21]	50%
2	62 MPa	DIN 53457 [8]	40%

Tensile Strength Yield:

No.	Value	Note
1	80 MPa	DIN 53457 [21]
2	93.8 MPa	[17]

Complex Moduli: The variation of tan δ with temp. [39] (-150–160°) and of G' and G'' with temp. [31] (-50–200°) and master curve for the complex shear modulus, G*, with frequency [30] has been reported

Impact Strength: Dart impact test. Impact energy 33–115 J m^{-1} (0.635–1.323 mm thick) [40]

Failure Properties General: Brittle at room temp. [13,40]. Critical crazing strain 1.59–1.95% (air, 23°, 20h) [17]

Izod Notch:

No.	Value	Notch	Note
1	3 J/m	Y	[8]
2	4 J/m	Y	DIN 53453 [21]

Electrical Properties:

Electrical Properties General: Dissipation power factor, tan δ, has been reported [4]

Complex Permittivity and Electroactive Polymers: Variation of ε'' with frequency [2,5,22] and temp. [4,22] has been reported

Optical Properties:

Refractive Index:

No.	Value	Note
1	1.546	20° [1,8,21]

Transmission and Spectra: Ir [21] and uv [41] spectral data have been reported

Polymer Stability:

Environmental Stress: Photolysed by light (254–365 nm). In a vacuum, photo-scission occurs, followed by decarboxylation and recombination to form cyclodienone-type compounds [41]. In the presence of oxygen, carbonate bond scission is followed by radical induced oxidation to form thermally stable hydroperoxides and secondary oxidation products [41]. These are further photo-oxidised to phenolic and quinone-methide groups. Only the extent of degradation varies with wavelength; at short wavelengths (254 nm) photo-oxidation occurs only on the surface (5–10 nm), at longer wavelengths (300–365 nm) photo-oxidation occurs deeper in the sample [41]. More stable than Bisphenol A polycarbonate [41]

Chemical Stability: Thermal oxidation is a two stage process. At temps. of 100-450°, thermo-oxidative pyrolysis occurs, with a predominance of endothermic scission reactions (leading to hydroperoxide formation) [41,42]. At higher temps., exothermic oxidation occurs [42]. Rate of decomposition temp. 444° (pyrolysis, heating rate 10° min^{-1}) [42]; 612° (oxidation, heating rate 10° min^{-1}) [42]. In N_2 or He only one thermal decomposition step is observed [42]

Hydrolytic Stability: Tetramethylpolycarbonate is highly resistant to hydrolysis [8,9], including boiling aq. NaOH [8]

Applications/Commercial Products:

Processing & Manufacturing Routes: Synth. by interfacial polycondensation. A large excess of phosgene is introduced into a suspension or soln. of tetramethyl Bisphenol A in aq. NaOH, 20–30° in CH_2Cl_2 at high pH and with high concentrations of tertiary amine catalyst [8,21]

Bibliographic References

[1] Kim, C.K. and Paul, D.R., *Macromolecules*, 1992, **25**, 3097, (compatibility, T$_g$, refractive index)
[2] Mansour, A.A. and Madbouly, S.A., *Polym. Int.*, 1995, **36**, 269, (compatibility, complex permittivity)
[3] Hellmann, E.H., Hellmann, G.P. and Rennie, A.R., *Colloid Polym. Sci.*, 1991, **269**, 343, (compatibility)
[4] Landry, C.J.T., Yang, H. and Machell, J.S., *Polymer*, 1991, **32**, 44, (compatibility, T$_g$, dissipation factor, complex permittivity)
[5] Mansour, A.A. and Madbouly, S.A., *Polym. Int.*, 1995, **37**, 267, (compatibility, complex permittivity)
[6] Guo, W. and Higgins, J.S., *Polymer*, 1990, **31**, 699, (compatibility)
[7] Kim, C.K. and Paul, D.R., *Polymer*, 1992, **33**, 1630, (compatibility, equation of state)
[8] Serini, V., Peters, H. and Morbitzer, L., *Polymer Blends Symposium 4*, Plastics and Rubber Inst., Chameleon Press, 1981, 9.1, (props, synth)
[9] Fernandes, A.C., Barlow, J.W. and Paul, D.R., *Polymer*, 1986, **27**, 1788, (props)
[10] Casper, R. and Morbitzer, L., *Angew. Makromol. Chem.*, 1977, **58/59**, 1, (compatibility)
[11] Kim, C.K. and Paul, D.R., *Polymer*, 1992, **33**, 2089, (compatibility, density)
[12] Fernandes, A.C., Barlow, J.W. and Paul, D.R., *Polymer*, 1986, **27**, 1799, (compatibility)
[13] Sommer, K., Batoulis, J., Jilge, W., Morbitzer, L. et al, *Adv. Mater.*, 1991, **3**, 590, (solubility parameter, T$_g$, mech failure)
[14] Muruganandam, N., Koros, W.J. and Paul, D.R., *J. Polym. Sci., Polym. Phys. Ed.*, 1987, **25**, 1999, (props)
[15] Kim, C.K. and Paul, D.R., *Polymer*, 1992, **33**, 4929, (solubility parameter)
[16] Kim, C.K. and Paul, D.R., *Polymer*, 1992, **33**, 4941, (solubility parameter, density)
[17] Kambour, R.P., *Polym. Commun.*, 1983, **24**, 292, (cohesive energy density, T$_g$, tensile props, mech failure)
[18] Muruganandam, N. and Paul, D.R., *J. Membr. Sci.*, 1987, **34**, 185, (density, T$_g$, gas permeability)
[19] Jordan, S.M. and Koros, W.J., *J. Membr. Sci.*, 1990, **51**, 233, (density, T$_g$, gas permeability)
[20] Fleming, G.K. and Koros, W.J., *J. Polym. Sci., Polym. Phys. Ed.*, 1990, **28**, 1137, (density, T$_g$)
[21] Serini, V., Freitag, D. and Vernaleken, H., *Angew. Makromol. Chem.*, 1976, **55**, 175, (props, synth)
[22] Katana, G., Kremer, F., Fischer, E.W. and Plaetschke, R., *Macromolecules*, 1993, **26**, 3075, (density, thermal expansion coefficient, T$_g$, complex permittivity)
[23] Kim, C.K., Aguilar-Vega, M. and Paul, D.R., *J. Polym. Sci., Polym. Phys. Ed.*, 1992, **30**, 1131, (density, gas permeability)
[24] Haraya, K. and Hwang, S.-T., *J. Membr. Sci.*, 1992, **71**, 13, (density, gas permeability)
[25] Kim, E., Kramer, E.J. and Osby, J.O., *Macromolecules*, 1995, **28**, 1979, (thermal expansion coefficient, T$_g$)
[26] Costello, L.M. and Koros, W.J., *J. Polym. Sci., Polym. Phys. Ed.*, 1994, **32**, 701, (thermal expansion coefficient, gas permeability)
[27] Brannock, G.R. and Sanchez, I.C., *Macromolecules*, 1993, **26**, 4970, (equation of state)
[28] Kim, E., Kramer, E.J., Osby, J.O. and Walsh, D.J., *J. Polym. Sci., Polym. Phys. Ed.*, 1995, **33**, 467, (equation of state)
[29] Floudas, G., Pakula, T., Stamm, M. and Fischer, E.W., *Macromolecules*, 1993, **26**, 1671, (T$_g$)
[30] Wisniewsky, C., Marin, G. and Monge, P., *Eur. Polym. J.*, 1984, **20**, 691, (T$_g$, complex moduli)
[31] Humme, G., Röhr, H. and Serini, V., *Angew. Makromol. Chem.*, 1977, **58/59**, 85, (Vicat temp, complex moduli)
[32] Freitag, D., Fengler, G. and Morbitzer, L., Angew. Chem., *Int. Ed. Engl.*, 1991, **30**, 1598, (γ relaxation)
[33] Horbach, A., Muller, H. and Bottenbruch, L., *Makromol. Chem.*, 1981, **82**, 2873, (dilute soln props)
[34] Hellums, M.W., Koros, W.J., Husk, G.R. and Paul, D.R., *J. Membr. Sci.*, 1989, **46**, 93, (gas permeability)
[35] Jordan, S.M., Fleming, G.K. and Koros, W.J., *J. Polym. Sci., Polym. Phys. Ed.*, 1990, **28**, 2305, (gas permeability)
[36] Moe, M.B., Koros, W.J. and Paul, D.R., *J. Polym. Sci., Polym. Phys. Ed.*, 1988, **26**, 1931, (gas permeability)
[37] Schmidhauser, J.C. and Longley, K.L., *J. Appl. Polym. Sci.*, 1990, **39**, 2083, (gas permeability, T$_g$)
[38] Koros, W.J., Coleman, M.R. and Walker, D.R.B., *Annu. Rev. Mater. Sci.*, 1992, **22**, 47, (gas diffusivity)
[39] Fischer, E.W., Hellmann, G.P., Spiess, H.W., Hörth, F.J., et al, Makromol. Chem., *Suppl.*, 1985, **12**, 189, (complex moduli)
[40] Fried, J.R., Zhang, C. and Liu, H.-C., *J. Polym. Sci., Polym. Lett. Ed.*, 1990, **28**, 7, (impact strength, mech failure)
[41] Rivaton, A. and Lemaire, J., *Polym. Degrad. Stab.*, 1988, **23**, 51, (uv, uv stability, thermal stability)
[42] Dobkowski, Z. and Rudnik, E., *J. Therm. Anal.*, 1992, **38**, 2211, (thermal stability, T$_g$)

Poly[2,2-propane bis[4-[2-methylphenyl]]carbonate]   P-468

Synonyms: *Poly[oxycarbonyloxy(2-methyl-1,4-phenylene)(1-methylethylidene)(3-methyl-1,4-phenylene)]. Poly(4,4'-dihydroxy-3,3'-dimethyldiphenyl-2,2-propane-co-carbonic dichloride)*
Monomers: 4,4'-Dihydroxy-3,3'-dimethyldiphenyl-2,2-propane, Carbonic dichloride
Material class: Thermoplastic
Polymer Type: polycarbonates (miscellaneous)
CAS Number:

CAS Reg. No.
26500-24-3

Molecular Formula: $(C_{18}H_{18}O_3)_n$
Fragments: $C_{18}H_{18}O_3$

Volumetric & Calorimetric Properties:
Density:

No.	Value	Note
1	d 1.22 g/cm^3	25° [3]
2	d 1.22 g/cm^3	20°, amorph. [4]
3	d 1.16 g/cm^3	[6]

Equation of State: Mark-Houwink parameters have been reported [1]
Melting Temperature:

No.	Value	Note
1	150–170°C	[3]
2	156–172°C	amorph. [4]

Glass-Transition Temperature:

No.	Value	Note
1	98°C	[5]
2	124°C	[6]
3	128°C	[7]
4	100°C	[8]
5	125°C	[1]

Surface Properties & Solubility:
Solubility Properties General: Forms miscible blends with Bisphenol A polycarbonate and tetramethylpolycarbonate; it forms immiscible blends with polymethyl methacrylate and polystyrene
Solvents/Non-solvents: Sol. MeCl, *m*-cresol, EtOAc, cyclohexanone, THF, C_6H_6, pyridine, DMF; swollen by Me_2CO [2]

Transport Properties:
Gas Permeability:

No.	Gas	Value	Note
1	O_2	0.015 cm^3 mm/(m^2 day atm)	2.3×10^{-4} cm^2 (s Pa)$^{-1}$, 25° [5]

Mechanical Properties:
Mechanical Properties General: Tensile stress-strain curve shows a viscoelastic region up to the yield point, followed by a stress drop and then necking behaviour (at approx. constant stress) until the stress increases to the break point [10]

Tensile (Young's) Modulus:

No.	Value	Note
1	1180 MPa	12000 kg cm^{-2}, amorph. [10]

Tensile Strength Break:

No.	Value	Note	Elongation
1	56 MPa	5.7 kg mm^{-2} [10]	116%

Tensile Strength Yield:

No.	Value	Note
1	38.2 MPa	3.9 kg mm^{-2} [10]

Complex Moduli: The variation of power dissipation (tan δ) with temp. (-150–240°) has been reported [11]

Electrical Properties:
Electrical Properties General: Dielectric constant and dissipation factor increase slightly with increasing temp. [3]
Dielectric Permittivity (Constant):

No.	Value	Frequency	Note
1	2.5	1 kHz	25° [3]
2	2.6	1 kHz	100° [3]

Dissipation (Power) Factor:

No.	Value	Frequency	Note
1	0.002	1 kHz	25° [3]
2	0.004	1 kHz	100° [3]

Optical Properties:
Refractive Index:

No.	Value	Note
1	1.5783	25° [3]
2	1.577	20° [7]
3	1.578	calc. [1]

Transmission and Spectra: Ir spectral data have been reported [2]

Applications/Commercial Products:
Processing & Manufacturing Routes: Synth. by two routes; phosgenation or ester interchange. Phosgenation is accomplished by interfacial polycondensation (most widely used) involving base-catalysed reaction of phosgene with an aromatic dihydroxy compound in CH_2Cl_2, with a catalyst to give high MW products [2,3], or by soln. polymerisation in pyridine [2] Ester exchange involves melt (150-300°) polymerisation of the aromatic dihydroxy compound with excess diphenyl carbonate in the absence of air. Reaction rate may be increased by use of alkaline catalysts or a medium-high vacuum in the final stages [2,3]

Bibliographic References:
[1] Kim, C.K. and Paul, D.R., *Macromolecules*, 1992, **25**, 3097, (miscibility, T_g, refractive index)
[2] Schnell, H., *Polym. Rev.*, 1964, **9**, 99, (solvents, ir, rev)

Poly[2,2-propanebis(4-phenyl)carbonate]-*block*-poly(dimethylsiloxane)

[3] Schnell, H., *Ind. Eng. Chem.*, 1959, **51**, 157, (props)
[4] Kozlov, P.V. and Perepelkin, A.N., *Polym. Sci. USSR (Engl. Transl.)*, 1967, **9**, 414, (density, T_m)
[5] Schmidhauser, J.C. and Longley, K.L., *J. Appl. Polym. Sci.*, 1990, **39**, 2083, (T_g, gas permeability)
[6] Cais, R.E., Nozomi, M., Kawai, M. and Miyake, A., *Macromolecules*, 1992, **18**, 4588, (T_g, density)
[7] Serini, V., Freitag, D. and Vernaleken, H., *Angew. Makromol. Chem.*, 1976, **55**, 175, (T_g, refractive index)
[8] Perepelkin, A.N. and Kozlov, P.V., *Polym. Sci. USSR (Engl. Transl.)*, 1966, **8**, 57, (T_g)
[9] Horbach, A., Muller, H. and Bottenbruch, L., *Makromol. Chem.*, 1981, **82**, 2873, (dilute solns.)
[10] Perepelkin, A.N. and Kozlov, P.V., *Polym. Sci. USSR (Engl. Transl.)*, 1968, **10**, 14, (tensile props.)
[11] Reding, F.P., Faucher, J.A. and Whitman, R.D., *J. Polym. Sci.*, 1961, **54**, 56, (complex moduli)

Poly[2,2-propanebis(4-phenyl)carbonate]-*block*-poly(dimethylsiloxane)   P-469

Synonyms: *Polycarbonate-block-poly(dimethylsiloxane). Polycarbonate-poly(dimethylsiloxane) block copolymer. Polycarbonate silicone copolymer. Silicone polycarbonate copolymer*
Related Polymers: Poly[2,2-propanebis(4-phenyl)carbonate], Poly(dimethylsiloxane)
Monomers: Bisphenol A, Carbonic dichloride, Dichlorodimethylsilane
Material class: Semithermoplastic, Copolymers, Synthetic Elastomers
Polymer Type: bisphenol A polycarbonate, dimethylsilicones
CAS Number:

CAS Reg. No.
64365-15-7

Molecular Formula: $[(C_{16}H_{14}O_3)_x.C_{15}H_{14}O.(C_2H_6OSi)_y]_n$
Fragments: $C_{16}H_{14}O_3$ $C_{15}H_{14}O$ C_2H_6OSi
General Information: The block copolymers are random alternating multi-block copolymers with polydisperse blocks. [8,9] They exhibit props. ranging from flexible elastomers to rigid engineering thermoplastics depending on composition [7,9]
Morphology: Block copolymers form separate microphases consisting of polycarbonate-rich domains in a polydimethylsiloxane-rich matrix. [7,8,10,11,12,13,14,15] Morphology is affected by the casting solvent (when present). [10,15,16] For example, casting from hexane-rich solvents causes aggregation of the polycarbonate blocks. [10] For polycarbonate content of 50% or less at about -73°, nearly all the polycarbonate is in rigid domains in a polydimethylsiloxane mobile matrix. [8] The rigid fraction decreases with increasing temp., especially as its T_g is approached. [8] At shorter polydimethylsiloxane block lengths (<20) increased mixing of the components is favoured [13]

Volumetric & Calorimetric Properties:
Density:

No.	Value	Note
1	d 1.147 g/cm^3	25% PDMS, 17.8 polycarbonate units [13]
2	d 1.104 g/cm^3	34% PDMS, 11 polycarbonate units [13]
3	d 1.097 g/cm^3	50% PDMS, 6 polycarbonate units [13]
4	d 1.052 g/cm^3	60% PDMS, 4.4 polycarbonate units [13]
5	d 1.04 g/cm^3	64% PDMS, 3.8 polycarbonate units [13]
6	d 1.05–1.09 g/cm^3	65% PDMS, total M_n 27800–96300 [16]

Thermal Expansion Coefficient:

No.	Value	Note	Type
1	9.56 × 10^{-5} K^{-1}	20°, 43% PDMS, MW 170000	L
2	0.000168 K^{-1}	20°, 50% PDMS, total MW 120000	L
3	0.000204 K^{-1}	20°, 55% PDMS, total MW 220000	L
4	0.000219 K^{-1}	20°, 65% PDMS, total MW 84000	L

Volumetric Properties General: Density varies with degree of polymerisation of the polydimethylsiloxane component. [13,16] The linear thermal expansion coefficients of the block copolymers lie between those of the homopolymers and increase with increasing polydimethylsiloxane content [10]
Melting Temperature:

No.	Value	Note
1	225°C	carbonate [1]

Glass-Transition Temperature:

No.	Value	Note
1	72°C	carbonate
2	-110°C	siloxane [1,2,7]
3	60–84°C	43% PDMS, total MW 170000 [10]
4	82–84°C	43% PDMS, total MW 170000 [15]
5	-85–-80°C	43% PDMS, total MW 170000 [15]
6	52°C	50% PDMS, total MW 120000 [15]
7	61–69°C	50% PDMS, total MW 120000 [15]
8	-90–-84°C	50% PDMS, total MW 120000 [15]

Transition Temperature General: T_g varies with degree of polymerisation and the PDMS content
Deflection Temperature:

No.	Value	Note
1	52°C	55% PDMS, total MW 220000 [10]
2	15.1°C	65% PDMS, total MW 84000 [10]
3	158°C	5% PDMS, end cap ratio 1.2 [9]
4	143°C	13% PDMS, end cap ratio 1.2 [9]
5	140°C	13% PDMS, end cap ratio 2 [9]
6	128°C	13% PDMS, end cap ratio 4 [9]
7	94°C	41% PDMS, end cap ratio 1.2 [9]
8	67°C	41% PDMS, end cap ratio 2 [9]
9	48°C	41% PDMS, end cap ratio 4 [9]
10	61°C	46% PDMS, end cap ratio 1.2 [9]
11	42°C	46% PDMS, end cap ratio 2 [9]
12	9°C	52% PDMS, end cap ratio 1.2 [9]
13	-19°C	57% PDMS, end cap ratio 1.2 [9]

Vicat Softening Point:

No.	Value	Note
1	70.5°C	43% PDMS, total MW 170000 [10]
2	54.4°C	50% PDMS, total MW 120000 [10]
3	100°C	55% PDMS, total MW 220000 [10]
4	66.3°C	65% PDMS, total MW 84000 [10]

Surface Properties & Solubility:

Cohesive Energy Density Solubility Parameters: δ 15.5 $J^{1/2}$ $cm^{-3/2}$ (7.5 $cal^{1/2}$ $cm^{-3/2}$, silicone), δ 20.5 $J^{1/2}$ $cm^{-3/2}$ (10 $cal^{1/2}$ $cm^{-3/2}$, polycarbonate) [7]

Solvents/Non-solvents: Sol. hexane/CH_2Cl_2 [10] CH_2Cl_2, [17] $CHCl_3$, THF, [18,19,20] chlorinated hydrocarbons, aromatic hydrocarbons. Mod. sol. ketones, esters, ethers. Insol. aliphatic hydrocarbons, alcohols, H_2O [2,5], MeOH [6,9,19]. Swollen by methylcyclohexane, isopropyl ether, decalin, silicone fluid, butanol, [21] heptane, hexane, cyclohexane [10,16]

Surface and Interfacial Properties General: Very high resistance to water. [4] Block copolymers show a distinct preference for polydimethylsiloxane at the surface over a wide range of compositions, [18,19,22] which is illustrated by the spreading of films on water. [23] Annealing of the polymer films at 180° for 24h *in vacuo* drives more polydimethylsiloxane into the surface area [18,19,22,24]

Wettability/Surface Energy and Interfacial Tension: Films containing 25–83% polydimethylsiloxane (M_n 6350) and polycarbonate (M_n 35000) spread on water at 24°. [23] The area occupied by the films at 2 dyne cm^{-1} is roughly proportional to the siloxane content; block lengths of polydimethylsiloxane (above 20 units) and polycarbonate seem unimportant [23]

Transport Properties:

Melt Flow Index:

No.	Value	Note
1	1.1 g/10 min	300°, total MW 10000 [12]

Permeability of Gases: Less gas permeable than (silicone rubber (heat vulcanised)) but relative permeability is similar. [1] Gas permeability depends upon temp., degree of polymerisation and amount of polydimethylsiloxane [13]

Permeability and Diffusion General: Diffusivity of propane depends upon the polydimethylsiloxane content and its degree of polymerisation. For a given degree of polymerisation, diffusivity generally increases with polydimethylsiloxane content [13]

Gas Permeability:

No.	Gas	Value	Note
1	Propane	96299 cm^3 mm/(m^2 day atm)	1.1×10^{-10} cm^2 $(s\ Pa)^{-1}$, 64% PDMS, degree of polymerisation 20, 50° [13]
2	Propane	122562 cm^3 mm/(m^2 day atm)	1.4×10^{-10} cm^2 $(s\ Pa)^{-1}$, 60% PDMS, degree of polymerisation 40, 50° [13]
3	Propane	201353 cm^3 mm/(m^2 day atm)	2.3×10^{-10} cm^2 $(s\ Pa)^{-1}$, 64% PDMS, degree of polymerisation 100, 50° [13]

Mechanical Properties:

Mechanical Properties General: Mech. props. depend on the end cap ratio. [9] Tensile strength increases and maximum elongation decreases as the percentage of polycarbonate increases [2,3,6,9,10]. Elastomer has high mech. strength. The variation of shear modulus with temp. (-100–270°) for polydimethylsiloxane content 25–75% and degrees of polymerisation of 20 and 40 has been reported [21]

Tensile Strength Break:

No.	Value	Note	Elongation
1	24 MPa	3500 psi, 50% silicone by weight [2,3,6]	270%
2	22.5–25.7 MPa	3270–3730 psi, 46% PDMS, end cap ratio 1.2, ASTM D412 [9]	360–370%
3	15.6–18.1 MPa	2270–2630 psi, 46% PDMS, end cap ratio 2, ASTM D412 [9]	320–380%
4	20–22.8 MPa	2900–3300 psi, 52% PDMS, end cap ratio 1.2, ASTM D412 [9]	510–570%
5	1.37 MPa	197 psi, no break, 51% PDMS, end cap ratio 4, ASTM D412 [9]	>950%
6	12–13.7 MPa	1740–1980 psi, 57% PDMS, end cap ratio 1.2, ASTM D412 [9]	690–890%
7	1.3–2.6 MPa	190–370 psi, no break, 57% PDMS, end cap ratio 2, ASTM D412 [9]	>950%
8	26.9 MPa	43% PDMS, total MW 170000 [10]	716%
9	13.3 MPa	50% PDMS, total MW 120000 [10]	729%
10	45 MPa	6500 psi, 15% PDMS [6]	110%
11	32 MPa	4600 psi, 44% PDMS [6]	110%
12	14.4 MPa	55% PDMS, total MW 229000 [10]	454%
13	18 MPa	2700 psi, 60% PDMS [6]	480%
14	12.5 MPa	65% PDMS, total MW 840000 [10]	1690%
15	43 MPa	6300 psi, 15% PDMS [6]	90%
16	28 MPa	4000 psi, 44% PDMS [6]	110%
17	16 MPa	2400 psi, 60% PDMS [6]	270%

Elastic Modulus:

No.	Value	Note
1	1900–1950 MPa	$1.9–1.95 \times 10^{10}$ dyne cm^{-2}, 25% PDMS, -196° [21]
2	2100 MPa	2.1×10^{10} dyne cm^{-2}, 50% PDMS, -196° [21]
3	2030 MPa	2.03×10^{10} dyne cm^{-2}, 65% PDMS, -196° [21]
4	1950 MPa	1.95×10^{10} dyne cm^{-2}, 60% PDMS, -196° [21]
5	1700 MPa	1.7×10^{10} dyne cm^{-2}, 75% PDMS, -196° [21]

Complex Moduli: The variation of loss tangent and loss modulus with temp. (-150–130°, 110 Hz) for block copolymers with 43–65% polydimethylsiloxane and degrees of polymerisation of 10 or 20 has been reported. [10,15] Props. are affected by the solvent from which the films are cast [10,15]

Viscoelastic Behaviour: Stress-strain curves have been reported. [10,16] As the polydimethylsiloxane content increases (45–65%) there is a transition to elastomeric behaviour. [6,9,10] The stress-strain curve for polydimethylsiloxane content of 43% shows a viscoelastic region up to the yield point followed by a slight stress drop and an increase in stress to the break point at 716% elongation. [10] At lower levels of polydimethylsiloxane the yield stress is more pronounced, and necking occurs. [6] At higher polydimethylsiloxane content no yield point is observed, and the stress increases to very large elongations (500–1500%). [10,16] Sonic modulus (room temp., 10 kHz) decreases with polydimethylsiloxane content and with decreasing end cap ratio. [9] Stress relaxation curves for block copolymers show relaxation modulus decreases with time and with increasing temp.; stress relaxation behaviour depends on the casting solvent [10]

Electrical Properties:
Electrical Properties General: Has good insulating props. [1] The variation of dielectric constant and dissipation factor with temp. (-160–160°, 50 Hz–100 kHz) has been reported [16,17]

Optical Properties:
Refractive Index:

No.	Value	Note
1	1.43–1.48	65% PDMS [16]

Transmission and Spectra: Transparent in the visible region [1]
Total Internal Reflection: Birefringence-strain curves for polydimethylsiloxane content 65% and degree of polymerisation 20–100 have been reported [16]
Volume Properties/Surface Properties: Clarity transparent (block copolymer films) [7,21]

Polymer Stability:
Thermal Stability General: Block copolymers have good thermal stability with decomposition occurring above 300°. [25]
Decomposition Details: Decomposes above 300° [5]. Decomposition is a two-step process. [25] In air, the first (lower temp.) reaction is thermooxidative and the second reaction is depolymerisation. In N_2, the second reaction has a lower activation energy than in air. [25] The appearance of the char (750°) is dark in N_2 but in air contains relatively large amounts of greyish oxides of silicon. The mechanism of decomposition is complicated as indicated by the higher levels of silicon in the char from copolymers with higher polycarbonate content. For polymers with polydimethylsiloxane degree of polymerisation 20, 2.5% weight loss in N_2 occurs at 407–521° (21.9% PDMS), 390–472° (47.8% PDMS), 415–537° (62.8% PDMS); in air at 384–514° (21.9% PDMS), 340–446° (47.8% PDMS), 364–462° (62.8% PDMS). Activation energy of decomposition in N_2 41–208 kJ mol^{-1} (117–164 kJ mol^{-1} in air) depending on polydimethylsiloxane content [25]
Flammability: No self-supporting combustion. [5] Limiting oxygen index as a function of polydimethylsiloxane content shows a max. of 38–40% (15–30% polydimethylsiloxane). Limiting oxygen index 30–40% (6–81% PDMS, degree of polymerisation 2–40, ASTM D2863-70), 50% (18% PDMS, degree of polymerisation 5, ASTM D2863-70). Limiting oxygen index is independent of siloxane block length [26]
Environmental Stress: High resistance to electric discharge [1]
Chemical Stability: Resistant to aliphatic hydrocarbons and alcohols; dissolved by chlorinated and aromatic hydrocarbons [5]
Hydrolytic Stability: Very resistant to hydrol. [4]
Recyclability: Not recyclable

Applications/Commercial Products:
Processing & Manufacturing Routes: Formed by reaction of hydroxy-terminated oligomers (made by phosgenation of Bisphenol A) with 'chlorine-terminated' poly(dimethylsiloxane) in the presence of tertiary amines. Reaction of dimethylamino-terminated poly(dimethylsiloxane) with the elimination of dimethylamine and thermal cleavage is also used. Block copolymers are synth in soln. in two stages; a polydisperse α,ω-dichlorosiloxane oligomer is treated with bisphenol A in pyridine/CH_2Cl_2 to give an end-capped oligomer. [6,8] Depending on the relative amounts of oligomer and bisphenol A, chain extension of the siloxane block may occur. [8] Secondly, the end-capped siloxane oligomer is treated with phosgene and bisphenol A to form the (polydisperse) polycarbonate block *in situ*. [6,8] The block copolymer is isol. by precipitation with MeOH or after solvent evaporation. [6] This procedure gives a range of block copolymers with polydimethylsiloxane degree of polymerisation 2–100 and polycarbonate 3–70. [6,8] The polycarbonate block size depends on the ratio of α,ω-dichlorosiloxane to bisphenol A in the end-capping reaction. [9] Block copolymers may also be synth. by interfacial polymerisation using a carboxylic acid terminated polydimethylsiloxane oligomer; this gives a SiC linkage [20]
Applications: Coatings for aluminium or concrete to prevent ice formation, in laminates and fillers, selective membranes, impact modifiers

Trade name	Details	Supplier
Copel	various grades	General Electric Plastics
LR 3320	43% polysiloxane	General Electric Plastics
LR 5630	65% polysiloxane	General Electric Plastics
Makrolon KUI-1198		Bayer Inc.
XD-11	25% polysiloxane, 75% PC	General Electric Plastics

Bibliographic References
[1] Noshay, A. and McGrath, J.E., *Block Copolym.*, Academic Press, 1977
[2] Allport, D.C., *Block Copolym.*, (eds. D.C. Allport and W.H. Jones), Applied Science, 1973, 532
[3] Hardman, B. and Torkelson, A., *Encycl. Polym. Sci. Eng.*, 2nd edn., (ed. J.I. Kroschwitz), John Wiley and Sons, 1985, **15**, 204
[4] Noll, W., *Chemistry and Technology of Silicones*, Academic Press, 1968
[5] Roff, W.J. and Scott, J.R., Fibres, Films, *Plastics and Rubbers*, Butterworths, 1971
[6] Vaughn, H.A., J. Polym. Sci., *Part B: Polym. Lett.*, 1969, **7**, 569
[7] Kambour, R.P., J. Polym. Sci., *Part B: Polym. Lett.*, 1969, **7**, 573
[8] Lind, A.C., *Macromolecules*, 1984, **17**, 300, (morphology, synth)
[9] Niznik, G.E. and LeGrand, D.G., J. Polym. Sci., *Polym. Symp.*, 1977, **60**, 97, (props, synth)
[10] Maung, W., Chua, K.M., Ng, T.H. and Williams, H.L., *Polym. Eng. Sci.*, 1983, **23**, 439, (props)
[11] LeGrand, D.G., J. Polym. Sci., *Part C: Polym. Lett.*, 1969, **7**, 579, (morphology)
[12] Matzner, M., Noshay, A. and McGrath, J.E., Polym. Prepr. (Am. Chem. Soc., *Div. Polym. Chem.*), 1973, **14**, 68, (morphology, melt flow index)
[13] Barrie, J.A., Williams, M.J.L. and Spencer, H.G., *J. Membr. Sci.*, 1984, **21**, 185, (morphology, density, gas permeability)
[14] LeGrand, D.G., J. Polym. Sci., *Part C: Polym. Lett.*, 1970, **8**, 195, (morphology)
[15] Maung, W. and Williams, H.L., *Polym. Eng. Sci.*, 1985, **25**, 113, (morphology, complex modulus)
[16] Magila, T.L. and LeGrand, D.G., *Polym. Eng. Sci.*, 1970, **10**, 349, (props)
[17] Kaqniskin, V.A., Kaya, A., Ling, A. and Shen, M., *J. Appl. Polym. Sci.*, 1973, **17**, 2695, (solvents, complex permittivity)
[18] Chen, X., Lee, H.F. and Gardella, J.A., *Macromolecules*, 1993, **26**, 4601, (solvents, surface props)
[19] Dwight, D.W., McGrath, J.E., Riffle, J.S., Smith, S.D. and York, G.A., *J. Electron. Spectrosc. Relat. Phenom.*, 1990, **52**, 457, (solvents, surface props)
[20] Riffle, J.S., Freelin, R.G., Banthia, A.K. and McGrath, J.E., J. Macromol. Sci., *Chem.*, 1981, **15**, 967, (solvents, synth)
[21] Narkis, M. and Tobolsky, A.V., J. Macromol. Sci., *Phys.*, 1970, **4**, 877, (solvents, elastic modulus, volume props)
[22] Schmitt, R.L., Gardella, J.A., Magill, J.H., Salvati, L. and Chin, R.L., *Macromolecules*, 1985, **18**, 2675, (surface props)
[23] Gaines, G.L., J. Polym. Sci., *Polym. Symp.*, 1971, **34**, 115, (surface props)
[24] Pertsin, A.J., Gorelova, M.M., Levin, V.Y. and Makarova, L.I., *J. Appl. Polym. Sci.*, 1992, **45**, 1195, (surface props)
[25] Grubbs, G.R., Kleppick, M.E. and Magill, J.H., *J. Appl. Polym. Sci.*, 1982, **27**, 601, (thermal stability)
[26] Kambour, R.P., Klopper, H.J. and Smith, S.A., *J. Appl. Polym. Sci.*, 1981, **26**, 847, (flammability)

Poly[2,2-propanebis(4-phenyl)carbonate]-*block*-poly(ethylene oxide)

P-470

Synonyms: Polycarbonate-block-poly(ethylene oxide). Block copolymer of 4,4'-(1-methylethylidene)bis[phenol] with α-hydro-ω-hydroxypoly(oxy-1,2-ethanediyl) and carbonic acid
Related Polymers: Poly[2,2-propanebis(4-phenyl)carbonate], Polyethylene oxide
Monomers: Bisphenol A, Carbonic dichloride, Ethylene oxide
Material class: Synthetic Elastomers, Copolymers

Polymer Type: polyethylene oxide, bisphenol A polycarbonate
CAS Number:

CAS Reg. No.
108795-16-0

Molecular Formula: $[(C_{16}H_{14}O_3).(C_2H_4O).CO]_n$
Fragments: $C_{16}H_{14}O_3$ C_2H_4O CO

Applications/Commercial Products:
Processing & Manufacturing Routes: Microporous membranes (5% poly(ethylene oxide) 4000). Haemodialysis membranes (20–40% PEG 6000 or 20,000). Used in the separation of polar and non-polar gases, ultrafiltration, microfiltration, electrophoresis, desalination of sea-water

Trade name	Supplier
Makrolon KL3-1013	Bayer Inc.

Poly[2,2-propanebis(4-phenyl)carbonate]-*block*-poly[4,4'-isopropylidenediphenoxydi(4-phenylene)-sulfone]   P-471

Synonyms: *Polycarbonate-block-polysulfone. Polycarbonate-polysulfone block copolymer*
Related Polymers: Poly[2,2-propanebis(4-phenyl)carbonate]
Material class: Thermoplastic, Copolymers
Polymer Type: polycarbonates (miscellaneous), polysulfone
Molecular Formula: $((C_{27}H_{22}O_4S).(C_{16}H_{14}O_3)_y)_n$
Fragments: $C_{27}H_{22}O_4S$ $C_{16}H_{14}O_3$

Poly[2,2-propanebis(4-phenyl)carbonate]-*block*-poly(methylmethacrylate)   P-472

Synonyms: *Polycarbonate-block-poly(methylmethacrylate). Polycarbonate-poly(methylmethacrylate) block copolymer. Block copolymer of 4,4'-(1-methylethylidene)bis[phenol] with carbonic acid and methyl 2-methyl-2-propenoate. Poly(4,4'-[1,4-phenylenebis(1-methylethylidene)]bisphenol-co-carbonic acid)-block-poly(1-methoxycarbonyl)-1-methylethylene)*
Related Polymers: Poly(methylmethacrylate), Poly[2,2-propanebis(4-phenyl)carbonate]
Monomers: Bisphenol A, Carbonic dichloride, Methyl methacrylate
Material class: Thermoplastic, Copolymers
Polymer Type: acrylic, bisphenol A polycarbonate
CAS Number:

CAS Reg. No.	Note
122873-25-0	source based: carbonic dichloride
125321-92-8	source: carbonic acid

Molecular Formula: $[(C_5H_8O_2).C_3H_4O_2S.(C_{16}H_{14}O_3)]_n$
Fragments: $C_5H_8O_2$ $C_{16}H_{14}O_3$

Poly[2,2-propanebis(4-phenyl)carbonate]-*block*-poly[2,2-propanebis[4-(2,6-dimethylphenyl)]carbonate]   P-473

Synonyms: *Polycarbonate-block-poly(tetramethyl Bisphenol A carbonate)*
Related Polymers: Poly[2,2-propanebis(4-phenyl)carbonate], Poly[2,2-propanebis[4-(2,6-dimethylphenyl)]carbonate]
Monomers: Bisphenol A, Carbonic dichloride, Tetramethyl Bisphenol A
Material class: Thermoplastic, Copolymers
Polymer Type: polycarbonates (miscellaneous), bisphenol A polycarbonate
CAS Number:

CAS Reg. No.	Note
132410-03-8	source: carbonic acid

Molecular Formula: $[(C_{20}H_{22}O_3).(C_{16}H_{14}O_3)]_n$
Fragments: $C_{20}H_{22}O_3$ $C_{16}H_{13}O_3$

Poly[2,2-propane[4-(2,6-dimethyl)phenyl]carbonate]/Polystyrene (High impact) blend   P-474

Related Polymers: Polystyrene (High Impact), Poly[2,2-propane[4-(2,6-dimethylphenyl)carbonate]
Monomers: Tetrabromobisphenol A, Styrene
Material class: Blends
Polymer Type: polycarbonates (miscellaneous), polystyrene
General Information: Tetramethylpolycarbonate is miscible in all proportions with polystyrene. [1,2,3,4,5,6]. However, phase separation occurs on heating (lower critical soln. temp. behaviour) [1,4,5,7,8]. The polymers are not miscible on a segmental level, but on a higher structural level [3]

Volumetric & Calorimetric Properties:
Density:

No.	Value	Note
1	d^{30} 1.055 g/cm^3	20% tetramethylpolycarbonate, polystyrene MW 10000–30000 [7]
2	d^{30} 1.062 g/cm^3	40% tetramethylpolycarbonate, polystyrene MW 10000–30000 [7]
3	d^{30} 1.071 g/cm^3	60% tetramethylpolycarbonate, polystyrene MW 10000–30000 [7]
4	d^{30} 1.076 g/cm^3	77% tetramethylpolycarbonate, polystyrene MW 10000–30000 [7]
5	d 1.02 g/cm^3	25% tetramethylpolycarbonate, 75% high impact polystyrene [6]
6	d 1.05 g/cm^3	50% tetramethylpolycarbonate, 50% high impact polystyrene [6]
7	d 1.07 g/cm^3	75% tetramethylpolycarbonate, 25% high impact polystyrene [6]

Glass-Transition Temperature:

No.	Value	Note
1	106°C	10% tetramethylpolycarbonate MW 67000, 90% polystyrene MW 200000 [2]

2	104–107°C	12.5% tetramethylpolycarbonate MW 40000, 87.5% polystyrene MW 250000 [3]
3	113°C	20% tetramethylpolycarbonate 80% polystyrene [6,7]
4	114°C	20% tetramethylpolycarbonate MW 330000, 80% polystyrene MW 330000 [5]
5	115°C	25% tetramethylpolycarbonate MW 67000, 75% polystyreneMW 200000 [2]
6	113–114°C	25% tetramethylpolycarbonate MW 40000, 75% polystyrene MW 250000 [3]
7	120°C	30% tetramethylpolycarbonate MW 67000, 70% polystyrene MW 200000 [2]
8	115°C	30% tetramethylpolycarbonate, MW 52600, 70% polystyrene, MW 289000 [4]
9	132°C	40% tetramethylpolycarbonate, 60% polystyrene [6]
10	124–125°C	40% tetramethylpolycarbonate, MW 33000, 60% polystyrene, MW 330000 [5,7]
11	128–134°C	50% tetramethylpolycarbonate 50% polystyrene dependent on MW of components [2,3,4,5]
12	135–152°C	60% tetramethylpolycarbonate 40% polystyrene,dependent on MW of components [5,6,7]
13	142–153°C	70% tetramethylpolycarbonate 30% polystyrene, dependent on MW of components [2,4]
14	152–156°C	75% tetramethylpolycarbonate, 25% polystyrene, dependent on MW of components [2,3,7]
15	173°C	80% tetramethylpolycarbonate 20% polystyrene [6]
16	159°C	80% tetramethylpolycarbonate MW 33000, 20% polystyrene, MW 330000 [5]
17	165–172°C	87.5% tetramethylpolycarbonate MW 40000, 12.5% polystyrene MW 250000 [3]
18	161°C	90% tetramethylpolycarbonate MW 52600, 10% polystyrene MW 289000 [4]
19	176°C	90% tetramethylpolycarbonate MW 67000, 10% polystyrene MW 200000 [2]

Vicat Softening Point:

No.	Value	Note
1	110°C	25% tetramethylpolycarbonate, 75% high-impact polystyrene B [6]
2	139°C	50% tetramethylpolycarbonate, 50% high-impact polystyrene B [6]
3	167°C	75% tetramethylpolycarbonate, 25% high-impact polystyrene B [6]

Surface Properties & Solubility:
Solubility Properties General: Interaction energy between tetramethylpolycarbonate [5] and polystyrene -0.75 – -0.71 J cm$_3$ (-0.18 – -0.17 cal cm^3) (30–80% tetramethylpolycarbonate) [5]. Interaction parameter 0.029 (tetramethylpolycarbonate (MW 67000), 50% polystyrene (MW 200000) [7]. This value is less negative for higher tetramethylpolycarbonate levels [7]
Solvents/Non-solvents: Sol. THF [2], CH$_2$Cl$_2$ [3,4,7], C$_6$H$_6$ [4]

Mechanical Properties:
Tensile (Young's) Modulus:

No.	Value	Note
1	1700 MPa	25% tetramethylpolycarbonate, 75% high-impact polystyrene [6]
2	2100 MPa	50% tetramethylpolycarbonate, 50% high-impact polystyrene [6]
3	2300 MPa	75% tetramethylpolycarbonate, 25% high-impact polystyrene [6]

Izod Notch:

No.	Value	Notch	Note
1	14 J/m	Y	25% tetramethylpolycarbonate, 75% high-impact polystyrene [6]
2	12 J/m	Y	50% tetramethylpolycarbonate, 50% high-impact polystyrene [6]
3	8 J/m	Y	75% tetramethylpolycarbonate, 25% high-impact polystyrene [6]

Electrical Properties:
Complex Permittivity and Electroactive Polymers: The variation of η'' with frequency (0.01Hz–100 kHZ) for several temps. (140–165°) for 25, 50 and 75% tetramethylpolycarbonate blends has been reported [3]

Optical Properties:
Volume Properties/Surface Properties: The appearance of the films can depend on the solvent from which they were cast [7]; films from CH$_2$Cl$_2$ are transparent [7]

Polymer Stability:
Thermal Stability General: The blends show phase separation on heating for 5 mins at 235–295°, depending on composition [4,5,8]. The process is reversible [5]

Applications/Commercial Products:
Processing & Manufacturing Routes: Synth. by casting from soln. [3,4,5,7]. A soln. containing 5% polymer in THF is cast onto a glass plate mounted on a heated block at 50–60° for 5 min [5]. Alternatively, CH$_2$Cl$_2$ may be used and the film left to dry at room temp. [3,4,7]. The films are then dried in a vacuum oven at 60–190° for 1–7 days [3,4,5,7]. Casting from toluene leads to crystallisation of the tetramethylpolycarbonate [7]
Applications: Suggested use as microwaveable dishes and car headlamp reflectors

Trade name	Supplier
Bayblend H	Bayer Inc.

Bibliographic References
[1] Kim, C.K. and Paul, D.R., *Macromolecules*, 1992, **25**, 3097, (morphology, thermal stability)
[2] Landry, C.J.T., Yang, H. and Machell, J.S., *Polymer*, 1991, **32**, 44, (morphology, solvent, T_g)
[3] Mansour, A.A. and Madbouly, S.A., *Polym. Int.*, 1995, **37**, 267, (morphology, solvent, T_g, complex permittivity, synth.)
[4] Guo, W. and Higgins, J.S., *Polymer*, 1990, **31**, 699, (morphology, thermal stability, solvent, synth.)
[5] Kim, C.K. and Paul, D.R., *Polymer*, 1992, **33**, 1630, (morphology, thermal stability, interaction energy, T_g, synth.)
[6] Serini, V., Peters, H. and Morbitzer, L., *Polymer Blends Symposium 4*, Plastics and Rubber Inst., Chameleon Press, 1981, **9**, 1, (props).
[7] Fernandes, A.C., Barlow, J.W. and Paul, D.R., *Polymer*, 1986, **27**, 1788, (props., synth.)
[8] Casper, R. and Morbitzer, L, *Angew. Makromol. Chem.*, 1977, **58/59**, 1, (thermal stability)

Poly(propene-*alt*-carbon monoxide-*alt*-ethene)

Synonyms: *1-Propene-carbon monoxide-ethene alternating copolymer. Aliphatic polyketones. Poly(propylene-alt-carbon monoxide-alt-ethylene)*
Related Polymers: Polypropylene, Polyethylene
Monomers: Propylene, Ethylene, Carbon monoxide
Material class: Thermoplastic, Copolymers
Polymer Type: polyethylene, polypropylenepolyketones
CAS Number:

CAS Reg. No.
121520-83-0

Molecular Formula: $[(C_3H_6).(C_2H_4).(CO)]_n$, $(C_6H_{10}O)_n$
Fragments: C_3H_6 C_2H_4 CO C_6H_{10}
Additives: Stabilisers (to improve thermal stability for melt processabilty and stability to uv). Glass fibre; flame retardants
General Information: A high-melting polymer with good impact props., good dimensional heat stability and good barrier props.
Morphology: α-form: orthorhombic a 0.701, b 0.507, c 0.759 [1] a 0.691, b 0.512, c 0.76 [4]. β form: orthorhombic, a 0.796, b 0.478, c 0.759 [1]

Volumetric & Calorimetric Properties:
Density:

No.	Value	Note
1	d 1.26 g/cm^3	3% propylene [1]
2	d 1.24 g/cm^3	[1]
3	d 1.48 g/cm^3	30% glass fibre [1]
4	d 1.4–1.6 g/cm^3	flame retardant grades [1]

Melting Temperature:

No.	Value	Note
1	220°C	6% propylene [1,2]
2	170°C	17% propylene [1]

Glass-Transition Temperature:

No.	Value	Note
1	15°C	[1]
2	17°C	7% propylene [2]

Transition Temperature General: A sub-T_g γ-transition has been reported [1]. T_m increases with increasing CO content [1]. T_g decreases with increasing levels of sorbed water [3]

Surface Properties & Solubility:
Solvents/Non-solvents: Sol. *m*-cresol, *o*-chlorophenol, trifluoroacetic acid (decomposes); Insol. common organic solvents

Transport Properties:
Melt Flow Index:

No.	Value	Note
1	6 g/10 min	240°, 2.16 kg, ISO 1133 [1]
2	5 g/10 min	240°, 2.16 kg, ISO 1133, 30% glass fibre [1]

Polymer Solutions Dilute: 1.8 dl g^{-1} (*m*-cresol, 30°, 6% propylene), 1.52 dl g^{-1} (*m*-cresol, 90°, 6% propylene), 2.3 dl g^{-1} (phenol/trichlorobenzene, 30°, 6% propylene), 1.75 dl g^{-1} (phenol/trichlorobenzene, 100°, 6% propylene) [1]. Huggins constants for polymer containing 6% propylene have been reported [1]
Permeability of Gases: Diffusivity He 8.46×10^{-6} cm^2 s^{-1} (25°, amorph.), O_2 9.58×10^{-8} cm^2 s^{-1} (25°, amorph.), CO_2 3.76×10^{-8} cm^2 s^{-1} (25°, amorph.) [2,3]. Permeability He 2.35×10^{-7} cm^2 (atm min)$^{-1}$, (25°, crystallinity 42%) [2,3], O_2 $5.03–6.43 \times 10^{-9}$ cm^2 (atm min)$^{-1}$ (25°, crystallinity 42%) [2,3], CO_2 $4.41–5.75 \times 10^{-8}$ cm^2 (atm min)$^{-1}$ (25°, crystallinity 42%) [2,3], N_2 1.23×10^{-9} cm^2 (atm min)$^{-1}$ (25°, crystallinity 42%) [2,3], CH_4 2.022×10^{-9} cm^2 (atm min)$^{-1}$ (25°, crystallinity 42%) [2,3], ethane 6.11×10^{-10} cm^2 (atm min)$^{-1}$, (25°, crystallinity 42%) [2,3]

Water Absorption:

No.	Value	Note
1	0.5 %	23°, 50% relative humidity ASTM D570 [1]
2	0.3 %	23°, 50% relative humidity, 30% glass fibre ASTM D570 [1]

Mechanical Properties:
Mechanical Properties General: The effect of shear controlled orientation in injection moulding (SCORIM) on the mechanical props. has been reported. Young's modulus and tensile strength both increase as a result of SCORIM processing [4]

Tensile (Young's) Modulus:

No.	Value	Note
1	1400 MPa	1400 N mm^{-2}, ISO/R 527 [1]
2	7300 MPa	7300 N mm^{-2}, 30% glass fibre, ISO/R 527 [1]
3	2500–6000 MPa	2500–6000 N mm^{-2}, flame retardant grades, ISO/R 527 [1]

Flexural Modulus:

No.	Value	Note
1	1400 MPa	1400 N mm^{-2}, ISO/R 527 [1]
2	7000 MPa	7000 N mm^{-2}, 30% glass fibre, ISO/R 527 [1]
3	2500 MPa	2500 N mm^{-2}, flame retardant grades, ISO/R 527 [1]

Tensile Strength Break:

No.	Value	Note	Elongation
1	55 MPa	55 N mm^{-2}, ISO/R 527 [1]	350%
2	120 MPa	120 N mm^{-2}, 30% glass fibre, ISO/R 527 [1]	3%
3	55–90 MPa	55–90 N mm^{-2}, flame retardant grades, ISO/R 527 [1]	2.5–9%

Tensile Strength Yield:

No.	Value	Elongation	Note
1	60 MPa	25%	60 N mm^{-2}, ISO/R 527 [1]

Miscellaneous Moduli:

No.	Value	Note	Type
1	1300 MPa	1300 N mm^{-2}, 1h, 0.5% elongation, ISO 899 [1]	creep modulus
2	900 MPa	900 N mm^{-2}, 1000h, 0.5% elongation, ISO 899 [1]	creep modulus

[9] Brydson, J.A., *Plast. Mater.*, 6th ed., Butterworth and Heinemann, 1995, 538
[10] Serini, V., Peters H., Morbitzer, L., *Polymer Blends Symposium 4*, 1981, 9.1

Hardness: Rockwell R105 (ASTM D785) [1]
Friction Abrasion and Resistance: Coefficient of dynamic friction 0.19 (against steel, ASTM D1894); 0.28 (against brass, ASTM D1894); 0.17 (against aluminium, ASTM D1894); 0.31 (against self, ASTM D1894). Taber abrasion resistance 12 mg (1000 cycles)$^{-1}$ (1 kg load, CS-17 wheel, ASTM D1044) [1]
Izod Notch:

No.	Value	Notch	Note
1	1500 J/m	Y	ISO 180/1A [1]
2	8000 J/m	Y	30% glass fibre ISO 180/1A [1]

Polymer Stability:
Decomposition Details: Thermodecomposition is a complex process [1]
Chemical Stability: Has good chemical resistance. Slight swelling can be caused by solvents which are capable of hydrogen-bonding or which have a chemical composition similar to that of the polymer. Aq. solns. of weak acids and bases have little effect; is attacked by strong acids and bases particularly at elevated temps. and with prolonged exposure [1]

Applications/Commercial Products:
Processing & Manufacturing Routes: Perfectly alternating terpolymers are obtained by using transition metal (e.g., palladium) catalysts [1]. Suitable for injection moulding, blow moulding, extrusion and rotational moulding
Applications: Applications due to good barrier props. in food packaging

Trade name	Supplier
Carilon	Shell Chemicals

Bibliographic References
[1] Sommazzi, A. and Garbassi, F., *Prog. Polym. Sci.*, 1997, **22**, 1547, (synth., mech. props.)
[2] Del Nobile, M.A., Mensitieri, G. and Sommazzi, A., *Polymer*, 1995, **36**, 4943, (gas permeability)
[3] Del Nobile, M.A., Mensitieri, G. Nicolais, L., Sommazzi, A. and Garbassi, F., *J. Appl. Polym. Sci.*, 1993, **50**, 1261, (gas permeability)
[4] Kalay, G and Bevis, M.J., *J. Polym. Sci., Part B: Polym. Phys.*, 1997, **35**, 415, (mechanical props.)

Poly(propene-*co*-carbon monoxide-*co*-ethene)   P-476

$$-[-CH_2CH_2CH_2-]_x[-\overset{O}{\underset{\|}{C}}-]_y[-CH_2CH_2-]_z-]_n$$

Synonyms: *1-Propene-carbon monoxide-ethene copolymer. Aliphatic polyketone. Poly(propylene-co-carbon monoxide-co-ethylene)*
Related Polymers: Polypropylene, Polyethylene, Poly(propene-*alt*-carbon monoxide-*alt*-ethene)
Monomers: Propylene, Carbon monoxide, Ethylene
Material class: Thermoplastic, Copolymers
Polymer Type: polyethylene, polypropylenepolyketones
CAS Number:

CAS Reg. No.
88995-51-1

Molecular Formula: $[(C_3H_6)_x(CO)_y(C_2H_4)_z]_n$
Fragments: C_3H_6 CO C_2H_4
General Information: A white highly crystalline solid with good impact props. Is dimensionally heat stable

Volumetric & Calorimetric Properties:
Transition Temperature General: T_m increases with increasing CO content [1]

Mechanical Properties:
Mechanical Properties General: Mechanical props. are dependent on the amount of CO [1]

Polymer Stability:
Environmental Stress: Non-alternating copolymers are photodegradable; at room temp. uv irradiation causes chain cleavage following a Norrish-II type process [1]

Applications/Commercial Products:
Processing & Manufacturing Routes: Can be synth. by free-radical polymerisation or by using γ-irradiation [1]

Trade name	Supplier
Carilon 87/014	Shell Chemicals

Bibliographic References
[1] Sommazzi, A. and Garbassi, F., *Prog. Polym. Sci.*, 1997, **22**, 1547

Poly(1-propenyl ether functional siloxane)   P-477
Related Polymers: PDMS [polymethyldimethylsiloxane fluid]
Material class: Synthetic Elastomers, Copolymers
Polymer Type: silicones
Molecular Formula: $[(C_2H_6OSi).(C_{11}H_{22}O_3Si)]_n$
Fragments: C_2H_6OSi $C_{11}H_{22}O_3Si$
Molecular Weight: MW 2000–3000
General Information: Poly functional siloxanes can be made from siloxane monomers and oligomers in order to prod. precisely determined products but use of polymethylhydrogen-*co*-dimethyl siloxane is likely to be commercially more important. Electron beam-induced cationic polymerisation takes place rapidly and at low radiation doses [1]
Miscellaneous: Incorporation of vinyl groups into the side chain allow ready polymerisation by UV or electron beam in presence of photoinitiator

Applications/Commercial Products:
Processing & Manufacturing Routes: Polymethyldimethylsiloxane fluid [1,2] is reacted with 1-allyloxy-4-(propyloxy)butane (PAB) in the presence of a noble metal catalyst for 1h at 65°. The product can be readily polymerised in the presence of an onium salt photoinitiator by UV radiation or an electron beam
Applications: Potentially useful in high speed coatings. Polyfunctional siloxane very rapidly polymerises with photoinitiator and UV or electron beam

Bibliographic References
[1] Crivello, J.V., Yang, B. and Kim, W.-G., *J. Macromol. Sci., Chem.*, 1996, **33**, 399
[2] Crivello, J.V., Yang, B. and Kim, W.-G., *J. Polym. Sci., Polym. Chem. Ed.*, 1995, **33**, 2415

Poly(propyl acrylate)   P-478
Synonyms: *Poly[1-(propoxycarbonyl)ethylene]. Poly(propyl 2-propenoate)*
Monomers: Propyl acrylate
Material class: Natural elastomers
Polymer Type: acrylic
CAS Number:

CAS Reg. No.
24979-82-6

Molecular Formula: $(C_6H_{10}O_2)_n$
Fragments: $C_6H_{10}O_2$
Molecular Weight: The MW can range from 1000–5000000
Additives: Soap-urea, activated thiol, soap-sulfur, lead thiourea, thiourea, diamine, carbon black, etc.

– **Polypropylene**

General Information: A soft, tacky polymer with low T_g, good thermal stability, low resistance to oils and non-polar solvents; it is only used commercially as copolymer. It forms copolymers with a variety of monomers (e.g. polyolefins, vinyl acetate, vinyl chloride) to give a wide range of products

Volumetric & Calorimetric Properties:
Density:

No.	Value
1	d_4^{25} 1.1 g/cm^3

Volumetric Properties General: Molar volume 103 cm^3 mol^{-1}
Melting Temperature:

No.	Value	Note
1	-37°C	
2	-25°C	emulsion or aq. soln. polymer, DSC 20° min^{-1}, mid point

Surface Properties & Solubility:
Cohesive Energy Density Solubility Parameters: δ 18.52 J$^{1/2}$ cm$^{-3/2}$. Molar cohesive energy 35.3 kJ mol^{-1}
Plasticisers: Thiokol TP-759, Plastolein 9720, Admex 760
Solvents/Non-solvents: Sol. C_6H_6, toluene, xylene, CH_2Cl_2, $CHCl_3$, dichloroethane, chlorobenzene, dioxane, 2-butanone, diisopropyl ketone, cyclohexanone, butyl acetate, isobutyl propionate, cyclohexyl acetate, cyclohexanol (hot), isobutanol (hot). Insol. aliphatic hydrocarbons, hydrogenated naphthalenes, Et_2O

Transport Properties:
Polymer Solutions Concentrated: Temporary contact between polymer chains results in the formation of gels which cannot be processed satisfactorily
Permeability of Gases: Has low permeability to gases and liqs.

Mechanical Properties:
Tensile Strength Break:

No.	Value	Note	Elongation
1	1.72–2.76 MPa	250–400 psi, pure gum, ASTM D412	450–750%
2	6.89–17.24 MPa	1000–2500 psi, black, ASTM D412	150–450%

Viscoelastic Behaviour: Has reasonable rebound when cold; very good when hot. Has a fair to good resistance to tear
Hardness: Durometer A40-A90
Friction Abrasion and Resistance: Has good abrasion resistance

Polymer Stability:
Polymer Stability General: Retains original props. under normal use conditions
Thermal Stability General: Stable to heat up to 250°. Decomposes slowly unless subjected to extreme heat conditions. Excellent resistance to heat ageing
Decomposition Details: Decomposition between 300–500° gives mainly propylene, propanol and CO_2
Flammability: Readily flammable
Environmental Stress: Unaffected by normal uv degradation, as the primary uv absorption occurs below solar spectrum of 290 nm. Uv stabilisers improve stability. Excellent resistance to oxidation
Chemical Stability: Undergoes cross-linking and degradation when heated in the presence of a free-radical initiator
Hydrolytic Stability: Stable to both acid and alkaline hydrol.

Applications/Commercial Products:
Processing & Manufacturing Routes: Commercially it is predominantly prepared by free radical emulsion polymerisation and soln. polymerisation. It is processed by injection moulding, blow moulding, extrusion and thermoforming. Stearic acid, TE-80 and Vanfore AP2 are added as processing aids
Applications: Exclusively used for applications in adhesives and coatings. It is used in paints, coatings and lacquers in the paper industry. It is also used in radiation curable systems, adhesives and sealing compounds

Bibliographic References
[1] *Ullmanns Encykl. Ind. Chem.*, 5th edn., (ed. W. Gerhartz), VCH, 1992
[2] *Encycl. Polym. Sci. Eng.*, 2nd edn., (eds. H.F. Mark, N.M. Bikales, C.G. Overberger and G. Menges), John Wiley and Sons, 1985, **1**
[3] *Kirk-Othmer Encycl. Chem. Technol.*, 4th edn., (ed. J.I. Kroschwitz), 1991, **1**
[4] Bovey, J., Abere, J.F., Rathmann, G.B. and Sandberg, C.L., *J. Polym. Sci.*, 1955, **15**, 520
[5] Riddle, E.H., *Monomeric Acrylic Esters*, Reinhold, 1954

Polypropylene P-479

$$-\!\!\left[CH_2CH(CH_3)\right]_n\!\!-$$

Synonyms: *Polypropene. PP. Poly(1-propene). Poly(methyl ethylene). Propene polymer. Propylene polymer*
Related Polymers: Polypropylene, isotactic, Polypropylene, atactic, Polypropylene, syndiotactic, Polypropylene Film, Polypropylene Fibres, Polypropylene powder/fibre filled, Polypropylene Blends, Polypropylene copolymers, Polypropylene (chlorinated), Polypropylene oxides
Monomers: Propylene
Material class: Thermoplastic
Polymer Type: polypropylene
CAS Number:

CAS Reg. No.	Note
9003-07-0	
25085-53-4	isotactic

Molecular Formula: $(C_3H_6)_n$
Fragments: C_3H_6
Molecular Weight: MW 25000–500000, MW 80000–500000 (commercial). M_w/M_n 2.1–8.0
Additives: Stabilisers: hindered phenol, Irganox 1010, Ethyl A0330, Topanol CA, Gooditte 3114, phosphites. Uv stabilisers: hindered piperidines, uv absorbents. Clarifier: Millad. Surface improver: PTS-CEL. Sarmalen
Tacticity: Commercial polypropylene is a complex mixt. of varying amounts of isotactic, stereoblock and atactic forms, each with a given MW distribution. In general, references tend to be to this product unless tacticity is specifically stated. Isotactic polypropylene is the major product from Ziegler-Natta and more recent order-specific catalysts. Syndiotactic polypropylene results from regular, alternating monomer insertion. Non-crystalline, atactic polypropylene exhibits random location of the pendant methyl groups. Stereoblock is defined as comprising segments of isotactic and atactic polypropylene. Commercial polypropylene is mainly isotactic. Tradenames given refer to commercial materials [1,2]
Identification: Thermoplastic [3]. Cl, N, S, P normally absent. Burns with bluish, non-smoky flame with flaring molten drops. Little or no residue, waxy, slightly acrid odour on extinction. On pyrolysis becomes transparent, melts at about 170° and decomposes at about 350° with evolution of low MW hydrocarbons. The isotactic index [4] is the fraction insoluble in boiling heptane though on a commercial basis, solubility in xylene or decalin is used to determine a measure of stereoregularity [5]. Because of the lack of chemical reactivity, physical tests are normally preferred. Polypropylene floats in water like polyethylene polyethylene (PE) and poly(4-methyl-1-pentene polymethylpentene) (PMP); floats in isophorone (distinct from PE) and sinks in 80/20 v/v. alcohol/water (distinct from PMP) [3]. Spot test: HgO in H_2SO_4 gives a yellow colour [45]

– Polypropylene

Miscellaneous: Polypropylene is prod. in a wide variety of MW, MW distributions and crystallinities in addition to modification by fillers, additives, co-polymerisation and blending. Hence hundreds of grades are commercially available [1,2]. Polypropylene use has increased at a higher rate than all other thermoplastics. Growth rate is such that it tends to be limited only by overall economic growth. Polymerisation propagation rate constant K_p 2000000 × exp (-32000/RT) (50–150°, radical telomerisation) [68,69]. Activation Energy of Propagation E_p 23.4 kJ mol^{-1}. (gas phase) [70]. It has been said that polypropylene is not a top grade plastic, but it has an attractive price to performance ratio, especially with respect to blends [50] see polypropylene blends

Volumetric & Calorimetric Properties:
Density:

No.	Value	Note
1	d 0.903 g/cm^3	ASTM 792A-2 [1]
2	d 0.9–0.91 g/cm^3	homopolymer
3	d 0.89–0.905 g/cm^3	copolymer
4	d 0.88–0.905 g/cm^3	impact modified [6,7]

Thermal Expansion Coefficient:

No.	Value	Note	Type
1	8.1 × 10^{-5}–0.0001 K^{-1}	ASTM D696	L
2	6.8 × 10^{-5}–9.5 × 10^{-5} K^{-1}	copolymer and impact modified copolymer [6]	L
3	6.5 × 10^{-5} K^{-1}	-30°-0°	L
4	0.000105 K^{-1}	0°-30°	L
5	0.000145 K^{-1}	30°-60°, ASTM D696 [7]	L

Equation of State: Tait and Spencer-Gilmore equation of state information has been reported [84,85,86,87,88].
Thermodynamic Properties General: Effects of isotacticity and crystallisation temp. on crystalline struct. and thermodynamic props. have been reported [54]. C_p values have been reported [82,91]. A detailed investigation of work on T_m, T_g, C_p, enthalpy, entropy and Gibbs function has been reported [82,83]
Latent Heat Crystallization: $\Delta H_{cryst.}$ 87–92 J g^{-1} [20]. ΔH_{fusion} 209 J g^{-1} (100% crystalline, 92–95% isotactic, MW range 22000–947000). $\Delta H_{cryst.}$ 97 J g^{-1} (approx.) (>95% isotactic, MW range 22000–947000), 40 J g^{-1} (approx., <50% isotactic, MW range 22000–947000) [46]. Values vary with T_m, tacticity and MW range [47]. Other $\Delta H_{cryst.}$ values have been reported [49]. Heat of polymerisation and entropy of polymerisation values have been reported [71]. ΔH 86.5 kJ mol^{-1}. $\Delta S°$ 167 J K^{-1} mol^{-1} (25°, monomer gas phase, polymer gas phase); ΔH 104 kJ mol^{-1}. $\Delta S°$ 205 J K^{-1} mol^{-1}, 25°, monomer gas phase, polymer crystalline/partly crystalline); ΔH 84 kJ mol^{-1}. $\Delta S°$ 113 J K^{-1} mol^{-1} (25°, monomer liquid phase, polymer condensed amorph. phase); ΔH 69 kJ mol^{-1} (-78°, monomer soln. phase butane, polymer condensed amorph. phase) [71]

Thermal Conductivity:

No.	Value	Note
1	0.117 W/mK	0.00028 cal (cm s °C)$^{-1}$, ASTM C177 [6]
2	0.25 W/mK	[14]

Specific Heat Capacity:

No.	Value	Note	Type
1	1.926 kJ/kg.C	[7,12]	P

Melting Temperature:

No.	Value	Note
1	168–175°C	crystalline, hompolymer and copolymer
2	160–168°C	crystalline, impact modified copolymer [6]

Glass-Transition Temperature:

No.	Value	Note
1	-20°C	amorph. [6,13]

Transition Temperature General: T_g varies with increasing M_n to an assymptotic value [13]. T_g dependant on tacticity and thermal history of the sample [7]. T_g is more MW dependent than T_m, hence effectiveness of low MW plasticisers [72]

Deflection Temperature:

No.	Value	Note
1	50–60°C	1.84 MPa, (264 psi), ASTM D648, homopolymer
2	96–120°C	225–250°F, 0.45 MPa, (66 psi), ASTM D648, hompolymer [6,7]
3	55–60°C	130–140°F, 1.84 MPa, (264 psi), ASTM D648, copolymer
4	85–105°C	185–220°F, 0.45 MPa, (66 psi), ASTM D648, copolymer [6]
5	45–55°C	115–135°F, 1.84 MPa, (264 psi), ASTM D648, impact modified
6	75–90°C	167–192°F, 0.45 MPa, (66 psi), ASTM D648, impact modified [6]

Vicat Softening Point:

No.	Value	Note
1	138–155°C	ASTM D1525 [7]

Brittleness Temperature:

No.	Value	Note
1	25°C	ASTM D746 [7]

Surface Properties & Solubility:
Solubility Properties General: Solubility varies greatly with tacticity
Cohesive Energy Density Solubility Parameters: δ 19.2 J$^{1/2}$ cm$^{-3/2}$ (calc.) [8,10]; δ 18.8 J$^{1/2}$ cm$^{-3/2}$ (25°) [9,10]
Internal Pressure Heat Solution and Miscellaneous: ΔH_s 31 J g^{-1} (18000 g mol^{-1}, C$_6$H$_6$, 25°) [55,56]; 6.4 J g^{-1} (18000 g mol^{-1}, CCl$_4$, 25° [55,57]; 3.9 J g^{-1}) (18000 g mol^{-1}, cyclohexane, 25°) [55,56]. Other heats of soln. values have been reported [58]
Plasticisers: Plasticisers used include esters of nonanoic acid (10–15%, by weight), which lower brittleness temp. and improve elongation at break [2,97], petrolatum which improves low temp. performance [2,98]; DOS (15% by weight) plus a cross-linking agent (5% by weight) which gives good low temp. props. without sacrificing room temp. strength, and bis(hexyl)azelate and bis(2-ethylhexyl)adipate which are used in food applications
Solvents/Non-solvents: Sol. hydrocarbons (above 80°), halogenated hydrocarbons (above 80°), higher aliphatic esters and ketones, diamyl ether [11]. Insol. all common organic solvents (room temp.), polar organic solvents, inorganic solvents [11]

Transport Properties:
Transport Properties General: Transport props. vary with composition (grade)

Polypropylene

Melt Flow Index:

No.	Value	Note
1	0.25–0.35 g/10 min	ASTM D1238L, pipes and sheets
2	1.5–2 g/10 min	ASTM D1238L, films
3	5–7 g/10 min	ASTM D1238L, injection moulding
4	7–10 g/10 min	ASTM D1238L, cast and tubular film[7]
5	0.3 g/10 min	MW 650000
6	1.2 g/10 min	MW 410000
7	5.3 g/10 min	MW 300000
8	13 g/10 min	MW 145000 [16]
9	0.25–70 g/10 min	range available according to grade [1]

Polymer Solutions Concentrated: K_θ values have been reported [53]

Polymer Melts: Polymer melts vary widely according to grade. Complex dynamic viscosity values have been reported [35,36]. η^* 470–4720 Pa s depending on M_n and low frequency storage modulus

Permeability of Gases: Permeability decreases with decreasing density and increasing crystallinity

Water Absorption:

No.	Value	Note
1	0.03 %	1/8" thick, 24h, ASTM D570 [6,7]

Gas Permeability:

No.	Gas	Value	Note
1	He	0.018 cm^3 mm/(m^2 day atm)	2.8 × 10^{-14} cm^2 (s Pa)$^{-1}$, 20°, 50% cryst. [51]
2	H$_2$	2.03 cm^3 mm/(m^2 day atm)	3.1 × 10^{-12} cm^2 (s Pa)$^{-1}$, 20°, 50% cryst. [51]
3	N$_2$	0.021 cm^3 mm/(m^2 day atm)	3.3 × 10^{-14} cm^2 (s Pa)$^{-1}$, 20°, 50% cryst. [51]
4	O$_2$	0.111 cm^3 mm/(m^2 day atm)	1.7 × 10^{-13} cm^2 (s Pa)$^{-1}$, 20°, 50% cryst. [51]
5	CO$_2$	0.45 cm^3 mm/(m^2 day atm)	6.9 × 10^{-13} cm^2 (s Pa)$^{-1}$, 20°, 50% cryst. [51]
6	H$_2$	3.34 cm^3 mm/(m^2 day atm)	5.1 × 10^{-12} cm^2 (s Pa)$^{-1}$, 20°, 50% cryst. [51]

Mechanical Properties:

Mechanical Properties General: Modifications to mech. props. have been reported [32,92,93]

Tensile (Young's) Modulus:

No.	Value	Note
1	1140–1550 MPa	165–225 kpsi, ASTM D638, homopolymer
2	895–1240 MPa	130–180 kpsi, ASTM D638, copolymer [6]
3	345–1035 MPa	50–150 kpsi, ASTM D638, impact modified
4	1032–1720 MPa	ASTM D638 [7]

Flexural Modulus:

No.	Value	Note
1	1172 MPa	ASTM D790 [7]
2	1172–1724 MPa	170–250 kpsi, 28° (73°F)
3	345 MPa	50 kpsi, 93°, (200°F)
4	241 MPa	35 kpsi, 121°, (250°F)
5	896–1379 MPa	130–200 kpsi, 28°, (73°F), ASTM D790, homopolymer [6]
6	276 MPa	40 kpsi, 93° (200°F)
7	207 MPa	30 kpsi, 121°, (250°F), ASTM D790, copolymer [6]
8	414–1103 MPa	60–160 kpsi, 28°

Tensile Strength Break:

No.	Value	Note	Elongation
1	31–41 MPa	4500–6000 psi, ASTM D638, homopolymer [6]	100–600%
2	28–38 MPa	4000–5500 psi, ASTM D638, copolymer [6]	200–500%
3	24–34 MPa	3500–5000 psi, ASTM D638, impact modified copolymer [6]	200–700%

Flexural Strength at Break:

No.	Value	Note
1	41–55 MPa	Flexural strength at break or yield, 6000–8000 psi, ASTM D790, homopolymer [6]
2	34–48 MPa	5000–7000 psi, ASTM D790, copolymer
3	28–41 MPa	4000–6000 psi, ASTM D790, impact modified copolymer [6]

Compressive Modulus:

No.	Value	Note
1	1034–2068 MPa	150–300 kpsi, ASTM D695 [6]

Elastic Modulus:

No.	Value	Note
1	1170–1720 MPa	170–250 kpsi, ASTM D790, general purpose
2	690–1380 MPa	100–290 kpsi, ASTM D790, high impact
3	1310–4200 MPa	190–600 kpsi, ASTM D790, flame retardant [22]

Poisson's Ratio:

No.	Value	Note
1	0.4	
2	-0.22	1.6% strain - a microporous auxetic PP has been reported [17]

Tensile Strength Yield:

No.	Value	Elongation	Note
1	29.3–38.6 MPa	11–15%	ASTM D638, mainly isotactic [7]

– Polypropylene

2	31–37 MPa		4500–5400 psi, ASTM D638, homopolymer [6]
3	22–30 MPa		3200–4300 psi, ASTM D638
4	11–28 MPa		1600–4000 psi, ASTM D638, impact modified copolymer [6]

Compressive Strength:

No.	Value	Note
1	38–55 MPa	5500–8000 psi, ASTM D695, homopolymer
2	24–55 MPa	3500–8000 psi, ASTM D695
3	24–41 MPa	3500–6000 psi, ASTM D695, impact modified copolymer [6]

Miscellaneous Moduli:

No.	Value	Note	Type
1	552 MPa	shear, d 0.91 g cm^{-3}	Creep modulus
2	526 MPa	shear, d 0.907 g cm^{-3}, injection	Creep modulus
3	1510 MPa	d 0.91 g cm^{-3}, tensile	E_c
4	1590 MPa	d 0.907 g cm^{-3}, tensile, injection moulding	E_c

Impact Strength: Notched impact strength improves with converted 60% β-form and by surface chlorination, see
Viscoelastic Behaviour: Viscoelastic behaviour has been reported [23,24,25,96]
Hardness: Shore D70–D80 (ASTM D1706, mainly isotactic); Rockwell R80–R102 (ASTM D785, commercial homopolymers); Rockwell R65–R96 (ASTM D785A); Shore D70–D73 (ASTM D2240, general commercial copolymers). Rockwell R50–R60 (ASTM D785A); Shore D45–D55 (ASTM D2240, impact modified copolymers). Rockwell C55–C60 (ASTM D785B); Rockwell R95 (ASTM D785A); Rockwell R80 (ASTM D785A, extrusion grade medium impact copolymer); Rockwell R60 (ASTM D785A, extrusion grade high-impact copolymer); Rockwell C60–C65 (ASTM D785B); Rockwell R99–R100 (ASTM D785A, general purpose, injection moulding homopolymer); Rockwell R98–R100 (ASTM D785A, thin complex part injection moulding homopolymer) [1,6,7]
Fracture Mechanical Properties: Fracture is brittle below about 30° for the homopolymer. Plasticising effect of ethylene additions suppresses yield stress and ductile/brittle transition temp. Fracture toughness, K_c 5.5–7.5 Mn m$^{-3/2}$ [34]. Can be altered by fillers [32,33] see polypropylene, filled and polypropylene polyethylene copolymer
Friction Abrasion and Resistance: Wear 0.03 mm^3 m^{-1} (max.) (5.5 N load, simulated marine slurries), 1.1 mm^3 m^{-1} (5.5 N load, 120 μm silicon carbide paper, dry, shows ductile flow at surface. The grade used is not stated) [18]. Particle erosion 17.72 mm^3 cm^{-2} (250 g of 120 grade silicon carbide at 0.123 g s^{-1}). Cavitation erosion 11–20 mm^3 cm^{-2} (5h). Abrasive wear 9.14 mm^3 cm^{-2} (sample, Shore D76) [19]
Izod Notch:

No.	Value	Notch	Note
1	20–320 J/m	Y	0.4–6 ft lb in^{-1}, isotactic, ASTM D256 [1,6,7]
2	5–40 J/m	Y	0.1–0.7 ft lb in^{-1}, isotactic, ASTM D256 [1,6,7]
3	80–130 J/m	Y	23°, extrusion grade [1,6,7]
4	530 J/m	Y	34°, medium impact [1,6,7]
5	640 J/m	Y	23°, high impact [1,6,7]
6	35–65 J/m	Y	23°, general purpose [1,6,7]
7	20–35 J/m	Y	23°, injection moulded [1,6,7]
8	130–530 J/m	Y	23°, general purpose, medium-high impact [1,6,7]
9	70 J/m	Y	23°, injection moulded [1,6,7]

Electrical Properties:
Electrical Properties General: Excellent electrical props., especially low electrical losses. May be modified by filler content where applicable [23]. Electrical conductivity [6] is proportional to exp. (inverse temp.). Plots have different slopes above and below the softening temp. (390K). Observed high activation energy at T >390K implies intrinsic conductivity, low activation energy at T <90K implies impurity conductivity. Conductivity increases with sample thickness. Variations with temp. have been reported [21]. Other electrical props. have been reported, see polypropylene films
Dielectric Permittivity (Constant):

No.	Value	Frequency	Note
1	2.2–2.3	1 kHz	ASTM D150 [7]
2	2.2–2.6	10 kHz–1 GHz	ASTM D1531

Dielectric Strength:

No.	Value	Note
1	24 kV.mm^{-1}	610 V mil^{-1}, ASTM D149 [6,7]
2	17 kV.mm^{-1}	120° [6,7]
3	20 kV.mm^{-1}	500 V mil^{-1}, ASTM D149, impact modified copolymer [6]

Arc Resistance:

No.	Value	Note
1	136–185s	ASTM D495 [7]

Static Electrification: If the concentration of light positive ions is greater than 500000 cm^{-3} (20°, 40% relative humidity) the surface charge density tends to zero [78]
Dissipation (Power) Factor:

No.	Value	Frequency	Note
1	0.0003	60 Hz	ASTM D510 [7]
2	0.001	100 MHz	ASTM D510 [7]
3	0.0005–0.0018	1 MHz	[7]
4	0.0002–0.0003	1 MHz	ASTM D150 [22]

Optical Properties:
Refractive Index:

No.	Value	Note
1	1.49	25°

Transmission and Spectra: Solid state nmr of atactic and isotactic polymer have been reported [80,81] (see also isotactic polypropylene and atactic polypropylene). FT-ir spectral data have been reported [90]. Other ir spectral data have been reported [91,95]
Molar Refraction: Concentration dependence of refractive index has been reported [61,62,63,64,65,66,67]
Volume Properties/Surface Properties: Translucent-opaque (ASTM D791) [22], yellow index 3–15% (ASTM D1925), gloss 48–51% (ASTM D523), luminosity 70 (min.) (ASTM D1635) [7]

Polymer Stability:
Polymer Stability General: Polymer stability is generally good

– Polypropylene

Upper Use Temperature:

No.	Value	Note
1	85°C	approx.
2	120°C	approx. [23]

Decomposition Details: Pyrolysis occurs at 469° and combustion occurs at 550°. Thermal decomposition reaction is predominantly 'back-biting' to the fifth or ninth carbon of the chain end
Flammability: Flammability has been reported [44,50,51,52,53,57]. Lowest flame spread rate 0.04–0.06 cm s^{-1} [54]. Flammability can be reduced by various additives phosphorus or halogen compds [56,58], antimony compounds [53,54,58], ammonium polyphosphate [59,60] and various inorganic fillers [61,62,63]. Limiting oxygen index 18.6% (ASTM D2863) [37]; 17.0% (ASTM D2863) [38]
Environmental Stress: Readily stress crazes [23]. Does not stress crack (ASTM D1693) [7]. Subject to uv degradation, with deterioration of tensile props., with maximum photochemical sensitivity at 370 nm [23,26]. Outdoor lifetime 0.2 y (50% loss of tensile and impact props.)
Chemical Stability: Attacked by oxidising agents [2,41,42,43]; photo-oxidative degradation leads to embrittlement. Antioxidants are always added to commercial grades. Decomposed by sulfuric acid above 100° [2] generating CO_2
Hydrolytic Stability: Excellent hydrolytic stability [2]
Recyclability: Recycled polypropylene can be blended with virgin material to give a 'monopolymer blend' [99]. Because PP undergoes rapid thermomechanical degradation, re-stabilising is necessary [99]

Applications/Commercial Products:

Processing & Manufacturing Routes: Early batch manufacture using $TiCl_3$ catalysts activated by $Al(C_2H_5)_2Cl$ in a hydrocarbon medium was replaced by continuous processes following basically the same reaction route. Older plants still use advanced $TiCl_3$ catalysts in the Solvay process, giving less catalyst residue and an atactic fraction low enough not to need separation. Counter-current washing of the polymer slurry with fresh monomer ensures high purity in the Sumitomo process. $MgCl_2$ - supported, high yield catalysts, introduced by Montedison and Mitsui Petrochemicals, give sufficiently low catalyst residue not to need removal but atactic fraction separation is still required. Current (simplified) plants based on these superactive supported catalysts have lower environmental impact due to reduced waste/recovery requirements and lower energy consumption. Metallocene catalyst processes are being commercialised by some producers. Recent developments have enabled polypropylene similar to commercial Ziegler-Natta materials to be prod. but with narrow MW distributions (M_w/M_n 2). High yield of isotactic fraction is obtained or, with an appropriate variation of catalyst, syndiotactic [1,28,29,30,31]. Injection moulding process temp. is traditionally 260–290° [6] but with current high flow rate polymers can be as low as 200°. Blow moulding is finding application with newer grades. Extrusion blow moulding uses lower melt temp. (205–215°); injection blow moulding is at higher temp. Stretch blow moulding causes biaxial orientation increasing stiffness, low temp. impact strength and clarity. In extrusion a temp. range of 205–260° is possible using low melt flow rate, high MW grades. Modifications, including graft co-polymerisation, are often used. Thermoforming is not a traditional way of processing polypropylene but newer grades and refined equipment have overcome the narrow temp. range required and the tendency to sag

Mould Shrinkage (%):

No.	Value	Note
1	1–2.5%	0.01–0.025 in/in linear, ASTM D955, homopolymer [6]
2	2–2.5%	0.02–0.025 in/in linear, ASTM D955, copolymer [6]
3	1–2.5%	0.01–0.025 in/in linear, ASTM D955, impact modified copolymer [6]

Applications: Because of the range of struct./property variations available, polypropylene grades are available to suit most polymer processing technologies. Injection moulded components are used as automotive parts (fascias, bumpers, batteries, etc.) appliances, consumer products (closures, integrally hinged boxes, child-proof caps, toys, cups, dishes, outdoor furniture), rigid packaging, medical products (syringes, pans, trays, utensils). Blow moulded products include bottles (high clarity) and automotive parts. Extrusion and thermoforming forms some pipes and profiles but mostly sheets which are then thermoformed for food containers, trays (esp. for microwavability). Fibres see polypropylene fibres used in carpeting, non-woven fabrics. Films see polypropylene films used in packaging

Trade name	Details	Supplier
"Standard Samples"		American Polymer Standards Corporation
3950X		Amoco Polymers
ACCPRO	various grades	Amoco Polymers
ACCTUF	various grades	Amoco Polymers
API	PP	American Polymers
APP/CPP	various grades	Ferro Corporation
Adell	various grades	Adell Plastics Inc.
Adpro	AP	Novacor Chemicals
Adpro AP		Genesis Polymer
Alkorprop		Alkor Plastics (UK) Ltd.
Amoco		Amoco Chemical Company
Appryl		Elf Atochem (UK) Ltd.
Appryl	isotactic	Elf Atochem (UK) Ltd.
Aqualoy	125/135/145	Comalloy Intl. Corp.
Arpro	expanded beads	ARCO Chemical Company
Astryn		Himont
Astryn	various grades	Montell
Avisun		Avisun Corporation
Azdel	P/PD/PH/PM	Azdel Inc.
BSP Polypropylene		Parkland Engineering Ltd.
Bapolene	various grades	Bamberger Polymers
Bynel	CXA	DuPont
C/CP/CRV/D/F/FF/FP/FT/L/LP/TI	various grades	Aristech Chemical
CTI	PP/PPX	CTI
Cadplen		Caduata sas di R. Fornasero and C.
Carlona	isotactic	Shell Chemicals
Carlona P		Hellyar Plastics Ltd.
Cefor	SRD4	Shell Chemical Co.
Celstran	PPG/PPS	Hoechst Celanese
Celstran	PPG	Polymer Composites
Colonial	various grades	Colonial
Comalloy	various grades	Comalloy Intl. Corp.
Comshield	101/105/106/110/160	Comalloy Intl. Corp.

Polypropylene

Comtuf	101/102/103/ 104/105/106	Comalloy Intl. Corp.
Conductomer	PPC-30	Synthetic Rubber
Daplen		PCD Polymere GmbH.
Daplen	isotactic	PCD Polymere GmbH.
Eastman		Eastman Chemical International AG
El Rexene		Rexall Drug and Chemical Corporation
Electrafil	various grades	DSM Engineering
Eltex P	isotactic	Solvay
Eltrex P		Solvay
Empee	PP	Monmouth Plastics
Encore	100% recycled	P. P. Payne Ltd.
Endura	various grades	Polymer Products
Enviro Plastic	PT-PP-150RCF	Planet Polymer
Epolene		Eastman Chemical Company
Escalloy	104/124/129	Comalloy Intl. Corp.
Escon		Enjay Chemical
Escorene	various grades	Exxon Chemical
Estraprop		Estra-Kunststoff GmbH.
Eucatherm		Van Raaijen Kunststoffen
FPC	various grades	Federal Plastics
FPP	40GR16NA	Ferro Corporation
Faralloy	PP-150	ICC Industries
Ferrene	RPE10HW	Ferro Corporation
Ferrex	various grades	Ferro Corporation
Ferrocon	EPP99GA01/02	Ferro Corporation
Ferropak	TPP20WL	Ferro Corporation
Fiberfil	various grades	DSM Engineering
Fiberstran	G-60	DSM Engineering
Fina	not exclusive	Fina Chemicals (Division of Petrofina SA)
Fortilene		Solvay Polymers
Fusabond	various grades	DuPont, Canada
Fusabond MZ		DuPont
G-60/G-63/J-60/ J-62/PF/PP60/ PP62	various grades	DSM Engineering
GXE	35	ICI, UK
Gapex	various grades	Ferro Corporation
HCPP	various grades	Chisso Corporation
Hiform	M_nB25B	Hitachi
Higlass	various grades	Montell
Hiloy	various grades	Comalloy Intl. Corp.
Hipol	various grades	Mitsui Petrochem
Hoechst Wax		Hoechst Celanese
Hostalen PP		Hoechst (UK) Ltd. Polymer Division
Hostaten PP	isotactic	Hoechst (UK) Ltd. Polymer Division
Isplen		Repsol Quimica
Isplen	isotactic	Repsol Quimica
LPP/MPP/NPP/ PPF/PPM/RPP/ TPP	various grades	Ferro Corporation
Lubricomp	MFL/ML	LNP Engineering
Lupareen		BASF
Magnacomp	ML/MM	LNP Engineering
Mapal		Technotherm Ltd.
Maplex		Alamo Polymer Corporation
Marlex	various grades	Phillips 66 Co.
Microthene	PP 6820-HU	Quantum Chemical
Mitsubishi		Mitsubishi Chemical
Mon Prop		Monoplas Industries Ltd.
Moplen		Himont
Moplen		Enimont Iberica SA
Moplen	various grades	Montell
Moplen	isotactic	Novamont
Multi-Pro	various grades	Multibase
Multi-flam	PP	Multibase
Napryl		Naphthachimie SA
Napryl	isotactic	Naphthachimie SA
Neopolen P	foam	BASF
Noblen		Mitsubishi Chemical
Noblen	various grades	Sumitomo Chem Co
Norchem	NPP	Quantum Chemical
Nortruff		Quantum Chemical
Nortruff		USI
Novolen		BASF
Novolen	isotactic	BASF
Nyloy	various grades	Nytex Composites
Oleform		Avisun Corporation
Oleform	various grades	Chisso Corporation
Olehard	R	Chisso Corporation
Optum	DPP	Ferro Corporation
P	various grades	Thermofil Inc.
PD	8020.62	A. Schulman Inc.
PD	8020.62	A. Schulman Inc.
PO/P	various grades	MA Industries
PP	E series	Epsilon
PP	various grades	Polycom Huntsman
PP	GF series	TP Composites
PP	various grades	Nan Ya Plastics
PP/APP	various grades	Washington Penn
PPH-200	10GA/20GA/ 30GA	MRC Polymers Inc.
PRO-FAX	various grades	Montell
Pemublend		Plastenterprises Pemú Ltd.

– Polypropylene

Petrothene PP		Quantum Chemical
Petrothene PP		USI
Plaspylene		Plastribution Ltd.
Plastecnics		Europlastecnics SA
Polifil	various grades	Polifil Inc.
Poly Pro		Chisso Corporation
Poly-Hi	PP	Poly-Hi
Polyfine	MF	Tokuyama Corp.
Polyflam	RPP	A.Schulman Inc.
Polyflam	RPP	A.Schulman Inc.
Polyfort	XP-1	Rototron Corp.
Polyfort	FPP/PP	A.Schulman Inc.
Polypro	isotactic	Chisso Corporation
Polypro	various grades	Tokuyama Corp.
Polypropylene	A12/PPA12	Albis Corp.
Polypropylene	various grades	Huntsman Chemical
Polypropylene	5C97	Shell Chemical Co.
Polypropylene	various grades	IPC/GWC
Procond	H/X	United Composites
Profax		Hercules
Profax	isotactic	Hercules
Prolen	various grades	Polibrasil
Propadex	recycled	Mainetti Recycling Ltd.
Propathene		ICI Acrylics
Propathene	isotactic	ICI Acrylics
Proplex		Royalite Plastics Ltd.
Propylex		Royalite Plastics Ltd. (Industrial Sheet Division)
Prostat	CLX/H	United Composites
Prylene		Société Normandde Mati&feres Plastique
RTP	various grades	RTP Company
RTP	various grades	RTP Company
RX5/SC5		Spartech
Repsol		Commercial Quimica Insular SA
Resinol	Type 0	Allied Resinous
Rexene		Rexene Products
Rotothon	XP-1	Rototron Corp.
Royalite 95		Royalite Plastics Ltd.
Rxloy	MG NPP00N-Q8012NA	Ferro Corporation
Salvage	233	MRC Polymers Inc.
Senolen		Senova Kunststoffe GmbH
Shell	not exclusive	Shell Chemicals
Shell	isotactic	Shell Chemicals
Shuman	various grades	Shuman Plastics
Snialene		Deutsche Snia Vertriebs GmbH
Stamyroid		DSM (UK) Ltd.
Stanylan P		DSM
Stat-Kon	various grades	LNP Engineering
Suplen		Sumica SA
Synthetic	ABS	Washington Penn
Tafmer	XR106L/107L	Mitsui Petrochem
Tamcin	PP-C2FR	Washington Penn
Tatren		Slovnaft Joint Stock Company
Tatren	isotactic	Slovnaft Joint Stock Company
Tecafine PP		Ensinger Ltd.
Technodur		Campos 1925 SA
Technopro		Mario Lombardini SRL
Tempalloy	105/114/120	Comalloy Intl. Corp.
Tenite		Eastman Chemical Company
Tenite		Albis (UK) Ltd.
Tenite	isotactic	Eastman Chemical Company
Thermocomp	M/MC/MF/MFX	LNP Engineering
Tipplen		Tiszai Vegyi Kombinát
Tipplen	isotactic	Tiszai Vegyi Kombinát
Tonen	RPP	Tonen Chemical
Tonen Polypro	various grades	Tonen Chemical
Torayfan	PC-1	Toray Industries
Transpalene		Neste
Trovidor 500	isotactic	Anbus Pares-Parfonry SA
Trovidur		HT (UK) Ltd.
UBE Polypro	various grades	UBE Industries Inc.
URT		Parkland Engineering Ltd.
Unipol	PP 5C06L/7C56	Shell Chemical Co.
Unite	MCP/MP	Aristech Chemical
VB	511	Washington Penn
Valtec		Enimont Iberica SA
Valtec	various grades	Montell
Verton	MFX	LNP Engineering
Vestolen	isotactic	Hüls AG
Vestolen P		Chemische Werke Huels AG
Vitralene		Stanley Smith and Company Plastics Ltd.
Volara	various grades	Voltek
Voloy	100/102/112/113	Comalloy Intl. Corp.
WRS7	300/301	Shell Chemical Co.

Bibliographic References
[1] *Kirk-Othmer Encycl. Chem. Technol.*, 4th edn., (eds. J.I. Kroschwitz and M. Howe-Grant), Wiley Interscience, 1993, **17**, 784
[2] *Encycl. Polym. Sci. Eng.*, 2nd edn., (ed. J.I. Kroschwitz), John Wiley and Sons, 1985, **13**, 474
[3] Roff, W.J. and Scott, J.R., Fibres, Films, *Plastics and Rubbers*, Butterworths, 1971

[4] Natta, G., Mazzanti, G., Crespi, G. and Moraglio, G., *Chim. Ind. (Milan)*, 1957, **39**, 275
[5] ASTM Method D5492, ASTM
[6] Guide to Plastics, *Property and Specification Charts*, McGraw-Hill, 1987
[7] Bai, F., Li, F., Calhoun, B.H., Quirk, R.P. and Cheng, S.Z.D., *Polym. Handb.*, 4th edn., (eds. J. Brandrup, E.H. Immergut and E.A. Grulke), John Wiley and Sons, 1999, V, 26
[8] Vocks, F., J. Polym. Sci., Part A-2, 1964, 5319
[9] Hayes, R.A., *J. Appl. Polym. Sci.*, 1961, **5**, 318
[10] Grulke, E.A., *Polym. Handb.*, 3rd edn., (eds. J. Brandrup and E.H. Immergut), Wiley Interscience, 1989
[11] Fuchs, O., *Polym. Handb.*, 3rd edn., (eds. J. Brandrup and E.H. Immergut), Wiley Interscience, 1989
[12] Wunderlich, B., *Polym. Handb.*, 3rd edn., (eds. J. Brandrup and E.H. Immergut), Wiley Interscience, 1989
[13] Cowrie, J.M.G., *Eur. Polym. J.*, 1973, **9**, 1041
[14] Branclas, J., Spieth, E. and Lekakau, C., *Polym. Eng. Sci.*, 1996, **36**, 49
[15] Ott, H.J., *Plast. Rubber Process. Appl.*, 1981, **1**, 9, (thermal conductivity)
[16] Chen, L.F., Wong, B. and Baker, W.E., *Polym. Eng. Sci.*, 1996, **36**, 1594
[17] Pickles, A.P., Alderson, K.L. and Evans, K.E., *Polym. Eng. Sci.*, 1996, **36**, 636
[18] Larsen-Basse, J. and Tadjvar, A., *Wear*, 1988, **122**, 135
[19] Bohm, H., Betz, S. and Ball, A., *Tribol. Int.*, 1990, **23**, 399
[20] Paukkerig, R. and Lehtinen, A., *Polymer*, 1993, **34**, 4075
[21] Singh, H.P. and Gupta, D., *Indian J. Pure Appl. Phys.*, 1986, **24**, 444
[22] *CRC Materials Science and Engineering Handbook*, 2nd edn., (eds. J.F. Shackleford, W. Alexander and J.S. Park), CRC Press, 1994
[23] *Engineered Materials Handbook*, (ASM International Handbook Committee), ASM International, 1995
[24] Kahl, R., *Principles of Plastics Materials Seminar Centre for Professional Advancement*, 1979
[25] Horsley, R., *Appl. Polym. Symp.*, (ed. O. Delatychi), 1971, **17**
[26] Hirt, R.C. and Searle, N.D., *Appl. Polym. Symp.*, 1967, **4**, 61
[27] Kojima, T., Mem. Natl. Def. Acad., Math., Phys., *Chem. Eng.*, 1989, **28**, 139
[28] Breslow, D.S. and Newburg, N.R., *J. Am. Chem. Soc.*, 1957, **79**, 5072, (metallocene catalysts)
[29] Ewen, J.A., *J. Am. Chem. Soc.*, 1984, **106**, 6355, (catalyst)
[30] Ewen, J.A., Jones, R.L., Razavi, A. and Ferrara, J.D., *J. Am. Chem. Soc.*, 1988, **110**, 6255, (catalyst)
[31] Schut, J.H., *Plast. Technol.*, 1992, **38**, 31, (metallocene catalyst)
[32] Fernando, P.L. and Williams, J.G., *Polym. Eng. Sci.*, 1980, **20**, 215
[33] Fernando, P.L., *Polym. Eng. Sci.*, 1988, **28**, 806
[34] Hashemi, S. and Williams, J.G., *J. Mater. Sci.*, 1984, **19**, 3746
[35] Ghodgaonkar, P.G. and Sundararaj, U., *Polym. Eng. Sci.*, 1996, **36**, 1656
[36] Levitt, L., Macosco, C.W. and Pearson, S.D., *Polym. Eng. Sci.*, 1996, **36**, 1647
[37] Price, D., Horrocks, A. and Tunc, M., *Chem. Br.*, 1987, **23**, 235, (flammability)
[38] Brydson, J.A., *Plast. Mater.*, 4th edn., Butterworth, 1982
[39] *Degrad. Stab. Polyolefins*, (ed. G. Geuskens), Applied Science, 1983, 63
[40] Garton, A., Carlsson, D.J. and Wiles, D.M., *Dev. Polym. Photochem.*, 1980, **1**, 93
[41] Razumovski, S.D. and Zaikov, G.E., *Dev. Polym. Stab.*, 1983, **6**, 239
[42] Carlsson, D.J., Garton, A. and Wiles, D.M., *Macromolecules*, 1976, **9**, 695
[43] Peeling, J. and Clark, D.T., J. Polym. Sci., *Polym. Chem. Ed.*, 1983, **21**, 2047
[44] Hirschler, M.M., *ACS Symp. Ser.*, 1990, 520
[45] *Encycl. Polym. Sci. Eng.*, 2nd edn., (ed. J.I. Kroschwitz), John Wiley and Sons, 1985, **3**, 390
[46] Paukkeri, R. and Lehtinen, A., *Polymer*, 1993, **34**, 4075, (props)
[47] Burfield, D.R., Loi, P.S.T., Doi, Y. and Mezjik, J., *J. Appl. Polym. Sci.*, 1990, **41**, 1095
[48] Ranby, B. and Rabek, J.F., Photodegradation, Photo-oxidation and Photostabilisation of Polymers, *Principles and Applications*, Wiley, 1975, 128
[49] Burfield, D.R., Loi, P.S.T., Dori, Y. and Menzik, J., *J. Appl. Polym. Sci.*, 1990, **41**, 1095
[50] Rudolph, H., Makromol. Chem., *Makromol. Symp.*, 1988, **16**, 57
[51] Naito, Y., Mizoguchi, K., Terada, K. and Kamiya, Y., J. Polym. Sci., *Part B: Polym. Phys.*, 1991, **29**, 457
[52] Pauly, S., *Polym. Handb.*, 3rd edn., (eds. J. Brandrup and E.H. Immergut), John Wiley and Sons, 1989
[53] Khan, H.U. and Bhattacharyya, K.K., *J. Macromol. Sci., Chem.*, 1987, **24**, 841
[54] Cheng, S.Z., Janimak, J.J., Zhang, A. and Hsieh, E.T., *Polymer*, 1991, **32**, 648
[55] Orwoll, R.A., *Polym. Handb.*, 3rd edn., (eds. J. Brandrup and E.H. Immergut), Wiley Interscience, 1989
[56] Ochiai, H., Ohashi, T., Tadokoro, Y. and Murakami, I., *Polym. J. (Tokyo)*, 1982, **14**, 457
[57] Ochiai, H., Nishihara, Y., Yamaguchi, S. and Murackami, I., J. Sci. Hiroshima Univ., Ser. A: Math., Phys., *Chem.*, 1977, **41**, 157
[58] Phuong-Nguyen, H. and Delmas, G., *Macromolecules*, 1979, **12**, 740, 746
[60] Schreiber, H.P. and Waldman, M.H., J. Polym. Sci., Part A-2, 1967, **5**, 555
[61] Huglin, M.B., *Polym. Handb.*, 3rd edn., (eds. J. Brandrup and Immergut), Wiley Interscience, 1989
[62] Elias, H.G. and Dietschy, H., *Makromol. Chem.*, 1967, **105**, 102
[63] Florian, S., Lath, D. and Manasek, Z., *Angew. Makromol. Chem.*, 1970, **13**, 43
[64] Kinsinger, J.B. and Hughes, R.E., *J. Phys. Chem.*, 1959, **63**, 2002
[65] Drott, E.E. and Mendelson, R.A., J. Polym. Sci., *Part B: Polym. Lett.*, 1964, **2**, 187
[66] Kotera, A., Saito, T., Matsuda, H. and Wada, A., *Rep. Prog. Polym. Phys. Jpn.*, 1965, **8**, 5
[67] Yamaguchi, K., *Makromol. Chem.*, 1969, **128**, 19
[68] Berger, K.C. and Meyerhoff, G., *Polym. Handb.*, 3rd edn., (eds. J. Brandrup and E.H. Immergut), Wiley Interscience, 1989
[69] Shostenko, A.G. and Myshkin, V.E., *Kinet. Katal.*, 1979, **20**, 781
[70] McKenna, T.F. and Hamielec, A.E., *Polym. Handb.*, 3rd edn., (eds. J. Brandrup and E.H. Immergut), Wiley Intescience, 1989
[71] Busfield, W.K., *Polym. Handb.*, 3rd edn., (eds. J. Brandrup and E.H. Immergut), Wiley Interscience, 1989
[72] Billmeyer, F.W., *Textbook of Polymer Science*, John Wiley and Sons, 1984
[73] Kwei, K.P. and Kwei, T.K., *J. Phys. Chem.*, 1962, **66**, 2146
[74] Allen, G., Booth, C., Gee, G. and Jones, M.N., *Polymer*, 1964, **5**, 367
[75] Brown, W.B., Gee, G. and Taylor, W.D., *Polymer*, 1964, **5**, 362
[76] Ochiai, H., Gekko, K. and Yamamura, H., J. Polym. Sci., Part A-2, 1971, **9**, 1629
[77] Di Paola-Baranyi, G., Braun, J.M. and Guillet, J.E., *Macromolecules*, 1978, **11**, 224
[78] Bigos, J., Skalny, J. and Dindosova, D., *CA*, 1990, **112**, 32027w
[79] Li, Z., He, Y., Shi, G., Zhang, J. and Cao, Y., *CA*, 1987, **106**, 103062h
[80] Koenig, J.L., *Chemtracts: Macromol. Chem.*, 1990, **1**, 356
[81] Schaefer, D., Speiss, H.W., Syter, U.W. and Flemming, W.W., *Macromolecules*, 1990, **23**, 3431
[82] Gaur, U. and Wunderlich, B., *J. Phys. Chem. Ref. Data*, 1981, **10**, 1051
[83] Varma-Nair, M. and Wunderlich, B., *J. Phys. Chem. Ref. Data*, 1991, **20**, 349
[84] Zoller, P., *Polym. Handb.*, 3rd edn., (eds. J. Brandrup and E.H. Immergut), Wiley Interscience, 1989
[85] Zoller, P., *J. Appl. Polym. Sci.*, 1979, **23**, 1057
[86] Foster, G.N., Waldman, N. and Griskey, R.G., *Polym. Eng. Sci.*, 1966, **6**, 131
[87] Zoller, P., J. Polym. Sci., *Polym. Phys. Ed.*, 1978, **16**, 1491
[88] Jain, R.K. and Simha, R., J. Polym. Sci., *Polym. Phys. Ed.*, 1979, **17**, 1929
[89] Pottiger, M.T. and Laurence, R.L., J. Polym. Sci., *Polym. Phys. Ed.*, 1984, **22**, 903
[90] Batra, N., Singh, S.R. and Seghal, V.N., *Indian J. Forensic Sci.*, 1987, **1**, 161
[91] Bu, H.-S., Aycock, W. and Wunderlich, B., *Polymer*, 1987, **28**, 1165
[92] Ramsteiner, F., Konig, G., Heckmann, W. and Gruber, W., *Polymer*, 1983, **24**, 365
[93] Fernando, P.L. and Williams, J.-G., *Polym. Eng. Sci.*, 1981, **21**, 1003
[94] Minoshima, W., White, J.L. and Spruiell, J.E., *Polym. Eng. Sci.*, 1980, **20**, 1166
[95] *The Infrared Spectra Atlas of Monomers and Polymers*, Sadtler Research Laboratories, 1980
[96] *U.S. Pat.*, 1965, 3 201 364
[97] *U.S. Pat.*, 1965, 3 178 386
[98] *U.S. Pat.*, 1966, 3 281 390
[99] Marrone, M. and La Mantia, F.P., *Polym. Recycl.*, 1996, **2**, 9, 17
[100] Rabello, M.S. and White, J.R., *Plast. Rubber Compos. Process. Appl.*, 1996, **25**, 237

Polypropylene, Aluminium filled

Synonyms: *PP, aluminium filled*
Related Polymers: Polypropylene, Isotactic Polypropylene, Polypropylene, Filled
Monomers: Propylene
Material class: Composites
Polymer Type: polypropylene, polyolefins
CAS Number:

CAS Reg. No.	Note
9003-07-0	
25085-53-4	isotactic

Additives: Aluminium (as powder, fibre or flake)
General Information: Although deleterious to mech. props. under some circumstances aluminium may be used to obtain a lightweight, electrically and thermally conductive composite
Morphology: Crystallinity calculated from DSC thermograms agrees with the estimate from x-ray studies. Crystallinity decreases with increasing Al powder content up to a content of 1.7%. Crystallinity then shows a small increase up to filler content of 3.3% [4]

Volumetric & Calorimetric Properties:
Volumetric Properties General: Thermal conductivity of polypropylene filled with aluminium fibre can be adequately predicted by Nielsen's theoretical development. An increase relative to the unfilled polymer of about 2 times at 12 vol.% Al and 8 times for 30 vol.% has been reported [1,2]. The gradual increase in thermal conductivity of PP filled with aluminium powder gives good agreement with several two-phase literature models. A non-linear increase in Vicat softening point with filler volume level occurs [4]

Transport Properties:
Polymer Melts: Melt viscosity of aluminium powder filled isotactic polymer shows power law dependence on shear stress at temps. 180–220°. Melt viscosity increases with aluminium content to a maximum at approx. 9% aluminium, decreasing with increasing aluminium content beyond this level to less than that of unfilled polymer at 26% aluminium. Die swell ratio and first normal stress difference decrease with increasing aluminium content to a minimum also at approx. 9% aluminium [4]

Mechanical Properties:
Mechanical Properties General: Tensile props. are dependent on crystallinity, decreasing with increasing aluminium powder content [4]. Tensile modulus shows a drastic decrease at 1.63% aluminium and only a slight further decrease at greater aluminium levels. Tensile strength decreases steadily as aluminium content increases, while elongation at break decreases sharply at 0.33% aluminium and then more gradually. Flexural modulus increases steadily with increasing aluminium content. Flexural strength increases up to 1% aluminium then decreases slightly up to 3% aluminium and increases again up to 7% aluminium, relative to unfilled isotactic polymer [3]. These trends are less marked if aluminium fibres are used, especially with polymer modified to give some polymer-filler adhesion [1]
Impact Strength: There is an initial increase in Izod impact strength with increasing aluminium powder content up to 5% aluminium but then a decrease with further increasing aluminium content [3]

Electrical Properties:
Electrical Properties General: Spherical (or irregular) shaped filler particles need to be present at >38 vol.% to obtain a conductive composite - defeating cost and density advantages. If aluminium fibres are used, resistivity as low as 200 Ω cm at 7.7 vol.% aluminium is possible. The greater the aspect ratio of the fibres (or flakes) the lower the filler content at which resistivity falls sharply [1]. The conductivity of polypropylene-aluminium composites prepared by mech. mixing or by polymerisation filling follows the percolation theory. Coefficients of conductivity anisotropy range from $0.1-10^{-12}$ depending on the volume fraction of filler [5]

Polymer Stability:
Thermal Stability General: Increasing aluminium filler content raises the activation energy of the composite and hence increases thermal stability [4]

Applications/Commercial Products:
Processing & Manufacturing Routes: Roll milling at approx. 160–200° [1,3]
Applications: EMI shielding, static electricity dissipation, thermal conductivity - especially where low density and/or reduced cost are important

Bibliographic References
[1] Bigg, D.M., *Composites*, 1979, **10**, 95
[2] Nielsen, L.E., *J. Appl. Polym. Sci.*, 1973, **17**, 3819
[3] Maiti, S.N. and Mahapatro, P.K., *Polym.-Plast. Technol. Eng.*, 1991, **30**, 559
[4] Maiti, S.N. and Mahapatro, P.K., *J. Polym. Mater.*, 1988, **5**, 179
[5] Borisov, Yu.V., Grinev, V.G., Kudinova, O.I., Novokshonova, L.A. and Tarasova, G.M., *Acta Polym.*, 1992, **43**, 131

Polypropylene blends

Synonyms: *Polypropene blends. PP blends. Polypropylene alloys. PAB's*
Related Polymers: Polypropylene, Polypropylene copolymers, Blends with polyethylene, Blends with propylene-ethylene copolymer, EPDM blends, Vulcanised, Blends with nylon, Blends with rubber, Blends with polycarbonate, Blends with styrenes, Blends with polyester, Isotactic polypropylene, Polypropylene films
Material class: Thermoplastic, Blends
Polymer Type: polypropylene
Additives: Compatibilisers: a copolymer or graft or by reactive blending
General Information: Blended polymers are rapidly increasing their market share. Polypropylene is blended with other polymers to improve toughness, impact strength, mech. props., chemical resistance, flame retardance, processability (very important), etc. Most commercially important are blends with polyethylene, propylene-ethylene random copolymer, ethylene-propylene elastomers (EPM, EPDM) and nylons. A recently expanding area is thermoplastic elastomer blends such as polypropylene with natural rubber, butyl rubber, butadiene rubber, modified butadiene rubbers, etc. Also significant but of lesser commercial importance as yet are blends with polycarbonate, styrenics and thermoplastic polyesters
Morphology: Almost all blends are immiscible and normally comprise a dispersion of one component within a matrix of the other. Blend props. are controlled by morphology, composition, compounding method and processing methods. With some compositions, the distinction between a blend and a copolymer becomes vague. Micromorphology depends on one of three situations; simultaneous crystallisation of both components (when T_c ranges overlap), crystallisation of one component in the presence of the liq. phase of the other or crystallisation of one component in the presence of a solid phase of the second (heterogeneous nucleation)
Miscellaneous: By the very nature of blends, large variation in props. occurs as a result of proportions of components, and the exact nature of components, method of mixing, as well as all the usual causes of variation. It follows that where props. are given, ranges may be very wide and not necessarily exhaustive. Blends of PP, rheology of such blends, melt mixing and applications have been reported [1]. Eastern Bloc work on polymer blends has developed rather differently from western countries

Bibliographic References
[1] Plochocki, A.P., *Polym. Blends*, (eds. D.R. Paul and S. Newman), Academic Press, 1978, **2**, (rev)

Polypropylene blends with polyethylene

Synonyms: *PP-PE blends. Polypropene-polyethene blends. Polyethylene blends with polypropylene*
Related Polymers: Polypropylene blends, Propylene-ethylene copolymers, Polyethylene
Monomers: Propylene, Ethylene
Material class: Thermoplastic, Blends
Polymer Type: polyethylene, polypropylene
CAS Number:

CAS Reg. No.
9003-07-0
9002-88-4

Additives: Compatibilisers may be added, usually a co-polymer such as EPR or EPDM, or a peroxide (for reactive blending)
General Information: Addition of PE to PP improves the low temp. impact props. and also the drawability of PP fibres. Most

– Polypropylene blends with polyethylene

common are PP blends with LDPE, LLDPE and HDPE. Less common are blends with ULDPE ($\rho = 890$ kg m^{-3}) and ultrahigh MW PE. Blend performance and props. are greatly affected by MW and MW distribution, component ratio, nucleating agents, compatibilisation and thermal or radiative history

Morphology: The blends are immiscible. Compatibilisation may be achieved by addition of a copolymer, reactive blending (using peroxides) or post blending treatment by chemical cross-linking, electron beam radiation or γ-radiation. Reactive compatibilisation or radiation lead to lower blend crystallinity. Micromorphology is controlled by composition, nucleating agents, compatibilisers, and thermal or radiative treatment. As little as 10% HDPE in PP gives a large reduction in the PP spherulite size. Finer spherulite size and better mech. props. are obtained if crystallisation is carried out at temps. less than or equal to 127° (T_c for PE). Blends have the most ordered struct. at 40–70% PP. PP-PE copolymers are more ordered than blends and are most ordered at about 95.5% PP. At up to approx. 50% PE, struct. is spherulitic within a PP matrix. At higher PE levels interpenetrating networks are observed [18,19]. There is no evidence of adhesion between the matrix and the dispersed phase at any composition. Morphology (and fracture props.) are greatly modified by the addition of small amounts of EPR [18,19]. Morphological props. of UHMWPE/LMWPP gel films which are fibrilous without apparent separate domains have been reported [17]. Considerable work has been done to relate mech. props. and rheological props. to morphology [19,31,32,33,34,35,36,37]

Volumetric & Calorimetric Properties:
Density:

No.	Value	Note
1	d 0.91–0.96 g/cm^3	PP/HDPE [8,16]
2	d 0.89–0.91 g/cm^3	approx. follows simple additive rule PP/ULDPE [24]

Latent Heat Crystallization: ΔH_f (PP) 85–102 J g^{-1} (max. at 30–40% ULDPE); ΔH_f (ULDPE) 51–54 J g^{-1} (monotonic increase with PP content); ΔH_f (total) 52–94 J.g^{-1} (slight devation-negative near pure components, positive for mid-range compositions from simple additive rule [24]

Melting Temperature:

No.	Value	Note
1	156–160°C	PP
2	130–136°C	HDPE, as PP content increases [8]
3	152.5°C	PP
4	105.5°C	LDPE, initial melting point, all compositions [20]

Transition Temperature General: For polypropylene blends with virtually all types of PE, two distinct melting, crystallisation (or other transition) temps. are observed. Slight changes may occur with composition (sometimes observed to be test method dependent [20]) but are so small as to not contradict the immiscibility of the components. T_m 161° (PP) increases to max. of 161.8° at 50% ULDPE then decreases to 161.4°; actual values depend on grade of ULDPE; similarly T_m (ULDPE) increases with ULDPE content to value depending on grade. T_c (PP) 112° increases to max. 119° then decreases (according to ULDPE grade); T_c (ULDPE) decreases with increasing ULDPE content to 59° [24]

Vicat Softening Point:

No.	Value	Note
1	60–150°C	PP-ULDPE, significant negative deviation from simple additive rule [24]

Surface Properties & Solubility:
Plasticisers: EPM [7,11,12,15]

Transport Properties:
Polymer Melts: Zero shear viscosity η_o 9.05 poise (180°, 25% HPPE/75% PP), 6.02 poise (200°, 25% HDPE/75% PP); 2.88 poise (180°, 50% HDPE, 50% PP); 2.12 poise (200°, 50% HDPE, 50% PP); 2.27 poise (180°, 75% HDPE, 25% PP); 1.77 poise (200°, 75% HDPE, 25% PP) [9,16]. Zero shear viscosity decreases with increasing temp. and increase in HDPE composition. Complex viscosity η^* 2.1–3.2 Pa s (PP/ULDPE) [24]. Values vary with composition. Other viscosity measurements have been reported [25]

Permeability of Gases: Water vapour transmission 0.128 g d^{-1} m^{-2} (60% PP, 40% PE, ASTM E96-66B); 0.054 g d^{-1} m^{-2} (33% PP, 60% PE, ASTM E96-66B); 0.63–0.8 g d^{-1} m^{-2} (10% PP, 90% PE, ASTM E96-66B) [16]

Mechanical Properties:
Mechanical Properties General: The nature and proportions of the components are not sufficient to determine the mech. props. Depending on the time, temp. and rate of mixing the 'same blend' can show different behaviour. The mixing procedure should be optimised [21,22,23]. Mech. props. can be greatly altered by the incorporation of a compatibilising agent. The extent of such effects is dependent on the chemical struct. and/or on the molecular mass of the added copolymer as well as on the blend composition [11,12,15]

Tensile (Young's) Modulus:

No.	Value	Note
1	735 MPa	max. at 80% PP [18]
2	686 MPa	7000 kg cm^{-2}, PP
3	1078 MPa	11000 kg cm^{-2}, HDPE
4	1034 MPa	150000 psi [30]
5	1379–1724 MPa	200000–250000 psi, injection moulded
6	620–724 MPa	90000–105000 psi, compression moulded
7	482–620 MPa	70000–90000 psi, max. at 25–50% PP, pressed and quenched [21]
8	90–896 MPa	13000–130000 psi [29]
9	240–1250 MPa	0.24–1.25 GPa [25]
10	2000–15000 MPa	draw ratio 20
11	15000–50000 MPa	min. at 70% PP, draw ratio 40
12	22000–50000 MPa	draw ratio 60
13	30000–100000 MPa	approx., linear, draw ratio 100 [17,30]

Flexural Modulus:

No.	Value	Note
1	130–840 MPa	linear [20]
2	245–1240 MPa	23°
3	1210–3310 MPa	-40° [20]
4	100–1500 MPa	[19]

Tensile Strength Break:

No.	Value	Note
1	22–35 MPa	max. at 75–90% PP
2	15.9–30 MPa	50% PP [8,16,18,28]

– Polypropylene blends with polyethylene

3	21 MPa	approx., max., PP/LLDPE [15,20,25]
4	300 MPa	draw ratio 10
5	500 MPa	PP
6	600 MPa	max. at 70% PP
7	400 MPa	PE, draw ratio 20
8	600–900 MPa	PP
9	1300–1500 MPa	PE, draw ratio 40
10	700 MPa	PP
11	1600 MPa	PE, draw ratio 60
12	1000 MPa	PP
13	2200 MPa	PE, draw ratio 100
14	3000 MPa	approx., draw ratio 300 [17,30]

Elastic Modulus:

No.	Value	Note
1	1000 MPa	approx., all compositions [12]

Tensile Strength Yield:

No.	Value	Elongation	Note
1	21–34 MPa		[8,18,28]
2	<25 MPa	20%	max., 50:50 PP/LDPE [31]

Miscellaneous Moduli:

No.	Value	Note	Type
1	827 MPa	PP	Secant modulus
2	896 MPa	max. at 90% PP	Secant modulus
3	669 MPa	PE [16]	Secant modulus
4	110–450 MPa	5% elongation [20,21]	5% modulus
5	50–220 MPa	linear [20]	Shear modulus

Impact Strength: Tensile impact strength 67 kJ m^{-2} (32 ft lb in^{-1}, 100% PP), 21 kJ m^{-2} (10 ft lb in^{-1}, 45% PP/55% LDPE), 189 kJ m^{-2} (90 ft lb in^{-1}, 100% LDPE) [15]

Fracture Mechanical Properties: Extent of stress whitening for impact test specimens, including those modified with, has been reported [11]

Izod Notch:

No.	Value	Notch	Note
1	53.4 J/m	Y	1 ft lb in^{-1}, 100% PP [26]
2	107 J/m	N	2 ft lb in^{-1}, 50% PP [26]
3	587 J/m	N	11 ft lb in^{-1}, 25% PP [26]
4	1070 J/m	N	20 ft lb in^{-1}, 100% HDPE [26]
5	60–87 J/m	Y	room temp., 90% PP/10% ULDPE [24]
6	21 J/m	Y	0°, 5–10% ULDPE [24]
7	16 J/m	Y	-20°, 5–10% ULDPE [24]
8	13–16 J/m	Y	-30°, 5–10% ULDPE [24]

Izod Area:

No.	Value	Notch	Note
1	4 kJ/m^2	Unnotched	100% PP [1]
2	2 kJ/m^2	Unnotched	75% PP/25% HDPE [1]
3	1.3 kJ/m^2	Unnotched	50% PP/50% HDPE [1]
4	2.2 kJ/m^2	Unnotched	25% PP/75% HDPE [1]
5	8.4 kJ/m^2	Unnotched	100% HDPE [1]
6	9 kJ/m^2	Unnotched	50% PP/50% LLDPE [29]
7	53 kJ/m^2	Notched	room temp., 95% PP/5% ULDPE [24]

Charpy:

No.	Value	Notch	Note
1	3.5	N	100% PP [20]
2	76	N	100% LDPE [20]

Optical Properties:
Transmission and Spectra: Ir absorption band ratios 720–731 cm^{-1} (PE). 973–998 cm^{-1} (PP) may be used to determine blend composition [23]
Total Internal Reflection: Birefringence 0.040–0.042 (25% PE); 0.043–0.048 (40% PE) (20° <T <170°, sharp fall in value at approx. 170°, UHMWPE/LMWPP Gel film) [17]

Polymer Stability:
Recyclability: Much of the interest in these blends has been generated by the fact that PP, HDPE, LDPE constitute a major proportion of plastic waste which is therefore a potential source [21]

Bibliographic References
[1] Polypropylene: Structure, *Blends and Composites: Copolymers and Blends,* (ed. J. Karger-Kocsis), Chapman and Hall, 1995, **2**
[2] Ledneva, O.A., Popov, A.A. and Zaikov, G.E., Vysokomol. Soedin., Ser. B, 1990, **32**, 785
[3] Tong, X., Zhang, Z., Xiang, Z., Lu, R. et al, Suliao, 1989, **18**, 11
[4] Sawatari, C., Satoh, S. and Matsuo, M., Polymer, 1990, **31**, 1456
[5] Wang, L. and Huang, B., J. Polym. Sci., Part B: Polym. Phys., 1990, **28**, 937
[6] Martuscelli, E., *Polym. Eng. Sci.,* 1984, **24**, 563
[7] Martuscelli, E., Silvestre, C. and Abate, G., *Polymer,* 1982, **23**, 229
[8] Greco, R., Mucciariello, G., Ragosta, G. and Martuscelli, E., *J. Mater. Sci.,* 1980, **15**, 845
[9] Alle, N. and Lyngaae-Jørgensen, J., *Rheol. Acta,* 1980, **19**, 94
[10] Alle, N. and Lyngaae-Jørgensen, J., *Rheol. Acta,* 1980, **19**, 104
[11] D'Orazio, L., Greco, R., Mancarella, C., Martuscelli, E. et al, Polym. Eng. Sci., 1982, **22**(9), 536, (HDPE/PP blends)
[12] D'Orazio, L., Greco, R., Martuscelli, E. and Ragosta, G., *Polym. Eng. Sci.,* 1983, **23**, 489, (mech props)
[13] Greco, R., Mucciariello, G., Ragosta, G. and Martuscelli, E., *J. Mater. Sci.,* 1981, **16**, 1001, (fibrous samples)
[14] Tang, M.R., Greco, R., Ragosta, G. and Cimmino, S., *J. Mater. Sci.,* 1983, **18**, 1031, (ultradrawn fibres)
[15] Nolley, E., Barlow, J.W. and Paul, D.R., *Polym. Eng. Sci.,* 1980, **20**, 364, (LDPE/PP blends)
[16] Noel, O.F. and Carley, J.F., *Polym. Eng. Sci.,* 1975, **15**, 117, (rev)
[17] Ogita, T., Kawahara, Y., Sawatari, C., Ozaki, F. and Matsuo, M., Polym. J. (Tokyo), 1991, **23**, 871
[18] Lovinger, A.J. and Williams, M.L., J. Appl. Polym. Sci., 1980, **25**, 1703
[19] Wenig, W. and Meyer, K., *Colloid Polym. Sci.,* 1980, **258**, 1009
[20] Teh, J.W., *J. Appl. Polym. Sci.,* 1983, **28**, 605
[21] Deanin, R.D. and Sansone, M.F., Polym. Prepr. (Am. Chem. Soc., *Div. Polym. Chem.),* 1978, **19**, 211
[22] Valenza, A., La Mantia, F.P. and Acierno, D., *Eur. Polym. J.,* 1984, **20**, 727
[23] Mirabella, F.M., *Adv. Chem. Ser.,* 1990, **227**
[24] Lee, Y.K., Jeong, Y.T., Kim, K.C., Jeong, H.M. and Kim, B.K., *Polym. Eng. Sci.,* 1991, **31**, 944
[25] Dumoulin, M.M., Farha, C. and Utracki, L.A., *Polym. Eng. Sci.,* 1984, **24**, 1319
[26] Bartlett, D.W., Barlow, J.W. and Paul, D.R., J. Appl. Polym. Sci., 1982, **27**, 2351
[27] Rizzo, G. and Spadaro, G., *Eur. Polym. J.,* 1988, **24**, 303
[28] Robertson, R.E. and Paul, D.R., *J. Appl. Polym. Sci.,* 1973, **17**, 2579
[29] Cheung, P., Suwanda, D. and Balke, S.T., *Polym. Eng. Sci.,* 1990, **30**, 1063
[30] Sawatari, C., Satah, S. and Matsuo, M., *Polymer,* 1990, **31**, 1456

[31] Bartozak, Z., Galeski, A. and Pracella, M., *Polymer*, 1986, **27**, 537
[32] Galeski, A., Pracella, M. and Martuscelli, E., *J. Polym. Sci., Polym. Phys. Ed.*, 1984, **22**, 739
[33] Martuscelli, E., Pracella, M., Della Volpe, G. and Greco, P., *Makromol. Chem.*, 1984, **185**, 1041
[34] Noel, O.F. and Carley, J.F., *Polym. Eng. Sci.*, 1984, **24**, 488
[35] Gallagher, G.A., Jakeways, R. and Ward, I.M., *J. Appl. Polym. Sci.*, 1991, **43**, 1399
[36] Long, Y., Stachurski, Z.H. and Shanks, R.A., *Polym. Int.*, 1991, **26**, 143
[37] Fujiyama, M. and Kawasaki, Y., *J. Appl. Polym. Sci.*, 1991, **42**, 467

Polypropylene/butadiene rubber blend P-483

Synonyms: *Polypropylene/polybutadiene blend. PP/BR blend. Polypropylene/cis-polybutadiene blend*
Related Polymers: Polypropylene/rubber blends
Monomers: Propylene, 1,3-Butadiene
Material class: Thermoplastic, Synthetic Elastomers
Polymer Type: polypropylene, polybutadiene
CAS Number:

CAS Reg. No.
9003-07-0
9003-17-2

Mechanical Properties:
Tensile Strength Break:

No.	Value	Note	Elongation
1	20.8 MPa	60:40 BR:PP [1]	258%

Viscoelastic Behaviour: Tension set 27% [1]

Bibliographic References
[1] Coran, A.Y., Patel, R.P. and Williams, D., *Rubber Chem. Technol.*, 1982, **55**, 116

Poly(propylene-co-butylene) P-484

$$\left[(CH_2CH(CH_3))_{\overline{x}}(CH_2CH(CH_2CH_3))_y \right]_n$$

Synonyms: *Poly(1-butene-co-1-propene). Poly(1-propene-co-1-butene). Butene, polymer with 1-propene*
Related Polymers: Polypropylene copolymers, Polypropylene-ethylene-butylene terpolymer
Monomers: 1-Butene, Propylene
Material class: Thermoplastic, Copolymers
Polymer Type: polypropylene, polyolefins
CAS Number:

CAS Reg. No.	Note
29160-13-2	1-butene, polymer with 1-propene
116257-98-8	block, isotactic
9019-30-1	butene, polymer with 1-propene
108645-83-6	1-butene, block polymer with 1-propene
133006-20-9	syndiotactic
63058-37-7	isotactic

Molecular Formula: $[(C_4H_8)(C_3H_6)]_n$
Fragments: C_4H_8 C_3H_6
General Information: Copolymers of propylene and butylene have low brittle points yet are comparable in strength and melting point to propylene homopolymers [1,2,3]
Morphology: It has been suggested that propylene/butylene copolymers are not completely random even when the monomers are mixed and processed using highly stereospecific $TiCl_3$ catalysts. For propylene between 13 and 76 mol%, two distinct crystalline phases form, one PP rich and one PB rich, suggesting poly[(propylene-co-butene-block-(butene-co-propylene)] as a general structural description

Volumetric & Calorimetric Properties:
Density:

No.	Value	Note
1	d 0.895 g/cm^3	approx., 48% wt. butene [9]

Volumetric Properties General: The apparent enthalpy of fusion, overall density and melting temp. all show a "eutectic type" variation with composition showing a minimum at about 48 wt.% butene [9]. Although melting temp. values differ, this trend has also been reported for homogenous samples prod. using metallocene catalysts [4]
Latent Heat Crystallization: Overall apparent enthalpy of fusion ΔH_f 9.84 kJ mol^{-1} (24 cal g^{-1}) (single crystals 100% PP); 4.92 kJ mol^{-1} (12 cal g^{-1}) (single crystals 100% PB); 2.05 kJ mol^{-1} (5 cal g^{-1}) (single crystals 52 wt.% propene/48 wt.% butene random copolymer) [9]
Melting Temperature:

No.	Value	Note
1	133°C	propylene/butylene 85:15 [8]
2	155°C	approx., 100% PP, melt crystallised
3	120°C	approx., 100% PB, melt crystallised
4	100°C	PP
5	65°C	50% propene/50% butene, melt crystallised random copolymer [9]

Transport Properties:
Melt Flow Index:

No.	Value	Note
1	3 g/10 min	85 wt.% propylene/15 wt.% butylene, T_m 133° [8]

Mechanical Properties:
Mechanical Properties General: Stiffness is greater than that of for comparable T_m and melt flow index [8]
Flexural Modulus:

No.	Value	Note
1	690 MPa	Secant, 85 wt.% propene/15 wt.% butene [8]

Optical Properties:
Transmission and Spectra: C-13 nmr has been used to determine comonomer sequences in the molecular struct. [7]

Polymer Stability:
Chemical Stability: 1.3% Hexane extractable material (50°); 7% xylene soluble - meets FDA requirements for food packaging materials [5]

Applications/Commercial Products:
Processing & Manufacturing Routes: Ziegler-Natta catalyst processes tend to give a heterogeneous copolymer [4]. Addition of alkoxysilane donors [5] or Lewis bases [6] during $MgCl_2$-supported Ziegler-Natta polymerisation can be used to modify the copolymer. Metallocene catalysts prod. homogenous copolymers over the whole composition range with more efficient uptake of the 1-butene [4]
Applications: Applications include food packaging

Bibliographic References

[1] Allport, D.C. and Janes, W.H., *Block Copolym.*, Applied Science, 1973
[2] Turner Jones, A., *Polymer*, 1966, **7**, 23
[3] Coover, H.W., McConnell, R.L., Joyner, F.B., Slonaker, D.F. et al, *J. Polym. Sci., Part A-1*, 1966, **4**, 2563
[4] Arnold, M., Henschke, O. and Knorr, J., *Macromol. Chem. Phys.*, 1996, **197**, 563
[5] Sacchi, M.C., Fan, Z.-Q., Forlini, F., Tritto, I. and Locatelli, P., *Macromol. Chem. Phys.*, 1994, **195**, 2805
[6] Sacchi, M.C., Shan, C., Forlini, F., Tritto, I. and Locatelli, P., *Makromol. Chem.*, *Rapid Commun.*, 1993, **14**, 231
[7] Acki, A. and Hayashi, T., *Macromolecules*, 1992, **25**, 155
[8] Ficker, H.K. and Walker, D.A., *Plast. Rubber Process. Appl.*, 1990, **14**, 103
[9] Cavallo, P., Martuscelli, E. and Pracella, M., *Polymer*, 1977, **18**, 42

Polypropylene/butyl rubber blends   P-485

Synonyms: *Polypropene/butyl rubber blends*
Related Polymers: Polypropylene/rubber blends
Material class: Thermoplastic, Synthetic Elastomers, Blends
Polymer Type: polypropylene

Volumetric & Calorimetric Properties:
Density:

No.	Value	Note
1	d 0.95–1 g/cm^3	[1]

Transport Properties:
Permeability of Gases: Relative air permeability 1.45 (35°, ASTM D1434, sample 0.76 mm thick) [2]

Mechanical Properties:
Tensile Strength Break:

No.	Value	Note	Elongation
1	21.6 MPa	60:40 IIR/PP, dimethylol phenolic cured [2,3]	380%

Viscoelastic Behaviour: Tension set 23% (60:40 IIR/PP, dimethylol phenolic cured) [2,3]
Hardness: Shore A40-A80 (ASTM D2240) [1]

Applications/Commercial Products:
Processing & Manufacturing Routes: Dynamically vulcanised during mixing
Applications: Applications due to low permeability and high damping

Trade name	Supplier
Sarlink 2000	Novacor Chem. Int.
Trefsin	Advanced Elastomer Systems

Bibliographic References

[1] *Kirk-Othmer Encycl. Chem. Technol.*, 4th edn., (eds. J.I. Kroschwitz and M. Howe-Grant), Wiley Interscience, 1994, **9**
[2] Coran, A.Y. and Patel, R.P., Polypropylene: Structure, *Blends and Composites: Copolymers and Blends*, (ed. J. Karger-Kocsis), Chapman and Hall, 1995, **2**
[3] Coran, A.Y., Patel, R.P. and Williams, D., *Rubber Chem. Technol.*, 1982, **55**, 116

Polypropylene, Calcium carbonate filled   P-486

Synonyms: *PP, Calcium carbonate filled. Polypropene, Calcium carbonate filled*
Related Polymers: Polypropylene, Polypropylene, Filled, Isotactic polypropylene
Monomers: Propylene
Material class: Thermoplastic, Composites
Polymer Type: polypropylene
CAS Number:

CAS Reg. No.
9003-07-0
25085-53-4

Molecular Formula: $(C_3H_6)_n$
Fragments: C_3H_6
Additives: Calcium carbonate (chalk, limestone, marble, dolomite, precipitated calcium carbonate, CCP) with or without coating
General Information: Addn. of chalk similar in effect to that of talc, but increases stiffness less and decreases impact strength less. Advantages (c.f. talc-filled): easy dispersal, high melt index, better UV stability, better surface finish, suitable for food use, lower abrasion

Volumetric & Calorimetric Properties:
Density:

No.	Value	Note
1	d 1.24 g/cm^3	40% CaCO$_3$ filled, ASTM D792 [1]

Thermal Expansion Coefficient:

No.	Value	Note	Type
1	$2.8 \times 10^{-5} – 5 \times 10^{-5}$ K^{-1}	40% CaCO$_3$ filled homopolymer, ASTM D696 [1]	L

Thermal Conductivity:

No.	Value	Note
1	0.29 W/mK	0.00069 cal cm^{-1} s^{-1} °C^{-1}, 40% CaCO$_3$ filled, ASTM C177 [1]

Deflection Temperature:

No.	Value	Note
1	65–77°C	150–170°F, 1.84 MPa (264 psi) [1]
2	99–121°C	210–250°F, 0.45 MPa (66 psi), 40% CaCO$_3$ filled, ASTM D648 [1]

Transport Properties:
Water Absorption:

No.	Value	Note
1	0.02–0.05 %	24h [1]
2	0.1 %	saturation, $\frac{1}{8}$" thick, 40% CaCO$_3$ filled, ASTM D570 [1]

Mechanical Properties:
Tensile (Young's) Modulus:

No.	Value	Note
1	2585–3450 MPa	375–500 kpsi, 40% CaCO$_3$, filled homopolymer [1]
2	24130 MPa	3500 kpsi, 40% CaCO$_3$, ASTM D638, filled copolymer [1]

— Polypropylene, Carbon filled

Flexural Modulus:

No.	Value	Note
1	2480–3100 MPa	360–450 kpsi, 28° (73°F)
2	2205 MPa	320 kpsi, 93° (200°F), 40% $CaCO_3$ filled homopolymer [1]
3	2070–2550 MPa	300–370 kpsi, 28° (73°F), 40% $CaCO_3$ filled copolymer, ASTM D790 [1]

Tensile Strength Break:

No.	Value	Note	Elongation
1	23–25 MPa	3400–3460 psi, 40% $CaCO_3$ filled homopolymer [1]	10–80%
2	19 MPa	2700 psi, 40% $CaCO_3$ filled copolymer, ASTM D638 [1]	40–50%

Flexural Strength at Break:

No.	Value	Note
1	38–48 MPa	5500–7000 psi, 40% $CaCO_3$ filled homopolymer [1]
2	30–45 MPa	4300–6500 psi, 40% $CaCO_3$ filled copolymer, rupture or yield, ASTM D790 [1]

Tensile Strength Yield:

No.	Value	Note
1	27 MPa	3850 psi, 40% $CaCO_3$ filled homopolymer [1]
2	26 MPa	3800 psi, 40% $CaCO_3$ filled copolymer, ASTM D638 [1]

Compressive Strength:

No.	Value	Note
1	21–50 MPa	3000–7200 psi, 40% $CaCO_3$ filled homopolymer, rupture or yield, ASTM D695 [1]

Hardness: Rockwell R78-R99 (homopolymer); R81 (40% $CaCO_3$ filled copolymer, ASTM D785) [1]

Izod Notch:

No.	Value	Notch	Note
1	32–53 J/m	N	0.6–1 ft lb in^{-1}, 40% $CaCO_3$ filled homopolymer [1]
2	37–53 J/m	N	0.7–1 ft lb in^{-1}, 1/8" thick, 40% $CaCO_3$ filled copolymer, ASTM D256A [1]

Electrical Properties:
Dielectric Strength:

No.	Value	Note
1	16.14–19.68 kV.mm^{-1}	410–500 V mil^{-1}, 1/8" thick, short time, 40% $CaCO_3$ filled, ASTM D149 [1]

Applications/Commercial Products:
Processing & Manufacturing Routes: Processing temp. range (injection moulding) 190–275° (375–525°F, 40% $CaCO_3$ filled homopolymer); 175–245° (350–470°F, 40% $CaCO_3$ filled copolymer). Moulding pressure range 55–138 MPa (8–20 kpsi, 40% $CaCO_3$ filled homopolymer); 103–138 MPa (15–20 kpsi, 40% $CaCO_3$ filled copolymer) [1]

Mould Shrinkage (%):

No.	Value	Note
1	0.7–1.4%	0.007–0.014 in/in, 40% $CaCO_3$ filled homopolymer
2	0.6–1.2%	0.006–0.012 in/in, 40% $CaCO_3$ filled copolymer (linear, ASTM D955) [1]

Applications: Used in food packaging, appliances and household goods

Trade name	Details	Supplier
ARP	Thermofill	Northern Industrial Plastics Ltd.
LPP	various grades	Ferro Corporation
PP5200 series	various grades homopolymer	Polycom Huntsman
PP7200 series	various grades copolymer	Polycom Huntsman
Polifil C	RMC - high impact	Plastics Group
Profil PP60CC		Wilson-Fiberfil
Thermofil P-[nn]CC		Thermofil
		Himont
		RTP Company
		A. Schulman Inc.
		Washington Penn

Bibliographic References
[1] Guide to Plastics, *Property and Specification Charts*, McGraw-Hill, 1987, (technical data)

Polypropylene, Carbon filled   P-487
Synonyms: *PP, Carbon filled. PP, Carbon fibre reinforced*
Related Polymers: Polypropylene, Polypropylene, filled
Monomers: Propylene
Material class: Thermoplastic, Composites
Polymer Type: polypropylene
CAS Number:

CAS Reg. No.
9003-07-0
25085-53-4

Additives: Carbon fibre or carbon powder
General Information: Carbon is a high cost filler generally used where high thermal conductivity, high electrical conductivity, and/or UV radiation protection are required. Carbon black is also used as a pigment

Volumetric & Calorimetric Properties:
Density:

No.	Value	Note
1	d 1.04 g/cm^3	30% PAN carbon fibre, conductive for EMI shielding [1]

– **Polypropylene chlorinated**

Deflection Temperature:

No.	Value	Note
1	118°C	245°F, 1.84 MPa (264 psi), 30% PAN carbon fibre, conductive for EMI shielding, ASTM D648 [1]

Transport Properties:
Water Absorption:

No.	Value	Note
1	0.12 %	24h, 1/8" thick, 30% PAN carbon fibre, conductive for EMI shielding, ASTM D570

Mechanical Properties:
Tensile (Young's) Modulus:

No.	Value	Note
1	12065 MPa	1750 kpsi, 30% PAN carbon fibre, conductive for EMI shielding, ASTM D638 [1]

Flexural Modulus:

No.	Value	Note
1	11375 MPa	1650 kpsi, 28° (73°F), 30% PAN carbon fibre, conductive for EMI shielding, ASTM D790 [1]

Tensile Strength Break:

No.	Value	Note	Elongation
1	47 MPa	6800 psi, 30% PAN carbon fibre, conductive for EMI shielding, ASTM D638 [1]	0.5%

Flexural Strength at Break:

No.	Value	Note
1	62 MPa	9000 psi, rupture or yield, 30% PAN carbon fibre, conductive for EMI shielding, ASTM D790 [1]

Izod Notch:

No.	Value	Notch	Note
1	58.7 J/m	N	1.1 ft lb in^{-1}, 1/8" thick, 30% PAN carbon fibre, conductive for EMI shielding, ASTM D256A [1]

Applications/Commercial Products:
Processing & Manufacturing Routes: Processing temp. range (injection moulding) 180–245° (360–470°F, 30% PAN carbon fibre, conductive for EMI shielding) [1]
Mould Shrinkage (%):

No.	Value	Note
1	0.1–0.3%	0.001–0.003 in/in, linear, 30% PAN carbon fibre, conductive for EMI shielding, ASTM D955 [1]

Trade name	Supplier
Cabelec	Cabot Plastics Ltd.
Electrafil	Akzo Engineering Plastics
	Wilson-Fiberfil

Bibliographic References
[1] Guide to Plastics, *Property and Specification Charts*, McGraw-Hill, 1987

Polypropylene chlorinated   P-488
Synonyms: *CPP. Poly(1-propene) chlorinated*
Related Polymers: Polypropylene
Monomers: Propylene
Material class: Synthetic Elastomers
Polymer Type: polypropylene
CAS Number:

CAS Reg. No.	Note
9003-07-0	polypropylene

Molecular Formula: $[(C_3H_5Cl).(C_3H_6)]_n$
Fragments: C_3H_5Cl C_3H_6
General Information: Chlorination up to about 50 wt.% is possible. Chlorination is reported to occur at a mixture of primary, secondary and tertiary carbons [1,4,19,22]. Industrial material has a heterogeneous struct. due to the presence of isotactic polypropylene [4]. White, odourless, non-flammable powder
Tacticity: Isotactic material has a tendency to chlorinate at the 2° carbon of the backbone ($CHClCHCH_3$) [1], and atactic at the 3° carbon (CH_2CClCH_3)
Morphology: On chlorination, the crystallinity of isotactic polypropylene decreases rapidly with increasing chloride content [1,2,6,7]. X-ray diffraction shows pronounced transformation from α to β PP when the chlorine content is between 10 and 17% [2]

Volumetric & Calorimetric Properties:
Volumetric Properties General: Density of isotactic polypropylene increases with increasing chlorine content [2]
Melting Temperature:

No.	Value	Note
1	132–197°C	25.1 wt% Cl
2	162–204°C	28 wt% Cl
3	132–236°C	35 wt% Cl
4	184–234°C	42.5 wt% Cl
5	204–252°C	53.5 wt% Cl [3]

Glass-Transition Temperature:

No.	Value	Note
1	160°C	[5]

Transition Temperature General: T_g increases sharply with increasing chlorine content [3,4]. Brittleness temp. varies with degree of chlorination; atactic polymer is brittle at room temp. if Cl content > 25 wt.% [3]. There is no sharp melting point because of chlorination at all three types of carbon [3]

Surface Properties & Solubility:
Cohesive Energy Density Solubility Parameters: Increasing polarity with increasing chlorination leads to increasing CED [3]
Wettability/Surface Energy and Interfacial Tension: Plasma surface chlorinated isotactic polymer has high surface tension; lower

values are obtained when other processes are used [8]
Adhesion General: Mechanism of adhesion promotion has been reported
Surface Tension:

No.	Value	Note
1	35.5 mN/m	$CHCl_3$ [8]

Transport Properties:
Transport Properties General: Intrinsic viscosity of chlorinated atactic polymer depends on the degree of chlorination and the synth. conditions [10]
Water Content: Chlorination of fibres by sodium hypochlorite increases water content (and affinity for cationic dyes) [9]

Mechanical Properties:
Mechanical Properties General: Tackiness of atactic polymer decreases with chlorine content and material is brittle above 25 wt% chlorine [3]. Strength of isotactic polymer decreases with increasing chlorine content [2]
Tensile Strength Break:

No.	Value	Note
1	20–27 MPa	at 20–30% Cl, low crystallinity, isotactic [2]

Impact Strength: Surface chlorination of isotactic polymer (e.g. by photochlorination with chlorine gas) increases impact strength by the formation of polyene chains [11]

Electrical Properties:
Electrical Properties General: Photochlorination of isotactic polymer in chlorine gas, followed by extensive dehydrochlorination can give conductive polyene chains [11]

Optical Properties:
Transmission and Spectra: Ir and C-13 nmr spectral data have been reported [3,4,19,22]
Volume Properties/Surface Properties: Chlorinated isotactic polymer with more than 4% chlorine gives transparent films [2]

Polymer Stability:
Polymer Stability General: Chlorination improves surface protection against photodegradation and weathering [11]
Thermal Stability General: Stability of chlorinated polymer is less than atactic polymer at above 30% chlorine; greater below 30% [15]
Decomposition Details: There are three steps to decomposition; 1. dehydrochlorination, 2. degradation of residual polypropylene segments, 3. further decomposition of the unsaturated chains from step 1 [11,15]
Flammability: Non-flammable (51% Cl) [12], high flame resistance and self extinguishing props. (Cl >30%) [2]. Limiting oxygen index 40–50% (48–60% Cl); 50–64% (64–72% Cl, which exceeds the oxygen index of PVC and fluoropolymers) [13,14]

Applications/Commercial Products:
Processing & Manufacturing Routes: Normally chlorinated with chlorine in suspension in aq. or organic solvents (e.g. CCl_4) in the presence of free radical initiators [8], e.g. with gaseous chlorine containing >0.1 vol% O or O_3 in the presence of water and $[Me_2C(CN)Ni]_2$ giving stable polymers with 5–75% Cl, at 50–60° [18]. Evolved HCl is catalytic and increased presence of isotactic polymer increases the reaction rate [19] but leads to the industrial product being heterogenous [4]. Thermal chlorination, its mechanism and the effect of parameters have been reported [3,20]. The rate of soln. photochlorination decreases with increasing degree of chlorination [4]. Metal salts accelerate chlorination of isotactic polymer but not atactic form [21]. Other production methods are surface photochlorination with Cl_2 gas [11], with $CHCl_3$ or CCl_4 plasma [8], or with sodium hypochlorite [9]
Applications: Atactic polymer used in surfactants, bonding of non-woven fabric, sizing of polypropylene yarn, fire-retardant adhesives, lubricating oil additives, antifouling coatings, primers, printing inks. Also used as a secondary plasticiser for PVC. Isotactic polymer is used as degradation and weathering protection, pre-dying treatment for fibres, film print promoter, coatings, laminates and rubber

Trade name	Supplier
Parlon P	Hercules

Bibliographic References

[1] Hawley's Condensed Chemical Dictionary, 11th edn., (eds. N.I. Sax and R.J. Lewis), Van Nostrand Reinhold, 1987
[2] Ronkin, G.M., Khromenkov, L.G. and Kolbasov, V.I., *Plast. Massy*, 1987, 20
[3] Mukherjee, A.K. and Patri, M., *J. Macromol. Sci., Chem.*, 1989, **26**, 213
[4] Li, W., Shi, L., Shen, D. and Zheng, J., *Shiyou Huagong*, 1989, **18**, 295
[5] Aubin, M. and Prud'homme, R.E., *Macromolecules*, 1988, **21**, 2945
[6] Mitani, T., Ogata, T. and Iwasaki, M., *J. Polym. Sci., Polym. Chem. Ed.*, 1974, **12**, 1653
[7] Keller, F., Pinther, P. and Hartmann, M., *Acta Polym.*, 1981, **32**, 82
[8] Joly, A.M., *Prog. Org. Coat.*, 1996, **28**, 209
[9] Shah, C.D. and Jain, D.K., *Text. Res. J.*, 1983, **53**, 274
[10] Mukherjee, A.K. and Patri, M., *Angew. Makromol. Chem.*, 1989, **171**, 131
[11] Ramelow, U.S., *Trends Inorg. Chem.*, 1991, **2**, 159
[12] Ronkin, G.M., *Lakokras. Mater. Ikh Primen.*, 1990, 23
[13] Ronkin, G.M., Serkov, B.B. and Izmailov, A.S., *Plast. Massy*, 1988, 41
[14] Aseeva, R.M., Ruban, L.V. and Lalayan, V.M., *Int. J. Polym. Mater.*, 1992, **16**, 289
[15] Li, W., Shi, L., Shen, D. and Luo, B., *Polym. Degrad. Stab.*, 1988, **22**, 375
[16] Sharma, Y.N., Satish, S. and Bhardwaj, I.S., *J. Appl. Polym. Sci.*, 1981, **26**, 3213
[17] Clemens, R.J., Batts, G.N., Lawniczak, J.E., Middleton, K.P. and Sass, C., *Prog. Org. Coat.*, 1994, **24**, 43
[18] *Jpn. Pat.*, 1967, 67 3 191, (Asahi Electrochemical)
[19] Voronin, N.I., Kulikova, V.A., Filimoshkin, A.G. and Bol'shakov, G.F., *Dokl. Akad. Nauk SSSR*, 1987, **295**, 870
[20] Mukherjee, A.K. and Patri, M., *Angew. Makromol. Chem.*, 1988, **163**, 23
[21] Puszynski, A. and Dwornicka, J., *Angew. Makromol. Chem.*, 1986, **139**, 123
[22] Li, W., Shi, L. and Shen, D., *Gaofenzi Xuebao*, 1989, 291

Polypropylene copolymers P-489
Synonyms: *Polypropene copolymers. PP copolymers*
Related Polymers: More general information on polypropylene, Isotactic polypropylene, Polypropylene copolymers with ethylene, Ethylene-propylene elastomers (EPM, EPDM, EPR), Polypropylene copolymers with butylene, Polypropylene Blends, Fillers and reinforcers - filled polypropylene, Polypropylene Films
Material class: Thermoplastic, Copolymers
Polymer Type: polypropylene
CAS Number:

CAS Reg. No.	Note
9010-79-1	propene-ethene copolymer
106565-43-9	
25895-47-0	
62849-76-7	

Additives: Stabilisers; fillers and reinforcers
General Information: Polypropylene copolymer (or copolymer polypropylene) frequently refers to propylene-ethylene (up to 10%) random copolymer. A wide range of co- and ter- (and even higher) polymers (alternating, random, block and graft) are of major importance. Depending on the manufacturing method and the resulting morphology, the difference between copolymers and blends may not be clearly defined

– Polypropylene - EPDM blends, Vulcanised

Applications/Commercial Products:

Trade name	Details	Supplier
Acctuf 3045	PP copolymer	Amoco Chemical Company
Astryn	PP copolymer	Himont
Fortilene	PP copolymer	Solvay Polymers
Fusabond MZ	PP copolymer	DuPont
Petrothene PP	PP copolymer	Quantum Chemical
Petrothene PP	PP copolymer	USI
Procond	PP copolymer electrically conductive	United Composites
Rexene	PP copolymer	Rexene Products
SD-376	PP copolymer	Himont

Polypropylene - EPDM blends, Vulcanised   P-490

Synonyms: *PP-EPDM thermoplastic vulcanisate*
Related Polymers: Polypropylene blends, Other blends with rubbers, Non-vulcanised blends with ethylene-propylene copolymers, EPDM
Material class: Thermoplastic, Synthetic Elastomers, Copolymers, Blends
Polymer Type: polyethylene, polypropylenepolyolefins
Additives: Compatibilisers may be present, otherwise, except for colourants additives tend to be detrimental
General Information: If polypropylene - EPDM mixtures are dynamically vulcanised during blending, the harder polypropylene phase is continuous while the rubbery EPDM phase is a uniform, fine, highly vulcanised dispersion. Props. improve as the EPDM particle size decreases
Identification: Distinct from unvulcanised/part-vulcanised thermoplastic elastomers, thermoplastic vulcanisates have less than 3% extractable rubber in cyclohexane at room temp.
Miscellaneous: Comparison of polypropylene/EPDM (vulcanised) blends with Polypropylene/ethylene-propylene copolymer blends has been reported [10]

Volumetric & Calorimetric Properties:
Density:

No.	Value	Note
1	d 0.88–1.04 g/cm^3	[4]

Vicat Softening Point:

No.	Value	Note
1	97–143°C	commercial grades, ASTM D1525

Brittleness Temperature:

No.	Value	Note
1	-63--34°C	commercial grades, ASTM D746

Surface Properties & Solubility:
Cohesive Energy Density Solubility Parameters: As δ for each phase is similar, mixing is good, giving smaller rubber particles and hence good mech. props. [3,9]
Solvents/Non-solvents: Thermoplastic vulcanised blends are swollen by a range of organic solvents, cyclohexane, DMF, propanol etc. and hot oil [8]

Transport Properties:
Polymer Melts: Rheology is highly non-Newtonian - viscosity varies greatly with shear rate. Log viscosity vs. log shear rate is linear over 3 orders of magnitude for each variable (viscosity high at low shear, low at high shear). Viscosity much less sensitive to temp. [3]
Permeability of Gases: Relative air permeability 4.44 (0.76 mm thick, 35°, ASTM D1434) [2]
Water Absorption:

No.	Value	Note
1	>0.8 %	min., 24h

Mechanical Properties:
Mechanical Properties General: Props. vary widely depending on the specific formulation. Commercial grades have increased variation because of the use of fillers (esp. carbon black) and extender oil. The diene (cross-linking agent) in the EPDM phase affects props. [2]. Fatigue performance is superior to natural rubber, EPDM, neoprene [3]. Tensile props. are anisotropic - a standard injection moulded plaque exhibits a strong direction and a weak direction

Tensile (Young's) Modulus:

No.	Value	Note
1	56–435 MPa	no additives [2,7]
2	7.3–235 MPa	ASTM D412, commercial grades [3]

Flexural Modulus:

No.	Value	Note
1	18.5–347 MPa	ASTM D790, commercial grades [3]

Tensile Strength Break:

No.	Value	Note	Elongation
1	9.1–28.8 MPa	no additives [2,7]	165–580%
2	4.5–31 MPa	[4]	150–600%
3	4.4–27.6 MPa	ASTM D412, commercial grades [3]	330–600%

Tensile Strength Yield:

No.	Value	Elongation	Note
1	11.4 MPa	50%	commercial grade at Shore D50 [3]

Miscellaneous Moduli:

No.	Value	Note	Type
1	3.9–13.6 MPa	no additives [2,7]	100% modulus
2	2–10 MPa	ASTM D412, commercial grades [3]	

Mechanical Properties Miscellaneous: Tension set 7–46% (no additives) [2,7]; 8–61% (ASTM D412, commercial grades, 100% strain) [3]; 16% (60–40 EPDM:PP) [8]. Stress at 5% compression 0.14–3.7 MPa (ASTM D695, commercial grades) [3]. Stress at 10% compression 0.27–>6.89 MPa (ASTM D695, commercial grades) [3]. Compression set 16–33% (23°, 22h, 25% deflection, ASTM D395B); 17–58% (100°, 22h, 25% deflection, ASTM

D395B, commercial grades) [3]. Bashore Resilience 38–51% (first rebound, ASTM D2632, commercial grades) [3]
Hardness: Shore A55-D50 (commercial grades) [3]; A45-D50 [5,7]
Failure Properties General: Tear strength 108–514 pli (23°, ASTM D624); 42–364 pli (100°, Die C, ASTM D624, commercial grades) [3]
Friction Abrasion and Resistance: Pico abrasion 82.9–544.1 (abrasion index, ASTM D2228), NBS abrasion 25–408 (abrasion index, ASTM D1630, commercial grades) [3]

Electrical Properties:
Electrical Properties General: Dielectric strength values have been reported [3]
Surface/Volume Resistance:

No.	Value	Note	Type
1	$10-20 \times 10^{15}$ Ω.cm	23° and 100°	S

Dielectric Permittivity (Constant):

No.	Value	Frequency	Note
1	2.7–3.1	60 Hz	ASTM D15081 [3]

Dissipation (Power) Factor:

No.	Value	Note
1	0.0021–0.0088	0.21–0.88%, 23° [3]
2	0.0015–0.0044	0.15–0.44%, 100° [3]

Polymer Stability:
Thermal Stability General: Heat stable up to 260° [3]
Upper Use Temperature:

No.	Value	Note
1	135°C	[3]

Flammability: Fl. p. 340° (min.). Limiting oxygen index 19% (max., general purpose grades); 23–28% (flame retardant grades) [3]
Environmental Stress: Good resistance to ionising radiation and fungi. Resistance to oxygen, ozone and atmospheric pollutants is less than for styrenic, olefinic blend and MPR thermoplastic elastomers [3]
Chemical Stability: Resistant to aq. media including conc. acids and alkalis, and polar solvents; as polarity decreases so does resistance - poor resistance to halocarbons [3]. Resistance to hydrocarbons only fair but better than Polypropylene/ethylene-propylene copolymer blends. Resistance to saturated hydrocarbons better than to aromatics [3]
Hydrolytic Stability: Resistant to aq. media including conc. acids and alkalis [3]
Recyclability: Recyclable [5]. Mech. props. decrease with each recycling [3]

Applications/Commercial Products:
Processing & Manufacturing Routes: Dynamically vulcanised during mixing. Components are mixed under conditions of intensive shear with cross-linking of the EPDM taking place during mixing (dynamic vulcanisation [6]) by the additon of a (non-peroxide) curing agent. Processing as normal for thermoplastics [4,5]
Applications: Better oil resistance, lower compression set and softer than non-vulcanised material. Applications in automotive industry (hose coverings, duct covers, gaskets, seals, convoluted boots, vibration dampers, ignition components, window seals, etc.); architecture and construction (expansion joints, roofing, flooring, weather seals); electrical and electronic industries (primary insulators, jacketing, combined primary insulation and jacketing, connectors, computers, telecommunications, office equipment), hose, tubing, sheet, medical and food-contact applications (FDA approved grades), mechanical rubber goods, impact modifiers for PP, PE (but not for more polar polymers)

Trade name	Supplier
Hifax XL	Himont
Santoprene	Advanced Elastomer Systems
Sarlink 3000	Novacor
Telprene	Teknor Apex

Bibliographic References
[1] Martuscelli, E., *Polym. Eng. Sci.*, 1984, **24**, 563
[2] Polypropylene: Structure, *Blends and Composites: Copolymers and Blends*, (ed. J. Karger-Kocsis), Chapman and Hall, 1995, **2**
[3] Rader, C.P., *Handb. Thermoplast. Elastomers*, 2nd edn. (eds. B.M. Walker and C.P. Rader), Van Nostrand Reinhold, 1988
[4] *Encycl. Polym. Sci. Eng.*, 2nd edn., (ed. J.I. Kroschwitz), John Wiley and Sons, 1986, **5**
[5] *Kirk-Othmer Encycl. Chem. Technol.*, 4th edn., (eds. J.I. Kroschwitz and M. Howe-Grant), Wiley Interscience, 1993, **9**
[6] *U.S. Pat.*, 1962, 3 037 954, (Esso)
[7] Coran, A.Y. and Patel, R., *Rubber Chem. Technol.*, 1980, **53**, 141
[8] Coran, A.Y., Patel, R. and Williams, D., *Rubber Chem. Technol.*, 1982, **55**, 116, 1063
[9] Barton, A.F.M., *CRC Handbook of Solubility Parameters and Other Cohesion Parameters*, CRC Press, 1983
[10] Abdou-Sabet, S. and Patel, R.P., *Rubber Chem. Technol.*, 1991, **64**, 769

Polypropylene-ethylene-butylene terpolymer   P-491

$$\text{-}[(CH_2CH(CH_3))_{\overline{x}}\text{-}(CH_2CH_2)_{\overline{y}}\text{-}(CH_2CH(CH_2CH_3))_{\overline{z}}]_n$$

Synonyms: *Poly(1-propene-co-ethene-co-1-butene)*
Related Polymers: Propylene-ethylene copolymers, Propylene-butylene copolymers
Monomers: Propylene, Ethylene, 1-Butene
Material class: Copolymers
Polymer Type: polyethylene, polypropylenepolyolefins
CAS Number:

CAS Reg. No.
25895-47-0
119008-27-4
133006-22-1

Molecular Formula: $[(C_4H_8).(C_2H_4).(C_3H_6)]_n$
Fragments: C_4H_8 C_2H_4 C_3H_6
General Information: Poly(propylene-ethylene-butylene) terpolymer is significant over a very wide range of monomer ratios. At certain compositions the terpolymers are elastomers without detectable crystallinity, demonstrating a random sequence distribution and thus being promising as rubbers [1]. Introduction of 1-butene into EPR decreases T_g and improves mech. props. Crystallinity and elasticity are determined by the distribution of monomer units [2]

Volumetric & Calorimetric Properties:
Density:

No.	Value	Note
1	d 0.86 g/cm^3	approx. [1]

– Polypropylene/ethylene-propylene copolymer blends

Melting Temperature:

No.	Value	Note
1	120–145°C	depending on composition; propylene 14–95 mol%, ethylene 3–75 mol%, 1-butene 2–23 mol% [1,3,4]

Surface Properties & Solubility:
Solvents/Non-solvents: Dependent on composition, polymer is insol. heptane [2]

Transport Properties:
Permeability of Gases: Gas permeability increases as ethylene content (density and crystallinity) decreases [1]

Gas Permeability:

No.	Gas	Value	Note
1	O_2	262.7–1543.2 cm^3 mm/(m^2 day atm)	4–23.5 × 10^{-10} cm^2 (s cmHg)$^{-1}$ [1]
2	N_2	65.7–394 cm^3 mm/(m^2 day atm)	1–6 × 10^{-10} cm^2 (s cmHg)$^{-1}$ [1]

Mechanical Properties:
Mechanical Properties General: Complex modulus, loss modulus and loss tangent as function of temp. have been reported [1]

Tensile Strength Break:

No.	Value	Note	Elongation
1	12.4 MPa	126 kg cm^{-2}, carbon filled, vulcanised [1]	492%

Miscellaneous Moduli:

No.	Value	Note	Type
1	8.7 MPa	89 kg cm^{-2}, carbon filled, vulcanised [1]	300% modulus

Viscoelastic Behaviour: Permanent deformation 16% (carbon filled, vulcanised) [1]
Hardness: Shore A66 (carbon filled, vulcanised) [1]

Optical Properties:
Transmission and Spectra: C-13 nmr spectral data have been reported [1]

Applications/Commercial Products:

Trade name	Supplier
Polypro	Chisso Corporation
Vestoplast	Hüls AG
Vestoplast	Shell Chemicals

Bibliographic References
[1] Sun, L. and Lin, S., *J. Polym. Sci., Part A: Polym. Chem.*, 1990, **28**, 1237
[2] Seidov, N.M., Guseinov, F.O., Abasov, A.I., Ibragimov, Kh.D. and Askerov, V.M., Vysokomol. Soedin., *Ser. B*, 1975, **17**, 2076, (Russian, English abstract)
[3] Kawamoto, N., Mori, H., Nitta, K., Yui, N. and Terano, M., *Macromol. Chem. Phys.*, 1996, **197**, 3523
[4] Guidetti, G.P., Busi, P., Giulianelli, I. and Zannetti, R., *Eur. Polym. J.*, 1983, **19**, 757

Polypropylene/ethylene-propylene copolymer blends
P-492

$$-[(CH_2CH(CH_3))_n]_p-[(CH_2CH(CH_3))_x-(CH_2CH_2)_y]_q-$$

Synonyms: *Polypropylene/EPR blend. Polypropylene/EPDM blends, Unvulcanised. PP/EPR TPO. PP/EPDM TPO*
Related Polymers: Polypropylene blends, Ethylene-propylene copolymers, Vulcanised blends, Polyethylene
Monomers: Propylene, Ethylene
Material class: Thermoplastic, Synthetic Elastomers, Blends
Polymer Type: polyethylene, polypropylene
Additives: Additives, which include fillers, reinforcing agents, lubricants, plasticisers, heat stabilisers, antioxidants, uv absorbers, colourants, flame retardants, flow modifiers, and processing aids are mostly present in the rubbery phase at room temp. Some migration to the polypropylene phase may occur at processing temp.
General Information: Blends of polypropylene with ethylene-propylene random copolymer or unvulcanised EPDM are treated as essentially similar. Continuity of both phases is present, resulting in a soft product
Miscellaneous: Comparison of polypropylene/ethylene-propylene copolymer blends with Polypropylene-EPDM blends, vulcanised, has been reported [10]

Volumetric & Calorimetric Properties:
Density:

No.	Value	Note
1	d 0.89–1.03 g/cm^3	[3]
2	d 1.95 g/cm^3	[3]

Thermal Expansion Coefficient:

No.	Value	Note	Type
1	5.4 × 10^{-5}–0.000117 K^{-1}	3 × 10^{-5}–6.5 × 10^{-5} F^{-1}	L

Melting Temperature:

No.	Value	Note
1	202°C	95 PP: 5 EPM
2	207°C	90 PP: 10 EPM
3	215°C	80 PP: 20 EPM [9]

Brittleness Temperature:

No.	Value	Note
1	-80--10°C	[3]

Surface Properties & Solubility:
Plasticisers: Paraffinic-type processing oils including Sunpar 150, Sunpar 2280 and Tufflo 6056 [7]. Plasticisers must have high fl. p. due to high processing temps.

Transport Properties:
Melt Flow Index:

No.	Value	Note
1	0.4–11.4 g/10 min	230°

Polymer Melts: Melt viscosity 220–365 Pa s [3]

Mechanical Properties:

Mechanical Properties General: Props. can be affected if this type of blend is formulated using EPDM, the specific diene, even if uncured. High ethylene content provides strength for the rubber component as some crystallinity is present. Equal ethylene/propylene contents gives a softer material with good impact props.

Tensile (Young's) Modulus:

No.	Value	Note
1	72 MPa	no additives [4]

Flexural Modulus:

No.	Value	Note
1	17.2–1558 MPa	2.5–226 kpsi, commercial grades [3]

Tensile Strength Break:

No.	Value	Note	Elongation
1	0.37–28.8 MPa	53–4177 psi, commercial grades [3]	50–1000%

Miscellaneous Moduli:

No.	Value	Note	Type
1	2.2 MPa	320 psi, TPR 5470 [3]	100%
2	9 MPa	TPR 5490 [3]	100%
3	117 MPa	17000 psi, -51°	Gehman
4	55 MPa	8000 psi, -43°	Gehman
5	24 MPa	3500 psi, -34°	Gehman
6	13.8 MPa	2000 psi, -26°	Gehman
7	6.9 MPa	1000 psi, -18°, commercial grades [3]	Gehman

Mechanical Properties Miscellaneous: Compression set 32–49% (22°, 22h, commercial grades) [3]
Hardness: Shore A66-D72 (commercial grades) [3]
Failure Properties General: Tear strength 80–900 pli (die C, commercial grades) [3]

Electrical Properties:

Dielectric Permittivity (Constant):

No.	Value	Frequency	Note
1	2.37	1 MHz	ASTM D150, filled [3]

Dielectric Strength:

No.	Value	Note
1	21.26–26 kV.mm^{-1}	540–660 V mil^{-1}, general
2	21.26 kV.mm^{-1}	540 V mil^{-1}, 0.08" thick, filled, ASTM D149 [3]

Arc Resistance:

No.	Value	Note
1	114s	filled (TPR 5595), ASTM D495 [3]

Polymer Stability:

Upper Use Temperature:

No.	Value	Note
1	105°C	general purpose formulations [3]
2	125°C	with stabilisers and antioxidants [3]
3	140°C	approx., depending on blend and/or components [3]

Environmental Stress: Weather resistance is good but non-fast grades may undergo a colour change. Stable to ozone [3]
Chemical Stability: Unaffected by H_2O and aq. reagents; acids and bases have little effect. Swollen and softened by hydrocarbon solvents especially hot solvents [3]

Applications/Commercial Products:

Processing & Manufacturing Routes: Processed on normal thermoplastic processing equipment-injection moulding, extrusion, vacuum forming, injection blow moulding, extrusion blow moulding etc. Melt viscosity may be relatively high; newer grades are available to overcome this. Machine parameter tables for processing have been reported [3]

Mould Shrinkage (%):

No.	Value	Note
1	0.6–1.7%	6.0–17 mil in^{-1} [3]

Trade name	Supplier
ETA	Republic Plastics
Ferroflex	Ferro Corporation
Hifax	Himont
Polytrope	Schulman
RPI	International
Ren-Flex	Dexter Plastics
Ren-Flex	Research Polymers
TPR	Advanced Elastomer Systems
TPR	BP Performance Polymers Inc
Telcar	Teknor Apex
Vistaflex	Exxon Chemical

Bibliographic References

[1] Martuscelli, E., *Polym. Eng. Sci.*, 1984, **24**, 563
[2] Varga, J. and Garzo, G., *Angew. Makromol. Chem.*, 1990, **180**, 15
[3] *Handb. Thermoplast. Elastomers*, 2nd edn., (eds. B.M. Walker and C.P. Rader), Van Nostrand Reinhold, 1988
[4] Polypropylene: Structure, *Blends and Composites: Copolymers and Blends*, (ed. J. Karger-Kocsis), Chapman and Hall, 1995
[5] *Kirk-Othmer Encycl. Chem. Technol.*, 4th edn., (eds. J.I. Kroschwitz and M. Howe-Grant), Wiley Interscience, 1993, **9**
[6] *Encycl. Polym. Sci. Eng.*, 2nd edn., (ed. J.I. Kroschwitz), John Wiley and Sons, 1986, **5**
[7] *Thermoplast. Elastomers Rubber-Plast. Blends*, (eds. S.K. De and A.K. Bhowmick), Horwood, 1990, (rev)
[8] Wolfe, J.R., *Thermoplastic Elastomers*, (ed. N.R. Legge), Hanser, 1987
[9] Martuscelli, E., Silvestre, C. and Abate, G., *Polymer*, 1982, **23**, 229
[10] Abdou-Sabet, S. and Patel, R.P., *Rubber Chem. Technol.*, 1991, **64**, 769

Polypropylene fibres

Synonyms: *Polypropene fibres. PP fibres*
Related Polymers: Polypropylene, Polypropylene copolymers, Isotactic polypropylene
Monomers: Propylene
Material class: Thermoplastic

— Polypropylene, Filled

Polymer Type: polypropylene
CAS Number:

CAS Reg. No.
9003-07-0
25085-53-4

Molecular Formula: $(C_3H_6)_n$
Fragments: C_3H_6
Molecular Weight: MW 20000–1000000. M_w/M_n 2–15
General Information: Circular cross section fibres are waxy. Fibres with non-circular cross section have texture and lustre. After melt drawing, fibres are cold drawn to give orientation. Dyeable fibres are copolymers containing a basic nitrogen atom

Surface Properties & Solubility:
Solvents/Non-solvents: Sol. aromatic and chlorinated hydrocarbons. Slightly sol. boiling hydrocarbons

Transport Properties:
Water Content: Effect of chlorination on water content,
Water Absorption:

No.	Value	Note
1	0.01 %	21°, 65% relative humidity [1,2,4]

Mechanical Properties:
Mechanical Properties General: Tensile props. are functions of MW, morphology, production and test conditions [1,2,5]

Tensile (Young's) Modulus:

No.	Value	Note
1	240–322 MPa	[1,4]

Tensile Strength Break:

No.	Value	Note	Elongation
1	160–440 MPa	[1,4]	20–200%

Viscoelastic Behaviour: Viscoelastic behaviour has been reported [3]

Polymer Stability:
Thermal Stability General: Thermal degradation temp. (290°) is lower than that of polyamides and polyesters [1,2]
Flammability: Use of flammability retardants often reduces light and temp. stability
Environmental Stress: Subject to uv degradation, stabilisers are usually radical scavengers
Chemical Stability: Highly sensitive to oxygen; highly substituted phenols are used as stabilisers. Resistant to most organic solvents at room temp., some swelling occurs in chlorinated hydrocarbons. Degraded by strong oxidising agents at high temp.
Hydrolytic Stability: Hydrophobic; resistant to acids and bases

Applications/Commercial Products:
Processing & Manufacturing Routes: Melt spinning is the commonest method of manufacture and involves extrusion, quenching, and take up, followed by drawing (to as much as 6 times original length), stress relieving heat treatment and texturising. The normal melt spin temp. is 240–310°. Also prod. by slit film (film-to-fibre) technology for fibres above 0.7 tex (6.6 den). Non wovens prod. by melt blown, spun bond and spurted fibre methods [1,2,3]
Applications: Used in carpet facing, carpet backing, upholstery, drapery, rope, geotextiles, non-wovens, filtration media, cordage, outdoor furniture webbing, bags and synthetic turf

Trade name	Supplier
Celmar	Royalite Plastics Ltd.
Daplen	PCD Polymere GmbH.
Escon A	Enjay Chemical
Fina	Fina Chemicals (Division of Petrofina SA)
Herculon	Hercules
Meraklon	Montecatini Edison SpA.
Monofil T	Monofil Technology Ltd.
Polysteen	Steen and Co. GmbH.
Tippfil	Tiszai Vegyi Kombinát
Tipptrex	Tiszai Vegyi Kombinát
Ulstron	ICI, UK

Bibliographic References
[1] *Kirk-Othmer Encycl. Chem. Technol.*, 4th edn., (eds. J.I. Kroschwitz and M. Howe-Grant), Wiley Interscience, 1993, **10**, 639
[2] *Encycl. Polym. Sci. Eng.*, 2nd edn., (ed. J.I. Kroschwitz), Wiley Interscience, 1987, **10**, 373
[3] Minoshima, W., White, J.L. and Spruiell, J.E., *Polym. Eng. Sci.*, 1980, **20**, 1166
[4] *Text. World*, 1984, **134**, 49
[5] Hall, I.M., *J. Polym. Sci.*, 1961, **54**, 505

Polypropylene, Filled P-494
Synonyms: *PP, Filled. Polypropene, Filled*
Related Polymers: Polypropylene, Talc filled, Calcium carbonate filled, Glass fibre filled, glass fibre reinforced, Mica filled, Carbon filled, Aluminium powder filled, Isotactic polypropylene
Monomers: Propylene
Material class: Thermoplastic, Composites
Polymer Type: polypropylene
CAS Number:

CAS Reg. No.
9003-07-0
25085-53-4

Molecular Formula: $(C_3H_6)_n$
Fragments: C_3H_6
Additives: Extenders (for cost reduction), e.g. calcium carbonate, talc. Reinforcement e.g. talc, mica, fibres. Conductivity e.g. carbon, aluminium. Fillers include glass fibres, glass spheres, asbestos, wood flour, Wollastonite, zinc oxide, polyamide fibres, polyester fibres, metallic powders, mica, talc and calcium carbonate
General Information: Solid fillers are added to reduce cost by increasing bulk or to modify mechanical, thermal or electrical props. The effect of a filler depends on its particle size, shape, physical props. and chemical composition. Fibres or powders are commonly used. The effects of ≤40% treated calcium carbonate and glass bead fillers on the props. of homopolymer and (ethylene) copolymer have been found to be increased density (linearly with the filler content), while T_m, tensile strength, elongation at break and enthalpies of fusion and crystallinity decrease. Hardness increased with glass bead filler and decreased with calcium carbonate filler content
Miscellaneous: The effects of various fillers on struct. and props. of polypropylene (including some blends and copolymers) have been reported [1]

Applications/Commercial Products:

Trade name	Details	Supplier
ARP	Thermofill	Northern Industrial Plastics Ltd.
Durapol		Monopol Plastics Compounding Ltd.

Elfix		Peter Sallach Elfix Kuststoffvertrieb
Ferrolene		Ferro Corporation
Hellyar		Hellyar Plastics Ltd.
Hostacom		Hoechst (UK) Ltd. Polymer Division
Hostalen PP		Hoechst (UK) Ltd. Polymer Division
Polyfill		Polykemi AB
Procom		ICI Acrylics
Profax		Hercules
Propathene		ICI Acrylics
Silvar		Carretier Robin
Starpylene		Ferro Corporation
Vestolen		Hüls AG

Bibliographic References

[1] Polypropylene: Structure, *Blends and Composites,* (ed. J. Karger-Kocsis), Chapman and Hall, 1995, **3**

Polypropylene films   P-495

Synonyms: *PP films. Polypropene films. OPP*
Related Polymers: Polypropylene, Polypropylene blends, Polypropylene copolymers, Isotactic polypropylene
Monomers: Propylene
Material class: Thermoplastic
Polymer Type: polypropylene
CAS Number:

CAS Reg. No.	Note
9003-07-0	
25085-53-4	isotactic

Molecular Formula: $(C_3H_6)_n$
Fragments: C_3H_6
General Information: Polypropylene films may be unoriented (cast) for sparkling optics and clarity, oriented for improved strength, durability and thermal props. or biaxially oriented for maximum strength and durability
Morphology: Morphological props. vary a little with MW distribution, and more so with draw temp. In general as distribution broadens, birefringence may increase, and amorph. orientation may increase (T > 165°), long period spacing decreases slightly; as draw temp. increases, crystalline fraction, crystalline orientation and especially long period spacing increase but birefringence and amorph. orientation decrease [3]

Transport Properties:
Permeability of Gases: Water vapour transmission rate 11.2–14.6 × 10^{-6} cm^3 cm^{-2} s^{-1} (0.5–0.65 nmol cm^{-2} s^{-1}, non-oriented, ASTM E96, 37°, 90% relative humidity); 6.72–8.96 × 10^{-6} cm^3 cm^{-2} s^{-1} (0.3–0.4 nmol cm^{-2} s^{-1}, oriented, ASTM E96, 37°, 97% relative humidity) [8,9]

Mechanical Properties:
Mechanical Properties General: Mech. props. vary with temp., draw ratio, melt flow and MW distribution. In general, as MW distribution broadens tensile strength increases, elongation decreases, tensile strength modulus increases and shrinkage decreases. As draw temp. increases tensile strength increases then decreases; elongation increases and tensile strength modulus decreases [3]

Tensile (Young's) Modulus:

No.	Value	Note
1	2500–4900 MPa	gauge 12–75 µm [4]

Tensile Strength Break:

No.	Value	Note	Elongation
1	50–275 MPa	biaxially oriented, ASTM D882 [9]	35–500%
2	172.3–206.8 MPa	oriented, ASTM D882 [8,9]	60–100%
3	20.7–62 MPa	non-oriented, ASTM D882 [8,9]	400–800%

Elastic Modulus:

No.	Value	Note
1	758–965 MPa	non-oriented
2	2206–2620 MPa	oriented [8,9]

Impact Strength: Impact strength 19–58 kJ m^{-1} (biaxially oriented) [7]. Impact strength 9.8–29.4 N cm (non-oriented); 49–147 N cm (oriented) [8,9]
Mechanical Properties Miscellaneous: Stiffness 4.9–10.8 N m^{-1} [4]
Fracture Mechanical Properties: Tear strength 15.7–129 N mm^{-1} (non-oriented); 1.57–2.35 N mm^{-1} (oriented) [8,9]; 1.2–3.9 N mm^{-1} (biaxially oriented) [7] (ASTM D1922)

Electrical Properties:
Electrical Properties General: Films exhibit good electrical and insulating props. Factors affecting props. include morphology, orientation, crystallinity, tacticity and MW. Dissipation factor increases with increasing irradiation dose; reversed by impregnation with monoisopropyl biphenyl
Dielectric Permittivity (Constant):

No.	Value	Frequency	Note
1	2–2.5	60 Hz–1 kHz	24° [1]

Dielectric Strength:

No.	Value	Note
1	315 $kV.mm^{-1}$	8000 V mil^{-1}, 24°, 60 Hz [1]
2	118 $kV.mm^{-1}$	3000 V mil^{-1}, 12–75 µm gauge [4]

Strong Field Phenomena General: Breakdown voltage (25.4 µm thick film) 8 kV (60 Hz) [1]; 16 kV (DC) [2]; 12 kV (60 Hz, monoisopropyl biphenyl impregnated) [3]; 16 kV (DC, monoisopropyl biphenyl impregnated) [3]
Static Electrification: Voltage-current characteristics, frequency dependence of conductance, capacitance and dielectric loss have been reported [3]
Dissipation (Power) Factor:

No.	Value	Frequency	Note
1	1 × 10^{-5}	10 kHz	approx.
2	0.002	50 Hz	

Optical Properties:
Optical Properties General: Crystallinity and the presence of spherulites give low transparency. Transparency is raised by quenching, addition of nucleating agents or destroying the spherulites by stretching. Quenching is preferred for food use due to lower contamination [8]. Transparent films are obtained from chlorinated polypropylene having 4% chlorine content
Refractive Index:

No.	Value	Note
1	1.5	[4]

– Polypropylene, Glass fibre reinforced

Transmission and Spectra: Turbidity 5 cm^{-1} (chill roll 60–80°); 0.35 cm^{-1} (quenched flowing water); 0.14 cm^{-1} (quenched, roll annealed) [6]
Total Internal Reflection: Birefringence 1090 (chill roll, 60–80°); 1750 (quenched in flowing water); 3290 (quenched, roll annealed) [6]
Volume Properties/Surface Properties: Haze 1–2% (12–75 μm gauge) [4]; 1.5–25% (biaxially oriented) [7]

Polymer Stability:
Thermal Stability General: Relatively high melting point allows sterilisation of PP film wrapped foods [8]
Upper Use Temperature:

No.	Value	Note
1	140°C	extruded
2	143°C	biaxially oriented [7]

Flammability: Processing conditions and environmental temp. influence burning characteristics and smoke generation. As extrusion temp. (or number of extrusions) is increased, Limiting Oxygen Index increases; linear burn rate decreases, mass burn rate increases; thin layers (flame retarded) offer greater resistance to burning [2]

Applications/Commercial Products:
Processing & Manufacturing Routes: Prod. by tenter-frame or double-bubble processes for oriented film, or by a blown process with water bath quench for unoriented film [6]. Heat seal range for unoriented film is 140–200° (oriented films require co-polymer, terpolymer or alloy layer for heat sealing) [4,7]
Applications: In the absence of other information commercial materials are assumed to be unoriented. Used in packaging of snack foods, baking items, cigarettes, sweets, overwraps, bottle labels, etc. Other uses include pressure-sensitive tape, sheet protectors, stationery products and labels. Electrical usage includes insulation for capacitors and power cables, especially in aerospace industries

Trade name	Details	Supplier
Biafol	oriented	Tiszai Vegyi Kombinát
Bolflex	for lamination	Bollore Technologies
Bolphane	shrinkable	Bollore Technologies
Borden		Northern Film Products Ltd.
Crispac	perforated	Cryovac Grace Packaging
Duplex	flexible foil	Tvornica Oplemenjenih Folija
Flexophan		Ciarolella Snc.
Forco-OPP	biaxially oriented	4P Folie Forchheim GmbH.
Ipelene	biaxially, oriented	IPEL Ltd.
Manucor	bioriented, coextruded	Manuli España SA
Mattflex	laminated, coated	Ace SA
Megofilm		Linsay and Williams Ltd.
Microflex	breathable, flexible	ASP Packaging
Moplefan	oriented	Moplefan (UK) Ltd.
Naprene	roll	Martyn Industrials
Propafilm	oriented	ICI Propafilm
Propex	bioriented, perforated	Borden
Radil	biaxially oriented	Radici Film SpA.
Rayopp	biaxially oriented	UCB Films
SPX	flexible, low temp.	Cairn Chemicals Ltd.
Satinflex	laminated, coated	Ace SA
Sekawrap	shrinkable	Amalgamated Kempner Group
Splendene		Nastrificio A. Bolis SpA.
Supermicro	perforated	Cryovac Grace Packaging
Taffaflex	laminated, coated	Ace SA
Tresaphan	biaxially oriented	Hoechst (UK) Ltd. Films Division
Ultralen	extruded	Lonza-Folien GmbH.
Velvaflex	laminated, coated	Ace SA
Vishene		Visual Packaging Ltd.
Walofilm	biaxially oriented, coated/metalised	Bayer Inc.
Walothen	biaxially, oriented	Bayer Inc.

Bibliographic References
[1] Hammoud, A.N., Laghari, J.R. and Krishnakumar, B., *IEEE Trans. Nucl. Sci.*, 1987, **34**, 1822
[2] McIlhagger, R. and Hill, B.J., *Plast. Rubber. Process.*, 1987, **7**, 179
[3] Flood, J.E. and Nulf, S.A., *Polym. Eng. Sci.*, 1990, **30**, 1504
[4] *Encycl. Polym. Sci. Eng.*, 2nd edn., (ed., J.I. Kroschwitz), John Wiley and Sons, 1987, **7**
[5] Zor, M. and Hogarth, C.A., *Phys. Status Solidi*, 1987, **104**, 761
[6] Shibayama, M., Katoh, K., Iwamoto, T., Takahashi, D. and Namura, S., *Polym. J. (Tokyo)*, 1991, **23**, 837
[7] *Kirk-Othmer Encycl. Chem. Technol.*, Vol. 10, 4th edn., (eds. J.I. Kroschwitz and M. Howe-Grant), Wiley Interscience, 1993, 761
[8] *Kirk-Othmer Encycl. Chem. Technol.*, Vol. 17, 4th edn., (eds. J.I.Kroschwitz and M. Howe-Grant), Wiley Interscience, 1993, 814
[9] *Packaging (Boston)*, 1986, **31**, 79

Polypropylene, Glass fibre reinforced P-496

Synonyms: PP, Glass fibre reinforced. Polypropene, Glass fibre reinforced. GRPP
Related Polymers: Polypropylene, Polypropylene, Filled, Isotactic polypropylene
Monomers: Propylene
Material class: Thermoplastic, Composites
Polymer Type: polypropylene
CAS Number:

CAS Reg. No.
9003-07-0
25085-53-4

Molecular Formula: $(C_3H_6)_n$
Fragments: C_3H_6
Additives: Glass fibres, short or long. Coupling agents
General Information: Gives high tensile strength over wide temp. range, high stiffness and improved creep behaviour. Use of coupling agents gives improved mechanical values, increased fatigue strength, good long-term props. and higher bend-creep modulus. Disadvantages of glass fibre reinforcement are reduced impact strength, matt surface and abrasion

Volumetric & Calorimetric Properties:
Density:

No.	Value	Note
1	d 1.04–1.13 g/cm^3	20–30% glass fibre reinforced homopolymer
2	d 1.22–1.23 g/cm^3	40% glass fibre reinforced homopolymer

– Polypropylene, Glass fibre reinforced

| 3 | d 0.98–1.04 g/cm³ | 10–20% glass fibre reinforced copolymer |
| 4 | d 1.11–1.21 g/cm³ | 30–40% glass fibre reinforced copolymer, ASTM D792 |

Thermal Expansion Coefficient:

No.	Value	Type
1	$2.7 \times 10^{-5} - 3.2 \times 10^{-5}$ K^{-1}	V

Thermal Conductivity:

No.	Value	Note
1	0.35–0.37 W/mK	0.00084–0.00088 cal cm^{-1} s^{-1} C^{-1}, ASTM C177, 40% glass fibre reinforced homopolymer

Melting Temperature:

No.	Value	Note
1	168°C	homopolymer
2	160–168°C	copolymer [1]

Deflection Temperature:

No.	Value	Note
1	132–138°C	270–280°F, 20–30% glass fibre reinforced homopolymer
2	149–166°C	300–330°F, 40% glass fibre reinforced homopolymer
3	127–138°C	260–280°F, 10–20% glass fibre reinforced copolymer
4	138°C	280°F, 1.84 MPa (264 psi), 30–40% glass fibre reinforced copolymer [1]
5	143–146°C	290–295°F, 20–30% glass fibre reinforced homopolymer
6	166°C	330°F, 40% glass fibre reinforced homopolymer
7	152°C	305°F, 10–20% glass fibre reinforced copolymer
8	154°C	310°F, 0.45 MPa (66 psi), 40% glass fibre reinforced copolymer, ASTM D648 [1]
9	121°C	250°F, 20–30% long glass fibre reinforced homopolymer
10	152°C	305°F, 20–30% long glass fibre reinforced homopolymer, ASTM D648 [1]

Transport Properties:
Water Absorption:

No.	Value	Note
1	0.05 %	20–30% glass reinforced homopolymer
2	0.05–0.06 %	40% glass reinforced homopolymer
3	0.01 %	1/8" thick, glass reinforced copolymer, 24h, ASTM D570
4	0.09–0.1 %	1/8" thick, 40% glass reinforced homopolymer, saturation, ASTM D570 [1]

Mechanical Properties:
Mechanical Properties General: Considerable variation in mech. props. occurs due to length, length variation and distribution uniformity of the fibres; degree of coupling and processing history [3]. Mech. props are improved by the addition of a β nucleating agent (e.g. adipic acid) [5]

Tensile (Young's) Modulus:

No.	Value	Note
1	4825–5515 MPa	700–800 kpsi, 20–30% glass fibre reinforced homopolymer
2	7580–10340 MPa	1100–1500 kpsi, 40% glass fibre reinforced homopolymer, ASTM D638 [1]
3	5170–6200 MPa	750–900 kpsi, 20–30% long glass fibre reinforced homopolymer [1]
4	3120 MPa	20°, short glass fibre reinforced
5	5090–5620 MPa	-43°
6	3000–3510 MPa	20°, long glass fibre reinforced, depending on processing [2]
7	16200 MPa	twill weave, long fibre reinforced sheet [4]

Flexural Modulus:

No.	Value	Note
1	3450–4140 MPa	500–600 kpsi, 28° (73°F), 20–30% glass fibre reinforced homopolymer
2	3550–6890 MPa	950–1000 kpsi, 28° (73°F), 40% glass fibre reinforced homopolymer
3	2450–3520 MPa	355–510 kpsi, 28° (73°F), 10–20% glass fibre reinforced copolymer
4	4895–6620 MPa	710–960 kpsi, 28° (73°F), 30–40% glass fibre reinforced copolymer
5	3790–4825 MPa	550–700 kpsi, 28° (73°F), 20–30% long glass fibre reinforced homopolymer [1]
6	6270 MPa	40% short glass fibre
7	4620 MPa	40% long glass fibre mat [6]

Tensile Strength Break:

No.	Value	Note	Elongation
1	45–48 MPa	6500–7000 psi, 20–30% glass fibre reinforced homopolymer	2.5–3.0%
2	58–103 MPa	8400–15000 psi, 40% glass fibre reinforced homopolymer	1.5–4.0%
3	48–55 MPa	7000–8000 psi, 10–20% glass fibre reinforced copolymer	3.0–3.4%
4	59–62 MPa	8500–9000, 30–40% glass fibre reinforced copolymer, - ASTM D638[1]	2.2–2.6%
5	52–55 MPa	7500–8000 psi, 20–30% long glass fibre reinforced, homopolymer [1]	2.1–2.2%
6	65.6 MPa	20° short fibre	
7	45.6 MPa	55°	

— Polypropylene, Glass fibre reinforced

8	45.4 MPa	20°	
9	24.8 MPa	55°, long fibre reducing with increasing processing [3]	
10	270 MPa	twill weave long fibre reinforced sheet	
11	275 MPa	non crimp fabric long fibre reinforced sheet [4]	

Flexural Strength at Break:

No.	Value	Note
1	48–59 MPa	7000–8500 psi, 20–30% glass fibre reinforced homopolymer
2	70–150 MPa	10500–22000 psi, 40% glass fibre reinforced - homopolymer
3	57–66 MPa	8200–9600 psi, 10–20% glass fibre reinforced copolymer
4	73–76 MPa	10600–11000 psi, rupture or yield, 30–40% glass fibre reinforced copolymer, ASTM D790 [1]
5	69–72 MPa	10000–10500 psi, 20–30% long glass fibre reinforced, homopolymer [1]
6	125.5 MPa	40% short glass fibre [6]
7	142.7 MPa	40% long glass fibre mat [6]

Compressive Strength:

No.	Value	Note
1	45–48 MPa	6500–7000 psi, 20–30% glass fibre reinforced homopolymer
2	61–68 MPa	8900–9800 psi, 40% glass fibre reinforced homopolymer
3	38–39 MPa	5500–5600 psi, 10–20% glass reinforced copolymer
4	37–39 MPa	5400–5700 psi, 30–40% glass fibre reinforced copolymer, ASTM D695 [1]

Impact Strength: 187–203 J m^{-1} (3.5–3.8 ft lb in^{-1}, $\frac{1}{8}$" thick, 20–30% long glass fibre reinforced homopolymer) [1]; 7 J m^{-1} (25°), 6 J m^{-1} (-30°, 40% short glass fibre, injection moulded), 63 J m^{-1} (25°), 76 J m^{-1} (-30°, 40% long glass fibre mat, sheet) [6]; 1110 J m^{-1} (twill weave long fibre reinforced sheet) [4]. Rheometrics instrumented rates have been reported [6]

Hardness: Rockwell R100-R115 (20–30% glass fibre reinforced homopolymer); R102-R111 (40% glass fibre reinforced homopolymer); R100-R103 (10–20% glass fibre reinforced copolymer); R104-R105, Shore D45-D55 (30–40% glass fibre reinforced copolymer, ASTM D785 (Rockwell), ASTM D2440 (Shore)) [1]

Izod Notch:

No.	Value	Notch	Note
1	53–64 J/m	N	1.0–1.2 ft lb in^{-1}, 20–30% glass fibre reinforced homopolymer [1]
2	75–107 J/m	N	1.4–2 ft lb in^{-1}, 40% glass fibre reinforced homopolymer [1]
3	51–53 J/m	N	0.95–1 ft lb in^{-1}, 10–20% glass fibre reinforced copolymer [1]
4	48–51 J/m	N	0.9–0.95 ft lb in^{-1}, $\frac{1}{8}$" thick, 30–40% glass fibre reinforced copolymer, ASTM D256A [1]

Electrical Properties:
Dielectric Strength:

No.	Value	Note
1	19.68–20.07 kV.mm^{-1}	500–510 V mil^{-1}, $\frac{1}{8}$" thick, 40% glass fibre reinforced homopolymer, ASTM D149 [1]

Applications/Commercial Products:
Processing & Manufacturing Routes: Processing temp. ranges (injection moulding) typically 180–225° (360–440°F, 20–30% glass fibre reinforced homopolymer); 230–290° (450–550°F, 40% glass fibre reinforced homopolymer; 175–245° (350–470°F, glass fibre reinforced copolymer)[1]. Moulding pressure range 70–170 MPa (10000–25000 psi, 40% glass fibre filled homopolymer) [1]
Mould Shrinkage (%):

No.	Value	Note
1	0.4–0.5%	0.004–0.005 in/in, 20–30% glass fibre reinforced homopolymer
2	0.3–0.4%	0.003–0.004 in/in, 20–30% long glass fibre reinforced homopolymer
3	0.3–0.5%	0.003–0.005 in/in, 40% glass fibre reinforced homopolymer
4	0.3–1%	0.003–0.01 in/in, 10–20% glass fibre reinforced copolymer
5	0.1–1%	0.001–0.01 in/in, 30–40% glass fibre reinforced copolymer, ASTM D955

Trade name	Supplier
ARP	Northern Industrial Plastics Ltd.
Adell ER	Adell Plastics Inc.
Celmar	Royalite Plastics Ltd.
Gapex RPP	Ferro Corporation
Higlass	Himont
Procom	ICI Acrylics
Profax	Hercules
Profil G	Wilson-Fiberfil
Profil J	Wilson-Fiberfil
Propathene	ICI, UK
Thecnoprene	Enichem
Thermocomp MF	LNP Engineering
Thermocomp MFX	LNP Engineering
Thermofil P-[nn] FG	Thermofil
	RTP Company
	A. Schulman Inc.

Bibliographic References
[1] Guide to Plastics, *Property and Specification Charts*, McGraw-Hill, 1987
[2] Gupta, V.B., Mittal, R.K., Sharma, P.K., Mennig, G. and Wolters, J., *Polym. Compos.*, 1989, **10,** 8, 16
[3] Hiscock, D.F. and Bigg, D.M., *Polym. Compos.*, 1989, **10,** 145
[4] Bigg, D.M., Hiscock, D.F., Preston, J.R. and Bradbury, E.J., *Polym. Compos.*, 1988, **9,** 222
[5] Ku, L., He, X., Song, J., Shou, Z. and Wang, X., *CA*, 1987, **106,** 120695j, (mech props)
[6] Silverman, E.M., *Polym. Compos.*, 1987, **8,** 8
[7] Denault, J. and Vu-Khank, T., *Polym. Compos.*, 1988, **9,** 360

Polypropylene, Mica filled   P-497

Synonyms: *PP, Mica filled. PP, Mica reinforced*
Related Polymers: Polypropylene, Polypropylene, Filled
Monomers: Propylene
Material class: Thermoplastic, Composites
Polymer Type: polypropylene
CAS Number:

CAS Reg. No.
9003-07-0
25085-53-4

Molecular Formula: $(C_3H_6)_n$
Fragments: C_3H_6
Additives: Mica; chlorinated paraffins
General Information: 40% Mica imparts the same stiffness as 30% glass fibre but at a lower cost level. Mica is generally used to enhance flexural modulus, heat deflection temp., dimensional stability and tensile strength

Volumetric & Calorimetric Properties:
Density:

No.	Value	Note
1	d 1.23 g/cm^3	40% mica filled, impact modified homopolymer, ASTM D792 [1]

Deflection Temperature:

No.	Value	Note
1	96°C	205°F, 1.84 MPa (264 psi), 40% mica filled, impact modified homopolymer, ASTM D648 [1]

Transport Properties:
Melt Flow Index:

No.	Value	Note
1	8–12 g/10 min	40% mica filled, ASTM D1238 [2]

Mechanical Properties:
Tensile (Young's) Modulus:

No.	Value	Note
1	4825 MPa	700 kpsi, 40% mica filled impact modified homopolymer, ASTM D638 [1]

Flexural Modulus:

No.	Value	Note
1	4136 MPa	600 kpsi, 28° (73°F), 40% mica filled impact homopolymer, ASTM D790 [1]

Tensile Strength Break:

No.	Value	Note	Elongation
1	31 MPa	4500 psi, 40% mica filled impact modified homopolymer, ASTM D638 [1]	4%

Flexural Strength at Break:

No.	Value	Note
1	48 MPa	7000 psi, rupture or yield, 40% mica filled impact modified homopolymer, ASTM D790 [1]

Hardness: Rockwell R97 (40% mica filled, ASTM D785) [2]. Shore D82 (40% mica filled, ASTM D2240)
Izod Notch:

No.	Value	Notch	Note
1	3.7 J/m	N	0.7 ft lb in^{-1}, 1/8" thick, 40% mica filled impact modified homopolymer, ASTM D256A [1]
2	261 J/m	N	4.9 ft lb in^{-1}, 40% mica filled, ASTM D256 [2]

Applications/Commercial Products:
Processing & Manufacturing Routes: Processing temp. range (injection moulding) 175–245° (350–470°F, 40% mica filled impact modified homopolymer) [1]
Mould Shrinkage (%):

No.	Value	Note
1	0.7–0.8%	0.007–0.008 in/in, linear, 40% mica filled impact modified homopolymer [1]

Applications: Used in automotive engine ducts and housings

Trade name	Details	Supplier
MPP	various grades	Ferro Corporation
Polifil M		Plastics Group
Profil PP60MI		Wilson-Fiberfil
Thermofil P	various grades	Thermofil

Bibliographic References
[1] Guide to Plastics, *Property and Specification Charts*, McGraw-Hill, 1987
[2] *Polyfil M Technical Datasheet*, Plastics Group of America, USA, 1995, (technical datasheet)

Polypropylene/natural rubber blends   P-498

Synonyms: *PP/NR blends. Polypropylene/polyisoprene blends. PP/IR blends. Polypropene/cis-1,4-polyisoprene rubber blends*
Related Polymers: Polypropylene/rubber blends
Material class: Thermoplastic, Natural elastomers, Blends
Polymer Type: polypropylene

Volumetric & Calorimetric Properties:
Density:

No.	Value	Note
1	d 0.99–1.04 g/cm^3	[1]

Brittleness Temperature:

No.	Value	Note
1	-50–-35°C	ASTM D746 [1]

Mechanical Properties:
Tensile Strength Break:

No.	Value	Note	Elongation
1	5–26.4 MPa	ASTM D412 [1,3]	300–620%

— Polypropylene/nitrile rubber blends

Miscellaneous Moduli:

No.	Value	Note	Type
1	2.1–10.5 MPa	ASTM D412 [1]	Rubber modulus

Viscoelastic Behaviour: Tension set 10–50% (ASTM D412) [1,3]. Compression set 24–45% (23°); 30–63% (100°, ASTM D395B) [1]
Hardness: Shore A60-D50 (ASTM D2240) [1,2]
Failure Properties General: Tear strength 22–98 kN m^{-1} (ASTM D624)

Polymer Stability:
Thermal Stability General: Good retention of tensile props. in hot air (100° for 1 month) [1]
Chemical Stability: Ozone resistance 10 (40°, 110 ppm O_3, ASTM D518) [1]

Applications/Commercial Products:
Processing & Manufacturing Routes: Dynamically vulcanised during mixing
Applications: Low cost

Trade name	Supplier
Vyram	Advanced Elastomer Systems

Bibliographic References
[1] Coran, A.Y. and Patel R.P., Polypropylene: Structure, *Blends and Composites: Copolymers and Blends*, (ed. J. Karger-Kocsis), Chapman and Hall, 1995, **2**
[2] *Kirk-Othmer Encycl. Chem. Technol.*, 4th edn. (eds. J.I. Kroschwitz and M. Howe-Grant), Wiley Interscience, 1994, **9**
[3] Coran, A.Y., Patel R.P. and Williams, D., *Rubber Chem. Technol.*, 1982, **55**, 116

Polypropylene/nitrile rubber blends   P-499
Synonyms: *Polypropylene/NBR blends*
Related Polymers: Polypropylene/rubber blends
Material class: Thermoplastic, Synthetic Elastomers, Blends
Polymer Type: polypropylene
Additives: Compatibiliser required
General Information: The blend is a compatibilised nitrile rubber fine dispersion in a polypropylene matrix

Volumetric & Calorimetric Properties:
Density:

No.	Value	Note
1	d 1–1.1 g/cm^3	[2]

Brittleness Temperature:

No.	Value	Note
1	-40–-33°C	ASTM D746 [3]

Mechanical Properties:
Mechanical Properties General: Since PP and NBR are grossly incompatible, mech. props. are largely dependent on compatibilisation. A given formulation may enhance one property and not another hence the wide ranges shown [1]

Tensile (Young's) Modulus:

No.	Value	Note
1	105–456 MPa	tensile tester, cross head speed 2.5 cm min^{-1}, up to 30% elongation [1]
2	36–118 MPa	5220–17100 psi, ASTM D412 [3]

Flexural Modulus:

No.	Value	Note
1	58–163 MPa	8460–23590 psi, ASTM D790 [3]

Tensile Strength Break:

No.	Value	Note	Elongation
1	8.2–26.7 MPa	tensile tester, 25 cm min^{-1} [1,4]	66–580%
2	9–171 MPa	true stress of break [1]	

Compressive Modulus:

No.	Value	Note
1	0.35–1.23 MPa	51–178 psi, ASTM D695, 5% compression
2	1.1–3.27 MPa	159–576 psi, ASTM D695, 10% compression

Miscellaneous Moduli:

No.	Value	Note
1	9.9–14.2 MPa	stress at 100% strain, tensile tester, 25 cm min^{-1} [1]
2	5.4 MPa	785 psi [3]

Viscoelastic Behaviour: Tension set 15–70% (ASTM D412-66) [1,3,4]. Compression set 17–19% (25% deflection, 22h, 23°, ASTM D395B); 33–48% (25% deflection, 22h, 100°, ASTM D395B) [3]
Hardness: Shore A70-D50 (ASTM D2240) [2,3]
Failure Properties General: Tear strength 250–425 pli (23°, ASTM D624, Die C); 110–230 pli (100°, ASTM D624, Die C) [3]

Polymer Stability:
Thermal Stability General: Service temp. range of -40–125° [3]
Upper Use Temperature:

No.	Value	Note
1	125°C	[3]

Chemical Stability: Excellent resistance to acids, alkalis and hot oils. Excellent resistance to hydrocarbons (less than C_{20}) at room temp. Unstable to diesel above 70° [3]
Hydrolytic Stability: Excellent resistance to aq. solns.

Applications/Commercial Products:
Processing & Manufacturing Routes: Dynamically vulcanised during mixing
Applications: Applications due to oil resistance

Trade name	Supplier
Geolast	Advanced Elastomer Systems

Bibliographic References
[1] Coran, A.Y. and Patel, R.P., *Rubber Chem. Technol.*, 1983, **56**, 1045
[2] *Kirk-Othmer Encycl. Chem. Technol.*, 4th edn., (eds. J.I. Kroschwitz and M. Howe-Grant), Wiley Interscience, 1991, **9**
[3] Rader, C.P., *Handb. Thermoplast. Elastomers*, 2nd edn., (eds. B.M. Walker and C.P. Rader), Van Nostrand Reinhold, 1988
[4] Coran, A.Y., Patel, R.P. and Williams, D., *Rubber Chem. Technol.*, 1982, **55**, 116

Polypropylene/Nylon blends P-500

Synonyms: *Polypropylene/polyamide blends. PP/PA blends*
Related Polymers: Polypropylene blends
CAS Number:

CAS Reg. No.
9003-07-0
25038-54-4
32131-17-2
63428-83-1

Additives: Compatibilisers are used such as maleated PP, maleated SBS, maleated SEBS or commercial formulations
General Information: Cited advantages of PP/PA blends are low moisture absorption, improved processability, good impact resistance and flexural modulus [8]
Morphology: Uncompatibilised blends are coarse mixtures with particles up to 50 μm which are readily deformable into fibrils. Compatibilisation reduces the particle size to about 1 μm [8,16]. Even in well dispersed, compatibilised blends, processing leads to flow segregation, coarsening of morphology and fibrillation. Changes in morphology with composition and compatibilisation correlate with variations in mech., rheological and other props. have been reported [9,10,11,12,13,17,18]

Volumetric & Calorimetric Properties:
Density:

No.	Value	Note
1	d 1.03 g/cm^3	PP 40%, PA-6 60% [3]

Latent Heat Crystallization: ΔH_{fusion} 55 kJ kg^{-1} (13.2 cal g^{-1}, from PP), 14 kJ kg^{-1} (3.3 cal g^{-1} from PA) (80% PP/90% PA-6); 20 kJ kg^{-1} (4.7 cal g^{-1}, from PP), 52 kJ kg^{-1} (12.5 cal g^{-1} from PA) (30% PP/70% PA-6) [1]
Melting Temperature:

No.	Value	Note
1	165.3°C	PP
2	220.6°C	PA-6, PP 40%/PA-6 60%, Orgallay R 6000 [2,3]

Deflection Temperature:

No.	Value	Note
1	66°C	75 PP/25 PA-6
2	64°C	25 PP/75 PA-6

Transition Temperature:

No.	Value	Note	Type
1	112.5°C	PP	T_c
2	185°C	PA-6, PP 40%/PA-6 60%, Orgalloy R 6000 [2]	T_c

Transport Properties:
Melt Flow Index:

No.	Value	Note
1	18 g/10 min	75 PP/25 PA-6
2	32 g/10 min	25 PP/75 PA-6 [10]

Polymer Melts: Melt flow props. depend on sample drying. There is a lack of agreement between steady state, dynamic and extensional viscosity data and between first normal stress difference and storage modulus. Elastic responses are larger than for single phase systems. Lowest viscosity occurs at 75% PP: 25% PA. Viscosity may increase as a result of compatibilisation mechanism [8,14]. In capillary flow η is insensitive to temp. between 225° and 250° giving poor agreement with η′ and η*; η$_e$ (elongation viscosity) measured is one order of magnitude less than calculated by the Cogswell relation. Therefore capillary flow is not suitable for characterisation [8]
Water Absorption:

No.	Value	Note
1	3.1–3.8 %	uncompatibilised [12]

Mechanical Properties:
Mechanical Properties General: Data on modified fibre have been reported [15]. Other values of tensile modulus have been reported. [12] Other values of storage modulus have been reported [1,4,6,7]. Loss tangent (tan δ) peaks have been reported [4,6]

Tensile (Young's) Modulus:

No.	Value	Note
1	1150 MPa	75 PP/25 PA-6
2	1500 MPa	25 PP/75 PA-6 [10]

Tensile Strength Break:

No.	Value	Note
1	44 MPa	25% PP/75% PA 1010 [9]

Flexural Strength at Break:

No.	Value	Note
1	33 MPa	75% PP/25% PA-6 [10]
2	58 MPa	25% PP/75% PA-6 [10]

Tensile Strength Yield:

No.	Value	Note
1	27 MPa	75 PP/25 PA-6 [10]
2	42 MPa	25 PP/75 PA-6 [10]

Miscellaneous Moduli:

No.	Value	Note	Type
1	1500–3500 MPa	0°, 40% PP, 60% PA-6, Orgalloy R6000	Storage modulus
2	40–100 MPa	[1,5,6,7]	Loss modulus

Hardness: Rockwell R84 (75% PP/25% PA-6); R95 (50% PP/50% PA-6); R104 (25% PP/75% PA-6) [10]
Fracture Mechanical Properties: Brittle fracture has been reported. [7,9] Charpy critical strain energy release rate, G_c 5800 J m^{-2} (transverse to machining direction); 10 kJ m^{-2} (PP 40%/PA-6 60%, Orgalloy R6000, specimen thickness 2mm along machining direction) [7]

– **Poly(propylene oxide)**

Izod Notch:

No.	Value	Notch	Note
1	14 J/m	N	75% PP/25% PA-6 [10]
2	37 J/m	N	25% PP/75% PA-6 [10]

Izod Area:

No.	Value	Notch	Note
1	2.8 kJ/m^2	Notched	ASTM D256-54T [1]

Applications/Commercial Products:
Processing & Manufacturing Routes: Production is typically by modification of the PP (normally by maleation in the presence of peroxide though other acids or anhydrides, (e.g. fumaric, acrylic and methacrylic), may be used followed by mixing with the PA [8]. A commercial compatibiliser is based on ethylene-acrylic ester-maleic anhydride-glycidyl methacrylate [8]. Because of the adverse effects of compatibilisation on crystallinity it must be optimised then closely controlled
Applications: Applications include structs. which require both rigid and flexible parts, e.g. automotive engine air intakes, compressed air lines

Trade name	Details	Supplier
Orgalloy		Ashland
Orgalloy	RS - rigid grades	Elf Atochem (UK) Ltd.
Orgalloy	LT - flexible grades	Elf Atochem (UK) Ltd.

Bibliographic References
[1] Ide, F. and Hasegawa, A., *J. Appl. Polym. Sci.*, 1974, **18**, 963
[2] Diaz, M.I.A. and Maeda, Y., *Rep. Prog. Polym. Phys. Jpn.*, 1990, **33**, 255
[3] Maeda, Y., Polym. Prepr. (Am. Chem. Soc., *Div. Polym. Chem.*), 1991, **32**, 281
[4] Liang, Z. and Williams, H.L., *J. Appl. Polym. Sci.*, 1992, **44**, 699
[5] Rogers, M. and Samurkas, T., *Polym. Eng. Sci.*, 1992, **32**, 1727
[6] Wippler, C., *Polym. Eng. Sci.*, 1993, **33**, 347
[7] Utracki, L.A. and Sammut, P., Plast., *Rubber Compos. Process. Appl.*, 1991, **16**, 221
[8] Utracki, L.A. and Dumoulin, M.M., Polypropylene: Structure, *Blends and Composites: Copolymers and Blends*, (ed. J. Karger-Kocsis), Chapman and Hall, 1995
[9] Zhang, X.-M. and Yin, J.-H., *Polym. Eng. Sci.*, 1997, **37**, 197
[10] Sathe, S.N., Spinivasa Rao, G.S., Rao, K.V. and Devi, S., *Polym. Eng. Sci.*, 1996, **36**, 2443
[11] Tchoudakov, R., Breuer, O., Narkis, M. and Siegmann, A., *Polym. Eng. Sci.*, 1996, **36**, 1336
[12] Lee, J.-D. and Yang, S.-M., *Polym. Eng. Sci.*, 1995, **35**, 1821
[13] Rosch, J., *Polym. Eng. Sci.*, 1995, **35**, 1917
[14] Liang, B.-R., White, J.L., Spruiell, J.E. and Goswami, B.C., *J. Appl. Polym. Sci.*, 1983, **28**, 2011
[15] Grof, I., Sain, M.M. and Durcova, O., *J. Appl. Polym. Sci.*, 1992, **44**, 1061
[16] Willis, J.M. and Favis, B.D., *Polym. Eng. Sci.*, 1988, **28**, 1416
[17] Van Gheluwe, P., Favis, B.D. and Chalifoux, J.-P., *J. Mater. Sci.*, 1988, **23**, 3910
[18] Park, S.J., Kim, B.K. and Jeong, H.M., *Eur. Polym. J.*, 1990, **26**, 131

Poly(propylene oxide) P-501

Synonyms: Polymethyloxirane. Polyoxypropylene. Polypropylene glycol. Poly[oxy(methane-1,2-ethanediyl)]. α-Hydro-ω-hydroxy-poly[oxy(methane-1,2-ethanediyl)]. Propylene glycol polyol. Poly(1,2-epoxypropane). PPO. Polypropylene oxide polyols. PO polyols
Related Polymers: Polypropylene, Oxetane polymers, Propylene oxide elastomers, Poloxamers
Monomers: Propylene oxide

Material class: Oligomers
Polymer Type: polyol, polypropylene oxide
CAS Number:

CAS Reg. No.	Note
25322-69-4	generic
25791-96-2	

Molecular Formula: $(C_3H_6O)_n$
Fragments: C_3H_6O
Molecular Weight: MW normally less than 7000, dependent on termination conditions, prepared via acid or base catalysis. Catalysts are reported that are capable of giving MW up to 1000000
Additives: Stabilisers, usually based on hindered phenols or aromatic amines are added. Antioxidants e.g. butylated hydroxytoluene (BHT)
General Information: Poly(propylene oxide) of MW less than 700 is soluble in water. Solubility decreases with increasing MW, giving rise to a cloud point
Tacticity: The common forms are atactic and isotactic. Syndiotactic is also possible as are optically active forms
Morphology: Conformational analysis shows a gauche oxygen effect [6,7]
Miscellaneous: End groups vary significantly according to the initiator/terminator and may give rise to diols, triols, and other polyols. Some catalysts, notably acidic, give "abnormal" head to head or tail to tail structs. in proportions up to about 45%. Information on blends has been reported [24]

Volumetric & Calorimetric Properties:
Density:

No.	Value	Note
1	d 1.157 g/cm^3	cryst.
2	d 1.002 g/cm^3	amorph.
3	d^{25} 1.005–1.02 g/cm^3	depending on functionality [14,21]

Latent Heat Crystallization: ΔH_f 8.4 kJ mol^{-1} (at T_m). ΔS_f 23.8 J mol^{-1} K^{-1} (at T_m) [14]
Thermal Conductivity:

No.	Value	Note
1	0.16 W/mK	MW 3000 [11]
2	0.15 W/mK	MW 5000 [11]

Specific Heat Capacity:

No.	Value	Note	Type
1	1.95 kJ/kg.C	25°	P
2	2.3 kJ/kg.C	150°, [11]	P

Melting Temperature:

No.	Value	Note
1	78°C	M_n 500000, isotactic sequence 55
2	82°C	multiple melting transition
3	70–75°C	[14]

Glass-Transition Temperature:

No.	Value
1	-83°C

– Poly(propylene oxide)

Transition Temperature General: T_g is independent of stereoregularity [14] and varies little with MW [11]. T_g for networks has been reported [28]
Transition Temperature:

No.	Value	Note	Type
1	-55°C	[14]	β dispersion
2	-150°C	[14]	γ dispersion

Surface Properties & Solubility:
Solubility Properties General: Soln. behaviour in H_2O has been reported [26,27]
Cohesive Energy Density Solubility Parameters: δ_d 15.3–17.3; δ_p 4.2–5.2; δ_h 6.9–7.9; δ 17.3–19.7 [15]
Solvents/Non-solvents: Sol. H_2O (polyols with MW <700), most organic solvents; insol. non-polar solvents (high hydroxyl content) [11]

Transport Properties:
Polymer Solutions Dilute: PPO undergoes conformational change at 30°. Below wt. fractions of 0.6 soln. behaviour is dominated by hydrophobic hydration [26,27]
Polymer Solutions Concentrated: Viscosity - temp. relationships at temps near to T_g for pure PPO 4000 and for polymer electrolytes based on PPO 4000 have been reported [23]
Polymer Melts: Poly (propylene oxide) polyols are liquid for 200 < MW <6000. Viscosity depends on functionality (inversely proportional to equivalent weight) and the initiator. Values may range from 600 mPa s (flexible foam use) to 10000–100000 mPa s (rigid foam use) to 1000000 mPa s (high functionality, sucrose initiated) [11,14]. Addition of dibenzylidene sorbitol (DBS) at amounts 1–2% by weight give highly connected rigid gel networks formed by strong interaction of DBS with all PPO repeat units [23]
Gas Permeability:

No.	Gas	Value	Note
1	H_2	0.0014 cm^3 mm/(m^2 day atm)	molar mass between crosslinks 425
2	H_2	0.0036 cm^3 mm/(m^2 day atm)	molar mass between crosslinks 725
3	H_2	0.009 cm^3 mm/(m^2 day atm)	molar mass between crosslinks 1025
4	H_2	0.014 cm^3 mm/(m^2 day atm)	molar mass between crosslinks 2000
5	H_2	0.022 cm^3 mm/(m^2 day atm)	molar mass between crosslinks 3000
6	CO	0.0003 cm^3 mm/(m^2 day atm)	molar mass between crosslinks 425
7	CO	0.003 cm^3 mm/(m^2 day atm)	molar mass between crosslinks 725
8	CO	0.0014 cm^3 mm/(m^2 day atm)	molar mass between crosslinks 1025
9	CO	0.004 cm^3 mm/(m^2 day atm)	molar mass between crosslinks 2000
10	CO	0.0062 cm^3 mm/(m^2 day atm)	molar mass between crosslinks 3000 [28]

Electrical Properties:
Electrical Properties General: Used as a constituent of polymer electrolytes [8,9,10]

Optical Properties:
Optical Properties General: Refractive index data have been reported [21]
Transmission and Spectra: Spectral data have been reported [22]
Molar Refraction: Specific refractivity 0.26879 cm^3 g^{-1} (MW 1000, T_g -83° (approx.)) [21]

Polymer Stability:
Thermal Stability General: Oxidative degradation is rapid at elevated temps. [11]
Decomposition Details: Oxidation products include acetaldehyde, Me_2CO, MeOH, H_2O, formic acid and AcOH. Two mechanisms have been proposed based on hydroperoxide formation at the secondary alkoxy radical [3,5] and/or the tertiary alkoxy radical [3,4]. Unsaturated hydrocarbons are not present - in contrast to purely thermal decomposition. Alkali metal salts strongly retard the formation of volatile products [29]
Flammability: Fl. p. is greater than 93° [11]
Environmental Stress: Poly(propylene oxides) and their precursors are not quantitatively eliminated by sewage treatment works [30]

Applications/Commercial Products:
Processing & Manufacturing Routes: Normal commercial production is by base catalysed anionic polymerisation. [11,16,17,18] e.g. Propylene oxide may be polymerised at 85° for 40h in the presence of tetrabutylammonium benzoate in dry acetonitrile [18] Other methods of synth. have been investigated [11,21]. Partially stereoregular polypropylene oxide may be prod. using the Pruitt-Baggett catalyst. [20]
Applications: Used as a soft polyether unit in polyurethane elastomers and foams; in polymer electrolytes; as surfactants (lubricants, dispersants, antistatic agents, foam control agents, in printing inks, as solubilisers); in hydraulic fluids, coolant compositions; in medical applications (protective bandages, drug delivery systems, organ preservation, dental compositions, fat substitute)

Trade name	Details	Supplier
Alkapol PPG		Rhone-Poulenc
Hodag PPG		Hodag
Jeffox		Texaco
Jeffox		Huntsman Chemical
Macol		PPG/Specialty Chem
Pluracol		BASF
Pluracol		BP Chemicals
Pluracol		Harcros
Pluracol		Miles
Pluracol		Olin
Pluracol		Hüls AG
Pluracol		Calgene
Polyglycol P		Dow
Unicol		UPI
Voranol		Dow
Witconol		Witco
Witconol		Shell Chemicals
Witconol		Akrochem
		ARCO Chemical Company
		Ashland

Bibliographic References
[3] Barton, Z., Kemp, T.J., Buzy, A. and Jennings, K.R., *Polymer*, 1995, **36**, 4927
[4] Griffiths, P.J.F., Hughes, J.G. and Park, G.S., *Eur. Polym. J.*, 1993, **29**, 437
[5] Lemaire, J. and Gauvin, P., *Makromol. Chem.*, 1987, **188**, 1815
[6] Sasanuma, Y., *Macromolecules*, 1995, **28**, 8629
[7] Sunghoe, Y., Ichikawa, K., MacKnight, W.J. and Hsu, S.L., *Macromolecules*, 1995, **28**, 4278
[8] *Solid Electrolytes Their Appl.*, (ed. E.C. Subbarao), Plenum, New York, 1980

[9] Watanabe, M. and Ogata, N., *Br. Polym. J.*, 1988, **20**, 181
[10] Mitani, K. and Adachi, K., *J. Polym. Sci., Polym. Phys. Ed.*, 1995, **33**, 947
[11] Gagnon, S.D., *Kirk-Othmer Encycl. Chem. Technol.*, 4th edn., (ed. J.I. Kroschwitz), Wiley Interscience, 1991, **19**
[12] Hähner, U., Habicher, W.D. and Schwetlick, K., *Polym. Degrad. Stab.*, 1991, **34**, 111, 119
[13] Szewczyk, P., *J. Appl. Polym. Sci.*, 1986, **31**, 1151
[14] Gagnon, S.D., *Encycl. Polym. Sci. Eng.*, 2nd edn., (ed. J.I. Kroschwitz), Wiley Interscience, 1985, **6**
[15] Mieczkowski, R., *Eur. Polym. J.*, 1991, **27**, 377
[16] Yu, G., Masters, A.J., Heatley, F., Booth, C. and Blease, T.G., *Macromol. Chem. Phys.*, 1994, **195**, 1517
[17] Yu, G., Heatley, F., Booth, C. and Blease, T.G., *J. Polym. Sci., Part A: Polym. Chem.*, 1994, **32**, 1131
[18] Chen, X. and Van De Mark, M., *J. Appl. Polym. Sci.*, 1993, **50**, 1923
[19] Kohjiya, S., Sato, T., Nakayama, T. and Yamashita, S., *Makromol. Chem., Rapid Commun.*, 1981, **2**, 231
[20] &kColak, N., *Polym.-Plast. Technol. Eng.*, 1996, **35**, 317
[21] Krbecek, H.H., Kupisch, W. and Pietralla, M., *Polymer*, 1996, **37**, 3483
[22] Trathnigg, B. and Maier, B., *Macromol. Symp.*, 1996, **110**, 231
[23] McLin, M.G. and Angell, C.A., *Polymer*, 1996, **37**, 4713
[24] Morales, E., Salmerón, M. and Acosta, J.L., *J. Polym. Sci., Part B: Polym. Phys.*, 1996, **34**, 2715
[25] Nuñez, C.M., Whitfield, J.K., Mercurio, D.J., Ilzhoefer, J.R., et al, *Macromol. Symp.*, 1996, **106**, 275
[26] Crowther, N.J. and Eagland, D., *J.C.S., Chem. Commun.*, 1994, 839
[27] Crowther, N.J. and Eagland, D., *J. Chem. Soc., Faraday Trans.*, 1996, **92**, 1859
[28] Andrady, A.L., Sefcik, M.D., *J. Polym. Sci., Part B: Polym. Phys.*, 1983, **21**, 2453
[29] Costa, L., Camino, G., Luda, M.P., Cameron, G.G and Qureshi, M.Y., *Polym. Degrad. Stab.*, 1996, **53**, 301
[30] Paxeus, N. and Schroeder, H.F., *World Surfactants Congr.*, *4th*, 1996, **4**, 463

Poly(propylene oxide-*co*-allyl glycidyl ether) P-502

Synonyms: *GPO. PO-AGE copolymer. Allyl glycidyl ether-propylene oxide synthetic rubber. Poly[methyloxirane-co-(2-propenyloxymethyl)oxirane]*
Related Polymers: Propylene oxide elastomers
Monomers: Propylene oxide, Allyl glycidyl ether
Material class: Synthetic Elastomers, Copolymers
Polymer Type: polyalkylene ether
CAS Number:

CAS Reg. No.
25104-27-2

Molecular Formula: $[(C_3H_6O).(C_6H_{10}O_2)]_n$
Fragments: C_3H_6O $C_6H_{10}O_2$
Additives: Curing/vulcanising systems: sulfur cure or sulfur-donor cure, not peroxide types which attack the backbone
General Information: GPO is the ASTM designation. Crystallinity is low, the pendant allyl group contributing to an amorph. state. Molar composition of PO ranges from about 65 to 90%. The copolymer is used in the vulcanised form

Volumetric & Calorimetric Properties:
Density:

No.	Value	Note
1	d 1.01 g/cm^3	[1]

Glass-Transition Temperature:

No.	Value	Note
1	-62°C	[1]

Brittleness Temperature:

No.	Value	Note
1	-90°C	uncompounded [2]
2	-61--58°C	vulcanised, commercial [1,2]
3	-56°C	94% PO, 6% AGE [3]

Surface Properties & Solubility:
Plasticisers: Plasticisers are required only in highly filled compounds or for softening or viscosity. Aromatic oils (e.g. Sundex 790 (Sun Oil)) and esters (e.g. DOP) may be used [2], also ethers (e.g. di(butoxyethoxyethyl) formal), polyesters (e.g. Paraplex G50), ether-esters (e.g. di(butoxyethoxyethyl)adipate) [1]
Solvents/Non-solvents: Unvulcanised polymers sol. aromatic hydrocarbons, aliphatic hydrocarbons [2]

Transport Properties:
Polymer Melts: η 21–26 (oscillating disc rheometry) [1]
Permeability and Diffusion General: Gas permeability is low [1]
Water Absorption:

No.	Value	Note
1	13 %	70h, 20°, vulcanised [2]
2	4 %	70h, 23°, vulcanised [1]
3	10 %	70h, 100°, vulcanised [1]

Mechanical Properties:
Mechanical Properties General: Mech. props. measured after vulcanisation (170°). Tensile strength and hardness are increased by reinforcing fillers. Plots of loading vs. parameter vs. filler have been reported. [2] Low temp. flexibility is excellent (to -65°) as is flex life [1] (slightly better than natural rubber)
Tensile Strength Break:

No.	Value	Note	Elongation
1	10.2 MPa		365%
2	10.6 MPa	air aged 70h, 125° [1]	80%
3	13.3 MPa		370%
4	10.9 MPa	air aged 70h, 150° [2]	210%
5	13.7 MPa	1980 psi, 94% PO, 6% AGE [3]	
6	19.2 MPa	2790 psi, MeOH purified [3]	

Miscellaneous Moduli:

No.	Value	Note	Type
1	3.4 MPa	[1]	100% Modulus
2	3.8 MPa	[2]	100% Modulus
3	6.2 MPa	air aged 70h, 125°	100% Modulus
4	4.6 MPa	air aged 70h, 150° [2]	100% Modulus

Viscoelastic Behaviour: Compression set 39% (70h, 100°); 66% (70h, 150°) [1]; 66% (70h, 125°); 67% (70h, 150°) [2]; 52% (94% PO, 6% AGE) [3]
Mechanical Properties Miscellaneous: Resilience 42% [1]; 65% (Lupke, 94% PO, 6% AGE) [3]
Hardness: Shore A68–A70 [1,2,3]; 76 (air aged, 70h, 125°) [1]; 73 (air aged, 70h, 150°) [2]; 90 (MeOH purified, 94% PO, 6% AGE) [3]

Failure Properties General: Tear strength 51 kN m^{-1} [1]; 42.7 kN m^{-1} (244 lb in^{-1}, 94% PO, 6% AGE) [3]
Friction Abrasion and Resistance: Wear resistance is poor [1]

Polymer Stability:
Polymer Stability General: Compounding with polycarbonate improves impact and chemical resistance. Copolymerising with styrene increases weather resistance
Thermal Stability General: Resistant to high temp. ageing [1]; further improved by addition of 2-mercaptobenzimidazole [2]
Environmental Stress: Resistance is moderate to poor. Volume change after 70h exposure 48% (ASTM fuel A, 23°); 134% (ASTM fuel B, 23°); 175% (ASTM fuel C, 23°); 11% (ASTM oil no. 1, 125°); 16% (ASTM oil no. 1, 150°); 92% (ASTM oil no. 3, 125°); 102% (ASTM oil no. 3, 150°) [1]
Chemical Stability: Ozone resistance 168h (min., 100 ppm, 49°) [1], further improved by addition of nickel dibutyldithiocarbamate [2] (see natural rubber. Chain backbone is sensitive to peroxides.

Applications/Commercial Products:
Processing & Manufacturing Routes: Normally, soln. polymerisation in aromatic or aliphatic hydrocarbon solvents (e.g. heptane) is carried out in a continuous process. Solvent, catalyst and monomers are fed to a controlled temp., back mixed reactor. Composition uniformity is maintained by adjusting catalyst feed rate to give a constant monomer concentration. Reactivity ratio for PO:AGE is 1. MW control is by organic acid anhydride and organic acid chloride addition [1,2]. Slurry systems have been reported [4,5]. GPO tends to crumble and does not bond well. Best results are obtained by starting mixing of the elastomer only in a cool mill/mixer. Ultimately 60–80° is recommended. Fillers are added incrementally followed by other dry ingredients (not vulcanising agent). Plasticisers follow and finally vulcanising agent (sulfur containing, not peroxide - which should only be added in very small amounts if breakdown is required) [1,2]
Applications: The cured rubber is used as engine mounts and suspension bushings. Applications due to vibration damping comparable to that of natural rubber, but where superior temp. resistance is required. Compounding with polycarbonate yields an impact and chemical resistant product. Copolymer with styrene has applications due to good weather resistance

Trade name	Supplier
Parel 58	Zeon

Bibliographic References
[1] Owens, K. and Kyllingstad, V.L., *Kirk-Othmer Encycl. Chem. Technol.*, 4th edn., (eds. J.I. Kroschwitz and M. Howe-Grant), John Wiley and Sons, 1993, **8**
[2] Body, R.W. and Kyllingstad, V.L., *Encycl. Polym. Sci. Eng.*, 2nd edn., (ed. J.I. Kroschwitz), John Wiley and Sons Inc., 1986, **6**
[3] Hendrickson, J.G., Gurgiolo, A.E. and Prescott, W.E., *Ind. Eng. Chem. Proc. Des. Dev.*, 1963, **2**, 199
[4] U.S. Pat., 1973, 3 776 863
[5] U.S. Pat., 1976, 3 957 697

Poly(propylene oxide-*co*-allyl glycidyl ether-*co*-epichlorohydrin)   P-503

$$\left[-(CH_2CHO)_x-(CH_2CHO)_y-(CH_2CHO)_z-\right]_n$$
$$CH_3CH_2ClCH_2OCH_2CH=CH_2$$

Synonyms: *Poly(epichlorohydrin-co-propylene oxide-co-allyl glycidyl ether). GPCO. ECH-PO-AGE terpolymer. Allyl glycidyl ether-epichlorohydrin-propylene oxide synthetic rubber*
Related Polymers: Propylene oxide elastomers
Monomers: Propylene oxide, Allyl glycidyl ether, Epichlorohydrin
Material class: Synthetic Elastomers
Polymer Type: polyalkylene ether, chlorinated polyether

CAS Number:

CAS Reg. No.
25213-15-4

Molecular Formula: $[(C_3H_6O).(C_3H_5ClO).(C_6H_{10}O_2)_2]_n$
Fragments: C_3H_6O C_3H_5ClO $C_6H_{10}O_2$
Additives: Curing/vulcanising systems (sulfur or peroxides) designed for use with conventional rubber. However, any system suitable for polyethers may be used
General Information: GPCO is the ASTM designation. Crystallinity is low and the struct. is generally amorph. Molar composition of PO ranges from about 65 to 90%

Volumetric & Calorimetric Properties:
Density:

No.	Value	Note
1	d 1.12 g/cm^3	[1]

Glass-Transition Temperature:

No.	Value	Note
1	-48°C	[1]

Brittleness Temperature:

No.	Value	Note
1	-46°C	ASTM 2137 [1]

Surface Properties & Solubility:
Solvents/Non-solvents: Uncured polymer sol. aromatic solvents, ketones [1]

Transport Properties:
Polymer Melts: η 60–80 (100°, Mooney viscosity) [1]
Permeability and Diffusion General: Gas permeability is low [1]

Mechanical Properties:
Mechanical Properties General: Mech. props. are reported for the vulcanised elastomer (170°). Low temp. flexibility is excellent but deteriorates with increasing epichlorohydrin content
Tensile Strength Break:

No.	Value	Note	Elongation
1	11.4 MPa	[1]	210%
2	11.8 MPa	air aged, 70h, 125° [1]	80%

Miscellaneous Moduli:

No.	Value	Note	Type
1	5.6 MPa		100% Modulus
2	0 MPa	air aged, 70h, 125° [1]	100% Modulus

Viscoelastic Behaviour: Compression set 39% (70h, 100°); 100% (70h, 150°)
Hardness: Shore A77; A90 (air aged, 70h, 125°) [1]
Failure Properties General: Tear strength 46 kN m^{-1} [1]

Polymer Stability:
Thermal Stability General: Exhibits heat resistance similar to epichlorohydrin-ethylene oxide copolymer
Environmental Stress: Fuel resistance is similar to polychloroprene and increases with increasing epichlorohydrin content. Volume change after 70h exposure 12% (ASTM fuel A, 23°); 65% (ASTM fuel B, 23°); 106% (ASTM fuel C, 23°); 0% (ASTM oil no. 1,

125°); 2% (ASTM oil no. 1, 150°); 40% (ASTM oil no. 3, 125°); 56% (ASTM oil no. 3, 150°) [1]. Ozone resistance 168h (min., 100 ppm, 49°) [1]

Applications/Commercial Products:
Processing & Manufacturing Routes: Manufacture and processing similar to polypropylene-co-allyl glycidyl ether. Reactivity ratio 1.5:1.5:1 hence background monomer ratio control is essential for product uniformity [1]
Applications: Applications in automotive industry as dust covers, fuel hose covers, suspension/transmission boots, antivibration mounts, cable covers. Also has high/low temp. high flex applications

Trade name	Supplier
Zeospan	Nippon Zeon

Bibliographic References
[1] Owens, K. and Kyllingstad, V.L., *Kirk-Othmer Encycl. Chem. Technol.*, 4th edn., (eds. J.I. Kroschwitz and M. Howe-Grant), John Wiley and Sons, 1993

Poly(propylene oxide-co-ethylene oxide) P-504
Synonyms: *PO-EO polyols. PO-EO copolymers. Propylene oxide-ethylene oxide copolymers. Methyloxirane-oxirane copolymer*
Related Polymers: Polypropylene oxide
Monomers: Propylene oxide, Ethylene oxide
Material class: Copolymers, Oligomers
Polymer Type: polyalkylene ether, polyethylene oxide
CAS Number:

CAS Reg. No.
9003-11-6

Molecular Formula: $[(C_2H_4O).(C_3H_6O)]_n$
Fragments: C_2H_4O C_3H_6O
Miscellaneous: Usually in the form of a low MW polyol

Polypropylene/polycarbonate blends P-505
Synonyms: *PP/PC blends*
Related Polymers: Polypropylene blends
Material class: Thermoplastic, Blends
Polymer Type: polypropylene, polycarbonates (miscellaneous)
CAS Number:

CAS Reg. No.
9003-07-0
24936-68-3

Additives: Compatibilisers especially SEBS, styrene-acrylonitrile grafted EPDM, maleated polypropylene and polybutyleneterephthalate
General Information: Addition of about 5 wt % polypropylene to polycarbonate or about 10 wt % polycarbonate to polypropylene results in improved processability, impact strength and modulus, while reducing notch sensitivity and mould shrinkage [1]
Morphology: Because of general immiscibility, morphology determines blend performance. Polycarbonate additions to polypropylene enhance crystallinity and slightly increase T_c. Compatibilisation increases the effect. Dispersion is dependent on the viscosity ratio. Processing effects are great, e.g. between the surface and centre of an extruded strand [2,3]. Most deformation and disintegration takes place during the first two minutes of mixing. Composition, viscosity ratio and torque ratio affect size, while viscosity ratio and elasticity ratio affect the shape of the dispersed phase. Some coalescence coarsening has been observed for 23% PC in PP. Shear rate and matrix shear stress have relatively little effect on the dispersed phase. The effect of composition is particularly great near the region of dual phase continuity. The die has a greater effect on the morphology than other aspects of twin-screw extrusion [2,3,4,5,6]

Volumetric & Calorimetric Properties:
Density:

No.	Value	Note
1	d 1.2 g/cm³	100% PC
2	d 1.06 g/cm³	100% PP [7]

Latent Heat Crystallization: ΔH_{fusion} 75 J g⁻¹ (PP 92%, PC 8%); 67 J g⁻¹ (PP 75%, PC 25%); 39 J g⁻¹ (PP 50%, PC 50%) [8]
Melting Temperature:

No.	Value	Note
1	158.5–159°C	[8]

Deflection Temperature:

No.	Value	Note
1	67–140°C	100% PC, 1.82 MPa, negative deviation [8]

Transition Temperature:

No.	Value	Note	Type
1	106–107.5°C	[8]	crystallisation temp.

Transport Properties:
Polymer Melts: Extrudate swell has a max. of 90% (20–25% PP/75–80% PC) [7]. Shear stress to shear rate relations are intermediate between those of the components

Mechanical Properties:
Mechanical Properties General: Small additions (less than 10%) of PC to PP increase hardness, modulus and impact strength but reduce notch sensitivity. Small additions (less than 5%) of PP to PC increase impact strength and modulus but reduce notch sensitivity [10]
Flexural Modulus:

No.	Value	Note
1	1170 MPa	100% PP [8]
2	1300 MPa	75% PP/25% PC [8]
3	900 MPa	60% PP/50% PC [8]
4	1700 MPa	100% PC [8]

Tensile Strength Break:

No.	Value	Note	Elongation
1	12–22 MPa	60% PC [8]	20–10%
2	44 MPa	100% PC [8]	

Tensile Strength Yield:

No.	Value	Note
1	29 MPa	100% PP [8]
2	46 MPa	100% PC [8]
3	12 MPa	50% PP/50% PC [8]

Electrical Properties:
Dielectric Permittivity (Constant):

No.	Value	Frequency	Note
1	2.6		90% PP/10% PC
2	2.72		60% PP/40% PC [10]
3	2.8		50% PP/50% PC
4	3.1		40% PP/60% PC
5	3.22	1–100 kHz	30–150°, 10% PP/90% PC

Dissipation (Power) Factor:

No.	Value	Note
1	0.0002–0.0005	60% PP/40% PC [10]
2	0.0005–0.0007	50% PP/50% PC [10]
3	0.0006–0.0007	50–150°, 40% PP/60% PC [10]

Polymer Stability:
Chemical Stability: PP containing 5–95% PC has excellent chemical resistance [9]

Bibliographic References
[1] Polypropylene: Structure, *Blends and Composites: Copolymers and Blends*, (ed. J. Karger-Kocsis), Chapman and Hall, 1995
[2] Favis, B.D. and Chalifoux, J.P., *Polymer*, 1988, **29**, 1761
[3] Favis, B.D., *J. Appl. Polym. Sci.*, 1990, **39**, 285
[4] Favis, B.D. and Therrieu, D., *Polymer*, 1991, **32**, 1474
[5] Favis, B.D. and Chalifoux, J.P., *Polym. Eng. Sci.*, 1987, **27**, 1591
[6] Favis, B.D. and Willis, J.M., *J. Polym. Sci., Polym. Phys. Ed.*, 1990, **28**, 2259
[7] Rudin, A. and Braithwaite, N.E., *Polym. Eng. Sci.*, 1984, **24**, 1312
[8] Liang, Z. and Williams, H.L., *J. Appl. Polym. Sci.*, 1991, **43**, 379
[9] Fisa, B., Favis, B.D. and Bourgeois, S., *Polym. Eng. Sci.*, 1990, **30**, 1051
[10] Utracki, L.A. and Dumoulin, M.M., Polypropylene: Structure, *Blends and Composites: Copolymers and Blends*, (ed. J. Karger-Kocsis), Chapman and Hall, 1995
[11] Pillai, P.K.C., Narula, G.K. and Tripathi, A.K., *Polym. J. (Tokyo)*, 1984, **16**, 575

Polypropylene/polyester blends   P-506
Synonyms: *PP/PEST*
Related Polymers: Polypropylene blends
Material class: Blends
Polymer Type: polypropylene
CAS Number:

CAS Reg. No.
9003-07-0

Additives: Compatibilisers, especially maleated polypropylene
General Information: PP/PEST blends are generally formulated within the dispersed phase region. PP/liq. cryst. polyester blends are a new development based on the orientation (uni- or bi-axial) of the polyester crysts. dispersed in the PP matrix

Volumetric & Calorimetric Properties:
Latent Heat Crystallization: ΔH_f 75.4–83.7 kJ kg^{-1} (18–20 cal g^{-1} PP/LCP, LCP $\leq$20%) [2]
Transition Temperature General: T_g tends to that for PET and is little affected by blending

Transport Properties:
Polymer Melts: Viscosity 4000–40000 Poise (400–4000 Pa s, shear rate 300-10 s^{-1}, 100% polypropylene); 1500–4000 Poise (150–400 Pa s, shear rate 300-10 s^{-1}, 80% polyproylene) [2]

Mechanical Properties:
Mechanical Properties General: Data refer primarily to fibres/filaments

Tensile (Young's) Modulus:

No.	Value	Note
1	10300 MPa	1500 kpsi, 30% PP/70% PET
2	8300 MPa	1200 kpsi, 50% PP/50% PET [1]
3	1000–5000 MPa	80% PP/20% LCP
4	1000–4000 MPa	100% PP [2]

Tensile Strength Break:

No.	Value	Note	Elongation
1	40–300 MPa	80% PP/20% sebacic acid, hydroxybenzoic acid, dihydroxybiphenyl	350–550%
2	50–400 MPa	100% PP [2]	400–700%

Mechanical Properties Miscellaneous: Plots of dynamic mech. props. for the PP/PET blends have been reported [1]. Tenacity 0.51 N tex^{-1} (32% elongation) (30% PP/70% PET). 0.5 N tex^{-1} (22% elongation) (50% PP/50% PET)

Bibliographic References
[1] Rudin, A., Loucks, D.A. and Goldwasser, J.M., *Polym. Eng. Sci.*, 1980, **20**, 741
[2] Ye, Y.-C., La Mantia, F.P., Valenza, A, Citta, V. et al, *Eur. Polym. J.*, 1991, **27**, 723
[3] Polypropylene: Structure, *Blends and Composites: Copolymers and Blends*, (ed. J. Karger-Kocsis), Chapman and Hall, 1995, **2**

Polypropylene/rubber blends   P-507
Synonyms: *Thermoplastic vulcanisates. PP/rubber blends. Thermoplastic elastomers*
Related Polymers: Polypropylene blends, Blends with EPDM (dynamically vulcanised), Blends with butadiene rubbers (polybutadiene), Blends with nitrile rubbers (NBR), Blends with butyl rubber, Blends with natural rubber, Blends with EPDM (unvulcanised) or EPR, Elastomers based on poly(propylene oxide), Propylene-ethylene elastomers
Material class: Thermoplastic, Synthetic Elastomers, Blends
Polymer Type: polypropylene
Additives: Except for colourants, use of additives within these blends is not recommended
General Information: Combination of a fine dispersion of a highly vulcanised rubber-like component within a thermoplastic matrix gives products comparable to thermoset rubbers but which, being thermoplastic, are re-processable. Blending polypropylene with other polymers improves some of its props. particularly its impact strength at low temps. Blends with rubbers are of particular interest as they can combine the processing and mechanical props. of polypropylene and the flexibility of rubbers. Polypropylene (PP)/rubber blends include: PP/ethylene-propylene-rubber (EPR); PP/ethylene-butene rubber (EBR); PP/EPR/polyethylene; PP/ethylene-propylene-diene (EPDM); PP/poly(*cis*-butadiene) (PCB); PP/natural rubber (NR); PP/nitrile rubber (NBR); PP/styrene-butadiene-styrene (SBS) triblock copolymer; PP/styrene-isoprene-styrene (SIS) triblock copolymer; PP/styrene-block-ethene-*co*-1-butene-block-styrene (SEBS). The most frequently encountered blends are PP/EPDM and PP/EPR; when these contain lower amounts of elastomer (5–20%) low temp. impact strength and toughness are improved with some maintenance of rigidity. Blends containing higher amounts of elastomer become semi-elastomeric and eventually fully elastomeric. PP may be blended with incompatible rubbers in the presence of compatibilisers or

– **Polypropylene/rubber blends**

emulsifiers. The potential performance of PP/rubber blends has been further improved by the development of dynamic vulcanisation process

Morphology: Domain morphology of the blends is closely related to their props. and depends on factors such as composition, preparation, vulcanisation/cross-linking, and viscosity ratio. The morphology of various blends has been studied extensively by SEM and optical microscopy [11,12,13,14,15,23,24,27,28,31,32,33]

Miscellaneous: Useful comparisons with extensive data have been reported [1,2,3]. Considerable information on many TPVs has been reported [5,6,7]

Volumetric & Calorimetric Properties:

Thermodynamic Properties General: The linear expansion coefficient of PP/EPR blends at -20° increases sharply as the elastomer content increases from approx. 5 wt% [12]

Transition Temperature General: Cross-linking in PP/EPDM blends lowers the crystallisation temp. (T_c. Increasing the amount of elastomer in the uncross-linked blend generally lowers T_c but after cross-linking the reverse trend is observed [21]. Addition of peroxide and co-agent has little effect on the melting point of PP in PP/NR and PP/EPR blends but T_c in the former is increased by addition of both. In PP/EPR blends increasing the co-agent also raises T_c significantly but the presence of peroxide can lower T_c significantly. T_m remains unchanged upon increasing viscosity ratio but T_c decreases [16,17]

Deflection Temperature:

No.	Value	Note
1	100–114°C	PP/EPR 58/31; varying MW of EPR, 10% talc [27]
2	100–106°C	PP/EBR 58/31; varying MW of EBR, 10% talc [27]

Vicat Softening Point:

No.	Value	Note
1	134–143°C	PP/EPDM 80/20; not cross-linked [22]
2	140–148°C	PP/EPDM 80/20; cross-linked [22]
3	135°C	PP/SBS 80/20; not cross-linked [22]
4	137°C	PP/SBS 80/20; not cross-linked [22]
5	134°C	PP/SIS 80/20; not cross-linked [22]
6	140°C	PP/SIS 80/20; not cross-linked [22]

Brittleness Temperature:

No.	Value	Note
1	<-60°C	max., PP/EPDM [2]
2	-24°C	compatibilised PP/NBR [2]
3	-47°C	PP/EPDM:compatibilised PP/NBR, 50:50 [2]

Transport Properties:

Transport Properties General: The melt flow rate of PP/EPR and PP/EPDM blends decreases with increasing amounts of elastomer and increases with cross-linking. Cross-linking of PP/SBS and PP/SIS blends results in a decrease in melt flow rate [12,20,22]. The addition of peroxide and co-agent to PP/NR blends has a significant effect on the melt index. At constant level of peroxide the addition of increasing amounts (0–0.1%) of co-agent (TMPTA) causes a slight increase in melt index. Melt index decreases sharply as the amount of co-agent rises to 0.4%. Above 0.4% co-agent, the melt index changes only slightly. At a constant level of peroxide the addition of up to 0.01% peroxide causes a sharp drop in the melt index. Amounts of peroxide between 0.01 and 0.02% cause only a slight decrease in the melt index and above 0.02% peroxide, the melt index increases slightly [16]

Melt Flow Index:

No.	Value	Note
1	100–510 g/10 min	PP/NR 90/10; varying levels of peroxide co-agent [16]
2	140–240 g/10 min	PP/EPDM; increasing EPDM; not cross-linked [20]
3	140–470 g/10 min	PP/EPDM; increasing EPDM; varying degress of cross-linking [20]
4	160 g/10 min	PP/EPDM 80/20; not cross-linked [22]
5	240 g/10 min	PP/EPDM 80/20; not cross-linked [22]
6	150 g/10 min	PP/SBS 80/20; not cross-linked [22]
7	70 g/10 min	PP/SBS 80/20; not cross-linked [22]
8	260 g/10 min	PP/SIS 80/20; not cross-linked [22]
9	130 g/10 min	PP/SIS 80/20; not cross-linked [22]

Polymer Melts: The effect of peroxide/co-agent addition on the melt viscosity of PP/EPR and PP/NR blends has been reported [16,17]. Changes in the rheological props. arising from the addition of glass fibre to PP/EPDM blends have been reported [19]. The melt rheology of PP/SBS blends has been studied with and without dynamic vulcanisation. Melt elasticity increases and melt fracture decreases with increasing SBS content and dynamic vulcanisation enhances these effects [30]

Mechanical Properties:

Mechanical Properties General: Effects of blend ratio and dynamic vulcanisation on mech. props. have been reported [5]. A range of thermoplastic vulcanisates with polypropylene has been reported [1]. In studies on the correlation between blend props. and characteristics of PP and elastomer it has been found that the best compositions are formed when there is a greater degree of crystallinity in PP, when there are higher entanglement densities in the elastomer and when the surface energy of the elastomer is similar to that of PP. The morphology and particle size of the rubber are important factors affecting the mechanical props. of blends, particularly impact strength and flexural modulus. These are improved with fibril formation. The effect of struct. on mechanical props. has been investigated for a number of blends including PP/EPR, PP/EPR/PE and PP/SEBS [24,33]. The props. of PP blends with elastomers in which the surface energies are different can be improved by compatibilisation. For example, PP/NBR blends may be compatibilised by the addition of EPDM or by maleic-modification of PP and the use of amine-terminated nitrile elastomer (NBR). PP/NBR blends may also be compatibilised by melt grafting using 2-isopropenyl-2-oxazoline or glycidyl methacrylate both of which allow chemical reaction between PP and the elastomer. By these means PP/NBR blends can be obtained which are comparable to PP/EPDM blends in their mechanical props.[2,25,26,31,32]. The variation of loss and storage moduli and loss tangent with temp. and composition has been reported for PP/NBR and PP/NR. The effect of compatibilisation and dynamic vulcanisation on the dynamic mechanical props. is also reported [14,29]. Yield strength and flexural modulus both decrease and the elongation at break increases as the elastomer content of a PP/EPDM blend is increased. The effect of cross-linking varies but generally gives rise to lowered yield strength and increases in elongation and flexural modulus. The impact strength and toughness are also improved as the level of elastomer increases. Cross-linking further improves the impact strength, the improvement being particularly effective at elastomer levels of approx. 20% and higher [20,21]. Increasing the extent of cross-linking improves storage modulus and decreases loss tangent values and has more effect in blends with a higher proportion of elastomer. Increasing the amount of elastomer reduces the storage modulus and increases the loss tangent

– Polypropylene/rubber blends

[10,21]. In PP/SBS blends the tensile strength, tensile modulus and impact strength decrease and the elongation increases as the amount of elastomer increases. Dynamic vulcanisation gives rise to a greater improvement in props. This improvement is relatively small at low levels of SBS but is dramatic at SBS amounts of 20% and more [24]. The effect of viscosity ratio and added peroxide/co-agent on mechanical props. has been reported for PP/NR, PP/NBR and PP/EPR [14,16,17]. PP blended with EPR/PE can have better impact strength than PP blended with EPR alone. [12,33]. EBR has been found to have a greater toughening effect in PP blends than EPR [27]. The effect of the addition of filler and extender oil on the mechanical props. of PP/rubber blends has been reported [2,10,23]. Tear strength 100.5 kN m^{-1} (30% natural rubber, not cross-linked), 115.2 kN m^{-1} (30% natural rubber, dynamically cross-linked), 22.1 kN m^{-1} (70% natural rubber, not cross-linked), 44.4 kN m^{-1} (70% natural rubber, dynamically cross-linked) [15]

Tensile (Young's) Modulus:

No.	Value	Note
1	97–103 MPa	PP/EPDM 66.7/100; decreasing particle size; static vulcanisation [2]
2	58 MPa	PP/EPDM 66.7/100; dynamic vulcanisation [2]
3	60–72 MPa	PP/EPDM 66.7/100; dynamic vulcanisation, decreasing cross-link density [2]
4	22–435 MPa	PP/EPDM 66.7/100; increasing levels of PP; dynamic vulcanisation [2]
5	691–1175 MPa	PP/SEBS; varying composition and MW [24]
6	691–998 MPa	PP/SEBS; 80/20 varying composition and MW [24]
7	843–1170 MPa	PP/SEBS; 90/10 varying composition and MW [24]
8	250 MPa	PP/NBR 70/30; no compatibilisation [31]
9	135 MPa	PP/NBR 50/50; no compatibilisation [31]
10	47 MPa	PP/NBR 30/70; no compatibilisation [31]

Flexural Modulus:

No.	Value	Note
1	1290–1340 MPa	23°, PP/EPR/HDPE 88/6/6 [11]
2	990–1060 MPa	23°, PP/EPR/HDPE 76/12/12 [11]
3	952 MPa	room temp., 9700 kg cm^{-2}, PP/EPR 80/20 [13]
4	991–1020 MPa	10100–10400 kg cm^{-2}, PP/EPR 80/20; increased viscosity ratio [17]
5	1148–1305 MPa	room temp., 11700–13300 kg cm^{-2}, PP/EPR/HDPE 80/10/10 [13]
6	1520–1560 MPa	PP/EPR 58/31, varying MW of EPR, 10% talc [27]
7	1560–1600 MPa	PP/EBR 58/31, varying MW of EBR, 10% talc [27]
8	961–1462 MPa	9800–14900 kg cm^{-2}, PP/EPDM increasing EPDM; not cross-linked [21]
9	932–1707 MPa	9500–17400 kg cm^{-2}, PP/EPDM; increasing EPDM; cross-linked [21]

Tensile Strength Break:

No.	Value	Note	Elongation
1	8.6–19.1 MPa	PP/EPDM (66.7/100); decreasing particle size; static vulcanisation [2]	165–480%
2	24.3 MPa	PP/EPDM (66.7/100); dynamic vulcanisation [2]	530%
3	4.9–18.2 MPa	PP/EPDM (66.7/100); dynamic vulcanisation; decreasing cross-link density [2]	
4	17.9–28.8 MPa	PP/EPDM (66.7/100); increasing levels dynamic vulcanisation [2]	470–580%
5	18.3 MPa	PP/NBR (70/30); no compatibilisation [31]	96%
6	9.2 MPa	PP/NBR (50/50); no compatibilisation [31]	43%
7	3.5 MPa	PP/NBR (30/70); no compatibilisation [31]	39%
8	24.3 MPa	PP/EPDM (40/60); dynamic vulcanisation [2]	530%
9	21.6 MPa	PP/butyl rubber (40/60); dynamic vulcanisation [2]	380%
10	22.7 MPa	PP/poly-transpentenamer (40/60); dynamic vulcanisation [2]	210%
11	26.4 MPa	PP/natural rubber (40/60); dynamic vulcanisation [2]	390%
12	20.8 MPa	PP/butadiene (40/60); dynamic vulcanisation [2]	258%
13	21.7 MPa	PP/styrene-butadiene rubber (40/60); dynamic vulcanisation [2]	128%
14	13 MPa	PP/chloroprene rubber (40/60); dynamic vulcanisation [2]	141%
15	17 MPa	PP/nitrile rubber (40/60); dynamic vulcanisation [2]	201%
16	8.6 MPa	PP/EPDM; dynamic vulcanisation [2]	415%
17	22.6 MPa	compatibilised PP/NBR; dynamic vulcanisation [2]	485%
18	15.9 MPa	50/50 mix of PP/EPDM and compatibilised PP/NBR; dynamic vulcanisation [2]	510%
19	32.4 MPa	room temp.; 330 kg cm^{-2}, PP/EPR(80/20) [13,17]	510%
20	20.1–30.4 MPa	room temp.; 205–310 kg cm^{-2}, PP/EPR/HDPE (80/10/10) [13]	470–610%

Tensile Strength Yield:

No.	Value	Note
1	23.5 MPa	room temp., 240 kg cm^{-2}, PP/EPR (80/20) [13]
2	28–30.4 MPa	room temp., 285–310 kg cm^{-2}, PP/EPR/HDPE (80/10/10) [13]

– Polypropylene/rubber blends

3	19.1–20.6 MPa	194–210 kg cm^{-2}, PP/NR (90/10), varying levels of cross-linking [16]
4	17.5–26.7 MPa	PP/SEBS; varying composition and MW [24]
5	17.5–23 MPa	PP/SEBS (80/20) varying MW [24]
6	20.9–26.7 MPa	PP/SEBS (90/10) varying MW [24]

Impact Strength: Falling weight impact strength 1.4–1.6 J (PP/EPR/HDPE (88/6/6)) [11], 12.6–12.9 (PP/EPR/HDPE (76/12/12)) [11]

Hardness: Shore D41–D43 (PP/EPDM (66.7/100); decreasing particle size; static vulcanisation), D42 (PP/EPDM (66.7/100); dynamic vulcanisation), D40–D22 (PP/EPDM (66.7/100); dynamic vulcanisation, decreasing cross-link density), D34–D59 (PP/EPDM (66.7/100), increasing levels of PP, dynamic vulcanisation) [2], D55 (PP/NBR (70/30); no compatibilisation), D45 (PP/NBR (50/50); no compatibilisation), D18 (PP/NBR (30/70); no compatibilisation) [31], D68 (PP/EPDM), D93 (compatibilised PP/NBR), D87 (50/50 mix of PP/EPDM and compatibilised PP/NBR [2]; Shore R61 (PP/EPR (80/20) [17]; Shore A95 (PP/NBR (70/30); no compatibilisation), A93 (PP/NBR (50/50); no compatibilisation), A83 (PP/NBR (30/70); no compatibilisation) [31]. Rockwell 75–76 (PP/EPR (58/31); varying, MW of EPR; 10% talc), 83–84 (PP/EBR (58/31); varying MW of EBR; 10% talc [27]

Failure Properties General: The effect of composition and dynamic cross-linking on the tear strength of PP/natural rubber blends has been studied. Examination of the fracture surfaces suggests that PP brittle fracture is changed to a ductile type on the addition of rubber. Dynamic cross-linking of the elastomer appears to result in restricted flow under tear pressure [15]. The toughening mechanisms of various blends with and without cross-linking have been reported [20,27]

Izod Notch:

No.	Value	Notch	Note
1	50 J/m	Y	23°, PP/EPR/HDPE (88/6/6) [11]
2	90–100 J/m	Y	23°, PP/EPR/HDPE (76/12/12) [11]
3		Y	room temp., 46.9 kg cm cm^{-1}, PP/EPR (80/20) [13]
4		Y	room temp., 7.3–31.6 kg cm cm^{-1} PP/EPR/HDPE (80/10/10) [13]
5		Y	-20°, 2.4–5 kg cm cm^{-1}, PP/NR (90/10) [16]
6		Y	25°, 6–30 kg cm cm^{-1} (PP/NR (90/10); varying levels of cross-linking [16]
7		Y	-10°, 47.3 kg cm cm^{-1} PP/SBS (80/20); not cross-linked [22]
8		Y	-10°, 74.1 kg cm cm^{-1} PP/SBS (80/20); cross-linked [22]
9		Y	-30°, 10.5 kg cm cm^{-1} PP/SBS (80/20); not cross-linked [22]
10		Y	-30°, 23.2 kg cm cm^{-1} PP/SBS (80/20); cross-linked [22]
11		Y	-10°, 59.3 kg cm cm^{-1} PP/SIS (80/20); not cross-linked [22]
12		Y	-10°, 75.8 kg cm cm^{-1} PP/SIS (80/20); cross-linked [22]
13		Y	-30°, 11.6 kg cm cm^{-1} PP/SIS (80/20); not cross-linked [22]
14		Y	-30°, 21.9 kg cm cm^{-1} PP/SIS (80/20); cross-linked [22]
15	700–750 J/m	N	-28°, PP/EPR/HDPE (88/6/6) [11]
16	1440–1550 J/m	N	-28°, PP/EPR/HDPE (76/12/12) [11]

Izod Area:

No.	Value	Notch	Note
1	3.6–38.4 kJ/m^2	Notched	PP/SEBS; varying composition and MW [24]
2	27.8–38.4 kJ/m^2	Notched	PP/SEBS (80/20); varying MW [24]
3	3.6–13.4 kJ/m^2	Notched	PP/SEBS (90/10); varying MW [24]

Polymer Stability:

Thermal Stability General: The thermal stability of PP is improved by the addition of rubber [29]
Decomposition Details: The main degradation of PP/natural rubber occurs in the range 400–425° [29]
Chemical Stability: PP/rubber blends have good resistance to hot oil [14]. Volume of swell in hot ASTM No.3 oil, 70h, 100°) 62.5 (PP/EPDM) 22 (compatibilised PP/NBR) 32.5 (50/50 mix of PP/EPDM and compatibilised PP/NBR) [2]

Applications/Commercial Products:

Processing & Manufacturing Routes: Prepared by melt mixing, soln. blending or latex mixing, melt mixing being the most usual. Banbury mixers, mixing extruders and twin-screw mixers are suitable. For dynamically vulcanised blends (TPVs), rubber and plastic plus any filler, plasticiser, lubricant, stabiliser, etc. are melt mixed. Curatives and cross-linkers are then added and mixing continued until mixing torque (or energy) passes through a maximum [7]. The thermoplastic vulcanisation of PP and EPDM in supercritical propane has been reported [28]. Sulfur-curing of PP/EPDM blends has been reported [18]. Dynamic vulcanisation produces elastomer particle sizes far smaller than those obtained using static process, i.e. precured, ground and melt-mixed. For example particle sizes of 1–2 μm are obtained by dynamic vulcanisation compared with 5–70 μm and above for static vulcanisation (PP/EPDM blends). Modification by melt grafting has been used to compatibilise PP blends with polymers such as NBR [25,26]. Blends can be processed by injection moulding or extrusion [2]

Applications: Applications in mechanical rubber goods (gaskets, seals, wheels, tubing, bumpers, suction cups, rollers, etc.); in automotive components (shock isolators, fuel line hose cover, body plugs, grommets, etc.); in hosing; in electrical applications (plugs, cable insulation, bushings, etc.); and in footwear and sports goods

Trade name	Details	Supplier
Santoprene	completely vulcanised PP/EPDM	Monsanto Chemical
TPR 1700	partially vulcanised PP/EPDM	Uniroyal

Bibliographic References

[1] *Handb. Thermoplast. Elastomers*, 2nd edn., (eds. B.M. Walker and C.P. Rader), Van Nostrand Reinhold, 1988
[2] *Thermoplastic Elastomers*, (ed. N.R. Legge), Hanser, 1987
[3] *Thermoplast. Elastomers Rubber-Plast. Blends*, (eds. S.K. De and A.K. Bhowmick), Horwood, 1990
[4] Coran, A.Y. and Patel, R.P., *Rubber Chem. Technol.*, 1980, **53**, 141, 781
[5] Coran, A.Y., Patel, R.P. and Williams, D., *Rubber Chem. Technol.*, 1982, **55**, 1063
[6] Coran, A.Y., Patel, R.P. and Williams-Headd, D., *Rubber Chem. Technol.*, 1985, **58**, 1014
[7] Coran, A.Y. and Patel, R.P., Polypropylene: Structure, *Blends and Composites: Copolymers and Blends*, (ed. J. Karger-Kocsis), Chapman and Hall, 1995, **2**
[8] Martuscelli, E., *Polym. Eng. Sci.*, 1984, **24**, 563

- Polypropylene/styrenic blends

[9] Axtell, F.-H., Autsawasatain, D.and Riyivanontapong, K., *Plast., Rubber Compos. Process. Appl.*, 1992, **18**, 47
[10] Kuriakose, B., De, S.K., Bhagawan, S.S., Sivaramkrishnan, R. and Athithan, S.K., *J. Appl. Polym. Sci.*, 1986, **32**, 5509, (dynamic mech props)
[11] Stehling, F.C., Huff, T., Speed, C.S. and Wissler, G., *J. Appl. Polym. Sci.*, 1981, **26**, 2693, (struct, props)
[12] Petrovic, Z.S., Budinski-Simendic, J., Divjakovic, V. and Skrbic, Z.., *J. Appl. Polym. Sci.*, 1996, **59**, 301
[13] Do, I.H., Yoon, L.K., Kim, B.K. and and Jeong, H.M., *Eur. Polym. J.*, 1996, **32**, 1387
[14] George, S., Neelakantan, N.R., Varughese, K.T. and Thomas, S., *J. Polym. Sci. Part B*, 1997, **35**, 2309, (dynamic mechanical props)
[15] Kuriakose, B., *J. Mater. Sci. Lett.*, 1985, **4**, 455, (failure props)
[16] Yoon, L.K., Choi, C.H. and Kim, B.K., *J. Appl. Polym. Sci.*, 1995, **56**, 239, (processing)
[17] Kim, B.K. and Do, I.H., *J. Appl. Polym. Sci.*, 1996, **61**, 439
[18] Sengupta, A., and Konar, B.B., *J. Appl. Polym. Sci.*, 1997, **66**, 1231
[19] Gupta, A.K., Kumar, P.K. and Ratnam, B.K., *J. Appl. Polym. Sci.*, 1991, **42**, 2595, (melt rheology)
[20] Ishikawa, M., Sugimoto, S. and Inoue, T., *J. Appl. Polym. Sci.*, 1996, **62**, 1495
[21] Inoue, T. and Suzuki, T., *J. Appl. Polym. Sci.*, 1995, **56**, 1113
[22] Inoue, T., *J. Appl. Polym. Sci.*, 1994, **54**, 723, (impact strength, mech props)
[23] Long, Y. and Shanks, R.A., *J. Appl. Polym. Sci.*, 1996, **61**, 1877, (mech props, struct)
[24] Saroop, M., and Mathur, G.N., *J. Appl. Polym. Sci.*, 1997, **65**, 2691, (mech props)
[25] Lui, N.C., and Baker, W.E., *Polymer*, 1994, **35**, 988
[26] Chen, L.F., Wong, B. and Baker, W.E., *Polym. Eng. Sci.*, 1996, **36**, 1594
[27] Yokoyama, Y. and Ricco, T., *Polymer*, 1998, **39**, 3675
[28] Han, S.J., Lohse, D.J., Radosz, M. and Sperling, L.H., *Macromolecules*, 1998, **31**, 5407, (synth, morphology)
[29] Choudhury, N.R., Chaki, T.K. and Bhowmick, A.K., *Thermochim. Acta*, 1991, **176**, 149
[30] Saroop, M. and Mathur, G.N., *J. Appl. Polym. Sci.*, 1997, **65**, 2703, (melt rheology)
[31] George, S., Joseph, R., Thomas, S. and Varughese, K.T., *Polymer*, 1995, **36**, 4405, (morphology, mech props)
[32] Phan, T.T.M., Denicola, J. and Schadler, L.S., *J. Appl. Polym. Sci.*, 1998, **68**, 1451, (morphology, mechanical props.)
[33] Kim, B.K. and Do, I., H., *J. Appl. Polym. Sci.*, 1996, **60**, 2207

Polypropylene/styrenic blends   P-508

Synonyms: *PP/PS blends*
Related Polymers: Polypropylene blends
Material class: Blends
Polymer Type: polypropylene, styrenes
CAS Number:

CAS Reg. No.
9003-07-0
9003-53-6

General Information: Primarily of academic interest, being good for blend morphology study. PP/PS blends have improved high temp. and grease resistance props. PP/HIPS (high impact) and PP/SBS show good mechanical props. and large strain at break. PP/SB/EPDM shows an exceptional balance of props. PP/poly(styrene-*co*-terpene) is useful for production of biaxially oriented, heat sealable film
Morphology: Photomicrographs showing struct. at various compositions and using different blending techniques have been reported [2]

Volumetric & Calorimetric Properties:
Density:

No.	Value	Note
1	d 0.8865 g/cm^3	100% PP
2	d 0.9149 g/cm^3	80% PP/20% PS
3	d 0.9304 g/cm^3	70% PP/30% PS
4	d 0.968 g/cm^3	50% PP/50% PS
5	d 0.9731 g/cm^3	30% PP/70% PS
6	d 0.9908 g/cm^3	20% PP/80% PS
7	d 1.075 g/cm^3	100% PS [1]

Volumetric Properties General: Fractional crystallinity, $_c$, 0.35 (100% PP)- 0.2 (>80% PS) [1]
Latent Heat Crystallization: Enthalpy of fusion, ΔH_f 75.4 kJ kg^{-1} (18 cal g^{-1}, 100% PP, 70% PP/30% PS); 74.5 J kg^{-1} (17.8 cal g^{-1}, 80% PP/20% PS); 74.9 kJ kg^{-1} (17.9 cal g^{-1}, 50% PP/50% PS); 58.6 kJ kg^{-1} (14 cal g^{-1}, 30% PP/70% PS); 54.8 kJ kg^{-1} (13.1 cal g^{-1}, 20% PP/80% PS) [1]
Melting Temperature:

No.	Value	Note
1	162°C	PP $\leq$50%
2	163°C	PP <50% [1]

Glass-Transition Temperature:

No.	Value	Note
1	103°C	80% $\leq$ PP $\leq$20% [1]

Surface Properties & Solubility:
Solubility Properties General: Equilibrium swelling props. in *n*-hexane have been reported [1,5]

Transport Properties:
Polymer Melts: A plot of melt viscosity versus blending ratio has been reported [2]

Mechanical Properties:
Mechanical Properties General: The effect of compatibilisers on mechanical props. has been reported [3,4]. Elongation falls sharply with increasing polystyrene content

Tensile (Young's) Modulus:

No.	Value	Note
1	686 MPa	7000 kg cm^{-2}, 100% PP
2	1863 MPa	19000 kg cm^{-2}, 100% PS [5]

Tensile Strength Break:

No.	Value	Note
1	32.4 MPa	4700 psi, 100% PP
2	34.5 MPa	5000 psi, 100% PS
3	20.7 MPa	3000 psi, 60% PP/40% PS, plunger machine
4	29.6 MPa	4300 psi, 75% PP/25% PS, plunger/static mixer [2]

Hardness: Shore 69 (100% PP); 80 (100% PS, ASTM D2240-68) [2]

Applications/Commercial Products:
Mould Shrinkage (%):

No.	Value	Note
1	1.4–1.8%	100% PP
2	0.5%	approx., 100% PS, ASTM D955-51 [2]

– Polypropylene, Talc filled

Bibliographic References
[1] Greco, R., Hopfenberg, H.B., Martuscelli, E., Ragosta, G. and Demma, G., *Polym. Eng. Sci.*, 1978, **18**, 654
[2] Han, C.D., Villamizar, C.A., Kim, Y.W. and Chen, S.J., *J. Appl. Polym. Sci.*, 1977, **21**, 353
[3] Barlow, J.W. and Paul, D.R., *Polym. Eng. Sci.*, 1984, **24**, 525
[4] Bartlett, D.W., Paul, D.R. and Barlow, J.W., *ANTEC*, 1981, **27**, 487
[5] Costagliola, M., Greco, R., Martuscelli, E. and Ragosta, G., *J. Mater. Sci.*, 1979, **14**, 1152
[6] Polypropylene: Structure, *Blends and Composites: Copolymers and Blends*, (ed. J. Karger-Kocsis), Chapman and Hall, 1995, **2**

Polypropylene, Talc filled   P-509

Synonyms: *PP, Talc filled. Polypropene, Talc filled*
Related Polymers: Polypropylene, Polypropylene, Filled, Isotactic polypropylene
Monomers: Propylene
Material class: Thermoplastic, Composites
Polymer Type: polypropylene
CAS Number:

CAS Reg. No.
9003-07-0
25085-53-4

Molecular Formula: $(C_3H_6)_n$
Fragments: C_3H_6
Additives: Talc (natural, hydrated magnesium silicate). Lamellar form 40%
General Information: Addn. of talc improves flexural and dimensional props. and gives high stiffness props. at the expense of tensile and impact strengths and thermal stability. Surface is matt and welding is impaired

Volumetric & Calorimetric Properties:
Density:

No.	Value	Note
1	d 1.23–1.27 g/cm^3	40% talc filled, homopolymer [1]
2	d 1.23–1.24 g/cm^3	40% talc filled, copolymer [1]

Thermal Expansion Coefficient:

No.	Value	Note	Type
1	4.2×10^{-5}–8×10^{-5} K^{-1}	homopolymer, 40% talc filled [1]	V

Thermal Conductivity:

No.	Value	Note
1	0.32 W/mK	0.00076 cal cm^{-1} s^{-1} °C^{-1}, ASTM C177, 40% talc filled [1]

Melting Temperature:

No.	Value	Note
1	156–168°C	homopolymer, 40% talc filled [1]

Deflection Temperature:

No.	Value	Note
1	79–82°C	175–180°F, homopolymer [1]
2	65–74°C	150–165°F, copolymer, 40% talc filled, 1.84 MPa (264 psi), ASTM D648 [1]
3	129–143°C	265–290°F, homopolymer
4	102–121°C	215–250°F, copolymer, 40% talc filled, 0.45 MPa (66 psi), ASTM D648 [1]

Transport Properties:
Water Absorption:

No.	Value	Note
1	0.01–0.03 %	1/8" thick, 24h, ASTM D570 [1]

Mechanical Properties:
Tensile (Young's) Modulus:

No.	Value	Note
1	3100–3960 MPa	450–575 kpsi, homopolymer, 40% talc filled, ASTM D638 [1]

Flexural Modulus:

No.	Value	Note
1	3100–4310 MPa	450–625 kpsi, 28° (73°F)
2	2756 MPa	400 kpsi, 93° (222°F), homopolymer
3	2345–2755 MPa	340–400 kpsi, 28° (73°F), copolymer, 40% talc filled, ASTM D790 [1]

Tensile Strength Break:

No.	Value	Note	Elongation
1	30–34 MPa	4300–5000 psi, homopolymer	3–8%
2	24 MPa	3500 psi, copolymer, 40% talc filled ASTM D638 [1]	20%

Flexural Strength at Break:

No.	Value	Note
1	43–63 MPa	7000–9200 psi, homopolymer
2	35 MPa	5100 psi, copolymer, rupture or yield, 40% talc filled, ASTM D790 [1]

Tensile Strength Yield:

No.	Value	Note
1	32 MPa	4600, homopolymer
2	21 MPa	3100 psi, copolymer, 40% talc filled, ASTM D638 [1]

Compressive Strength:

No.	Value	Note
1	52 MPa	7500 psi, rupture or yield, homopolymer, 40% talc filled, ASTM D695 [1]

Hardness: Rockwell R94-R110 (homopolymer); R83 (copolymer, 40% talc filled, ASTM D785) [1]

Izod Notch:

No.	Value	Notch	Note
1	21–32 J/m	N	0.4–0.6 ft lb in^{-1}, 1/8" thick, homopolymer [1]
2	43–214 J/m	N	0.8–4 ft lb in^{-1}, 1/8" thick, copolymer, 40% talc filled, ASTM D256A [1]

– Polypropylene, Wood-filled

Electrical Properties:
Dielectric Strength:

No.	Value	Note
1	19.68 kV.mm^{-1}	500 V mil^{-1}, 1/8" thick, homopolymer, 40% talc filled, ASTM D149 [1]

Applications/Commercial Products:
Processing & Manufacturing Routes: Processing temp. range (40% talc filled) 175–290° (350–550°F), homopolymer, injection moulding); 175–245° (350–470°F, injection moulding), 220–245° (425–475°F, extrusion grade, copolymer) [1]. Moulding pressure range (40% talc filled) 70–140 MPa (10–20 kpsi), homopolymer); 100–140 MPa (15–20 kpsi, copolymer) [1]

Mould Shrinkage (%):

No.	Value	Note
1	0.8–1.5%	linear (0.008–0.015 in/in, homopolymer)
2	0.9–1.2%	0.009–0.12 in/in, copolymer (40% talc filled, ASTM D955) [1]

Applications: Applications include food containers and various structural parts, appliances, packaging and other household goods. Also used in automotive engine housings and ducts

Trade name	Details	Supplier
ARP	Thermofill	Northern Industrial Plastics Ltd.
Escorene PP		Exxon Chemical
PP 5100 series	Various grades	Polycom Huntsman
Polifil T	RMT high impact	Plastics Group
Profil PP60TC		Wilson-Fiberfil
TPP	Various grades	Ferro Corporation
Tenite		Eastman Chemical Company
Thermofil P-TC		Thermofil
Vestolen		Hüls AG
		ARCO Chemical Company
		Washington Penn
		Himont
		RTP Company
		A. Schulman Inc.
		Bamberger

Bibliographic References
[1] Guide to Plastics, *Property and Specification Charts*, McGraw-Hill, 1987, (technical data)

Polypropylene, Wood-filled   P-510

Synonyms: *Polypropene. PP. Poly(1-propene). Poly(methyl ethylene). Propene polymer.* Propylene polymer

Related Polymers: Polypropylene, Isotactic Polypropylene, Atactic Polypropylene, Polypropylene, Filled, Polypropylene, Aluminium filled, Polypropylene, Calcium carbonate filled, Polypropylene, Talc filled, Polypropylene, Glass fibre reinforced, Polypropylene, Mica filled, Polypropylene, Carbon filled, Polypropylene, Glass fibre reinforced
Monomers: Propylene
Material class: Thermoplastic
Molecular Formula: $(C_3H_6)n$
Fragments: C_3H_6
General Information: Extrusion and injection molding grades available.

Volumetric & Calorimetric Properties:
Density:

No.	Value	Note
1	d 0.966–1.145 g/cm^3	(ASTM D792, 20–60% filled, molding grade) [1]
2	d 0.951–1.142 g/cm^3	(ASTM D792, 20–60% filled, extrusion grade) [1]

Deflection Temperature:

No.	Value	Note
1	71–94°C	ASTM D648, 20–60% filled, extrusion grade [1]
2	61–98°C	ASTM D648, 20–60% filled, molding grade [1]

Vicat Softening Point:

No.	Value	Note
1	153–158°C	ASTM D1525, 20–60% filled, extrusion grade [1]

Transport Properties:
Transport Properties General: Processing characteristics often better than MFI suggests [1].
Melt Flow Index:

No.	Value	Note
1	5 g/10 min	(value less than) ASTM D1238, at 190°C, 20–60% filled, extrusion grade [1]
2	0.5–7.4 g/10 min	(min. value less than) ASTM D1238, at 190°C, 20–60% filled, molding grade [1]

Mechanical Properties:
Tensile (Young's) Modulus:

No.	Value	Note
1	2730–6010 MPa	ASTM D638, 20–60% filled, extrusion grade [1]
2	2500–6200 MPa	ASTM D638, 20–60% filled, molding grade [1]

Flexural Modulus:

No.	Value	Note
1	2480–4880 MPa	ASTM D790, 20–60% filled, extrusion grade [1]
2	2200–4900 MPa	ASTM D790, 20–60% filled, molding grade [1]

– Poly(n-propyl methacrylate)

Tensile Strength Break:

No.	Value	Note	Elongation
1	21–26.6 MPa	ASTM D638, 40–60% filled, extrusion grade [1]	1.2–6.2%
2	19.7–25.3 MPa	ASTM D638, 20–60% filled, molding grade [1]	1.2–9.2%

Tensile Strength Yield:

No.	Value	Note
1	30.9–32.3 MPa	ASTM D638, 20–40% filled, extrusion grade [1]
2	26.3–27.4 MPa	ASTM D638, 20–40% filled, molding grade [1]

Flexural Strength Yield:

No.	Value	Note
1	41.5–56.4 MPa	ASTM D790, 20–60% filled, extrusion grade [1]
2	39.1–47.6 MPa	ASTM D790, 20–60% filled, molding grade [1]

Izod Notch:

No.	Value	Notch	Note
1	23.3–46.4 J/m	Y	ASTM D256, 20–60% filled, extrusion grade [1]
2	49–196 J/m	N	ASTM D256, 20–60% filled, molding grade [1]
3	22.3–29.5 J/m	Y	ASTM D256, 20–60% filled, extrusion grade [1]
4	40–126 J/m	N	ASTM D256, 20–60% filled, molding grade [1]

Applications/Commercial Products:
Processing & Manufacturing Routes: Injection molding processing:
Dryer inlet temp. 104°C
Dew point -29°C
Drying time 2–4 hours
Processing temps.: rear <182°C, middle <188°C, front <193°C, nozzle <193°C
Residence time <15 min.
Blowing agents low temp.
Screw type conventional
Extrusion processing:
Dryer inlet temp. 104°C
Dew point -29°C
Drying time 2–4 hours
Temp. settings.: zone 1 <149–171°C, zone 2 <164–188°C, zone 3 <171–193°C, zone 4 <182–193°C, adapter <182–204°C, die <182–204°C
Residence time <15 min.
Blowing agents low temp.
Screen packs <20 mesh
Draw down 0–5%
Screw type conventional

Applications/Commercial Products:
Mould Shrinkage (%):

No.	Value	Note
1	0.18–0.87%	ASTM D955, 20–60% filler, molding grade [1]

Applications:

Trade name	Supplier
Tipwood	Tipco Industries Ltd.
	North Wood Plastics Inc.

Bibliographic References

[1] North Wood Plastics Ltd. http://www.northwoodplastics.com, 2006

Poly(n-propyl methacrylate) P-511

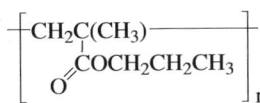

Synonyms: Poly(propyl-2-methyl-2-propenoate). Poly[1-(propoxycarbonyl)-1-methyl ethylene]
Monomers: n-Propyl methacrylate
Material class: Thermoplastic
Polymer Type: acrylic
CAS Number:

CAS Reg. No.
25609-74-9

Molecular Formula: $(C_7H_{12}O_2)_n$

Volumetric & Calorimetric Properties:
Density:

No.	Value	Note
1	d^{20} 1.085 g/cm^3	[1,2]

Thermal Expansion Coefficient:

No.	Value	Note	Type
1	0.00058 K^{-1}	uv polymerisation, above T_g [3]	V
2	0.000315 K^{-1}	uv polymerisation, below T_g [3]	V

Volumetric Properties General: Molar volumes: Van der Waal's, V_w 76.6 cm^3 mol (amorph. polymer, 25°); glassy polymer, V_g 118.7 cm^3 mol (amorph. polymer, 25°)
Melting Temperature:

No.	Value	Note
1	33°C	refractometric method [4]

Glass-Transition Temperature:

No.	Value	Note
1	35–43°C	[2]
2	35°C	Uv polymerisation, dilatometric method [3]

Transition Temperature General: γ-transition, $T_{2\gamma}$ -150° [5]
Vicat Softening Point:

No.	Value	Note
1	55°C	[1]

Poly(pyrrole)

Brittleness Temperature:

No.	Value	Note
1	35°C	refractometric method [4]

Transition Temperature:

No.	Value	Note	Type
1	-150°C	[5]	γ-transition temp.

Surface Properties & Solubility:
Cohesive Energy Density Solubility Parameters: 36.8 $J^{1/2}\ cm^{-3/2}$ (8.8 $cal^{1/2}\ cm^{-3/2}$) [6,7]; 36.4–40.2 $J^{1/2}\ cm^{-3/2}$ (8.7–9.6 $cal^{1/2}\ cm^{-3/2}$) [8]
Solvents/Non-solvents: Sol. cyclohexane (hot), gasoline (hot), turpentine, castor oil (hot), linseed oil (hot), CCl_4, EtOH, Et_2O, Me_2CO [9]. Insol. formic acid
Surface and Interfacial Properties General: Surface tension, γ 33.3 $mN\ m^{-1}$ (20°, MV 8500) [10], 24.7 $mN\ m^{-1}$ (150°, MV 8500) [10], 21.5 $mN\ m^{-1}$ (200°, MV 8500) [10], $-d\gamma/dT$ 0.065 $mN\ m^{-1}$, $°C^{-1}$ (MV 8500) [10]
Wettability/Surface Energy and Interfacial Tension: Contact angle 65° (advancing, H_2O) [11]

Transport Properties:
Polymer Solutions Dilute: Polymer-liq. interaction parameters [12], 0.82 (hexane), 0.63 (2-butanone), 1.42 (acetonitrile), 0.01 ($CHCl_3$), 0.98 (n-propanol), 0.82 (Me_2CO), 0.65 (EtOAc)
Permeability of Liquids: Diffusion coefficient [H_2O] $18.4 \times 10^{-8}\ cm^2\ s^{-1}$ (37°) [11]

Mechanical Properties:
Hardness: Vickers 7 (5 kg load) [13]

Electrical Properties:
Electrical Properties General: Refractive index μ_n 1.484 (20°) [2,13]
Dielectric Permittivity (Constant):

No.	Value	Note
1	3.1	[2]

Polymer Stability:
Decomposition Details: Decomposes at 250° yields monomer [13]

Bibliographic References
[1] Van Krevelen, D.W., *Properties of Polymers: Their Correlation with Chemical Structure* Elsevier, 1976, 79
[2] Rogers, S.S. and Mandelkern, L., *J. Phys. Chem.*, 1957, **61**, 985
[3] Wiley, R.H. and Brauer, G.M., *J. Polym. Sci.*, 1948, **3**, 647, (transition temp)
[4] Plazek, D.J. and Ngai, K.L., *Physical Properties of Polymers Handbook*, (ed. J.E. Mark), AIP Press, 1996, 139
[5] Frank, C.W., and Gashgari, M.A., *Macromolecules*, 1979, **12**, 163
[6] Hughes, L.J. and Britt, G.E., *J. Appl. Polym. Sci.*, 1961, **5**, 337
[7] Ahmed, H. and Yaseen, M., *Polym. Eng. Sci.*, 1979, **19**, 858
[8] Strain, D.E., Kennelly, R.G. and Dittmar, H.R., *Ind. Eng. Chem.*, 1939, **31**, 382
[9] Wu, S., *Polymer Interface and Adhesion*, Marcel Dekker, 1982, (surface tension)
[10] Kalachandra, D.J. and Kusy, R.G., *Polymer*, 1991, **32**, 2428
[11] Walsh, D.J. and McKeown, J.G, *Polymer*, 1980, **21**, 1335
[12] Crawford, J.W.C., *J. Soc. Chem. Ind., London*, 1949, **68**, 201

Poly(pyrrole) P-512

Synonyms: *PPY. Pyrrole black. Pyrrole red*
Related Polymers: Poly(*N*-methylpyrrole)
Monomers: Pyrrole
Polymer Type: polypyrrole
CAS Number:

CAS Reg. No.
30604-81-0

Molecular Formula: $(C_4H_3N)_n$
Fragments: C_4H_3N
General Information: Props. of poly(pyrrole) are determined by the method of synth. [1,2,3,4,38]. Electrochemical synth. has been used to form durable films of conducting polymer [39]
Morphology: Cryst. a 0.82, b 0.735, c 0.682, α, β 90° and γ 117° [5]

Volumetric & Calorimetric Properties:
Density:

No.	Value	Note
1	d 1.4 g/cm^3	[7,8,9]

Thermal Conductivity:

No.	Value	Note
1	0.62–1.12 W/mK	7° [10]

Thermal Diffusivity: 0.0028–0.0066 $cm^2\ s^{-1}$ [24]; 0.0091–0.0093 $cm^2\ s^{-1}$ (11–59°) [10]
Specific Heat Capacity:

No.	Value	Note	Type
1	0.8–1.4 kJ/kg.C	[10]	P

Transition Temperature:

No.	Value	Note	Type
1	60–80°C	[6]	Decomposition temp.

Surface Properties & Solubility:
Solvents/Non-solvents: Sol. C_6H_6, $CHCl_3$, CH_2Cl_2, dioxane, EtOH [6]. Insol. conc. sulfuric acid, 40% NaOH [8], H_2O, pentane [6]

Mechanical Properties:
Tensile Strength Break:

No.	Value	Note	Elongation
1	18–33 MPa	[11,12,13,14]	2–410%

Electrical Properties:
Electrical Properties General: Electrical conductivity [40,41,42,43] depends on the counter-ion [15,16,17,18,19]. Doping increases conductivity [20,21]; conductivity decreases on ageing [9,22]
Surface/Volume Resistance:

No.	Value	Note	Type
1	6.6×10^{15} Ω.cm	[8]	S

Dielectric Permittivity (Constant):

No.	Value	Note
1	2.031–8.035	[7,23]

Complex Permittivity and Electroactive Polymers: Conductivity values σ 51.3 (3h); 35.1 (14 days) [9]; 59.2 Ω^{-1} cm^{-1} [24]
Dissipation (Power) Factor:

No.	Value	Frequency	Note
1	0.78–1.766	10 GHz	[7,23]

Optical Properties:
Transmission and Spectra: Ir [6,8,25,26], Raman [27], H-1 nmr [6,28], epr [25,29,30], uv [27] and x-ray photoelectron [31] spectral data have been reported

Polymer Stability:
Decomposition Details: Poly(pyrrole) is oxidised in the presence of oxygen. At temps. above 230° this oxidation is a bulk phenomenon leading to decomposition. At temps. below 230° the oxidation is a slow, gradual process [22]

Applications/Commercial Products:
Processing & Manufacturing Routes: Prod. by chemical oxidative polymerisation of pyrrole [29,32,33,34]; or by electrochemical polymerisation of pyrrole [35,36,37,39]
Applications: Lightweight secondary batteries, electromagnetic interference shielding, transistors, and many potential applications as electrode materials, e.g. in enzyme electrodes for studying enzyme-catalysed kinetics

Trade name	Supplier
Lutamer	BASF

Bibliographic References
[1] Otero, T.F. and DeLarreta, E., *Synth. Met.*, 1988, **26**, 79
[2] Schneider, O. and Schwitzgebel, G., *Synth. Met.*, 1993, **55**, 1406
[3] Saunders, B.R., Fleming, R.J. and Murray, K.S., *Chem. Mater.*, 1995, **7**, 1082
[4] *Ullmanns Encykl. Ind. Chem.*, (ed. B. Elners), VCH, 1992, **A21**, 435
[5] Geiss, R.H., Street, G.B., Volksen, W. and Economy, J., *IBM J. Res. Dev.*, 1983, **27**, 321
[6] Armour, M., Davies, A.G., Upadhay, J. and Wassermann, A., *J. Polym. Sci., Part A-1*, 1967, **5**, 1527
[7] Truong, V.T., Codd, A.R. and Forsyth, M., *J. Mater. Sci.*, 1994, **29**, 4331
[8] Kanazawa, K.K., Diaz, A.F. and Krombi, M., *J. Polym. Sci., Polym. Lett. Ed.*, 1982, **20**, 187
[9] Ribo, J.M., Dicko, A., Valles, M.A., Ferrer, N. et al, *Synth. Met.*, 1989, **33**, 403
[10] Lunn, B.A., Unsworth, J., Booth, N.G. and Innis, P.C., *J. Mater. Sci.*, 1993, **28**, 5092, (thermal props, electrical props)
[11] Bates, N., Cross, M., Lines, R. and Walton, D., *J.C.S., Chem. Commun.*, 1985, 871, (mech props)
[12] Buckley, L.J., Roylance, D.K. and Wrek, G.E., *J. Polym. Sci., Part B: Polym. Phys.*, 1987, **25**, 2179
[13] Cvetko, B.F., Brungs, M.P., Burford, R.P. and Skyllas-Kazacos, M., *J. Mater. Sci.*, 1988, **23**, 2102
[14] Ansari, R. and Wallace, G.G., *Polymer*, 1994, **35**, 2372
[15] Bittihn, R., Ely, G. and Woeffler, F., *Makromol. Chem., Makromol. Symp.*, 1987, **8**, 51
[16] McNeill, R., Siudak, R., Wardlow, J.H. and Weiss, D.E., *Aust. J. Chem.*, 1963, **16**, 1056
[17] Naarmann, H., *Makromol. Chem., Makromol. Symp.*, 1987, **8**, 1
[18] Kaeriyama, K. and Masuda, H., *Synth. Met.*, 1991, **41**, 389
[19] Bakshi, A.K., Ladik, J. and Seel, M., *Phys. Rev. B: Condens. Matter*, 1987, **35**, 704
[20] Phillips, G., Suresh, R., Chen, J.I., Kumar, J. et al, *Mol. Cryst. Liq. Cryst.*, 1990, **190**, 27
[21] Bloor, D., Monkman, A.P., Stevens, G.C., Cheung, K.M. and Pugh, S., *Mol. Cryst. Liq. Cryst.*, 1990, **187**, 231
[22] Thieblemont, J.C., Brun, A., Marty, J., Planche, M.F. and Calo, P., *Polymer*, 1995, **36**, 1605
[23] Kaynak, A., Unsworth, J., Beard, G.E. and Clout, R., *Mater. Res. Bull.*, 1993, **28**, 1109
[24] Tsutsumi, N., Ishida, S. and Kiyotsukuri, T., *J. Polym. Sci., Polym. Phys. Ed.*, 1994, **32**, 1899
[25] Kaplin, D.A. and Qutubuddin, S., *Polymer*, 1995, **36**, 1275
[26] Zerbi, G., Veronelli, M., Martina, S., Schlüter, A.D. and Wegner, G., *J. Chem. Phys.*, 1994, **100**, 978
[27] Furukawa, Y., Tazawa, S., Fujii, Y. and Haruda, I., *Synth. Met.*, 1988, **24**, 329
[28] Baker, P., Matthews, D. and Hope, A., *Aust. J. Chem.*, 1994, **47**, 1
[29] Khulbe, K.C., Mann, R.S. and Khulbe, C.P., *J. Polym. Sci., Polym. Chem. Ed.*, 1982, **20**, 1089
[30] Kang, E.T., Neoh, K.G., Matsuyama, T. and Yamaoka, H., *Polym. Commun.*, 1988, **29**, 201
[31] Dujordin, S., Lazzaroni, R., Rigo, L., Riga, J. and Verbist, J.J., *J. Mater. Sci.*, 1986, **21**, 4342
[32] Wei, Y., Tian, J. and Yang, D., *Makromol. Chem., Rapid Commun.*, 1991, **12**, 617
[33] Sak-Bosnar, M., Budimer, M.V., Kovac, S., Kukulj, D. and Duic, L., *J. Polym. Sci., Part A: Polym. Chem.*, 1992, **30**, 1609
[34] Chen, X., Devaux, J., Issi, J.P. and Billard, D., *Polym. Eng. Sci.*, 1995, **35**, 642
[35] Rau, J.R., Chen, S.C. and Liu, P.H., *J. Electroanal. Chem.*, 1991, **307**, 269
[36] Diaz, A.F., *Chem. Scr.*, 1981, **17**, 145
[37] Nakazawa, Y., Ebine, T., Kusunoki, T., Nishizawa, H. et al, *Jpn. J. Appl. Phys., Part 1*, 1988, **27**, 1304
[38] *Handbook of Conducting Polymers*, Vol. 1 (ed. T.A. Skotheim), Marcel Dekker, 1986, 265
[39] Diaz, A.F., Kanazawa, K.K. and Gardini, G.P., *J.C.S., Chem. Commun.*, 1979, 635
[40] Diaz, A.F., Castillo, J.I., Logan, J.A. and Lee W.-Y., *J. Electroanal. Chem.*, 1981, **129**, 115, (electrochem)
[41] Wegner, G., Wernet, W., Glatzhofer, D.T., Ulanski, J. et al, *Synth. Met.*, 1987, **18**, 1, (conductivity)
[42] Yamaura, M., Hagiwara, T. and Iwata, K., *Synth. Met.*, 1988, **26**, 209, (conductivity)
[43] Warren, L.F., Walker, J.A., Anderson, D.P., Rhodes, C.G. and Buckley, L.J., *J. Electrochem. Soc.*, 1989, **136**, 2286

Poly(3-pyrroleacetic acid)

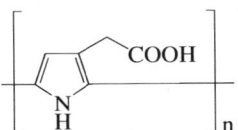

Synonyms: *1H-Pyrrole-3-acetic acid homopolymer*
Monomers: 1*H*-Pyrrole-3-acetic acid
Material class: Thermoplastic
Polymer Type: polypyrrole
CAS Number:

CAS Reg. No.
134215-74-0

Molecular Formula: $(C_6H_5NO_2)n$
Fragments: $C_6H_5NO_2$
General Information: Conductive polymer with sensitivity towards pH [1]. Prod. electrochemically as blue-black film.

Electrical Properties:
Electrical Properties General: Conductivity as high as 37 S cm^{-1} has been reported for film, with a proposed self-doping mechanism [1]

Optical Properties:
Transmission and Spectra: Ftir [2], xps and ms [3] spectral data have been reported.

Applications/Commercial Products:
Processing & Manufacturing Routes: Prod. by electropolymerisation of 3-pyrroleacetic acid in a suitable solvent such as MeCN [2,3] or propylene carbonate [4] with a supporting electrolyte such as lithium perchlorate [2,4] or tetrabutylammonium tetrafluoroborate [3].
Applications:

Bibliographic References
[1] Delabouglise, D. and Garnier, F., *New J. Chem.*, 1991, **15**, 233
[2] Ryder, K.S., Morris, D.G. and Cooper, J.M., *Langmuir*, 1996, **12**, 5681

Poly(resorcinol isophthalate)   P-514

Synonyms: *Poly(metaphenylene isophthalate). Poly(oxy-m-phenyleneoxyisophthaloyl). Poly(oxy-1,3-phenyleneoxycarbonyl-1,3-phenylenecarbonyl)*
Monomers: 1,3-Benzenediol, 1,3-Benzenedicarboxylic acid
Material class: Thermoplastic
Polymer Type: unsaturated polyester
CAS Number:

CAS Reg. No.	Note
26618-61-1	from isophthalic acid
28211-79-2	from isophthaloyl chloride
26637-46-7	repeating unit

Molecular Formula: $(C_{14}H_8O_4)_n$
Fragments: $C_{14}H_8O_4$

Volumetric & Calorimetric Properties:
Density:

No.	Value	Note
1	d^{25} 1.34 g/cm^3	[3]

Melting Temperature:

No.	Value	Note
1	196–243°C	[1]
2	209°C	[2]
3	247°C	520K [3]
4	234°C	[6]
5	230–245°C	[11]

Glass-Transition Temperature:

No.	Value	Note
1	155°C	[2]
2	132°C	405K [3]
3	124.4°C	397.6K, calc. [7]
4	130°C	[9]

Transition Temperature:

No.	Value	Note	Type
1	312°C	[4]	Polymer melt temp.
2	245°C	[12]	Polymer melt temp.

Surface Properties & Solubility:
Solvents/Non-solvents: Sol. hot *m*-cresol, hot *o*-chlorophenol, hot DMF, hot DMSO, hot 1,1,2,2-tetrachloroethane, hot *N*-methyl-2-pyrrolidone [6]. Insol. H$_2$O [5], Me$_2$CO [5], CHCl$_3$, formic acid [6], chlorobenzene [6]. Solubility in other solvents has been reported [11]
Wettability/Surface Energy and Interfacial Tension: Surface free energy, γ_s 41.7 mJ m^{-2}, γ_d 37.4 mJ m^{-2}, γ_p 4.3 mJ m^{-2}; γ_p/γ_s 0.1 [6]

Transport Properties:
Polymer Solutions Dilute: $[\eta]$ 0.31 dl g^{-1} [1]; $[\eta]_{inh}$ 0.29 dl g^{-1} (*m*-cresol, 25°) [2]; $[\eta]_{inh}$ 1.32 dl g^{-1} (phenol/tetrachloroethane, (60:40), 25°) [3]; $[\eta]_{inh}$ 0.25 dl g^{-1} (phenol/tetrachloroethane, (60:40), 30°) [4]; $[\eta]_{inh}$ 0.14 dl g^{-1} (1-methyl-2-pyrrolidinone, 30°) [5]; $[\eta]$ 0.19 dl g^{-1} (phenol/tetrachloroethane, (60:40), 30°) [6]; $[\eta]_{inh}$ 1.38 dl g^{-1} (phenol/tetrachloroethane, (60:40), 30°) [12]

Mechanical Properties:
Mechanical Properties General: Dynamic elastic modulus has been reported [3]

Optical Properties:
Optical Properties General: Exhibits optical anisotropy [6] in the melting state
Transmission and Spectra: Ir [1], nmr [3] and mass spectral [5] data have been reported

Polymer Stability:
Thermal Stability General: Heat resistance has been reported [8]
Decomposition Details: Decomposition temp. 540° [2], 490° [5]. Thermal degradation details have been reported [3,6]. 10% Weight loss occurs at 474° [6]. Activation energy for thermal degradation 190 kJ mol^{-1} (45 kcal mol^{-1}) [10]

Applications/Commercial Products:
Processing & Manufacturing Routes: Synth. by the interfacial polymerisation of isophthaloyl chloride with 1,3-benzenediol in CHCl$_3$/H$_2$O at 20° [1,2,4,5,12]

Bibliographic References
[1] Bellus, D., Manasek, Z., Hrdlovic, P. and Slama, P., *J. Polym. Sci., Part C: Polym. Lett.*, 1966, **16**, 267, (synth, ir)
[2] Ehlers, G.F.L., Evers, R.C. and Fisch, K.R., *J. Polym. Sci., Part A-1*, 1969, **7**, 3413, (synth, transition temps)
[3] Frosini, V., Levita, G., Landis, J. and Woodward, A.E., *J. Polym. Sci., Polym. Phys. Ed.*, 1977, **15**, 239, (nmr, density, mech props)
[4] Higashi, F., Kubota, K., Sekizuko, M. and Goto, M., *J. Polym. Sci., Polym. Lett. Ed.*, 1980, **18**, 385, (synth, polymer melt temp)
[5] Foti, S., Giuffrida, M., Maravigna, P. and Montaudo, G., *J. Polym. Sci., Polym. Chem. Ed.*, 1984, **22**, 1201, (ms, thermal degradation, synth)
[6] Kiyotsukuri, T., Tsutsumi, N., Okada, K. and Asai, K., *J. Polym. Sci., Part A: Polym. Chem.*, 1988, **26**, 225, (solubility, surface free energy, optical anisotropy)
[7] Hopfinger, A.J., Koehler, M.G., Peartstein, R.A. and Tripathy, S.K., *J. Polym. Sci., Part B: Polym. Phys.*, 1988, **26**, 2007, (transition temp)
[8] Korshak, V.V., Vinogradova, S.V., Danilov, V.G. and Salazkin, S.N., *Dokl. Akad. Nauk SSSR*, 1972, **202**, 1076, (heat resistance)
[9] Lee, W.L. and Stagg, B., *Tech. Rep. - R. Aircr. Establ. (G.B.)*, 1971, TR71233, (transition temp)
[10] Goldfarb, I.J., Bain, D.R., McGuchan, R. and Meeks, A.C., *Am. Chem. Soc., Div. Org. Coat. Plast. Chem., Pap*, 1971, **31**, 130, (thermal degradation)
[11] Korshak, V.V. and Vinogradova, S.V., *Izv. Akad. Nauk SSSR, Otd. Khim. Nauk*, 1958, 614, (solubility)
[12] Eareckson, W.M., *J. Polym. Sci.*, 1959, **40**, 399, (synth, soln props)

Poly(selenocarbonothioylseleno-1,4-phenylene)   P-515

Synonyms: *Poly(p-phenylenediselenothiocarbonate). PPSSe$_2$*
CAS Number:

CAS Reg. No.	Note
182067-75-0	Poly(selenocarbonothioylseleno-1,4-phenylene)

Molecular Formula: $(C_7H_4SSe_2)n$
Morphology: Dopant agent produces a high regularity in the polymer structure, with a corresponding increase in conductivity [2].

Identification: Elemental analysis: %C calc. 30.2 found 31.2; %H calc. 1.4 found 2.0; %Se calc. 56.8 found 55 [1].

Surface Properties & Solubility:
Solubility Properties General: PPSSe$_2$ has low solubility

Electrical Properties:
Electrical Properties General: When the polymer was doped with FeCl$_3$, AlCl$_3$ and SbF$_5$, the conductivity value was not increased satisfactorily as tabulated [1]. The iodine-doped compound exhibits semiconducting properties. The small conductivity of poly(p-thioselenocarbonate) after doping with iodine has been attributed to the localisation of a positive charge on the sulfur [1]. Tends to form dimers through polar bonds that cause the decrease in conductivity [1].

Optical Properties:
Transmission and Spectra: IR: C-Se 490 cm^{-1}; p-arom. 805 cm^{-1}; C-S 1030 cm^{-1} [1].

Applications/Commercial Products:
Processing & Manufacturing Routes:
Applications: Iodine doped polymers exhibit semiconducting properties [1,2].

Bibliographic References
[1] Diaz F.R., Godoy A., Tagle L.H., Valdebenito N., and Bernede J.C., *Eur. Polym. J.*, 1996, **32(9)**, 1155, (em, ir, nmr, pes, props, syn, use)
[2] Sandman D.J., Stark J., Rubner M., Hamill G., Acampora L., McGrath M., and Allen, Proc. 4th Int. Conf. on the Organic Chemistry of Selenium and Tellurium Eastman Kodak Corp, IBM, Birmingham, 1983

Poly(selenocarbonylseleno-1,4-phenylene)   P-516

Synonyms: *Poly(p-phenylenediselenocarbonate)*. PPSe$_2$
CAS Number:

CAS Reg. No.	Note
182067-71-6	Poly(selenocarbonylseleno-1,4-phenylene)

Molecular Formula: (C$_7$H$_4$OSe$_2$)n
General Information: Bright red powder in 75% yield.
Morphology: Morphology studied using scanning electron microscopy. The PPSe$_2$ chains are observed with a globular and laminar phase with great disorder. When the compound contains selenocarbonate, the structure is more laminar, with a high degree of disorder and bigger spacing between the sheets. The morphology of this structure is in agreement with low measured conductivities [1]. For poly-p-phenyleneselenide (PPSe), the dopant agent produces a high regularity in the polymer structure, with a corresponding increase in conductivity [2].
Identification: Elemental analysis: %C calc. 30.7 found 27.9; %H calc. 1.7 found 1.6; %Se calc. 67.5 found 68.8 [1].

Surface Properties & Solubility:
Solubility Properties General: PPSe$_2$ has low solubility.

Electrical Properties:
Electrical Properties General: Doping with iodine was carried out according to the Gutierrez method. When the products were doped with FeCl$_3$, AlCl$_3$ and SbF$_5$, the conductivity values were not increased satisfactorily as tabulated [1].

Optical Properties:
Transmission and Spectra: Ir : C-Se 490 cm^{-1}; p-arom. 800 cm^{-1} [1]

Applications/Commercial Products:
Processing & Manufacturing Routes:
Applications: Iodine doped polymers exhibit semiconducting properties [1,2].

Bibliographic References
[1] Diaz F.R., Godoy A., Tagle L.H., Valdebenito N., and Bernede J.C., *Eur. Polym. J.*, 1996, **32(9)**, 1155
[2] Sandman D.J., Stark J., Rubner M., Hamill G., Acampora L., McGrath M., and Allen, Proc. 4th Int. Conf. on the Organic Chemistry of Selenium and Tellurium Eastman Kodak Corp, IBM, Birmingham, 1983

Poly(silane)   P-517

$\bigl[\text{SiH}_2\bigr]_n$

Synonyms: *Poly(silylene)*. PSA. PS. Psi. TP
Related Polymers: Polycarbosilanes
Polymer Type: polysilanes
CAS Number:

CAS Reg. No.
32028-95-8

Molecular Formula: (H$_2$Si)n
Fragments: H$_2$Si
General Information: Si analogue of polyethylene. Predictions of theoretical studies was an unusually high torsional mobility for the PS chains, owing to the relatively long Si-Si bonds and, consequently, low barriers for rotation [6]
Morphology: Initially reported as an air-sensitive orange solid. *Ab initio* crystal orbital calculations reveal that the *trans*-conformer is the ground state of polysilane (and not gauche-polysilane as originally thought). The gauche-polysilane (GP) is 0.15 kcal mol^{-1} per unit above the *trans*-polysilane (TP). TP has a smaller band gap, lower ionisation potential and greater electron delocalisation than GP. The effective mass of the hole at the valence band edge is ultimately greater in GP than in TP [6,7,13]
Miscellaneous: Conformationally more flexible than polysilaethylene [8]

Volumetric & Calorimetric Properties:
Glass-Transition Temperature:

No.	Value	Note
1	140°C	predicted [8]

Optical Properties:
Transmission and Spectra: Ir [10,11,12], uv [10], nmr [11], epr [11] and pes [9] spectra reported.
Total Internal Reflection: In the early stages of Si deposition at a substrate temperature of 200°C, SiH$_2$ is the major surface species although (SiH$_2$)n chains and SiH$_3$ also exist. At room temperature, SiH$_3$ and (SiH$_2$)n chains are the dominant surface products. At -95°C, polymerisation reactions among adsorbates proceed on the surface to form polysilanes consisting of (SiH$_2$)n chains terminated with SiH$_3$ [2].

Applications/Commercial Products:
Processing & Manufacturing Routes: A polysilane -(SiHn)x- where n is 1 or 2 and x is 10, 20, or more is manufactured by treating SiH$_m$X$_{4-m}$, where m is 1–3 and X is a halogen, with Li in a liquid that is inert to the reagents and a nonsolvent for the polysilane [1]. Reacting Si surfaces in contact with a SiH$_4$ glow discharge - in the early stages of Si deposition at a substrate temperature of 200°, SiH$_2$ is the major surface species [2,4]
Applications: Polysilane-yttria hybrid thin films [3]. Active layers of electroluminescent devices [4]

– Polysiloxanes

Bibliographic References
[1] *U.K. Pat.*, 1981, 2 077 710
[2] Miyazaki, S., Shin, H., Miyoshi, Y. and Hirose, M., *Jpn. J. Appl. Phys., Part 1*, 1995, **34**, 787
[3] Matsuura, Y. and Matsukawa, K., *Kagaku to Kogyo (Osaka)*, 2005, **79**, 126
[4] *U.S. Pat.*, 2005, 05 238 910
[5] *U.S. Pat.*, 2004, 04 211 458
[6] Interrante, L.V., Wu, H.J, Apple, T., Shen, Q., Ziemann, B. and Narsavaget, D.M., *J. Am. Chem. Soc.*, 1994, **116**, 12085
[7] Teramae, H. and Takeda, K., *J. Am. Chem. Soc.*, 1989, **111**, 1281
[8] Bharadwaj, R.K., Berry, R.J. and Farmer, B.L., *Polym. Mater. Sci. Eng.*, 2001, **85**, 408
[9] Kurokawa, Y., Nomura, S., Takemori, T. and Aoyagi, Y., Prog. Theor. Phys. Suppl. Computational Physics and Related Topics, 2000, **138**, 147
[10] Perpete, E.A., Andre, J.M. and Champagne, B., *J. Chem. Phys.*, 1998, **109**, 4624
[11] Onischuk, A.A., Strunin, V.P., Samoilova, R.I., Nosov, A.V., Ushakova, M.A. and Panfilov, V.N., *J. Aerosol Sci.*, 1997, **28**, 1425
[12] John, P., Odeh, I.M., Thomas, M.J.K. and Wilson, J.I.B., *J. Phys., Colloq.*, 1981, **C4**,Pt.2, 651
[13] Crespo, R., Piqueras, M.C. and Tomas, F., *J. Chem. Phys.*, 1994, **100**, 6953

Polysiloxanes   P-518

Synonyms: *Polyorganosiloxanes. Silicones*
Related Polymers: Polydimethylsiloxane, Cyclic siloxane, Resins (methyl silicon resin), Polycarbonate silicone copolymer, Silicone polyetherimide copolymer, Silicone phenol formaldehyde copolymer, Silicone epoxy copolymer
General Information: Commercial development of silicone products began in 1937 at the Corning Glass Works (later Dow Corning) and in 1938 at General Electric. Early research was devoted to producing methyl and phenyl silicone resins as possible electrical insulators, but methyl silicone fluids were the first siloxanes to be made commercially. Methyl silicone fluids were valued for high temp. stability and the low temperature-dependence of viscosity. Before the end of the second world war silicone rubber became available for use at temps. exceeding that of other elastomers and silicone resins for use as insulating materials at high temps. became available. Silicone polymers are more expensive than most other polymers. Commercial importance has always been based upon their capacity to function under extreme conditions. By using copolymers of methyl and phenyl siloxanes the low and high temp. performance of silicone fluids and rubbers can be improved compared to dimethyl silicones. Substitution of fluorine or nitrile containing groups allows drastic improvement of solvent resistance. The incorporation of small amounts of vinyl groups into silicone rubber allows for easier vulcanisation by organic peroxide and better elastic props. at high temps. These discoveries resulted in a wide range of silicone products becoming available for many high and low temp. applications particularly where a wide range of temps. would be encountered during use. Room temp. vulcanising (RTV) silicone rubbers became commercially available in the 60's. Whereas heat vulcanising silicone rubber gum is a high MW very high viscosity material which requires crosslinking at elevated temps. with organic peroxide, RTV silicone rubber gum is a relatively low MW fluid with reactive silanol end groups. These end groups can be crosslinked at room temp. by a reactive cross-linker to produce the cured rubber. There is currently considerable interest in the props. and applications of polysiloxane liq. crysts. Such liq. crysts. are likely to have a significant impact in the field of electrooptics and liq. cryst. displays. Polymers of siloxanes and other materials have been developed in a wide variety of forms, resulting in fluids and resins with special props. for particular applications. Fluid copolymers of polysiloxanes and polyethers are commercially important for surfactant props.; other copolymers have found uses as specialist fluids. A very large number of cocondensation products of silicones with organic resins have been described. Copolymers with polyesters are commercially the most important, being harder, more adhesive and more solvent resistant than silicone resins as well as cheaper and more compatible with pigments and dyes. If the siloxane comprises at least 50% of the copolymer then the heat resistance, water repellency and weather resistance typical of silicone resins are largely retained. Silicone production reached half a million metric tonnes a year in the late 80's and continues to rise [1,2,3,4,5,6]

Surface Properties & Solubility:
Solvents/Non-solvents: Insol. H_2O, alcohols [2,3]
Surface and Interfacial Properties General: Generally low surface adhesion [3]

Electrical Properties:
Electrical Properties General: Excellent insulating and dielectric props. [3]

Polymer Stability:
Polymer Stability General: Stable material with wide range of uses. Although silicones are non-toxic their stability and non-biodegradability has caused concern about environmental accumulation [1]
Thermal Stability General: Usable from -100–300° (temp. range varies with particular silicone product)
Flammability: Low flammability [1]
Environmental Stress: Resistant to oxidative degradation, H_2O, weathernig, ozone and corona. Unaffected by uv radiation [3]. Liq. silicones may form very thin films capable of being broken down by oixdative degradation
Hydrolytic Stability: Stable in H_2O and aq. solns. at room temp. Damaged by high pressure steam. Fluids in very thin films undergo hydrolytic degradation [1,3]
Biological Stability: Non-biodegradable [2]
Recyclability: Can often be recycled by depolymerisation using steam or acid catalysis [3]. This does not occur on a large scale

Applications/Commercial Products:
Processing & Manufacturing Routes: The development of new methods of curing and vulcanising silicones has increased their versatility and usefulness. Modified silicone resins can be crosslinked at room temp. by methyl triacetoxy silane to produce flexible elastomers similar to RTV silicone rubbers. The liq. silicone rubber system; in which low viscosity vinyl containing silicone gums are rapidly cured by hydrosilation at no more than 100°; allows easier and more economical moulding and extrusion of high quality silicone elastomers
Applications: Because of the low toxicity, bio-stability and water repellency of silicone polymers, silicone fluids have found a role in cosmetics. Silicone rubbers have been widely used in cosmetic surgery particularly for breast implants, but there are currently grave doubts about the long term safety of such procedures. Silicone rubbers generally have low adhesion to most materials. "Non-sticking" characteristics have become particularly important in the manufacture of moulds for the plastics industry. The resistance of silicones to weathering, sunlight and high humidity has proved particularly important in the use of silicone resins as paints or varnishes and as resin films to protect masonry. Low surface tension and high water repellency of silicone fluids makes them useful in polishes, and because of their high compressibility silicone fluids are used for vibration damping. Thermal stability and high thermal conductivity makes them valuable as heat-transfer media. The most important single application of silicones is in the electrical and electronics industry, where high electrical resistance, good dielectric props., and low flammability have made them invaluable. Silicone fluids are used as dielectric media in transformers; silicone resins are used as silicone glass laminates in spacers and wedges in electrical units and as moulding resins for connectors, terminal boards and enclosures. Heat cured silicone rubber is widely used in the manufacture of coatings for wires and cables, providing high reliability over a wide range of temps.; while RTV silicone rubbers are used as protective encapsulants for fragile circuitry. 50% of silicones are sold as fluids or derivatives, 40% as elastomers and 10% as resins or derivatives. Over 1000 products are available commercially

Bibliographic References
[1] Hardman, B. and Torkelson, A., *Encycl. Polym. Sci. Eng.*, 2nd edn., (ed. J.I. Kroschwitz), John Wiley and Sons, 1985, **15**, 204
[2] Roff, W.J. and Scott, J.R., Fibres, Films, *Plastics and Rubbers*, Butterworths, 1971

[3] Noll, W., *Chemistry and Technology of Silicones*, Academic Press, 1968
[4] Gair, T.S., *Silicone Technology*, (ed. P.F. Bruins), John Wiley and Sons, 1970, 1
[5] Liebhafsky, H.A., Silicones under the Monograph John Wiley and Sons, 1978
[6] Clarson, S.J. and Semlyen, J.A., *Siloxane Polymers*, 1993

Poly[silylene(methylene)] — P-519

$$-[\text{SiH}_2-\text{CH}_2]_n-$$

Synonyms: *Poly(silaethylene)*. PSM. PSE. SP4000
Related Polymers: Polycarbosilanes, Poly[(methylphenylsilylene)methylene], Poly[(dimethylsilylene)(methylene)]
Polymer Type: carbosilane polymers
CAS Number:

CAS Reg. No.
74056-94-3

Molecular Formula: $(CH_4Si)_n$
Fragments: CH_4Si
Molecular Weight: MW 33,000; M_n 12,300 [1,3,4]
General Information: PSM with its 1:1 Si:C ratio is an attractive candidate for a SiC precursor to silicon carbide ceramics [1,5]. Obtained as a viscous liquid at room temperature upon isolation from hydrocarbon solvents. On cooling to just below room temperature, however, it forms a translucent white solid [3].
Tacticity: The preference for the *trans*-state, which is well known in PE, is weakened in PSM [4].
Morphology: Reported to be a linear polycarbosilane with a regularly alternating Si-C backbone structure [1,2,6,10]; Simulation C-Si bond, 0.189 nm [4]
Identification: Calc: C, 27.27; H, 9.09; Found: C, 27.21; H, 9.09 [13]

Volumetric & Calorimetric Properties:
Thermodynamic Properties General: Exhibits a melting transition which is ca. 100°C lower than that of PE [3]
Glass-Transition Temperature:

No.	Value	Note
1	-140--135°C	Unusually low value for T_g suggests that the uncrystallised portion of PSM remains mobile down to extremely low temperatures [3]

Surface Properties & Solubility:
Solubility Properties General: Soluble in hydrocarbon solvents [3]

Mechanical Properties:
Mechanical Properties General: The modulus of elasticity and the nanohardness of the SiC films were measured with the aid of a nanoindenter at various depths, which did not exceed 25% of the film thickness. The average nanohardness at indentation depths of approx. 10% of the film thickness was measured up to 13.4GPa [7]

Optical Properties:
Optical Properties General: Specific peaks in the nmr spectra arise either from -SiH-(Me)- branch sites or the CH_3 and SiH_3 end groups [1].
Transmission and Spectra: Ir [3,13], pmr, cmr and Si-29 nmr spectra reported [1,3,13].

Polymer Stability:
Thermal Stability General: Thermal analysis in nitrogen gave a remarkably high ceramic yield (calc. 90.9; found 87). Weight loss started at approx. 100°C, and there was almost no weight loss after 600°C [13]. Soft-hard binary polymer blends consisting of amorphous polysilylenemethylenes and crystalline poly(diphenylsilylenemethylene) have been studied [6]

Decomposition Details: The observed evolution of D_2 from the deuterio-deriv. of PSE, as the primary gaseous product at 250-400°C, where crosslinking of the polymer occurs, suggests that loss of H_2 from the Si is a key step in the crosslinking process. A reaction pathway is postulated for the crosslinking and pyrolysis of PSE in which both 1,1-H_2 elimination and intramol. H-transfer reactions lead to highly reactive silylene intermediates; these insert into Si-H bonds of neighbouring polymer chains forming Si-Si bonds which rapidly rearrange to Si-C bonds at these temps. to form Si-C interchain cross-links. The cross-links prevent extensive fragmentation of the polycarbosilane network as the temp. is increased further to the range (420°C) where homolytic bond cleavage occurs at an appreciable rate, leading to free radicals. These free radical processes are presumably the main mechanisms at higher temps. (>475°C) where extensive rearrangement of the Si/C network structure is evidenced by solid-state NMR spectroscopy. Further heating of the polymer to 1000°C leads to the formation of SiC in high yield (approx. 85%) [1,10]

Applications/Commercial Products:
Processing & Manufacturing Routes: Prepared by the ring-opening polymerisation of 1,3-disilacyclobutane [2,10]. SiC films have been successfully deposited on various substrates by oligomer thermal CVD from a novel, halogen-free, oligomer precursor family of polysilylenemethylenes [7]
Applications: Composite materials containing fine particles [8]. Method for forming carbide material on substrates [9]. Porous ceramics [12]

Bibliographic References
[1] Shen, Q.H. and Interrante, L.V., *Macromolecules*, 1996, **29**, 5788
[2] *U.S. Pat.*, 1986, 4 631 179
[3] Interrante, L.V., Wu, H.-J., Apple, T., Shen, Q. and Smith, K., *J. Am. Chem. Soc.*, 1994, **116**, 12085
[4] Helfer, C.A., Mattice, W.L. and Chen, D., *Polyhedron*, 2004, **45**, 1297
[5] Mattice, W.L., Helfer, C.A. and Farmer, B.L., *Polym. Prepr. (Am. Chem. Soc., Div. Polym. Chem.)*, 2005, **93**, 834
[6] Ogawa, T. and Murakami, M., *J. Polym. Sci., Part A*, 2003, **41**, 257
[7] Futschik, U., Efstathiadis, H., Castracane, J., Kaloyeros, A.E., Mcdonald, L., Hayes, S. and Fountzoulas, C., *Mater. Res. Soc. Symp. Proc.*, 2002, **697**, 153
[8] *Eur. Pat.*, 2000, 973 049
[9] *U.S. Pat.*, 1999, 5 952 046
[10] Liu, Q., Wu, H.-J., Lewis, R., Maciel, G.E. and Interrante L.V., *Chem. Mater.*, 1999, **11**, 2038
[11] Auner, N. and Grobe, J., *J. Organomet. Chem.*, 1980, **188**, 151
[12] *Jpn. Pat.*, 2005, 05 145 776
[13] Wu, H.-J. and Interrante, L.V., *Macromolecules*, 1992, **25**, 1840

Poly[sodium 4-(3-pyrrolyl)butanesulfonate] — P-520

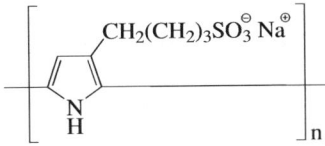

Synonyms: *1H-Pyrrole-3-butanesulfonic acid monosodium salt homopolymer*. PPyBS
Material class: Thermoplastic
Polymer Type: polypyrrole
CAS Number:

CAS Reg. No.
111984-81-7

Molecular Formula: $(C_8H_{10}NNaO_3S)_n$
Fragments: $C_8H_{10}NNaO_3S$
General Information: Water-soluble conductive polypyrrole. Prod. as blue-black film [1,2].
Morphology: Amorphous films prod. electrochemically [1,2].

— Polystyrene

Surface Properties & Solubility:
Solvents/Non-solvents: Sol. H_2O, MeOH [1,2]. Insol. MeCN [1].

Electrical Properties:
Electrical Properties General: Conductive form of polymer is self-doped with zwitterion formation and protonation of pyrrole rings [1]. Maximum conductivity of 0.5 S cm^{-1} recorded [2]

Optical Properties:
Optical Properties General: UV absorption band with λ_{max} ca. 400nm for film reported [1,2,3]. In aqueous solution there are bands with λ_{max} ca. 390 and 580nm [6].
Transmission and Spectra: Pmr [2], ir [4,6] and uv [1,2,3,6] spectral data have been reported..

Polymer Stability:
Thermal Stability General: Weight loss in major decomposition stage at ca. 340°C on heating in nitrogen [4]

Applications/Commercial Products:
Processing & Manufacturing Routes: Prod. either by electrochemical polymerisation of sodium 4-(3-pyrrolyl)butanesulfonate in MeCN [1,2,5] or $NaNO_3$ soln. [3], or chemically by oxidation of the monomer with $FeCl_3$ in water [6]. The protonated polyacid polymer has also been produced [6].
Applications:

Bibliographic References
[1] Havinga, E.E., van Horssen, L.W., ten Hoeve, W., Wynberg, H. and Meijer, E.W., *Polym. Bull. (Berlin)*, 1987, **18**, 277
[2] Havinga, E.E., ten Hoeve, W., Meijer, E.W. and Wynberg, H., *Chem. Mater.*, 1989, **1**, 650
[3] Innis, P.C., Chen, Y.C., Ashraf, S. and Wallace, G.G., *Polymer*, 2000, **41**, 4065
[4] Li, H.C. and Khor, E., *Macromol. Chem. Phys.*, 1995, **196**, 1801
[5] Huang, B.J., Shieh, D.T., Chieh, W.C., Liaw, D.-J. and Li, L.-J., *Thin Solid Films*, 1997, **301**, 175
[6] Chang, E.-C., Hua, M.-Y. and Chen, S.-A., *J. Polym. Res.*, 1998, **5**, 249

Polystyrene

P-521

$$-[CH_2CH(Ph)]_n-$$

Synonyms: *Poly(1-phenylethylene). Poly(vinyl benzene). Poly(phenyl ethene). Poly(phenyl ethylene). Poly(1-vinyl benzene). Poly(ethenyl benzene). PS*
Related Polymers: High-impact polystyrene, Polystyrene, Foamed, Polystyrene latex
Monomers: Styrene
Material class: Thermoplastic
Polymer Type: polystyrene
CAS Number:

CAS Reg. No.	Note
9003-53-6	atactic
25086-18-4	isotactic
28325-75-9	syndiotactic

Molecular Formula: $(C_8H_8)_n$
Fragments: C_8H_8
Molecular Weight: Typically n = 1000 styrene units. A wide range of MW is possible depending on the polymerisation conditions. General purpose polystyrene has MW 100000–300000
Additives: Various additives may be used depending on product application e.g. uv screening agents, antioxidants, flame retardants, anti-static agents. The physical props. of polystyrene may be modified by glass reinforcement. Polystyrene may be filled with glass fibres or glass beads/spheres

General Information: Atactic polystyrene is a crystal clear thermoplastic material, hard, rigid and free of odour and taste [4]. Isotactic polystyrene is an opaque material. General purpose polystyrene is available in the following commercial grades: easy flow and medium flow are used in injection moulding; high heat is used in extrusion applications. Speciality grades include fast cycle resins, food contact grade resins (low residual) and low MW resins [2]
Tacticity: Commercially prod. polystyrene is atactic [8,9] and non-stereospecific. It is an amorph. material. Isotactic polystyrene is prod. on a small scale and is not important commercially. The amorph. isotactic form has similar props. to the atactic material. A partially crystalline isotactic form may also be prepd. This has a degree of crystallinity in the range 45–60%. The crystallisation rate of the isotactic form is very slow. The low crystallinity and rate of crystallisation are the main obstacles to commercial application of isotactic polystyrene as a general purpose plastic. The syndiotactic form has also been prepared on a small scale and may become commercially important in the future. It is a highly crystalline material with a high rate of crystallisation. Many physical props. of isotactic and syndiotactic forms differ significantly from the props. of the amorph. material
Identification: Isotactic and syndiotactic (crystalline) polystyrene may be distinguished from atactic, amorph. polystyrene by their lack of solubility in ketones and their higher melting temps. Tacticity and steric composition can be identified from ir spectra, H-1 nmr and C-13 nmr spectra or X-ray diffraction patterns. Polystyrene softens and burns fairly readily with an extremely sooty flame. The characteristic odour of styrene noticeable on extinction

Volumetric & Calorimetric Properties:
Density:

No.	Value	Note
1	d 1.05 g/cm^3	standard, atactic, ASTM D792 [3]
2	d 1.08 g/cm^3	isotactic, crystalline [8]

Thermal Expansion Coefficient:

No.	Value	Note	Type
1	0.0002085 K^{-1}	T < Tg [20]	V
2	0.000565 K^{-1}	T > Tg [20]	V
3	6.95 × 10^{-5} K^{-1}	T < Tg [20]	L
4	0.0001883 K^{-1}	T > Tg [20]	L

Volumetric Properties General: The density of polystyrene depends on the temp. This temp. dependence has been calculated [21]. Density also depends on the tacticity and crystallinity. It is higher for isotactic crystalline polystyrene than for the common atactic form. [8]. Glass fibre reinforcement of polystyrene also increases its density [3]. The density of a polystyrene product may also be affected by processing conditions. Foamed polystyrene products have a much lower density than that of General purpose polystyrene. The volumetric coefficient of thermal expansion has also been fitted to an equation for the temp. range 100–322° [21]. For an oriented thermoformed polystyrene part, the coefficients of expansion parallel and perpendicular to the draw direction are expected to be different [22]. The coefficient of thermal expansion may also be affected by crystallinity (this may be significant for isotactic and syndiotactic forms), cross-linking and the use of fillers. The latter two factors usually reduce the thermal expansion coefficient [23]
Equation of State: Equation of state information has been reported [26,27,28,29]
Thermodynamic Properties General: Thermal conductivity increases with temp. until the T_g is reached; it then decreases slightly. At higher temps. it again increases slightly [2]. The large

discrepancies for polystyrene samples at any given temp. and the varying temp. dependence have been attributed to polymer characteristics such as MW or MW distribution [33]. Thermal conductivity has also been found to depend on sample thickness; however, heat pretreatment of the polystyrene sample is found to reduce this effect and increase the thermal conductivity [33]. DSC based methods for measuring thermal conductivity are most suitable for small sized samples and may be used for polymer melts [32,34]. Heat capacity data for amorph. atactic and isotactic polystyrene have been reviewed [35]. The heat capacity of amorph. polystyrene is found to be a smooth function of temp. from -273–37°. The heat capacity of molten polystyrene is a linear function of temp. from T_g to 327°. The heat capacity of polystyrene from -203° to temps. just below T_g is practically independent of crystallinity and tacticity. The change in heat capacity at T_g does not fit the rule of constant change in heat capacity and this is similar to deviations observed for other polymers containing phenyl structural units [35]

Latent Heat Crystallization: ΔH 19.4 kJ mol^{-1}, ΔS 133.7 J mol^{-1} K^{-1} (25°, amorph.) [35]; ΔH 30.2 kJ mol^{-1}, ΔS 166 J mol^{-1} K^{-1} (100°, amorph.) [35] ΔH_f 10 kJ mol^{-1} (isotactic) [38]; ΔS_f 17.6 J K^{-1} (isotactic) [8]; ΔH_f 5.53 kJ mol^{-1} (53.2 J g^{-1}) (syndiotactic) [39]

Thermal Conductivity:

No.	Value	Note
1	0.123 W/mK	-173°, ASTM C177 [2,31]
2	0.141 W/mK	-73°, ASTM C177 [2,31]
3	0.154 W/mK	7°, ASTM C177 [2,31]
4	0.16 W/mK	127°, ASTM C177 [2,31]
5	0.164 W/mK	227°, ASTM C177 [2,31]
6	0.14 W/mK	DSC method [32]
7	0.165 W/mK	DSC method with pretreatment by grease and high temp. [33]

Thermal Diffusivity: a 24.52 × 10^{-8} (-173°); 16.42 × 10^{-8} (-73°); 12.35 × 10^{-8} (7°); 8.28 × 10^{-8} (127°); 8.02 × 10^{-8} m^2 s^{-1} (227°) [2,31]

Specific Heat Capacity:

No.	Value	Note	Type
1	127.5 kJ/kg.C	25°, glassy state [25]	P
2	163.4 kJ/kg.C	100°, at T_g [35]	P

Melting Temperature:

No.	Value	Note
1	210–220°C	isotactic [36]
2	270°C	syndiotactic [9]
3	243°C	isotactic [38]

Glass-Transition Temperature:

No.	Value	Note
1	100°C	100°
2	85–110°C	[5]

Transition Temperature General: General purpose polystyrene is an amorph. material and its glass-transition temp. is its most important thermal property. The glass-transition temp. of polystyrene depends on M_n and this is shown in Fig. 2.1 [24]. The greatest change in T_g is for low values of M_n and there is little change when M_n >85000. The T_g is decreased by the presence of diluents and hence materials such as mineral oil or paraffin oil may be added to lower the T_g and the processing temp. of polystyrene polymers. Diluents can be reactants, solvents or by-products forming during the polymer synth. (e.g. styrene monomer, ethylbenzene) or additives, e.g. stabilisers, flame-retardants and antioxidants. Different diluents lower the T_g by various amounts. Ethylbenzene is more effective on a weight basis at lowering the T_g than paraffin oil [2]. Cross-linking may either increase or decrease T_g depending on whether compositional changes occur [2,23]. The T_g of isotactic polystyrene is in the same range as that for the common, atactic form [5,37]. Amorph. polymers such as common, atactic polystyrene do not have a melting transition temp. However, a sharp melting transition may be observed for crystalline, isotactic samples and also for syndiotactic polystyrene [9]

Deflection Temperature:

No.	Value	Note
1	103°C	1.84 MPa, ASTM D648, injection moulded samples, MW 300000, M_n 130000 [2]
2	84°C	1.84 MPa, 264 psi, ASTM D648, injection moulded samples, MW 225000, M_n 92000 [2]
3	77°C	1.84 MPa, 264 psi, ASTM D648, injection moulded samples, MW 218000, M_n 74000 [2]
4	85°C	0.45 MPa, ASTM D648, unreinforced [3]
5	93°C	0.45 MPa, ASTM D648, 20% w/w glass reinforced [3]
6	101°C	0.45 MPa, 66 psi, ASTM D648, 30% w/w glass reinforced [3]
7	104°C	0.45 MPa, 66 psi, ASTM D648, 40% w/w glass reinforced [3]

Vicat Softening Point:

No.	Value	Note
1	108°C	ASTM D1525, injection moulded, MW 300000, M_n 130000 [2]
2	102°C	ASTM D1525, injection moulded sample, MW 225000, M_n 92000 [2]
3	88°C	ASTM D1525, injection moulded sample, MW 218000, M_n 74000 [2]

Transition Temperature:

No.	Value	Note	Type
1	-240°C	atactic [35]	δ relaxation
2	27°C	atactic [6]	β relaxation

Surface Properties & Solubility:

Solubility Properties General: Polystyrene is compatible with poly(tetramethyl bisphenol polycarbonate), poly(vinyl methyl ether), poly(2-vinylnaphthalene), poly(3-bromo-2,6-dimethyl-1,4-phenylene oxide-co-2,6-dimethyl 1,4-phenylene oxide), poly(2-chlorostyrene), poly(2-chlorostyrene-co-4-chlorostyrene), poly(2,6-diethyl-1,4-phenylene oxide), poly(2,6-dimethyl-1,4-phenylene oxide), poly(2,6-dimethyl-1,4-phenylene oxide) (carboxylated) [5], poly(2,6-dipropyl-1,4-phenylene oxide), poly(2-ethyl-6-propyl-1,4-phenylene oxide), poly(6-ethyl-2-methyl-1,4-phenylene oxide), poly(2-methyl-6-propyl-1,4-phenylene oxide), poly(4-bromostyrene-co- styrene), poly(butadiene-co-styrene), poly(n-butyl methacrylate-co-styrene), poly(3-bromo-2,6-dimethyl-1,4-phenylene oxide), poly(n-butyl methacrylate), poly(cis-1,4-isoprene), poly(methyl methacrylate), poly(methyl methacrylate-co-styrene), poly(α-methyl styrene), poly(butadiene)

Cohesive Energy Density Solubility Parameters: δ 18.6 [59]. Polystyrene is generally sol. in solvents for which the difference between solubility parameters of the polymer and the solvent is less than 1.8 MPa$^{1/2}$

Internal Pressure Heat Solution and Miscellaneous: Internal spreading pressure Π_e 16 mN m^{-1} (polystyrene-water interface, 25°) [61]. There is some variation in the values of contact angles because polymers are not ideal surfaces. Contact angle hysteresis may occur if the surface is not smooth and homogeneous. The measured contact angles may not be equilibrium contact angles [62]. θ 84–95° (H$_2$O) [60,61,65], θ 10–15° (α-bromonaphthalene) [61,65], θ 35° (methyl iodide) [65], θ 74° (formamide) [65], θ 80° (glycerol) [65]. Effective molar energy of adhesion 1314.6 J mol^{-1} [66]

Plasticisers: Plasticisers, such as mineral oil, phthalate and adipate esters, are only used for specialist applications as they are detrimental to mech. props. [1,4]

Solvents/Non-solvents: Isotactic and syndiotactic: Insol. ketones, esters, alkanes [8]. Atactic: Sol. butyl acetate, CCl$_4$, ethylbenzene, toluene, EtOAc, tributyl phosphate, C$_6$H$_6$, 2-butanone, CHCl$_3$, styrene, THF, chlorobenzene, methyl isobutyl ketone, DMF, CH$_2$Cl$_2$, cyclohexanone, tetrahydrodimethylfuran, diethyl phthalate, dioxane, CS$_2$, dioxolane, butyrolactone, 1,2-dichlorobenzene, 1-nitropropane, cyclopentanone, 1,2-dibromoethane, pyridine, glycol formal, dimethyl phthalate, PCl$_3$, cyclohexane/Me$_2$CO, decahydronaphthalene/diethyl oxalate, phenol/Me$_2$CO, tetralin [4]; slightly sol. methylcyclohexane, Me$_2$CO, nitroethane; spar. sol. cyclohexane. Insol. pentane, Et$_2$O, heptane, 1,2,3,4-tetrafluorobenzene, tricresyl phosphate, dibutyl phthalate, isoamyl alcohol, 4-methyl-2-pentanol glycol ethers, AcOH, 2-ethyl-1-butanol, 2-butanol, 1-butanol, MeCN, 1-propanol, ethylene chlorohydrin, nitromethane, EtOH, 1,2-propylene carbonate, propiolactone, MeOH, ethylene carbonate, H$_2$O [2,5], trichloroethyl phosphate [5]

Surface and Interfacial Properties General: Surface tension depends on the solvent (or solvent pair) used for the experimental measurements [65]. The spreading pressure Π_e needs to be included in the calculations for accurate results [61]. The experimental error is found to be largest for the γ_s^p calculations [65]

Wettability/Surface Energy and Interfacial Tension: Polarity 0.168 [5]. - (dt/dT) 0.072 mN (mK)$^{-1}$ (MV 44000). [5] Interfacial tension between polystyrene and other polymers [5] polystyrene-poly(chlorophene) 0.7 mN m^{-1} (20°), 0.5 mN m^{-1} (150°), 0.4 mN m^{-1} (200°) (dt$_{12}$/dT) 0.0014 mN (mK)$^{-1}$; polystyrene-poly(methyl acrylate) 3.2 mN m^{-1} (150°); polystyrene-poly(ethyl acrylate) 1.4 mN m^{-1} (150°); polystyrene-poly(n-butyl acrylate) 1.4 mN m^{-1} (150°); polystyrene-poly(2-ethyl hexyl acrylate) 2.5 mN m^{-1} (150°); polystyrene-poly(vinyl acetate) 4.2 mN m^{-1} (20°), 3.7 mN m^{-1} (150°), 3.4 mN m^{-1} (200°) (-dt$_{12}$/dT) 0.0044 mN (mK)$^{-1}$; polystyrene-poly(methyl methacrylate) 3.2 mN m^{-1} (20°), 1.5 mN m^{-1} (150°), 0.8 mN m^{-1} (200°) (-dt$_{12}$/dT) 0.013 mN (mK)$^{-1}$ [5]; polystyrene-linear polyethylene 8.3 mN m^{-1} (20°), 5.7 mN m^{-1} (150°), 4.7 mN m^{-1} (200°) (-dt$_{12}$/dT) 0.02 mN (mK)$^{-1}$ [5]; polystyrene-branched polyethylene 5.6 mN m^{-1} (20°), 5 mN m^{-1} (150°), 4.8 mN m^{-1} (200°) (-dt$_{12}$/dT) 0.0046 mN (mK)$^{-1}$ [5]. Polystyrene-water interfacial tension 25.6 mN m^{-1} [60] (calculated from contact angle measurements at 25°). The surface energy (solid surface tension) may be calculated from contact angle measurements [61,65]. For polystyrene t_s 41 mN m^{-1} [61] (calculated from the contact angle with H$_2$O). For a range of different solns. the average value of γ_s is 42 mN m^{-1} [65]. Different values of γ_s may be obtained when other calculation methods are used [65]. Other values of γ_s 40.7 mN m^{-1} (polymer melt method) [64], 43 mN m^{-1} (equation of state method) [64], 45.2 mN m^{-1} [66,67]

Surface Tension:

No.	Value	Note
1	40.7 mN/m	20°, MV 44000 [5]
2	31.4 mN/m	150°, MV 44000 [5]
3	27.8 mN/m	200°, MV 44000 [5]
4	33–43 mN/m	[63]
5	34–39.6 mN/m	[61,65]
6	3.8–7 mN/m	[61,65]

Transport Properties:

Transport Properties General: The melt flow index depends on MW and storage temp. [2]

Melt Flow Index:

No.	Value	Note
1	1.6 g/10 min	ASTM D1238, condition G, MW 300000, M$_n$ 130000 [2]
2	7.5 g/10 min	ASTM D1238, condition G, MW 225000, M$_n$ 92000 [2]
3	16 g/10 min	ASTM D1238, condition G, MW 218000, M$_n$ 74000 [2]
4	9 g/10 min	ASTM D1238, condition G, MW 195000, M$_n$ 95000 [2]
5	9 g/10 min	DIN 53735, standard [8]
6	1 g/10 min	DIN 53735, 30% glass reinforced [8]

Polymer Solutions Dilute: Theta temps. for various solvents have been reported [5,104] θ -49° (isoamyl acetate), -9° (n-butyl formate), 34–35° (cyclohexane), 20–154.2° (cyclopentane), 22° (bis(2-ethylhexyl)phthalate), -44–139° (EtOAc), 70° (ethyl cyclohexane), 60–70.5° (methyl cyclohexane), 28.0–30.6° (1-phenyldecane), -80–178° (n-propyl acetate), -46° (isobutyl acetate), 6° (1-chlorodecane), 58.6° (1-chlorododecane), 79–87.8° (cyclohexanol), 31–35.9° (diethyl malonate), 51.5–59.6° (diethyl oxalate), 108.5° (ethyl acetoacetate), 43–114° (methyl acetate), 8.0° (3-methyl cyclohexanol), -27–107° (isopropyl acetate). Theta temps. have also been reported for solvent mixtures [5,104]. The different concentration regimes for rheology have been reported for different MWs [2,106]. The intrinsic viscosity of dilute solns. is relatively insensitive to temp. over the usual measurement range (25–55°) [2,107] and depends on the value of MW and also the solvent quality [108]. The effect of MW on intrinsic viscosity in toluene (10% w/w, ASTM D445) has been investigated [109]. Various values for the Mark-Houwink equation parameters have been reported for different solvents [5,105,108,110]

Polymer Solutions Concentrated : For more conc. polymer solns., the zero shear viscosity depends on both the concentration and the average MW [106,111]. This relationship has been investigated for polystyrene solns. in toluene and trans-decalin at 25° and n-butylbenzene at 30° [112]. The zero shear viscosity at polymer concentrations of 50–100% (w/w) and 60–225° in ethylbenzene have been studied [113]

Polymer Melts: A mathematical relationship between the melt viscosity and MW has been reported [1,5,114,115]. In the melt, polystyrene exhibits both linear and non-linear viscoelastic behaviour. The flow characteristics of the polymer melt using capillary rheometry have been reported [118]. Rheological studies in the melt and in soln. have been reported [2,119,120]

Permeability and Diffusion General: The oxygen permeabilities of styrene polymers decrease as the temp. decreases, but this effect is smaller than that observed for other polymers [2]. The gas and vapour diffusion coefficients of many small molecules have been determined [96]. The permeation of CO$_2$ depends on M$_n$ but a similar effect is not observed for the diffusion coefficient [97]. The diffusion coefficient of O$_2$ at 25° is constant at high MW (M$_n \geq$ 20000) but varies at lower MWs [98]. The diffusion coefficient of organic vapours, e.g. Me$_2$CO, is slightly dependent on M$_n$ [96]. Organic vapour transmission rates have been measured for over pure liquids [MeOH] 0.14–0.84 nmol ms^{-1} (23°) [2]; [EtOH] 0.1 nmol ms^{-1} (23°) [2]; [formaldehyde] 0.6–0.8 nmol ms^{-1} (23°) [2]; [C$_6$H$_6$] 70 nmol ms^{-1} (35°) [2]. Diffusion coefficient for styrene [99] 8×10^{-15} m^2 s^{-1} (1100 ppm, 25°); 2×10^{-15} m^2 s^{-1} (300 ppm, 25°). Activation energy for diffusion [100,101] [N$_2$] 42.3 kJ mol^{-1}; [EtOH] 41 kJ mol^{-1} (T < Tg); 87.9 kJ mol^{-1} (T > Tg); [CH$_2$Cl$_2$] 41.8 kJ mol^{-1} (T < Tg); 100.4 kJ mol^{-1} (T > Tg); [Ethyl bromide] 51.9 kJ mol^{-1} (T < Tg); 108.8 kJ mol^{-1} (T > Tg). The activation energy for diffusion has also been measured for other gases and organic compounds [96]. The diffusion props. of CHCl$_3$ [103],

– Polystyrene

2-butanone [103], Me$_2$CO [103] and alkanes [102,103] have been reported. The gas permeability of isotropic and biaxially oriented polystyrene film has been reviewed [116]. Uv radiation increases the permeability of polystyrene films to air [117]
Water Content: Water vapour transmission rate 0.5–2.5 nmol ms^{-1} (24–38°, 90% relative humidity) [2]
Water Absorption:

No.	Value	Note
1	0.03–0.05 %	24h, ASTM D570, general purpose polystyrene [3]
2	0.07 %	24h, ASTM D570, 20% w/w glass reinforced [3]
3	0.05 %	24h, ASTM D570, 30% w/w glass reinforced and 40% w/w glass reinforced [3]

Gas Permeability:

No.	Gas	Value	Note
1	O$_2$	116.2–154.9 cm^3 mm/(m^2 day atm)	600–800 nmol (m s GPa)$^{-1}$ [2]
2	O$_2$	166.3 cm^3 mm/(m^2 day atm)	1.9 × 10^{-13} cm^3 cm^{-2} (s Pa)$^{-1}$, 25° [5,95]
3	N$_2$	7.7–9.7 cm^3 mm/(m^2 day atm)	40–50 nmol (m s GPa)$^{-1}$ [2]
4	CO$_2$	387.4–581 cm^3 mm/(m^2 day atm)	2000–3000 nmol (m s GPa)$^{-1}$ [2]
5	CO$_2$	65.6–525.2 cm^3 mm/(m^2 day atm)	0.75–6 × 10^{-13} cm^3 cm^{-2} (s Pa)$^{-1}$, 25°, ultradrawn [5]

Mechanical Properties:
Mechanical Properties General: Polystyrene is generally hard and rigid but brittle. The mech. props. depend on the MW of the sample and the processing conditions. They may be altered by rubber modification or by glass reinforcement. The relation between tensile strength and elongation has been reported in a graphical form [3]. There are several general reviews of mech. props. [126,127]. A graph of flexural strength against tensile strength has been reported [2]. Variation of Young's modulus with temp. has been reported [5,122]. Stress-strain curves have been reported [1]. The dependence of tensile strength on MW has been reported [2,153]

Tensile (Young's) Modulus:

No.	Value	Note
1	2800–3500 MPa	ASTM D638, various grades [3]
2	2400–3200 MPa	ASTM D638, various grades [119]
3	7200 MPa	23°, ASTM D638, 20% w/w glass reinforced [3]
4	8900 MPa	23°, ASTM D638, 30% w/w glass reinforced [3]
5	11300 MPa	23°, ASTM D638, 40% w/w glass reinforced [3]
6	3100 MPa	23°, ASTM D638, unreinforced [3]
7	3340 MPa	ASTM D638, MW 300000, M$_n$ 130000, injection moulded [2]
8	3240 MPa	ASTM D638, MW 300000, M$_n$ 130000, compression moulded [2]
9	2450 MPa	ASTM D638, MW 225000, M$_n$ 92000 [2]
10	3100 MPa	ASTM D638, MW 218000, M$_n$ 74000 [2]
11	3170 MPa	ASTM D638, MW 195000, M$_n$ 95000 [2]
12	3240 MPa	ASTM D638, MW 300000, M$_n$ 130000, low residual [2]

Flexural Modulus:

No.	Value	Note
1	2800–3200 MPa	20°, ASTM D790, various grades [3]
2	3100 MPa	ASTM D790, unreinforced [3]
3	4500 MPa	ASTM D790, 20% w/w glass reinforced [3]
4	8200 MPa	ASTM D790, 30% w/w glass reinforced [3]
5	10300 MPa	ASTM D790, 40% w/w glass reinforced [3]
6	3155 MPa	ASTM D790, MW 300000, M$_n$ 130000, injection moulded [2]
7	3170 MPa	ASTM D790, MW 225000, M$_n$ 92000 [2]
8	3100 MPa	ASTM D790, MW 195000, M$_n$ 95000 [2]
9	3155 MPa	ASTM D790, MW 300000, M$_n$ 130000 [2]

Tensile Strength Break:

No.	Value	Note	Elongation
1	44 MPa	23°, ASTM D638, unreinforced [3]	
2	72 MPa	23°, ASTM D638, 20% w/w glass reinforced [3]	
3	86 MPa	23°, ASTM D638, 30% w/w glass reinforced [3]	
4	99 MPa	23°, ASTM D638, 40% w/w glass reinforced [3]	
5	56.6 MPa	ASTM D638, MW 300000, M$_n$ 130000, injection moulded [2]	2.4%
6	44.1 MPa	ASTM D638, MW 300000, M$_n$ 130000, compression moulded [2]	1.4%
7	44.8 MPa	ASTM D638, MW 225000, M$_n$ 92000 [2]	2.0%
8	35.9 MPa	ASTM D638, MW 218000, M$_n$ 74000 [2]	1.6%
9	40 MPa	ASTM D638, MW 195000, M$_n$ 95000 [2]	1.3%
10	44.1 MPa	ASTM D638, MW 300000, M$_n$ 130000 [2]	1.4%
11	35–84 MPa	20°, ASTM D638, various grades [3,119]	1.0–4.5%

Flexural Strength at Break:

No.	Value	Note
1	83–118 MPa	20°, ASTM D790, various grades [3]
2	58 MPa	ASTM D790, unreinforced [3]
3	103 MPa	ASTM D790, 20% w/w glass reinforced [3]

— Polystyrene

4	111 MPa	ASTM D790, 30% w/w glass reinforced [3]
5	120 MPa	ASTM D790, 40% w/w glass reinforced [3]
6	82.8 MPa	ASTM D638, MW 225000 and 300000, M_n 92000 and 130000 [2]
7	51 MPa	ASTM D638, MW 195000, M_n 95000 [2]

Compressive Modulus:

No.	Value	Note
1	3000 MPa	ASTM D695, unoriented [1,5]

Elastic Modulus:

No.	Value	Note
1	1200 MPa	[121]
2	3000 MPa	[121]
3	3330–3660 MPa	[2,123]

Poisson's Ratio:

No.	Value	Note
1	0.33	general purpose grade [5,6,121]

Tensile Strength Yield:

No.	Value	Elongation	Note
1	42 MPa	1.8% elongation	ASTM D638, unreinforced [1]
2	131 MPa	1.5% elongation	ASTM D638, glass filled [1]

Compressive Strength:

No.	Value	Note
1	77–112 MPa	ASTM D695 [3]

Impact Strength: Pigments and other fillers typically lower the Izod impact strength [124]. Glass reinforcement increases Izod impact strength, depending on fibre length [2]. Izod impact toughness shows slight temp. dependence for toughened polystyrene [3]. Izod impact strength is affected by MW and processing conditions [2]. Dart drop impact strength of general purpose grade is very low [1] as this polymer is completely brittle under this test [2]. For glass-filled polystyrene the dart drop impact strength is medium high [1]

Viscoelastic Behaviour: Polystyrene polymers exhibit linear viscoelastic behaviour in the melt and in conc. soln. [2]. Viscoelastic props. are characterised by storage and loss moduli under shear deformation, relaxation time, the steady state shear compliance and the zero shear viscosity. The storage and loss moduli depend on the polymer MW and MW distribution [2]; data have been reported [128,129,130,131,132,133]. The zero shear viscosity depends on the weight average MW of the polystyrene and this effect has been reported [2,131,134,135]. The steady state shear compliance is less dependent on MW than the zero shear viscosity but is more sensitive to MW distribution [2]. Compliance data have been reported [136,137]. The storage modulus changes with temp. [2] and the data obtained may be used to determine the glass-transition temp. and other transitions [138]. Changes in viscoelastic behaviour with temp. are characterised by the shift factor α_T which is obtained using the Williams, Landel and Ferry equation [2]. Values of the shift factor for samples at 150–260° have been reported [128,139,140]. The Williams-Landel-Ferry equation constants used to calculate the shift factors for polystyrene samples of various MWs have been reported [119] as have the experimental and calculation methods used [119,128,141,142,143,144,145,146,147,148]. Viscoelastic props. are also affected by pressure and the shift for polystyrene has been reported [149,150]. Non-linear viscoelastic props. are observed for molten and conc. polystyrene solns. under conditions of large strain or high shear rates [2]. A master plot of viscosity and melt elasticity as a function of shear rate [2,139,140] illustrates this non-linear behaviour. Non-linear behaviour may also affect the steady state elongational and shear viscosities at high tensile or shear stresses [151]. Conversion graphs for melt flow rate (MFR) to apparent melt viscosity at various shear rates for commercial polystyrene resins have been reported [152]

Hardness: Rockwell M65-M90 (various grades). [3] Variation of hardness with MW has been reported [2]. Ball indentation hardness 150 N mm^{-2} (DIN 53 456; increases for a 30% w/w glass fibre reinforced sample) [13]

Failure Properties General: Creep tests involving the measurement of deformation as a function of time at constant stress or load have been reported [1,154,155,156]. Fatigue testing of polystyrene polymers in air and harsh environments has been reported [1,157,158]

Fracture Mechanical Properties: Fracture toughness 0.7–1.1 J cm^{-3} m$^{1/2}$ [119]. Fracture surface energy 250–1000 J m^{-2} [125]

Izod Notch:

No.	Value	Notch	Note
1	0.0013–0.0034 J/m	N	20°, ASTM D256 [3]
2	13–25 J/m	N	ASTM D256 [119]
3	16 J/m	N	ASTM D256 [2]
4	21 J/m	N	ASTM D256 [1]
5	0.0048 J/m	N	20% w/w glass reinforced, ASTM D256 [3]
6	0.0053 J/m	N	30% w/w glass reinforced, ASTM D256 [3]
7	0.0064 J/m	N	40% w/w glass reinforced, ASTM D256 [3]
8	24 J/m	N	MW 300000, M_n 130000, injection moulded [2]
9	14 J/m	N	MW 300000, M_n 130000, compression moulded [2]
10	16 J/m	N	MW 225000, M_n 92000 [2]
11	19 J/m	N	MW 218000, M_n 74000 [2]

Electrical Properties:

Electrical Properties General: Polystyrene has good electrical insulation props. It has a very low power factor and very low permittivity in high frequency electric fields [3]. The dissipation factor of pure, atactic polystyrene is relatively unaffected by frequency [2,91]. Dielectric constant decreases with increasing temp. [2]. This effect is most marked near T_g [2,90]

Surface/Volume Resistance:

No.	Value	Note	Type
1	0.00084×10^{15} Ω.cm	dry	S
2	0.02×10^{15} Ω.cm	93% relative humidity [88]	S

Dielectric Permittivity (Constant):

No.	Value	Frequency	Note
1	2.49–2.55		amorph. [5,6]
2	2.61		crystalline [5,6]
3	2.45–3.1	60 Hz	[3]

– Polystyrene

Dielectric Strength:

No.	Value	Note
1	20–28 kV.mm^{-1}	[3]

Arc Resistance:

No.	Value	Note
1	60–140s	[3]

Strong Field Phenomena General: Electrical breakdown has been reported [88]. Intrinsic voltage breakdown strength 600 kV mm^{-1} (20°), 720 kV mm^{-1} (-200°). Thermal breakdown - limiting stress 16 kV mm^{-1} (60 Hz). Tracking resistance 2.5 kV/50 mm spacing (ASTM D2303) [88].
Static Electrification: Triboelectric charge -5.5 kV (25°, 60% relative humidity) [92]. The half decay time at 5 kV and -5 kV is greater than 1h [92]. Percentage decay has been reported [92]. The charge transfer from metals to polymers has been reported [93]. Charge transfer -0.04–0.05 pC (gold) [93], -0.01–0.03 pC (platinum) [93], -0.04–-0.03 pC (aluminium) [93], -0.08–-0.04 pC (magnesium) [93]. The charge transfer to polystyrene is smaller than that for other polymers, but the value increases when the polystyrene samples are oxidised [93]
Magnetic Properties: The magneto-optical props. of transparent polystyrene materials have been reported [94]
Dissipation (Power) Factor:

No.	Value	Frequency	Note
1	0.00015		amorph. [5,6]
2	0.0003		cryst. [5,8,9]
3	0.0001–0.0006	60 Hz	[3]

Optical Properties:

Optical Properties General: Atactic polystyrene is transparent to visible light [2]. Polystyrene has a relatively high refractive index compared to other polymers because of the high polarisability of the phenyl groups [2]
Refractive Index:

No.	Value	Note
1	1.59	ASTM D542 [2]
2	1.59–1.6	[3,5,6]

Transmission and Spectra: Ir spectra of atactic and isotactic forms have been reported [78,79,80,81,82]. Uv absorption bands at 260, 194 and 80 nm, *in vacuo* [5,83]. Other uv absorption spectra have been reported [84,85]. H-1 nmr spectral data have been reported [5,86,87]
Molar Refraction: Abbé value 30.8 [68]. Dispersion 0.0325 [68]; dη/dC 0.226, 0.217, 0.215 (2-butanone, 25°) [69], dη/dC 0.214, 0.203, 0.201 (ethyl butyrate, 25°) [69], dη/dC 0.184, 0.172, 0.17 (dioxane, 25°) [69], dη/dC 0.167, 0.155, 0.153 (CHCl$_3$, 25°) [69], dη/dC 0.161, 0.149, 0.147 (CCl$_4$, 25°) [69], dη/dC 0.116, 0.112, 0.112 (toluene, 25°) [69], dη/dC 0.086, 0.085, 0.084 (chlorobenzene, 25°) [69]. The temp. and MW dependence of refractive index have been reported [70]
Total Internal Reflection: Birefringence can be induced by stressing the polymer below T_g (photoelastic effect); by orientation in the melt above T_g; or by orientation in soln. (flow birefringence) [2]. Birefringence occurs when the refractive indices parallel and perpendicular to the applied stress are different and indicates the level of orientation in a transparent polymer article [2]. Intrinsic birefringence -0.10 ± 0.05 (amorph.) [71,72]. Birefringence -0.001–-0.002 (melt drawn) [2,73]. Birefringence of polymer solns. [74] -1.5 × 10^{-8} s cm^3 g^{-1} (MW 10000) -2.95 × 10^{-7} s cm^3 g^{-1} (MW 390000). Optical polarisability -0.65 × 10^{-23} cm^3 (MW 10000) [74], -1.48 × 10^{-23} cm^3 (MW 390000) [74]. Birefringence dispersion has been reported [2,75]. Stress optical coefficient C -4700 × 10^{-12} Pa^{-1} (atactic PS in the melt) [76], -2942 × 10^{-12} Pa^{-1} (dilute soln., 25°, MW 10000) [74], -6695 × 10^{-12} Pa^{-1} (dilute soln., 25°, MW 390000) [74], 10.1 × 10^{-12} Pa^{-1} (glassy monofilament) [2,77], 9.5 × 10^{-12} Pa^{-1} (extruded sheet) [2,77], 8.7 × 10^{-12} Pa^{-1} (compression moulded, MW 145000) [2,77], 8.3 × 10^{-12} Pa^{-1} (compression moulded, MW 225000) [2,77]. The optical stress coefficient changes sign for glassy polystyrene [2]. Optical strain coefficients for glassy polystyrene 29600 (monofilament) [77], 30500 (extruded sheet) [77], 31000 (compression moulded, MW 145000) [77], 26400 (compression moulded, MW 225000) [77]
Volume Properties/Surface Properties: Unmodified polystyrene is normally colourless [2]. Degradation may cause yellowing. Atactic polystyrene is transparent [2]. Isotactic polystyrene is not transparent because crystallites in the material scatter light [2]

Polymer Stability:

Thermal Stability General: Polystyrene is easy to fabricate thermally into a wide variety of shapes without significant decomposition. However, conditions do exist where polystyrene will degrade thermally in the absence of oxygen or in the presence of oxygen by thermal-oxidative degradation
Upper Use Temperature:

No.	Value	Note
1	105°C	standard [3]

Decomposition Details: Decomposes to monomer, dimer and trimer with small amounts of toluene, α-methyl styrene and 1,3-diphenylpropane [2,46,47]. Degradation mechanism involves depropagation and first order termination [2,44]. Initial decomposition temp. 300–330° [2]. Rate of degradation depends on the method of synth. [44]. Anionically initiated polystyrene is more stable thermally than that prod. thermally or from free-radical initiators, e.g. benzoyl peroxide or azoisobutyronitrile (AIBN) [44]. After about 10% weight loss all polystyrene polymers degrade at about the same rate [45]. E_{act} 188 kJ mol^{-1} (dec.) in vacuum or flowing nitrogen. Oxygen plays a very important role in degradation. The activation energy for the thermal degradation of polystyrene above 350° decreases to 90 kJ mol^{-1} with an excess of oxygen [2]. Degradation mechanism involves depropagation. Thermal-oxidative products include benzaldehyde, benzoic acid, acetophenone, phenol, benzyl alcohol, cinnamaldehyde, styrene, styrene dimer (diphenyl butene) and styrene trimer (triphenyl hexene). Antioxidants retard oxidative degradation by scavenging peroxy radicals as they are formed [2]. Oxidation during moulding or extrusion may be reduced by removing absorbed oxygen from the resin before melting. Complete oxygen removal can be accomplished by storage under vacuum or in an inert atmosphere [2]
Flammability: Limiting oxygen index 17.6–18.3% (23°, ISO 4589/ASTM D2863) [23,40]. The flammability of styrene polymers has been reduced by incorporating fire resistant comonomers or flame retardant additives. Effective flame retardant compounds generally contain phosphorus or halogen. Synergists and diluents (or fillers) are also significant. Monochlorostyrene (comonomer), dibromostyrene (comonomer), vinyl bromide, bis(dibromopropyl)fumarate, pentabromophenyl methacrylate (comonomer) and ammonium bromide are used as flame retardants [41]. The effect of systems containing phosphorus, nitrogen and bromine on flammability has been reported [48]. Phosphorus decreases flammability but increases smoke formation [48]
Environmental Stress: Uv radiation of wavelength 254 nm is strongly absorbed resulting in significant physical and chemical changes. [49] Polystyryl radicals are formed from the scission of the tertiary CH bond. Cross-linking of two polystyryl radicals results in rapid embrittlement of the polymer [50]. Cross-linking is the dominant reaction under short wavelength irradiation in the absence of oxygen [50]. In the presence of oxygen (or under vacuum) short wavelength uv radiation causes yellowing. Experimental evidence shows that yellowing is primarily due to the formation of conjugated polyene structs. Weathering may cause chain scission, yellowing and oxidation [2]. Impurities in the

polymer, e.g. monomer residues, acetophenone end groups, in-chain peroxides (formed by copolymerisation with oxygen in free radical polymerisation) and hydroperoxides are thought to initiate uv degradation reactions. In outdoor weathering both uv degradation and oxidation reactions occur in competition. Weatherability is improved by using stabilisers, e.g. hindered amine light stabilisers (0.1–1.0% w/w) [52] and benzotriazoles/benzophenones (0.25% w/w as uv stabilisers) [1]. Pigments such as TiO_2 can contribute to polymer stability by screening the damaging wavelengths. Pigments can also mask colour changes making yellowing less noticeable [2]. Polystyrene is susceptible to environmental stress cracking and crazing by oils, esters, higher alcohols, ketones, aromatics and chlorinated hydrocarbons [3]. The critical stress required for the onset of craze and crack formation generally depends on the difference in solubility parameters between the polymer and the solvent. This critical stress is low when the solubility parameters of polymer and solvent are close together [53]. The critical stress can also be lowered by decreasing the interfacial tension [54]. Environmental stress cracking is affected by MW, MW distribution and composition (for copolymers) [2]. The environmental stress crack resistance increases with MW and narrowing MW distribution for polystyrene and n-butanol [55]. Environmental stress cracking is also affected by orientation (see oriented film polystyrene) and rubber modification (see high-impact polystyrene). Environmental stress cracking may also be induced by gases. Helium, nitrogen, carbon dioxide and hydrogen have been shown to be very mild craze growth enhancers [2,56,57]. Trichlorofluoromethane is a very aggressive cracking agent [2]

Chemical Stability: Affected by conc. acids but has some chemical resistance when the acid is diluted. Unaffected by conc. or dilute alkali [3]. Dissolution, swelling and environmental stress cracking occurs for a range of chemicals; aromatic and halogenated hydrocarbons, ethers, ketones, esters and higher alcohols [23]

Hydrolytic Stability: Has low water absorption [2,23]. Stability is generally unaffected by dilute aq. solns.

Biological Stability: Resistant to biodeterioration [58]. Microorganisms and insects have little effect, although surface growths of fungi can impair electrical props. [4]

Recyclability: Suitable for recycling because there is a large, dependable source of material, it is easy to sort and separate, the recycled resins are relatively high in value and there are acceptable (non-food) applications for the recycled material [42]. Industrial scrap (regrind) from moulding, blowing, extrusion and thermoforming is usually melted down and reprocessed which satisfies both economic and environmental considerations [2,18]. Polystyrene foam products and other polystyrene resins may be recycled [42,43]

Stability Miscellaneous: High resistance to x-rays and γ-rays due to the protective action of the phenyl groups [2,4] which dissipate excitation energy without decomposing [51]. In the absence of air flexural strength and tensile strength are not decreased by γ-irradiation at doses up to 500 Mrads. These mech. props. decrease at this radiation level in the presence of oxygen. Even in the presence of oxygen, changes in props. and discoloration are negligible at γ-radiation levels of 2–3 Mrads (the dose range of γ-radiation used for sterilising medical supplies) [2]

Applications/Commercial Products:

Processing & Manufacturing Routes: Polystyrene can be prod. via anionic, cationic, Ziegler-Natta and free radical mechanisms using appropriate initiators. The free radical process is the most important commercially and typically uses peroxide, persulfate or azo initiators [1,2]. Free radical polystyrene can be prod. by bulk (or mass), soln., suspension or emulsion processes. The division between bulk and soln. processes is ill-defined; however, bulk polymerisation is usually defined as a system where pure styrene monomer is polymerised and soln. polymerisation as a system where the styrene monomer is diluted with solvent. Bulk polymerisation of styrene monomer may be accomplished by thermal initiation alone, without an added initiator [17]. Bulk, soln., suspension and emulsion polymerisation reactions may use a batch process. In a batch process all reagents are added to the reactor initially and remain in the reactor for the same length of time. Batch processes enable a high conversion of monomers to polymer. However, there may be difficulties with heat control and in maintaining batch-to-batch uniformity. For bulk and soln. polymerisation reactions, a process where reagents are fed continuously into the reactor and waste products are removed is preferred for polystyrene manufacture. Batch processes are however used for emulsion polymerisation and suspension polymerisation reactions. Emulsion polymerisation is used on a semi-commercial scale for the production of polystyrene latex. It is not attractive commercially for the production of general purpose polystyrene due to the high recovery costs of washing and drying the polymer in order to remove the emulsion stabilisers, pelleting of dried polymer and treatment of waste water. Suspension polymerisation is used almost exclusively to produce expandable polystyrene. Because a blowing agent can be introduced during polymerisation, residual monomer requirements are not severe and the spherical particle form prod. in suspension can be used directly in the product. Products for injection moulding and extrusion applications are usually extruded to the pellet form. The suspension polymerisation process is declining in importance for general purpose polystyrene. The main commercial route for production is the soln. (or bulk-soln.) continuous process. Main technical challenges are controlling the exothermic heat of polymerisation and handling the highly viscous partially converted polymer. The addn. of ethylbenzene solvent reduces the viscosity. Temp. control and mixing (stirring) are important factors in controlling the product. For thermal initiation the typical temp. range is 120–180°; for chemical initiation the temp. may be reduced to 90°. A chain transfer agent may be added to control the MW. Anionic polymerisation of styrene is limited to the production of narrow MW distribution polystyrenes in laboratory quantities. Typical initiators are organolithium compounds. Cationic polymerisation is used to produce low MW resins. Typical initiators are strong protonic acids and Lewis Acids. Polystyrene obtained with any heterogeneous Ziegler-Natta catalyst is a mixture of products with different tacticities usually atactic and isotactic. The isotactic and atactic products may be separated by utilising their different solubility props. The syndiotactic form can be prod. with some homogeneous Ziegler-Natta catalysts. Typical catalyst systems contain tetrabenzyltitanium or tetrabenzylzirconium and methylaluminoxane in aromatic solvents [9,10,11,12]. Injection moulding is used to make polystyrene plastics for a wide variety of applications. Easy-flow resins are used for injection moulding of thin-walled parts. 3–4% Mineral oil is added to the resin to decrease the melt viscosity. Medium flow resins containing 1–2% added mineral oil are also used for injection moulding applications. Colour concentrates and other additives are added to the resin at the extrusion stage which takes place before injection moulding [2]. Extrusion methods are used for processing styrene polymers and copolymers. PS resins are easy to plasticate and have good melt strength. They are readily fabricated into sheet, film, rod, pipe, profiles, tubing and monofilament products

Mould Shrinkage (%):

No.	Value	Note
1	0.2–0.6%	unreinforced [3]
2	0.05–0.1%	glass reinforced [3]

Applications: Non-foamed polystyrene are used in appliances and consumer electronics (air conditioners, refrigerators and freezers, small appliances, cassettes and related products, cabinets, light fittings, furniture and toilet seats, toys and novelties e.g. used in advertising and photographic accessories). They are used in the building and construction industry but the use of foamed materials is more common. Non-foamed products are used in medical and cosmetic applications as table phials and tablet packaging. Polystyrene is used in a wide range of packaging applications including rigid packaging, closures and produce baskets. It is also used for disposable products such as tumblers and cocktail glasses, flatware and cutlery, dishes, bowls and cups. Heat resistant products are used for vending machine products

– Polystyrene

and hot food and drink containers. Other applications include prismatic reflectors and diffusers. Narrow MW distribution polystyrenes are used for instrument calibration. Low MW resins are used for coatings and glues. Injection moulding is used to produce furniture, toys, radio, television and computer cabinets, automotive parts, medical ware, houseware, bottle caps and containers. Blow moulding is used to form bottles, containers, furniture and automotive parts. Extrusion is used to manufacture film products (multilayer and oriented), profiles, light diffusers and wall coverings. Extrusion and thermoforming are used to make refrigerator and freezer parts; luggage; food containers, both solid and formed; disposable cups and dinnerware; and large automotive parts. Coextruded PS composites include dairy containers, two-colour dinnerware and drink cups, and sheet for outdoor use

Trade name	Details	Supplier
A/CSO/H/NSC/SE/SH/Y	various grades	Thermofil Inc.
AIM	4800/4810/4900	Dow Chemical
API	various grades	American Polymers
Amoco	unfilled	Amoco Chemical Company
Arloy	1000/1100	ARCO Chemical Co.
Arprylene	glass-reinforced	Thermofil
Arpylene	glass-reinforced	TPA
Bapolan	6050/6445/6450	Bamberger Polymers
Carinex	unfilled	Shell Chemicals
Comalloy	various grades	Comalloy Intl. Corp.
Conductomer	PSC-20/SC-26	Synthetic Rubber
DC-1007		Spartech
Dart	PS	Dart Polymers Inc.
Dylark	polystyrene and styrene-maleic anhydride copolymer	ARCO Chemical Company
Dylene	glass-reinforced and unfilled products	ARCO Chemical Company
Dylite	R2595B EPS	ARCO Chemical Co.
EA/EB/EC/EH	various grades	Chevron Chemical
EMI-X	PDX-C-89603	LNP Engineering
EPS	54/84/86	Huntsman Chemical
ES	various grades	BASF
Edistir	unfilled and uv stabilised	Montedison
Electrafil	J-30/PS-31	DSM Engineering
Emiclear	EC-1300 PS	Toshiba Chem. Prod.
Empee	Jenahrd/PS	Monmouth Plastics
Esbrite	various grades	Sumitomo Chem. Co.
Eslorite	unfilled product	Sumitomo Chem Co
FPC	various grades	Federal Plastics
Fiberstran	G-30/G-35	DSM Engineering
Fina	various grades	Fina Oil and Chem.
Fostalite	unfilled	Foster Grant
Fostarene	unfilled	Foster Grant
G31/35		DSM Engineering
Gedex		Cd F Chemie
Gordon (Super-dense)	unfilled	Hammond
Gordon (Super-flow)	unfilled	Hammond
HIPS	0014-L	Multibase
Hostyren	unfilled, uv stabilised	Hoechst
Huntsman	unfilled, uv stabilised	Huntsman Chemical
J-30	CF/30	DSM Engineering
Kaofulex	GPS/HPS	Kaofu Chemical
Kaofulon	various grades	Kaofu Chemical
Kaofulux	ABS	Kaofu Chemical
Kaofusan	SAN	Kaofu Chemical
Lacqrene	miscellaneous and high-impact grades	Atochem Polymers
Lastirol	glass-reinforced	Freeman
Lubricomp	CL series	LNP Engineering
Lucky	PS series	Lucky
Lustrex	unfilled, uv stabilised	Monsanto Chemical
MA/MB/MC/MD	various grades	Chevron Chemical
Marvaloy	750	Marval Indust.
Metastyrene	miscellaneous grades	BP Chemicals
Metastyrol		BP Chemicals
Mobil	various grades	Mobil Chem. Co.
Multi-Alloy	3006/4006	Multibase
Novalite		Cole
PS	various grades	BASF
PS	2835	Multibase
PS	various grades	Huntsman Chemical
PS	H23MF-1/H63CC-1	Washington Penn
PS	GF20 IM	TP Composites
Piccolastic	low MW resin	Hercules
Polyflam	RMS/RS	A.Schulman Inc.
Polyrex	P series	Chi Mei Corporation
Polystyrene	various grades	Novacor Chemicals
Polystyrene	various grades	Ricard Group/RRT
Polystyrene High Impact		IPC/GWC
Polystyrol	miscellaneous grades	BASF
RTP	glass-reinforced	Fiberite (Vigilant)
Replay	RP series	Huntsman Chemical
Resoglass	glass-reinforced	BP Chemicals
Resoglaz	glass-reinforced	BP Chemicals
Restirolo		SIR
Retain	PS Series	Dow Chemical
Rexene		Rexene Products
SC2F-1090		Spartech
Shuman	various grades	Shuman Plastics
Stat-Kon	C	LNP Engineering
Sternite	moulding powder	Sterling

– Polystyrene

Stilletex		Cole
Stycond	E	United Composites
Styrafil	flame retardant and glass-reinforced	Fiberfil
Styrol	various grades	Denka
Styron	unfilled and uv stabilised products	Dow
Styronol		Allied Resinous
Stystat	Q	United Composites
Styvex	22000/22001	Ferro Corporation
Superstyrex		BP Chemicals
Thermacomp	glass-reinforced, PTFE filled, glass bead/sphere filled	LNP Engineering
Toporex		Mitsui Petrochemicals
Trolitul		Dynamit Nobel
Valtra	various grades	Chevron Chemical
Vestypor		Hüls AG
Vestyron	polystyrene and styrene-butadiene copolymer products	Hüls AG
Vistron		BP Chemicals
XU	70016.01/70029.00	Dow Chemical

Bibliographic References

[1] *Kirk-Othmer Encycl. Chem. Technol.*, 3rd edn., (eds. M. Grayson and D. Eckreth), Wiley Interscience, 1984, **21**, 801
[2] *Encycl. Polym. Sci. Eng.*, 2nd edn. (eds. N.M. Bikales, H.F. Mark, G. Menges and C.G. Overberger), John Wiley and Sons, 1989, **16**, 1
[4] Roff, W.J. and Scott, J.R., Fibres, Films, *Plastics and Rubbers*, 2nd edn., Butterworths, 1971
[5] Schrader, D., *Polym. Handb.*, 4th edn., (eds. J. Brandrup, E.H. Immergut and E.A. Grulke), John Wiley and Sons, 1999, **V**, 91
[6] Styrene, Its Polymers, Copolymers and Derivatives, (eds. R.H. Boundy and R.F. Boyer), Reinhold Corp. (New York), 1952
[7] Campbell, I.M., *Introduction to Synthetic Polymers*, (eds. P.W. Atkins, J.S.E. Holker and A.K. Holliday), OUP, 1994
[8] *Encycl. Polym. Sci. Eng.*, 2nd edn. (eds. N.M. Bikales, H.F. Mark, G. Menges and C.G. Overberger), John Wiley and Sons, 1989, **15**, 656, 763, (isotactic form)
[9] *Encycl. Polym. Sci. Eng.*, 2nd edn. (eds. N.M. Bikales, H.F. Mark, G. Menges and C.G. Overberger), John Wiley and Sons, 1989, **15**, 660, 766, (syndiotactic form)
[10] Ishihara, N., Seiyama, T., Kuramoto, M. and Uoi, M., *Macromolecules*, 1986, **19**, 2464, (syndiotactic form)
[11] Pellecchia, C., Longo, P., Grossi, A., Ammendola, P. and Zambelli, A., Makromol. Chem., *Rapid Commun.*, 1987, **8**, 277, (syndiotactic form)
[12] Oliva, L., Pellecchia, C., Cinquina, P. and Zambelli, A., *Macromolecules*, 1989, **22**, 1642, (syndiotactic form, synth)
[13] *New Horizons in Plastics*, (ed. J. Murphy), WEKA, 1991
[14] Gardner, W., Cooke, E. and Cooke, R., *Chemical Synonyms and Tradenames*, 8th edn., Technical Press Ltd., 1978
[15] *Chemical Industry Directory*, Tonbridge, Benn Business Information Services, 1988
[16] *UK Kompass Register*, Reed Information Services, 1996
[17] Amos, J., *Polym. Eng. Sci.*, 1974, **14**, 1, (synth)
[18] *Encycl. Polym. Sci. Eng.*, 2nd edn., (eds. N.M. Bikales, H.F. Mark, G. Menges and C.G. Overberger), John Wiley and Sons, 1989, **2**, 512, (polymerisation)
[19] *Encycl. Polym. Sci. Eng.*, 2nd edn., (eds. N.M. Bikales, H.F. Mark, G. Menges and C.G. Overberger), John Wiley and Sons, 1989, **9**, 502, (applications)
[20] Aharoni, S.M., *J. Appl. Polym. Sci.*, 1973, **42**, 795, (thermal expansion coefficient)
[21] Hocker, H., Blake, G.J. and Flory, P.J., *Trans. Faraday Soc.*, 1971, **67**, 2251, (density-temp dependence)
[22] Wang, L.H., Choy, C.L. and Porter, R.S., *J. Polym. Sci., Polym. Phys. Ed.*, 1982, **20**, 633, (thermal expansion coefficient)
[23] Birley, A.W., Haworth, B. and Batchelor, J., Physics of Plastics: Processing, *Properties and Materials Engineering*, Hanser/Oxford University Press, 1991
[24] Claudy, P., Letoffe, J.M., Chamberlain, Y. and Pascault, J.P., *Polym. Bull. (Berlin)*, 1983, **9**, 208, (transition temps)
[25] Domalski, E.S. and Hearing, E.D., *J. Phys. Chem. Ref. Data*, 1990, **19**, 881, (heat capacity)
[26] Hartmann, B. and Haque, M.A., *J. Appl. Polym. Sci.*, 1985, **30**, 1553, (equation of state)
[27] Zoller, P., *J. Appl. Polym. Sci.*, 1977, **21**, 3129, (equation of state)
[28] Simha, R., Wilson, P.S. and Olabisi, O., *Kolloid Z. Z. Polym.*, 1973, **251**, 402, (equation of state)
[29] Cho, B., *Polym. Eng. Sci.*, 1985, **25**, 1139, (equation of state)
[30] Quach, A. and Simha, R., *J. Appl. Phys.*, 1971, **42**, 4592, (equation of state)
[31] Ho, C.Y., Desai, P.D., Wu, K.Y., Havill, T.N. and Lee, Y., *Proc. Symp. Thermophys. Prop.*, 1977, **7**, 198, 217, (thermal conductivity)
[32] Chiu, J. and Fair, P.G., *Thermochim. Acta*, 1979, **34**, 4592, (thermal conductivity)
[33] Hall, J.A., Ceckler, W.H. and Thompson, E.V., *J. Appl. Polym. Sci.*, 1987, **33**, 2029, (thermal conductivity)
[34] Khanna, Y.P., Taylor, T.J. and Chomyn, G., *Polym. Eng. Sci.*, 1988, **28**, 1034, (thermal conductivity)
[35] Gaur, U. and Wunderlich, B., *J. Phys. Chem. Ref. Data*, 1982, **11**, 313, (heat capacity, entropy, enthalpy)
[36] Maruselli, E., Demma, G., Driol, E., Nicolais, L. et al, *Polymer*, 1979, **20**, 571, (transition temps)
[37] Boon, J., Challa, G. and van Krevelen, D.W., *J. Polym. Sci., Part A-2*, 1968, **6**, 1791, (transition temps)
[38] Wunderlich, B., *Polym. Eng. Sci.*, 1978, **18**, 431, (transition temps)
[39] Pasztor, A.J., Landes, B.G. and Karjala, P.J., *Thermochim. Acta*, 1991, **177**, 187, (thermal props, syndiotactic form)
[40] *Flammability Handbook for Plastics*, 3rd edn., (ed. C.J. Hilado), Technomic, 1982
[41] Tesoro, G.C., *J. Polym. Sci., Part D: Macromol. Rev.*, 1978, **13**, 283, (flame retardants)
[42] Kline, C.H., *Chem. Ind. (London)*, 1989, **14**, 440, (recycling)
[43] *The World of Plastics*, British Plastics Federation, 1986, (recycling)
[44] Cameron, G.G., Bryce, W.A. and McWalter, I.T., *Eur. Polym. J.*, 1984, **20**, 563, (thermal stability)
[45] Cameron, C.G. and McWalter, I.T., *Eur. Polym. J.*, 1982, **18**, 1029, (thermal stability)
[46] Guaita, M., *Br. Polym. J.*, 1986, **18**, 226, (thermal stability)
[47] Daoust, D., Bormann, S., Legras, R. and Mercier, J.P., *Polym. Eng. Sci.*, 1981, **21**, 721, (thermal stability)
[48] Cullis, C.F., Hirschler, M.M. and Tao, Q.M., *Eur. Polym. J.*, 1991, **27**, 281, (flammability)
[49] Ranby, B. and Rabek, J.F., Photodegradation, Photo-oxidation and Photostabilisation of Polymers, *Principles and Applications*, Wiley, 1975, 165, (polymer stability)
[50] David, C., Baeyens-Volant, D., Delaunois, G., Lu Vinh, Q. et al, *Eur. Polym. J.*, 1978, **14**, 501, (weathering)
[51] Bowmer, T.N., Cowen, L.K., O'Donnell, J.H. and Winzor, D.J., *J. Appl. Polym. Sci.*, 1979, **24**, 425, (weathering)
[52] Loelinger, H. and Gilg, B., *Angew. Makromol. Chem.*, 1985, **137**, 163, (weathering)
[53] Yaffe, M.B. and Kramer, E.J., *J. Mater. Sci.*, 1981, **16**, 2130, (environmental stress cracking)
[54] Brown, H.R. and Kramer, E.J., *Polymer*, 1981, **22**, 687, (environmental stress cracking)
[55] Rudd, J.F., *J. Polym. Sci., Polym. Lett. Ed.*, Part B, 1963, **1**, 1, (environmental stress cracking)
[56] Wang, W.-C.V., Kramer, E.J. and Sachse, W.H., *J. Polym. Sci., Polym. Phys. Ed.*, 1982, **20**, 1371, 1383, (environmental stress cracking)
[57] Wang, W.-C.V. and Kramer, E.J., *Polymer*, 1982, **23**, 1667, (environmental stress cracking)
[58] Seal, K.J. and Morton, L.H.G., *Biotechnology*, 1986, **8**, 583, (biological stability)
[59] Suh, K.W. and Clarke, D.H., *J. Polym. Sci., Part A-1*, 1967, **5**, 1671, (solubility parameter)
[60] Absolom, D.R. and Neumann, A.W., *Colloids Surf.*, 1988, **30**, 25, (surface props)
[61] Busscher, H.J., Van Pelt, A.W.J., De Jong, H.P. and Arends, J., *J. Colloid Interface Sci.*, 1983, **95**, 23, (surface props)
[62] Ward, C.A. and Neumann, A.W., *J. Colloid Interface Sci.*, 1974, **49**, 286, 291, (surface props)
[63] Good, R.J., *J. Colloid Interface Sci.*, 1977, **59**, 398, (surface props)
[64] Wu, S., *Polymer Interface and Adhesion*, Marcel Dekker, 1982, (surface props)
[65] Dalal, E.N., *Langmuir*, 1987, **3**, 1009, (surface props)
[66] Pritykin, L.M., *J. Colloid Interface Sci.*, 1989, **112**, 539, (surface props)
[67] Yagnyatinskaya, S.M., Voyutsky, S.S. and Kaplunova, L.Y., *Russ. J. Phys. Chem. (Engl. Transl.)*, 1970, **44**, 1445, (surface props)

[68] *Encycl. Polym. Sci. Eng.*, 2nd edn., (eds. N.M. Bikales, H.F. Mark, G. Menges and C.G. Overberger), John Wiley and Sons, 1989, **10**, 494, (optical props)
[69] Huglin M.B., O'Donohue, S.J. and Rudson, M.A., *Eur. Polym. J.*, 1989, **25**, 543, (refractive index)
[70] Krause, S. and Lu, Z.-H., *J. Polym. Sci., Polym. Phys. Ed.*, 1981, **19**, 1925, (refractive index)
[71] Lefebvre, D., Jasse, B. and Monnerie, L., *Polymer*, 1982, **23**, 706, (birefringence)
[72] Lefebvre, D., Jasse, B. and Monnerie, L., *Polymer*, 1983, **24**, 1240, (birefringence)
[73] Gurnee, E.F., *J. Appl. Phys.*, 1954, **25**, 1232, (birefringence)
[74] Lodge, T.P., Miller, J.W. and Schrag, J.L., *J. Polym. Sci., Polym. Phys. Ed.*, 1982, **20**, 1409, (optical props)
[75] Gurnee, E.F., *J. Polym. Sci., Part A-2*, 1967, **5**, 817, (birefringence)
[76] Muller, R. and Froelich, D., *Polymer*, 1985, **26**, 1477, (stress optical coefficient)
[77] Rudd, J.F. and Andrews, R.D., *J. Appl. Phys.*, 1960, **31**, 818, (stress optical coefficient)
[78] Krimm, S., *Adv. Polym. Sci.*, 1960, **2**, 51, (ir)
[79] Painter, P. and Koenig, J., *J. Polym. Sci., Polym. Phys. Ed.*, 1977, **15**, 1885, (ir)
[80] *Aldrich Library of Infra Red Spectra*, (ed. C.J. Poucher), Aldrich Chemical Company, 1981
[81] Jasse, B., *Polym. Sci. Technol. (Plenum)*, 1987, **36**, 245, (ir)
[82] Nyquist, R.A., *Appl. Spectrosc.*, 1984, **38**, 264, (ir)
[83] Partridge, R.H., *J. Chem. Phys.*, 1967, **47**, 4223, (uv)
[84] Ramelow, U. and Baysal, B.M., *J. Appl. Polym. Sci.*, 1986, **32**, 5865, (uv)
[85] Soares, B.G. and de Souza Gomes, A., *Polym. Bull. (Berlin)*, 1988, **20**, 543, (uv)
[86] Bovey, F.A., Tiers, G.V.D. and Filopovich, G., *J. Polym. Sci.*, 1959, **38**, 73, (nmr)
[87] Live, D. and Kent, S.B.H., *ACS Symp. Ser.*, 1982, **193**, 501, (nmr)
[88] *Encycl. Polym. Sci. Eng.*, 2nd edn., (eds. N.M. Bikales, H.F. Mark, G. Menges and C.G. Overberger), John Wiley and Sons, 1989, **5**, 446, 523, (electrical props)
[89] Illers, K.-H., *Z. Elektrochem.*, 1961, **65**, 679, (electrical props)
[90] Broens, O. and Muller, F.H., *Kolloid Z. Z. Polym.*, 1955, **140**, 121, (electrical props)
[91] Hippel, A.R. and Wesson, L.G., *Ind. Eng. Chem.*, 1946, **38**, 1121, (electrical props)
[92] Uyama, Y. and Ikada, Y., *J. Appl. Polym. Sci.*, 1990, **41**, 619, (electrostatic props)
[93] Lowell J. and Akanda, A.R., *J. Phys. D: Appl. Phys.*, 1988, **21**, 125, (electrostatic props)
[94] Muto, S., Ichikawa, S.-I., Nagata, T., Matsuzaki, A. and Ito, H., *J. Appl. Phys.*, 1989, **66**, 3912, (magneto-optical props)
[95] Salame, M., *J. Polym. Sci., Polym. Symp.*, 1973, **41**, 1, (gas permeability)
[96] Berens, A.R. and Hopfenberg, H.B., *J. Membr. Sci.*, 1982, **10**, 283, (diffusion coefficient)
[97] Toi, K., Ohori, Y., Maeda, Y. and Tokuda, T., *J. Polym. Sci., Polym. Phys. Ed.*, 1980, **18**, 1621, (gas permeability)
[98] Weir, N.A., *J. Macromol. Sci., Phys.*, 1975, **11**, 553, (diffusion coefficient)
[99] Miltz, J. and Rosen-Doody, V., *J. Food Process. Preserv.*, 1984, **8**, 151, (diffusion coefficient)
[100] *Phys. Glassy Polym.*, (ed. R.N. Haward), Applied Science, 1973, 504, 522, (diffusion coefficient)
[101] *Polym. Permeability*, (ed. J. Comyn), Elsevier Applied Science, 1985, (diffusion props)
[102] Hopfenberg, H.B. and Frisch, H.L., *J. Polym. Sci.*, 1969, **7**, 405, (diffusion props)
[103] Hopfenberg, H.B., Holley, R.H. and Stannett, V.T., *Polym. Eng. Sci.*, 1970, **10**, 376, (diffusion props)
[104] Saeki, S., Konno, S., Kuwahara, N., Nakata, M. and Kanebo, M., *Macromolecules*, 1974, **7**, 521, (theta solvents)
[105] Boon, J., Challa, G. and Hermans, P.H., *Makromol. Chem.*, 1964, **74**, 129, (dilute soln rheology)
[106] Graessley, W.W., *Polymer*, 1980, **21**, 1323, (polymer soln rheology)
[107] Sun, S.F., *Polymer*, 1987, **28**, 283, (dilute soln rheology)
[108] Einaga, Y., Miyari, Y. and Fujita, H., *J. Polym. Sci., Polym. Phys. Ed.*, 1979, **17**, 2103, (dilute soln rheology)
[109] Keskkula, H. and Taylor, W.C., *J. Polym. Sci., Part B: Polym. Lett.*, 1970, **8**, 867, (dilute soln rheology)
[110] Aharoni, S., *J. Appl. Polym. Sci.*, 1977, **21**, 1323, (dilute soln rheology)
[111] Kulicke, W.M. and Kniewske, R., *Rheol. Acta*, 1984, **23**, 75, 82, (conc soln rheology)
[112] Graessley, W.W., Hazleton, R.L. and Lindemann, L.R., *Trans. Soc. Rheol.*, 1967, **11**, 267, (conc soln rheology)
[113] Mendelson, R.A., *J. Rheol. (N.Y.)*, 1980, **24**, 765, (conc soln rheology)
[114] Fox, T.E. and Flory, P.J., *J. Polym. Sci.*, 1954, **14**, 315, (melt rheology)
[115] Boon, J., Challa, G. and Hermans, P.H., *Makromol. Chem.*, 1964, **74**, 129, (melt rheology)
[116] Burmester, A.F., Manial, T.A., McHattie, J.S. and Wessling, R.A., *Polym. Prepr. (Am. Chem. Soc., Div. Polym. Chem.)*, 1986, **27**, 414, (gas permeability)
[117] Kaminska, A., *Polimery (Warsaw)*, 1980, **25**, 442, (permeability)
[118] Cogswell, F., *Polymer Melt Rheology*, John Wiley and Sons, 1981, 17, (melt rheology)
[119] *Physical Properties of Polymers Handbook*, (ed. J.E. Mark), AIP Press, 1996
[120] Fetters, L.J., Hajichristidis, N., Lindner, J.S. and Mays, J.W., *J. Phys. Chem. Ref. Data*, 1994, **23**, 619, (rheological and molecular props)
[121] *Properties of Polymers: Their Correlation with Chemical Structure*, 3rd edn., (ed. D.W. van Krevelen), Elsevier, 1990
[122] Rudd, J.F. and Gurnee, E.F., *J. Appl. Phys.*, 1957, **28**, 1096, (Young's modulus)
[123] Gilmour, I., Trainor, A. and Haward, R.N., *J. Polym. Sci., Polym. Phys. Ed.*, 1974, **12**, 1939, (bulk modulus)
[124] *Plastics Additives Handbook*, (eds. R. Gaechter and H. Mueller), Macmillan, 1985, 397, (Izod impact strength)
[125] *Polymer Toughening*, (ed. C.B. Arends), Marcel Dekker, 1996, (fracture surface energy)
[126] Martin, J.R., Johnson, J.F. and Cooper, A.R., *J. Macromol. Sci., Rev. Macromol. Chem. Phys.*, 1972, **8**, 57, (mech props)
[127] Nunes, R.W., Martin, J.R. and Johnson, J.F., *Polym. Eng. Sci.*, 1982, **22**, 205, (mech props)
[128] Onogi, S., Masuda, T. and Kitagawa, K., *Macromolecules*, 1970, **3**, 109, (viscoelastic props)
[129] Martin, G., Montford, J.P., Arman, J. and Monge, P., *Rheol. Acta*, 1979, **18**, 629, (viscoelastic props)
[130] Schausberger, A., Schindlauer, G. and Janeschitz-Kriegel, H., *Rheol. Acta*, 1985, **24**, 220, (viscoelastic props)
[131] Prest, W.M. and Porter, R.S., *Polym. J. (Tokyo)*, A2, 1972, **4**, 154, (viscoelastic props)
[132] Watanabe, H., Sakamoto, T. and Kotaka, T., *Macromolecules*, 1985, **18**, 1008, (viscoelastic props)
[133] Wu, S., *Polym. Eng. Sci.*, 1985, **25**, 122, (viscoelastic props)
[134] Friedman, E.M. and Porter, R.S., *Trans. Soc. Rheol.*, 1975, **18**, 103, (viscoelastic props)
[135] Bernard, D.A. and Noolandi, J., *Macromolecules*, 1982, **15**, 1553, (viscoelastic props)
[136] Graessley, W.W., *Adv. Polym. Sci.*, 1974, **16**, 14, (viscoelastic props)
[137] Montford, J.P., Marin, G. and Monge, P., *Macromolecules*, 1986, **19**, 1979, (viscoelastic props)
[138] Chung, C.I. and Sauer, J.A., *J. Polym. Sci., Part A-2*, 1971, **9**, 1097, (viscoelastic props)
[139] Chee, K.K. and Rudin, A., *Trans. Soc. Rheol.*, 1974, **18**, 103, (viscoelastic props)
[140] Foster, R.W. and Lindt, J.T., *Polym. Eng. Sci.*, 1987, **27**, 1292, (viscoelastic props)
[141] Plazek, D.J. and O'Rourke, M.V., *J. Polym. Sci., Part A-2*, 1971, **9**, 209, (viscoelastic props)
[142] Kaufman, S., Wefling, S., Schaeffer, D. and Spiess, H.W., *J. Chem. Phys.*, 1990, **93**, 197, (viscoelastic props)
[143] Inoue, T., Hayashihara, H., Okamoto, H. and Osaki, K., *J. Polym. Sci., Polym. Phys. Ed.*, 1992, **30**, 409, (viscoelastic props)
[144] Ferry, J.D., *Viscoelastic Properties of Polymers*, 3rd edn., Wiley, 1980, (viscoelastic props)
[145] Plazek, D.J., *J. Phys. Chem.*, 1965, **69**, 3480, (viscoelastic props)
[146] Schwarzl, F.R. and Zahradnik, F., *Rheol. Acta*, 1980, **19**, 137, (viscoelastic props)
[147] Barlow, A.J., Erginsay, A. and Lamb, J., *Proc. R. Soc. London*, Part A, 1967, **298**, 481, (viscoelastic props)
[148] Nemoto, N., *Polym. J. (Tokyo)*, 1970, **1**, 485, (viscoelastic props)
[149] Hellwege, K.H., Knappe, W., Paul, P. and Semjonow, V., *Rheol. Acta*, 1967, **6**, 165, (viscoelastic props)
[150] Cogswell, F.N. and McGowan, J.C., *Br. Polym. J.*, 1972, **4**, 183, (viscoelastic props)
[151] Munstedt, H., *J. Rheol. (N.Y.)*, 1980, **24**, 847, (viscoelastic props)
[152] Shenoy, A.V. and Saini, D.R., *Rheol. Acta*, 1984, **23**, 368, (melt flow rate viscosity)
[153] Hauss, A., *Angew. Makromol. Chem.*, 1969, **8**, 73, (tensile strength)
[154] Bergen, R.L. and Wolstenholme, W.E., *Soc. Plast. Eng., Tech. Pap.*, 1960, **16**, 1235, (creep tests)
[155] Jackson, G.B. and McMillan, J.L., *Soc. Plast. Eng., Tech. Pap.*, 1963, **19**, 203, (creep tests)
[156] Turley, S.G. and Keskkula, H., *Polym. Eng. Sci.*, 1967, **7**, 1, (creep tests)
[157] *Test. Polym.*, (eds. J.V. Schmitz and W.E. Brown), Wiley Interscience, 1969, **4**, 237, (creep tests)
[158] Haslett, W.H. and Cohen, L.A., *Soc. Plast. Eng., Tech. Pap.*, 1964, **20**, 246, (fatigue testing)

Poly(styrene-*block*-butadiene-*block*-vinylpyridine)

$-[CH_2CH]-[CH_2CH=CHCH_2]-[CH_2CH]-$
    Ph   (pyridyl)

Synonyms: *Poly(ethylbenzene-co-1,3-butadiene-co-ethylenepyridine). Styrene-butadiene-vinylpyridine terpolymer. PS/PBD/PVP. Styrene-vinylpyridine-butadiene triblock polymer. Poly(styrene-butadiene-vinylpyridine) triblock copolymer. SBP block polymer. Poly(1,3-butadiene-co-ethenylbenzene-co-ethenylpyridine). Styrene/butadiene/vinylidene terpolymer rubbers*
Monomers: Styrene, 1,3-Butadiene, 2-Vinylpyridine, 4-Vinylpyridine
Material class: Copolymers, Synthetic Elastomers
Polymer Type: polybutadiene, polystyrenepolyvinyls
CAS Number:

CAS Reg. No.	Note
130196-29-1	4-vinylpyridine; block/graft
122720-06-3	4-vinylpyridine, block
135819-00-0	2-vinyl pyridine, graft
26658-72-0	4-vinylpyridine
9019-71-0	vinyl pyridine
130193-99-6	4-vinylpyridine; graft
25053-48-9	2-vinyl pyridine
109881-14-3	2-vinyl pyridine, block

Molecular Formula: $[(C_8H_8)_x(C_4H_6)_y(C_7H_7N)_z]_n$
Fragments: C_8H_8 C_4H_6 C_7H_7N
General Information: Commercial materials may be poly(styrene-*block*-butadiene-*block*-4-vinylpyridine) or Poly(styrene-*block*-butadiene-*block*-2-vinylpyridine). Commercial materials may contain either 2-vinylpyridine or 4-vinylpyridine [5,6] as terpolymer component, or a derivative such as 2-methylvinylpyridine. Block and graft copolymers (terpolymers) [10,11,12] are also available. Terpolymers used mainly as latices (40–42% total solids) and typically have the composition styrene/butadiene/vinylpyridine (15/70/15). May be regarded as styrene butadiene (30/70) rubber in which half of the styrene units have been replaced by more polar vinylpyridine units [7]
Morphology: Electron microscopy of graft (4-vinylpyridine) terpolymer shows a microphase-separated domain struct. [11]

Volumetric & Calorimetric Properties:
Density:

No.	Value	Note
1	d 0.98 g/cm^3	typical polymer latex; 3–10% of vinylpyridine replacing styrene; 40–42% total solids; solids 0.95–0.98 g cm^{-2} [5]

Surface Properties & Solubility:
Solvents/Non-solvents: Sol. CHCl$_3$; insol. MeOH (4-vinylpyridine graft copolymer) [11,12]
Surface Tension:

No.	Value	Note
1	48–54 mN/m	48–54 dyne cm^{-1}, typical latex formulation; 40–42% total solids; styrene/butadiene/vinylpyridine (15/70/15) [6]

Transport Properties:
Polymer Solutions Dilute: Viscosity 30–150 cP s (typical latex formulations, 40–42% total solids; styrene/butadiene/vinylpyridine (15/70/15)) [6]
Permeability of Gases: Permeability of (4-vinyl pyridine) graft copolymers, i.e. 'modified SBS triblock copolymers' has been reported; such modifications reduce permeability to both O$_2$ and N$_2$ [10,12]. The effects of the degree of grafting and operating temp. have also been reported [12]. Gas permeability behaviour is related to morphological struct. [10]

Mechanical Properties:
Mechanical Properties General: Mechanical props of 4-vinylpyridine graft copolymers have been reported; correlations between modulus and tensile strength with grafting ratio have been reported [11]

Optical Properties:
Transmission and Spectra: Ir spectral data for 4-vinylpyridine graft copolymers have been reported [11]

Polymer Stability:
Polymer Stability General: Polymer latices (40–42% total solids) have excellent chemical, mechanical, storage and thermal (freeze-thaw) stability [6]

Applications/Commercial Products:
Processing & Manufacturing Routes: Terpolymer latices are produced by emulsion copolymerisation at 50–60°, using reaction systems similar to those used for SBR latices [7,8,9]. Typically potassium peroxodisulfate is used as initiator [7,8,9]. Copolymerisation takes place more rapidly than corresponding emulsion polymerisation of styrene and butadiene under similar conditions [7,9]. Graft copolymers containing 4-vinylpyridine are synth. by reaction of latter with SBS triblock copolymer using γ-irradiation; rate of grafting depends on irradiation dose [11,12]
Applications: Terpolymer latices have been developed principally for applications as rubber-to-textile adhesives, in particular for tyre cords; latices are used either alone or mixed with styrene/butadiene latices to give stronger and more fatigue-resistant rubber/textile bond than latter latex alone. Particularly useful for nylon and polyester cords and for belts requiring good rubber/textile adhesion, e.g. V-belts and conveyor belting [6], hoses [6] and adhesive systems with rayon, cotton, fibreglass and aramid fabrics. Other applications (mainly as latex formulations) include use in impact-resistant moulding compositions, printing inks, composites, laminates, fast-curing rubber blends; (glass and carbon) fibre compositions, (pigmented) coatings. Graft copolymers (2-or 4-vinylpyridine) produced by γ-irradiation are used for ion-exchange and gas-separation membranes, and various coatings

Trade name	Details	Supplier
Gen-Tac	latex formulation; (40–42% solids)	Gencorp
Pliocord	LVP-4668; latex formulation (42% solids); S/BD/VP (15/70/15). LVP-5263; latex formulation (41% solids); S/BD/VP (19/70/11). LVP-5622 latex formulation; (40–42% solids)	Gencorp
Pyratex	J1904; latex formulation (40–41.5% solids); terpolymer containing 2-vinyl pyridine	Bayer Inc.
VP-100		Goodyear

Bibliographic References
[1] *Macromolecules*, 1993, **26**, No. 24, 6338–6345
[2] J. Polym. Sci., *Part B: Polym. Phys.*, 1993, **31**, No. 9, 1075–1081
[3] *Macromolecules*, 1980, **13**, No. 6, 1670–1678
[4] *Macromolecules*, 1984, **17**, No. 11, 2325–2332
[5] Roff, W.J. and Scott, J.R., Fibres, Films, *Plastics and Rubbers*, 1971, 394, (uses, volumetric props., surface props, transport props.)

- Polystyrene, Foamed

[6] Ash, M. and Ash, I., Handbook of Plastic Compounds, *Elastomers and Resins*, VCH, 1992, 547, 604, (stability, transport props, surface props, uses)
[7] Blackley, D.C., *Polymer Latices: Science and Technology*, 1997, 320, (synth., uses)
[8] Tsailingold, V.L. et al, *Soviet Rubber Technology*, 1959, **18**(3), 6, (synth.)
[9] Frank, R.L. et al, *Ind. Eng. Chem.*, 1948, **40**, 879, (synth.)
[10] Yang, J.-M. and Hsiue, G.-H., *Angew. Makromol. Chem.*, 1995, **231**, 1, (graft copolymer, synth., transport props., uses)
[11] Hsiue, G.H. and Yeh, Z.K., *Angew. Makromol. Chem.*, 1987, **153**, 33, (graft copolymer, synth., microstruct., ir, mechanical props, surface props.)
[12] Yang, J.-M. and Hsiue, G.-H., *J. Appl. Polym. Sci.*, 1990, **41**, 1141, (graft copolymer, synth., transport props., uses)

Polystyrene, Foamed   P-523

Synonyms: *Polystyrene, Cellular. Polystyrene, Expanded. Polystyrene, Expandable. EPS*
Related Polymers: Polystyrene General information, Polystyrene Latex, Polystyrene Oriented Film and Sheet, High-impact polystyrene
Monomers: Styrene
Material class: Fibres and Films, Thermoplastic
Polymer Type: polystyrene
CAS Number:

CAS Reg. No.
9003-53-6

Molecular Formula: $(C_8H_8)_n$
Fragments: C_8H_8
Molecular Weight: For extruded products high MW polystyrene is used MW 320000. For expanded foam products medium MW polystyrene is used MW 250000 [3]
Additives: For building applications flame retardants are added which are typically organic bromine compds. e.g. hexabromocyclododecane and pentabromomonochlorocyclohexane [1,2]
General Information: Polystyrene foam is a rigid plastic material [3]. There are several types depending on the manufacturing conditions and physical props. such as foam density and thickness. Extrusion processes are used to produce extruded board and sheet products. Extruded board has a typical thickness of 25 mm and density in the range 27–40 kgm^{-3}, whereas extruded sheet is in the range 0.4–6.5 mm with density of 40–160 kgm^{-3}
Miscellaneous: Expanded polystyrene beardboard is prod. by lowmoulding of expandable polystyrene beads. This product has densities in the range 13–48 kgm^{-3}. Expanded PS loosefill has a bulk density of 4.0–4.8 kgm^{-3} and is usually prepared by a blow moulding process [1,3]. Polystyrene structural foam may be prepared by injection moulding or extrusion

Volumetric & Calorimetric Properties:
Density:

No.	Value	Note
1	d 0.032–0.16 g/cm^3	extruded foam sheet [3]
2	d 0.016–0.16 g/cm^3	moulded, expandable PS [3]
3	d 0.027–0.066 g/cm^3	extruded foam board [3]
4	d 0.008–0.03 g/cm^3	expanded, loose-fill [3]
5	d 0.7 g/cm^3	injection-moulded structural foam [4]
6	d 0.2–0.4 g/cm^3	PS bead foam [14]
7	d 0.7 g/cm^3	PS structural foam [4,37]
8	d 0.84 g/cm^3	PS structural foam, 20% (w/w) glass reinforced [4,37]

Thermal Expansion Coefficient:

No.	Value	Note	Type
1	6.3×10^{-5} K^{-1}	extruded plank, density 0.035 g cm^{-3} and 0.053 g cm^{-3}, ASTM D696 [4,9,24]	L
2	5.4×10^{-5} – 7.2×10^{-5} K^{-1}	expanded plank, densities 0.016 and 0.08 g cm^{-3}, ASTM D696 [4,9,31]	L
3	5×10^{-5} K^{-1}	PS bead foam, density 0.2–0.4 g cm^{-3} [14]	L
4	0.0001 K^{-1}	PS structural foam [5]	L

Volumetric Properties General: Density depends on processing conditions
Thermodynamic Properties General: Linear thermal expansion coefficient varies with density [4,24]. Thermal conductivity depends on the density of the foam [29,30], the thermal conductivity of the gas used as the blowing agent in the foam manufacture, the temp., the foam thickness, the amount of absorbed water [4,9,54] and the cell struct. [4]. The plasticising effect of the blowing agent on the polymer affects the thermal props. in the same way as it affects the unfoamed polymer [4]. Other thermal conductivity data have been reported [32]. Thermal conductivity is related to insulation props [4,9,54,56], and increases with time [56] and moisture absorption [54]

Thermal Conductivity:

No.	Value	Note
1	0.03 W/mK	extruded plank, density 0.035 g cm^{-3}, ASTM C177 [4,9,24]
2	0.035 W/mK	extruded sheet, density 0.096 g cm^{-3}, ASTM C177 [4,9,24]
3	0.035 W/mK	extruded sheet, density 0.16 g cm^{-3}, ASTM C177 [4,9,24]
4	0.035 W/mK	expanded plank, density 0.032 g cm^{-3} and 0.08 g cm^{-3}, ASTM C177 [4,9,31]
5	0.037 W/mK	expanded plank, density 0.016 g cm^{-3}, ASTM C177 [4,9,31]

Specific Heat Capacity:

No.	Value	Note	Type
1	1.1 kJ/kg.C	extruded plank, ASTM C351 [4,9]	P

Transition Temperature General: The max. service temp. of polymer foams cannot be defined precisely because polymer foams gradually decrease in modulus as the temp. increases as opposed to undergoing a sharp change in props. [4]. The upper value of T_{use} depends strongly on the plastic phase. When a polymer foam is manufactured this process imposes a stress on the polymer phase. This stress may tend to relax at a temp. below the heat distortion temp. of the unfoamed polymer [4]

Deflection Temperature:

No.	Value	Note
1	82–83°C	Contin, ISO 175A [16]
2	88–90°C	Marten, DIN 53458, ISO 175B [16]
3	80°C	PS bead foam [14]
4	85°C	foam sheet/PS bead board (moulded) [1]

Polystyrene, Foamed

Vicat Softening Point:

No.	Value	Note
1	90°C	DIN 53460 [16]

Transport Properties:

Permeability of Gases: The diffusion of dichlorodifluoromethane, pentane, N_2 and O_2 in extruded foams has been reported [57]. The permeability of water vapour in expanded foam and particle board has been reported [60]. Moisture vapour transmission rate 35 g m^{-1} s^{-1} G Pa^{-1} (extruded plank, density 0.035 g cm^{-3}, ASTM E96) [4,9,24], <120 g m^{-1} s^{-1} G Pa^{-1} (expanded plank, density 0.016 g cm^{-3}, ASTM E96) [4,9,31], 35–120 g m^{-1} s^{-1} G Pa^{-1} (expanded plank, density 0.032 g cm^{-3}, ASTM E96) [4,9,31], 23–35 g m^{-1} s^{-1} G Pa^{-1} (expanded plank, density 0.08 g cm^{-3}, ASTM E96) [4,9,31], 86 g m^{-1} s^{-1} G Pa^{-1} (extruded sheet, density 0.096 g cm^{-3}, ASTM E96) [4,9], 56 g m^{-1} s^{-1} G Pa^{-1} (extruded sheet, density 0.016 g cm^{-3}, ASTM E96) [4,9]. Other gas diffusion data have been reported [61]

Water Absorption:

No.	Value	Note
1	0.02 %	extruded plank, density 0.035 g cm^{-3}, ASTM C272 [4,9,24]
2	0.05 %	extruded plank, density 0.053 g cm^{-3}, ASTM C272 [4,9,24]
3	1–4 %	expanded plank, ASTM C272 [4,9,31]
4	0.2–0.27 %	24h, 23°, foam, DIN 53495 [16]
5	0.2 %	5 years, extruded [4,9,54]
6	8–30 %	5 years, moulded [4,9,54]

Gas Permeability:

No.	Gas	Value	Note
1	Chlorotrifluoromethane	426.1 cm^3 mm/(m^2 day atm)	2200 nmol (m s G Pa)$^{-1}$, 24° [3,58]

Mechanical Properties:

Mechanical Properties General: The mechanical props. of rigid polymer foams, e.g. polystyrene, are different to those of flexible foams [4]. The compressive strength and modulus, and the shear strength and modulus are affected by plastic phase composition, density, cell struct. and plastic state; the effect of gas composition is small [4,35]. PS foams do not exhibit a definite yield point, but show an increased deviation from Hooke's Law as the compressive load is increased [4,33]. The effect of density and polymer composition on the compressive props. has been reported [4,24]. Strengths and moduli and shear props. increase with decreasing temp. [4]. The tensile strength and modulus of rigid foams vary with density in a similar manner to the compressive strength and modulus [4,18,23,24]. Tensile props. are affected by the polymer composition, density and cell shape [4]. Flexural strength and modulus of PS foams are also affected by density [4]. Variations in the strength props. of expanded PS caused by the block moulding process have been reported [50]. The compressive modulus of PS bead foam depends on temp. and is very low at high temps. [14]

Tensile (Young's) Modulus:

No.	Value	Note
1	10–28 MPa	PS bead foam [14]

Flexural Modulus:

No.	Value	Note
1	41 MPa	extruded plank, density 0.035 g cm^{-3}, ASTM D790 [4,9,24]
2	9–26 MPa	expanded plank, density 0.032 g cm^{-3}, ASTM D790 [4,9,31]
3	1400 MPa	PS structural foam [4,37]
4	5200 MPa	PS structural foam, 20% glass reinforced [4,37]
5	1600 MPa	20°, PS structural foam [5]

Tensile Strength Break:

No.	Value	Note	Elongation
1	2.1–9.1 MPa	ASTM D882 [31,36]	2–8%
2	2.07–6.89 MPa	[14]	1–2%
3	0.517 MPa	extruded plank, density 0.035 g cm^{-3}, ASTM D1623 [4,9,24]	
4	0.145–0.193 MPa	expanded plank, density 0.016 g cm^{-3}, ASTM D1623 [4,9,31]	
5	0.31–0.379 MPa	expanded plank, density 0.032 g cm^{-3}, ASTM D1623 [4,9,31]	
6	1.02–1.186 MPa	expanded plank, density 0.08 g cm^{-3}, ASTM D1623 [4,9,31]	
7	2.07–3.45 MPa	extruded sheet, density 0.096 g cm^{-3}, ASTM D1623 [4,9]	
8	4.137–6.9 MPa	extruded sheet, density 0.16 g cm^{-3}, ASTM D1623 [4,9]	
9	12.4 MPa	PS structural foam [4,37]	
10	34.5 MPa	PS structural foam, 20% glass reinforced [4,37]	
11	15–17 MPa	DIN 53455 [16]	14–21%
12	22 MPa	20°, PS structural foam [5]	2%

Flexural Strength at Break:

No.	Value	Note
1	1.138 MPa	extruded plank, density 0.035 g cm^{-3}, ASTM D790 [4,9,24]
2	0.379–0.517 MPa	expanded plank, density 0.032 g cm^{-3}, ASTM D790 [4,9,31]
3	34–37 MPa	DIN 53452 [16]
4	31 MPa	PS structural foam [4,37]
5	58.6 MPa	PS structural foam, 20% glass reinforced [4,37]

Compressive Modulus:

No.	Value	Note
1	10.3 MPa	extruded plank, density 0.035 g cm^{-3}, ASTM D1621 [4,9,24]
2	3.4–14 MPa	expanded plank, density 0.032 g cm^{-3}, ASTM D1621 [4,9,31]

– **Polystyrene, Foamed**

Elastic Modulus:

No.	Value	Note
1	10.3 MPa	extruded plank, density 0.035 g cm^{-3}, ASTM C273 [4,9,24]
2	7.6–11 MPa	expanded plank, density 0.032 g cm^{-3}, ASTM C273 [4,9,31]

Compressive Strength:

No.	Value	Note
1	0.31 MPa	10% deflection, extruded plank, density 0.035 g cm^{-3}, ASTM D1621 [4,9,24]
2	0.862 MPa	10% deflection, extruded plank, density 0.053 g cm^{-3}, ASTM D1621 [4,9,24]
3	0.09–0.124 MPa	10% deflection, expanded plank, density 0.016 g cm^{-3}, ASTM D1621 [4,9,31]
4	0.207–0.276 MPa	10% deflection, expanded plank, density 0.032 g cm^{-3}, ASTM D1621 [4,9,31]
5	0.586–0.896 MPa	10% deflection, expanded plank, density 0.08 g cm^{-3}, ASTM D1621 [4,9,31]
6	2.07–8.27 MPa	max., PS bead foam, density 0.2–0.4 g cm^{-3} [14]
7	0.29 MPa	10% deflection, extruded sheet, density 0.096 g cm^{-3}, ASTM D1621 [4,9]
8	0.469 MPa	10% deflection, extruded sheet, density 0.16 g cm^{-3}, ASTM D1621 [4,9]

Impact Strength: The relationship between the impact strength of the foam and that of the polymer has been reported [4,44]. The impact characteristics of foamed plastics have been reported [45]
Mechanical Properties Miscellaneous: The hydrostatic compression curves for PS foams show three stages of deformation; bending of the cell walls (low pressure); buckling of the walls; contact between the cell walls (high pressure) [62]
Hardness: Rockwell M30 (PS structural foam) [5]
Failure Properties General: Shear strength 241 kPa (extruded plank, density 0.035 g cm^{-3}, ASTM C273) [4,9,24]. The min. load required to cause long-term creep in moulded PS increases with increasing foam density 50 kPa (density 0.016 g cm^{-3}); 455 kPa (density 0.16 g cm^{-3}) [4,38]. Shear strength 241 kPa (expanded plank, density 0.032 g cm^{-3}, ASTM C273) [4,9,31]. Preloading of rigid foams reduces tensile modulus and affects the dependency of absorbed energy on the compressive stress [39]. The relationship between the fatigue and creep behaviour of PS foams and those of the polymer has been reported [4,44]. The stress relaxation of PS foams and its relation to the cushioning behaviour of the foams has been reported [45,46,47]
Fracture Mechanical Properties: The creep characteristic of polymer foams is important in structural applications. The percentage deformation against time under various static loads has been reported [4,38]. Fracture toughness of PS bead foams depends on the processing conditions and increases with moulding time, steam pressure, temp. and decreasing foam density (increasing bead fusion) [48]. However, brittle interbead failure also increases with decreasing toughness [49]
Izod Notch:

No.	Value	Notch	Note
1	30 J/m	Y	20°, PS structural foam [5]

Izod Area:

No.	Value	Notch	Note
1	11–16 kJ/m^2	Unnotched	DIN 53453 [16]

Electrical Properties:
Electrical Properties General: PS foams have good electrical insulation props. [4]
Dielectric Permittivity (Constant):

No.	Value	Frequency	Note
1	1.02		expanded plank, ASTM D1673 [4,9,31]
2	1.27		extruded sheet, density 0.096 g cm^{-3}, ASTM D1673 [4,9]
3	1.28		extruded sheet, density 0.16 g cm^{-3}, ASTM D1673 [4,9]
4	<1.05		max., extruded plank, ASTM D1673 [4,9,24]
5	3	1 kHz	PS structural foam [5]

Dielectric Strength:

No.	Value	Note
1	0.1 kV.mm^{-1}	PS structural foam [5]

Dissipation (Power) Factor:

No.	Value	Frequency	Note
1	0.0007		expanded plank [4,9,31]
2	0.00011		extruded sheet, density 0.096 g cm^{-3} [4,9]
3	0.00014		extruded sheet, density 0.16 g cm^{-3} [4,9]
4	<0.0004		max., extruded plank [4,9,24]
5	0.0005	1 kHz	PS structural foam [5]

Optical Properties:
Optical Properties General: The penetration of visible light through PS foam approximates to the Beer-Lambert Law [4,24]
Transmission and Spectra: Ir spectral data have been reported [32]
Molar Refraction: Complex refractive index for PS foams has been reported [32]
Volume Properties/Surface Properties: Colour white

Polymer Stability:
Thermal Stability General: Polymer beads collapse at approx. 110–120°. The collapsed beads melt at 160° and start to vaporise above 275° (approx.). Complete volatilisation occurs at 460–500°. The heat of degradation of expanded PS is estimated to be 912 J g^{-1} [51]
Upper Use Temperature:

No.	Value	Note
1	74°C	max. service temp., extruded plank [4,9,24]
2	74–80°C	max. service temp., expanded plank [4,9,31]
3	77–80°C	max. service temp., extruded sheet [4,9]

Decomposition Details: Decomposition temp. 265°; 50% weight loss occurs at 355° [72]. For flame retardant material decomposition temp. 320°; 50% weight loss at 381° [72]

Flammability: Continuous ingress of oxygen into PS foams is made easier by their open, cellular struct. and burning tends to be more rapid than in solid materials [63]. The fire resistance of PS foams can be improved by incorporating appropriate chemical additives and/or fire retardant interlayers [2,18,64,65,66,67,68,69,70], e.g. halogenated aliphatic, alicyclic and aromatic compounds, phosphorus compounds and inorganic diluents and fillers [65]. The combustion rate of hexabromocyclohexane treated PS foam brands is 100–160 mm min^{-1} and the extinguishing time 10–50 s [68]. The smoke and gas evolution rates of PS foams have also been reported [71]. Flammability tests for PS foams have been reported [73]

Environmental Stress: Has poor resistance to uv radiation [31,36]. Outdoor exposure of expanded, extruded PS foam results in rapid discoloration and embrittlement of the surface, increased uv absorbance, and the formation of a possibly protective yellow surface layer, the thickness of which varies with time. Yellowing is accompanied by marked decreases in the average MW and tensile strength of the material. The protective yellow surface layer decreases, but does not eliminate, the effects of degradation in the underlying foam layers. Although initially the rate of deterioration is faster in air, in the long-term the rate of deterioration is greater in sea water, which is due to the continuous loss of the yellowed surface [52,53]. Outdoor exposure of PS insulation increases its thermal conductivity [54,56]

Chemical Stability: Has good resistance to strong acids and strong alkalis. Has fair resistance to greases, and poor resistance to organic solvents [31,36]

Biological Stability: The resistance of cellular polymers to rot, mildew and fungus can be related to the amount of moisture that can be absorbed by the foam [55]. Open-celled foams are much more likely to support growth than are closed cell-foams [4,9]. Very high temps. and humidity are needed for the growth of microbes on plastic foam [4,9]

Recyclability: PS foam products can be recycled [40,41,42,43]. The critical props. of PS foams are unaffected by the recycling process [43]

Applications/Commercial Products:

Processing & Manufacturing Routes: Extruded foam sheet is prepared using polystyrene resin, blowing agent, nucleating agent, colour concentrate (optional) and regrind. A high MW unplasticised PS resin with MW $\leq$ 320000 is used [3]. The most common blowing agents are aliphatic hydrocarbons (potential fire hazard) and partially halogenated chlorofluorocarbons. Carbon dioxide and nitrogen may also be used [27]. For aliphatic hydrocarbon blowing agents, a blend of citric acid and sodium bicarbonate is typically used as nucleating agent and for chlorofluorocarbons a fine particle size talc. For food applications all ingredients must meet the apllicable government regulations. Rolls of PS foam sheet are stored and allowed to age for 2–5 days. During ageing blowing agent diffuses out of the cells at a slower rate than air diffuses into the cells. The increased cell pressure allows the proper concentration of blowing agent and air in the cells for optimum postexpansion of the sheet when it is reheated before thermoforming. The PS foamed sheet is thermoformed to produce the desired products. Polystyrene extruded foam board is prod. in a similar way to extruded foam sheet. Additives to control cell nucleation, extrusion performance, colour and flammability behaviour are used. Halogenated aromatic, aliphatic and cycloaliphatic hydrocarbons are used to increase the foams resistance to ignition. Particulate inorganics e.g. silica or clay or organics e.g. indigotin may be used as nucleating agents. Typical blowing agents are CFC's, hydrocarbons, halocarbons or mixtures of these. Expanded PS beadboard is manufactured from preformed expandable beads or granules prepd. by suspension polymerisation. The polymer in the expandable beads is a styrene polymer or copolymer with MW $\leq$ 250000 containing a minimal amount of volatile material or additives. Polystyrene, styrene-acrylonitrile copolymer and styrene-maleic anhydride copolymers may be used. Blowing agents for expandable foam are hydrocarbons and chlorofluorocarbons. Other additives may be used to change the flammability of the final expanded objects; to control the cell size of the foam, and to control the cohesion of the expanded beads to themselves in the pre expansion step. Expanded PS loose-fill packaging materials are prod. by suspension polymerisation with blowing agent incorporated into the polymer during the polymerisation or by extrusion followed by multiple steam expansions. Expansion with steam produces a low-density foam in the shape of an S, 8 or a hollow shell. Pentane, CFC-11 or a mixture of these may be used as a blowing agent. Two or three expansions are carried out using steam with at least one day of ageing after each expansion. Stabilisation is accomplished by cooling the polymer phase below its glass-transition temp. during the expansion process. Structural foam may be prod. by injection moulding or by extrusion. The process reduces the density of the solid polymer by 25 to 30%. Decomposable chemical blowing agents e.g. azodicarbonamide give a high density foamed product. Medium-density products can be prod. with a physical or chemical blowing agent, or a mixture of both. Most PS structural foam is prod. from impact modified polystyrene or high-impact polystyrene. Composite syntactic foam [1,19], using expandable polystyrene or styrene-acrylonitrile copolymer particles (in either the unexpanded or prefoamed state) are mixed with a resin or a resin containing a blowing agent which has a large exotherm during curing. The mixture is then placed in a mould and the exotherm from the resin cure causes the expandable particles to foam and squeeze the resin or foamed matrix to the surface of the moulding

Applications: Extruded foam board is used as an insulating material in building construction, due to high insulation value per unit thickness, high dimensional stability, excellent resistance to moisture and freeze-thaw deterioration, excellent compression strength and good handling props. (non-flammable and non-toxic). The main construction applications are residential and industrial sheathing and roofing in commercial buildings. The use of extruded foam as a highway underlayment to prevent frost damage to roads and pavements has been successfully tried out in the USA, Canada and Japan [1]. Extruded foam products may be used as perimeter and floor insulation under concrete and combined plaster base and insulation for walls. It may also be used for floral, decorative and buoyancy products. Foam sheet is primarily used as a packaging material. It is often thermoformed into diposable packaging (egg cartons, produce trays; fast-food containers, dinnerware and drinking cups) Foam sheet has the advantages for disposable packaging, reduced weight, high flexural modulus, low water absorption and low cost. Expanded foam may be used as billets in the building and construction industry and for buoyancy and decorative applications, due to good insulating value, light weight, low water pick up and low cost. Expanded foam can also be used as backing e.g. on aluminium siding to provide heat and sound insulation. Expanded foam has been used in Europe in the design of sound-absorbing floors but cost, flammability and cleaning difficulties have prevented market penetration in this area. Expanded shapes are used as loose-fill packaging due to good shock absorbency, excellent resiliency, light weight and reuseable. Common shapes include S or 8 shapes and rounded discs. Expanded beads may be mixed with concrete for light-weight masonry structs. Expanded spheres may also be used as stabilisers for explosives. Foam insulation is used in refrigeration for large commercial refrigerators, freezers and cold storage areas including cryogenic equipment and gas tanks. Foam is used in load bearing sandwich panels for low temp. applications, as solid propellant for rocket motors and for low frequency electrical insulation. Foam is used where cost and moisture resistance are important. Expandable bead-foam is used for encapsulation of electronic packages. The advantages of this material are low density, low modulus, good vibration and thermal shock resistance and it is solvent removable, but props. depend on the degree of bead fusion and the residual solvent content. A high temp. polystyrene bead foam is also used for encapsulation. Injection moulded structural foam is used widely for high-density articles (picture frames, furniture appliances, housewares, utensils, toys, pipes and fittings). Impact modified polystyrene (HIPS) is the most common structural foam, and is used as a replacement for wood, metals and solid plastics. Injection moulded foam may also be used for display and novelty pieces. Expandable polystyrene can be steam-moulded

– Polystyrene, Foamed

into useful, light-weight, low cost, closed cell foam shapes (beverage cups, packaging components, picnic chests, ice-buckets and insulation board). Insulation prepd. from expanded polystyrene beads has different props. to that prepd. from extruded foam board. It has inferior insulation value, moisture resistance and other physical props., though the cost of expanded PS beadboard is less than extruded PS foam board. Polystyrene composite syntactic foam (Voraspan, an expandable PS in flexible polyurethane foam matrix) is used for cushioning appliactions such as automobile crash pad and moulded furniture

Trade name	Details	Supplier
Certifoam	extruded foam insulation	Minnesota Diversified Products Inc.
Denistyr		Vamp-Technologies S.p.A.
Dylite	construction and insulation foam	ARCO Chemical Company
Edistir	sheet and film	Enichem
Foamulak	extruded foam insulation	US Gypsum and Condec Corp.
Huntsman	sheet and film	Huntsman Chemical
Koplen	PS cellular board and expanded polystyrene	Div. Plasty Kauouk SP
Krasten		Kaucuk A.S.
Lacqrene		Elf Atochem (UK) Ltd.
Ladene		SABIC Marketing Ltd.
Lastirol		LATI Industria Thermoplastics S.p.A.
Lubricomp		LNP Engineering
Pelaspan Pac	expanded loose-fill packaging (S-shaped)	Dow
Poly-Star		Idemitsu Chemicals
Polystyrene		BP Chemicals
Polystyrol		BASF
Poron	expanded polystyrene	Metal Closures Rosslite Ltd.
RTP 400		RTP Company
Solarene		Dongbu Corp.
Stat-Kon		LNP Engineering
Styrocell	expandable PS	Shell Chemicals
Styrofil	structural foam	Wilson-Fiberfil
Styrofoam	extruded foam insulation	Dow
Styron	general purpose and high impact	Dow
Styropoc	expanded beads	BASF
Thermo-comp		LNP Engineering
Tyril Foam	copolymer foam used in buoyancy products	Dow
Vampstyr		Vamp-Technologies S.p.A.
Vestyron	expandable foam	Hüls AG
Voraspan	expandable polystyrene in flexible polyurethane foam matrix	Dow
	light weight construction materials extruded foam	Kubota Kk
	composite building materials containing foam	Niioka Yoshio
	fire retardant foam containing polystyrene	Dainippon Ink and Chemicals
	blow-moulded foam	Dau Kako Kk
	expandable foam	Dyfoam
	sheet and film	Foam Plus Ltd.
	roof insulation	Arms Technical Engineering
	expanded polystyrene	Mitsubishi Chemical
	foamable pellets	Dainippon Ink and Chemicals
	construction and insulation foam	Polymers Inc.
	expandable foam	Koppers Co. Ltd.
	foam heat insulation panels	Kanegafuchi Chemical Ind.
	fireproofed foam mouldings	Hoechst
	composite	Jsp Corp.
	foamed, laminated sheets	Mitsubishi Plastics Ltd.
	foamed sheet	Diacel Chem. Ind.
	cellular PS sheet laminated with polyolefin film (HDPE)	Sekisui Plastics

Bibliographic References

[1] *Kirk-Othmer Encycl. Chem. Technol.*, 3rd edn., vol. 21 (ed. M. Grayson), Wiley Interscience, 1978, 801
[2] Kuryla, W. and Papa, A., *Flame Retard. Polym. Mater.*, Marcel Dekker, 1973, (flame retardants)
[3] *Encycl. Polym. Sci. Eng.*, 2nd edn., vol. 16 (ed. J.I. Kroschwitz), John Wiley and Sons, 1985, 193
[4] *Eur. Pat. Appl.*, 1995, 665 195, (foam concrete)
[5] Ravindrarajah, R.S. and Tuck, A.J., *Cem. Concr. Compos.*, 1994, **16**, 273, (foam concrete)
[6] *Jpn. Pat.*, 1995, 07 144 990
[7] Fossey, D.J. and Smith, C.H., *Electron. Packag. Prod.*, 1984, **24**, 110, (use)
[8] *New Horizons in Plastics*, (ed. J. Murphy), WEKA, 1991
[9] Kennedy, R.N., *Handbook of Foamed Plastics*, (ed. R.J. Bender), Lake Publishing Co., 1965
[10] *Plast. Foams*, (eds. K.C. Frisch and J.H. Saunders), Vol. 1, Marcel Dekker, 1973, 538
[11] Brookes, J.B. and Rey, L.G.J., *J. Cell. Plast.*, 1973, **9**, 232
[12] Rubens, L.C., *J. Cell. Plast.*, 1965, **1**, 3
[13] Martino, R., *Mod. Plast.*, 1978, **55**, 34
[14] Burt, J.G., *J. Cell. Plast.*, 1979, **15**, 158
[15] Benning, C., *Plast. Foams*, Vol. 1, John Wiley and Sons, 1969
[16] Griffin, J.D. and Skochpole, R.E., *Engineering Design for Plastics*, (ed. E. Baer), Reinhold Publishing Co., 1964
[17] *Struct. Foam*, (ed. B. Wendle), The Society of the Plastics Industry, Structural Foams Div., 1982
[18] *Mod. Plast.*, (ed. D. Brownbill), 1986, **63**, 64, (uses)
[19] Fehse, H.F., *FCKW - Ausstieg Wohin?*, *Beitr. Dechema-Fachgespraechs Umweltschutz*, 8th edn., (eds. D. Behrens and J. Wiesner,), 1990, 41
[20] *U.S. Pat.*, 1993, 22 180, (use)
[21] *Encycl. Polym. Sci. Eng.*, 2nd edn., (ed. J.I. Kroschwitz), John Wiley and Sons, 1985, **3**, 38
[22] *The Materials Selector*, 2nd edn., (eds. N.A. Waterman and M.F. Ashby), Chapman and Hall, 1997, (uses, processing, tradenames)
[23] Gardner, W., Cooke, E. and Cooke R., *Chemical Synonyms and Tradenames*, 8th edn., Technical Press Ltd., 1978

[24] *Chemical Industry Directory,* Tonbridge, Benn Business Information Services, 1988
[25] *UK Kompass Register,* Reed Information Services, 1996
[26] *Kirk-Othmer Encycl. Chem. Technol.,* 3rd edn., (ed. M. Grayson), Wiley Interscience, 1978, **11**, 86, (foam, sound insulation)
[27] *Kirk-Othmer Encycl. Chem. Technol.,* 3rd edn., (ed. M. Grayson), Wiley Interscience, 1978, **18**, 101, (foam, thermal insulation)
[28] *Encycl. Polym. Sci. Eng.,* 2nd edn., (ed. J.I. Kroschwitz), John Wiley and Sons, 1985, **5**, 804, (use)
[29] Skochdopole, R.E., *Chem. Eng. Prog.,* 1961, **57**, 55, (thermal props.)
[30] Guenther, F.O., *SPE Trans.,* 1962, **2**, 243, (thermal props.)
[31] Agranoff, J., Guide to Plastics, *Property and Specification Charts,* (ed. W.A. Kaplan), McGraw Hill, 1977
[32] Kuhn, J., Ebert, H.-P., Arduini-Schuster, M.C., Buttner, D. and Fricke, J., *Int. J. Heat Mass Transfer,* 1992, **35**, 1795, (thermal props.)
[33] Phillips, T.L. and Lannon, D.A., *Br. Plast.,* 1961, **34**, 236, (mechanical props.)
[34] Nussinovitch, A., Cohen, G. and Peleg, M., *J. Cell. Plast.,* 1991, **27**, 527, (compression characteristics)
[35] Meinecke, E.A. and Clark, R.C., *Mechanical Properties of Polymeric Foams,* Technomic Publishing Co., 1972, (mechanical props.)
[36] *Kirk-Othmer Encycl. Chem. Technol.,* 3rd edn., Vol.10, (ed. M. Grayson), Wiley Interscience, 1978, 216
[37] Eakin, J.L., *Plast. Eng. (N.Y.),* 1978, **34**, 56, (mechanical props.)
[38] Brown, W.B., *Plastics Progress,* Iliffe, 1959, 149, (mechanical props.)
[39] Ozbul, M.H. and Mark, J.E., *Polym. Eng. Sci.,* 1994, **34**, 794, (mechanical props.)
[40] Kline, C.H., *Chem. Ind. (London),* 1989, **14**, 440, (recycling)
[41] Wiedmann, W. and Wohlfahrt-Laymann, H., *Kunststoffe,* 1992, **82**, 934, (recycling)
[42] Hohwiller, F., *Betonwerk + Fertigteil-Tech.,* 1992, **58**, 86, (recycling)
[43] Kampouris, E.M., Papaspyrides, C.D. and Lekakou, C., *Polym. Eng. Sci.,* 1988, **28**, 534, (recycling)
[44] *Engineering Guide to Structural Foams,* (ed. B.C. Wendle), Technomic Publishing Co., 1976
[45] Jemian, W.A., Jang, B.Z. and Chan, J.S., *ASTM Spec. Tech. Publ.,* 1987, **936**, 117, (mechanical props.)
[46] Ramon, O. and Miltz, J., *J. Appl. Polym. Sci.,* 1990, **40**, 1683, (mechanical props.)
[47] Ramon, O. and Miltz, J., *Polym. Eng. Sci.,* 1990, **30**, 129, (mechanical props.)
[48] Stupak, P.R. and Donovan, J.A., *Mater. Res. Soc. Symp. Proc.,* 1991, **207**, 157, (mechanical props.)
[49] Stupak, P.R., Frye, W.O. and Donovan, J.A., *J. Cell. Plast.,* 1991, **27**, 484, (mechanical props.)
[50] Sarlin, J., Jarvella, P., Jarvela, P. and Tormala, P., *Plast. Rubber Process. Appl.,* 1987, **7**, 207, (mechanical props.)
[51] Mehta, S., Biederman, S. and Shivkumar, S., *J. Mater. Sci.,* 1995, **30**, 2944, (mechanical props.)
[52] Andrady, A.L. and Pegram, J.E., *Polym. Fiber Sci.: Recent Adv.,* 1992, 287, (weathering)
[53] Andrady, A.L. and Pegram, J.E., *J. Appl. Polym. Sci.,* 1991, **42**, 1589, (weathering)
[54] Dechow, F.J. and Epstein, K.A., *ASTM Spec. Tech. Publ.,*(Thermal Transmission Measurements of Inuslation), ASTM, 1978, (insulation props.)
[55] Cooper, A., Plast. Inst., *Trans.,* 1958, **26**, 299, (biological stability)
[56] Yarbrough, D.W., Graves, R.S. and Christian J.E., *ASTM Spec. Tech. Publ.,* (Insul. Mater. Test Appl.), **1116**, 214, (insulation props.)
[57] Schwartz, N.V., Bomberg, M.T. and Kumaran, M.K., *J. Therm. Insul.,* 1989, **13**, 48, (gas permeability)
[58] Ol'Shevskii, M.V., Kukharskii, Y.M., Mitrofanov, A.D. and Mamonto, V.M., *Khim. Khim. Tekhnol.,* 1987, **30**, 89, (gas permeability)
[59] Gigna, G., Merlotti, M. and Castellani, L., *Cell. Polym.,* 1986, **5**, 241, (permeability)
[60] Galbraith, G.H., Comm. Eur. Communities, *(Rep.),* EUR, **14349**, (permeability)
[61] Shankland, I.R., *ASTM Spec. Tech. Publ.,* **1030**, 174, (diffusion)
[62] Fortes, M.A., Fernandes, J.J., Serralheiro, I. and Rosa, M.E., *J. Test. Eval.,* 1989, **17**, 67, (hydrostatic compression)
[63] Birley, A.W., Haworth, B. and Batchelor, J., Physics of Plastics: Processing, Properties and Materials Engineering Engineering, Hanser/Oxford University Press, 1991
[64] *Fire and Cellular Polymers,* (eds. J. Buist, S.J. Grayson and W.D. Woolley), Applied Science, 1987, (flammability)
[65] Tesoro, G.C., J. Polym. Sci., Part D: Macromol. Rev., 1978, **13**, 283, (flame retardants)
[66] Katz, H.S. and Milewski, J.V., *Handb. Fillers Reinf. Plastics,* H.S. Katz and J.V. Milewski, Van Nostrand Reinhold, 1978
[67] Gouininlock, E.V., Porter, J.F. and Hindersinn, R.R., *J. Fire Flammability,* 1971, **2**, 206, (flammability)
[68] Korzhova, I.T. and Sokolova, N.V., *Plast. Massy,* 1990, **11**, 79, (flammability)
[69] Guttmann, H. and Sarsour, N., Polym. Prepr. (Am. Chem. Soc., Div. Polym. Chem.), 1989, **30**, 526, (flame retardants)
[70] Ku, P.L., *Adv. Polym. Technol.,* 1989, **9**, 57, (flame retardants)
[71] Babrauskas, V., *Fire Saf. J.,* 1989, **14**, 135, (flame retardants)
[72] Kalousstian, J., Arfi, C., Pauli, A.M. and Pastor, J., *J. Therm. Anal.,* 1993, **39**, 447, (flammability, thermal anal.)
[73] Kim, A., *ASTM Spec. Tech. Publ.,* (Fire and Flammability of Furnishings and Contents of Buildings), **1233**, 186, (flammability)

Polystyrene, Glass filled P-524

Synonyms: *Poly(1-phenylethylene). Poly(vinyl benzene). Poly(-phenyl ethene). Poly(phenyl ethylene). Poly(1-vinyl benzene). Poly(ethenyl benzene).* PS
Related Polymers: Polystyrene, High-impact polystyrene, Polystyrene, wood filled
Monomers: Styrene
Material class: Thermoplastic
Polymer Type: polystyrene
Additives: Flame-retardant, anti-static and flow-enhanced grades are available.
General Information: Glass fibre and beads are used as fillers. High-impact grades also available.

Volumetric & Calorimetric Properties:
Density:

No.	Value	Note
1	d 1.1–1.27 g/cm^3	10–30% filler, ASTM D792 [1]

Mechanical Properties:
Tensile (Young's) Modulus:

No.	Value	Note
1	2965–9998 MPa	10–30% filler, ASTM D638 [1]
2	4137–8274 MPa	10–30% filler, high-impact grade, ASTM D638 [1]

Flexural Modulus:

No.	Value	Note
1	4826–9308 MPa	10–30% filler, ASTM D790 [1]
2	3792–7584 MPa	10–30% filler, high-impact grade, ASTM D790 [1]

Tensile Strength Break:

No.	Value	Note	Elongation
1	59–79 MPa	10–30% filler, ASTM D638 [1]	1.0–2.5%
2	34–53 MPa	10–30% filler, high-impact grade, ASTM D638 [1]	1.0–2.5%

Flexural Strength Yield:

No.	Value	Note
1	97–124 MPa	10–30% filler, ASTM D790 [1]
2	50–76 MPa	10–30% filler, high-impact grade, ASTM D790 [1]

Izod Notch:

No.	Value	Notch	Note
1	48 J/m	Y	10–30% filler, ASTM D256, 3.2 mm section [1]

2	160–187 J/m	N	10–30% filler, ASTM D4812, 3.2 mm section [1]
3	53–80 J/m	Y	10–30% filler, high-impact grade, ASTM D256, 3.2 mm section [1]
4	160 J/m	N	10–30% filler, high-impact grade, ASTM D4812, 3.2 mm section [1]

Polymer Stability:
Flammability: HB at 1.5mm (ASTM D635) [1]

Applications/Commercial Products:
Processing & Manufacturing Routes: Processing for injection molding [1]:
Injection pressure: 69–103 Mpa
Melt temperature: 210–249°C
Mold temperature: 38–66°C
Drying: 2 hrs at 82°C

Applications/Commercial Products:
Mould Shrinkage (%):

No.	Value	Note
1	0.1–0.4%	10–30% filler, ASTM D955 [1]

Applications:

Trade name	Details	Supplier
Dylark		Nova Chemicals
Hiloy		A. Schulman Inc.
Lastirol		Lati
RTP	401 (10% filler), 405 (30% filler)	RTP Co.
Thermocomp		LNP (GE Plastics)
	401HI (10% filler, high impact)	
	415HI (30% filler, high impact)	

Bibliographic References
[1] *RTP Co. datasheet*, http://www.rtpcompany.com/info/data/index.htm, 2006

Poly(styrene-*block*-isoprene-*block*-2-vinylpyridine)

P-525

Synonyms: *Styrene-isoprene-2-vinylpyridine triblock copolymer. ISP triblock copolymer. Poly(ethenylbenzene-block-2-ethenylpyridine-block-2-methyl-1,3-butadiene). Poly(ethenylbenzene-block-2-methyl-1,3-butadiene-block-2-ethenylpyridine)*
Monomers: 2-Methyl-1,3-butadiene, Isoprene, Styrene, 2-Vinylpyridine, 4-Vinylpyridine, 2-Vinylpyridine
Material class: Copolymers
Polymer Type: polydienes, polybutadienepolystyrene
CAS Number:

CAS Reg. No.	Note
143106-01-8	Block
53895-47-9	Styrene

Morphology: The variation of morphology with composition has been reported. Four different morphologies are identified in a series of isoprene-styrene-vinylpyridine polymers where the end block polymer volume fractions are kept constant and the middle block (polystyrene) is varied: a three-phase, four-layer lamellar struct., an ordered tricontinuous double diamond struct., a cylindrical struct. and a spherical struct. The ordered tricontinuous double-diamond struct. is also found in styrene-isoprene-vinylpyridine polymers. More recent work, however, indicates that the double-diamond morphology of isoprene-styrene-vinylpyridine terpolymers is actually gyroid [1,2]

Applications/Commercial Products:
Processing & Manufacturing Routes: Can be prepared by anionic polymerisation [1]

Bibliographic References
[1] Mogi, Y.; Kotsuji, H.; Kaneko, Y.; Mori, K. *et al*, *Macromolecules*, 1992, **25**, 5408, 5412, (synth, morphology)
[2] Matsen, M.W., *J. Chem. Phys.*, 1998, **108**, 785, (struct)

Polystyrene latex

P-526

Synonyms: *Polystyrene emulsion polymers. Polystyrene polymer colloids*
Related Polymers: Polystyrene-General information, Styrene-butadiene copolymer, Styrene-acrylonitrile copolymer, (SAN), High-impact polystyrene, ABS
Monomers: Styrene
Material class: Emulsions and Gels, Thermoplastic
Polymer Type: polystyrene
CAS Number:

CAS Reg. No.
9003-53-6

Molecular Formula: $(C_8H_8)_n$
Fragments: C_6H_8
Molecular Weight: Polystyrene prod. by emulsion polymerisation has high MW. The size or diameter of the latex particles is usually used to identify latex products
Additives: Polystyrene latex may be stabilised by surface active agents (surfactants) or by an adsorbed layer of a polymeric stabiliser
General Information: Polystyrene latex is generally a milky white suspension of small, discrete particles of polymer dispersed in aq. soln. When the diameter of the particles is a similar order of magnitude to the wavelength of light, the scattering of white light gives an irridescent appearance [2]. When the latex particle diameter is very small (approx. 30 nm or less) a coloured dispersion may be observed [2]
Miscellaneous: The PS latex particle size affects both its physical props. and stability. The PS latex particle size may be determined by electron microscopy and light scattering techniques [12]. Large particles, with diameters greater than 1 μm may also have their particle sizes measured by optical microscopy and electronic counting techniques e.g. the Coulter Counter [10]. PS latex particles are observed as solid spheres using electron microscopy, but the particle consists of many chains of entangled polystyrene polymer with charged end groups, arising from initiator fragments, located at the particle surface. These initiator fragments may be anionic or cationic [11,12], depending on the type of initiator used. Persulfate initiator systems give rise to sulfate end groups which are liable to hydrol. producing carboxyl and hydroxyl groups in addition to sulfate. The initiator 4,4′-azobis-4-cyanopentanoic acid gives carboxyl groups which are stable to hydrol. Cationic surface groups include amidine derivs. from 2,2′-azobisisobutyramidine and 2,2′-azobis-N,N'-dimethyleneisobutyramidine initiators. The solids content of polystyrene latex prepd. by emulsion polymerisation may be up to 50% (w/w) for latex particle sizes in the range 0.1 μm–0.5 μm. However when PS latex is prepd. by emulsifier-free polymerisation the solids content is lower (typically 10% w/w) [12]

Applications/Commercial Products:

Trade name	Supplier
Estaper	Dow
	Rhone-Poulenc
	National Bureau of Standards
	Seradyn Inc.
	Interfacial Dynamics Corp.
	Polymer Laboratories
	Polysciences Inc
	Duke Scientific Corp.

Bibliographic References

[1] Bovey, F.A., Kolthoff, I.M., Medalia, A.I. and Meehan, E.J., *Emulsion Polym.*, Wiley Interscience, 1955, (general reference on emulsion polymerisation)
[2] Vanderhoff, J.W., van den Hul H.J., Tausk, R.J.M. and Overbeek, J.Th.G., *Clean Surf.: Their Prep. Charact. Interfacial Stud.*, (ed. G. Goldfinger), Marcel Dekker, 1970, (general reference on emulsion polymerisation)
[3] *Polym. Colloids*, (ed. R.M. Fitch), Plenum Press, 1971, (general reference on emulsion polymerisation)
[4] Polymer Colloids 2 (Two), *(Proc. Symp. Phys. Chem. Prop. Colloidal Part.)*, (ed. R.M. Fitch), Plenum Press, 1980, (general reference on emulsion polymerisation)
[5] NATO ASI Ser., Ser. E, No. 138, 1987, (general reference on emulsion polymerisation)
[6] *Emulsion Polym.*, (ed. I. Piirma), Academic Press, 1982, (general reference on emulsion polymerisation)
[7] Bassett, D.R. and Hamielec, A.E., *ACS Symp. Ser.*, 1981, **165**, (general reference on emulsion polymerisation)
[8] *Emulsion Polym.*, (ed. D.C. Blackley), Applied Science, 1975, (general reference on emulsion polymerisation)
[9] Eliseeva, S.S., Ivanchev, S., Kuchanov, S.I. and Lebedev, A.V., *Emulsion Polymerization and Its Applications in Industry*, Consultants Bureau, 1981, (general reference on emulsion polymerisation)
[10] Chung-Li, Y., Goodwin, J.W. and Ottewill, R.H., *Prog. Colloid Polym. Sci.*, 1976, **60**, 173, (large PS latex particles)
[11] Goodwin, J.W., Ottewill, R.H., Pelton, R., Vainello, G. and Yates, D.E., *Br. Polym. J.*, 1978, **10**, 170, (polystyrene latex review)
[12] Hearn, J., Wilkinson, M.C. and Goodall, A.R., *Adv. Colloid Interface Sci.*, 1981, **14**, 173, (polymer Colloid review)
[15] Singer, J.M., NATO ASI Ser., Ser. E, No. 138, 1987, (immunomicrospheres)
[16] Ortega-Vinuesa, J.L., Gálvez-Riuz, M.J. and Hidalgo-Alvarez, R., *Prog. Colloid Polym. Sci.*, 1995, **98**, 233, (immunomicrospheres)
[17] Quali, L., Pefferkorn, E., Elaissari, A., Pichot, C., and Mandrand, B., *J. Colloid Interface Sci.*, 1995, **171**, 276, (Latex in pregnancy test)
[18] Ugelstad, J., M&oork, P.C., Kaggerud, K.H. and Berge, A., *Adv. Colloid Interface Sci.*, 1980, **13**, 101, (large latex particles)
[19] Vanderhoff, J.W., El-Aasser, M.S. and Tseng, C.M., *J. Dispersion Sci. Technol.*, 1984, **5**, 231, (large latex particles)
[20] Sun, F. and Ruckenstien, E., *J. Appl. Polym. Sci.*, 1993, **48**, 1279, (large latex particles)
[21] El-Aasser, M.S., NATO ASI Ser., Ser. E, 1983, (Latex Cleaning Review)
[23] Vanderhoff, J.W., NATO ASI Ser., Ser. E, 1983, (polymer Colloid reference)
[24] *Encycl. Polym. Sci. Eng.*, 2nd edn. (eds. N.M. Bikales, H.F. Mark, G. Menges and C.G. Overberger), John Wiley and Sons, 1989, **10**, 779, (paper filling)
[25] Rossin, E.H., *Pulp Pap.*, 1974, **48**, 57, (paper filling)
[26] U.S. Pat., 1981, 4 282 060, (paper filling)
[27] U.S. Pat., 1981, 4 235 982
[28] Kotera, A., Furusawa, K. and Takedo, Y., *Kolloid Z. Z. Polym.*, 1970, **239**, 677, (surfactant free, polymerisation)
[29] Kotera, A., Furusawa, K. and Kudo, K., *Kolloid Z. Z. Polym.*, 1970, **240**, 837, (suractant free, polymerisation)
[30] Goodwin, J.W., Hearn, J., Ho, C.C. and Ottewill, R.H., *Br. Polym. J.*, 1973, **5**, 347, (surfactant free polymerisation)
[31] Goodwin, J.W., Hearn, J., Ho, C.C. and Ottewill, R.H., *Colloid Polym. Sci.*, 1976, **60**, 173, (surfactant free polymerisation)
[33] Vanderhoff, J.W., El-Aasser, M.S., Micale, F.J., Sudol, E.D., Tseng, C.-M., Sheu, H.-R. and Cornfield, R., *Polym. Prepr. (Am. Chem. Soc., Div. Polym. Chem.)*, 1987, **28 (2)**, 455, (space processing)
[35] Koopal, C.G.J. and Nolte, R.J.M., *Enzyme Microb. Technol.*, 1994, **16**, 402, (glucose biosensor)
[36] Kim, K.J. and Ruckenstein, E., *J. Appl. Polym. Sci.*, 1989, **38**, 441, (latex for controlled release of herbicides)
[37] U.S. Pat., 1992, 5 154 749, (latex for controlled release of herbicides)
[38] Gombooz, E., Tietz, D., Hurtt, S.S. and Chrombach, A., Electrophoresis (Weinheim, Fed. Repub. Ger.), 1987, **8**, 261, (size standards)
[39] Styring, M.G., Honing, J.A.J., Hamielec, A.E., *Liq. Chromatogr. Polym. Relat. Mater.*, 1986, **9**, 3505, (size standards)
[40] Wilkinson, M.C., Hearn, J., Karpowicz, F. and Chainey, M., *Part. Charact.*, 1986, **3**, 56, (size standards storage and handling)
[41] *Jpn. Pat.*, 1991, 03 211 565, (electrostatographic liq developer)
[42] *Encycl. Polym. Sci. Eng.*, 2nd edn. (eds. N.M. Bikales, H.F. Mark, G. Menges and C.G. Overberger), John Wiley and Sons, 1989, **2**, 512, (polymerisation)
[43] *Austrian Pat.*, 1982, 526, 106, (products containing latex for the protection of shorn sheep)
[44] Floyd, F., *Org. Coat. Plast. Chem.*, 1980, **43**, 31, (polymer coatings containing latex)
[45] U.S. Pat., 1992, 5 114 479, (newsprint inks)
[46] U.S. Pat., 1991, 5 075 358, (cement binder)
[47] U.S. Pat., 1990, 1 599 396, (manufacture of asbestos linerboard)
[48] Eur. Pat. Appl., 1987, 214 626, (hair setting compositions)
[49] PCT Int. Appl., 1993, 93 18 126, (defoamer for liq detergent during storage)

Poly(styrene oxide)   P-527

Synonyms: Poly(oxy(1-phenyl)ethylene). Poly[oxy(1-phenyl-1,2-ethanediyl)]. Poly(phenyloxirane). Poly[(epoxyethyl)benzene]
Monomers: Styrene oxide
Material class: Thermoplastic
Polymer Type: polyalkylene ether
CAS Number:

CAS Reg. No.	Note
25189-69-9	homopolymer
105390-15-6	isotactic
105761-00-0	(R)-form
101062-46-8	
105369-63-9	homopolymer, isotactic

Molecular Formula: $(C_8H_8O)_n$
Fragments: C_8H_8O
Molecular Weight: M_w/M_n 1.23; M_n typically 1600, 1450
General Information: Bright yellow crystalline solid [1,2]

Volumetric & Calorimetric Properties:
Density:

No.	Value	Note
1	d 1.118 g/cm^3	[2]
2	d^{25} 1.147–1.151 g/cm^3	[3]

Melting Temperature:

No.	Value	Note
1	149°C	[2,3]
2	162°C	[2,3]
3	179°C	[2,3]

Glass-Transition Temperature:

No.	Value	Note
1	40–41°C	[2,3]

Surface Properties & Solubility:
Solvents/Non-solvents: Sol. C_6H_6, toluene, $CHCl_3$, CCl_4 [2]. Insol. paraffins, Me_2CO, MeOH, 2-butanone

Mechanical Properties:
Failure Properties General: Max. shear stress 3 g cm^{-1} s^{-2} [2]

Optical Properties:
Molar Refraction: $[M]^{15}_D$ 121.2 (c.1 in dioxan, *S*-form). $[M]^{15}_D$ -127.2 (c.0.9 in dioxan, *R*-form) [4]

Bibliographic References
[1] Takata, T., Takuma, K. and Endo, T., Makromol. Chem., *Rapid Commun.*, 1993, **14**, 203, (general)
[2] Allen, G., Booth, C. and Hurst, S.J., *Polymer*, 1967, **8**, 385, 406, 414, (general)
[3] Lal, J. and Trick, G.S., *J. Polym. Sci., Part A-1*, 1970, **8**, 2339, (transition temps, density)
[4] Jedliński, Z., Kasperczyk, J. and Dworak, A., *Eur. Polym. J.*, 1983, **19**, 899, (mean residue rotation values)

Polystyrene/styrene-butadiene block copolymers blends   P-528

Synonyms: *Blends of polystyrene and styrene-butadiene di-block copolymers*
Related Polymers: Styrene-butadiene block copolymers
Monomers: Styrene, 1,3-Butadiene
Material class: Blends
Polymer Type: polybutadiene, polystyrene

Bibliographic References
[1] *Polymer*, 1994, **35**, No. 23, 5051-5056
[2] *Polymer*, 1987, **28**, No. 2, 244-250
[3] *Macromolecules*, 1991, **24**, No.19, 5324-5329
[4] *Macromolecules*, 1993, **26**, No.11, 2983-2986

Poly(styrene-*co*-2-vinylpyridine)   P-529

Synonyms: *Poly(2-ethenylpyridine-co-ethenylbenzene)*. *PS/P2VP*. *PS/PV2P*. *P2VPS*
Related Polymers: Poly(2-vinylpyridine)
Monomers: 2-Vinylpyridine, Styrene
Material class: Copolymers
Polymer Type: polyolefins, polystyrene
CAS Number:

CAS Reg. No.	Note
24980-54-9	styrene-*co*-2-vinylpyridine
108614-86-4	block
27476-03-5	dimer
96526-40-8	hexamer
9019-70-9	styrene polymer with vinylpyridine
176663-42-6	
113810-78-9	
129863-88-3	
143731-12-8	
118677-53-5	
129572-32-3	

Molecular Formula: $[(C_8H_8).(C_7H_7N)]_n$
Fragments: C_8H_8 C_7H_7N
Morphology: Block copolymer exhibits a mesomorphic struct. in presence of a preferential or selective solvent for one of the blocks [9]. Depending on polymer and solvent the structural element is lamellar or cylindrical [9,19]. Morphology investigated by near-field optics [25]

Volumetric & Calorimetric Properties:
Thermodynamic Properties General: Soln. deposited film surface density has been reported [2]. Thermal expansion of adsorbed diblock copolymers on mica has been reported [16]

Melting Temperature:

No.	Value	Note
1	125–175°C	random copolymer [17]
2	150–240°C	block copolymer [17]

Glass-Transition Temperature:

No.	Value	Note
1	85–100°C	linear block copolymer [6]
2	101–106°C	macrocyclic block copolymer [6]
3	102°C	random copolymer [17]
4	105°C	block copolymer [17]

Surface Properties & Solubility:
Solvents/Non-solvents: Sol. $CHCl_3$/1,4-dioxane (6:4) [4], DMF [20], pyridine [21,22], THF [21], 1,1,2-trichloroethane [21]. Slightly sol. 2-butanone [22], C_6H_6 [22]
Surface and Interfacial Properties General: Contact angle of water on deposited film has been reported. Wetting behaviour of polystyrene on PS-block-P2VP films [27] has been studied by AFM, neutron scattering and SIMS.

Surface Tension:

No.	Value	Note
1	1.2 mN/m	block copolymer [7]

Transport Properties:
Polymer Solutions Dilute: Zero shear viscosities and values for the Mark-Houwink-Sakurada equation for pyridine, 2-butanone and C_6H_6 for the block copolymer have been reported [22]. Critical concentration for microphase separation in pyridine has been reported [22]. Flory interaction parameters for block copolymers are temp. dependent [16]

Mechanical Properties:
Compressive Modulus:

No.	Value	Note
1	0.1 MPa	1000000 dyn cm^{-2}, block copolymer, calc. [7]

Miscellaneous Moduli:

No.	Value	Note	Type
1	0.3 MPa	3000000 dyn cm^{-2}, block copolymer, calc. [7]	Splay modulus

Mechanical Properties Miscellaneous: Forces between layers of adsorbed block copolymer have been reported [5,8]. Maximum stretch of 40% in lamellar configuration on silicon (block copolymer). [7] Forces between layers of adsorbed block copolymer [5,8] have been reported. Maximum stretch of 40% in lamellar configuration on silicon. [7] (block copolymer)Forces between layers of adsorbed block copolymer [5,8] have been reported. Maximum stretch of 40% in lamellar config. on silicon [7] (block copolymer)

Fracture Mechanical Properties: The fracture toughness of the glass-deutero PS-P2VP interface has been investigated [24]; the mechanism of interface failure has been assigned to chain scission or crazing followed by chain scission, depending on the areal chain densities [24]

Electrical Properties:

Electrical Properties General: Electrical anisotropy of cast films has been reported. Conductivity of quaternised film and iodine doped block copolymers has been reported. Conductivity increases parallel to film orientation [4]

Surface/Volume Resistance:

No.	Value	Note	Type
1	0.0018×10^{15} Ω.cm	quaternised film [4]	S

Optical Properties:

Transmission and Spectra: Electron micrographs of graft ABB polymers have been reported [15]. SAXS diffraction of graft ABB polymers has been reported [15]. Ir spectral data (star copolymer) have been reported [18]. Atomic force microscopy has been reported [26]. H-1 nmr spectral data (star copolymer) have been reported [18]. Transmission electron microscopy [20] (block copolymer) has been reported [21]. Electron paramagnetic resonance (epr) of charge-transfer complexes [23]. Near-field optical microscopy has been reported [25]. XPS data have been reported [1,2,3]. FTIR and SIMS spectral data have been reported [3]. Transmission electron microscopy of films has been reported [4]. Atomic force microscopy of block copolymer has been reported [7]. Cryo-electron microscopy of block copolymer has been reported [10]. Scanning force microscopy (SFM) on block copolymer has been reported [13]. Laser-induced desorption photoelectron ionisation LD/PEI mass spectrum of copolymer [14]. TEM of block copolymer has been reported [7]

Polymer Stability:

Decomposition Details: Decompostion temp. 399–407° (linear block copolymer) [6]. Decomposition temp. 372–406° (macrocyclic block copolymer) [6]

Bibliographic References

[1] Fabish, T.J. and Thomas, H.R., *Macromolecules*, 1980, **13**, 1487, (XPS spectra)
[2] Parsonage, E. Tirrell, M., Watanabe, H. and Nuzzo, R.G., *Macromolecules*, 1991, **24**, 1987, (XPS, surface density of deposited film, contact angle)
[3] Lee, H.F. and Gardella, J.A., *Polymer*, 1992, **33**, 4250, (surface analysis)
[4] Ishizu, K., Yamada, Y., Saito, R., Yamamoto, T. and Kanbara, T., *Polymer*, 1992, **33**, 1816, (electrical props)
[5] Patel, S. and Tirrell, M., Polym. Prepr. (Am. Chem. Soc., *Div. Polym. Chem.*), 1987, **28**, 42, (block copolymers)
[6] Gan, Y., Zoller, J. and Hogen-Esch, T.E., Polym. Prepr. (Am. Chem. Soc., *Div. Polym. Chem.*), 1993, **34**, 69, (synth)
[7] Liu, Y., Rafailovich, M.H., Sokolov, J., Schwarz, S.A. and Bahal, S., *Macromolecules*, 1996, **29**, 899, (block copolymer dislocations)
[8] Watanabe, H. and Tirrell, M., *Macromolecules*, 1993, **26**, 6455, (diblock copolymer)
[9] Grosius, P., Gallot, Y. and Skoulious, A., *Eur. Polym. J.*, 1970, **6**, 355, (synth)
[10] Oostergetel, G.T., Esselink, F.J. and Hadziioannou, G., *Langmuir*, 1995, **11**, 3721, (cryo-electron microscopy)
[11] Webber, R.M., van der Linden, C.C. and Anderson, J.L., *Langmuir*, 1996, **12**, 1040, (thermal expansion)
[12] Moller, M. and Lenz, R.W., *Makromol. Chem.*, 1989, **190**, 1153, (iodine complex, conductivity)
[13] Grim, P.C.M., Brouwer, H.J., Seyger, R.M., Oostergetel, G.T., et al, *Makromol. Chem., Makromol. Symp.*, 1992, **62**, 141, (scanning force microscopy)
[14] Schriemer, D.L. and Li, L., *Anal. Chem.*, 1996, **68**, 250, (laser desorption photoelectron ionisation mass spectroscopy)
[15] Matsushita, Y., Watanabe, J., Katano, F., Yoshida, Y. and Noda, I., *Polymer*, 1996, **37**, 321, (graft ABB polymers)
[16] Dai, K.H. and Kramer, E.J., *Polymer*, 1994, **35**, 157, (Flory interaction parameter)
[17] Noel, C., *Bull. Soc. Chim. Fr.*, 1967, **10**, 3733, (thermal anal)
[18] Khan, I.M., Gao, Z., Khougaz, K. and Eisenberg, A., *Macromolecules*, 1992, **25**, 3002, (star copolymer)
[19] Grosius, P., Gallot, Y. and Skoulious, A., *Makromol. Chem.*, 1969, **127**, 94, (molecular struct)
[20] Kunz, M., Moller, M., Heinrich, U.-R. and Cantow, H.-J., Makromol. Chem., *Makromol. Symp.*, 1989, **23**, 57, (electron spectroscopic studies)
[21] Kunz, M., Moller, M. and Cantow, H.-J., Makromol. Chem., *Rapid Commun.*, 1987, **8**, 401, (block copolymer)
[22] Yamaguchi, M., Maeda, N., Takahashi, Y., Matsushita, Y. and Noda, I., *Polym. J. (Tokyo)*, 1991, **23**, 227, (viscosity)
[23] Palaniappan, S. and Sathyanarayana, D.N., *Synth. Met.*, 1990, **38**, 53, (esr)
[24] Calistri-Yeh, M., Park, E.J., Kramer, E.J., Smith, J.W. and Sharma, R., *J. Mater. Sci.*, 1995, **30**, 5953, (fracture toughness)
[25] Williamson, R.L., Miles, M.J. and Jandt, K.D., *Proc. SPIE-Int. Soc. Opt. Eng.*, **2535**, 82, (optical microscopy)
[26] Stocker, W., Bickmann, B., Magonov, S.N., Cantow, H.-J., et al, *Ultramicroscopy*, 1992, **42-44**, 1141, (atomic force microscopy)
[27] Liu, Y., Rafailovich, M.H., Sokolov, J., Schwarz, S.A., et al, *Phys. Rev. Lett.*, 1994, **73**, 440, (wetting behaviour)

Polystyrene, Wood filled

Synonyms: *Poly(1-phenylethylene). Poly(vinyl benzene). Poly(phenyl ethene). Poly(phenyl ethylene). Poly(1-vinyl benzene). Poly(ethenyl benzene). PS*
Related Polymers: Polystyrene, High-impact polystyrene, Polystyrene, glass filled
Monomers: Styrene
Material class: Thermoplastic
Polymer Type: polystyrene
General Information: Molding and extrusion grades available.

Volumetric & Calorimetric Properties:
Density:

No.	Value	Note
1	d 1.074–1.156 g/cm^3	20–40% filler, ASTM D792 [1]

Deflection Temperature:

No.	Value	Note
1	75–81°C	at 264 psi, ASTM D648, 20–40% filler [1]

Transport Properties:
Melt Flow Index:

No.	Value	Note
1	0.5–1.6 g/10 min	at 190°C, 20–40% filler, ASTM D1238 [1]

Mechanical Properties:
Tensile (Young's) Modulus:

No.	Value	Note
1	4000–5800 MPa	20–40% filler, ASTM D638 [1]

Flexural Modulus:

No.	Value	Note
1	3200 MPa	20–40% filler, ASTM D790 [1]

Polysulfone

Tensile Strength Break:

No.	Value	Note	Elongation
1	19.9–28.7 MPa	20–40% filler, ASTM D638 [1]	1.5–5.8%

Tensile Strength Yield:

No.	Value	Note
1	22.5–28.9 MPa	20–40% filler, ASTM D638 [1]

Flexural Strength Yield:

No.	Value	Note
1	48.3–56.3 MPa	20–40% filler, ASTM D790 [1]

Izod Notch:

No.	Value	Notch	Note
1	20.3–40.5 J/m	Y	20–40% filler, ASTM D256 [1]
2	50–80 J/m	N	20–40% filler, ASTM D256 [1]

Applications/Commercial Products:
Processing & Manufacturing Routes: Injection molding [1]:
Dryer inlet temperature: 87.8°C
Dew point: -28.9°C
Drying time: 2–4 hours
Processing temperatures: rear <182°C, middle <188°C, front <193°C, nozzle °193°C
Residence time: <15 mins
Extrusion [1]:
Temperature settings [1]:
Zone 1: <149–171°C
Zone 2: <160–188°C
Zone 3: <171–193°C
Zone 4: <182–193°C
Adapter and die: <182–204°C
Screen packs: <20 mesh
Draw down: 0–5%

Applications/Commercial Products:
Mould Shrinkage (%):

No.	Value	Note
1	0.08–0.26%	20–40% filler, ASTM D955 [1]

Supplier
North Wood Plastics Inc.

Bibliographic References
[1] North Wood Plastics Inc. Data sheet, http://www.northwoodplastics.com, 2006

Polysulfone   P-531

Synonyms: *Polysulfone unreinforced. Bisphenol A polysulfone. PSF. PSU. Poly[oxy-1,4-phenylenesulfonyl-1,4-phenyleneoxy-1,4-phenylene(1-methylethylidene)-1,4-phenylene]*
Related Polymers: Polysulfone, 30% Glass fibre reinforced, Polysulfone, 30% Carbon fibre reinforced, Polysulfone, 15% PTFE lubricated, Polysulfone, Mineral filler modified, ABS/PSU blend, Crystalline modified polysulfone with glass fibre, Polysulfone-based proprietary blend
Monomers: 4,4′-Dichlorodiphenyl sulfone, Bisphenol A
Material class: Thermoplastic
Polymer Type: polysulfone
CAS Number:

CAS Reg. No.	Note
25135-51-7	Udel

Molecular Formula: $(C_{27}H_{22}O_4S)_n$
Fragments: $C_{27}H_{22}O_4S$
Molecular Weight: M_n 20400 and MW 40000 (determined by osmometry and light scattering on unfractionated polymer)
Additives: Fillers include carbon fibre, glass, PTFE and minerals
General Information: Polysulfone is a tough, rigid, high-strength amorph. thermoplastic for moulding and extrusion. Its principal features are good high temp. performance, resistance to creep, transparency and self-extinguishing characteristics. It maintains its props. over a wide temp. range (-101° to above 149°). Its excellent physical and electrical props. are virtually unaffected by long-term thermal ageing at 149°. Hydrolytically stable, it has exceptional steam resistance. On a cost to performance basis, it replaces metals and glass. Polysulfone is available in transparent and opaque forms in both moulding and extrusion grades (unfilled). A special medical grade is certified for contact with blood and other body parts. Also available are polysulfone compounds with fillers such as glass fibre, carbon fibre and Teflon. Glass-fibre reinforced polysulfone grades offer materials which can be used for strength purposes and can compete with metal die castings. Strength, stiffness and dimensional stability are increased with glass-fibre reinforcement

Volumetric & Calorimetric Properties:
Density:

No.	Value	Note
1	d 1.24 g/cm³	ASTM D792 [1]

Thermal Expansion Coefficient:

No.	Value	Note	Type
1	5.6×10^{-5} K^{-1}	ASTM D696 [1]	L

Volumetric Properties General: Has a higher coefficient of linear thermal expansion compared to reinforced grades
Thermal Conductivity:

No.	Value	Note
1	0.26 W/mK	ASTM C177 [1]

Specific Heat Capacity:

No.	Value	Note	Type
1	1 kJ/kg.C	23° [1]	P

Melting Temperature:

No.	Value	Note
1	310–390°C	[2]

– Polysulfone

Glass-Transition Temperature:

No.	Value	Note
1	185°C	[3]

Transition Temperature General: Heat deflection temp. increases significantly with thermal ageing [6]

Deflection Temperature:

No.	Value	Note
1	174°C	1.8 MPa, ASTM D648 [3]
2	181°C	0.5 MPa, ASTM D648 [3]

Vicat Softening Point:

No.	Value	Note
1	188°C	ASTM D1525 [3]

Transition Temperature:

No.	Value	Note	Type
1	-100°C	[1]	Second order (β)

Surface Properties & Solubility:
Cohesive Energy Density Solubility Parameters: δ 21.8 $J^{1/2}$ $cm^{-3/2}$ [1]
Solvents/Non-solvents: Sol. polar solvents, DMF, *N*-methylpyrrolidinone, chlorinated aliphatic hydrocarbons, CH_2Cl_2, $CHCl_3$, trichloroethylene [1]

Transport Properties:
Melt Flow Index:

No.	Value	Note
1	6.5 g/10 min	0.3 MPa, 343°, ASTM D1238 [6]

Polymer Solutions Dilute: Weight average MW intrinsic viscosity relationships for polysulfone have been reported [9]. Theta temp. θ - 105.5° [9]

Water Absorption:

No.	Value	Note
1	0.3 %	24h, 23°, 3.2 mm, ASTM D570 [6]

Mechanical Properties:
Mechanical Properties General: Rigid and tough with practical engineering strength and stiffness. Exhibits ductile yielding over a wide range of temps. and deformation rates. Continues to exhibit useful mech. props. at temps. up to 160° under prolonged or repeated thermal exposure [3,8]

Tensile (Young's) Modulus:

No.	Value	Note
1	2480 MPa	ASTM D638 [6]

Flexural Modulus:

No.	Value	Note
1	2690 MPa	ASTM D790 [6]

Tensile Strength Break:

No.	Note	Elongation
1	ASTM D638 [3]	50–100%

Compressive Modulus:

No.	Value	Note
1	2579 MPa	ASTM D695 [3]

Poisson's Ratio:

No.	Value	Note
1	0.37	0.5% strain [3]

Tensile Strength Yield:

No.	Value	Note
1	70.3 MPa	ASTM D638 [3]

Flexural Strength Yield:

No.	Value	Note
1	106 MPa	ASTM D790 [3]

Compressive Strength:

No.	Value	Note
1	96 MPa	ASTM D695 [3]

Miscellaneous Moduli:

No.	Value	Note	Type
1	917 MPa	[3]	Shear

Complex Moduli: Tensile creep modulus 2500 MPa (1000h, 23°, Dehnung ≤0.5%, ISO/IEC 899) [7]
Impact Strength: Tensile impact strength 420 kJ m^{-2} (ASTM D1822) [3]
Viscoelastic Behaviour: Creep under continuous load is unusually low, even at elevated temps. [6]. Creep modulus under ambient conditions and applied stress of 28 MPa is 2.413 MPa (1h), 2200 MPa (100h) and 2137 MPa (1000h) [15]
Hardness: Rockwell M69, R120 (ASTM D785) [3]
Failure Properties General: Shear strength 41.4 MPa (yield). Shear strength, 62.1 MPa (ultimate ASTM D732) [3]
Fracture Mechanical Properties: Fatigue endurance limit (10000000 cycles 1800 cycles min^{-1}) at room temp. is approx. equivalent to the performance of bisphenol A polycarbonate but lower than that of most crystalline engineering thermoplastics [1]
Friction Abrasion and Resistance: Abrasion resistance 20 mg (1000 cycles)$^{-1}$ (ASTM D1040, Taber CS-17 test) [3]

Izod Notch:

No.	Value	Notch	Note
1	64 J/m	Y	6.35 mm, ASTM D256 [3]
2	3200 J/m	N	3.18 mm, ASTM D256 [3]

Electrical Properties:
Electrical Properties General: Offers excellent electrical props., which are maintained over a wide temp. range and after immersion in water

Polysulfone

Surface/Volume Resistance:

No.	Value	Note	Type
2	30×10^{15} Ω.cm	ASTM D257 [1]	S

Dielectric Permittivity (Constant):

No.	Value	Frequency	Note
1	3.07	60 Hz	ASTM D150 [6]
2	3.06	1 kHz	ASTM D150 [6]
3	3.03	1 MHz	ASTM D150 [6]

Dielectric Strength:

No.	Value	Note
1	16.6 kV.mm^{-1}	3.2 mm thick, ASTM D149 [1]

Dissipation (Power) Factor:

No.	Value	Frequency	Note
1	0.0008	60 Hz	ASTM D150 [6]
2	0.001	1 kHz	ASTM D150 [6]
3	0.003	1 MHz	ASTM D150 [6]

Optical Properties:
Optical Properties General: Exhibits optical transparency
Refractive Index:

No.	Value	Note
1	1.63	ASTM D1505 [3]

Transmission and Spectra: Light transmittance 80% [3]
Volume Properties/Surface Properties: Haze <7% (3.1 mm thick, ASTM D1004). Light yellow colour [3]

Polymer Stability:
Thermal Stability General: Excellent thermal stability [3] is reflected in long-term retention of electrical props. at elevated temp.
Upper Use Temperature:

No.	Value	Note
1	160°C	UL 746 [1]

Decomposition Details: Degrades gradually *in vacuo* above 400°. Rapid decomposition begins above 460° [10]
Flammability: Flammability rating V0 (6.1 mm thick, UL 94) [1]. Limiting oxygen index 30% (ASTM D286 [1]. Exhibits excellent resistance to burning, and requires no flame-retardant additives [1]. Smoke density 90 at 1.5 mm (ASTM E662) [3]
Environmental Stress: High degree of oxidative stability. Poor resistance to uv light. Absorbance in the uv region occurs with attendant discoloration and losses in mechanical props. due to polymer degradation. For outdoor applications, protective lacquers or grades of the material containing carbon black are recommended [3]. Resistant to most forms of radiation [12,13,14]
Chemical Stability: Highly resistant to mineral acids, alkali and salt solns. Resistance to detergents, oils and alcohols is good. Attacked by polar organic solvents [6]. Is crazed, swollen or dissolved by chlorinated hydrocarbons, aromatic hydrocarbons, esters and ketones [1]
Hydrolytic Stability: Exceptional hydrolytic stability and steam resistance. In contaminated steam, its useful life is calculated to be at least 12 years [6]

Biological Stability: Good resistance to bacterial and fungal attack [15]
Stability Miscellaneous: Good dimensional stability [3]

Applications/Commercial Products:
Processing & Manufacturing Routes: Polysulfone is made by treating the sodium salt of bisphenol A with 4,4'-dichlorodiphenyl sulfone in a mixed solvent of chlorobenzene and DMSO. The polymerisation is carried out at 130–160°. At these temps. unacceptably high reaction rates and MW of 250000 (after 1h) occur. Terminators regulate chain growth to a range that facilitates commercial melt processing. A comonomer mole ratio of 1 results in the highest MW. Contamination with air has to be avoided and all traces of water must be removed. Polysulfone resin can be melt-processed on conventional equipment. Injection moulding is the most common fabrication technique used for all natural, coloured, filled and otherwise modified grades of the resin. Blow moulding, thermoforming, structural-foam moulding, soln. casting, and continuous-fibre reinforced polysulfone pre-pregging have been carried out. The polymer has a high melt viscosity and melt processing should be performed under the recommended conditions in order to produce parts of low residual stress levels. Conditions include the use of mould temp. of 95–150°, depending on part configuration. The resin's viscosity is quite insensitive to shear and its excellent thermal and oxidative stability within the broad melt-temp. processing range of 330–400° makes possible the use of melt temp. as the primary moulding parameter. Prior to melt processing, the resin must be dried to reduce the absorbed atmospheric moisture to below 0.05% wt. Once formed, parts made of polysulfone can be annealed. Polysulfones are extruded on conventional extrusion equipment as sheet, film, pipe or wire coating. Melt temps. of the extruded stock are in the range of 315–375°. Extrusion-grade polysulfones have excellent drawdown props. for the production of thin film due to their high melt strengths [1,3,10]
Mould Shrinkage (%):

No.	Value	Note
1	0.7%	ASTM D955 [6]

Applications: The primary uses of polysulfones include appliances and cookware. Also used in chemical, food and beverage processing; electrical and electronic components; medical products and some aerospace and automotive components. Polysulfone is suitable for use in articles intended for repeated use in contact with foods, particularly in microware cookware due to the polymer's outstanding transparency to microwave radiation. Electrical and electronic applications include automotive fuses, integrated-circuit carriers, connectors, switch housings and printed circuit boards. The resistance of polysulfones to chemical attack has resulted in their use in chemical processing equipment. Examples of components in this area are corrosion-resistant pipes, filter modules, support plates and tower packing. One unique application area for PSF is in membrane separation uses. Asymmetric PSF membranes are used in ultrafiltration, reverse osmosis and ambulatory hemodialysis (artificial kidney) units. The PRISM (Monsanto) gas-separation system is based on a polysulfone coating applied to a hollow-fibre support. Glass-reinforced grades of polysulfone have substantially enhanced tensile and flexural props. as well as resistance to cracking in chemical environments. Proprietary blends based on polysulfone are sold under the Mindel trademark. They offer maximum cost-effectiveness while maintaining key props.. Polysulfone meets U.S. Food and Drug Administration (FDA) requirements for direct food contact

Trade name	Details	Supplier
CTI PSUL		Compounding Technology
Comalloy PSUL		ComAlloy International

– Polysulfone-based proprietary blend

Electrafil	J-1500/CF/20	DSM Engineering
Fiberfil	J-1500	DSM Engineering
Fiberstran	G-1500/10/20	DSM Engineering
Hyvex	various grades	Ferro Corporation
J-1500/1505		DSM Engineering
Lubricomp	various grades	LNP Engineering
MPSL	various grades	Modified Plastics
Mindel	various grades	Amoco Polymers
Plastalloy	J-1505/30	DSM Engineering
RTP	various grades	RTP Company
Radel	R-5000/5100/7000A	Amoco Polymers
S	various grades	Thermofil Inc.
SU-1500	TF/15	DSM Engineering
Stat-Kon	GC-1003	LNP Engineering
Thermocomp	various grades	LNP Engineering
Thermofil PSUL		Thermofil
Udel P	includes 1700- standard moulding grade, unreinforced and 3500 MW series for extrusion	Amoco Performance Products
Ultrason S	1010, 2010, 3010, easy flow, general purpose injection moulding product, high viscosity product for injection moulding/extrusion	BASF

Bibliographic References

[1] *Encycl. Polym. Sci. Eng.*, Vol. 13, 2nd edn., (ed. J.I. Kroschwitz), John Wiley and Sons, 1985
[2] *The Plastics Compendium, Key Properties and Sources*, Vol. 1, (ed. R. Dolbey), Rapra Technology, Ltd., 1995
[3] *Kirk-Othmer Encycl. Chem. Technol.*, 4th edn., (eds. J.I. Kroschwitz and M. Howe-Grant), Wiley Interscience, 1993, **19**, 945
[4] Brydson, J.A., *Plast. Mater.*, 5th edn., Butterworths, 1988
[5] Guide to Plastics, *Property and Specification Charts*, McGraw-Hill, 1993
[6] *Engineering Plastics for Performance and Value*, Amoco Performance Products, Inc., (technical datasheet)
[7] Ultrason (Polyethersulfone/Polysulfone PES/PSU) Range Chart, Features Applications, *Typical Values*, BASF, (technical datasheet)
[8] Attwood, T.E., Cinderey, M.B. and Rose, J.B., *Polymer*, 1993, **34**, 1322
[9] Allen, G., McAinsh, J. and Strazielle, C., *Eur. Polym. J.*, 1969, **5**, 319, (soln props)
[10] Rose, J.B., *Polymer*, 1974, **15**, 456, (synth, props, struct)
[11] Johnson, R.N., Farnham, A.G., Glendinning, R.A., Hale W.F. and Merriam, C.N., *J. Polym. Sci., Part A-1*, 1967, **5**, 2375, 2399, (synth, props)
[12] Davis, A., Gleaves, M.H., Golden, J.H. and Huglin, M.B., *Makromol. Chem.*, 1969, **129**, 63
[13] Brown, J.R. and O'Donnel, J.H., *J. Polym. Sci., Part B: Polym. Lett.*, 1970, **8**, 121
[14] Lyons, A.R., Symons, M.C.R. and Yandel, J.K., *Makromol. Chem.*, 1972, **157**, 103
[15] Daniels, C.A., *Polymers: Structure and Properties*, Technomic, 1989
[16] Koros, W.J. and Chern, R.T., *Handbook of Separation Process Technology*, (ed. R.W. Rousseau), John Wiley and Sons, 1987

Polysulfone-based proprietary blend P-532

Synonyms: *Polysulfone-based proprietary resin*
Related Polymers: Polysulfone
Monomers: 4,4′-Dichlorodiphenyl sulfone, Bisphenol A
Material class: Thermoplastic
Polymer Type: polysulfone

CAS Number:

CAS Reg. No.	Note
25135-51-7	Udel

Molecular Formula: $(C_{27}H_{22}O_4S)_n$
Fragments: $C_{27}H_{22}O_4S$
General Information: Polysulfone-based proprietary resin is developed to fit a cost/performance niche in medical and food service applications. It has chemical resistance comparable to polysulfone, steam and hot water resistance and autoclavability with excellent electrical and mech. props.

Volumetric & Calorimetric Properties:
Density:

No.	Value	Note
1	d 1.23 g/cm^3	ASTM D792 [1]

Thermal Expansion Coefficient:

No.	Value	Note	Type
1	5.7×10^{-5} K^{-1}	ASTM D696 [1]	L

Melting Temperature:

No.	Value	Note
1	315–345°C	[1]

Deflection Temperature:

No.	Value	Note
1	149°C	1.8 MPa, ASTM D648 [1]

Transition Temperature:

No.	Value	Note	Type
1	149°C	[1]	High heat deflection

Surface Properties & Solubility:
Solvents/Non-solvents: Sol. polar solvents, DMF, N-methylpyrrolidinone, chlorinated aliphatic hydrocarbons, CH_2Cl_2, $CHCl_3$ and trichloroethylene [2]. Attacked by some polar organic solvents

Transport Properties:
Melt Flow Index:

No.	Value	Note
1	13 g/10 min	0.3 MPa, 320°, ASTM D1238 [1]

Water Absorption:

No.	Value	Note
1	0.2 %	3.2 mm, 24h, 23°, ASTM D570 [1]

Mechanical Properties:
Mechanical Properties General: Has excellent mech. props. [1]

Tensile (Young's) Modulus:

No.	Value	Note
1	2400 MPa	ASTM D638 [1]

– Polysulfone, 30% carbon fibre reinforced

Flexural Modulus:

No.	Value	Note
1	2630 MPa	ASTM D790 [1]

Tensile Strength Break:

No.	Note	Elongation
1	ASTM D638 [1]	50% (min.)

Tensile Strength Yield:

No.	Value	Note
1	66 MPa	ASTM D638 [1]

Flexural Strength Yield:

No.	Value	Note
1	97 MPa	ASTM D790 [1]

Impact Strength: Tensile impact strength 273 kJ m^{-2}
Izod Notch:

No.	Value	Notch	Break	Note
1	85 J/m	Y		ASTM
2		N	Y	ISO

Izod Area:

No.	Value	Notch	Note
1	9.9 kJ/m^2	Notched	ISO

Electrical Properties:
Electrical Properties General: Has excellent electrical props., which are maintained over a wide temp. range and after immersion in water [1]
Dielectric Permittivity (Constant):

No.	Value	Frequency	Note
1	3.12	60 Hz	ASTM D150 [1]
2	3.1	1 kHz	ASTM D150 [1]
3	3.5	1 MHz	ASTM D150 [1]

Dielectric Strength:

No.	Value	Note
1	20 kV.mm^{-1}	3.2 mm, ASTM D149 [1]

Dissipation (Power) Factor:

No.	Value	Frequency	Note
1	0.001	1 kHz	ASTM D150 [1]
2	0.007	1 MHz	ASTM D150 [1]

Optical Properties:
Volume Properties/Surface Properties: Opaque [1]

Polymer Stability:
Polymer Stability General: Stable over a wide temp. range

Thermal Stability General: Excellent thermal stability
Upper Use Temperature:

No.	Value	Note
1	105°C	UL 746B, provisional rating [1]
2	160°C	UL 746 [1]

Flammability: Flammability rating V0 (3.3 mm thick, UL 94) [1]
Environmental Stress: High degree of oxidative stability [3]
Chemical Stability: Chemical resistance comparable to polysulfone [1]
Hydrolytic Stability: Superior hydrolytic stability and steam resistance [1]

Applications/Commercial Products:
Processing & Manufacturing Routes: Melt temp. range is 315–345° and mould temp. range is 90–140° for injection moulding. Prior to melt processing, the resin must be dried for 3h at 135° [1]
Mould Shrinkage (%):

No.	Value	Note
1	0.66%	ASTM D955 [1]

Applications: Food processing application, has FDA approval. Used in medical devices, for biological contact and institutional feeding systems. Mindel S-1000 resin provides strength and sterilisability to components of medical analysers

Trade name	Details	Supplier
Mindel S	20% Glass fibre grade available	Amoco

Bibliographic References
[1] *Engineering Plastics for Performance and Value*, Amoco Performance Products, Inc., (technical datasheet)
[2] Harris, J.E. and Johnson, R.N., *Encycl. Polym. Sci. Eng.*, Vol. 13, 2nd edn., (ed. J.I. Kroschwitz), John Wiley and Sons, 1985, 196
[3] *Kirk-Othmer Encycl. Chem. Technol.*, Vol. 19, 4th edn., (eds. J.I. Kroschwitz and M. Howe-Grant), Wiley Interscience, 1993, 945

Polysulfone, 30% carbon fibre reinforced P-533
Synonyms: *PSU, 30% carbon fibre reinforced. PSF, 30% carbon fibre reinforced*
Related Polymers: Polysulfone
Monomers: 4,4'-Dichlorodiphenyl sulfone, Bisphenol A
Material class: Thermoplastic
Polymer Type: polysulfone
CAS Number:

CAS Reg. No.	Note
25135-51-7	Udel

Molecular Formula: $(C_{27}H_{22}O_4S)_n$
Fragments: $C_{27}H_{22}O_4S$
Additives: Carbon fibre
General Information: A tough rigid engineering thermoplastic. Compared with unreinforced grades it has superior tensile strength, creep resistance and rigidity, low coefficient of thermal expansion and very good mould skrinkage. It is expensive and has poor electrical insulation characteristics. Due to the incorporation of carbon fibre it offers full resistance to uv light.

Volumetric & Calorimetric Properties:
Density:

No.	Value	Note
1	d 1.37 g/cm^3	ASTM D792 [1]

– Polysulfone, 30% carbon fibre reinforced

Thermal Expansion Coefficient:

No.	Value	Note	Type
1	1.9×10^{-5} K^{-1}	ASTM D696 [3]	L

Volumetric Properties General: Low coefficient of thermal expansion compared with unreinforced PSU [1]

Thermal Conductivity:

No.	Value	Note
1	0.45 W/mK	ASTM C177 [3]

Melting Temperature:

No.	Value	Note
1	310–390°C	[1]

Glass-Transition Temperature:

No.	Value	Note
1	190°C	[2]

Deflection Temperature:

No.	Value	Note
1	197°C	0.45 MPa [1]
2	185°C	1.8 MPa, ASTM D648 [1]

Transition Temperature:

No.	Value	Note	Type
1	-100°C	[4]	Second order (β)

Surface Properties & Solubility:

Solvents/Non-solvents: Sol. polar solvents such as DMF and *N*-methylpyrrolidinone, chlorinated aliphatic hydrocarbons such as CH_2Cl_2, $CHCl_3$ and trichloroethylene [4]

Transport Properties:

Water Absorption:

No.	Value	Note
1	0.15–0.25 %	1/8" thick, 24h, ASTM D570 [2]

Mechanical Properties:

Mechanical Properties General: Superior tensile strength, creep resistance and rigidity compared with unreinforced PSU [1]

Tensile (Young's) Modulus:

No.	Value	Note
1	14822–19303 MPa	2150–2800 kpsi, ASTM D638 [2]

Flexural Modulus:

No.	Value	Note
1	15500 MPa	ASTM D790 [3]

Tensile Strength Break:

No.	Value	Note
1	160 MPa	ASTM D638 [3]

Flexural Strength at Break:

No.	Value	Note
1	220 MPa	ASTM D790 [3]

Tensile Strength Yield:

No.	Value	Elongation	Note
1	1 MPa	2%	ASTM D638 [3]

Compressive Strength:

No.	Value	Note
1	140 MPa	ASTM D695 [3]

Hardness: Rockwell M96 (ASTM D785) [3]
Failure Properties General: Shear strength 105 MPa (ASTM D732) [3]
Friction Abrasion and Resistance: Coefficient of friction (static) 0.17 (LNP SOP). Coefficient of friction (dynamic) 0.14 (LNP SOP) [3]

Izod Notch:

No.	Value	Notch	Note
1	90 J/m	Y	ASTM D256 [3]
2	400 J/m	N	ASTM D256 [3]

Electrical Properties:

Electrical Properties General: Due to the incorporation of carbon fibre this material is conductive. Surface and volume resistivities are in the range 10–100 Ω cm (ASTM D257) [3]

Polymer Stability:

Polymer Stability General: Stable over a wide temp. range (-100° to above 149°)
Thermal Stability General: Excellent thermal stability
Upper Use Temperature:

No.	Value	Note
1	160°C	UL 746 [3]

Flammability: Flammability rating V0 (UL 94) [1]. Limiting oxygen index 32% [1]. Excellent resistance to burning and, in most applications, requires no flame-retardant additive [4]
Environmental Stress: High degree of oxidative stability. Resistant to uv light [5]
Chemical Stability: Highly resistant to aq. mineral acid, alkali and salt solns. Resistant to aliphatic hydrocarbons and most alcohols, but is crazed, swollen or dissolved by chlorinated hydrocarbons, aromatic hydrocarbons, esters and ketones [4]
Hydrolytic Stability: Excellent hydrolytic stability. Withstands many hours exposure to hot water and steam [4]
Biological Stability: Good resistance to bacterial and fungal attack [6]

Applications/Commercial Products:

Processing & Manufacturing Routes: Mould temp. of 120–160° and melt temp. of 310–390° are recommended for injection moulding. Prior to melt processing, the resin is dried for 5h at 120° to prevent bubbling, surface streaks, and splash marks in moulded parts [1]

Polysulfone, 30% Glass fibre reinforced

Mould Shrinkage (%):

No.	Value	Note
1	0.15%	ASTM D-955, very good mould shrinkage compared with unreinforced PSU, moulded components can be highly anisotropic [1]

Applications: Automotive applications for under the bonnet components, aricraft interior and exterior components. Chemical plant pumps and valves, process pipe and plumbing components, pump impellers, switch devices, reflectors, knobs and buttons

Trade name	Details	Supplier
CTI PSUL		Compounding Technology
ComAlloy PSUL		ComAlloy International
Grafil GG		Courtaulds
RTP	Also used to refer to other materials	RTP Company
Thermocomp GC-1006	30% carbon fibre reinforced. Also used to refer to other materials other grades available	LNP Engineering
Thermocomp LC		LNP Engineering
Thermofil PSUL		Thermofil
Udel		Amoco Performance Products

Bibliographic References

[1] The Plastics Compendium, *Key Properties and Sources*, (ed. R. Dolbey), Rapra Technology Ltd., 1995, **1**
[2] Guide to Plastics, *Property and Specification Charts*, McGraw-Hill, 1993
[3] LNP Engineering Plastics, *Product Data Book*,
[4] Harris, J.E. and Johnson, R.N., *Encycl. Polym. Sci. Eng.*, Vol. 13, 2nd edn., (Ed. Kroschwitz, J.I.), John Wiley and Sons, 1985, 196
[5] *Kirk-Othmer Encycl. Chem. Technol.*, Vol. 19, 4th edn., (eds. Kroschwitz, J.I. and Howe-Grant, M.), Wiley Interscience, 1993, 945
[6] Daniels, C.A., *Polymers: Structure and Properties*, Technomic, 1989

Polysulfone, 30% Glass fibre reinforced   P-534

Synonyms: *PSU, 30% Glass fibre reinforced. PSF, 30% Glass fibre reinforced*
Related Polymers: Polysulfone
Monomers: 4,4′-Dichlorodiphenylsulfone, Bisphenol A
Material class: Thermoplastic
Polymer Type: polysulfone
CAS Number:

CAS Reg. No.	Note
25135-51-7	Udel

Molecular Formula: $(C_{27}H_{22}O_4S)_n$
Fragments: $C_{27}H_{22}O_4S$
Additives: Glass fibre
General Information: A tough rigid engineering thermoplastic with good high temp. performance. Compared with unreinforced grades it has excellent creep resistance, greater strength and rigidity, good mould shrinkage, better flammability characteristics, but lower thermal expansion and reduced elongation at break. It is used in harsh chemical environments for enhanced resistance and long service life

Volumetric & Calorimetric Properties:
Density:

No.	Value	Note
1	d 1.45 g/cm^3	ASTM D792 [1]

Thermal Expansion Coefficient:

No.	Value	Note	Type
1	2×10^{-5} K^{-1}	ASTM D696 [4]	L

Volumetric Properties General: Lower coefficient of linear thermal expansion compared with unreinforced
Thermal Conductivity:

No.	Value	Note
1	0.2 W/mK	ASTM 177 [3]

Melting Temperature:

No.	Value	Note
1	340–400°C	[4]

Glass-Transition Temperature:

No.	Value	Note
1	189–190°C	[2]

Transition Temperature General: Slightly improved heat deflection temp. compared with unreinforced [1]
Deflection Temperature:

No.	Value	Note
1	191°C	0.45 MPa [1]
2	185°C	1.8 MPa, ASTM D648 [1]

Vicat Softening Point:

No.	Value	Note
1	197°C	VST/A/50, ISO/IEC 306 [5]

Transition Temperature:

No.	Value	Note	Type
1	-100°C	[6]	Second order (β)

Surface Properties & Solubility:
Solubility Properties General: δ 21.8 J$^{1/2}$ cm$^{-3/2}$ [1]
Solvents/Non-solvents: Sol. polar solvents such as DMF and N-methylpyrrolidinone, chlorinated aliphatic hydrocarbons such as CH_2Cl_2, $CHCl_3$ and trichloroethylene [6]

Transport Properties:
Melt Flow Index:

No.	Value	Note
1	7–8 g/10 min	ASTM D1238 [2]

Polysulfone, 30% Glass fibre reinforced

Water Absorption:

No.	Value	Note
1	0.3 %	1/8" thick, 24h, ASTM D570 [2]

Mechanical Properties:

Mechanical Properties General: Tensile strength, flexural strength, notched Izod impact strength and creep resistance are superior to those of unreinforced. Elongation at break is significantly reduced compared with unreinforced [1]

Tensile (Young's) Modulus:

No.	Value	Note
1	9300–10000 MPa	1350–1450 kpsi, ASTM D638 [2]
2	7380 MPa	ASTM D638 [4]

Flexural Modulus:

No.	Value	Note
1	8500 MPa	ASTM D790 [3]

Tensile Strength Break:

No.	Value	Note
1	130 MPa	ASTM D638 [3]

Flexural Strength at Break:

No.	Value	Note
1	150 MPa	ASTM D790 [3]

Tensile Strength Yield:

No.	Elongation	Note
1	2–3%	ASTM D638 [3]

Compressive Strength:

No.	Value	Note
1	110 MPa	ASTM D695 [3]

Complex Moduli: Tensile creep modulus 8300 MPa (23°, 1000h, Dehnung $\leq 0.5\%$, ISO/IEC 899) [5]
Impact Strength: Tensile impact strength 109 kJ m^{-2} (ASTM D1822) [4]
Viscoelastic Behaviour: Significantly increased creep resistance compared with unreinforced PSU [7]
Hardness: Rockwell M96 (ASTM D785) [3]
Failure Properties General: Shear strength 75 MPa (ASTM D732) [3]
Friction Abrasion and Resistance: Coefficient of friction (static) 0.24 (LNP SOP). Coefficient of friction (dynamic) 0.22 (LNP SOP) [3]
Izod Notch:

No.	Value	Notch	Note
1	85 J/m	Y	ASTM D256 [3]
2	350 J/m	N	ASTM D256 [3]

Electrical Properties:

Electrical Properties General: Has excellent electrical props. which are maintained over a wide temp. range and after immersion in water [4]

Surface/Volume Resistance:

No.	Value	Type
1	0.1×10^{15} Ω.cm	S

Dielectric Permittivity (Constant):

No.	Value	Frequency	Note
1	3.5	60 Hz	ASTM D150 [3]
2	3.5	1 MHz	ASTM D150 [3]

Dielectric Strength:

No.	Value	Note
1	19 kV.mm^{-1}	ASTM D149 [3]

Dissipation (Power) Factor:

No.	Value	Frequency	Note
1	0.001	60 Hz	ASTM D150 [3]
2	0.007	1 MHz	ASTM D150 [3]

Optical Properties:
Volume Properties/Surface Properties: Opaque [4]

Polymer Stability:
Polymer Stability General: Stable over a wide temp. range (-101° to above 149°)
Thermal Stability General: Excellent thermal stability. A high proportion of its initial props. are retained upon exposure to temps. of 149° to 177° during continual or long-term intermittent service [4]

Upper Use Temperature:

No.	Value	Note
1	160°C	UL 746 [3]

Flammability: Flammability rating V0 (UL 94) [1]. Limiting oxygen index 35% [1]. Excellent resistance to burning. Requires no flame-retardant additives [6]
Environmental Stress: High degree of oxidative stability. Poor resistance to uv light. Improved stress cracking resistance compared with unreinforced [12]. Resistant to many forms of ionising radiation [9,10,11,12]
Chemical Stability: Highly resistant to mineral acids, alkali and salt solns. Resistance to detergents and hydrocarbon oils is good. Attacked by polar organic solvents [4]
Hydrolytic Stability: Exceptional hydrolytic stability and steam resistance. In contaminated steam, its useful life is calculated to be at least 12 years [4]
Biological Stability: Good resistance to bacterial and fungal attack [13]
Stability Miscellaneous: Improved dimensional stability compared with unreinforced [8]

Applications/Commercial Products:
Processing & Manufacturing Routes: Moulded in modern injection-moulding equipment. A mould temp. of 120–160° and a melt temp. of 340–400° are recommended. Prior to melt processing, the resin is dried for 4h at 150° to prevent bubbling, surface streaks, and splash marks in moulded parts [4]

– Polysulfone, Mineral filler modified

Mould Shrinkage (%):

No.	Value	Note
1	0.25%	ASTM D955, good mould shrinkage compared with unreinforced. Mech. props. of moulded components can be anisotropic [1]

Applications: Electrical components - connectors, circuit boards, TV components, hairdryer parts, oven, projector and fan heater components. Pumps and valves for use in chemical and petrochemical industries, suitable also for use with hot water. Appliance housings, process equipment and food service items

Trade name	Details	Supplier
CTI PSUL		Compounding Technology
Comalloy PSUL		ComAlloy International
Lasulf		LATI Industria Thermoplastics S.p.A.
Starglass PSU		Ferro Corporation
Thermocomp GF	Also used to refer to other materials	LNP Engineering
Thermofil PSUL		Thermofil
Udel	GF-120, 20% Glass fibre reinforced, GF-110 and -130 available	Amoco Performance Products
Ultrason S2010G2	10% Glass fibre reinforced general purpose injection moulding grade other grades available	BASF

Bibliographic References

[1] The Plastics Compendium, *Key Properties and Sources*, (ed. R. Dolbey), Rapra Technology Ltd., 1995, **1**
[2] Guide to Plastics, *Property and Specification Charts*, McGraw-Hill, 1993
[3] LNP Engineering Plastics, *Product Data Book*,
[4] *Engineering Plastics for Performance and Value*, Amoco Performance Products, Inc.,
[5] Ultrason (Polyethersulfone/Polysulfone PES/PSU) Range Chart, Features Applications, *Typical Values*, BASF, (technical datasheet)
[6] *Encycl. Polym. Sci. Eng.*, Vol. 13, 2nd edn., (ed. J.I. Kroschwitz), John Wiley and Sons, 1985
[7] *Plast. Mater.*, 5th edn., Butterworths, 1988
[8] Harper, C.A., *Handb. Plast. Elastomers*, (ed. C.A. Harper), McGraw-Hill, 1975
[9] Davis, A., Gleaves, M.H., Golden, J.H. and Huglin, M.B., *Makromol. Chem.*, 1969, **129**, 63
[10] Brown, J.R. and O'Donnel, J.H., *J. Polym. Sci., Part B: Polym. Lett.*, 1970, **8**, 121
[11] Lyons, A.R., Symons, M.C.R. and Yandel, J.K., *Makromol. Chem.*, 1972, **157**, 103
[12] *Kirk-Othmer Encycl. Chem. Technol.*, Vol. 19, 4th edn., (eds. J.I. Kroschwitz and M. Howe-Grant), Wiley Interscience, 1993, 945
[13] Daniels, C.A., *Polymers: Structure and Properties*, Technomic, 1989

Polysulfone, Mineral filler modified P-535

Synonyms: *PSF, Mineral filler modified. PSU, Mineral filler modified*
Related Polymers: Polysulfone
Monomers: 4,4'-Dichlorodiphenyl sulfone, Bisphenol A
Material class: Thermoplastic
Polymer Type: polysulfone

CAS Number:

CAS Reg. No.	Note
25135-51-7	Udel

Molecular Formula: $(C_{27}H_{22}O_4S)_n$
Fragments: $C_{27}H_{22}O_4S$
Additives: Mineral fillers
General Information: Polysulfones modified with mineral filler are formulated for maximum cost-effectiveness while maintaining key props. They have special props. such as improved environmental stress crack resistance, dimensional stability, creep resistance, toughness and mouldability.

Volumetric & Calorimetric Properties:
Density:

No.	Value	Note
1	d 1.48 g/cm^3	ASTM D792 [1]

Thermal Expansion Coefficient:

No.	Value	Note	Type
1	3.4×10^{-5}–0.00039 K^{-1}	ASTM D696 [1]	L

Glass-Transition Temperature:

No.	Value	Note
1	190°C	[1]

Deflection Temperature:

No.	Value	Note
1	174–179°C	345–354°F, 1.8 MPa, ASTM D648 [1]

Transition Temperature:

No.	Value	Note	Type
1	100°C	[3]	Second order (β)

Surface Properties & Solubility:
Solvents/Non-solvents: Sol. polar solvents such as DMF and *N*-methylpyrrolidinone, chlorinated aliphatic hydrocarbons such as CH_2Cl_2, $CHCl_3$ and trichloroethylene [3]

Transport Properties:
Melt Flow Index:

No.	Value	Note
1	7–8.5 g/10 min	ASTM D238 [1]

Water Absorption:

No.	Value	Note
1	0.5–0.6 %	1/8" thick, saturation, ASTM D570 [1]

Mechanical Properties:
Mechanical Properties General: Good creep resistance, toughness and mouldability [2]

Polysulfone, 15% PTFE lubricated

Tensile (Young's) Modulus:

No.	Value	Note
1	3791–4481 MPa	550–650 kpsi, ASTM D638 [1]

Flexural Modulus:

No.	Value	Note
1	4136–5170 MPa	22.8°, ASTM D638 [1]
2	3929–4894 MPa	93.3°, ASTM D638 [1]
3	3792–4757 MPa	121.1°, ASTM D638 [1]

Tensile Strength Break:

No.	Value	Note
1	65.5–67.6 MPa	9500–9800 psi, ASTM D638 [1]

Flexural Strength Yield:

No.	Value	Note
1	98.6–106 MPa	14300–15400 psi, ASTM D790 [1]

Viscoelastic Behaviour: Creep resistant [2]
Hardness: Rockwell M70-M74 (ASTM D785) [1]
Izod Notch:

No.	Value	Notch	Note
1	34.6–53.1 J/m	N	0.65-1 ft lb in^{-1}, $\frac{1}{8}$" thick, ASTM D256A [1]

Electrical Properties:

Electrical Properties General: Has excellent electrical props., which are maintained over a wide temp. range and after immersion in water. Props. are virtually unaffected by long-term ageing at 149° [2]

Dielectric Permittivity (Constant):

No.	Value	Frequency	Note
1	3.7–3.8	60 Hz	ASTM D150

Dielectric Strength:

No.	Value	Note
1	17 kV.mm^{-1}	ASTM D149 [1]

Arc Resistance:

No.	Value	Note
1	125s	ASTM D495

Dissipation (Power) Factor:

No.	Value	Frequency	Note
1	0.003	60 Hz	ASTM D150

Optical Properties:

Volume Properties/Surface Properties: Provides a smooth surface [2]

Polymer Stability:

Polymer Stability General: Stable over a wide temp. range (-101° to above 149°) [2]
Thermal Stability General: Excellent thermal stability
Flammability: Flammability rating V0 (UL 94) [2]. Excellent resistance to burning. Requires no flame-retardant additives [4]
Environmental Stress: Improved environmental stress crack resistance compared with unreinforced [2]. High degree of oxidative stability
Chemical Stability: Highly resistant to mineral acids, alkali and salt solns. Resistance to detergents and hydrocarbon oil is good. Attacked by polar organic solvents [2]
Hydrolytic Stability: Has excellent hydrolytic stability and has demonstrable long-term resistance to steam [2]
Stability Miscellaneous: Has long-term dimensional stability

Applications/Commercial Products:

Processing & Manufacturing Routes: Mould temp. range is 357–413° in injection moulding
Mould Shrinkage (%):

No.	Value	Note
1	0.4–0.5%	ASTM D955 [1]

Applications: Applications in electrical devices, food handling equipment, and medical equipment. Qualified for FDA approved devices

Trade name	Details	Supplier
Mindel M	mineral filled resins	Amoco Performance Products

Bibliographic References

[1] Guide to Plastics, *Property and Specification Charts*, McGraw-Hill, 1993
[2] *Engineering Plastics for Performance and Value*, Amoco Performance Products, Inc., (technical datasheet)
[3] *Encycl. Polym. Sci. Eng.*, Vol. 13, 2nd edn., (ed. J.I. Kroschwitz), John Wiley and Sons, 1985
[4] The Plastics Compendium, *Key Properties and Sources*, Vol. 1, (ed. R. Dolbey), Rapra Technology Ltd., 1995

Polysulfone, 15% PTFE lubricated   P-536

Synonyms: *PSU, 15% PTFE lubricated. PSF, 15% PTFE lubricated*
Related Polymers: Polysulfone
Monomers: 4,4′-Dichlorodiphenyl sulfone, Bisphenol A
Material class: Thermoplastic
Polymer Type: polysulfone
CAS Number:

CAS Reg. No.	Note
25135-51-7	Udel

Molecular Formula: $(C_{27}H_{22}O_4S)_n$
Fragments: $C_{27}H_{22}O_4S$
Additives: PFTE as lubricant
General Information: A tough rigid engineering thermoplastic with high resistance to creep, good high temp. performance and self-extinguishing characteristics. Good chemical resistance and hydrolytic stability. Compared with unreinforced grades it has low coefficient of friction and lower tensile strength

Volumetric & Calorimetric Properties:

Density:

No.	Value	Note
1	d 1.32 g/cm^3	ASTM D792 [1]

– Polysulfone, 15% PTFE lubricated

Thermal Expansion Coefficient:

No.	Value	Note	Type
1	5.5×10^{-5} K^{-1}	ASTM D696 [3]	L

Thermal Conductivity:

No.	Value	Note
1	0.17 W/mK	ASTM C177 [3]

Melting Temperature:

No.	Value
1	310–390°C

Transition Temperature General: Slightly better heat deflection temp. compared with unreinforced polysulfone [1]
Deflection Temperature:

No.	Value	Note
1	260°C	0.45 MPa [1]
2	177°C	1.8 MPa, ASTM D648 [1]

Transition Temperature:

No.	Value	Note	Type
1	-100°C	[4]	Second-order (β)

Surface Properties & Solubility:
Solvents/Non-solvents: Sol. polar solvents such as DMF and *N*-methylpyrrolidinone, chlorinated aliphatic hydrocarbons such as CH_2Cl_2, $CHCl_3$ and trichloroethylene [4]

Transport Properties:
Water Absorption:

No.	Value	Note
1	0.25 %	24h, ASTM D570 [3]
2	0.6 %	saturation, ASTM D570 [3]

Mechanical Properties:
Mechanical Properties General: Lower tensile strength, elongation at break and notched Izod impact strength compared with unreinforced polysulfone [1]
Flexural Modulus:

No.	Value	Note
1	2600 MPa	ASTM D790 [3]

Tensile Strength Break:

No.	Value	Note
1	55 MPa	ASTM D638 [3]

Flexural Strength at Break:

No.	Value	Note
1	90 MPa	ASTM D790

Tensile Strength Yield:

No.	Elongation	Note
1	10–20%	ASTM D638 [3]

Compressive Strength:

No.	Value	Note
1	60 MPa	ASTM D695 [3]

Hardness: Rockwell M81 (ASTM D785) [3]
Failure Properties General: Shear strength 40 MPa (ASTM D732)
Friction Abrasion and Resistance: Coefficient of friction (static) 0.09 (LNP SOP). Coefficient of friction (dynamic) 0.14 (LNP SOP) (low coefficient of friction compared with unreinforced polysulfone) [3]
Izod Notch:

No.	Value	Notch	Note
1	65 J/m	Y	
2	350 J/m	N	ASTM D256 [3]

Electrical Properties:
Electrical Properties General: Has excellent electrical props. which are maintained over a wide temp. range and after immersion in water. Unaffected by long-term ageing at 149° [2]. Arc resistance values have been reported [3]
Dielectric Permittivity (Constant):

No.	Value	Frequency	Note
1	3.15	60 Hz	ASTM D150 [3]
2	3.1	1 MHz	ASTM D150 [3]

Dielectric Strength:

No.	Value	Note
1	17 kV.mm^{-1}	ASTM D149 [3]

Dissipation (Power) Factor:

No.	Value	Frequency	Note
1	0.0011	60 Hz	ASTM D150 [3]
2	0.005	1 MHz	ASTM D150 [3]

Polymer Stability:
Polymer Stability General: Stable over a wide temp. range (-101° to above 149°) [2]
Thermal Stability General: Excellent thermal stability. A high proportion of its initial props. are retained upon exposure to temps. of 149° to 177° during continual or long-term intermittent service [2]
Upper Use Temperature:

No.	Value	Note
1	160°C	UL 746 [3]

Flammability: Flammability rating V1 (UL 94) [1]. Limiting oxygen index 32% [1]. Excellent resistance to burning. Requires no flame-retardant additives [4]
Environmental Stress: High degree of oxidative stability. Poor resistance to uv light [5]

Chemical Stability: Highly resistant to mineral acids, alkali and salt solns. Resistance to detergents and hydrocarbon oil is good. Attacked by polar organic solvents [2]. Is crazed, swollen or dissolved by chlorinated hydrocarbons, aromatic hydrocarbons, esters and ketones [4]
Hydrolytic Stability: Has excellent hydrolytic stability and has demonstrable long term resistance to steam [2]
Biological Stability: Good resistance to bacterial and fungal attack [6]

Applications/Commercial Products:
Processing & Manufacturing Routes: Melt temp. range of 310–390° and mould temp. range of 90–150° are recommended for injection moulding. Prior to melt processing, the resin is dried for 5h at 120° to prevent bubbling, surface streaks, and splash marks in moulded parts [1]
Mould Shrinkage (%):

No.	Value	Note
1	0.55%	3mm section, ASTM D955 [3]

Applications: Used in the manufacture of electrical components - coil bobbins, connectors, terminal blocks, cooking appliances, milking machine parts, process and sanitary pipes, brushes for compact hair stylers, housing for electrical appliances, pumps and valves and medical equipment that requires sterilisation

Trade name	Details	Supplier
CTI PSUL		Compounding Technology
ComAlloy PSUL		ComAlloy International
Lubricomp GL	GL-4030	LNP Engineering
RTP 905 TFE	905 TFE 15 also used to refer to other materials	RTP Company
Thermofil PSUL		Thermofil
Udel		Amoco Performance Products

Bibliographic References
[1] The Plastics Compendium, *Key Properties and Sources*, (ed. R. Dolbey), Rapra Technology Ltd., 1995, **1**
[2] *Engineering Plastics for Performance and Value*, Amoco Performance Products, Inc., (technical datasheet)
[3] LNP Engineering Plastics, *Product Data Book*,
[4] *Encycl. Polym. Sci. Eng.*, 2nd edn., (ed. J.I. Kroschwitz), John Wiley and Sons, 1985
[5] *Kirk-Othmer Encycl. Chem. Technol.*, Vol. 19, 4th edn., (eds. Kroschwitz, J.I. and Howe-Grant, M.), Wiley Interscience, 1993, 945
[6] Daniels, C.A., *Polymers: Structure and Properties*, Technomic, 1989

Poly[sulfonyl bis(4-phenyl)carbonate]   P-537

Synonyms: *Poly[oxycarbonyloxy-1,4-phenylenesulfonyl-1,4-phenylene]*. *4,4′-Dihydroxydiphenylsulfone polycarbonate*. *Poly(4,4′-sulfonylbisphenol-co-carbonic dichloride)*
Monomers: 4,4′-Sulfonylbisphenol, Carbonic dichloride
Material class: Thermoplastic
Polymer Type: polycarbonates (miscellaneous)

CAS Number:

CAS Reg. No.
28930-33-8

Molecular Formula: $(C_{13}H_8O_5S)_n$
Fragments: $C_{13}H_8O_5S$

Volumetric & Calorimetric Properties:
Melting Temperature:

No.	Value	Note
1	200–210°C	[3]
2	233°C	cryst. [4]

Glass-Transition Temperature:

No.	Value	Note
1	132°C	[2]
2	195°C	cryst. [4]

Surface Properties & Solubility:
Solvents/Non-solvents: Sol. *m*-cresol, cyclohexanone, THF, pyridine, DMF [1], *N*-methylpyrrolidinone, DMSO [2], phenol (above 60°); insol. CH_2Cl_2, EtOAc, C_6H_6, Me_2CO [1], THF, MeOH, $CHCl_3$, 1,4-dioxane [2]. Reported to be both sol. and insol. in THF [1,2]
Wettability/Surface Energy and Interfacial Tension: Contact angle 78° (H_2O in air, 25°, 65% relative humidity) [2]

Optical Properties:
Transmission and Spectra: Ir spectral data have been reported [2]

Polymer Stability:
Decomposition Details: Weight loss of 10%, at 380° [2]. Residual mass at 500°, 45.3% [2]
Flammability: Limiting Oxygen Index 34 (ASTM D2863-77) [2]

Applications/Commercial Products:
Processing & Manufacturing Routes: Synth. by transesterification, in which the dihydroxy compound and a slight excess of diphenyl carbonate are mixed together in the melt (150–300°) in the absence of O_2, with elimination of phenol. Reaction rates are increased by catalysts (basic salts such as zinc acetate) and the use of a medium-high vacuum in the final stages [1,2,3]

Bibliographic References
[1] Schnell, H., *Polym. Rev.*, 1964, **9**, 99, (solvents, synth., rev)
[2] Liaw, D.-J. and Chang, P., *Polymer*, 1996, **37**, 2857, (props., synth.)
[3] Schnell, H., *Ind. Eng. Chem.*, 1959, **51**, 157, (T_m, synth.)
[4] Kim, C.K. and Paul, D.R., *Macromolecules*, 1992, **25**, 3097, (T_m, T_g)

Polytetrachloroethylene   P-538

Synonyms: *Polytetrachloroethene*
Monomers: Tetrachloroethene
Polymer Type: polyhaloolefins
CAS Number:

CAS Reg. No.
25135-99-3

Applications/Commercial Products:

Processing & Manufacturing Routes: Hydrophobic films synth. by radiofrequency glow-discharge (Ar plasma) polymerisation of monomer, often on supporting layer of another polymer, e.g. PET [10,11]. Conducting films can be produced by electrolytic reaction polymerisation, in methyl cyanate, of monomer in presence of $Bu_4N\ BF_4$ [7]

Applications: Used as component of binding agents in electro-thermal materials in anti-corrosion coatings (flexible), e.g. for batteries; in laminates/filters for gases as hydrophobic filter papers for spot tests in membranes for sensors, reflection spectroscopy boards, paints; adhesives and sorbents. Plasma-produced polymer films have potential medical applications (diagnostic assays, immunoassays, catheters, vascular graft and vascular assist devices) also have potential application as chromatography packing material. Conducting films have potential applications in batteries, semiconductor batteries, semiconductor devices and switching elements

Poly(1-tetracosene)   P-539

$$\mathrm{-[CH_2CH]_n\!-}$$
$$\mathrm{\ \ |}$$
$$\mathrm{CH_2(CH_2)_{19}CH_2CH_3}$$

Synonyms: *Poly(1-docosyl-1,2-ethanediyl). Poly(1-docosylethylene)*
Monomers: 1-Tetracosene
Material class: Thermoplastic
Polymer Type: polyolefins
CAS Number:

CAS Reg. No.	Note
31259-31-1	atactic

Molecular Formula: $(C_{24}H_{48})_n$
Fragments: $C_{24}H_{48}$

Applications/Commercial Products:
Processing & Manufacturing Routes: Synth. from 1-tetracosene using $AlEt_3$ at 198–200° and 2 mm pressure [1]

[1] *Brit. Pat.*, 1966, 1 048 984, (synth)

Poly(1-tetradecene)   P-540

$$\mathrm{-[CH_2CH]_n\!-}$$
$$\mathrm{\ \ |}$$
$$\mathrm{CH_2(CH_2)_{10}CH_3}$$

Synonyms: *Poly(1-dodecylethylene). Poly(1-dodecyl-1,2-ethanediyl)*
Monomers: 1-Tetradecene
Material class: Thermoplastic
Polymer Type: polyolefins
CAS Number:

CAS Reg. No.	Note
25608-58-6	atactic
26746-87-2	isotactic

Molecular Formula: $(C_{14}H_{28})_n$
Fragments: $C_{14}H_{28}$
General Information: Stiff rubber [2]
Morphology: Three polymorphic forms: form I orthorhombic, a 0.75, b 5.6, c 0.67, helix 4_1; form II orthorhombic, a 0.75, b 5.60, c 0.67 [5,7,8,9,17]

Volumetric & Calorimetric Properties:
Density:

No.	Value	Note
1	d 0.882 g/cm^3	[10]
2	d^{25} 0.848 g/cm^3	[13]

Melting Temperature:

No.	Value	Note
1	57°C	[2,3,4,13,17]
2	54°C	327 K [6]
3	56.3°C	[10]
4	52–55°C	[9,14,16]

Glass-Transition Temperature:

No.	Value	Note
1	10°C	[2]
2	-32°C	[12]

Transition Temperature:

No.	Value	Note	Type
1	-130°C	450 Hz [2]	Tβ
2	33°C	[10]	T_m
3	3°C	276 K, isotactic	T_1
4	14°C	287 K, isotactic	T_2
5	-14°C	110 Hz, isotactic	Tα
6	-36°C	110 Hz, isotactic	Tα
7	-48°C	110 Hz, isotactic	Tα
8	-160°C	110 Hz, isotactic [14]	Tβ

Surface Properties & Solubility:
Solvents/Non-solvents: Sol. heptane. Insol. Me_2CO [12]

Transport Properties:
Polymer Solutions Dilute: [η] 2.45 dl g^{-1} [13]. [η] 1.00–1.94 dl g^{-1} [16]

Mechanical Properties:
Mechanical Properties General: Dynamic tensile storage and loss moduli have been reported [1]
Viscoelastic Behaviour: Viscoelastic props. have been reported [15]

Optical Properties:
Transmission and Spectra: Ir and Raman spectral data have been reported [3,4,5,11]
Total Internal Reflection: Stress optical coefficient -6980–-7210 [16]

Applications/Commercial Products:
Processing & Manufacturing Routes: Synth. using Ziegler-Natta catalysts such as $TiCl_3$-$AlEt_3$ at 40° in heptane [3,4,10,12,13,16]

Bibliographic References
[1] Takayanagi, M., *Pure Appl. Chem.*, 1970, **23**, 151, (mech props)
[2] Clark, K.J., Jones, A.T. and Sandiford, D.J.H, *Ind. Chem.*, 1962, 2010, (transition temps)
[3] Holland-Moritz, K., *Colloid Polym. Sci.*, 1975, **253**, 922, (ir, Raman, synth)
[4] Modric, I., Holland-Moritz, K. and Hummel, D.O., *Colloid Polym. Sci.*, 1976, **254**, 342, (ir, Raman, synth)
[5] Fraser, G.V., Hendra, P.J., Chalmers, J.M., Cudby, M.E.A. and Willis, H.A., *Makromol. Chem.*, 1973, **173**, 195, (ir, Raman, struct)

[6] Trafara, G., Koch, R., Blum, K. and Hummel, D., *Makromol. Chem.*, 1976, **177**, 1089, (transition temps)
[7] Privalko, V.P., *Macromolecules*, 1980, **13**, 370, (struct)
[8] Trafara, G., *Makromol. Chem.*, 1980, **181**, 969, (struct)
[9] Blum, K. and Trafara, G., *Makromol. Chem.*, 1980, **181**, 1097, (struct)
[10] Soga, K., Lee, D.H., Shiono, T. and Kashiwa, N., *Makromol. Chem.*, 1989, **190**, 2683, (synth)
[11] Holland-Moritz, K., Modric, I., Heinen, K.U. and Hummel, D.O., *Kolloid Z. Z. Polym.*, 1973, **251**, 913, (ir, Raman)
[12] Natta, G., Danusso, F. and Moraglio, G., *CA*, 1958, **52**, 17789e, (synth, solubilities)
[13] *Neth. Pat.*, 1965, 6 411 790, (synth)
[14] Manabe, S. and Takayanagi, M., *Kogyo Kagaku Zasshi*, 1970, **73**, 1581, (transition temps)
[15] Manabe, S. and Takayanagi, M., *Kogyo Kagaku Zasshi*, 1970, **73**, 1595, (viscoelastic props)
[16] Philippoff, W. and Tomquist, E., *J. Polym. Sci., Part C: Polym. Lett.*, 1966, **23**, 881, (stress optical coefficient)
[17] Turner-Jones, A., *Makromol. Chem.*, 1964, **71**, 1, (struct)

Poly(tetrafluoroethylene-co-ethylene), Glass fibre filled   P-541

Synonyms: ETFE, Glass fibre filled. ETFE copolymer, Glass fibre filled. Tetrafluoroethylene-ethylene copolymer, Glass fibre filled
Related Polymers: Poly(tetrafluoroethylene-co-ethylene), Polytetrafluoroethylene
Monomers: Tetrafluoroethylene, Ethylene
Material class: Thermoplastic, Composites, Copolymers
Polymer Type: polyethylene, PTFE
Additives: Glass fibre
General Information: Glass fibre filler improves high temp. props.

Volumetric & Calorimetric Properties:
Density:

No.	Value	Note
1	d 1.86 g/cm^3	Tefzel [1]
2	d 1.89 g/cm^3	30% glass fibre [3]

Thermal Expansion Coefficient:

No.	Value	Note	Type
1	3.06×10^{-5} K^{-1}	Tefzel [1]	L
2	3.1×10^{-5} K^{-1}	30% glass fibre [2]	L

Thermal Conductivity:

No.	Value	Note
1	0.24 W/mK	0.000572 cal (s cm °C)$^{-1}$, Tefzel [1]

Melting Temperature:

No.	Value	Note
1	271°C	Tefzel [1]

Deflection Temperature:

No.	Value	Note
1	210°C	1.8 MPa, Tefzel [1]
2	238°C	1.8 MPa, 30% glass fibre [2]

Transport Properties:
Water Absorption:

No.	Value	Note
1	0.02 %	max., 24h, 30% glass fibre [2]

Mechanical Properties:
Tensile (Young's) Modulus:

No.	Value	Note
1	8267 MPa	84300 kg cm^{-2}, Tefzel [1]

Flexural Modulus:

No.	Value	Note
1	6541 MPa	66700 kg cm^{-2}, Tefzel [1]
2	7200 MPa	30% glass fibre [2]

Tensile Strength Break:

No.	Value	Note	Elongation
1	82.7 MPa	843 kg cm^{-2}, Tefzel [1]	8%
2	96 MPa	30% glass fibre [2]	4–5%

Compressive Strength:

No.	Value	Note
1	68.9 MPa	703 kg cm^{-2}, Tefzel [1]

Hardness: Rockwell R74 (Tefzel) [1,2]
Friction Abrasion and Resistance: Wear rate increases at high speeds
Izod Notch:

No.	Value	Notch	Note
1	480.4 J/m	Y	9 ft lb in^{-1}, room temp., Tefzel [1]
2	400 J/m	Y	30% glass fibre [2]

Electrical Properties:
Dielectric Permittivity (Constant):

No.	Value	Frequency	Note
1	3.4	1 MHz	Tefzel [1]

Dissipation (Power) Factor:

No.	Value	Frequency	Note
1	0.003	1 MHz	Tefzel [1]

Optical Properties:
Volume Properties/Surface Properties: Clarity opaque (30% glass fibre) [2]

Polymer Stability:
Chemical Stability: Attacked by chlorinated solvents

Applications/Commercial Products:
Mould Shrinkage (%):

No.	Value
1	0.2–0.3%

Trade name	Supplier
Tefzel HT-2004	DuPont

Bibliographic References

[1] *Tefzel*, DuPont, 1997, (technical datasheet)
[2] *The Materials Selector*, 2nd edn., (eds. N.A. Waterman and M.F. Ashby), Chapman and Hall, 1997, **3**, 221, (rev)

Poly(tetrafluoroethylene-*co*-hexafluoropropylene) P-542

Synonyms: FEP. Tetrafluoroethylene-hexafluoropropylene copolymer. FEP fluorocarbon. Fluorinated ethylene-propylene copolymer. Poly(tetrafluoroethene-co-1,1,2,3,3,3-hexafluoro-1-propene)
Related Polymers: Polytetrafluoroethylene, Polyhexafluoropropylene copolymers
Monomers: Tetrafluoroethylene, Hexafluoropropylene
Material class: Thermoplastic, Copolymers
Polymer Type: PTFE, FEP
CAS Number:

CAS Reg. No.
25067-11-2

Molecular Formula: $[(C_2F_4).(C_3F_6)]_n$
Fragments: C_2F_4 C_3F_6
Molecular Weight: M_n 52000–260000; 248000–629000
General Information: This copolymer has a lower melt temp. than PTFE and is thus easier to process using conventional techniques. The perfluoromethyl side groups reduce crystallinity to approx. 30–45%. The copolymer is random. Lower service temp. than virgin PTFE

Volumetric & Calorimetric Properties:
Density:

No.	Value	Note
1	d 2.14–2.17 g/cm^3	ASTM D792-50 [3]
2	d^{23} 2.136 g/cm^3	10.7 mol% hexafluoropropene, 49% cryst. [6]

Thermal Expansion Coefficient:

No.	Value	Note	Type
1	0.000135 K^{-1}	0–100°, ASTM E831, Teflon 100 [4]	L
2	0.000208 K^{-1}	100–150°, ASTM E831, Teflon 100 [4]	L
3	0.000266 K^{-1}	150–200°, ASTM E831, Teflon 100 [4]	L
4	9.3 × 10^{-5} K^{-1}	T > 23°, ASTM D696-44 [4]	V
5	5.7 × 10^{-5} K^{-1}	T < 23°, ASTM D696-44 [4]	V

Volumetric Properties General: Latent heat of fusion 24.3 kJ kg^{-1} (DSC) [1]
Thermal Conductivity:

No.	Value	Note
1	2.4 W/mK	-129–182°, Cenco Fitch [3]
2	1.4 W/mK	-253°, Cenco Fitch [3]
3	0.209 W/mK	23°, ASTM C177 [1]

Specific Heat Capacity:

No.	Value	Note	Type
1	1.17 kJ/kg.C	100° [1]	P
2	1.09 kJ/kg.C	20°, DSC [1]	P

Melting Temperature:

No.	Value	Note
1	253–282°C	[1]
2	259–306.9°C	[7]

Glass-Transition Temperature:

No.	Value	Note
1	126°C	15 mol% hexafluoropropene [2]

Transition Temperature General: α and γ relaxation temps. increase with increasing hexafluoropropene content [6]
Deflection Temperature:

No.	Value	Note
1	70–77°C	0.455 MPa [1]
2	51–57°C	1.82 MPa, ASTM D648 [1]

Brittleness Temperature:

No.	Value	Note
1	-250°C	approx.

Transition Temperature:

No.	Value	Note	Type
1	70–126°C	0–15% mol% hexafluoropropene [2]	α transition temp.
2	-70–-10°C	[2]	β transition temp.
3	-11–29°C	[6]	γ transition temp.
4	127–197°C	[6]	α transition temp.

Surface Properties & Solubility:
Solvents/Non-solvents: Absorption of organic solvents is less than 1% at elevated temps. Insol. most chemicals and solvents, even at elevated temps. and pressures [4]

Transport Properties:
Transport Properties General: Low permeability to liquids and gases compared with many other polymers
Melt Flow Index:

No.	Value	Note
1	7 g/10 min	Teflon 100 [4,9]
2	3 g/10 min	Teflon 140 [4,9]
3	1.5 g/10 min	Teflon 160 [4,9]

Polymer Melts: Melt fracture occurs approx. 100 s^{-1} [11]. Melt viscosity 8–50 kPa s (capillary rheometer) [1]
Water Absorption:

No.	Value	Note
1	0.01 %	max. [3,4]

Gas Permeability:

No.	Gas	Value	Note
1	O_2	3620 cm^3 mm/(m^2 day atm)	18.69 × 10^{-15} mol (m s Pa)$^{-1}$ [4,13]

– Poly(tetrafluoroethylene-co-hexafluora

2	He	21979 cm^3 mm/(m^2 day atm)	113.47 × 10^{-15} mol (m s Pa)$^{-1}$ [4,13]
3	N$_2$	1181.5 cm^3 mm/(m^2 day atm)	6.1 × 10^{-15} mol (m s Pa)$^{-1}$ [4,13]
4	H$_2$	7777 cm^3 mm/(m^2 day atm)	40.15 × 10^{-15} mol (m s Pa)$^{-1}$ [4,13]
5	CH$_4$	614 cm^3 mm/(m^2 day atm)	3.17 × 10^{-15} mol (m s Pa)$^{-1}$ [4,13]

Mechanical Properties:

Mechanical Properties General: Good flexibility at low temps. Has similar props to PTFE

Tensile (Young's) Modulus:

No.	Value	Note
1	340 MPa	[9]
2	1500 MPa	[10]

Flexural Modulus:

No.	Value	Note
1	655 MPa	23°, ASTM D790 [4]

Tensile Strength Break:

No.	Value	Note	Elongation
1	40 MPa	max. [4,9,10]	300–350%

Flexural Strength at Break:

No.	Note
1	no break [1]

Poisson's Ratio:

No.	Value	Note
1	0.48	Teflon 100 [9]

Tensile Strength Yield:

No.	Value	Note
1	12 MPa	23°, ASTM D638 [4]

Flexural Strength Yield:

No.	Value	Note
1	18 MPa	23°, ASTM D790 [4]

Compressive Strength:

No.	Value	Note
1	15 MPa	23°, ASTM D695 [1,4]
2	21 MPa	23° [9]

Mechanical Properties Miscellaneous: Tear strength 174.49 kN m^{-1} (ASTM D1004) [9]
Hardness: Durometer D59 (23°, ASTM D2240-T) [4]. Rockwell R25 [1]

Friction Abrasion and Resistance: Taber abrasion 7.5 g 1000000 cycles^{-1} (100 g load, CS17 wheel). [4] Coefficient of friction 0.27 (metal/film, ASTM D1894, Teflon 100) [4]
Izod Notch:

No.	Notch	Break	Note
1	Y	Y	ASTM D256 [1]

Electrical Properties:

Electrical Properties General: Excellent electrical props., widely used as an insulator
Surface/Volume Resistance:

No.	Value	Note	Type
2	>1 × 10^{15} Ω.cm	min., ASTM D257 [4,9]	S

Dielectric Strength:

No.	Value	Note
1	80 kV.mm^{-1}	0.25 mm film, 23°, ASTM D149 [4]
2	20 kV.mm^{-1}	3.2 mm sheet, 23°, ASTM D149 [4]

Arc Resistance:

No.	Value	Note
1	>300s	min. [1]
2	165s	ASTM D495 [9]

Dissipation (Power) Factor:

No.	Value	Frequency	Note
1	0.0003	100 kHz	ASTM D1531 [4,9]
2	0.0007	1 MHz	ASTM D1531 [4,9]

Optical Properties:

Optical Properties General: Transmits more uv, visible light and ir than ordinary window glass [4]
Refractive Index:

No.	Value	Note
1	1.341–1.347	[4,9]

Volume Properties/Surface Properties: Gloss 70–90% (ASTM D2457) [10]. Haze 2.0% (ASTM D1003) [10]. Light transmittance 94–96% (ASTM E424, 1 mm) [10]

Polymer Stability:

Polymer Stability General: Excellent weatherability, chemical resistance and thermal stability
Thermal Stability General: Mech. props. retained in continuous service at 200° [4]
Upper Use Temperature:

No.	Value	Note
1	-250–206°C	[1]
2	200°C	

Decomposition Details: Very low weight loss at elevated temps.; hourly weight loss 0.0004% (230°), 0.3% (370°) [4]
Flammability: Non-flammable (ASTM D635) [3]; flammability rating VE0 (UL 94) [1]. Limiting Oxygen Index 34% (ASTM D2863) [9]

Environmental Stress: After 15 years' outdoor exposure in Florida tensile strength, light transmission and clarity remained unchanged (25 μm film) [4]

Chemical Stability: Resistant to most chemicals but does react with fluorine, molten alkali metal, and molten sodium hydroxide [3]

Hydrolytic Stability: Unaffected by aq. solns.

Stability Miscellaneous: Processed at approx. 370°; long hold up times can cause degradation [4]. Ten times more resistant to radiation in air than PTFE

Applications/Commercial Products:

Processing & Manufacturing Routes: Synth. by aq. and non-aq. dispersion polymerisation. Processed by injection moulding, compression moulding and extrusion. Films can be heat sealed and vacuum formed

Applications: Used in bearings, seals, coatings, linings, chemically resistant mouldings, wire insulation and other electrical and electronic applications

Trade name	Supplier
Korton	Norton
Neoflon	Daikin Kogyo
Teflex	Niitechim
Teflon FEP	DuPont
Teflon N	DuPont

Bibliographic References

[1] Kerbow, D.L. and Sperati, C.A., *Polym. Handb.*, 4th edn., (eds. J. Brandrup, E.H. Immergut and E.A. Grulke), John Wiley and Sons, 1999, **V**, 41, (mech props)
[2] McCrum, N.G., Read, B.E. and Williams, G., *Anelastic and Dielectric Effects in Polymeric Solids* Wiley, 1967
[3] *Kirk-Othmer Encycl. Chem. Technol.*, Wiley Interscience, 1980, **11**, 24
[4] *Encycl. Polym. Sci. Eng.*, Wiley Interscience, 1989, **16**, 601
[5] Starkweather, H.W., Zoller, P. and Jones, G.A., *J. Polym. Sci., Polym. Phys. Ed.*, 1984, **22**, 1431
[6] Eby, R.K. and Wilson, F.C., *J. Appl. Phys.*, 1962, **33**, 2951
[7] Colson, J.P. and Eby, R.K., *J. Appl. Phys.*, 1966, **37**, 3511
[8] Wu, S., *Macromolecules*, 1985, **18**, 2023
[9] Teflon 100, Materials Data Sheet DuPont, 1992, (technical datasheet)
[10] Korton, Materials Datasheet Norton, 1995, (technical datasheet)
[11] Brydson, J.A., *Plast. Mater.*, 5th edn., Butterworth-Heinemann, 1989, 353
[12] Weeks, J.J., Sanchez, I.C., Eby, R.K. and Poser, C.I., *Polymer*, 1980, **21**, 325
[13] *J. Teflon*, 1970, **11**, 8

Poly(tetrafluoroethylene-*co*-perfluoropropyl vinyl ether) P-543

Synonyms: *Tetrafluoroethylene-perfluoropropyl vinyl ether copolymers. Poly(1,1,1,2,2,3,3-heptafluoro-3-[(trifluoroethenyl)oxy]propane-co-tetrafluoroethene)*

Related Polymers: Polytetrafluoroethylene
Monomers: Tetrafluoroethylene, Perfluoro(propyl vinyl ether)
Material class: Thermoplastic, Copolymers
Polymer Type: PTFE, fluorocarbon (polymers)
CAS Number:

CAS Reg. No.
26655-00-5

Molecular Formula: $[(C_2F_4).(C_5F_{10}O)]_n$
Fragments: C_2F_4 $C_5F_{10}O$
Additives: Stable inorganic pigments at 0.1–1.0%
General Information: Polymer contains approx. 65–75% crystallinity

Applications/Commercial Products:

Processing & Manufacturing Routes: For aq. polymerisation, water soluble initiators and a perfluorinated emulsifying agent are used. For non-aq. copolymerisation, fluorinated acyl peroxides that are soluble in the medium are used as initiators. Processed by injection moulding, transfer moulding, dispersion processing, powder coating

Applications: Electrical insulation, mechanical parts, coatings and linings, speciality tubing, injection moulded articles

Trade name	Supplier
Hostaflon TFA	Hoechst Celanese
Neoflon AP	Daikin Kogyo
Teflon PFA	DuPont

Poly(tetrafluoroethylene-*co*-vinylidene fluoride) P-544

$$\left[\left[CH_2CF_2 \right]_x \left[CF_2CF_2 \right]_y \right]_n$$

Synonyms: *Poly(vinylidene fluoride-co-perfluoroethylene). Poly(vinylidene fluoride-co-tetrafluoroethylene). Tetrafluoroethylene-vinylidene fluoride copolymer. PTFE-PVdF copolymer. Poly(1,1-difluoroethene-co-tetrafluoroethene). PVDF-co-PTFE*

Related Polymers: Polyvinylidene fluoride, Polytetrafluoroethylene
Monomers: Tetrafluoroethylene, Vinylidene fluoride
Material class: Thermoplastic, Copolymers
Polymer Type: PVDF, PTFE
CAS Number:

CAS Reg. No.
25684-76-8

Molecular Formula: $[(C_2H_2F_2)_x(C_2F_2)]_n$
Fragments: C_2F_2 CH_2F_4
Additives: Fillers (alkaline earth metals oxides)
General Information: Copolymers can be produced across the entire composition range (very close reactivity ratios) [17] Addition of further comonomer, e.g. hexafluoropropylene (10 wt%), gives terpolymers with increased solubility in organic solvents, plus reduction in degree of crystallinity and T_m; chlorotrifluoroethylene also used as additional comonomer. Such formulations are used in paint preparations which dry at room temp. [17]. Is of interest owing to its dielectric, piezoelectric and pyroelectric props.

Morphology: Copolymers exhibit high degree of crystallinity, which increases with increasing TFE content over wide composition range [8,12]. X-ray studies show struct. is very similar to that of PVDF, form I [1]. Semicrystalline films can be made from copolymer containing 27 wt% TFE. The degree of crystallinity may be 35–50%, depending on rate of cooling from melt. Crystallites are surrounded by amorph. regions; addition of TFE to VDF induces crystallisation into an all-*trans* conformn. of PVDF (β-phase) hexagonal struct. to within 0.1% [3]. Vinylidene fluoride blocks in the copolymer have same struct. as polyvinylidene fluoride [11]. Copolymer (containing 27 wt% TFE) crystallises directly from the melt into cryst. phase analogous to the β-phase of PVDF [13] Degree of crystallinity 20–40 mol% (39–34 mol% TFE) [14]. The presence of tetrafluoroethylene units favours crystalline transformation of disordered γ-form into ordered phase; head-to-head struct. favours γ-phase formation and transformation of disordered γ-form into ordered phase [30]. The effects of mechanical and electrical stresses on struct. have been reported [7]. Addition of tetrafluoroethylene comonomer does not introduce new chemical species (as with other fluorinated comonomers e.g. trifluoroethylene), but merely increases head-to-head defect content [24]. The relationship between composition and solid-state transformations have been reported [25,26]. Isothermal crystallisation from the melt first produces paraelectric crystal phase. This then undergoes a crystal-crystal transition into a ferroelectric phase [26]. Crystallisation and transition kinetics are extremely temp. dependent [26]. The morphology during stretching has been reported [40]

Identification: Elemental anal. of such insol., partially hydrogenated, partially fluorinated polymers is difficult [9]

Miscellaneous: Reviews of physical props, mech. props., compounding, processing and uses have been reported [16,17]. At a TFE content of less than 30%, copolymer shows β-PVDF type packing. At a TFE content above 30%, copolymer displays X-ray patterns analogous to those of PTFE. Extensive isodimorphism appears to occur between monomeric units in general [12]

Volumetric & Calorimetric Properties:

Density:

No.	Value	Note
1	d 1.865–1.896 g/cm^3	17.8–30.7 mol% TFE [4]
2	d 1.88 g/cm^3	average value, commercial material; 20 wt% TFE, ISOR 1183D [17]
3	d 1.79–1.95 g/cm^3	4–39 mol% TFE [14]
4	d 1.849–1.865 g/cm^3	slow and fast cooling after heat-pressing, 17.8 mol% tetrafluoroethylene [4]
5	d 1.89–1.94 g/cm^3	slow and fast cooling after heat-pressing, 25.1 mol% tetrafluoroethylene [4]
6	d 1.896 g/cm^3	no difference between cooling methods, 30.7 mol% tetrafluoroethylene [4]

Latent Heat Crystallization: ΔH_f 13–21 J g^{-1} (average value, commercial material, 20 wt% TFE) [17]

Melting Temperature:

No.	Value	Note
1	126°C	27 wt% TFE [13]
2	132–155°C	16–4 mol% TFE [8]
3	243–272°C	55 mol% TFE [7]
4	122–126°C	average values, commercial material, 20 wt% TFE, ISO 12086 [17]
5	120°C	approx., 20 mol% TFE [17]
6	136.4°C	commercially available copolymer, 19 wt% tetrafluoroethylene; slow cooled from melt [24]
7	124–325°C	dependent on vinylidene fluoride content
8	410–420°C	[36]

Glass-Transition Temperature:

No.	Value	Note
1	-50°C	amorph. regions, 27 wt% TFE [3]
2	-50--44°C	16–4 mol% TFE [8]

Transition Temperature General: Copolymer containing 19% TFE displays conformational change at 110–115° ascribed to Curie transition [4,15]. For copolymers containing 17–30 mol% TFE, small endotherms occurring below T_m can be correlated with Curie transitions [4]. T_m increases with increasing TFE content [17]. T_m increases with decreasing vinylidene fluoride content [37]. Compositions with a vinylidene fluoride content greater than 82% show no phase transitions between room temp. and the melting point. Curie transitions are observed at vinylidene fluoride content of 70–82% decreasing in temp. with decreasing vinylidene fluoride content. When the vinylidene fluoride content is less than 70% more gradual transitions are observed [36,38]. Mechanical relaxations have been reported [39]. The effects of poling on the Curie temp. have been reported [36,38]

Deflection Temperature:

No.	Value	Note
1	39–43°C	average values, commercial material, 20 wt% TFE, ISO 75 [17]

Transition Temperature:

No.	Value	Note	Type
1	72–176°C	30.7–17.8 wt% TFE [3,4]	Curie temp.
2	120.4°C	commercially available copolymer, 19 wt% tetrafluoroethylene, slow cooled from melt [24]	Curie temp.

Surface Properties & Solubility:

Solubility Properties General: Solubility in several solvents is better than that of PVDF [17]

Solvents/Non-solvents: Commercial material (20 wt% TFE) sol. Me$_2$CO, 2-butanone, EtOAc, dimethyl acetamide, DMF, isophorone; spar. sol. isobutylmethylketone, cyclohexanone, THF, dioxane; insol. hydrocarbons, alcohols [17]

Transport Properties:

Transport Properties General: The relationship between melt index and molecular characteristics such as MW and MW distribution has been reported. There is an inverse correlation between high-MW MW distribution component and melt index [5]

Melt Flow Index:

No.	Value	Note
1	2.3 g/10 min	6 mol% tetrafluoroethylene [40]

Polymer Melts: Melt flow index increases with comonomer content (TFE) and temp. [21]

Permeability of Gases: Permeability of copolymer films to ammonia increases with time and thermal treatment [28]. Permeability to N$_2$, O$_2$, CO$_2$ and He decreases after exposure to fluoride and fluorine/hydrogen fluoride mixtures [34]

Permeability and Diffusion General: Diffusion of hydrogen chloride and effect on service life have been reported [33]

Mechanical Properties:

Mechanical Properties General: Tensile strength increases with decreasing temp. down to -253°, and then decreases down to temp. of -268.8°. Breaking elongation decreases with decreasing temp. Addition of fillers gives increased tensile strength at room temp., but lower values at low temps. [10] Tensile strength and elongation increase with increasing high MW component (in MW distribution curve) [5]. Copolymer fibres may be strengthened by stretching six or seven times at 90–135° [27]. Mechanical props are significantly affected by exposure to fluorine and fluorine/hydrogen fluoride mixtures

Tensile (Young's) Modulus:

No.	Value	Note
1	2360 MPa	20°, isotropic [40]
2	4350 MPa	20°, cold-stretched [40]
3	3600 MPa	20°, isometric annealing [40]
4	2700 MPa	20°, free annealing [40]

– Poly(tetrafluoroethylene-co-vinylidene fluoride)

5	38300 MPa	-160°, cold stretched [40]
6	40000 MPa	-160°, isometric annealing [40]
7	26200 MPa	-160°, free annealing [40]

Flexural Modulus:

No.	Value	Note
1	600–700 MPa	23°, commercial material, 20 wt% TFE, ISO 178 [17]

Tensile Strength Break:

No.	Value	Note	Elongation
1	32–45 MPa	23°, commercial material, 20 wt% TFE, ISO R527-2/ISO 12086 [17]	500–800%

Tensile Strength Yield:

No.	Value	Note
1	15–19 MPa	23°, commercial material, 20 wt% TFE, ISO R 527-2/ISO 12086 [17]

Viscoelastic Behaviour: Longitudinal speed of sound 1300 m s^{-1} (200 kHz, 20°, isotropic), 1550 m s^{-1} (200 kHz, 20°, cold-stretched), 1410 m s^{-1} (200 kHz, 20°, isometric annealing), 1220 m s^{-1} (200 kHz, 20°, free annealing), 2430 m s^{-1} (200 kHz, -160°, isotropic), 4600 m s^{-1} (200 kHz, -160°, cold-stretched), 4700 m s^{-1} (200 kHz, -160°, isometric annealing), 3800 m s^{-1} (200 kHz, -160°, free annealing) [40]

Mechanical Properties Miscellaneous: Copolymers containing 17–30 mol% TFE show relaxations that can be correlated with Curie transitions [4]

Izod Notch:

No.	Value	Notch	Note
1	1100 J/m	Y	23°, commercial material, 20 wt% TFE, ISO 180 [17]

Electrical Properties:
Electrical Properties General: Dielectric loss angle (tan δ_{max}) decreases with increasing tetrafluoroethylene content and decreasing degree of branching [31]. Dielectric constant as a function of temp. for copolymer containing 27 wt% TFE has been reported [13]

Dielectric Permittivity (Constant):

No.	Value	Frequency	Note
1	7.2	1 Hz	23°, commercial material, 20 wt% TFE [17]
2	1.09		quenched, degree of crystallinity 43%, 80% vinylidene fluoride) [41]
3	0.95		crystallised at 120°, degree of crystallinity 60%, 80% vinylidene fluoride [41]
4	0.76		crystallised at 250°, degree of crystallinity 80%, 80% vinylidene fluoride [41]

Complex Permittivity and Electroactive Polymers: The dielectric props., charge distribution and change transport props. of thin films in combination with Si in FET structs. have been reported [3]. Ferroelectric props. of copolymer (27 wt% TFE) have been reported. Ferroelectricity results from each CH_2CF_2 unit and copolymer crystallising into polar (β) crystal phase. Exhibits reversible remnant polarisation [13]. Undergoes a ferroelectric transition to a paraelectric state [15,20], which is induced by electron irradiation [15]. Brillouin scattering experiments have also been reported [19]. Copolymers (4–39 mol% TFE) have piezoelectric d-constants that are dependent on degree of crystallinity and film compliance [14]. A basic relationship exists between charging behaviour and bulk/surface polymer struct. [32]. Ferroelectric/paraelectric transitions have been reported [26]

Magnetic Properties: Ferroelectric, piezoelectric and pyroelectric props., and with pyroelectric and piezoelectric constants for drawn and undrawn copolymer have been measured [41,42]. Piezoelectric stress constants 3.7 kC m^{-2} (quenched; degree of crystallinity 43%, 80% vinylidene fluoride, 4.6 kC m^{-2} (crystallised at 120°, degree of crystallinity 60%, 80% vinylidene fluoride), 10.2 kC m^{-2} (crystallised at 250°, degree of crystallinity 80%, 80% vinylidene fluoride) [41]

Dissipation (Power) Factor:

No.	Value	Frequency	Note
1	0.013	1 Hz	23°, commercial material, 20 wt% TFE [17]

Optical Properties:
Refractive Index:

No.	Value	Note
1	1.59	25 wt% TFE [1]
2	1.4	commercial material, 20 wt% TFE [17]

Transmission and Spectra: Wide-line F-19 nmr, [9,12] far-ir (15–400 cm^{-1}) [1,7,40], esr, [6] Raman [24] and secondary emission mass [11] spectral data have been reported

Total Internal Reflection: Stretched samples show high dichroism [7]. H-1 nmr, C-13 nmr and F-19 nmr spectral data have been reported [35]

Polymer Stability:
Polymer Stability General: Addition of fillers (e.g. alkaline earth oxides), reportedly gives up to 25-fold improvement in both service life and chemical stability [33]

Decomposition Details: The products of thermal decomposition have been evaluated [22]

Flammability: Fire resistance is inferior to that of the chlorine-substituted analogue. Oxygen index 60.5–69.0% (approx.) for copolymer containing 26 wt% TFE (various MWs) [2]. Fl. pt., ignition temp. and self-ignition temp. are dependent on comonomer ratio [23]. Limiting oxygen index 44% (average value; commercial material; 20 wt% TFE ISO 4589) [17]. Flammability rating V0 (average value, commercial material (0.8mm thick), 20 wt% TFE, UL94) [17]

Chemical Stability: Reacts with ammonia; two types of bonds are formed: thermally unstable ammonium salt bonds and thermally stable bonds ($CF-NH_2$, $CF-NH-CF$, $C=NH$) [28]. Is resistant to AcOH [29]

Stability Miscellaneous: Stability to γ-irradiation has been reported [43]

Applications/Commercial Products:
Processing & Manufacturing Routes: Typical synth. involves aq. emulsion copolymerisation of mixture of comonomers, e.g. using ammonium persulfate as free-radical initiator at 80° and 8 atm [12], or at 90° and 8–11 atm [8]. Suspension and soln. polymerisation are also used [17]. Melt processed by standard extrusion and moulding techniques at 190–260° [17]. Film, sheet, tube, pipe, cable jackets and other standard or special shapes produced by extrusion or moulding [17]

Applications: Industrial applications in general limited to two compositions, i.e. 5% and 20 wt% TFE, the latter being the most important. Applications due to solubility and relatively low T_m in paint industry (coatings). May often be formulated with acrylic resins. Also used as components of high-performance greases, cellular and porous plastics and membranes. Applications due to piezoelectric and pyroelectric props. in microphones (and other acoustic transducers), temperature sensors. Has potential uses (doped with Si) in field-effect transistors. Other applications include use in pipelines, pumps and valves (for use in harsh environments, e.g. wood chemistry industry), membranes for separation of gas mixtures, and in electrophotography

Trade name	Details	Supplier
Ftorlon	42, 42A, 2M, non-exclusive; used for other VDF-based polymers; former USSR	
Kynar	various grades; non-exclusive; used for other VDF-based polymers	Pennwalt Chemical Corporation
Kynar	7200/SL (200 wt% TFE); 7201 (30 wt% TFE, semi-crystalline)	Atochem Polymers
Neoflon	non-exclusive, used for other VDF-based polymers	Daikin Kogyo

Bibliographic References

[1] Latour, M. and Moreira, R.L., *J. Polym. Sci., Polym. Phys. Ed.*, 1987, **25**, 1913, (ir, optical props., conformn.)
[2] Pavlenkova, G.A., Borisov, E.M. and Mizerovskii, L.N., *Plast. Massy*, 1990, 88, (flammability)
[3] Langlois, J.-M., Noirhomme, B., Fillon, A., Rambo, A. and Caron, L.G., *J. Appl. Phys.*, 1987, **61**, 5360, (electrical props., morphology, uses)
[4] Murata, Y. and Koizumi, N., *Polym. J. (Tokyo)*, 1985, **17**, 1071, (volumetric props., electrical props., thermal props., mechanical props.)
[5] Madorskaya, L.Ya., Otradine, G.A., Andreeva, A.I., Zaichenko, Yu.A., *Plast. Massy*, 1988, 31, (MW, transport props., mechanical props.)
[6] Noël, C., et al, *Plast. Massy*, 1986, **19**, 201, (esr)
[7] Latour, M. and Moreira, R.L., *J. Polym. Sci., Polym. Phys. Ed.*, 1987, **25**, 1913, (ir, struct., conformn., morphology)
[8] Moggi, G., Bonardelli, P., Monti, C. and Bart, J.C.J., *Polym. Bull. (Berlin)*, 1984, **11**, 35, (synth., thermal props., morphology)
[9] Lutringer, G., Meurer, B. and Weill, G., *Makromol. Chem.*, 1989, **190**, 2815, (F-19 nmr, anal.)
[10] Novikov, N.M., Ill'ichev, V.Ya. and Drozdov, V.V., *Nizkotemp. Vak. Materialoved.*, 1973, **3**, 71, (mechanical props.)
[11] Tantsyrev, G.D., Povolotskaya, M.I. and Kleimenov, N.A., *Vysokomol. Soedin., Ser. A*, 1977, **19**, 2057, (ms, microstruct.)
[12] Moggi, G., Bonardelli, P. and Bart, J.C.J., *J. Polym. Sci., Polym. Phys. Ed.*, 1984, **22**, 357, (morphology, struct., synth., F-19 nmr)
[13] Hicks, J.C., Jones, T.E. and Logan, J.C., *J. Appl. Phys.*, 1978, **49**, 6093, (ferroelectric props., dielectric props., morphology)
[14] Stefanou, H., *J. Appl. Phys.*, 1979, **50**, 1486, (piezoelectric props., volumetric props., morphology)
[15] Lovinger, A.J., *Macromolecules*, 1983, **16**, 1486, (ferroelectric props., morphology, struct., thermal props.)
[16] Schroeder, H., *Rubber Technology*, (ed. M. Morton), 3rd edn., Van Nostrand Reinhold, 1987, 410, (rev.)
[17] *Modern Fluoropolymers*, (ed. J. Scheirs), John Wiley, 1997, (rev.)
[18] Murata, Y. and Koizumi, N., *Ferroelectrics*, 1989, **92**, 47, (ferroelectric props)
[19] Liu, Z. and Schmidt, V.H., *Ferroelectrics*, 1990, **112A**, 237, (ferroelectric props)
[20] Legrand, J.F. et al, *Ferroelectrics*, 1987/88, **78**, 151, (ferroelectric props)
[21] Loginova, N.N., Podlesskaya, N.K., Kochkina, L.G., Bronov, M.V. and Madorskaya, L.Ya., *Plast. Massy*, 1988, 24, (transport props., thermal props.)
[22] Ryabikova, V.M., Ivanova, T.L., Ziegel, A.N., Pirozhnaya, L.N. and Popova, G.S., *Zh. Anal. Khim.*, 1988, **43**, 1093, (thermal stability, thermal decomposition, ir, anal)
[23] Korzhova, I.T., Madorskaya, L.Ya., Makeenko, T.G., Sokolova, N.V. and Alekseeva, N.I., *Plast. Massy*, 1991, 58, (flammability)
[24] Green, J. and Rabolt, J.F., *Macromolecules*, 1987, **20**, 456, (ferroelectric props, thermal props, Raman)
[25] Lovinger, A.J., Davis, D.D., Cais, R.E. and Kometani, J.M., *Macromolecules*, 1988, **21**, 78, (morphology, struct, ferroelectric props)
[26] Marand, H. and Stein, R.S., *Macromolecules*, 1989, **22**, 444, (struct, crystallisation, thermal props, ferroelectric props)
[27] Poddubnyi, V.I., El'yashevich, G.K., Strel'tses, B.V. and Bezprozvannykh, A.V., *Khim. Volokna*, 1990, 31, (mech props)
[28] Kostrov, Yu.A., Velikanova, I.M., Litovchenka, G.D., Reibarkh, F.E. et al, *Polym. Sci. USSR (Engl. Transl.)*, 1987, **29**, 2328, (chemical stability, permeability)
[29] Zotova, L.M., Kurilova, M.M., Efimov, A.A., Druzhinin, V.F. and Ivanou, V.N., *Gidroliz. Lesokhim. Prom.-St.*, 1986, 11, (chemical stability, uses)
[30] Wen, J. and Lu, W., *Gaofenzi Cailiao Kexue Yu Gongcheng*, 1988, **4**, 29, (morphology)
[31] Matveev, V.K., Smirnova, N.A., Madorskaya, L.Ya., Loginova, N.N. and Milinchuk, V.K., *Vysokomol. Soedin., Ser. B*, 1990, **32**, 762, (dielectric props)
[32] De Reggi, A.S., *CA*, 1986, **106**, 157208d, (ferroelectric props, morphology, struct)
[33] Shterenzon, A.L., Mulin, Yu.A. and Gemusova, I.B., *Plast. Massy*, 1988, 35, (chemical stability, transport props)
[34] Semenova, S.I. et al, *Plast. Massy*, 1991, 55, (permeability, chemical stability, mech props)
[35] Lutringer, G., Meurer, B. and Weill, G., *Polymer*, 1992, **33**, 4920, (F-19 nmr)
[36] Murata, Y. and Koizumi, N., *Polym. J. (Tokyo)*, 1985, **17**, 1071, (transition temps)
[37] Tashiro, K., Kaito, H. and Kobayashi, M., *Polymer*, 1992, **33**, 2915, (T_m)
[38] Murata, Y., *Polym. J. (Tokyo)*, 1987, **19**, 337
[39] Abe, Y., Kakizaki, M. and Hideshima, T., *J. Appl. Phys.*, 1985, **24**, 208
[40] Kochervinskii, V.V., Glukhov, V.A., Romadin, V.F., Sokolov, V.G. and Lokshin, B.V., Polym. Sci., Ser. A Ser. B (Transl. Vysokomol. Soedin., *Ser. A Ser. B)*, 1988, **30**, 2037, (morphology, mech props)
[41] Tasaka, S. and Miyata, S., *J. Appl. Phys.*, 1985, **57**, 906, (ferroelectric props)
[42] Wen, J.X., *Polym. J. (Tokyo)*, 1985, **17**, 399, (piezoelectric props)
[43] Khlyabich, P.P., Otradina, G.A., Budtov, V.P. and Sirota, V.P., Polym. Sci., Ser. A Ser. B (Transl. Vysokomol. Soedin., *Ser. A Ser. B)*, 1989, **31**, 503

Poly(tetramethylene adipate)   P-545

Synonyms: *PBA. PTA. Poly(butylene adipate). Poly(oxytetramethyleneoxyadipoyl). Poly(oxytetramethylene oxycarbonyl tetramethylene carbonyl). Poly(butylene glycol adipate). Poly[oxy-1,4-butanediyloxy(1,6-dioxo-1,6-hexanediyl)]. PTMA. Polybutilate. Poly(1,4-butylene adipate)*
Monomers: Hexanedioic acid, 1,4-Butanediol
Material class: Thermoplastic
Polymer Type: saturated polyester
CAS Number:

CAS Reg. No.	Note
25103-87-1	Poly(tetramethylene adipate)
24936-97-8	

Molecular Formula: $[(C_{10}H_8O_4).(C_{10}H_{16}O_4)]_n$
Fragments: $C_{10}H_8O_4$ $C_{10}H_{16}O_4$
Morphology: Two polymorphic forms [1,9], α-form monoclinic a 0.670, b 0.800, c 1,420, β 45.5° [1], β-form orthorhombic a 0.505, b 0.736, c 1.44 [1], α-form monoclinic a 0.673, b 0.794, c 1.420, β 45.5° [2], β-form orthorhombic a 0.5062, b 0.7352, c 1.467 [2]. α-form monoclinic a 0.677, b 0.811, c 1.466, β 43° 47' [3], a 0.561, b 0.727, c 1.465, β 115.5° [3], a 1.005, b 0.811, c 1.466, β 90.0° [3]. The struct. has been studied [5]

Volumetric & Calorimetric Properties:
Thermodynamic Properties General: Thermodynamic props have been reported [24,26]
Latent Heat Crystallization: ΔH_f 50.5 J g^{-1} [5]. ΔH_f 75.4−80.4 J g^{-1} [9]. ΔH_f 95.3 kJ kg^{-1} (high MW). ΔH_f 104.1 kJ kg^{-1} (low MW) [28]. ΔH_m 24.8 kJ mol^{-1} [26]. S_0 24 J mol^{-1} K^{-1} [26]. $\Delta H_{combustion}$ -5564 kJ mol^{-1}, ΔH_f -973 kJ mol^{-1} [26]. C_p values over temp. range -193° to -197° have been reported [11,12,26]

Poly(tetramethylene adipate)

Melting Temperature:

No.	Value	Note
1	60°C	[5]
2	63–65°C	[7]
3	57.2–60.2°C	330.4–333.4 K [9]
4	54°C	327 K [13]
5	58°C	α-form [1]
6	48°C	β-form [1]
7	58°C	[16]
8	55.6°C	328.8 K [26]
9	60°C	[33]

Glass-Transition Temperature:

No.	Value	Note
1	-59°C	[5]
2	-61°C	212 K [13]
3	-74°C	199 K [26]
4	-68°C	[32]

Surface Properties & Solubility:
Cohesive Energy Density Solubility Parameters: Energy of cohesion has been reported [21]
Internal Pressure Heat Solution and Miscellaneous: Phase diagram with 1,4-dioxane has been reported [16]
Solvents/Non-solvents: Sol. $CHCl_3$. Insol. MeOH [5]
Wettability/Surface Energy and Interfacial Tension: Surface tension values have been reported [7,21]

Transport Properties:
Polymer Solutions Dilute: Soln. props. have been reported [6,14]. θ 77° (cyclohexanol) [31]. [η] 0.288 dl g^{-1} ($CHCl_3$, 30°, MW 6700). [η] 0.488 dl g^{-1} ($CHCl_3$, 30°, MW 14500). [η] 0.322 dl g^{-1} (25°, C_6H_6) [29]. Theta temp. θ 25° (C_6H_6/hexane 83.3:16.7, chlorobenzene/hexane 75.9:24.1, hexane, nitrobenzene 43.3:56.7) [6]. Other viscosity values have been reported [19]
Permeability of Liquids: Diffusion coefficient has been reported [18]. D_o 367.9–550.8 μm^2 s^{-1} (octadecane, 130–145°) [29]
Water Content: Water absorption has been reported [18]

Mechanical Properties:
Tensile Strength Break:

No.	Value	Note
1	66 MPa	[28]

Electrical Properties:
Electrical Properties General: Other dielectric constants have been reported [30]
Dielectric Permittivity (Constant):

No.	Value	Note
1	2.19	
2	2.21	[6]

Optical Properties:
Transmission and Spectra: Ir [4,22,28], C-13 nmr [8,27,28], H-1 nmr [20,27,28], Raman [22] and mass spectral data [15,25] have been reported

Molar Refraction: dn/dc 0.1 (50°, AcOH), 0.0784 (35°, THF), 0.084 (35°, diethyl carbonate), 0.091 (70°, diethyl carbonate), 0.0518 (35°, dioxane) [6]

Polymer Stability:
Decomposition Details: Thermal degradation [25] at 500° produces a mixture of esters of the monomer, whereas at 700° adipic acid is a major constituent [17]
Biological Stability: Biodegradation details have been reported [23]. The high MW material degrades more slowly than the low MW material [28]

Applications/Commercial Products:
Processing & Manufacturing Routes: Prod. by condensation polymerisation of dimethyl adipate with 1,4-butanediol using Ti(OBu)$_4$ as catalyst at 150–210° [14,4,5,8]
Applications: Plasticisers. Used in polyurethane industry for manufacturing of pre-polymers, thermoplastic elastomers, coatings and adhesives, dispersing agents, and cast elastomers

Trade name	Supplier
Formrex 44	Witco

Bibliographic References
[1] Minke, R. and Blackwell, J., *J. Macromol. Sci., Phys.*, 1979, **16**, 407, (cryst struct)
[2] Minke, R. and Blackwell, J., *J. Macromol. Sci., Phys.*, 1980, **18**, 233, (cryst struct)
[3] Kim, C.W., *PhD Thesis*, University of Delaware, 1971, (cryst struct)
[4] Davison, W.H.T. and Corish, P.J., *J. Chem. Soc.*, 1955, 2428, (ir, synth)
[5] Gilbert, M. and Hybart, F.J., *Polymer*, 1974, **15**, 407, (synth, transition temps)
[6] Knecht, M.R. and Elias, H.G., *Makromol. Chem.*, 1972, **157**, 1, (dielectric constant, soln props)
[7] Mayrhofer, R.C. and Sell, P.J., *Angew. Makromol. Chem.*, 1971, **20**, 153, (surface tension)
[8] Kricheldorf, H.R., *Makromol. Chem.*, 1978, **179**, 2133, (C-13 nmr, synth)
[9] Szekely-Pecsi, Z., Vancso-Szmercsanyi, I., Cser, F., Varga, J. and Belina, K., *J. Polym. Sci., Polym. Phys. Ed.*, 1981, **19**, 702, (thermodynamic props)
[10] Liau, W.B. and Boyd, R.H., *Macromolecules*, 1990, **23**, 1531, (struct)
[11] Nistratov, V.P., Shuetsova, K.G. and Babinkov, A.G., *Termodin. Org. Soedin.*, 1981, 32, (heat capacity)
[12] Lim, S. and Wunderlich, B., *Polymer*, 1987, **28**, 777, (heat capacity)
[13] Jutier, J.J., Lernieux, E. and Prud Homme, R.E., *J. Polym. Sci., Part B: Polym. Phys.*, 1988, **26**, 1313, (transition temps)
[14] Munari, A., Manaresi, P., Chiorboli, E. and Chiolle, A., *Eur. Polym. J.*, 1992, **28**, 101, (synth, soln props)
[15] Kim, Y.L. and Hercules, D.M., *Macromolecules*, 1994, **27**, 7855, (ms)
[16] Carbonnel, L., Rosso, J.C. and Ponge, C., *Bull. Soc. Chim. Fr.*, 1972, **3**, 941, (phase diagram)
[17] Messier, F. and De Jongh, D.C., *Can. J. Chem.*, 1977, **55**, 2732, (thermal degradation)
[18] Willmott, P.W.T. and Billmeyer, F.W., *CA*, 1963, **60**, 679b, (water absorption, diffusion coefficient)
[19] Orszagh, A. and Fejgin, J., *Vysokomol. Soedin.*, 1963, **5**, 1861, (soln props)
[20] Urman, Y.G., Khramova, T.S., Gorbunova, V.G., Barshtein, R.S. and Slonim, I.Y., *Vysokomol. Soedin., Ser. A*, 1970, **12**, 160, (H-1 nmr)
[21] Sewell, J.H., *Rev. Plast. Mod.*, 1971, **22**, 1367, (energy of cohesion, surface tension)
[22] Holland-Moritz, K., *Kolloid Z. Z. Polym.*, 1973, **251**, 906, (ir, Raman)
[23] Fields, R.D. and Rodriguez, F., *CA*, 1975, **87**, 53708, (biodegradation)
[24] Yagfarov, M.S., Vlasov, V.V. and Yagfarova, T.T., *CA*, 1976, **87**, 118252, (thermodynamic props)
[25] Plage, B. and Schulten, H.R., *J. Anal. Appl. Pyrolysis*, 1988, **15**, 197, (ms, thermal degradation)
[26] Rabinovich, I.B., Nistratov, V.P., Babinkov, A.G., Shuetsava, K.G. and Larina, V.N., *Vysokomol. Soedin., Ser. A*, 1984, **26**, 743, (thermodynamic props)
[27] Yu, H., Chen, X. and Ren, L., *CA*, 1990, **114**, 208417, (H-1 nmr, C-13 nmr)
[28] Albertsson, A.C. and Ljungquist, O., *J. Macromol. Sci., Chem.*, 1986, **23**, 393, (ir, H-1 nmr, C-13 nmr, biodegradation)
[29] Uriarte, C., Alfagerne, J., Etxeberria, A. and Iruin, J.J., *Eur. Polym. J.*, 1995, **31**, 609, (soln props, diffusion coefficient)
[30] Wurstlin, F., *Kolloid Z. Z. Polym.*, 1948, **110**, 71, (dielectric constant)

[31] Zavaglia, E.A. and Billmeyer, F.W., *CA*, 1964, **60**, 15996h, (soln props)
[32] Grieveson, B.M., *Polymer*, 1960, **1**, 499, (transition temp)
[33] Miller, G.W. and Saunders, J.H., *J. Appl. Polym. Sci.*, 1969, **13**, 1277, (transition temps)

Polytetramethylene ether P-546

Synonyms: *Polytetramethylene glycol. PTMEG*
Monomers: Tetrahydrofuran
Material class: Oligomers
Polymer Type: polyalkylene ether, polyol
CAS Number:

CAS Reg. No.
25190-06-1

Molecular Formula: $(C_4H_8O)_n$
Fragments: C_4H_8O
Additives: Antioxidants
Miscellaneous: The standard commercial grades available have MW of 650, 1000 and 2000. Used in the formation of elastomeric materials, these form the "soft segments" of a polymer chain

Applications/Commercial Products:
Processing & Manufacturing Routes: Tetrahydrofuran is polymerised using fluorosulfonic acid catalyst (HSO_3F)
Applications: Used in polyester and polyamide elastomers and polyurethanes. Used in Spandex fibres

Trade name	Supplier
Polymeg	QO Chemicals
Terathane	DuPont

Poly(tetramethylene isophthalate) P-547

Synonyms: *PBI. PBIP. Poly(butylene isophthalate). Poly(oxytetramethyleneoxyisophthaloyl). Poly(oxy-1,4-butanediyloxycarbonyl-1,3-phenylenecarbonyl). PTMI*
Monomers: 1,4-Butanediol, Isophthalic acid
Material class: Thermoplastic
Polymer Type: unsaturated polyester
CAS Number:

CAS Reg. No.	Note
28087-45-8	repeating unit
26615-64-5	from isophthalic acid
59199-72-3	from isophthaloyl chloride

Molecular Formula: $(C_{12}H_{12}O_4)_n$
Fragments: $C_{12}H_{12}O_4$
Morphology: c 2.6 [4]

Volumetric & Calorimetric Properties:
Density:

No.	Value	Note
1	d^{25} 1.268 g/cm^3	amorph. [1]
2	d^{25} 1.309 g/cm^3	cryst. [1]
3	d^{152} 1.14 g/cm^3	[1]

Thermal Expansion Coefficient:

No.	Value	Note	Type
1	4×10^{-5} K^{-1}	0° [13]	L

Volumetric Properties General: ΔC_p 5.4 J mol^{-1} K^{-1} (1.3 cal mol^{-1} °C^{-1}) [6], 0.32 J g^{-1} K^{-1} [19], 85 J mol^{-1} K^{-1} [25]
Latent Heat Crystallization: ΔH_f 42.3 kJ mol^{-1} (10.1 kcal mol^{-1}); ΔS_f 99.2 J mol^{-1} K^{-1} [1]; ΔH_f 27.2 J g^{-1} [4]; ΔH_f 9.84 kJ mol^{-1} (2.351 cal mol^{-1}) [7]; ΔH_f 39.4 J g^{-1} (9.4 cal g^{-1}) [18]; ΔH_f 8.1 kJ mol^{-1} [25], $\Delta H_f°$ 121.7 J g^{-1}; $\Delta S_f°$ 0.278 J g^{-1} K^{-1} [19]. Melt flow activiation energy 63 kJ mol^{-1} (15 kcal mol^{-1}) [24]
Specific Heat Capacity:

No.	Value	Note	Type
1	234 kJ/kg.C	56 cal mol^{-1} °C^{-1}, 34° [6]	P

Melting Temperature:

No.	Value	Note
1	152.5°C	[1]
2	150°C	[4,8]
3	135°C	[6]
4	141°C	[7]
5	137°C	410K [9]
6	128°C	[13]
7	147°C	[18]
8	146°C	[20]
9	145°C	418K [25]

Glass-Transition Temperature:

No.	Value	Note
1	17°C	[3,18]
2	34°C	[6]
3	24°C	[8,19]
4	23–26°C	[40]
5	26°C	[25]

Transition Temperature General: Other transition temps. have been reported [9]
Deflection Temperature:

No.	Value	Note
1	35°C	[13]

Vicat Softening Point:

No.	Value	Note
1	144°C	[13]

Transition Temperature:

No.	Value	Note	Type
1	85°C	[1]	Softening temp.
2	161°C	[19]	Equilbrium melt temp.
3	60°C	[6]	$T_{\beta max}$

Surface Properties & Solubility:
Cohesive Energy Density Solubility Parameters: Constant of solvent interaction parameters -3 cal cm^{-3} (α-methylnaphthalene), -3.1 cal cm^{-3} (isophorone), -1 cal cm^{-3} (benzonitrile) [1]. Energy of cohesion has been reported [21]

Solvents/Non-solvents: Sol. CH_2Cl_2 [10]. Insol. MeOH [10]
Wettability/Surface Energy and Interfacial Tension: Critical surface tension 33.7 mJ m^{-2} (33.7 dynes cm^{-1}) [5]. Other critical surface tension values have been reported [21,22]

Transport Properties:
Polymer Solutions Dilute: $[\eta]$ 0.39 dl g^{-1} (phenol/tetrachloroethane, (60:40)) [1]; $[\eta]_{inh}$ 1.01 dl g^{-1} (phenol/tetrachloroethane, (60:40), 25°) [3]; $[\eta]$ 0.61 dl g^{-1} (phenol/tetrachloroethane, (60:40), 25°) [7]; $[\eta]_{inh}$ 0.31 dl g^{-1} (phenol/tetrachloroethane, (60:40), 30°) [11]; $[\eta]$ 1.14 dl g^{-1} (phenol/tetrachloroethane, (60:40), 30°) [13]. Calculated viscosity values have been reported [17].
Polymer Melts: Melt flow viscosity has been reported [24]

Mechanical Properties:
Mechanical Properties General: Dynamic tensile storage and loss moduli (-150–140°) have been reported [20]

Tensile (Young's) Modulus:

No.	Value	Note
1	940 MPa	[15]

Viscoelastic Behaviour: Stress relaxation modulus (-34–35°) has been reported [14]. Stress-strain curves [15], stress relaxation [16] and creep [16] have been reported. Viscoelastic props have been reported [23]
Mechanical Properties Miscellaneous: T-peel strength 7 lb in^{-1} (23°) [3]

Electrical Properties:
Electrical Properties General: Variation of tan δ with temp. has been reported in graphical form [6,20]

Optical Properties:
Transmission and Spectra: C-13 nmr [10], mass [11] and fluorescence spectral [12] data have been reported

Polymer Stability:
Decomposition Details: Thermal degradation has been studied by ms [11]. Polymer decomposition temp. 410° [11]

Applications/Commercial Products:
Processing & Manufacturing Routes: Synth. by transesterification of dimethyl isophthalate or isophthaloyl chloride with 1,4-butanediol at 240° followed by condensation polymerisation at 240° under vacuum [1,2,4,7,9,10,17]

Bibliographic References
[1] Conix, A. and Van Kerpel, R., *J. Polym. Sci.*, 1959, **40**, 521, (synth, solubility, thermodynamic props)
[2] Ogata, N. and Okamoto, S., *J. Polym. Sci., Polym. Chem. Ed.*, 1973, **11**, 2537, (synth)
[3] Jackson, W.J., Gray, T.F. and Caldwell, J.R., *J. Appl. Polym. Sci.*, 1970, **14**, 685, (mech props, soln props)
[4] Gilbert, M. and Hybart, F.J., *Polymer*, 1972, **13**, 327, (synth, struct)
[5] Sewell, J.H., *J. Appl. Polym. Sci.*, 1973, **17**, 1741, (surface tension)
[6] Yip, H.K. and Williams, H.L., *J. Appl. Polym. Sci.*, 1976, **20**, 1217, (heat capacity, transition temps)
[7] Yip, H.K. and Williams, H.L., *J. Appl. Polym. Sci.*, 1976, **20**, 1205, (synth, enthalpy of fusion)
[8] Gilbert, M. and Hybart, F.J., *Polymer*, 1974, **15**, 407, (transition temps)
[9] Uralil, F., Sederel, W., Anderson, J.M. and Hiltner, A., *Polymer*, 1979, **20**, 51, (transition temps)
[10] Kricheldorf, H.R., *Makromol. Chem.*, 1978, **179**, 2133, (C-13 nmr, synth, solubility)
[11] Foti, S., Giuffrida, M., Maravigna, P. and Montaudo, G., *J. Polym. Sci., Polym. Chem. Ed.*, 1984, **22**, 1217, (thermal degradation, ms)
[12] Mendicuti, F., Patel, B., Waldeck, D.H. and Mattice, W.L., *Polymer*, 1989, **30**, 1680, (fluorescence spectrum)
[13] Ng, T.H. and Williams, H.L., *Makromol. Chem.*, 1981, **182**, 3323, (transition temps, expansion coefficients)
[14] Ng, T.H. and Williams, H.L., *Makromol. Chem.*, 1981, **182**, 3331, (stress relaxation)
[15] Ng, T.H. and Williams, H.L., *J. Appl. Polym. Sci.*, 1986, **32**, 4883, (mech props)
[16] Ng, T.H. and Williams, H.L., *J. Appl. Polym. Sci.*, 1987, **33**, 739, (stress relaxation, creep)
[17] Pilati, F., Munari, P., Manaresi, P., Milani, G. and Bonova, V., *Eur. Polym. J.*, 1987, **23**, 265, (synth, soln props)
[18] Chang, S.J. and Tsai, H.B., *J. Appl. Polym. Sci.*, 1994, **51**, 999, (transition temps)
[19] Phillips, R.A., McKenna, J.M. and Cooper, S.L., *J. Polym. Sci., Polym. Phys. Ed.*, 1994, **32**, 791, (thermodynamic props)
[20] Castles, J.L., Vallance, M.A., McKenna, J.M. and Cooper, S.L., *J. Polym. Sci., Polym. Phys. Ed.*, 1985, **23**, 2119, (mech props)
[21] Sewell, J.H., *Rev. Plast. Mod.*, 1971, **22**, 1367, (energy of cohesion, surface tension)
[22] Kasemura, T., Kondo, T. and Hata, T., *Kobunshi Ronbunshu*, 1979, **36**, 815, (surface tension)
[23] Chiorboli, E. and Pizzoli, M., *Polym. Bull. (Berlin)*, 1989, **21**, 77, (viscoelastic props)
[24] Munari, A., Pezzin, G. and Pilati, F., *Rheol. Acta*, 1990, **29**, 469, (melt flow viscosity)
[25] Righetti, M.C., Pizzoli, M. and Munari, A., *Macromol. Chem. Phys.*, 1994, **195**, 2039, (thermodynamic props)

Poly[thio bis(4-phenyl)carbonate]   P-548

Synonyms: *Poly[oxycarbonyloxy-1,4-phenylenethio-1,4-phenylene]*. *Poly(4,4'-dihydroxydiphenylsulfide-co-carbonic dichloride)*
Related Polymers: Copolymers of Thiobisphenol and Bisphenol A, Poly[1,1-cyclohexane bis(4-(2,6-dichlorophenyl)carbonate)]
Monomers: 4,4'-Thiobisphenol, Carbonic dichloride
Material class: Thermoplastic
Polymer Type: polycarbonates (miscellaneous)
CAS Number:

CAS Reg. No.
28930-32-7
94411-09-3

Molecular Formula: $(C_{13}H_8O_3S)_n$
Fragments: $C_{13}H_8O_3S$

Volumetric & Calorimetric Properties:
Melting Temperature:

No.	Value	Note
1	220–240°C	[2]

Glass-Transition Temperature:

No.	Value	Note
1	156°C	[3]

Surface Properties & Solubility:
Solvents/Non-solvents: Sol. *m*-cresol, cyclohexanone, pyridine, DMF; swollen by THF; insol. CH_2Cl_2, EtOAc, C_6H_6, Me_2CO [1]

Optical Properties:
Transmission and Spectra: Ir spectral data have been reported [1]

Applications/Commercial Products:
Processing & Manufacturing Routes: Synth. by transesterification in view of its poor solubility in many solvents The dihydroxy compound and a slight excess of diphenyl carbonate are mixed together in the melt (150–300°) in the absence of O_2, with elimination of phenol. Reaction rates are increased by alkaline catalysts and the use of a medium-high vacuum in the final stages [1,2]

Bibliographic References

[1] Schnell, H., *Polym. Rev.*, 1964, **9**, 99, (solvents, ir, synth., rev)
[2] Schnell, H., *Ind. Eng. Chem.*, 1959, **51**, 157, (T_m, synth.)
[3] Kim, C.K. and Paul, D.R., *Macromolecules*, 1992, **25**, 3097, (T_g)

Poly[[thio bis(4-phenyl)carbonate]-*co*-[2,2-propane bis(4-phenyl)carbonate]]   P-549

Synonyms: *Thiobisphenol-Bisphenol A copolycarbonates. Poly[oxycarbonyloxy-1,4-phenylenethio-1,4-phenyleneoxycarbonyloxy-1,4-phenylene(1-methylethylidene)-1,4-phenylene]. Poly(4,4′-thiobisphenol-co-4,4′-(1-methylethylidene)bisphenol-co-carbonic dichloride). Copolycarbonates*
Related Polymers: Poly[thiobis(4-phenyl)carbonate], Poly[2,2-propanebis(4-phenyl)carbonate]
Monomers: 4,4′-Thiobisphenol, Bisphenol A, Carbonic dichloride
Material class: Thermoplastic, Copolymers
Polymer Type: polycarbonates (miscellaneous), bisphenol A polycarbonate
CAS Number:

CAS Reg. No.	Note
109759-92-4	
84287-08-1	Merlon T85

Molecular Formula: $[(C_{13}H_8O_3S).(C_{16}H_{14}O_3)]_n$
Fragments: $C_{13}H_8O_3S$ $C_{16}H_{14}O_3$

Applications/Commercial Products:

Trade name	Supplier
Merlon T	Mobay

Poly(thiophene)   P-550

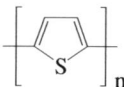 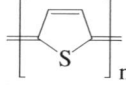

Benzoid form   Quinoid form

Synonyms: *PT*
Monomers: Thiophene
Polymer Type: polythiophene
CAS Number:

CAS Reg. No.
25233-34-5

Molecular Formula: $(C_4H_2S)_n$
Fragments: C_4H_2S
General Information: Polymer struct. has two forms, an aromatic benzenoid and a quinoid form [1,2]
Morphology: Orthorhombic; a 7.80, b 5.55, c 8.03; monoclinic; a 7.83, b 6.55, c 8.20, β 96° [7]
Miscellaneous: Major product on polymerisation is a trimer [3,4,5,6]

Volumetric & Calorimetric Properties:
Glass-Transition Temperature:

No.	Value	Note
1	187°C	approx. [8]

Surface Properties & Solubility:
Solvents/Non-solvents: Sol. C_6H_6, CCl_4, $CHCl_3$, CH_2Cl_2, nitrobenzene. Insol. pentane, EtOH, H_2O [9]

Transport Properties:
Polymer Solutions Dilute: [η] 0.028 (C_6H_6, 25°) [9]

Mechanical Properties:
Tensile (Young's) Modulus:

No.	Value	Note
1	2000–3300 MPa	room temp., neutral film [10]

Tensile Strength Break:

No.	Value	Note
1	30–82 MPa	room temp., neutral film [10]

Mechanical Properties Miscellaneous: Stress-strain curves have been reported [10]

Electrical Properties:
Electrical Properties General: Electrical conductivity increases through doping with various chemicals [14,15,16,17,28]
Complex Permittivity and Electroactive Polymers: Conductivity 2×10^{-8} Ω^{-1} cm^{-1} [12,13]

Optical Properties:
Transmission and Spectra: Ir [13,18,19,20], Raman [21,22], uv [13], H-1 nmr [9], C-13 nmr [8] and mass [4] spectral data have been reported

Polymer Stability:
Thermal Stability General: Polymer is thermally stable up to 200°; weight loss occurs at 400° [8]

Applications/Commercial Products:
Processing & Manufacturing Routes: Prod. by chemical or electrochemical oxidn. polymerisation of thiophene with various catalysts [24,25]; by oxidn. polymerisation of bithiophenes [26] or redox elimination of poly(5,5′-bithiophenediyl-*p*-acetoxybenzylidene, [27]
Applications: Potential applications in photoelectric solar cells, membranes, semiconductors, solid electrolytes, diodes and transistors

Bibliographic References

[1] Karpfen, A. and Kertesz, M., *J. Phys. Chem.*, 1991, **95**, 7680
[2] Hong, S.Y. and Marynick, D.S., *Macromolecules*, 1992, **25**, 4652
[3] Margosian, D. and Kovacic, P., J. Polym. Sci., *Polym. Chem. Ed.*, 1979, **17**, 3695
[4] Kovacic, P. and McFarland, K.N., J. Polym. Sci., *Polym. Chem. Ed.*, 1979, **17**, 1963
[5] Curtis, R.F., Jones, D.M., Ferguson, G., Hawley, D.M. et al, J.C.S., *Chem. Commun.*, 1969, 165
[6] Kovacic, P., Margosian, D. and McFarland, K.N., Polym. Prepr. (Am. Chem. Soc., Div. Polym. Chem.), 1979, **20**, 863
[7] Mo, Z., Lee, K., Moon, Y., Kobayashi, M. et al, *Macromolecules*, 1985, **18**, 1972
[8] Erdmann, K., Czerwinski, W., Gerstein, B.C. and Pruski, M., J. Polym. Sci., *Polym. Phys. Ed.*, 1994, **32**, 1799
[9] Armour, M., Davies, A.G., Upadhyay, J. and Wassermann, A., J. Polym. Sci., Part A-1, 1967, **5**, 1527
[10] Ito, M., Tsuruno, A., Osawa, S. and Tanaka, K., *Polymer*, 1988, **29**, 1161
[11] Bradley, A. and Hammes, J.P., *J. Electrochem. Soc.*, 1963, **110**, 5
[12] Kaneto, K., Kohno, Y., Yoshino, K. and Inuishi, Y., J.C.S., *Chem. Commun.*, 1983, 382
[13] Wang, S., Tanaka, K. and Yamabe, T., *Synth. Met.*, 1989, **32**, 141
[14] Waltman, R.J., Bargon, J. and Diaz, A.F., *J. Phys. Chem.*, 1983, **87**, 1459
[15] Hotta, S., Soga, M. and Sonoda, N., *J. Phys. Chem.*, 1989, **93**, 4994
[16] Springborg, M., *J. Phys.: Condens. Matter*, 1992, **4**, 101
[17] Kaneto, K., Yoshino, K. and Inuishi, Y., Jpn. J. Appl. Phys., Part 2, 1982, **21**, 567

[8] Hasoon, S., Galtier, M. and Sauvajol, J.C., *Synth. Met.*, 1989, **28**, C317
[9] Louam, G., Mevellec, J.Y., Buisson, J.P. and Lefrant, S., *Synth. Met.*, 1993, **55**, 587
[20] Poussigue, G., Benoit, C., Sauvajol, J.L., Lene-Porte, J.P. and Chowo, C., *J. Phys.: Condens. Matter*, 1991, **3**, 8803
[21] Botta, C., Luzzati, S., Dellepiane, J. and Taliani, C., *Synth. Met.*, 1989, **28**, D331
[22] Lene-Porte, J.P., Sauvájol, J.L., Hasoon, S., Chenouri, D. et al, *Mol. Cryst. Liq. Cryst.*, 1987, **161**, 223
[23] Hatta, S., *Synth. Met.*, 1987, **22**, 103
[24] Ramsey, J.S. and Kovacic, P., *J. Polym. Sci., Part A-1*, 1969, **7**, 127
[25] Bazzaoui, E.A., Aeiyach, S. and Lacaze, P.C., *J. Electroanal. Chem.*, 1994, **364**, 63
[26] Krische, B., Zagorska, M. and Hellberg, J., *Synth. Met.*, 1993, **58**, 295
[27] Jenekhe, S.A., Polym. Prepr. (Am. Chem. Soc., *Div. Polym. Chem.*), 1986, **27**, 74
[28] *Handbook of Conducting Polymers*, Vol. 1, (ed. T.A. Skotheim), Marcel Dekker, 1986, 293

Poly(3-thiopheneacetic acid)   P-551

Synonyms: *3-Thiopheneacetic acid homopolymer. PTAA. P3TA*
Monomers: 3-Thiopheneacetic acid
Material class: Thermoplastic
Polymer Type: polythiophene
CAS Number:

CAS Reg. No.
114815-74-6

Molecular Formula: $(C_6H_4O_2S)_n$
Fragments: $C_6H_4O_2S$
General Information: Conductive polymer with electrochromic and photoluminescent properties. Prod. as red or black film.

Surface Properties & Solubility:
Solubility Properties General: Solubility of polymer increases in some solvents such as THF when protonated [1].
Solvents/Non-solvents: Sol. aqueous alkali, DMSO, DMF, THF; insol. $CHCl_3$ [1,2,3]

Electrical Properties:
Electrical Properties General: Polymer in undoped form has typical conductivity of 1.9×10^{-4} S cm^{-1} [4]. A self-doping mechanism can be excluded because of the weak acidity of the carboxyl group [5]. Photovoltaic properties are exhibited particularly in multilayer films [6] or as sensitiser for titanium dioxide devices [7]. Polymer doped with perchloric acid exhibits electrorheological props. [8].

Optical Properties:
Optical Properties General: In aqueous NaCl solutions the UV absorption band is very sensitive to pH in the region 5–6, with a sharp increase of λ_{max} from ca. 405 to 445nm, corresponding to a change in conformation with more extended conjugation [2,9]. Photoluminescence band recorded with λ_{max} 545nm [10]. Polymer exhibits electrochromism, changing colour from red or yellow in reduced state to blue or black in oxidised form [1,11].
Transmission and Spectra: Pmr [2,8], ir [1,2,4,8] and uv [3,8,10,11,12] spectral data have been reported.

Polymer Stability:
Thermal Stability General: Thermal degradation at 253°C and 542°C corresponds to side chain and backbone breakdown [8].

Applications/Commercial Products:
Processing & Manufacturing Routes: Prod. chemically by polymerisation of the methyl ester [2,13] or ethyl ester [3] of 3-thiopheneacetic acid with $FeCl_3$ and subsequent hydrolysis of the ester polymer. Direct polymerisation of acid monomer has also been used [1]. Electrochemical synthesis from 3-thiopheneacetic acid in acetonitrile or chlorobenzene with tetrabutylammonium perchlorate or lithium perchlorate as supporting electrolyte is an alternative method [4,5,11,12].

Bibliographic References
[1] Giglioti, M., Trivinho-Strixino, F., Matsushima, J.T., Bulhões, L.O.S. and Pereira, E.C., *Sol. Energy Mat. and Sol. Cells*, 2004, **82**, 413
[2] Kim, B., Chen, L., Gong, J. and Osada, Y., *Macromolecules*, 1999, **32**, 3964
[3] Royappa, A.T. and Rubner, M.F., *Langmuir*, 1992, **8**, 3168
[4] Liaw, O.-J., Liaw, B.-Y., Gong, J.-P. and Osada, Y., *Synth. Met.*, 1999, **99**, 53
[5] Hou, C.-N., Hua, M.-Y. and Chen, S.-A., *Mater. Chem. Phys.*, 1994, **36**, 359
[6] Kawai, T., Yamane, T., Tada, K., Onoda, M., Jin, S.-H., Choi, S.-K. and Yoshino, K., *Jpn. J. Appl. Phys.*, 1996, **36**, L741
[7] Kim, Y.-G., Walker, G., Samuelson, L.A. and Kumar, J., Polym. Prepr. (Am. Chem. Soc., *Div. Polym. Chem.*), 2002, **43**, 576
[8] Chotpattanonant, D., Sirivat, A. and Jamieson, A.M., *Colloid Polym. Sci.*, 2004, **282**, 357
[9] Kim, Y.-G., Samuelson, L.A., Kumar, J. and Tripathy, S.K., *J. Macromol. Sci., Part A: Chem.*, 2002, **39**, 1127
[10] Youk, J.H., Locklin, J., Prussia, A. and Advincula, R., *Langmuir*, 2003, **19**, 8119
[11] Albery, W.J., Li, F. and Mount, A.R., *J. Electroanal. Chem.*, 1991, **310**, 239
[12] Hara, K., Sayama, K. and Arakawa, H., *Bull. Chem. Soc. Jpn.*, 2000, **73**, 583
[13] Senadeera, G.K.R., *Curr. Sci. (India)*, 2005, **88**, 145

Poly(thiophene vinylene)   P-552

Synonyms: *PTV. Poly(2,5-thiophenediyl-1,2-ethenediyl)*
Related Polymers: Poly(thiophene), Poly(phenylene vinylene)
Monomers: 5-Methyl-2-thiophenecarboxaldehyde
Material class: Thermoplastic
Polymer Type: polythiophene
CAS Number:

CAS Reg. No.	Note
26498-02-2	
144240-81-3	(*E*)-form

Molecular Formula: $(C_6H_4S)_n$
Fragments: C_6H_4S

Mechanical Properties:
Tensile (Young's) Modulus:

No.	Value	Note
1	1100 MPa	[1]

Tensile Strength Break:

No.	Value	Note
1	500 MPa	[1]

Electrical Properties:
Electrical Properties General: Chemical doping increases conductivity [2,3,4,5,6,7]. Band gaps are lowered compared to poly(thiophene) [8]
Complex Permittivity and Electroactive Polymers: Conductivity for doped materials has been reported, 60 Ω^{-1} cm^{-1} [4]

Optical Properties:
Refractive Index:

No.	Value	Note
1	2.42–3.56	700–2500 nm [9]

Transmission and Spectra: Ir [10,11,12,13,14], Raman [12,13,14,15], Mössbauer [16], epr [2], uv [10,18] and mass [17] spectral data have been reported

Applications/Commercial Products:
Processing & Manufacturing Routes: Prod. by elimination of a precursor polymer derived from a bissulfonium salt [1,2,7,19]; or by polycondensation of 5-methyl-2-thiophenecarboxaldehyde when reacted with strong base [10]
Applications: Development of applications as electrical conductors and in transistors

Bibliographic References
[1] Tokito, S., Smith, P. and Heeger, A.J., *Synth. Met.*, 1990, **36**, 183
[2] Jen, K.Y., Jow, R., Shacklette, L.W., Maxfield, M. et al, *Mol. Cryst. Liq. Cryst.*, 1988, **160**, 69
[3] Jen, K.Y., Maxfield, M., Shacklette, L.W. and Elsenbaumer, R.L., J.C.S., *Chem. Commun.*, 1987, 209
[4] Yamada, S., Tokito, S., Tsutsui, T. and Saito, S., J.C.S., *Chem. Commun.*, 1987, 1449
[5] Onoda, M., Moita, S., Iwasa, T., Nakayama, H. and Yoshino, K., *J. Chem. Phys.*, 1991, **95**, 8584
[6] Murase, I., Ohnishi, T., Noguchi, T. and Hirooka, M., *Polym. Commun.*, 1985, **26**, 362
[7] Tokito, S., Momii, T., Murata, H., Tsutsui, T. and Saito, S., *Polymer*, 1990, **31**, 1137
[8] Pellegrin, E., Fink, J., Martens, J.H.F., Bradley, D.D.C. et al, *Synth. Met.*, 1991, **41**, 1353
[9] Yang, C.J. and Jenekhe, S.A., *Chem. Mater.*, 1995, **7**, 1276
[10] Kossmehl, G. and Yaridjianian, A., *Makromol. Chem.*, 1981, **182**, 3419
[11] Foster, C.M., Kim, Y.H., Votani, N. and Heeger, A.J., *Synth. Met.*, 1989, **29**, E135
[12] Furukawa, Y., Sakamoto, A. and Tasumi, M., *J. Phys. Chem.*, 1989, **93**, 5354
[13] Mevellec, J.Y., Buisson, J.P., Lefrant, S. and Eckhardt, H., *Synth. Met.*, 1991, **41**, 283
[14] Buisson, J.P., Lefrant, S., Louarn, G., Mevellec, J.Y. et al, *Synth. Met.*, 1992, **49**, 305
[15] Furukawa, Y., Ohta, H., Sakamoto, A. and Tasumi, M., Spectrochim. Acta, Part A, 1991, **47**, 1367
[16] Briers, J., Eeuers, W., De Wit, M., Geise, H. et al, *J. Phys. Chem.*, 1995, **99**, 12971
[17] Holzmann, G. and Kossmehl, G., *Org. Mass Spectrom.*, 1980, **15**, 336
[18] Tsutsui, T., Murata, H., Momii, T., Yoshiura, K. et al, *Synth. Met.*, 1991, **41**, 327
[19] Murase, I., Ohnishi, T., Noguchi, T. and Hirooka, M., *Polym. Commun.*, 1987, **28**, 229

Poly(1-tridecene) P-553

—[CH$_2$CH]$_n$—
      |
   CH$_2$(CH$_2$)$_9$CH$_3$

Synonyms: *Poly(1-undecylethylene). Poly(1-undecyl-1,2-ethanediyl)*
Monomers: 1-Tridecene
Material class: Thermoplastic
Polymer Type: polyolefins
CAS Number:

CAS Reg. No.	Note
29758-00-7	atactic
59284-10-5	Isotactic

Molecular Formula: $(C_{13}H_{26})_n$
Fragments: $C_{13}H_{26}$
Morphology: Three forms: form I monoclinic, a 0.59, b 5.88, c 1.34, β 96° [5]

Volumetric & Calorimetric Properties:
Thermal Expansion Coefficient:

No.	Value	Note	Type
1	0.00073 K^{-1}	[3]	V

Latent Heat Crystallization: ΔH_m 48 J g^{-1} (form I); 15 J g^{-1} (form III) [5]
Melting Temperature:

No.	Value	Note
1	48°C	321 K [4]
2	50°C	323 K, Form I [5]
3	37°C	310 K, Form II [5]
4	-9°C	264 K, Form III [5]

Transport Properties:
Polymer Solutions Dilute: Theta temp. has been reported [1]
Polymer Melts: Flow activation energy 33.5 kJ mol^{-1} (8 kcal mol^{-1}) [2,3,7]. Zero shear viscosity η_0 820000 [6]

Optical Properties:
Transmission and Spectra: Ir and Raman spectral data have been reported [4]

Applications/Commercial Products:
Processing & Manufacturing Routes: Synth. using Ziegler-Natta catalysts, such as TiCl$_3$-AlEt$_2$Cl [4,5]

Bibliographic References
[1] Wang, J.S., Porter, R.S. and Knox, J.R., *Polym. J. (Tokyo)*, 1978, **10**, 619, (expansion coefficient, soln props)
[2] Wang, J.S., Porter, R.S. and Knox, J.R., J. Polym. Sci., Part B: *Polym. Lett.*, 1970, **8**, 671, (melt flow activation energy)
[3] Porter, R.S., Chuah, H.H. and Wang, J.S., *J. Rheol. (N.Y.)*, 1995, **39**, 649, (expansion coefficient, flow activation energy)
[4] Modric, I., Holland-Moritz, K. and Hummel, D.D., *Colloid Polym. Sci.*, 1976, **254**, 342, (ir, Raman)
[5] Trafara, G., Makromol. Chem., *Rapid Commun.*, 1980, **1**, 319, (struct, melting points)
[6] Wang, J.S., Knox, J.R. and Porter, R.S., J. Polym. Sci., *Polym. Phys. Ed.*, 1978, **16**, 1709, (shear modulus, zero shear viscosity)
[7] Wang, J.S. and Porter, R.S., *Rheol. Acta*, 1995, **34**, 496, (melt flow activation energy)

Poly[1,1-(3,3,5-trimethylcyclohexane) bis(4-phenyl)carbonate] P-554

Synonyms: *Poly[oxycarbonyloxy-1,4-phenylene(3,3,5-trimethylcyclohexylidene)-1,4-phenylene]. Poly(4,4'-dihydroxydiphenyl-1,1-(3,3,5-trimethylcyclohexane)-co-carbonic dichloride). Bisphenol TMC polycarbonate*
Related Polymers: Copolymers of Bisphenol TMC and Bisphenol A
Monomers: Bisphenol TMC, Carbonic dichloride
Material class: Thermoplastic
Polymer Type: polycarbonates (miscellaneous)

CAS Number:

	CAS Reg. No.
	129510-27-6

Molecular Formula: $(C_{22}H_{24}O_3)_n$
Fragments: $C_{22}H_{24}O_3$

Volumetric & Calorimetric Properties:
Density:

No.	Value	Note
1	d 1.107 g/cm^3	[1]

Glass-Transition Temperature:

No.	Value	Note
1	239°C	[2,3]
2	233°C	[1]

Transition Temperature:

No.	Value	Note	Type
1	-108°C	[4]	Gamma relaxation temp.

Transport Properties:
Permeability of Gases: The polycarbonate from trimethylcyclohexanone shows high gas permeabilities and diffusivities, but relatively low selectivities [1]. The permeability of CO_2 is found to decrease with increasing pressure

Gas Permeability:

No.	Gas	Value	Note
1	CO_2	2456 cm^3 mm/ (m^2 day atm)	2.8 × 10^{-12} cm^2 (s Pa)$^{-1}$, 35°, 2 atm [1]
2	CH_4	147 cm^3 mm/ (m^2 day atm)	1.68 × 10^{-13} cm^2 (s Pa)$^{-1}$, 35°, 2 atm [1]
3	O_2	497 cm^3 mm/ (m^2 day atm)	5.69 × 10^{-13} cm^2 (s Pa)$^{-1}$, 35°, 2 atm [1]
4	N_2	113 cm^3 mm/ (m^2 day atm)	1.29 × 10^{-13} cm^2 (s Pa)$^{-1}$, 35°, 2 atm [1]
5	He	2723 cm^3 mm/ (m^2 day atm)	3.111 × 10^{-12} cm^2 (s Pa)$^{-1}$, 35°, 2 atm [1]

Mechanical Properties:
Impact Strength: Possesses good impact strength. Impact strength test results mainly no break or 150 kJ m^{-2} (ISO 180) [3]

Optical Properties:
Volume Properties/Surface Properties: Transparent [1,3]

Polymer Stability:
Thermal Stability General: Short term exposure to air at 400° results in minimal discolouration [4]
Environmental Stress: Shows excellent uv stability [3]

Applications/Commercial Products:
Processing & Manufacturing Routes: Synth. by interfacial polycondensation. Phosgene is introduced into a soln. of the monomer in aq. NaOH at 20–25°, in an inert solvent (CH$_2$Cl$_2$/chlorobenzene). N-Ethylpiperidine as a catalyst results in high MWs [3]

Bibliographic References
[1] Hägg, M.-B., Koros, W.J. and Schmidhauser, J.C., J. Polym. Sci., Polym. Phys. Ed., 1994, **32**, 1625, (density, T$_g$, gas permeability, volume props.)
[2] Sommer, K., Batoulis, J., Jilge, W., Morbitzer, L. et al, Adv. Mater., 1991, **3**, 590, (T$_g$)
[3] Freitag, D. and Westeppe, U., Makromol. Chem., Rapid Commun., 1991, **12**, 95, (T$_g$, impact strength, volume props., uv stability, synth.)
[4] Freitag, D., Fengler, G. and Morbitzer, L., Angew. Chem., Int. Ed. Engl., 1991, **30**, 1598, (transition temp., oxidation)

Poly[[1,1-(3,3,5-trimethylcyclohexane) bis(4-phenyl)carbonate]-co-[2,2-propane bis(4-phenyl) carbonate]]

P-555

Synonyms: Bisphenol TMC-Bisphenol A copolycarbonates. Poly[oxycarbonyloxy-1,4-phenylene(3,3,5-trimethylcyclohexylidene)-1,4-phenyleneoxycarbonyloxy-1,4-phenylene(1-methylethylidene)-1,4-phenylene]. Poly[4,4'-(1-methylethylidene)bisphenol-co-4,4'-(3,3,5-trimethylcyclohexylidene)bisphenol-co-carbonic acid]
Related Polymers: Poly[1,1-(3,3,5-trimethylcyclohexane)bis(4-phenyl)carbonate], Poly[2,2-propanebis(4-phenyl)carbonate]
Monomers: Bisphenol A, Bisphenol TMC, Carbonic acid
Material class: Thermoplastic, Copolymers
Polymer Type: polycarbonates (miscellaneous), bisphenol A polycarbonate
CAS Number:

	CAS Reg. No.
	132721-26-7

Molecular Formula: $[(C_{22}H_{24}O_3).(C_{16}H_{14}O_3)]_n$
Fragments: $C_{22}H_{23}O_3$ $C_{16}H_{14}O_3$
Morphology: Amorph

Volumetric & Calorimetric Properties:
Thermal Expansion Coefficient:

No.	Value	Note	Type
1	7.5 × 10^{-5} K^{-1}	Apec HT9351 [2]	L

Thermal Conductivity:

No.	Value	Note
1	0.21 W/mK	DIN 5612, Apec HT 9351

Specific Heat Capacity:

No.	Value	Note	Type
1	1 kJ/kg.C	DIN 52612, Apec HT9351	P

Glass-Transition Temperature:

No.	Value	Note
1	174°C	20 mol%
2	187°C	35 mol%
3	205°C	55 mol%

Transition Temperature General: Tg increases with increasing bisphenol TMC content [1]
Deflection Temperature:

No.	Value	Note
1	174°C	0.45 Mpa DIN 53461, Apec HT9351
2	162°C	1.8 MPa, DIN 53461, Apec HT9351

Surface Properties & Solubility:
Solubility Properties General: Sol. CH_2Cl_2/chlorobenzene (45:55) [1]

Transport Properties:
Melt Flow Index:

No.	Value	Note
1	5–8 g/10 min	320°, Apec HT9351

Water Absorption:

No.	Value	Note
1	0.32 %	DIN 53495, Apec HT9351

Mechanical Properties:
Tensile (Young's) Modulus:

No.	Value	Note
1	2250 MPa	DIN 53457, Apec HT9351

Flexural Modulus:

No.	Value	Note
1	2200 MPa	DIW 53457, Apec HT9351

Tensile Strength Break:

No.	Value	Note	Elongation
1	60 MPa	DIN 53455, Apec HT9351	70%

Flexural Strength at Break:

No.	Value	Note
1	95 MPa	DIN 43457, Apec HT9351

Tensile Strength Yield:

No.	Value	Elongation	Note
1	65 MPa	7%	DIN 53455, Apec HT9351

Impact Strength: No break (0.2–0.55 bisphenol TMC, Iso 180
Hardness: Ball Indentation hardness 115 Mmm^{-2} (DIN 53458, Apec HT9351)

Electrical Properties:
Surface/Volume Resistance:

No.	Value	Note	Type
1	10×10^{15} Ω.cm	Dry, Apec HT9351 [2]	S
2	10×10^{15} Ω.cm	Wet, Apec HT9351 [2]	S

Dielectric Permittivity (Constant):

No.	Value	Frequency	Note
1	2.9	50 Hz–1 MHz	IEC 250, Apec HT9351

Dielectric Strength:

No.	Value	Note
1	35 $kV.mm^{-1}$	Dry IEC 243, Apec HT9351
2	32 $kV.mm^{-1}$	Wer IEC 243, Apec HT9351

Dissipation Power Factor: 0.0014 (50 Hz) (IEC 250, Apec HT93510; 0.001 (1KHz) (IEC 250, Apec HT9351); 0.01 (1MHz) (IEC 250, Apec HT9351)
Arc Resistance:

No.	Value	Note
1	110s	ASTM D495, Apec HT9351

Optical Properties:
Refractive Index:

No.	Value	Note
1	1.572	Apec HT9351

Transmission and Spectra: Light transmission 90%

Polymer Stability:
Thermal Stability General: Exhibits outstanding heat resistance
Flammability: Flammability rating (UL94, 1.6–3.2 mm, Apec HT9351) [2] Limiting oxygen index 24 % (ASTM D2863, Apec HT9351 [2]
Chemical Stability: Excellent u.v stability

Applications/Commercial Products:
Processing & Manufacturing Routes: Synth. by interfacial polycondensation. Phosgene is introduced into a solution of Bisphenol TMC and Bisphenol A in aqueous NaOH at 20–25° in an inert solvent. the catalyst, N-ethylpiperidene, is used to achieve high MW [1]. May be pprocessed by injection moulding and extrusion.
Mould Shrinkage (%):

No.	Value	Note
1	0.85%	Apec HT9351

Applications: Car headlights, lamps, household appliances and in medical applications that require superheated steam sterilisation.

Trade name	Supplier
Apec HT 9351	Bayer Inc.

Poly(tripropargylamine) P-556

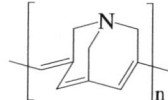

Synonyms: *Poly(TPA)*
Monomers: Tripropargylamine
Polymer Type: polyacetylene

CAS Number:

CAS Reg. No.
177743-09-8

Molecular Formula: $(C_9H_9N)_n$
Fragments: C_9H_9N
Morphology: Black or occasionally brown. Calculated identity period 11.97Å. Quasi-helical trans-gauche three-step struct. (TG_3) [1]

Surface Properties & Solubility:
Solvents/Non-solvents: Insol. almost all organic solvents [1]

Electrical Properties:
Electrical Properties General: Highly conjugated polymer struct. may confer electrical conduction props.

Optical Properties:
Transmission and Spectra: C-13 nmr and ir spectral data have been reported [1]

Applications/Commercial Products:
Processing & Manufacturing Routes: Polymerised by the use of transition-metal chlorides, namely $PdCl_2$, $PtCl_2$ and $RuCl_3$ [1]
Applications: Has potential use due to electrically conducting props.

Bibliographic References
[1] Gal, Y.-S., Lee, W.-C., Choi, S.-K., Kim, Y.-C. and Jung, B., *Eur. Polym. J.*, 1996, **32**, 579, (synth, spectroscopy)

Poly(1-undecene)   P-557

$$\left[CH_2CH\right]_n$$
$$CH_2(CH_2)_7CH_3$$

Synonyms: *Poly(1-nonylethylene)*. *Poly(1-nonyl-1,2-ethanediyl)*
Monomers: 1-Undecene
Material class: Thermoplastic
Polymer Type: polyolefins
CAS Number:

CAS Reg. No.	Note
29250-41-7	atactic
59284-09-2	Isotactic

Molecular Formula: $(C_{11}H_{22})_n$
Fragments: $C_{11}H_{22}$
Morphology: Three forms: form I monoclinic, a 0.59, b 5.19, c 1.34, β 96° [4]

Volumetric & Calorimetric Properties:
Thermal Expansion Coefficient:

No.	Value	Note	Type
1	0.0008 K^{-1}	[2]	V

Latent Heat Crystallization: ΔH_m 40 J g^{-1} (form I); 8 J g^{-1} (form III) [4]
Melting Temperature:

No.	Value	Note
1	36°C	309 K [3]
2	38°C	311 K, form I [4]
3	27°C	300 K, form II [4]
4	-38°C	235 K, form III [4]

Glass-Transition Temperature:

No.	Value	Note
1	-37°C	[6]

Transport Properties:
Polymer Solutions Dilute: Theta temp. in anisole/cyclohexanone has been reported [1]. [η] 4.3 dl g^{-1} (30°, CCl_4) [6]
Polymer Melts: Melt flow activation energy 33.5 kJ mol^{-1} (8 kcal mol^{-1}) [2]

Mechanical Properties:
Mechanical Properties General: Shear modulus values have been reported in graphical form [5]

Optical Properties:
Transmission and Spectra: Ir and Raman spectral data have been reported [3]

Applications/Commercial Products:
Processing & Manufacturing Routes: Synth. using Ziegler-Natta catalysts, such as $TiCl_3$-$AlEt_2Cl$ [3,4]

Bibliographic References
[1] Wang, J.S., Porter, R.S. and Knox, J.R., *Polym. J. (Tokyo)*, 1978, **10**, 619, (soln props)
[2] Porter, R.S., Chuah, H.H. and Wang, J.S., *J. Rheol. (N.Y.)*, 1995, **39**, 649, (expansion coefficient, melt flow activation energy)
[3] Modric, I., Holland-Moritz, K. and Hummel, D.D., *Colloid Polym. Sci.*, 1976, **254**, 342, (ir, Raman)
[4] Trafara, G., Makromol. Chem., *Rapid Commun.*, 1980, **1**, 319, (struct, melting points)
[5] Wang, J.S., Knox, J.R. and Porter, R.S., *J. Polym. Sci., Polym. Phys. Ed.*, 1978, **16**, 1709, (shear modulus)
[6] Natta, G., *CA*, 1958, **52**, 17789e, (transition temp)

Polyurethane, Adhesive applications   P-558

Related Polymers: Polyurethane General information
Polymer Type: polyurethane
General Information: Two component polyurethane adhesive which has good resistance to water, oil and fuel. The struct. is unknown

Volumetric & Calorimetric Properties:
Thermal Conductivity:

No.	Value	Note
1	0.122 W/mK	0.122 J (m s °C)$^{-1}$ [1]

Mechanical Properties:
Tensile Strength Break:

No.	Value	Note	Elongation
1	1.03 MPa	[1]	265%

Viscoelastic Behaviour: Compression set 14% (ASTM D395-55)
Hardness: Shore D95

Electrical Properties:
Dielectric Strength:

No.	Value	Note
1	9 kV.mm^{-1}	kV mm^{-1} [1]
2	8.5 kV.mm^{-1}	kV mm^{-1}, 48h, in H_2O [1]

Polyurethane, Diisocyanate deficient

P-559

Applications/Commercial Products:
Processing & Manufacturing Routes: The two compounds are mixed and applied with a spatula or trowel. Working time 25–45 mins.; cure time 24h at 25°
Applications: Used as a contact adhesive for repairing rubber rolls, rubber hoses and belting. Also used to bond metals (e.g. iron, steel, aluminium), and wood, glass and canvas

Trade name	Details	Supplier
Bostik	2064, 3206	Bostik
Flexane	30, 60, 95	Devcon Ltd.

Bibliographic References
[1] *Flexane*, Devcon Ltd, (technical datasheet)

Polyurethane, Diisocyanate deficient

P-559

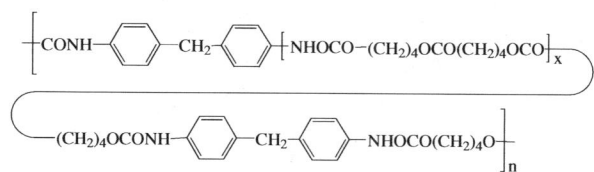

Related Polymers: Polyurethanes General Information
Monomers: Hexanedioic acid, Ethanediol, 1,5-Naphthalene diisocyanate, Ethanolamine
Material class: Synthetic Elastomers, Gums and resins
Polymer Type: polyurethane
General Information: A diisocyanate deficient gum that is stable over relatively long time periods. Full cure is affected by adding further diisocyanate, often during processing. One of the first urethane elastomers to make a commercial entry in Europe and U.S.

Mechanical Properties:
Tensile (Young's) Modulus:

No.	Value	Note
1	6 MPa	880 psi, 300% elongation [1]

Tensile Strength Break:

No.	Value	Note	Elongation
1	24.8 MPa	3600 psi [1]	730%

Hardness: Shore A65 [1]

Applications/Commercial Products:
Processing & Manufacturing Routes: Final curing of the prepolymer is achieved by mixing a small amount of diisocyanate on a mill roll and completing the cure in the mould by oven heating
Applications: Uses are restricted by relatively high cost and are thus used only when high performance is the main consideration; in oil seals, shoe soles, forklift truck tyres, diaphragms, chute linings, fabric coatings resistant to dry cleaning, bicycle tyres, traction drive wheels, tail wheels of carrier based aircraft

Trade name	Details	Supplier
Chemigum	SL (Neothane)	Goodyear
Genthane	S	General Tyre and Rubber Co.

Vulcaprene	A	ICI Chemicals and Polymers Ltd.
Vulkollan	non-exclusive	Bayer Inc.

Bibliographic References
[1] Mueller, E., Bayer, S., Peterson, S., Piepenbrink, H.F. *et al*, *Angew. Chem.*, 1952, **64**, 523

Polyurethane elastomer

P-560

Related Polymers: Polyurethanes General Information
Monomers: 1,6-Hexanedioic acid, 1,4-Butanediol, 4,4'-Diphenylmethane diisocyanate
Material class: Thermoplastic, Synthetic Elastomers
Polymer Type: polyurethane
Molecular Weight: M_n 36000
Additives: Fillers, pigments, lubricants and certain plasticisers
General Information: Shows good extensibility and high elasticity. Considered to be 'virtually cross-linked'; does not require vulcanising agents

Volumetric & Calorimetric Properties:
Density:

No.	Value	Note
1	d_4^{23} 1.21 g/cm^3	ASTM D1505-57T [2]

Vicat Softening Point:

No.	Value	Note
1	67°C	ASTM D1525-58T [2]

Brittleness Temperature:

No.	Value	Note
1	-65°C	ASTM D746-57T [2]

Transport Properties:
Melt Flow Index:

No.	Value	Note
1	1 g/10 min	ASTM D1238-52T, 0.3 MPa [2]

Mechanical Properties:
Tensile Strength Break:

No.	Value	Note	Elongation
1	39.9 MPa	4560 psi, ASTM D638-58T [2]	415%

Hardness: Shore A85 (durometer, ASTM D676-55T) [2]

Applications/Commercial Products:
Processing & Manufacturing Routes: Processed by extrusion, injection moulding [3] and calendering. Can also be soln. cast.

Mould Shrinkage (%):

No.	Value	Note
1	0.9%	[3]

Applications: Are relatively expensive so are used where their excellent toughness and abrasion resistance, resistance to lubricating and fuel oils, and wide range of operating temps. are required. Used in automotive brushes and bearings. As injection moulded gears, bearings and diaphragms for hydraulic systems. In moulded shoe soles, wear resistant moulded heel pieces for ladies' shoes, and in ski boots. Previously used as a biomedical material but abandoned due to degradation *in vivo*

Trade name	Details	Supplier
Estane	5740X1(general purpose); 5740X2 (low modulus for soln. applications); 5740X7 (high modulus)	B.F. Goodrich

Bibliographic References

[1] Saunders, J.H. and Frisch, K.C., Polyurethanes Chemistry and Technology Part II: Technology, John Wiley, 1964, 378
[2] Bonatto, S. and Walton, R.K., *Mod. Plast.*, 1963, **40(a)**, 143
[3] Waugaman, C.A., *Mod. Plast.*, 1961, **39**, 146
[4] Mirkovitch, V.A., Akutsu, T. and Kolff, W.J., *Trans. Am. Soc. Artif. Intern. Organs*, 1962, **8**, 79

Polyurethane, Liquid prepolymer   P-561

$$HNOCO[CH_2CH_2CH_2CH_2O]_xCONH$$

(structure with two toluene rings bearing CH_3 and NCO groups)

Related Polymers: Polyurethanes General Information
Monomers: 2,4-Diisocyanatotoluene, 3,4-Diisocyanatotoluene, 1,4-Butanediol, 2,6-Diisocyanatotoluene
Material class: Fluids
Polymer Type: polyurethane
Molecular Weight: MW 2000
Additives: Carbon black, fillers and plasticisers
General Information: A liq. prepolymer cured by reaction with polyols, diamines, or water. Cross-linked by catalysts such as potassium acetate. Physical props. are dependent upon type and conc. of curing agent, mixing and curing temp. and time

Volumetric & Calorimetric Properties:
Density:

No.	Value	Note
1	d 1.05 g/cm^3	[2]

Thermal Expansion Coefficient:

No.	Value	Note	Type
1	7.1×10^{-5} K^{-1}	[2]	L

Mechanical Properties:
Tensile (Young's) Modulus:

No.	Value	Note
1	3.7 MPa	540 psi, 10% elongation, cured, ASTM D638-52T [1]

Tensile Strength Break:

No.	Value	Note	Elongation
1	18.9 MPa	2750 psi, cured, ASTM D412-51T [1]	565%

Hardness: Shore A77 (cured) [1]

Electrical Properties:
Dielectric Strength:

No.	Value	Note
1	11.8 kV.mm^{-1}	300 V mil^{-1}, 125 mil thick ASTM D149 [2]

Applications/Commercial Products:
Processing & Manufacturing Routes: Processed by a liq. casting technique. Chain extension and cross-linking are controlled by the type and conc. of curing agent as well as the temp. and time of cure. Curing agents may be polyols, diamines or water. Metal salt catalysts can also be used to promote cross-linking
Applications: Used in shoe heels and soles. Friction drives and belts; in washing machines, tape recorders, record players, oil seals, rings and diagphragms. Also used in encapsulation of electrical components

Trade name	Details	Supplier
Adiprene	L100	E.I. DuPont de Nemours & Co., Inc.

Bibliographic References

[1] Quant, A.J., Soc. Plast. Eng., *Tech. Pap.*, 1959, **15**, 298
[2] Saunders, J.H. and Frisch, K.C., Polyurethanes Chemistry and Technology Part II: Technology, John Wiley, 1964, 349

Polyurethane, Peroxide cured   P-562

$$\{HN(CH_2)_6NHOCNH\text{-}\phi\text{-}CH_2\text{-}\phi\text{-}NHOC[O(CH_2)_2OCO(CH_2)_4OC]_x$$
$$[OCH_2CHCH_3OCO(CH_2)_4OC]_y O(CH_2)_2OCONH\text{-}\phi\text{-}CH_2\text{-}\phi\text{-}NHCO\}_n$$

Related Polymers: Polyurethanes General information
Monomers: 1,6-Hexanediamine, 1,6-Hexanedioic acid, 4,4'-Diphenylmethane diisocyanate, 1,2-Propanediol, Ethanediol
Material class: Synthetic Elastomers
Polymer Type: polyurethane
Additives: Carbon black
General Information: Is designed to be cured by peroxides

Mechanical Properties:
Tensile (Young's) Modulus:

No.	Value	Note
1	20 MPa	2900 psi, 300% elongation, cured, carbon black filled [1]

Tensile Strength Break:

No.	Value	Note	Elongation
1	28.9 MPa	4200 psi, cured, carbon black filled [1]	400%

Hardness: Shore A65 (cured, carbon black filled) [1]

Applications/Commercial Products:
Processing & Manufacturing Routes: The polymer gum is compounded with either dicumyl peroxide or 2,5-dimethyl-2,5-di-*tert*-butylperoxyhexane to yield a stable compound. Curing is effected by heating (3 min., 177°)

Applications: Used in oil seals, tyres, drive wheels, fabric coatings and diaphragms

Trade name	Supplier
Vibrathane	US Rubber Co.

Bibliographic References
[1] Urs, S.V., *Ind. Eng. Chem. Proc. Des. Dev.*, 1962, **1**, 199

Polyurethanes General Information   P-563

Related Polymers: Polyurethane, elastomers, Spandex fibres, Urethane coating, Polyurethane, diisocyanate deficient, Polyurethane, liquid prepolymer, Urethane rubber, multrathane based, Polyurethane, adhesive, Polyurethane, peroxide cured

General Information: Polyurethanes are a family of heterogeneous polymers which are the product of the reaction of isocyanates (NCO) with a hydroxy compound RNCO + HOR′ → RNHCOOR′. They contain the urethane linkage RHNC(O)OR′ and may be considered as esters of carbamic acid, or ester amides of carbonic acid. When a diisocyanate and a diol react together, a linear polyurethane is obtained, whilst a diisocyanate and a polyol lead to a cross-linked polymer. The urethane group does not usually constitute the majority of the functional groups within a polyurethane, and it is the ability to incorporate other functional groups into the polymer network that contributes to the range of props. shown by polyurethane materials. Polyurethanes range from rigid thermosetting materials to soft elastomers. Interest in polyurethanes stems from Otto Bayer's work on the commercial production of urethane polymers in 1937 [1]. This was followed in 1938 by the first US patent application by Rinke [2]. The patent application for Lycra(R) was filed in 1954 [3]

Morphology: Composed of short alternating blocks of soft and hard segments. The soft segment is typically a low T_g polyether, polyester or polyalkanediol, and the hard segment is usually a high T_g aromatic diisocyanate, linked with a low MW chain extender. Polyurethanes are generally synth. with chain extenders consisting of low MW diols or diamines which produce additional urethane or urethane-urea segments respectively. A degree of immiscibility between the hard urethane segments and the soft polyol segments means that on a microscopic level polyurethanes are not structurally homogeneous. The structure may be considered as hard segment domains dispersed in a soft segment matrix

Identification: Polyols: The earliest polyurethanes were based on aliphatic diols (glycols) with 1,4-butanediol being most frequently used. However, now most polyurethane production involves polymeric hydroxy compounds, mainly polyesters and polyethers. Approx. 90% of commercial polyurethanes are based on polyethers, while polyesters are used in the remainder. Diisocyanates: Isocyanates are commonly prepared by the reaction of phosgene and primary amines: $RNH_2 + COCl_2 \rightarrow RNCO + 2HCl$ The most commonly used diisocyanates are tolylene diisocyanate (TDI) (as a mixture of isomers), diphenylmethane diisocyanate(MDI), naphthylene 1,5-diisocyanate (NDI), hexamethylene diisocyanate (HDI) and isophorone diisocyanate (IPDI)

Applications/Commercial Products:
Processing & Manufacturing Routes: Polyurethanes are often prepared in two stages with the production of a prepolymer followed by a chain extension reaction. This type of process is normally used in the production of elastomeric-type polymers. During the first step the diisocyanate is reacted with a short chain polyether or polyester (e.g. poly(ethylene adipate), poly(η-caprolactone)) with MW 1000–3000. This product is then chain-extended by subsequent reaction with a short chain diol (e.g. 1,4-butanediol, ethylene glycol) or a diamine. Using a diamine gives a urea linkage and a poly(urethane-*co*-urea) struct. which can also react with other isocyanate groups, leading to the formation of Biuret links. Cross-linked systems can also be obtained by using multifunctional monomers. For the production of rigid and flexible foams the chain extender molecule can be omitted, and polyols with more than 3 hydroxy groups are used. The reaction is base catalysed by tertiary amines or organotin compounds and the reaction mixture must also include a blowing agent. This is often achieved by adding controlled quantities of H_2O to the system. The H_2O reacts with an isocyanate group to form an unstable carbamic acid which then decomposes to produce the corresponding amine and CO_2 gas. The released gas forms bubbles which then expand through the polymer matrix to form a foam struct. Volatile liqs. such as fluorocarbons can be used as alternative blowing agents. In the past CFC-11 (1,1,1,2,2-pentachloro-2-fluoroethane) was extensively used in the production of thermal insulation foams, but the worldwide ban on ozone-depleting CFCs has led to the search for alternative blowing agents (e.g. HCFC-141b, 1,1-dichlorofluoroethane, bp 32°, pentane, bp 36°) as well as modifying to urethane-foaming processes. For flexible foams, high MW polyoxypropylene triols are used with toluene diisocyanate. To produce rigid foams polymeric methylene diphenylisocyanates are used in conjunction with low MW high-functionality polyols

Applications: Elastomers: Solid polyurethane elastomers may be divided into 3 categories- cast, millable and thermoplastic. (a) Cast elastomers are produced by pouring a liq. reaction mixture comprising low MW material into a heated mould wherein the material is converted to a solid high MW elastomer. A wide range of physical props. can be obtained by appropriate formulation but in general cast elastomers are characterised by high tensile strength, abrasion resistance and tear strength. These props. have lead to such uses as solid tyres, bearings and rollers. (b) Millable elastomers are rubber-like gums which may undergo mill pounding and vulcanisation. Their props. are similar to those of cast elastomers, but with a narrower use range. (c) Thermoplastic elastomers have functional props. of conventional vulcanised rubbers but may be processed as normal thermoplastics. Their props. are relatively similar to those of cast polyurethanes but the absence of cross-linking results in higher compression set and a more pronounced loss of strength with increasing temp. A major use is in the production of ski boots. They may also be produced in fibre form to give elastic fibres (spandex fibres). These fibres are extensively used in the clothing industry in sportswear, swimwear, underwear and foundation garments. Foams: these can be classified as rigid and flexible foams. Rigid foams are closed cell structures which are highly cross-linked polymers with good resistance to organic solvents and inorganic substances. The major interest in rigid foams is for thermal insulation. They are also widely used in the fabrication of composite structures for aircraft components such as doors, bulkheads, and wing tips, and dinghy hulls, packaging and shock absorption. Flexible foams are open cell structures and highly resistant to detergents and solvents. and inert to oxidation. They find use in the clothing industry (shoulder pads, interlinings), upholstery (cushions, pillows, seating) and packaging. Coatings: several polyurethanes are used in coating applications, and can be divided into 2 broad categories: one-component systems, which cure by reaction with atmospheric O_2, moisture or heat, and two-component systems which cure through interaction of two materials on mixing. They find use as wood varnishes, quick-drying enamels, lacquers, electrical wire coatings and floor paints. Adhesives: Three general categories of polyurethane adhesives may be distinguished (a) Isocyanate-polyol adhesives are two-component systems similar to those used for surface coatings and may be used for bonding wood, plastics and metals. They are low in viscosity and tack. (b) Soluble elastomers may be used as adhesives provided the MW is low enough to permit solubility in organic solvents. Such adhesives give flexible bonds and are used in the footwear industry (c) Polyisocyanates, especially triphenylmethane 4,4′,4″-triisocyanate, are used for bonding of rubber to rubber, metals, glass and synthetic fibres

Polyvanillylidenecycloalkanone phosphoramide esters

P-564

Bibliographic References

[1] *Ger. Pat.*, 1937, 728981
[2] *U.S. Pat.*, 1938, 2511544
[3] *U.S. Pat.*, 1954, 2692893, (Lycra, synth.)
[4] *Polyurethane Handbook*, (ed., G. Oertel) 2nd edn., Hanser/Gardner Publications Inc., 1993
[5] Lamba, N.M.K., Woodhouse, K.A. and Cooper, S.L., *Polyurethanes in Biomedical Applications*, CRC Press, 1998, (rev)
[6] Szycher, M., *Szycher's Handbook of Polyurethanes*, CRC Press, 1999
[7] *Kirk-Othmer Encycl. Chem. Technol.*, Vol.10, (ed. J.I. Kroschwitz), John Wiley, 1993, 628
[8] *Encycl. Polym. Sci. Eng.*, Vol.6, (ed. J.I. Kroschwitz), John Wiley, 1996, 739

Polyvanillylidenecycloalkanone phosphoramide esters P-564

x = 0 (cyclopentanone)
x = 1 (cyclohexanone)

Y = H, OMe, CH_3, Cl, Br

Monomers: 2,5-Bis(4-hydroxy-3-methoxybenzylidene)cyclopentanone, 2,6-Bis(4-hydroxy-3-methoxybenzylidene)cyclohexanone
Molecular Formula: $(C_{21}H_{20}O_5)_n (C_{22}H_{22}O_5)_n$
Fragments: $C_{21}H_{20}O_5$ $C_{22}H_{22}O_5$

Volumetric & Calorimetric Properties:
Melting Temperature:

No.	Value	Note
1	172–182°C	cyclopentanone derivative [1]
2	168–174°C	cyclohexanone derivative [1]

Glass-Transition Temperature:

No.	Value	Note
1	80–87°C	cyclopentanone derivative
2	80–85°C	cyclohexanone derivative [1]

Surface Properties & Solubility:
Solvents/Non-solvents: Sol. CH_2Cl_2, $CHCl_3$, Me_2CO, THF, DMF, dimethyl acetamide and DMSO. Insol. C_6H_6, toluene, pentane [1]

Optical Properties:
Transmission and Spectra: H-1 nmr, C-13 nmr and ir spectral data have been reported [1]

Polymer Stability:
Polymer Stability General: Cyclohexanone-based polymers have increased thermal stability (less ring strain). Halogens in the phosphoramide also increase thermal stability [1]
Thermal Stability General: 290–330° (10% wt. loss); 430–520° (50% wt. loss) (20°, N_2) [1]

Applications/Commercial Products:
Processing & Manufacturing Routes: Prepared by interfacial condensation using a phase transfer catalyst (hexadecyltrimethyl ammonium bromide) [1]
Applications: Applications in photo-resists for integrated circuits, printing plates and inks, photocurable coatings and energy exchange systems

Bibliographic References

[1] Kannan, P. and Murugavel, S.C., *Polym. Int.*, 1996, **40**, 287, (polymerisation)

Poly(vinyl acetal) P-565

Related Polymers: Polyvinyl formal, Polyvinyl butyral, Polyvinyl alcohol
Monomers: Vinyl acetate, Acetaldehyde
Material class: Thermoplastic
Polymer Type: polyvinyls
Molecular Formula: $[(C_6H_{10}O_2)_x(C_2H_4O)_y(C_4H_6O_2)_z]_n$
Fragments: $C_6H_{10}O_2$ C_2H_4O $C_4H_6O_2$
Additives: Weak alkalis act as stabilisers
General Information: Props. depend on degree of acetalisation and free hydroxyl groups present in terpolymer. [1]

Volumetric & Calorimetric Properties:
Melting Temperature:

No.	Value	Note
1	81°C	10% free OH

Transition Temperature General: Increase in hydroxyl group content increases T_m, softening and brittle temps
Vicat Softening Point:

No.	Value	Note
1	97°C	12% free OH [1]
2	82°C	10% free OH [1]

Brittleness Temperature:

No.	Value	Note
1	36°C	12% OH [1]
2	27°C	10% OH [1]

Surface Properties & Solubility:
Internal Pressure Heat Solution and Miscellaneous: Sol. MeOH, EtOH, AcOH, dioxane. Insol. Me_2CO, EtOAc, C_6H_6. [1] Increased free hydroxyl content increases solubility in polar solvents. High acetate content increases solubility in less polar solvents
Plasticisers: Addition of dibutyl phthalate decreases T_m, softening and brittle temps. and decreases mech. props markedly [1]

Mechanical Properties:
Mechanical Properties General: High hydroxyl content increases mech. props, which decrease with increasing temp.
Tensile Strength Break:

No.	Value	Note	Elongation
1	78.6 MPa	11.4 kpsi, 9% OH [1]	6%

Elastic Modulus:

No.	Value	Note
1	3585 MPa	520 kpsi, 0°, 12% OH [1]

2	3102 MPa	450 kpsi, 40°, 12% OH [1]
3	2688 MPa	390 kpsi, 0°, 12% OH [1]
4	2316 MPa	336 kpsi, 40°, 10% OH [1]

Polymer Stability:
Environmental Stress: Oxidation my cause cleavage of chain leading to acid autocatalytic decomposition. Should not be stored indefinitely or without stabilisers
Hydrolytic Stability: Unstable to acids and strong base

Applications/Commercial Products:
Processing & Manufacturing Routes: Prod. by treating acid hydrolysed polyvinyl acetate in EtOH with acetadehyde at 60–75°
Applications: Now of little commercial importance

Bibliographic References
[1] Fitzhugh, A.F and Crozier, R.N., *J. Polym. Sci.*, 1952, **8**, 225, (synth, props)

Poly(vinyl alcohol)   P-566

$$-[CH_2CH]_n-$$
$$\quad\ \ |$$
$$\quad\ OH$$

Synonyms: *Poly(1-hydroxyethylene). Poly(hydroxyethylene). Poly(1-hydroxyethene). Poly(hydroxyethene). Poly(ethenol). PVA. PVOH. PVAL*
Related Polymers: Poly(vinyl alcohol) fibres, Poly(vinyl alcohol) films, Poly(vinyl alcohol) foamed material
Monomers: Vinyl acetate
Material class: Gums and resins
Polymer Type: polyvinyl alcohol
CAS Number:

CAS Reg. No.	Note
9002-89-5	
25067-41-8	isotactic
27101-67-3	syndiotactic

Molecular Formula: $(C_2H_4O)_n$
Fragments: C_2H_4O
Molecular Weight: There are four MW ranges which are commercially important; low grade M_n 25000; intermed. grade 40000; medium grade 60000; high grade 100000
General Information: Pure PVA is a light, white, powdery solid. Degree of hydrol. is important in controlling polymer props. The three main grades available are fully hydrolysed (98–98.8 mol % hydrolysed); intermed. (95–97 mol %) and partially hydrolysed (87–89 mol %). Additional grades are super (>99.3 mol %) and low (79–81 mol %). Polymers with other degrees of hydrol. are prod. but have a smaller market value
Tacticity: Tacticity of PVA is determined by the struct. and props. of the polymer used in preparation. Most commercially available PVAs are called atactic, consisting of a random mixture of isotactic and syndiotactic sequences. The nomenclature as regards tacticity is not applied as literally to PVA as it is to other polyolefins because PVAs with syndiotacticity values close to 100 or 0 % have not been obtained [5]
Morphology: PVA can be amorph. or polycrystalline depending on mechanical treatment [5]
Identification: Fully hydrolysed PVA forms a complex with iodine which has a blue colour (λ_{max} 620 nm). The colour of this complex varies with the degree of hydrol.

Volumetric & Calorimetric Properties:
Density:

No.	Value	Note
1	d 1.345 g/cm^3	cryst. [11]
2	d 1.269 g/cm^3	amorph. [11]
3	d 1.27–1.31 g/cm^3	[1,2,3,4,5]
4	d^{25} 1.02 g/cm^3	10% w/w aq. soln. [5]

Thermal Expansion Coefficient:

No.	Value	Note	Type
1	7×10^{-5}–0.00012 K^{-1}	[1]	L
2	0.0001 K^{-1}	[3]	L
3	2.9×10^{-5}–0.0005 K^{-1}	[2]	V

Volumetric Properties General: Degree of crystallinity increases with heat treatment and the degree of hydrol. [1]. Density increases with the degree of crystallinity, and is affected by moisture content [3] and tacticity [12]. The relationship between density and concentration has been reported [2]
Equation of State: Mark-Houwink constants have been reported [4,16,18,82,83,84,85,89]
Thermodynamic Properties General: Thermal conductivity of PVA solns. increases with temp. and decreases with increasing PVA concentration [14]. Thermal diffusity increases with temp. and decreases with PVA concentration [14]. C_v of aq. solns. of PVA increases with temp. but decreases with PVA concentration [14]
Latent Heat Crystallization: Heat of fusion 78.7–98.3 J g^{-1} (18.8–23.6 cal g^{-1}, degree of crystallinity 0.55–0.58) [12]. Heat of formation 6.28 kJ mol^{-1} (1.5 kcal mol^{-1}, cryst., per segment) [2]. Entropy of fusion 13.8 J K^{-1} mol^{-1} (3.3 cal °C^{-1} mol^{-1}, cryst., 1 mole of segments) [2]
Thermal Conductivity:

No.	Value	Note
1	0.2 W/mK	[1,5]
2	0.02–0.4 W/mK	[14]

Thermal Diffusivity: 0.08–1.4 $\times 10^{-7}$ m^2 s^{-1} [14]
Specific Heat Capacity:

No.	Value	Note	Type
1	1.04 kJ/kg.C	45.9 J mol^{-1} K^{-1}, cryst. [15]	P
2	1.67 kJ/kg.C	[1]	P
3	1.5 kJ/kg.C	[5]	P
4	0.2674 kJ/kg.C	11.78 J mol^{-1} K^{-1}, solid	P
5	0.7187 kJ/kg.C	31.66 J mol^{-1} K^{-1}, solid	P
6	1.185 kJ/kg.C	52.21 J mol^{-1} K^{-1}, solid	P
7	1.546 kJ/kg.C	68.11 J mol^{-1} K^{-1}, solid [16,17,18]	P

Melting Temperature:

No.	Value	Note
1	230°C	unplasticised, 98–99% hydrolysed [1,5]
2	180°C	unplasticised, 87–89% hydrolysed [1,5]
3	220–267°C	fully hydrolysed [1,22,23,24,25]
4	212–235°C	isotactic

5	228–240°C	atactic
6	230–267°C	syndiotactic [4,22]

Glass-Transition Temperature:

No.	Value	Note
1	85°C	high MW, 98–99% hydrolysed [1]
2	58°C	87–89% hydrolysed, depends on the degree of polymerisation [1]
3	85°C	[18,19,20,21]
4	75–85°C	[5]

Transition Temperature General: Determination of T_m is difficult because decomposition occurs above 130° [1]. Second-order transition temps. have been reported [2]. T_m depends on tacticity [4] and crystallinity [12]. T_g is reduced by the presence of plasticisers [5] and by absorption of H_2O. Absorption of 6% w/w H_2O lowers T_g to room temp. [2]

Surface Properties & Solubility:

Solubility Properties General: Solubility in H_2O is controlled by the degree of hydrol. and the degree of polymerisation. Fully hydrolysed PVA must be dissolved in hot to boiling H_2O but remains in soln. on cooling [5]. The optimum hydrol. range for solubility in both cold and hot H_2O is 87–89% [5]. The temp. for H_2O solubility decreases as the degree of hydrol. decreases. At 75–80% hydrol., PVA is fully sol. cold H_2O but precipitates upon heating above 40° [1,2,5]. PVA solubility is improved by decreasing the particle size and the polymer MW but is decreased by heat treatment [1] or increasing crystallinity [12]. The heat of soln. of PVA in H_2O has a negative value [1,2]
Cohesive Energy Density Solubility Parameters: δ 7.26 $J^{1/2} cm^{-3/2}$ (5.78 $MPa^{1/2}$) [16,18,27]
Plasticisers: H_2O [3,5]. Water sol. polyhydric alcohols, e.g. glycerol, poly(ethylene glycol), glycol, triethylene glycol, sorbitol; hydroxy-esters (glyceryl lactate), amides (urea), and calcium chloride [3,5]. Pentaerythritol and 1,2,6-hexanetriol are used in high temp. processing [5]
Solvents/Non-solvents: All commercial PVA samples sol. H_2O. Atactic sol. glycols (hot), glycerol (hot), piperazine, formamide, H_2O, triethylenediamine, hexamethylphosphoric triamide [18], DMF, mixtures of lower alcohols and H_2O (up to 50% w/w) [5], acetamide [1]; mod sol. DMSO; insol. hydrocarbons, chlorinated hydrocarbons, lower alcohols, THF, ketones, carboxylic acids, esters, conc. aq. salt solns [18], gasoline, kerosene, C_6H_6, xylene, MeOH, trichloroethylene, CCl_4, Me_2CO, methyl acetate [5]. Syndiotactic sol. H_2O (above 160°), propanediol (above 160°) [18]
Surface and Interfacial Properties General: The surface tension of PVA solns. in DMSO/H_2O mixtures has been reported [48]. The surface tension of aq. PVA solns. varies with concentration and degree of hydrol., and is at its lowest with a degree of hydrol. of 78–81% [1]. Surface tension decreases slightly with decreasing MW. Surface tension increases linearly with degree of hydrol. from 88–98% but then increases sharply for fully hydrolysed samples [5]
Wettability/Surface Energy and Interfacial Tension: γ_c 37 mN m^{-1} [16,28,29], γ_d 42 [16,29,30], γ_a 0 [16,29,30]. Has exceptional adhesion to cellulosic surfaces and exceptional binding power in cement formulations. All PVA samples show good adhesion to hydrophilic materials but fully hydrolysed products have the best adhesive props. Partially hydrolysed PVA is the only grade that gives good adhesion to hydrophobic materials, e.g. glass, metal and many plastics [5]. The adhesive props. of PVA films have been reported [5]. Contact angle 22° (H_2O, receding), 54° (H_2O, advancing) [2]. γ_{LV} 50 mN m^{-1} (50 dyne cm^{-1}, 25°, 1% w/w PVA soln. in H_2O) [2]. Other values of contact angle have been reported [37]

Transport Properties:

Polymer Solutions Dilute: Theta temps. 97° (H_2O) [16,31], 25° (EtOH/H_2O 41.5/58.5% (w/w)), 25° (MeOH/H_2O 41.7/58.3% (w/w)), 25° (1-propanol/H_2O 35.1/64.9% (w/w)), 25° (2-propanol/H_2O 39.4/60.6% (w/w)), 25° (tert-butanol/H_2O 32/68% (w/w)) [16,32]. Soln. viscosity increases with concentration and MW (degree of polymerisation) and decreases with increasing temp. [1,5]
Polymer Solutions Concentrated: The degree of hydrol. does not affect the soln. viscosity as strongly but the viscosity is proportional to the degree of hydrol. at constant MW and concentration [1,5]. The concentration of PVA solns. is limited by the viscosity rather than the polymer solubility. The practical concentration limits are 30% (w/w) for low MW, 15 and 20% (w/w) for high MW [5]. The stability of the soln. viscosity depends on the PVA concentration, degree of hydrol., storage temp. and dissolution temp. [1]. The viscosities of partially hydrolysed PVA solns. remain constant over a wide range of concentrations if the solns. are stored at high temp. [5]. The viscosities of concentrated solns. of fully hydrolysed PVA gradually increase over several days' storage at room temp. Gelation occurs in products containing less than 1 mol % acetate groups. This viscosity increase or gelation can be reversed by reheating. The effect of ageing on the soln. viscosity of PVA has been reported [22,75,76,77,78,79,80]. The viscosity of PVA solns. may be stabilised by adding small amounts of lower aliphatic alcohols [81], urea or salts, e.g. thiocyanates [1]. Intrinsic viscosity - MW relationships for PVA have been reported [4,18]. The concentration dependence of intrinsic viscosity has been reported [87]. The effect of temp. on intrinsic viscosity and the Mark-Houwink parameters for PVA in H_2O has been studied from 20–80° [4,88]. Intrinsic viscosity decreases steadily as the temp increases [1]. Intrinsic viscosity - MW relationship for PVA samples with different degrees of hydrol. has been reported [1,4,40,90,91]. Intrinsic viscosity decreases as the acetate content increases [4,40,91]. The effect of temp. on the intrinsic viscosity of PVA samples with different acetate contents has been reported [91]
Permeability of Gases: The oxygen barrier props. of PVA at low humidity are greater than those of any other synthetic resin [5]. Fully hydrolysed PVA has transmission rates that are two to four times lower than those of poly(vinylidene chloride). However PVA, as a water sensitive polymer, exhibits rapid loss of barrier performance above 50% relative humidity [1,5,74]. The gas barrier performance of PVA increases with increasing degree of hydrol. and rapidly diminishes below a degree of hydrol. of 98% [1,5]. High sodium acetate content also increases the gas transmission rates and must be less than 0.1% w/w for optimum performance [5]. PVA is highly impermeable to most organic vapours, inert gases and H_2. It is relatively impermeable to alcohol vapour but permeable to water vapour and NH_3 [3]
Permeability and Diffusion General: Permeability of H_2O vapour at 20–30° has been reported [3]. The permeability of H_2O vapour depends on relative humidity and temp. [2] and is about 100000 times greater than that for O_2 and CO_2
Water Content: 4% (approx., room temp., 50% relative humidity) [5], 8% (80% relative humidity) [5], 6–9% (65% relative humidity) [3]. For samples of relatively low crystallinity, the amount of water absorbed per unit volume is proportional to the external vapour pressure of H_2O up to 50% relative humidity. The coefficient of absorption at 25° is 0.0039 (moles H_2O) (cm^3 polymer)$^{-1}$ (cm Hg pressure of H_2O vapour)$^{-1}$ [2]. At 40° this absorption coefficient is reduced to 0.0019 [2]. At relative humidities higher than 50% the amount absorbed increases greatly with increasing external water vapour pressure but amount absorbed is no longer proportional to the vapour pressure [2]
Water Absorption:

No.	Value	Note
1	25–30 %	100% relative humidity [94]

Mechanical Properties:

Mechanical Properties General: Most of the mechanical props. reported in the literature are for films and fibres. Mechanical props. of extruded PVA [3] and PVA gels [92,93,95,96,97,98,99,100,101] have been reported. Mechanical props. are sensitive to moisture (humidity) [3] and are also affected by temp. and processing conditions [1]

Poly(vinyl alcohol)

Tensile Strength Break:

No.	Value	Note	Elongation
1	34 MPa	3.5 kgf mm^{-2}, extruded material fully hydrolysed [3]	225% extension

Viscoelastic Behaviour: PVA gels formed from concentrated soln. (in H_2O or aq. solvent mixtures) have viscoelastic props. The effects of temp., MW, concentration and other additives on viscoelastic props. have been reported [92,95,96,97,98,99,100,101]
Failure Properties General: At high humidity, PVA samples creep under load [3]. Force-extension curves for PVA gels have been reported [93]

Electrical Properties:
Electrical Properties General: The electrical props. of PVA are very limited because of the absorption of H_2O under normal atmospheric conditions [3]. Most measurements of the electrical props. of PVA have been carried out on films [2,3]. Volume resistivity 31–38 M ohm^{cm} [1,5]. Dielectric constant increases with increasing water content [2]
Dielectric Permittivity (Constant):

No.	Value	Note
1	40	room temp., dry PVA in bulk [2]

Strong Field Phenomena General: Intrinsic breakdown strength 300 kV mm^{-1} (20°), 1500 kV mm^{-1} (-180°) [38]
Static Electrification: The electrostatic charge on PVA samples induced by contact with metals has been reported [73]

Optical Properties:
Optical Properties General: The refractive index of aq. PVA solns. increases linearly with concentration with a discontinuity at a (critical) PVA concentration [2,4,36]. The critical concentration depends on MW and the soln. viscosity [2,4]
Transmission and Spectra: Ir [2,4,22,41,42,43], C-13 nmr [16,43,44,45,46] and H-1 nmr [47] spectral data have been reported. Pure PVA should show no uv/visible absorption [2,4] Uv spectral data regarding films have been reported [102]
Volume Properties/Surface Properties: Colour white (solid)

Polymer Stability:
Polymer Stability General: The presence of hydroxyl groups in PVA gives the polymer a range of chemical reactions [1,2,22]
Thermal Stability General: The thermal decomposition of PVA in the absence of oxygen is a two stage process: dehydration at 200° accompanied by the formation of volatile products [1,2,49,50,51,52,53]; decomposition at 400–500° to yield carbon and hydrocarbons [1,2,53]. The residue from the first stage is predominantly polyenes [1,53]. The decomposition products may depend on the manufacturing conditions of the PVA [1]. In the presence of oxygen, oxidation of the unsaturated polymeric residue introduces ketone groups into the polymer chain [1,26]. Thermal decomposition is inhibited by carbon dioxide. [2] Dehydration at high temps. is catalysed by dil. acids and alkalis [2]. Other studies of the thermal degradation have been reported [56,57,58]
Decomposition Details: $T_{1/2}$ 268°; monomer yield 0% [16]. Initial decomposition temp. 100° [3]. Activation energy of decomposition 14–18 kJ mol^{-1} (60–75 kcal mol^{-1}) (in the absence of O_2) [2], 8–13 kJ mol^{-1} (35–55 kcal mol^{-1}) (in the presence of O_2) [2]. Decomposition products include H_2O, MeOH, Me_2CO, EtOH, propanol, AcOH, crotonaldehyde, C_6H_6, acetaldehyde, benzaldehyde, acetophenone and carbon monoxide [1,2]
Flammability: Limiting oxygen index 22.5% [16,54]. Burns in a similar manner to paper [1,5]. Fire retardant props. have been reported [55]
Environmental Stress: Has excellent resistance to sunlight [1,5], showing only slight loss in strength after prolonged exposure [3]. Oxidised by air, oxygen or ozone to form ketone groups along the chain. Air oxidation at 60–100° in the presence of alkali during manufacture produces impurities [2]

Chemical Stability: Oxidised by hydrogen peroxide, acidified potassium dichromate, ceric nitrate and periodate [2]. Swollen by polyhydric alcohols, especially when hot [3]. Stable to hydrocarbons (including petrol), chlorinated hydrocarbons, alcohol, Me_2CO, carboxylic acid esters, greases, animal and vegetable oils and fats [1,3,5]. Resistance to organic solvents increases with increasing hydrol. [1,5]
Hydrolytic Stability: Small concentrations of strong inorganic acids and bases do not precipitate PVA from soln. However, for partially hydrolysed PVA in aq. soln., addition of strong acid or base can cause an increase in the degree of hydrol. over a period of several days. Under extreme conditions of pH, the hydrol. reaction continues to completion. Fully hydrolysed PVA samples are essentially unaffected by pH [5]. PVA can absorb large amounts of H_2O. It needs to be stored in bags with moisture barriers to keep the polymer dry and prevent caking [5]. Reacts with aq. alkali to form gels [60]
Biological Stability: Aq. solns. of PVA must be protected from bacterial and mould growth by the addition of a biocide [3,5]. PVA can be effectively biodegraded in acclimatised, activated sludge wastewater systems [1,5,65,66]. The presence of PVA does not appear to interfere with the treatment of other biodegradable materials normally present in waste water [5]. PVA is biodegradable under both aerobic [67,68,69] and anaerobic conditions [70] using bacteria isol. from soil and water. The effect of MW and struct. on biodegradability has been reported [71]
Stability Miscellaneous: γ-Irradiation in air at room temp. induces oxidation of hydroxyl groups to carbonyl groups. Chain scission with the production of carboxyl and carbonyl end groups occurs, and cross-linking occurs particularly at elevated temps. [2]. The irradiation of aq. PVA solns. causes a substantial amount of cross-linking [2], which may be prevented by the addition of thiourea [4]. The effect of x-rays and γ-rays on the stability of PVA in aq. soln. has been reported [72]. γ-Irradiation also affects thermal props. owing to an increase in MW [14]

Applications/Commercial Products:
Processing & Manufacturing Routes: Poly(vinyl alcohol) is most commonly synth. from poly(vinyl acetate) by acid or base catalysed alcoholysis. NaOH is most common catalyst and MeOH is the most common alcohol used. Reaction is terminated by neutralising or removing the catalyst to give degree of hydrol. required [2]. Fully hydrolysed PVA is obtained if the reaction is allowed to go to completion. PVA can also be synth. via the hydrol. of other poly(vinyl esters) and poly(vinyl ethers). These reactions are not commercially significant. Syndiotactic PVA may be synth. on a laboratory scale by the hydrol. of poly(vinyl trifluoroacetate). Isotactic poly(vinyl alcohol) may be synth. on a laboratory scale from the saponification of isotactic poly(vinyl benzyl ether)
Applications: Used as an intermed. in the production of poly(vinyl acetals), e.g. poly(vinyl butyral) and poly(vinyl formal) and in the preparation of poly(vinyl sulfates), poly(vinyl phosphates), poly(vinyl carbonates) and poly(vinyl nitrates). PVA is used in adhesives. All grades show good adhesion to hydrophilic materials where fully hydrolysed PVA is the best adhesive. Partially hydrolysed PVA is the only grade that gives good adhesion to hydrophobic substrates e.g. glass, metal and many plastics. PVA is also used as a modifier for other aq. adhesive systems to improve film forming props. PVA based adhesives are used in paper laminating, the manuf. of plasterboard, re-moistenable adhesive for envelopes and stamps, binder for ceramics and non-woven fabrics and as a concrete additive. PVA may be used as a stabiliser for vinyl monomers during emulsion polymerisation. PVA may be used as an emulsion stabiliser in the photographic industry (silver halide emulsions); paints; printing inks; cosmetics (hair cream, sun-tan cream and skin conditioners). Also used as a thickening agent in cosmetics. PVA may be used as a coating. PVA-amylose mixture makes paper greaseproof. It can also be used as a paper sizer prior to the paper being coated with wax. PVA coatings have excellent grease, oil and oxygen barrier performance. PVA coated paper is used to package food and chemicals. PVA soln. may be used as a blood plasma extender to maintain the normal volume and osmotic pressure of the systematic blood. The largest worldwide application is warp sizing of textiles. On changing

– Poly(vinyl alcohol)

degree of hydrol. PVA can be used for natural fibres e.g. cotton (fully hydrolysed PVA) to synthetic fibres and fibre-blends (partially hydrolysed PVA). PVA may be used as a textile finisher so fibres slip over each other on weaving and are used as wash and wear finishes for knit and woven fabrics. PVA is also used as a binder for non-woven fabrics, important when solvent resistance is required. Thin films have potential applications in microelectronics and optical engineering

Trade name	Details	Supplier
Alcotex		Revertex
Alcotex		Harlow Chem. Co. Ltd. (U.K.)
CCP		Chang Chun
Elvanol		DuPont
Gelvatol	prior to 1985	Monsanto Chemical
Gohsenol Gohsenal		Nippon Gohsei
Gohseran Gohsefimer		Nippon Gohsei
Moviol		Hoechst
Mowiol		Hoechst
Polybond	adhesive	Polybond Ltd.
Polyviol		Wacker
Poral		Kuraray-Mitsui Toatsu
Poval		Kuraray-Mitsui Toatsu
Rhodoviol		Rhone-Poulenc
Vinol Airvol		Air Products
		Wako Pure Chem. Co.
		Kaso Chem. Co.
		Neste

Bibliographic References
[1] *Encycl. Polym. Sci. Eng.*, 2nd edn., (ed. J.I. Kroschwitz), John Wiley and Sons, 1985, **17**, 167
[2] Pritchard, J.G., *Polyvinyl alcohol,* Polymer Monographs Series, MacDonald, 1970
[3] Roff, W.J. and Scott, J.R., Fibres, Films, *Plastics and Rubbers,* 2nd edn., Butterworths, 1971, 72
[4] *Water Soluble Synthetic Polymers,* (ed. P. Molyneux), CRC Press, 1984
[5] *Kirk-Othmer Encycl. Chem. Technol.,* 3rd edn., (ed. M. Grayson), Wiley Interscience, 1978, **23**, 848
[6] Tanigami, T., Shirai, Y., Yamura, K. and Matsuzawa, S., *Polymer,* 1994, **35**, 1970
[7] Shields, J., *Adhesives Handbook,* Butterworth, SIRA, 1970
[8] *Chemical Industry Directory,* Tonbridge, Benn Business Information Services, 1988
[9] Matsuzawa, S., Yamaura, K., Tanigami, T., Somura, T. and Nakata, M., *Polym. Commun.,* 1987, **28**, 105
[10] *UK Kompass Register,* Reed Information Services, 1996
[11] Sakurada, I., Nuhunshina, Y. and Sone, Y., *Kobunshi Kagaku,* 1955, **12**, 506, (density)
[12] Nakamae, K., Nishino, T., Ohkubo, H., Matsuzawa, S. and Yamaura, K., *Polymer,* 1992, **33**, 2581, (tacticity)
[13] Bronnikov, S.V., Vettegren, V.I. and Frenkel, S.Y., *Polym. Eng. Sci.,* 1992, **32**, 1204, (thermal props, mech props)
[14] Kanawy, M.A., Dakroury, A.Z. and Osman, M.B.S., *J. Appl. Polym. Sci.,* 1991, **43**, 1393, (thermal props)
[15] Domalski, E.S. and Hearing, E.D., *J. Phys. Chem. Ref. Data,* 1990, **19**, 881, (heat capacity)
[16] *Physical Properties of Polymers Handbook,* (ed. J.E. Mark), AIP Press, 1996
[17] Gaur, U., Lau, S.F., Wunderlich, B.B. and Wunderlich, B., *J. Phys. Chem. Ref. Data,* 1983, **12**, 29, (heat capacity)
[18] *Polym. Handb.,* 3rd edn., (eds. J. Brandrup and E.H. Immergut), John Wiley and Sons, 1989
[19] Gillham, J.K. and Schwenker, R.F., *Appl. Polym. Symp.,* 1966, **2**, 59, (glass transition temp)
[20] Frosini, V., Butta, E. and Calamia, M., *J. Appl. Polym. Sci.,* 1967, **11**, 527, (glass transition temp)
[21] Clark, J.B., *Polym. Eng. Sci.,* 1967, **7**, 137, (glass transition temp)
[22] *Polyvinyl alcohol,* (ed. C.A. Finch), Wiley, 1973
[23] Sakurada, I., Nakajima, A. and Takida, H., *Kobunshi Kagaku,* 1955, **12**, 21, (melting temp)
[24] Osugi, T., Man-Made Fibers, *Sci. Technol.,* 1968, **3**, 245, (melting temp)
[25] Hamada, F. and Nakajima, A., *Kobunshi Kagaku,* 1966, **23**, 395, (melting temp)
[26] *Polyvinyl alcohol,* (ed. C.J. Finch), Monograph no. 30, Soc. of Chem. Ind., 1968
[27] Shvarts, A.G., *Kolloidn. Zh.,* 1956, **18**, 755, (solubility parameter)
[28] Ray, B.R., Anderson, J.R. and Scholz, *J. Phys. Chem.,* 1958, **62**, 1220, (surface props)
[29] van Oss, C.J., Interfacial Forces in Aqueous Media Marcel Dekker, 1994, (surface props)
[30] van Oss, C.J., Chaudhury, M.K. and Good, R.J., *Adv. Colloid Interface Sci.,* 1987, **28**, 35, (surface props)
[31] Dieu, H.A., *J. Polym. Sci.,* 1954, **12**, 417, (solvent props)
[32] Wolfram, E., *Kolloid Z. Z. Polym.,* 1968, **227**, 86, (solvent props)
[33] Noro, K., *Br. Polym. J.,* 1970, **2**, 128, (surface props)
[34] Hayashi, S., Nakano, C. and Motoyama, T., *Kobunshi Kagaku,* 1963, **20**, 303, (surface props)
[35] Hayashi, S., Nakano, C. and Motoyama, T., *Kobunshi Kagaku,* 1964, **21**, 300, (surface props)
[36] Matsumoto, M. and Ohyanagi, Y., *J. Polym. Sci.* 1958, **31**, 225, (surface props, refractive index)
[37] Uyama, Y., Inoue, H. Ito, K., Kishida, A. and Ikada, Y., *J. Colloid Interface Sci.,* 1991, **141**, 275, (surface props)
[38] *Encycl. Polym. Sci. Eng.,* 2nd edn., (ed. J.I. Kroschwitz), John Wiley and Sons, 1985, **5**, 523, (electrical props)
[39] Klenin, V.J., Klenina, O.V., Shvartsburd, B.I. and Frenkel, S.Y., *J. Polym. Sci., Polym. Symp.,* 1974, **44**, 131, (refractive index)
[40] Beresniewicz, A., *J. Polym. Sci.,* 1959, **39**, 63, (refractive index)
[41] Nyquist, R.A., Infra Red Spectra of Plastics, *Polymers and Resins,* 2nd edn., Dow Chemical Co., 1961, (ir)
[42] Haslam, J., Willis, H.A. and Squirrel, D.C.M., *Identification and Analysis of Plastics,* 2nd edn., I.L.I.F.F.E., Butterworths, 1972, (ir)
[43] Bovey, F.A. and Jelinski, L.W., *J. Phys. Chem.,* 1985, **89**, 571, (C-13 nmr)
[44] Laupretre, F., Noel, C. and Monnerie, L., *J. Polym. Sci., Polym. Phys. Ed.,* 1977, **15**, 2143, (C-13 nmr)
[45] Horii, F., Hu, S., Ito, T., Odani, H. *et al, Polymer,* 1992, **33**, 2299, (C-13 nmr)
[46] Hu, S., Tsuji, M. and Horii, F., *Polymer,* 1994, **35**, 2516, (C-13 nmr)
[47] Horii, F., Hu, S., Deguchi, K., Sugisawa, H. *et al, Macromolecules,* 1996, **29**, 3330, (H-1 nmr)
[48] Matsuzawa, S. and Ueberreiter, K., *Colloid Polym. Sci.,* 1978, **256**, 490, (surface tension)
[49] Vasile, C., Cascaval, C.N. and Barbu, P., J. Polym. Sci., *Polym. Chem. Ed.,* 1981, **19**, 907, (thermal stability)
[50] Vasile, C., Patachia, S.F. and Dumitrascu, V., J. Polym. Sci., *Polym. Chem. Ed.,* 1983, **21**, 329, (thermal stability)
[51] Vasile, C., Odochian, L., Patachia, S.F. and Popoutanu, M., J. Polym. Sci., *Polym. Chem. Ed.,* 1985, **23**, 2579, (thermal stability)
[52] Ettre, K. and Varadi, P.F., *Anal. Chem.,* 1963, **35**, 69, (thermal stability)
[53] Tsuchiya, Y. and Sumi, K., J. Polym. Sci., Part A-1, 1969, **7**, 3151, (thermal stability)
[54] Cullis, C.F. and Hirschler, M.M., *The Combustion of Organic Polymers,* Clarendon Press, 1981, (flammability)
[55] Lomakin, S.M., Artsis, M.I. and Zaikov, G.E., *Flammability Polym. Mater.,* 1994, 89, (flammability)
[56] Lomakin, S.M., Artsis, M.I. and Zaikov, G.E., *Int. J. Polym. Mater.,* 1994, **26**, 187, (thermal degradation)
[57] Gilman, J.W. and van der Hart, D.L., *ACS Symp. Ser.,* 1995, **599**, 161, (thermal degradation)
[58] Anders, H. and Zimmerman, H., *Polym. Degrad. Stab.,* 1987, **18**, 111, (thermal degradation)
[59] Shibayama, M., Sato, M., Kimura, Y., Fujiwara, H. and Nomura, S., *Polymer,* 1988, **29**, 336, (props)
[60] Uragami, T. and Sugihara, M., *Angew. Makromol. Chem.,* 1977, **57**, 123, (props)
[61] Hirai, T., Okazaki, A. and Hayashi, S., *J. Appl. Polym. Sci.,* 1986, **32**, 3919, (props)
[62] Scholtens, B.J.R. and Bijsterbosch, B.H., J. Polym. Sci., *Polym. Phys. Ed.,* 1979, **17**, 1771, (props)
[63] Hayashi, S., Hirai, T., Shimomichi, S. and Hojo, N., J. Polym. Sci., *Polym. Chem. Ed.,* 1982, **20**, 839, (props)
[64] Hirai, T., Okazaki, A., Ohno, S. and Hayashi, S., J. Polym. Sci., *Polym. Lett. Ed.,* 1988, **26**, 299, (props)
[65] Casey, J.P. and Manley, D.G., Proc. Int. Biodegrad. Symp., *3rd,* Applied Science, 1975, (biological stability)
[66] Nishikawa, H. and Fujita, Y., *Chem. Econ. Eng. Rev.,* 1975, **7**, 33, (biological stability)

[67] Suzuki, T., Ichihara, Y., Yamada, M. and Tonomura, K., *Agric. Biol. Chem.*, 1973, **37**, 747, (biological stability)
[68] Sakazawa, C., Shimao, M., Taniguchi, Y. and Kato, N., *Appl. Environ. Microbiol.*, 1981, **42**, 261, (biological stability)
[69] Shimao, M., Saimoto, H., Kato, N. and Sakazawa, C., *Appl. Environ. Microbiol.*, 1983, **46**, 605, (biological stability)
[70] Matsumura, S., Kurita, H. and Shimoke, H., *Biotechnol. Lett.*, 1994, **15**, 749, (biological stability)
[71] Matsumura, S., Shimura, Y., Toshima, K., Tsuji, M. and Hatanaka, T., *Macromol. Chem. Phys.*, 1995, **196**, 3437, (biological stability)
[72] Alexander, P. and Charlesby, A., *J. Polym. Sci.*, 1957, **23**, 355, (irradiation)
[73] Lowell, J. and Akanda, A.R., *J. Phys. D: Appl. Phys.*, 1988, **21**, 125, (electrostatic props)
[74] Toyoshima, K., *Polyvinyl alcohol*, (ed. C.A. Finch), Wiley, 1973, (gas barrier props)
[75] Prokopova, E., Stern, P. and Quadrat, O., *Colloid Polym. Sci.*, 1985, **263**, 899, (viscosity)
[76] Stern, P., Prokopova, E. and Quadrat, O., *Colloid Polym. Sci.*, 1987, **265**, 234, (viscosity)
[77] Prokopova, E., Stern, P. and Quadrat, O., *Colloid Polym. Sci.*, 1987, **265**, 903, (viscosity)
[78] Vercauteren, F.F., Donners, W.A.B., Smith, R., Crowther, N.J. and Eagland, D., *Eur. Polym. J.*, 1987, **23**, 711, (viscosity)
[79] Fujishige, S., *J. Colloid Interface Sci.*, 1958, **13**, 193, (viscosity)
[80] Amaya, K. and Fujishiro, R., *Bull. Chem. Soc. Jpn.*, 1956, **29**, 361, (viscosity)
[81] Eagland, D., Vercauteren, F.F., Scholte, T.G., Donners, W.A.B. et al, *Eur. Polym. J.*, 1986, **22**, 351, (viscosity)
[82] Matsumoto, M. and Ohyanagi, Y., *Kobunshi Kagaku*, 1960, **17**, 191, (viscosity)
[83] Beresniewicz, A., *J. Polym. Sci.*, 1959, **35**, 321, (viscosity)
[84] Nakajima, A. and Furatate, K., *Kobunshi Kagaku*, 1949, **6**, 460, (viscosity)
[85] Elias, H.G., *Makromol. Chem.*, 1962, **54**, 78, (viscosity)
[86] Yamaura, K., Hirata, K., Tamura, S. and Matsuzawa, S., *J. Polym. Sci., Polym. Phys. Ed.*, 1985, **23**, 1703, (viscosity)
[87] Matsumoto, M. and Imai, K., *J. Polym. Sci.*, 1957, **24**, 125, (viscosity)
[88] Kuroiwa, T., *Kobunshi Kagaku*, 1952, **9**, 253, (viscosity)
[89] Matsuo, T. and Inagaki, H., *Makromol. Chem.*, 1962, **55**, 150, (viscosity)
[90] Lankveld, J.M.G. and Lyklema, J., *J. Colloid Interface Sci.*, 1972, **41**, 454, (viscosity)
[91] Sakurada, I., Nakajima, A. and Takita, H., *Kobunshi Kagaku*, 1955, **12**, 15, (viscosity)
[92] Nishinari, K., Watase, M. and Tanaka, F., *J. Chim. Phys. Phys. - Chim. Biol.*, 1996, **93**, 880, (viscoelastic props)
[93] Nishinari, K., Watase, M., Ogino, K. and Nambu, M., *Polym. Commun.*, 1983, **24**, 345
[94] *Kirk-Othmer Encycl. Chem. Technol.*, 3rd edn., (ed. M. Grayson), Wiley Interscience, 1978, **20**, 210
[95] Watase, M. and Nishinari, K., *Makromol. Chem.*, 1988, **189**, 871, (viscoelastic props)
[96] Watase, M. and Nishinari, K., *J. Polym. Sci. Polym. Phys. Ed.*, 1985, **23**, 1803, (viscoelastic props)
[97] Nagura, M., Hanano, T. and Ishikawa, H., *Polymer*, 1989, **30**, 762, (viscoelastic props)
[98] Nishinari, K. and Watase, M., *Polym. J. (Tokyo)*, 1993, **25**, 463, (viscoelastic props)
[99] Watase, M. and Nishinari, K., *Polym. Commun.*, 1983, **24**, 270, (viscoelastic props)
[100] Lozinsky, V.I., Vainerman, E.S., Domotenko, L.V., Mamtsis, A.M. et al, *Colloid Polym. Sci.*, 1986, **264**, 19, (viscoelastic props)
[101] Lazinsky, V.I., Zubov, A.L., Kulalova, V.K., Titova, E.F. and Rogozhin, S.V., *J. Appl. Polym. Sci.*, 1992, **44**, 1423, (viscoelastic props)
[102] Wang D.-H., *Thin Solid Films*, 1997, **293**, 270, (uv)

Poly(vinyl alcohol), Fibres   P-567

Synonyms: *Vinal. Vinylon*
Related Polymers: Poly(vinyl alcohol) general information, Poly(vinyl alcohol) films, Poly(vinyl alcohol) foamed material, Poly(vinyl alcohol)-poly(vinyl chloride) fibre blend
Monomers: Vinyl acetate
Polymer Type: polyvinyl alcohol
CAS Number:

CAS Reg. No.
9002-89-5

Molecular Formula: $(C_2H_4O)_n$
Fragments: C_2H_4O
Additives: To make fibres water insol. cross-linking agents (formaldehyde, acetaldehyde, benzaldehyde) may be used. For gel spinning process, boric acid is added to poly(vinyl alcohol). A plasticiser (glycerol, low MW polyethylene glycol) may also be used for extrusion process
General Information: PVA fibre is prod. extensively in Japan, China and the Far East, Russia and Eastern European countries e.g. Hungary, Romania. The high cost of production (compared to other fibres) has limited its use in the UK, USA and Europe [1,2,3,4,5]
Miscellaneous: The US Federal Trade Commission defines vinal fibres as having the following composition: at least 50% vinyl alcohol units and at least 85% total vinyl alcohol and acetal units

Volumetric & Calorimetric Properties:
Density:

No.	Value	Note
1	d 1.348 g/cm^3	cryst. [8]
2	d 1.345 g/cm^3	cryst. [10,11]
3	d 1.265 g/cm^3	amorph. [8]
4	d 1.269 g/cm^3	amorph. [10,11]
5	d 1.26 g/cm^3	[7]
6	d 1.26–1.32 g/cm^3	depending on the degree of crystallinity [8]
7	d 1.34 g/cm^3	[16]

Thermal Expansion Coefficient:

No.	Value	Note	Type
1	1.7×10^{-5} K^{-1}	[23]	L

Melting Temperature:

No.	Value	Note
1	220–230°C	[7]
2	246°C	[12]
3	237°C	[13]
4	235°C	[26]
5	222°C	[1]
6	250°C	[16]

Glass-Transition Temperature:

No.	Value	Note
1	50°C	[18]

Transition Temperature General: T_m increases with increasing crystallinity [26]. T_g is lower than for isotropic PVA, which may be due to the presence of plasticiser [18]. A secondary transition is observed at low temps. [17]
Transition Temperature:

No.	Value	Note	Type
1	-3°C	[23]	characteristic temp.

Surface Properties & Solubility:
Solubility Properties General: PVA fibres can be rendered insol. H_2O by heat treatment and chemical cross-linking [1,3,8,9]. PVA fibres are compatible with cement [6]

Solvents/Non-solvents: Water insol. fibres: sol. formic acid; sl. sol. *m*-cresol [8]. Insolubilised fibres are slightly swollen by H$_2$O [8]
Wettability/Surface Energy and Interfacial Tension: Fibres containing boric acid are hydrophilic and have excellent adhesion to hard materials such as cement and slate [2]

Transport Properties:
Water Content: Water retention 30–35% (staple fibres) [8]
Water Absorption:

No.	Value	Note
1	1.3–1.8 %	20–25°, 20% relative humidity [8]
2	4.5–5 %	20–25°, 65% relative humidity [8]
3	10–12 %	20–25°, 95% relative humidity [8]
4	3.5–5 %	equilib. [7]

Mechanical Properties:
Mechanical Properties General: Mechanical props. depend on the processing conditions (drawing [12,21], heat setting temp., moisture absorption [5,13,27] and the amount of cross-linking) [7]. Tensile strength at break 0.41 N tex^{-1} (max., draw ratio 7.6) [19], 0.79 N tex^{-1} (10–20% elongation) [7]. Young's modulus 8.562 N tex^{-1} (max.) [19], 8.8 N tex^{-1} [7]. The tensile modulus and tensile strength of PVA fibres increase with increasing birefringence and decrease with decreasing hydrol. [18]. Tensile modulus increases with draw ratio [3,10,13,14,19,20,26] and is slightly dependent on crystallinity [14,19,20,26]. Tensile strength also increases with draw ratio [10,14,19] and this increase is greatest for low MW PVA [13]. Elongation at break decreases with increasing draw ratio [19,21]. The wet strength of PVA fibres is approx. 80% of the dry air strength [8]. Tensile modulus decreases rapidly as the temp. exceeds the glass-transition temp. [3,23]. Insolubilised PVA fibres retain 70% of their normal strength at 100° [8]. Although it is expected that fibre from high MW PVA has high tensile modulus and strength [24,25] it has been found that the increase in draw ratio with MW (degree of polymerisation) is small [20]

Tensile (Young's) Modulus:

No.	Value	Note
1	4800 MPa	amorph. [14]
2	12000–18000 MPa	max. [18]
3	<70000 MPa	approx., max. [20]
4	76000 MPa	65% relative humidity, degree of polymerisation 3420, draw temp. 190°, max.draw ratio 21.5 [13]
5	6300 MPa	draw ratio 4 [21]
6	14000–19800 MPa	draw ratio 14.0–19.8 [21]
7	90000 MPa	mean value, 20–22°, 50% relative humidity [22]
8	<120000 MPa	max. [22]
9	3.92 MPa	40 kg mm^{-2} [3]
10	29.4–117.7 MPa	300–1200 kg f mm^{-2}, 25–100 gf den^{-1}, 20–25°, 65% relative humidity [8]
11	<196.2 MPa	max., 2000 kg f mm^{-2}, 180 gf den^{-1}, 20–25°, 65% relative humidity [8]
12	<60000 MPa	max.[10]
13	51760 MPa	dry [27]
14	46000 MPa	[16]
15	49000 MPa	[28]
16	2200 MPa	longitudinal [16]
17	4200 MPa	transverse [16]

Tensile Strength Break:

No.	Value	Note	Elongation
1	900–1100 MPa	max. [18]	
2	<2300 MPa	max. [20]	
3	400 MPa	draw ratio 4 [21]	59.0%
4	2000–2500 MPa	draw ratio 14.0–19.8 [21]	3.9–5.2%
5	5000 MPa	max., 20–22°, 50% relative humidity [22]	
6	4.91–7.36 MPa	50–75 kg f mm^{-2} [8]	15–26%
7	<10.3 MPa	max., 105 kg f mm^{-2}, 9 gf den^{-1} [8]	approx. 10% elongation
8	<2800 MPa	max. [10]	
9	1820 MPa	dry [27]	
10	1800 MPa	[16]	
11	2500 MPa	[28]	4.3%
12	690 MPa	longitudinal [16]	3.2%
13	10 MPa	transverse [16]	0.24%

Elastic Modulus:

No.	Value	Note
1	7.85–245 MPa	80–2500 kg mm^{-2}, 8% strain, 0.0001–2 s [3]
2	2550070 MPa	2550000 kg cm^{-2} [25]
3	240000 MPa	[10]
4	1900 MPa	[16]

Complex Moduli: Dynamic modulus 90600 MPa (20°), 26600 MPa (200°, draw ratio 13.6, 62.2% crystallinity) [15], 115000 MPa (max., room temp.), 42000 MPa (max., 200°) [12]. Dynamic modulus decreases with increasing temp. and increases with processing [12,15]. The temp. dependence of the dynamic modulus and loss factor at 1 Hz has been reported [17]. The effect of applied stress at constant temp. on the dynamic modulus showsan increase in the dynamic modulus of 30000 MPa over a stress range 50–900 MPa [17]
Viscoelastic Behaviour: Elastic recovery of PVA fibres 75% (1% short-term strain), 30% (5% short-term strain), 20% (10% short-term strain), 70–80% (3% short-term strain, special fibres), 45–60% (5% short-term strain, special fibres) [8]. Stress relaxation of PVA fibres at 0.5, 1 and 2.0% strain at 30°, 50° and 70° has been reported [17]. Stress relaxation is less time dependent at high loads [17]. The amorph. regions in PVA fibre display strong viscoelastic effects upon loading at room temp. [17]. The non-linear viscoelastic behaviour of PVA fibres in relation to its fatigue props. has been reported [28]
Failure Properties General: Creep props. and their relation to brittle failure and viscoelastic props. have been reported [16]. Fatigue behaviour using zone non-linear viscoelastic analysis has been reported [28]. Shear strength 16 MPa [16]
Friction Abrasion and Resistance: The abrasion resistance of PVA fibres is at least as good as that of cotton [8]

Optical Properties:
Optical Properties General: Refractive index depends on the fibre type [8]
Refractive Index:

No.	Value	Note
1	1.532–1.55	parallel to draw [8]
2	1.505–1.526	perpendicular to draw [8]

Transmission and Spectra: Ir spectral data have been reported [23]
Total Internal Reflection: Birefringence 0.022–0.037 [8], 0.04 (max., highly oriented fibres) [8], 0.047 (crystallinity 62.2%, draw ratio 13.6) [15]. Intrinsic birefringence 0.0079 (amorph.), 0.0052 (cryst.) [14]. Birefringence increases with increasing draw ratio [14,15], attributed to an increase in chain orientation in the amorph. phase of the fibre [14]. Birefringence is also affected by crystallinity [15]

Polymer Stability:
Thermal Stability General: On heating, PVA fibres shrink [8,26]. In addition to the decomposition products for PVA, formaldehyde is prod. from the cross-links for insol. fibres [8]. Dehydration of PVA fibres occurs at temps. above about 170°, causing discolouration. However, dehydrated PVA fibres retain most of their normal strength [3,8]. PVA fibres can be completed carbonised to give graphite fibres under prolonged heating conditions [3]
Flammability: Has good fire resistance [6]
Chemical Stability: Decomposed by conc. mineral acids [8] and warm, dil. nitric acid. Warming with 20–25% sulfuric acid produces formaldehyde in 3% yield from the cross-links. Commercial PVA fibre is resistant to alkalis [2]. Insolubilised PVA fibres have good chemical resistance [1,6]
Hydrolytic Stability: Insolubilised fibres show only slight shrinkage in boiling H_2O [8]. Insolubilised PVA fibres are unaffected by immersion in sea water [6,8]
Biological Stability: Water insol. PVA fibres are resistant to biological attack and are unaffected by burial in soil [1,8]

Applications/Commercial Products:
Processing & Manufacturing Routes: Water sol. PVA fibre may be formed by wet spinning of poly(vinyl alcohol) from hot water into a coagulating bath containing salt soln. e.g. sodium sulfate, sodium carbonate, ammonium sulfate or an organic solvent e.g. ethanol, propanol or acetone. The fibres are drawn under wet or dry conditions (or both) to develop orientation and crystallinity. Heat treatment at temps. up to 220° induces further crystallisation and improves water resistance. Water insol. fibres are prod. by cross-linking fibres e.g. with formaldehyde under acid conditions or with formalin and heat. The processing conditions used drawing and heat setting and degree of cross-linking affect the mechanical and physical props. Superhydrolysed grade poly(vinyl alcohol) is used for water-insol. fibres. High MW fibres may be prepared by gel-spinning PVA containing boric acid and drying/drawing at high temps. PVA fibres have also been prepared by extrusion
Applications: PVA fibres may be blended with cotton, wool and other fibres and used to produce knitted and woven fabrics. PVA fibres are used to make Japanese fishing nets, in making artificial fur, plush pile, carpeting, tarpaulins, fire hose and coarse canvas-like cloth. PVA fibre is used in the building industry to form reinforced slate and mortar cement, and is used as an asbestos substitute in insulating board due to its fire resistance. Plastics reinforced with PVA fibres have improved impact and weathering props. PVA fibres are used in bristles and sewing thread (including medical sutures). Suitably sized fibres can be used to make filter cloths and pads which are effective in trapping fine particles (e.g. bacteria), and fibres may be used in waste water treatment. Fibres are water absorbent and may be used in nappies and sanitary products. Fibres can be used as a base in the fabrication of good quality paper, and fibrous PVA can also be used as a binder in the manufacture of conventional paper from wood pulp

Trade name	Details	Supplier
Mewlon		Toray Industries
Tac board	PVA reinforced plasterboard	Taconic Plastics Ltd.
Vinal	USA	Allied Inc (USA)
Vinylon	Japan	Kuroshiki Rayon Co (Japan)
		Kuraray-Mitsui Toatsu
		Nichibo Co (Japan)

Bibliographic References
[1] *Kirk-Othmer Encycl. Chem. Technol.*, 3rd edn., Vol. 10, (ed. M. Grayson), Wiley Interscience, 1978, 148
[2] Hongu, T. and Phillips, G., *New Fibers*, Ellis Horwood, 1990
[3] Pritchard, J.G., *Polyvinyl alcohol*, Polymer Monographs Series, Macdonald, 1970
[4] Sakurada, I., *Fibre Chem. (Engl. Transl.)*, (eds. M. Lewin and E.M. Pearce), Marcel Dekker Inc., 1985
[5] Sakurada, I., Polyvinyl alcohol Fibres, Marcel Dekker 1985
[6] *The Materials Selector*, 2nd edn., (eds. N.A. Waterman and M.F. Ashby), Chapman and Hall, 1997, (applications, processing, tradenames)
[7] *Encycl. Polym. Sci. Eng.*, 2nd edn., Vol. 6, (ed. J.I. Kroschwitz), John Wiley and Sons, 1985, 723
[8] Roff, W.J. and Scott, J.R., Fibres, Films, *Plastics and Rubbers*, 2nd edn., Butterworths, 1971
[9] *Kirk-Othmer Encycl. Chem. Technol.*, 3rd edn., Vol. 23, (ed. M. Grayson), Wiley Interscience, 1978, **23**, 848
[10] Sawatari, C., Yamamoto, Y., Yanagida, N. and Matsuo, M., *Polymer*, 1993, **34**, 956, (synth)
[11] Sakurada, I., Nukushina, Y. and Sone, Y., *Kobunshi Kagaku*, 1955, **12**, 506, (density)
[12] Kunugi, T., Kawasumi, T. and Ito, T., *J. Appl. Polym. Sci.*, 1990, **40**, 2101, (synth, props)
[13] Schellekens, R. and Ketels, H., *Polym. Commun.*, 1990, **31**, 212, (mech and thermal props)
[14] van Gurp, M., *Int. J. Polym. Mater.*, 1993, **22**, 219, (birefringence, mech props)
[15] Suzuki, A., Kawasaki, S. and Kunugi, T., *Kobunshi Ronbunshu*, 1994, **51**, 201, (birefringence, mech props)
[16] Peijs, T., van Vught, R.J.M. and Govaert, L.E., *Composites*, 1995, **26**, 83, (mech props)
[17] Govaert, L.E. and Peijs, T., *Polymer*, 1995, **36**, 3589, (dynamic mech props)
[18] Tanigami, T., Zhu, L.-H., Yamaura, K. and Matsuzawa, S., *Sen'I Gakkaishi*, *Sen'i Gakkaishi*, 1994, **50**, 53, (mechanical props)
[19] Lin, C.-A., Hwang, K.-S. and Lin, C.-H., *J. Polym. Res.*, 1994, **1**, 215, (mechanical props.)
[20] Schellekens, R. and Bastiaansen, C., *J. Appl. Polym. Sci.*, 1991, **43**, 2311, (mechanical props.)
[21] Hong, P.D. and Miyasaka, K., *Polymer*, 1994, **35**, 1369, (mechanical props.)
[22] Yamaura, K., Tanigami, T., Hayashi, N., Kosuda, K.-I. et al, *J. Appl. Polym. Sci.*, 1990, **40**, 905, (mechanical props.)
[23] Bronnikov, S.V., Vettegren, V.I. and Frenkel, S.Y., *Polym. Eng. Sci.*, 1992, **32**, 1204, (mechanical props., thermal props.)
[24] Yamamoto, T., Seki, S., Fukae, R., Sangen, O. and Kamachi, M., *Polym. J. (Tokyo)*, 1990, **22**, 567
[25] Sakurada, I., Ito, T. and Nakamae, K., J. Polym. Sci., Part C: Polym. Lett., 1966, **15**, 75, (mechanical props.)
[26] Cebe, P. and Grubb, D., *J. Mater. Sci.*, 1985, **20**, 4465, (synth., props.)
[27] Lin, C.-A., Hwang, K.-S. and Lin, C.-H., *Text. Res. J.*, 1995, **65**, 278, (mechanical props.)
[28] Liang, T., Takahara, A., Saito, K. and Kajiyama, T., *Polym. J. (Tokyo)*, 1996, **28**, 801, (fatigue props.)

Poly(vinyl alcohol), Films

Related Polymers: Poly(vinyl alcohol), Ethylene-vinyl alcohol copolymer films
Monomers: Vinyl acetate
Material class: Fibres and Films
Polymer Type: polyvinyl alcohol
CAS Number:

CAS Reg. No.	Note
9002-89-5	atactic
27101-67-3	syndiotactic

Molecular Formula: $(C_2H_4O)_n$
Fragments: C_2H_4O
Additives: A plasticiser needs to be added when films are produced by extrusion. Plasticisers used are typically high boiling, water sol. compounds containing hydroxyl groups e.g. glycerol and low MW poly(ethylene glycol). Water is the best plasticiser but produces a foamy product. High boiling methylol compounds e.g. pentaerythritol and 1,2,6-hexanetriol are preferred for high temp. processing

Poly(vinyl alcohol), Films

General Information: Water solubility of the films can be controlled by changing the degree of hydrolysis of the poly(vinyl alcohol) used. The cost of producing film is high and vinyl films only have a very small share of the market [3]

Tacticity: As well as atactic PVA films, syndiotactic PVA films have been prepared on a laboratory scale

Miscellaneous: The physical and mechanical props. of vinyl films (including PVA) are highly dependent on the method of manufacture. Can retain fragrances and aromas [15]

Volumetric & Calorimetric Properties:
Density:

No.	Value	Note
1	d 1.26 g/cm^3	[2]
2	d 1.302 g/cm^3	[16]
3	d 1.294 g/cm^3	dry [6]
4	d 1.22 g/cm^3	200 mol% H$_2$O [6]

Volumetric Properties General: The density of PVA films increases with increasing crystallinity [11,12]

Latent Heat Crystallization: ΔH_f 0.071 J kg^{-1} (17 cal g^{-1}) [16], ΔH_f 0.1 J kg^{-1} (24 cal g^{-1}) [17]

Melting Temperature:

No.	Value	Note
1	220°C	428°F [15]
2	227°C	442°F [15]
3	225°C	[16]

Transition Temperature General: Transition temps. are affected by the polymer MW (degree of polymerisation), degree of hydrol., crystallinity and tacticity. T_g is affected by the presence of plasticisers, especially H$_2$O. T_g is also reduced by iodine absorption [18]

Transition Temperature:

No.	Value	Note	Type
1	260–275°C		Heat sealing temp.
2	240–265°C	20 μm film thickness, depending on manufacturing conditions [15]	Heat sealing temp.

Surface Properties & Solubility:
Solvents/Non-solvents: Sol. H$_2$O [2,38,39]

Wettability/Surface Energy and Interfacial Tension: The adhesion props. of PVA have been expressed as peel strength of PVA film on polyester film [1] and plotted against the degree of hydrol. The peel strength sharply decreases above 95% hydrol., showing that partially hydrolysed PVA gives the best adhesion to hydrophobic surfaces [1]. MW does not significantly affect PVA adhesion to a substrate [1]. PVA shows good adhesion to hydrophilic materials with fully hydrolysed materials having the best adhesive props. [1]. PVA adhesives can also be formed by modifying the PVA with boric acid [1,6]. The adhesive props. of PVA are related to its film forming ability [31,40]

Transport Properties:
Permeability of Gases: Other gas permeability data have been reported [2,4,7,30]. Dry PVA films have excellent gas barrier props. and form barriers to H$_2$, He, O$_2$, N$_2$, CO$_2$, SO$_2$ and H$_2$S [15,31]. However, the gas barrier props. of PVA decrease at high humidity (greater than 50% relative humidity) when the permeability of films to CO$_2$ and O$_2$ increases rapidly [1,3,30]

Permeability of Liquids: PVA is permeable to H$_2$O and NH$_3$ vapour [7,31]. The H$_2$O and NaCl transport props. of untreated and heat treated PVA film membranes have been reported [33]. Permeabilities of H$_2$O and NaCl both decrease as a result of heat treatment but this effect is greater for NaCl permeability [33]. Radiation cross-linking of PVA films does not give a significant improvement over untreated PVA with regard to the selective transport of H$_2$O and NaCl [34]. The hydraulic permeability of H$_2$O in PVA membranes increases with increasing degree of hydration [35]. The diffusion coefficients for H$_2$O and EtOH in PVA membranes with different degrees of crystallinity have been reported [36]. The diffusion coefficient increases exponentially with the average H$_2$O content in the membrane. The limiting diffusion coefficient decreases as the crystallinity of the PVA increases and is greater for H$_2$O than for EtOH [36]

Permeability and Diffusion General: Permeability of H$_2$O, alcohols and aq. alcohols through syndiotactic PVA films has been reported [4]. For very thin films (0.3 μm), the permeability of alcohols decreases with increasing MW of the alcohol and the permeability of aq. alcohol mixtures is lower than those of pure H$_2$O and alcohols [4]. The temp. dependence of the H$_2$O permeability for syndiotactic PVA films has also been reported [5]. The sorption and permeation of sodium chloride, Congo red dye and Sunset Yellow dye in PVA membranes and the relationship between the degree of hydration, permeability and the diffusion coefficient has been reported [37]

Water Content: Water-sol. plasticisers e.g., propylene glycol, glycerol and polyethylene glycol decrease water resistance [39]. PVA films are hygroscopic [6] and the amount of H$_2$O absorbed depends on the film thickness, degree of crystallinity, temp. and whether the polymer has been heat-treated or plasticised [11]. The presence of syndiotactic PVA in the films lowers their hygroscopicity [38]

Gas Permeability:

No.	Gas	Value	Note
1	H$_2$O	39.4 cm^3 mm/(m^2 day atm)	0.04 m^3 m m^{-2} day^{-1} (PPa)$^{-1}$, 23°, 0% relative humidity [30]
2	H$_2$O	21.4 cm^3 mm/(m^2 day atm)	0.3 cm^3 m^{-2} day^{-1} atm^{-1}, 20°, 0% relative humidity, 14 μm thick [15]
3	H$_2$O	428.6 cm^3 mm/(m^2 day atm)	6 cm^3 m^{-2} day^{-1} atm^{-1}, 20°, 92% relative humidity, 14 μm thick [15]
4	H$_2$O	600 cm^3 mm/(m^2 day atm)	3 cm^3 m^{-2} day^{-1} atm^{-1}, 20°, 0% relative humidity, 25 μm thick [15]
5	CO$_2$	67.8 cm^3 mm/(m^2 day atm)	0.95 cm^3 m^{-2} day^{-1} atm^{-1}, 20°, 0% relative humidity, 14 μm thick [15]
6	N$_2$	2.85 cm^3 mm/(m^2 day atm)	0.04 cm^3 m^{-2} day^{-1} atm^{-1}, 20°, 0% relative humidity, 14 μm thick [15]
7	H$_2$O	2720474 cm^3 mm/(m^2 day atm)	14045 × 10^{-15} mol (m s Pa)$^{-1}$, 25°, 100% relative humidity [32]

Mechanical Properties:
Mechanical Properties General: Mechanical props. are affected by the relative humidity, method of synth. [5,17,44] and temp. Tensile strength depends on the degree of hydrol., MW and relative humidity; it decreases linearly with increasing humidity and increases with degree of hydrol. and degree of polymerisation (MW) [1]. Tensile strength yield 26 m^2 (kg mm)$^{-1}$ (ASTM D2103) [2]. The tensile response to MW is non-linear [1]. Heat treatment of PVA films increases their tensile strength [3]. The addition of plasticisers reduces the tensile strength of PVA films in proportion to the amount added. Tensile elongation increases from 10–400% (dry - 80% relative humidity) with increasing humidity and the addition of plasticiser [1,3]. Elongation of PVA films is relatively independent of degree of hydrol. but is proportional to MW.

– Poly(vinyl alcohol), Films

Storage modulus is affected by the solvent composition [8]. Young's modulus depends on the preparation method, draw ratio, polymer concentration and film thickness [5]. Young's modulus increases with increasing draw ratio [43]. The tensile strength of PVA films is higher than that of other water sol. polymers [31]

Tensile (Young's) Modulus:

No.	Value	Note
1	12300 MPa	25°, atactic [42]
2	27800 MPa	25°, 55.2% syndiotactic [42]
3	17500 MPa	25°, 63.0% syndiotactic [42]
4	166.7–196.1 MPa	1700–2000 kg cm^{-2}, 20°, 65% relative humidity, film thickness 25 μm [15]
5	6668.5 MPa	68000 kg cm^{-2}, 20°, 65% relative humidity, film thickness 14 μm [15]
6	<62000 MPa	max., gel film
7	<50300 MPa	max., cast film
8	6600–23200 MPa	degree of polymerisation 1570, film thickness 180 nm [5]
9	2400–30000 MPa	degree of polymerisation 1970, film thickness 180 nm [5]
10	3500–31400 MPa	film thickness 26–116 nm [5]
11	4903 MPa	500 kg mm^{-2} [17]

Tensile Strength Break:

No.	Value	Note	Elongation
1	27.6–69 MPa	ASTM D882 [2]	180–600%
2	67–110 MPa	98–99% hydrolysed [3]	0–300%
3	24–79 MPa	87–89% hydrolysed [3]	0–300%
4	55–69 MPa	50% relative humidity, fully hydrolysed, unplasticised [1]	
5	0.39–1.2 MPa	4–12 kg mm^{-2}, cast film with 2% H$_2$O [7]	2–15%
6	0.2–0.49 MPa	2–5 kg mm^{-2}, with 10% H$_2$O [7]	100–280%
7	30–35 MPa	80% relative humidity, fully hydrolysed [1]	
8	0.04–0.12 MPa	0.4–1.2 kg mm^{-2}, with 25% H$_2$O [7]	200–550%
9	0.98–1.5 MPa	10–15 kg mm^{-2}, dry unplasticised film [7]	
10	53.9 MPa	550 kg cm^{-2}, max., 20°, 65% relative humidity, film thickness 25 μm [15]	290–330%
11	2098 MPa	2100 kg cm^{-2}, max., 20°, 65% relative humidity, film thickness 14 μm [15]	50–55%
12	472 MPa	[16]	10.6%
13	190–610 MPa	dependent on preparation conditions and film thickness [5]	6.1–11.0%
14	100–1500 MPa	dependent on preparation conditions and film thickness [5]	1.3–20.0%

Elastic Modulus:

No.	Value	Note
1	12600 MPa	[16]
2	333.4 MPa	3400 kg mm^{-2} [44]

Miscellaneous Moduli:

No.	Value	Note	Type
1	19400–31400 MPa	degree of polymerisation 4400 [8]	Storage modulus
2	22000–28500 MPa	degree of polymerisation 2000 [8]	Storage modulus
3	240000 MPa	[8]	Crystal lattice modulus

Impact Strength: Pendulum impact strength 300–500 kg cm^{-1} (3.2 mm thick) [2], 2.56 kg m mm^{-1} (20°, 65% relative humidity) [15]. The impact strength of PVA films is high [7]

Mechanical Properties Miscellaneous: Tear strength 60 g μm^{-1} (min., 20°, 65% relative humidity, film thickness 25 μm) [15], 1.4–1.5 g μm^{-1} (20°, 65% relative humidity, film thickness 14 μm) [15], 250–800 g mm^{-1} [2]. Tear strength increases with increasing relative humidity or with the addition of small amounts of plasticiser [1,3]. Tear resistance increases with increasing degree of hydrol. and with increasing MW (viscosity) [31]. Pinhole strength 300 g (20°, 65% relative humidity, film thickness 14 μm) [15]

Hardness: Hardness increases with MW and varies inversely with moisture content [7]

Failure Properties General: Is resistant to repeated flexure [7]. PVA film is flexible and flexibility increases with increasing degree of hydrol. and increasing MW [1,3,31]. Plasticised PVA remains flexible at -40° or lower [7]

Friction Abrasion and Resistance: Coefficient of friction 0.5 μs (against self, 20°, 65% relative humidity, film thickness 14 μm) [15]. Abrasion resistance is good [7]

Electrical Properties:

Electrical Properties General: Electrical props. are affected by H$_2$O absorption under normal atmospheric conditions [7], and may also depend on the film thickness and the presence of doping materials such as iodine. The electrical conduction of very thin PVA films has been reported [19,20]. Electrical conductivity increases for films doped with iodine [20]. Specific surface resistance 24–35 G Ω (20°, 65% relative humidity) [15]. Specific surface resistance increases with decreasing film thickness [15]

Dielectric Permittivity (Constant):

No.	Value	Frequency	Note
1	3.5–10	1 kHz	65% relative humidity [7]

Dielectric Strength:

No.	Value	Note
1	>40 kV.mm^{-1}	min., dry
2	0.4 kV.mm^{-1}	under humid conditions [7]

Strong Field Phenomena General: Electric field breakdown 30 MV cm^{-1} [19], 3000 kV cm^{-1} (room temp., film thickness 0.001–0.006 cm) [6]

Static Electrification: Static electricity charging voltage 12 V (20°, 65% relative humidity, film thickness 25 μm), 95 V (20°, 65% relative humidity, film thickness 14 μm) [15]. Static electricity half-value period 1 s (20°, 65% relative humidity, film thickness 25 μm), 1.9s (20°, 65% relative humidity, film thickness 14 μm) [15]

– Poly(vinyl alcohol), Films

Dissipation (Power) Factor:

No.	Value	Frequency	Note
1	0.03–0.1	1 kHz	65% relative humidity [7]

Optical Properties:
Optical Properties General: Refractive index depends on the temp. and the amount of absorbed H_2O. The refractive index - temp. correlation has been used to measure the T_g of PVA [11,21]. Transmittance is affected by the addition of a polarising agent e.g., iodine or dichroic dye [20,22]

Refractive Index:

No.	Value	Note
1	1.55	20° [1,5]
2	1.51	[11,13]
3	1.5–1.53	depending on moisture content [7]
4	1.483–1.485	25° [14]

Transmission and Spectra: Uv permeability 75% (λ 200–400 nm, 50 μm thick film) [15], ir permeability 0% (λ 600–1700 cm^{-1}) [15]. The relationship between the transmittance at 3300 cm^{-1} in the ir spectra and film thickness for syndiotactic PVA films has been reported [5]. The optical absorption of PVA film is very low, decreasing from 8% (λ 300 nm) to 3% (λ 700 nm) with no absorption maximum [20]. Ir [6,11,13,23,24,25], C-13 nmr [25,26,27,28] and H-1 nmr [29] spectral data have been reported

Total Internal Reflection: Birefringence 0.00404–0.00481 (20°, degree of polymerisation 4400), 0.00205–0.00462 (20°, degree of polymerisation 2000) [8]. Birefringence is affected by degree of polymerisation, the concentration of PVA soln. used to form a gel and the composition of solvent (aq. DMSO) used. Solvent composition has the greatest effect on the physical props. Birefringence has a maximum value at a $DMSO/H_2O$ composition of 70:30 [8]

Volume Properties/Surface Properties: Transparency 90.5–91.5% (decreases with increasing film thickness, ASTM D1003-61) [15]. Haze 1.5–2.5% (increases with film thickness, ASTM D1003-61) [15]. Gloss high [13,15]. Clarity high [13,15] to excellent [1]

Polymer Stability:
Polymer Stability General: Plasticised films are heat sealable at approx. 150° [7]

Thermal Stability General: Heat resistance of PVA film 206° (film thickness 25 μm, ASTM D1637) and 220° (film thickness 14μm, ASTM D1637) [15]. PVA film is not heat shrinkable [2]

Upper Use Temperature:

No.	Value	Note
1	120–140°C	[2,7]

Decomposition Details: Undergoes slow degradation above 100° [7]

Flammability: Burn rate 2.5 cm s^{-1} [2]

Environmental Stress: Has excellent resistance to uv radiation (ASTM D1435) [2,41]. Has good weathering resistance [15]. Mechanical loading of PVA films during photo- and γ-irradiation-induced degradation increases the rate of formation of oxide groups. A similar effect is observed for thermal degradation [45]

Chemical Stability: Has excellent resistance to greases and oils (ASTM D722) [2,31,41]. Has poor resistance to strong acids (ASTM D534) and strong bases (ASTM D534) [2,41]. Has good to excellent resistance to organic solvents (ASTM D543) [2,15,31,41]

Hydrolytic Stability: Has poor resistance to high relative humidity (ASTM D756) [41]

Stability Miscellaneous: Dimensional stability -3.2% (100°) [2]

Applications/Commercial Products:
Processing & Manufacturing Routes: PVA films are made by casting from soln. One method of film preparation involves gelation/crystallisation from DMSO/water solns. by drawing the film in an oven at 125° under nitrogen. Films can also be prepared by extrusion. This is difficult because the melting point and decomposition temps. are close together and a plasticiser must be added. Soln. casting is the most common method of preparation, and the film is then biaxially stretched

Applications: Water-sol. films are used for disposables such as detergent packages, dyes, composting bags and hospital laundry bags; may be used in food packaging or as a semi-permeable membrane e.g. for osmotic pressure measurements and ion exchange. Film is used as a temporary protective coating for metals, plastics and ceramics. The coating reduces damage from mechanical or chemical agents during manufacture, transport and storage. The protective films can be removed by peeling or washing with water. Film may be used as packaging for pesticides and agrochemical products, and as a laminate because of its gas barrier props. PVA films made from PVA-iodine complexes may be used as polarisers

Trade name	Details	Supplier
Bovlon		Nippon Synthetic Chemical Industry Co
Pevafix	adhesive for Pevalon	May and Baker
Pevafix		Enak Ltd (UK)
Pevafix		Rotalac Plastics Ltd (UK)
Pevafix		Toray Industries
Pevafix		W R Grace
Pevafix		Kuraray Co
Pevalon		May and Baker
	films with addded iodine (polarisers)	Nippon Synthetic Chemical Industry Co
		Shin-Etsu Polymer Co Ltd
		Kanebo Ltd
		Sliontec Corp

Bibliographic References
[1] *Kirk-Othmer Encycl. Chem. Technol.*, 3rd edn., (ed. M. Grayson), Wiley Interscience, 1978, **23**, 848
[2] *Kirk-Othmer Encycl. Chem. Technol.*, 3rd edn., (ed. M. Grayson), Wiley Interscience, 1978, **10**, 216
[3] *Encycl. Polym. Sci. Eng.*, 2nd edn., (ed. J.I. Kroschwitz), John Wiley and Sons, 1985, **17**, 167
[4] Yamaura, K., Kirikawa, S., Sakuma, N., Tanigami, T. and Matsuzawa, S., *J. Appl. Polym. Sci.*, 1991, **41**, 2453, (syndiotactic)
[5] Yamaura, K., Ikeda, K.-I., Fuji, M., Yamada, R. et al, *J. Appl. Polym. Sci.*, 1987, **34**, 989, (syndiotactic)
[6] Pritchard, J.G., *Polyvinyl alcohol*, Polymer Monographs Series, Macdonald, 1970
[7] Roff, W.J. and Scott, J.R., Fibres, Films, *Plastics and Rubbers*, 2nd edn., Butterworths, 1971
[8] Sawatari, C., Yamamoto, Y., Yanagida, N. and Matsuo, M., *Polymer*, 1993, **34**, 956, (synth)
[9] Gardiner, W., Cooke, E. and Cooke, R, *Chemical Synonyms and Tradenames*, 8th edn., Technical Press Ltd, 1978
[10] *UK Kompass Register*, Reed Information Services, 1996
[11] *Water Soluble Synthetic Polymers*, (ed. P. Molyneux), CRC Press, 1984
[12] Sakurada, I., Nukushina, Y. and Sone, Y., *Kobunshi Kagaku*, 1955, **12**, 506, (density)
[13] *Polyvinyl alcohol*, (ed. C.A. Finch), Wiley, 1973
[14] *Physical Properties of Polymers Handbook*, (ed. J.E. Mark), AIP Press, 1996
[15] Moroi, H., *Br. Polym. J.*, 1988, **20**, 335, (props)
[16] Nishino, T., Takano, K. and Nakamae, K., *Polymer*, 1995, **36**, 959
[17] *Jpn. Pat.*, 1989, 01 93 325

[18] Choi, Y.S., Oishi, Y. and Miyasaka, K., *Technol. Rep. Osaka Univ.*, 1988, **37**, 2613, (electrical props, T_g)
[19] Gu, H.B., Yoshino, K., Akiya, T., Yamamura, K. and Matsuzawa, S., *Technol. Rep. Osaka Univ.*, 1987, **37**, 105, (electrical props)
[20] Kulanthaisami, S., Mangalaraj, D. and Narayandass, S.K., *Eur. Polym. J.*, 1995, **10**, 969, (electrical and optical props)
[21] Packter, A. and Nerurkar, M.S., *Eur. Polym. J.*, 1968, **4**, 685, (refractive index)
[22] *Eur. Pat. Appl.*, 1989, 297 927
[23] Nyquist, R.A., Infra Red Spectra of Plastics, *Polymers and Resins*, 2nd edn., Dow Chemical Co., 1961, (ir)
[24] Haslam, J., Willis, H.A. and Squirrel, D.C.M., *Identification and Analysis of Plastics*, 2nd edn., Butterworths, 1972, (ir)
[25] Bovey, F.A. and Jelinski, L.W., *J. Phys. Chem.*, 1985, **89**, 571, (nmr)
[26] Laupretre, F., Noel, C. and Monnerie, L., *J. Polym. Sci., Polym. Phys. Ed.*, 1977, **15**, 2143, (nmr)
[27] Horii, F., Hu, S., Ito, T., Odani, H. et al, *Polymer*, 1992, **33**, 2299, (nmr)
[28] Hu, S., Tsuji, M. and Horii, F., *Polymer*, 1994, **35**, 2516, (nmr)
[29] Horii, F., Hu, S., Deguchi, K., Sugisawa, H. et al, *Macromolecules*, 1996, **29**, 3330, (nmr)
[30] *Kirk-Othmer Encycl. Chem. Technol.*, 3rd edn., (ed. M. Grayson), Wiley Interscience, 1978, **3**, 480
[31] *Encycl. Polym. Sci. Eng.*, 2nd edn., (ed. J.I. Kroschwitz), John Wiley and Sons, 1985, **17**, 757
[32] *Polym. Permeability*, (ed. J. Comyn), Elsevier Applied Science, 1985, (permeability)
[33] Katz, M.G. and Wydeven, T., *J. Appl. Polym. Sci.*, 1982, **27**, 79, (permeability, diffusion)
[34] Katz, M.G. and Wydeven, T., *J. Appl. Polym. Sci.*, 1981, **26**, 2935, (permeability, diffusion)
[35] Kojima, Y., Furuhata, K. and Miyasaka, K., *J. Appl. Polym. Sci.*, 1983, **28**, 2401, (permeability, diffusion)
[36] Perrin, L., Quang, T.N., Clement, R. and Neel, J., *Polym. Int.*, 1996, **39**, 251, (permeability, diffusion)
[37] Kojima, Y., Furuhata, K. and Miyasaka, K., *J. Appl. Polym. Sci.*, 1984, **29**, 533, (permeability, diffusion)
[38] Matsuzawa, S., Yamaura, K., Nagura, M. and Fukata, T., *J. Appl. Polym. Sci.*, 1988, **35**, 1661, (solubility props)
[39] Lim, L.Y. and Wan, L.S.C., *Drug. Dev. Ind. Pharm.*, 1994, **20**, 1004, (water resistance)
[40] Shields, J., *Adhesives Handbook*, Butterworth, 1970, (adhesive props)
[41] *Encycl. Polym. Sci. Eng.*, 2nd edn., (ed. J.I. Kroschwitz), John Wiley and Sons, 1985, **17**, 86, (stability)
[42] Nakamae, K., Nishino, T., Ohkubo, H., Matsuzawa, S. and Yamaura, K., *Polymer*, 1992, **33**, 2581, (elastic modulus)
[43] Yamaura, K., Tanigami, T., Hayashi, N., Kosuda, K.-I. et al, *J. Appl. Polym. Sci.*, 1990, **40**, 905
[44] *Jpn. Pat.*, 1990, 02 266 914
[45] Baimuratov, E., Saidov, D.S. and Kalontarov, I.Y., *Polym. Degrad. Stab.*, 1993, **39**, 35, (degradation)

Poly(vinyl alcohol), Foamed P-569

Related Polymers: Poly(vinyl alcohol) general information
Monomers: Vinyl acetate
Material class: Gums and resins
Polymer Type: polyvinyl alcohol
CAS Number:

CAS Reg. No.
9002-89-5
26876-25-5

Molecular Formula: $(C_2H_4O)_x$
Fragments: C_2H_4O
Additives: Formalin (formaldehyde/methanol mixture) is used as a cross-linking agent

Applications/Commercial Products:
Processing & Manufacturing Routes: A concentrated, very viscous solution of PVA in acidified formalin is impregnated with air bubbles to form a foam. The foam is treated to about 60° to form a sponge via cross-linking [1]
Applications: Poly(vinyl alcohol) sponge has medicinal uses and may also be used as a water-absorbent material

Supplier
DuPont
Hoechst
Sekisui Chem. Co Ltd
H.G. Hammon

Bibliographic References
[1] Pritchard, J., *Polyvinyl alcohol*, Polymer Monographs Series, Macdonald, 1970
[2] *Encycl. Polym. Sci. Eng.*, 2nd edn., John Wiley and Sons, 1989, **17**, 167

Poly(vinyl alcohol)/poly(vinyl chloride), Fibre blend P-570

Synonyms: *PVA/PVC fibre blend. Polychlal fibres*
Related Polymers: Poly(vinyl alcohol) fibres
Monomers: Vinyl chloride, Vinylacetate
Material class: Blends, Fibres and Films
Polymer Type: PVC, polyvinyl alcohol
CAS Number:

CAS Reg. No.
25822-51-9

Molecular Formula: $[(C_2H_4O).(C_2H_3Cl)]_n$
Fragments: C_2H_4O C_2H_3Cl
Additives: The fibres are treated with formaldehyde (cross-linking agent)
General Information: These fibres are unique to Japan and are closely related to poly(vinyl alcohol) and poly(vinyl chloride) fibres. Production in Japan occurs on a modest scale

Applications/Commercial Products:
Processing & Manufacturing Routes: Polychlal fibres are prod. from an emulsion of poly(vinyl chloride) and a matrix of poly(vinyl alcohol). This is wet spun into aq. sodium sulfate and treated with formaldehyde
Applications: Used in the manufacture of flame retardant apparel

Trade name	Details	Supplier
Polychlal	Japan	Nissin Spinning
Polychlal		Kojin KK

Bibliographic References
[1] *Kirk-Othmer Encycl. Chem. Technol.*, 3rd edn., Wiley Interscience, 1984, **10**, 159
[2] *Jpn. Pat.*, 1993, 05 321 140
[3] *Jpn. Pat.*, 1994, 06 02 239

Poly(vinyl benzoate) P-571

$$\begin{array}{c}\text{—}[\text{CH}_2\text{—CH}]_n\text{—}\\|\\\text{OCOPh}\end{array}$$

Synonyms: *Poly(benzoyloxyethylene). PVB*
Monomers: Vinyl benzoate
Material class: Synthetic Elastomers
Polymer Type: polyvinyls
CAS Number:

CAS Reg. No.
24991-32-0

– Poly(vinyl bromide)

Molecular Formula: $(C_9H_8O_2)_n$
Fragments: $C_9H_8O_2$
Tacticity: Very little is given on specific tacticities; an isotactic preparation has been reported [13]

Volumetric & Calorimetric Properties:
Density:

No.	Value	Note
1	d^{20} 1.198 g/cm^3	[1]
2	d^{25} 1.198 g/cm^3	[7]

Glass-Transition Temperature:

No.	Value	Note
1	71°C	[1]

Transition Temperature:

No.	Value	Note	Type
1	-70°C	100 Hz [2]	β

Surface Properties & Solubility:
Solubility Properties General: Miscible with poly(phenyl acrylate) in all proportions [9]
Solvents/Non-solvents: Sol. EtOAc, CHCl$_3$, Me$_2$CO, dioxane and CCl$_4$ [14]

Transport Properties:
Transport Properties General: Gas permeability measurements suggest that the permeability of heavier (larger) gases is greater for PVB than for PVAC; for the lighter (smaller) gases the inverse is true [11]
Polymer Solutions Dilute: [η] 0.29 dl g^{-1} (MW 133000) [5]
Permeability of Gases: [O$_2$] 8.05 × 10^{-8} (below T$_g$); 3.68 × 10^{-4} (above T$_g$); [N$_2$] 1.38 × 10^{-7} (below T$_g$); 1.68 × 10^{-4} (above T$_g$); [CO$_2$] 2.66 × 10^{-8} (below T$_g$); 9.79 × 10^{-5} cm^2 (s cmHg)$^{-1}$ (above T$_g$) [11]

Electrical Properties:
Dielectric Permittivity (Constant):

No.	Value	Frequency	Note
1	3.3	1 kHz	room temp.; monolayer [22]

Optical Properties:
Refractive Index:

No.	Value	Note
1	1.513–1.516	[6,15]
2	1.57	30° [1]

Transmission and Spectra: Main area of interest 10–15μ region of ir spectrum. Bands of significance at 11.8 and 12.48μ, 13.25μ. Vinyl group deformations are shifted, relative to established position - a characteristic of vinyl esters [16]

Polymer Stability:
Environmental Stress: Frost resistant but may coagulate if aq. phase freezes [12]
Biological Stability: Unprotected emulsions may be susceptible to attack from microorganisms [12]

Applications/Commercial Products:
Processing & Manufacturing Routes: Polymerisation of vinyl benzoate [10,12] using a typical peroxide initiator [14,3] or other radical sources e.g. azobisisobutyronitrile [4] may be stabilised with water-soluble protective colloids, anionic or non-ionic emulsifiers.

Can also be prod. heterogeneously by emulsion polymerisation [8]. A number of catalysts have been developed for the preparation of polyvinyl benzoate. These include amine oxides [17], organic polyperoxides [18] and a three-component initiator system, comprising an oxidising agent, metal chelate and electron donors [19,20]. Metal-halide silane cocatalysts have also been used [21]

Bibliographic References
[1] Magagnini, P.L., *Chim. Ind. (Milan)*, 1967, **49**, 1041, (refractive indices, glass transition temp)
[2] Matsuoka, S. and Ishida, Y., *J. Polym. Sci., Part C: Polym. Lett.*, 1966, **14**, 247, (glass transition temp)
[3] Pizzirani, G., Magagnini, P. and Guisti, P., *J. Polym. Sci., Part A-2*, 1971, **9**, 1133, (polymerisation)
[4] Kinoshita, M., Irie, T. and Imoto, M., *Makromol. Chem.*, 1967, **110**, 47, (radical polymerisation)
[5] Magagnini, P.L. and Frosini, V., *Eur. Polym. J.*, 1966, **2**, 139
[6] Morrison, E.D., Gleason, E.H. and Stannett, V., *J. Polym. Sci.*, 1959, **36**, 267
[7] Frosini, V., Marchetti, A. and Butta, E., *Chim. Ind. (Milan)*, 1977, **59**, 415
[8] Plavljanić, B. and Janović, Z., *J. Polym. Sci., Polym. Chem. Ed.*, 1981, **19**, 1795, (polymerisation)
[9] Maiti, N., Dutta, S., Bhattacharyya, S.N. and Mandal, B.M., *Polym. Commun.*, 1988, **29**, 363
[10] Bevington, J.C. and Johnson, M., *Eur. Polym. J.*, 1968, **4**, 373, (radical polymerisation)
[11] Hirose, T., Mizoguchi, K. and Kamiya, Y., *J. Appl. Polym. Sci.*, 1985, **30**, 410
[12] *Ullmanns Encykl. Ind. Chem.*, 5th edn. (ed. W. Gerhartz), VCH, 1985, (synth)
[13] Magagnini, P.L. and Frosini, V., *Chim. Ind. (Milan)*, 1966, **48**, 137
[14] Han, G.E. and Ringwald, E.L., *J. Polym. Sci.*, 1952, **8**, 91, (polymerisation)
[15] Burnett, G.M. and Wright, W.W., *Trans. Faraday Soc.*, 1953, **49**, 1108, (polymerisation studies)
[16] Smets, G. and Hertoghe, A., *Makromol. Chem.*, 1956, **17**, 189, (ir)
[17] *Brit. Pat.*, 1947, 585 396
[18] *Brit. Pat.*, 1948, 604 580
[19] *Brit. Pat.*, 1968, 1 136 326
[20] *Fr. Pat.*, 1966, 1 492 940
[21] *U.S. Pat.*, 1966, 3 285 895
[22] Scala, L.C. and Handy, R.M., *J. Appl. Polym. Sci.*, 1965, **9**, 3111

Poly(vinyl bromide)

$$-[CF_2-CH(Br)]_n-$$

Synonyms: *Poly(bromoethene)*
Monomers: Vinyl bromide
Polymer Type: polyhaloolefins
CAS Number:

CAS Reg. No.
25951-54-6

Polymer Stability:
Thermal Stability General: Has poor thermal stability and is substantially less stable than PVC [1,3,4,12]
Decomposition Details: Decomposition begins at approx. 100°, dehydrobromination is the dominant reaction. Kinetics studies of decomposition shows evidence for autocatalytic reactions [1]. Activation energy of decomposition is 17 KJ mol^{-1} less than the corresponding value for PVC. Decomposition products include HBr, and aromatic compounds (e.g. C$_6$H$_6$ and naphthalene) [2,9]. Thermal degradation leads to approx. 100% of available bromine as HBr via chain-stripping reaction, after which resultant polyene undergoes degradation [15]. Mechanism of decomposition appears to be 'gradual zip growth' type; initiation step for this is not autocatalytic. Activation energy of decomposition 153 KJ mol^{-1} (for random elimination of HBr). Long polyene sequences are formed during dehydrobromination reaction (longer than those formed during corresponding dehydrochlorination reactions of PVC) [16]
Environmental Stress: May darken on exposure to uv light [24]

Poly(vinyl butyral) P-573

Synonyms: *PVB*
Related Polymers: Polyvinyl formal, Polyvinyl acetal, Polyvinyl alcohol
Monomers: Vinyl acetate, Butyraldehyde
Material class: Thermoplastic
Polymer Type: polyvinyls
Molecular Formula: $[(C_8H_{14}O_2)_x(C_2H_4O)_y(C_4H_6O_2)_z]_n$
Fragments: $C_8H_{14}O_2$ C_2H_4O $C_4H_6O_2$
Molecular Weight: 30000–250000
General Information: Atactic, isotactic and syndiotactic forms known. Tacticity depends on that of poly (vinyl acetate) percursor. Isotactic chain portions acetalise more rapidly than syndiotactic. [1]. In general, commercial resins have low acetate content. Increasing acetate groups decreases hydrophobicity, lowers distortion temp. and decreases toughness. The inherent terpolymer struct. gives excellent adhesion props. May cross-link in presence of acid
Tacticity: Props. of polymer depend on MW and degree of acetalisation. Manuf. process yields a terpolymer struct, consisting of acetal, free OH and acetate groups. May contain up to 10–20% acetate and 10% OH groups. [2,3]

Volumetric & Calorimetric Properties:
Density:

No.	Value	Note
0	d 1.083 g/cm^3	Butvar B76 [4]
1	d 1.1 g/cm^3	Butvar B72 [4]

Glass-Transition Temperature:

No.	Value	Note
1	62.72°C	Butvar B79 [4]
2	72–78°C	Butvar B72 [4]

Deflection Temperature:

No.	Value	Note
1	50–54°C	Butvar B79, ASTM D648 [4]
2	56–60°C	Butvar B72, ASTM D648 [4]

Brittleness Temperature:

No.	Value	Note
1	-40°C	18% OH [3]
2	-32°C	25% OH [3]

Surface Properties & Solubility:
Solubility Properties General: Compatible with epoxy resins, phenolics, polyurethanes and other plasticising resins [9]
Plasticisers: Platicisers include dibutyl phthalate, dibutyl sebacate, triethylcitrate, dihexyl adipate, castor oil, phosphates, polyethylene glycols, triethylene glycol-2,2-diethyl butyrate. Compatibility with plasticisers has been reported [6]
Solvents/Non-solvents: Sol. MeOH (25% OH), EtOH, AcOH, Me_2CO (12% OH), dioxane, EtOAc (12% OH). Insol. C_6H_6 (12% OH), Me_2CO (25% OH), EtOAc (25% OH), C_6H_6 (25% OH). High degree free-hydroxyl group increases solubility in polar solvents, decreases solubility in non-polar solvents [3]
Surface and Interfacial Properties General: Adhesion increases with increasing free hydroxyl groups. [8] Potassium and lithium salts are added to reduce film adhesion.

Transport Properties:
Polymer Solutions Dilute: Viscosity 170–260 cP (MW 170000–250000) [4]; 6.0–9 (MW 40000–70000) [4]
Water Absorption:

No.	Value	Note
1	0.3–0.5 %	24h, Butvar, ASTM D570 [4]

Mechanical Properties:
Tensile Strength Break:

No.	Value	Note	Elongation
1	68.9 MPa	10 kpsi, 25% OH [3]	8%
2	59.3 MPa	8.6 kpsi, 12% OH [3]	10%
3	48.2 MPa	7–8 kpsi, Butvar B72, ASTM D638	70%

Elastic Modulus:

No.	Value	Note
1	2620 MPa	380 kpsi, 0°, 25% OH [3]
2	2068 MPa	300 kpsi, 40°, 25% OH [3]
3	2275 MPa	330 kpsi, 0°, 12% OH [3]
4	1654 MPa	240 kpsi, 40°, 12% OH [8]
5	1930 MPa	280–340 kpsi, Butvar, ASTM D638 M[4]

Tensile Strength Yield:

No.	Value	Elongation	Note
1	46.9–53.8 MPa	8%	6.8–7.8 kpsi, Butvar B72, ASTM D638

Flexural Strength Yield:

No.	Value	Note
1	7239–8963 MPa	1.050–13 kpsi, Butvar, ASTM D790 [4]

Hardness: Rockwell M100–M115, Butvar, ASTM D785 [4]; E5-E20, Butvar, ASTM D785
Izod Notch:

No.	Value	Notch	Note
1	58.7 J/m	Y	1.1 ft lb in^{-1}, Butvar B72 [4]
2	42.7 J/m	Y	0.8 ft lb in^{-1}, Butvar B79 [4]

Electrical Properties:
Electrical Properties General: Increasing butyral content decreases dielectric permittivity, increases dielectric strength
Dielectric Permittivity (Constant):

No.	Value	Frequency	Note
1	3.2	50 Hz	Butvar B72, ASTM D150 [4]
2	2.7	50 Hz	Butvar B79, ASTM D150 [4]
3	2.6	1 kHz	Butvar B72, ASTM D150 [4]
4	2.8	1 MHz	Butvar B79, ASTM D150 [4]
5	2.6	1 MHz	Butvar B79, ASTM D150 [4]
6	2.6	1 kHz	Butvar B79, ASTM D150 [4]

Poly(N-vinylcarbazole)

Dielectric Strength:

No.	Value	Note
1	15.74–18.89 kV.mm^{-1}	400–480 Vmil^{-1}, short time, Butvar, ASTM D

Dissipation (Power) Factor:

No.	Value	Frequency	Note
1	0.005	50 Hz	Butvar B79, ASTM D150 [4]
2	0.0039	1 kHz	Butvar B79, ASTM D150 [4]
3	0.013	1 MHz	Butvar B79, ASTM 150 [4]
4	0.0064	50 Hz	Butvar B72, ASTM D150 [4]
5	0.0062	1 kHz	Butvar B72, ASTM 150 [4]
6	0.0027	1 MHz	Butvar B72, ASTM D150 [4]

Optical Properties:
Refractive Index:

No.	Value	Note
1	1.485–1.49	Butvar, ASTM D542 [4]

Transmission and Spectra: Nmr spectral data have been reported [4]

Polymer Stability:
Polymer Stability General: Colours with ageing and thermal degradation
Upper Use Temperature:

No.	Value	Note
1	93°C	200°F

Flammability: Fl. pt 370° (min.). Burns smoothly and completely
Chemical Stability: Stable to aromatic hydrocarbons, aliphatic hydrocarbons, esters, ketones; attacked by alcohols. High butyral content decreases solvent resistance
Hydrolytic Stability: Good resistance to weak and strong bases (ASTM D543) [4] and to weak acids
Biological Stability: Non toxic

Applications/Commercial Products:
Processing & Manufacturing Routes: Produced by 2 methods. Poly(vinyl acetate) is saponified in MeOH with NaOH, the alcohol purified and dissolved in water and created with butyraldehyde and hydrochloric acid at 0° - hetergeneous acetalisation. [10] Alternatively, the saponified poly(vinyl acetate) in MeOH is treated with butyraldehyde and hydrochloric acid at -60° - homogeneous acetalisation [11]
Applications: Major use of film is as a laminate in safety glass for automotive use eg windscreens. Also used in bullet-proof and other glass products. Other uses are as ceramic binders. Surface coatings for wood, structural adhesives (combined with other resins) in printed circuits, inks and dyes, paints; solder masks and hot melt adhesives. Improves water and stain resistance of textiles

Trade name	Supplier
Butvar	Solutia
Mowibal B	Hoechst
Pioloform B	Wacker
Rhovinal B	Rhone-Poulenc
S-lec	Sekisui Chem. Co Ltd
Vinylite	Union Carbide

Bibliographic References
[1] Bolomstron, T.P., *Encycl. Polym. Sci. Eng.*, 2nd edn (ed. J.I. Kroschwitz), 1985, **117**, 136, (rev)
[2] Flory, P.J., *J. Am. Chem. Soc.*, 1939, **61**, 1518, (synth)
[3] Fitzhugh, A.F. and Crozier, R.N., *J. Polym. Sci.*, 1952, **8**, 9, 225, (synth, MW, viscosity)
[4] *Butvar resin*, Solutia Inc., 1997, (technical datasheet)
[5] Bruch, M.D. and Bonesteel, J.K., *Macromolecules*, 1986, **19**, 1622, (2D nmr)
[6] Tyazhlo, N.I., Uspenskaya, Z.R., Piastro, O.V., Trofimova, N.V. et al, *Zh. Prikl. Khim. (Leningrad)*, 1974, **47**, 2285, (plasticiser)
[7] Lavin, E. and Snelgrove, J.A., *Kirk-Othmer Encycl. Chem. Technol.*, 3rd edn., (eds M.Grayson and D. Eckroth), John Wiley and Sons, 1983, **23**, 798, (rev)
[8] Svoboda, J. and Mabys, V., *Plasty Kauc.*, 1975, **12**, 131, (adhesion)
[9] *Pioloform B*, Wacker Chemie, 1996, (technical datasheet)
[10] *Ger. Pat.*, IG Farber, 1939, 891 745, (manuf)
[11] *Ger. Pat.*, Hoechst, 1953, 899 864, (manuf)

Poly(N-vinylcarbazole) P-574

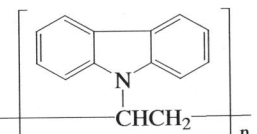

Synonyms: *Poly(9-ethenyl-9H-carbazole). PVK. PVCZ. Poly[1-(N-carbazolyl)ethylene]. PNVK. PNVC. PVCA*
Monomers: N-Vinylcarbazole
Material class: Thermoplastic
Polymer Type: polyvinyls
CAS Number:

CAS Reg. No.	Note
25067-59-8	homopolymer

Molecular Formula: $(C_{14}H_{11}N)_n$
Fragments: $C_{14}H_{11}N$
Molecular Weight: M_n 11000 (film)
Additives: Addition of additive changes wavelength of radiation which produces photoconduction [2,3]. Small amounts of styrene (0.1–2%) as a copolymer increase workability [34]
General Information: A brittle photoconductive polymer. The polymer has some toxicity due to residual monomer being present
Morphology: Unit cell of PVK is orthorhombic with pseudo hexagonal packing of the polymer chains. Orthorhombic unit cell a 21.6Å, b 12.5Å [15] (melt extruded PVK). Previously assigned as hexagonal unit cell $a_1=a_2=a_3$ 12.3Å [2,15]. X-ray spacings 10.8Å, 21.5Å, 8.44Å, 6.15Å [15] (melt extruded PVK)
Miscellaneous: First known organic photoconducting polymer; conducts on exposure to uv wavelengths [2,3]. A green conducting form of PNVK is formed by electropolymerisation of N-vinylcarbazole [12]

Volumetric & Calorimetric Properties:
Density:

No.	Value	Note
1	d 1.184 g/cm^3	amorph. [15]
2	d 1.191 g/cm^3	approx., extruded
3	d 1.193 g/cm^3	crystallisation [2,15]
4	d 1.2 g/cm^3	Polectron, Luvican [2,34]
5	d 1.19 g/cm^3	DIN 1306 [35]

Poly(N-vinylcarbazole)

Thermal Expansion Coefficient:

No.	Value	Note	Type
1	$4.5 \times 10^{-5} - 5.5 \times 10^{-5}$ K^{-1}	Polectron [2,34]	L
2	5.7×10^{-5} K^{-1}	20–100°, Luvican [2,34]	L

Equation of State: Mark-Houwink equation of state parameters have been reported [17,18,26]

Thermodynamic Properties General: Specific heat values have been reported [2,34]

Thermal Conductivity:

No.	Value	Note
1	2.51 W/mK	0.006 cal cm^{-1} s^{-1} K^{-1}, Luvican [34]
2	0.125 W/mK	0.0003 cal cm^{-1} s^{-1} K^{-1}, 20° [35]
3	0.164 W/mK	0.00039 cal cm^{-1} s^{-1} K^{-1}, 173° [35]

Specific Heat Capacity:

No.	Value	Note	Type
1	1.6 kJ/kg.C	160° [14]	P

Melting Temperature:

No.	Value	Note
1	290°C	[34]

Glass-Transition Temperature:

No.	Value	Note
1	227°C	amorph. [2]
2	195–200°C	values in this range contain approx. 4% carbazole and other impurities in the polymer [9]
3	225°C	[21,22]

Transition Temperature General: Flow temp. 270° (Luvican) [2]. Shape stability temp. 160–170° (Martens); 200° (Vicat); 150° (ASTM) [35].

Deflection Temperature:

No.	Value	Note
1	100–150°C	ASTM [2,34]

Vicat Softening Point:

No.	Value	Note
1	150°C	Martens [2], Luvican [34]
2	190°C	Vicat [2], Luvican [34]

Transition Temperature:

No.	Value	Note	Type
1	80°C	carbazole wagging motion [21,22]	β
2	-60°C	rotational libration of pendant carbazole group, only in presence of oxygen [21,22]	γ
3	-160°C	carbazole localised backbone motion [21,22]	δ

Surface Properties & Solubility:

Plasticisers: Plasticisers to reduce brittleness include benzyl tetralin, amyl naphthalene, phenanthrene, diamylbiphenyl terphenyl, tritolyl phosphate, diphenyl and polyglycols [34]

Solvents/Non-solvents: Sol. C_6H_6 [34], dioxane [10], THF, CHCl$_3$, tetrachloroethane [17], aromatic hydrocarbons [34], 1,2-dichloroethane, nitrobenzene, chlorobenzene [26]. Mod. sol. cyclohexanone [17]. Insol. MeOH, hexane [9], aliphatic alcohols, esters, ethers, ketones [34]

Transport Properties:

Polymer Solutions Dilute: Theta temp. θ -36.5° (chlorobenzene) [28], θ -20.4° (nitrobenzene), θ 37° (toluene) [18]. Table of interaction parameter (B) for chlorobenzene and nitrobenzene at 25–45° have been reported [26]. Dipole moment vs. weight fraction of solute in 1,4-dioxan has been reported [34]. Kerr constants of PVK in 1,4-dioxan [32] have been reported. Viscosity coefficient 40–100 (Luvican) [34]

Water Absorption:

No.	Value	Note
1	0.1 %	Polectron [2,34]
2	0.012 %	1 week, Luvican [2,34]

Mechanical Properties:

Mechanical Properties General: Strain failure 1.1% (extruded) [14]; 0.3% (amorph.) [14]

Tensile (Young's) Modulus:

No.	Value	Note
1	4120 MPa	597 kpsi, extruded PNVK [14]
2	4000 MPa	580 kpsi, amorph., Polectron [2,14,34]
3	3240 MPa	470 kpsi, Luvican [2,34]

Tensile Strength Break:

No.	Value	Note	Elongation
1	12.41 MPa	1800 psi, Polectron [2,34]	0.32%
2	14.13 MPa	2050 psi, Luvican [2,34]	
3	124–138 MPa	18–20 kpsi, oriented fibres [2,34]	

Flexural Strength at Break:

No.	Value	Note
1	10.34–12.41 MPa	1500–1800 psi, Polectron [34]
2	37.9–47.5 MPa	5500–6900 psi, Luvican [34]

Compressive Strength:

No.	Value	Note
1	33 MPa	4800 psi, Luvican [2,34]

Impact Strength: Impact strength 9.8 J m^{-1} (0.19 ft lb in^{-1}, Polectron) [2]; 117 J m^{-1} (2.2 ft lb in^{-1}, Luvican) [2,34]. Impact bending stress 10–15 cm kg^{-1} cm^{-1} (Luvican) [2,34]

Hardness: Rockwell R113–R125 (Polectron) [2,34]

Fracture Mechanical Properties: Shear strength 24.13 MPa (3500 psi, Polectron) [34]

Electrical Properties:

Electrical Properties General: Dipole moment and dielectric permittivity values have been reported. [10,11,13] DC dark

– Poly(*N*-vinylcarbazole)

conductivity has been reported [31]. $\Delta\epsilon \simeq 0.39$ eV. Threshold ionisation potential 5.85 eV [20]. Conductivity of green film prod. by electrochemical polymerisation of *N*-vinylcarbazole 3×10^{-5} Ω^{-1} cm^{-1} [19]

Dielectric Permittivity (Constant):

No.	Value	Frequency	Note
1	4.96	100 kHz	AIBN polymerised [14]
2	4.6	1 MHz	AIBN polymerised [14]
3	2.99	10 Hz	[10]

Dielectric Strength:

No.	Value	Note
1	11–86 kV.mm^{-1}	25–150° [34]
2	40 kV.mm^{-1}	2.5 mm thick samples [34]
3	120 kV.mm^{-1}	0.225 mm thick samples [34]

Arc Resistance:

No.	Value	Note
1	20s	0.15 in. thick, ASTM D495-42 [34]

Dissipation (Power) Factor:

No.	Value	Frequency	Note
1	0.0027	100 kHz	[14]
2	0.0022	1 MHz	[14]

Optical Properties:
Refractive Index:

No.	Value	Note
1	1.68	[23]
2	1.69	20° [2,34]

Transmission and Spectra: H-1 nmr [5,6,19,34], C-13 nmr [7,8,13,25], ir [12,15,30], Raman [12], x-ray [15,16], electron diffraction [16], uv [23,35] and fluorescence [35] spectral data have been reported

Volume Properties/Surface Properties: Electrochemically polymerised PNVK has a green colour due to bicarbazoyl groups [19]. The 'normal' PNVK is white [19]

Polymer Stability:
Thermal Stability General: Degradation studies have been reported [4,14,24]

Upper Use Temperature:

No.	Value	Note
1	120°C	Luvican [2,34]
2	150–200°C	[34]
3	300°C	approx.

Decomposition Details: Below 260° minimal weight loss occurs, above 300° noticeable weight loss due to depolymerisation occurs and above 410° pyrolysis is complete after 18h [4]. Polymer becomes cross-linked and insol. when heated up to 410° [4,14]

Environmental Stress: Sunlight causes slight discoloration [34]

Chemical Stability: Decomposes to carbazole when soln. is passed through silica gel [4]. Attacked by conc. sulfuric acid or conc. nitric acid [34]. Resistant to alkalis (60% sodium hydroxide at 160°), dilute acids (including hydrofluoric acid) [34]. Unaffected by paraffinic compounds and mineral oils [34]

Hydrolytic Stability: Rapidly attacked by 0.5% sodium chlorite soln. at 80°. [34] Resistant to water

Stability Miscellaneous: Low temp. mech. destruction leads to radical products. Room temp. destruction gives more stable products [33]

Applications/Commercial Products:
Processing & Manufacturing Routes: Thin films can be obtained by vapour deposition polymerisation of *N*-vinylcarbazole using a heated filament [39] or a low pressure mercury lamp. [40] The moulded polymer may be turned, sawn, milled and polished by machines with cutting angles set for brass [34]

Applications: Used as a capacitor dielectric, in switch parts, cable connectors, coil formers, cable spacers, stand-off insulators. Has photoconducting applications in electric photographic processes such as photocopying. Has also found applications in microlithography, holographic recording, blue-green electrochromic displays and optical switching devices. Thin films used as layer in organic electroluminescent devices for flat optical displays

Trade name	Supplier
Luvican	BASF
Polectron	GAF

Bibliographic References

[1] Biswas, M. and Chakravarty, D., *J. Macromol. Sci., Rev. Macromol. Chem. Phys.*, 1972, **C7**, 189, (rev)
[2] Penwell, R.C., Ganguly, B.N. and Smith, T.W., *Macromol. Rev.*, 1978, **13**, 63, (rev)
[3] Pearson, J.M. and Stolka, M., Polymer Monographs, Vol. 6: Poly(*N*-vinylcarbazole), Gordon and Breach, 1981, (rev)
[4] Chu, J.Y.C. and Stolka, M., *J. Polym. Sci., Part A: Polym. Chem.*, 1975, **13**, 2867, (thermal degradation)
[5] Yoshimoto, S., Akana, Y., Kimura, A., Hirata, H. *et al*, *J.C.S., Chem. Commun.*, 1969, 987, (nmr)
[6] Williams, D.J., *Macromolecules*, 1970, **3**, 602, (nmr)
[7] Kawamura, T. and Matsuzaki, K.V., *Makromol. Chem.*, 1978, **179**, 1003, (C-13 nmr)
[8] Natansohn, A., *J. Polym. Sci., Part A: Polym. Chem.*, 1989, **27**, 4257, (nmr)
[9] Bergfjord, J.A., Penwell, R.C. and Stolka, M., *J. Polym. Sci., Part B: Polym. Phys.*, 1979, **17**, 711, (glass transition)
[10] Molina, M.S., Barrales-Rienda, J.M., Riande, E. and Saiz, E., *Macromolecules*, 1984, **17**, 2728, (dipole moments, dielectric constant)
[11] Baysal, B.M. and Aras, L., *Macromolecules*, 1985, **18**, 1693, (dipole moment)
[12] Sacak, M., Akbulut, U., Cheng, C. and Batchelder, D.N., *Polymer*, 1994, **35**, 2495, (ir and Raman)
[13] Mumby, S.J. and Beevers, M.S., *Polymer*, 1985, **26**, 2014, (nmr)
[14] Haque, S.A. and Biswas, M., *Polym. Commun.*, 1985, **26**, 122, (dielectric constant)
[15] Penwell, R.C. and Prest, W.M., *Polymer*, 1978, **19**, 537, (melt extruded PNVK)
[16] Sundararajan, P.R. and Zamin, J., *Polymer*, 1979, **20**, 1567, (melt extruded PNVK)
[17] Sitaramaiah, G. and Jacobs, D., *Polymer*, 1970, **11**, 165, (soln props)
[18] Kuwahara, N., Higashida, S., Nakata, M. and Kaneko, M., *J. Polym. Sci., Part B: Polym. Phys.*, 1969, **7**, 285, (unperturbed dimensions)
[19] Geissler, U., Hallensleben, M.L. and Toppare, L., *Synth. Met.*, 1993, **55**, 1662, (conductivity)
[20] Shirota, Y., Matsumoto, Y., Mikawa, H., Seki, K. and Inokuchi, H., *Chem. Phys. Lett.*, 1983, **97**, 57, (ionisation potential)
[21] Pochan, J.M., Hinman, D.F. and Nash, R., *J. Appl. Phys.*, 1975, **46**, 4115, (dielectric relaxation)
[22] Froix, M.F., Williams, D.J. and Goedde, A.O., *J. Appl. Phys.*, 1975, **46**, 4166, (nmr)
[23] Klopffer, W., *J. Chem. Phys.*, 1969, **50**, 2337, (uv)
[24] Pielichowski, K., *J. Therm. Anal.*, 1995, **43**, 509
[25] Siove, A. and Ades, D., *Eur. Polym. J.*, 1992, **28**, 1583, (nmr)
[26] Leon, L.M., Katime, I. and Rodriguez, M., *Eur. Polym. J.*, 1979, **15**, 29, (soln props)
[27] Leon, L.M., Katime, I., Gonzalez, M. and Figueruelo, J., *Eur. Polym. J.*, 1978, **14**, 671, (soln props)

[28] Leon, L.M., Galaz, J., Garcia, L.M. and Anasagasti, M.S., *Eur. Polym. J.*, 1980, **16**, 921, (soln props)
[29] *Encycl. Polym. Sci. Eng.*, 2nd edn., (ed. J.I. Kroschwitz), John Wiley and Sons, 1985, **3**, 334, 340
[30] Papež, V., Inganäs, O., Cimrová, V. and Neηspůrek, S., *J. Electroanal. Chem.*, 1990, **282**, 123, (PNVC)
[31] Santos-Lémus, S. and Luna, M., *Polym. Eng. Sci.*, 1993, **33**, 501, (conductivity)
[32] Beevers, M.S. and Mumby, S.J., *Polym. Commun.*, 1984, **25**, 173, (PNVK)
[33] Tino, J., Szöcs, F. and Hlousková, Z., *Polymer*, 1982, **23**, 1443, (PNVK)
[34] Cornish, E.H., *Plastics (London)*, 1963, **28**, 61, (applications)
[35] Klöpffer, W., *Kunststoffe*, 1971, **61**, 533, (rev)
[36] *Photocond. Relat. Phenom.*, (eds. J. Mort and D.M. Pai), Elsevier, 1976, (rev)
[37] Stolka, M. and Pai, D.M., *Adv. Polym. Sci.*, 1978, **29**, 1, (rev)
[38] *Electron. Prop. Polym.*, (eds. J. Mort and G.P. Pfister), Wiley-Interscience, 1982, (rev)
[39] Tamada, M., Omichi, H. and Okui, N., *Thin Solid Films*, 1995, **268**, 18
[40] Tamada, M., Koshikawa, H. and Omichi, H., *Thin Solid Films*, 1997, **292**, 164

Poly(vinyl chloride-*co*-acrylonitrile)   P-575

$$-[(CH_2CH(Cl))_x-(CH_2CH(CN))_y]_n-$$

Synonyms: *Vinyl chloride-acrylonitrile copolymer. Poly(chloroethene-co-2-propenenitrile). VCA. VC/AN copolymer. Modacrylic fibres*
Related Polymers: PVC Copolymers, Polyvinyl chloride, unplasticised, Polyacrylonitrile
Monomers: Vinyl chloride, Acrylonitrile
Material class: Thermoplastic, Copolymers
Polymer Type: PVC, acrylonitrile and copolymerspolyhaloolefins
CAS Number:

CAS Reg. No.
9003-00-3
30642-67-2

Molecular Formula: $[(C_2H_3Cl).(C_3H_3N)]_n$
Fragments: C_2H_3Cl C_3H_3N
Additives: Plasticisers; heat stabilisers
General Information: Copolymers form the basis of useful fibres with low flammability; typical compositions often contain vinyl chloride as minor constituent (10–15%); original Dynel modacrylic fibres, however, had copolymer composition of 60/40 vinyl chloride/acrylonitrile; cheaper than materials with higher acrylonitrile content, and PAN homopolymer [6]. Fibres show good dyeing characteristics [3]
Morphology: Fibres have low degree of crystallinity; no significant changes after stretching or heating [3]

Volumetric & Calorimetric Properties:
Density:

No.	Value	Note
1	d 1.3 g/cm³	approx., fluxed resin [3]

Glass-Transition Temperature:

No.	Value	Note
1	90°C	40 wt.% acrylonitrile [2,3]

Transition Temperature General: Softening points are higher than PVC homopolymer; lower than PAN homopolymer (fibres readily shrink unless stabilised) (40 wt.% acrylonitrile) [6]; softening points are increased after prolonged exposure to elevated temps. [3]

Deflection Temperature:

No.	Value	Note
1	85°C	approx., 40 wt.% acrylonitrile, moulded resin, ASTM D648 [3]

Surface Properties & Solubility:
Solubility Properties General: Resins become markedly less sol. in organic solvents as a result of exposure to elevated temps. [3]; copolymers show improved solubilities in organic solvents, particularly in Me_2CO, compared with both homopolymers
Plasticisers: *o*- and *p*-toluene ethyl sulfonamides; tetrahydrofurfuryl phthalate [3]
Solvents/Non-solvents: Sol. Me_2CO, cyclohexanone, DMF. Spar. sol. higher aliphatic ketones [3]

Transport Properties:
Transport Properties General: Flow characteristics of copolymer resins are poor compared to those of PVC/PVA, [3]
Polymer Solutions Dilute: η_{sp} 0.25–0.3 (40 wt.% acrylonitrile, 0.2% soln. in cyclohexanone) [3]
Polymer Melts: Has higher melt viscosity than PVC homopolymer
Gas Permeability:

No.	Gas	Value	Note
1	[O_2]	43.34 cm³ mm/ (m² day atm)	0.66×10^{-10} cm² (s cm Hg)⁻¹, 23°, 40 wt% acrylonitrile, 20% cryst. [1]
2	[CO_2]	131.33 cm³ mm/ (m² day atm)	2.0×10^{-10} cm² (s cm Hg)⁻¹, 23@0, 40 wt% acrylonitrile, 20% cryst. [1]

Mechanical Properties:
Mechanical Properties General: Fibres have slightly lower strengths than those of higher acrylonitrile copolymers or PAN homopolymer [6]; outdoor ageing leads to decline in strength [3]

Optical Properties:
Refractive Index:

No.	Value	Note
1	1.53	40 wt.% acrylonitrile [6]

Total Internal Reflection: Birefringence 0.003 (approx., 40 wt.% acrylonitrile) [6]
Volume Properties/Surface Properties: Colour light amber to dark brown (fluxed resin, 40 wt.% acrylonitrile) [3]; fibres display high gloss

Polymer Stability:
Flammability: Fibres have low flammability [6]
Environmental Stress: Fibres display good resistance to sunlight [6]; however, bleaching occurs after exposure to strong sunlight [3]
Chemical Stability: Moderately resistant to acids and alkalis, but attacked by hot alkalis [6]; more readily swollen or dissolved by common organic solvents, as compared to PAN homopolymer [6]
Hydrolytic Stability: Fibres display good resistance to water (e.g. moisture regain 0.5–1% at 65% relative humidity) [6]

Applications/Commercial Products:
Processing & Manufacturing Routes: Polymerisation of comonomer mixture occurs by free-radical initiation (e.g. azo-compounds, peroxides, persulfates, etc.) in aq. emulsions or organic solvents (e.g. Me_2CO, ethylene carbonate, THF, etc.). Polymer soln. can be spun directly into hot H_2O, followed by stretching, to give 'modacrylic' fibres [6]. Graft copolymers are prepared by the reaction of acrylonitrile [4] (or acrylonitrile/ethyl acrylate [5]) in the presence of PVC. Resin can also be compression moulded but has comparatively poor flow characteristics compared to PVC/PVA

Applications: Fibres find applications due to flameproof props. and low cost (cf. PAN homopolymer fibres) in blankets, carpets, curtains/draperies; filter cloths and media (e.g. dust collection bags); work clothes; children's nightwear; mouldable fabrics. Miscellaneous uses of fibres as membranes in ultra filtration - e.g. will preferentially extract partially sol. alcohols from H_2O; in felts, as electrode binders and absorbers, and vented cell separators in secondary batteries/alkaline cells; in triboelectric valves (for air pollution control). Terpolymer (with ethyl acrylate) used for flooring

Trade name	Details	Supplier
Dynel	Fibres; also used to refer to other acrylic-based materials	Union Carbide
Dynel NYGL	Fibres; AN/VC copolymer	Union Carbide
Vinyon N	Fibres; AN/VC copolymer; 'Vinyon' also used to refer to other vinyl-based materials	Carbide and Carbon Chem.

Bibliographic References

[1] *Sci. Technol. Polym. Films,* (ed. O.J. Sweeting), Wiley, 1971, **II**, 114, (permeability)
[2] *Kirk-Othmer Encycl. Chem. Technol.,* 3rd edn., (eds. M. Grayson and D. Eckroth), John Wiley and Sons, 1983, **23**, 443, 896, 911, (thermal processing, graft copolymers, uses)
[3] Rugeley, E.W. et al, *Ind. Eng. Chem.,* 1948, **40**, 1724, (thermal processing)
[4] *Fr. Pat.,* 1957, 1 147 722, (graft copolymers)
[5] *U.S. Pat.,* 1966, 3 254 049, (graft copolymers)
[6] Roff, W.J. and Scott, J.R., Fibres, Films, *Plastics and Rubbers,* Butterworths, 1971, 101, 105, (processing, optical props, stability)
[7] *Kirk-Othmer Encycl. Chem. Technol.,* 3rd edn., (eds. M. Grayson and D. Eckroth), John Wiley and Sons, 1978, **3**, 605, 625, 626, (uses)

Poly(vinyl chloride), Chlorinated   P-576

$$-[-[CH(Cl)CH(Cl)]_x-[CH_2CH(Cl)]_y-]_n$$

Synonyms: *CPVC. PVC-C. Post chlorinated PVC. Chlorinated PVC. Chlorinated poly(vinyl chloride). Perchlorovinyl. Per CV. Pe Ce (fibres)*
Related Polymers: Polyvinyl chloride, Polyvinyl chloride, unplasticised
Monomers: Chloroethylene
Material class: Thermoplastic
Polymer Type: PVC, polyhaloolefins
CAS Number:

CAS Reg. No.
54242-36-3
69865-59-4
107248-35-1

Molecular Formula: $[(C_2H_2Cl_2).(C_2H_3Cl)]_n$
Fragments: $C_2H_2Cl_2$ C_2H_3Cl
General Information: Higher softening point and better solubility than PVC. Obtained by post-chlorinating PVC to a chlorine content of 62–70%; typical commercial materials contain 66–67% chlorine [1,2]. Can be regarded as hypothetical 'copolymer' of vinyl chloride and symmetrical dichloroethylene [1]. Fibres (62–65% chlorine) have x = y in struct. unit; mainly CHCl-CHCl, but with some CH_2-CCl_2 [1]. Heat-resistant chlorinated form ("high-temp. PVC") has x > y in struct. unit (high chlorine content) [1]

Volumetric & Calorimetric Properties:
Density:

No.	Value	Note
1	d 1.54 g/cm^3	[1]
2	d 1.44 g/cm^3	fibres, 62–65% chlorine content [1]

Thermal Expansion Coefficient:

No.	Value	Note	Type
1	5.7×10^{-5}–8×10^{-5} K^{-1}	25–50° [1]	L
2	9.4×10^{-5} K^{-1}	50–70° [1]	L
3	4×10^{-5}–6.9×10^{-5} K^{-1}	70–100° [1]	L

Thermodynamic Properties General: C_p/C_v 0.29 [1]
Thermal Conductivity:

No.	Value	Note
1	0.039 W/mK	0.00039 cal cm^{-1} s^{-1} °C^{-1} [1]

Glass-Transition Temperature:

No.	Value	Note
1	110°C	approx., typical commercial material, 66–67% Cl content [2]

Deflection Temperature:

No.	Value	Note
1	98–105°C	1.8 MPa, 0.185 kg f mm^{-2}, normal grades [1]
2	83.2°C	1.8 MPa, 0.185 kg f mm^{-2}, transparent grade [1]
3	90.6°C	0.45 MPa, 0.046 kg f mm^{-2}, transparent grade [1]

Vicat Softening Point:

No.	Value	Note
1	110–128°C	normal grades [1]
2	96°C	transparent grade [1]

Transition Temperature:

No.	Value	Note	Type
1	100°C	approx., typical commercial material, 66–67% Cl content [2]	Softening temp.
2	102–107°C	normal grades [1]	Softening temp.
3	87°C	transparent grades [1]	Softening temp.

Surface Properties & Solubility:
Solubility Properties General: Forms miscible blends with PVC at 62.2% chlorine; multiphase system produced above 67.5% chlorine [4,5]
Cohesive Energy Density Solubility Parameters: Solubility parameter, δ 19 J$^{1/2}$ cm$^{-3/2}$ (9.3 cal$^{1/2}$ cm$^{-3/2}$, 25°) [3]
Solvents/Non-solvents: Displays better solubility than 'ordinary' PVC. Sol. aromatic hydrocarbons, $CHCl_3$, chlorobenzene, THF, dioxane, Me_2CO, cyclohexane, cyclohexanone, butyl acetate, tetrachloroethylene, nitrobenzene, DMF, DMSO (63% chlorine

– Poly(vinyl chloride), Chlorinated

content) [1,6]. Swollen by butanol, conc. hydrochloric acid, nitric acid [1]. Insol. aliphatic hydrocarbons, cycloaliphatic hydrocarbons, CCl_4, methyl acetate, nitromethane, organic acids, dil. inorganic acids, conc. sulfuric acid, dil. alkalis, glycerol, petroleum [1,6]

Transport Properties:
Polymer Melts: Effects of temp., frequency and strain amplitude on dynamic viscosity for varying chlorinated compositions have been reported; results obtained compare with those of PVC. Strong temp. dependence of rheological behaviour reflects significant structural changes [7]

Water Absorption:

No.	Value	Note
1	0.048–0.08 %	0.75" pipe, 24h, 23° [1]
2	0.49 %	28 days, 23°
3	14.3 %	100° [1]

Mechanical Properties:
Tensile (Young's) Modulus:

No.	Value	Note
1	2840–3236 MPa	290–330 kg f mm^{-2} [1]

Compressive Modulus:

No.	Value	Note
1	1863–1961 MPa	190–200 kg f mm^{-2} [1]

Tensile Strength Yield:

No.	Value	Elongation	Note
1	58.8–65.7 MPa	8.9%	6.0–6.7 kg f mm^{-2} [1]

Compressive Strength:

No.	Value	Note
1	80.4–86.2 MPa	8.6–9.1% compression, 8.2–8.8 kg f mm^{-2} [1]

Hardness: Rockwell R117-R119 [1]
Izod Notch:

No.	Value	Notch	Note
1	3.92–10.8 J/m	Y	0.4–1.1 kg f mm^{-1}, impact modified [1]

Charpy:

No.	Value	Notch	Note
1	4.7–14.7	Y	0.48–1.5 kg f mm^{-1}, impact modified [1]

Electrical Properties:
Dielectric Permittivity (Constant):

No.	Value	Frequency	Note
1	2.93	60 Hz	30°, 55% relative humidity [1]
2	2.92	10 kHz	30°, 55% relative humidity [1]
3	2	1 MHz	20°, 55% relative humidity [1]

Dielectric Strength:

No.	Value	Note
1	39 kV.mm^{-1}	30° [1]

Dissipation (Power) Factor:

No.	Value	Frequency	Note
1	0.0109	60 Hz	30°, 55% relative humidity [1]
2	0.011	1 kHz	30°, 55% relative humidity [1]
3	0.009	1 MHz	30°, 55% relative humidity [1]

Optical Properties:
Molar Refraction: dn/dc 0.045–0.73 ml g^{-1} (56–66% Cl, 25°, tetrachloroethane) [8]

Polymer Stability:
Thermal Stability General: Longitudinal heat shrinkage 0–3% (80–140°, transparent sheet) [1]. Transverse heat shrinkage 0.5–1% (80–140°, transparent sheet) [1]. Thermal stability during processing is inferior to that of ordinary PVC, [1]
Hydrolytic Stability: Slightly hygroscopic

Applications/Commercial Products:
Processing & Manufacturing Routes: Prod. by post-chlorinating PVC to a chlorine content of 62–70% PVC dissolved in chlorinated solvent; can be carried out at low temps.; photo-chemical methods can be used. Products can be moulded (injection or compression), extruded, thermo-formed or calendered into sheets. Fibres for textiles are stretch-spun into hot H_2O from dil. (30%) Me_2CO solns. [1,2]. Drying at 90–100° for 1–2h recommended before processing/fabrication [1]
Applications: Hot water pipes, containers (heat tolerant), adhesives. Applications due to high softening point and superior chemical resistance. Moulded products include waste and soil systems (passing hot water); chemical plant. Calendered sheet used in hot filling techniques (e.g. jam packing). Fibres used in filter fabrics; tarpaulins; belting; nets (fishing). Miscellaneous uses include adhesives and lacquers

Trade name	Details	Supplier
Boltaron 4125	non-exclusive	Gencorp
Genclor S		ICI, UK
Gendor		ICI Chemicals and Polymers Ltd.
Geon	non-exclusive	B.F. Goodrich
Kane Ace		Kaneka
Lucalor		Atochem Polymers
Lucalor		Rhone-Poulenc
Rhenoflex		Dynamit Nobel
Rhenoflex 63		Dynamit Nobel
Solvitherm	Fibres	Solvay
Welvic	Heat-resistant form, non-exclusive	ICI, UK

Bibliographic References
[1] Roff, W.J. and Scott, J.R., Fibres, Films, *Plastics and Rubbers*, Butterworths, 1971, 116, (general props., struct., volumetric props., thermal props.)
[2] Brydson, J.A., *Plast. Mater.*, 6th edn., Butterworth-Heinemann, 1995, 346, (thermal props.)

[3] Tillaev, R.S., Khasaukhanova, M., Tashmukhamedov, S.A. and Usmanov, Kh.U., *J. Polym. Sci., Part C: Polym. Lett.*, 1972, **39**, 107, (solubility)
[4] *Kirk-Othmer Encycl. Chem. Technol.*, Vol. 18, 3rd edn., (eds M. Grayson and D. Eckroth), Wiley-Interscience, 1982, 461, (blends)
[5] Olabisi, O., Robeson, L.M. and Shaw, M.T., *Plast. Compd.*, 1980, **3**, 51, (rev.)
[6] Bloch, D.R., *Polym. Handb.*, 4th edn., (eds. J. Brandrup, E.H. Immergut and E.A. Grulke), John Wiley and Sons, 1999, **VII**, 505, (solubility)
[7] Bonnebat, C. and De Vries, A.J., *Polym. Eng. Sci.*, 1978, **18**, 824, (viscosity)
[8] Michielsen, S., *Polym. Handb.*, 4th edn., (eds. J. Brandrup, E.H. Immergut and E.A. Grulke), John Wiley and Sons, 1999, **VII**, 591, (optical props.)

Poly(vinyl chloride), Crystalline P-577

Synonyms: *PVC, Crystalline. Crystalline PVC. Crystalline poly(vinyl chloride)*
Related Polymers: Polyvinyl chloride, unplasticised, Polyvinyl chloride, Plasticised
Monomers: Vinyl chloride
Material class: Thermoplastic
Polymer Type: PVC, polyhaloolefins
Molecular Formula: $(C_2H_3Cl)_n$
Fragments: C_2H_3Cl
Molecular Weight: Can be produced in a wide range of MWs; chain-transfer reaction suppressed at low temps., with greater control over MW by adjustment of initiator concentrations; very high MW material (K-values ≥ 130) can be easily prepared [3]
Additives: Plasticisers (esters); stabilisers; toughening agents and impact modifiers (EVA polymers most effective)
General Information: The effects of polymerisation temp., degree of branching and stereoregularity affect the struct. and physical props.; syndiotacticity is the major contributing factor. Development of stereoregular polyolefins and polydienes arising from discovery of Ziegler-Natta-type catalysts stimulated attempts to produce vinyl chloride polymers with increased stereoregularity. Usual Ziegler-Natta catalysts cannot be used for vinyl chloride as they react with both monomer and polymer. Production of highly crystalline PVC can be achieved by low-temp. free-radical polymerisation using γ-irradiation and active substances such as alkyl boranes. Major differences when compared to conventional PVC, are considerable change in microstruct., i.e. degree of crystallinity and syndiotacticity increased; solubility (and swelling) reduced; resistance to solvents increased; increase in heat distortion temp.; tensile strength and creep resistance increased; T_g raised (consequence of decreased polymerisation temp.); T_m increased; Vicat softening point increased; increased brittleness (major disadvantage) as direct consequence of increased crystallinity. Changes in physical props. are caused primarily by differences in stereoregularity
Tacticity: Syndiotactic propagation is favoured over isotactic propagation in radical polymerisation of vinyl chloride if polymerisation temp. is reduced [4]; difference in activation enthalpy between two forms is approx. 600 cal mol^{-1} [4,5,6]; degree of syndiotacticity approx. 0.65 for material polymerised at -50° (measured by high-resolution pmr and ir spectroscopy) [1,4]
Morphology: Level of crystallinity 25% (polymerisation temp. -60°) [3]; 20% (polymerisation temp. -50° [1,3,4,6,7]. Degree of branching (from ir spectroscopy) reduced when compared to conventional PVC, (reduction in polymerisation temp.) [4]. Generally agreed that only syndiotactic sequences in PVC are capable of crystallisation [4]; x-ray cryst. struct.: orthorhombic, a 10.6, b 5.4, c 5.1 Å [9,10]; crystallites have small dimensions, of the order of 50–100 Å, or even less; order is poor along chain length [8]. Preparation of (apparently) completely syndiotactic polymer, by the x-ray irradiation of vinyl chloride-urea canal complex at -78°, has been reported. Highly stereoregular struct. indicated by ir spectroscopy and reduced solubilities in common organic solvents [11]
Miscellaneous: Extensive details on props. of crystalline form have been reported [3,4]. Plasticised materials can be produced, although mixing temps. of up to 190° are necessary [1]; some evidence that degree of crystallinity slightly higher in plasticised compositions, with consequent increase in brittleness for such materials [2]

Volumetric & Calorimetric Properties:

Volumetric Properties General: Shrinkage of oriented material can be much lower than that of conventional PVC; e.g. 10% shrinkage of fibres at 150° [4]. More dense than conventional PVC
Melting Temperature:

No.	Value	Note
1	273°C	syndiotactic, calc. [4,7]
2	310°C	77% syndiotactic, 84% crystalline [2]

Glass-Transition Temperature:

No.	Value	Note
1	81°C	polymerised at 50° [4]
2	100°C	polymerised at -30° [4]
3	106°C	polymerised at -50° [4]

Transition Temperature General: T_m increases with increasing stereoregularity. T_g increases with a decrease in polymerisation temp.; consequence of both stereoregularity and degree of crystallinity [4]; influenced by additives and processing conditions. Vicat softening point increases with a decrease in polymerisation temp. [15]; consequence of both stereoregularity and degree of crystallinity [4]; 15° higher than conventional PVC; influenced by additives and processing conditions [4]; has effect on processing, e.g. critical processing temp. 210° (min.) required [3]
Deflection Temperature:

No.	Value	Note
1	92°C	1.88 MPa (264 lb ft in^{-2}), polymerised at -40° [4]

Vicat Softening Point:

No.	Value	Note
1	100°C	approx., polymerised at -40° [4]

Surface Properties & Solubility:
Solvents/Non-solvents: Solubility in organic solvents much lower than that of conventional PVC; sol. THF, cyclohexanone (130°); insol. THF, cyclohexanone (room temp.) [4]

Mechanical Properties:
Mechanical Properties General: Has higher tensile strengths (at break) and Young's moduli than conventional PVC [17], plus greater resistance to creep [4]. Mech. props. are greatly influenced by processing conditions [4]. Plasticised compositions display different mech. props. to those of equivalent plasticised conventional PVC materials [4]; elasticity modulus is 3 times as large, with rupture strain being usually much smaller [4]. Fibres made from crystalline PVC have high tensile strength [4]; Young's modulus above softening point much higher than that of conventional PVC [4]
Impact Strength: Has lower impact strengths than conventional PVC [4]; typical value 3.92 kJ m^{-2} (4 kgf cm cm^{-2}) [18]; improved by addition of EVA 11.8 kJ m^{-2} (12 kgf cm cm^{-2}) [18]
Mechanical Properties Miscellaneous: Dynamic mech. props. (such as storage modulus) are very similar to conventional PVC from -150–50° [4,19]; above 50°, show significant differences; at 140°, storage modulus improved by factor of 4 [4]; such differences also maintained in plasticised compositions, e.g. creep compliance can be 4–6 times lower than that of normal PVC up to relatively high temps. [4]. Higher crystallinities lead to increased brittleness; further increased by presence of plasticisers [2]; harder and stiffer products produced [2]

Polymer Stability:
Thermal Stability General: Does not differ markedly from that of conventional polymer; consequently has effect on processing conditions [16]
Hydrolytic Stability: Is resistant to hot H_2O [4]

Applications/Commercial Products:
Processing & Manufacturing Routes: Synth. by free-radical polymerisation of vinyl chloride at low temps. (-80°) using γ-irradiation or very active reagents such as boron or aluminium alkyls [2,3], in conjunction with oxygen, halogens or peroxides [3]; conventional redox catalysts [3] and uranyl nitrate (photopolymerisation) [10] have also been used as initiators. Preparation of crystalline material has been the subject of extensive investigation [12,13,14]. Ziegler-Natta catalysts cannot be used as both monomer and polymer react with them [1,3]
Applications: Although of great potential interest, e.g. increased heat distortion temp. and greater tensile strength and creep resistance than conventional polymer, no appreciable commercial development of crystalline material appears to have taken place. Increased temps. needed for thermoplastic processing and relatively brittle, expensive products obtained which are rather expensive, preclude any significant developments

Bibliographic References

[1] Brydson, J.A., *Plast. Mater.*, 6th edn., Butterworth-Heinemann, 1995, 345, (synth, morphology, tacticity, thermal props, mech props, solubility)
[2] Matthews, G., *Vinyl Allied Polym.*, (ed. P.D. Ritchie), Iliffe Books (Butterworths Group), 1972, **2**, 42, 57, (synth, morphology, tacticity, thermal props, mech props, solubility)
[3] Bockman, O.C., *Br. Plast.*, 1965, **38**, 364, (rev)
[4] Pezzin, G., *Plast. Polym.*, 1969, **37**, 295, (rev)
[5] Fordham, J.W.L., Burleigh, P.H. and Sturm, C.L., *J. Polym. Sci.*, 1959, **41**, 73, (tacticity)
[6] Talamini, G. and Vidotto, G., *Makromol. Chem.*, 1967, **100**, 48, (tacticity, morphology, crystallinity)
[7] Kockott, D., *Kolloid Z. Z. Polym.*, 1964, **198**, 17, (morphology, crystallinity, thermal props)
[8] Natta, G. and Corradini, P., *J. Polym. Sci.*, 1956, **20**, 251, (morphology)
[9] Smith, R.W. and Wilkes, C.E., *J. Polym. Sci., Polym. Lett. Ed.*, 1967, **5**, 433, (morphology)
[10] Nakajima, A., Hamada, H. and Hayashi, S., *Makromol. Chem.*, 1966, **95**, 40, (synth, thermal props, tacticity, morphology)
[11] White, D.M., *J. Am. Chem. Soc.*, 1960, **82**, 5678, (synth, morphology, tacticity)
[12] *Jpn. Pat.*, 1965, 4 347, (synth)
[13] *Jpn. Pat.*, 1965, 4 348, (synth)
[14] *Jpn. Pat.*, 1965, 4 349, (synth)
[15] Reding, F.P., Walter, E.R. and Welch, F.J., *J. Polym. Sci.*, 1962, **56**, 225, (thermal props)
[16] Crosato-Arnaldi, A., Palma, G., Peggion, E. and Talamini, G., *J. Appl. Polym. Sci.*, 1964, **8**, 747, (thermal stability)
[17] Bier, G., *Kunststoffe*, 1965, **55**, 694, (mech props)
[18] Trautvetter, W., *Makromol. Chem.*, 1967, **101**, 214, (mech props)
[19] Pezzin, G., Arjroldi, G. and Garbuglio, C., *J. Appl. Polym. Sci.*, 1967, **11**, 2553, (mech props)

Poly(vinyl chloride-*co*-N-cyclohexylmaleimide) P-578

Synonyms: *Vinyl chloride-N-cyclohexylmaleimide copolymer. Vinyl chloride-cyclohexylmaleimide copolymer*
Related Polymers: Polyvinyl chloride, unplasticised, PVC copolymers
Monomers: Vinyl chloride, N-Cyclohexylmaleimide
Material class: Thermoplastic, Copolymers
Polymer Type: PVC, polyhaloolefins
CAS Number:

CAS Reg. No.
28210-06-2

Molecular Formula: $[(C_2H_3Cl).(C_{10}H_{13}NO_2)]_n$
Fragments: C_2H_3Cl $C_{10}H_{13}NO_2$

General Information: Improved heat deformation props. compared to PVC without affecting processing or clarity. Incorporation of approx. 5% cyclohexylmaleimide as comonomer improves softening point of PVC homopolymer; no detrimental effects on processing temps./conditions and hence stability; clarity also maintained [1]
Morphology: Degree of crystallinity increases with polymerisation temp. [5]

Volumetric & Calorimetric Properties:
Density:

No.	Value	Note
1	d 0.47 g/cm^3	approx., bulk, DIN 53468 [4]

Glass-Transition Temperature:

No.	Value	Note
1	90°C	transparent film, increases with polymerisation temp. [4]

Transition Temperature General: Introduction of 5% comonomer increases Vicat and heat distortion temps. by 7° [3], and significantly increases T_g [3]. Copolymer displays higher heat distortion temp. than polyvinyl chloride, good workability and high resistance to heat fatigue [4]
Vicat Softening Point:

No.	Value	Note
1	87°C	5% comonomer [4]

Transport Properties:
Polymer Melts: [η] 8.8 dl g^{-1} (approx., 5% N-cyclohexylmaleimide, DIN 53726) [4]

Applications/Commercial Products:

Trade name	Details	Supplier
Hostalit LP HT 5060	Hostalit used to refer to other PVC homopolymer and copolymer materials	Hoechst
Hostalit SC 3060/5	Hostalit used to refer to other PVC homopolymer materials	Hoechst

Bibliographic References

[1] Brydson, J.A., *Plast. Mater.*, 6th edn., Butterworth-Heinemann, 1995, 347, (thermal props, use)
[2] Whelan, T., *Polymer Technology Dictionary*, Chapman and Hall, 1994, 475, (thermal props, use)
[3] Titow, W.V., *PVC Plastics: Properties, Processing, and Applications*, Elsevier Applied Science, 1990, 74, (thermal props)
[4] Kühne, G. *et al*, *Kunststoffe*, 1973, **63**, 139, (synth, thermal props, uses)

Poly(vinyl chloride-*co*-ethylene) P-579

$$\left[(CH_2CH_2)_x-(CH_2CH(Cl))_y\right]_n$$

Synonyms: *VC/E copolymer. Vinyl chloride-ethylene copolymer. Poly(chloroethene-co-ethene). VCE*
Related Polymers: PVC Copolymers, Polyvinyl chloride, unplasticised
Monomers: Vinyl chloride, Ethylene
Material class: Thermoplastic, Copolymers
Polymer Type: polyethylene, PVCpolyhaloolefins

Poly(vinyl chloride-co-ethylene)

CAS Number:

CAS Reg. No.
25037-78-9

Molecular Formula: $[(C_2H_3Cl).(C_2H_4)]_n$
Fragments: C_2H_3Cl C_2H_4
Molecular Weight: MW decreases with increasing ethylene content. M_n 26000, 50000. MW 109000 (84.3 mol% vinyl chloride copolymer, reductive dechlorination of PVC); MW 38000 (13.6 mol% vinyl chloride copolymer, reductive dechlorination of PVC)
Additives: Plasticisers; stabilisers; lubricants
General Information: Copolymers contain relatively small amounts of ethylene comonomers; softer, easier flowing material than homopolymer; can be processed at lower melt temps. [1]. Copolymers have generally similar props. and uses to those of propylene copolymers. Limited commercial use, as considerably more expensive than poly(vinyl chloride)-co-vinyl acetate
Miscellaneous: Random copolymer produced, with broad compositional distribution, using Ziegler-Natta catalyst system (Al/Ti <2.5 (molar)); block copolymer (rich in ethylene), along with random copolymer (rich in vinyl chloride), produced when Al/Ti >3 (molar) [1]

Volumetric & Calorimetric Properties:
Density:

No.	Value	Note
1	d 0.43–0.58 g/cm^3	bulk density; dependent on copolymer composition [26]

Glass-Transition Temperature:

No.	Value	Note
1	38°C	20% PE [3,7]
2	60°C	10% PE [3,7]

Surface Properties & Solubility:
Plasticisers: Ethylene as internal plasticiser is 2–3 times more efficient than bis(2-ethylhexyl) phthalate; crystallinity still retained in all copolymers examined [21]

Transport Properties:
Transport Properties General: Ease of melt flow increases with increasing ethylene content [5]
Melt Flow Index:

No.	Value	Note
1	>0.5 g/10 min	min., 2–10% PE [8]
2	70.4 g/10 min	4.7% PE, typical value; resin blended with stabiliser [8]
3	17 g/10 min	suspension polymerisation, 6.3 wt% ethylene, ASTM D1238-57T-F [10]

Polymer Solutions Dilute: Reduced specific viscosity η_{sp} 0.1734–0.4234 (66–77 mol% vinyl chloride, THF) [9]. Inherent viscosity η_{inh} 1.17 dl g^{-1} (1 wt% ethylene), 1.18 dl g^{-1} (3.7 wt% ethylene), 1 dl g^{-1} (4 wt% ethylene), 0.94 dl g^{-1} (5.5 wt% ethylene) [21]
Polymer Melts: Intrinsic viscosity decreases with increasing PE content; typical values 0.76 dl g^{-1} (2.5% PE) [6]; 0.67 dl g^{-1} (2.7% PE) [6]; 0.6 dl g^{-1} (4.7% PE; resin blended with stabiliser) [8]; 0.8 dl g^{-1} (suspension polymerisation, 6.3 wt% ethylene) [10]; 0.5–1.5 dl g^{-1} (2–10% PE) [8]

Mechanical Properties:
Mechanical Properties General: Presence of comonomer has comparatively little adverse effect on mech. props. of PVC homopolymer (in contrast to PVCA) [5]

Tensile (Young's) Modulus:

No.	Value	Note
1	>6864.6 MPa	min., 7000 kg mm^{-2}, 'apparent' values; 2–10% PE at 40–75° [8]

Izod Notch:

No.	Value	Notch	Note
1	21.2 J/m	Y	0.4 ft lb in^{-1}., 4.7% PE; typical value; resin blended with stabiliser [8]

Optical Properties:
Transmission and Spectra: Ir 750 cm^{-1} (THF soluble fraction; random copolymer) [9]. C-13 nmr spectral data have been reported for series of copolymers (97.6–14.7 mol% ethylene) [22]. Comprehensive study of ir spectral data of full range of copolymers (97.6–14.7 mol% ethylene) has been reported [19,20]

Polymer Stability:
Thermal Stability General: Presence of comonomer has comparatively little adverse effect on thermal stability of PVC homopolymer [5]
Flammability: Inherently flame retardant (emulsion/latex forms) [2]
Chemical Stability: Displays excellent resistance to alcohols and greases (emulsion/latex forms) [2]
Hydrolytic Stability: Displays excellent resistance to water (emulsion/latex forms) [2]

Applications/Commercial Products:
Processing & Manufacturing Routes: Free-radical polymerisation of mixture of comonomers; emulsion and suspension polymerisation most commonly used; melt processing of latter using injection moulding and vacuum moulding techniques [2,4]. Copolymerisation reactivity parameters have been reported [3]. Offers outstanding plasticiser absorption, quick gelation and excellent flow behaviour [2]. Copolymerisation can be effected by using a Ziegler-Natta catalyst system of AlEtCl$_2$/Ti(n-OBu)$_4$; mixture of block and random copolymers produced, dependent on Al/Ti ratio [9]. Synth. of copolymers by partial reduction of PVC using tributyltin hydride has also been reported [22]. Copolymerisation occurs by the following methods: suspension polymerisation to produce materials with high melt indexes [10]; bulk polymerisation (or in inert solvent) with reaction stopped at 5–20% conversion by addition of Lewis base. Semi-rigid copolymers produced, exhibiting improved transparencies and surface smoothness [11]; emulsion polymerisation to produce coatings [15], pastos (under pressure) [12], and non-sticky, transparent elastic films (graft polymers) [18]; production of elastomeric materials for use as self-extinguishing mastics, adhesives or coatings [17], and for mouldings with enhanced impact strength and elongation [13]
Applications: Applications of moulded components: phonographic discs; electrical components; sheets; formed and rigid profiles; vacuum moulded plates Film applications highly elastic forms; flexible and rigid transparent forms. Miscellaneous uses in blown bottles; flexible tubes; leather (shoes, boots); crash pads; wire and cable insulation. Applications due to flexibility and water resistance in caulks, mastics and barrier coats (building industry); non-conductive and dielectric products; carpet backings; surfactant stabilisation (emulsion polymerisation). Applications due to flame retardancy as non-woven binder for fibrefill and high loft stock; binder for flame-retardant fabrics and heat-sensitive non-wovens; filters; textiles (latex). Also used as cross-linkable flexible binder and saturant for paper and paperboard applications (latex)

Trade name	Details	Supplier
Airflex 4500	Latex	Air Products

— Poly(vinyl chloride), Plasticised

Airflex 4514	Latex	Air Products
Airflex 4530	Latex	Air Products
Airflex 4814	Latex	Air Products
Airflex 742-BP	Emulsion	Air Products
Airflex 7522 DEV	Emulsion	Air Products
Airflex RB-35	Emulsion. Airflex used for other vinyl ethene systems	Air Products
Airflex RB-40	Emulsion	Air Products
Bakelite	QSQH, QSQL, QSQM	Bakelite
Ryuron	E-430, E-650, E-800, E-1050, E-1300, E-2800. Ryuron also used for other PVC materials	Tosoh
Vinnol LL 352	Used to refer to other PVC-based materials (homo-, co-, and terpolymers)	Wacker-Chemie

Bibliographic References

[1] Whelan, T., *Polymer Technology Dictionary*, Chapman and Hall, 1994, 475, (processing/production)
[2] Ash, M. and Ash, I., Handbook of Plastic Compounds, *Elastomers and Resins*, (eds. M.B. Ash and I.A. Ash), Wiley-VCH, 1992, 114, (props., stability, polymerisation, processing, uses)
[3] *Kirk-Othmer Encycl. Chem. Technol.*, 3rd edn., (eds. M. Grayson and D. Eckroth), John Wiley and Sons, 1983, **23**, 896, 910, (thermal copolymerisation reactions)
[4] Titow, W.V., *PVC Technol.*, 4th edn., Elsevier Applied Science, 1984, 41, (polymerisation)
[5] Titow, W.V., PVC Plastics: Properties, Processing, *and Applications*, Elsevier Applied Science, 1990, 79, (transport props)
[6] Heiberger, C.A. et al, *Polym. Eng. Sci.*, 1969, **9**, 445, (processing, transport props)
[7] Reding, F.P. et al, *J. Polym. Sci.*, 1962, **57**, 483, (thermal props)
[8] *Neth. Pat.*, 1966, 6 510 714, (processing, transport props, mech props)
[9] Misono, A. et al, *Bull. Chem. Soc. Jpn.*, 1966, **39**, 1822, (synth, ir, transport props, struct)
[10] *Neth. Pat.*, 1966, 6510712, (synth, use)
[11] *Neth. Pat.*, 1965, 6507034, (synth, use)
[12] *Fr. Pat.*, 1965, 1 409 600, (synth)
[13] *Brit. Pat.*, 1966, 1 035 339, (synth, use)
[14] *Fr. Pat.*, 1966, 1 423 164, (synth, use)
[15] *Belg. Pat.*, 1964, 642345, (synth, use)
[16] *Fr. Pat.*, 1964, 1 379 842, (synth)
[17] *Fr. Pat.*, 1965, 1 386 575, (synth, use)
[18] *Ger. Pat.*, 1966, 1 211 395, (synth, use)
[19] Koenig, J.L., *Spectroscopy of Polymers*, ACS, 1991, 81, (ir)
[20] Bowmer, T.N. and Tonelli, A.E., J. Polym. Sci., *Polym. Phys. Ed.*, 1986, **24**, 1681, (ir, microstructure)
[21] Hanna, R.J. and Fields, J.W., *J. Vinyl Technol.*, 1982, **4**, 57, (plasticisation, viscosity)
[22] Schilling, F.C., Tonelli, A.E. and Valenciano, M., *Macromolecules*, 1985, **18**, 356, (synth, nmr, microstructure, MW)

Poly(vinyl chloride), Plasticised

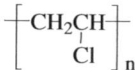

Synonyms: *Plasticised PVC. Plasticised poly(vinyl chloride). Plasticised poly(chloroethene). PPVC. PVC-P*
Related Polymers: Polyvinyl chloride, unplasticised, Poly(vinyl chloride-*co*-vinyl acetate), PVC copolymers
Monomers: Vinyl chloride
Material class: Thermoplastic
Polymer Type: PVC, polyhaloolefins

CAS Number:

CAS Reg. No.
9002-86-2

Molecular Formula: $(C_2H_3Cl)_n$
Fragments: C_2H_3Cl
Additives: Thermal stabiliser and plasticiser required. Sometimes used: fire retardants, blowing agents, lubricants. Lubricants include lead stearate, stearic acid, 12-hydroxystearic acid, calcium stearate, paraffin wax, stearamide, cetyl/stearyl acetate, cetyl/stearyl alcohol, polyethylene waxes. Plasticisers include phthalates, phosphates, adipic esters, azelaic esters, sebacic esters, epoxidised soybean oil, chlorinated paraffins and trimellitates. Blowing agents include azodicarbonamide and sodium bicarbonate. Thermal stabilisers include lead salts, barium/cadmium, barium/zinc, calcium/zinc, epoxidised oils, polyols and organic phosphates. Flame retardants include antimony oxide, aluminium trihydrate, zinc borate and chlorinated paraffins
General Information: Plasticisers are added to PVC compounds in order to confer flexibility, softness and ease of processing to the resulting material
Morphology: There are significant differences between orientation distributions for crystalline and non-crystalline regions of various samples; amorph. chains behave in a rubber-like way during orientation; behaviour dependent on plasticiser content and draw temp. [26,27,28]
Identification: For general identification and distinction from polychloroprene, see polyvinyl chloride, unplasticised

Volumetric & Calorimetric Properties:
Density:

No.	Value	Note
1	d 1.1–1.7 g/cm^3	dependent on plasticiser [2]

Thermal Expansion Coefficient:

No.	Value	Note	Type
1	0.00017–0.000175 K^{-1}	$T > T_g$ [3]	L
2	0.000246–0.000537 K^{-1}	$T < T_g$, extruded rods, phthalate plasticisers [3]	V
3	0.000504–0.00055 K^{-1}	$T > T_g$, extruded rods, phthalate plasticisers [3]	V
4	0.00025 K^{-1}	approx., typical value [2]	L

Volumetric Properties General: At 20–25% crystallinity isotactic or crystalline PVC has a higher softening temp. and reduced solubility compared to commercial PVC. Crystalline PVC has not been developed as a commercial product. A PVC with a high degree of plasticiser becomes stiff and brittle. Changes in various physical props. (e.g. density, degree of crystallinity, T_m and T_g) with variations in polymerisation temp. attributed principally to corresponding changes in stereoregularity, rather than degree of branching, particularly for samples polymerised at low temps. [14] Degree of branching decreases with decreasing polymerisation temp. of resin; [14] with corresponding increase in syndiotacticity over same temp. range. [14] Degree of crystallinity increases with decreasing polymerisation temp. of resin
Transition Temperature:

No.	Value	Note	Type
1	66–88°C	[4]	Softening temp.

Surface Properties & Solubility:
Plasticisers: Di-2-ethylhexyl phthalate, tricresyl phosphate, di-2-ethyl adipate. Colour tests for plasticisers include fluorescence test

(phthalates); Gibbs iodophenol; Liebermanns test; diazo test; phenolphthalein (aryl esters, tritolyl phosphate) [2]; ammonium molybdate/benzidine test (phosphate plasticisers) [2]
Solvents/Non-solvents: Swollen by aromatic and chlorinated hydrocarbons, nitroparaffins, Ac_2O, aniline, Me_2CO [2]. Sol. plasticisers leached out by aliphatic hydrocarbons [2]
Wettability/Surface Energy and Interfacial Tension: Values for contact angles (e.g. with H_2O, methylene iodide) increase on addition of plasticisers (see poly(vinyl chloride), unplasticised) [1]

Transport Properties:
Polymer Melts: Zero shear viscosity studies of PVC containing di-2-ethylhexylphthalate and dioctyltin stabiliser (also functions as plasticiser) have been reported. η_o is dependent on MW (24000–100000) in simple shearing flow experiments; monomolecular melts follow same struct.-viscosity relationships as found for other linear melts in viscometric flow [19]. Studies of the primary (first) normal stress difference coefficient (σ_{11}-σ_{22}) have been reported; complex behaviour is exhibited [19,22,23,24]. Effect of plasticiser content on melt flow activation energy has been reported; increases with amount of plasticiser for $T < T_b$, but decreases with amount of plasticiser for $T > T_b$ (T_b, break temp., di-2-ethylhexylphthalate plasticiser) [24]
Permeability of Gases: Effect of plasticiser content on gas permeability has been reported [13]
Water Content: Surface absorption 200 g m^{-2} (max., 32 days, 20°) [2]

Water Absorption:

No.	Value	Note
1	0.6 %	48h, 50°, without filler [2]
2	11 %	max., 48h, 50°, with mineral fillers [2]

Gas Permeability:

No.	Gas	Value	Note
1	H_2O	16.41 cm^3 mm/(m^2 day atm)	0.25×10^{-10} cm^2 (s cm Hg)$^{-1}$, 30°, 25% tritolyl phosphate [2]
2	H_2O	72.2 cm^3 mm/(m^2 day atm)	1.1×10^{-10} cm^2 (s cm Hg)$^{-1}$, 30°, 25% dioctylphthalate [2]
3	H_2	122.6–236 cm^3 mm/(m^2 day atm)	$1.4–2.7 \times 10^{-13}$ cm^2 (s Pa)$^{-1}$, 5–40% tricresyl triphosphate [13]
4	CO	1.05–24.6 cm^3 mm/(m^2 day atm)	$0.012–0.28 \times 10^{-13}$ cm^2 (s Pa)$^{-1}$, 5–40% tricresyl triphosphate [13]

Mechanical Properties:
Tensile Strength Break:

No.	Value	Note	Elongation
1	19.6 MPa	2 kg f mm^{-2}, 60% plasticiser content [2]	200–450%

Viscoelastic Behaviour: Stress at 100% elongation 2.7–18.6 MPa (0.28–1.9 kg f mm^{-2}) [2]
Hardness: Shore A50–A100 [2]
Friction Abrasion and Resistance: Abrasion resistance is better for suitably compounded plasticised polymers at room temp. than vulcanised rubber; however, soon becomes too soft at elevated temps. [2]. Coefficient of friction μ_{stat}, 0.350–0.797 (25.9–39.4% plasticiser); μ_{dyn}, 0.719–0.925 (25.9–39.4% plasticiser); di-2-ethylhexylphthalate, diisodecylphthalate, n-octyl, n-decylphthalate as plasticisers [11]. μ_{stat} increases with plasticiser content; may pass through a max. [2]: μ_{dyn} 0.45 (grey cotton yarn; running at 1 m s^{-1}, 65% relative humidity) [2]

Electrical Properties:
Electrical Properties General: Certain electrical props., e.g. dielectric constant, power factor, volume resistivity, etc., are highly dependent on plasticiser type, concentration and test temp. [2]. Dielectric props., i.e. dielectric constant and dielectric loss factor have been reported, -110–110° and 20 Hz to 100 kHz; effects of type of resin, plasticiser (di-2-ethylhexylphthalate) content and ageing have been studied [20]. Where electrical props. are of importance in PVC formulation, essential to use suspension or bulk polymer and not emulsion polymer [2]

Optical Properties:
Refractive Index:

No.	Value	Note
1	1.52	40% dioctylphthalate [8]
2	1.55	40% tricresylphosphate [8]
3	1.55–1.6	[2]

Transmission and Spectra: CPMAS C-13 nmr spectral data have been reported; discrimination by contact time allows identification of plasticiser [26]. Raman spectral data of drawn samples containing 0–20 pph dioctyl sebacate plasticiser have been reported; CCl stretching region examined; [27,28] data used in conjunction with birefringence measurements in studies of molecular orientation [27]
Total Internal Reflection: Effect of added plasticiser (e.g. phthalate, adipate) on stress optical coefficient of PVC homopolymer has been reported, positive values found, even at lowest loads, except at low temps.; large positive values at $T > T_g$; increases with increasing plasticiser content [17] Birefringence measurements have been reported for drawn samples containing 0–20 pph dioctyl sebacate plasticiser; data used to complement Raman spectroscopy results in studies of molecular orientation [27]

Polymer Stability:
Polymer Stability General: Stability almost entirely controlled by type and amount of plasticiser and stabiliser present
Flammability: Not easy to ignite; has self-extinguishing props. [4]; when ignited, burns with yellow and smoky flame; biting, acidic odour; melts, forming burning droplets [4]. Limiting oxygen index (ca. 45% for unplasticised material) considerably reduced by presence of plasticiser [21]
Environmental Stress: Under certain circumstances, low MW plasticisers can diffuse out of plasticised PVC formulations; of particular concern with food packaging; although no evidence that such plasticisers are toxic, systems involving polymeric plasticisers with lower rates of diffusion have been developed as precaution [1]. Major problems are loss of plasticiser by volatilisation or exudation on exposure to uv-radiation [5]; extenders (used in conjunction with plasticisers), such as chlorinated paraffins, particularly prone to exude under effects of uv-radiation [5]
Chemical Stability: Has good chemical resistance in general
Hydrolytic Stability: Unaffected by H_2O and salt solns.; dil. acids and alkalis have little effect at room temp., but some hydrolysis and extraction of plasticiser may occur at elevated temps.; conc. acids and alkalis hydrolyse plasticisers slowly when cold, but more rapidly on heating; most organic solvents extract plasticisers; such 'leaching out' leads to compound hardening [4]
Biological Stability: Plasticisers such as adipates, azelates, ricinolates, and sebacates are prone to microbial attack; phosphates and phthalates are usually resistant; biological attack can lead to stiffening and consequently cracking of material [5]
Stability Miscellaneous: γ-Irradiation leads to cross-linking of plasticised PVC [9]. Addition of small quantities of tetrafunctional monomer, e.g. ethylene glycol dimethacrylate, is reported to reduce radiation dose required for cross-linking and improves thermal stability; graft cross-linking process [9]

– Poly(vinyl chloride), Plasticised

Applications/Commercial Products:
Processing & Manufacturing Routes: Polymerisation of vinyl chloride carried out by using bulk (mass), suspension or emulsion techniques [2,14,15]; suspension and bulk polymer contain porous particles (100 µm in size); readily absorb plasticisers (and other additives). Premixing of polymer mixing usually carried out by employing trough mixer at room temp., or dry blending (with vigorous stirring and/or gentle heating); latter product known as easy processing (EP) resin [2]. Processed using injection moulding, compression moulding, blow moulding, extrusion, casting, rotational moulding, calendering and laminating techniques [2]; calendered sheet can be prod. with excellent surface finish and/or with remarkable transparency; such sheet can be rapidly and strongly welded into large and complex shapes using process of high or radio-frequency welding. Emulsion polymer contains fine particles (1 µm in size); much of this type of polymer used to make pastes (plastisols); dispersed with stabilisers etc. in plasticiser to form creamy paste (spreadable and free-flowing); plasticisation achieved by heating to >150–160°; such gel form is similar to plasticised polymer prod. by other means, although inferior in some physical props. [2]. In addition to plastisols (no diluents), other pastes, with different viscosities and processing considerations, can be prod.: organosols (with diluents, reduced viscosities); aerogels (high viscosities); rigisols (incorporate polymerisable plasticiser - give hard mouldings on gelation); plastigels (incorporate thickening agents; thick putty-like form) [2,4,5]. Pastes processed by using spread-coating, rotational casting, dipping and spraying techniques [2,16]. Paste injection-moulding techniques have also been developed [14]. Lattices prod. from emulsion polymer by coagulation with conc. acids, polyvalent cations and by dehydration with water-miscible liquids; plasticisers, fillers, thickening agents, etc. added and processed for coating applications using similar techniques to those employed for pastes [14]. Cellular form can be prod. by incorporating blowing agents [2]

Mould Shrinkage (%):

No.	Value	Note
1	1.5–5%	0.015–0.05 in in [4]

Applications: PVC leathercloth, cable coating, packaging, clothing, household goods, toys. Injection moulded components used for toys; watch straps; cable grommets/ends; gaskets; washers; handle coverings; footwear (beach shoes, sandals); heel tags; electrical shields; plugs and other electrical components; knobs; components for medical industry (prosthetics). Extruded components used in cable sheathing and domestic wiring insulation (low frequency); decorative trim; flexible hose and ducting; sachets for household items. Calendered sheet used in protective coverings; flooring (vinyl and vinyl/asbestos); fabric-backed wall coverings; curtaining; cover-sheets and tarpaulins; clothing (aprons, raincoats); hats; galoshes; inflammable toys; air-beds; fancy goods; 'transparent' windows in flexible doors; swimming pool linings. Lattices used in paper and textile impregnation and coatings. Pastes: spreading techniques used in fabric-backed leathercloth and other coated substrates for upholstery in transport, furniture; handbags; clothing (heavy duty); shoes; wall coverings; carpet backing; flooring; automobile parts; tool handles; inks; sealants; rotational casting and slush moulding techniques used for toys; balls; arm rests; dip-forming and dip-coating techniques can be used where degree of cushioning and anti-corrosion is required for gloves; handle-bar grips; spouts; wire-basket coverings; anti-corrosion sealants; spraying techniques used for tank linings. Pastes also used as binding resins for viscosity-building, low-gloss finishes. Cellular form can be bonded to woven or knitted fabrics: garment and upholstery uses

Trade name	Details	Supplier
Aisloplastic		Aiscondel SA
Alpha 661/3X-88		Dexter Plastics
BVC		Bayshore Vinyl
Begra		Begra
Belcon	modified material	
Bellotex		Oesterr. Linoleum-Wachstuch-u. Kunstlederfabriken
Benecor		J.H. Benecke
Benefol		J.H. Benecke
Benelit	non-exclusive, used for other PVC-based materials	J.H. Benecke
Benequick		J.H. Benecke
Benova		J.H. Benecke
Biodrak	non-exclusive	
Blacar		
Bocato		J.H. Benecke
Boltaflex		General Tyre and Rubber Co.
Bondene		BIP
Breon		British Geon
Breon M		British Geon
Breon S		British Geon
Carina		Shell Chemicals
Clean Ace		Mitsubishi Monsanto Chem. Co
Coloflor		NL7 Krommenie BV
Colorplast		VEB Piesteritz, Germany
Colorvinyl		NL7 Krommenie BV
Conrex		Continental Gummi-Werke
Contan		Continental Gummi-Werke
Contimatic		Continental Gummi-Werke
Contiplast		Continental Gummi-Werke
Contiplastex		Continental Gummi-Werke
Dural	modified material	Alpha
Ekalit		VEB Piesteritz, Germany
Ekalon	non-exclusive	VEB Piesteritz, Germany
Elkavyl	non-exclusive, used for other PVC-based materials	Ugine Kuhlmann SA Div. Plastiques
Elkavyl	non-exclusive	Ugine Kuhlmann SA Div. Plastiques
Excelon		
FG 7108		Georgia-Gulf
Foamex	foamed composites and adhesives	
Franklin		Franklin
Geon	non-exclusive, used for other vinyl-based materials	B.F. Goodrich
Geon 121	a 'plastisol'	B.F. Goodrich
Geon 575 × 43	contains 25% phosphate plasticiser	B.F. Goodrich
Geon HTX		BF Goodrich Chemical
Geon RX		BF Goodrich Chemical

– Poly(vinyl chloride), Plasticised

Goeppinger plastics		Goeppinger Kaliko and Kunstleder Werke
Gorlex	non-exclusive	Polymer Gorla
Grabiol		Gyoer/Raab
Graboplast		
Greizella		Grand Pacific
HM 2230		Georgia-Gulf
Howelon		Konrad Hornschuh AG
Juvogress		Julius Votteler's Nachf.
Juvotecto		Julius Votteler's Nachf.
Kanaflex		Techno-Chemie
Kanaflex		Kessler Co. GmbH
Kanevinyl	non-exclusive, used for other PVC-based materials	Kanegafuchi Chemical Ind.
Kanevinyl Paste PSL-10	plasticised paste	Kanegafuchi Chemical Ind.
Ledron		
Ledron		
Letex		
Letex		
Lonfix		Kawaguchi Rubber
Lonfix		Kawaguchi Rubber
Lonleum		Kawaguchi Rubber
Lonleum		Kawaguchi Rubber
Lucavil		Chimica Lucana S.p.A
Lucolene		Rhone-Poulenc Polymers
Lucolene		Pechiney-St Gobain
Lucomix		Rhone-Poulenc Polymers
Lucovyl		Pechiney-St Gobain
Marleyflex		Marlex-Werke
Marleyflor		Marlex-Werke
Marvelon		Konrad Hornschuh AG
Marvelon		
Marvinol	non-exclusive	Uniroyal Chemical
Marvinol		Naugatuch Chemical Div.
Marvylan		DSM
MayFair		Commercial Plastic
Mipofolie	non-exclusive	Alfred Schwarz GmbH
Mipolam		Dynamit Nobel
Mipoplast		Dynamit Nobel
Nairn		
Norvinyl DX550		Hydro Plast
Novilon		NL7 Krommenie BV
Opalon	non-exclusive	Monsanto Chemical
Or-on		
Orit		
Oxy		OxyChem
Oxy 225PG		OxyChem
Oxy 225SG		OxyChem
Pe Va Clair		Sonobat SA
Pevikon 737	Plastisol, non-exclusive, used for other PVC-based materials	Kema-Nord AB
Polytherm		4P Folie Forchheim GmbH.
Polytron SM50		BF Goodrich Chemical
Polytuf 200	modified material	Diamond Shamrock
Polyvin		A. Schulman Inc.
Primavil		Giovanni Crespi
Ravinil		European Vinyls Corporation Ltd.
Rhenofol		Brass u. Co. GmbH, Kunstoffwerke Schildkroet
Rhenoglas		Brass u. Co. GmbH, Kunstoffwerke Schildkroet
Rhodopas X		Rhone-Poulenc
SP 7107		Georgia-Gulf
Solvic		Laporte
Superkleen 3003-85		Dexter Plastics
Sylphane		UCB Films
Symadur		
Symalit		
Symkanal		
Telon		Oy Nars AB
Telonette		Oy Nars AB
Thermaflo		Evode Plastics
Trosiplast	non-exclusive, used for other PVC-based materials	Hüls AG
Trosiplast 7804	Phthalate-ester plasticised/$CaCO_3$ filled	Hüls AG
Trosiplast 8008	Phthalate-ester plasticised	Hüls AG
Trovidur	non-exclusive, general	Hüls AG
Trovipor		Dynamit Nobel
Ultron	non-exclusive	Monsanto Chemical
Ultryl		Philips Petroleum International
Unichem		Colorite
VC-47B		Borden
Velbex		BIP
Velkor	non-exclusive	Alkor Plastics (UK) Ltd.
Vestolit		Hüls AG
Vestolit B7021		Hüls AG
Vestolit E		Hüls AG
Vestolit P1330K		Hüls AG

Vinatex	non-exclusive (general)	Vinatex
Vinelle		General Tyre and Rubber Co.
Vinifol		Aismalibar S.A.
Vinika	non-exclusive, used for other PVC-based materials	Mitsubishi Monsanto Chem. Co
Vinilen	non-exclusive, used for PVC-based materials	Vinisa
Viniplast		Aismalibar S.A.
Vinoflex		BASF
Vinoflex KR		BASF
Vinoflex P		BASF
Vinoflex S		BASF
Vinol	non-exclusive, used for other polyvinyl materials	Gummifabriken Gislaved AB
Vinolfex compound	non-exclusive, general	BASF
Vinophane		British Cellophane Ltd.
Vinybel		Caselith National SA
Vinylchlore		Saplast SA
Viplast	non-exclusive	Montedison
Vista		Vista
Vista 5305		Vista
Vitafilm		Goodyear
Vocavyn		
Vygen		Vygen Corporation
Vygen 301		Vygen Corporation
Vynella	non-exclusive	Adretta-Werke
Vynide		ICI, UK
Welvic	non-exclusive, used for other PVC-based materials	ICI, UK

Bibliographic References

[1] Bowers, R.C., Jarvis, N.L. and Zisman, W.A., *Ind. Eng. Chem. Proc. Des. Dev.*, 1965, **4**, 86, (surface props., mechanical props.)
[2] Roff, W.J. and Scott, J.R., Fibres, Films, *Plastics and Rubbers*, Butterworths, 1971, 108, 230, 304, 521, (props.)
[3] Dunlap, L., *J. Polym. Sci., Part C: Polym. Lett.*, 1968, **30**, 561, (volumetric props.)
[4] Whelan, T., *Polymer Technology Dictionary*, Chapman and Hall, 1994, 308, 534, 538, (flammability, thermal props., chemical stability, processing)
[5] Matthews, G., *Vinyl Allied Polym.*, (ed. P.D. Ritchie), Iliffe Books, (Butterworth), 1972, **2**, 242, 299, (biological stability)
[6] Sheldon, R.P. and Lane, K., *Polymer*, 1965, **6**, 77, (thermal conductivity)
[7] Eierman, K., *Kunststoffe*, 1961, **51**, 512, (thermal conductivity)
[8] Seferis, J.C., *Polym. Handb.*, 4th edn., (eds. J. Brandrup, E.H. Immergut and E.A. Grulke), John Wiley and Sons, 1999, **VI**, 581, (optical props.)
[9] Miller, A.A., *Ind. Eng. Chem.*, 1959, **51**, 1271, (radiation stability, thermal props.)
[10] Hoffmann, R. and Knappe, W., *Kolloid Z. Z. Polym.*, 1970, **240**, 784, (volumetric props.)
[11] Collins, E.A., Daniles, C.A. and Witenhafer, D.E., *Polym. Handb.*, 4th edn., (eds. J. Brandrup, E.H. Immergut and E.A. Grulke), John Wiley and Sons, 1999, **V**, 67, (mechanical props.)
[12] Sefcik, M.D., Schaefer, J., May, F.L., Racher, D. and Dub, S.M., *J. Polym. Sci., Polym. Phys. Ed.*, 1983, **21**, 1041, (transport props.)
[13] Brydson, J.A., *Plast. Mater.*, 6th edn., Butterworth-Heinemann, 1995, 298, (synth., processing, applications)
[14] *Polymer Yearbook - 5*, (ed. R.A. Pethrich), Harwood Academic, 1989, 4, (synth., processing, applications)
[15] Ash, M. and Ash, I., Handbook of Plastic Compounds, *Elastomers and Resins*, (eds. M.B. Ash and I.A. Ash), Wiley-VCH, 1992, 512, (processing, applications)
[16] Andrews, R.D. and Kazama, Y., *J. Appl. Phys.*, 1968, **39**, 4891, (optical props., mechanical props.)
[17] Heydemann, P. and Guicking, H.D., *Kolloid Z. Z. Polym.*, 1963, **193**, 16, (volumetric props.)
[18] Lyngaae-Jørgensen, J., *J. Appl. Polym. Sci.*, 1976, **20**, 2497, (transport props.)
[19] Utracki, L.A. and Jukes, J.A., *J. Vinyl Technol.*, 1984, **6**, 85, (electrical props.)
[20] Alger, M.S.M., *Polymer Science Dictionary*, Elsevier Applied Science, 1989, 386, (flammability)
[21] Utracki, L.A., Bakerdjian, Z. and Kamal, M.R., *Trans. Soc. Rheol.*, 1975, **19**, 173, (transport props.)
[22] Utracki, L.A., *Polym. Eng. Sci.*, 1974, **14**, 308, (transport props.)
[23] Utracki, L.A., *J. Polym. Sci., Polym. Phys. Ed.*, 1974, **12**, 563, (transport props.)
[24] *Encycl. Polym. Sci. Eng.: Supplement Volume*, (eds. H.F. Mark, N.M. Bikales, C.G. Overberger and G. Menge), 2nd edn., Wiley-Interscience, 1989, 889, (environmental stability)
[25] *Engineering Properties of Thermoplastics*, (ed. R.M. Ogorkiewicz), Wiley, 1970, 252, (applications)
[26] *Polymer Spectroscopy*, (ed. A.H. Fawcett), John Wiley, 1995, 130, 183, (nmr, Raman)
[27] Robinson, M.E.R., Bower, D.I. and Maddams, W.F., *J. Polym. Sci., Polym. Phys. Ed.*, 1978, **16**, 2115, (Raman, birefringence, orientation)
[28] Bower, D.I., King, J. and Maddams, W.F., *J. Macromol. Sci., Phys.*, 1981, **20**, 305, (Raman, orientation)

Poly(vinyl chloride-*co*-propylene)   P-581

$$\left[\begin{array}{c} CH_2CH \\ | \\ Cl \end{array}\right]_x \left[\begin{array}{c} CH_2CH \\ | \\ CH_3 \end{array}\right]_y \Big]_n$$

Synonyms: *PVC-PP copolymer. VCP. VC/P copolymer. Vinyl chloride-propylene copolymer. Poly(chloroethene-co-1-propene)*
Related Polymers: Poly (vinyl chloride), unplasticised, PVC copolymers
Monomers: Vinyl chloride, Propylene
Material class: Thermoplastic, Copolymers
Polymer Type: polypropylene, PVC polyhaloolefins
CAS Number:

CAS Reg. No.
72027-02-2
25119-90-8

Molecular Formula: $[(C_2H_3Cl).(C_3H_6)]_n$
Fragments: C_2H_3Cl C_3H_6
Molecular Weight: MW decreases with increasing propylene content (at given copolymerisation reaction temp.)
Additives: Stabilisers; processing aids; waxes; lubricants; impact-modifiers
General Information: Typically contains 2-10% propylene. Reduced tendency to dehydrochlorination compared with PVC, leading to improved heat stability. Copolymer more stable than PVC homopolymer; has lower melt viscosity, thus requires lower processing temps and hence less need for stabilisers; latter fact improves rigidity at high temps, and lowers permeability [1]. Restricted commercial use of copolymers, as considerably more expensive than PVC/PVA [3]
Tacticity: Low-propylene-content (<15 mol%) materials contain long uninterrupted runs of vinyl chloride units. Copolymers exhibit reduced sensitivity to stereosequencing [14,15] Theoretical conformational model has been reported [16] based on rotational isomeric state (RIS) theory; calculations for obtaining random-coil dimensions (mean-square unperturbed dimensions), $[r^2]_0$ and dipole moments, $[\mu^2]$

Poly(vinyl chloride-co-propylene)

Miscellaneous: A comprehensive review of the struct. and physical props. of a series of propylene copolymers, and the processing and uses of such copolymer resins developed for rigid applications has been reported [17]

Volumetric & Calorimetric Properties:
Density:

No.	Value	Note
1	d 1.286–1.445 g/cm^3	[1,12]

Volumetric Properties General: Density decreases with increasing propylene content [1,12]

Glass-Transition Temperature:

No.	Value	Note
1	60–80°C	[2]

Transition Temperature:

No.	Value	Note	Type
1	60–80°C	[2]	Degradation temp.
2	178–181°C	2.7 wt% propylene [5]	Equiviscous temp.

Surface Properties & Solubility:
Plasticisers: Polypropylene acts as internal (permanent) plasticiser [7]

Transport Properties:
Transport Properties General: Ease of melt flow increases with increasing propylene content (for the same MW) [3,11]; may be varied by 100-fold according to composition [9]

Melt Flow Index:

No.	Value	Note
1	0.75–180 g/10 min	1.9%–7.6% propylene ASTM D 1238 [10]
2	3.8 g/10 min	3.5 wt% propylene, ASTM D1238-57T-F [13]

Polymer Melts: Intrinsic viscosity of commercially available resins decreases with increasing propylene content [10,11]; [η] 1.393–1.356 dl g^{-1} (1.9–7.6 wt% propylene) [10]. Melt viscosity 69000P (simple formulation (minimum additives), shear rate 50 s^{-1}); 2400P (simple formulation, shear rate 250 s^{-1}); 52000P (blow-moulded formulation (various additives), shear rate 50 s^{-1}); 17000P (blow-moulded formulation, shear rate 250 s^{-1}) (2.7 wt% propylene) [5]

Permeability of Gases: [O$_2$] 45 cm^3(m$_2$ day cm^3 atm)$^{-1}$ 250°, film; typical value [8]

Permeability of Liquids: [H$_2$O] 4.3g m^2 mil, 95% relative humidity, 37.7° (100°F), film; typical value [8]

Water Absorption:

No.	Value	Note
1	0.05–0.25 %	24h, 3mm H$_2$ [2]

Mechanical Properties:
Mechanical Properties General: Presence of comonomer has comparatively little adverse effect on mechanical props., compared to PVC homopolymer [3]

Tensile (Young's) Modulus:

No.	Value	Note
1	2353–3138 MPa	240–320 kgf mm^{-2} [1]

Flexural Modulus:

No.	Value	Note
1	2647–3206 MPa	270–327 kgf mm^{-2} [1]

Tensile Strength Break:

No.	Value	Note	Elongation
1	29.4–48 MPa	3–4.9 kgf mm^{-2}	100–200%

Flexural Strength at Break:

No.	Value	Note
1	59.8–76 MPa	6.1–7.75 kgf mm^{-2} [1]

Tensile Strength Yield:

No.	Value	Note
1	46–59.8 MPa	4.7–6.1 kgf mm^{-2}

Compressive Strength:

No.	Value	Note
1	54–78.4 MPa	550–800 kg cm^{-2} [2]

Hardness: Rockwell R105-R118 [1]

Izod Area:

No.	Value	Notch	Note
1	16–1600 kJ/m^2	Unnotched	0.3–30 ft lb in^{-1}, 23° [1]

Electrical Properties:
Dielectric Permittivity (Constant):

No.	Value	Note
1	2.8–3.5	[2]

Dissipation (Power) Factor:

No.	Value	Frequency	Note
1	0.01–0.02	1 MHz	[2]

Optical Properties:
Optical Properties General: Copolymers show lower light absorbance and 'yellowness' index than PVC homopolymer; (ASTM D1925-637) [5] commercial significance; e.g. formulations for blow-moulded products [4,5]

Transmission and Spectra: High resolution C-13 nmr spectral data have been reported [15]

Polymer Stability:
Polymer Stability General: Has better thermal stability than PVC homopolymer; presence of propylene units, randomly distributed in small amounts (also terminate polymer chains), break up zipper dehydrochlorination (decomposition) reactions; need less heat stabiliser in processing [1,6,7]; importance both in application and during processing - greater ease of moulding [9]

Flammability: Self-extinguishing [2]

Applications/Commercial Products:
Processing & Manufacturing Routes: Synth. by free-radical suspension [1] polymerisation of comonomer mixture; melt processing involving injection-moulding (150–165°; mould temp

40–60°) and extrusion (150–175°) techniques; lower melt-extrusion range than PVC homopolymer (190–210°) [6]; ease of moulding is an advantage; can also be blow-moulded. Also synth. by suspension polymerisation at 56° involving intermittent addition of vinyl chloride to reaction mixture. The product, containing 3.5 wt% propylene, has good mech. props. and high melt index [13]. In the copolymerisation reaction propylene acts as a degradative chain-transfer reagent; thus both polymerisation rate and resulting copolymer(s) MW are lower than those of PVC homopolymer; decrease with increasing propylene content. The copolymerisation process is kinetically controlled by propylene [18]

Applications: Injection and blow-moulded products used in gramophone records; sheets; (glass-clear) blown bottles; coatings; flame-resistant films; thermoformed packaging; extended films for packaging/food wrapping (fresh meat); rigid/transparent compositions used for moulded/extended components in medical applications

Trade name	Details	Supplier
Airco	400 series	Air Products

Bibliographic References

[1] Roff, W.J. and Scott, J.R., *Fibres, Films, Plastics and Rubbers* Butterworths, 1971, 121, (processing, mech. props)
[2] *Polymer Yearbook - 5* (ed. R.A. Pethrick), Harwood Academic, 1989, 24, (thermal props, mech. props, electrical props)
[3] Titow, W.V., *PVC Plastics: Properties, Processing, and Applications,* Elsevier Applied Science, 1990, **79**, 483, 781, (applications)
[4] Titow, W.V., *PVC Technol.* 4th edn., Elsevier Applied Science, 1984, 795, (stability)
[5] Taylor, W. and King, L.F, *Polym. Eng. Sci.,* 1970, **10**, 204, (optical processing props, transport props, stability)
[6] *Kirk-Othmer Encycl. Chem. Technol.,* 3rd edn. (eds. M. Grayson and D. Eckroth), John Wiley and Sons, 1983, 910, (processing, applications, thermal stability, props)
[7] Whelan, T., *Polymer Technology Dictionary,* Chapman and Hall, 1994, 475, (applications, stability)
[8] *Sci. Technol. Polym. Films,* (ed. O.J. Sweeting), Wiley, 1971, 321, (permeability)
[9] Brydson, J.A., *Plast. Mater.,* 6th edn., Butterworth-Heinemann, 1995, 347, (mech. props, applications)
[10] Cantow, M.J.R., et al, *Mod. Plast.,* 1969, **126**, 46, (polymerisation mechanism/reactions, processing, transport props)
[11] Heiberger, C.A. et al, *Polym. Eng. Sci.,* 1969, **9**, 445, (processing, transport props)
[12] Ravey, M., *J. Polym. Sci., Polym. Chem. Ed.,* 1975, **13**, 2635, (density, polymer composition)
[13] *Neth. Pat.,* 1966, 6510712, (synth, mech. props, transport props)
[15] Tonelli, A.E. and Schilling, F.C., *Macromolecules,* 1984, **17**, 1946, (C-13 nmr, tacticity, sequence distribution)
[16] Mark, J.E., *J. Polym. Sci., Polym. Phys. Ed.,* 1973, **11**, 1375, (stereochemistry)
[17] Heiberger, C.A. and Phillips, R., *Rubber Plast. Age,* 1967, **48**, 636, (rev)
[18] Ravey, M., Waterman, J.A., Shorr, L.M. and Kramer, M., *J. Polym. Sci., Polym. Chem. Ed.,* 1976, **14**, 1609, (synth, polymerisation reaction)

Poly(vinyl chloride-*co*-trifluoro chloroethylene)

P-582

$$-[(CH_2CH(Cl))_x-(CF(Cl)CF_2)_y]_n-$$

Synonyms: *Poly(vinyl chloride-co-chlorotrifluoroethylene). Vinyl chloride-trifluorochloroethylene copolymer. Poly(chloroethene-co-chlorotrifluoroethene). VC/TFCE copolymer. VC/CTFE copolymer. Poly(vinyl chloride-co-(1-chloro-1,2,2-trifluoroethylene)). Vinyl chloride-chlorotrifluoroethylene copolymer*
Related Polymers: Polyvinyl chloride, unplasticised, PVC copolymers
Monomers: Vinyl chloride, Chlorotrifluoroethylene
Material class: Thermoplastic, Copolymers
Polymer Type: PVC, PCTFEfluorocarbon (polymers)polyhaloolefins

CAS Number:

CAS Reg. No.
24937-97-1

Molecular Formula: $[(C_2H_3Cl)_x.(C_2ClF_3)_y]_n$
Fragments: C_2H_3Cl C_2ClF_3
Additives: Stabilisers include Cd, Ba, Zn, Ca compounds
Morphology: Studies of monomer sequence distribution have been reported [1]
Miscellaneous: Finds main application as surface coatings; has good solubility characteristics and particularly good service props.; displays poor adhesion to metals unless stoved at high temp. after application [19]. Struct. and physical props. have been reviewed [1]. Graft copolymers have been reported [14,22]; possess good heat, impact and abrasion resistance [22]; flexible, elastomeric, melt-processable systems [14]

Volumetric & Calorimetric Properties:
Density:

No.	Value	Note
1	d 1.7 g/cm^3	FPC 461 [23]

Melting Temperature:

No.	Value	Note
1	100°C	approx., 50% chlorotrifluoroethylene [1]

Transition Temperature General: Minimum melting point observed at 50% chlorotrifluoroethylene content; linear decrease in T_g observed with increasing chlorotrifluoroethylene content [1]; heat resistance decreases with increasing chlorotrifluoroethylene content [1]

Surface Properties & Solubility:
Solubility Properties General: Solubility characteristics (with respect to coatings applications) have been reviewed in comparison with PVC homopolymer, terpolymers and other copolymers [26]
Solvents/Non-solvents: (FPC 461) sol. toluene, xylene, ketones, chlorinated hydrocarbons, esters [23]

Transport Properties
Polymer Solutions Dilute: η_{rel} 1.5 (FPC 461) [23]; η 210 cP (80% chlorotrifluoroethylene; 20% 2-butanone, 20°) [27]
Polymer Melts: Viscosity decreases sharply with increase in chlorotrifluoroethylene content; related to decrease in chain length of copolymer [1]

Mechanical Properties:
Impact Strength: Notched impact strength is independent of chlorotrifluoroethylene content; values remain constant [1]
Hardness: Hardness decreases with increasing chlorotrifluoroethylene content [1]; Pencil hardness HB (film, 80% chlorotrifluoroethylene) [27]

Optical Properties:
Transmission and Spectra: XPES and ion scattering spectroscopy studies (thin film coatings) have been reported [17,18]. H-1 nmr and F-19 nmr electron-nuclear double resonance (ENDOR) spectral data have been reported [2]

Polymer Stability:
Polymer Stability General: Displays good heat and light stability (FPC 461) [23]
Thermal Stability General: Thermal stability (plus processability) increases with increase in chlorotrifluoroethylene content and MW of copolymer [1]
Decomposition Details: Rate of hydrogen halide cleavage decreases with increase in chlorotrifluoroethylene content [1]; blackening and carbonisation occur after heating at 195° for 10 min. [24]

Chemical Stability: Has good resistance to chemicals (80% chlorotrifluoroethylene) [27]; acid/alkali resistance 2% (80% chlorotrifluoroethylene) (JIS K-5400) [27]

Applications/Commercial Products:
Processing & Manufacturing Routes: Synth. by suspension [20] or emulsion [19] polymerisation of a mixture of comonomers. Dry processing material fabricated by moulding, extrusion, calendering or coating techniques [23,24]. Graft copolymers prod. by radical suspension polymerisation in alkaline solns. [22]
Applications: Main applications as surface coatings and films, thin coatings (5–10 Å) on plastic-bonded and other explosives; in precision detonator compositions, e.g. pentaerythritaltetranitrate (PETN); typically 1.5/1 vinyl chloride/chlorotrifluoroethylene molecular ratios. Major use in electrophotographic toner/developer compositions (xerography), including formulations with PMMA and liq. cryst. polymers; colour toner systems also reported: such developers provide excellent and stable electrostatic behaviour, good chargeability, high durability, and good environmental stability. Used in blend compositions with other fluorinated polymers (e.g. PTFE), with copolymer as minor component, as lubricants for various substrates; gives improved lubricity characteristics and abrasion resistance; lowers friction coefficients of substrates such as photographic films and magnetic surfaces and tapes. Miscellaneous uses include compositions of magnetic cleaning systems; release agents for heat shrinkable films; mould release agents for epoxy (thermosetting) resins. Medically acceptable polymer (non-thrombogenic material) used in conjunction with amphiphilic polyethers or polysulfides, with latter bound to copolymer surface; potential applications as artificial arteries, heart valves, oxygenation membranes for heart-lung machines for cardiovascular surgery, dialysis shunts, intravenous or intraarterial catheters

Trade name	Details	Supplier
Exon 46	Non-exclusive, used for other PVC-based materials	Firestone
FPC-461		Occidental Chemical Corp.
Oxy		Occidental Chemical Corp.

Bibliographic References
[1] Ulbricht, J. and Raessler, K., *Plaste Kautsch.*, 1976, **23**, 487, (rev, thermal props., transport props., struct., stability, mechanical props.)
[2] Veregin, R.P., Harbour, J.R., Kotake, Y. and Janzon, E.G., *J. Colloid Interface Sci.*, 1989, **132**, 542, (H-1 nmr, F-19 nmr)
[3] *Ger. Pat.*, 1989, 3 730 797, (uses)
[4] *Jpn. Pat.*, 1989, 01 120 566, (uses)
[5] *Jpn. Pat.*, 1989, 01 120 567, (uses)
[6] *U.S. Pat.*, 1989, 4 876 169, (uses)
[7] *Jpn. Pat.*, 1990, 02 22 673, (uses)
[8] *U.S. Pat.*, 1990, 4 891 293, (uses)
[9] *U.S. Pat.*, 1990, 4 912 005, (uses)
[10] *Jpn. Pat.*, 1990, 02 203 356, (applications)
[11] *U.S. Pat.*, 1990, 4 937 167, (uses)
[12] Vannet, M.D. and Ball, G.L., *J. Energ. Mater.*, 1987, **5**, 35, (uses)
[13] *Jpn. Pat.*, 1987, 62 09 358, (uses)
[14] *Ger. Pat.*, 1987, 3 631 789, (processing)
[15] *Jpn. Pat.*, 1988, 63 30 864, (uses)
[16] Worley, C.M., *Energy Res. Abstr.*, 1986, **11**, no. 38516, (uses)
[17] Wang, P.S. and Wittberg, T.N., *J. Energ. Mater.*, 1984, **2**, 167, (uses)
[18] Worley, C.M., Vannet, M.D., Ball, G.L. and Moddeman, W.E., SIA, *Surf. Interface Anal.*, 1987, **10**, 273, (uses, surface studies, spectroscopy)
[19] Titow, W.V., *PVC Technol.*, 4th edn., Elsevier Applied Science, 1984, **23**, 1049, (synth, uses)
[20] *U.S. Pat.*, 1976, 3 998 939, (synth)
[21] *Eur. Pat. Appl.*, 1984, 123 044, (synth)
[22] *U.S. Pat.*, 1986, 4 574 141, (synth)
[23] Ash, M. and Ash, I., Handbook of Plastic Compounds, *Elastomers and Resins*, (eds. M.B. Ash and I.A. Ash), Wiley-VCH, 1992, 517, (processing, stability, solubility)
[24] *Jpn. Pat.*, 1985, 60 155 211, (processing, thermal stability)
[25] *U.S. Pat.*, 1975, 3 862 860, (uses)
[26] Park, R.A., Am. Chem. Soc., Div. Org. Coat. Plast. Chem., *Pap*, 1971, **31**, 515, (uses, solubility)
[27] *Jpn. Pat.*, 1974, 74 60 391, (processing, transport props., mechanical props., stability)

Poly(vinyl chloride), Unplasticised P-583

$$-[CH_2CH(Cl)]_n-$$

Synonyms: *Vinyl chloride polymer. Chloroethene polymer. Chloroethylene polymer. Unplasticised PVC. PVC. UPVC. PVC-U. Poly(chloroethene). Unplasticised poly(vinyl chloride). Poly(1-chloroethylene). RPVC. Poly(monochlorethylene). Rigid PVC*
Related Polymers: Polyvinyl chloride, Plasticised, Poly(vinyl chloride-*co*-vinyl acetate), Poly(vinyl chloride-*co*-vinyl acetate-*co*-2-hydroxypropyl acrylate), Poly(vinyl chloride-*co*-N-cyclohexyl-maleimide), Poly(vinyl chloride-*co*-vinyl acetate-*co*-maleic acid), Poly(vinyl chloride-*co*-vinylidene chloride), Poly(vinyl chloride-*co*-propylene), Poly(vinyl chloride-*co*-vinyl acetate-*co*-vinyl alcohol), Chlorinated polyvinyl chloride, PVC copolymers
Monomers: Vinyl chloride
Material class: Thermoplastic
Polymer Type: PVC
CAS Number:

CAS Reg. No.	Note
37331-71-8	
8050-05-3	
88813-65-4	
9002-86-2	
26793-37-3	isotactic
74192-23-7	

Molecular Formula: $(C_2H_3Cl)_n$
Fragments: C_2H_3Cl
Molecular Weight: MW 100000–200000, MW 45000–64000. (MW can range from as low as 40000 to as high as 450000)
Additives: Stabilisers, plasticisers, extenders, lubricants, fillers, pigments, processing aids, impact modifiers, fire-retardants, blowing agents. Process aids can include PMMA or SAN. Heat stabilisers: basic lead carbonate (white lead), tribasic lead sulfate, dibasic lead phosphite (also acts as light stabiliser), and dibasic lead phthalate; cadmium-barium mixed systems, of stearates, ricinoleates, palmitates and octoates; organotin systems; epoxide resins and oils; amines, diphenylurea, 2-phenylindole and aminocrotonates. Antioxidants include trisnonyl phenyl phosphite; hindered phenols. Uv absorbers: modified benzophenones and benzotriazoles. Extenders (used in conjunction with plasticisers): chlorinated paraffin waxes and liq. paraffinic fractions; oil extracts. Lubricants: (external) calcium stearate, stearic acid, normal and dibasic lead stearate; (internal) wax derivatives glyceryl esters and long-chain esters. Fillers: china clay; various forms of calcium carbonate; talc; baryter; silicas; silicates. Impact modifiers/processing aids: ABS graft terpolymers; MBS terpolymers; chlorinated polyethylene; EVA-PVC graft polymers; polyacrylates. Fire-retardants: antimony oxide. Blowing agents: azodicarbonamide; azodiisobutyronitrile. Antistatic agents: quaternary ammonium compounds. Viscosity depressants: poly(ethylene glycol) derivatives
General Information: Major thermoplastic material; as a consequence of all-round generally good props. and exceptionally wide range of additives that may be employed to modify these props., may be regarded as the most widely used plastic material [1]. Polymer is characterised by extreme flexibility, resiliency and toughness, has excellent resistance to water and low flammability. PVC has limited thermal stability and requires the addn. of

Poly(vinyl chloride), Unplasticised

stabilisers before or during processing. Typically chlorine content: 57%

Tacticity: Approx. 55% syndiotactic with the rest largely atactic. This results in some short-range ordering of the molecules, corresponding to approx. 10% 'crystallinity' [1] typically head-to-tail linkages; molecules contain slight branching; branching has been estimated as approx. 1.5 branches per 100 monomer units [2]

Morphology: Typically head to tail linkages. Substantially amorph. Typical commercial material contains approx. 5% crystallinity. Molecules contain slight branching. Branching content as function of polymerisation temp. has been reported. [17,71,72] Formation of anomalous structs., e.g. those containing labile (tertiary) chlorine groups, has significant effect on thermal stability. [71] Various branch structs. contribute to mechanism of chain transfer to monomer during polymerisation process. [72] Orthorhombic; space group, Pacm; 2 monomers per unit cell: a 10.6, b 5.40, c 5.1 (commercial material, polymerised at 50–60°) [17,28]; a 10.4, b 5.30, c 5.1 (solution blended high-MW, low-cryst. polymer; low-MW, high cryst. polymer) [17,91]; a 10.32, b 5.32 (single crystals, polymerised at -75°) [17,92]; a 10.24, b 5.24, c 5.08 (single crystals, low-MW, polymer produced in n-butyraldehyde). [17,93] Readily forms thermoreversible gels on cooling solns. of polymer in a variety of solvents, e.g. dioxane, [45,46,47,48] dichloroethane, [46] bromobenzene, [45,46,47,48] nitrobenzene, [45,47,48] 1,2-dichlorobenzene, [47,48] butanone, [45,47,48] DMF, [45] diisopropyl ketone, [47,48] etc. Various PVC/dialkyl phthalate gel systems also reported [46,47,48,49]; gelation believed to be associated with crystallisation; [48] has important technological implications concerning plasticisation. [47] Crystalline regions in commercial materials have syndiotactic structure; repeat distance 5.1 Å [28,55]

Identification: Distinguished from polychloroprene ($\leq 40\%$ chlorine) as polychloroprene decomposes in conc. HNO_3 or boiling 80% H_2SO_4, while PVC is unaffected. Polychloroprene fairly rapidly decolorises dil. soln. of I_2 (e.g. 0.02%) in CCl_4, while PVC remains ineffective (subject to type of plasticiser present). Various colour tests are available for plasticisers [13]

Miscellaneous: At 20–25% crystallinity isotactic or crystalline PVC has a higher softening temp. and reduced solubility compared to commercial PVC. Crystalline PVC has not been developed as a commercial product. A PVC with a high degree of plasticiser becomes stiff and brittle. Changes in various physical props. (e.g. density, degree of crystallinity, T_m and T_g) with variations in polymerisation temp. attributed principally to corresponding changes in stereoregularity, rather than degree of branching, particularly for samples polymerised at low temps. [14] Degree of branching decreases with decreasing polymerisation temp. of resin, [14] with corresponding increase in syndiotacticity over same temp. range. [14] Degree of crystallinity increases with decreasing polymerisation temp. of resin; [14,17] low-temperature values are greatly overestimated owing to assumptions made in calculations [17]

Volumetric & Calorimetric Properties:

Density:

No.	Value	Note
1	d 1.385 g/cm³	x-ray density, amorph. [28]
2	d 1.44 g/cm³	[28]
3	d 1.34 g/cm³	high impact [13]
4	d 1.4 g/cm³	fibres [13]
5	d 1.39 g/cm³	carboxylated powder, normal impact [1,13]

Thermal Expansion Coefficient:

No.	Value	Note	Type
1	0.00017–0.000175 K^{-1}	$T > T_g$ [15]	L
2	$6.6 \times 10^{-5} - 7.5 \times 10^{-5}$ K^{-1}	$T < T_g$, before annealing [15]	L
3	6.9×10^{-5} K^{-1}	$T < T_g$, after annealing [15]	L

Volumetric Properties General: Heat capacity components related to various modes of vibration, on basis of literature spectral data or Debye functions, have been reported. [24] Effect of thermal history on C_p for PVC samples of different stereoregularities has been reported; partially syndiotactic materials have higher C_p values than atactic samples; differences reduced after annealing. [56] Reviews of C_p measurements have been reported. [24,40,59] Comparative studies of commercially available material (suspension-, emulsion-, and bulk-polymerised; high-impact PVC) have been reported. [61,62] Activation energies for free-radical polymerisation of vinyl chloride E_{prop} 16 kJ mol^{-1} [32,57]; E_{ter} 17.6 kJ mol^{-1}. [32,51] Heat of polymerisation -96 - -109 kJ mol^{-1}. [33,34,35] Temp. variation of C_p for amorph. polymer 3.38–26 cal deg^{-1} mol^{-1} (60–420 K). [40] Density increases with decreasing polymerisation temp. of resin; [14,21] dependence on thermal history has also been reported. [17,22] Isothermal compressibility studies of uniaxially stretched PVC have been reported [17,73,74]

Equation of State: Tait and Simha-Somcynsky equation of state parameters have been reported [42,43]

Thermodynamic Properties General: Heat of polymerisation -95.88–131.88 kJ mol^{-1}. [33,34,35] Other thermodynamic props. have been reported [24]. Thermal conductivities not significantly influenced by MW, polymerisation temp., and syndiotacticity, except for cases where tacticity leads to crystallisation. [17] Thermal conductivity increases with temp. up to T_g and then decreases. [18]

Latent Heat Crystallization: Enthalpy and entropy functions of amorph. PVC have been reported. [59] Entropy as a function of pressure and temp. has been reported. [60] Heat of fusion of 2.76–12.65 kJ mol^{-1} [17] ΔH_u 3.28 kJ mol^{-1} (785 cal mol^{-1}) appears to be most probable value [14,17]

Thermal Conductivity:

No.	Value	Note
1	0.1588 W/mK	20° [20]
2	0.129–0.165 W/mK	-170–100° [16]

Thermal Diffusivity: 0.00144–0.000706 cm² s^{-1} (-73–207°) [16]

Specific Heat Capacity:

No.	Value	Note	Type
1	1.046–1.255 kJ/kg.C	0.25–0.3 cal (g k)$^{-1}$, $T < T_g$	P
2	1.757 kJ/kg.C	0.42 cal (g k)$^{-1}$, $T > T_g$ [21,24]	P

Melting Temperature:

No.	Value	Note
1	273°C	syndiotactic, estimated [53]
2	155–300°C	polymerisation temp. 125–80° [14,17,21,23]

Glass-Transition Temperature:

No.	Value	Note
1	81°C	can be increased to 98° on increasing syndiotactic content [58]
2	68–110°C	polymerisation temp. 125–-60° [21,23,25,26]
3	81°C	approx., pure commercial resins [23,27]

Transition Temperature General: Heat distortion temp. increases with increasing syndiotactic contact. [53] Gel melting (gel-sol transition) points have been reported for a wide range of PVC/organic solvents and PVC/dialkyl phthalate systems [49,50]; T_m increases with increasing size of phthalate [49]. Variation of T_g with polymerisation temp., [17,56] MW, [17] degree of crystallinity (syndiotactic content) [56,57] and thermal history (annealing) has been reported. [56,57,58] Partially syndiotactic materials have

– Poly(vinyl chloride), Unplasticised

higher T_g values than atactic samples; differences reduced after annealing. [56] Effect of pressure on T_g has been reported. [17] T_g displays an approx. linear increase with decreasing polymerisation temp. of resin. [21] T_m cannot be measured directly because of thermal instability of PVC resin; usual method is to determine from T_m^* of polymer-diluent (stabiliser) systems using Flory relationship; [14,21] increases linearly with decreasing polymerisation temp. of resin [17,21]

Deflection Temperature:

No.	Value	Note
1	75–80°C	[53]

Vicat Softening Point:

No.	Value	Note
1	70–80°C	[54]
2	100°C	syndiotactic

Transition Temperature:

No.	Value	Note
1	>165°C	min., critical processing temp. [53]
2	>210°C	min., syndiotactic, polymerisation at -40° [53]

Surface Properties & Solubility:

Cohesive Energy Density Solubility Parameters: Solubility parameters δ_h 44.35–46.44 $J^{1/2}$ cm$^{-3/2}$ (poor solvent); [63] δ_h 38.91–41.42 $J^{1/2}$ cm$^{-3/2}$ (moderate solvent); [63] 0 (strong solvent) [63] δ 19.19–22.1 $J^{1/2}$ cm$^{-3/2}$ (9.38–10.8 cal$^{1/2}$ cm$^{-3/2}$) [63,64]

Internal Pressure Heat Solution and Miscellaneous: Values for heats of solution (and dilution) in THF and cyclohexanone at 30° have been reported [3]

Plasticisers: Plasticisers used in PVC more than other polymer; typical amounts 50–70 phr; can be as high as 85 wt.%; often incorporated at high temps. [1,13]. [13,31] Primary plasticisers include phthalates, phosphates, sebacates and various fatty acid derivatives; polymeric species such as sebacates, adipates and azelates of propylene glycol. Phthalates used as general-purpose plasticisers, aliphatic esters for low-temp. applications, phosphates used in products for good flame-resistance. Most common are dioctyl (or 2-ethylhexyl) and diisooctyl phthalates, 'dialphanyl' (C_7–C_9) phthalate, trixylyl phosphate, dioctyl sebacate, polypropylene sebacate and epoxidised soya bean oil. Phthalates prepared from C_8–C_{10} alcohols are the most important class; approx. 75% of plasticisers used. Secondary plasticisers (limited compatibility) include esters of fatty acids, petroleum residues, alkyl or aryl hydrocarbons and nitrated or halogenated derivatives, always used in conjunction with primary plasticisers. Special-purpose plasticisers include trimellitates (high-temp. applications); epoxidised oils (light-stability); solid ethylene-vinyl acetate modified polymers (biodegradation resistance); diallyl phthalate (adhesives/metal finishing). Internal (permanent) plasticisation occurs when vinyl chloride is copolymerised with certain monomers, e.g. propylene and vinyl acetate, where the comonomers function as plasticisers; plasticiser cannot be lost or extracted [7]

Solvents/Non-solvents: High MW sol. THF, DMF, DMSO, nitrobenzene, methyl ethyl ketone, cyclopentanone, cyclohexanone, Me_2CO/CCl_4, C_6H_6, EtOAc/CCl_4, trichloroethylene/nitromethane. [13,44] Swollen by aromatic hydrocarbons and chlorinated hydrocarbons, nitroparaffins, Ac_2O, aniline, Me_2CO. [13,44] Insol. aliphatic hydrocarbons, esters, alcohols, carboxylic acids, carbon disulfide, vinyl chloride, H_2O, conc. alkalis, non-oxidising acids, oils. [13,44] Low MW sol. warm tetrahydrofurfuryl alcohol, Me_2CO, dioxane, ethylene dichloride, o-dichlorobenzene; diisopropyl ketone, mesityl oxide, isophorone, hexamethylphosphorustrisamide, warm toluene [13,44]

Wettability/Surface Energy and Interfacial Tension: Contact angles 75° (H_2O); 20° (methylene iodide); 0° (hexadecane) (bulk polymer, 25°). [1] Values increase on addition of plasticiser [4]

Surface Tension:

No.	Value	Note
1	419 mN/m	polarity 0.146, 20° [5,6]

Transport Properties:

Polymer Solutions Dilute: Mark-Houwink parameters have been reported for cyclohexanone, cyclopentanone and THF. [85] Second virial coefficients for cyclohexanone have been reported. [90] Unperturbed dimensions are reported for cyclohexane, THF and benzyl alcohol. [89,95,96] Values for Huggins constant, K_H (20–25°) in wide range of solvents have been reported. [82,88] Comprehensive studies of the thermodynamics of solns. in cyclohexanone, [83,89] cyclopentanone, [84] benzyl alcohol, [89] chlorobenzene [89] and THF, [85,89] have been reported

Polymer Melts: η_{inh} 0.51–1.4 ('research' grades) [1]. Relative viscosity 2.23–3.04 ('research' grades). [1] Dynamic viscosity studies have been reported: effects of temp., frequency and strain amplitude studied. [80] Rheological characterisation, using capilliary and parallel plate rheometers, has been reported; results obtained from oscillatory shear experiments show importance of thermomechanical history and time-independent viscoelastic response; capilliary experiments indicate importance of stabilisers and lubricants on flow behaviour. [81] Dependence of melt-flow activation energy on polymerisation temp., [100] MW and shear-rate has been reported [17,97,99]

Permeability of Gases: Other gas permeability and diffusivity values have been reported [65,66,67]

Water Content: Water absorption increases with increasing temp.; greater for emulsion than suspension polymerised material. [13] Moisture retention 8% (fibres, 60% relative humidity), [13] regain 0.1–0.15% [13]

Water Absorption:

No.	Value	Note
1	0.95 %	0.4 mm, 32d, 20° [13]
2	11.7 %	0.4 mm, 32d, 60° [13]
3	0.4–0.6 %	emulsion polymerised [13]
4	0.2–0.3 %	suspension polymerised [13]

Gas Permeability:

No.	Gas	Value	Note
1	He	131 cm^3 mm/(m^2 day atm)	1.5 × 10^{-13} cm^2 (s Pa)$^{-1}$, 25° [65,66]
2	H_2	113.8 cm^3 mm/(m^2 day atm)	1.3 × 10^{-13} cm^2 (s Pa)$^{-1}$, 25° [65,66]
3	Ne	25.4 cm^3 mm/(m^2 day atm)	2.9 × 10^{-14} cm^2 (s Pa)$^{-1}$, 25° [65,66]
4	N_2	0.78 cm^3 mm/(m^2 day atm)	8.9 × 10^{-16} cm^2 (s Pa)$^{-1}$, 25° [65,66]
5	O_2	2.98 cm^3 mm/(m^2 day atm)	3.4 × 10^{-15} cm^2 (s Pa)$^{-1}$, 25° [65,66]
6	CO_2	10.5 cm^3 mm/(m^2 day atm)	1.2 × 10^{-14} cm^2 (s Pa)$^{-1}$, 25° [65,66]
7	CH_4	1.84 cm^3 mm/(m^2 day atm)	2.1 × 10^{-15} cm^2 (s Pa)$^{-1}$, 25° [65,66]
8	H_2O	18034 cm^3 mm/(m^2 day atm)	2.06 × 10^{-11} cm^2 (s Pa)$^{-1}$, 25° [65,66]
9	NH_3	324 cm^3 mm/(m^2 day atm)	3.7 × 10^{-13} cm^2 (s Pa)$^{-1}$, 20° [67]
10	CO	1.66 cm^3 mm/(m^2 day atm)	1.99 × 10^{-15} cm^2 (s Pa)$^{-1}$, 27° [68]

– Poly(vinyl chloride), Unplasticised

Mechanical Properties:

Mechanical Properties General: Young's modulus and tensile strength fall with increasing temp. Fibre wet tensile strength is same as dry strength. [13] Mech. props. vary with MW, polymerisation temp. and additives

Tensile (Young's) Modulus:

No.	Value	Note
1	7584 MPa	-196° [17,75,94]
2	276 MPa	70° [17,75,94]
3	2745 MPa	280 kg f mm^{-2} [13]
4	4900 MPa	approx., 500 kg f mm^{-2}, filled fibres [13]
5	980 MPa	approx., 100 kg f mm^{-2}, fibres [13]

Tensile Strength Break:

No.	Value	Note	Elongation
1	19.6 MPa	2 kg f mm^{-2}, 60° [13]	
2	14.7 MPa	1.5 kg f mm^{-2}, 70° [13]	
3	41.6–54.9 MPa	approx., 4.25–5.6 kg f mm^{-2} [13]	5–25%
4	333 MPa	34 kg f mm^{-2}, filled fibres [13]	20%
5	147 MPa	15 kg f mm^{-2}, relaxed fibres [13]	180%

Flexural Strength at Break:

No.	Value	Note
1	93.2 MPa	9.5 kg f mm^{-2} [13]
2	83.4 MPa	8.5 kg f mm^{-2}, elastomer modified [13]

Poisson's Ratio:

No.	Value	Note
1	0.38	[29]

Compressive Strength:

No.	Value	Note
1	54.9–65.7 MPa	5.6–6.7 kg f mm^{-2} [13]

Viscoelastic Behaviour: Deformational behaviour, tensile creep curves and creep recovery have been reported. [75] Room temp. viscoelastic props. may vary with additives; at higher temp. thermal history becomes a factor [76]
Mechanical Properties Miscellaneous: Dynamic mech. props., elastic moduli, shear moduli and loss factor have been reported. [17,80,81,100] Deformation props. at high pressure have been reported, brittle to ductile transition at 10 MPa. [39] Elastic recovery of fibres 100% (1% extension, filled), 95% (2% extension), 85% (4% extension), 75% (9% extension) [13]
Hardness: Rockwell R110-R120. [13] Vickers diamond pyramid 14.3 (bulk polymer, 25°) [4]
Fracture Mechanical Properties: Factors affecting fatigue resistance in UPVC pipe compositions have been reported; effects of processing on crack propagation and lifetime of specimens have important practical implications both for designing and processing rigid polymers. [77] Effects of MW and presence of modifiers on fatigue, crack propagation response have been reported: addition of 6-14 phr MBS enhances resistance of PVC; increases significantly with increasing MW (at given level of modifier) [78]
Friction Abrasion and Resistance: Coefficient of friction of bulk polymer μ_{stat} (polymer/steel): 0.5 (as moulded, 25°); 0.39 (abraded, 25°); 0.46 (reheated, 25°); [4] μ_{dyn} (polymer/steel): 0.42 (as moulded, 25°); 0.34 (abraded, 25°); 0.37 (reheated, 25°). [4] Values decrease on addition of plasticisers [4]. μ_{stat} 0.4–0.45; [13] μ_{dyn} 0.23 (grey cotton yarn, running at 1 m s^{-1}, 65% relative humidity [13]

Izod Notch:

No.	Value	Notch	Note
1	42.5 J/m	Y	0.8 ft lb in^{-1} [13]
2	799 J/m	Y	max. ,15 ft lb in^{-1}, impact modified [13]

Electrical Properties:

Electrical Properties General: Behaves as typical polymer insulator; additives influence max. retention of insulation props. Where electrical props. are of importance, essential to use suspension or bulk polymer, not emulsion polymer. [13] Other values for dielectric constant and loss factor have been reported [79]

Dielectric Permittivity (Constant):

No.	Value	Frequency	Note
1	3	800 Hz	23° [13]
2	3.3	800 Hz	high impact, 23° [13]
3	3.5	60 Hz	25° [17]
4	3.39	1 kHz	25° [17]
5	3.29	10 kHz	25° [17]
6	11.76	60 Hz	140° [17]
7	11.27	1 kHz	140° [17]
8	10.94	10 kHz	140° [17]

Dielectric Strength:

No.	Value	Note
1	14–16 kV.mm^{-1}	[13]

Static Electrification: Has good resistance to tracking [13]
Dissipation (Power) Factor:

No.	Value	Frequency	Note
1	0.02	800 Hz	23° [13]

Optical Properties:

Optical Properties General: Specific refractive index increment values have been reported for dil. solns. in a wide range of solvents and plasticisers

Refractive Index:

No.	Value	Note
1	1.53–1.56	[13]
2	1.54	fibres [13]
3	1.54–1.55	[19]
4	1.54806	486 nm [15,17]
5	1.54151	589 nm [15,17]
6	1.53843	656 nm [15,17]

Transmission and Spectra: Ir, [30] H-1 nmr [36] and C-13 nmr [38] spectral data have been reported
Total Internal Reflection: Birefringence essentially zero (fibres). [13] Stress optical coefficient -6.5 (-200–50°); becomes positive at T_g [17,69,70]
Volume Properties/Surface Properties: Transparent, colourless; becomes translucent or opaque with appropriate additives

– Poly(vinyl chloride), Unplasticised

Polymer Stability:
Polymer Stability General: Decomposition may be caused by excessive shear, such as in resin processing
Thermal Stability General: Anomalous structs. within bulk polymer may promote thermal instability, e.g. tertiary or allylic chlorine. [71] Thermal degradation decreases mech. props.
Upper Use Temperature:

No.	Value	Note
1	60°C	[7]
2	90°C	modified [7]

Decomposition Details: Decomposition of PVC is a complex chemical mechanism. [1,13,31] Dehydrochlorination occurs above 300° up to 200°, leading to conjugated polymers above 300°, hydrocarbons also released, and above 400° benzene and other aromatics are detected. [8,9] Free radical mechanism postulated. [8] HCl liberated may autocatalyse decomposition, ionic complexes may also be involved [10] with recombination to double bonds. Decomposition in oxygen leads to deepening of colour, and oxygen may also be responsible for scission and cross-linking reactions. Stabilisers added to scavenge HCl. Under photodegradation oxidation may occur first, then HCl evolution
Flammability: Limiting oxygen index 45% (reduced by plasticisers). [1] Has self extinguishing props.; [7] burns with yellow flame, acidic odour, and softens and blackens. [7] Burning releases CO_2 and HCl, no monomer, phosgene or Cl_2 [102] contrary to popular opinion. [103] PVC dust not reported to be explosive [102]
Environmental Stress: Stable to Uv radiation in moderate climates for outdoor use; may be stabilised by Pb. Uv radiation may cause loss of plasticisers. [11] Residual monomer unconverted in polymerisation process is environmental hazard [102] as are heat stabilisers (Pb, cadmium salts) and low MW plasticisers [102]
Chemical Stability: Resistant to salt, oxidising agents, reducing agents, aliphatic hydrocarbons, detergents, oils, fats, petrol. Attacked by conc. oxidising agents, accelerated by metals. Unsuitable for use with chlorinated hydrocarbons, aromatics, ketones, esters, ethers. Attacked by bromine and fluorine at room temp. [7]
Hydrolytic Stability: Stable to water, dil. acid and conc. alkali, and to conc. acids below 60°. Hot acids e.g. nitric acid and sulfuric acid, may cause oxidation
Biological Stability: Generally resistant to microbial attack. Certain plasticisers and lubricants can be attacked by microbes, leading to stiffening and cracking. [11] Organotin species inhibit microbial attack
Recyclability: May be recycled and reformulated when material is free from other polymers; if not, then separation must take place. Used as fuel in incinerators [102] HCl gas is major decomposition product so must be removed from flue gases [102]
Stability Miscellaneous: γ-Radiation leads to cross-linking [11] with loss of chlorine and increase of mech. props., [52] shear and impact strength. Copolymers are degraded; colour deepens [52]

Applications/Commercial Products:
Processing & Manufacturing Routes: Vinyl chloride can be polymerised [12,13,31] by high-energy radiation, or in commercial practice by free-radical initiators; currently used initiators include peroxydicarbonates, *tert*-butylperpivalate, azobis(2,4-dimethylvaleronitrile) and acetyl cyclohexylsulfonyl peroxide; polymerisation can be effected by bulk (mass), suspension, emulsion or solution techniques, although latter rarely used; suspension and emulsion methods are most common heterogeneous polymerisation, as polymer is insol. in its monomer; low polymerisation temps. used when high MW material required. In bulk polymerisation, vinyl chloride, liquefied under pressure, is polymerised with initiators such as benzoyl peroxide or azobisisobutyronitrile; polymer is precipitated from the monomer; purest product obtained by this method; e.g. using benzoyl peroxide, 80% conversion achieved after 17h at 55°. Suspension polymerisation employs water-sol. suspension agents, e.g. poly(vinyl alcohol) or methylcellulose, and monomer-sol. initiators, e.g. benzoyl or lauroyl peroxide; polymerisation is effected by heating, and stirred to maintain suspension; product removed by filtration. Emulsion polymerisation is a similar process to that used for suspension technique, except that emulsifying agents are employed. Initiator system is initially contained in aq. phase, and polymer is obtained in very finely divided state, e.g. particle diameters $<10^{-3}$ mm, cf. approx. 0.1 mm diameter in suspension process; higher MW materials produced by emulsion process, and this coupled with greater surface area/weight ratios allows greater compatibility with plasticisers; more expensive process, however, and presence of residual emulsifying agents (which cannot be completely removed) leads to inferior water resistance and electrical props. In addition, 'highly crystalline' (20–25% crystalllinity) polymer with increased stereoregularity can be produced by using low temp. free-radical polymerisation techniques using γ-irradiation and active species such as alkyl boranes. Ziegler-Natta catalyst systems cannot be used, due to reaction with both vinyl chloride monomer and polymer. Cannot be processed successfully in the absence of stabilisers, severe discoloration and progressive deterioration result from oxidation and elimination of hydrogen chloride during heating at high temps.; in use, action of heat and light can have same effects. [12] Pre-mixing of polymer with stabilisers and other additives achieved by using dry-blending techniques. [31] Emulsion-, suspension- and bulk-polymerised materials are processed by using injection moulding, blow moulding, compression moulding, extrusion, vacuum- and thermo-forming, calendering, high-pressure laminating and coating techniques. [13] Films produced by solvent casting, calendering (emulsion polymer) or extrusion (latter being usual method). Fibres can be wet-spun from, e.g. THF or cyclohexanone, or dry-spun from mixed solvents. [13] Cellular polymer (foam) can be produced by incorporating suitable blowing agents, e.g. azodicarbonamide and azodiisobutyronitrile, into formulations [31]

Mould Shrinkage (%):

No.	Value	Note
1	0.2–0.4%	0.002–0.004 in/in or mm/mm [7]

Applications: Versatility and low cost leads to extensive use in a variety of applications where toughness, abrasion and chemical resistance, and self-extinguishing props. are required. Injection moulded components to pipe fittings for rainwater and irrigation projects (large market); housings for electrical equipment (e.g. televisions, computers); textile bobbins; electrical conduit fittings; bottles; junction boxes; door handles; telephone handsets; chair backs and seats; water filtration pressure tanks; ventilation grilles; corrugated roofing; general chemical engineering applications. Extruded components used as pipes for potable and non-potable water supplies, chemical plant and waste and soil systems (largest market); profiles for building industry, e.g. window and door frames (large market), shiplap wall cladding, fencing and bench-type seating; rainwater goods; curtain rails; computer housings; tube; wire; rod. Extruded films: in-stove packaging; thin (shrink-film) for foods, e.g. cooked meats, cheese and other produce; packing of toys, games, stationery, cosmetics, confectionery, pharmaceuticals; magnetic tapes. Blow moulded articles used in tubing; bottles and containers for consumable liquids (e.g. fruit squashes), liquids for household use (e.g. detergents, disinfectants, cosmetics and toiletries, pharmaceuticals). Calendered (rigid) sheets used in internal wall-cladding for chemical plant, food stores and tunnels; light fittings, reflectors and signs; fume ducting; tank linings; food processing equipment; general chemical engineering applications. Cellular forms: cushioning; mattresses, pillows etc.; battery separators; thermal insulation. Fibres used in garments (also used in blends with wool). Foils used in decorative signs and displays; refrigerator components; liners; trays; packaging; liners for concrete shuttering. Powders used as coatings for metal protection; pyrotechnics

Trade name	Details	Supplier
Afcodur	non-exclusive	Armosig
Afcoplast		Pechiney-St Gobain

Poly(vinyl chloride), Unplasticised

Name	Grade	Manufacturer
Afcovyls		Pechiney-St Gobain
Aismadur		Aismalibar S.A.
Aron	non-exclusive	Toa Gosei
Begra		Begra
Belcon	'modified' form	Maruto Chem.
Benedur		Benecke
Benelit	non-exclusive	J.H. Benecke
Benvic		Solvay
Boltaron		Gencorp
Boltaron	non-exclusive	General Tyre and Rubber Co.
Bonyl		N. Lundbergs Fabriks
Brandur		Hoehn
Breon	non-exclusive	BP Chemicals
Colorit		Gehr-Kunststoffwerk KG
Colorite		Unichem
Colorite		Gehr-Kunststoffwerk KG
Conticel	modified form	Continental Gummi-Werke
Conticell		Continental
Continol		Continental Gummi-Werke
Corvic		European Vinyls Corporation Ltd.
Crinovyl	fibres	Soc. Rhovyl
Darvic		ICI Chemicals and Polymers Ltd.
Densite	non-exclusive	Tenneco Chem.
Ekadur		VEB Elektrochem. Kombinat
Ekalon	non-exclusive	VEB Elektrochem. Kombinat
Ekavin		VEB Elektrochem. Kombinat
Ekavyl	non-exclusive	Ugine Kuhlmann SA Div. Plastiques
Ekavyl	SDF58, SD2, SK55, SK60, SK64, SK66, SL66	Ugine Kuhlmann SA Div. Plastiques
Ekazell		VEB Elektrochem. Kombinat
Envyloid		Daicel Chem.
Evilon		Dietzel
Exon	non-exclusive	Firestone
Exon	605, 6337, 640, 654, 9269, 9269A, 9290, 965	Firestone
FPC	non-exclusive	Hooker
Gedevyl		Cd F Chemie
Geon		BF Goodrich Chemical
Goelzalit		VEB Goelzaplast
Gorlex	non-exclusive	Polymer Gorla
Grabokid	71C	Kunstlederfabrik
Grabolak	RV71, 03, 04, 40, 41	Gyoer/Raab
Grabona	C402, C472	Gyoer/Raab
Gracon		W R Grace
Grancon		W R Grace
Hostalit	homopolymer; modified	Hoechst Celanese
Howeflex		Hornschurch
Howesol		Hornschurch
Howetex		Hornschurch
Igelit		BASF
Isovyl	fibres	Rhoyl
Iztavil		Polimeros De Mexico
Jushi Glass		Mitsubishi Chemical
Kaneace KVC 745		Kanegafuchi Chemical Ind.
Kanevinyl	non-exclusive	Kanegafuchi Chemical Ind.
Koonkote		Plastic Coatings
Krene	non-exclusive	Union Carbide
Lacovyl		Alpha
Lacovyl		Elf Atochem (UK) Ltd.
Lacqvyl		ATO Chimie
Laistell		Lavorazione Materie Plastiche
Leavil		Montecatini Edison SpA
Leavil	fibres	Montecatini Edison SpA
Leavin		Montecatini Edison SpA
Leben		Day Nippon Togyo. Co.
Longlass		Kawaguchi Rubber
Lonplate		Kawaguchi Rubber
Lucalor		Atochem Polymers
Lucoflex		Armosig
Lucolene		Pechiney-St Gobain
Lucolene	non-exclusive; other PVC-based materials	Rhone-Poulenc Polymers
Lucorex		Elf Atochem (UK) Ltd.
Lucovyl	non-exclusive	Rhone-Poulenc Polymers
Lucovyl		Pechiney-St Gobain
Lutofan	non-exclusive	BASF
Marvinol	non-exclusive	Uniroyal
Marvinol		Naugatuch Chemical Div.
Marvylflo		LVM
Mipofolie	non-exclusive	Alfred Schwarz GmbH
Movil		Montecatini Edison SpA.

Poly(vinyl chloride), Unplasticised

Movinyl 100		Montecatini Edison SpA.
Nipolit	non-exclusive	Chisso
Norvinyl	non-exclusive	Norsk Hydro
Novablend		Novatec
Novitex		Benecke
Opalon	non-exclusive	Monsanto Chemical
Opalon		Monsanto Chemical
Organit	"normal" and modified forms	Unitecta Oberflaechenschutz
Oriex		Tenneco Chem.
Pekema		Punda
Pekevic		Neste
Pevikon	non-exclusive	Norsk Hydro
Ravinil	non-exclusive	Anic
Resilon		Canadian General-Tower
Rhovyl	fibres	Soc. Rhovyl
Rovil		Soc. Rhovyl
Ryuron	non-exclusive	Tekkosha
Scon		Vinatex
Selectophore	High MW available; non-exclusive	Fluka
Sicovinil	non-exclusive	Mazzucchelli Celuloide SpA
Sicron		Montedison
Sicron	non-exclusive	Montecatini Edison SpA.
Solvic		Solvay
Somvyl		Deutsche Sommer
Sonwood		Sonesson
Soplasco		Southern Plastics Co., USA
Standard samples	Carboxylated (1.8% carboxyl content)	Aldrich
Standard samples		Fluka
Standard samples		Aldrich
Telcovin		Telcon Plastics Ltd
TempRite		BF Goodrich Chemical
Teviron		Teijin Chem. Ltd.
Thermovyl	heat-stable fibres	Rhone-Poulenc Textiles
Trosiplast	non-exclusive	Hüls AG
Trovidur	non-exclusive	Dynamit Nobel
Trovilon		Dynamit Nobel
Trovitherm		Dynamit Nobel
Trulon		Olin
Ultron	non-exclusive	Monsanto Chemical
Ultryl 6800		Philips Petroleum International
Varian		DSM
Varlan		DSM
Velkor	non-exclusive	Alkor Plastics (UK) Ltd.
Vestolit		Hüls AG
Vestolit M		Hüls AG
Vestolit S		Hüls AG
Vinatex	non-exclusive	Vinatex
Vinex		Air Products
Vinidur	non-exclusive	BASF
Vinika	non-exclusive	Mitsubishi Monsanto Chem. Co
Vinilen	non-exclusive (other PVC-based materials)	Vinisa
Vinnol	non-exclusive (other PVC-based materials)	Wacker-Chemie
Vinoflex	non-exclusive	BASF
Vinopac		Rhone Poulenc Inc.
Vinychlon		Mitsui Toatsu Chemicals
Vinyclair		Rhone-Poulenc
Vinydur		Saplast SA
Vinyfoil		Mitsubishi Plastics Ltd.
Vinylaz		Soc. Belge de L'Azote
Vinylcel		Johns-Mauville
Vinylite	non-exclusive	Carbide and Carbon Chem.
Vinylon 39	non-exclusive	Synthetic Fiber
Vipla		European Vinyls Corporation Ltd.
Viplast	non-exclusive	Montedison
Viplast		Montecatini Edison SpA.
Vipolit		Lonza-Folien GmbH.
Vista		Vista
Vitacel		British Vita Co., UK
Vitalpol		Vitafoam
Volkaril	modified	Plastimer
Vybak	non-exclusive	BIP
Vynella	non-exclusive	Adretta-Werke
Vynoid	non-exclusive	Plastic Coatings
Wehodur		W. & H. Chem.
Welvic	non-exclusive	ICI Chemicals and Polymers Ltd.

Bibliographic References

[1] Alger, M.S.M., *Polymer Science Dictionary*, Elsevier Applied Science, 1989, 386, (rev)
[2] Cotman, J.D., *Ann. N. Y. Acad. Sci.*, 1953, **57**, 417, (morphology, tacticity)
[3] Maron, S.H. and Filisko, F.E., *J. Macromol. Sci., Phys.*, 1972, **6**, 413, (solubility, props)
[4] Bowers, R.C., Jarvis, N.L. and Zisman, W.A., *Ind. Eng. Chem. Proc. Des. Dev.*, 1965, **4**, 86, (mechanical props, surface props)
[5] Wu, S., *J. Polym. Sci., Part C: Polym. Lett.*, 1971, **34**, 19, (surface props)
[6] Wu, S., *J. Adhes.*, 1973, **5**, 39, (surface props)

[7] Whelan, A., *Polymer Technology Dictionary*, Chapman and Hall, 1994, 207, (flammability, chemical stability, processing, rev)
[8] Stromberg, R.R., Straus, S. and Achhammer, B.G., *J. Polym. Sci.*, 1959, **35**, 355, (thermal props, stability)
[9] Ilda, T., Nakanishi, M. and Gotō, K., *J. Polym. Sci., Polym. Chem. Ed.*, 1974, **12**, 737, (thermal props, stability)
[10] Guyot, A. and Bert, M., Polym. Prepr. (Am. Chem. Soc., *Div. Polym. Chem.*), 1971, **12**, 303, (thermal studies)
[11] Matthews, G., *Vinyl Allied Polym.*, (ed. P.D. Ritchie), Iliffe Books, 1972, **2**, 299, (biol. stability)
[12] *Polymer Yearbook - 5*, (ed. R.A. Pethrich), Harwood Academic, 1989, 4, (volumetric props, synth, uses)
[13] Roff, W.J. and Scott, J.R., Fibres, Films, *Plastics and Rubbers*, Butterworths, 1971, 108, (rev)
[14] Nakajima, A., Hamada, H. and Hayashi, S., *Makromol. Chem.*, 1966, **95**, 40, (volumetric props)
[15] Dannis, M.L., *Modern Plast. Intern.*, 1954, **31**(7), 120, (volumetric props)
[16] Ho, C.Y., Desai, P.D., Wy, K.Y., Havill, T.N. and Lee, T.Y., *NBS Report No. NBS-GCR-77-83*, 1977, (thermal props)
[17] Collins, E.A., Daniels, C.A. and Witenhafer, D.E., *Polym. Handb.*, 4th edn., (eds. J. Brandrup, E.H. Immergut and E.A. Grulke), John Wiley and Sons, 1999, **V**, 72, (rev)
[18] Sheldon, R.P. and Lane, K., *Polymer*, 1965, **6**, 77, (thermal props)
[19] Seferis, J.C., *Polym. Handb.*, 4th edn., (eds. J. Brandrup, E.H. Immergut and E.A. Grulke), John Wiley and Sons, 1999, **VI**, 576, (optical props)
[20] Eiermann, K., *Kunststoffe*, 1961, **51**, 512, (thermal conductivity)
[21] *Kirk-Othmer Encycl. Chem. Technol.*, Vol. 23, 3rd edn., (eds. M. Grayson and D. Eckroth), Wiley-Interscience, 1983, 896, (thermal props)
[22] Illers, K.-H., *Makromol. Chem.*, 1969, **127**, 1, (volumetric props)
[23] Reding, F.P., Walter, E.R. and Welch, F.J., *J. Polym. Sci.*, 1962, **56**, 225, (thermal props)
[24] Lebedev, B.V., Rabinovich, I.B. and Budarina, V.A., Vysokomol. Soedin., Ser. A, 1967, **9**, 488, (volumetric props)
[25] Pezzin, G., *Plast. Polym.*, 1969, **37**, 295, (thermal props)
[26] Garbuglio, C., Rodella, A., Borsini, G.C. and Gallinella, E., *Chim. Ind. (Milan)*, 1964, **46**, 166, (thermal props, morphology, tacticity)
[27] Keavney, J.J. and Eberlin, E.C., *J. Appl. Polym. Sci.*, 1960, **3**, 47, (thermal props)
[28] Natta, G. and Corradini, P., *J. Polym. Sci.*, 1956, **20**, 251, (volumetric props)
[29] Raghava, R., Caddell, R.M. and Yeh, G.S.Y., *J. Mater. Sci.*, 1973, **8**, 225, (mechanical props)
[30] Krimm, S., Folt, V.L., Shipman, J.J. and Berens, A.R., J. Polym. Sci., Part A: Polym. Chem., 1963, **1**, 2621, (ir)
[31] Brydson, J.A., *Plast. Mater.*, 6th edn., Butterworth-Heinemann, 1995, 298, (rev)
[32] Burnett, G.M. and Wright, W.W., *Proc. R. Soc. London A*, 1954, **222**, 41, (thermal props)
[33] Sinko, G.C. and Stull, D.R., *J. Phys. Chem.*, 1958, **62**, 397, (thermal props)
[34] Joshi, R.M., *Indian J. Chem.*, 1964, **2**, 125, (thermal props)
[35] Joshi, R.M. and Zwolinski, B.J., J. Polym. Sci., *Part B: Polym. Lett.*, 1965, **3**, 779, (thermal props)
[36] Heatley, F. and Bovey, F.A., *Macromolecules*, 1969, **2**, 241, (H-1 nmr, tacticity)
[37] Pham, Q.T., Millan, J.-L. and Madruga, E.L., *Makromol. Chem.*, 1974, **175**, 945, (C-13 nmr, tacticity)
[38] Carman, C.J., *Macromolecules*, 1973, **6**, 725, (C-13 nmr, tacticity)
[39] Yuan, J., Hiltner, A. and Baer, E., *J. Mater. Sci.*, 1983, **18**, 3063, (mechanical props)
[40] Wunderlich, B. and Baur, H., *Adv. Polym. Sci.*, 1970, **7**, 335, (volumetric props)
[41] White, D.M., *J. Am. Chem. Soc.*, 1960, **82**, 5678, (synth, morphology)
[42] Zoller, P., *Polym. Handb.*, 3rd edn., (eds J. Brandrup and E.H. Immergut), Wiley-Interscience, 1989, **VI**, 475, (volumetric props)
[43] Simha, R., Wilson, P.S. and Olabisi, O., *Kolloid Z. Z. Polym.*, 1973, **251**, 402, (volumetric props)
[44] Fuchs, O., *Polym. Handb.*, 3rd edn., (eds. J. Brandrup and E.H. Immergut), Wiley-Interscience, 1989, **VII**, 384, (solubility)
[45] Harrison, M.A., Morgan, P.H. and Park, G.S., *Eur. Polym. J.*, 1972, **8**, 1361, (gelation)
[46] Yang, Y.C. and Geil, P.H., *J. Macromol. Sci., Phys.*, 1983, **22**, 463, (gelation)
[47] Guerrero, S.J. and Keller, A., *J. Macromol. Sci., Phys.*, 1981, **20**, 167, (gelation)
[48] Guerrero, S.J., Keller, A., Soni, P.L. and Geil, P.H., J. Polym. Sci., *Polym. Phys. Ed.*, 1980, **18**, 1533, (gelation)
[49] Leharne, S.A. and Park, G.S., *Eur. Polym. J.*, 1985, **21**, 383, (gelation)
[50] Hiltner, A., *Polym. Handb.*, 4th edn., (eds. J. Brandrup, E.H. Immergut and E.A. Grulke), John Wiley and Sons, 1999, **VII**, 767, (gelation)
[51] McKenna, T.F. and Santos, A.M., *Polym. Handb.*, 4th edn., (eds. J. Brandrup, E.H. Immergut and E.A. Grulke), John Wiley and Sons, 1999, **II**, 420, (thermal props)
[52] Byrne, J., Costikyan, T.W., Handford, C.B., Johnson, D.L. and Mann, W.L., *Ind. Eng. Chem.*, 1953, **45**, 2549, (radiation stability)
[53] Bockman, O.C., *Br. Plast.*, 1965, **38**, 364, (morphology, thermal props)
[54] *Sci. Technol. Polym. Films*, (ed. O.J. Sweeting), Wiley, 1971, 307, (thermal props)
[55] Fuller, C.S., *Chem. Rev.*, 1940, **26**, 143, (morphology)
[56] Ceccorulli, G., Pizzoli, M. and Pezzin, G., *J. Macromol. Sci., Phys.*, 1977, **14**, 499, (thermal props)
[57] Singh, P. and Lyngaae-Jørgensen, J., *J. Macromol. Sci., Phys.*, 1981, **19**, 177, (thermal props)
[58] Andrews, R.J. and Grulke, E.A., *Polym. Handb.*, 4th edn., (eds. J. Brandrup, E.H. Immergut and E.A. Grulke), John Wiley and Sons, 1999, **VI**, 193, (thermal props)
[59] Gaur, U., Wunderlich, B.B. and Wunderlich, B., *J. Phys. Chem. Ref. Data*, 1983, **12**, 29, (volumetric props, thermodynamic props)
[60] Griskey, R.G. and Waldman, N., *Mod. Plast.*, 1966, **43**(7), 119, (thermodynamic props)
[61] Grewer, T. and Wilski, H., *Kolloid Z. Z. Polym.*, 1968, **226**, 46, (volumetric props)
[62] Chang, S.-S., *J. Res. Natl. Bur. Stand. (U.S.)*, 1977, **82**, 9, (volumetric props)
[63] Grulke, E.A., *Polym. Handb.*, 3rd edn., (eds. J. Brandrup and E.H. Immergut), Wiley-Interscience, 1989, **VII**, 549, 554, (transport props, solubility props)
[64] Di Benedetto, A.T., J. Polym. Sci., *Part A: Polym. Chem.*, 1963, **1**, 3459, (solubility props)
[65] Pauly, S., *Polym. Handb.*, 4th edn., (eds. J. Brandrup, E.H. Immergut and E.A. Grulke), John Wiley and Sons, 1999, **VI**, 550, (transport props)
[66] Tikhomirov, B.P., Hopfenberg, H.B., Stannett, V.T. and Williams, J.L., *Makromol. Chem.*, 1968, **118**, 177, (transport props)
[67] Braunisch, V.L. and Lenhart, H., *Kolloidn. Zh.*, 1961, **177**, 24, (transport props)
[68] Sefcik, M.D., Schaefer, J., May, F.L., Racher, D. and Dub, S.M., J. Polym. Sci., *Polym. Phys. Ed.*, 1983, **21**, 1041, (transport props)
[69] Andrews, R.D. and Chatre, V., *J. Appl. Phys.*, 1969, **40**, 4266, (optical props, mechanical props)
[70] Andrews, R.D. and Kazama, Y., *J. Appl. Phys.*, 1968, **39**, 4891, (optical props, mechanical props)
[71] Hjertberg, T. and Sorvik, E.M., *ACS Symp. Ser.*, 1985, **280**, 259, (morphology, thermal stability)
[72] Starnes, W.H., Schilling, F.C., Plitz, I.M., Cais, R.E., et al, *Macromolecules*, 1983, **16**, 790, (morphology, polymerisation mechanism)
[73] Hennig, J., *Kolloid Z. Z. Polym.*, 1965, **202**, 127, (volumetric props)
[74] Heydemann, P. and Guicking, H.D., *Kolloid Z. Z. Polym.*, 1963, **193**, 16, (volumetric props)
[75] *Engineering Properties of Thermoplastics*, (ed. R.M. Ogorkiewicz), Wiley-Interscience, 1970, 247, (mechanical props, optical props)
[76] Turner, S., *Br. Plast.*, 1964, **37**, 682, (mechanical props)
[77] Gotham, K.V. and Hitch, M.J., *Br. Polym. J.*, 1978, **10**, 47, (mechanical props)
[78] Skibo, M.D., Manson, J.A., Webler, S.M., Hertzberg, R.W. and Collins, E.A., *ACS Symp. Ser.*, 1979, **95**, 311, (mechanical props)
[79] Utracki, L.A. and Jukes, J.A., *J. Vinyl Technol.*, 1984, **6**, 85, (electrical props)
[80] Bonnebat, C. and De Vries, A.J., *Polym. Eng. Sci.*, 1978, **18**, 824, (transport props, mechanical props)
[81] Rangel-Nafaile, C., Garcia Leon, A., *J. Vinyl Technol.*, 1984, **6**, 63, (transport props, mechanical props)
[82] Utracki, L.A., *Polym. J. (Tokyo)*, 1972, **3**, 551, (viscosity)
[83] Maron, S.H. and Lee, M.-S., *J. Macromol. Sci., Phys.*, 1973, **7**, 29, (viscosity)
[84] Maron, S.H. and Lee, M.-S., *J. Macromol. Sci., Phys.*, 1973, **7**, 47, (viscosity)
[85] Maron, S.H. and Lee, M.S., *J. Macromol. Sci., Phys.*, 1973, **7**, 61, (viscosity)
[86] Berens, A.R. and Hopfenberg H.B., *J. Membr. Sci.*, 1982, **10**, 283, (diffusion)
[87] Michielsen, S., *Polym. Handb.*, 4th edn., (eds. J. Brandrup, E.H. Immergut and E.A. Grulke), John Wiley and Sons, 1999, **VII**, 590, (optical props)
[88] Schoff, C.K., *Polym. Handb.*, 4th edn., (eds. J. Brandrup, E.H. Immergut and E.A. Grulke), John Wiley and Sons, 1999, **VII**, 275, (viscosity)
[89] Kurata, M. and Tsunashima, Y., *Polym. Handb.*, 4th edn., (eds. J. Brandrup, E.H. Immergut and E.A. Grulke), John Wiley and Sons, 1999, **VII**, 18, 53, (viscosity, soln props)
[90] Lechner, M.D., Nordmeier, L. and Steinmeier, D.G., *Polym. Handb.*, 4th edn., (eds. J. Brandrup, E.H. Immergut and E.A. Grulke), John Wiley and Sons, 1999, **VII**, 178, (soln props)

[91] Natta, G., Bassi, I.W. and Corradini, P., *Rend. Accad. Naz. Lincei.*, 1961, **31**(**1**,2), 1, (morphology)
[92] Nakajima, A. and Hayashi, S., *Kolloid Z. Z. Polym.*, 1969, **229**, 12, (morphology)
[93] Wilkes, C.E., Folt, V.L. and Krimm, S., *Macromolecules*, 1973, **6**, 235, (morphology)
[94] Dyment, J. and Ziebland, H., *J. Appl. Chem.*, 1958, **8**, 203, (mechanical props)
[95] Utracki, L.A., Bakerdjian, Z. and Kamal, M.R., *Trans. Soc. Rheol.*, 1975, **19**, 173, (transport props)
[96] Utracki, L.A., *Polym. Eng. Sci.*, 1974, **14**, 308, (transport props)
[97] Collins, E.A., *Pure Appl. Chem.*, 1977, **49**, 581, (transport props)
[98] Lyngaae-Jørgensen, J., *J. Appl. Polym. Sci.*, 1976, **20**, 2497, (transport props)
[99] Utracki, L.A., *J. Polym. Sci., Polym. Phys. Ed.*, 1974, **12**, 563, (transport props)
[100] Utracki, L.A., *J. Vinyl Technol.*, 1985, **7**, 150, (transport props, mechanical props)
[101] Haslam, J. and Hall, J.I., *Analyst (London)*, 1958, **83**, 196, (anal)
[102] *Encycl. Polym. Sci. Eng.: Supplement Volume*, (eds. H.F. Mark, N. Bikales, C.G. Overberger and G. Menges), 2nd edn., Wiley-Interscience, 1989, 884, (stability, toxicity, flammability, recyclability, haz)
[103] Hirschler, M., *Fire Prevention*, 1987, **204**, 19, (flammability, haz)

Poly(vinyl chloride-co-vinyl acetate)   P-584

$$\left[\left[CH_2CH\right]_x \left[CH_2CH\right]_y\right]_n$$
$$\qquad\ \ |\qquad\qquad\ |$$
$$\qquad\ Cl\qquad\quad\ OCCH_3$$
$$\qquad\qquad\qquad\quad\ \|$$
$$\qquad\qquad\qquad\quad\ O$$

Synonyms: *Poly(ethenyl acetate-co-chloroethene). Poly(chloroethene-co-ethenyl acetate). VC/VAC. VC/VA. PVCA. Poly(vinyl chloride-co-ethenyl acetate)). PVC/PVAc*
Related Polymers: PVC copolymers, Poly(vinyl chloride), unplasticised, Poly(vinyl chloride), plasticised, Epoxidised poly(vinyl chloride-co-vinyl acetate), Carboxylated poly(vinyl chloride-co-vinyl acetate), Hydroxylated poly(vinyl chloride-co-vinyl acetate), Poly(vinyl chloride-co-vinyl acetate-co-2-hydroxypropyl acrylate), Poly(vinyl chloride-co-vinyl acetate-co-maleic acid), Poly(vinyl chloride-co-vinyl acetate-co-vinyl alcohol), Poly(vinyl chloride-co-ethylene-co-vinyl acetate)
Monomers: Vinyl chloride, Vinyl acetate
Material class: Gums and resins, Copolymers, Thermoplastic
Polymer Type: PVC, polyvinyl acetatepolyhaloolefins
CAS Number:

CAS Reg. No.
9003-22-9
59740-43-1
25034-86-0

Molecular Formula: $[(C_2H_3Cl).(C_4H_6O_2)]_n$
Molecular Weight: M_n 12000–44000
Additives: Polymers, stabilisers, plasticisers, extenders, lubricants, fillers (e.g. carbon black), pigments, processing aids, impact modifiers, fire retardants, blowing agents. Stabilisers include solid calcium/zinc complex, barium/cadmium/zinc complex, epoxidised soybean oil. Lubricants include stearic acid
General Information: The most important vinyl chloride copolymer; typically contains 20 wt% vinyl acetate; it is softer and more easy flowing than the homopolymer; can be processed at lower melt temps. Useful range: 5–20% vinyl acetate content although can be as high as 40%; 48.4% Cl for copolymer containing 15% vinyl acetate (cf. 56.7% Cl in PVC). Variation in vinyl acetate content and degree of polymerisation leads to range of copolymers for which physical properties such as tensile strength, elongation, impact strength, and solubility may be varied over wide range
Tacticity: Presence of vinyl acetate comonomer interrupts regular polymer sequences in PVC homopolymer; syndiotacticity reduced by increase in vinyl acetate content (from ir measurements); degree of crystallinity reduced in a similar way; interference with crystallisability of polymer caused by both large size of acetate group (cf. Cl) as well as interruption of syndiotactic sequences; ΔH(syndio-iso) approx. 0.3 kcal mol^{-1} [7]
Morphology: Typically head-to-tail linkages; molecules contain slight degree of branching; estimated as >1.5 branches per 100 monomer units; higher than PVC homopolymer. [12] Crystallinity reduced with increasing vinyl acetate content (from transmission electron microscopy and electron diffraction techniques): e.g. non-uniform structure, i.e. large crystallinites plus amorph. regions for copolymer containing 3 wt% acetate, small crystallites in amorph. matrix for material with 10 wt% acetate, while copolymer with 17 wt% acetate shows only amorph. structure [28]; essentially amorph. even when stretched [18]
Identification: Vinyl acetate content determined by hydrolysis of acetate groups and back-titration (ISO 1159). [33]
Miscellaneous: Copolymers are colourless, water- and flame-resistant, and flexible (as thin films); in addition, are non-toxic, odourless and tasteless, and thus find uses in food-contact applications. [13,37] Initial attempts were made to combine properties of poly(vinyl chloride) and poly(vinyl acetate) by blending, but they show poor compatibility and no worthwhile properties. [37] Often copolymerised in the presence of a third comonomer to give range of terpolymers poly(vinyl chloride-co-vinyl acetate-co-maleic acid) and poly(vinyl chloride-co-vinyl acetate-co-vinyl alcohol); find major uses in coating applications. US Food and Drug Administration have published details of maximum levels of vinyl chloride monomer allowed in food-contact applications; established under 'The Federal Food, Drug and Cosmetic Act'; levels of 5–50 ppb (based on vinyl chloride component weight of material), depending on polymer and specific use [35]

Volumetric & Calorimetric Properties:
Density:

No.	Value	Note
1	d 1.36 g/cm^3	10% vinyl acetate [18]
2	d 1.39 g/cm^3	15% vinyl acetate [19]
3	d 1.34–1.36 g/cm^3	fibres [13]
4	d 1.34–1.45 g/cm^3	high vinyl acetate [37]

Thermal Expansion Coefficient:

No.	Value	Note	Type
1	7×10^{-5} K^{-1}	[13]	L

Volumetric Properties General: C_p/C_v 0.2–0.5 [13]
Thermal Conductivity:

No.	Value	Note
1	0.146–0.167 W/mK	0.00035–0.0004 cal (cm s °C)$^{-1}$ [13]

Specific Heat Capacity:

No.	Value	Note	Type
1	0.96 kJ/kg.C	0.23 cal (g °C)$^{-1}$ [37]	P

Glass-Transition Temperature:

No.	Value	Note
1	63°C	5% vinyl acetate [1]
2	59°C	10% vinyl acetate [1]
3	68°C	12% vinyl acetate; M_n 12000 [18]

Poly(vinyl chloride-co-vinyl acetate)

No.	Value	Note
4	72°C	14% vinyl acetate; M_n 22000–27000 [18]
5	72°C	15% vinyl acetate; M_n 26000 [19]
6	66.3°C	DSC [29]
7	71.7°C	TMA [29]
8	72.3°C	3 wt% acetate [29]
9	66.2°C	calc., DSC [29]
10	66.9°C	TMA [29]
11	67.9°C	calc., 10 wt% acetate [29]
12	66.1°C	DSC [21]
13	60.1°C	TMA [29]
14	65.1°C	calc., 17 wt% acetate [29]

Transition Temperature General: T_g of copolymers is an approximately linear function of composition (approx. 80°, PVC homopolymer - approx. 30°, PVAc homopolymer). [26] Plasticised material flexible to very low temp., -50°; unplasticised material becomes rubbery in boiling water, and fluid at 125–150°; fibres shrink at 60–65°; becomes tacky at 85°; fluid at 135–150° [13]

Deflection Temperature:

No.	Value	Note
1	60–65°C	[37]

Transition Temperature:

No.	Value	Note
1	70–125°C	softening temp. [37]

Surface Properties & Solubility:

Solubility Properties General: Solubility characteristics and coating properties (cf. applications as coatings) have been reviewed. [14] Copolymer much more soluble in wider range of solvents and solvent mixtures than PVC homopolymer

Cohesive Energy Density Solubility Parameters: Solubility parameter δ 43.51–44.35 $J^{1/2}$ cm$^{-3/2}$ (13 wt% vinyl acetate) [6,10,36]; 41.76 $J^{1/2}$ cm$^{-3/2}$ (14 wt% vinyl acetate) [36] δ_h 38.91–46.44 $J^{1/2}$ cm$^{-3/2}$ (14 wt%vinyl acetate) [36] 32.64–55.65 $J^{1/2}$ cm$^{-3/2}$ (14 wt% vinyl acetate, moderate solvent); [36] 0 $J^{1/2}$ cm$^{-3/2}$ (14 wt% vinyl acetate, strong solvent); 38.91–46.44 $J^{1/2}$ cm$^{-3/2}$ (12 wt% vinyl acetate, poor solvent); [36] 32.64–55.23 $J^{1/2}$ cm$^{-3/2}$ (12 wt% vinyl acetate, moderate solvent); [36] 0 $J^{1/2}$ cm$^{-3/2}$ (12 wt% vinyl acetate, strong solvent) [36]

Plasticisers: Vinyl acetate comonomer acts as an internal (permanent) plasticiser, it cannot be lost or extracted, and will not migrate; [1,11] comonomer has softening effect on copolymerisation with vinyl chloride; internal plasticisation means that less additional plasticiser is needed than for pure PVC. Other plasticisers include dioctyl and dibutyl phthalates, dibutyl cellosolve phthalate, cellosolve ricinoleates, glyceryl esters, tricresyl phosphate, camphor and triacetin [37]

Solvents/Non-solvents: Sol. THF, cyclohexanone, Me_2CO, methyl ethyl ketone, methyl isobutyl ketone, benzene, toluene, methylene chloride, esters (general). Insol. alcohols, hydrocarbons, Me_2CO (<5 wt% vinyl acetate) [13]

Transport Properties:

Transport Properties General: Viscosity is dependent on vinyl acetate content, concentration of soln., plus presence of other additives, e.g. plasticisers

Polymer Solutions Dilute: Viscosity 200–5000 cP (25% soln) [15]. Solution viscosity (15% solids in methyl ethyl ketone/toluene (1/1), 25°) 16 cP ("medium/low" MW, 12–13 wt% vinyl acetate); [33] 36 cP ("medium" MW, 13–16 wt% vinyl acetate); [33] 57 cP ("medium" MW, 14–15 wt% vinyl acetate); [33] 82 cP ("medium/high" MW, approx. 10 wt% vinyl acetate); [33] 175 cP (M_n 22000, 14 wt% vinyl acetate, 25% solids, methyl ethyl ketone/toluene (1/2); 200 cP (MW 27000, 14 wt% vinyl acetate, 20% solids, methyl ethyl ketone/toluene (1/1); 250 cP (MW 44000, 10 wt% vinyl acetate, 15% solids, methyl ethyl ketone/toluene (2/1); 350 cP (MW 12000, 12 wt% vinyl acetate, 30% solids, methyl ethyl ketone/toluene (1/2) [20]

Polymer Melts: η_{inh} 0.28 (M_n 12000, 12 wt% vinyl acetate); [18] 0.4 (M_n 22000, 14 wt% vinyl acetate); [18] 0.5 (M_n 27000, 14 wt% vinyl acetate) [18]

Permeability of Gases: Diffusion coefficients for a range of gases have been reported [37]

Permeability of Liquids: Water absorption as a function of vinyl acetate content has been reported, related to morphology of copolymers; absorption increases with increasing acetate content (cf. reduction in crystallinity) [28]

Permeability and Diffusion General: Permeability of O_2 and iron salts through membranes has been studied; permeability reduced when TiO_2 used as filler; found that mechanical degradation increases, hence anticorrosive props. diminish use. [22]

Water Content: Water retention of fibres approx. 12%; [13] moisture regain of fibres, approx. 0.1% (65% relative humidity) [13]

Water Absorption:

No.	Value	Note
1	0.23 %	50h, 3 wt% vinyl acetate [28]
2	0.27 %	50h, 10 wt% vinyl acetate [28]
3	0.3 %	50h, 17 wt% vinyl acetate [28]
4	0.07–0.1 %	24h, 25°, 1/8" thick [37]

Gas Permeability:

No.	Gas	Value	Note
1	O_2	51.2 cm^3 mm/(m^2 day atm)	7.8×10^{-11} cm^2 (s cm Hg)$^{-1}$, 25°, 75 wt% vinyl acetate [37]
2	O_2	23.6 cm^3 mm/(m^2 day atm)	3.6×10^{-11} cm^2 (s cm Hg)$^{-1}$, 25°, 50 wt% vinyl acetate [37]
3	O_2	7.87 cm^3 mm/(m^2 day atm)	1.2×10^{-11} cm^2 (s cm Hg)$^{-1}$, 25°, 20 wt% vinyl acetate [37]
4	He	55.8 cm^3 mm/(m^2 day atm)	8.5×10^{-11} cm^2 (s cmHg)$^{-1}$, 25°, 40 wt% vinyl acetate [37]
5	H_2	60.4 cm^3 mm/(m^2 day atm)	9.2×10^{-11} cm^2 (s cmHg)$^{-1}$, 25°, 40 wt% vinyl acetate [37]
6	N_2	3.94 cm^3 mm/(m^2 day atm)	6×10^{-12} (s cmHg)$^{-1}$, 25°, 40 wt% vinyl acetate [37]
7	CO_2	105 cm^3 mm/(m^2 day atm)	1.06×10^{-10} cm^2 (s cmHg)$^{-1}$, 25°, 40 wt% vinyl acetate [37]

Mechanical Properties:

Mechanical Properties General: Wet strength of fibres is almost the same as air-dry strength [13]

Tensile (Young's) Modulus:

No.	Value	Note
1	2353–4168 MPa	240–425 kgf mm^{-2}, plastics [13]
2	3432–4412 MPa	approx. 350–450 kgf mm^{-2}, fibres, 30–40 gf den^{-1} [13]

— Poly(vinyl chloride-*co*-vinyl acetate)

Tensile Strength Break:

No.	Value	Note	Elongation
1	41.4–48.3 MPa	[37]	
2	51.9–58.8 MPa	5.3–6 kgf mm^{-2}, plasticised [13]	200–450%
3	245–294 MPa	25–30 kgf mm^{-2}, fibres containing filler, 2.5 gf den$^-$ [13]	20–30%
4	83.4 MPa	8.5 kgf mm^{-2}, staple, 0.7 gf den^{-1} [13]	100%

Flexural Strength at Break:

No.	Value	Note
1	83.4 MPa	8.5 kgf mm^{-2} [13]

Mechanical Properties Miscellaneous: Elastic recovery 70–90% (continuous filament, 1–2% extension); 30% (approx., continuous filament, >5% extension); 10% (staple, >5% extension); 5–10% (staple, >10% extension) [13]
Hardness: Brinell hardness: 15 (approx., 25 kg load, 2.5 mm ball) (unplasticised material); softened by plasticisers and increasing vinyl acetate content [13]
Izod Notch:

No.	Value	Notch	Note
1	26.68–53.37 J/m	Y	0.5–1 ft lb in^{-1} [13]

Electrical Properties:
Electrical Properties General: Dielectric constant increased by presence of plasticisers [13]
Dielectric Permittivity (Constant):

No.	Value	Frequency	Note
1	3–3.5	50 Hz–1 MHz	[13]
2	3.2–3.3	60 Hz	[37]
3	3.1–3.2	1 kHz	[37]
4	3–3.1	1 MHz	[37]

Dielectric Strength:

No.	Value	Note
1	55 kV.mm^{-1}	unplasticised [13]
2	16–24 kV.mm^{-1}	plasticised [13]
3	16.7 kV.mm^{-1}	425 V mil^{-1}, short time, ⅛" thick [37]

Dissipation (Power) Factor:

No.	Value	Frequency	Note
1	0.01–0.02	50 Hz–1 MHz	[13]

Optical Properties:
Refractive Index:

No.	Value	Note
1	1.525–1.535	average values; 5–10% vinyl acetate [3]
2	1.52–1.55	20° [13,37]
3	1.53	fibres [13]

Transmission and Spectra: Ir spectral data for copolymers, unplasticised and containing plasticiser (tricresyl phosphate), have been reported; presence of plasticiser has little effect on carbonyl peak. [25] X-ray photoelectron spectroscopy has been used to examine the surfaces of copolymers moulded against metal substrates [34]
Total Internal Reflection: Birefringence nil [13]

Polymer Stability:
Thermal Stability General: Thermal analysis of coating materials has been reported. [27] Thermal stability is less than each of comonomers; [2] minimum stability occurs for compositions containing 40–50 mol% vinyl acetate. [2,5] Dependent on content of internal unsaturated groups and blocks of polyconjugated double bonds; independent of MW and total number of double bonds; stability reduced when terpolymers formed [17]
Decomposition Details: Unplasticised material decomposes above 150°. [13] Bulk degradation studies using thermal volatilisation analysis have been reported; evolution of AcOH and hydrogen chloride with relative proportions remaining equivalent to copolymer composition; measurements made over ambient-500°. [2,5] Corresponding experiments carried out in tritolyl phosphate plasticised solns. at 180° under isothermic degradation conditions; similar behaviour to bulk polymer exhibited; minimum stability occurs for compositions containing 30–40 mol% vinyl acetate; [4,5] uv spectroscopy used to examine development of conjugation along polymer chain; longer sequences found (at given extent of degradation for copolymers containing 20–30 mol% vinyl acetate. [4] Other thermal degradation studies have been reported, both for bulk polymer [9] and in solution (DMF). [24] At 153°, both dehydrochlorination and elimination of the comonomer units occurs (analysis by using ir and Raman spectroscopy); comonomer units randomly distributed in small amounts and break up 'zipper' dehydrochlorination reaction of PVC homopolymer; [9] conjugated polyenes formed which are significantly longer than average PVC block length, by elimination of vinyl acetate moieties [24]
Flammability: Is self-extinguishing [37]
Environmental Stress: Resistance to uv irradiation reduced on addition of comonomer; approx. in proportion to vinyl acetate content. [16] Darkens on prolonged exposure to sunlight; [37] less stable to light than PVC homopolymer. Low water absorption results in excellent dimensional stability under all kinds of weather conditions [37]
Chemical Stability: Virtually unaffected by most chemical reagents. Displays excellent resistance to acids, alcohols, alkalis and oils of all kinds; swollen by aromatic hydrocarbons; poor resistance to aq. ammonia, ketones, esters, aldehydes etc.; decomposed by alcoholic alkalis and hot oxidising acids. [13,37] Chemical degradation studies using potassium-*tert*-butoxide/DMF have been reported: essentially complete dehydrochlorination occurs, plus hydrolysis to vinyl alcohol; latter groups remain intact and block any extension of conjugated polyene sequences; ir and Raman spectroscopy show only short polyene sequences to be formed [24]
Hydrolytic Stability: Displays excellent resistance to water and salt solns.

Applications/Commercial Products:
Processing & Manufacturing Routes: Synth. by free-radical emulsion polymerisation of mixture of comonomers; typical initiators include peroxides (benzoyl, cyclohexanone and hydroxyheptyl), AIBN (often with UV light) and tri-*n*-butyl boron [7,8,13]
Applications: Flooring, records, sheet, film, packaging. In general, finds applications for products where better melt flow and fusion is required than is easily achievable with homopolymer, such as pipes, fittings, coatings and films

Trade name	Details	Supplier
Corvic	non-exclusive	ICI Chemicals and Polymers Ltd.
Exon	450,454,470, 481,760	Firestone
FPC 6338		OxyChem

FPC 6338		Hooker
Lucovyl	GA 8502, MA 6028, MA 6035, PA 1208, PA 1302, SA 6001	Rhone-Poulenc
Nipolit (MH,MR)		Chisso
Norvinyl	P6	Norsk Hydro
Opalon 400	non-exclusive	Monsanto Chemical
Pevikon	C870	Kema-Nord AB
Rhodopas	30/10 85/15	Rhone-Poulenc
Rhodopas AX		Rhone-Poulenc
Ryuron	710B	Tekkosha
Ucar VYHD		Union Carbide Corporation
Ucar VYHH		Union Carbide Corporation
Ucar VYLF		Union Carbide Corporation
Ucar VYNS-3		Union Carbide Corporation
VC-113R		Borden
VC-168		Borden
VC-171C		Borden
Vestolit B7090		Hüls AG
Vinidur		BASF
Vinnol		Wacker-Chemie
Vinnol 50		Wacker-Chemie
Vinnol 50/25C		Wacker-Chemie
Vinylite R	non-exclusive	Carbide and Carbon Chem.
Vinyon		Carbide and Carbon Chem.

Bibliographic References

[1] Simril, V.L., *J. Polym. Sci.*, 1947, **2**, 142, (thermal props)
[2] Grassie, N., McLaren, I.F. and McNeill, I.C., *Eur. Polym. J.*, 1970, **6**, 679, (thermal stability, degradation)
[3] Seferis, J.C., *Polym. Handb.*, 4th edn., (eds. J. Brandrup, E.H. Immergut and E.A. Grulke), John Wiley and Sons, 1999, **VI**, 581, (optical props)
[4] Grassie, N., McLaren, I.F. and McNeill, I.C., *Eur. Polym. J.*, 1970, **6**, 865, (thermal stability, degradation)
[5] Liggat, J., *Polym. Handb.*, 4th edn., (eds. J. Brandrup, E.H. Immergut and E.A. Grulke), John Wiley and Sons, 1999, **II**, 460, (thermal stability, degradation)
[6] DiBenedetto, A.T., *J. Polym. Sci., Part A-1*, 1963, **1**, 3459, (solubility parameters)
[7] Rosen, I. and Marshall, W.E., *J. Polym. Sci.*, 1962, **56**, 501, (stereoregularity, morphology, tacticity synth)
[8] Fordham, J.W.L., Burleigh, P.H. and Sturm, C.L., *J. Polym. Sci.*, 1959, **41**, 73, (synth)
[9] Weintraub, L., Zufall, J. and Heiberger, C.A., *Polym. Eng. Sci.*, 1968, **8**, 64, (thermal stability)
[10] Burrell, H., *J. Paint Technol.*, 1968, **40**, 197, (solubility parameters)
[11] Whelan, A., *Polymer Technology Dictionary* Chapman & Hall, 1994, 207, 475, (plasticisation)
[12] Cotman, J.D., *Ann. N. Y. Acad. Sci.*, 1953, **57**, 417, (branching, ordering)
[13] Roff, W.J. and Scott, J.R., *Fibres, Films, Plastics and Rubbers*, Butterworths, 1971, 118, (synth, stability, thermal props, volumetric props, optical props, electrical props, mech. props, uses)
[14] Park, R.A., *Am. Chem. Soc., Div. Org. Coat. Plast. Chem., Pap*, 1971, **31**, 513, (solubility, uses)
[15] Matthews, G., *Vinyl Allied Polym.*, Vinyl Chloride and Vinyl Acetate Polymers, Iliffe Books, 1972, **72**, 279, (stability, transport props, solubility)
[16] Dürselen, L.F.J., Wegmann, D., May, K. Oesch, U. and Simon, W., *Anal. Chem.*, 1988, **60**, 1455, (uses)
[17] Minsker, K.S. *et al*, *Plast. Massy*, 1982, 16, (thermal stability)
[18] *Catalogue Handbook of Fine Chemicals*, Aldrich, 1996–1997, 1274, (volumetric props, thermal props, transport props, MW)
[19] Oishi, T. and Prausnitz, J.M., *Ind. Eng. Chem. Fundam.*, 1978, **17**, 109, (chemical identification, analysis, health/safety)
[20] Ash, M. and Ash, I., Handbook of Plastic Compounds, Elastomers and Resins VCH, 1992, 527, (MW, transport props, volumetric props)
[21] *Jpn. Pat.*, 1990, 0288229, (uses)
[22] Zobov, E.V. and Mashina, Z.V., *Deposited Doc.*, 1975, **VINITI**, 2253, (transport porps)
[23] Gibb, T.B. and Wolfe, P.H., *J. Chromatogr. Sci.*, 1982, **20**, 471, (chemical identification, analysis, health/safety)
[24] Bowley, H.J., Gerrard, D.L. and Maddams, W.F., *Makromol. Chem.*, 1985, **186**, 715, (chemical/thermal stability)
[25] Newey, H.A., Busso, C.J., Wedgewood, A.R. and Beck, T.R., *J. Appl. Polym. Sci.*, 1985, **30**, 675, (ir)
[26] Reding, F.P., Faucher, J.A. and Whitman, R.D., *J. Polym. Sci.*, 1962, **57**, 483, (thermal props)
[27] Porowska, E. and Stareczek, T., *Plaste Kautsch.*, 1973, **20**, 643, (thermal analysis)
[28] Panwar, V.S. *et al*, *J. Mater. Sci. Lett.*, 1988, **7**, 572, (morphology, transport props)
[29] Panwar, V.S. *et al*, *J. Mater. Sci. Lett.*, 1988, **7**, 203, (thermal props)
[30] Ginsberg, T. and Stevens, J.J., *Paint Varn. Prod.*, 1974, **64**, 19, (uses)
[31] *Jpn. Pat.*, 1980, 8081811, (uses)
[32] *Czech. Pat.*, 1983, 215160, (uses)
[33] Titow, W.V., *PVC Technol.*, 4th Edn., Elsevier Applied Science, 1984, 22, 50, 1049, 1060, (transport props, uses, chemical identification, processing)
[34] Pennings, J.M.F., *Physiochem. Aspects Polym. Surf. (Proc. Int. Symp.)*, (Ed. K.L. Mittal), Plenum, 1983, 1199, (XPS, surface analysis)
[35] *Fed. Regist.*, 1986, **51**, 4177, (chemical identification, health/safety)
[36] Grulke, E.A., Polym. Handb. 3rd Edn., (Eds. J. Brandrup and E.H. Immergut), Wiley Interscience, 1989, **VII-550**,VII-554, (solubility parameters)
[37] *Sci. Technol. Polym. Films*, Vol. II (Ed. O.J. Sweeting), Wiley Interscience, 1971, **91**, 115, 309, (transport props, mech. props, stability electrical props, thermal props, volumetric props, optical props)
[38] Polymeric Materials Encyclopedia Vol. 11, (Ed. J.C. Salamone), CRC Press, 1996, 8569, (MW, processing, uses)

Poly(vinyl chloride-*co*-vinyl-acetate), Carboxylated

$$\left[\left[\begin{matrix} CH_2CH \\ | \\ Cl \end{matrix} \right]_w \left[\begin{matrix} CH_2CH \\ | \\ OCCH_3 \\ \| \\ O \end{matrix} \right]_x \left[\begin{matrix} CH_2CH \\ | \\ OH \end{matrix} \right]_y \left[\begin{matrix} CH_2CH \\ | \\ OCR \\ \| \\ O \end{matrix} \right]_z \right]_n$$

R = poly (carboxylic acid) containing side chain

Synonyms: *Carboxylated VC/VA. Carboxylated PVC/PVAc. Poly(vinyl chloride-co-vinyl acetate), carboxyl-modified. Carboxylated vinyl chloride-vinyl acetate copolymer. Carboxyl-modified vinyl chloride-vinyl acetate copolymer.' COOH containing ' vinyl chloride-vinyl acetate copolymer*
Related Polymers: Poly(vinyl chloride-*co*vinyl acetate), PVC copolymers, Poly(vinyl chloride-*co*-vinyl acetate-*co*-maleic acid), Poly(vinyl chloride-*co*-vinyl acetate), hydroxylated
Monomers: Vinyl chloride, Vinyl acetate
Material class: Gums and resins, Copolymers
Polymer Type: PVC, polyvinyl acetatepolyvinylspolyhaloolefins
CAS Number:

CAS Reg. No.	Note
39423-34-2	aq. soln.

Molecular Formula: $[(C_2H_3Cl).(C_4H_6O_2).(C_2H_4O).(C_3H_3O_2R)]_n$
Molecular Weight: M_n 4000–23000 (polycarboxylated side-chain copolymers)
Identification: Hydroxyl equivalent weight 740 (polycarboxylated-side-chain copolymer, M_n 4000); 1000 (polycarboxylated-side-chain copolymer, M_n 23000) [7]. Hydroxyl content 2.3% (polycarboxylated-side-chain copolymer, M_n 4000); 1.7% (polycarboxylated-side-chain copolymer, M_n 23000) [7]. Acid number 75 (polycarboxylated-side-chain copolymer, M_n 4000–23000) [7]
Miscellaneous: Presence of COOH group improves adhesion when used in coatings applications (particularly to metals) and improves solubility in relatively weak solvent systems. A specific example is poly(vinyl chloride-co-vinyl acetate-co-maleic acid), which is most well studied carboxylated form. Structure and props. of series of copolymers with polycarboxylate side chains have been reported [7]; hydroxyl groups randomly distributed on backbone; side chains become extremely hydrophobic in form of amine salts; waterborne vinyl dispersions have been formulated for use as inks and paints [7] presence of hydroxyl and carboxyl side-groups imparts improved adhesion to many plastic substrates and provides sites for optional cross-linking [7]. Has environmental and health aspects, dispersions of polycarboxylated-side-chain copolymers in aq. systems used for waterborne ink and paint formulations; reductions in volatile organic content during handling and drying operations [7]

Volumetric & Calorimetric Properties:
Density:

No.	Value	Note
1	d 1.26 g/cm^3	polycarboxylated-side-chain copolymer; M_n 4000 [7]
2	d 1.32 g/cm^3	polycarboxylated-side-chain copolymer; M_n 23000 [7]

Glass-Transition Temperature:

No.	Value	Note
1	67°C	polycarboxylated-side-chain copolymer; M_n 4000 [7]
2	81°C	polycarboxylated-side-chain copolymer; M_n 23000 [7]

Transport Properties:
Polymer Solutions Dilute: Viscosity 2000 cP (polycarboxylated-side-chain copolymers, M_n 4000, aq. dispersion, 31% solids) [7]; 1900 cP (polycarboxylated-side-chain copolymers, M_n 23000, aq. dispersion, 36% solids) [7]

Mechanical Properties:
Mechanical Properties General: Low MW material is relatively brittle; however, displays excellent combination of toughness, adhesion and hardness in formulated thermoplastic systems (e.g. in inks); toughness, e.g. impact strength, increases with both MW and cross-linking [7]

Polymer Stability:
Polymer Stability General: Waterborne dispersions of polycarboxylate-side-chain copolymers (pH 5–9) display good shelf lives and freeze/thaw stability: typically, can be stored for e.g. 4 weeks at 50° and at least 1 year at 20° [7]
Chemical Stability: Good solvent and chemical resistance shown by amino-cured films of polycarboxylated-side-chain copolymers [7]

Applications/Commercial Products:

Trade name	Details	Supplier
Cevian A678	aq. solution	Diacel Chem. Ind.
Ucar AW 100	low MW aq. dispersion	Union Carbide Corporation
Ucar AW 850	high MW aq. dispersion	Union Carbide Corporation

Bibliographic References
[1] *Jpn. Pat.*, 1973, 73 50 812, (use)
[2] *Jpn. Pat.*, 1980, 80 84 035, (use)
[3] Del Amo, B., Caprari, J.J., Chiesa, M.J. and Ingeniero, R.D., *An. - CIDEPINT*, 1981, 1, (use)
[4] *Jpn. Pat.*, 1983, 58 179 642, (use)
[5] *Eur. Pat. Appl.*, 1985, 139 244, (use)
[6] *U.S. Pat.*, 1988, 4 767 644, (use)
[7] Burns, R.J. and Mayer, W.P., *J. Water Borne Coat.*, 1987, **10**, 3, (use, MW, chemical identification, environmental aspects, transport props, thermal props, mech props, stability)
[8] Jung, C., Raghavan, S. and Mathur, M.C.A., *IEEE Trans. Magn.*, 1992, **28**, 2371, (use)

Poly(vinyl chloride-co-vinyl acetate), Epoxidised   P-586

Synonyms: *Epoxidised VC/VA. Epoxidised PVC/PVAc. Poly(vinyl chloride-co-vinyl acetate), epoxy-modified. Epoxidised vinyl chloride-vinyl acetate copolymer. Epoxy-modified vinyl chloride-vinyl acetate copolymer*
Related Polymers: Poly(vinyl chloride-co-vinyl acetate), PVC copolymers, Poly(vinyl chloride-co-vinyl acetate), hydroxylated, Poly(vinyl chloride-co-vinyl acetate-co-(2-hydroxypropyl acetate)
Monomers: Vinyl chloride, Vinyl acetate
Material class: Gums and resins, Copolymers
Polymer Type: PVC, polyvinyl acetatepolyhaloolefins

Molecular Weight: 15000 (82% vinyl chloride, 9% vinyl acetate 9% epoxy-containing monomer)
General Information: Epoxy-modified copolymers can crosslink to carboxyl-containing polymers after stoving, thus improving toughness and solvent resistance. Solns. in methyl ethyl ketone/toluene (3/2) (40% solids) commercially available; used, often in conjunction with solns. of acid-modified (carboxylated) terpolymers such as poly(vinyl chloride-co-vinyl acetate-co-maleic acid), to produce high-solid baked coatings, having very good adhesion to metal, hardness, and gloss [5]

Transport Properties:
Polymer Solutions Dilute: Viscosity 1000 cP (vinyl resin soln. in methyl ethyl ketone/toluene (3:2); 40% solids, 82% vinyl chloride, 9% vinyl acetate, 9% epoxy-containing monomer) [1]

Applications/Commercial Products:

Trade name	Details	Supplier
Ucar VERR-40	vinyl resin soln. 40% solids in methyl ethyl ketone/toluene (3/2)	Union Carbide Corporation

Bibliographic References
[1] Ash, M. and Ash, I., Handbook of Plastic Compounds, *Elastomers and Resins*, VCH, 1992, 527, (transport props, MW)
[2] *Ger. Pat.*, 1975, 2 453 101, (use)
[3] *Rom. Pat.*, 1975, 59 098, (use)
[4] *Jpn. Pat.*, 1989, 01 163 040, (use)
[5] Titow, W.V., *PVC Technol.*, Elsevier Applied Science, 4th edn., 1984, 1059, (use)

Poly(vinyl chloride-co-vinyl acetate), Hydroxylated   P-587

Synonyms: *Hydroxylated VC/VA. Poly(vinyl chloride-co-vinyl acetate), hydroxy-modified. Hydroxylated vinyl chloride-vinyl acetate copolymer. Hydroxy-modified vinyl chloride-vinyl acetate copolymer.' OH-containing' vinyl chloride-vinyl acetate copolymer*

Related Polymers: Poly(vinyl chloride-*co*-vinyl acetate), PVC copolymers, Poly(vinyl chloride-*co*-vinyl acetate-*co*-vinyl alcohol), Poly(vinyl chloride-*co*-vinyl acetate-*co*-(2-hydroxypropyl acrylate)
Monomers: Vinyl chloride, Vinyl acetate
Material class: Gums and resins, Copolymers
Polymer Type: PVC, polyvinyl acetatepolyhaloolefins
CAS Number:

CAS Reg. No.	Note
25822-51-9	17% hydroxy functionality

Molecular Weight: MW 3500-33000 (soln. vinyl resins containing hydroxyalkyl acrylates)
General Information: General class: examples include terpolymers containing poly (vinyl alcohol), poly(vinyl chloride-*co*-vinyl acetate-*co*-vinyl alcohol) and various hydroxyalkyl acrylates, such as 2-hydroxypropyl acrylate, Poly (vinyl chloride-*co*-vinyl acetate-*co*-(2-hydroxypropyl acrylate))
Miscellaneous: Modified copolymers display improved compatibility with other film-forming resins; in addition, they can be cured by cross-linking with isocyanates or melamine-formaldehyde resins. Solutions in methyl ethyl ketone, methyl ethyl ketone/toluene and isopropyl acetate (20-74% solids) commercially available; used for coatings [6]. Terpolymers often used blended with polyurethanes, e.g. in applications in magnetic recording media

Volumetric & Calorimetric Properties:
Density:

No.	Value	Note
1	d 1.36-1.37 g/cm^3	81% vinyl chloride, 4% vinyl acetate, 15% hydroxyalkyl acrylate [6]
2	d 1.32 g/cm^3	67% vinyl chloride, 11% vinyl acetate, 22% hydroxyalkyl acrylate [6]
3	d 1.32 g/cm^3	60% vinyl chloride 32% vinyl acetate, 8% hydroxyalkyl acrylate [6]
4	d 1.27 g/cm^3	58% vinyl chloride 10% vinyl acetate, 32% hydroxyalkyl acrylate [6]

Surface Properties & Solubility:
Solvents/Non-solvents: Sol. THF. Insol. MeOH
Transport Properties:
Polymer Solutions Dilute: Viscosity (terpolymers containing hydroxyalkyl acrylates): 171-1000 cP (20-50% solids, in methyl ethyl ketone/toluene) [6]; 1100 cP (74% solids, methyl ethyl ketone) [6]; 1500 cP (40% solids, isopropyl acetate) [6]

Optical Properties:
Transmission and Spectra: Ir spectrum shows broad OH signal at 3400 cm^{-1} (copolymer with 17% hydroxyl functionality) [1]

Applications/Commercial Products:
Processing & Manufacturing Routes: Synth. by partial hydrolysis of poly(vinyl chloride-*co*-vinyl acetate) using NaOH in MeOH [1] or by copolymerisation of poly(vinyl chloride-*co*-vinyl acetate) with hydroxyl-containing monomers in a free radical polymerisation process
Applications: Applications as components of binder resins, e.g. for use in magnetic recording media; formed by reactions of isocyanates with OH groups of copolymers, magnetic particles wetted, also used in abrasive tapes for polishing magnetic (tape) heads. Additional applications include components of waterproof fillers for decorative building boards and in the preparation of membranes for ion-selective electrodes

Trade name	Details	Supplier
Ucar	VAGC, VAGF, VROH, VYES, VYES-4, VYNC, VP-200; solns. of resins in isopropyl acetate, methyl ethyl ketone or methyl ethyl ketone/toluene mixts	Union Carbide Corporation

Bibliographic References
[1] Durselen, L.F.J., Wegmann, D., May, K., Oesch, U. and Simon, W., *Anal. Chem.*, 1988, **60**, 1455, (use, synth)
[2] Hellner, A.H.C. and Masson, C., *Res. Discl.*, 1979, **178**, 57, (use)
[3] *Jpn. Pat.*, 1980, 80 84 035, (use)
[4] *U.S. Pat.*, 1988, 4 767 644, (use)
[5] *Jpn. Pat.*, 1990, 02 167 956, (use)
[6] Ash, M. and Ash, I., Handbook of Plastic Compounds, *Elastomers and Resins*, VCH, 1992, 526, (transport props, MW, volumetric props, use)
[7] Jung, C., Rybicki, E., Raghavan, S. and Mathur, M.C.A., *Colloids Surfaces A: Physiochem. Eng. Aspects*, 1993, **80**, 77, (use)
[8] Jung, C., Raghavan, S. and Mathur, M.C.A., *IEEE Trans. Magn.*, 1992, **28**, 2371, (use)

Poly(vinyl chloride-*co*-vinyl acetate-*co*-2-hydroxypropyl acrylate)

$$\mathrm{-[\!\!-CH_2CH\!\!-]_{\mathit{x}}[\!\!-CH_2CH\!\!-]_{\mathit{y}}[\!\!-CH_2CH\!\!-]_{\mathit{z}}\!-]_{\mathit{n}}}$$
Cl, OCCH$_3$(=O), C(=O)OCH$_2$CHCH$_3$ with OH

Synonyms: *Poly(vinyl chloride-co-vinyl acetate-co-(2-hydroxypropyl propenoate). Poly(chloroethene-co-ethenyl acetate-co-(2-hydroxypropyl-2-propenoate)). Poly(chloroethene-co-ethenyl acetate-co-(β-hydroxypropyl-2-propenoate)). Poly(vinyl chloride-co-vinyl acetate-co-(β-hydroxypropylacrylate)*
Related Polymers: Polyvinyl chloride, Poly(vinyl chloride-*co*-vinyl acetate), PVC copolymers, Poly(vinyl chloride-*co*-vinyl acetate), hydroxylated, Poly(vinyl chloride-*co*-vinyl acetate), epoxidised
Monomers: Vinyl chloride, Vinyl acetate, 2-Hydroxypropyl acrylate
Material class: Thermoplastic, Copolymers
Polymer Type: PVC, polyvinyl acetateacrylic terpolymerspolyhaloolefins
CAS Number:

CAS Reg. No.	Note
41618-91-1	
123814-61-9	graft copolymer
114237-66-0	phosphate, P$_2$O$_5$

Molecular Formula: [(C$_2$H$_3$Cl).(C$_4$H$_6$O$_2$).(C$_6$H$_{10}$O$_3$)]$_n$
Fragments: C$_2$H$_3$Cl C$_4$H$_6$O$_2$ C$_6$H$_{10}$O$_3$
Molecular Weight: M$_n$ 5500-3000 (research grade), M$_n$ 10000-12000 (commercial material)
General Information: Is available in research quantities
Identification: Vinyl chloride content 58-81%; vinyl acetate 4-34%; hydroxypropyl acrylate 8-22%: commercial materials usually contain 80-81 wt% vinyl chloride 4-5% vinyl acetate, 15% acrylate [7,12]. Hydroxyl number 99-59 (22-15 wt% acrylate; M$_n$ 5500-33000) [12]

Volumetric & Calorimetric Properties:
Density:

No.	Value	Note
1	d 1.35 g/cm^3	80% vinyl chloride, 5% vinyl acetate, 15% acrylate [7]
2	d 1.2 g/cm^3	estimated: 58% vinyl chloride, 34% vinyl acetate, 8% acrylate [7]

Glass-Transition Temperature:

No.	Value	Note
1	63°C	80% vinyl chloride, 5% vinyl acetate, 15% acrylate, M_n 10000; 58% vinyl chloride, 34% vinyl acetate, 8% acrylate, M_n 12000 [7],
2	65–70°C	M_n 15000–33000, 81% vinyl chloride, 4% vinyl acetate, 15% acrylate [12]
3	53°C	M_n 5500, 67% vinyl chloride, 11% vinyl acetate, 22% acrylate [12]

Surface Properties & Solubility:
Solvents/Non-solvents: Sol. cyclohexane [2]

Transport Properties:
Polymer Solutions Dilute: η 0.15–0.56 (M_n 5500–33000) [12]; 0.222 (52% vinyl chloride, 30% vinyl acetate, 18% acrylate) [2]; 0.34 (59% vinyl chloride, 32.1% vinyl acetate, 8.9% acrylate) [5]

Polymer Stability:
Stability Miscellaneous: Devolatilisation in environmental and health studies has been reported [7]

Applications/Commercial Products:
Processing & Manufacturing Routes: Synth. by free-radical initiated (diisopropyl peroxydicarbonate) polymerisation of co-monomer mixt. in Me$_2$CO at 73° (41–69 psig) [2]. Coatings compositions prepared by reaction with neopentyl glycol diacrylate, and cured by photocrosslinking using Hg or Ar irradiation [2]; thermosetting coating compositions have also been reported [3,4,6]

Applications: Major applications as coatings, layers and binders: coating compositions for wood (radiation curable); binders for magnetic recording media (e.g. tapes) cured with diisocyanates; various thermosetting coating compositions, furniture polishes and lacquers (with nitrocellulose); as polyurethane prepolymers for high-strength mouldings. Graft copolymers used in production of sheet and foam laminates. Phosphate (containing P$_2$O$_5$) derivative used in compositions for magnetic recording media

Trade name	Details	Supplier
Standard samples	M_n 5500–33000; 22–15 wt% acrylate	Aldrich

Bibliographic References
[1] U.S. Pat., 1976, 3 983 302, (use)
[2] U.S. Pat., 1976, 3 943 103, (use, synth, processing, transport props, solubility)
[3] Ger. Pat., 1974, 2 337 681, (processing, use)
[4] Ger. Pat., 1973, 2 256 393, (use)
[5] U.S. Pat., 1973, 3 755 271, (use, transport props)
[6] U.S. Pat., 1978, 4 093 575, (use, processing)
[7] Oishi, T. and Prausnitz, J.M., Ind. Eng. Chem. Fundam., 1978, **17**, 109, (chem identification, thermal props, volumetric props, stability, environmental props)
[8] U.S. Pat., 1976, 3 996 394, (use)
[9] Jpn. Pat., 1986, 61 126 180, (use)
[10] Jpn. Pat., 1989, 01 163 040, (graft copolymers, use)
[11] Jpn. Pat., 1987, 62 187 748, (phosphate, use)

Poly(vinyl chloride-co-vinyl acetate-co-maleic acid)

Synonyms: *Poly(chloroethene-co-(ethenyl acetate)-co-((Z)-2-butenedioic acid)). 'Carboxylated' vinyl chloride-vinyl acetate copolymers. 'Carboxyl-modified' vinyl chloride-vinyl acetate copolymers. 'Acid-modified' vinyl chloride-vinyl acetate copolymers*
Related Polymers: PVC copolymers, Carboxylated poly(vinyl chloride-co-vinyl acetate)
Material class: Thermoplastic, Copolymers
Polymer Type: polyvinyls, polyhaloolefins
CAS Number:

CAS Reg. No.
9005-09-8

Molecular Weight: M_n 15000–27000; M_n 25400; MW 61900 (86% vinyl chloride, 13% vinyl acetate, 1% maleic acid)
Identification: Typical compositions: 79–88% vinyl chloride; 11.8–18.9% vinyl acetate; 0.8–2.5% maleic acid: most generally used copolymers contain 86% vinyl chloride, 13% vinyl acetate and 1% maleic acid (acid number 10, 83% vinyl chloride, 16% vinyl acetate and 1% maleic acid (acid number 10), and 81% vinyl chloride, 17% vinyl acetate and 2% maleic acid (acid number 19) [10]
Miscellaneous: Specific example of carboxylated vinyl chloride-vinyl acetate copolymer. Addition of maleic acid improves air-dry adhesion to metals, paper and other substrates; important in coating and adhesive applications; often used in conjunction with cross-linking systems such as epoxidised vinyl chloride-vinyl acetate copolymers. [9] Sometimes referred to as maleic anhydride terpolymer, as anhydride used in synthesis as third comonomer. Comparative study of polydispersity for terpolymers prepared by both solution and suspension-polymerisation techniques has been carried out; similar values found for polydispersity, 1.80–1.94. [28] United States Food and Drug Administration have published data on maximum levels of vinyl chloride monomer that may be tolerated in terpolymers used for food-contact applications (Federal Food, Drug and Cosmetic Act); levels of 5–50 ppb reported, depending on polymer composition and use [31] may be determined by gc [13]

Volumetric & Calorimetric Properties:
Density:

No.	Value	Note
1	d 1.34–1.35 g/cm^3	[11]

Glass-Transition Temperature:

No.	Value	Note
1	70–74°C	2-1 wt % maleic acid [10]

Transition Temperature General: Thermal analysis of terpolymer coating materials (films and lacquers) has been reported [6]

Surface Properties & Solubility:
Solubility Properties General: Solubility characteristics and coating properties (in connection with use as coatings) have been reported [12]
Cohesive Energy Density Solubility Parameters: Solubility parameters δ 41.8 J$^{1/2}$ cm$^{-3/2}$; δ$_h$ 44.35–46.44 J$^{1/2}$ cm$^{-3/2}$ (poor solvent).[3] 32.64–50.63 J$^{1/2}$ cm$^{-3/2}$ (moderate solvent) [3], 0 (poor solvent) [3]
Solvents/Non-solvents: Sol. THF [7], 2-butanone [8], (some) ketones, esters, chlorinated hydrocarbons, aromatic hydrocarbons, chlorinated hydrocarbons [9]. Solubility in weak solvent systems improved by increasing carboxyl groups. V. sol. ketone/aromatic solvent mixtures [9]
Surface and Interfacial Properties General: 2.1–2.5% maleic acid content used in adhesives [9]

Mechanical Properties:

Friction Abrasion and Resistance: Adhesive strengths of terpolymer coating formulations have been reported; with respect to MW, polarity and flexibility of polymer chains [32]

Polymer Stability:

Environmental Stress: Effect of environment on adhesion to steel of various paint formulations containing terpolymer has been reported; water environments have most disruptive effect [23]
Stability Miscellaneous: Sterilisation by uv radiation or H_2O_2 of heat sealable packaging material formulations (for pharmaceutical and food products) reduces seam (tear) strength [35]

Applications/Commercial Products:

Processing & Manufacturing Routes: Synth. by free-radical-initiated copolymerisation of vinyl chloride/vinyl acetate/maleic anhydride mixtures

Applications: Major applications as coatings and adhesives; solutions in methyl ethyl ketone/toluene (25/75 and 50/50) containing 20–30% solids are used either alone or in combination with other resins e.g. cross-linking systems such as epoxidised vinyl chloride-vinyl acetate copolymers for coatings systems. Coatings consisting of interpenetrating networks of terpolymers with various urethanes have also been reported. Applications include wetting binders for magnetic and optical recording media; heat-sealable packaging films for pharmaceutical and food products; plastisol compositions for sealing (welded) joints and sound-insulation coatings; heat-transferable and chemical and abrasion-resistant labels and coatings for glass and metals; preparation of membranes for ion-selective electrodes; food contact applications; release agents; leather susbstitutes (e.g. footwear); inks for thermal-transfer printing; primers for metal surfaces; anticorrosive sheet steel (e.g. for cans); binders and fire-resistant coatings for textiles; fluoroelastomer film compositions; modifiers for polymer fillers; electrophotographic coatings; microlithography/photoresists (e.g in highly sensitive negative electron-beam resist compositions); waterproof coating materials for building applications (e.g. on concretes); in blends with heparinised urethane rubbers for medical applications such as catheters. Used as anti-corrosive paint and varnish

Trade name	Details	Supplier
Bakelite	VMCH, VMCC, VMCA	Union Carbide
Rhodopas	AXCM, AXCM2, AXCM3	Rhone-Poulenc Polymers
Standard samples		Scientific Polymer Products Inc.
Standard samples		Aldrich
Ucar	VMCH, VMCA, VMCC 1–2% maleic acid	Union Carbide
Vinylite	VHCC, VMCA, VMCC, VMCH	Carbide and Carbon Chem.

Bibliographic References

[1] *Rom. Pat.*, 1975, 59098, (uses)
[2] *Ger. Pat.*, 1975, 2 453 101, (uses)
[3] Grulke, E.A., *Polym. Handb.*, 3rd edn., (eds. J. Brandrup and E.H. Immergut), Wiley-Interscience, 1989, **VII**, 550, 554, (solubility props.)
[4] Dürselen, L.F.J., Wegmann, D., May, K., Oesch, U. and Simon, W., *Anal. Chem.*, 1988, **60**, 1455, (uses)
[5] Brezinski, J.J., Kolerko, J.V. and Potter, G.H., *Congr. FATIPEC*, 1972, **11**, 327, (transport props.)
[6] Porowska, E. and Stareczek, T., *Plaste Kautsch.*, 1973, **20**, 643, (thermal props.)
[7] Kandon, K., Kitahara, A. and Kon-No, K., *J. Colloid Interface Sci.*, 1984, **99**, 455, (uses)
[8] Jung, C., Rybicki, E., Raghavan, S., and Mathur, M.C.A., *Colloids Surfaces A: Physiochem. Eng. Aspects*, 1993, **80**, 77, (uses)
[9] Titow, W.V., *PVC Technol.*, 4th edn., Elsevier Applied Science, 1984, 1049, (applications, synth., solubility, transport props.)
[10] Ash, M. and Ash, I., Handbook of Plastic Compounds, *Elastomers and Resins*, 1992, 527, (MW, transport props., volumetric props.)
[12] Park, R.A., Am. Chem. Soc., Div. Org. Coat. Plast. Chem., *Pap*, 1971, **31**, 513, (solubility, uses)
[13] Gibb, T.B. and Wolfe, P.H., *J. Chromatogr. Sci.*, 1982, **20**, 471, (chem identification)
[14] *Jpn. Pat.*, 1991, 03 53 984, (use)
[15] Zobov, E.V., Shinik, G.M. and Russu, I.V., *Zashch. Met.*, 1990, **26**, 400, (permeability)
[16] Eloy, R. et al, *Thromb. Res.*, 1987, **45**, 223, (use)
[17] *Jpn. Pat.*, 1989, 01 295 843, (use)
[18] *Jpn. Pat.*, 1989, 01 97 684, (use)
[19] Xiao, H.X., Frisch, K.C. and Patsis, A., *Polym. Mater. Sci. Eng.*, 1987, **56**, 551, (use)
[20] Namaste, Y.M.N., Oberndorf, S.K. and Rodriguez, F., J. Vac. Sci. Technol., *B*, 1988, **6**, 2245, (use)
[21] *Jpn. Pat.*, 1989, 01 297 482, (use)
[22] *Czech. Pat.*, 1988, 256 051, (use)
[23] Ruggeri, R.T. and Beck, T.R., *Adhesive Aspects of Polymeric Coatings (Conf. Proc.)*, (ed. K.L. Mittal), Plenum, 1983, 329, (stability, use)
[24] *U.S. Pat.*, 1982, 4 347 268, (use)
[25] *Czech. Pat.*, 1982, 215 160, (use)
[26] *Eur. Pat. Appl.*, 1981, 38 878, (use)
[27] *Jpn. Pat.*, 1983, 58204075, (use)
[28] Kronman, A.G. et al, *Plast. Massy*, 1984, **Part 9**, 14, (MW, polydispersity)
[29] *Jpn. Pat.*, 1984, 59 146 602, (use)
[30] *Ger. Pat.*, 1985, 3 418 000, (use)
[31] *CA*, 1986, **104**, 147304b, (toxicity, safety, use)
[32] Smekhov, F.M., *Lakokras. Mater. Ikh Primen.*, 1986, **Part 1**, 23, (mech. props. use)
[33] Zobov, E.V. and Mishina, Z.V., *CA*, 1977, **87**, 86406t, (transport props)
[34] Perera, D.Y. and Vanden Eynde, D., *J. Coat. Technol.*, 1979, **51**, 74, (transport props, use)
[35] Cerny, G. and Schricker, G., *Aluminium (Dusseldorf)*, 1978, **54**, 316, (stability)
[36] Tolstaya, S.N., *Faserforsch. Textiltech.*, 1978, **29**, 19, (use)
[37] Tolstaya, S.N., *CA*, 1977, **87**, 102961c, (use)
[38] Pila, S., *Ind. Vernice*, 1973, **27**, 25, (use)
[39] Medard, M., *Double Liaison*, 1973, **20**, 236, (use)
[40] Ginsberg, T., *FATIPEC Congr.*, 1974, **12**, 497, (transport props)

Poly(vinyl chloride-co-vinyl acetate-co-vinyl alcohol)   P-590

$$\left[\left[CH_2CH \atop Cl \right]_x \left[CH_2CH \atop OCCH_3 \atop \| \atop O \right]_y \left[CH_2CH \atop OH \right]_z \right]_n$$

Synonyms: *Poly(chloroethene-co-ethenyl acetate-co-ethenol)*. 'Hydroxyl-modified' vinyl chloride-vinyl acetate copolymers. 'Hydroxylated' vinyl chloride-vinyl acetate copolymers. Hydrolysed vinyl chloride-vinyl acetate copolymers
Related Polymers: Polyvinyl chloride, Poly(vinyl chloride-co-vinyl acetate), PVC copolymers, Hydroxylated Poly(vinyl chloride-co-vinyl acetate)
Monomers: Vinyl alcohol, Vinyl acetate, Vinyl chloride
Material class: Thermoplastic, Copolymers
Polymer Type: PVC, polyvinyl acetatepolyvinyl alcohol
CAS Number:

CAS Reg. No.
25086-48-0
51990-36-4

Molecular Formula: $[(C_2H_3Cl).(C_2H_4O).(C_4H_6O_2)]_n$
Molecular Weight: M_n 22000–27000; M_n 32900. MW 69500 (91.5% vinyl chloride, 3% vinyl acetate, 5.5% vinyl alcohol)
Additives: Plasticisers (phthalates, phosphates)
Morphology: Thin films cast from isobutyl methyl ketone have

Poly(vinyl chloride-*co*-vinyl acetate-*co*-vinyl alcohol)

completely amorph. struct. (from electron microscopic and electron diffraction studies) [15]; ir spectroscopic studies and temp. dependence investigations indicate the presence of intermolecular hydrogen bonding [18]

Identification: Typical composition: 90–91.5% vinyl chloride; 3–4% vinyl acetate; 5.5–6% vinyl alcohol: most generally used copolymers contain 90% vinyl chloride, 4% vinyl acetate, 6% vinyl alcohol (hydroxyl number, 76). A headspace gas-chromatographic method, for the routine determination of low-level (ppb) amounts of residual vinyl chloride in various vinyl chloride terpolymer systems, has been reported [24]; rapid method reported for identifying small particles of terpolymer by using light microscopy and dispersion staining techniques [40]

Miscellaneous: Specific example of hydrolysed vinyl chloride-vinyl acetate copolymer. Hydrolysed copolymers have increased adhesion to organic substrates, compatibility with wide range of modifying resins and reactive sites for cross-linking and modification; cross-linking is possible of OH groups by reaction with amino resins (urea-formaldehyde or melamine-formaldehyde) or isocyanates. Presence of OH groups also improves compatibility with e.g. nitrocellulose and alkyd resins in surface-coating formulations. Toxicological studies of terpolymer dust have been reported: shows no skin resorption and fibrogenic props.; recommended max. permissible concentration in working environment 10 mg m^{-3} [39]

Volumetric & Calorimetric Properties:
Density:

No.	Value	Note
1	d 1.39 g/cm^3	[1,2]

Glass-Transition Temperature:

No.	Value	Note
1	76–77°C	
2	78.1°C	TMA, 91% vinyl chloride, 3% vinyl acetate, 6% vinyl alcohol [19]
3	81°C	differential scanning calorimetry, 91% vinyl chloride, 3% vinyl acetate, 6% vinyl alcohol [18]
4	82.5°C	calc., 91% vinyl chloride, 3% vinyl acetate, 6% vinyl alcohol [18]

Transition Temperature General: Thermal analysis of terpolymer coating materials (films and lacquers) has been reported [11]

Surface Properties & Solubility:
Solubility Properties General: Solubility reduces with increasing MW [14]. Solubility characteristics and coating props. (in connection with use as coatings) have been reported [22]

Cohesive Energy Density Solubility Parameters: Solubility parameter δ 42.47 J$^{1/2}$ cm$^{-3/2}$ [3]; δ_h 44.35–46.44 J$^{1/2}$ cm$^{-3/2}$ (poor solvent) [3]; 32.64–41.42 J$^{1/2}$ cm$^{-3/2}$ (moderate solvent) ; 0 (strong solvent) [3]

Solvents/Non-solvents: Sol. THF [13], (some) ketones, esters and chlorinated hydrocarbons [14], isobutyl methyl ketone

Transport Properties:
Polymer Solutions Dilute: Hydrodynamic props. (viscosity parameters, α and k, etc) have been reported for both dilute and conc. solns [5]. Viscosity 350–400 cP (20–25%, methyl ethyl ketone/toluene, Ucar) [2]; 17–87 cP (15% solns. methyl ethyl ketone/toluene) [14]. Comprehensive study of rheology of terpolymer solns. in methyl ethyl ketone and dichloroethane has been reported. Displays an inverse dependence of intrinsic viscosity on critical concentration [26]

Polymer Melts: Inherent viscosity η_{inh} 0.44–0.53

Permeability and Diffusion General: Effects of type and content of different plasticisers, e.g. phthalates and phosphates, on permeability and diffusion of (artificial) seawater on terpolymer (91 wt% vinyl chloride, 3 wt% vinyl acetate, 6 wt% vinyl alcohol) have been reported [17]. Water absorption of terpolymer (91% vinyl chloride, 3% vinyl acetate, 6% vinyl alcohol) greater than that of poly(vinyl chloride-*co*-vinyl acetate) (3–17% vinyl acetate); indications of hydrogen-bonding in terpolymer, and effect of amorph. struct. [15]

Mechanical Properties:
Mechanical Properties Miscellaneous: Thermomechanical spectra have been reported, including systems cross-linked with diisocyanates, effect of OH content and degree of cross-linking, on T_g [38]

Electrical Properties:
Static Electrification: Low frequency dielectric relaxation, relative permittivity (ϵ') and loss (ϵ'') have been reported [29,30] for terpolymer (91% vinyl chloride, 3% vinyl acetate, 6% vinyl alcohol). Strong α and weak β relaxations are observed

Optical Properties:
Transmission and Spectra: Ir spectral data: temp. dependence for plasticised and unplasticised terpolymer has been reported [18,34,35]. XPES have been reported [37]

Polymer Stability:
Thermal Stability General: Thermal dehydrochlorination studies, including effects of accelerated uv-ageing, have been reported [7]. Thermal stability is reduced with respect to poly(vinyl chloride-*co*-vinyl acetate, possibly due to formation of conjugated double bonds [33,36]

Decomposition Details: Degradation involves dehydrochlorination and formation of conjugated polyene, initiated at several sites in polymer unzipping reaction

Applications/Commercial Products:

Trade name	Details	Supplier	
Cordelan	graft copolymer; fibres	Kojin KK	Aldrich

Bibliographic References

[1] Sarvetnick, H.A., *Polyvinyl Chloride*, Van Nostrand Reinhold, 1969, 21, (uses, synth, volumetric props)
[2] Ash, M. and Ash, I., Handbook of Plastic Compounds, *Elastomers and Resins*, VCH, 1992, 526, (MW, transport props, volumetric props)
[3] Grulke, E.A., *Polym. Handb.*, 3rd edn., (eds. J.Brandrup and E.H. Immergut), Wiley-Interscience, 1989, 550, 554, (solubility props)
[4] Durselen, L.F.J., Wegmann, D., May, K., Oesch, U. and Simon, W., *Anal. Chem.*, 1988, **60**, 1455, (uses)
[5] Brezinski, J.J., Kolesko, J.V. and Potter, G.H., *FATIPEC Congr.*, 1972, **11**, 327, (transport props)
[6] Bonnet, L., *Res. Discl.*, 1975, **135**, 74, (uses)
[7] Juhass, K. and Kiss, W., *FATIPEC Congr.*, 1972, **11**, 251
[8] Ginsberg, T. and Stevens, J.J., *J. Paint Varnish Products*, 1974, **64**, 19, (uses)
[9] Lonrenz, J., *J. Oil Colour Chem. Ass.*, 1973, **56**, 369, (uses)
[10] Mearns, R.D., *J. Oil Colour Chem. Ass.*, 1973, **56**, 353, (uses)
[11] Porowska, E. and Stareczek, T., *Plaste Kautsch.*, 1973, **20**, 643, (thermal props)
[12] Sumiya, K., Hirayama, N., Hayama, F. and Matsumoto, T., *IEEE Trans. Magn.*, 1984, **20**, 745, (uses)
[13] Kandori, K., Kitahara, A. and Kon-No, K., *J. Colloid Interface Sci.*, 1984, **99**, 455, (uses, MW)
[14] Titow, W.V., *PVC Technol.*, 4th edn., Elsevier, 1984, 1049, (uses, synth, solubility, transport props)
[15] Panwar, V.S., Singh, R., Malhotra, G.L. Sharma, S.K. et al, *J. Mater. Sci. Lett.*, 1988, **7**, 572, (morphology, transport props)
[16] Battilotti, M., Colapicchioni, C., Giannini, I., Porcelli, F. et al, *Anal. Chim. Acta*, 1989, **221**, 157, (uses)
[17] Di Sarli, A.R., Schwiderke, E.E. amd Podesta, J.J., *J. Chem. Technol. Biotechnol.*, 1989, **45**, 29, (permeability, diffusion)
[18] Panwar, V.S., Singh, R., Mehendru, P.C. and Gupta, N.P., *Thin Solid Films*, 1990, **192**, 157, (ir, struct)
[19] Panwar, V.S., Singh, R., Mathur, R.B., Mehendru, P.C. and Gupta, N.P., *J. Mater. Sci. Lett.*, 1988, **7**, 203, (thermal props)
[20] Tu, R.S., ACS Symp. Ser. Polym. Inf. Storage Technol., 1989, 345, (uses, reactions)
[21] *Jpn. Pat.*, 1980, 80 81 811, (uses)
[22] Park, R.A., *Am. Chem. Soc., Div. Org. Coat. Plast. Chem., Pap*, 1971, **31**, 513

[23] Safdy, M.E., *J. Appl. Polym. Sci.*, 1982, **27**, 4753, (uses)
[24] Gibb, T.B. and Wolfe, P.H., *J. Chromatogr. Sci.*, 1982, **20**, 471, (chem identification)
[25] *Jpn. Pat.*, 1977, 77 150 474, (uses)
[26] Klark, L.N., Kronman, A.G. and Emel'yanov, D.N., *Fiz.-Khim. Osn. Sint. Pererab. Polim.*, 1978, **3**, 72, (transport props)
[27] Franceschini, A., Fidilio, G. and Actis, G., *Desalination*, 1978, **26**, 37, (uses)
[28] *Neth. Pat.*, 1978, 77 14 429, (uses)
[29] Mehendru, P.C., Singh, R., Panwar, V.S. and Gupta, N.P., *Thin Solid Films*, 1984, **120**, L83, (electrical props)
[30] Mehendru, P.C., Singh, R., Panwar, V.S. and Gupta, N.P., *J. Phys. D: Appl. Phys.*, 1984, **17**, L61, (electrical props)
[31] *Neth. Pat.*, 1985, 83 03 139, (blends, uses)
[32] *Neth. Pat.*, 1985, 83 03 140, (blends, uses)
[33] Misker, K.S. et al, *Plast. Massy*, 1982, **Part 1**, 16, (thermal stability)
[34] Pomerantseva, E.G., Kronman, A.G., Chervyokova, G.N. and Ivanova, L.F., *Plast. Massy*, 1983, **Part 9**, 39, (ir, chem identification)
[35] Newey, H.A., Busso, C.J., Wedgewood, A.R. and Bech, T.R., *J. Appl. Polym. Sci.*, 1985, **30**, 675, (ir)
[36] Bowley, H.J., Gerrard, D.L. and Maddams, W.F., *Makromol. Chem.*, 1985, **186**, 715, (ir, Raman, chem stability, thermal stability)
[37] Pennings, J.F.M., *Physiochem. Aspects Polym. Surf. (Proc. Int. Symp.)*, 1983, **2**, 1199, (surface props, XPS)
[38] Ma, D., Zhu, Q., Zhou, Y., Chen, S., Luo, X. and Hou, J., *Gaodeng Xuexiao Huaxue Xuebao*, 1982, **3**, 533, (thermomechanical spectra)
[39] Kogai, A.M., *Gig. Sanit.*, 1982, **Part 8**, 89, (tox)
[40] Skirius, S.A., *Microscope*, 1986, **34**, 29, (identification)

Poly(vinyl chloride-*co*-vinyl bromide) P-591

$$\left[\left[CH_2CH\atop Cl\right]_x\left[CH_2CH\atop Br\right]_y\right]_n$$

Synonyms: *Vinyl chloride-vinyl bromide copolymer. Poly(chloroethene-co-bromoethene). VC/VB copolymer*
Related Polymers: Poly(vinyl chloride), unplasticised, PVC copolymers
Monomers: Vinyl chloride, Vinyl bromide
Material class: Thermoplastic, Copolymers
Polymer Type: PVC, polyhaloolefins
CAS Number:

CAS Reg. No.
30875-00-4

Molecular Formula: $[(C_2H_3Cl)_x(C_2H_3Br)_y]_n$
Molecular Weight: MW 30000–32400 (AIBN/uv, 25°, 7.6–1.7% vinyl bromide), 40500 (AIBN/uv, -30°, 6.8% vinyl bromide), 63900–148400 (degree of polymerisation 510–1190, suspension polymerisation, cyclohexyl percarbonate initiator, 40°, 13.87–0.49 wt% bromine)
General Information: Presence of vinyl bromide as comonomer increases copolymer's flame-retardant props. Has been used as model for PVC homopolymer with chain irregularities and variable tacticities (latter by ir) [6,9,10]; comonomer vinyl bromide content <10% [6,10]. Have relatively low MW
Tacticity: Tacticity decreases with increase in vinyl bromide content. [10] Analysis of tacticity and sequence distribution facilitated by reductive debromination, thus eliminating ionisation interferences; approximate Bernoullian statistics exhibited for copolymerisation reactions [1,2]

Surface Properties & Solubility:
Solvents/Non-solvents: Sol. 1,2,4-trichlorobenzene (100°) [1]; DMF [3]. Insol. MeOH, H_2O [3]

Transport Properties:
Polymer Solutions Dilute: [η] 8.0–26.5 (THF, 25°, soln. polymerisation; cyclohexyl percarbonate initiator, 40°, increases with increasing vinyl chloride content). [3] [η] 73.2–138.8 (THF, 25°, suspension polymerisation; cyclohexyl percarbonate initiator, 40°, increases with increasing vinyl chloride content) [3]

Optical Properties:
Transmission and Spectra: High-resolution H-1 nmr and C-13 nmr spectral data have been reported. [1,2] Ir spectral data have been reported [6]

Polymer Stability:
Thermal Stability General: Used as model for investigations of thermal dehydrochlorination reaction of PVC homopolymer. [3,9,10,11]
Decomposition Details: Rate of decomposition at 180° greater than for PVC; lower activation energy of hydrogen halide elimination. Thermal elimination of hydrogen chloride starts from comonomer unit (in N_2 atmosphere); first step in decomposition requires sequence of at least two neighbouring comonomer units. [11] Ozonolysis studies at 0–20° on suspension polymerised copolymers (1.32 and 12.45 mol.% vinyl bromide) have been reported. Ionic degradation with LiCl and LiBr also examined; polymer degrades by autooxidation at 0–70°; used in determination of internal double bond content of PVC homopolymer. It is reported that suspension polymerised PVC contains 0.05 double bonds for each 1000 vinyl chloride units; presence of Bu_2S during ozonolysis protects double bonds. [9] Dehydrobromination reaction is not selective; isolated double bonds not produced even at low temps. [3]

Applications/Commercial Products:
Processing & Manufacturing Routes: Synth. by bulk [1,6,10], solution (dichloroethane) [3], suspension [3] or emulsion [7,8] polymerisation of comonomers. Free radical initiators are used, e.g. AIBN (60°) [1], AIBN + uv radiation (-30 to 25°) [6,10], cyclohexyl percarbonate (40°) [3] or γ-irradiation (0°) [1]
Applications: Has applications due to fireproofing characteristics that include its compositions with poly(vinyl alcohol), e.g. transparent fireproof films (30% vinyl bromide content, emulsion polymerised) and other fire-resistant films, fibres and coatings; component of paddings in fire-retardant blend compositions for fireproofing textiles, cotton/polyester blends and 50/50 polyester/cotton wovens [8]; emulsions with amides and phosphonium salts; flame-resistant finish durable up to 50 washing cycles, with no yellowing during curing

Bibliographic References
[1] Cais, R.E., Kometani, J.M. and Salzman, N.H., *Macromolecules*, 1986, **19**, 1006, (H-1 nmr, C-13 nmr, tacticity)
[2] Cais, R.E., Kometani, J.M. and Salzman, N.H., *Polym. Mater. Sci. Eng.*, 1984, **50**, 285, (nmr, tacticity)
[3] Michel, A., Schmidt, G. and Guyot, A., *J. Macromol. Sci., Chem.*, 1973, **7**, 1279, (stability, sequence distribution, synth, transport props)
[4] *Jpn. Pat.*, 1973, 7 343 443, (uses)
[5] *Jpn. Pat.*, 1975, 7 551 149, (uses)
[6] Milán, J. and Guzmán-Perote, J., *Rev. Plast. Mod.*, 1974, **28**, 227, (synth., struct., tacticity)
[7] *U.S. Pat.*, 1975, 618 188, (uses)
[8] Donaldson, D.J., Normand, F.L., Drake, G.L. and Reeves, W.A., *Am. Dyest. Rep.*, 1975, **64**, 30, (uses)
[9] Michel, A., Schmidt, G. and Guyot, A., *Polym. Prepr. (Am. Chem. Soc., Div. Polym. Chem.)*, 1973, **14**, 665, (stability, sequence distribution)
[10] Guzmán Perote, J. and Millán, J., *Eur. Polym. J.*, 1976, **12**, 295, 299, (synth, tacticity, thermal stability)
[11] Braun, D. and Thallmaier, M., *J. Polym. Sci., Part C: Polym. Lett.*, 1967, **16**, 2351, (thermal stability)

Poly(vinyl chloride-*co*-vinylidene chloride) P-592

$$\left[(CH_2CH(Cl))_x-(CH_2C(Cl)_2)_y\right]_n$$

Poly(vinyl chloride-co-vinylidene chloride)

Synonyms: *Vinyl chloride-vinylidene chloride copolymer. High-vinylidene copolymer. Poly(vinylidene chloride) copolymer. VDC/VC. Poly(1,1-dichloroethene-co-1-chloroethene)*
Related Polymers: Polyvinyl chloride, unplasticised, Poly(vinylidene chloride), PVC copolymers
Monomers: Vinyl chloride, Vinylidene chloride
Material class: Thermoplastic, Copolymers
Polymer Type: PVC, polyvinylidene chloridepolyhaloolefins
CAS Number:

CAS Reg. No.
9011-06-7

Molecular Formula: $[(C_2H_3Cl).(C_2H_2Cl_2)]_n$
Fragments: C_2H_3Cl $C_2H_2Cl_2$
Molecular Weight: MW 200000 (emulsion copolymers); MW 44250 (degree of polymerisation 200, 38 wt.% vinyl chloride); MW 28530 (degree of polymerisation 130, 68 wt.% vinyl chloride); MW 29340 (degree of polymerisation 150, 76% vinyl chloride); MW 43190 (degree of polymerisation 200, 95% vinyl chloride, free-radical (AIBN) initiated suspension polymerisation)
Additives: Stabilisers (chemical; light), plasticisers, antioxidants
General Information: Has improved permeability compared to PVC. Although it has many valuable props., e.g. resistance to solvents and low gas permeability, PVDC not used to any great extent on its own, owing to difficulties in fabrication and processing; high-vinylidene copolymers used - typically contain 10–15% vinyl chloride as comonomer. In addition, copolymers are available, at other end of range, containing approx. 95% vinyl chloride. Terpolymers (e.g. vinylidene chloride/vinyl chloride/acrylonitrile (85/13/2) are also available commercially
Morphology: High-vinylidene copolymers prepared as aq. dispersions are usually amorph. but can become crystalline when converted to films [2]. Cold drawing leads to highly oriented (highly crystalline) material, e.g. films and fibres [2]. X-ray diffraction studies show morphology of PVDC homopolymer retained in copolymers (0.4–95% vinyl chloride) [5,6]; crystallinity 20–34% (40–90% vinyl chloride, cold drawn and annealed fibres) [5,6]; SAXS studies on cold drawn fibres indicate microstructure of fringe-micelle type, with little folded-chain-lamellar struct. [5,7]. PVDC homopolymer has higher degree of crystallinity (lamellar particles); structural regularity is reduced as vinyl chloride content is increased, until at vinyl chloride $\geq$ 30%, copolymer becomes amorph. (spherical particles) [5,9]
Miscellaneous: Fibres have characteristic pale straw colour [12]

Volumetric & Calorimetric Properties:
Density:

No.	Value	Note
1	d 1.69 g/cm^3	granules, 200–260 μm [1]
2	d 1.6 g/cm^3	approx., flexible film [4]
3	d 1.65–1.72 g/cm^3	typical range, monofilament [5]

Melting Temperature:

No.	Value	Note
1	163°C	granules, 200–260 μm [1]
2	164°C	powder [1]
3	135°C	10 wt.% vinyl chloride [13]
4	183–192°C	vinyl chloride mole fraction 0.44–0.095, cold drawn and annealed fibres [5,6]

Glass-Transition Temperature:

No.	Value	Note
1	-17°C	[13]

Transition Temperature General: T_g varies with copolymer composition; can be approximated using a modified Gordon-Taylor equation [5,10]. High-vinylidene copolymers have lower T_m values than PVDC homopolymer [11]
Vicat Softening Point:

No.	Value	Note
1	115–140°C	typical range, monofilament [5]

Transition Temperature:

No.	Value	Note	Type
1	77°C	intermittant, typical value, monofilament [5]	heat resistance temp.
2	100°C	continuous, shrinks, typical value, monofilament [5]	heat resistance temp.

Surface Properties & Solubility:
Solubility Properties General: Modified forms of PVC (e.g. 95 wt% vinyl chloride) have greater solubility than PVC homopolymer; solubility increases with increasing vinyl chloride content [2]; copolymers, in particular those of low crystallinity, are much more soluble in common organic solvents at room temp. than PVDC homopolymer [11]
Plasticisers: Alkyl esters of adipic, sebacic and citric acids (e.g. acetyltributyl citrate); dioctyl phthalate; polymeric plasticisers, e.g. poly(ethylene-co-vinyl acetate) and polyurethane rubbers. Most of common PVC plasticisers can be used, although compatibility with high-vinylidene copolymers is sometimes poor [5]
Solvents/Non-solvents: Sol. THF (vinylidene chloride:vinyl chloride (90:10)) [2], ketone/aromatic hydrocarbon mixtures (vinylidene chloride:vinyl chloride (85:15)) [2], ketones (low vinylidene chloride content materials) [2]
Wettability/Surface Energy and Interfacial Tension: Critical contact angle 39–40° (15 wt% vinyl chloride, 25°) [5,8]. Contact angle 80° (H$_2$O), 31° (methylene iodide), 0° (hexadecane, 25°, 15 wt% vinyl chloride) [5,8]

Transport Properties:
Transport Properties General: High vinylidene copolymers are highly impermeable to most gases and vapours; transmission increases as vinylidene chloride content is reduced, and with increasing temp. Vinylidene chloride polymers more impermeable to wider range of gases and liquids than many other polymers due to a combination of high density and high crystallinity [11]; permeabilities increased by presence of plasticisers [11]
Polymer Solutions Dilute: [η] (after gel permeation chromatography, in THF) 0.23 dl g^{-1} (38 wt% vinyl chloride); 0.44 dl g^{-1} (68 wt% vinyl chloride); 0.5 dl g^{-1} (76 wt% vinyl chloride); 0.875 dl g^{-1} (95 wt% vinyl chloride) [16]
Polymer Melts: Melt viscosity 5300 P (175°, 100 s^{-1}, granules, 200–260 μm) [1]; 9100 P (185°, 100 s^{-1}, powder) [1]
Permeability of Gases: 0.015 Barrer, (20°, 40 wt% vinyl chloride) [17]
Permeability of Liquids: [H$_2$O] 1.38–7.38 × 10^{13} cm^3 cm (cm^2 s Pa)$^{-1}$ (30°, 0.5–7.2 wt.% acetyltributyl citrate plasticiser; vinylidene chloride/vinyl chloride (88/12), (no relative humidity variation effects of any significance, but permeability increases markedly with temp. between 30° and 45° [3]

– Poly(vinyl chloride-co-vinylidene chloride)

Water Absorption:

No.	Value	Note
1	0.1 %	27h, max., typical value, 0.01 in diameter filament, [5]

Gas Permeability:

No.	Gas	Value	Note
1	[O_2]	0.33 cm^3 mm/ (m^2 day atm)	0.005×10^{-10} cm^2 (s cmHg)$^{-1}$ [2]
2	[N_2]	0.06 cm^3 mm/ (m^2 day atm)	0.0009×10^{-10} cm^2 (s cmHg)$^{-1}$ [2>
3	[CO_2]	1.97 cm^3 mm/ (m^2 day atm)	0.03×10^{-10} cm^2 (s cmHg)$^{-1}$ [2]
4	[H_2O]	0.03 cm^3 mm/ (m^2 day atm)	0.0004×10^{-10} g (cm s cmHg)$^{-1}$ [2] (typical values, high vinylidene copolymers);
5	[N_2]	0.08 cm^3 mm/ (m^2 day atm)	0.0012 Barrer, (25°, moulded sheets, low to moderate crystallinity (<50%)
6	[CO_2]	0.19 cm^3 mm/ (m^2 day atm)	0.003 Barrer, (20°, 15 wt% vinyl chloride) [17]
7	[CO_2]	0.46 cm^3 mm/ (m^2 day atm)	0.007 Barrer, (20°, 30 wt% vinyl chloride) [17]
8	[CO_2]	0.98 cm^3 mm/ (m^2 day atm)	0.015 Barrer, (20°, 40 wt% vinyl chloride) [17]

Mechanical Properties:
Mechanical Properties General: Copolymers are non-linear viscoelastic materials; mechanical props. are very dependent on sample (thermal) history and morphology [5]. Presence of vinyl chloride units in high-vinylidene copolymers results in lowering of dynamic modulus (reduction in crystallinity); corresponding rise in T_g; softening effect observed at room temp. accompanied by increased brittleness at lower temps.; 2–10% plasticiser added to avoid this [11]. Increase in tensile strength with crystallinity, with corresponding falls in toughness and elongation; orientation improves all of these props. [11]

Tensile (Young's) Modulus:

No.	Value	Note
1	350–400 MPa	10–15 wt% vinyl chloride, 25 μm film [13]

Tensile Strength Break:

No.	Value	Note	Elongation
1	235–330 MPa	[11,14]	33.1–16.2%
2	>70 MPa	min., flexible film [4]	
3	275 MPa	40 kpsi, typical value, monofilament [5]	15–25%

Flexural Strength at Break:

No.	Value	Note
1	103–117 MPa	15–17 kpsi, typical range, monofilament [5]

Mechanical Properties Miscellaneous: Yield 23 m^2 kg^{-1} (high-vinylidene copolymer, 10–15 wt.% vinyl chloride, 25 μm film) [13]. Shrinkage 2–60% (possible range from typical filament; dependent on formulation and after-treatment) [5]. Tensile strength 1.4–1.7 kN m^{-1} (35–25% extension, high vinylidene copolymer, 25 μm film) [13]
Hardness: Vickers diamond pyramid (136°) 16.7 (3 × 100 g, 3 × 500 g; moulded discs, 15 wt.% vinyl chloride) [8]
Friction Abrasion and Resistance: High-vinylidene copolymers display good abrasion resistance; μ_{stat} (polymer/polymer) 0.49; μ_{dyn} (polymer/steel sphere) 0.46 (15 wt.% vinyl chloride) [5,8]

Optical Properties:
Transmission and Spectra: Crystalline materials show strong doublet at 1046 and 1071 cm^{-1}, plus signals in CCl stretching region (500–700 cm^{-1}) [5]

Polymer Stability:
Polymer Stability General: In melt-processing of high-vinylidene copolymers, great care must be taken to avoid degradation caused, for example, by overheating or contact with metals, such as Fe, Cu, steel etc.
Thermal Stability General: High-vinylidene copolymers have greater stability than PVDC homopolymers cf. greater stability of CCl bond on vinyl chloride unit (secondary halides less reactive than tertiary halides) [5]
Upper Use Temperature:

No.	Value	Note
1	75–95°C	intermittently, high-vinylidene copolymers [2]

Decomposition Details: Thermal decomposition occurs above 150° and involves a two-stage dehydrochlorination reaction; rate of latter falls progressively with increasing vinyl chloride content (cf. change in morphology) [5]. Slow (overall) degradation rate observed [11]. Hydrogen chloride is product of decomposition, and is itself a catalyst for further decomposition; stabilisers (e.g. alkaline-earth and heavy metal oxide phosphites) which absorb or combine with evolved HCl are used [11]
Flammability: High-vinylidene copolymers are flame-resistant [2]; self-extinguishing [5]
Environmental Stress: Fibres are resistant to light [12]
Biological Stability: Resistant to bacterial and insect attack (fibres) [12]
Stability Miscellaneous: High-energy irradiation, e.g. γ-rays, causes copolymers to undergo cross-linking (PVC units) and chain scission [11]; behaviour dependent on morphology and composition [11,15]

Applications/Commercial Products:
Processing & Manufacturing Routes: Prepared by conventional free-radical copolymerisation of mixture of comonomers [18]; initiators used include potassium peroxysulfates, organic peroxides, peroxycarbonates, azo compounds, H_2O_2, organic hydroperoxides, peroxyborates etc.; redox pairs often used. Emulsion and suspension polymerisation are preferred industrial processes; continuous addition process employed to ensure constant composition. Emulsion polymerisation - resulting latex is often used directly, plus various stabilising additives, as coating vehicle for application to specific substrates. Suspension polymerisation used for producing moulding and extrusion resins: fabrication by injection, compression and transfer moulding techniques. Extrusion is main fabrication technique for filaments, films, rods, tubing or pipe. Manufacture of filaments: stretching produces orientation in single direction; highly crystalline materials. Oriented films produced in similar fashion. Fibres can also be produced by melt-spinning and then stretching. Low-density, fine-celled foams made by extrusion in presence of blowing agent. Copolymerisation can also be achieved by ionic initiation (Ziegler catalysis)

Applications: Used in the manufacture of fibres and films. Modified forms of PVC (approx. 95% vinyl chloride) are used for mouldings and in lacquers and impregnating compositions. Typical applications include unplasticised calendered sheets, mouldings, extruded films for packaging, viscosity-reducing polymer in pastes (suspension polymerisation); coatings and finishes (emulsion polymerisation). High-vinylidene copolymers are useful for films with high durability and chemical resistance, and as films and coatings with low gas and water vapour permeability. Major applications are adhesion coatings for paper and board; packaging of food products; pipes for chemical processing equipment; upholstery, seat covers, fibres, bristles, latex coatings. Such barrier applications utilise copolymer films with 10–15% vinyl chloride content. Fibres (Saran); high-vinylidene copolymers; both staple fibres and filament yarns produced, latter in both monofilament and multifilament forms; applications in upholstery, filter cloths and fishnets. Applications due to excellent chemical resistance, barrier props. and extended service life. Moulded parts as gas filters, valves, pipe fittings, containers, chemical processing equipment. Extruded monofilaments are widely used in textile field, e.g. furniture and automobile upholstery; drapery fabric; outdoor furniture; filter cloths. Biaxially oriented extruded films used in packaging, transparent barriers and shrink film for food products. Multilayer films and sheets, e.g. with HDPE, used for perishable food packaging and rigid barrier containers. Lacquer resins used as coatings for films, board and paper; interior coatings for fuel storage tanks; packaging of foodstuffs; binders for magnetic (audio, video and computer) tapes. Applications due to flame-suppression props., e.g. in fibres. Miscellaneous uses include rubber-modified products and blends; improved toughness and other mech. props.

Trade name	Details	Supplier
Breon	CS 100/30, 202	BP Chemicals
Cryovac		WR Grace
Daran	CR 6795H	WR Grace
Geon	222, 652, 650 X17	Aku-Goodrich
Ixan	SGA 1	Solvay
Kurehalon	AO, SOA	Kureha Chemical Industry Co.
Saran	Films and fibres, B and HB series; high-vinylidene copolymers include 506, 683, 746, 872, 925	Dow
Saran	Film	Ashley Polymers
Standard samples		Aldrich
Velon		Firestone

Bibliographic References
[1] *Catalogue Handbook of Fine Chemicals*, Aldrich, 1996, 1275, (thermal props, transport props)
[2] Roff, W.J. and Scott, J.R., Fibres, Films, *Plastics and Rubbers*, Butterworths, 1971, 121, 123, 125, 126, (uses, morphology, solubility, transport props, stability)
[3] Delassus, P.T. and Grieser, D.J., *J. Vinyl Technol.*, 1980, **2**, 195, (permeability)
[4] *Kirk-Othmer Encycl. Chem. Technol.*, 3rd edn., (eds. M. Grayson and D. Eckroth), John Wiley and Sons, 1980, **11**, 181, (volumetric props, mech props)
[5] Wessling, R.A., Polymer Monographs, *Vol. 5: Polyvinylidene chloride*, Gordon and Breach, 1977, (props, morphology)
[6] Okuda, K., *J. Polym. Sci., Part A: Polym. Chem.*, 1964, **2**, 1749, (thermal props, morphology)
[7] Okuda, K. *et al*, *Kogyo Kagaku Zasshi*, 1970, **73**, 1398, (morphology)
[8] Bowers, R.C. *et al*, *Ind. Eng. Chem. Proc. Des. Dev.*, 1965, **4**, 86, (mech props, surface props)
[9] Titow, W.V., PVC Plastics: Properties, Processing, *and Applications*, Elsevier Applied Science, 1990, 74, 76, (morphology, processing, uses)
[10] Wessling, R.A. *et al*, *Appl. Polym. Symp.*, 1974, **25**, 83, (thermal props)
[11] *Kirk-Othmer Encycl. Chem. Technol.*, 3rd edn., (eds. M. Grayson and D. Eckroth), John Wiley and Sons, 1983, **23**, 772, (transport props, thermal props, solubility, stability, mech props, processing, uses)
[12] *Kirk-Othmer Encycl. Chem. Technol.*, 3rd Edn., Vol. 10, (eds. M. Grayson and D. Eckroth), John Wiley and Sons, 1980, 158, (fibres, uses)
[13] Oswin, C.R., *Plastic Films and Packaging*, Applied Science, 1975, **2**, 72, (thermal props, mech props)
[14] Serdensky, E.D., *Man-Made Fibers*, (eds. H. Mark, E. Cernia and S.M. Atlas), John Wiley and Sons, 1968, 319, (mech props)
[15] Harmer, D.E. and Raab, J.A., *J. Polym. Sci.*, 1961, **55**, 821, (stability, γ-irradiation)
[16] Revilon, A. *et al*, *J. Polym. Sci., Polym. Chem. Ed.*, 1976, **14**, 2263, (MW, fractionation, viscosity)
[17] *Sci. Technol. Polym. Films*, (ed. O.J. Sweeting), Wiley, 1971, **II**, 96, 101, 123, 124, (permeability)
[18] *Encycl. Polym. Sci. Eng.*, 2nd Edn., Vol. 17, (eds. H.F. Mark, N.M. Bikales, C.G. Overberger and G. Menges), John Wiley and Sons, 1989, 516, (processing, uses)

Poly(vinyl chloroacetate)   P-593

Synonyms: *Poly(ethenyl chloroacetate). Poly[(chloroacetoxy)ethylene]*
Related Polymers: Poly(vinyl fluoroacetate)
Monomers: Vinyl chloroacetate
Material class: Thermoplastic
Polymer Type: polyvinyl acetate
CAS Number:

CAS Reg. No.
24991-33-1

Molecular Formula: $(C_4H_5ClO_2)_n$
Fragments: $C_4H_5ClO_2$
Molecular Weight: Typically MW 153000, M_n 57000

Volumetric & Calorimetric Properties:
Density:

No.	Value	Note
1	d^{20} 1.45 g/cm^3	amorph.

Glass-Transition Temperature:

No.	Value	Note
1	23°C	[1,2]

Surface Properties & Solubility:
Solvents/Non-solvents: Sol. $CHCl_3$, EtOAc, chlorobenzene, pyridine, dioxane, cyclohexanone. Insol. saturated hydrocarbons. Swollen by Me_2CO

Transport Properties:
Polymer Solutions Dilute: [η] 0.28–0.83 (2-butanone) [3]; 0.25–0.49 (Me_2CO, 20°) [4]

Electrical Properties:
Dielectric Permittivity (Constant):

No.	Value	Frequency	Note
1	24.73	60 Hz	70° [6]

Optical Properties:
Refractive Index:

No.	Value	Note
1	1.5131	20°
2	1.504–1.512	30°

Transmission and Spectra: X-ray [5] spectral data have been reported

Applications/Commercial Products:
Processing & Manufacturing Routes: Prod. by free radical polymerisation of vinyl chloroacetate
Applications: Lacquer resin, adhesive, electrophotographic binder

Bibliographic References
[1] Kästner, S. and Schlosser, E., *Makromol. Chem.*, 1978, **179**, 2467
[2] Dubey, K.S., Ramachandrarao, P. and Lele, S., *Polymer*, 1987, **28**, 1341
[3] Cooper, W., Johnston, F.R. and Vaughan, G., *J. Polym. Sci., Part A: Polym. Chem.*, 1963, **1**, 1509
[4] Kelley, D.J., *J. Polym. Sci.*, 1962, **59**, 56
[5] Fordham, J.W.L., McGain, G.H. and Alexander, L.E., *J. Polym. Sci.*, 1959, **39**, 335
[6] Mead, D.J., *J. Am. Chem. Soc.*, 1941, **63**, 2832, (electrical props)

Poly(vinyl cyclobutane)   P-594

$$\mathrm{-\!\!-\!\!\!\left[CH_2CH\right]_n\!\!-\!\!-}$$

Synonyms: *Poly(ethenylcyclobutane). Poly(1-cyclobutyl-1,2-ethanediyl)*
Monomers: Vinylcyclobutane
Material class: Thermoplastic
Polymer Type: polyolefins
CAS Number:

CAS Reg. No.
66969-20-8

Molecular Formula: $(C_6H_{10})_n$
Fragments: C_6H_{10}
Morphology: Tetragonal, a 3.412, b 3.412, c 0.66, helix 4_1 [1]

Volumetric & Calorimetric Properties:
Transition Temperature:

No.	Value	Note	Type
1	228°C	[2]	Softening point

Surface Properties & Solubility:
Solvents/Non-solvents: Sol. hot decahydronaphthalene [2]. Insol. C_6H_6, isopropanol [2]

Applications/Commercial Products:
Processing & Manufacturing Routes: Synth. using Ziegler-Natta catalysts such as $TiCl_4$-$Al(^iBu)_3$ in heptane at 75° [2]

Bibliographic References
[1] Noether, H.D., *J. Polym. Sci., Part C: Polym. Lett.*, 1967, **16**, 725, (struct)
[2] Overberger, C.G., Kaye, H. and Walsh, G., *J. Polym. Sci., Part A: Polym. Chem.*, 1964, **2**, 755, (synth, solubility)

Poly(vinyl cycloheptane)   P-595

$$\mathrm{-\!\!-\!\!\!\left[CH_2CH\right]_n\!\!-\!\!-}$$

Synonyms: *Poly(ethenylcycloheptane). Poly(1-cycloheptyl-1,2-ethanediyl)*
Monomers: Vinylcycloheptane
Material class: Thermoplastic
Polymer Type: polyolefins
CAS Number:

CAS Reg. No.	Note
66906-09-0	
30847-54-2	isotactic

Molecular Formula: $(C_9H_{16})_n$
Fragments: C_9H_{16}
Morphology: Tetragonal, a 2.34–2.35, b 2.34–2.35, c 0.65, helix 4_1 [1]

Volumetric & Calorimetric Properties:
Density:

No.	Value	Note
1	d 0.926 g/cm³	[1]

Melting Temperature:

No.	Value	Note
1	>300°C	min. [2]

Surface Properties & Solubility:
Solvents/Non-solvents: Sol. C_6H_6 [2]. Insol. Me_2CO, Et_2O [2]

Transport Properties:
Polymer Solutions Dilute: [η] 0.39 dl g⁻¹ (C_6H_6, 30°) [2]

Applications/Commercial Products:
Processing & Manufacturing Routes: Synth. using Ziegler-Natta catalysts such as $TiCl_4$-$Al(^iBu)_3$ in heptane at 72° [2]

Bibliographic References
[1] Noether, H.D., *J. Polym. Sci., Part C: Polym. Lett.*, 1967, **16**, 725, (struct)
[2] Overberger, C.G., Kaye, H. and Walsh, G., *J. Polym. Sci., Part A: Polym. Chem.*, 1964, **2**, 755, (synth, melting point, solubility)

Poly(vinyl cyclohexane)   P-596

$\mathrm{-[CH_2-CH(C_6H_{11})]_n-}$

Synonyms: *PVCH. Hydrogenated polystyrene. Poly(cyclohexylethylene). Poly(ethenylcyclohexane). Poly(1-cyclohexyl-1,2-ethanediyl)*
Monomers: Vinylcyclohexane
Material class: Thermoplastic
Polymer Type: polyolefins
CAS Number:

CAS Reg. No.	Note
25498-06-0	atactic
26951-20-2	isotactic
123880-54-6	syndiotactic

Molecular Formula: $(C_8H_{14})_n$
Fragments: C_8H_{14}
General Information: Fine, white powder [3]
Tacticity: Atactic, isotactic and syndiotactic forms known
Morphology: Tetragonal, a 2.19, b 2.19, c 0.65, helix 4_1 [4]; tetragonal, a 2.176, b 2.176, c 0.650, helix 4_1 [5]; form I tetragonal, a 2.199, b 2.199, c 0.643, helix 4_1 [6]; form II tetragonal, a 2.048, b 2.048, c 4.458, helix 24_7 [6]; triclinic, a 1.16, b 0.78, c 0.66 or 1.32, α 92$\pm$5°, β 108$\pm$5°, γ 98$\pm$5° [7]

Volumetric & Calorimetric Properties:
Density:

No.	Value	Note
1	d^{25} 0.9467 g/cm^3	[8]
2	d^{90} 0.9272 g/cm^3	[8]
3	d 0.9475 g/cm^3	[6]
4	d 0.945 g/cm^3	[7]
5	d 0.94 g/cm^3	[31]

Volumetric Properties General: α_v values have been reported [9]
Equation of State: Thermodynamic functions have been measured (0–400K) [68]
Thermodynamic Properties General: Activation energy of polymerisation 21.8 kJmol^{-1} (5.2 kcalmol^{-1}) [10], 52.3 kJmol^{-1} (12.5 kcalmol^{-1}) [11]. Heat of combustion has been reported [68]. Cp values have been reported [12,13,14,68]
Latent Heat Crystallization: Equilibrium thermodynamic props. of formation have been reported [12,68]. Thermodynamic props. of melting have been reported [68]
Melting Temperature:

No.	Value	Note
1	280–285°C	[10]
2	300°C	[15]
3	305°C	[16]
4	320–325°C	[3,17,18,19]
5	340°C	[20]
6	365°C	[6]
7	370–375°C	[2,22,23]
8	382°C	[24]
9	383°C	[25]
10	351°C	(syndiotactic) [1]

Glass-Transition Temperature:

No.	Value	Note
1	90°C	[23,26]
2	118–128.3°C	[8,24]
3	120°C	[27,67]
4	135°C	[28]
5	80°C	atactic [12]
6	127–150°C	atactic [21]
7	138°C	atactic [29]
8	127°C	isotactic [29]
9	133°C	isotactic [21]
10	88°C	head-head [29]

Transition Temperature General: Additional Tg values have been reported [68]
Deflection Temperature:

No.	Value	Note
1	142.5°C	1.84 MPa (264 psi), ASTM D648) [31]

Vicat Softening Point:

No.	Value	Note
1	140°C	[30]

Transition Temperature:

No.	Value	Note	Type
1	405°C	Equilibrium melting point (isotactic) [12]	
2	-130°C	[27,67]	β
3	200°C	Softening range [32]	
4	95°C	Softening point [15]	
5	28°C	Polymer melt temp. [16]	
6	-90°C	Relaxation temp. (183K) [36]	
7	223°C	High temp. transition [24]	

Surface Properties & Solubility:
Solubility Properties General: Solvent-polymer interaction parameter 0.463–0.468 (toluene, 37°) [8]
Cohesive Energy Density Solubility Parameters: (CED 285 Jcm^{-3} (0.000285 J^{-3}); δ 16.9 J$^{1/2}$cm$^{3/2}$ [28]. Energy of molecular cohesion [9] has been reported
Solvents/Non-solvents: Sol. toluene, C_6H_6, cyclohexane, carbon disulfide, tetrahydronaphthalene; [3,7,18,22] CHCl$_3$ [33, 34]; trichloroethylene [34]; boiling *p*-xylene, boiling chlorobenzene, boiling trichlorobenzene, boiling methylcyclohexane, boiling

– Poly(vinyl cyclohexane)

ethylbenzene, boiling decahydronaphthalene, boiling o-dichlorobenzene [10]. Mod. sol. xylene, decahydronaphthalene [3,18]; n-heptane, nonane [10]. Insol. Me_2CO [22]; butanone [10,33]; dioxane [33]; Et_2O [7]; nitrobenzene, CH_2Cl_2 [10]. Atactic, sol. hexane, heptane, cyclohexane, decahydronaphthalene, C_6H_6, toluene, xylene, chlorobenzene, bromobenzene, $CHCl_3$, CCl_4 1,2-dichloroethylene-cis, 1,2-dichloroethylene-trans, THF [35]. Atactic, insol. dioxane, diisopropylether, Et_2O, MeOH, EtOH, propanol, 2-propanol, cyclohexanol, benzyl alcohol, MeOAc, EtOAc, methyl propionate, Me_2CO, butanone, 2-pentanone, 2-heptanone, 3-methyl-2-heptanone [35]. THF is a theta solvent [33]

Transport Properties:
Polymer Solutions Dilute: Other T_θ values have been reported [37]. $[\eta]_{inh}$ 1.05 dl g^{-1} (decahydronaphthalene, 135°) [32]. $[\eta]_{inh}$ 2.4 dl g^{-1} (trichloroethylene) [34]. $[\eta]_{sp}$ 0.61 dl g^{-1} (cyclohexane) [31]. $[\eta]$ 0.36 dl g^{-1} (C_6H_6, 30°, atactic) [29]. $[\eta]$ 0.85 dl g-1 (tetrahydronaphthalene, 90°, isotactic) [29]. $[\eta]$ 0.38 dl g^{-1} (C_6H_6, 30°, head-head) [29]. Other $[\eta]$ values have been measured for various solvents [3,10,15,18,19,20,21,22,31,33,36,37]

Mechanical Properties:
Mechanical Properties General: Thermomechanical prop. curves have been reported [18,24]. Mech. props. have been reported [38]. Dynamic mech. props. have been reported [36]

Tensile Strength Break:

No.	Value	Note
1	16.67–19.61 MPa	170–200 kgm cm^{-2} [39]
2	26.9 MPa	3900 psi, 25° [31]
3	13.1 MPa	1900 psi, 60° [31]
4	10.3 MPa	1500 psi, 80° [31]
5	9.7 MPa	1400 psi, 100° [31]
6	8.3 MPa	1200 psi, 120° [31]

Flexural Strength at Break:

No.	Value	Note
1	19.61–29.42 MPa	200–300 kg cm^{-2} [39]

Elastic Modulus:

No.	Value	Note
1	2900 MPa	[28]

Complex Moduli: Dynamic tensile storage modulus and loss modulus (110Hz, -150–180°, atactic) [21], (110Hz, -160–320°, isotactic) [21], (-180–230°) have been reported [27]
Impact Strength: Impact strength 5–8 kg cm cm^{-2} [39]. Dynstat impact strength 2.9 kg cm cm^{-2} (½″ unnotched) [31]
Viscoelastic Behaviour: Viscoelastic behaviour [40,41] has been reported
Mechanical Properties Miscellaneous: Critical strain 0.35% [28]
Hardness: Rockwell R125.5 [31]

Electrical Properties:
Electrical Properties General: Specific volume electrical resistance (180–160°, 50–1000000Hz) [42]. Dielectric constant of atactic and isotactic forms has been reported. [43] Other dissipation factor values have been reported [42,43]

Dielectric Permittivity (Constant):

No.	Value	Frequency	Note
1	2.1–2.4	50–1000000Hz	-160–220° [20]
2	2.56		[31]

Static Electrification: Conductivity has been reported [43]. Dielectric permeability coefficient -0.2% (glass phase); -0.6% (100° approx.) [42]

Dissipation (Power) Factor:

No.	Value	Frequency	Note
1	0.0006		up to 200° [19]
2	0.0005–0.0008	50 Hz-1 MHz	-10–220° [20]

Optical Properties:
Transmission and Spectra: Ir [3,5,11,18,21,22,25,29,32,36,44,45], H-1 nmr [29,32], C-13 nmr [46,47], esr [48,49,50], epr (20°, 160°) [24] and photoelectron [51] spectral data have been reported. Light transmission 88% (450 nm), 90% (550 nm), 91% (600 nm) [31]
Molar Refraction: dn/dC 0.131 cm^3 g^{-1} (THF, 25°), 0.17 cm^3 (hexane, 25°), 0.101 cm^3 g^{-1} (cyclohexane, 25°) [33]

Polymer Stability:
Decomposition Details: Decomposition temp., T_{dec} [39], 280–530° [7,24,39,60], 368° (atactic) [29], 379° (isotactic) [2,29], 400° (syndiotactic) [2], 366° (head-head) [29]. Weight loss 0.02–0.17% (100°) [8], 0.71–0.93% (150°) [8]. 10% (320°) [24], 50% (390°) [24], 50% (369°) [61]. Temp.-weight loss curves have been reported [12,39,62]. Thermal degradation has been reported [60,62,63,64,65,66]. Activation energy of decomposition 236.6 kJmol^{-1} (56.5 kcal mol^{-1}, 390–420°, argon) [60]. Activation energy of decomposition 184.2 kJmol^{-1} (44 kcal mol^{-1}, atactic), 138.2 kJmol^{-1} (33 kcalmol^{-1}, isotactic), 184.2 kJmol^{-1} (44 kcalmol^{-1}, head-head) [29]. Other values for activation energy of decomposition have been reported [61,62]
Environmental Stress: Effects of antioxidants [39,54,55], ozonolysis [48,49,50,56,57,58], and oxidative thermal degradation [55,59,60] have been reported. Oxidation onset 220° (approx.) [24], 240° (approx.) [60], 260° (approx.) [19] and ozonolysis degradation kinetics have been reported [53]. Activation energy of oxidation 20.5 kJmol^{-1} (4.9 kcalmol^{-1}) [39]. Activation energy of oxidation 129.8 kJmol^{-1} (approx.) (31 kcalmol^{-1} (approx.), 240–320°, air) [60]

Applications/Commercial Products:
Processing & Manufacturing Routes: Synth. using Ziegler-Natta catalysts [47,52], such as $TiCl_3$-$AlMe_3$ in toluene at 50°, or by hydrogenation of polystyrene [21,47]

Bibliographic References
[1] Eur. Pat. Appl. EP, 1989, 322 731, (syndiotactic)
[2] Soga, K., *Jpn. Pat.*, 1990, **02 18 404**, (syndiotactic)
[3] Mushima, E.A, Perelman, A.I., Topchiev, A.V. and Krentsel, B.A., *J. Polym. Sci.*, 1961, **52**, 199, (solubilities, transition temps)
[4] Natta, G., Corradini, P. and Bassi, I.W., *Makromol. Chem.*, 1959, **33**, 247, (cryst struct)
[5] Natta, G., *Makromol. Chem.*, 1960, **35**, 94, (cryst struct)
[6] Noether, H.D., J. Polym. Sci., *Part C: Polym. Lett.*, 1967, **16**, 725, (cryst struct)
[7] Overberger, C.G., Borchert, A.E. and Katchman, A., *J. Polym. Sci.*, 1960, **44**, 491, (cryst struct)
[8] Taylor, G.L. and Davidson, S., J. Polym. Sci., *Part B: Polym. Lett.*, 1968, **6**, 699, (solubilities, density, thermal decomposition)
[9] Eskin, V.E. and Korotkina, Q.Z., *Vysokomol. Soedin., Ser. A*, 1974, **16**, 41, (expansion coefficient, energy of molecular cohesion)
[10] McCarty, W.H. and Parravano, G., J. Polym. Sci., *Part A: Polym. Chem.*, 1965, **3**, 4029, (solubilities)
[11] Topchiev, A.V., Mushina, E.A, Perelman, A.I. and Shishkina, M.V., *Neftekhimiya*, 1964, **4**, 735, (ir)
[12] Grebowicz, J.S., *Polym. Eng. Sci.*, 1992, **32**, 1228, (transition temps)
[13] Eskin, V.E. and Kipper, A.I., *Vysokomol. Soedin., Ser. B*, 1977, **19**, 883, (specific heat capacity)

- Poly(vinyl cyclopentane)

[14] Kipper, A.I. and Eskin, V.E., *Vysokomol. Soedin.*, Ser. B, 1985, **27**, 307, (heat capacity)
[15] Dunham, K.R., Vandenberghe, J., Faber, J.W.H. and Contois, L.E., *J. Polym. Sci., Part A: Polym. Chem.*, 1963, **1**, 751, (transition temps)
[16] Campbell, T.W. and Haven, A.C., *J. Appl. Polym. Sci.*, 1959, **1**, 73, (transition temps)
[17] Topchiev, A.V., Mushima, E.A, Perelman, A.I. and Krentsel, B.A., *Dokl. Akad. Nauk SSSR*, 1960, **130**, 344, (melting point)
[18] Mushina, E.A., Perelman, A.I., Topchiev, A.V. and Krentsel, B.A., *Mezhdunar. Simp. Makromol. Khim.*, Dokl. Avtoreferaty, 1960, **1**, 118, (ir, solubility)
[19] Perelman, A.I., Mushina, E.A. and Topchiev, A.V., *Plast. Massy*, 1964, 3, (electrical props, transition temps.)
[20] *Brit. Pat.*, 1970, 1 212 558, (electrical props)
[21] Abe, A. and Hama, T., *J. Polym. Sci., Part B: Polym. Lett.*, 1969, **7**, 427, (synth, mech props)
[22] Natta, G. and Sianesi, D., *Atti Accad. Naz. Lincei, Rend., Cl. Sci. Fis., Mat. Nat., Rend.*, 1959, **26**, 418, (ir, solubilities)
[23] *High Polymers*, (eds. R.A.V. Raff and K.W. Doak), Interscience, 1965, (transition temps)
[24] Kharas, G.B., Kleiner, V.I., Stotskaya, L.L., Karpacheva, G.P. et al, *J. Polym. Sci., Polym. Chem. Ed.*, 1977, **15**, 755, (thermal decomposition, epr spectra, transition temps)
[25] Ketley, A.D. and Ehrig, R.J., *J. Polym. Sci., Part A: Polym. Chem.*, 1964, **2**, 4461, (ir, melting point)
[26] Reding, F.P., Faucher, J.A. and Whitman, R.D., *J. Polym. Sci.*, 1962, **57**, 483, (transition temp)
[27] Seefield, C.G. and Koleske, J.V., *J. Polym. Sci., Polym. Phys. Ed.*, 1976, **14**, 663, (transition temp, mech props)
[28] Kambour, R.P., *Polym. Commun.*, 1983, **24**, 292, (mech props)
[29] Helbig, M., Inoue, H. and Vogl, O., *J. Polym. Sci., Polym. Symp.*, 1978, **63**, 329, (ir, H-1 nmr)
[30] *Ger. Pat.*, 1962, 1 131 885, (density, vicat temp)
[31] Pendleton, J.F., Hoeg, D.F. and Goldberg, E.P., *Adv. Chem. Ser.*, 1973, **129**, 27, (mech props, optical props, electrical props)
[32] Kennedy, J.P., Elliott, J.J. and Naegele, W., *J. Polym. Sci., Part A: Polym. Chem.*, 1964, **2**, 5029, (ir, H-1 nmr)
[33] Elias, H.G. and Etter, O.J., *J. Makromol. Chem.*, 1966, **1**, 431, (soln props)
[34] Hutchison, J.D., *J. Polym. Sci., Part A: Polym. Chem.*, 1965, **3**, 2710, (cryst struct, solubility)
[35] Elias, H.G. and Etter, O., *J. Macromol. Sci., Chem.*, 1967, **A1(5)**, 943, (solubility)
[36] Frasini, V., Magagnini, P., Butta, E. and Baccanedda, M., *Kolloid Z. Z. Polym.*, 1966, **213**, 115, (ir, mech props)
[37] Kipper, A.I., Korottzina, O.Z. and Eskin V.E., *Vysokomol. Soedin.*, Ser. B, 1973, **3**, 190, (soln props)
[38] Zhorin, V.A., Malkin, A, Y. and Enikolopyan, N.S., *Vysokomol. Soedin.*, Ser. A, 1979, **21**, 820, (mech props)
[39] Zinin, E.F., Ataitin, M.S. and Andrianov, B.V., *Tr. Inst. - Mosk. Khim.-Tekhnol. Inst. im. D.I. Mendeleeva*, 1969, **61**, 233, (mech props)
[40] Zhorin, V.A., Usichenko, V.M., Budnitskii, Y.U., Akutin, M.S. and Enikolopyan, N.S., *Vysokomol. Soedin.*, Ser. A, 1982, **24**, 1889, (mech props)
[41] Kelchner, R.E. and Aklonis, J.J., *J. Polym. Sci., Part A-2*, 1970, **8**, 799, (visoselastic behaviour)
[42] Barsamyan, S.T., Apresyan, A.S., Kleiner, V.I. and Stotskaya, L.L., *Plast. Massy*, 1968, 13, (electrical props)
[43] Talykov, V.A., Kleiner, V.I. and Stotskaya, L.L, *Vysokomol. Soedin.*, Ser. A, 1972, **14**, 432, (electrical props)
[44] Borchert, A.E. and Overberger, C.G., *J. Polym. Sci.*, 1960, **44**, 483, (ir)
[45] Kharas, G.B., Yu, V.K., Kleiner, V.I., Krentsel, B.A., Stotskaya, L.L., Zakharyan, R.Z., *Eur. Polym. J.*, 1973, **9**, 315, (ir)
[46] Weill, G. and Vogl, O., *Polym. Bull. (Berlin)*, 1978, **1**, 181, (C-13 nmr)
[47] Ammendola, P., Tancredi, T. and Zambelli, A., *Macromolecules*, 1986, **19**, 307, (synth, C-13 nmr)
[48] Pokholok, T.V., Vikhlyaev, R.M., Karpukhin, O.N. and Razumouskii, S.D., *Vysokomol. Soedin.*, Ser. B, 1969, **11**, 692, (ozonolysis, ESR)
[49] Rakovskii, S.K., Razumouskii, S.D. and Zaikou, G.E., *Izv. Akad. Nauk SSSR, Ser. Khim.*, 1976, **3**, 701, (ozonolysis, ESR)
[50] Gapanova, I.S., Goldberg, V.M., Zaikov, G.V., Kefeli, A.A., Pariiskii, G.B., Razumouskii, S.D. and Toptygin, D.Y., *Vysokomol. Soedin.*, Ser. A, 1978, **20**, 2038, (ozonolysis, ESR)
[51] Pireaux, J.J., Riga, J., Caudano, R., Verbist, J.J. et al, *Phys. Scr.*, 1977, **16**, 329, (photoelectron spectrum)
[52] Endo, K. and Otsu, T., *J. Polym. Sci., Part A: Polym. Chem.*, 1992, **30**, 679, (synth)
[53] Razumouskii, S.D., Kefeli, A.A. and Zaikov, G.E., *Eur. Polym. J.*, 1971, **7**, 275, (ozonolysis)
[54] Ozyubina, M.A., Kuzmina, G.N., Bakunin, V.N. and Parenago, O.P., *Neftekhimiya*, 1994, **34**, 268, (antioxidants)
[55] Zinin, E.F., Akutin, M.S., Kovarskaya, B.M., Blyumenfeld, A.B. et al, *Plast. Massy*, 1970, **1**, 32, (antioxidants, thermal decomposition)
[56] Razumouskii, S.D., Kefeli, A.A. and Zaikov, G.E., *Kinet. Mech. Polyreactions, Int. Symp. Macromol. Chem., Prepr.*, 1969, **5**, 285, (ozonolysis)
[57] Gapanova, I.S., Gol'dberg, V.M., Zaikov, G.E., Kefali, A.A. et al, *Vysokomol. Soedin.*, Ser. B, 1978, **20**, 699, (ozonolysis)
[58] Gapanova, I.S., Kefeli, A.A., Pariiskii, G.B., Razumouskii, S.D. and Toptygin, D.Y., *Dokl. Akad. Nauk SSSR*, 1977, **234**, 592, (ozonolysis)
[59] Hansen, R.H., Pascale, J.V., DeBenedictus, T. and Rentzepis, P.M., *J. Polym. Sci., Part A: Polym. Chem.*, 1965, **3**, 2205, (oxidation)
[60] Kleiner, V.I., Nschitailo, N.A. and Stotskaya, L.L, *Plast. Massy*, 1971, **11**, 48, (chemical decomposition)
[61] Achammer, B.G., Tryon, M. and Kline, G.M., *Kunststoffe*, 1959, **49**, 600, (thermal decomposition)
[62] Madorsky, S.L., *J. Polym. Sci.*, 1953, **11**, 491, (thermal decomposition)
[63] Konovalov, V.V., Blyumenfel'd, A.B., Zinin, E.F., Akatin, M.S. et al, *Vysokomol. Soedin.*, Ser. A, 1974, **16**, 323, (thermal decomposition)
[64] Luederwald, I. and Vogl, O., *Makromol. Chem.*, 1979, **180**, 2295, (thermal decomposition)
[65] Madorsky, S.L., *J. Res. Natl. Bur. Stand., Sect. A*, 1954, **53**, 361, (pyrolysis)
[66] Straus, S. and Madorsky, S.L., *J. Res. Natl. Bur. Stand., Sect. A*, 1953, **50**, 165, (pyrolysis)
[67] Cowie, J.M.G., *J. Macromol. Sci., Phys.*, 1980, **B18**, 569, (transition temps)
[68] Lebedev, B.V., Smirnova, N.N., Kiparisova, E.G., Vasil'ev, V.G. et al, *Vysokomol. Soedin.*, Ser. A, 1993, **35**, 767, (thermodynamic props)

Poly(vinyl cyclopentane) P-597

Synonyms: *PVCP. Cyclopentane-ethenyl homopolymer. Poly(ethenylcyclopentane). Poly(1-cyclopentyl-1,2-ethanediyl)*
Related Polymers: Poly(vinylcyclopropane)
Monomers: Vinylcyclopentane
Material class: Thermoplastic
Polymer Type: polyolefins
CAS Number:

CAS Reg. No.	Note
30229-22-2	
30847-53-1	isotactic

Molecular Formula: $(C_7H_{12})_n$
Fragments: C_7H_{12}
Morphology: Form I tetragonal, a 2.014, b 2.014, c 0.650, helix 4_1 [1]; form II tetragonal, a 2.011, b 2.011, c 1.95, helix 12_3 [1]; form III tetragonal, a 3.73, b 3.73, c 1.98, helix 10_3 [1]; triclinic, a 1.05, b 0.74, c 0.66, α $92 \pm 5°$, β $108 \pm 5°$, γ $99 \pm 5°$ [3]

Volumetric & Calorimetric Properties:
Density:

No.	Value	Note
1	d 0.965 g/cm^3	[3]
2	d 0.95 g/cm^3	form III [1]

Melting Temperature:

No.	Value	Note
1	270°C	[7]
2	292°C	[5]

Glass-Transition Temperature:

No.	Value	Note
1	70°C	[2,8]
2	75°C	[6,7]

Transition Temperature:

No.	Value	Note	Type
1	-140°C	[2,8]	Relaxation temp.

Surface Properties & Solubility:
Solvents/Non-solvents: Sol. C_6H_6 [4], $CHCl_3$ [4], trichloroethylene [4]. Slightly sol. pentane [5]. Insol. Et_2O [4,6].

Transport Properties:
Polymer Solutions Dilute: [η] 0.56 dl g^{-1} [3]

Mechanical Properties:
Mechanical Properties General: Variation of dynamic loss modulus with temp. has been reported in graphical form [2]

Optical Properties:
Transmission and Spectra: Ir [3,5] and H-1 nmr [5] spectral data have been reported

Polymer Stability:
Thermal Stability General: Decomposition temp. 260° (min.) [3]

Applications/Commercial Products:
Processing & Manufacturing Routes: Synth. using Ziegler-Natta catalysts such as $TiCl_4$-$AlBu_3$ in heptane at 68° [2,3,5]

Bibliographic References
[1] Noether, H.D., *J. Polym. Sci., Part C: Polym. Lett.*, 1967, **16**, 725, (struct)
[2] Seefried, C.G. and Koleske, J.V., *J. Polym. Sci., Polym. Phys. Ed.*, 1976, **14**, 663, (synth, transition temps)
[3] Borchert, A.E. and Overberger, C.G., *J. Polym. Sci.*, 1960, **44**, 483, 491, (ir, struct, synth, decomposition temp)
[4] Jamison, S.E., *Text. Res. J.*, 1970, **40**, 955, (solubilties)
[5] Ketley, A.D. and Ehrig, R.J., *J. Polym. Sci., Part A: Polym. Chem.*, 1964, **2**, 4461, (ir, H-1 nmr, synth, melting point)
[6] Reding, F.P., Faucher, J.A. and Whitman, R.D., *J. Polym. Sci.*, 1962, **57**, 483, (transition temps)
[7] High Polymers Part 1, (eds. R.A.V. Raff and K.W. Doak), Interscience, 1965, **20**, (transition temps)
[8] Cowie, J.M.G., *J. Macromol. Sci., Phys.*, 1980, **18**, 569, (transition temps)

Poly(vinyl cyclopropane)   P-598
Synonyms: *PVCP. Poly(ethenylcyclopropane). Poly(1-cyclopropyl-1,2-ethanediyl)*
Related Polymers: Poly(vinylcyclopentane)
Monomers: Vinylcyclopropane
Material class: Thermoplastic
Polymer Type: polyolefins
CAS Number:

CAS Reg. No.	Note
26617-60-7	
26617-61-8	isotactic

Molecular Formula: $(C_5H_8)_n$
Fragments: C_5H_8
General Information: Powder [1]
Morphology: Form I hexagonal, a 1.36, b 1.36, c 0.648, helix 3_1 [2]; form II tetragonal, a 1.521, b 1.521, c 2.085, helix 10_3 [2] (c 0.65, helix 3_1) [3]

Volumetric & Calorimetric Properties:
Density:

No.	Value	Note
1	d 0.9752 g/cm^3	[1]
2	d 0.975 g/cm^3	form I [2]
3	d 0.9805 g/cm^3	form I [2]
4	d 0.926 g/cm^3	form II [2]

Melting Temperature:

No.	Value	Note
1	228–230°C	[1,3]

Surface Properties & Solubility:
Solvents/Non-solvents: Insol. C_6H_6 [3,4], cyclohexane, CCl_4, carbon disulfide [3]

Transport Properties:
Polymer Solutions Dilute: [η] 1.3 dl g^{-1} (135°, tetrahydronaphthalene) [3]

Optical Properties:
Transmission and Spectra: Ir [1,3,5,6] and H-1 nmr [1,5] spectral data have been reported

Polymer Stability:
Decomposition Details: Decomposition temp. 260° (min.) [4]

Applications/Commercial Products:
Processing & Manufacturing Routes: Synth. using Ziegler-Natta catalysts such as $TiCl_3$-$AlEt_2Cl$ in heptane at 25° [1,3,4,7]

Bibliographic References
[1] Overberger, C.G. and Halek, G.W., *J. Polym. Sci., Part A-1*, 1970, **8**, 359, (synth, ir, H-1 nmr, melting point)
[2] Noether, H.D., Overberger, C.G. and Halek, G.W., *J. Polym. Sci., Part A-1*, 1969, **7**, 201, (struct)
[3] Natta, G., Sianesi, D., Morero, D., Bassi, I.W. and Caporiccio, G., *CA*, 1960, **55**, 9940i, (synth, solubility, melting point)
[4] Overberger, C.G., Borchert, A.E. and Katchman, A., *J. Polym. Sci.*, 1960, **44**, 491, (synth, decomposition temp)
[5] Ketley, A.D., Berlin, A.J. and Fisher, L.P., *J. Polym. Sci., Part A-1*, 1967, **5**, 227, (ir, H-1 nmr)
[6] Borchert, A.E. and Overberger, C.G., *J. Polym. Sci.*, 1960, **44**, 483, (ir)
[7] Halek, G.W., *Diss. Abstr. Int. B*, 1967, **28**, 876, (synth)

Poly(vinyl difluoroacetate)   P-599
Synonyms: *Poly[(difluoroacetoxy)ethylene]. Poly(ethenyl difluoroacetate)*
Monomers: Vinyl difluoroacetate
Material class: Thermoplastic
Polymer Type: polyvinyl acetate
CAS Number:

CAS Reg. No.
105596-75-6

Molecular Formula: $(C_4H_4F_2O_2)_n$
Fragments: $C_4H_4F_2O_2$
General Information: Reported to have similar props. to poly(vinyl trifluoroacetate)

Bibliographic References
[1] *U.S. Pat.*, 1948, 2 436 144

Poly(vinyl ethers)　　　　　　　　　P-600

$$\mathrm{-[CH_2CH]_n}$$
$$\mathrm{\quad\quad |}$$
$$\mathrm{\quad\quad OR}$$

where R = Me, Et, etc

Synonyms: *Poly(vinyl alkyl ethers). Poly(alkyl vinyl ethers)*
Related Polymers: Poly(vinyl methyl ether-*co*-maleic anhydride), Poly(vinyl methyl ether), Poly(vinyl octadecyl ether), Poly(vinyl propyl ether), Poly(vinyl hexyl ether), Poly(vinyl methyl ether-*co*-maleic acid)
Monomers: For vinyl ether derivatives
Material class: Synthetic Elastomers
Polymer Type: polyvinyls
General Information: The physical props. of polymers based upon alkyl vinyl ethers depend upon MW, the alkyl group, polymerisation initiator, stereospecificity and crystallinity. Materials range from viscous liqs., through to sticky liqs., rubbery or brittle solids. Long alkyl chains prod. low MW waxy materials. Copolymers with maleic anhydride are the most significant commercial class of vinyl ether copolymers. Methyl vinyl ether-maleic anhydride copolymers dominate the commercial market [1,2]

Applications/Commercial Products:
Processing & Manufacturing Routes: Homopolymers prod. by cationic initiation with Friedel-Crafts catalysts. Copolymers prod. by free radical initiated copolymerisation. Vinyl ether polymerisation rates decrease rapidly with solvent polarity [3]. Many initiator systems have been reported to give stereoregular homopolymers [4,5]
Applications: Adhesives, surface coatings, lubricants, greases, elastomers, moulding compounds, fibres, films. Also used as plasticisers. Waxy materials are used as polishing and waxing agents

Trade name	Details	Supplier
Gantrez		GAF
Gantrez B	isobutyl ether	GAF
Gantrez M	methyl ether	GAF
Lutonal		BASF
Lutonal A	ethyl ether	BASF
Lutonal I	isobutyl ether	BASF
Lutonal M	methyl ether	BASF
PVEE	ethyl ether elastomeric solid	Union Carbide
V-Wax	octadecyl ether, low MW	BASF

Bibliographic References
[1] Schröder, G., Ullmanns Encykl. Ind. Chem. VCH, 1993, 11, (general)
[2] Kirk-Othmer Encycl. Chem. Technol. John Wiley and Sons, 3rd edn., (ed. M. Grayson), 1978, **23**, 937, (general)
[3] Hsieh, W.C., Kubota, H., Squire, V.R. and Stannett, V., *J. Polym. Sci., Polym. Chem. Ed.*, 1980, **18**, 2773, (polymerisation)
[4] Schildknecht, C.E., Zoss, A.O. and McKinley, C., *Ind. Eng. Chem.*, 1947, **39**, 180, (stereoregular polymerisation)
[5] Schildknecht, C.E., Gross, S.T. and Zoss, A.O., *Ind. Eng. Chem.*, 1949, **41**, 1998, (stereoregular polymerisation)

Poly(vinyl ethyl ether)　　　　　　　　　P-601

$$\mathrm{-[\quad]_n}$$
$$\mathrm{EtO}$$

Synonyms: *Poly(ethoxyethylene). Poly(ethoxyethene). Poly(1-ethoxyethylene). Poly(ethyl vinyl ether). PVE. PVEE*
Related Polymers: Poly(vinyl ethers), Poly(vinyl ethyl ether-*co*-carbon dioxide), Poly(vinyl ethyl ether-*co*-maleic anhydride)
Monomers: Ethyl vinyl ether
Material class: Gums and resins
Polymer Type: polyvinyls
CAS Number:

CAS Reg. No.	Note
25104-37-4	polyethoxyethene
31550-08-0	polyethoxyethylene

Molecular Formula: $(C_4H_8O)_n$
Fragments: C_4H_8O
Tacticity: Atactic and isotactic forms are known

Volumetric & Calorimetric Properties:
Density:

No.	Value	Note
1	d^{20} 0.957 g/cm^3	[15]
2	d^{27} 0.968 g/cm^3	
3	d^{30} 0.951 g/cm^3	[4]

Thermal Expansion Coefficient:

No.	Value	Note	Type
1	0.000242 K^{-1}	T > -33° [17,19]	L
2	0.000101 K^{-1}	-90 < T < -33°	L
3	8.9×10^{-5} K^{-1}	-178° < T < -90° [17,19]	L
4	4.9×10^{-5} K^{-1}	T < -178°	L

Thermodynamic Properties General: Thermodynamic props. have been reported [18,26]
Latent Heat Crystallization: ΔH_f 9.39 kJ mol^{-1} (2240 cal mol^{-1}) [7]. ΔS_f 0.015 kJ mol^{-1} (3.5 cal mol^{-1}) [7]
Melting Temperature:

No.	Value	Note
1	86°C	[1,27]

Glass-Transition Temperature:

No.	Value	Note
1	-33°C	[26]
2	-43°C	[4]

Transition Temperature General: $T_{gg}(1)$ -90° [17]. $T_{gg}(2)$ -222° [17]
Vicat Softening Point:

No.	Value	Note
1	57–68°C	[16]

Transition Temperature:

No.	Value	Note	Type
1	-23°C	[4]	Inflection temp.
2	-21°C	[4]	Inflection temp.
3	-20°C	[4]	Low temp. stiffening
4	-10°C	[4]	Low temp. stiffening

Surface Properties & Solubility:
Cohesive Energy Density Solubility Parameters: δ 31.59 $J^{1/2}$ $cm^{-3/2}$ (7.54 $cal^{1/2}$ $cm^{-3/2}$) [4], 38.21 $J^{1/2}$ $cm^{-3/2}$ (9.12 $cal^{1/2}$ $cm^{-3/2}$) [29]. Other solubility parameters have been reported [29]
Solvents/Non-solvents: Sol. CH_2Cl_2, $CHCl_3$, EtOH, MeOH [8], C_6H_6, Et_2O, Me_2CO, vinyl alcohol [15]. Insol. H_2O [8], MeOH, heptane, hexane, ethylene glycol, Et_2O

Transport Properties:
Polymer Solutions Dilute: [η] 1.45 dl g^{-1} [16]. Absolute viscosity 0.8015 centipoise. $η_{inh}$ 1.9 dl g^{-1} [4]

Mechanical Properties:
Hardness: Shore A15 [16], A20 (24h) [30]

Optical Properties:
Refractive Index:

No.	Value	Note
1	1.454	30° [4,5]
2	1.452	20° [15]
3	1.3761	20° [16]

Transmission and Spectra: Nmr [21,25] and ir [31] spectral data have been reported
Molar Refraction: Molar refraction 20.53 [4]

Polymer Stability:
Polymer Stability General: Oxidatively degraded by air
Stability Miscellaneous: Stability to γ-irradiation has been reported 0.36 G scissions $(100\ eV)^{-1}$ [18]

Applications/Commercial Products:
Processing & Manufacturing Routes: Stereoregular polymerisation takes place in toluene with ferric oxide as initiator at -20° giving approx. 97% conversion [6,7]. The atactic form may be obtained by using heterogeneous aluminium sulfate-sulfuric acid as catalyst [9]. A considerable number of catalysts and synthetic routes have been developed and studied. Catalysts include hydrosilicates [10], inorganic acids [11], ferric sulfate [12] (which gives high MW stereoregular polymers), and a variety of organic peroxides [13]. Low MW polymers can be prepared by the use of zeolites [14]. Ionic polymerisations [15], and cationic polymerisations initiated by antimonate salts [21,23,24] have been developed along with the use of Grignard reagents [20] and photopolymerisation techniques [22]
Applications: Used as a plasticiser for cellulose nitrate and cellulose resin lacquers, as a base for pressure-sensitive adhesives in films and tapes, especially in surgical applications. Also used for non-irritating surgical casts requiring moisture permeability and to impart dry-film flexibility and increased viscosity of photo-chemical resist-coating compositions

Trade name	Supplier
Lutonal A	BASF
PVEE	Union Carbide

Bibliographic References
[1] Nielsen, L.E. and Landel, R.F., *Mechanical Properties of Polymers and Composites,* 2nd edn., Marcel Dekker, 1993, (melting point, glass transition temp)
[2] Schmeider, K. and Wolf, K., *Kolloid Z. Z. Polym.,* 1953, **134**, 149, (glass transition temp)
[3] Jenckel, E., *Kolloid Z. Z. Polym.,* 1962, **100**, 163, (glass transition temp)
[4] Lal, J. and Trick, G.S., *J. Polym. Sci., Part A-2,* 1964, **2**, 4559, (glass transition temp, refractive indices)
[5] Andrews, R.J. and Grulke, E.A., *Polym. Handb.,* 4th edn., (eds. J. Brandrup, E.H. Immergut and E.A. Grulke), John Wiley and Sons, 1999, **VI**, 214, (refractive index, glass transition temp)
[6] *Encycl. Polym. Sci. Eng.,* 2nd edn., (ed. J.I. Kroschwitz), John Wiley and Sons, 1985, **17**, 446, (synth)
[7] Hino, M. and Arata, K, *Chem. Lett.,* 1980, **8**, 1963, (stereoregular polymerisation)
[8] *Polymeric Materials Encyclopedia,* (ed. J.C. Salamone), CRC Press, 1996, (uses, solubility)
[9] Rapoport, V.O., Portyanskii, A.E., Gurevich, V.R. and Maksimova, M.V., *Vysokomol. Soedin., Ser. A,* 1970, **12**, 1958, (polymerisation)
[10] *Brit. Pat.,* 1932, 379 674, (catalysts)
[11] *Brit. Pat.,* 1932, 378 544, (polymerisation)
[12] *Brit. Pat.,* 1959, 846 690, (catalyst)
[13] *U.S. Pat.,* 1961, 2 967 203, (catalysts)
[14] Barrer, R.M. and Oei, T.T., *J. Catal.,* 1974, **34**, 19, (polymerisation)
[15] Shostakovsky, M.F., Mikhantyer, B.I. and Orchinnikova, N.N., *Bull. Acad. Sci. USSR, Div. Chem. Sci. (Engl. Transl.),* 1953, 939, (polymerisation)
[16] Fishbein, L. and Crowe, B.F., *Makromol. Chem.,* 1961, **48**, 221, (props)
[17] Schell, W.J., Simha, R. and Aklonis, J.J., *J. Macromol. Sci., Chem.,* 1969, **3**, 1297, (transition temps)
[18] Suzuki, Y., Rooney, J.M. and Stannett, V., *J. Macromol. Sci., Chem.,* 1978, **12**, 1055, (degradation studies)
[19] Haldon, R.A., Schell, W.J. and Simha, R., *J. Macromol. Sci., Phys.,* 1967, **1**, 759, (transition temp)
[20] Bruce, J.M. and Farrer, D.W., *Polymer,* 1963, **4**, 407, (polymerisation)
[21] Harris, J.J. and Temin, S.C., *J. Polym. Sci., Part A-1,* 1972, **10**, 1165, (polymerisation)
[22] Kaeriyama, K. and Shimura, Y., *J. Polym. Sci., Polym. Chem. Ed.,* 1972, **10**, 2833, (photopolymerisation)
[23] Chung, Y.J., Rooney, J.M., Squire, D.R. and Stannett, V., *Polymer,* 1975, **16**, 527, (polymerisation)
[24] Ledwith, A., Lockett, E. and Sherrington, D.C., *Polymer,* 1975, **16**, 31, (props)
[25] Ramey, K.C., Field, N.D. and Borchert, A.E., *J. Polym. Sci., Part A: Polym. Chem.,* 1965, **3**, 2885, (nmr)
[26] Aharoni, S.M., *Polym. Prepr. (Am. Chem. Soc., Div. Polym. Chem.),* 1973, **14**, 334
[27] Vandenburg, E.J., Heck, R.F. and Breslow, D.S., *J. Polym. Sci.,* 1959, **41**, 519
[28] Hopfinger, A.J., Pearlstein, R.A., Taylor, P.L. and Boyle, F.P., *J. Macromol. Sci., Phys.,* 1987, **26**, 359
[29] Ahmad, H. and Yaseen, M., *Polym. Eng. Sci.,* 1979, **19**, 858, (solubility parameters)
[30] Schildknecht, C.E., *Polym. Prepr. (Am. Chem. Soc., Div. Polym. Chem.),* 1973, **13**, 1071
[31] Haque, S.A., Uryu, T. and Okawa, H., *Makromol. Chem.,* 1987, **188**, 2523

Poly(vinyl ethyl ether-*co*-carbon dioxide) P-602

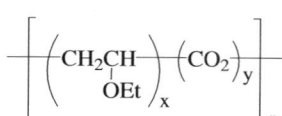

Related Polymers: Poly(vinyl ethers)
Monomers: Ethyl vinyl ether, Carbon dioxide
Material class: Thermoplastic, Copolymers
Polymer Type: polyester, polyvinyls
CAS Number:

CAS Reg. No.
55993-87-8

Molecular Formula: $[(C_4H_8O).(CO_2)]_n$
Fragments: C_4H_8O CO_2

Optical Properties:
Transmission and Spectra: Ir spectral data have been reported [1]

Applications/Commercial Products:
Processing & Manufacturing Routes: Prod. by cationic polymerisation with Lewis Acid catalyst

Bibliographic References
[1] Soga, K., Hosoda, S., Tazuke, Y. and Ikeda, S., *J. Polym. Sci., Polym. Lett. Ed.,* 1975, **13**, 265

Poly(vinyl fluoride) P-603

$$\mathrm{\left[CH_2CH \atop F \right]_n}$$

Synonyms: *PVF. Vinyl fluoride polymer. Poly(fluoroethene)*
Monomers: Vinyl fluoride
Material class: Thermoplastic
Polymer Type: PVF, fluorocarbon (polymers)
CAS Number:

CAS Reg. No.
24981-14-4

Molecular Formula: $(C_2H_3F)_n$
Fragments: C_2H_3F
Molecular Weight: M_n 76000–234000, MS 143000–654000, MS $\cong$ MW
Additives: Pigments, stabilisers, flame retardants, plasticisers.
Tacticity: Commercial material is usually atactic
Morphology: Polyvinyl fluoride is a semicrystalline polymer with a planar zig-zag chain configuration. Amount of crystallinity will depend on the polymerisation method

Volumetric & Calorimetric Properties:
Density:

No.	Value	Note
1	d 1.38–1.57 g/cm^3	[5]
2	d 1.45 g/cm^3	[3]

Thermal Expansion Coefficient:

No.	Value	Note	Type
1	5×10^{-5} K^{-1}	30 min., air [5]	L
2	9×10^{-5} K^{-1}	[3]	L

Thermodynamic Properties General: ΔH 120–240 kJ mol^{-1} (upper T_g value); ΔH 160 kJ mol^{-1} (lower T_g value) [5]
Thermal Conductivity:

No.	Value	Note
1	0.14 W/mK	-30 [5]
2	0.17 W/mK	60° [5]

Melting Temperature:

No.	Value	Note
1	197–235°C	dependent on degree of crystallinity [1,15]

Glass-Transition Temperature:

No.	Value	Note
1	40–45°C	(upper value) relaxation in the amorph. region under restraint by crystallinity [5]
2	-20°C	(lower value) relaxation in the amorph. region essentially free from restraint
3	40–50°C	(upper value) relaxation in the amorph. region under restraint by crystallinity [8,9,10,11]
4	-20–-15°C	(lower value) relaxation in the amorph. region, essentially free from restraint [7]

Transition Temperature General: T_m increases with decreasing polymerisation temp. and with increasing degree of crystallinity [15]
Deflection Temperature:

No.	Value	Note
1	121°C	0.45 Mpa
2	82°C	1.8 MPa [3]

Transition Temperature:

No.	Value	Note	Type
1	-70°C	[3]	Minimum service temp.
2	-80°C	[1]	Short chain relaxation in the amorph. region
3	150°C	[1]	Premelting intracrystalline relaxation

Surface Properties & Solubility:
Solubility Properties General: Low solubility in all solvents below 100°. Solubility may be increased by addition of 0.1% 2-propanol as modifier. δ 38.9 J cm^{-3} (9.3 cal cm^{-3}, CHCl$_3$, 23°); 53.2 J cm^{-3} (12.7 cal cm^{-3}, EtOH, 23°); 41.4 J cm^{-3} (9.9 cal^{-3}, Me$_2$CO, 23°) [12]
Cohesive Energy Density Solubility Parameters: δ 50.2–50.7 J$^{1/2}$ cm$^{-3/2}$ (12.0–12.1 cal$^{1/2}$ cm$^{-3/2}$) [1]
Solvents/Non-solvents: Slightly sol. DMF, dimethyl acetamide, γ-butyrolactone, propylene carbonate (above 100°) [1]

Transport Properties:
Transport Properties General: Has good barrier props.
Permeability of Gases: [CO$_2$] 1.7, [He] 22.9, [H$_2$] 8.9, [N$_2$] 0.04, [O$_2$] 0.5 cm^3 (m^2 day kPa)$^{-1}$ (98 kPa, 23.5°, ASTM D1434) [5]
Permeability and Diffusion General: Vapour permeability g (m^2 day kPa)$^{-1}$, 23.5°), AcOH 10, Me$_2$CO 2400, C$_6$H$_6$ 22, CCl$_4$ 12, EtOAc 240, hexane 13 [5], H$_2$O 40–50 g (m^2 d)$^{-1}$ (7 kPa, 39.5°, ASTM E96-58T)
Water Absorption:

No.	Value	Note
1	>0.5 %	min., 23° [5]

Mechanical Properties:
Tensile (Young's) Modulus:

No.	Value	Note
1	1700–2600 MPa	23°, ASTM D882 Method A [5]

Flexural Modulus:

No.	Value	Note
1	1400 MPa	1.4 GN m^{-2} [3]

Tensile Strength Break:

No.	Value	Note	Elongation
1	48–120 MPa	23°, ASTM D882 Method A [5]	100% elongation min.

Tensile Strength Yield:

No.	Value	Elongation	Note
1	33–41 MPa	100% elongation min.	23°, ASTM D882 Method A [5]

Impact Strength: Impact strength 10–22 kJ m^{-1} (DuPont pneumatic tester) [5]
Mechanical Properties Miscellaneous: Ultimate elongation 115–250% (ASTM D882 Method A, 100% min., 23°) [5]. Tear strength 4.6–39 kJ m^{-1}, (Elmendorf propagated) [5]. Tear strength 174–239 kJ m^{-1}, (ASTM 1004 initial value). Bursting strength, 131–482 kPa (Mullen, ASTM D774) [5]
Hardness: Shore D80 [3]
Friction Abrasion and Resistance: Good abrasion resistance [3]. Coefficient of friction 0.15–0.3 (film/metal, Instron)
Izod Notch:

No.	Value	Notch	Note
1	1800 J/m	Y	[3]

Electrical Properties:
Dielectric Permittivity (Constant):

No.	Value	Frequency	Note
1	8.5–9.9	1 kHz	23°, ASTM D150 [5]

Dielectric Strength:

No.	Value	Note
1	130 kV.mm^{-1}	ASTM D150 [5]
2	20 kV.mm^{-1}	
3	140 kV.mm^{-1}	[3]

Strong Field Phenomena General: Corona endurance 40 V μm^{-1} (60 Hz, ASTM D2275) [5]
Dissipation (Power) Factor:

No.	Value	Frequency	Note
1	1.6	1 MHz	ASTM D150 [3]

Optical Properties:
Refractive Index:

No.	Value	Note
1	1.46	30°, ASTM D542 [5]

Transmission and Spectra: Transparent to radiation in the uv, visible and near ir regions of the solar spectrum, transmitting over 90% of the radiation at 350–2000 nm. It absorbs radiation of 7000–12000 nm [1]. Other details of transmission and spectra have been reported [14]

Polymer Stability:
Polymer Stability General: Has excellent weatherability and strength over a wide range of temps.
Upper Use Temperature:

No.	Value	Note
1	107–175°C	[5]
2	105–120°C	ASTM D150 [3]
3	105–120°C	[3]

Decomposition Details: Degradation in air is a two step process with loss of hydrogen fluoride followed by backbone cleavage at 350 and 450° respectively. Char also forms which resists further degradation until 650° [1]. The main products of pyrolysis and non-flaming oxidative degradation at 450° are hydrogen fluoride and methane [13]

Flammability: Self ignition temp. 390° (ASTM D1929) [5]. Limiting oxygen index 22.6% [1]
Chemical Stability: Inert to a wide variety of chemicals, corrosive and staining agents [1]
Stability Miscellaneous: Relatively unstable at processing temps.

Applications/Commercial Products:
Processing & Manufacturing Routes: Polymerisation requires high pressure and high temp. The use of Ziegler-Natta catalysts lowers polymerisation temp. and pressure. Bulk, suspension and emulsion techniques may be used. Thin film may be produced by extrusion and dispersion. The polymer may be applied to substrates using dispersion or powder coating
Applications: PVF finds wide use as protective or decorative coatings, either as preformed film in a laminating step, or from dispersion. Uses include highway sound barriers, automobile trim, aircraft cabin interiors, solar panel covers, greenhouses, bags for insulating mats and release sheets in plastics processing.

Trade name	Supplier
Dalvor	Diamond Shamrock
Tedlar PVF film	DuPont

Bibliographic References
[1] *Encycl. Polym. Sci. Eng.*, Wiley Interscience, 1989, **17**
[2] Brydson, J.A., *Plast. Mater.*, 5th edn., Butterworth Heinemann Ltd., 1989
[3] *The Materials Selector*, Elsevier Applied Science, 1991, **3**
[4] *Encyclopedia of Advanced Materials*, (eds. D. Bloor, R.J. Brook, M.C. Flemings, S. Mahajan and R.W. Cahn), Pergamon Press, 1994, **3**
[5] *Kirk-Othmer Encycl. Chem. Technol.*, Wiley Interscience, 1980, **11**
[6] Kerbow, D.L. and Sperati, C.A., *Polym. Handb.*, 4th edn., (eds. J. Brandrup, E.H. Immergut and E.A. Grulke), John Wiley and Sons, 1999, **V**, 48
[7] Boyer, R.F., *J. Polym. Sci., Polym. Symp.*, 1975, **50**, 189
[8] Enns, J.B. and Simha, R., *J. Macromol. Sci., Phys.*, 1977, **13**, 11
[9] Fujii, M., Stannett, V. and Hopfenberg, H.B., *J. Macromol. Sci., Phys.*, 1978, **15**, 421
[10] Ziegel, K.D., Frensdorf, H.I. and Blair, D.E., *J. Polym. Sci., Part A: Polym. Chem.*, 1969, **7**, 809
[11] Gorlitz, M., Minke, R., Trautvetter, W. and Weisberger, G., *Angew. Makromol. Chem.*, 1973, **29/30**, 137
[12] Chapira, A., Mankowski, Z. and Schmitt, N., *J. Polym. Sci., Polym. Chem. Ed.*, 1982, **20**, 1791
[13] Chatfield, D.A., *J. Polym. Sci., Polym. Chem. Ed.*, 1983, **21**, 1681
[14] Petitt, K.B., *Sol. Energy Mater.*, 1979, **1**, 125
[15] Sianesi, D. and Caporricio, D., *J. Polym. Sci., Part A: Polym. Chem.*, 1968, **6**, 335

Poly(vinyl fluoroacetate)   P-604
Synonyms: Poly[(fluoroacetoxy)ethylene]
Monomers: Vinyl fluoroacetate
Material class: Thermoplastic
Polymer Type: polyvinyl acetate
Molecular Formula: $(C_4H_5FO_2)_n$
Fragments: $C_4H_5FO_2$
General Information: Reported to have similar props. to poly(vinyl trifluoroacetate)

Bibliographic References
[1] *U.S. Pat.*, 1948, 2 436 144

Poly(vinyl formal)   P-605

Poly(vinyl formal)

Synonyms: *PVF*
Related Polymers: Polyvinyl butyral, Polyvinyl acetal, Polyvinyl alcohol
Monomers: Vinyl acetate, Formaldehyde
Material class: Thermoplastic
Polymer Type: polyvinyl acetate, polyvinyls
CAS Number:

CAS Reg. No.
9003-33-2

Molecular Formula: $[(C_5H_8O_2)_x(C_2H_4O)_y(C_4H_6O_2)_z]$
Fragments: $C_5H_8O_2$ C_2H_4O $C_4H_6O_2$
Molecular Weight: 15000–45000
Additives: Antioxidants improve thermal stability
General Information: Props. of resin depend on MW and degree of acetalisation. Manufacturing process yields a terpolymer struct. [2]. May contain up to 10–20% acetate and 5–10% hydroxyl groups. Statistical degree of acetalisation is 87% (approx.) [3]. Hydroxyl content is independent of reaction time and temp. Increased polyvinyl acetate content contributes to softer resins, lower heat distortion temp. and reduced tensile and impact strength
Tacticity: Atactic, isotactic and syndiotactic forms known, related to the tacticity of starting material, polyvinyl acetate. Isotactic polymer chains undergo acetalisation more rapidly than syndiotactic portions [1]

Volumetric & Calorimetric Properties:
Density:

No.	Value	Note
1	d 1.2 g/cm^3	[5]

Glass-Transition Temperature:

No.	Value	Note
1	103–113°C	ASTM D1043, Formvor [6]

Transition Temperature General: MW change little influences thermal props. Increasing polyvinyl acetate content lowers distortion temp. [6]
Deflection Temperature:

No.	Value	Note
1	88–93°C	Formvor 7/95 [5]
2	50–60°C	7% OH, 40–50% acetate [5]

Vicat Softening Point:

No.	Value
1	140–170°C

Surface Properties & Solubility:
Solubility Properties General: Compatible with phenolic and epoxy resins
Plasticisers: Plasticisers include phthalates, phosphates and adipate esters
Solvents/Non-solvents: Sol. THF, phenol, dimethylacetamide, DMSO, *N*-methyl pyrrolidone. Formvor 7/95 (5% OH, 15% PVA): sol. AcOH, aniline, CHCl$_3$, dioxane, pyridine, tetrachloroethane, nitropropane, furfural, DMF; insol. Me$_2$CO, EtOH, MeOH, cyclohexanone, EtOAc, hydrocarbons, methyl cellosolve, 2-butanone [3]. Formvor 12/85 (5% OH, 22–30% PVA): sol. AcOH, aniline, CHCl$_3$, cyclohexanone, tetrachloroethane, dioxane, isophorone, nitropropane, methyl cellosolve, pyridine, DMF; insol. MeOH, EtOH, Me$_2$CO, hydrocarbons, EtOAc, 2-butanone [3]. Increasing polyvinyl acetate content increases solubility in less polar solvents; increasing free hydroxyl content increases H$_2$O solubility

Transport Properties:
Polymer Solutions Dilute: Soln. viscosity 500–600 cP (MW 26000–34000, 22–30% PVA, 5% OH), 100–200 cP (MW 10000–15000, 9–13% PVA, 5–6% OH), 300–500 cP (MW 16000–20000, 9–13% PVA, 5–6% OH) [6], 40–60 cP (20°, dichloroethane) [5]. Viscosity increases with increasing MW and increasing poly(vinyl acetate) content
Water Absorption:

No.	Value	Note
1	1 %	24h, 22–30% PVA, 5% OH [5]
2	1.2 %	24h, 9–13% PVA, 5% OH [5]

Mechanical Properties:
Mechanical Properties General: Strength, rigidity, elongation and impact strength decrease with increasing polyvinyl acetate content [5]
Tensile Strength Break:

No.	Value	Note	Elongation
1	68.9–75.8 MPa	10–11 kpsi, ASTM D638, Formvor [5]	7–50%

Elastic Modulus:

No.	Value	Note
1	2413–4136 MPa	350–600 kpsi, ASTM D1043, Formvor [5]

Tensile Strength Yield:

No.	Value	Elongation	Note
1	59–66 MPa	7%	ASTM D638, Formvor [6]

Flexural Strength Yield:

No.	Value	Note
1	115–124 MPa	ASTM D790, Formvor [6]

Hardness: Rockwell M150-M155 (ASTM D785) [6], E65-E75 (ASTM D785, Formvor) [6]
Friction Abrasion and Resistance: Has good abrasion resistance
Izod Notch:

No.	Value	Notch	Note
1	70 J/m	Y	ASTM D256, Formvor [6]
2	3202 J/m	N	600 ft lb in^{-1}, ASTM D256 [5]

Electrical Properties:
Dielectric Permittivity (Constant):

No.	Value	Frequency	Note
1	3.2–3.4	50 Hz	ASTM D150, Formvor [6]
2	3–3.3	1 kHz	ASTM D150, Formvor [6]
3	2.8–3.1	1 MHz	ASTM D150, Formvor [6]

– Poly(vinyl formate)

Dielectric Strength:

No.	Value	Note
1	12–13 kV.mm^{-1}	short-time, ASTM D149, Formvor [6]

Dissipation (Power) Factor:

No.	Value	Frequency	Note
1	0.0081–0.0087	50 Hz	ASTM D150, Formvor [6]
2	0.01	1 kHz	ASTM D150, Formvor [6]
3	0.021	1 MHz	ASTM D150, Formvor [6]

Optical Properties:
Refractive Index:

No.	Value	Note
1	1.5	25° [5]

Transmission and Spectra: Ir [4,8] and H-1 nmr [7,8] spectral data have been reported
Volume Properties/Surface Properties: Colourless

Polymer Stability:
Polymer Stability General: High degree of acetalisation implies good thermal stability
Thermal Stability General: Stable up to 320° in N_2. Polymer darkens to deep brown colour at elevated temp.
Decomposition Details: Decomposition in air occurs via oxidative cleavage of acetal ring producing formaldehyde [10]
Flammability: Fl. p. 370° (min.)
Chemical Stability: Has good resistance to oils and hydrocarbons
Hydrolytic Stability: Stable to base, but sensitive to oxidation in base. Hydrolysed by strong acid

Applications/Commercial Products:
Processing & Manufacturing Routes: Synth. [11] by *in situ* hydrolysis of poly(vinyl acetate) in AcOH by sulfuric acid with formaldehyde at 65–90°. Polymer is produced after neutralisation. MW depends on MW of starting acetate, and composition is determined by quantities of AcOH, H_2O and formaldehyde added. Also produced by acid catalysed acetalisations of poly(vinyl alcohol)
Applications: Primarily used as oil resistant coating for electrical wires in combination with phenolic resins. Also used as an adhesive, surface coating agent, binder in alkali batteries and in rocket casings

Trade name	Supplier
Formvor	Monsanto Chemical
Pioloform F	Wacker
Vinilex	Shin Nippon

Bibliographic References
[1] Blomstrom, T.P., *Encycl. Polym. Sci. Eng.*, 2nd edn., (ed. J.I. Kroschwitz), Vol. 17, John Wiley and Sons, 1985, 136, (rev)
[2] Flory, P.J., *J. Am. Chem. Soc.*, 1939, **61**, 1518, (synth)
[3] Raghavendrachar, P. and Chanda, M., *Eur. Polym. J.*, 1983, **19**, 391, (kinetics)
[4] Chanda, M. and Kumar, V., *Angew. Makromol. Chem.*, 1977, **62**, 229, (kinetics, synth., ir)
[5] Fitzhugh, A.F., Lavin, E. and Morrison, G.O., *J. Electrochem. Soc.*, 1953, **100**, 351, (rev, props, synth.)
[6] Lavin, E. and Snelgrove, J.A., *Kirk-Othmer Encycl. Chem. Technol.*, 3rd edn., Vol. 23 (ed. M. Grayson), Wiley Interscience, 1978, 798, (rev, props)
[7] Shibatani, K., Fujii, K., Oyanagi, Y., Ukidu, J. and Matsumoto, M., *J. Polym. Sci., Part C: Polym. Lett.*, 1968, **23**, 647, (H-1 nmr)
[8] Shibatani, K., Fujiwara, Y. and Fujii, K., *J. Polym. Sci., Part A: Polym. Chem.*, 1970, **8**, 1693, (ir, H-1 nmr)
[9] Beachell, H.C., Fotis, P. and Hucks, J., *J. Polym. Sci.*, 1951, **7**, 353, (degradation)
[10] Chanda, M., Kumar, W.S.J. and Raghavendrachar, P., *J. Appl. Polym. Sci.*, 1979, **23**, 755, (thermal degradation)
[11] *U.S. Pat.*, 1934, 2 168 827, (manuf)

Poly(vinyl formate) P-606
Synonyms: *Poly(ethenyl formate)*. PVF. PVFO
Monomers: Vinyl formate
Material class: Thermoplastic
Polymer Type: polyvinyls
CAS Number:

CAS Reg. No.	Note
25567-89-9	homopolymer
27101-53-7	syndiotactic

Molecular Formula: $(C_3H_4O_2)_n$
Fragments: $C_3H_4O_2$
Tacticity: Syndiotactic and isotactic forms are known as well as a stereoblock form
Morphology: Rhombohedral, a 15.9Å, b 15.9Å, c 6.55Å [5]

Volumetric & Calorimetric Properties:
Density:

No.	Value	Note
1	d^{30} 1.34 g/cm^3	[5]

Thermodynamic Properties General: ΔH_{act} for polymerisation has been reported [4]
Glass-Transition Temperature:

No.	Value	Note
1	30.5°C	[5]
2	37°C	60% syndiotactic [1]
3	33°C	50% syndiotactic [1]

Transition Temperature General: T_g varies considerably with polymerisation temp. [5] and the extent of syndiotacticity [1,5]

Surface Properties & Solubility:
Solvents/Non-solvents: Sol. $CHCl_3$, H_2O, dioxane [11] and Me_2CO [12]. Syndiotactic [5] sol. acetonitrile. Isotactic sol. formic acid, DMSO [5]. Isotactic insol. acetonitrile, methyl formate [5]

Optical Properties:
Refractive Index:

No.	Value	Note
1	1.4757	20° [1]

Transmission and Spectra: Visible absorption [11] and nmr spectral data have been reported [6]. Important characteristic ir band at 1026 cm^{-1} in the syndiotactic [5] and stereoblock forms [9]. This band is absent from the isotactic ir spectrum

Polymer Stability:
Environmental Stress: Frost resistant but may coagulate if aq. phase freezes [3]
Biological Stability: Unprotected emulsions susceptible to attack from microorganisms [3]

Applications/Commercial Products:
Processing & Manufacturing Routes: Prod. by radical polymerisation of vinyl formate using typical peroxide initiators [2,3,10]. May be stabilised with anionic or non-ionic emulsifiers or water-sol.

protective colloids. Syndiotactic propagation is achieved using AIBN [4] or triethylboron catalyst [6,8]. Final polymerisation result is very dependent upon polymerisation medium [5,7]
Applications: Major applications in the area of coatings

Bibliographic References

[1] Serefis, J.C., *Polym. Handb.*, 4th edn., (eds. J. Brandrup, E.H. Immergut and E.A. Grulke), John Wiley and Sons, 1999, **VI**, 581, (refractive index)
[2] Bevington, J.C. and Johnson, M., *Eur. Polym. J.*, 1968, **4**, 373, (radical polymerisation)
[3] *Ullmanns Encykl. Ind. Chem.*, 5th edn., (ed. W. Gerhartz), VCH, 1985, (stability, synth)
[4] Nozakura, S.-I., Sumi, M., Uoi, M., Okamoto, T. and Murahashi, S., *J. Polym. Sci., Polym. Chem. Ed.*, 1973, **11**, 279
[5] Fujii, K., Mochizuki, T., Imoto, S., Ukida, J. and Matsumoto, M., *J. Polym. Sci., Part A: Polym. Chem.*, 1964, **2**, 2327
[6] Ramey, K.C., Lini, D.C. and Stratton, G.L., *J. Polym. Sci., Part A: Polym. Chem.*, 1967, **1**, 257, (nmr)
[7] Fujii, K., Imoto, S., Ukida, J. and Matsumoto, M., *J. Polym. Sci., Part B: Polym. Lett.*, 1963, **1**, 497
[8] *Jpn. Pat.*, 1961, 14 039
[9] Fujii, K., Nagoshi, K., Ukida, J. and Matsumoto, M., *Makromol. Chem.*, 1963, **65**, 81
[10] Elias, H.-G., Riva, M. and Göldi, P., *Makromol. Chem.*, 1971, **145**, 163, (free-radical polymerisation)
[11] Rosen, I., McCain, G.H., Endrey, A.L. and Sturm, C.L., *J. Polym. Sci., Part A: Polym. Chem.*, 1963, **1**, 951
[12] Vansheydt, A.A. and Chelpanora, L.F., *J. Gen. Chem. USSR (Engl. Transl.)*, 1950, **20**, 2353
[13] *U.S. Pat.*, 1944, 2 360 308, (polymerisation)
[14] *U.S. Pat.*, 1949, 2 492 929, (polymerisation)
[15] *U.S. Pat.*, 1965, 3 214 419, (ionic polymerisation)
[16] *U.S. Pat.*, 1966, 3 239 494, (ionic polymerisation)
[17] *Ger. Pat.*, 1941, 703 075, (gas phase polymerisation)

Poly(vinyl hexadecyl ether-*co*-vinyl octadecyl ether-*co*-styrene-*co*-maleic anhydride)   P-607

Synonyms: *Poly(hexadecyloxyethene-co-octadecyloxyethene-co-ethenylbenzene-co-2,5-furandione)*
Related Polymers: Poly(vinyl ethers)
Monomers: Styrene, Maleic anhydride
Material class: Synthetic Elastomers, Copolymers
Polymer Type: polystyrene, polyvinyls
CAS Number:

CAS Reg. No.
40472-29-5

Molecular Formula: $[(C_{18}H_{36}O).(C_{20}H_{40}O).(C_8H_8).(C_4H_2O_3)]_n$
Fragments: $C_{18}H_{36}O$ $C_{20}H_{40}O$ C_8H_8 $C_4H_2O_3$

Applications/Commercial Products:
Processing & Manufacturing Routes: Prod. by free radical copolymerisation of a mixture of maleic anhydride, styrene, a vinyl ether of Alfal 2.8 and AIBN heated at 65° for 21h
Applications: Aq. soln. thickening agent for heavy-duty cleaners and ammonia-based fertilisers

Bibliographic References
[1] *U.S. Pat.*, 1973, 3 723 375, (use)

Poly(vinyl hexadecyl ether-*co*-vinyl octadecyl ether-*co*-N-vinyl pyrrolidinone)   P-608

Synonyms: *Poly(hexadecyloxyethene-co-octadecyloxyethene-co-1-ethenyl-2-pyrrolidinone)*
Related Polymers: Poly(vinyl ethers)
Monomers: *N*-Vinylpyrrolidinone
Material class: Thermoplastic, Copolymers
Polymer Type: polyvinyls
CAS Number:

CAS Reg. No.
29351-75-5

Molecular Formula: $[(C_{18}H_{36}O).(C_{20}H_{40}O).(C_6H_9NO)]_n$
Fragments: $C_{18}H_{36}O$ $C_{20}H_{40}O$ C_6H_9NO

Bibliographic References
[1] *U.S. Pat.*, 1974, 3 839 310, (use)

Poly(vinyl hexyl ether)   P-609

Synonyms: *Poly(hexyl vinyl ether). Poly(hexoxyethene)*
Related Polymers: Poly(vinyl ethers)
Monomers: Vinyl hexyl ether
Material class: Synthetic Elastomers
Polymer Type: polyvinyls
CAS Number:

CAS Reg. No.
25232-88-6

Molecular Formula: $(C_8H_{16}O)_n$
Fragments: $C_8H_{16}O$

Volumetric & Calorimetric Properties:
Density:

No.	Value	Note
1	d^{30} 0.902 g/cm^3	[1]
2	d^{27} 0.925 g/cm^3	

Thermodynamic Properties General: Molar volume V_r 138.6 cm^3 mol^{-1} [3]. Thermodynamic props. have been reported [2]
Glass-Transition Temperature:

No.	Value	Note
1	-74°C	[3]
2	-77°C	[1]

– Poly(vinylidene chloride)

Transition Temperature General: Boiling point 145° [4]
Vicat Softening Point:

No.	Value	Note
1	49–57°C	[4]

Transition Temperature:

No.	Value	Note	Type
1	-61°C	[1]	Inflection temp.

Surface Properties & Solubility:
Cohesive Energy Density Solubility Parameters: δ 32.72 $J^{1/2}$ $cm^{3/2}$ (7.81 $cal^{1/2}$ $cm^{3/2}$) [1]

Optical Properties:
Refractive Index:

No.	Value	Note
1	1.4591	30° [1]

Molar Refraction: Molar refraction 38.86 [1]

Applications/Commercial Products:
Processing & Manufacturing Routes: Prod. by cationic polymerisation with Friedel-Crafts catalysts (BF_3, $AlCl_3$, $SnCl_4$ or their complexes)

Bibliographic References
[1] Lal, J. and Trick, G.S., *J. Polym. Sci., Part A: Polym. Chem.*, 1964, **2**, 4559
[2] Aharoni, S.M., *Polym. Prepr. (Am. Chem. Soc., Div. Polym. Chem.)*, 1973, **14**, 334
[3] Van Krevelen, D.W., *Properties of Polymers: Their Correlation with Chemical Structure*, Elsevier, 1972, 44
[4] Fishbein, L. and Crowe, B.F., *Makromol. Chem.*, 1961, **48**, 221

Poly(vinylidene chloride) P-610

$$\text{-}[CH_2CCl_2]_n\text{-}$$

Synonyms: *Poly(1,1-dichloroethene). Poly(1,1-dichloroethylene). PVDC. PVdC*
Related Polymers: Poly(vinyl chloride-*co*-vinylidene chloride), Poly(vinylidene chloride-*co*-acrylonitrile), Poly(vinylidene chloride-*co*-butadiene-*co*-co styrene, Vinylidene chloride copolymers
Monomers: Vinylidene chloride
Material class: Thermoplastic
Polymer Type: polyvinylidene chloride
CAS Number:

CAS Reg. No.
9002-85-1
98846-23-2

Molecular Formula: $(C_2H_2Cl_2)_n$
Fragments: $C_2H_2Cl_2$
Molecular Weight: Mw 9100–126000 Mv 8500–117000
Additives: Heat and light stabilisers acid-acceptor type e.g. amines, phenyl glycidyl ether, certain organometallics plasticisers, dyes pigments lubricants
General Information: Due to its high thermal instability, the homopolymer has little commercial importance. Principal applications are with copolymers
Morphology: Hard, tough, translucent to transparent material sometimes faintly yellow in colour [8] Stereoregular polymer [9, 28] Symmetrical molecular structure result in high crystallinity [27]
Identification: Cl present (Ca 70%), N, P, S absent [8] Colour tests darkens when left for some days in contact with conc. NH4OH [8] brown/black colour with pyridine/methanolic KOH soln [6] brown/black ppt with pyridine/methanolic NaOH soln (c.f. PVC homopolymer . Where brown/ black colouration is initially observed, followed by slow formalin of brown precipitate [6], immersion in morpholine causes both sample and liquid to blacken within a few hours [6, 8] latter two methods allow identification of PVDC even in presence of PVC

Polymer Stability:
Thermal Stability General: Displays low thermal stability [7], may darken an lose strength (especially in presence Cu, Fe, etc); deterioration not serious with properly compounded materials [8]. Min. use temp. 0° (film) [7]
Upper Use Temperature:

No.	Value	Note
1	143°C	approx. commercial film [7]

Decomposition Details: TGA Studies show some decomposition beginning at 140° [6]; decomposition occurs as a result of dehydrochlorination reaction [8], formation of conjugated unsaturated units gives yellow or brown. Coloration (particularly under alkaline conditions) [8]. Decomp. temp. 210–225 [7], 200° (film) [8], 250° (cryst.)
Flammability: Flammability rating V0 (commercial vinyl-based resin) [28]. Excellent fire-resistance props. as a consequence of high chlorine content [9]; does not readily burn, self-extinguishing [7,8], softens, melts, leaves swollen carbonaceous residue [8]
Environmental Stress: Displays poor resistance to sunlight and uv-irradiation; darkens on exposure [7,8] and may lose strength [8]
Chemical Stability: Swollen by oxygen-containing liquids, e.g. dioxane, cyclohexanone, dimethylformamide and some chlorination hydrocarbons [8]; relatively unaffected by hydrocarbons (mineral oils), alcohols, phenols, many common organic solvents, moderately concentrated acids and alkalis, except ammonia and strong amines) [8]; exceptional resistance to water [8]. Chemical resistance is reduced with increases in temp. [8] decomposed by prolonged contact with ammonia, strong amines and related compounds [8]; darkening observed, some yellowing observed in presence of other alkalis [8]; attacked by chlorine, and slowly by cocentrated sulfuric acid [8], copper and iron catalyse decomposition of polymer [8]
Biological Stability: Unaffected by bacteria, fungi and insects [8]
Stability Miscellaneous: High chlorine content can lead to rapid degradation at normal process temps.; risk of splitting off of chlorine during processing, thus only non-ferrous materials should be allowed to come into contact with melt [8,27]; however, compromise/balance must be made, as high crystallinity necessitates use of high tool temps. [27]

Applications/Commercial Products:

Trade name	Details	Supplier
Daran		W R Grace
Daran X		W R Grace
Ixan		Laporte
Saran		Dow Plastics
Trovidur	300	Huls America
Tygan	coated material	Coulthards Advanced Materials (Holdings) Ltd
Unocal 76 Res 5001		Unocal
Unocal 76 Res 5527		Unocal
Unocal 76 Res 5581		Unocal
Unocal 76 Res 701		Unocal

Poly(vinylidene chloride-co-acrylonitrile)   P-611

$$\left[\left(CH_2C\genfrac{}{}{0pt}{}{Cl}{Cl}\right)_x\left(CH_2CH\genfrac{}{}{0pt}{}{}{CN}\right)_y\right]_n$$

Synonyms: *Poly (1,1-dichloroethene-co-2-propenenitrile). Poly(vinylidene chloride-co-vinyl cyanide)*
Related Polymers: Vinylidene chloride copolymers, Poly acrylonitrile
CAS Number:

CAS Reg. No.	Note
9010-76-8	
124322-63-0	block copolymer

Fragments: C_2H_3Cl C_3H_3N
Additives: Anti-blocking agents (films, coatings)
General Information: Copolymers usually contain 5–15% acrylonitrile. Copolymers are more easily processed than the homopolymer. Copolymer exhibits excellent barrier props.
Morphology: Unperturbed chain dimension parameters derived from viscosity/molecular weight data; A = 0.664 a.u., Af = 0.313 a.u; steric factor (A/AP), 2.12; copolymers with low AN content used as model to make comparisons with polyvinyl chloride homopolymer; results indicate that latter has more extended chain, as a consequence of increased hindering potential about C-C bond through addition of second Cl atom [1] Modacrylic fibres contain 35–84% acrylonitrile content; flame-proof ; good resistance to sunlight, water acids and alkalis; cheaper than materials containing higher proportions of acrylonitrile, although not resistant to org.solvents, soften at lower temps and have slightly lower strengths [10]
Miscellaneous: The terms polymer and copolymer are used interchangeably with regard to Poly(vinylidene chloride) copolymers; and care must be take to avoid ambiguities

Volumetric & Calorimetric Properties:
Density:

No.	Value	Note
1	d 1.34 g/cm^3	modacrylic fibre [10]

Transition Temperature General: Modacrylic fibres soften at lower temps than other commercial products containing higher proportions of acrylonitrile; readily shrink unless stabilised [10]

Transport Properties:
Transport Properties General: Display excellent barrier props
Polymer Solutions Dilute: Viscosity/molecular weight relationship study of copolymer samples η 0.177–0.568 dl g^{-1} (MW 31300–189800; M_n 27600–139700, 9wt% acrylonitrile, TMF) Mark-Howink relationships, and unperturbed chain dimensions have been reported [1]
Permeability of Gases: Gas permeability increases with acrylonitrile content and decreasing crystallininity
Water Absorption:

No.	Value	Note
1	0.5–1 %	65% relative humidity [10]

Mechanical Properties:
Mechanical Properties General: Modacrylic fibres have lower strengths than other commercial products containing higher proportoons of acrylonitrile [10]

Friction Abrasion and Resistance: Coefficient of friction/μstat (polymer/poymer) 0.8; μdyn (polymer/steel sphere) 0.65 [4]
Optical Properties:
Transmission and Spectra: IR spectrum of copolymer containing 20% acrylonitrile has been reported [7, 10] Characteristic absorption at 2240 cm^{-1}
Polymer Stability:
Flammability: Modacrylic fibres are flame-proof [10]
Environmental Stress: Resistant to sunlight (modacrylic fibres) [10]
Chemical Stability: Modacrylic fibres resistant to water; moderately resistant to acids and cold alkalis; attacked by hot alkalis; fibres swollen by some org. solvs. [10]
Hydrolytic Stability:

Applications/Commercial Products:

Trade name	Supplier
Saran	Dow
Teklan	Courtaulds
Viclan	Dow

Poly(vinylidene fluoride)   P-612

$$+\!\!\left[CH_2CF_2\right]\!\!+_n$$

Synonyms: *PVDF. Vinylidene fluoride polymer. Poly(1,1-difluoroethene). PVF_2. PVdF*
Related Polymers: Poly(vinylidene fluoride-co-hexafluoroisobutylene), Poly(vinylidene fluoride-co-hexafluoropropylene, Poly(vinylidene fluoride-co-tetrafluoroethylene)
Monomers: Vinylidene fluoride
Material class: Thermoplastic
Polymer Type: PVDF, fluorocarbon (polymers)
CAS Number:

CAS Reg. No.
24937-79-9

Molecular Formula: $(C_2H_2F_2)_n$
Fragments: $C_2H_2F_2$
Additives: Carbon, carbon fibre
General Information: Partially crystalline polymer containing 59.4% w/w fluorine and 3% hydrogen. Typical commercial material contains 3.5–6 mol% of head-to-head and tail-to-tail units
Morphology: PVdF crystallises in three different forms: Form I (β), form II (α) and form III (γ). Form I is monoclinic with a trans-gauche-trans-gauche (TGTG) conformn.; form II is orthorhombic with a trans-trans (TT) or planar zig-zag conformn., and form III is monoclinic with planar zig-zag TT conformn. The most commonly available are the polar form I and the electrically inactive form II. Commercial bulk polymer usually consists of form II. For high degrees of electrical activity, a significant amount of form I crystalline material is necessary. The degree of electrical activity will depend on the polarity, i.e. the degree to which the dipoles can be made parallel to one another. A common method of achieving this is to apply a high potential across oriented sheet
Miscellaneous: PVdF is an electronically-active polymer, the piezoelectric and pyroelectric responses of PVdF rival those of ceramics. (PVdF is known as a ferroelectric material)

Volumetric & Calorimetric Properties:
Density:

No.	Value	Note
1	d 1.75–1.8 g/cm^3	ASTM D792 [3]

Poly(vinylidene fluoride)

Thermal Expansion Coefficient:

No.	Value	Note	Type
1	9×10^{-5}–0.00012 K^{-1}	ASTM D696 [2]	L

Latent Heat Crystallization: Heat of fusion 29.5 kJ mol^{-1} (46.1 kJ kg^{-1}) [5]. ΔS 74.7 J K^{-1} mol^{-1} (17°) [8]. ΔH 111.7 kJ mol^{-1} (17°) [8]. ΔG 104.9 kJ mol^{-1} (17°)

Thermal Conductivity:

No.	Value	Note
1	0.1–0.13 W/mK	[3]

Specific Heat Capacity:

No.	Value	Note	Type
1	1.255–1.425 kJ/kg.C	[5]	P
2	1.217 kJ/kg.C	77.9 J K^{-1} mol^{-1}, -3° [8]	P

Melting Temperature:

No.	Value	Note
1	155–192°C	ASTM D3418 [3]

Glass-Transition Temperature:

No.	Value	Note
1	-40°C	[3]

Deflection Temperature:

No.	Value	Note
1	150°C	0.455 MPa [2]
2	100°C	1.82 MPa [2]

Brittleness Temperature:

No.	Value	Note
1	-64–-62°C	ASTM D746 [3]

Surface Properties & Solubility:
Solubility Properties General: Has good compatibility with a range of oxygen containing polymers [2]
Solvents/Non-solvents: Sol. dimethylacetamide, DMF, hexamethyl phosphoramide and DMSO [5]. Some highly polar solvents such as ketones and ethers are absorbed

Transport Properties:
Melt Flow Index:

No.	Value	Note
1	5–180 g/10 min	265°, ASTM D1238 [2]

Volatile Loss and Diffusion of Polymers: Thermogravimetric weight loss 0.5% (1h, 300°) [2]

Water Absorption:

No.	Value	Note
1	0.03–0.06 %	ASTM D570 [5]

Mechanical Properties:
Mechanical Properties General: Has excellent mech. props., with good impact strength and excellent ability to withstand fatigue crack propagation. Tensile strengths up to 290 MPa can be achieved by orientation [1]

Tensile (Young's) Modulus:

No.	Value	Note
1	1200–1600 MPa	ASTM D638 [2]

Flexural Modulus:

No.	Value	Note
1	1200–1800 MPa	ASTM D790 [2]

Tensile Strength Break:

No.	Value	Note	Elongation
1	30–60 MPa	25° [2]	25–500%
2	20–30 MPa	100°, ASTM D638 [2]	300–600%

Flexural Strength at Break:

No.	Value	Note
1	55–75 MPa	ASTM D790 [2]

Compressive Modulus:

No.	Value	Note
1	1000–1400 MPa	ASTM D695 [2]

Poisson's Ratio:

No.	Value	Note
1	0.383	Kynar 730

Tensile Strength Yield:

No.	Value	Note
1	30–60 MPa	25° [2]
2	20–40 MPa	25°

Compressive Strength:

No.	Value	Note
1	55–100 MPa	ASTM D695

Mechanical Properties Miscellaneous: Tear strength 250–320 N mm^{-1} (40 μm thick, oriented film, machine direction), 160–200 N mm^{-1} (25 μm thick, oriented film, machine direction), 170–245 N mm^{-1} (9 μm thick, oriented film, machine direction) [1]
Hardness: Durometer D70-D80 (ASTM D2240) [2]. Shore R110 [6]
Friction Abrasion and Resistance: Tabor abrasion 7.0–10 mg (1000 cycles)$^{-1}$ (Wheel CS-17, 1 kg) [3]; 17.6 mg (1000 cycles)$^{-1}$ (Wheel CS-17, 0.5 kg) [4]. Sand abrasion 4 m^3 mm^{-1} (ASTM D968) [3]

Poly(vinylidene fluoride)

Izod Notch:

No.	Value	Notch	Note
1	75–235 J/m	Y	25°, ASTM D256 [4]
2	700–2300 J/m	N	25°, ASTM D256 [4]

Electrical Properties:
Electrical Properties General: Exhibits ferroelectric behaviour after alignment of crystal dipoles in an electric field. This makes PVdF a very useful material in piezoelectric and pyroelectric applications

Dielectric Permittivity (Constant):

No.	Value	Frequency	Note
1	8.4–13.5	60 Hz	ASTM D150 [2]
2	7.4–13.2	1 kHz	ASTM D150 [2]
3	6–7.6	1 MHz	ASTM D150 [2]

Dielectric Strength:

No.	Value	Note
1	10.23–37.4 kV.mm^{-1}	260–950 V mil^{-1}, ASTM D149 [5]

Arc Resistance:

No.	Value	Note
1	50–60s	ASTM D495 [2]

Complex Permittivity and Electroactive Polymers: Pyroelectric coefficient 24–28 μC m^{-2} K^{-1} [1]; 27–40 μC m^{-2} K^{-1} [2,5]. Piezoelectric coefficient d_{31} 21–28, d_{32} 2.3, d_{33} -32 [2]. Piezoelectric coefficient g^{31} 0.12–0.14, g^{32} 0.18–0.022 Vm N^{-1} [1]. Piezoelectric strain constant 20–50 pm V^{-1} [3]. Piezoelectric stress constant (pressure) 0.23 V m^{-1} Pa^{-1} [3]. Piezoelectric constants, d 20 × 10^{-12} mV^{-1}, e 6 × 10^{-2} N Vm^{-1}, g 0.174 Vm N^{-1}, h 530000000 Vm^{-1} [9]

Dissipation (Power) Factor:

No.	Value	Frequency	Note
1	0.012–0.019	60 Hz	ASTM D150
2	0.013–0.019	1 kHz	ASTM D150
3	0.15–0.22	1 MHz	ASTM D150

Optical Properties:
Refractive Index:

No.	Value	Note
1	1.41–1.42	25°, ASTM D542 [2]

Transmission and Spectra: Exhibits high transmittance (80–90%) at 100 μm thickness in the visible region of the electromagnetic spectrum making it suitable for solar applications

Volume Properties/Surface Properties: Clarity transparent to translucent [6]

Polymer Stability:
Upper Use Temperature:

No.	Value	Note
1	148°C	[6]
2	150–190°C	

Decomposition Details: The predominant method of thermal degradation is dehydrofluorination. The second major product is trifluorobenzene, which is prod. by chain scission and cyclisation. The onset of rapid thermal degradation occurs at 390°. After a weight loss of 70%, the charred PVdF is stabilised and no further gaseous evolution occurs [2]

Flammability: Flammability rating V0 (UL 94). Limiting oxygen index 43% (ASTM D2863, Kynar 730) [5]

Environmental Stress: Has superb weathering characteristics, with no deterioration of mech. props. after years of outdoor exposure [2]

Chemical Stability: Resistant to most inorganic and organic chemicals at room temp. It is, however, especially sensitive to attack from amines [5]

Stability Miscellaneous: Has exceptional resistance to γ-radiation [2]. At 100 Mrad there is essentially no loss in tensile strength

Applications/Commercial Products:
Processing & Manufacturing Routes: Free radical suspension, emulsion or soln. polymerisation. Processed by extrusion or injection moulding

Mould Shrinkage (%):

No.	Value	Note
1	2.5–3%	[6]

Applications: Long lasting metal finishes, wire and cable insulation, fluid handling systems, packaging, chemical processing equipment. Transducers, ultrasonics, hydrophones, loudspeakers, headphones, microphones and pressure sensors

Trade name	Details	Supplier
Dyflor		Dynamit Nobel
Foraflon		Atochem Polymers
Foraflon HD	HD, LD, VLD grades	Atochem Polymers
KF		Kureha Chemical Industry Co.
Kynar		Pennwalt Chemical Corporation
Kynar		Atochem Polymers
Nesflon		Daikin Kogyo
Solef		Solvay & Cie
Vidar		Solvay & Cie

Bibliographic References

[1] Holmes-Siedle, A.G., Wilson, P.D. and Verrall, A.P., *Mater. Des.*, 1984, **4**, 910
[2] Lovinger, A.G., *Dev. Cryst. Polym.*, (ed. D.C. Bassett), Applied Science, 1982
[3] *Kirk-Othmer Encycl. Chem. Technol.*, Vol. 11, 4th edn., (ed. J.I. Kroschwitz), 1980, **11**, 65
[4] *Encycl. Polym. Sci. Eng.*, Vol. 17, 2nd edn., (ed. J.I. Kroschwitz), Wiley Interscience, 1989, 532
[5] Kerbow, D.L. and Sperati, C.A., *Polym. Handb.*, 4th edn., (eds. J. Brandrup, E.H. Immergut and E.A. Grulke), John Wiley and Sons, 1999, **V**, 48, (mech props)
[6] *The Materials Selector*, 2nd edn., (eds. N.A. Waterman and M.F. Ashby), Chapman and Hall, 1997, **3**, 1714
[7] Basics of Design Engineering: Materials: Plastics, *Mach. Des.*, 1995, **67**, 79
[8] Lee, W.K. and Choy, C.L., *J. Polym. Sci., Polym. Phys. Ed.*, 1975, **13**, 619
[9] Sussner, H., *Ultrason. Symp. Proc.*, 1979, 491

Poly(vinylidene fluoride-*co*-chloro trifluoroethylene)

P-613

$$\left[\left[CF_2CF_2 \right]_x \left[\begin{array}{c} CFCF_2 \\ | \\ Cl \end{array} \right]_y \right]_n$$

Synonyms: *Poly(1,1-difluoroethene-co-chlorotrifluoroethene). PVDF-co-PCTFE*
Related Polymers: Poly(vinylidene fluoride)
Monomers: Vinylidene fluoride, Chlorotrifluoroethylene
Material class: Thermoplastic, Copolymers, Synthetic Elastomers
Polymer Type: PVDF, PCTFEfluorocarbon (polymers)
CAS Number:

CAS Reg. No.	Note
9010-75-7	
123236-28-2	graft

Molecular Formula: $[(C_2H_2F_2)_x(C_2ClF_3)_y]_n$
Fragments: $C_2H_2F_2$ C_2ClF_3
Molecular Weight: MW 750000–1000000; MW 840000; M_n 480000; MW 75000; M_n 29000 (23–28 mol% vinylidene fluoride); MW 447000 (vinylidene fluoride:CTFE 1:30); MW 322000 (vinylidene fluoride:CTFE 1:12); MW 996000 (vinylidene fluoride:CTFE 1:9); MW 279000 (vinylidene fluoride:CTFE 2:1) MW 416000 (vinylidene fluoride:CTFE 4:1); MW 75700 (commercial resin, 22 wt% CTFE); M_n 29400 (commercial resin, 22 wt% CTFE)
Additives: Plasticisers/softeners; vulcanising (cross-linking) agents; thermal stabilisers ('Gossypol', phenyl ketone derivatives, hydroquinone derivatives, $CaCl_2$, methyl cellulose); fillers (silica, silicates, carbon blacks, Cu powder); modifiers (low MW chlorofluorocarbons); anti-oxidants (phenyl ketone derivatives); anti-caking agents (polyolefins, alkaline earth salts); processing aids; release agents and colours (occasionally)
General Information: Are speciality, high-performance fluorocarbon copolymers available either as thermoplastics (containing less than 15 mol% CTFE and known as 'flexible PVDF') or as fluoroelastomers, which are commercially available as 30/70 and 50/50 CTFE/vinylidene fluoride compositions (most commonly used/well known); fluoroelastomers are usually tough transparent to translucent gums [36,37]. Copolymers containing low amounts of vinylidene fluoride have excellent optical, electrical, chemical, thermal and mechanical props. They find uses where good moisture barrier and optical transparency props. are needed: such props. are dependent on morphology [31]. Physical and mechanical props., compounding, processing and applications have been reviewed [36,37]
Tacticity: Atactic fluorine rubber [6]
Morphology: Has a hexagonal struct. (helical arrangement) (3 mol% vinylidene fluoride) [41]. Copolymer containing 25–70 mol% vinylidene fluoride is amorph. [41]. Cryst. struct. of 95 wt% vinylidene fluoride copolymer before and after poling has been reported [46,52]. Monoclinic, a 0.496, b 0.965, c, 0.462 [46,52]; these dimensions are v. close to those of PVDF α-form, plus very small amount of PVDF β-form [46,52]. There is little change in struct. after poling although ir spectral data reveal decrease in α-phase component and increase in β-form component [46,52]. Positron annihilation studies of the microdefect struct. have been reported [30]. Structural assignments of chemical shifts in F-19 high-resolution (338.7 MHz) nmr spectra have been correlated with 5-carbon (pentad) sequences in copolymer backbone [8,29]. Methods to monitor crystallinity of copolymers containing 0.72–3.4 wt% vinylidene fluoride have been reported; degree of crystallinity (films), 51–16% (from dsc), 64–11% (from xrd), 77–13% (from density measurements) [31]. Morphology of block copolymers consists of hard (PVF) and soft segments [48]. Copolymers containing 96–97 mol% CTFE have a crystalline struct. (hexagonal, helical arrangement). Increasing vinylidene fluoride content 25–70 mol% gives amorph. materials [41]. Amorph. fluoroelastomer has globular struct; same struct. both in films and bulk. It is resistant to corrosive liqs., milling, and unchanged by vulcanisation and is altered only by high pressure [42]. Ms studies of microstruct. reveal comonomers linked by intermediate $-CH_2CFCl-$ and $-CF_2CF_2-$ units [45]. Far ir measurements have been used to investigate degree of ordering/chain conform.; order reduced by increasing amounts of CTFE [13]. X-ray measurements show struct. is similar to that of PVDF (Form II) [13]. Typical commercially available copolymer, e.g. CTFE:VDF (75:25) is partially crystalline. The extent of crystallinity depends on thermal history. Annealing at temps. $T_g < T < T_m$ (also under strain) produces slow crystallisation resulting in a rather low ultimate degree of crystallinity, as randomly distributed ribbon-like lamellae. Thermal and mechanical (tensile) behaviour following annealing have also been reported [15]. The crystallisation rate is also dependent on processing history; DSC studies have revealed 'memory effect' in some copolymer systems [34]. X-ray diffraction anal. of semi-crystalline polymer shows unusually large crystallites (less than 500Å) depending on processing conditions; apparent crystallite size can be regarded as measure of crystalline perfection [16]. Molecular motions of copolymer have been investigated by esr spectroscopy and fluorescence polarisation experiments [55]. Motions of graft (VDF on CTFE) copolymer have been studied by pulsed pmr spectroscopy; the struct. consists of a ribbon-like (mobile) component and a crystalline (immobile) component; motions correlate with mechanical props. [33]. Commercial elastomer is amorph. at temps. down to -40°, but partially crystallises at 300% elongation [58]. Small amounts of CTFE reduce crystallinity of polyvinylidene fluoride. Greater than 30 mol% vinylidene fluoride can be incorporated into helical struct. of CTFE homopolymer before crystallinity and struct. are significantly affected. 50/50 Compositions have a fair amount of disorder along chain direction and are of random struct. with low degree of alternation, head-to-head, head-to-tail [51]
Identification: Fluorine determined by burning in presence of urethane rubber, PE and $NaNO_3$, using distilled water as absorbing soln.; fluoride ions in solns. containing urotropin (pH 5–5.2) are titrated with $La(NO_3)_3$ soln., using a fluoride-selective electrode [18]. Controlled pyrolysis- electron impact mass spectrometry has been reported as diagnostic method for identifying commercial material [54]. Atom ion emission- ms anal. for qualitative and quantitative evaluation has the particular advantages of high accuracy, efficiency and the need for only small sample sizes [39]. H-1 nmr and C-13 nmr spectroscopy is used for identification/anal. of impurities in copolymer resin and its lacquers: F-19 nmr is used to identify impurities and emulsifiers (e.g. perfluorodecanoates) from polymerisation process [57]. Anal. of esca data gives compositional results to within 2% of those obtained from elemental anal. [43]. CTFE content in copolymers can be monitored by the intensity of the 840 cm^{-1} band in the vibrational (ir) spectrum; intensity decreases with increasing CTFE content [56]
Miscellaneous: Graft copolymers are available [23,24,26] e.g. with PVDF, vinylidene fluoride grafted on to copolymer [23]. Block copolymers have been reported [48]. Blends of copolymer with PTFE, EPDM, ethylene-propylene copolymers, natural rubber, 'reclaimed' rubber, EPR, PVC, ethylene-norbornene (rubber), acrylic polymers and rubbers, graft fluoropolymers and (oligomeric) PE have been reported

Volumetric & Calorimetric Properties:
Density:

No.	Value	Note
1	d 2–2.04 g/cm^3	commercial resin, 25 wt% vinylidene fluoride [60]
2	d 2.09–2.12 g/cm^3	3 mol% vinylidene fluoride, higher crystallinity, hexagonal (helical) arrangement [41]
3	d 2.02 g/cm^3	25 mol% vinylidene fluoride, amorph. [41]

4	d 1.85 g/cm^3	50 mol% vinylidene fluoride, amorph. [41]
5	d 1.85 g/cm^3	70 mol% vinylidene fluoride, amorph. [41]
6	d 1.85 g/cm^3	68 wt% CTFE, commercial elastomers [37,58]
7	d 2.093–2.144 g/cm^3	3.84–0.72 wt% vinylidene fluoride [31]

Volumetric Properties General: C_P 26–100 J mol^{-1} K^{-1}, -193°- 67°, 70 wt% vinylidene fluoride; 28–107 J mol^{-1} K^{-1}, -193°-67°, 56 wt% vinylidene fluoride [38]

Thermodynamic Properties General: Thermal conductivity increases with increasing temp.; no change is observed at temps. greater than T_g [21]

Melting Temperature:

No.	Value	Note
1	197°C	approx., 3–4 mol% vinylidene fluoride, high crystallinity, hexagonal (helical) arrangement [41]
2	92–164°C	85–99 mol% vinylidene fluoride [47]
3	186–215°C	3.84–0.72 wt% vinylidene fluoride [31]

Glass-Transition Temperature:

No.	Value	Note
1	52°C	approx., 3–4 mol% vinylidene fluoride, high crystallinity, hexagonal (helical) arrangement [41]
2	0°C	50 mol% vinylidene fluoride, amorph. [41]
3	-15°C	70 mol% vinylidene fluoride, amorph. [41]
4	-44--26°C	99–85 mol% vinylidene fluoride [47]
5	-13°C	amorph., 70 wt% vinylidene fluoride [49,55]
6	28°C	25 wt% vinylidene fluoride [15]
7	-2°C	amorph., 50 wt% vinylidene fluoride [49,55]
8	-18°C	unknown composition [30]
9	20–40°C	VDF:CTFE 1:3 [9]

Transition Temperature General: Transition temps. are dependent on thermal and mechanical history of sample [49]

Brittleness Temperature:

No.	Value	Note
1	-40°C	25 wt% CTFE, commercial vulcanisate [37]

Surface Properties & Solubility:

Plasticisers: Fluorodiethers, fluorosiloxanes, chlorofluorocarbons, fluoroelastomers. Use of (oligo)fluorosiloxane liqs. (5–7 phr) leads to products with high strength, good heat resistance, and improved processability [19]. Addition of fluorinated species (optimal conc. 50 parts) such as H(CF$_2$)$_n$CH$_2$O-CH$_2$OCH$_2$(CF$_2$)$_n$H (where n=4 or 6; n=6 most effective) gives products displaying reduced stress at 100% elongation, high breaking elongation and high cold resistance (to -50°) [17]. Fluoroacrylate polymers and cyanoethyl ethers also used

Solvents/Non-solvents: Sol. certain ketones and esters [37]; THF (22 wt% CTFE, commercial resin) [25]. Insol. MeCN, MeOH; (22 wt% CTFE, commercial resin). 50/50 Copolymer sol. DMF [10,53]

Transport Properties:

Transport Properties General: Rheological props. and their dependence on MW and MW distribution have been reported. Melt index displays inverse correlation to high MW component [20]. Rheological behaviour (entanglement network and segmental mobility) and its effect on electrification has been reported [6]. Commercial vulcanisate has low water absorption props. [58]

Polymer Solutions Dilute: Intrinsic viscosity and Mark-Houwink constants of vinylidene fluoride:CTFE (1:4) copolymer in EtOAc at 25° have been reported; 0.0004–0.0005 poise (graft copolymer with vinylidene fluoride) [33]; melt viscosity 0.45–0.55 cSt (commercial resin, 25 wt% vinylidene fluoride) [60]. Mooney viscosity 127 (170°, commercial elastomer, 68% CTFE) [37]. Mooney viscosity of elastomer depends on MW amd MW distribution [20]

Permeability of Liquids: Permeability to solns. of hydrochloric acid has been reported. The lowest permeabilities are found for copolymers with highest T_g values [12] Has low permeability to moisture [9]

Gas Permeability:

No.	Gas	Value	Note
1	[O$_2$]	1.313 cm^3 mm/(m^2 day atm)	0.02×10^{-10} cm^2 (s cm Hg)$^{-1}$, 25°, 97% CTFE [14]
2	[N$_2$]	0.657 cm^3 mm/(m^2 day atm)	0.01×10^{-10} cm^2 (s cm Hg)$^{-1}$, 25°, 97% CTFE [14]
3	[CO$_2$]	5.253 cm^3 mm/(m^2 day atm)	0.08×10^{-10} cm^2 (s cm Hg)$^{-1}$, 25°, 97% CTFE [14]
4	[N$_2$]	1.97 cm^3 mm/(m^2 day atm)	0.03×10^{-10} cm^2 (s cm Hg)$^{-1}$, 25°, 75% CTFE [14]
5	[CO$_2$]	11.82 cm^3 mm/(m^2 day atm)	0.18×10^{-10} cm^2 (s cm Hg)$^{-1}$, 25°, 75% CTFE [14]
6	[O$_2$]	27.58 cm^3 mm/(m^2 day atm)	0.42×10^{-10} cm^2 (s cm Hg)$^{-1}$, 25°, 68% CTFE [14]
7	[N$_2$]	10.51 cm^3 mm/(m^2 day atm)	0.16×10^{-10} cm^2 (s cm Hg)$^{-1}$, 25°, 68% CTFE [14]
8	[CO$_2$]	177.29 cm^3 mm/(m^2 day atm)	2.7×10^{-10} cm^2 (s cm Hg)$^{-1}$, 25°, 68% CTFE [14]
9	[O$_2$]	61.07 cm^3 mm/(m^2 day atm)	0.93×10^{-10} cm^2 (s cm Hg)$^{-1}$, 25°, 50% CTFE [14]
10	[CO$_2$]	262.66 cm^3 mm/(m^2 day atm)	4.0×10^{-10} cm^2 (s cm Hg)$^{-1}$, 25°, 50% CTFE [14]
11	[O$_2$]	2.76–37.43 cm^3 mm/(m^2 day atm)	$0.042–0.57 \times 10^{-10}$ cm^2 (s cm Hg)$^{-1}$, 25°, 25% CTFE [14]
12	[N$_2$]	0.79–0.98 cm^3 mm/(m^2 day atm)	$0.012–0.015 \times 10^{-10}$ cm^2 (s cm Hg)$^{-1}$, 25°, 25% CTFE [14]
13	[CO$_2$]	6.3–15.76 cm^3 mm/(m^2 day atm)	$0.096–0.24 \times 10^{-10}$ cm^2 (s cm Hg)$^{-1}$, 25°, 25% CTFE [14]
14	[H$_2$]	308.63 cm^3 mm/(m^2 day atm)	4.7×10^{-10} cm^2 (s cm Hg)$^{-1}$, 33°, 20% CTFE [14]
15	[He]	22.33 cm^3 mm/(m^2 day atm)	0.34×10^{-10} cm^2 (s cm Hg)$^{-1}$, 25°, 20% CTFE [14]
16	[CH$_4$]	0.55 cm^3 mm/(m^2 day atm)	0.0084×10^{-10} cm^2 (s cm Hg)$^{-1}$, 25°, 20% CTFE [14]

Mechanical Properties:

Mechanical Properties General: Tensile strength (and corresponding elongation) increases with increasing MW [20]. Tensile strength increases with decrease in temp. from 20° to -253°, and then shows a further increase from -253° to -268.8°. Breaking elongation decreases with decreasing temp. The presence of fillers gives increased tensile strength at room temp., with lower strengths recorded at lower temps [40]. Compounding with carbon black lends reinforcement by breaking down the original globular struct. Mechanical props. of filled rubber vulcanisates improve with increasing filler/rubber interface area; effects of moisture also reported [62]. Tensile behaviour following annealing has been reported for commercially available (75% CTFE) copolymer [15]. Elasticity and good tensile strength are attributable to the presence of CH_2 units in normally rigid highly fluorinated chain [58]

Tensile Strength Break:

No.	Value	Note	Elongation
1	28–30 MPa	graft copolymer with vinylidene fluoride [33]	520–563%

Tensile Strength Yield:

No.	Value	Elongation	Note
1	10.3–20.7 MPa	250–350%	1500–3000 psi, commercial resin, 25% vinylidene fluoride [60]

Miscellaneous Moduli:

No.	Value	Note	Type
1	2.1–5.5 MPa	25 wt% CTFE, commercial vulcanisate [37]	100% modulus

Impact Strength: Impact experiments on thin layers of copolymers with thin layers of intermediate explosives (e.g. PETN and other aminotetryls) have been reported [3]. The copolymer has a sensitising effect on the explosive. Dynamic compression experiments carried out at 500–1000 MPa are related to ignition behaviour by impact [3]
Viscoelastic Behaviour: Viscoelastic props. have been reported. Temp. dependence of both filled and unfilled vulcanisates appears to be determined by stress-relaxation process [35]
Mechanical Properties Miscellaneous: Copolymer exhibits changes in its mechanical behaviour upon storage at room temp. (or at temps. just above) which can lead to problems with respect to end-use applications [15]. Compression set 40% at 200° (70h) [Method B], (25 wt% CTFE, commercial vulcanisate) [37]. Shear storage modulus, relaxation strength and loss tangent have been reported. Mechanical props. vary with degree of crystallinity which is of relevance to applications as sensitisers for explosives, e.g. PETN [15]. Mechanical relaxation spectra have been reported for copolymers containing 30–97 mol% CTFE for isochronal measurements at 3.5–110 Hz as function of temp. (-90° to highest temp. before onset of flow) [41]
Hardness: Shore D64 (commercial resin, 25% vinylidene fluoride) [60]. Durometer 53–85 (25 wt% CTFE, commercial vulcanisate) [37]
Failure Properties General: Failure appears to be determined by stress-relaxation process; maximum tensile strength (and elongation) occurs at -22°, with maximum at 120°. The strength of filled copolymer is up to twice as that of unfilled copolymer [35]
Friction Abrasion and Resistance: Rubber vulcanisates filled with Cu powder show a reduction in friction coefficient of 20–60% and 2–6 fold increase in wear resistance [44]
Izod Notch:

No.	Notch	Break	Note
1	Y	N	commercial resin, 25% vinylidene fluoride [60]

Electrical Properties:
Electrical Properties General: Tracking erosion and arc resistance have been reported [4]
Complex Permittivity and Electroactive Polymers: Electrical conductivity of soln.-grown thin films of copolymer (50% fluoride) as a function of temp. and field has been reported. Current is strongly temp. dependent. Richardson-Schottky effect suggested as dominant mechanism of conduction at high fields [53]. Substantial piezoelectric activity is observed on poling a copolymer film (5% CTFE). Near ir spectral anal. shows increase in β-phase (with corresponding decrease in γ-phase) on poling, although x-ray structural anal. shows no change on poling; α-phase of PVDF predominates, plus small amount of β-phase [46,52]. Electrical characteristics (dielectric props., charge distribution, charge-transport props.) and related potential applications of thin films in combination with silicon in FET structures have been reported [10]
Static Electrification: Static electrification study of copolymer extruded above T_g through metal and dielectric ducts has been reported; negatively charged surfaces are produced [6]

Optical Properties:
Refractive Index:

No.	Value	Note
1	1.416	commercial resin, 25 wt% vinylidene fluoride [60]
2	1.62	91 wt% vinylidene fluoride [13]

Transmission and Spectra: Nmr spectral data of graft copolymer (with vinylidene fluoride) have been reported [33]. Esr spectral data have been reported [55]. Mass [45], fluorescence [55], F-19 nmr [51] spectral data have been reported [51]. Near (400–4000 cm^{-1})- and far (50–500 cm^{-1}) ir spectral data of copolymer containing 95 wt% vinylidene fluoride have been reported [13,46,52]
Total Internal Reflection: Mechanical stretching results in highly dichroic material (highly anisotropic) [13]

Polymer Stability:
Polymer Stability General: As T_g of copolymers is close to ambient temp., both physical ageing and annealing effects can occur during its end use [9]
Thermal Stability General: Vulcanisate is thermally stable up to approx. 125° [58]. Heat resistance has been reported [37]. Thermal anal. of semicrystalline polymer has been reported [15]
Upper Use Temperature:

No.	Value	Note
1	175°C	25 wt% CTFE, commercial vulcanisate [37]

Decomposition Details: Pyrolysis of the copolymer causes depolymerisation, with elimination of HF, and chain cleavage accompanied by H-transfer. Major pyrolysis products are CF_2CH_2, CF_2CFCl, C_3F_5Cl, HCl, $C_3F_5H_2^{\oplus}$, $C_3F_4H_2Cl^{\oplus}$, $C_5F_6H^{\oplus}$, plus various allylic ions [7]. Hydrogen halide gases are liberated above 230°. Mechanism of decomposition has been reported [54]. Presence of stabiliser ('Gossipol', 15%) increases degradation induction period and reduces weight loss. Degradation reactions and mechanisms for various copolymers compositions have been reported [28]; decomposition products include CO_2, CTFE, VDF, C_3F_5Cl, HF and HCl [28]. Chain unzipping is the predominant mechanistic route [28] Decomposition is promoted by presence of chlorine in copolymer chain [28]. Themal degradation proceeds via free-radical mechanism and is a single stage process, starting at approx. 400° for pure elastomer. Decomposition temp. is greatly reduced for compounds, but increases again on vulcanisation, although it is still lower than that of the pure elastomer. Degradation rates and activation energies have also been reported [5]. Thermal degradation of a range of copolymers produces the constituent comonomers plus some hydrogen chloride which results from elimination reactions

involving adjoining -CF_2CFCl- and -CH_2CF_2- units in the backbone [32].
Flammability: Elastomers cured with hexamethylenediamine carbamate or bisphenol A have oxygen indices 48–60%. Is non-flammable [58,60]. Flammability rating V0 (UL94) [36]. Fire resistance is greatly increased by introduction of chlorine into the copolymer macromolecule. Oxygen index 52% (11 wt% CTFE); 85% (33 wt% CTFE); 88% (50 wt% CTFE); 97% (80 wt% CTFE); 100% (approx., 86 wt% CTFE) [11]
Chemical Stability: Vulcanisate is completely destroyed by contact with potassium hydroxide (specific gravity 1.4) at 90° [61]. Unaffected by hexane; quite stable in aliphatic hydrocarbons, aromatic hydrocarbons, chlorinated hydrocarbons, alcohols and primary aromatic amines. Swollen by ketones, carboxylic acids, ethers, esters, tertiary amines and DMF [50]. Resists oxidising acids [37,58,59] 90% H_2O_2, O_2 and petroleum products [58]. Displays good general resistance to most chemicals [9]. Coatings show satisfactory resistance to acids and bases [2]

Applications/Commercial Products:

Processing & Manufacturing Routes: Usually synth. by aq. emulsion polymerisation of comonomer mixture using potassium or ammonium persulfate as free-radical initiator [32,37,51], or by use of redox system with or without chain-transfer agents (e.g. CCl_4, alkyl esters or halogen salts) [37]. Graft copolymers with vinylidene fluoride synth. by reaction of vinylidene fluoride, CTFE and unsaturated peroxide (e.g. CH_2CHCH_2OCOOO-*tert*-butyl) at low temps.; the intermediate (base) polymer, containing peroxide side chains, is heated to give required polymer [33]. Polymer has a saturated backbone and therefore cannot be vulcanised in conventional sense. It can, however, be cross-linked by use of poly(isocyanates), polyamines or organic peroxides to form elastomeric materials [58, 59]. 'Vulcanisation' can also be achieved using γ-irradiation. Formulation/compounding/processing [59] with various additives can be carried out on 2-roll rubber mill at 77–88° [37]. Films (e.g. containing 5% CTFE) are prepared by chill-roll extrusion process [46,52]. Resin (e.g. containing 70 mol% CTFE) can be moulded into thin, clear sheets that are flexible and stress-free at ambient temps. [60]
Applications: Used in lacquers, paints, putties, propellants, lubricants, adhesives, acid-proofing compounds, laminates; as corrosion-resistant coatings and radiation-protective coatings. A major application is as coatings (elastomeric polymeric binder) for plastic-bonded explosives and propellants, e.g. with PETN and other aminotetryls (sensitiser), and 1,3,5-triamino-2,4,6-trinitro-benzene (TATB) (sensitiser). Also used in gaskets, oil seals, high-barrier films and membranes, and in fibres (with other fluoropolymers). Has wide range of applications in optics and electronics, including moisture-resistant and weather-resistant coatings (insulators) for electrical braid, wire, cable and harnesses, fibre optics, piezoelectric components, electrets, microcircuitry, LCD devices, cathodes and batteries, thin films for FETS (with Si) [10] and in electrophotographic and photothermographic imaging systems. Other applications include ion-exchange membranes, HPLC column packings (with aminopropyl silica) [25] and in dentistry. Block copolymers are used in optical device materials. Graft copolymers are used as leather substitutes, owing to their resistance to stains, chemicals and weathering, as laminated sheets and in flexible layers for various pharmaceutical applications. Copolymer films containing 0.5–3.5 wt% vinylidene fluoride find applications where good moisture-barrier and optical transparency props. are required

Trade name	Details	Supplier
Aclar	22 (96 wt% CTFE), 22A, 22S, 33C. Non-exclusive also used to refer to other PCTFE materials	Allied Chemical
Kel-F	5 wt% CTFE; thin (20µ) films	Solvay
Kel-F	various grades 25–97 wt% CTFE non-exclusive; also used to refer to other VDF and CTFE polymer materials	3M Industrial Chemical Products
P994	Fluorine rubber	Atochem Polymers
SKF-32	fluorine rubber	
Voltalef	3700 (30 wt% CTFE) and 5500 (50 wt% CTFE), amorph., copolymers; non-exclusive; also used to refer to other PCTFE and rubber materials	Atochem Polymers

Bibliographic References

[1] Savinkina, N.P. and El'kina, L.V., *Bum. Prom-St.*, 1987, 27, (uses)
[2] Shigorina, I.I., Egorov, B.N., Timofeeva, L.P. and Zvyagintseva, N.V., *Lakokras. Mater. Ikh Primen.*, 1987, 55, (uses, stability)
[3] Swallowe, G.M., Field, J.E. and Hutchinson, C.D., *Shock Waves Condens. Matter*, (ed. Y.M. Gupta), Plenum, 1986, 891, (uses, mechanical props.)
[4] Kuznetsova, A.V., Solov'ev, E.P., Sazhin, B.I. and Truskova, L.I., Izv. Vyssh. Uchebn. Zaved., Energ., 1986, 21, (electrical props.)
[5] Jaroszynska, D. and Kleps, T., *J. Therm. Anal.*, 1986, 31, 955, (stability, thermal degradation, additvies)
[6] Vinogradov, G.V., Dreval, V.E. and Protasov, V.P., *Proc. R. Soc. London A*, 1987, 409, 249, (electrical props.)
[7] Lonfei, J., Jingling, W. and Shuman, X., *J. Anal. Appl. Pyrolysis*, 1986, 10, 99, (thermal stability, degradation, ms)
[8] Dec, S.F., Wind, R.A. and Maciel, G.E., *Macromolecules*, 1987, 20, 2754, (F-19 nmr)
[9] Siegmann, A., Cohen, G. and Baraam, Z., *J. Appl. Polym. Sci.*, 1989, 37, 1567, (morphology, thermal props, mechanical props.)
[10] Andry, P., Filion, A.Y., Blain. S., Rambo, A. and Caron, L.G., *J. Appl. Phys.*, 1991, 69, 2644, (electrical props.)
[11] Pavlenkova, G.A., Borisov, E.M. and Mizerovskii, L.N., *Plast. Massy*, 1990, 88, (flammability)
[12] Kazantseva, T.V., Gemusova, I.B., Shterenzon, A.L., Mulin, Y.A. and Maksimova, E.I., *Plast. Massy*, 1989, 25, (permeability)
[13] Latour, M. and Moreira, R.L., J. Polym. Sci., *Polym. Phys. Ed.*, 1987, 25, 1913, (ir, optical props., conformn.)
[14] *Sci. Technol. Polym. Films*, (ed. O.J. Sweeting), Wiley Interscience, 1971, 2, 118, (permeability)
[15] Hoffman, D.M., Matthews, F.M. and Pruneda, C.O., *Thermochim. Acta*, 1989, 156, 365, (morphology, thermal props., mechanical props., uses)
[16] Murthy, N.S. and Minor, H., *Polymer*, 1990, 31, 996, (morphology)
[17] Evchik, V.S., Leshchenko, V.I., Markova, L.A. and Sokolova, G.A., *Kauch. Rezina*, 1988, 26, (plasticisers)
[18] Kimstach, T.B., Kupreeva, G.S. and Lapshova, A.A., *Kauch. Rezina*, 1989, 34, (anal.)
[19] Evchik, V.S., Leshchenko, V.I., Markova, L.A. and Sokolova, G.A., *Kauch. Rezina*, 1989, 25, (plasticisers)
[20] Madorskaya, L.Ya., Otradina, G.A., Andreeva, A.I., Zaichenko, Yu.A. and Kornilova, T.V., *Plast. Massy*, 1988, 31, (transport props., MW, mechanical props.)
[21] Rozov, I.A., Truskova, L.I. and Belomutskaya, O.K., *Plast. Massy*, 1988, 42, (thermal conductivity)
[22] Mirvaliev, Z.Z., Khakimov, R., Dzhalilou, A.T., Asamov, M.K. and Fatkhullaev, E., *Plast. Massy*, 1990, 71, (stability, thermal props, additives)
[23] *Jpn. Pat.*, 1989, 01 22 547, (graft copolymers, uses)
[24] *Jpn. Pat.*, 1990, 02 274 534, (graft copolymers, uses)
[25] Danielson, N.D., Wangsa, J. and Shamsi, S.A., *J. Liq. Chromatogr.*, 1995, 18, 2579, (uses)
[26] *Jpn. Pat.*, 1995, 07 33 646, (graft copolymers, uses)
[27] Luo, S. and Ji, G., *Hanneng Cailiao*, 1993, 1, 31, (transport props, viscosity)
[28] Rizvi, M., Munir, A., Zulfiqar, S. and Zulfiqar, M., *J. Therm. Anal.*, 1995, 45, 1597, (stability, thermal degradation)
[29] Dec, S.F., Wind, R.A. and Maciel, G.E., *J. Magn. Reson.*, 1986, 70, 355, (F-19 nmr)
[30] Arifov, P., Vasserman, S. and Tishin, S., *Phys. Status Sol. A*, 1987, 102, 565, (struct)
[31] Murthy, N.S., Khanna, Y.P. and Signorelli, A.J., *Polym. Eng. Sci.*, 1994, 36, 1254, (morphology, thermal props, volumetric props)
[32] Zulfiqar, S. Zulfiqar, S., Rizvi, M., Munir, A. and McNeill, I.C., *Polym. Degrad. Stab.*, 1994, 43, 423, (synth, stability, thermal props, thermal degradation)
[33] Katoh, E., Kawashima, C. and Ando, I., *Polym. J. (Tokyo)*, 1995, 27, 645, (graft copolymers, nmr, synth, transport props, mech props, ordering)
[34] Khanna, Y.P., Kumar, R. and Reimschuessel, A.C., *Polym. Eng. Sci.*, 1988, 28, 1612, (morphology, thermal props)
[35] Zobina, M.V., Kurlyand, S.K., Akopyan, L.A. and Burtseva, M.A., *Kauch. Rezina*, 1988, 25, (mech. props., viscoelasticity)

[36] Modern Fluoropolymers John Wiley, (ed. J. Scheirs), 1997, (rev)
[37] Schroeder, H., Rubber Technology 3rd edn., (ed. M. Morton), Van Nostrand Reinhold, 1987, 410, (rev)
[38] Wong, K.C., Chen, F.C. and Choy, C.L., *Polymer*, 1975, **16**, 858, (heat capacity)
[39] Tantsyrev, G.D. and Kleimenov, N.A., *Dokl. Akad. Nauk SSSR*, 1973, **213**, 649, (anal)
[40] Novikov, N.M., Il'chev, V.Ya. and Drozdov, V.V., *Nizkotemp. Vak. Materialoved.*, 1973, **3**, 71, (mech. props)
[41] Kalfoglou, N.K. and Williams, H.L., *J. Appl. Polym. Sci.*, 1973, **17**, 3367, (mech. props, thermal props, morphology)
[42] Litvinova, T.V., Ivanova, R.P. and Titov, B.G., *Kauch. Rezina*, 1974, **10**, (morphology, struct)
[43] Clark, D.T., Feast, W.J., Kilcast, D. and Musgrave, W.K.R., *J. Polym. Sci., Polym. Chem. Ed.*, 1973, **11**, 389, (esca, anal)
[44] Lebedev, V.M., Asheichik, A.A., Khachatryan, G.R. and Ratner, B.V., *Kauch. Rezina*, 1981, **23**, (tribology)
[45] Tantsyrev, G.D., Povolotskaya, M.I. and Kleimenov, N.A., *Vysokomol. Soedin., Ser. A*, 1977, **19**, 2057, (ms, microstruct)
[46] Latour, M. and Dorra, H.A., *Ferroelectrics*, 1982, **44**, 197, (struct, morphology, ir, electrical props)
[47] Moggi, G., Bonardelli, P., Monti, C. and Bart, J.C.J., *Polym. Bull. (Berlin)*, 1984, **11**, 35, (synth, thermal props)
[48] Jpn. Pat., 1983, 58164611, (block copolymers, synth, optical props, struct)
[49] Leonard, C., Halary, J.L., Monnerie, L. and Micheron, F., *Polym. Bull. (Berlin)*, 1984, **11**, 195, (thermal props)
[50] Lytkina, N.I., Vansyatskaya, I.N., Mizerovskii, L.N. and Shcheglova, O.V., *Kauch. Rezina*, 1984, **13**, (solvents, solubility, chemical stability)
[51] Moggi, G., Bonardelli, P. and Bart, J.C.J., *J. Polym. Sci., Polym. Phys. Ed.*, 1984, **22**, 357, (morphology, struct, synth, F-19 nmr)
[52] Latour, M., *Ferroelectrics*, 1984, **60**, 71, (struct, ir, electrical props, morphology)
[53] Mehendru, P.C., Sharma, D.C., Chand, S. and Gupta, N.P., *Indian J. Pure Appl. Phys.*, 1985, **23**, 337, (electrical props)
[54] Pidduck, A.J., *J. Anal. Appl. Pyrolysis*, 1985, **7**, 215, (thermal degradation, anal, ms)
[55] Noël, C., et al, *Macromolecules*, 1986, **19**, 201, (esr)
[56] Benedetti, E., D'Alessio, A., Vergamini, P., Bonardelli, P. and Moggi, G., *Conv. Ital. Sci. Macromol.*, (Atti) 6th, 1983, **2**, 125, (ir, anal)
[57] Rutenberg, A.C., *Energy Res. Abstr.*, Abstract No. 46854, 10, (H-1 nmr, C-13 nmr, F-19 nmr, anal)
[58] Conry, M.E., Honn, F.J., Robb, L.E. and Wolf, D.R., *Rubber Age (N.Y.)*, 1955, **76**, 543, (rev)
[59] Griffis, C.B. and Montermoso, J.C., *Rubber Age (N.Y.)*, 1955, **77**, 559, (additives, compounding, vulcanisation, stability)
[60] Ash, M. and Ash, I, Handbook of Plastic Compounds, Elastomers and Resins VCH, 1992, 131, (uses, transport props, mech. props)
[61] Goikhman, Ts.M., et al, *Issled. Pbl. Fiz. Khim. Rezin*, 1973, **2**, 60, (chemical stability)
[62] Dontsov, A., Ivanov, A.B., Treshchalov, V.I. and Chulyukina, A.V., *Kauch. Rezina*, 1976, **17**, (mech. props, additives)

Poly(vinylidene fluoride-*co*-hexafluoro isobutylene)

P-614

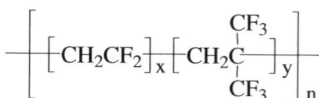

Synonyms: Poly(vinylidene fluoride-co-hexafluoroisobutene). Poly(HFIB/VF$_2$). Poly(1,1-difluoroethene-co-(3,3,3-trifluoro-2-(trifluoromethyl)-1-propene)). PVDF/HFIB. Poly(vinylidene fluoride-co-hexafluoro(γ-butylene)
Related Polymers: Poly(vinylidene fluoride)
Monomers: Vinylidene fluoride, Hexafluoroisobutylene
Material class: Thermoplastic, Copolymers
Polymer Type: PVDF, fluorocarbon (polymers)
CAS Number:

CAS Reg. No.
34149-71-8

Molecular Formula: $[(C_2H_2F_2).(C_4H_2F_6)]_n$
Fragments: $C_2H_2F_2$ $C_4H_2F_6$

Additives: Phosphites, bis(carboxylates) of bis(hydroxyalkyl), oxamides; salts and/or oxides of divalent metals, including ZnO, CaO as stabilisers
General Information: Fluorocarbon copolymer exhibiting many of the desirable props. of PTFE but with additional advantage that it is melt processable and film forming [4]. The polymer has the same crystalline melting point as PTFE but a much lower density. The copolymer may be injection moulded at approx. 380°. Commercial material is available as a 50:50 mixt. and is hard, rigid and highly crystalline, displaying good resistance to chemicals, good electrical props., anti-stick characteristics, and is non-flammable [4]. Reviews of the synth., processing, applications and physical props. have been reported [1,3]
Morphology: Orthorhombic; a 1.064, b 1.837, c 0.783 [2]; helix 2_1, with approx. *trans-trans-gauche-gauche-* backbone dihedral angle sequence. Severe intrachain steric crowding gives a rigid chain, with some backbone bond angles being highly extended [2]. Ir and Raman spectroscopy show head-to-tail struct., ie. $C(CF_3)_2CH_2CF_2 CH_2)_n$. Evaluation of CH_2 bending region supports such a conclusion. There is more than one monomer per repeat unit [7]
Miscellaneous: MW values have not been reported, partly due to non-solubility of copolymer in organic solvents [7,8]

Volumetric & Calorimetric Properties:
Density:

No.	Value	Note
1	d 1.88 g/cm^3	50/50 copolymer [1,2,3]

Thermal Expansion Coefficient:

No.	Value	Note	Type
0	4.7×10^{-5} K^{-1}	149–204° [1]	1
1	3.7×10^{-5} K^{-1}	-45–24° [1]	1

Melting Temperature:

No.	Value	Note
1	327°C	50/50 copolymer [2,3,7,9]
2	326°C	[6]
3	303°C	64% VDF [15]

Glass-Transition Temperature:

No.	Value	Note
1	84–98°C	[6]

Transition Temperature General: Melting point increases with increasing isobutylene content [15], T_g of high MW material difficult to measure by DSC [10]. Copolymer does not display any reversible crystalline (relaxation) transition temp. [5,6,11]
Deflection Temperature:

No.	Value	Note
1	220°C	1.82 MPa [2,3,7]
2	220°C	1.89 MPa [1]

Transition Temperature:

No.	Value	Note	Type
1	132°C	approx. [10]	T_α
2	-68°C	approx. [10]	T_β

Surface Properties & Solubility:

Solvents/Non-solvents: Insol. all common organic solvents [7,10]; swollen by EtOAc [8]
Surface and Interfacial Properties General: Has low surface energy [20] and tension [8,24]
Surface Tension:

No.	Value	Note
1	19.3 mN/m	19.3 dyne cm^{-1} [2,3]

Transport Properties:

Transport Properties General: A study of the effect of metal oxide stabilisers on melt flow index values has been reported [9,14]
Melt Flow Index:

No.	Value	Note
1	3.07 g/10 min	no stabilisers
2	1.04 g/10 min	1.5 wt% ZnO, extrusion at 345° [9]
3	1.31 g/10 min	CaO, extrusion [14]
4	1–3 g/10 min	ASTM D2116 [2,3]
5	0.8 g/10 min	powder, 5 kg load [7]
6	13 g/10 min	film, after compression, 2.16 kg load, moulding at 350° [7]
7	15 g/10 min	350°, 2.16 kg load [8]
8	0.042 g/10 min	no stabilisers
9	0.005 g/10 min	stabilised [9,14]
10	9 g/10 min	345°, 2.16 kg load [5]

Permeability and Diffusion General: Permeability to water vapour is approx. 4 times lower than that of PTFE. The copolymer has one of the lowest known values of water vapour permeability [1]

Mechanical Properties:

Mechanical Properties General: Displays high moduli and low creep even at temps. above 150° [20]. Its mechanical props. e.g. tensile strength and modulus are superior to those of PTFE, particularly at elevated temps. [1,2]. Its impact strength is inferior to that of PTFE [1]

Tensile (Young's) Modulus:

No.	Value	Note
1	3792 MPa	550 kpsi [2,3]
2	3800 MPa	380 N mm^{-2}, 23° [1,2,3]
3	758 MPa	110 kpsi [2,3]
4	690 MPa	690 N mm^{-2}, 200° [1,2,3]

Flexural Modulus:

No.	Value	Note
1	4481 MPa	650 kpsi, 23° [2,3]

Tensile Strength Break:

No.	Value	Note	Elongation
1	37.9 MPa	5.5 kpsi [2,3]	
2	38 MPa	38 N mm^{-2} [1]	2%, 23° [1,2,3]
3	20.7 MPa	3000 psi [2,3]	
4	21 MPa	21 N mm^{-2}	200%, 200° [1,2,3]
5	15.8 MPa	2300 psi [2,3]	
6	16 MPa	16 N mm^{-2}	220%, 300° [1,2,3]

Flexural Strength at Break:

No.	Value	Note
1	36.5 MPa	5300 psi, 23° [2,3]

Impact Strength: Impact strength 0.4 ft lb in^{-1}, notched, 23° [2,3]
Mechanical Properties Miscellaneous: Dielectric and dynamic mechanical relaxation studies have been reported [10]
Hardness: Rockwell R111 [2,3]. Pencil hardness 8H (coatings containing metal oxides) [16]. R116 (surface) [1]
Fracture Mechanical Properties: Creep strain 1.4% (1hr, 100°, 2 kpsi); strain 1.7% (after 150h) [2,3]
Friction Abrasion and Resistance: Displays anti-stick, non-wetting characteristics, and good abrasion and scratch resistance [1,4]. Shows good adhesion to metal substrates [4] and unprimed metals [16]

Electrical Properties:

Electrical Properties General: Has good electrical props.[4,15], low electrical loss props. [12] and high electrical resistance [8]

Optical Properties:

Transmission and Spectra: Ir and Raman [7,8]; epr [12] H-1 nmr, and F-19 nmr [7,10] spectral data have been reported
Total Internal Reflection: Birefringent, pseudosphenilitic struct. revealed by polarised light microscopy studies [10]. Film drawn at 300° to only 200% elongation displays high degree of dichroism [8]
Volume Properties/Surface Properties: Colour yellow-brown (un-stabilised extrudate); grey (ZnO stabilised) [9]

Polymer Stability:

Thermal Stability General: Has good thermal stability [15]
Upper Use Temperature:

No.	Value	Note
1	288°C	[4]
2	280°C	[1,2,3]

Decomposition Details: Weight loss of 50/50 copolymer in N_2 0.03%, 0.22% and 53% at 350°, 400° and 500°, respectively [15]
Flammability: Non-flammmable [4]. Has high oxygen index [20]
Chemical Stability: Has good resistance to chemicals (in general) [1,4,8,15,20]. Inert to strong bases, halogens and metal salt solns. [1]. Shows slight weight gain in ketones and esters [1]

Applications/Commercial Products:

Processing & Manufacturing Routes: Synth. by free-radical suspension or emulsion polymerisation of mixture of comonomers under pressure [1]. Typical initiators include trichloroacetyl peroxide [12,15], trifluoroacetyl peroxide [12], *tert*-butyl acetyl peroxide [12] and di-sec-butyl peroxydicarbamate [12]. Processed by compression moulding at 350°, injection moulding at 370–380°, or powding coating by fluidised bed, electrostatic spray or plasma spray techniques [1]
Applications: Aplications where superior mechanical props. (particularly at elevated temps.) are required in mouldings, coatings, cellular and porous plastic materials. Aqueous dispersions are used to produce organosols. Other uses include mould release agents, non-stick coatings for industrial rolls, cookware, chemical process equipment, fuel cells, dust precipitators industrial fabrics. Often used in blends with other polymers; examples include porous films; non-wetting, flexible coatings, bearing seals, components with high compressive strength (with PTFE); sliding parts (with PEEK); components with good non-stick props.,

low-friction coefficients and good abrasion resistance (with polyimides). Potential application as sorbent in chromatography

Trade name	Supplier
CM-1 fluoropolymer	Allied Chemical
CM-X fluoropolymer	Allied Chemical

Bibliographic References
[1] Elias, H.-G., *Polym. News,* 1977, **4**, 78, (rev.)
[2] Weinhold, S., Litt, M.H. and Lando, J.B., *J. Polym. Sci., Polym. Phys. Ed.,* 1982, **20**, 535, (morphology, mechanical props.)
[3] Minhas, P.S. and Petrucelli, F., *Plast. Eng. (N.Y.),* 1977, **33**, 60, (rev.)
[4] Petrucelli, F., Am. Chem. Soc., Div. Org. Coat. Plast. Chem., Pap, 1975, **35**, 107, (mechanical props., processing, uses, stability)
[5] Aharoni, S.M. and Sibilia, J.P., *J. Appl. Polym. Sci.,* 1979, **23**, 133, (processing, extrusion, thermal props., conformational behaviour)
[6] Aharoni, S.M. and Sibilia, J.P., *Polym. Eng. Sci.,* 1979, **19**, 450, (processing, extrusion, thermal props., conformational behaviour)
[7] Hsu, S.L., Sibilia, J.P., O'Brien, K.P. and Snyder, R.G., *Macromolecules,* 1978, **11**, 990, (ir, ordering and orientation)
[8] *Ger. Pat.,* 1977, 2528519, (processing, additives, stability)
[9] Pochan, J.M., Hinman, D.F., Froix, M.F. and Davidson, T., *Macromolecules,* 1977, **10**, 113, (mechanical props, electrical props., thermal props., optical props.)
[10] Aharoni, S.M., and Sibilia, J.P., *American Chem. Soc. Div. Polym. Chem. Preprints,* 1978, **19**, 350, (morphology, processing, thermal props., conformational behaviour)
[11] Yue, H., Nelson, G., O'Brien, K., Chandrasekaran, S. and Sibilia, J., *J. Polym. Sci., Polym. Phys. Ed.,* 1980, **18**, 1393, (epr, ordering and orientation, synth.)
[12] *U.S. Pat.,* 1976, 3962373, (uses)
[13] *U.S. Pat.,* 1975, 3903045, (processing, additives, stability)
[14] *Ger. Pat.,* 1971, 2117654, (synth., stability, thermal props.)
[15] *U.S. Pat.,* 1976, 3936569, (uses)
[16] *Ger. Pat.,* 1976, 2545897, (uses)
[17] *U.S. Pat.,* 1987, 4655945, (uses, processing)
[18] *U.S. Pat.,* 1988, 4775709, (processing additives, stability)
[19] Robertson, A.B., Chandrasekaran, S. and Chen, C.S., *Int. SAMPE Tech. Conf.,* 1986, **18**, 490, (uses)
[20] *Eur. Pat. Appl.,* 1990, 373588, (processing, uses)
[21] *Jpn. Pat.,* 1993, 05279506, (uses)
[22] *Jpn. Pat.,* 1993, 0542530, (uses)
[23] Ulinskaya, N.N., Shadrina, N.E., Korsakov, V.G. and Ivanchev, S.S., *Zh. Fiz. Khim.,* 1992, **66**, 2235, (surface props., uses)
[24] *Jpn. Pat.,* 1992, 0415257, (uses)
[25] *Jpn. Pat.,* 1992, 0415254, (uses)

Poly(vinylidene fluoride-*co*-hexafluoro propene-*co*-tetrafluoroethylene)   P-615

$$\left[\left[CH_2CF_2\right]_x\left[\begin{array}{c}CF_2CF\\|\\CF_3\end{array}\right]_y\left[CF_2CF_2\right]_z\right]_n$$

Synonyms: *Vinylidene fluoride-hexafluoropropylene-tetrafluoroethylene terpolymer. Fluoroelastomer. Fluorine-containing rubber*
Related Polymers: Polyvinylidene fluoride, Polytetrafluoroethylene, Fluoroelastomers, Poly(vinylidene fluoride-*co*-hexafluoropropylene)
Monomers: Tetrafluoroethylene, Vinylidene fluoride, Hexafluoropropene
Material class: Synthetic Elastomers, Copolymers
Polymer Type: PVDF, PTFEfluorocarbon (polymers)
CAS Number:

CAS Reg. No.
25190-89-0

Molecular Formula: $[(C_2H_2F_2).(C_3F_6).(C_2F_4)]_n$
Fragments: $C_2H_2F_2$ C_3F_6 C_2F_4

Additives: Magnesium oxide, calcium hydroxide, fillers, carbon black, cross-linking agents, accelerators, process aids
Miscellaneous: The terpolymer has better resistance to swelling in oils, bettter long term heat resistance and better resistance to chemical degradation than vinylidene fluoride/hexafluoropropylene copolymers

Applications/Commercial Products:
Processing & Manufacturing Routes: Typically prepared by high pressure, free radical, aq. emulsion polymerisation. Polymer is processed further using typical rubber compounding techniques such as milling

Trade name	Supplier
Fluorel	3M Industrial Chemical Products
Tecnoflon	Ausimont
Viton B	DuPont

Poly(vinylidene fluoride-*co*-hexafluoro propylene)   P-616

$$\left[\left[CH_2CF_2\right]_x\left[\begin{array}{c}CF_2CF\\|\\CF_3\end{array}\right]_y\right]_n$$

Synonyms: *Vinylidene fluoride-hexafluoropropylene copolymer. FKA Synthetic Rubber. Poly(1,1-difluoroethene-co-1,1,2,3,3,3-hexafluoro-1-propene). Perfluoropropene-PVDF copolymer. Poly(vinylidene fluoride-co-perfluoropropylene). PVdF-HFP copolymer. Fluoroelastomer. Fluorine containing rubber*
Related Polymers: Polyvinylidene fluoride, Poly(vinylidene fluoride-*co*-hexafluoropropylene-*co*-tetrafluoroethylene)
Monomers: Vinylidene fluoride, Hexafluoropropene
Material class: Synthetic Elastomers, Copolymers
Polymer Type: PVDF, fluorocarbon (polymers)
CAS Number:

CAS Reg. No.	Note
9011-17-0	
107221-31-8	block

Molecular Formula: $[(C_2H_2F_2).(C_3F_6)]_n$
Fragments: $C_2H_2F_2$ C_3F_6
Additives: Magnesium oxide, calcium hydroxide, fillers, carbon black, cross-linking agents, accelerators, process aids. Cure agents include diamines, bisphenols and peroxides [1,2,4]
General Information: Several versions of this copolymer have been commercialised. Props. may be elastomeric depending on the mol% of hexafluoropropylene. Typical values are in the region of 30–40 mol%
Miscellaneous: The copolymer is generally used as an elastomer after vulcanisation. Typical fluorine contents of elastomers are 68–69%

Volumetric & Calorimetric Properties:
Density:

No.	Value	Note
1	d 1.8 g/cm^3	[1]

Glass-Transition Temperature:

No.	Value	Note
1	-9.5°C	TGA, unvulcanised [2]
2	-22°C	vulcanisate [4]

Brittleness Temperature:

No.	Value	Note
1	-50--18°C	ASTM D746, dependent on formulation [1]

Transition Temperature:

No.	Value	Note	Type
1	-84°C	unvulcanised [1]	T_β

Surface Properties & Solubility:
Cohesive Energy Density Solubility Parameters: δ 8.7 $J^{1/2}$ $cm^{-3/2}$ [3]
Solvents/Non-solvents: Sol. 2-butanone, other ketones and esters. (Raw elastomer (non-cross-linked)). The cross-linked material is swollen by compatible solvents. Degree of swelling depends on the formulation and the extent of cross-links [2,3]

Transport Properties:
Transport Properties General: Very low gas permeability
Polymer Solutions Dilute: Mooney viscosity 30–120 (100°, ML 1+10, ASTM D1646) [1,4]; 74 (100°, ML 1+4, ASTM D1646) [2]
Polymer Melts: Mooney scorch, min. viscosity 42 (121°), time to a 10 unit rise 26 min. (Viton A, DuPont) [4]

Mechanical Properties:
Mechanical Properties General: The exact values depend on the formulation and compounding of the elastomer
Tensile Strength Break:

No.	Value	Note	Elongation
1	8.96–18.62 MPa	[1]	100–500%
2	14.4 MPa	moulding compound [1]	265%
3	15 MPa	O-ring compound [1]	200%

Miscellaneous Moduli:

No.	Value	Note
1	2.07–15.17 MPa	100% Modulus [1]

Viscoelastic Behaviour: Compression set 9–16% (70h, 25°); 10–30% (70h, 200°), 50–70% (1000h, 200°) (ASTM D395, Method B for 5 mm O-ring) [1]. Compression set 20% (70h, 200°, ASTM D395, Method B, 1.27 cm disk, moulding compound) [1]. Compression set 15% (70h, 200°, ASTM D395, Method B, O-ring) [1]
Hardness: Shore A50-A95 [1]. Shore A74 (moulding compound) [1]. Shore A75 (O-ring compound) [1]
Friction Abrasion and Resistance: Abrasion resistance is good and is satisfactory for most uses [1]

Polymer Stability:
Polymer Stability General: Has excellent resistance to a range of fuels and oils, and good resistance to solvents, acids and alkalis
Thermal Stability General: Has good thermal stability 50% of the initial tensile strength is retained after 1 year at 205° [1]
Upper Use Temperature:

No.	Value	Note
1	315°C	
2	230°C	3000h
3	260°C	1000h
4	285°C	240h
5	315°C	48h [5]

Decomposition Details: Thermal degradation temp. 400–550° [1]
Flammability: Self extinguishing or non-burning [1]
Environmental Stress: Has excellent resistance to ozone and weathering. Sample unaffected after 200h exposure to 150 ppm ozone [1]
Chemical Stability: Low volume change after exposure to various solvents has been reported [2,4]
Stability Miscellaneous: Has fair to good resistance to radiation

Applications/Commercial Products:
Processing & Manufacturing Routes: Typically prepared by high pressure, free radical, aq. emulsion polymerisation. Polymer is processed further using typical rubber compounding techniques such as milling. Additives such as cross-linking agents are added at this stage. The material can be processed using compression, transfer and injection moulding, as well as extrusion. Material is vulcanised in press cure and post cure stages. Typical curing temps. are in the ranges 175–205° press cure and 204–260° post cure (16–24h)
Mould Shrinkage (%):

No.	Value
1	2.6–3.2%

Applications: Components for ground transportation including, engine oil seals and fuel system components such as hoses, O-rings and drive train seals. Also used for industrial hydraulic and pneumatic applications

Trade name	Supplier
Dai-el	Daikin Kogyo
Fluorel	3M Industrial Chemical Products
K-31	Bacon
K-50	Bacon
Kynar Flex	Atochem Polymers
Tecnoflon	Ausimont
Viton	DuPont

Bibliographic References
[1] *Encycl. Polym. Sci. Eng.*, (ed. J.I. Kroschwitz), Wiley Interscience, 1989, **7**, 257
[2] Moggi, G., Geri, S., Flabbi, L. and Arjoldi, G., Rubber Conf., Programme Pap., *10th*, 1979, 1015
[3] Myers, M.E. and Abu-Isa, I.A., *J. Appl. Polym. Sci.*, 1986, **32**, 3515
[4] Viton E-60C, Fluoroelastomer, Product Information, *VT-225.E60C*, DuPont, (technical datasheet)
[5] Viton Fluoroelastomer, Product Information, H-22833 DuPont, (technical datasheet)

Poly(*N*-vinylimidazole)   P-617

Synonyms: *1-Ethenyl-1H-imidazole homopolymer. PVI*
Monomers: 1-Vinylimidazole
Material class: Thermoplastic
CAS Number:

CAS Reg. No.
25232-42-2

Molecular Formula: $(C_5H_6N_2)n$
Fragments: $C_5H_6N_2$
General Information: Weakly basic polymer which forms complexes with a wide range of metal ions such as Cu^{2+}, Ag^+, Co^{2+}, Cd^{2+}, Zn^{2+} and Pb^{2+} [1,2,3]. It also forms interpolymer complexes with polyacrylic acid and other hydroxy-containing polymers [4] and with silicic acid [5]. Prod. as white powder or hygroscopic solid.
Tacticity: The effect of temperature and nature of polymerisation medium on tacticity of the polymer has been studied [6].

Volumetric & Calorimetric Properties:
Melting Temperature:

No.	Value	Note
1	340–374°C	[8]

Glass-Transition Temperature:

No.	Value	Note
1	109–167°C	Different values of T_g for different samples [4,7]. Doping with iodine and complexation with copper increases T_g [9]

Surface Properties & Solubility:
Solvents/Non-solvents: Sol. water, MeOH, EtOH, isopropanol, butanol [10], DMF [4]. Insol. THF [4]. Water is a theta solvent for the polymer [10]

Electrical Properties:
Electrical Properties General: Polymer alone is insulator (conductivity less than 10^{-10} S cm^{-1}). On doping with iodine, conductivity increases to 10^{-7} S cm^{-1}. Quaternisation with methyl iodide and iodine doping increases conductivity to 10^{-4} S cm^{-1} [9].

Optical Properties:
Optical Properties General: UV absorption bands with λ_{max} in region of 205 and 245nm reported [2]
Transmission and Spectra: Pmr [6,11,12], cmr [6], ftir [5,7,9,12,13], Raman [13] and UV [2] spectral data have been reported.

Polymer Stability:
Thermal Stability General: Gradual weight loss (23%) on heating to 330°C then rapid decomposition [8].

Applications/Commercial Products:
Processing & Manufacturing Routes: Prod. by free-radical solution or bulk polymerisation of *N*-vinylimidazole. A common solvent is benzene [1,7,10], although other solvents such as absolute EtOH have been used [4,6]. The free-radical initiator used is normally azobisisobutyronitrile (AIBN) [1,4,6,7,8,10].
Applications: Possible applications for extraction of metal ions from different media. Has been used as anticorrosion agent for copper at high temperatures [14].

Supplier
GFS Chemicals

Bibliographic References
[1] Savin, G., Burchard, W., Luca, C. and Beldie, C., *Macromolecules*, 2004, **37**, 6565
[2] Pekel, N. and Güven, O., *Colloid Polym. Sci.*, 1999, **277**, 570
[3] Miyajima, T., Nishimura, H., Kodama, H. and Ishiguro, H., *React. Funct. Polym.*, 1998, **38**, 183
[4] Luo, X., Goh, S.H. and Lee, S.Y., *Macromol. Chem. Phys.*, 1999, **200**, 399
[5] Annenkov, V.V., Danilovtseva, E.N., Filina, E.A. and Likhoshway, Y.V., *J. Polym. Sci., Part A: Polym. Chem.*, 2006, **44**, 820
[6] Dambatta, B.B., Ebdon, J.R. and Huckerby, T.N., *Eur. Polym. J.*, 1984, **20**, 645
[7] Özyalçin, M. and Küçükyavuz, Z., *Synth. Met.*, 1997, **87**, 123
[8] Skvortsova, G.G., Domnina, Ye. S., Glazkova, N.P., Ivlev, Yu.N. and Chipanina, N.N., *Polym. Sci. USSR (Engl. Transl.)*, 1972, **14**, 660
[9] Küçükyavuz, Z., Küçükyavuz, S., and Abbasnejad, N., *Polymer*, 1996, **37**, 3215
[10] Tan, J.S. and Sochor, A.R., *Macromolecules*, 1981, **14**, 1700
[11] Henrichs, P.M., Whitlock, L.R., Sochor, A.R. and Tan, J.S., *Macromolecules*, 1980, **13**, 1375
[12] Pekel, N., Rzaev, Z.M.O. and Güven, O., *Macromol. Chem. Phys.*, 2004, **205**, 1088
[13] Lippert, J.L., Robertson, J.A., Havens, J.R. and Tan, J.S., *Macromolecules*, 1985, **18**, 63
[14] Eng, F.P. and Ishida, H., *J. Appl. Polym. Sci.*, 1986, **32**, 5035

Poly(vinyl isobutyl ether-*co*-monoethyl maleate)

Synonyms: *Poly(vinyl 2-methyl propane ether-co-monoethyl maleate)*
Related Polymers: General information
Monomers: Vinyl isobutyl ether, Maleic anhydride
Material class: Thermoplastic, Copolymers
Polymer Type: polyvinyls
CAS Number:

CAS Reg. No.
36572-31-3

Molecular Formula: $[(C_6H_{12}O).(C_6H_6O_4)]_n$
Fragments: $C_6H_{12}O$ $C_6H_6O_4$

Applications/Commercial Products:
Processing & Manufacturing Routes: Prod. by dissolving poly(isobutyl vinyl ether-*co*maleic anhydride) in ethanol. Poly(isobutyl vinyl ether-*co*-maleic anhydride) formed by free-radical initiated copolymerisation
Applications: Application as an aerosal hairspray resin

Poly(vinyl isobutyl ether-*co*-vinyl chloride)

Synonyms: *Poly(vinyl-2-methyl propane ether-co-vinyl chloride)*
Related Polymers: General information
Monomers: Vinyl isobutyl ether, Vinyl chloride
Material class: Synthetic Elastomers, Copolymers
Polymer Type: PVC, polyvinyls
CAS Number:

CAS Reg. No.
25154-85-2

Molecular Formula: $[(C_6H_{12}O).(C_2H_3Cl)]_n$
Fragments: $C_6H_{12}O$ C_2H_3Cl

Applications/Commercial Products:
Processing & Manufacturing Routes: Prod. by free-radical initiated copolymerisation e.g. AIBN, benzoyl peroxide
Applications: Used in marine paints, electrostatic image-developing powders i.e. toner, binding resin, adhesive tape for joining and sealing PVC single-ply roofing membranes

Trade name	Supplier
Caroflex MP-45	BASF

Poly(vinyl *l*-menthyl ether-*co*-indene) P-620

Synonyms: *Poly[(5-methyl-2-(1-methylethyl)cyclohexyloxyethene-co-indene)*
Related Polymers: Poly(vinyl ethers)
Monomers: Vinyl *l*-menthyl ether, Indene, Levomenthol
Material class: Thermoplastic
Polymer Type: polyvinyls
CAS Number:

CAS Reg. No.
69860-91-9

Molecular Formula: $[(C_{12}H_{22}O).(C_9H_8)]_n$
Fragments: $C_{12}H_{22}O$ C_9H_8

Volumetric & Calorimetric Properties:
Melting Temperature:

No.	Value	Note
1	158–169°C	[1]

Optical Properties:
Transmission and Spectra: Uv and ir spectral data have been reported [1]
Total Internal Reflection: $[\alpha]_D^{25}$ -3.4–26.1 (C_6H_6) [1]

Applications/Commercial Products:
Processing & Manufacturing Routes: Prod. by asymmetric induction radical copolymerisation with AIBN at 50°

Bibliographic References
[1] Kurokawa, M. and Minoura, Y., *J. Polym. Sci., Polym. Chem. Ed.*, 1979, **17**, 3297

Poly(vinyl methyl ether) P-621

Synonyms: *Poly(methyl vinyl ether)*. PVME. *Poly(methoxyethene)*
Related Polymers: Poly(vinyl ethers), Poly(vinyl methyl ether-*co*-maleic anhydride)
Monomers: Methyl vinyl ether
Material class: Synthetic Elastomers
Polymer Type: polyvinyls
CAS Number:

CAS Reg. No.	Note
9003-09-2	homopolymer
27082-59-3	isotactic

Molecular Formula: $(C_3H_6O)_n$
Fragments: C_3H_6O
Additives: Antioxidants generally added
General Information: The polymer can vary from a viscous, readily water-sol. liq. to a stiff rubber depending on MW. Compatible with aq. solns.
Tacticity: Isotactic form known

Volumetric & Calorimetric Properties:
Density:

No.	Value	Note
1	d^{27} 1.037 g/cm^3	[1]

Thermal Expansion Coefficient:

No.	Value	Note	Type
1	0.00047 K^{-1}	[2,3]	l

Equation of State: Tait, [4] Dee-Walsh, [5] van der Waals, [13] and Sanchez-Lacombe equation of state information has been reported [14]
Thermodynamic Properties General: ΔC_p 0.533 J g^{-1} K^{-1}. Specific heat capacity expressions have been reported [6]
Latent Heat Crystallization: ΔH_m 13.64 kJ mol^{-1}, ΔS_m 14.7 J mol^{-1} K^{-1} [10]. E_a 7.54 kJ mol^{-1} (1.8 kcal mol^{-1}) [11]. ΔH_a 31.5–38.5 [3]. ΔH_c -30.5 kJ g^{-1} (specific heat of combustion) [12]
Specific Heat Capacity:

No.	Value	Note	Type
1	2.53 kJ/kg.C	320–420K [9]	P

Melting Temperature:

No.	Value	Note
1	147°C	[7]
2	144°C	Crystalline form [19,31]

Glass-Transition Temperature:

No.	Value	Note
1	-31°C	[1]
2	-22°C	[3]
3	-13°C	[1,19]

Vicat Softening Point:

No.	Value	Note
1	55–70°C	[8]

Transition Temperature:

No.	Value	Note	Type
1	-20°C	[1]	Inflection temp.
2	-10°C	[1]	Low temp. stiffening
3	-31°C	[1]	Low temp. stiffening

Surface Properties & Solubility:
Solubility Properties General: Compatible with chlorinated rubber [27]; also with poly *p*-(2-hydroxyhexafluoroisopropyl)styrene-*co*-styrene [21]; novolac resin [22]; phenoxy resin [29]; polystyrene [20,25,26,27,28,29,33]; polyvinylpropionate [30]; polyvinylbutanoate [30] and the poly(hydroxyether) of Bisphenol A [23]

Cohesive Energy Density Solubility Parameters: δ 37.78 $J^{1/2}$ $cm^{-3/2}$ (8.13 $cal^{1/2}$ $cm^{-3/2}$) [2]; 33.27 $J^{1/2}$ $cm^{-3/2}$ (7.94 $cal^{1/2}$ $cm^{-3/2}$) [1]
Solvents/Non-solvents: Sol. EtOAc, Me_2CO, cold H_2O, MeOH, C_6H_6, CH_2Cl_2, $CHCl_3$, EtOH [30,34]. Insol. hexane, heptane, decane, ethylene glycol, Et_2O, hot H_2O. Polymer absorbs solvents to a greater or lesser degree depending on solubility [15]
Wettability/Surface Energy and Interfacial Tension: γ_s 0.00029 mN m^{-1} (29 dyn cm^{-1})

Transport Properties:
Polymer Solutions Dilute: [η] 1.4 dl g^{-1} [2]; 1.8 dl g^{-1} [1]. Viscosity constant K, 76000 ml g^{-1} (30°, light scattering) [2]. η_{sp} 0.47 dl g^{-1}
Permeability of Gases: Permeability of gases such as Ar and He has been reported [2]

Mechanical Properties:
Tensile (Young's) Modulus:

No.	Value	Note
1	82.73 MPa	12000 psi, 10 mil film [15]
2	56.53 MPa	8200 psi, 23°, 10 mil film, 25% H_2O absorbed [15]
3	158.6 MPa	23000 psi, 23°, 10 mil film, oriented [15]

Tensile Strength Break:

No.	Value	Note	Elongation
1	20.68 MPa	3000 psi, 23°, 10 mil film [15]	450%
2	59.98 MPa	8700 psi, 23°, 10 mil film, oriented [15]	46%
3	8.96 MPa	1300 psi, 33°, 10 mil film, 23°, 25% H_2O absorbed [15]	180%
4	18.99 MPa	2754 psi [16]	370%
5	1.01 MPa	146 psi [16]	525%

Tensile Strength Yield:

No.	Value	Elongation	Note
1	0.68 MPa	525%	98 psi [16]
2	9.24 MPa	370%	1340 psi [16]

Miscellaneous Moduli:

No.	Value	Note	Type
1	2.6 MPa	2600000 dyn cm^{-2} [17]	Plateau modulus

Mechanical Properties Miscellaneous: Steady state compliance 3×10^{-5} MPa (3×10^{-6} dyn cm^{-2}) [17]
Hardness: Shore A70 [8]

Optical Properties:
Refractive Index:

No.	Value	Note
1	1.47	30° [1]

Transmission and Spectra: Ir spectral data have been reported [18]
Molar Refraction: Molar refractivity 15.54 [1]

Polymer Stability:
Polymer Stability General: Exposure to oxygen, heat and light can lead to chain cleavage, cross-linking and oxidation [30,34]
Decomposition Details: Decomposition occurs by a free-radical pathway [34]
Flammability: Fl. p. 22° (Tagliabue open cup)

Chemical Stability: Stability to γ-radiation has been reported 0.31 G scissions (100 $eV)^{-1}$ [11]
Hydrolytic Stability: Very resistant to hydrol. [34]

Applications/Commercial Products:
Processing & Manufacturing Routes: Prod. by bulk or soln. polymerisation with a Lewis acid catalyst such as BF_3 or $AlCl_3$. May also be prepared by etherification of polyvinyl alcohol. Isotactic form synth. by using aluminium sulfate/sulfuric acid initiator in toluene
Applications: Rubber plasticiser and tackifier, raw material for adhesives and paints, heat sensitiser. Used for pressure sensitive adhesives for tapes and labels. Promotes adhesion on glass and metal. Also used to treat reverse osmosis membranes; to impart dry-film flexibility and to increase viscosity of photochemical-resist coatings

Trade name	Supplier
Gantrez M	GAF
Lutonal M	BASF

Bibliographic References

[1] Lal, J. and Trick, G.S., *J. Polym. Sci., Part A: Polym. Chem.*, 1964, **2**, 4559, (physical data)
[2] Allen, S.M., Stannett, V. and Hopfenberg, H.B., *Polymer*, 1981, **22**, 912
[3] Aharoni, S.M., *Polym. Prepr. (Am. Chem. Soc., Div. Polym. Chem.)*, 1973, **14**, 334
[4] Rodgers, P.A., *J. Appl. Polym. Sci.*, 1993, **48**, 1061, (equations of state)
[5] Dee, G.T. and Walsh, D.J., *Macromolecules*, 1988, **21**, 815
[6] Corrie, J.M.G. and Ferguson, R., *Macromolecules*, 1989, **22**, 2307
[7] Vandenberg, E.J., Heck, R.F. and Breslow, D.S., *J. Polym. Sci.*, 1959, **41**, 519
[8] Fishbein, L. and Crowe, B.F., *Makromol. Chem.*, 1961, **48**, 221, (physical data)
[9] Uriarte, C., Eguiazábal, J.I., Llanoz, M., Iribarren, J.I. and Iruin, J.J., *Macromolecules*, 1987, **20**, 3038
[10] Hopfinger, A.J., Pearlstein, R.A., Taylor, P.L. and Boyle, F.P., *J. Macromol. Sci., Phys.*, 1987, **26**, 359
[11] Suzuki, Y., Rooney, J.M. and Stannett, V., *J. Macromol. Sci., Chem.*, 1978, **12**, 1055
[12] Ng, S.C. and Chee, K.K., *Polymer*, 1993, **34**, 3870
[13] Brannock, G.R. and Sanchez, I.C., *Macromolecules*, 1993, **26**, 4970
[14] Rudolf, B., Schneider, H.A. and Cantow, H.J., *Polym. Bull. (Berlin)*, 1995, **34**, 109
[15] Vandenberg, E.J., *J. Polym. Sci., Part C: Polym. Lett.*, 1963, **1**, 207
[16] Kern, R.J., Hawkins, J.J. and Calfee, J.D., *Makromol. Chem.*, 1963, **66**, 126
[17] Takahashi, Y., Suzuki, H., Nakagawa, Y., Yamaguchi, M. and Noda, I., *Polym. J. (Tokyo)*, 1991, **23**, 1333
[18] Nakano, S., Iwasaki, K. and Fukutani, H., *J. Polym. Sci., Part A: Polym. Chem.*, 1963, **1**, 3277
[19] Nielsen, L.E. and Landel, R.F., *Mechanical Properties of Polymers and Composites*, 2nd edn., Marcel Dekker, 1993, (glass transition temp)
[20] Bank, M., Leffingwell, J. and Thies, C., *Macromolecules*, 1971, **4**, 43, (compatibility)
[21] Seferis, J.C., *Polym. Handb.*, 4th edn., (eds. J. Brandrup, E.H. Immergut and E.A. Grulke), John Wiley and Sons, 1999, **VI**, 581, (refractive indices)
[22] Fahrenholtz, S.R. and Kwei, T.K., *Macromolecules*, 1981, **14**, 1076, (novolac resins)
[23] Robesan, L.M., Hale, W.F. and Mermain, C.N., *Macromolecules*, 1981, **14**, 1644, (miscibility)
[24] McMaster, L.P., *Macromolecules*, 1973, **6**, 760, (thermodynamics)
[25] Nishi, T., Wang, T.T. and Kwei, T.K., *Macromolecules*, 1975, **8**, 227, (separation)
[26] Kwei, T.K., Nishi, T. and Roberts, R.F., *Macromolecules*, 1974, **7**, 667, (compatibility)
[27] Nishi, T. and Kwei, T.K., *Polymer*, 1975, **16**, 285, (compatibility)
[28] Bank, M., Leffingwell, J. and Thies, C., *J. Polym. Sci., Part A-2*, 1972, 1097, (phase separation)
[29] Corrie, J.M.G. and Saeki, S., *Polym. Bull. (Berlin)*, 1981, **6**, 75, (compatibility with polystyrene)
[30] Dutta, S., Chakraborty, S.S., Mandal, B. and Bhattacharyya, S.N., *Polymer*, 1993, **34**, 3499, (compatibility with various polyvinyl esters)
[31] *Encycl. Polym. Sci. Eng.*, 2nd edn., (ed. J.I. Kroschwitz), John Wiley and Sons, 1985, (manufacture)
[32] Chiellini, E., Montagnoli, G. and Pirio, P., *J. Polym. Sci., Part B: Polym. Lett.*, 1969, **7**, 121, (polymerisation)

[33] Lu, F.J., Benedetti, E. and Hsu, S.L., *Macromolecules*, 1983, **16**, 1525, (spectroscopic studies)
[34] Barnich, J. and Muller, H.W.J., *Polymeric Materials Encyclopedia*, (ed. J.C. Salamone), CRC Press, 1996, 8572, (stability, production and applications)
[35] *Eur. Pat. Appl.*, 1982, 46 371, (synth)

Poly(vinyl methyl ether-*co*-disodium maleate) P-622

$$\left[\left(CH_2CH\right)_x\left(\begin{array}{c}CHCH\\|\\COONa\\COONa\end{array}\right)_y\right]_n$$
$\quad\quad$ OMe

Synonyms: *Poly(vinyl methyl ether-co-disodium 1,3-propenoate). Poly(disodium (Z)-2-butenedioate-co-methoxyethene). Poly(-methoxyethene-co-disodium (Z)-2-butenedioate)*
Related Polymers: Poly(vinyl ethers), Poly(vinyl methyl ether-*co*-maleic acid), Poly(vinyl methyl ether-*co*-maleic anhydride)
Monomers: Methyl vinyl ether, Maleic anhydride
Material class: Synthetic Elastomers, Copolymers
Polymer Type: polyvinyls
CAS Number:

CAS Reg. No.
9019-25-4

Molecular Formula: $[(C_3H_6O).(C_4H_2O_4Na_2)]_n$
Fragments: $C_3H_6O \quad C_4H_2O_4Na_2$

Applications/Commercial Products:
Processing & Manufacturing Routes: Prod. by cation exchange of free acid derived from the hydrol. of PVM-MA. PVM-MA formed by free radical copolymerisation using AIBN or benzoyl peroxide initiators
Applications: Phosphate replacement in detergents

Trade name	Supplier
Gantrez DS 1935	GAF
Sakalan CP-2	BASF

Poly(vinyl methyl ether-*co*-maleic acid) P-623

$$\left[\left(\begin{array}{c}CH_2CH\\|\\OMe\end{array}\right)_x\left(\begin{array}{c}CH_2CH_2\\|\\COOH\\COOH\end{array}\right)_y\right]_n$$

Synonyms: *Poly(vinyl methyl ether-co-butenedioic acid). Poly((Z)-2-butenedioic acid-co-methoxyethene). Poly(methoxyethene-co-(Z)-2-butenedioic acid)*
Related Polymers: Poly(vinyl ethers), Poly(vinyl methyl ether-*co*-monoethyl maleate), Poly(vinyl methyl ether-*co*-monoisopropyl maleate), Poly(vinyl methyl ether-*co*-disodium maleate), Poly(vinyl methyl ether-*co*-maleic anhydride)
Monomers: Maleic anhydride, Butenedioic acid, Butenedioic acid, Vinyl methyl ether
Material class: Synthetic Elastomers, Copolymers
Polymer Type: polyvinyls
CAS Number:

CAS Reg. No.	Note
25153-40-6	
36906-62-4	ammonium salt
111263-13-9	alternating
26300-19-6	sodium salt
54578-91-5	butyl ester
54018-18-7	monobutyl ester
62386-95-2	calcium, sodium salt
133222-53-4	calcium, strontium salt
133222-51-2	calcium, zinc salt
133222-52-3	calcium, zinc, sodium salt

Molecular Formula: $[(C_3H_6O).(C_4H_2O_2)]_n$
Fragments: $C_3H_6O \quad C_4H_2O_2$

Volumetric & Calorimetric Properties:
Thermodynamic Properties General: Molar volume V_r 36.4 cm^3 mol^{-1} [3]
Glass-Transition Temperature:

No.	Value	Note
1	144°C	[1]

Transport Properties:
Polymer Solutions Dilute: [η] 1.56 dl g^{-1} (25°, sodium salt) [2]

Electrical Properties:
Complex Permittivity and Electroactive Polymers: Soln. dielectric props. have been reported [4]

Optical Properties:
Transmission and Spectra: Ir spectral data have been reported [1,2]

Applications/Commercial Products:
Processing & Manufacturing Routes: Prod. by hydrolysis of PVM-MA. PVM-MA formed by free-radical initiated copolymerisation e.g. AIBN, benzoyl peroxide
Applications: Thickener, protective colloid, inhibitor of tartar and antibacterial agent in toothpastes

Trade name	Details	Supplier
Gantrex AT 795		GAF
Gantrez S	The S series appears to refer to this material	GAF

Bibliographic References
[1] Chung, K.H., Wu, C.S. and Malawer, E.J., *J. Appl. Polym. Sci.*, 1990, **41**, 793
[2] Kokufata, E., *Polymer*, 1980, **21**, 177
[3] Zana, R., *J. Polym. Sci., Polym. Phys. Ed.*, 1980, **18**, 121
[4] Minakata, A., Nishio, T. and Nakamura, H., *Ferroelectrics*, 1988, **86**, 7, (dielectric props.)

Poly(vinyl methyl ether-*co*-maleic anhydride) P-624

$$\left[\left(\begin{array}{c}CH_2CH\\|\\OMe\end{array}\right)_x\left(\begin{array}{c}\\O\diagup\diagdown O\end{array}\right)_y\right]_n$$

Synonyms: *PVM-MA. Methyl vinyl ether-maleic anhydride copolymer. Poly(2,5-furandione-co-methoxyethene). Poly(methoxyethene-co-2,5-furandione)*
Related Polymers: Poly(vinyl ethers), Poly(vinyl methyl ether-*co*-butenedioic acid)
Monomers: Vinyl methyl ether, Maleic anhydride
Material class: Synthetic Elastomers, Copolymers
Polymer Type: polyvinyls

CAS Number:

CAS Reg. No.	Note
9011-16-9	
52229-50-2	alternating

Molecular Formula: $[(C_3H_6O).(C_4H_4O_3)]_n$
Fragments: C_3H_6O $C_4H_4O_3$
Additives: Water sol. uv absorbers and antioxidants such as thiourea may be added to Gantrez
General Information: Copolymers of maleic anhydride are the most significant commercial class of vinyl ether copolymers and the methyl vinyl ether copolymer is by far the most dominant

Volumetric & Calorimetric Properties:
Glass-Transition Temperature:

No.	Value	Note
1	153°C	[1,3]
2	133°C	[2]

Transport Properties:
Polymer Solutions Dilute: η_{inh} 2.43 dl g^{-1} (0.5% in DMF, 30°) [1]

Optical Properties:
Transmission and Spectra: H-1 nmr spectral data have been reported [1]

Polymer Stability:
Chemical Stability: Preferentially degrades rather than cross-links upon exposure to γ-radiation resulting in greatly reduced viscosity

Applications/Commercial Products:
Processing & Manufacturing Routes: Prod. by free radical initiated polymerisation
Applications: Raw material for adhesives, coatings and pharmaceuticals. FDA approved for food packaging adhesives

Trade name	Details	Supplier
Gantrez	The AN series appears to refer to this material	GAF
Lupasol		BASF
Luviform		BASF
Sokulan		BASF
Viscofras		ICI, UK

Bibliographic References
[1] Fu, X., Bassett, W. and Vogl, O., *J. Polym. Sci., Polym. Chem. Ed.*, 1983, **21**, 891
[2] Kaneko, R., *Kobunshi Kagaku*, 1967, **24**, 272
[3] Chung, K.H., Wu, C.S. and Malamer, E.G., *J. Appl. Polym. Sci.*, 1990, **41**, 793
[4] Pahl, K.U. and Rodriguez, F., Polym. Prepr. (Am. Chem. Soc., Div. Polym. Chem.), 1984, **25**, 320

Poly(vinyl methyl ether-*co*-monobutyl maleate) P-625

Synonyms: *Poly(vinyl methyl ether-co-3-(butyloxycarbonyl)propenoic acid)*
Related Polymers: General information
Monomers: Vinyl methyl ether, Maleic anhydride
Material class: Synthetic Elastomers, Copolymers
Polymer Type: polyvinyls
CAS Number:

CAS Reg. No.
54578-91-5

Molecular Formula: $[(C_3H_6O).(C_8H_{10}O_4)]_n$
Fragments: C_3H_6O $C_8H_{10}O_4$

Applications/Commercial Products:
Processing & Manufacturing Routes: Prod. by action of PVM-MA in butanol. PVM-MA formed by free-radical initiated copolymerisation e.g. AIBN, benzoyl peroxide
Applications: Adhesive, cosmetic and pharmaceutical formulations, hairsprays, manufacture of speciality coatings, thermographic copying material

Trade name	Supplier
Gantrez-ES-425	GAF

Poly(vinyl methyl ether-*co*-monoethyl maleate) P-626

$$\left[\left[\begin{array}{c} CH_2CH \\ | \\ OMe \end{array} \right]_x \left[\begin{array}{c} CHCH \\ | \quad | \\ COOEt \\ COOH \end{array} \right]_y \right]_n$$

Synonyms: *Poly(vinyl methyl ether-co-3-(ethyloxycarbonyl)propenoic acid). Poly(ethyl hydrogen ((Z)-2-butenedioate-co-methoxyethene). Poly(methoxyethene-co-ethyl hydrogen (Z)-2-butenedioate)*
Related Polymers: Poly(vinyl ethers), Poly(vinyl methyl ether-*co*-maleic acid), Poly(vinyl methyl ether-*co*-maleic anhydride)
Monomers: Methyl vinyl ether, Maleic anhydride
Material class: Synthetic Elastomers, Copolymers
Polymer Type: polyvinyls
CAS Number:

CAS Reg. No.	Note
50935-57-4	
67724-93-0	maleic anhydride ethyl ester

Molecular Formula: $[(C_3H_6O).(C_6H_6O_4)]_n$
Fragments: C_3H_6O $C_6H_6O_4$

Applications/Commercial Products:
Processing & Manufacturing Routes: Prod. by action of EtOH on PVM-MA. PVM-MA is formed by free-radical initiated copolymerisation of vinyl methyl ether and maleic anhydride
Applications: Hairsprays, adhesive, cosmetic and pharmaceutical formulations, manufacture of speciality coatings

Trade name	Supplier
Gantrez SP 215	GAF

Bibliographic References
[1] Bonelli, D., Clementi, S. Ebert, C., Lovrecich, M. and Rubessa, F., *Drug. Dev. Ind. Pharm.*, 1989, **15**, 1375, (applications)

Poly(vinyl methyl ether-*co*-monoisopropyl maleate) P-627

$$\left[\left[\begin{array}{c} CH_2CH \\ | \\ OMe \end{array} \right]_x \left[\begin{array}{c} CHCH \\ | \quad | \\ COOCH(CH_3)_2 \\ COOH \end{array} \right]_y \right]_n$$

Synonyms: *Poly(vinyl methyl ether-co-3-(isopropyloxycarbonyl)-propenoic acid). Poly(vinyl methyl ether-co-3-(2-propyloxycarbonyl)propenoic acid). Poly(1-methylethyl hydrogen ((Z)-2-*

butenedioate-co-methoxyethene). Poly(methoxyethene-co-1-methylethyl hydrogen (Z)-2-butenedioate)
Related Polymers: Poly(vinyl ethers), Poly(vinyl methyl ether-co-maleic acid), Poly(vinyl methyl ether-co-maleic anhydride)
Monomers: Methyl vinyl ether, Maleic anhydride
Material class: Synthetic Elastomers, Copolymers
Polymer Type: polyvinyls
CAS Number:

CAS Reg. No.	Note
31307-95-6	
54077-45-1	maleic anhydride isopropyl ester
56091-51-1	Gantrez ES 335I

Molecular Formula: $[(C_3H_6O).(C_7H_8O_4)]_n$
Fragments: C_3H_6O $C_7H_8O_4$

Applications/Commercial Products:
Processing & Manufacturing Routes: Prod. by action of 2-propanol on PVM-MA. PVM-MA formed by free-radical initiated copolymerisation using AIBN or benzoyl peroxide initiator
Applications: Hairsprays, adhesive, cosmetic and pharmaceutical formulations, manufacture of speciality coatings

Trade name	Details	Supplier
Gantrez	ES 335, 335I	GAF

Poly(vinyl octadecyl ether)   P-628

$$\left[\begin{array}{c} CH_2CH \\ | \\ OCH_2(CH_2)_{16}CH_3 \end{array}\right]_n$$

Synonyms: *Poly(octadecyl vinyl ether). Poly(octadecyloxyethene)*
Related Polymers: General information, Poly(vinyl octadecyl ether-co-maleic anhydride)
Monomers: Octadecyl vinyl ether
Material class: Synthetic Elastomers
Polymer Type: polyalkylene ether
Molecular Formula: $(C_{20}H_{40}O)_n$
Fragments: $C_{20}H_{40}O$

Volumetric & Calorimetric Properties:
Vicat Softening Point:

No.	Value	Note
1	48–54°C	[1]

Transport Properties:
Polymer Solutions Dilute: [η] 0.19 dl g^{-1} [1]; 0.046–0.132 dl g^{-1} [2]

Applications/Commercial Products:
Processing & Manufacturing Routes: Prod. by cationic polymerisation with Lewis acid catalysts
Applications: Polishing and waxing agent

Trade name	Supplier
Luwax	BASF
VWax	BASF

Bibliographic References
[1] Fishbein, L. and Crowe, B.F., *Makromol. Chem.*, 1961, **48**, 221
[2] Fee, J.G., Port, W.S. and Witnauer, L.P., *J. Polym. Sci.*, 1958, **33**, 95

Poly(vinyl octadecyl ether-co-maleic anhydride)   P-629

$$\left[\begin{array}{c} (CH_2CH)_x \\ | \\ OCH_2(CH_2)_{16}CH_3 \end{array} \begin{array}{c} \\ O \\ \diagup \diagdown \\ O \quad O \end{array}_y \right]_n$$

Synonyms: *Poly(1-(ethenyloxy)octadecane-co-2,5-furandione). Poly(2,5-furandione-co-1-(ethenyloxy)octadecane)*
Related Polymers: Poly(vinyl ethers)
Monomers: Vinyl octadecyl ether, Maleic anhydride
Material class: Thermoplastic, Copolymers
Polymer Type: polyvinyls
CAS Number:

CAS Reg. No.
28214-64-4

Molecular Formula: $[(C_{20}H_{40}O).(C_4H_2O_3)]_n$
Fragments: $C_{20}H_{40}O$ $C_4H_2O_3$

Volumetric & Calorimetric Properties:
Melting Temperature:

No.	Value	Note
1	140°C	[1]

Electrical Properties:
Static Electrification: Conductance 9.1–22.9 mS cm^{-2} (25°, 3.9 nm film); 12.5–21.7 mS cm^{-2} (140°, 3.9 nm film) [1]

Applications/Commercial Products:
Processing & Manufacturing Routes: Prod. by free-radical initiated copolymerisation with AIBN, benzoyl peroxide etc
Applications: Antiblocking agents, release coatings. Potential applications in dentistry

Trade name	Supplier
Gantrez AN 8194	GAF

Bibliographic References
[1] Tredgold, R.H. and Winter, C.S., *J. Phys. D: Appl. Phys.*, 1982, **15**, L55

Poly(vinyl octyl ether)   P-630

$$\left[\begin{array}{c} CH_2CH \\ | \\ OCH_2(CH_2)_6CH_3 \end{array}\right]_n$$

Synonyms: *Poly(octyl vinyl ether). Poly(octoxyethene)*
Related Polymers: Poly(vinyl ethers)
Monomers: Vinyl octyl ether
Material class: Synthetic Elastomers
Polymer Type: polyvinyls
CAS Number:

CAS Reg. No.
25232-89-7

Molecular Formula: $(C_{10}H_{20}O)_n$
Fragments: $C_{10}H_{20}O$

Poly(vinyl propionate)

Volumetric & Calorimetric Properties:
Density:

No.	Value	Note
1	d^{30} 0.893 g/cm^3	
2	d^{27} 0.914 g/cm^3	[1]

Thermodynamic Properties General: Molar volume V_r 171 cm^3 mol^{-1} [2]

Glass-Transition Temperature:

No.	Value	Note
1	-79°C	[1]

Transition Temperature:

No.	Value	Note	Type
1	-62°C	[1]	Inflection temp.

Surface Properties & Solubility:
Cohesive Energy Density Solubility Parameters: δ 32.98 J$^{1/2}$ cm$^{-3/2}$ (7.87 cal$^{1/2}$ cm$^{-3/2}$) [1]

Optical Properties:
Refractive Index:

No.	Value	Note
1	1.4613	30° [1]

Molar Refraction: Molar refraction 48.03 [1]

Applications/Commercial Products:
Processing & Manufacturing Routes: Prod. by cationic polymerisation with Friedel-Crafts catalysts (BF$_3$, AlCl$_3$, SnCl$_4$ or their complexes)

Bibliographic References
[1] Lal, J. and Trick, G.S., *J. Polym. Sci., Part A: Polym. Chem.*, 1964, **2**, 4559
[2] Van Krevelen, D.W., *Properties of Polymers: Their Correlation with Chemical Structure*, Elsevier, 1972, 44

Poly(vinyl propionate) P-631

OCOEt

Synonyms: *PVPr. Poly(propionyloxyethylene)*
Monomers: Vinyl propionate
Material class: Synthetic Elastomers
Polymer Type: polyvinyls
CAS Number:

CAS Reg. No.
25035-84-1

Molecular Formula: (C$_5$H$_8$O$_2$)$_n$
Fragments: C$_5$H$_8$O$_2$

Volumetric & Calorimetric Properties:
Thermodynamic Properties General: ΔH_p -86.5 kJ mol^{-1} (-20.6 kcal mol^{-1}) [8]

Glass-Transition Temperature:

No.	Value	Note
1	10°C	[1]

Surface Properties & Solubility:
Solubility Properties General: Miscible with poly(vinyl methyl ether) [3] and poly(ethyl acrylate) [11]. Limited miscibility with poly(vinyl acetate-*co*-ethylene) and poly(vinyl acetate-*co*-stearate) [6]

Optical Properties:
Refractive Index:

No.	Value	Note
1	1.4665	20° [2]

Polymer Stability:
Decomposition Details: Photodegradation leads mainly to the formation of propanoic acid (70%) and propanol (19%). Small volatiles account for the mass balance [7]
Environmental Stress: Frost resistant but may coagulate if aq. phase freezes [5]
Chemical Stability: More stable towards alkalis than polyvinyl acetate [5]
Biological Stability: Emulsions may be prone to attack from microorganisms unless protected frequently [5]
Stability Miscellaneous: Photodegradable by the action of a mercury lamp in vacuum at 20° [7]

Applications/Commercial Products:
Processing & Manufacturing Routes: Manufactured by emulsion polymerisation of vinyl propionate using typical peroxide initiators [5]. May be stabilised with water-sol. protective colloids, anionic emulsifiers or non-ionic emulsifiers. Poly vinyl propionate may be prepared by radical polymerisation [4] using AIBN [9] or in the presence of lithium [10]. Polymerisation giving polymer beads from a monomer suspension has been reported [16], as have a number of catalysts [12,13,14,15], some of which are used for poly(vinyl benzoate) manufacture [12,13]
Applications: Main applications include emulsion paints, coatings and adhesives. Used in anti-noise coatings, chewing gum bases, as a concrete additive and in soil stabilisation. Also used as glass fibre sizing material and in binders for non-woven fabrics, pigment printing pastes, as coatings for food and the undersides of carpets, cheese rinds and paper, and as a fibrous leather substitute

Trade name	Supplier
Propiofan	BASF

Bibliographic References
[1] van Hoorn, H., *Rheol. Acta*, 1971, **10**, 208, (glass transition temp)
[2] Serefis, J.C., *Polym. Handb.*, 4th edn., (eds. J. Brandrup, E.H. Immergut and E.A. Grulke), John Wiley and Sons, 1999, **VI**, 581, (refractive index)
[3] Dutta, S., Chakraborty, S.S., Mandal, B.M and Bhattacharyya, S.N., *Polymer*, 1993, **34**, 3499
[4] Bevington, J.C. and Johnson, M., *Eur. Polym. J.*, 1968, **4**, 373, (radical polymerisation)
[5] *Ullmanns Encykl. Ind. Chem.*, Vol. 10, 5th edn., (ed. W. Gerhartz), VCH, 1985, 395, (synth)
[6] Braun, D., Leiss, D., Bergmann, M.J. and Hellmann, G.P., *Eur. Polym. J.*, 1993, **29**, 225
[7] Buchanan, K.J. and McGill, W.J., *Eur. Polym. J.*, 1980, **16**, 309, 313, 319, (decomp)
[8] Hopfinger, A.J., Pearlstein, R.A., Mabilia, M. and Tripathy, S.K., *Pure Appl. Chem.*, 1988, **60**, 271

[9] Burnett, G.M. and Wright, W.W., *Trans. Faraday Soc.*, 1953, **49**, 1108, (radical polymerisation)
[10] Kelley, D.J., *J. Polym. Sci.*, 1962, **59**, S6, (polymerisation)
[11] Bhattacharyya, C., Bhattacharyya, S.N. and Mandal, B.M., *J. Indian Chem. Soc.*, 1986, **63**, 157, (miscibility)
[12] *Brit. Pat.*, 1947, 585 396, (amine oxide catalysts)
[13] *Brit. Pat.*, 1948, 604 580, (organic polyperoxide catalysts)
[14] *U.S. Pat.*, 1962, 3 052 661, (aluminium-alkyl ether catalysts)
[15] *U.S. Pat.*, 1962, 3 053 822, (aluminium-alkyl ketone catalysts)
[16] *Ger. Pat.*, 1960, 1 088 716, (polymerisation)

Poly(vinyl propyl ether) P-632

$$\left[\begin{array}{c} CH_2CH \\ | \\ OCH_2CH_2CH_3 \end{array} \right]_n$$

Synonyms: *Poly(propyl vinyl ether). Poly(propoxyethene). PVPr*
Related Polymers: Poly(vinyl ethers)
Monomers: Propyl vinyl ether
Material class: Synthetic Elastomers
Polymer Type: polyvinyls
CAS Number:

CAS Reg. No.	Note
25585-50-6	homopolymer
28209-42-9	isotactic

Molecular Formula: $(C_5H_{10}O)_n$
Fragments: $C_5H_{10}O$

Volumetric & Calorimetric Properties:
Density:

No.	Value	Note
1	d 0.9368 g/cm^3	[4]

Melting Temperature:

No.	Value	Note
1	76°C	[1]

Glass-Transition Temperature:

No.	Value	Note
1	-49°C	[3,5]

Surface Properties & Solubility:
Cohesive Energy Density Solubility Parameters: δ cohesive energy 37.54 J$^{1/2}$ cm$^{-3/2}$ (8.96 J$^{1/2}$ cm$^{-3/2}$) [2]
Solvents/Non-solvents: Sol. $CHCl_3$, Et_2O, C_6H_6, propanol. Insol. heptane, Me_2CO [1]

Transport Properties:
Polymer Solutions Dilute: Absolute viscosity (0.8010 centipoise) [2] has been reported [4]

Optical Properties:
Refractive Index:

No.	Value	Note
1	1.4528	[4]

Applications/Commercial Products:
Processing & Manufacturing Routes: Prod. by cationic polymerisation with Friedel-Crafts catalysts (BF_3, $AlCl_3$, $SnCl_4$). May also be prepared in good yield by using ferric chloride catalyst [4]. The isotactic form can be obtained by using ferric sulfate catalyst [6]
Applications: Used as a diesel fuel additive to reduce particulates

Bibliographic References
[1] Vandenberg, E.J., Heck, R.F. and Breslow, D.S., *J. Polym. Sci.*, 1959, **41**, 519, (synth)
[2] Ahmad, H. and Yaseen, M., *Polym. Eng. Sci.*, 1979, **19**, 858, (solubility parameters)
[3] Mikhailev, G.P. and Sheyelev, V.A., *Polym. Sci. USSR (Engl. Transl.)*, 1967, **9**, 2762, (glass transition temp)
[4] Shostakovsky, M.F., Mikhantyev, B.L. and Orchimnikova, N.N., *Bull. Acad. Sci. USSR, Div. Chem. Sci. (Engl. Transl.)*, 1953, 939, (polymerisation)
[5] Lal, J. and Scott, W., *J. Polym. Sci., Part C: Polym. Lett.*, 1965, **9**, 113, (physical props)
[6] *Brit. Pat.*, 1960, 846 690, (polymerisation)

Poly(2-vinylpyridine) P-633

$$\left[\begin{array}{c} \text{(2-pyridyl)} \\ | \\ CHCH_2 \end{array} \right]_n$$

Synonyms: *Poly(2-ethenylpyridine). PVP*
Monomers: 2-Vinylpyridine
Material class: Thermoplastic
Polymer Type: polyolefins
CAS Number:

CAS Reg. No.	Note
9003-47-8	Polyethenylpyridine
25014-15-7	homopolymer
25585-16-4	isotactic
28601-69-6	syndiotactic
80123-18-8	trimer
80123-19-9	tetramer
96526-39-5	pentamer
96526-40-8	hexamer
54579-18-9	syndiotactic homopolymer ion(1-)
36604-92-9	homopolymer radical ion(1-) caesium
71114-01-7	tetramer isotactic
36604-93-0	homopolymer radical ion(2-) disodium
59914-48-6	2-ethenyl-2,2-d$_2$ pyridine homopolymer
146666-38-8	2-ethenyl-2,2-d$_2$ pyridine isotactic homopolymer
91499-22-8	homopolymer ion(1-), lithium
27476-03-5	dimer
53125-47-6	homopolymer ion(1-), sodium

Molecular Formula: $(C_7H_7N)_n$
Fragments: C_7H_7N
Molecular Weight: MV 400000 (isotactic, organo hafnium catalyst)
Tacticity: Isotactic and syndiotactic forms known
Morphology: Threefold helix, hexagonal unit cell a=b 15.49Å, c 6.56Å (fibre axis) [3,17], space group P3$_1$ or P3$_2$ [17]
Miscellaneous: Side reactions occur during the preparation and handling (e.g. in soln.) of polymers of 2-vinylpyridine [6]

Volumetric & Calorimetric Properties:
Thermodynamic Properties General: Heat of fusion per unit

volume values have been reported [3]. Thermal expansion coefficient values have been reported [27]

Melting Temperature:

No.	Value	Note
1	212.5°C	isotactic [3]

Glass-Transition Temperature:

No.	Value	Note
1	84°C	isotactic [5]

Transition Temperature General: Up to three melting endotherms can be present [3]. Other T_g values have been reported [4,24]

Surface Properties & Solubility:

Solubility Properties General: Acts as an electron donor due to pyridine nitrogen of polymer and hence is more sol. in acidic solvents

Cohesive Energy Density Solubility Parameters: Solubility parameter 21.26 $J^{1/2}$ cm$^{-3/2}$ (approx., 10.4 cal$^{1/2}$ cm$^{-3/2}$) [10]. Other solubility parameter values have been reported [10]

Solvents/Non-solvents: Sol. MeOH, ethylene glycol, nitroethane, EtOH aq., CHCl$_3$, heptanol, 2-butanol, pentanol, butanol, propanol, pyridine [3,10,29]. Insol. hexane, H$_2$O [10,29], DMF, THF, 2-propanol, chlorobenzene, C$_6$H$_6$ [29], dioxane, 2-butanone and CH$_2$Cl$_2$ [29]

Surface Tension:

No.	Value	Note
1	39.5 mN/m	[3]
2	3.2 mN/m	perpendicular to direction of growing crystal [3]
3	34–36.2 mN/m	varies with MW [14]

Transport Properties:

Polymer Solutions Dilute: Various calculated viscosity values and Mark-Houwink parameters have been reported for pyridine [31], CHCl$_3$/EtOH [36,38], MeOH [11,31], DMF [11], C$_6$H$_6$ [11] and dioxane [11]. Critical miscibility temps. of soln. have been calculated from Flory-Fox [12] and Kurata-Stockmayer equations [12]

Permeability and Diffusion General: Diffusion of C$_6$H$_6$, CH$_2$Cl$_2$, MeOH, H$_2$O vapours has been reported; [29] C$_6$H$_6$ and CH$_2$Cl$_2$ vapours behave in a non-Fickian manner [29]

Mechanical Properties:

Mechanical Properties General: Sound velocity 11.1 kHz [32]. Variation of damping factor vs. temp. has been reported [32]

Electrical Properties:

Electrical Properties General: Electrical behaviour of P2VP has been studied [28]. Activation energy for conduction 1.23 eV [28]. Conductivity doped with iodine σ 10^{-4} Ω^{-1} cm^{-1} (M_n 40000) [32]

Surface/Volume Resistance:

No.	Value	Note	Type
1	1×10^{15} Ω.cm	conductivity value [32]	S

Optical Properties:

Optical Properties General: Molecular optical anisotropy has been reported [34,35]

Transmission and Spectra: Ir [1,5,16], x-ray [16,17], H-1 nmr [13,21,22], C-13 nmr [13], epr [19,24], ms [2,8,25,26], uv [1], xps [18], esca [19], EELS [15], Auger [7], carbon core 1s electron spectrum [15], and laser desorption spectral data [8] have been reported

Polymer Stability:

Decomposition Details: Decomposition by depolymerisation occurs at 396° leaving no char [2]. Complexation with copper(II) chloride lowers decomposition temp. to 320° and leaves 26–41% by wt. of char [2]. Activation energy of decomposition 250 kJ mol^{-1} (60 kcal mol^{-1})

Environmental Stress: Irradiation at wavelengths above 300 nm produces vinyl pyridinyl radicals [9]

Stability Miscellaneous: Laser desorption degrades polymer to give 2-vinylpyridine monomer [8]. γ-Irradiation causes cross-linking [19,20]. P2VP is 5–9 times less stable to ionising radiation than styrene

Bibliographic References

[1] De Bruyne, A., Delplancke, J.-L. and Winand, R., *J. Appl. Electrochem.*, 1995, **25**, 284, (electropolymerisation, uv, ir)
[2] Lyons, A.M., Pearce, E.M. and Mujsce, A.M., *J. Polym. Sci., Part A: Polym. Chem.*, 1990, **28**, 245, (thermal decomposition)
[3] Alberda van Ekenstein, G.O.R., Tan, Y.Y. and Challa, G., *Polymer*, 1985, **26**, 283, (crystallisation, isotactic)
[4] Gan, Y., Dong, D. and Hogen-Esch, T.E., *Macromolecules*, 1995, **28**, 383, (T_g)
[5] McKenna, W.P. and Apai, G., *J. Phys. Chem.*, 1992, **96**, 5902, (spectra)
[6] Schmitz, F.P., Hilger, S.H. and Gemmel, B.V., *Makromol. Chem.*, 1990, **191**, 1033
[7] Rye, R.R., Kelber, J.A., Kellog, G.E., Nebesny, K.W. and Lichtenberger, D.L., *J. Chem. Phys.*, 1987, **86**, 4375, (Auger)
[8] Schriemer, D.C. and Li, L., *Anal. Chem.*, 1996, **68**, 250, (laser desorption)
[9] Quaegebeur, J.P., Lablache-Combier, A. and Chachaty, C., *Makromol. Chem.*, 1977, **178**, 1507, (irradiation)
[10] Arichi, S., Matsuura, H., Tanimoto, Y. and Murata, H., *Bull. Chem. Soc. Jpn.*, 1966, **39**, 434, (solubility)
[11] Arichi, S., *Bull. Chem. Soc. Jpn.*, 1966, **39**, 439, (intrinsic viscosities)
[12] Arichi, S., Tanimoto, Y. and Murata, H., *Bull. Chem. Soc. Jpn.*, 1963, **41**, 1296, (soln data)
[13] Brigodiot, M., Cheraderne, H., Fontanille, M. and Vairon, J.P., *Polymer*, 1976, **17**, 254, (nmr)
[14] Vitt, E. and Shull, K.R., *Macromolecules*, 1995, **28**, 6349
[15] Ritsko, J.J. and Bigelow, R.N., *J. Chem. Phys.*, 1978, **69**, 4162
[16] Arichi, S., *Bull. Chem. Soc. Jpn.*, 1968, **41**, 244, (x-ray, ir)
[17] Puterman, M., Kolpak, F.J., Blackwell, J. and Lando, J.B., *J. Polym Sci., Part B: Polym. Phys.*, 1977, **15**, 805, (x-ray)
[18] Fabishi, T.J. and Thomas, H.R., *Macromolecules*, 1980, **13**, 1487, (XPS)
[19] David, C., Verhasselt, A. and Geuskens, G., *J. Polym. Sci., Part C: Polym. Lett.*, 1967, **16**, 2181, (γ-irradiation)
[20] David, C., Verhasselt, A. and Geuskens, G., *Polymer*, 1964, **5**, 544, (γ-irradiation)
[21] Weill, G. and Hermann, G., *J. Polym. Sci., Part B: Polym. Phys.*, 1967, **5**, 1293, (nmr)
[22] Matsuzaki, K. and Sugimoto, T., *J. Polym. Sci., Part B: Polym. Phys.*, 1967, **5**, 1320, (nmr)
[23] Toreki, W.M., Hogen-Esch, T.E. and Butler, G.B., *Polym. Prepr. (Am. Chem. Soc., Div. Polym. Chem.)*, 1987, **28**, 343, (macrocyclic P2VP)
[24] Sundararajan, J. and Hogen-Esch, T.E., *Polym. Prepr. (Am. Chem. Soc., Div. Polym. Chem.)*, 1991, **32**, 63, (macrocyclic P2VP)
[25] Fletcher, R.A. and Fatiadi, A.J., *J. Trace Microprobe Tech.*, 1986, **4**, 215, (ms)
[26] Hook, K.J., Hook, T.J., Wandass, J.H. and Gardella, J.A., *Appl. Surf. Sci.*, 1990, **44**, 29, (SIMS)
[27] Zanten, J.H. van, Wallace, W.E. and Wu, W.-L., *Phys. Rev. E: Stat. Phys., Plasmas, Fluids, Relat. Interdiscip. Top.*, 1996, **53**, R2053, (thermal expansion)
[28] Chohan, M.H., *J. Mater. Sci. Lett.*, 1994, **13**, 6
[29] Odani, H., Uchikura, M. and Kurata, M., *Bull. Inst. Chem. Res., Kyoto Univ.*, 1984, **62**, 188, (diffusion)
[30] Fréchet, J.M.J., *Polym. Prepr. (Am. Chem. Soc., Div. Polym. Chem.)*, 1983, **24**, 340, (applications)
[31] Arichi, S., *J. Sci. Hiroshima Univ., Ser. A-2*, 1965, **29**, 97, (solvent parameters)
[32] Baccaredda, M., Butta, E., Frosini, V. and De Petris, S., *Mater. Sci. Eng.*, 1968, **3**, 157, (dynamic mech props)
[33] Moeller, M., Capistran, J. and Lenz, R.W., *Polym. Prepr. (Am. Chem. Soc., Div. Polym. Chem.)*, 1983, **24**, 342, (conductivity)
[34] Fourche, G. and Tourenne, C., *Eur. Polym. J.*, 1976, **12**, 663, (molecular optical anisotropy)
[35] Fourche, G. and Lemaire, B., *Eur. Polym. J.*, 1976, **12**, 677, (molecular optical anisotropy)
[36] Dondos, A. and Beno&git, H., *Eur. Polym. J.*, 1970, **6**, 1439, (intrinsic viscosity)

[37] Menger, F.M., Shinozaki, H. and Lee, H.-C., *J. Org. Chem.*, 1980, **45**, 2724, (polymer supported reducing agent)
[38] Dondos, A., *Polymer*, 1978, **19**, 1305, (viscosity)
[39] Howard, G.J. and Leung, W.M., *Colloid Polym. Sci.*, 1981, **259**, 1031, (use)
[40] Clark, D.T. and Thomas, H.R., *J. Polym. Sci., Part A: Polym. Chem.*, 1978, **16**, 791, (ESCA)

Poly(2-vinylpyridine)-borane complex   P-634

Synonyms: Poly(2-ethenylpyridine-co-borane)
Related Polymers: Poly(2-vinylpyridine)
Monomers: 2-Vinylpyridine
Material class: Copolymers
Polymer Type: polyolefins
CAS Number:

CAS Reg. No.
72514-01-3

Molecular Formula: $[(C_7H_7N).BH_3]_n$
Fragments: C_7H_7N BH_3

Surface Properties & Solubility:
Solvents/Non-solvents: Insol. THF, C_6H_6, hexane [1]. Swells in refluxing C_6H_6

Applications/Commercial Products:

Supplier
Aldrich

Bibliographic References
[1] Menger, F.M., Shinozaki, H. and Lee, H.-C., *J. Org. Chem.*, 1980, **45**, 2724, (synth)

Poly(vinylpyrrolidinone-co-(2-dimethyl aminoethyl)methacrylate)   P-635

Synonyms: Poly(1-ethenyl-2-pyrrolidinone-co-2-(dimethylamino)ethyl-2-methyl-2-propenoate). PVP-dimethylaminoethyl methacrylate copolymer. PVP-DMAEMA. (DAM-VP) copolymer
Related Polymers: Poly N-vinylpyrrolidone
Monomers: N-Vinylpyrrolidinone, 2-Dimethylaminoethyl methacrylate
Material class: Copolymers
Polymer Type: acrylic, polyvinyls
CAS Number:

CAS Reg. No.	Note
30581-59-0	
172284-56-9	graft copolymer
53633-54-8	
55008-57-6	
99716-87-7	

Molecular Formula: $[(C_6H_9NO).(C_8H_{15}NO_2)]_n$
Fragments: C_6H_9NO $C_8H_{15}NO_2$
Tacticity: Tacticity has been examined by nmr [2]

Surface Properties & Solubility:
Solubility Properties General: Molal compressibility and molal volume of the copolymer in MeOH, dioxane and water have been reported [4]
Solvents/Non-solvents: Sol. DMF [2,6], tetramethylene sulfone [2], EtOH [3] (copolymer), MeOH (copolymer) [4], dioxane (copolymer) [4], H_2O (copolymer (insol. when quarternised with C12 or C16 octyl bromides) [2] [4], 2-propanol, n-butanol, 2-butanol [6], glycerine [6], propylene glycol [6], benzyl alcohol [6]

Transport Properties:
Polymer Solutions Dilute: Intrinsic viscosity of quarternised polymer in a pH 7 buffer, and Mark-Houwink constants have been reported [1]. Rheological props. of quarternised PVPDMAEMA in aq. solns. have been studied [2]. Second viral coefficient (A_2), intrinsic viscosity and the Huggins parameter have been investigated for copolymers and quarternised copolymers [3]

Optical Properties:
Transmission and Spectra: Ir (copolymer) [2,4] and H-1 nmr (copolymer) [3,4] spectral data have been reported

Applications/Commercial Products:

Trade name	Details	Supplier
Copolymer 845		International Speciality Products
Copolymer 937	film forming resin	International Speciality Products
Copolymer 958	ethyl alcohol	International Speciality Products
Gafquat	quarternised product	International Speciality Products
PVPDMAEMA	734 low MW	International Speciality Products
	755 high MW	International Speciality Products
	755N high MW, neutralised copolymer	International Speciality Products

Bibliographic References
[1] Wu, C.S. and Senak, L., *J. Liq. Chromatogr.*, 1990, **13**, 851, (size exclusion chromatography)
[2] Deboudt, K., Delporte, M. and Loucheux, C., *Macromol. Chem. Phys.*, 1995, **196**, 279, 291, 303, (synth, characterisation)
[3] Ushakova, V.N., Kipper, A.I., Afanakina, N.A., Samarova, O.E. *et al*, *Vysokomol. Soedin., Ser. B*, 1995, **37**, 933, (copolymer, quarternised copolymers)
[4] Roy-Chowdhury, P. and Kirtiwar, M.S., *J. Appl. Polym. Sci.*, 1982, **27**, 1883, (solvent effect)
[5] Spitsina, N.I., Aidarova, S.B. and Musabekov, K.B., *Kolloidn. Zh.*, 1984, **46**, 806, (surface tension)
[6] Bronnsack, A.H., *Riechst., Aromen, Kosmet.*, 1979, **29**, 206, (use)

Poly(vinylpyrrolidinone)-iodine complex   P-636

Synonyms: PVP-I. Povidone-iodine
Monomers: N-Vinylpyrrolidinone
Polymer Type: polyvinyls
CAS Number:

CAS Reg. No.
25655-41-8
30849-75-3

Molecular Formula: $[(C_6H_9NO).I_2]_n$
Fragments: C_6H_9NO I_2
Additives: Iodine
General Information: Povidone-iodine is an antibacterial, antiviral, antiseptic and an antifungal agent. These props. are due to the 10–12% iodine in the complex

Surface Properties & Solubility:
Solvents/Non-solvents: V. sol. H_2O [1]. Sol. EtOH [1], $CHCl_3$. Spar. sol. Et_2 [1], Me_2CO, C_6H_6, CCl_4 [1]

Optical Properties:
Transmission and Spectra: Uv [1], x-ray [2] and ir [3] spectral data have been reported

Applications/Commercial Products:

Trade name	Supplier
Betadine	Napp Laboratories Ltd.
Efo-Dine	E. Fougera & Co. Inc.
Isodine	Blair Laboratories Inc.
PVP-iodine	BASF
Pevidine	Berk Pharmaceuticals Ltd.
Ultradine	West Agro
Videne	3M Industrial Chemical Products
	ISP
	Aldrich

Bibliographic References
[1] Runti, C., Univ. Studi Trieste, Ist. Chim., (Pubbl.), 1957, No. 16, 11 pages, (spectroscopic features)
[2] Gazarov, R.A., Sukiasyan, A.N., Limanov, V.E. and Zevin, L.S., Zh. Prikl. Khim. (Leningrad), 1977, **50**, 1424, (x-ray)
[3] Mokhnach, V.O. and Propp, L.N., Iodinol Med. Vet., 1967, 21, (ir)
[4] Schwarz, W. and Schenk, H.U., PVP-Jod Oper. Med., (eds. G. Hierholzer and G. Goertz), Springer, 1984, 1, (rev)
[5] Proc. Int. Symp. Povidone, (eds. G.A. Digenis and J. Ansell), University of Kentucky, 1983, (microbiology, struct, function)

Poly(vinylpyrrolidinone-co-vinylacetate) P-637

Synonyms: *PVP-VA copolymer. Poly(1-ethenyl-2-pyrrolidinone-co-ethenylacetate). SVAP*
Related Polymers: Poly *N*-vinylpyrrolidone
Monomers: Vinyl acetate, Vinyl pyrrolidinone
Material class: Copolymers
Polymer Type: polyvinyl acetate, polyvinyls
CAS Number:

CAS Reg. No.	Note
25086-89-9	
116219-49-9	block
119786-14-0	graft

Molecular Formula: $[(C_6H_9NO).(C_4H_6O_2)]_n$
Fragments: C_6H_9NO $C_4H_6O_2$

Volumetric & Calorimetric Properties:
Density:

No.	Value
1	d 0.91 g/cm^3

Glass-Transition Temperature:

No.	Value	Note
1	55°C	10% vinyl acetate [2]
2	40°C	35% vinyl acetate [2]

Surface Properties & Solubility:
Internal Pressure Heat Solution and Miscellaneous: Enthalpy of soln. of PVP-VA (60% VA) in water at various pressures has been reported [4]. $\Delta H°$ -73.9 cal mol^{-1} (1 atm.) [4]. ΔH -38.3 cal mol^{-1} (2000 atm.). The volume changes on dissolution of the same copolymer in water at various temps. has also been reported [4]. ΔV increases with temp. increase
Solvents/Non-solvents: Sol. Me_2CO [4], H_2O [4,7], EtOH [7], 2-propanol [7]. Insol. naphtha

Transport Properties:
Polymer Solutions Dilute: Viscosities of copolymers (initiated by radiation) are reported in DMF [2], EtOH, H_2O [7]. Surface pressure versus vinyl acetate subphase area, in soln., has been examined [3]

Optical Properties:
Transmission and Spectra: H-1 nmr [5] and ir [5] spectral data have been reported

Polymer Stability:
Thermal Stability General: Pyrolysed products have been identified [2]
Decomposition Details: Decomposition temp. at 60% weight loss for radiation polymerised copolymer 432° (10% vinyl acetate) [2], 435° (35% vinyl acetate)

Applications/Commercial Products:

Trade name	Details	Supplier
Agrimer VA		International Speciality Products
Kolima		GAF
Luviskol VA	range of compositions in EtOH or aq. soln.	BASF
Nasuna B		Henkel Canada
PVP/VA	solns. in EtOH, H_2O and 2-propanol	International Speciality Products
		Aldrich

Bibliographic References
[1] Seymour, R.B. and Stahl, G.A., Polym. Prepr. (Am. Chem. Soc., Div. Polym. Chem.), 1975, **16**, 2, 65, (copolymer)
[2] Ramakrishna, M.S., Dhal, P.K., Deshpande, D.D. and Babu, G.N., J. Polym. Sci., Part A: Polym. Chem., 1986, **24**, 2107, (thermal props)
[3] Zatz, J.L. and Knowles, B., J. Colloid Interface Sci., 1972, **40**, 475, (surface pressure, graft copolymer, props)
[4] Taniguchi, Y., Suzuki, K. and Enomoto, T., J. Colloid Interface Sci., 1974, **46**, 511, (cloud point)
[5] Peppas, N.A. and Gehr, T.W.B., J. Appl. Polym. Sci., 1979, **24**, 2159, (analysis, copolymer)
[6] Kabanova, T.A. and Kamenskaya, E.V., Zh. Vses. Khim. O.va im. D.I. Mendeleeva, 1987, **32**, 117, (electronmicroscopy)
[7] Wu, C., Curry, J.F. and Senak, L., ACS Symp. Ser., (ed. T. Provder), 1993, **521**, 292, (size exclusion chromatography)
[8] Kadosaka, T. and Makita, K., Kanzei Chuo Bunsekishoho, 1981, **22**, 115, (C-13 nmr)
[9] Feng, R., Lu, D., Liang, G. and Mo, B., Zhongshan Daxue Xuebao, Ziran Kexueban, 1995, **34**, 52, (C-13 nmr)

Poly(*N*-vinyl-2-pyrrolidone) P-638

Synonyms: *PVP. PNVP. Polyvidone. Poly(1-ethenyl-2-pyrrolidinone). Poly(VPD). PVPyr. Poly[1-(2-oxo-1-pyrrolidinyl)ethylene]. PVPD*
Monomers: *N*-Vinylpyrrolidinone
Material class: Thermoplastic
Polymer Type: polyvinyls

Poly(*N*-vinyl-2-pyrrolidone)

CAS Number:

CAS Reg. No.	Note
9003-39-8	homopolymer
89231-03-8	isotactic
41227-10-5	dimer

Molecular Formula: $(C_6H_9NO)_n$
Fragments: C_6H_9NO
General Information: A water soluble white hygroscopic powder [33]. Cast films are dry, very hard and clear (may have slight yellow coloration). [33] The polymer is insol. in water when cross-linked
Miscellaneous: Also obtainable as an aq. soln. [33]

Volumetric & Calorimetric Properties:
Density:

No.	Value	Note
1	d 1.25 g/cm^3	25° [35]

Thermodynamic Properties General: Thermodynamic props. for a range of aq. solns. have been reported [27]. Relationship of specific heat capacity with temp. has been reported [26,27]. Molar volume 31.1 cm^3 mol^{-1} (calc.) [17]
Glass-Transition Temperature:

No.	Value	Note
1	175°C	[8]

Transition Temperature General: Transition temp. varies with water content of bulk polymer [18,37]

Surface Properties & Solubility:
Cohesive Energy Density Solubility Parameters: Cohesive energy density parameters have been reported [17]
Solvents/Non-solvents: Very sol. H_2O [8]. Solubility in H_2O limited by viscosity of soln. [9,33,34,37]. Sol. alcohols, e.g. MeOH, EtOH, cyclohexanol, phenol, 1,3-butanediol [8], organic acids, AcOH, CH_2Cl_2, $CHCl_3$, amines, Freons, aniline, diethanolamine, butylamine, morpholine. Insol. EtOAc, hydrocarbons, C_6H_6, petrol, hexane, Et_2O, CCl_4, Me_2CO, THF [8]

Transport Properties:
Transport Properties General: Melt viscosity is impractically high [33]. Decomposition occurs before onset of flow [8]
Polymer Solutions Dilute: Viscosity values calc. from Stockmayer-Fixman-Burchard equation parameters have been reported [3,23,31,33]. Adiabatic compressibility of aq. solns. and ultrasonic velocity have been reported [28]. Dielectric response function is of the Kohlrausch type in aq. soln. [15], and behaviour and constants have been reported [36]. Heats of soln. -16.62 kJ mol^{-1} (-3.97 kcal mol^{-1}, H_2O) [20]; -8.37 kJ mol^{-1} (-2 kcal mol^{-1}, $CHCl_3$). Theta temp. 130° (H_2O) [6]. Hydrodynamic parameters, temp. coefficient of intrinsic viscosity have been reported [16]
Permeability of Gases: Diffusion coefficients vary with the thickness of polymer film and relative humidity [24]. Permeation coefficient [O_2] 31 × 10^{10} cm^2 s^{-1} [24]
Water Content: Absorbance 0.5 mol H_2O per monomer unit. Equilib. water content is 33% (approx.) of relative humidity. [33] Moisture sorption capability of thin films has been reported [44]
Water Absorption:

No.	Value	Note
1	18 %	50% relative humidity [33]
2	8 %	28% relative humidity [33]
3	57 %	89% relative humidity

Mechanical Properties:
Mechanical Properties General: Apparent plasticity 0.0369 cm N^{-1} (62% relative humidity) [13]. Adhesive force f_{50} 0.01-0.0001 N (40-80% relative humidity) [13]

Electrical Properties:
Electrical Properties General: Dielectric behaviour of various aq. solns. has been reported [30]. Poole-Frenkel type conduction mechanisms have been reported [5]. Conductivity of unannealed films decreases during annealing cycles [5]
Surface/Volume Resistance:

No.	Value	Note	Type
1	2.56 × 10^{15} Ω.cm	conductivity value, room temp., annealed film [5]	S
2	0.00145 × 10^{15} Ω.cm	conductivity value, unannealed film [5]	S

Optical Properties:
Refractive Index:

No.	Value	Note
1	1.53	25° [35]

Transmission and Spectra: H-1 nmr [2,37], C-13 nmr [2,37], ir [8,19,37,38], ms [10,11], Raman [12,25], ESCA [21] and uv [38] spectral data have been reported
Molar Refraction: Refractive index parameters from the Sellmeyer dispersion [32] equation and at various T_m [28] values have been reported

Polymer Stability:
Polymer Stability General: Cross-linking occurs at 150° [33]
Thermal Stability General: Weight loss below 100° due to loss of absorbed moisture [1]
Decomposition Details: Decomposition studies have been reported [1]
Environmental Stress: Aq. solns. degrade on exposure to ozone [23]
Chemical Stability: Conc. nitric acid causes an increase in viscosity of solns. and formation of a gel. Precipitation from soln. occurs with strong caustic solns. [33]. Treatment with ammonium peroxydisulfate and heating at 90° causes cross-linking [33]. Exposure to light in presence of diazo compds. or oxidising agents can cause gel formation [33]
Biological Stability: Solns. are stable if protected from mould [33]
Stability Miscellaneous: Ultrasound causes decomposition in aq. soln., yielding CO, CO_2, CH_2O, CH_3CHO, CH_4, C_2H_4, C_2H_6 [3]. These products are attributed to pyrolysis rather than hydrodynamic shear forces [3]. Rate of ultrasonic decomposition 4 × 10^{-5} M min^{-1} [3]. Ultrasonic degradation rate 0.032 min^{-1} [7]. PVP degrades under high speed stirring (30000 rpm) in aq. soln. [22]

Applications/Commercial Products:

Trade name	Details	Supplier
Agrimer		ISP
Albigen A		BASF
Ganex		ISP
Ganex		GAF
Kollidon		BASF
Neocompensan		Heilmittel Werke
PVP		ISP
PVP		GAF
Peregal		GAF
Peregal ST		ISP

Periston		Bayer Inc.
Plasdone		ISP
Plasdone		GAF
Plasdone C	pharmaceutical grade blood plasma volume expander	GAF
Plasdone K	pharmaceutical grade tablet manufacture	GAF
Polyclar		ISP
Polyclar	(L, AT) cross-linked beverage clarification	GAF
Polyclar		GAF
Polyplas-done		ISP
Polyplas-done		GAF
Povidone		ISP
Povidone		BASF
Sokalan		BASF
Tri-K 30		Tri-K Industries Inc.
Vinisil		Abbott Laboratories

Bibliographic References

[1] Radic, D., Tagle, L.H., Opazo, A., Godoy, A. and Bargallo, L., *J. Therm. Anal.*, 1994, **41**, 1007, (thermal data)
[2] Hunter, T.C. and Price, G.C., *Polymer*, 1994, **35**, 3530, (nmr)
[3] Gavara, R., Campos, A. and Figueruelo, J.E., *Makromol. Chem.*, 1990, **191**, 1915
[4] Gutiérrez, M. and Henglein, A., *J. Phys. Chem.*, 1988, **92**, 2978, 3705, (ultrasonic degradation)
[5] Narasimha Rao, V.V.R. and Kalpalatha, A., *Polymer*, 1987, **28**, 648, (electrical conduction)
[6] Sakellariou, P., *Polymer*, 1992, **33**, 1339, (theta temp)
[7] Koda, S., Mori, H., Matsumoto, K. and Nomura, H., *Polymer*, 1994, **35**, 30, (ultrasonic degradation)
[8] Haaf, F., Sanner, A. and Straub, F., *Polym. J. (Tokyo)*, 1985, **17**, 143, (synth, uses)
[9] Linke, W. and Vogel, F.G.M., *Polym. News*, 1987, 12, 232, (rev)
[10] Ohnishi, J., Hashimoto, K., Izushi, T. and Aoki, M., *Bunseki Kagaku*, 1991, **40**, 705, (ms)
[11] Thomson, B., Suddaby, K., Rudin, A. and Lajoie, G., *Eur. Polym. J.*, 1996, **32**, 239, (ms)
[12] Tanaka, N., *Macromol. Chem. Phys.*, 1994, **195**, 3369, (Raman)
[13] Iida, K., Otsuka, A., Danjo, K. and Sunada, H., *Chem. Pharm. Bull.*, 1992, **40**, 189, (adhesive force)
[14] Ali, S. and Ahmad, N., *Br. Polym. J.*, 1982, **14**, 113, (soln props)
[15] Miura, N., Skinyashiki, N. and Mashimo, S., *J. Chem. Phys.*, 1992, **97**, 8722, (dielectric relaxation)
[16] Pavlov, G.M., Panarin, E.F., Korneeva, E.V., Kurochkin, C.V. et al, *Makromol. Chem.*, 1990, **191**, 2889, (hydrodynamic props)
[17] Eldin, S.H., *J. Appl. Polym. Sci.*, 1986, **32**, 3971, (adhesive props)
[18] Tan, Y.Y. and Challa, G., *Polymer*, 1976, **17**, 739, (T_g)
[19] Léonard-Stibbe, E., Lécayon, G., Deniau, G., Viel, P. et al, *J. Polym. Sci., Part A: Polym. Chem.*, 1994, **32**, 1551, (ir)
[20] Meza, R. and Gargallo, L., *Eur. Polym. J.*, 1977, **13**, 235
[21] Clark, D.T. and Thomas, H.R., *J. Polym. Sci., Part A: Polym. Chem.*, 1978, **16**, 791, (ESCA)
[22] Nakano, A. and Minoura, Y., *J. Appl. Polym. Sci.*, 1977, **21**, 2877, (degradation)
[23] Suzuki, J., Taumi, N. and Suzuki, S., *J. Appl. Polym. Sci.*, 1979, **23**, 3281, (degradation)
[24] Petrak, K. and Pitts, E., *J. Appl. Polym. Sci.*, 1980, **25**, 879, (permeability)
[25] Viras, F. and King, T.A., *J. Non-Cryst. Solids*, 1990, **119**, 65, (Raman)
[26] Wakabayashi, T. and Franks, F., *Cryo-Lett.*, 1986, **7**, 361, (heat capacity)
[27] Dakroury, A.Z., Osman, M.B.S. and El-Sharkawy, A.W.A., *Int. J. Thermophys.*, 1990, **11**, 515, (thermal props)
[28] Rajulu, A.V., Rao, K.C. and Naidu, S.V., *Acustica*, 1991, **75**, 213, (ultrasonic studies)
[29] Vogel, F.G.M., *Soap, Cosmet., Chem. Spec.*, 1989, **65**(4), 42, 46, 128, (rev)
[30] Weyt, D.A. and Neale, S.M., *Dielectrics*, 1963, **1**, 22, (dielectric behaviour)
[31] Reppe, W., Monographien zu "Angewandte Chemie" und "Chemie-Ingenier-Technik" Number 66, 1954, (rev)
[32] Warnecke, A.J. and LoPresti, P.J., *IBM J. Res. Dev.*, 1973, **17**, 256, (refractive index)
[33] Blecher, L., Lorenz, D.H., Lowd, H.L., Wood, A.S. and Wyman, D.P., *Handb. Water-Soluble Gums Resins*, (ed. R.L. Davidson), McGraw-Hill, Chapter 21, 1980, (rev)
[34] *Water. Compr. Treatise*, (ed. F. Franks), Plenum Press, 1972, **4**, 649, (rev)
[35] Remond, J., *Rev. Prod. Chim.*, 1956, **59**, 127, 260, (rev)
[36] Grant, E.H., McClean, V.E.R., Nightingale, N.R.V., Sheppard, R.J. and Chapman, M.J., *Bioelectromagnetics (N.Y.)*, 1986, **7**, 151, (dielectric behaviour)
[37] Adeyeye, C.M. and Barabas, E., *Anal. Profiles Drug Subst.*, 1993, **22**, 555, (rev)
[38] Oster, G. and Immergut, E.H., *J. Am. Chem. Soc.*, 1954, **76**, 1393, (ir, uv)
[39] PVP: An Annotated Bibliography, 1951-66, General Aniline and Film Corp., N.Y., 1967, (rev)
[40] PVP: Polyvinylpyrrolidone: Physical, Chemical, *Physiological and Functional Properties*, 1964, GAF Corp., N.Y., (technical datasheet)
[41] Wang, L., *Xiandai Huagong*, 1995, **15**, 21, (rev)
[42] *Proc. Int. Symp. Povidone*, (eds. G.A. Digenis and J. Ansell), University of Kentucky, Levington, 1983, April 17-20, (rev)
[43] *PVP Polyvinylpyrrolidone*, General Aniline and Film Corp, 1951, (rev)
[44] Ranucci, E., Ferruti, P., Opelli, P., Ferrari, V. et al, *Sens. Mater.*, 1994, **5**, 221, (moisture sorption)

Poly(*N*-vinylpyrrolidone-*co*-acrylic acid)   P-639

Synonyms: *Poly(1-ethenyl-2-pyrrolidinone-co-2-propenoic acid)*
Related Polymers: Poly(*N*-vinylpyrrolidone)
Monomers: *N*-Vinylpyrrolidinone, Acrylic acid
Material class: Copolymers
Polymer Type: acrylic, polyvinyls
CAS Number:

CAS Reg. No.	Note
28062-44-4	
127383-17-9	graft
153990-36-4	sodium salt
150198-94-0	
53608-09-6	
102685-02-9	

Molecular Formula: $[(C_6H_9NO).(C_3H_4O_2)]_n$
Fragments: C_6H_9NO $C_3H_4O_2$
General Information: White free flowing powders (Acrylidone R) [6]. Copolymer is a weak acid [6]. Increasing the pH of soln. causes the polymer chains to extend due to electrostatic repulsions, thus altering the rheological props. [6]. The copolymer is amphiphilic due to pyrrolidone and carboxylic groups [6]
Morphology: Crystallinity is due to interpolymer adduct, by hydrogen bonding, between vinylpyrrolidinone and acrylic acid [2]
Miscellaneous: The copolymer contains dissociated ionogenic groups [3]

Volumetric & Calorimetric Properties:
Glass-Transition Temperature:

No.	Value	Note
1	116-185°C	varies with MW and composition [1,6]

Surface Properties & Solubility:
Solubility Properties General: Molal compressibility and molal volume of the copolymer in MeOH and water have been reported [8]
Solvents/Non-solvents: Sol. *N*-ethyl-2-pyrrolidinone, *m*-pyrrole, *N*-cyclohexyl-2-pyrrolidinone, glycerol, ethylene glycol, propylene glycol, 1,4-butanediol, DMF, DMSO, dimethyl acetamide, EtOH/H_2O, *N*-methyl-2-pyrrolidinone [6], alkali solns. Insol. butyrolactone, toluene, heptane, xylene, $CHCl_3$, CH_2Cl_2, CCl_4, EtOAc,

MeOH, EtOH, isopropanol, butanol, propanol, Me$_2$CO, 2-butanone, H$_2$O
Wettability/Surface Energy and Interfacial Tension: Surface tension values have been reported [6]

Transport Properties:
Polymer Solutions Dilute: Reduced viscosity varies according to neutralisation of the ionogenic macromolecule [3]. Viscosimetric titration [3]. Viscosity of copolymer, partially neutralised and fully neutralised copolymer have been reported [6]. Potentiometric titration of copolymer follows the Henderson-Hasselbach equation [7]. Viscosimetric titration of copolymer has been performed [7], the curves are typical for ionogenic polymers [7] and some constants have been determined [7]
Water Content: Depends on MW and composition ratio 5–14% water content [1,6] (0–100% acrylic acid). Water resistance is enhanced if unneutralised copolymer is cast from 2-propanol/H$_2$O or from ammonium neutralised solns. [6]. Water resistance is also enhanced by cross-linking with ammonium zirconium carbonate or methylated melamine formaldehyde [6]

Mechanical Properties:
Mechanical Properties General: Flexibility can be increased by treatment with plasticisers (e.g. dioctyl phthalate, glycerine or polyethyleneglycol) [6]

Optical Properties:
Transmission and Spectra: X-ray [2], C-13 nmr [4], ir [8] and H-1 nmr [8] spectral data have been reported

Polymer Stability:
Polymer Stability General: Copolymer is precipitated from soln. by strong acids or divalent and trivalent salt solns. [6]
Thermal Stability General: Thermally decomposes by dehydration and decarboxylation releasing H$_2$O and CO$_2$ respectively [5]. Copolymer is approx. 4 times as slow in releasing H$_2$O compared to polyacrylic acid at 200° [5]. The copolymer decarboxylates more readily than polyacrylic acid homopolymer [5]
Stability Miscellaneous: Compatibility in soln. with various resins and gums, and inorganic salts has been reported [6]

Applications/Commercial Products:

Trade name	Details	Supplier
ACP		ISP
Acrylidone	range of grades	ISP

Bibliographic References
[1] Shih, J.S., Chuang, J.C. and Login, R.B., *Polym. Mater. Sci. Eng.*, 1992, **67**, 266, (random copolymers)
[2] Subramanian, R., Natarajan, P. and Parthasarathi, V., Makromol. Chem., *Rapid Commun.*, 1980, **1**, 47, (x-ray)
[3] Roman, A. and Scondac, I., *Makromol. Chem.*, 1968, **113**, 171, (viscosimetric titration)
[4] Khodzhaev, S.G. and Mushraipov, R., Vysokomol. Soedin., *Ser. A*, 1990, **32**, 1321, (C-13 nmr)
[5] Bolyachevskaya, K.I., Litmanovich, A.A., Markov, S.V., Izvolenskii, V.V. and Papisov, I.M., Vysokomol. Soedin., *Ser. A*, 1993, **35**, 1449, (thermolysis)
[6] Hornby, J.C., Pelesko, J.D. and Jon, D., Soap, Cosmet., *Chem. Spec.*, 1993, **69**, 6, 40, 66, (applications)
[7] Dima, M., Scondac, I. and Roman, A., *Rev. Roum. Chim.*, 1966, **11**, 965, 1229, 1333, (potentiometric titrations)
[8] Roy-Chowdhury, P. and Kirtiwar, M.S., *J. Appl. Polym. Sci.*, 1982, **27**, 1883, (solvent effects)
[9] Izvolenskii, V.V., Semchikov, Y.D., Sveshnikova, T.G. and Shalin, Sk., Vysokomol. Soedin., *Ser. A*, 1992, **34**, 53, (copolymerisation)
[10] Ushakova, V.N., Denisov, V.M., Kol'tsov, A.I., Panarin, E.F. and Khachaturov, A.S., Vysokomol. Soedin., *Ser. B*, 1989, **31**, 345, (C-13 nmr)

Poly(*N*-vinylpyrrolidone-*co*-α-eicosene) P-640
Synonyms: *Poly(1-ethenyl-2-pyrrolidinone-co-1-eicosene)*
Related Polymers: Poly *N*-vinylpyrrolidone
Monomers: 1-Eicosene, *N*-Vinylpyrrolidinone
Material class: Copolymers
Polymer Type: polyolefins, polyvinyls
CAS Number:

CAS Reg. No.	Note
28211-18-9	PVP-1-eicosene
157002-67-0	PVP-eicosene
77035-98-4	PVP-eicosene
151081-15-1	PVP-eicosene, graft
92815-93-5	
121602-88-8	
116921-54-1	
117803-73-3	

Molecular Formula: [(C$_6$H$_9$NO).(C$_{16}$H$_{32}$)]$_n$
Fragments: C$_6$H$_9$NO   C$_{16}$H$_{32}$

Volumetric & Calorimetric Properties:
Density:

No.	Value
1	d 0.947 g/cm^3

Transition Temperature:

No.	Value	Type
1	56–64°C	melting point

Surface Properties & Solubility:
Solvents/Non-solvents: Insol. H$_2$O

Electrical Properties:
Electrical Properties General: Charge decay of Antaron V220 coated phosphor has been reported. Charge to mass ration (q/m) of Antaron V220 coated phosphor 0.005 C kg^{-1} [1] (50% relative humidity, 20°)
Surface/Volume Resistance:

No.	Value	Note	Type
1	0.0009 × 10^{15} Ω.cm	Antaron V220 coating on calcium halophosphate phosphor [1], 50% relative humidity, 20°	S

Polymer Stability:
Chemical Stability: Unstable to strong reducing agents. Prod. nitrous oxides in contact with oxidising agents
Biological Stability: Unlikely to be bioavailable due to insol. in water

Applications/Commercial Products:

Trade name	Supplier
Antaron V220	International Speciality Products
Ganex V220	International Speciality Products

Bibliographic References
[1] Chandrasekhar, R., *Powder Technol.*, 1992, **71**, 81, (electrical props)
[2] *Fed. Regist.*, United States Environmental Protection Agency, 1991, **56**, 32514, (dispersant props)

Poly(N-vinylpyrrolidone-co-styrene) P-641

Synonyms: *PVP-styrene copolymer. Poly(1-ethenyl-2-pyrrolidinone-co-ethenylbenzene). PSV. SVP*
Related Polymers: Poly N-vinylpyrrolidone
Monomers: Styrene, N-Vinylpyrrolidinone
Material class: Copolymers
Polymer Type: polystyrene, polyvinyls
CAS Number:

CAS Reg. No.	Note
25086-29-7	
116219-50-2	block copolymer
123924-78-7	graft

Molecular Formula: $[(C_8H_8).(C_6H_9NO)]_n$
Fragments: C_8H_8 C_6H_9NO
General Information: Copolymer (Antara) is tough and clear [5]

Surface Properties & Solubility:
Solvents/Non-solvents: Sol. N-methylpyrrolidone [1], C_6H_6 [4], dichloroethane [4], $CHCl_3$ [4], benzyl alcohol [4], acetylacetone [4], DMF/2-propanol (9:1) [4], dioxane [2]. Insol. H_2O, cyclohexane (hot); cyclohexane-heptane [2]

Transport Properties:
Polymer Solutions Dilute: Soln. props. (intrinsic viscosity, second viral coefficient A_2, average radius of gyration of the polymer coils $(R^{-2})^{1/2}$) in C_6H_6 and dichloroethane for various ratios of polystyrene and overall MW have been reported. Intrinsic viscosity and second viral coefficient have been determined in $CHCl_3$, DMF, dioxane and the temp. dependance of the intrinsic viscosity has been reported [2]
Permeability of Liquids: Flow rate of water through polymer versus time has been reported [5] (antara). Permeate flux of ultrafiltration membrane has been reported [1] (100000 cm^3 m^{-2} h^{-1})
Water Absorption:

No.	Value	Note
1	9 %	antara [5]

Mechanical Properties:
Mechanical Properties General: Bursting strength of ultrafiltration membrane (styrene grafted onto PVP) has a maximum of approx. 5.5 kPa (85% w/w grafted styrene) [1]. The bursting pressure also depends on the strength of the casting soln. (ideally >6% polymer) [1]

Optical Properties:
Transmission and Spectra: SEM data have been reported [1]

Applications/Commercial Products:

Trade name	Details	Supplier
Antara 130		Anchor Chemical Co
Polectron	430, 450	GAF

Bibliographic References
[1] Uliana, C., Vigo, F. and Traverso, P., *Sep. Sci. Technol.*, 1994, **29**, 1621, (ultrafiltration membranes)
[2] Eskin, V.Y., Grigor'ev, A.I., Baranovskaya, I.A. and Rudkovskaya, G.D., Vysokomol. Soedin., *Ser. A*, 1978, **20**, 55, (conformational props)
[3] Negulescu, I., Feldman, D. and Simionescu, Cr., *Polymer*, 1972, 13, 149, (copolymerisation)
[4] Eskin, V.Y. and Korotkina, O.Z., *Polym. Sci. USSR (Engl. Transl.)*, 1970, **12**, 2511, (soln behaviour)
[5] Maconochie, G., Sharples, A. and Thompson, G., *Eur. Polym. J.*, 1971, **7**, 499, (reverse osmosis membranes)
[6] Ueda, A., Agari, Y., Hayanshi, K. and Nagai, S., *Kagaku Kogyo*, 1995, **69**, 364, (block copolymer)

Poly(vinyl trifluoroacetate) P-642

Synonyms: *Poly(ethenyl trifluoroacetate). Poly[(trifluoroacetoxy)ethylene]*
Monomers: Vinyl trifluoroacetate
Material class: Thermoplastic
Polymer Type: polyvinyl acetate
CAS Number:

CAS Reg. No.	Note
25748-85-0	
27101-55-9	syndiotactic

Molecular Formula: $(C_4H_3F_3O_2)_n$
Fragments: $C_4H_3F_3O_2$
Molecular Weight: Typically MW 306000, M_n 111000
General Information: A hard and fibrous, clear and tough material [1,2]

Volumetric & Calorimetric Properties:
Density:

No.	Value	Note
1	d^{20} 1.633 g/cm^3	cryst.

Latent Heat Crystallization: ΔS 4 [3]. ΔH_m 7.5 kJ mol^{-1} (approx.)
Melting Temperature:

No.	Value	Note
1	165°C	[3]

Glass-Transition Temperature:

No.	Value	Note
1	>75°C	min. [2]

Vicat Softening Point:

No.	Value	Note
1	172°C	[1]

Surface Properties & Solubility:
Cohesive Energy Density Solubility Parameters: δ 8.25 (bulk) [5]
Solvents/Non-solvents: Sol. Me_2CO, pyridine, cyclohexanone, butyl acetate, diethylenetriamine [1], isobutyl methyl ketone, n-propyl methyl ketone, isopropyl methyl ketone, 2-butanone [4]. Spar. sol. methyl formate, methyl acetate, methyl propionate, methyl n-butyrate [4]. Insol. C_6H_6, hexane, MeOH, γ-butyrolactone, $CHCl_3$, formic acid [1], nonane, octane, heptane, hexane, pentane, vinyl trifluoroacetate [4]

Transport Properties:
Polymer Solutions Dilute: [η] 1.15−1.31 [1]

Mechanical Properties:
Mechanical Properties Miscellaneous: Tenacity 0.8 g den^{-1} (7%) [3]

Optical Properties:
Refractive Index:

No.	Value	Note
1	1.383	[1]
2	1.375	[2]

Transmission and Spectra: Ir [1], H-1 nmr [6] and x-ray [1,7] spectral data have been reported
Total Internal Reflection: Optical birefringence -0.0015 [3]

Polymer Stability:
Flammability: Non flammable
Hydrolytic Stability: Hydrolyses quickly to poly(vinyl alcohol) [1]

Applications/Commercial Products:
Processing & Manufacturing Routes: Prod. by free radical polymerisation of vinyl trifluoroacetate

Bibliographic References
[1] Haas, H.C., Emerson, E.S. and Schuler, N.W., *J. Polym. Sci.*, 1956, **22**, 291
[2] Reid, T.S., Codding, D.W. and Bovey, F.A., *J. Polym. Sci.*, 1955, **18**, 417
[3] Bohn, C.R., Schaefgen, J.R. and Statton, W.O., *J. Polym. Sci.*, 1961, **55**, 531
[4] Matsuzawa, S., Yamaura, K., Noguchi, H. and Hayashi, H., *Makromol. Chem.*, 1973, **165**, 217
[5] Nozakura, S.I., Sumi, M., Uoi, M., Okamato, T. and Murahashi, S., *J. Polym. Sci., Polym. Chem. Ed.*, 1973, **11**, 279
[6] Ramey, K.C. and Field, N.D., *J. Polym. Sci., Part B: Polym. Lett.*, 1965, **3**, 63
[7] Briggs, D. and Beamson, G., *Anal. Chem.*, 1993, **65**, 1517
[8] U.S. Pat., 1948, 2 436 144

Poly(*m*-xylenediamine-*co*-*p*-xylenediamine-*co*-trimethylhexamethylene terephthalamide-*co*-terephthalic acid)   P-643

Synonyms: *Poly(m-xylenediamine-co-wp-xylenediamine-co-nylon 6(3)T-co-terephthalic acid. MXD/PXD/6(3)T/IT*
Related Polymers: Nylon 6(3)T
Monomers: *m*-Xylenediamine, *p*-Xylenediamine, Terephthalic acid
Material class: Thermoplastic, Copolymers
Polymer Type: polyamide
Molecular Formula: $[(C_8H_{10}N_2).(C_{17}H_{26}N_2O_2).(C_8H_4O_2)]_n$
Fragments: $C_8H_{10}N_2$ $C_{17}H_{26}N_2O_2$ $C_8H_4O_2$
Molecular Weight: 15000–40000
General Information: Is amorph.

Volumetric & Calorimetric Properties:
Glass-Transition Temperature:

No.	Value	Note
1	150°C	7.5–15% *m/p*-xylenediamine

Deflection Temperature:

No.	Value	Note
1	150–180°C	7.5–15% *m/p*-xylenediamine

Vicat Softening Point:

No.	Value	Note
1	156–159°C	[1]

Surface Properties & Solubility:
Plasticisers: *p*-Toluenesulfonamide
Solvents/Non-solvents: Sol. sulfuric acid, formic acid, dimethylacetamide, *N*-methylpyrrolidone (containing 1–2% boric acid)

Mechanical Properties:
Flexural Modulus:

No.	Value	Note
1	2500–3000 MPa	20°, 7.5–15% *m/p*-xylenediamine [1]

Tensile Strength Break:

No.	Value	Note
1	91–93 MPa	20°, 7.5–15% *m/p*-xylenediamine [1]

Impact Strength: Falling weight impact strength 50 J (min., 2mm disc, 0% failure rate, 7.5–15% *m/p*-xylenediamine
Hardness: Rockwell R130 (7.5–15% *m/p*-xylenediamine)

Electrical Properties:
Electrical Properties General: Has very good electrical insulation props., particularly when dry

Polymer Stability:
Upper Use Temperature:

No.	Value	Note
1	110°C	
2	150°C	500 h

Decomposition Details: Volatilisation of the diamines occurs at above 300°
Flammability: Limiting oxygen index 25% (approx.)
Environmental Stress: Ageing in air above 80° causes yellowing
Chemical Stability: Injection moulded polymer is resistant to C_6H_6, paraffin, oil, CCl_4; slightly cracked by petroleum ether, aq. alkali, hydrochloric acid, sulfuric acid and salt soln. Severe cracking caused by Me_2CO, lower alcohols, THF and brake fluid. Is resistant to most aliphatic and aromatic hydrocarboons.
Hydrolytic Stability: Is resistant to boiling H_2O for over 500h

Applications/Commercial Products:
Processing & Manufacturing Routes: Polymer containing up to 80 mol% terephthalic acid is synth. by aq. melt polymerisation of nylon, *m* and *p*-xylenediamine and terephthalic acid salts at up to 275° for 3h
Applications: Potential applications due to excellent impact toughness and general performance. Uses have not been investigated commercially

Bibliographic References
[1] Dolden, J.G., *Polymer*, 1976, **17**, 875, (rev.)
[2] *Encycl. Polym. Sci. Eng.*, Vol.11 2nd edn., (ed. J.I. Kroschwitz), John Wiley and Sons, 1986, 372, (rev.)

Poly(*m*-xylenediamine-*co*-terephthalic acid-*co*-isophthalic acid)   P-644

Synonyms: *MXD/I/T copolymer*
Monomers: Terephthalic acid, Isophthalic acid, *m*-Xylenediamine
Material class: Copolymers

– Poly(*m*-xylylene adipamide)

Molecular Formula: $[(C_8H_{10}N_2).(C_8H_4O_2)]_n$
Fragments: $C_6H_{14}N_2$ $C_8H_4O_2$
Molecular Weight: 15000–40000
Miscellaneous: Amorph. when the terephthalic acid content is less than 70 mol%

Surface Properties & Solubility:
Plasticisers: *p*-Toluenesulfonamide
Solvents/Non-solvents: Sol. H_2SO_4, formic acid, dimethylacetamide, *N*-methylpyrrolidone (containing 1–2% boric acid)

Transport Properties:
Gas Permeability:

No.	Gas	Value	Note
1	[O_2]	0.02–0.08 cm^3 mm/(m^2 day atm)	$0.3–1.3 \times 10^{-13}$ cm^2 (s cm Hg)$^{-1}$, isophthalic acid : terephthalic acid (1:1) [1]

Electrical Properties:
Electrical Properties General: Has good electrical insulation props., particularly when dry

Polymer Stability:
Decomposition Details: Volatilisation of diamines occurs above 300°
Flammability: Limiting oxygen index 25% (approx.)
Environmental Stress: Ageing in air above 80° causes yellowing
Chemical Stability: Injection moulded material is resistant to C_6H_6, paraffin, oil, CCl_4; slight cracking caused by petroleum ether, aq. alkali, hydrochloric acid, sulfuric acid and salt soln. Severe cracking caused by Me_2CO, lower alcohols, THF and brake fluid. Is resistant to most aliphatic and aromatic hydrocarbons
Hydrolytic Stability: Boiling H_2O causes crystallisation after 24h

Applications/Commercial Products:
Processing & Manufacturing Routes: Polymer containing up to 80 mol% terephthalic acid may be synth. by aq. melt polymerisation of *m*-xylylenediamine, isophthalic acid and terephthalic acid salts at up to 275° for 3h [1]
Applications: Potential applications due to low gas permeability have been restricted by cost

Bibliographic References
[1] *Encycl. Polym. Sci. Eng.*, 2nd edn., (eds John Wiley and Sons), 1985, **11**, 372, (rev.)

Poly(*m*-xylylene adipamide)   P-645

Synonyms: *Nylon MXD-6. Poly(m-xylylene diamine-co-adipic acid). Poly(iminomethylene-1,3-phenylenemethyleneimino(1,6-dioxo-1,6-hexanediyl)]. Poly(meta-xylylene adipamide)*
Related Polymers: Poly(meta-xylylene adipamide), 50% Glass fibre filled
Monomers: Hexanedioic acid, 1,3-Bis(aminomethyl)benzene
Material class: Thermoplastic
Polymer Type: aromatic polyamides
CAS Number:

CAS Reg. No.
25805-74-7

Molecular Formula: $C_{16}H_{18}N_2O_2$
Fragments: $C_{16}H_{18}N_2O_2$
Additives: Normally sold with 30% glass fibre reinforcement to improve thermal and mech. props.
General Information: A semi-crystalline semi-aromatic nylon. The pure polymer has a very high resistance to gas and water vapour permeation. Glass reinforced polymer maintains remarkably high stiffness and strength up to near its melting point and compares favourably in mech. and thermal props. with wholly aromatic liq. cryst. polyesters, PEEK, Kevlar etc. Coefficients of linear expansion are comparable to those of many metals and alloys

Volumetric & Calorimetric Properties:
Density:

No.	Value	Note
1	d 1.22 g/cm^3	amorph. [6]
2	d 1.22–1.25 g/cm^3	crystalline [6]

Thermal Expansion Coefficient:

No.	Value	Note	Type
1	1.1×10^{-5} K^{-1}	ASTM D696	L

Melting Temperature:

No.	Value	Note
1	243°C	[2,3]
2	245°C	[6]
3	236°C	

Glass-Transition Temperature:

No.	Value	Note
1	95°C	[5]

Surface Properties & Solubility:
Solvents/Non-solvents: Sol. strong acids, e.g. H_2SO_4, formic acid, and polar solvents such as DMF and *N*-methylpyrrolidone

Transport Properties:
Permeability of Gases: Maintains excellent oxygen impermeability under very moist conditions
Water Content: 4.0% (saturated) [2,3]
Gas Permeability:

No.	Gas	Value	Note
1	O_2	0.03 cm^3 mm/(m^2 day atm)	0.5×10^{-13} cm^2 (s cmHg)$^{-1}$, 0% relative humidity, ASTM D3982-81 [4]
2	O_2	0.06 cm^3 mm/(m^2 day atm)	0.9×10^{-13} cm^2 (s cmHg)$^{-1}$, 100% relative humidity, ASTM D3982-81 [4]

Mechanical Properties:
Mechanical Properties General: The unreinforced polyamide does not have a particularly useful range of physical props. and is used as an engineering polymer only when reinforced with glass fibre. However, limited quantities of unfilled material are sold into the barrier market where the excellent oxygen impermeability of MXD-6 is utilised by blending with main stream barrier polymers to enhance performance. Tenacity 0.26–0.44 N tex^{-1} [3]

Polymer Stability:
Polymer Stability General: Poor dimensional stability above 75° unless reinforced with glass fibre
Thermal Stability General: Similar to other semi-aromatic nylons. Stiffness and creep resistance drop dramatically above T_g. Susceptible to oxidation above 80°
Upper Use Temperature:

No.	Value	Note
1	75–80°C	[1]

Flammability: Limiting oxygen index 25% (ASTM D2863)

– Poly(*m*-xylylene adipamide), 50% Glass fibre filled

Chemical Stability: Resistant to aliphatic and aromatic hydrocarbons, chlorinated solvents, ketones, esters, ethers. Attacked by these solvents above 60°. Resistant to many aq. salt or weak base solns. Degraded by strong mineral acids, oxidising agents and bases
Hydrolytic Stability: Sensitive to aq. chloride solns. and strong acid solns.

Applications/Commercial Products:
Processing & Manufacturing Routes: Believed to be manufactured from the MXD-6 salt in an autoclave under pressure in temp. range 200–300° to prepolymer, followed by final conversion in a thin film extrusion process. Up to 50% of 3 mm glass fibre, treated with e.g. silane to give better adhesion to the polymer, may be incorporated in a separate extrusion blending stage. Similarly, grades containing up to 50% inorganic filler are prod. by extrusion blending

Mould Shrinkage (%):

No.	Value	Note
1	0.1–0.5%	direction of flow [1]

Applications: Used as a barrier material. High price dictates its use, in blending with other nylons to increase oxygen impermeability. Used in multi-extruded layers for packaging. Filled material used in engineering components such as pulleys, shafts and gears

Trade name	Supplier
Ixef	Solvay
Ixef	Laporte
Reny	Mitsubishi Plastics Ltd.

Bibliographic References
[1] Ixef Polyarylamide Solvay, (technical datasheet)
[2] Carlston, E.F. and Lum, F.G., *Ind. Eng. Chem.*, 1957, **49**, 1239
[3] Langbottom, R.W., *Mod. Text.*, 1968, **49**, 19
[4] 1991, BP Chemicals
[5] Dolden, J.G., *Polymer*, 1976, **17**, 883
[6] Van Krevelen, D.W., *Properties of Polymers: Their Correlation with Chemical Structure*, 3rd edn., Elsevier, 1990

Poly(*m*-xylylene adipamide), 50% Glass fibre filled P-646

Synonyms: *MXD-6, 50% Glass fibre filled. Poly(metaxylylene adipamide)*
Related Polymers: Poly(*m*-xylylene adipamide)
Monomers: Hexanedioic acid, 1,3-Bis(aminomethyl)benzene
Material class: Thermoplastic
Polymer Type: polyamide
CAS Number:

CAS Reg. No.
25805-74-7

Molecular Formula: $(C_{16}H_{18}N_2O_2)_n$
Fragments: $C_{16}H_{18}N_2O_2$
Additives: 50% Glass fibre by weight

Volumetric & Calorimetric Properties:
Density:

No.	Value
1	d 1.64 g/cm^3

Thermal Expansion Coefficient:

No.	Value	Note	Type
1	1.1×10^{-5} K^{-1}	ASTM D696	L

Thermal Conductivity:

No.	Value	Note
1	0.55 W/mK	ASTM C177

Deflection Temperature:

No.	Value	Note
1	232°C	1.8 MPa, ASTM D648 [1]

Transition Temperature:

No.	Value	Note	Type
1	249–282°C	[1]	Processing temp.
2	121–149°C	[1]	Moulding temp.

Transport Properties:
Transport Properties General: Very impermeable to oxygen, CO_2 and water vapour, being superior to most other polymers and on a par with polyvinyl alcohol. It absorbs water much more slowly than nylons 6 or 6,6
Permeability of Liquids: Coefficient of diffusion D [H_2O] 1×10^{-16} cm^2 s^{-1} (20°), 200×16^{-10} cm^2 s^{-1} (80°)
Water Content: 1.7% (65% relative humidity, ASTM D570)
Water Absorption:

No.	Value	Note
1	0.17 %	24h, 20°, ASTM D570

Mechanical Properties:
Tensile (Young's) Modulus:

No.	Value	Note
1	20000 MPa	ISO R527

Flexural Modulus:

No.	Value	Note
1	17700 MPa	ASTM D790

Tensile Strength Break:

No.	Value	Note	Elongation
1	255 MPa	ISO R527 [1]	1.9%

Flexural Strength at Break:

No.	Value	Note
1	340 MPa	ASTM D790

Compressive Strength:

No.	Value	Note
1	230 MPa	ASTM D695

Miscellaneous Moduli:

No.	Value	Type
1	12.2 MPa	Specific modulus

Hardness: Rockwell M110 (ASTM D785)
Friction Abrasion and Resistance: Taber abrasion 16 mg $(1000h)^{-1}$ (CS-17, ASTM D1044)
Izod Notch:

No.	Value	Notch	Note
1	107 J/m	Y	[1,2]
2	900 J/m	N	[1,2]

Electrical Properties:
Electrical Properties General: Relatively unchanged at 50% relative humidity compared to dry or moulded values
Surface/Volume Resistance:

No.	Value	Note	Type
1	0.9×10^{15} Ω.cm	ASTM D257	S

Dielectric Permittivity (Constant):

No.	Value	Note
1	4.2	ASTM D150

Dielectric Strength:

No.	Value	Note
1	3200 kV.mm^{-1}	ASTM D149

Arc Resistance:

No.	Value	Note
1	129s	ASTM D495

Polymer Stability:
Thermal Stability General: Processed at temps. up to 280°. Thermally stable up to 300° in N_2
Upper Use Temperature:

No.	Value	Note
1	138°C	[1]

Flammability: Limiting oxygen index 25% (ASTM D2863). Flammability rating HB (3.2 mm, UL 94)
Chemical Stability: Unaffected up to 60° by aliphatic and aromatic hydrocarbons, chlorinated solvents, ketones, esters, ethers. Also resistant to many aq. salts and weak bases. Resistant to engine oil at 120° for 2000h and gasoline at 40°. Degraded by strong mineral acids, oxidising agents and bases
Hydrolytic Stability: Sensitive to aq. chloride solns. and strong acid solns.

Applications/Commercial Products:
Processing & Manufacturing Routes: Glass fibre is incorporated into resin in melt extrusion process
Mould Shrinkage (%):

No.	Value	Note
1	0.44%	ASTM D955

Applications: Glass reinforced grades are suitable for injection moulding with a 'hot' mould to allow gradually cooling crystallisation which imparts vacuum property benefits. It replaces metals in many applications due to low coefficient of expansion, high stiffness and heat distortion temp. Examples of use include continuous use at 120°, fishing reels, bell housings for cars, one shot mouldings of frames e.g. chart recorders, nails, scraber vice for precision engineering. Other uses include electronics (connectors, housings and boxes for video recorder lids, tape recorder cases etc.); electrotechnology (bases for equipment, circuit breakers, electrical induction etc.); domestic electrical appliances (parts for electric can openers, irons, vacuum cleaner motors, refrigerator hinges, sewing machines etc.); mechanical applications (gear wheels, axles, housings, vices, nails, air pumps for soldering irons, cable clamps, display shelving); motor vehicles (petrol pumps, oil filter bodies, parcel shelf supports, rear mirror adjusters, window opening mechanisms, door handles etc.)

Trade name	Supplier
Ixef 1022	Solvay

Bibliographic References
[1] *J. Plastics Technology,* 1991, **37**, 14, (mech props)
[2] Ixef 1022 Technical Datasheet Solvay, (technical datasheet)
[3] *J. Plast. Ind. News,* 1989, **35**, (heat resistance)
[4] *J. High Perform. Plast.,* 1990, **Dec**, 4, (mech props)

Poly(*p*-xylylene sebacamide) P-647

$$\left[-NHCH_2-\underset{}{\bigcirc}-CH_2NH\overset{O}{\overset{\|}{C}}CH_2(CH_2)_6CH_2\overset{O}{\overset{\|}{C}}\right]_n$$

Synonyms: *Poly(p-xylylene diamine-co-sebacic acid). Poly[iminomethylene-1,4-phenylenemethyleneimino(1,10-dioxo-1,10-decanediyl)]*
Monomers: 1,4-Bis(aminomethyl)benzene, Decanedioic acid
Material class: Thermoplastic
Polymer Type: polyamide
CAS Number:

CAS Reg. No.
31711-07-6

Molecular Formula: $(C_{18}H_{26}N_2O_2)_n$
Fragments: $C_{18}H_{26}N_2O_2$

Volumetric & Calorimetric Properties:
Density:

No.	Value	Note
1	d <1.14– <1.14 g/cm^3	max., amorph.
2	d >1.168	min., cryst.

Melting Temperature:

No.	Value
1	168–200°C

Glass-Transition Temperature:

No.	Value
1	115°C

PPP-OR11

P-648

Synonyms: PPP(EO)₃. Poly[2,5-bis[2-[2-(2-methoxyethoxy)ethoxy]ethoxy][1,1'-biphenyl]-4,4'-diyl]
Material class: Thermoplastic
CAS Number:

CAS Reg. No.
187754-90-1

Molecular Formula: $(C_{26}H_{36}O_8)n$
Fragments: $C_{26}H_{36}O_8$
General Information: Soluble poly(p-phenylene) polymer with photoluminescent and electroluminescent properties.
Morphology: Layers of parallel main chains are separated by side chains that are mostly crystalline [1,2].

Volumetric & Calorimetric Properties:
Transition Temperature General: The endothermic transition between 18 and 65°C corresponds to side chain melting. Above 120°C there is a broad endotherm due to polymer backbone melting with a sharp peak at 216°C, which is probably due to melting of the high molecular-weight fraction [2].

Surface Properties & Solubility:
Solvents/Non-solvents: Sol. $CHCl_3$ and chlorinated hydrocarbons; sparingly sol. THF [1,2].

Electrical Properties:
Electrical Properties General: Polymer film exhibits electroluminescence in the blue region with λ_{max} in region of 420nm [4,5]. Oxidative degradation can cause a fairly rapid shift to the green region in LED devices, although blue emission is stable in light-emitting electrochemical cells [5]. When combined with lanthanide complexes, in particular ytterbium tetraphenylporphyrin [4] and β-diketonate complexes [6], the blue emission is quenched and Förster energy transfer to the Yb complex takes place with sharp emission at 977nm in the near-IR region [4,6,7]. The wavelength of emission can be tuned with the use of other lanthanide complexes, for example, with erbium emission is at 1570nm [7]. Turn-on voltage for LED devices is ca. 4v [4].

Optical Properties:
Optical Properties General: For polymer film, uv absorption band exhibited with λ_{max} ca. 365nm and photoluminescence band with λ_{max} ca. 415nm [3,4]. For polymer solutions, the absorption and photoluminescence bands occur at slightly shorter wavelength [4].
Transmission and Spectra: In addition to the UV and photoluminescence [3,4,5,8] and electroluminescence [4,5,7,8] spectral data, the electron diffraction spectrum [2] has been reported.

Applications/Commercial Products:
Processing & Manufacturing Routes: Prod. by Suzuki coupling of the oligo(ethylene oxide) derivative 2,5-dibromo-1,4-bis[2-(2-methoxyethoxy)ethoxy]ethoxy]benzene and bis(1,3-propanediyl) 1,4-benzenediboronate [1].
Applications: Potential application for organic light-emitting diodes, particularly near-infrared devices in blends with lanthanide compounds.

Bibliographic References
[1] Lauter, U., Meyer, W.H. and Wegner, G., *Macromolecules*, 1997, **30**, 2092
[2] Lauter, U., Meyer, W.H., Enkelmann, V. and Wegner, G., *Macromol. Chem. Phys.*, 1998, **199**, 2129
[3] Wenzl, F.P., Mauthner, G., Collon, M., List, E.J.W., Suess, C., Haase, A., Jakopic, G., Somitsch, D., Knoll, P., Bouguettaya, M., Reynolds, J.R. and Leising, G., *Thin Solid Films*, 2003, **433**, 287
[4] Harrison, B.S., Foley, T.J., Knefely, A.S., Mwaura, J.K., Cunningham, G.B., Kang, T.-S., Bouguettaya, M., Boncella, J.M., Reynolds, J.R. and Schanze, K.S., *Chem. Mater.*, 2004, **16**, 2938
[5] Mauthner, G., Collon, M., List, E.J.W., Wenzl, E.P., Bouguettaya, M. and Reynolds, J.R., *J. Appl. Phys.*, 2005, **97**, 063508
[6] Schanze, K.S. and Reynolds, J.R., *Adv. Funct. Mater.*, 2003, **13**, 205
[7] Schanze, K.S., Reynolds, J.R., Boncella, J.M., Harrison, B.S., Foley, T.J., Bouguettaya, M. and Kang, T.-S., *Synth. Met.*, 2003, **137**, 1013
[8] Schanze, K.S., Reynolds, J.R., Boncella, J.M., Harrison, B.S., Foley, T.J., Bouguettaya, M. and Kang, T.-S., *Polym. Prepr. (Am. Chem. Soc., Div. Polym. Chem.)*, 2002, **43**, 499

Propylene-ethylene copolymer

P-649

Synonyms: Propylene copolymer. Polypropylene impact copolymer. Polypropylene copolymer. Poly(1-propene-co-ethene). Poly(1-propene-block-ethene)
Related Polymers: More general information on polypropylene, More general information on polypropylene copolymers, Ethylene-propylene elastomers (EPM, EPDM), Polypropylene-isotactic, Polypropylene-atactic, Ethylene-propylene-butylene terpolymer, Polyethylene, Blends of polypropylene and polyethylene
Monomers: Propylene, Ethylene
Material class: Thermoplastic, Copolymers
Polymer Type: polyethylene, polypropylene
CAS Number:

CAS Reg. No.	Note
9010-79-1	
106565-43-9	block

Molecular Formula: $[(C_3H_6).(C_2H_4)]_n$
Fragments: C_3H_6 C_2H_4
Molecular Weight: MW 3000–1000000. $M_w/M_n \leq 50$. MZ/MW ≤ 20. MV/MW ≥ 0.85
General Information: Propylene with up to about 10% ethene (actual amount depends on the process) polymerises to a random copolymer with improved elongation and toughness compared to polypropylene. The effect is similar to, but more efficient than, an increase in the atactic to isotactic ratio. At higher ethylene levels up to about 40%, *block* copolymer is formed, comprising blocks of highly isotactic polypropylene bonded to blocks of random ethylene-propylene copolymer. Still higher ethylene levels give ethylene-propylene elastomers
Miscellaneous: Polypropylene, impact copolymer is a term often used to describe the low (<10%) ethylene copolymer. Propylene-ethylene copolymers with very small percentages of ethylene (<10%) may be referred to as ethylene-propylene random copolymer or as copolymer polypropylene which in general is discussed under Polypropylene. Propylene/ethylene copolymer reactivity ratios of r_1 0.620; r_2 0.32 have been measured [6]

Surface Properties & Solubility:
Internal Pressure Heat Solution and Miscellaneous: Heats of mixing at infinite dilution have been reported for ethylene: propylene (33:67) ratio in a range of aliphatic solvents [1]

Propylene oxide elastomers – PTFE

Transport Properties:
Permeability of Gases: Permeability of gases [He] 1.93×10^{-7}; [Ne] 4.89×10^{-7}; [Ar] 46.7×10^{-7}; [N$_2$] 167×10^{-7} cm^2 (s Pa)$^{-1}$ [4,5]

Mechanical Properties:
Mechanical Properties General: Mech. props. have been reported [2,3]. At low ethylene content, the material is normally referred to as polypropylene copolymer

Applications/Commercial Products:

Trade name	Details	Supplier
Amoco [nnnn]		Amoco Chemical Company
Buplen	random	Hercules
Daplen	random	PCD Polymere GmbH.
El Rexene	random	Rexall Drug and Chemical Corporation
Eltex	random, block	Solvay
Fina	random	Fina Chemicals (Division of Petrofina SA)
Hostalen	random, block	Hoechst (UK) Ltd. Polymer Division
Moplen	random, block	Enimont Iberica SA
Naprene		Martyn Industrials
Napryl	random	Naphthachimie SA
Noblen	random, block	Mitsubishi Chemical
Novatec	random	Hoechst (UK) Ltd
Novolen	block	BASF
Profax	random	Hercules
Propathene	random, block	ICI Acrylics
Shell	random	Shell Chemicals
Shoallomer	random, block	Hoechst (UK) Ltd
Snialene		Deutsche Snia Vertriebs GmbH.
Stamylan	random	DSM
Technodur		Campos 1925 SA
Technopro		Mario Lombardini SRL
Technopro		Tiszai Vegyi Kombinát
Tenite	random, block	Eastman Chemical Company
Tipplen	random	Tiszai Vegyi Kombinát
Trespaphan	random	Hoechst (UK) Ltd. Films Division
Vestolen	random, block	Hüls AG

Bibliographic References
[1] Phuong-Nguyen, H. and Delmas, G., *Macromolecules*, 1979, **12**, 740, 746
[2] Fernando, P.L. and Williams, J.G., *Polym. Eng. Sci.*, 1980, **20**, 215
[3] Ramsteiner, F., Kanig, G., Heckmann, W. and Gruber, W., *Polymer*, 1983, **24**, 365
[4] Pauly, S., *Polym. Handb.*, 3rd edn., (eds. J. Brandrup and E.H. Immergut), John Wiley and Sons, 1989, 435
[5] Paul, D.R. and DiBenedetto, A.T., *J. Polym. Sci., Part C: Polym. Lett.*, 1965, **10**, 17
[6] Mortimer, G.A., *J. Polym. Sci., Part B: Polym. Lett.*, 1965, **3**, 343

Propylene oxide elastomers   P-650

Synonyms: *Synthetic elastomers - polyether. 1,2-Epoxypropane elastomers*
Related Polymers: Other polymers of propylene oxide, Propylene oxide, allyl glycidyl ether copolymer (GPO), Propylene oxide, allyl glycidyl ether, epichlorohydrin terpolymer (GPCO), Propylene-ethylene copolymers, Polypropylene-rubber blends
Material class: Synthetic Elastomers
Polymer Type: polyalkylene ether
CAS Number:

CAS Reg. No.	Note
25104-27-2	GPO
25213-15-4	GPCO

Additives: Reinforcing fillers (e.g. carbon black, silica or alumina); non-reinforcing fillers; stabilisers/antioxidants (especially nickel dibutyldithiocarbamate for air and ozone resistance and 2-mercaptobenzimidazole for heat resistance). Plasticisers may be added if the elastomer is highly filled or for softening or viscosity requirements. Ether and/or esters are suitable, hydrocarbon and peroxidised plasticisers are not suitable
General Information: Because of its low cost, propylene oxide is an attractive monomer to include in polyether elastomers. The homopolymer is not readily vulcanised and so has not been commercialised. The copolymers GPO and GPCO however are used as an alternative to natural rubber where the improvements in high and low temp. resistance and in resistance to air and ozone are required

Proteins   P-651

Synonyms: *Polypeptides. Peptides*
Related Polymers: Casein (protein), Collagen (protein), Gelatin (protein), Silk (protein), Wool (protein), Poly(L-alanine) fibre, Poly(γ-methylL-glutamate) fibre

PTFE   P-652

$$-[CF_2CF_2]_n-$$

Synonyms: *Polytetrafluoroethylene. Teflon. Polytetrafluoroethene*
Related Polymers: Poly(tetrafluoroethylene-*co*-hexafluoropropylene), Poly(tetrafluoroethylene-*co*-ethylene), Poly(tetrafluoroethylene-*co*-perfluoropropyl vinyl ether), Polytetrafluoroethylene Bronze filled, Polytetrafluoroethylene Glass fibre filled, Polytetrafluoroethylene Graphite filled
Monomers: Tetrafluoroethylene
Material class: Thermoplastic
Polymer Type: PTFE, fluorocarbon (polymers)
CAS Number:

CAS Reg. No.
9002-84-0

Molecular Formula: $(C_2F_4)_n$
Fragments: C_2F_4
Molecular Weight: MW 400000–9000000
Additives: Cadmium compounds, iron oxides, ultramarines, glass, asbestos, alumina, silica and lithia, graphite, titanium dioxide, bronze
Morphology: Virgin PTFE has a crystallinity of approx. 92–98% with chains containing little branching. The fluorine atoms are 'too large' to allow a planar zig-zag struct., therefore chains are rigid

Volumetric & Calorimetric Properties:
Density:

No.	Value	Note
1	d 2.18 g/cm^3	[1]

– PTFE

Thermal Expansion Coefficient:

No.	Value	Note	Type
1	0.00012 K^{-1}	23–60°, ASTM D696 [1,5]	L

Latent Heat Crystallization: Heat of formation 813 kJ mol^{-1}. Heat of fusion 8.2 kJ mol^{-1} (82 kJ kg^{-1}). Entropy of melting 477 J K^{-1} mol^{-1} [6]

Thermal Conductivity:

No.	Value	Note
1	0.24 W/mK	4.6 mm thickness, Cenco fitch [1]

Melting Temperature:

No.	Value	Note
1	327°C	[3]
2	342°C	

Glass-Transition Temperature:

No.	Value	Note
1	127°C	[3]

Deflection Temperature:

No.	Value	Note
1	120°C	0.45 MPa [3]
2	132°C	0.455 MPa [6]
3	60°C	1.82 MPa, ASTM D648 [6]

Brittleness Temperature:

No.	Value	Note
1	-269°C	PTFE remains ductile in compression [2]

Transition Temperature:

No.	Value	Type
1	79°C	1st order
2	30°C	1st order, cryst. disordering
3	80–110°C	1st order, stress relaxation
4	-110–-73°C	2nd order, amorph. region
5	-30°C	2nd order, amorph. region >

Surface Properties & Solubility:
Solvents/Non-solvents: Sol. high boiling perfluorocarbons e.g. perfluorokerosene at 350° [4]. Insol. almost all industrial solvents and chemicals at room temp.
Surface and Interfacial Properties General: Has very low surface friction
Wettability/Surface Energy and Interfacial Tension: Critical surface tension, γ_c, 18.5 mN m^{-1}. The coefficients of static and dynamic friction are numerically equal to wet ice on wet ice [4]

Transport Properties:
Polymer Melts: Viscosity 10 GPa (380°). Critical shear rate for melt fracture 0.0001 s^{-1} (380°) [6]
Permeability of Gases: Permeability constants [CO_2] 0.93 × 10^{15} mol m^{-1} s^{-1} Pa^{-1}; [N_2] 0.18 × 10^{15}; [He] 2.47 × 10^{15}; [HCl] <0.01 × 10^{15}. Other permeability constants have been reported [12]

Permeability and Diffusion General: Permeability to vapours at 23° [acetophenone] 8.61 g m^{-2} d^{-1}, [C_6H_6] 5.54 g m^{-2} d^{-1}, [CCl_4] 0.92 g m^{-2} d^{-1}, [EtOH] 1.99 g m^{-2} d^{-1}, [HCl (20%)] <0.15 g m^{-2} d^{-1}, [piperidine] 1.08 g m^{-2} d^{-1}, [Skydrol hydraulic fluid] 0.92 g m^{-2} d^{-1}, [NaOH (50%)] 0.0008 g m^{-2} d^{-1}, [H_2SO_4] 0.0008 g m^{-2} d^{-1} (ASTM 96-35T at vapour pressure for 25.4 μm film thickness) [7]

Water Absorption:

No.	Value	Note
1	0.005 %	ASTM D570 [2]

Mechanical Properties:
Mechanical Properties General: PTFE is a tough, non-resilient material with moderate tensile strength [1,2]

Tensile (Young's) Modulus:

No.	Value	Note
1	340 MPa	23°, ASTM D638 [9]

Flexural Modulus:

No.	Value	Note
1	350–630 MPa	23°, ASTM D747, granular material [1]
2	280–630 MPa	fine powder [9]

Elastic Modulus:

No.	Value	Note
1	410–550 MPa	tensile elastic modulus [3]

Poisson's Ratio:

No.	Value	Note
1	0.46	Teflon, DuPont

Tensile Strength Yield:

No.	Value	Note
1	7–28 MPa	23°, ASTM D638, granular material [1]
2	17.5–24.5 MPa	fine powder

Compressive Strength:

No.	Value	Note
1	4.2 MPa	23°, 1% deformation, ASTM D695, Compressive stress
2	7 MPa	23°, 1% offset, Compressive stress

Miscellaneous Moduli:

No.	Value	Note	Type
1	186 MPa	100h, 0.6895 MPa, 23°, ASTM D695 [9]	Compressive creep modulus

Mechanical Properties Miscellaneous: Elongation 100–200% (23°, granular) [1,30], 300–600% (23°, fine powder), 350% (ASTM D638). Deformation under load 2.4% (6.86 MPa, 24h, 26°), 15% (13.72 MPa, 24h, 26°) [1]. Creep modulus 2 kN m^{-1} [5]
Hardness: Durometer D50–D65 (ASTM D1706) [1]
Friction Abrasion and Resistance: Static coefficient of friction with

— PTFE

polished steel 0.05–0.08 [1]. Wear factor 3000 × 10^{-17} Pa^{-1} [8]
Izod Notch:

No.	Value	Notch	Note
1	106 J/m	N	2 ft lb in^{-1}, 23°, ASTM D256 [2]
2	160 J/m	N	[15]

Electrical Properties:
Electrical Properties General: Has excellent electrical insulating props.
Surface/Volume Resistance:

No.	Value	Note	Type
1	>10 × 10^{15} Ω.cm	min., 100% relative humidity [1,10]	S

Dielectric Permittivity (Constant):

No.	Value	Frequency	Note
1	2.1	60 Hz-2 GHz	ASTM D150 [10]

Dielectric Strength:

No.	Value	Note
1	23.6 $kV.mm^{-1}$	short time, 2 mm thick [1]
2	55 $kV.mm^{-1}$	1397 V mil^{-1}, Hostaflon

Arc Resistance:

No.	Value	Note
1	>300s	min., ASTM D495 [1,10]

Dissipation (Power) Factor:

No.	Value	Frequency	Note
1	0.0003	60 Hz-2 GHz	ASTM D150 [1,10]

Optical Properties:
Refractive Index:

No.	Value	Note
1	1.35	ASTM D542 [3]

Volume Properties/Surface Properties: Clarity opaque [1,4]

Polymer Stability:
Polymer Stability General: Has excellent stability to heat
Thermal Stability General: PTFE degrades to monomer under vacuum. Degradation begins at 440°, peaks at 540° and continues until 590° [1]
Upper Use Temperature:

No.	Value	Note
1	230–260°C	[3]
2	260°C	intermittent
3	230–260°C	continuous [3]
4	500°C	gaskets in totally enclosed system

Decomposition Details: Fine powder degrades more readily than granular material. At temp. 232–371° degradation ranges from 0.0001–0.03% per hour [1]

Flammability: Non-flammable
Environmental Stress: Exceptional weathering resistance. Samples exposed in Florida for 10 years showed little change in physical props. [2]
Chemical Stability: Chemically inert to almost all industrial chemicals and solvents. It will react with molten alkali metals, fluorine, strong fluorinating agents and sodium hydroxide above 300°. The surface can be etched by solns. of sodium naphthalenide or sodium in liq. ammonia [1,3]
Hydrolytic Stability: Resistant to aq. solns. [2]
Biological Stability: Has excellent resistance to fungi and bacteria [3]
Stability Miscellaneous: Degraded by high energy radiation. Exposure to a dose of 70 Mrad halves the tensile strength. [2] Degradation due to uv radiation has been reported [16]

Applications/Commercial Products:
Processing & Manufacturing Routes: Manufactured by suspension polymerisation, which produces a granular resin. Emulsion polymerisation produces a fine dispersion. Processed by pre-forming the powder, sintering and then cooling. Commonly machined and coined. Specialist extrusion with sintering zone. Used for impregnation of fabrics
Applications: Seals, tapes, gaskets, valves, pump parts and laboratory equipment

Trade name	Details	Supplier
75Y-P84/25Y		Lenzing AG/Pacrim
Aflon	various grades	Asahi Glass
Algaflon		Montedison
Algoflon		Ausimont
Algoflon		Montedison
Allied	CM-X	Allied Signal Corporation
EGC Alloy	various grades	EGC Corporation
Fluon		ICI, UK
Fluorocomp		LNP Engineering
Fluorogold		Furon
Fluoroloy	various grades	Furon
Fluorosint	507/500 (mica filled)	Polymer Corp.
Foraflon		Atochem Polymers
Halon		Allied Chemical
Halon		Ausimont
Hostaflon TF		Hoechst Celanese
IPC	5508	IPC
Klingerflon		Klinger
Polycomp	various grades	ICI Americas Inc.
Polyflon		Daikin Kogyo
RT/duroid	D series	Rogers Corp.
Soreflon		Atochem Polymers
Stat-Kon	KCL-4022 (carbon fibre filled)	LNP Engineering
Teflon		DuPont
Teflon	TFE	Furon
Tetraflon		Nitto Chemical Industry Co. Ltd
Tetrafluor	TFC series	Tetrafluor Inc.
Tetralon	various grades	Tetrafluor Inc.
Thermocomp	FC-102CF (carbon fibre filled)	LNP Engineering

– PTFE, Bronze filled		
Thermocomp	FC-446CS	LNP Engineering
Ultralon		ICI Advanced Materials
Whitcon		ICI Advanced Materials

Bibliographic References

[1] *Encycl. Polym. Sci. Eng.*, 2nd edn., (ed. J.I. Kroschwitz), Wiley Interscience, 1989, **3**
[2] Brydson, J.A., *Plast. Mater.*, 5th edn., Butterworth-Heinemann, 1989
[3] *The Materials Selector*, 2nd edn., (eds. N.A. Waterman and M.F. Ashby), Chapman and Hall, 1997, **3**
[4] *Encyclopedia of Advanced Materials*, (eds. D. Bloor, R.J. Brook, M.C. Flemings, S. Mahajan and R.W. Cahn), Pergamon Press, 1994, **3**
[5] *Kirk-Othmer Encycl. Chem. Technol.*, Vol. 11, 4th edn., (ed. J.I. Kroschwitz), Wiley Interscience, 1980
[6] Sperati, C.A. and Starkweather, H.W., *Adv. Polym. Sci.*, 1961, **2**, 465, (deflection temp)
[7] *Encyclopedia of Advanced Materials*, (eds. D. Bloor, R.J. Brook, M.C. Flemings, S. Mahajan and R.W. Cahn), Pergamon Press, 1994, **2**
[8] Lewis, R.B., *J. Eng. Ind.*, 1986, **1**, 1, (wear factor)
[9] Kerbow, D.L. and Sperati, C.A., *Polym. Handb.*, 4th edn., (eds. J. Brandrup, E.H. Immergut and E.A. Grulke), John Wiley and Sons, 1999, **V**, 31
[10] Mechanical Design Data, *Teflon Fluorocarbon Resins*, bulletin, E.I. DuPont de Nemours and Co. Inc., 1964
[11] Clark, E.S. and Muus, L.T., *Paper presented at the 133rd Meeting of the American Chemical Society*, Sept., 1957, New York
[12] Araki, Y., *J. Appl. Polym. Sci.*, 1965, **9**, 3585
[13] McCrum, N.G., *J. Polym. Sci.*, 1959, **34**, 355
[14] *J. Teflon*, DuPont, 1970, **11**, 8
[15] Banks, R.E. et al, *Fluoropolymers '92*, RAPRA Technology, 1992
[16] Ferry, L. et al, *Polym. Adv. Technol.*, 1996, **7**, 493, (uv degradation)

PTFE, Bronze filled

P-653

Synonyms: *Polytetrafluoroethylene, Bronze filled*
Related Polymers: Polytetrafluoroethylene
Monomers: Tetrafluoroethylene
Material class: Thermoplastic, Composites
Polymer Type: PTFE

Volumetric & Calorimetric Properties:
Density:

No.	Value	Note
1	d 3.85 g/cm^3	Hostaflon [1]
2	d 3.12 g/cm^3	Fluorocomp [3]
3	d 3.9 g/cm^3	Hostaflon, ISO 1183 [1]

Thermal Expansion Coefficient:

No.	Value	Note	Type
1	0.000129 K^{-1}	Hostaflon [1]	L
2	0.00011 K^{-1}	Fluorocomp [3]	L

Thermodynamic Properties General: Higher thermal conductivity than other filled materials
Thermal Conductivity:

No.	Value	Note
1	0.69 W/mK	0.00165 cal (s cm °C)$^{-1}$, Hostaflon [1]
2	0.62 W/mK	0.00148 cal (s cm °C)$^{-1}$, Fluorocomp [3]

Melting Temperature:

No.	Value	Note
1	330°C	Hostaflon [1]

Mechanical Properties:
Mechanical Properties General: Elongation at break 150% (Hostaflon, 2 mm thick) [1]
Flexural Modulus:

No.	Value	Note
1	965 MPa	9840 kg cm^{-2}, Fluorocomp [3]

Tensile Strength Break:

No.	Value	Note	Elongation
1	17 MPa	173 kg cm^{-2}, Hostaflon, 200 μm film [1]	
2	16.2 MPa	165 kg cm^{-2}, Fluorocomp, ASTM D1457 [3]	200%

Flexural Strength Yield:

No.	Value	Note
1	9.65 MPa	98.4 kg cm^{-2}, Fluorocomp, 1% strain [3]

Compressive Strength:

No.	Value	Note
1	7.58 MPa	77.3 kg cm^{-2}, Fluorocomp, 1% strain [3]

Hardness: Shore D60 (Fluorocomp) [3]

Electrical Properties:
Surface/Volume Resistance:

No.	Value	Note	Type
3	0.0001 × 10^{15} Ω.cm	Hostaflon, IEC 93 [1]	S

Polymer Stability:
Chemical Stability: Poor resistance to 50% sulfuric acid, conc. hydrochloric and nitric acids and conc. NH$_3$. Poor resistance to mercury [2]

Applications/Commercial Products:
Processing & Manufacturing Routes: Processed by extrusion or compression moulding
Mould Shrinkage (%):

No.	Value	Note
1	2%	Hostaflon [1]

Applications: Unsuitable for any application involving contact with foodstuffs. Used in seals, bearings and gaskets. Potential applications in the electrical industry

Trade name	Details	Supplier
FC-446CS		LNP Engineering
Fluorocomp FC-144	40% bronze	ICI Fluoropolymers
Hostaflon TF4406	60% bronze	Hoechst Celanese
Klingerflon	60% bronze	Klinger
Rulon 142		Dixon
Rulon 142		Furon

PTFE, Carbon and graphite filled

Bibliographic References
[1] *Hostaflon*, Hoechst AG, 1997, (technical datasheet)
[2] *The Materials Selector*, 2nd edn., (eds. N.A. Waterman and M.F. Ashby), Chapman and Hall, 1997, **3**, 221
[3] *Fluorocomp*, ICI Fluoropolymers, 1997, (technical datasheet)

PTFE, Carbon and graphite filled — P-654

Synonyms: *Polytetrafluoroethylene, Carbon and Graphite filled*
Monomers: Tetrafluoroethylene
Material class: Composites
Polymer Type: PTFE
General Information: Reduces coefficient of friction; good deformation and wear resistance

Mechanical Properties:
Friction Abrasion and Resistance: Wear factor K 4 (26.3% filled) [1]

Applications/Commercial Products:

Trade name	Supplier
Fluorocomp 192HE	LNP Engineering

Bibliographic References
[1] *The Materials Selector*, 2nd edn., (eds. N.A. Waterman and M.F. Ashby), Chapman and Hall, 1997, **3**, 221

PTFE, Glass fibre filled — P-655

Synonyms: *Polytetrafluoroethylene, Glass fibre filled*
Related Polymers: Polytetrafluoroethylene
Monomers: Tetrafluoroethylene
Material class: Thermoplastic, Composites
Polymer Type: PTFE
General Information: Glass filler improves creep resistance and chemical resistance
Miscellaneous: Grades available with 15% and 25% glass fibre. May be filled up to 36% by volume, with 22% being optimal [2]

Volumetric & Calorimetric Properties:
Density:

No.	Value	Note
1	d 2.21 g/cm^3	Hostaflon, 15% filled, ISO 1133 [1]
2	d 2.24 g/cm^3	Hostaflon, 25% filled, ISO 1133 [1]
3	d 2.23 g/cm^3	Hostaflon, 25% filled [1]
4	d 2.17 g/cm^3	Fluorocomp, 5% filled, ASTM D1457 [3]
5	d 2.25 g/cm^3	30% filled [2]

Thermal Expansion Coefficient:

No.	Value	Note	Type
1	0.000166 K^{-1}	Hostaflon, 15% filled [1]	L
2	0.00015 K^{-1}	Hostaflon, 25% filled [1]	L
3	0.000131 K^{-1}	Fluorocomp, 5% filled [3]	L
4	$6 \times 10^{-5} - 8 \times 10^{-5}$ K^{-1}	30% filled [2]	L

Thermal Conductivity:

No.	Value	Note
1	0.35 W/mK	0.000836 cal (s cm °C)$^{-1}$, Hostaflon, 15% filled [1]
2	0.41 W/mK	0.000979 cal (s cm °C)$^{-1}$, Hostaflon, 25% filled [1]
3	0.3 W/mK	0.000725 cal (s cm °C)$^{-1}$, Fluorocomp, 5% filled [3]
4	0.36 W/mK	30% filled [2]

Melting Temperature:

No.	Value	Note
1	330°C	Hostaflon, 15% and 25% filled [1]

Mechanical Properties:
Flexural Modulus:

No.	Value	Note
1	827.7 MPa	8440 kg cm^{-2}, Fluorocomp, 5% filled [3]
2	1650 MPa	30% filled [2]

Tensile Strength Break:

No.	Value	Note	Elongation
1	17 MPa	173 kg cm^{-2}, Hostaflon, 15% filled, 500 μm skived film [1]	
2	13 MPa	133 kg cm^{-2}, Hostaflon, 25% filled, 500 μm skived film [1]	350%
3	20 MPa	204 kg cm^{-2}, Fluorocomp, 5% filled, ASTM D1457 [3]	320%
4	12–20 MPa	room temp., 30% filled [2]	200–300%

Flexural Strength Yield:

No.	Value	Note
1	7.8 MPa	79.4 kg cm^{-2}, Fluorocomp, 5% filled, 1% strain [3]

Compressive Strength:

No.	Value	Note
1	5.9 MPa	60.5 kg cm^{-2}, Fluorocomp, 5% filled, 1% strain [3]

Viscoelastic Behaviour: Creep resistance of glass-filled grades may be improved by inert gas sintering [2]
Hardness: Shore D54 (Fluorocomp, 5% filled) [3], D60–D75 (30% filled) [2]
Friction Abrasion and Resistance: Wear factor K 5.5 [2]. Coefficient of friction 0.12–0.15 (30% filled) [2]
Izod Notch:

No.	Value	Notch	Note
1	220 J/m	Y	30% filled [2]

Electrical Properties:
Surface/Volume Resistance:

No.	Value	Note	Type
2	$>1 \times 10^{15}$ Ω.cm	min., Hostaflon, 15% and 25% filled, IEC 93 [1]	S

– PTFE, Graphite filled

Dielectric Permittivity (Constant):

No.	Value	Frequency	Note
1	2.2	50 Hz-1 MHz	Hostaflon, 15% filled, IEC 250 [1]
2	2.3	50 Hz-1 MHz	Hostaflon, 25% filled, IEC 250 [1]
3	2.35	1 MHz	95% relative humidity, 30% filled [2]
4	2.35	1 MHz	dry, 30% filled [2]

Dielectric Strength:

No.	Value	Note
1	40 kV.mm^{-1}	Hostaflon, 15% and 25% filled, IEC 243-1 [1]

Complex Permittivity and Electroactive Polymers: Comparative tracking index 600 V (Hostaflon, 15% and 25% filled, IEC 112) [1]

Dissipation (Power) Factor:

No.	Value	Frequency	Note
1	5×10^{-5}	50 Hz	Hostaflon, 15% and 25% filled, ISO 1325 [1]
2	7×10^{-5}	1 MHz	Hostaflon, 15% and 25% filled, ISO 1325 [1]

Applications/Commercial Products:

Processing & Manufacturing Routes: Processed by extrusion or compression moulding

Mould Shrinkage (%):

No.	Value	Note
1	2.5%	Hostaflon, 15% filled [1]
2	2%	Hostaflon, 25% filled [1]
3	2%	30% filled [2]

Applications: In bearings, components and seals

Trade name	Details	Supplier
Fluorocomp FC-101		ICI Fluoropolymers
Hostaflon	TF4103, TF4105	Hoechst Celanese
Klingerflon	25% filled	Klinger
TLY-3		Taconic Plastics Ltd.
TLY-5		Taconic Plastics Ltd.

Bibliographic References

[1] *Hostaflon*, Hoechst AG, 1997, (technical datasheet)
[2] *The Materials Selector*, 2nd edn., (eds. N.A. Waterman and M.F. Ashby), Chapman and Hall, 1997, **3**, 221
[3] *Fluorocomp*, ICI Fluoropolymers, 1997, (technical datasheet)

PTFE, Graphite filled   P-656

Synonyms: *Polytetrafluoroethylene, Graphite filled*
Related Polymers: Polytetrafluoroethylene
Monomers: Tetrafluoroethylene
Material class: Thermoplastic, Composites
Polymer Type: PTFE
General Information: Graphite filler improves wear resistance and lowers coefficient of friction

Volumetric & Calorimetric Properties:

Density:

No.	Value	Note
1	d 2.17 g/cm^3	Algoflon, 15% filled, ASTM D792 [1]
2	d 2.15 g/cm^3	Hostaflon, 15% filled, ISO 1183 [2]
3	d 2.11 g/cm^3	15% filled [3]

Thermal Expansion Coefficient:

No.	Value	Note	Type
1	0.00019 K^{-1}	Algoflon, 15% filled [1]	L
2	0.00012 K^{-1}	15% filled [3]	L
3	0.000144 K^{-1}	Hostaflon, 15% filled [2]	L

Thermal Conductivity:

No.	Value	Note
1	0.78 W/mK	0.00186 cal (s cm °C)$^{-1}$, Hostaflon, 15% filled [2]

Melting Temperature:

No.	Value	Note
1	330°C	Hostaflon, 15% filled [2]

Deflection Temperature:

No.	Value	Note
1	120°C	0.45 MPa, 15% filled [3]
2	95°C	1.8 MPa, 15% filled [3]

Transport Properties:

Water Absorption:

No.	Value	Note
1	0.07 %	24h, 15% filled [3]

Mechanical Properties:

Mechanical Properties General: Elongation break at 270% (Hostaflon, 15% filled, 200 μm film) [2]

Flexural Modulus:

No.	Value	Note
1	1030 MPa	15% filled [3]

Tensile Strength Break:

No.	Value	Note	Elongation
1	27 MPa	room temp., 15% filled [3]	240%
2	18 MPa	184 kg cm^{-2}, Hostaflon, 15% filled, 2 mm thick [2]	

Tensile Strength Yield:

No.	Value	Elongation	Note
1	24 MPa	230%	Algoflon, 15% filled, ASTM D638 [1]

Hardness: Shore D57 (Algoflon, 15% filled) [1], D63 (15% filled) [3]
Friction Abrasion and Resistance: Coefficient of friction is lower than that of other filled grades [3]
Izod Notch:

No.	Value	Notch	Note
1	140 J/m	Y	15% filled [3]

Electrical Properties:
Electrical Properties General: Volume resistivity 10^7 Ω cm (Algoflon, 15% filled, ASTM D257) [1]
Surface/Volume Resistance:

No.	Value	Note	Type
2	0.01×10^{15} Ω.cm	Hostaflon, 15% filled, IEC 93 [2]	S

Dissipation (Power) Factor:

No.	Value	Frequency	Note
1	0.0007	1 kHz	dry, 15% filled [3]

Applications/Commercial Products:
Processing & Manufacturing Routes: May be processed by extrusion and compression moulding
Mould Shrinkage (%):

No.	Value	Note
1	3%	Hostaflon, 15% filled [2]

Applications: Piston rings, bearings, ball valve seats, gaskets, seals, insulators, electrical components

Trade name	Details	Supplier
Algoflon	15GR	Ausimont
Hostaflon TF4303		Hoechst Celanese

Bibliographic References
[1] *Algoflon*, Ausimont, 1997, (technical datasheet)
[2] *Hostaflon*, Hoechst AG, 1997, (technical datasheet)
[3] *The Materials Selector*, 2nd edn., (eds. N.A. Waterman and M.F. Ashby), Chapman and Hall, 1997, **3**, 221, (rev)

PTFE, PPS filled   P-657

Synonyms: *Polytetrafluoroethylene, Polyphenylene sulfide filled*
Monomers: Tetrafluoroethylene
Material class: Composites
Polymer Type: PTFE, fluorocarbon (polymers)polyphenylene sulfide
General Information: Has good high temp. mech. props.; is tough, and has good wear and chemical resistance
Miscellaneous: Has limited availability commercially [1]

Volumetric & Calorimetric Properties:
Density:

No.	Value	Note
1	d 2.14 g/cm³	[1]

Thermal Expansion Coefficient:

No.	Value	Note	Type
1	7.2×10^{-5} K^{-1}	[1]	L

Thermal Conductivity:

No.	Value	Note
1	0.28 W/mK	[1]

Deflection Temperature:

No.	Value	Note
1	>260°C	min., 1.8 MPa [1]

Transport Properties:
Water Absorption:

No.	Value	Note
1	0.03 %	24h, max. [1]

Mechanical Properties:
Flexural Modulus:

No.	Value	Note
1	1170 MPa	[1]

Tensile Strength Yield:

No.	Value	Elongation	Note
1	45 MPa	6%	room temp. [1]

Hardness: Shore D70 [1]
Friction Abrasion and Resistance: Coefficient of friction 0.08 [1]
Izod Notch:

No.	Value	Notch	Note
1	221 J/m	Y	[1]

Applications/Commercial Products:

Trade name	Details	Supplier
Alton		International Polymer Corp.
Polycomp	139, 142, 148, 149, 158	ICI Advanced Materials

Bibliographic References
[1] *The Materials Selector*, 2nd edn., (eds. N.A. Waterman and M.F. Ashby), Chapman and Hall, 1997, **3**, 221, (rev)

PTFE, Stainless steel filled   P-658

Synonyms: *Polytetrafluoroethylene, Stainless steel filled*
Monomers: Tetrafluoroethylene
Material class: Composites
Polymer Type: PTFE, fluorocarbon (polymers)

Volumetric & Calorimetric Properties:
Density:

No.	Value	Note
1	d 0.8 g/cm³	Algoflon, bulk, 50% filled, ASTM D4894 [1]
2	d 3.38 g/cm³	Algoflon, 50% filled, ASTM D792 [1]

– PTFE, TFE filled

Thermal Expansion Coefficient:

No.	Value	Note	Type
1	0.00014 K^{-1}	Algoflon, 50% filled, ASTM D696 [1]	L

Mechanical Properties:
Tensile Strength Yield:

No.	Value	Elongation	Note
1	20 MPa	220%	Algoflon, 50% filled, ASTM D638 [1]

Hardness: Shore D63 (Algoflon, 50% filled) [1]

Applications/Commercial Products:
Processing & Manufacturing Routes: Processed by compression moulding
Applications: Piston rings, bearings, ball valve seats, gaskets, seals, insulators, electrical components

Trade name	Details	Supplier
Algoflon	50 INOX	Ausimont
FC-60% SS		LNP Engineering

Bibliographic References
[1] *Algoflon*, Ausimont, 1997, (technical datasheet)

PTFE, TFE filled P-659
Synonyms: *Polytetrafluoroethylene, Tetrafluoroethylene filled*
Monomers: Tetrafluoroethylene
Material class: Composites
Polymer Type: PTFE
General Information: Available as tubes, rods and sheets

Volumetric & Calorimetric Properties:
Density:

No.	Value	Note
1	d 2.03 g/cm^3	Fluoroloy K [1]
2	d 2.2 g/cm^3	Fluoroloy S [1]
3	d 2.21 g/cm^3	Fluoroloy X [1]
4	d 2.22 g/cm^3	Fluoroloy Y [1]
5	d 2.24 g/cm^3	Fluoroloy A [1]
6	d 2.04 g/cm^3	Fluoroloy B [1]
7	d 0.941 g/cm^3	Fluoroloy G [1]
8	d 2.09 g/cm^3	Fluoroloy H [1]

Thermal Expansion Coefficient:

No.	Value	Note	Type
1	3.6 × 10^{-5} K^{-1}	Fluoroloy H [1]	L
2	6.48 × 10^{-5} K^{-1}	Fluoroloy K [1]	L
3	5.04 × 10^{-5} K^{-1}	Fluoroloy S [1]	L
4	7.38 × 10^{-5} K^{-1}	Fluoroloy X and Y [1]	L
5	0.000198 K^{-1}	Fluoroloy A [1]	L
6	0.000135 K^{-1}	Fluoroloy B [1]	L
7	0.000145 K^{-1}	Fluoroloy C [1]	L

Thermal Conductivity:

No.	Value	Note
1	0.26 W/mK	0.000614 cal (s cm K)$^{-1}$, Fluoroloy X [1]
2	0.28 W/mK	0.000665 cal (s cm K)$^{-1}$, Fluoroloy Y [1]
3	1.21 W/mK	0.00289 cal (s cm K)$^{-1}$, Fluoroloy H [1]
4	0.25 W/mK	0.000593 cal (s cm K)$^{-1}$, Fluoroloy K [1]
5	0.21 W/mK	0.00051 cal (s cm K)$^{-1}$, Fluoroloy S [1]
6	0.39 W/mK	0.000921 cal (s cm K)$^{-1}$, Fluoroloy A [1]
7	0.63 W/mK	0.0015 cal (s cm K)$^{-1}$, Fluoroloy B [1]
8	0.632 W/mK	0.00151 cal (s cm K)$^{-1}$, Fluoroloy C [1]

Mechanical Properties:
Tensile (Young's) Modulus:

No.	Value	Note
1	619.78 MPa	6320 kg cm^{-2}, Fluoroloy G [1]

Flexural Modulus:

No.	Value	Note
1	696.3 MPa	7100 kg cm^{-2}, Fluoroloy S [1]
2	990.5 MPa	10100 kg cm^{-2}, Fluoroloy X and Y [1]

Tensile Strength Yield:

No.	Value	Elongation	Note
1	10.98 MPa	50–200%	112 kg cm^{-2}, Fluoroloy A, Fluoroloy B [1]
2	12.36 MPa	65%	126 kg cm^{-2}, Fluoroloy C [1]
3	37.17 MPa	450%	379 kg cm^{-2}, Fluoroloy G [1]
4	14.4 MPa	190%	147 kg cm^{-2}, Fluoroloy X [1]
5	13.04 MPa	170%	133 kg cm^{-2}, Fluoroloy Y [1]
6	6.9 MPa	4%	70.3 kg cm^{-2}, Fluoroloy H [1]
7	16.5 MPa	205%	168 kg cm^{-2}, Fluoroloy K [1]
8	28.2 MPa	320%	288 kg cm^{-2}, Fluoroloy S [1]

Flexural Strength Yield:

No.	Value	Note
1	6.2 MPa	63.2 kg cm^{-2}, Fluoroloy S and X [1]
2	4.8 MPa	49.2 kg cm^{-2}, Fluoroloy Y [1]

Compressive Strength:

No.	Value	Note
1	12.36 MPa	126 kg cm^{-2}, Fluoroloy A [1]
2	11.67 MPa	119 kg cm^{-2}, Fluoroloy B [1]

No.	Value	Note
3	10.98 MPa	112 kg cm^{-2}, Fluoroloy C [1]
4	42 MPa	428 kg cm^{-2}, Fluoroloy H [1]

Hardness: Shore D60 (Fluoroloy A) [1], D66 (Fluoroloy B) [1], D65 (Fluoroloy C) [1]. Rockwell R62 (Fluoroloy G) [1]. Shore D58 (Fluoroloy H), D59 (Fluoroloy K), D57 (Fluoroloy S and X), D55 (Fluoroloy Y) [1]

Izod Notch:

No.	Value	Notch	Note
1	960.8 J/m	Y	18 ft lb in^{-1}, room temp., Fluoroloy G [1]
2	320.3 J/m	Y	6 ft lb in^{-1}, room temp., Fluoroloy H [1]

Electrical Properties:
Dielectric Permittivity (Constant):

No.	Value	Frequency	Note
1	2.3	1 MHz	Fluoroloy G [1]

Dielectric Strength:

No.	Value	Note
1	27.9 kV.mm^{-1}	Fluoroloy G [1]
2	19.6 kV.mm^{-1}	Fluoroloy H [1]

Dissipation (Power) Factor:

No.	Value	Frequency	Note
1	0.0002	1 MHz	Fluoroloy G [1]

Applications/Commercial Products:

Trade name	Details	Supplier
Fluoroloy	A,B,C,D,G,H,K,S,X,Y	Furon

Bibliographic References
[1] Fluoroloy Furon, 1996, (technical datasheet)

Pullulan
P-660

Related Polymers: Cellulose I, Cellulose II
Monomers: Glucose
Material class: Polysaccharides
Polymer Type: cellulosics
CAS Number:

CAS Reg. No.
9057-02-7

Molecular Formula: $(C_6H_{10}O_5)_n$
Fragments: $C_6H_{10}O_5$
Molecular Weight: MW 15000–810000. MW 200000 (commercial material)
General Information: Polysaccharide extracted from the fungus *Aureobasidium pullulans*. Contains repeating triads of 1-4,1-4 and 1-6-linked α-D-glucose monomers. The dried product is a water-sol. powder, which can be formed into films, fibres or compression mouldings

Volumetric & Calorimetric Properties:
Thermodynamic Properties General: Heat of combustion 2705.4 kJ mol^{-1} (16.7 kJ g^{-1}) [1]

Transport Properties:
Permeability and Diffusion General: Films ≤0.01 mm thick are impermeable to O_2 and grease [2]

Optical Properties:
Total Internal Reflection: $[\alpha]_D$ +180–190° (H_2O) [1]

Applications/Commercial Products:

Trade name	Supplier
Pollulan	Hayashibara

Bibliographic References
[1] Yuen, S., *Process Biochem.*, 1974, **7**, 22
[2] *Encycl. Polym. Sci. Eng.*, 2nd edn., (eds. H.F. Mark, N.M. Bikales, C.G. Overberger and C. Menges), John Wiley and Sons, 1985, **13**
[3] *Chem. Eng. News*, Dec., 1973, **51**, 40

PURET
P-661

Synonyms: *Poly[2-(3-thienyl)ethanol-n-butoxycarbonylmethylurethane]*. *N-[[2-(3-Thienyl)ethoxy]carbonyl]glycine butyl ester homopolymer*
Material class: Thermoplastic
Polymer Type: polythiophene
CAS Number:

CAS Reg. No.
182691-54-9

Molecular Formula: $(C_{13}H_{17}NO_4S)n$
Fragments: $C_{13}H_{17}NO_4S$
General Information: Urethane-substituted polythiophene with electroluminescent properties and good solubility in a range of organic solvents.
Morphology: Amorphous polymer structure according to x-ray diffraction studies [1]. Polymer has a non-planar coil-like structure with inter- and intramolecular hydrogen bonding [2].

Volumetric & Calorimetric Properties:
Glass-Transition Temperature:

No.	Value	Note
1	18°C	[1]

Surface Properties & Solubility:
Solvents/Non-solvents: Sol. $CHCl_3$, THF, dioxan, toluene, DMF, *N*-methylpyrrolidone, DMSO, trifluoroacetic acid [1,3].

Electrical Properties:

Electrical Properties General: Conductivity of films doped with I_2 or $FeCl_3$ of the order of 1 S cm^{-1} [1]. Polymer exhibits electroluminescence in single-layer LED devices with emission of orange-red light with λ_{max} in the region of 600nm [3,4]. Electroluminescence intensity is enhanced by the addition of europium or ruthenium complexes [5]. On doping with the laser dye DCM there is increase in EL intensity and voltage tunable emission, with a red shift on increasing the voltage [6]. Addition of tris(8-hydroxyquinoline)aluminium causes a red shift in EL emission [7]. On increasing the temperature there is a blue shift in EL emission and an increase in intensity [2].

Optical Properties:

Optical Properties General: Uv absorption band occurs with λ_{max} in the region of 465–480nm [1,4] for film and 420–430nm for solutions in various solvents [1]. The optical band gap is ca. 2eV [4,5]. Photoluminescence with emission of orange-red light occurs with λ_{max} in the region of 570–600nm [1,4,5]. Increasing the temperature causes a blue shift in the uv absorption band [8] and in photoluminescence with an increase in intensity [2].

Transmission and Spectra: Pmr [1], ftir [1], uv and photoluminescence [1,2,4,5,8], electroluminescence [2,3,4,5,6,7], xps [3,8] and ups [8] spectral data have been reported.

Polymer Stability:

Thermal Stability General: Polymer is stable up to 220°C in nitrogen [1]. The side chain decomposes in the region 200–300°C, while the backbone is stable above 400°C [8].

Stability Miscellaneous: Polymer backbone decomposes slowly at r.t on exposure to light at wavelength 397nm [8].

Applications/Commercial Products:

Processing & Manufacturing Routes: Prod. by polymerisation of the monomer (2-(3-thienyl)ethanol n-butoxycarbonylmethylurethane) with $FeCl_3$ in $CHCl_3$ [1,3]. Films can be obtained by solution casting [1].

Applications:

Bibliographic References

[1] Chittibabu, K.G., Balusubramanian, S., Kim, W.H., Cholli, A.L., Kumar, J. and Tripathy, S.K., *J. Macromol. Sci., Part A: Pure Appl. Chem.*, 1996, **33**, 1283
[2] Jung, S.-D., Hwang, D.-H., Zyung, T., Kim, W.H., Chittibabu, K.G., and Tripathy, S.K., *Synth. Met.*, 1998, **98**, 107
[3] Seung, H.Y. and Whitten, J.E., *Synth. Met.*, 2000, **114**, 305
[4] Jung, S.-D., Zyung, T., Kim, W.H., Chittibabu, K.G. and Tripathy, S.K., *Polym. Prepr. (Am. Chem. Soc., Div. Polym. Chem.)*, 1997, **38**, 425
[5] Cazeca, M.J., Chittibabu, K.G., Kim, J., Kumar, J., Jain, A., Kim, W. and Tripathy, S.K., *Synth. Met.*, 1998, **98**, 45
[6] Kaur, A., Cazeca, M.J., Sengupta, S.K., Kumar, J. and Tripathy, S.K., *Synth. Met.*, 2002, **126**, 283
[7] Jung, S.-D., Zyung, T., Kim, W.H., Lee, C.J. and Tripathy, S.K., *Synth. Met.*, 1999, **100**, 223
[8] Herrera, G.J. and Whitten, J.E., *Synth. Met.*, 2002, **128**, 317

PVC copolymers   P-662

Synonyms: *Poly(vinyl chloride) copolymers*
Related Polymers: PVC, Vinyl chloride-acrylonitrile copolymer, Vinylidene chloride-vinyl chloride copolymer, Vinyl chloride-propylene copolymer, Vinyl chloride-chlorotrifluoroethylene copolymer, Cinyl chloride-ethylene copolymer, Polyvinyl chloride, unplasticised, Poly(vinyl chloride-*co*-vinyl acetate), Poly(vinyl chloride-*co*-vinyl acetate), epoxidised, Poly(vinyl chloride-*co*-vinyl acetate), carboxylated, Poly(vinyl chloride-*co*-vinyl acetate-*co*-vinyl alcohol), Poly(vinyl chloride-*co*-vinyl acetate-*co*-2-hydroxypropyl acrylate), Poly(vinyl chloride-*co*-vinyl acetate-*co*-maleic acid), Poly(vinyl chloride-*co*-acrylonitrile), Poly(vinyl chloride-*co*-chlorotrifluoroethylene), Poly(vinyl chloride-*co*-ethylene), Poly(vinyl chloride-*co*-propylene), Poly(vinyl chloride-*co*-N-cyclohexylmaleimide), Poly(vinyl chloride-*co*-vinylidene chloride), Poly(vinyl chloride-*co*-vinyl bromide)

General Information: The presence of a significant amount of comonomer leads, in general, to the following; processing temp. is reduced; reduction in T_g; decrease in heat stability; reduction in softening temp.; reduction in temp. of deflection under load; hardness is decreased; extensibility is increased [1]. Copolymers are also more readily soluble, in general, than homopolymer [1]

Bibliographic References

[1] Titow, W.V., *PVC Technol.*, Elsevier Applied Science, 1984, 20, (props)

PVC, EVA filled   P-663

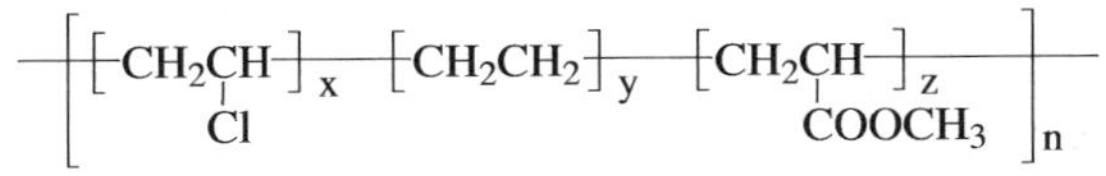

$$\left[\left[CH_2CH\underset{Cl}{|}\right]_x\left[CH_2CH_2\right]_y\left[CH_2CH\underset{COOCH_3}{|}\right]_z\right]_n$$

Synonyms: *Polyvinyl chloride, Ethylene-vinyl acetate copolymer filled. Poly(chloroethene-co-ethene-co-ethenyl acetate). VC/EVA. Vinyl chloride-ethylene-vinyl acetate terpolymer. PVC-EVA. Ethylene-vinyl acetate-vinyl chloride graft copolymer. Grafted VC/EVA copolymer. EVA/VC graft copolymer*
Related Polymers: Poly(vinyl chloride), unplasticised, Poly(vinyl chloride-*co*-vinyl acetate), Poly(vinyl chloride-*co*-ethylene), Poly(ethylene-*co*-vinyl acetate), PVC copolymers
Material class: Thermoplastic, Synthetic Elastomers, Copolymers, Composites
Polymer Type: polyethylene, PVCpoly(vinyl acetate)polyhaloolefins
CAS Number:

CAS Reg. No.	Note
25085-46-5	(terpolymer)
24937-78-8	
55599-81-0	

Molecular Formula: $[(C_2H_3Cl).(C_2H_4).(C_4H_6O_2)]_n$
Molecular Weight: MW 250000 (approx., main chain terpolymer).
K-value: 67 (high-impact graft copolymer, 88% vinyl chloride); 78–79 (high-impact graft copolymer, 50% vinyl chloride
Additives: Stabilisers (organotin compounds, Ba/Cd soaps, Ca/Zn compounds, Pb compounds); co-stabilisers (phosphites); lubricants; plasticisers; blowing-agents; pigments; fillers
General Information: Both main-chain co(ter)polymers and graft copolymers are available; for latter, most usual form is where vinyl chloride is grafted onto ethylene-vinyl acetate backbone, [18] although other forms have been described, e.g. grafting of ethylene to vinyl chloride-vinyl acetate backbone. [29] Graft copolymer displays interesting combination of physical props., i.e. plastic rigidity (PVC component) and toughness of a saturated rubber (ethylene-vinyl acetate rubber component); [27] finds most important application as an impact modifier, e.g. for PVC homopolymer mouldings; typical composition 27.5/22.5/50 (ethylene/vinyl acetate/vinyl chloride) with 50–70% EVA, can act as plasticiser or processing aid (e.g. lubricant) for (U)PVC homopolymer; vinyl chloride component promotes excellent compatibility with PVC resins; graft copolymer can act as solid plasticiser of exceptionally high permanence, in amounts as high as 80 phr; [5, 31] graft copolymers more expensive, in general, than main-chain copolymers [13]
Morphology: Electron microscopy studies of graft copolymers[21,27] with varying EVA content (5,20,50%) reveal that EVA phase forms network structure which separates PVC matrix into particles (approx. 0.5μ diameter; 5% EVA); mesh of network decreases and thread becomes thicker with increasing EVA content; 50% EVA sample can be best described as EVA matrix containing dispersed PVC particles, i.e. an EVA rubber 'filled' with PVC particles; such a structure helps to explain improved impact strength and flow properties (cf. rigid PVC homopolymer); network of EVA phase containing PVC particles undergoes phase inversion when subjected to high temp., pressure and shear; such phenomena generally applicable to most 'impact' grade PVC-2 phase systems; reduction in impact resistance observed after such treatment. [17] Weight fraction of grafted chains for solution-grafted copolymer (75% MeOH) is greater than that of

bulk-grafted (electron-irradiated) material for equivalent grafted monomer contents [24]
Identification: Pyrolysis-gas chromatography has been used for simple and rapid identification/analysis of terpolymer [14]
Miscellaneous: Anionic and non-ionic latex systems (52% solids), terpolymer emulsions (52% solids) and terpolymer dispersions (approx. 50% solids) are commercially available. [32] Properties and uses of graft copolymers have been reviewed [3,12]

Volumetric & Calorimetric Properties:
Density:

No.	Value	Note
1	d 1.13 g/cm^3	graft copolymer
2	d 1.38 g/cm^3	engineered terpolymer [32]
3	d 0.56–0.69 g/cm^3	bulk density, graft copolymer [32]
4	d 0.88 g/cm^3	8.8 lb gal^{-1}, terpolymer anionic latex system component [32]
5	d 0.91 g/cm^3	9.1 lb gal^{-1}, terpolymer emulsion system component [32]
6	d 1.17–1.18 g/cm^3	high-impact graft copolymer, 50% vinyl chloride [17]
7	d 1.4 g/cm^3	high-impact graft copolymer, 95% vinyl chloride [17]
8	d 1.35 g/cm^3	high-impact graft copolymer, 88% vinyl chloride [17]

Transition Temperature General: Variation of Vicat softening point (to VDE 302) as a function of admixture content (5–20% ethylene-vinyl acetate) has been reported [18]
Deflection Temperature:

No.	Value	Note
1	82°C	1 kg load, Vicat, DIN 53 460, 10 pbw, [4,31] 70°
2	70°C	5 kg load, Vicat, DIN 53460, 10 pbw [4,31]
3	85°C	1 kg load, Vicat, DIN 53460, 15 pbw [4,31]
4	69°C	5 kg load, Vicat, DIN 53460, 15 pbw [4,31]

Vicat Softening Point:

No.	Value	Note
1	81°C	min., Vicat/B, high-impact PVC formulation [32]
2	76–82°C	high-impact graft copolymer, 95% vinyl chloride [17]
3	76°C	min., high-impact graft copolymer, 94% vinyl chloride [17]

Brittleness Temperature:

No.	Value	Note
1	-55°C	high-impact graft copolymer, 50% vinyl chloride [17]

Surface Properties & Solubility:
Plasticisers: Phthalate esters, e.g. ethyl hexyl and di-n-butyl phthalates. VC/EVA graft copolymers (containing 50–70% EVA) themselves can act as plasticisers for (U)PVC compositions; excellent compatibility with PVC resins; low extractability, volatility and migration; true ('internal') plasticiser; high concentrations possible ($\leq$ 80 phr), and high degree of resistance to biodegradation [5,31]
Solvents/Non-solvents: Graft copolymers sol. cyclohexane

Transport Properties:
Transport Properties General: Flow properties of graft copolymer considerably improved (cf. rigid PVC homopolymer); can be understood by considering network struct. of rubber phase in plastic matrix, as revealed by electron microscopy studies; EVA phase (possibly partially contaminated with PVC) serves as lubricant at high temps., while acting as binder for PVC phases at room temp.; [27] high melt flow displayed by graft copolymer; [32] melt flow index of solution-grafted copolymer lower than that of bulk-grafted form [24]
Polymer Solutions Dilute: η_{rel} 2.26 (high-impact graft copolymer, 88% vinyl chloride, cyclohexanone (1%, 25°), K-value 67) [17]
Polymer Solutions Concentrated: Viscosity 50–12 000 mPa s (terpolymer dispersions, 50% solids); [32] 100–600 cps (self-crosslinking anionic terpolymer system, 52% solids); [32] 100–500 cps (anionic terpolymer latex system, 52% solids); [32] 100–500 cps, (terpolymer emulsion system, 52% solids) [32]
Polymer Melts: $\eta_{intrinsic}$ 153 cm^3 g^{-1} (high-impact graft copolymer, 50% vinyl chloride, K-value 79) [17]

Mechanical Properties:
Mechanical Properties General: Changes from a semirigid PVC to a soft PVC on increasing EVA content of graft copolymer; mechanical props. of latter form comparable to those of plasticised PVC. [27] Comparative study of mechanical properties of bulk (electron irradiation at 0° and 20°) and soln. (75% MeOH) grafted copolymers has been reported: yield strength increases with increasing graft content for both forms; yield strength for electron-irradiated graft copolymer prepared at 20° is greater than that of material produced at 0° for equivalent graft contents [24]. Mech. props. of a range of EVA/PVC systems have been reported [18]

Tensile (Young's) Modulus:

No.	Value	Note
1	2500 MPa	high-impact graft copolymer, 95% vinyl chloride [17]

Flexural Modulus:

No.	Value	Note
1	2400 MPa	high-impact graft copolymer, 95% vinyl chloride [17]
2	2500 MPa	high-impact graft copolymer, 94% vinyl chloride [17]

Tensile Strength Break:

No.	Value	Note	Elongation
1	12.75 MPa	1850 psi, engineered terpolymer [32]	250%
2	4.5 MPa	terpolymer dispersion material [32]	550–700%

Tensile Strength Yield:

No.	Value	Elongation	Note
1	45 MPa	<5% elongation	min., high-impact PVC formulation, 94% vinyl chloride [17,32]
2	14.5 MPa	250% elongation	high-impact graft copolymer, 50% vinyl chloride [17]
3	37 MPa	60% elongation	high-impact graft copolymer, 95% vinyl chloride [17]

– PVC, EVA filled

Flexural Strength Yield:

No.	Value	Note
1	80 MPa	high-impact graft copolymer, 95% vinyl chloride [17]

Impact Strength: Impact strength of graft copolymer is greatly improved cf. rigid PVC homopolymer, e.g. by factor of 10 for 5% EVA content. [27] The notched Charpy impact strength (DIN 53 453, 23° and -36°) of unplasticised PVC, containing EVA/VC copolymer as modifier increases with increasing modifier content. Comparative study of impact strengths of bulk (electron irradiation at 0° and 20°) and soln. (75% MeOH) grafted copolymers has been reported: reduction with increasing graft for bulk-grafted material, while increases with increasing content for solution-grafted product; in former case, impact strength for material prepared at 0° greater than that of material prepared at 20° for equivalent graft contents. [24] Notched impact tensile strength 170 kJ m^{-2} (high-impact graft copolymer, 94–95% vinyl chloride). [17] Maximum impact strength of graft copolymers is achieved with EVA rubbers having 45% vinyl acetate content.[18] Impact strengths 40 kJ m^{-2} (high-impact graft copolymer, 95% vinyl chloride); [17] 50 kJ m^{-2} (high-impact graft copolymer, 92% vinyl chloride); [17] higher concentrations of EVA rubber give no further improvement; [17] reduced to low levels (e.g. 10–15 kJ m^{-2}) when subjected to conditions of high temp., pressure and shear (e.g. in roll-compounding mill); related to phase inversion of structure [17]

Hardness: Shore A90 (10 s, engineered terpolymer) [32]; Shore D78 (high-impact PVC formulation, graft copolymers, 94–95% vinyl chloride) [17,32]

Friction Abrasion and Resistance: Engineered terpolymer has good abrasion props.; [32] good abrasion resistance displayed by terpolymer latex component [32]

Charpy:

No.	Value	Notch	Note
1	640	y	23°, 90% PVC, 10% EVA/VC copolymer [4,31]
2	820	y	23°, 85% PVC, 15% EVA/VC copolymer [4,31]
3	480	y	-36°, 90% PVC, 10% EVA/VC copolymer [4,31]
4	490	y	-36°, 85% PVC, 15% EVA/VC copolymer [4,31]
5	20	y	min., 23°, high-impact graft copolymer, 95% vinyl chloride [17]
6	8	y	min., 23°, high-impact graft copolymer, 94% vinyl chloride [17]
7	30	y	min., 23°, high-impact graft copolymer, 94% vinyl chloride [17]
8	5	y	min., 0°, high-impact graft copolymer, 94% vinyl chloride [17]

Electrical Properties:
Dielectric Permittivity (Constant):

No.	Value	Frequency	Note
1	3	1 MHz	high-impact graft copolymer, 95% vinyl chloride [17]

Dielectric Strength:

No.	Value	Note
1	43 kV.mm^{-1}	high-impact graft copolymer, 50% vinyl chloride [17]
2	28 kV.mm^{-1}	high-impact graft copolymer, 95% vinyl chloride [17]

Static Electrification: Thermodielectric loss measurements (using DTA) have been reported for graft copolymer; contour maps produced show similar shapes to those obtained for incompatible polymer mixtures (blends) (cf. two distinct polymer phases of graft copolymer) [26]

Dissipation (Power) Factor:

No.	Value	Frequency	Note
1	0.0185	1 MHz	high-impact graft copolymer, 95% vinyl chloride [17]

Optical Properties:
Transmission and Spectra: Wide-line nmr spectral data of graft copolymers have been reported: comparative studies with blends of PVC and EVA used to examine phase separation [28]

Polymer Stability:
Polymer Stability General: Study of effect of heat and light on degradation processes and structural changes has been reported: presence of pigments in addition to impact-modifier; surface examination by electron microscopy shows degradation by escaping pigment; high pigment concentrations lead to low weather resistance [23]

Thermal Stability General: Engineered terpolymer displays good resistance to high temps. [32]; graft copolymer (high-impact form, 88% vinyl chloride) displays good thermal stability at elevated temps. [17]. Good stability at normal extrusion temps. (100 min.), [19] compared to PVC. Other thermal stability studies have been reported [29]

Decomposition Details: Pyrolysis-gas chromatography studies show production of dimers and trimers on pyrolysis [14]. Raman spectroscopy has been used to evaluate polyene sequences in heat-treated graft copolymers (50% vinyl chloride content) over range 60–145°; engineered terpolymer shows low HCl generation on heating [25,32]

Flammability: Engineered terpolymer has low flammability and low smoke generation; [32] terpolymer (anionic) latex component exhibits flame retardancy props. [32] Limiting oxygen index 23.5% (high-impact graft copolymer, 50% vinyl chloride) [17]

Environmental Stress: Engineered terpolymer displays good weathering props. [30,31,32,34]

Chemical Stability: Terpolymer emulsion component shows good resistance to alkalis; [32] terpolymer displays greater resistance to alkalis than vinyl chloride-vinyl acetate copolymers, [34]

Hydrolytic Stability: Has good resistance to water [32]

Biological Stability: Has high degree of resistance to biodegradation [5]

Applications/Commercial Products:
Processing & Manufacturing Routes: Main-chain terpolymers synth. by emulsion or suspension polymerisation of mixture of comonomers; free-radical initiators used include peroxydicarbonates, [20] alkali peroxydisulfates, [33] organic hydroperoxides [33] and lauroyl peroxide; [28] typical emulsion polymerisation involves treating a mixture of vinyl chloride-vinyl acetate (1:9) at 68–70° with ethylene (17–20 atm., excess pressure) (vinyl chloride:ethylene 1:2–3:1) using $K_2S_2O_8$ as initiator. [8] Graft

– PVC, EVA filled

copolymers typically made by grafting PVC chains onto ethylene-*co*-vinyl acetate backbone by suspension polymerisation of EVA rubber dissolved in vinyl chloride monomer. [27,30] Graft copolymers usually suspension-grade materials; [18] grafting can be carried out in bulk (e.g. by electron irradiation at 0–20°) or in soln. (e.g. 75% MeOH); [24] typical synthesis involves grafting of 0.2–2 parts of vinyl chloride to 1 part of EVA copolymer (containing 15–75% ethylene) below 80°. [15] Some graft copolymers used immediately for thermoplastic processing, while others serve as mixing components for achieving certain props. in formulations, e.g. impact-modification, plasticisation, lubrication, etc.; [18,30] processed by injection moulding or extrusion [30]

Applications: Terpolymer emulsions (water-based dispersions/pressure-polymer emulsions) used as components of various adhesives, paints, textiles, paper and board coatings, wallpapers and binder formulations. Uses include adhesives for bonding plastics, foam, tiles, flooring, walls; binders for pigments, fillers and mortar, textured finishes, thermal insulation systems, fabrics (non-woven, fire-resistant), glass fibre, textile auxiliaries, and wipes, cover stock and medical surgical applications; house and masonry emulsion paints; surface coatings and finishing agents (paper and paperboard), foam-forming (fire-resistant) applications for use as electrical insulators; high strength cements with improved waterproofing props., chemical resistance and mechanical strength; shoe-sole formulations and other leather substitutes (often in blends with PVC); flexible foams, with improved processability and mechanical strength. Graft copolymer (extruded material) has major application as profiles, either on its own or in compositions (blends), with UPVC in window frames (≤ 10% grafted polymer); impact-modifier, offering good combination of impact resistance and tensile strength, with good weatherability (at low temps.): extruded or injection moulded products used for rigid pipes (drainpipes), roof guttering, profiles for benches and fences, light panels, road-sign posts, panels for facade cladding, ventilation systems, chemical plant, waste, gas and drainage pipes and cable conduit, thermoforming sheet, fittings, covers, bottles and containers. Graft copolymers containing high concentrations of EVA behave as internally (permanently) plasticised materials; used for plasticiser-free sheet roofing or artificial leather; also used as processing aids for PVC, as lubricants

Trade name	Details	Supplier
Airflex 120	anionic; self-cross-linking	Air Products
Airflex 456	anionic; terpolymer latex	Air Products
Airflex 728	terpolymer emulsion	Air Products
Baymod L		Bayer Inc.
Denkovinyl	non-exclusive	Denki Kagaku Kogyo Kabushiki Kaisha
Elvaloy	741, 742; non-exclusive	DuPont
Ethyl	non-exclusive	Ethyl Corporation
Graftmer	GR5, R, R3, R5	Nippon Zeon
Hostalit	non-exclusive	Hoechst
Krene	non-exclusive	Union Carbide
Laroflex	non-exclusive	BASF
Levapren	VC 45/50; non-exclusive	Bayer Inc.
Lutofan	non-exclusive	BASF
Organil		Solvay
Pantalast	L; 50% vinyl chloride graft 60% EVA; W; 12% EVA graft vinyl chloride	Pantasote Inc.
Povikon	A6710; non-exclusive	Kema-Nord AB
Ryuron - Graft	H110, H120, H130	Tekkosha
Solvic	non-exclusive	Solvay
Trosiplast	non-exclusive	Dynamit Nobel
Varlan	non-exclusive	DSM
Vestolit	BAU; 'high-impact' form; HIS; graft copolymer	Hüls AG
Vinamul	R34075, R34078, R34083, 3400, 3401, 3405; non-exclusive	Vinyl Products
Vinidur	non-exclusive	BASF
Vinnapas Dispersion - CE	216, F10; dispersion containing protective colloid	Wacker-Chemie
Vinnol	K, K510/68, VK801, K550/79, K505/68; CE-35 (terpolymer dispersion); non-exclusive	Wacker-Chemie

Bibliographic References

[1] Bier, G., *Kunststoffe,* 1965, **55**, 694, (rev)
[2] Göbel, W., Bartl, H., Hardt, D. and Reischl, A., *Kunststoffe,* 1965, **55**, 329, (processing)
[3] Edser, M.H. and Bulezuik, B.W., *Polym. Paint Colour J.,* 1974, **164**, 1051, 1056, 1090-1, 1093-4, (uses, rev)
[4] Sahajpal, V.K., *Kunststoffe,* 1976, **66**, 18, (processing, uses, thermal props, mechanical props)
[5] Titow, W.V., *Developments in PVC Production and Processing - 1,* (eds. A. Whelan and J.L. Craft), Applied Science, 1977, 64, (uses, stability)
[6] *Ger. Pat.,* 1972, 2 107 287, (uses)
[7] *Jpn. Pat.,* 1974, 74 329 32, (uses)
[8] *U.S. Pat.,* 1974, 3 816 363, (uses, synth)
[9] *Jpn. Pat.,* 1973, 73 050 11, (uses)
[10] *East Ger. Pat.,* 1974, 110 283, (uses)
[11] *Jpn. Pat.,* 1974, 74 98 830, (uses)
[12] Waki, I., *Jpn. Plast. Age,* 1971, **9**, 29, (rev)
[13] *Jpn. Pat.,* 1976, 76 86 569, (uses)
[14] Okumoto, T. and Takeuchi, T., *Nippon Kagaku Kaishi,* 1972, **71**, (identification, analysis, thermal stability/decomposition)
[15] *Jpn. Pat.,* 1972, 72 385 50, (synthesis, uses)
[16] Brydson, J.A., *Plast. Mater.,* 6th edn., Butterworth-Heinemann, 1995, 346-7, (uses, synth)
[17] Elias, H.-G. and Vohwinkel, F., *New Commercial Polymers,* 2nd edn., Gordon and Breach, 1986, 58, (volumetric props, mechanical props, stability, electrical props, thermal props, transport props, molecular weight)
[18] Domininghaus, H., Plastics for Engineers: Materials, Properties, Applications, Hanser, 1993, 173, (synth, processing, uses, thermal props)
[19] Sieglaff, C.L. and Harris, R.L., *SPE J.,* 1972, **28(6)**, 48, (thermal stability, transport props, processing)
[20] *Czech. Pat.,* 1974, **152**, 770, (synth)
[21] Matsuo, M. and Sagaye, S., *Colloidal Morphology Behaviour of Block and Graft Copolymers: Proceedings of American Chemical Society Symposium,* (ed. G.E. Molau), Plenum, 1971, **1**, (morphology)
[22] *Ger. Pat.,* 1972, 221 454 6, (uses)
[23] Menzel, G., Nat. Kuenstliche Alterung Kunstst., Donaulaendergespraech 7th, Fac. Verlag, 1975, 167, (stability)
[24] Sasuga, T., Shimizu, Y., Sasaki, T., Tamaki, H. and Araki, K., *Kobunshi Kagaku,* 1973, **30**, 761, (synth, mechanical props, structure)
[25] Peitscher, G. and Holtrup, W., *Angew. Makromol. Chem.,* 1975, **47**, 111, (thermal stability/decomposition; Raman)
[26] Akiyama, S., Komatsu, Y. and Kaneko, R., *Polym. J. (Tokyo),* 1975, **7**, 172, (electrical props)
[27] Matsuo, M. and Sagaye, S., Polym. Prepr. (Am. Chem. Soc.), Div. Polym. Chem.), 1970, **11**, 384, (transport props, mechanical props, morphology, synth)
[28] Elmqvist, C. and Svanson, S.E., *Eur. Polym. J.,* 1975, **11**, 789, (nmr, synth)
[29] Sasuga, T., Araki, K. and Kuriyama, I., *Kobunshi Ronbunshu,* 1975, **32**, 83, (thermal anal, morphology, synth)

[30] Titow, W.V., *PVC Plastics: Properties, Processing, and Applications*, Elsevier Applied Science, 1990, 77, 477, (synthesis, processing, uses)
[31] Titow, W.V., *PVC Technol.*, 4th edn, Elsevier Applied Science, 1984, 20, 367, 394, (uses, mechanical props, thermal props, stability)
[32] Ash, M. and Ash, I., Handbook of Plastic Compounds, *Elastomers and Resins*, VCH, 1992, 47, 52, 57, 113, 529, (stability, mech props, thermal props, volumetric props, electrical props, uses, transport props)
[33] Farmer, D.B. and Edser, M.H., *Chem. Ind. (London)*, 1983, 228, (uses, processing, synth)
[34] *Polymeric Materials Encyclopedia*, (ed. J.C. Salamone), CRC Press, 1996, 11, 8570-71, (uses, mechanical props)

PVDF, Carbon fibre filled   P-664

Synonyms: *Polyvinylidene fluoride, Carbon fibre filled*
Related Polymers: PVDF, Mineral fibre filled, PVDF, Mica filled
Monomers: Vinylidene fluoride
Material class: Thermoplastic, Composites
Polymer Type: PVDF, fluorocarbon (polymers)

Volumetric & Calorimetric Properties:
Density:

No.	Value	Note
1	d 1.76 g/cm^3	15% filled [1]
2	d 1.78 g/cm^3	17% filled [2]
3	d 1.76 g/cm^3	20% filled [3]

Thermal Expansion Coefficient:

No.	Value	Note	Type
1	3.6×10^{-5} K^{-1}	17% filled [2]	L
2	6×10^{-5} K^{-1}	20% filled [3]	L

Thermal Conductivity:

No.	Value	Note
1	0.22 W/mK	0.000524 (cm s°C)$^{-1}$, 17% filled [2]

Deflection Temperature:

No.	Value	Note
1	134°C	1.82 MPa, 17% filled [2]
2	150°C	0.45 MPa, 20% filled [3]
3	140°C	1.82 MPa, 20% filled [3]

Transport Properties:
Water Absorption:

No.	Value	Note
1	0.05 %	24h, 17% filled [2]
2	0.04 %	24h, 20% filled [3]

Mechanical Properties:
Tensile (Young's) Modulus:

No.	Value	Note
1	6002 MPa	61200 kg cm^{-2}, 17% filled [2]

Flexural Modulus:

No.	Value	Note
1	7855 MPa	80100 kg cm^{-2}, 15% filled [1]
2	6002 MPa	61200 kg cm^{-2}, 17% filled [2]
3	5500 MPa	20% filled [3]

Tensile Strength Break:

No.	Value	Note	Elongation
1	931 MPa	9490 kg cm^{-2}, 17% filled [2]	1%

Tensile Strength Yield:

No.	Value	Elongation	Note
1	90.9 MPa	2%	927 kg cm^{-2}, 15% filled [1]
2	93 MPa		949 kg cm^{-2}, 17% filled [2]
3	100 MPa	6%	room temp., 20% filled [3]

Flexural Strength Yield:

No.	Value	Note
1	122.5 MPa	1250 kg cm^{-2}, 15% filled [1]
2	171 MPa	1740 kg cm^{-2}, 17% filled [2]

Compressive Strength:

No.	Value	Note
1	96 MPa	977 kg cm^{-2}, 17% filled [2]

Hardness: Shore D82 (17% filled) [2], D90 (20% filled) [3]
Izod Notch:

No.	Value	Notch	Note
1	53.3 J/m	Y	1 ft lb in^{-1}, room temp., 15% filled [1]
2	120 J/m	Y	20% filled [3]

Electrical Properties:
Electrical Properties General: Thermocomp FP-VC-1003 passes NFPA code 56A for static decay [1]. Volume resistivity 3000 Ω cm(15% filled) [1], 1000 Ω m (20% filled) [3]
Dielectric Permittivity (Constant):

No.	Value	Frequency	Note
1	6.1	1 MHz	15% filled [1]

Dielectric Strength:

No.	Value	Note
1	10.2 kV.mm^{-1}	15% filled [1]

Dissipation (Power) Factor:

No.	Value	Frequency	Note
1	0.159	1 MHz	15% filled [1]

PVDF, Filled

Polymer Stability:
Polymer Stability General: Has excellent resistance to ageing [2]
Upper Use Temperature:

No.	Value	Note
1	130°C	continuous service, 20% filled [3]

Flammability: Has good fire resistance [2]
Environmental Stress: Has excellent uv stability [2]

Applications/Commercial Products:
Processing & Manufacturing Routes: Processed by extrusion or injection moulding
Mould Shrinkage (%):

No.	Value	Note
1	0.7%	20% filled [3]

Applications: Potential applications in the chemical, nuclear and pharmaceutical industries

Trade name	Details	Supplier
Dyflor		Huls America
Fluoromelt FP-VC-1003		LNP Engineering
Grafil PVDF		Grafil
Solef 8808	discontinued material	Solvay Polymers
Stat-Kon FP-VC-1003		LNP Engineering
Thermocomp FP-VC-1003	15% filled; discontinued material	LNP Engineering

Bibliographic References

[1] 1996, LNP Engineering, (technical datasheet)
[2] 1996, Solvay Polymers, (technical datasheet)
[3] *The Materials Selector*, 2nd edn., (eds. N.A. Waterman and M.F. Ashby), Chapman & Hall, 1997, **3**, 232

PVDF, Filled

Synonyms: *Polyvinylidene fluoride, Filled*
Related Polymers: PVDF, Carbon fibre filled PVDF, Mineral fibre filled PVDF, Mica filled PVDF, PTFE and carbon-filled PVDF, Graphite filled PVDF
Monomers: Vinylidene fluoride
Material class: Composites
Polymer Type: PVDF, fluorocarbon (polymers)

Applications/Commercial Products:

Trade name	Details	Supplier
Electrafil	J-1405/CF/30	DSM Engineering
Foraflon	various grades	Atochem Polymers
Hylar	320/460/700 series	Ausimont USA Inc.
Kureha	KF	Kreha Corp. of America
Kynar	various grades	Atochem Polymers
Poly-Hi PVDF		Poly-Hi
Solef	various grades	Solvay Polymers
Thermocomp	FP series	LNP Engineering

PVDF, Graphite filled

Synonyms: *Polyvinylidene fluoride, Graphite filled*
Monomers: Vinylidene fluoride
Material class: Composites
Polymer Type: PVDF
Morphology: Semi-crystalline

Volumetric & Calorimetric Properties:
Density:

No.	Value	Note
1	d 1.85 g/cm^3	Kynar 320 [1]
2	d 1.86 g/cm^3	Kynar 370 [1]

Thermal Expansion Coefficient:

No.	Value	Note	Type
1	0.00013 K^{-1}	Kynar 320 [1]	L
2	3.96×10^{-5} K^{-1}	Kynar 370 [1]	L

Melting Temperature:

No.	Value	Note
1	168°C	Kynar 320, 370

Deflection Temperature:

No.	Value	Note
1	48°C	1.82 MPa, Kynar 320 [1]
2	118°C	1.82 MPa, Kynar 370 [1]

Transport Properties:
Water Absorption:

No.	Value	Note
1	0.05 %	24h, Kynar 370 [1]

Mechanical Properties:
Tensile (Young's) Modulus:

No.	Value	Note
1	4138 MPa	42200 kg cm^{-2}, Kynar 370 [1]

Flexural Modulus:

No.	Value	Note
1	6208 MPa	63300 kg cm^{-2}, Kynar 370 [1]

Tensile Strength Break:

No.	Value	Note	Elongation
1	45 MPa	459 kg cm^{-2}, Kynar 370 [1]	32%
2	46 MPa	475 kg cm^{-2}, Kynar 370 [1]	12%

Compressive Strength:

No.	Value	Note
1	57 MPa	586 kg cm^{-2}, Kynar 370 [1]

Hardness: Shore D77 (Kynar 320, 370) [1]
Izod Notch:

No.	Value	Notch	Note
1	4.3 J/m	Y	0.08 ft lb in^{-1}, room temp., Kynar 320 [1]
2	53 J/m	Y	1 ft lb in^{-1}, room temp., Kynar 370 [1]

Electrical Properties:
Dielectric Permittivity (Constant):

No.	Value	Frequency	Note
1	0.047	1 MHz	Kynar 320 [1]
2	0.078	1 MHz	Kynar 370 [1]

Dissipation (Power) Factor:

No.	Value	Frequency	Note
1	0.21	1 MHz	Kynar 320 [1]
2	28.8	1 MHz	Kynar 370 [1]

Applications/Commercial Products:
Mould Shrinkage (%):

No.	Value	Note
1	1%	Kynar 320, 370 [1]

Trade name	Details	Supplier
Kynar	320, 370	Atochem Polymers

Bibliographic References
[1] *Kynar Range*, Atochem Polymers, 1996, (technical datasheet)

PVDF, Mica filled   P-667
Synonyms: *Polyvinylidene fluoride, Mica filled*
Related Polymers: Filled PVDF, PVDF, carbon fibre filled, PVDF, mineral fibre filled
Monomers: Vinylidene fluoride
Material class: Composites, Thermoplastic
Polymer Type: PVDF, fluorocarbon (polymers)
General Information: Has excellent chemical and ageing resistance and excellent mechanical and self-extinguishing props.

Volumetric & Calorimetric Properties:
Density:

No.	Value	Note
1	d 1.84 g/cm^3	[1]

Transport Properties:
Water Absorption:

No.	Value	Note
1	0.04 %	24h [1]

Mechanical Properties:
Tensile (Young's) Modulus:

No.	Value	Note
1	4197 MPa	42800 kg cm$_{-2}$ [1]

Flexural Modulus:

No.	Value	Note
1	4687 MPa	47800 kg cm$_{-2}$ [1]

Tensile Strength Break:

No.	Value	Note	Elongation
1	46.8 MPa	478 kg cm^{-2} [1]	6%

Tensile Strength Yield:

No.	Value	Note
1	48.9 MPa	499 kg cm^{-2} [1]

Flexural Strength Yield:

No.	Value	Note
1	80.6 MPa	822 kg cm^{-2} [1]

Hardness: Shore D70 [1]

Applications/Commercial Products:
Processing & Manufacturing Routes: Processed by injection moulding

Trade name	Details	Supplier
Solef 8908	discontinued material	Solvay Polymers

Bibliographic References
[1] 1996, Solvay Polymers, (technical datasheet)

PVDF, Mineral fibre filled   P-668
Synonyms: *Polyvinylidene fluoride, Mineral fibre filled*
Related Polymers: PVDF, Carbon fibre filled
Monomers: Vinylidene fluoride
Material class: Thermoplastic, Composites
Polymer Type: PVDF, fluorocarbon (polymers)

Volumetric & Calorimetric Properties:
Density:

No.	Value	Note
1	d 1.9–1.93 g/cm^3	25% filled [1]
2	d 1.88 g/cm^3	15% filled [1]

Thermal Conductivity:

No.	Value	Note
1	0.36 W/mK	0.000862 cal (s cm°C)$^{-1}$, 25% filled [1]
2	0.35 W/mK	0.000828 cal (s cm°C)$^{-1}$, 15% filled [1]

Melting Temperature:

No.	Value	Note
1	160°C	15–25% filled [1]

Deflection Temperature:

No.	Value	Note
1	121–146°C	1.82 MPa, 25% filled [1]
2	110°C	1.82 MPa, 15% filled [1]

– PVDF, PTFE and Carbon filled

Mechanical Properties:
Flexural Modulus:

No.	Value	Note
1	2569–3717 MPa	26200–37900 kg cm^{-2}, 25% filled [1]
2	5374 MPa	54800 kg cm^{-2}, 15% filled [1]

Tensile Strength Break:

No.	Value	Note
1	38–46 MPa	386–468 kg cm^{-2}, 25% filled, 8–14% elongation [1]
2	43 MPa	435 kg cm^{-2}, 15% filled, 5% elongation [1]

Flexural Strength Yield:

No.	Value	Note
1	59–65 MPa	604–660 kg cm^{-2}, 25% filled [1]
2	62 MPa	632 kg cm^{-2}, 15% filled [1]

Izod Notch:

No.	Value	Notch	Note
1	48 J/m	Y	0.9 ft lb in^{-1}, room temp., 25% filled [1]
2	53 J/m	Y	1 ft lb in^{-1}, room temp., 15% filled [1]

Electrical Properties:
Dielectric Permittivity (Constant):

No.	Value	Frequency	Note
1	6.5	1 MHz	25% filled [1]
2	7	1 MHz	15% filled [1]

Dielectric Strength:

No.	Value	Note
1	7.08–7.87 kV.mm^{-1}	25% filled [1]
2	8.66 kV.mm^{-1}	15% filled [1]

Dissipation (Power) Factor:

No.	Value	Frequency	Note
1	0.175	1 MHz	25% filled [1]
2	0.184	1 MHz	15% filled [1]

Polymer Stability:
Upper Use Temperature:

No.	Value	Note
1	121°C	continuous service, 15–25% filled [1]

Applications/Commercial Products:
Processing & Manufacturing Routes: Processed by extrusion or injection moulding

Mould Shrinkage (%):

No.	Value	Note
1	0.2–2%	25% filled [1]
2	2.2%	15% filled [1]

Applications: In electronic parts

Trade name	Details	Supplier
Fluoromelt FP-VM-2550		LNP Engineering
Fluoromelt FP-VM-3830		LNP Engineering
Fluoromelt FP-VM-3850		LNP Engineering
Thermocomp FP-VM-2550	25% filled; discontinued material	LNP Engineering
Thermocomp FP-VM-3830	15% filled; discontinued material	LNP Engineering
Thermocomp FP-VM-3850	25% filled; discontinued material	LNP Engineering

Bibliographic References
[1] *Thermocomp Range*, LNP Engineering, 1996, (technical datasheet)

PVDF, PTFE and Carbon filled P-669
Synonyms: *Polyvinylidene fluoride, Polytetrafluoroethylene and carbon filled*
Monomers: Tetrafluoroethylene, Vinylidene fluoride
Material class: Composites
Polymer Type: PVDF, PTFEfluorocarbon (polymers)

Volumetric & Calorimetric Properties:
Density:

No.	Value	Note
1	d 1.81 g/cm^3	10% PTFE, 20% carbon fibre [1]

Thermal Conductivity:

No.	Value	Note
1	0.35 W/mK	0.000828 cal (cm s °C)$^{-1}$, 10% PTFE, 20% carbon fibre [1]

Melting Temperature:

No.	Value	Note
1	160°C	10% PTFE, 20% carbon fibre [1]

Deflection Temperature:

No.	Value	Note
1	173°C	1.82 MPa, 10% PTFE, 20% carbon fibre [1]

Mechanical Properties:
Flexural Modulus:

No.	Value	Note
1	11376 MPa	116000 kg cm^{-2}, 10% PTFE, 20% carbon fibre [1]

– PVDF, PTFE and Carbon filled

Tensile Strength Break:

No.	Value	Note
1	65 MPa	667 kg cm^{-2}, 10% PTFE, 20% carbon fibre [1]

Flexural Strength Yield:

No.	Value	Note
1	73 MPa	745 kg cm^{-2}, 10% PTFE, 20% carbon fibre [1]

Izod Notch:

No.	Value	Notch	Note
1	59 J/m	Y	1.1 ft lb in^{-1}, room temp., 10% PTFE, 20% carbon fibre [1]

Electrical Properties:
Electrical Properties General: Volume resistivity 1000 Ω cm (10% PTFE, 20% carbon fibre) [1]

Polymer Stability:
Upper Use Temperature:

No.	Value	Note
1	121°C	continuous service, 10% PTFE, 20% carbon fibre [1]

Applications/Commercial Products:
Processing & Manufacturing Routes: Processed by extrusion or injection moulding
Mould Shrinkage (%):

No.	Value	Note
1	0.4%	10% PTFE, 20% carbon fibre [1]

Applications: In electronic parts

Trade name	Details	Supplier
Fluoromelt FP-VCL-4024		LNP Engineering
Thermocomp FP-VCL-4024	discontinued material	LNP Engineering

Bibliographic References
[1] 1996, LNP Engineering, (technical datasheet)

Resorcinol formaldehyde resins R-1

Synonyms: *Resorcinol resins. 1,3-Benzenediol resins. RF resins*
Related Polymers: Phenolic Resins
Monomers: Resorcinol, Formaldehyde
Material class: Thermosetting resin
Polymer Type: phenolic
Molecular Weight: MW 400 (uncured)
General Information: In resorcinol OH groups activate the 2,4 and 6 ring positions. The effect of this activation is that the resin system will cure at much lower temps. than for phenol formaldehyde resins. Resorcinol is a relatively high-cost phenol and hence is only used in high-value applications where its cold setting props. are significant. Resorcinol can be co-condensed with phenol to form a range of resorcinol phenol formaldehyde (RPF) resins which are more reactive than simple PF resins. This is significant when designing cold-curing adhesives, and is the usual commercial practice

Volumetric & Calorimetric Properties:
Density:

No.	Value	Note
1	d1.364 g/cm^3	uncured [2]

Melting Temperature:

No.	Value	Note
1	70–105°C	uncured, dependent on synth. [1]

Vicat Softening Point:

No.	Value	Note
1	89–125°C	uncured, dependent on synth. [1]

Surface Properties & Solubility:
Cohesive Energy Density Solubility Parameters: δ 6.68 J$^{1/2}$ cm$^{-3/2}$ (10.66 cal$^{1/2}$ cm$^{-3/2}$, uncured) [2]
Solvents/Non-solvents: Uncured sol. Me$_2$CO, 2-butanone, EtOAc, EtOH; insol. petroleum, C$_6$H$_6$, turpentine, toluene, CCl$_4$, CHCl$_3$ [1]

Transport Properties:
Polymer Solutions Dilute: Viscosity of resins for laminated wood beams 60–160 mPa s (20°, 53–54% solids RF30); 3000–8000 mPa s (20°, 58–59%, solids, WATEX-1410, 1400)

Mechanical Properties:
Miscellaneous Moduli:

No.	Value	Note	Type
1	1000 MPa	10^{10} dyne cm^{-2}, 20–150°, cured [5]	Storage modulus

Optical Properties:
Transmission and Spectra: C-13 nmr spectral data have been reported [3,4]
Volume Properties/Surface Properties: Colour 7–10 (Gardner scale, uncured, dependent on synth.) [1]

Polymer Stability:
Chemical Stability: Uncured resin is incompatible with linseed oil, tung oil and castor oil. By altering conditions of synth., uncured resin can be compatible with double-boiled linseed oil [1]

Applications/Commercial Products:
Processing & Manufacturing Routes: Synth. in a manner analogous to that of phenol-formaldehyde resins i.e. reaction of resorcinol with aq. formaldehyde in the presence of an acid (e.g. oxalic acid) catalyst. The reaction rate of resorcinol with formaldehyde is considerably higher than that of phenol. Phenolic resins prepared with some resorcinol as comonomer have greater reactivity. Resorcinol formaldehyde adhesives are curable at room temp. and at neutral pH. They require the addition of a curing agent and solid paraformaldehyde is often used. Resorcinol formaldehyde adhesives confer greater moisture resistance in outdoor structural applications
Applications: Structural wood bonding methods have been used to construct loadbearing structs. for both indoor and outdoor applications, e.g. boat building. Resorcinol or resorcinol-formaldehyde prepolymers can be used as accelerating compounds for curing phenolic resins. The addition of 3–10% permits shorter curing cycles in particle board and grinding wheel production. Resorcinol formaldehyde resins are curable under neutral conditions enabling the adhesives to be used with acid-sensitive systems such as fibre-reduced cement. Shaped refractory products can be made using resorcinol resins by a cold-mixing technique. Resorcinol formaldehyde resoles are used as tyre-cord adhesives and as high performance binders for wood composites

Trade name	Details	Supplier
Aerodux	resorcinol resins	Dyno
Dynosol		Dyno
Lauxite	PP-1597, wheatflour-filled	Huntsman Chemical
Rütaphen HL	cold bonding adhesives	Bakelite
RF-10	RPF resin for laminated wood	Neste
RF-30L	RPF resin for laminated wood	Neste

Bibliographic References
[1] Ahisanuddin, Saksena, S.C., Panda, H. and Rakhshinda, *Paintindia*, 1982, **32**, 3, (synth., solubility)
[2] Shvarts, A.G., Chefranova, E.K. and Iotkovskaya, L.A., *Kolloidn. Zh.*, 1970, **32**, 603, (solubility parameter)
[3] Kim, M.G., Ames, L.W. and Barnes, E.E., *J. Polym. Sci., Part A: Polym. Chem.*, 1993, **31**, 1871, (C-13 nmr)
[4] De Breet, A.J.J., Dankelman, W., Huysmans, W.G.B. and De Wit, J., *Angew. Makromol. Chem.*, 1977, **62**, 7, (C-13 nmr)
[5] Yamada, M., Taki, K., *Mokuzai Gakkaishi*, 1991, **37**, 529

Resorcinol resins, Wood applications R-2

Related Polymers: Resorcinol Formaldehyde Resin
Monomers: Resorcinol, Formaldehyde
Material class: Thermosetting resin
Polymer Type: phenolic
General Information: In resorcinol -OH groups activate the 2,4 and 6 ring positions. The effect of this activation is that the resin system will cure at much lower temps. than for phenol formaldehyde resins

Applications/Commercial Products:
Processing & Manufacturing Routes: Bonded wood beams (88%) are used with resin (2%) and 10% moisture. Resorcinol resin adhesives with curing agents (paraformaldehyde) are curable at ambient temps. (>15°) and are used where enhanced moisture resistance is required
Applications: Structural wood bonding methods have been used to construct loadbearing structs. for both indoor and outdoor applications, e.g. boat building

— Rubber hydrochloride

Trade name	Details	Supplier
Aerodux	resorcinol resins	Dyno
Dynosol	resorcinol resins	Dyno
RF-10	RPF resin for laminated wood	Neste
RF-30L	RPF resin for laminated wood	Neste

Rubber hydrochloride
Related Polymers: Natural rubber
Monomers: 2-Methyl-1,3-butadiene
Material class: Synthetic Elastomers
Polymer Type: polybutadiene
Additives: Plasticisers: dibutylphthalate, tritolyl phosphate
General Information: Hydrohalogenated rubber

SAN   S-1

$$\left[\left[CH_2CH(Ph)\right]_x \left[CH_2CH(CN)\right]_y\right]_n$$

Synonyms: *Styrene-acrylonitrile copolymer. Poly(styrene-co-acrylonitrile) copolymer. ANS. Poly(2-propenenitrile-co-ethenylbenzene)*
Related Polymers: SAN, general purpose, SAN, injection moulding unfilled, SAN, injection moulding glass fibre filled, SAN, extrusion unfilled, SAN, glass reinforced, SAN, flame retardant injection moulding glass fibre filled, SAN, blow moulding, Polyacrylonitrile, Polystyrene
Monomers: Styrene, Acrylonitrile
Material class: Thermoplastic, Copolymers
Polymer Type: styrene-acrylonitrile (SAN), acrylonitrile and copolymers
CAS Number:

CAS Reg. No.	Note
9003-54-7	
106209-33-0	alternating
107440-17-5	block copolymer
110902-08-4	graft copolymer

Molecular Formula: $(C_8H_8)_x(C_3H_3N)_y$
Fragments: C_8H_8 C_3H_3N
General Information: SAN resins are rigid, hard and transparent. A typical composition is 76% styrene and 24% acrylonitrile [14]

Volumetric & Calorimetric Properties:
Density:

No.	Value	Note
1	d 1.07–1.08 g/cm^3	[14]

Equation of State: Equation of state information [35,36] for the alternating copolymer [37] has been reported
Thermodynamic Properties General: Heat capacity as a function of temp. (10–240°) has been reported [39]. Heat of combustion has been reported [40]
Glass-Transition Temperature:

No.	Value	Note
1	127°C	[30]
2	104°C	amorph. [38]

Transition Temperature General: T_g varies non-linearly with acrylonitrile content, with a maxima at 50% wt. acrylonitrile [10]
Deflection Temperature:

No.	Value	Note
1	72°C	5.5% wt. acrylonitrile [10]
2	88°C	27% wt. acrylonitrile [10]
3	101–104°C	1.8 MPa, 264 psi [14]

Surface Properties & Solubility:
Solubility Properties General: Miscible with poly(*N*-vinyl-2-pyrrolidone), poly (*N*-methyl-*N*-vinylacetamide) and poly(*N,N*-dimethylacrylamide) provided the acrylonitrile content of SAN exceeds 62% [1]. Miscible with bisphenol A carbonate cyclic oligomers [2]

Cohesive Energy Density Solubility Parameters: Solubility parameters have been reported [22,23]
Solvents/Non-solvents: Softened and swollen by toluene. Unaffected by immersion in 2-propanol, petrol, motor oil, hexane, detergent or chlorine bleach [47]

Transport Properties:
Melt Flow Index:

No.	Value	Note
1	1.6 g/10 min	10 min, 220°, ASTM D1238 [11]

Polymer Solutions Dilute: η 0.68 (25°, C_6H_6) (alternating; styrene:acrylonitrile 1:2) [3]; η 0.75 dl g^{-1} (30°, 2-butanone, 26% wt. acrylonitrile) [5]. Dilute soln. props. have been reported [17,27]
Polymer Melts: Relationship between limiting viscosity and MW, and calculation of the Flory constant, K_θ, have been reported [6]. Limiting viscosity number decreases with increasing temp. [7]. The viscosity of polymer melts at high shear rate has been reported [11]
Permeability of Gases: Permeability to oxygen [19] and water vapour [19,20] has been reported
Water Content: Water content data have been reported [29]

Mechanical Properties:
Mechanical Properties General: Mechanical props. are dependent on the acrylonitrile content

Tensile (Young's) Modulus:

No.	Value	Note
1	0.0033–0.0038 MPa	0.475–0.56 psi [14]

Tensile Strength Break:

No.	Value	Note	Elongation
1	68.9–75.8 MPa	10–11 kpsi [14]	2–3%
2	42.27 MPa	5.5% wt. acrylonitrile [10]	1.6%
3	72.47 MPa	27% wt. acrylonitrile [10]	3.2%

Tensile Strength Yield:

No.	Value	Elongation	Note
1	39.2 MPa	1.8%	400 kg cm^{-2}, 40° [4,6]

Miscellaneous Moduli:

No.	Value	Note	Type
1	2059.4 MPa	21000 kg cm^{-2}, 23°, 10000h [46]	creep modulus

Impact Strength: Impact strength 26.6 J m^{-1} (notched, 5.5% wt. acrylonitrile), 27.1 J m^{-1} (27% wt. acrylonitrile) [10]; 13.7 J m^{-1} (0.4–0.6 ft lb in^{-1}). Charpy impact strength behaviour has been reported [18].
Viscoelastic Behaviour: The dynamic viscoelastic behaviour of SAN random copolymers has been reported [15]. Dynamic viscoelastic props. are dependent on the method of synth. [30]
Mechanical Properties Miscellaneous: Critical shear rate for injection moulding has been reported [28]
Fracture Mechanical Properties: SAN resins with acrylonitrile content of 25, 30 and 51.2% have greater fatigue fracture resistance than unmodified polystyrene. Fatigue resistance of resins increases with increasing acrylonitrile content up to a max. of 30% [13]

Optical Properties:

Optical Properties General: 30% Styrene content in the polymer film leads to a deterioration in optical props. [41]. The relationship between refractive index and struct. has been reported [45]

Transmission and Spectra: Ir [42,43] and C-13 nmr [44] spectral data have been reported

Polymer Stability:

Thermal Stability General: Deteriorates with increasing acrylonitrile content [9]

Decomposition Details: Initial temp. of weight loss (alternating copolymer) 323° [3]. Activation energy of decomposition 142 kJ mol^{-1} [3]. Max. rate of decomposition occurs at 425° with NH_3 and HCN among the major products [8]. Weight loss is smaller when heating occurs in a N_2 rather than air environment, with the least amount of degradation occurring in the presence of 0.15–0.5% moisture [12]. Pyrolysis at 500° under N_2 produces HCN, acetonitrile, acrylonitrile, styrene and toluene among the products. The amount of HCN is proportional to the acrylonitrile content [34]

Flammability: Ignition temp. 542° [4]. Ignition limiting oxygen indices 16.9 (550°), 5.4 (600°), 4.5 (650°) [4]

Environmental Stress: Outdoor exposure reduces impact strength by 75% and bending strength by 50% after 1 year [21]. Other weathering data have been reported [25]

Chemical Stability: SAN resins are resistant to non-oxidising acids, aliphatic hydrocarbons, alkalis, except KOH, vegetable oils, foodstuffs and detergents. Attacked by aromatic and chlorinated hydrocarbons, esters and ketones [14]

Biological Stability: Biodegradation details have been reported [26]

Stability Miscellaneous: γ-Irradiation causes embrittlement and degradation of phys. props. [24]

Applications/Commercial Products:

Processing & Manufacturing Routes: Synth. comly. by emulsion, suspension or continuous bulk polymerisation [9,14]. The latter method is preferred for applications requiring high optical quality and gives better colour and haze [9]. Bulk polymerisation in toluene or acetonitrile can show a penultimate unit effect [31]. May also be synth. by free radical polymerisation using a $ZnCl_2$ catalyst without initiator in soln. at 40° [32]. Processed by injection moulding, blow moulding, extrusion, thermoforming and casting [9]. The mechanism of the $ZnCl_2$-catalysed polymerisation has been reported [33]

Applications: Has applications in household appliances and goods - washing machine parts, refrigerator shelves, blenders, mixers, food trays, bottles and as bristle in brooms and brushes; in the automotive industry (batteries, lenses, interior trim)

Trade name	Details	Supplier
B	various grades	Thermofil Inc.
Lubricomp	BBL/BFL/BL/BML	LNP Engineering
Lucky	SAN	Lucky
Lupan	GP/HF/HR	Lucky
Lustran	SAN	Monsanto Chemical
RTP	500 series	RTP Company
RX6	6090	Spartech
SAN	various grades	Multibase/Network Polymers
Sangel	D/I	Proquigel
Suprel	various grades	Vista Chemical Co.
Taitalac	1400/1500	Taita Chem. Co.
Thermocomp	B/BF	LNP Engineering
Toyolac	ASG20	Toray Industries

Bibliographic References

[1] Yeo, Y.T., Goh, S.H. and Lee, S.Y., *Polym. Polym. Compos.*, 1996, **4**, 235, (miscibility)
[2] Kambour, R.P., Nachlis, W.L. and Carbeck, J.D., *Polymer*, 1994, **35**, 209, (miscibility)
[3] Singh, P.K., Mathur, A.B. and Mathur, G.N., *J. Therm. Anal.*, 1982, **25**, 387, (thermal behaviour)
[4] Morimoto, T., Mori, T. and Enomoto, S., *J. Appl. Polym. Sci.*, 1978, **22**, 1911
[5] Khan, H.U., Gupta, V.K. and Bhargava, G.S., J. Macromol. Sci., Chem., 1985, **22**, 1593, (soln viscosity)
[6] Venkataramana Reddy, G., Srnivasan, K.S.V. and Santappa, M., J. Macromol. Sci., *Chem.*, 1977, **11**, 2123, (soln props)
[7] Reddy Rami, C. and Kalpagam, V., Indian J. Chem., *Sect. A*, 1977, **15**, 389, (viscosity)
[8] Luda di Cortemiglia, M.P., Camino, G., Costa, L. and Guaita, M., *Thermochim. Acta*, 1985, **93**, 187, (thermal degradation)
[9] *Encycl. Polym. Sci. Eng.*, Vol. 1, 2nd edn., (eds. H.F. Mark, N.M. Bikales, C.G. Overberger and G. Menges), 1985, 452
[10] Hanson, A.W. and Zimmerman, R.L., *Ind. Eng. Chem.*, 1957, **49**, 1803, (mech props)
[11] Kurihara, F. and Kimura, S., *Polym. J. (Tokyo)*, 1985, **17**, 863
[12] Jabarin, S.A. and Lofgren, E.A., *Polym. Eng. Sci.*, 1988, **28**, 1152, (thermal degradation)
[13] Sauer, J.A. and Chen, C.C., *Polym. Eng. Sci.*, 1984, **24**, 786, (fatigue behaviour)
[14] Ku, P.L., *Adv. Polym. Technol.*, 1988, **8**, 177, (synth, props)
[15] Lomellini, P. and Lavagnini, L., *Rheol. Acta*, 1992, **31**, 175, (viscoelastic behaviour)
[16] Severini, F., Gallo, R. and Pegoraro, M., *Polym. Degrad. Stab.*, 1985, **13**, 351, (degradation)
[17] Kashyap, A.K. and Kalpagam, V., *J. Sci. Ind. Res.*, 1981, **40**, 113, (soln props)
[18] Hirose, H., Kobayashi, T. and Kohno, Y., *Kobunshi Ronbunshu*, 1981, **38**, 377, (impact strength)
[19] Neitzert, W.A., *Plastverarbeiter*, 1979, **30**, 773, (permeability)
[20] Kelleher, P.G. and Boyle, D.J., *Rev. Plast. Mod.*, 1980, **40**, 473, (permeability)
[21] Dolezel, B., *Koroze Ochr. Mater.*, 1973, **17**, 92, (weathering)
[22] Seymour, R.B. and Wood, H.A., Struct.-Solubility Relat. Polym., *(Proc. Symp.)*, (Proc. Symp.), (eds. F.W. Harris and R.B. Seymour), Academic Press, 1976, 99, (solubility parameters)
[23] Imoto, M., *Secchaku*, 1972, **16**, 436, (solubility parameters)
[24] Popovic, B., *SPE J.*, 1970, **26**, 54
[25] Putman, W.J., McGreer, M. and Conrad, M., *Annu. Tech. Conf. - Soc. Plast. Eng.*, 1995, **3**, 3322, (weathering)
[26] Mlinac-Misak, M., *Polimeri (Zagreb)*, 1994, **15**, 214, (biodegradation)
[27] Khan, H.U. and Goel, P.K., *Pop. Plast. Packag.*, 1993, **38**, 53, (soln props)
[28] Serrano, M. and Little, J., *Annu. Tech. Conf. - Soc. Plast. Eng.*, 1995, **1**, 357
[29] Bozzelli, J.W., Furches, B.J. and Janicki, S.L., *Mod. Plast.*, 1988, **65**, 78, 80, 82, (water content)
[30] Perepechenko, I.I. and Tekut'eva, Z.E., *Nauchn. Tr.-Kursk. Gos. Pedagog. Inst.*, 1980, **203**, 63, (viscoelastic props)
[31] Hill, D.J.T., Lang, A.P., Munro, P.D. and O'Donnell, J.H., *Eur. Polym. J.*, 1992, **28**, 391
[32] Wang, H., Chu, G., Srisiri, W., Padias, A.B. and Hall, H.K., *Acta Polym.*, 1994, **45**, 26, (synth)
[33] Srivastava, N. and Rai, J.S.P., *Polym. Int.*, 1992, **27**, 315
[34] Braun, D. and Disselhoff, R., *Angew. Makromol. Chem.*, 1972, **23**, 103, (pyrolysis)
[35] Can@fe, F. and Capaccioli, *Eur. Polym. J.*, 1978, **14**, 185, (eqn of state)
[36] Hino, T., Song, Y. and Prausnitz, J.M., *Macromolecules*, 1985, **28**, 5717, (eqn of state)
[37] Cincu, C., *Eur. Polym. J.*, 1974, **10**, 29, (eqn of state)
[38] Kelleher, P.G., *Adv. Polym. Technol.*, 1990, **10**, 219, (glass transition temp)
[39] Fischer, F., Gummi, Asbest, *Kunstst.*, 1974, **27**, 430, 432, 434, 436, (heat capacity)
[40] Rabinovich, I.B., Sheiman, M.S. and Selivanov, V.D., Vysokomol. Soedin., *Ser. B*, 1972, **14**, 24, (heat of combustion)
[41] Mamedov, A.A., Babaev, P.I., Buniyat-Zade, A.A., Aliguliev, R.M. et al, *Azerb. Khim. Zh.*, 1987, 73
[42] Nyquist, R.A., *Appl. Spectrosc.*, 1987, **41**, 797, (ir)
[43] Wolfram, L.E., Grasselli, J.G. and Koenig, J.L., *Appl. Polym. Symp.*, 1974, **25**, 27, (ir)
[44] Barron, P.F., Hill, D.J.T., O'Donnell, J.H. and O'Sullivan, P.W., *Macromolecules*, 1984, **17**, 1967, (C-13 nmr)
[45] Ayano, S. and Murakawa, T., *Kobunshi Kagaku*, 1972, **29**, 723, (refractive index)
[46] Wilmes, K., *Kunststoffe*, 1968, **58**, 133
[47] Loveless, H.S. and McWilliams, D.E., *Polym. Eng. Sci.*, 1970, **10**, 139

SAN, Blow moulding   S-2

Related Polymers: SAN
Monomers: Styrene, Acrylonitrile
Material class: Thermoplastic, Copolymers
Polymer Type: styrene-acrylonitrile (SAN), acrylonitrile and copolymers
CAS Number:

CAS Reg. No.
9003-54-7

Polymer Stability:
Chemical Stability: Has excellent resistance to corn oil, hydrochloric acid, formaldehyde, petrol and ammonia solns. [1]

Applications/Commercial Products:

Trade name	Supplier
Tyril 880B	Dow

Bibliographic Reference
[1] *Tyril SAN Resins*, Dow Chemical Co., 1996, (technical datasheet)

SAN, Extrusion, Unfilled   S-3

Related Polymers: SAN
Monomers: Styrene, Acrylonitrile
Material class: Thermoplastic, Copolymers
Polymer Type: styrene-acrylonitrile (SAN), acrylonitrile and copolymers
CAS Number:

CAS Reg. No.
9003-54-7

Volumetric & Calorimetric Properties:
Density:

No.	Value	Note
1	d 1.08 g/cm^3	Luran 388S, ASTM D792 [1]

Melting Temperature:

No.	Value	Note
1	220–270°C	Luran 388S [1]

Deflection Temperature:

No.	Value	Note
1	102.8°C	0.45 MPa, 66 psi, Luran 388S, ASTM D648 [1]
2	98.9°C	1.8 MPa, 264 psi, Luran 388S, ASTM D648, [1]

Transport Properties:
Polymer Melts: Melt volume rate 8 cm^3 (10 min.)$^{-1}$ (Luran 388S, ASTM D1238), 7 cm^3 (10 min.)$^{-1}$ (Luran 388S, ASTM D1238) [1]
Water Absorption

No.	Value	Note
1	0.3%	23°, saturated, ASTM D570 [1]

Mechanical Properties:
Tensile (Young's) Modulus:

No.	Value	Note
1	3899 MPa	565.5 kpsi, Luran 388S, ASTM D638 [1]

Tensile Strength Yield:

No.	Value	Note
1	82 MPa	11.89 kpsi, Luran 388S, ASTM D638 [1]

Izod Notch:

No.	Value	Notch	Note
1	16.2 J/m	Y	0.47 ft lb in^{-1}, 23°, ASTM D256) [1]

Electrical Properties:
Surface/Volume Resistance:

No.	Value	Note	Type
2	0.1 × 10^{15} Ω.cm	Luran 388S, ASTM D257 [1]	S

Dielectric Permittivity (Constant):

No.	Value	Frequency	Note
1	2.8	1 MHz	Luran 388S, ASTM D150 [1]

Complex Permittivity and Electroactive Polymers: Comparative tracking index 475 (Luran 388S, ASTM D3638) [1]

Polymer Stability:
Upper Use Temperature:

No.	Value	Note
1	85°C	continuous, few hrs., Luran 388S [1]

Flammability: Flammability rating HB (Luran 388S, UL 94) [1] RTI mechanical with impact 50° (Luran 388S, UL 746B) [1]. RTI electrical 50° (Luran 388S, UL 746B) [1]

Applications/Commercial Products:
Mould Shrinkage (%):

No.	Value	Note
1	0.3–0.7%	Luran 388S [1]

Trade name	Supplier
Luran 388S	BASF
Luran KR 2556	BASF
Lustran SAN 49	Monsanto Chemical

Bibliographic Reference
[1] *Luran SAN Copolymer*, BASF, 1995, (technical datasheet)

SAN, Flame retardant, Injection moulding, Glass fibre filled   S-4

Related Polymers: SAN
Monomers: Styrene, Acrylonitrile
Material class: Composites, Copolymers

– SAN, General purpose

Polymer Type: styrene-acrylonitrile (SAN), acrylonitrile and copolymers
CAS Number:

CAS Reg. No.
9003-54-7

Volumetric & Calorimetric Properties:
Density:

No.	Value	Note
1	d 1.31 g/cm^3	30% filled [1]

Thermal Expansion Coefficient:

No.	Value	Note	Type
1	3.2×10^{-5} K^{-1}	30% filled [1]	L

Thermal Conductivity:

No.	Value	Note
1	0.29 W/mK	30% filled [1]

Deflection Temperature:

No.	Value	Note
1	105°C	1.81 MPa, 30% filled [1]
2	110°C	0.45 MPa, 30% filled [1]

Transport Properties:
Water Absorption:

No.	Value	Note
1	0.08%	24 h [1]

Mechanical Properties:
Flexural Modulus:

No.	Value	Note
1	10200 MPa	30% filled [1]

Tensile Strength Break:

No.	Value	Note	Elongation
1	130 MPa	30% filled [1]	1–2%

Flexural Strength at Break:

No.	Value	Note
1	155 MPa	30% filled [1]

Compressive Strength:

No.	Value	Note
1	100MPa	30% filled [1]

Hardness: Rockwell M93 (30% filled) [1]
Failure Properties General: Shear strength 65 Mpa (30% filled) [1]
Friction Abrasion and Resistance: Wear factor 220 (30% filled) [1].

Static coefficient of friction 0.28; dynamic coefficient of friction 0.33 (30% filled) [1]
Izod Notch:

No.	Value	Notch	Note
1	40 J/m	Y	30% filled [1]
2	185 J/m	N	30% filled [1]

Electrical Properties:
Surface/Volume Resistance:

No.	Value	Note	Type
2	0.1×10^{15} Ω.cm	30% filled [1]	S

Dielectric Permittivity (Constant):

No.	Value	Frequency	Note
1	3.4	60 Hz	30% filled [1]
2	3.1	1 MHz	30% filled [1]

Complex Permittivity and Electroactive Polymers: Tracking resistance 600 kC (30% filled) [1]
Dissipation (Power) Factor:

No.	Value	Frequency	Note
1	0.007	60 Hz	30% filled [1]
2	0.01	1 MHz	30% filled [1]

Polymer Stability:
Upper Use Temperature:

No.	Value	Note
1	60°C	continuous service, 30% filled [1]

Flammability: Flammability rating HB (3.2 mm sample, 30% filled, UL 94) [1]

Applications/Commercial Products:
Mould Shrinkage (%):

No.	Value	Note
1	0.15–0.25%	3 mm sample [1]

Trade name	Supplier
SN-30GF/000FR	Compounding Technology
Thermocomp BF-1006FR	LNP Engineering

Bibliographic Reference
[1] LNP Engineering Plastics, *Product Data Book*, (technical datasheet)

SAN, General purpose S-5
Related Polymers: SAN
Monomers: Styrene, Acrylonitrile
Material class: Thermoplastic, Copolymers
Polymer Type: styrene-acrylonitrile (SAN), acrylonitrile and copolymers
CAS Number:

CAS Reg. No.
9003-54-7

SAN, General purpose

Volumetric & Calorimetric Properties:

Density:

No.	Value	Note
1	d 1.08 g/cm^3	[1]

Thermal Expansion Coefficient:

No.	Value	Note	Type
1	6.2×10^{-5} K^{-1}	[1]	L

Thermal Conductivity:

No.	Value	Note
1	0.23 W/mK	[1]

Deflection Temperature:

No.	Value	Note
1	98°C	1.81 MPa [1]
2	102°C	0.45 MPa [1]

Vicat Softening Point:

No.	Value	Note
1	104°C	ASTM D1525 [2]

Transport Properties:

Melt Flow Index:

No.	Value	Note
1	2.8 g/10 min	ASTM D1238 [2]

Water Absorption:

No.	Value	Note
1	0.1%	24h [1]

Mechanical Properties:

Flexural Modulus:

No.	Value	Note
1	4000 MPa	ASTM D790 [1]

Tensile Strength Break:

No.	Value	Note	Elongation
1	75 MPa	ASTM D638 [1]	3%

Flexural Strength at Break:

No.	Value	Note
1	125 MPa	ASTM D790 [1]

Compressive Strength:

No.	Value	Note
1	60 MPa	ASTM D695 [1]

Hardness: Rockwell M83 (ASTM D785) [1,2]
Failure Properties General: Shear strength 30 Mpa [1] (ASTM D732)
Friction Abrasion and Resistance: Wear factor 3000 [1]. Coefficient of friction 0.28 (static), 0.33 (dynamic) [1]

Izod Notch:

No.	Value	Notch	Note
1	20 J/m	Y	ASTM D256 [1]
2	180 J/m	N	ASTM D256 [1]

Electrical Properties:

Surface/Volume Resistance:

No.	Value	Note	Type
2	0.1×10^{15} Ω.cm	ASTM D257 [1]	S

Dielectric Permittivity (Constant):

No.	Value	Frequency	Note
1	3	60 Hz	ASTM D150 [1]
2	2.7	1 MHz	ASTM D150 [1]

Complex Permittivity and Electroactive Polymers: Tracking resistance 425 kC (DIN IEC 112) [1]

Dissipation (Power) Factor:

No.	Value	Frequency	Note
1	0.004	60 Hz	ASTM D150 [1]
2	0.007	1 MHz	ASTM D150 [1]

Polymer Stability:

Upper Use Temperature:

No.	Value	Note
1	55°C	continuous service [1]

Flammability: Flammability rating HB (3.2 mm, UL 94) [1,2]

Applications/Commercial Products:

Mould Shrinkage (%):

No.	Value	Note
1	0.4–0.6%	3 mm sample [1]

Trade name	Details	Supplier
Kibisan		Chi Mei Corporation
Luran		BASF
Lustran DMC	alternating	Monsanto Chemical
Lustran SAN		Monsanto Chemical
Thermocomp	B-1000 series	LNP Engineering
Thermocomp		LNP Engineering
Tyvil	various grades	Dow

SAN, Glass reinforced S-6

Related Polymers: SAN
Monomers: Styrene, Acrylonitrile
Material class: Composites, Copolymers
Polymer Type: styrene-acrylonitrile (SAN), acrylonitrile and copolymers
CAS Number:

CAS Reg. No.
9003-54-7

General Information: Glass fibre content is usually 20% or 40%

Volumetric & Calorimetric Properties:
Density:

No.	Value	Note
1	d 1.22–1.4 g/cm^3	20% filled [1]

Volumetric Properties General: Thermal expansion coefficient has been reported [7]
Transition Temperature General: Heat deflection temp. has been reported [5]
Deflection Temperature:

No.	Value	Note
1	99–110°C	264 psi, 20% filled [1]

Mechanical Properties:
Mechanical Properties General: Addition of glass fibre improves toughness, modulus and hardness [1]. Studies on the tensile modulus of Acrylafil have been reported [2]. Props. of 20% filled material have been reported [4]. Tensile modulus 1.45 GN m^{-2} [7]

Tensile (Young's) Modulus:

No.	Value	Note
1	0.008–0.012 MPa	1.2–1.7 psi, 20% filled [1]

Tensile Strength Break:

No.	Value	Note	Elongation
1	73.8 MPa	10.7–12 kpsi	1.4–1.9%
2	82.7 MPa	20% glass fibre [1]	

Poisson's Ratio:

No.	Value	Note
1	0.36	105.6° [7]

Tensile Strength Yield:

No.	Value	Elongation	Note
1	39.2 MPa	0.6%	40°, 400 kg cm^{-2} [6]

Miscellaneous Moduli:

No.	Value	Note	Type
1	7355 MPa	75000 kg cm^{-2}, 23°, 10000h [6]	creep modulus

Izod Notch:

No.	Value	Notch	Note
1	86.1–103.1 J/m	Y	2.5–3 ft lb in^{-1} [1]

Polymer Stability:
Flammability: A 0.125" sample cont. long fibres passes the UL 484 test whereas a 0.065" sample fails [8]. UL 94 self-extinguishing (0.25" bar) [8]
Chemical Stability: Chemical resistance and hydrolytic stability are improved by the addition of glass fibre [3]

Applications/Commercial Products:

Trade name	Details	Supplier
Acrylafil		Fiberfil
SN	Various GF series	Compounding Technology

Bibliographic References
[1] Ku, P.L., *Adv. Polym. Technol.*, 1988, **9**, 177, (mech props)
[2] Lees, J.K., *Polym. Eng. Sci.*, 1968, **8**, 186, 195, (tensile strength)
[3] Theberge, J.E. and Arkles, B., *Mater. Prot. Perform.*, 1972, **11**, 25, (stability)
[4] Carvalho, A., *Plast. Rev.*, 1977, **16**, 22, 24, 28, 30, (props)
[5] Fienhold, G. and Schmiedel, H., *Oesterr. Kunstst.-Z.*, 1990, **21**, 210, (heat deflection temp)
[6] Wilmes, K., *Kunststoffe*, 1968, **58**, 133
[7] Kajiyama, T., Yoshinaga, T. and Takayenagi, M., *J. Polym. Sci., Polym. Phys. Ed.*, 1977, **15**, 1557, (mech props)
[8] Hattori, K., *Tech. Pap., Reg. Tech. Conf. - Soc. Plast. Eng.*, 1969, 9, (flammability)

SAN, Injection moulding, Glass fibre filled S-7

Related Polymers: SAN
Monomers: Styrene, Acrylonitrile
Material class: Composites, Copolymers
Polymer Type: styrene-acrylonitrile (SAN), acrylonitrile and copolymers
CAS Number:

CAS Reg. No.
9003-54-7

Mechanical Properties:
Flexural Modulus:

No.	Value	Note
1	6205 MPa	900 kpsi, 20% filled [1]
2	1034.2 MPa	1500 kpsi, 40% filled [1]

Tensile Strength Break:

No.	Value	Note
1	124.8 MPa	18.1 kpsi, short term, 40% filled
2	44.1 MPa	6400 psi, long term, 40% filled [1]
3	103.4 MPa	15 kpsi, short term, 20% filled
4	37.2 MPa	5400 psi, long term, 20% filled [1]

Bibliographic References
[1] LNP Engineering Plastics, *Product Data Book*, LNP Engineering, (technical datasheet)
[2] *Thermoplastics Directory and Databook*, Chapman and Hall, 1997

– SAN, Injection moulding, Unfilled

Flexural Strength Yield:

No.	Value	Note
1	130.3 MPa	18.9 kpsi, 40% filled [1]
2	126.9 MPa	18.4 kpsi, 20% filled [1]

Compressive Strength:

No.	Value	Note
1	117.9 MPa	17.1 kpsi, 40% filled [1]
2	119.3 MPa	17.3 kpsi, 20% filled [1]

Izod Notch:

No.	Value	Notch	Note
1	44.7 J/m	Y	1.3 ft lb in^{-1}, 40% filled [1]
2	44.7 J/m	Y	1.3 ft lb in^{-1}, 20% filled [1]

Applications/Commercial Products:

Trade name	Details	Supplier
Luran 378 PG7		BASF
RTP	500 series	RTP Company
SN-40GF/000		Compounding Technology
Styvex 32002 NA		Ferro Corporation
Thermocomp BF-1006		LNP Engineering

Bibliographic Reference
[1] Loveless, H.S. and McWilliams, D.E., *Polym. Eng. Sci.*, 1970, **10**, 139, (mech props)

SAN, Injection moulding, Unfilled   S-8
Related Polymers: SAN
Monomers: Styrene, Acrylonitrile
Material class: Thermoplastic, Copolymers
Polymer Type: styrene-acrylonitrile (SAN), acrylonitrile and copolymers
CAS Number:

CAS Reg. No.
9003-54-7

Volumetric & Calorimetric Properties:
Density:

No.	Value	Note
1	d 1.08 g/cm^3	Luran 358N, ASTM D792 [3]

Thermal Expansion Coefficient:

No.	Value	Note	Type
1	3.8 × 10^{-5} K^{-1}	Lustran, ASTM D696 [4]	L

Melting Temperature:

No.	Value	Note
1	200–250°C	Luran 358N [3]
2	200–250°C	Luran 378P [3]
3	220–270°C	Luran 368R [3]

Deflection Temperature:

No.	Value	Note
1	101.7°C	0.45 MPa, 66 psi, ASTM D648 [3]
2	97.8°C	1.8 MPa, 264 psi, ASTM D648 [3]

Vicat Softening Point:

No.	Value	Note
1	111°C	Lustran, ASTM D1525 [4]

Transport Properties:
Melt Flow Index:

No.	Value	Note
1	5.1 g/10 min	10 min.$^{-1}$, Lustran [4]

Polymer Melts: Melt volume rate 27 cm^3 (10 min.)$^{-1}$ (ASTM D1238), 22 cm^3 (10 min.)$^{-1}$ (ASTM D1238) [3]
Water Absorption:

No.	Value	Note
1	0.2%	23°, saturated, ASTM D570 [3]

Mechanical Properties:
Tensile (Young's) Modulus:

No.	Value	Note
1	3699 MPa	536.5 kpsi, ASTM D638 [3]

Flexural Modulus:

No.	Value	Note
1	3585 MPa	520 kpsi, ASTM D790 [2]

Tensile Strength Break:

No.	Value	Note
1	68.3 MPa	9900 psi, short term, ASTM D638 [2]
2	28.9 MPa	4200 psi, long term [2]

Tensile Strength Yield:

No.	Value	Note
1	93 MPa	13.5 kpsi, ASTM D790 [2]
2	72 MPa	10.44 kpsi, ASTM D638 [3]

Compressive Strength:

No.	Value	Note
1	120 MPa	17.4 kpsi, ATM D695 [2]

Mechanical Properties Miscellaneous: RTI mechanical with impact 50° (0.063", UL 746B) [3]
Hardness: Rockwell M83 (ASTM D785) [4]

Izod Notch:

No.	Value	Notch	Note
1	10.16 J/m	Y	0.3 ft lb in^{-1}, [2]
2	12.74 J/m	Y	0.37 ft lb in^{-1}, 23° [3]

Electrical Properties:
Electrical Properties General: RTI electrical 50° (0.063", UL 746B) [3]
Surface/Volume Resistance:

No.	Value	Note	Type
2	0.1×10^{15} Ω.cm	ASTM D257 [3]	S

Dielectric Permittivity (Constant):

No.	Value	Frequency	Note
1	2.7	1 MHz	ASTM D150 [3]

Complex Permittivity and Electroactive Polymers: Comparative tracking index CTI 400, ASTM D3638 [3]

Optical Properties:
Refractive Index:

No.	Value	Note
1	1.57	Lustran, ASTM D542 [4]

Volume Properties/Surface Properties: Haze 1% (Lustran, ASTM D1003) [4]

Polymer Stability:
Upper Use Temperature:

No.	Value	Note
1	85°C	few hrs. [3]

Flammability: Flammability rating HB (UL 94)
Recyclability: Luran regrind can be recycled if it has not been contaminated or thermally degraded. Mixing regrind with virgin pellets may alter the feed and flow characteristics and affect demoulding, shrinkage and mech. props. Therefore only virgin materials recommended for high performance engineering parts [3]

Applications/Commercial Products:
Mould Shrinkage (%):

No.	Value	Note
1	0.3–0.7%	[3]

Trade name	Details	Supplier
Luran	358N, 378P, KR 2556	BASF
Lustran	29-2040	Bayer Inc.
Tyril		Dow

Bibliographic References
[1] Daly, J.H., Guest, M.J. Hayward, D. and Pethrick, R.A., *Polym. Commun.*, 1990, **31**, 325, (dielectric props)
[2] Loveless, H.S. and McWilliams, D.E., *Polym. Eng. Sci.*, 1970, **10**, 139, (mech props)
[3] *Lutan SAN Copolymer*, BASF, 1995, (technical datasheet)
[4] *Lustran*, Bayer Corporation, 1996, (technical datasheet)

SBS S-9
Synonyms: *Poly(styrene-block-butadiene-block-styrene) triblock copolymer. Styrene-butadiene-styrene block copolymer*
Related Polymers: Styrene-butadiene copolymers, Epoxidised SBS
Monomers: Styrene, 1,3-Butadiene
Material class: Thermoplastic, Synthetic Elastomers
Polymer Type: polybutadiene, polystyrene
CAS Number:

CAS Reg. No.
9003-55-8
106107-54-4

Applications/Commercial Products:

Trade name	Supplier
C-Flex R	Concept
Kraton	Shell Chemicals

Silicone acrylate S-10

Related Polymers: PDMS fluid
Material class: Synthetic Elastomers
Polymer Type: silicones
Molecular Formula: [C_2H_6OSi]$_n$
Fragments: C_2H_6OSi
Molecular Weight: MW 5000–25000
General Information: End groups are readily cross-linked by uv radiation in the presence of photoinitiatiors and the absence of oxygen

Transport Properties:
Polymer Melts: Viscosity values have been reported 10 Poise (25°, before cross-linking) [3]

Mechanical Properties:
Mechanical Properties General: Has good elastomeric props.

Applications/Commercial Products:
Processing & Manufacturing Routes: Acrylate groups can be attached to silicones in various ways. A major route involves reaction of polydimethylsiloxane, bearing terminal SiH groups, with 3-chloro-1-propene. The product is treated with acrylic acid to give the acrylate end group polysiloxane [1,2,3,4]
Applications: Rapid cure makes silicone acrylate commercially valuable for release coatings on paper and film. Has applications in fibre optics

Trade name	Supplier
Photomer 7020	Harcros
Tego Silicone Acrylate	Goldschmidt AG

Bibliographic References
[1] Clarson, S.J. and Semlyen, J.A., *Siloxane Polymers*, Prentice Hall, 1993
[2] *U.S. Pat.*, 1982, 4 348 454
[3] *Jpn. Pat.*, 1981, 86 922
[4] *Ger. Pat.*, 1980, 2 948 708

Silicone epoxy copolymer S-11

Synonyms: *Epoxysilicone copolymer*
Related Polymers: Polysiloxanes, Epoxy resins
Monomers: Dichlorodimethylsilane, Epichlorohydrin, Diphenylolpropane
Material class: Gums and resins
Molecular Formula: $[(C_2H_6OSi).(C_{18}H_{20}O_3)]_n$
Fragments: C_2H_6OSi $C_{18}H_{20}O_3$
General Information: The struct. is very complex; the structural unit illustrated is a simplified version

Surface Properties & Solubility:
Surface and Interfacial Properties General: Good adhesive props., particularly to copper [1,3]

Electrical Properties:
Electrical Properties General: Has good insulating props. [2]

Polymer Stability:
Polymer Stability General: Greater resistance to chemicals than polysiloxanes. Usable at higher temps. than epoxy resins [1,2]
Thermal Stability General: Usable above 200° [1,2]
Environmental Stress: Highly weather resistant [1,2]
Chemical Stability: Resistant to most chemicals and solvents; decomposed by conc. acids [2]
Hydrolytic Stability: SiOC bonds are susceptible to hydrolysis at elevated temps. [1]
Recyclability: Not recyclable

Applications/Commercial Products:
Processing & Manufacturing Routes: Polydimethylsiloxane with reactive end groups is mixed with epoxy resin, which is obtained by the condensation of epichlorohydrin and diphenylolpropane. Under moderate heating co-condensation occurs. Resin can be cured at room temp. [1,3,4]
Applications: Can be used for tough resistant coatings and varnishes used particularly to coat metals

Bibliographic References
[1] Noll, W., *Chemistry and Technology of Silicones*, Academic Press, 1968
[2] Roff, W.J. and Scott, J.R., Fibres, Films, *Plastics and Rubbers*, Butterworths, 1971
[3] *Ger. Pat.*, 1956, 937 554
[4] *U.S. Pat.*, 1962, 3 055 858

Silicone phenol formaldehyde copolymer S-12

Synonyms: *Silicone phenolic resin copolymer. Phenolic resin silicone copolymer*
Related Polymers: Polysiloxanes, Phenolic resins
Monomers: Dichlorodimethylsilane, Formaldehyde, Phenol
Material class: Gums and resins
Molecular Formula: $[(C_2H_6OSi).(C_7H_6O)]_n$
Fragments: C_2H_6OSi C_7H_6O
General Information: The struct. is very complex. The structural unit illustrated is a simplified version

Electrical Properties:
Electrical Properties General: Has good insulating props. [1]

Polymer Stability:
Thermal Stability General: Usable above 200° [1,2]
Environmental Stress: Strongly resistant to weathering and exposure [1,2,3]
Chemical Stability: Decomposed by conc. oxidising acids and hot alkalis [1]
Hydrolytic Stability: SiOC bonds are susceptible to hydrolysis at elevated temps. [1,2]
Recyclability: Not recyclable

Applications/Commercial Products:
Processing & Manufacturing Routes: Polydimethylsiloxane with reactive end groups is heated with glycerol under dehydrating conditions. The product is mixed with phenol formaldehyde resin in an organic solvent and heated under dehydrating conditions [2,3]
Applications: Used as a varnish where resistance to weathering is important

Bibliographic References
[1] Roff, W.J. and Scott, J.R., Fibres, Films, *Plastics and Rubbers*, Butterworths, 1971
[2] Noll, W., *Chemistry and Technology of Silicones*, Academic Press, 1968
[3] *U.S. Pat.*, 1955, 2 718 507

Silicone polycarbinol copolymer S-13

Synonyms: *Silicone polyallyl alcohol copolymer*
Related Polymers: PDMS fluid
Monomers: Dichlorodimethylsilane, Allyl alcohol
Material class: Fluids
Polymer Type: silicones
Molecular Formula: $[(C_2H_6OSi).(C_{(5+3m)}H_{(12+6m)}O_{(2+m)}Si)]_n$
Fragments: C_2H_6OSi $C_{(5+3m)}H_{(12+6m)}O_{(2+m)}Si$
Molecular Weight: M_n 1000–20000

Volumetric & Calorimetric Properties:
Transition Temperature:

No.	Value	Note	Type
1	-62°C	-80°F [1,2]	Pour point

Surface Properties & Solubility:
Solvents/Non-solvents: Sol. aliphatic hydrocarbons, chlorinated hydrocarbons, aromatic hydrocarbons [1,2]

Transport Properties:
Polymer Melts: Viscosity 2–5 St [1,2]

Polymer Stability:
Polymer Stability General: SiC bonds make this a relatively stable copolymer
Flammability: Fl. p. 120° (250°F) [1,2]
Hydrolytic Stability: Stable to H_2O at room temp.; hydrolysed by steam [4]
Recyclability: Not recyclable

Applications/Commercial Products:
Processing & Manufacturing Routes: Polydimethylsiloxane is heated with allyl alcohol and peroxide [3,5]
Applications: Surfactant for organic systems and urethanes

Trade name	Details	Supplier
Dow Corning	1248 Fluid, Q-28026 Fluid, Q4-3557 Fluid	Dow Corning STI

Bibliographic References

[1] Ash, M. and Ash, I., Handbook of Plastic Compounds, *Elastomers and Resins*, (eds. M.B. Ash and I.A. Ash), Wiley-VCH, 1992
[2] Ash, M. and Ash, I., *Condensed Encyclopedia of Surfactants*, Edward Arnold, 1989
[3] Ceresa, R.J., *Block Graft Copolym.*, Butterworth, 1962
[4] Noll, W., *Chemistry and Technology of Silicones*, Academic Press, 1968
[5] *Brit. Pat.*, 1957, 766 528

Silicone polyetherimide copolymer   S-14

Synonyms: *Polyetherimide silicone copolymer*
Related Polymers: Polysiloxanes
Material class: Synthetic Elastomers, Copolymers
Polymer Type: polyimide, polysilanes
Molecular Formula: $[(C_2H_6OSi).(C_{37}H_{24}N_2O_6)]_n$
Fragments: C_2H_6OSi $C_{37}H_{24}N_2O_6$

Volumetric & Calorimetric Properties:
Deflection Temperature:

No.	Value	Note
1	200°C	[1]

Surface Properties & Solubility:
Solvents/Non-solvents: Insol. aliphatic hydrocarbons, aromatic hydrocarbons, chlorinated hydrocarbons, esters, H_2O [1,2]

Mechanical Properties:
Tensile Strength Break:

No.	Value	Note	Elongation
1	37 MPa	5500 psi [4]	47%

Electrical Properties:
Electrical Properties General: Has excellent insulating props. [3]

Optical Properties:
Transmission and Spectra: Low light transmission [1]

Polymer Stability:
Polymer Stability General: Highly resistant to heat, chemicals and flames [1,3]
Thermal Stability General: Usable up to 200° [1]
Upper Use Temperature:

No.	Value	Note
1	200°C	[1]

Flammability: Flame retardant [1,2]
Environmental Stress: Highly resistant to weathering [1,2]
Chemical Stability: Unaffected by aliphatic, aromatic and chlorinated hydrocarbons or by esters [1,2,3]
Hydrolytic Stability: High hydrolytic stability [1,2]
Recyclability: Non-recyclable

Applications/Commercial Products:
Processing & Manufacturing Routes: Polydimethylsiloxane with aminoalkyl end groups is treated with polyetherimide formed by reaction of a bis(ether anhydride) with diaminobenzene [1,2,3]
Applications: Used for wire and cable manufacture where fire resistance is important

Trade name	Supplier
Siltem	General Electric Plastics
Ultem D9000	General Electric Plastics

Bibliographic References

[1] Clagett, D.C., *Encycl. Polym. Sci. Eng.*, 2nd edn., (ed. J.I. Kroschwitz), John Wiley and Sons, 1985, **6**, 94
[2] Hardman, B. and Torkelson, A., *Encycl. Polym. Sci. Eng.*, Vol. 15, 2nd edn., (ed. J.I. Kroschwitz), John Wiley and Sons, 1985, 204
[3] Ash, M. and Ash, I., Handbook of Plastic Compounds, *Elastomers and Resins*, (eds. M.B. Ash and I.A. Ash), Wiley-VCH, 1992
[4] *Can. Pat.*, 1993, 2 050 182

Silicone resin, Glass filled   S-15

Synonyms: *Silicone resin, Glass reinforced*
Related Polymers: Methyl silicone resin, Methyl phenyl silicone resin
Monomers: Trichloromethylsilane, Trichlorophenylsilane, Dichlorodimethylsilane, Dichloromethylphenylsilane
Material class: Gums and resins, Copolymers
Polymer Type: silicones, dimethylsilicones
CAS Number:

CAS Reg. No.	Note
159011-85-5	methyl silicone
159968-66-8	methyl phenyl silicone

Molecular Formula: $[(C_2H_6OSi)(C_7H_8OSi)]_n$
Fragments: C_2H_6OSi C_7H_8OSi
Additives: Glass fibre used as filler to increase toughness and mechanical strength

Volumetric & Calorimetric Properties:
Density:

No.	Value	Note
1	d 1.85 g/cm³	[4,6]

Thermal Expansion Coefficient:

No.	Value	Note	Type
1	7.5×10^{-5} K^{-1}	laminate [1]	L
2	4×10^{-5} K^{-1}	glass filled [4,6]	L

Thermal Conductivity:

No.	Value	Note
1	0.21 W/mK	0.0005 cal (cm s °C)$^{-1}$, laminate [1]
2	0.45 W/mK	0.0011 cal (cm s °C)$^{-1}$, glass filled [4]

Specific Heat Capacity:

No.	Value	Note	Type
1	1.05 kJ/kg.C	0.25 Btu lb^{-1}°F^{-1}, laminate [1]	P

Deflection Temperature:

No.	Value	Note
1	260°C	[6]
2	300°C	570°F [1,3]

– Silicone resin, Glass filled

Transport Properties:

Water Absorption:

No.	Value	Note
1	0.3%	laminate
2	0.07%	glass filled [1]

Mechanical Properties:

Mechanical Properties General: Silicone resin laminates are much stiffer than glass fibre filled resins

Tensile (Young's) Modulus:

No.	Value	Note
1	15000 MPa	2200 kpsi, laminate [1]

Flexural Modulus:

No.	Value	Note
1	9000 MPa	13500 kpsi, glass filled [4]
2	18000 MPa	2700 kpsi, laminate [1]

Tensile Strength Break:

No.	Value	Note	Elongation
1	28 MPa	4000 psi, glass filled [1,4]	
2	200 MPa	30000 psi, laminate [1]	1.5%

Flexural Strength at Break:

No.	Value	Note
1	62 MPa	9000 psi, glass filled [1,4]
2	240 MPa	35000 psi, laminate [1]

Compressive Strength:

No.	Value	Note
1	82 MPa	12000 psi, glass filled [1,4]
2	100 MPa	15000 psi, laminate [1]

Hardness: Rockwell M70–M90 [4]

Izod Notch:

No.	Value	Notch	Note
1	16–550 J/m	N	0.3–10 ft lb in^{-1}, glass filled [1,4]
2	270–1360 J/m	N	5–25 ft lb in^{-1}, laminate [1]

Electrical Properties:

Electrical Properties General: Good insulating props.

Dielectric Permittivity (Constant):

No.	Value	Note
1	3.9	[2,3,6]

Dielectric Strength:

No.	Value	Note
1	14 kV.mm^{-1}	340 V mil^{-1}, ⅛" thick) [1]

Arc Resistance:

No.	Value	Note
1	200–300s	[3,4]

Dissipation (Power) Factor:

No.	Value	Frequency	Note
1	0.002		glass filled [2,3,4]
2	0.005	1 kHz	laminate [1]
3	0.00015	1 MHz	laminate

Polymer Stability:

Polymer Stability General: Stable over a wide temp. range
Thermal Stability General: Usable from -100–250° [1,2,3]
Upper Use Temperature:

No.	Value
1	250°C

Flammability: Non-burning [4]
Environmental Stress: Outstanding weather resistance. Unaffected by sunlight or water; excellent radiation resistance [3,5]
Chemical Stability: Attacked by strong acids; unaffected by alkali or weak acid. Degraded by chlorinated hydrocarbons, aromatic hydrocarbons, ketones and esters. Alcohols and aliphatic hydrocarbons have little effect [3,5]
Hydrolytic Stability: Fully stable to cold water; poor resistance to steam [5]
Biological Stability: Non-biodegradable [2]
Recyclability: Not recyclable

Applications/Commercial Products:

Processing & Manufacturing Routes: Silicone resin is prepared as [2,4] for methyl silicone resin or methyl phenyl silicone resin. It is mixed with glass fibre and partially cured, then moulded and given a final cure. Glass filler can be regularly distributed as sheets, rather than randomly in fibres, to prod. silicone resin glass laminates

Mould Shrinkage (%):

No.	Value	Note
0	0.003%	[6]

Applications: Used as tough heat resistant moulding compound and laminate particularly in the electrical and electronics industry. Valuable in environments where large temp. variations are encountered

Trade name	Supplier
SC	Bacon

Bibliographic References

[1] Meals, R.N. and Lewis, F.M., *Silicones*, Reinhold, 1959
[2] Roff, W.J. and Scott, J.R., Fibres, Films, *Plastics and Rubbers*, Butterworths, 1971
[3] McGregor, R.R., *Silicones and their Uses*, McGraw-Hill, 1954
[4] Elliott, E.C., *Silicone Technology*, (ed. P.F. Bruins), John Wiley and Sons, 1970, 121
[5] Noll, W., *Chemistry and Technology of Silicones*, Academic Press, 1968

Silicone resin, Room temperature vulcanising S-16

Synonyms: *Flexible silicone resin. Silicone encapsulant*
Related Polymers: Methylsilicone resin, Methylphenylsilicone resins
Monomers: Trichloromethylsilane, Trichlorophenylsilane, Dichlorodimethylsilane, Dichloromethylphenylsilane
Material class: Gums and resins, Copolymers
Polymer Type: silicones, dimethylsilicones
CAS Number:

CAS Reg. No.	Note
159011-85-5	methyl silicone
159968-66-8	methyl phenyl silicone

Molecular Formula: $[(C_2H_6OSi)(C_7H_8OSi)]_n$
Fragments: C_2H_6OSi C_7H_8OSi

Volumetric & Calorimetric Properties:
Density:

No.	Value	Note
1	d 1.23 g/cm^3	[2]

Thermal Expansion Coefficient:

No.	Value	Note	Type
1	0.0009 K^{-1}	[2]	V

Thermal Conductivity:

No.	Value	Note
1	0.3 W/mK	0.0007 cal (cm s °C)$^{-1}$ [2]

Transition Temperature:

No.	Value	Note	Type
1	-80°C	approx. [2]	Stiffening temp.

Transport Properties:
Polymer Melts: Viscosity values for the pre-cure resin have been reported 1000 Poise [3]
Water Absorption:

No.	Value	Note
1	0.1%	[2]

Mechanical Properties:
Mechanical Properties General: Mech. props. are similar to room temp. vulcanised methyl silicone rubber. Resins remain flexible down to low temps. unlike unmodified low functionality soft resins
Tensile Strength Break:

No.	Value	Note	Elongation
1	3.4 MPa	490 psi [1,3]	180%

Hardness: Shore A22–A70 [1,2]
Fracture Mechanical Properties: Tear strength 0.3–1.7 J cm^{-2} (15–100 ppi) [2]

Electrical Properties:
Dielectric Permittivity (Constant):

No.	Value	Note
1	2.9	[1,2,3]

Dielectric Strength:

No.	Value	Note
1	22 kV.mm^{-1}	550 V mil^{-1} [2]

Arc Resistance:

No.	Value	Note
1	115–130s	[2]

Dissipation (Power) Factor:

No.	Value	Note
1	0.007	[1,2]

Polymer Stability:
Polymer Stability General: Stable material which is particularly valuable in high temp., high humidity conditions
Thermal Stability General: Usable from -100–250° [4]
Upper Use Temperature:

No.	Value
1	250°C

Flammability: Self-extinguishing [2]
Environmental Stress: Outstanding weather resistance. Unaffected by sunlight or water. Excellent resistance to radiation, electrical fields and currents [4]
Chemical Stability: Attacked by strong acids; unaffected by alkali or weak acid. Degraded by chlorinated hydrocarbons, aromatic hydrocarbons, ketones and esters. Alcohols and aliphatic hydrocarbons have little effect [1,4]
Hydrolytic Stability: Fully stable to cold water; poor resistance to steam [2,4]
Biological Stability: Non-biodegradable
Recyclability: Not recyclable

Applications/Commercial Products:
Processing & Manufacturing Routes: Silicone resin is prepared as [4,5] methyl silicone resin or (methyl phenyl silicone resin) and the free silanol groups are modified by reaction with a diorganobisacyloxysilane. The resulting resin can be cured at room temp. by an organotrisacyloxysilane in the presence of atmospheric moisture
Mould Shrinkage (%):

No.	Value	Note
1	0.15%	[2]

Applications: Used as sealant encapsulant and adhesive particularly in the electrical and electronics industry

Trade name	Details	Supplier
Amicon SC		Emerson & Cuming
Eccosil		Emerson & Cuming
RTV	various grades, not exclusive	General Electric Silicones

Sylgard		Dow Corning STI
Thermoset		Thermoset

Bibliographic References
[1] *The Materials Selector*, 2nd edn., (eds. N.A. Waterman and M.F. Ashby), Chapman and Hall, 1997, **3**, 455
[2] Hainline, A.N., *Silicone Technology*, (ed. P.F. Bruins), John Wiley and Sons, 1970, 17
[3] Ash, M. and Ash, I., Handbook of Plastic Compounds, *Elastomers and Resins*, (eds. M.B. Ash and I.A. Ash), Wiley-VCH, 1992
[4] Noll, W., *Chemistry and Technology of Silicones*, Academic Press, 1968
[5] *U.S. Pat.*, 1962, 3 032 529

Silicone rubber, Conductive S-17

Synonyms: *Conductive silicone rubber*
Related Polymers: Pre-crue gum, More general information
Monomers: Dichlorodimethylsilane, Hexamethylcyclotrisiloxane
Material class: Synthetic Elastomers
Polymer Type: dimethylsilicones
CAS Number:

CAS Reg. No.	Note
9016-00-6	
63394-02-5	elastomer

Molecular Formula: $(C_2H_6OSi)_n$
Fragments: C_2H_6OSi
Additives: Carbon block used as a filler to prod. an electrically conducting elastomer

Mechanical Properties:
Tensile Strength Break:

No.	Value	Note	Elongation
1	4.8 MPa	700 psi [2]	200%

Hardness: Shore A60 [2]
Fracture Mechanical Properties: Tear strength 1 J cm^{-2} (60 ppi) [2]

Electrical Properties:
Electrical Properties General: Is electrically conducting. Volume resistivity 5 Ω cm [1,2]

Optical Properties:
Volume Properties/Surface Properties: Colour black, owing to presence of carbon filler

Applications/Commercial Products:
Processing & Manufacturing Routes: The gum is prepared as in polydimethylsiloxane gum. Organic peroxide (typically benzoyl peroxide) is used at 1–2 parts per 100 rubber to ease the gum to an elastomer at 110–200°, under pressure typically for less than an hour. It is often post cured for several hours at temps. at 200–250°
Applications: Used in [1,2,3] electrical and electronic industry as antistatic material and for low amperage electric contacts

Trade name	Supplier
COHRlastic EC	CHR Industries
Eccoshield	Emerson & Cuming

Bibliographic References
[1] Noll, W., *Chemistry and Technology of Silicones*, Academic Press, 1968
[2] Ash, M. and Ash I., Handbook of Plastic Compounds, *Elastomers and Resins*, VCH, 1992

Silicone rubber, Glass fibre filled S-18

Synonyms: *Silicone elastomer, Glass fibre filled. Dimethylsiloxane rubber, Glass fibre filled. Poly[oxy(dimethylsilylene)]*
Related Polymers: Methylsilicone rubber, Heat vulcanised, Polydimethylsiloxane gum
Monomers: Dichlorodimethylsilane, Hexamethylcyclotrisiloxane
Material class: Synthetic Elastomers
Polymer Type: dimethylsilicones
CAS Number:

CAS Reg. No.	Note
63394-02-5	elastomer
31900-57-9	PDMS

Molecular Formula: $[C_2H_6OSi]_n$
Fragments: C_2H_6OSi
Additives: 10–46% Glass fibre

Mechanical Properties:
Mechanical Properties General: Very high tear strength; silicone rubber elongation at break 10% (max) [1]
Hardness: Shore A70–A80 [1]
Fracture Mechanical Properties: Tear strength 5 J cm^{-2} (300 ppi) [1]

Electrical Properties:
Dielectric Strength:

No.	Value	Note
1	20 kV.mm^{-1}	500 V mil^{-1} [1]

Dissipation (Power) Factor:

No.	Value	Note
1	0.04	[2]

Applications/Commercial Products:
Processing & Manufacturing Routes: Organic peroxide, typically benzoyl peroxide, is used at 1–2 parts per 100 rubber to cure the gum to an elastomer. The rubber is cured at 110–200°, under pressure, typically for less than an hour. Post cured often for several hours at temps. of 200–250°
Applications: Used as solid rubber where high tear strength is essential. Used as flexible insulating material in the electrical industry

Trade name	Details	Supplier
COHR lastic	3300, 4000, 4400 series	CHR Industries

Bibliographic References
[1] Ash, M. and Ash, I., Handbook of Plastic Compounds, *Elastomers and Resins*, (eds. M.B. Ash and I.A. Ash), Wiley-VCH, 1992
[2] Noll, W., *Chemistry and Technology of Silicones*, Academic Press, 1968
[3] Johnson, P., *Silicones*, (ed. S. Fordham), 1960, 213

Silk S-19

(Gly.Ala.Gly.Ala.Gly.Ser)$_n$

Synonyms: *Fibroin*
Related Polymers: Proteins
Monomers: Amino acids
Material class: Proteins and polynucleotides
Molecular Weight: 350000, 370000 (subunits of 170000 and 25000). 60000–150000 (regenerated)
General Information: Complex natural fibre secreted by the silkworm *Bombyx mori*, related insects and some spiders. Highly orientated crystalline fibre forming antiparallel pleated sheet struct. [3,4]. Has poor thermoplasticity [5]

Volumetric & Calorimetric Properties:

Density:

No.	Value	Note
1	d 1.31 g/cm^3	[1]
2	d 1.3 g/cm^3	[5]

Thermal Conductivity:

No.	Value	Note	
1	0.092 W/mK	0.00022 cal (cm s °C)$^{-1}$	[1]

Surface Properties & Solubility:

Solubility Properties General: Amphoteric, absorbing acids and alkalis. Isoelectric point in H$_2$O pH 5 [2]
Solvents/Non-solvents: Sol. sulfuric acid, sodium hypochlorite. Insol. AcOH, Me$_2$CO, DMF, hydrochloric acid

Transport Properties:

Water Absorption:

No.	Value	Note
1	10%	20°, 60% relative humidity [2], 18.7% swelling in diameter
2	11%	[1,5]

Mechanical Properties:

Mechanical Properties General: Tenacity 0.34–0.39 N tex^{-1} (20% elongation) [5]

Tensile (Young's) Modulus:

No.	Value	Note	
1	9800 MPa	1000 kg mm^{-2}	[1]

Impact Strength: Impact strength 1300000000 erg cm^{-3} [1]
Friction Abrasion and Resistance: Elongation (standard conditions) 18–25% (15–60% wet) [2]

Electrical Properties:

Electrical Properties General: Has high electrical resistance. It is electrostatic when dry [2]

Optical Properties:

Refractive Index:

No.	Value	Note
1	0.06	[1]

Polymer Stability:

Decomposition Details: Decomposes above 150°. Weight loss occurs above 175°, with charring at 250° [2]
Flammability: Combustible but self-extinguishing [1]
Environmental Stress: Slowly yellows upon exposure to light. Degrades rapidly in sunlight [2]
Chemical Stability: Discoloured by oxidants [2]. Stable to strong alkali (16–18% sodium hydroxide at low temp.) [2]

Bibliographic References

[1] *Polymeric Materials Encyclopedia,* (ed. J.C. Salamone), CRC Press, 1996, **10**, 7711
[2] *Ullmanns Encykl. Ind. Chem.,* VCH, Weinheim, 1993, **A24**, 95
[3] Frazer, R.D.B., MacRae, T.P. Stewart, F.H.C.S. and Suzuki, E., *J. Mol. Biol.,* 1965, **11**, 706
[4] Takahashi, Y., Gehoh, M. and Yuzuriha, K., J. Polym. Sci., *Polym. Phys. Ed.,* 1991, **29**, 889
[5] *Encycl. Polym. Sci. Eng.,* 2nd edn., (eds. H.F. Mark, N.M. Bikales, C.G. Overberger and G. Menges), John Wiley and Sons, 1985, **15**, 309
[6] Takahashi, Y., *Silk Polym.,* (eds. D. Kaplan, W.W. Adams, B. Former and C. Viney), ACS Symposium No. 544, ACS Washington, 1994

Silly putty S-20

Synonyms: *Bouncing putty*
Related Polymers: PDMS fluid
Monomers: Dichlorodimethylsilane, Hexamethylcyclotrisiloxane
Material class: Synthetic Elastomers
Polymer Type: dimethylsilicones
CAS Number:

CAS Reg. No.	Note
31900-57-9	PDMS

Molecular Formula: [C$_2$H$_6$OSi]$_n$
Fragments: C$_2$H$_6$OSi
Molecular Weight: MW 1000–10000
Additives: Fillers are often incorporated to produce desired mech. props.
General Information: Contains (SiOBOSi) linkages

Surface Properties & Solubility:

Solvents/Non-solvents: Sol. esters, ethers, ketones, alcohols, chlorinated hydrocarbons, aromatic hydrocarbons. Slightly sol. H$_2$O, aliphatic hydrocarbons [4]

Mechanical Properties:

Mechanical Properties General: The incorporation of boron atoms into a linear dimethylpolysiloxane allows temporary cross-linking between chains. In the short term the putty behaves as a cross-linked rubber but in the long term as a fluid gum. Behaves as very viscous fluid on storage in a vessel but shows elastic rebound on hard surfaces, brittleness on sudden impact and plastic flow under moderate tension
Viscoelastic Behaviour: 80% Rebound from sharp impact [3,5]

Polymer Stability:

Polymer Stability General: Not particularly stable
Chemical Stability: Dissolved by Me$_2$CO, alcohols and Et$_2$O [4]. Swollen by chlorinated or aromatic hydrocarbons
Hydrolytic Stability: Slowly disintegrated by H$_2$O [4]
Recyclability: Not recyclable

Applications/Commercial Products:

Processing & Manufacturing Routes: Silicone fluid prepared as in PDMS fluid. The fluid is heated with boric oxide and ferric chloride for some hours at 150–250°
Applications: Used as a novelty toy, as core for golf balls and as an exerciser for weak arm muscles

Bibliographic References

[1] *U.S. Pat.,* 1947, 2 431 878
[2] *U.S. Pat.,* 1951, 2 541 851
[3] Freeman, G.G., Silicones The Plastics Institute, 1962
[4] McGregor, R.R., Silicones and their Uses McGraw Hill, 1954
[5] Roff, W.J. and Scott, J.R., Fibres, Films, *Plastics and Rubbers,* Butterworths, 1971
[6] Noll, W., *Chemistry and Technology of Silicones,* Academic Press, 1968

SIS S-21

Synonyms: *Poly(styrene-block-isoprene-block-styrene).* SIS polymer. *Poly(ethenylbenzene-block-2-methyl-1,3-butadiene-block-Ethenylebenzene). Styrene-isoprene-styrene triblock copolymers*
Related Polymers: Styrene-isoprene copolymers, Styrene-isoprene star block copolymers
Monomers: Styrene, Isoprene
Material class: Copolymers, Thermoplastic, Natural elastomers
Polymer Type: polydienes, polybutadienepolystyrene

CAS Number:

CAS Reg. No.	Note
25038-32-8	
105729-79-1	block

Additives: Carbon fibre
General Information: Is probably the most important styrene-isoprene copolymer
Morphology: Contains a varying number of phases depending on MW and composition [3]. Lamellae, spherical and cylindrical microdomains have been identified for copolymers of different compositions. A mobile polystyrene-polyisoprene junction point has been identified by dielectric spectroscopy; the associated chain dynamics have been reported [4,8,9]

Volumetric & Calorimetric Properties:
Density:

No.	Value	Note
1	d 0.92 g/cm^3	14% styrene [1]

Melting Temperature:

No.	Value	Note
1	153°C	[10]

Glass-Transition Temperature:

No.	Value	Note
1	-47--37°C	[1]
2	-90.5--62.5°C	dependent on composition [3]

Transition Temperature General: There may be one, two or three glass-transition temps. depending on the composition [3]
Transition Temperature:

No.	Value	Note	Type
1	220°C	14.3% styrene, total MW 140000 [2,7,8]	Order-disorder transition
2	180°C	13% styrene, total MW 113800 [4,7,8]	Order-disorder transition
3	90°C	30% styrene, total MW 39000 [4,7,8]	Order-disorder transition
4	105°C	52% styrene, total MW 27700 [4,7,8]	Order-disorder transition
5	166°C	13% styrene, total MW 149000 [4,7,8]	Order-disorder transition

Transport Properties:
Polymer Solutions Concentrated: In an isoprene-selective solvent (tetradecane) solns. show rubbery, plastic and viscous behaviour at low, intermediate and high temps. Homogeneous mixing of styrene and isoprene blocks occurs in the viscous region but in the rubbery and plastic regions styrene blocks are segregated to form spherical domains and the isoprene blocks have either loop or bridge conformns. [5]
Polymer Melts: Melt elasticity is independent of temp. and total MW for copolymers in the disordered state with constant block length ratio. The melt viscosity, however, is dependent on both total MW and block length ratio. In the ordered state props. would depend on the morphological state of the microdomains [4]

Mechanical Properties:
Mechanical Properties General: At similar compositions the lower MWs tend to give better tensile resistance. At similar MWs tensile strength increases with polyisoprene content. Optimum shear strength is found with MWs of 30000–50000 and a polyisoprene content of 20–40%. Shear strength can be improved by chemical modification, e.g. by reaction with small amounts of maleic anhydride [9]
Tensile Strength Break:

No.	Value	Note	Elongation
1	15.7 MPa	14% styrene [1]	167%
2	12.2–12.4 MPa	14% styrene, 5 phr carbon fibre [1]	2025–2100%
3	10.8–11.7 MPa	14% styrene, 10 phr carbon fibre [1]	2025–2100%
4	4.4–4.6 MPa	14% styrene, 20 phr carbon fibre [1]	1860–1880%

Miscellaneous Moduli:

No.	Value	Note	Type
1	0.25 MPa	14% styrene [1]	10% Modulus
2	0.25 MPa	14% styrene, 5 phr carbon fibre [1]	10% Modulus
3	0.35–0.51 MPa	14% styrene, 10 phr carbon fibre [1]	10% Modulus
4	0.59–0.62 MPa	14% styrene, 20 phr carbon fibre [1]	10% Modulus
5	0.38 MPa	14% styrene [1]	25% Modulus
6	0.41 MPa	14% styrene, 5 phr carbon fibre [1]	25% Modulus
7	0.6–0.67 MPa	14% styrene, 10 phr carbon fibre [1]	25% Modulus
8	0.1–1.15 MPa	14% styrene, 20 phr carbon fibre [1]	25% Modulus
9	0.5 MPa	14% styrene [1]	50% Modulus
10	0.51–0.52 MPa	14% styrene, 5 phr carbon fibre [1]	50% Modulus
11	0.7–0.75 MPa	14% styrene, 10 phr carbon fibre [1]	50% Modulus
12	1.02–1.1 MPa	14% styrene, 20 phr carbon fibre [1]	50% Modulus

Viscoelastic Behaviour: The ratio of loops to bridges of the polyisoprene midblock is a major factor influencing the viscoelastic props. Viscoelastic props. can be very sensitive to the presence of styrene-isoprene diblocks and homopolymer. The effects of added diblock and homopolymer on the props. of a copolymer with spherical domain morphology have been reported [6,9]
Hardness: Shore A30 (14% styrene); A33 (14% styrene; 5 phr carbon fibre); A35 (14% styrene, 10 phr carbon fibre), A37 (14% styrene, 20 phr carbon fibre) [1]
Failure Properties General: Tear strength 28.2 N mm^{-1} (14% styrene), 20.30–21.6 N mm^{-1} (14% styrene, 5 phr carbon fibre), 19.90–20.6 N mm^{-1} (14% styrene, 10 phr carbon fibre), 15.00–6.2 N mm^{-1} (14% styrene, 20 phr carbon fibre) [1]. The failure props. of adhesives containing triblock copolymers have been reported [9]

Optical Properties:
Transmission and Spectra: Ir spectral data have been reported [2]

Polymer Stability:
Polymer Stability General: Ageing tests show that oxidation of surface molecules initially occurs slowly when stored at 95°. After

– Sodium carboxymethyl cellulose

approx. 90 min., however, oxidation proceeds much more rapidly; the MW distribution shifts to low values and there is a significant loss of mech. props. During degradation the surface energy of the polymer changes from nonpolar to polar [2]

Applications/Commercial Products:
Processing & Manufacturing Routes: May be synth. using a three-stage anionic polymerisation in C_6H_6 at 45° using n-butyl lithium as initiator [3]
Applications: Used in adhesives, and as reinforcing agents

Trade name	Details	Supplier
Cariflex TR 1107		Philips Petroleum International
Kraton	D-1107	Shell Chemicals

Bibliographic References

[1] Roy, D., Bhowmick, A.K. and De, S.K., *J. Appl. Polym. Sci.*, 1993, **49**, 263, (dynamic mech props)
[2] Harrison, D.J.P.; Johnson, J.F. and Yates, W.R, *Polym. Eng. Sci.*, 1982, **22**, 865, (thermal stability)
[3] Meyer, G.C. and Widmaier, J.M., *J. Polym. Sci., Part B: Polym. Lett.*, 1982, **20**, 389, (T_g)
[4] Han, C.D.; Baek, D.M.; Kim J.K. and Chu, S.G., *Polymer*, 1992, **33**, 294, (rheology)
[5] Watanabe, H.; Sato, T.; Osaki, K; Yao, M.-L. and Yamagishi, A., *Macromolecules*, 1997, **30**, 5877, (rheology)
[6] Berglund, C.A. and McKay, K.W., *Polym. Eng. Sci.*, 1993, **33**, 1195, (viscoelastic props)
[7] Winter, H.H., Scott, D.B. Gronski, W., Okamoto, S. and Hashimoto, T., *Macromolecules*, 1993, **26**, 7236, (ordering)
[8] Adams, L.L.; Quiram, D.J.; Graessley, W.W; Register, R.A. and Marchand, G.R., *Macromolecules*, 1996, **29**, 2929, (ordering)

Sodium carboxymethyl cellulose S-22

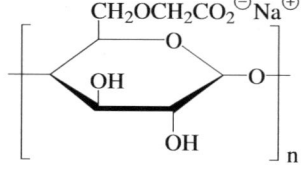

Synonyms: *CMC. Cellulose carboxymethyl ether sodium salt. Croscarmellose sodium*
Related Polymers: Cellulose
Monomers: Base monomer unit 6-carboxymethyl glucose
Material class: Polysaccharides
Polymer Type: cellulosics
CAS Number:

CAS Reg. No.	Note
9000-11-7	carboxymethyl cellulose
9004-32-4	

Molecular Formula: $(C_8H_{11}O_7Na)_n$
Fragments: $C_8H_{11}O_7Na$
Molecular Weight: MW 90000–700000 (typically 200 to 1000 'monomer' units)
General Information: Degree of substitution 0.65–1.45 [4]; sodium content 7.0–12.0% [4]

Volumetric & Calorimetric Properties:
Density:

No.	Value
1	d 1.067 g/cm^3
2	d 0.55–1 g/cm^3

Volumetric Properties General: Film density 1.59 g m^{-1} [2,3]
Melting Temperature:

No.	Value	Note
1	>300°C	min. [1]

Transition Temperature:

No.	Value	Note	Type
1	227°C	[2,3]	Browning
2	252°C	[2,3]	Charring

Surface Properties & Solubility:
Plasticisers: 10–30% glycerol (film) [4]. Diglycerol, 2-methyl-2-nitro-1,3-propanediol, propylene glycol [4], ethanolamines, ethylene glycol, glycerol alphamonomethyl ether, glycerol monochlorohydrin, 2,3-butylene glycol, 1,2,6-hexanetriol
Solvents/Non-solvents: Sol. H_2O; sl. sol. EtOH, Me_2CO; insol. organic solvents [4]
Surface Tension:

No.	Value	Note
1	71000 mN/m	1% soln. [3]

Transport Properties:
Water Content: 8.0% (max., commercial product) [2,3]; 26–32% (80% relative humidity, 25°) [4]
Mechanical Properties:
Viscoelastic Behaviour: Viscosity ranges 1000–4500 mPa s (1%, 25°); 25–3100 mPa s (2%, 25°) [4]
Electrical Properties:
Complex Permittivity and Electroactive Polymers: Zeta potential 3.4 mV (multilayer in $10^{-4}N$ KCl) [5]
Optical Properties:
Refractive Index:

No.	Value	Note
1	1.515	film [2,3]
2	1.3355	25°, 2% soln. [2,3]

Polymer Stability:
Chemical Stability: Maximum stability occurs between pH 7–9
Applications/Commercial Products:
Processing & Manufacturing Routes: Alkali cellulose (the complexes obtained by treating cellulose with aq. sodium hydroxide) is treated with sodium chloroacetate in alcoholic solvent
Applications: Food applications as an emulsifier; to control ice cryst. growth; a protein stabiliser; texture modifier; thickener and rapid solubiliser. Used in personal care products as a thickener; stabiliser and viscosity regulator; in pharmaceuticals, tablet disintegrant, as a binder, viscosifier and emulsifier. Also used in adhesives; paints; paper coatings; textile sizing; drilling muds; films and detergents

Trade name	Details	Supplier
Aquasorb	Suberabsorbent CMC	Hercules
Blanose	Blanose refined CMC	Hercules
Cellofas		ICI, UK
Cellogen		Dai-Ichi
Edifas		ICI, UK

Gabrosa		Montedison
Hertcopac		Hercules
Tylose C		Hoechst
Walocel C	Registered trademark	Wolff

Bibliographic References
[1] Xu, Q., *Fenxi Huaxue*, 1985, **13**, 170, (melting point)
[2] Cellulose Gum, Chemical and Physical Properties Hercules Inc., Wilmington, Delaware, U.S.A., 1980, (technical datasheet)
[3] Just, E.K. and Majewicz, T.G., *Encycl. Polym. Sci. Eng.*, 2nd edn., (eds. H.F. Mark, N.M. Bikales, C.G. Overberger and G. Menges), John Wiley and Sons, 1985, **3**, 239
[4] Blanose, *Cellulose Gum*, Technical Bulletin 87.502-E7, Hercules Inc., Wilmington, Delaware, U.S.A., (technical datasheet)
[5] Onabe, F., *J. Appl. Polym. Sci.*, 1979, **23**, 2999

Spandex fibres S-23

Related Polymers: Polyurethanes General information
Monomers: Ethylenediamine, 1,4-Butanediol, 3,4-Diisocyanatotoluene, 2,6-Diisocyanatotoluene, 4,4′-Diphenylmethane diisocyanate, 2,4-Diisocyanatotoluene
Material class: Fibres and Films
Polymer Type: polyurethane
General Information: Spandex (FTC) is the generic name given to synthetic elastomeric fibres that contain at least 85% of segmented polyurethane. The term elastane (ISO) is most common in Europe. May be used to replace natural rubber thread in many applications. Fibres are lightweight, and have high tensile and tear strength and excellent abrasion resistance. Melt or soln. spun fibres of a wide variety of denier sizes are available

Volumetric & Calorimetric Properties:
Density:

No.	Value	Note
1	d 1 g/cm^3	[1]

Melting Temperature:

No.	Value	Note
1	250°C	[1]

Transition Temperature General: Sticking temp. 175° [1]
Mechanical Properties:
Mechanical Properties General: Tenacity 0.7g den^{-1} 550% elongation [1]
Polymer Stability:
Thermal Stability General: Yellows and degrades above 150° [1]. Stable to a wide range of chemicals including dry-cleaning fluids, soaps, detergents, oils, perspiration, acids and alkalis [1]
Applications/Commercial Products:
Processing & Manufacturing Routes: Usually prepared by a two-stage process using an isocyanate terminated prepolymer that is chain extended with a low MW diol, diamine or hydrazine. Several methods of spinning: dry spinning from soln., reaction spinning, wet spinning, hot-melt extrusion. The most important process at present is dry spinning from DMF soln. Lycra (DuPont) and Dorlastan (Bayer) are made by this process. Lycra is the first and most sucessful spandex fibre [2] and is based on poly(oxytetramethylene)glycol reacted with excess diisocyanate, HMDI and chain extended in soln. with aliphatic diamines. Dorlastan is based on MDI with soft segment consisting of hydrophobic polyester chains derived from adipic acid and aliphatic glycols e.g.1,6-hexanediol. Much work has been done on melt-spinning to avoid the use of expensive solvents and solvent recovery but this process is commercially less important than dry-spinning because of limitations on elastomer melting point
Applications: Spandex fibres have established a substantial and competitive market in the clothing industry replacing natural rubber threads. They can be produced as fine yarns and have better resistance to oxidation, sunlight and dry cleaning fluids than natural rubber threads and may be dyed using acid, basic or disperse dyes. Inferior to natural rubber in stress retention and some grades are less resistant to hot detergent wash treatment. Degraded by solns. of bleach and chlorine but generally satisfactory in low chlorine levels. Used in supportive clothing (belts, girdles, corsets, brassieres, garters, surgical stockings). In swimwear, cycling shirts, cycling shorts and various sportswear applications. Other uses in upholstery and carpets

Trade name	Supplier
Acelan	Tae Kwang
Cleerspan	
Dorlastan	Bayer Inc.
Elastan	Chemitex Cellviskoza
Enkaswing	Le Seda de Barcelona
Espa	Toyobo
Fujibo Spandex	Fuji Chemical
Gomelast	Spandhaven
Kane/Loobell	Kanebo Ltd
Likra	Nylmex
Likra	DuPont
Lycra	DuPont
Lynel/Lyneltex	Fillatice
Mobilon	Nissin Spinning
Opelon	Toray
Opelon	DuPont
Roica	Asahi Chem. Co Ltd
Spantel	Kuraray Co
Spanzelle	Courtaulds
Vispan	Elastofibre

Bibliographic References
[1] Marshall, T.B., *Textile Ind.*, 1961, **125**, 75
[2] U.S. Pat., 1954, 2 692 893, (synth.)
[3] *Kirk-Othmer Encycl. Chem. Technol.*, 4th edn., Vol.10 (ed. J.I.Kroschwitz), John Wiley and Sons, 1993, 628
[4] *Encycl. Polym. Sci. Eng.*, 2nd ed., Vol.6, (ed. J.I. Kroschwitz) John Wiley and Sons, 1986, 739
[5] *Encyclopedia of Advanced Materials*, Vol. (A-E), (eds., Bloor, Brook, Flemmings and Mahajan) Elsevier Science, 1994, 676

Starch S-24

Synonyms: *Amylum. Polyamylose*
Related Polymers: Amylose, Amylopectin

Monomers: Base monomer unit glucose
Material class: Polysaccharides
Polymer Type: cellulosics
CAS Number:

CAS Reg. No.
9005-84-9

Molecular Formula: $(C_6H_{10}O_5)_n$
Fragments: $C_6H_{10}O_5$
General Information: Starch is a polysaccharide polymer found in seeds and tubers of many plants. It comprises two related components: **Amylose**, a mainly linear polymer of $(1\rightarrow4)$-α-D-glucose comprising 200–2000 glucose residues, and Amylopectin, a branched polymer of $(1\rightarrow4)$-α-D-glucose, containing both 1–4 and 1–6 bonds, in which each branch may contain 20–30 sugar residues. The degree of polymerisation can be up to ten times greater than for amylose [1]. Starch forms into granules with diameter range 2–150μm. Typical values are 20–22μm in wheat seeds and 25–50μm in potato tubers. Differences in the ratio of the two polymers and the size of granules, sometimes rather small, lead to changes in props. that are important in processing and performance. The amylose component is usually found as a linear polymer, though occasional branching is found in samples from some sources. Amylose usually comprises 18–28% of the mixt. in starch, though in some seeds it appears to make up ca. 85% of the total. In parallel with starch, amylose has been identified in four forms, A, B, C and V, though detailed conformational information is scarce. Polymer chains appear to have a left-handed helical struct., containing six glucose residues per turn [3]. This struct. accounts for the ability of amylose to absorb water (up to 10%) and to form clathrates with a wide range of molecules. Complexation with butanol is important in its isolation and purification. In soln. amylose is thought to form random coils, consisting of helical sections separated by disordered areas of the polymer [4]. Amylopectin is one of the largest natural polymers, its MW range is variously quoted in the range 1000000–500000000, depending on the source [5,6]. Branches occur at every tenth-twelfth glucose residue, usually by reaction at C-6, though C-3 bonds are also found. Each side-chain comprises 20–30 glucose residues, which may also be non-linear. Thus the struct. of amylopectin is variable and complex
Morphology: Granules generally contain crystalline regions which make up 25-45% of the total. X-ray data suggest four types of macromolecular arrangements are possible. Types A and B are found in cereals and tubers respectively. Type C is present in smooth peas and some beans, while type V is generated when starch is precipitated from soln. Samples high in type A material may contain only 10–15% crystalline materials [2]
Polymer Stability:
Polymer Stability General: Moderately explosive when exposed to flame

Bibliographic References
[1] *Modif. Starches: Prop. Uses*, (ed. O.B. Wurzburg), CRC Press, 1987
[2] Zobel, H.F., *Starch/Staerke*, 1988, **40**, 1
[3] Blackwell, J., Sarko, R.H. and Marchessault, R.H., *J. Mol. Biol.*, 1969, **42**, 379
[4] Greenwood, C.T., *Carbohydr. Res.*, 1968, **7**, 349
[5] Banks, W. and Greenwood, C.T., *Starch and its Components*, Edinburgh University Press, 1975
[6] Zimm, B.H. and Thurmond, C.D., *J. Am. Chem. Soc.*, 1952, **74**, 1111

Styrene-butadiene block copolymers   S-25
Synonyms: *Styrene-butadiene di-block copolymers. Poly(styrene-block-butadiene). Poly(ethenylbenzene-block-1,3-butadiene)*
Related Polymers: Styrene-butadiene copolymers, K-resin, Styrene-butadiene radial block copolymers, Blends of polybutadiene and styrene-butadiene block copolymers, Blends of polystyrene and styrene-butadiene block copolymers
Monomers: Styrene, 1,3-Butadiene
Material class: Synthetic Elastomers
Polymer Type: polybutadiene, polystyrene
CAS Number:

CAS Reg. No.	Note
9003-55-8	
106107-54-4	block

Applications/Commercial Products:

Trade name	Supplier
Buna BL 6533	Bayer Inc.
Buna BL 6533	Mobay
Ricon 500	Advanced Resins
Solprene 1205	Housemex
Stereon 840A	Firestone
Styrolux	BASF

Styrene-butadiene copolymers   S-26
Synonyms: *Poly(styrene-co-butadiene). SB Polymer*
Related Polymers: Polybutadiene, Styrene-butadiene rubber, Styrene-butadiene block copolymers, High styrene resins, SBS
Monomers: Styrene, 1,3-Butadiene
Material class: Synthetic Elastomers
Polymer Type: polybutadiene, polystyrene
CAS Number:

CAS Reg. No.
9003-55-8

Applications/Commercial Products:

Trade name	Details	Supplier
Ashlene	various grades	Ashley Polymers
Cevian	N	Daicel Chem.
Comalloy	240-3020/3030/3040	Comalloy Intl. Corp.
Darex		W R Grace
Darex		Organics
Duranit		Hüls AG
Europrene		Enichem Elastomers Ltd.
FPC	various grades	Federal Plastics
Fiberfil	G-40	DSM Engineering
Fiberstran	G-40	DSM Engineering
Finaclear	520/530	Fina Oil and Chem.
Finaprene	various grades	Fina Oil and Chem.
GPC Delta	D series	Grand Pacific
K-Resin	various grades	Phillips 66 Co.
Kibisan	PN	Chi Mei Industrial
Kostil	B26/B36	Enichem America
Kraton	D	Shell Chemical Co.
Luran	various grades	BASF
Pliolite		Goodyear

RSA2OKEO1BL		Ferro Corporation
Stereon	840A/881/900	Firestone
Styvex	32000/32002	Ferro Corporation
Tyril	various grades	Dow Chemical
Vector	various grades	Dexco Polymers

Styrene butadiene radial block copolymers   S-27

Synonyms: *Branched butadiene-styrene block copolymers. Radial teleblock butadiene-styrene copolymers. Radial multichain styrene-butadiene block copolymer. Star styrene butadiene block copolymers. Star-branched butadiene-styrene block copolymers*
Related Polymers: Styrene butadiene block copolymers
Monomers: Styrene, 1,3-Butadiene
Material class: Copolymers, Thermoplastic
Polymer Type: polybutadiene, polystyrene
CAS Number:

CAS Reg. No.
9003-55-8
106107-54-4
61789-96-9

Molecular Formula: $[(C_8H_8).(C_4H_6)]_n$
Molecular Weight: MW 190 000 (20% styrene); M_n 150 000 (20% styrene); MW 115000–136000 (30% styrene, trichain); MW 182 000 (30% styrene, tetrachain); MW 115 000 (40% styrene, trichain); MW 182000 (40% styrene, tetrachain); MW 77000–216000 (30% wt% styrene; trichain; terminating in butadiene blocks); MW 63000–155000 (30 wt% styrene; trichain; terminating in butadiene blocks); MW 135000–177000 (30 wt% styrene; trichain; terminating in styrene blocks); MW 111000–139000 (30 wt% styrene; trichain; terminating in styrene blocks); MW 156000–173000 (30 wt% styrene; tetrachain; terminating in styrene blocks); M_n 127000–148000 (30 wt% styrene; tetrachain; terminating in styrene blocks)
Additives: Fillers (carbon black, clay, silica, whitings); resins (polymers, e.g. polythene or polystyrene or other polymers (as blends); TiO_2; stearic acid; thermal stabilisers (e.g. zinc dibutyl dithiocarbamate); antioxidants (phenolic based, e.g. phenyl-2-naphthylamine); pigments; light (uv) stabilisers (e.g. benzotriazoles and benzophenones)
General Information: May be represented by general formula $(SB)_n$ where n = 3 (trichain (Y-shaped) molecule) or n = 4 (tetrachain (X-shaped) molecule) [8]; radial copolymers uniquely combine effects of long-chain branching with aggregation phenomenon of terminal block polystyrene-containing materials [8] Radial block copolymers belong to class of 'themoplastic elastomers', i.e. they combine various desirable props. of thermoplastic and elastomeric materials without the need for vulcanisation [8,10]. They possess high coefficients of friction and good flexibility at low temps.
Morphology: Polystyrene segments associate with one another to form microdomains which effectively cross-link molecules into network struct. [6] Reinforcement of rubbery segments in this way leads to high tensile strengths [10]
Miscellaneous: Reviews describing the physical and mech. props., processing and compounding, and applications of radial block copolymers have been reported [8,10]

Volumetric & Calorimetric Properties:
Density:

No.	Value	Note
1	d 0.92–0.94 g/cm^3	30 wt% styrene [16]
2	d 0.94–0.95 g/cm^3	30–40 wt% styrene [10]

Glass-Transition Temperature:

No.	Value	Note
1	-87°C	lower, 30% styrene [8]
2	110°C	higher, 30% styrene [8]
3	-92°C	lower, 40% styrene [8]
4	102°C	upper, 40% styrene [8]

Transition Temperature General: Incompatability of blocks causes raw polymers to exhibit distinct T_g values for each segment [6,10]. Such block copolymers can have T_g values well below those of corresponding random copolymers, e.g. systems containing 40% styrene [10]

Surface Properties & Solubility:
Solubility Properties General: Can be blended with both elastomers and thermoplastics (e.g. polystyrene, polythene, polybutadiene, polyurethanes, neoprene, EPDM) to give combinations of props. desirable in specific applications [4,5,6]
Plasticisers: Naphthenic oil, paraffin oil
Solvents/Non-solvents: Sol. toluene [5,6], THF [5], cyclohexane [12], hexane [6] Insol. isopropanol [12]; EtOH [14] Generally block copolymers are sol. in solvents in which corresponding homopolymers are insoluble. This is a consequence of block struct and 'non-permanence' of network struct, which lead to unique solubility characteristics [6]
Wettability/Surface Energy and Interfacial Tension: Adhesive props. of various formulations containing radial block copolymers have been reported [5,6,7]. The reduction in viscosity with an increase in radiality (n = 2–4, linear to tetrachain) [5], and the effects of this parameter, MW and copolymer composition on shear resistance and adhesion and tack have been reported [1,4]. Shear resistance increases as a function of radiality [5,6] while tack shows little change [1]. Tack of formulations as a function of styrene content has been reported [6,7]. Peel strength 2.1–4.6 lb in^{-1} (15–40 wt% styrene, ASTM D 1000) [6,7,8]. Shear adhesion 0.03–4.5 h (to failure, 15–40 wt% styrene, 90°, slippage of sample bonded vertically to stainless steel) [6,7]. Tack, peel strength and shear adhesion as functions of MW, degree of branching and monomer distribution in blocks have been reported [6,7,8]

Transport Properties:
Transport Properties General: Block copolymers behave as thermorheologically complex materials [11]. Radial block copolymers display markedly different viscosity props to those of linear analogues [5,12]. Rheological behaviour is dominated by their two-phase domain struct., which persists in the melt, i.e. at temps. T_g (polystyrene domains) [12]. Viscous flow is significantly affected by long-chain branching [15]
Melt Flow Index:

No.	Value	Note
1	< 1 g/10 min	max., commercial materials; 30% styrene [16]
2	30 g/10 min	commercial material; 44% styrene; plasticised [16]
3	2–4 g/10 min	180°, 5 kg; 40% styrene, MW 150 000 [8]

Polymer Solutions Dilute: Solution viscosity measurements (17% soln. in toluene/ naptha at 25°) of comparable block copolymer systems show a reduction in viscosity with increasing radiality (n = 2–4, linear to tetrachain) (e.g. for copolymer containing 30% styrene) [5] Viscosity [12] 0.73–1.66 dl g^{-1} (THF, 25°, 30% styrene, trichain, terminating in butadiene blocks); 0.99–1.49 dl g^{-1} (THF, 25°, 30% styrene, trichain, terminating in styrene blocks); 1.13–1.14 dl g^{-1} (THF, 25°, 30% styrene, tetrachain, terminating in styrene blocks) $[\eta]_{inh}$ 1.27–1.01 dl g^{-1} (toluene, 15–40 wt% styrene) [6,7]. The effect of MW degree of branching and monomer distribution in blocks on inherent viscosity has also been reported [6,7] Other viscosity data have been reported [16]

Polymer Melts: Mooney viscosity of comparable block copolymer systems decreases with increasing radiality ($n = 2$–4, linear to tetrachain) [5]. Corresponding effects are also observed in measurements of melt flow (e.g. for copolymers containing 30% styrene) [5]. At constant MW, the viscosity of a polymer terminating in S block is the greatest regardless of branching. Mooney viscosity has been reported as a function of styrene content [6,7,8]

Mechanical Properties:
Mechanical Properties General: Thermoplastic elastomers possess high coefficients of friction and good flexibility at low temps.: depending on MW and styrene content, block copolymers can exhibit relatively high tensile strength [6], which increases with styrene content and MW [8,10]. Tensile strength decreases with increasing temp.; may become low as softening point is approached [8,10]

Tensile Strength Break:

No.	Value	Note	Elongation
1	4.5–27.5 MPa	4000–650 psi, 26–60°, MW 130 000 [4]	
2	12.4–26.2 MPa	3800–1800 psi, 26–60° MW 250 000 [4]	
3	1.72–27.6 MPa	250–4000 psi, 30% styrene [16]	600–820%
4	9.2–10.7 MPa	1330–1500 psi, 30–46% styrene, plasticised, commercial material [16]	1100–1200%
5	11.7–33.8 MPa	1700–4900 psi, 15–40 wt% styrene [6,7]	
6	19.3–27.6 MPa	2800–4000 psi, 30–40 wt% styrene [8]	700–750%

Miscellaneous Moduli:

No.	Value	Note	Type
1	19.3–27.6 MPa	2800–4000 psi, 30–40 wt% styrene [8]	300% modulus

Viscoelastic Behaviour: Viscoelastic behaviour is governed by the length of the terminal block and not by MW [12]
Mechanical Properties Miscellaneous: Tensile strength as a function of styrene content, MW, degree of branching and monomer distribution in blocks has been reported [6,7]
Hardness: Shore A increases with increasing styrene content [10]; Shore A61–A75 (30% styrene, commercial material) [16]; A46–A82 (30–40% styrene, commercial material, plasticised) [16]; Shore A42–A98 25°, 10–50 wt% styrene) [10]. Copolymers can display a wide range in hardness values [8,10]. Shore A80–A92 (30–40% styrene) [10]. Hardness decreases with increasing temp; may become low as softening point is reached [10]
Friction Abrasion and Resistance: Resistance to abrasion and cracking improves with increasing MW [8,10]

Polymer Stability:
Thermal Stability General: An increase in temp. causes loss of tensile strength thereby limiting serviceable use at higher temps. [8,10]
Environmental Stress: Presence of unsaturated polybutadiene block renders copolymers susceptible to attack by ozone; resistance can be improved by compounding techniques [10]

Applications/Commercial Products:
Processing & Manufacturing Routes: Synth. by addition of butadiene to styrene and butyl lithium initiator in cyclohexane at 70°; polymerisation is allowed to proceed for approx. 30 min., and then a coupling agent is added, e.g. PCl_3 for trichain ($n = 3$) or $SiCl_4$ for tetrachain ($n = 4$). The reaction is terminated after a further 30 min. by the addition of isopropanol [12]. Many variations on this general route have been reported [13,14,15]. Thermoplastic-elastomeric nature allows processing using conventional equipment for both thermoplastics and rubbers, e.g. injection moulding, compression moulding and extrusion [8,10]. Deposition from solvents is also used to form products
Applications: As an elastomeric component of pressure-sensitive adhesives; as an asphalt modifier (to give optimun high-temp. performance); in contact cements and hot melts. There is a wide range of elastomeric applications including use in shoe soles and heels, hose and tubing, toys, cove bases, mats, weatherstrip and in various extruded goods and moulded parts. Used in blends with other rubbers and plastics to modify props. Examples include addition in small quantities to improve impact strength/resistance in plastics or green strength in rubbers. Blending with other rubbers e.g. with EPDM, can also give improved resistance to ozone, etc [10]

Trade name	Details	Supplier
Europrene Sol T	non-exclusive; used for other butadiene polymer systems	Anic
Kraton	D 11884 non-exclusive; used for other styrene and butadiene rubber systems branched S/B copolymer; 30% styrene	Shell Chemical Co.
Solprene	various grades; butadiene-styrene radial teleblock co-polymers (30–40% styrene)	Philips Petroleum International
Solprene	non-exclusive; used for other butadiene/styrene/isoprene polymer and copolymer systems styrene-butadiene radial block co-polymers (40:60%)	Philips Petroleum International
	D 1188X SBS radial block copolymer (80% diblock; 30% styrene	Shell Chemical Co.
	D 458 branched S/B copolymer; 30% styrene; contains 33% plasticiser oil	Shell Chemical Co.
	D 4230 branched S/B copolymer; 46% styrene; contains 23% plasticiser oil	Shell Chemical Co.
	D 4240 branched S/b copolymer; 44% styrene; contains 465 plasticiser oil	Shell Chemical Co.

Bibliographic References
[1] *Polymer*, 1990, **6**, No. 6, 447-454
[2] *J. Rheol. (N.Y.)*, 1987, **31**, No. 8, 711-724
[3] *Polymer*, 1988, **20**, No. 4, 293-305
[4] *J. Rheol. (N.Y.)*, 1989, **33**, No. 3, 469-480
[5] Marrs, O.L., Zelinski, R.P. and Doss, R.C., *J. Elastomers Plast.*, 1974, **6**, 246, (transport props., surface props., adhesion, uses)
[6] Marrs, O.L., Naylor, F.E. and Edmonds, L.O., *Recent Advances in Adhesion (Proc. Am. Chem. Soc. Symp.)*, (ed., L.-H., Lee), Gordon and Breach, 1973, 213, (transport props., surface props., mechanical props., adhesion, uses)
[7] Marrs, O.L., Naylor, F.E. and Edmonds, L.O., *J. Adhes.*, 1972, **4**, 211, (transport props., surface props., mechanical props., adhesion, uses)
[8] Haws, J.R., *ACS Symp. Ser.*, 1974, **4**, 1, (rev.)
[9] Haws, J.R. and Middlebrook, *Rubber World*,
[10] Haws, J.R. and Middlebrook, T.C., *Rubber World*, 1973, **167**, 27, (rev.)
[11] Kraus, G. and Rollmann, K.W., J. Polym. Sci., *Polym. Phys. Ed.*, 1977, **15**, 385, (blends, microstruct.)

- Styrene-butadiene rubber

[12] Kraus, G., Naylor, F.E. and Rollmann, K.W., *J. Polym. Sci., Part A-2*, 1971, **9**, 1839, (synth., transport, props., rheology)
[13] *U.S. Pat.*, 1961, 2 975 160, (synth.)
[14] *U.S. Pat.*, 1966, 3 251 905, (synth.)
[15] Zelinski, R.P. and Wofford, C.F., *J. Polym. Sci., Part A-1*, 1965, **3**, 93, (synth.)
[16] Ash, M. and Ash, I., Handbook of Plastic Compounds, *Elastomers and Resins*, 1992, 633, (transport, props., mechanical props.)

Styrene-butadiene rubber S-28
Synonyms: *SBR. GR-S. Poly(ethenylbenzene-co-1,3-butadiene)*
Related Polymers: Polybutadiene, Styrene-butadiene copolymers, Polystyrene, High-*trans*-styrene butadiene rubber
Monomers: Styrene, 1,3-Butadiene
Material class: Synthetic Elastomers
Polymer Type: polybutadiene, polystyrene
CAS Number:

CAS Reg. No.
9003-55-8

Additives: Vulcanised by sulfur. Fillers include carbon black and extender oil. Mineral fillers such as alumina and lead silicate
General Information: Usually contains 25% styrene to 75% butadiene. Styrene and butadiene may be randomly dispersed mixts. or blocks or grafts. Does not break down on mastication. Synthetic polymer is more uniform in quality than natural rubber, and contains less contaminants
Morphology: Amorph.: variation of catalyst alters diene microstructure. Does not crystallise at high strain unlike natural rubber

Volumetric & Calorimetric Properties:
Density:

No.	Value	Note
1	d 0.933 g/cm^3	unvulcanised [1]
2	d 0.98 g/cm^3	pure gum vulcanisate
3	d 1.15 g/cm^3	vulcanisate, 50 phv carbon black [2]
4	d 0.94 g/cm^3	pure gum vulcanisate [25]

Thermal Expansion Coefficient:

No.	Value	Note	Type
1	0.00066 K^{-1}	unvulcanised [1]	V
2	0.00066 K^{-1}	pure gum vulcanisate [2,3]	V

Thermodynamic Properties General: Variation of thermodynamic props. with concentration of vulcanising sulfur and carbon black has been reported. [139,140] Thermal conductivity and coefficient of thermal expansion values have been reported for a range of SBR/butyl rubber blends [69,110]
Thermal Conductivity:

No.	Value	Note
1	0.19–0.25 W/mK	pure gum vulcanisate [7,8]
2	0.3 W/mK	vulcanisate, 50 phv carbon black [8]
3	0.246 W/mK	pure gum vulcanisate [69]
4	0.19 W/mK	SBR/BR blend (50/50) pure gum vulcanisate [69]

Thermal Diffusivity: 9×10^{-8} m^2 s^{-1} (pure gum vulcanisate) [9]

Specific Heat Capacity:

No.	Value	Note	Type
1	1.89 kJ/kg.C	unvulcanised [5]	P
2	1.83 kJ/kg.C	pure gum vulcanisate [6]	P
3	1.5 kJ/kg.C	vulcanisate, 50 phv carbon black [6]	P

Glass-Transition Temperature:

No.	Value	Note
1	-108 – -54°C	SBR/butyl rubber blends [89]
2	-64 – -59°C	unvulcanised [1]
3	-52°C	pure gum vulcanisate [4]
4	-52°C	vulcanisate, 50 phv carbon black [4]

Transition Temperature General: T_g increases with increasing styrene content [25] and variation with hydrocarbon content has been reported. [110] Random emulsion polymerisation is reported to produce single T_g value [78,111]. Blends of SBR with butyl rubber before vulcanisation give two distinct T_gs for the separate components and only one T_g after vulcanisation [89]
Brittleness Temperature:

No.	Value	Note
1	-60°C	vulcanisate, carbon black filled [25]

Surface Properties & Solubility:
Solubility Properties General: The compatibility of SBR with other tyre elastomers, with and without carbon black, cured and uncured has been investigated by DTA [112]. The miscibility of SBR with other tyre elastomers has been investigated by examining the morphology of blends after curing. Immiscible with butyl rubber, chlorobutyl rubber, EPDH and CR. Miscibility with natural rubber, and polybutadiene is borderline [119]
Cohesive Energy Density Solubility Parameters: Solubility parameter δ 17.39 J$^{1/2}$ cm$^{-3/2}$ (butadiene:styrene 85:15) [33]
Plasticisers: The effectiveness of pitch as a plasticiser for SBR and blends compared with oil has been reported. The 300% modulus for pitch, 5.6 Mpa, is higher than for oil, 2.6 Mpa, and elongation is similar (800–900%) but there is more heat build up for pitch [99]
Solvents/Non-solvents: The feasibility of using supercritical fluids to extract extender oil, organic acids and antioxidants from SBR has been reported [137]. Unvulcanised sol. hydrocarbon solvents
Surface and Interfacial Properties General: The distribution of carbon black between phases in blends of SBR with natural rubber and polybutadiene has been characterised [42,46]
Surface Tension:

No.	Value	Note
1	31.3 mN/m	PTFE mould, pure gum vulcanisate, 23% styrene [40]
2	35.6 mN/m	PET mould, pure gum vulcanisate, 23% styrene [40]
3	39.6 mN/m	stainless-steel mould, pure gum vulcanisate, 23% styrene [40]
4	33 mN/m	critical surface tension spreading [73]

Transport Properties:
Polymer Solutions Dilute: θ temp. 21° (2-pentanone, M_n 150000–860000, butadiene:styrene 76:24) [31]. 21° (octane, M_n 40000–800000, butadiene:styrene 75:25) [32]. Viscosity is affected by

amount of carbon black and MW [90]. Effect of shear rate, temp. and blend ratio on rheological props. of polymer lattices has been reported [55,143]. The distribution of carbon black between phases and effect of vulcanisation in NR/SBR blends has been determined [51,57].

Polymer Melts: Extensional viscosity data have been reported for SBR melt with 30–80 phv carbon black [74,75]. The effect of carbon black type and loading on the extensional viscosity of SBR melt and also the effect of other fillers - clay, calcium carbonate and silica - have been investigated [75]. The viscosity of SBR melt with different levels of silica and carbon black has been measured at shear rates from 3 to 3000 s^{-1} [98]. Mark-Houwink constants K 0.0525 ml g^{-1}, a 0.66 (C_6H_6, 25°) [34]; 0.0316 ml g^{-1}, a 0.7 (cyclohexane, 30°) [35]; 0.00525 ml g^{-1}, a 0.66 (toluene, 25°) [36]

Permeability of Gases: Permeability of gases increases with increasing temp. [157]. Relative gas permeabilities 1.09 (He/CH_4, 23°); 2.71 (O_2:N_2; 23–25°); 1.91 (H_2:CH_4); 5.9 (CO_2:CH_4) [30]. More gas permeable than polybutadiene. Diffusion coefficients of a range of hydrocarbons have been reported for a range of temp. and vulcanisation conditions [110,116,124,147,148,150,153]

Gas Permeability:

No.	Gas	Value	Note
1	air	34.8 cm^3 mm/(m^2 day atm)	24° [157]
2	air	2048 cm^3 mm/(m^2 day atm)	66° [157]
3	air	4346 cm^3 mm/(m^2 day atm)	94° [157]

Mechanical Properties:

Mechanical Properties General: Gum vulcanisates have low mech. strength, vastly improved using carbon black fillers [59,72,88,113,125,133]. Has been used as an additive in oriented fibres [104]. Other values for tensile strength have been reported [84,86] including blends with other elastomers [77]. Does not crystallise under high strain, unlike natural rubber, so mechanical props. are adversely affected

Tensile (Young's) Modulus:

No.	Value	Note
1	1–2 MPa	pure gum vulcanisate [37]
2	14–19 MPa	vulcanisate, 50 phv carbon black [37]
3	1.6 MPa	pure gum vulcanisate [4,16]
4	3–6 MPa	vulcanisate, 50 phv carbon black [15]

Tensile Strength Break:

No.	Value	Note	Elongation
1	16–23 MPa	carbon black filled; SBR/BR blends [88]	
2	24–28 MPa	carbon black filled [82]	415–500%
3	7 MPa	pure gum vulcanisate [26]	600%
4	14 MPa	vulcanisate, carbon black filled [26]	600%
5	1.4–3 MPa	pure gum vulcanisate [14,15,25,37]	400–600%
6	17–28 MPa	vulcanisate, 50 phv carbon black [14,15]	400–600%
7	14–24 MPa	vulcanisate, 50 phv carbon black [25,37]	300–700%

Elastic Modulus:

No.	Value	Note
1	1890 MPa	unvulcanised [11]
2	1960 MPa	vulcanised [11]
3	2500 MPa	vulcanised, 50 phv carbon black [11]

Miscellaneous Moduli:

No.	Value	Note	Type
1	0.53 MPa	pure gum vulcanisate [4]	shear modulus
2	2 MPa	vulcanisate, 50 phv carbon black [17]	shear modulus

Complex Moduli: In-phase shear modulus, G′, out-of-phase shear modulus, G″, and tan δ have been measured over a wide range of amplitude for SBR gum and vulcanisate and for blends with other elastomers [93]. Complex dynamic shear modulus for unvulcanised SBR [18,19,20], pure gum vulcanisate [18,19,22,23] and vulcanisate with 50 phv carbon black [2,17,19] has been reported. Dynamic mechanical props. have been reported for carbon black and other filled systems [55,62,70,81,82,100,101,102,114,115]. Effect of cross-linking on dynamic mechanical props. has been reported [92]. Storage modulus and loss modulus values have been reported [19,20,22,60,114]. log G′ 5.82 (log Pa) (unvulcanised) [20]; 6.94 (log Pa) (vulcanisate, 50 phv carbon black) [19]; log G″ 4.94 (log Pa) (unvulcanised) [20]; 6.28 (log Pa) (vulcanisate, 50 phv carbon black) [19]. Storage modulus is increased on addition of octadecylamine diluent; hydrocarbons have little effect [130]

Viscoelastic Behaviour: Constants in Williams-Landel-Ferry equation for temperature shift factor for viscous flow have been reported [19,56]. Dynamic elastic and loss moduli, and tan δ have been reported for cured SBR with different types of carbon black from 0.01 to 10 Hz [50]. Resilience [77] 65% (pure gum vulcanisate) [15,24,158]; 40% (vulcanisate, 50 phv carbon black) [24]; 45–61% (carbon black filled) [88]. Effects of microstructure and styrene content on rebound values have been reported [111]. Tack data have been reported [29]. Heat build up occurs readily. Values of 53–66° have been reported for a range of filled polymers [47,88]. Creep behaviour of carbon-filled polymer is characterised by temp. jump experiments [82]. Viscoelastic props. are affected by particle size and loading of carbon black fillers [83]. Creep activation energy increases with increase in carbon filler 54 kJ mol^{-1} (unfilled); 167 kJ mol^{-1} (80 phv carbon black) [61]. Extrudate swell [121,125] is dependent on volume fraction of carbon black, but not on shear rate, wall slip [149] or rubber-carbon network [121]. Green strength [44] 1.7 kN m^{-1}; tack 1.56 kN m^{-1} [72]. Peel strength 1.2–1.5 kN m^{-1} (bonded to UHMWPE) [43]

Mechanical Properties Miscellaneous: Longitudinal bulk wave velocity 1485 m s^{-1} (pure gum vulcanisate); 1510 m s^{-1} (vulcanisate, 50 phv carbon black) [12]. Longitudinal strip velocity 73 m s^{-1} (pure gum vulcanisate) [13]; 161 m s^{-1} (vulcanisate, 50 phv carbon black) [12]. Compressibility 530/TPa (unvulcanised) [11]; 510/TPa (pure gum vulcanisate) [12]; 400/TPa (vulcanisate, 50 phv carbon black) [11]. Retraction after extrusion has been reported [72]

Hardness: Shore A40–A100 (vulcanisate, carbon black filled) [25]; A40–A90 [26]; A65–A71 (carbon black filled) [82]. Other hardness values have been reported [111]

Failure Properties General: Crack growth [80] is slower in vacuum than in air, but these effects are less marked under high loads and tearing rates [45,47,108]. Threshold tear strength [77,127,128] 60 J m^{-2} (Young's modulus 2.25) [63]. Tearing energy [76,77,111] 10–26 kJ m^{-2} (carbon filled); 26–100 kJm^{-2} (SBR elastomer blends) [88].

Friction Abrasion and Resistance: Coefficient of friction [57] 0.6–1.5 [54], 2 [40], 2.2 [40], 2.3 [40]; values increase when moulded against surfaces with increasing surface energy. Carbon black filled grades [136] have lower values [94,135]. Skid resistance has been reported for wet ice [53] and correlated to phase behaviour [103,111]. Has better abrasion resistance and resistance to tearing

[56] but lower resilience (causing heat build-up) [27] than natural rubber. Abradability is affected by amount of carbon black and other fillers [41,49,105,151,155]. Wear by metal is much higher in nitrogen than in air [79]

Electrical Properties:
Electrical Properties General: Volume resistivity 10^5 Ω cm (pure gum vulcanisate) [25]. Electrical props. of mica filled [52,151], carbon black filled [48,141,152] and sulfur vulcanised [139] polymer have been reported

Dielectric Permittivity (Constant):

No.	Value	Frequency	Note
1	2.5	1 kHz	unvulcanised [10]
2	3	1 kHz	pure gum vulcanisate [10]

Dissipation (Power) Factor:

No.	Value	Note
1	0.0009	unvulcanised [10]
2	0.0009	pure gum vulcanisate [10]

Optical Properties:
Refractive Index:

No.	Value	Note
1	1.5345	unvulcanised [1]
2	1.5368	[112]

Transmission and Spectra: Nmr spectral data have been reported [66,91]. Sequence distribution [71,117] has been reported and H-1 nmr magic angle spinning spectral data indicate loosely and tightly bound monomer segments on the carbon black surface [132]

Polymer Stability:
Upper Use Temperature:

No.	Value	Note
1	110°C	vulcanisate, carbon black filled [14,25]
2	90°C	[26]

Decomposition Details: Weight loss occurs at 442–449° (23–40% styrene) and 261–309° (23–40% styrene) [25]. Volatilisation reported to occur at 330–440° [87]. Decomposition onset 350–390° (N_2, gum) [93]. Carbon black filler has no effect on pyrolysis
Flammability: Limiting oxygen index 16.9% (foam) [38]; 18.5%; 24% (53 wt.% alumina trihydrate) [95]. Smoke evolution has been reported [87]; addition of $CaCO_3$ reduces soot emissions by up to 60% [95]. Flammability rating 3.9 (Matel) [107]; 1.8 (Mil 417) [107]
Environmental Stress: Vulcanisate tears more rapidly in air than in vacuum, and tearing is increased by thiophenol, probably due to prevention of broken chain radicals recombining [47]. Electron irradiation in air liberates H_2, and methane (87%) plus higher mass products [67]. Has good resistance to oxidation and γ-radiation [122,125]; oxygen is reponsible for stress relaxation between 75–115° [97]. Oxidative hardening also has been reported [100]. Carbon black filled vulcanisate has poor ozone resistance; the rate of crack propagation has been reported [109]. Ageing at 70° has been reported [28]. Fate of airborne particles has been reported [65,142]
Chemical Stability: Has poor resistance to gasoline, oil and hydrocarbons [2,26,64]
Biological Stability: Has been monitored for mutagenic props. [44]
Recyclability: Pyrolysis kinetics have been reported; potential for energy recovery from scrap [126]

Applications/Commercial Products:
Processing & Manufacturing Routes: Originally produced by emulsion polymerisation of styrene and butadiene at 50° - hot rubber; also produced at 5° - cold rubber. Broader MW distribution obtained from hot rubber process, more suitable for adhesives. 'Cold' product has more *trans* isomer (72%). May also be produced by solution polymerisation using anionic catalysis, with more control over molecular structure. Suitable initiators determine microstructure, and monomer concentration controls distribution of monomer segments
Applications: Main application is in tyre tread, also used for conveyor belts. Not used in thick sections due to excessive heat generation. Other applications include electrical sheathing, rollers, moulded and extruded articles. May also be blended with natural rubber and other elastomers

Trade name	Details	Supplier
Ameripol Synpol 1009		Ameripol Synpol
Buna Hüls EM		Hüls AG
Buna SL		Bayer Inc.
COPO 1500, 1712		Copolymer Rubber
Cariflex	styrene-butadiene rubber	Shell Chemicals
Cariflex S		Shell Chemicals
Krylene		Bayer Inc.
Plioflex		Goodyear
Pliolite LPF-2108		Goodyear

Bibliographic References
[1] Wood, L.A., Physical Chemistry of Synthetic Rubbers (ed. G.S. Whitby), John Wiley, 1954
[2] Wildschut, A.J., *Technological and Physical Investigations on Natural and Synthetic Rubbers*, Elsevier, 1946
[3] Prettyman, I.B., *Handbook of Chemistry and Physics*, 44th edn., CRC Press, 1962, 1564
[4] Wood, L.A. and Roth, F.L., *Rubber Chem. Technol.*, 1963, **36**, 611
[5] Rands, R.D., Ferguson, W.J. and Prather, J.L., *J. Res. Natl. Bur. Stand. (U.S.)*, 1944, **33**, 63
[6] Hamill, W.H., Mrowca, B.A. and Anthony, R.L., *Rubber Chem. Technol.*, 1946, **19**, 622
[7] Payne, A.R. and Scott, J.R., *Engineering Design with Rubber*, Interscience, 1960
[8] Schilling, H., *Kautsch. Gummi Kunstst.*, 1963, **16**, 84
[9] Rehner, J., *Rubber Chem. Technol.*, 1948, **21**, 82
[10] McPherson, A.T., *Rubber Chem. Technol.*, 1963, **36**, 1230
[11] Naunton, W.H.S. et al, *Rubber Engineering*, Ministry of Supply, 1945
[12] Cramer, W.S. and Silver, I., 1951, U.S. Naval Ordnance Lab.
[13] Witte, R.S., Mrowca, B.A. and Guth, E., *Rubber Chem. Technol.*, 1950, **23**, 163
[14] Ball, J.M. and Maasen, G.C., ASTM, *Symp. Appl. Synth. Rubbers*, 1944, 27
[15] Boomstra, B.B.S.T., Elastomers. Their Chemistry, *Physics and Technology*, (ed. R. Houwink), Elsevier, 1948, **3**
[16] Martin, G.M., Roth, F.L. and Stickler, R.D., *Rubber Chem. Technol.*, 1957, **30**, 876
[17] Philipoff, W., *J. Appl. Phys.*, 1953, **24**, 685
[18] Mancke, R.G. and Ferry, J.D., *Trans. Soc. Rheol.*, 1968, **12**, 335
[19] Payne, A.R., The Physical Properties of Polymers, *S.C.I. Monograph No.5*, Society of Chemical Industry, London, 1959, 273
[20] Zapas, L.J., Shufler, S.L. and de Witt, T.W., *Rubber Chem. Technol.*, 1956, **29**, 725
[21] Fletcher, W.P. and Gent, A.N., *Br. J. Appl. Phys.*, 1957, **8**, 194
[22] Ivey, D.G., Mrowca, B.A. and Guth, E., *Rubber Chem. Technol.*, 1950, **23**, 172
[23] Boomstra, B.B.S.T., *Rubber Chem. Technol.*, 1951, **24**, 199
[24] Ferry, J.D., Mancke, R.G., Maekawa, E., Oyanagi, M. and Dickie, R.A., *J. Phys. Chem.*, 1964, **68**, 3414
[25] Lattime, R.R., *Kirk-Othmer Encycl. Chem. Technol.*, 4th edn., (ed. J.I. Kroschwitz), John Wiley & Sons, 1997, **22**, 994, 997

[26] *The Materials Selector,* 2nd edn., (eds. N.A. Waterman and M.F. Ashby), Chapman & Hall, 1997, **3**, 482
[27] Morton, M., Kirk-Othmer Encycl. Chem. Technol. (ed. J.I.K. Kroschwitz), Vol. 8, 4th Ed., John Wiley & Sons, 1993, 913
[28] Lattime, R.R., Kirk-Othmer Encycl. Chem. Technol. (ed. J.I.K. Kroschwitz), Vol. 22, 4th Ed., John Wiley & Sons, 1997, 994
[29] Struick, L.C.E., Encycl. Polym. Sci. Eng. 2nd edn., Vol. 1 (ed. J.I.K. Kroschwitz), John Wiley & Sons, 1985, 605
[30] Shieh, C.-H. and Hamed, G.R., *J. Polym. Sci., Polym. Phys. Ed.,* 1983, **21**, 1415
[31] Cabasso, I., Encycl. Polym. Sci. Eng. 2nd edn., Vol. 2, (ed. J.I. Kroschwitz), John Wiley & Sons, 1985, 200
[32] Homma, T., Kawahara, K., Fujita, H., *J. Appl. Polym. Sci.,* 1964, **8**, 2853
[33] Poddubnyi, I.Ya., Grechanovskii, V.A. and Podulinskii, A.V., *J. Polym. Sci., Part C: Polym. Lett.,* 1968, **16**, 3109
[34] Lautout, M. and Magat, M., *Z. Phys. Chem. (Munich),* 1958, **16**, 292
[35] French, D.M. and Ewart, R.H., *Anal. Chem.,* 1947, **19**, 165
[36] Altares, T., Wyman, D.P. and Allen, V.R., *J. Polym. Sci., Part A-2,* 1964, 4533
[37] Scott, R.L., Carter, W.C., Magat, M., *J. Am. Chem. Soc.,* 1949, **71**, 220
[38] Brostow, W., Kubát, J., Kubát, M.M., *Physical Properties of Polymers Handbook,* (ed. J.E. Mark), AIP Press, 1996, 333
[39] Tewarson, A., *Physical Properties of Polymers Handbook,* (ed. J.E. Mark), AIP Press, 1996, 587
[40] Billmeyer, F.W., *Textbook of Polymer Science,* Wiley-Interscience, 1984, **136**, 361
[41] Mori, K., Kaneda, S., Kanae, K., Hirahara, H. et al, *Rubber Chem. Technol.,* 1994, **67**, 797
[42] Thavamani, P. and Bhowmick, A.K., *Rubber Chem. Technol.,* 1994, **67**, 129
[43] Hess, W.M., Herd, C.R. and Vegvari, P.C., *Rubber Chem. Technol.,* 1993, **66**, 329
[44] Hamed, G.R. and Dweik, H.S., *Rubber Chem. Technol.,* 1993, **66**, 92
[45] Cho, P.L. and Hamed, G.R., *Rubber Chem. Technol.,* 1992, **65**, 475
[46] Gent, A.N., Liu, G.L. and Sueyasu, T., *Rubber Chem. Technol.,* 1991, **64**, 96
[47] Ayala, J.A., Hess, W.M., Ristler, F.D. and Joyce, G.A., *Rubber Chem. Technol.,* 1991, **64**, 19
[48] Gent, A.N. and Hindi, M., *Rubber Chem. Technol.,* 1988, **61**, 892
[49] Burton, L.C., Hwang, K. and Zhang, T., *Rubber Chem. Technol.,* 1989, **62**, 838
[50] Goldberg, A., Leseur, D.R. and Patt, J., *Rubber Chem. Technol.,* 1989, **62**, 272, 288
[51] Funt, J.M., *Rubber Chem. Technol.,* 1988, **61**, 842
[52] Cotton, G.R. and Murphy, L.J., *Rubber Chem. Technol.,* 1988, **61**, 601
[53] Debnath, S., De, P.P. and Khastgir, D., *Rubber Chem. Technol.,* 1988, **61**, 555
[54] Ahagon, A., Kobeyashi, T. and Misawa, M., *Rubber Chem. Technol.,* 1988, **61**, 14
[55] Lazeration, J.J., *Rubber Chem. Technol.,* 1987, **60**, 966
[56] Nakajima, N., Sobbo, J.J. and Harrell, E.R., *Rubber Chem. Technol.,* 1987, **60**, 761
[57] Stacer, R.G. and Kelley, F.N., *Rubber Chem. Technol.,* 1985, **58**, 421, 913, 924
[58] Inoue, T., Shomura, F., Ongizawa, T. and Miyasaka, K., *Rubber Chem. Technol.,* 1985, **58**, 873
[59] Kannabiran, R., *Rubber Chem. Technol.,* 1984, **57**, 1001
[60] Hess, W.M., Swor, R.A. and Micer, E.J., *Rubber Chem. Technol.,* 1984, **57**, 959
[61] Isono, Y. and Ferry, J.D., *Rubber Chem. Technol.,* 1984, **57**, 925
[62] Thiole, S.L. and Cohen, R.E, *Rubber Chem. Technol.,* 1983, **56**, 465
[63] Hess, W.M. and Klamp, W.K., *Rubber Chem. Technol.,* 1983, **56**, 390
[64] Bhowmick, A.K., Gent, A.N. and Pulford, C.T.R., *Rubber Chem. Technol.,* 1983, **56**, 226
[65] Clamroth, R. and Ruitz, L., *Rubber Chem. Technol.,* 1983, **56**, 31
[66] Bohm, G.G.A. and Tveerem, J.O., *Rubber Chem. Technol.,* 1982, **55**, 575
[67] Copely, B.C., *Rubber Chem. Technol.,* 1982, **55**, 416
[68] Sircar, A.K. and Wells, J.L., *Rubber Chem. Technol.,* 1982, **55**, 191
[69] Meinecke, E.A. and Makcin, S., *Rubber Chem. Technol.,* 1981, **54**, 857
[70] Tanaka, Y., Sato, H., Saito, K. and Miyashita, K., *Rubber Chem. Technol.,* 1981, **54**, 685
[71] Hamed, G.R., *Rubber Chem. Technol.,* 1981, **54**, 403, 576
[72] Coran, A.Y. and Patel, R., *Rubber Chem. Technol.,* 1981, **54**, 91
[73] Cotten, G.R., *Rubber Chem. Technol.,* 1979, **52**, 181
[74] Kadir, A. and Thomas, A.G., *Rubber Chem. Technol.,* 1981, **54**, 151
[75] Bhowmick, A.K. and De, S.K., *Rubber Chem. Technol.,* 1980, **53**, 960
[76] Brazier, D.W., *Rubber Chem. Technol.,* 1980, **53**, 437
[77] Gent, A.N.and Pulford, C.T.R., *Rubber Chem. Technol.,* 1980, **53**, 176
[78] Southern, E. and Thomas, A.G., *Rubber Chem. Technol.,* 1979, **52**, 1008
[79] Medalia, A.I., *Rubber Chem. Technol.,* 1978, **51**, 437
[80] Caruthers, J.M. and Cohen, R.E., *Rubber Chem. Technol.,* 1976, **49**, 1076
[81] Kraus, G., *Rubber Chem. Technol.,* 1978, **51**, 297
[82] Smith, T.L., *Rubber Chem. Technol.,* 1978, **51**, 225
[83] Gent, A.N. and Kim, H.J., *Rubber Chem. Technol.,* 1978, **51**, 35
[84] Nakajima, N. and Collins, E.A., *Rubber Chem. Technol.,* 1976, **49**, 52
[85] Fabris, H.J. and Sommer, J.G., *Rubber Chem. Technol.,* 1977, **50**, 523
[86] Hess, W.M. and Chirico, V.E., *Rubber Chem. Technol.,* 1977, **50**, 301
[87] Ghijsols, A., *Rubber Chem. Technol.,* 1977, **50**, 278
[88] Abu-Isa, I.A., *Rubber Chem. Technol.,* 1983, **56**, 135
[89] Harwood, H.J., *Rubber Chem. Technol.,* 1982, **55**, 620, 769
[90] Cotten, G.R., Thiele, J.L., *Rubber Chem. Technol.,* 1978, **51**, 749
[91] Turetzky, S.B., Van Bushkirk, P.R. and Gunberg, P.F., *Rubber Chem. Technol.,* 1976, **49**, 1
[92] Katritzky, A.R. and Weiss, D.E., *Rubber Chem. Technol.,* 1975, **48**, 1055
[93] Thirion, P., *Rubber Chem. Technol.,* 1975, **48**, 981
[94] Sircar, A.K. and Lamond, T.G., *Rubber Chem. Technol.,* 1972, **45**, 293
[95] Gnörich, W. and Grosch, K.A., *Rubber Chem. Technol.,* 1975, **48**, 527
[96] Lawson, D.F., Kay, E.L. and Roberts, D.T., *Rubber Chem. Technol.,* 1975, **48**, 124
[97] Pierson, W.R. and Brachaczek, W.W., *Rubber Chem. Technol.,* 1974, **47**, 1275
[98] Kalfayan, S.H., Pakutis, R. and Silver, R.H., *Rubber Chem. Technol.,* 1974, **47**, 1265
[99] Derringer, G.C., *Rubber Chem. Technol.,* 1974, **47**, 825
[100] Deviney, M.L., Weaver, E.J., Wade, W.H. and Gardner, J.E., *Rubber Chem. Technol.,* 1974, **47**, 837
[101] Studebaker, M.L. and Beatty, J.R., *Rubber Chem. Technol.,* 1973, **46**, 450
[102] Voet, A. and Morawski, J.C., *Rubber Chem. Technol.,* 1974, **47**, 765
[103] Ulmer, J.D., Hess, W.M. and Chirico, V.E., *Rubber Chem. Technol.,* 1974, **47**, 729
[104] Corish, P.J. and Powell, B.D.W., *Rubber Chem. Technol.,* 1974, **47**, 481
[105] Coran, A.Y., Boustany, K. and Hamed, P., *Rubber Chem. Technol.,* 1974, **47**, 396
[106] Dizon, E.S., Hicks, A.E. and Chirico, V.E., *Rubber Chem. Technol.,* 1974, **47**, 231
[107] Deviney, M.L., Whittington, L.E., Good, C.W. and Sperley, R.J., *Rubber Chem. Technol.,* 1974, **47**, 127
[108] Trexler, H.E., *Rubber Chem. Technol.,* 1973, **46**, 1114
[109] Lake, G.J., *Rubber Chem. Technol.,* 1972, **45**, 309
[110] Veith, A.G., *Rubber Chem. Technol.,* 1972, **45**, 293
[111] Corman, B.G., Deviney, M.C. and Whittington, L.E., *Rubber Chem. Technol.,* 1972, **45**, 278
[112] Kieule, R.N., Dizon, E.S., Brett, T.J. and Eckert, C.F., *Rubber Chem. Technol.,* 1971, **44**, 996
[113] Callan, J.E., Hess, W.M. and Scott, C.E., *Rubber Chem. Technol.,* 1971, **44**, 814
[114] Sambrook, R.W., *Rubber Chem. Technol.,* 1971, **44**, 728
[115] Mayer, D.A. and Sommer, J.G., *Rubber Chem. Technol.,* 1971, **44**, 258
[116] Sircar, A.K., Voet, A. and Cook, F.R, *Rubber Chem. Technol.,* 1971, **44**, 175
[117] Deviney, M.L. and Whittington, L.E., *Rubber Chem. Technol.,* 1971, **44**, 87
[118] Moehel, V.D. and Johnson, B.L., *Rubber Chem. Technol.,* 1970, **43**, 1138
[119] Harwood, J.A.C., Payne, A.R. and Smith, J.F., *Rubber Chem. Technol.,* 1970, **43**, 687
[120] Gardiner, J.B., *Rubber Chem. Technol.,* 1970, **43**, 370
[121] Waldrop, M.A. and Kraus, G., *Rubber Chem. Technol.,* 1969, **42**, 1155
[122] Freakley, P.K. and Sirisinha, C., *J. Appl. Polym. Sci.,* 1997, **65**, 305
[123] Abdel Aziz, M.M. and Gwaily, S.E., *Polym. Degrad. Stab.,* 1997, **55**, 269
[124] Chen, K.S., Zeh, R.Z. and Chang, Y.R., *Combustion and Flame,* 1997, **108**, 408
[125] George, S.C., Thomas, S. and Ninan, K.N., *Polymer,* 1996, **37**, 5839
[126] Abdel Aziz, M.M., Youssef, H.A., Elmiligy, A.A., Yoshii, F. and Makuuchi, K., *Polym. Polym. Compos.,* 1996, **4**, 259
[127] Lin, J.P., Chang, C.Y. and Wu, C.H., *J. Chem. Technol. Biotechnol.,* 1996, **66**, 7
[128] Kumar, R.P. and Thomas, S., *Polym. Int.,* 1995, **88**, 173
[129] Kumar, R.P., Amma, M.L.G. and Thomas, S., *J. Appl. Polym. Sci.,* 1995, **58**, 597
[130] Hamed, G.R. and Wu, P.S., *Rubber Chem. Technol.,* 1995, **68**, 248
[131] Hamed, G.R. and Roberts, G.D., *Rubber Chem. Technol.,* 1995, **68**, 212
[132] Hamed, G.R. and Roberts, G.D., *J. Adhes.,* 1994, **47**, 95
[133] Dutta, N.K., Choudhury, N.R., Haidar, B., Vidal, A. and Donnet, J.B., *Polymer,* 1994, **35**, 4293
[134] Furtaco, C.R.G., Nunes, R.C.R. and Filko, A.S.D., *Eur. Polym. J.,* 1994, **30**, 1151
[135] Khairy, S.A. and Ateia, E., *J. Phys. D: Appl. Phys.,* 1993, **26**, 2272

[136] Thavaman, P. and Bhowmick, A.K., Plast., *Rubber Compos. Process. Appl.*, 1993, **20**, 239
[137] Neogi, C., Bhowmick, A.K. and Basse, S.P., *J. Elastomers Plast.*, 1992, **24**, 96
[138] Sekinger, J.K., Ghebremeskel, G.N. and Concienne, L.H., *Rubber Chem. Technol.*, 1996, **69**, 851
[139] Ward, J.B., Ammenheuser, M.M., Whorton, E.B., Bechfold, W.E. et al, *Toxicology*, 1996, **113**, 84
[140] Badawy, M.M., *Polym. Test.*, 1996, **15**, 507
[141] Nasr, G.M., Badawy, M.M., Gwaily, S.E. and Attia, G., *Polym. Int.*, 1995, **38**, 249
[142] Karasek, L., Meissner, B., Asai, S. and Sumita, M., *Polym. J. (Tokyo)*, 1996, **28**, 121
[143] Yamaguchi, T., Yamazaki, H., Yamauchi, A. and Kakiuchi, Y., *Japanese Journal of Toxicology and Environmental Health*, 1995, **41**, 155
[144] Varkey, J.T., Rao, S.S. and Thomas, S., Plast., *Rubber Compos. Process. Appl.*, 1995, **23**, 249
[145] Varkey, J.T., Thomas, S. and Rao, S.S., *J. Appl. Polym. Sci.*, 1995, **56**, 451
[146] Matheson, M.J., Wampler, T.P. and Simonsick, W.J., *J. Anal. Appl. Pyrolysis*, 1994, **29**, 129
[147] Little, J.C., Hodgson, A.T. and Gadgil, A.J., *Atmospheric Environment*, 1994, **28**, 227
[148] Aminabhavi, T.M. and Munnolli, R.S., *Polym. Int.*, 1994, **34**, 59
[149] Iwai, Y., Miyamoto, S., Ikada, H., Arai, Y. et al, *Polym. Eng. Sci.*, 1993, **33**, 322
[150] Mourniae, P., Agassant, J.F. and Vergnes, B., *Rheol. Acta*, 1992, **31**, 565
[151] Harogoppad, S.B. and Aminabhavi, T.M., *J. Appl. Polym. Sci.*, 1991, **42**, 2329
[152] Helaly, F.M., Badran, A.S. and Ramadan, A.M., *J. Elastomers Plast.*, 1991, **23**, 301
[153] Hassan, H.H., Khairy, S.A., Elguiziri, S.B. and Abdelmonein, H.M., *J. Appl. Polym. Sci.*, 1991, **42**, 2879
[154] Harogoppad, S.B. and Aminobhavi, T.M., *Macromolecules*, 1991, **24**, 2598
[155] Debnath, S., *Int. J. Polym. Mater.*, 1989, **12**, 225
[156] Debnath, S., Khastgir, D.K. and Dutta, D., *Polym. Test.*, 1987, **7**, 371
[157] Rodriguez, E.L. and Filisko, F.E., *Polymer*, 1986, **27**, 1943
[158] *Permeability and other Film Properties of Plastics and Elastomers*, Plastic Design Library, 1995, 415, (air permeability)

Styrene butadiene rubber, High-*trans* S-29

Synonyms: Poly(styrene-co-1,3-butadiene), high-trans. High trans styrene butadiene rubber. High trans SBR
Related Polymers: Styrene butadiene rubber, Polystyrene, Polybutadiene, ABS, SBS, Styrene butadiene copolymers
Monomers: Styrene, 1,3-Butadiene
Material class: Synthetic Elastomers, Copolymers
Polymer Type: polydienes, polystyrene
CAS Number:

CAS Reg. No.
9003-55-8

Molecular Formula: $[(C_8H_8)_x(C_4H_6)_y]_n$
Additives: Carbon black
General Information: High *trans* polymer contains greater than 80% *trans* butadiene segments with respect to *cis* segments

Volumetric & Calorimetric Properties:
Melting Temperature:

No.	Value	Note
1	-11°C	76% 1,4 *trans* [1]
2	30°C	88% 1,4 *trans* [1]

Transition Temperature General: T_m increases with increasing *trans* content

Mechanical Properties:
Mechanical Properties General: Carbon black-filled vulcanisates show high levels of abrasion resistance and resistance to oxidation and crack initiation and when blended with natural rubber, show remarkably low rates of crack growth with high green strength and tack strength

Tensile (Young's) Modulus:

No.	Value	Note
1	1.3 MPa	100% elongation [1]
2	5 MPa	300% elongation [1]

Tensile Strength Break:

No.	Value	Note
1	16.4 MPa	[1]

Mechanical Properties Miscellaneous: The green strength of high *trans* SBR has been reported as the variation of stress with elongation for different levels of *trans* content. [1] Stress relaxation of vulcanisates with carbon black has been measured for high *trans* SBR and compared with soln. SBR and natural rubber [1]
Failure Properties General: Fatigue life has been reported for different degrees of elongation and compared with natural rubber. [1] Tack strength 0.29 MPa (uncured); green strength 2 MPa (uncured) [1]
Friction Abrasion and Resistance: Pico abrasion index for high *trans* SBR in tyre tread vulcanisate 194

Applications/Commercial Products:
Processing & Manufacturing Routes: Produced by soln. polymerisation using anionic catalysts

Bibliographic Reference
[1] Fabris, H.J., Hargis, I.G., Livigni, R.A., Aggarwal, S.L., *Rubber Chem. Technol.*, 1987, **60**, 721

Styrene-isoprene block copolymers S-30

Synonyms: *Styrene-isoprene diblock copolymers. Poly(styrene-block-isoprene). Poly(ethylbenzene-block-2-methyl-1,3-butadiene)*
Related Polymers: Styrene-isoprene copolymers, SIS, Styrene-isoprene star block copolymers
Monomers: Styrene, Isoprene
Material class: Copolymers
Polymer Type: polyolefins, polydienespolybutadienepolystyrenestyrenes
CAS Number:

CAS Reg. No.	Note
25038-32-8	
105729-79-1	block

General Information: The props. of these polymers can vary greatly; for example, SIS triblock copolymers at room temp. have many props. associated with vulcanised rubbers whereas the diblock copolymers are weak materials. Perhaps the most important group commercially are the SIS copolymers
Morphology: Diblock copolymers undergo several phase transitions. These transitions are listed together with transition temp. and any other information for diblock copolymers with varying compositions given in parentheses: lamellae to disordered 250° (42.3% isoprene, M_n 30300) or 239° (40.7% isoprene, M_n 30800); hexagonally perforated layers to bicontinuous cubic (Ia3d) 167° (39.3% isoprene, M_n 31500) or 175° (37.6% isoprene, M_n 31800); bicontinuous cubic (Ia3d) to disordered (39.3% isoprene, M_n 31500) or 234° (37.6% isoprene, M_n 31800); hexagonally packed cylinders to bicontinuous cubic (Ia3d) 182° (35.9% isoprene, M_n 33800); bicontinuous cubic (Ia3d) to disordered 245° (35.9% isoprene, M_n 33800); hexagonally packed cylinders to disordered 285° (33.1% isoprene, M_n 36800). [7] Lamellar spacing remains essentially constant up to 180–200°. Above this temp. the lamellar thickness increases significantly and irreversibly. Linear multiblock copolymers $(SI)_n$ exhibit lamellar morphologies in which the

microdomain periodicity decreases with n. [7,10] Morphology of ideal and tapered block copolymers has been reported [4]

Volumetric & Calorimetric Properties:
Melting Temperature:

No.	Value	Note
1	180–200°C	lamellar, MW 25000–100000 [7]

Glass-Transition Temperature:

No.	Value	Note
1	15°C	ideal block, 51% styrene, M_n 97000 [4]
2	39–40°C	tapered block, 47–48% styrene, M_n 43000–36000 [4]
3	82.5–97.7°C	diblock, varying MW and composition [8]
4	-44–40°C	tetrablock, varying composition [2]

Transition Temperature General: T_g increases with increasing M_n [8]. The order-disorder transition temp. of a multiblock copolymer is lower than that of the corresponding diblock copolymer of the same composition and total MW. The order-disorder transition temp. for tapered block copolymers is lower than that of ideal block copolymers [2,4,7,10,17]

Transition Temperature:

No.	Value	Note	Type
1	50°C	approx., diblock, MW 45000, varying styrene content [2,4]	Order-disorder transition
2	230–285°C	diblock [7]	Order-disorder transition
3	111°C	asymmetric diblock, 13% styrene, MW 62600 [17]	Order-disorder transition
4	159°C	asymmetric diblock, 13.5% styrene, MW 78400 [17]	Order-disorder transition
5	200°C	asymmetric diblock, 13.1% styrene, MW 90600 [17]	Order-disorder transition
6	<25°C	max., tetrablock, MW 45000, varying styrene content [2,4]	Order-disorder transition

Surface Properties & Solubility:
Solvents/Non-solvents: Sol. $CHCl_3$, THF, toluene, cyclohexane, methyl cyclohexane, decalin [3,5]

Transport Properties:
Polymer Solutions Dilute: Intrinsic viscosity 0.4 dl g^{-1} (THF, diblock, MW 117000), 1.61 dl g^{-1} (toluene, 33 wt% styrene, MW 293000), 1.4 dl g^{-1} (toluene, 59 wt% styrene, MW, 189000), 0.69 dl g^{-1} (isobutyl methyl ketone, 33 wt% styrene, MW 263000), 0.6 dl g^{-1} (isobutyl methyl ketone, 78 wt% styrene, MW 263000), 1.41 dl g^{-1} (cyclohexane 33 wt% styrene, MW 268000), 0.86 dl g^{-1} (cyclohexane, 78 wt% styrene, MW 269200), 0.29 dl g^{-1} (cyclohexane, tapered long blocks, MW 47800), 0.44 dl g^{-1} (cyclohexane, tapered short blocks, MW 62500). [1,3] Theta temps. [1] 44° (butanone), -88° (toluene, MW 256000), -80° (toluene, MW 221000), -93° (toluene, MW 259000), -81° (cyclohexane, MW 256000), -24° (cyclohexane, MW 221000), 6° (cyclohexane, MW 259000). Second virial coefficients have been reported [1]

Polymer Melts: The use of large amplitude oscillatory shear to induce lamellae alignment in diblock copolymer melts has been reported. Lamellae orient parallel to the shearing surfaces at high frequency or low temp. and perpendicular at low frequency/high temp. Mixed parallel and perpendicular morphologies have been identified at intermediate temps. [14,15]

Permeability of Gases: The sorption/desorption of toluene vapour has been used to study the order-disorder transition of the diblock copolymer [16]

Permeability and Diffusion General: Self-diffusion coefficients have been reported for tetrablock copolymers over a range of temps. Self-diffusion coefficient increases with isoprene content and becomes more temp. dependent as the styrene content increases [2]

Mechanical Properties:
Mechanical Properties General: For $(AB)_n$ block copolymers the yield stress and elastic modulus increase with n. [11] The optimum shear strength of SI copolymers occurs at approx. 50% polyisoprene. Star copolymers may have higher shear strengths than triblock copolymers of the same MW range; exceptions occur at the extremes of polyisoprene content. In the case of the star copolymers it is the MW of the branches rather than the total number of branches which exerts the greater influence on shear strength [18]

Viscoelastic Behaviour: The viscoelastic props. of tetrablock copolymers of varying compositions and Willams-Landel-Fleming parameters have been reported [2]

Electrical Properties:
Electrical Properties General: Block copolymers can be made electrically conductive by doping with iodine [9]

Optical Properties:
Optical Properties General: The variation of refractive index increment with composition for various block copolymers in several solvents has been reported; a linear relationship independent of chain struct. is found [3]

Transmission and Spectra: Diblock copolymers with MW greater than 1000000 are visibly coloured above a critical concentration in C_6H_6. The critical concentration decreases as MW increases; e.g. critical concentration for MW 470000 is approx. 10% weight and the soln. is blue. Optical props. of these polymers are related to the size of the domains present as a result of phase separation [11]

Applications/Commercial Products:
Processing & Manufacturing Routes: Block copolymers with varying block lengths and sequence can be prepared using anionic living polymerisation. It is necessary to exclude water, oxygen or any impurity which may react with the active propagating species. The different copolymers are produced by varying the solvents, initiators and polar modifiers. Sequential living anionic polymerisation has also been used to prepare $(SI)_n$ copolymers with varying styrene content [2,3,10,12]

Applications: Used in pressure-sensitive adhesives, and as reinforcing agents

Trade name	Details	Supplier
SI(11-10)	diblock	Pressure Chemical Co.

Bibliographic References
[1] Cramond, D.N. and Urwin, J.R., *Eur. Polym. J.*, 1969, **5**, 35, 45, (soln props)
[2] Apman, B.R., Hamersky, M.W., Milhaupt, J.M., Kostelecky, C. and Lodge, T.P., *Macromolecules*, 1998, **31**, 4562, (struct.)
[3] Velichkova, R, Toncheva, V., Getova, C., Pavlova, S. et al, *J. Polym. Sci., Part A-1*, 1991, **29**, 1107, (synth, soln props)
[4] Hashimoto, T., Tsukahara, Y. and Kawai, H., *Polym. J. (Tokyo)*, 1983, **15**, 699, (struct, props)
[5] Urwin, J.R. and Girolamo, M, *Makromol. Chem.*, 1971, **150**, 179, (phase transition)
[6] Forster, S., Khandpur, A.K., Zhao, J., Bates, F.S. et al, *Macromolecules*, 1994, **27**, 6922
[7] Hadziioannou, G. and Skoulios, A., *Macromolecules*, 1982, **15**, 271, (T_m)
[8] Spontak, R.J., Smith, S.D. and Ashraf, A, *Macromolecules*, 1993, **26**, 956, (morphology)
[9] Lehrke, J., Cramer, K., Luchow, H. and Gronski, W., *Makromol. Chem., Rapid Commun.*, 1990, **11**, 495, (electrical conductivity)
[10] Smith, S.D., Spontak, R.J, Satkowski, M.M, Ashraf, A., *Polymer*, 1994, **35**, 4527

[11] Bywater, S. and Toporowski, P.M., *Polymer*, 1981, **22**, 29, (optical props)
[12] Meyer, G.C. and Widmaier, J.M, *J. Polym. Sci., Part B: Polym. Lett.*, 1982, **20**, 389, (T_g)
[13] Berglund, C.A. and McKay, K.W., *Polym. Eng. Sci.*, 1993, **33**, 1195, (viscoelastic props)
[14] Patel, S.S., Larson, R.F., Winey, K.I. and Watanabe, H., *Macromolecules*, 1995, **28**, 4313, (rheology)
[15] Pinheiro, B.S. and Winey, K.I., *Macromolecules*, 1998, **31**, 4447, (morphology, viscoelastic props)
[16] Hong, S.-U., Laurer, J.H., Zielinski, J.M., Samseth, J. et al, *Macromolecules*, 1998, **31**, 2174, (anal)
[17] Adams, L.L., Quiram, D.J., Graessley, W.W., Register, R.A. and Marchand, G.R., *Macromolecules*, 1996, **29**, 2929, (ordering)
[18] Goodman, I., *Developments in Block Copolymers*, Elsevier, 1985, **2**, 132

Styrene-isoprene copolymers   S-31

$$\left[\left[CH_2CH\right]_x\left[CH_2CH\right]_y\right]_n$$
(Ph on first unit; isopropenyl on second)

Synonyms: *Poly(styrene-co-isoprene). Poly(ethenylbenzene-co-2-methyl-1,3-butadiene). Poly(vinylbenzene-co-2-methyl-1,3-butadiene)*
Related Polymers: Styrene-isoprene star block copolymers, Styrene-isoprene block copolymers, SIS
Monomers: Styrene, 2-Methyl-1,3-butadiene
Material class: Copolymers
Polymer Type: styrenes
CAS Number:

CAS Reg. No.
25038-32-8

Molecular Formula: $[(C_8H_8)_x(C_5H_8)_y]_n$
Fragments: C_8H_8 C_5H_8
General Information: Styrene and isoprene may copolymerise to give a variety of products including random, diblock, triblock and star copolymers. The props. of these polymers can vary greatly; e.g. SIS triblock copolymers, at room temp., have many props., associated with vulcanised rubbers whereas the diblock copolymers are weak materials. Perhaps the most important group commercially are the SIS copolymers.

Transport Properties:
Polymer Solutions Dilute: Intrinsic viscosity 0.56 dl g^{-1} (THF, random copolymer, MW 75200, 0.56 dl g^{-1} (cyclohexane, random copolymer, MW 83000)

Styrene-isoprene star block copolymers   S-32
Synonyms: *Styrene-isoprene radial block copolymers. Poly(ethenylbenzene-block-2-methyl-1,3-butadiene)*
Related Polymers: Styrene-isoprene copolymers, SIS, Styrene-isoprene block copolymers
Monomers: Styrene, 2-Methyl-1,3-butadiene
Material class: Copolymers
Polymer Type: polydienes, polybutadienepolystyrene
CAS Number:

CAS Reg. No.	Note
25038-32-8	
105729-79-1	block

General Information: There are several types of styrene/isoprene star copolymers: $(AB)_n$, A_nB_n, A_mB_n and gradient-modulus stars
Morphology: Morphology of these polymers has been extensively investigated. Body-centred cubic, a mixture of body-centred and face-centred cubic and face-centred cubic structs. have been reported for $(AB)_n$ stars depending on the concentration. The microphase-separated structs. of stars with diblock arms have been reported. [1,2,5,6] Polystyrene outer blocks in the arms may exist in the form of spheres, cylinders, lamellae or a double-diamond struct. The morphology may depend on the number of arms, the amount of polystyrene in the arm, and the MW of polystyrene. The microphase-separated morphologies of several types of miktoarm polymers has been reported: e.g. morphology is lamellar for star type A_2B (55 wt% styrene, total M_n 70400) or for star type A_3B (57 wt% styrene, total M_n 104000) [9]

Volumetric & Calorimetric Properties:
Transition Temperature:

No.	Value	Note	Type
1	112°C	$(AB)_n$ star, 44% styrene, 4 arms, polystyrene MW 6300 [2,10]	lamellar/disorder
2	155°C	$(AB)_n$ star, 29% styrene, 4 arms, polystyrene MW 7000 [3,12]	PS spheres/disorder
3	97°C	A_3B star, 41% styrene, total M_n 77200 [2,10]	PS cylinder/disorder
4	88°C	A_3B star, 50% styrene, total M_n 73000 [2,10]	PS cylinder/disorder

Transport Properties:
Polymer Solutions Dilute: The micellisation behaviour of miktoarm polymers in comparison to that of linear diblock copolymers has been reported [11]. Studies on the dilute soln. props. of gradient-modulus stars show that they behave as soft rather than hard spheres and that they form unimolecular micelles [12]. Intrinsic viscosity 2.29 dl g^{-1} (toluene, 35°, 3-miktoarm $PS(PI)_2$, 32% styrene, MW 25500), 2.38 dl g^{-1} (dioxane, 26.5°, 3-miktoarm $PS(PI)_2$, 32% styrene, MW 25500), 1.83 dl g^{-1} (cyclohexane, 34.5°, 3-miktoarm $PS(PI)_2$, 32% styrene, MW 25500), 4.83–11.19 d lg^{-1} (dioxane, 26.5°, 16-miktoarm $(PS)_8(PI)_8$, 48–41% styrene, MW 330000–894000) [4,11]. Huggins and second virial coefficients have been reported [13]

Mechanical Properties:
Mechanical Properties General: Tensile and shear strengths of star copolymers are higher than those of linear block copolymers. The MW of the branches has a greater influence on the strength of the polymer than the total MW. The lowest strengths are found at the extremes of polyisoprene content [13]
Tensile Strength Break:

No.	Value	Note
1	0.5–1.5 MPa	4–17 triblock arms, MW 220000–780000 [13]

Fracture Mechanical Properties: Shear strength 0.9–13.4 MPa (4–17 triblock arms, MW 220000–780000) [13]

Optical Properties:
Transmission and Spectra: H-1 nmr and C-13 nmr spectral data have been reported [3]

Applications/Commercial Products:
Processing & Manufacturing Routes: $(AB)_n$ Star polymers are prepared by cross-linking poly(styrene-block-isoprene) diblocks with divinylbenzene or chlorosilanes [5]. Copolymers with two polystyrene arms are two polyisoprene arms (A_2B_2) have been synth. by various routes. [7,8] The synth. of various miktoarm copolymers by anionic polymerisation has been reported [4,9,11]. Gradient-modulus star copolymers are prepared by anionic copolymerisation of tapered block copolymer monoanions with divinylbenzene [12]

– Styrene/methacrylate copolymers

Applications: Used in adhesives

Trade name	Supplier
Kratom D 1320X	Shell Chemicals

Bibliographic References

[1] Herman, D.S., Kinning, D.J., Thomas, E.L. and Fetters, L.J., *Macromolecules*, 1986, **19**, 215, (morphology)
[2] Floudas, G., Paraskeva, S., Hadjichristidis, N., Fytas, G. et al, *J. Chem. Phys.*, 1997, **107**, 5502
[3] Pitsikalis, M. and Hadjichristidis, N., *Macromol. Chem. Phys.*, 1995, **196**, 2767, (nmr)
[4] Iatrou, H., Siakali-Kioulafa, E., Hadjichristidis, N., Roovers, J. and Mays, J., *J. Polym. Sci., Part B: Polym. Lett.*, 1995, **33**, 1925, (hydrodynamic props)
[5] Ishizu, K. and Uchida, S., *Polymer*, 1994, **35**, 4712
[6] Ishizu, K. and Uchida, S., *J. Colloid Interface Sci.*, 1995, **175**, 293, (struct)
[7] Ishizu, K. and Kuwahara, K., *J. Polym. Sci., Part A-1*, 1993, **31**, 661, (synth)
[8] Wright, S.J.; Young, R.N. and Croucher, T.G., *Polym. Int.*, 1994, **33**, 123, (synth)
[9] Tselikas, Y.; Iatrou, H.; Hadjichristidis, N.; Liang, K.S. et al, *J. Chem. Phys.*, 1996, **105**, 2456, (morphology)
[10] Floudas, G., Hadjichristidis, N., Tselikas, Y. and Erukhimovich, I., *Macromolecules*, 1997, **30**, 3090, (struct)
[11] Pispas, S.; Poulos, Y. and Hadjichristidis, N., *Macromolecules*, 1998, **31**, 4177
[12] Ishizu, K.; Sunahara, K.; Asai, S.-I., *Polymer*, 1998, **39**, 953, (synth, soln props)
[13] Goodman, I., *Developments in Block Copolymers*, 1985, **2**, 132

Styrene/methacrylate copolymers S-33

$$\left[\left[-CH_2C(CH_3)- \atop \underset{O}{\overset{\|}{C}}-OR \right]_x \left[-CH_2CH- \atop Ph \right]_y \right]_n$$

Where R = Me (MMA), Et (ethylmethacrylate),
$CH_2CH_2CH_2CH_3$ (butylmethacrylate),
$CH_2CH(CH_3)_2$ (isobutylmethacrylate), $CH_2(CH_2)_4CH_3$ (hexyl),
$CH_2(CH_2)_6(CH_3)$ (octyl)

Synonyms: *Poly(methyl methacrylate-co-styrene). Poly(ethyl methacrylate-co-styrene). Poly(butyl methacrylate-co-styrene). Poly(isobutyl methacrylate-co-styrene). Poly(hexyl methacrylate-co-styrene). Poly(octyl methacrylate-co-styrene). Poly(isooctyl methacrylate-co-styrene). Poly(methyl methacrylate-co-acrylonitrile-co-styrene)*
Related Polymers: Polymethacrylates, General, Poly(methyl methacrylate), Poly(ethyl methacrylate), Poly(butyl methacrylate), Poly(isobutyl methacrylate), Poly(hexyl methacrylate), Poly(octyl methacrylate), Poly(isooctyl methacrylate), Polyacrylonitrile, Polystyrene, Poly(methyl methacrylate-co-butadiene-co-styrene), Poly(methyl methacrylate-co-acrylonitrile-co-butadiene-co-styrene)
Monomers: Styrene, Methyl methacrylate, Ethyl methacrylate, Butyl methacrylate, Isobutyl methacrylate, Hexyl methacrylate, Octyl methacrylate, Isooctyl methacrylate, Acrylonitrile, Butadiene
Material class: Thermoplastic
Polymer Type: acrylic copolymers
CAS Number:

CAS Reg. No.	Note
25034-86-0	MMA-*co*-styrene
108266-99-5	alternating
106911-77-7	block
107741-20-8	graft
26634-88-8	ethyl methacrylate-*co*-styrene
25213-39-2	butyl methacrylate-*co*-styrene
107493-06-1	alternating copolymer
107391-68-4	block copolymer
27136-15-8	MMA-*co*-butyl acrylate-*co*-styrene

Morphology: Copolymers with the same composition have the same microstruct. regardless of the solvent used in their preparation [1]. Comb-like graft copolymers can be made with short polystyrene sidechains [2]

Volumetric & Calorimetric Properties:
Density:

No.	Value	Note
1	d 1.09 g/cm^3	MMA-*co*-styrene, injection-moulding grade [3]
2	d 1.16 g/cm^3	MMA-*co*-styrene, extrusion grade [3]
3	d^{100} 0.974 g/cm^3	hexyl methacrylate-*co*-styrene, 26 mol% styrene [4]
4	d^{100} 0.984 g/cm^3	hexyl methacrylate-*co*-styrene, 41 mol% styrene [4]
5	d^{25} 1.022 g/cm^3	hexyl methacrylate-*co*-styrene, 26 mol% styrene [4]
6	d^{25} 1.031 g/cm^3	hexyl methacrylate-*co*-styrene, 41 mol% stryrene [4]

Thermal Expansion Coefficient:

No.	Value	Note	Type
1	6×10^{-5}–8×10^{-5} K^{-1}	MMA-*co*-styrene, ASTM D696 [5]	L

Thermodynamic Properties General: The heat capacities of alternating, statistical and diblock copolymers are additive irrespective of sequence length distribution. The glass-transition temp., however, is dependent on sequence length distribution as well as tacticity and composition [6]
Thermal Conductivity:

No.	Value	Note
1	0.189 W/mK	0.0004–0.0005 cal (s cm °C)$^{-1}$, MMA-*co*-styrene, ASTM C177 [5]

Specific Heat Capacity:

No.	Value	Note	Type
1	1.42 kJ/kg.C	0.34 cal (g°C)$^{-1}$, MMA-*co*-styrene [5]	P

Glass-Transition Temperature:

No.	Value	Note
1	111.5°C	MMA-*co*-styrene, statistical copolymer, 29.4% styrene, MW 447000 [6]
2	102.5°C	MMA-*co*-styrene, statistical copolymer, 55.3% styrene, MW 137000 [6]
3	101.1°C	MMA-*co*-styrene, statistical copolymer, 70.3% styrene, MW 270000 [6]
4	97.1°C	MMA-*co*-styrene, alternating copolymer, 50.4% styrene, MW 233000 [6]

5	94.5°C	MMA-co-styrene, alternating copolymer, 49.0% styrene, MW 186000 [6]
6	101.8°C	MMA-co-styrene, diblock copolymer, 47.0% styrene, MW 828000 [6]
7	173°C	MMA-co-styrene, diblock copolymer, 47% styrene, MW 828000 [6]
8	78°C	ethyl methacrylate-co-styrene, statistical copolymer, 52.6% styrene [8,9]
9	75°C	ethyl methacrylate-co-styrene, alternating copolymer, 50.2% styrene [8,9]
10	43°C	butyl methacrylate-co-styrene, alternating copolymer, 47.4% and 52.1% styrene [8,9]
11	65°C	isobutyl methacrylate-co-styrene [9]
12	4°C	hexyl methacrylate-co-styrene, 26 mol% styrene [4]
13	14°C	hexyl methacrylate-co-styrene, 41 mol% styrene [4]
14	-1°C	octyl methacrylate-co-styrene, alternating copolymer, 51.4% styrene [8]
15	-21–77°C	octyl methacrylate-co-styrene, statistical copolymer, 15.6–92.1% styrene [8]

Transition Temperature General: T_g at the surface of a diblock copolymer is much lower than that of the bulk copolymer. T_g decreases as depth decreases. MW dependence of T_g is more pronounced than in the bulk sample [7]. T_g varies with styrene content [8]

Deflection Temperature:

No.	Value	Note
1	98°C	1.82 MPa, MMA-co-styrene, annealed, injection-moulding grade, ASTM D648 [3]
2	99°C	1.82 MPa, MMA-co-styrene, annealed, extrusion grade, ASTM D648 [3]

Surface Properties & Solubility:

Cohesive Energy Density Solubility Parameters: δ 15.14 $J^{1/2}$ $cm^{-3/2}$ (7.41 $cal^{1/2}$ $cm^{-3/2}$, butyl methacrylate-co-styrene, 58% wt styrene) [10], 15.14 $J^{1/2}$ $cm^{-3/2}$ (7.41 $cal^{1/2}$ $cm^{-3/2}$, isobutyl methacrylate-co-styrene, 80% wt styrene) [10]

Solvents/Non-solvents: MMA-co-styrene sol. CCl_4, toluene, o-xylene, C_6H_6, 2-butanone, chlorobenzene, CH_2Cl_2, Me_2CO, isopropyl methyl ketone, o-dichlorobenzene, cyclopentanone, pyridine, o-toluidine.; insol. Freon-215, isopentane, isooctane, pentane, hexane, octane, methylcyclohexane, cyclohexane, propanol, EtOH, MeOH; isobutyl methacrylate-co-styrene sol. white spirit, aromatics, esters, chlorinated hydrocarbons. Other solubility data have been reported [11]

Transport Properties:

Transport Properties General: At moderate concentrations in solvents that are selective (i.e., solvent for one block only), the block copolymers form multinuclear micelles. Addition of a selective non-solvent for one component to a soln. of a block copolymer, e.g., acetonitrile to C_6H_6 soln., forms a colloidal dispersion with polystyrene at the core of the aggregated chains and poly(methyl methacrylate) at the surface [14,15,16]

Melt Flow Index:

No.	Value	Note
1	0.2 g/10 min	190°, 10 kg, MMA-co-styrene, extrusion grade, ASTM D1238 [3]
2	0.7 g/10 min	230°, 3.8 kg, MMA-co-styrene, extrusion grade, ASTM D1238 [3]
3	0.13 g/10 min	230°, 1.2 kg, MMA-co-styrene, extrusion grade, ASTM D1238 [3]
4	4.3 g/10 min	190°, 10 kg, MMA-co-styrene, injection-moulding grade, ASTM D1238 [3]

Polymer Solutions Dilute: Intrinsic viscosity (MMA-co-styrene, random copolymer) 0.93 dl g^{-1} (C_6H_6, 35% wt styrene, MW 250000), 0.96 dl g^{-1} (chlorobenzene), 0.94 dl g^{-1} (dioxane), 1.18 dl g^{-1} ($CHCl_3$), 0.95 dl g^{-1} (THF), 0.72 dl g^{-1} (p-xylene) [12]; (MMA-co-styrene, diblock copolymer) 0.227 dl g^{-1} (30°, toluene, 48% wt styrene, MW 581000) [13], 1.53 dl g^{-1} (C_6H_6, 52% wt styrene, MW 603000) [12], 1.42 dl g^{-1} (dioxane, 52% wt styrene, MW 603000) [12], 1.78 dl g^{-1} ($CHCl_3$, 52% wt styrene, MW 603000) [12], 1.36 dl g^{-1} (THF, 52% wt styrene, MW 603000) [12], 0.168 dl g^{-1} (diethyl malonate, 48% wt styrene, MW 581000) [13], 0.192 dl g^{-1} (p-xylene, 48% wt styrene, MW 581000) [13], 0.167 dl g^{-1} (30°, 1-chlorobutane, 48% wt styrene, MW 581000) [13]. Theta temps. 25° (butanone/MeOH, 99.4/0.6, isooctyl methacrylate-co-styrene, 80% styrene) [17]; (MMA-co-styrene): 20° (C_6H_6/hexane, 44/56, 76.3% styrene) [18], 63° (cyclohexanol, 70.2% styrene) [19], 58.4° (2-ethoxyethanol, 56.2% styrene) [19], 20° (C_6H_6/isopropanol, 48/52, 42.3% styrene) [18], 60.8° (cyclohexanol, 50% styrene, alternating copolymer) [20], 68.2° (cyclohexanol, 28.5% styrene, random copolymer) [21], 72.8° (2-ethoxyethanol, 69.4% styrene, random copolymer) [21], 84° (cyclohexanol, 84.6% styrene, block copolymer) [21], 69.5° (2-ethoxyethanol, 35.9% styrene, block copolymer) [21]

Polymer Melts: Viscoelastic coefficients have been reported [22]

Water Absorption:

No.	Value	Note
1	0.15%	24h, MMA-co-styrene, injection-moulding grade, ASTM D570 [3]
2	0.17%	24h, MMA-co-styrene, extrusion grade, ASTM D570 [3]

Mechanical Properties:

Mechanical Properties General: The tensile strength and hardness of the MMA/butyl acrylate/styrene copolymer increase with increasing polystyrene or methyl methacrylate content but the elongation decreases. Tensile strength decreases as the MW of the polystyrene macromonomer increases

Tensile (Young's) Modulus:

No.	Value	Note
1	3500 MPa	MMA-co-styrene, injection-moulding grade, ASTM D638 [3]
2	3300 MPa	MMA-co-styrene, extrusion grade, ASTM D638 [3]
3	2960 MPa	430 kpsi, MMA-co-styrene, ASTM D638 [5]
4	330 MPa	MMA-co-styrene, alternating copolymer [6]

Flexural Modulus:

No.	Value	Note
1	3300 MPa	MMA-co-styrene, extrusion grade, ASTM D790 [3]
2	3500 MPa	MMA-co-styrene, injection-moulding grade, ASTM D790 [3]
3	1790–2620 MPa	260-380 kpsi, MMA-co-styrene, ASTM D790 [5]

Tensile Strength Break:

No.	Value	Note	Elongation
1	57.2 MPa	MMA-co-styrene, injection-moulding grade, ASTM D638 [3]	2%
2	68.2 MPa	MMA-co-styrene, extrusion grade, ASTM D638 [3]	5%
3	68.9 MPa	10 kpsi, MMA-co-styrene, ASTM D638 [5]	3%
4	11.3 MPa	MMA-co-styrene, alternating copolymer [6]	3.4%

Flexural Strength at Break:

No.	Value	Note
1	103 MPa	MMA-co-styrene, injection-moulding grade, ASTM D790 [3]
2	116 MPa	MMA-co-styrene, extrusion grade, ASTM D790 [3]

Compressive Modulus:

No.	Value	Note
1	1650–2550 MPa	240–370 kpsi, MMA-co-styrene, ASTM D695 [5]

Compressive Strength:

No.	Value	Note
1	75.8–103.3 MPa	11–15 kpsi, MMA-co-styrene, ASTM D695 [5]

Viscoelastic Behaviour: Williams-Landel-Ferry constants of MMA-co-styrene [22] and hexyl methacrylate-co-styrene [4] for temp. superposition have been reported
Hardness: Rockwell M64 (MMA-co-styrene, injection-moulding grade, ASTM D785) [3], M80 (MMA-co-styrene, extrusion grade, ASTM D785) [3]
Izod Notch:

No.	Value	Notch	Note
1	20 J/m	Y	MMA-co-styrene, MEC, ASTM D256 [3]
2	16 J/m	Y	0.3 ft lb in^{-1}, MMA-co-styrene, ASTM D256 [5]

Optical Properties:
Refractive Index:

No.	Value	Note
1	1.53	MMA-co-styrene, extrusion grade, ASTM D542 [3]
2	1.56	MMA-co-styrene, injection-moulding grade, ASTM D542 [3]
3	1.567	MMA-co-styrene, ASTM D542 [5]

Transmission and Spectra: Light transmission 90% (MMA-co-styrene, MEC) [3]
Volume Properties/Surface Properties: Haze 2% (MMA-co-styrene, extrusion grade) [3]

Polymer Stability:
Upper Use Temperature:

No.	Value	Note
1	82–93°C	180–200°F, MMA-co-styrene, continuous [5]

Flammability: Burning rate of MMA-co-styrene is slow (1.3 in min^{-1}) (ASTM D635) [5]
Environmental Stress: MMA-co-styrene is unaffected by sunlight [5]
Chemical Stability: MMA-co-styrene is unaffected by weak acids, weak or strong alkalis. Attacked by strong oxidising acids and organic solvents (ketones, esters, aromatic hydrocarbons and chlorinated hydrocarbons) [5]
Stability Miscellaneous: Most of the volatile products from the γ-irradiation of the copolymers result from fragmentation of the MMA units [23]

Applications/Commercial Products:
Processing & Manufacturing Routes: Processed by compression or injection moulding
Mould Shrinkage (%):

No.	Value	Note
1	0.2–0.6%	MMA-co-styrene [5]

Applications: May be used to strengthen the interfaces between immiscible homopolymers; random copolymers are more effective than block copolymers. The copolymers are added to homopolymer blends to improve adhesion and hence the mech. props., e.g. the fracture toughness of the interface can be increased with the copolymer efficiency increasing in the order: block > graft > random. Styrene/MMA block copolymers are used as coupling agents between PS and PMMA or between PPO and PMMA in blends where miscibility is a problem. Strong interfaces result when the MW of the copolymer is ≥ 84000 (for PS/PMMA blends) and 80000 (PPO/PMMA blends). Other applications of the copolymers are as binders in lithographic and other coatings; as toners; in photographic paper; in optical discs and in adhesives

Trade name	Supplier
Almitrex	Mitsui Petrochemicals
Centrex	C.E. Plastics
Geloy	C.E. Plastics
Luran	BASF
NAS	Novacor
NAS	Monsanto Chemical
Neocryl	Zeneca
Noan	Monsanto Chemical
Noan	Novacor
Styrocel	Shell Chemicals
Zerlon	Dow

Bibliographic References
[1] Davis, T.P., *Polym. Commun.*, 1990, **31**, 442
[2] Magarik, S.Y., Pavlov, G.M. and Fomin, G.A., *Macromolecules*, 1978, **11**, 294, (hydrodynamic props., optical props.)
[3] *Encycl. Polym. Sci. Eng.*, Vol. 16, 2nd edn., (ed. J.I. Kroschwitz), Wiley Interscience, 1985
[4] Abralkhodja, S., Bi, L.K., Wong, C.-P., Schrag, J.L. et al, *J. Polym. Sci., Part A-2*, 1970, **8**, 1927, (dynamic mechanical props.)
[5] Guide to Plastics, *Property and Specification Charts*, (ed. W.A. Kaplan), McGraw-Hill,
[6] Suzuki, H., Nishia, Y., Kimura, N., Mathot, V.B.F. et al, *Polymer*, 1994, **35**, 3698

[7] Tanaka, K., Takahara, A. and Kajiyama, T., *Acta Polym.*, 1995, **46**, 476
[8] Podesva, J. and Biros, J., *Makromol. Chem.*, 1981, **182**, 3341, (glass transition temps.)
[9] *Neocryl Range*, Zeneca, 1997, (technical datasheet)
[10] Dipaola-Baranyi, G., *Macromolecules*, 1982, **15**, 622, (solubility)
[11] Barton, A.F.M., *CRC Handbook of Polymer-Liquid Interaction Parameters and Solubility Parameters*, CRC Press, 1990, 265
[12] Dondos, A., Rempp, P. and Benoit, H., *Makromol. Chem.*, 1969, **130**, 233
[13] Tanaka, T., Kotaka, T., Ban, K. and Hattori, M., *Macromolecules*, 1977, **10**, 960
[14] Newman, S., *J. Appl. Polym. Sci.*, 1962, **6**, S15
[15] Gallot, Y., Franta, E., Rempp, P. and Benoit, H., *J. Polym. Sci., Part C: Polym. Lett.*, 1963, **4**, 473
[16] Krause, S., *J. Phys. Chem.*, 1964, **68**, 1948, (dilute soln. props.)
[17] Kalfus, M. and Mitus, J., *J. Polym. Sci., Part A-1*, 1966, **4**, 953, (theta temps.)
[18] Elias, H.G. and Gruber, U., *Makromol. Chem.*, 1965, **86**, 168, (theta temps.)
[19] Kotaka, T., Ohnuma, H. and Murakami, Y., *J. Phys. Chem.*, 1966, **70**, 4099, (theta temps.)
[20] Kotaka, T., Tanaka, T., Ohnuma, H., Murakami, Y. and Inagaki, H., *Polym. J. (Tokyo)*, 1970, **1**, 245, (theta temps.)
[21] Kotaka, T., Ohnuma, H. and Inagaki, H., *Polymer*, 1969, **10**, 517, (theta temps.)
[22] Halary, J.L., Oultache, A.K., Louyot, J.F., Jasse, B. *et al*, *J. Polym. Sci., Part B: Polym. Phys.*, 1991, **29**, 933, (viscoelastic props.)
[23] Busfield, W.K., O'Donnell, J.H. and Smith, C.A., *Polymer*, 1982, **23**, 431, (degradation)

Syndiotactic polypropylene   S-34

Synonyms: *Polypropene-syndiotactic. PP-syndiotactic. Poly(1-propene) syndiotactic*
Related Polymers: For more general information including other stereoisomers see Polypropylene.
Monomers: Propylene
Material class: Thermoplastic
Polymer Type: polypropylene
CAS Number:

CAS Reg. No.
9003-07-0

Molecular Formula: $(C_3H_6)_n$
Fragments: C_3H_6
General Information: Only crystalline syndiotactic poly(1-alkene) known at present. Not as readily prepared by normal commercial processes as the other stereoisomers and so tends to be of limited application

Volumetric & Calorimetric Properties:
Density:

No.	Value	Note
1	d 0.91–0.989 g/cm^3	cryst.
2	d 0.856 g/cm^3	amorph., calc.

Latent Heat Crystallization: Entropy of polymerisation $\Delta S°$ 191 J K^{-1} mol^{-1} (25°, monomer gas phase, polymer crystalline/partly crystalline) [1,4]

Transport Properties:
Polymer Solutions Dilute: Theta temp. θ 42° (isoamyl acetate, by virial coefficients) [1,2]

Optical Properties:
Transmission and Spectra: Ir spectral data have been reported [5]
Molar Refraction: dη/dc 0.1077 (30°, 436 nm, heptane) [1,2]

Bibliographic References
[1] Elias, H.G., *Polym. Handb.*, 3rd edn., (eds. J. Brandrup and E.H. Immergut), Wiley Interscience, 1989
[2] Inagaki, H., Miyamoto, T. and Ohta, S., *J. Phys. Chem.*, 1966, **70**, 3420
[3] *Encycl. Polym. Sci. Eng.*, 2nd edn., (ed. J.I. Kroschwitz), John Wiley and Sons, 1985
[4] Gee, D.R. and Melia, T.P., *Makromol. Chem.*, 1968, **116**, 122
[5] *Wunderlich Macromolecular Physics*, Academic Press, 1973

(2,2′,6,6′-Tetrabromobisphenol A diglycidyl ether) resin T-1

Synonyms: *[2,2′-[(1-Methylethylidene)bis(2,6-dibromo-4,1-phenylene)oxymethylene]]bisoxirane resin*
Monomers: 2,2′,6,6′-Tetrabromobisphenol A, (Chloromethyl)oxirane
Material class: Thermosetting resin
Polymer Type: epoxy
CAS Number:

CAS Reg. No.	Note
33294-14-3	
107632-52-0	(copolymer with methylenedianiline)
105186-73-0	(copolymer with bisphenol A dicyanate)
87036-50-8	(copolymer with phthalic anhydride)

Molecular Formula: $(C_{21}H_{20}Br_4O_4)_n$

Volumetric & Calorimetric Properties:
Density:

No.	Value	Note
1	d 1.646 g/cm^3	phthalic anhydride cured, 0.67 mol phthalic anhydride [13]

Transition Temperature General: T_g of methylenedianiline cured resin decreases with increasing water content [10]
Deflection Temperature:

No.	Value	Note
1	82°C	1.82 MPa, DER 542, anhydride cured [14]
2	192°C	dry, bisphenol A dicyanate cured [11]
3	172°C	wet, bisphenol A dicyanate cured [11]

Transport Properties:
Polymer Solutions Dilute: Viscosity has a minima at approx. 130° and depends upon the degree of B-staging of the resin; the lower the degree of B-staging, the lower the minimum viscosity [8]
Water Content: Cured resin shows very little water sorption compared to its non-brominated analogue [10]
Water Absorption:

No.	Value	Note
1	0.8%	64h, 92°, >95% relative humidity, bisphenol A dicyanate cured [11]

Mechanical Properties:
Mechanical Properties General: Elastic props. have been reported [1]
Flexural Modulus:

No.	Value	Note
1	3585 MPa	520 kpsi, bisphenol A dicyanate cured [11]
2	3589 MPa	36600 kg cm^{-2}, DER 542, anhydride cured [14]

Tensile Strength Break:

No.	Value	Note	Elongation
1	40.6 MPa	414 kg cm^{-2}, DER 542, anhydride cured [14]	0%

Flexural Strength at Break:

No.	Value	Note
1	124.8 MPa	18.1 kpsi, bisphenol A dicyanate cured [11]

Flexural Strength Yield:

No.	Value	Note
1	73 MPa	745 kg cm^{-2}, DER 542, anhydride cured [14]

Compressive Strength:

No.	Value	Note
1	229.5 MPa	2340 kg cm^{-2}, DER 542, anhydride cured [14]

Viscoelastic Behaviour: Acoustic behaviour has been reported [6]
Hardness: Rockwell M106 (DER 542, anhydride cured) [14]
Izod Notch:

No.	Value	Notch	Note
1	12.28 J/m	Y	0.23 ft lb in^{-1}, room temp., DER 542, anhydride cured [14]

Electrical Properties:
Electrical Properties General: Dielectric loss factor as a function of temp. has been reported [9]
Dielectric Permittivity (Constant):

No.	Value	Frequency	Note
1	3.4	1 MHz	DER 542, anhydride cured [14]

Dissipation (Power) Factor:

No.	Value	Frequency	Note
1	0.038	1 MHz	DER 542, anhydride, cured [14]

Optical Properties:
Refractive Index:

No.	Value	Note
1	1.614	20°, phthalic anhydride cured, 0.67 mol phthalic anhydride [13]

Transmission and Spectra: H-1 nmr spectral data have been reported [7]

Polymer Stability:
Polymer Stability General: Degradation studies have been reported [2]
Decomposition Details: Details of thermal degradation have been reported. [3] Pyrolysis produces hydrogen bromide gas, with max. liberation at 400° [5]
Flammability: Inherently flame retarding owing to a lowering of temp. through degradation and the formation of hydrogen bromide. [5] Flammability rating V0 (bisphenol A dicyanate cured, UL94) [11]

Environmental Stress: The effects of irradiation with uv have been reported [4]

Applications/Commercial Products:
Processing & Manufacturing Routes: Solventless curing may be undertaken using bisaspartimide-diamine compounds. [12] Processed by casting [14]
Applications: Main use in printed circuit boards. Other applications where flame retardancy required

Trade name	Details	Supplier
DER 542	Anhydride; discontinued material	Dow
EpiRez	5163	Interez

Bibliographic References
[1] Ramakrishna, K., Chen, W.T., Thiel, G.H., Steinwall, J.E. and Niu, T.M., *AMD*, 1994, **195**, 1, (elastic props.)
[2] Li, M.J., Gohari, K. and Pecht, M., *Int. SAMPE Electron. Conf.*, 1994, **7**, 446, (degradation)
[3] Yuan, J. and Paczkowski, M.A, Proc. - Electron. Compon. Technol. Conf., **43rd**, 1993, 330, (thermal degradation)
[4] Anderson, J.E., Adams, K.M. and Troyk, P.R., *Proc. - Electrochem. Soc.*, 1993, 93, (uv irradiation)
[5] Nara, S. and Matsuyama, K., *J. Macromol. Sci., Chem.*, 1971, **5**, 1205, (thermal degradation)
[6] Thompson, C.M. and Ting, R.Y., *Org. Coat. Appl. Polym. Sci. Proc.*, 1981, **46**, 661, (acoustic props)
[7] Gulino, D., Galy, J., Pascault, J.-P. and Pham, Q.T., *Makromol. Chem.*, 1984, **185**, 2269, (H-1 nmr)
[8] Martin, G.C. and Tungare, A.V., *Polym. Mater. Sci. Eng.*, 1988, **59**, 980, (rheological props)
[9] Gotro, J. and Yandrasits, M., *Polym. Eng. Sci.*, 1989, **29**, 278, (dielectric loss)
[10] Burton, B.L., *Int. SAMPE Tech. Conf.*, 1986, **18**, 124, (water sorption)
[11] Shimp, D.A., Hudock, F.A. and Ising, S.J., *Int. SAMPE Symp. Exhib.*, 1988, **33**, 754, (props)
[12] Kumar, D. and Gupta, A.D., *J. Macromol. Sci., Chem.*, 1985, **22**, 1101
[13] Fridman, Yu. B. and Shchurov, A.F., *Vysokomol. Soedin.*, *Ser. A*, 1983, **25**, 1473, (refractive index)
[14] *DER542 Anhydride*, Dow Chemical Co., 1996, (technical datasheet)

Tetrafluoroethylene-perfluoromethylvinyl ether copolymer   T-2

$$\left[\left[CF_2CF_2\right]_x\left[\begin{array}{c}CF_2CF\\|\\OCF_3\end{array}\right]_y\right]_n$$

Synonyms: PFA. *Poly(tetrafluoroethene-co-trifluoro(trifluoromethoxy)ethene)*. *Poly(tetrafluoroethylene-co-trifluoromethyl trifluorovinyl ether)*. *Poly(tetrafluoroethylene-co-trifluorovinyl trifluoromethyl ether)*. *Poly(tetrafluoroethene-co-perfluoro(methyl vinyl) ether)*
Related Polymers: Carbon-filled PFA, Glass filled PFA, PTFE filled PFA, Mineral and mica filled PFA, PTFE and carbon-filled PFA
Monomers: Tetrafluoroethylene, Perfluoro(methyl vinyl ether)
Material class: Copolymers, Thermoplastic
Polymer Type: PFA, fluorocarbon (polymers)
CAS Number:

CAS Reg. No.	Note
26425-79-6	
112652-06-9	alternating

Molecular Formula: $[(C_2F_4)_x(C_3F_6O)_y]_n$
Fragments: C_2F_4 C_3F_6O
Morphology: Chain flexibility is increased relative to PTFE alone [23]

Miscellaneous: The characteristic props. and uses of the Kalrez elastomer have been reported [8,20]. Props. of ECD-006 have been reported [21]

Volumetric & Calorimetric Properties:
Density:

No.	Value	Note
1	d 2.12–2.19 g/cm^3	[1]
2	d 2.14 g/cm^3	Hyflon [2]
3	d 2.15 g/cm^3	Teflon PFA [3]
4	d 2.13 g/cm^3	Teflon PFA 200 LP [25]

Thermal Expansion Coefficient:

No.	Value	Note	Type
1	0.00012–0.000199 K^{-1}	[1]	L
2	0.000288 K^{-1}	Hyflon [2]	L

Melting Temperature:

No.	Value	Note
1	300°C	Hyflon [2]
2	306°C	Teflon PFA [3]

Glass-Transition Temperature:

No.	Value	Note
1	-12°C	17.5 mol% perfluoro(methyl vinyl) ether [13]
2	-12°C	PTFE/perfluoro(methyl vinyl) ether 60:40 mol% [26]

Transition Temperature General: Low-temp. transitions are observed only in copolymers where the perfluoro(methyl vinyl) ether content is below 4 mol% [11]
Transition Temperature:

No.	Value	Note	Type
1	-60°C	1 kHz, 17.5 mol% perfluoro(methyl vinyl) ether [13]	T_γ
2	-179°C	1 kHz, 17.5 mol% perfluoro(methyl vinyl) ether [13]	T_δ
3	-245°C	1 kHz, 17.5 mol% perfluoro(methyl vinyl) ether [13]	T_ϵ

Surface Properties & Solubility:
Cohesive Energy Density Solubility Parameters: Solubility parameters of the rubber have been reported [18,19]
Solvents/Non-solvents: ECD-006 sol. highly fluorinated solvents; insol. most organic solvents, acids, bases [26]

Transport Properties:
Transport Properties General: Melt flow index is only slightly dependent upon the time of thermal ageing [6]
Permeability of Gases: ECD-006 has low permeability to gases [26]
Permeability of Liquids: Permeability to moisture has been reported [12]
Water Absorption:

No.	Value	Note
1	0.03%	24h [1]

Gas Permeability:

No.	Gas	Value	Note
1	H$_2$O	985 cm^3 mm/(m^2 day atm)	1.5 × 10^{-10} cm^2 (s torr)$^{-1}$, Kalrez [17]

Mechanical Properties:
Mechanical Properties General: ECD-006 has poor mech. props. at high temps. [26]
Flexural Modulus:

No.	Value	Note
1	690 MPa	[1]
2	624.7 MPa	6370 kg cm^{-2}, Hyflon [2]

Tensile Strength Break:

No.	Value	Note	Elongation
1	29 MPa	room temp. [1]	200%
2	24.8 MPa	253 kg cm^{-2}, Hyflon [2]	300%

Tensile Strength Yield:

No.	Value	Note
1	13.8–18.6 MPa	141–190 kg cm^{-2}, Teflon PFA [3]

Hardness: Shore D60 [1,2], A74 (Kalrez 1050) [22]
Izod Notch:

No.	Notch	Break	Note
1	Y	No break	[1,2]

Electrical Properties:
Electrical Properties General: Dielectric props. at low temps. have been reported [16]. The dissipation factor spectrum shows only a small γ-transition peak [25]
Dielectric Permittivity (Constant):

No.	Value	Frequency	Note
1	2.1	60–1 kHz	[1]
2	2.1	1 MHz	dry [1,2]

Dielectric Strength:

No.	Value	Note
1	20 kV.mm^{-1}	[1]

Complex Permittivity and Electroactive Polymers: Stability to charge storage can be improved by constant current corona charging [4,15]
Dissipation (Power) Factor:

No.	Value	Frequency	Note
1	0.00415	50 Hz	Hyflon [2]

Optical Properties:
Transmission and Spectra: F-19 nmr spectral data have been reported [7,10]
Volume Properties/Surface Properties: Clarity transparent to translucent [1]

Polymer Stability:
Thermal Stability General: ECD-006 is stable to heating in air for one month at 300° [26]
Upper Use Temperature:

No.	Value	Note
1	260°C	[1,2,3]

Decomposition Details: The major products of pyrolysis include tetrafluoroethene, hexafluoroethane and perfluoro(methyl vinyl) ether [14]. Heating at 275° for up to 80h gives only low yields (approx. 0.0002 mol g^{-1} of polymer) of hydrogen fluoride [24]
Flammability: Flammability is dependent on struct. but not on MW [9]
Chemical Stability: Kalrez 1050 is swollen by 91.6% by dichlorotetrafluoroethane after 21 d at 177° and by 26.2% under the same conditions in geothermal brine. Essentially unaffected, apart from slight soln. discoloration, after immersion under the same conditions in Mobil lubricating oil [22]. ECD-006 is stable to inorganic acids, bases, oxidising and reducing agents, and is not swollen by most common solvents
Stability Miscellaneous: Effect of γ-irradiation upon physico-chemical props., including dielectric loss, has been reported [5]. Thermal annealing produces two melting peaks which depend strongly on annealing temp. and time [11]

Applications/Commercial Products:
Processing & Manufacturing Routes: Processed by compression, injection or extrusion moulding. May be synth. by aq. emulsion polymerisation at 40–100° under pressure for a few hours using ammonium persulfate initiator [26]
Mould Shrinkage (%):

No.	Value	Note
1	4%	[1]
2	5%	Hyflon [2]

Applications: Potential applications as coating for electrical wires; in hoses, pipes and tubes. ECD-006 has potential applications in electrical insulation and seals for aircraft engines and scientific equipment

Trade name	Details	Supplier
Aflon	P range	Asahi Glass
ECD	006	DuPont
Hyflon	PFA 420/450	Ausimont
Hyflon	PFA 420/450	Ausimont USA Inc.
Kalrez	1050	DuPont
Stat-Kon	FP-P	LNP Engineering
Teflon	C series/310/340/350	DuPont
Teflon	PFA	Furon
Teflon PFA	510/560/580/200 LP	DuPont
Thermocomp	FP series	LNP Engineering

Bibliographic References
[1] *The Materials Selector*, 2nd edn., (eds. N.A. Waterman and M.F. Ashby), Chapman and Hall, 1997, **3**
[2] *Hyflon*, Ausimont USA, Inc., (technical datasheet)
[3] Teflon PFA DuPont,(technical datasheet)
[4] Xia, Z., *Tongji Daxue Xuebao*, 1991, **19**, 315
[5] Matveev, V.K., Smirnova, E.A., Kochkina, L.G., Loginova, N.N. and Milinchuk, V.K., Vysokomol. Soedin., *Ser. B*, 1991, **33**, 351, (γ-irradiation)
[6] Loginova, N.N., Podlesskaya, N.K., Kochkina, L.G., Bronov, M.V. and Madorskaya, L.Ya., *Plast. Massy*, 1988, **24**, (rheological props)

[7] Sokolov, Yu.P., Filippov, N.N., Veretennikov, N.V. and Konshin, A.I., Issled. Stroeniya Makromolekul Metodom YAMR Vysok. Razresheniya, *M.*, 1983, 45, (F-19 nmr)
[8] Muto, I., *Nikko Materiaru*, 1985, **3**, 72, (props, uses)
[9] Zhevlakov, A.F., Ermakova, I.S., Loginova, N.N., Madorskaya, L.Ya. et al, *Plast. Massy*, 1980, 62, (flammability)
[10] Lovchikov, V.A., Sass, V.P., Konshin, A.I., Dolgopol'skii, I.M. and Sokolov, S.V., Vysokomol. Soedin., *Ser. B*, 1975, **17**, 622, (F-19 nmr)
[11] Pucciariello, R., *J. Polym. Sci., Part B: Polym. Phys.*, 1996, **32**, 1771
[12] Ma, C., Verma, N., Shero, E., Gilbert, S.L. and Shadman, F., *Proc. - Inst. Environ. Sci.*, 1994, **1**, 193, (moisture permeability)
[13] Starkweather, H.W., Avakian, P., Fontanella, J.J. and Wintersgill, M.C., *J. Therm. Anal.*, 1996, **46**, 785
[14] Jiang, L., Wu, J. and Xue, S., *J. Anal. Appl. Pyrolysis*, 1986, **10**, 99, (pyrolysis)
[15] Xia, Z., Ding, H., Yang, G., Lu, T. and Sun, X., *IEEE Trans. Electr. Insul.*, 1991, **26**, 35
[16] Meyer, W., *Proc. Int. Cryog. Eng. Conf.*, 1976, **6**, 367, (dielectric props)
[17] Ma, C., Shero, E., Verma, N., Gilbert, S.L. and Shadman, F., *J. IES*, 1995, **38**, 43, (permeability)
[18] Kirillova, T.L., Safanov, A.V. and Frenkel, R.Sh., Vysokomol. Soedin., *Ser. B*, 1992, **34**, 7, (solubility parameters)
[19] Safanov, A.V., Frenkel, R.Sh. and Kirillova, T.L., *Kauch. Rezina*, 1991, **10**, (solubility parameters)
[20] Knox, J.B., *Chem. Rundsch.*, 1977, **30**, 13, (props)
[21] Maskornick, M.J., Kalb, G.H. and Graff, R.S., Soc. Plast. Eng., *Tech. Pap.*, 1974, **20**, 675, (props)
[22] Thibeau, R.J., *Ind. Eng. Chem. Proc. Des. Dev.*, 1983, **22**, 127
[23] Villani, V., Pucciariello, R. and Fusco, R., *Colloid Polym. Sci.*, 1991, **269**, 477, (conform)
[24] Knight, G.J. and Wright, W.W., *Thermochim. Acta*, 1983, **60**, 187, (thermal degradation)
[25] Sacher, E., *J. Macromol. Sci., Phys.*, 1981, **19**, 109, (dielectric props)
[26] Thompson, D.C., *DuPont Innovation*, 1973, **4**, 8, (props)

Tetrafluoroethylene-propylene copolymer   T-3

$$-[-CF_2CF_2-]_x[-CH_2CH(CH_3)-]_y-]_n$$

Synonyms: *TFE/P copolymer. PTFE/PP copolymer. Poly(tetrafluoroethene-co-1-propene)*
Monomers: Tetrafluoroethylene, Propylene
Material class: Copolymers
Polymer Type: polypropylene, PTFEfluorocarbon (polymers)
CAS Number:

CAS Reg. No.
27029-05-6

Polymer Stability:
Polymer Stability General: Vulcanised elastomer displays excellent heat and chemical resistance, owing to high stability of cross-links, in addition to inherent thermal and chemical inertness
Thermal Stability General: When vulcanised is stable at temps. of approx. 230° for months
Upper Use Temperature:

No.	Value	Note
1	300°C	approx., short time

Decomposition Details: Onset of thermal decomposition in both air and N_2 occurs at approx. 415°
Chemical Stability: Swollen to some extent by trichoroethylene, $CHCl_3$, Me_2CO, CCl_4 and C_6H_6 vulcanised (peroxide-cured) elastomer displays excellent resistance to acids, bases, water, steam and oils (ASTM D471-68). Has excellent resistance to amine-corrosion inhibitors and bleach. A typical (peroxide-cured) vulcanisate approx. 55 mol% TFE) is unaffected by conc. strong acids and bases, even at high temp. Is fairly resistant to oxidising and reducing agents and lubricating oils, has limited resistance to aromatic hydrocarbons (relatively low fluorine content). Resistant to ketones, and display exceptional resistance to dehydrofluorination and embrittlement by organic bases such as amines, relatively highly swollen by AcOH

Applications/Commercial Products:

Trade name	Supplier
Aflas 100H	3M Industrial Chemical Products
Aflas FA 100H	3M Industrial Chemical Products
Aflas FA 100S	3M Industrial Chemical Products
Aflas FA 150E	3M Industrial Chemical Products
Aflas FA 150L	3M Industrial Chemical Products
Aflas FA 150P	3M Industrial Chemical Products

Ultem 1000 U-1

Synonyms: *Poly[(1,3-dihydro-1,3-dioxo-2H-isoindole-5,2-diyl)-1,3-phenylene(1,3-dihydro-1,3-dioxo-2H-isoindole-2,5-diyl)oxy-1,4-phenylene(1-methylethylidene)-1,4-phenyleneoxy]*
Related Polymers: Ultem 5001
Monomers: Bisphenol A
Material class: Thermoplastic
Polymer Type: polyether-imide
CAS Number:

CAS Reg. No.
61128-24-3

Molecular Formula: $(C_{37}H_{24}N_2O_6)_n$
Fragments: $C_{37}H_{24}N_2O_6$
Molecular Weight: M_n 8490–15000. MW 14700–28800. M_w/M_n 2.95. MW 35700
Additives: Glass fibre may be used to increase mech. props.
Morphology: Amorph.
Miscellaneous: Performance characteristics have been reported [5]. Reviews of props. have been reported [8,9,10,11,40]

Volumetric & Calorimetric Properties:
Density:

No.	Value	Note
1	d1.28 g/cm^3	1 mil thick film [18]
2	d1.285 g/cm^3	[24]
3	d1.278 g/cm^3	compression moulded [32]
4	d1.27 g/cm^3	ASTM D792 [36,41]

Thermal Expansion Coefficient:

No.	Value	Note	Type
1	5.6×10^{-5} K^{-1}	mould direction, ASTM D696 [36]	L

Thermal Conductivity:

No.	Value	Note
1	0.22 W/mK	ASTM C177 [36,41]

Thermal Diffusivity: 1.39×10^{-7} m^2 s^{-1} (40°, 50 μm thick). Thermal diffusivity decreases with increasing temp. [14]
Glass-Transition Temperature:

No.	Value	Note
1	215°C	[13,28]
2	231°C	[14,26]
3	215°C	[18]
4	218.7°C	[22]
5	230°C	[27]
6	217°C	[29]
7	220°C	[32,34]

Transition Temperature General: The temp. range of physical ageing extends from T_g to temps. within the T_β range [34]
Deflection Temperature:

No.	Value	Note
1	200°C	1.8 MPa, ASTM D648 [29,36,41]
2	210°C	0.45 MPa, ASTM D648) [29,36,41]

Vicat Softening Point:

No.	Value	Note
1	219°C	Method B, ASTM D1525 [36,41]

Transition Temperature:

No.	Value	Note	Type
1	82°C	approx. [22]	Tβ
2	74°C	approx. [34]	Tβ
3	-113°C	[22]	Tγ
4	-111°C	[34]	Tγ

Surface Properties & Solubility:
Solubility Properties General: Miscibility with poly(aryl ether ketones) is dependent on the struct. of both components. Poly(aryl ether ketones) based on bisphenol A or a sulfonyl diphenol struct. are immiscible with Ultem [4]. Miscible with poly(ether ether ketone) in the amorph. state [20]
Cohesive Energy Density Solubility Parameters: δ 28.8 J$^{1/2}$ cm$^{-3/2}$ [24]
Plasticisers: CO_2 [22]
Solvents/Non-solvents: Sol. DMF, $CHCl_3$, CH_2Cl_2, *N*-methylpyrrolidone, dimethylacetamide [28]. Insol. DMSO

Transport Properties:
Melt Flow Index:

No.	Value	Note
1	8.9 g/10 min	ASTM D1238 [41]

Polymer Solutions Dilute: Rheological props. have been reported [17]. η 0.39–0.56 dl g^{-1} (DMF, 25°, dependent on method of synth.) [27], 0.28 dl g^{-1} (CH_2Cl_2) [27], 0.5 dl g^{-1} (*m*-cresol) [28]
Polymer Melts: Melt rheology has been reported [35]
Permeability of Gases: The permeability of CO_2 has been reported [1]. Diffusion coefficients (1 mil film, 35°) [CO_2] 1.14×10^{-8} cm^2 s^{-1}, [CH_4] 0.113×10^{-8} cm^2 [N_2] 0.572×10^{-8} cm^2 s^{-1} [18]. Diffusion coefficient of O_2 has been reported [31]
Water Absorption:

No.	Value	Note
1	0.25%	24 h, 23°, ASTM D570 [36,41]
2	1.25%	equilibrium, 23°, ASTM D570 [36,41]

Mechanical Properties:
Mechanical Properties General: Ultimate elongation 60% (4 mm thick) [32], (Type I, 0.125" thick) [41]

Tensile (Young's) Modulus:

No.	Value	Note
1	3000 MPa	4 mm thick [32]
2	3800 MPa	20°, compression moulded [32]

– Ultem 1000

Flexural Modulus:

No.	Value	Note
1	2068.2 MPa	300 kpsi, 180° [29]

Tensile Strength Break:

No.	Value	Note
1	41.4 MPa	6 kpsi, 180° [29]

Poisson's Ratio:

No.	Value	Note
1	0.36	ASTM D638 [41]

Viscoelastic Behaviour: Stress-strain relationship under load is linear below the proportional limit [29]
Hardness: Rockwell M109 [33] (ASTM D785) [36,41]
Fracture Mechanical Properties: Shear strength 103.41 MPa (ASTM D732) [41]. K_{Ic} 2.93 Mpa m$^{1/2}$ (20°, 4 mm thick, 0.00016 s^{-1} strain rate) [32] G_{Ic} 2042 J m^{-2} (20°, 4 mm thick, 0.00016 s^{-1} strain rate) [32]. Stress endurance limit 24 MPa, which increases with orientation [32]

Electrical Properties:
Surface/Volume Resistance:

No.	Value	Note	Type
1	100×10^{15} Ω.cm	24°, ASTM D257 [29]	S
2	0.17×10^{15} Ω.cm	150°, ASTM D257 [29]	S

Dielectric Permittivity (Constant):

No.	Value	Frequency	Note
1	3.08	60 Hz	150°, ASTM D150 [29]
2	3.06	1 kHz	150°, ASTM D150 [29]
3	3.05	1 MHz	150°, ASTM D150 [29]
4	3.15	60 Hz	24°, ASTM D150 [29]
5	3.15	1 kHz	24°, ASTM D150 [29,36,41]
6	3.13	1 MHz	24°, ASTM D150 [29]
7	3.15	100 Hz	ASTM D150 [41]

Dielectric Strength:

No.	Value	Note
1	24 kV.mm^{-1}	24°, ASTM D149 [29]
2	28 kV.mm^{-1}	0.0625" thick, in oil, ASTM D149 [36,41]
3	33 kV.mm^{-1}	0.0625" thick, in oil, ASTM D149 [36,41]

Arc Resistance:

No.	Value	Note
1	128s	ASTM D495 [36]

Complex Permittivity and Electroactive Polymers: Comparative tracking index 160 V (24°, ASTM D3638) [29]

Optical Properties:
Optical Properties General: Complex refractive index data have been reported [2]

Refractive Index:

No.	Value	Note
1	1.662	22° [21]

Polymer Stability:
Thermal Stability General: Thermal stability under a flow of N_2 has been reported [6]
Upper Use Temperature:

No.	Value	Note
1	170°C	[36]

Decomposition Details: Heating at 380° *in vacuo* for 6 min. causes cross-linking. 5% Weight loss occurs at 514°. 30% Weight loss at 572° [26]. 5% Weight loss in air at 499° (at 508° under N2)[28]
Flammability: Self-ignition temp. 535° (ASTM D1929) [29]; fl. p. 520° (ASTM D1929) [29]. Flammability rating V0 (0.025" thick, UL94) [29,36,41], 5V (0.075" thick, UL94) [29,36,41]. Limiting oxygen index 47% (ASTM D2863) [29,36,41]. Other details on flammability have been reported [38]
Environmental Stress: Exposure to atomic oxygen for 4 h leads to mass loss and deterioration in flexural props. [16] Good resistance to uv, with only slight surface degradation after 2500 h exposure. [29] Has greater resistance to environmental stress cracking and crazing than do most other amorph. thermoplastics [30]
Chemical Stability: Has greater stability than most amorph. thermoplastics. [29] Attacked in the long term by aliphatic amines, NH_3 and hydrazine, and aq. solns. with pH greater than 8 [29]. Disintegrates on prolonged exposure to 10% sodium hydroxide at 96° [39]
Hydrolytic Stability: Stability to hot H_2O and steam has been reported [37] Exhibits internal crack formation upon both cyclic and continuous exposure to H_2O at 96°. The cracks form 0.0625" from the injection moulded surface [39]. Steam sterilisation causes only minimal internal crack formation and increases tensile strength. Tensile impact strength is only reduced significantly after 100 cycles of steam sterilisation [39]
Stability Miscellaneous: Electron-beam irradiation causes deterioration of tensile props. only at exposure levels of 100 Mrad [3]. Electron-beam irradiation at 1 Grad causes cross-linking and embrittlement due to dehydrogenation of the methyl groups and cleavage of ether linkages [7]. Good stability to γ-irradiation [29]

Applications/Commercial Products:
Processing & Manufacturing Routes: May be synth. by imidisation of poly(amic acid) salts at 250–300° [12]. Also synth. by polyimidisation by treating dianhydrides with diamines [15]. Processed by injection moulding [23] or extrusion [41]. Synth. by reaction of the disodium salt of bisphenol A with a nitro-substituted bisimine in DMSO/toluene at 40–50° under N_2; displacement of the nitro groups is achieved to form the polymer [27,28]
Mould Shrinkage (%):

No.	Value	Note
1	0.5–0.7%	[36]

Applications: Potential use in underground trains, shipboards and in wiring of airframes. Other usage includes microwave oven trays, iron compounds and medical products which require repeated sterilisation

Trade name	Supplier
Standard sample	Mitsubishi Chem.
Ultem	General Electric

Bibliographic References

[1] Tanaka, K., Yokoshi, O., Kita, H. and Okamoto, K., *Kogakubu Kenkyu Hokoku (Yamaguchi Daigaku)*, 1988, **38**, 277, (gas permeability)
[2] Stead, M. and Simonis, G.J., *Proc. SPIE-Int. Soc. Opt. Eng.*, 1988, 1039, (refractive index)
[3] Kiefer, R.L. and Orwoll, R.A., *NASA Contract. Rep.*, 1987, 31, (irradiation)
[4] Hedrick, J.C., Arnold, C.A., Zumbrum, M.A., Ward, T.C. and McGrath, J.E., *Int. SAMPE Symp. Exhib.*, 1990, **35**, 82, (miscibility)
[5] Okamura, M., *Kobunshi Kako*, 1986, **35**, 503
[6] Yokota, R., Sakino, T. and Mita, I., *Kobunshi Ronbunshu*, 1990, **47**, 207, (thermal stability)
[7] Long, S.A.T. and Long, E.R., *IEEE Trans. Nucl. Sci.*, 1984, **NS-31**, 1293, (irradiation)
[8] Dazai, T., *Baruka Rebyu*, 1993, **37**, 9, (props, use)
[9] Ma, T.C. and Shia, H.C., *Hua Hsueh*, 1986, **44**, A1, (props, use)
[10] Okamura, M., *Gosei Jushi*, 1985, **31**, 7, (props)
[11] Suematsu, K., *Nikko Materiaru*, 1984, **2**, 53, (props, use)
[12] Sensenich, C.L., Facinelli, J.V., Dong, L., Gardner, S.L. et al, *Polym. Prepr. (Am. Chem. Soc., Div. Polym. Chem.)*, 1996, **37**, 400, (synth)
[13] Higuchi, A., Nakajima, T., Morisato, A., Ando, M. et al, *J. Polym. Sci., Part B: Polym. Phys.*, 1996, **34**, 2153, (gas permeability)
[14] Morikawa, J., Tan, J. and Hashimoto, T., *Polymer*, 1995, **36**, 4439, (thermal diffusivity)
[15] Davies, M., Hay, J.N. and Woodfine, B., *High Performance Polym.*, 1993, **5**, 37, (synth)
[16] Stancil, P.C., Long, E.R., Long, S.A.T. and Harries, W.L., *Polym. Prepr. (Am. Chem. Soc., Div. Polym. Chem.)*, 1991, **32**, 644
[17] Lin, W.L., Hsiue, L.T., Hu, J.T. and Ma, C.C.M., *MRL Bull. Res. Dev.*, 1987, **1**, 9, (rheological props)
[18] Barbari, T.A., Koros, W.J. and Paul, D.R., *J. Polym. Sci., Part B: Polym. Phys.*, 1988, **26**, 709, (gas permeability)
[19] Schmidhauser, J.C. and Longley, K.L., *Polym. Prepr. (Am. Chem. Soc., Div. Polym. Chem.)*, 1989, **30**, 13, (gas permeability)
[20] Crevecoeur, G. and Groeninckx, G., *Macromolecules*, 1991, **24**, 1190, (miscibility)
[21] Philipp, H.R., Le Grand, D.G., Cole, H.S. and Liu, Y.S., *Polym. Eng. Sci.*, 1989, **29**, 1574, (optical props)
[22] Fried, J.R., Liu, H.-C. and Zhang, C., *J. Polym. Sci., Part C: Polym. Lett.*, 1989, **27**, 385
[23] Wagner, A.H., Yu, J.S. and Kalyon, D.M., *Adv. Polym. Technol.*, 1989, **9**, 17
[24] Okamoto, K.-I., Tanaka, K., Hidetoshi, K., Nakamura, A. and Kusuki, Y., *J. Polym. Sci., Part B: Polym. Phys.*, 1989, **27**, 2621, (gas permeability)
[25] Nied, H.F., Stokes, V.K. and Ysseldyke, D.A., *Polym. Eng. Sci.*, 1987, **27**, 101, (strain behaviour)
[26] Kuroda, S.-I., Terauchi, K., Nogami, K. and Mita, I., *Eur. Polym. J.*, 1989, **25**, 1, (thermal degradation)
[27] White, D.M., Takeroshi, T., Williams, F.J., Relles, H.M. et al, *J. Polym. Sci., Polym. Chem. Ed.*, 1981, **19**, 1635, (synth)
[28] Takekoshi, T., Kochanowski, J.E., Manello, J.S. and Webber, M.J., *J. Polym. Sci., Polym. Symp.*, 1986, **74**, 93, (synth, thermal degradation)
[29] Johnson, R.O. and Burlhis, H.S., *J. Polym. Sci., Polym. Symp.*, 1983, **70**, 129, (synth, props)
[30] White, S.A., Weissman, S.R., Kambour, R.P., *J. Appl. Polym. Sci.*, 1982, **27**, 2675, (stress cracking)
[31] Alger, M.M. and Stanley, T.J., *J. Appl. Polym. Sci.*, 1988, **36**, 1501, (gas permeability)
[32] Trotignon, J.P., Verdu, J., Martin, Ch. and Morel, E., *J. Mater. Sci.*, 1993, **28**, 2207, (fatigue behaviour)
[33] Bijwe, J., Logani, C.M. and Tewari, U.S., *Wear*, 1990, **138**, 77, (abrasion)
[34] Brennan, A.B. and Feller, F., *J. Rheol. (N.Y.)*, 1995, **39**, 453
[35] Kalyon, D.M., Yu, D.-W. and Yu, J.S., *J. Rheol. (N.Y.)*, 1988, **32**, 789, (melt rheology)
[36] Relles, H.M., *Contemp. Top. Polym. Sci.*, 1984, **5**, 261, (props)
[37] Rosato, D.V., *Med. Device Diagn. Ind.*, 1985, **7**, 48, 50, 85, (hydrolytic stability)
[38] Bridgman, A.L. and Nelson, G.L., *J. Fire Flammability*, 1982, **13**, 114, (flammability)
[39] Robeson, L.M., Dickinson, B.L. and Crisafulli, S.T., *Polym. News*, 1986, **11**, 359, (hydrolytic stability)
[40] Serfaty, I.W., Polyimides: Synth., Charact., *Appl. (Proc. Tech. Conf. Polyimides)*, 1982, **1**, 149, (props)
[41] 1997, GE Plastics, (technical datasheet)

Ultem 5001

Synonyms: *Poly[(1,3-dihydro-1,3-dioxo-2H-isoindole-5,2-diyl)-1,4-phenylene(1,3-dihydro-1,3-dioxo-2H-isoindole-2,5-diyl)oxy-1,4-phenylene(1-methylethylidene)-1,4-phenyleneoxy]*
Related Polymers: Ultem 1000
Monomers: Bisphenol A
Material class: Thermoplastic
Polymer Type: polyether-imide
CAS Number:

CAS Reg. No.
61128-25-4

Molecular Formula: $(C_{37}H_{24}N_2O_6)_n$
Fragments: $C_{37}H_{24}N_2O_6$
Miscellaneous: The Ultem grade CRS5001 has improved resistance to strong acids, strong bases, aromatics and ketones. It complies with FDA and USP class VI regulations in colour 1000 [2]

Volumetric & Calorimetric Properties:
Density:

No.	Value	Note
1	d1.28 g/cm^3	ASTM D792 [2]

Deflection Temperature:

No.	Value	Note
1	208.8°C	1.82 MPa, 0.25 in. thick, unannealed, ASTM D648 [2]

Surface Properties & Solubility:
Cohesive Energy Density Solubility Parameters: δ 14.14 [1]
Solvents/Non-solvents: Sol. CCl_4, $CHCl_3$, CH_2Cl_2, dichloroethane, THF, cyclohexanone, Me_2CO, 1,4-dioxane, N,N-dimethylacetamide, DMF, pyridine [1]. Insol. hexane, cyclohexane, xylene, toluene, C_6H_6, Et_2O, pentyl acetate, EtOAc, 2-butanone, butanol, isopropyl alcohol, propanol, EtOH, AcOH, formic acid, MeOH, glycol, glycerol

Transport Properties:
Melt Flow Index:

No.	Value	Note
1	4.2 g/10 min	ASTM D1238 [2]

Permeability of Liquids: The diffusion of distilled H_2O through a membrane of the polymer at 28 psi decreases with an increase in polymer concentration but increases in the presence of additives, provided the additive concentration does not exceed 6% [1]
Water Absorption:

No.	Value	Note
1	0.16%	24 h, 23°, ASTM D570 [2]

Mechanical Properties:
Mechanical Properties General: Elongation at break of Ultem Type I material 70% (0.125" thick, ASTM D638) [2]

Tensile (Young's) Modulus:

No.	Value	Note
1	2895.5 MPa	420 kpsi, 0.125" thick, Ultem Type I, ASTM D638 [2]

Ultra high molecular weight PE

Synonyms: *UHMWPE*
Related Polymers: Polyethylenes, HDPE, Polyethylene film, Polyethylene fibre
Monomers: Ethylene
Material class: Thermoplastic
Polymer Type: polyethylene, polyolefins
CAS Number:

CAS Reg. No.	Note
9002-88-4	used for all density types

Molecular Formula: $(C_2H_4)_n$
Fragments: C_2H_4
Molecular Weight: 3000000–6000000 (ASTM D4020), MW as low as 1000000 have been reported
Additives: Graphite fibre, powdered metal, glass fibre, glass bead fillers used to increase stiffness and heat deflection temp
General Information: High degree of linearity and very long chain length lead to low crystallinity (ca. 40–45%) and fairly low density, [1,2,3] giving rise to anomalous props.
Morphology: Drawing into fibres gives almost perfect orientations. This results in predominantly orthorhombic crystallisation (with traces of non-orthorhombic crystals and a low amorph. content) unit cell dimensions a 7.36, b 4.89. Thermal expansion coefficient is higher in 'a' direction than in 'b' direction giving a transformation to a pseudohexagonal form [4]
Miscellaneous: Props. can be substantially improved using a melt/solid phase compressive deformation process [24]

Volumetric & Calorimetric Properties:
Density:

No.	Value	Note
1	d 0.93–0.94 g/cm³	ASTM D742 [1,2,3,5]

Thermal Expansion Coefficient:

No.	Value	Note	Type
1	0.00013 K^{-1}	-30–30° ASTM D696 [2,5]	L
2	0.0002 K^{-1}	30–60°, ASTM D696 [2,5]	L

Melting Temperature:

No.	Value	Note
1	125–135°C	cryst. [2,5,6]

Transition Temperature General: T_m and crystallinity vary with thermal history of sample, size and batch variation [12]. Heating above T_m changes appearance from opaque to clear [2,5,6]
Deflection Temperature:

No.	Value	Note
1	75°C	70–80°, 0.45 MPa, ASTM D648 [2,5]
2	43–50°C	110–120°F, 1.84 MPa (264 psi) [5]

Vicat Softening Point:

No.	Value	Note
1	135°C	[1,2]

Brittleness Temperature:

No.	Value	Note
1	<-200 °C	max., ASTM D746 [1]

Flexural Modulus:

No.	Value	Note
1	3102.3 MPa	450 kpsi 0.25 in. thick, Ultem, ASTM D790 [2]

Tensile Strength Yield:

No.	Value	Note
1	99.96 MPa	14.5 kpsi 0.125 in. thick, Ultem Type I, ASTM D638 [2]

Flexural Strength Yield:

No.	Value	Note
1	137.9 MPa	20 kpsi, 0.25" thick, Ultem, ASTM D790 [2]

Hardness: Rockwell R123 (Ultem, ASTM D785) [2]
Friction Abrasion and Resistance: Taber abrasion 10 mg (1000 cycles)$^{-1}$ (CS-17, 1 kg, ASTM D1044) [2]
Izod Notch:

No.	Value	Notch	Note
1	1281 J/m	N	24 ft lb in^{-1}, Ultem, ASTM D4812 [2]
2	53.3 J/m	Y	1 ft lb in^{-1}, Ultem, ASTM D256 [2]

Electrical Properties:
Surface/Volume Resistance:

No.	Value	Note	Type
2	58 × 10^{15} Ω.cm	ASTM D257 [2]	S

Dielectric Permittivity (Constant):

No.	Value	Frequency	Note
1	3.12	100 Hz	ASTM D150 [2]

Dielectric Strength:

No.	Value	Note
1	17.95 kV.mm^{-1}	456 V mil^{-1}, in oil, 125 mils, ASTM D149

Dissipation (Power) Factor:

No.	Value	Frequency	Note
1	0.0017	100 Hz	ASTM D150 [2]

Polymer Stability:
Flammability: Flammability rating V0 (0.063 in. thick, UL94) [2]

Applications/Commercial Products:
Processing & Manufacturing Routes: Processed by extrusion or injection moulding [2]

Trade name	Details	Supplier
Standard sample		Shanghai Synthetic Resin Institute
Ultem	5000	General Electric

Bibliographic References
[1] Dong, B. and Zhu, K., *J. Membr. Sci.*, 1991, **60**, 63, (props.)
[2] 1977, GE Plastics, (technical datasheet)

Transport Properties:

Melt Flow Index:

No.	Value	Note
1	<0.01 g/10 min	max., ASTM D1238 [1]

Water Absorption:

No.	Value	Note
1	0.01%	max. 24 h, 1/8 in thick, ASTM D570 [5]

Mechanical Properties:

Mechanical Properties General: Stiffness and yield strength are less than for HDPE, but impact toughness (even at cryogenic temps) and abrasion resistance are the highest of all thermoplastics. ESCR creep resistance, cyclical fatigue resistance and corrosion resistance are also high. Friction coefficient is low. Under tension there is no necking but uniform stretching. Orientation can change tensile props by a factor of 15 [6]. Graphite fibre, powdered metal, glass fibre, glass bead fillers all increase stiffness and heat deflection temp. and decrease deformation under load, but all except graphite and glass fibre reduce abrasion resistance and impact strength [1,2] Poisson's ratio may be negative. [6] Sterilisation for biomedical applications results in a loss in fracture toughness of approx. 50% and a decrease in tearing modulus of approx. 30% [25]

Tensile (Young's) Modulus:

No.	Value	Note
1	650 MPa	[6]
2	10000 MPa	biaxially oriented [6]
3	150000 MPa	max, gel spun fibres [1]

Flexural Modulus:

No.	Value	Note
1	896–965 MPa	130–140 kpsi, 23°, (73°F), ASTM D790 [5]

Tensile Strength Break:

No.	Value	Note	Elongation
1	21–42 MPa	ASTM D638 [1,2,5,6]	200–500%
2	330 MPa	biaxially oriented [6]	17%

Tensile Strength Yield:

No.	Value	Elongation	Note
1	20–28 MPa	4%	ASTM D638 [2,5]

Impact Strength: Tensile impact resistance 893 kJ m^{-2} (425 ft lb in^{-2} 23°, 2.54 mm thick, ASTM D1822) [5]
Hardness: Shore D67 (ASTM D2240) [1]; Rockwell R50–R52 (ASTM D785) [2,5]. Scratch Hardness 50–150 MPa (V 0.0026 mm s^{-1}); 380–550 MPa (v 2.6 mm s^{-1} 21°) [21]
Fracture Mechanical Properties: Fracture toughness (critical J integral method) has been reported [23]
Friction Abrasion and Resistance: Coefficient of friction μ_{stat} 0.20–0.25; μ_{dyn} 0.15–0.2 [2]. Abrasion resistance is highest of all thermoplastics [1]. Wear props are of biomedical importance (use of UHMWPE for implants) [7,8,9] as well as other areas. [10,11] Effects of γ-irradiation on wear volume, surface area and coefficient of friction have been reported [27]. Wear rates of aged and non-aged γ-irradiated and gas-plasma sterilised materials have been reported [28]

Izod Notch:

No.	Value	Notch	Break	Note
1	1335–1708 J/m	N		double notch; 23° [2]
2		N	No break	single notch [5]

Electrical Properties:

Dielectric Permittivity (Constant):

No.	Value	Frequency	Note
1	2.3	1 kHz	ASTM D150 [5]

Dielectric Strength:

No.	Value	Note
1	27.95 kV.mm^{-1}	710 V mil^{-1} 1/8 in thick, short time, ASTM D149 [5]
2	51.18 kV.mm^{-1}	1300 V mil^{-1}, 0.01 mil thick, ASTM D 149 [5]

Dissipation (Power) Factor:

No.	Value	Frequency
1	0.00023	1 kHz

Polymer Stability:

Polymer Stability General: Stabilisers not normally required (the long chains are resistant to scission) [2]. However, mechano-oxidative degradation (induced by microtaming) has been reported. [22]
Environmental Stress: Possesses excellent environmental stress crack resistance [1,2,3]
Chemical Stability: Chemical stability is as good as or better than other types of polyethylene
Stability Miscellaneous: Radiation resistant; has highest neutron resistance of thermoplastics [1] Effects of γ-irradiation have been reported [25,26,27,28]

Applications/Commercial Products:

Processing & Manufacturing Routes: Normally manufactured by slurry process but soln or bulk processes can be used. (see HDPE, LLDPE). Ziegler-Natta catalysts can be used. Processed by compression moulding at 204–260° (400–500°F) [5]; also by ram extrusion, forging or die stamping at ca. 150° or by gel spinning (MW 2000000–3000000) at 160–200° in decalin, 5–10 wt.% soln, passed through spinarette, cooled fibre stretched 10–40 times [1,2,13,14]
Applications: Used in chemical processing, food and drink industries, foundries, timber industry, electrical industry; for medical implants, mining, mineral processing, sewage treatment equipment, paper manufacture, recreational equipment, transport, textiles (inc. bullet proof vests), composites (both as fibre and matrix). Speciality film

Trade name	Details	Supplier
CV	non exclusive	Cestidur Industries
Cadco (UHMW)	also other polyolefins	Cadillac Plastic and Chemical
Cestidur		DSM
Cestidur		Cestidur Industries
Cestilene	non exclusive	DSM
Cestilene	non exclusive	Cestidur Industries

Cestilite	antistatic	DSM Engineering Plastics
Cestilite	antistatic	Cestidur Industries
Ertalene	non exclusive	Cestidur Industries
Hostalen	non exclusive	Hoechst
Hostalen	non exclusive	Hoechst
Hostalen	non exclusive	Tinto Plastics
Hunz		Poly Hi Solidur
Murtfeldt S		Industria Engineering Products
PAS	non exclusive	Faigle Kunststoffe
Primax	modified	Air Products
Ramex	non exclusive	Sym Plastics
Solidur	non exclusive	Poly Hi Solidur
Stamylan	non exclusive	DSM
Stamylan	non exclusive	Macrodan A/S
Stamylan	non exclusive	Ashland Plastics
Ulpolen		Polikin Polimer Ve Kinya Sanayii
Ultradur	sheet	S.W. Plastics
Ultradur	sheet	BASF

Bibliographic References

[1] Kissin, Y.V., *Kirk-Othmer Encycl. Chem. Technol.*, 4th edn. (eds. J.I. Kroschwitz and M. Howe-Grant) Wiley Interscience, 1996, **17**, 724
[2] Coughlan, J.J. and Hug, D.P., *Encycl. Polym. Sci. Eng.*, 2nd edn. (ed. J.I. Kroschwitz) Wiley Interscience, 1986, **6**, 490
[3] Alger, M., *Polymer Science Dictionary*, 2nd edn., Chapman and Hall, 1997
[4] You-Lo Hsieh and Xiao-Ping Hu, J. Polym. Sci., *Polym. Phys. Ed.*, 1997, **35**, 623
[5] Guide to Plastics, *Property and Specification Charts*, McGraw-Hill, 1987, **74**,92, 150
[6] Prins, A.J. and Kortschott, M.T., *Polym. Eng. Sci.*, 1997, **37**, 261
[7] Halm, D.W., Wolfarth, D.L. and Parks, N.L., *J. Biomed. Mater. Res.*, 1997, **35**, 31
[8] Lewis, G., *J. Biomed. Mater. Res.*, 1997, **38**, 55
[9] King, R., *Plastics News (Detroit)*, 1996, **8**, 5
[10] Zhang, R., Qi, S., Walter, R., and Haeger, A.M., *J. Mater. Sci.*, 1996, **31**, 5191
[11] Briscoe, B.J., Pelillo, E., and Sinha, S.K., *Polym. Eng. Sci.*, 1996, **36**, 2996
[12] Pascand, R.S., Evans, W.T., McCollogh, P.J.J., and Fitzpatrick, D., *J. Biomed. Mater. Res.*, 1996, **32**, 619
[13] Smith, P., and Lemstra, P.J., *J. Mater. Sci.*, 1980, **15**, 505
[14] Lemstra, P.J., *Dev. Oriented Polym.*, Elsevier Science Publishing, 1987, 99
[15] Wilson, L. and co-workers, *Proceedings of SPE Polyolefins: VIII International Conference*, 1993, 494
[16] U.S. Pat., 1985, 4 650 710
[17] 1985, Bulletin No. 500-667, HIMONT USA, (technical datasheet)
[18] Kerber, M.L., Panomarev, I.N., Lapshova, O.A., and Grinenko, E.S., Polym. Sci., Ser. A Ser. B (Transl. Vysokomol. Soedin., Ser. A Ser. B), 1996, **38**, 867
[19] *High Performance Textiles*, Elsevier Sequoia, 1996, **10**,12
[20] *J. Mater. Eng.*, 1996, **5**, 278
[21] Briscoe, B.J., Pelillo, E., and Sinha, S.K., *Polym. Eng. Sci.*, 1996, **36**, 2996
[22] Costa, L., Luda, M.P., and Trossarelli, L., *Polym. Degrad. Stab.*, 1997, **55**, 329
[23] Pascand, R.S., Evans, W.T., and McCullogh, P.J.J., *Polym. Eng. Sci.*, 1997, **37**, 11
[24] Prins, A.J., Kortschot, M.T., and Woodhams, R.T., *Polym. Eng. Sci.*, 1997, **37**, 261
[25] Pascaud, R.S. et al, *Biomaterials*, 1997, **18**, 727
[26] Cornwall, G.B. et al, *J. Mater. Sci.: Mater. Med.*, 1997, **8**, 303
[27] Oonishi, H. et al, *J. Mater. Sci.: Mater. Med.*, 1997, **8**, 11
[28] Fisher, J. et al, *J. Mater. Sci.: Mater. Med.*, 1997, **8**, 375

Urea formaldehyde resins

Synonyms: *Urea resins*
Related Polymers: Urea Resins, Wood applications, Urea Resin, Moulding powders, Cellulose filled
Monomers: Urea, Formaldehyde, Paraformaldehyde
Material class: Thermosetting resin
Polymer Type: urea formaldehyde
Miscellaneous: Urea is a tetra-functional monomer, which readily forms a tetra-methylol deriv. Melamine can also be co-condensed with urea to form a range of melamine urea formaldehyde (MUF) resins

Applications/Commercial Products:

Processing & Manufacturing Routes: Urea-formaldehyde resins are made by the condensation reaction of urea with formaldehyde, generally 37% aq. soln. The condensation generally proceeds with basic catalyst and a molar excess of formaldehyde. The resins are generally used in aq. soln., but may also be used if alcoholic solvents are added. They cure by application of heat or acid
Applications: Urea-formaldehyde resins have various applications. See the relevant cross references for further details

Trade name	Details	Supplier
Aerolite	UF resins	Dyno
Beetle	amine resins	BIP
Blagden LE	urea resins	Blagden
Blagden MPI	urea resins	Blagden
Blagden UF	urea resins	Blagden
Borden UL	urea resins	Borden
Dynorit	UF resins	Dyno
Granular Compound		Sun Coast
Perstorp	UF	Perstorp Compounds
Urea Compounds		Patent Plastics
Urex	urea resins	Neste

Urea resin, Moulding powders, Cellulose filled

Related Polymers: Urea Formaldehyde Resins, Melamine Formaldehyde Resins, Melamine Resin, Moulding Compounds, Cellulose Filled
Monomers: Urea, Melamine, Formaldehyde
Material class: Thermosetting resin
Polymer Type: urea formaldehyde
Additives: Additives include cellulose filler, pigments and a curing catalyst

Volumetric & Calorimetric Properties:

Density:

No.	Value	Note
1	d 1.47–1.52 g/cm^3	ASTM D792 [1]

Thermal Expansion Coefficient:

No.	Value	Note	Type
1	2.2×10^{-5}–3.6×10^{-5} K^{-1}	ASTM D896 [1]	L

Thermal Conductivity:

No.	Value	Note
1	0.00423 W/mK	ASTM C177 [1]

Deflection Temperature:

No.	Value	Note
1	130°C	1.8 MPa, ASTM D648 [1]

Transport Properties:
Water Absorption:

No.	Value	Note
1	0.48%	24h, 3.2 mm thick, ASTM D590

Mechanical Properties:
Tensile (Young's) Modulus:

No.	Value	Note
1	9000–9700 MPa	ASTM D638 [1]

Flexural Modulus:

No.	Value	Note
1	9700–10300 MPa	ASTM D9700 [1]

Tensile Strength Break:

No.	Value	Note
1	38–48 MPa	ASTM D638 [1]

Flexural Strength at Break:

No.	Value	Note
1	70–124 MPa	ASTM D790 [1]

Tensile Strength Yield:

No.	Elongation	Note
1	0.5–1%	ASTM D638 [1]

Hardness: Rockwell M110–M120 (ASTM D785) [1]

Electrical Properties:
Dielectric Permittivity (Constant):

No.	Value	Frequency	Note
1	0.034–0.043	60 Hz	ASTM D149 [1]

Arc Resistance:

No.	Value	Note
1	80–100s	ASTM D495 [1]

Polymer Stability:
Upper Use Temperature:

No.	Value	Note
1	77°C	[1]

Flammability: Flammability rating V0 (UL 94) [1]

Applications/Commercial Products:
Processing & Manufacturing Routes: Urea formaldehyde moulding materials combine a virtually unlimited range of colours and excellent colour retention with good electrical props., solvent resistance, high flame retardance, rigidity, surface hardness, scratch resistance and anti-static props. For the compression moulding process preheating of the resin powder pellets or granules is required. Injection moulding grades are preheated efficiently both by the injection machine and by the frictional heat generated as the material flows through the sprue, runner or gate

Mould Shrinkage (%):

No.	Value	Note
1	0.64–0.7%	compression moulding
2	0.073–0.78%	injection moulding

Applications: Urea formaldehyde materials give general purpose mouldings for all types of articles, particularly electrical accessories. Wet-granulated grades are used for thick section mouldings - such as toilet seats. Injection moulded material is used for household articles such as electrical accessories

Trade name	Details	Supplier
Beetle GXT	urea materials	BIP
Beetle ZXT	urea materials	BIP
Scarab	urea materials	BIP

Bibliographic Reference
[1] *Amino Moulding Powders*, BIP Plastics.

Urea resins, Wood applications   U-6
Related Polymers: Urea Formaldehyde Resins, Melamine Resins, Wood Applications
Monomers: Urea, Melamine, Formaldehyde
Material class: Thermosetting resin
Polymer Type: urea formaldehyde
General Information: Relatively poor moisture resistance. For superior moisture resistance for outdoor applications refer to melamine urea resins (MUF) and also resorcinol (RF) or resorcinol-phenol resins (RPF)

Urethane coating   U-7

– Urethane rubber, Multrathane based

Synonyms: *Urethane, polyether*
Related Polymers: Polyurethanes General Information
Monomers: 3,4-Diisocyanatotoluene, 2,6-Diisocyanatotoluene, 3,3-Dihydroxymethylbutanol, 2,4-Diisocyanatotoluene, 1,3-Butanediol
Material class: Fibres and Films, Fluids
Polymer Type: polyurethane
General Information: An isocyanate-terminated prepolymer. A one component polyether urethane coating cured by moisture

Transport Properties:
Polymer Solutions Dilute: Brookfield viscosity 1200 cP s (25°) [1]. Gardner-Holdt viscosity W
Permeability of Liquids: [H_2O] 130 g m^{-2} day^{-1} 10 mil film [1]

Mechanical Properties:
Tensile Strength Break:

No.	Value	Note	Elongation
1	34.5 MPa	5000 psi [1]	40%

Hardness: Sward 60; pencil hardness H [1]
Fracture Mechanical Properties: Taber abrasion resistance 18 mg (1000 cycles)$^{-1}$ (CS-10, 0.5 kg), 75 mg (1000 cycles)$^{-1}$ (CS-10, 1 kg) [1]
Friction Abrasion and Resistance: Has excellent resistance to abrasion and wear, excellent adhesion and resistance to chipping

Optical Properties:
Volume Properties/Surface Properties: Gloss 98% (Hunter) [1]. Colour 4 (max., Gardner 1953 scale) [1]

Applications/Commercial Products:
Processing & Manufacturing Routes: Is applied as a soln. The solvent rapidly evaporates and moisture diffuses into the remaining film effecting the final cure
Applications: Used as a coating for wood, metal, concrete and electrical apparatus

Trade name	Supplier
ECD-154	DuPont
PR-942	Wyandotte Chemical Corp.

Bibliographic Reference
[1] Saunders, J.H., Frisch, K.C., *Polyurethanes Chemistry and Technology*, Part II: Technology, Interscience, 1964, 479, 480

Urethane rubber, Multrathane based U-8
Related Polymers: Polyurethanes General Information
Monomers: 1,4-Butanediol, 4,4′-Diphenylmethanediisocyanate
Material class: Synthetic Elastomers
Polymer Type: polyurethane
General Information: A cast polyester urethane elastomer with improved low temp. flexibility. Based on a specifically designed polyester, multrathane F144. The struct. of the polymer is unknown

Mechanical Properties:
Mechanical Properties General: Has improved low temp. props.

Tensile (Young's) Modulus:

No.	Value	Note	Elongation
1	5.5 MPa	800 psi [1]	100%

Tensile Strength Break:

No.	Value	Note	Elongation
1	45 MPa	6500 psi [1]	550%

Hardness: Shore A85 [1]
Fracture Mechanical Properties: Shows solenoid brittleness at -70°

Electrical Properties:
Surface/Volume Resistance:

No.	Value	Note	Type
1	$5 \times 10^{-5} \times 10^{15}$ Ω.cm	[2]	S
2	$9 \times 10^{-6} \times 10^{15}$ Ω.cm	24h in H_2O [2]	S

Dielectric Permittivity (Constant):

No.	Value	Frequency	Note
1	7.6	800 Hz	[2]
2	7.1	1 MHz	[2]

Dissipation (Power) Factor:

No.	Value	Frequency	Note
1	0.016	800 Hz	[2]
2	0.058	1 MHz	[2]

Applications/Commercial Products:
Processing & Manufacturing Routes: Cast elastomer is processed in the fluid state. Prepolymer is mixed with diol curing agent and heated in the casting machine to effect cure. Can be gravity or centrifugally cast or compression or transfer moulded
Applications: Has applications in oil seals, drive wheels and diaphragms

Trade name	Details	Supplier
Multrathane	F66	Mobay

Bibliographic References
[1] Pigott, K.A., Cote, R.J., Ellegast, K., Frye, B.F. *et al*, *Rubber Age (N.Y.)*, 1962, **90**, 629
[2] Saunders, J.H., Frisch, K.C., Polyurethanes Chemistry and Technology Part II Technology, Interscience, 1964, 412

Vinyl methyl phenyl silicone rubber, Heat vulcanised

Synonyms: *Vinyl phenyl methyl silicone rubber, Heat vulcanised. Methyl phenyl vinyl silicone rubber, Heat vulcanised. Phenyl methyl vinyl silicone rubber, Heat vulcanised. Vinyl methyl phenyl based silicone elastomer*
Related Polymers: Methylsilicone rubber, Heat vulcanised, Vinyl methyl silicone rubber, Heat vulcanised, Methylphenylsilicone rubber, Heat vulcanised, Phenylsilicone rubber, Heat vulcanised
Monomers: Dichlorodimethylsilane, Dichloroethenylmethylsilane, Dichloromethylphenylsilane, Hexamethylcyclotrisiloxane
Material class: Synthetic Elastomers, Copolymers
Polymer Type: silicones, dimethylsilicones
CAS Number:

CAS Reg. No.
158195-32-5

Molecular Formula: $[(C_2H_6OSi).(C_3H_6OSi).(C_7H_8OSi)]_n$
Fragments: C_2H_6OSi C_3H_6OSi C_7H_8OSi
Additives: Finely divided silica is incorporated as a reinforcer to improve mech. props.
General Information: Vinyl groups are present at less than 1%. They cause the gum to cure more readily with lower activity peroxides

Volumetric & Calorimetric Properties:
Density:

No.	Value	Note
1	d 1.15 g/cm^3	[5]

Thermal Expansion Coefficient:

No.	Value	Note	Type
1	0.00075 K^{-1}	approx. 60% dimethyl siloxane, 40% methyl phenyl siloxane [1]	V
2	0.0009 K^{-1}	approx. 90% dimethyl siloxane, 10% methyl phenyl siloxane [1]	V

Thermal Conductivity:

No.	Value	Note
1	0.25–0.3 W/mK	0.0006–0.0007 cal cm^{-2} (s K)$^{-1}$ [12,13]

Specific Heat Capacity:

No.	Value	Note	Type
1	1.3 kJ/kg.C	0.3 Btu lb^{-1}°F^{-1} [5]	P

Melting Temperature:

No.	Value	Note
1	-77°C	approx. 60% dimethyl siloxane, 40% methyl phenyl siloxane [1,10]

Glass-Transition Temperature:

No.	Value	Note
1	-96°C	approx. 60% dimethyl siloxane, 40% methyl phenyl siloxane [1]
2	-84°C	approx. 60% dimethyl siloxane, 40% methyl phenyl siloxane [10]
3	-112°C	approx 90% dimethyl siloxane, 10% methyl phenyl siloxane [1]

Transition Temperature General: For transitions of high content silicone rubber see (phenylsilicone rubber (heat-vulcanised)). For transitions of low phenyl content silicone rubber see (methyl phenyl silicone rubber, heat vulcanised))
Transition Temperature:

No.	Value	Note	Type
1	-70°C	approx. 60% dimethyl siloxane, 40% methyl phenyl siloxane [1]	Stiffening
2	-109°C	approx. 90% dimethyl siloxane, 10% methyl phenyl siloxane [1]	Stiffening

Surface Properties & Solubility:
Cohesive Energy Density Solubility Parameters: δ 16–17 J$^{1/2}$ cm$^{-3/2}$ (7.7–8.2 cal$^{1/2}$ cm$^{-3/2}$). Value increases with increasing phenyl substitution [8]
Solvents/Non-solvents: Sol. aliphatic hydrocarbons, chlorinated hydrocarbons, aromatic hydrocarbons, ketones, esters, ethers; insol. alcohols, H$_2$O [7,8,9]
Surface and Interfacial Properties General: Generally low surface adhesion [3]

Transport Properties:
Polymer Solutions Concentrated: Viscosity 100000 St ((approx.) pre-cure gum)
Permeability of Gases: For permeability of gas see (methyl phenyl silicone rubber, heat vulcanised). Gas permeability falls sharply as phenyl substitution rises [4]

Mechanical Properties:
Mechanical Properties General: Mech. props. are improved by incorporation of vinyl groups; phenyl groups reduce high temp. performance
Tensile Strength Break:

No.	Value	Note	Elongation
1	6.4 MPa	65 kg cm^{-2}, 25°, filled [3,6]	400%

Viscoelastic Behaviour: Compression set 30% ((max.) 150° 22h compression for high phenyl filled rubber) [7]

Electrical Properties:
Electrical Properties General: Excellent insulating props. though inferior to dimethyl silicone rubber
Dielectric Permittivity (Constant):

No.	Value	Note
1	3	approx. [2,11]

Dielectric Strength:

No.	Value	Note
1	20 kV.mm^{-1}	500 V mil^{-1} [5]

Optical Properties:
Transmission and Spectra: No transmission below 290 nm [3,7]

Polymer Stability:
Thermal Stability General: Usable over very wide temp. range [7]. -100° to 315° low temp. rubber *ca.* 10% methyl phenyl silicone; high temp. rubber >40% methyl phenyl silicone

– Vinyl methyl silicone rubber, Heat vulcanised

Upper Use Temperature:

No.	Value	Note
1	315°C	high phenyl rubber

Decomposition Details: Oxidation temp. 350° (dry air, 55% methyl phenyl silicone) [7]
Environmental Stress: Resistant to oxidative degradation, water, weathering, ozone and corona. Unaffected by uv. Very resistant to high energy radiation (radioactivity) [3]
Chemical Stability: Swollen by toluene, aliphatic hydrocarbons and CCl_4. Disintegrates in conc. H_2SO_4. Relatively unaffected by weak alkali, weak acids and alcohols [9]
Hydrolytic Stability: Stable in H_2O and aq. solns. at room temp. Damaged by high pressure steam
Biological Stability: Non-biodegradable. If properly vulcanised it is not subject to biological attack
Recyclability: Can be recycled by depolymerisation using steam, or by acid catalysis [3]
Stability Miscellaneous: Stability to solvents is reduced by phenyl incorporation [3]

Applications/Commercial Products:
Processing & Manufacturing Routes: Dimethylmethylphenyl and methylvinyl cyclic polysiloxanes are equilibrated and polymerised together at 150–200° with an alkaline catalyst. Methyl vinyl cyclic polysiloxane is only present in small amounts producing a linear polymer with less than 1% methyl vinyl siloxane. Traces of hexamethyldisiloxane are added to terminate the siloxane chains. Elastomer is readily cured by low activity organic peroxides (typically 1–2 parts per 100 rubber) at temps. of over 110°. Several hours post cure treatment is required to remove residual peroxide
Applications: Used in low and very high temp. applications; as encapsulant coating, insulation for electric cables, and in pads and belts

Trade name	Supplier
C-154	Wacker Silicones

Bibliographic References
[1] Polmanteer, K.G. and Hunter, M.J., *J. Appl. Polym. Sci.*, 1959, **1**, 3, (transition temps)
[2] Stark, F.O., Fallender, J.R. and Wright, A.P., *Comprehensive Organometallic Chemistry*, (eds. G. Wilkinson, F.G.A. Stone and E.W. Abel), Pergamon Press, 1982, **2**, 305
[3] Noll, W., *Chemistry and Technology of Silicones*, Academic Press, 1968
[4] Robb, W.L., *Ann. N. Y. Acad. Sci.*, 1986, **146**, 119, (gas permeability)
[6] Meals, R.N. and Lewis, F.M., Silicones Reinhold, 1959
[7] Hardman, B. and Torkelson, A., *Encycl. Polym. Sci. Eng.*, Vol. 15, 2nd edn, (Ed. Kroshwitz, J.I.), John Wiley and Sons, 1985, 204
[8] Yerrick, K.B. and Beck, H.N., *Rubber Chem. Technol.*, 1964, **37**, 261, (solubility)
[9] Roff, W.J. and Scott, J.R., Fibres, Films, *Plastics and Rubbers*, Butterworths, 1971
[10] Borisov, M.F., *Soviet Rubber Technology*, 1966, **25**, 5, (transition temps)
[11] Vincent, G.A., Feuron, F.W.G. and Orbech, T., *Annu. Rep., Conf. Electr. Insul. Dielectr. Phenom.*, 1972, 17
[12] Ames, J., *J. Sci. Instrum.*, 1958, **35**, 1
[13] Jamieson, D.T. and Irving, J.B., *Proc. Int. Conf. Therm. Conduct.*, 279, (thermal conductivity)

Vinyl methyl silicone rubber, Heat vulcanised V-2
Synonyms: *Methyl vinyl silicone rubber, Heat vulcanised. Vinyl methyl based silicone elastomer*
Related Polymers: Methylsilicone rubber, Heat vulcanised, Vinyl methyl phenyl silicone rubber, Heat vulcanised
Monomers: Dichlorodimethylsilane, Dichloroethenylmethylsilane, Hexamethylcyclotrisiloxane
Material class: Synthetic Elastomers, Copolymers
Polymer Type: silicones, dimethylsilicones

CAS Number:

CAS Reg. No.
155665-02-4

Molecular Formula: $[(C_2H_6OSi).(C_3H_6OSi)]_n$
Fragments: C_2H_6OSi C_3H_6OSi
Additives: Finely divided silica incorporated as a reinforcer to improve mech. props.
General Information: Vinyl groups are present at less than 1%. They cause the gum to cure more readily with lower activity peroxides

Volumetric & Calorimetric Properties:
Density:

No.	Value	Note
1	d 1.1 g/cm^3	[12]

Thermal Expansion Coefficient:

No.	Value	Note	Type
1	0.00095 K^{-1}	[6]	V

Thermal Conductivity:

No.	Value	Note
1	0.3 W/mK	0.0007 cal cm^{-1} cm^{-2} s^{-1} [11]

Specific Heat Capacity:

No.	Value	Note	Type
1	1300 kJ/kg.C	0.3 Btu (lb °F)$^{-1}$ [12]	P

Melting Temperature:

No.	Value	Note
1	-54°C	[6]

Glass-Transition Temperature:

No.	Value	Note
1	-123°C	[6,15]

Transition Temperature General: Stiffening temp. is related to T_m rather than to T_g [2,3,4,5,6]
Brittleness Temperature:

No.	Value	Note
1	-65°C	-85°F [13]

Transition Temperature:

No.	Value	Note
1	-38°C	stiffening temp. [6]

Surface Properties & Solubility:
Cohesive Energy Density Solubility Parameters: δ 15.5 J$^{1/2}$ cm$^{-3/2}$ (7.5–7.6 cal$^{1/2}$ cm$^{-3/2}$); δ$_d$ 15.9; δ$_p$ 0.0; δ$_h$ 4.1 J$^{1/2}$ cm$^{-3/2}$ [3,7,8,9] (see methylsilicone rubber heat vulcanised for more details)

Solvents/Non-solvents: Sol. aliphatic hydrocarbons, chlorinated hydrocarbons, aromatic hydrocarbons [3,7,8]. Mod. sol. ketones, esters, ethers. Insol. alcohols, H_2O [3,7,8]
Surface and Interfacial Properties General: Generally low surface adhesion [4]

Transport Properties:
Transport Properties General: Silicone rubber is ten times more permeable to gases than other rubbers. [14] The solubility of gases in polysiloxanes is similar to that for other polymers but the diffusion coefficients are much higher
Polymer Melts: Viscosity 100000 St (approx., pre-cure gum) [3], 100–1000 St (pre-cure liq. silicone rubber) [16]

Mechanical Properties:
Mechanical Properties General: The vinyl groups sharply reduce compression set at elevated temps. [1]. Silicone rubbers preserve their mechanical props. better up to higher temps. compared to other elastomers. Requires vulcanisation with a reinforcing filler (usually silica) in order to possess acceptable mechanical props. [1]
Tensile Strength Break:

No.	Value	Note	Elongation
1	6.5 MPa	65 kg cm^{-2}, 25°, filled [4]	300%
2	3.4 MPa	500 psi, 200°, filled [10]	150%
3	2.7 MPa	400 psi, 250°, filled [10]	100%
4	5 MPa	0.5 kg f mm^{-2}, 25°, high vinyl content, unfilled [3]	200%
5	5.5 MPa	liq. silicone rubber [16]	520%

Viscoelastic Behaviour: Compression set 10% at 175° and 22h compression (filled) [1]
Hardness: Shore A55–A60 [12]. Shore A32 (liq. silicone rubber) [16]
Fracture Mechanical Properties: Tear strength 1.6 J cm^{-2} (90 pse, 25°, filled) [13]. Tear strength 1.4 J cm^{-2} (14 kN m^{-1}, liq. silicone rubber) [16]

Electrical Properties:
Electrical Properties General: Silicone rubber has excellent insulating props. [4]
Dielectric Permittivity (Constant):

No.	Value	Note
1	2.5–4.5	depends on filler [4]

Dielectric Strength:

No.	Value	Note
1	21 kV.mm^{-1}	[4]

Arc Resistance:

No.	Value	Note
1	200–250s	[17]

Dissipation (Power) Factor:

No.	Value	Frequency	Note
1	0.005	800 Hz	[4]

Optical Properties:
Refractive Index:

No.	Value	Note
1	1.404	[3]

Transmission and Spectra: Uv absorption at 290 nm [4]

Polymer Stability:
Polymer Stability General: Stable material with a wide range of uses (see (vinylmethyl phenyl silicone rubber (heat vulcanised)) [2]
Thermal Stability General: Usable from -60 to 260°. Some varieties are usable at slightly higher temps.
Upper Use Temperature:

No.	Value
1	260°C

Decomposition Details: Oxidation temp. dry air 290° (CO produced) [2,16]
Flammability: Limiting oxygen index 20% (approx.) [2]
Environmental Stress: Resistant to oxidative degradation, water, weathering, ozone and corona. Unaffected by uv [4]
Chemical Stability: Swollen by toluene, aliphatic hydrocarbons and CCl_4. Disintegrates in conc. sulfuric acid. Relatively unaffected by weak alkalis, weak acids, alcohols and oils (except silicone oil) [3]
Hydrolytic Stability: Stable in H_2O and aq. solns. at room temp. Damaged by high pressure steam
Biological Stability: Non-biodegradable. If properly vulcanised it is not subject to biological attack
Recyclability: Can be recycled by depolymerisation using steam or acid catalysis [4]

Applications/Commercial Products:
Processing & Manufacturing Routes: Dimethyl and methyl vinyl cyclic polysiloxanes are polymerised together at 150–200° with an alkaline catalyst. Less than 1% of the methyl vinyl siloxane monomer and small quantities of hexamethyldisiloxane are added to terminate the siloxane chains. The elastomer is readily cured by low activity organic peroxides (typically 1–2 parts per 100 rubber) at temps. of over 110°. Several hours post cure treatment are required to remove residual peroxide. Liq. silicon process involves low viscosity (<1000 St) vinyl methyl gums which can be vulcanised as a liq. in the presence of (polymethyl hydrogen siloxane) as crosslinker and platinum catalyst at temps. no higher than 100°. Cure takes only a few minutes [2,16]
Mould Shrinkage (%):

No.	Value
1	0.025%

Applications: Used where good temp. props. and stability are important particularly where temps. of >200° may be encountered. Used for coatings, electrical cables, encapsulants, pads and belts. Used in aircraft engines, gas turbines, autoclaves, irons and ovens. Important use as electrical insulant and in aircraft and the aerospace industry

Trade name	Details	Supplier
B-179	elastomer	Wacker Silicones
Blensil		General Electric
C-155	gum	Wacker Silicones
COHR lastic	various grades not exclusive	CHR Industries
Rhodosil RS	various grades not exclusive	Rhone-Poulenc
SE	various grades not exclusive	General Electric Silicones
SWS 720 series	elastomer	Wacker Silicones
SWS 7800 series		Wacker Silicones
Silastic 590 series	liq. silicone rubber	Dow Corning STI

Silastic GP		Dow Corning STI
Siloprene VS	elastomer	Bayer Inc.

Bibliographic References

[1] Dunham, M.L., Bailey, D.L. and Mixer, R.Y., *Ind. Eng. Chem.*, 1957, **49**, 1373
[2] Hardman, B. and Torkelson, A., *Encycl. Polym. Sci. Eng.*, 2nd edn., (ed. J.I. Kroschwitz), John Wiley and Sons, 1985, **15**, 204
[3] Roff, W.J. and Scott, J.R., Fibres, Films, *Plastics and Rubbers*, Butterworths, 1971
[4] Noll, W., *Chemistry and Technology of Silicones*, Academic Press, 1968
[5] Stark, F.O., Fallender, J.R. and Wright, A.P., *Comprehensive Organometallic Chemistry*, (eds. G. Wilkinson, F.G.A. Stone and E.W. Abel), Pergamon Press, 1982, **2**, 305
[6] Polmanteer, K.E. and Hunter, M.J., *J. Appl. Polym. Sci.*, 1959, **1**, 3, (transition temps)
[7] Yerrick, K.B. and Beck, H.N., *Rubber Chem. Technol.*, 1964, **37**, 261, (solubility)
[8] Baney, R.H., Voigt, C.E. and Mentele, J.W., Struct.-Solubility Relat. Polym., *(Proc. Symp.)*, (eds. F.W. Harris and R.B. Seymour), Academic Press, 1977, 225, (solubility)
[9] Hauser, R.L., Walker, C.A. and Kilbourne, F.L., *Ind. Eng. Chem.*, 1956, **48**, 1202, (solubility)
[10] Ames, J., *Silicones*, (ed. S. Fordham), 1960, 154
[11] Ames, J., *J. Sci. Instrum.*, 1958, **35**, 1
[12] Ash, M. and Ash, I., Handbook of Plastic Compounds, *Elastomers and Resins*, (eds. M.B. Ash and I.A. Ash), Wiley-VCH, 1992
[13] Hamilton, S.B., *Silicone Technology*, (ed. P.F. Bruins), John Wiley and Sons, 1970, 17
[14] Robb, W.L., *Ann. N. Y. Acad. Sci.*, 1968, **146**, 119, (gas permeability)
[15] Yim, A. and St Pierre, L.E., J. Polym. Sci., *Part B: Polym. Lett.*, 1969, **7**, 237, (transition temps)
[17] McGregor, R.R., Silicones and their Uses McGraw Hill, 1954

Vinylcyclohexene dioxide resin   V-3

Synonyms: *Poly(3-oxiranyl-7-oxabicyclo[4.1.0]heptane). Poly(1,2-epoxy-4-epoxyethylcyclohexane). Poly(1-epoxyethyl-3,4-epoxycyclohexane)*
Monomers: 4-Vinylcyclohexene dioxide
Material class: Thermosetting resin
Polymer Type: epoxy
CAS Number:

CAS Reg. No.	Note
25086-25-3	
123698-11-3	copolymer with diaminodiphenylsulfone

Molecular Formula: $(C_8H_{12}O_2)_n$
Fragments: $C_8H_{12}O_2$
General Information: Uncured ERL-4206 is a low viscosity liq. with a characteristic odour [9]. Props. of cured resin depend on curing agent, stoichiometry, cure time and cure temp.

Volumetric & Calorimetric Properties:
Density:

No.	Value	Note
1	d^{25} 1.08–1.1 g/cm^3	ERL-4206, uncured [8]
2	d 1.1 g/cm^3	ERL-4206, uncured [9]

Transition Temperature General: Heat distortion temp. of Unox 206 cured with Sylkyd 50-H_3BO_3 increases with decreasing curing agent content [6]. Bp 227° (ERL-4206, uncured) [8,9]
Deflection Temperature:

No.	Value	Note
1	95°C	1.82 MPa, borosiloxane polymer (30% wt) cured [1,6]

Transport Properties:
Polymer Solutions Dilute: Viscosity 20 cP (25°, RD-4), 8 cP (25°, ERL-4206)
Water Absorption:

No.	Value	Note
1	0.36%	7 d, 25°, borosiloxane polymer (30% wt) cured [1,6]
2	1.9%	30 d, 25°, borosiloxane polymer (30% wt) cured [1,6]
3	0.5%	7 d, 100°, borosiloxane polymer (30% wt) cured [1,6]

Mechanical Properties:
Mechanical Properties General: Flexural strength of Unox 206 cured with Sylkyd 50-H_3BO_3 increases on ageing at 135° or 160°, but decreases on ageing at 180° [6]
Flexural Strength at Break:

No.	Value	Note
1	56.5 MPa	8.2 kpsi, borosiloxane polymer (30% wt) cured [1,6]
2	77.9 MPa	11.3 kpsi, borosiloxane polymer (30% wt) cured, 135°, 30 d [1,6]

Electrical Properties:
Electrical Properties General: Dissipation factor of Unox 206 cured with Sylkyd 50-H_3BO_3 generally decreases after heat ageing in air [6]
Dissipation (Power) Factor:

No.	Value	Frequency	Note
1	0.005	60 Hz	25°, Sylkyd 50-H_3BO_3 (50% wt) cured, Unox 206 [6]
2	0.002	60 Hz	25°, Sylkyd 50-H_3BO_3 (20% wt) cured, Unox 206 [6]
3	0.014	60 Hz	100°, Sylkyd 50-H_3BO_3 (20–50% wt) cured, Unox 206 [6]

Optical Properties:
Volume Properties/Surface Properties: Colour 1 (max., 1933 Gardner, ERL-4206, uncured) [8]

Polymer Stability:
Thermal Stability General: Thermal stability can depend on the method of polymerisation used. The least stable polymers result from electroinitiation or use of BF_3-etherate as initiator [4]
Decomposition Details: Borosiloxane polymer (30% wt) cured material shows 3.5% weight loss after 30 d at 180° [1,6]. Diaminodiphenylsulfone cured material shows approx. 35% weight loss in air after 2h at approx. 290° [2]. Activation energy for 5–10% weight loss 148 kJ mol^{-1} (diaminodiphenylsulfone cured) [2]

Applications/Commercial Products:

Processing & Manufacturing Routes: 4-Vinylcyclohexene dioxide can be cationically polymerised in bulk using coumarin/onium salt as initiator [3]; by bulk polymerisation in the presence of a free radical donor and cationic salt [5]. Soluble or cross-linked (depending on radiation dose) product may be obtained by γ-ray-induced polymerisation of 4-vinylcyclohexene dioxide [7]

Applications: Used to improve the thermal stability of polyamide resins. Potential application in electron microscopic embedding. May also be used as a reactive diluent in epoxy resin formulations

Trade name	Details	Supplier
ERL	4206	Union Carbide
ERL	4206	Boehringer Ingelheim Bioproducts
ERL	4206	Electron Microscopy Sciences
ERL	4206	SpI Chem.
RD-4		Ciba-Geigy Corp.
Unox	206	Union Carbide

Bibliographic References

[1] Lee, S.M., *Epoxy Resins: Chemistry and Technology,* 2nd edn., (ed. C.A. May), Marcel Dekker, 1988, 783
[2] Knight, G.J. and Wright, W.W., *Br. Polym. J.,* 1989, **21**, 303, (thermal stability)
[3] Zhu, Q.Q. and Schnabel, W., *Polymer,* 1996, **37**, 4129, (synth.)
[4] Hacaloglu, J., Yalçin, T. and &cOnal, A.M., *J. Macromol. Sci., Pure Appl. Chem.,* 1995, **32**, 1167, (thermal stability)
[5] Yağci, Y., Hizal, G. and Aydoğan, A.C., *Eur. Polym. J.,* 1985, **21**, 25, (synth.)
[6] Markovitz, M., Am. Chem. Soc., Div. Org. Coat. Plast. Chem., *Pap,* 1970, **30**, 254, (props.)
[7] Onal, A.M., Usanmaz, A., Akbulut, U. and Toppare, L., *Br. Polym. J.,* 1983, **15**, 187
[8] ERL-4206, Union Carbide, (technical datasheet)
[9] 1997, SPI-CHEM, (technical datasheet)

VLDPE V-5

Synonyms: *Very low-density polyethylene. PE-VLD. Ultra low-density polyethylene. ULDPE*
Related Polymers: polyethylene, LLDPE, High-density polyethylene
Monomers: Ethylene, 1-butene, 1-octene, 4-methyl-1-pentene, 1-hexene
Material class: Thermoplastic
Polymer Type: polyethylene, polyolefins
CAS Number:

CAS Reg. No.	Note
9002-88-4	all density types
25087-34-7	1-butene comonomer
25213-02-9	1-hexene
26221-73-8	1-octene comonomer
25213-96-1	4-methyl-1-pentene comonomer
60785-11-7	1-hexene, 1-butene comonomers
74746-95-5	4-methyl-1-pentene 1-butene comonomers

Molecular Formula: $(C_2H_4)_n$
Fragments: C_2H_4
Molecular Weight: 50000–200000
General Information: Two types of PE with density less than 0.915 g cm^{-3}; PE plastomers of low crystallinity (ca. 10–20%) and amorph. PE elastomers. [1] Elastomers approach ethylene-propene elastomers in prop., and the plastomers compete with poly(ethene -*co*-vinyl acetate), as flexible film material. In general, VLDPE may be regarded as a modification of LLDPE. Struct. consists of high number of branches from high proportion of comonomers.

Morphology: Crystallinity is low, typically 25% (10–20% for VLDPE plastomers; 10% (max) for ULDPE-elastomers) but exceptions (a resin with crystallinity 42.9%) do exist. Crystallinity decreases with an increase in α olefin content; branching uniformity and chain length of α olefin. [1] Mech. and optical props vary with crystallinity.

Volumetric & Calorimetric Properties:
Density:

No.	Value	Note
1	d 0.9–0.915 g/cm^3	VLDPE, plastomers
2	d 0.86–0.9 g/cm^3	ULDPE, elastomers [1]

Volumetric Properties General: Density is affected by comonomer, comonomer content and branching uniformity [1]
Melting Temperature:

No.	Value	Note
1	125–128°C	non-uniform branching [1,3]
2	110°C	max, uniform branching [1]

Transition Temperature General: T_m is dependent on α-olefin content
Brittleness Temperature:

No.	Value	Note
1	<-100°C	max. [1]

Transport Properties:
Melt Flow Index:

No.	Value	Note
1	1–3.8 g/10 min	[1]

Polymer Melts: Introduction of long-chain branches (using metallocene catalysts in the soln process) increases shear thinning, thereby decreasing viscosity.

Mechanical Properties:
Mechanical Properties General: Higher values of moduli and lower values of strain recovery are associated with higher α olefin content and with non-uniformity of branching. [1]

Tensile (Young's) Modulus:

No.	Value	Note
1	25–150 MPa	[1]

Elastic Modulus:

No.	Value	Note
1	60 MPa	8% butene, 25% crystallinity, non-uniform branching
2	1.3 MPa	18% butene, uniform branching [1]

Miscellaneous Moduli:

No.	Value	Note	Type
1	35–310 MPa	1 Hz [1]	Dynamic modulus

Mechanical Properties Miscellaneous: Strain recovery 63–98% [1]

Optical Properties:
Volume Properties/Surface Properties: Haze 4–50%. Increases with higher α olefin content and non-uniformity of branching [1]

Applications/Commercial Products:
Processing & Manufacturing Routes: Manufactured using Ziegler, metallocene or chromium based catalysts in gas phase, solution or slurry processes as for LLDPE

Applications: Used for clear film (used for uniformly branched mVLDPE); heavy duty film [5]; mVLDPE is extensively used for blending with HDPE, commodity grade LLDPE and PP. Other uses in speciality flexible tubing; wire and cable insulation.

Trade name	Supplier
Attane	Dow Plastics
Clearflex	Enichem
Clearflex	Ashland Plastics
Flexirene	Enichem
Flexirene	Ashland Plastics
Teamex	DSM

Bibliographic References
[1] Kissin, Y.V., *Kirk-Othmer Encycl. Chem. Technol.*, 4th edn., (eds. J.I. Kroschwitz and M. Howe-Grant) Wiley Interscience, 1996, **17,** 756
[2] Alger, M., *Polymer Science Dictionary*, 2nd edn., Chapman and Hall, 1997
[3] Kashiwa, N., Proceedings of the Worldwide Metallocene Conference, *MetCon '93,* Houston, Texas, 1993, 237
[4] *U.S. Pat.*, 1978, 4 105 842
[5] Krider, G.A., *Encycl. Polym. Sci. Eng.*, 2nd edn., (J.I. Kroschwitz), Wiley Interscience, 1986, **6,** 571
[6] *Plastics Extrusion Technology Handbook,* (ed. S. Levy), Industrial Press, 1981
[7] Yu, T.C. and Wagner, G.J., *Proceedings of SPE Polyolefins: VIII International Conference,* 1993, 539

Wool

Synonyms: *Keratin*
Related Polymers: Proteins
Monomers: Amino acids
Material class: Proteins
General Information: Natural staple fibre from the fleece of sheep, alpaca, vicuna and some goats. Composed of keratin, which is assembled into cross-linked and complex multistrand fibres. Partly crystalline

Volumetric & Calorimetric Properties:

Density:

No.	Value	Note
1	d 1.32 g/cm^3	[3]
2	d 1.304 g/cm^3	[4]

Glass-Transition Temperature:

No.	Value	Note
1	170°C	dry [1]
2	-10°C	wet [1]

Transport Properties:

Water Content: 15% (65% relative humidity) [1,6], 16% (21.2°, 65% relative humidity) [3]

Mechanical Properties:

Tensile (Young's) Modulus:

No.	Value	Note
1	3550 MPa	3.55 × 10^{10} dynes cm^{-2}, 65% relative humidity [9]
2	1400 MPa	1.4 × 10^{10} dynes cm^{-2}, 20° [10]

Tensile Strength Break:

No.	Value	Note
1	152–207.9 MPa	1550–2120 kg cm^{-2}, 65% relative humidity, 22° [7]

Miscellaneous Moduli:

No.	Value	Note	Type
1	1400 MPa	[2]	longitudinal modulus

Electrical Properties:

Electrical Properties General: Mass specific resistance 10000000000 Ω g cm^{-2} (65% relative humidity) [8]

Optical Properties:

Refractive Index:

No.	Value	Note
1	1.553–1.555	along fibre axis
2	1.542–1.546	transverse [5]

Polymer Stability:

Decomposition Details: Decomposes above 126° with scorching at 204° [3]
Flammability: Combustible
Chemical Stability: Stable in moderately acidic media (except sulfuric acid). Decomposed by alkalis and hypochlorite bleach

Applications/Commercial Products:

Processing & Manufacturing Routes: The wool is cleaned and then spun
Applications: Applications include clothing and insulating material

Bibliographic References

[1] *Polymeric Materials Encyclopedia,* (ed. J.C. Salamone), CRC Press, 1996
[2] Feughelman, M. and Robinson, M.S., *Text. Res. J.,* 1971, **41,** 469
[3] *Hawley's Condensed Chemical Dictionary,* 11th edn., (eds. N.I. Sax and R.J. Lewis), Van Nostrand Reinhold, New York, 1987
[4] King, A.T., *J. Text. Inst.,* 1926, **17,** T53
[5] Bunn, C.W., *Fibre Sci.,* Chapter 10, (ed. J.M. Preston), Textile Institute, Manchester, 1949
[6] Speakman, J.B., *J. Soc. Chem. Ind., London,* 1930, **49,** 209T
[7] von Bergen, W. and Matthews, *Text. Fibers,* 6th edn., Chapter 11, Wiley, New York, 1954
[8] Hearle, J.W.S., *J. Text. Inst.,* 1957, **48,** 40
[9] *The British Wool Manual,* 2nd edn., (ed. H. Spibey), Columbine Press, Buxton, 1969
[10] Feughelman, M. and Robinson, M.S., *Text. Res. J.,* 1971, **41,** 6

Xylenol formaldehyde resins X-1

Synonyms: *Dimethylphenol formaldehyde resins*
Related Polymers: Phenol Aldehyde Resins, Phenolic Resoles, Phenolic Novolaks, Cresol Formaldehyde Resins
Monomers: Formaldehyde, 2,4-Xylenol, 2,3-Xylenol, 2,5-Xylenol, 3,5-Xylenol, 3,4-Xylenol
Material class: Thermosetting resin
Polymer Type: phenolic
CAS Number:

CAS Reg. No.	Note
25053-90-1	2,5-xylenol-formaldehyde copolymer
25053-93-4	2,6-xylenol-formaldehyde copolymer
25053-87-6	3,4-xylenol-formaldehyde copolymer
25086-35-5	3,5-xylenol-formaldehyde copolymer
25053-91-2	2,3-xylenol-formaldehyde copolymer
25053-94-5	2,4-xylenol-formaldehyde copolymer

Molecular Formula: $(C_9H_{10}O.CH_2O)_n$
Fragments: $C_9H_{10}O$ CH_2O
Molecular Weight: M_n 1043 (3,4-xylenol resin), 488 (3,5-xylenol resin)
General Information: Xylenol isomers have different functionality in the condensation reaction with formaldehyde and this influences the degree of cross-linking possible. Of the possible isomers, 2,4-xylenol is monofunctional; 2,3- 3,4- and 2,5-xylenol are difunctional; the only trifunctional isomer is 3,5-xylenol. The methyl groups tend to activate the phenolic molecule and 3,5-xylenol is a reactive cross-linking component

Volumetric & Calorimetric Properties:
Melting Temperature:

No.	Value	Note
1	178–191°C	3,4-xylenol resin [1]
2	136–142°C	3,5-xylenol resin [1]

Polymer Stability:
Decomposition Details: The major pyrolysis products of 3,5-xylenol-formaldehyde resin (hexa cured) at 600° are 2,3,5-trimethylphenol (35.3 mol%) and 3,5-dimethylphenol (17.8 mol%) [2]

Applications/Commercial Products:
Processing & Manufacturing Routes: May be synth. by reaction of the appropriate dimethylphenoxymagnesium bromide with paraformaldehyde in dry C_6H_6 for 12 h at 80° followed by acidic quenching [1]. This procedure can be used to synth. "high-*ortho*" novolaks [3] Xylenol-formaldehyde resins are made by the condensation reaction of molten xylenols with formaldehyde, generally 37% aq. soln. With basic catalyst and a molar excess of formaldehyde, a Resole is formed. With an acidic catalyst and excess xylenol, a Novolak resin is formed
Applications: 3,4 and 3,5-xylenol resins have potential applications as antimicrobial agents. High-*ortho* 3,4-xylenol resin has potential application as an antioxidant in lard

Bibliographic References
[1] Uchibori, T., Kawada, K., Watanabe, S., Asakura, K. *et al*, *Bokin Bobai*, 1990, **18**, 215, (synth. use)
[2] Blazsó, M., and Tóth, T., *J. Anal. Appl. Pyrolysis*, 1986, **10**, 41, (thermal decomposition)
[3] Abe, Y., Matsumura, S., Asakawa, K. and Kasama, H., *Yukagaku*, 1986, **35**, 751

Zein Z-1

$$\left[-NHCHC(=O)- \right]_n$$
where R is the side chain

Synonyms: *Vicara. Maize protein*
Related Polymers: Groundnut protein fibre, Proteins
Material class: Amino acid
Polymer Type: protein
General Information: A naturally coiled protein polymer isol. from maize meal [1,2]

Volumetric & Calorimetric Properties:
Density:

No.	Value
1	d 1.25 g/cm^3

Melting Temperature:

No.	Value	Note
1	240°C	[1]

Surface Properties & Solubility:
Solvents/Non-solvents: Sol. hot alkalis. Insol. organic solvents

Transport Properties:
Water Absorption:

No.	Value	Note
1	10%	moisture regain [1]
2	40%	imbibition max.

Mechanical Properties:
Mechanical Properties General: Tensile strength at break 0.75–12 g den^{-1}. [2] Tenacity 1060 N tex^{-1} (dry); 574 N tex^{-1} (wet)
Tensile Strength Break:

No.	Value	Note
1	120.1–133 MPa	1225–1365 kg cm^{-2} [1]

Tensile Strength Yield:

No.	Value	Note
1	25–35 MPa	dry [1]
2	30–45 MPa	wet

Polymer Stability:
Decomposition Details: Decomposes at 185° [1]
Flammability: Has low flammability
Environmental Stress: Undergoes some deterioration on prolonged exposure to sunlight
Chemical Stability: Very resistant to acids. Sol. hot alkali with deterioration.

Applications/Commercial Products:
Processing & Manufacturing Routes: Corn meal protein is dissolved in aq. 2-propanol. The extracted Zein is dissolved in alkali and spun into an acidic soln. The resulting fibre can be stabilised by cross-linking with formaldehyde [1,2]
Applications: Used in fabrics as a blend with nylon, rayon, cotton and wool

Trade name	Details	Supplier
Vicara	discontinued	Virginia-Carolina Chemical Co.

Bibliographic References
[1] Gordon Cook, J., *Handbook of Textile Fibers,* 5th edn., Merrow, 1984, 2
[2] Moncrieff, R.W, *Man-Made Fibers,* 6th edn., Newnes-Butterworths, 1975

Zenite Z-2
Material class: Polymer liquid crystals
Polymer Type: polyester
CAS Number:

CAS Reg. No.
177772-13-3
177772-14-4
177772-15-5

Volumetric & Calorimetric Properties:
Thermal Expansion Coefficient:

No.	Value	Note	Type
1	1.3 × 10^{-5} K^{-1}	25–150°, flow direction, Zenite 6130 [1]	l
2	8 × 10^{-6} K^{-1}	25–150°, flow direction, Zenite 6330 [1]	L
3	1.4 × 10^{-5} K^{-1}	25–150°, flow direction, Zenite 7130 [1]	L
4	3.7 × 10^{-5} K^{-1}	25–150°, transverse direction, Zenite 6130 [1]	L
5	2.2 × 10^{-5} K^{-1}	25–150°, transverse direction, Zenite 6330 [1]	L
6	3.6 × 10^{-5} K^{-1}	25–150°, flow direction, Zenite 7130 [1]	L

Thermal Conductivity:

No.	Value	Note
1	0.32 W/mK	ASTM C177, Zenite 7130 [1]
2	0.27 W/mK	ASTM C177, Zenite 6130 [1]

Specific Heat Capacity:

No.	Value	Note	Type
1	0.8 kJ/kg.C	25°, Zenite 6130 [1]	P
2	1.1 kJ/kg.C	125°, Zenite 6130 [1]	P
3	1.3 kJ/kg.C	225°, Zenite 6130	P
4	1.4 kJ/kg.C	325°, Zenite 6130 [1]	P

Melting Temperature:

No.	Value	Note
1	335°C	Zenite 6130 [1]
2	335°C	Zenite 6330 [1]
3	352°C	Zenite 7130 [1]
4	350°C	end point, ASTM D3418 [1]

Zenite

Glass-Transition Temperature:

No.	Value
1	120°C

Deflection Temperature:

No.	Value	Note
1	265°C	Zenite 6130 [1]
2	245°C	Zenite 6330 [1]
3	295°C	Zenite 7130 [1]

Transport Properties:
Polymer Melts: Melt viscosity 40 Pa s (500 s^{-1}); 30 Pa s (1000 s^{-1}); 25 Pa s (2000 s^{-1}); 20 Pa s (3000 s^{-1}) (Zenite 6130) [2]. 150 Pa s (500 s^{-1}); 100 Pa s (1000 s^{-1}); 65 Pa s (2000 s^{-1}); 40 Pa s (5000 s^{-1}) (Zenite 7130) [2]

Water Absorption:

No.	Value	Note
1	0.002%	24h, 23°, ASTM D570, Zenite 6130 [1]
2	0.05%	6 month immersion, ASTM D570, Zenite 6130 [1]

Mechanical Properties:
Tensile (Young's) Modulus:

No.	Value	Note
1	29000 MPa	-40°, Zenite 6130 [1]
2	21000 MPa	23°, Zenite 6130 [1]
3	12000 MPa	120–150°, Zenite 6130 [1]
4	10000 MPa	200°, Zenite 6130 [1]
5	4000 MPa	250°, Zenite 6130 [1]
6	23000 MPa	-40°, Zenite 7130 [1]
7	18000 MPa	23°, Zenite 7130 [1]
8	14000 MPa	120°, Zenite 7130 [1]
9	9000 MPa	150–250°, Zenite 7130 [1]

Flexural Modulus:

No.	Value	Note
1	15000 MPa	-40°, Zenite 6130 [1]
2	12000 MPa	23°, Zenite 6130 [1]
3	6200 MPa	120°, Zenite 6130 [1]
4	5500 MPa	150°, Zenite 6130 [1]
5	2800 MPa	250°, Zenite 6130 [1]
6	9600 MPa	23°, Zenite 6330 [1]
7	16000 MPa	-40°, Zenite 7130 [1]
8	13000 MPa	23°, Zenite 7130 [1]
9	8000 MPa	120–150°, Zenite 7130 [1]

Tensile Strength Break:

No.	Value	Note
1	185 MPa	-40°, Zenite 6130 [1]
2	250 MPa	23°, Zenite 6130 [1]
3	60 MPa	120°, Zenite 6130 [1]
4	50 MPa	150°, Zenite 6130 [1]
5	14 MPa	250°, Zenite 6130 [1]
6	110 MPa	23°, Zenite 6330, ASTM D790 [1]
7	230 MPa	-40°, Zenite 7130 [1]
8	174 MPa	23°, Zenite 7130 [1]
9	78 MPa	120°, Zenite 7130 [1]
10	64 MPa	150°, Zenite 7130 [1]
11	30 MPa	250°, Zenite 7130 [1]

Flexural Strength at Break:

No.	Value	Note
1	245 MPa	-40°, Zenite 6130 [1]
2	170 MPa	23°, Zenite 6130 [1]
3	62 MPa	120°, Zenite 6130 [1]
4	33 MPa	200°, Zenite 6130 [1]
5	17 MPa	250°, Zenite 6130 [1]
6	270 MPa	-40°, Zenite 7130 [1]
7	174 MPa	23°, Zenite 7130 [1]
8	78 MPa	120°, Zenite 7130 [1]
9	48 MPa	200°, Zenite 7130 [1]
10	30 MPa	250°, Zenite 7130 [1]

Compressive Modulus:

No.	Value	Note
1	6900 MPa	Zenite 6130, ASTM D621 [1]
2	4100 MPa	Zenite 6330, ASTM D621 [1]
3	5300 MPa	Zenite 7130, ASTM D621 [1]

Compressive Strength:

No.	Value	Note
1	105 MPa	ASTM D621, Zenite 6130 [1]
2	7130 MPa	ASTM D621, Zenite 7130 [1]

Mechanical Properties Miscellaneous: Isochronous stress-strain 20 MPa stress, 0.6% strain (10 g) [1]
Hardness: Rockwell M61–M63 [1], R108–R110 [1]
Failure Properties General: Flexural fatigue 60 MPa (10000 h); 40 MPa (10000000 h)
Friction Abrasion and Resistance: Static coefficient of friction μ_{stat} 0.09 (steel, Zenite 6130); 0.1 (steel, Zenite 7130) [1]; 0.29 (self, Zenite 7130) [1]. Dynamic coefficient of friction μ_{dyn} 0.12 (steel, Zenite 6130) [1]; 0.1 (steel, Zenite 7130) [1]; 0.23 (self, Zenite 7130) [1]. Taber abrasion 49 mg (1000 cycles)$^{-1}$ (CS-17, Zenite 6130); 63 mg (1000 cycles)$^{-1}$ (CS-17, Zenite 7130) [1]

Izod Notch:

No.	Value	Notch	Note
1	110 J/m	Y	-40°, ASTM D256, Zenite 6130 [1]
2	120 J/m	Y	23°, ASTM D256, Zenite 6130 [1]
3	160 J/m	Y	23°, ASTM D256, Zenite 6330 [1]

Zenite

4	185 J/m	Y	-40°, ASTM D256, Zenite 6130 [1]
5	225 J/m	Y	23°, ASTM D256, Zenite 7130 [1]
6	440 J/m	N	-40°, ASTM D4812, Zenite 6130 [1]
7	655 J/m	N	23°, ASTM D4812, Zenite 6130 [1]
8	555 J/m	N	-40°, ASTM D4812, Zenite 7130 [1]
9	740 J/m	N	23°, ASTM D4812, Zenite 7130 [1]

Electrical Properties:

Surface/Volume Resistance:

No.	Value	Note	Type
1	10×10^{15} Ω.cm	ASTM D257, Zenite 6130 [1]	S
2	1×10^{15} Ω.cm	ASTM D257, Zenite 7130 [1]	S

Dielectric Permittivity (Constant):

No.	Value	Frequency	Note
1	4.4	1 kHz	23°, Zenite 6130 [1]
2	5	1 kHz	120–200°, Zenite 6130 [1]
3	3.9	1 MHz	23°, Zenite 6130 [1]
4	4.8–4.9	1 MHz	120–200°, Zenite 6130 [1]
5	4.3	1 GHz	23°, Zenite 6130 [1]
6	4.4–4.5	1 GHz	120–200°, Zenite 6130 [1]
7	4.3	1 kHz	23°, Zenite 7130 [1]
8	4.9–5	1 kHz	120–200°, Zenite 7130 [1]
9	3.8	1 MHz	23°, Zenite 7130 [1]
10	4.5–4.9	1 MHz	120–200°, Zenite 7130 [1]
11	4.3	1 GHz	23°, Zenite 6130 [1]
12	4.4–4.7	1 GHz	120–200°, ASTM D150, Zenite 7130 [1]

Dielectric Strength:

No.	Value	Note
1	28–29 kV.mm^{-1}	23–120°, ASTM D149 [1]
2	27 kV.mm^{-1}	150–200°, ASTM D149 [1]
3	24–26 kV.mm^{-1}	23°, step by step [1]

Strong Field Phenomena General: Comparative Tracking Index 175 V (ASTM D3636, Zenite 6130) [1]; 167 V (ASTM D3636, Zenite 7130) [1]

Dissipation (Power) Factor:

No.	Value	Frequency	Note
1	0.013	1 kHz	23°, ASTM D150, Zenite 6130 [1]
2	0.027	1 MHz	23°, ASTM D150, Zenite 6130 [1]
3	0.004	1 GHz	23°, ASTM D2520, Zenite 6130 [1]
4	0.013	1 kHz	23°, ASTM D150, Zenite 7130 [1]
5	0.029	1 MHz	23°, ASTM D150, Zenite 7130 [1]
6	0.004	1 GHz	23°, ASTM D2520, Zenite 7130 [1]

Polymer Stability:

Polymer Stability General: Is extremely stable
Thermal Stability General: Resins have excellent thermal stability even in melt above 300°. Physical and electrical props. retained at high temp. over long periods [1]
Flammability: Very inflammable
Chemical Stability: Resistant to a range of automotive fluids, unstable to antifreeze at 100°
Hydrolytic Stability: Very stable to most alkalis and acids and to salt solns. ($ZnCl_2$)
Recyclability: A very high retention level of physical props. is maintained when reusing resin (as regrind) even after six passes. All major props. retained in excess of 80% of original level

Applications/Commercial Products:

Processing & Manufacturing Routes: Normally synth. from the diacetate of aromatic hydroxyl compound and aromatic acid. Acetates may be formed in situ or separately by treatment with glacial AcOH using 1% sulfuric acid catalyst. Stoichiometric quantities of diacids and diacetates/acetoxy aromatic acids are mixed in the melt with distillation of AcOH from the mixture. Polymers may be formed in a continuous reactor and extruded to form oriented fibrous material. Processing is carried out by injection moulding (melt temp. 350–370°, injection pressure 20–40 MPa. Mould temp. 60–110°). Melt viscosity falls dramatically with shear rate. Very high fill rates can be utilised over the whole injection stroke. Fast injection rates also contribute to anisotropic props., including high physical props., low thermal expansion and shrinkage in the flow direction.

Mould Shrinkage (%):

No.	Value	Note
1	-0.07%	flow direction, Zenite 6130
2	0.5%	transverse direction, 23°, Zenite 6130
3	0%	flow direction, 80°, Zenite 6130
4	0.05%	flow direction, 100°, Zenite 6130
5	0.08%	flow direction, 140°, Zenite 6130
6	0.5%	transverse direction, 23°, Zenite 7130
7	0%	50°, Zenite 7130
8	0.08%	flow direction, 100°, Zenite 7130

Applications: Suitable for applications requiring high temp. performance with retention of props. over a wide temp. range, dimensional stability, chemical resistance and excellent electrical props. Main applications include: electrical/electronic; uses such as surface mount components, sockets, connectors, chip carriers, bobbins, ceramic replacements, electric motor insulation components, fibre optic connectors, guides, closures, fuse holders. Automotive; sensors, lamp sockets, coil forms, chip carriers, transmission system components, pump components, ignition system components. Aerospace; all electrical components, imaging and optoelectric components, sensor devices, composite materials

Trade name	Details	Supplier
Zenite	Several grades	DuPont

Bibliographic References

[1] *Zenite LCP Product and Properties Guide,* DuPont, 1996, (technical datasheet)
[2] *Zenite LCP Moulding Guide,* DuPont, 1996, (technical datasheet)
[3] Huber, H.J., *Kunststoffe,* 1996, **86,** 530

Name and Synonym Index

A-Stage Resin, P-21
AAS, A-39
ABS Flame retardant, A-4
ABS General purpose, A-5
ABS Polymers, A-17
ABS, Blow moulding, A-1
ABS, Carbon reinforced, A-2
ABS, Extrusion, Unfilled, A-3
ABS, Glass fibre reinforced, A-6
ABS, Heat Resistant, A-7
ABS, High impact, A-8
ABS, Injection moulding, Glass fibre reinforced, A-9
ABS, Injection moulding, Unfilled, A-10
ABS, Plating, A-12
ABS, Stainless steel filled, Flame retardant, A-21
ABS, Stainless steel filled, General purpose, A-22
ABS/Nylon blends, Injection moulding, A-11
ABS/PA, A-11
ABS/PC blend, A-13
ABS/PC blends, Flame retardant, A-15
ABS/PC blends, A-14, A-16
ABS/PMMA blends, P-342
ABS/Polyamide Alloy, A-11
ABS/Polycarbonate alloy, A-13
ABS/Polycarbonate alloys, Flame retardant, A-15
ABS/Polycarbonate alloys, A-14, A-16
ABS/Polycarbonate blend, General purpose injection moulding, A-13
ABS/Polycarbonate blends, extrusion unreinforced, A-14
ABS/Polycarbonate blends, Flame retardant, A-15
ABS/Polycarbonate blends, Glass fibre reinforced, A-16
ABS/polyurethane blends, General purpose, A-18
ABS/Polyvinylchloride blends, A-19
ABS/PSO blend, A-20
ABS/PSU blend, A-20
ABS/PVC Alloy, A-19
ABS/PVC Blends, A-19
Acemannan, A-23
Acetal copolymer, A-24
Acetal, P-432
Acetate, C-9
'Acid-modified' vinyl chloride-vinyl acetate copolymers, P-589
Acrylamide copolymer emulsion, A-25
Acrylamide gels, P-61
Acrylates-PVP copolymer, P-230
Acrylic fibres, A-26
Acrylic-styrene-acrylonitrile terpolymer, A-39
Acrylic/methacrylic copolymers, A-27
Acrylic, P-117, P-381
Acrylics, M-8
Acrylonitrile-butadiene copolymer, N-5
Acrylonitrile-butadiene-isoprene terpolymer, A-28
Acrylonitrile-butadiene-styrene copolymer, A-2, A-3, A-4, A-5, A-7, A-8, A-9, A-12, A-17, A-21, A-22
Acrylonitrile-butadiene-styrene terpolymer and polyvinylchloride blends, A-19
Acrylonitrile-butadiene-styrene terpolymer/polyamide blends, A-11
Acrylonitrile-butadiene-styrene terpolymer/polycarbonate blend, A-13
Acrylonitrile-butadiene-styrene terpolymer/polycarbonate blends, Flame retardant, A-15
Acrylonitrile-butadiene-styrene terpolymer/polycarbonate blends, A-14, A-16
Acrylonitrile-butadiene-styrene terpolymer, A-2, A-3, A-4, A-5, A-7, A-8, A-9, A-12, A-17, A-21, A-22
Acrylonitrile-butadiene-styrene, Glass fibre reinforced, A-6
Acrylonitrile-butadiene-styrene/polysulfone blend, A-20

Acrylonitrile-methyl acrylate copolymer, Injection moulding, A-30
Acrylonitrile-methyl acrylate copolymer, A-29
Acrylonitrile-styrene-acrylate terpolymer, A-39
Adipic acid-ethylene glycol copolymer, P-250
AES, Impact modified, A-31
Agar-Agar, A-32
Agar, A-32
Agarose, A-32
Alginate, A-33
Alginic acid, A-33
Aliphatic polykeone, P-476
Aliphatic polyketones, P-475
Alkene-sulfur dioxide copolymers, O-2
Alkylfluoroalkoxyphosphazene elastomers, F-12
Alkyne-sulfur dioxide copolymers, O-2
Allyl cyanoacrylate polymer, P-68
Allyl glycidyl ether-epichlorohydrin-propylene oxide synthetic rubber, P-503
Allyl glycidyl ether-propylene oxide synthetic rubber, P-502
Amorphous Nylon, A-34
Amylopectin, A-35
Amylose, A-36
α-Amylose, A-36
Amylum, S-24
Aniline black, P-72
ANS, S-1
Araban, A-37
Arabinan, A-37
Aramids, A-38
Artificial Horn, C-5
Arylfluoroalkoxyphosphazene elastomers, F-12
ASA, A-39
ASA terpolymer, A-39
ASA, Blow moulding, A-40
ASA, Extrusion unfilled, A-41
ASA, High impact, A-42
ASA, Injection moulding, Unfilled, A-43
ASA/PBT blends, P-3
ASA/PC blend, P-130
ASA/PVC blend, A-44
Atactic Polypropylene, A-45
Avimid K-III, A-46

B/SB, P-95
BADGE, B-3
Balata rubber, N-1
Barex 210, A-30
Barex, A-29
BEH-PPV, P-81
1,4-Benzenediamine, N-phenyl-, homopolymer, P-454
1,3-Benzenedicarboxylic acid polymer with [1,1'-biphenyl]-3,3',4,4'-tetramine, P-2
1,3-Benzenediol resins, R-1
Biaxially oriented polystyrene sheet and film, O-3
BIIR, B-21
1-(N,N-Bis(2,3-epoxypropyl)amino)-4-(2,3-epoxyprop-1-oxy)benzene) resin, B-2
2,2'-Bis(3,4-dicarboxyphenyl)hexafluoropropane dianhydride-co-m-phenylenediamine-co-p-phenylenediamine, F-6
Bis(2,3-epoxycyclopentyl)ether resin, B-1
Bis(4-butylphenyl)dichlorosilane homopolymer, P-76
Bisphenol A Diglycidyl ether resin, B-3
Bisphenol A polycarbonate, Carbon fibre reinforced, B-6
Bisphenol A Polycarbonate, Glass fibre reinforced, B-7
Bisphenol A polycarbonate, Glass filled, B-8
Bisphenol A polycarbonate, Metal filled, B-9
Bisphenol A polycarbonate, PTFE lubricated, B-10
Bisphenol A polycarbonate, Structural Foam, B-11
Bisphenol A polycarbonate, B-5

Bisphenol A polysulfone, P-531
Bisphenol A-1,3-benzenedicarboxylic acid polyester carbonate, B-4
Bisphenol A-1,4-benzenedicarboxylic acid polyester carbonate, B-13
Bisphenol A-isophthalic acid polyester carbonate, B-4
Bisphenol A-terephthalic acid polyester carbonate, B-13
Bisphenol A-terephthalic acid-isophthalic acid polyester carbonate, B-12
Bisphenol A-Tetrabromobisphenol A copolycarbonates, P-465
Bisphenol ACP polycarbonate, P-452
Bisphenol AF polycarbonate, B-14
Bisphenol B polycarbonate, P-97
Bisphenol F diglycidyl ether resin, B-15
Bisphenol G polycarbonate, B-16
Bisphenol P polycarbonate, B-17
Bisphenol TMC polycarbonate, P-554
Bisphenol TMC-Bisphenol A copolycarbonates, P-555
Bisphenol Z polycarbonate, P-157
Bisphenol Z-Bisphenol A copolycarbonates, P-158
Bisphenol-dichloroethylene-Bisphenol A copolycarbonates, P-174
Blended polyethylenes, P-252
Blends of polybutadiene and styrene-butadiene di-block copolymers, P-95
Blends of polystyrene and styrene-butadiene di-block copolymers, P-528
Block copolymer of 4,4'-(1-methylethylidene)bis[phenol] with α-hydro-ω-hydroxypoly(oxy-1,2-ethanediyl) and carbonic acid, P-470
Block copolymer of 4,4'-(1-methylethylidene)bis[phenol] with carbonic acid and methyl 2-methyl-2-propenoate, P-472
Bouncing putty, S-20
BPDA-pPDA, B-20
BPDA-ODA, B-19
BR, L-7
Branched butadiene-styrene block copolymers, S-27
Brominated isobutylene-isoprene elastomer, B-21
Bromobutyl rubber, B-21
BTDA-1,3-BABB, L-1
BTDA-3,3'-DABP, L-2
BTDA-4-BDAF, B-22
BTDA-mPDA/ODA, B-28
BTDA-mPDA, B-27
BTDA-DAPI, B-23
BTDA-diaminophenylindane, B-23
BTDA-MDA, B-24
BTDA-MPD/MDA, B-25
BTDA-ODA, B-26
Butadiene rubber, L-7, P-87
Butadiene-acrylonitrile copolymers, B-29
Butadiene-pentadiene copolymers, B-30
Butadiene-styrene-vinylidene chloride terpolymer, B-31
1,4-Butanediol diglycidyl ether resin, B-32
Butene, polymer with 1-propene, P-484
Butyl glycoyl ether resin, B-33
para-tert-Butyl Phenolic resins, B-34
Butyl rubber, B-35
Butyl rubber brominated, B-21
Butyl rubber chlorinated, B-36

C12-C5, P-220
C2-C5, P-277
C4-C5, P-122
C5-C6, P-304
C5-C8, P-424
Calcium caseinate, C-4
9H-Carbazole homopolymer, P-129
Carbonaceous char, P-35

Carbonaceous materials, P-35
Carboxyl-modified vinyl chloride-vinyl acetate copolymer, P-585
'Carboxyl-modified' vinyl chloride-vinyl acetate copolymers, P-589
Carboxylated acrylonitrile-butadiene copolymer, N-6
Carboxylated NBR, N-6
Carboxylated nitrile rubber, N-6
Carboxylated poly(acrylonitrile-co-butadiene), N-6
Carboxylated PVC/PVAc, P-585
Carboxylated VC/VA, P-585
Carboxylated vinyl chloride-vinyl acetate copolymer, P-585
'Carboxylated' vinyl chloride-vinyl acetate copolymers, P-589
6-Carboxymethyl-2'-hydroxyethyl cellulose, C-1
Carboxymethylhydroxyethyl cellulose, C-1
Cardanol Resins, C-2
Carrageenan, C-3
Carrisyn, A-23
Casein Formaldehyde, C-5
Casein Plastic, C-5
Casein, C-4
Cashew nut shell liquid resins, C-2
Celloidin, C-12
Cellophane, C-8
Celluloid, C-12
Cellulon, C-7
Cellulose acetate butyrate, C-10
Cellulose acetate propionate, C-11
Cellulose acetate, C-9
Cellulose carboxymethyl ether sodium salt, S-22
Cellulose ethyl-2-hydroxyethyl ether, E-18
Cellulose I, C-7
Cellulose II, C-8
Cellulose nitrate, C-12
Cellulose triacetate, C-13
Cellulose, C-6
α-Cellulose, C-7
CF, C-5
Chicle, C-14
Chitin, C-15
Chitosan, C-16
Chloral polycarbonate, P-173
Chlorinated isobutylene-isoprene copolymer/elastomer, B-36
Chlorinated poly(vinyl chloride), P-576
Chlorinated PVC, P-576
Chlorobutyl rubber, B-36
Chloroethene polymer, P-583
Chloroethylene polymer, P-583
Chloroprene rubber, P-149
Chloroprene rubber, Vulcanised, P-150
Chlorotrifluoroethylene-ethylene copolymer, E-4
CIIR, B-36
CMC, S-22
CMHEC, C-1
CN-PPV6, C-17
CNSL resins, C-2
Collagen, C-18
Collodion cotton, C-12
Collodion wool, C-12
Colloxylin, C-12
Conductive silicone rubber, S-17
'COOH containing ' vinyl chloride-vinyl acetate copolymer, P-585
Copolycarbonates, P-549
CPP, P-488
CPVC, P-576
CR, P-149
Cresol Novolaks, C-21
Cresol Resoles, C-20
Cresol-formaldehyde Novolaks, C-21
Cresol-formaldehyde Resins, C-19
Cresol-formaldehyde Resoles, C-20
Cresylic Novolaks, C-21
Cresylic Resoles, C-20
Croscarmellose sodium, S-22
Crystalline modified polysulfone with glass fibre, C-22
Crystalline poly(vinyl chloride), P-577
Crystalline PVC, P-577

CS, C-5
CTFE, P-154
Cyclic dimethyl polysiloxane, C-23
Cyclic dimethyl siloxane, C-23
Cyclohexane-1-ethenyl-3-methyl homopolymer, P-415
Cyclomethicone, C-23
Cyclopentane-ethenyl homopolymer, P-597
Cyclosiloxane, C-23

(DAM-VP) copolymer, P-635
Decamethylcyclopentasiloxane, C-23
Deoxyribonucleic acid, D-1
DGEBPA, B-3
2,7-Dibromo-9,9-bis(2-ethylhexyl)-9H-fluorene homopolymer, P-79
1,4-Dibutoxy-2,5-diethynylbenzene homopolymer, P-171
Dibutyldichlorosilane homopolymer, P-172
2,5-Dichlorobenzophenone homopolymer, P-75
Dichlorobutylpentylsilane homopolymer, sru, P-122
Dichlorodihexylsilane homopolymer, P-179
Dichlorodipentylsilane homopolymer, P-204
Dichloroheptylhexylsilane homopolymer, P-288
Dichlorohexyloctylsilane homopolymer, P-302
Dichloromethylpropylsilane homopolymer, P-406
Dichlorooctylpentylsilane homopolymer, P-424
Dichloropoly(methylphenylsilane), D-2
α,ω-Dichloropolymethylphenylsilane, D-2
Dichloropolyphenylmethylsilane, D-2
Dichloropolyphenylmethylsilylene, D-2
Dicyclopentadiene dioxide resin, D-3
Diethylsilane homopolymer, sru, P-176
Diglycidyl aniline resin, D-4
Diglycidyl ether of Bisphenol A, B-3
1,3-Dihexyl-1,3-dimethyl-1,3-disilacyclobutane homopolymer, P-301
2,3-Dihydro-6H-1,4-dioxino[2,3-c]pyrrole homopolymer, P-256
2,5-Dihydroxy-1,4-benzenedicarboxylic acid polymer with 2,3,5,6-pyridinetetramine, P-50
4,4'-Dihydroxydiphenylsulfone polycarbonate, P-537
Dimethicone, P-192
Dimethicone polyol, D-5
Dimethicones fluid, P-7
Dimethiconol, D-5
Dimethyl polysiloxane, P-192
1,3-Dimethyl-1,3-diphenyl-1,3-disilacyclobutane homopolymer, SRU, P-403
4-(1,1-Dimethylethyl)phenol based resins, B-34
Dimethylphenol formaldehyde resins, X-1
Dimethylsilicone, P-192
Dimethylsilicone fluid, P-7
Dimethylsilicone gum, P-192a
Dimethylsiloxane rubber, Glass fibre filled, S-18
9,9-Dioctylfluorene 2,1,3-benzothiadiazole copolymer, P-197
9,9-Dioctylfluorene-2,2'-bithiophene copolymer, P-199
1,4-Dioxane-2,5-dione homopolymer, P-283
Diphenylamine homopolymer, P-440
DIVEMA, P-216
DMOS-PPV, P-186
DMPPO, P-188
DNA, D-1
Dodecamethylcyclohexasiloxane, C-23
(DodecMeSi)n, P-219
DOO-PPV, P-84

E-Bonite, E-1
EAA copolymer, P-248
Easy flow polycarbonate, B-5
ECH-PO-AGE terpolymer, P-503
ECO, P-255
Ecolyte PS, E-3
Ecolyte, E-2
ECTFE, E-4
ECTFE, Carbon fibre filled, E-5
ECTFE, Glass fibre filled, E-6
EHEC, E-18

EHO-OPPE, P-80
Emeraldine, P-72
Emulsion polymerised polybutadiene, P-87
ENR, E-7
EO/PO block polymer, P-55
EPDM, E-17
EPDM Rubber, E-17
EPM, E-17
EPM Rubber, E-17
Epoxidised natural rubber, E-7
Epoxidised poly(styrene-block-butadiene-block-styrene) triblock copolymer, E-8
Epoxidised PVC/PVAc, P-586
Epoxidised SBS, E-8
Epoxidised styrene-butadiene-styrene block copolymer, E-8
Epoxidised VC/VA, P-586
Epoxidised vinyl chloride-vinyl acetate copolymer, P-586
Epoxy Resins, General Information, E-9
Epoxy Resins, E-9
Epoxy-modified vinyl chloride-vinyl acetate copolymer, P-586
2-(3,4-Epoxycyclohexyl-5,5-spiro-3,4-epoxy)cyclohexane metadioxide resin, E-10
3',4'-Epoxycyclohexylmethyl-3,4-epoxycyclohexane carboxylate resin, E-11
1,2-Epoxypropane elastomers, P-650
Epoxysilicone copolymer, S-11
EPR, E-17
EPS, P-523
EPT, E-17
ESBS, E-8
ETFE, E-12
ETFE copolymer, Glass fibre filled, P-541
ETFE copolymer, E-12
ETFE, Carbon fibre filled, E-13
ETFE, Glass fibre filled, E-14
ETFE, Glass fibre filled, P-541
2,2',2'',2'''-[1,2-Ethanediylidenetetrakis (phenyleneoxymethylene)]tetrakisoxirane, P-282
1-Ethenyl-1H-imidazole homopolymer, P-617
Ethenylbenzenesulfonic acid homopolymer compound with 2,3-dihydrothieno[3,4-b]-1,4-dioxin homopolymer, P-258
Ethyl cyanoacrylate polymer, P-246
Ethylcellulose, E-15
Ethylene copolymer with carbon monoxide, P-255
Ethylene copolymers with acrylates, E-16
Ethylene glycol succinate polyester, P-270
Ethylene oxide-trioxane copolymer, A-24
Ethylene-acrylic elastomer, P-263
Ethylene-chlorotrifluoroethylene copolymer, Carbon fibre filled, E-5
Ethylene-chlorotrifluoroethylene copolymer, Glass fibre filled, E-6
Ethylene-propylene copolymer, E-17
Ethylene-propylene elastomers, E-17
Ethylene-propylene rubber, E-17
Ethylene-propylene terpolymer, E-17
Ethylene-propylene-diene polymer, E-17
Ethylene-propylene-diene rubber, E-17
Ethylene-tetrafluoroethylene copolymer, Carbon fibre filled, E-13
Ethylene-tetrafluoroethylene copolymer, Glass fibre filled, E-14
Ethylene-vinyl acetate-vinyl chloride graft copolymer, P-663
Ethylhydroxyethyl cellulose, E-18
α-(3-Ethynylphenyl)-ω-[5-[1-[2-(3-ethynylphenyl)-2,3-dihydro-1,3-dioxo-1H-isoindol-5-yl]-2,2,2-trifluoro-1-(trifluoromethyl)ethylidene]-1,3-dihydro-1,3-dioxo-2H-isoindol-2-yl]poly[(1,3-dihydro-1,3-dioxo-2H-isoindole-2,5-diyl)[2,2,2-trifluoro-1-(trifluoromethyl)ethylidene](1,3-dihydro-1,3-dioxo-2H-isoindole-5,2-diyl)-1,3-phenyleneoxy-1,3-phenyleneoxy-1,3-phenylene], F-7
ETVA, P-273
Eucommia ulmodia gum, E-19
EVA, P-273
EVA/VC graft copolymer, P-663
EVAL, P-273

Name and Synonym Index

EVAL films, P-274
EVOH, P-273
EVOH films, P-274
Extrusion polycarbonate, B-5

F6, P-178
F8, P-196
F8BT, P-197
F8T2, P-199
6FDA-3,3'-6FDA, F-3
6FDA-4,4'-6FDA, F-4
6FDA-4-BDAF, F-1
6FDA-mPDA/pPDA, F-6
6FDA-pTeMPD, F-2
6FDA-durene, F-2
6FDA-ODA, F-5
6FDA-TPE Addition cured (acetylenic end-caps), F-7
6FDE-pPDA Addition cured (nadic end-groups), P-53
FEP fluorocarbon, P-542
FEP, Carbon fibre filled, F-8
FEP, Filled, F-9
FEP, Glass fibre filled, F-10
FEP, Polytetrafluoroethylene filled, F-11
FEP, PTFE filled, F-11
FEP, P-542
Fibre fleece materials, P-45
Fibroin, S-19
FKA Synthetic Rubber, P-616
Flexible silicone resin, S-16
Fluorinated ethylene-propylene copolymer, Carbon fibre filled, F-8
Fluorinated ethylene-propylene copolymer, Filled, F-9
Fluorinated ethylene-propylene copolymer, Glass fibre filled, F-10
Fluorinated ethylene-propylene copolymer, P-542
Fluorine containing rubber, P-616
Fluorine-containing rubber, P-615
Fluoroalkoxyphosphazene elastomers, F-12
Fluoroelastomer, P-615, P-616
Fluoromethyl silicone rubber, Heat vulcanised, F-13
Fluoromethyl silicone rubber, Room temp. vulcanised, F-14
Fluoromethylsilicone fluid, F-15
Fluorosilicone oil, F-15
Fluorosilicone rubber, F-13
Fluorosilicone rubber, Room temp. vulcanised, F-14
Furan resins, P-281
Furan resins, Foundry applications, F-16

Galalith, C-5
Gelatin, G-1
Glass reinforced Nylon 6, N-49
Glassy carbon, P-35
GPCO, P-503
GPO, P-502
GR-I, B-35
GR-S, S-28
Grafted VC/EVA copolymer, P-663
Groundnut protein fibre, G-2
GRPP, P-496
4GT, P-103
Guayule rubber, G-3
Gun cotton, C-12
Gutta, G-4
Gutta-percha, G-4

Hard rubber, E-1
HBA/HNA, P-311
HDPE, P-261
Heparin, H-1
Heparinic acid, H-1
Hevea guianensis, N-2
Hexafluoropropylene polymer, P-290
α-[4-(1,3,3aα,4α,7α,7aα-Hexahydro-1,3-dioxo-4,7-methano-2H-isoindol-2-yl)phenyl]-ω-(1,3,3aα,4α,7α,7aα-hexahydro-1,3-dioxo-4,7-methano-2H-isoindol-2-yl)poly[(1,3-dihydro-aα,4α,7α,7aα-hexahydro-1,3-dioxo-4,7-methano-2H-isoindol-2-yl)poly[(1,3-dihydroH-isoindole-5,2-diyl)-1,4-phenylene], P-53
α-[4-[[4-(1,3,3a,4,7,7a-Hexahydro-1,3-dioxo-4,7-methano-2H-isoindol-2-yl)phenyl]methyl]phenyl]-ω-(1,3,3a,4,7,7a-hexahydro-1,3-dioxo-4,7-methano-2H-isoindol-2-yl)poly[(1,3-dihydro-1,3-dioxo-2H-isoindole-2,5-diyl)carbonyl(1,3-dihydro-1,3-dioxo-2H-isoindole-5,2-diyl)-1,4-phenylenemethylene-1,4-phenylene], P-54
Hexamethylcyclotrisiloxane, C-23
(HexMeSi)n, P-300
Hexylmethylsilane homopolymer, sru, P-300
High trans SBR, S-29
High trans styrene butadiene rubber, S-29
High density polyethylene, P-261
High impact polystyrene/poly(2,6-dimethyl-1,4-phenylene oxide) blends, H-3
High impact polystyrene/poly(phenylene ether) blends, H-3
High impact polystyrene/poly(phenylene oxide) blends, H-3
High impact polystyrene/polyoxyphenylene blend, H-3
High impact polystyrene, H-2
High pressure polyethylene, L-4
High styrene resins, H-4
High styrene rubbers, H-4
High temperature nylon, N-54
High-vinylidene copolymer, P-592
Highly saturated nitrile elastomer, N-9
HIPS/DMPPO blends, H-3
HIPS/PPE blends, H-3
HIPS/PPO blends, H-3
HIPS, H-2
HNBR, N-9
HPPE, L-4
HSN, N-9
α-Hydro-ω-hydroxypoly[oxy(methane-1,2-ethanediyl)], P-501
α-Hydro-w-hydroxy-poly[oxy(dimethylsilylene)], M-17
α-Hydro-w-hydroxypoly[oxy(dimethylsilylene)], D-5, M-19
Hydrogenated acrylonitrile-butadiene copolymer, N-9
Hydrogenated nitrile rubber, N-9
Hydrogenated poly(acrylonitrile-co-butadiene), N-9
Hydrogenated polystyrene, P-596
Hydrolysed collagen, G-1
Hydrolysed ethylene-vinyl acetate films, P-274
Hydrolysed ethylene-vinyl alcohol copolymer, P-273
Hydrolysed vinyl chloride-vinyl acetate copolymers, P-590
Hydroxy-modified vinyl chloride-vinyl acetate copolymer, P-587
Hydroxyalkyl methacrylate copolymers, H-5
Hydroxyethyl cellulose, H-6
'Hydroxyl-modified' vinyl chloride-vinyl acetate copolymers, P-590
Hydroxylated VC/VA, P-587
Hydroxylated vinyl chloride-vinyl acetate copolymer, P-587
'Hydroxylated' vinyl chloride-vinyl acetate copolymers, P-590
Hydroxypropyl cellulose, H-7

IDPA-mPDA, L-3
IIR, B-35
Impact polystyrene, H-2
Impact resistant polystyrene, H-2
1H-Indole homopolymer, P-319
1H-Indole-5-carboxylic acid homopolymer, P-144
Ionomer, P-350
6/IPD//I/T copolymer, P-293
IR, P-326
Isobutyl cyanoacrylate polymer, P-321
Isotactic polypropylene, I-1
ISP triblock copolymer, P-525

Jelutong, J-1

K-Resin, K-1
Keratin, W-1

Langley Research Center Crystalline Polyimide, L-1
Langley Research Center Isomeric Thermoplastic Polyimide, L-3
Langley Research Center Thermoplastic Polyimide, L-2
LARC-CPI, L-1
LaRC-I-TPI, L-3
LaRC-TPI, L-2
LDPE, L-4
Leucoemeraldine, P-72
Linear low density polyethylene, L-5
LLDPE film, P-9
LLDPE, L-5
Long chain Poly(alkyl methacrylates), L-6
Low cis-1,4-polybutadiene, L-7
Low cis-1,4-poly(1,3-butadiene), L-7
Low density polyethylene, L-4
Low pressure polyethylene, P-261

MABS, P-377
Maize protein, Z-1
MDMO-PPV, P-356
MDPE, M-1
ME12, P-219
ME4, P-363
ME6, P-300
ME8, P-389
Medium density polyethylene, M-1
MEH-PPV, P-357
Melamine formaldehyde resins, M-2
Melamine resin, Moulding powders, Cellulose filled, M-3
Melamine resins, M-2
Melamines, M-2
Mercerised cellulose, C-8
Meroxapol, M-4
Methacrylate amide/imide copolymers, M-5
Methacrylate-urethane polymers and copolymers, M-7
Methacrylate/olefin copolymers, M-6
Methacrylates, General, M-8
Methacrylic acid copolymers, M-9
Methacrylic acid ionomers, M-9
3-Amino-4-methoxybenzenesulfonic acid homopolymer, P-355
Methyl cyanoacrylate polymer, P-365
Methyl hydrogen based silicone elastomer, M-10
Methyl hydrogen rubber, Heat vulcanised, M-10
Methyl oxirane polymers, P-55
Methyl phenyl based silicone elastomer, M-15
Methyl phenyl silicone resin, M-11
Methyl phenyl siloxane resin, M-11
Methyl phenyl vinyl silicone rubber, Heat vulcanised, V-1
Methyl rubber, P-182
Methyl silicone resin, M-12
Methyl silicone rubber, One part room temp. vulcanised, M-13
Methyl siloxane resin, M-12
Methyl vinyl ether-maleic anhydride copolymer, P-624
Methyl vinyl silicone rubber, Heat vulcanised, V-2
Methylacrylate polymer, P-361
Methylcellulose, M-14
N-(1-Methylethyl)-N'-phenyl-1,4-benzenediamine homopolymer, P-334
[2,2'-[(1-Methylethylidene)bis(2,6-dibromo-4,1-phenylene)oxymethylene]]bisoxirane resin, T-1
Methylhydrogen dimethylsiloxane copolymer, P-374
Methylnitrile silicone rubber, Room temp. vulcanised, N-12
Methyloxirane-oxirane copolymer, P-504
Methylphenol-formaldehyde Novolaks, C-21
Methylphenol-formaldehyde Resoles, C-20

Name and Synonym Index

Methylphenyl rubber, Room temp. vulcanised, M-18
Methylphenylsilicone rubber, Heat vulcanised, M-15
Methylsilicone gum, P-192a
Methylsilicone rubber, Heat vulcanised, M-16
Methylsilicone rubber, Room temp. vulcanised, M-17
Methylsilicone rubber, Room temp. vulcanised, M-18
Methylsilicone rubber, Two part room temp. vulcanised, M-19
Methylsilicone, P-192
Methyltrifluoropropyl silicone rubber, Room temp. vulcanised, F-14
Milk protein, C-4
MO-PPV, P-180
Modacrylic fibres, M-20
Modacrylic fibres, P-575
Modified hydrogenated acrylonitrile-butadiene copolymer, P-90
Modified PPO, P-188
mPEK, P-235
MPS-PPV, P-360
MXD-6, 50% Glass fibre filled, P-646
6/MXD//I/T copolymer, P-295
MXD//I/T copolymer, P-644
MXD/PXD/6(3)T//T, P-643
n-Butylhexyldichlorosilane homopolymer, P-116

N3, N-17
Natural balata, N-1
Natural cellulose, C-7
Natural rubber/PMMA blend, N-3
Natural rubber/PMMA/*graft*-natural rubber blend, N-3
Natural rubber/poly(methyl methacrylate) inter-penetrating network, N-3
Natural rubber, N-2
NBR, carbon black filled, P-89
NBR, mineral filled, P-92
NBR/PMMA/PVC blends, P-345
NBR/PVC blends, P-93
NBR, N-5
NEW-TPI, N-4
Nigraniline, P-72
Nitrile butadiene rubber, N-5
Nitrile rubber modifier, N-11
Nitrile rubber, carbon black filled, P-89
Nitrile rubber, Carboxylated, N-6
Nitrile rubber, Extrusion unfilled, N-7
Nitrile rubber, General purpose, N-8
Nitrile rubber, Hydrogenated, N-9
Nitrile rubber, Injection moulding, Unfilled, N-10
Nitrile rubber, mineral filled, P-92
Nitrile rubber/polyvinyl chloride blends, P-93
Nitrile rubber/PVC blends, P-93
Nitrile rubber, N-5
Nitrilemethyl silicone rubber, Room temp. vulcanised, N-12
Nitrocellulose, C-12
NNPheBz, P-205
para-Nonyl Phenolic resins, N-13
4-Nonylphenol based resins, N-13
Novolaks, P-19, P-20
Novoloid fibres, N-37
NR, carbon black filled, P-89
NR, mineral filled, P-92
NR/PVC blends, P-93
Nucleic acid, N-14
Nylon 10,10, N-37
Nylon 10,9, N-36
Nylon 10, N-35
Nylon 11, N-38
Nylon 12 copolymers with cycloaliphatic and aromatic comonomers, N-45
Nylon 12 cycloaliphatic/aromatic copolymers, N-45
Nylon 12-*block*-poly(tetramethylene glycol), N-56
Nylon 12, N-39
Nylon 13, N-40
Nylon 3, N-17
Nylon 4,6 (30% glass fibre reinforced), N-48

Nylon 4,6, N-19
Nylon 4, N-18
Nylon 5,6, N-21
Nylon 5, N-20
Nylon 6 T/Me(5)/PET copolymer, N-54
Nylon 6(3)T, N-61
Nylon 6, glass reinforced, N-49
Nylon 6,10, N-26
Nylon 6,12, N-27
Nylon 6,6 Aramid fibre filled, N-42
Nylon 6,6 Carbon fibre filled, N-44
Nylon 6,6 Mineral filled, N-51
Nylon 6,6 Stainless steel fibre filled, N-58
Nylon 6,6, Glass reinforced, N-50
Nylon 6,6, N-23
Nylon 6,8, N-24
Nylon 6,9, N-25
Nylon 6,I, N-16
Nylon 6,T, N-59
Nylon 6-12, glass and carbon reinforced, N-47
Nylon 6-EEA/EEA-MA blend, N-46
Nylon 6-Nylon 6T (Semi-Aromatic) copolymer, N-52
Nylon 6.12 copolyamide, P-127
Nylon 6/Ethylene ethacrylate copolymer ethylene ethacrylate-maleic anhydride blend, N-46
Nylon 6, N-22
Nylon 7,7, N-29
Nylon 7, N-28
Nylon 8.8, N-31
Nylon 8, N-30
Nylon 9,6, N-33
Nylon 9,9, N-34
Nylon 9, N-32
Nylon Acrylic Blends, N-41
Nylon Barrier blends, N-43
Nylon HT, P-446
Nylon MPD-I, P-446
Nylon MXD-6, N-645
Nylon polyester copolymer glass fibre filled, N-54
Nylon powder coatings, N-57
Nylon PPD-T, P-450
Nylon-6, T/nylon-6, I copolymer, N-294
Nylon-PC blends, N-53
Nylon-Polycarbonate blends, N-53
Nylon/ABS Alloy, A-11
Nylon/polyolefin blends, N-55
Nylon, N-15

O-OPPE, P-201
OC1C10-PPV, P-356
Octamethylcyclotetrasiloxane, C-23
para-Octyl Phenolic resins, O-1
4-Octylphenol based resins, O-1
p-tert-Octylphenol resins, O-1
'OH-containing' vinyl chloride-vinyl acetate copolymer, P-587
Olefin - sulfur dioxide colpolymers, O-2
Oligo[(methylsilylene)(methylene)], P-408
OMSM, P-408
One part room temp. vulcanised methyl silicone rubber, M-13
OPP, P-495
OPS, O-3
Orgalloy RS 6000, N-55
Oriented polystyrene sheet and film, O-3
Oxetane polymers, O-4
Oxirane polymers with methyl oxirane (block, triblock), P-55
Oxirane polymers with methyloxirane (block, triblock), M-4
Oxirane, (methoxymethyl), homopolymer, 9CI, P-358

P 3-HT, P-305
Pα-MS, P-411
P2VPS, P-529
P3BT, P-123
P3DDT, P-222
P3DT, P-168, P-222
P3HT:PCBM, P-306
P3M1P, P-392

P3MH, P-370
P3TA, P-551
P4MP, P-393
P61S, P-300
PA12-*block*-PTMG, N-56
PA12, N-39
PAA, P-62
PAB's, P-481
PACM-12, P-1
PAN, P-63
PANI, P-72
PAO, P-70
1,2-pb, P-56, P-56
PBA, P-545
PBCMO, O-4
PBES, P-115
PBHS, P-116
PBI, P-2
PBI, P-547
PBIP, P-547
PBMA, P-117
PBPS, P-76, P-122
PBT, P-103, P-443
PBT - PHE, P-109
PBT, Flame retardant, P-4
PBT, High impact, Impact modified, P-5
PBT, injection moulding carbon fibre filled, P-107
PBT, injection moulding glass fibre filled, P-6
PBT, injection-moulding mineral filled, P-108
PBT-PEE, P-104
PBT-poly(oxytetramethylene) copolymer, P-104
PBT-PTMEG, P-104
PBT/ASA blends, P-3
PBT/PET blends, P-110
PBTF, P-199
PBZT, P-443
PC/ABS alloy, A-13
PC/ABS blend, A-13
PC/ASA blend, P-130
PC/PBT blend, P-135
PC/PET blends, P-139
PC/PMMA blends, P-343
PCHDMT, P-159
PCHDT, P-159
PCI, B-4
PCL, P-128
PCT, B-13, P-159
PCTFE, P-154
PD10S, P-175
PD12S, P-221
PDB, P-310
PDBOPA, P-171
PDBS, P-172
PDCT, P-159
PdDecSi, P-175
PDDS, P-175
PDEO, O-4
PDES, P-176
PDGEBD, B-32
PDHepS, P-177
PDHepSi, P-177
PDHF, P-178
PDHS, P-179
PDMA, P-164, P-181
PDMB, P-391
PDMO, O-4
PDMS fluid, P-7
PDMS gum, P-192a
PDMS-polyether copolymer, P-8
PDMS, P-191, P-192
PDMSi, P-191
PDMSM, P-194
PDmTSM, P-82
PDMV, P-180
PdnHS, P-177
PDNS, P-195
PdnTDS, P-213
PDOctS, P-202
PDOctSi, P-202
PDOF, P-196
PDOS, P-202
PDPA, P-440
PDPhSM, P-210

PDPrS, P-211
PDPrSM, P-212
PDPS, P-204
PDPSM, P-212
PDpTSM, P-83
PDS, P-165
PDTS, P-213
Pe Ce (fibres), P-576
PE fibres, P-260
PE film, P-9
PE-HD, P-261
PE-LD, L-4
PE-LLD, L-5
PE-MD, M-1
PE-VLD, V-5
PE/PS blends, P-253
PE7, P-251
PE, P-247
PEA, P-244, P-250
Peanut protein fibre, G-2
PEB, P-268
PEBS, P-115
PECA, P-246
PEDOP, P-256
PEDOT-PSS, P-258
PEDOT/PSS, P-258
PEDOT:PSS, P-258
PEDOT, P-257
PEDP, P-256
PEDT-PSS, P-258
PEDT, P-257
PEEK, P-231
PEEK, 30% Carbon fibre filled, P-232
PEEK, 30% Glass fibre filled, P-233
PEEKK, P-234
PEG, P-267
PEK, P-235
PEKEKK, P-236
PEKK, P-237
PEKK, carbon-filled, P-238
PEKK, glass-filled, P-239
PEMA, P-275
PEN-2,6, P-265, P-266
PEN, P-265, P-266
PEO, P-267
PEOB, P-268
PEPS, P-277
Peptides, P-651
Per CV, P-576
Perchlorovinyl, P-576
Perethylcellulose, E-15
Perfluoropropene-PVDF copolymer, P-616
Permethylcellulose, M-14
Pernigraniline, P-72
PES, 20% Glass fibre reinforced, P-242
PES, 30% Carbon fibre reinforced, P-241
PES, 30% Glass fibre reinforced, P-243
PES, P-240, P-269
PET/PC blends, P-139
PET, P-271
PEtCA, P-246
PF Novolaks, P-19
PF resins, P-20, P-33
PF Resoles, P-21
PF2/6, P-79
PFA, T-2
PFA, Carbon filled, P-10
PFA, Glass filled, P-11
PFA, Mineral and mica filled, P-12
PFA, PTFE and carbon filled, P-13
PFA, PTFE filled, P-14
PFB, P-198
PFO, P-196
PGA, P-283
PGE, P-453
PGLA, P-337
PGLY, P-283
PH, P-297
PHA, P-308
PHAc, P-283
PHB, P-317
PHBA, P-310
PHBV, P-318

PHE, P-297
Phenol Aralkyl Resins, Glass Filled, P-16
Phenol Aralkyl resins, Mica filled, P-17
Phenol Aralkyl Resins, P-15
Phenol-α,α′-dimethoxy-p-xylene resins, Mica filled, P-17
Phenol-α,α′-dimethoxy-p-xylene resins, Glass filled, P-16
Phenol-α,α′-dimethoxy-p-xylene resins, P-15
Phenol-1,4-bis(methoxymethyl)benzene resins, Glass filled, P-16
Phenol-1,4-bis(methoxymethyl)benzene resins, Mica filled, P-17
Phenol-1,4-bis(methoxymethyl)benzene resins, P-15
Phenol-butanal resins, P-18
Phenol-Butyraldehyde Resins, P-18
Phenol-formaldehyde foam, P-24
Phenol-formaldehyde Novolaks, P-19
Phenol-formaldehyde resins, P-20
Phenol-formaldehyde resins, P-33
Phenol-formaldehyde Resoles, P-21
Phenolic dispersions, P-22
Phenolic foam, P-24
Phenolic laminating Resins, P-23
Phenolic microballoons, P-43
Phenolic microspheres, P-43
Phenolic resin foam, P-24
Phenolic resin silicone copolymer, S-12
Phenolic resin Spheres, P-43
Phenolic Resin, Insulation Wool Binders, P-25
Phenolic resin, Moulding powders, Mica filled, P-30
Phenolic resin, Moulding powders, Cellulose filled, P-26
Phenolic resin, Moulding powders, Cotton filled, P-27
Phenolic resin, Moulding powders, Glass fibre filled, P-28
Phenolic resin, Moulding powders, Graphite filled, P-29
Phenolic resin, Moulding powders, Mineral filled, P-31
Phenolic resin, Moulding powders, Woodflour filled, P-32
Phenolic resins high-ortho, P-40
Phenolic resins, Abrasive Materials, P-34
Phenolic resins, Carbon materials, P-35
Phenolic resins, Cast resin, P-36
Phenolic resins, Fibres, P-37
Phenolic resins, Friction materials, P-38
Phenolic resins, Glass-reinforced plastic, P-39
Phenolic Resins, Ion exchangers, P-41
Phenolic resins, Oil Recovery, P-42
Phenolic resins, Rubber adhesives, P-44
Phenolic resins, Textile fleece binders, P-45
Phenolic resins, P-33
Phenolic Resins, P-20
Phenolic Resoles, Coatings, P-46
Phenolic Resoles, Foundry applications, P-47
Phenolic Resoles, Wood applications, P-48
Phenolic Resoles, P-21
Phenoplasts, P-20, P-21
Phenyl based silicone elastomer, P-49
Phenyl methyl silicone rubber, Heat vulcanised, M-15
Phenyl methyl vinyl silicone rubber, Heat vulcanised, V-1
N-phenylbenzenamine homopolymer, P-440
p-Phenylenediamine homopolymer, P-169
Phenylmethylsilicone rubber, Room temp. vulcanised, M-18
Phenylpolysilane, P-455
Phenylsilicone rubber, Heat vulcanised, P-49
PHEP, P-287
PHEX, P-297
PHFP, P-290
PHHS, P-288
PHMS, P-300
PHMT, P-296
PHOS, P-302
PHT, P-296
Phthalocyaninatopoly(siloxane), P-433

PIBVE, P-323, P-323
PIPD, P-50
PIPVE, P-335
PLA fibre, P-64
PLA-PGA, P-337
PLA, P-336
Plasticised poly(chloroethene), P-580
Plasticised poly(vinyl chloride), P-580
Plasticised PVC, P-580
PMA, P-349, P-361
PMAAm, P-338
PMAM, P-338
PMAN, P-353
PMAS, P-355
PMBSM, P-121
PMCN, P-353
PMDA-3,3′-BAPB, N-4
PMDA-4-BDAF, P-51
PMDA-ODA, P-52
pMeBuSi, P-363
PMHE, P-371
PMHSM, P-301
PMI, P-352
PMIPK, P-375
PMMA/ABS blends, P-342
PMMA/NBR/PVC blends, P-345
PMMA/PC blends, P-343
PMMA/PEO blends, P-344
PMMA/polybutadiene interpenetrating networks, P-342
PMMA/PVC blends, P-345
PMMA/PVDF blends, P-348
PMMA/PVF2 blends, P-348
PMMA, P-381
PMOS, P-389
PMPD, P-391
PMPSM, P-403
PMR-15, Unfilled, P-54
PMR-II (Second generation) Resins, P-53
PMR-II, P-53
PMSM, P-408
PMVAc, P-413
PMVK, P-417
PNIPAM, P-330
PNMP, P-407
PNVC, P-574
PNVK, P-574
PNVP, P-638
PO polyols, P-501
PO-AGE copolymer, P-502
PO-EO copolymers, P-504
PO-EO polyols, P-504
POD, P-427
PODZ, P-427
POHS, P-302
Poliglusam, C-15
Poloxamers, P-55
Polyacetal, P-432
Polyacetal copolymer, A-24
Poly(acetaldehyde), P-57
Poly[(1→4)-β-2-acetamido-2-deoxy-D-glucose], C-15
Poly(acetyl ethylene), P-417
Poly(acetylene), P-58
Poly(acetylenes), P-59
Polyacrylamide gels, P-61
Polyacrylamide, P-60
Poly(acrylic acid), P-62
Polyacrylonitrile, P-63
Poly(acrylonitrile-co-butadiene), N-5
Poly(acrylonitrile-co-butadiene-co-isoprene), A-28
Poly(acrylonitrile-co-butadiene-co-styrene), A-17
Poly(acrylonitrile-co-methyl acrylate), A-29
Poly(adamantyl methacrylate), P-347
Poly(L-alanine) fibre, P-64
Poly(allyl cyanoacrylate), P-68
Poly(alkyl 2-cyanoacrylates), P-65
Poly(alkyl 2-methyl-2-propenoate), L-6
Poly(alkyl ether methacrylate), P-67
Poly(alkyl methacrylate-co-ethene), M-6
Poly(alkyl methacrylate-co-ethylene), M-6
Poly(alkyl vinyl ethers), P-600
Polyalkylene dicarboxylates, general, P-66

Name and Synonym Index

Poly(alkylene sulfone), O-2
Poly(1-allyloxy-2,3-epoxypropane), P-69
Poly(alpha olefin), P-70
Polyalphaolefin oligomer, P-70
Poly[(1→4)-2-amino-2-deoxy-β-D-glucose], C-16
Poly(p-aminoaniline), P-169
Poly(4-aminobutyric acid), N-18
Poly(10-aminocapric acid), N-35
Poly(8-aminocaprylic acid), N-30
Poly(10-aminodecanoic acid), N-35
Poly(4-aminodiphenylamine), P-454
Poly(12-aminododecanoic acid), N-39
Poly(7-aminoheptanoic acid), N-28
Poly(6-aminohexanoic acid], N-22
Poly(9-aminononanoic acid), N-32
Poly(8-aminooctanoic acid), N-30
Poly(9-aminopelargonic acid), N-32
Poly(5-aminopentanoic acid), N-20
Poly(1-(4-aminophenyl)-2,3-dihydro-1,3,3-trimethyl-1H-inden-5-amine-alt-5,5′-carbonylbis[1,3-isobenzofurandione], B-23
Poly(3-(4-aminophenyl)-2,3-dihydro-1,1,3-trimethyl-1H-inden-5-amine-alt-5,5′-carbonylbis(1,3-isobenzofurandione), B-23
Poly(13-aminotridecanoic acid), N-40
Poly(11-aminoundecanamide), N-38
Poly(ω-aminoundecanoic acid), N-38
Poly(5-aminovaleric acid), N-20
Poly(amyl methacrylate), P-437
Polyamylose, S-24
Poly(anhydrides), P-71
Polyaniline (emeraldine salt), P-169
Poly(aniline), P-72
Poly(aryl methacrylates), P-74
Poly(aryl 2-methyl-2-propenoate), P-74
Poly(arylalkyl methacrylates), P-73
Poly(arylalkyl 2-methyl-2-propenoate), P-73
Polyaryletheretherketone, P-231
Polyaryletheretherketone, 30% Carbon fibre filled, P-232
Polyaryletheretherketone, 30% Glass fibre filled, P-233
Polyaryletheretherketoneketone, P-234
Polyaryletherketoneketone, P-237
Poly(azacyclotridecan-2-one-co-4,4′-methylenebis(2-methylcyclohexanamine-co-1,3-benzenedicarboxylic acid)), N-45
Poly(azacycloundecan-2-one), N-35
Poly(BEMO), O-4
Poly(benzenamine), P-72
Poly(1,3-benzenediamine-co-5,5′-carbonylbis[1,3-isobenzofurandione]-co-4,4′-oxybisbenzenamine), B-28
Poly(1,4-benzenediamine-alt-[5,5′-biisobenzofuran]-1,1′,3,3′-tetrone), B-20
Poly(1,3-benzenedicarboxylic acid-co-1,4-benzenedicarboxylic acid-co-1,6-hexanediamine), A-34
Poly(1,3-benzenedicarboxylic acid-co-1,1′-biphenyl-4,4′-diol-co-4-hydroxybenzoic acid), P-312
Poly(1,4-benzenedicarboxylic acid-co-1,1′-biphenyl-4,4′-diol-co-4-hydroxybenzoic acid), P-313
Poly(1,4-benzenedicarboxylic acid-co-1,4-butanediol), P-103
Poly(1,4-benzenedicarboxylic acid-co-1,4-cyclohexane dimethanol), P-162
Poly(1H,3H-benzo[1,2-c:4,5-c′]difuran-1,3,5,7-tetrone-alt-3,3′-bis(4-aminophenoxy) biphenyldiamine), N-4
Poly(1H,3H-benzo[1,2-c:4,5-c′]difuran-1,3,5,7-tetrone-alt-4,4′-oxybisbenzenamine], P-52
Poly(benzo[1,2-d:4,5-d′]bisthiazole-2,6-diyl-1,4-phenylene), P-443
Poly[1H,3H-benzo[1,2-c:4,5-c′]difuran-1,3,5,7-tetrone-alt-4,4′-[(1-methylethylidene)bis[(2,6-dichloro-4,1-phenylene)oxy]]bis[benzenamine]], A-46
Poly[1H,3H-benzo[1,2-c:4,5-c′]difuran-1,3,5,7-tetrone-alt-4,4′-[[2,2,2-trifluoro-1-(trifluoromethyl)ethylidene]bis(4,1-phenyleneoxy)bis[benzenamine]], P-51

Poly(2,5-benzophenone), P-75
Poly(3,3′,4,4′-benzophenonetetracarboxylic dianhydride-alt-bis(4-aminophenyl)methane), B-24
Poly(3,3′,4,4′-benzophenonetetracarboxylic dianhydride-alt-m-phenylenediamine), B-27
Poly(3,3′,4,4′-benzophenonetetracarboxylic dianhydride-alt-2,2-bis[4-(4-aminophenoxy)phenyl]hexafluoropropane), B-22
Poly(3,3′,4,4′-benzophenonetetracarboxylic dianhydride-alt-4,4′-methylenedianiline), B-24
Poly(3,3′,4,4′-benzophenonetetracarboxylic dianhydride-alt-bis(4-aminophenyl) ether), B-26
Poly(3,3′,4,4′-benzophenonetetracarboxylic dianhydride-co-m-phenylene-4,4′-oxydianiline), B-28
Poly(3,3′,4,4′-benzophenonetetracarboxylic dianhydride-co-2-methyl-1,3-phenylenediamine-co-4,4′-methylenedianiline), B-25
Poly(3,3′,4,4′-benzophenonetetracarboxylic dianhydride-alt-4,4′-oxydianiline), B-26
Poly(3,3′,4,4′-benzophenonetetracarboxylic dianhydride-1,3-bis(4-aminophenoxy-4′-benzoyl)benzene, L-1
Poly(3,3′,4,4′-benzophenonetetracarboxylic dianhydride-alt-diaminophenylindane), B-23
Poly[2,1,3-benzothiadiazole-4,7-diyl(9,9-dioctyl-9H-fluorene-2,7-diyl)], P-197
Poly(benzoyl-1,4-phenylene), P-75
Poly(benzoyl-p-phenylene), P-75
Poly(benzyl methacrylate), P-73
Poly(benzoyloxyethylene), P-571
Poly([5,5′-bi-1H-benzimidazole]-2,2′-diyl-1,3-phenylene), P-2
Poly(bicyclo[2.2.1]hept-2-yl methacrylate), P-347
Poly([5,5′-biisobenzofuran]-1,1′,3,3′-tetrone-alt-1,4-benzenediamine), B-20
Poly([5,5′-biisobenzofuran]-1,1′,3,3′-tetrone-alt-4,4′-oxybisbenzenamine), B-19
Poly(biphenyl methacrylate), P-74
Poly(1-(4-[1,1′-biphenyl])-1,2-ethanediyl), P-458
Poly(3,3′,4,4′-biphenyltetracarboxylic dianhydride-alt-bis(4-aminophenyl)ether), B-19
Poly(3,3′,4,4′-biphenyltetracarboxylic dianhydride-alt-4,4′-oxydianiline), B-19
Poly(3,3′,4,4′-biphenyltetracarboxylic dianhydride-alt-p-phenylenediamine), B-19
Poly[9,9-bis(2-ethylhexyl)-9H-fluorene-2,7-diyl], P-79
Poly[(2,5-bis(2-ethylhexyloxy)-1,4-phenylene ethynylene)-co-(2,5-dioctyloxy-1,4-phenylene ethynylene)], P-80
Poly[2,5-bis(2′-ethylhexyloxy)-1,4-phenylene vinylene], P-81
Poly(2,2′-bis(3,4-dicarboxyphenyl)hexafluoropropane dianhydride-alt-1,3-bis(3-aminophenoxy)benzene) Addition cured (acetylenic end-caps), F-7
Poly(2,2′-bis(3,4-dicarboxyphenyl)hexafluoropropane dianhydride-alt-2,3,5,6-tetramethyl-1,4-benzenediamine), F-2
Poly(2,2′-bis(3,4-dicarboxyphenyl)hexafluoropropane dianhydride-alt-2,2-bis[4-(4-aminophenoxy)phenyl]hexafluoropropane), F-1
Poly(2,2′-bis(3,4-dicarboxyphenyl)hexafluoropropane dianhydride bis(4-aminophenyl)ether), F-5
Poly(2,2′-bis(3,4-dicarboxyphenyl)hexafluoropropane dianhydride-4,4′-oxydianiline), F-5
Poly(2,5-bis(3-sulfopropoxy)-1,4-phenylene ethynylene)-(1,4-phenylene ethynylene) disodium salt], P-85
Poly(bis(4-aminocyclohexyl)methane 1,10-decanedicarboxamide), P-1
Poly(bis(4-aminocyclohexyl)methane-co-dodecanoic acid), P-1
Poly(bis(4-butylphenyl)silylene), P-76
Poly[bis(4-butylphenyl)silane), P-76
Poly[2,2-bis(chloromethyl)trimethylene-3-oxide], P-77
Poly(bis(chloromethyl)oxetane), O-4
Poly[3,3-bis(chloromethyl)oxetane], P-77

Poly[3,3-bis(chloromethyl)oxacyclobutane], P-77
Poly(1,2-bis(2,3-epoxyprop-1-oxy)benzene), P-78
Poly[3,3-bis(exothymethyl)oxetane], O-4
Poly[2,5-bis(octyloxy)-1,4-phenylene vinylene], P-84
Poly[2,5-bis[2-[2-(2-methoxyethoxy)ethoxy]ethoxy][1,1′-biphenyl]-4,4′-diyl], P-648
Poly(bisphenol F bis(oxiranylmethyl) ether), B-15
Poly[[2,2′-bithiophene]-5,5′-diyl(9,9-dioctyl-9H-fluorene-2,7-diyl)], P-199
Polyblend, P-95
Poly(bornyl methacrylate), P-347
Poly(1-bromo-4-ethenylbenzene), P-86
Poly(bromoethene), P-572
Poly(2-bromoethyl methacrylate), P-285
Poly(1-(4-bromophenyl)-1,2-ethanediyl), P-86
Poly(p-bromostyrene), P-86
Poly(4-bromostyrene), P-86
Poly(1,3-butadiene-co-acrylonitrile-co-ethylene), P-90
Poly(1,3-butadiene-co-1,1-dichloroethene-co-ethenylbenzene), B-31
Poly(1,3-butadiene-co-1,3-pentadiene), B-30
Poly(1,3-butadiene-2-methyl), N-2
Poly(1,3-butadiene-co-acrylonitrile-co-2-methyl-1,3-butadiene), P-91
Poly(1,3-butadiene-co-acrylonitrile), carbon black filled, P-89
Poly(1,3-butadiene-co-acrylonitrile), mineral filled, P-92
Poly(1,3-butadiene-co-acrylonitrile)/polyvinyl chloride blends, P-93
Poly(1,3-butadiene-co-acrylonitrile), carboxylated, N-6
Poly(1,3-butadiene-co-acrylonitrile), hydrogenated, N-9
Poly(1,3-butadiene-co-acrylonitrile-co-isoprene), P-91
Poly(1,3-butadiene-co-ethenylbenzene-co-ethenylpyridine), P-522
Poly(1,3-butadiene-co-propenenitrile-co-2-methyl-1,3-butadiene), P-91
Poly(butadiene)/poly(methyl methacrylate) blends, P-342
Polybutadiene, Carbon black filled, P-94
Polybutadiene/styrene-butadiene block copolymer blends, P-95
Polybutadiene/PMMA blends, P-342
Polybutadiene, P-87
1,2-Poly(1,3-butadiene), P-56
1,2-Polybutadiene, P-56
cis-1,4-Polybutadiene, P-88
cis-Polybutadiene, P-88
Poly(1,3-butadiene), P-87
Poly(1,3-butadiene-co-propenenitrile-co-ethene), P-90
Poly(butadiene-co-acrylonitrile), B-29, N-5
Poly(1,1-(1,3-butadiyne-1,4-diyl)bisbenzene), P-206
Poly[1,1-butane bis(4-phenyl)carbonate], P-96
Poly[2,2-butane bis(4-phenyl)carbonate], P-97
Poly(2,2′-[1,4-butanediylbis(oxymethylene)] bisoxirane), B-32
Poly(3-buten-2-one), P-417
Poly(1-butene-co-1-propene), P-484
Poly((trans-2-butene oxide), P-98
Poly((Z)-2-butenedioic acid-co-methoxyethene), P-623
Poly(1-butenylene-graft-1-phenylethylene-co-cyanoethylene), A-17
Poly(1-butenylene-co-1-cyanoethylene), N-5
Polybutilate, P-545
Poly(sec-butoxy ethene), P-125
Poly(3-butoxy-1,2-epoxypropane), P-99
Poly(1-butoxy-2,3-epoxypropane), B-33, P-99
Poly[1-(2-butoxycarbonyl)ethylene], P-100
Poly[1-(butoxycarbonyl)-1-methylethylene], P-117
Poly(tert-butoxyethene), P-126
Poly(butoxyethene), P-124
Poly(2-butoxyethyl methacrylate), P-67
Poly((butoxymethyl) oxirane), B-33
Poly[(butoxymethyl)oxirane], P-99

Poly(3-butoxypropylene oxide), P-99
Poly(sec-butyl 2-methyl-2-propenoate), P-118
Poly(butyl 2-methyl-2-propenoate), P-117
Poly(tert-butyl 2-methyl-2-propenoate), P-119
Poly(butyl 2-propenoate), P-100
Poly(butyl acrylate), P-100
Poly(butyl methacrylate)/nitrile rubber/poly(vinyl chloride), P-345
Poly(butyl methacrylate-co-vinyl chloride), P-120
Poly(tert-butyl methacrylate), P-119
Poly(sec-butyl methacrylate), P-118
Poly(butyl methacrylate)/poly(vinyl chloride-co-vinyl acetate), P-345
Poly(butyl methacrylate-co-methacrylamide), M-5
Poly(butyl methacrylate-co-vinylidene chloride), M-6
Poly(butyl methacrylate-co-isobutyl methacrylate), A-27
Poly(n-butyl methacrylate), P-117
Poly(butyl methacrylate-co-styrene), S-33
Poly(sec-butyl vinyl) ether, P-125
Poly(tert-butyl vinyl) ether, P-126
Poly(butyl vinyl ether), P-124
Poly(tert-butyl vinyl ether), P-126
Poly(3-butyl-2,5-thiophenediyl), P-123
Poly(1-butyl-1,2-ethanediyl), P-297
Poly(tert-butylaminoethyl methacrylate), P-340
Polybutylene terephthalate-polyethylene terephthalate blends, P-110
Polybutylene terephthalate, P-103
Poly(1,4-butylene adipate), P-545
Poly(butylene adipate), P-545
Poly(butylene glycol adipate), P-545
Poly(butylene glycol diglycidyl ether), B-32
Poly(butylene isophthalate), P-547
Polybutylene succinate adipate, P-102
Polybutylene succinate, P-101
Polybutylene terephthalate, Films/Sheets, P-106
Polybutylene terephthalate, injection moulding glass fibre filled, P-6
Polybutylene terephthalate, Structural foam, P-113
Poly(butylene terephthalate-co-tetrahydrofuran), P-104
Polybutylene terephthalate, Silicone lubricated, P-112
Polybutylene terephthalate, injection moulding mineral filled, P-108
Polybutylene terephthalate, phenoxy blends, P-109
Polybutylene terephthalate, Uv stabilised, P-114
Poly(butylene terephthalate-co-butylene oxide), P-104
Poly(butylene terephthalate/acrylonitrile-styrene-acrylic) blends, P-3
Polybutylene terephthalate poly(hydroxy ether) of bisphenol A blends, P-109
Poly(butylene terephthalate-co-butylene ether glycol terephthalate), P-104
Polybutylene terephthalate, PTFE lubricated, P-111
Polybutylene terephthalate, fibres, P-105
Polybutylene terephthalate, injection moulding carbon fibre filled, P-107
Poly(butylethylsilane), P-115
Poly(1-butylethylene), P-297
Poly(butylethylsilylene), P-115
Poly(butylglycidyl ether), P-99
Poly(butylhexylsilane), P-116
Poly(butylhexylsilylene), P-116
Poly(butylmethylsilylene), P-363
Poly[(butylmethylsilylene)methylene], P-121
Poly[(butylmethylsilane)methylene], P-121
Poly(butyloxirane), P-298
Poly(butylpentylsilane), P-122
Poly(butylpentylsilylene), P-122
Poly(tert-butylphenyl methacrylate), P-74
Poly(3-butylthiophene), P-123
Poly(γ-butyrolactam), N-18
Poly(caproamide-co-terephthalic acid-co-isophthalic acid), P-294
Polycaproamide, N-22
Poly(ε-caprolactam), N-22

Poly(ε-caprolactam-co-hexamethylene diamine-co-terephthalic acid), N-52
Poly(caprolactam-co-laurolactam), P-127
Poly(ε-caprolactone), P-128
Poly(caprolactone), P-128
Poly(caprylene), P-421
Poly(capryllactam), N-30
Poly(2-(9H-carbazol-9-yl)ethyl methacrylate), P-346
Poly(2-(9H-carbazol-9-yl)hexyl methacrylate), P-346
Poly(2-(9H-carbazol-9-yl)pentyl methacrylate), P-346
Poly(2-(9H-carbazol-9-yl)propyl methacrylate), P-346
Poly(2-(9H-carbazol-9-yl)undecyl methacrylate), P-346
Polycarbazole, P-129
Poly[1-(N-carbazolyl)ethylene], P-574
Poly(4-carbomethoxyphenyl methacrylate), P-74
Polycarbomethylsilane, P-408
Polycarbonate/ABS alloy, A-13
Polycarbonate foam, Glass fibre reinforced, P-132
Polycarbonate silicone copolymer, P-469
Polycarbonate film, B-5
Polycarbonate fast cycling resins, B-5
Polycarbonate, Flame retardant, B-5
Polycarbonate, Glass fibre and carbon fibre reinforced, P-133
Polycarbonate, Glass fibre reinforced, PTFE lubricated, P-134
Polycarbonate, Structural foam, B-11
Polycarbonate, Uv stabilised, B-5
Polycarbonate-block-poly(methylmethacrylate), P-472
Polycarbonate-block-poly(dimethylsiloxane), P-469
Polycarbonate-block-poly(ethylene oxide), P-470
Polycarbonate-poly(methylmethacrylate) block copolymer, P-472
Polycarbonate-polysulfone block copolymer, P-471
Polycarbonate/ABS blend, A-13
Polycarbonate/polybutylene terephthalate alloy, Glass reinforced, P-136
Polycarbonate/polyurethane alloy, P-141
Polycarbonate Z, P-157
Polycarbonate, carbon fibre reinforced, PTFE lubricated, P-131
Polycarbonate-block-poly(tetramethyl Bisphenol A carbonate), P-473
Polycarbonate-block-polysulfone, P-471
Polycarbonate/acrylic alloy, P-343
Polycarbonate/acrylic blend, P-343
Polycarbonate/acrylic-styrene-acrylonitrile alloy, P-130
Polycarbonate/ASA alloy, P-130
Polycarbonate/polybutylene terephthalate alloy, P-135
Polycarbonate/polybutylene terephthalate blend, P-135
Polycarbonate/polyester alloys, P-137
Polycarbonate/polyester blend, P-137
Polycarbonate/polyethylene alloy, P-138
Polycarbonate/polyethylene blend, P-138
Polycarbonate/polyethylene terephthalate alloy, P-139
Polycarbonate/polyethylene terephthalate alloy, Glass fibre reinforced, P-140
Polycarbonate/polyethylene terephthalate blend, P-139
Polycarbonate/polyurethane blend, P-141
Polycarbonate/SMA alloy, P-142
Polycarbonate/styrene-maleic anhydride alloy, P-142
PolycarbonatePolycarbonate-poly(dimethylsiloxane) block copolymer, P-469
Poly[4,4′-carbonylbis-1,2-benzenedicarboxylic acid-co-1,3-diisocyanatomethylbenzene-co-1,1′-methylenebis(4-isocyanatobenzene)], B-25
Poly(5,5′-carbonylbis[1,3-isobenzofurandione]-alt-1,3-phenylenebis[[4-(4-aminophenoxy)phenyl]methanone]), L-1
Poly(5,5′-carbonylbis[1,3-isobenzofurandione]-alt-bis(3-aminophenylmethanone), L-2

Poly(5,5′-carbonylbis[1,3-isobenzofurandione]-co-1,3-benzenediamine-co-4,4′-oxybisbenzenamine), B-28
Poly[5,5′-carbonylbis(1,3-isobenzofurandione)-alt-1,3-benzenediamine], B-27
Poly[5,5′-carbonylbis(1,3-isobenzofurandione)-alt-4,4′-methylene(bisbenzenamine)], B-24
Poly(5,5′-carbonylbis[1,3-isobenzofurandione]-co-1,3-diisocyanatomethylbenzene-co-1,1′-methylenebis(4-isocyanatobenzene)), B-25
Poly[5,5′-carbonylbis-1,3-isobenzofurandione-alt-4,4′-oxybis(benzenamine)], B-26
Polycarbosilanes, P-143
Poly[1-(carboxy)ethylene], P-62
Poly(5-carboxyindole), P-144
Poly(cetyl methacrylate), L-6
Polychlal fibres, P-570
Poly(1-chloro-2-ethenylbenzene), P-151
Poly(1-chloro-3-ethenylbenzene), P-152
Poly(1-chloro-3-vinylbenzene), P-152
Poly(1-chloro-4-vinylbenzene), P-153
Poly(2-chloro-1,3-butadiene), Vulcanised, P-150
Poly(2-chloro-1,3-butadiene), P-149
Poly[(chloroacetoxy)ethylene], P-593
Poly(chloroethene), P-583
Poly(chloroethene-co-ethene), P-579
Poly(chloroethene-co-ethenyl acetate), P-584
Poly(chloroethene-co-(ethenyl acetate)-co-((Z)-2-butenedioic acid)), P-589
Poly(chloroethene-co-1-propene), P-581
Poly(chloroethene-co-2-propenenitrile), P-575
Poly(chloroethene-co-bromoethene), P-591
Poly(chloroethene-co-chlorotrifluoroethene), P-582
Poly(chloroethene-co-ethene-co-ethenyl acetate), P-663
Poly(chloroethene-co-ethenyl acetate-co-(β-hydroxypropyl-2-propenoate)), P-588
Poly(chloroethene-co-ethenyl acetate-co-(2-hydroxypropyl-2-propenoate)), P-588
Poly(chloroethene-co-ethenyl acetate-co-ethenol), P-590
Poly[(2-chloroethoxy)ethene-co-1-ethenyl-2-pyrrolidinone], P-148
Poly[(2-chloroethoxy)ethene-co-2,2′-(2,5-cyclohexadiene-1,4-diylidene)bispropanedinitrile], P-147
Poly(2-chloroethoxyethene), P-145
Poly(2-chloroethyl methacrylate), P-285
Poly(1-chloroethyl methacrylate), P-285
Poly(2-chloroethyl vinyl ether-co-N-vinylpyrrolidinone), P-148
Poly(2-chloroethyl vinyl ether), P-145
Poly(2-chloroethyl vinyl ether-co-isobutylene-co-butyl acrylate), P-146
Poly(2-chloroethyl vinyl ether-co-tetracyanoquinodimethane), P-147
Poly(1-chloroethylene), P-583
Poly(chlorohexafluoroisopropyl methacrylate), P-285
Poly(1-(2-chlorophenyl)-1,2-ethanediyl), P-151
Poly(1-(4-chlorophenyl)-1,2-ethanediyl), P-153
Poly(2-chlorophenyl methacrylate), P-74
Poly(4-chlorophenyl methacrylate), P-74
Poly(1-(3-chlorophenyl)-1,2-ethanediyl), P-152
Poly(1-o-chlorophenylethyl methacrylate), P-73
Polychloroprene, Vulcanised, P-150
Polychloroprene, P-149
Polychloroprene, Unvulcanised, P-149
Poly(m-chlorostyrene), P-152
Poly(o-chlorostyrene), P-151
Poly(p-chlorostyrene, P-153
Poly(2-chlorostyrene), P-151
Poly(3-chlorostyrene), P-152
Poly(4-chlorostyrene), P-153
Poly(chlorotrifluoroethene-co-ethylene), E-4
Polychlorotrifluoroethylene, P-154
Polychlorotrifluoroethene, P-154
Poly(2-cyanoethyl methacrylate), P-285
Poly(4-cyanomethylphenyl methacrylate), P-74
Poly(4-cyanophenyl methacrylate), P-74
Poly(cycloaliphatic methacrylates), P-155
Poly(1-cyclobutyl-1,2-ethanediyl), P-594
Poly(1-cycloheptyl-1,2-ethanediyl), P-595

Name and Synonym Index

Poly[[1,1-cyclohexane bis(4-phenyl)carbonate]-co-[2,2-propane bis(4-phenyl)carbonate]], P-158
Poly[1,1-cyclohexane bis[4-(2,6-dichlorophenyl)]carbonate], P-156
Poly(1,4-cyclohexylene dimethylene terephthalate), P-159
Poly[1,1-cyclohexane bis(4-phenyl)carbonate], P-157
Poly(cyclohexyl dimethylene terephthalate), P-159
Poly(1-cyclohexyl-1,2-ethanediyl), P-596
Poly(4-cyclohexyl-1-butene), P-160
Poly(cyclohexylene dimethylene terephthalate), P-159
Poly(cyclohexylenedimethylene terephthalate), P-162
Poly(1-(2-cyclohexylethyl)-1,2-ethanediyl), P-160
Poly(cyclohexylethylene), P-596
Poly[4,4'-(cyclohexylidene)bisphenol-co-4,4'-(1-methylethylidene)-co-diphenyl carbonate], P-158
Poly(1,4-cyclohexylidene dimethylterephthalate), P-162
Poly(cyclohexylphenyl methacrylate), P-74
Poly[1,1-cyclopentane bis(4-phenyl)carbonate], P-163
Poly(1-cyclopentyl-1,2-ethanediyl), P-597
Poly(1-cyclopropyl-1,2-ethanediyl), P-598
Poly(decamethylene azelamide), N-36
Poly(decamethylene adipate), P-164
Poly(decamethylene diamine-co-azelaic acid), N-36
Poly(decamethylene diamine-co-sebacic acid), N-37
Poly(decamethylene sebacamide), N-37
Poly(decamethylene sebacate), P-165
Poly(decamethylene terephthalate), P-166
Poly(1,10-decanediol-co-1,4-benzenedicarboxylic acid), P-166
Poly(1,10-decanediol-co-hexanedioic acid), P-164
Poly(1-decene) oligomeric, P-70
Poly(1-decene), P-167
Poly(1-decyl-1,2-ethanediyl), P-218
Poly(3-decyl-2,5-thiophenediyl), P-168
Poly(1-decylethylene), P-218
Poly(3-decylthiophene), P-168
Poly(di-n-decylsilane), P-175
Poly(1,4-diaminobenzene), P-169
Poly(p-diaminophenylene), P-169
Poly(2,6-dibromophenylene oxide), P-170
Poly(2,3-dibromopropyl methacrylate), P-285
Poly(2,5-dibutoxy-1,4-phenylene ethynylene), P-171
Poly[(2,5-dibutoxy-1,4-phenylene)-1,2-ethynediyl], P-171
Poly(dibutylsilane), P-172
Poly(dibutylsilylene), P-172
Poly[(1,1-dichloroethene)-co-(1,4-butadiene)-co-ethenylbenzene], B-31
Poly(1,1-dichloroethene), P-610
Poly(1,1-dichloroethene-co-1-chloroethene), P-592
Poly(1,1-dichloroethene-co-2-propenenitrile), P-611
Poly[[1,1-dichloroethylene bis(4-phenyl)carbonate]-co-[2,2-propane bis(4-phenyl)carbonate], P-174
Poly[1,1-dichloroethylene bis(4-phenyl) carbonate], P-173
Poly(1,1-dichloroethylene), P-610
Poly(dicyclopentadiene-ethylene-propene) rubber, E-17
Poly(didecylsilane), P-175
Poly(didecylsilylene), P-175
Poly(1,2:5,6-diepoxyhexahydro-4,7-methanoindan), D-3
Poly(diethylaminoethyl methacrylate), P-340
Poly(diethylsilane), P-176
Poly(diethylsilylene), P-176
Poly[(difluoroacetoxy)ethylene], P-599
Poly(1,1-difluoroethene), P-612
Poly(1,1-difluoroethene-co-1,1,2,3,3,3-hexafluoro-1-propene), P-616
Poly(1,1-difluoroethene-co-chlorotrifluoroethene), P-613
Poly(1,1-difluoroethene-co-tetrafluoroethene), P-544

Poly(1,1-difluoroethene-co-(3,3,3-trifluoro-2-(trifluoromethyl)-1-propene)), P-614
Poly(diheptylsilane), P-177
Poly(diheptylsilylene), P-177
Poly(9,9-dihexyl-9H-fluorene-2,7-diyl), P-178
Poly(9,9-dihexylfluorene), P-178
Poly(dihexylsilane), P-179
Poly(dihexylsilylene), P-179
Poly[(1,3-dihydro-1,3-dioxo-2H-isoindole-2,5-diyl)[2,2,2-trifluoro-1-(trifluoromethyl)ethylidene](1,3-dihydro-1,3-dioxo-2H-isoindole-5,2-diyl)(2,3,5,6-tetramethyl-1,4-phenylene)], F-2
Poly[(1,3-dihydro-1,3-dioxo-2H-isoindole-2,5-diyl)carbonyl(1,3-dihydro-1,3-dioxo-2H-isoindole-5,2-diyl)-1,4-phenyleneoxy-1,4-phenylene[2,2,2-trifluoro-1-(trifluoromethyl)ethylidene]-1,4-phenyleneoxy-1,4-phenylene], B-22
Poly[(1,3-dihydro-1,3-dioxo-2H-isoindole-2,5-diyl)-1,3-phenylene(1,3-dihydro-1,3-dioxo-2H-isoindole-2,5-diyl)oxy-1,4-phenylene(1-methylethylidene)-1,4-phenyleneoxy], U-1
Poly[(1,3-dihydro-1,3-dioxo-2H-isoindole-2,5-diyl)carbonyl(1,3-dihydro-1,3-dioxo-2H-isoindole-5,2-diyl)-1,4-phenyleneoxy-1,4-phenylene], B-26
Poly[(1,3-dihydro-1,3-dioxo-2H-isoindole-2,5-diyl)[2,2,2-trifluoro-1-(trifluoromethyl)ethylidene](1,3-dihydro-1,3-dioxo-2H-isoindole-5,2-diyl)-1,4-phenyleneoxy-1,4-phenylene[2,2,2-trifluoro-1-(trifluoromethyl)ethylidene]-1,4-phenyleneoxy-1,4-phenylene], F-1
Poly[(1,3-dihydro-1,3-dioxo-2H-isoindole-2,5-diyl)-1,3-phenylene(1,3-dihydro-1,3-dioxo-2H-isoindole-2,5-diyl)carbonyl-1,3-phenylenecarbonyl], L-3
Poly[(1,3-dihydro-1,3-dioxo-2H-isoindole-2,5-diyl)carbonyl(1,3-dihydro-1,3-dioxo-2H-isoindole-5,2-diyl)-1,3-phenylenecarbonyl-1,3-phenylene], L-2
Poly[(1,3-dihydro-1,3-dioxo-2H-isoindole-2,5-diyl)carbonyl(1,3-dihydro-1,3-dioxo-2H-isoindole-5,2-diyl)-1,3-phenylene], B-27
Poly[(1,3-dihydro-1,3-dioxo-2H-isoindole-2,5-diyl)carbonyl(1,3-dihydro-1,3-dioxo-2H-isoindole-5,2-diyl)-1,4-phenylenemethylene-1,4-phenylene], B-24
Poly[(1,3-dihydro-1,3-dioxo-2H-isoindole-2,5-diyl)carbonyl(1,3-dihydro-1,3-dioxo-2H-isoindole-5,2-diyl)-1,4-phenyleneoxy-1,4-phenylenecarbonyl-1,3-phenylenecarbonyl-1,4-phenyleneoxy-1,4-phenylene], L-1
Poly[(1,3-dihydro-1,3-dioxo-2H-isoindole-2,5-diyl)[2,2,2-trifluoro-1-(trifluoromethyl)ethylidene](1,3-dihydro-1,3-dioxo-2H-isoindole-5,2-diyl)-1,4-phenyleneoxy-1,4-phenylene], F-5
Poly[(1,3-dihydro-1,3-dioxo-2H-isoindole-2,5-diyl)[2,2,2-trifluoro-1-(trifluoromethyl)ethylidene](1,3-dihydro-1,3-dioxo-2H-isoindole-5,2-diyl)phenylene], F-6
Poly[(1,3-dihydro-1,3-dioxo-2H-isoindole-2,5-diyl)[2,2,2-trifluoro-1-(trifluoromethyl)ethylidene](1,3-dihydro-1,3-dioxo-2H-isoindole-5,2-diyl)[2,2,2-trifluoro-1-(trifluoromethyl)ethylidene]-1,3-phenylene], F-3
Poly[(1,3-dihydro-1,3-dioxo-2H-isoindole-2,5-diyl)[2,2,2-trifluoro-1-(trifluoromethyl)ethylidene](1,3-dihydro-1,3-dioxo-2H-isoindole-5,2-diyl)[2,2,2-trifluoro-1-(trifluoromethyl)ethylidene]-1,4-phenylene], F-4
Poly[(1,3-dihydro-1,3-dioxo-2H-isoindole-5,2-diyl)-1,4-phenylene(1,3-dihydro-1,3-dioxo-2H-isoindole-2,5-diyl)oxy-1,4-phenylene(1-methylethylidene)-1,4-phenyleneoxy], U-2
Poly[(5,7-dihydro-1,3,5,7-tetraoxobenzo[1,2-c:4,5-c']dipyrrole-2,6(1H,3H)-diyl)-1,3-phenyleneoxy[1,1'-biphenyl]-4,4'-diyloxy-1,3-phenylene], N-4
Poly[(5,7-dihydro-1,3,5,7-tetraoxobenzo[1,2-c:4,5-c']dipyrrole-2,6(1H,3H)-diyl)-1,4-phenyleneoxy(2,6-dichloro-1,4-phenylene)(1-methylethylidene)(3,5-dichloro-1,4-phenylene)oxy-1,4-phenylene], A-46

Poly[(5,7-dihydro-1,3,5,7-tetraoxobenzo[1,2-c:4,5-c']dipyrrole-2,6(1H,3H)-diyl)-1,4-phenyleneoxy-1,4-phenylene], P-52
Poly[(5,7-dihydro-1,3,5,7-tetraoxobenzo[1,2-c:4,5c']dipyrrole-2,6(1H,3H)-diyl)-1,4-phenylene[2,2,2-trifluoro-1-(trifluoromethyl)ethylidene]-1,4-phenyleneoxy-1,4-phenylene], P-51
Poly[(1,4-dihydrodiimidazo[4,5-b:4',5'-e]pyridine-2,6(diyl)(2,5-dihydroxy-1,4-phenylene)], P-50
Poly(4,4'-dihydroxy-3,3',5,5'-tetrabromodiphenyl-2,2-propane-co-carbonic dichloride), P-464
Poly(4,4'-dihydroxy-3,3',5,5'-tetrachlorodiphenyl-2,2-propane-co-carbonic dichloride), P-466
Poly(4,4'-dihydroxy-3,3',5,5'-tetramethyldiphenyl-2,2-propane-co-carbonic dichloride), P-467
Poly(4,4'-dihydroxy-3,3'-dichlorodiphenyl-2,2-propane-co-carbonic dichloride), P-463
Poly(4,4'-dihydroxy-3,3'-dimethyldiphenyl-2,2-propane-co-carbonic dichloride), P-468
Poly(4,4'-dihydroxydiphenyl-1,1-cyclohexane-co-carbonic dichloride), P-157
Poly(4,4'-dihydroxydiphenyldiphenylmethane-co-carbonic dichloride), P-207
Poly(4,4'-dihydroxydiphenylsulfide-co-carbonic dichloride), P-548
Poly(4,4'-dihydroxydiphenyl-1,1-(3,3,5-trimethylcyclohexane)-co-carbonic dichloride), P-554
Poly(4,4'-dihydroxydiphenyl-1,1-butane-co-carbonic dichloride), P-96
Poly(4,4'-dihydroxydiphenyl-1,1-cyclopentane-co-carbonic dichloride), P-163
Poly(4,4'-dihydroxydiphenyl-1,1-dichloroethylene-co-carbonic dichloride), P-173
Poly(4,4'-dihydroxydiphenyl-1,1-ethane-co-carbonic dichloride), P-228
Poly(4,4'-dihydroxydiphenyl-1,1-isobutane-co-carbonic dichloride), P-405
Poly(4,4'-dihydroxydiphenyl-2,2-butane-co-carbonic dichloride), P-97
Poly(4,4'-dihydroxydiphenyl-2,2-pentane-co-carbonic dichloride), P-435
Poly(4,4'-dihydroxydiphenyl-2,2-propane-co-carbonic dichloride), B-5
Poly(4,4'-dihydroxydiphenyl-4,4-heptane-co-carbonic dichloride), P-286
Poly(4,4'-dihydroxydiphenylmethane-co-carbonic dichloride), P-354
Poly(4,4'-dihydroxydiphenylphenylmethylmethane-co-carbonic dichloride), P-452
Poly(2,5-dimethoxy-1,4-phenylene-1,2-ethenediyl), P-180
Poly(p-2,5-dimethoxyphenylene vinylene), P-180
Poly(2,3-dimethyl-1,3-butadiene), P-182
Poly(2,6-dimethyl-1,4-phenylene oxide), P-188
Poly(1,3-dimethyl-1,3-butadiene), P-391
Poly(4,4-dimethyl-1-hexene), P-184
Poly(4,4-dimethyl-1-pentene), P-187
Poly(N,N-dimethyl-2-propenamide), P-181
Poly(3,3-dimethyl-2-oxetanone), P-190
Poly(α,α-dimethyl-β-propiolactone), P-190
Poly(dimethyl)(methylphenyl)siloxane) fluid, P-402
Poly(2,6-dimethyl-1,4-phenylene ether), P-188
Poly(N,N-dimethylacrylamide), P-181
Poly(dimethylaminoethyl methacrylate), P-340
Polydimethylbutadiene, P-182
Poly(3,3-dimethylbutyl methacrylate), P-299
Poly(1,3-dimethylbutyl methacrylate), P-299
Poly(1-(2,2-dimethylbutyl)-1,2-ethanediyl), P-184
Polydimethyldiphenylsiloxane fluid, P-183
Poly(1,1-dimethylethyl methacrylate), P-119
Poly(dimethylglycolic acid), P-336
Poly(2,2-dimethylhydracrylic acid), P-190
Poly[[(dimethyloctylsilyl)-1,4-phenylene]-1,2-ethenediyl], P-186
Poly[[2-(3,7-dimethyloctyl)oxy]-5-methoxy-1,4-phenylene]-1,2-ethenediyl], P-356

Poly[[[(3,7-dimethyloctyl)oxy]methoxy-1,4-
  phenylene]-1,2-ethenediyl], P-356
Poly[2-(dimethyloctylsilyl)-1,4-phenylene
  vinylene], P-186
Poly(3,3-dimethyloxetane), O-4
Poly(*trans*-2,3-dimethyloxirane), P-98
Poly(2,6-dimethylphenol), P-188
Poly(2,6-dimethylphenyl methacrylate), P-74
Poly(2,6-dimethylphenylene oxide)/polystyrene
  blends, P-189
Poly(2,2-dimethylpropyl vinyl) ether, P-126
Poly(3,3′-dimethylpropiolactam), N-17
Poly(4,4′-dimethylpropiolactam), N-17
Poly[(dimethylsilylene)methylene], P-194
Poly(dimethylsilane), P-191
Polydimethylsilicone fluid, P-7
Polydimethylsilicone, P-192
Poly(dimethylsiloxane)diol, D-5
Poly(dimethylsiloxanes) with liquid crystal
  sidechains, P-193
Poly[dimethylsiloxane(oxyethylene
  cooxypropylene)], P-8
Polydimethylsiloxane rubber, M-16
Polydimethylsiloxane, P-192
Polydimethylsiloxane gum, P-192a
Poly(dimethylsilylene), P-191
Poly(dinonylsilane), P-195
Poly(dinonylsilylene), P-195
Poly(9,9-dioctyl-9*H*-fluorene-2,7-diyl), P-196
Poly(9,9-dioctylfluorene), P-196
Poly(9,9-dioctylfluorene-*co*-N-(4-(1-
  methylpropyl)phenyl)diphenylamine), P-200
Poly(9,9-dioctylfluorene-*co*-2,1,3-
  benzothiadiazole), P-197
Poly(9,9-dioctylfluorene-*co*-2,2′-
  bithiophene), P-199
Poly(9,9-dioctylfluorene-*co*-bis-N,N′-(4-
  butylphenyl)-N,N′-diphenyl-1,4-
  benzenediamine), P-198
Poly(2,5-dioctyloxy-1,4-phenylene
  ethynylene), P-201
Poly(dioctylsilane), P-202
Poly(dioctylsilylene), P-202
Poly[(1,3-dioxo-2,5-isoindolinediyl)carbonyl(1,3-
  dioxo-5,2-isoindolinediyl)-*p*-
  phenylenemethylene-*p*-phenylene], B-24
Poly[3,3′-(4,4′-dioxyphenyl)diphenylene
  pyromellitimide], N-4
Poly(dipentene dioxide), P-203
Poly(dipentylsilane), P-204
Poly(dipentylsilylene), P-204
Poly(2,6-diphenyl-1,4-phenylene ether), P-209
Poly(2,6-diphenyl-1,4-phenylene oxide), P-209
Poly(diphenylamine), P-440
Poly(N,N′-diphenylbenzidine), P-205
Poly(diphenylcarbosilane), P-210
Poly(diphenyldiacetylene), P-206
Polydiphenyldimethylsiloxane fluid, P-183
Poly(1,2-diphenylethyl methacrylate), P-73
Poly[diphenylmethane
  bis(4-phenyl)carbonate], P-207
Poly(diphenylmethane-4,4′-dicarboxylic acid),
  P-430
Poly(diphenylmethyl methacrylate), P-73
Poly(2,6-diphenylphenylene oxide), P-208
Poly[(diphenylsilylene)(methylene)], P-210
Poly(diphenylsilylmethylene), P-210
Poly(dipropylsilane), P-211
Poly[(dipropylsilylene)methylene], P-212
Poly(dipropylsilylene), P-211
Poly(disodium (Z)-2-butenedioate-*co*-
  methoxyethene), P-622
Poly(ditetradecylsilylene), P-213
Poly(ditetradecylsilane), P-213
Poly(divinyl ether), P-215
Poly(divinyl ether-maleic anhydride), P-216
Poly(divinyl-1,4-butanediyl ether-*co*-maleic
  anhydride), P-214
Poly(1-docosene), P-217
Poly(docosyl methacrylate), L-6
Poly(1-docosyl-1,2-ethanediyl), P-539
Poly(1-docosylethylene), P-539
Poly(dodecafluoroheptyl methacrylate), P-285

Poly(ω-dodecanolactam), N-39
Poly(1-dodecene), P-218
Poly(dodecyl methacrylate), L-6
Poly(3-dodecyl-2,5-thiophenediyl), P-222
Poly(1-dodecyl-1,2-ethanediyl), P-540
Poly(1-dodecylethylene), P-540
Poly(dodecylmethylsilylene), P-219
Poly(*n*-dodecylmethylsilane), P-219
Poly(dodecylpentylsilylene), P-220
Poly(dodecylpentylsilane), P-220
Poly(dodecylsilane), P-221
Poly(dodecylsilylene), P-221
Poly(3-dodecylthiophene), P-222
Poly(DPB), P-205
Poly(DPBz), P-205
Poly(1-eicosene), P-223
Poly(1-eicosyl-1,2-ethanediyl), P-217
Poly(1-eicosylethylene), P-217
Poly(ω-enanthamide), N-28
Poly(enantholactam), N-28
Poly(endo-1,7,7-trimethylbicyclo[2.2.1]hept-2-yl
  methacrylate), P-347
Poly(epichlorohydrin-*co*-propylene oxide-*co*-allyl
  glycidyl ether), P-503
Poly(3,4-epoxy-6-methylcyclohexylmethyl-
  3,4-epoxy-6-methylcyclohexane
  carboxylate), P-225
Poly(1,2-epoxy-3-hexyloxypropane), P-303
Poly(1,2-epoxy-3-methoxypropane), P-358, P-367
Poly(1,2-epoxy-3-phenoxypropane), P-226
Poly(1,2-epoxy-3-phenoxypropane), P-438, P-453
Poly(1,2-epoxy-4-epoxyethylcyclohexane), V-3
Poly(*trans*-2,3-epoxybutane), P-98
Poly(*trans*-epoxybutane), P-98
Poly(2-(2,3-epoxycyclopentyl)phenyl glycidyl
  ether), P-224
Poly[(epoxyethyl)benzene], P-527
Poly(1-epoxyethyl-3,4-epoxycyclohexane), V-3
Poly(1,2-epoxyhexane), P-298
Poly(1,2-epoxyoctane), P-422
Poly(1,2-epoxypropane), P-501
Poly[2-[*o*-(2,3-epoxypropoxy)phenyl]-6-
  oxabicyclo[3.1.0]hexane], P-224
Polyester carbonates, P-227
Poly(ethanal), P-57
Poly[1,1-ethane bis(4-phenyl)carbonate], P-228
Poly(1,2-ethanediol-*co*-decanedioic acid), P-269
Poly(1,2-ethanediol-*co*nonanedioic acid), P-251
Poly(1,2-ethanediylbis(oxy-2,1-ethanediyl)
  methacrylate), P-284
Polyethene, P-247
Poly(ethenol), P-566
Poly(ethenol-*co*-ethene), P-273
Poly(ethenyl acetate-*co*-chloroethene), P-584
Poly(ethenyl benzene), P-521, P-524, P-530
Poly(ethenyl chloroacetate), P-593
Poly(ethenyl difluoroacetate), P-599
Poly(ethenyl formate), P-606
Poly(ethenyl trifluoroacetate), P-642
Poly(4-ethenyl-1,1′-biphenyl), P-458
Poly(9-ethenyl-9*H*-carbazole), P-574
Poly(N-ethenyl-N-methylacetamide), P-413
Poly(1-ethenyl-2-methylcyclohexane), P-414
Poly(1-ethenyl-2-pyrrolidinone), P-638
Poly(1-ethenyl-2-pyrrolidinone-*co*-ethenyl acetate-
  co-ethenyl propanoate), P-229
Poly(1-ethenyl-2-pyrrolidinone-*co*-[ethyl
  2-methyl-2-propenoate]-*co*-2-methyl-2-
  propenoic acid), P-230
Poly(1-ethenyl-2-pyrrolidinone-*co*-1-eicosene),
  P-640
Poly(1-ethenyl-2-pyrrolidinone-*co*-2-
  (dimethylamino)ethyl-2-methyl-2-
  propenoate), P-635
Poly(1-ethenyl-2-pyrrolidinone-*co*-2-propenoic
  acid), P-639
Poly(1-ethenyl-2-pyrrolidinone-*co*-
  ethenylacetate), P-637
Poly(1-ethenyl-2-pyrrolidinone-*co*-
  ethenylbenzene), P-641
Poly(1-ethenyl-3-methylcyclohexane), P-415
Poly(1-ethenyl-4-methylcyclohexane), P-416

Poly(ethenylbenzene-*block*-2-methyl-1,3-
  butadiene), S-30, S-32
Poly(ethenylbenzene-*block*-2-ethenylpyridine-
  block-2-methyl-1,3-butadiene), P-525
Poly(ethenylbenzene-*block*-1,3-butadiene), S-25
Poly(ethenylbenzene-*block*-2-methyl-1,3-butadiene-
  block-2-ethenylpyridine), P-525
Poly(ethenylbenzene-*block*-2-methyl-1,3-butadiene-
  block-Ethenylebenzene), S-21
Poly(ethenylbenzene-*co*-1,3-butadiene), H-4, S-28
Poly(ethenylbenzene-*co*-2-methyl-1,3-
  butadiene), S-31
Poly(ethenylcyclobutane), P-594
Poly(ethenylcycloheptane), P-595
Poly(ethenylcyclohexane), P-596
Poly(ethenylcyclopentane), P-597
Poly(ethenylcyclopropane), P-598
Poly(1-(ethenyloxy)butane), P-124
Poly[1-(ethenyloxy)-2-methylpropane-*co*-2,5-
  furandione), P-324
Poly(1-(ethenyloxy)octadecane-*co*-2,5-
  furandione), P-629
Poly(2-(ethenyloxy)butane), P-125
Poly(2-ethenylpyridine-*co*-borane), P-634
Poly(2-ethenylpyridine-*co*-ethenylbenzene), P-529
Poly(2-ethenylpyridine), P-633
Polyether Block Amide, N-56
Polyetheretherketone, 30% Glass
  fibre filled, P-233
Polyetheretherketone, P-231
Polyetheretherketone, 30% Carbon
  fibre filled, P-232
Polyetheretherketoneketone, P-234
Polyetherimide silicone copolymer, S-14
Polyetherketone, P-235
Polyetherketoneetherketoneketone, P-236
Polyetherketoneketone, glass-filled, P-239
Polyetherketoneketone, carbon-filled, P-238
Polyetherketoneketone, P-237
Polyethers (oxetane polymers), O-4
Polyethersulfone, 30% Carbon fibre
  reinforced, P-241
Polyethersulfone, 20% Glass fibre
  reinforced, P-242
Polyethersulfone, 30% Glass fibre
  reinforced, P-243
Polyethersulfone, P-240
Poly[1-(ethoxycarbonyl)-1-methyl ethylene], P-275
Poly[1-(ethoxycarbonyl)ethylene], P-244
Poly(ethoxyethene), P-601
Poly(ethoxyethene-*co*-2,5-furandione), P-278
Poly(2-ethoxyethyl methacrylate), P-67
Poly(1-ethoxyethylene), P-601
Poly(ethoxyethylene), P-601
Poly(ethyl 2-cyano-2-propenoate), P-246
Poly(ethyl 2-propenoate), P-244
Poly(ethyl acrylate), P-244
Poly(ethyl-*n*-butylsilylene), P-115
Poly(ethylbenzene-*co*-1,3-butadiene-*co*-
  ethylenepyridine), P-522
Polyethylbutadiene, P-245
Poly(2-ethylbutyl methacrylate), P-299
Poly(ethyl cyanoacrylate), P-246
Poly(ethyl hydrogen ((Z)-2-butenedioate-*co*-
  methoxyethene), P-626
Poly(ethyl methacrylate-*co*-acrylamide), M-5
Poly(ethyl methacrylate-*co*-styrene), S-33
Poly(ethyl methacrylate)/nitrile rubber/poly(vinyl
  chloride), P-345
Poly(ethyl methacrylate)/poly(vinyl chloride-*co*-
  vinyl acetate), P-345
Poly(ethyl methacrylate), P-275
Poly(ethyl methacrylate-*co*-chloroprene), M-6
Poly(ethyl methacrylate-*co*-ethyl acrylate), A-27
Poly(ethyl vinyl ether-*co*-maleic anhydride), P-278
Poly(ethyl vinyl ether), P-601
Poly(2-ethyl-1,3-butadiene), P-245
Poly(ethyl-2-methyl-2-propenoate, P-275
Poly(ethylene adipate), P-250
Polyethylene, L-4
Polyethylene, P-247
Polyethylene alloys, P-252
Poly(ethylene azelate), P-251

Name and Synonym Index

Poly(ethylene benzoate), P-268
Polyethylene blends with polypropylene, P-482
Polyethylene blends with polystyrenes, P-253
Polyethylene blends with rubbers, P-254
Polyethylene blends with styrenics, P-253
Polyethylene blends, P-252
Polyethylene fibres, P-260
Polyethylene film, P-9
Polyethylene glycol succinate, P-270
Poly(ethylene glycol adipate), P-250
Poly(ethylene glycol dimethacrylate), P-284
Poly(ethylene glycol methacrylate), P-284
Poly(ethylene glycol trimethacrylate), P-284
Poly(ethylene glycol), P-267
Poly(ethylene isophthalate), P-262
Poly(ethylene naphthalene-1,5-dicarboxylate), P-264
Poly(ethylene oxide) polymethacrylate, P-284
Poly(ethylene oxide), P-267
Poly(ethylene oxybenzoate), P-268
Poly(ethylene sebacate), P-269
Poly(ethylene sebacinate), P-269
Poly(ethylene succinate), P-270
Poly(ethylene terephthalate), P-271
Polyethylene, High density, P-261
Poly(ethylene-1,5-naphthalate), P-264
Poly(ethylene-2,6-naphthalenedicarboxylate), P-265, P-266
Poly(ethylene-2,6-naphthalate) films, P-265
Poly(ethylene-2,6-naphthalate) general purpose, P-266
Poly(ethylene-ethylidenenorbornene-propene) rubber, E-17
Poly(ethylene-hexadiene-propene) rubber, E-17
Poly(ethylene-methylidenenorbornene-propene) rubber, E-17
Polyethylene-polypropylene glycols, P-55
Poly(ethylene-propylene monomer) rubber, E-17
Poly(ethylene-propylene-diene monomer), E-17
Poly(ethylene-co-acrylic acid), P-249
Poly(ethylene-co-acrylic acid), P-248
Poly(ethylene-co-carbon monoxide), P-255
Poly(ethylene-co-methyl acrylate), P-263
Poly(ethylene-co-vinyl alcohol), films, P-274
Poly(ethylene-co-vinyl alcohol), P-273
Poly(ethylene-co-vinylacetate), P-272
Poly(ethylene-co-chlorotrifluoroethylene), Carbon fibre filled, E-5
Poly(ethylene-co-chlorotrifluoroethylene), Glass fibre filled, E-6
Poly(ethylene-co-ethyl acrylate), P-259
Poly(ethylene-co-methyl 2-propenoate), P-263
Poly(ethylene-co-tetrafluoroethylene), Carbon fibre filled, E-13
Poly(ethylene-co-tetrafluoroethylene), Glass fibre filled, E-14
Poly(ethylene-co-tetrafluoroethylene), E-12
Poly(ethylene-p-hydroxybenzoate), P-268
Poly(3,4-ethylenedioxypyrrole), P-256
Poly(3,4-ethylenedioxythiophene), P-257
Poly(3,4-ethylenedioxythiophene)/poly(4-styrenesulfonate) blend, P-258
Poly(p-(ethyleneoxy)benzoate), P-268
Poly[[2,5-bis[(2-ethylhexyl)oxy]-1,4-phenylene]-1,2-ethynediyl[2,5-bis(octyloxy)-1,4-phenylene]-1,2-ethynediyl], P-80
Poly[[2,5-bis[(2-ethylhexyl)oxy]-1,4-phenylene]-1,2-ethenediyl], P-81
Poly[[[(2-ethylhexyl)oxy]methoxy-1,4-phenylene]-1,2-ethenediyl], P-357
Poly[[2-[(2-ethylhexyl)oxy]-5-methoxy-1,4-phenylene]-1,2-ethenediyl], P-357
Poly[(ethylmethylsilane)(methylene)], P-276
Poly[(ethylmethylsilylene)(methylene)], P-276
Poly(ethylpentylsilane), P-277
Poly(ethylpentylsilylene), P-277
Poly((ethylthio)ethyl methacrylate), P-340
Poly(ethyne), P-58
Poly(ethynylbenzene), P-439
Poly(exo-1,7,7-trimethylbicyclo [2.2.1]hept-2-yl methacrylate), P-347
Poly(2-ferrocenylethyl methacrylate), P-346
Poly(2-ferrocenylmethyl methacrylate), P-346

Poly(9H-fluorene), P-279
Poly(9H-fluorene-2,7-diyl), P-279
Poly(fluorene), P-279
Poly(4-fluoro-1-vinyl-benzene), P-280
Poly[(fluoroacetoxy)ethylene], P-604
Poly(fluoroethene), P-603
Poly(1-(4-fluorophenyl)-1,2-ethanediyl), P-280
Poly(4-fluorostyrene), P-280
Poly(p-fluorostyrene), P-280
Polyformaldehyde, P-432
Poly(2,5-furandione-co-methoxyethene), P-624
Poly(2,5-furandione-co-1-(ethenyloxy)octadecane), P-629
Poly(2,5-furandione-co-ethoxyethane), P-278
Poly[2,5-furandione-co-1,1'-oxybisethene], P-216
Poly(2-furanmethanol), P-281
Poly(furfuryl alcohol), P-281
Poly(furfuryl alcohol-co-formaldehyde), P-281
Poly(furfuryl methacrylate), P-346
Poly[(1→4)-α-D-glucose], A-36
Polyglycidyl ether of Tetraphenolethane, P-282
Poly(glycidyl methacrylate-co-chloroprene), M-6
Poly(glycol methacrylates), P-284
Poly(glycolic acid), P-283
Polyglycolide, P-283
Poly(glycollic acid), P-283
Poly(glycollic ester), P-283
Polyglycollide, P-283
Polyguluronic acid, A-33
Poly(haloalkyl methacrylate), P-285
Poly(heptadecafluorooctyl methacrylate), P-285
Poly(1-heptadecyl-1,2-ethanediyl), P-418
Poly(1-heptadecylethylene), P-418
Poly(1,1,1,2,2,3,3-heptafluoro-3-[(trifluoroethenyl)oxy]propane-co-tetrafluoroethene), P-543
Poly(heptafluorobutyl methacrylate), P-285
Poly(heptafluoroisopropyl methacrylate), P-285
Poly(heptamethylene pimelamide), N-29
Poly(heptamethylenediamine-co-pimelic acid), N-29
Poly[4,4-heptane bis(4-phenyl)carbonate], P-286
Poly(1-heptene), P-287
Poly(1-heptyl-1,2-ethanediyl), P-419
Poly(1-heptylethylene), P-419
Poly(heptylhexylsilane), P-288
Poly(heptylhexylsilylene), P-288
Poly(1-hexadecene), P-289
Poly(hexadecyl methacrylate), L-6
Poly(hexadecyloxyethene-co-octadecyloxyethene-co-1-ethenyl-2-pyrrolidinone), P-608
Poly(hexadecyloxyethene-co-octadecyloxyethene-co-ethenylbenzene-co-2,5-furandione), P-607
Poly(1,1,2,3,3,3-hexafluoro-1-propene), P-290
Poly(4,4'-(hexafluoroisopropylidene)bis(phthalic anhydride)-1,3-isobenzofurandione-alt-1,3-bis(3-aminophenoxy)benzene) Addition cured (acetylenic end-caps), F-7
Poly(1H-hexafluoroisopropyl methacrylate), P-285
Poly(4,4'-(hexafluoroisopropylidene)bis(phthalic anhydride)-alt-2,2-bis[4-(4-aminophenoxy)phenyl]hexafluoropropane), F-1
Poly(4,4'-(hexafluoroisopropylidene)bis(phthalic anhydride)-alt-2,3,5,6-tetramethyl-1,4-benzenediamine, F-2
Poly(4,4'-(hexafluoroisopropylidene)bis(phthalic anhydride)-co-m-phenylenediamine-co-p-phenylenediamine), F-6
Polyhexafluoropropene, P-290
Poly(hexafluoropropene oxide), P-292
Polyhexafluoropropylene, P-290
Poly(hexafluoropropylene oxide), P-292
Poly(hexamethylene adipamide), N-23, N-42, N-44, N-50, N-51, N-58
Poly(hexamethylene azelamide), N-25
Poly(hexamethylene diamine-co-isophthalic acid, N-16
Poly(hexamethylene diamine-co-suberic acid), N-24
Poly(hexamethylene diamine-co-terephthalic acid), N-59
Poly(hexamethylene dodecanamide), N-27

Poly(hexamethylene isophthalamide), N-16
Poly(hexamethylene sebacamide), N-26
Poly(hexamethylene suberamide), N-24
Poly(hexamethylene terephthalate), N-59
Poly(hexamethylene terephthalate), P-296
Poly(hexamethylenediamine-co-dodecanedioic acid), N-27
Poly(hexamethylenediamine-co-m-xylylenediamine-co-isophthalic acid-co-terephthalic acid), P-295
Poly(hexamethylenediamine-co-adipic acid), N-23, N-42, N-44, N-50, N-51, N-58
Poly(hexamethylenediamine-co-azelaic acid), N-25
Poly(hexamethylenediamine-co-isophorone diamine-co-isophthalic acid-co-terephthalic acid), P-293
Poly(hexamethylenediamine-co-sebacic acid), N-26
Poly(hexamethylenediamine-co-terephthalic acid-co-isophthalic acid), P-294
Poly(1,2-hexane oxide), P-298
Poly(hexane terephthalate), P-296
Poly(1,6-hexanediyl terephthalate), P-296
Poly(hexene oxide), P-298
Poly(1-hexene), P-297
Poly(1-hexenylene), P-297
Poly(3-n-hexoxy-1,2-epoxypropane), P-303
Poly(hexoxyethene), P-609
Poly(hexyl 2-methyl-2-propenoate), P-299
Poly(n-hexyl methacrylate), P-299
Poly(hexyl methacrylate-co-styrene), S-33
Poly(hexyl methacrylates), P-299
Poly(hexyl vinyl ether), P-609
Poly(1-hexyl-1,2-ethanediyl), P-421
Poly(1-hexylethylene), P-421
Poly(hexylglycidyl ether), P-303
Poly[(hexylmethylsilane)(methylene)], P-301
Poly(hexylmethylsilane), P-300
Poly[(hexylmethylsilylene)(methylene)], P-301
Poly(hexylmethylsilylene), P-300
Poly(hexyloctylsilane), P-302
Poly(hexyloctylsilylene), P-302
Poly[[2,5-bis(hexyloxy)-1,4-phenylene](1-cyano-1,2-ethenediyl) [2,5-bis(hexyloxy)-1,4-phenylene](2-cyano-1,2-ethenediyl)], C-17
Poly(hexyloxirane), P-422
Poly[(hexyloxymethyl)oxirane], P-303
Poly(3-hexyloxypropylene oxide), P-303
Poly(hexylpentylsilane), P-304
Poly(hexylpentylsilylene), P-304
Poly(3-hexylthiophene), P-305
Poly(3-hexylthiophene)/PCBM blend, P-306
Poly(3-hexylthiophene)/[6,6]-phenyl-C61-butyric acid methyl ester blend, P-306
Poly(HFIB/VF2), P-614
Poly(4-hybe), P-310
Poly(hydrogenmethylsiloxanes) with liquid crystal sidechains, P-307
Poly(3-hydroxy-2,2-dimethylpropanoic acid), P-190
Poly(α-hydroxy-ω-hydroxy oxy-1,2-ethanediyl), P-267
Poly(hydroxyacetic acid), P-283
Poly(hydroxyacetic ester), P-283
Poly(hydroxyalkanoate), P-308
Poly(hydroxyalkyl methacrylate), P-309
Poly(4-hydroxybenzoate-co-2-oxy-6-naphthoyl), P-311
Poly(4-hydroxybenzoic acid-co-6-hydroxy-2-naphthoic acid), glass fibre filled, P-315
Poly(4-hydroxybenzoic acid-co-2-oxy-6-naphthoic acid), mineral filled, P-316
Poly(4-hydroxybenzoate-co-2-oxy-6-naphthoyl), glass fibre filled, P-315
Poly(4-hydroxybenzoate-co-6-hydroxy-2-naphthalenecarboxylate, P-311
Poly(4-hydroxybenzoate-co-isophthalic acid-co-4,4'-dihydroxydiphenylether), P-312
Poly(4-hydroxybenzoate-co-terephthalic acid-co-4,4'-dihydroxydiphenyl ether), P-313
Poly(p-hydroxybenzoate), P-310
Poly(4-hydroxybenzoic acid), P-310
Poly(4-hydroxybenzoic acid-co-6-hydroxy-2-naphthoic acid) carbon fibre filled, P-314

Poly(4-hydroxybenzoic acid-co-2-oxy-6-naphthoic acid), carbon fibre filled, P-314
Poly(4-hydroxybenzoic acid-co-6-hydroxy-2-naphthoic acid), mineral filled, P-316
Poly(3-hydroxybutanoate-co-3-hydroxypentanoate), P-318
Poly(3-hydroxybutanoate), P-317
Poly(3-hydroxybutyrate), P-317
Poly(3-hydroxybutyrate-co-3-hydroxyvalerate), P-318
Poly(α-hydroxybutyrate), P-317
Poly(α-hydroxybutyrate-co-α-hydroxyvalerate), P-318
Poly(1-hydroxyethene), P-566
Poly(hydroxyethene), P-566
Poly(4-(2-hydroxyethoxy)benzoic acid), P-268
Poly(hydroxyethyl methacrylate-co-acrylic acid), H-5
Poly(hydroxyethyl methacrylate-co-butyl acrylate), H-5
Poly(hydroxyethyl methacrylate-co-butyl methacrylate), H-5
Poly(hydroxyethyl methacrylate-co-dimethylaminoethyl methacrylate), H-5
Poly(hydroxyethyl methacrylate-co-ethyl methacrylate), H-5
Poly(hydroxyethyl methacrylate-co-ethylene glycol dimethacrylate), H-5
Poly(hydroxyethyl methacrylate-co-hydroxyethoxyethyl methacrylate), H-5
Poly(hydroxyethyl methacrylate-co-hydroxypropyl acrylate), H-5
Poly(hydroxyethyl methacrylate-co-methacrylic acid), H-5
Poly(hydroxyethyl methacrylate-co-methyl methacrylate), H-5
Poly(2-hydroxyethyl methacrylate), P-309
Poly(2-hydroxyethyl methacrylate-co-N,N-dimethylacrylamide), H-5
Poly(2-hydroxyethyl methacrylate-co-N-vinylpyrrolidone), H-5
Poly(2-hydroxyethyl methacrylate-co-styrene), H-5
Poly(2-hydroxyethyl methacrylate-graft-propylene), H-5
Poly(1-hydroxyethylene), P-566
Poly(hydroxyethylene), P-566
Poly(hydroxypivalic acid), P-190
Poly(2-hydroxypropanoic acid), P-336
Poly(hydroxypropyl methacrylate-co-methacrylic acid), H-5
Poly(2-hydroxypropyl methacrylate), P-309
Poly(imino adipoyl iminononamethylene), N-33
Poly[imino(1,10-dioxo-1,10-decanediyl)imino-1,10-decanediyl], N-37
Poly[imino(1,6-dioxo-1,6-hexanediyl)imino-1,6-hexanediyl], N-50, N-58
Poly(imino(1,6-dioxo-1,6-hexanediyl)imino-1,6-hexanediyl], N-23, N-42, N-44, N-51
Poly[imino(1,6-dioxo-1,6-hexanediyl)imino-1,9-nonanediyl], N-33
Poly[imino(1,7-dioxo-1,7-heptanediyl)imino-1,7-heptanediyl], N-29
Poly[imino(1,8-dioxo-1,8-octanediyl)imino-1,8-octanediyl], N-31
Poly(imino(1,9-dioxo-1,9-nonanediyl)imino-1,10-decanediyl], N-36
Poly(imino(1,9-nonanediyl)imino-1,9-nonanediyl], N-34
Poly[imino(1-oxo-1,10-decanediyl)], N-35
Poly[imino(1-oxo-1,11-undecanediyl)], N-38
Poly[imino(1-oxo-1,12-dodecanediyl)], N-39
Poly[imino(1-oxo-1,4-butanediyl)], N-18
Poly[imino(1-oxo-1,6-hexanediyl)], N-22
Poly[imino(1-oxo-1,7-heptanediyl)], N-28
Poly[imino(1-oxo-1,9-nonanediyl)], N-32
Poly[imino(1-oxo-1,3-propanediyl)], N-17
Poly[imino(1-oxo-1,8-octanediyl)], N-30
Poly(imino-1,3-phenyleneiminocarbonyl-1,3-phenylenecarbonyl), P-446
Poly[imino-1,4-butanediylimino(1,6-dioxo-1,6-hexanediyl)], N-48
Poly(imino-1,4-butanediylimino(1,6-dioxo-1,6-hexanediyl)), N-19

Poly[imino-1,4-cyclohexanediylmethylene-1,4-cyclohexanediylimino(1,12-dioxo-1,12-dodecanediyl)], P-1
Poly[imino-1,4-phenyleneiminocarbonyl-1,4-phenylenecarbonyl), P-450
Poly[imino-1,5-pentanediylimino(1,6-dioxo-1,6-hexanediyl)], N-21
Poly[imino-1,6-hexanediylimino(1,10-dioxo-1,10-decanediyl)], N-26
Poly[imino-1,6-hexanediylimino(1,12-dioxo-1,12-dodecanediyl)], N-27
Poly[imino-1,6-hexanediylimino(1,9-dioxo-1,9-nonanediyl)], N-25
Poly[imino-1-oxopentamethylene), N-20
Poly[imino-1-oxotridecamethylene), N-40
Poly[iminocarbonyl-1,4-phenylenecarbonylimino(trimethyl-1,6-hexanediyl)], N-61
Poly[iminocarbonyl-1,3-phenylenecarbonylimino-1,6-hexanediyl], N-16
Poly[iminocarbonyl-1,4-phenylenecarbonylimino-1,6-hexanediyl], N-59
Poly(iminohexamethylene iminoadipoyl), N-23, N-42, N-44, N-51, N-58
Poly[iminohexamethyleneiminoazelaoyl], N-25
Poly[iminohexamethyleneiminododecanedioyl], N-27
Poly(iminohexamethyleneiminosuberoyl), N-24
Poly[iminomethylene-1,3-phenylenemethyleneimino(1,6-dioxo-1,6-hexanediyl)], P-645
Poly[iminomethylene-1,4-phenylenemethyleneimino(1,10-dioxo-1,10-decanediyl)], P-647
Poly(iminopentamethyleneadipoyl), N-21
Poly(5-indanyl methacrylate), P-74
Polyindole, P-319
Poly(4-iodo-1-vinyl-styrene), P-320
Poly(1-(4-iodophenyl)-1,2-ethanediyl), P-320
Poly(4-iodostyrene), P-320
Poly(p-iodostyrene), P-320
Poly[9,9-bis(2-ethylhexyl)fluorene], P-79
Poly(isobornyl methacrylate), P-347
Poly(isobutoxyethene), P-323
Poly(isobutyl 2-methyl-2-propenoate), P-322
Poly(isobutyl cyanoacrylate), P-321
Poly(isobutyl methacrylate-co-styrene), S-33
Poly(isobutyl methacrylate), P-322
Poly(isobutyl vinyl ether-co-maleic anhydride), P-324
Poly(isobutyl vinyl ether), P-323
Poly(isobutylene-co-isoprene), B-35
Poly(isodecyl methacrylate), L-6
Poly(isooctyl methacrylate-co-styrene), S-33
Poly(isophthalic anhydride), P-431
Poly(4,4′-isophthaloyldiphthalic anhydride-alt-m-phenylenediamine), L-3
Polyisoprene star polymers, P-329
1,2-Polyisoprene, P-325
3,4-Polyisoprene, P-328
cis-1,4-Polyisoprene, P-326
trans-1,4-Polyisoprene, P-327
cis-1,4-Polyisoprene, G-3, N-2
cis-Polyisoprene, P-326
trans-1,4-Polyisoprene, E-19, G-4, N-1
trans-Polyisoprene, G-4, N-1, P-327
Polyisoprene, G-3, P-325, P-328
Poly(isopropyl acrylate), P-331
Poly(N-isopropylacrylamide), P-330
Poly(isopropyl methacrylate), P-333
Poly(isopropyl vinyl ether), P-335
Poly(2-isopropyl-1,3-butadiene), P-332
Poly(N-isopropyl-N′-phenyl-p-phenylenediamine), P-334
Polyisopropylbutadiene, P-332
Polyisopropylethylene, P-362
Poly(lactic acid), P-336
Poly(lactic acid-co-glycolic acid), P-337
Poly(lactide), P-336
Poly(lactide-co-glycolide), P-337
Poly(lactilic acid), P-336
Poly(lauryl methacrylate-co-N,N-dimethylacrylamide), M-5

Poly(lauryl methacrylate), L-6
Poly(lauryl methacrylate-co-glycidyl methacrylate), A-27
Poly(limonene dioxide), P-203
Polymannoacetate, A-23
Polymannuronic acid, A-33
Polymerisation of Monomeric Reactants Second generation resins, P-53
Poly(meta-xylylene adipamide), P-645
Poly(metaphenylene isophthalate), P-514
Poly(metaxylylene adipamide), P-646
Polymethacrylamide, P-338
Poly(methacrylate-co-butadiene-co-styrene), P-339
Poly(methacrylate ionomers), P-341
Polymethacrylate/polycarbonate blends, P-343
Polymethacrylate/polybutadiene blends, P-342
Polymethacrylate/polyvinyl blends, P-345
Polymethacrylate/polyethylene oxide blends, P-344
Polymethacrylate/vinylidene halide blends, P-348
Polymethacrylate esters of substituted alkyls, P-340
Polymethacrylates with fused-ring side chains, P-347
Polymethacrylates with cyclic sidechains containing heteroatoms, P-346
Poly(methacrylic acid-co-N-vinylpyrrolidone), M-9
Poly(methacrylic acid-co-styrene), M-9
Poly(methacrylic acid salt-co-ethylene), P-350
Poly(methacrylic acid), salts, P-351
Poly(methacrylic acid), P-349
Poly(methacrylic acid-co-urethane), M-9
Poly(methacrylimide), P-352
Polymethacrylonitrile, P-353
Poly(methacrylourethane), M-7
Poly[methane bis(4-phenyl)carbonate], P-354
Poly(3-methoxy-1,2-epoxypropane), P-358
Poly[[2-methoxy-5-(3-sulfopropoxy)-1,4-phenylene]-1,2-ethenediyl], P-360
Poly[1-(methoxy carbonyl)ethylene], P-361
Poly(4-methoxy-1-ethenyl-benzene), P-359
Poly[2-methoxy-5-(2′-ethylhexyloxy)-1,4-phenylene vinylene], P-357
Poly[2-methoxy-5-(3′,7′-dimethyloctyloxy)-1,4-phenylene vinylene], P-356
Poly[2-methoxy-5-(3-sulfopropoxy)-1,4-phenylene vinylene], P-360
Poly(2-methoxyaniline-5-sulfonic acid), P-355
Poly[1-(methoxycarbonyl)-1-methyl ethylene], P-381
Poly(methoxyethene-co-(Z)-2-butenedioic acid), P-623
Poly(methoxyethene), P-621
Poly(methoxyethene-co-1-methylethyl hydrogen (Z)-2-butenedioate), P-627
Poly(methoxyethene-co-2,5-furandione), P-624
Poly(methoxyethene-co-disodium (Z)-2-butenedioate), P-622
Poly(methoxyethene-co-ethyl hydrogen (Z)-2-butenedioate), P-626
Poly(2-methoxyethyl methacrylate), P-67
Poly[(methoxymethyl)oxirane], P-358, P-367
Poly(1-(4-methoxyphenyl)-1,2-ethanediyl), P-359
Poly(3-methoxypropylene oxide), P-358
Poly(3-methoxypropylene oxide), P-367
Poly(p-methoxystyrene), P-359
Poly(4-methoxystyrene), P-359
Poly(methyl 2-methyl-2-propenoate), P-381
Poly(methyl 2-propenoate), P-361
Poly(methyl 4-(2-hydroxyethoxy)benzoate), P-268
Poly(methyl acrylate), P-361
Poly(methyl cyanoacrylate), P-365
Poly(methyl ethylene), P-479, P-510
Poly(methyl isopropenyl ketone-co-styrene), E-3
Poly(methyl isopropenyl ketone), P-375
Poly(methyl methacrylate-co-butadiene), M-6
Poly(methyl methacrylaPoly(methyl methacrylate-co-butyl acrylate)/poly(vinyl chloride), P-345
Poly(methyl methacrylate), cast sheet, rods, and tubes, P-379
Poly(methyl methacrylate-co-vinyl alcohol), P-386
Poly(methyl methacrylate-co-acrylonitrile-co-α-methylstyrene), P-378
Poly(methyl methacrylate-co-methyl itaconate), P-384

Poly(methyl methacrylate-co-N,N-dimethylacrylamide), M-5
Poly(methyl methacrylate-co-tert-butylphenyl methacrylate), A-27
Poly(methyl methacrylate-co-ethyl acrylate), A-27
Poly(methyl methacrylate), extruded sheet, P-380
Poly(methyl methacrylate), General, P-381
Poly(methyl methacrylate-co-vinyl chloride), P-387
Poly(methyl methacrylate-co-butyl acrylate), A-27
Poly(methyl methacrylate-co-butyl methacrylate), A-27
Poly(methyl methacrylate-co-ethyl methacrylate-co-ethyl acrylate), A-27
Poly(methyl methacrylate-graft-natural rubber), M-6
Poly(methyl methacrylate-co-acrylonitrile-co-butadiene-co-styrene), P-377
Poly(methyl methacrylate-co-itaconic acid), P-382
Poly(methyl methacrylate-co-N-phenylmaleimide), M-5
Poly(methyl methacrylate-co-N-tolylmaleimide), M-5
Poly(methyl methacrylate-co-dodecyl methacrylate), A-27
Poly(methyl methacrylate-co-EPDM-graft-2-vinylnaphthalene), M-6
Poly(methyl methacrylate-co-EPDM-graft-styrene), M-6
Poly(methyl methacrylate-co-methyl acrylate)/poly(vinyl chloride), P-345
Poly(methyl methacrylate-co-vinylidene fluoride), M-6
Poly(methyl methacrylate)/ABS blends, P-342
Poly(methyl methacrylate)/nitrile rubber/poly(vinyl chloride), P-345
Poly(methyl methacrylate)/poly(vinyl chloride), P-345
Poly(methyl methacrylate-co-N-vinylpyrrolidone), P-388
Poly(methyl methacrylate-co-itaconic anhydride), P-383
Poly(methyl methacrylate-co-acrylonitrile), P-376
Poly(methyl methacrylate-co-EPDM-co-acrylonitrile), M-6
Poly(methyl methacrylate-co-ethyl acrylate)/poly(vinyl chloride), P-345
Poly(methyl methacrylate-co-ethyl methacrylate), A-27
Poly(methyl methacrylate-co-ethylhexyl acrylate)/poly(vinyl chloride), P-345
Poly(methyl methacrylate-co-methyl acrylate), A-27
Poly(methyl methacrylate-co-styrene), S-33
Poly(methyl methacrylate-co-tetrafluoroethylene), M-6
Poly(methyl methacrylate), Moulding and Extrusion Compounds, P-385
Poly(methyl methacrylate)/poly(vinyl chloride-co-vinyl acetate), P-345
Poly(methyl methacrylate)/polycarbonate blends, P-343
Poly(methyl methacrylate-co-acrylonitrile-co-styrene), S-33
Poly(methyl methacrylate-co-ethylene), M-6
Poly(N-methylpyrrole), P-407
Poly(methyl vinyl ether), P-621
Poly(methyl vinyl ketone), P-417
Poly(2-methyl-1,3-butadiene), N-2
Poly(2-methyl-1,3-pentadiene), P-391
Poly(2-methyl-1,4-pentadiene), P-391
Poly(3-methyl-1-butene), P-362
Poly(4-methyl-1-hexene), P-371
Poly(5-methyl-1-heptene), P-368
Poly(6-methyl-1-heptene), P-369
Poly(3-methyl-1-hexene), P-370
Poly(5-methyl-1-hexene), P-372
Poly(methyl-1-methylvinyl ketone), P-375
Poly(4-methyl-1-pentene), Injection moulding, Mineral filled, P-398
Poly(4-methyl-1-pentene) rubber, P-400
Poly(4-methyl-1-pentene), Extrusion, Unfilled, P-394
Poly(4-methyl-1-pentene), Moulded, P-399

Poly(3-methyl-1-pentene), P-392
Poly(4-methyl-1-pentene) films, P-395
Poly(4-methyl-1-pentene), P-393
Poly(4-methyl-1-pentene), Glass reinforced, P-396
Poly(4-methyl-1-pentene), Injection moulding, Glass filled, P-397
Poly(1-methyl-1-phenyl-1,2-ethanediyl), P-411
Poly(1-methyl-1-silylenemethylene), P-408
Poly(4-methyl-1-vinyl-benzene), P-409
Poly(4-methyl-1-vinyl-benzene), P-410
1,2-Poly(2-methyl-1,3-butadiene), P-325
3,4-Poly(2-methyl-1,3-butadiene), P-328
cis-1,4-Poly(2-methyl-1,3-butadiene), P-326
trans-1,4-Poly(2-methyl-1,3-butadiene), P-327
Poly(2-methyl-1,3-butadiene) star polymers, P-329
Poly(2-methyl-2-propenenitrile), P-353
Poly(2-methyl-2-propenamide), P-338
Poly(2-methyl-2-propenoic acid), salts, P-351
Poly(2-methyl-2-propenoic acid), P-349
Poly(3-methyl-3-buten-2-one), P-375
Poly(3-methyl-3-buten-2-one-co-ethenylbenzene), E-3
Poly(4-methyl-L-glutamic acid) fibre, P-366
Poly(γ-methyl-L-glutamate) fibre, P-366
Poly(N-methyl-N-vinylacetamide), P-413
Poly(methyl-n-octylsilane), P-389
Poly(methyl-n-butylsilane), P-363
Poly[(4-methyl-7-oxabicyclo[4.1.0]hept-3-yl)methyl 4-methyl-7-oxabicyclo[4.1.0]heptane-3-carboxylate], P-225
Poly[(5-methyl-2-(1-methylethyl)cyclohexyloxy-ethene-co-indene), P-620
Poly[1-methyl-4-(2-methyloxiranyl)-7-oxabicyclo[4.1.0]heptane], P-203
Polymethylalkylsiloxane fluid, P-390
Poly(1-(1-methylbutyl)-1,2-ethanediyl), P-370
Poly(2-methylbutyl methacrylate), P-118
Poly(1-(2-methylbutyl)-1,2-ethanediyl), P-371
Poly(1-(3-methylbutyl)-1,2-ethanediyl), P-372
Poly[(2-methylbutyl)undecylsilylene], P-364
Poly(1-(3-methylcyclohexyl)-1,2-ethanediyl), P-415
Poly(1-(2-methylcyclohexyl)-1,2-ethanediyl), P-414
Poly(1-(4-methylcyclohexyl)-1,2-ethanediyl), P-416
Poly(methylene carboxylate), P-283
Poly(methylene ethyl methyl ketone), P-375
Poly-4,4′-methylenedicyclohexylene dodecanediamide, P-1
Poly(4,4′-methylenebisbenzenamine-alt-5,5′-carbonylbis[1,3-isobenzofurandione]), B-24
Poly(4,4′-methylenebisbenzoic acid), P-430
Poly[(1-methylethenyl)benzene], P-411
Poly(1-(1-methylethyl)-1,2-ethanediyl), P-362
Poly(1-methylethyl 2-propenoate), P-331
Poly(1-methylethyl hydrogen ((Z)-2-butenedioate-co-methoxyethene), P-627
Poly(N-(1-methylethyl)-2-propenamide), P-330
Poly[4,4′-(1-methylethylidene)bisphenol-co-4,4′-(3,3,5-trimethylcyclohexylidene)bisphenol-co-carbonic acid], P-555
Poly[4,4′-(1-methylethylidene)bis[phenol]-co-1,4-benzenedicarbonyldichloride-co-1,3-benzenedicarbonyldichloride-co-carbonic dichloride], B-12
Poly(4,4′-(1-methylethylidene)bisbenzoic acid), P-429
Poly[4,4′-(1-methylethylidene)bisphenol-co-(chloromethyl)oxirane], B-3
Poly[4,4′-(1-methylethylidene)bis[2-(1-methylethyl)phenol]-co-carbonic dichloride), B-16
Poly[4,4′-(1-methylethylidene)bis[phenol]-co-1,3-benzenedicarboxylic acid-co-carbonic acid], B-4
Poly[4,4′-(1-methylethylidene)bis[phenol]-co-1,4-benzenedicarboxylic acid-co-1,3-benzenedicarboxylic acid-co-carbonic acid], B-12
Poly[4,4′-(1-methylethylidene)bis[phenol]-co-1,4-benzenedicarboxylic acid-co-carbonic acid], B-13
Poly(4,4′-(1-methylethylidene)bis(2,6-dibromophenol)co-4,4′-(1-methylethylidene)bisphenol-co-carbonic acid], P-465

Poly(methylglycidyl ether), P-367
Poly(methylglycidyl ether), P-358
Polymethylhydrogen siloxane, P-373
Poly(methylhydrogensiloxane-co-dimethylsiloxane) fluid, P-374
Polymethyloctylsiloxane fluid, P-390
Poly(methyloctylsilylene), P-389
Poly[methyloxirane-co-(2-propenyloxymethyl)oxirane], P-502
Polymethyloxirane, P-501
Poly(1-(3-methylpentyl)-1,2-ethanediyl), P-368
Poly(2-methylpentadiene), P-391
Poly[(3-methylpentyl)ethylene], P-368
Poly(1-methylpentyl methacrylate), P-299
Poly(1-(4-methylpentyl))-1,2-ethanediyl), P-369
Poly(1-(1-methylphenyl)-1,2-ethanediyl), P-409
Poly(methylphenyl)silicone rubber, P-49
Poly[[bis(4-methylphenyl)silylene]methylene], P-83
Poly(1-(4-methylphenyl)-1,2-ethanediyl), P-410
Poly[[bis(3-methylphenyl)silylene]methylene], P-82
Poly(2-methylphenyl methacrylate), P-74
Poly[(methylphenylsilylene)(methylene)], P-403
Poly[(methylphenylsilylene)methylene], P-404
Poly(methylphenylcarbosilane), P-403
Poly(methylphenylsilylene), P-401
Poly(methylphenylsilane), P-401
Polymethylphenylsiloxane fluid, P-402
Polymethylphenylsiloxane fluid, P-456
Poly(1-(1-methylpropyl)-1,2-ethanediyl), P-392
Poly[[[4-(1-methylpropyl)phenyl]imino]-1,4-phenylene(9,9-dioctyl-9H-fluorene-2,7-diyl)-1,4-phenylene], P-200
Poly[1,1-(2-methylpropane) bis(4-phenyl)carbonate], P-405
Poly(2-methylpropoxyethene), P-323
Poly(2-methylpropyl methacrylate), P-322
Poly(2-methylpropyl vinyl ether), P-125, P-323
Poly(1-(2-methylpropyl)-1,2-ethanediyl), P-187
Poly(methylpropylsilane), P-406
Poly(methylpropylsilylene), P-406
Polymethylsiloxane, P-373
Poly[(methylsilylene)(methylene)], P-408
Poly(4-methylstyrene), P-410
Poly(o-methylstyrene), P-409
Poly(p-methylstyrene), P-410
Poly(α-methylstyrene), P-411
Poly(2-methylstyrene), P-409
Poly(3-methylthiophene), P-412
Poly(4-methylvinylcyclohexane), P-416
Poly(2-methylvinylcyclohexane), P-414
Poly(3-methylvinylcyclohexane), P-415
Poly(monochlorethylene), P-583
POLY MVK, P-417
Poly(myristyl methacrylate), L-6
Poly(naphthyl methacrylate), P-74
Poly(NBR-co-ethylene), P-90
Poly(NBR-co-isoprene), P-91
Poly(2-nitratoethyl methacrylate), P-340
Poly(nitrile rubber-co-ethylene), P-90
Poly(nitrile rubber-co-isoprene), P-91
Poly(1-nonadecene), P-418
Poly(tert-nonafluorobutyl methacrylate), P-285
Poly(nonamethylene azelaimide), N-34
Poly(nonamethylenediamine-co-azelaic acid), N-34
Poly(1-nonene), P-419
Poly(1-nonyl-1,2-ethanediyl), P-557
Poly(1-nonylethylene), P-557
Poly(norbornyl methacrylate), P-347
Poly(NR-co-ethylene), P-90
Poly(NR-co-isoprene), P-91
Poly(1-octadecene), P-420
Poly(octadecyl methacrylate), L-6
Poly(octadecyl vinyl ether), P-628
Poly(1-octadecyl-1,2-ethanediyl), P-223
Poly(1-octadecylethylene), P-223
Poly(octadecyloxyethene), P-628
Poly(octafluoropentyl methacrylate), P-285
Poly(octahydro-2,4-methano-2H-indeno[1,2-b:5,6-b′]bisoxirene), D-3
Poly(octamethylene suberamide), N-31
Poly(octamethylenediamine-co-suberic acid), N-31
Poly(1,2-octane oxide), P-422
Poly(octene oxide), P-422

Poly(1-octene), P-421
Poly(octoxyethene), P-630
Poly(octyl 2-methyl-2-propenoate), P-423
Poly(octyl methacrylate-co-styrene), S-33
Poly(octyl methacrylate), P-423
Poly(octyl vinyl ether), P-630
Poly(1-octyl-1,2-ethanediyl), P-167
Poly(n-octyl-n-hexylsilane), P-302
Poly(1-octylethylene), P-167
Poly[[2,5-bis(octyloxy)-1,4-phenylene]-1,2-ethenediyl], P-84
Poly[[2,5-bis(octyloxy)-1,4-phenylene]-1,2-ethynediyl], P-201
Poly(octylpentylsilylene), P-424
Poly(3-octylthiophene), P-425
Polyolefin rubber, synthetic, E-17
Poly(olefin sulfone), O-2
Poly(α-olefin), P-70
Polyorganosiloxanes, P-518
Poly(7-oxabicyclo[4.1.0]hept-3-ylmethyl 7-oxabicyclo[4.1.0] heptane-3-carboxylate), E-11
Poly[2-(7-oxabicyclo[4.1.0]hept-3-yl)spiro[1,3-dioxane-5,3'-[7]oxabicyclo[4.1.0]heptane]], E-10
Poly(1,3,4-oxadiazole-2,5-diyl-1,4-phenylene), P-427
Poly(1,3,4-oxadiazole-2,5-diyl-1,3-phenylene), P-426
Poly(oxadiazoles), P-428
Poly(oxetane), O-4
Poly(3-oxiranyl-7-oxabicyclo[4.1.0]heptane), V-3
Poly[N-[4-(oxiranylmethoxy)phenyl]-N-(oxiranylmethyl)oxiranemethanamine], B-2
Poly[N-(oxiranylmethyl)-N-phenyloxiranemethanamine], D-4
Poly[1-(2-oxo-1-pyrrolidinyl)ethylene], P-638
Poly[oxy(1,2-dimethyl-1,2-ethanediyl)], P-98
Poly[oxy(1,2-dimethylethylene)], P-98
Poly[oxy(1,6-dioxo-1,6-hexanediyl)oxy-1,10-decanediyl], P-164
Poly[oxy(1-oxo-1,2-ethanediyl)], P-283
Poly[oxy(1,1'-biphenyl)-4,4'-diyloxy-1,4-phenylenesulfonyl-1,4-phenylene], P-459
Poly[oxy(2,2-bis(chloromethyl)-1,3-propanediyl]], P-77
Poly[oxy(2,6-dimethyl-1,4-phenylene)], P-188
Poly[oxy(butyl-1,2-ethanediyl)], P-298
Poly[oxy(dimethylsilylene)], M-16, P-7, P-192, P-192a, S-18
Poly[oxy(dibromophenylene)], P-170
Poly[oxy(hexyl-1,2-ethanediyl)], P-422
Poly[oxy(methoxymethyl)-1,2-ethanediyl], P-358
Poly[oxy(methyloctylsilylene)], P-390
Poly[oxy(methane-1,2-ethanediyl)], P-501
Poly[oxy(methylphenylsilylene)], P-456
Poly[oxy(methyl(3,3,3-trifluoropropyl)silylene)], F-15
Poly[oxy(methylsilylene)], P-373
Poly[oxy(phenoxymethyl)-1,2-ethanediyl], P-438
Poly(oxy-1,2-ethanediyloxycarbonyl-1,3-phenylenecarbonyl), P-262
Poly(oxy-1,2-ethanediyloxycarbonyl-1,4-phenylenecarbonyl), P-271
Poly(oxy-1,2-ethanediyloxycarbonyl-1,5-naphthalenediylcarbonyl), P-264
Poly(oxy-1,2-ethanediyloxy-1,4-phenylenecarbonyl), P-268
Poly(oxy-1,2-ethanediyloxycarbonyl-2,6-naphthalenediylcarbonyl), P-265, P-266
Poly(oxy-1,3-phenylenecarbonyl-1,4-phenylene), P-235
Poly(oxy-1,3-phenyleneoxycarbonyl-1,3-phenylenecarbonyl), P-514
Poly(oxy-1,3-propanediyl), O-4
Poly(oxy-1,4-butanediyloxycarbonyl-1,3-phenylenecarbonyl), P-547
Poly(oxy-1,4-butanediyloxycarbonyl-1,4-phenylenecarbonyl), P-103
Poly(oxy-1,4-phenylenecarbonyl-1,4-phenylene), P-235
Poly(oxy-1,4-phenylenecarbonyl-1,4-phenyleneoxy-1,4-phenylenecarbonyl-1,3-phenylenecarbonyl-1,4-phenylene), P-236
Poly(oxy-1,4-phenylenecarbonyl), P-310

Poly(oxy-1,4-phenylenecarbonyl-1,4-phenyleneoxy-1,4-phenylenecarbonyl-1,4-phenylenecarbonyl-1,4-phenylene), P-236
Poly(oxy-1,4-phenylenecarbonyl-1,4-phenyleneoxy-1,4-phenylenecarbonyl), P-237
Poly(oxy-1,4-phenyleneoxy-1,4-phenylenecarbonyl-1,4-phenylenecarbonyl-1,4-phenylene), P-234
Poly(oxy-1,4-phenyleneoxy-1,4-phenylenecarbonyl-1,4-phenylene), P-231
Poly(oxy-1,4-phenylenesulfonyl-1,4-phenylene), P-240
Poly[oxy-1,4-phenylenesulfonyl-1,4-phenyleneoxy-1,4-phenylene(1-methylethylidene)-1,4-phenylene], P-531
Poly[oxy-1,4-phenylenesulfonyl-1,4-phenyleneoxy-1,4-phenylene-(1-methylethylidene)-1,4-phenylene], C-22
Poly[oxy-1,10-decanediyloxy(1,10-dioxo-1,10-decanediyl)], P-165
Poly[oxy-1,2-ethanediyloxy(1,10-dioxo-1,10-decanediyl)], P-269
Poly[oxy-1,2-ethanediyloxy(1,6-dioxo-1,6-hexanediyl)], P-250
Poly[oxy-1,2-ethanediyloxy(1,9-dioxo-1,9-nonanediyl)], P-251
Poly(oxy-1-oxoethylene), P-283
Poly[oxy-1,2-ethanediyloxy(1,4-dioxo-1,4-butanediyl)], P-270
Poly[oxy-1,4-butanediyloxy(1,6-dioxo-1,6-hexanediyl)], P-545
Poly(oxy-2,6-diphenyl-1,4-phenylene), P-208
Poly(oxy-m-phenyleneoxyisophthaloyl), P-514
Poly(oxy [1,1':3',1''-terphenyl]-2',5'-diyl), P-209
Poly(oxy(1-phenyl)ethylene), P-527
Poly(oxy(2,2-dimethyl-1-oxo-1,3-propanediyl)), P-190
Poly(oxy-m-terphenyl-2',5'-ylene), P-209
Poly[oxy(1-phenyl-1,2-ethanediyl)], P-527
Poly[oxy[(butoxymethyl)-1,2-ethanediyl]], P-99
Poly[oxy[(hexyloxymethyl)-1,2-ethanediyl]], P-303
Poly[oxy[2,2-bis(chloromethyl)trimethylene]], P-77
Polyoxyacetyl, P-283
Poly(oxyadipoyloxydecamethylene), P-164
Poly(p-oxybenzoate), P-310
Poly(poxybenzoyl), P-310
Poly[4,4'-oxybis(benzenamine)-alt-5,5'-carbonylbis-1,3-isobenzofurandione], B-26
Poly[2,2'-oxybis(6-oxabicyclo[3.1.0]hexane)], B-1
Poly(4,4'-oxybisbenzenamine-co-1,3-benzenediamine-co-5,5'-carbonylbis[1,3-isobenzofurandione]), B-28
Poly[4,4'-oxybis[benzenamine]-alt-[5,5'-biisobenzofuran]-1,1',3,3'-tetrone), B-19
Poly[4,4'-oxybis[benzenamine]-alt-5,5'-[2,2,2-trifluoro-1-(trifluoromethyl)ethylidene]bis-1,3-isobenzofurandione), F-5
Poly[oxycarbonyloxy(2,6-dichloro-1,4-phenylene)(1-methylethylidene)(3,5-dichloro-1,4-phenylene)], P-466
Poly[oxycarbonyl-1,4-phenylene(1-methylethylidene)-1,4-phenylene], B-5
Poly[oxycarbonyl-1,4-phenylene(3,3,5-trimethylcyclohexylidene)-1,4-phenyleneoxycarbonyloxy-1,4-phenylene(1-methylethylidene)-1,4-phenylene], P-555
Poly[oxycarbonyloxy(2,6-dibromo-1,4-phenylene)(1-methylethylidene)(3,5-dibromo-1,4-phenylene)oxycarbonyloxy-1,4-phenylene(1-methylethylidene)-1,4-phenylene], P-465
Poly[oxycarbonyl-1,4-phenylene(1-methylbutylidene)-1,4-phenylene], P-435
Poly[oxycarbonyloxy-1,4-phenylene(dichloroethenylidene)-1,4-phenyleneoxycarbonyloxy-1,4-phenylene(1-methylethylidene)-1,4-phenylene], P-174
Poly[oxycarbonyl-1,4-phenylenecyclohexylidene-1,4-phenylene], P-157
Poly[oxycarbonyloxy(2,6-dibromo-1,4-phenylene)(1-methylethylidene)(3,5-dibromo-1,4-phenylene)], P-464
Poly[oxycarbonyloxy-1,4-phenylene(1-methylpropylidene)-1,4-phenylene], P-97

Poly(oxycarbonyl(1,1-dimethylethylene)), P-190
Poly(oxycarbonyl-1,3-phenylene carbonyl), P-431
Poly(oxycarbonyl-1,4-phenylenecarbonyloxy-1,6-hexanediyl), P-296
Poly(oxycarbonyl-1,4-phenylenecarbonyloxymethylene-1,4-cyclohexanediylmethylene), P-159
Poly(oxycarbonyl-1,4-phenylenecarbonyloxymethylene-1,4-cyclohexanediyl methylene), P-162
Poly(oxycarbonyl-1,4-phenyleneisopropylidene-1,4-phenylenecarbonyl), P-429
Poly(oxycarbonyl-1,4-phenylenemethylene-1,4-phenylene carbonyl), P-430
Poly[oxycarbonyloxy-1,4-phenylene(1-methylethylidene)-1,4-phenylene(1-methylethylidene)-1,4-phenylene], B-17
Poly[oxycarbonyloxy-1,4-phenylene(1-propylbutylidene)-1,4-phenylene], P-286
Poly[oxycarbonyloxy-1,4-phenylene(3,3,5-trimethylcyclohexylidene)-1,4-phenylene], P-554
Poly[oxycarbonyloxy-1,4-phenylene(dichloroethenylidene)-1,4-phenylene], P-173
Poly[oxycarbonyloxy-1,4-phenylene(diphenylmethylene)-1,4-phenylene], P-207
Poly[oxycarbonyloxy-1,4-phenylenebutylidene-1,4-phenylene], P-96
Poly[oxycarbonyloxy-1,4-phenylenecyclohexylidene-1,4-phenyleneoxycarbonyloxy-1,4-phenylene(1-methylethylidene)-1,4-phenylene], P-158
Poly[oxycarbonyloxy-1,4-phenylenemethylene-1,4-phenylene], P-354
Poly[oxycarbonyloxy-1,4-phenylenethio-1,4-phenylene], P-548
Poly(oxycarbonylethylidene), P-336
Poly[oxycarbonyloxy(2,6-dimethyl-1,4-phenylene)(1-methylethylidene)(3,5-dimethyl-1,4-phenylene)], P-467
Poly[oxycarbonyloxy-1,4-phenylenethio-1,4-phenyleneoxycarbonyloxy-1,4-phenylene(1-methylethylidene)-1,4-phenylene], P-549
Poly[oxycarbonyloxy-1,4-phenylene[2,2,2-trifluoro-1-(trifluoromethyl)ethylidene]-1,4-phenylene], B-14
Poly(oxycarbonyloxy[2-(1-methylethyl)-1,4-phenylene](1-methylethylidene)[3-(1-methylethyl)-1,4-phenylene]), B-16
Poly[oxycarbonyloxy(2-methyl-1,4-phenylene)(1-methylethylidene)(3-methyl-1,4-phenylene)], P-468
Poly[oxycarbonyloxy-1,4-phenylene(2-methylpropylidene)-1,4-phenylene], P-405
Poly[oxycarbonyloxy-1,4-phenyleneethylidene-1,4-phenylene], P-228
Poly[oxycarbonyloxy-1,4-phenylenesulfonyl-1,4-phenylene], P-537
Poly[oxycarbonyloxy(2,6-dichloro-1,4-phenylene)cyclohexylidene(3,5-dichloro-1,4-phenylene)], P-156
Poly[oxycarbonyloxy(2-chloro-1,4-phenylene)(1-methylethylidene)(3-chloro-1,4-phenylene)], P-463
Poly[oxycarbonyloxy-1,4-phenylene(1-phenylethylidene)-1,4-phenylene], P-452
Poly[oxycarbonyloxy-1,4-phenylenecyclopentylidene-1,4-phenylene], P-163
Poly(oxydecamethylene oxycarbonyl tetramethylene carbonyl), P-164
Poly(oxydecamethyleneoxysebacoyl), P-165
Poly(oxydecamethyleneoxycarbonyloctamethylenecarbonyl), P-165
Polyoxydin, P-267
Poly(4,4'-oxydiphenylene pyromellitimide), P-52
Poly(4,4'-oxydiphenylene biphenyltetracarboximide), B-19
Poly(4,4'-oxydiphenylene benzophenonetetracarboximide), B-26
Poly(oxyethylene-block-oxypropylene), P-55
Poly(oxyethylene oxyterephthaloyl), P-271
Poly(oxyethyleneoxy-1,4-phenylenecarbonyl), P-268

Polyoxyethylene glycol, P-267
Polyoxyethylene-polyoxypropylene block copolymer, P-55
Poly(oxyethyleneoxyadipoyl), P-250
Poly(oxyethyleneoxycarbonyl-1,5-naphthalenecarbonyl), P-264
Poly(oxyethyleneoxyazelaoyl), P-251
Poly(oxyethyleneoxycarbonyl-2,6-naphthalenecarbonyl), P-265, P-266
Poly(oxyethyleneoxyisophthaloyl), P-262
Poly(oxyethyleneoxysuccinoyl), P-270
Poly(oxyethyleneoxysebacoyl), P-269
Poly(oxyethyleneoxysuccinyl), P-270
Poly(oxyisophthaloyl), P-431
Poly(oxymethylene-1,4-cyclohexylenemethylene-oxyterephthaloyl), P-159
Poly(oxymethylene-co-ethylene oxide), A-24
Polyoxymethylene, P-432
Poly(oxyphenylene), P-447
Polyoxypropylene, P-501
Polyoxypropylene-polyoxyethylene block copolymer, M-4
Poly(oxysiliconphthalocyanin), P-433
Poly(oxyterephthaloyl oxymethylene-1,4-cyclohexylenemethylene), P-159
Poly(oxyterephthaloyloxydecamethylene), P-166
Poly(oxyterephthaloyloxyhexamethylene), P-296
Poly(oxytetramethylene oxycarbonyl tetramethylene carbonyl), P-545
Poly(oxytetramethyleneoxyadipoyl), P-545
Poly(oxytetramethyleneoxyisophthaloyl), P-547
Poly(oxytetramethyleneoxyterephthaloyl), P-103
Poly(oxy[2,6-diphenyl-1,4-phenylene]), P-209
Poly(palmityl methacrylate), L-6
Poly(paraphenylene), P-442
Poly(ω-pelargonamide), N-32
Poly(pentachlorophenyl methacrylate), P-74
Poly(1H,1H-pentadecafluorooctyl methacrylate), P-285
Poly(1-pentadecene), P-434
Poly(pentamethylene diamine-co-adipic acid, N-21
Polypentamethylenecarbonamide, N-22
Poly[2,2-pentane bis(4-phenyl)carbonate], P-435
Poly(1-pentene), P-436
Poly(pentyl 2-methyl-2-propenoate), P-437
Poly(pentyl methacrylate), P-437
Poly(1-pentyl-1,2-ethanediyl), P-287
Poly(1-pentylethylene), P-287
Poly(pentylhexylsilane), P-304
Polypeptides, P-651
Polyperfluoropropylene, P-290
Poly[(phenoxymethyl)oxirane], P-226, P-438, P-453
Poly(3-phenoxypropylene oxide), P-438
Poly(3-phenoxypropylene oxide), P-453
Poly(phenyl ethene), P-521, P-524, P-530
Poly(phenyl ethylene), P-521, P-524, P-530
Poly(phenyl glycidyl ether), P-226
Poly(phenyl methacrylate), P-74
Poly(phenyl methacrylate)/polycarbonate blends, P-343
Poly[[thio bis(4-phenyl)carbonate]-co-[2,2-propane bis(4-phenyl)carbonate]], P-549
Poly(4-phenyl-1-butene), P-441
p-Polyphenyl, P-442
Poly(N-phenyl-p-phenylenediamine), P-454
Poly(phenylacetylene), P-439
Poly(N-phenylaniline), P-440
Poly(4-phenylbutyl methacrylate), P-73
Poly(p-phenylene biphenyltetracarboximide), B-20
Polyphenylene ether, P-188
Polyphenylene oxide, P-188
Poly[1,4-phenylene bis[(2,2-propane)(4-phenyl)]carbonate], B-17
Poly(p-phenylene diamine-co-terephthalic acid), P-450
Poly(1,4-phenylene ethynylene), P-445
Poly(m-phenylene isophthalamide), P-446
Poly(phenylene oxide), P-447
Poly(phenylene sulfide), 40% Glass fibre filled, P-449
Poly(phenylene sulfide), P-448
Poly(p-phenylene terephthalamide), P-450
Poly(p-phenylene vinylene), P-451

Poly(p-phenylene xylidine), P-445
Poly(1,4-phenylene-1,2-ethynediyl), P-445
Poly(1,4-phenylene-1,2-ethenediyl), P-451
Poly(p-phenylene-1,3,4-oxadiazole), P-427
Poly(p-phenylene), P-442
Poly(m-phenylene-1,3,4-oxadiazole), P-426
Poly(p-phenylenebenzobisthiazole), P-443
Poly(2,2'-(1,2-phenylenebis(oxymethylene) bisoxirane)), P-78
Poly(4,4'-[1,4-phenylenebis(1-methylethylidene)]bisphenol-co-carbonic acid)-block-poly(1-methoxycarbonyl)-1-methylethylene), P-472
Poly(4,4'-[1,4-phenylenebis(1-methylethylidene)]bisphenol-co-carbonic dichloride), B-17
Polyphenylenediselenothiocarbonate, P-444
Poly(p-phenylenediamine), P-169
Poly(m-phenylenediamine-co-isophthalic acid), P-446
Poly[5,5'-(1,3-phenylenedicarbonyl)bis-1,3-isobenzofurandione-alt-1,3-benzenediamine], L-3
Poly(p-phenylenediselenocarbonate), P-516
Poly(p-phenylenediselenothiocarbonate), P-515
Poly(phenylenediselenocarbonate), P-444
Poly(p-phenyleneethynylene), P-445
Poly[1,1-(1-phenylethane) bis(4-phenyl)carbonate], P-441
Poly(1-(2-phenylethyl)-1,2-ethanediyl), P-441
Poly(2-phenylethyl methacrylate), P-73
Poly(1-phenylethyl methacrylate), P-73
Poly(1-phenylethylene), P-521, P-524, P-530
Poly(phenylglycidyl ether), P-453
Poly(phenylglycidyl ether), P-438
Poly(phenylmethylsilane), P-401
Poly(phenyloxirane), P-527
Poly(phenylsilane), P-455
Polyphenylsiloxane fluid, P-456
Poly[(phenylsilylene)methylene], P-457
Poly(phenylsilylene), P-455
Poly(4-phenylstyrene), P-458
Polyphenylsulfone, Unreinforced, P-459
Poly(phenylthiol methacrylate), P-74
Polyphosphazenes, P-460
Polyphthalamide, P-461
Poly(phthalocyaninato silicone oxo), P-433
Poly(piperazine-co-sebacic acid), P-462
Poly(piperazine sebacamide), P-462
Poly[1,4-piperazinediyl(1,10-dioxo-1,10-decanediyl)], P-462
Poly(β-pivalolactone), P-190
Poly(pivalolactone), P-190
Poly[2,2-propane bis[4-(2,6-dibromophenyl))carbonate]-co-(2,2-propane bis(4-phenyl)carbonate], P-465
Poly[2,2-propane bis[4-(2,6-dibromophenyl))carbonate], P-464
Poly[2,2-propane bis[4-(2,6-dichlorophenyl)]carbonate], P-466
Poly[2,2-propane bis[4-[2,6-dimethylphenyl]]carbonate], P-467
Poly[2,2-propane bis[4-[2-methylphenyl]]carbonate], P-468
Poly[2,2-propane bis[4-(2-(1-methylethyl)phenyl]carbonate], B-16
Poly[2,2-propane bis[4-(2-chlorophenyl)]carbonate], P-463
Poly[2,2-propanebis(4-phenyl)carbonate]-block-poly(ethylene oxide), P-470
Poly[2,2-propanebis(4-phenyl)carbonate]-block-poly(methylmethacrylate), P-472
Poly[2,2-propanebis(4-phenyl)carbonate], Carbon fibre reinforced, B-6
Poly[2,2-propanebis(4-phenyl)carbonate], Glass fibre reinforced, B-7
Poly[2,2-propanebis(4-phenyl)carbonate], Glass filled, B-8
Poly[2,2-propanebis(4-phenyl)carbonate], PTFE lubricated, B-10
Poly[2,2-propanebis(4-phenyl)carbonate]-block-poly[2,2-propanebis[4-(2,6-dimethylphenyl]carbonate], P-473

Poly[2,2-propanebis(4-phenyl)carbonate], B-5
Poly[2,2-propanebis(4-phenyl)carbonate], Metal filled, B-9
Poly[2,2-propanebis(4-phenyl)carbonate]-block-poly[4,4'-isopropylidenediphenoxydi(4-phenylene)sulfone], P-471
Poly[2,2-propanebis(4-phenyl)carbonate]-block-poly(dimethylsiloxane), P-469
Poly[2,2-propane[4-(2,6-dimethyl)phenyl]carbonate]/Polystyrene (High impact) blend, P-474
Poly(propene-co-carbon monoxide-co-ethene), P-476
Poly(2-propenamide), P-60
Polypropene blends, P-481
Polypropene copolymers, P-489
Polypropene fibres, P-493
Polypropene films, P-495
Poly(1-propene) atactic, A-45
Poly(1-propene) chlorinated, P-488
Poly(1-propene) isotactic, I-1
Poly(1-propene) syndiotactic, S-34
Poly(1-propene), P-479, P-510
Polypropene, Calcium carbonate filled, P-486
Polypropene, Filled, P-494
Polypropene, Glass fibre reinforced, P-496
Polypropene, Talc filled, P-509
Polypropene-atactic, A-45
Polypropene-isotactic, I-1
Polypropene-polyethene blends, P-482
Polypropene-syndiotactic, S-34
Polypropene/cis-1,4-polyisoprene rubber blends, P-498
Polypropene/butyl rubber blends, P-485
Polypropene, P-479, P-510
Poly(1-propene-block-ethene), P-649
Poly(1-propene-co-1-butene), P-484
Poly(1-propene-co-ethene), P-649
Poly(1-propene-co-ethene-co-1-butene), P-491
Poly(propene-alt-carbon monoxide-alt-ethene), P-475
Poly(2-propenenitrile), P-63
Poly(2-propenenitrile-co-(methyl 2-propenoate), A-29
Poly(2-propenenitrile-co-1,3-butadiene), N-5
Poly(2-propenenitrile-co-1,3-butadiene-co-2-methyl-1,3-butadiene), A-28
Poly(2-propenenitrile-co-1,3-butadiene-co-ethenylbenzene), A-2, A-3, A-4, A-5, A-7, A-8, A-9, A-12, A-17, A-21, A-22
Poly(2-propenenitrile-co-ethenylbenzene), S-1
Poly(2-propenoic acid), P-62
Poly(2-propenoic acid-co-ethenylbenzene-co-2-propenenitrile), A-39
Poly(1-propenyl ether functional siloxane), P-477
Poly[[(2-propenyloxy)methyl]oxirane], P-69
Polypropiolactam, N-17
Poly(propionyloxyethylene), P-631
Poly[1-(propoxycarbonyl)ethylene], P-478
Poly[1-(propoxycarbonyl)-1-methyl ethylene], P-511
Poly[1-(2-propoxycarbonyl)ethylene], P-331
Poly(propoxyethene), P-632
Poly(2-propoxyethene), P-335
Poly(2-propoxyethyl methacrylate), P-67
Poly(propyl 2-propenoate), P-478
Poly(propyl acrylate), P-478
Poly(n-propyl methacrylate), P-511
Poly(2-propyl methacrylate), P-333
Poly(2-propyl vinyl ether), P-335
Poly(propyl vinyl ether), P-632
Poly(propyl-2-methyl-2-propenoate), P-511
Poly(1-propyl-1,2-ethanediyl), P-436
Poly(propylene-co-carbon monoxide-co-ethylene), P-476
Polypropylene - EPDM blends, Vulcanised, P-490
Poly(propylene oxide-co-allyl glycidyl ether-co-epichlorohydrin), P-503
Polypropylene alloys, P-481
Polypropylene blends with polyethylene, P-482
Polypropylene blends, P-481
Polypropylene chlorinated, P-488
Polypropylene copolymer, P-649

Polypropylene copolymers, P-489
Polypropylene films, P-495
Polypropylene fibres, P-493
Polypropylene glycol, P-501
Polypropylene impact copolymer, P-649
Poly(propylene oxide-co-ethylene oxide), P-504
Poly(propylene oxide), P-501
Polypropylene oxide polyols, P-501
Poly(propylene oxide-co-allyl glycidyl ether), P-502
Polypropylene polyethylene glycol, M-4
Polypropylene, Calcium carbonate filled, P-486
Polypropylene, Carbon filled, P-487
Polypropylene, Filled, P-494
Polypropylene, Glass fibre reinforced, P-496
Polypropylene, Mica filled, P-497
Polypropylene, Talc filled, P-509
Polypropylene, wood-filled, P-510
Polypropylene, Aluminium filled, P-480
Polypropylene-ethylene-butylene terpolymer, P-491
Polypropylene/cis-polybutadiene blend, P-483
Polypropylene/natural rubber blends, P-498
Polypropylene/NBR blends, P-499
Polypropylene/butadiene rubber blend, P-483
Polypropylene/butyl rubber blends, P-485
Polypropylene/EPDM blends, Unvulcanised, P-492
Polypropylene/EPR blend, P-492
Polypropylene/ethylene-propylene copolymer blends, P-492
Polypropylene/nitrile rubber blends, P-499
Polypropylene/Nylon blends, P-500
Polypropylene/polybutadiene blend, P-483
Polypropylene/polycarbonate blends, P-505
Polypropylene/polyester blends, P-506
Polypropylene/polyisoprene blends, P-498
Polypropylene/polyamide blends, P-500
Polypropylene/rubber blends, P-507
Polypropylene/styrenic blends, P-508
Poly(propylene-co-butylene), P-484
Poly(propylene-alt-carbon monoxide-alt-ethylene), P-475
Polypropylene, P-479
Poly(1-propylethylene), P-436
Poly(di-n-propylsilylenemethylene), P-212
Poly(PrPPD), P-334
Poly(pyromellitic dianhydride-alt-4,4'-oxydianiline), P-52
Poly(pyromellitic dianhydride-alt-2,2-bis[4-(4-aminophenoxy)phenyl]hexafluoropropane), P-51
Poly(pyromellitic dianhydride-alt-3,3'-bis(4-aminophenoxy)biphenyldiamine), N-4
Poly(pyromellitic dianhydride-alt-bis(4-aminophenyl)ether), P-52
Poly(pyrrole), P-512
Poly(3-pyrroleacetic acid), P-513
Polypyrrolidone, N-18
Poly(resorcinol isophthalate), P-514
Poly(sarcolatic acid), P-336
Poly(sebacic acid-co-decanediol), P-165
Poly(selenocarbonothioylseleno-1,4-phenylene), P-515
Poly(selenocarbonylseleno-1,4-phenylene), P-516
Poly(silaethylene), P-519
Poly(silane), P-517
Polysilapropylene, P-408
Polysiloxanes, P-518
Poly[silylene(methylene)], P-519
Poly(silylene), P-517
Poly[sodium 4-(3-pyrrolyl)butanesulfonate], P-520
Poly(stearyl methacrylate), L-6
Polystyrene emulsion polymers, P-526
Polystyrene film, O-3
Polystyrene latex, P-526
Poly(styrene oxide), P-527
Polystyrene polymer colloids, P-526
Polystyrene, Cellular, P-523
Polystyrene, Expanded, P-523
Polystyrene, Expandable, P-523
Polystyrene, Foamed, P-523
Polystyrene, glass filled, P-524
Polystyrene, High impact, H-2
Polystyrene, Oriented film, O-3
Polystyrene, wood filled, P-530

Poly(styrene-butadiene-vinylpyridine) triblock copolymer, P-522
Polystyrene/poly(phenylene oxide) blends, High impact, H-3
Polystyrene/styrene-butadiene block copolymers blends, P-528
Polystyrene, P-521
Poly(styrene-block-butadiene-block-vinylpyridine), P-522
Poly(styrene-block-isoprene-block-2-vinylpyridine), P-525
Poly(styrene-block-butadiene), S-25
Poly(styrene-block-butadiene-block-styrene) triblock copolymer, S-9
Poly(styrene-block-isoprene), S-30
Poly(styrene-block-isoprene-block-styrene), S-21
Poly(styrene-co-1,3-butadiene), high-trans, S-29
Poly(styrene-co-2-vinylpyridine), P-529
Poly(styrene-co-acrylonitrile) copolymer, S-1
Poly(styrene-co-butadiene), S-26
Poly(styrene-co-isoprene), S-31
Poly(styrene-butadiene), H-4
Polysulfone unreinforced, P-531
Polysulfone, 15% PTFE lubricated, P-536
Polysulfone, 30% carbon fibre reinforced, P-533
Polysulfone, 30% Glass fibre reinforced, P-534
Polysulfone, Mineral filler modified, P-535
Polysulfone-based proprietary blend, P-532
Polysulfone-based proprietary resin, P-532
Polysulfone, P-531
Poly[sulfonyl bis(4-phenyl)carbonate], P-537
Poly(4,4'-sulfonylbisphenol-co-carbonic dichloride), P-537
Poly[[2,5-bis(3-sulfopropoxy)-1,4-phenylene]-1,2-ethynediyl-1,4-phenylene-1,2-ethynediyl disodium salt], P-85
Poly(3,3',5,5'-tetrachloro-4,4'-dihydroxydiphenyl-1,1-cyclohexane-co-carbonic dichloride), P-156
Polytetrachloroethene, P-538
Polytetrachloroethylene, P-538
Poly(1-tetracosene), P-539
Poly(1-tetradecene), P-540
Poly(1-tetradecyl-1,2-ethanediyl), P-289
Poly(tetradecyl methacrylate), L-6
Poly(1-tetradecylethylene), P-289
Poly(tetraethylene glycol methacrylate), P-284
Poly(tetraethylene glycol dimethacrylate), P-284
Polytetrafluoroethylene, Stainless steel filled, P-658
Polytetrafluoroethene, P-652
Polytetrafluoroethylene, Polyphenylene sulfide filled, P-657
Polytetrafluoroethylene, P-652
Polytetrafluoroethylene, Bronze filled, P-653
Polytetrafluoroethylene, Carbon and Graphite filled, P-654
Polytetrafluoroethylene, Glass fibre filled, P-655
Polytetrafluoroethylene, Tetrafluoroethylene filled, P-659
Poly(tetrafluoroethylene-co-trifluoromethylperfluorovinyl ether), PTFE and carbon filled, P-13
Polytetrafluoroethylene, Graphite filled, P-656
Poly(tetrafluoroethene-co-hexafluoropropene), Carbon fibre filled, F-8
Poly(tetrafluoroethene-co-1-propene), T-3
Poly(tetrafluoroethene-co-trifluoro(trifluoromethoxy)ethene), Glass filled, P-11
Poly(tetrafluoroethene-co-trifluoro(trifluoromethoxy)ethene), PTFE filled, P-14
Poly(tetrafluoroethene-co-trifluoro(trifluoromethoxy)ethene), T-2
Poly(tetrafluoroethene-co-1,1,2,3,3,3-hexafluoro-1-propene), P-542
Poly(tetrafluoroethene-co-hexafluoropropylene), PTFE filled, F-11
Poly(tetrafluoroethene-co-hexafluoropropylene), PTFE filled, F-11
Poly(tetrafluoroethene-co-perfluoro(methyl vinyl) ether), T-2

Poly(tetrafluoroethene-co-trifluoro(trifluoromethoxy)ethene), Mineral and mica filled, P-12
Poly(tetrafluoroethene-co-trifluoro(trifluoromethoxy)ethene), PTFE and carbon filled, P-13
Poly(tetrafluoroethylene-co-trifluoromethylperfluorovinyl ether), Glass filled, P-11
Poly(tetrafluoroethylene-co-trifluoromethylperfluorovinyl ether), PTFE filled, P-14
Poly(tetrafluoroethylene-co-trifluorovinyl trifluoromethyl ether), T-2
Poly(tetrafluoroethylene-co-trifluoromethylperfluorovinylether), Mineral and mica filled, P-12
Poly(tetrafluoroethylene-co-hexafluoropropylene), Glass fibre filled, F-10
Poly(tetrafluoroethylene-co-trifluoromethyl trifluorovinyl ether), T-2
Poly(tetrafluoroethylene-co-hexafluoropropylene), Carbon fibre filled, F-8
Poly(tetrafluoroethylene-co-ethylene), Glass fibre filled, P-541
Poly(tetrafluoroethylene-co-hexafluoropropylene), P-542
Poly(tetrafluoroethylene-co-perfluoropropyl vinyl ether), P-543
Poly(tetrafluoroethylene-co-vinylidene fluoride), P-544
Poly(tetrafluoroethylene-co-ethylene), E-12
Poly(tetrafluoroethylene-co-hexafluoropropene), Glass fibre filled, F-10
Poly(tetrafluoroethylene-co-hexafluoropropylene), Filled, F-9
Poly(1H-tetrafluoroisopropyl methacrylate), P-285
Poly(tetrafluoropropyl methacrylate), P-285
Poly(tetrahydro-4H-2-pyranyl methacrylate), P-346
Poly[(1,1',3,3'-tetrahydro-1,1',3,3'-tetraoxo[5,5'-bi-2H-isoindole]-2,2'-diyl)-1,4-phenyleneoxy-1,4-phenylene], B-19
Poly(tetrahydrofurfuryl methacrylate), P-346
Poly(1,1,3,3-tetramethyl-1,3-disilacyclobutane), SRU, P-194
Polytetramethylene succinate adipate, P-102
Poly(tetramethylbutylphenyl methacrylate), P-74
Polytetramethylene glycol, P-546
Polytetramethylene succinate, P-101
Poly(tetramethylene adipamide), N-19
Poly(tetramethylene adipate), P-545
Poly(tetramethylene isophthalate), P-547
Poly(tetramethylene terephthalate), P-103
Poly(tetramethylene terephthalate-co-tetramethylene oxide), P-104
Polytetramethylene ether, P-546
Poly(tetramethylenediamine-co-adipic acid), N-19, N-48
Poly[(1,1',3,3'-tetraoxo[5,5'-bi-2H-isoindole]-2,2'-diyl)-1,4-phenylene], B-20
Polythene, L-4, P-247
Poly[2-(3-thienyl)ethanol-n-butoxycarbonylmethylurethane], P-661
Poly[thio bis(4-phenyl)carbonate], P-548
Poly(4,4'-thiobisphenol-co-4,4'-(1-methylethylidene)bisphenol-co-carbonic dichloride), P-549
Poly(thiophene vinylene), P-552
Poly(thiophene), P-550
Poly(3-thiopheneacetic acid), P-551
Poly(2,5-thiophenediyl-1,2-ethenediyl), P-552
Poly(thiophenylene), P-448
Poly[(di-p-tolylsilylene)methylene], P-83
Poly[(di-m-tolylsilylene)methylene], P-82
Poly(TPA), P-556
Poly(2,2,2-trichloroethyl methacrylate), P-285
Poly(2,4,5-trichlorophenyl methacrylate), P-74
Poly(tricyclo[3.3.3.1]dec-1-yl methacrylate), P-347
Poly(1-tridecene), P-553
Poly(1-tridecyl-1,2-ethanediyl), P-434
Poly(tridecyl methacrylate), L-6
Poly(1-tridecylethylene), P-434
Poly(triethylene glycol dimethacrylate), P-284

Poly(triethylene glycol methacrylate), P-284
Poly[1,1-(2,2,2-trifluoro-1-(trifluoromethyl)ethane bis(4-phenyl)carbonate], B-14
Poly(trifluoro(trifluoromethyl)oxirane), P-292
Poly(4,4'-[2,2,2-trifluoro-1-(trifluoromethyl)ethylidene]bisphenol-co-carbonic dichloride), B-14
Poly(4,4'-[[2,2,2-trifluoro-1-(trifluoromethyl)ethylidene]bis(4,1-phenyleneoxy)]bisbenzenamine-alt-5,5'-carbonylbis[1,3-isobenzofurandione]), B-22
Poly(5,5'-[2,2,2-trifluoro-1-(trifluoromethyl)ethylidene]bis-1,3-isobenzofurandione-alt-1,3-bis(3-aminophenoxy)benzene) Addition cured (acetylenic end-caps), F-7
Poly(5,5'-[2,2,2-trifluoro-1-(trifluoromethyl)ethylidene]bis-1,3-isobenzofurandione-alt-2,2-bis[4-(4-aminophenoxy)phenyl]hexafluoropropane), F-1
Poly(5,5'-[2,2,2-trifluoro-1-(trifluoromethyl)ethylidene]bis-1,3-isobenzofurandione-alt-2,3,5,6-tetramethyl-1,4-benzenediamine), F-2
Poly(5,5'-[2,2,2-trifluoro-1-(trifluoromethyl)ethylidene]bis-1,3-isobenzofurandione-alt-3,3'-[2,2,2-trifluoro-1-(trifluoromethyl)ethylidene]bisbenzenamine), F-3
Poly(5,5'-[2,2,2-trifluoro-1-(trifluoromethyl)ethylidene]bis-1,3-isobenzofurandione-alt-4,4'-oxybisbenzenamine), F-5
Poly(5,5'-[2,2,2-trifluoro-1-(trifluoromethyl)ethylidene]bis-1,3-isobenzofurandione-alt-4,4'-[2,2,2-trifluoro-1-(trifluoromethyl)ethylidene]bisbenzenamine), F-4
Poly(5,5'-[2,2,2-trifluoro-1-(trifluoromethyl)ethylidene]bis-1,3-isobenzofurandione-co-m-phenylenediamine-co-p-phenylenediamine, F-6
Poly[(trifluoroacetoxy)ethylene], P-642
Poly(2,2,2-trifluoroethyl methacrylate), P-285
Poly(1,1,1-trifluoroisopropyl methacrylate), P-285
Poly(triglycidyl-p--aminophenol), B-2
Poly(2,2,4-trimethyl-1,6-hexanediamine-co-2,4,4-trimethyl-1,6-hexanediamine-co-1,4-benzenedicarboxylic acid), N-61
Poly[1,1-(3,3,5-trimethylcyclohexane) bis(4-phenyl)carbonate], P-554
Poly[[1,1-(3,3,5-trimethylcyclohexane) bis(4-phenyl)carbonate]-co-[2,2-propane bis(4-phenyl)carbonate]], P-555
Poly(trimethylene oxide), O-4
Poly(3,5,5-trimethylhexyl methacrylate), L-6
Poly(1,2,2-trimethylpropyl methacrylate), P-299
Poly(trimethylsilyl methacrylate), P-340
Poly(1,3,5-trioxane-co-oxirane), A-24
Poly(triphenylmethoxyethyl methacrylate), P-73
Poly(triphenylmethyl methacrylate), P-73
Poly(tripropargylamine), P-556
Poly(1-undecene), P-557
Poly(1-undecyl-1,2-ethanediyl), P-553
Poly(1-undecylethylene), P-553
Polyurethane elastomer, P-560
Poly(urethane methacrylate), M-7
Polyurethane, adhesive applications, P-558
Polyurethane, diisocyanate deficient, P-559
Polyurethane, liquid prepolymer, P-561
Polyurethane, peroxide cured, P-562
Polyurethanes General Information, P-563
Poly(δ-valerolactam), N-20
Polyvanillylidenecycloalkanone phosphoramide esters, P-564
Polyvidone, P-638
Poly(vinyl 2,2-dimethylpropyl ether), P-126
Poly(vinyl 2-butyl ether), P-125
Poly(vinyl 2-chloroethyl ether), P-145
Poly(vinyl 2-methyl propane ether-co-monoethyl maleate), P-618
Poly(vinyl 2-methylpropyl ether), P-323

Poly(vinyl 2-methylpropyl ether-co-maleic anhydride), P-324
Poly(vinyl 2-propyl ether), P-335
Poly(vinyl alcohol), foamed, P-569
Poly(vinyl alcohol)/poly(vinyl chloride), fibre blend, P-570
Poly(vinyl alcohol), P-566
Poly(vinyl alcohol), films, P-568
Poly(vinyl alkyl ethers), P-600
Poly(1-vinyl benzene), P-521, P-524, P-530
Poly(vinyl benzene), P-521, P-524, P-530
Poly(vinyl benzoate), P-571
Polyvinyl bromide, P-572
Poly(vinyl butyl ether), P-124
Poly(vinyl butyral), P-573
Polyvinyl chloride, Ethylene-vinyl acetate copolymer filled, P-663
Poly(vinyl chloride), Crystalline, P-577
Poly(vinyl chloride), Plasticised, P-580
Poly(vinyl chloride-co-ethylene), P-579
Poly(vinyl chloride-co-propylene), P-581
Poly(vinyl chloride-co-trifluoro chloroethylene), P-582
Poly(vinyl chloride-co-vinyl acetate), epoxidised, P-586
Poly(vinyl chloride-co-vinyl acetate-co-maleic acid), P-589
Poly(vinyl chloride-co-vinyl acetate-co-vinyl alcohol), P-590
Poly(vinyl chloride-co-chlorotrifluoroethylene), P-582
Poly(vinyl chloride-co-ethenyl acetate)), P-584
Poly(vinyl chloride-co-vinyl acetate), carboxyl-modified, P-585
Poly(vinyl chloride-co-vinyl acetate), epoxy-modified, P-586
Poly(vinyl chloride-co-vinyl acetate-co-(2-hydroxypropyl propenoate), P-588
Poly(vinyl chloride) copolymers, P-662
Poly(vinyl chloride) unplasticised, P-583
Poly(vinyl chloride), chlorinated, P-576
Poly(vinyl chloride-co-N-cyclohexylmaleimide), P-578
Poly(vinyl chloride-co-vinyl acetate), hydroxylated, P-587
Poly(vinyl chloride-co-vinyl acetate-co-2-hydroxypropyl acrylate), P-588
Poly(vinyl chloride-co-vinyl bromide), P-591
Poly(vinyl chloride-co-vinyl-acetate), carboxylated, P-585
Poly(vinyl chloride-co-vinylidene chloride), P-592
Poly(vinyl chloride-co-(1-chloro-1,2,2-trifluoroethylene)), P-582
Poly(vinyl chloride-co-acrylonitrile), P-575
Poly(vinyl chloride-co-vinyl acetate), hydroxy-modified, P-587
Poly(vinyl chloride-co-vinyl acetate-co-(β-hydroxypropylacrylate), P-588
Poly(vinyl chloroacetate), P-593
Poly(vinyl difluoroacetate), P-599
Poly(vinyl ethers), P-600
Poly(vinyl ethyl ether), P-601
Poly(vinyl ethyl ether-co-carbon dioxide), P-602
Poly(vinyl fluoride), P-603
Poly(vinyl fluoroacetate), P-604
Polyvinyl formal, P-605
Poly(vinyl formate), P-606
Poly(vinyl hexadecyl ether-co-vinyl octadecyl ether-co-N-vinyl pyrrolidinone), P-608
Poly(vinyl hexadecyl ether-co-vinyl octadecyl ether-co-styrene-co-maleic anhydride), P-607
Poly(vinyl hexyl ether), P-609
Poly(vinyl isobutyl ether-co-monoethyl maleate), P-618
Poly(vinyl isobutyl ether-co-vinyl chloride), P-619
Poly(vinyl isobutyl ether), P-323
Poly(vinyl isopropyl ether), P-335
Poly(vinyl methyl ether-co-butenedioic acid), P-623
Poly(vinyl methyl ether-co-maleic anhydride), P-624
Poly(vinyl methyl ether-co-3-(isopropyloxycarbonyl)propenoic acid), P-627

Poly(vinyl methyl ether-co-monobutyl maleate), P-625
Poly(vinyl methyl ether-co-maleic acid), P-623
Poly(vinyl methyl ether-co-monoethyl maleate), P-626
Poly(vinyl methyl ether-co-monoisopropyl maleate), P-627
Poly(vinyl methyl ether-co-3-(2-propyloxycarbonyl)propenoic acid), P-627
Poly(vinyl methyl ether-co-3-(butyloxycarbonyl)propenoic acid), P-625
Poly(vinyl methyl ether-co-3-(ethyloxycarbonyl)propenoic acid), P-626
Poly(vinyl methyl ether), P-621
Poly(vinyl methyl ether-co-disodium maleate), P-622
Poly(vinyl methyl ether-co-disodium 1,3-propenoate), P-622
Poly(vinyl octadecyl ether), P-628
Poly(vinyl octadecyl ether-co-maleic anhydride), P-629
Poly(vinyl octyl ether), P-630
Poly(vinyl propionate), P-631
Poly(vinyl propyl ether), P-632
Poly(vinyl-2-methyl propane ether-co-vinyl chloride), P-619
Poly(N-vinyl-2-pyrrolidone), P-638
Poly(1-vinyl-2-pyrrolidinone-co-ethyl methacrylate-co-methacrylic acid), P-230
Poly(1-vinyl-2-pyrrolidinone-co-vinyl acetate-co-vinyl propanoate), P-229
Poly(N-vinyl-N-methylacetamide), P-413
Poly(vinyl l-menthyl ether-co-indene), P-620
Poly(vinyl sec-butyl ether), P-125
Poly(vinyl acetal), P-565
Poly(vinyl alcohol), fibres, P-567
Poly(vinylbenzene-co-2-methyl-1,3-butadiene), S-31
Poly(2-vinylbutane), P-392
Poly(N-vinylcarbazole), P-574
Polyvinylchloride/acrylic-styrene-acrylonitrile terpolymer blend, A-44
Poly(vinyl chloride-co-vinyl acetate), P-584
Poly(vinylcyclobutane), P-594
Poly(vinylcycloheptane), P-595
Poly(vinylcyclohexane), P-596
Poly(vinylcyclopentane), P-597
Poly(vinylcyclopropane), P-598
Polyvinylidene fluoride, Mica filled, P-667
Polyvinylidene fluoride, Polytetrafluoroethylene and carbon filled, P-669
Poly(vinylidene chloride-co-acrylonitrile), P-611
Poly(vinylidene chloride-co-vinyl cyanide), P-611
Poly(vinylidene chloride) copolymer, P-592
Poly(vinylidene chloride), P-610
Polyvinylidene fluoride, Filled, P-665
Polyvinylidene fluoride, Graphite filled, P-666
Poly(vinylidene fluoride-co-perfluoropropylene), P-616
Poly(vinylidene fluoride-co-hexafluoro propene-co-tetrafluoroethylene), P-615
Poly(vinylidene fluoride-co-hexafluoro(γ-butylene), P-614
Poly(vinylidene fluoride-co-hexafluoroisobutene), P-614
Poly(vinylidene fluoride-co-perfluoroethylene), P-544
Poly(vinylidene fluoride-co-tetrafluoroethylene), P-544
Poly(vinylidene fluoride-co-chloro trifluoroethylene), P-613
Poly(vinylidene fluoride-co-hexafluoro isobutylene), P-614
Poly(vinylidene fluoride-co-hexafluoro propylene), P-616
Poly(vinylidene fluoride), P-612
Polyvinylidene fluoride, Carbon fibre filled, P-664
Polyvinylidene fluoride, Mineral fibre filled, P-668
Poly(N-vinylimidazole), P-617
Poly(2-vinylpropane), P-362
Poly(2-vinylpyridine)-borane complex, P-634
Poly(2-vinylpyridine), P-633
Poly(vinylpyrrolidinone)-iodine complex, P-636

Poly(vinylpyrrolidinone-co-(2-dimethyl aminoethyl)methacrylate), P-635
Poly(vinylpyrrolidinone-co-vinylacetate), P-637
Poly(N-vinylpyrrolidone-co-α-eicosene), P-640
Poly(N-vinylpyrrolidone-co-acrylic acid), P-639
Poly(N-vinylpyrrolidone-co-styrene), P-641
Poly(vinyltrifluoroacetate), P-642
Poly(VPD), P-638
Poly(m-xylenediamine-co-p-xylenediamine-co-trimethylhexamethylene terephthalamide-co-terephthalic acid), P-643
Poly(m-xylenediamine-co-terephthalic acid-co-isophthalic acid), P-644
Poly(m-xylene adipamide), P-645
Poly(m-xylene adipamide), 50% Glass fibre filled, P-646
Poly(m-xylene diamine-co-adipic acid), P-645
Poly(p-xylene diamine-co-sebacic acid), P-647
Poly(p-xylene sebacamide), P-647
Poly(m-xylylenediamine-co-wp-xylylenediamine-co-nylon 6(3)T-co-terephthalic acid, P-643
Poly(p-xylylidene), P-451
POM, P-432
Post chlorinated PVC, P-576
Povidone-iodine, P-636
PP, P-479, P-510
PP blends, P-481
PP copolymers, P-489
PP fibres, P-493
PP films, P-495
PP, aluminium filled, P-480
PP, Calcium carbonate filled, P-486
PP, Carbon fibre reinforced, P-487
PP, Carbon filled, P-487
PP, Filled, P-494
PP, Glass fibre reinforced, P-496
PP, Mica filled, P-497
PP, Mica reinforced, P-497
PP, Talc filled, P-509
PP-atactic, A-45
PP-EPDM thermoplastic vulcanisate, P-490
PP-isotactic, I-1
Pp-MS, P-410
PP-PE blends, P-482
PP-syndiotactic, S-34
PP/BR blend, P-483
PP/EPDM TPO, P-492
PP/EPR TPO, P-492
PP/IR blends, P-498
PP/NR blends, P-498
PP/PA blends, P-500
PP/PC blends, P-505
PP/PEST, P-506
PP/PS blends, P-508
PP/rubber blends, P-507
PP1, P-436
PPA, P-461
PPD, P-169
PPDA, P-169
PPE, P-188, P-436, P-436
PPE-SO3 ⊖, P-85
PPEN, P-436
PPESO3, P-85
PPHS, P-304
PPO, P-447, P-501
PPO-M, P-188
PPP, P-442
PPP(EO)3, P-648
PPP-OR11, P-648
PPPD, P-454
PPS, P-448
PPSe2, P-516
PPSF, P-459
PPSi, P-455
PPSM, P-457
PPSSe2, P-515
PPV, P-451
PPVC, P-580
PPY, P-512
PPyBS, P-520
2,2-Propanebis[4-phenol]-1,4-benzenedicarboxylic acid-1,3-benzenedicarboxylic acid polyester carbonate, B-12

Propene polymer, P-479, P-510
1-Propene-carbon monoxide-ethene alternating copolymer, P-475
1-Propene-carbon monoxide-ethene copolymer, P-476
Propylene copolymer, P-649
Propylene glycol polyol, P-501
Propylene oxide elastomers, P-650
Propylene oxide-ethylene oxide copolymers, P-504
Propylene polymer, P-479, P-510
Propylene-ethylene copolymer, P-649
Proteins, P-651
PS/P2VP, P-529
PS/PBD/PVP, P-522
PS/PPE blends, H-3
PS/PPO blends, H-3
PS/PV2P, P-529
PS, P-517, P-521, P-524, P-530
PSA, P-517
PSE, E-3, P-519
PSF, P-531
PSF, 15% PTFE lubricated, P-536
PSF, 30% carbon fibre reinforced, P-533
PSF, 30% Glass fibre reinforced, P-534
PSF, Mineral filler modified, P-535
Psi, P-517
PSM, P-519
PSU, P-531
PSU, 15% PTFE lubricated, P-536
PSU, 30% carbon fibre reinforced, P-533
PSU, 30% Glass fibre reinforced, P-534
PSU, Mineral filler modified, P-535
PSV, P-641
PT, P-550
PTA, P-545
PTAA, P-551
PtBuVE, P-126
PtBVE, P-126, P-126
PTDSi, P-213
PTFE, P-652
PTFE, Bronze filled, P-653
PTFE, Carbon and Graphite filled, P-654
PTFE, Glass fibre filled, P-655
PTFE, Graphite filled, P-656
PTFE, PPS filled, P-657
PTFE, Stainless steel filled, P-658
PTFE, TFE filled, P-659
PTFE-PVdF copolymer, P-544
PTFE/PP copolymer, T-3
PTGAP, B-2
PTMA, P-545
PTMEG, P-546
PTMI, P-547
PTMT, P-103
PTO, O-4
PTV, P-552
Pullulan, P-660
PURET, P-661
PVA/PVC fibre blend, P-570
PVA, P-566
PVAL, P-566
PVB, P-571, P-573
PVC, P-583
PVC copolymers, P-662
PVC, Crystalline, P-577
PVC, EVA filled, P-663
PVC-C, P-576
PVC-EVA, P-663
PVC-P, P-580
PVC-PP copolymer, P-581
PVC-U, P-583
PVC/ABS Blends, A-19
PVC/ASA blend, A-44
PVC/NBR blends, P-93
PVC/NBR/PMMA blends, P-345
PVC/PMMA blends, P-345
PVC/PVAc, P-584
PVCA, P-574, P-584
PVCH, P-596
PVCP, P-597, P-598
PVCZ, P-574
PVDC, P-610, P-610
PVDF, P-612

PVDF, Carbon fibre filled, P-664
PVDF, Filled, P-665
PVDF, Graphite filled, P-666
PVDF, Mica filled, P-667
PVDF, Mineral fibre filled, P-668
PVDF, PTFE and Carbon filled, P-669
PVDF-co-PCTFE, P-613
PVDF-co-PTFE, P-544
PVdF-HFP copolymer, P-616
PVDF/HFIB, P-614
PVDF/PMMA blends, P-348
PVE, P-601
PVEE, P-601
PVF, P-603, P-605, P-606
PVF2/PMMA blends, P-348
PVF2, P-612
PVFO, P-606
PVI, P-617
PVK, P-574
PVL, P-190
PVM-MA, P-624
PVME, P-621
PVnBE, P-124
PVOH, P-566
PVP, P-633, P-638
PVP-dimethylaminoethyl methacrylate copolymer, P-635
PVP-DMAEMA, P-635
PVP-ethyl methacrylate-methacrylic acid copolymer, P-230
PVP-I, P-636
PVP-styrene copolymer, P-641
PVP-VA copolymer, P-637
PVPD, P-638
PVPr, P-631, P-632
PVPyr, P-638
PVsBE, P-125, P-125
Pyran polymer, P-216
Pyrocellulose, C-12
1H-Pyrrole-3-acetic acid homopolymer, P-513
Pyrrole black, P-512
1H-Pyrrole-3-butanesulfonic acid monosodium salt homopolymer, P-520
Pyrrole red, P-512

Radial multichain styrene-butadiene block copolymer, S-27
Radial teleblock butadiene-styrene copolymers, S-27
Rayon, C-8
Regenerated cellulose, C-8
Resoles, P-20, P-21
Resorcinol formaldehyde resins, R-1
Resorcinol resins, wood applications, R-2
Resorcinol resins, R-1
RF resins, R-1
Rigid PVC, P-583
Rohacell, P-352
RPVC, P-583
Rubber, N-2
Rubber hydrochloride, R-3
Rubber modified polystyrene, H-2
Rubber reinforced polystyrene, H-2

SAN, S-1
SAN, Blow moulding, S-2
SAN, Extrusion, Unfilled, S-3
SAN, Flame retardant, Injection moulding, Glass fibre filled, S-4
SAN, General purpose, S-5
SAN, Glass reinforced, S-6
SAN, Injection moulding, Glass fibre filled, S-7
SAN, Injection moulding, Unfilled, S-8
SB Polymer, S-26
SBP block polymer, P-522
SBR, S-28
SBS, S-9
Side chain liquid crystalline polydimethylsiloxane copolymers, P-193
Sidechain liquid crystalline polymethylhydrogensiloxane copolymers, P-307
Silicone acrylate, S-10

Name and Synonym Index

Silicone elastomer, Glass fibre filled, S-18
Silicone elastomer, One part room temp. vulcanised, M-13
Silicone elastomer, Room temp. vulcanised, M-17
Silicone elastomer, Two part room temp. vulcanised, M-19
Silicone elastomer, M-16
Silicone encapsulant, S-16
Silicone epoxy copolymer, S-11
Silicone glycol copolymer, P-8
Silicone phenol formaldehyde copolymer, S-12
Silicone phenolic resin copolymer, S-12
Silicone polyallyl alcohol copolymer, S-13
Silicone polycarbinol copolymer, S-13
Silicone polycarbonate copolymer, P-469
Silicone polyetherimide copolymer, S-14
Silicone resin, Glass filled, S-15
Silicone resin, Glass reinforced, S-15
Silicone resin, Room temperature vulcanising, S-16
Silicone rubber, Conductive, S-17
Silicone rubber, Glass fibre filled, S-18
Silicones, P-518
Silk, S-19
Silly putty, S-20
SIS polymer, S-21
SIS, S-21
SMiPK, E-3
Sodium carboxymethyl cellulose, S-22
SP4000, P-519
Spandex Fibres, S-23
Star styrene butadiene block copolymers, S-27
Star-branched butadiene-styrene block copolymers, S-27
Star-branched polymer of polyisoprene, P-329
Starch, S-24
Styrene butadiene radial block copolymers, S-27
Styrene butadiene rubber, high-*trans*, S-29
Styrene-acrylonitrile copolymer, S-1
Styrene-butadiene block copolymer, K-1
Styrene-butadiene block copolymers, S-25
Styrene-butadiene copolymers, S-26
Styrene-butadiene di-block copolymers, S-25
Styrene-butadiene polymer, K-1
Styrene-butadiene rubber, S-28
Styrene-butadiene-styrene block copolymer, S-9
Styrene-butadiene-vinylpyridine terpolymer, P-522
Styrene-epoxidised butadiene-styrene block copolymers, E-8
Styrene-isoprene block copolymers, S-30
Styrene-isoprene copolymers, S-31
Styrene-isoprene diblock copolymers, S-30
Styrene-isoprene radial block copolymers, S-32
Styrene-isoprene star block copolymers, S-32
Styrene-isoprene-2-vinylpyridine triblock copolymer, P-525
Styrene-isoprene-styrene triblock copolymers, S-21
Styrene-vinylpyridine-butadiene triblock polymer, P-522
Styrene/butadiene/vinylidene terpolymer rubbers, P-522
Styrene/methacrylate copolymers, S-33
SVAP, P-637
SVP, P-641
Syndiotactic Polypropylene, S-34
Synthetic elastomers - polyether, P-650

6//T/I copolymer, P-294
Teflon, P-652
(2,2′,6,6′-Tetrabromobisphenol A diglycidyl ether) resin, T-1
Tetrabromobisphenol A polycarbonate, P-464
Tetrachloro Bisphenol A polycarbonate, P-466
Tetrafluoroethylene perfluoromethyl vinyl ether copolymer, Polytetrafluoroethylene and carbon filled, P-13
Tetrafluoroethylene-ethylene copolymer, Glass fibre filled, P-541
Tetrafluoroethylene-ethylene copolymer, E-12
Tetrafluoroethylene-hexafluoropropylene copolymer, Carbon fibre filled, F-8, F-10
Tetrafluoroethylene-hexafluoropropylene copolymer, Filled, F-9
Tetrafluoroethylene-hexafluoropropylene copolymer, Glass fibre filled, F-10
Tetrafluoroethylene-hexafluoropropylene copolymer, P-542
Tetrafluoroethylene-perfluoro(methyl vinyl) ether copolymer, Glass filled, P-11
Tetrafluoroethylene-perfluoromethyl vinyl ether copolymer, Polytetrafluoroethylene filled, P-14
Tetrafluoroethylene-perfluoromethylvinyl ether copolymer, Carbon filled, P-10
Tetrafluoroethylene-perfluoromethylvinyl ether copolymer, Mineral and mica filled, P-12
Tetrafluoroethylene-perfluoromethylvinyl ether copolymer, T-2
Tetrafluoroethylene-perfluoropropyl vinyl ether copolymers, P-543
Tetrafluoroethylene-propylene copolymer, T-3
Tetrafluoroethylene-vinylidene fluoride copolymer, P-544
Tetrakis (4-glycidyloxyphenyl)ethane, P-282
1,1,2,2-Tetrakis[4-(2,3-epoxypropoxy)phenyl]ethane, P-282
Tetramethyl Bisphenol A polycarbonate, P-467
4-(1,1,3,3-Tetramethylbutyl)phenol resins, O-1
1,1,3,3-Tetraphenyl-1,3-disilacyclobutane homopolymer, sru, P-210
TFB, P-200
TFE/P copolymer, T-3
Thermoplastic elastomers, P-507
Thermoplastic vulcanisates, P-507
N-[[2-(3-Thienyl)ethoxy]carbonyl]glycine butyl ester homopolymer, P-661
Thiobisphenol-Bisphenol A copolycarbonates, P-549
3-Thiopheneacetic acid homopolymer, P-551
Thymus nucleic acid, D-1
TP, P-517
Tree rubber, N-2
Trifluoropropylmethyl silicone fluid, F-15
Two part room temp. vulcanised methylsilicone rubber, M-19

UHMWPE, U-3
ULDPE, V-5
Ultem 1000, U-1
Ultem 5001, U-2
Ultra high molecular weight PE, U-3
Ultra low density polyethylene, V-5
Ultramid T, N-52
Unplasticised poly(vinyl chloride), P-583
Unplasticised PVC, P-583
Unvulcanised polychloroprene, P-149
UPVC, P-583
Urea formaldehyde resins, U-4
Urea resin, Moulding powders, Cellulose filled, U-5
Urea resins, Wood applications, U-6
Urea resins, U-4

Urethane coating, U-7
Urethane rubber, Multrathane based, U-8
Urethane, polyether, U-7

VC/AN copolymer, P-575
VC/CTFE copolymer, P-582
VC/E copolymer, P-579
VC/EVA, P-663
VC/P copolymer, P-581
VC/TFCE copolymer, P-582
VC/VA, P-584
VC/VAC, P-584
VC/VB copolymer, P-591
VCA, P-575
VCE, P-579
VCP, P-581
VDC/VC, P-592
Very low density polyethylene, V-5
Vicara, Z-1
Vinal, P-567
Vinyl chloride polymer, P-583
Vinyl chloride-N-cyclohexylmaleimide copolymer, P-578
Vinyl chloride-acrylonitrile copolymer, P-575
Vinyl chloride-chlorotrifluoroethylene copolymer, P-582
Vinyl chloride-cyclohexylmaleimide copolymer, P-578
Vinyl chloride-ethylene copolymer, P-579
Vinyl chloride-ethylene-vinyl acetate terpolymer, P-663
Vinyl chloride-propylene copolymer, P-581
Vinyl chloride-trifluorochloroethylene copolymer, P-582
Vinyl chloride-vinyl bromide copolymer, P-591
Vinyl chloride-vinylidene chloride copolymer, P-592
Vinyl fluoride polymer, P-603
Vinyl methyl based silicone elastomer, V-2
Vinyl methyl phenyl based silicone elastomer, V-1
Vinyl methyl phenyl silicone rubber, Heat vulcanised, V-1
Vinyl methyl silicone rubber, Heat vulcanised, V-2
Vinyl phenyl methyl silicone rubber, Heat vulcanised, V-1
Vinyl-polyisoprene, P-325
Vinyl-type butadiene, P-56
Vinylcyclohexene dioxide resin, V-3
Vinylidene fluoride polymer, P-612
Vinylidene fluoride-hexafluoropropylene copolymer, P-616
Vinylidene fluoride-hexafluoropropylene-tetrafluoroethylene terpolymer, P-615
Vinylon, P-567
Viscose, C-8
VLDPE, V-5
Vulcanised polychloroprene, P-150
Vulcanite, E-1

Wool, W-1

Xylenol formaldehyde resins, X-1
Xylok resins, P-15
Xylok resins, Glass filled, P-16
Xylok resins, Mica filled, P-17

Zein, Z-1
Zenite, Z-2